CHILTON'S GUIDE TO

VACUUM DIAGRAMS 1980-86 DOMESTIC CARS

Vacuum diagrams for 1980-86 domestic cars

CHILTON BOOK COMPANY
Chilton Way, Radnor, PA 19089

Manufactured in USA

ISBN 0-8019-7821-1
Library of Congress Card Catalog No. 87-47941
234567890 765432109

VACUUM CIRCUITS

CONTENTS

SAFETY NOTICE

Proper service and repair procedures are vital to the safe, reliable operation of all motor vehicles, as well as the personal safety of those performing repairs. This manual outlines procedures for servicing and repairing vehicles using safe effective methods. The procedures contain many NOTES, CAUTIONS and WARNINGS which should be followed along with standard safety procedures to eliminate the possibility of personal injury or improper service which could damage the vehicle or compromise its safety.

It is important to note that repair procedures and techniques, tools and parts for servicing motor vehicles, as well as the skill and experience of the individual performing the work vary widely. It is not possible to anticipate all of the conceivable ways or conditions under which vehicles may be serviced, or to provide cautions as to all of the possible hazards that may result. Standard and accepted safety precautions and equipment should be used when handling toxic or flammable fluids, and safety goggles or other protection should be used during cutting, grinding, chiseling, prying, or any other process that can cause material removal or projectiles.

Some procedures require the use of tools specially designed for a specific purpose. Before substituting another tool or procedure, you must be completely satisfied that neither your personal safety, nor the performance of the vehicle will be endangered.

American Motors
Vacuum Circuits

INDEX

1980 VACUUM CIRCUITS

1981 VACUUM CIRCUITS

1982 VACUUM CIRCUITS

1983 VACUUM CIRCUITS

1984—85 VACUUM CIRCUITS

1986 VACUUM CIRCUITS

VACUUM CIRCUITS

(© American Motors Corp.)

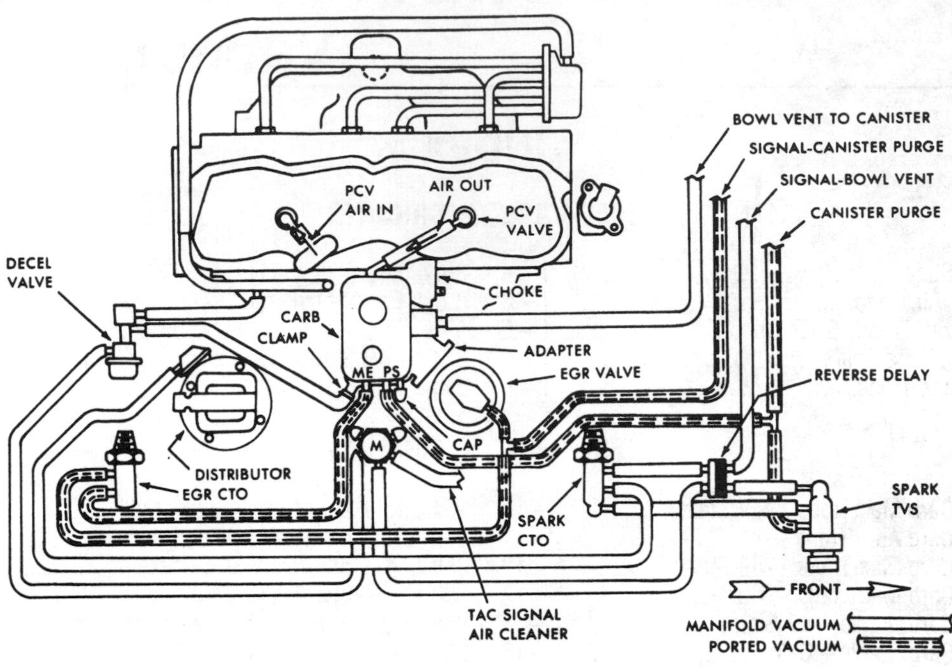

1980—4 CYL. ENG.-49 STATES

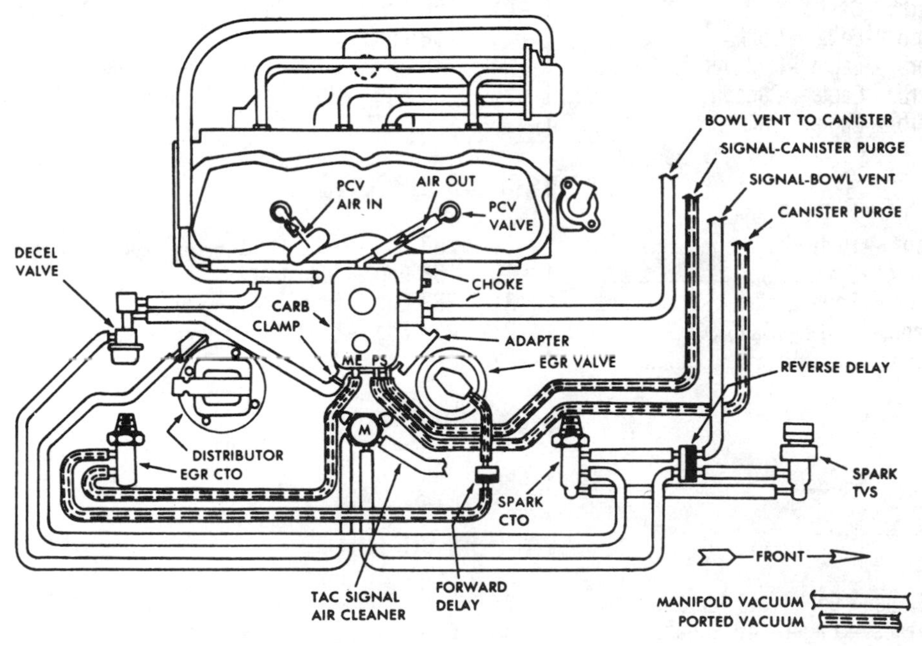

1980—4 CYL. ENG.-W/AUTO. TRANS.-WO/AIR COND.-49 STATES

VACUUM CIRCUITS

(© American Motors Corp.)

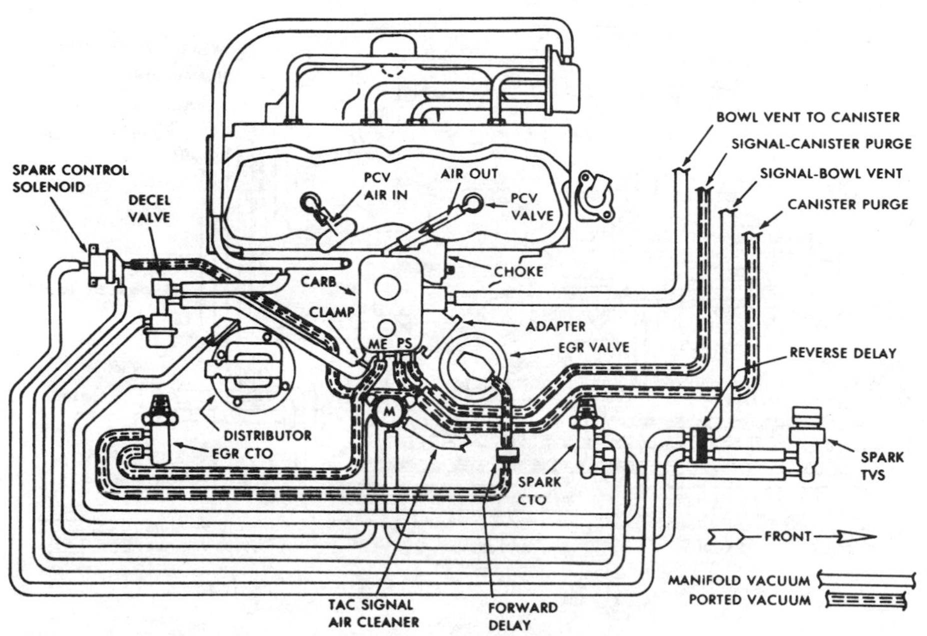

1980—4 CYL. ENG.-W/AUTO. TRANS. & AIR COND.-49 STATES

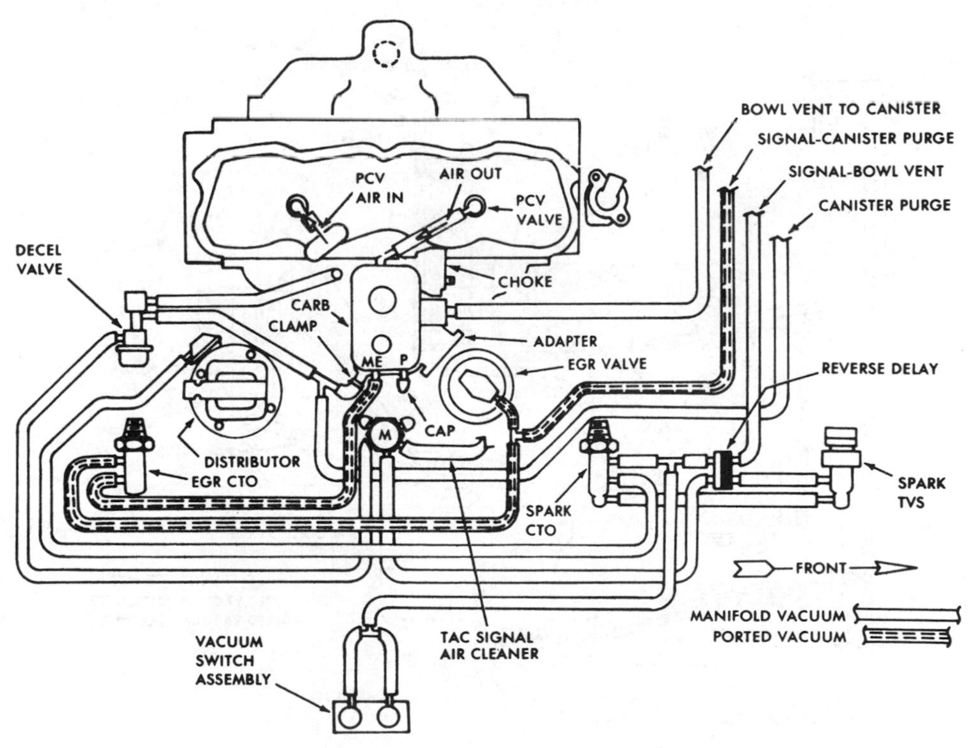

1980—4 CYL. ENG.-W/MAN. TRANS.-W & WO/AIR COND.-CALIFORNIA

VACUUM CIRCUITS

(© American Motors Corp.)

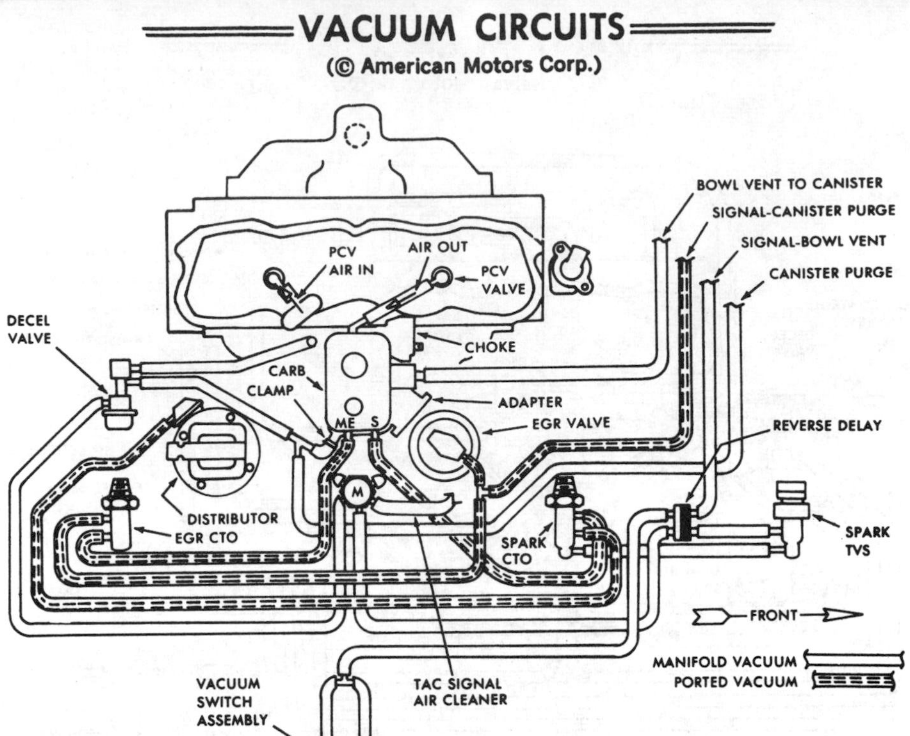

1980—4 CYL. ENG.-W/AUTO. TRANS.-W & WO/AIR COND.-CALIFORNIA

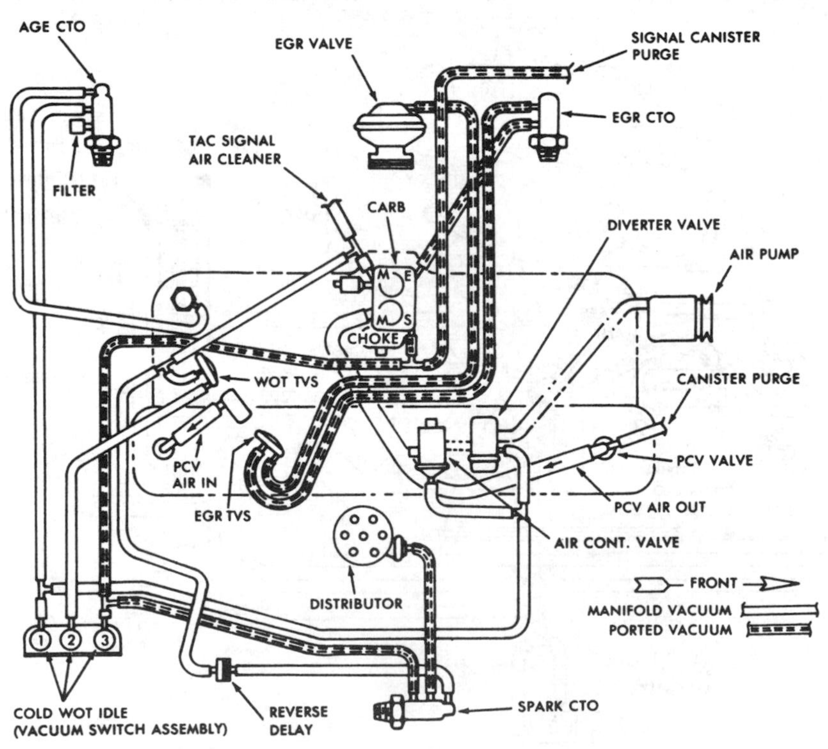

1980—6 CYL. ENG.-W/AUTO. TRANS.-49 STATES

VACUUM CIRCUITS

(© American Motors Corp.)

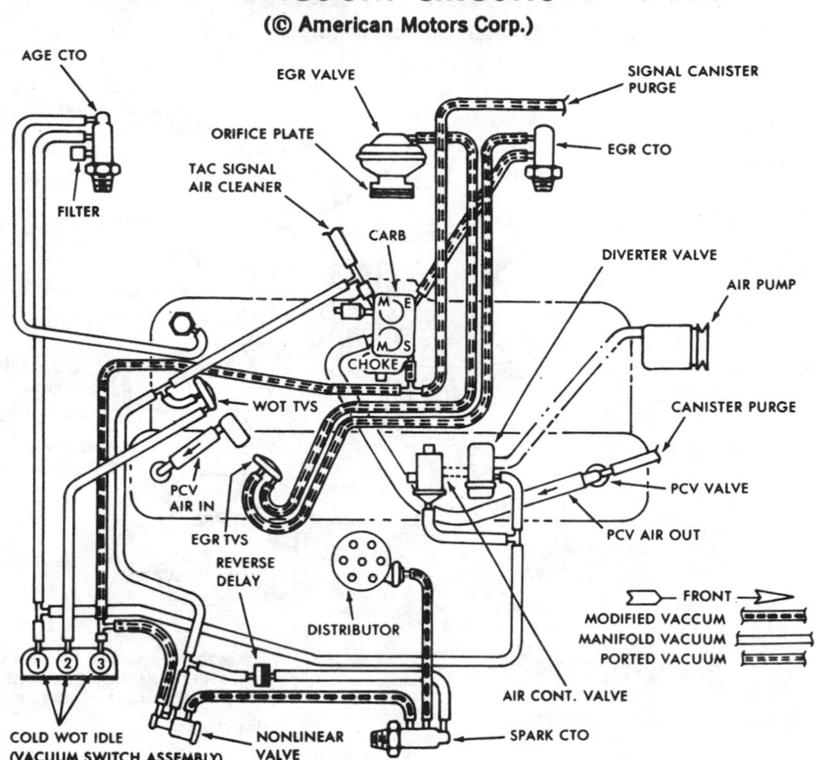

1980—6 CYL. ENG.-W/MAN. TRANS.-49 STATES

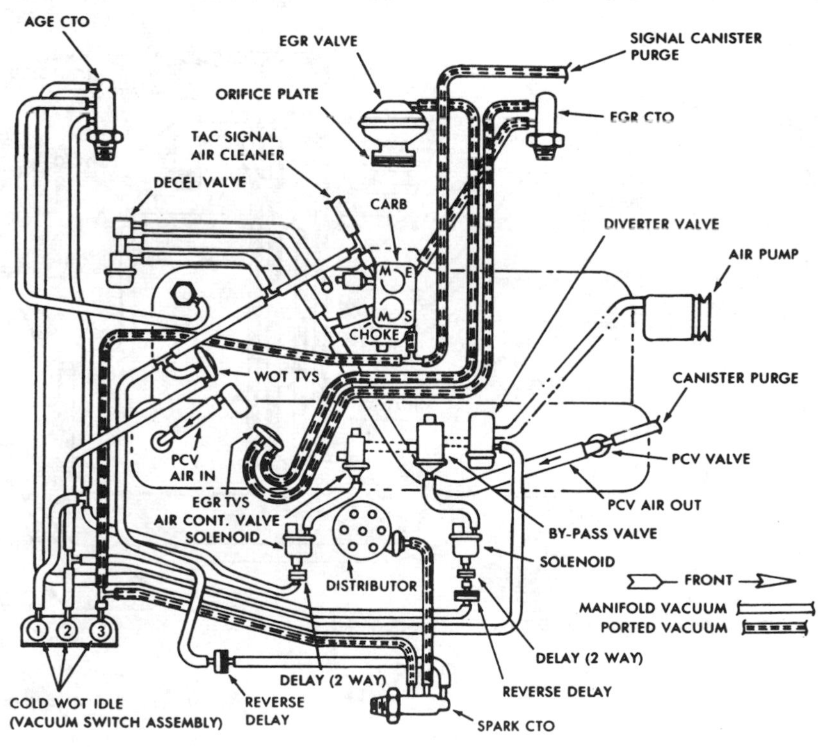

1980—6 CYL. ENG.-W/AUTO. TRANS.-CALIFORNIA

VACUUM CIRCUITS

(© American Motors Corp.)

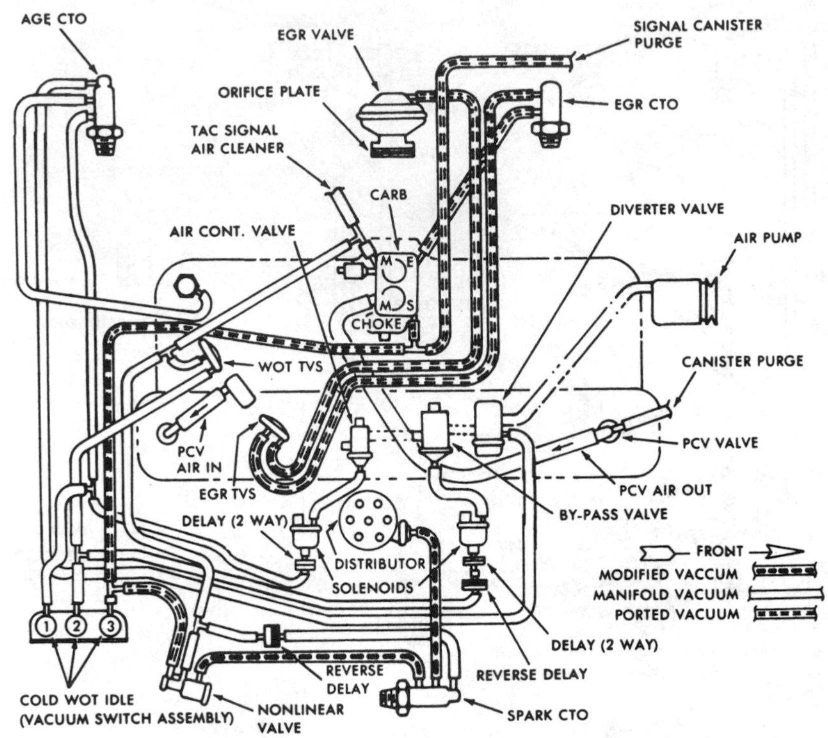

1980—6 CYL. ENG.-W/AUTO. TRANS.-CALIFORNIA

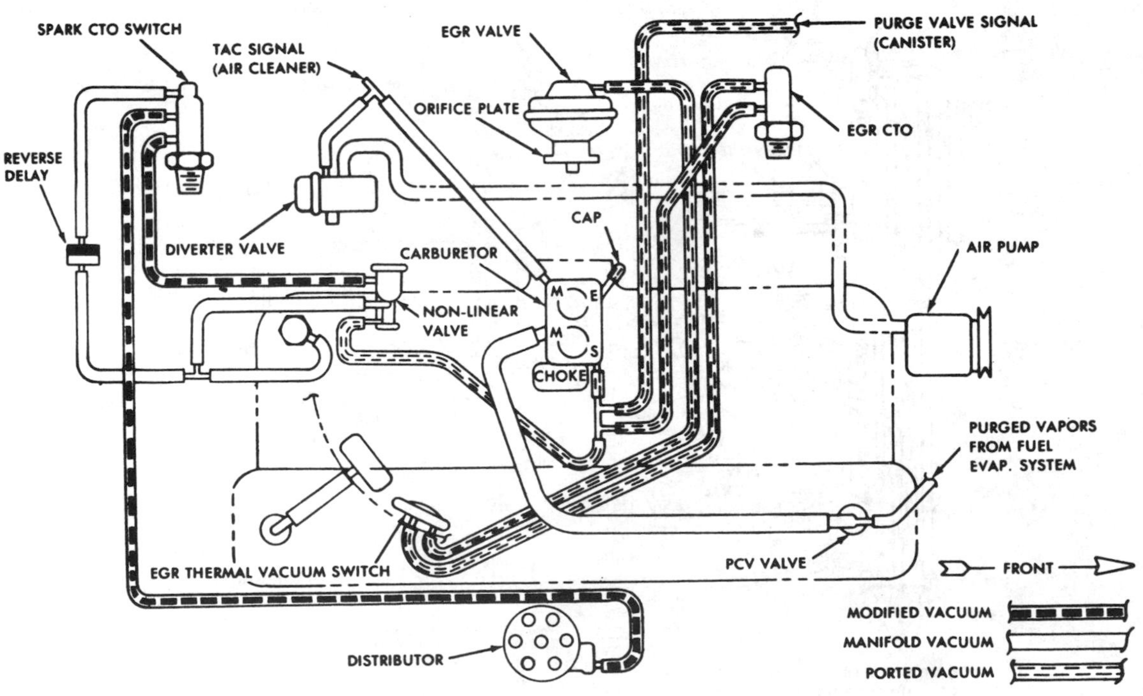

1980 EAGLE—W/3.08 AXLE-49 STATES

VACUUM CIRCUITS

(© American Motors Corp.)

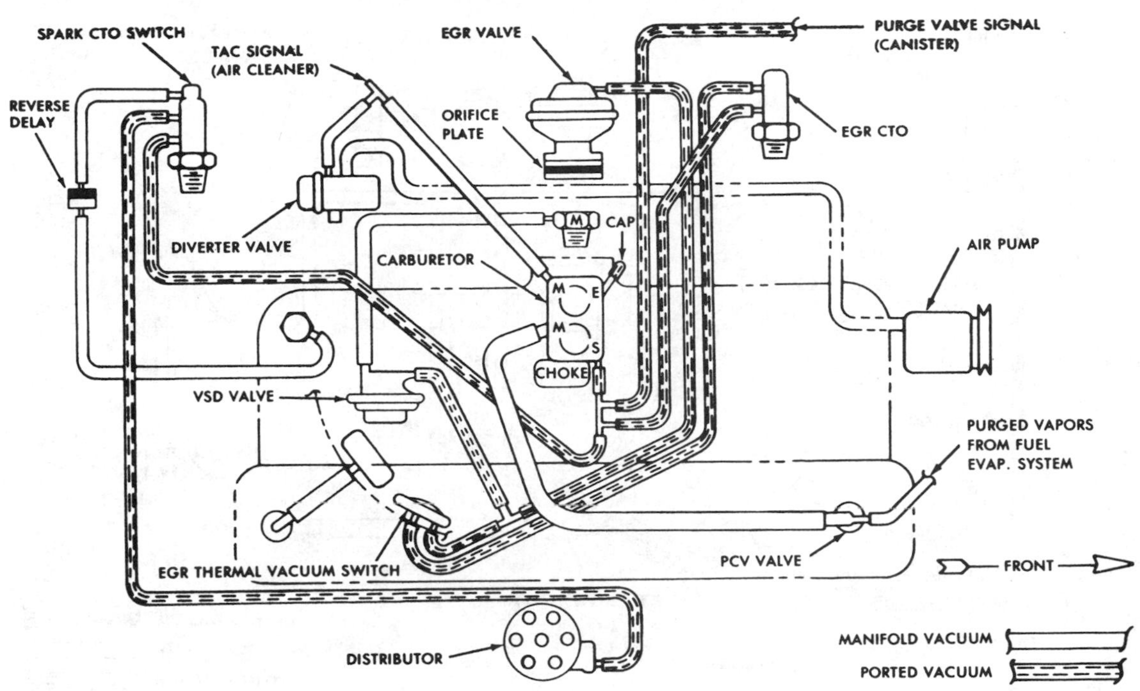

1980 EAGLE—W/3.54 AXLE-49 STATES

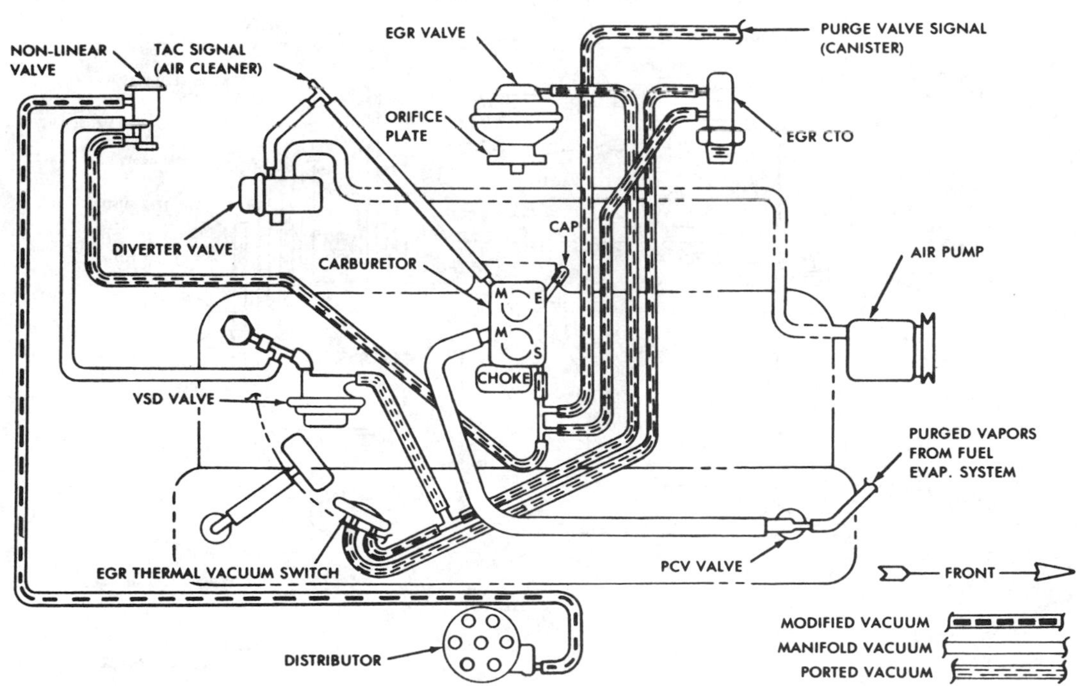

1980 EAGLE—HILLY TERRAIN

VACUUM CIRCUITS

(© American Motors Corp.)

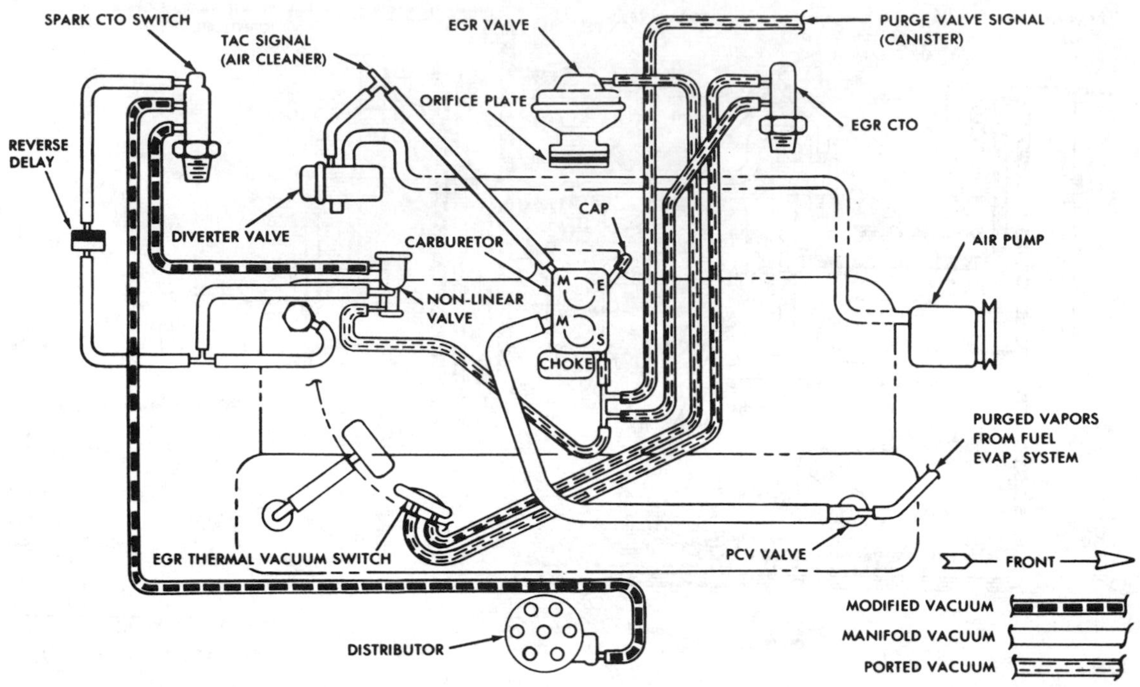

1980 EAGLE—CALIFORNIA

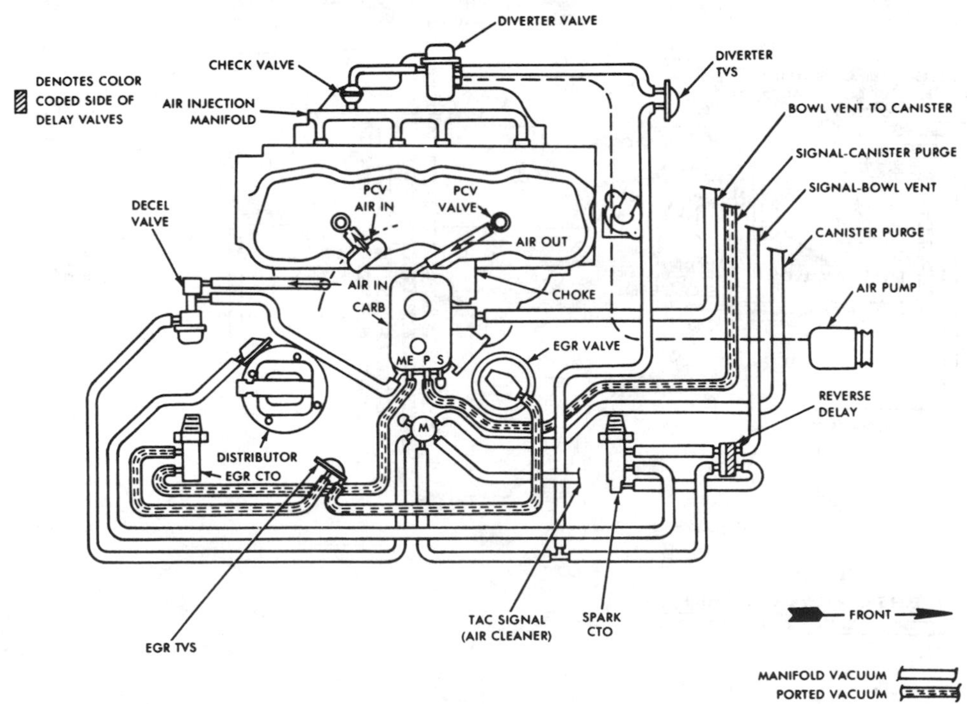

1981—151 ENG.-2BBL CARB.-W/MAN. TRANS.-WO/AIR COND.-SPIRIT-CONCORD-49 STATES

VACUUM CIRCUITS

(© American Motors Corp.)

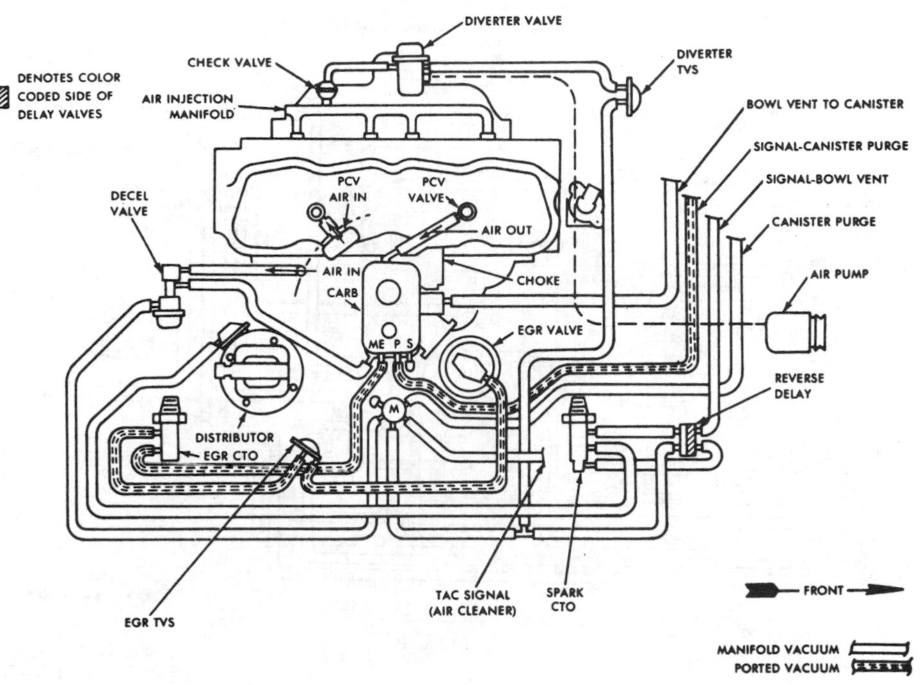

1981—151 ENG.-2BBL CARB.-W/AUTO. TRANS.-WO/AIR COND.-SPIRIT-CONCORD-49 STATES

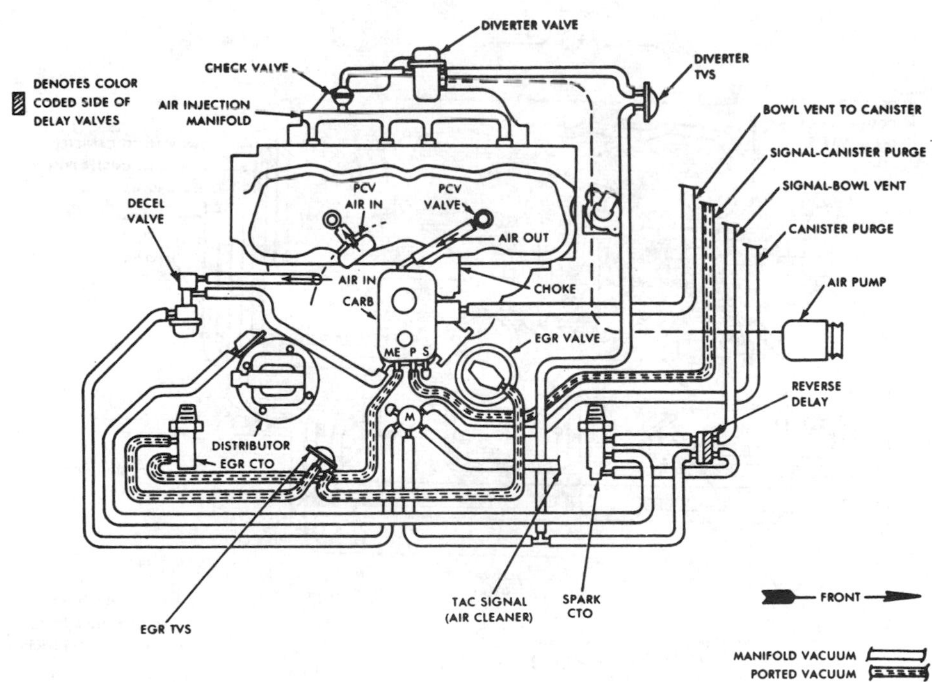

1981—151 ENG.-2BBL CARB.-W/MAN. TRANS. & W/AIR COND.-SPIRIT-CONCORD-49 STATES

VACUUM CIRCUITS

(© American Motors Corp.)

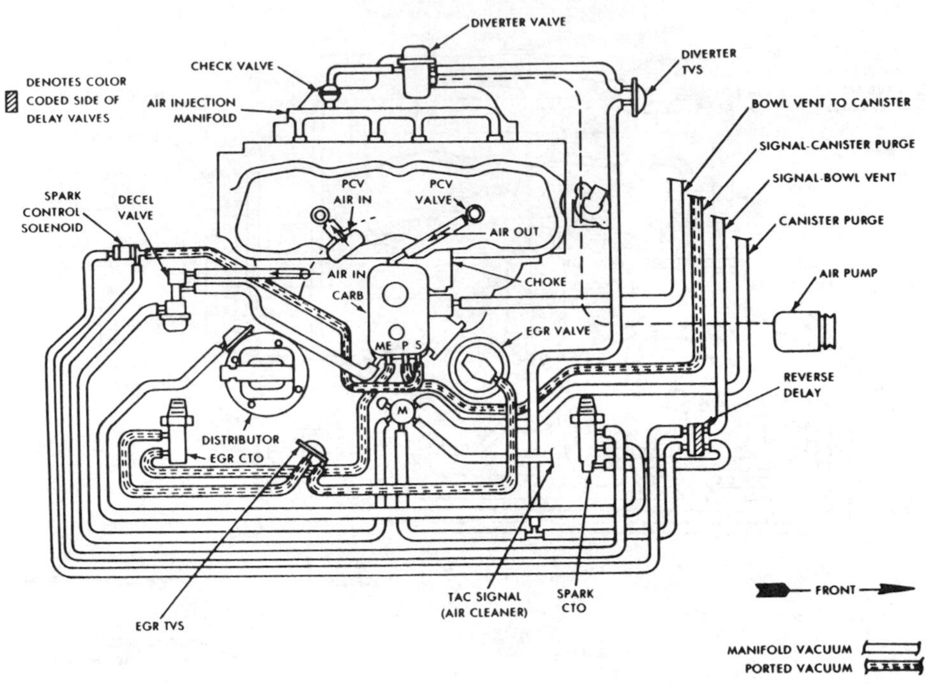

1981—151 ENG.-2BBL CARB.-W/AUTO. TRANS. & W/AIR COND.-SPIRIT-CONCORD-49 STATES

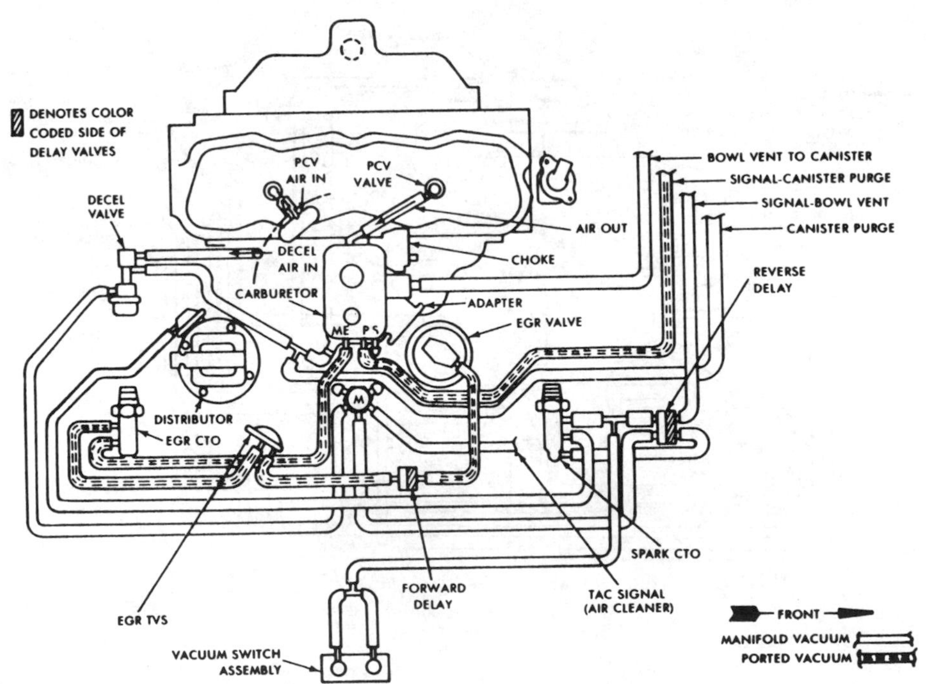

1981—151 ENG.-2BBL CARB.-W/MAN. TRANS.-SPIRIT-CONCORD-CALIFORNIA

VACUUM CIRCUITS

(© American Motors Corp.)

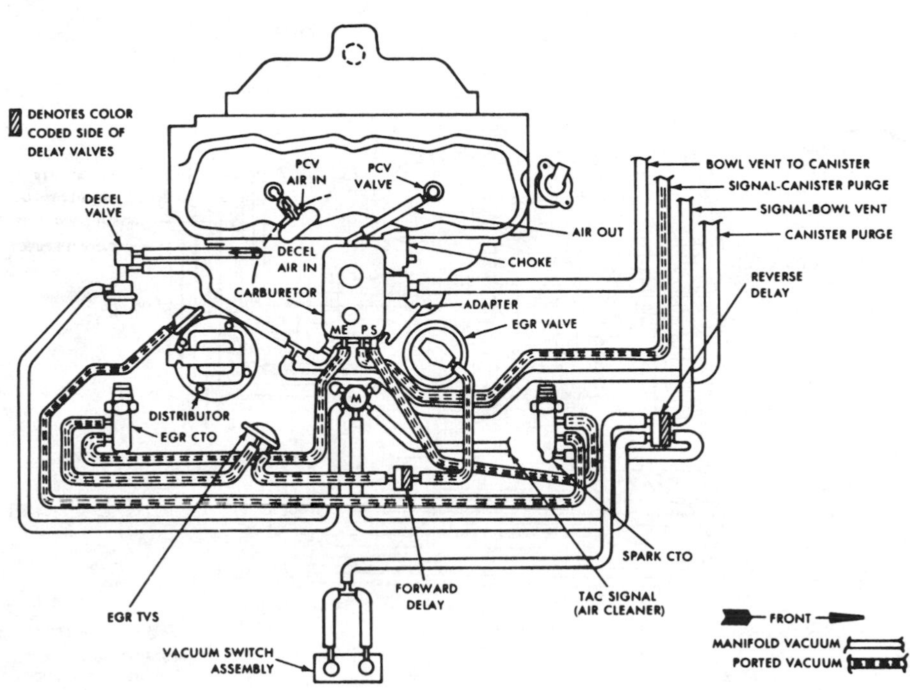

1981—151 ENG.-2BBL CARB.-W/AUTO. TRANS.-SPIRIT-CONCORD-CALIFORNIA

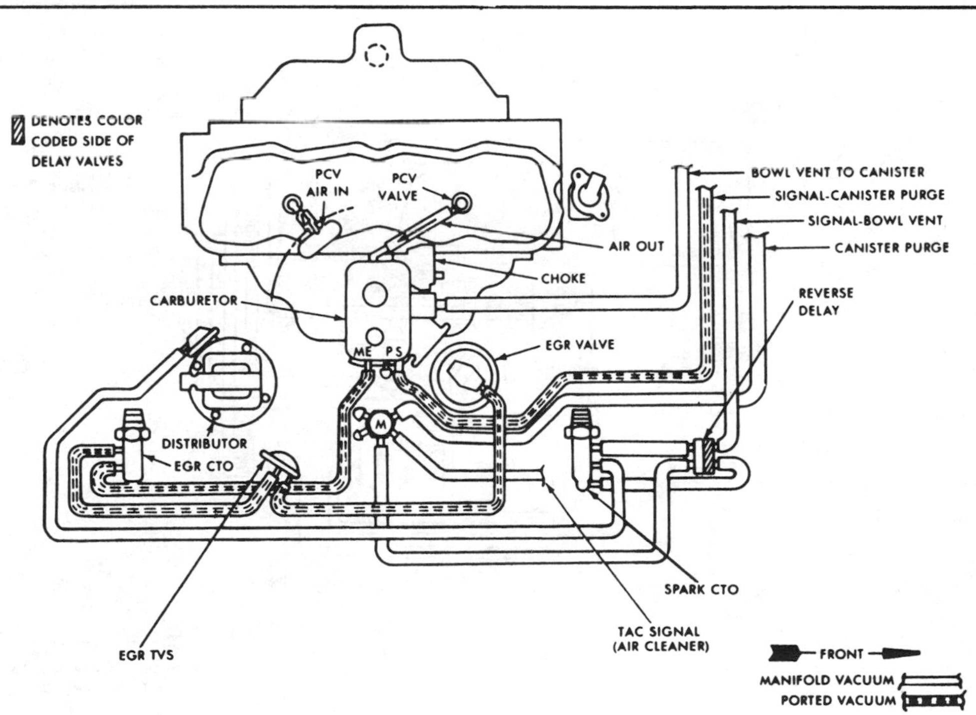

1981—151 ENG.-2BBL CARB.-W/MAN. TRANS.-WO/AIR COND.-EAGLE-49 STATES

VACUUM CIRCUITS

(© American Motors Corp.)

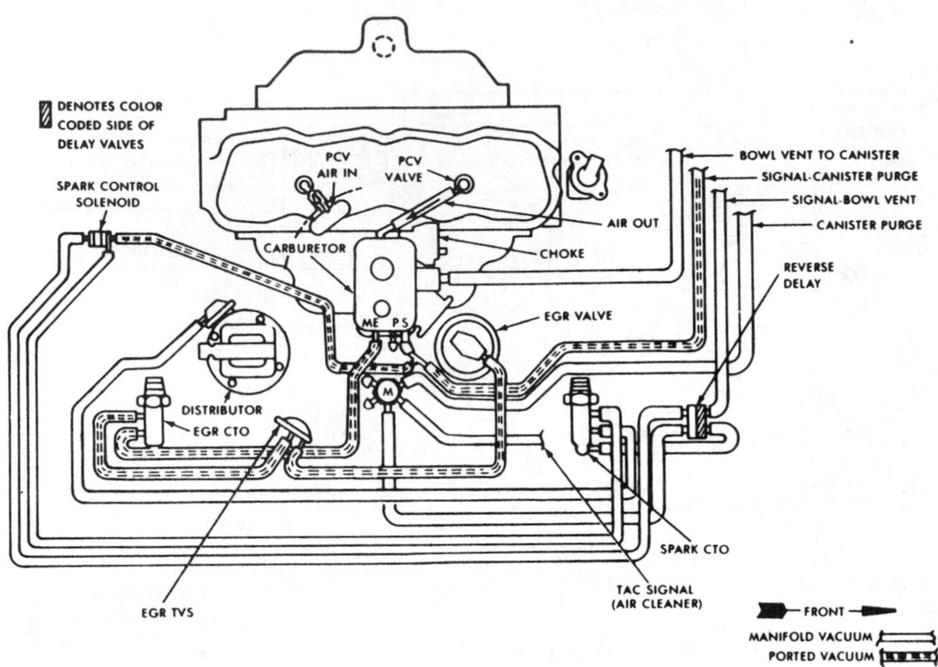

1981—151 ENG.-2BBL CARB.-W/MAN. TRANS. & W/AIR COND.-EAGLE-49 STATES

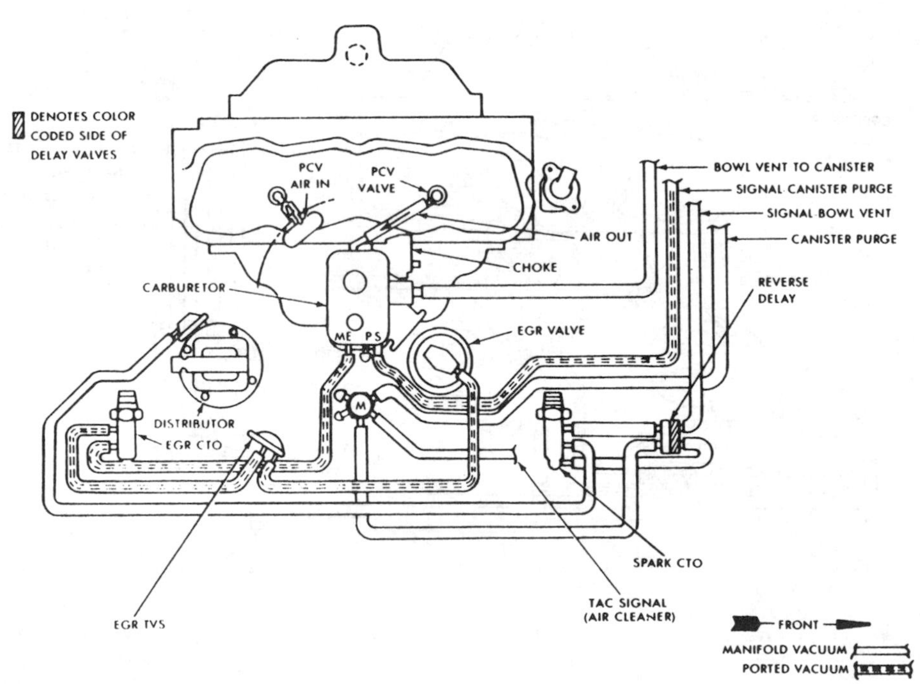

1981—151 ENG.-2BBL CARB.-W/AUTO. TRANS.-WO/AIR COND.-EAGLE-49 STATES

VACUUM CIRCUITS

(© American Motors Corp.)

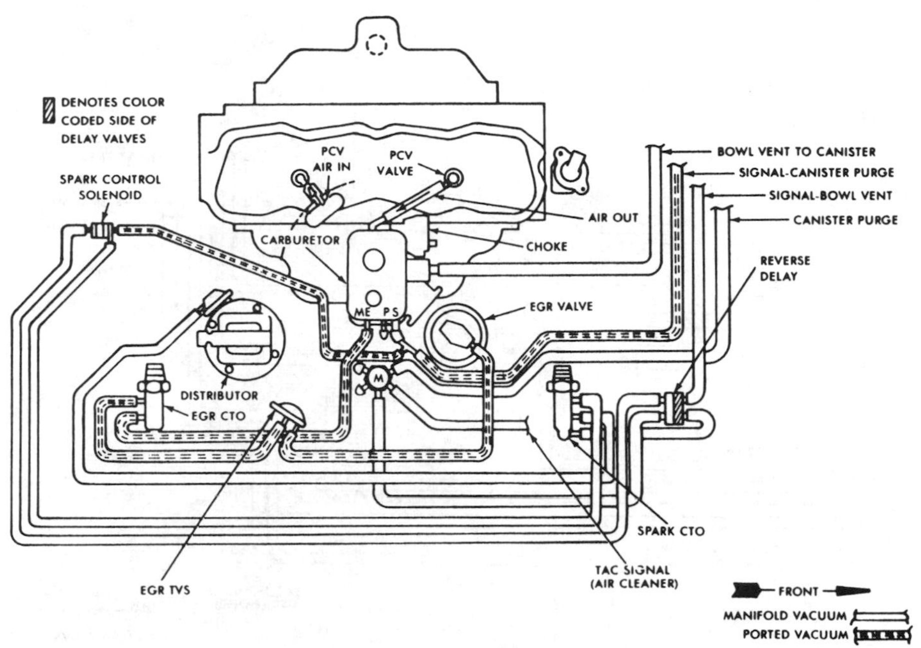

1981—151 ENG.-2BBL CARB.-W/AUTO. TRANS.-WO/AIR COND.-EAGLE-49 STATES

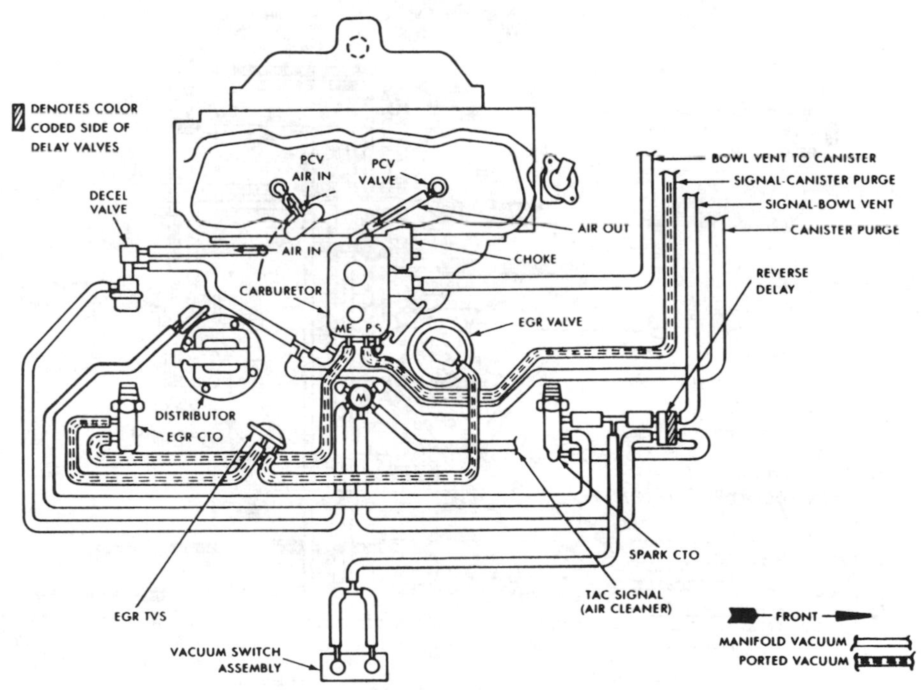

1981—151 ENG.-2BBL CARB.-W/MAN. TRANS.-EAGLE-CALIFORNIA

VACUUM CIRCUITS

(© American Motors Corp.)

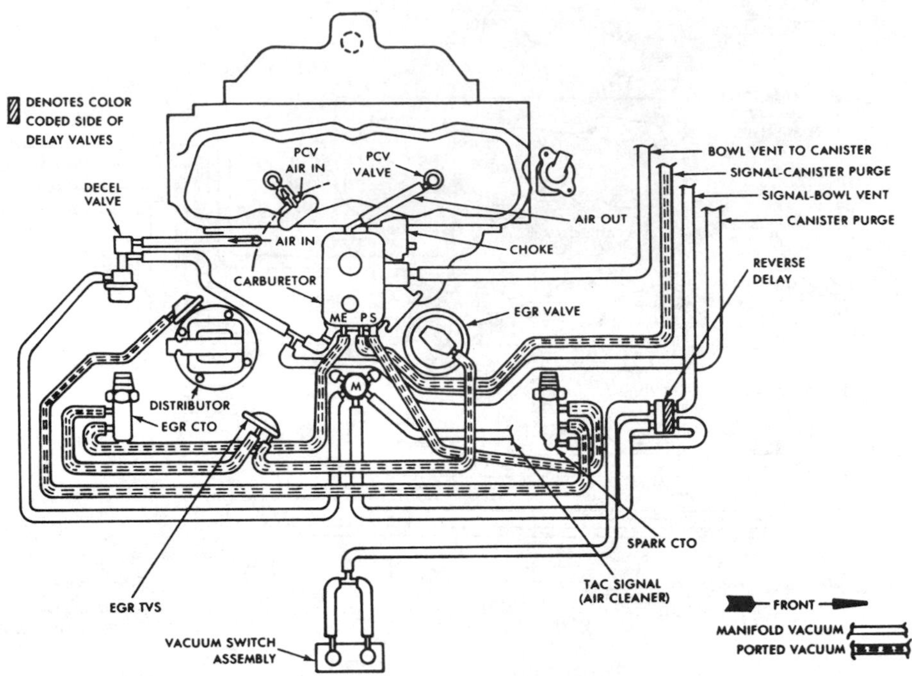

1981—151 ENG.-2BBL CARB.-W/AUTO. TRANS.-EAGLE-CALIFORNIA

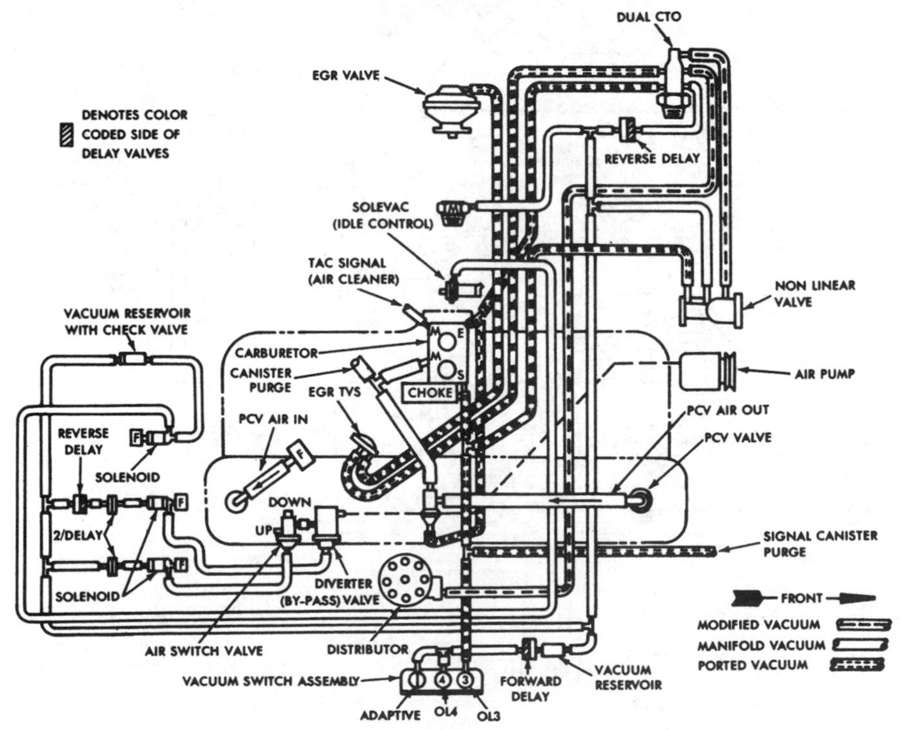

1981—258 ENG.-W/MAN. TRANS.-SPIRIT-CONCORD-50 STATES

VACUUM CIRCUITS

(© American Motors Corp.)

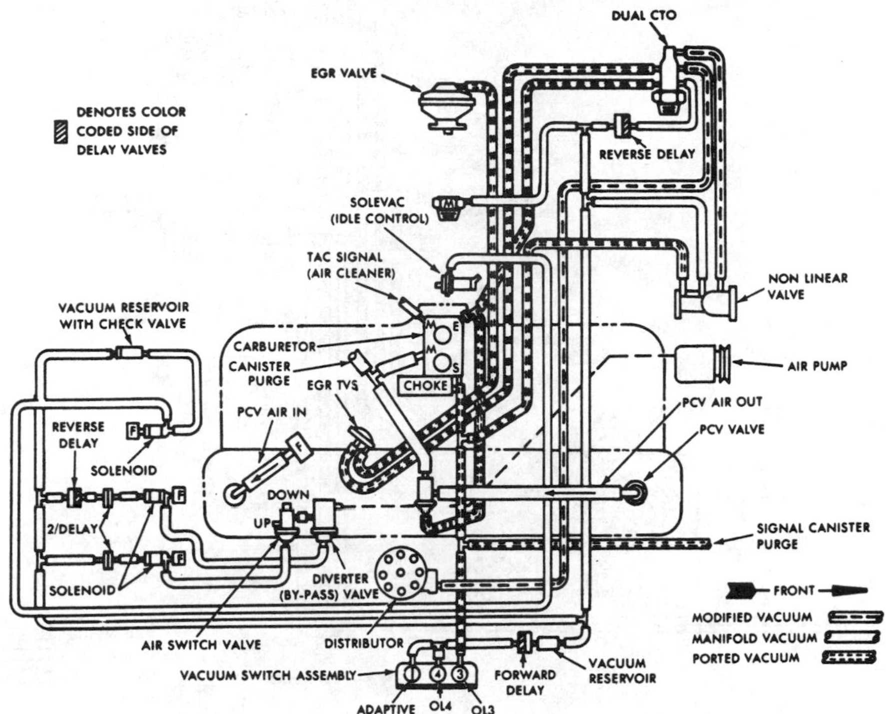

1981—258 ENG.-W/AUTO. TRANS.-SPIRIT-CONCORD-50 STATES

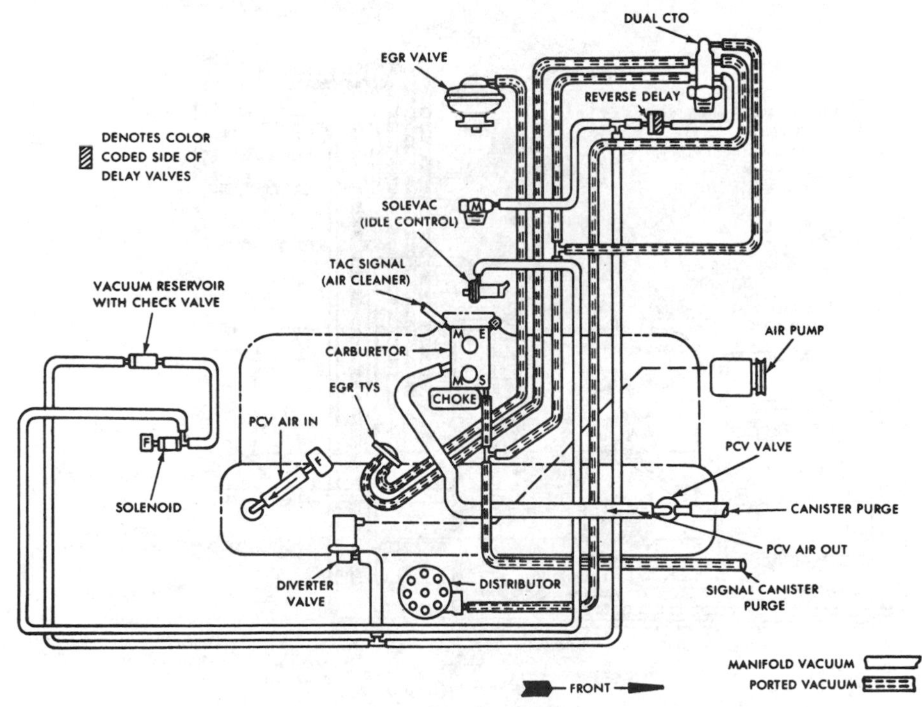

1981—258 ENG.-W/MAN. TRANS.-EAGLE-49 STATES

VACUUM CIRCUITS

(© American Motors Corp.)

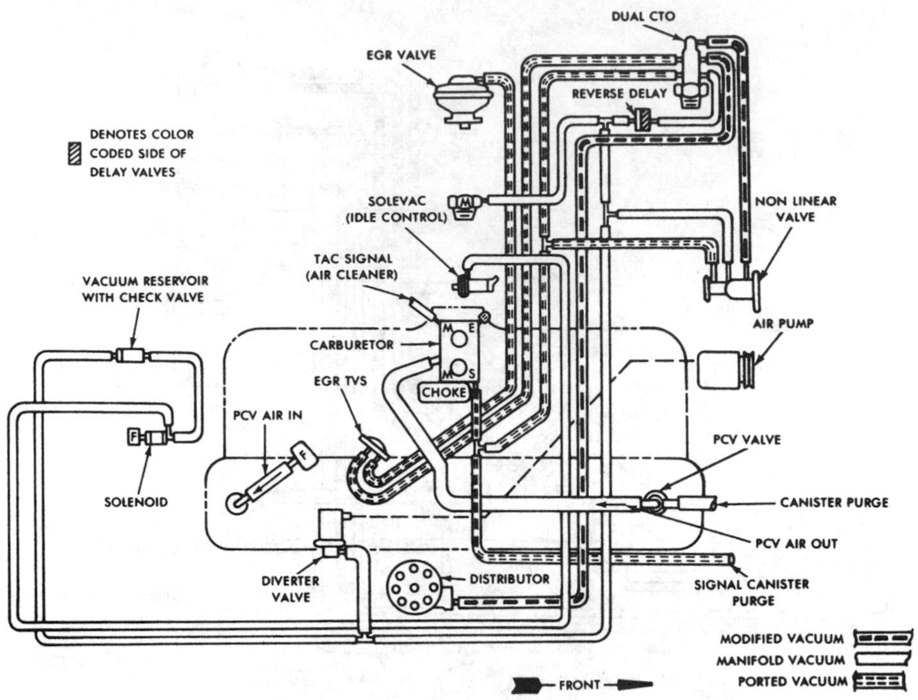

1981—258 ENG.-W/AUTO. TRANS.-EAGLE-49 STATES

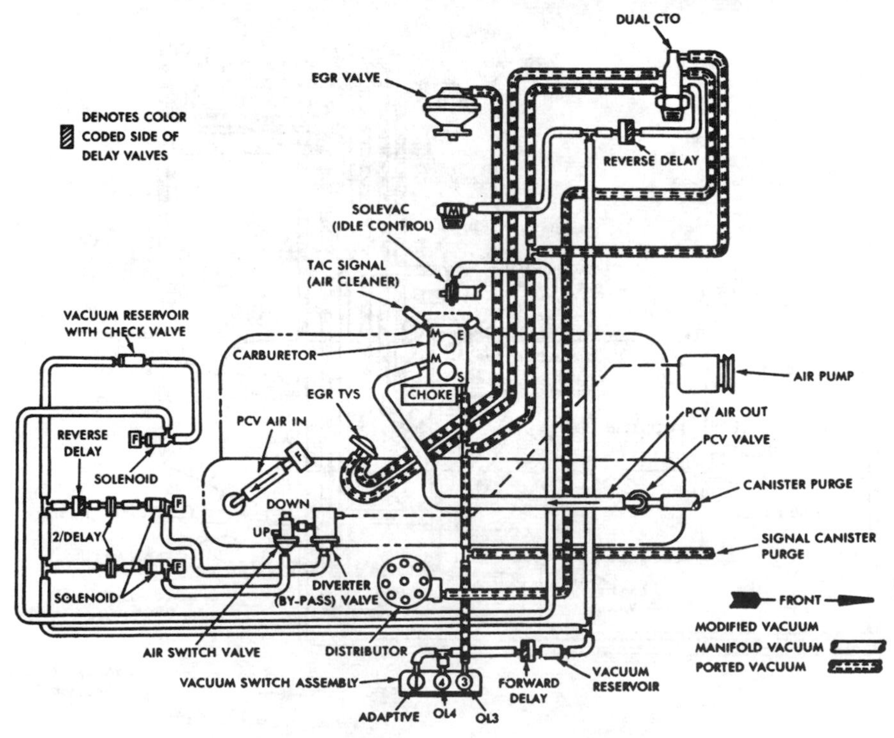

1981—258 ENG.-2BBL CARB.-W/MAN. TRANS.-EAGLE-CALIFORNIA

VACUUM CIRCUITS

(© American Motors Corp.)

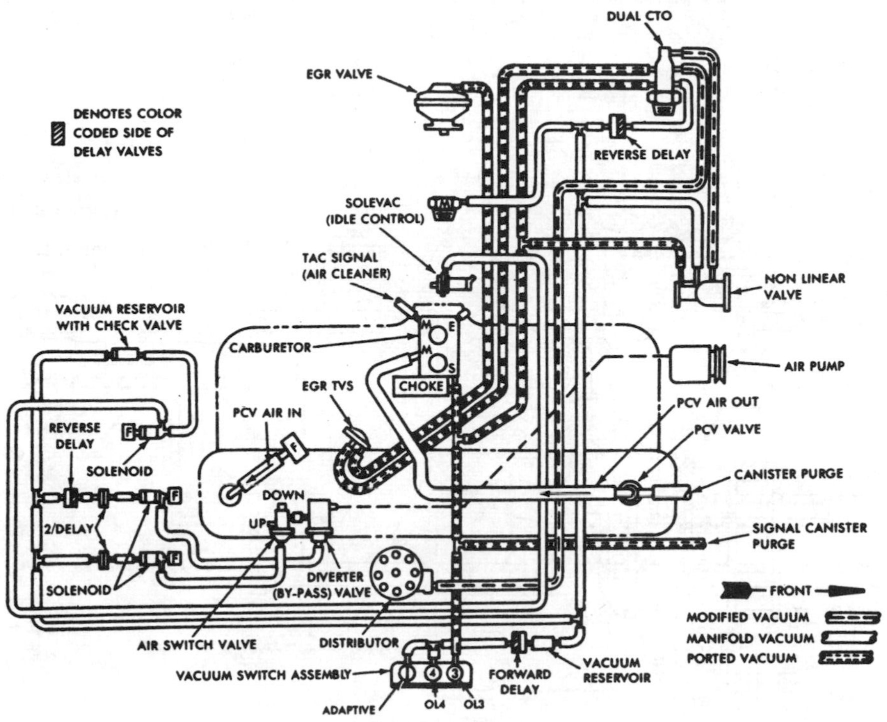

1981—258 ENG.-2BBL CARB.-W/AUTO. TRANS.-EAGLE-CALIFORNIA

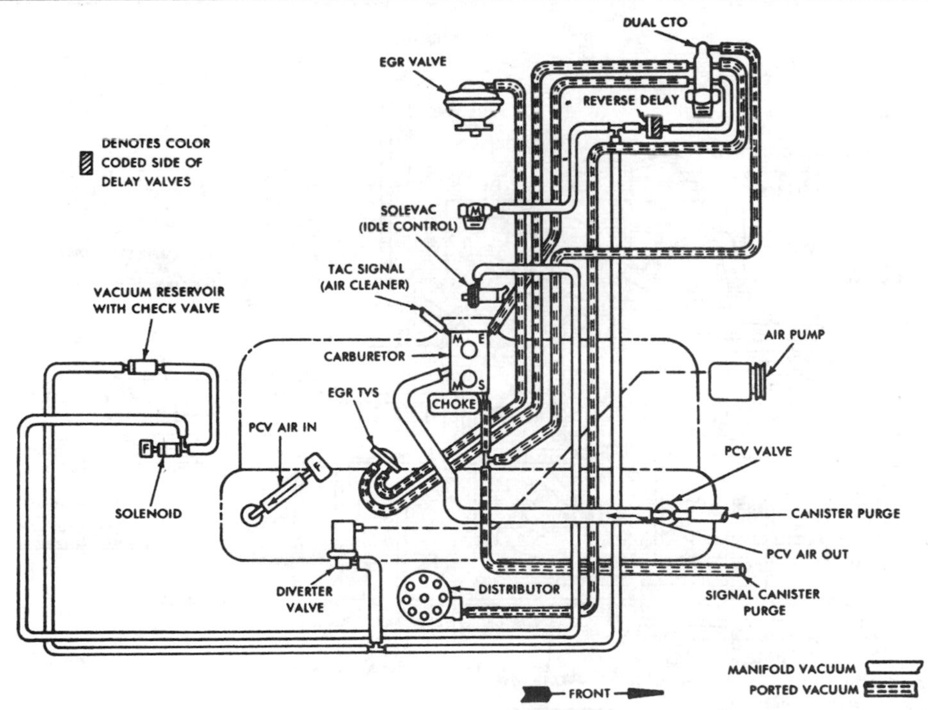

1981—258 ENG.-W/AUTO. TRANS.-EAGLE-HIGH ALT.

VACUUM CIRCUITS

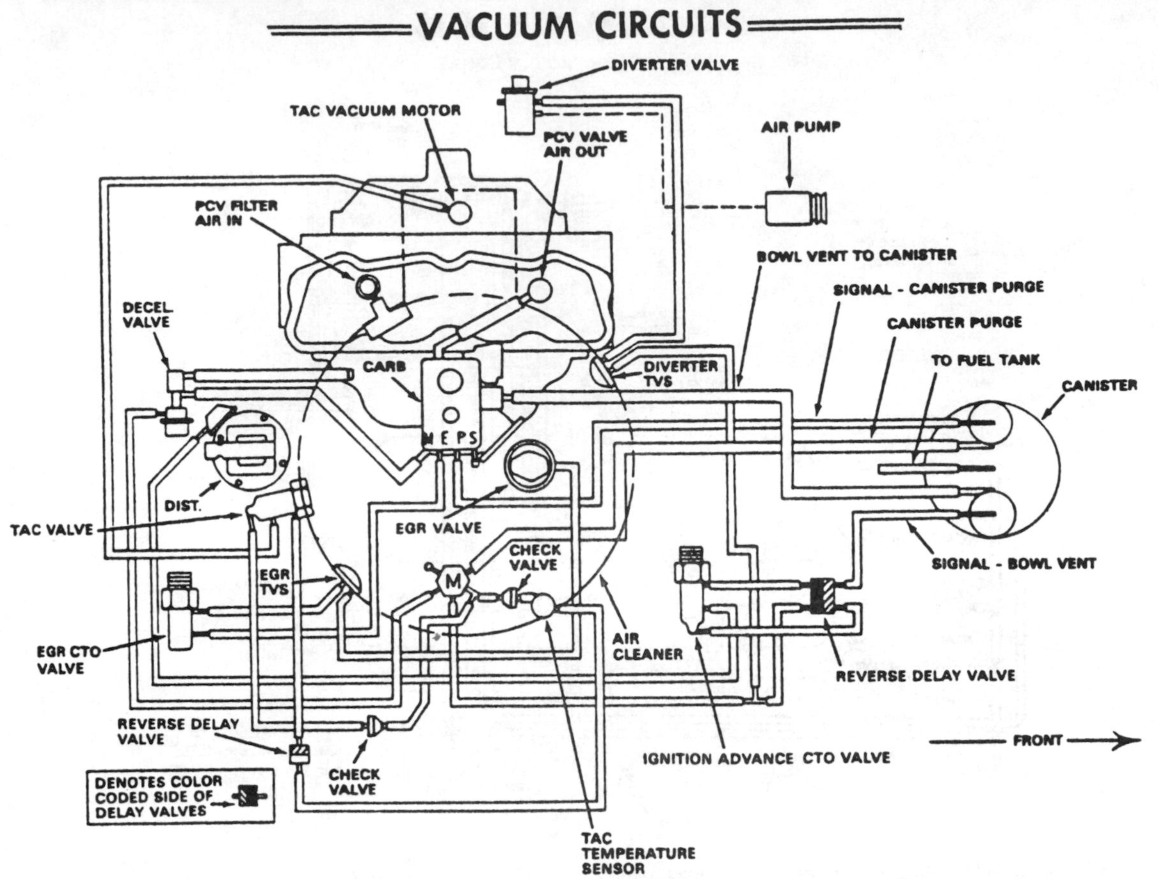

1982—151 4 CYL. ENG.—2 BBL CARB.—W/MAN. TRANS.—WO/A.C. SPIRIT, CONCORD—49-STATE, CANADA, HIGH ALTITUDE

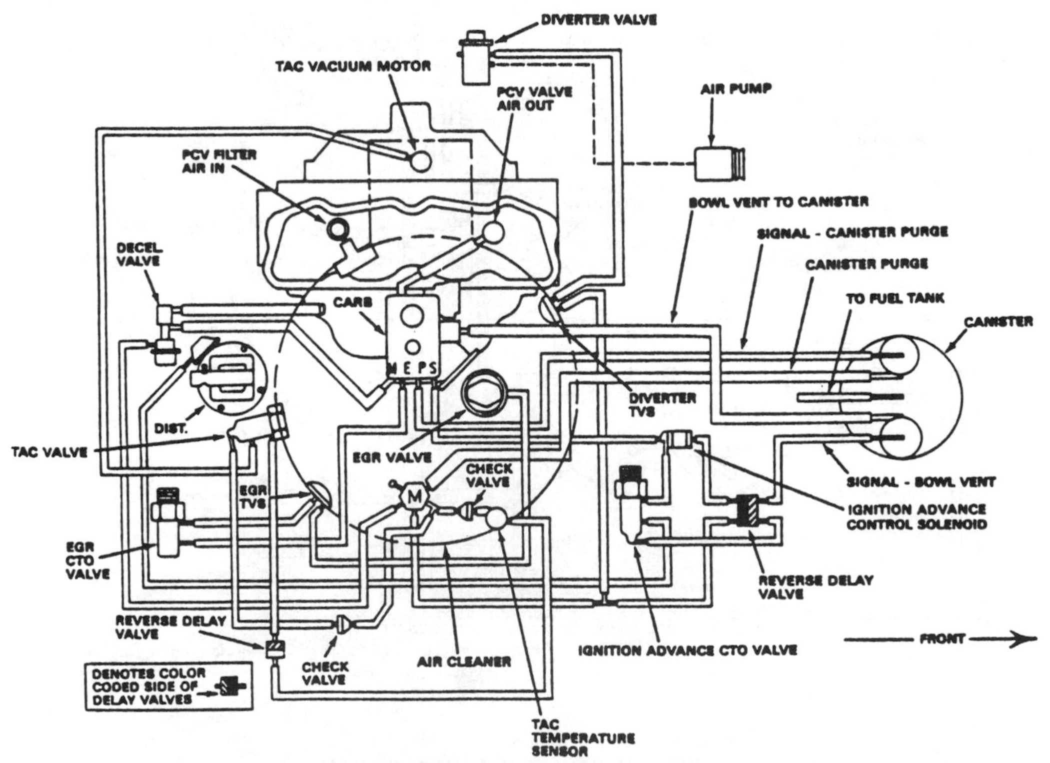

1982—151 4 CYL. ENG.—2 BBL CARB.—W/MAN. TRANS.—W/A.C. SPIRIT, CONCORD—49-STATE, CANADA, HIGH ALTITUDE

VACUUM CIRCUITS

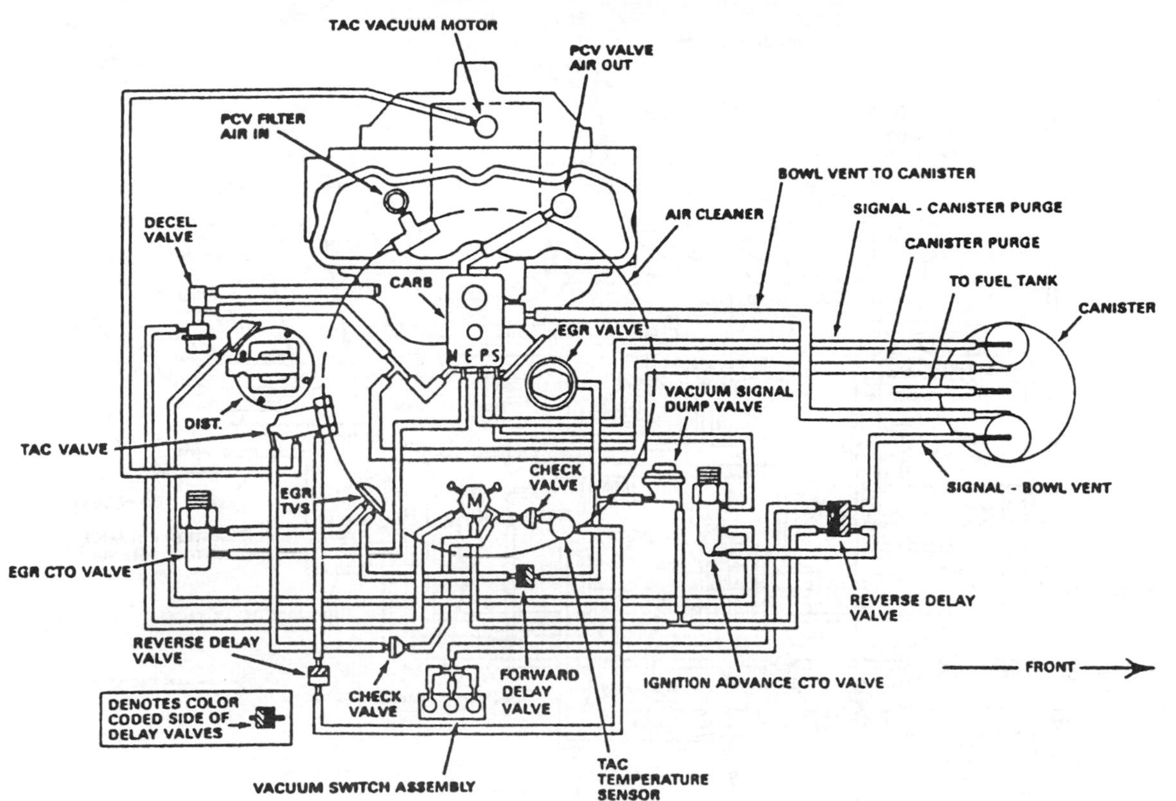

1982—151 4 CYL. ENG.—2 BB CARB.—W/AUTO TRANS.—WO/A.C. SPIRIT, CONCORD—49 STATE, CANADA

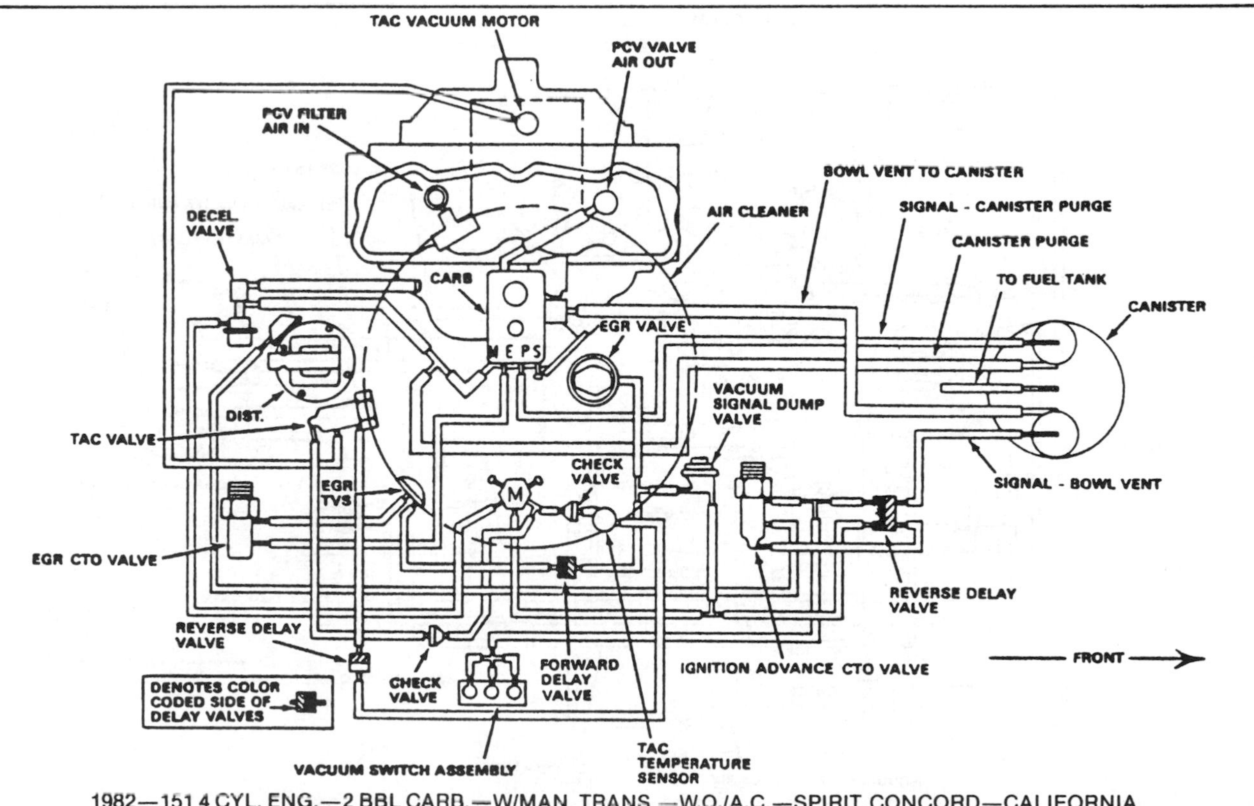

1982—151 4 CYL. ENG.—2 BBL CARB.—W/MAN. TRANS.—W.O./A.C.—SPIRIT, CONCORD—CALIFORNIA

VACUUM CIRCUITS

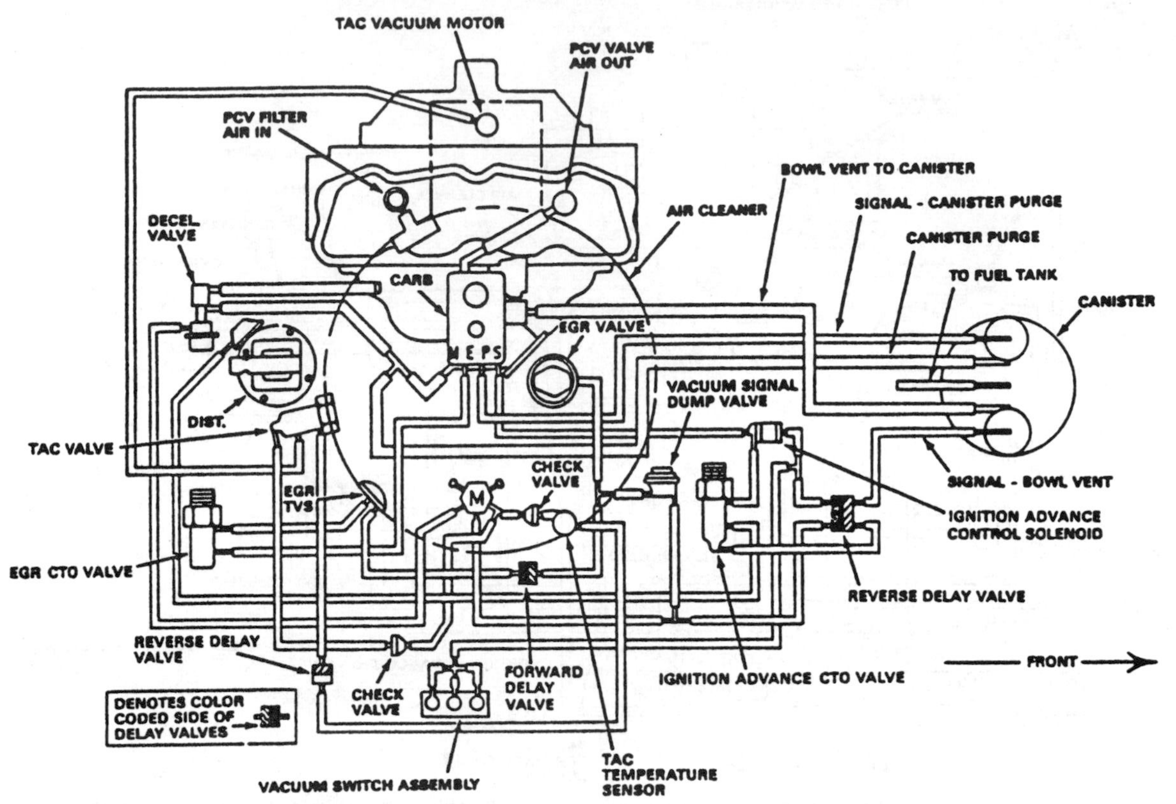

1982—151 4 CYL. ENG.—2 BBL CARB.—W/MAN. TRANS.—W./A.C.—SPIRIT, CONCORD-CALIFORNIA

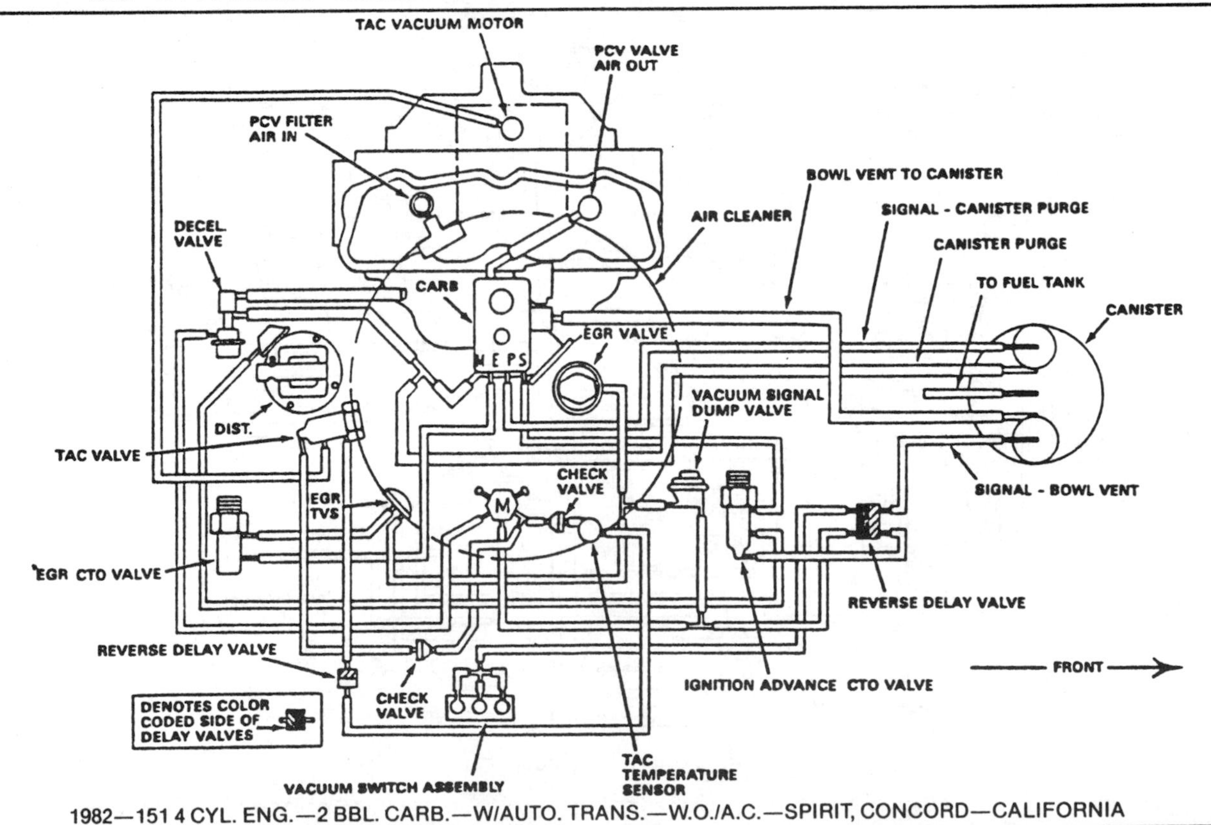

1982—151 4 CYL. ENG.—2 BBL. CARB.—W/AUTO. TRANS.—W.O./A.C.—SPIRIT, CONCORD—CALIFORNIA

VACUUM CIRCUITS

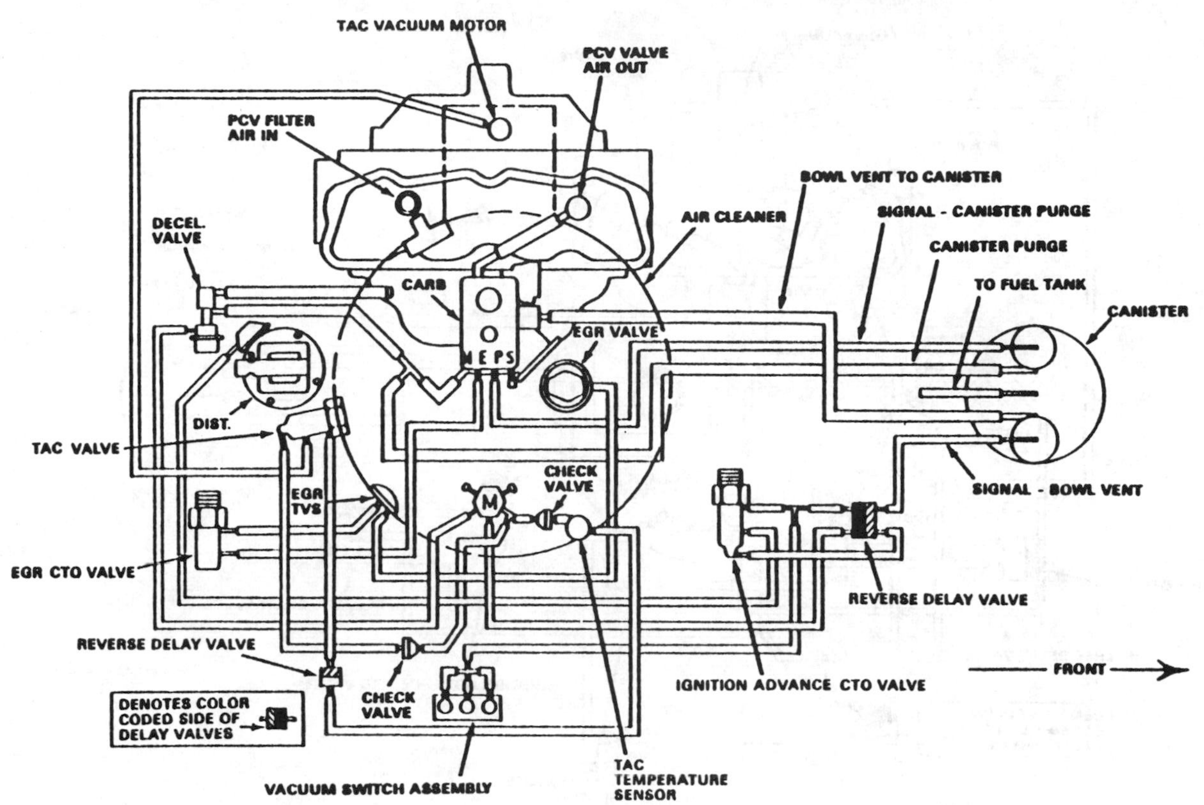

1982—151 4 CYL. ENG.—2 BBL. CARB.—W/MAN. TRANS.—WO./A.C.—EAGLE, EAGLE SX4—CALIFORNIA

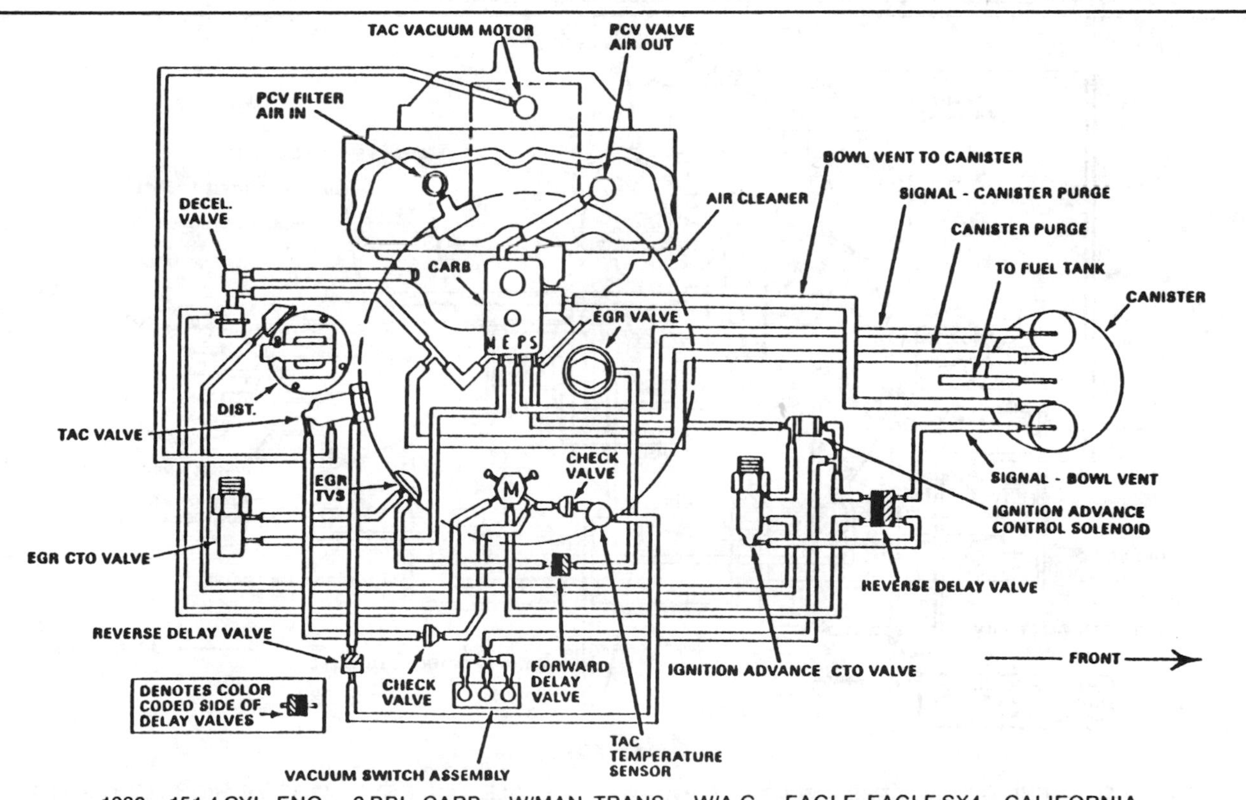

1982—151 4 CYL. ENG.—2 BBL. CARB.—W/MAN. TRANS.—W/A.C.—EAGLE, EAGLE SX4—CALIFORNIA

VACUUM CIRCUITS

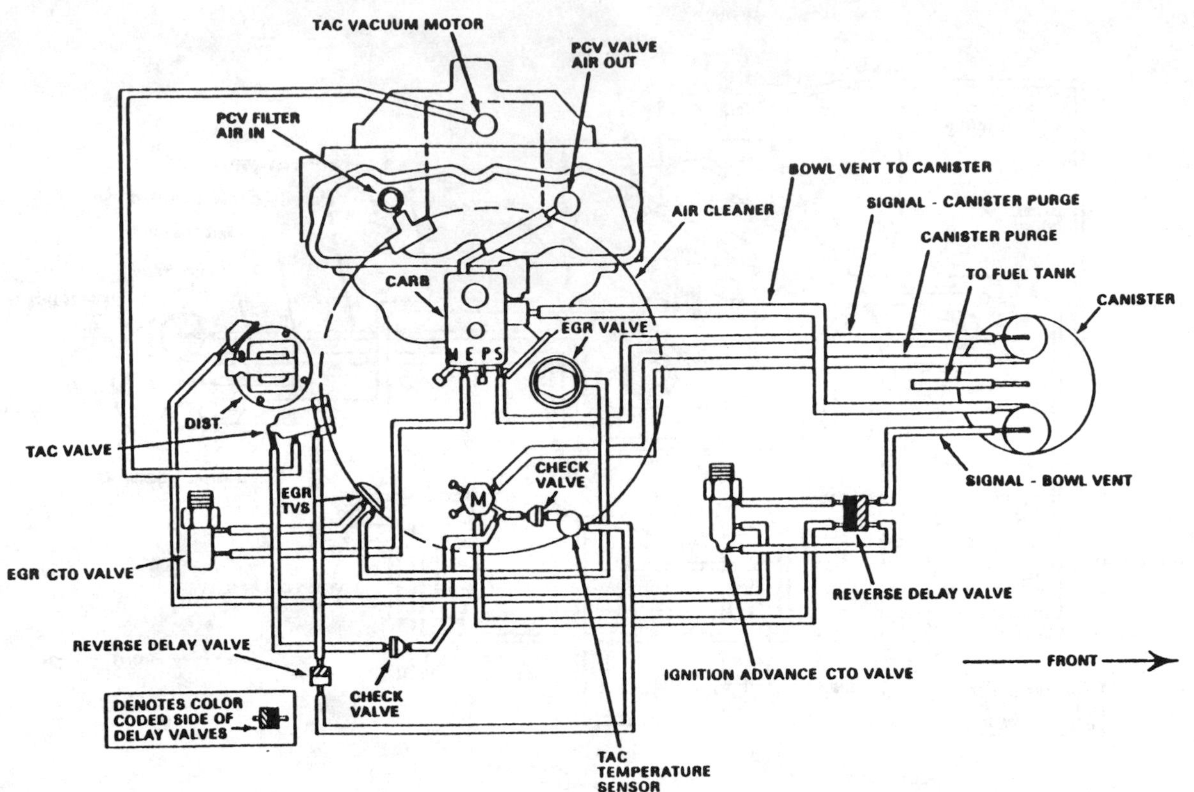

1982—151 4 CYL. ENG.—2 BBL CARB.—W/MAN. TRANS.—WO/A.C.—EAGLE, EAGLE SX4—49 STATE, HIGH ALT.
1982—151 4 CYL. ENG.—2 BBL CARB—W/AUTO. TRANS.—WO/AC.—EAGLE SX4—50 STATE

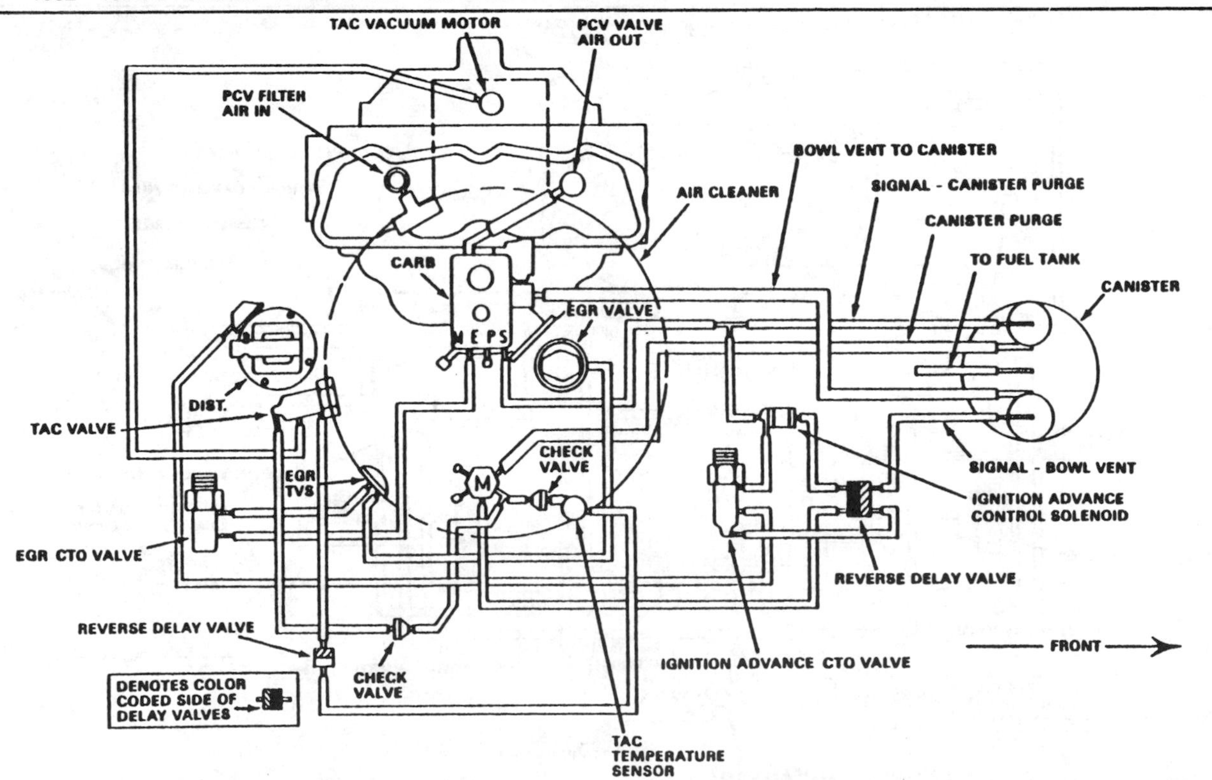

1982—151 4 CYL. ENG.—2 BBL. CARB.—W/MAN. TRANS.—W/AC—EAGLE, EAGLE SX4—49 STATE, HIGH ALT., CANADA.
1982—151 4 CYL. ENG.—2 BBL. CARB.—W/AUTO. TRANS.—W/A.C.—EAGLE SX4—49 STATE, CANADA

VACUUM CIRCUITS

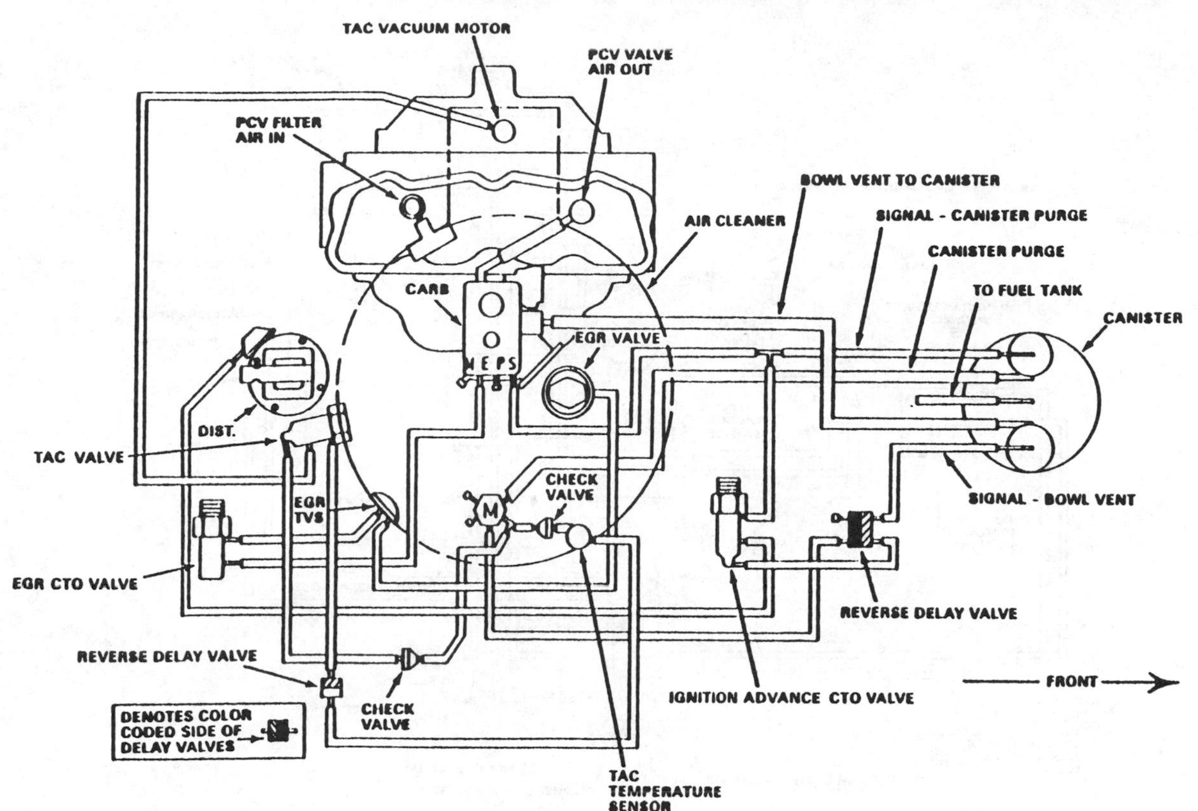

1982—151 4 CYL. ENG.—2 BBL CARB.—W/MAN. TRANS.—WO/A.C.—EAGLE, EAGLE SX4—CANADA.
1982—151 4 CYL. ENG.—2 BBL. CARB.—W/AUTO. TRANS.—WO/A.C.—EAGLE SX4—CANADA.

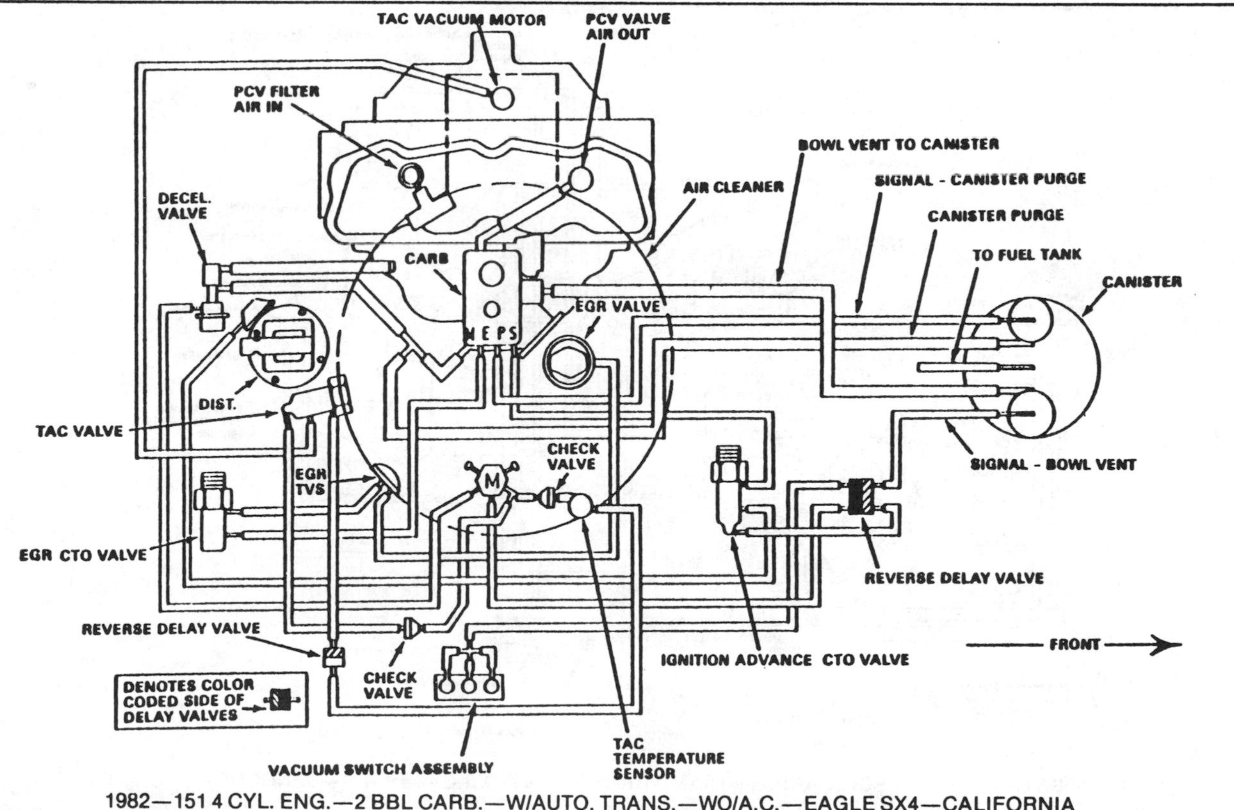

1982—151 4 CYL. ENG.—2 BBL CARB.—W/AUTO. TRANS.—WO/A.C.—EAGLE SX4—CALIFORNIA

VACUUM CIRCUITS

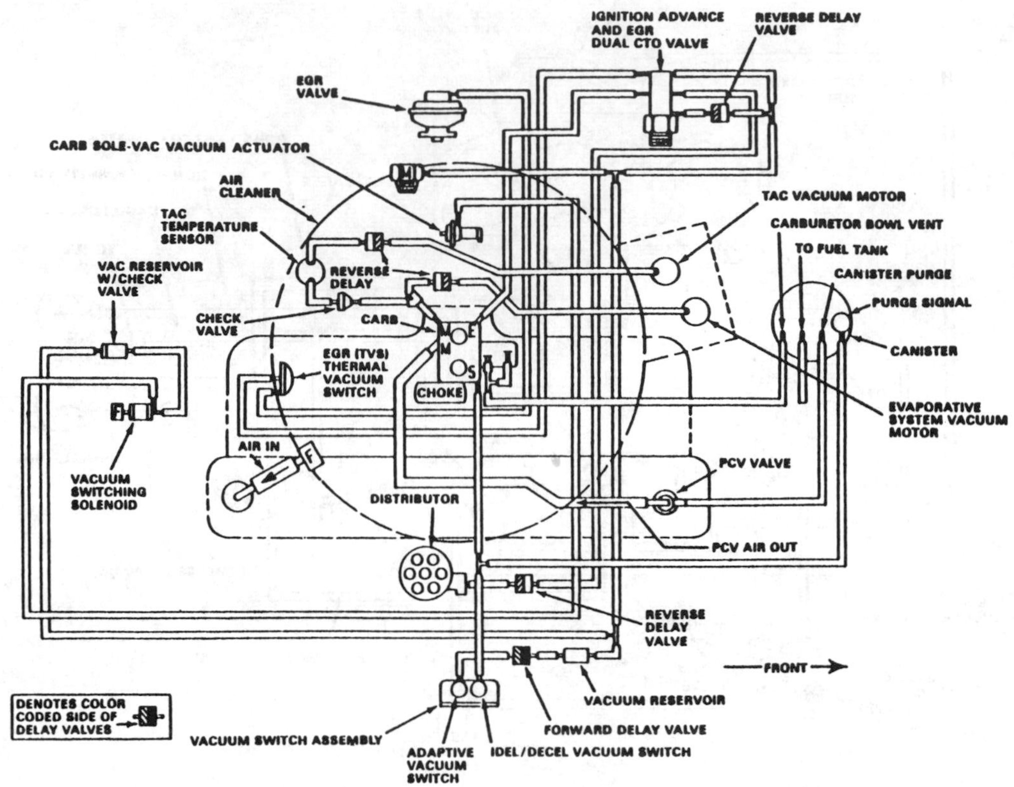

1982—258 6 CYL. ENG.—2 BBL CARB—W/MAN. TRANS.—W & WO/A.C.—EAGLE, EAGLE SX4—CANADA

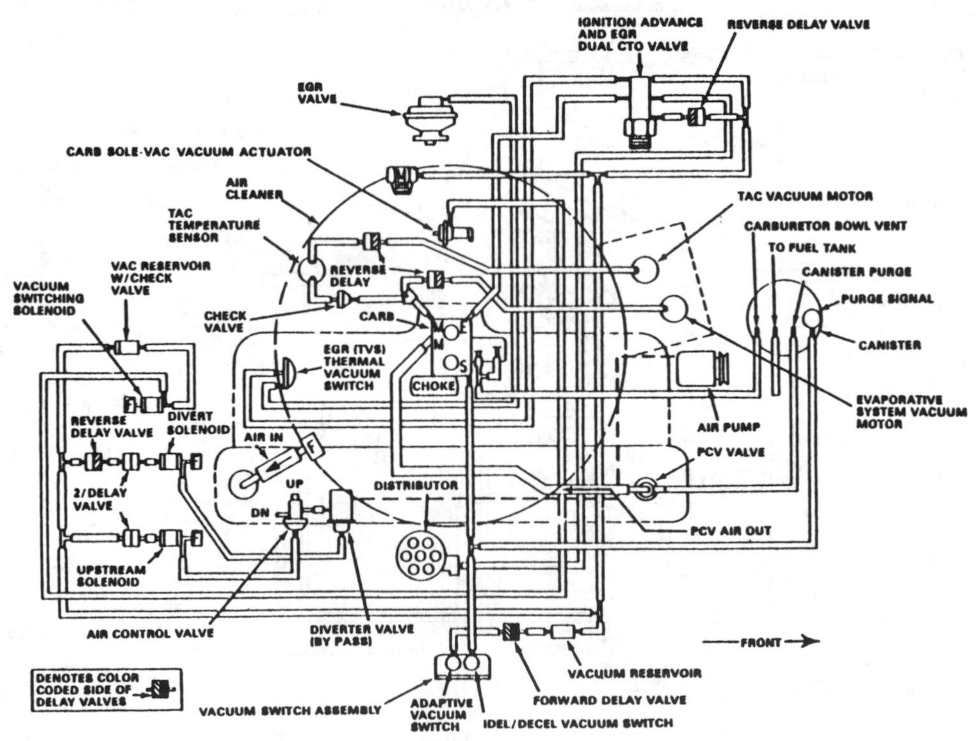

1982—258 6 CYL. ENG.—2 BBL. CARB.—W/MAN. TRANS.—W & WO/A.C.—SPIRIT—CONCORD—50 STATE—
EAGLE, EAGLE SX4—HIGH ALTITUDE.

VACUUM CIRCUITS

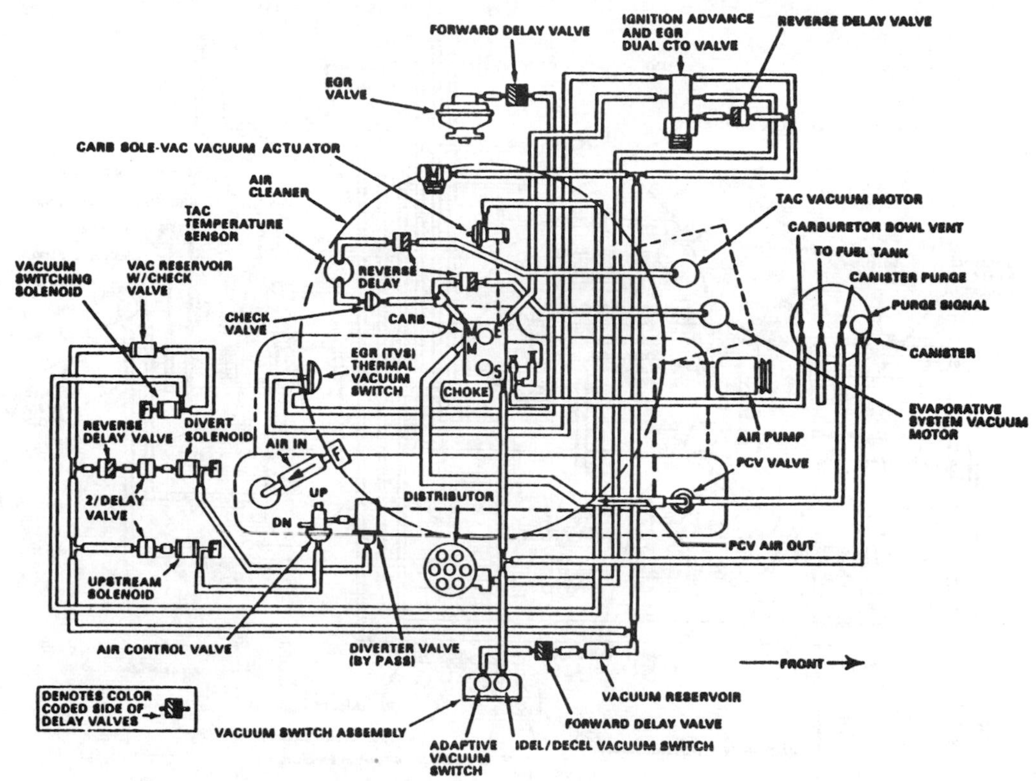

1982—258 6 CYL. ENG.—2 BBL CARB.—W/MAN. TRANS.—W & WO/A.C.—EAGLE, EAGLE SX4—49 STATE

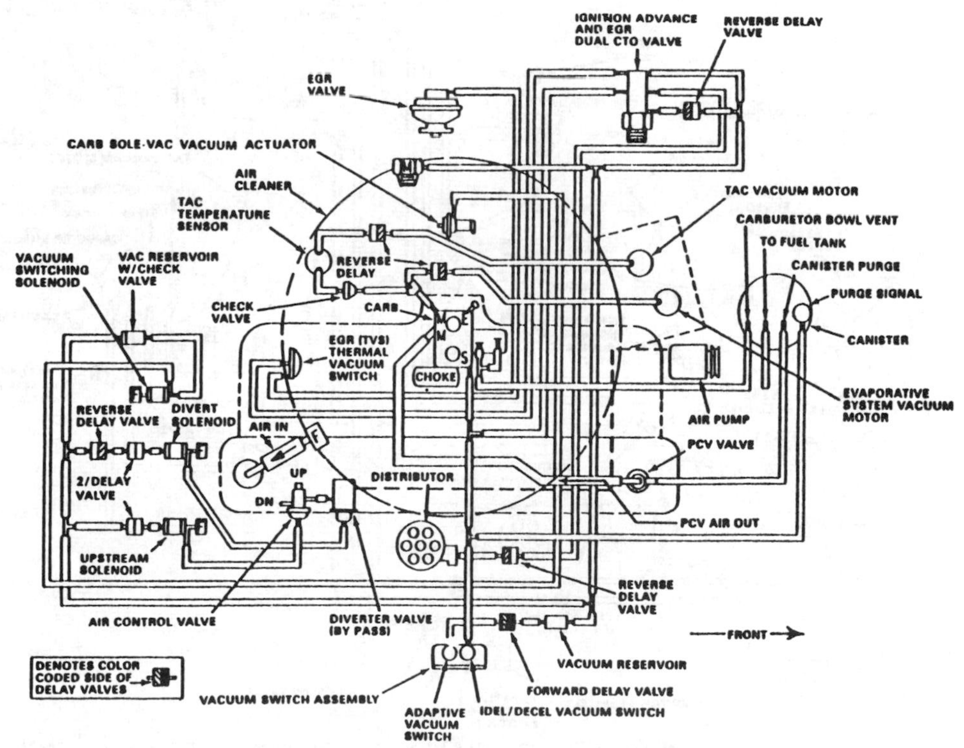

1982—258 6 CYL. ENG.—2 BBL. CARB.—W/AUTO. TRANS.—W & WO/A.C.—SPIRIT, CONCORD—CALIFORNIA

VACUUM CIRCUITS

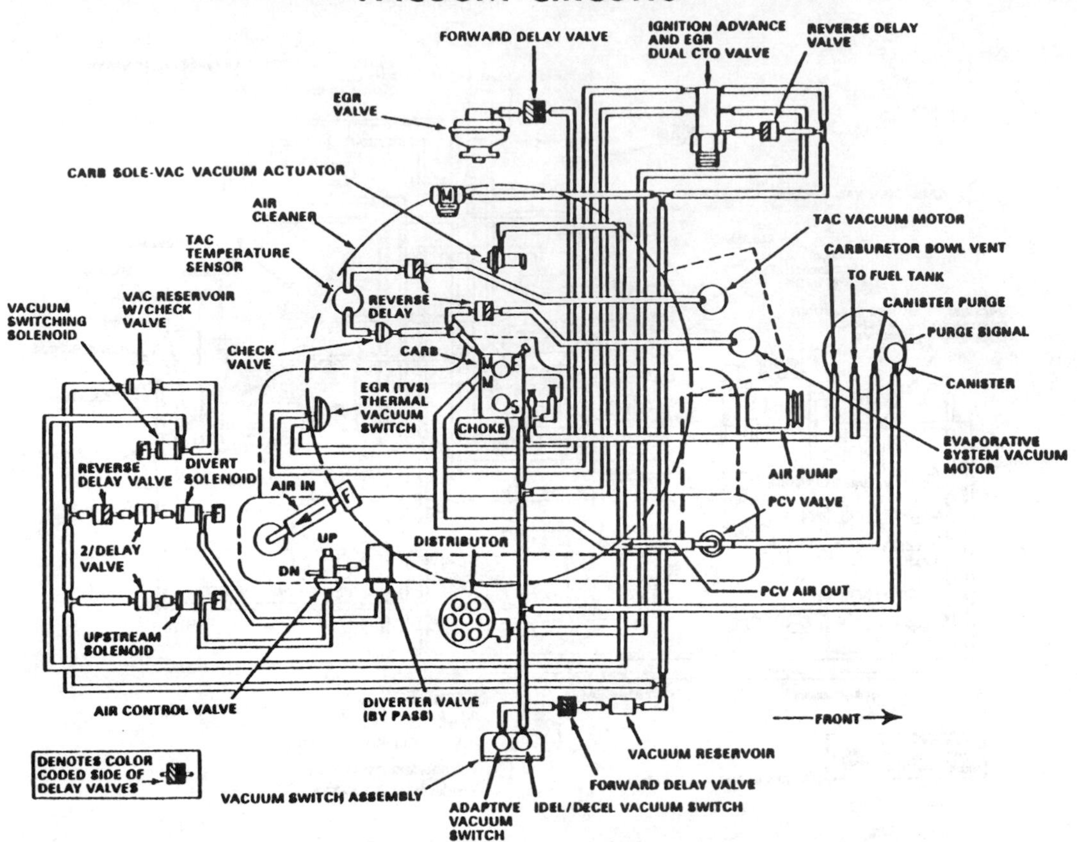

1982—258 6 CYL. ENG.—2 BBL CARB.—W/AUTO. TRANS.—W & WO/A.C.—EAGLE, EAGLE SX4—CALIFORNIA

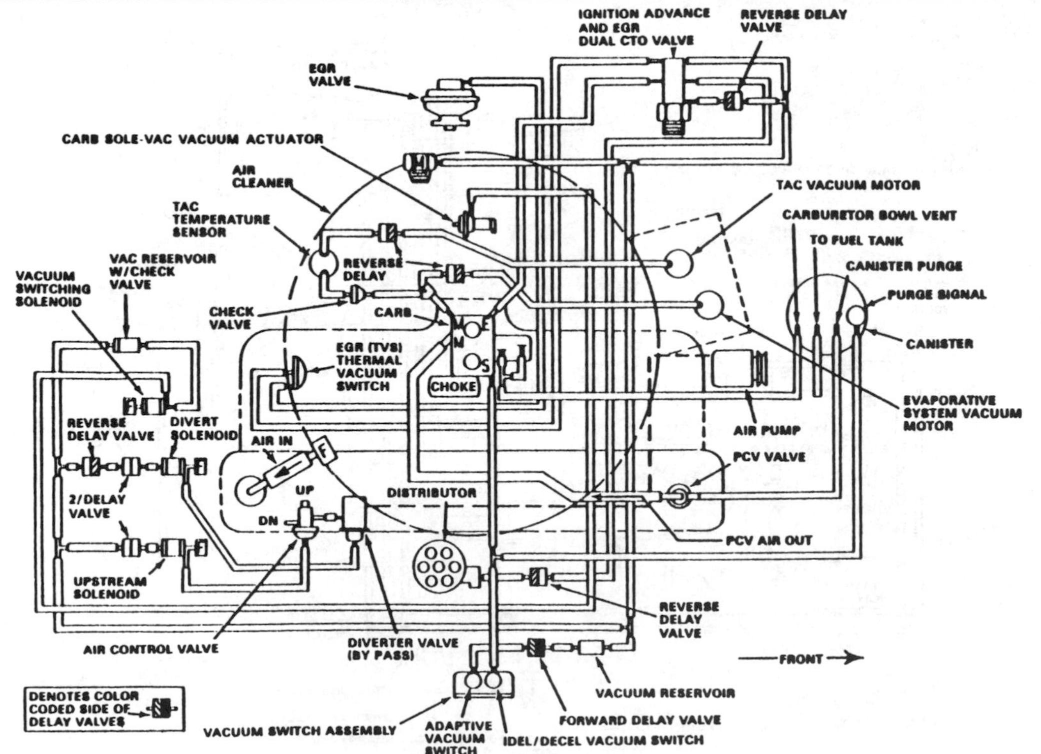

1982—258 6 CYL. ENG.—2 BBL CARB.—W/AUTO. TRANS.—W & WO/AC—SPIRIT, CONCORD, EAGLE, EAGLE SX4—49 STATE, CANADA, HIGH ALT.—EAGLE, EAGLE SX4—CALIFORNIA

VACUUM CIRCUITS

(© American Motors Corp.)

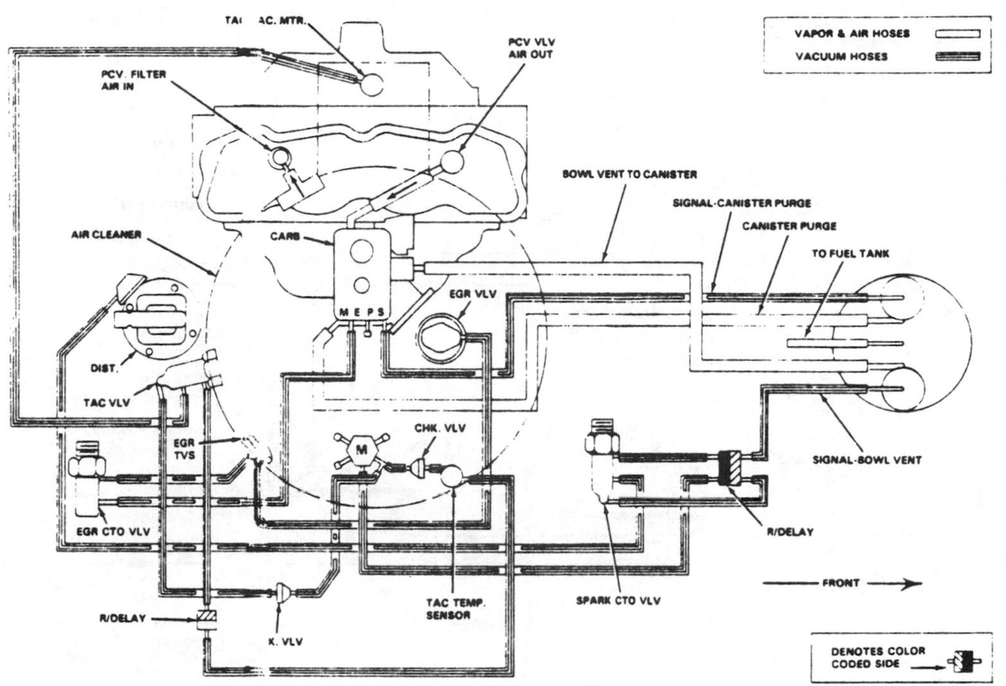

151 4-CYL. 2V, MANUAL TRANSMISSION, W/O AIR CONDITIONING, 49-STATE–CANADA–HIGH ALTITUDE, EAGLE-EAGLE SX/4

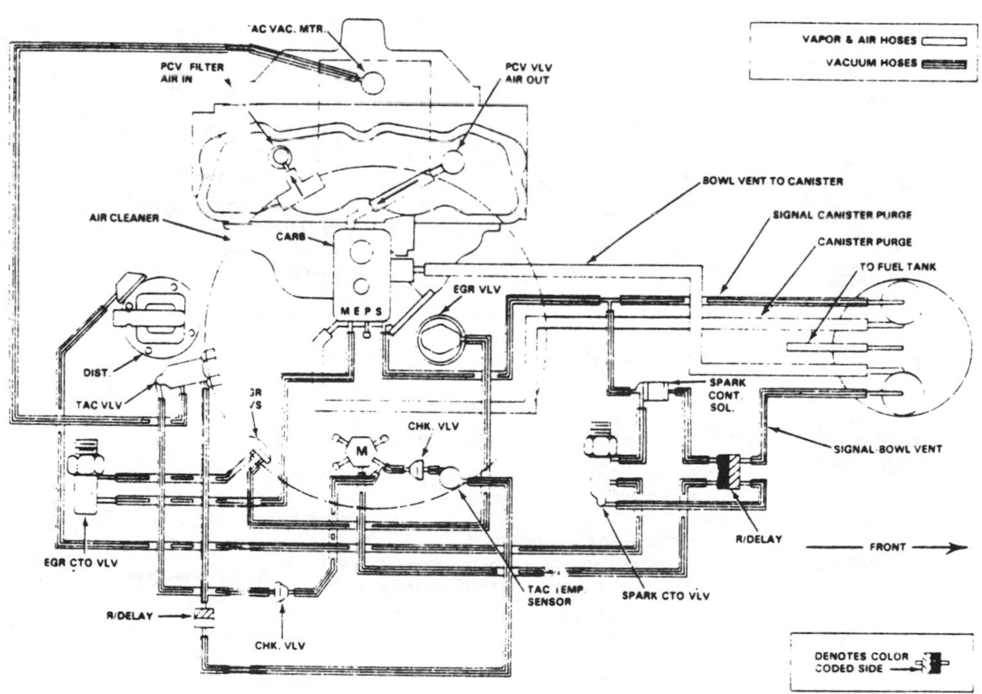

151 4-CYL. 2V, MANUAL TRANSMISSION, W/ AIR CONDITIONING, 49-STATE–CANADA–HIGH ALTITUDE, EAGLE-EAGLE SX/4

VACUUM CIRCUITS

(© American Motors Corp.)

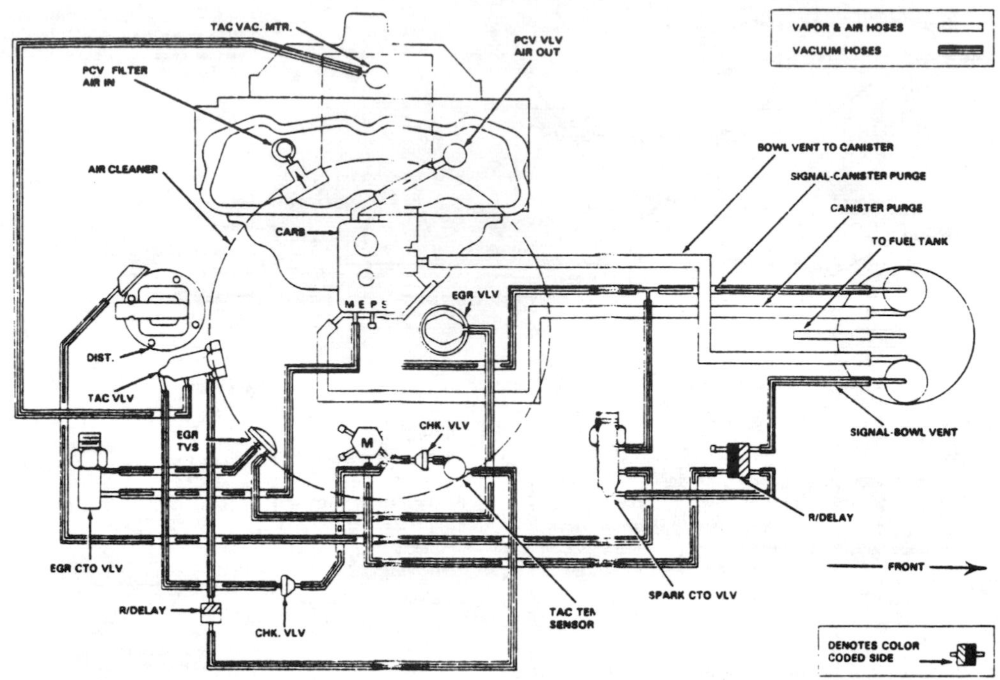

151 4-CYL. 2V, MANUAL & AUTOMATIC TRANSMISSION, W/ & W/O AIR CONDITIONING, 49-STATE—CANADA—HIGH ALTITUDE, EAGLE-EAGLE SX/4

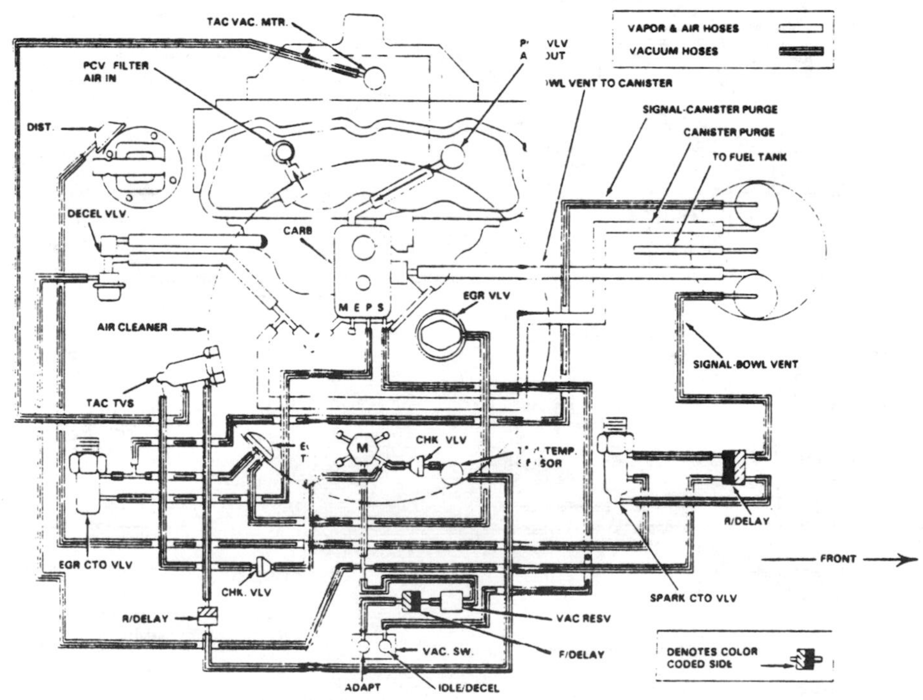

151 4-CYL. 2V, MANUAL TRANSMISSION, W/O AIR CONDITIONING, CALIFORNIA, EAGLE-EAGLE SX/4

VACUUM CIRCUITS

(© American Motors Corp.)

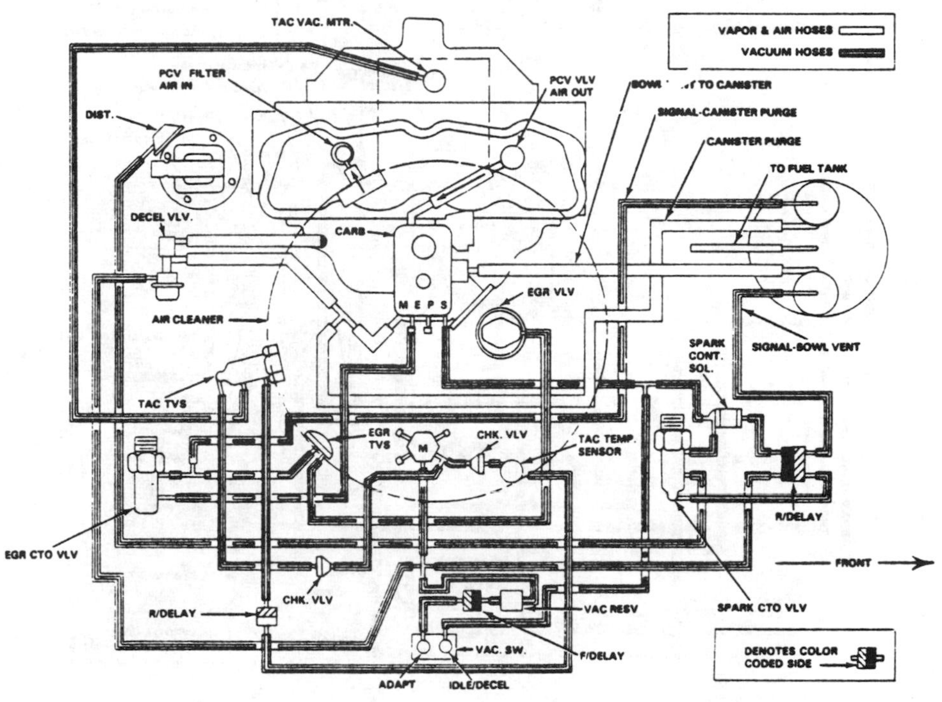

151 4-CYL. 2V, MANUAL TRANSMISSION, W/ AIR CONDITIONING, CALIFORNIA, EAGLE-EAGLE SX/4

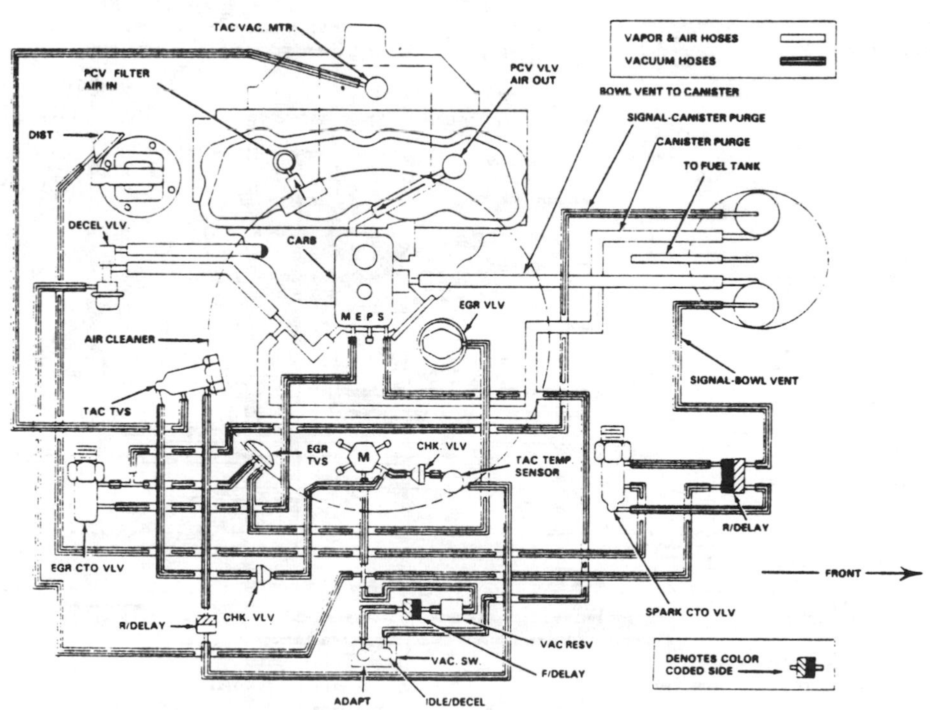

151 4-CYL. 2V, AUTOMATIC TRANSMISSION, W/O AIR CONDITIONING, CALIFORNIA, EAGLE SX/4

VACUUM CIRCUITS

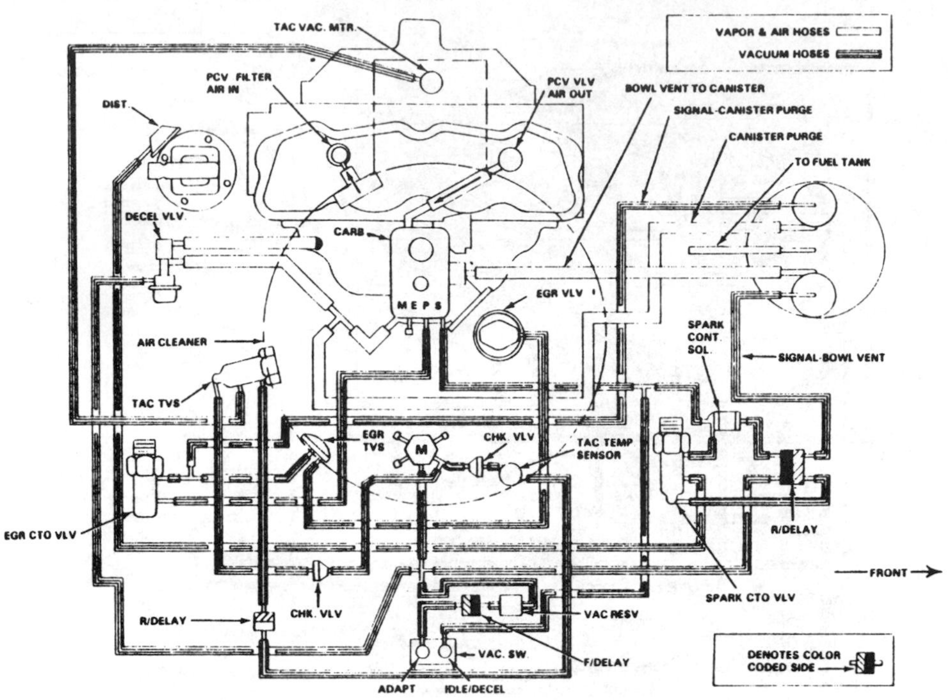

151 4-CYL. 2V, AUTOMATIC TRANSMISSION, W/ AIR CONDITIONING, CALIFORNIA, EAGLE SX/4

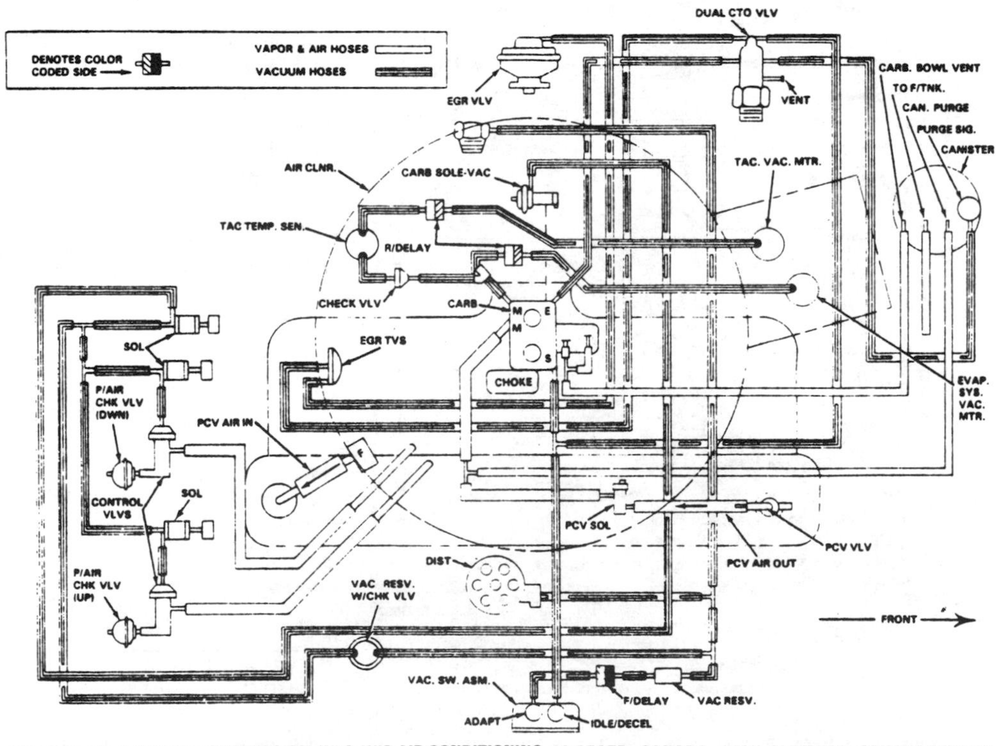

258 6-CYL. 2V, MANUAL TRANSMISSION, W/ & W/O AIR CONDITIONING, 49-STATE—CANADA—HIGH ALTITUDE, SPIRIT-CONCORD

VACUUM CIRCUITS

(© American Motors Corp.)

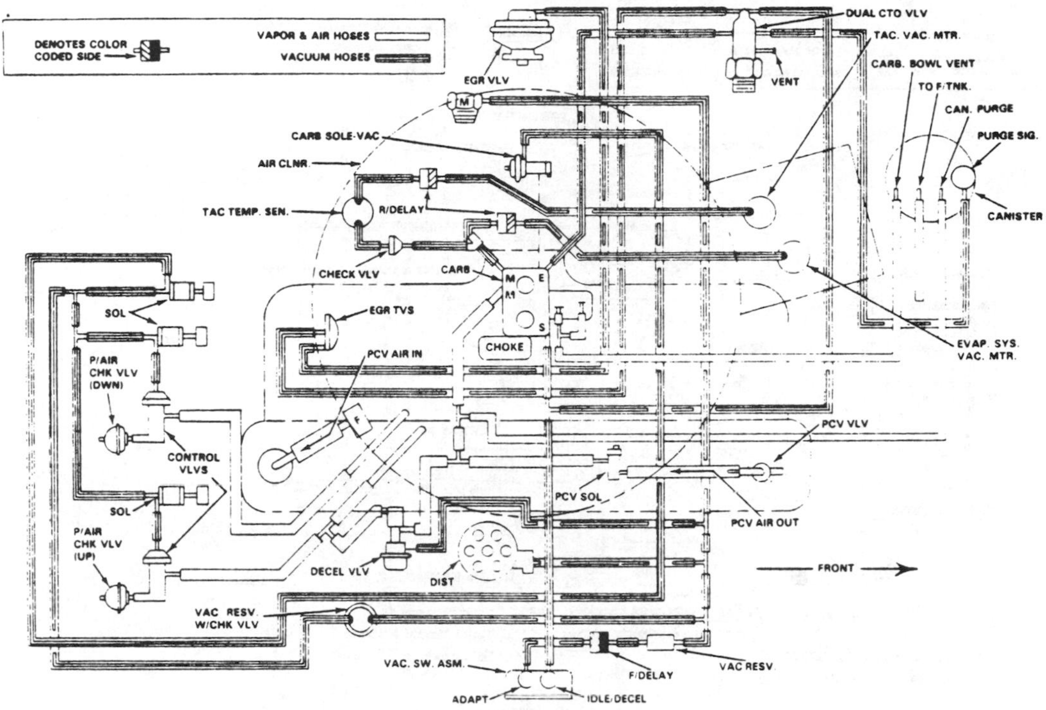

258 6-CYL. 2V, AUTOMATIC TRANSMISSION, W/ & W/O AIR CONDITIONING, 49-STATE—CANADA—HIGH ALTITUDE, SPIRIT-CONCORD

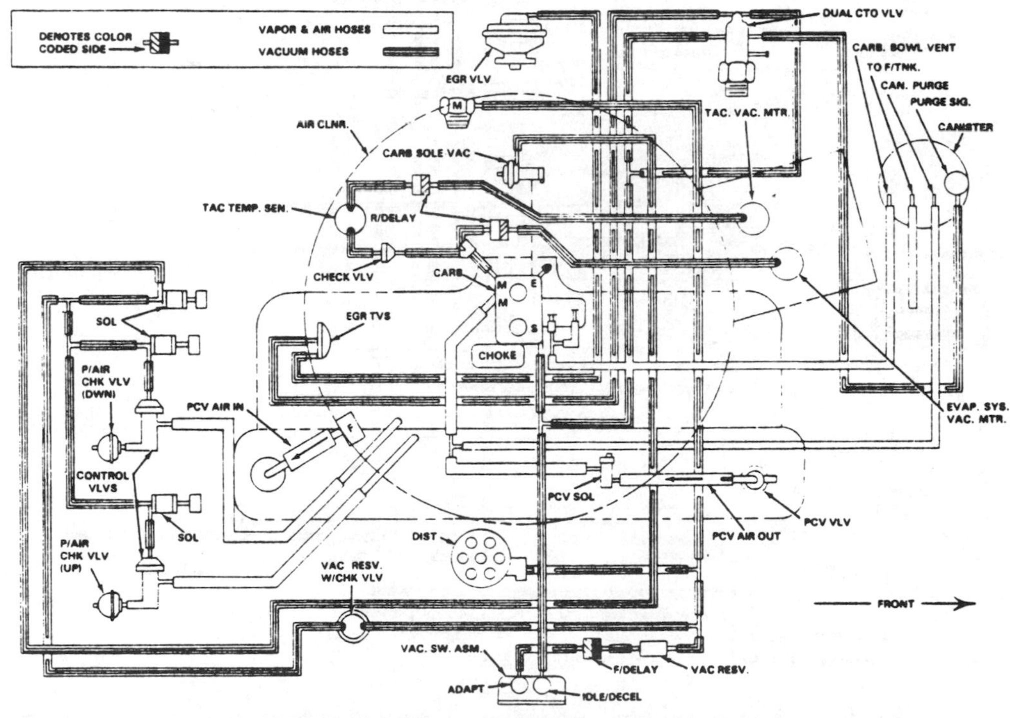

258 6-CYL. 2V, AUTOMATIC TRANSMISSION, W/ & W/O AIR CONDITIONING, CALIFORNIA, EAGLE-EAGLE SX/4 – SPIRIT-CONCORD

VACUUM CIRCUITS

(© American Motors Corp.)

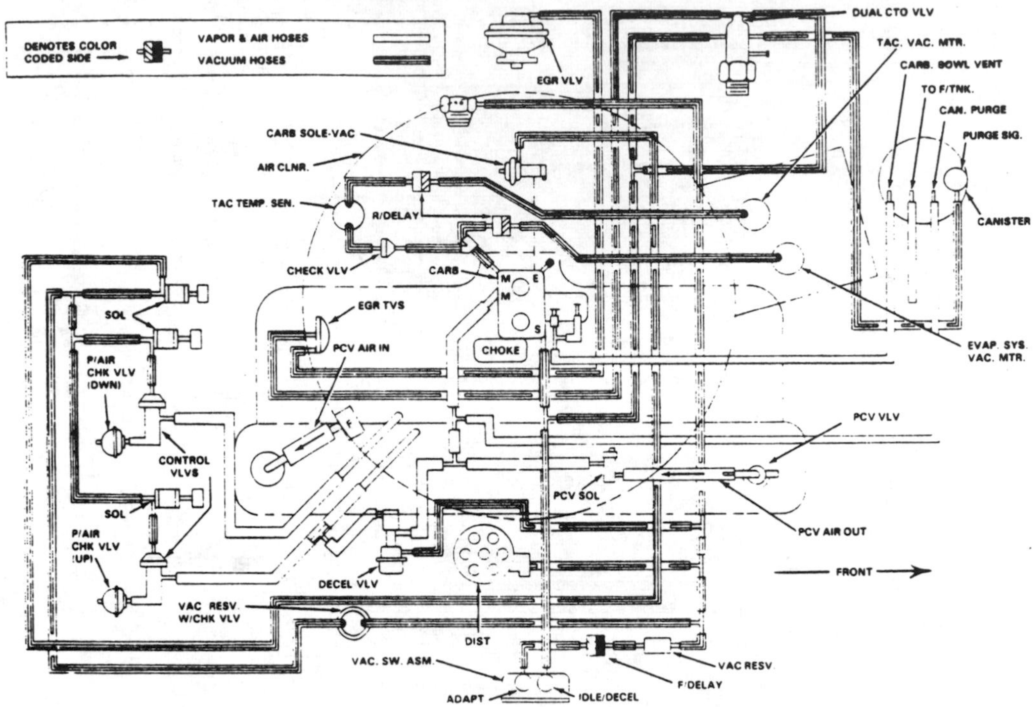

258 6-CYL. 2V, MANUAL TRANSMISSION, W/ & W/O AIR CONDITIONING, CALIFORNIA, EAGLE-EAGLE SX/4 – SPIRIT-CONCORD

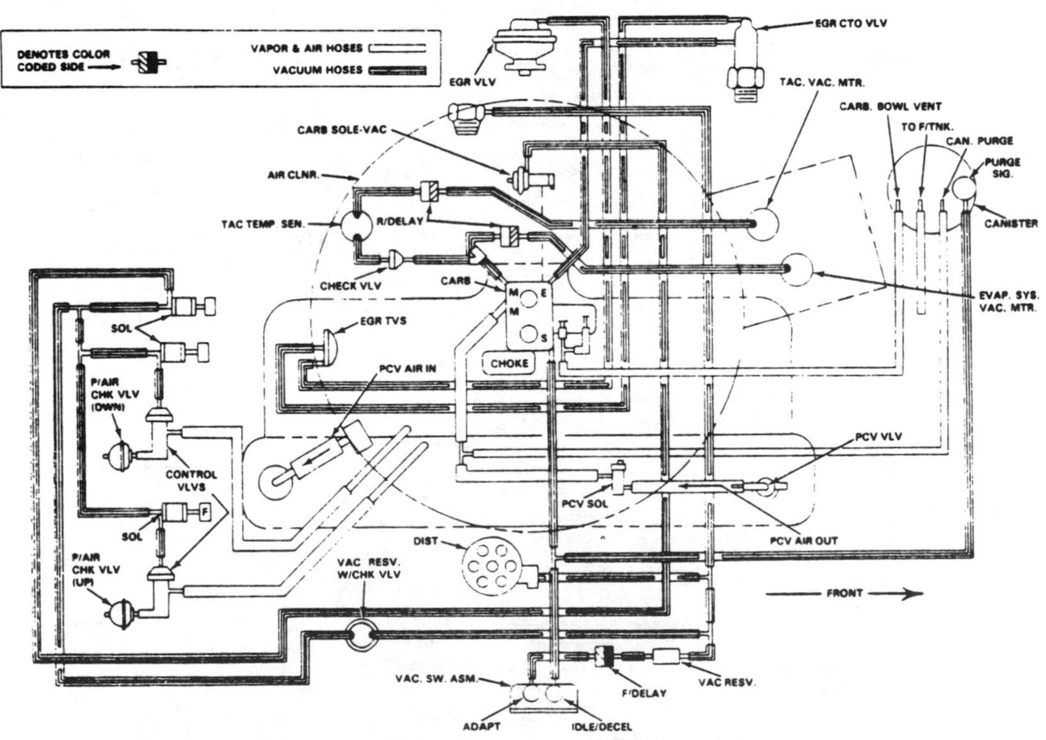

258 6-CYL. 2V, MANUAL & AUTOMATIC TRANSMISSION, W/ & W/O AIR CONDITIONING, 49-STATE – HIGH ALTITUDE, EAGLE-EAGLE SX/4

VACUUM CIRCUITS

(© American Motors Corp.)

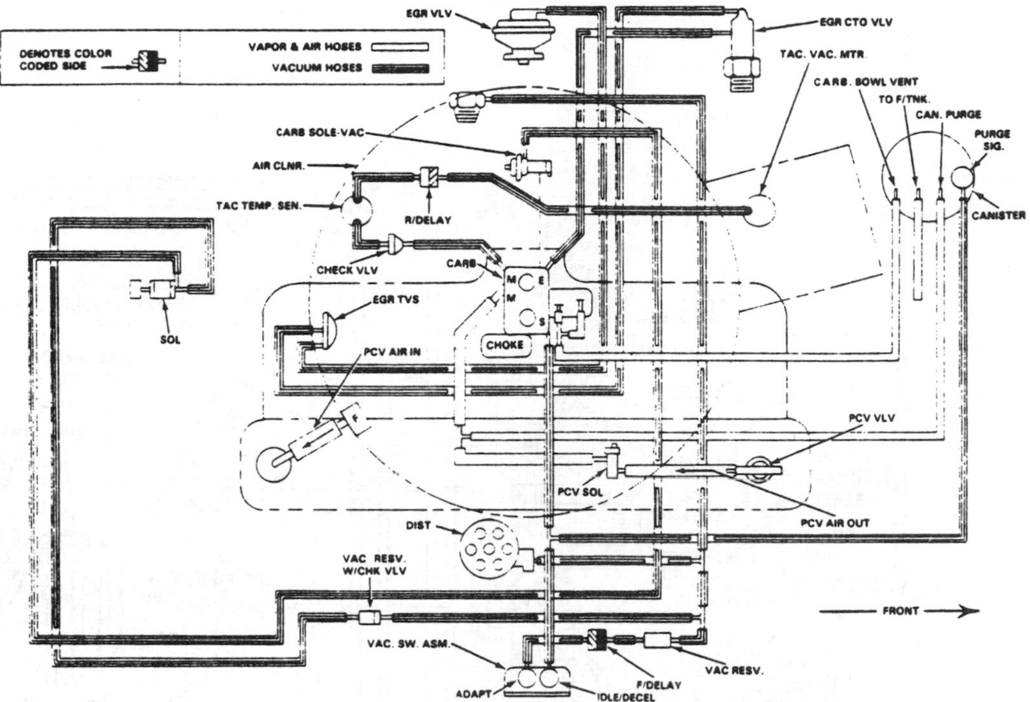

258 6-CYL. 2V, MANUAL AND AUTOMATIC TRANSMISSION, W/ & W/O AIR CONDITIONING, CANADA, EAGLE-EAGLE SX/4

VACUUM CIRCUITS

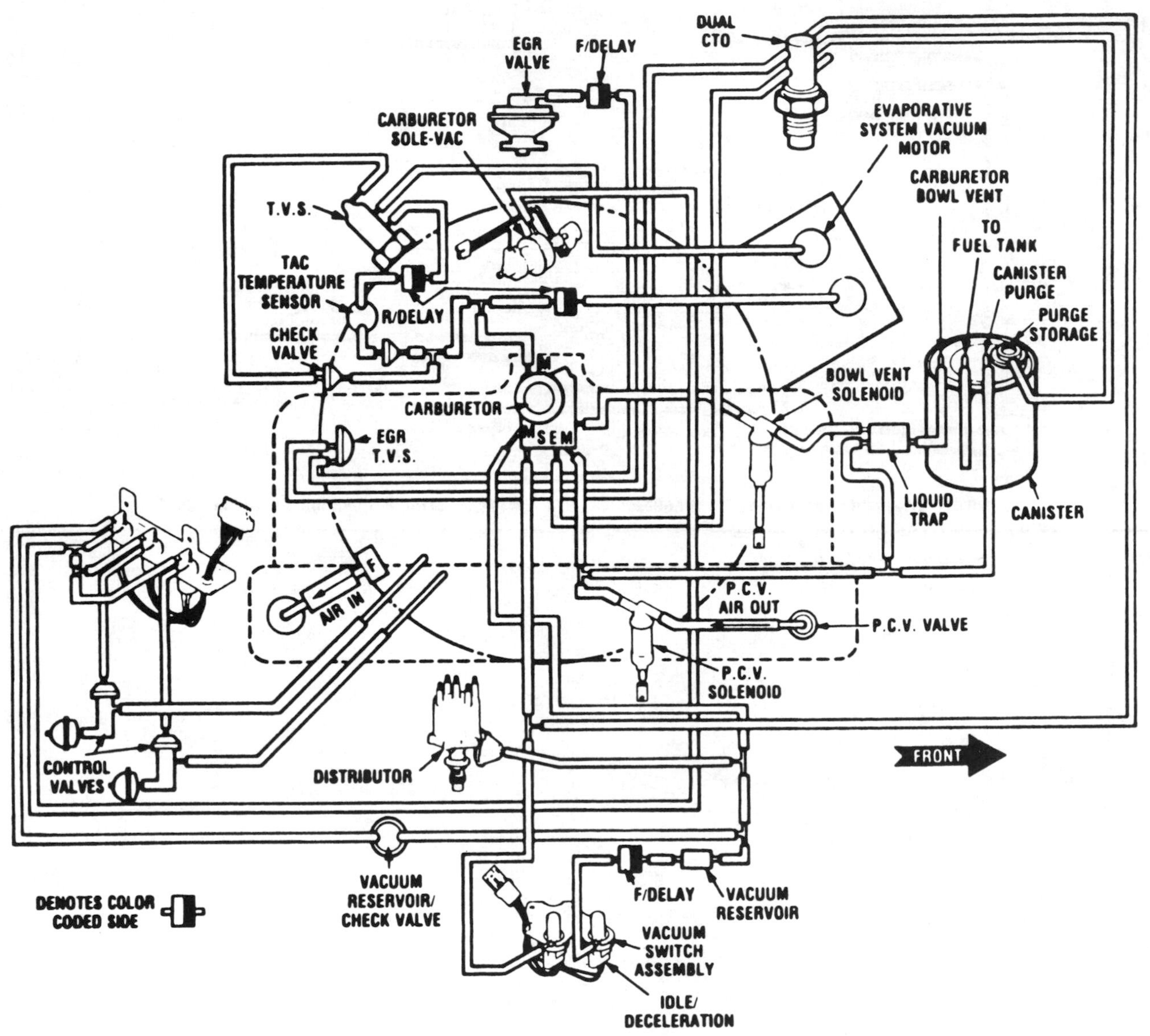

1984–85 4 CYL. SJ EAGLE

VACUUM CIRCUITS

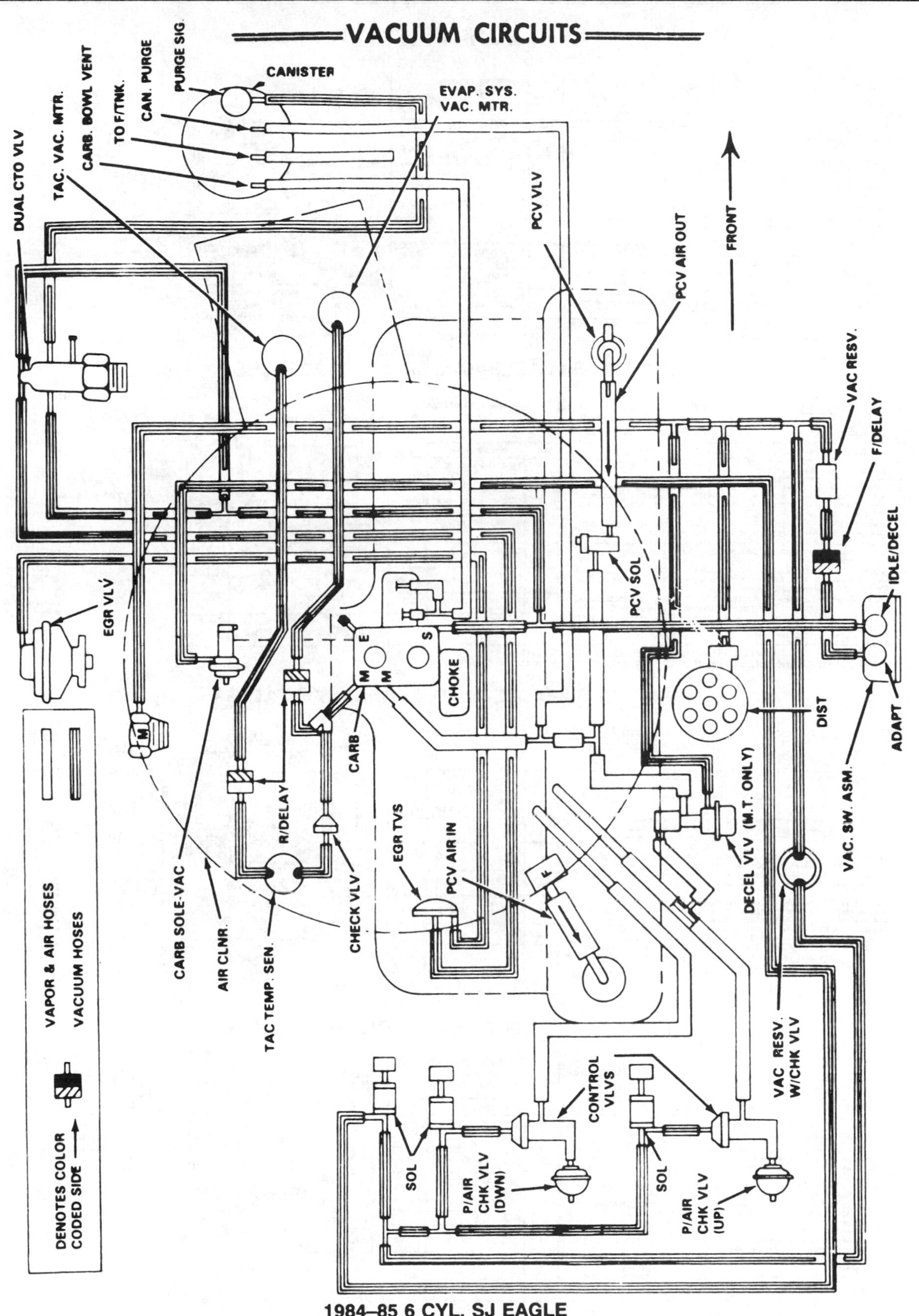

1984–85 6 CYL. SJ EAGLE

VACUUM CIRCUITS

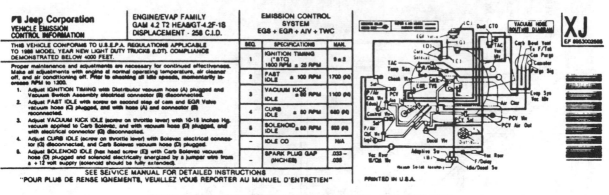

Jeep Corporation
VEHICLE EMISSION CONTROL INFORMATION

ENGINE/EVAP FAMILY
GAM 4.2 T2 HEA8/GT-4.2F-1S
DISPLACEMENT - 258 C.I.D.

EMISSION CONTROL SYSTEM
EGS + EGR + AIV + TWC

THIS VEHICLE CONFORMS TO U.S.E.P.A. REGULATIONS APPLICABLE TO 1986 MODEL YEAR NEW LIGHT DUTY TRUCKS (LDT). COMPLIANCE DEMONSTRATED BELOW 4000 FEET.

Proper maintenance and adjustments are necessary for continued effectiveness. Make all adjustments with engine at normal operating temperature, air cleaner off, and air conditioning off. Prior to checking all idle speeds, momentarily increase RPM to 1200.

1. Adjust IGNITION TIMING with Distributor vacuum hose (A) plugged and Vacuum Switch Assembly electrical connector (B) disconnected.
2. Adjust FAST IDLE with screw on second step of cam and EGR Valve vacuum hose (C) plugged, and with hose (A) and connector (B) reconnected.
3. Adjust VACUUM KICK IDLE (screw on throttle lever) with 10-15 inches Hg. vacuum applied to Carb Solevac, and with vacuum hose (D) plugged, and with electrical connector (G) disconnected.
4. Adjust CURB IDLE (screw on throttle lever) with Solevac electrical connector (G) disconnected, and Carb Solevac vacuum hose (D) plugged.
5. Adjust SOLENOID IDLE (hex head screw (E)) with Carb Solevac vacuum hose (D) plugged and solenoid electrically energized by a jumper wire from a + 12 volt supply (solenoid should be fully extended).

SEQ.	SPECIFICATIONS	MAN.
1	IGNITION TIMING (°BTC) 1600 RPM ± 25 RPM	9 ± 2
2	FAST IDLE ± 100 RPM	1700 (M)
3	VACUUM KICK IDLE ± 50 RPM	1100 (M)
4	CURB IDLE ± 50 RPM	680 (M)
5	SOLENOID IDLE ± 50 RPM	800 (M)
-	IDLE CO	N/A
-	SPARK PLUG GAP (INCHES)	.033 - .038

SEE SERVICE MANUAL FOR DETAILED INSTRUCTIONS
"POUR PLUS DE RENSE IGNEMENTS, VEUILLEZ VOUS REPORTER AU MANUEL D'ENTRETIEN"

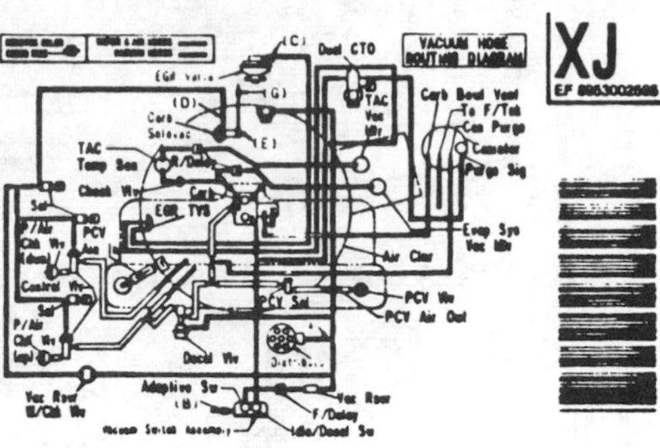

1986 6 CYL.-SJ EAGLE-MAN. TRANS.—49 STATES

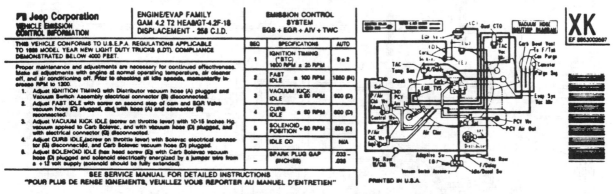

Jeep Corporation
VEHICLE EMISSION CONTROL INFORMATION

ENGINE/EVAP FAMILY
GAM 4.2 T2 HEA8/GT-4.2F-1S
DISPLACEMENT - 258 C.I.D.

EMISSION CONTROL SYSTEM
EGS + EGR + AIV + TWC

THIS VEHICLE CONFORMS TO U.S.E.P.A. REGULATIONS APPLICABLE TO 1986 MODEL YEAR NEW LIGHT DUTY TRUCKS (LDT). COMPLIANCE DEMONSTRATED BELOW 4000 FEET.

Proper maintenance and adjustments are necessary for continued effectiveness. Make all adjustments with engine at normal operating temperature, air cleaner off, and air conditioning off. Prior to checking all idle speeds, momentarily increase RPM to 1200.

1. Adjust IGNITION TIMING with Distributor vacuum hose (A) plugged and Vacuum Switch Assembly electrical connector (B) disconnected.
2. Adjust FAST IDLE with screw on second step of cam and EGR Valve vacuum hose (C) plugged, and with hose (A) and connector (B) reconnected.
3. Adjust VACUUM KICK IDLE (screw on throttle lever) with 10-15 inches Hg. vacuum applied to Carb Solevac, and with vacuum hose (D) plugged, and with electrical connector (G) disconnected.
4. Adjust CURB IDLE (screw on throttle lever) with Solevac electrical connector (G) disconnected, and Carb Solevac vacuum hose (D) plugged.
5. Adjust SOLENOID IDLE (hex head screw (E)) with Carb Solevac vacuum hose (D) plugged and solenoid electrically energized by a jumper wire from a + 12 volt supply (solenoid should be fully extended).

SEQ	SPECIFICATIONS	AUTO
1	IGNITION TIMING (°BTC) 1600 RPM ± 25 RPM	9 ± 2
2	FAST IDLE ± 100 RPM	1850 (N)
3	VACUUM KICK IDLE ± 50 RPM	800 (D)
4	CURB IDLE ± 50 RPM	600 (D)
5	SOLENOID POSITION + 50 RPM	800 (D)
-	IDLE CO	N/A
-	SPARK PLUG GAP (INCHES)	.033 - .038

SEE SERVICE MANUAL FOR DETAILED INSTRUCTIONS
"POUR PLUS DE RENSE IGNEMENTS, VEUILLEZ VOUS REPORTER AU MANUEL D'ENTRETIEN"

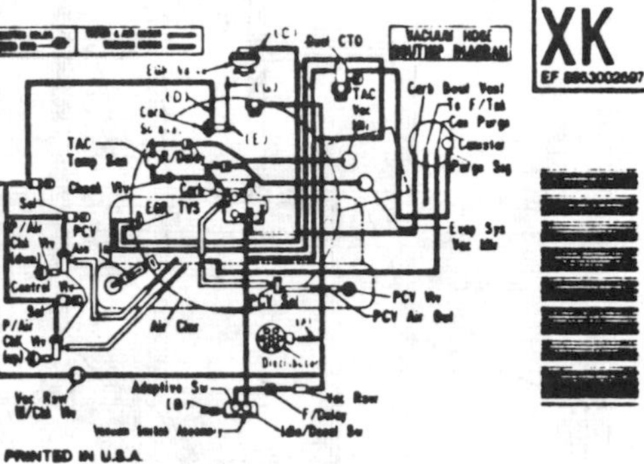

1986 6 CYL.-SJ EAGLE-AUTO. TRANS.—49 STATES

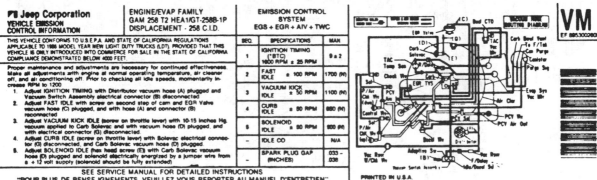

Jeep Corporation
VEHICLE EMISSION CONTROL INFORMATION

ENGINE/EVAP FAMILY
GAM 258 T2 HEA1/GT-258B-1P
DISPLACEMENT - 258 C.I.D.

EMISSION CONTROL SYSTEM
EGS + EGR + AIV + TWC

THIS VEHICLE CONFORMS TO U.S.E.P.A. AND STATE OF CALIFORNIA REGULATIONS APPLICABLE TO 1986 MODEL YEAR NEW LIGHT DUTY TRUCKS (LDT), PROVIDED THAT THIS VEHICLE IS ONLY INTRODUCED INTO COMMERCE FOR SALE IN THE STATE OF CALIFORNIA. COMPLIANCE DEMONSTRATED BELOW 4000 FEET.

Proper maintenance and adjustments are necessary for continued effectiveness. Make all adjustments with engine at normal operating temperature, air cleaner off, and air conditioning off. Prior to checking all idle speeds, momentarily increase RPM to 1200

1. Adjust IGNITION TIMING with Distributor vacuum hose (A) plugged and Vacuum Switch Assembly electrical connector (B) disconnected
2. Adjust FAST IDLE with screw on second step of cam and EGR Valve vacuum hose (C) plugged, and with hose (A) and connector (B) reconnected
3. Adjust VACUUM KICK IDLE (screw on throttle lever) with 10-15 inches Hg. vacuum applied to Carb Solevac and with vacuum hose (D) plugged, and with electrical connector (G) disconnected.
4. Adjust CURB IDLE (screw on throttle lever) with Solevac electrical connector (G) disconnected, and Carb Solevac vacuum hose (D) plugged.
5. Adjust SOLENOID IDLE (hex head screw (E)) with Carb Solevac vacuum hose (D) plugged and solenoid electrically energized by a jumper wire from a + 12 volt supply (solenoid should be fully extended)

SEQ.	SPECIFICATIONS	MAN
1	IGNITION TIMING (°BTC) 1600 RPM ± 25 RPM	9 ± 2
2	FAST IDLE ± 100 RPM	1700 (M)
3	VACUUM KICK IDLE ± 50 RPM	1100 (M)
4	CURB IDLE ± 50 RPM	680 (M)
5	SOLENOID IDLE ± 50 RPM	900 (M)
-	IDLE CO	N/A
-	SPARK PLUG GAP (INCHES)	.033 - .038

SEE SERVICE MANUAL FOR DETAILED INSTRUCTIONS
"POUR PLUS DE RENSE IGNEMENTS, VEUILLEZ VOUS REPORTER AU MANUEL D'ENTRETIEN"

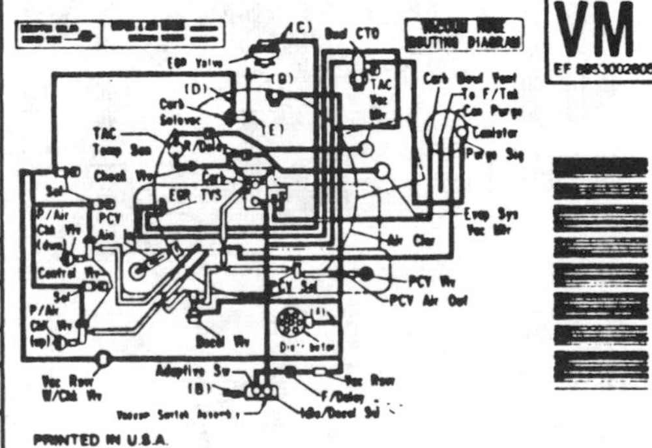

1986 6 CYL.-SJ EAGLE-MAN. TRANS.—CALIFORNIA

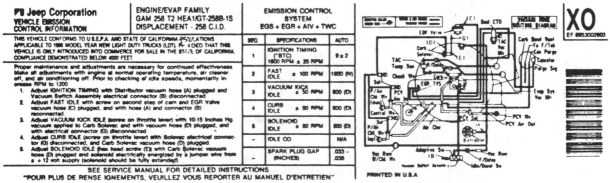

Jeep Corporation
VEHICLE EMISSION CONTROL INFORMATION

ENGINE/EVAP FAMILY
GAM 258 T2 HEA1/GT-258B-1S
DISPLACEMENT - 258 C.I.D.

EMISSION CONTROL SYSTEM
EGS + EGR + AIV + TWC

THIS VEHICLE CONFORMS TO U.S.E.P.A. AND STATE OF CALIFORNIA REGULATIONS APPLICABLE TO 1986 MODEL YEAR NEW LIGHT DUTY TRUCKS (LDT), PROVIDED THAT THIS VEHICLE IS ONLY INTRODUCED INTO COMMERCE FOR SALE IN THE STATE OF CALIFORNIA. COMPLIANCE DEMONSTRATED BELOW 4000 FEET.

Proper maintenance and adjustments are necessary for continued effectiveness. Make all adjustments with engine at normal operating temperature, air cleaner off, and air conditioning off. Prior to checking all idle speeds, momentarily increase RPM to 1200

1. Adjust IGNITION TIMING with Distributor vacuum hose (A) plugged and Vacuum Switch Assembly electrical connector (B) disconnected
2. Adjust FAST IDLE with screw on second step of cam and EGR Valve vacuum hose (C) plugged, and with hose (A) and connector (B) reconnected
3. Adjust VACUUM KICK IDLE (screw on throttle lever) with 10-15 inches Hg. vacuum applied to Carb Solevac and with vacuum hose (D) plugged, and with electrical connector (G) disconnected.
4. Adjust CURB IDLE (screw on throttle lever) with Solevac electrical connector (G) disconnected, and Carb Solevac vacuum hose (D) plugged.
5. Adjust SOLENOID IDLE (hex head screw (E)) with Carb Solevac vacuum hose (D) plugged and solenoid electrically energized by a jumper wire from a + 12 volt supply (solenoid should be fully extended)

SEQ	SPECIFICATIONS	AUTO
1	IGNITION TIMING (°BTC) 1600 RPM ± 25 RPM	9 ± 2
2	FAST IDLE ± 100 RPM	1850 (N)
3	VACUUM KICK IDLE ± 50 RPM	800 (D)
4	CURB IDLE ± 50 RPM	600 (D)
5	SOLENOID IDLE ± 50 RPM	800 (D)
-	IDLE CO	N/A
-	SPARK PLUG GAP (INCHES)	.033 - .038

SEE SERVICE MANUAL FOR DETAILED INSTRUCTIONS
"POUR PLUS DE RENSE IGNEMENTS, VEUILLEZ VOUS REPORTER AU MANUEL D'ENTRETIEN"

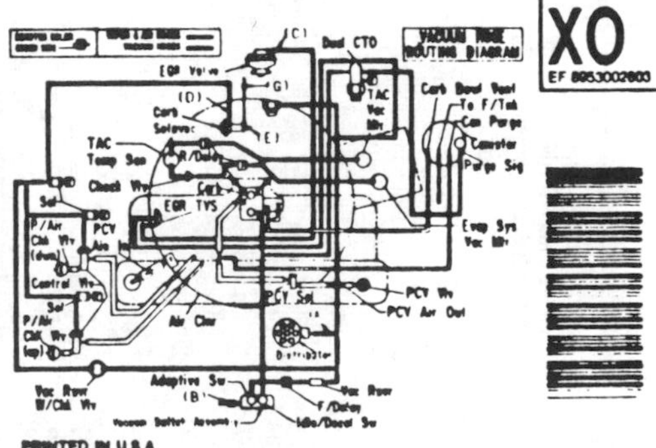

1986 6 CYL.-SJ EAGLE-AUTO. TRANS.—CALIFORNIA

VACUUM CIRCUITS

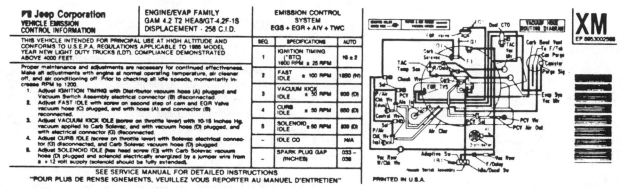

Jeep Corporation
VEHICLE EMISSION CONTROL INFORMATION

ENGINE/EVAP FAMILY
GAM 4.2 T2 HEA8/GT-4.2F-1S
DISPLACEMENT - 258 C.I.D.

EMISSION CONTROL SYSTEM
EGS + EGR + AIV + TWC

THIS VEHICLE INTENDED FOR PRINCIPAL USE AT HIGH ALTITUDE AND CONFORMS TO U.S.E.P.A. REGULATIONS APPLICABLE TO 1986 MODEL YEAR NEW LIGHT DUTY TRUCKS (LDT). COMPLIANCE DEMONSTRATED ABOVE 4000 FEET.

Proper maintenance and adjustments are necessary for continued effectiveness. Make all adjustments with engine at normal operating temperature, air cleaner off, and air conditioning off. Prior to checking all idle speeds, momentarily increase RPM to 1200.

1. Adjust IGNITION TIMING with Distributor vacuum hose (A) plugged and Vacuum Switch Assembly electrical connector (B) disconnected.
2. Adjust FAST IDLE with screw on second step of cam and EGR Valve vacuum hose (C) plugged, and with hose (A) and connector (B) reconnected.
3. Adjust VACUUM KICK IDLE (screw on throttle lever) with 10-15 inches Hg. vacuum applied to Carb Solevac, and with vacuum hose (D) plugged, and with electrical connector (G) disconnected.
4. Adjust CURB IDLE (screw on throttle lever) with Solevac electrical connector (G) disconnected, and Carb Solevac vacuum hose (D) plugged.
5. Adjust SOLENOID IDLE (hex head screw (E)) with Carb Solevac vacuum hose (D) plugged and solenoid electrically energized by a jumper wire from a + 12 volt supply (solenoid should be fully extended).

SEQ.	SPECIFICATIONS	MAN.
1	IGNITION TIMING (°BTC) 1600 RPM ± 25 RPM	16 ± 2
2	FAST IDLE ± 100 RPM	1700 (N)
3	VACUUM KICK IDLE ± 50 RPM	1100 (N)
4	CURB IDLE ± 50 RPM	700 (D)
5	SOLENOID IDLE ± 50 RPM	800 (N)
–	IDLE CO	N/A
–	SPARK PLUG GAP (INCHES)	.033 – .038

SEE SERVICE MANUAL FOR DETAILED INSTRUCTIONS
"POUR PLUS DE RENSE IGNEMENTS, VEUILLEZ VOUS REPORTER AU MANUEL D'ENTRETIEN"

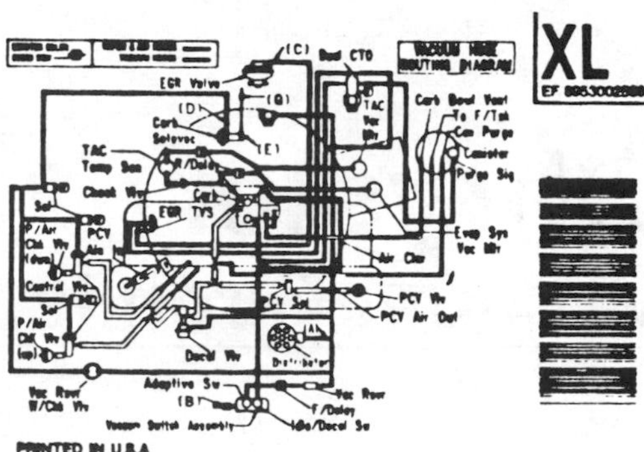

XL
EF 8953002588

1986 6 CYL.-SJ EAGLE-MAN. TRANS.—HIGH ALT.

Jeep Corporation
VEHICLE EMISSION CONTROL INFORMATION

ENGINE/EVAP FAMILY
GAM 4.2 T2 HEA8/GT-4.2F-1S
DISPLACEMENT - 258 C.I.D.

EMISSION CONTROL SYSTEM
EGS + EGR + AIV + TWC

THIS VEHICLE INTENDED FOR PRINCIPAL USE AT HIGH ALTITUDE AND CONFORMS TO U.S.E.P.A. REGULATIONS APPLICABLE TO 1986 MODEL YEAR NEW LIGHT DUTY TRUCKS (LDT). COMPLIANCE DEMONSTRATED ABOVE 4000 FEET.

Proper maintenance and adjustments are necessary for continued effectiveness. Make all adjustments with engine at normal operating temperature, air cleaner off, and air conditioning off. Prior to checking all idle speeds, momentarily increase RPM to 1200.

1. Adjust IGNITION TIMING with Distributor vacuum hose (A) plugged and Vacuum Switch Assembly electrical connector (B) disconnected.
2. Adjust FAST IDLE with screw on second step of cam and EGR Valve vacuum hose (C) plugged, and with hose (A) and connector (B) reconnected.
3. Adjust VACUUM KICK IDLE (screw on throttle lever) with 10-15 inches Hg. vacuum applied to Carb Solevac, and with vacuum hose (D) plugged, and with electrical connector (G) disconnected.
4. Adjust CURB IDLE (screw on throttle lever) with Solevac electrical connector (G) disconnected, and Carb Solevac vacuum hose (D) plugged.
5. Adjust SOLENOID IDLE (hex head screw (E)) with Carb Solevac vacuum hose (D) plugged and solenoid electrically energized by a jumper wire from a + 12 volt supply (solenoid should be fully extended).

SEQ.	SPECIFICATIONS	AUTO
1	IGNITION TIMING (°BTC) 1600 RPM ± 25 RPM	16 ± 2
2	FAST IDLE ± 100 RPM	1850 (N)
3	VACUUM KICK IDLE ± 50 RPM	900 (D)
4	CURB IDLE ± 50 RPM	650 (D)
5	SOLENOID IDLE ± 50 RPM	800 (D)
–	IDLE CO	N/A
–	SPARK PLUG GAP (INCHES)	.033 – .038

SEE SERVICE MANUAL FOR DETAILED INSTRUCTIONS
"POUR PLUS DE RENSE IGNEMENTS, VEUILLEZ VOUS REPORTER AU MANUEL D'ENTRETIEN"

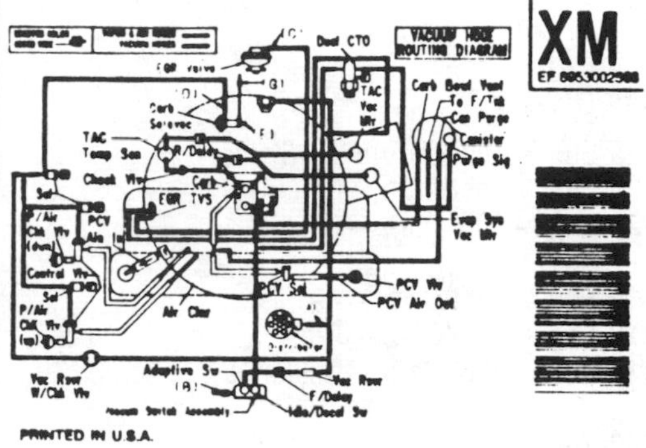

XM
EF 8953002589

1986 6 CYL.-SJ EAGLE-AUTO. TRANS.—HIGH ALT.

Jeep Corporation
VEHICLE EMISSION CONTROL INFORMATION

ENGINE/EVAP FAMILY
GAM 5.9 T2 HLE3/GT-5.9F-1P
DISPLACEMENT-360 C.I.D.

EMISSION CONTROL SYSTEM
EGR + AIP + TWC

THIS VEHICLE CONFORMS TO U.S.E.P.A. REGULATIONS APPLICABLE TO 1986 MODEL YEAR NEW LIGHT DUTY TRUCKS (LDT). COMPLIANCE DEMONSTRATED BELOW 4000 FEET.

Proper maintenance and adjustments are necessary for continued effectiveness. Make all adjustments with engine at normal operating temperature, air cleaner off, and air conditioning off.

1. Adjust IGNITION TIMING with Distributor vacuum hose (A) plugged.
2. Adjust FAST IDLE with screw on second step of cam and EGR Valve vacuum hose (B) plugged.
3. Adjust CURB IDLE by turning hex head screw (C) on Solenoid carriage.
4. Adjust BASE IDLE (screw on carb base flange) with solenoid electrical connector (D) disconnected.

SEQ	SPECIFICATIONS	AUTO
1	IGNITION TIMING (°BTC) 600 RPM +0 −50	12 ± 2
2	FAST IDLE ± 100 RPM	1600(N)
3	CURB IDLE ± 50 RPM	600(D)
4	BASE IDLE +0 −50 RPM	500(N)
–	IDLE CO	N/A
–	SPARK PLUG GAP (INCHES)	.033-.038

SEE SERVICE MANUAL FOR DETAILED INSTRUCTIONS
"POUR PLUS DE RENSEIGNEMENTS, VEUILLEZ VOUS REPORTER AU MANUEL D'ENTRETIEN"

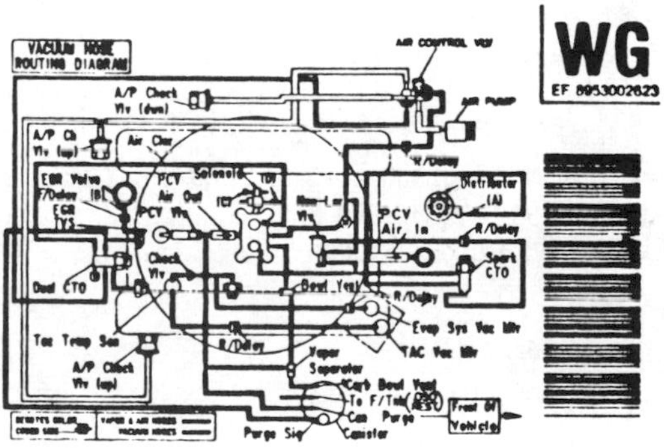

WG
EF 8953002623

1986 V8-SJ EAGLE-AUTO. TRANS.—50 STATES

Jeep Corporation
VEHICLE EMISSION CONTROL INFORMATION

ENGINE/EVAP FAMILY
GAM 5.9 T2 HLE3/GT-5.9F-1P
DISPLACEMENT 360 C.I.D.

EMISSION CONTROL SYSTEM
EGR + AIP + TWC

THIS VEHICLE INTENDED FOR PRINCIPAL USE AT HIGH ALTITUDE AND CONFORMS TO U.S.E.P.A. REGULATIONS APPLICABLE TO 1986 MODEL YEAR NEW LIGHT DUTY TRUCKS (LDT). COMPLIANCE DEMONSTRATED ABOVE 4000 FEET.

Proper maintenance and adjustments are necessary for continued effectiveness. Make all adjustments with engine at normal operating temperature, air cleaner off, and air conditioning off.

1. Adjust IGNITION TIMING with Distributor vacuum hose (A) plugged.
2. Adjust FAST IDLE with screw on second step of cam and EGR Valve vacuum hose (B) plugged.
3. Adjust CURB IDLE by turnigg hex head screw (C) on Solenoid carriage.
4. Adjust BASE IDLE (screw on carb base flange) with solenoid electrical connector (D) disconnected.

SEQ.	SPECIFICATIONS	AUTO
1	IGNITION TIMING (°BTC) 600 RPM +0 −50	19 ± 2
2	FAST IDLE ± 100 RPM	1600(N)
3	CURB IDLE ± 50 RPM	600(D)
4	BASE IDLE +0 −50 RPM	500(N)
–	IDLE CO	N/A
–	SPARK PLUG GAP (INCHES)	.033-.038

SEE SERVICE MANUAL FOR DETAILED INSTRUCTIONS
"POUR PLUS DE RENSEIGNEMENTS, VEUILLEZ VOUS REPORTER AU MANUEL D'ENTRETIEN"

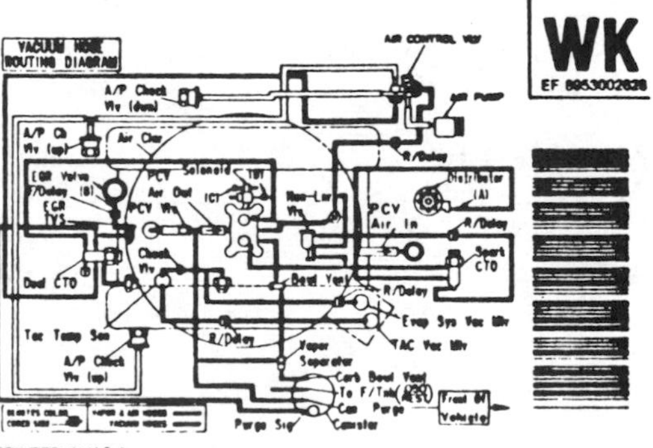

WK
EF 8953002628

1986 V8-SJ EAGLE-AUTO. TRANS.—HIGH ALT.

Chrysler
Vacuum Circuits

INDEX

1980 VACUUM CIRCUITS

1981—82 VACUUM CIRCUITS

1983 VACUUM CIRCUITS

VACUUM CIRCUITS

(© Chrysler Corp.)

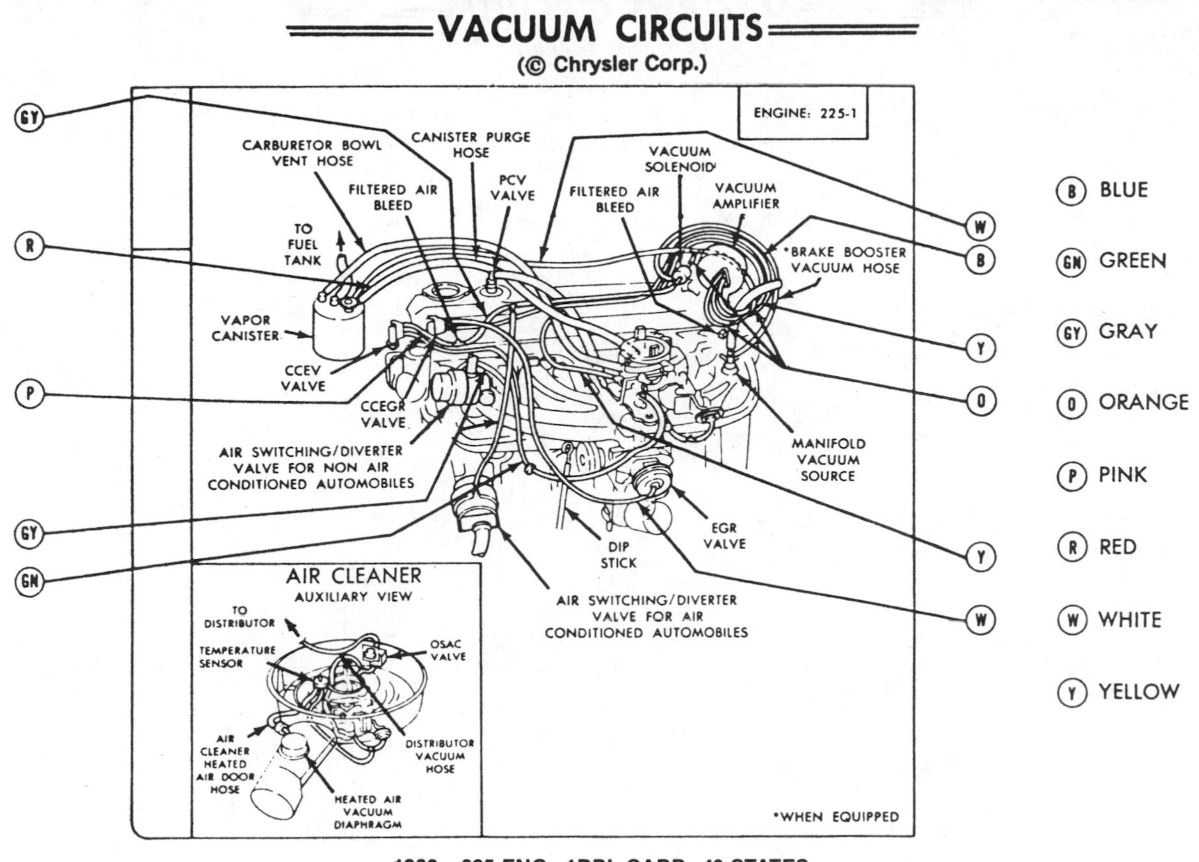

1980—225 ENG.-1BBL CARB.-49 STATES

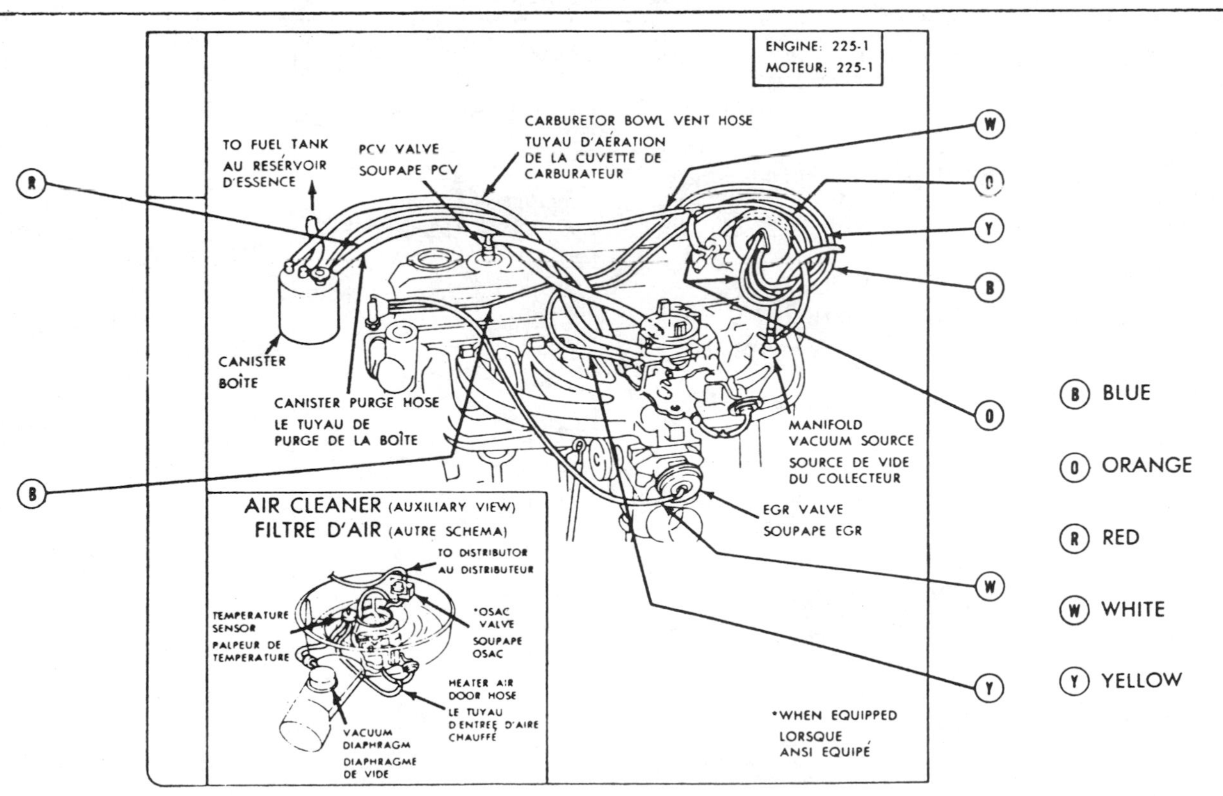

1980—225 ENG.-1BBL CARB.-CANADA

VACUUM CIRCUITS

(© Chrysler Corp.)

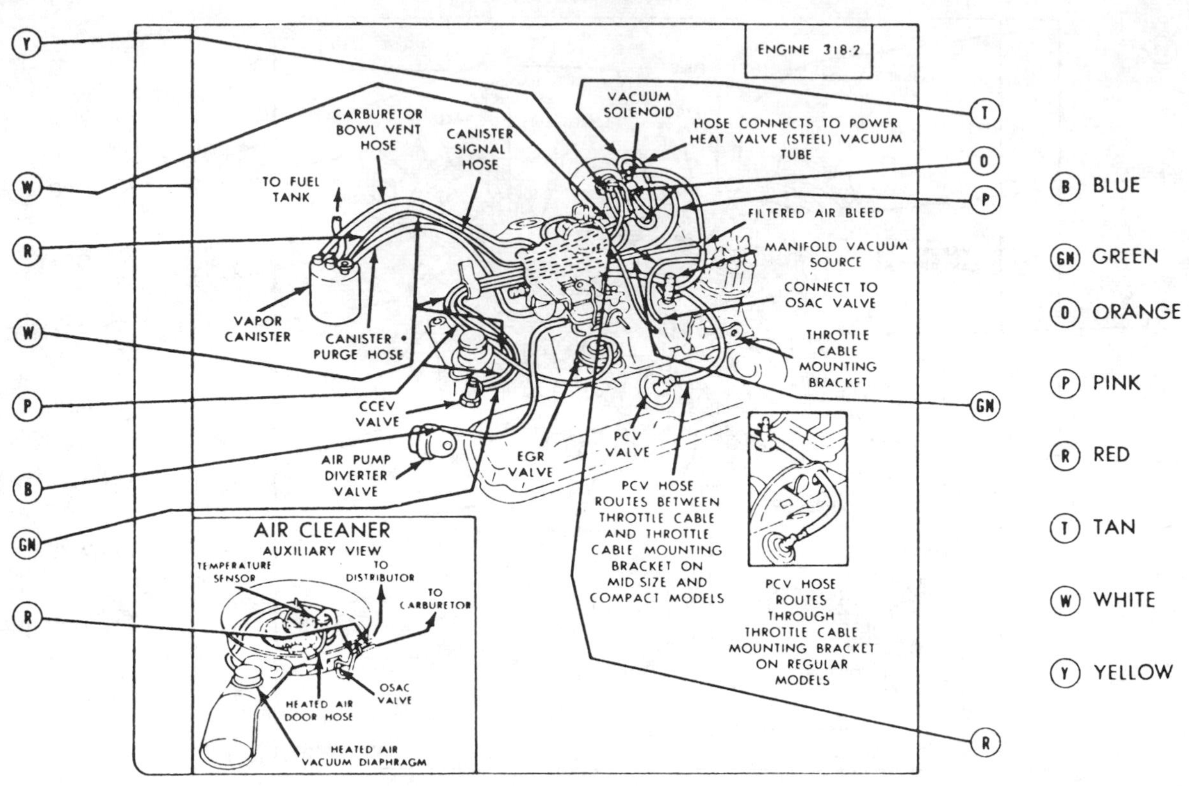
ENGINE 318-2
VACUUM SOLENOID
HOSE CONNECTS TO POWER HEAT VALVE (STEEL) VACUUM TUBE
CARBURETOR BOWL VENT HOSE
CANISTER SIGNAL HOSE
TO FUEL TANK
FILTERED AIR BLEED
MANIFOLD VACUUM SOURCE
CONNECT TO OSAC VALVE
VAPOR CANISTER
CANISTER • PURGE HOSE
THROTTLE CABLE MOUNTING BRACKET
CCEV VALVE
AIR PUMP DIVERTER VALVE
EGR VALVE
PCV VALVE
PCV HOSE ROUTES BETWEEN THROTTLE CABLE AND THROTTLE CABLE MOUNTING BRACKET ON MID SIZE AND COMPACT MODELS
PCV HOSE ROUTES THROUGH THROTTLE CABLE MOUNTING BRACKET ON REGULAR MODELS
AIR CLEANER
AUXILIARY VIEW
TEMPERATURE SENSOR
TO DISTRIBUTOR
TO CARBURETOR
OSAC VALVE
HEATED AIR DOOR HOSE
HEATED AIR VACUUM DIAPHRAGM
Y
W
R
W
P
B
GN
R
T
O
P
GN
R
B BLUE
GN GREEN
O ORANGE
P PINK
R RED
T TAN
W WHITE
Y YELLOW

1980—318 ENG.-2BBL CARB.-49 STATES

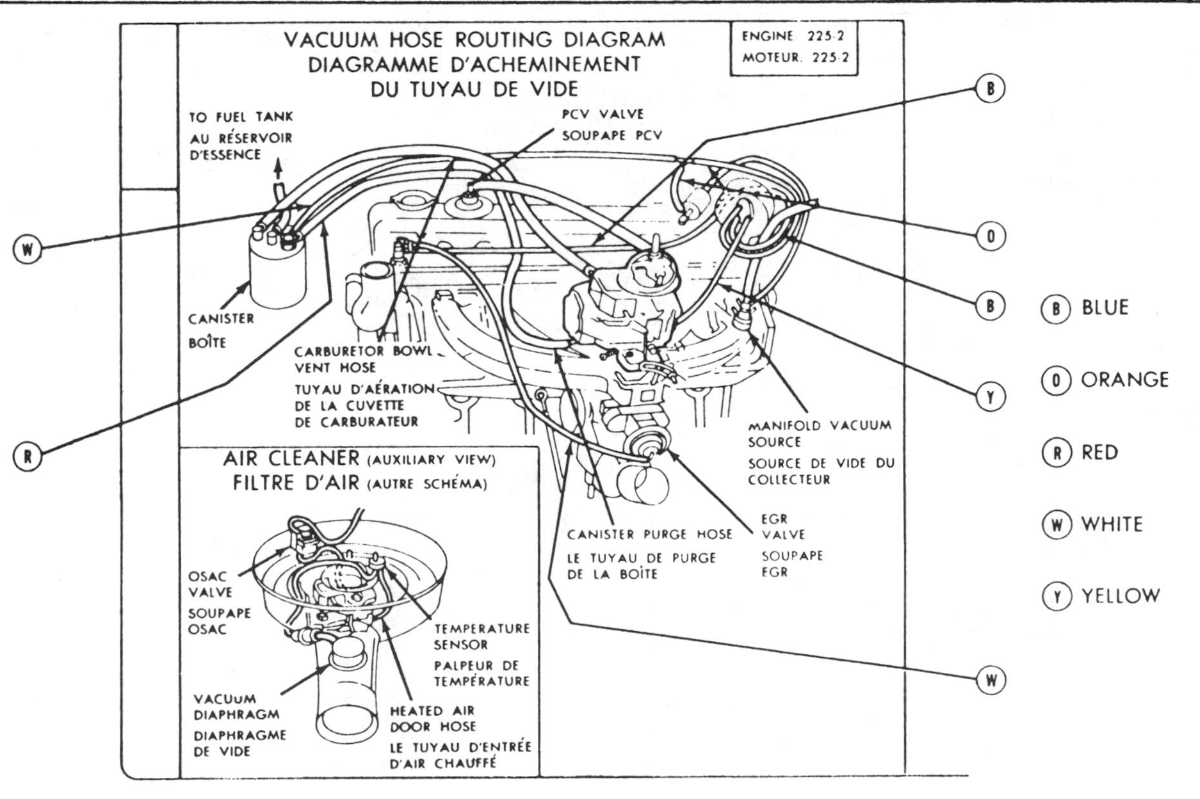
VACUUM HOSE ROUTING DIAGRAM
DIAGRAMME D'ACHEMINEMENT
DU TUYAU DE VIDE
ENGINE 225-2
MOTEUR 225-2
TO FUEL TANK
AU RÉSERVOIR D'ESSENCE
PCV VALVE
SOUPAPE PCV
CANISTER
BOÎTE
CARBURETOR BOWL VENT HOSE
TUYAU D'AÉRATION DE LA CUVETTE DE CARBURATEUR
MANIFOLD VACUUM SOURCE
SOURCE DE VIDE DU COLLECTEUR
CANISTER PURGE HOSE
LE TUYAU DE PURGE DE LA BOITE
EGR VALVE
SOUPAPE EGR
AIR CLEANER (AUXILIARY VIEW)
FILTRE D'AIR (AUTRE SCHÉMA)
OSAC VALVE
SOUPAPE OSAC
TEMPERATURE SENSOR
PALPEUR DE TEMPÉRATURE
VACUUM DIAPHRAGM
DIAPHRAGME DE VIDE
HEATED AIR DOOR HOSE
LE TUYAU D'ENTRÉE D'AIR CHAUFFÉ
W
R
B
O
B
Y
W
B BLUE
O ORANGE
R RED
W WHITE
Y YELLOW

1980—360 ENG.-2BBL CARB.-CANADA

VACUUM CIRCUITS

(© Chrysler Corp.)

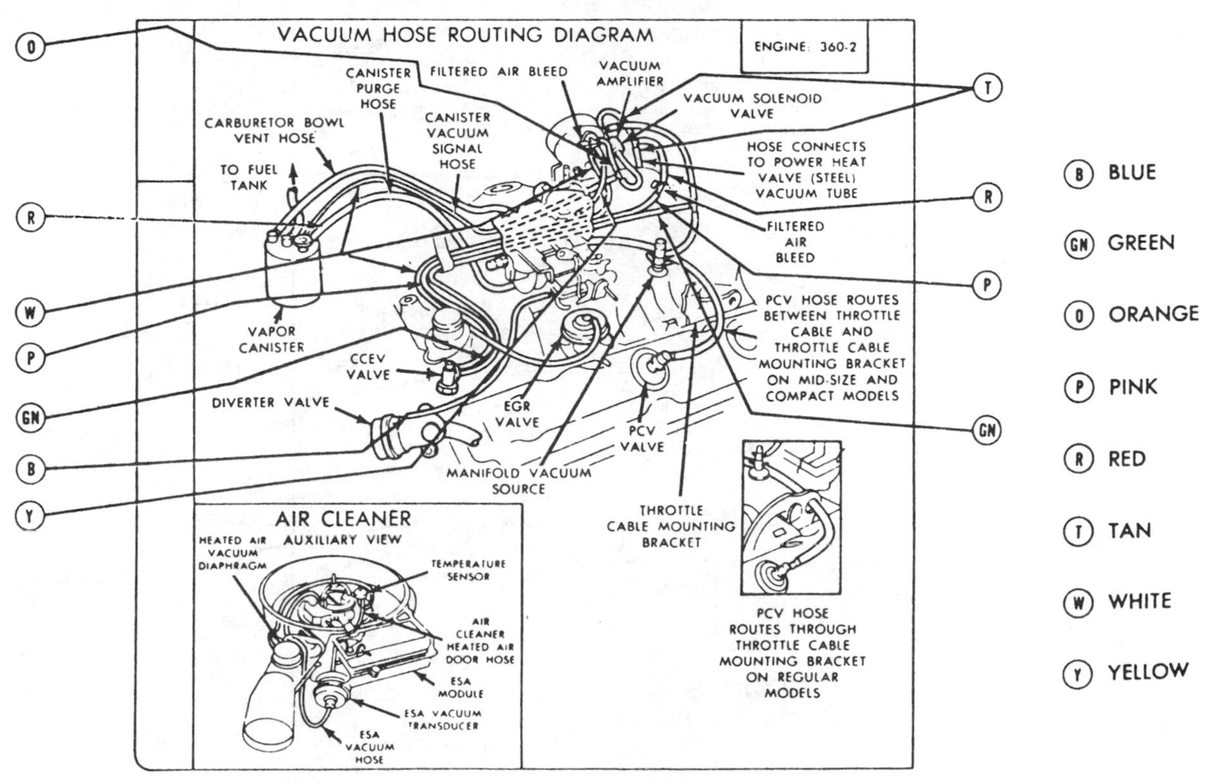
VACUUM HOSE ROUTING DIAGRAM
ENGINE: 360-2
CANISTER PURGE HOSE
FILTERED AIR BLEED
VACUUM AMPLIFIER
VACUUM SOLENOID VALVE
CARBURETOR BOWL VENT HOSE
CANISTER VACUUM SIGNAL HOSE
TO FUEL TANK
HOSE CONNECTS TO POWER HEAT VALVE (STEEL) VACUUM TUBE
FILTERED AIR BLEED
PCV HOSE ROUTES BETWEEN THROTTLE CABLE AND THROTTLE CABLE MOUNTING BRACKET ON MID-SIZE AND COMPACT MODELS
VAPOR CANISTER
CCEV VALVE
DIVERTER VALVE
EGR VALVE
PCV VALVE
MANIFOLD VACUUM SOURCE
THROTTLE CABLE MOUNTING BRACKET
PCV HOSE ROUTES THROUGH THROTTLE CABLE MOUNTING BRACKET ON REGULAR MODELS
AIR CLEANER AUXILIARY VIEW
HEATED AIR VACUUM DIAPHRAGM
TEMPERATURE SENSOR
AIR CLEANER HEATED AIR DOOR HOSE
ESA MODULE
ESA VACUUM TRANSDUCER
ESA VACUUM HOSE
O
T
R
W
P
GN
B
Y
B BLUE
GN GREEN
O ORANGE
P PINK
R RED
T TAN
W WHITE
Y YELLOW

1980—360 ENG.-2BBL CARB.-49 STATES

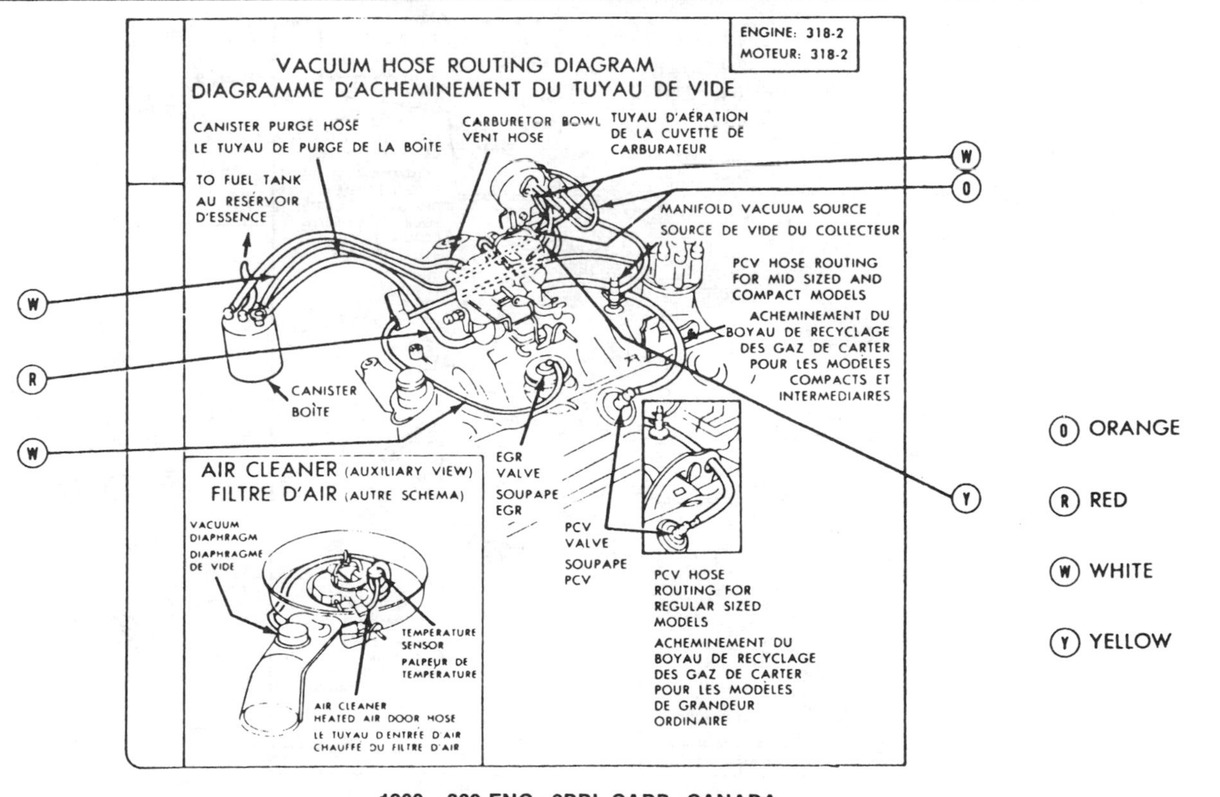
ENGINE: 318-2
MOTEUR: 318-2
VACUUM HOSE ROUTING DIAGRAM
DIAGRAMME D'ACHEMINEMENT DU TUYAU DE VIDE
CANISTER PURGE HOSE
LE TUYAU DE PURGE DE LA BOITE
CARBURETOR BOWL VENT HOSE
TUYAU D'AÉRATION DE LA CUVETTE DE CARBURATEUR
TO FUEL TANK
AU RESÉRVOIR D'ESSENCE
MANIFOLD VACUUM SOURCE
SOURCE DE VIDE DU COLLECTEUR
PCV HOSE ROUTING FOR MID SIZED AND COMPACT MODELS
ACHEMINEMENT DU BOYAU DE RECYCLAGE DES GAZ DE CARTER POUR LES MODELES COMPACTS ET INTERMEDIAIRES
CANISTER
BOÎTE
EGR VALVE
SOUPAPE EGR
PCV VALVE
SOUPAPE PCV
PCV HOSE ROUTING FOR REGULAR SIZED MODELS
ACHEMINEMENT DU BOYAU DE RECYCLAGE DES GAZ DE CARTER POUR LES MODÈLES DE GRANDEUR ORDINAIRE
AIR CLEANER (AUXILIARY VIEW)
FILTRE D'AIR (AUTRE SCHEMA)
VACUUM DIAPHRAGM
DIAPHRAGME DE VIDE
TEMPERATURE SENSOR
PALPEUR DE TEMPERATURE
AIR CLEANER HEATED AIR DOOR HOSE
LE TUYAU D'ENTRÉE D'AIR CHAUFFÉ DU FILTRE D'AIR
W
O
R
Y
O ORANGE
R RED
W WHITE
Y YELLOW

1980—360 ENG.-2BBL CARB.-CANADA

VACUUM CIRCUITS

(© Chrysler Corp.)

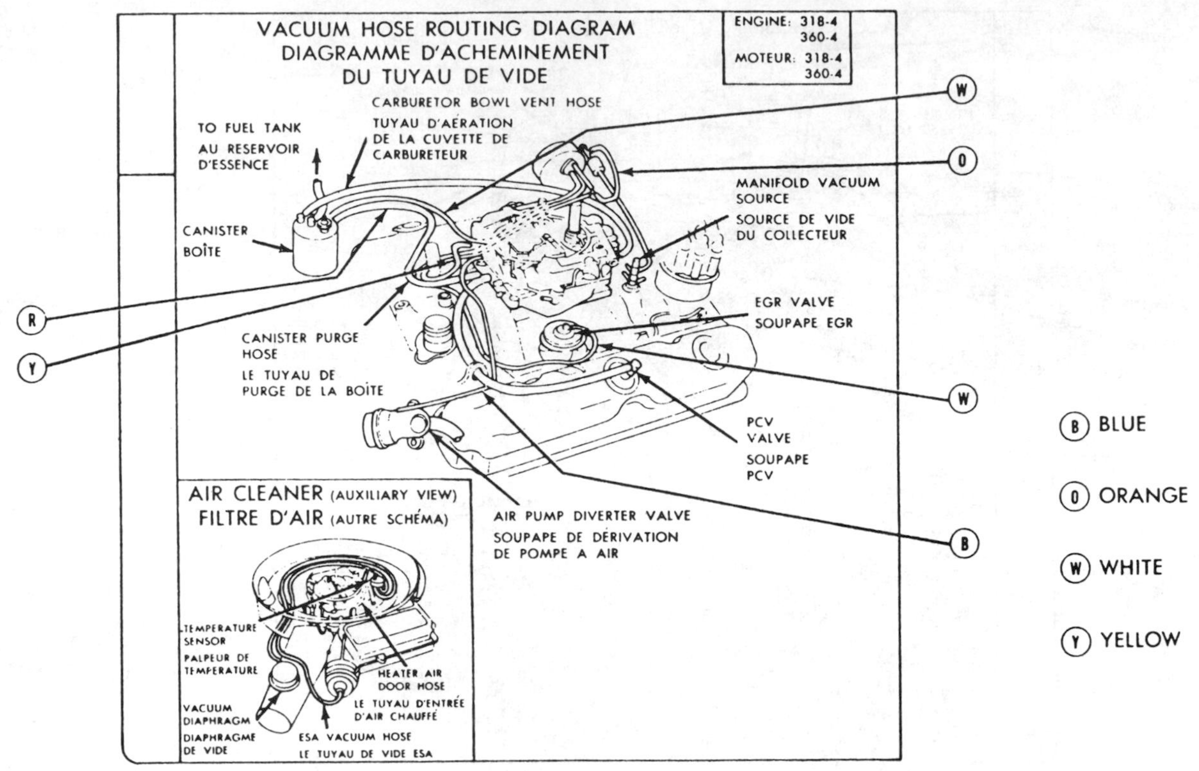
VACUUM HOSE ROUTING DIAGRAM
DIAGRAMME D'ACHEMINEMENT
DU TUYAU DE VIDE
ENGINE: 318-4
360-4
MOTEUR: 318-4
360-4
CARBURETOR BOWL VENT HOSE
TUYAU D'AÉRATION
DE LA CUVETTE DE
CARBURETEUR
TO FUEL TANK
AU RESERVOIR
D'ESSENCE
CANISTER
BOÎTE
MANIFOLD VACUUM
SOURCE
SOURCE DE VIDE
DU COLLECTEUR
EGR VALVE
SOUPAPE EGR
CANISTER PURGE
HOSE
LE TUYAU DE
PURGE DE LA BOÎTE
PCV
VALVE
SOUPAPE
PCV
AIR CLEANER (AUXILIARY VIEW)
FILTRE D'AIR (AUTRE SCHÉMA)
AIR PUMP DIVERTER VALVE
SOUPAPE DE DÉRIVATION
DE POMPE A AIR
TEMPERATURE
SENSOR
PALPEUR DE
TEMPERATURE
HEATER AIR
DOOR HOSE
LE TUYAU D'ENTRÉE
D'AIR CHAUFFÉ
VACUUM
DIAPHRAGM
DIAPHRAGME
DE VIDE
ESA VACUUM HOSE
LE TUYAU DE VIDE ESA
W
O
R
Y
W
B
B BLUE
O ORANGE
W WHITE
Y YELLOW

1980—318 & 360 ENGINE-4BBL CARB.-LeBARON, GRAN FURY & HORIZON-CANADA

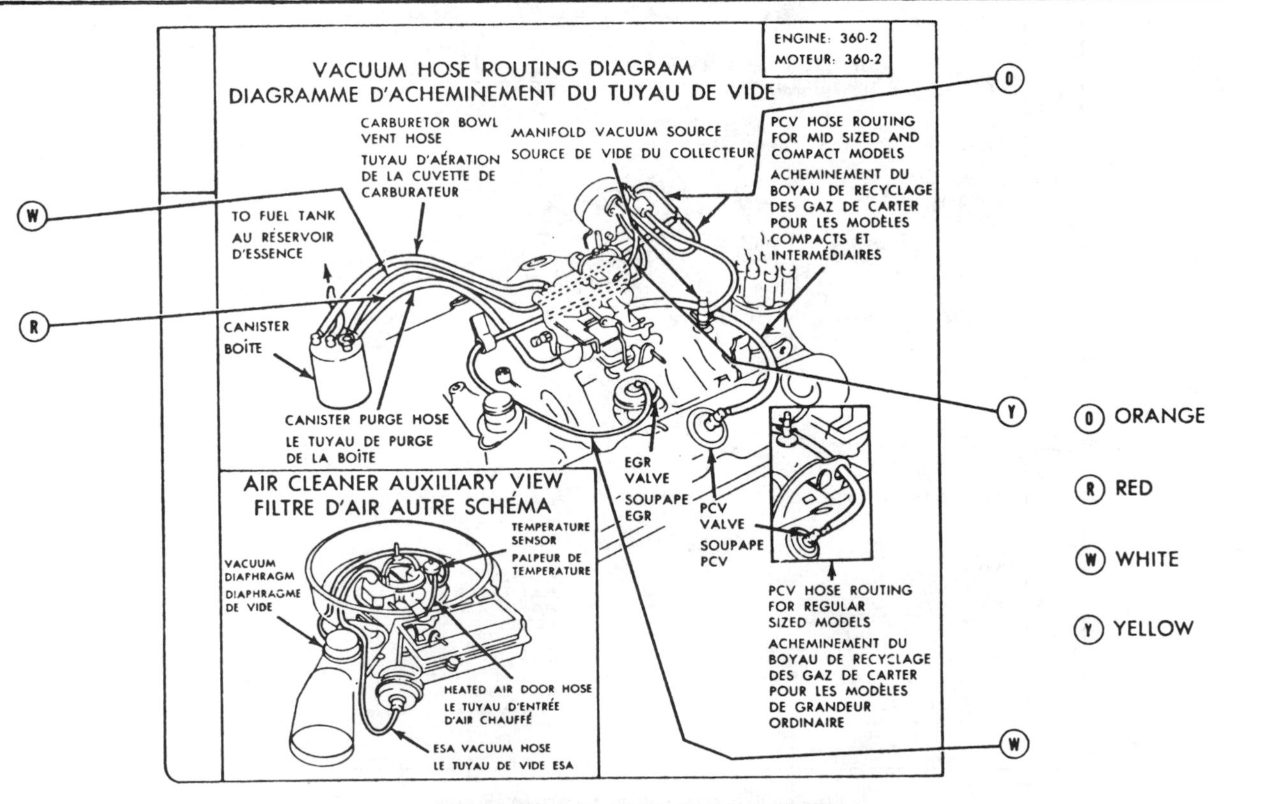
ENGINE: 360-2
MOTEUR: 360-2
VACUUM HOSE ROUTING DIAGRAM
DIAGRAMME D'ACHEMINEMENT DU TUYAU DE VIDE
CARBURETOR BOWL
VENT HOSE
TUYAU D'AÉRATION
DE LA CUVETTE DE
CARBURATEUR
MANIFOLD VACUUM SOURCE
SOURCE DE VIDE DU COLLECTEUR
PCV HOSE ROUTING
FOR MID SIZED AND
COMPACT MODELS
ACHEMINEMENT DU
BOYAU DE RECYCLAGE
DES GAZ DE CARTER
POUR LES MODÈLES
COMPACTS ET
INTERMÉDIAIRES
TO FUEL TANK
AU RESERVOIR
D'ESSENCE
CANISTER
BOÎTE
CANISTER PURGE HOSE
LE TUYAU DE PURGE
DE LA BOÎTE
EGR
VALVE
SOUPAPE
EGR
PCV
VALVE
SOUPAPE
PCV
AIR CLEANER AUXILIARY VIEW
FILTRE D'AIR AUTRE SCHÉMA
TEMPERATURE
SENSOR
PALPEUR DE
TEMPERATURE
VACUUM
DIAPHRAGM
DIAPHRAGME
DE VIDE
HEATED AIR DOOR HOSE
LE TUYAU D'ENTRÉE
D'AIR CHAUFFÉ
ESA VACUUM HOSE
LE TUYAU DE VIDE ESA
PCV HOSE ROUTING
FOR REGULAR
SIZED MODELS
ACHEMINEMENT DU
BOYAU DE RECYCLAGE
DES GAZ DE CARTER
POUR LES MODÈLES
DE GRANDEUR
ORDINAIRE
O
W
R
Y
W
O ORANGE
R RED
W WHITE
Y YELLOW

1980—360 ENG.-2BBL CARB.-CANADA

VACUUM CIRCUITS

(© Chrysler Corp.)

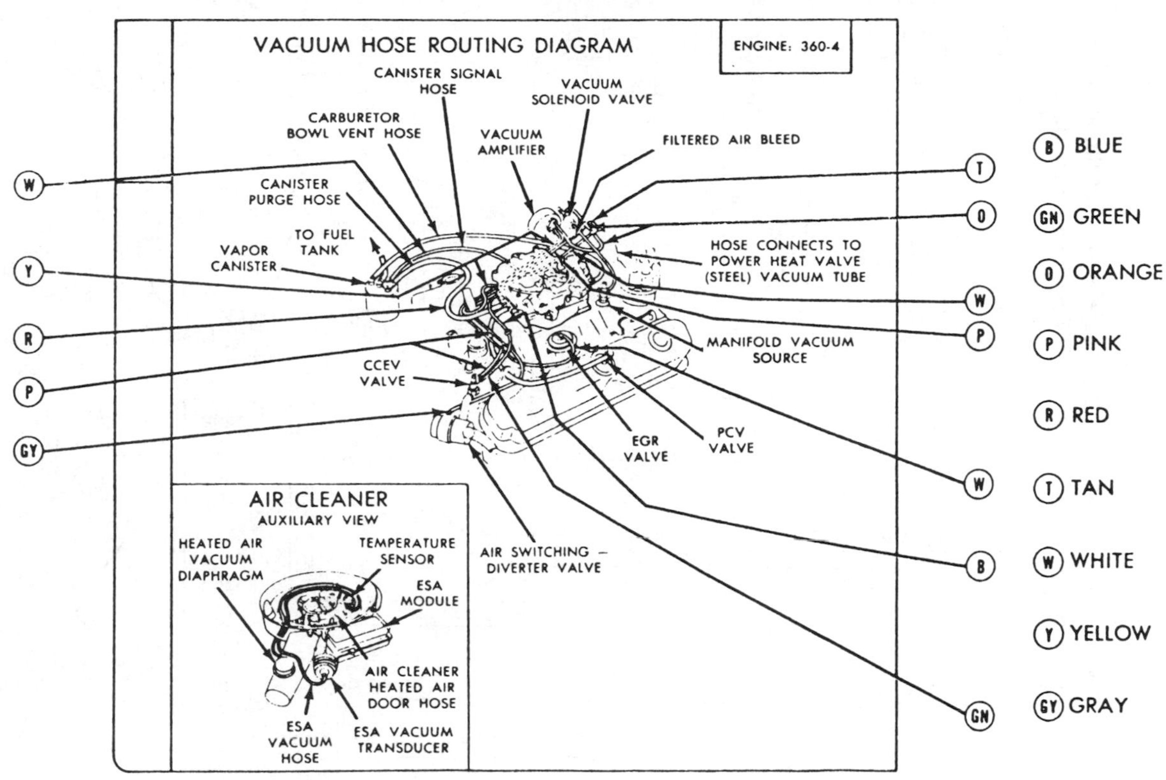
VACUUM HOSE ROUTING DIAGRAM
ENGINE: 360-4
CANISTER SIGNAL HOSE
VACUUM SOLENOID VALVE
CARBURETOR BOWL VENT HOSE
VACUUM AMPLIFIER
FILTERED AIR BLEED
CANISTER PURGE HOSE
TO FUEL TANK
VAPOR CANISTER
HOSE CONNECTS TO POWER HEAT VALVE (STEEL) VACUUM TUBE
MANIFOLD VACUUM SOURCE
CCEV VALVE
EGR VALVE
PCV VALVE
AIR SWITCHING – DIVERTER VALVE
AIR CLEANER
AUXILIARY VIEW
HEATED AIR VACUUM DIAPHRAGM
TEMPERATURE SENSOR
ESA MODULE
AIR CLEANER HEATED AIR DOOR HOSE
ESA VACUUM HOSE
ESA VACUUM TRANSDUCER
B BLUE
GN GREEN
O ORANGE
P PINK
R RED
T TAN
W WHITE
Y YELLOW
GY GRAY

1980—360 ENG.-4BBL CARB.-LeBARON, GRAN FURY & HORIZON-49 STATES

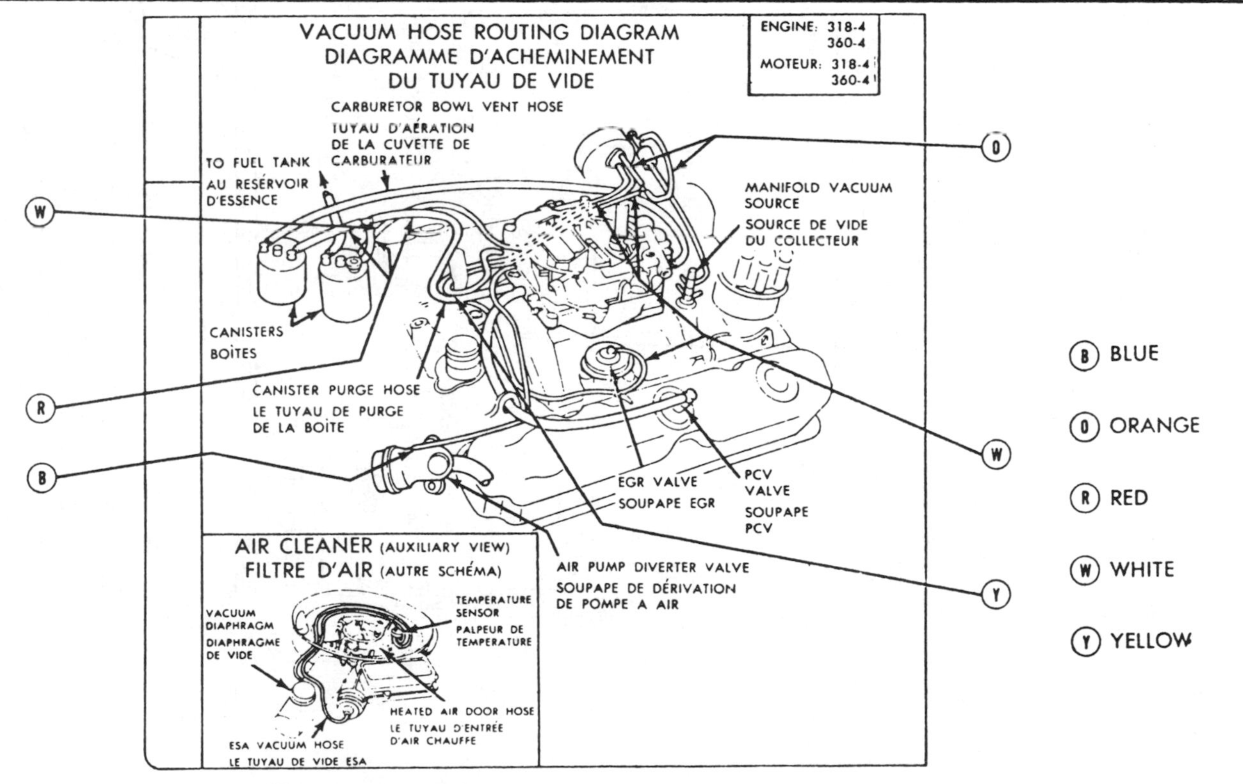
VACUUM HOSE ROUTING DIAGRAM
DIAGRAMME D'ACHEMINEMENT DU TUYAU DE VIDE
ENGINE: 318-4 360-4
MOTEUR: 318-4 360-4
CARBURETOR BOWL VENT HOSE
TUYAU D'AÉRATION DE LA CUVETTE DE CARBURATEUR
TO FUEL TANK
AU RESÉRVOIR D'ESSENCE
MANIFOLD VACUUM SOURCE
SOURCE DE VIDE DU COLLECTEUR
CANISTERS
BOITES
CANISTER PURGE HOSE
LE TUYAU DE PURGE DE LA BOITE
EGR VALVE
SOUPAPE EGR
PCV VALVE
SOUPAPE PCV
AIR PUMP DIVERTER VALVE
SOUPAPE DE DÉRIVATION DE POMPE A AIR
AIR CLEANER (AUXILIARY VIEW)
FILTRE D'AIR (AUTRE SCHÉMA)
VACUUM DIAPHRAGM
DIAPHRAGME DE VIDE
TEMPERATURE SENSOR
PALPEUR DE TEMPERATURE
HEATED AIR DOOR HOSE
LE TUYAU D'ENTRÉE D'AIR CHAUFFE
ESA VACUUM HOSE
LE TUYAU DE VIDE ESA
B BLUE
O ORANGE
R RED
W WHITE
Y YELLOW

1980—318 & 360 ENG.-4BBL CARB.-FURY-CANADA

VACUUM CIRCUITS

(© Chrysler Corp.)

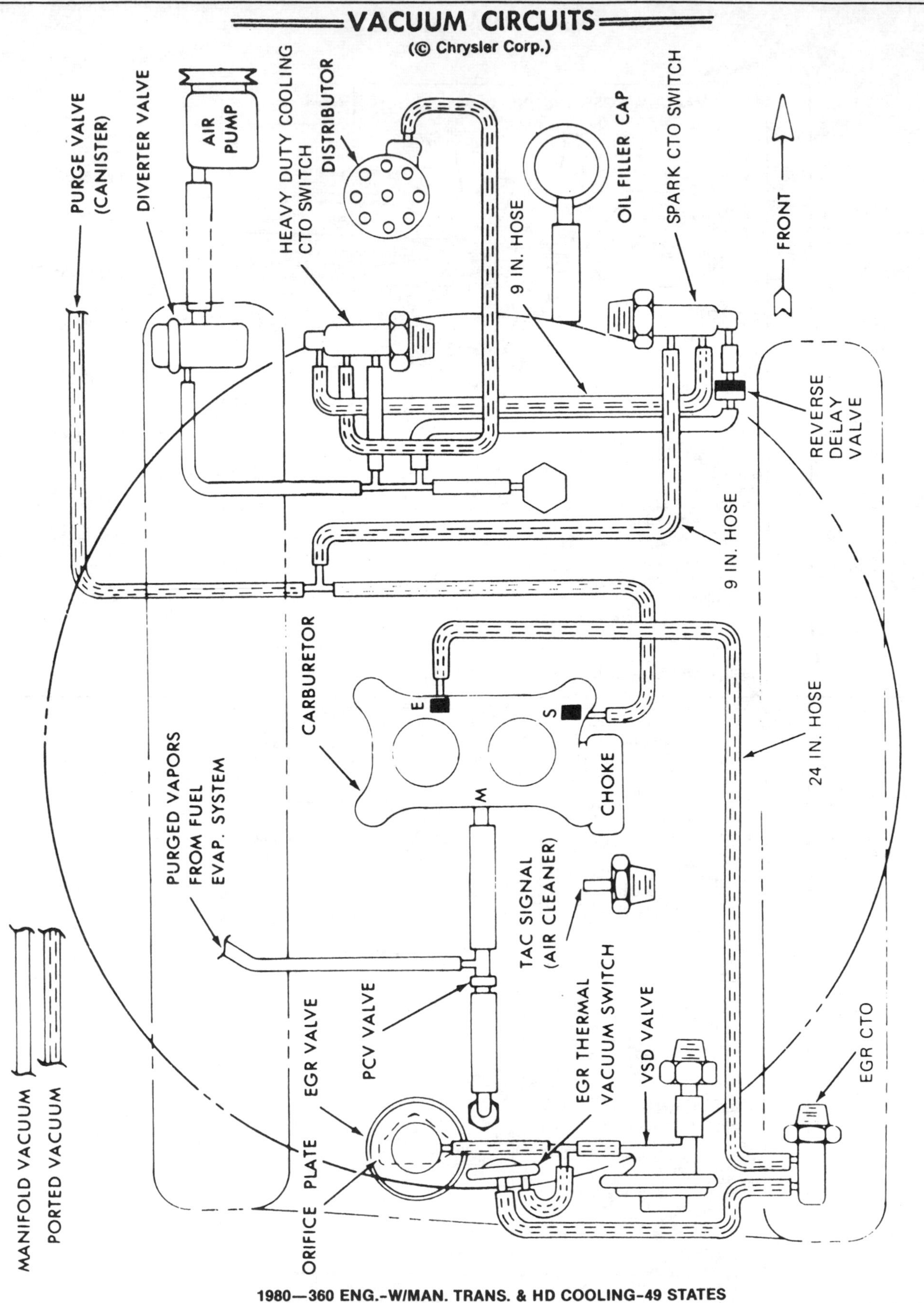

1980—360 ENG.-W/MAN. TRANS. & HD COOLING-49 STATES

VACUUM CIRCUITS

(© Chrysler Corp.)

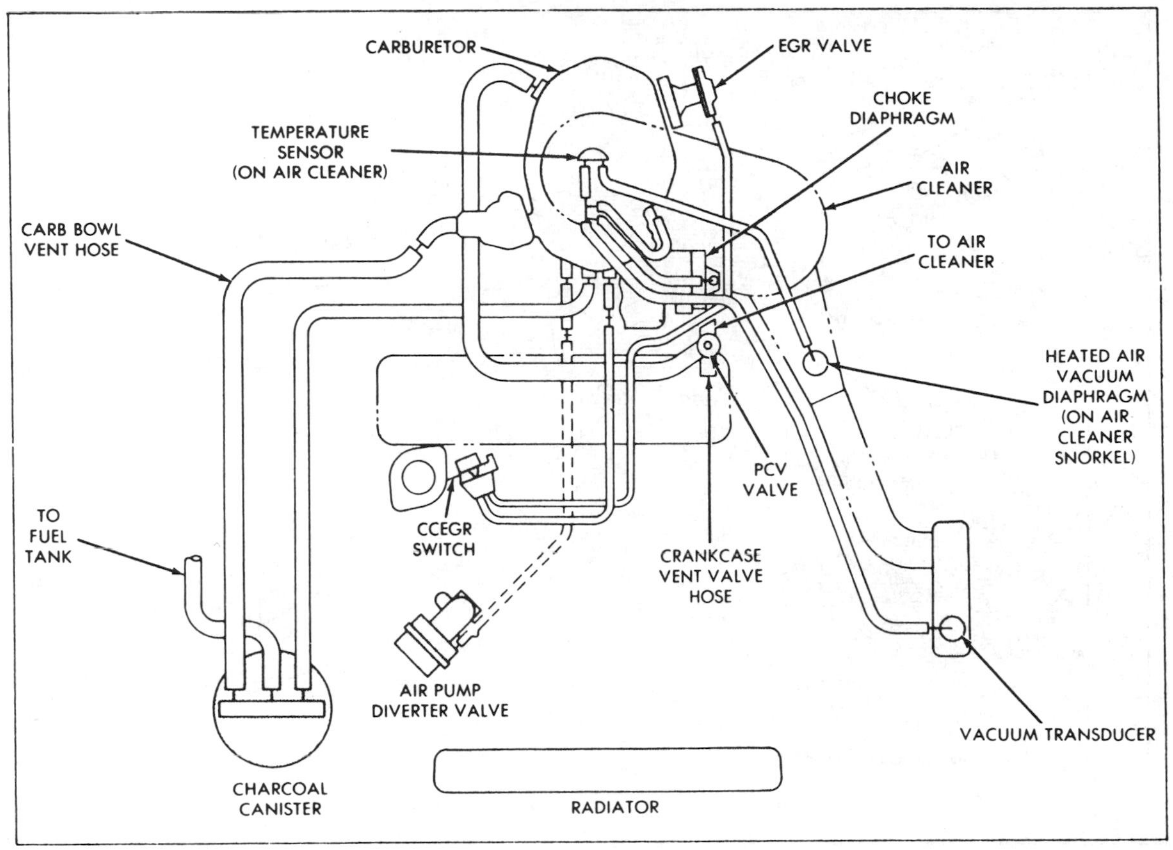

1980—OMNI & HORIZON

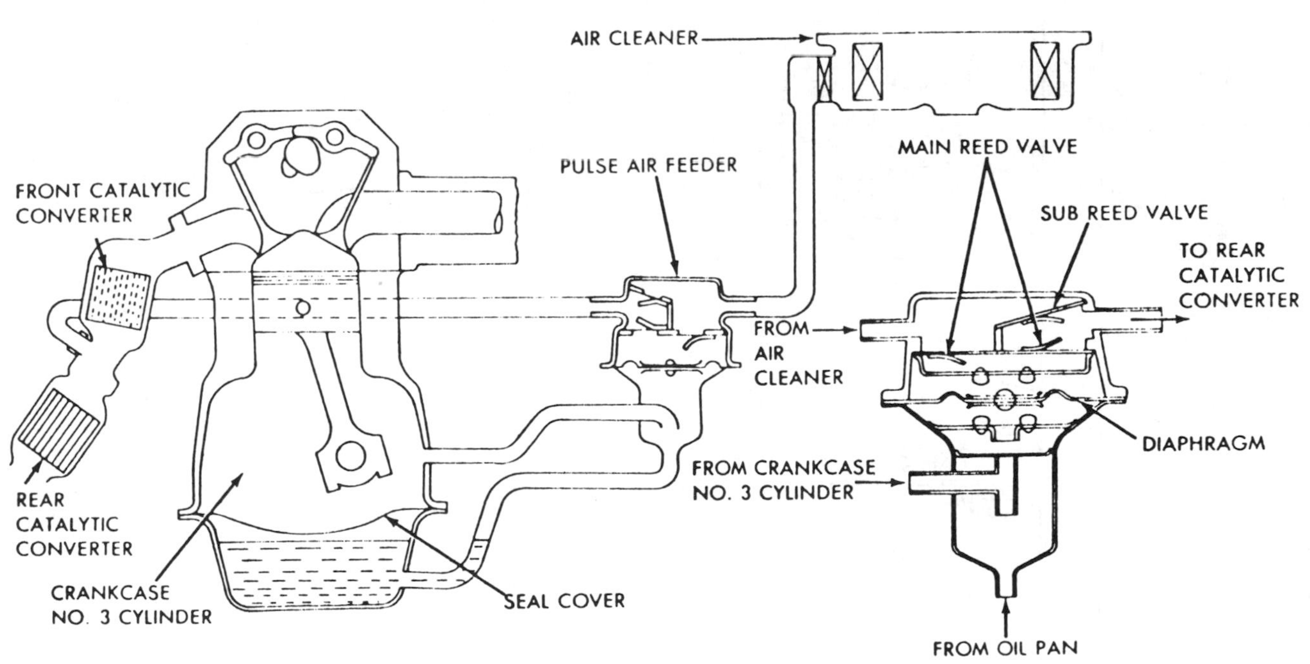

1980—PULSE AIR FEEDER SYSTEM

VACUUM CIRCUITS

(© Chrysler Corp.)

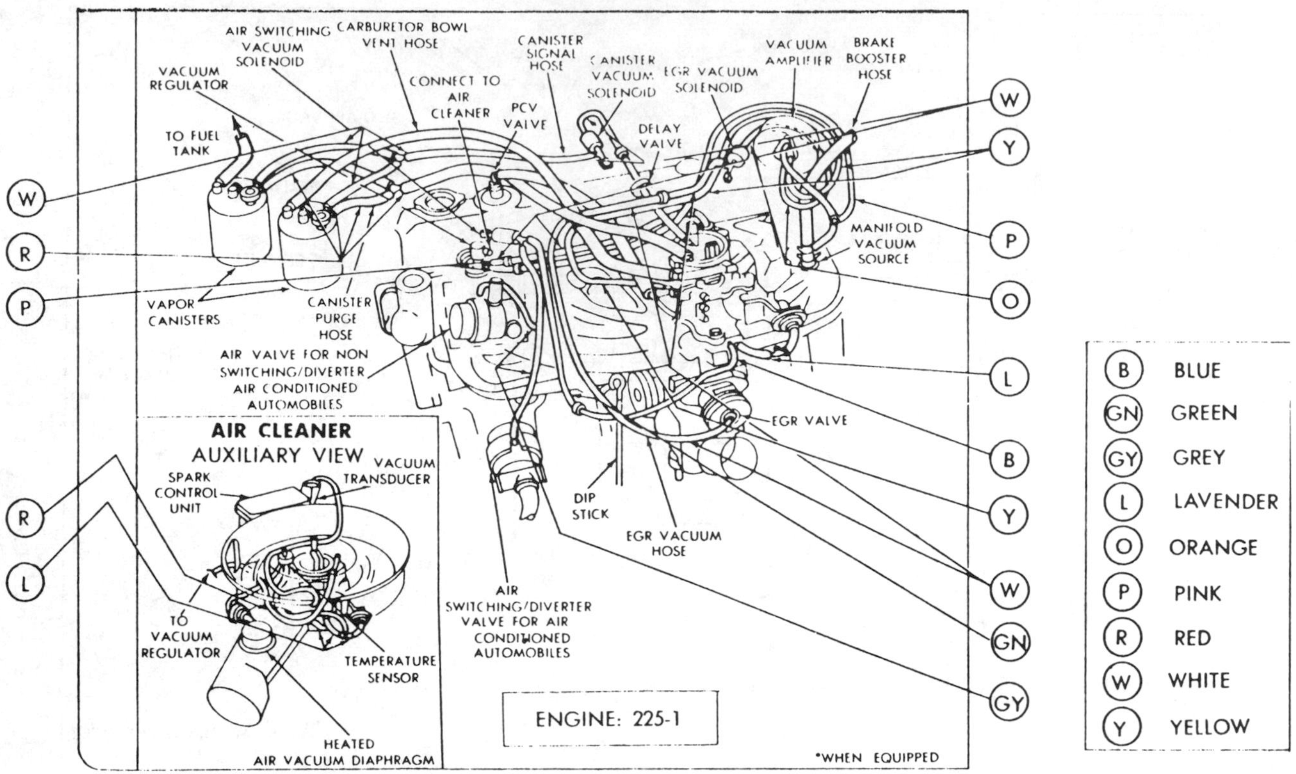

1980—225 ENG.-1BBL CARB.-CALIFORNIA

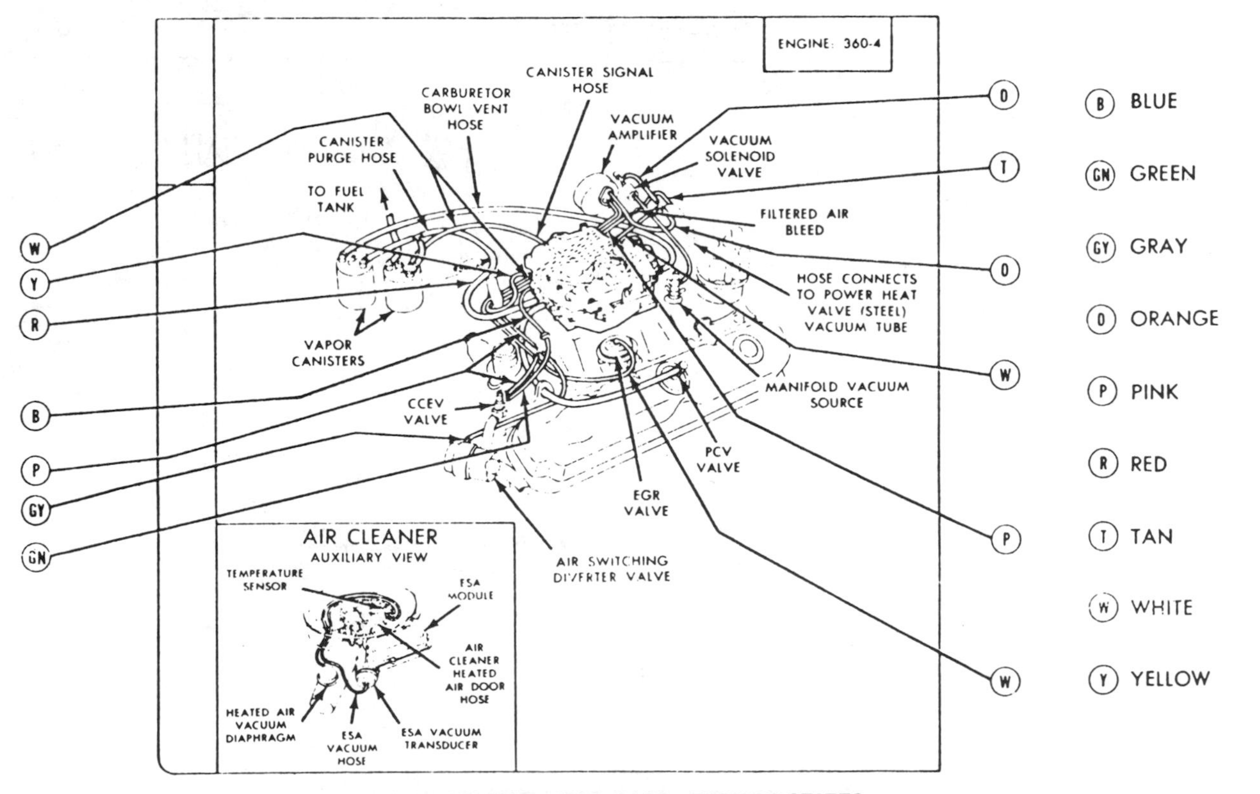

1980—360 ENG.-4BBL CARB.-FURY-49 STATES

VACUUM CIRCUITS

(© Chrysler Corp.)

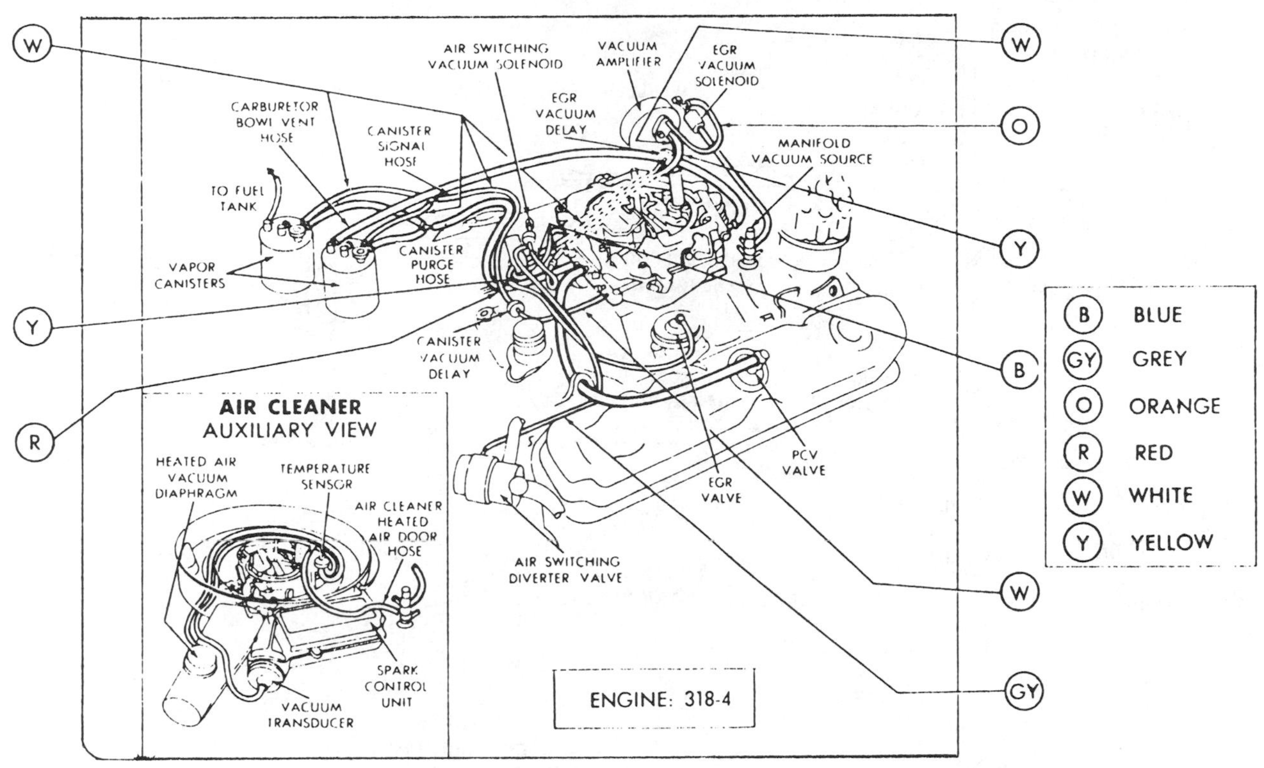

1980—318 ENG.-4BBL CARB.-LeBARON, GRAN FURY & HORIZON-CALIFORNIA

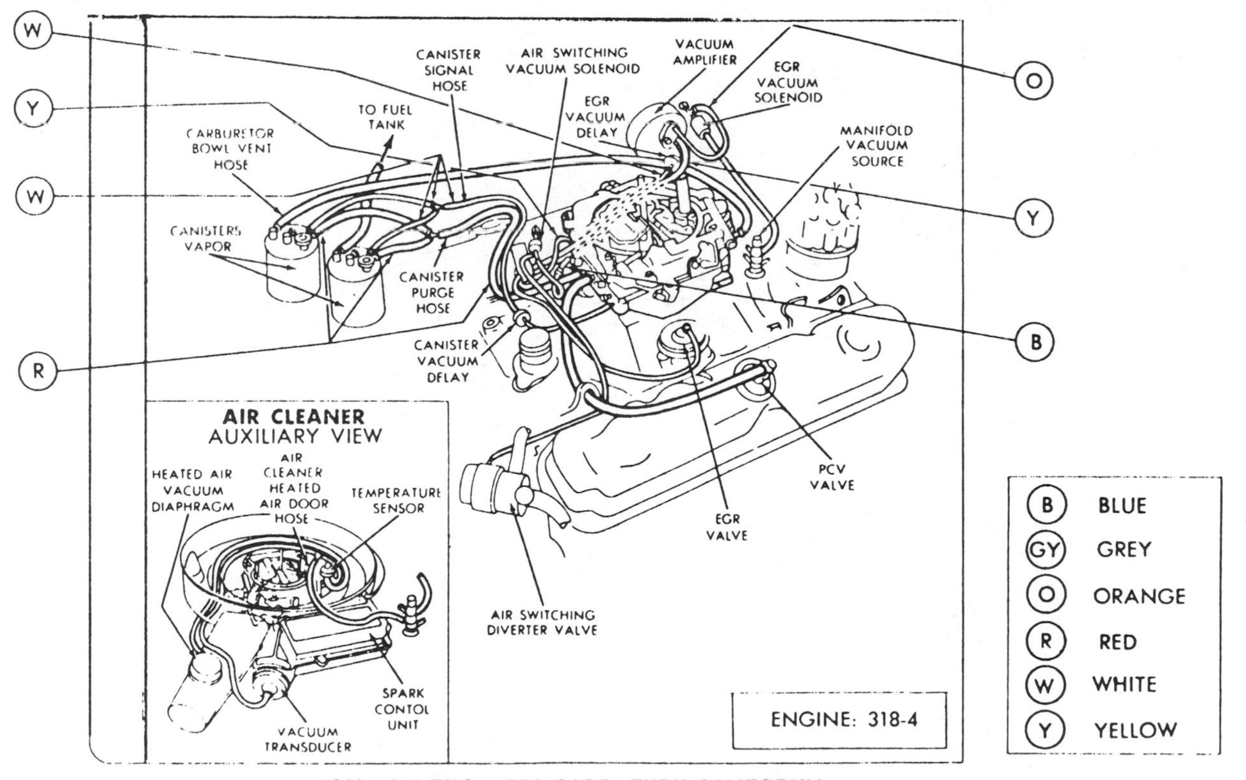

1980—318 ENG.-4BBL CARB.-FURY-CALIFORNIA

VACUUM CIRCUITS

(© Chrysler Corp.)

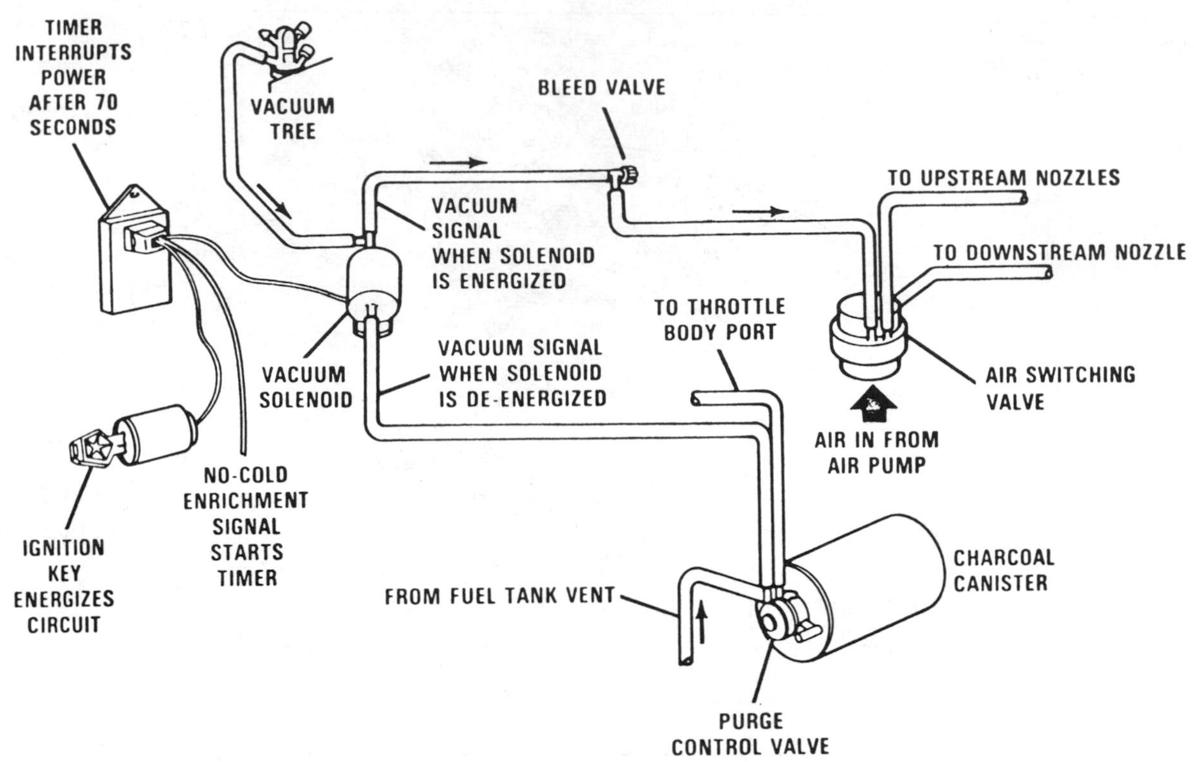

1981-82—AIR SWITCHING/CANISTER PURGING CONTROL CIRCUITS

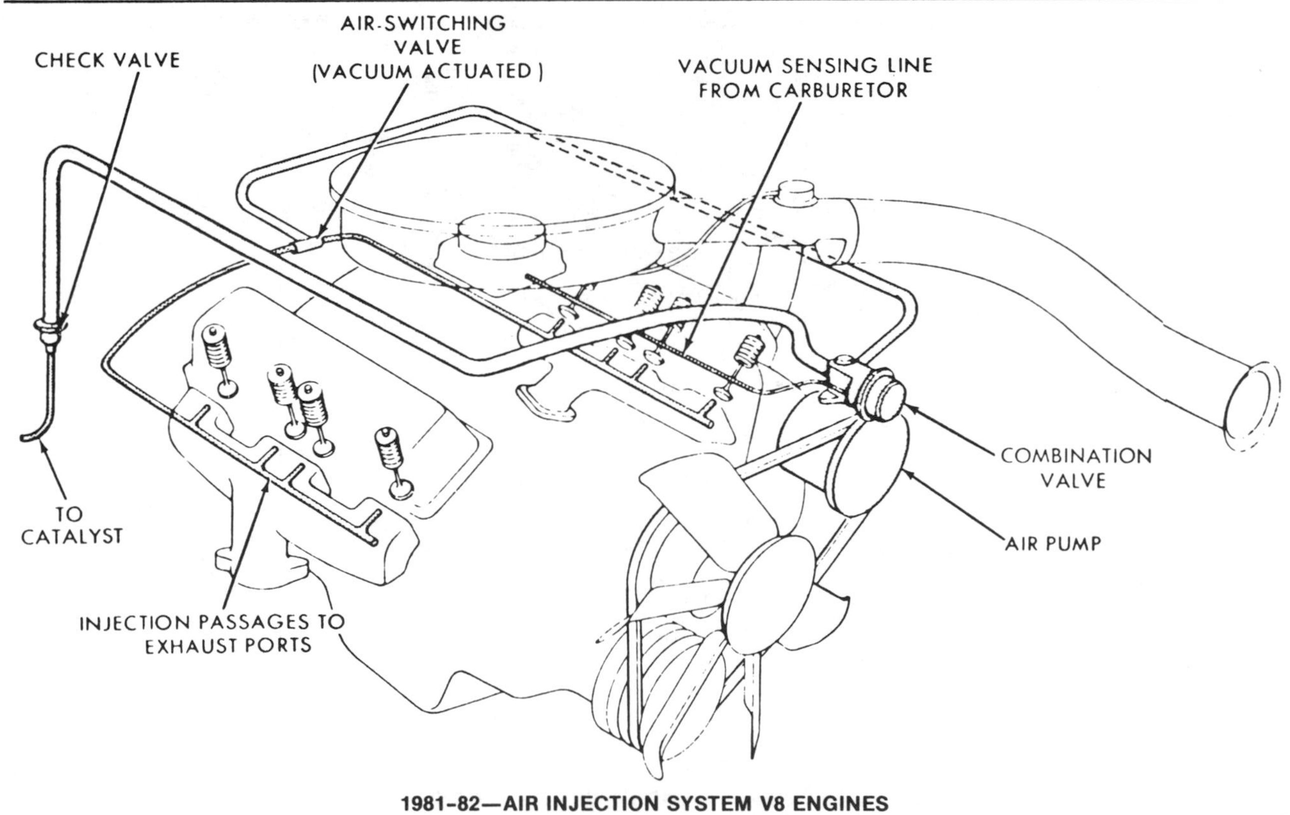

1981-82—AIR INJECTION SYSTEM V8 ENGINES

VACUUM CIRCUITS

(© Chrysler Corp.)

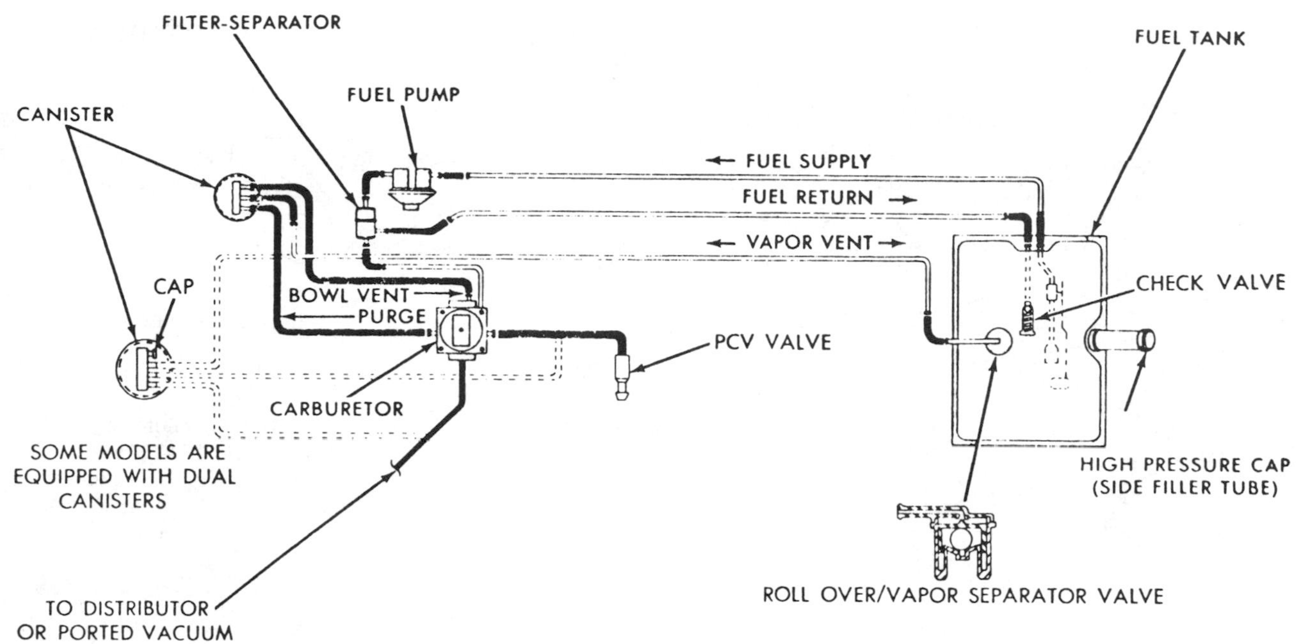

1981-82—EVAPORATIVE CONTROL SYSTEM

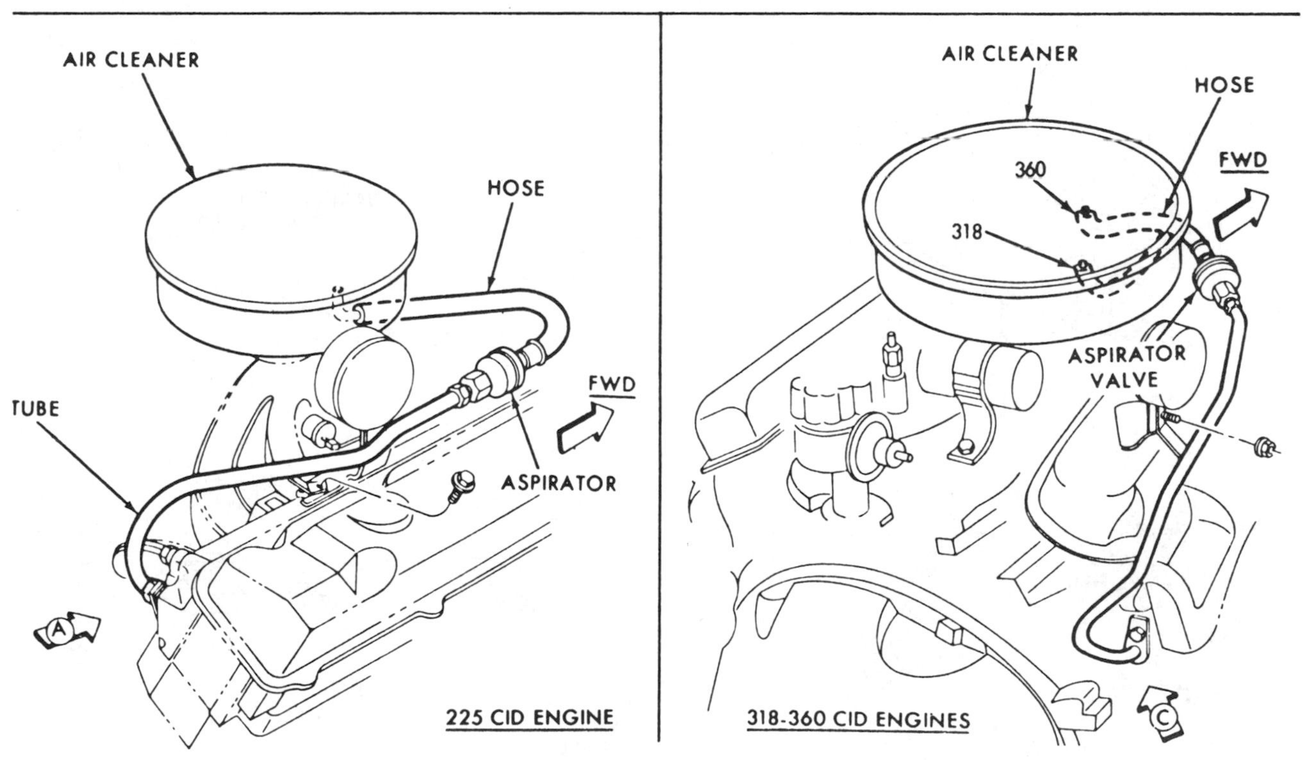

1981-82—ASPIRATOR SYSTEM—CANADA

VACUUM CIRCUITS

(© Chrysler Corp.)

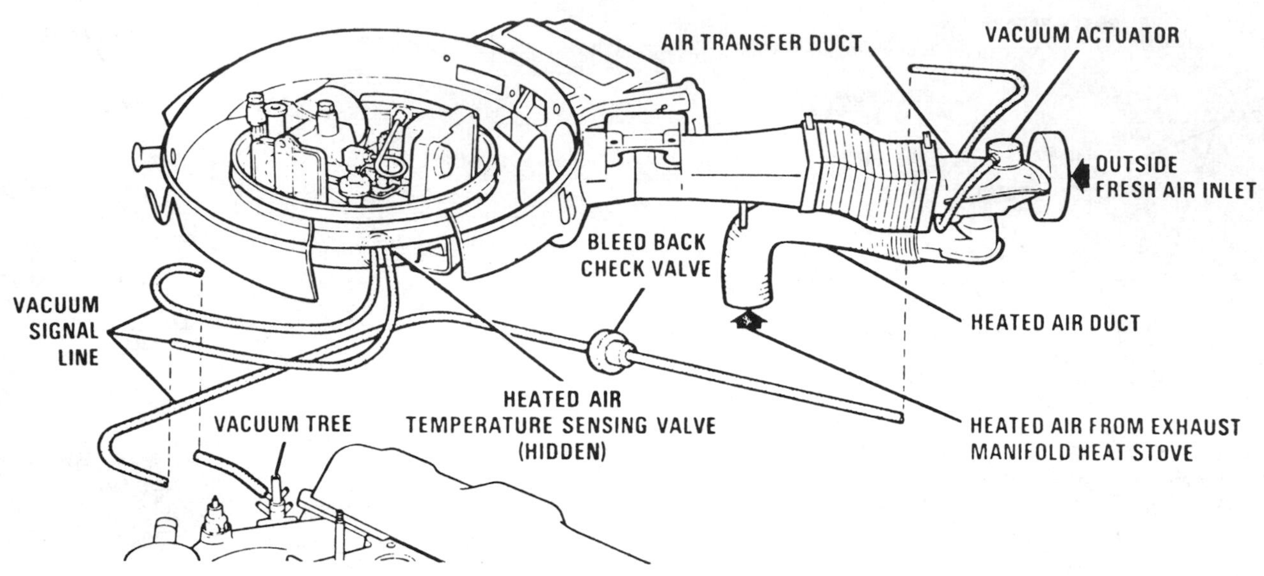

1981-82—FRESH/HEATED AIR MIXING ASSEMBLY

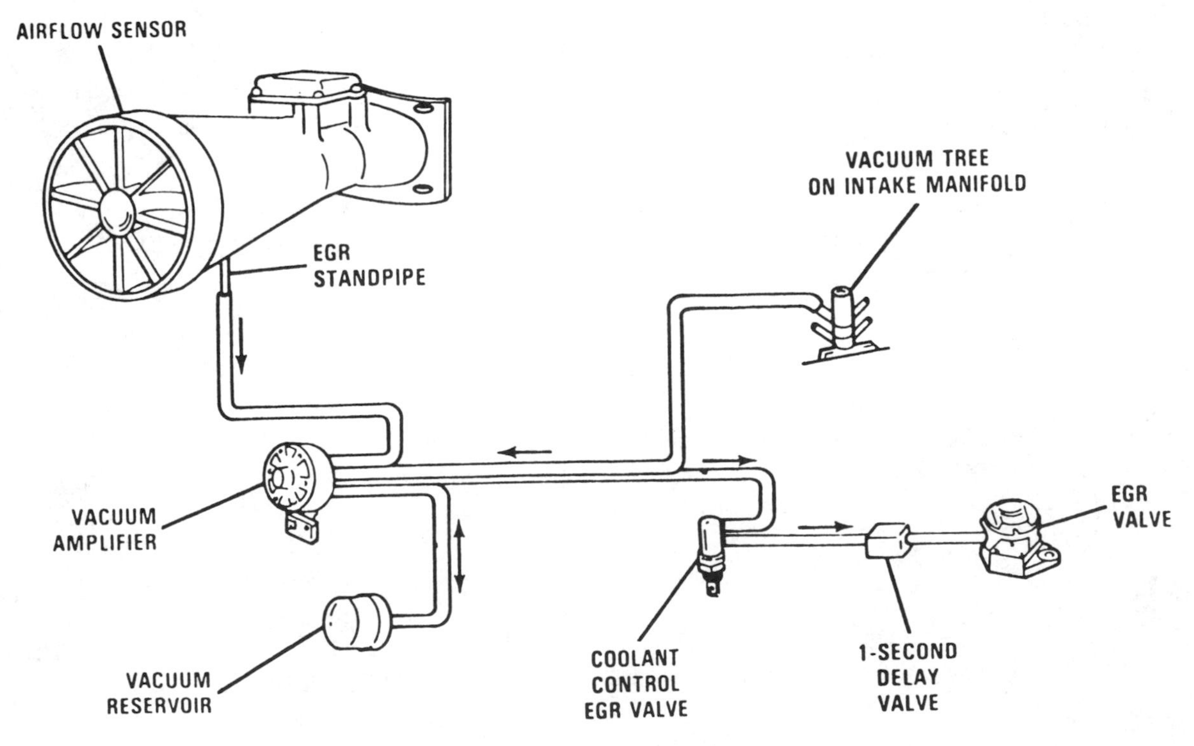

1981-82—EXHAUST GAS RECIRCULATION CONTROL CIRCUIT

VACUUM CIRCUITS

(© Chrysler Corp.)

1983

VEHICLE EMISSION CONTROL INFORMATION (VECI) LABEL

All vehicles are equipped with a **VECI** label. which is located in the engine compartment. This label is permanently attached and cannot be removed without destroying it. The specifications shown on the label are correct for the vehicle the label is mounted on. If any difference exists between the specifications shown on the label and those shown in the service manual, those shown on the label should be used.

4288351
CATALYST
CHRYSLER CORPORATION
VEHICLE EMISSION CONTROL INFORMATION
THIS VEHICLE CONFORMS TO U.S. EPA REGULATIONS APPLICABLE TO 1984 MODEL YEAR NEW MOTOR VEHICLES.
1.6L ECR1 6V2HDF7 ECRVA
SPARK PLUGS .035 IN. GAP RN12YC
ADJUSTMENTS MADE BY OTHER THAN APPROVED SERVICE MANUAL PROCEDURES MAY VIOLATE FEDERAL AND STATE LAWS

IDLE SETTINGS	MAN	AUTO
TIMING BTC	12°	12°
IDLE RPM	850	1000
FAST IDLE RPM	1400	1600

UNDER 300 MILES REDUCE ALL IDLE SPEEDS 75 RPM

COLD ENGINE VALVE LASH	
INTAKE	0.010 IN
EXHAUST	0.012 IN

- THE CARBURETOR IDLE MIXTURE HAS BEEN PRESET AT THE FACTORY. ADJUSTMENTS SHOULD NOT BE MADE DURING ROUTINE TUNE-UP.
- SEE SERVICE MANUAL FOR ADDITIONAL INFORMATION
- CHECK BASIC TIMING W/ESA MODULE VACUUM HOSE DISCONNECTED. BE SURE TO RECONNECT ESA HOSE AFTER SETTING BASIC TIMING.
- THIS VEHICLE IS EXEMPT FROM COMPLYING WITH HIGH ALTITUDE EMISSION STANDARD. VEHICLE PERFORMANCE IS DEEMED UNSUITABLE FOR PRINCIPLE USE AT HIGH ALTITUDE.
- CHECK IN NEUTRAL, LIGHTS AND ACCESSORIES OFF

Typical Federal VECI Label

4288357
CATALYST
EGR, OX, TWC, AP, CL
CHRYSLER CORPORATION
VEHICLE EMISSION CONTROL INFORMATION
2.2L ECR2 2V2HCL8 ECRVA
SPARK PLUGS .035 IN. GAP RN12YC
CHECK BASIC TIMING W/ESA MODULE VACUUM HOSE DISCONNECTED. BE SURE TO RECONNECT ESA HOSE AFTER SETTING BASIC TIMING.

- THE CARBURETOR IDLE MIXTURE HAS BEEN PRESET AT THE FACTORY. ADJUSTMENTS SHOULD NOT BE MADE DURING ROUTINE TUNE-UP.
- ADJUSTMENTS MADE BY OTHER THAN APPROVED SERVICE MANUAL PROCEDURES MAY VIOLATE FEDERAL AND STATE LAWS
- SEE SERVICE MANUAL FOR ADDITIONAL INFORMATION

THIS VEHICLE CONFORMS TO U.S. EPA AND STATE OF CALIFORNIA REGULATIONS APPLICABLE TO 1984 MODEL YEAR NEW MOTOR VEHICLES PROVIDED THAT THIS VEHICLE IS ONLY INTRODUCED INTO COMMERCE FOR SALE IN THE STATE OF CALIFORNIA.

IDLE SETTINGS	MAN	AUTO
TIMING BTC	10°	10°
IDLE RPM	800	900
FAST IDLE RPM	1500	1600

UNDER 300 MILES REDUCE ALL IDLE SPEEDS 75 RPM
CHECK IN NEUTRAL, LIGHTS AND ACCESSORIES OFF

Typical California VECI Label

REGLAGES DU RALENTI	MAN	AUTO
DISTRIBUTION AV P.H.	6°	6°
REGLAGE DU RALENTI EN TR/MIN	850	*700
RALENTI ACCELERE EN TR/MIN	N.A	N.A

CHRYSLER CANADA LTEE
RENSEIGNEMENTS RELATIFS AU CONTROLE DES EMISSIONS
CONSULTER LE MANUEL D'ENTRETIEN POUR DE PLUS AMPLES RENSEIGNEMENTS. LES REGLAGES QUI SERAIENT FAITS SELON D'AUTRES PROCEDURES QUE CELLES DUE MANUEL D'ENTRETIEN POURRAIENT ENFREINDRE LES LOIS FEDERALES ET PROVINCIALES.
TAILLE DU MOTEUR 2.2 LITRE
BOUGIES 0.9 ECARTEMENT X MM RN12YC
VERIFIER EN POSITION DE POINT MORT, LES ECLAIRAGES ET LES ACCESSOIRES ETANT A L'ARRET
*VERIFIER AVEC LA BOITE DE VITESSES EN MARCHE

4288365
CHRYSLER CANADA LTD
VEHICLE EMISSION CONTROL INFORMATION
SEE SERVICE MANUAL FOR ADDITIONAL INFORMATION. ADJUSTMENTS MADE BY OTHER THAN APPROVED SERVICE MANUAL PROCEDURES MAY VIOLATE FEDERAL AND PROVINCIAL LAWS.
ENGINE SIZE 2.2 LITRE
SPARK PLUGS 0.9 MM. GAP RN12YC
CHECK IN NEUTRAL, LIGHTS AND ACCESSORIES OFF
*CHECK IN DRIVE.

IDLE SETTINGS	MAN	AUTO
TIMING BTC	6°	6°
IDLE R/MIN	850	*700
FAST IDLE R/MIN	NA	N.A

Typical Canadian VECI Label

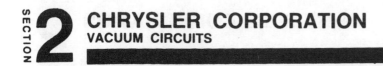

VACUUM CIRCUITS

(© Chrysler Corp.)

1983 VACUUM HOSE ROUTING LABEL

All vehicles are equipped with a vacuum hose routing label which is located in the engine compartment. This label is permanently attached and cannot be removed without destroying it. All hoses must be connected and routed as shown on the label.

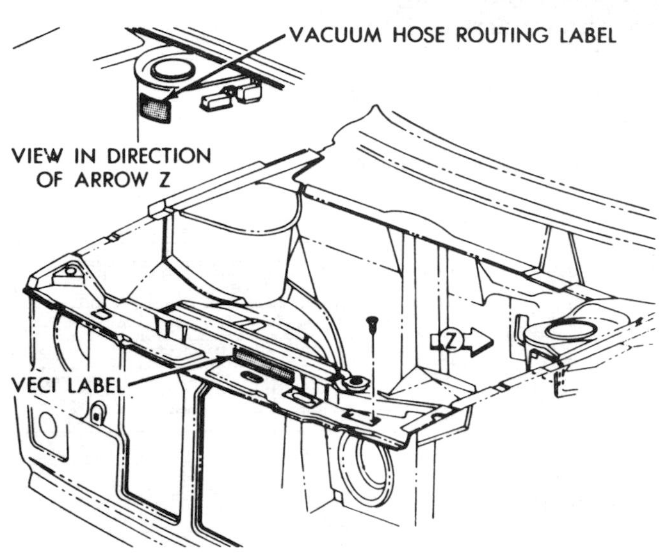

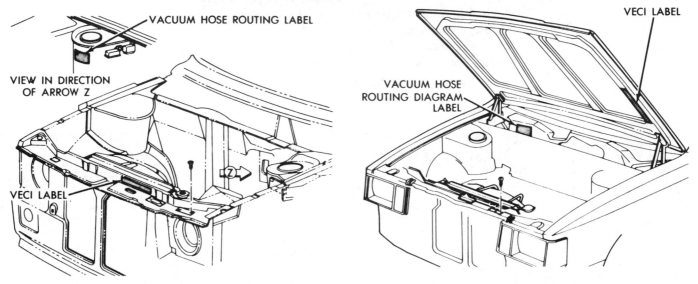

Underhood Label Location MZ

Underhood Label Location—Typical KEG

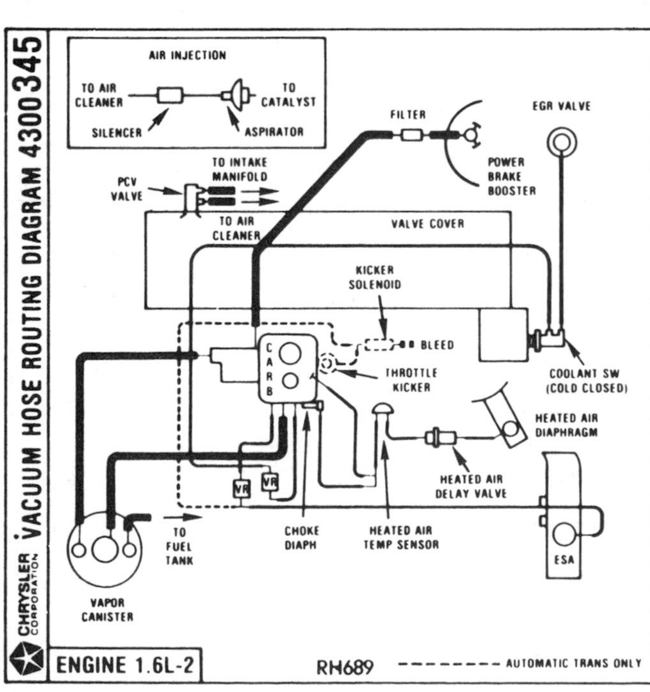

Hose Routing Label 1.6L Federal/California

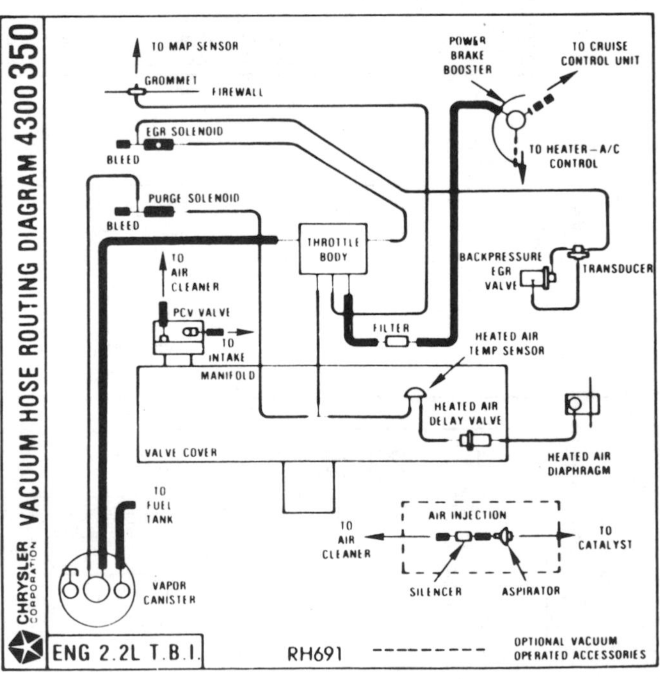

Hose Routing Label 2.2L EFI

VACUUM CIRCUITS

(© Chrysler Corp.)

1983

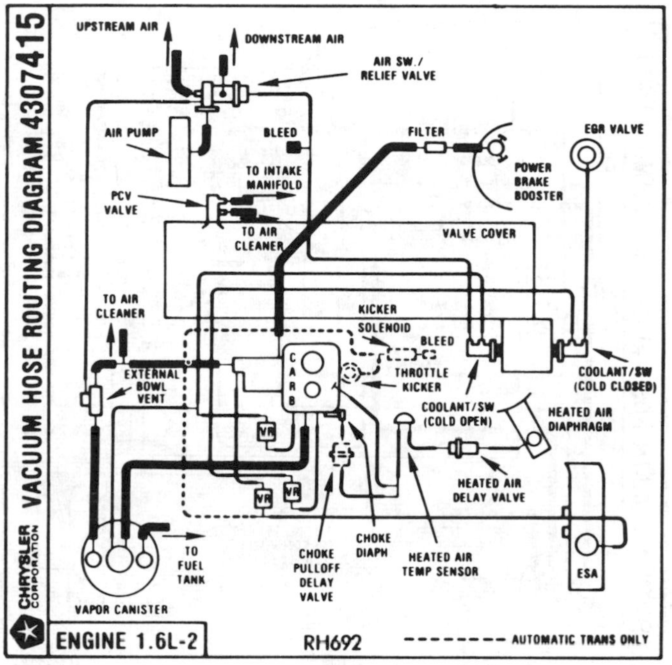

Hose Routing Label 1.6L Canada

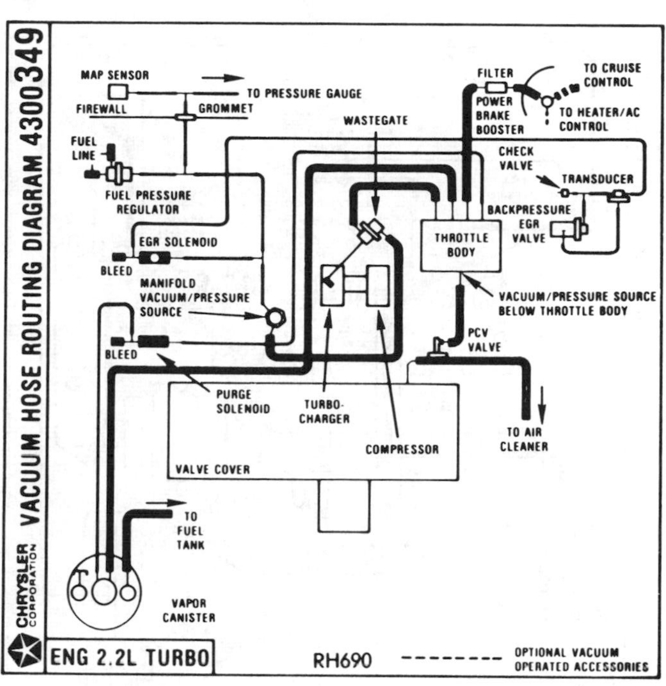

Hose Routing Label 2.2L Turbo

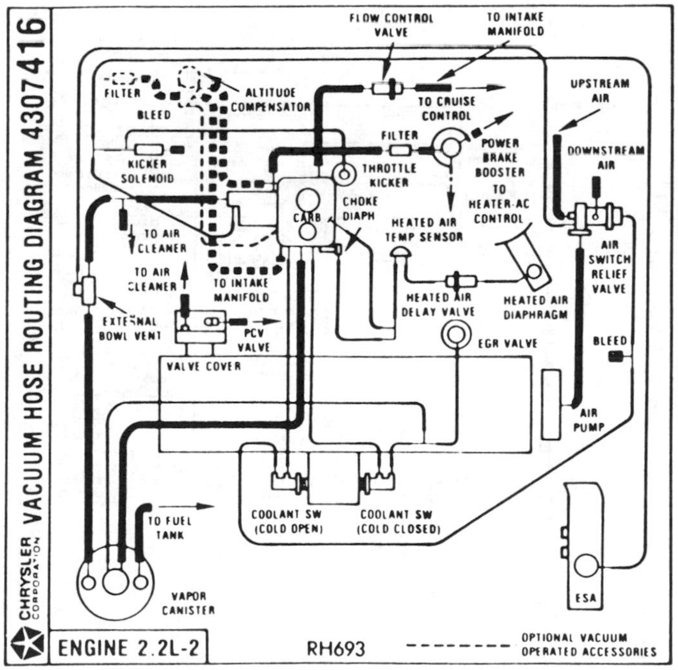

Hose Routing Label 2.2L Federal

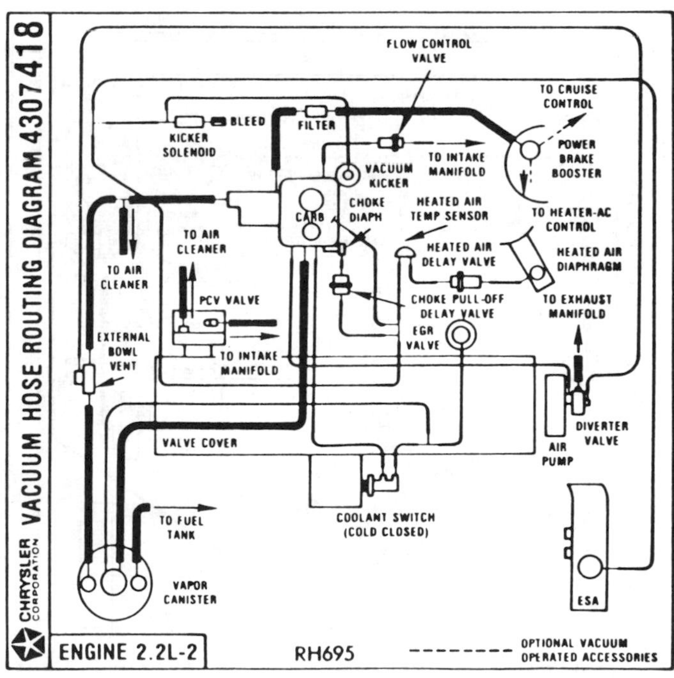

Hose Routing Label 2.2L Canada

VACUUM CIRCUITS

(© Chrysler Corp.)

1983

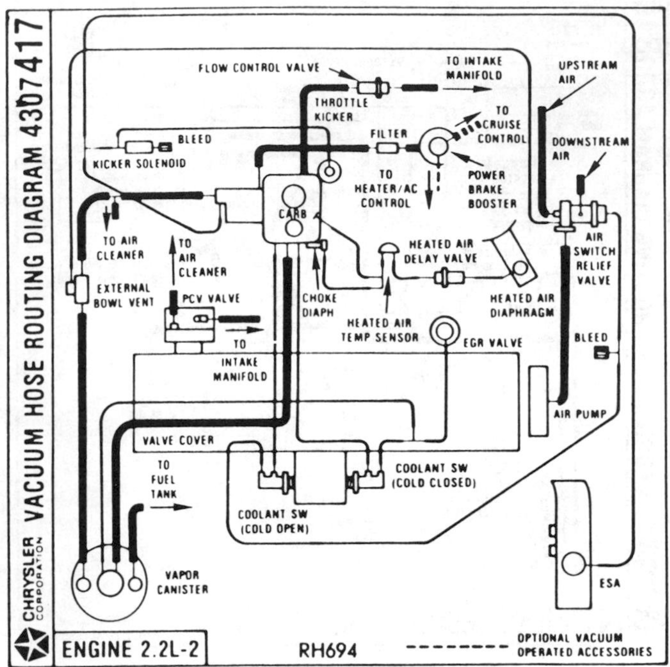

Hose Routing Label 2.2L California

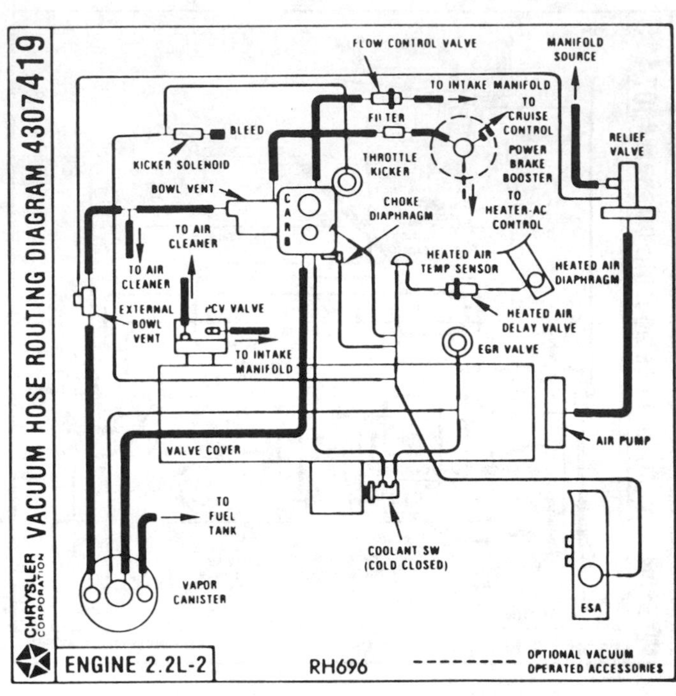

Hose Routing Label 2.2L High Altitude

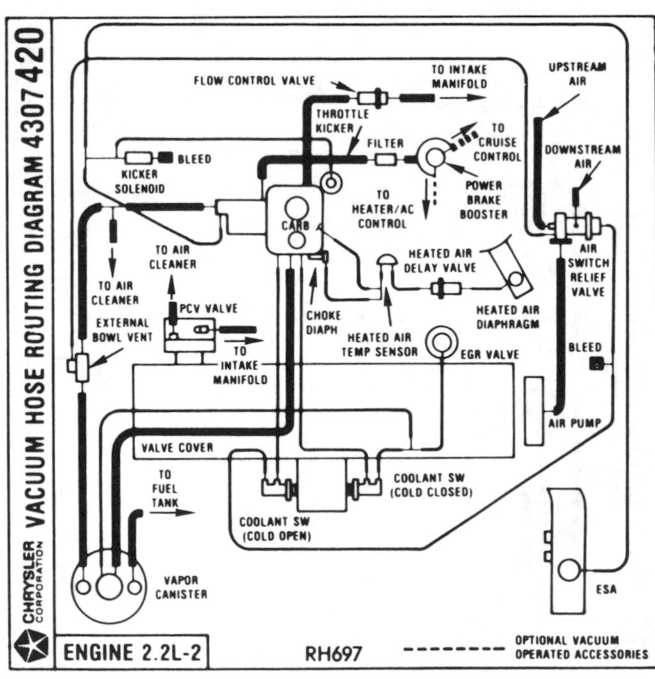

Hose Routing Label Z 28 Federal/Canada

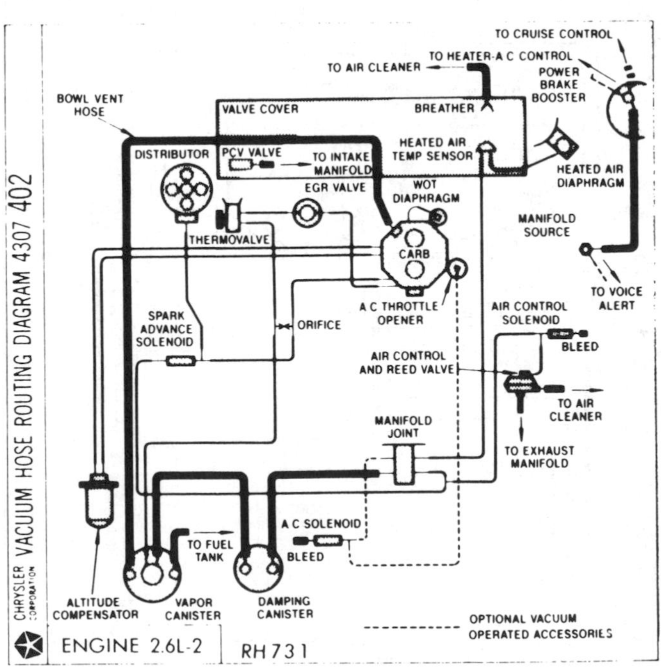

Hose Routing Label 2.6L Federal

VACUUM CIRCUITS

(© Chrysler Corp.)

1983

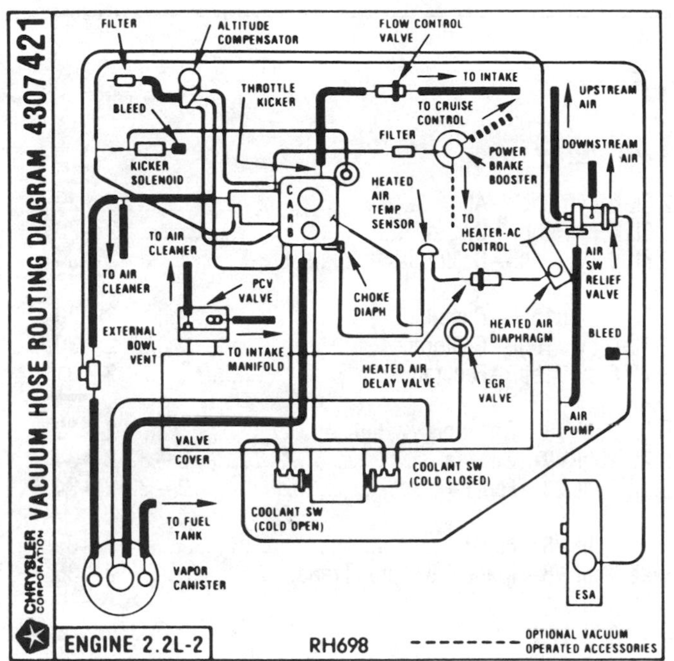

Hose Routing Label Z 28 California

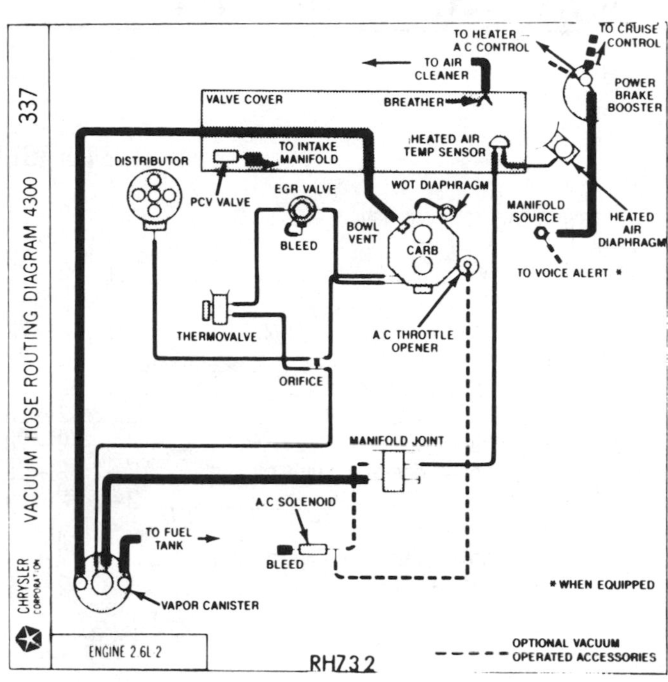

Hose Routing Label 2.6L Canadian

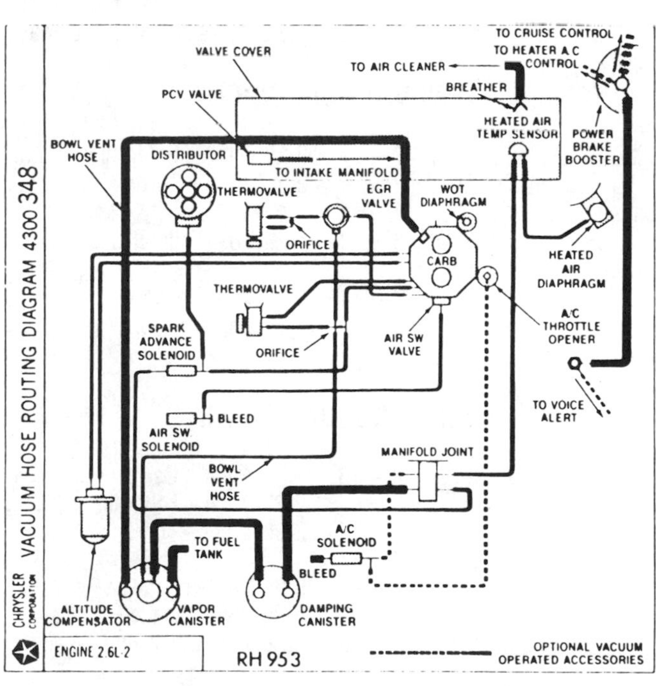

Hose Routing Label 2.6L California

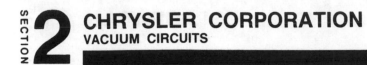

Chrysler
Vacuum Circuits

INDEX

1984 VACUUM CIRCUITS

1985 VACUUM CIRCUITS

1986 VACUUM CIRCUITS

1987 VACUUM CIRCUITS

VACUUM CIRCUITS

(© Chrysler Corp.)

1984

VEHICLE EMISSION CONTROL INFORMATION (VECI) LABEL

All vehicles are equipped with a **VECI** label. which is located in the engine compartment. This label is permanently attached and cannot be removed without destroying it. The specifications shown on the label are correct for the vehicle the label is mounted on. If any difference exists between the specifications shown on the label and those shown in the service manual, those shown on the label should be used.

4288351
CATALYST

CHRYSLER CORPORATION
VEHICLE EMISSION CONTROL INFORMATION

THIS VEHICLE CONFORMS TO U.S. EPA REGULATIONS APPLICABLE TO 1984 MODEL YEAR NEW MOTOR VEHICLES.

1.6L
ECR1.6V2HDF7
ECRVA

SPARK PLUGS
035 IN. GAP
RN12YC

ADJUSTMENTS MADE BY OTHER THAN APPROVED SERVICE MANUAL PROCEDURES MAY VIOLATE FEDERAL AND STATE LAWS.

IDLE SETTINGS	MAN	AUTO
TIMING BTC	12°	12°
IDLE RPM	850	1000
FAST IDLE RPM	1400	1600

UNDER 300 MILES REDUCE ALL IDLE SPEEDS 75 RPM.

COLD ENGINE VALVE LASH

INTAKE	0.010 IN.
EXHAUST	0.012 IN.

- THE CARBURETOR IDLE MIXTURE HAS BEEN PRESET AT THE FACTORY. ADJUSTMENTS SHOULD NOT BE MADE DURING ROUTINE TUNE-UP.
- SEE SERVICE MANUAL FOR ADDITIONAL INFORMATION.
- CHECK BASIC TIMING W/ESA MODULE VACUUM HOSE DISCONNECTED. BE SURE TO RECONNECT ESA HOSE AFTER SETTING BASIC TIMING.
- THIS VEHICLE IS EXEMPT FROM COMPLYING WITH HIGH ALTITUDE EMISSION STANDARC VEHICLE PERFORMANCE IS DEEMED UNSUITABLE FOR PRINCIPLE USE AT HIGH ALTITUDE.
- CHECK IN NEUTRAL, LIGHTS AND ACCESSORIES OFF.

Typical Federal VECI Label

4288357
CATALYST
EGR,OX,TWC,AP,CL

CHRYSLER CORPORATION
VEHICLE EMISSION CONTROL INFORMATION

2.2L
EGR2.2V2HCLB
ECRVA

SPARK PLUGS
.035 IN. GAP
RN12YC

CHECK BASIC TIMING W/ESA MODULE VACUUM HOSE DISCONNECTED. BE SURE TO RECONNECT ESA HOSE AFTER SETTING BASIC TIMING.

- THE CARBURETOR IDLE MIXTURE HAS BEEN PRESET AT THE FACTORY. ADJUSTMENTS SHOULD NOT BE MADE DURING ROUTINE TUNE-UP.
- ADJUSTMENTS MADE BY OTHER THAN APPROVED SERVICE MANUAL PROCEDURES MAY VIOLATE FEDERAL AND STATE LAWS.
- SEE SERVICE MANUAL FOR ADDITIONAL INFORMATION.

THIS VEHICLE CONFORMS TO U.S. EPA AND STATE OF CALIFORNIA REGULATIONS APPLICABLE TO 1984 MODEL YEAR NEW MOTOR VEHICLES PROVIDED THAT THIS VEHICLE IS ONLY INTRODUCED INTO COMMERCE FOR SALE IN THE STATE OF CALIFORNIA.

IDLE SETTINGS	MAN	AUTO
TIMING BTC	10°	10°
IDLE RPM	800	900
FAST IDLE RPM	1500	1600

UNDER 300 MILES REDUCE ALL IDLE SPEEDS 75 RFM.

CHECK IN NEUTRAL, LIGHTS AND ACCESSORIES OFF.

Typical California VECI Label

REGLAGES DU RALENTI	MAN	AUTO
DISTRIBUTION AV P.H.	6°	6°
REGLAGE DU RALENTI EN TR/MIN.	850	*700
RALENTI ACCELERE EN TR/MIN.	N.A	N.A.

CHRYSLER CANADA LTEE
RENSEIGNEMENTS RELATIFS AU CONTROLE DES EMISSIONS

CONSULTER LE MANUEL D'ENTRETIEN POUR DE PLUS AMPLES RENSEIGNEMENTS. LES REGLAGES QUI SERAIENT FAITS SELON D'AUTRES PROCEDURES QUE CELLES DUE MANUEL D'ENTRETIEN POURRAIENT ENFREINDRE LES LOIS FEDERALES ET PROVINCIALES.

TAILLE DU MOTEUR
2.2 LITRE

BOUGIES
0.9 ECARTEMENT X MM.
RN12YC

VERIFIER EN POSITION DE POINT MORT, LES ECLAIRAGES ET LES ACCESSOIRES ETANT A L'ARRET.
*VERIFIER AVEC LA BOITE DE VITESSES EN MARCHE

4288365

CHRYSLER CANADA LTD
VEHICLE EMISSION CONTROL INFORMATION

SEE SERVICE MANUAL FOR ADDITIONAL INFORMATION. ADJUSTMENTS MADE BY OTHER THAN APPROVED SERVICE MANUAL PROCEDURES MAY VIOLATE FEDERAL AND PROVINCIAL LAWS.

ENGINE SIZE
2.2 LITRE

SPARK PLUGS
0.9 MM. GAP
RN12YC

CHECK IN NEUTRAL, LIGHTS AND ACCESSORIES OFF.
*CHECK IN DRIVE.

IDLE SETTINGS	MAN	AUTO
TIMING BTC	6°	6°
IDLE R/MIN.	850	*700
FAST IDLE R/MIN.	NA.	N.A

Typical Canadian VECI Label

VACUUM CIRCUITS

(© Chrysler Corp.)

1984—1.6L (98 CID) 4 CYL. engines

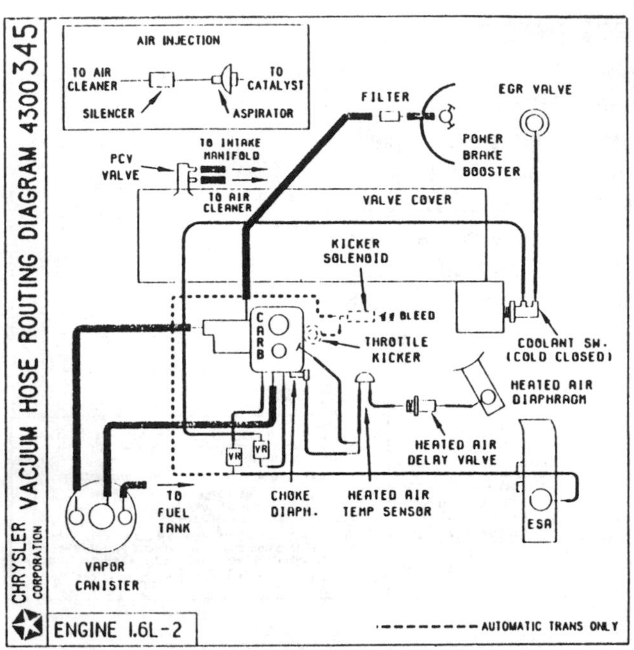

1984—4 CYL.-1.6L-2BBL, AUTO TRANS.—ALL

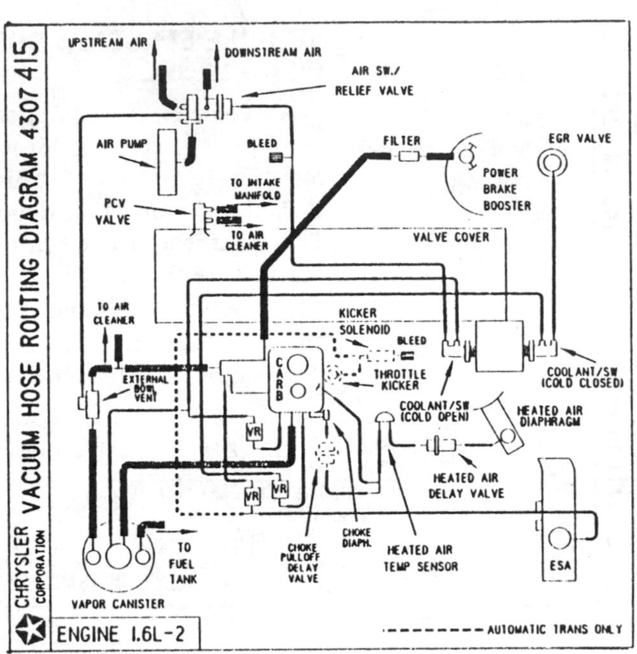

1984—4 CYL.-1.6L-2BBL, AUTO TRANS.—ALL

1984—2.2L (135 CID) 4 CYL. engines

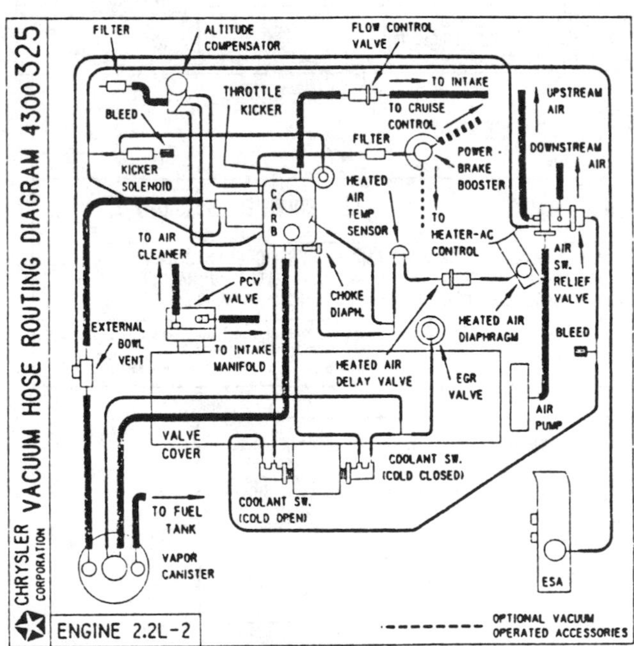

1984—4 CYL.-2.2L-2BBL—ALL

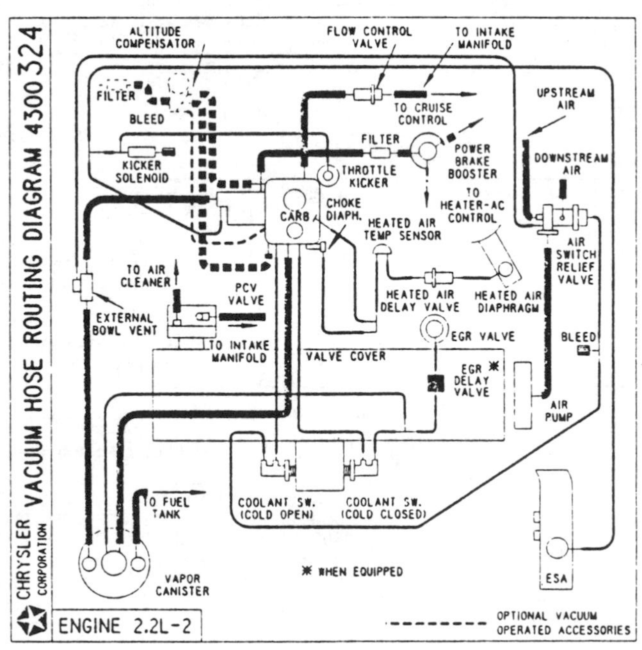

1984—4 CYL.-2.2L-2BBL—FEDERAL

VACUUM CIRCUITS

(© Chrysler Corp.)

1984—2.2L (135 CID) 4 CYL. engines

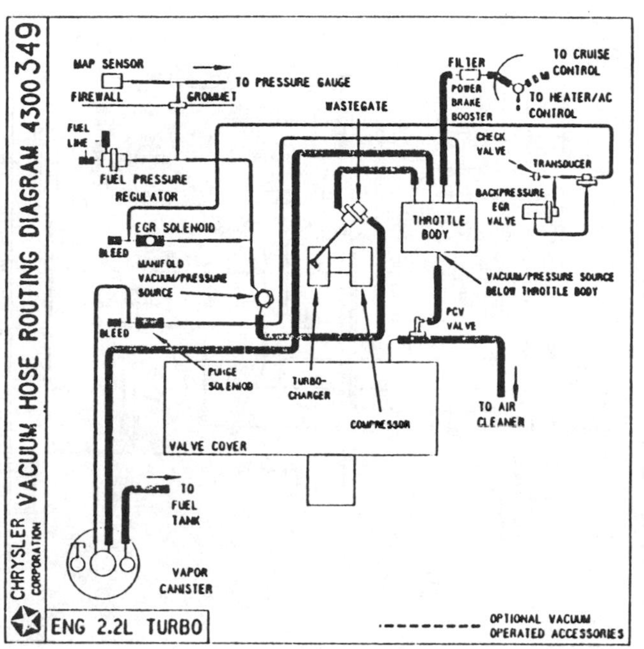

1984—4 CYL.-2.2L-T.B.I., TURBO—ALL

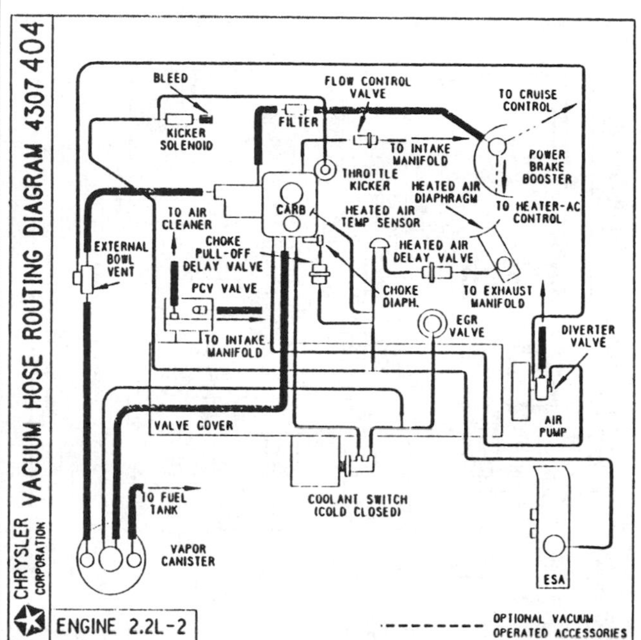

1984—4 CYL.-2.2L-2BBL—CANADA & EXPORT

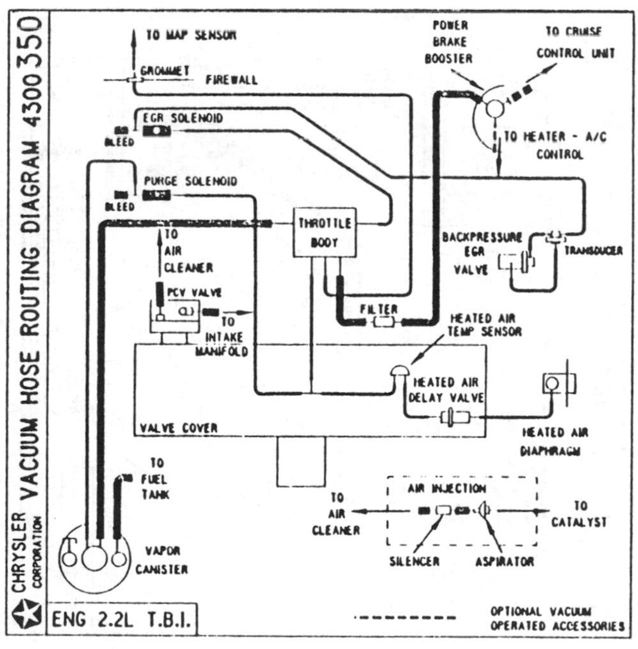

1984—4 CYL.-2.2L-T.B.I.—ALL

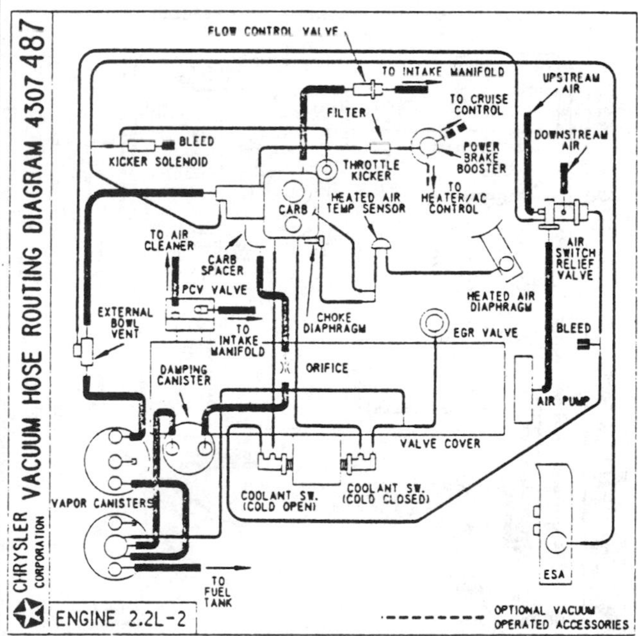

1984—4 CYL.-2.2L-2BBL, E.S.A.—CANADA

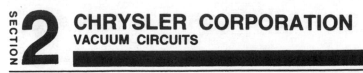

VACUUM CIRCUITS

(© Chrysler Corp.)

1984—2.2L (135 CID) 4 CYL. engines

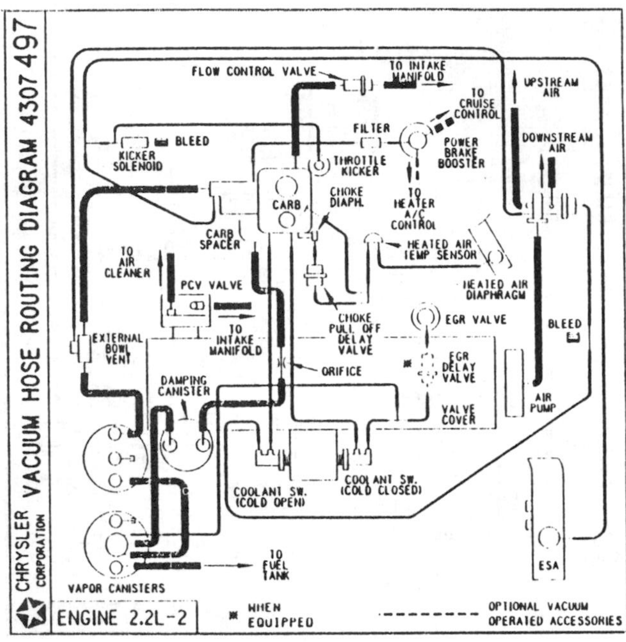

1984—4 CYL.-2.2L-2BBL—CALIFORNIA

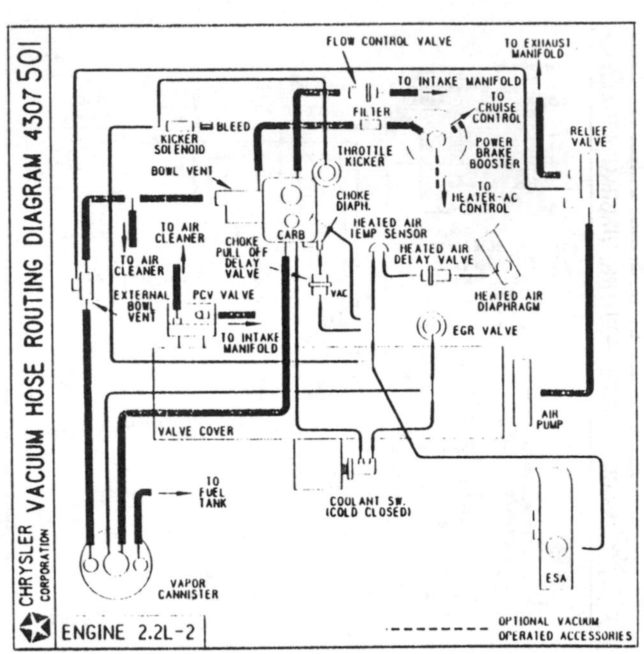

1984—4 CYL.-2.2L-2BBL—FEDERAL & HIGH ALT.

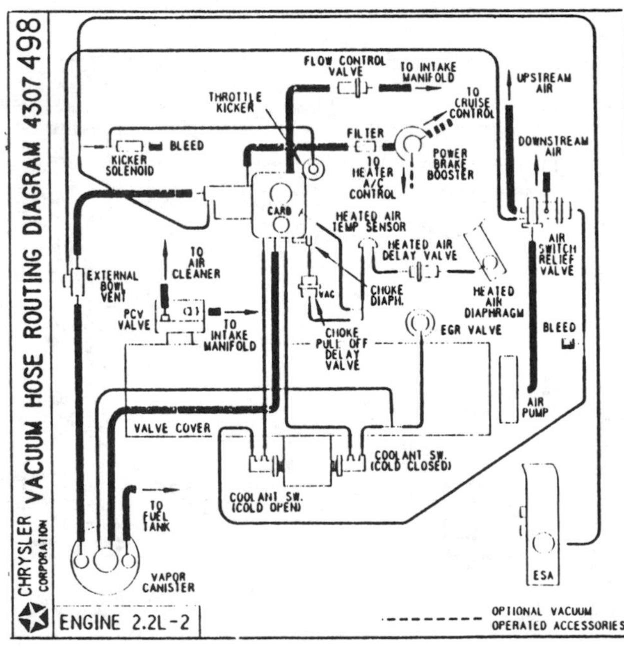

1984—4 CYL.-2.2L-2BBL—CALIFORNIA

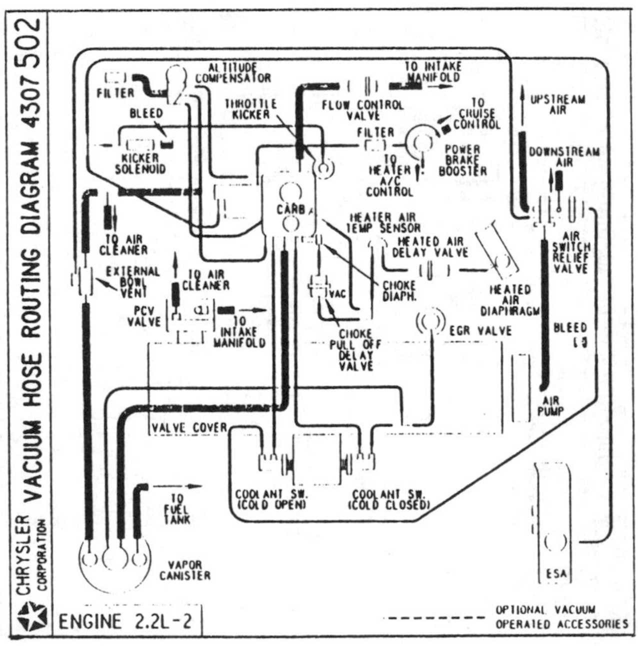

1984—4 CYL.-2.2L-2BBL, E.S.A.—FEDERAL

VACUUM CIRCUITS

1984—2.2L (135 CID) 4 CYL. engines

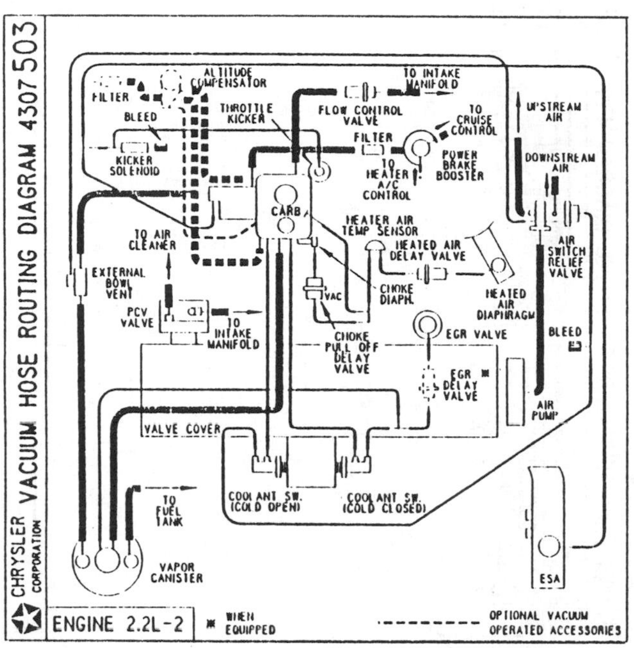

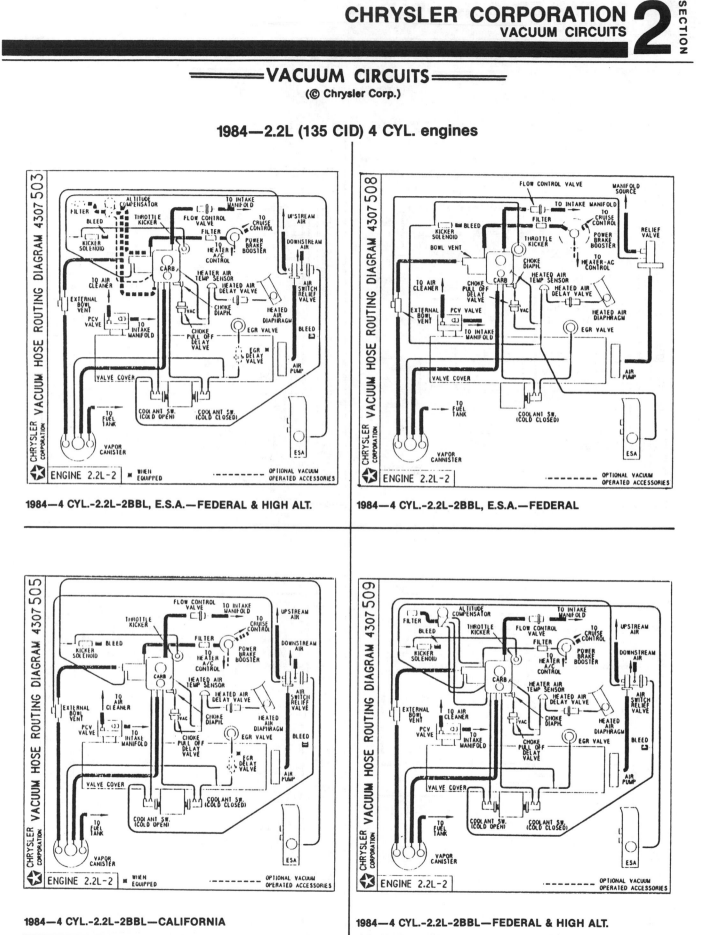

1984—4 CYL.-2.2L-2BBL, E.S.A.—FEDERAL & HIGH ALT.

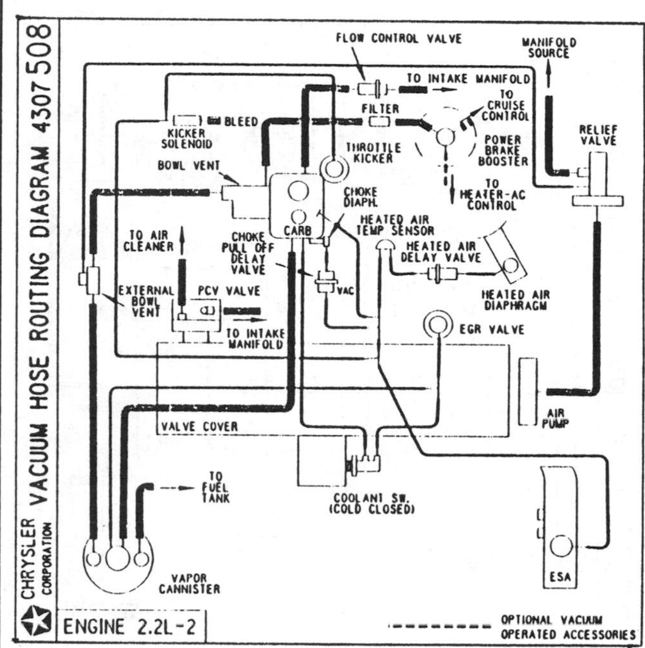

1984—4 CYL.-2.2L-2BBL, E.S.A.—FEDERAL

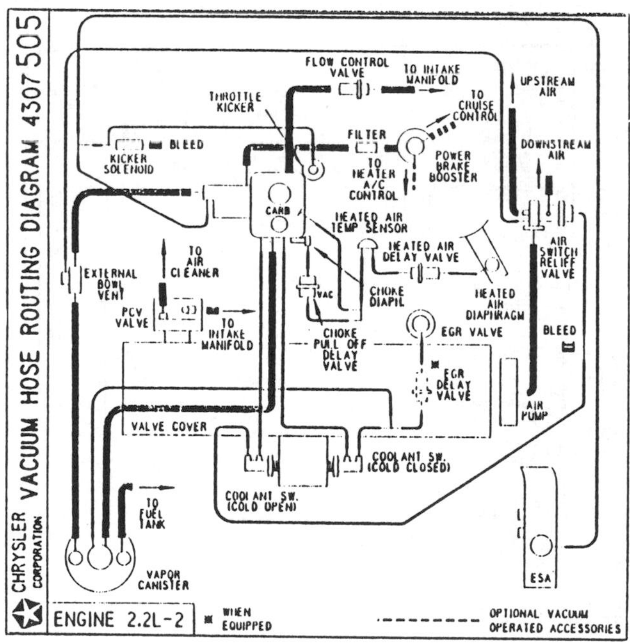

1984—4 CYL.-2.2L-2BBL—CALIFORNIA

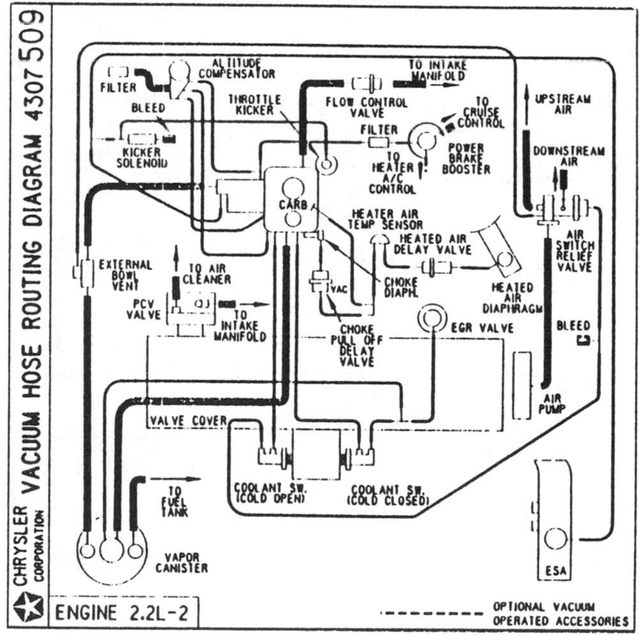

1984—4 CYL.-2.2L-2BBL—FEDERAL & HIGH ALT.

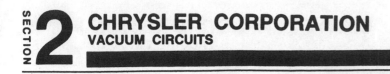

VACUUM CIRCUITS

(© Chrysler Corp.)

1984—2.2L (135 CID) 4 CYL. engines

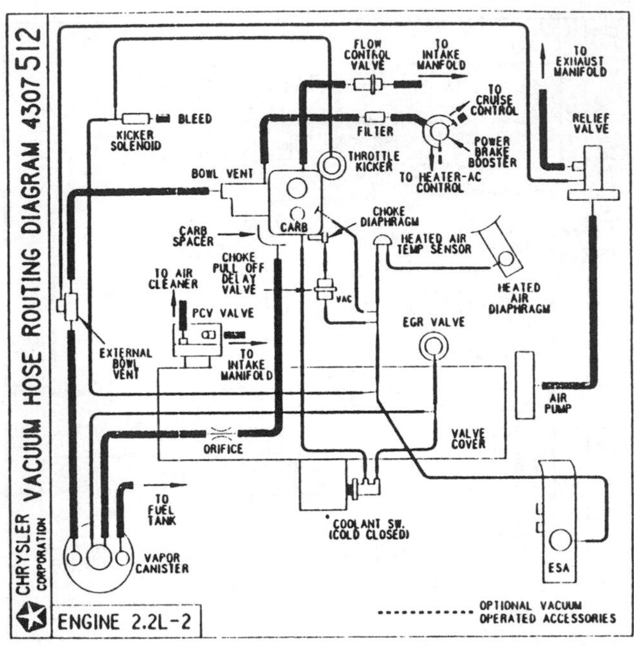

1984—4 CYL.-2.2L-2BBL, E.S.A.—CANADA

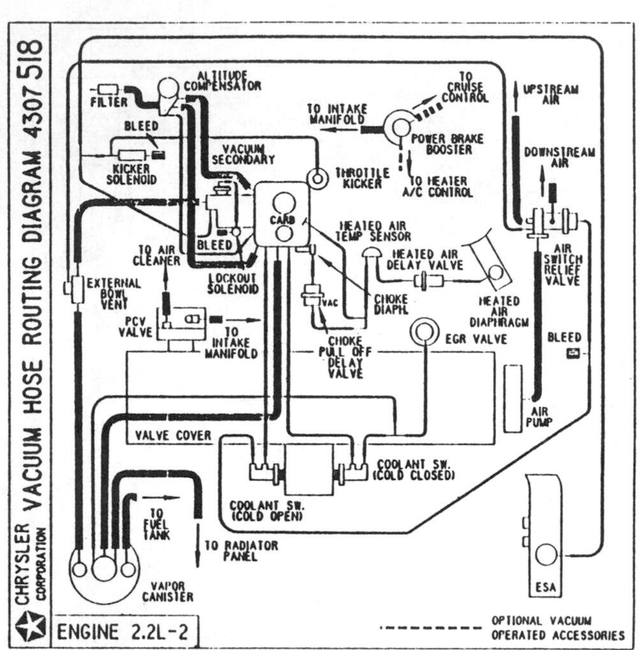

1984—4 CYL.-2.2L-2BBL—FEDERAL & HIGH ALT.

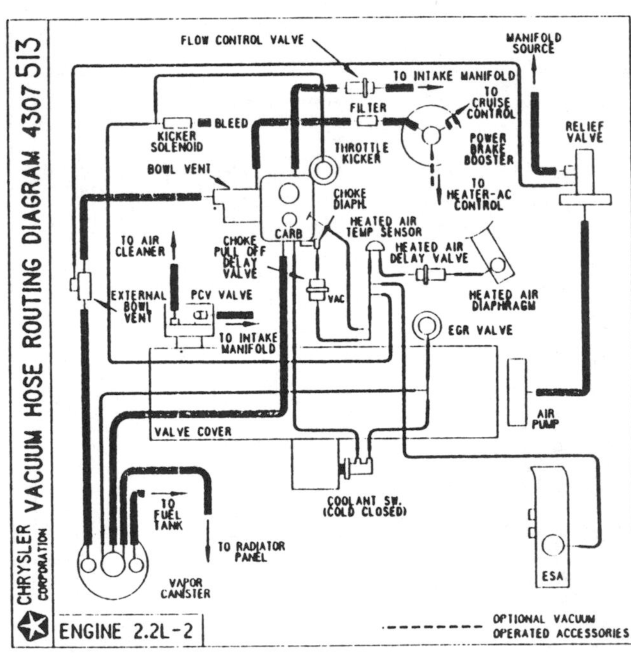

1984—4 CYL.-2.2L-2BBL, E.S.A.—FEDERAL

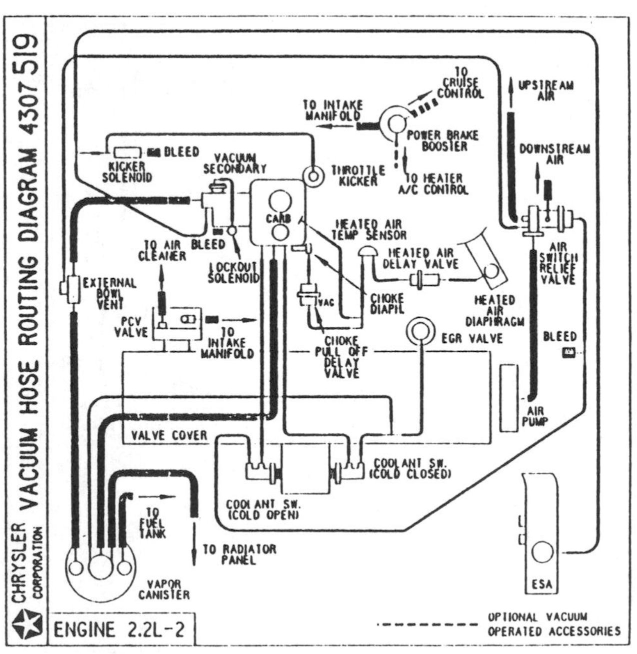

1984—4 CYL.-2.2L-2BBL—CALIFORNIA

VACUUM CIRCUITS

(© Chrysler Corp.)

1984—2.2L (135 CID) 4 CYL. engines

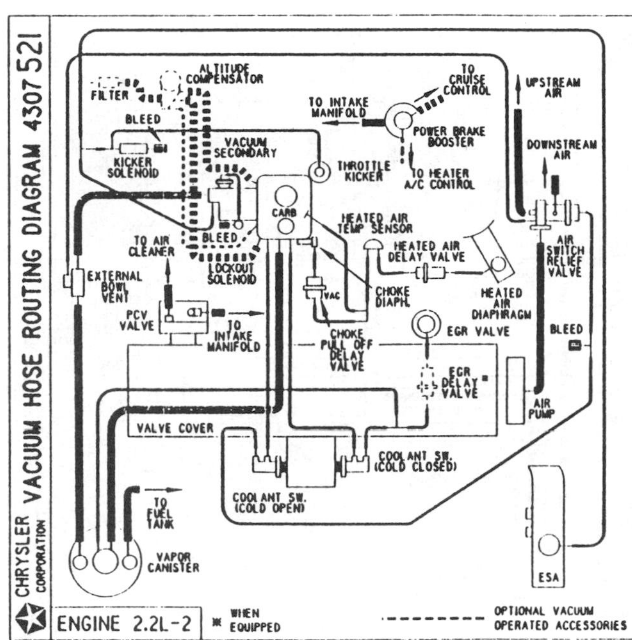

1984—4 CYL.-2.2L-2BBL, E.S.A.—MAN. TRANS.—FED. & HIGH ALT.

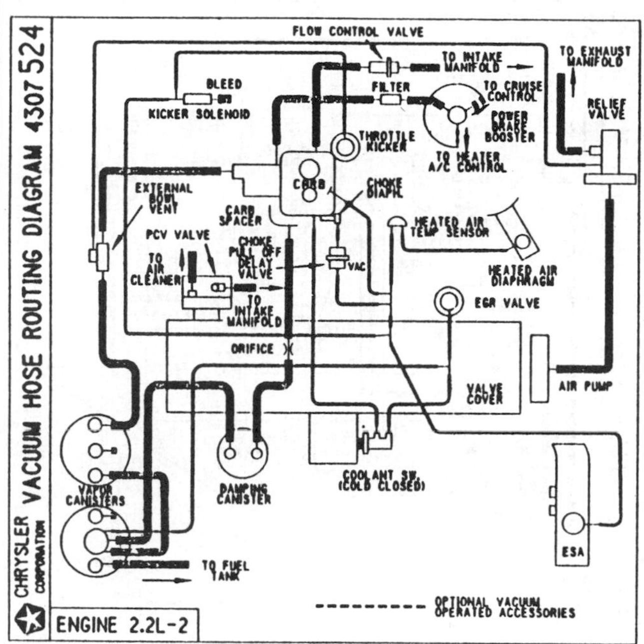

1984—4 CYL.-2.2L-2BBL, E.S.A.—FEDERAL

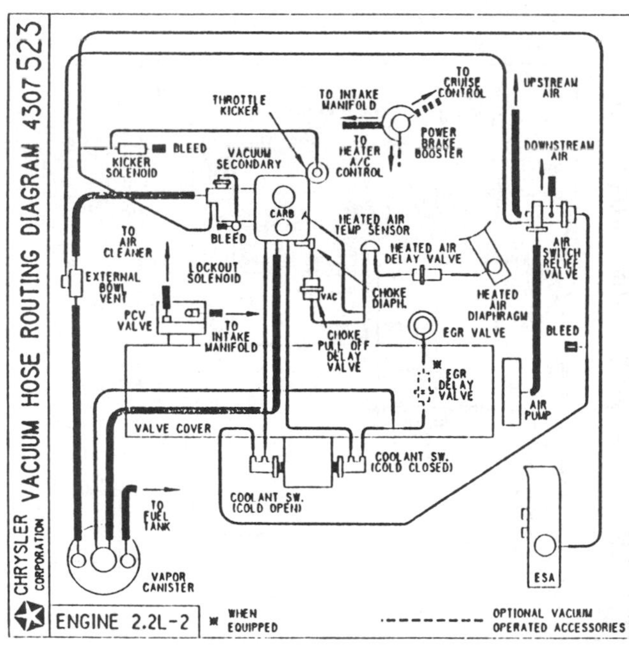

1984—4 CYL.-2.2L-2BBL—CALIFORNIA

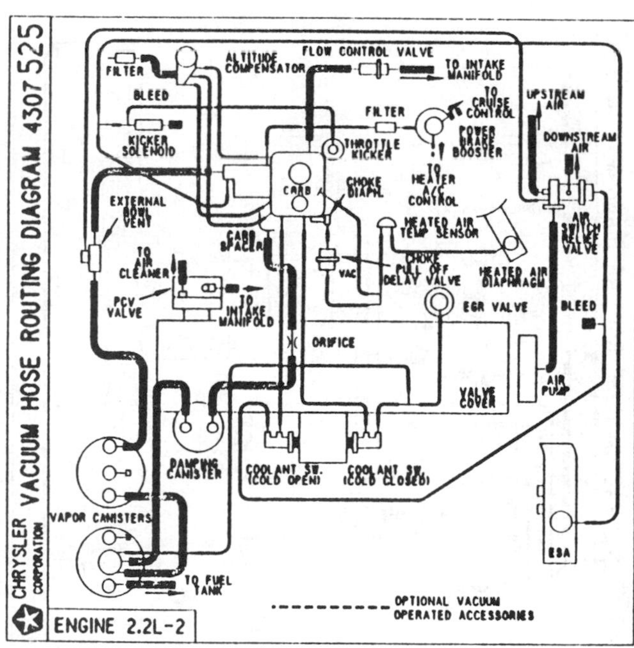

1984—4 CYL.-2.2L-2BBL—FEDERAL & HIGH ALT.

VACUUM CIRCUITS

(© Chrysler Corp.)

1984—2.2L (135 CID) 4 CYL. engines

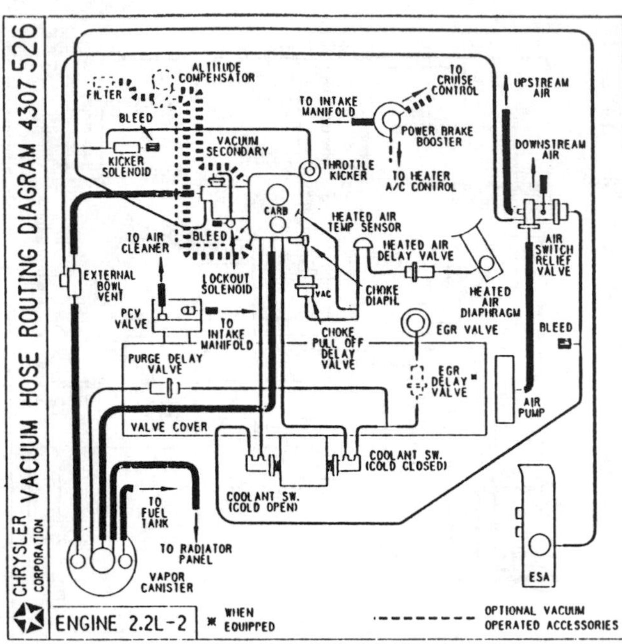

1984—4 CYL.-2.2L-2BBL, E.S.A.—FEDERAL & HIGH ALT.

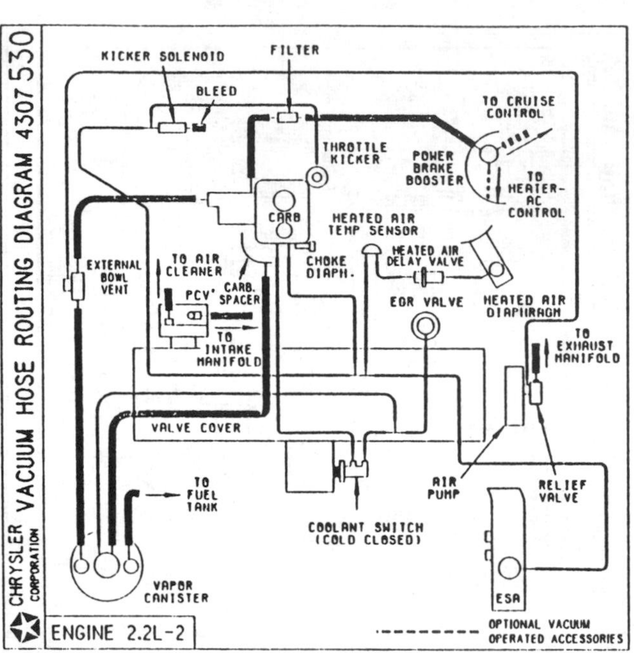

1984—4 CYL.-2.2L-2BBL, E.S.A.—EXPORT

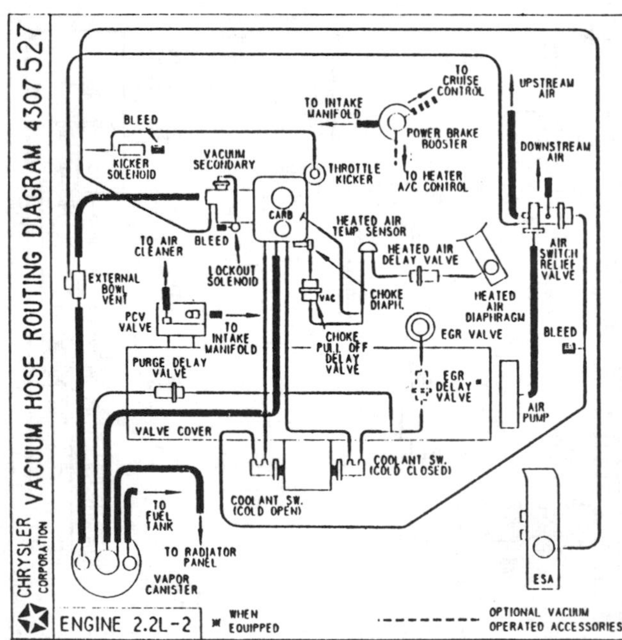

1984—4 CYL.-2.2L-2BBL—CALIFORNIA

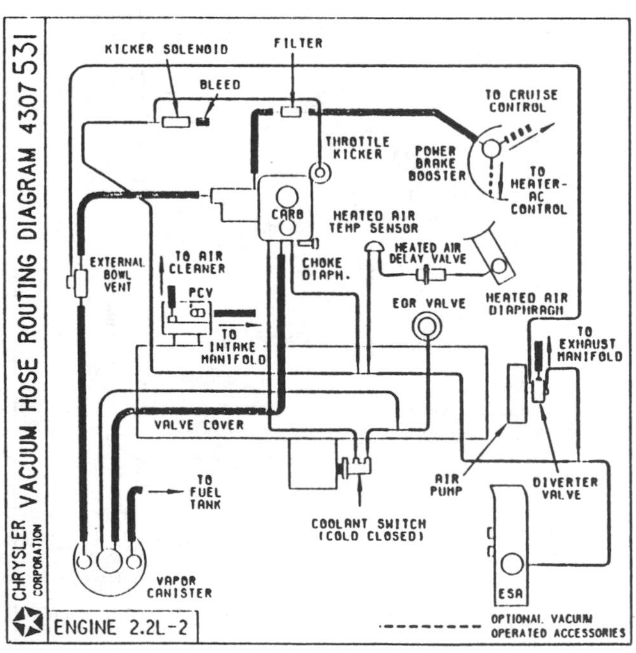

1984—4 CYL.-2.2L-2BBL, E.S.A.—EXPORT

VACUUM CIRCUITS

(© Chrysler Corp.)

1984—2.6L (156 CID) 4 CYL. engines

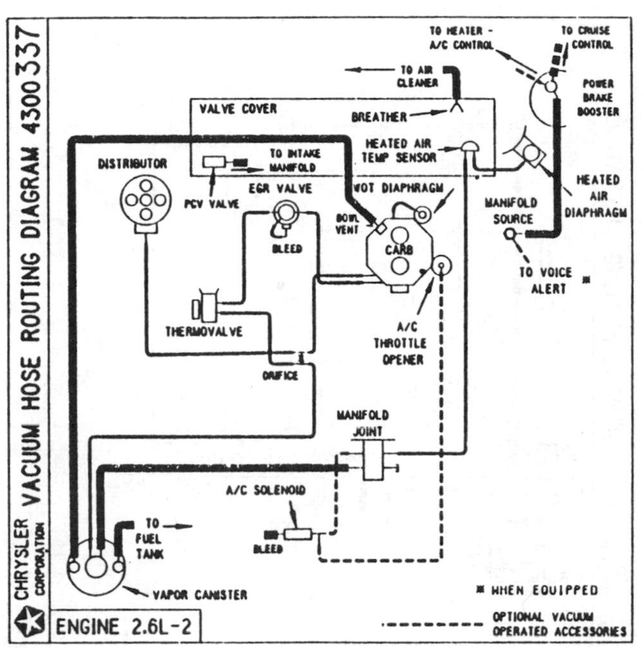

1984—4 CYL.-2.6L-2BBL—CANADA & EXPORT

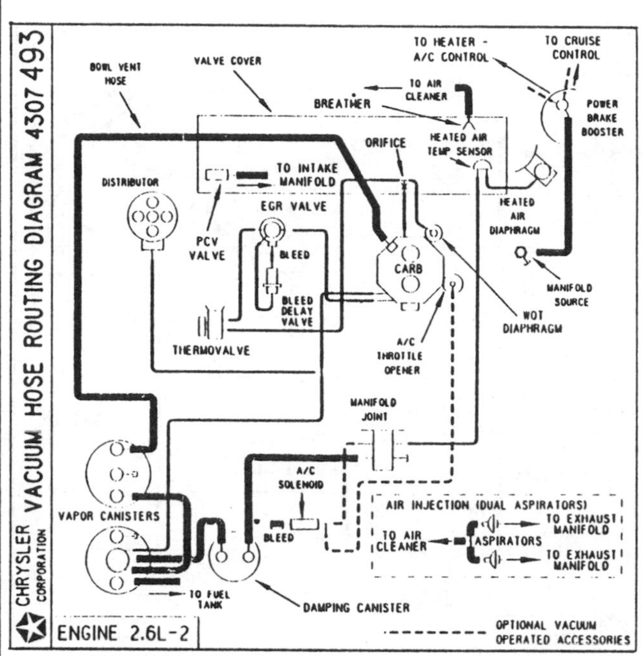

1984—4 CYL.-2.6L-2BBL—FEDERAL

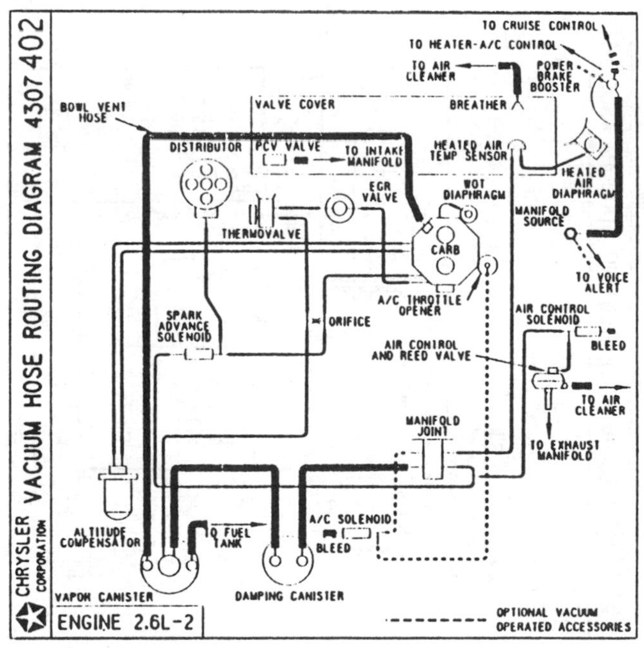

1984—4 CYL.-2.6L-2BBL—ALL

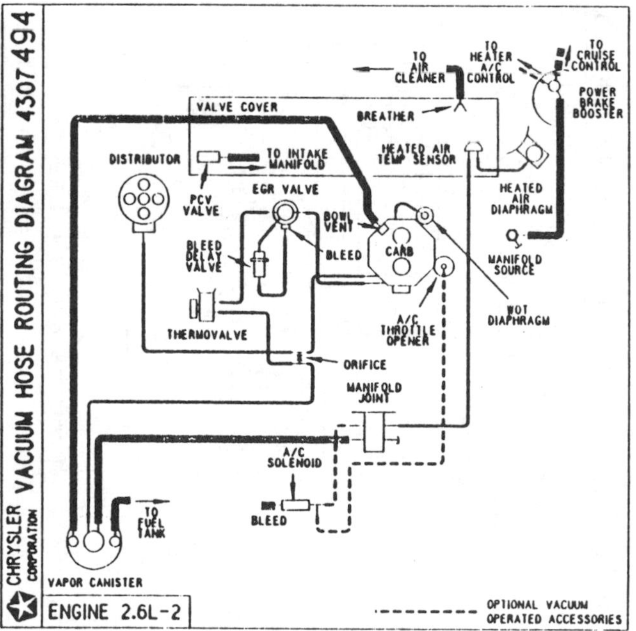

1984—4 CYL.-2.6L-2BBL—CANADA

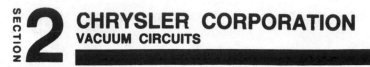

VACUUM CIRCUITS

(© Chrysler Corp.)

1984—2.6L (156 CID) 4 CYL. engines

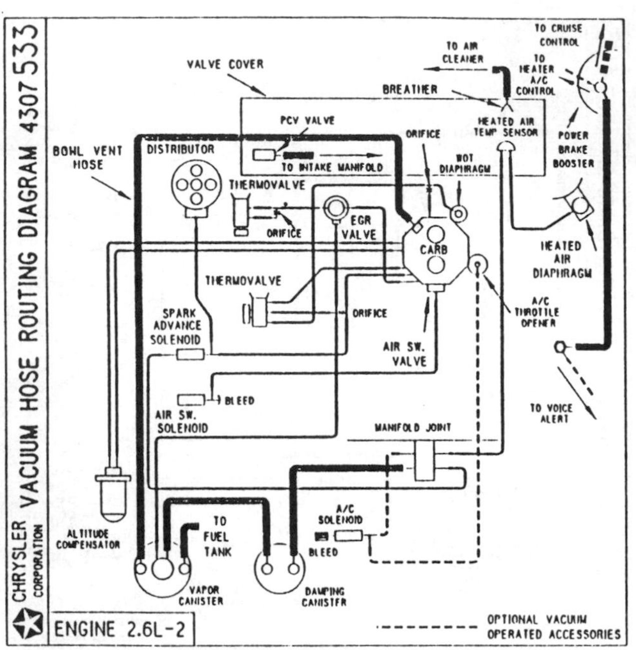

1984—4 CYL.-2.6L-2BBL—CALIFORNIA

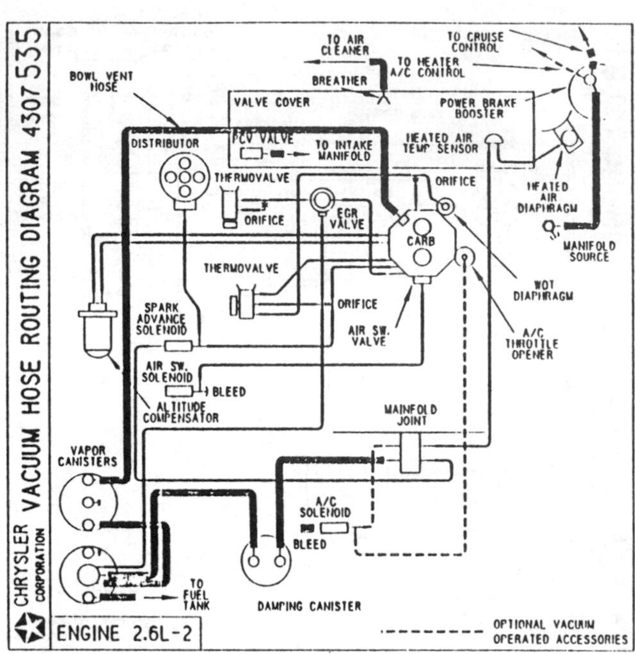

1984—4 CYL.-2.6L-2BBL—CALIFORNIA

1984—3.7L (225 CID) 6 CYL. engines

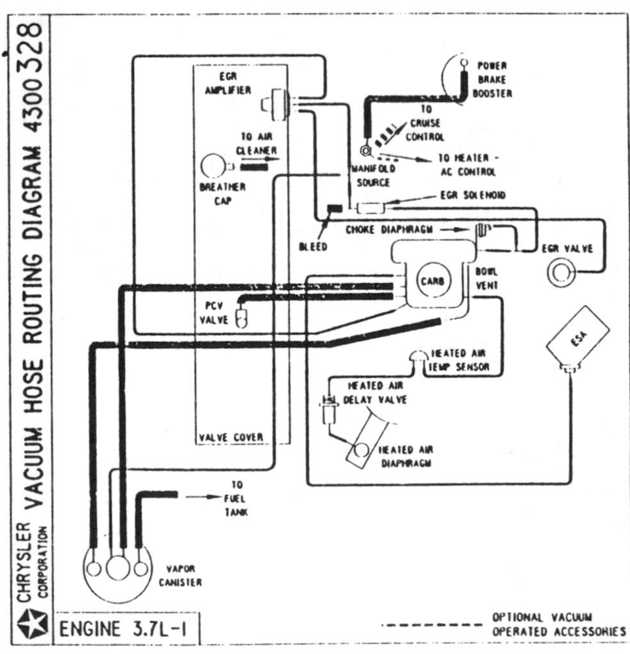

1984—6 CYL.-3.7L-1BBL, E.S.A.—CANADA

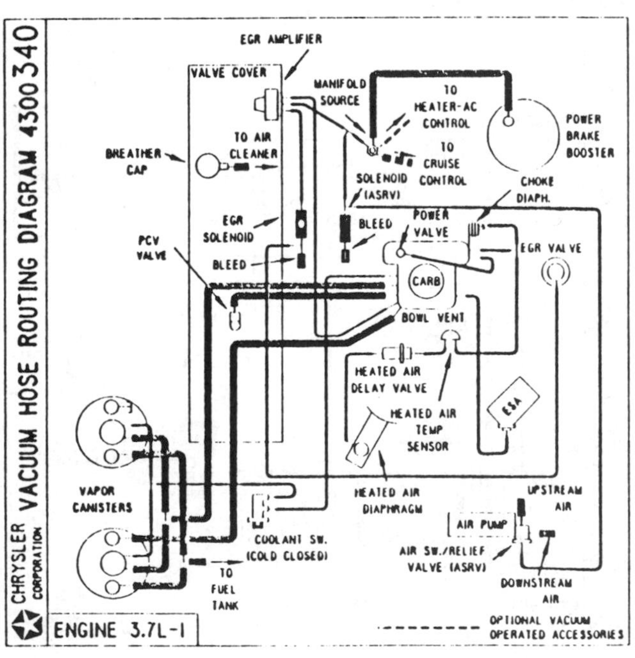

1984—6 CYL.-3.7L-1BBL—CALIFORNIA

VACUUM CIRCUITS

(© Chrysler Corp.)

1984—3.7L (225 CID) 6 CYL. engines

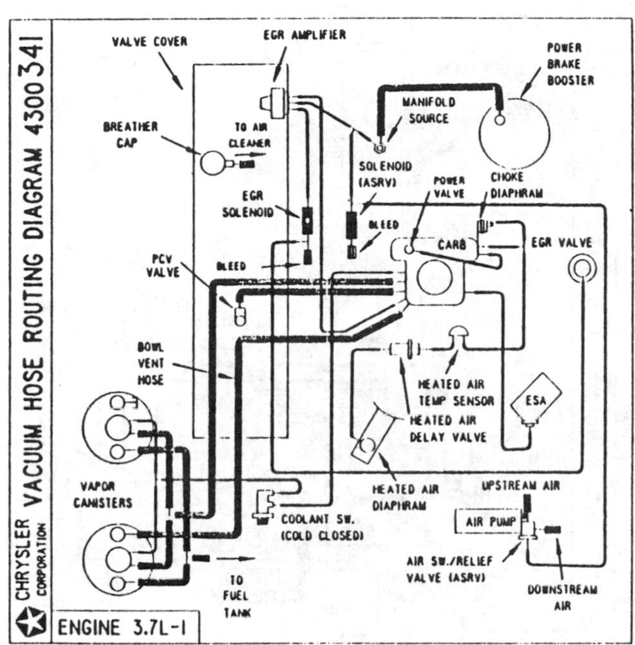

1984—6 CYL.-3.7L-1BBL, E.S.A.—FEDERAL

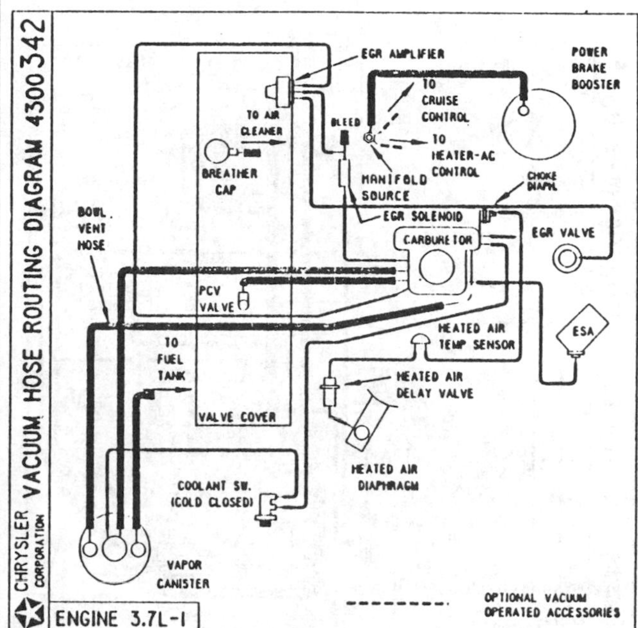

1984—6 CYL.-3.7L-1BBL, AUTO. TRANS., E.S.A.—FEDERAL

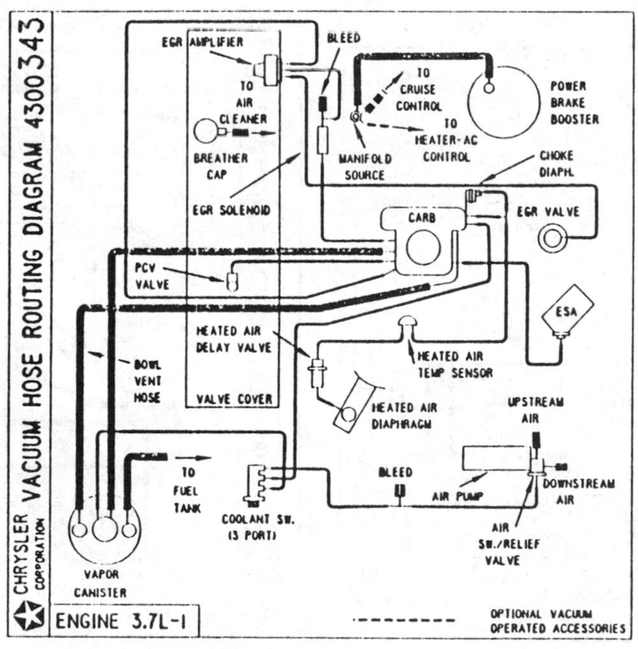

1984—6 CYL.-3.7L-1BBL, MAN. TRANS., E.S.A.—FEDERAL

VACUUM CIRCUITS

(© Chrysler Corp.)

1984—5.2L & 5.9L (318 & 360 CID) V8 engines

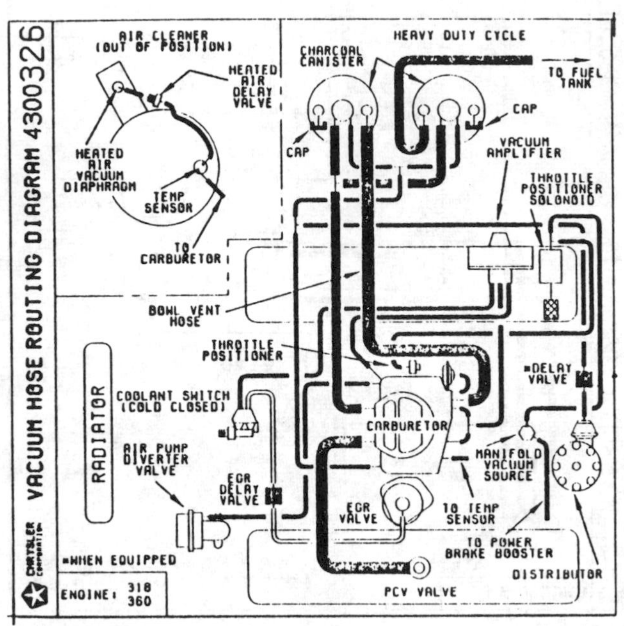

1984—V8-318, 360—CALIFORNIA

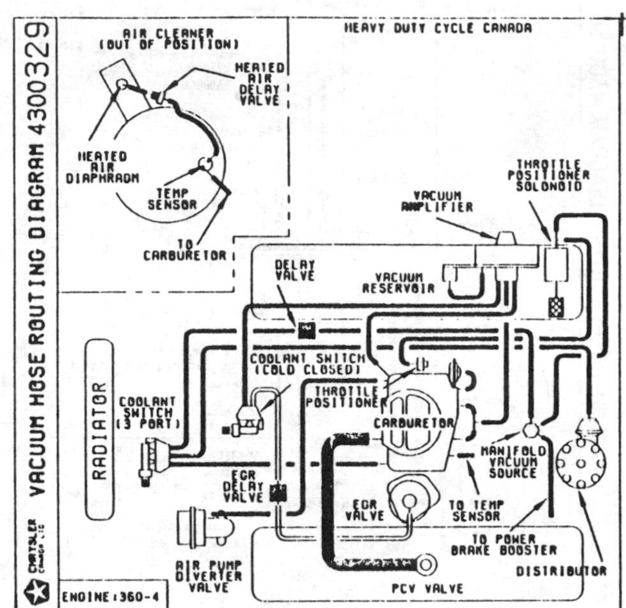

1984—V8-360—CANADA

1984—5.2L & 5.9L (318 & 360 CID) V8 engines

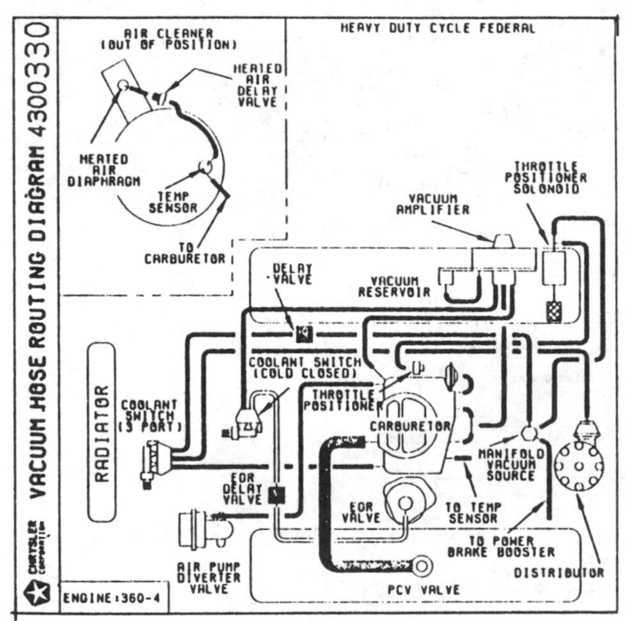

1984—V8-360—FEDERAL

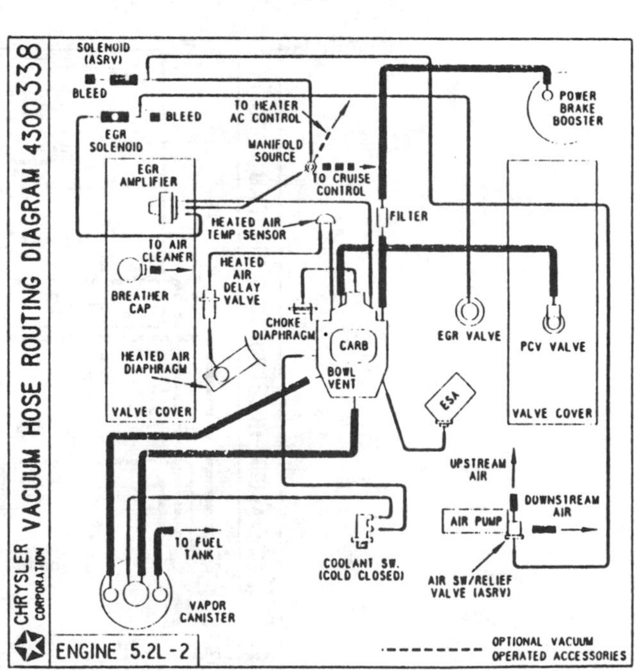

1984—V8-5.2L-2BBL—CALIFORNIA & HIGH ALT.

VACUUM CIRCUITS

(© Chrysler Corp.)

1984—5.2L (318 CID) V8 engines

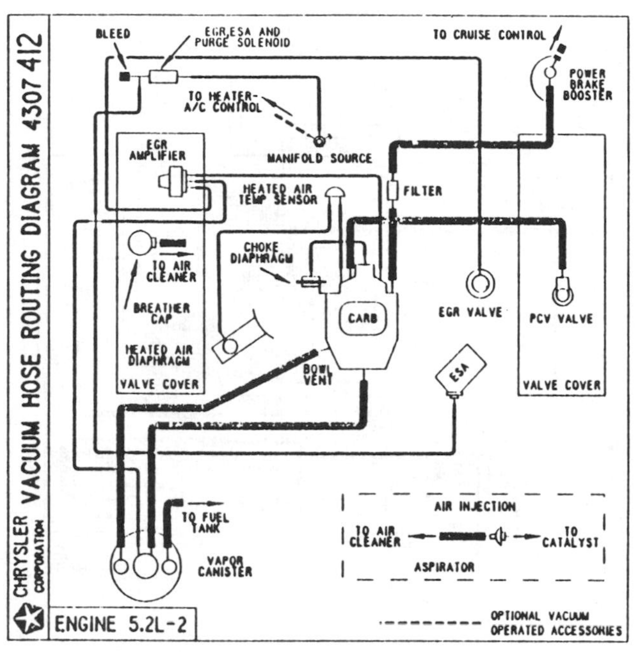

1984—V8-5.2L-2BBL, E.S.A.—CANADA

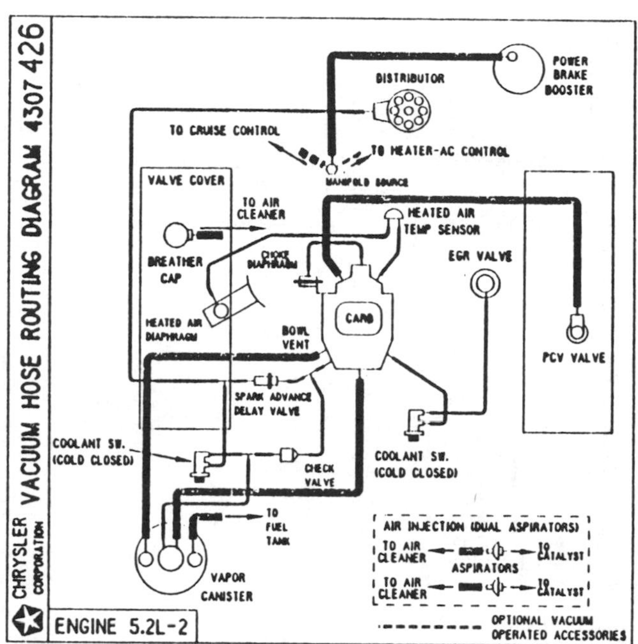

1984—V8-5.2L-2BBL, AUTO. TRANS. CANADA

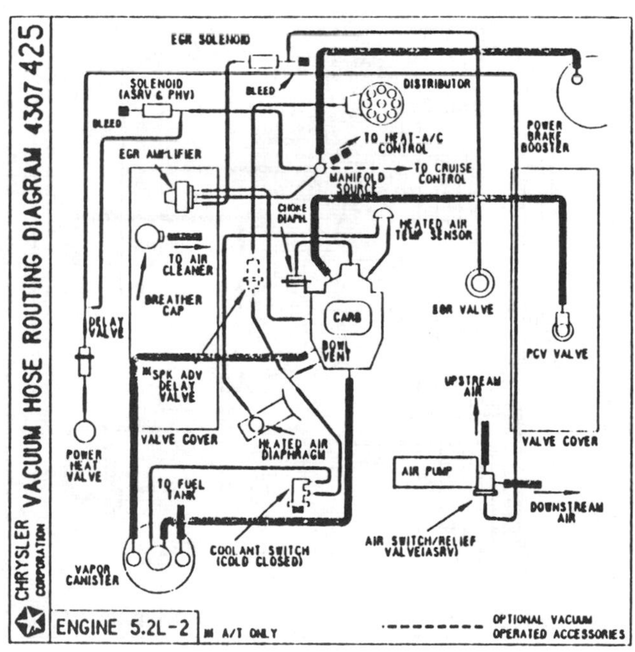

1984—V8-5.2L-2BBL, MAN. TRANS.—FEDERAL

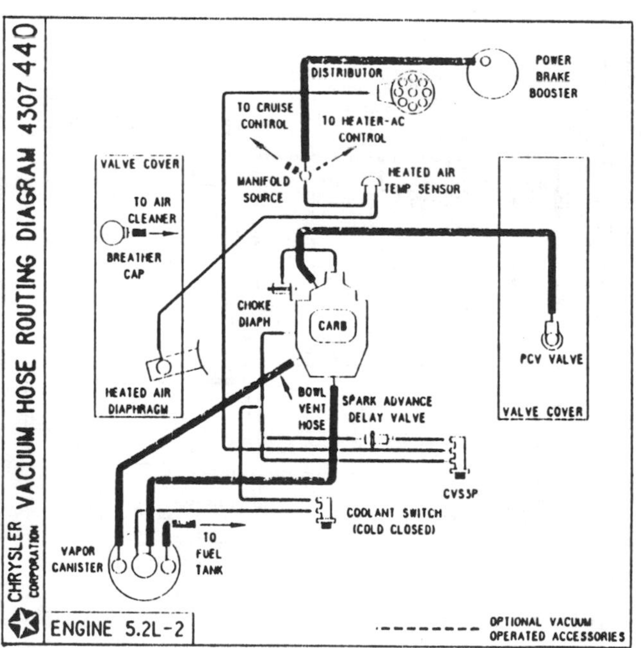

1984—V8-5.2L-2BBL, MAN. TRANS.

VACUUM CIRCUITS

(© Chrysler Corp.)

1984—5.2L (318 CID) V8 engines

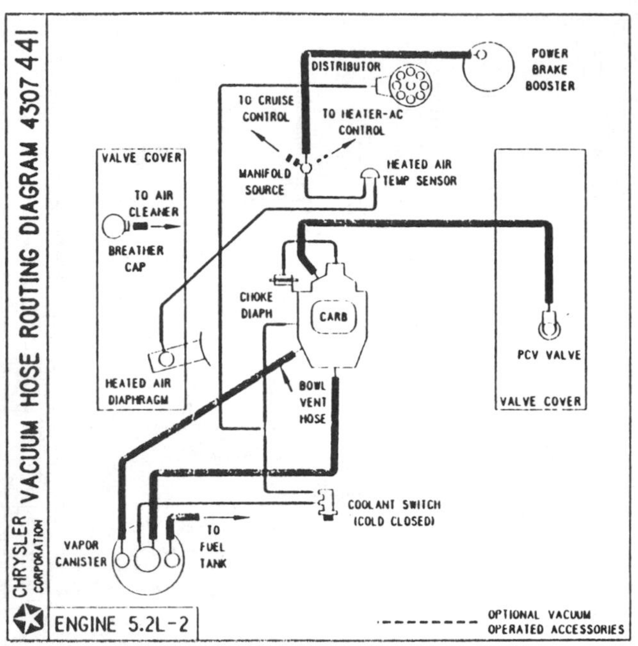

1984—V8-5.2L-2BBL, AUTO. TRANS.

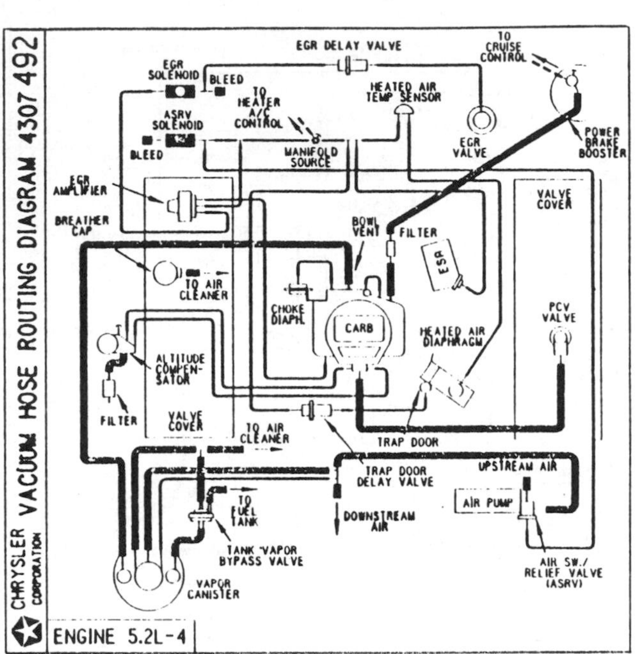

1984—V8-5.2L-4BBL, E.S.A.—FED. & CANADA

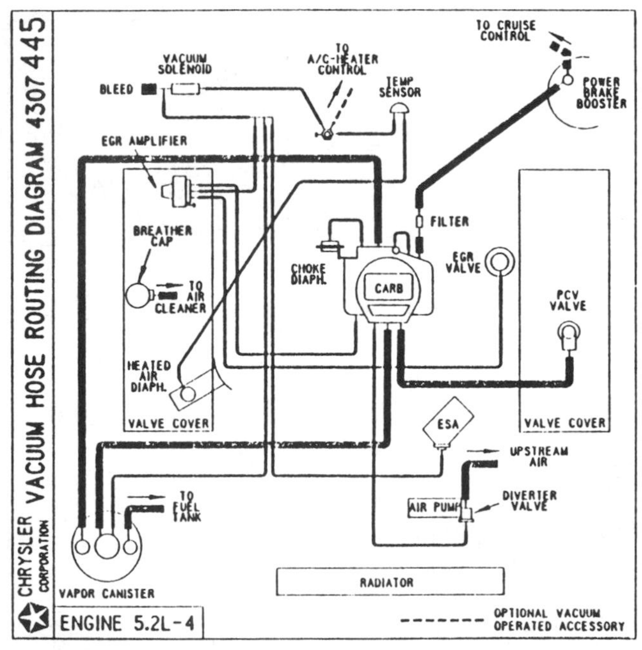

1984—V8-5.2L-4BBL, E.S.A.—ALL

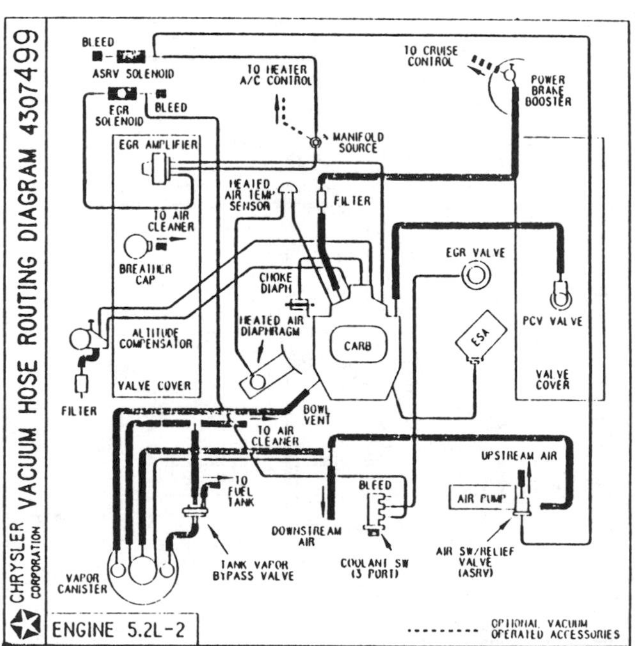

1984—V8-5.2L-2BBL, E.S.A.—FED. & HIGH ALT.

VACUUM CIRCUITS

(© Chrysler Corp.)

1984—5.2L (318 CID) V8 engines

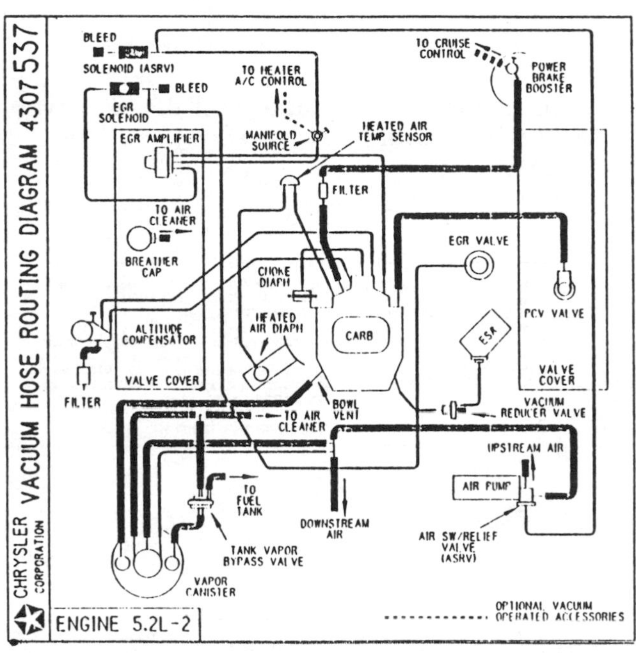

1984—V8-5.2L-2BBL, CALIFORNIA

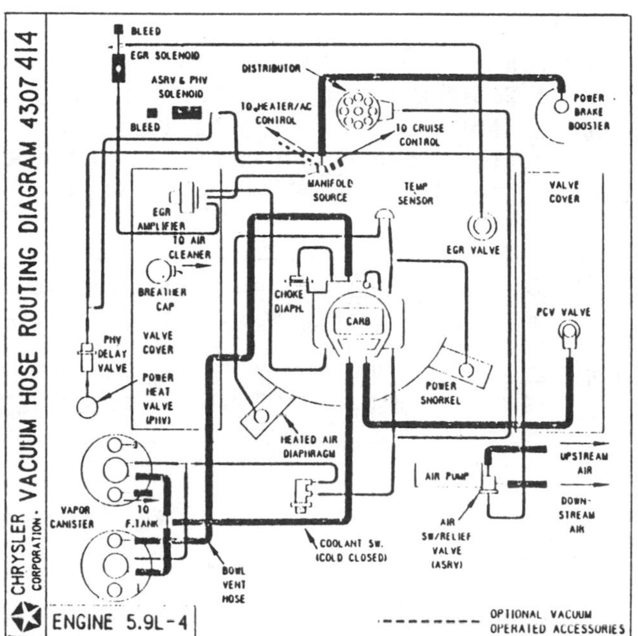

1984—V8-5.9L-4BBL, AUTO TRANS.—FEDERAL

1984—5.9L (360 CID) V8 engines

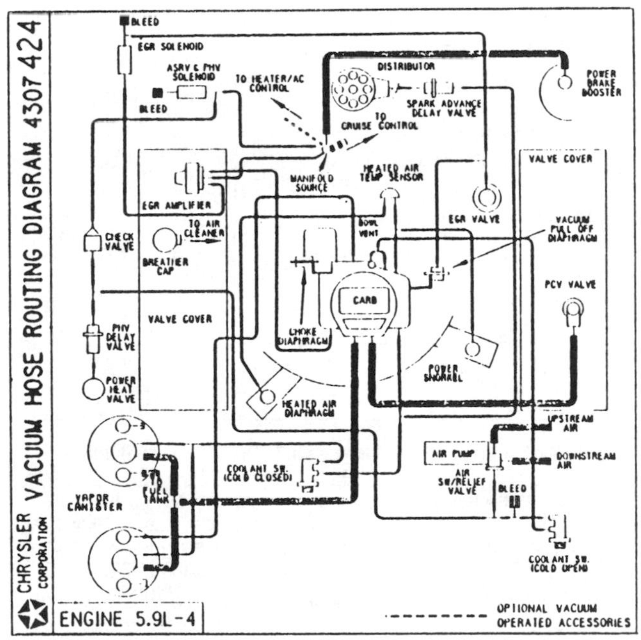

1984—V8-5.9L-4BBL, FEDERAL

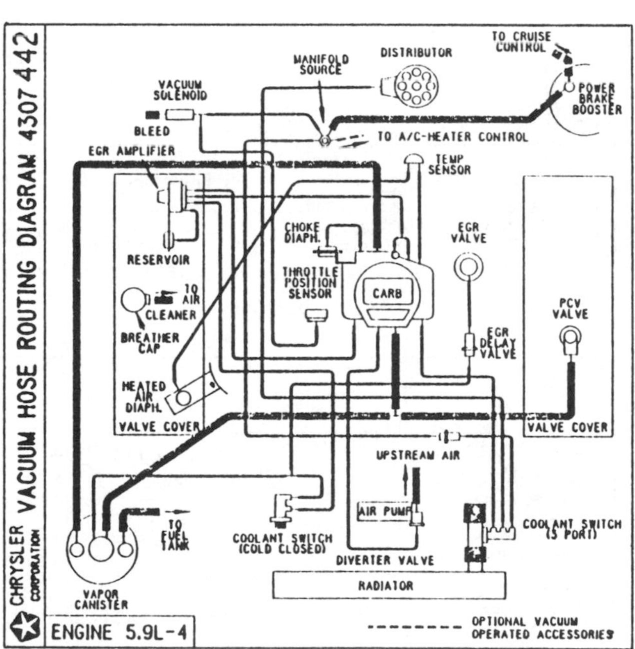

1984—V8-5.9L-4BBL, MAN. & AUTO. TRANS.

1985 Vacuum Circuit Diagrams

VEHICLE EMISSION CONTROL INFORMATION (VECI) LABEL

All vehicles are equipped with a **VECI** label (Figs. 1, 2, and 3) which is located in the engine compartment (Figs. 4, 5 and 6). This label is permanently attached and cannot be removed without defacing information and destroying it.

The specifications shown on the label are correct for the vehicle the label is mounted on. Propane idle speeds are shown on the VECI label. If any difference exists between the specifications shown on the label and those shown in the service manual, those shown on the label should be used.

4288 803
CATALYST
CHRYSLER CORPORATION
VEHICLE EMISSION CONTROL INFORMATION
THIS VEHICLE CONFORMS TO U.S. EPA REGULATIONS APPLICABLE TO 1985 MODEL YEAR NEW MOTOR VEHICLES AT ALL ALTITUDES.
2.2 LITER FCR2.2V2HDH6 FCRVA
SPARK PLUGS .036 IN. GAP RN12YC
ADJUSTMENTS MADE BY OTHER THAN APPROVED SERVICE MANUAL PROCEDURES MAY VIOLATE FEDERAL AND STATE LAWS.

IDLE SETTINGS	MAN	AUTO
TIMING BTC	10°	10°
IDLE RPM	800	900
FAST IDLE RPM	1700	1880

UNDER 300 MILES REDUCE ALL IDLE SPEEDS 75 RPM.

- THE CARBURETOR IDLE MIXTURE HAS BEEN PRESET AT THE FACTORY. ADJUSTMENTS SHOULD NOT BE MADE DURING ROUTINE TUNE-UP.
- SEE SERVICE MANUAL FOR ADDITIONAL INFORMATION.

CHECK BASIC TIMING W/ESA MODULE VACUUM HOSE DISCONNECTED. BE SURE TO RECONNECT ESA HOSE AFTER SETTING BASIC TIMING.

- CHECK IN NEUTRAL, LIGHTS AND ACCESSORIES OFF.

Fig. 1—Typical Federal VECI Label

4288 812
CATALYST
EGR, OX, TWC, AP, CL
CHRYSLER CORPORATION
VEHICLE EMISSION CONTROL INFORMATION
1.6 LITER FCR1.6V2HDJ3 FCRVA
SPARK PLUGS .036 IN. GAP RN12YC
CHECK BASIC TIMING W/ESA MODULE VACUUM HOSE DISCONNECTED. BE SURE TO RECONNECT ESA HOSE AFTER SETTING BASIC TIMING.

- THE CARBURETOR IDLE MIXTURE HAS BEEN PRESET AT THE FACTORY. ADJUSTMENTS SHOULD NOT BE MADE DURING ROUTINE TUNE-UP.
- ADJUSTMENTS MADE BY OTHER THAN APPROVED SERVICE MANUAL PROCEDURES MAY VIOLATE FEDERAL AND STATE LAWS.
- SEE SERVICE MANUAL FOR ADDITIONAL INFORMATION.

THIS VEHICLE CONFORMS TO U.S. EPA AND STATE OF CALIFORNIA REGULATIONS APPLICABLE TO 1985 MODEL YEAR NEW MOTOR VEHICLES PROVIDED THAT THIS VEHICLE IS ONLY INTRODUCED INTO COMMERCE FOR SALE IN THE STATE OF CALIFORNIA.

IDLE SETTINGS	MAN
TIMING BTC	12°
IDLE RPM	850
FAST IDLE RPM	1400

UNDER 300 MILES REDUCE ALL IDLE SPEEDS 75 RPM.

COLD ENGINE VALVE LASH	
INTAKE	0.010 IN
EXHAUST	0.010 IN.

CHECK IN NEUTRAL, LIGHTS AND ACCESSORIES OFF.

Fig. 2—Typical California VECI Label

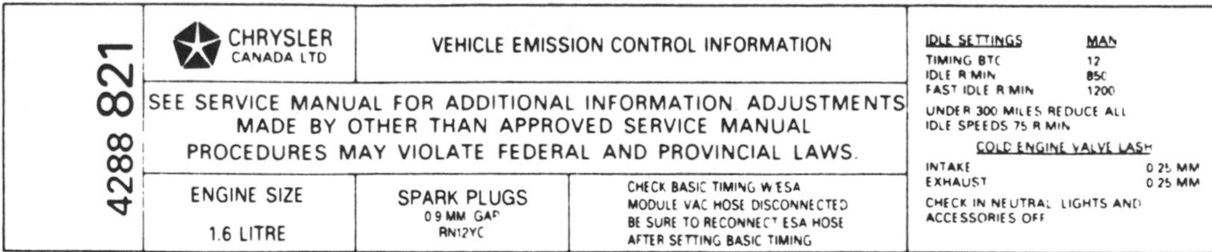
4288 821
CHRYSLER
CANADA LTD
VEHICLE EMISSION CONTROL INFORMATION
SEE SERVICE MANUAL FOR ADDITIONAL INFORMATION. ADJUSTMENTS MADE BY OTHER THAN APPROVED SERVICE MANUAL PROCEDURES MAY VIOLATE FEDERAL AND PROVINCIAL LAWS.
ENGINE SIZE
1.6 LITRE
SPARK PLUGS
0.9 MM GAP
RN12YC
CHECK BASIC TIMING W/ESA MODULE VAC HOSE DISCONNECTED BE SURE TO RECONNECT ESA HOSE AFTER SETTING BASIC TIMING
IDLE SETTINGS MAN
TIMING BTC 12
IDLE R/MIN 850
FAST IDLE R/MIN 1200
UNDER 300 MILES REDUCE ALL IDLE SPEEDS 75 R/MIN
COLD ENGINE VALVE LASH
INTAKE 0.25 MM
EXHAUST 0.25 MM
CHECK IN NEUTRAL LIGHTS AND ACCESSORIES OFF

REGLAGES DU RALENTI	MAN
DISTRIBUTION AV P.H	12
REGLAGE DU RALENTI EN TR/MIN	850
RALENTI ACCELERE EN TR/MIN	1200

SI MOINS QUE 500 KW REDUIRE TOUS LES RALENTIS DE 75 TR/MIN

JES DES SOUPAPES-MOTEUR FROID	
ADMISSION	0.25 MM
ECHAPPEMENT	0.25 MM

VERIFIER EN POSITION DE POINT MORT LES ECLAIRAGES ET LES ACCESSOIRES ETANT A L'ARRÊT

CHRYSLER CANADA LTEE
RENSEIGNEMENTS RELATIFS AU CONTRÔLE DES ÉMISSIONS
CONSULTER LE MANUEL D'ENTRETIEN POUR DE PLUS AMPLES RENSEIGNEMENTS. LES RÉGLAGES QUI SERAIENT FAITS SELON D'AUTRES PROCÉDURES QUE CELLES DUE MANUEL D'ENTRETIEN POURRAIENT ENFREINDRE LES LOIS FÉDÉRALES ET PROVINCIALES.
TAILLE DU MOTEUR 1.6 LITRE
BOUGIES 0.9 ECARTEMENT X MM RN12YC
VÉRIFIER LE REGLAGE DE BASE LE TUYAU DE DEPRESSION DU MODULE ESA ETANT DEBRANCHE NE PAS OUBLIER DE REBRANCHER LE TUYAU ESA APRÉS AVOIR EFFECTUE LE REGLAGE DE BASE

Fig. 3—Typical Canadian VECI Label

VACUUM CIRCUITS

(© Chrysler Corp.)

Front Wheel Drive

VACUUM HOSE ROUTING LABEL

All vehicles are equipped with a vacuum hose routing label which is located in the engine compartment (Figs. 4, 5 and 6).
This label is permanently attached and cannot be removed without destroying it. All hoses must be connected and routed as shown on the label.

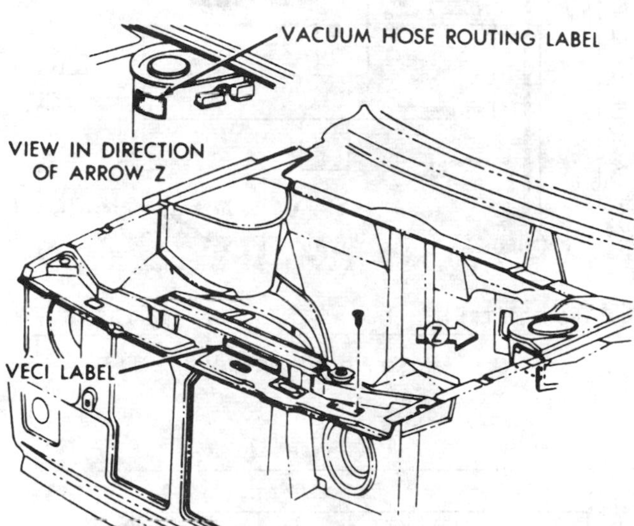

Fig. 4—Underhood Label Location L/MZ

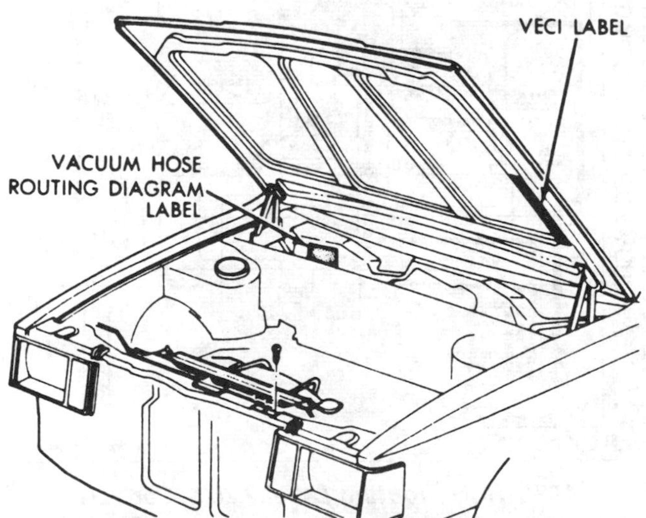

Fig. 5—Underhood Label Location—KEH

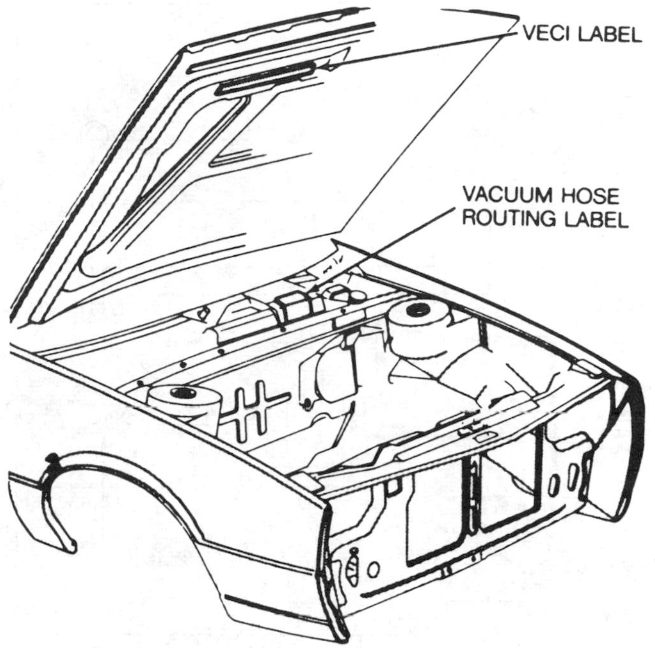

Fig. 6—Underhood Label Location G/CV

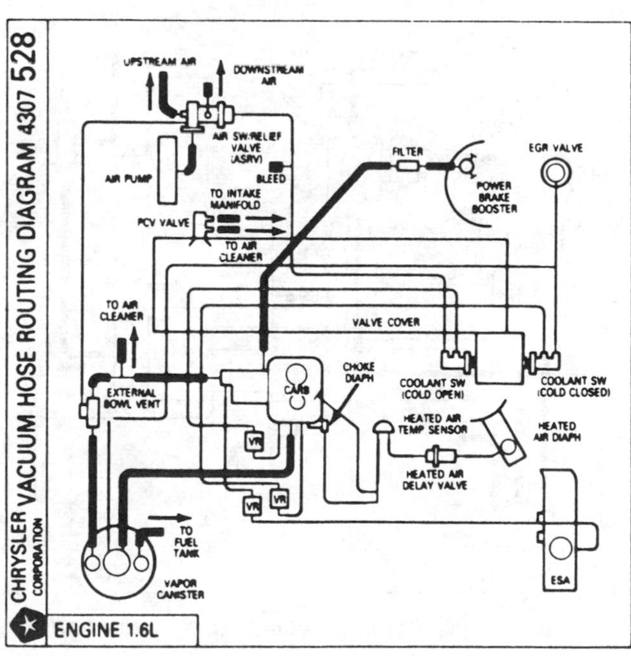

1985-*Hose Routing Label 1.6L Federal/California*

VACUUM CIRCUITS

(© Chrysler Corp.)

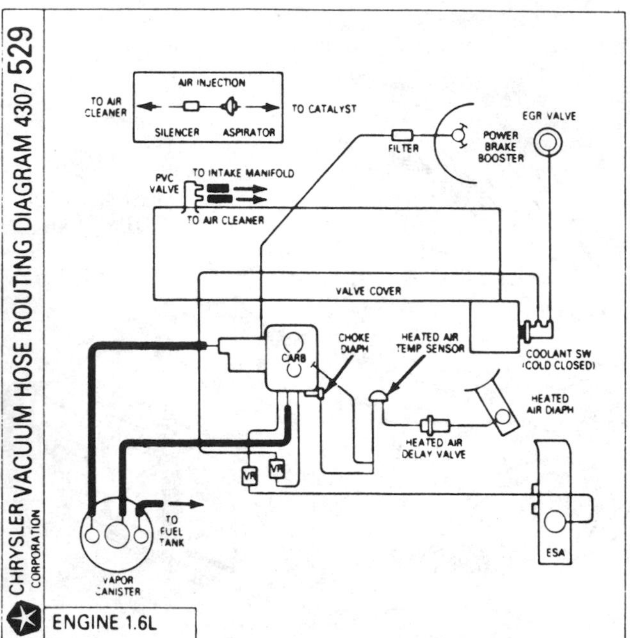

1985–Hose Routing Label 1.6L Canadian

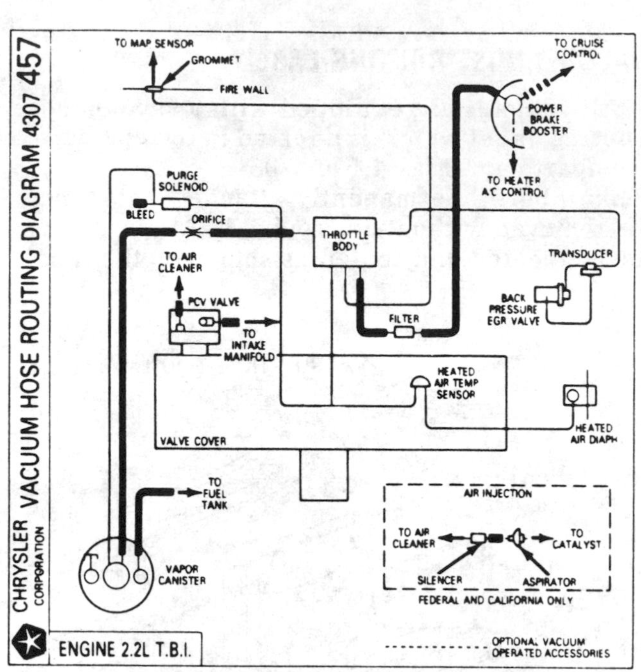

1985–Hose Routing Label 2.2L EFI

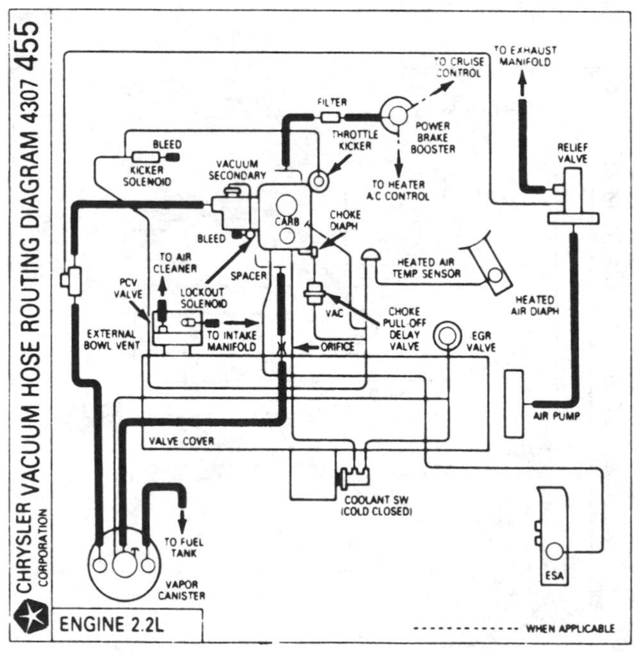

1985–Hose Routing Label 2.2L K, Canadian

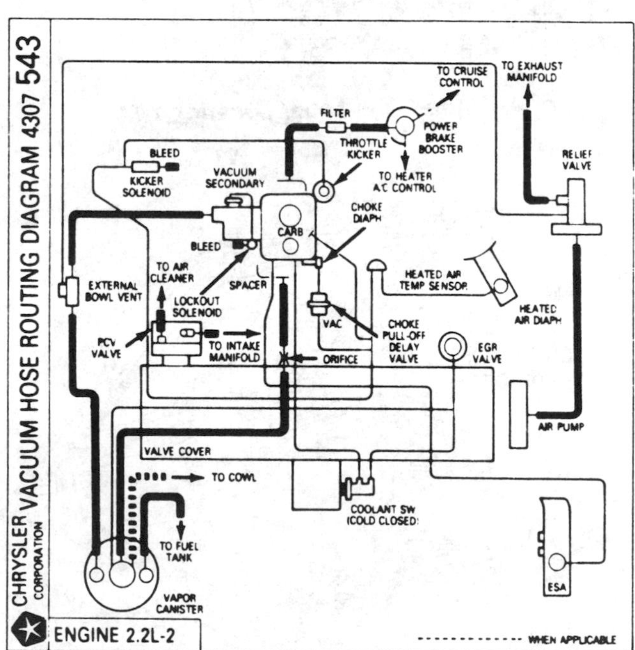

1985–Hose Routing Label 2.2L L, Canadian

VACUUM CIRCUITS

(© Chrysler Corp.)

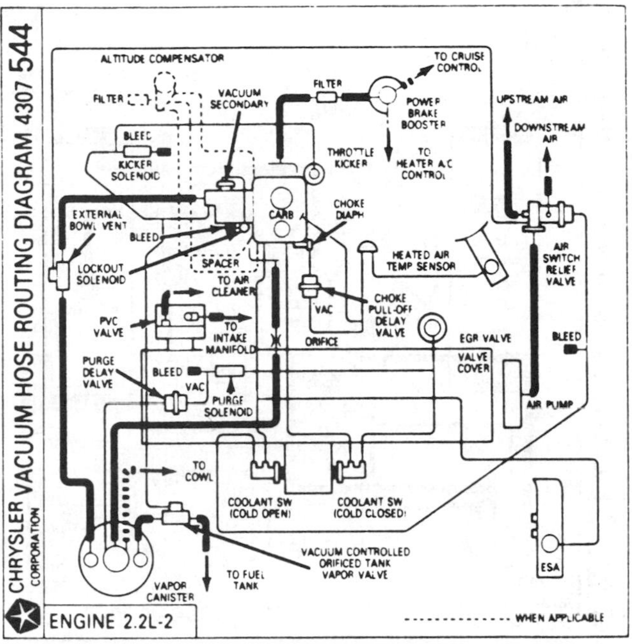

1985–Hose Routing Label 2.2L K,L Federal

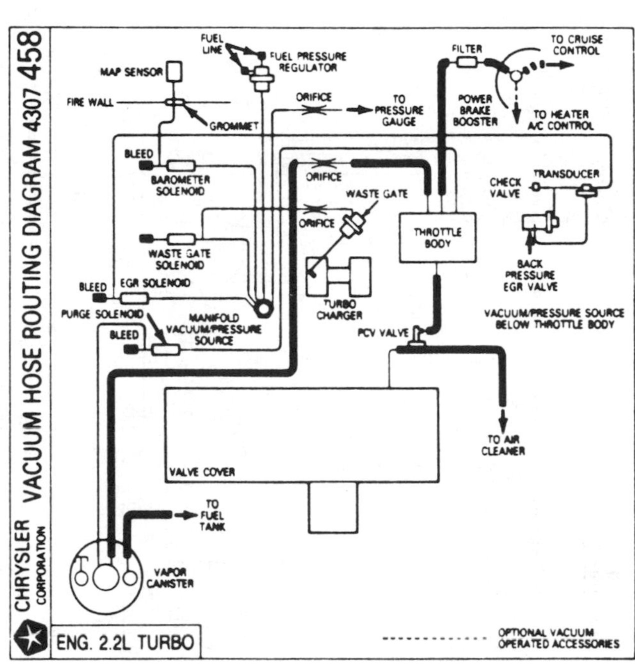

1985–Hose Routing Label 2.2L Turbo

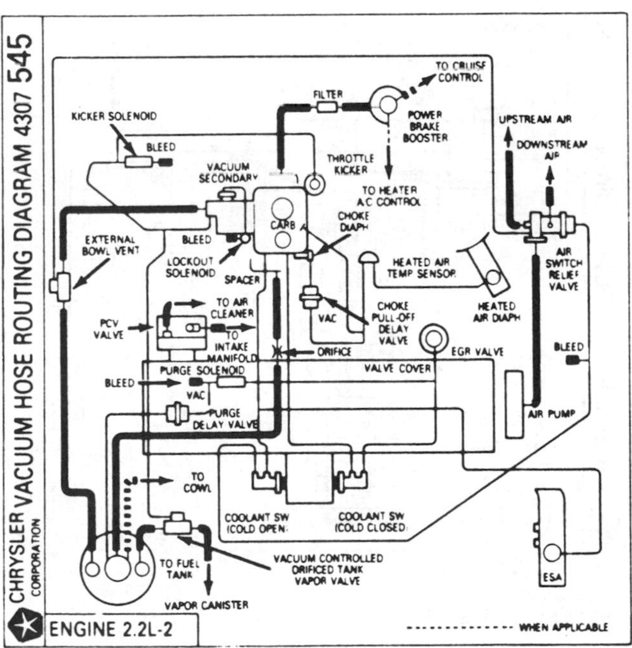

1985–Hose Routing Label 2.2L K,L California

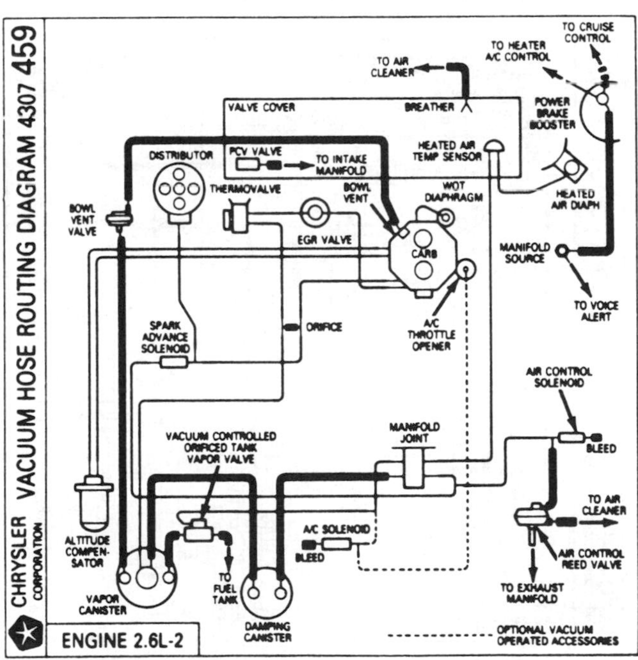

1985–Hose Routing Label 2.2L K,E Federal California and Canadian

VACUUM CIRCUITS

(© Chrysler Corp.)

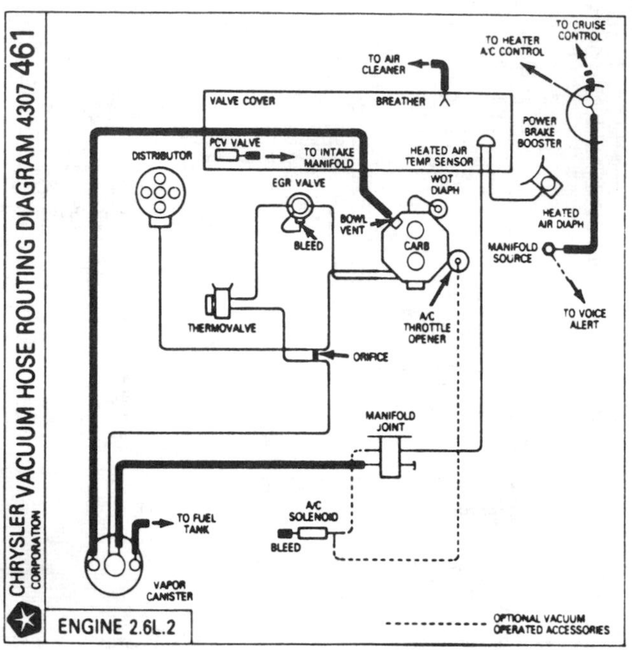

1985–*Hose Routing Label 2.6L Canadian*

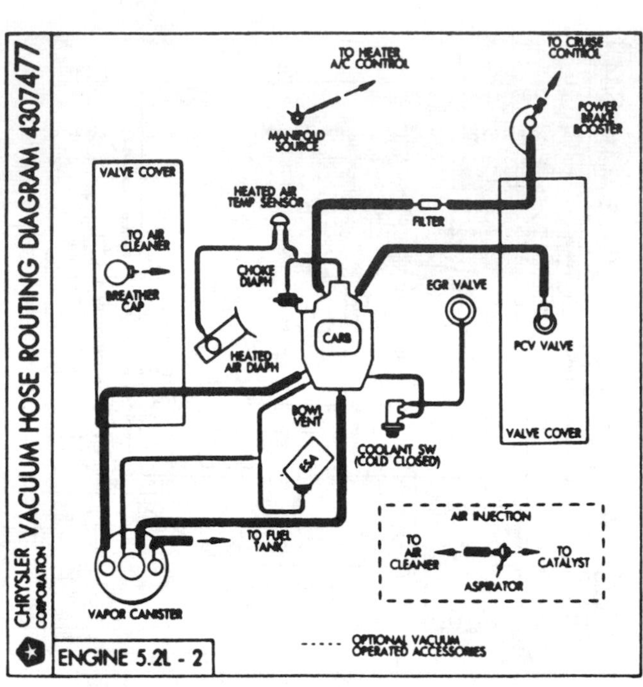

1985–*Canadian Vacuum Hose Routings 5.2L-2*

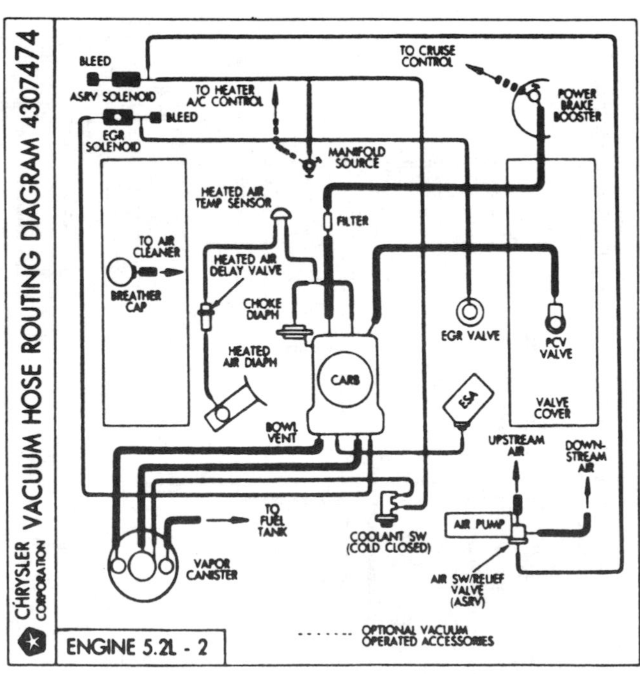

1985–*Federal/California Vacuum Hose Routings 5.2L-2*

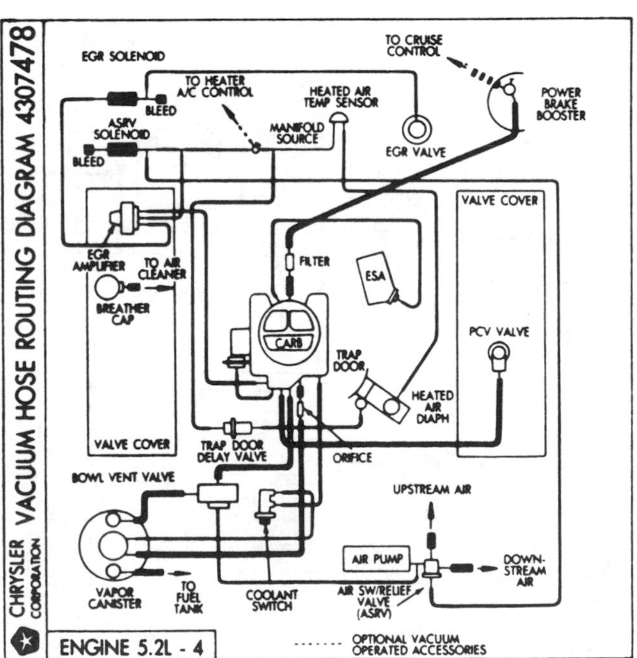

1985–*Federal/Canada Vacuum Hose Routings 5.2L-4*

1986 Vacuum Circuit Diagrams

VACUUM HOSE ROUTING LABEL

All vehicles are equipped with a vacuum hose routing label (Figs. 7 through 13) which is located in the engine compartment (Figs. 4, 5 and 6). This label is permanently attached and cannot be removed without destroying it. All hoses must be connected and routed as shown on the label.

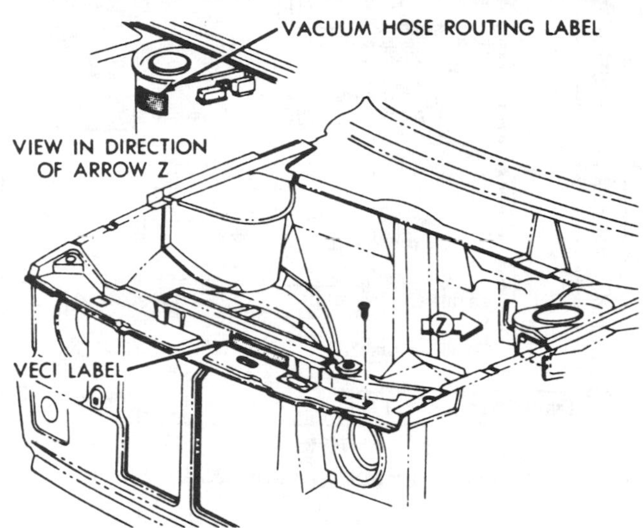

Fig. 4—Underhood Label Location L/MZ

CHRYSLER RWD

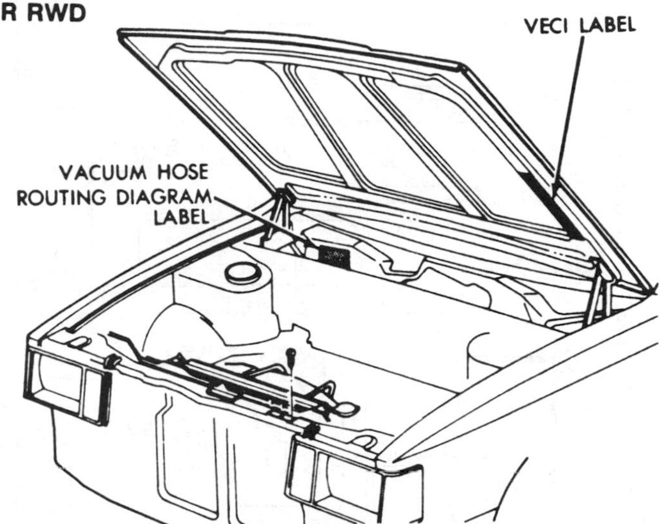

Fig. 5—Underhood Label Location—KEH

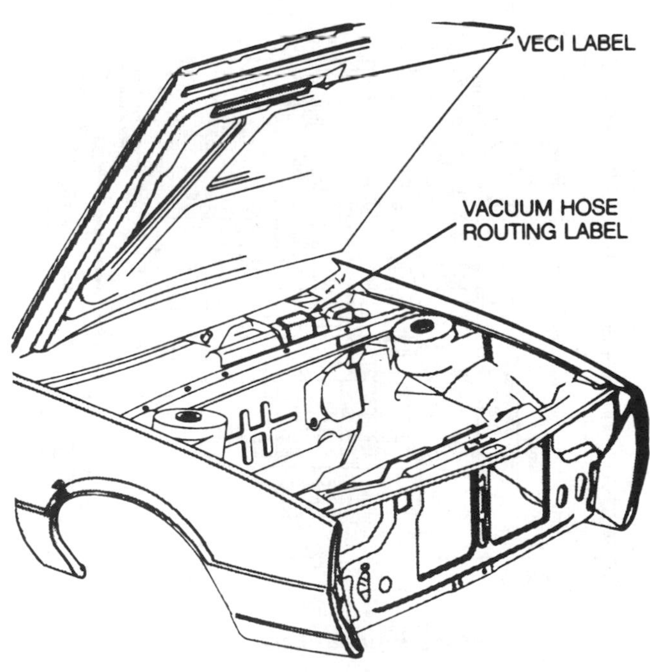

Fig. 6—Underhood Label Location G/CV

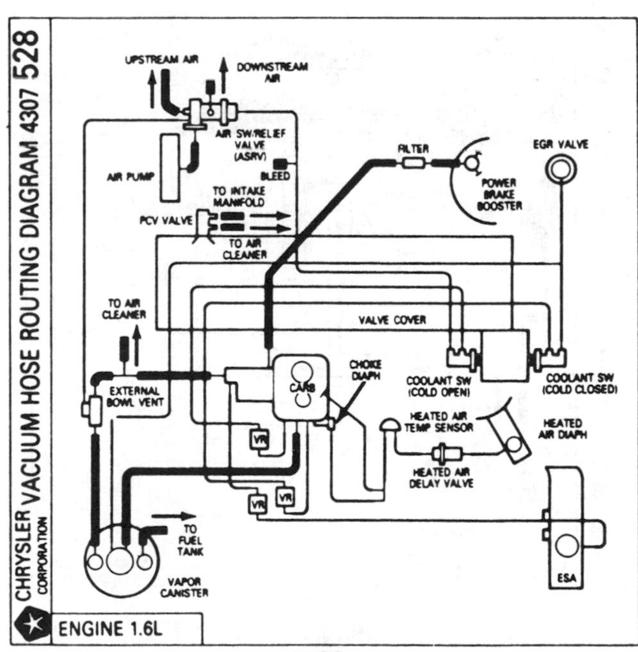

1986

Fig. 7—Hose Routing Label 1.6L Federal/California

VACUUM CIRCUITS

(© Chrysler Corp.)

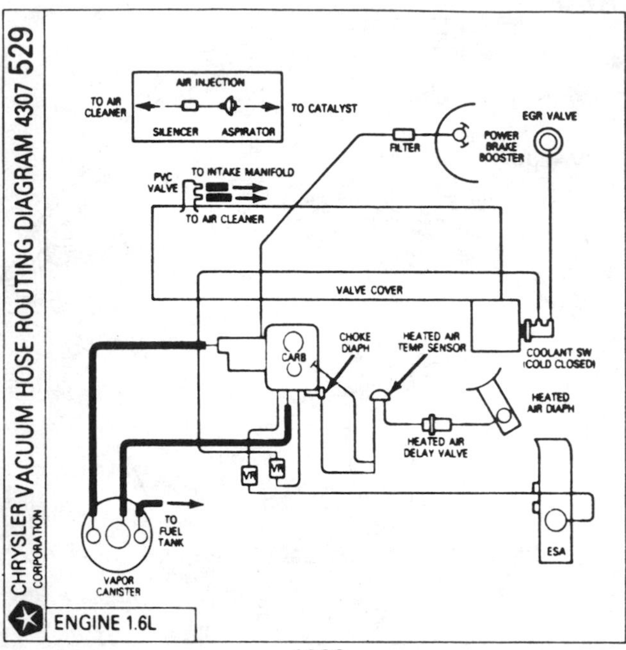

1986

Fig. 8—Hose Routing Label 1.6L Canadian

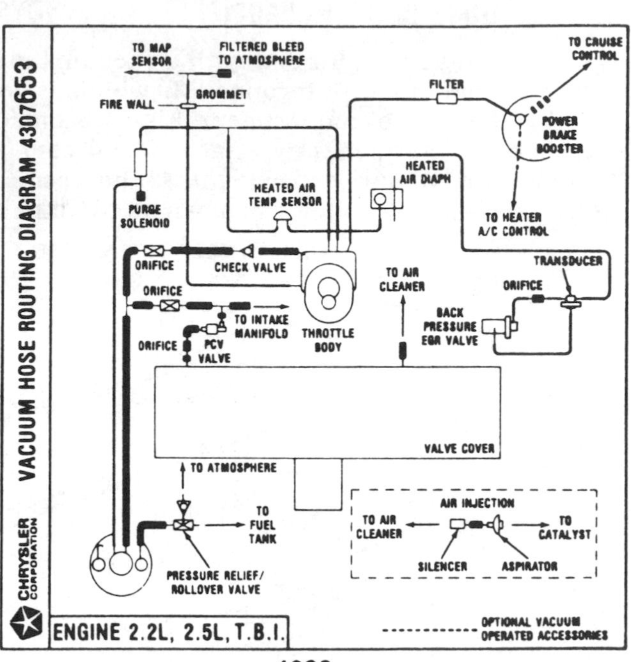

1986

Fig. 9—Hose Routing Label 2.2L 2.5L EFI

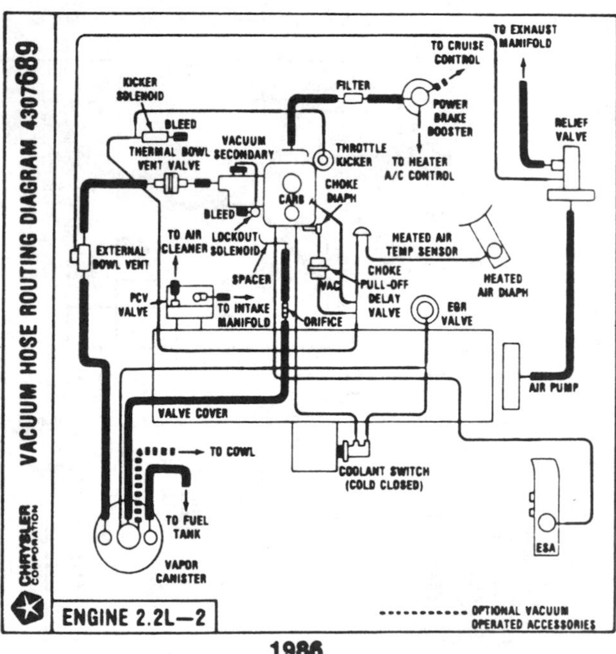

1986

Fig. 10—Hose Routing Label 2.2L L, Canadian

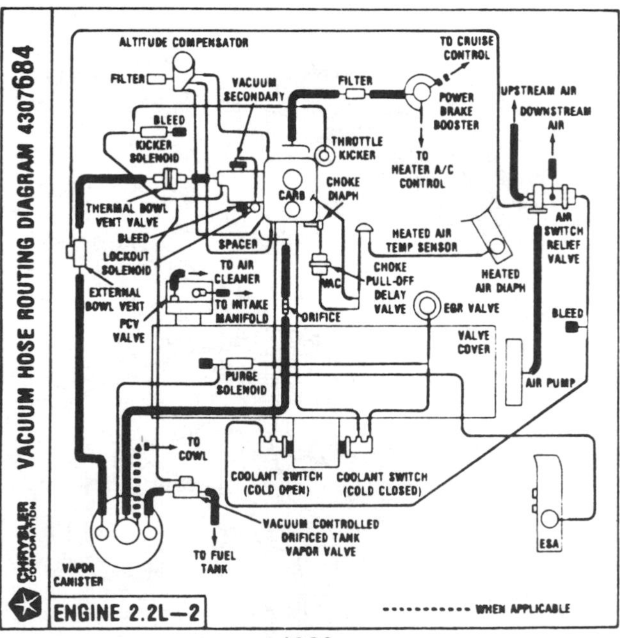

1986

Fig. 11—Hose Routing Label 2.2L L Federal

VACUUM CIRCUITS

(© Chrysler Corp.)

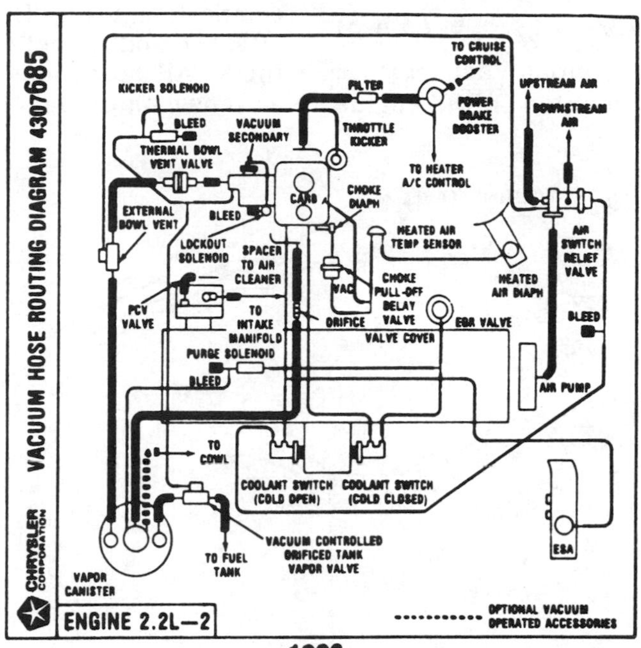

1986
Fig. 12—Hose Routing Label 2.2L L California

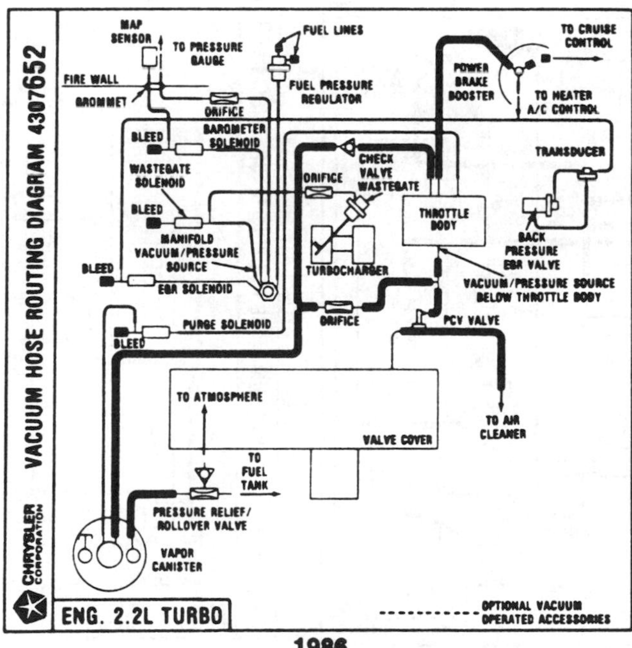

1986
Fig. 13—Hose Routing Label 2.2L Turbo

VACUUM CIRCUITS

(© Chrysler Corp.)

VACUUM HOSE ROUTING LABEL (Figs, 5, 6, 7 and 8)

All vehicles are equipped with a Vacuum Hose Routing Label which is mounted in the engine compartment (Fig. 1). This label is permamently attached and cannot be removed without destroying it. All hoses must be routed and connected as shown on the label.

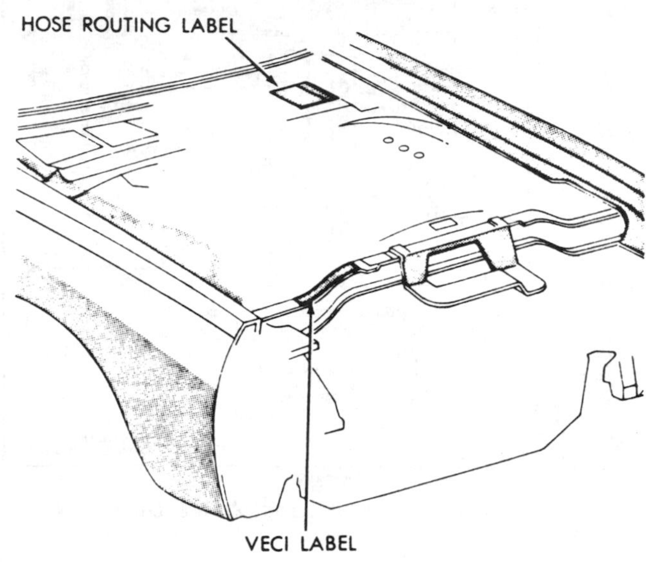

Underhood Label Locations

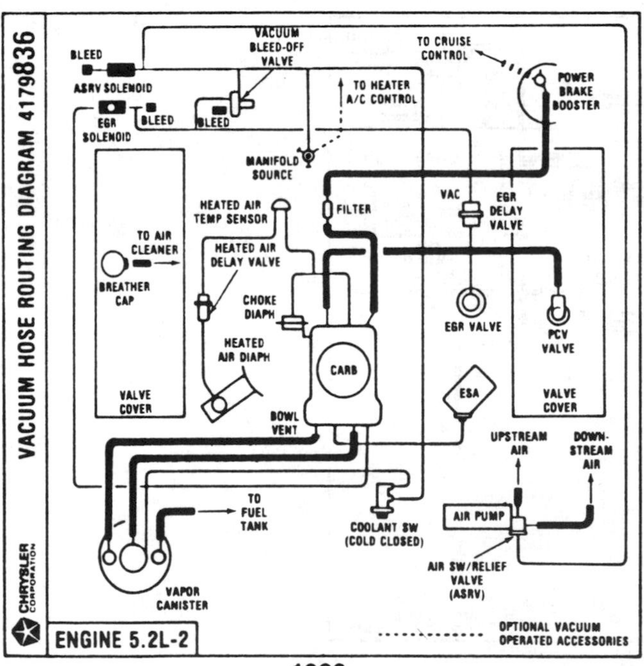

1986

Fig. 5—Federal 5.2L-2 Hose Routing Label

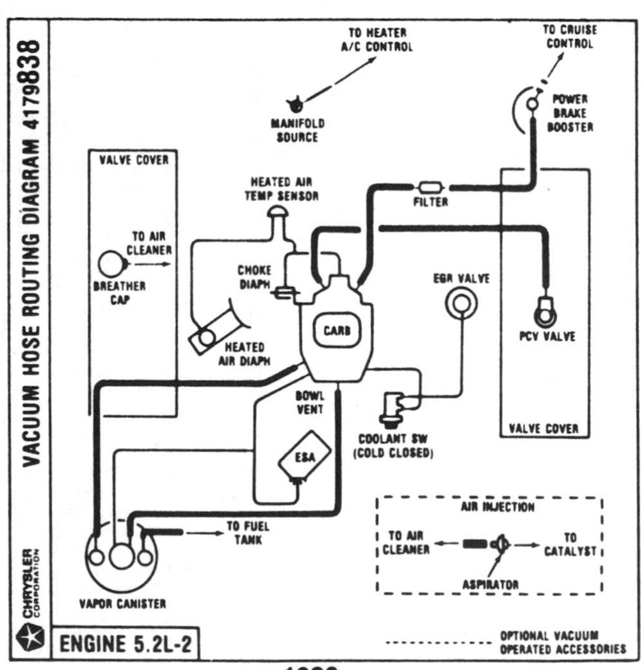

1986

Fig. 6—Canadian 5.2L-2 Hose Routing Label

VACUUM CIRCUITS

(© Chrysler Corp.)

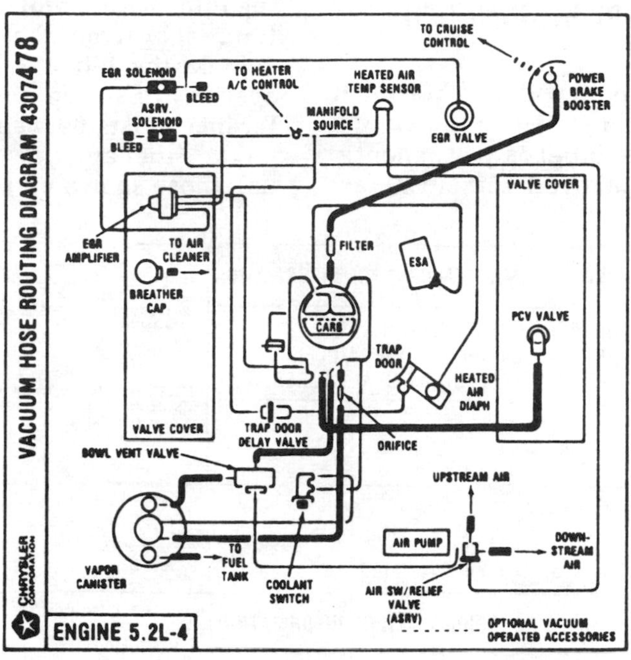

1986

Fig. 7—Federal/Canada 5.2L-4 Hose Routing Label

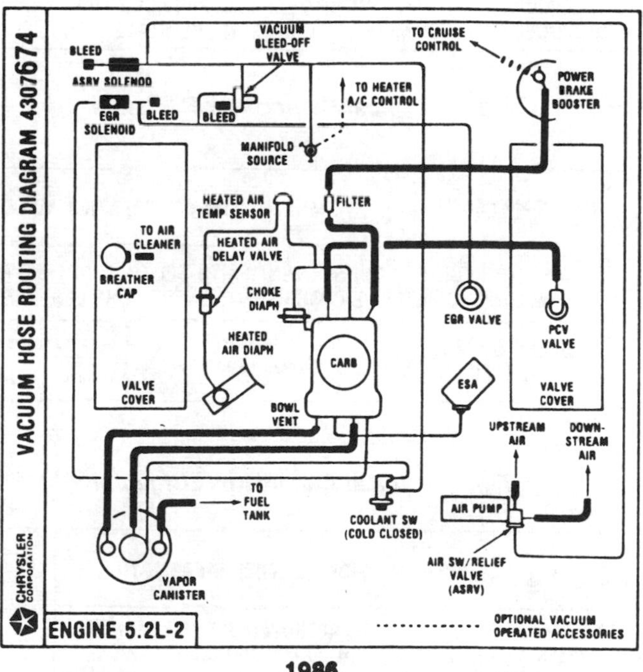

1986

Fig. 8—California Vacuum Hose Routings 5.2L-2

1987 Vacuum Circuit Diagrams

VEHICLE EMISSION CONTROL INFORMATION (VECI) LABEL

All vehicles are equipped with a **VECI** label (Figs. 1, 2, and 3) which is located in the engine compartment (Fig. 4). This label is permanently attached and cannot be removed without defacing information and destroying it. The specifications shown on the label are correct for the vehicle the label is mounted on. Propane idle speeds are shown on the VECI label. If any difference exists between the specifications shown on the label and those shown in the service manual, those shown on the label should be used.

4288953
CATALYST

CHRYSLER CORPORATION
VEHICLE EMISSION CONTROL INFORMATION

THIS VEHICLE CONFORMS TO U.S. EPA REGULATIONS APPLICABLE TO 1986 MODEL YEAR NEW MOTOR VEHICLES AT ALL ALTITUDES.

2 2 LITER
GCR2 2V5FAAX
GCRVA

SPARK PLUGS
035 IN GAP
RN12YC

*BASIC IGNITION TIMING AND FUEL INJECTION MIXTURE HAVE BEEN PRESET AT THE FACTORY. SEE THE SERVICE MANUAL FOR PROPER PROCEDURES AND OTHER ADDITIONAL INFORMATION.

• ADJUSTMENTS MADE BY OTHER THAN APPROVED SERVICE MANUAL PROCEDURES MAY VIOLATE FEDERAL AND STATE LAWS.

IDLE SETTINGS*	MAN	AUTO
TIMING BTC	12°	12°
IDLE RPM	900 (N)	800 (D)
PROPANE RPM	N A	N A

(N) = NEUTRAL
(D) = DRIVE

Fig. 1—Typical Federal VECI Label

4288961
CATALYST
EGR, TWC, EGS, EFI, AIV, CL

CHRYSLER CORPORATION
VEHICLE EMISSION CONTROL INFORMATION

2 2 LITER
GCR2 5V5FAM9
GCRVB

SPARK PLUGS
035 IN GAP
RN12YC

*BASIC IGNITION TIMING AND FUEL INJECTION MIXTURE HAVE BEEN PRESET AT THE FACTORY. SEE THE SERVICE MANUAL FOR PROPER PROCEDURES AND OTHER ADDITIONAL INFORMATION.

• ADJUSTMENTS MADE BY OTHER THAN APPROVED SERVICE MANUAL PROCEDURES MAY VIOLATE FEDERAL AND STATE LAWS

THIS VEHICLE CONFORMS TO U.S. EPA AND STATE OF CALIFORNIA REGULATIONS APPLICABLE TO 1986 MODEL YEAR NEW MOTOR VEHICLES PROVIDED THAT THIS VEHICLE IS ONLY INTRODUCED INTO COMMERCE FOR SALE IN THE STATE OF CALIFORNIA.

IDLE SETTINGS*	MAN	AUTO
TIMING BTC	12°	12°
IDLE RPM	900 (N)	700 (D)

Fig. 2—Typical California VECI Label

RÉGLAGES DU RALENTI *	MAN.	AUTO
DISTRIBUTION AV P.H.	12 °	12 °
RÉGLAGE DU RALENTI EN TR/MIN.	900	700(D)

*CONSULTEZ LE MANUEL D'ENTRETIEN POUR LES BONNES MÉTHODES DE DIAGNOSTIC ET AUTRES RENSEIGNEMENTS PERTINENTS.

CHRYSLER CANADA LTÉE
RENSEIGNEMENTS RELATIFS AU CONTRÔLE DES ÉMISSIONS

LES RÉGLAGES QUI SERAIENT FAITS SEL-ON D'AUTRES PROCÉDURES QUE CELLES DUE MANUEL D'ENTRETIEN POURRAIENT ENFREINDRE LES LOIS FÉDÉRALES ET PROVINCIALES.

TAILLE DU MOTEUR
2.2 LITRE

BOUGIES
0.9 ÉCARTEMENT X mm.
RN12YC

LE CALAGE DE L'ALLUMAGE INITIAL ET LE MÉLANGE D'INJECTION D'ESSENCE ONT ÉTÉ PRÉRÉGLES A L'USINE. N'EFFECTUEZ AUCUN RÉGLAGE LORS DE TRAVAUX D'ENTRETIEN RÉGULIERS.

Fig. 3—Typical Canadian VECI Label

4288968

CHRYSLER CANADA LTD.
VEHICLE EMISSION CONTROL INFORMATION

ADJUSTMENTS MADE BY OTHER THAN APPROVED SERVICE MANUAL PROCEDURES MAY VIOLATE FEDERAL AND PROVINCIAL LAWS.

ENGINE SIZE
2.2 LITRE

SPARK PLUGS
0.9 mm GAP
RN12YC

BASIC IGNITION TIMING AND FUEL INJECTION MIXTURE HAVE BEEN PRESET AT THE FACTORY. ADJUSTMENTS SHOULD NOT BE MADE DURING ROUTINE SERVICE.

IDLE SETTINGS *	MAN.	AUTO
TIMING BTC	12 °	12 °
IDLE R/MIN.	900	700(D)

*CONSULT THE SERVICE MANUAL FOR PROPER DIAGNOSTIC PROCEDURES AND OTHER ADDITIONAL INFORMATION.

VACUUM CIRCUITS

(© Chrysler Corp.)

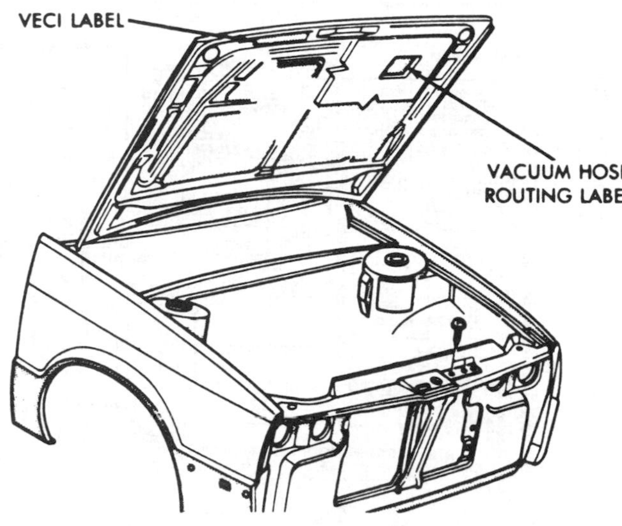

Fig. 4—Underhood Label Location

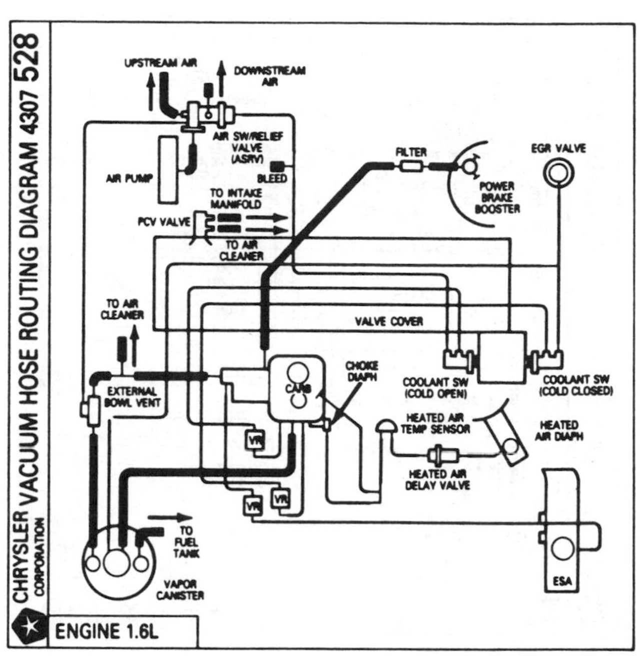

1987 *-Hose Routing Label 1.6L Federal/California*

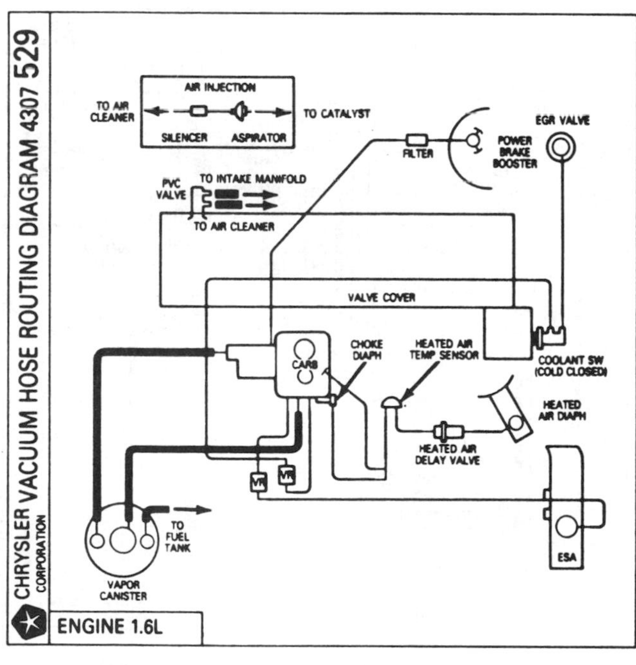

1987*-Hose Routing Label 1.6L Canadian*

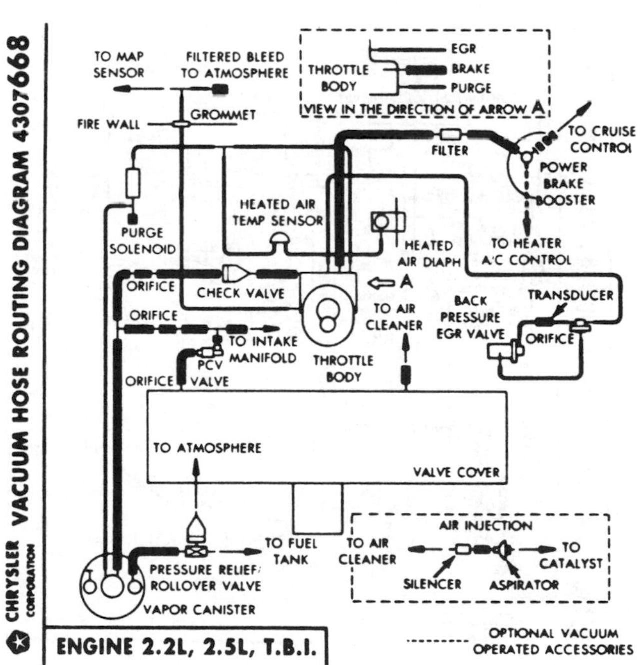

1987*–Hose Routing Label 2.2L E.F.I.*

VACUUM CIRCUITS

(© Chrysler Corp.)

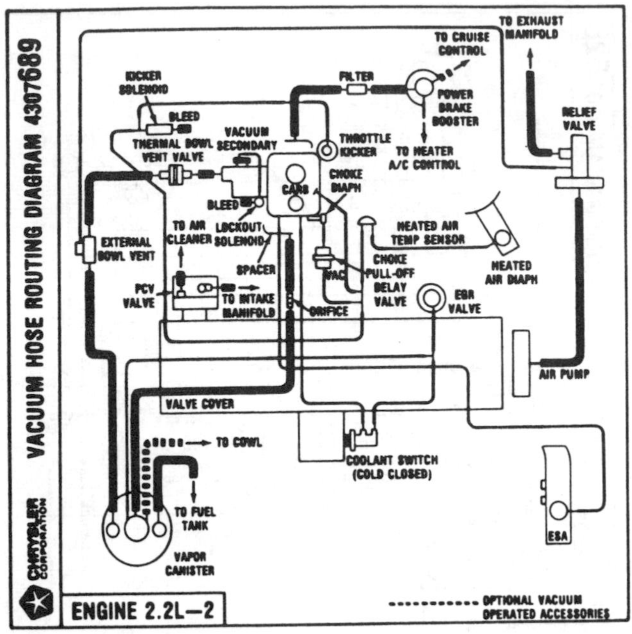

1987 *Hose Routing Label 2.2L L, Canadian*

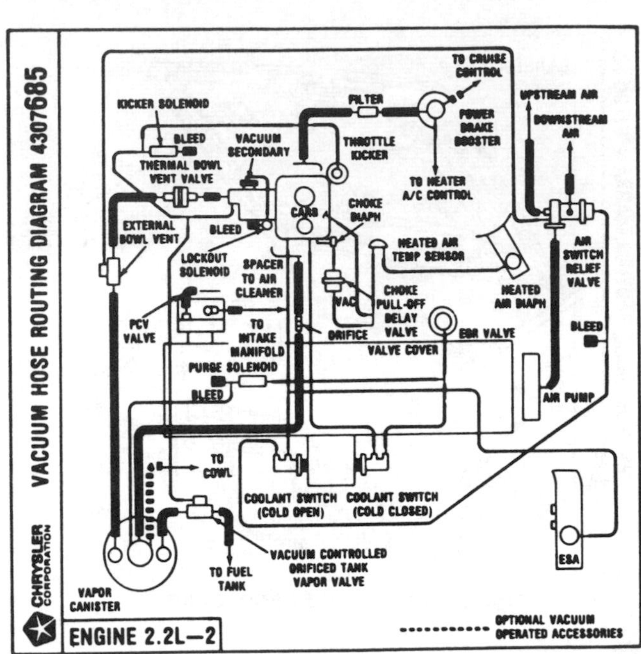

1987 *Hose Routing Label 2.2L L California*

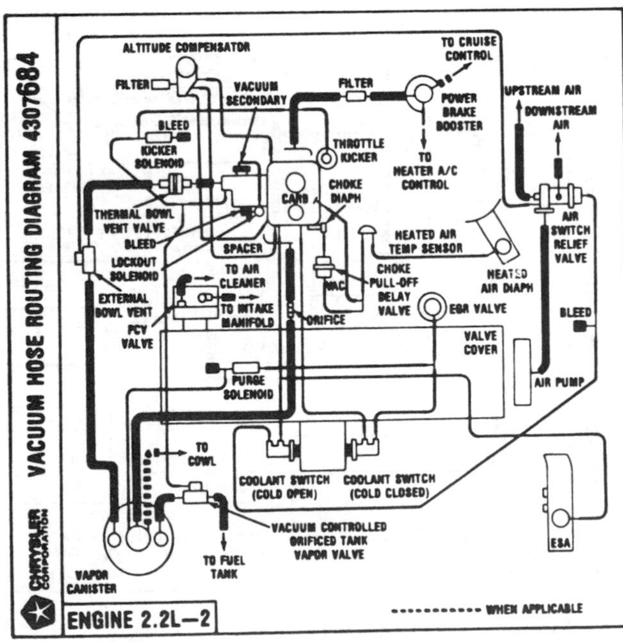

1987 *Hose Routing Label 2.2L L Federal*

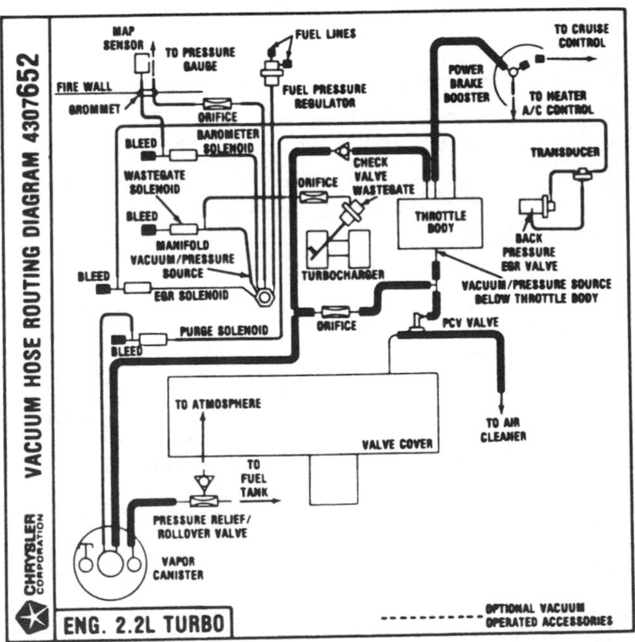

1987 *Hose Routing Label 2.2L Turbo*

VACUUM CIRCUITS

(© Chrysler Corp.)

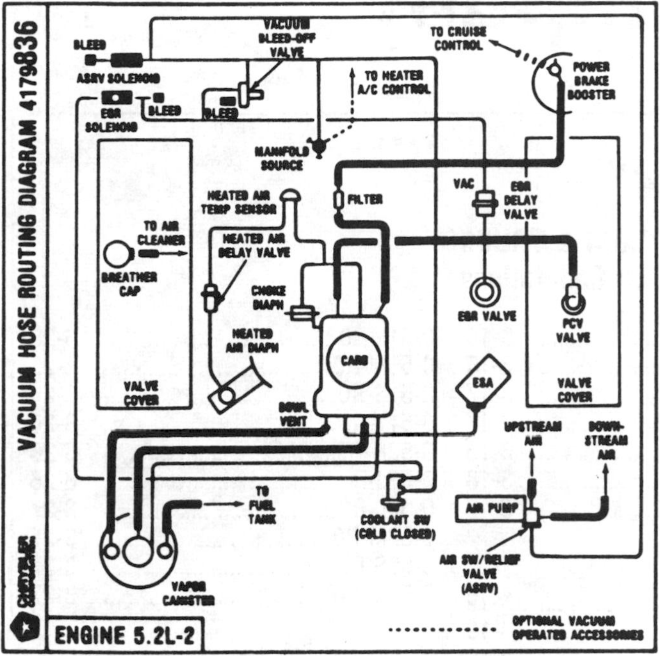

1987 Federal vacuum schematic—5.2L-2-BBL

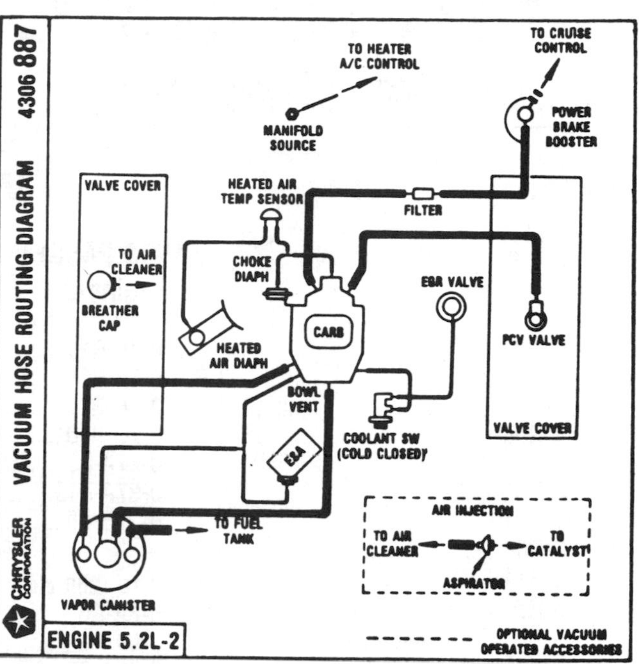

1987 Canadian vacuum schematic—5.2L-2-BBL

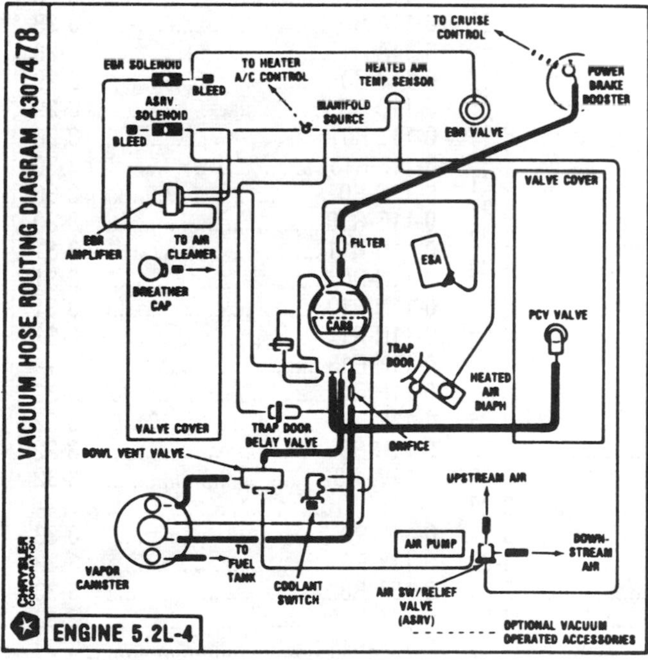

1987 Federal/Canada vacuum schematic—5.2L-4-BBL

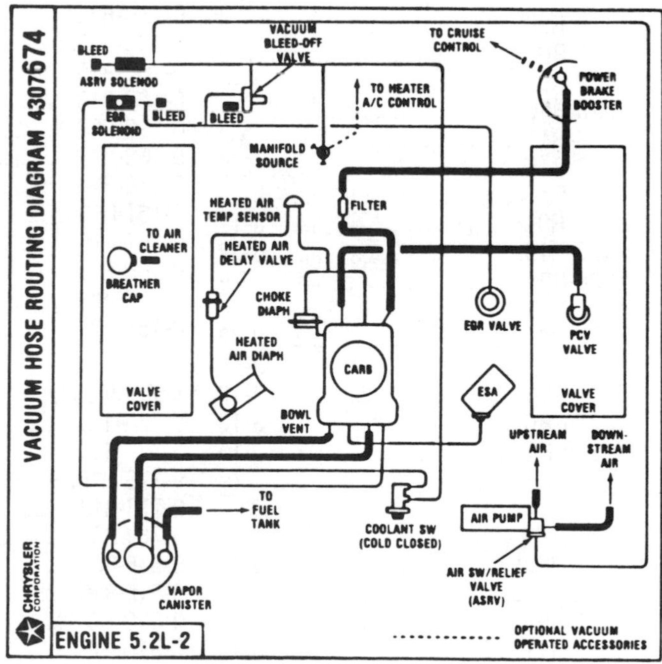

1987 California vacuum schematic—5.2L-2-BBL

Ford Motor Company
Vacuum Circuits

INDEX

1980 PASSENGER CARS & TRUCKS
Engine-Car Emission Calibrations

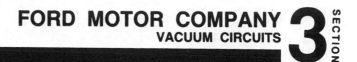

Ford
Vacuum Circuits

INDEX

1980 PASSENGER CARS & TRUCKS
Engine-Car Emission Calibrations

1980 VACUUM CIRCUITS

HOW TO DETERMINE VEHICLE CALIBRATION

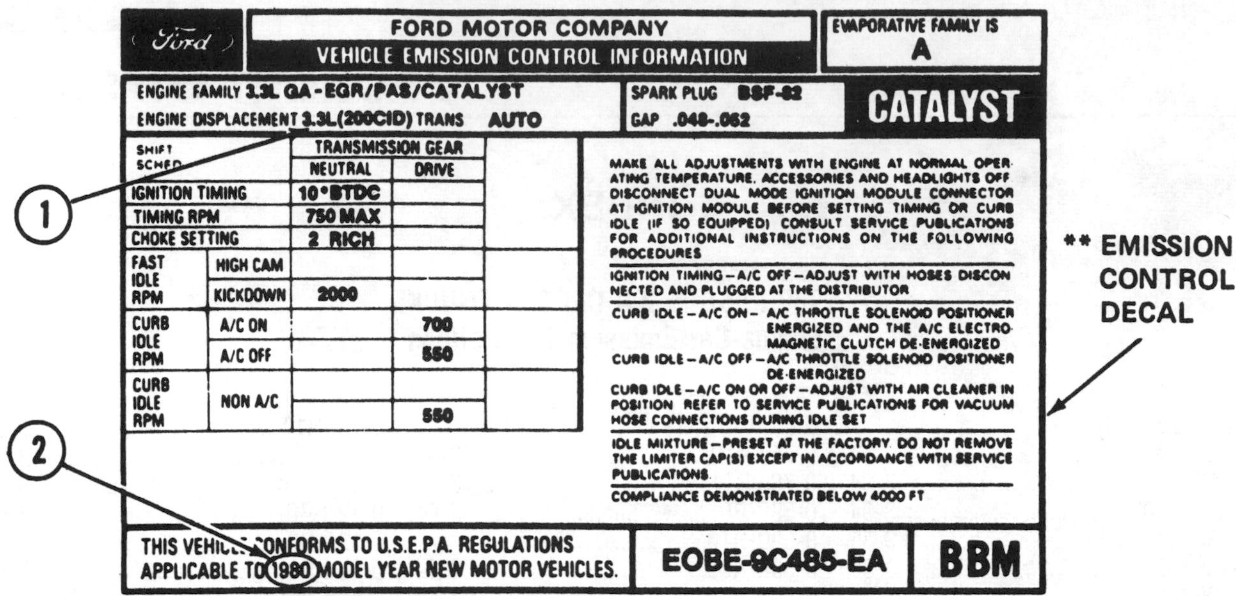

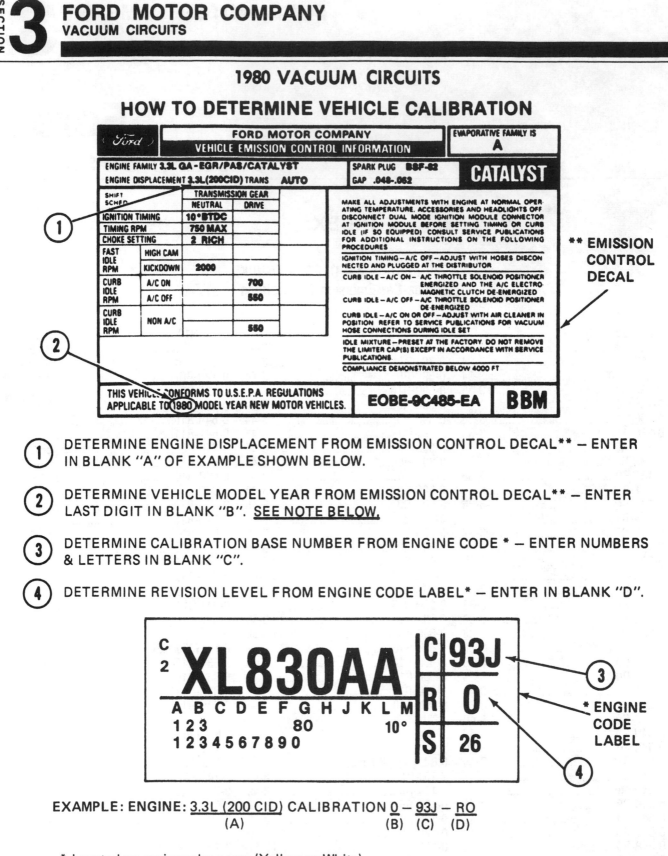

(1) DETERMINE ENGINE DISPLACEMENT FROM EMISSION CONTROL DECAL** — ENTER IN BLANK "A" OF EXAMPLE SHOWN BELOW.

(2) DETERMINE VEHICLE MODEL YEAR FROM EMISSION CONTROL DECAL** — ENTER LAST DIGIT IN BLANK "B". SEE NOTE BELOW.

(3) DETERMINE CALIBRATION BASE NUMBER FROM ENGINE CODE * — ENTER NUMBERS & LETTERS IN BLANK "C".

(4) DETERMINE REVISION LEVEL FROM ENGINE CODE LABEL* — ENTER IN BLANK "D".

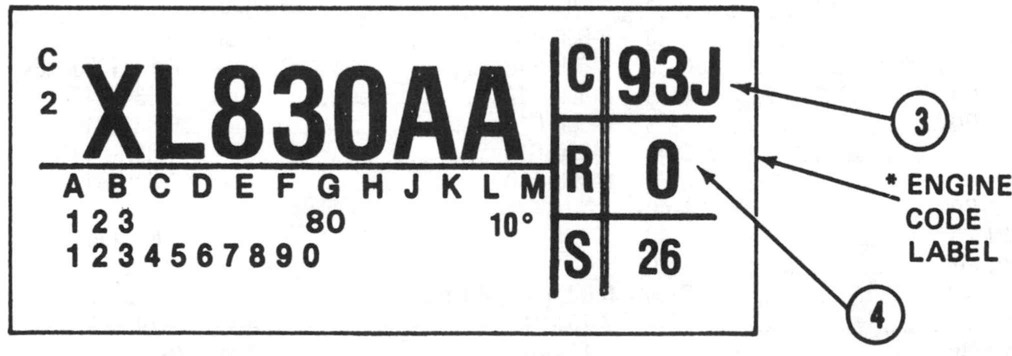

EXAMPLE: ENGINE: 3.3L (200 CID) (A) CALIBRATION 0 (B) – 93J (C) – RO (D)

* Located on engine valve cover (Yellow or White)

** Located on engine valve cover, radiator support bracket, underside of hood or fresh air inlet tube, (Gold or Silver), or windshield washer bottle.

NOTE: Some calibrations are carryover. If you cannot locate calibrations using "0" prefix, check listing under "9", "8", or "7" prefixes.

VACUUM SCHEMATIC PART NAME ABBREVIATIONS LIST

PART NAME	ABBRV.	PART NAME	ABBRV
Carburetor	CARB	Ported Pressure Switch	PPS
TSP-Vacuum Operated Throttle Modulator	TSP-VOTM	Vacuum Regulator Valve (3 Port)	VRV
Vacuum Operated Throttle Modulator	VOTM	Air Control Valve (Thermactor)	ACV
Spark Port	S	Idle Vacuum Valve	IVV
EGR Port	E	Vent Valve Vacuum	VVVac
Bowl Vent Port	BV	Vacuum Restrictor	VRest
Control Valve	CV	Fuel Decel Valve	FDV
Air Cleaner TVS Valve	TVS	Vacuum Retard Delay Valve	VRDV
Thermactor Air By-Pass Valve	Air BPV		
Anti-Backfire Valve	ANTI-BFV	Vacuum Control Valve	VCV
EGR Valve Actuator	EGR VA		
Purge Control Valve	Purge CV	Vacuum Controlled Switch	VCS
Solenoid Valve-Carb Bowl Vent	SV-CBV	Vacuum Controlled Switch (Cold Temp)	VCS-CT
Solenoid Vent Valve	SVV	Vacuum Controlled Switch (Decel Idle)	VCS-DI
Vacuum Regulator/Solenoid	VR/S	Vacuum Delay Valve	VDV
EGR Valve	EGR	Vacuum Check Valve	V Ck V
		Vacuum Retard Delay Valve	VRDV
Vacuum Control Valve	VCV		
		Vacuum Vent Valve	VacVV
Solenoid Valve	Sol V	Delay Valve - Two Way	DV-TW
EGR Valve	EGR	Ignition Timing Vacuum Switch	ITVS
		Ignition Pressure Switch	IPS
Air Cleaner Duct & Valve Vacuum Motor	A/CL DV	Barometric & Manifold Absolute Pressure Sensor	BMAP
Differential Vacuum Control Valve	DVCV	Manifold Absolute Pressure Sensor	MAP
Venturi Vacuum Amplifier	VVA		
Vacuum Reservoir	VReser	Distributor	DIST
Separator Assy.-Fuel Vacuum	SA-FV		
Thermal Vent Valve	TVV	Distributor	DIST
Air Cleaner Bi-Metal Sensor	A/CL Bi Met		
Air Cleaner Cold Weather Modulator	A/CL CWM		
Load Control Valve	LCV		
EGR B/P Transducer	B/P		
Signal Conditioner	SC		

DIAGNOSTIC SPECIFICATION ABBREVIATIONS

Calib.	Calibration number	N	Neutral
D	Drive transmission position	N/R	Not required
KD	Kickdown step	PPM	Parts per million
max	Maximum	TBD	Specification to be supplied later
min	Minimum	CALIF. ONLY	California Only
A/C ONLY	Air Conditioning Only	A/T	Automatic transmission
A/C & NON A/C	Air Conditioning & Non Air Conditioning	M/T	Manual transmission
49S	49 States	NON A/C	Non Air Conditioning
50S	50 States	A/C ONLY x—x—x	Shows Air Conditioning Only
MM	Millimeters	NON A/C —ıı—ıı—ı	Shows Non Air Conditioning Only

4 STEPS
For Using Your 1980 ESSDS

1. Determine vehicle CALIBRATION by the example on the following page.
2. Examine the SCHEMATIC. Check hose routing to see the vehicle hose routing matches the schematic.
3. Examine the PARTS LIST. Verify that the correct parts are installed.
4. USE SPECIFICATIONS during tune-up and normal adjustment.

VACUUM CIRCUITS

(© Ford Motor Co)

1980—(1.6 L) engines

49S A/C & NON A/C

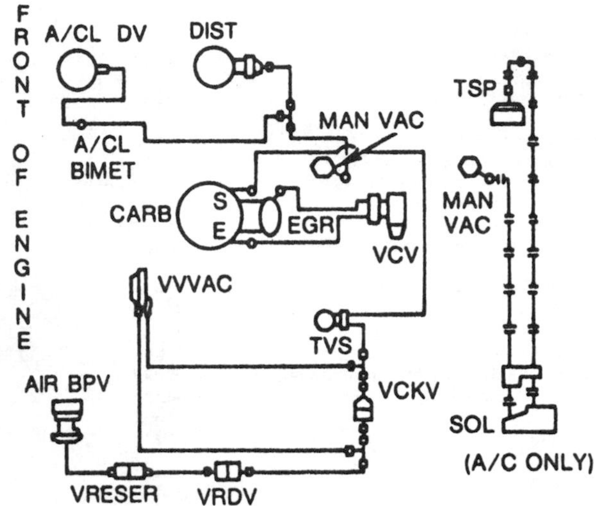

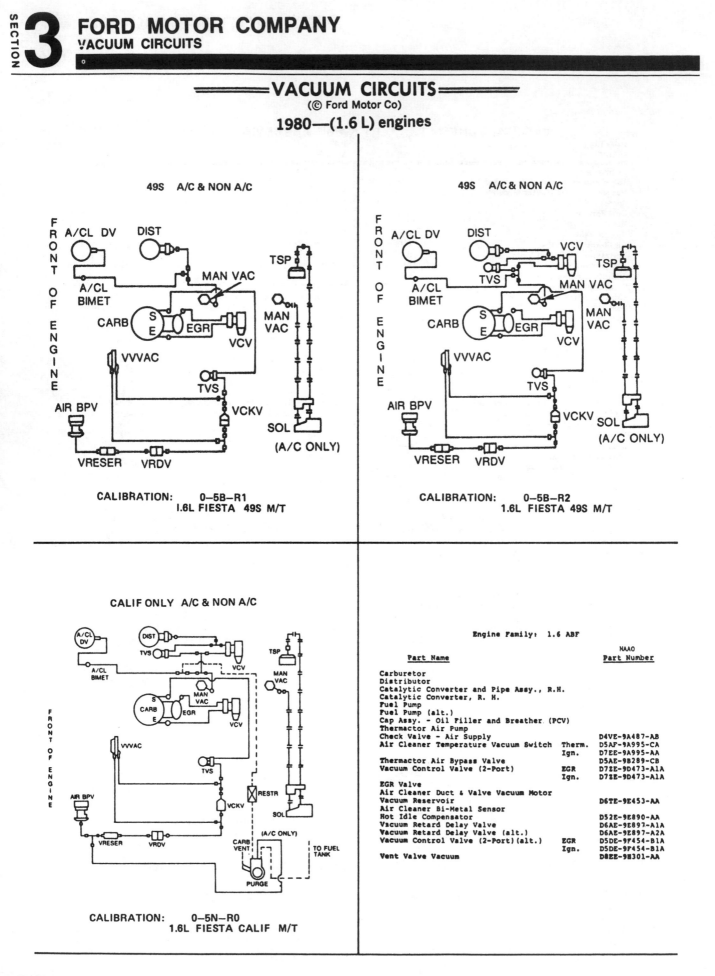

CALIBRATION: 0–5B–R1
1.6L FIESTA 49S M/T

49S A/C & NON A/C

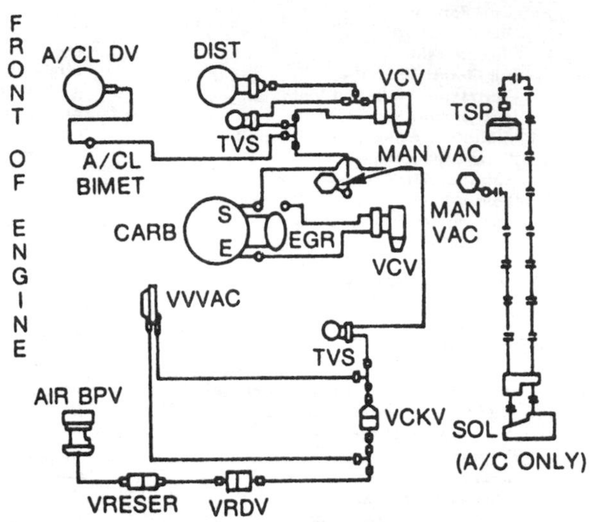

CALIBRATION: 0–5B–R2
1.6L FIESTA 49S M/T

CALIF ONLY A/C & NON A/C

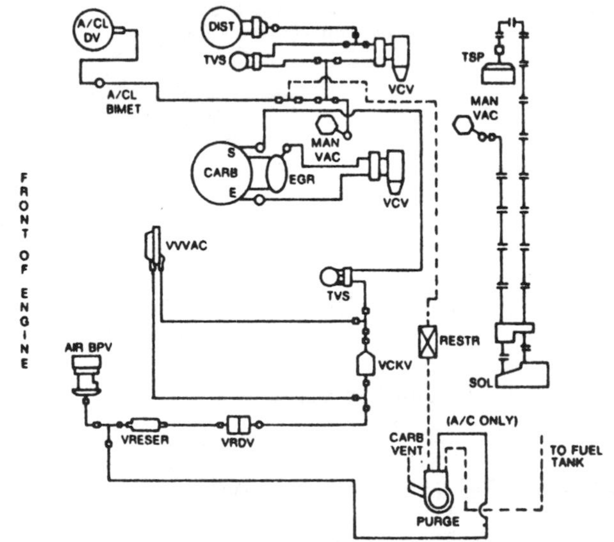

CALIBRATION: 0–5N–R0
1.6L FIESTA CALIF M/T

Engine Family: 1.6 ABF

Part Name		NAAO Part Number
Carburetor		
Distributor		
Catalytic Converter and Pipe Assy., R.H.		
Catalytic Converter, R. H.		
Fuel Pump		
Fuel Pump (alt.)		
Cap Assy. - Oil Filler and Breather (PCV)		
Thermactor Air Pump		
Check Valve - Air Supply		D4VE-9A487-AB
Air Cleaner Temperature Vacuum Switch	Therm.	D5AF-9A995-CA
	Ign.	D7EE-9A995-AA
Thermactor Air Bypass Valve		D5AE-9B289-CB
Vacuum Control Valve (2-Port)	EGR	D7ZE-9D473-A1A
	Ign.	D7ZE-9D473-A1A
EGR Valve		
Air Cleaner Duct & Valve Vacuum Motor		
Vacuum Reservoir		D6TE-9E453-AA
Air Cleaner Bi-Metal Sensor		
Hot Idle Compensator		D52E-9E890-AA
Vacuum Retard Delay Valve		D6AE-9E897-A1A
Vacuum Retard Delay Valve (alt.)		D6AE-9E897-A2A
Vacuum Control Valve (2-Port)(alt.)	EGR	D5DE-9F454-B1A
	Ign.	D5DE-9F454-B1A
Vent Valve Vacuum		D8EE-9H301-AA

VACUUM CIRCUITS

(© Ford Motor Co.)

1980—140 CID (2.3 L) engines

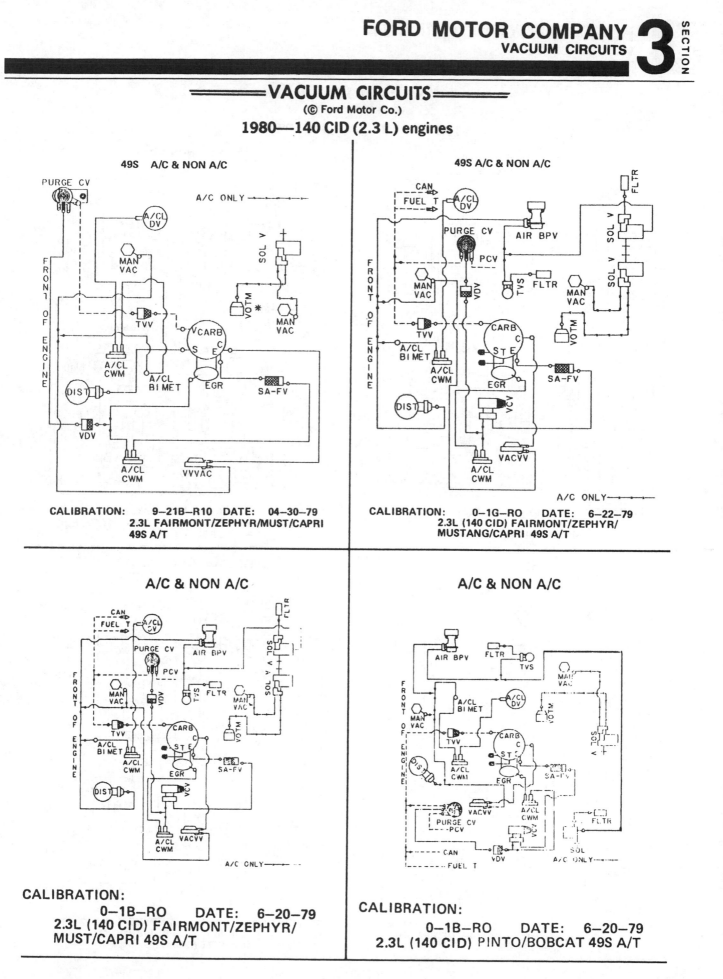

49S A/C & NON A/C

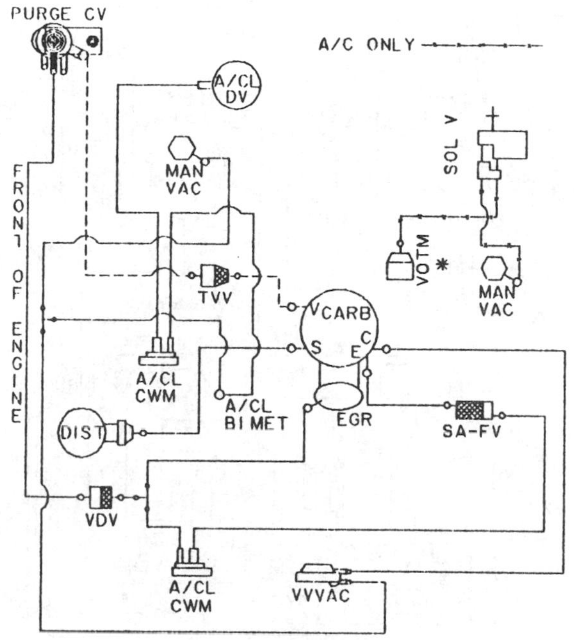

CALIBRATION: 9–21B–R10 DATE: 04–30–79
2.3L FAIRMONT/ZEPHYR/MUST/CAPRI
49S A/T

49S A/C & NON A/C

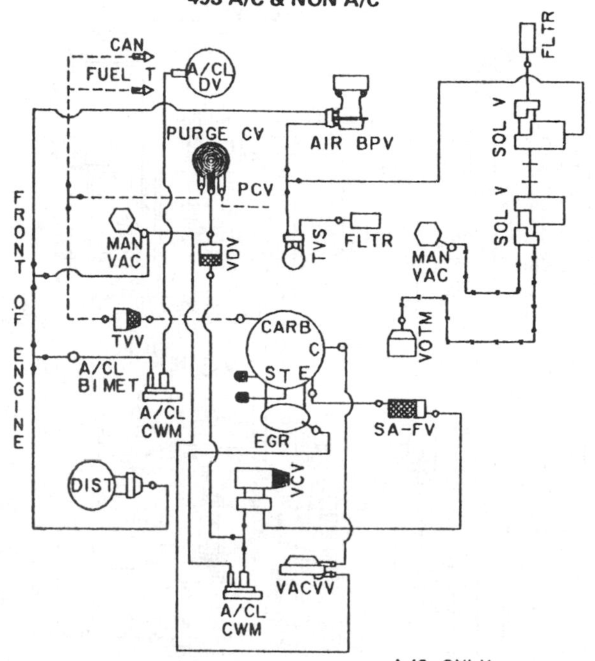

CALIBRATION: 0–1G–RO DATE: 6–22–79
2.3L (140 CID) FAIRMONT/ZEPHYR/
MUSTANG/CAPRI 49S A/T

A/C & NON A/C

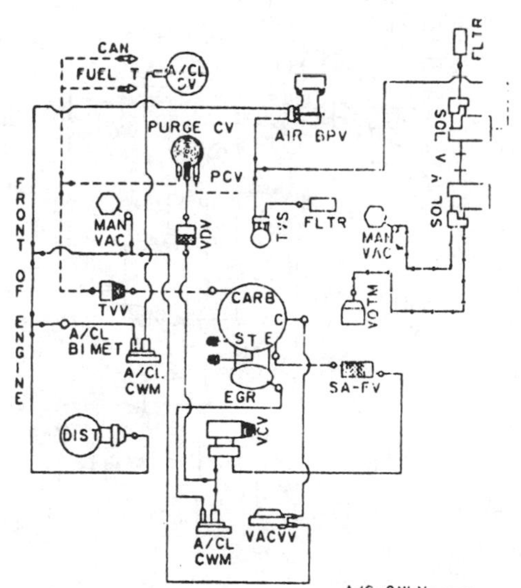

CALIBRATION:

0–1B–RO DATE: 6–20–79
2.3L (140 CID) FAIRMONT/ZEPHYR/
MUST/CAPRI 49S A/T

A/C & NON A/C

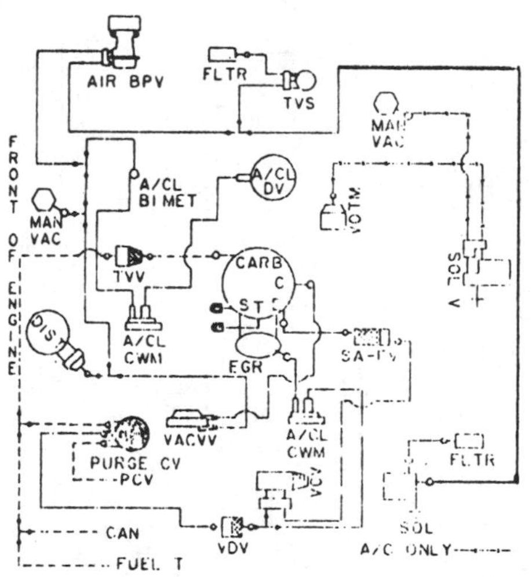

CALIBRATION:

0–1B–RO DATE: 6–20–79
2.3L (140 CID) PINTO/BOBCAT 49S A/T

VACUUM CIRCUITS

(© Ford Motor Co.)

1980—140 CID (2.3 L) engines

49S A/C & NON A/C

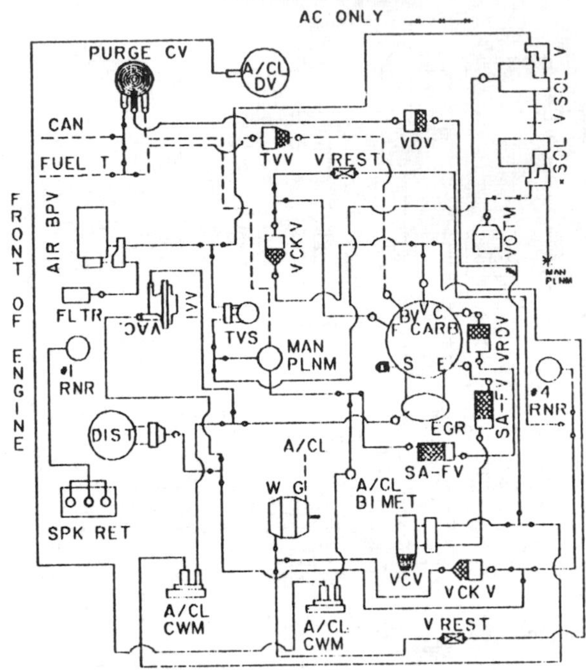

CALIBRATION: 0–01H–R10 DATE: 09–17–79
2.3L MUST/CAPRI/FAIRMONT/ZEPHYR
49S TURBO A/T

49S A/C & NON A/C

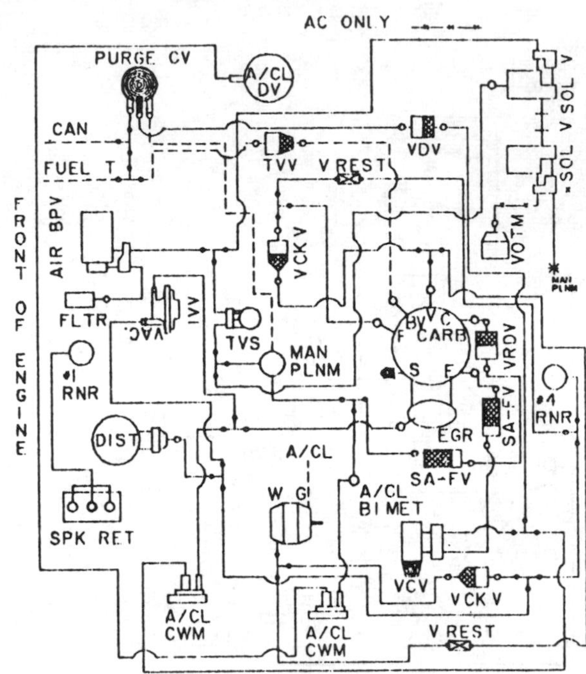

CALIBRATION: 0–1H–R15 DATE: 12–18–79
2.3L MUST/CAPRI/FAIRMONT/ZEPHYR
49S TURBO A/T

49S A/C & NON A/C

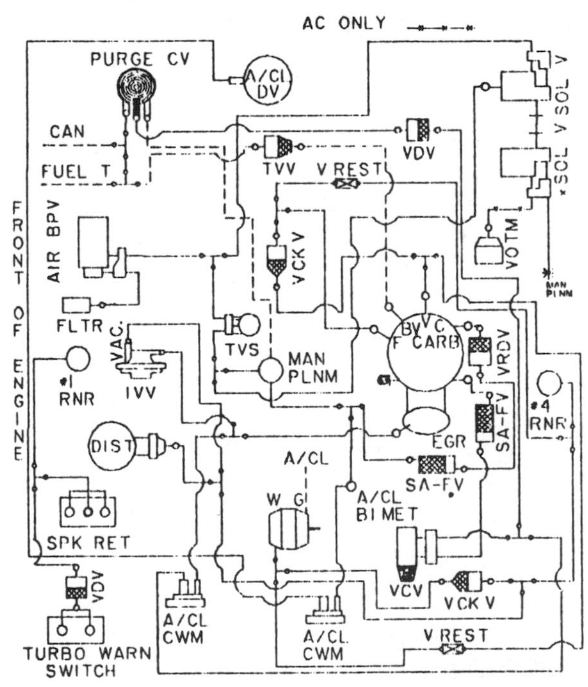

CALIBRATION: 0–1H–R19 DATE: 2–19–80
2.3L MUSTANG/CAPRI/FAIRMONT/
ZEPHYR 49S A/T

CALIF ONLY · A/C & NON A/C

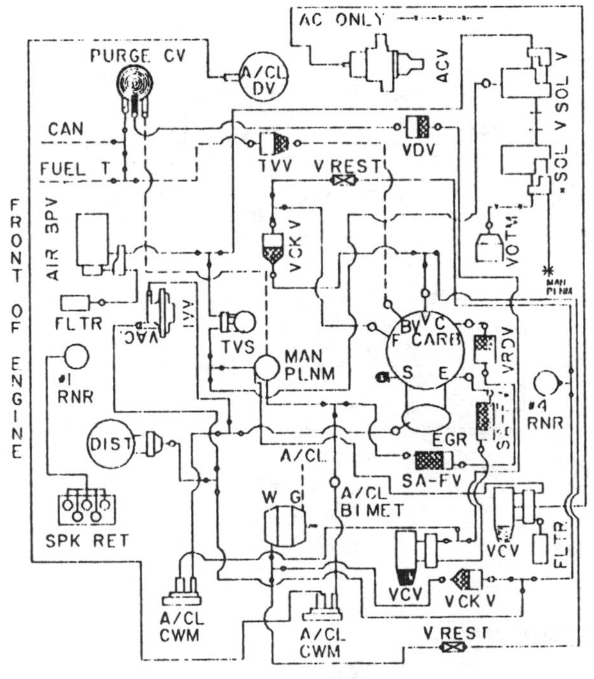

CALIBRATION: 0–1S–R11 DATE: 12–12–79
2.3L MUST/CAPRI/FAIRMONT/ZEPHYR
CALIF TURBO A/T

VACUUM CIRCUITS

1980—140 CID (2.3 L) engines

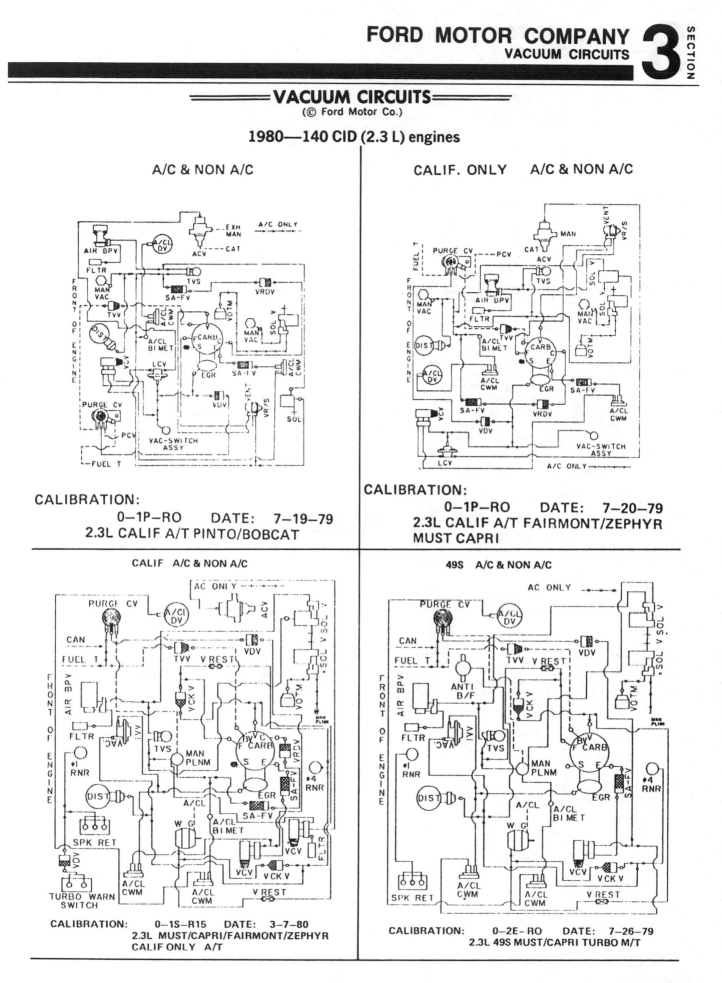

A/C & NON A/C

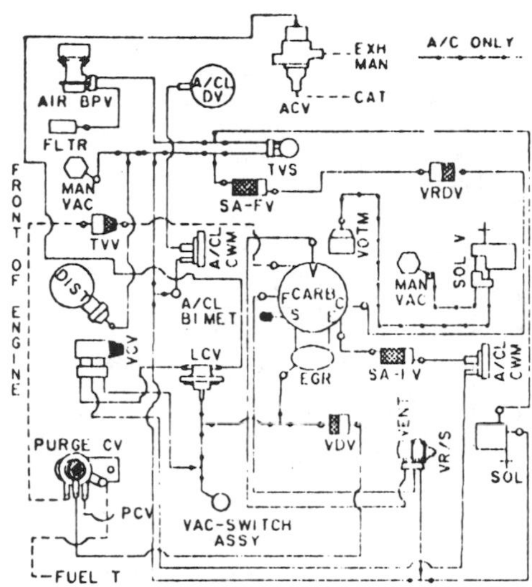

CALIBRATION:
0–1P–RO DATE: 7–19–79
2.3L CALIF A/T PINTO/BOBCAT

CALIF. ONLY A/C & NON A/C

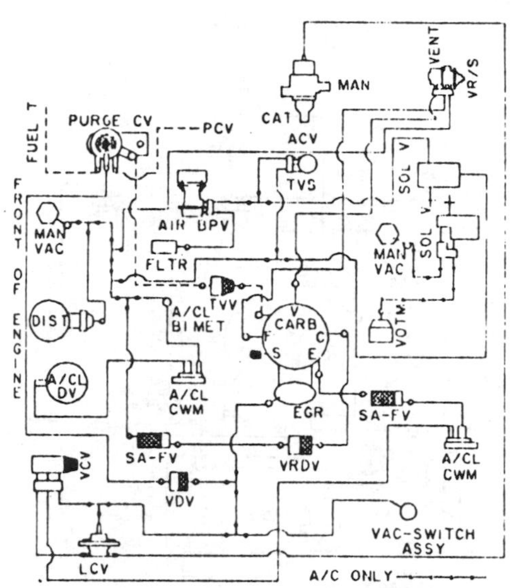

CALIBRATION:
0–1P–RO DATE: 7–20–79
2.3L CALIF A/T FAIRMONT/ZEPHYR
MUST CAPRI

CALIF A/C & NON A/C

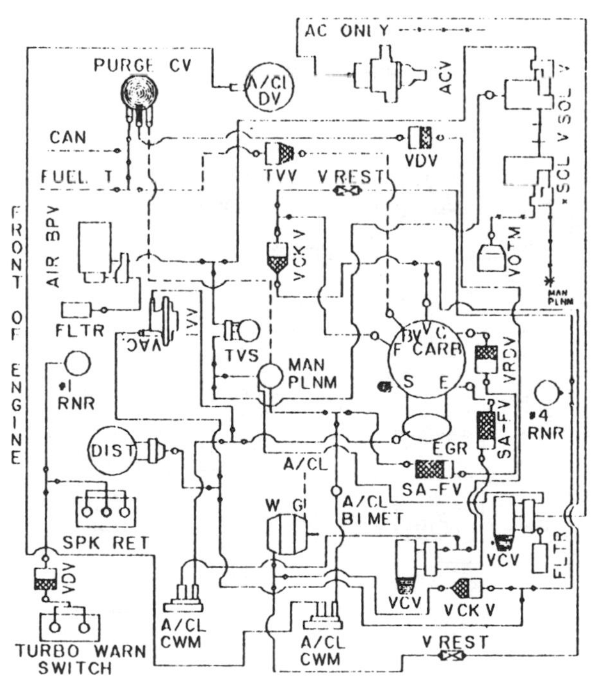

CALIBRATION: 0–1S–R15 DATE: 3–7–80
2.3L MUST/CAPRI/FAIRMONT/ZEPHYR
CALIF ONLY A/T

49S A/C & NON A/C

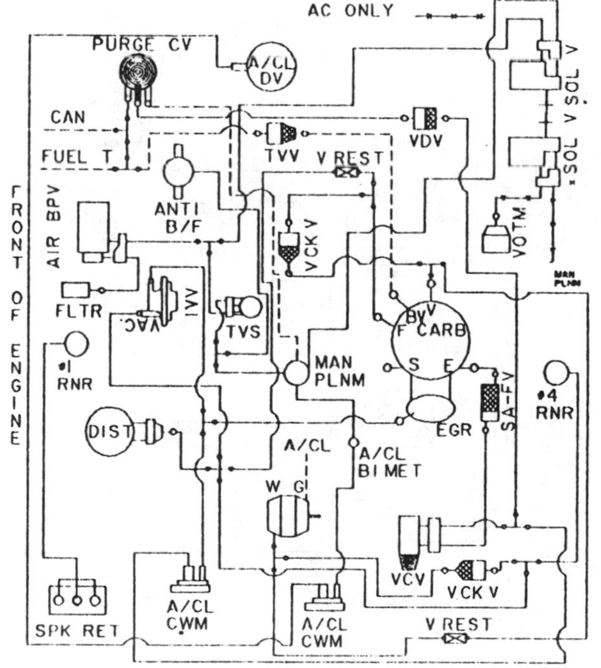

CALIBRATION: 0–2E–RO DATE: 7–26–79
2.3L 49S MUST/CAPRI TURBO M/T

VACUUM CIRCUITS

(© Ford Motor Co)

1980—140 CID (2.3 L) engines

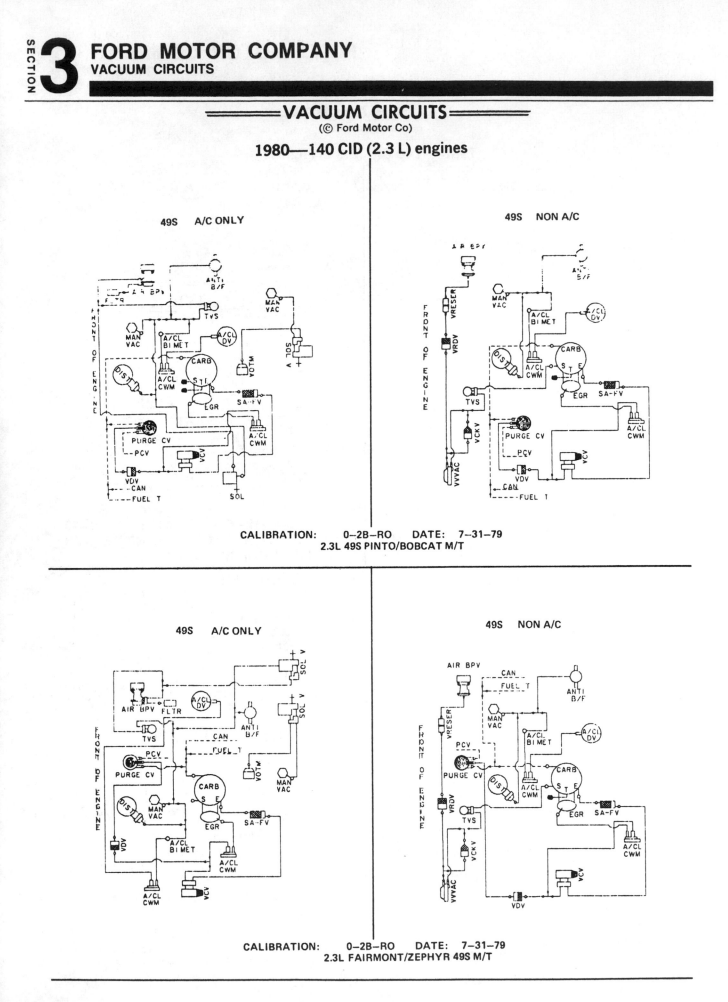

49S A/C ONLY

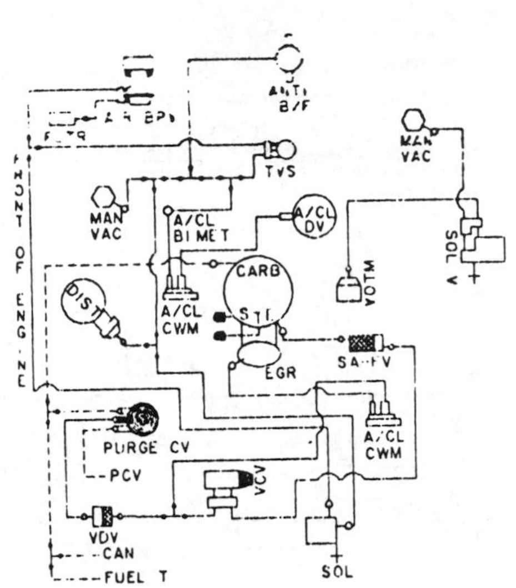

49S NON A/C

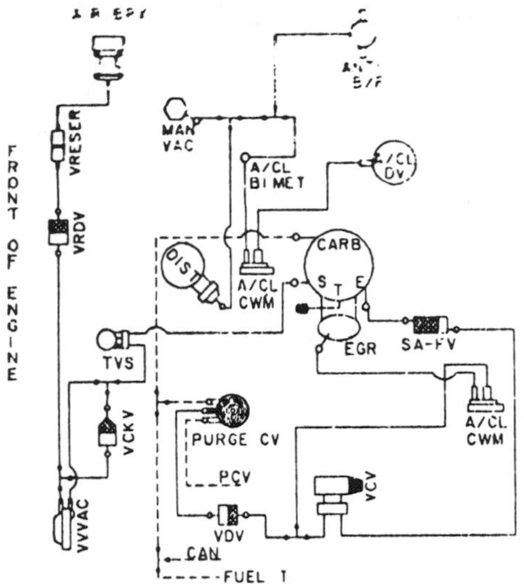

AIR BPV
ANTI B/F
VRESER
MAN VAC
A/CL BIMET
A/CL DV
FRONT OF ENGINE
VRDV
CARB
S T E
DIST
A/CL CWM
TVS
EGR
SA-FV
VCKV
PURGE CV
PCV
A/CL CWM
VCV
VVVAC
VDV
CAN
FUEL T

CALIBRATION: 0–2B–RO DATE: 7–31–79
2.3L 49S PINTO/BOBCAT M/T

49S A/C ONLY

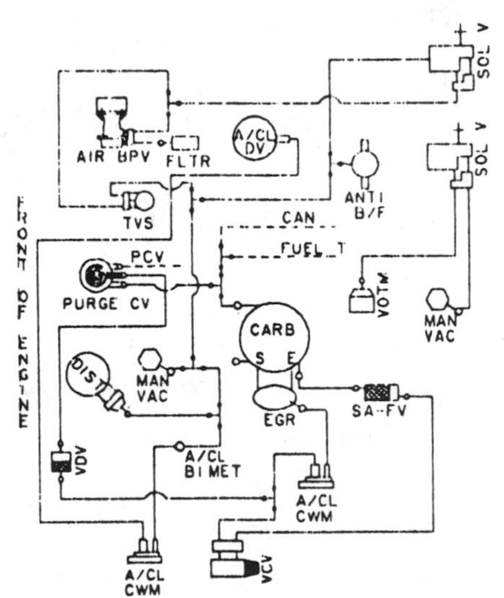

49S NON A/C

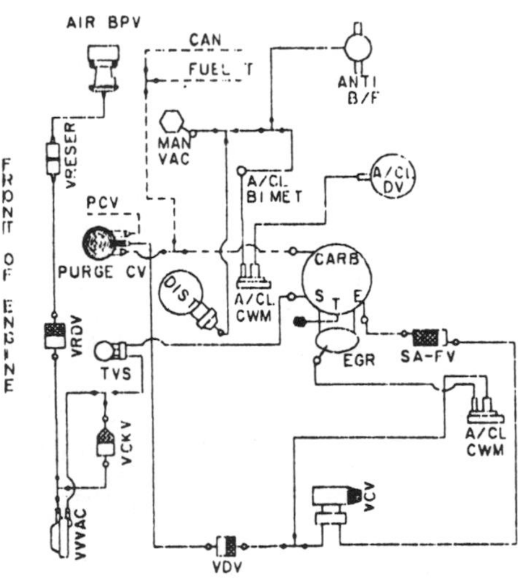

CALIBRATION: 0–2B–RO DATE: 7–31–79
2.3L FAIRMONT/ZEPHYR 49S M/T

VACUUM CIRCUITS

(© Ford Motor Co.)

1980—140 CID (2.3 L) engines

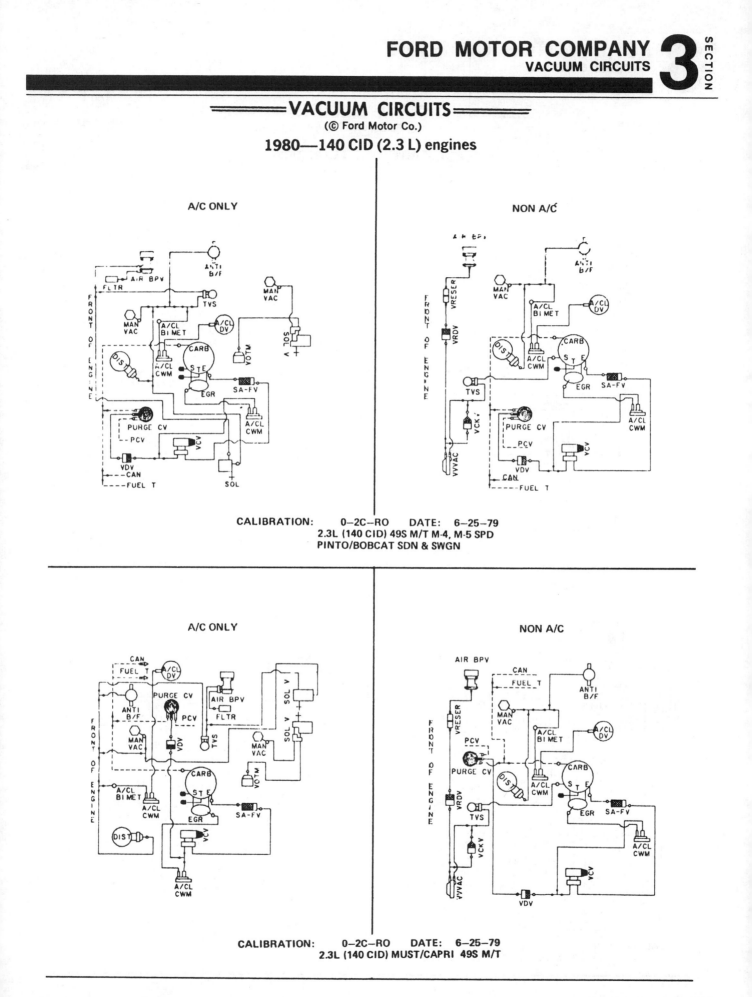

A/C ONLY

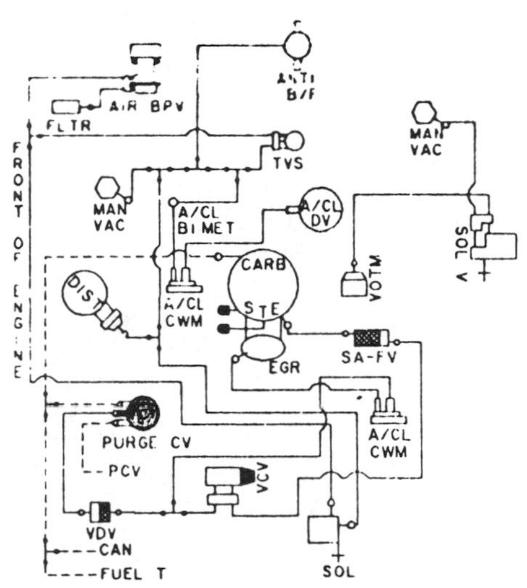

NON A/C

CALIBRATION: 0–2C–RO DATE: 6–25–79
2.3L (140 CID) 49S M/T M-4, M-5 SPD
PINTO/BOBCAT SDN & SWGN

A/C ONLY

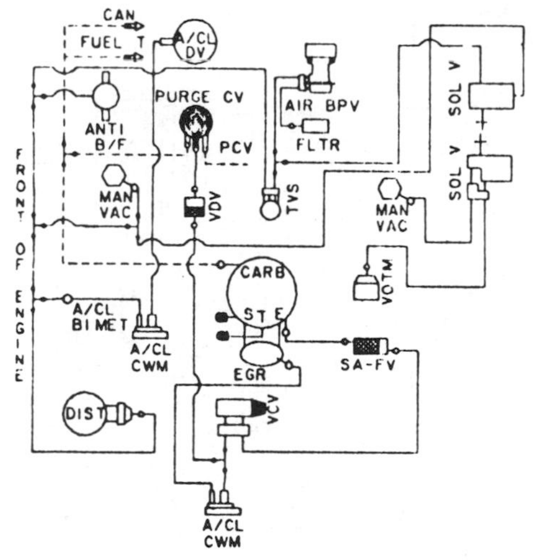

NON A/C

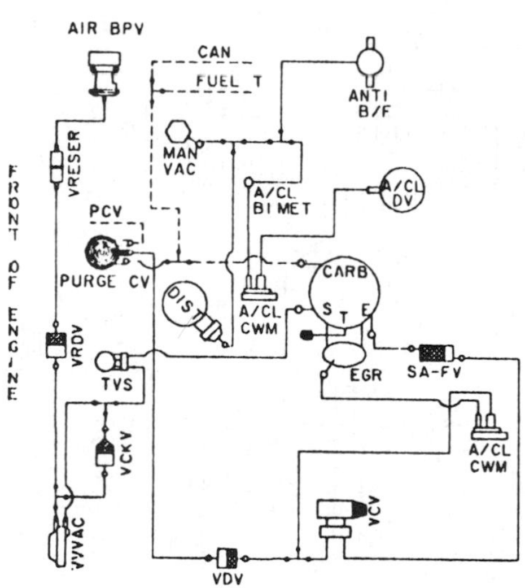

CALIBRATION: 0–2C–RO DATE: 6–25–79
2.3L (140 CID) MUST/CAPRI 49S M/T

VACUUM CIRCUITS

(© Ford Motor Co)

1980—140 CID (2.3 L) engines

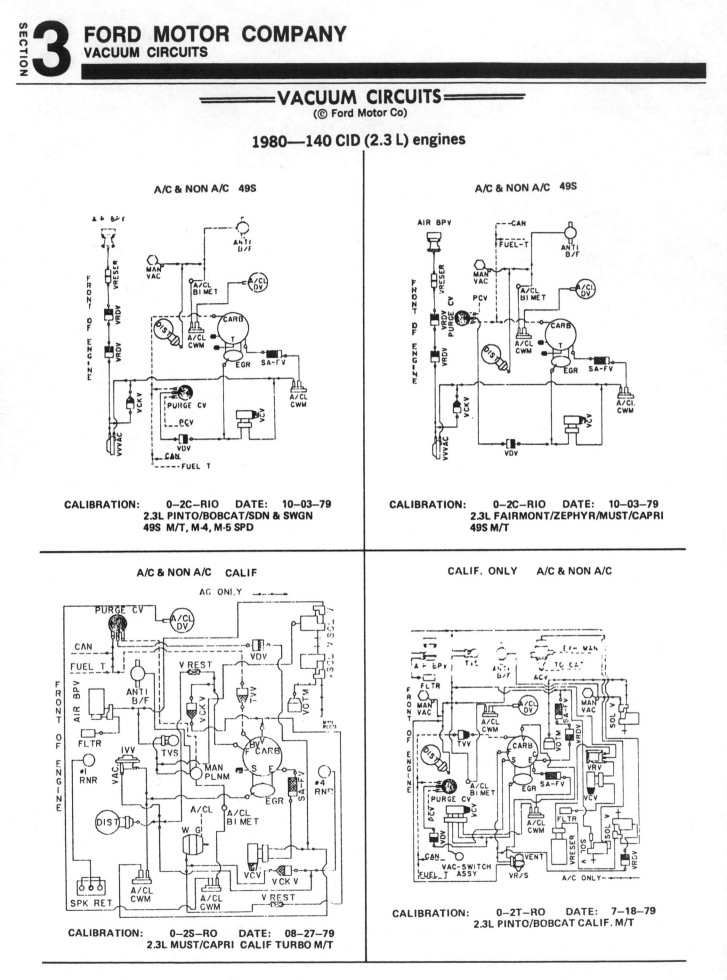

A/C & NON A/C 49S

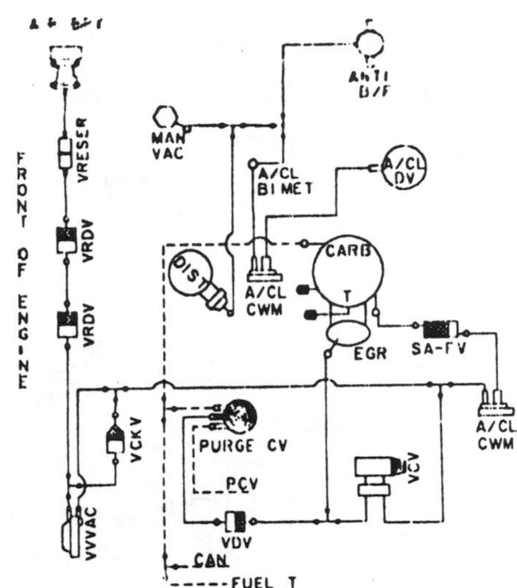

CALIBRATION: 0–2C–RIO DATE: 10–03–79
2.3L PINTO/BOBCAT/SDN & SWGN
49S M/T, M-4, M-5 SPD

A/C & NON A/C 49S

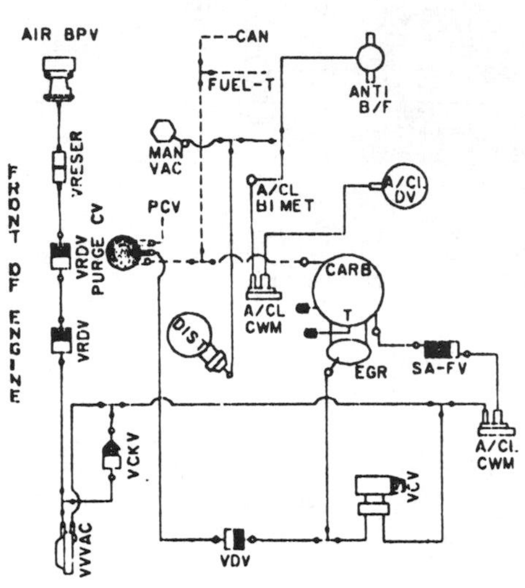

CALIBRATION: 0–2C–RIO DATE: 10–03–79
2.3L FAIRMONT/ZEPHYR/MUST/CAPRI
49S M/T

A/C & NON A/C CALIF

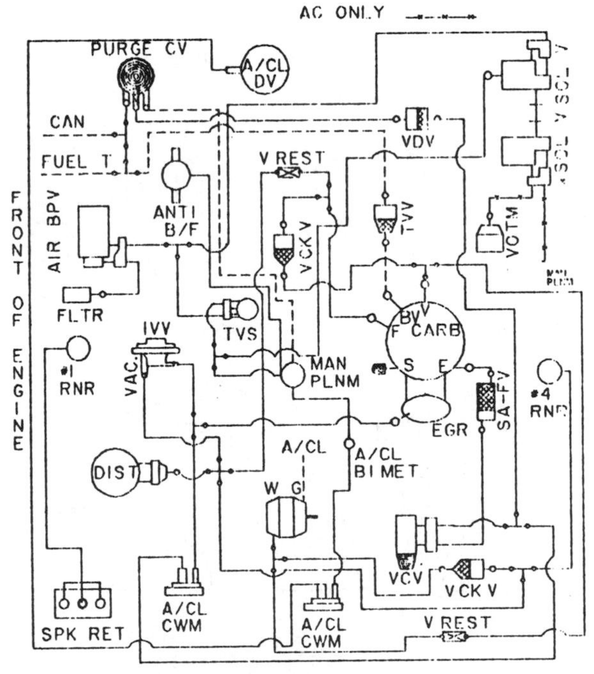

CALIBRATION: 0–2S–RO DATE: 08–27–79
2.3L MUST/CAPRI CALIF TURBO M/T

CALIF. ONLY A/C & NON A/C

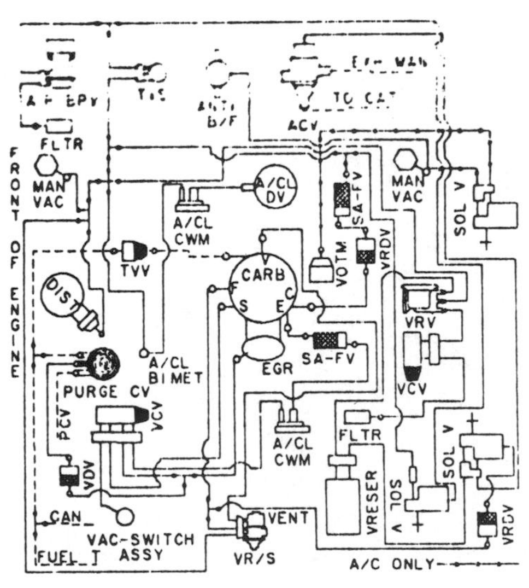

CALIBRATION: 0–2T–RO DATE: 7–18–79
2.3L PINTO/BOBCAT CALIF. M/T

VACUUM CIRCUITS

(© Ford Motor Co.)

1980—140 CID (2.3 L) engines

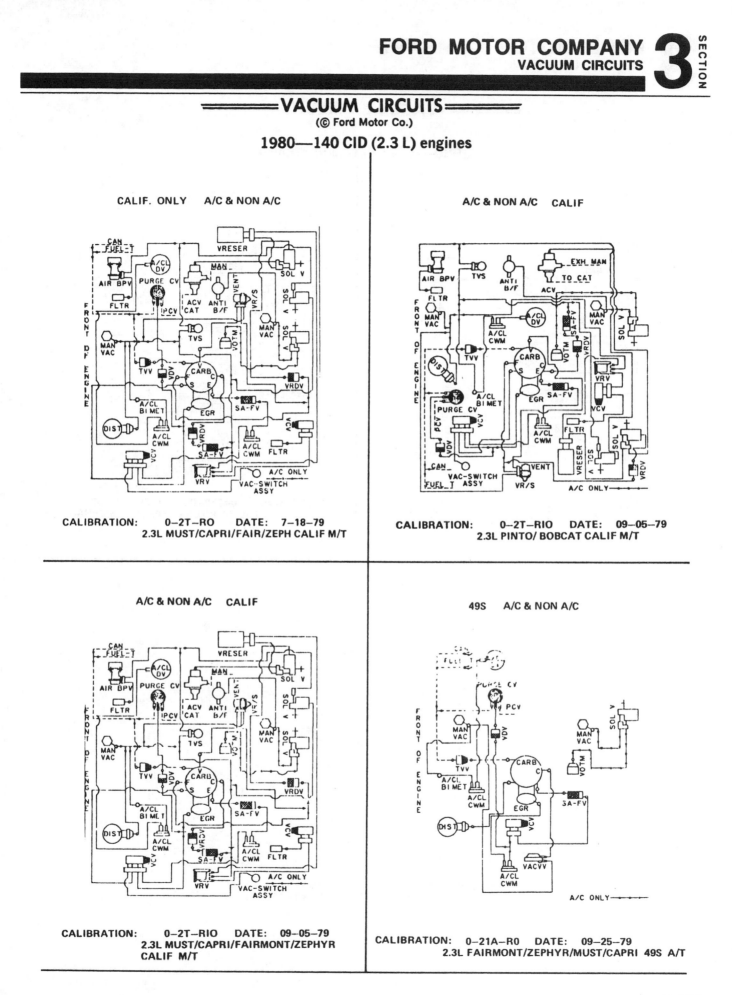

CALIF. ONLY A/C & NON A/C

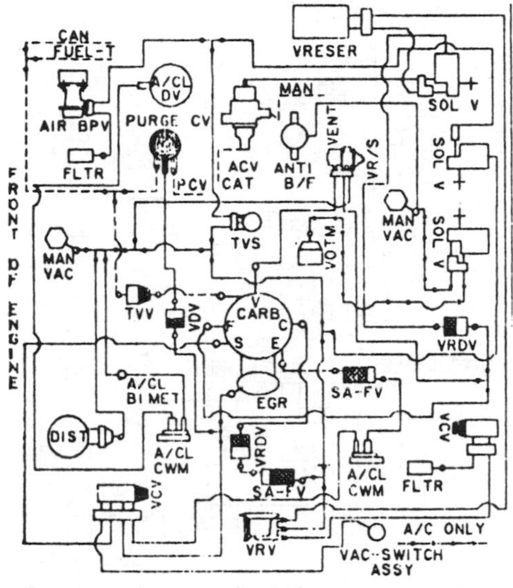

CALIBRATION: 0–2T–RO DATE: 7–18–79
2.3L MUST/CAPRI/FAIR/ZEPH CALIF M/T

A/C & NON A/C CALIF

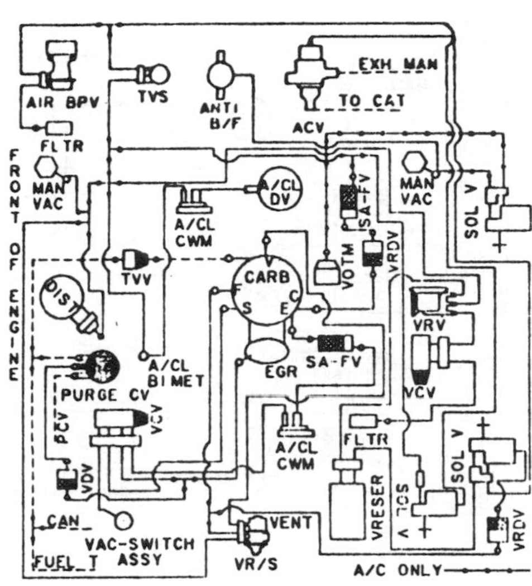

CALIBRATION: 0–2T–RIO DATE: 09–05–79
2.3L PINTO/ BOBCAT CALIF M/T

A/C & NON A/C CALIF

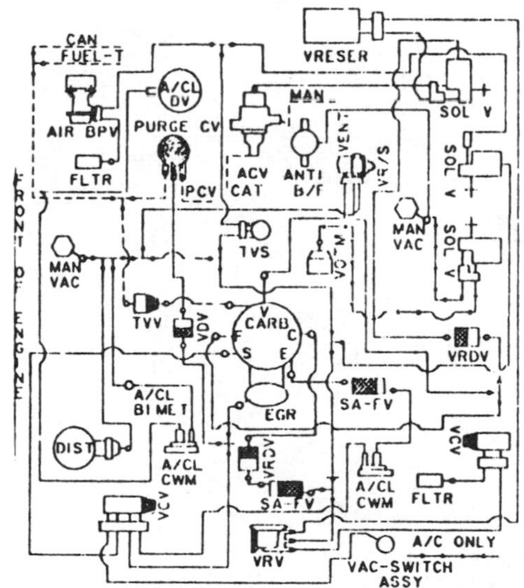

CALIBRATION: 0–2T–RIO DATE: 09–05–79
2.3L MUST/CAPRI/FAIRMONT/ZEPHYR
CALIF M/T

49S A/C & NON A/C

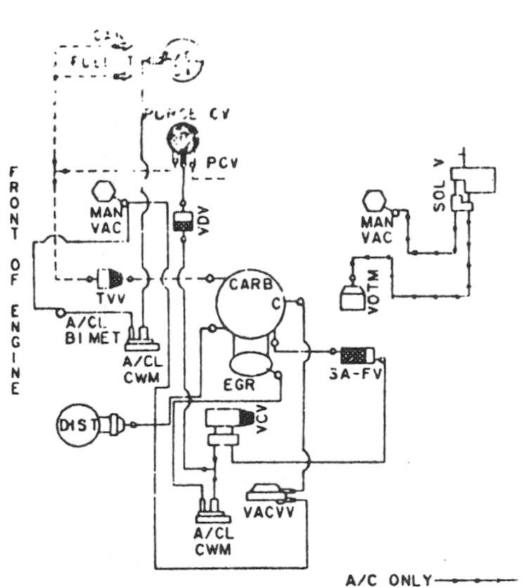

CALIBRATION: 0–21A–R0 DATE: 09–25–79
2.3L FAIRMONT/ZEPHYR/MUST/CAPRI 49S A/T

VACUUM CIRCUITS

(© Ford Motor Co.)

1980—140 CID (2.3 L) engines

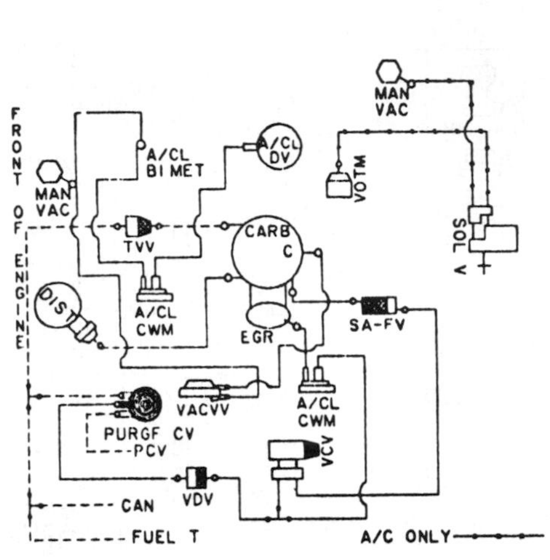

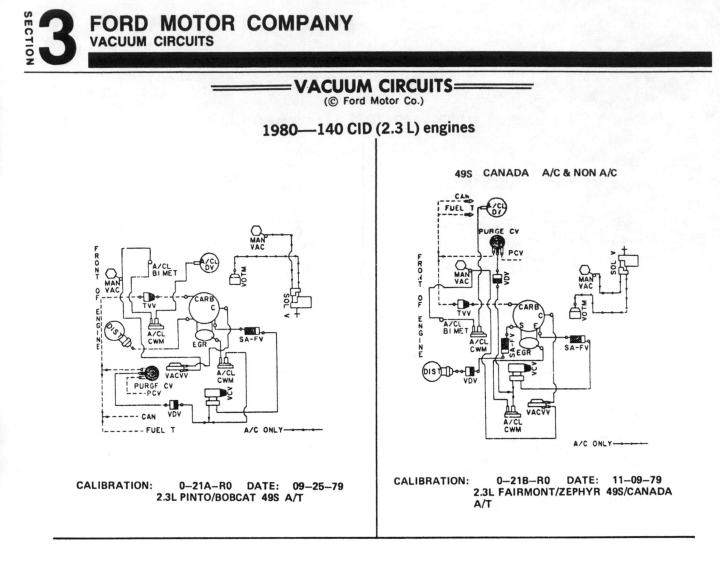

CALIBRATION: 0–21A–R0 DATE: 09–25–79
2.3L PINTO/BOBCAT 49S A/T

49S CANADA A/C & NON A/C

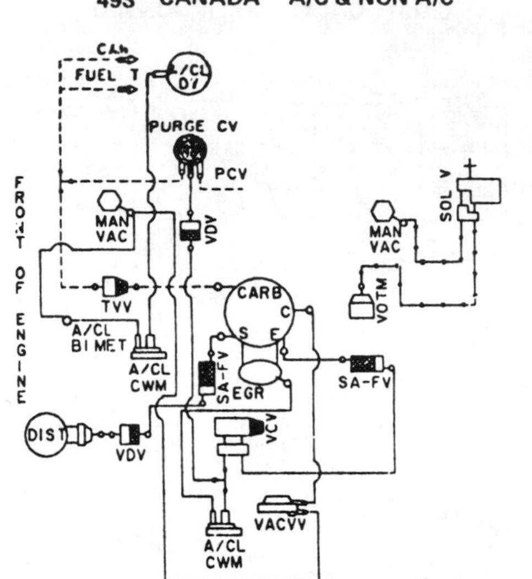

CALIBRATION: 0–21B–R0 DATE: 11–09–79
2.3L FAIRMONT/ZEPHYR 49S/CANADA
A/T

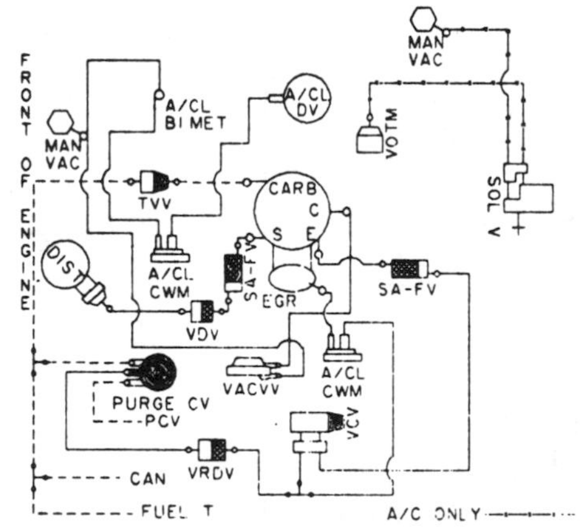

CALIBRATION: 0–21B–R0 DATE: 11–09–79
2.3L PINTO/BOBCAT 49S/CANADA A/T

VACUUM CIRCUITS

(© Ford Motor Co)

1980—200 CID (3.3 L) engines

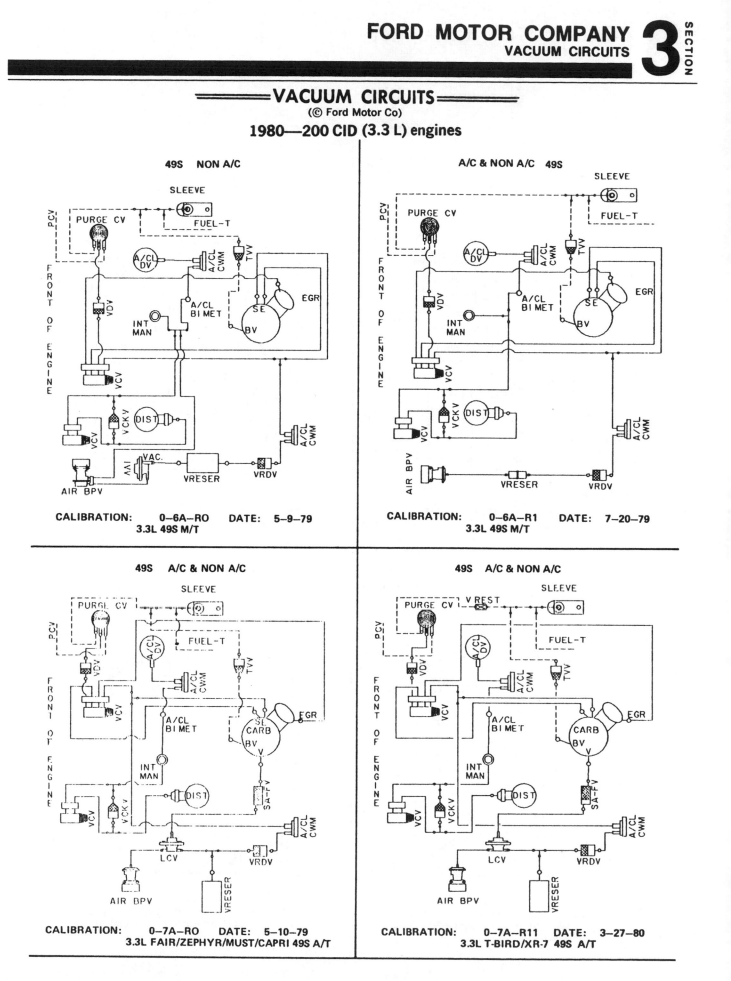

49S NON A/C

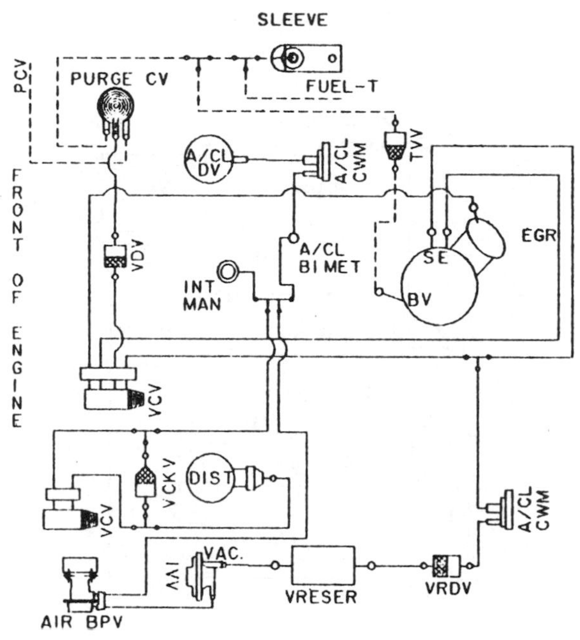

CALIBRATION: 0–6A–RO DATE: 5–9–79
3.3L 49S M/T

A/C & NON A/C 49S

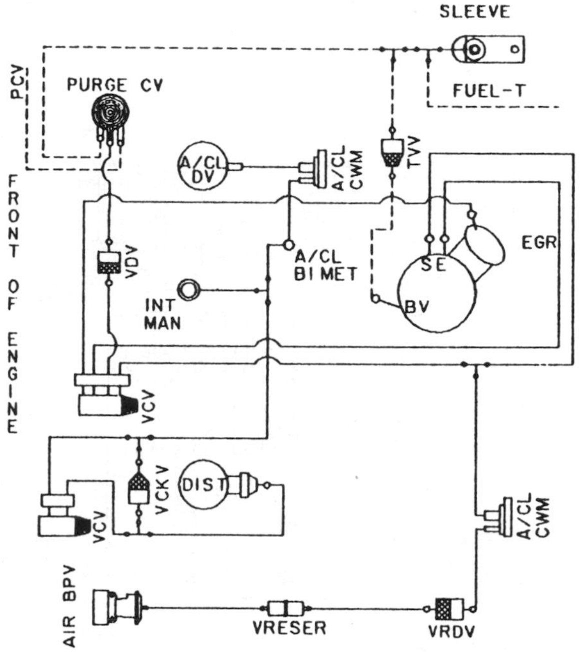

CALIBRATION: 0–6A–R1 DATE: 7–20–79
3.3L 49S M/T

49S A/C & NON A/C

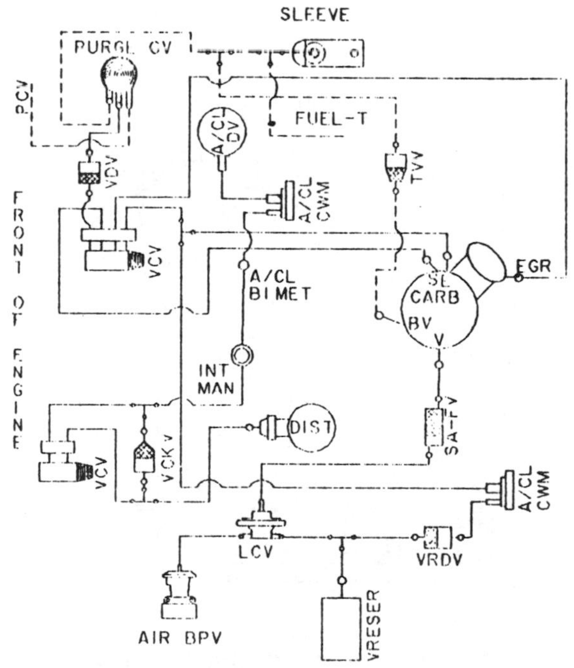

CALIBRATION: 0–7A–RO DATE: 5–10–79
3.3L FAIR/ZEPHYR/MUST/CAPRI 49S A/T

49S A/C & NON A/C

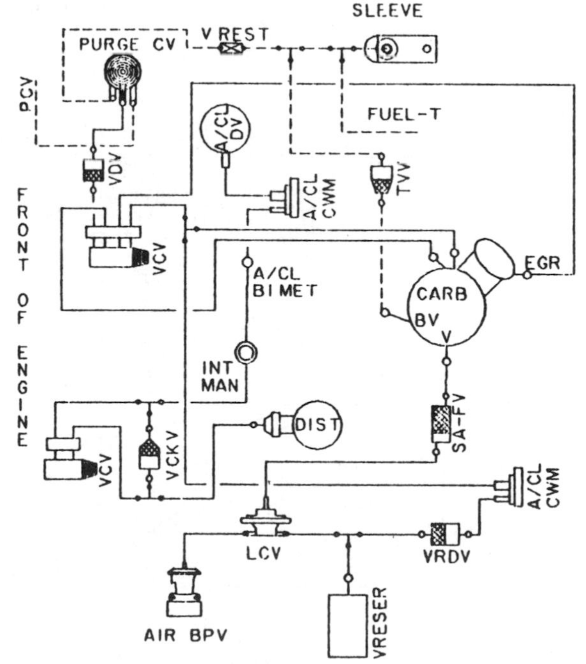

CALIBRATION: 0–7A–R11 DATE: 3–27–80
3.3L T-BIRD/XR-7 49S A/T

VACUUM CIRCUITS

(© Ford Motor Co.)

1980—200 CID (3.3 L) engines

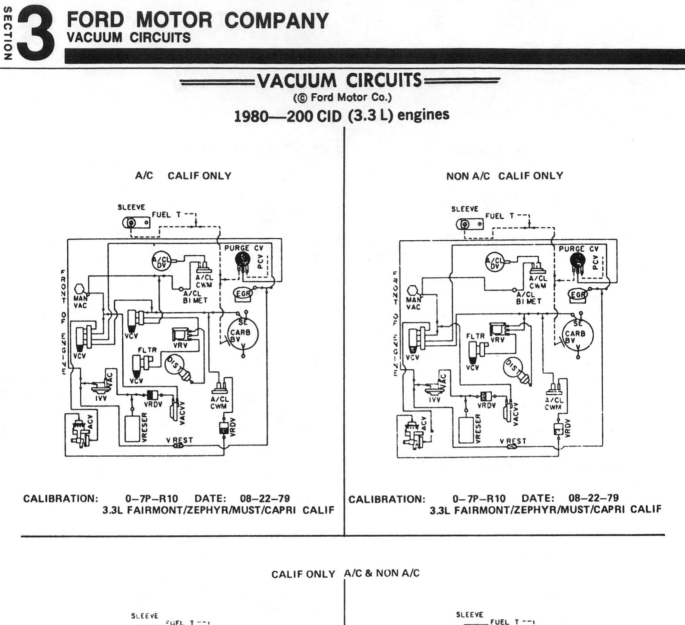

A/C CALIF ONLY

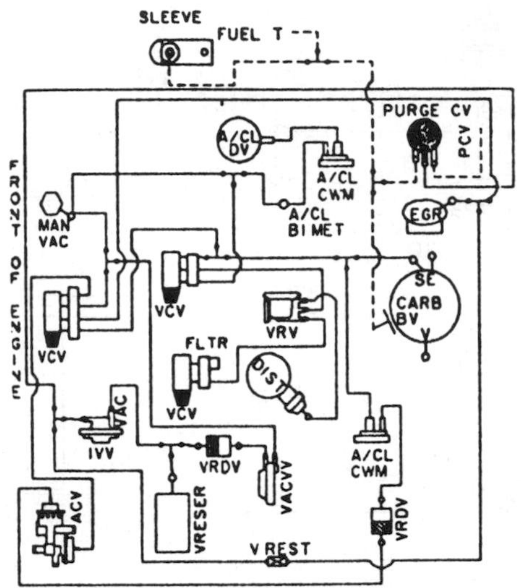

CALIBRATION: 0–7P–R10 DATE: 08–22–79
3.3L FAIRMONT/ZEPHYR/MUST/CAPRI CALIF

NON A/C CALIF ONLY

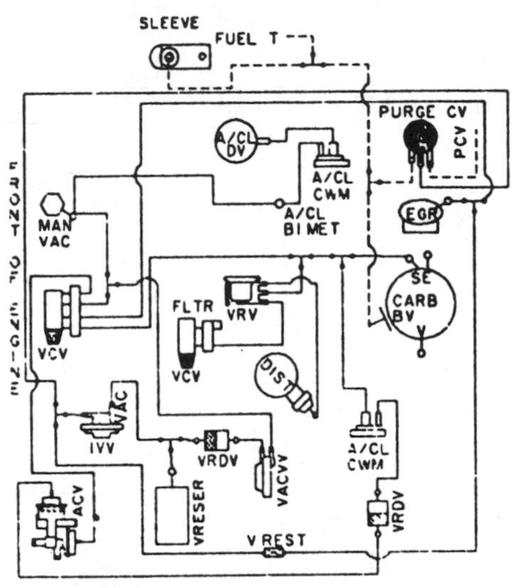

CALIBRATION: 0–7P–R10 DATE: 08–22–79
3.3L FAIRMONT/ZEPHYR/MUST/CAPRI CALIF

CALIF ONLY A/C & NON A/C

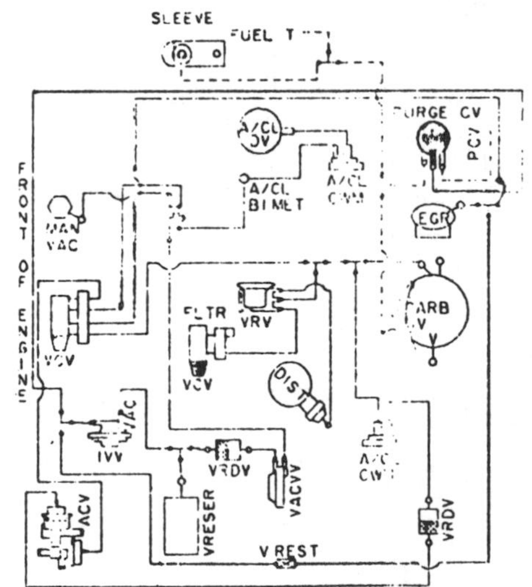

CALIBRATION: 0–7P–R11 DATE: 11–05–79
3.3L FAIRMONT/ZEPHYR/MUST/CAPRI
CALIF

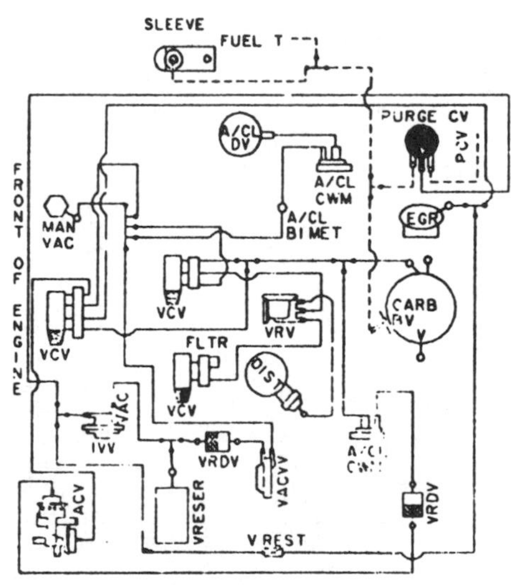

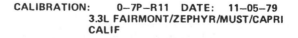

CALIBRATION: 0–7P–R11 DATE: 11–05–79
3.3L FAIRMONT/ZEPHYR/MUST/CAPRI
CALIF

VACUUM CIRCUITS

(© Ford Motor Co)

1980—200 CID (3.3 L) engines

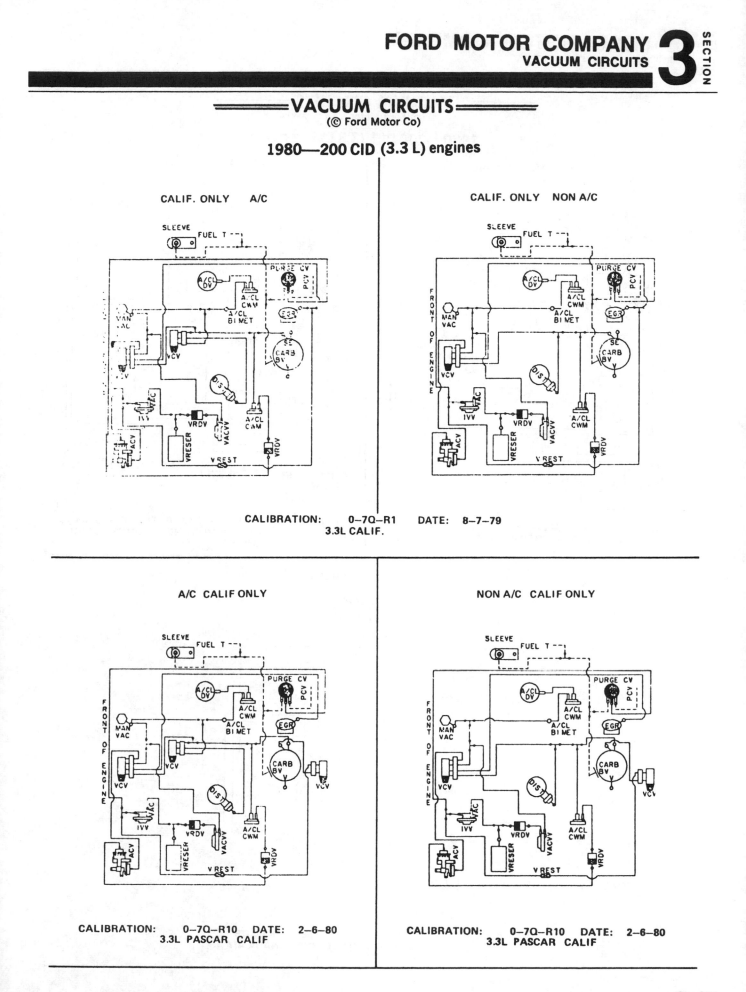

CALIF. ONLY A/C

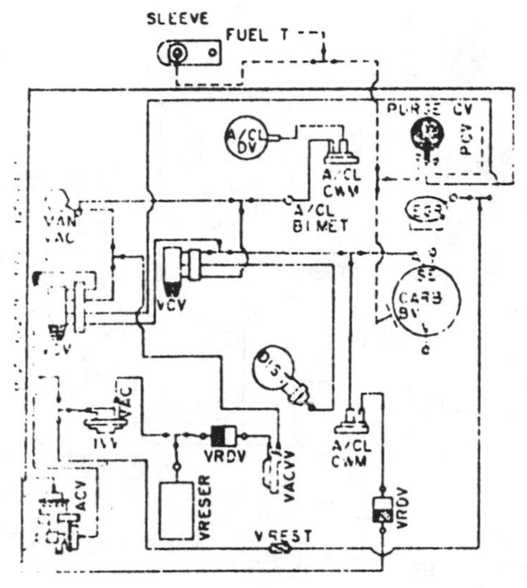

CALIF. ONLY NON A/C

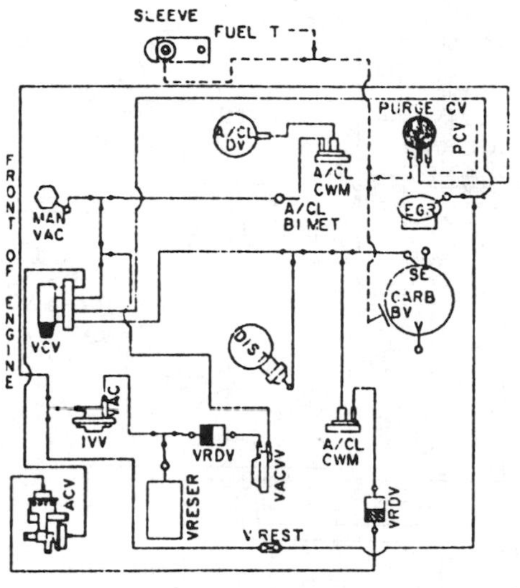

CALIBRATION: 0–7Q–R1 DATE: 8–7–79
3.3L CALIF.

A/C CALIF ONLY

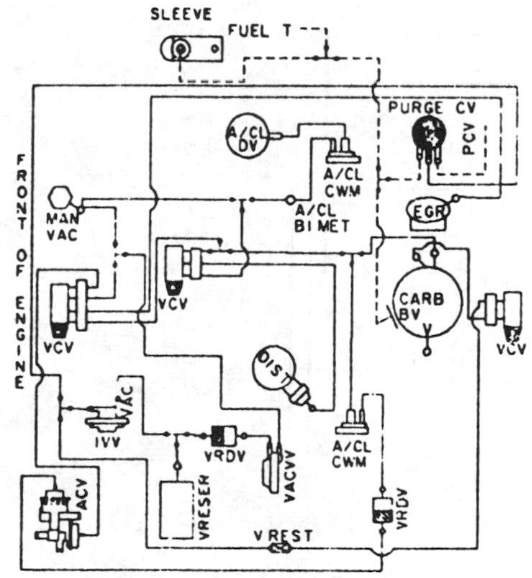

CALIBRATION: 0–7Q–R10 DATE: 2–6–80
3.3L PASCAR CALIF

NON A/C CALIF ONLY

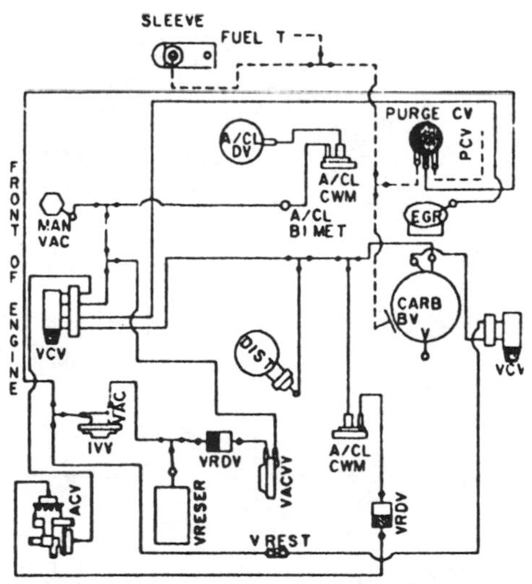

CALIBRATION: 0–7Q–R10 DATE: 2–6–80
3.3L PASCAR CALIF

VACUUM CIRCUITS

(© Ford Motor Co.)

1980—200 CID (3.3 L) engines

49S A/C & NON A/C

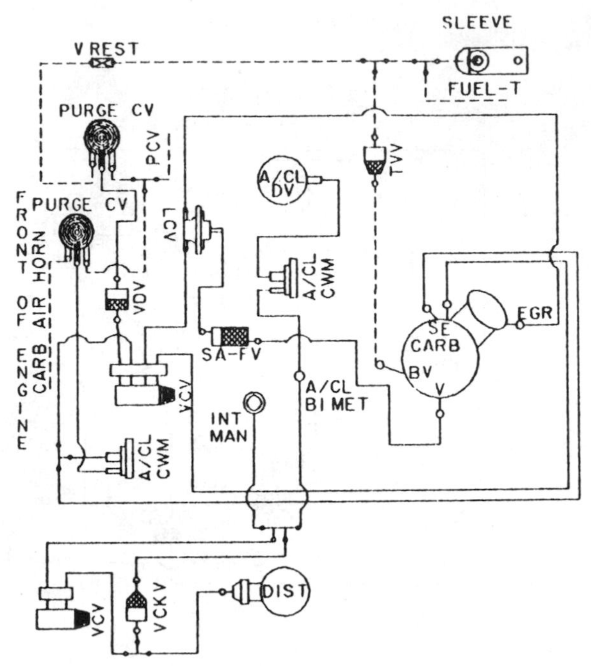

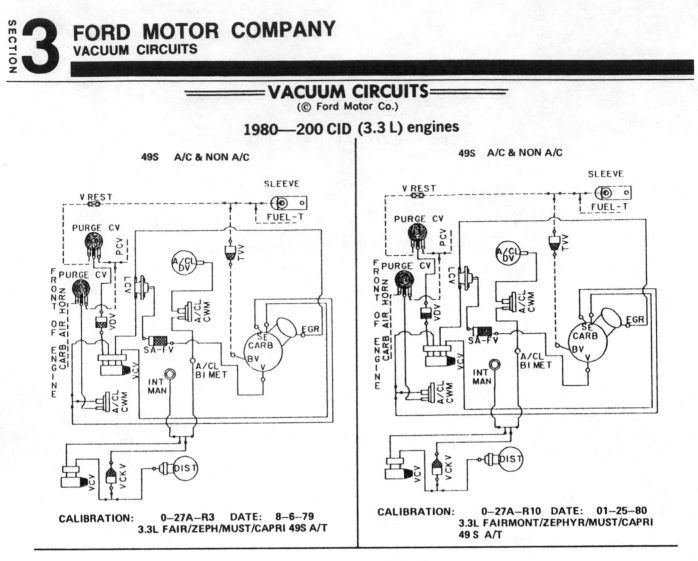

CALIBRATION: 0–27A–R3 DATE: 8–6–79
3.3L FAIR/ZEPH/MUST/CAPRI 49S A/T

49S A/C & NON A/C

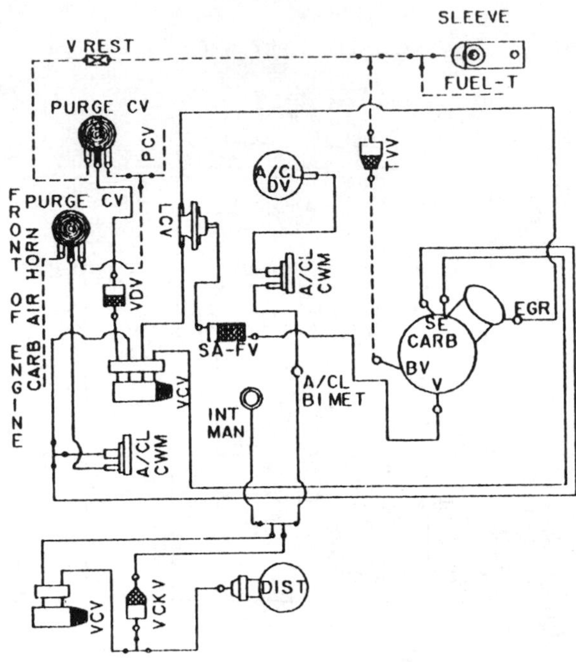

CALIBRATION: 0–27A–R10 DATE: 01–25–80
3.3L FAIRMONT/ZEPHYR/MUST/CAPRI
49 S A/T

CANADA A/C & NON A/C

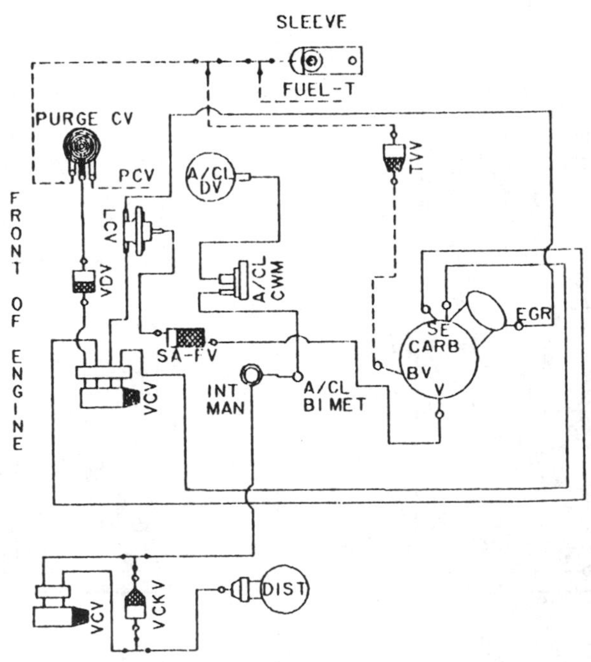

CALIBRATION: 0–27G–R0 DATE: 08–09–79
3.3L FAIRMONT/ZEPHYR CANADA A/T

VACUUM CIRCUITS

(© Ford Motor Co.)

1980—250 CID (4.1 L) engines

49S A/C & NON A/C

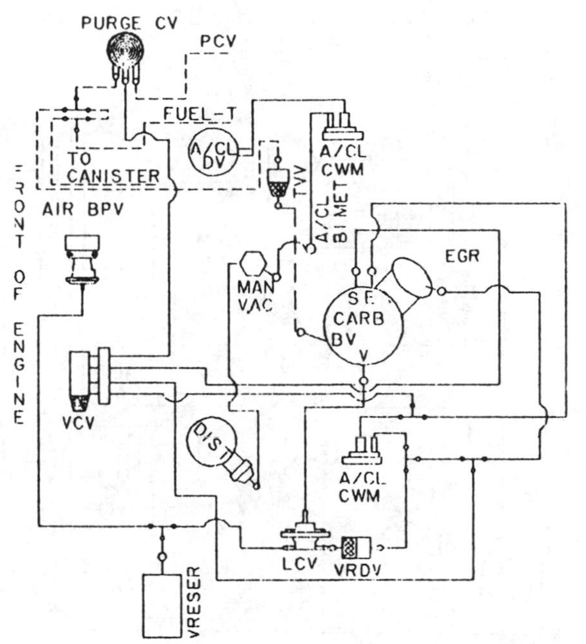

CALIBRATION: 0–8A–RO **DATE:** 6–1–79
4.1L GRANADA/MONARCH 49S M/T

49S A/C & NON A/C

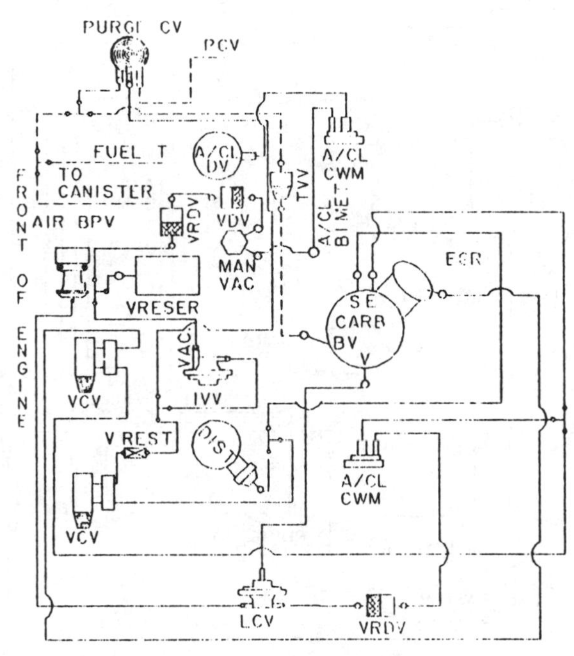

CALIBRATION: 0–9A–RO **DATE:** 6–1–79
4.1L GRANADA/MONARCH 49S A/T

49S CANADA A/C & NON A/C

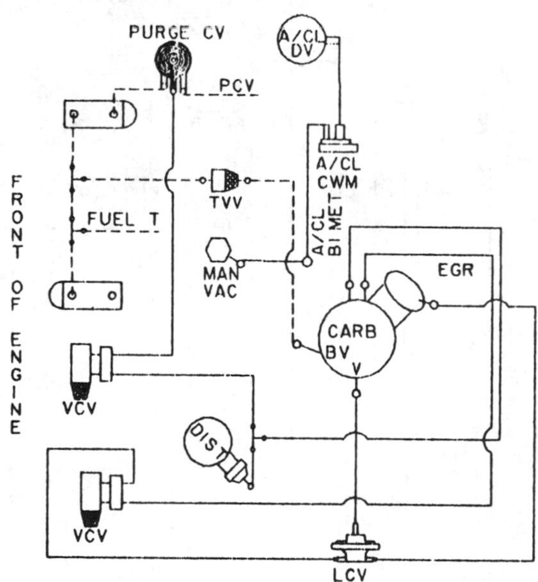

CALIBRATION: 0–29A–R0 **DATE:** 10–1–79
4.1L GRANADA/MONARCH 49S/ CANADA M/T

VACUUM CIRCUITS

(© Ford Motor Co)

1980—255 CID (4.2 L) engines

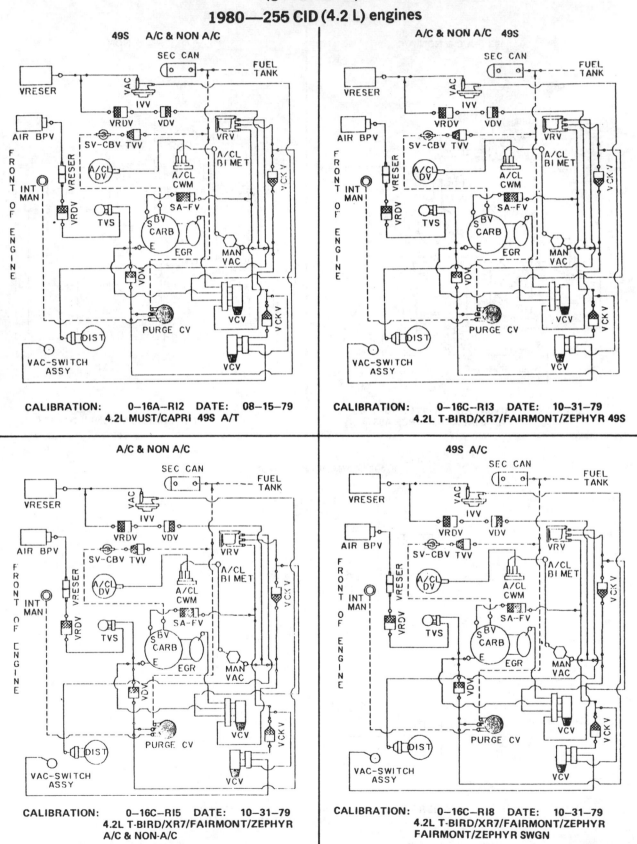

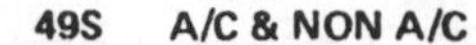

49S A/C & NON A/C

SEC CAN
FUEL TANK
VRESER
VAC
IVV
VRDV
VDV
AIR BPV
SV-CBV TVV
VRV
A/CL BI MET
A/CL DV
A/CL CWM
VRESER
VCKV
FRONT OF ENGINE
INT MAN
SA-FV
VRDV
TVS
S BV CARB E
EGR
MAN VAC
VDV
VCV
VCKV
PURGE CV
DIST
VAC-SWITCH ASSY
VCV

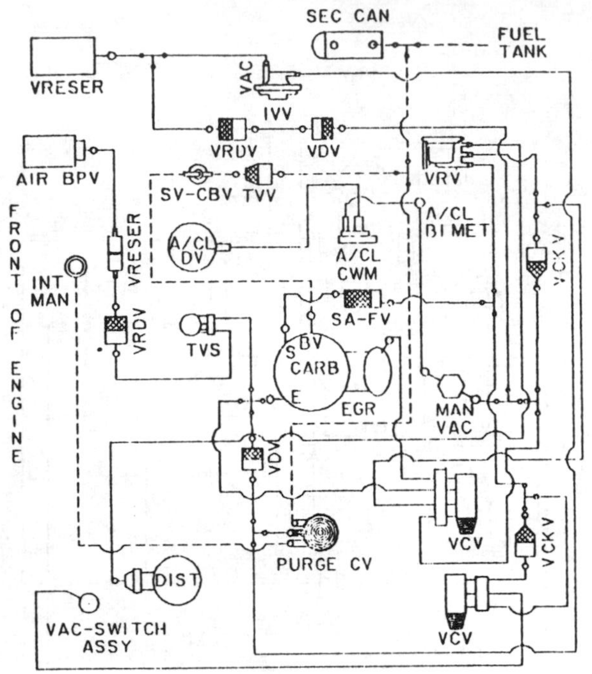

CALIBRATION: 0–16A–RI2 DATE: 08–15–79
4.2L MUST/CAPRI 49S A/T

A/C & NON A/C 49S

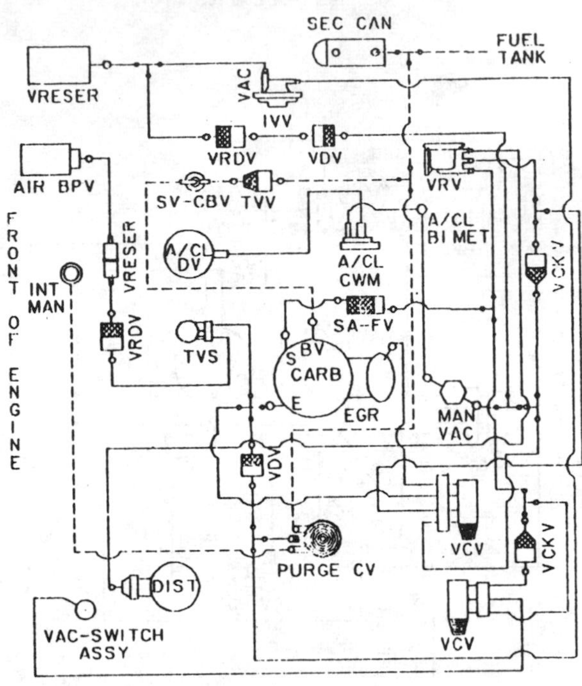

SEC CAN
FUEL TANK
VRESER
VAC
IVV
VRDV
VDV
AIR BPV
SV-CBV TVV
VRV
A/CL BI MET
A/CL DV
A/CL CWM
VRESER
VCKV
FRONT OF ENGINE
INT MAN
SA-FV
VRDV
TVS
S BV CARB E
EGR
MAN VAC
VDV
VCV
VCKV
PURGE CV
DIST
VAC-SWITCH ASSY
VCV

CALIBRATION: 0–16C–RI3 DATE: 10–31–79
4.2L T-BIRD/XR7/FAIRMONT/ZEPHYR 49S

A/C & NON A/C

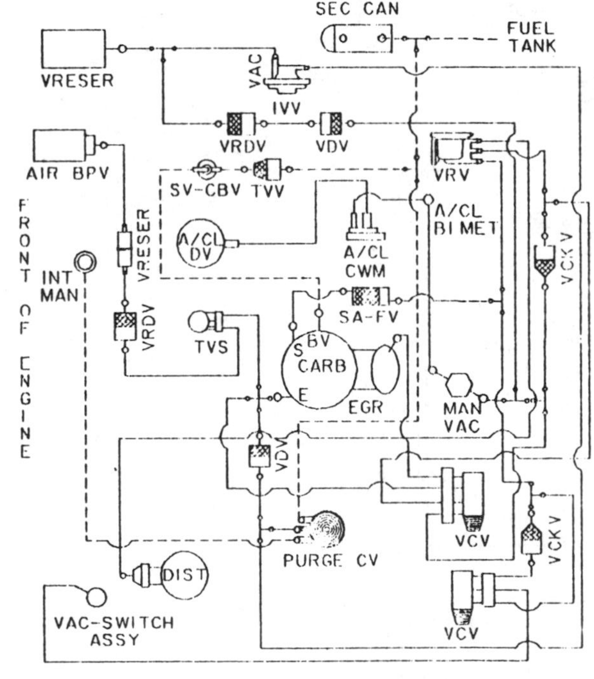

SEC CAN
FUEL TANK
VRESER
VAC
IVV
VRDV
VDV
AIR BPV
SV-CBV TVV
VRV
A/CL BI MET
A/CL DV
A/CL CWM
VRESER
VCKV
FRONT OF ENGINE
INT MAN
SA-FV
VRDV
TVS
S BV CARB E
EGR
MAN VAC
VDV
VCV
VCKV
PURGE CV
DIST
VAC-SWITCH ASSY
VCV

CALIBRATION: 0–16C–RI5 DATE: 10–31–79
4.2L T-BIRD/XR7/FAIRMONT/ZEPHYR
A/C & NON-A/C

49S A/C

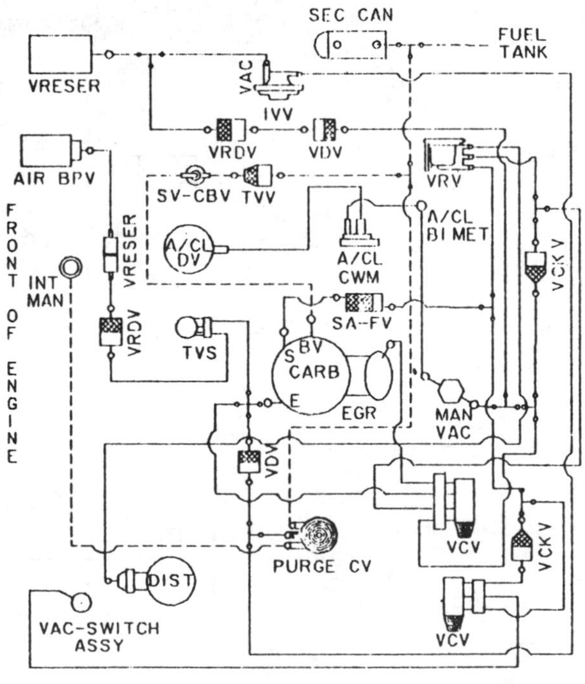

SEC CAN
FUEL TANK
VRESER
VAC
IVV
VRDV
VDV
AIR BPV
SV-CBV TVV
VRV
A/CL BI MET
A/CL DV
A/CL CWM
VRESER
VCKV
FRONT OF ENGINE
INT MAN
SA-FV
VRDV
TVS
S BV CARB E
EGR
MAN VAC
VDV
VCV
VCKV
PURGE CV
DIST
VAC-SWITCH ASSY
VCV

CALIBRATION: 0–16C–RI8 DATE: 10–31–79
4.2L T-BIRD/XR7/FAIRMONT/ZEPHYR
FAIRMONT/ZEPHYR SWGN

VACUUM CIRCUITS

(© Ford Motor Co.)

1980—255 CID (4.2 L) engines

A/C & NON A/C

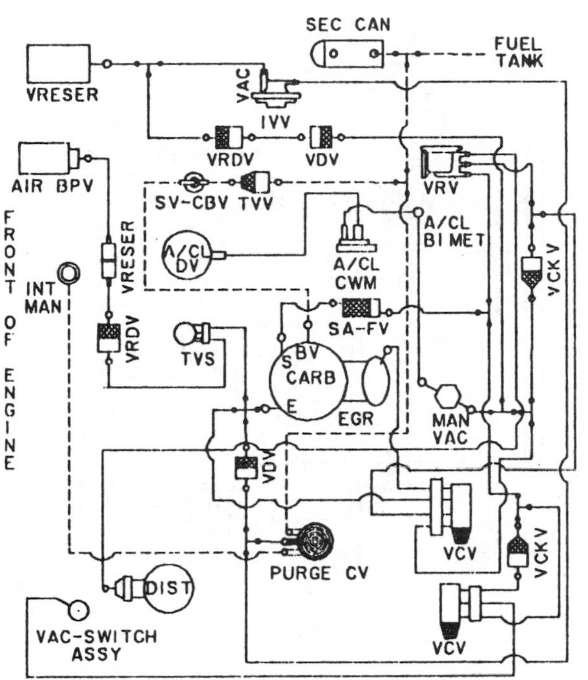

CALIBRATION: 0–16C–R19 **DATE:** 10–31–79
4.2L T-BIRD/XR-7

A/C & NON A/C

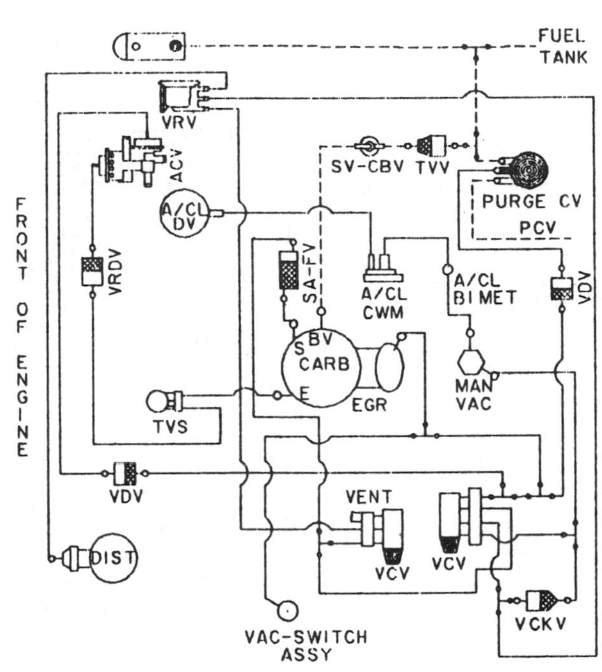

CALIBRATION: 0–16S–R14 **DATE:** 09–20–79
4.2L T-BIRD/XR7/FAIRMONT/ZEPHYR A/T

A/C & NON A/C

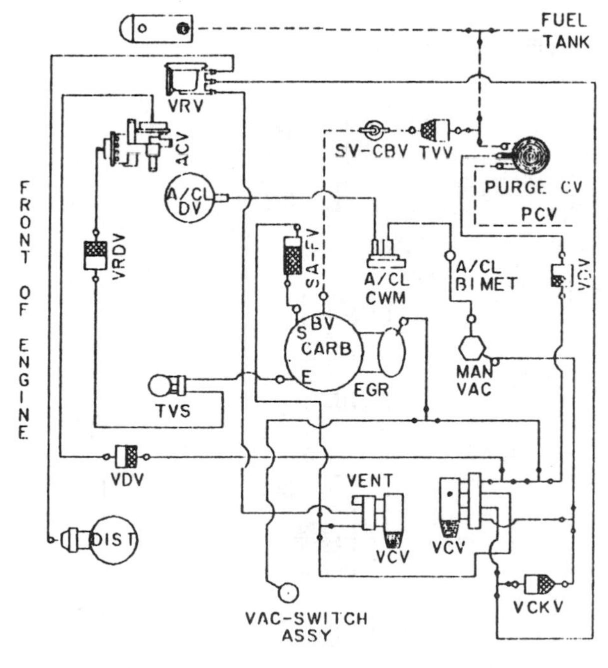

CALIBRATION: 0–16S–R16 **DATE:** 12–05–79
4.2L T-BIRD/XR7/FAIRMONT/ZEPHYR A/T

CALIF ONLY A/C & NON A/C

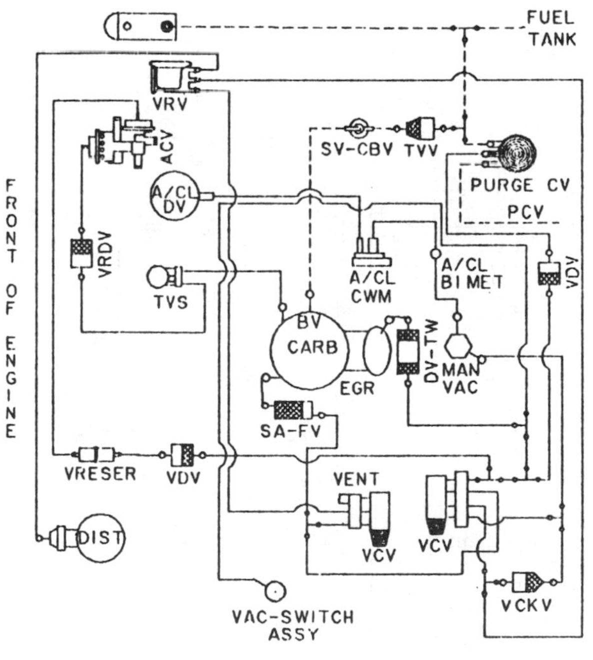

CALIBRATION: 0–16T–R13 **DATE:** 11–19–79
4.2L MUST/CAPRI CALIF

VACUUM CIRCUITS

(© Ford Motor Co)

1980—255 CID (4.2 L) engines

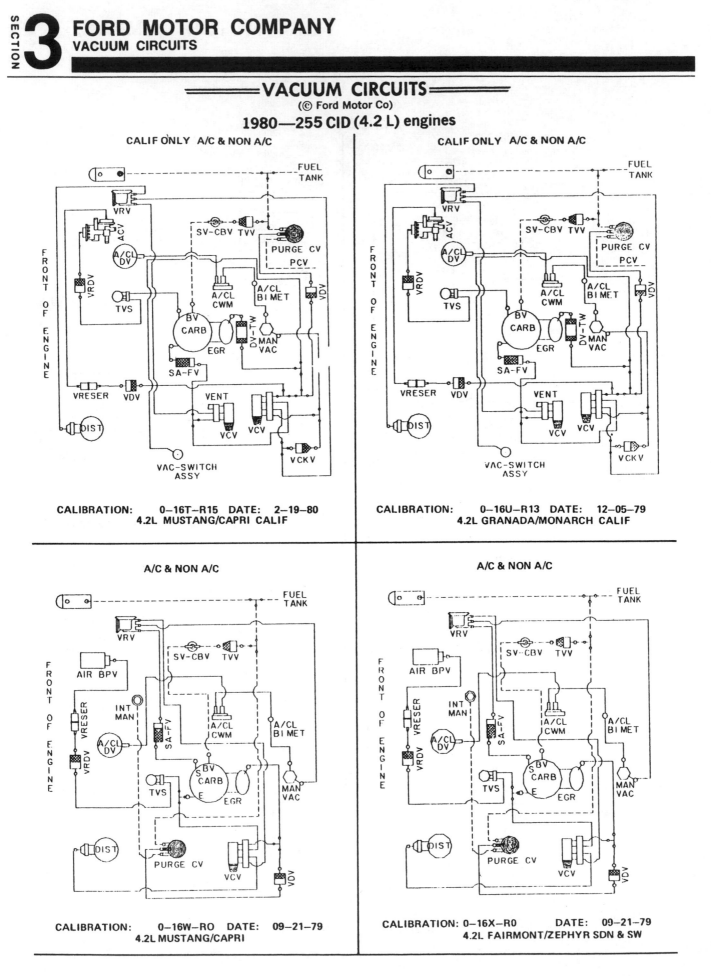

CALIF ONLY A/C & NON A/C

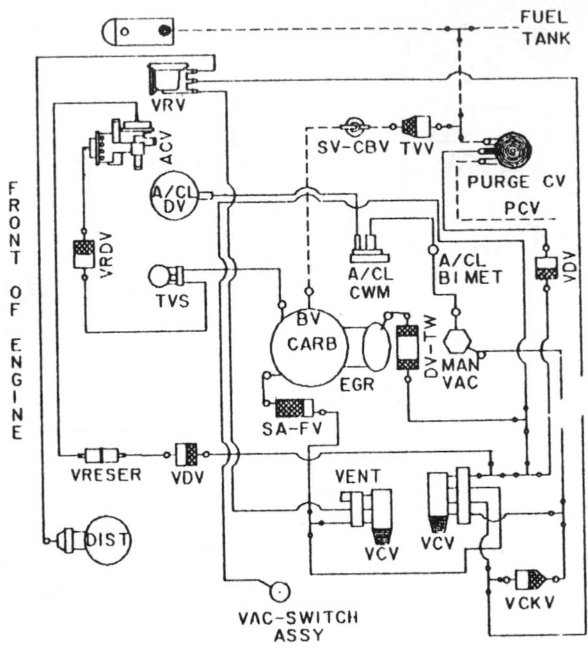

CALIBRATION: 0–16T–R15 DATE: 2–19–80
4.2L MUSTANG/CAPRI CALIF

CALIF ONLY A/C & NON A/C

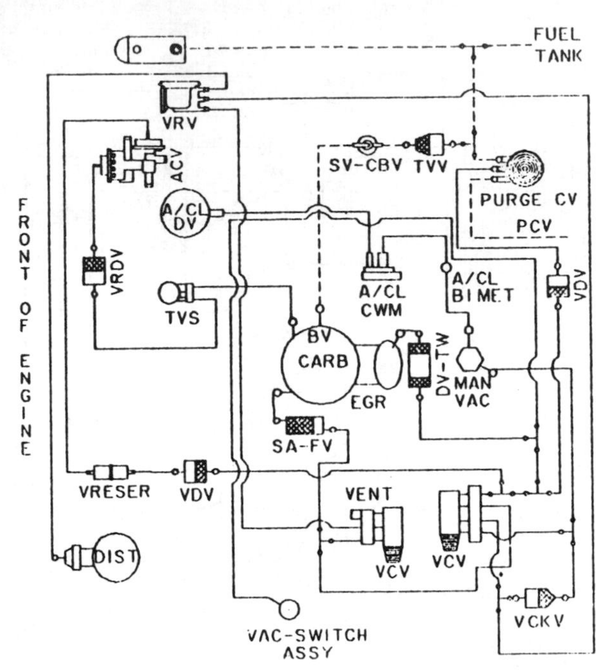

CALIBRATION: 0–16U–R13 DATE: 12–05–79
4.2L GRANADA/MONARCH CALIF

A/C & NON A/C

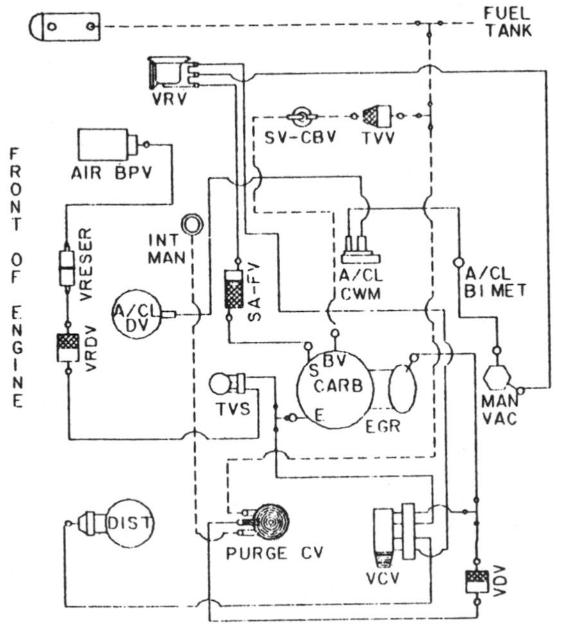

CALIBRATION: 0–16W–RO DATE: 09–21–79
4.2L MUSTANG/CAPRI

A/C & NON A/C

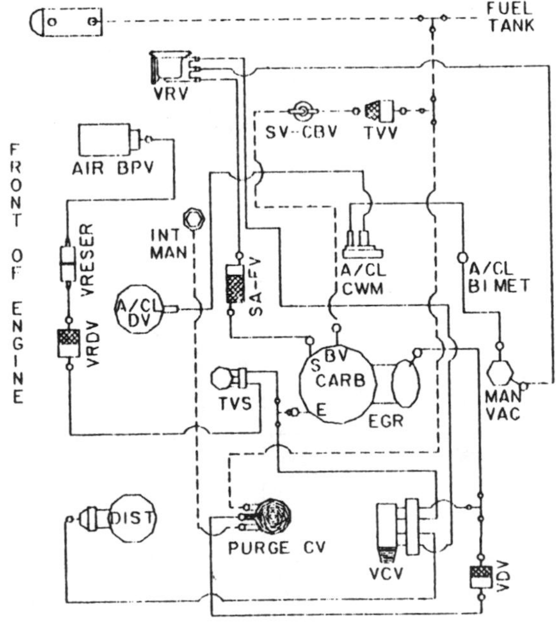

CALIBRATION: 0–16X–R0 DATE: 09–21–79
4.2L FAIRMONT/ZEPHYR SDN & SW

VACUUM CIRCUITS

(© Ford Motor Co.)

1980—300 CID (4.9 L) engines

CANADA

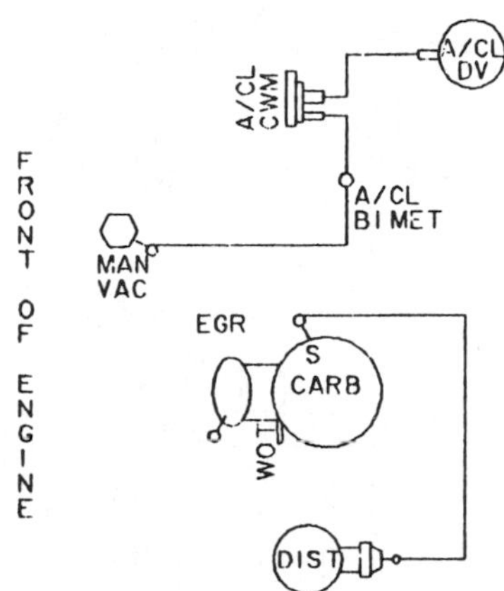

CALIBRATION: 5–77–R1 DATE: 6–23–80
4.9L E & F 150/250/350 1981 CANADA M/T

CANADA

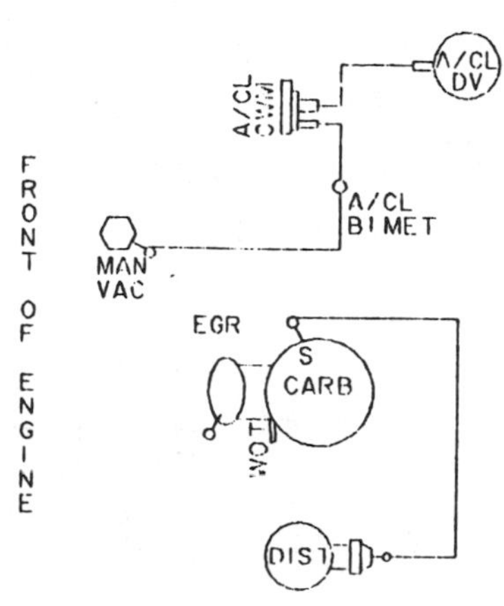

CALIBRATION: 5–78–R1 DATE: 6–23–80
4.9L E & F 150/250/350 1981 CANADA M/T

CALIF ONLY

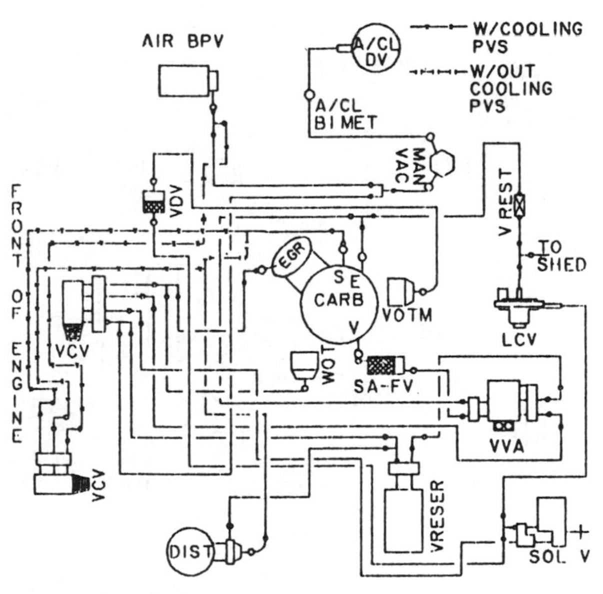

CALIBRATION: 9–77S–R10 DATE: 6–23–80
4.9L F-350 CALIF M/T

49S A/C & NON A/C

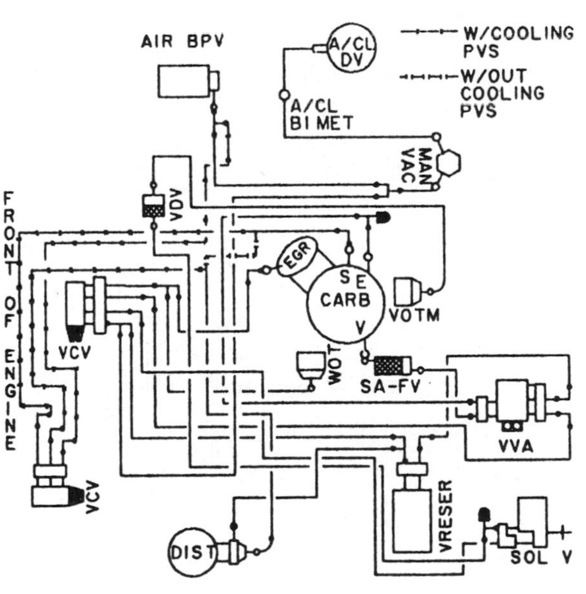

CALIBRATION: 9–77J–R12 DATE: 6–23–80
4.9L F-350 49S M/T

VACUUM CIRCUITS

(© Ford Motor Co.)

1980—300 CID (4.9 L) engines

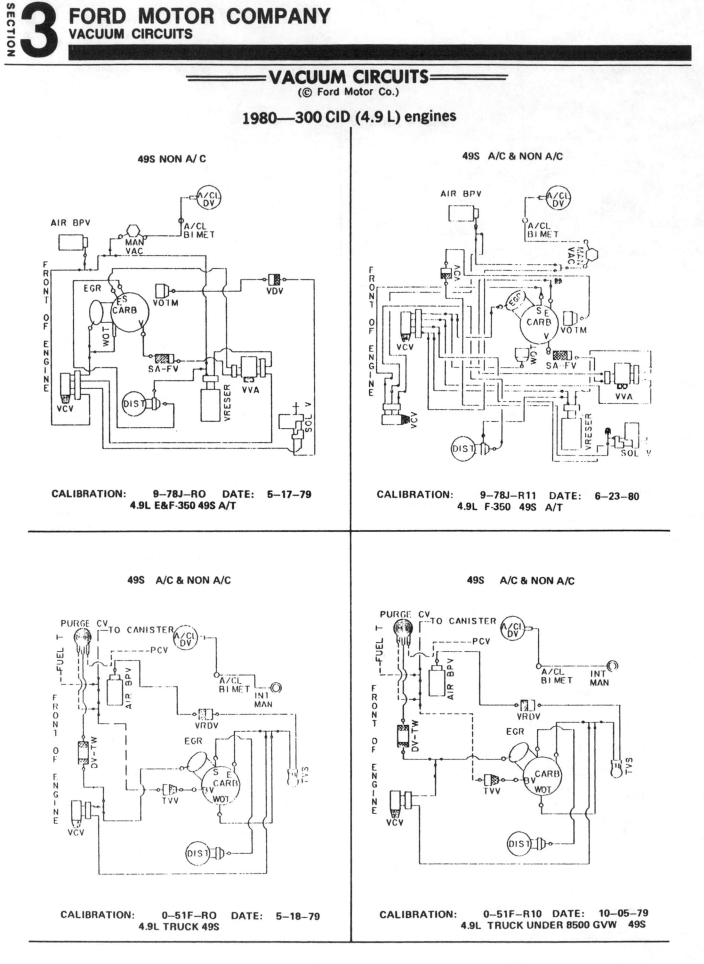

49S NON A/C

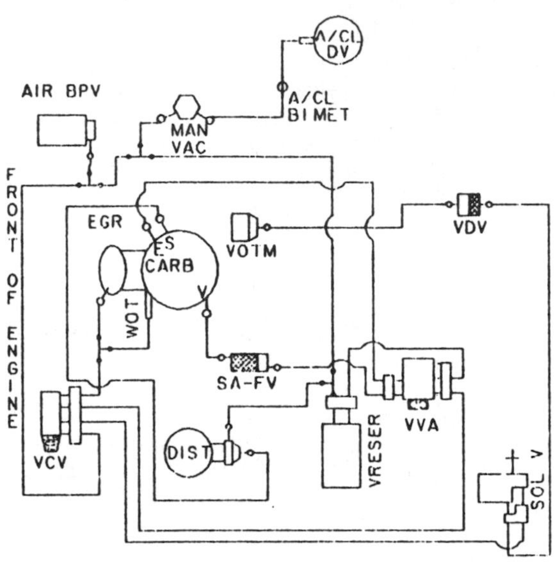

CALIBRATION: 9–78J–RO DATE: 5–17–79
4.9L E&F-350 49S A/T

49S A/C & NON A/C

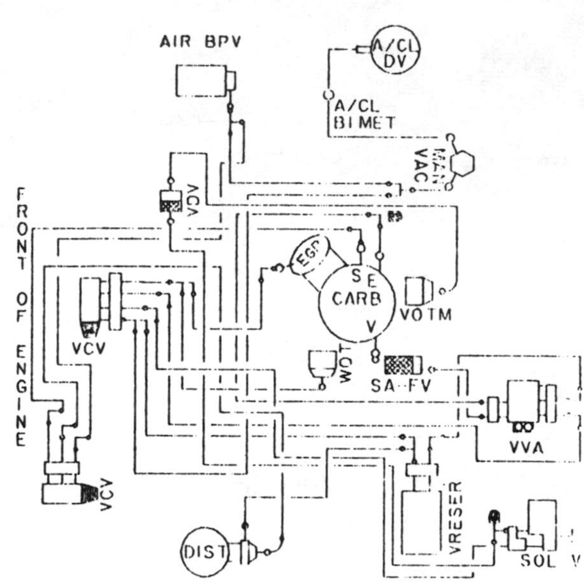

CALIBRATION: 9–78J–R11 DATE: 6–23–80
4.9L F-350 49S A/T

49S A/C & NON A/C

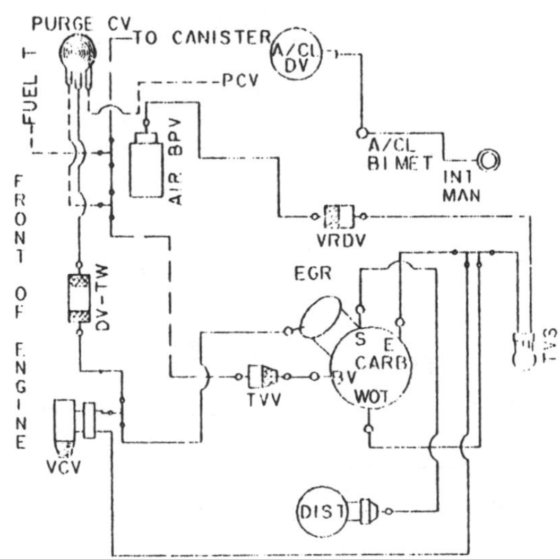

CALIBRATION: 0–51F–RO DATE: 5–18–79
4.9L TRUCK 49S

49S A/C & NON A/C

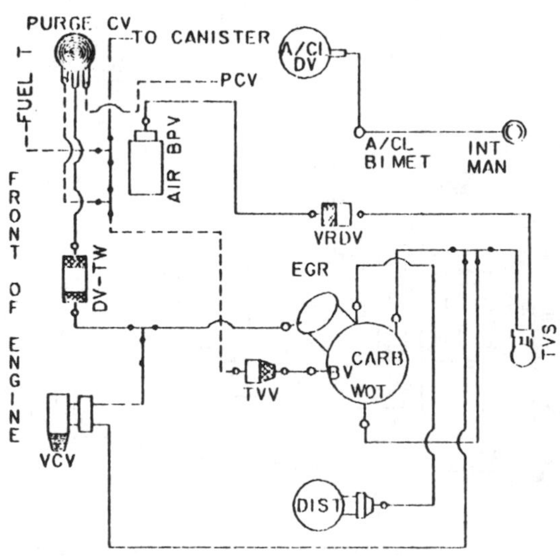

CALIBRATION: 0–51F–R10 DATE: 10–05–79
4.9L TRUCK UNDER 8500 GVW 49S

VACUUM CIRCUITS

(© Ford Motor Co)

1980—300 CID (4.9 L) engines

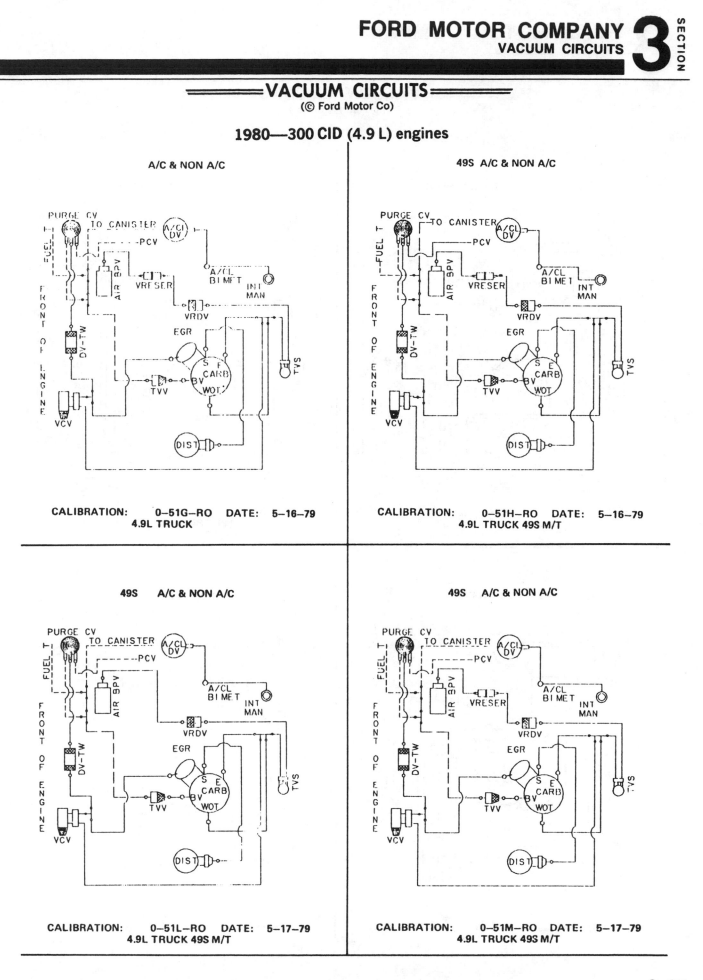

A/C & NON A/C

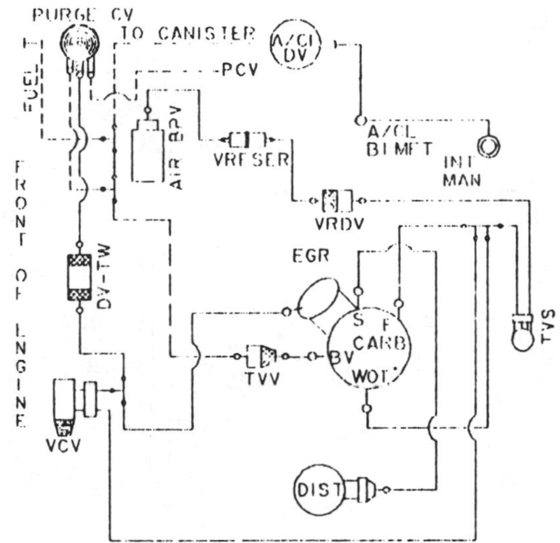

CALIBRATION: 0–51G–RO DATE: 5–16–79
4.9L TRUCK

49S A/C & NON A/C

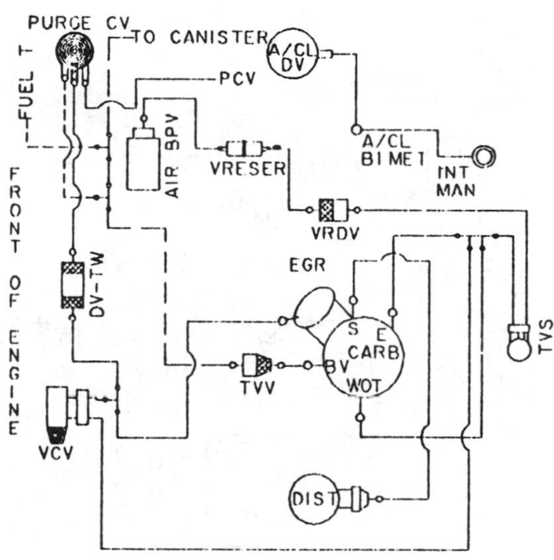

CALIBRATION: 0–51H–RO DATE: 5–16–79
4.9L TRUCK 49S M/T

49S A/C & NON A/C

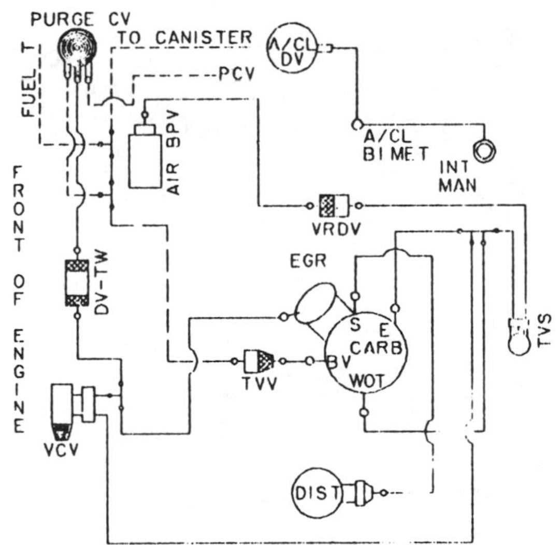

CALIBRATION: 0–51L–RO DATE: 5–17–79
4.9L TRUCK 49S M/T

49S A/C & NON A/C

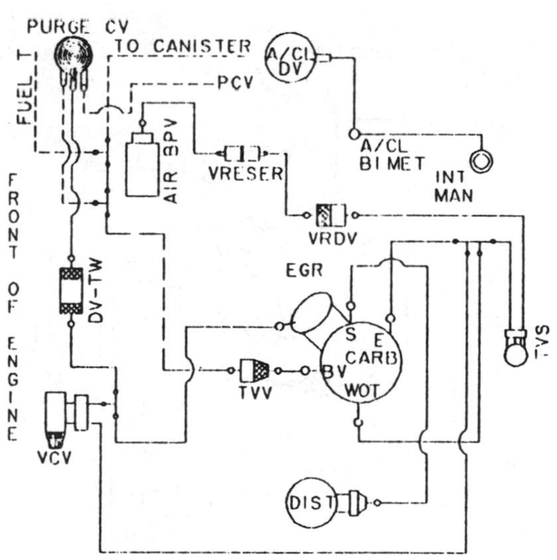

CALIBRATION: 0–51M–RO DATE: 5–17–79
4.9L TRUCK 49S M/T

VACUUM CIRCUITS

(© Ford Motor Co.)

1980—300 CID (4.9 L) engines

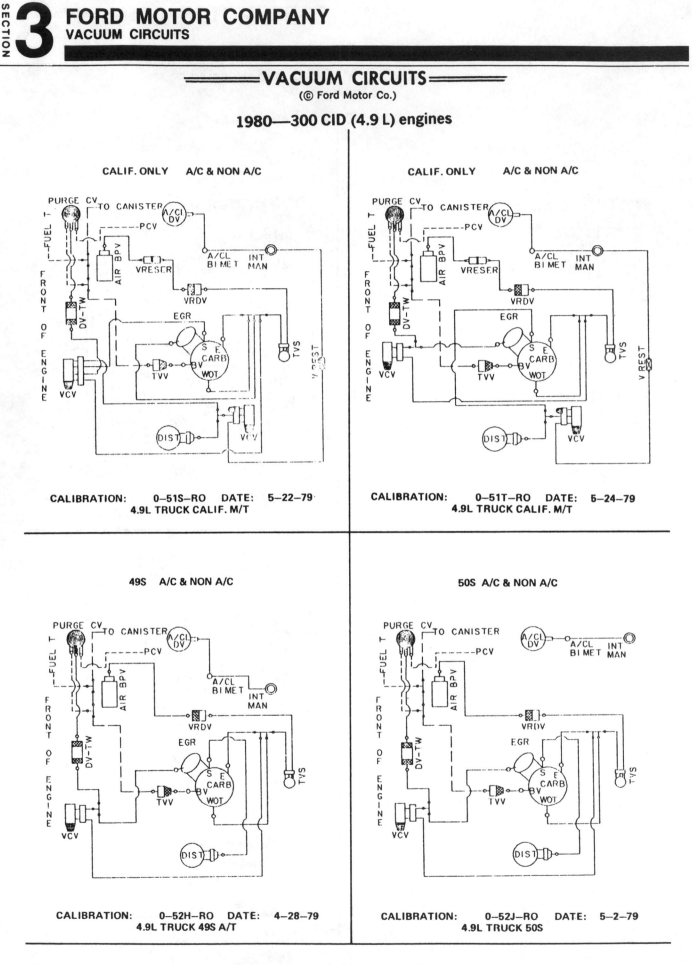

CALIF. ONLY A/C & NON A/C

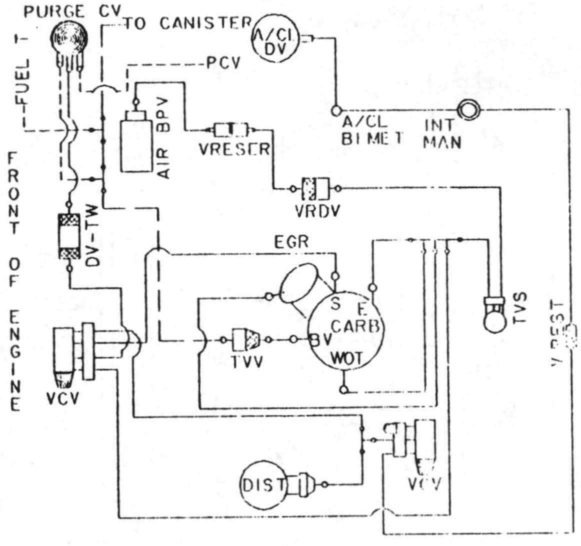

CALIBRATION: 0–51S–RO DATE: 5–22–79
4.9L TRUCK CALIF. M/T

CALIF. ONLY A/C & NON A/C

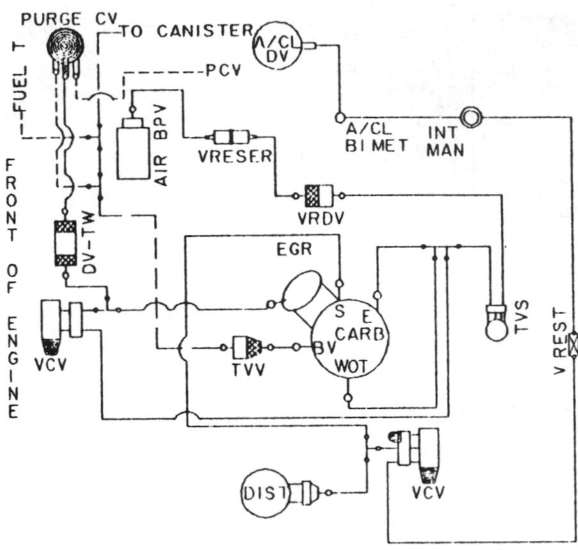

CALIBRATION: 0–51T–RO DATE: 5–24–79
4.9L TRUCK CALIF. M/T

49S A/C & NON A/C

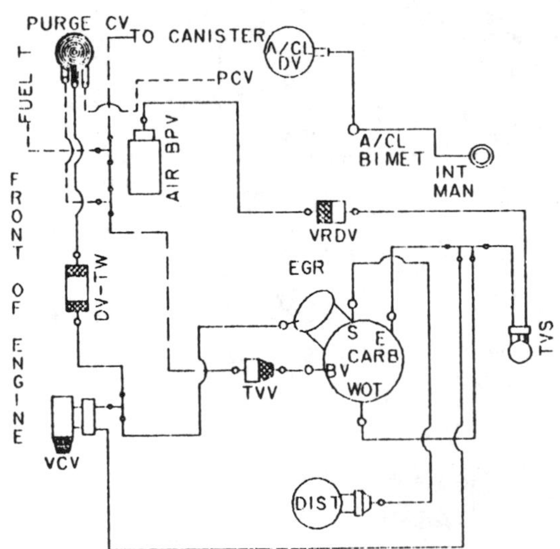

CALIBRATION: 0–52H–RO DATE: 4–28–79
4.9L TRUCK 49S A/T

50S A/C & NON A/C

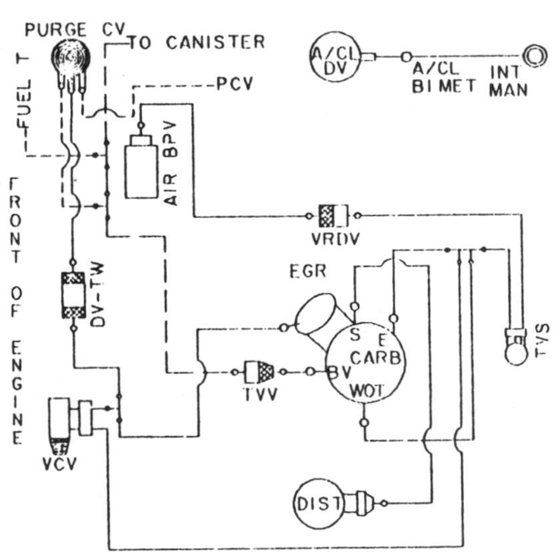

CALIBRATION: 0–52J–RO DATE: 5–2–79
4.9L TRUCK 50S

VACUUM CIRCUITS

(© Ford Motor Co)

1980—300 CID (4.9 L) engines

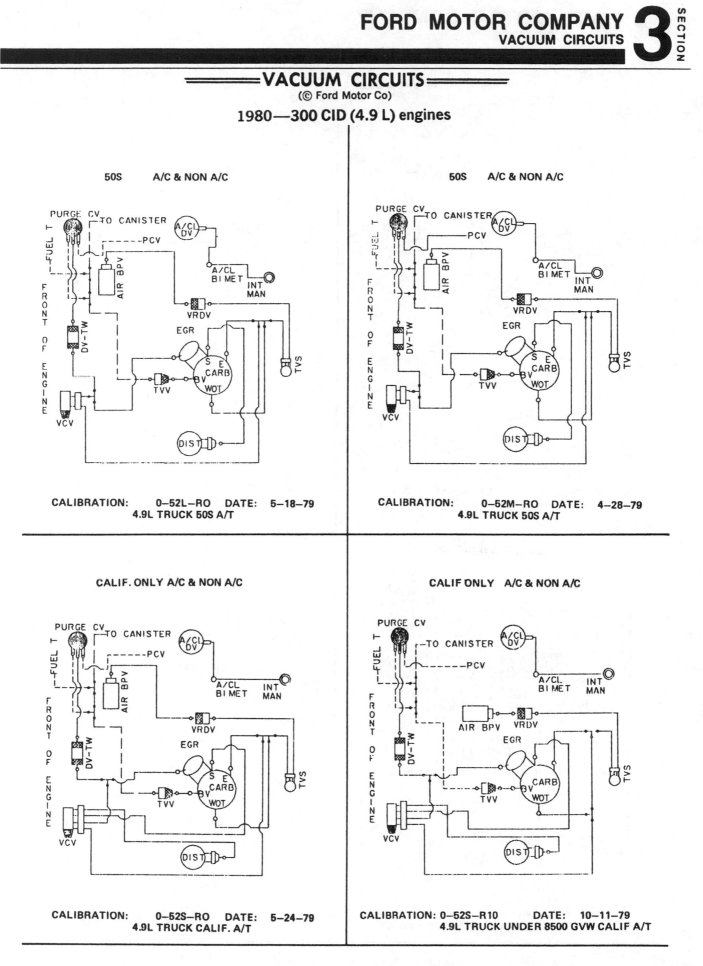

50S A/C & NON A/C

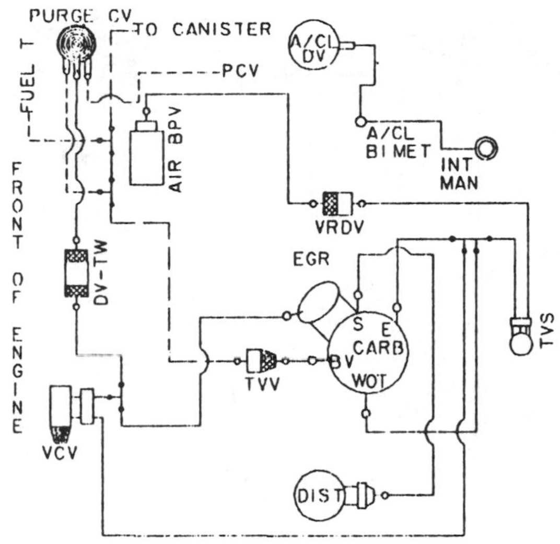

CALIBRATION: 0–52L–RO DATE: 5–18–79
4.9L TRUCK 50S A/T

50S A/C & NON A/C

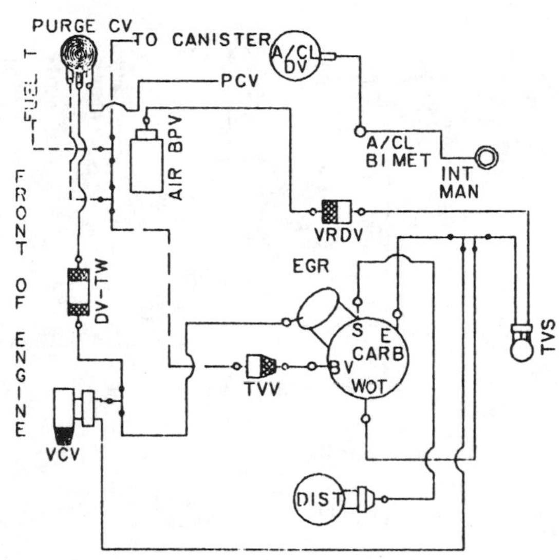

CALIBRATION: 0–52M–RO DATE: 4–28–79
4.9L TRUCK 50S A/T

CALIF. ONLY A/C & NON A/C

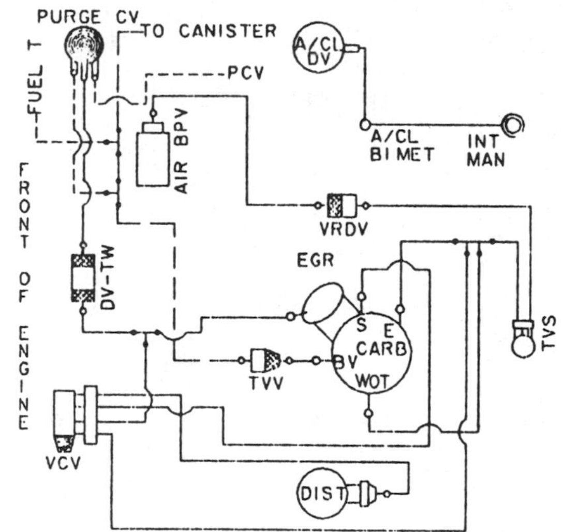

CALIBRATION: 0–52S–RO DATE: 5–24–79
4.9L TRUCK CALIF. A/T

CALIF ONLY A/C & NON A/C

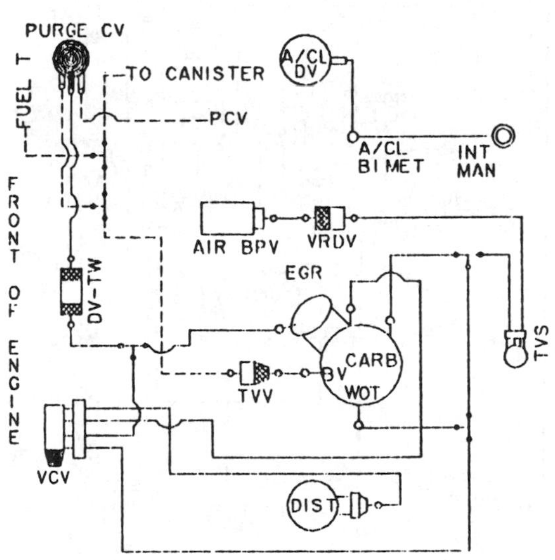

CALIBRATION: 0–52S–R10 DATE: 10–11–79
4.9L TRUCK UNDER 8500 GVW CALIF A/T

VACUUM CIRCUITS

(© Ford Motor Co.)

1980—302 CID (5.0 L) engines

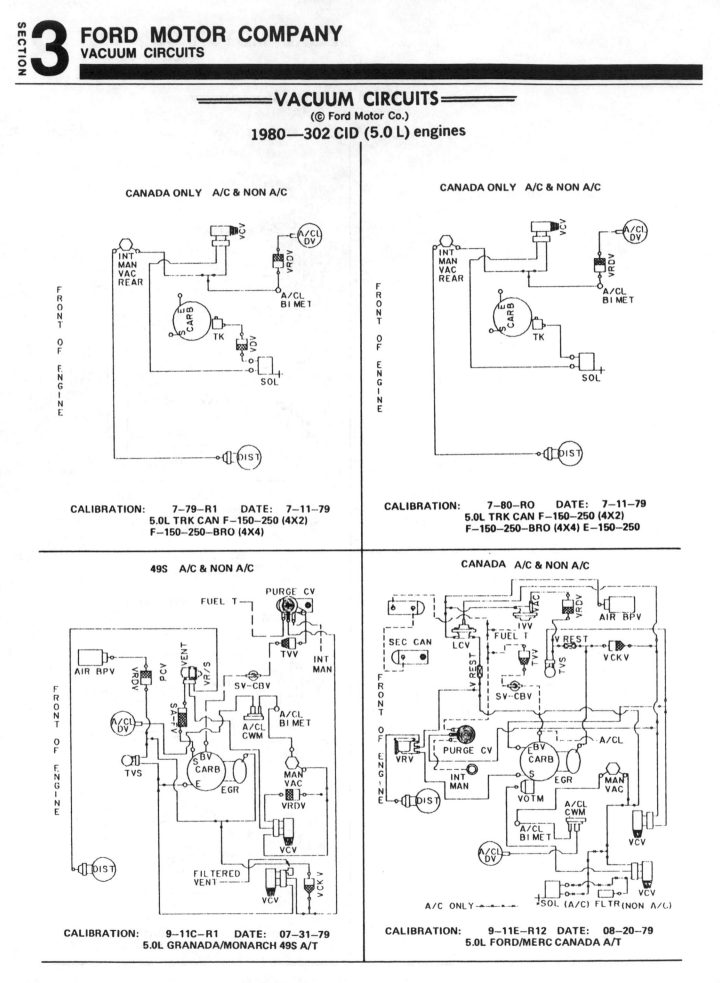

CANADA ONLY A/C & NON A/C

VCV
A/CL DV
INT MAN VAC REAR
VRDV
A/CL BI MET
FRONT OF ENGINE
E
CARB
S
TK
VDV
SOL
DIST

CALIBRATION: 7–79–R1 DATE: 7–11–79
5.0L TRK CAN F–150–250 (4X2)
F–150–250–BRO (4X4)

CANADA ONLY A/C & NON A/C

VCV
A/CL DV
INT MAN VAC REAR
VRDV
A/CL BI MET
FRONT OF ENGINE
E
CARB
S
TK
SOL
DIST

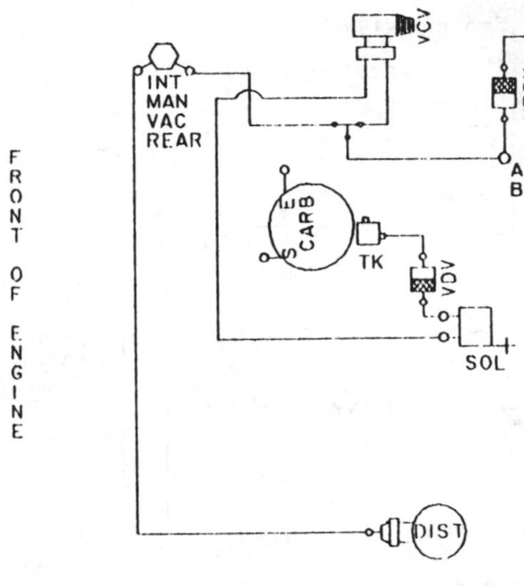

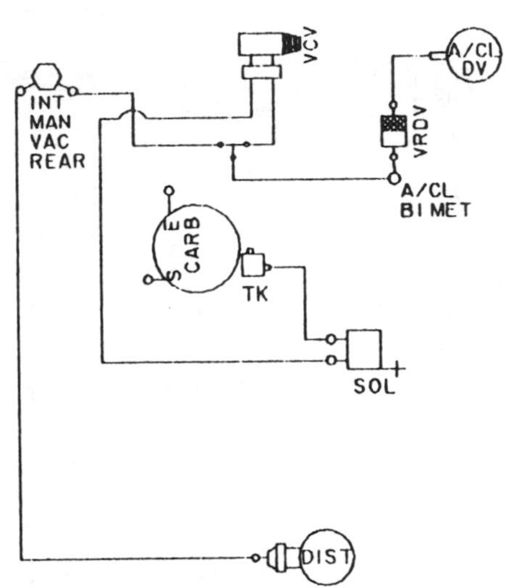

CALIBRATION: 7–80–RO DATE: 7–11–79
5.0L TRK CAN F–150–250 (4X2)
F–150–250–BRO (4X4) E–150–250

49S A/C & NON A/C

PURGE CV
FUEL T
TVV
INT MAN
AIR BPV
VRDV
PCV
VENT
VR/S
SV–CBV
FRONT OF ENGINE
A/CL DV
SA–FV
A/CL CWM
A/CL BI MET
TVS
BV
S
CARB
E
EGR
MAN VAC
VRDV
VCV
DIST
FILTERED VENT
VCV
VCKV

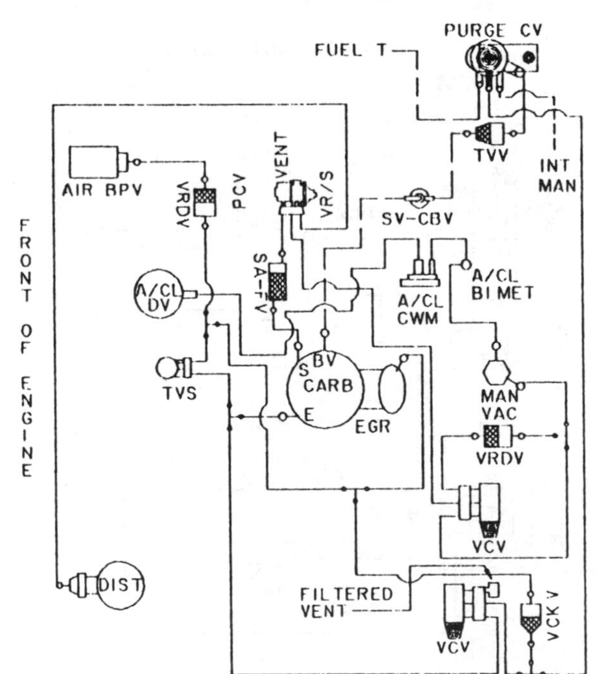

CALIBRATION: 9–11C–R1 DATE: 07–31–79
5.0L GRANADA/MONARCH 49S A/T

CANADA A/C & NON A/C

VAC
VRDV
AIR BPV
IVV
SEC CAN
LCV
FUEL T
V REST
TVV
TVS
VCKV
V REST
SV–CBV
FRONT OF ENGINE
A/CL
VRV
PURGE CV
BV
E
CARB
S
INT MAN
EGR
MAN VAC
DIST
VOTM
A/CL CWM
A/CL BI MET
VCV
A/CL DV
VCV
A/C ONLY
SOL (A/C)
FLTR (NON A/C)

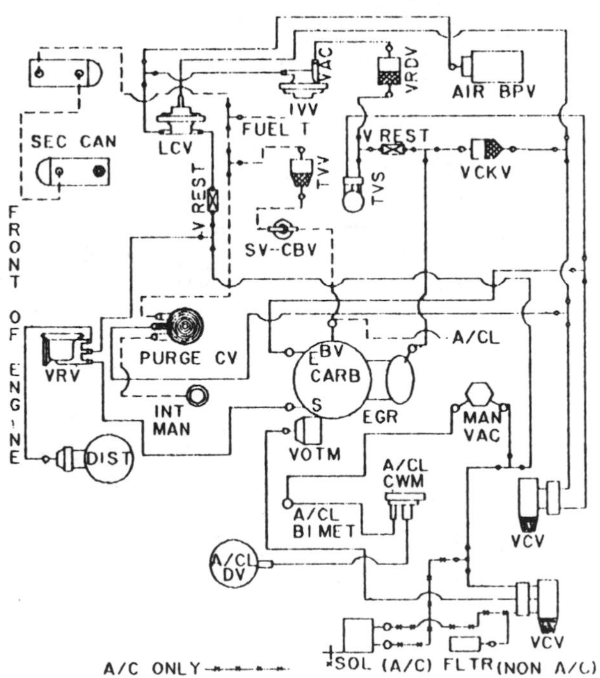

CALIBRATION: 9–11E–R12 DATE: 08–20–79
5.0L FORD/MERC CANADA A/T

VACUUM CIRCUITS

(© Ford Motor Co.)

1980—302 CID (5.0 L) engines

A/C & NON A/C

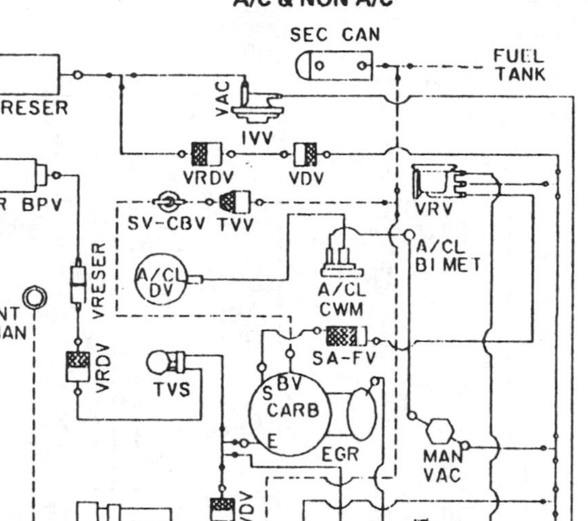

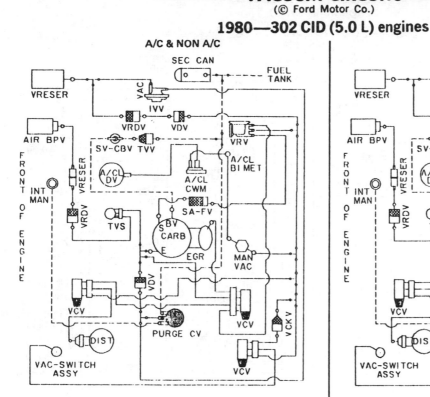

CALIBRATION: 0–11A–RO DATE: 8–8–79
5.0L T-BIRD/XR-7

A/C

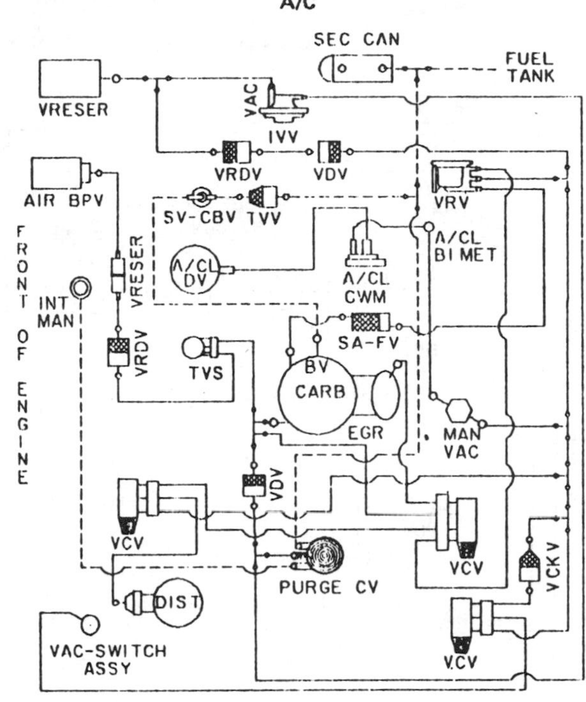

CALIBRATION: 0–11A–R10 DATE: 10–23–79
5.0L T–BIRD/XR7

A/C & NON A/C

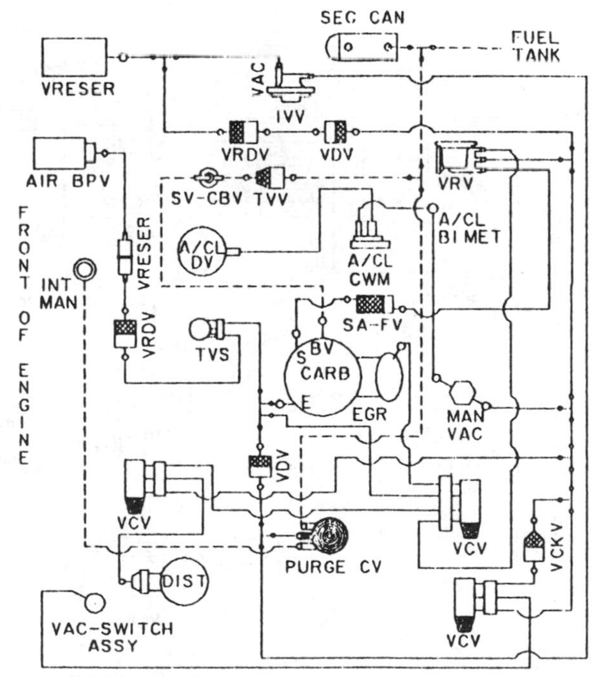

CALIBRATION: 0–11B–RO DATE: 8–8–79
5.0L T-BIRD/XR-7

A/C & NON A/C

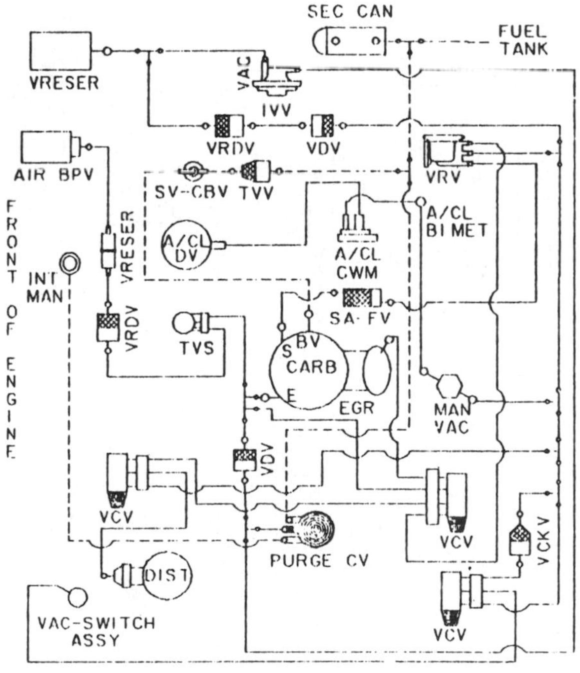

CALIBRATION: 0–11D–RO DATE: 8–8–79
5.0L GRANADA/MONARCH

VACUUM CIRCUITS

(© Ford Motor Co)

1980—302 CID (5.0 L) engines

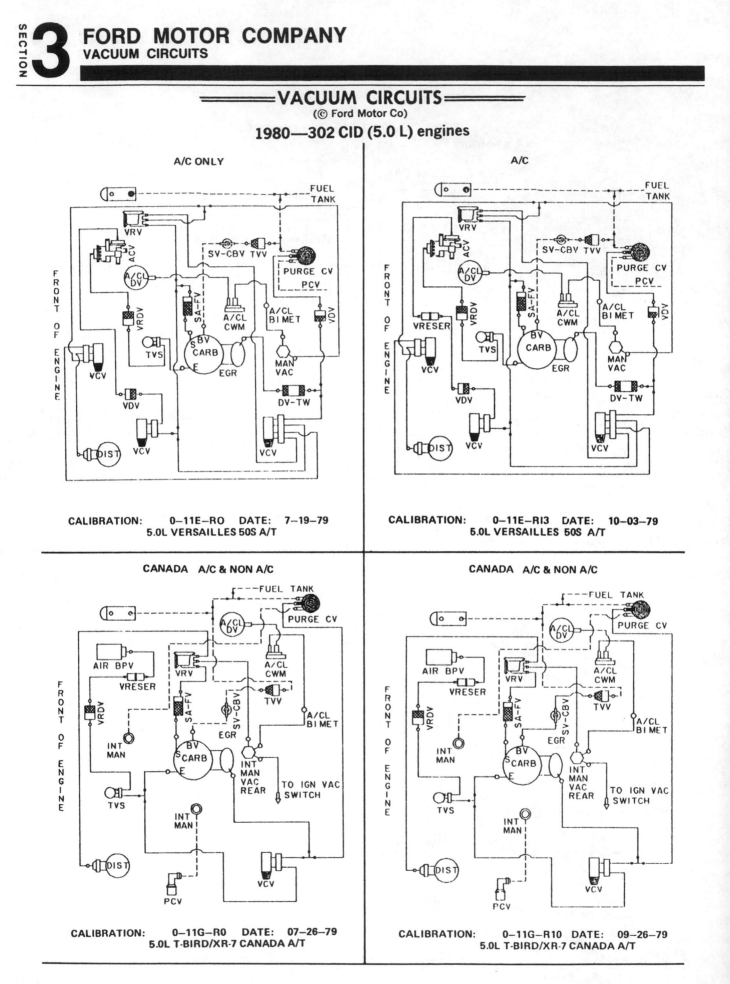

A/C ONLY

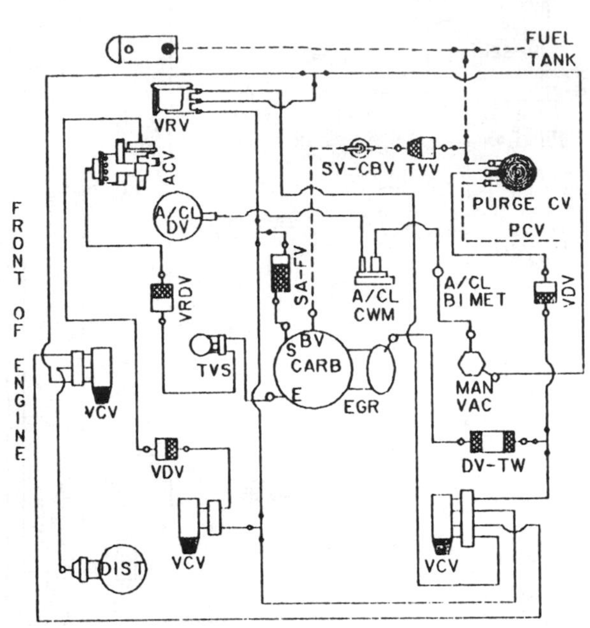

CALIBRATION: 0–11E–RO DATE: 7–19–79
5.0L VERSAILLES 50S A/T

A/C

CALIBRATION: 0–11E–RI3 DATE: 10–03–79
5.0L VERSAILLES 50S A/T

CANADA A/C & NON A/C

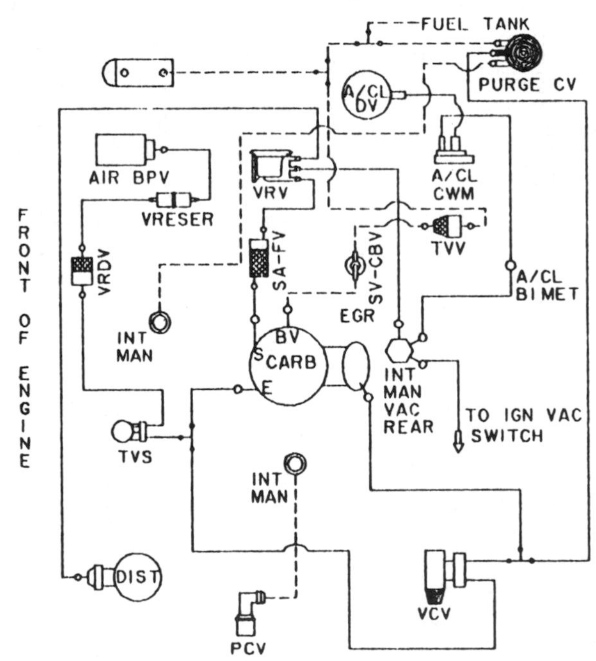

CALIBRATION: 0–11G–R0 DATE: 07–26–79
5.0L T-BIRD/XR-7 CANADA A/T

CANADA A/C & NON A/C

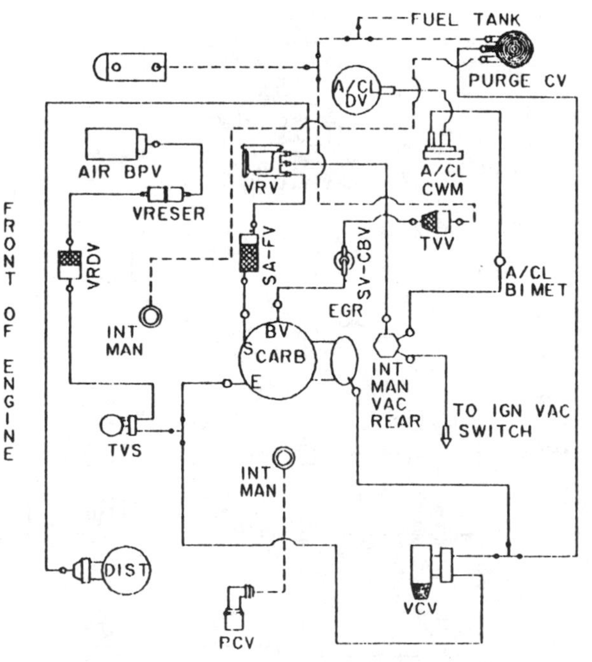

CALIBRATION: 0–11G–R10 DATE: 09–26–79
5.0L T-BIRD/XR-7 CANADA A/T

VACUUM CIRCUITS

(© Ford Motor Co.)

1980—302 CID (5.0 L) engines

CANADA A/C & NON A/C

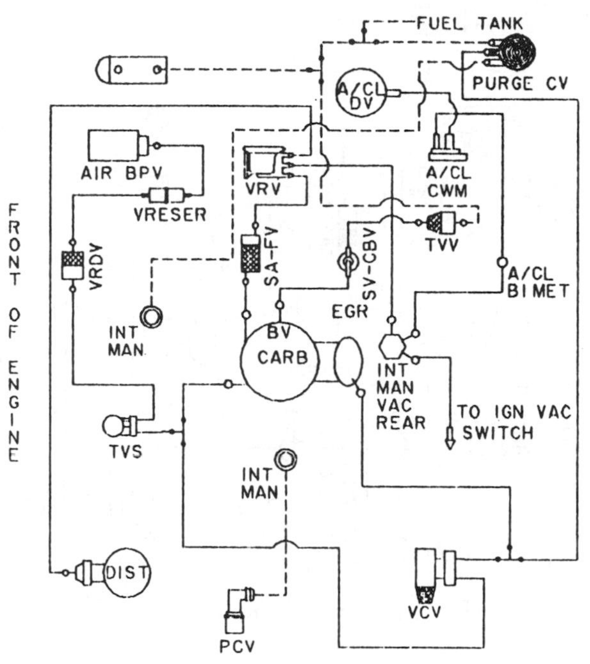

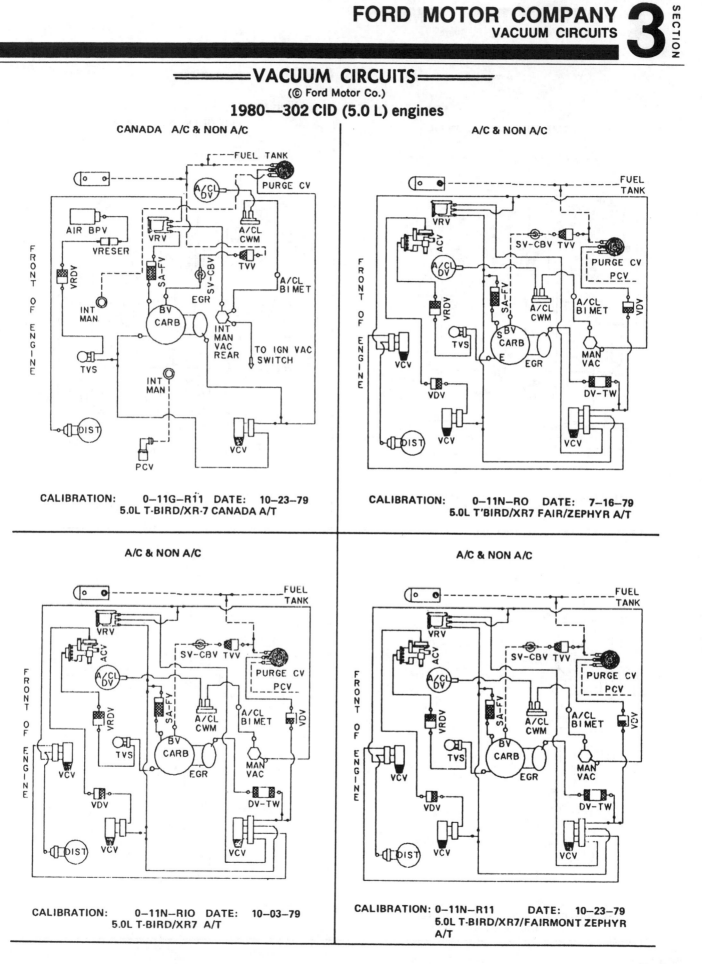

CALIBRATION: 0–11G–R11 DATE: 10–23–79
5.0L T-BIRD/XR-7 CANADA A/T

A/C & NON A/C

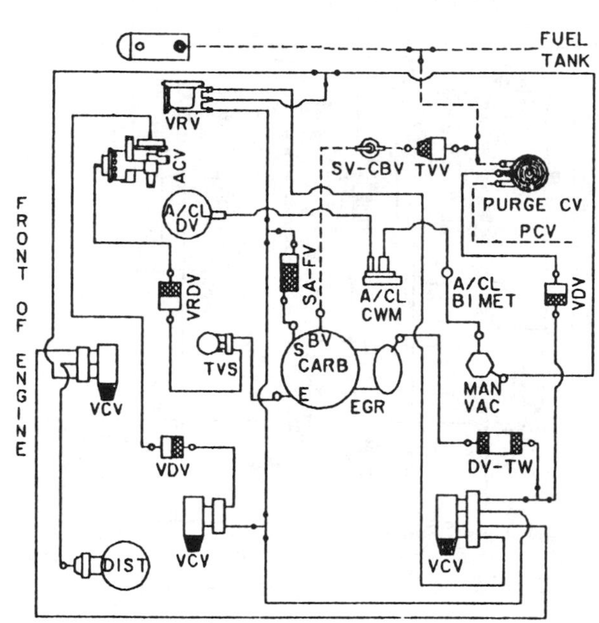

CALIBRATION: 0–11N–RO DATE: 7–16–79
5.0L T'BIRD/XR7 FAIR/ZEPHYR A/T

A/C & NON A/C

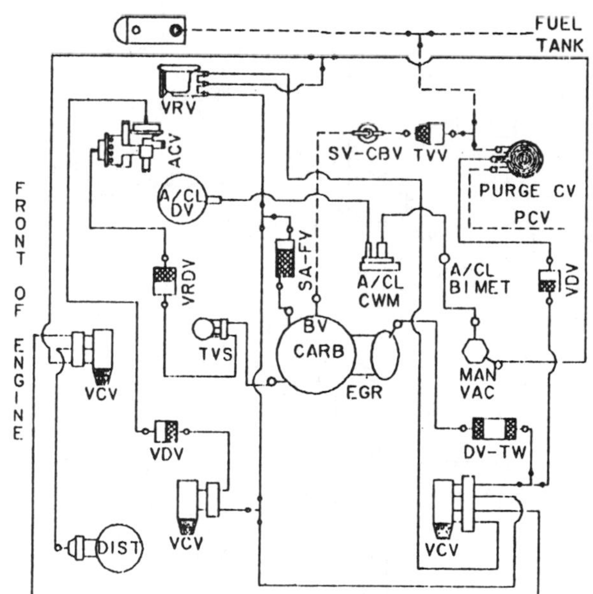

CALIBRATION: 0–11N–RIO DATE: 10–03–79
5.0L T-BIRD/XR7 A/T

A/C & NON A/C

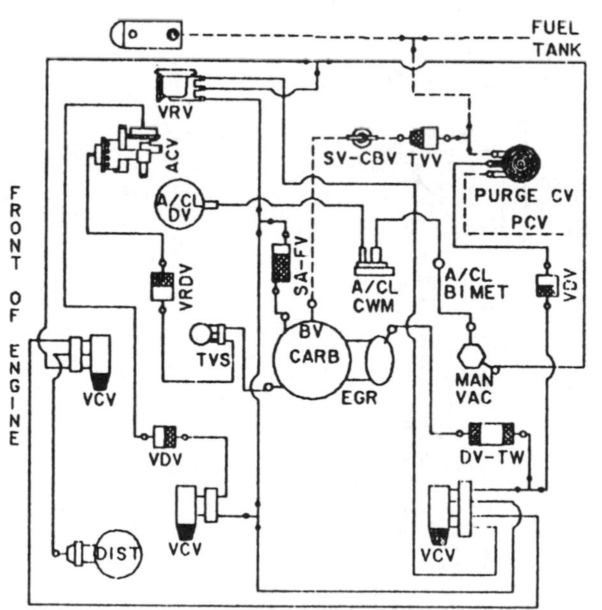

CALIBRATION: 0–11N–R11 DATE: 10–23–79
5.0L T-BIRD/XR7/FAIRMONT ZEPHYR
A/T

VACUUM CIRCUITS

1980—302 CID (5.0 L) engines

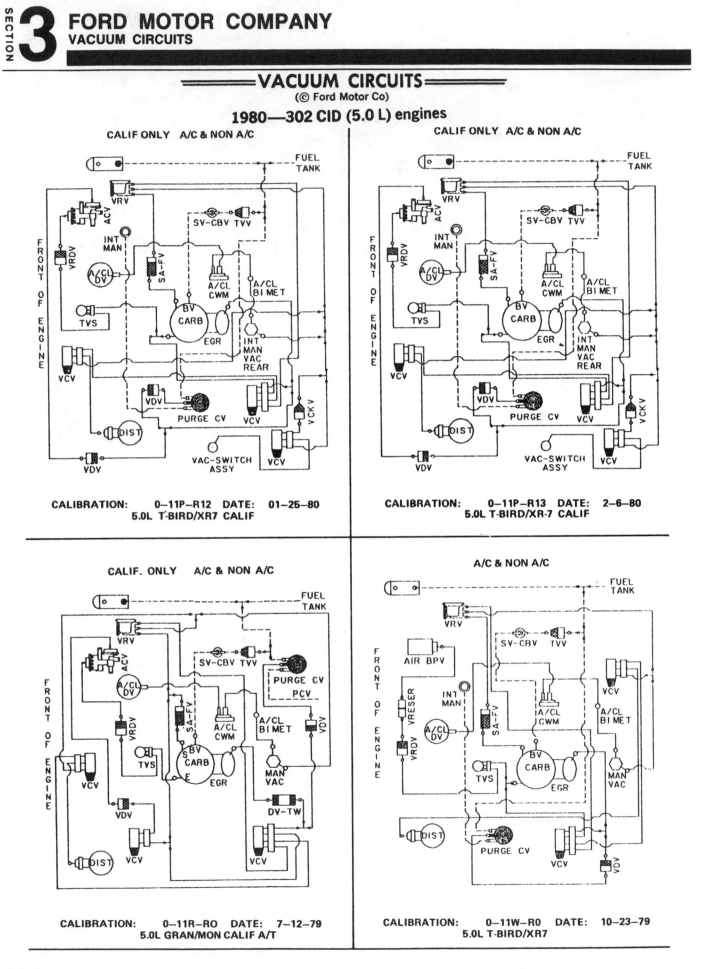

CALIF ONLY A/C & NON A/C

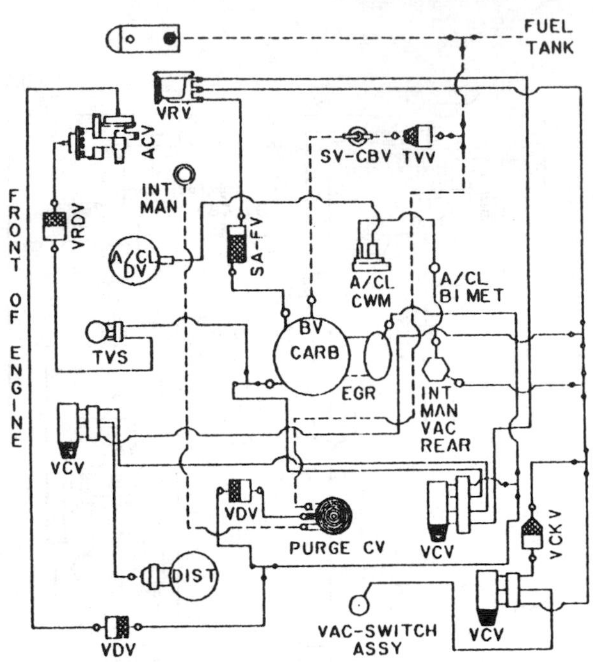

CALIBRATION: 0–11P–R12 DATE: 01–25–80
5.0L T-BIRD/XR7 CALIF

CALIF ONLY A/C & NON A/C

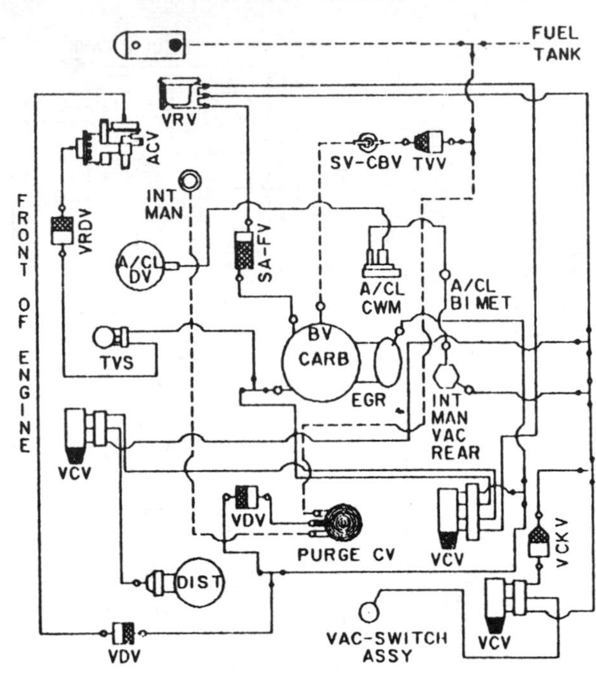

CALIBRATION: 0–11P–R13 DATE: 2–6–80
5.0L T-BIRD/XR-7 CALIF

CALIF. ONLY A/C & NON A/C

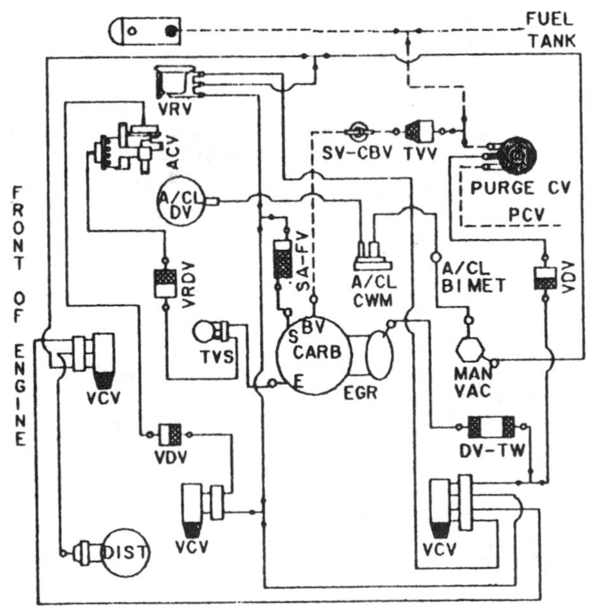

CALIBRATION: 0–11R–RO DATE: 7–12–79
5.0L GRAN/MON CALIF A/T

A/C & NON A/C

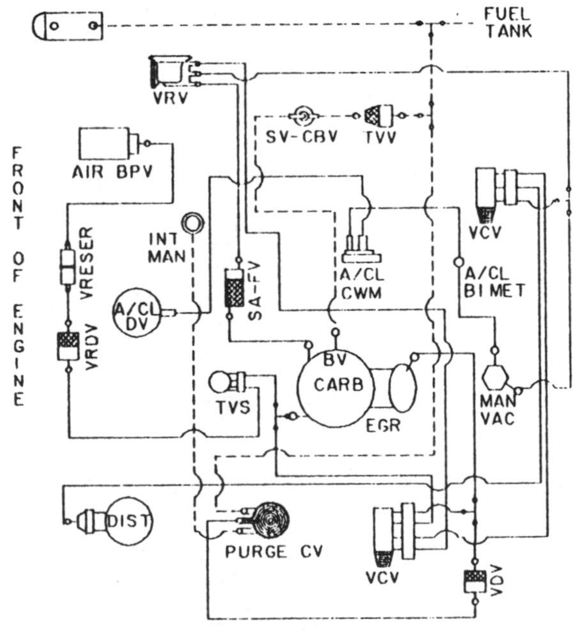

CALIBRATION: 0–11W–RO DATE: 10–23–79
5.0L T-BIRD/XR7

VACUUM CIRCUITS

(© Ford Motor Co.)

1980—302 CID (5.0 L) engines

A/C & NON A/C

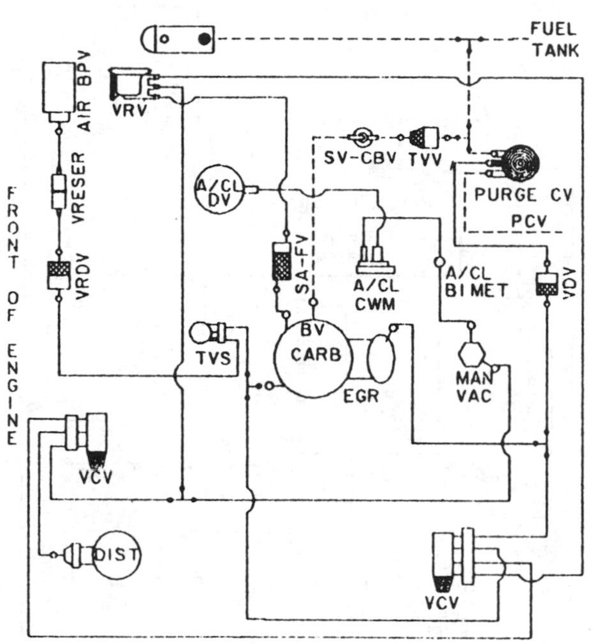

CALIBRATION: 0–11X–R0 DATE: 09–24–79
5.0L GRANADA/MONARCH A/T

A/C ONLY

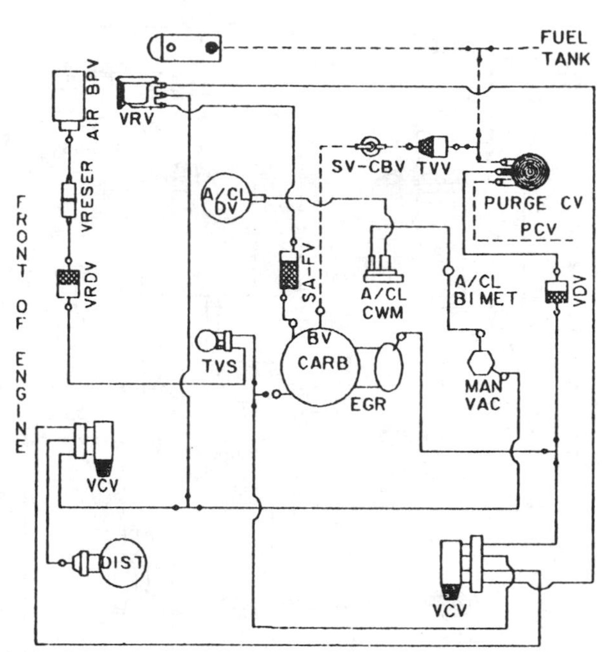

CALIBRATION: 0–11Y–R0 DATE: 9–24–79
5.0L VERSAILLES A/T

49S A/C & NON A/C

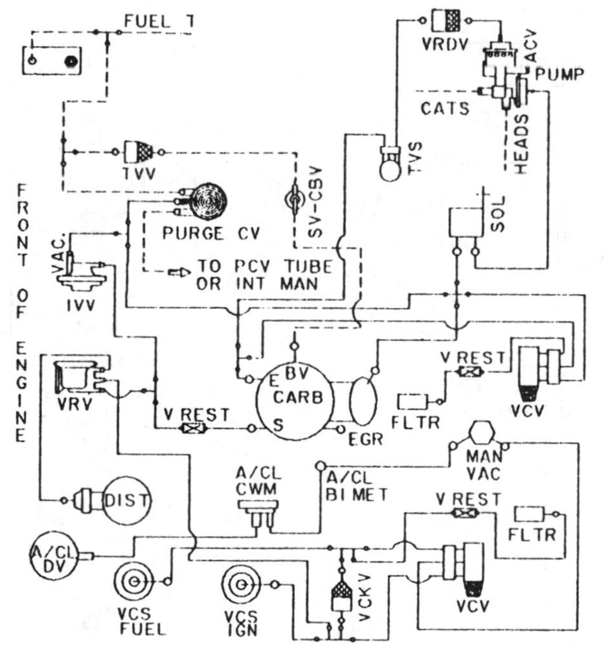

CALIBRATION: 0–13A–RO DATE: 7–18–79
5.0L 49S FORD–MERC SED A/T

49S A/C & NON A/C

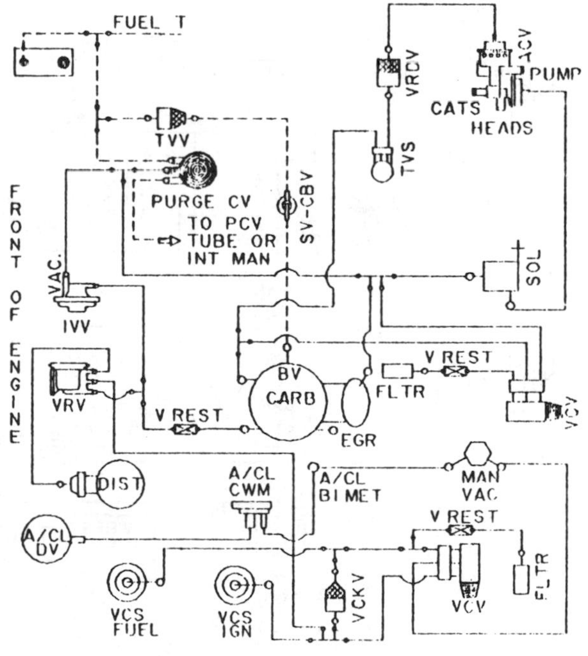

CALIBRATION: 0–13A–RIO DATE: 10–11–79
5.0L FORD/MERC SWNG 49S A/T

VACUUM CIRCUITS

(© Ford Motor Co)

1980—302 CID (5.0 L) engines

49S A/C & NON A/C

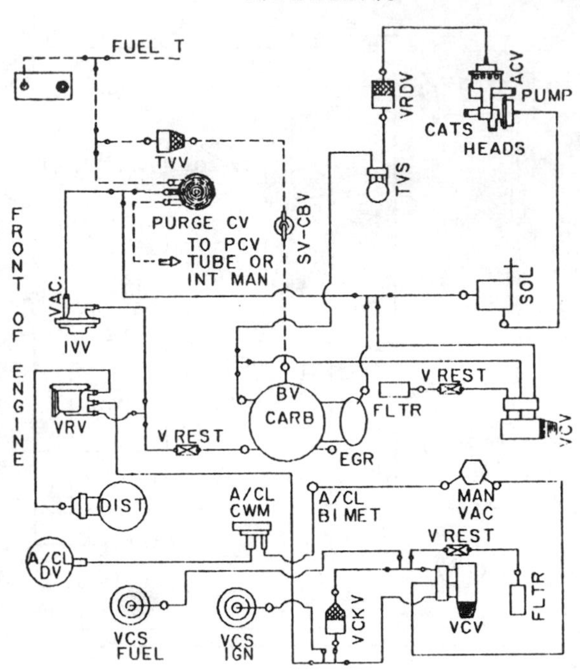

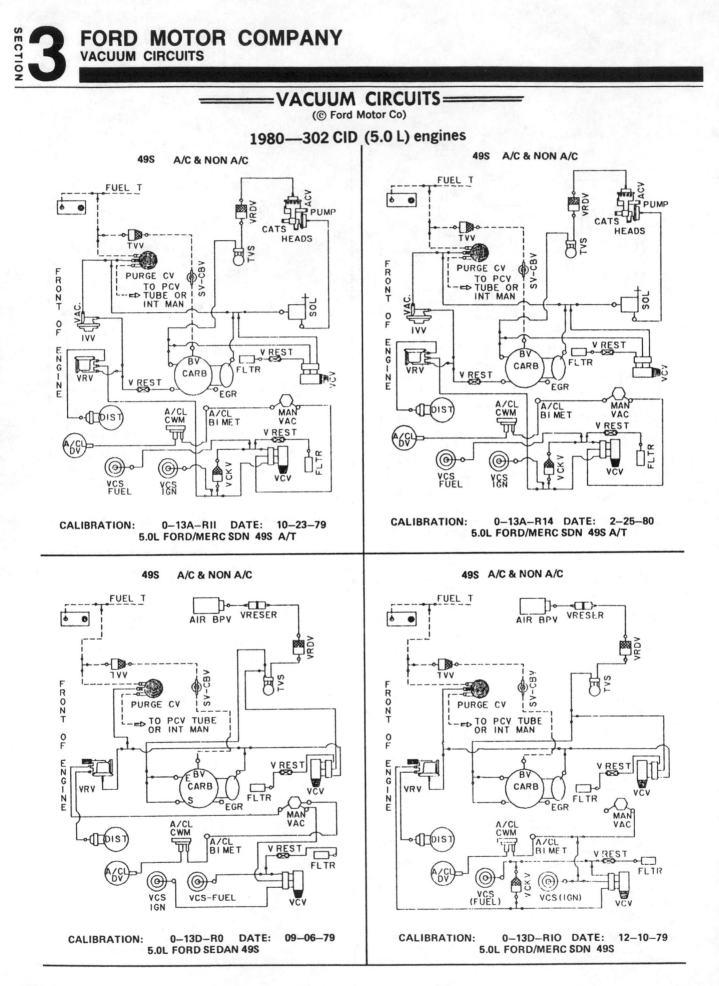

CALIBRATION: 0–13A–RII DATE: 10–23–79
5.0L FORD/MERC SDN 49S A/T

49S A/C & NON A/C

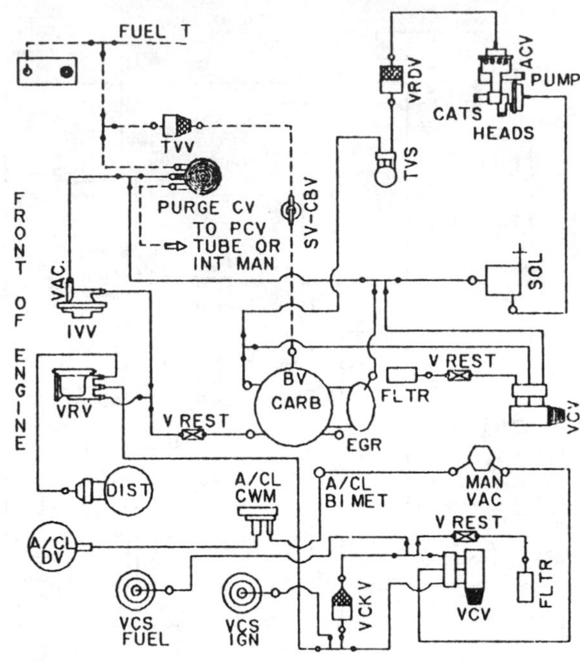

CALIBRATION: 0–13A–R14 DATE: 2–25–80
5.0L FORD/MERC SDN 49S A/T

49S A/C & NON A/C

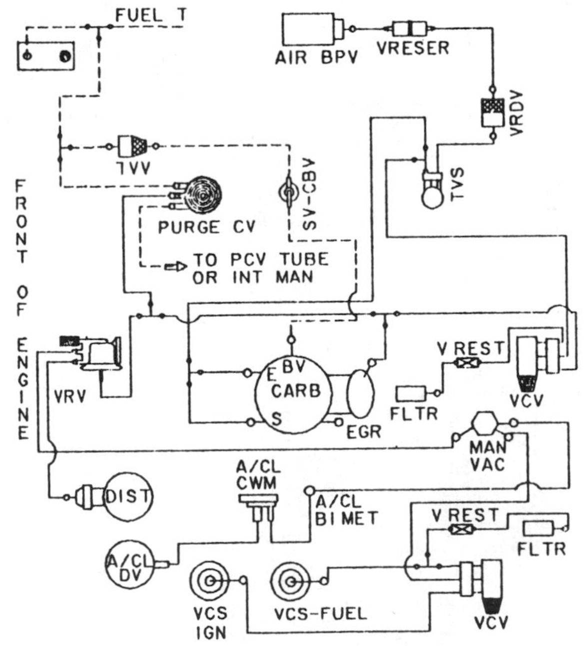

CALIBRATION: 0–13D–R0 DATE: 09–06–79
5.0L FORD SEDAN 49S

49S A/C & NON A/C

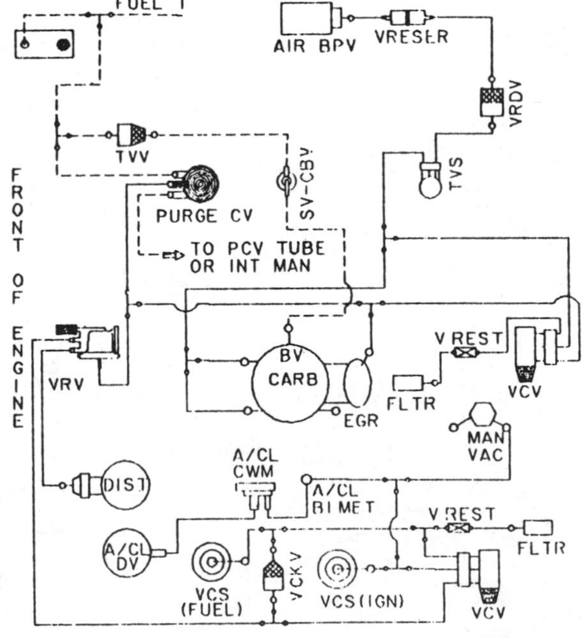

CALIBRATION: 0–13D–RIO DATE: 12–10–79
5.0L FORD/MERC SDN 49S

VACUUM CIRCUITS

(© Ford Motor Co.)

1980—302 CID (5.0 L) engines

49S A/C & NON A/C

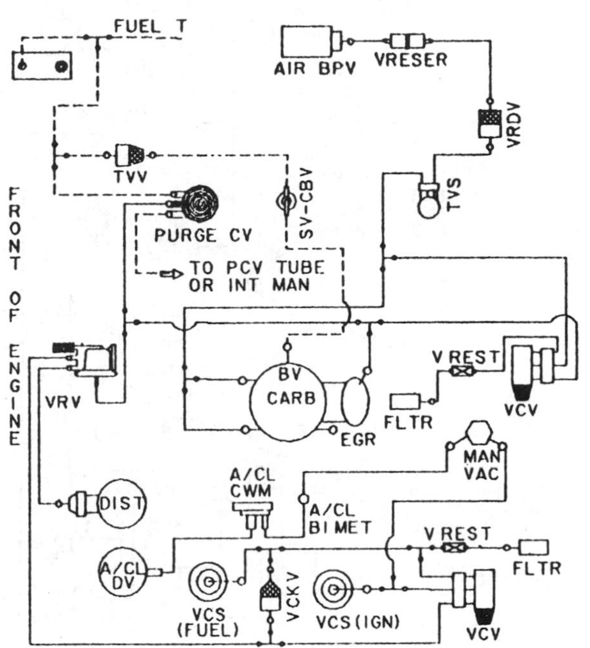

CALIBRATION: 0–13D–R11 DATE: 3–4–80
5.0L FORD & MERC SDN 49S

49S A/C & NON A/C

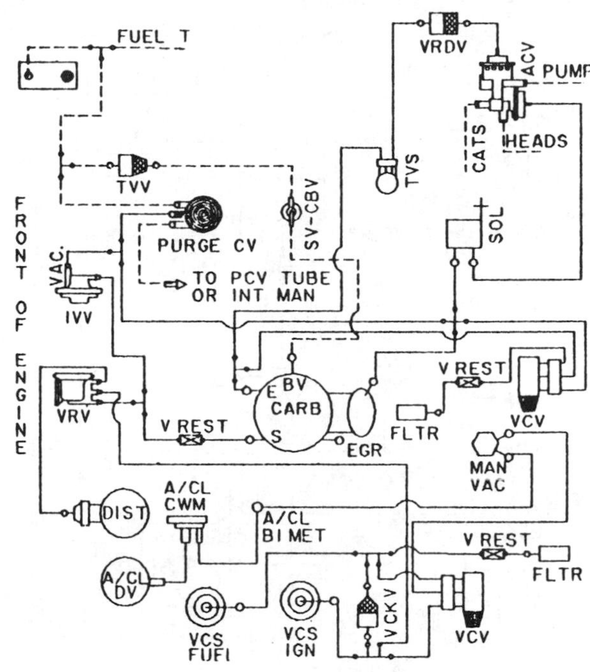

CALIBRATION: 0–13F–RO DATE: 7–18–79
5.0L 49S FORD–MERC S/W A/T

49S A/C & NON A/C

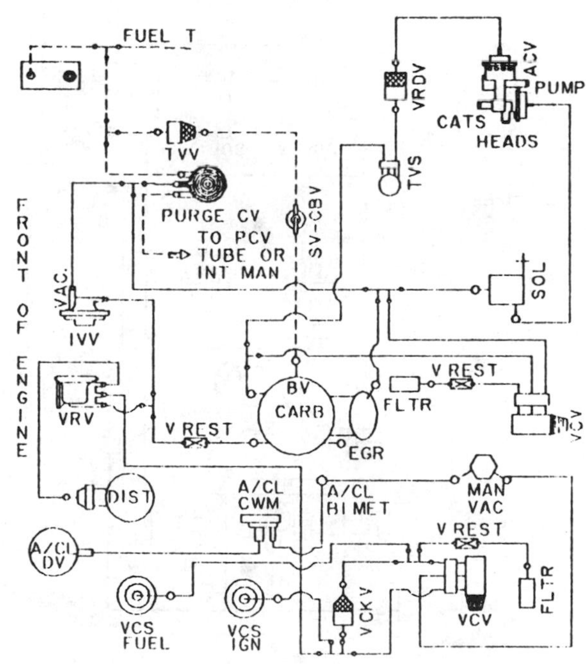

CALIBRATION: 0–13F–R11 DATE: 10–23–79
5.0L FORD/MERC SDN 49S A/T

49S A/C & NON A/C

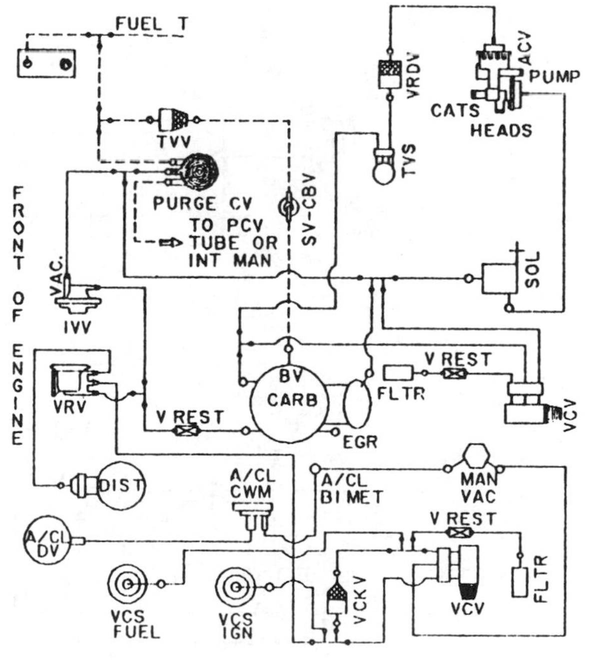

CALIBRATION: 0–13F–R12 DATE: 2–25–80
5.0L FORD & MERC S.W. 49S A/T

VACUUM CIRCUITS

(© Ford Motor Co.)

1980—302 CID (5.0 L) engines

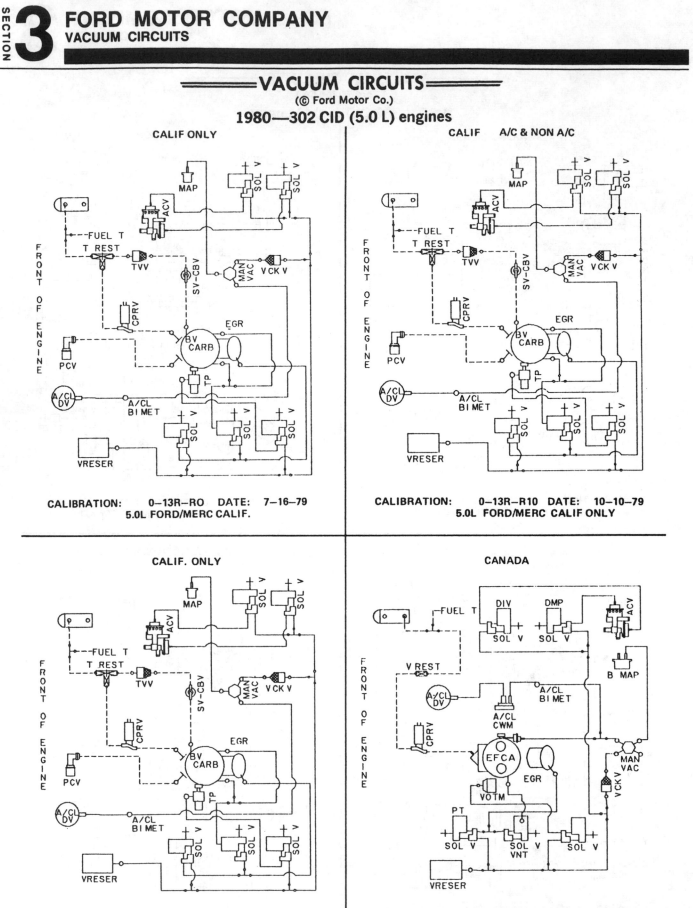

CALIF ONLY

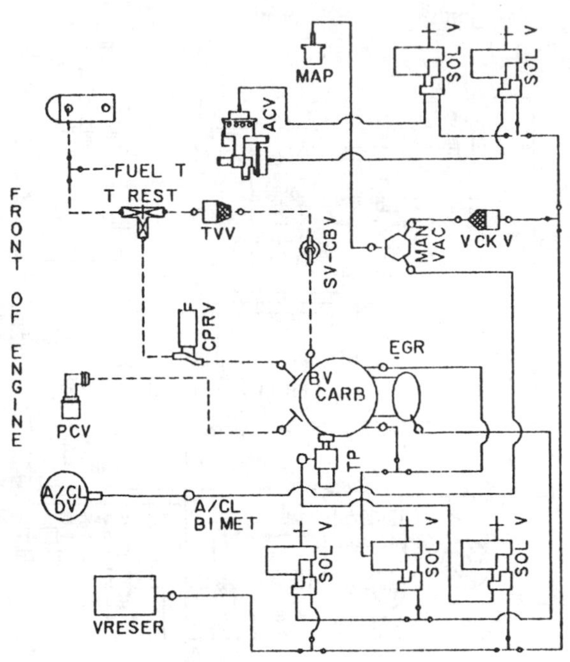

CALIBRATION: 0–13R–RO DATE: 7–16–79
5.0L FORD/MERC CALIF.

CALIF A/C & NON A/C

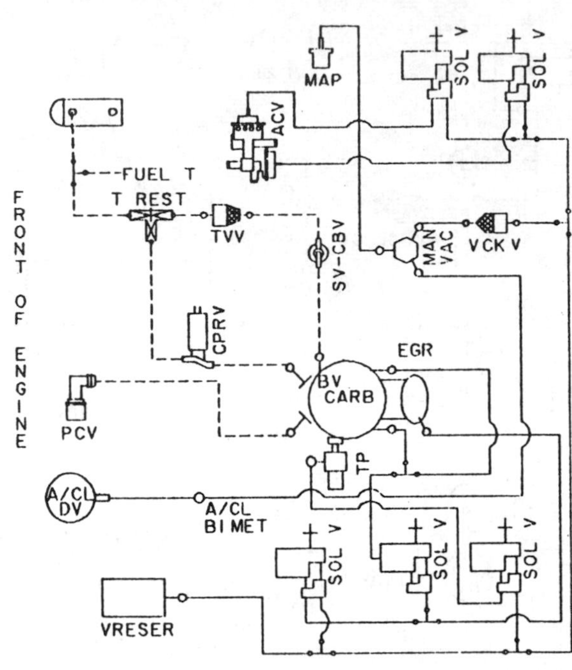

CALIBRATION: 0–13R–R10 DATE: 10–10–79
5.0L FORD/MERC CALIF ONLY

CALIF. ONLY

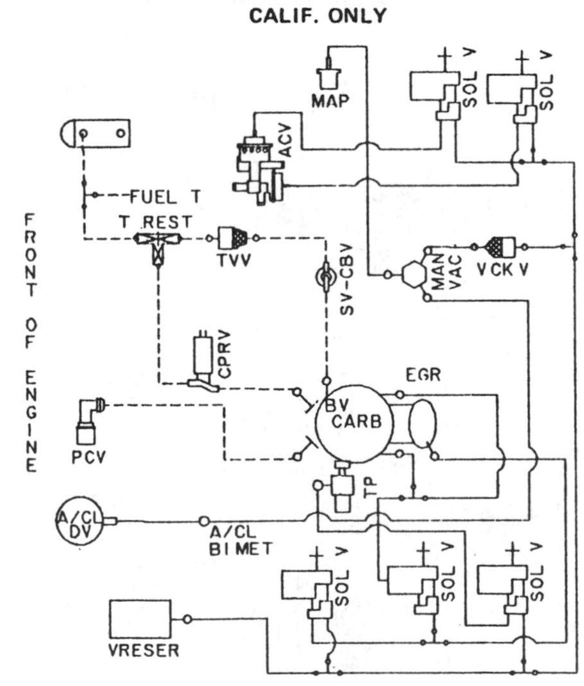

CALIBRATION: 0–13T–RO DATE: 7–16–79
5.0L FORD/MERC CALIF.

CANADA

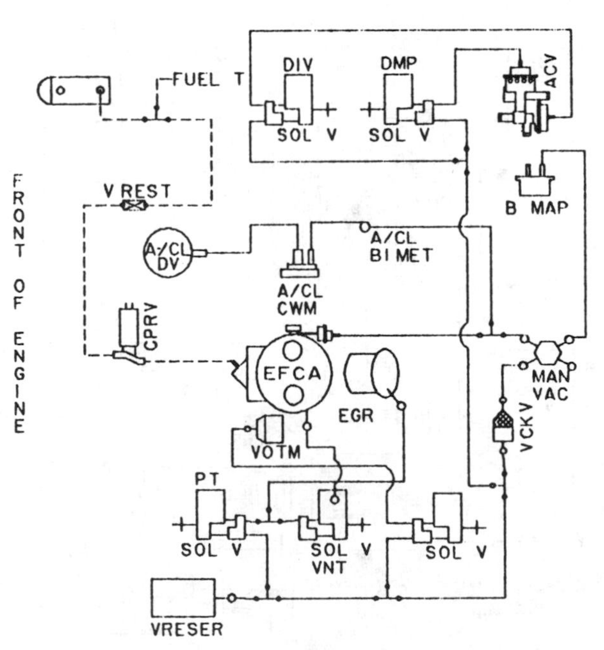

CALIBRATION: 0–14A–RO DATE: 7–17–79
5.0L 49S/CANADA LINC/MARK 4250
IW FORD/MERC F10D

VACUUM CIRCUITS

(© Ford Motor Co)

1980—302 CID (5.0 L) engines

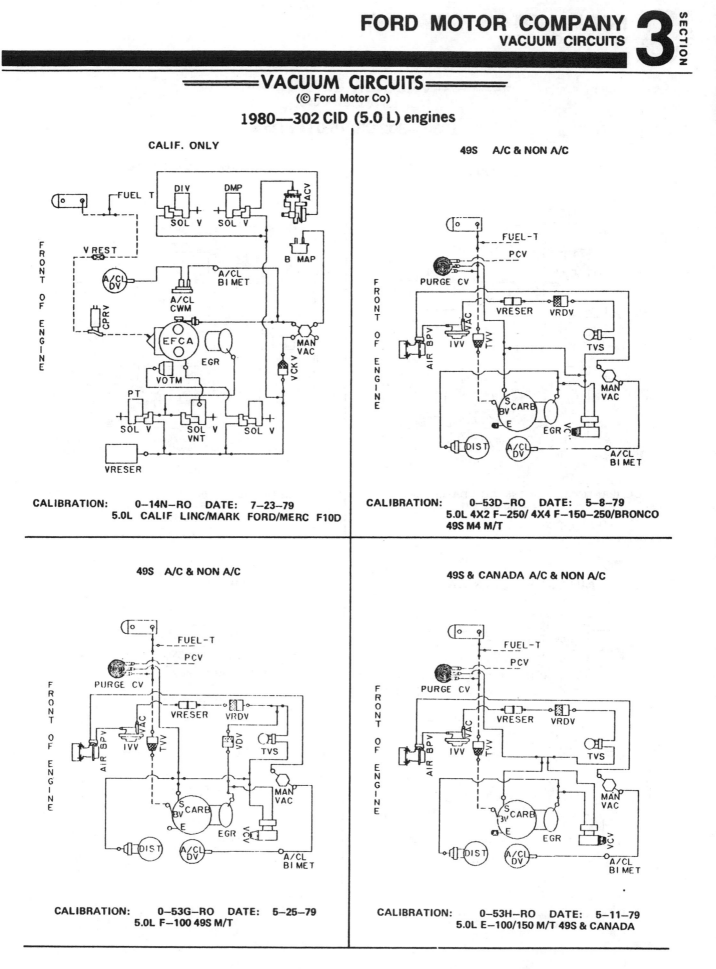

CALIF. ONLY

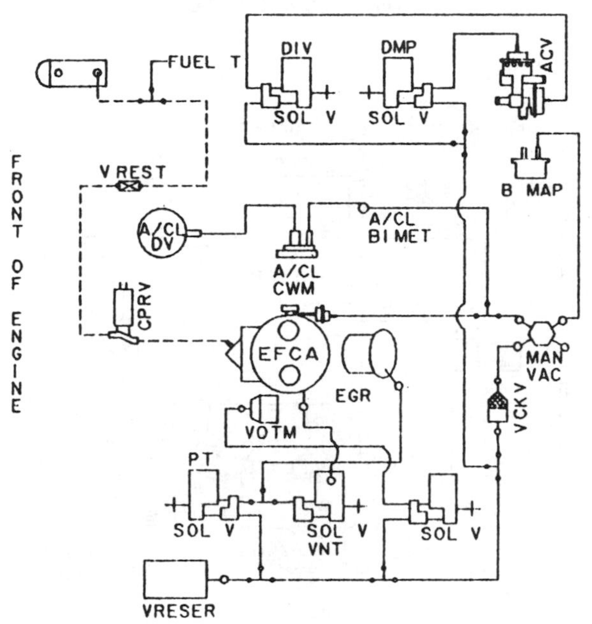

CALIBRATION: 0–14N–RO DATE: 7–23–79
5.0L CALIF LINC/MARK FORD/MERC F10D

49S A/C & NON A/C

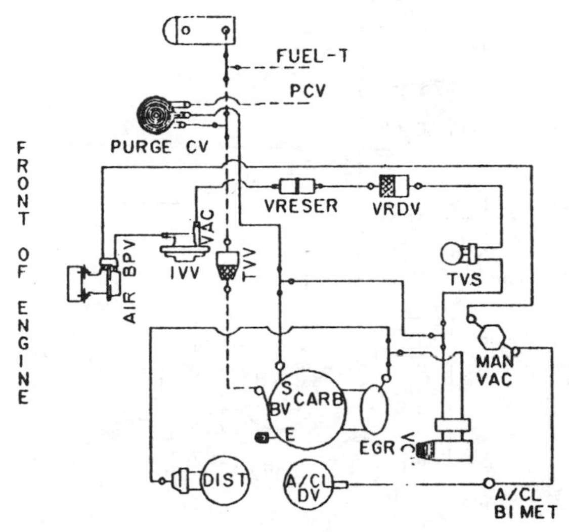

CALIBRATION: 0–53D–RO DATE: 5–8–79
5.0L 4X2 F–250/ 4X4 F–150–250/BRONCO
49S M4 M/T

49S A/C & NON A/C

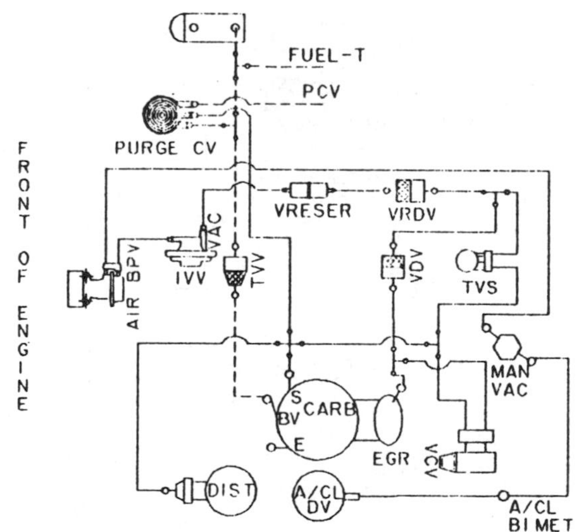

CALIBRATION: 0–53G–RO DATE: 5–25–79
5.0L F–100 49S M/T

49S & CANADA A/C & NON A/C

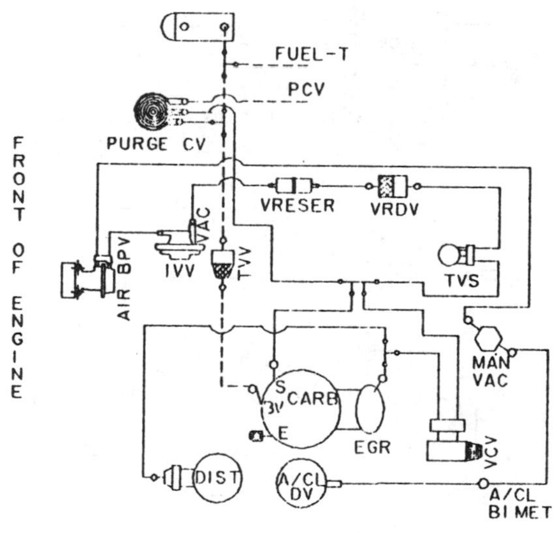

CALIBRATION: 0–53H–RO DATE: 5–11–79
5.0L E–100/150 M/T 49S & CANADA

VACUUM CIRCUITS

1980—302 CID (5.0 L) engines

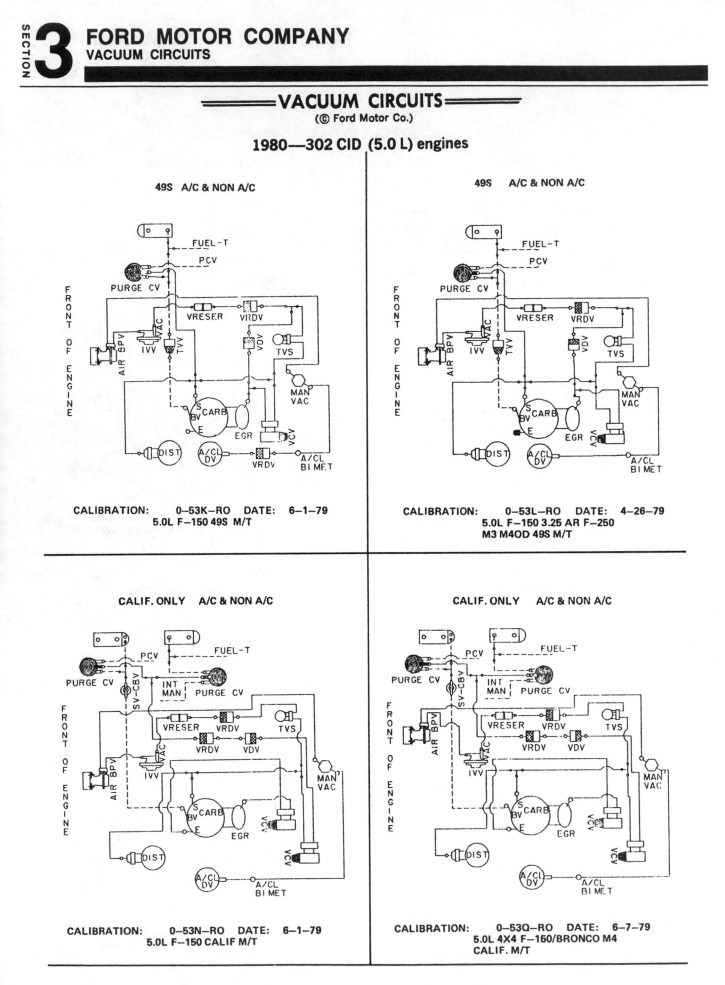

49S A/C & NON A/C

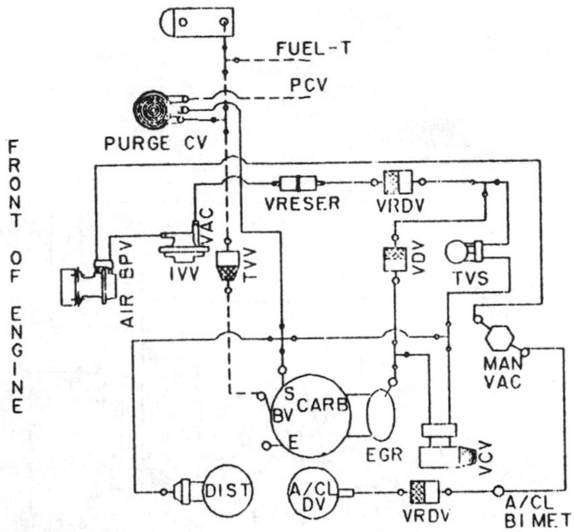

CALIBRATION: 0–53K–RO DATE: 6–1–79
5.0L F–150 49S M/T

49S A/C & NON A/C

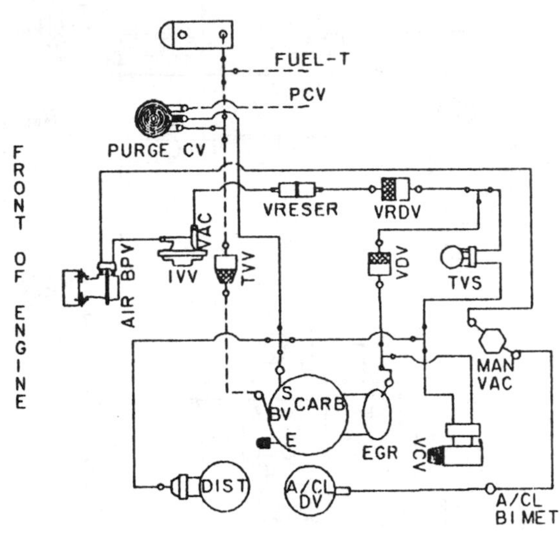

CALIBRATION: 0–53L–RO DATE: 4–26–79
5.0L F–150 3.25 AR F–250
M3 M4OD 49S M/T

CALIF. ONLY A/C & NON A/C

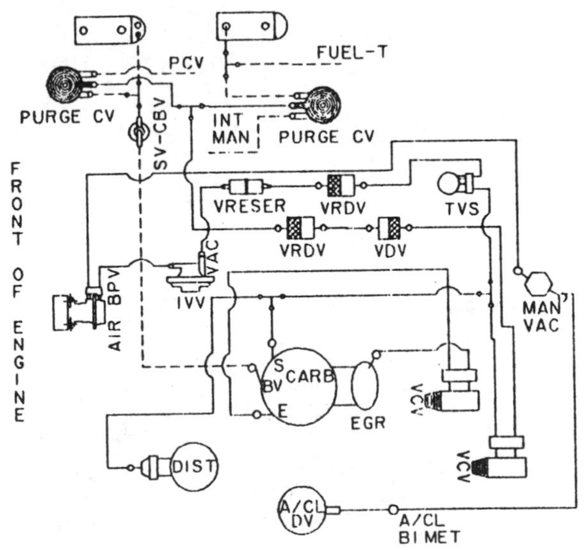

CALIBRATION: 0–53N–RO DATE: 6–1–79
5.0L F–150 CALIF M/T

CALIF. ONLY A/C & NON A/C

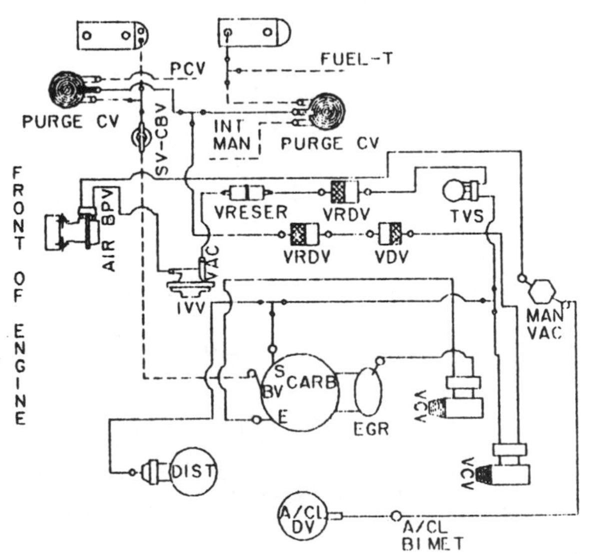

CALIBRATION: 0–53Q–RO DATE: 6–7–79
5.0L 4X4 F–150/BRONCO M4
CALIF. M/T

VACUUM CIRCUITS

(© Ford Motor Co.)

1980—302 CID (5.0 L) engines

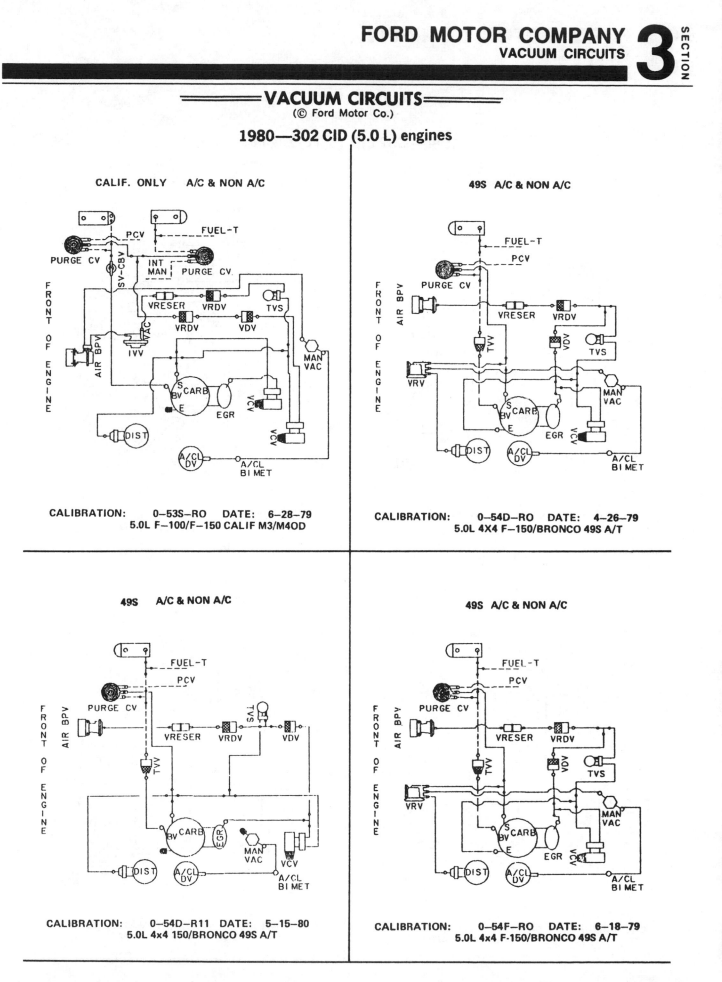

CALIF. ONLY A/C & NON A/C

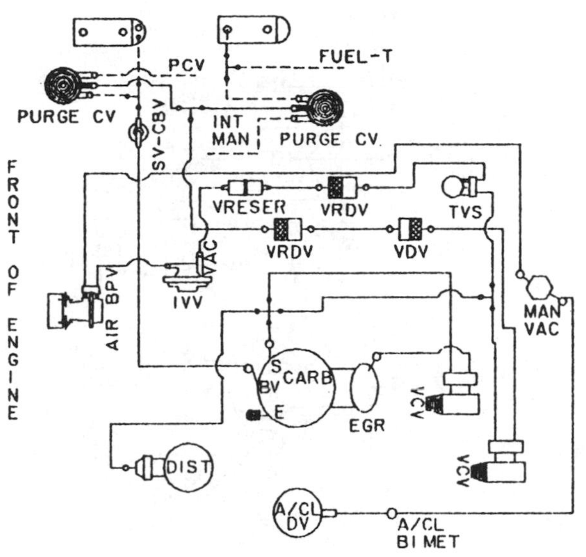

CALIBRATION: 0–53S–RO DATE: 6–28–79
5.0L F–100/F–150 CALIF M3/M4OD

49S A/C & NON A/C

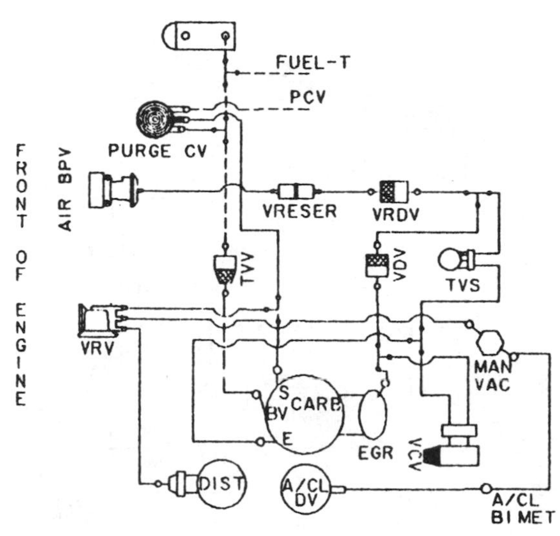

CALIBRATION: 0–54D–RO DATE: 4–26–79
5.0L 4X4 F–150/BRONCO 49S A/T

49S A/C & NON A/C

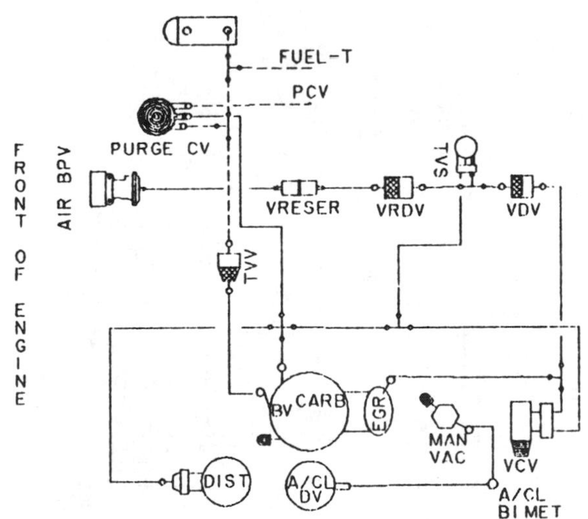

CALIBRATION: 0–54D–R11 DATE: 5–15–80
5.0L 4x4 150/BRONCO 49S A/T

49S A/C & NON A/C

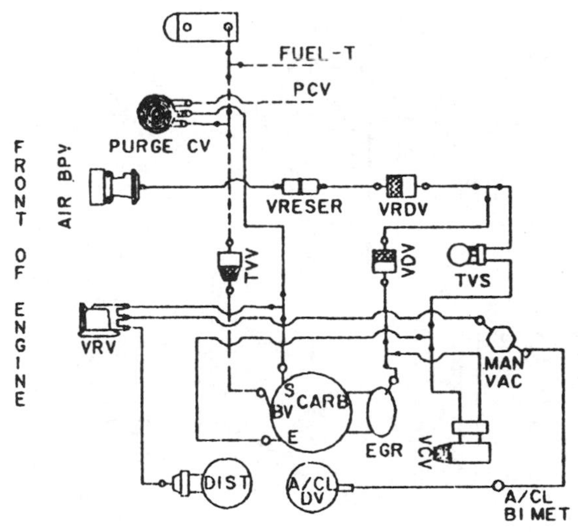

CALIBRATION: 0–54F–RO DATE: 6–18–79
5.0L 4x4 F-150/BRONCO 49S A/T

VACUUM CIRCUITS

(© Ford Motor Co.)

1980—302 CID (5.0 L) engines

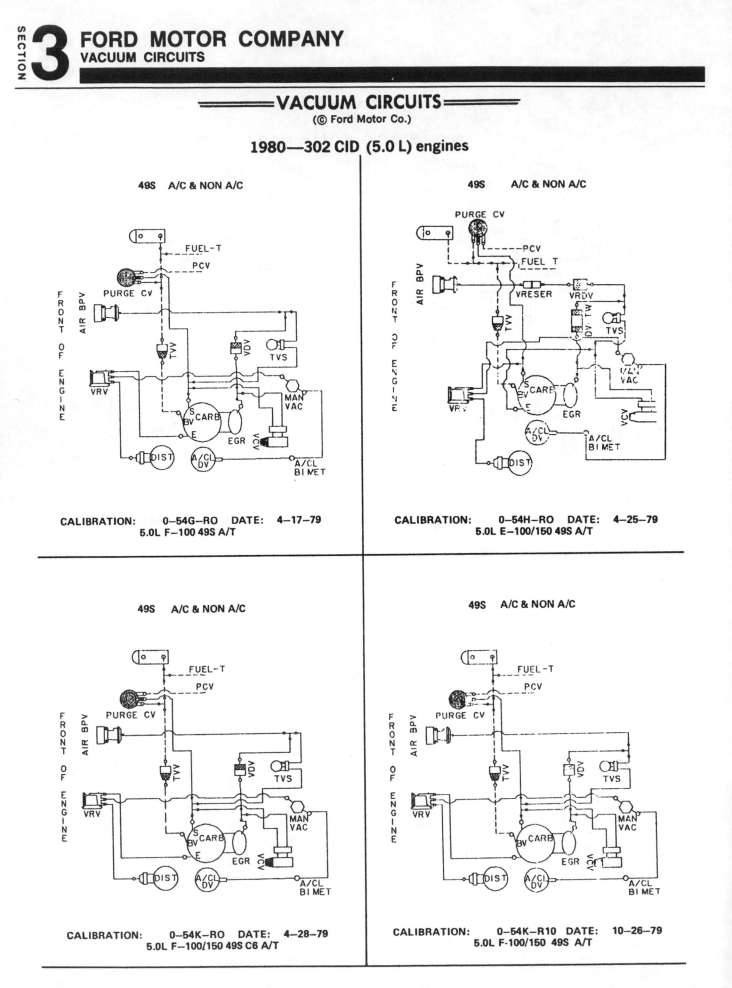

49S A/C & NON A/C

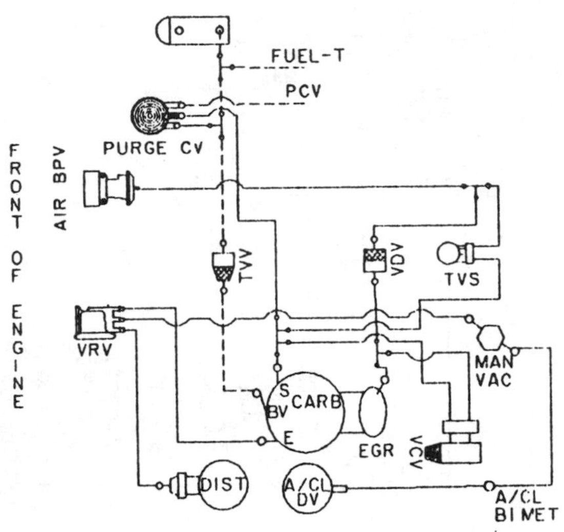

CALIBRATION: 0–54G–RO DATE: 4–17–79
5.0L F–100 49S A/T

49S A/C & NON A/C

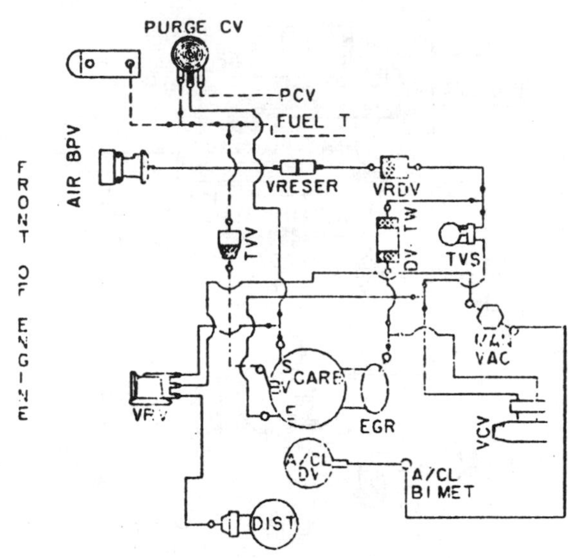

CALIBRATION: 0–54H–RO DATE: 4–25–79
5.0L E–100/150 49S A/T

49S A/C & NON A/C

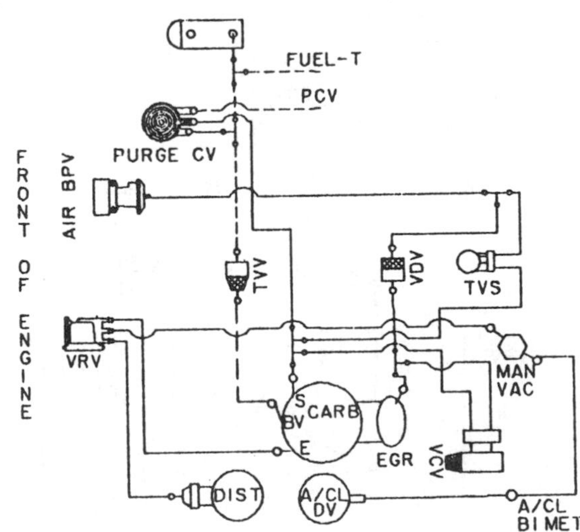

CALIBRATION: 0–54K–RO DATE: 4–28–79
5.0L F–100/150 49S C6 A/T

49S A/C & NON A/C

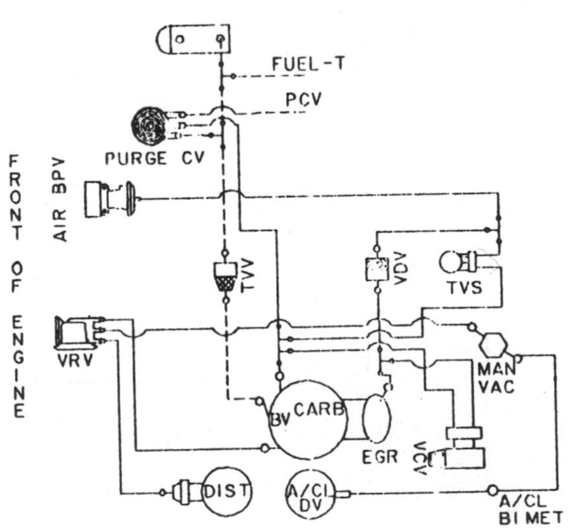

CALIBRATION: 0–54K–R10 DATE: 10–26–79
5.0L F-100/150 49S A/T

VACUUM CIRCUITS

(© Ford Motor Co)

1980—302 CID (5.0 L) engines

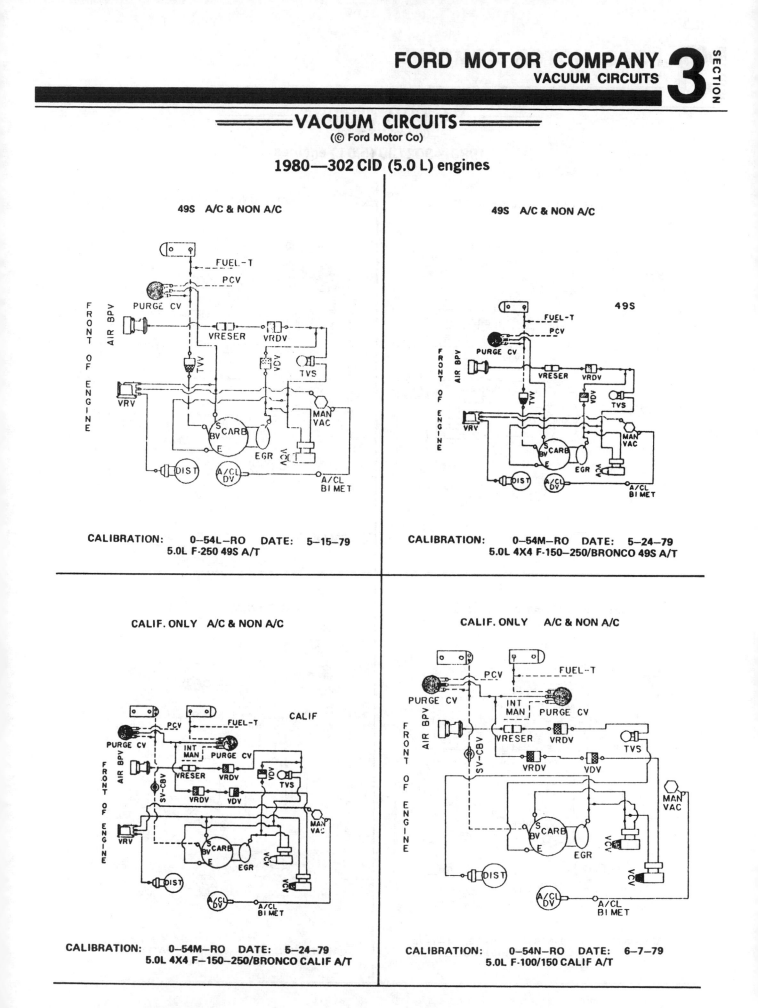

49S A/C & NON A/C

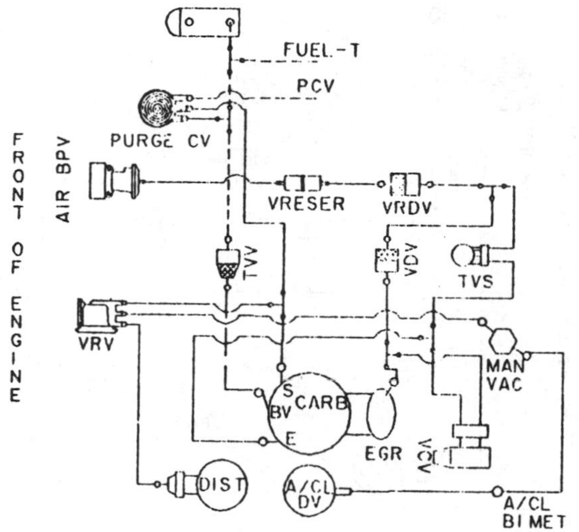

CALIBRATION: 0–54L–RO DATE: 5–15–79
5.0L F-250 49S A/T

49S A/C & NON A/C

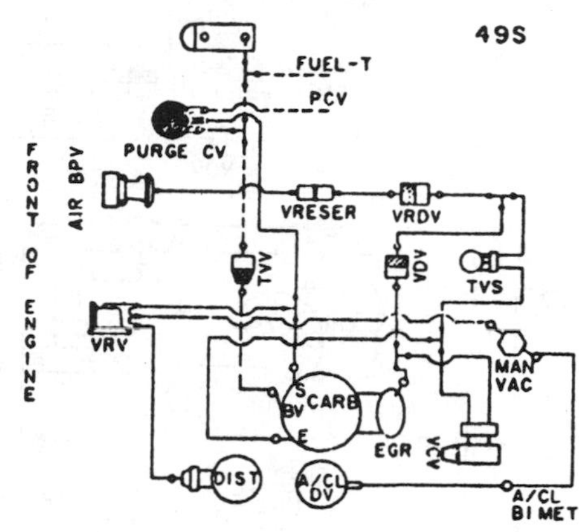

CALIBRATION: 0–54M–RO DATE: 5–24–79
5.0L 4X4 F-150–250/BRONCO 49S A/T

CALIF. ONLY A/C & NON A/C

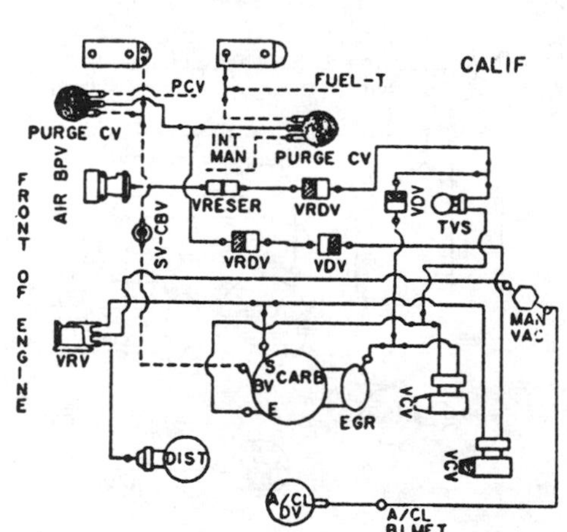

CALIBRATION: 0–54M–RO DATE: 5–24–79
5.0L 4X4 F–150–250/BRONCO CALIF A/T

CALIF. ONLY A/C & NON A/C

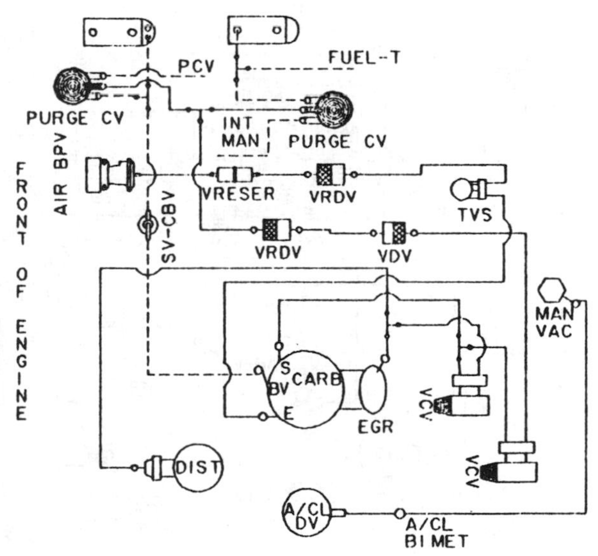

CALIBRATION: 0–54N–RO DATE: 6–7–79
5.0L F-100/150 CALIF A/T

VACUUM CIRCUITS

(© Ford Motor Co)

1980—302 CID (5.0 L) engines

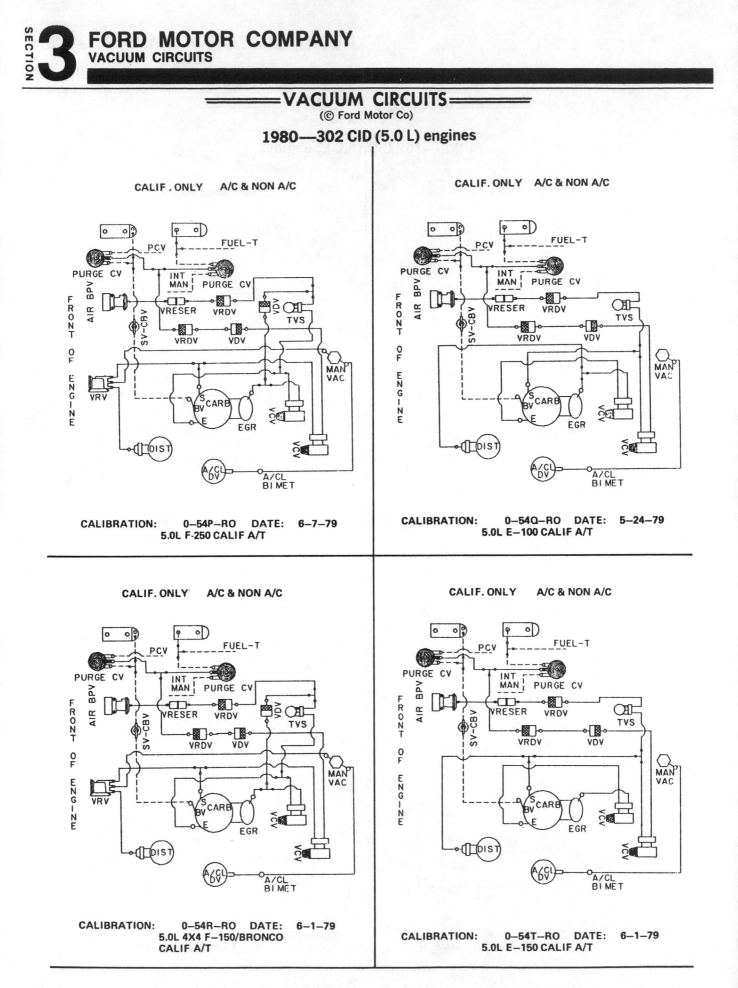

CALIF. ONLY A/C & NON A/C

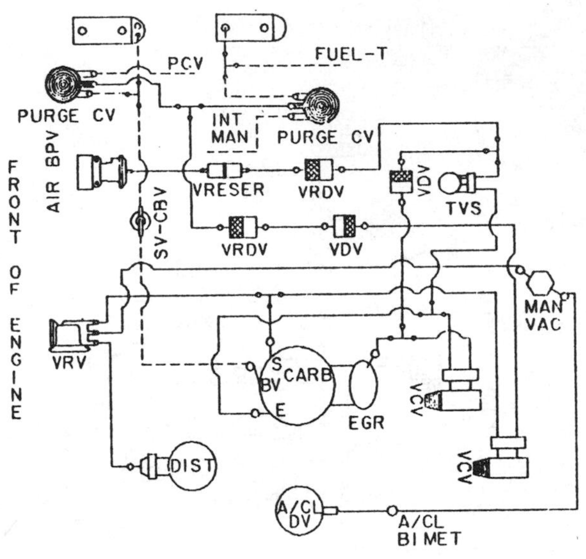

CALIBRATION: 0–54P–RO DATE: 6–7–79
5.0L F-250 CALIF A/T

CALIF. ONLY A/C & NON A/C

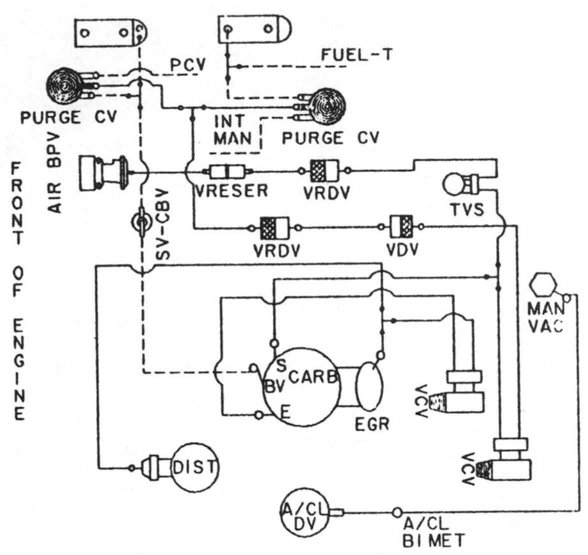

CALIBRATION: 0–54Q–RO DATE: 5–24–79
5.0L E–100 CALIF A/T

CALIF. ONLY A/C & NON A/C

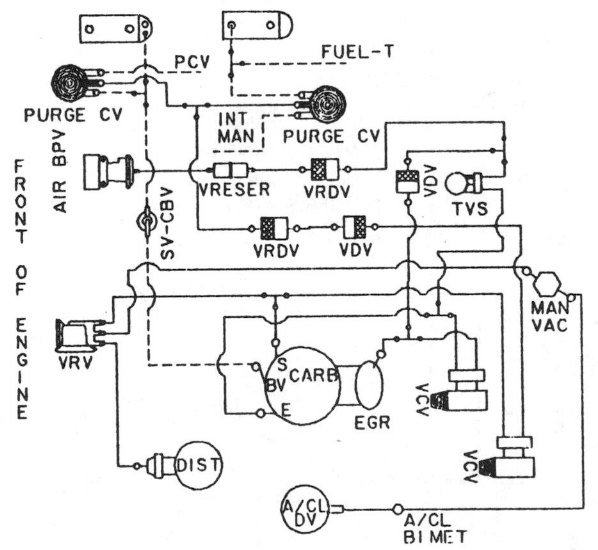

CALIBRATION: 0–54R–RO DATE: 6–1–79
5.0L 4X4 F–150/BRONCO
CALIF A/T

CALIF. ONLY A/C & NON A/C

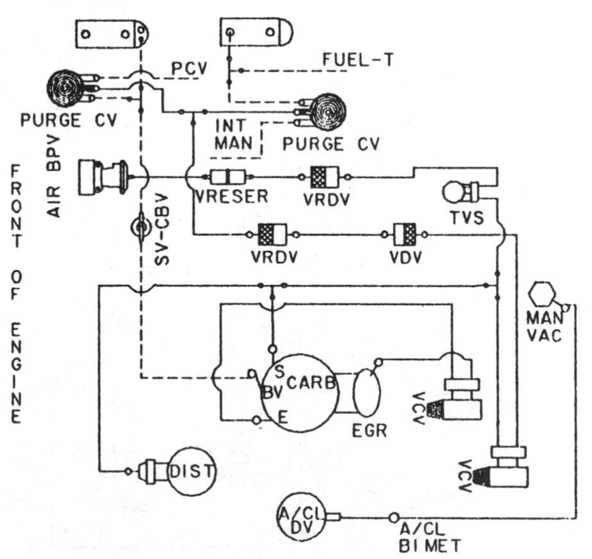

CALIBRATION: 0–54T–RO DATE: 6–1–79
5.0L E–150 CALIF A/T

VACUUM CIRCUITS

(© Ford Motor Co.)

1980—302 CID (5.0 L) engines

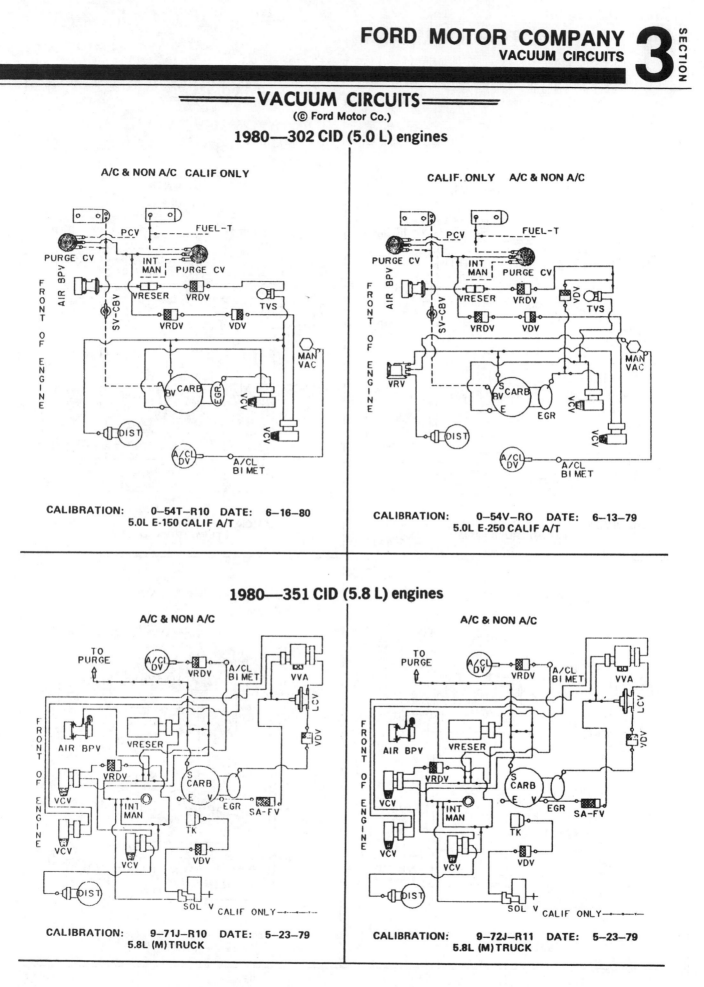

A/C & NON A/C CALIF ONLY

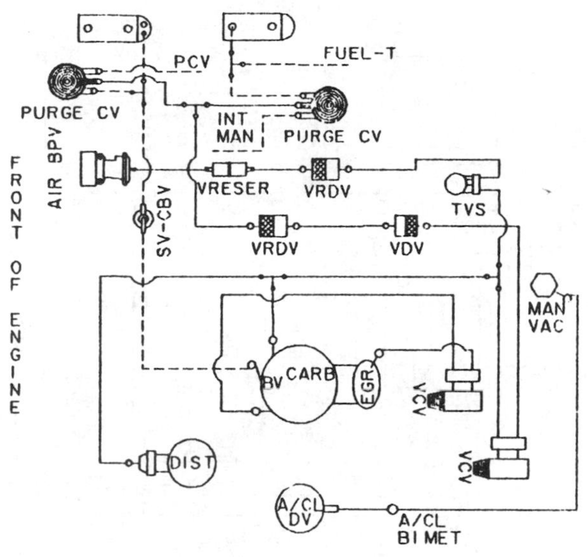

CALIBRATION: 0–54T–R10 DATE: 6–16–80
5.0L E-150 CALIF A/T

CALIF. ONLY A/C & NON A/C

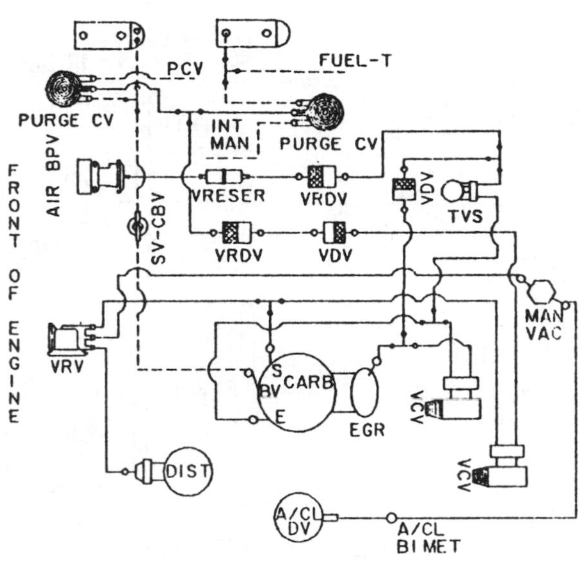

CALIBRATION: 0–54V–RO DATE: 6–13–79
5.0L E-250 CALIF A/T

1980—351 CID (5.8 L) engines

A/C & NON A/C

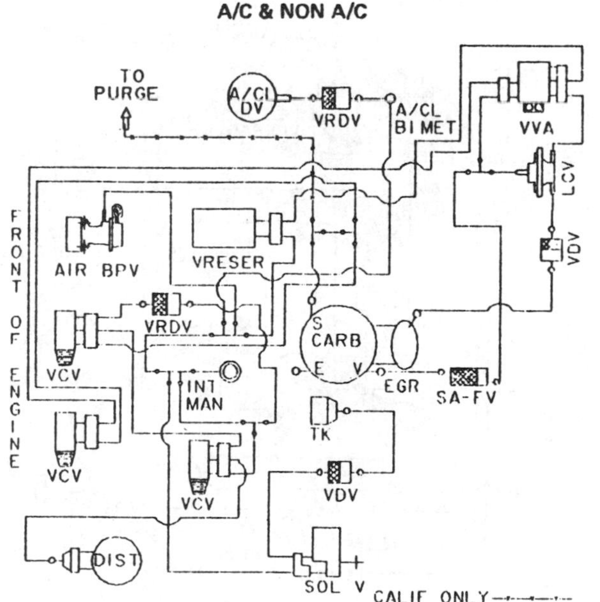

CALIBRATION: 9–71J–R10 DATE: 5–23–79
5.8L (M) TRUCK

A/C & NON A/C

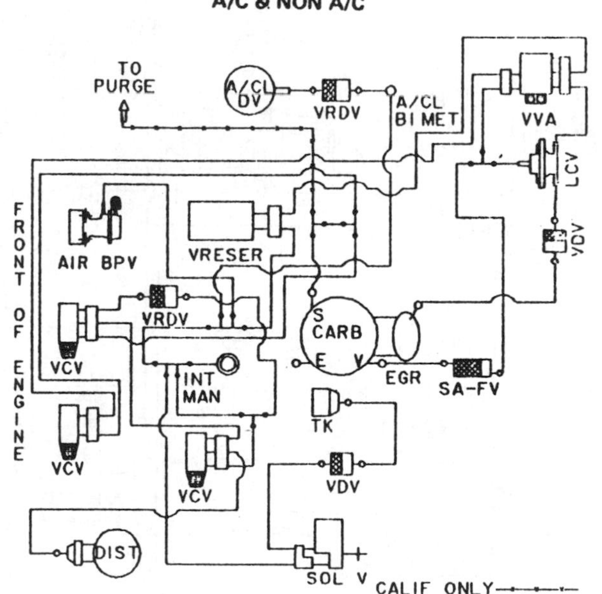

CALIBRATION: 9–72J–R11 DATE: 5–23–79
5.8L (M) TRUCK

VACUUM CIRCUITS

1980—351 CID (5.8 L) engines

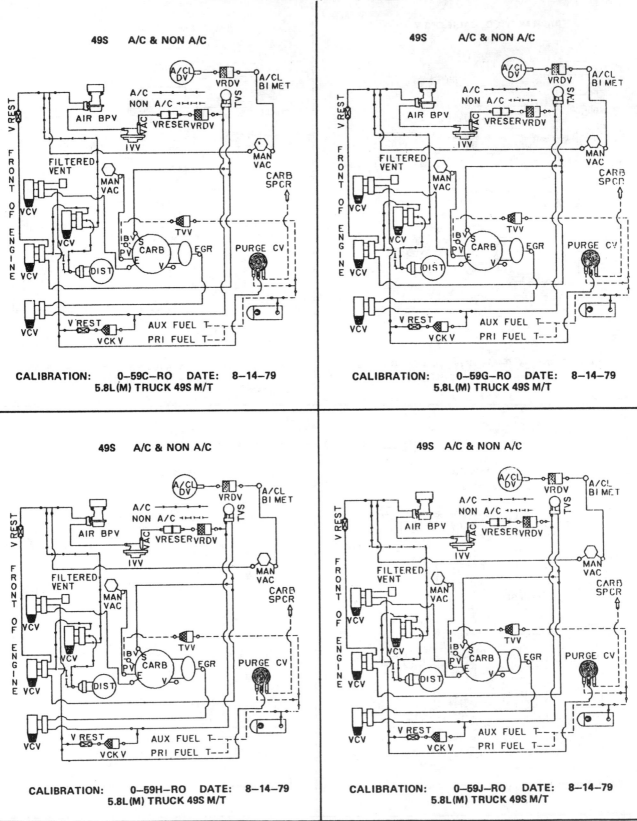

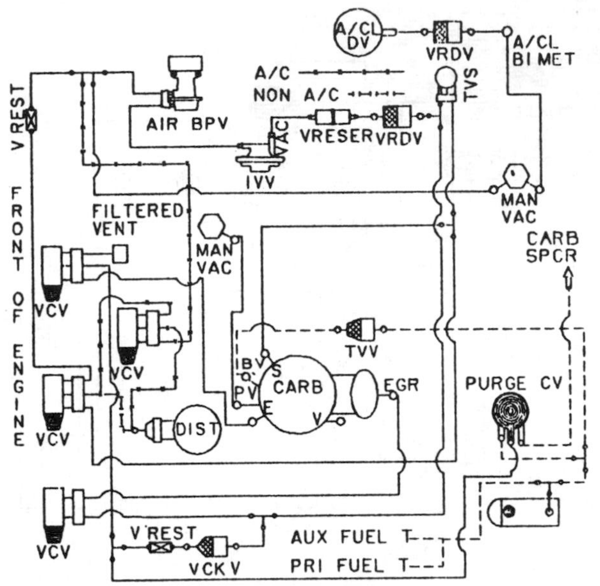

CALIBRATION: 0–59C–RO DATE: 8–14–79
5.8L(M) TRUCK 49S M/T

49S A/C & NON A/C

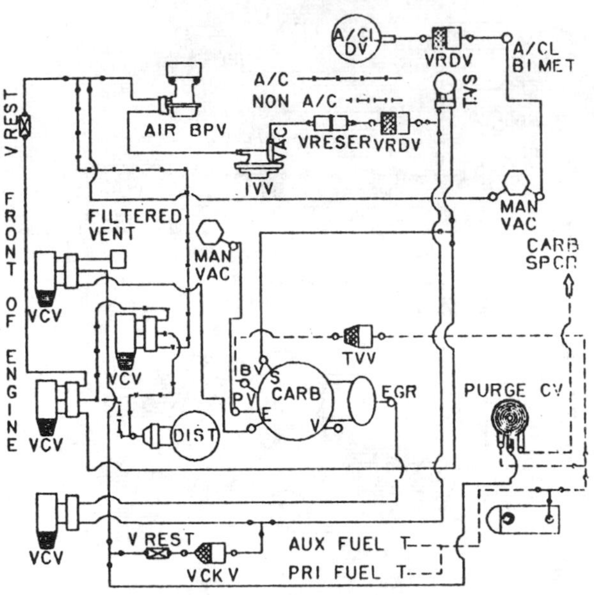

CALIBRATION: 0–59G–RO DATE: 8–14–79
5.8L(M) TRUCK 49S M/T

49S A/C & NON A/C

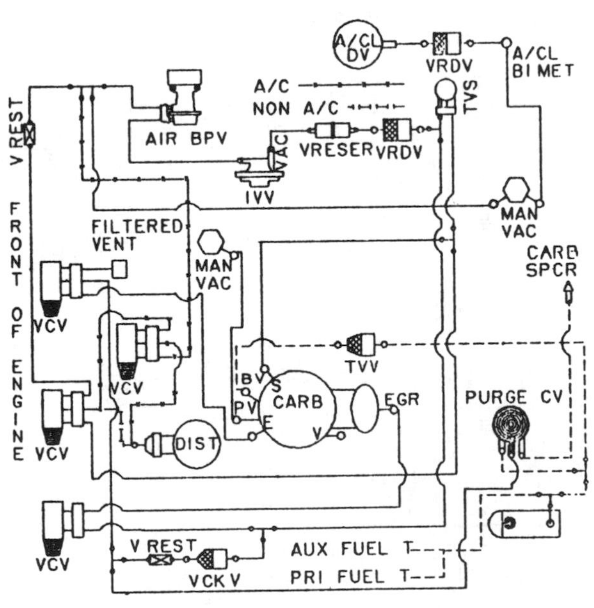

CALIBRATION: 0–59H–RO DATE: 8–14–79
5.8L(M) TRUCK 49S M/T

49S A/C & NON A/C

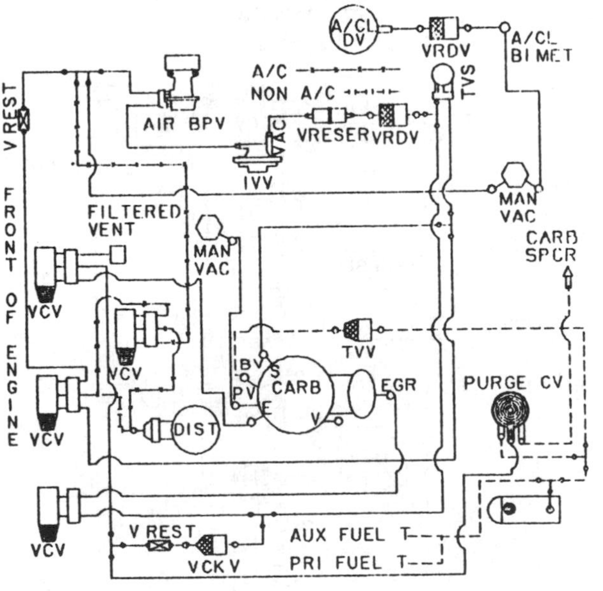

CALIBRATION: 0–59J–RO DATE: 8–14–79
5.8L(M) TRUCK 49S M/T

VACUUM CIRCUITS

1980—351 CID (5.8 L) engines

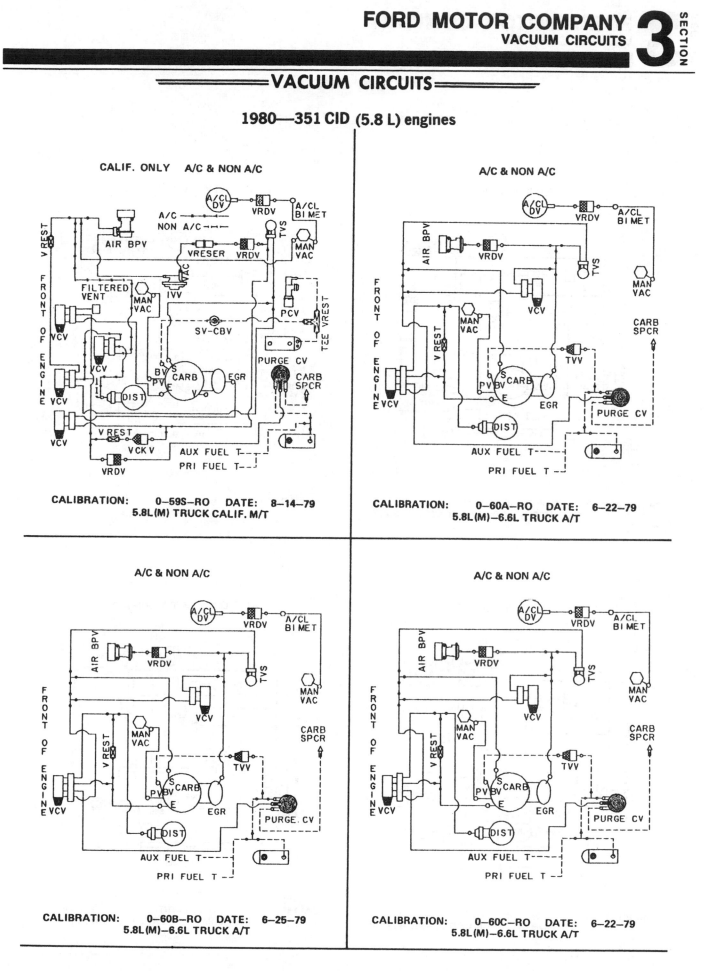

CALIF. ONLY A/C & NON A/C

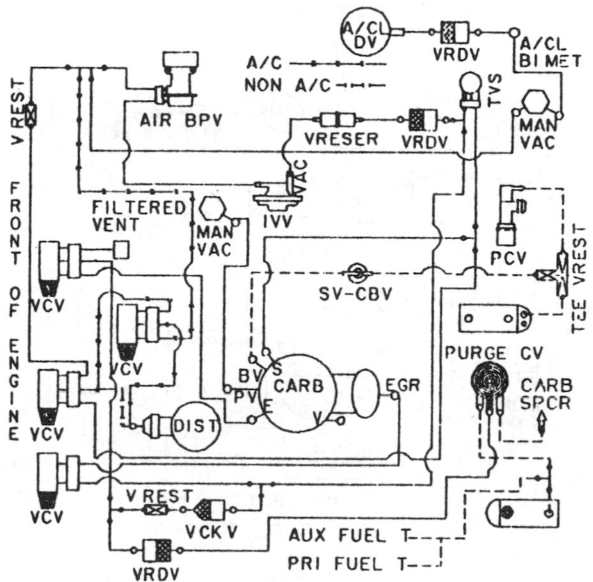

CALIBRATION: 0–59S–RO DATE: 8–14–79
5.8L(M) TRUCK CALIF. M/T

A/C & NON A/C

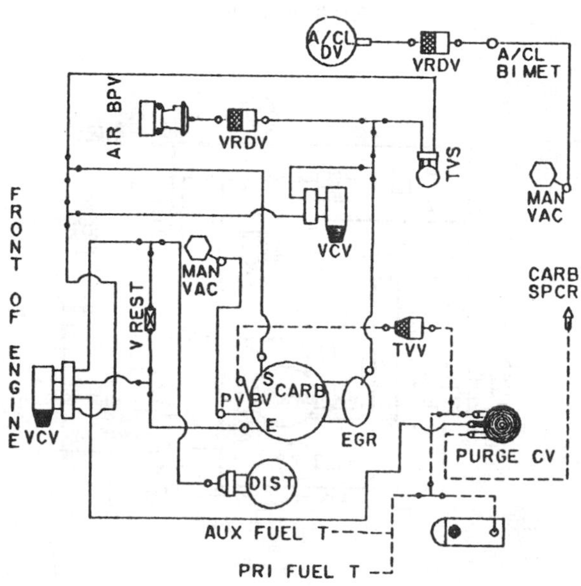

CALIBRATION: 0–60A–RO DATE: 6–22–79
5.8L(M)–6.6L TRUCK A/T

A/C & NON A/C

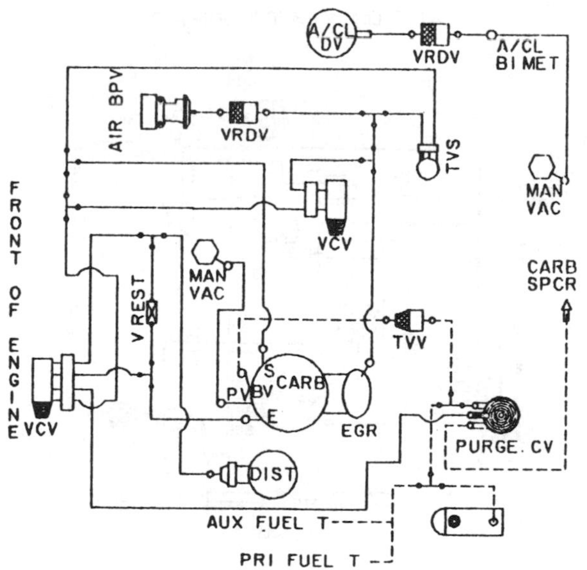

CALIBRATION: 0–60B–RO DATE: 6–25–79
5.8L(M)–6.6L TRUCK A/T

A/C & NON A/C

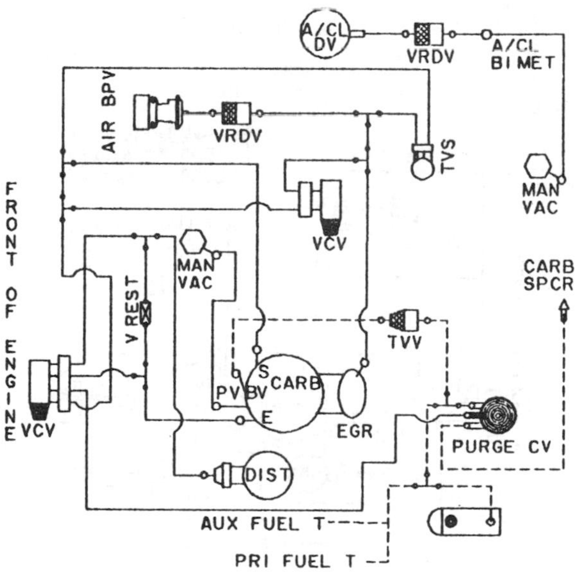

CALIBRATION: 0–60C–RO DATE: 6–22–79
5.8L(M)–6.6L TRUCK A/T

VACUUM CIRCUITS

1980—351 CID (5.8 L) engines

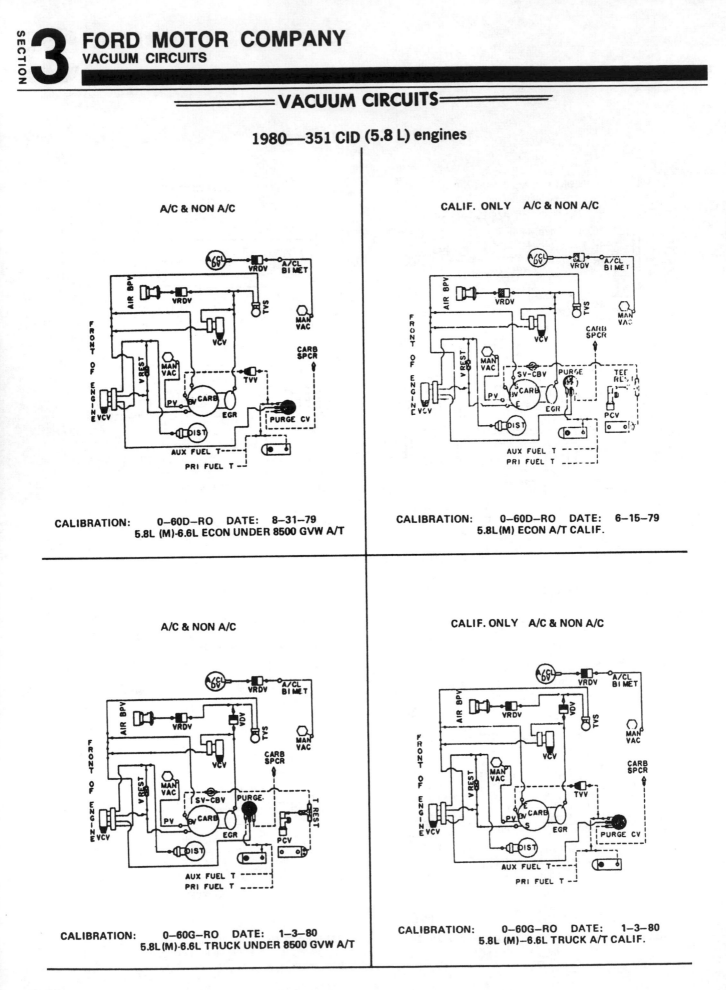

A/C & NON A/C

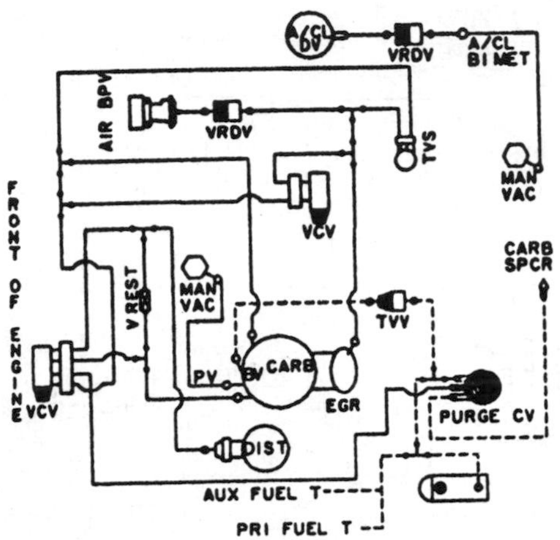

CALIBRATION: 0–60D–RO DATE: 8–31–79
5.8L (M)-6.6L ECON UNDER 8500 GVW A/T

CALIF. ONLY A/C & NON A/C

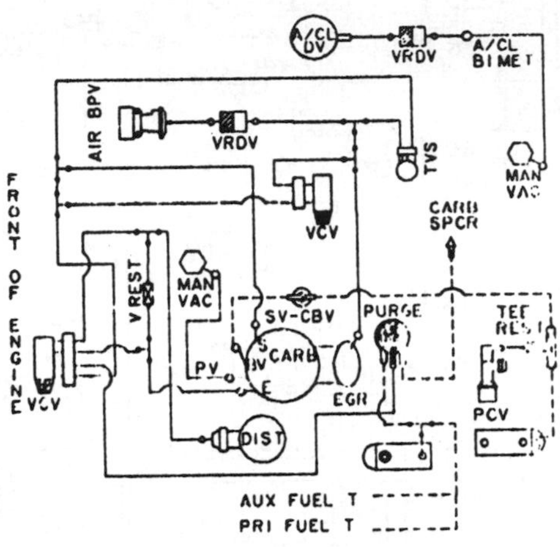

CALIBRATION: 0–60D–RO DATE: 6–15–79
5.8L(M) ECON A/T CALIF.

A/C & NON A/C

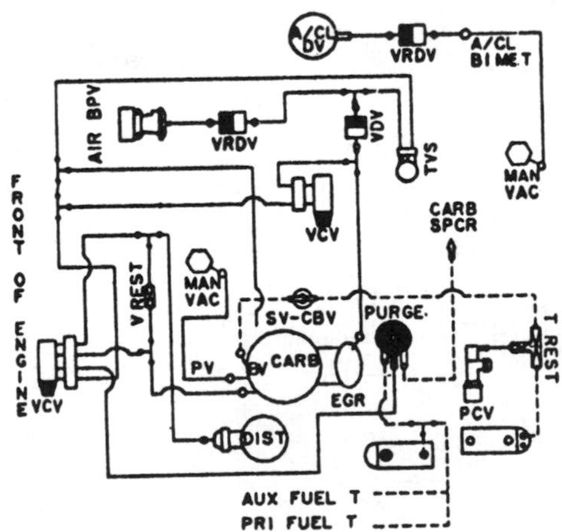

CALIBRATION: 0–60G–RO DATE: 1–3–80
5.8L(M)-6.6L TRUCK UNDER 8500 GVW A/T

CALIF. ONLY A/C & NON A/C

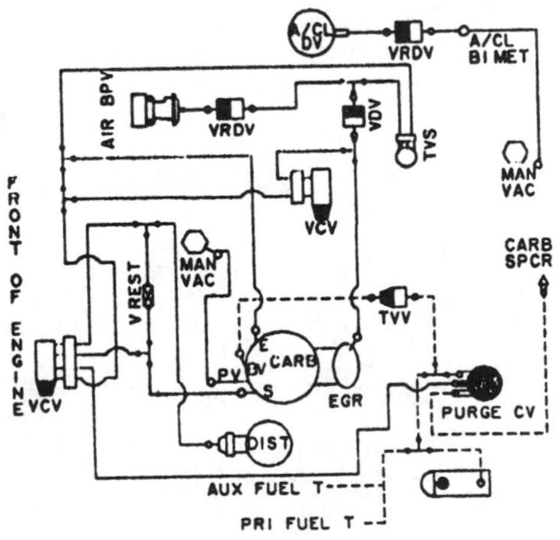

CALIBRATION: 0–60G–RO DATE: 1–3–80
5.8L (M)–6.6L TRUCK A/T CALIF.

VACUUM CIRCUITS

1980—351 CID (5.8 L) engines

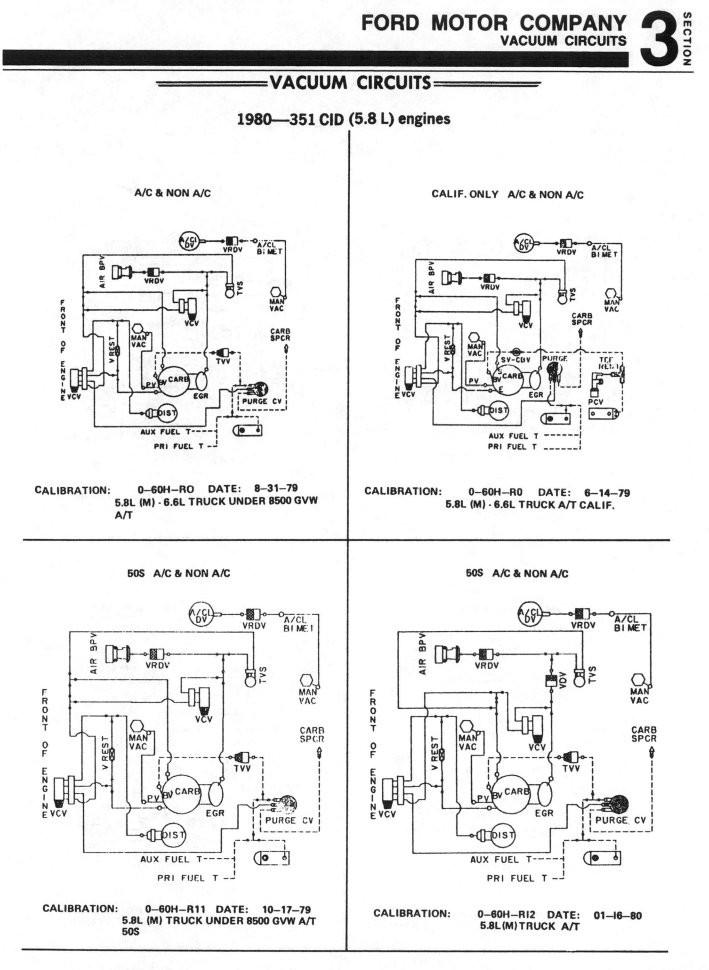

A/C & NON A/C

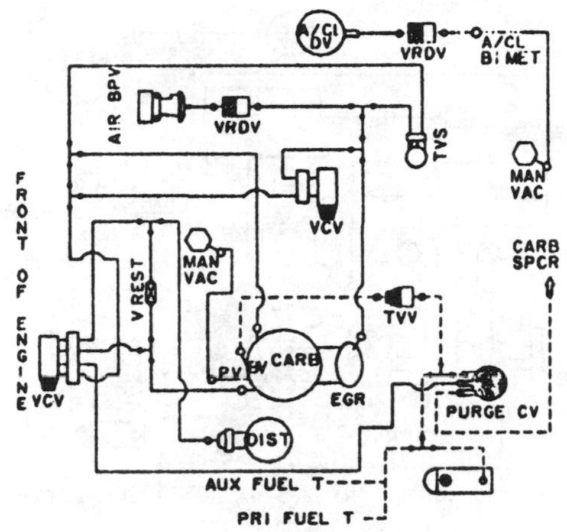

CALIBRATION: 0–60H–RO DATE: 8–31–79
5.8L (M) - 6.6L TRUCK UNDER 8500 GVW A/T

CALIF. ONLY A/C & NON A/C

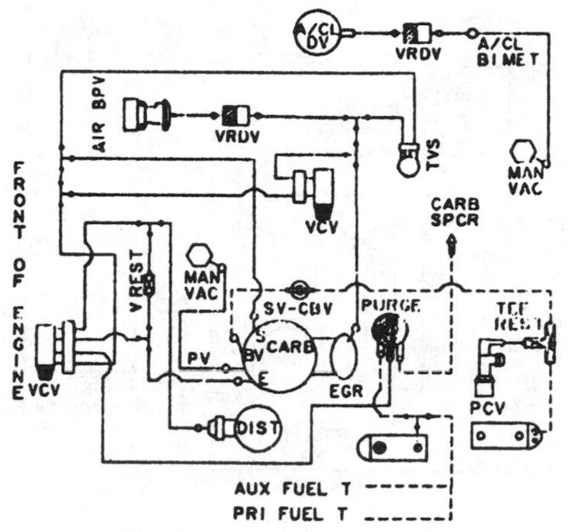

CALIBRATION: 0–60H–R0 DATE: 6–14–79
5.8L (M) - 6.6L TRUCK A/T CALIF.

50S A/C & NON A/C

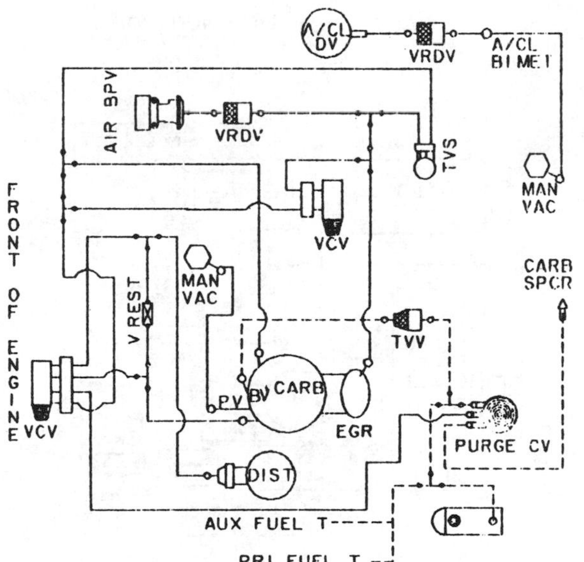

CALIBRATION: 0–60H–R11 DATE: 10–17–79
5.8L (M) TRUCK UNDER 8500 GVW A/T 50S

50S A/C & NON A/C

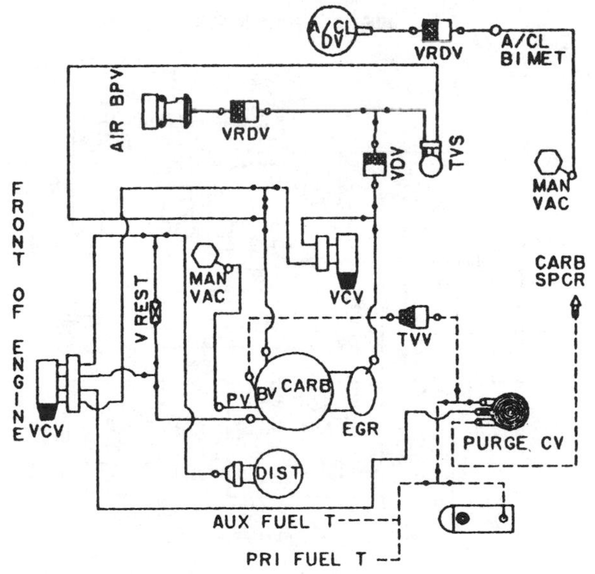

CALIBRATION: 0–60H–RI2 DATE: 01–16–80
5.8L(M) TRUCK A/T

VACUUM CIRCUITS

1980—351 CID (5.8 L) engines

49S A/C & NON A/C

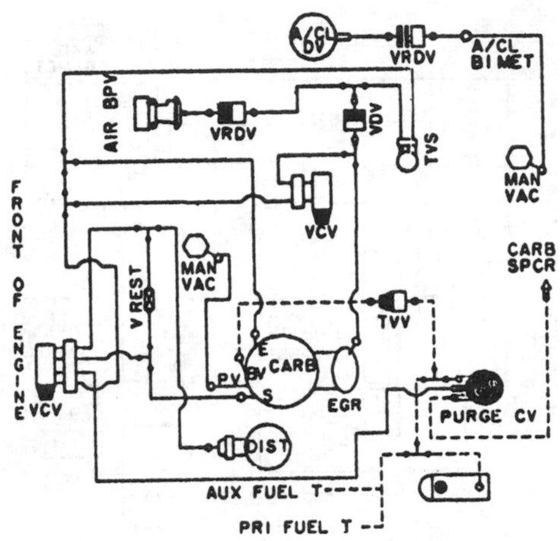

CALIBRATION: 0–60J–RO DATE: 1–03–80
5.8L (M)–6.6L TRUCK A/T 49S

CALIF. ONLY A/C & NON A/C

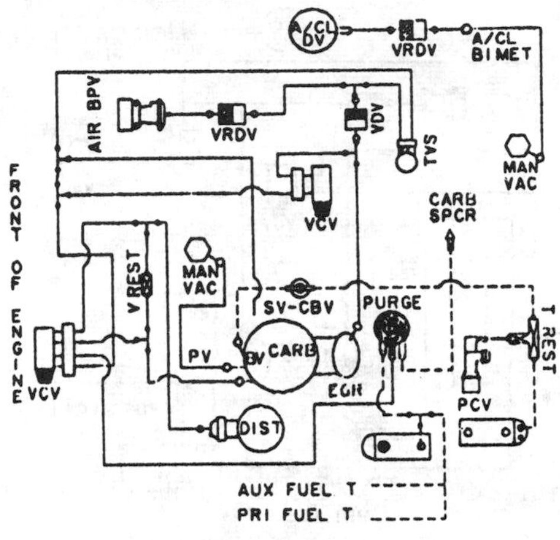

CALIBRATION: 0–60J–RO DATE: 1–03–80
5.8L (M)-6.6L TRUCK UNDER 8500 GVW A/T

49S A/C & NON A/C

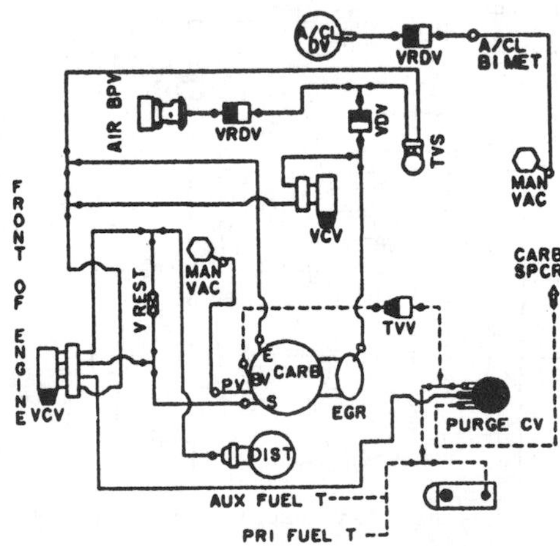

CALIBRATION: 0–60K–RO DATE: 1–3–80
5.8L (M)-6.6L TRUCK UNDER 8500 GVW A/T

CALIF. ONLY A/C & NON A/C

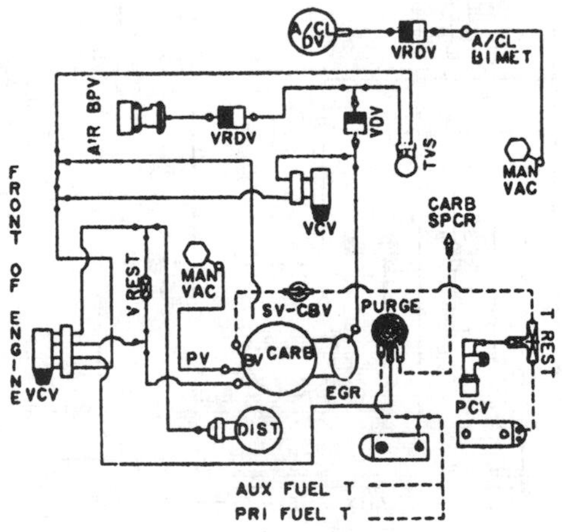

CALIBRATION: 0–60K–RO DATE: 1–3–80
5.8L (M)–6.6L TRUCK A/T CALIF.

VACUUM CIRCUITS

1980—351 CID (5.8 L) engines

50S A/C & NON A/C

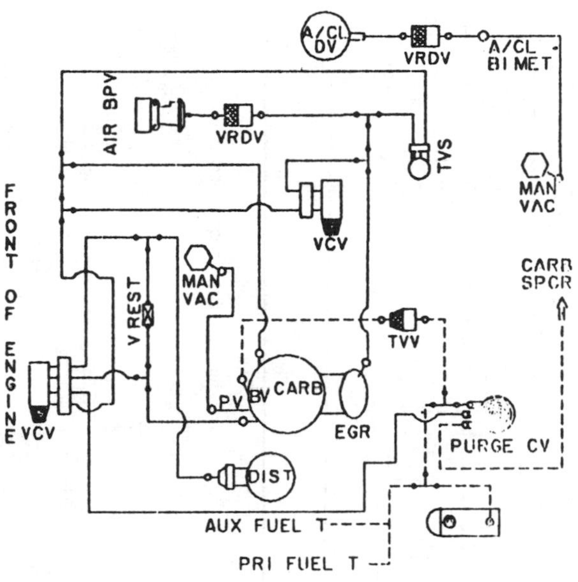

CALIBRATION: 0–60K–R11 DATE: 10–18–79
5.8L (M) TRUCK UNDER 8500 GVW A/T
50S

A/C & NON A/C

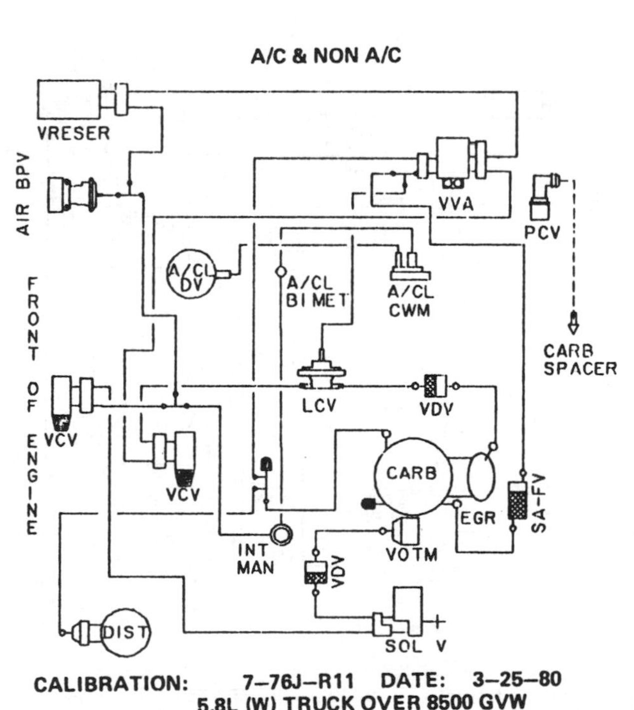

CALIBRATION: 7–76J–R11 DATE: 3–25–80
5.8L (W) TRUCK OVER 8500 GVW
49S A/T

49S A/C & NON A/C

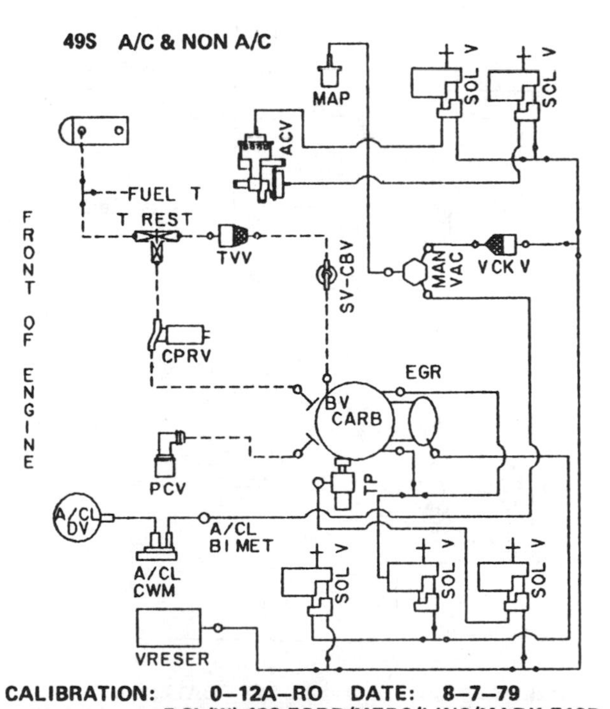

CALIBRATION: 0–12A–RO DATE: 8–7–79
5.8L(W) 49S FORD/MERC/LINC/MARK F10D

VACUUM CIRCUITS

1980—351 CID (5.8 L) engines

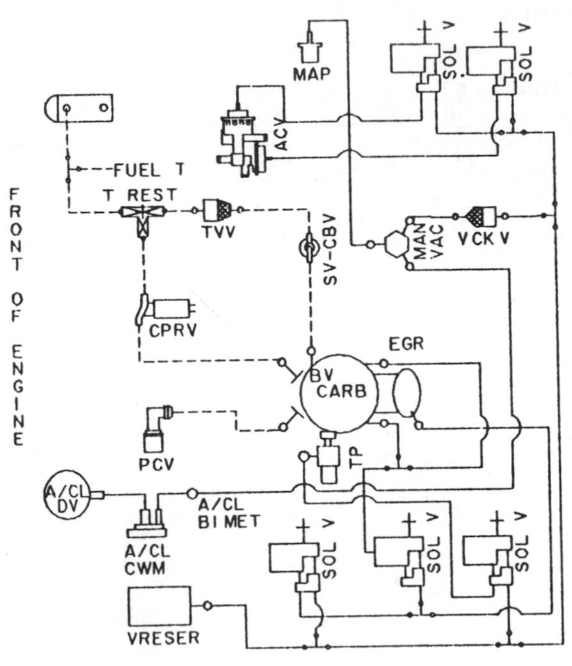

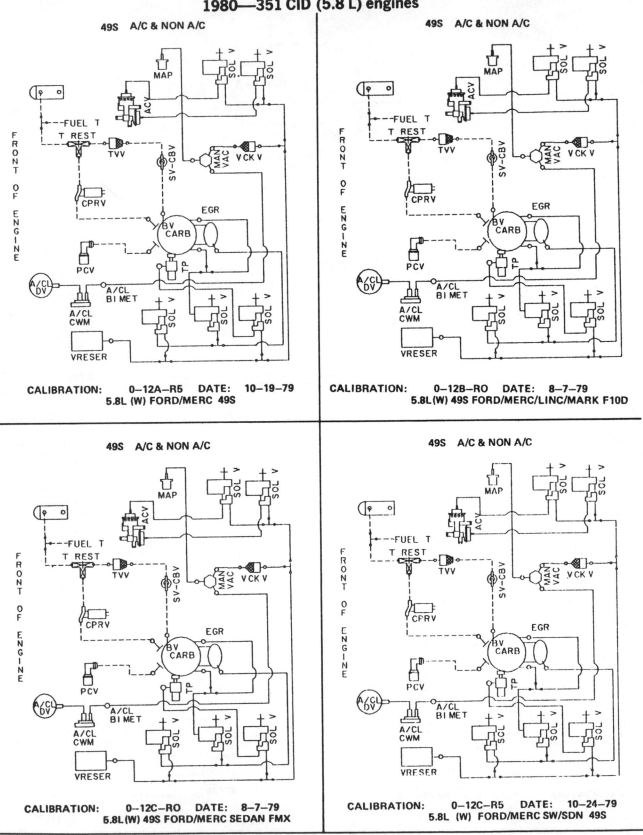

CALIBRATION: 0–12A–R5 **DATE:** 10–19–79
5.8L (W) FORD/MERC 49S

49S A/C & NON A/C

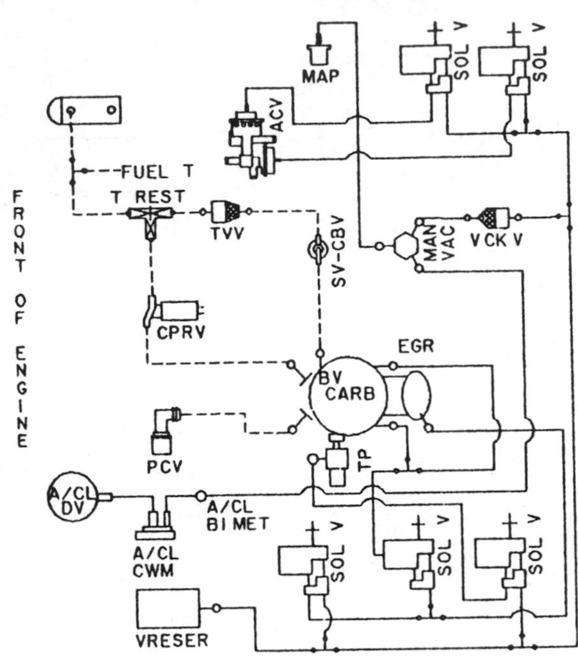

CALIBRATION: 0–12B–RO **DATE:** 8–7–79
5.8L(W) 49S FORD/MERC/LINC/MARK F10D

49S A/C & NON A/C

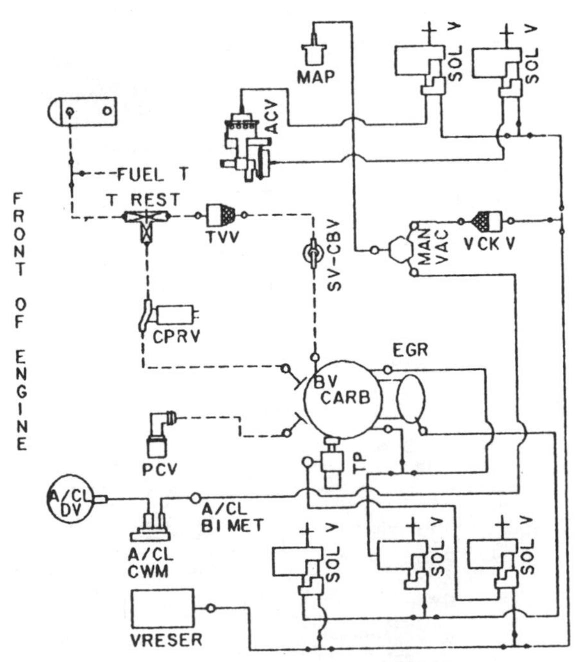

CALIBRATION: 0–12C–RO **DATE:** 8–7–79
5.8L (W) 49S FORD/MERC SEDAN FMX

49S A/C & NON A/C

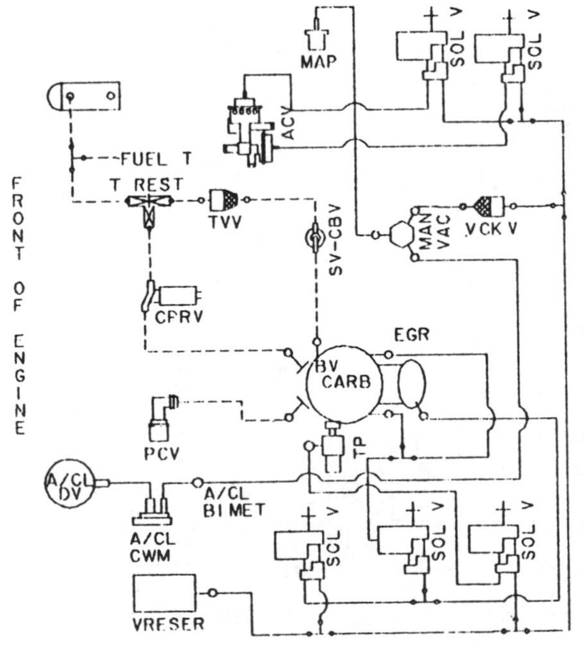

CALIBRATION: 0–12C–R5 **DATE:** 10–24–79
5.8L (W) FORD/MERC SW/SDN 49S

VACUUM CIRCUITS

1980—351 CID (5.8 L) engines

CANADA A/C & NON A/C

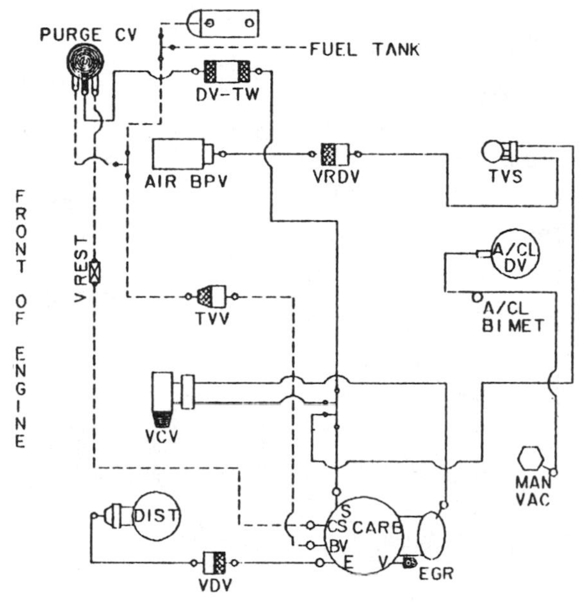

CALIBRATION: 0–12G–R0 DATE: 08–23–79
5.8L (W) LINCOLN/MARK CANADA

CANADA A/C & NON A/C

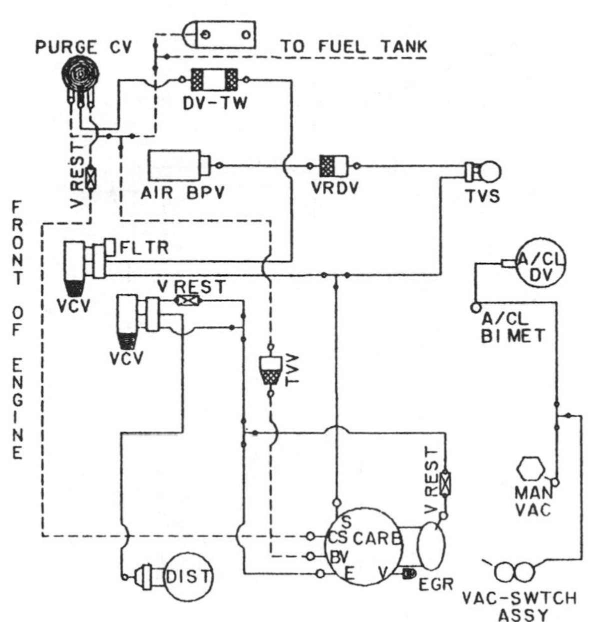

CALIBRATION: 0–12H–ROO DATE: 08–29–79
5.8L (W) FORD/FORD SWGN/MERC/
MERC SWGN

A/C & NON A/C

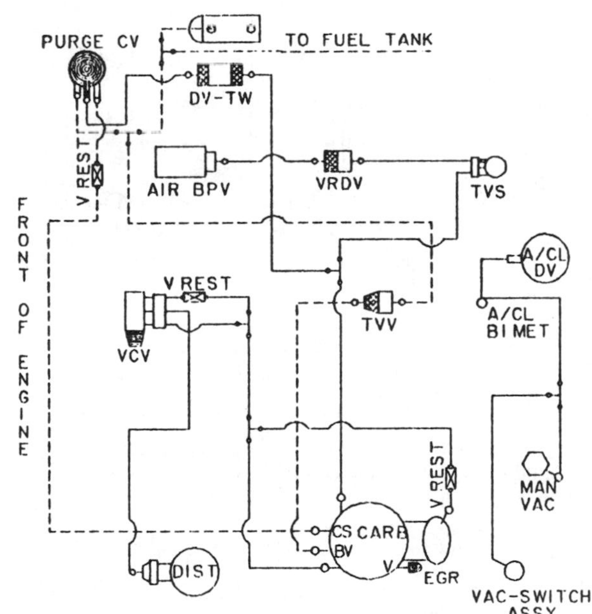

CALIBRATION: 0–12H–R10 DATE: 10–09–79
5.8L (W) FORD/FORD S.W./MERC/MERC
S.W.

A/C & NON A/C

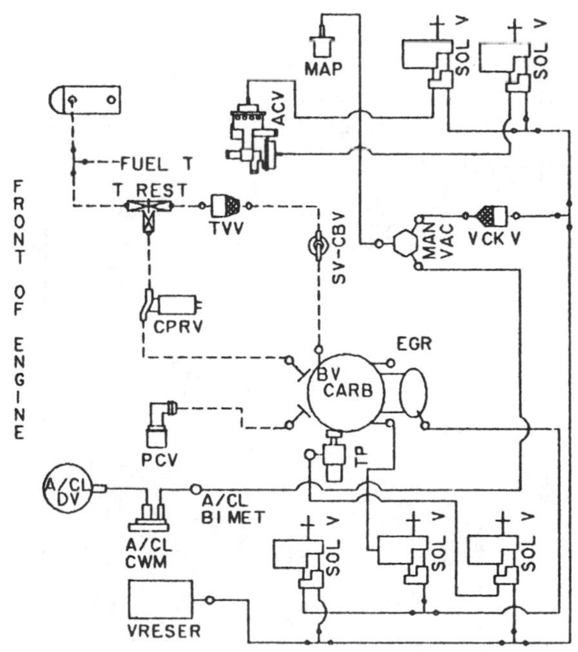

CALIBRATION: 0–12I–RIO DATE: 01–17–80
5.8L FORD/MERC H.O.

VACUUM CIRCUITS

1980—351 CID (5.8 L) engines

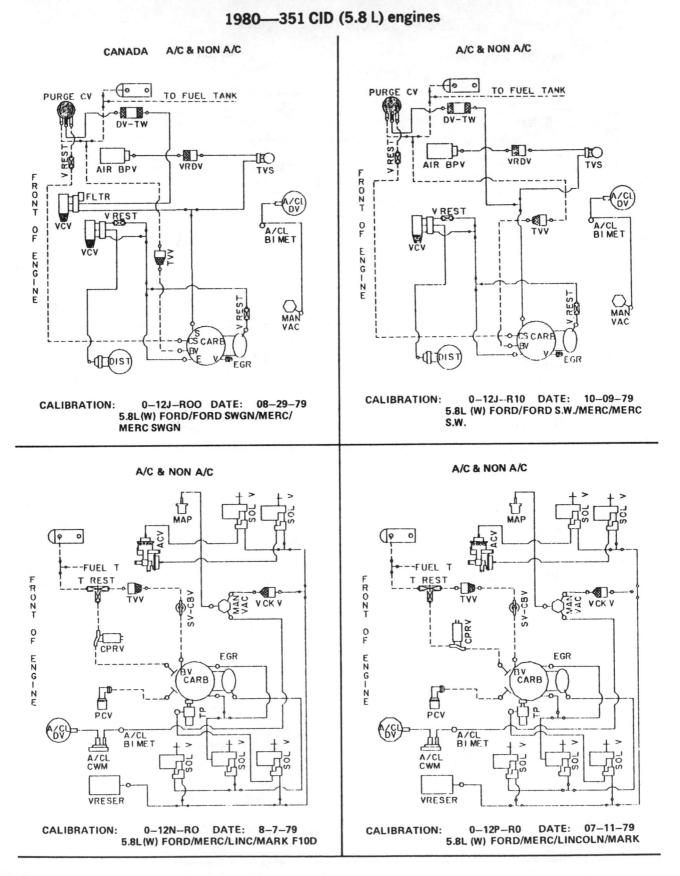

CANADA A/C & NON A/C

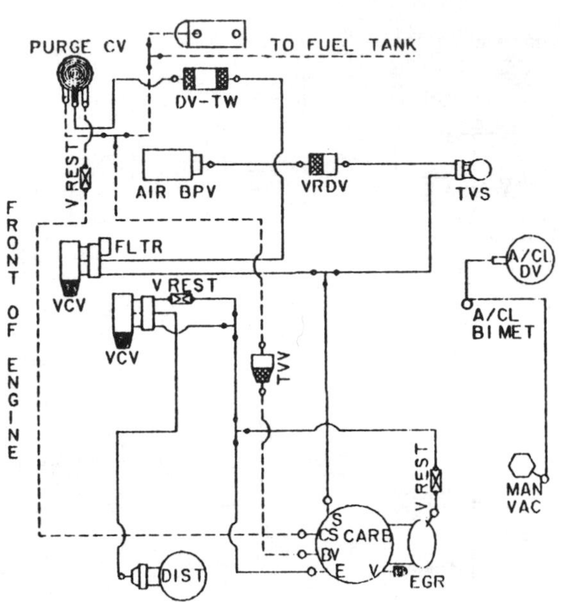

CALIBRATION: 0–12J–ROO DATE: 08–29–79
5.8L (W) FORD/FORD SWGN/MERC/
MERC SWGN

A/C & NON A/C

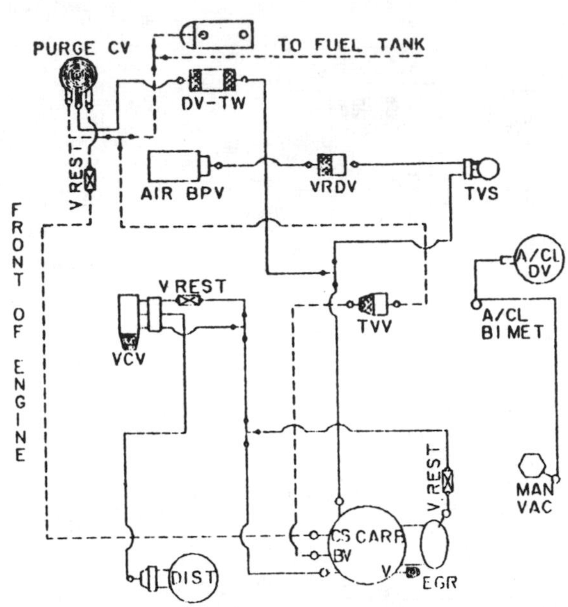

CALIBRATION: 0–12J–R10 DATE: 10–09–79
5.8L (W) FORD/FORD S.W./MERC/MERC
S.W.

A/C & NON A/C

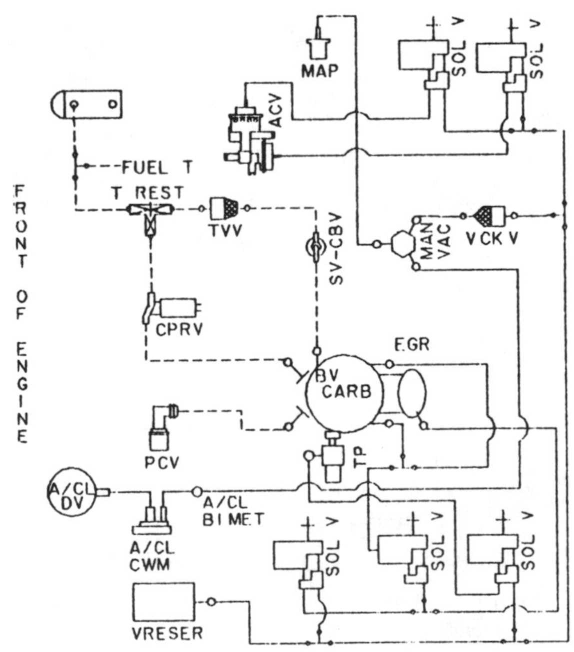

CALIBRATION: 0–12N–RO DATE: 8–7–79
5.8L (W) FORD/MERC/LINC/MARK F10D

A/C & NON A/C

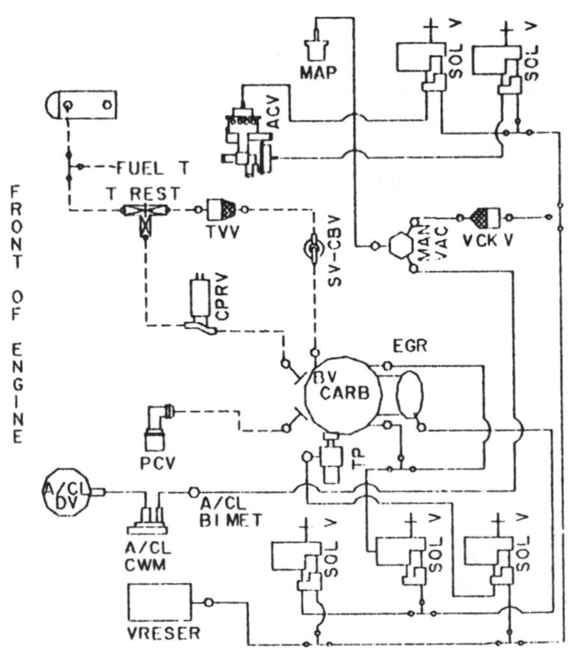

CALIBRATION: 0–12P–RO DATE: 07–11–79
5.8L (W) FORD/MERC/LINCOLN/MARK

VACUUM CIRCUITS

1980—351 CID (5.8 L) engines

49S A/C & NON A/C

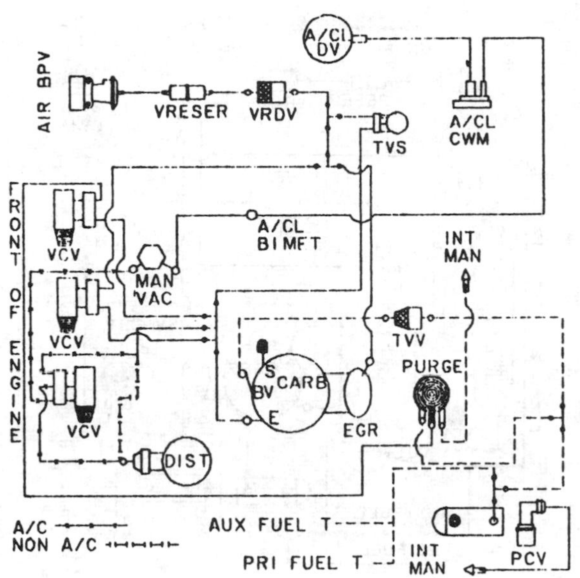

CALIBRATION: 0–64A–RO DATE: 6–22–79
5.8L (W) 49S ECON A/T

49S A/C & NON A/C

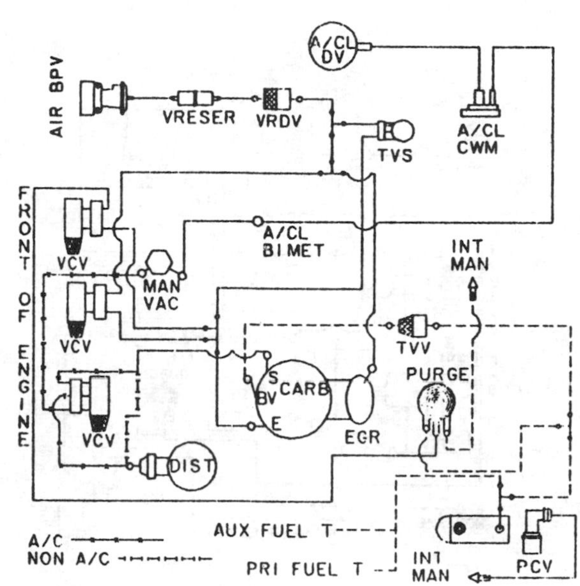

CALIBRATION: 0–64B–RO DATE: 6–21–79
5.8L (W) 49S ECON A/T

49S

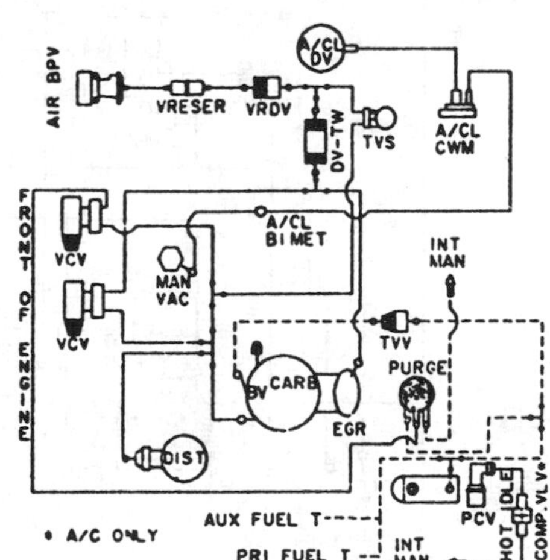

CALIBRATION: 0–64G–RO DATE: 6–5–79
5.8L (W) 49S ECON A/T

CALIF. ONLY A/C & NON A/C

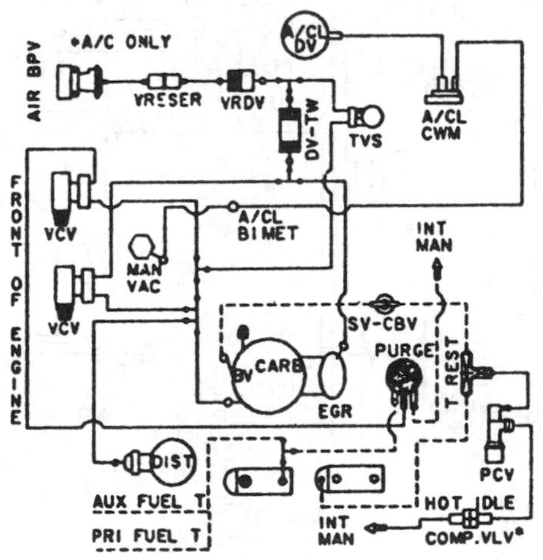

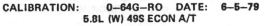

CALIBRATION: 0–64G–RO DATE: 6–5–79
5.8L (W) CALIF ECON A/T

VACUUM CIRCUITS

1980—351 CID (5.8 L) engines

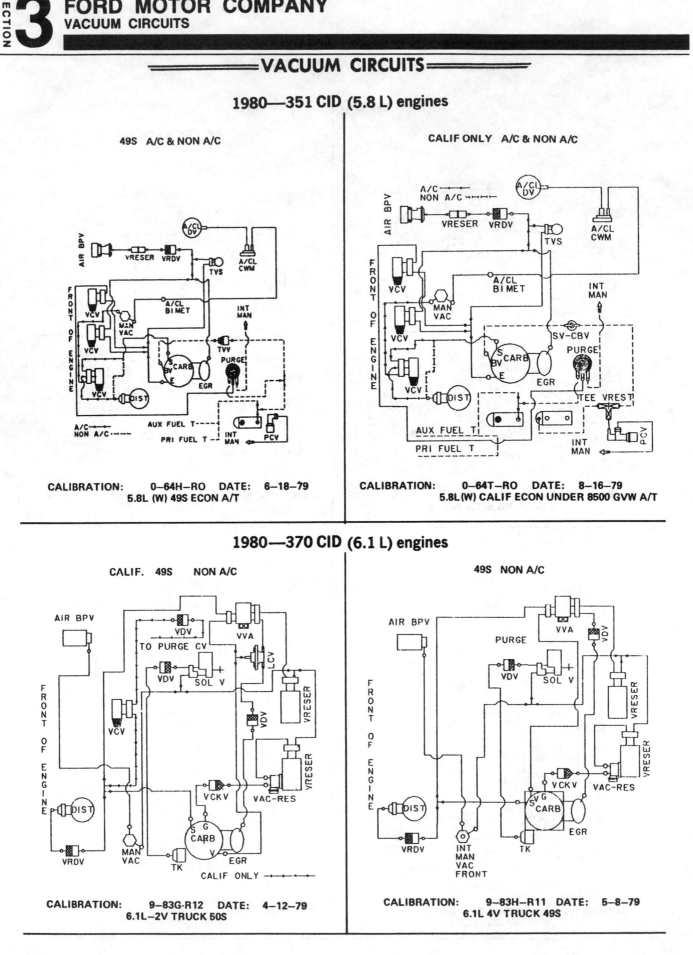

49S A/C & NON A/C

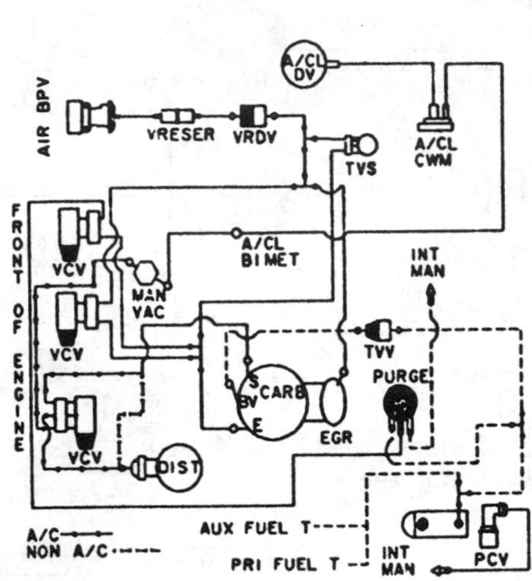

CALIBRATION: 0–64H–RO DATE: 6–18–79
5.8L (W) 49S ECON A/T

CALIF ONLY A/C & NON A/C

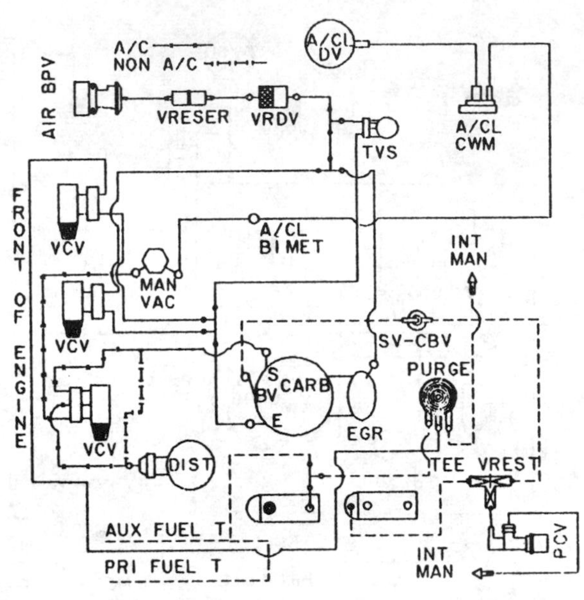

CALIBRATION: 0–64T–RO DATE: 8–16–79
5.8L (W) CALIF ECON UNDER 8500 GVW A/T

1980—370 CID (6.1 L) engines

CALIF. 49S NON A/C

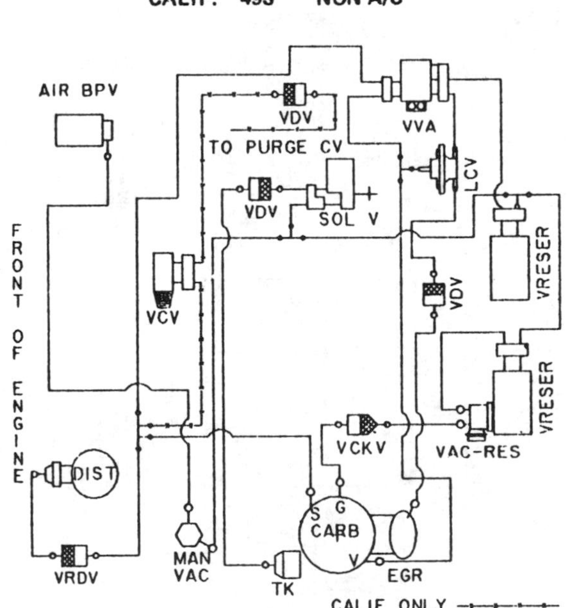

CALIBRATION: 9–83G-R12 DATE: 4–12–79
6.1L–2V TRUCK 50S

49S NON A/C

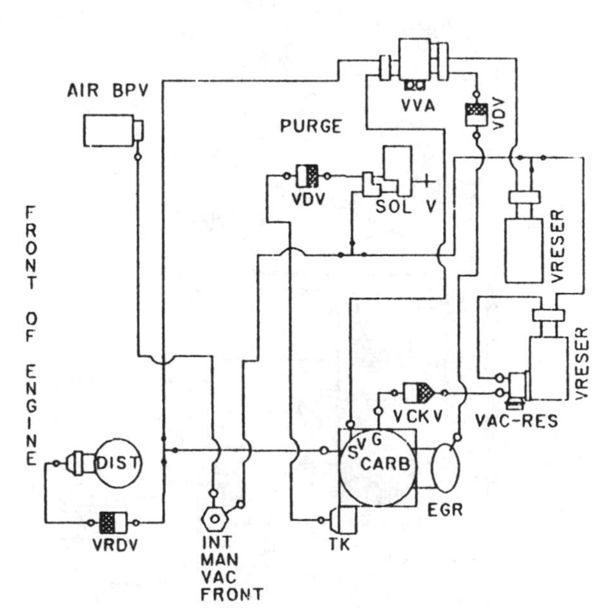

CALIBRATION: 9–83H–R11 DATE: 5–8–79
6.1L 4V TRUCK 49S

VACUUM CIRCUITS

1980—370 CID (6.1 L) engines

CALIF. ONLY NON A/C

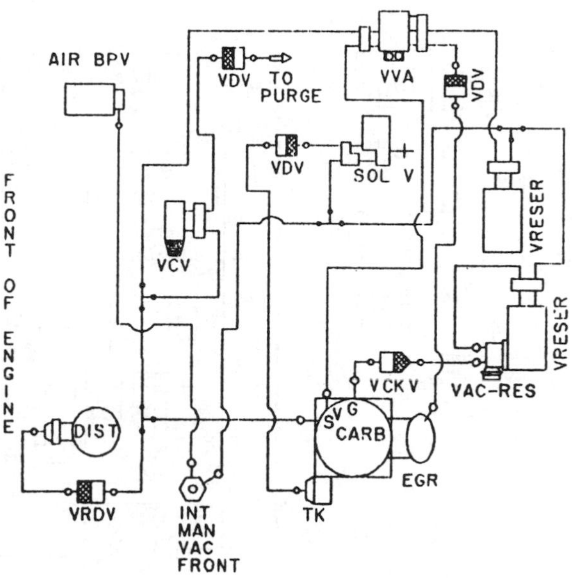

CALIBRATION: 9–83H–R14 DATE: 5–14–79
6.1L 4V TRUCK CALIF.

1980—400 CID (6.6 L) engines

49S A/C & NON A/C

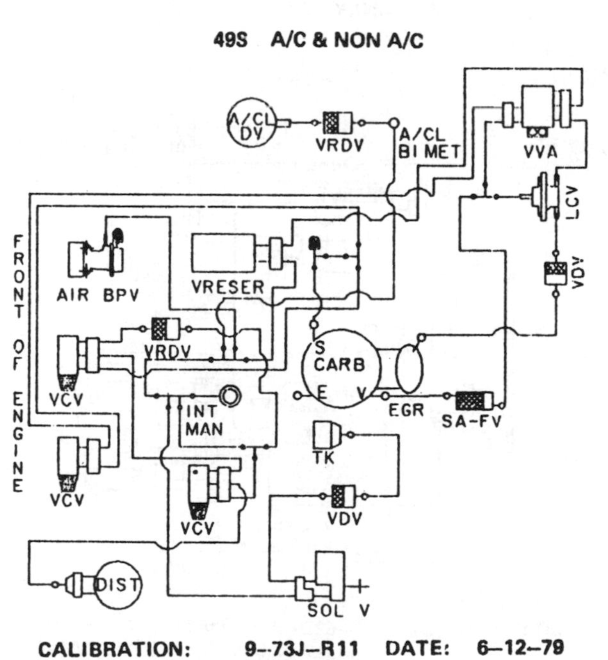

CALIBRATION: 9–73J–R11 DATE: 6–12–79
6.6L TRUCK 49S

CALIF. ONLY A/C & NON A/C

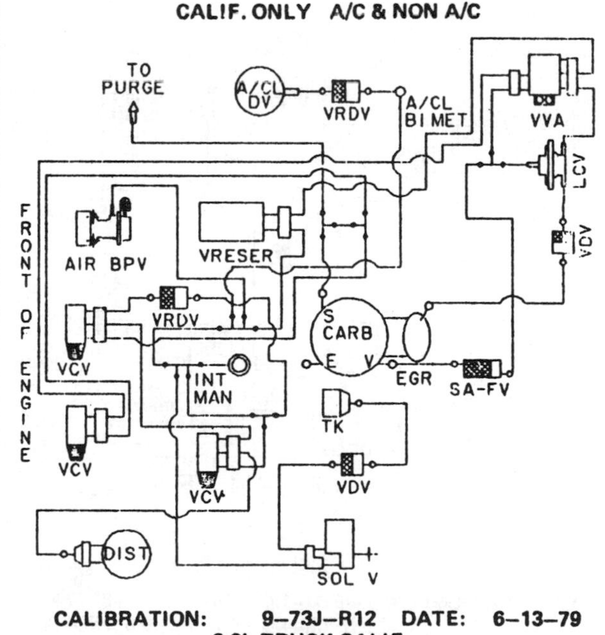

CALIBRATION: 9–73J–R12 DATE: 6–13–79
6.6L TRUCK CALIF.

VACUUM CIRCUITS

1980—400 CID (6.6 L) engines

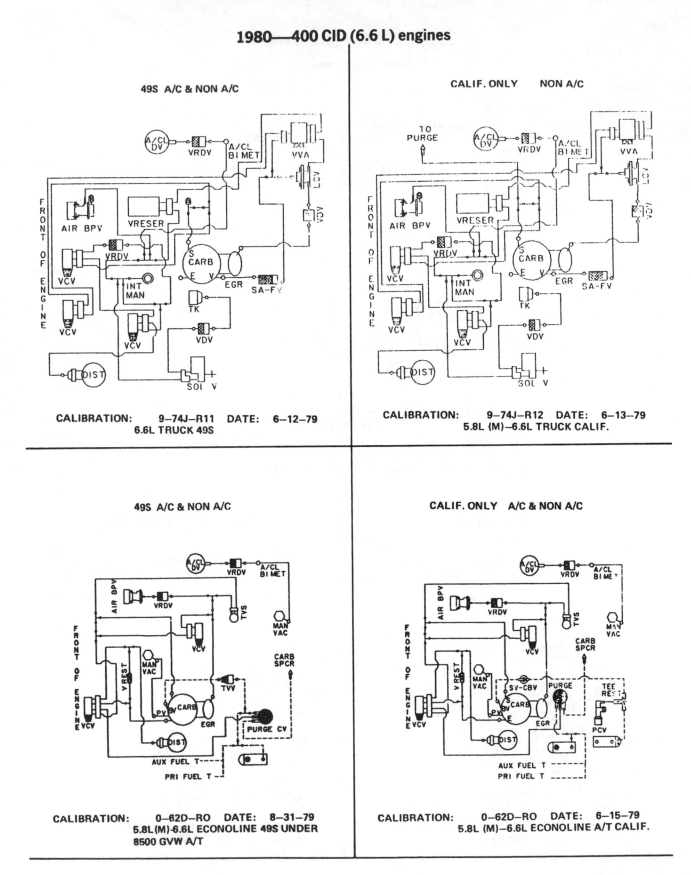

49S A/C & NON A/C

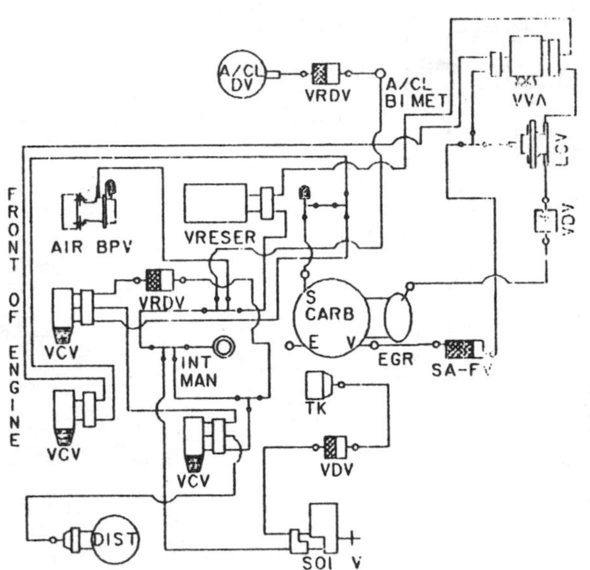

CALIBRATION: 9–74J–R11 DATE: 6–12–79
6.6L TRUCK 49S

CALIF. ONLY NON A/C

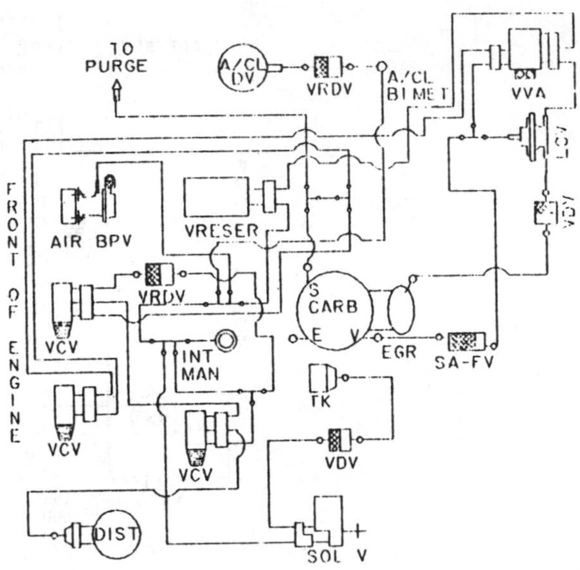

CALIBRATION: 9–74J–R12 DATE: 6–13–79
5.8L (M)–6.6L TRUCK CALIF.

49S A/C & NON A/C

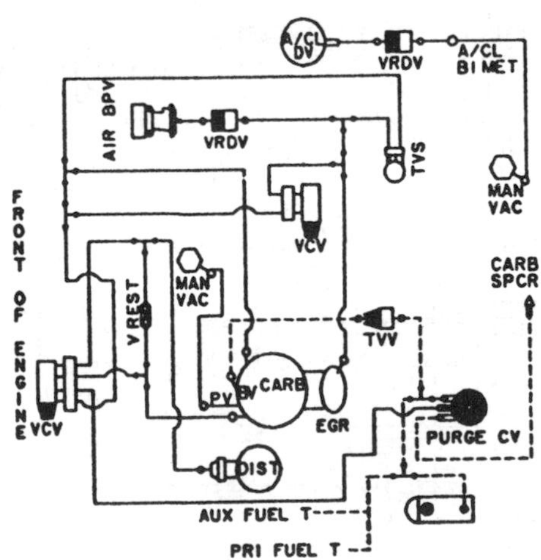

CALIBRATION: 0–62D–RO DATE: 8–31–79
5.8L (M)-6.6L ECONOLINE 49S UNDER
8500 GVW A/T

CALIF. ONLY A/C & NON A/C

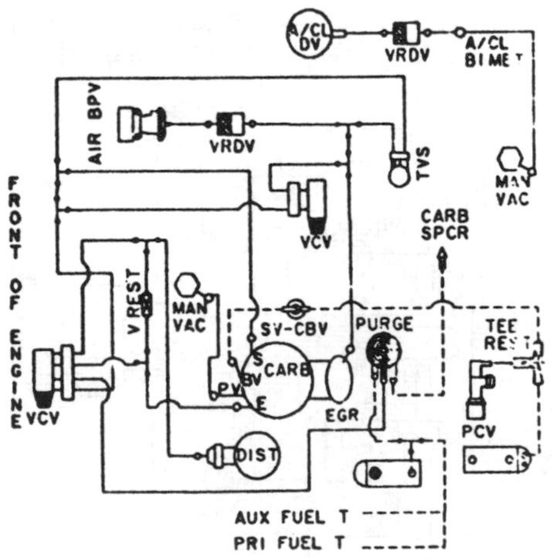

CALIBRATION: 0–62D–RO DATE: 6–15–79
5.8L (M)–6.6L ECONOLINE A/T CALIF.

VACUUM CIRCUITS

1980—400 CID (6.6 L) engines

49S A/C & NON A/C

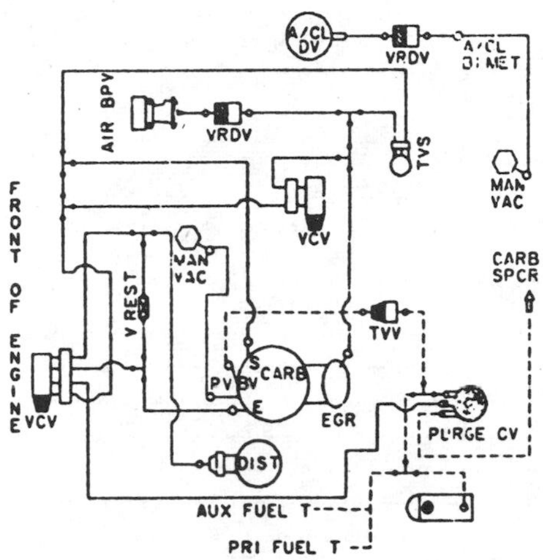

CALIBRATION: 0–62L–RO DATE: 6–14–79
5.8L (M)–6.6L TRUCK A/T 49S

CALIF. ONLY A/C & NON A/C

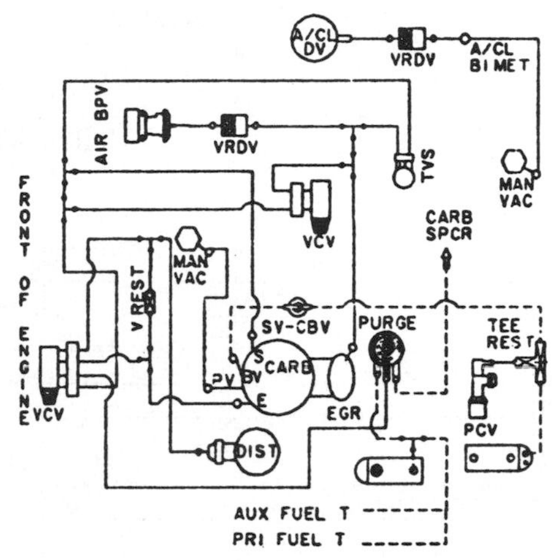

CALIBRATION: 0–62L–RO DATE: 6–14–79
5.8L (M)–6.6L TRUCK A/T CALIF.

VACUUM CIRCUITS

1980—429 CID (7.0 L) engines

50S A/C & NON A/C

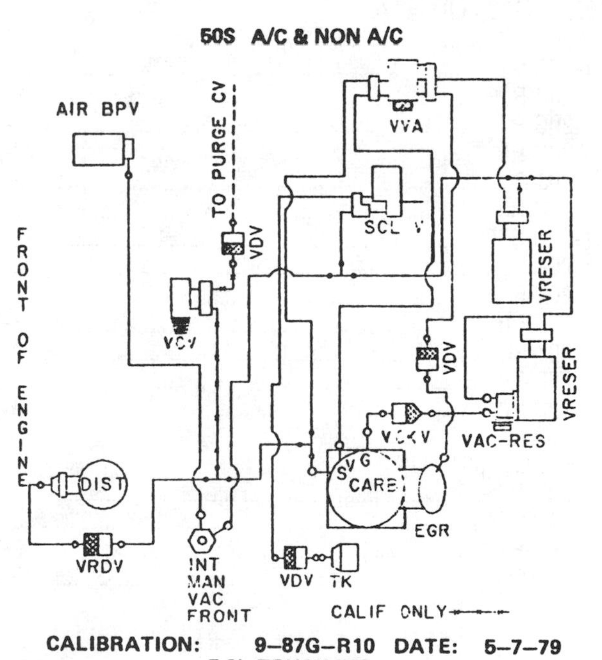

CALIBRATION: 9–87G–R10 DATE: 5–7–79
7.0L TRUCK 50S

50S NON A/C

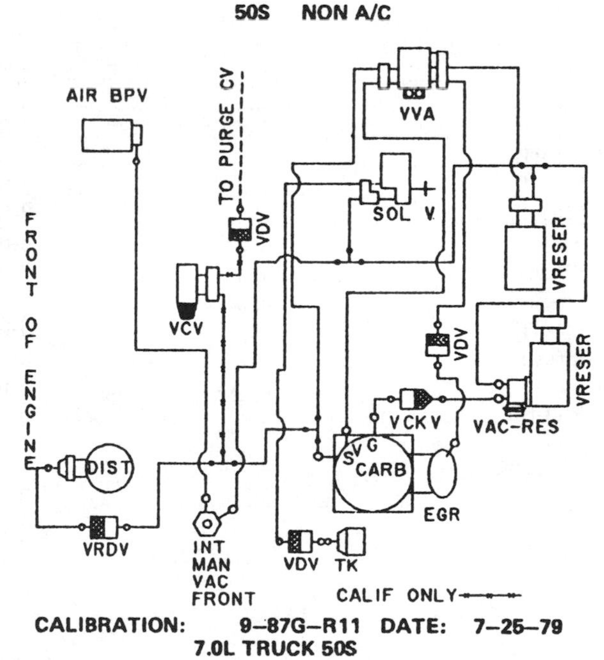

CALIBRATION: 9–87G–R11 DATE: 7–25–79
7.0L TRUCK 50S

Ford
Vacuum Circuits

INDEX

1981 PASSENGER CAR
Engine-Car Emission Calibrations

VACUUM CIRCUITS

(© Ford Motor Co.)

1981—1.6L engines

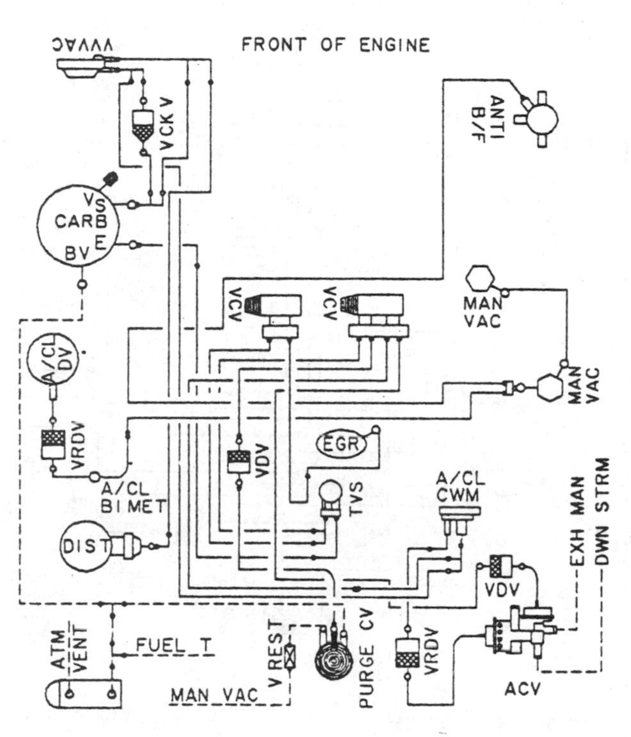

CALIBRATION: 1—3A—R1 DATE: 6-25-80

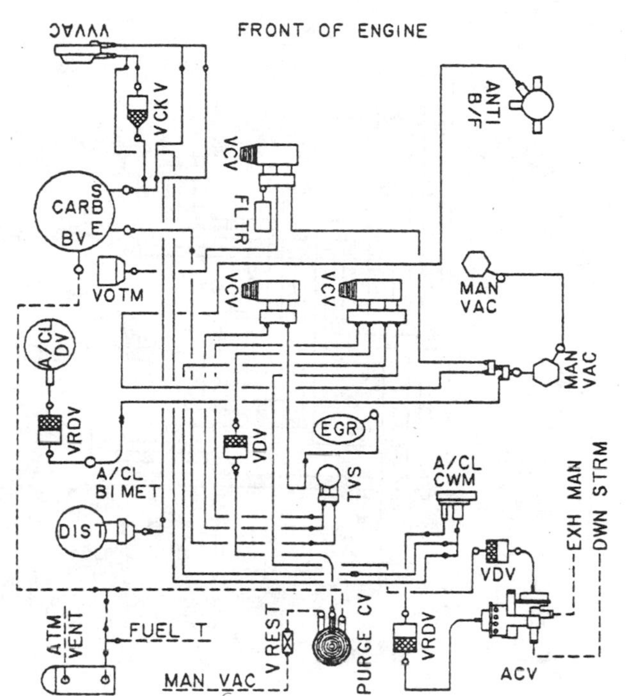

CALIBRATION: 1—3A—R10 DATE: 11-13-80

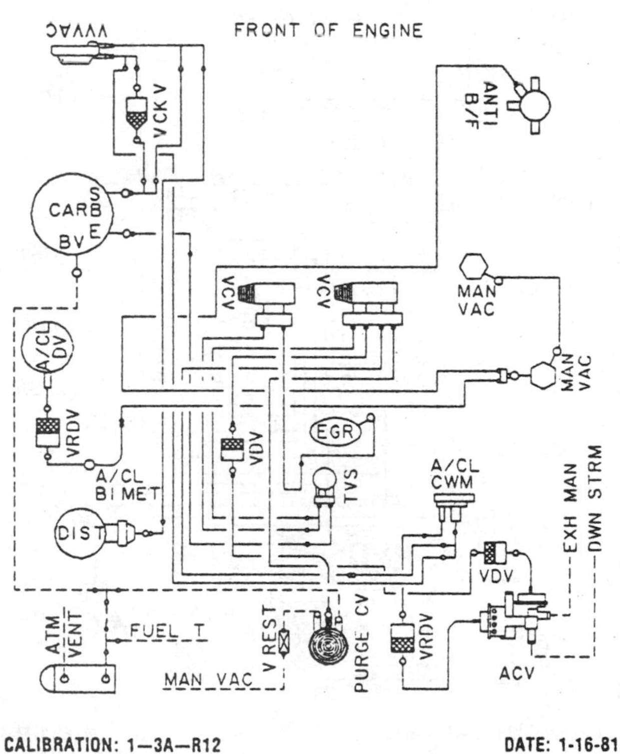

CALIBRATION: 1—3A—R12 DATE: 1-16-81

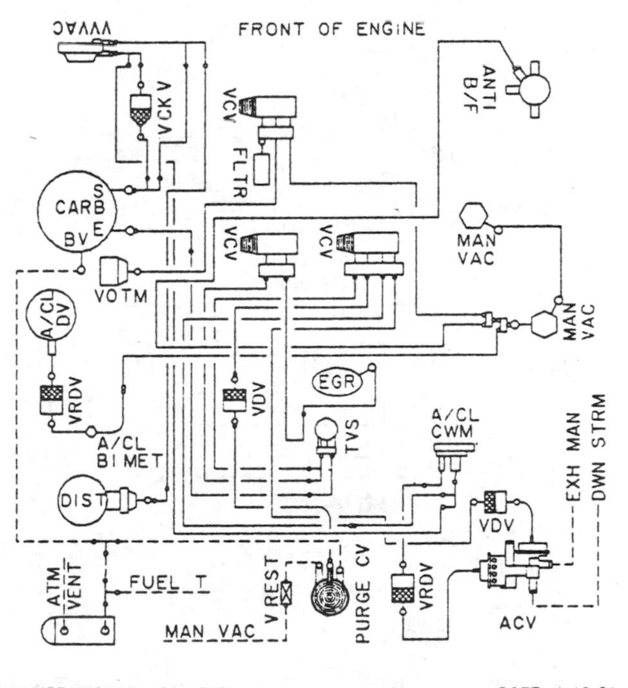

CALIBRATION: 1—3A—R13 DATE: 1-16-81

VACUUM CIRCUITS

(© Ford Motor Co.)

1981—1.6L engines

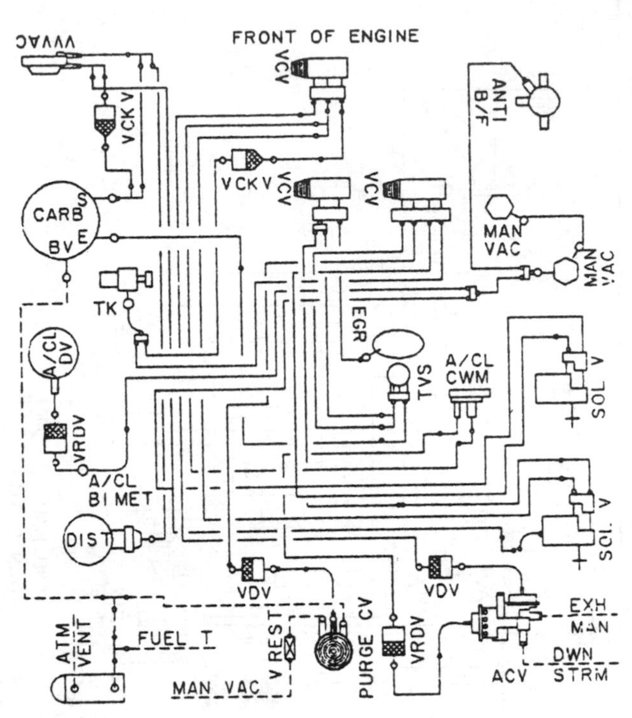

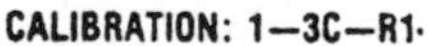

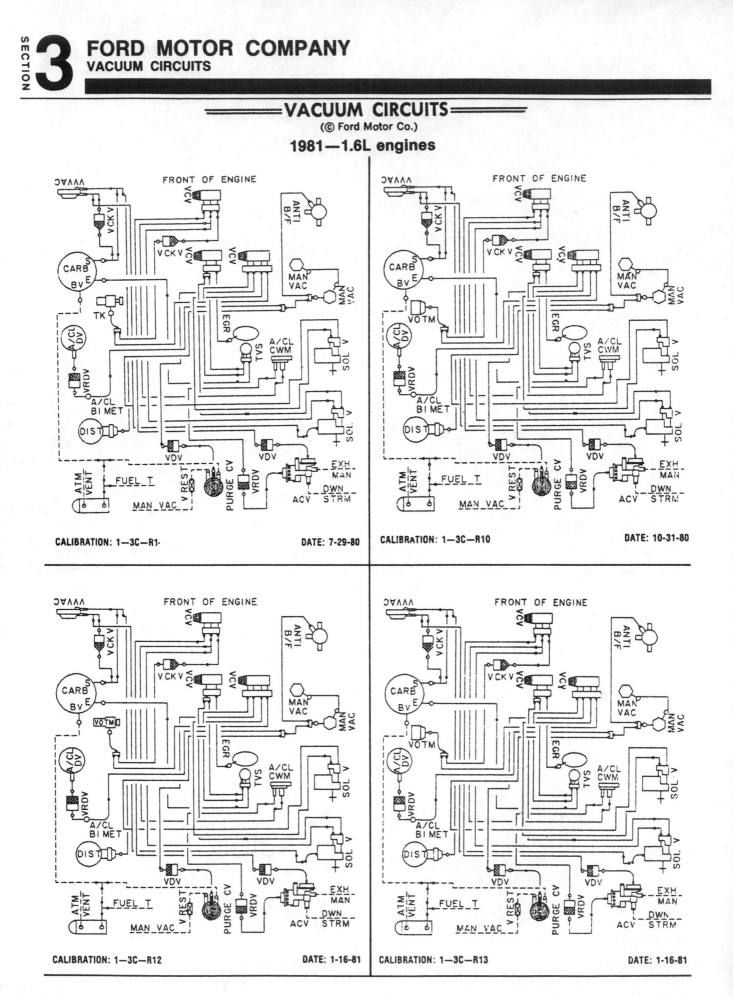

CALIBRATION: 1—3C—R1· DATE: 7-29-80

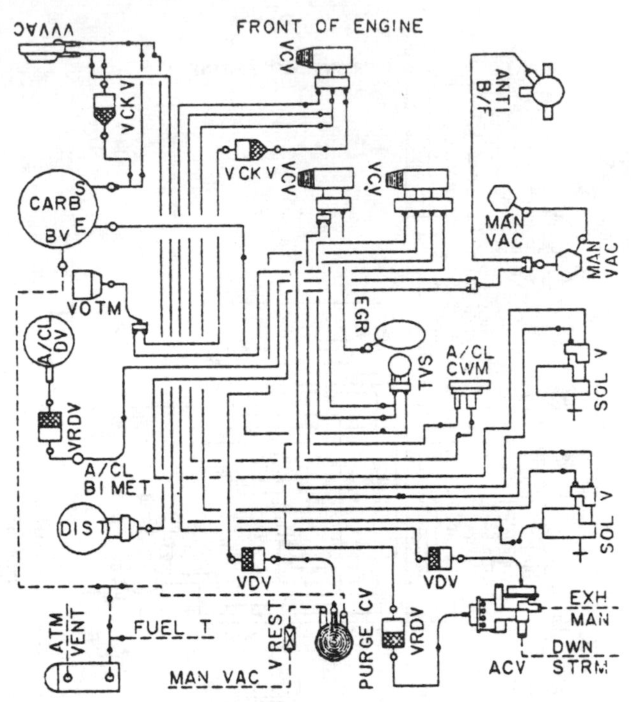

CALIBRATION: 1—3C—R10 DATE: 10-31-80

CALIBRATION: 1—3C—R12 DATE: 1-16-81

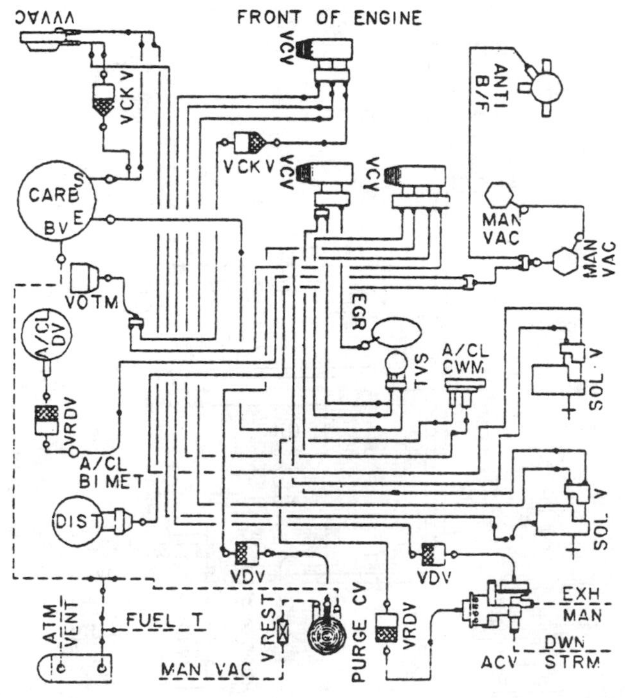

CALIBRATION: 1—3C—R13 DATE: 1-16-81

VACUUM CIRCUITS

(© Ford Motor Co)

1981—1.6L engines

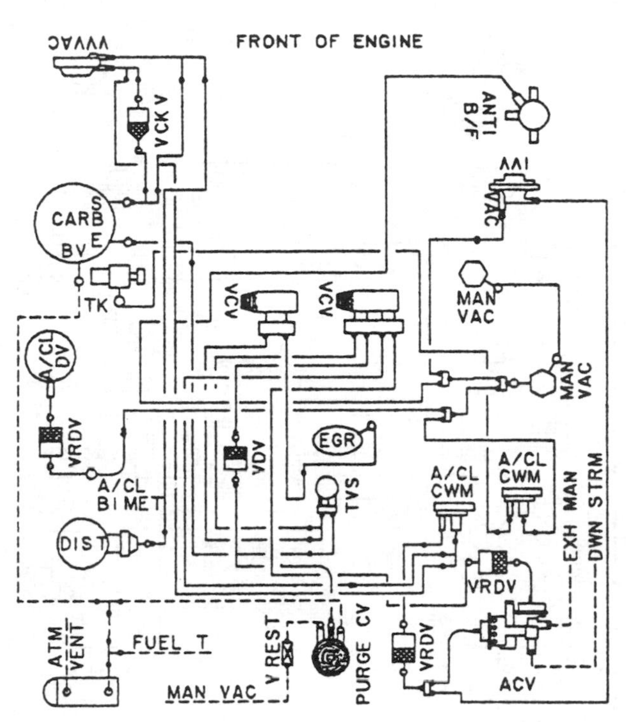

CALIBRATION: 1—3D—R10 DATE: 7-29-80

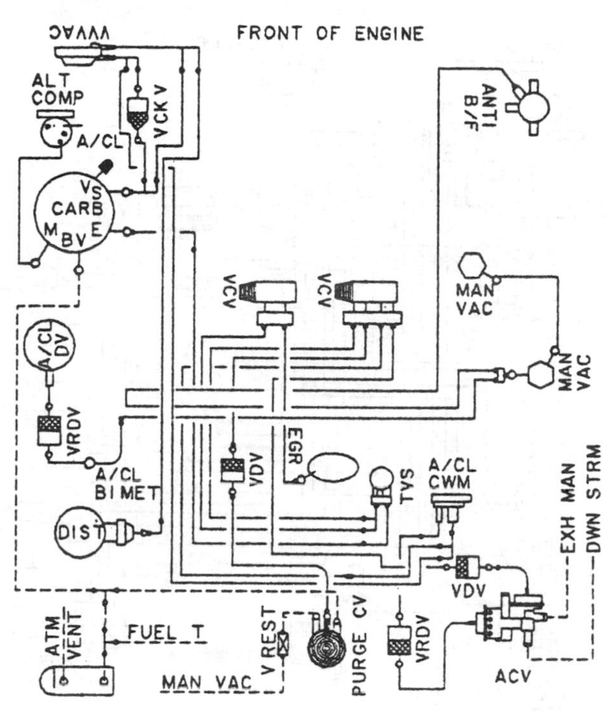

CALIBRATION: 1—3W—R0 DATE: 7-23-80

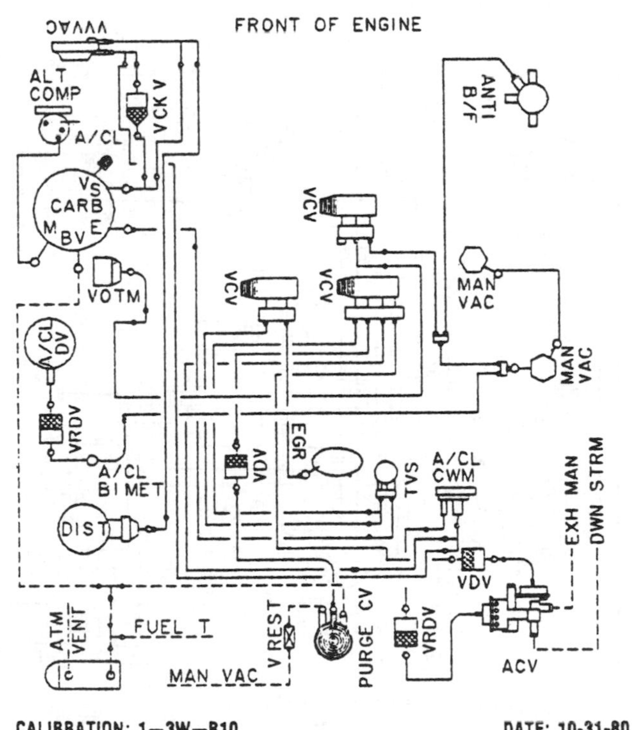

CALIBRATION: 1—3W—R10 DATE: 10-31-80

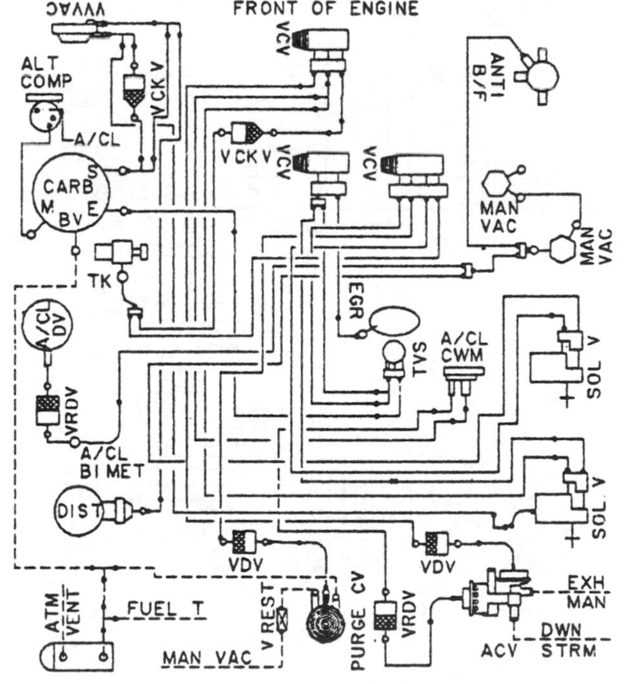

CALIBRATION: 1—3X—R0 DATE: 7-10-80

VACUUM CIRCUITS

(© Ford Motor Co)

1981—1.6L engines

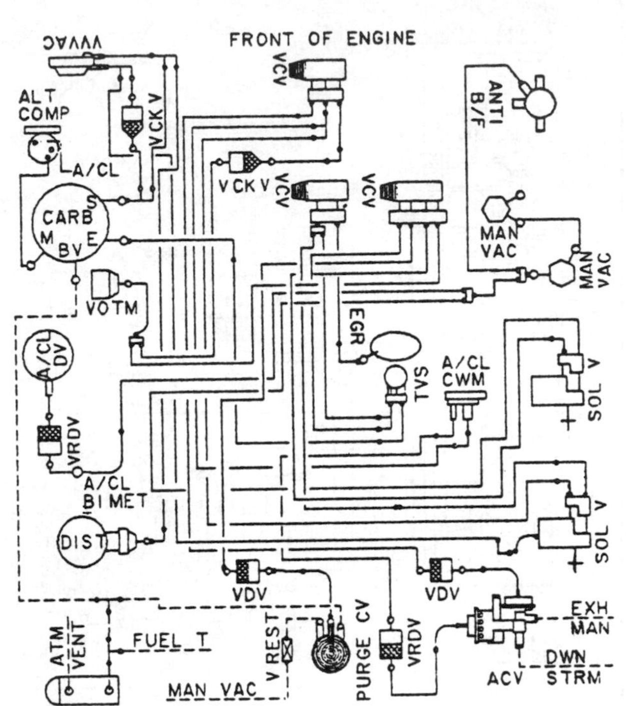

CALIBRATION: 1—3X—R10 DATE: 10-31-80

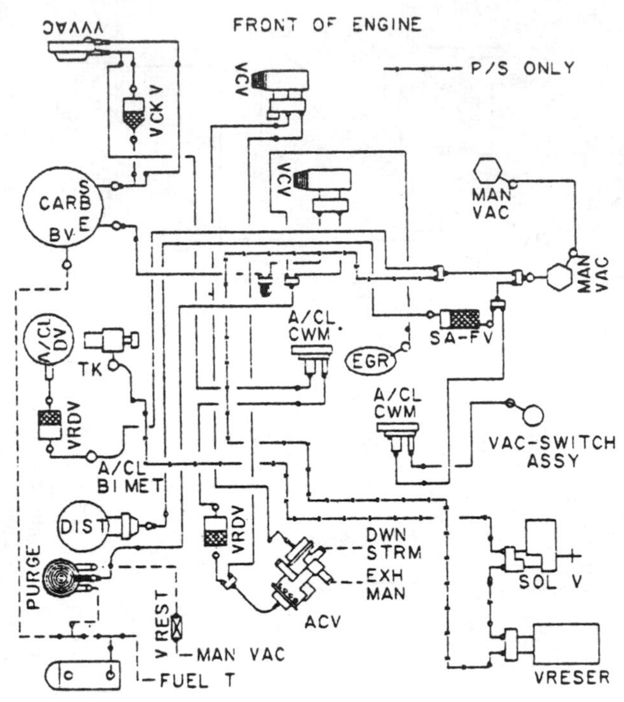

CALIBRATION: 1—4A—R12 DATE: 7-29-80

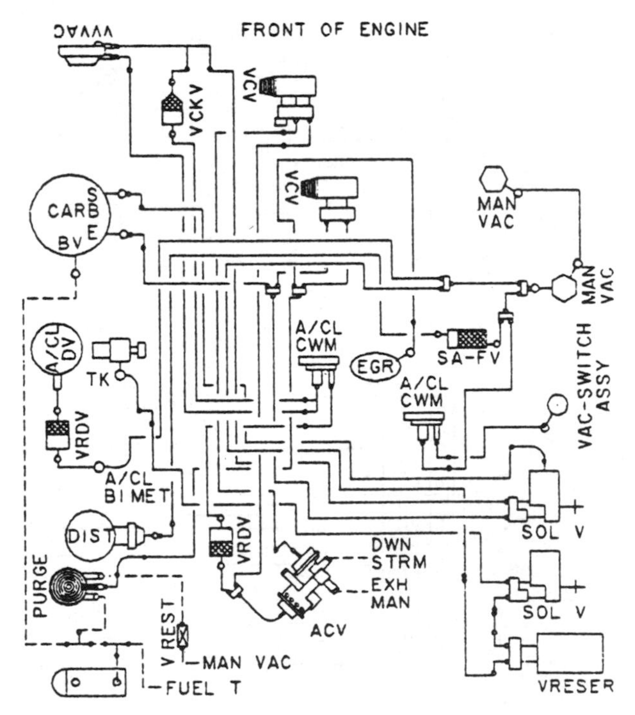

CALIBRATION: 1—4C—R10 DATE: 7-29-80

A/C

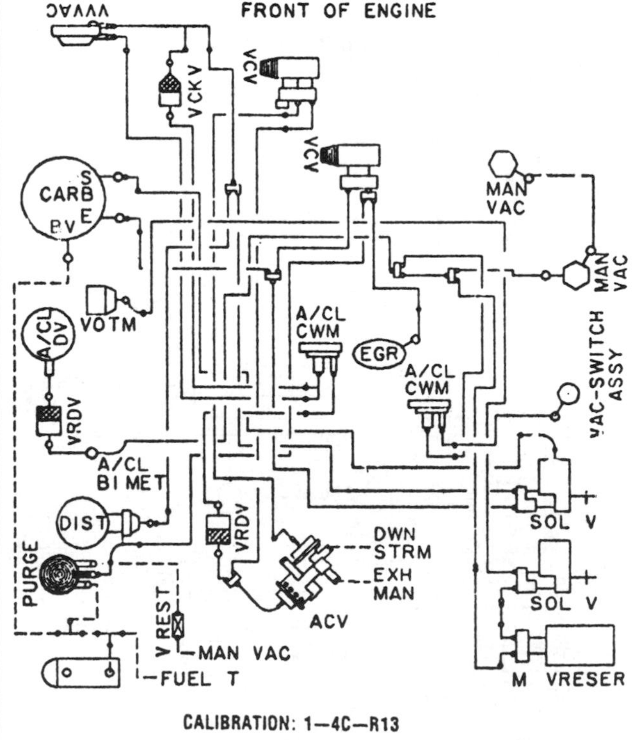

CALIBRATION: 1—4C—R13
DATE: 11-4-80

VACUUM CIRCUITS

(© Ford Motor Co.)

1981—1.6L engines

NON A/C

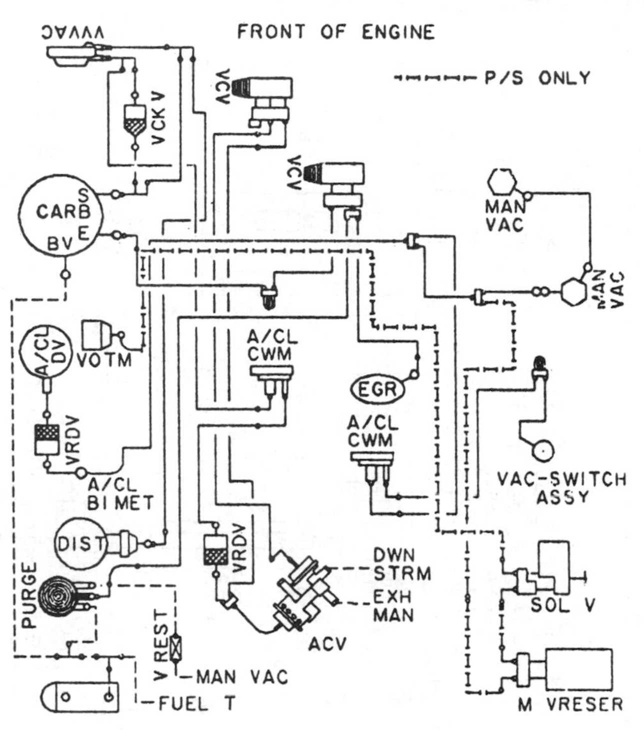

CALIBRATION: 1—4C—R13
DATE: 11-4-80

A/C

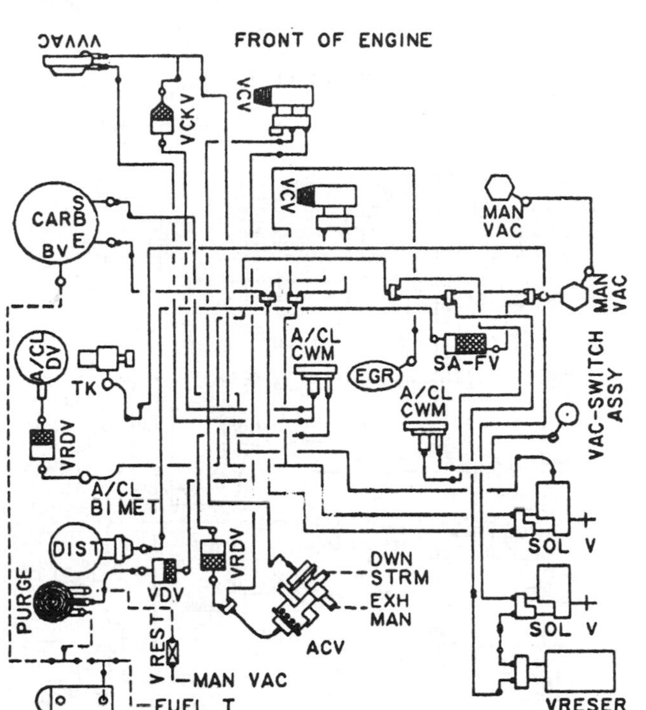

CALIBRATION: 1—4Q—R1
DATE: 8-27-80

NON A/C

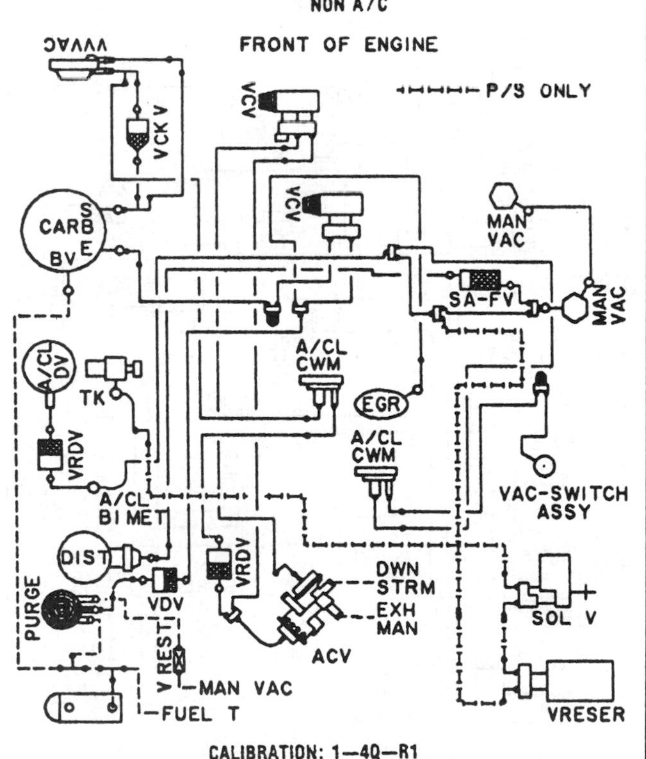

CALIBRATION: 1—4Q—R1
DATE: 10-9-80

A/C

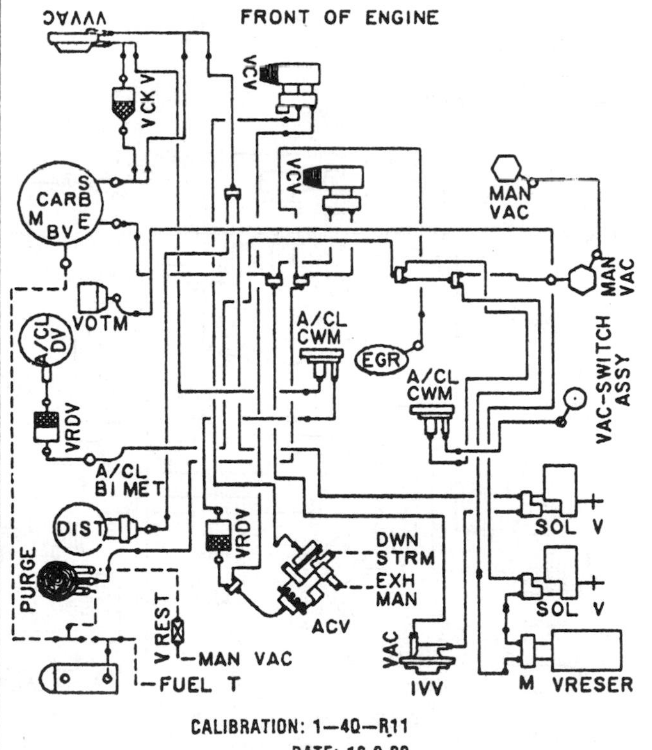

CALIBRATION: 1—4Q—R11
DATE: 12-2-80

VACUUM CIRCUITS

(© Ford Motor Co.)

1981—1.6L engines

NON A/C

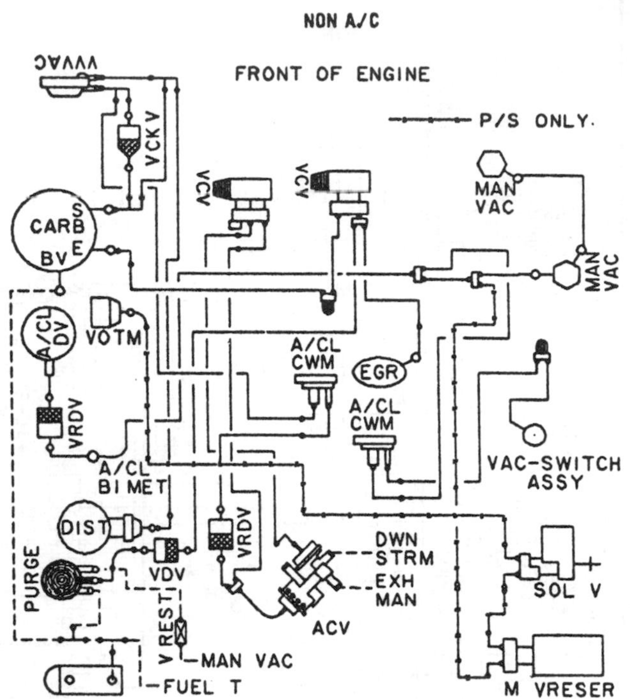

CALIBRATION: 1—4Q—R11
DATE: 10-31-80

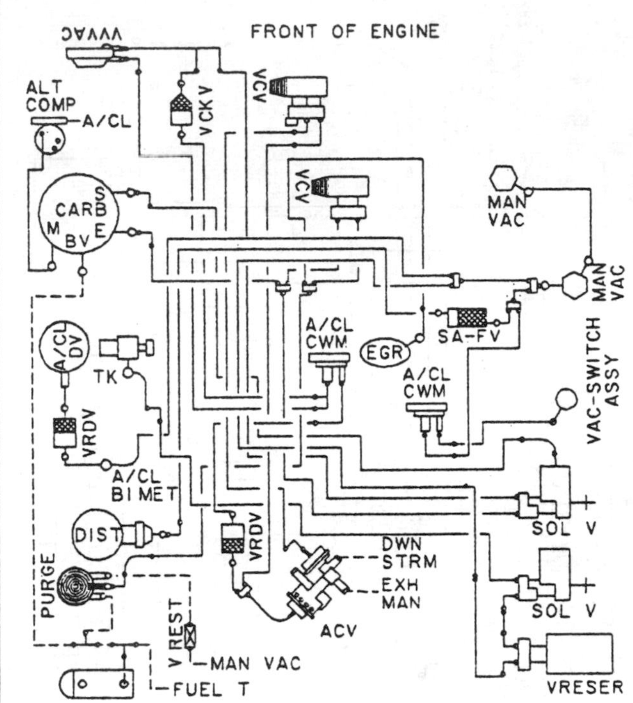

CALIBRATION: 1—4X—R10
DATE: 7-30-80

A/C

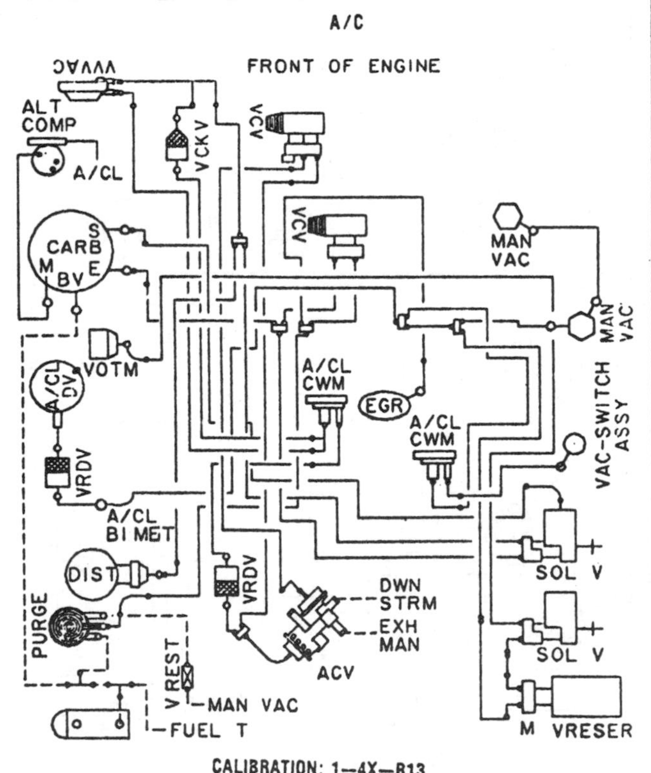

CALIBRATION: 1—4X—R13
DATE: 10-30-80

NON A/C

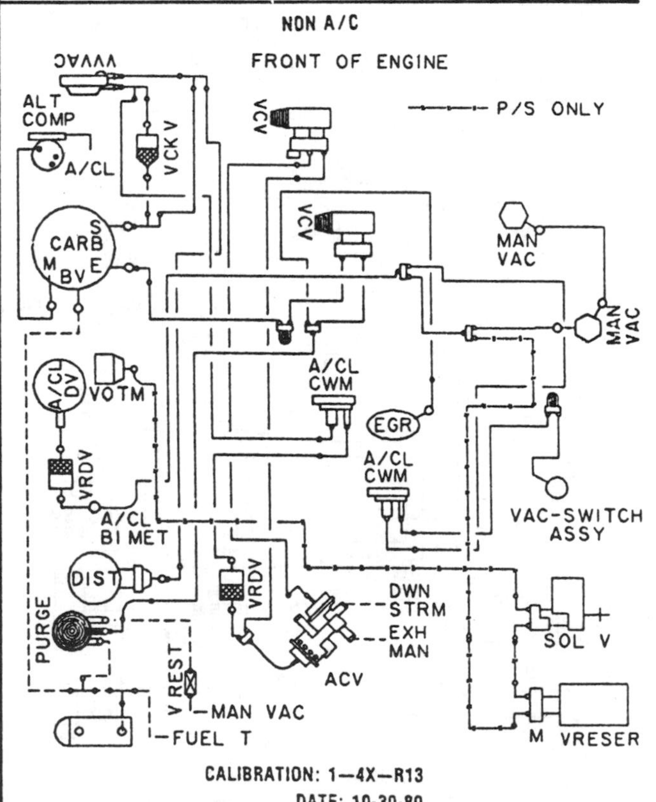

CALIBRATION: 1—4X—R13
DATE: 10-30-80

VACUUM CIRCUITS

(© Ford Motor Co)

1981—2.3L engines

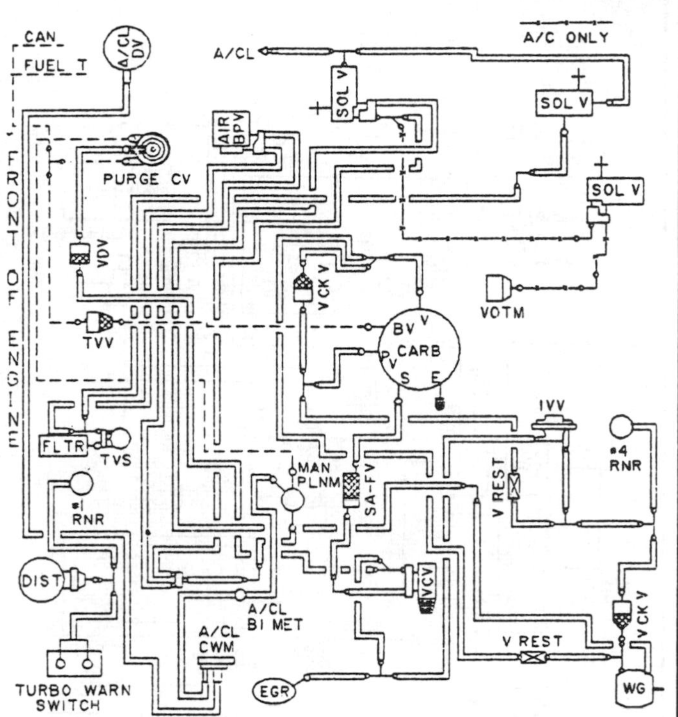

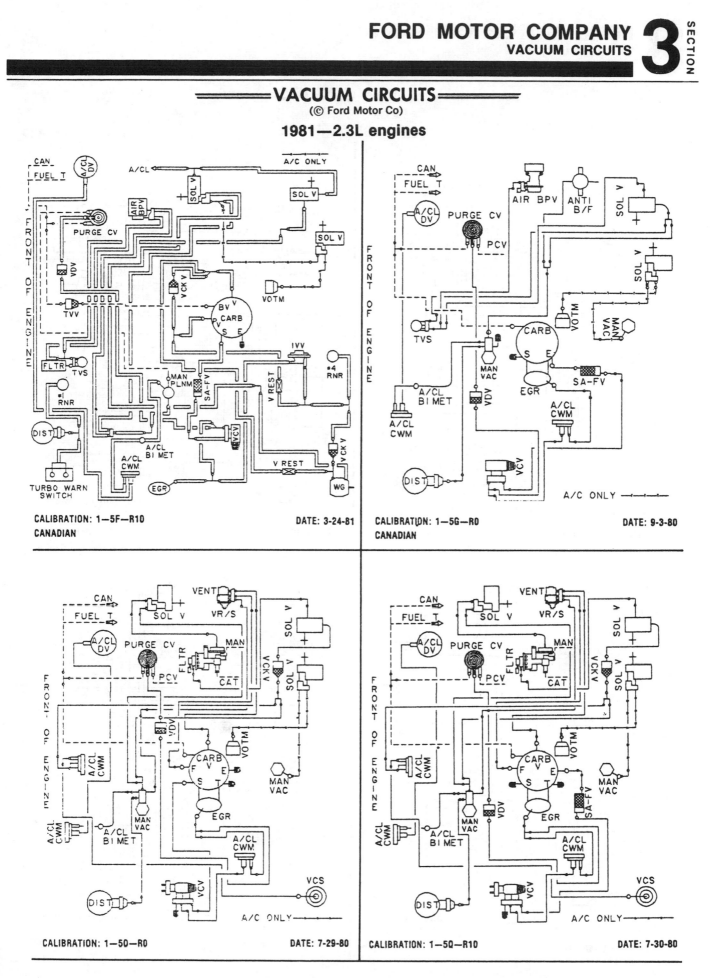

CALIBRATION: 1—5F—R10
CANADIAN

DATE: 3-24-81

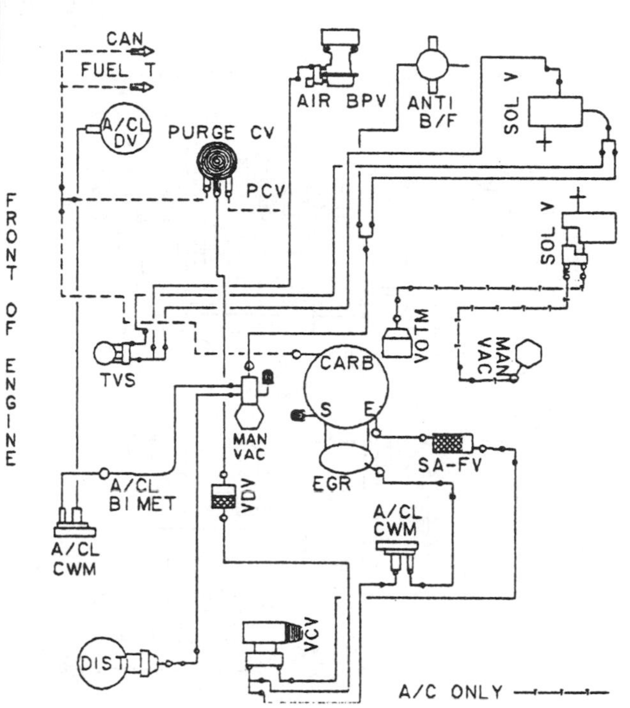

CALIBRATION: 1—5G—R0
CANADIAN

DATE: 9-3-80

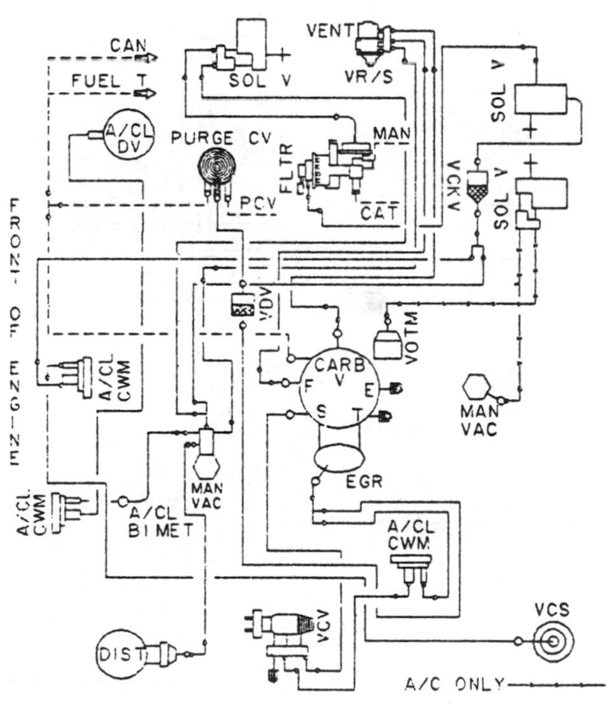

CALIBRATION: 1—5O—R0

DATE: 7-29-80

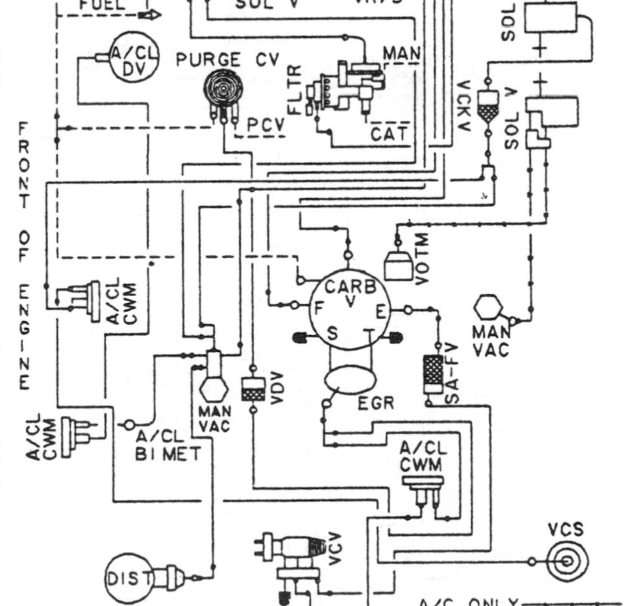

CALIBRATION: 1—5Q—R10

DATE: 7-30-80

VACUUM CIRCUITS

1981—2.3L engines

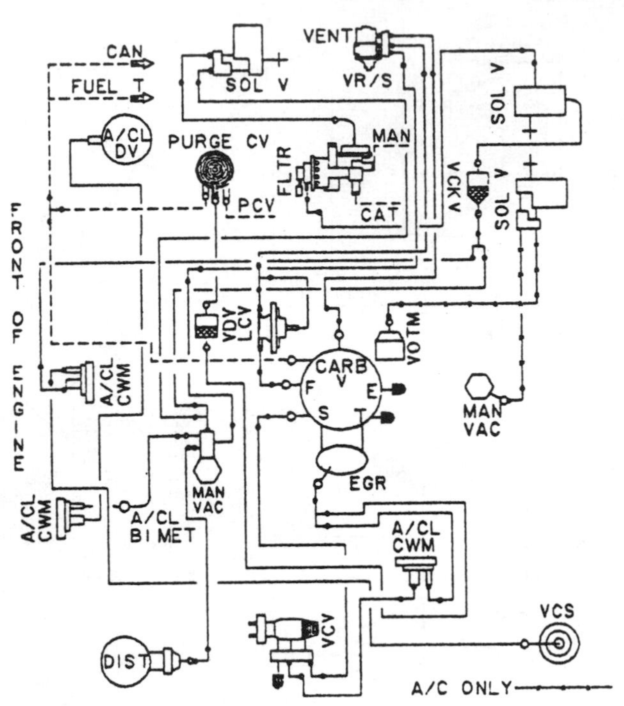

CALIBRATION: 1—5R—R10 DATE: 8-6-80

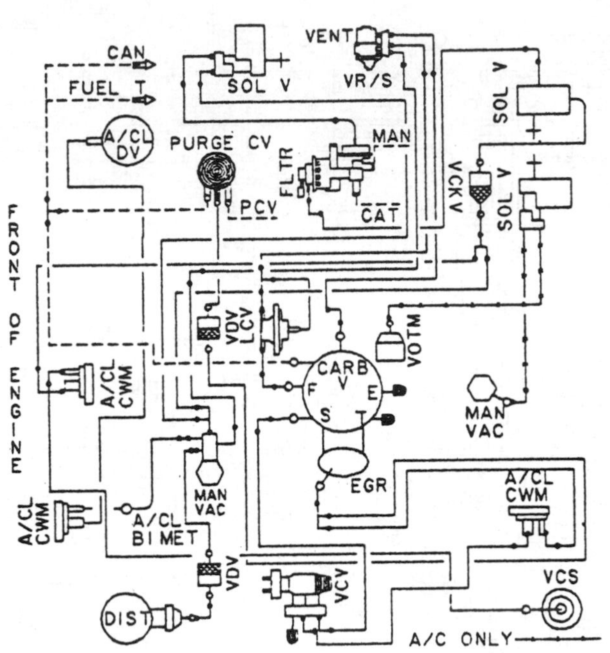

CALIBRATION: 1—5R—R11 DATE: 9-24-80

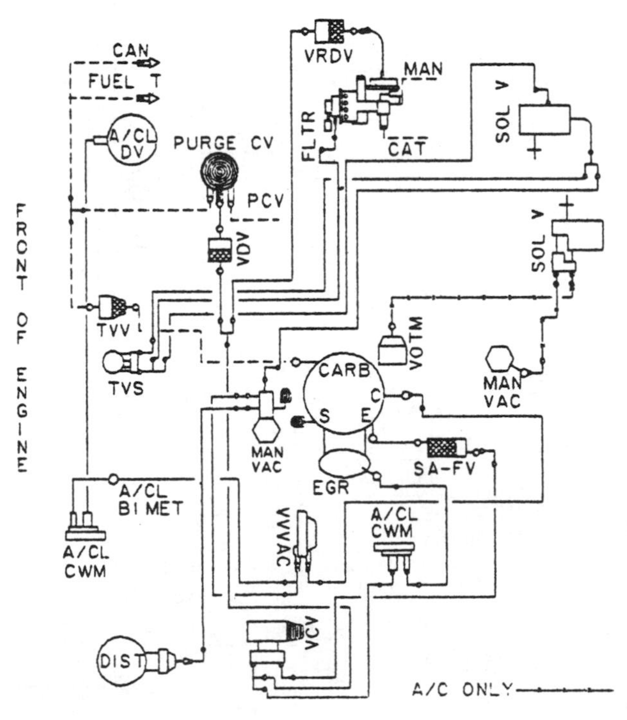

CALIBRATION: 1—6A—R10 DATE: 12-3-80

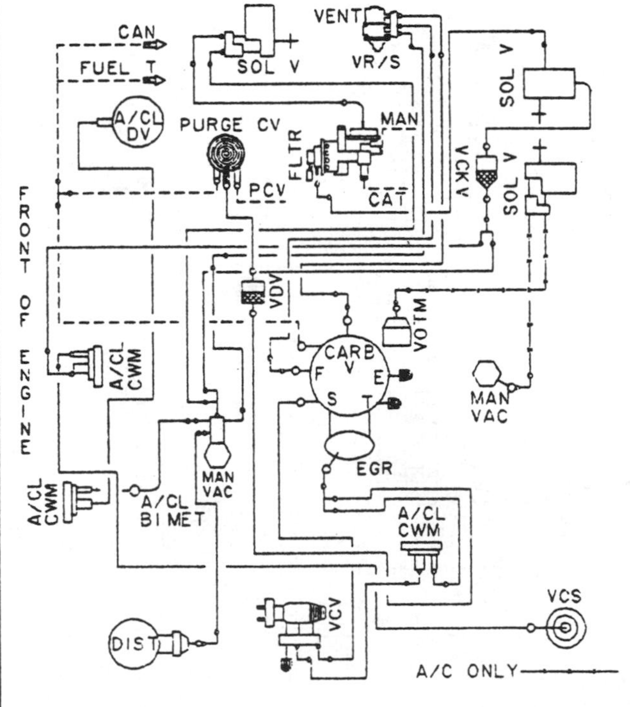

CALIBRATION: 1—6N—R0 DATE: 7-29-80

VACUUM CIRCUITS

(© Ford Motor Co)

1981—2.3L engines

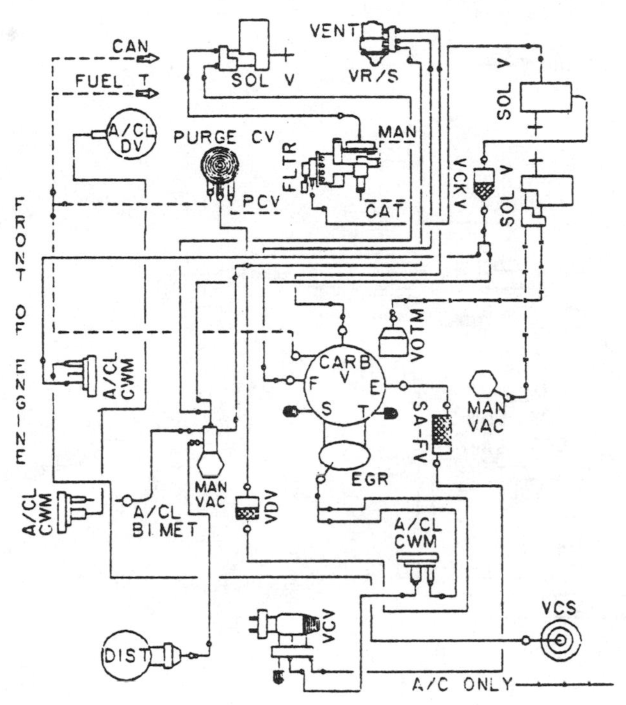

CALIBRATION: 1—6N—R10 DATE: 7-30-80

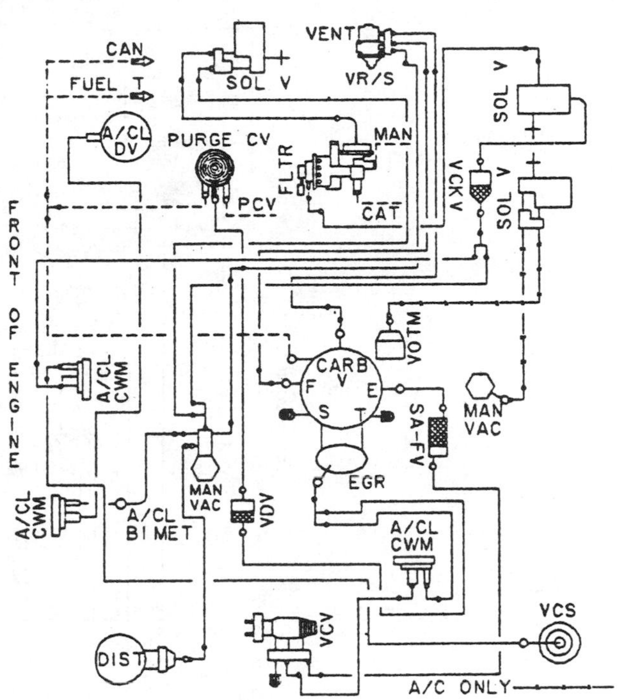

CALIBRATION: 1—6N—R12 DATE: 11-17-80

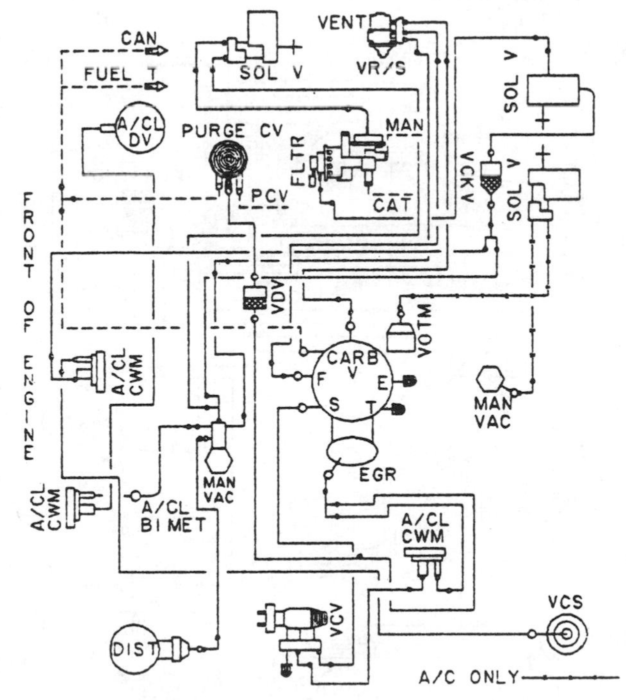

CALIBRATION: 1—6V—R0 DATE: 7-29-80

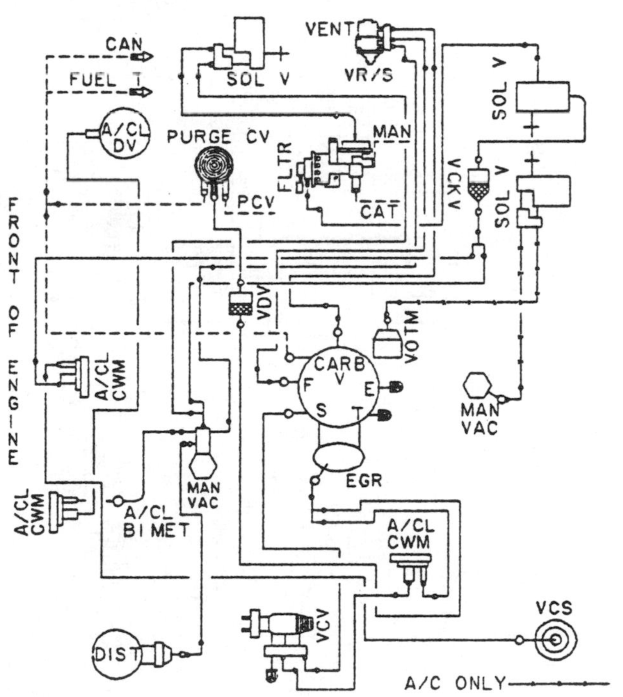

CALIBRATION: 1—6V—R12 DATE: 11-17-80

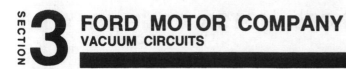

VACUUM CIRCUITS

(© Ford Motor Co.)

1981—2.3L engines

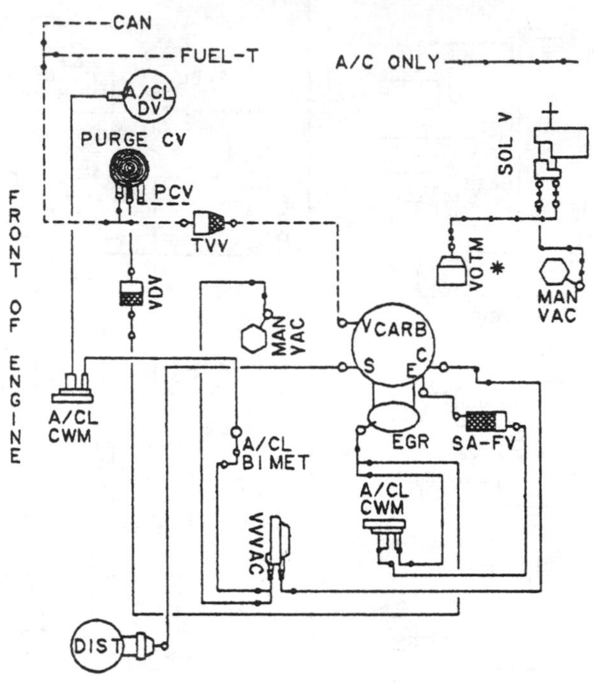

CALIBRATION: 1—26B—R0
CANADIAN

DATE: 7-14-80

1981—200 CID (3.3L) engines

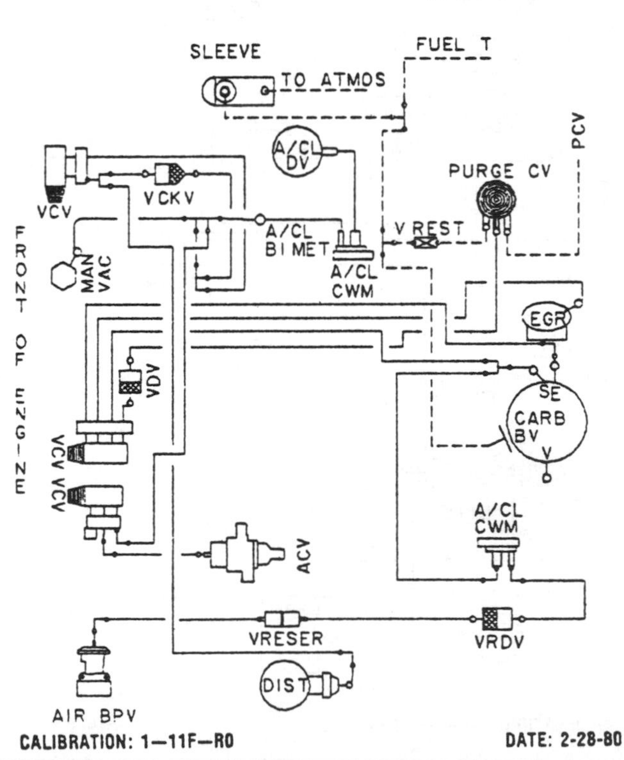

CALIBRATION: 1—11F—R0

DATE: 2-28-80

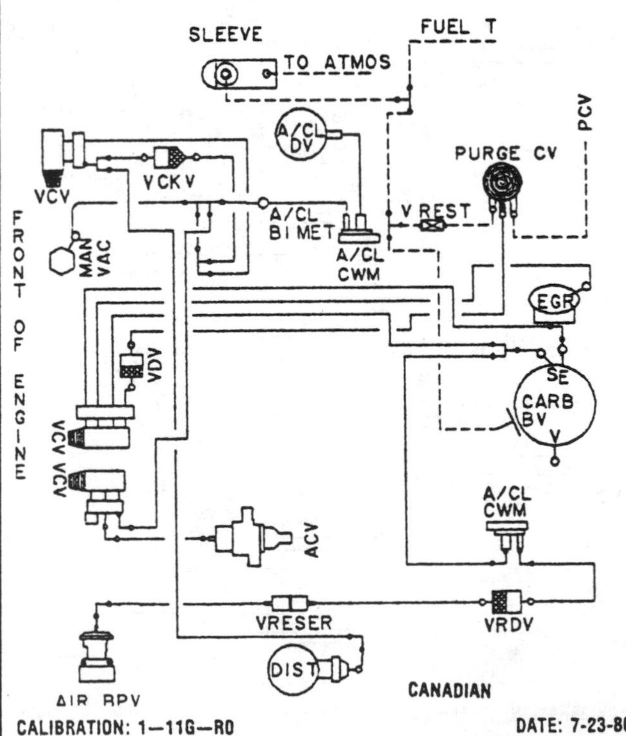

CANADIAN

CALIBRATION: 1—11G—R0

DATE: 7-23-80

VACUUM CIRCUITS

(© Ford Motor Co.)

1981—200 CID (3.3L) engines

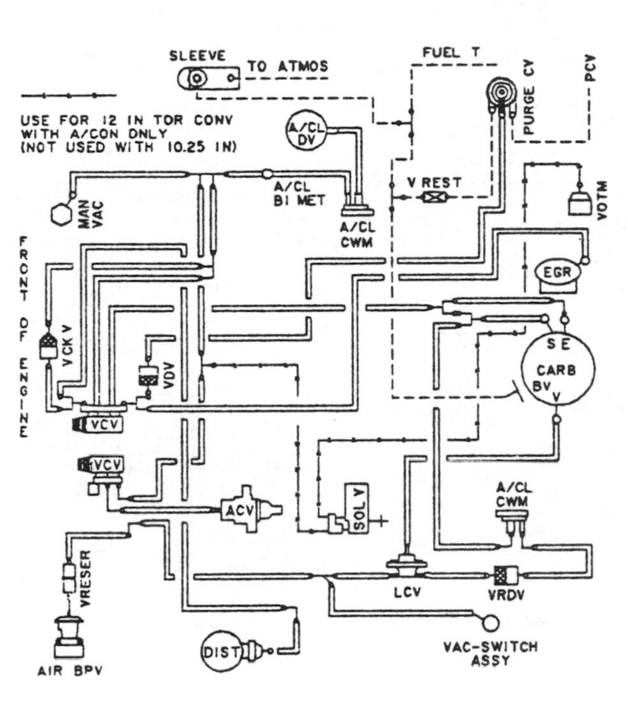

CALIBRATION: 1—12B—R0 DATE: 11-7-80

SLEEVE
TO ATMOS
FUEL T
PCV
A/CL DV
PURGE CV
VCV
VCKV
FRONT OF ENGINE
MAN VAC
A/CL BI MET
A/CL CWM
EGR
VDV
TVV
SE
CARB
BV
V
VCV
LCV
DIST

CALIBRATION: 1—32G—R0
CANADIAN
DATE: 8-12-80

1981—255 CID (4.2L) engines

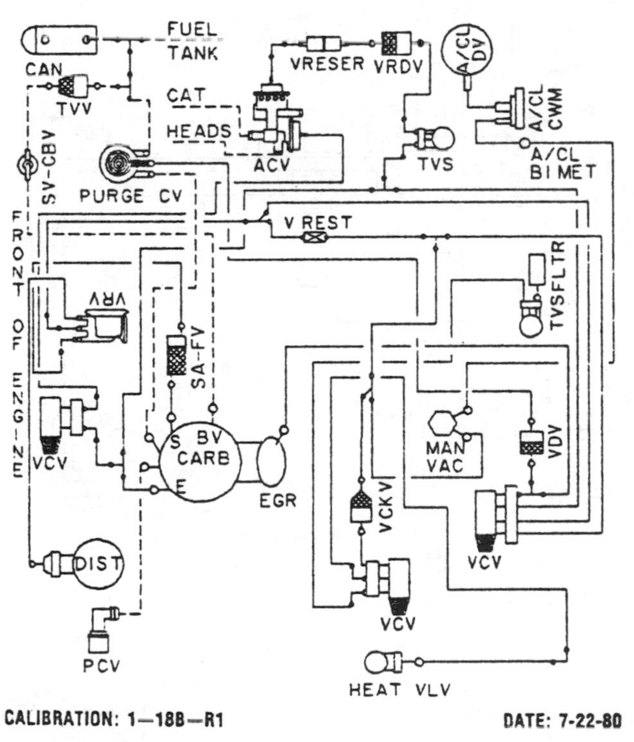

CALIBRATION: 1—18B—R1 DATE: 7-22-80

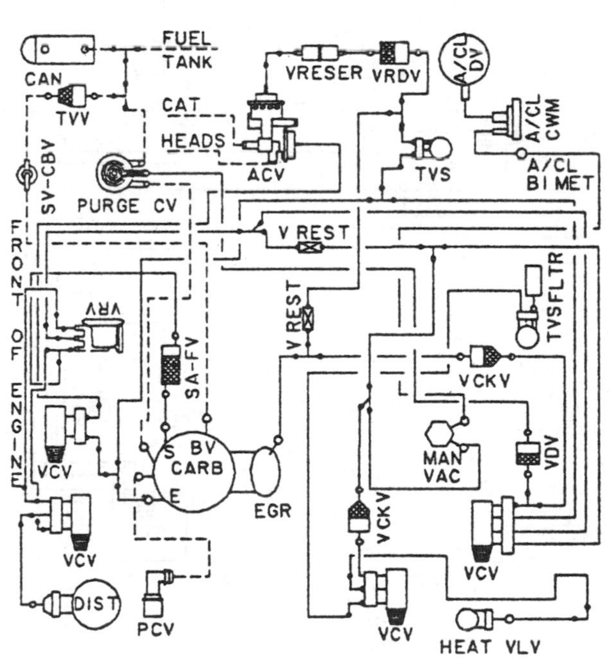

CALIBRATION: 1—18C—R11 DATE: 10-1-80

VACUUM CIRCUITS

(© Ford Motor Co)

1981—255 CID (4.2L) engines

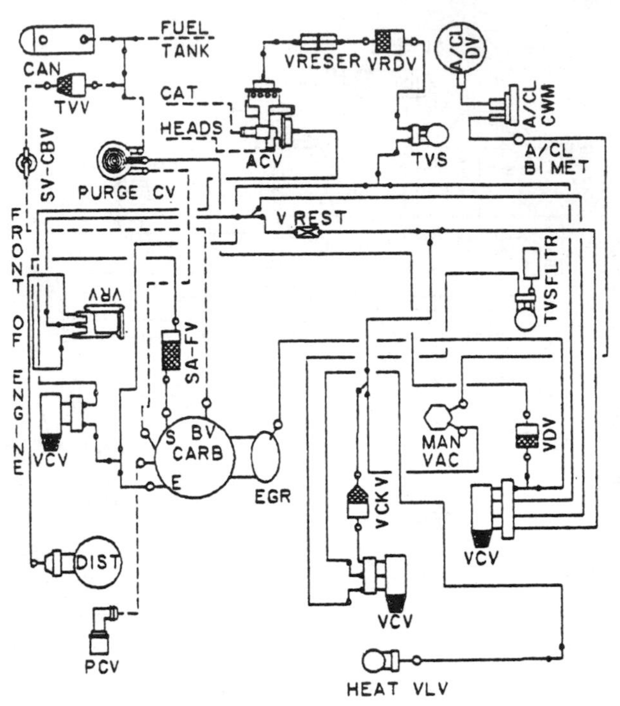

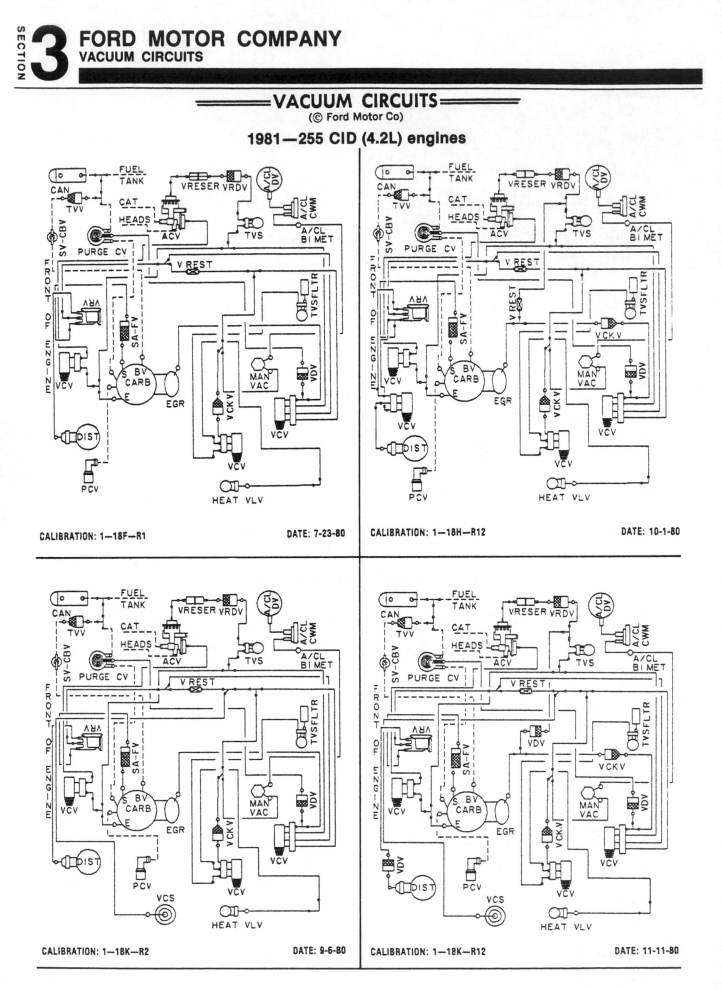

CALIBRATION: 1—18F—R1 **DATE: 7-23-80**

CALIBRATION: 1—18H—R12 **DATE: 10-1-80**

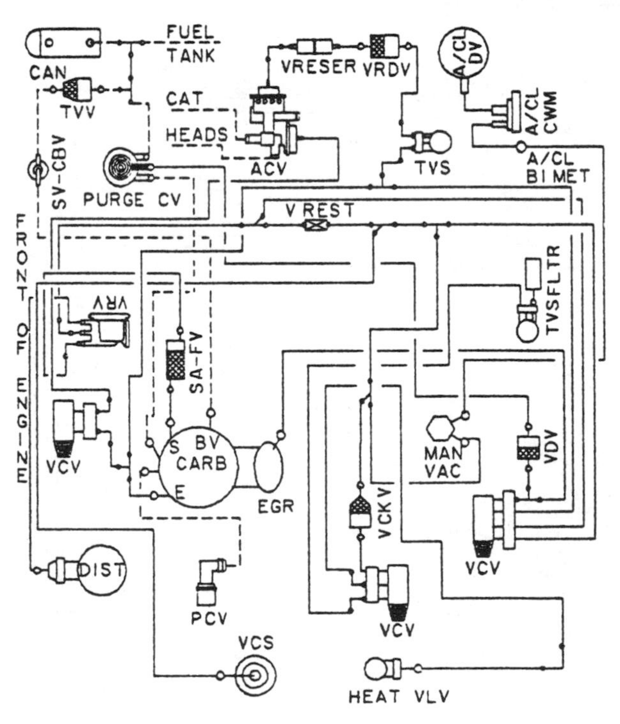

CALIBRATION: 1—18K—R2 **DATE: 9-6-80**

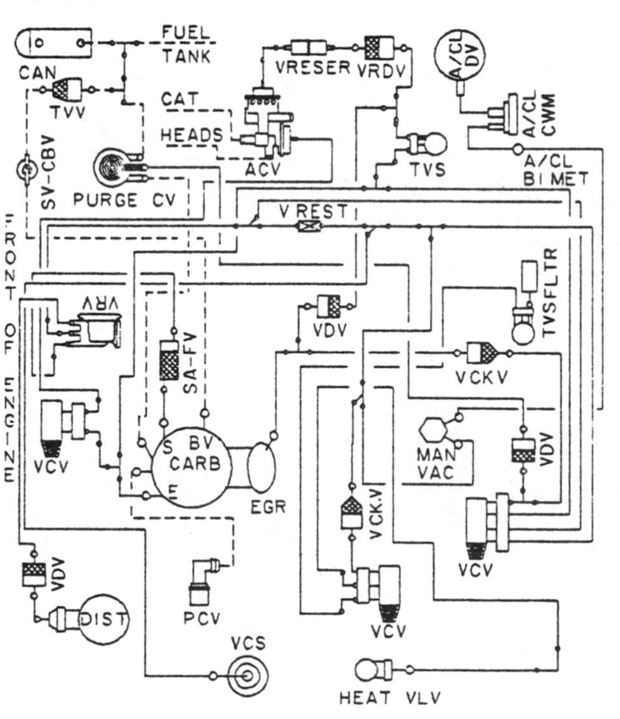

CALIBRATION: 1—18K—R12 **DATE: 11-11-80**

VACUUM CIRCUITS

(© Ford Motor Co.)

1981—255 CID (4.2L) engines

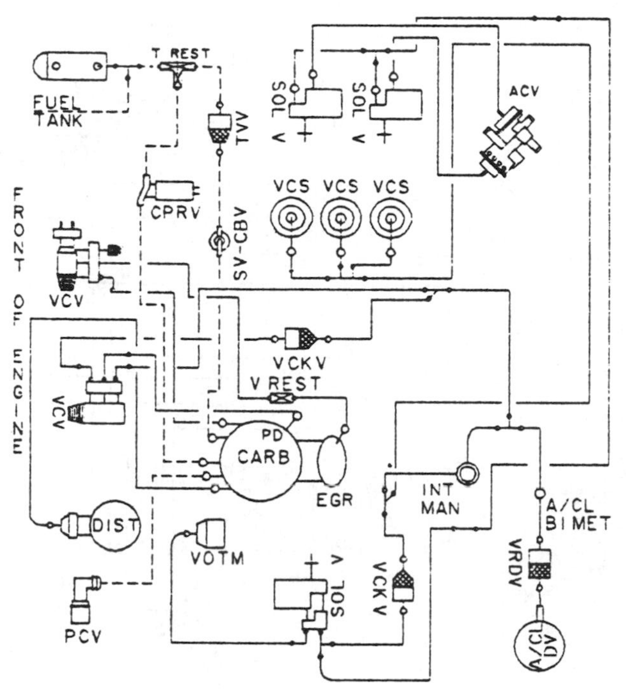

CALIBRATION: 1—18M—R1 DATE: 8-22-80

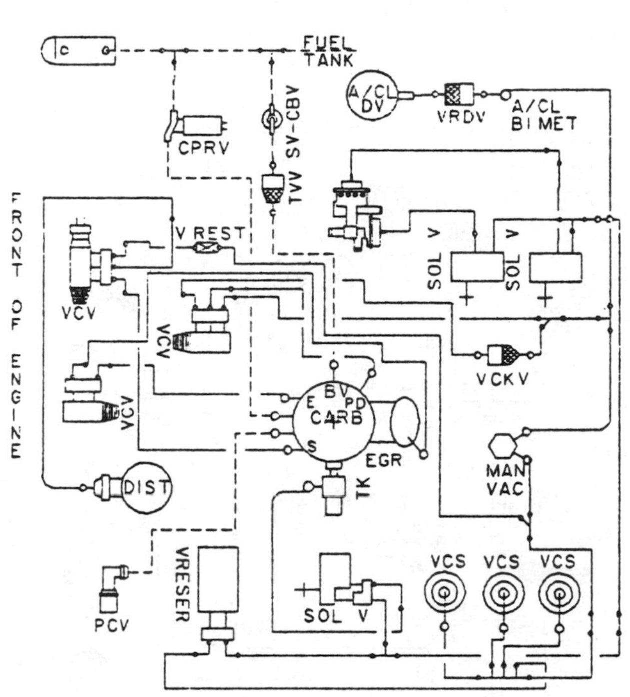

CALIBRATION: 1—18N—R0 DATE: 7-10-80

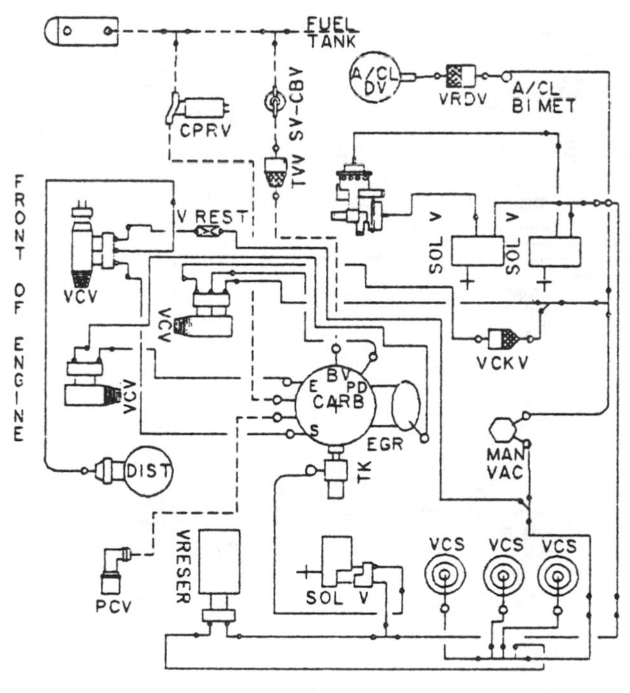

CALIBRATION: 1—18R—R0 DATE: 7-10-80

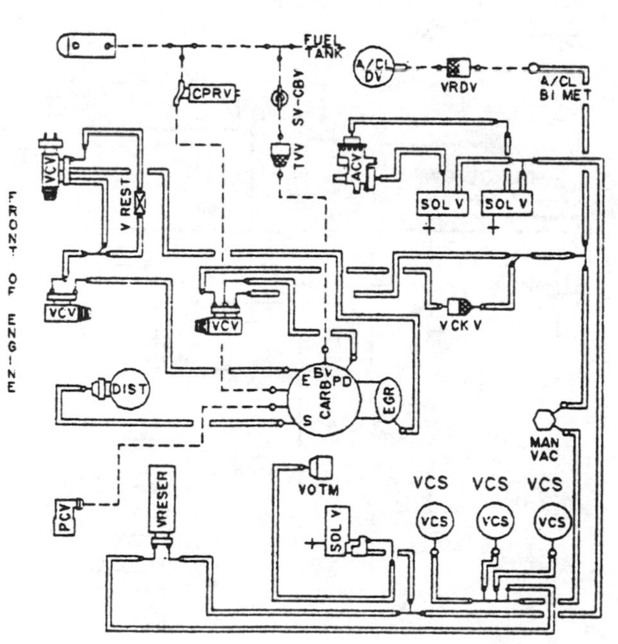

CALIBRATION: 1—18T—R0 DATE: 11-17-80

VACUUM CIRCUITS

(© Ford Motor Co)

1981—255 CID (4.2L) engines

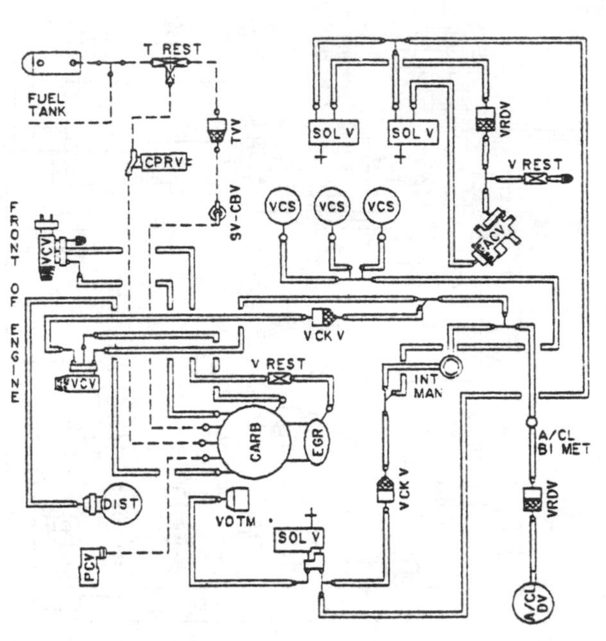

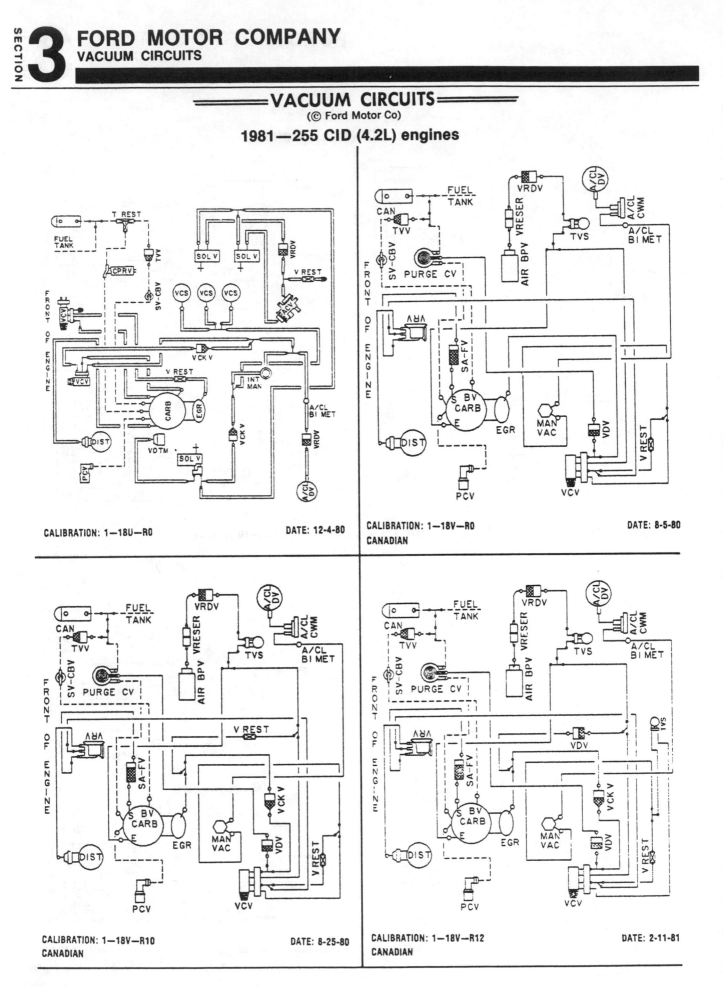

CALIBRATION: 1—18U—R0 DATE: 12-4-80

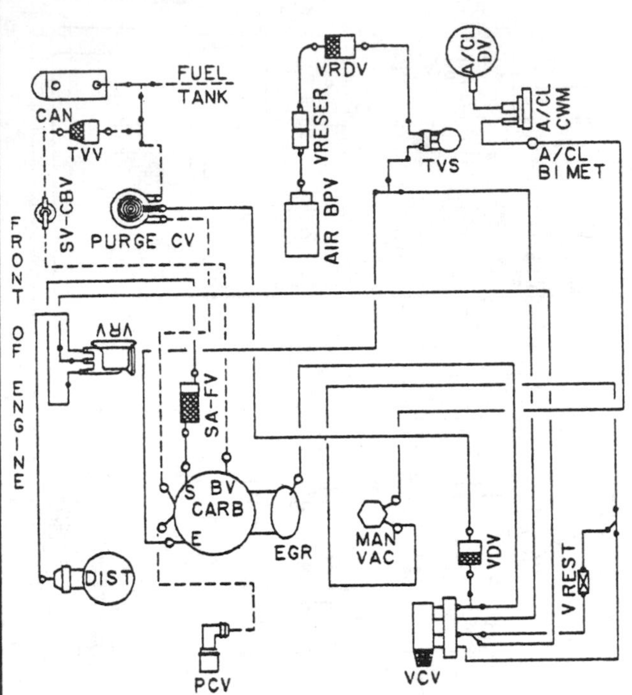

CALIBRATION: 1—18V—R0 DATE: 8-5-80
CANADIAN

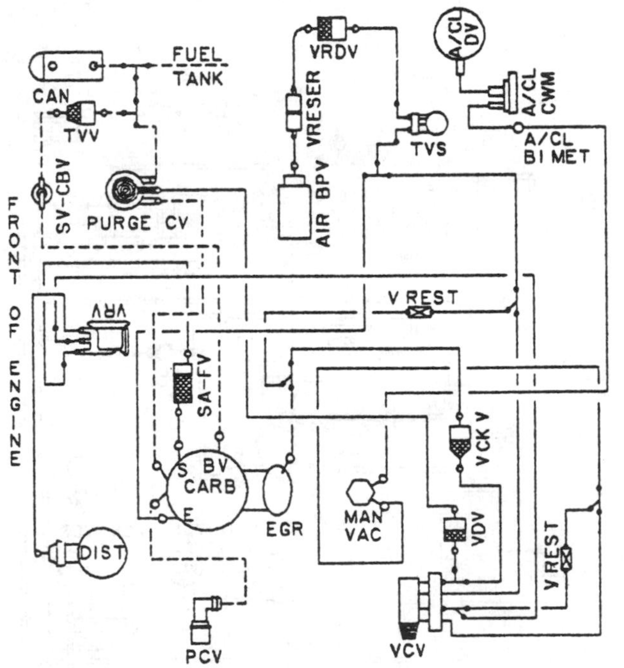

CALIBRATION: 1—18V—R10 DATE: 8-25-80
CANADIAN

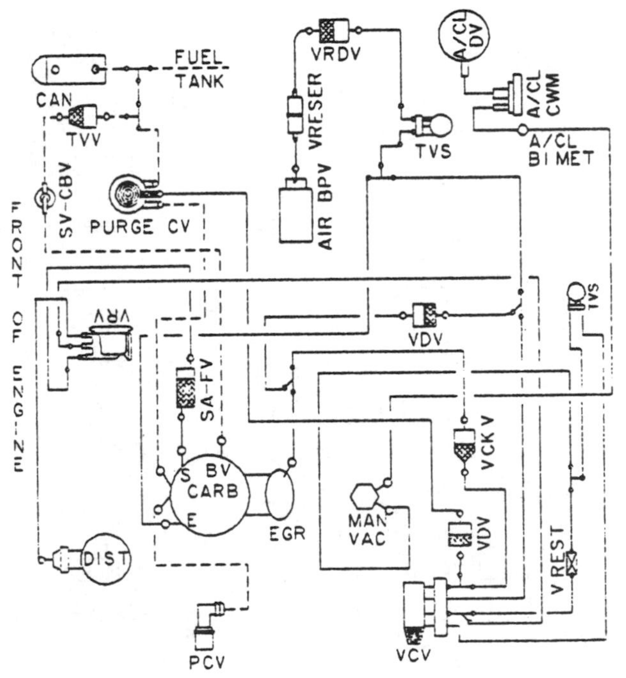

CALIBRATION: 1—18V—R12 DATE: 2-11-81
CANADIAN

VACUUM CIRCUITS

(© Ford Motor Co.)

1981—255 CID (4.2L) engines

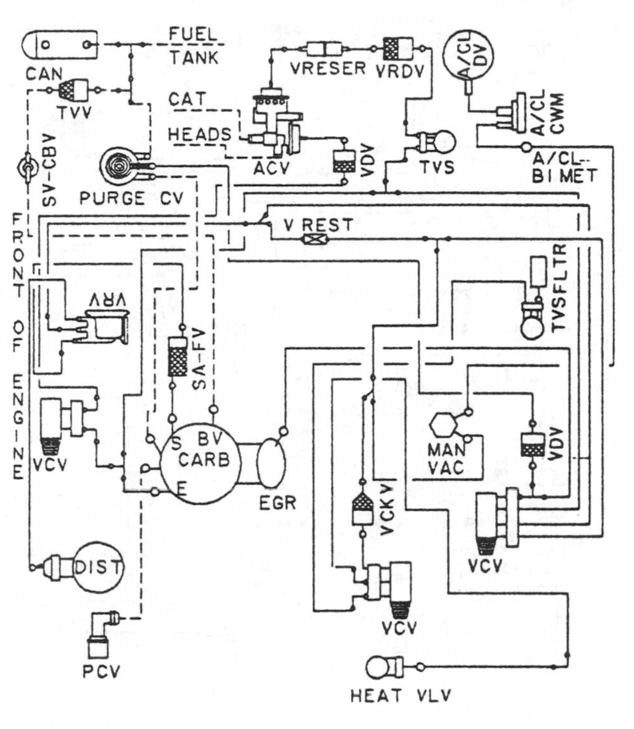

CALIBRATION: 1—18X—R0 DATE: 10-24-80

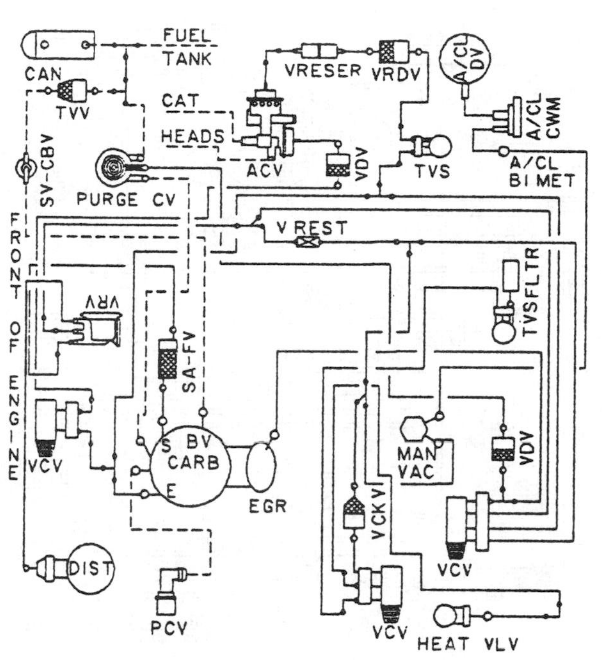

CALIBRATION: 1—18Z—R0 DATE: 10-24-80

1981—302 CID (5.0L) engines

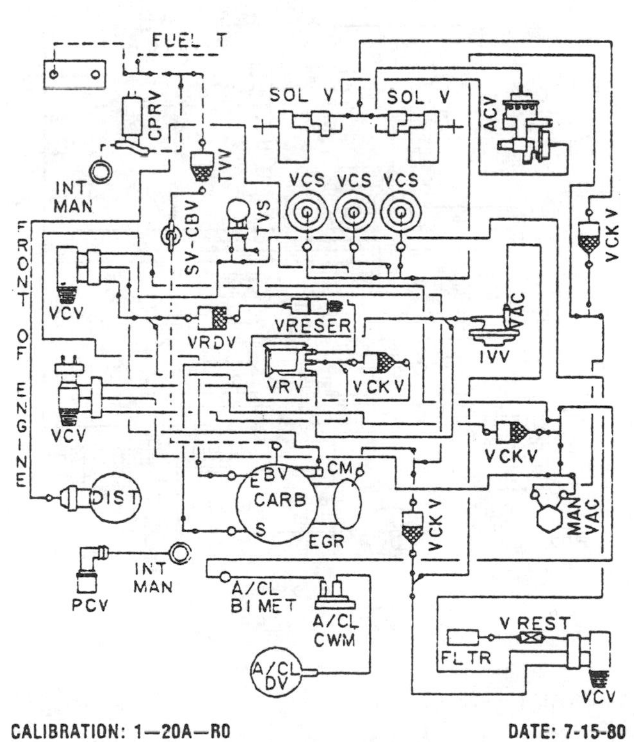

CALIBRATION: 1—20A—R0 DATE: 7-15-80

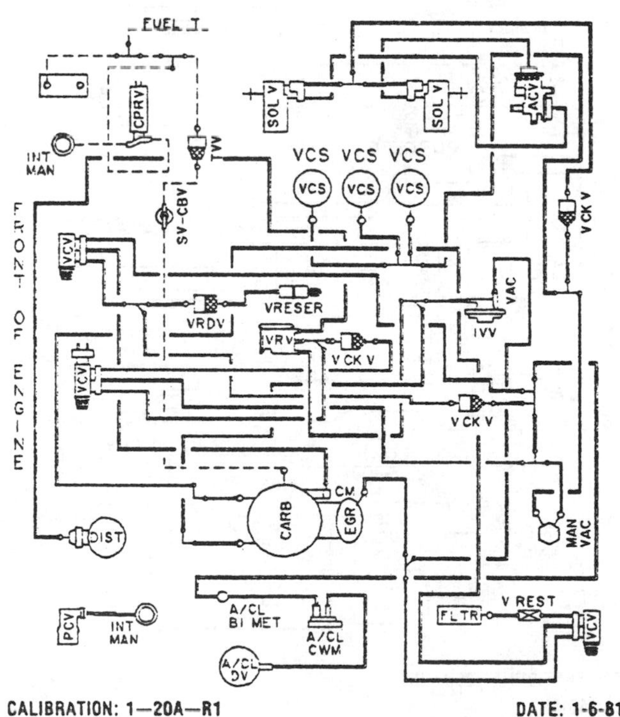

CALIBRATION: 1—20A—R1 DATE: 1-6-81

VACUUM CIRCUITS

(© Ford Motor Co)

1981—302 CID (5.0L) engines

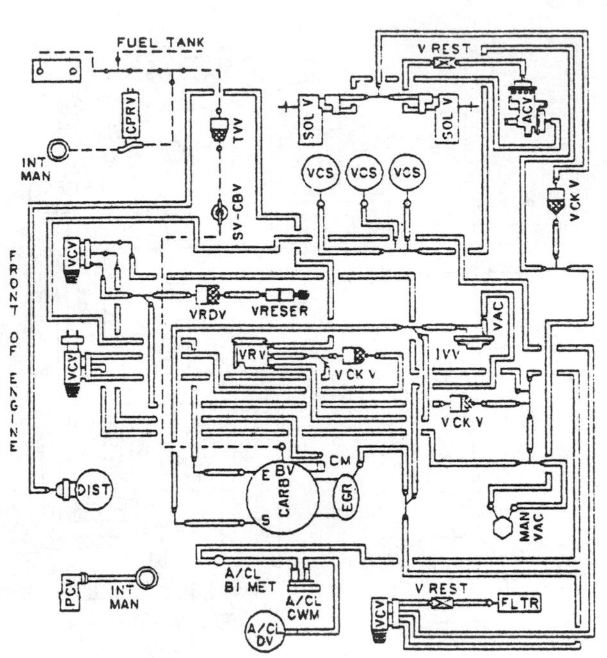

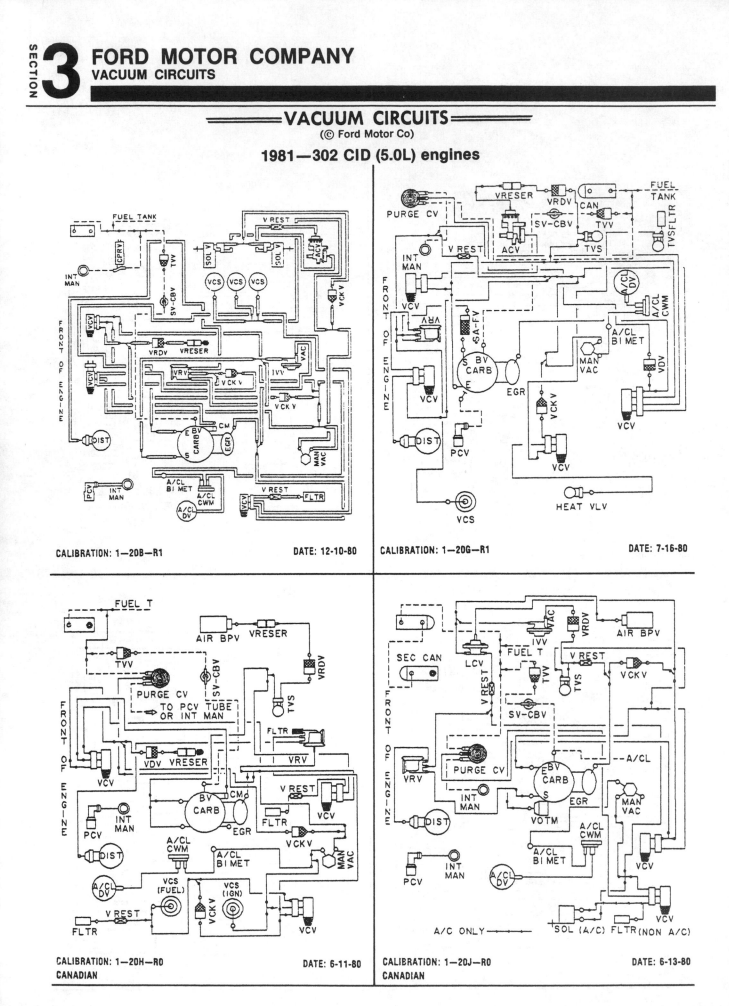

CALIBRATION: 1—20B—R1 DATE: 12-10-80

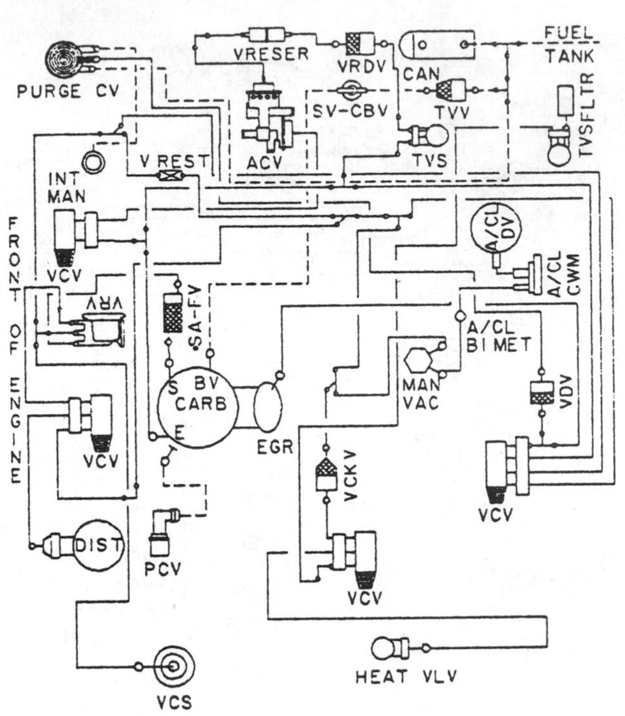

CALIBRATION: 1—20G—R1 DATE: 7-16-80

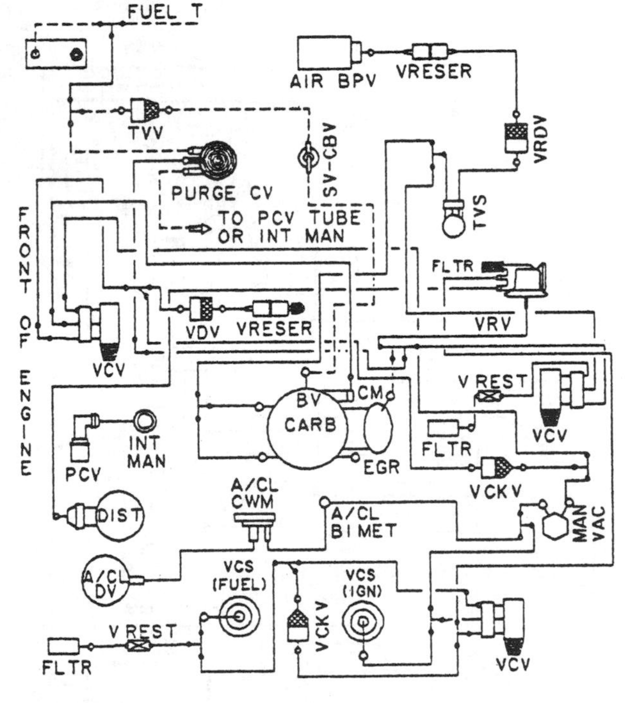

CALIBRATION: 1—20H—R0 DATE: 6-11-80
CANADIAN

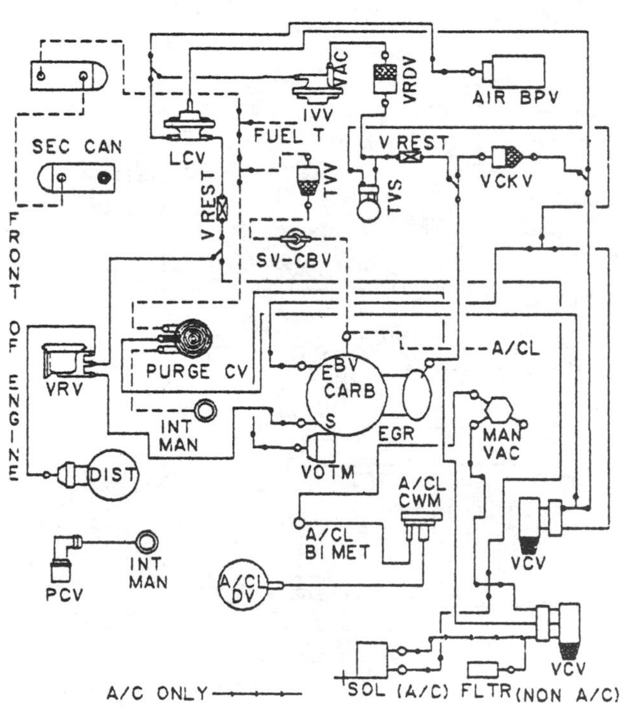

CALIBRATION: 1—20J—R0 DATE: 6-13-80
CANADIAN

VACUUM CIRCUITS

(© Ford Motor Co.)

1981—302 CID (5.0L) engines

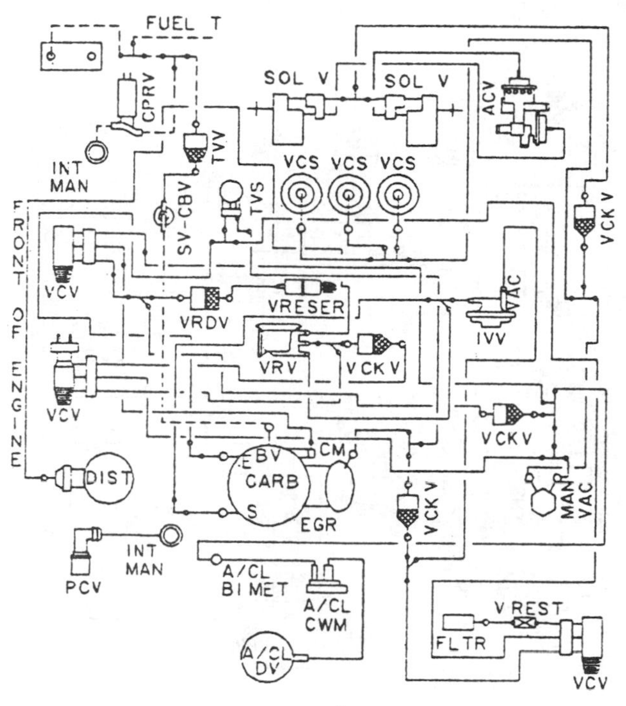

CALIBRATION: 1—20N—R0 DATE: 9-5-80

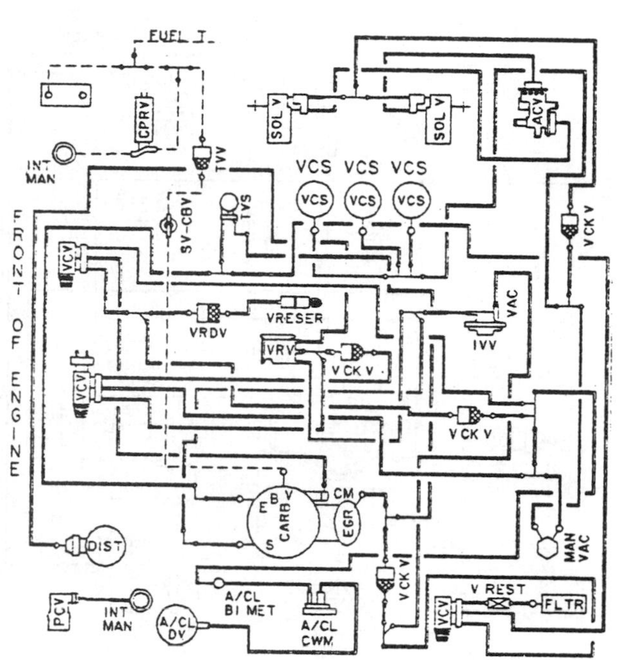

CALIBRATION: 1—20N—R10 DATE: 1-13-81

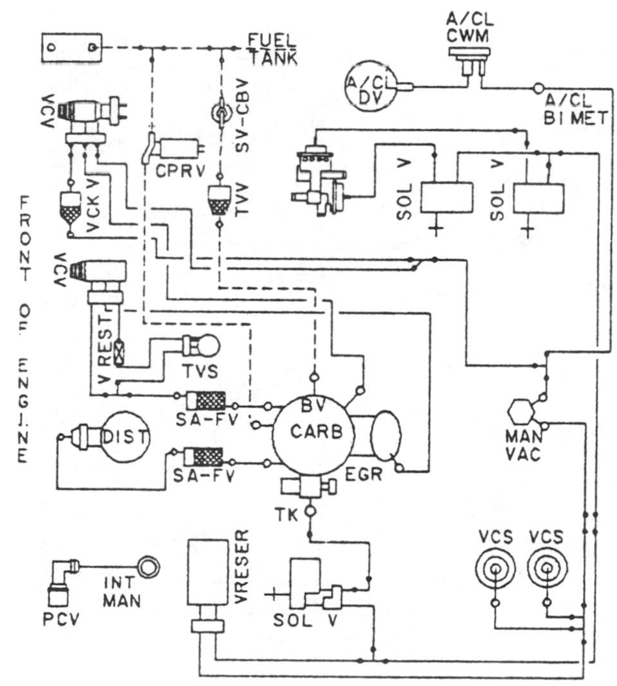

CALIBRATION: 1—20T—R0 DATE: 6-30-80

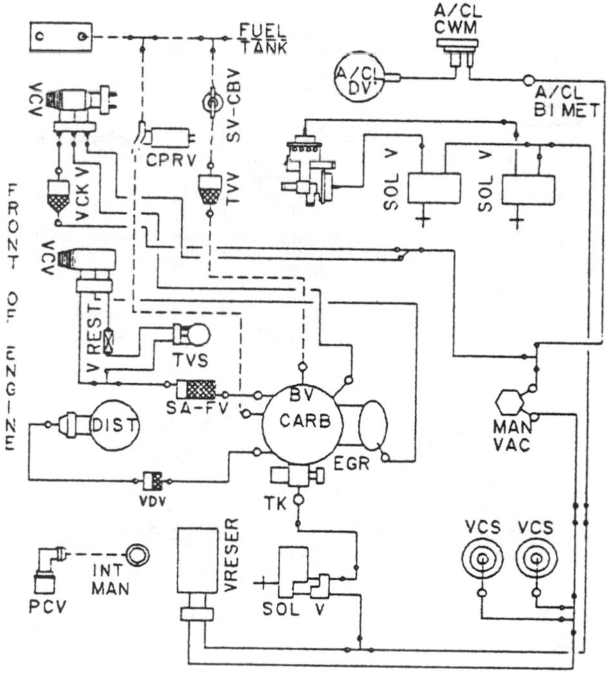

CALIBRATION: 1—20T—R11 DATE: 1-27-81

VACUUM CIRCUITS

(© Ford Motor Co)

1981—302 CID (5.0L) engines

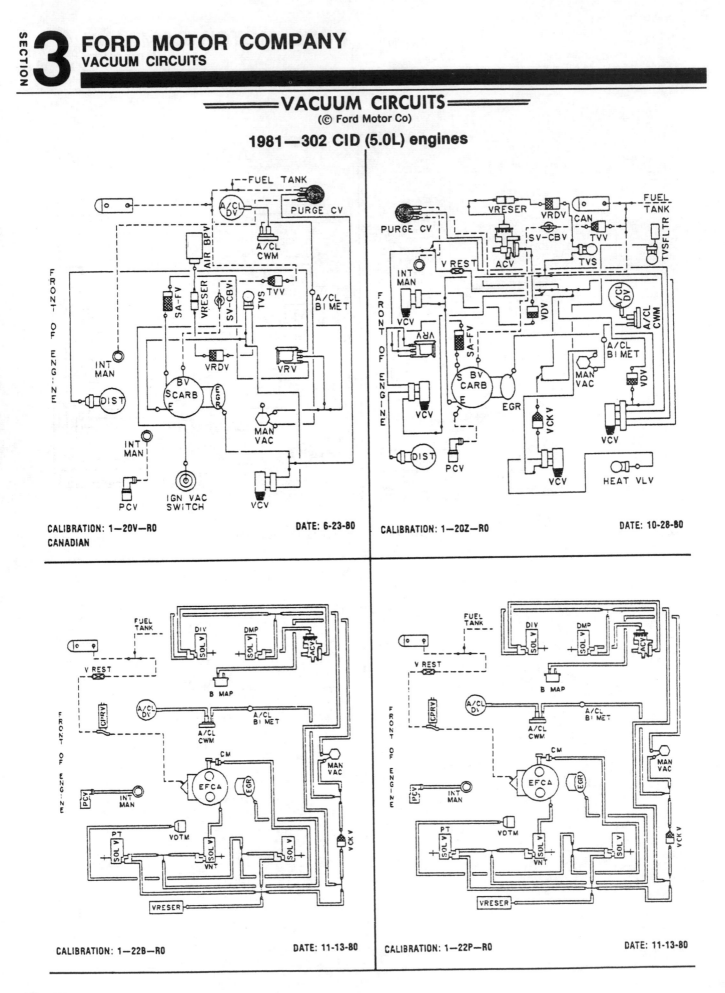

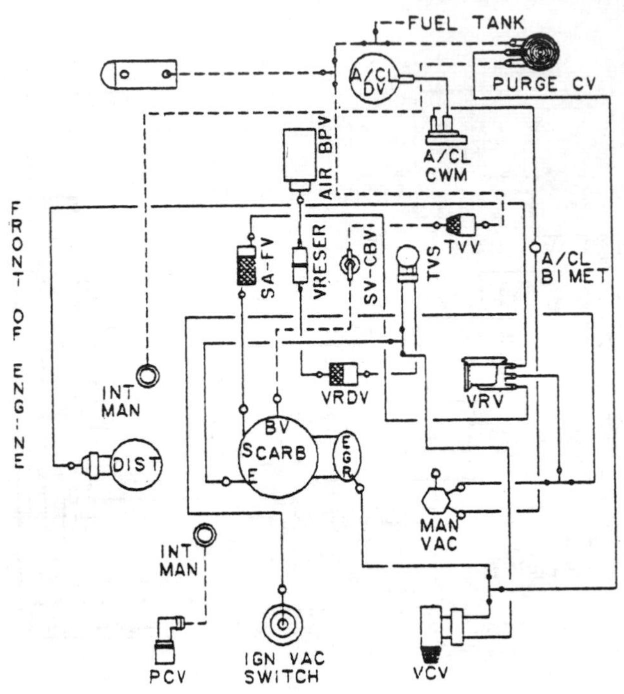

CALIBRATION: 1—20V—R0
CANADIAN

DATE: 6-23-80

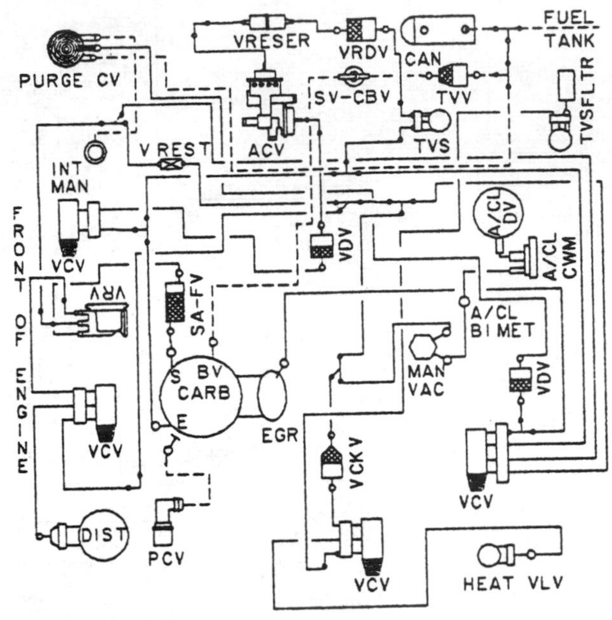

CALIBRATION: 1—20Z—R0

DATE: 10-28-80

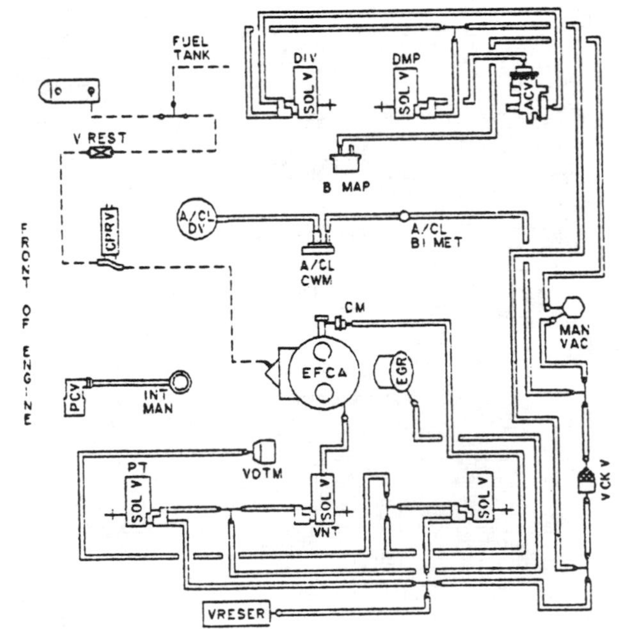

CALIBRATION: 1—22B—R0

DATE: 11-13-80

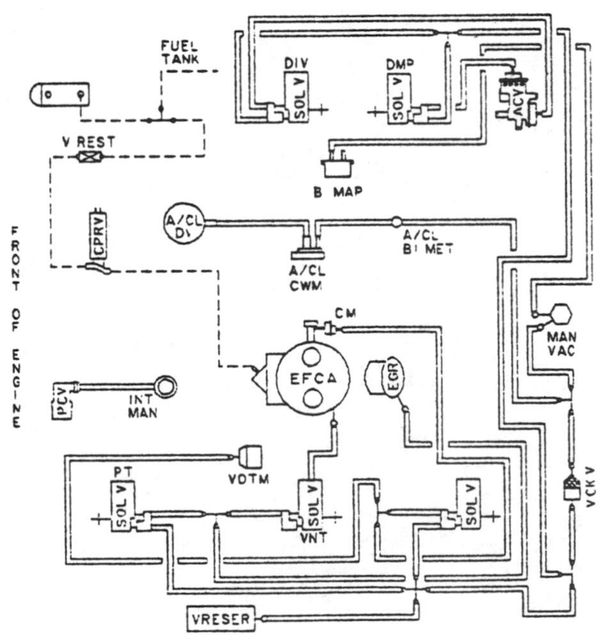

CALIBRATION: 1—22P—R0

DATE: 11-13-80

VACUUM CIRCUITS

(© Ford Motor Co.)

1981—351 CID (5.8L) engines

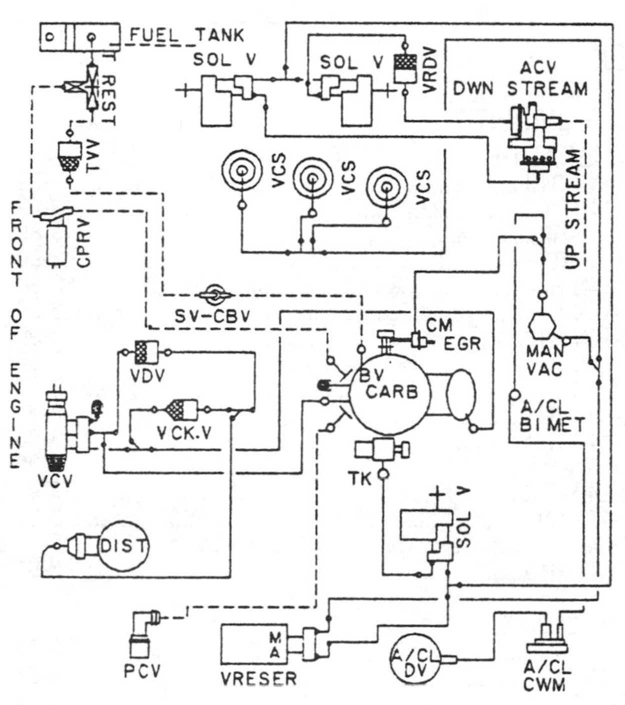

CALIBRATION: 1—24P—R0
CANADIAN

DATE: 8-8-80

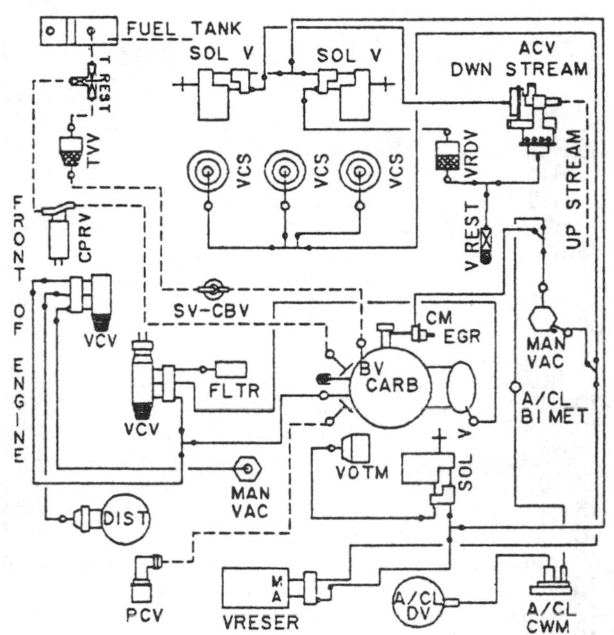

CALIBRATION: 1—24P—R21

DATE: 11-11-80

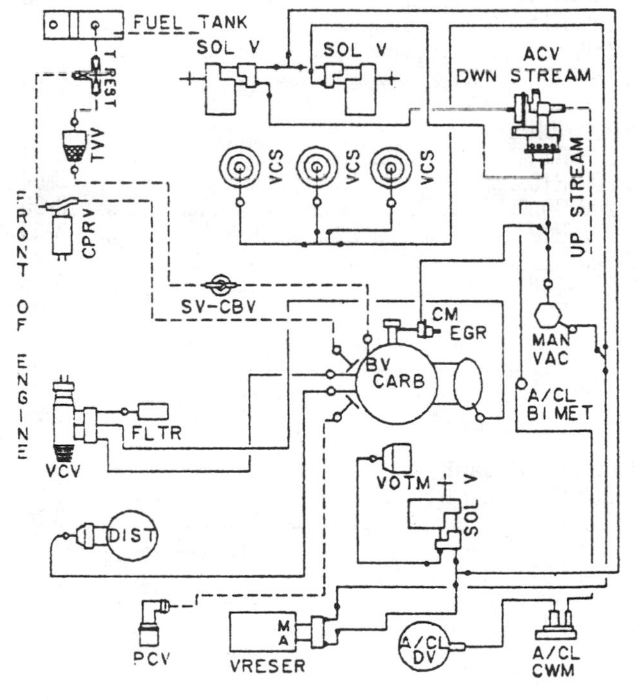

CALIBRATION: 1-24P-R22
CANADIAN

DATE: 11-26-80

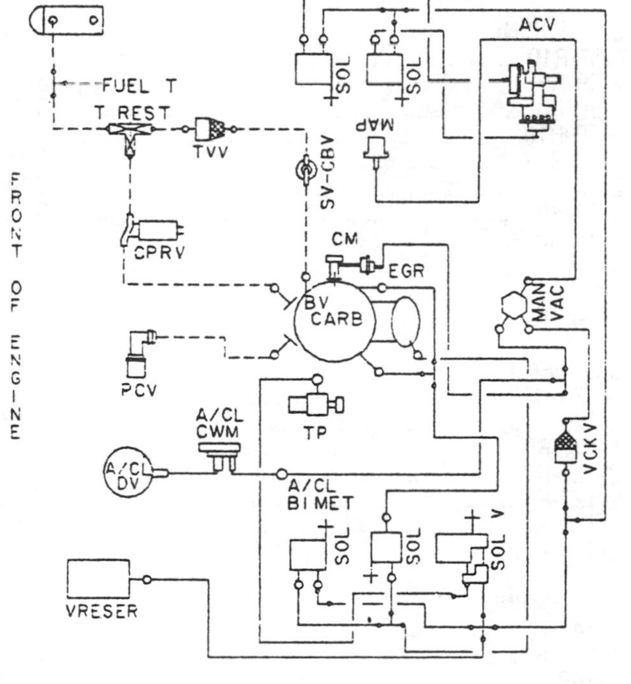

CALIBRATION: 1—24R—R0

DATE: 7-30-80

Ford
Vacuum Circuits

INDEX

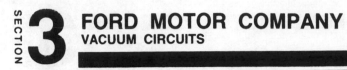

1982 FORD VACUUM CIRCUITS
ENGINE AND CALIBRATION

VACUUM CIRCUITS

(© Ford Motor Co)

HOW TO DETERMINE VEHICLE CALIBRATION

ITEM A—Vehicle Emission Decal

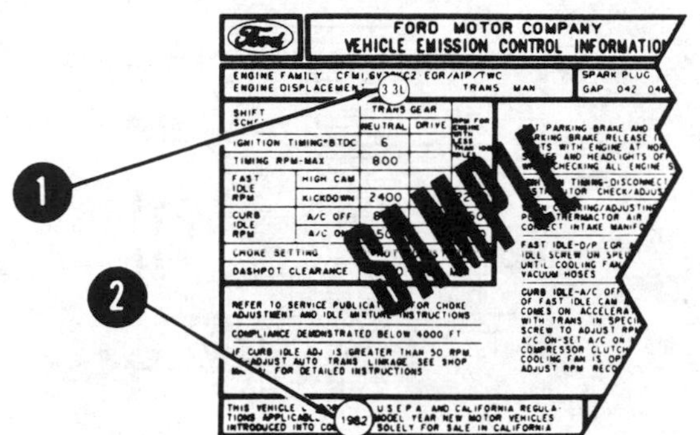

1982 MODELS

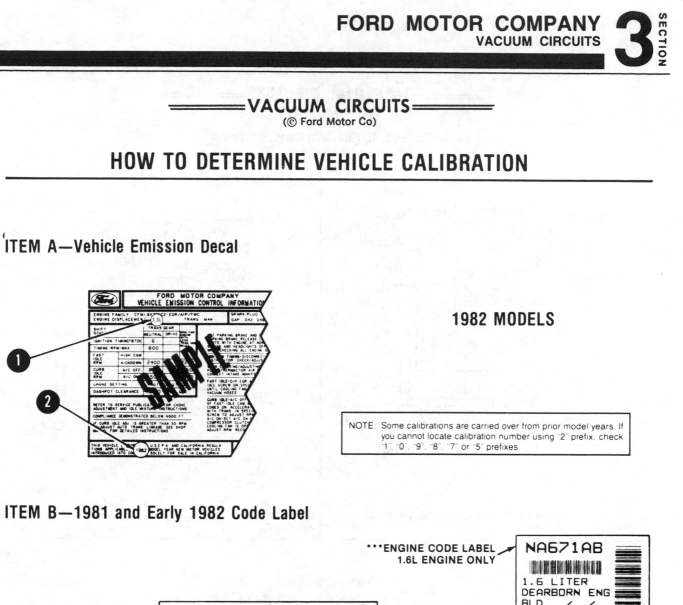

NOTE: Some calibrations are carried over from prior model years. If you cannot locate calibration number using "2" prefix, check "1", "0", "9", "8", "7" or "5" prefixes.

ITEM B—1981 and Early 1982 Code Label

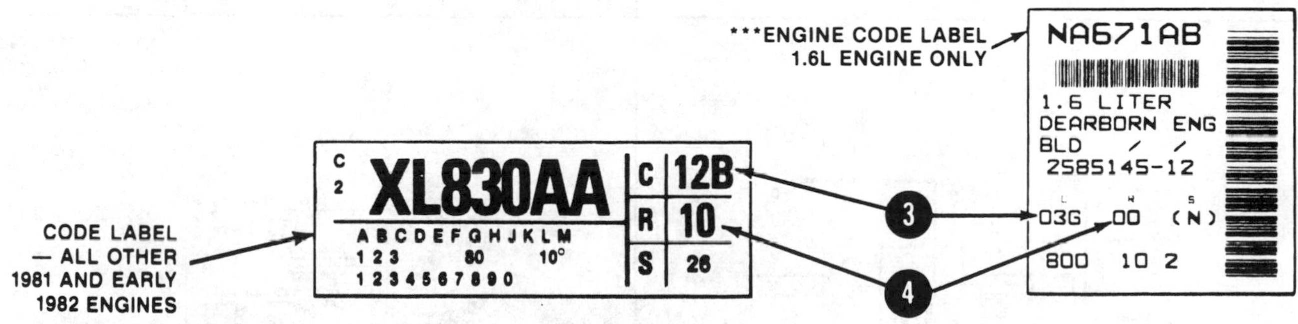

1. DETERMINE ENGINE DISPLACEMENT FROM EMISSION CONTROL DECAL**—ENTER IN BLANK "A" OF EXAMPLE SHOWN ON NEXT PAGE.
2. DETERMINE VEHICLE MODEL YEAR FROM EMISSION CONTROL DECAL**—ENTER LAST DIGIT IN BLANK "B." SEE EXAMPLE ABOVE.
3. DETERMINE CALIBRATION BASE NUMBER FROM ENGINE CODE LABEL*—ENTER NUMBERS AND LETTERS IN BLANK "C."
4. DETERMINE REVISION LEVEL FROM ENGINE CODE LABEL*—ENTER IN BLANK "D."

EXAMPLE: ENGINE: 3.3L (200 CID) (A) CALIBRATION 2 (B) – 12B (C) – R10 (D)

VACUUM CIRCUITS

(© Ford Motor Co)

1982—1.6L (98 CID) 4 cyl. engines

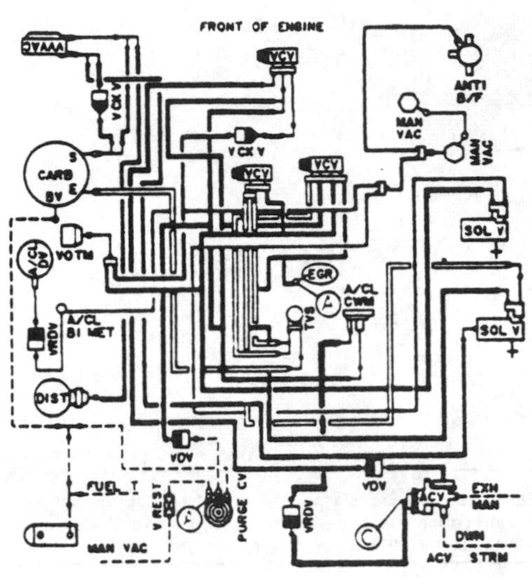

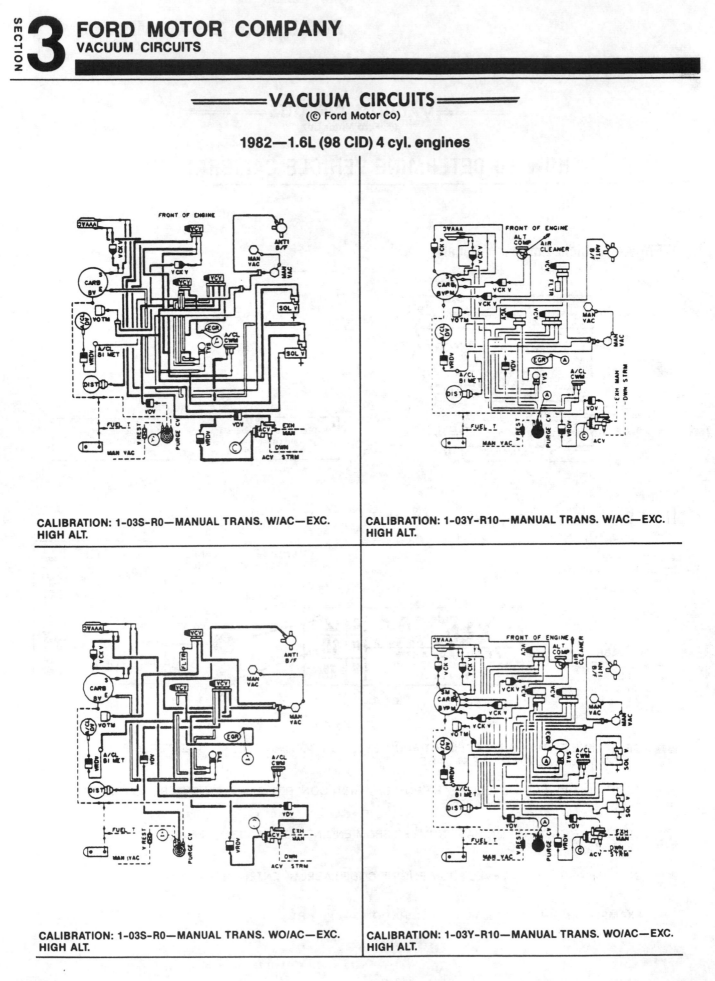

CALIBRATION: 1-03S-R0—MANUAL TRANS. W/AC—EXC. HIGH ALT.

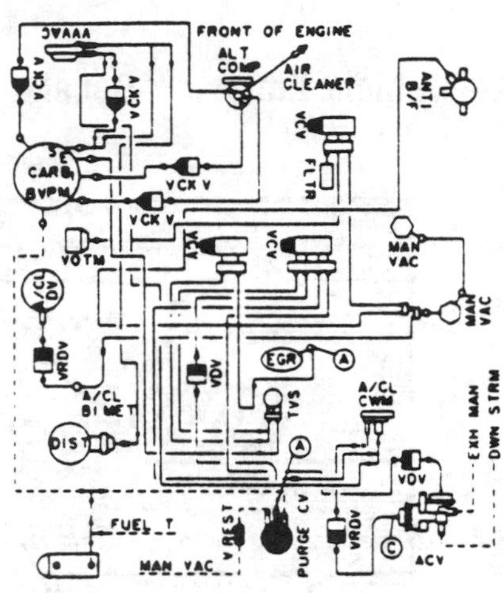

CALIBRATION: 1-03Y-R10—MANUAL TRANS. W/AC—EXC. HIGH ALT.

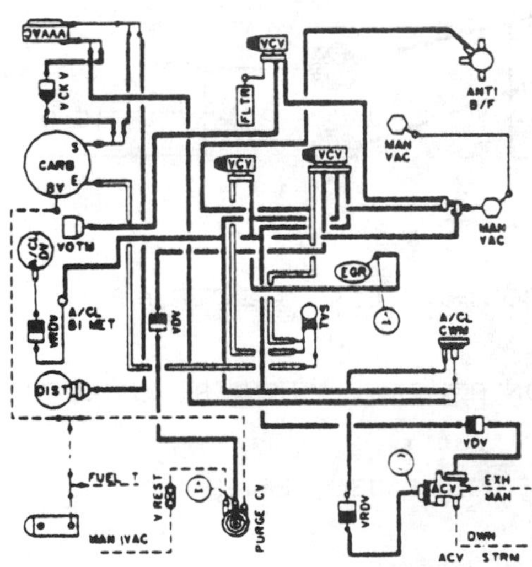

CALIBRATION: 1-03S-R0—MANUAL TRANS. WO/AC—EXC. HIGH ALT.

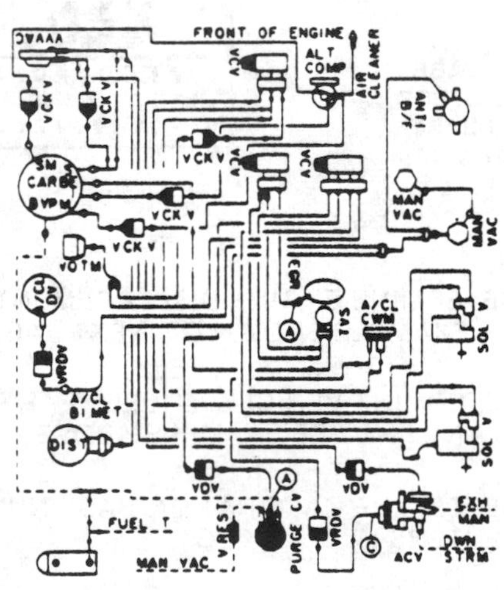

CALIBRATION: 1-03Y-R10—MANUAL TRANS. WO/AC—EXC. HIGH ALT.

VACUUM CIRCUITS

(© Ford Motor Co)

1982—1.6L (98 CID) 4 cyl. engines

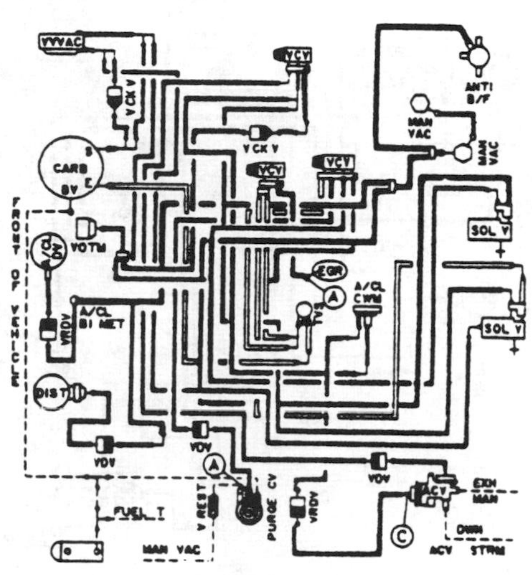

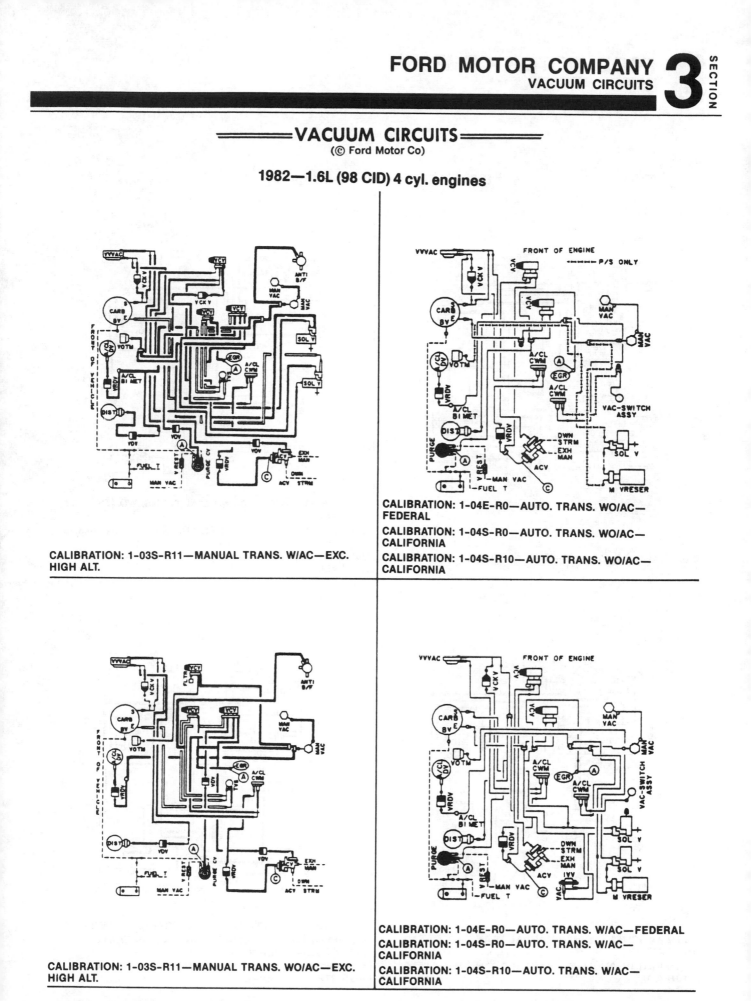

CALIBRATION: 1-03S-R11—MANUAL TRANS. W/AC—EXC. HIGH ALT.

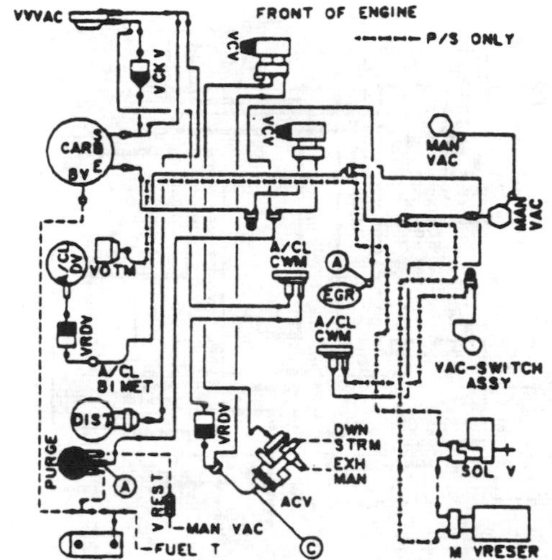

CALIBRATION: 1-04E-R0—AUTO. TRANS. WO/AC—FEDERAL

CALIBRATION: 1-04S-R0—AUTO. TRANS. WO/AC—CALIFORNIA

CALIBRATION: 1-04S-R10—AUTO. TRANS. WO/AC—CALIFORNIA

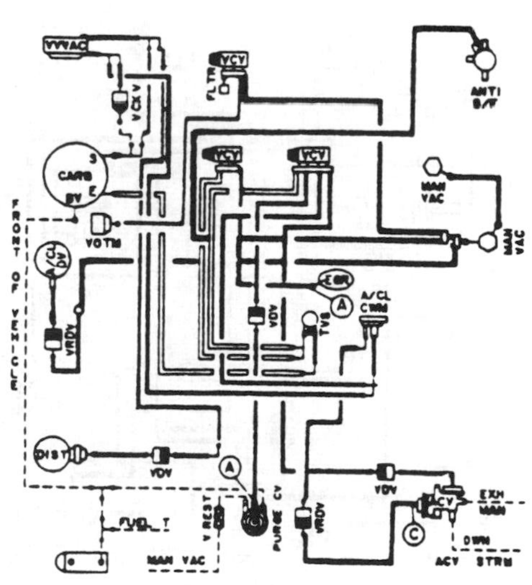

CALIBRATION: 1-03S-R11—MANUAL TRANS. WO/AC—EXC. HIGH ALT.

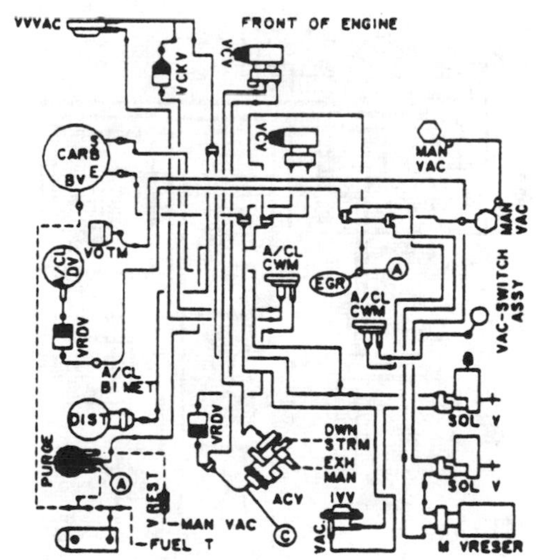

CALIBRATION: 1-04E-R0—AUTO. TRANS. W/AC—FEDERAL

CALIBRATION: 1-04S-R0—AUTO. TRANS. W/AC—CALIFORNIA

CALIBRATION: 1-04S-R10—AUTO. TRANS. W/AC—CALIFORNIA

VACUUM CIRCUITS

(© Ford Motor Co.)

1982—1.6L (98 CID) 4 cyl. engines

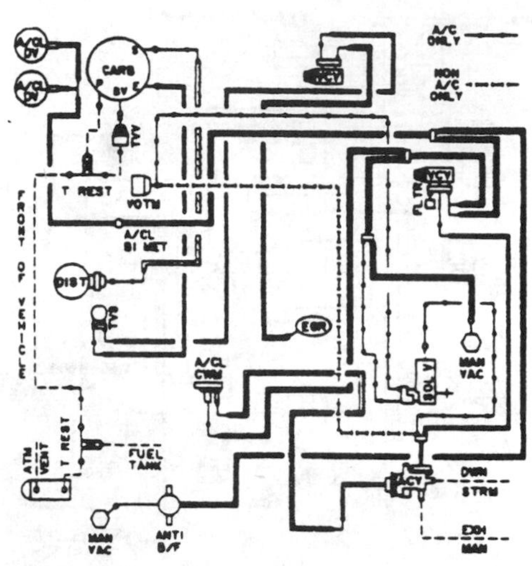

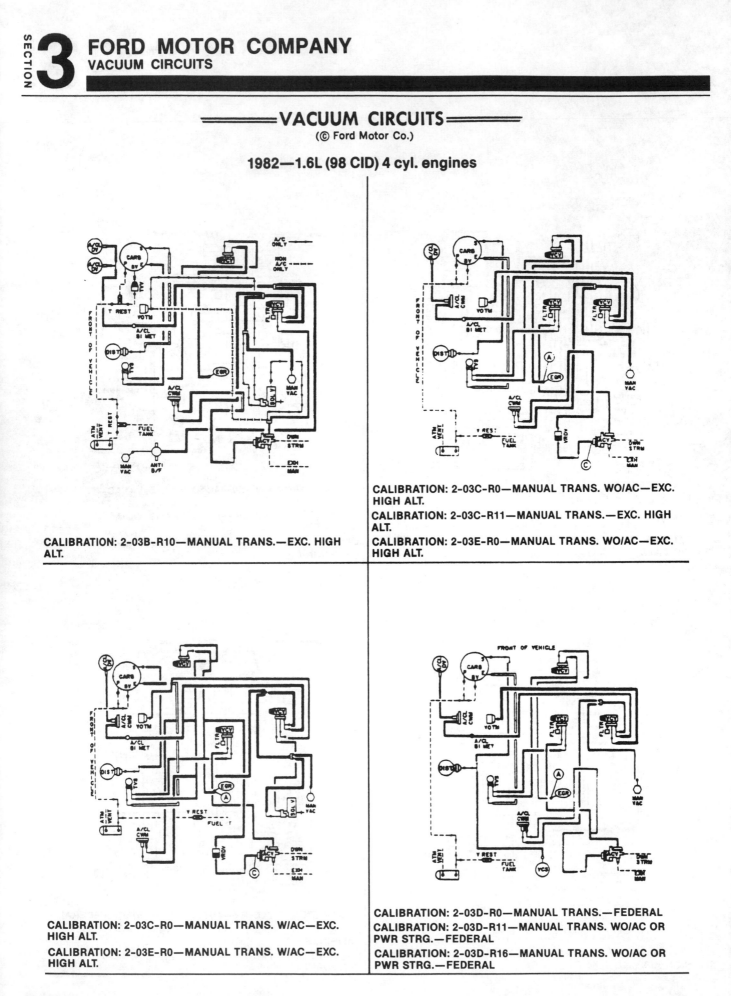

CALIBRATION: 2-03B-R10—MANUAL TRANS.—EXC. HIGH ALT.

CALIBRATION: 2-03C-R0—MANUAL TRANS. WO/AC—EXC. HIGH ALT.

CALIBRATION: 2-03C-R11—MANUAL TRANS.—EXC. HIGH ALT.

CALIBRATION: 2-03E-R0—MANUAL TRANS. WO/AC—EXC. HIGH ALT.

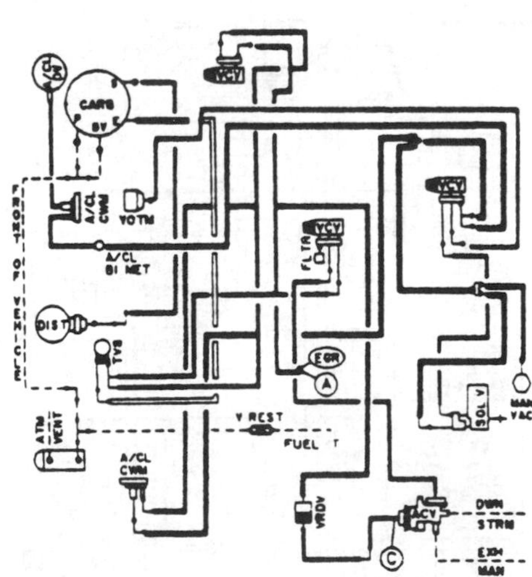

CALIBRATION: 2-03C-R0—MANUAL TRANS. W/AC—EXC. HIGH ALT.

CALIBRATION: 2-03E-R0—MANUAL TRANS. W/AC—EXC. HIGH ALT.

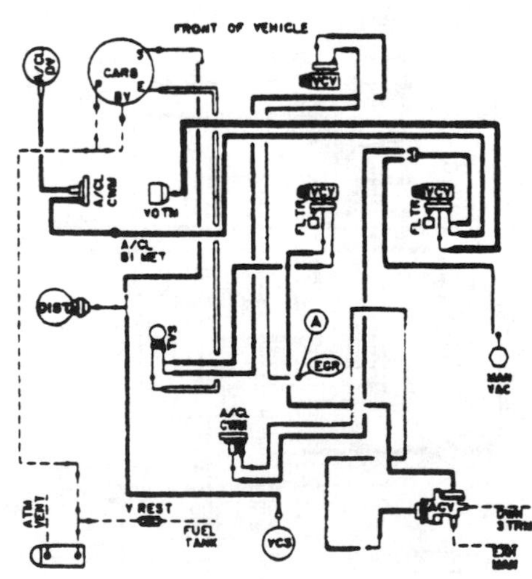

CALIBRATION: 2-03D-R0—MANUAL TRANS.—FEDERAL

CALIBRATION: 2-03D-R11—MANUAL TRANS. WO/AC OR PWR STRG.—FEDERAL

CALIBRATION: 2-03D-R16—MANUAL TRANS. WO/AC OR PWR STRG.—FEDERAL

VACUUM CIRCUITS

(© Ford Motor Co.)

1982—1.6L (98 CID) 4 cyl. engines

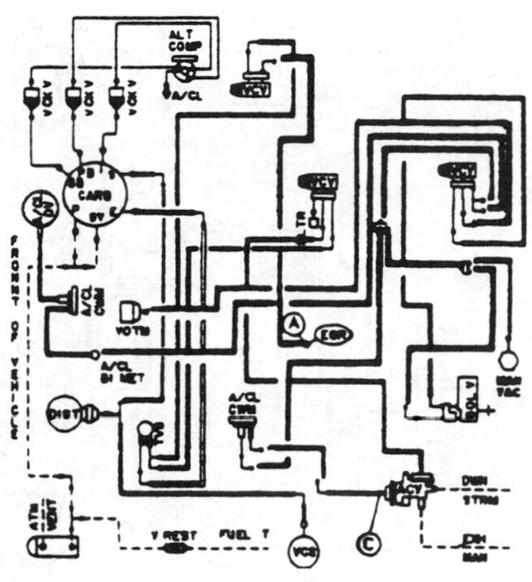

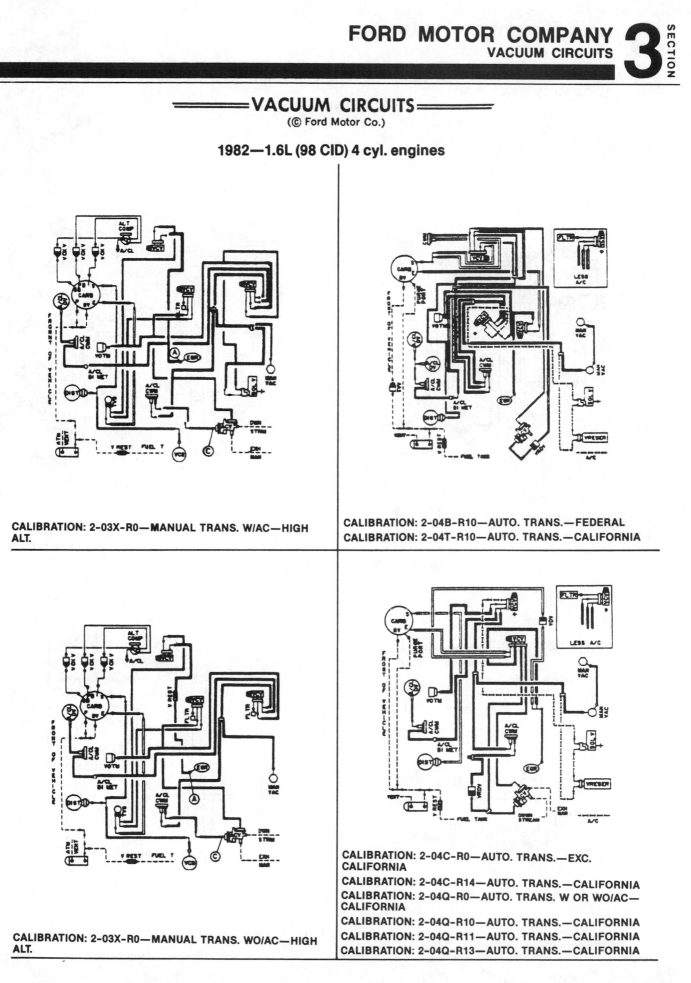

CALIBRATION: 2-03X-R0—MANUAL TRANS. W/AC—HIGH ALT.

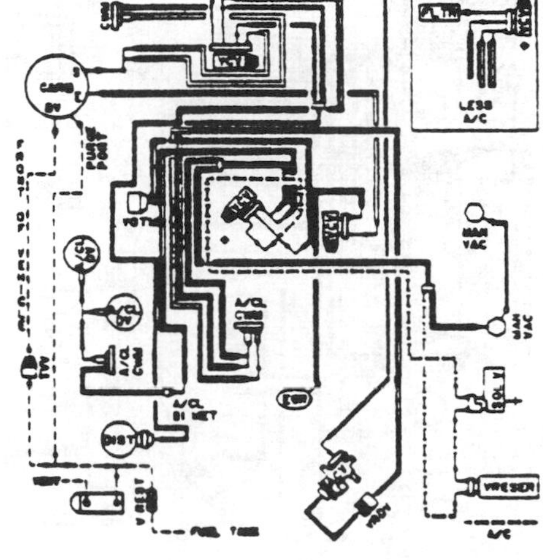

CALIBRATION: 2-04B-R10—AUTO. TRANS.—FEDERAL

CALIBRATION: 2-04T-R10—AUTO. TRANS.—CALIFORNIA

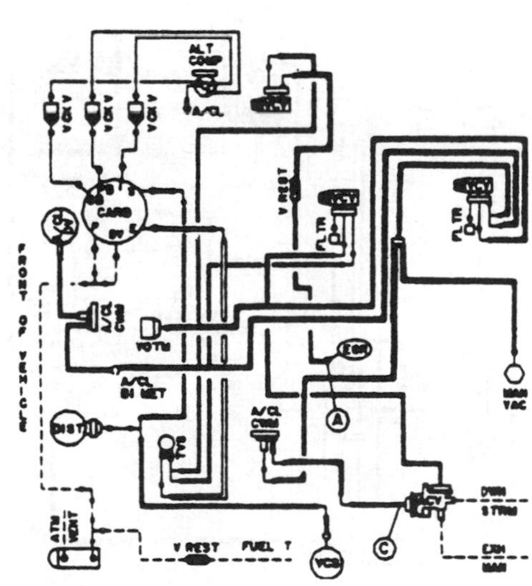

CALIBRATION: 2-03X-R0—MANUAL TRANS. WO/AC—HIGH ALT.

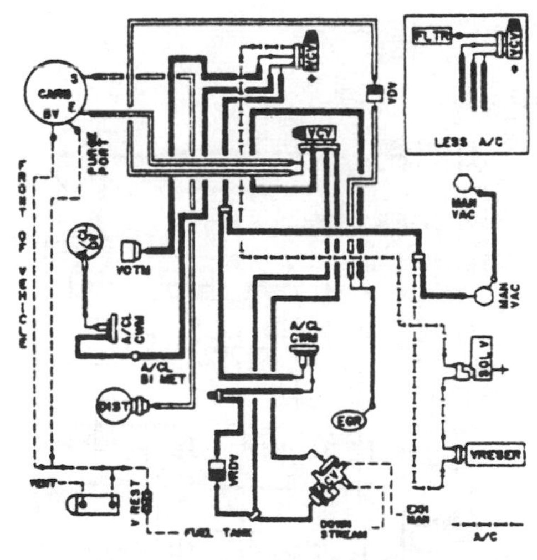

CALIBRATION: 2-04C-R0—AUTO. TRANS.—EXC. CALIFORNIA

CALIBRATION: 2-04C-R14—AUTO. TRANS.—CALIFORNIA

CALIBRATION: 2-04Q-R0—AUTO. TRANS. W OR WO/AC—CALIFORNIA

CALIBRATION: 2-04Q-R10—AUTO. TRANS.—CALIFORNIA

CALIBRATION: 2-04Q-R11—AUTO. TRANS.—CALIFORNIA

CALIBRATION: 2-04Q-R13—AUTO. TRANS.—CALIFORNIA

VACUUM CIRCUITS

1982—1.6L (98 CID) 4 cyl. engines

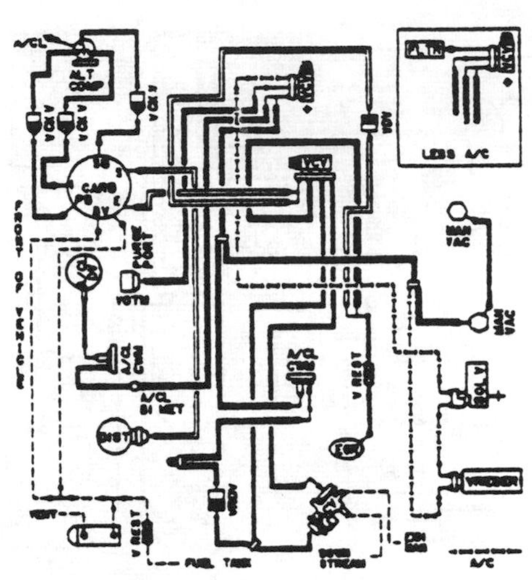

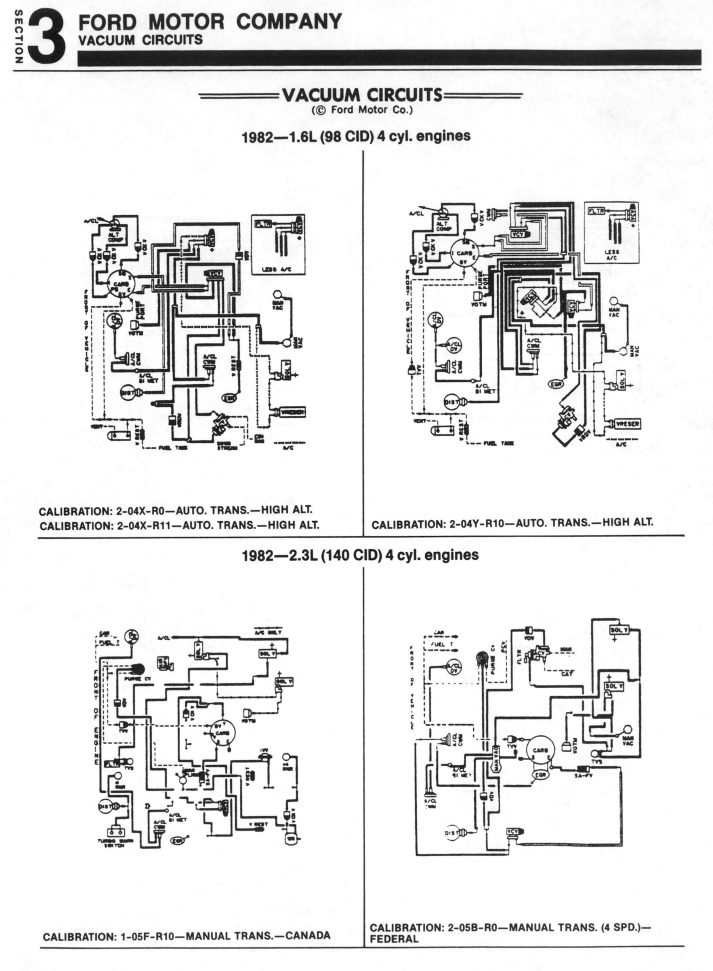

CALIBRATION: 2-04X-R0—AUTO. TRANS.—HIGH ALT.
CALIBRATION: 2-04X-R11—AUTO. TRANS.—HIGH ALT.

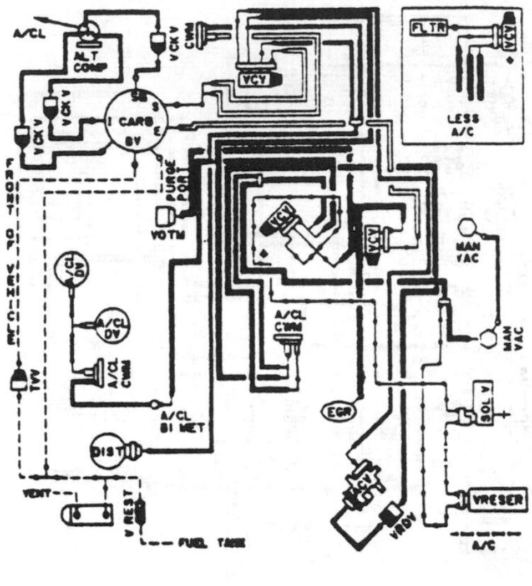

CALIBRATION: 2-04Y-R10—AUTO. TRANS.—HIGH ALT.

1982—2.3L (140 CID) 4 cyl. engines

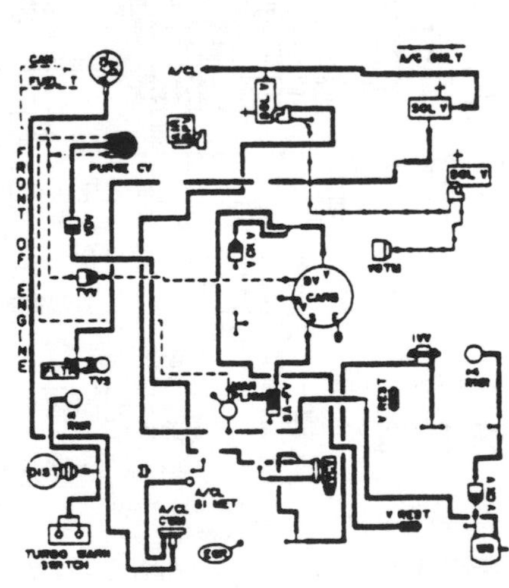

CALIBRATION: 1-05F-R10—MANUAL TRANS.—CANADA

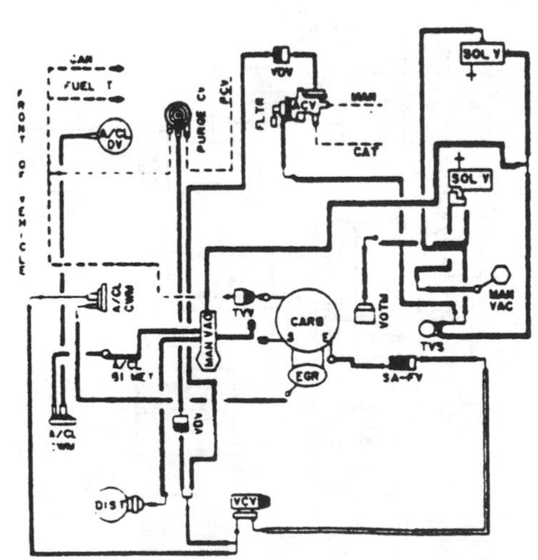

CALIBRATION: 2-05B-R0—MANUAL TRANS. (4 SPD.)—FEDERAL

VACUUM CIRCUITS

(© Ford Motor Co.)

1982—2.3L (140 CID) 4 cyl. engines

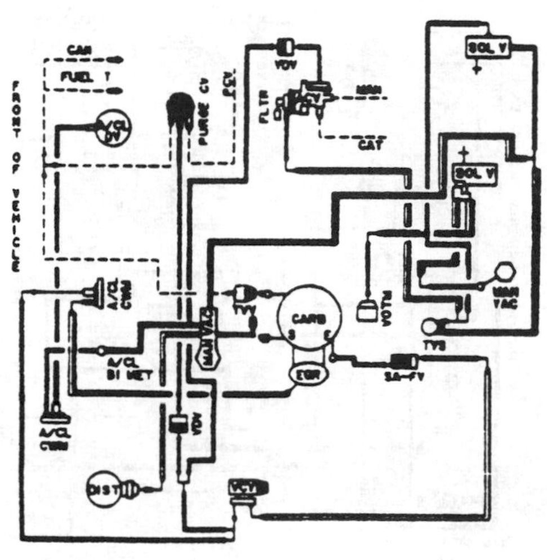

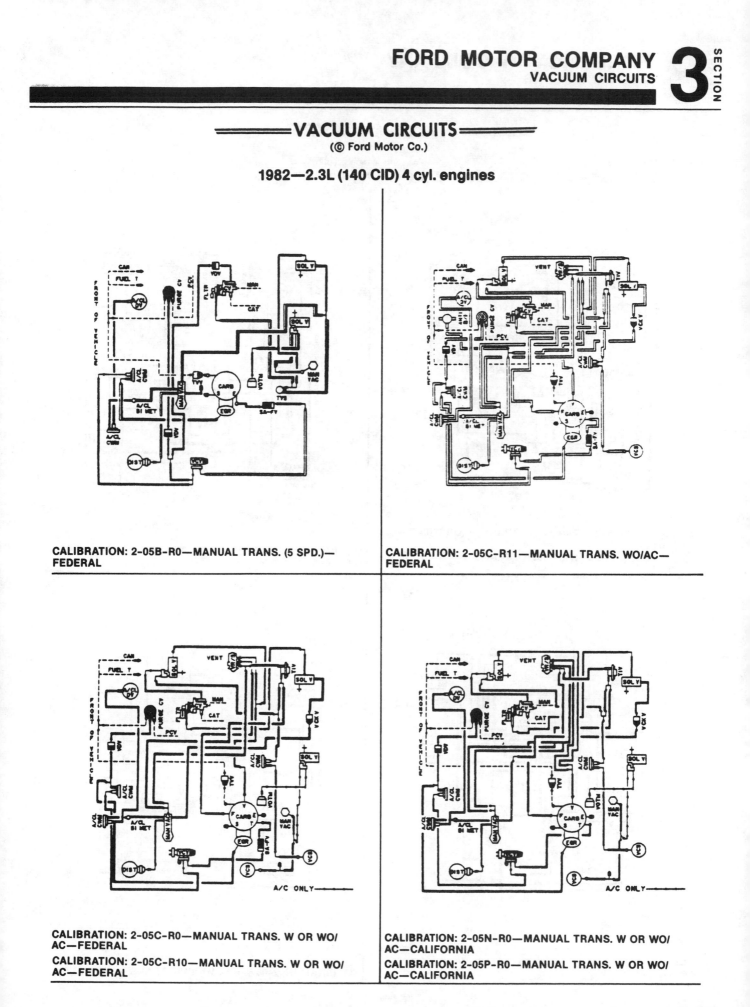

CALIBRATION: 2-05B-R0—MANUAL TRANS. (5 SPD.)—FEDERAL

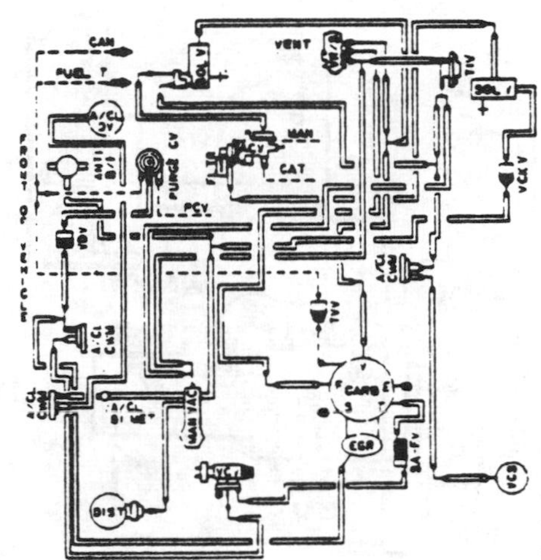

CALIBRATION: 2-05C-R11—MANUAL TRANS. WO/AC—FEDERAL

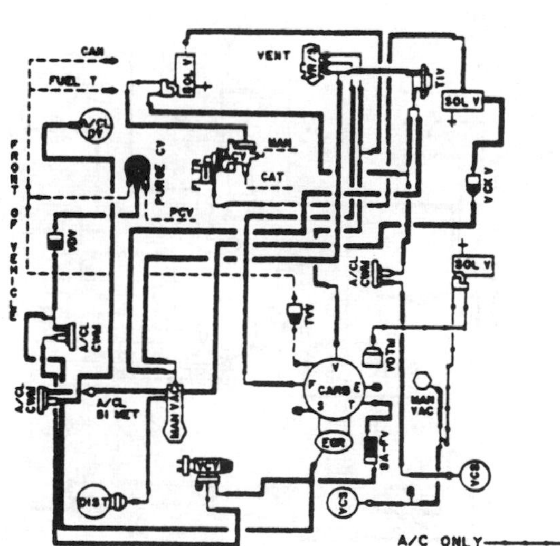

CALIBRATION: 2-05C-R0—MANUAL TRANS. W OR WO/AC—FEDERAL

CALIBRATION: 2-05C-R10—MANUAL TRANS. W OR WO/AC—FEDERAL

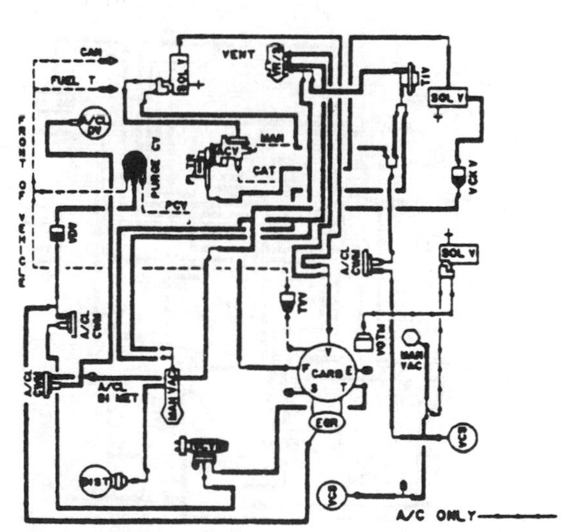

CALIBRATION: 2-05N-R0—MANUAL TRANS. W OR WO/AC—CALIFORNIA

CALIBRATION: 2-05P-R0—MANUAL TRANS. W OR WO/AC—CALIFORNIA

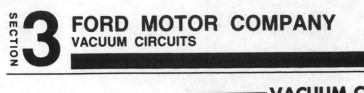

VACUUM CIRCUITS

(© Ford Motor Co.)

1982—2.3L (140 CID) 4 cyl. engines

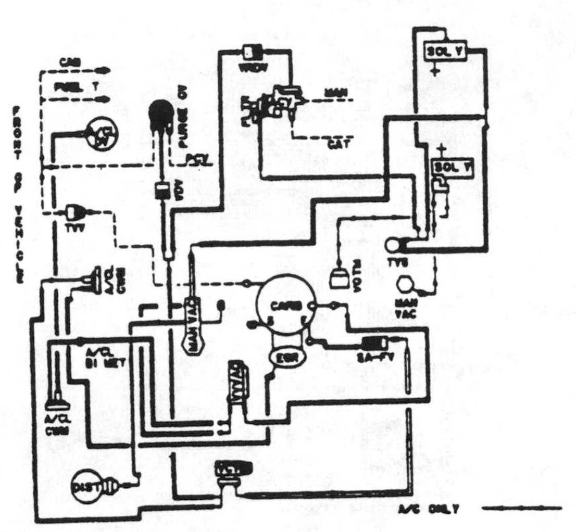

CALIBRATION: 2-06A-R0—AUTO. TRANS. W OR WO/AC—FEDERAL
CALIBRATION: 2-06A-R10—AUTO. TRANS. W OR WO/AC—FEDERAL

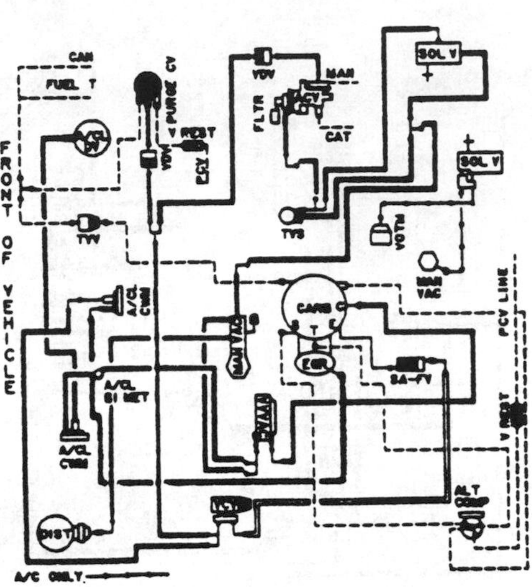

CALIBRATION: 2-06W-R0—AUTO. TRANS. W OR WO/AC—HIGH ALT.

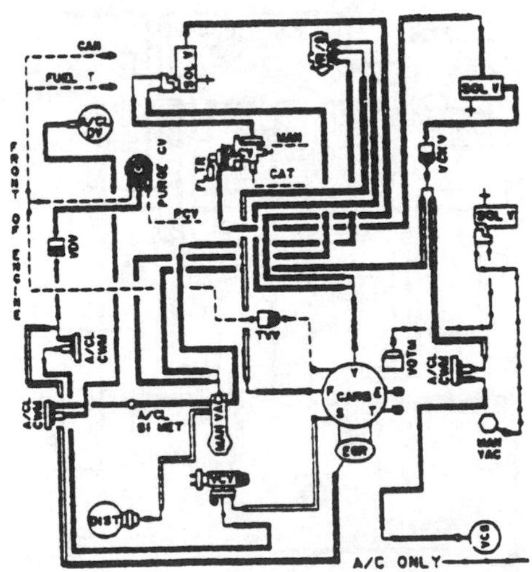

CALIBRATION: 2-06N-R0—AUTO. TRANS. W OR WO/AC—CALIFORNIA

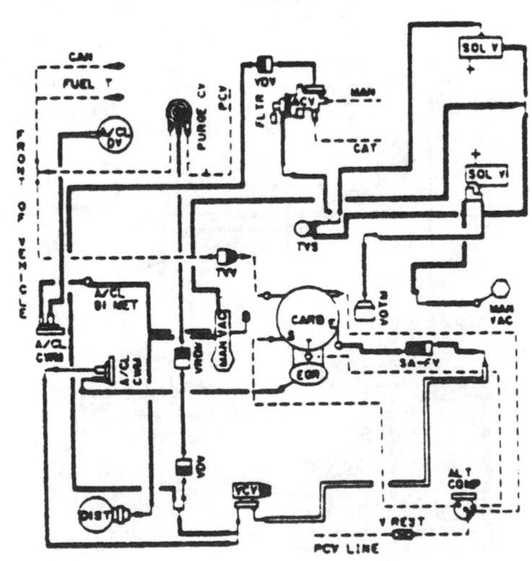

CALIBRATION: 2-05W-R0—MANUAL TRANS.—HIGH ALT.

VACUUM CIRCUITS

(© Ford Motor Co)

1982—2.3L (140 CID) 4 cyl. engines

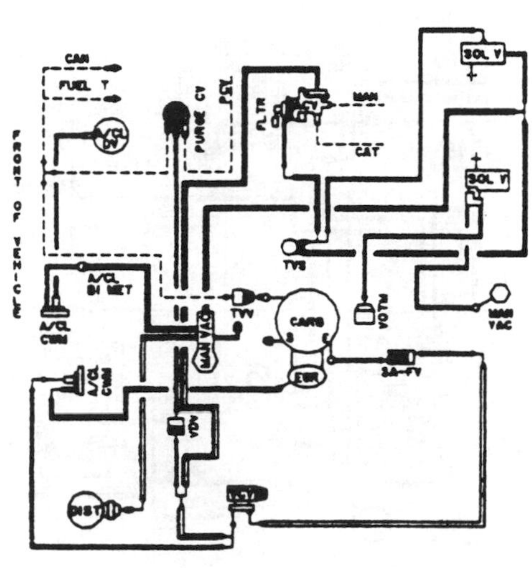

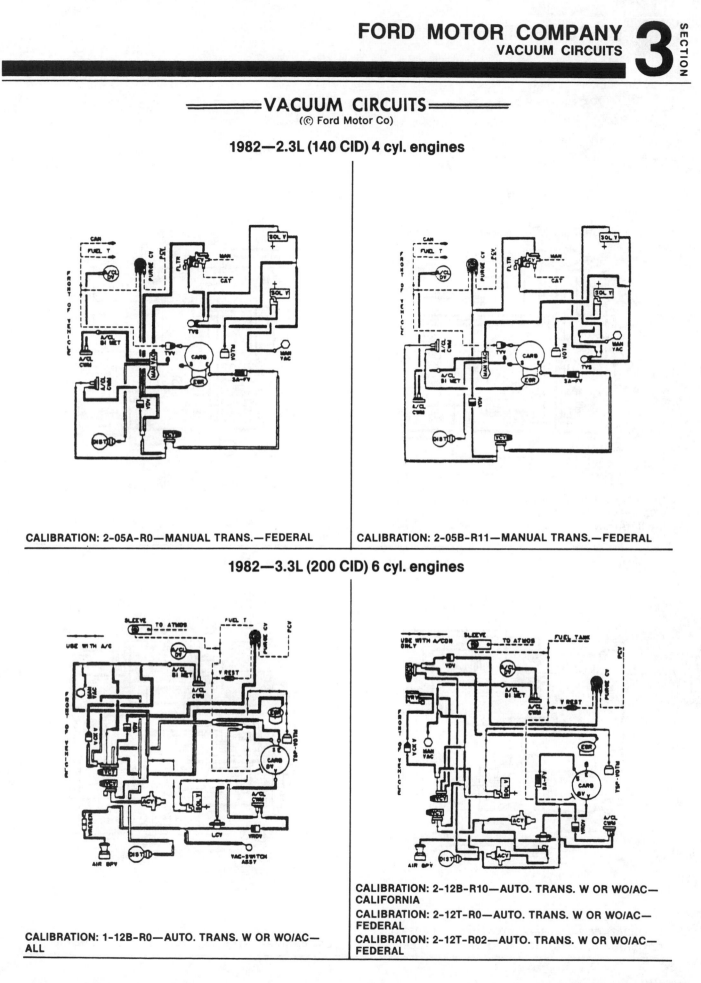

CALIBRATION: 2-05A-R0—MANUAL TRANS.—FEDERAL

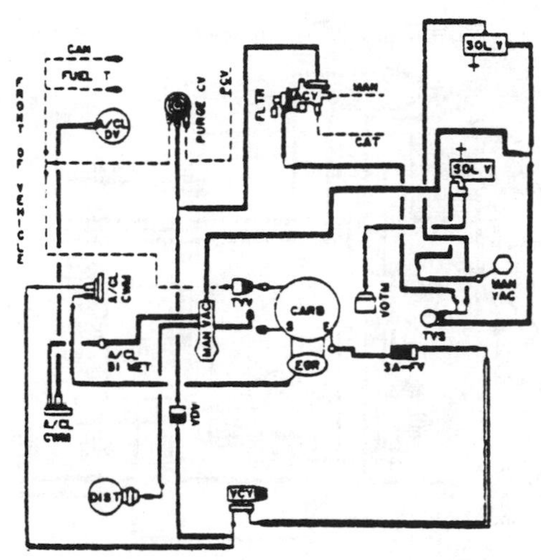

CALIBRATION: 2-05B-R11—MANUAL TRANS.—FEDERAL

1982—3.3L (200 CID) 6 cyl. engines

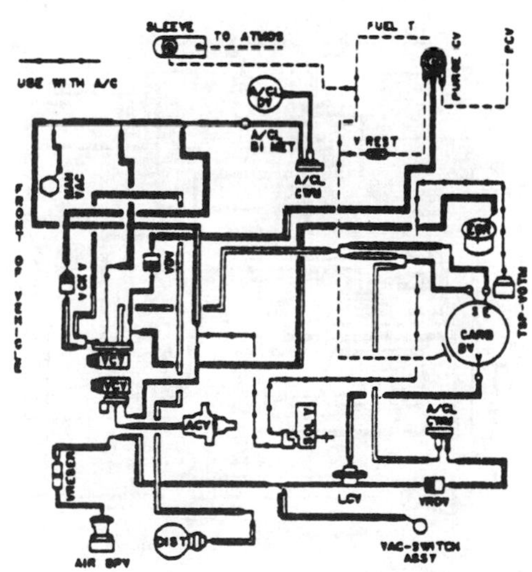

CALIBRATION: 1-12B-R0—AUTO. TRANS. W OR WO/AC—ALL

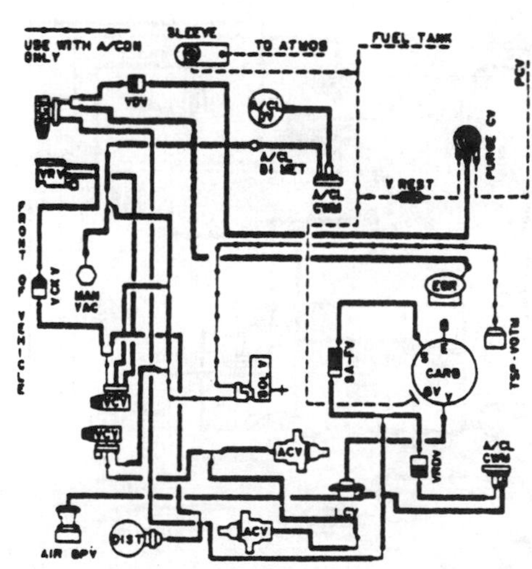

CALIBRATION: 2-12B-R10—AUTO. TRANS. W OR WO/AC—CALIFORNIA

CALIBRATION: 2-12T-R0—AUTO. TRANS. W OR WO/AC—FEDERAL

CALIBRATION: 2-12T-R02—AUTO. TRANS. W OR WO/AC—FEDERAL

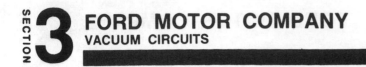

VACUUM CIRCUITS

(© Ford Motor Co.)

1982—3.3L (200 CID) 6 cyl. engines

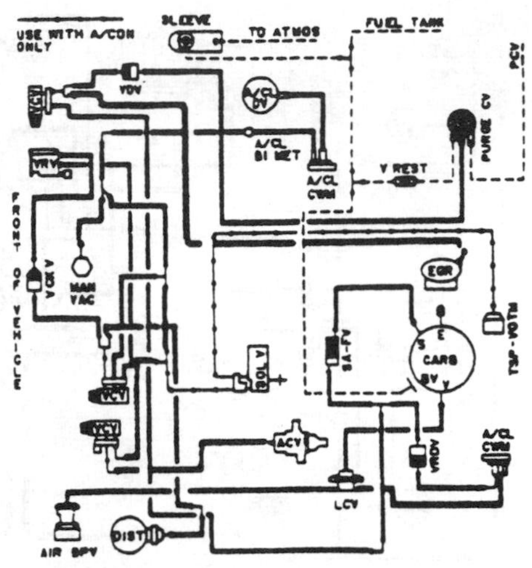

CALIBRATION: 2-12T-R04—AUTO. TRANS. W OR WO/AC—EXC. CALIF.

CALIBRATION: 2-12T-R15—AUTO. TRANS. W OR WO/AC—FEDERAL

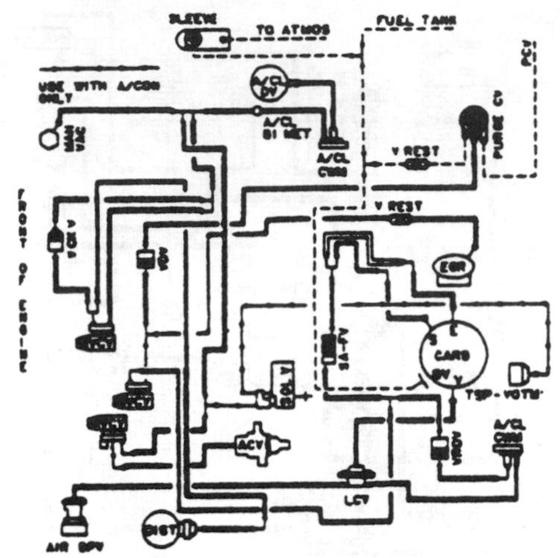

CALIBRATION: 2-12W-R0—AUTO. TRANS. W OR WO/AC—HIGH ALT.

1982—3.8L (230 CID) 6 cyl. engines

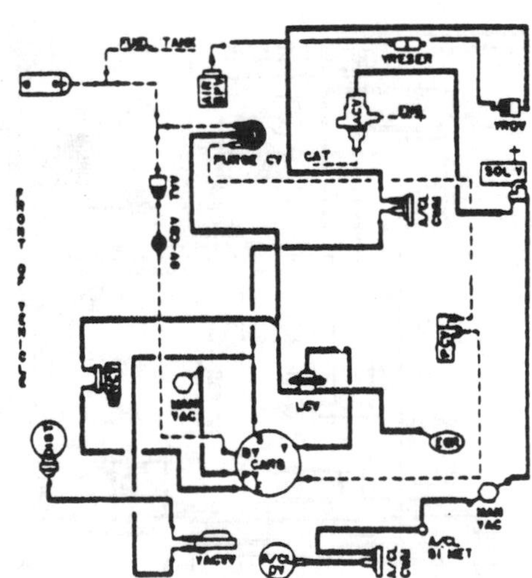

CALIBRATION: 2-14A-R0—AUTO. TRANS.—FEDERAL

CALIBRATION: 2-14A-R13—AUTO. TRANS.—FEDERAL

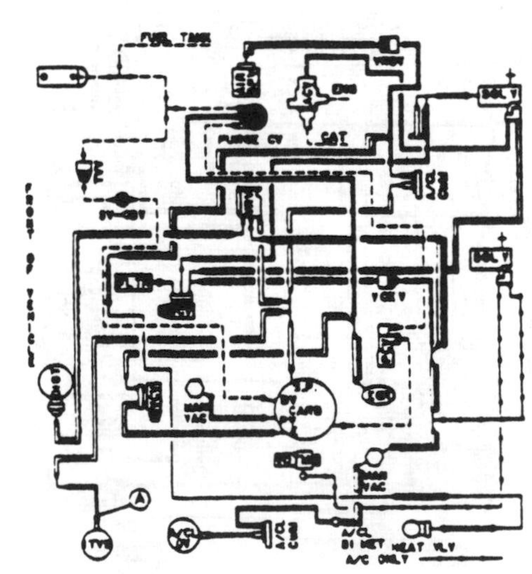

CALIBRATION: 2-14B-R1 —AUTO. TRANS.—FEDERAL

VACUUM CIRCUITS

(© Ford Motor Co.)

1982—3.8L (230 CID) 6 cyl. engines

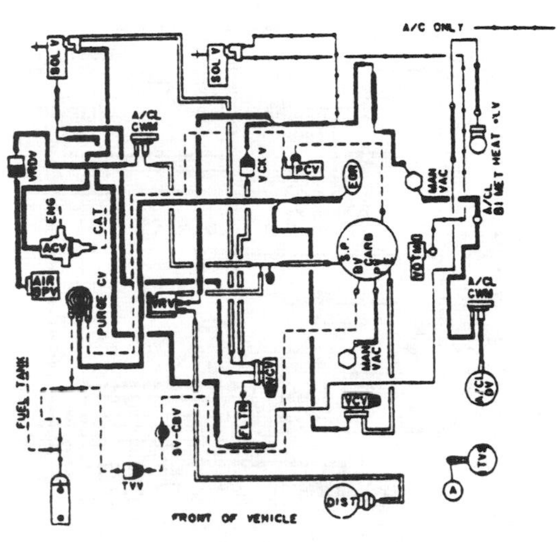

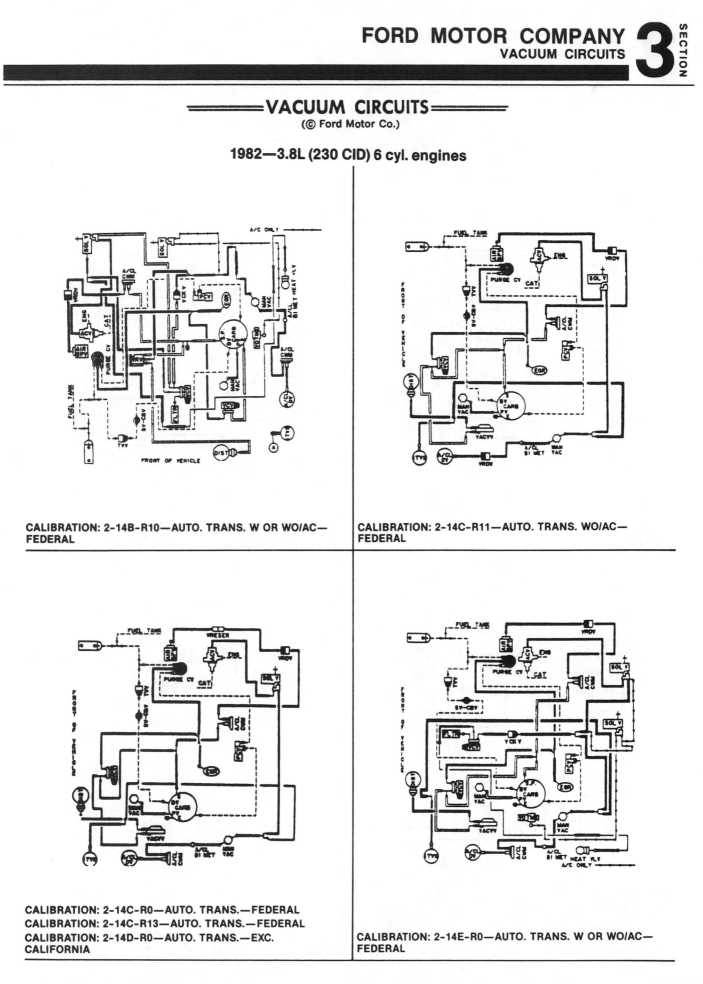

CALIBRATION: 2-14B-R10—AUTO. TRANS. W OR WO/AC—FEDERAL

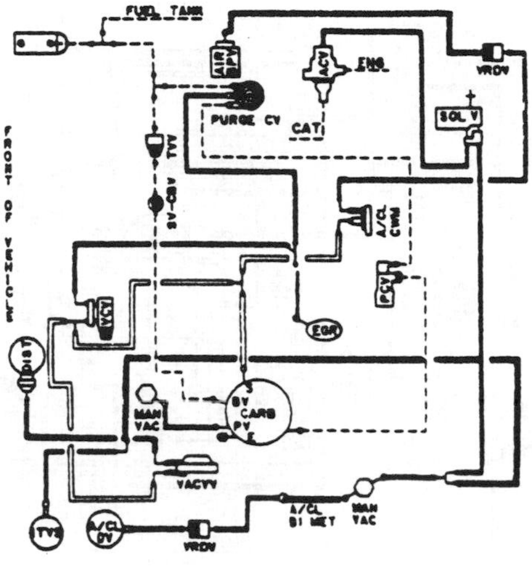

CALIBRATION: 2-14C-R11—AUTO. TRANS. WO/AC—FEDERAL

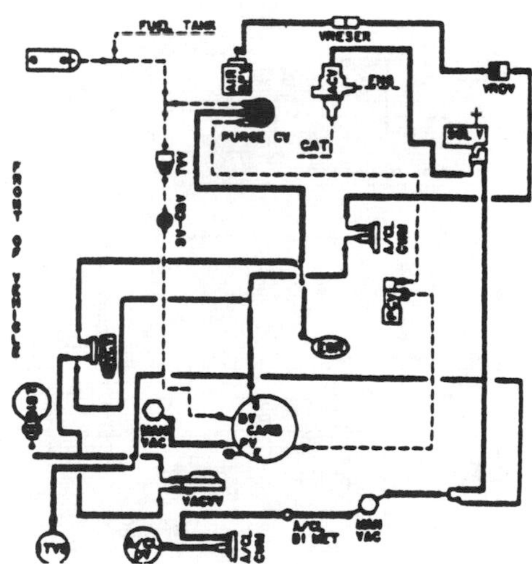

CALIBRATION: 2-14C-R0—AUTO. TRANS.—FEDERAL
CALIBRATION: 2-14C-R13—AUTO. TRANS.—FEDERAL
CALIBRATION: 2-14D-R0—AUTO. TRANS.—EXC. CALIFORNIA

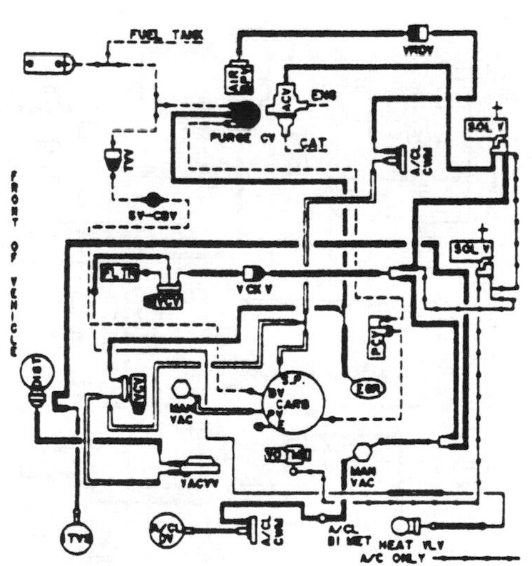

CALIBRATION: 2-14E-R0—AUTO. TRANS. W OR WO/AC—FEDERAL

VACUUM CIRCUITS

1982—3.8L (230 CID) 6 cyl. engines

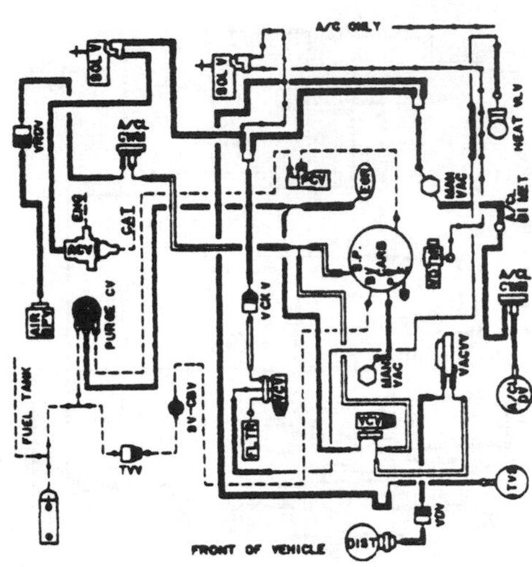

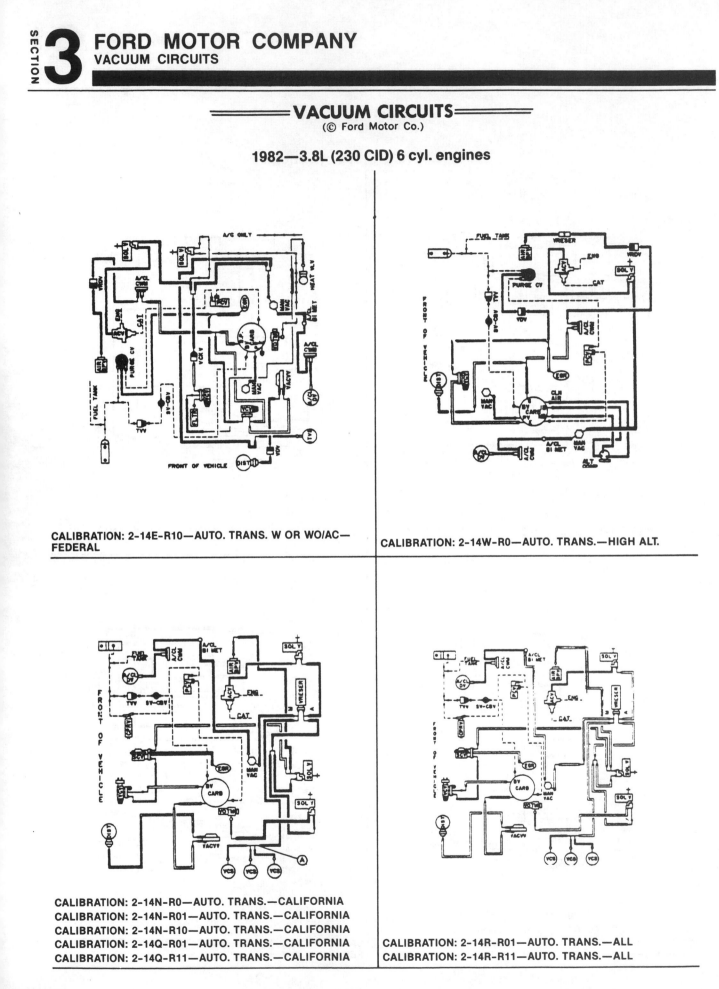

CALIBRATION: 2-14E-R10—AUTO. TRANS. W OR WO/AC—FEDERAL

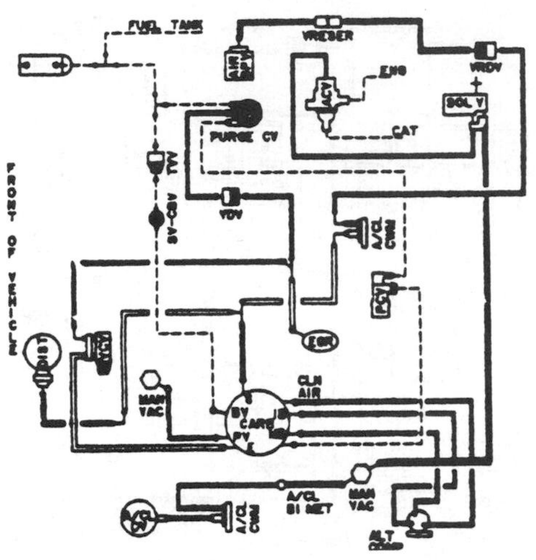

CALIBRATION: 2-14W-R0—AUTO. TRANS.—HIGH ALT.

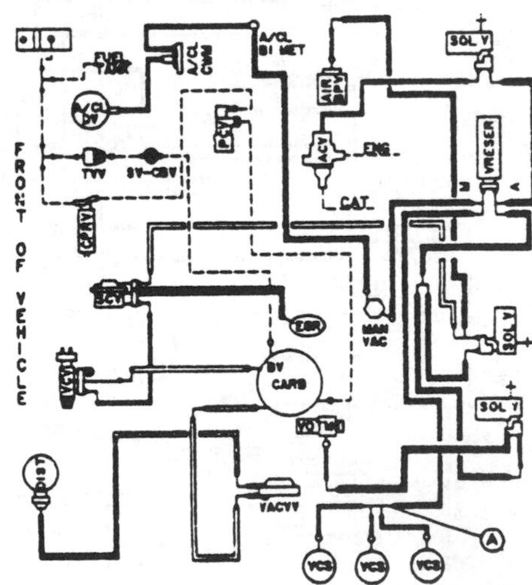

CALIBRATION: 2-14N-R0—AUTO. TRANS.—CALIFORNIA
CALIBRATION: 2-14N-R01—AUTO. TRANS.—CALIFORNIA
CALIBRATION: 2-14N-R10—AUTO. TRANS.—CALIFORNIA
CALIBRATION: 2-14Q-R01—AUTO. TRANS.—CALIFORNIA
CALIBRATION: 2-14Q-R11—AUTO. TRANS.—CALIFORNIA

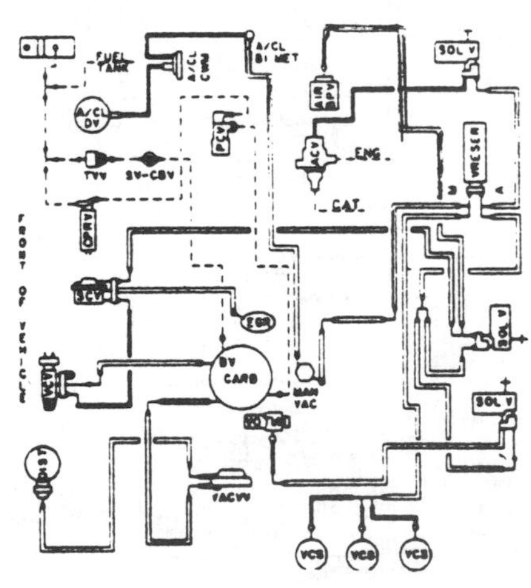

CALIBRATION: 2-14R-R01—AUTO. TRANS.—ALL
CALIBRATION: 2-14R-R11—AUTO. TRANS.—ALL

VACUUM CIRCUITS

(© Ford Motor Co)

1982—3.8L (230 CID) 6 cyl. engines

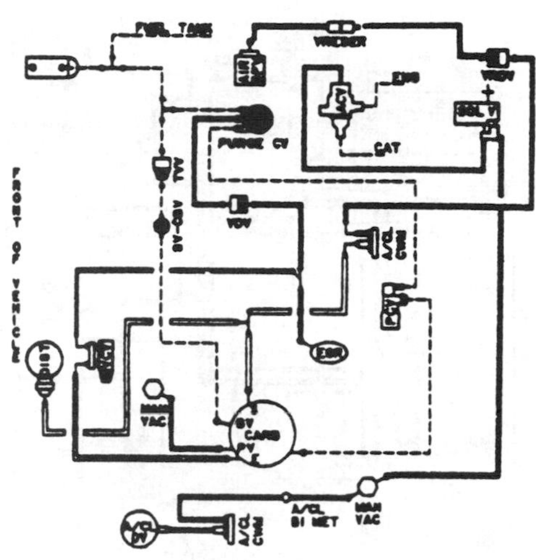

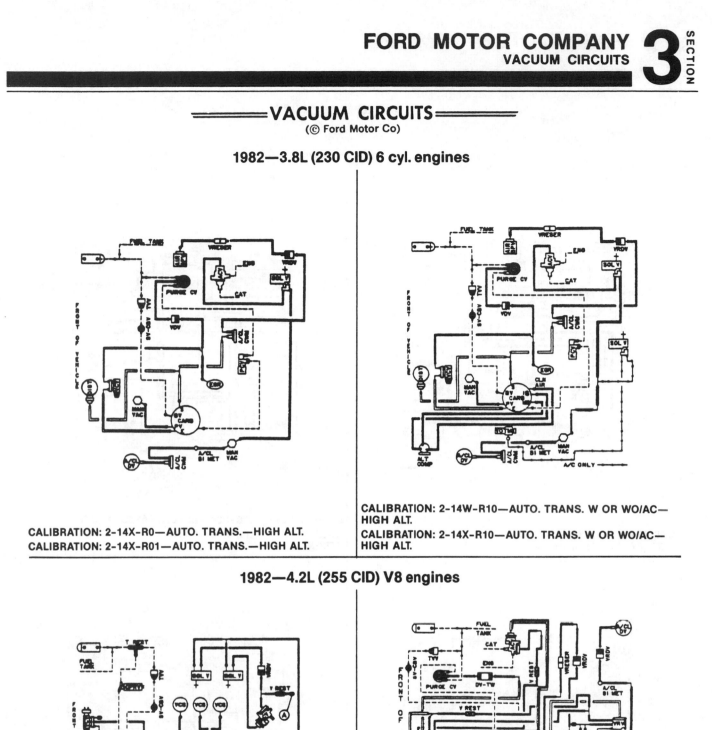

CALIBRATION: 2-14X-R0—AUTO. TRANS.—HIGH ALT.
CALIBRATION: 2-14X-R01—AUTO. TRANS.—HIGH ALT.

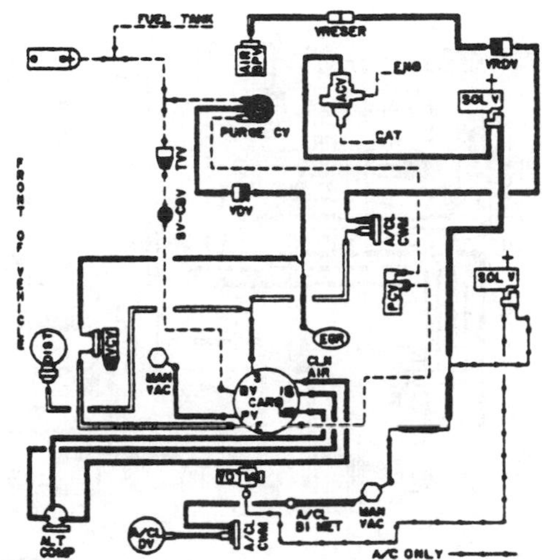

CALIBRATION: 2-14W-R10—AUTO. TRANS. W OR WO/AC—HIGH ALT.
CALIBRATION: 2-14X-R10—AUTO. TRANS. W OR WO/AC—HIGH ALT.

1982—4.2L (255 CID) V8 engines

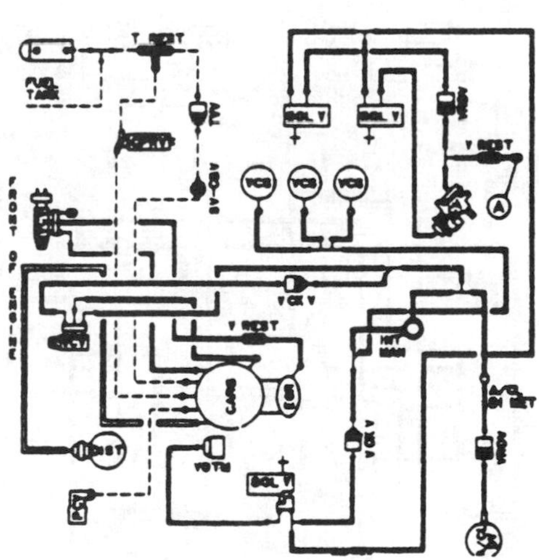

CALIBRATION: 1-18U-R0—AUTO. TRANS.—EXC. HIGH ALT.
CALIBRATION: 2-18U-R10—AUTO. TRANS.—EXC. HIGH ALT.

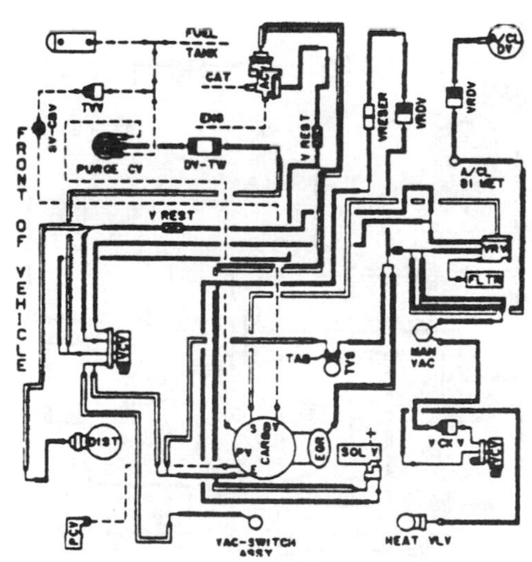

CALIBRATION: 2-18B-R0—AUTO. TRANS.—EXC. HIGH ALT.—FAIRMONT & ZEPHYR

VACUUM CIRCUITS

(© Ford Motor Co)

1982—4.2L (255 CID) V8 engines

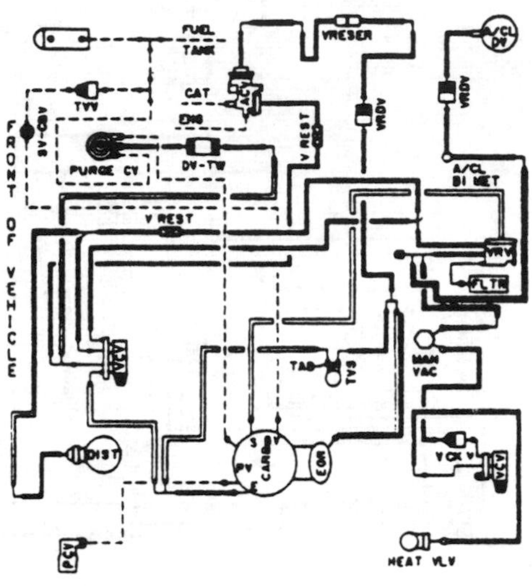

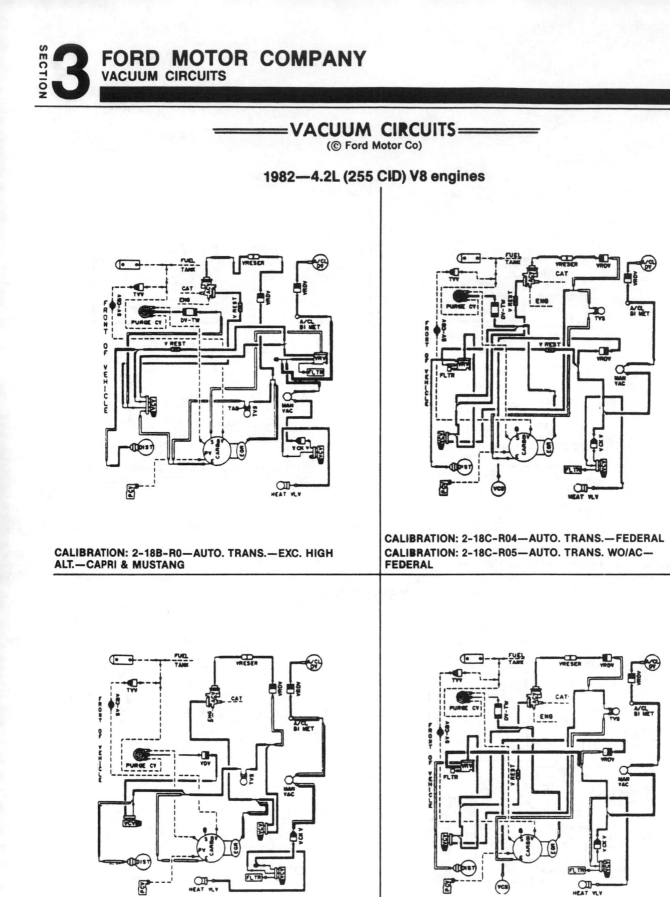

CALIBRATION: 2-18B-R0—AUTO. TRANS.—EXC. HIGH ALT.—CAPRI & MUSTANG

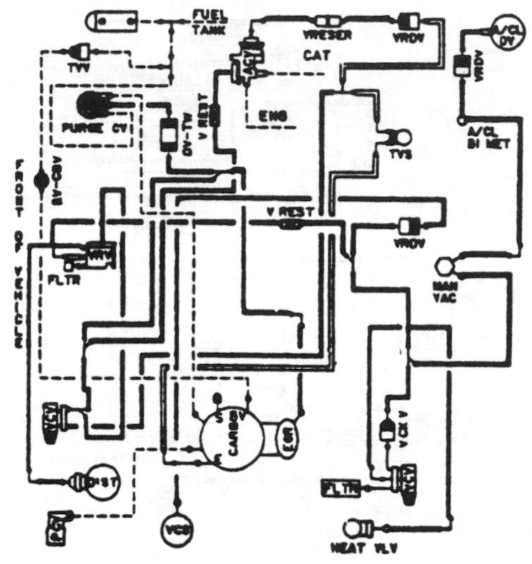

CALIBRATION: 2-18C-R04—AUTO. TRANS.—FEDERAL
CALIBRATION: 2-18C-R05—AUTO. TRANS. WO/AC—FEDERAL

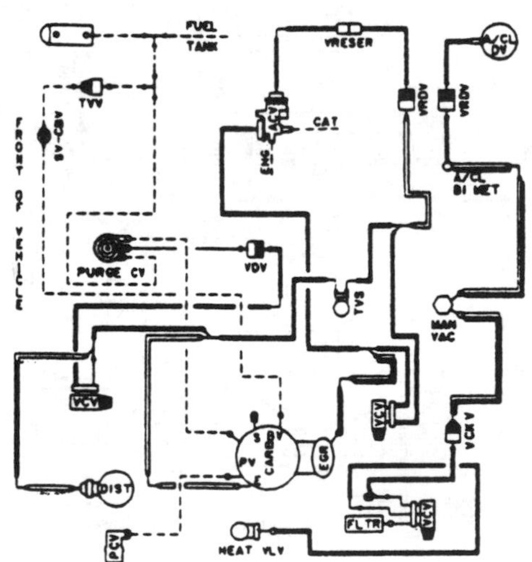

CALIBRATION: 2-18B-R13—AUTO. TRANS.—EXC. HIGH ALT.

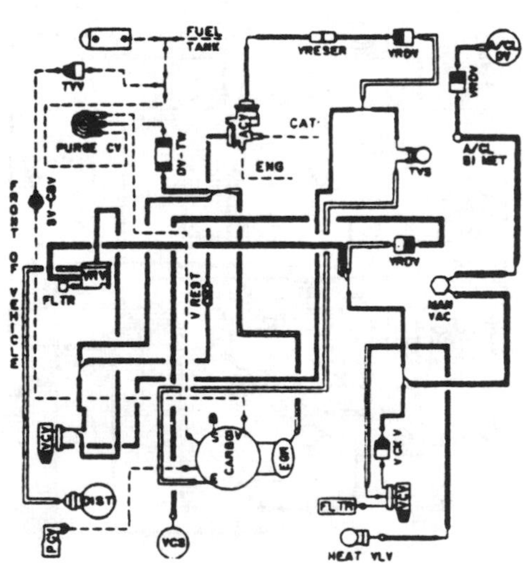

CALIBRATION: 2-18C-R11—AUTO. TRANS.—FEDERAL

VACUUM CIRCUITS

(© Ford Motor Co.)

1982—4.2L (255 CID) V8 engines

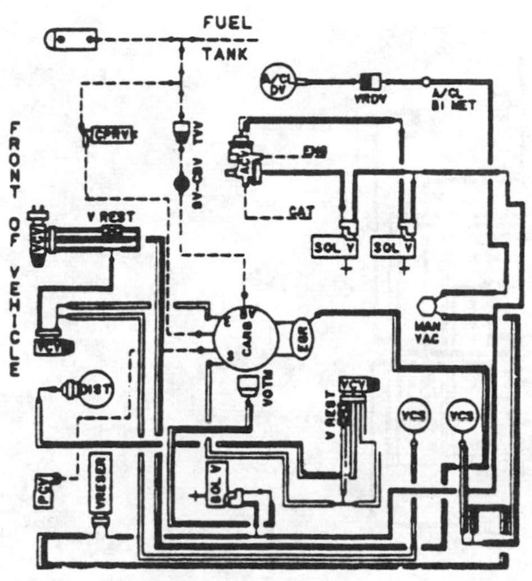

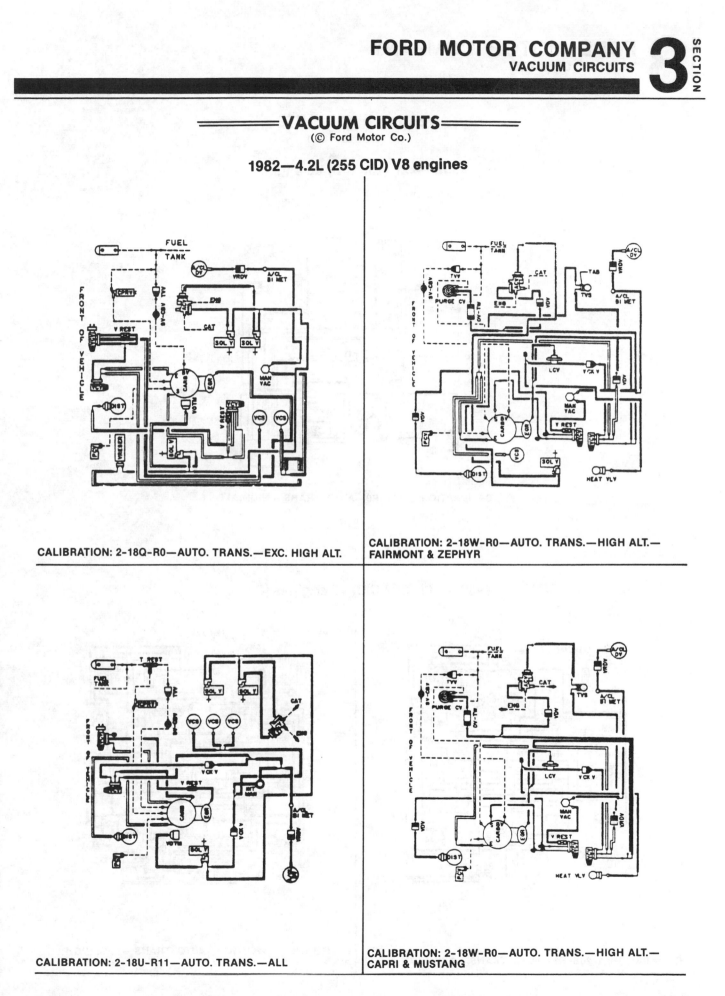

CALIBRATION: 2-18Q-R0—AUTO. TRANS.—EXC. HIGH ALT.

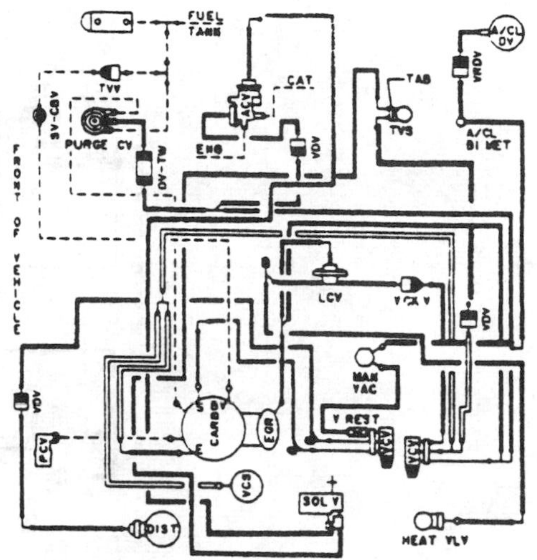

CALIBRATION: 2-18W-R0—AUTO. TRANS.—HIGH ALT.—FAIRMONT & ZEPHYR

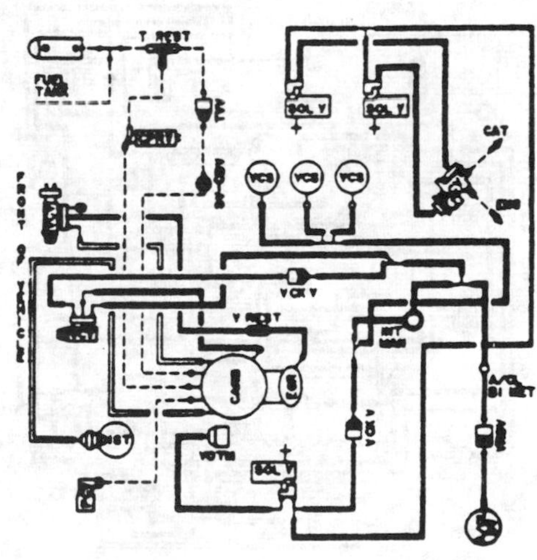

CALIBRATION: 2-18U-R11—AUTO. TRANS.—ALL

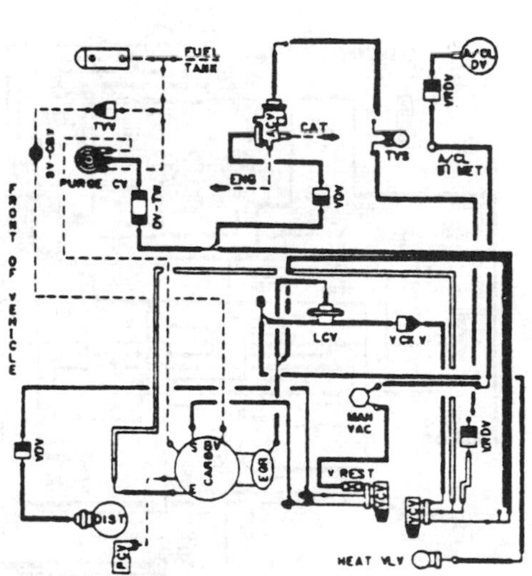

CALIBRATION: 2-18W-R0—AUTO. TRANS.—HIGH ALT.—CAPRI & MUSTANG

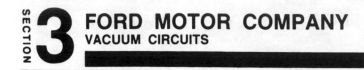

VACUUM CIRCUITS

(© Ford Motor Co.)

1982—4.2L (255 CID) V8 engines

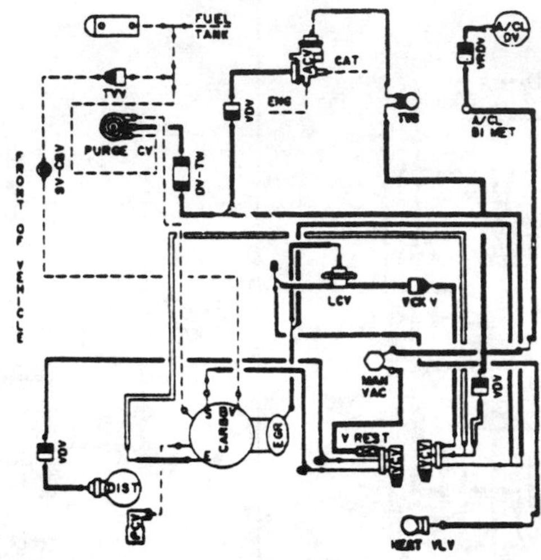

CALIBRATION: 2-18X-R0—AUTO. TRANS.—HIGH ALT.

1982—5.0L (302 CID) V8 engines (Carb.)

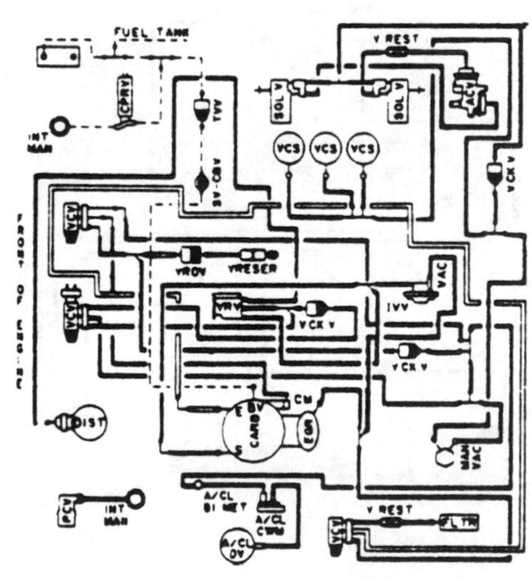

CALIBRATION: 1-20B-R1—AUTO. TRANS.—EXC. HIGH ALT.

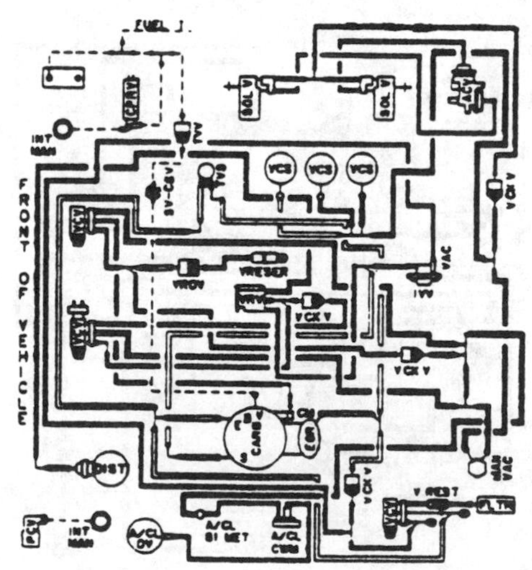

CALIBRATION: 1-20N-R10—AUTO. TRANS.—EXC. HIGH ALT.

VACUUM CIRCUITS

(© Ford Motor Co.)

1982—5.0L (302 CID) V8 engines (Carb.)

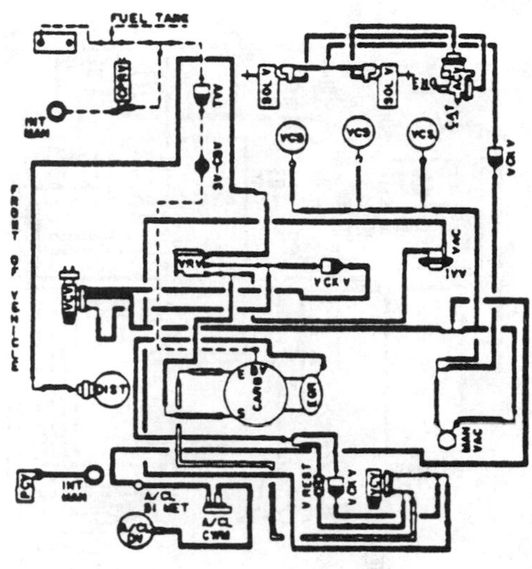

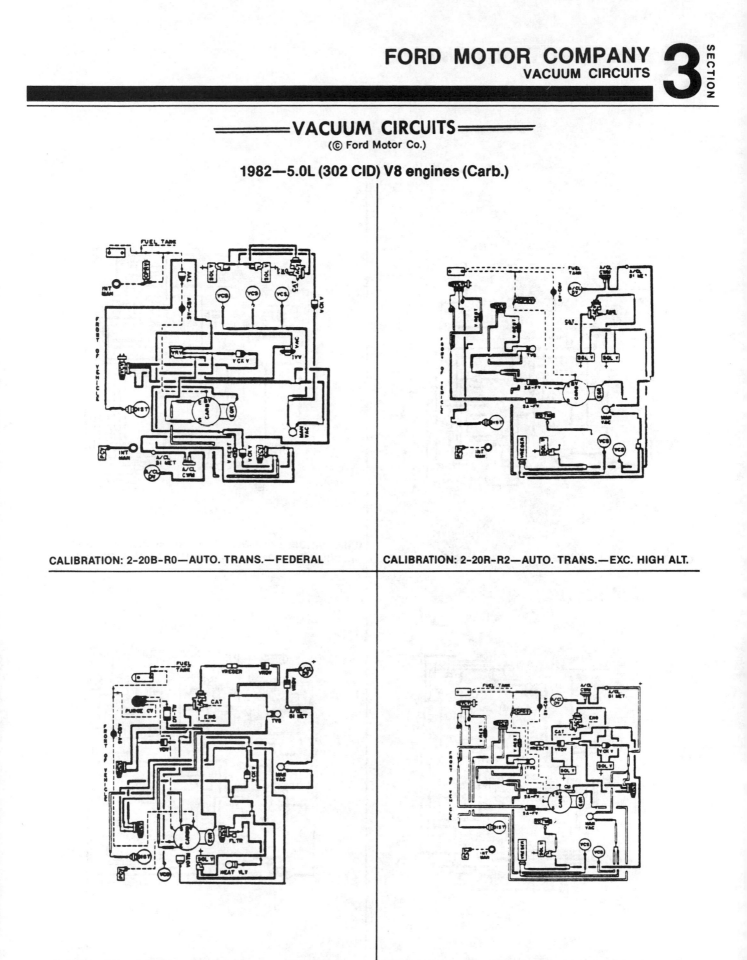

CALIBRATION: 2-20B-R0—AUTO. TRANS.—FEDERAL

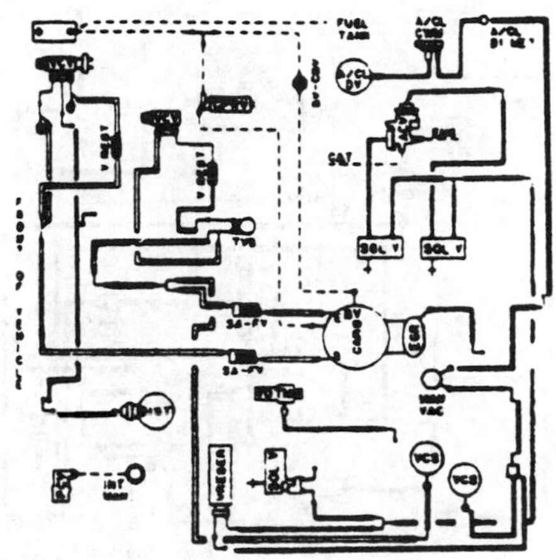

CALIBRATION: 2-20R-R2—AUTO. TRANS.—EXC. HIGH ALT.

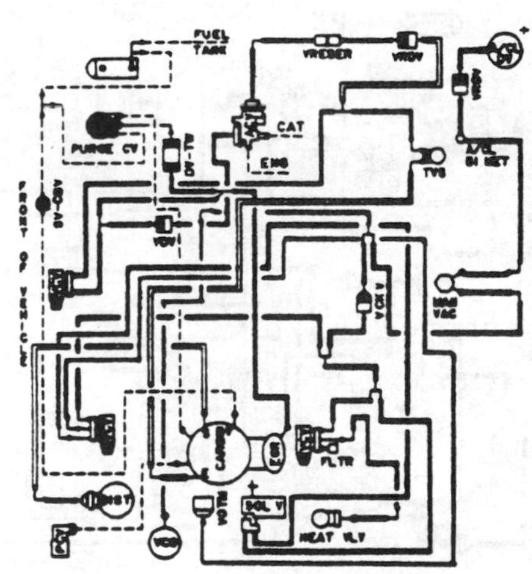

CALIBRATION: 2-20D-R1—AUTO. TRANS.—EXC. HIGH ALT.

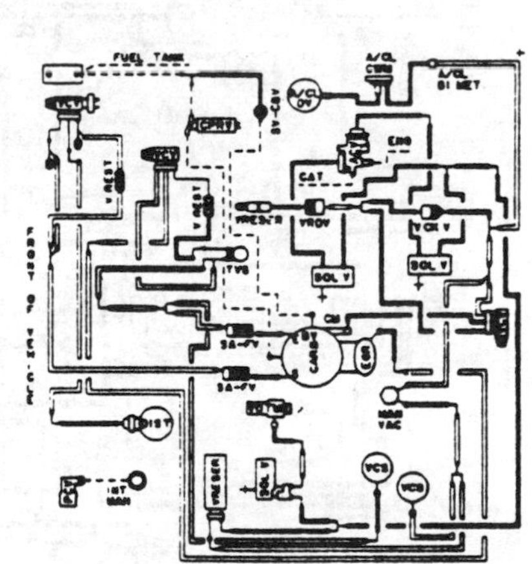

CALIBRATION: 2-20R-R16—AUTO. TRANS.—EXC. HIGH ALT.

VACUUM CIRCUITS

1982—5.0L (302 CID) V8 engines (Carb.)

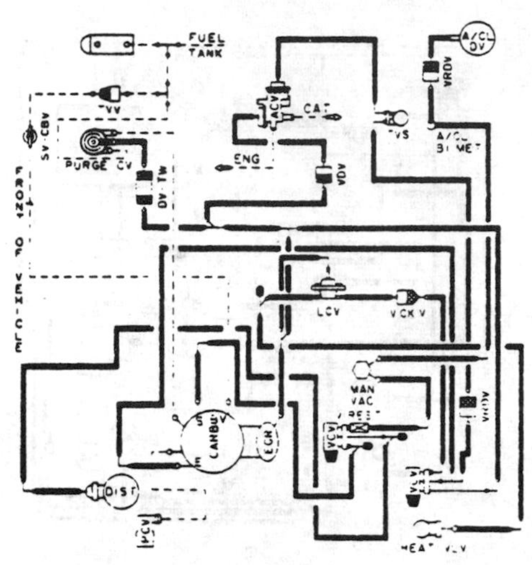

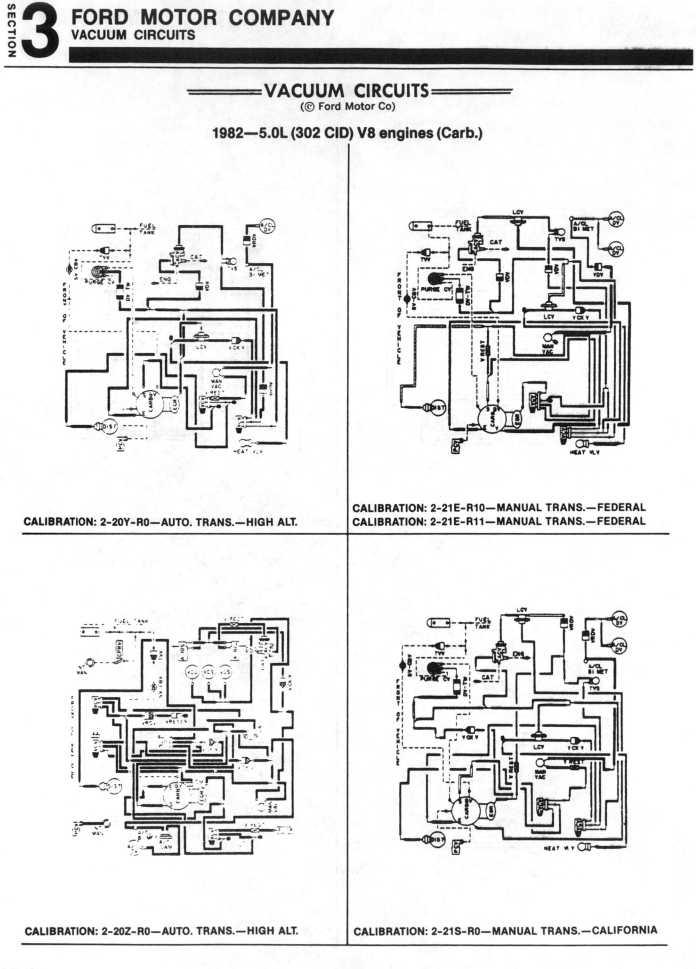

CALIBRATION: 2-20Y-R0—AUTO. TRANS.—HIGH ALT.

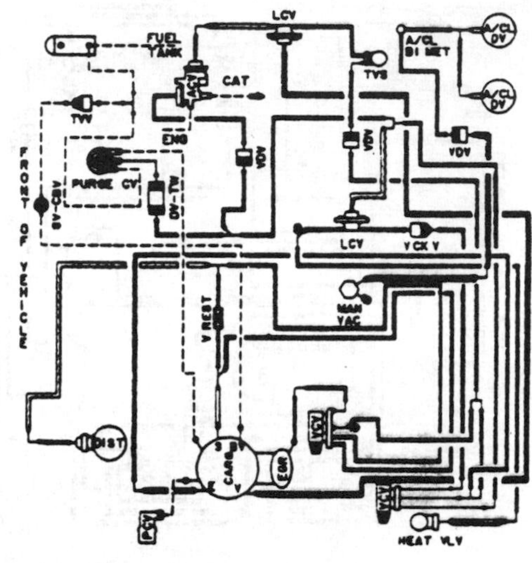

CALIBRATION: 2-21E-R10—MANUAL TRANS.—FEDERAL
CALIBRATION: 2-21E-R11—MANUAL TRANS.—FEDERAL

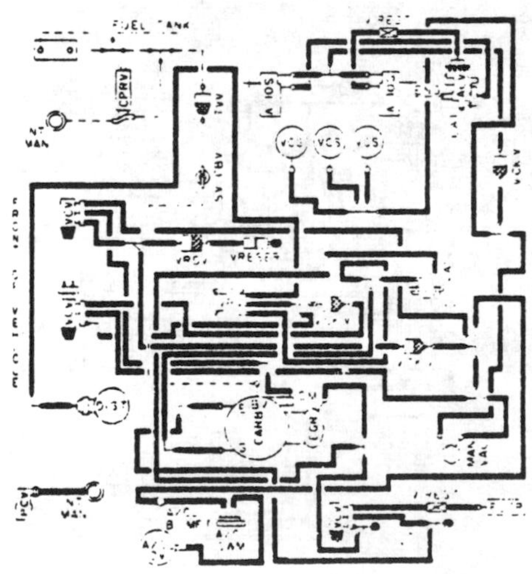

CALIBRATION: 2-20Z-R0—AUTO. TRANS.—HIGH ALT.

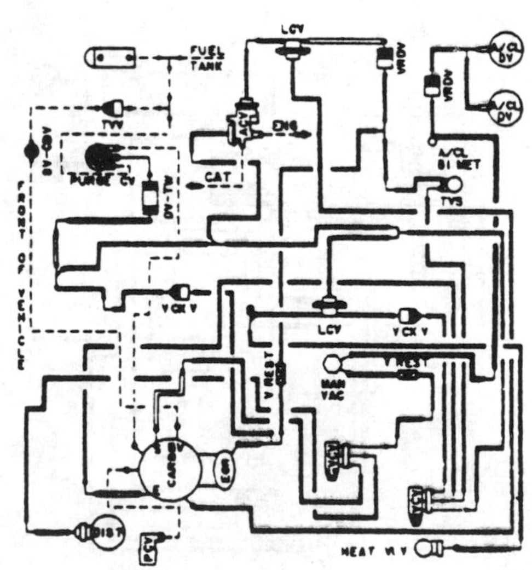

CALIBRATION: 2-21S-R0—MANUAL TRANS.—CALIFORNIA

VACUUM CIRCUITS

(© Ford Motor Co)

1982—5.0L (302 CID) V8 engines (Carb.)

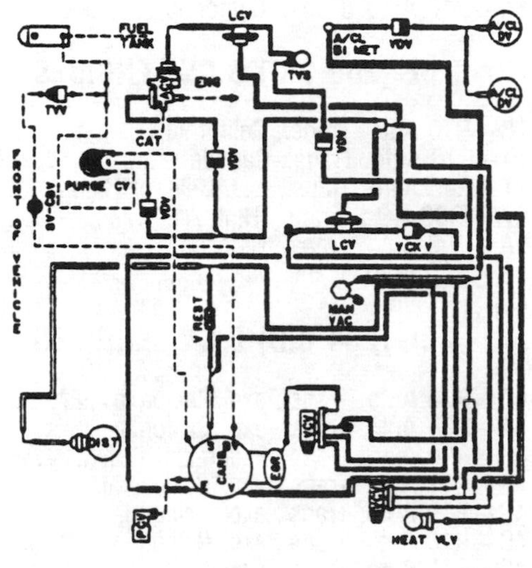

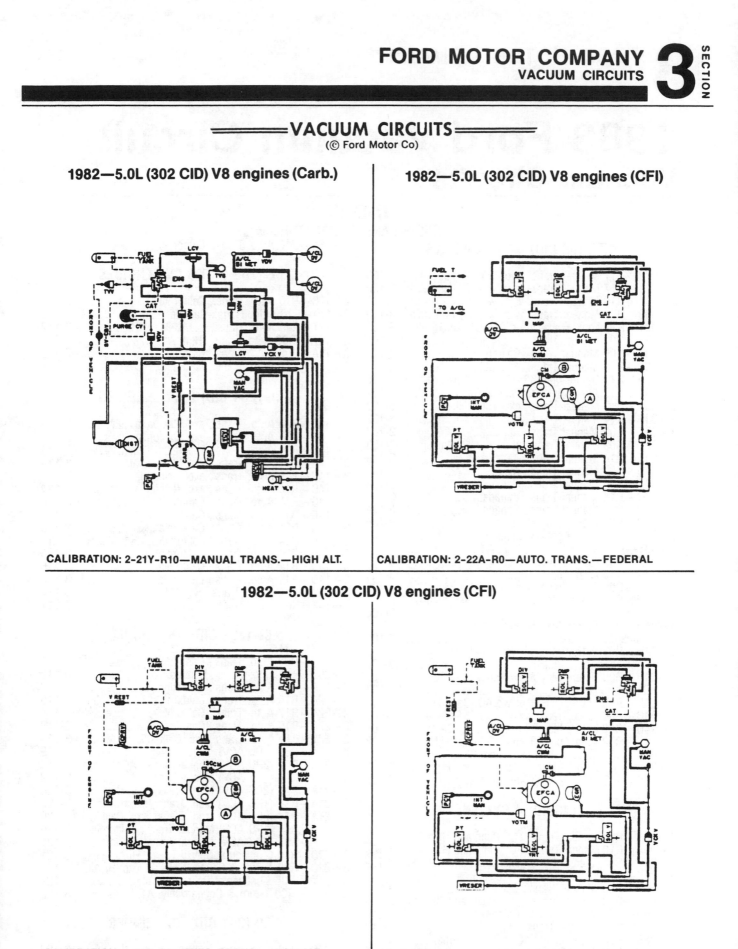

CALIBRATION: 2-21Y-R10—MANUAL TRANS.—HIGH ALT.

1982—5.0L (302 CID) V8 engines (CFI)

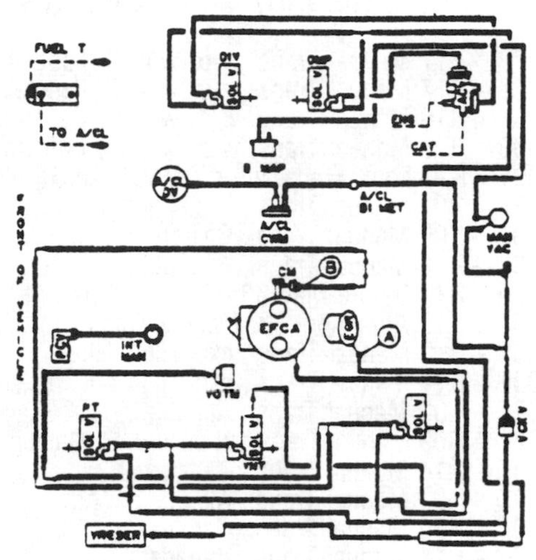

CALIBRATION: 2-22A-R0—AUTO. TRANS.—FEDERAL

1982—5.0L (302 CID) V8 engines (CFI)

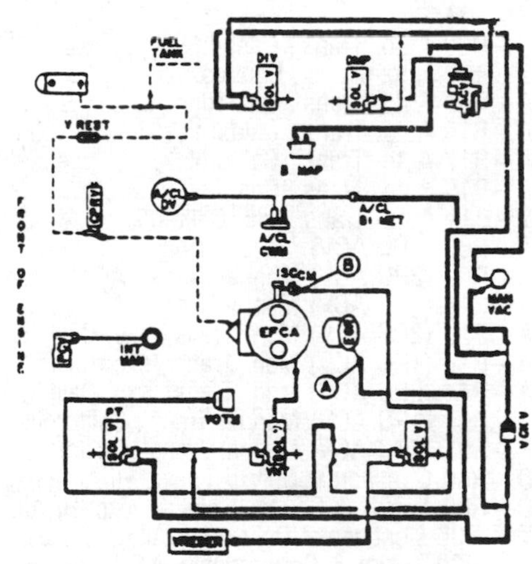

CALIBRATION: 1-22B-R0—AUTO. TRANS.—HIGH ALT.
CALIBRATION: 1-22P-R0—AUTO. TRANS.—CALIFORNIA

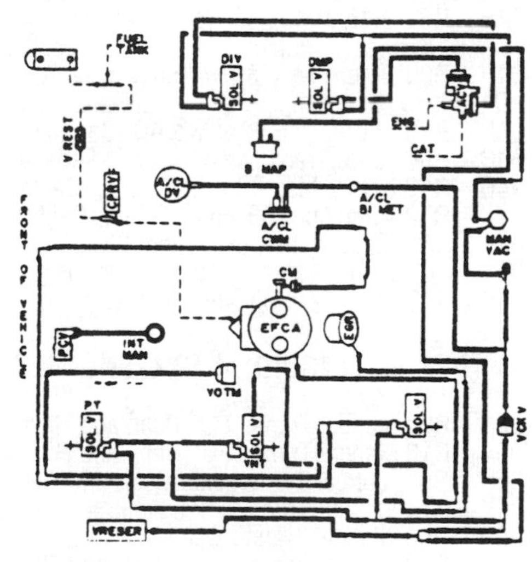

CALIBRATION: 2-22Y-R11—AUTO. TRANS.—EXC. CALIFORNIA

1983 Ford Vacuum Circuit

Vacuum Circuits

INDEX

ENGINE AND CALIBRATION

1.6L (98 CID) 4 CYL ENGINES

2.3L (140CID) 4 CYL ENGINES

3.3L (200 CID) 6 CYL ENGINES

3.8L (230 CID) 6 CYL ENGINES

5.0L (302 CID) V8 ENGINES

5.8L (351 CID) V8 ENGINES

HOW TO DETERMINE VEHICLE CALIBRATION

Attached to all 1983 production engines is an Engine Code Information Label containing among other pertinent data—the engine calibration number. The size and shape of the decal is dependent upon the engine displacement. To determine the engine calibration, look at Step 1A. You **do not** have to refer to the Vehicle Emission decal for calibration numbers as in prior years.

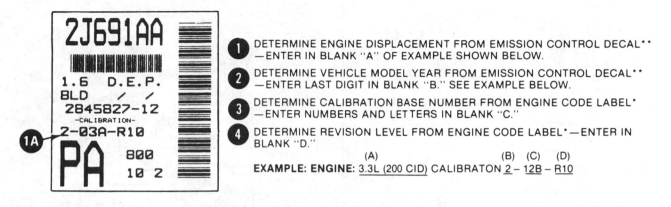

1. DETERMINE ENGINE DISPLACEMENT FROM EMISSION CONTROL DECAL**—ENTER IN BLANK "A" OF EXAMPLE SHOWN BELOW.
2. DETERMINE VEHICLE MODEL YEAR FROM EMISSION CONTROL DECAL**—ENTER LAST DIGIT IN BLANK "B." SEE EXAMPLE BELOW.
3. DETERMINE CALIBRATION BASE NUMBER FROM ENGINE CODE LABEL*—ENTER NUMBERS AND LETTERS IN BLANK "C."
4. DETERMINE REVISION LEVEL FROM ENGINE CODE LABEL*—ENTER IN BLANK "D."

EXAMPLE: ENGINE: 3.3L (200 CID) (A) CALIBRATON 2 (B) – 12B (C) – R10 (D)

* Located on engine valve cover (Yellow or White).

** Located on engine valve cover, radiator support bracket, underside of hood or fresh air inlet tube (Gold or Silver), or windshield washer bottle.

HOW TO READ VEHICLE EMISSION CONTROL INFORMATION (VECI) DECALS

The following callouts identify the various information areas of the VECI decals.

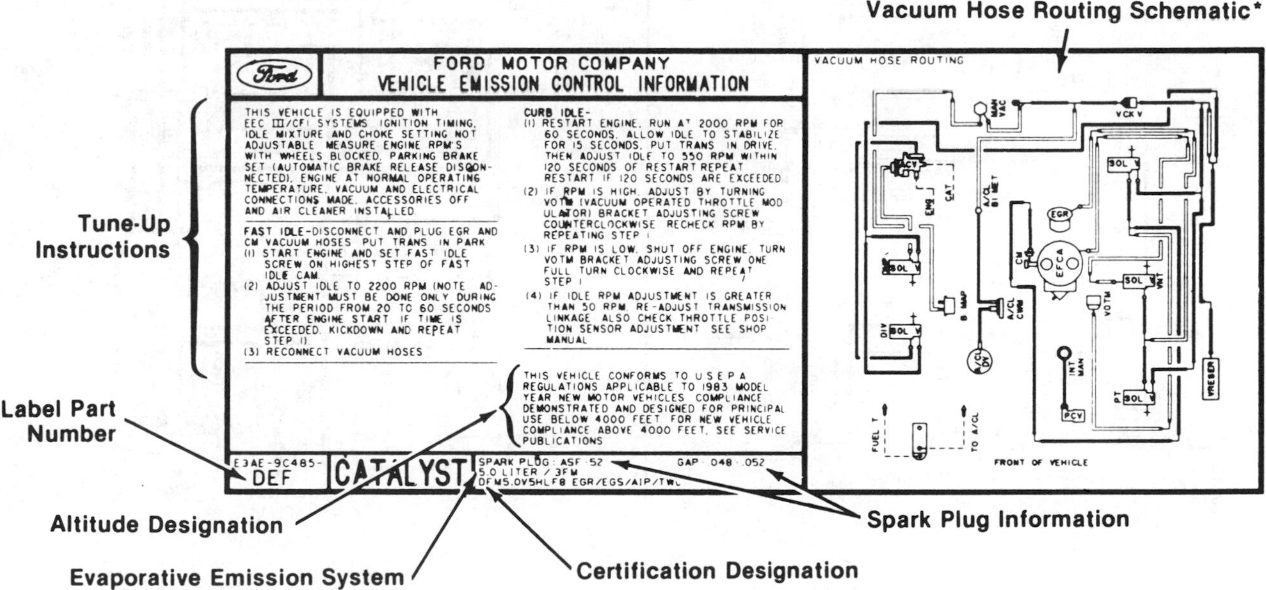

* The schematic refers only to emission systems (except exhaust). For vehicle vacuum systems, refer to Electrical and Vacuum Troubleshooting Manuals (EVTM).

VACUUM CIRCUITS

(© Ford Motor Co)

1983—1.6L (98 CID) 4 cyl. engines

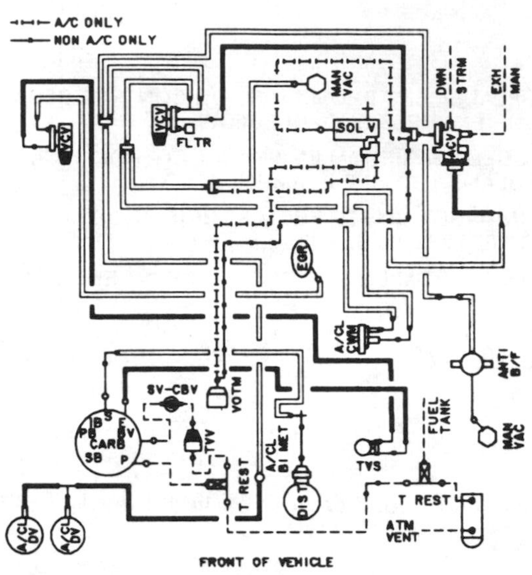

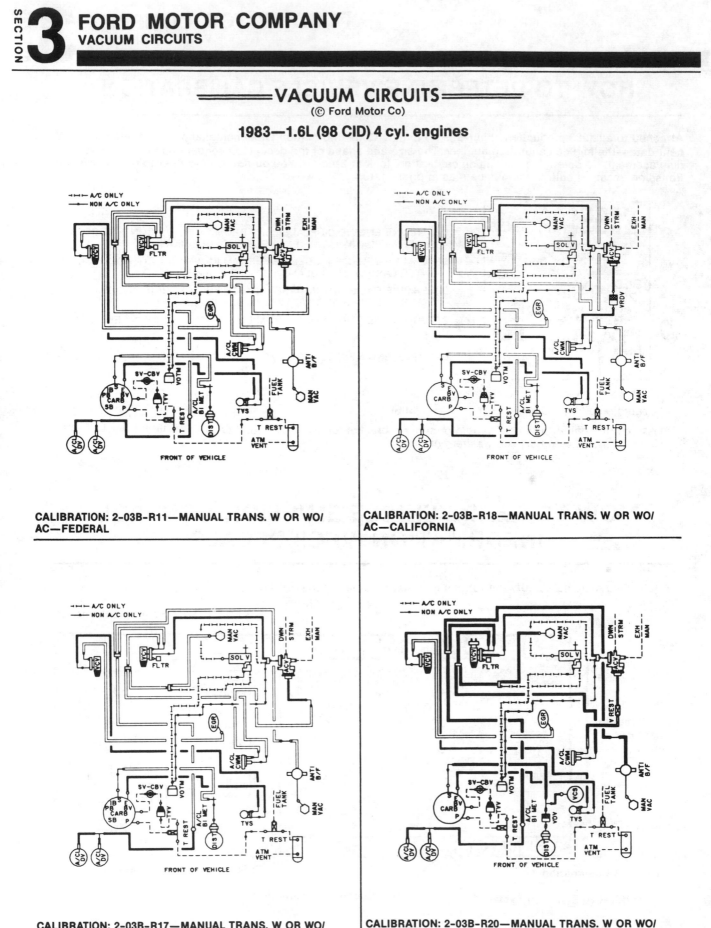

CALIBRATION: 2-03B-R11—MANUAL TRANS. W OR WO/ AC—FEDERAL

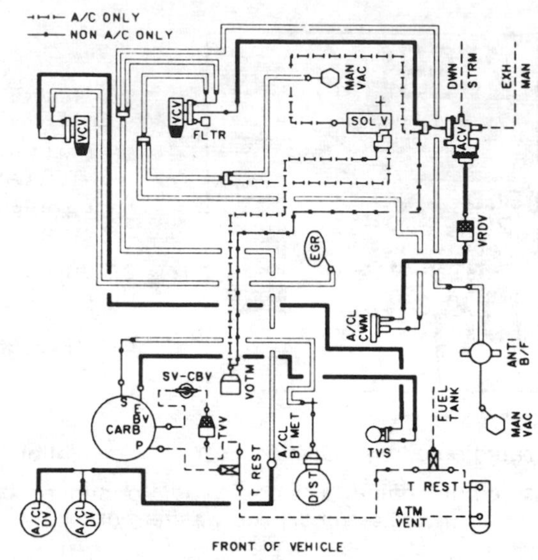

CALIBRATION: 2-03B-R18—MANUAL TRANS. W OR WO/ AC—CALIFORNIA

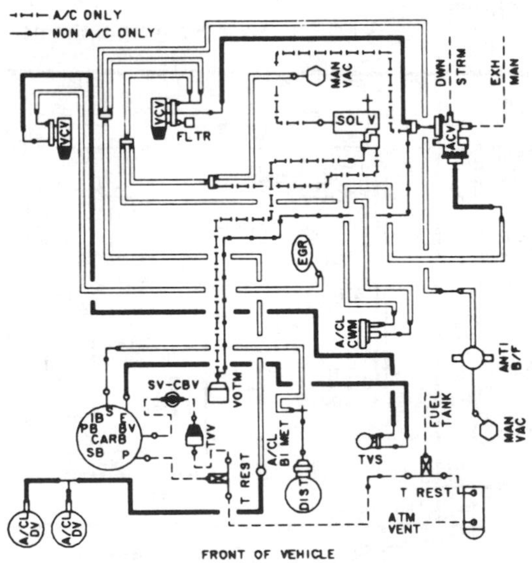

CALIBRATION: 2-03B-R17—MANUAL TRANS. W OR WO/ AC—FEDERAL

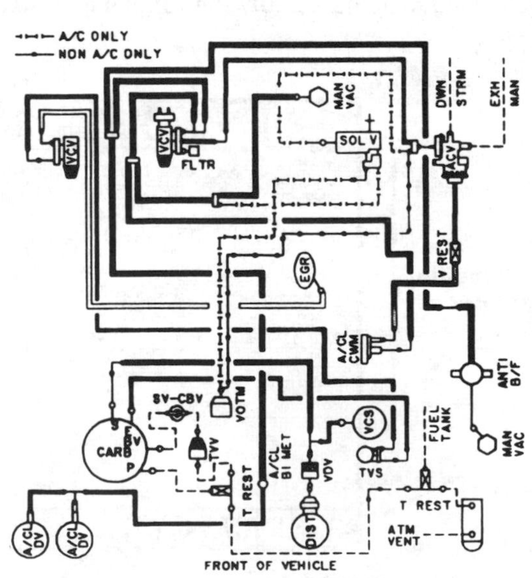

CALIBRATION: 2-03B-R20—MANUAL TRANS. W OR WO/ AC—CALIFORNIA

VACUUM CIRCUITS

(© Ford Motor Co)

1983—1.6L (98 CID) 4 cyl. engines

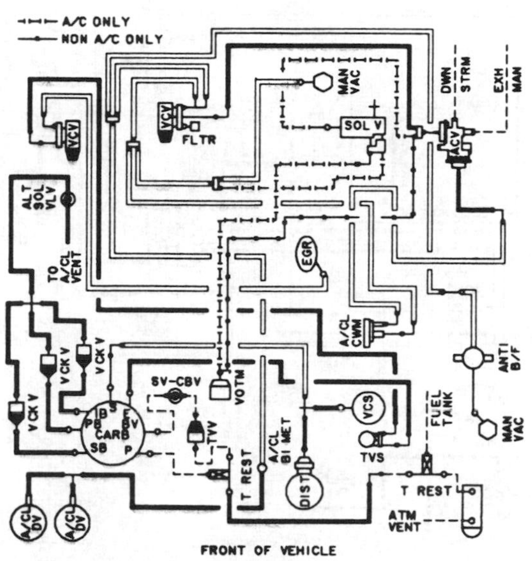

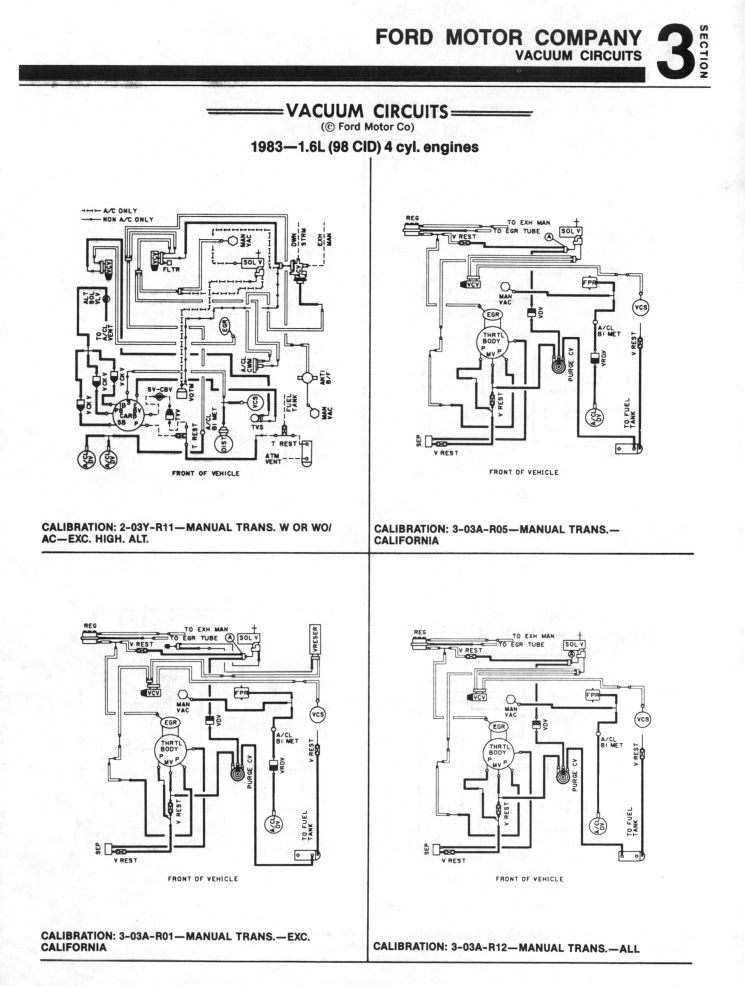

CALIBRATION: 2-03Y-R11—MANUAL TRANS. W OR WO/ AC—EXC. HIGH. ALT.

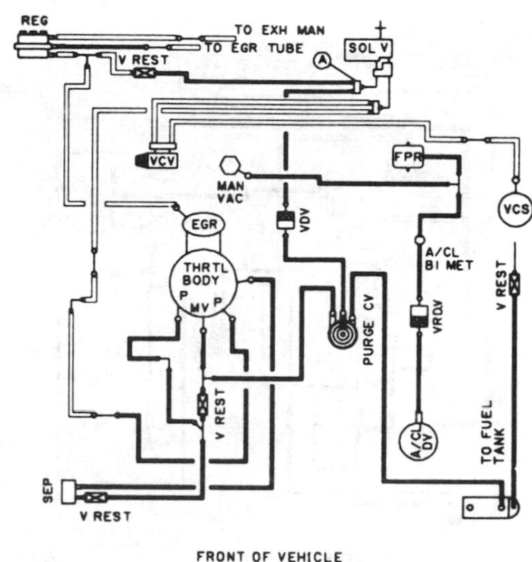

CALIBRATION: 3-03A-R05—MANUAL TRANS.—CALIFORNIA

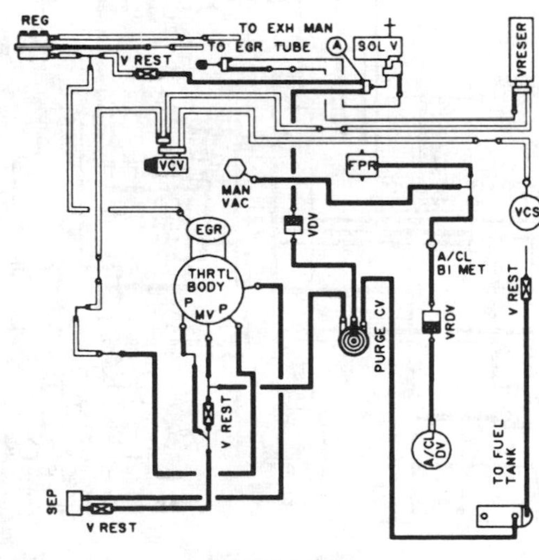

CALIBRATION: 3-03A-R01—MANUAL TRANS.—EXC. CALIFORNIA

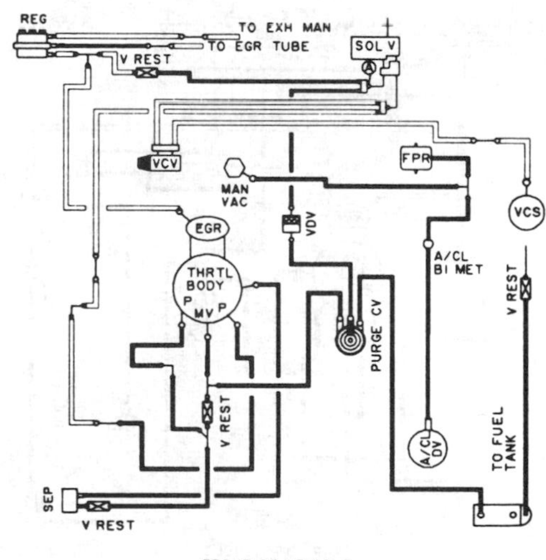

CALIBRATION: 3-03A-R12—MANUAL TRANS.—ALL

VACUUM CIRCUITS

(© Ford Motor Co)

1983—1.6L (98 CID) 4 cyl. engines

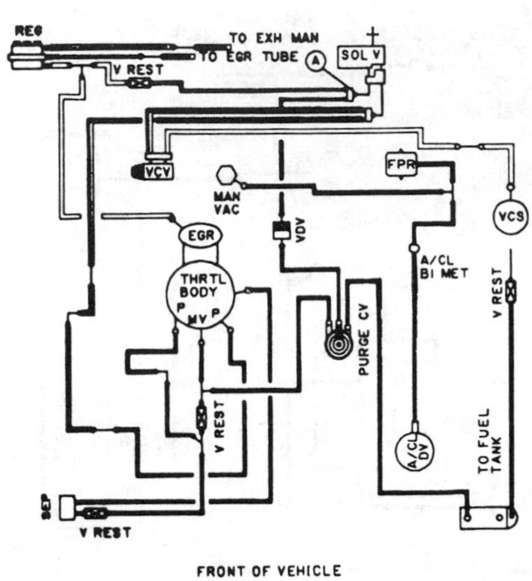

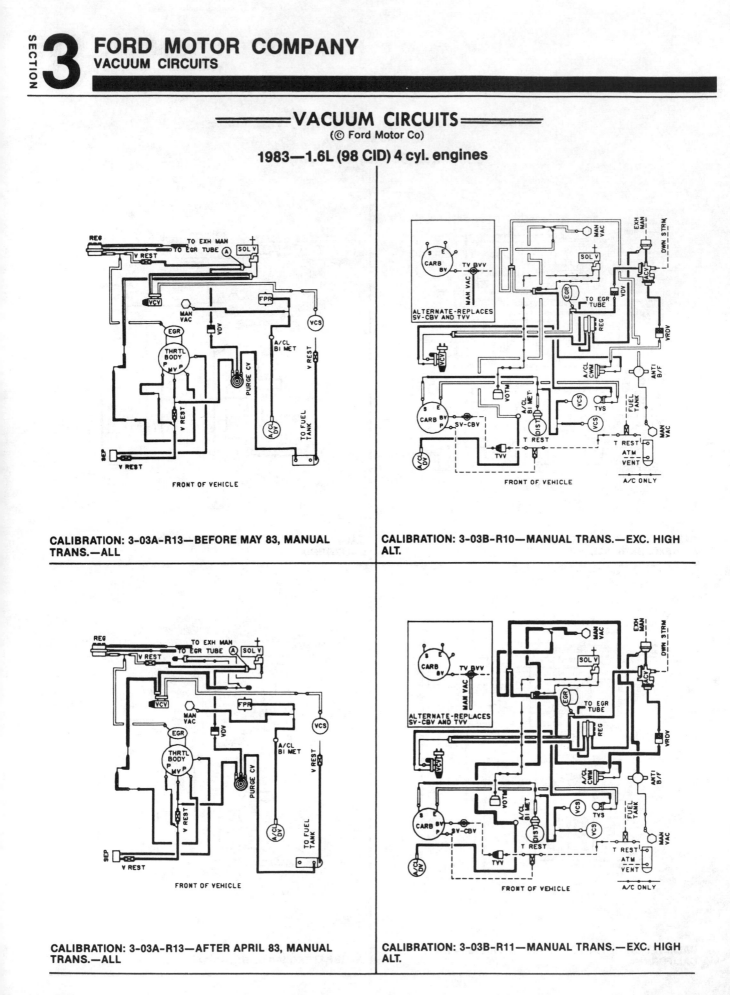

CALIBRATION: 3-03A-R13—BEFORE MAY 83, MANUAL TRANS.—ALL

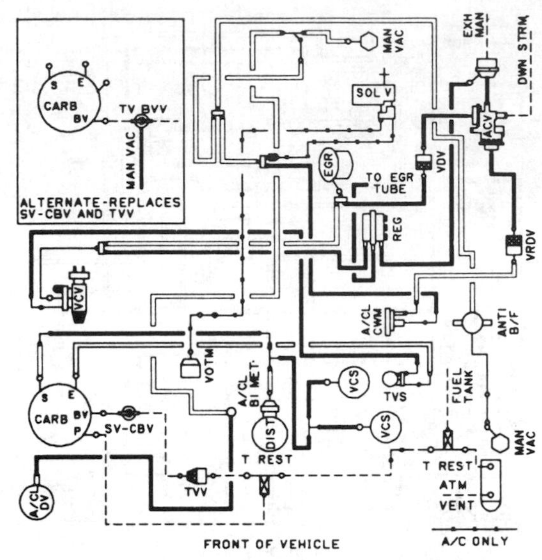

CALIBRATION: 3-03B-R10—MANUAL TRANS.—EXC. HIGH ALT.

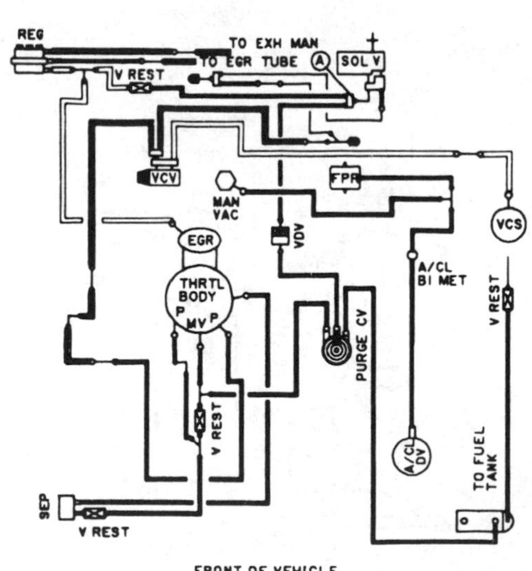

CALIBRATION: 3-03A-R13—AFTER APRIL 83, MANUAL TRANS.—ALL

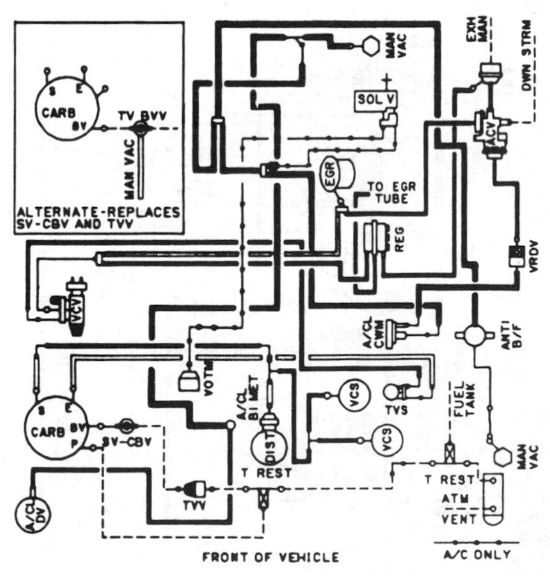

CALIBRATION: 3-03B-R11—MANUAL TRANS.—EXC. HIGH ALT.

VACUUM CIRCUITS

(© Ford Motor Co)

1983—1.6L (98 CID) 4 cyl. engines

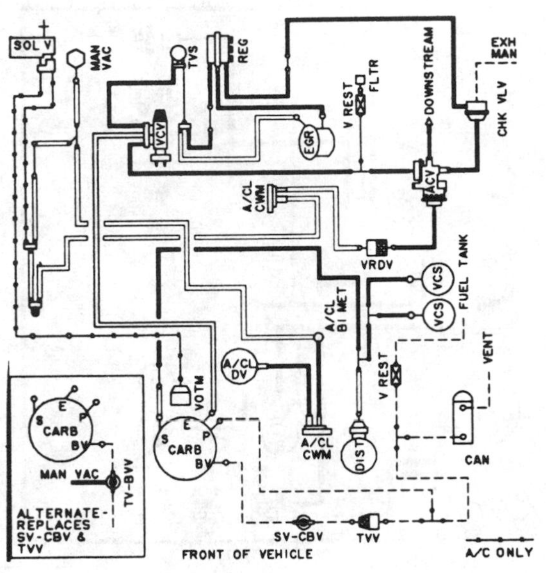

CALIBRATION: 3-03C-R00—MANUAL TRANS.—ALL

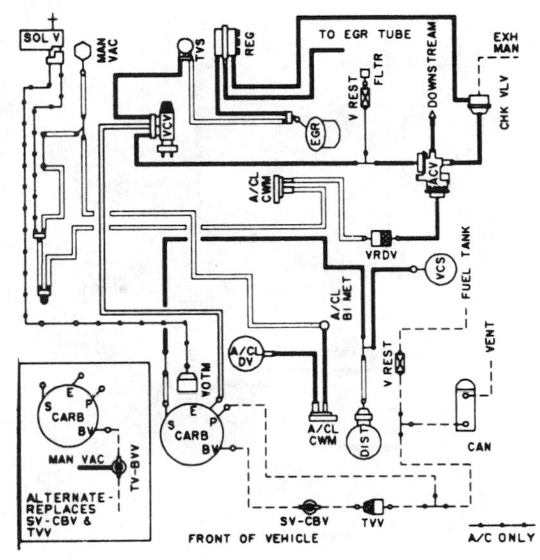

CALIBRATION: 3-03C-R14—MANUAL TRANS.—ALL

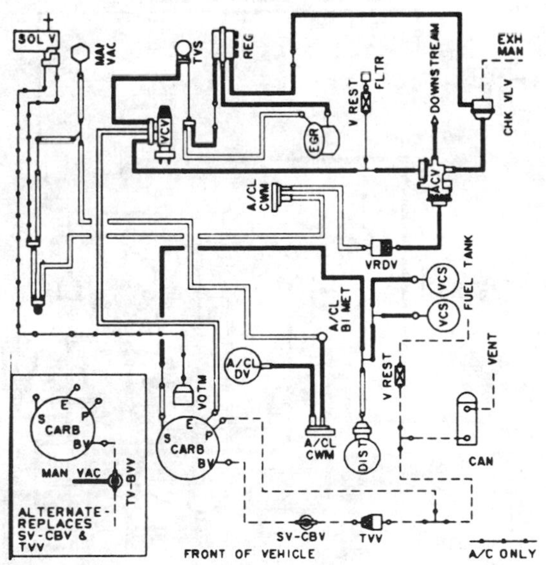

CALIBRATION: 3-03C-R09—MANUAL TRANS. W OR WO/ AC—CALIFORNIA

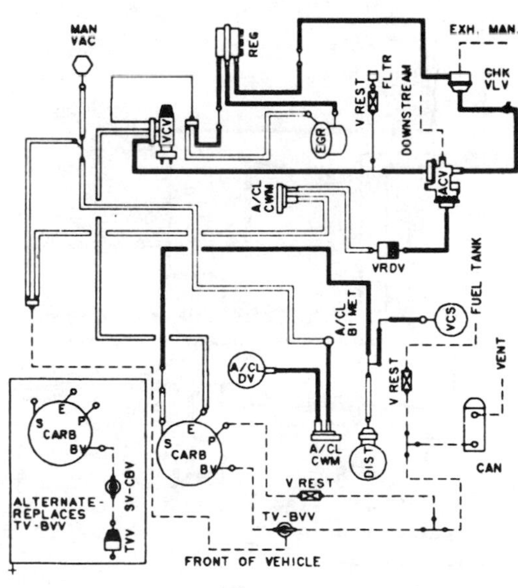

CALIBRATION: 3-03D-R01—MANUAL TRANS.—ALL

VACUUM CIRCUITS

(© Ford Motor Co)

1983—1.6L (98 CID) 4 cyl. engines

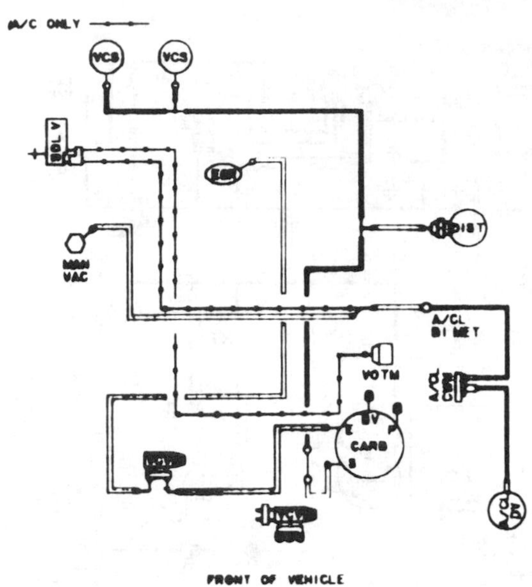

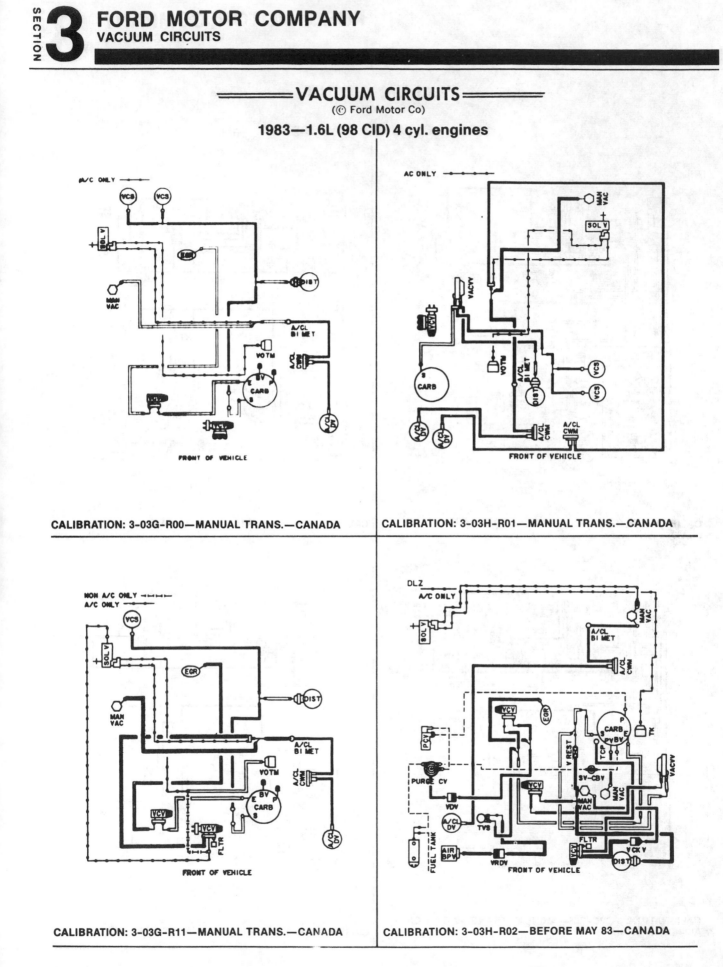

CALIBRATION: 3-03G-R00—MANUAL TRANS.—CANADA

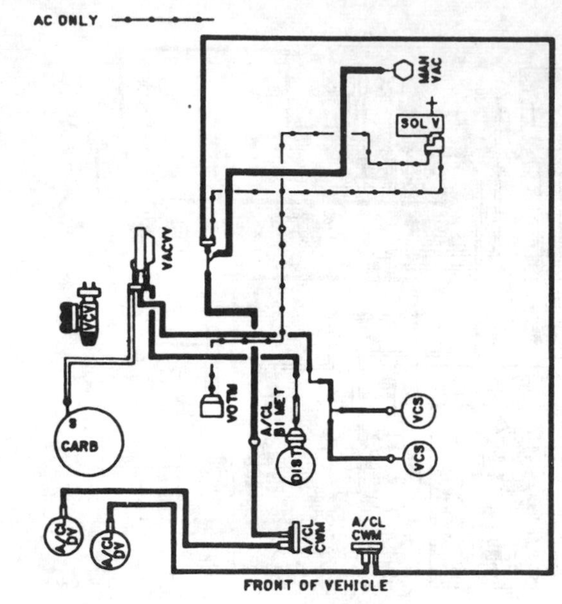

CALIBRATION: 3-03H-R01—MANUAL TRANS.—CANADA

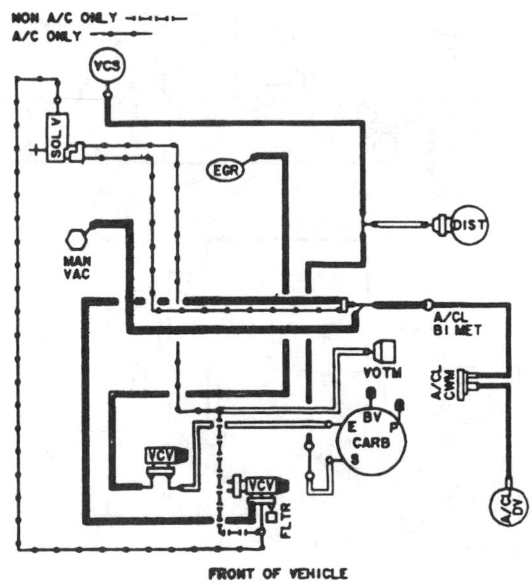

CALIBRATION: 3-03G-R11—MANUAL TRANS.—CANADA

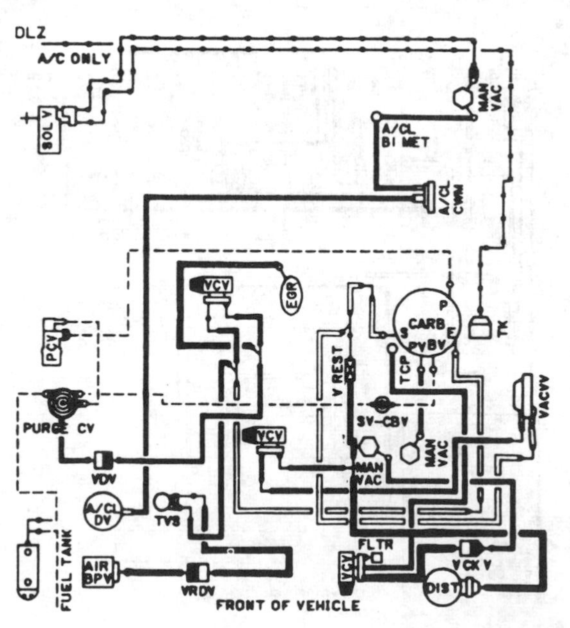

CALIBRATION: 3-03H-R02—BEFORE MAY 83—CANADA

VACUUM CIRCUITS

(© Ford Motor Co.)

1983—1.6L (98 CID) 4 cyl. engines

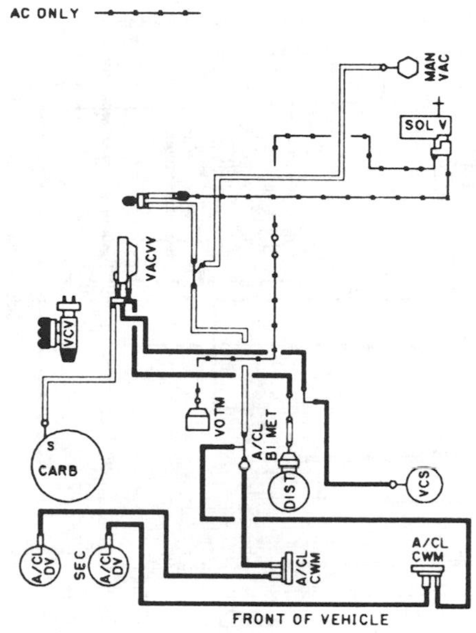

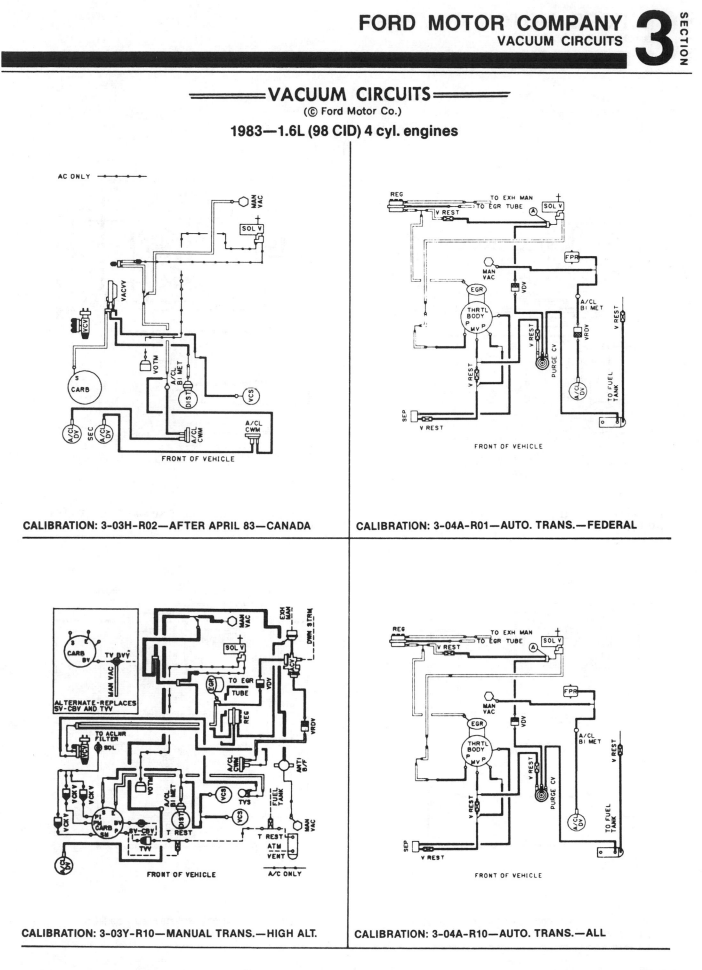

CALIBRATION: 3-03H-R02—AFTER APRIL 83—CANADA

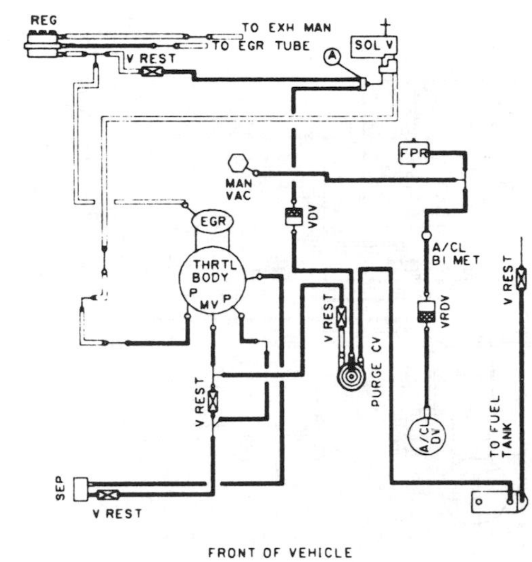

CALIBRATION: 3-04A-R01—AUTO. TRANS.—FEDERAL

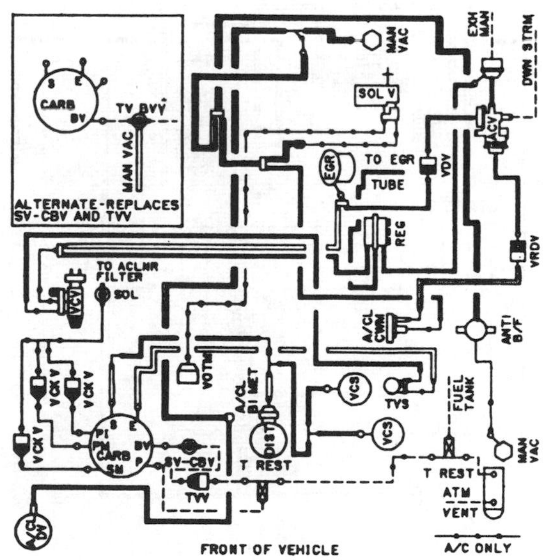

CALIBRATION: 3-03Y-R10—MANUAL TRANS.—HIGH ALT.

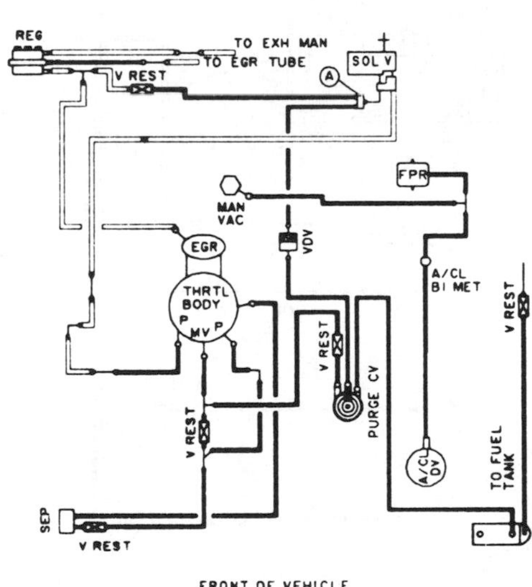

CALIBRATION: 3-04A-R10—AUTO. TRANS.—ALL

VACUUM CIRCUITS

(© Ford Motor Co.)

1983—1.6L (98 CID) 4 cyl. engines

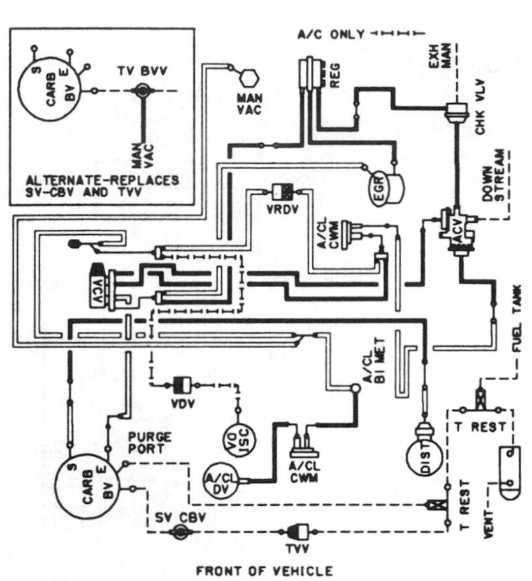

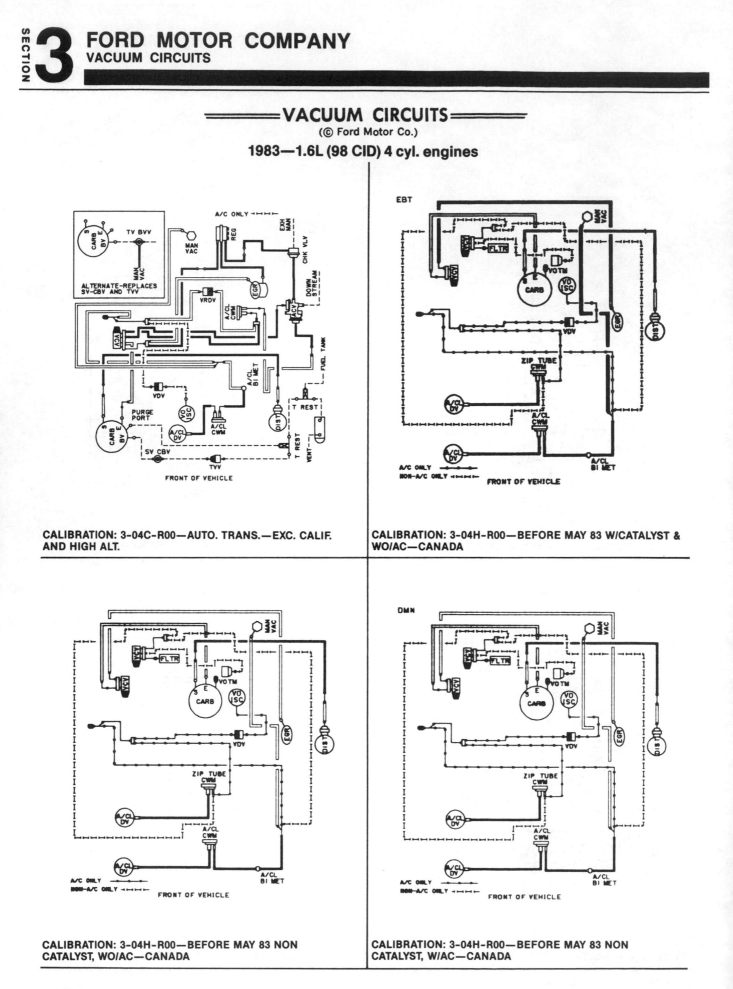

CALIBRATION: 3-04C-R00—AUTO. TRANS.—EXC. CALIF. AND HIGH ALT.

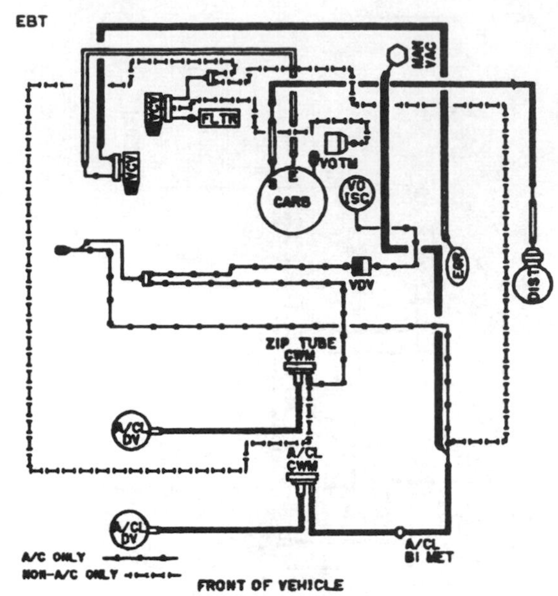

CALIBRATION: 3-04H-R00—BEFORE MAY 83 W/CATALYST & WO/AC—CANADA

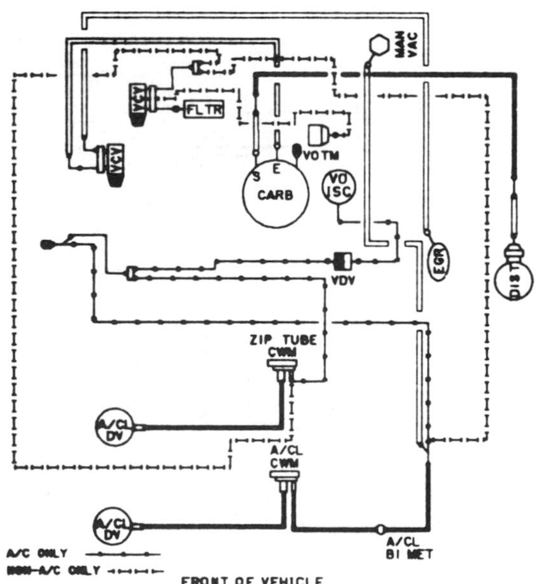

CALIBRATION: 3-04H-R00—BEFORE MAY 83 NON CATALYST, WO/AC—CANADA

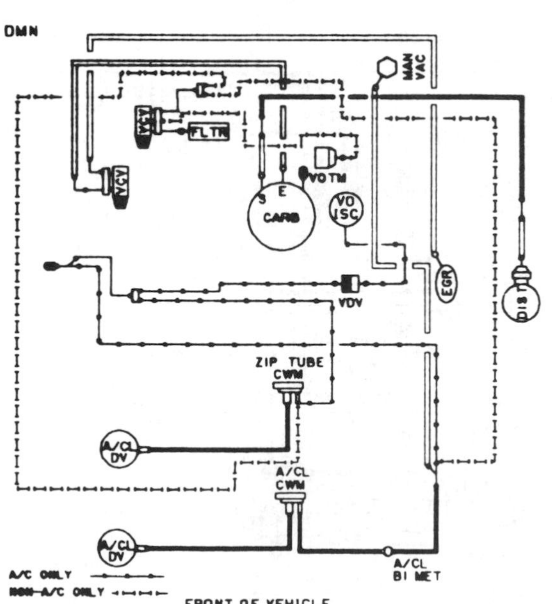

CALIBRATION: 3-04H-R00—BEFORE MAY 83 NON CATALYST, W/AC—CANADA

VACUUM CIRCUITS

(© Ford Motor Co.)

1983—1.6L (98 CID) 4 cyl. engines

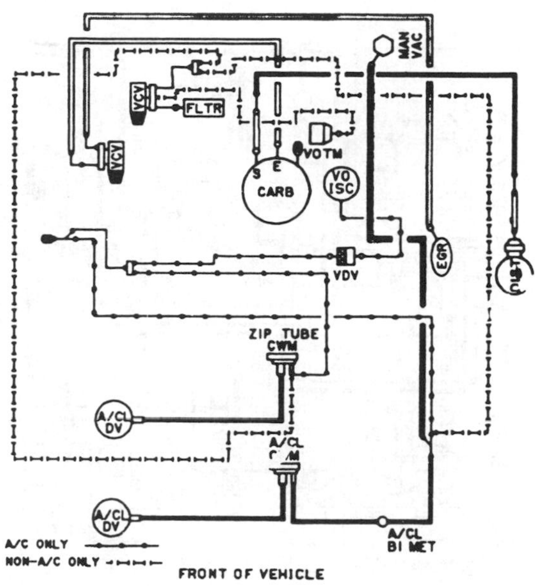

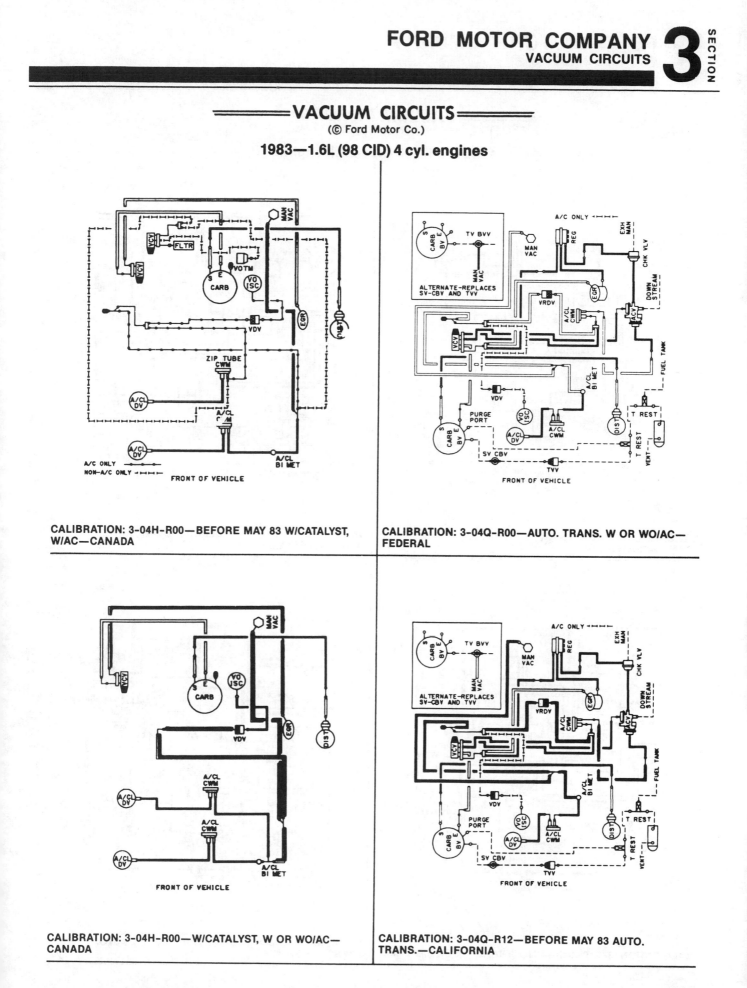

CALIBRATION: 3-04H-R00—BEFORE MAY 83 W/CATALYST, W/AC—CANADA

CALIBRATION: 3-04Q-R00—AUTO. TRANS. W OR WO/AC—FEDERAL

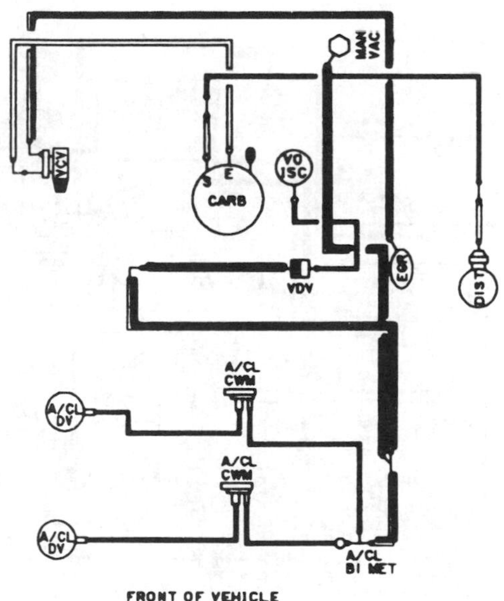

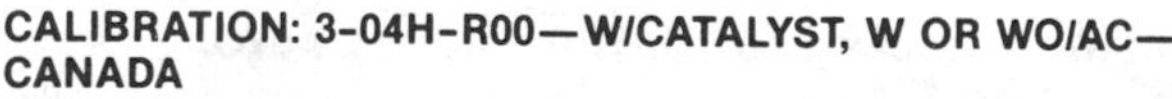

CALIBRATION: 3-04H-R00—W/CATALYST, W OR WO/AC—CANADA

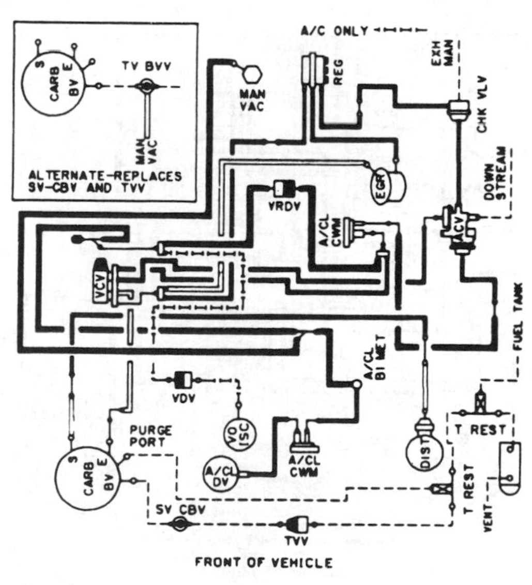

CALIBRATION: 3-04Q-R12—BEFORE MAY 83 AUTO. TRANS.—CALIFORNIA

VACUUM CIRCUITS

(© Ford Motor Co.)

1983—1.6L (98 CID) 4 cyl. engines

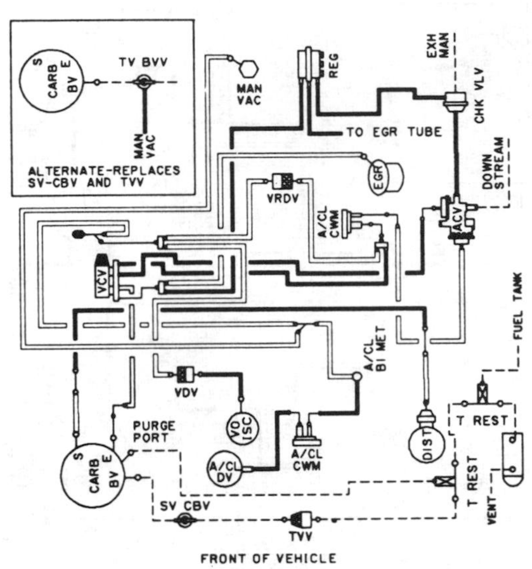

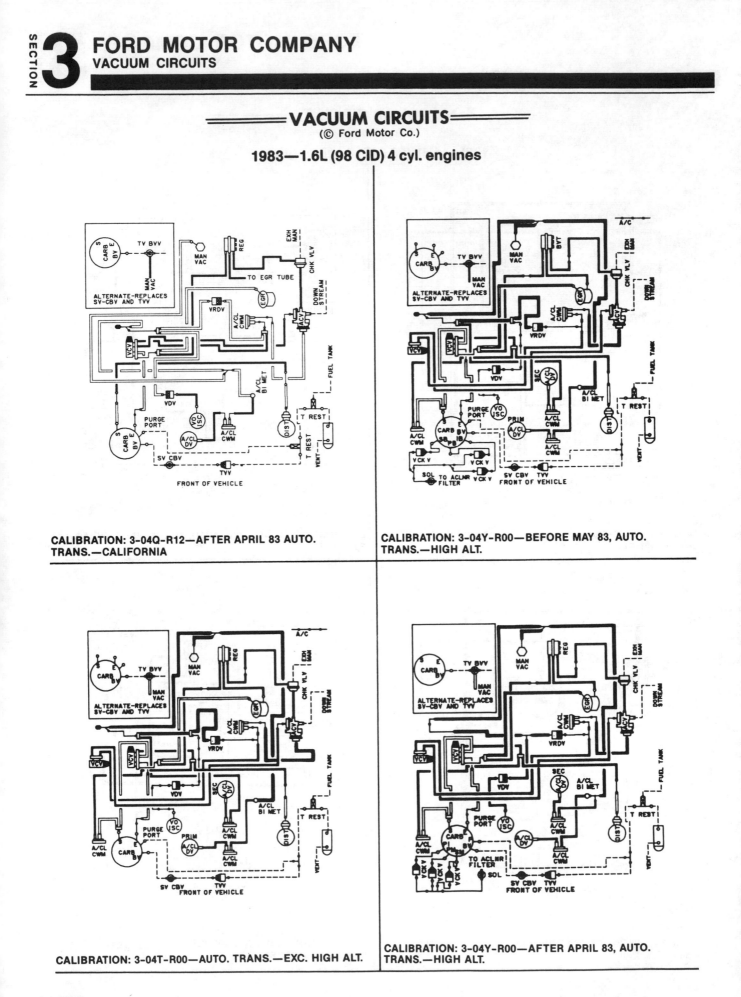

CALIBRATION: 3-04Q-R12—AFTER APRIL 83 AUTO. TRANS.—CALIFORNIA

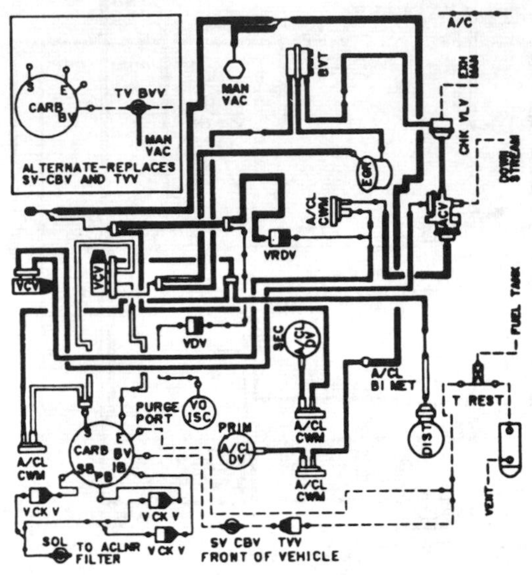

CALIBRATION: 3-04Y-R00—BEFORE MAY 83, AUTO. TRANS.—HIGH ALT.

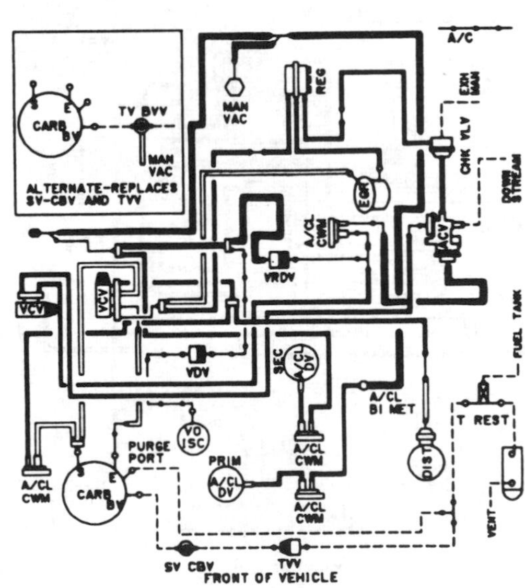

CALIBRATION: 3-04T-R00—AUTO. TRANS.—EXC. HIGH ALT.

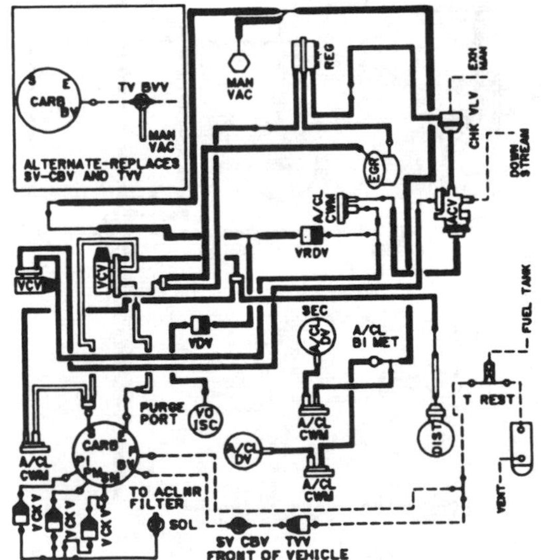

CALIBRATION: 3-04Y-R00—AFTER APRIL 83, AUTO. TRANS.—HIGH ALT.

VACUUM CIRCUITS

(© Ford Motor Co.)

1983—2.3L (140 CID) 4 cyl. engines

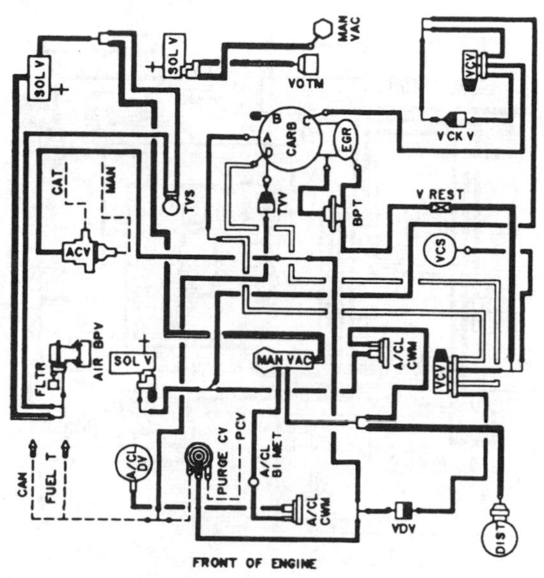

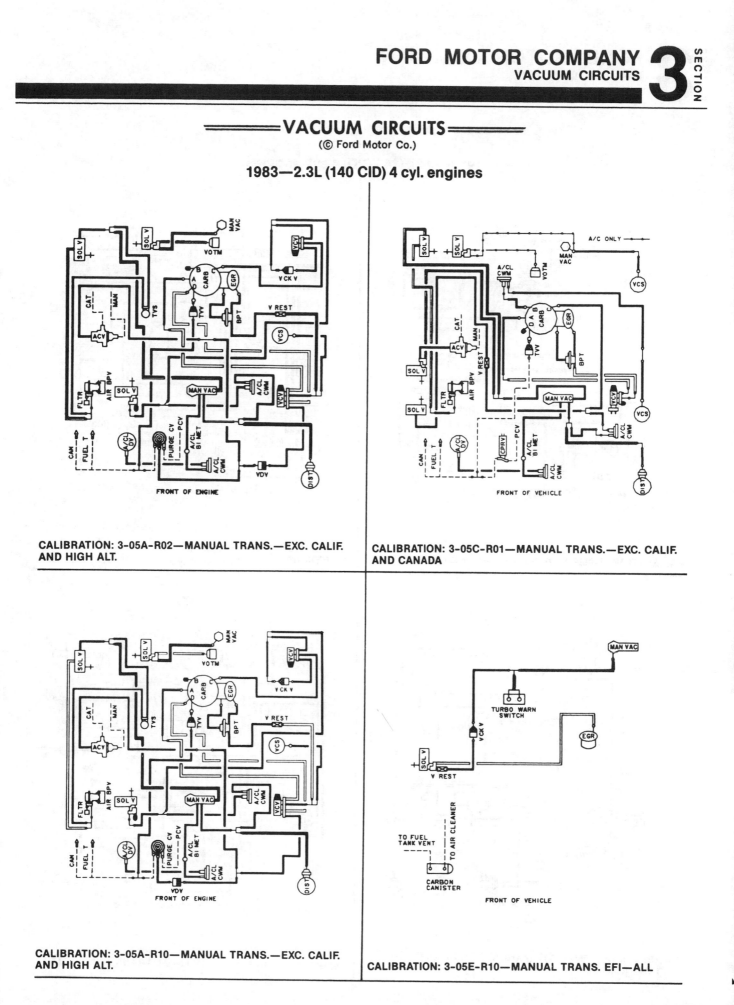

CALIBRATION: 3-05A-R02—MANUAL TRANS.—EXC. CALIF. AND HIGH ALT.

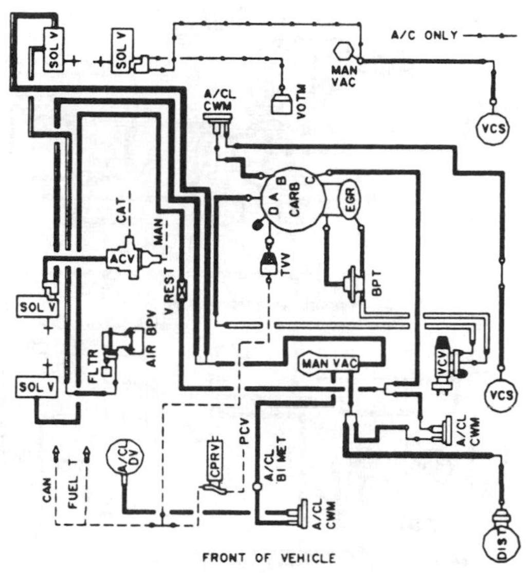

CALIBRATION: 3-05C-R01—MANUAL TRANS.—EXC. CALIF. AND CANADA

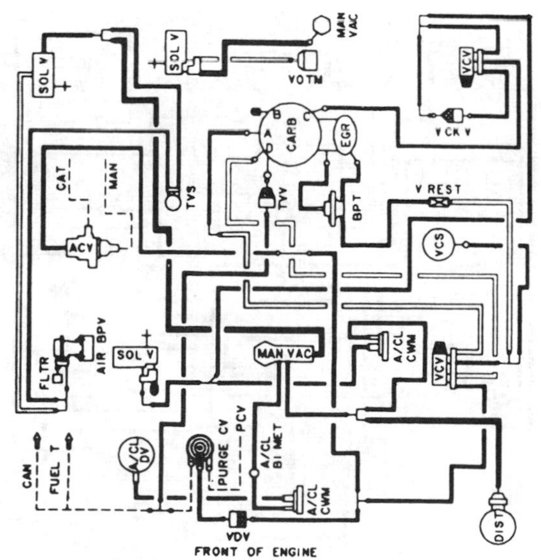

CALIBRATION: 3-05A-R10—MANUAL TRANS.—EXC. CALIF. AND HIGH ALT.

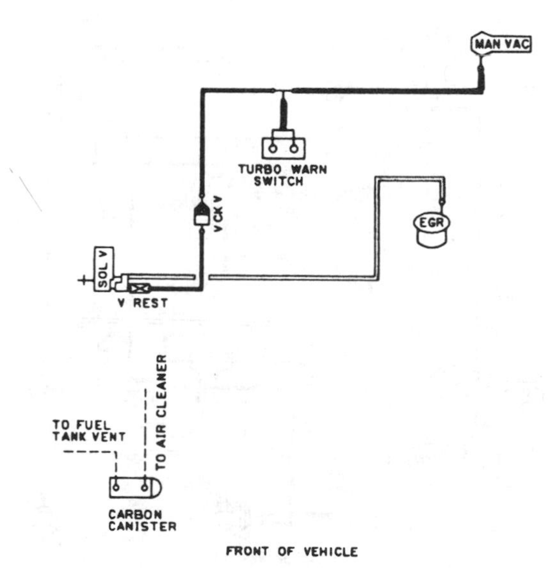

CALIBRATION: 3-05E-R10—MANUAL TRANS. EFI—ALL

VACUUM CIRCUITS

(© Ford Motor Co)

1983—2.3L (140 CID) 4 cyl. engines

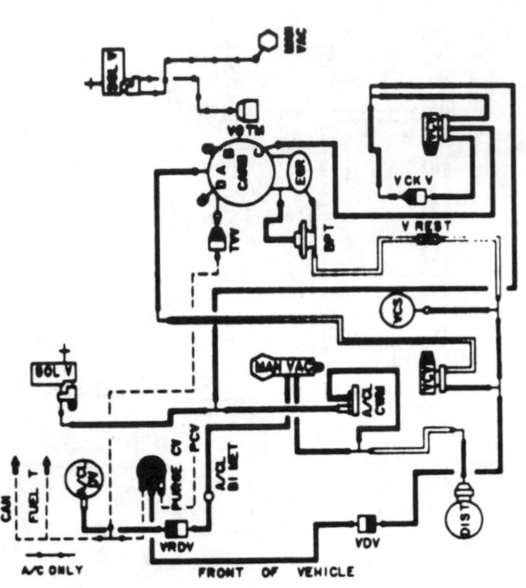

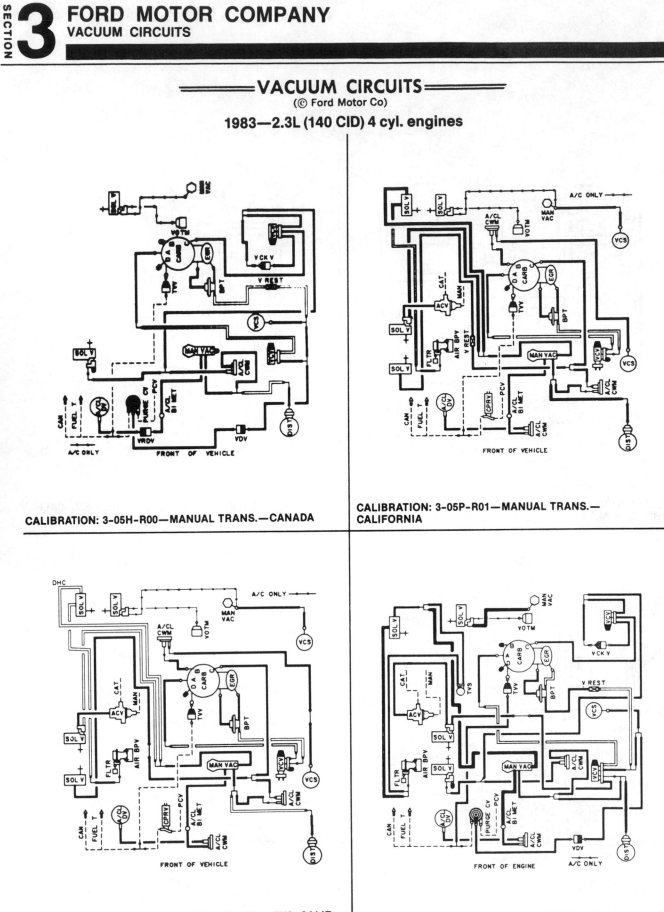

CALIBRATION: 3-05H-R00—MANUAL TRANS.—CANADA

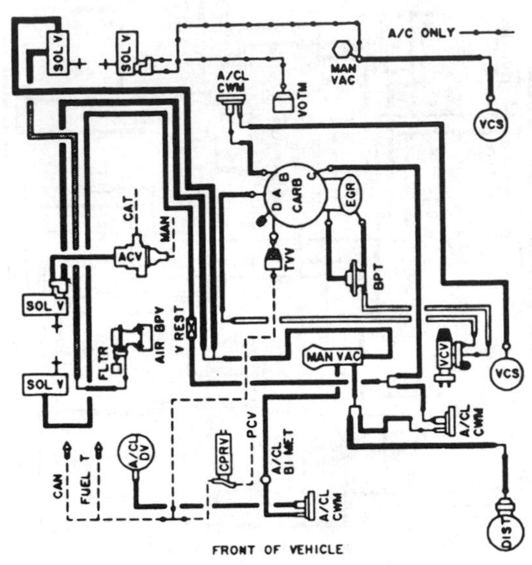

CALIBRATION: 3-05P-R01—MANUAL TRANS.—CALIFORNIA

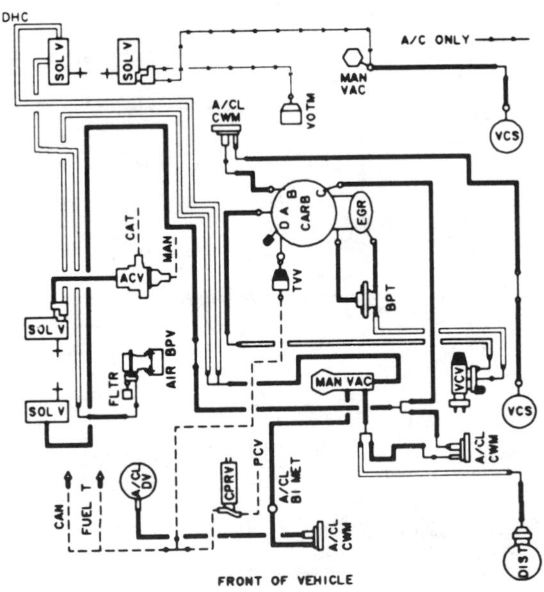

CALIBRATION: 3-05N-R00—MANUAL TRANS.—EXC. CALIF. AND CANADA

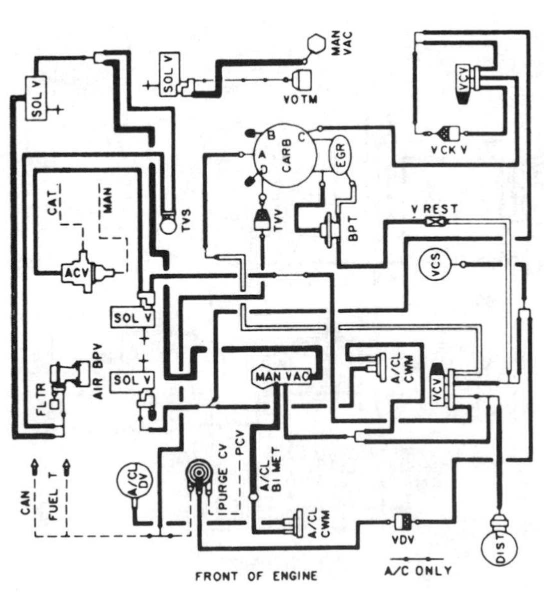

CALIBRATION: 3-05W-R00—MANUAL TRANS.—HIGH ALT.

VACUUM CIRCUITS

(© Ford Motor Co.)

1983—2.3L (140 CID) 4 cyl. engines

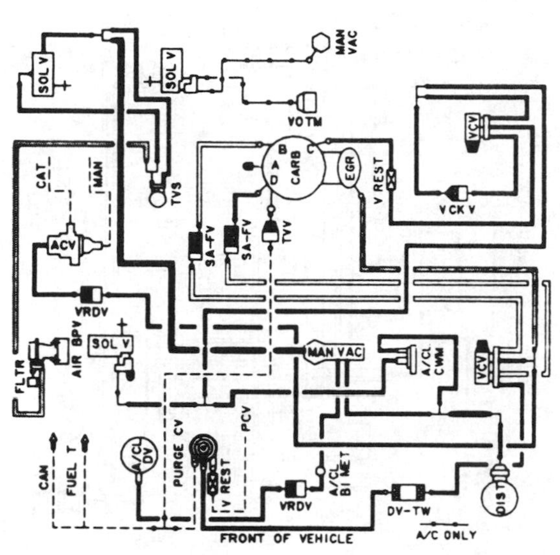

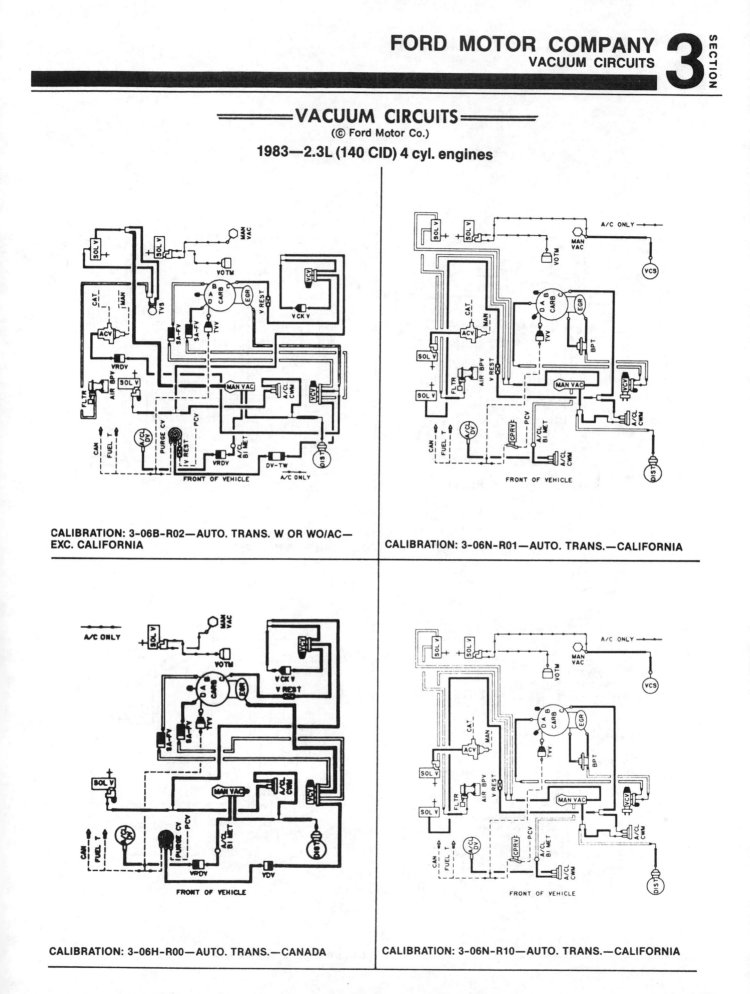

CALIBRATION: 3-06B-R02—AUTO. TRANS. W OR WO/AC—EXC. CALIFORNIA

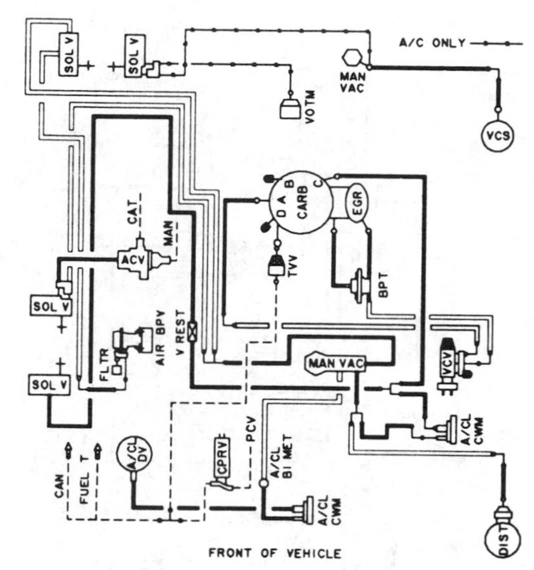

CALIBRATION: 3-06N-R01—AUTO. TRANS.—CALIFORNIA

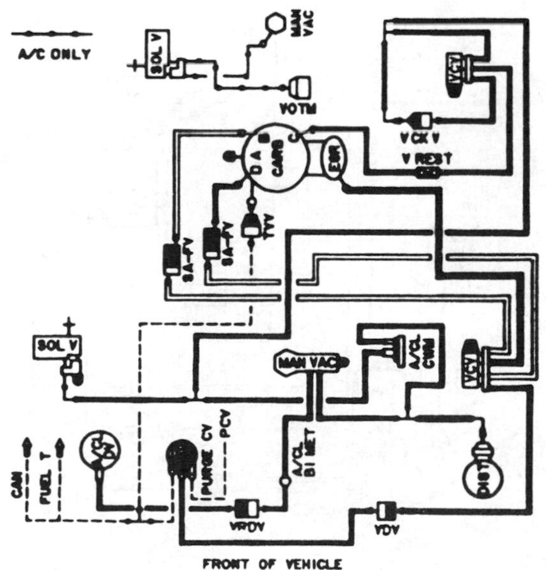

CALIBRATION: 3-06H-R00—AUTO. TRANS.—CANADA

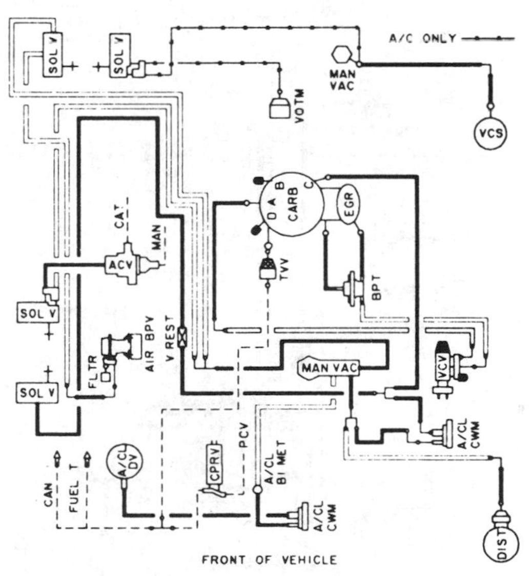

CALIBRATION: 3-06N-R10—AUTO. TRANS.—CALIFORNIA

VACUUM CIRCUITS

(© Ford Motor Co)

1983—3.3L (200 CID) 6 cyl. engines

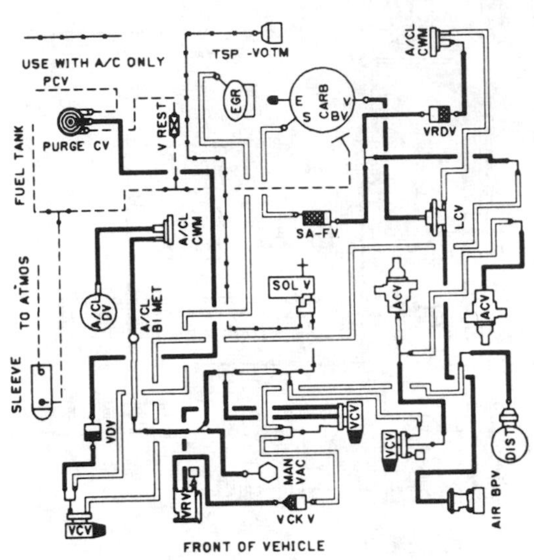

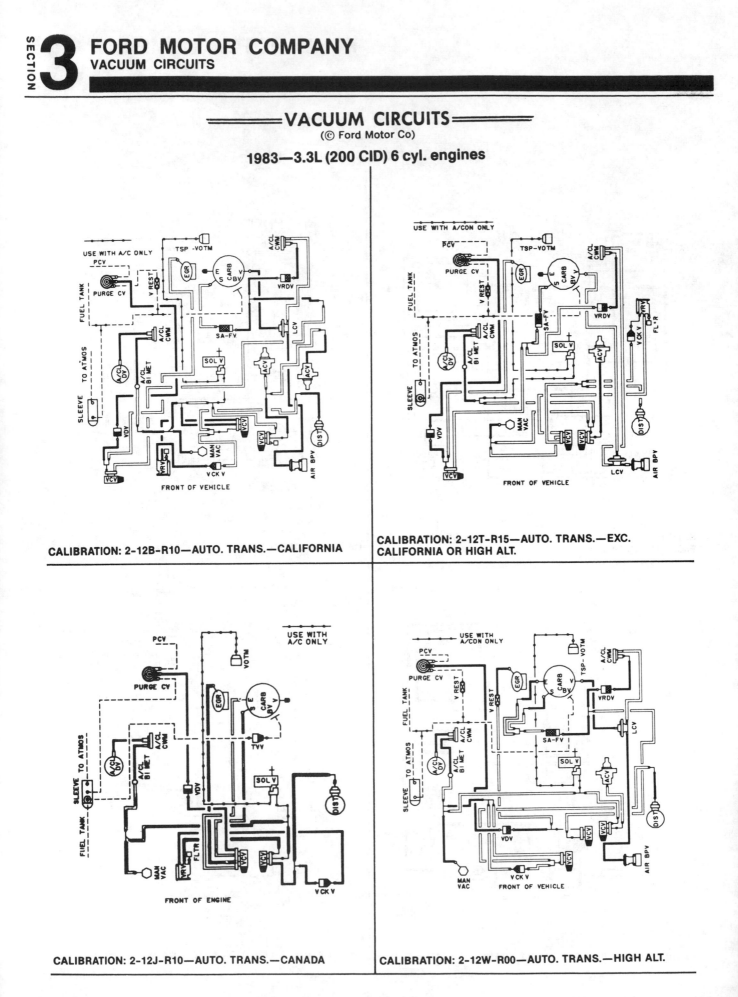

CALIBRATION: 2-12B-R10—AUTO. TRANS.—CALIFORNIA

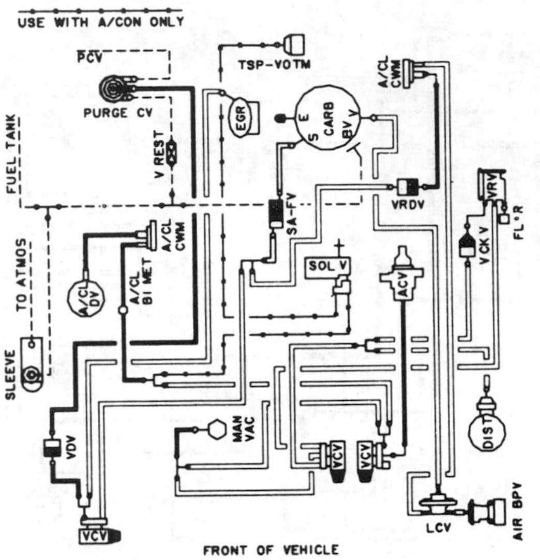

CALIBRATION: 2-12T-R15—AUTO. TRANS.—EXC. CALIFORNIA OR HIGH ALT.

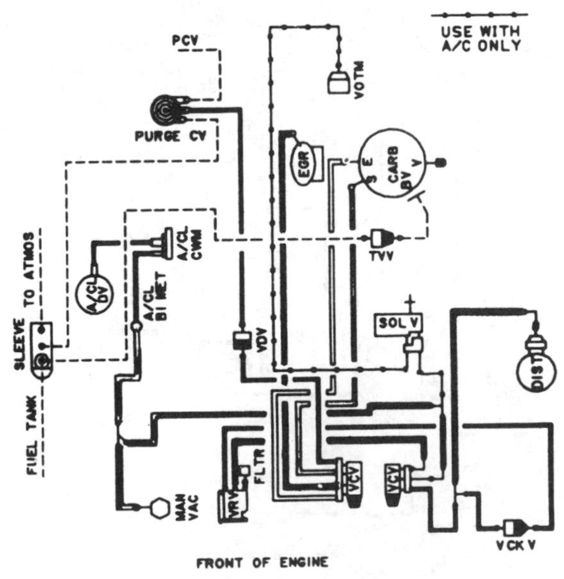

CALIBRATION: 2-12J-R10—AUTO. TRANS.—CANADA

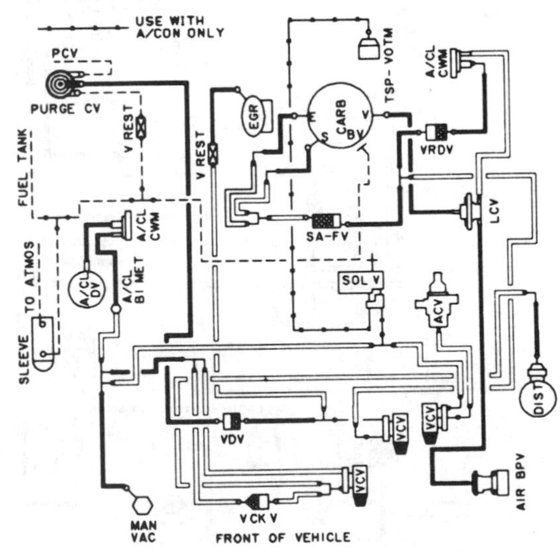

CALIBRATION: 2-12W-R00—AUTO. TRANS.—HIGH ALT.

VACUUM CIRCUITS

(© Ford Motor Co.)

1983—3.3L (200 CID) 6 cyl. engines

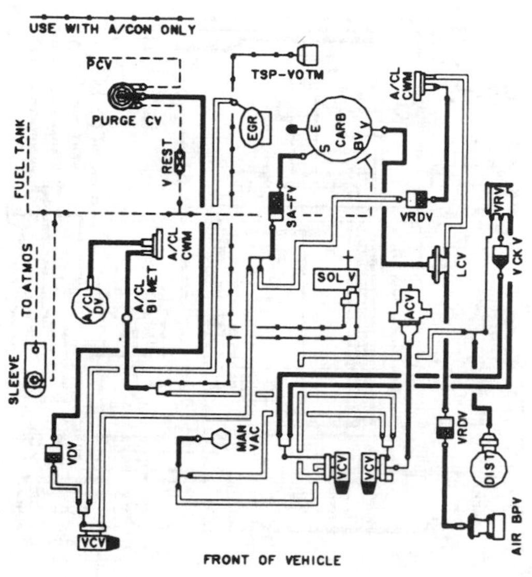

CALIBRATION: 3-12A-R00—AUTO. TRANS.—HIGH ALT.

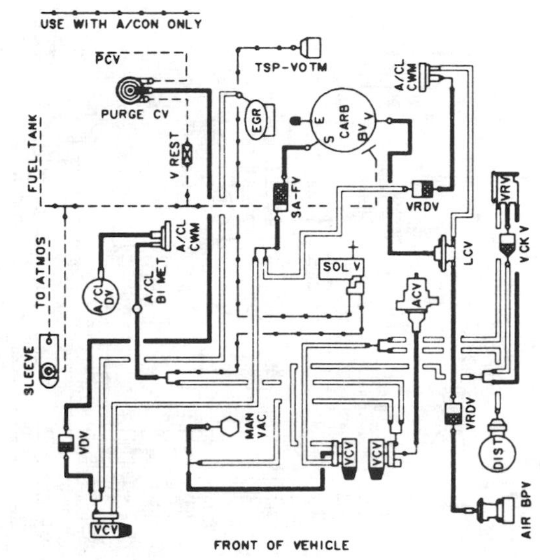

CALIBRATION: 3-12Q-R00—AUTO. TRANS.—ALL

1983—3.8L (230 CID) 6 cyl. engines

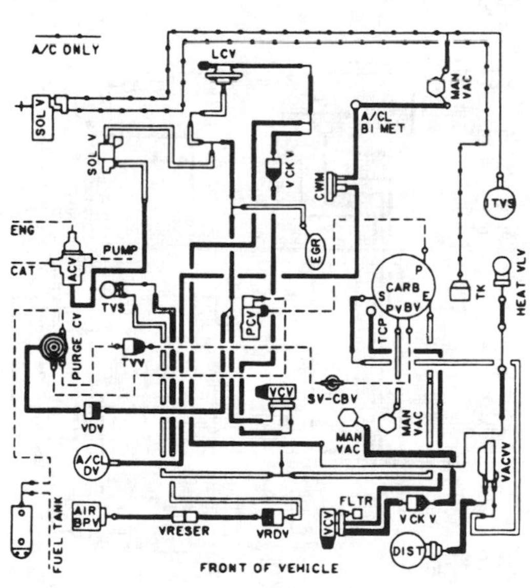

CALIBRATION: 3-14A-R00—AUTO. TRANS.—EXC. CALIFORNIA

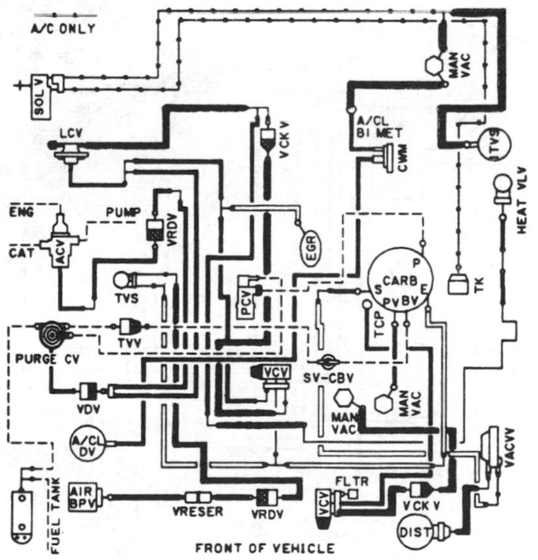

CALIBRATION: 3-14A-R04—AUTO. TRANS.—EXC. CALIFORNIA

VACUUM CIRCUITS

(© Ford Motor Co)

1983—3.8L (230 CID) 6 cyl. engines

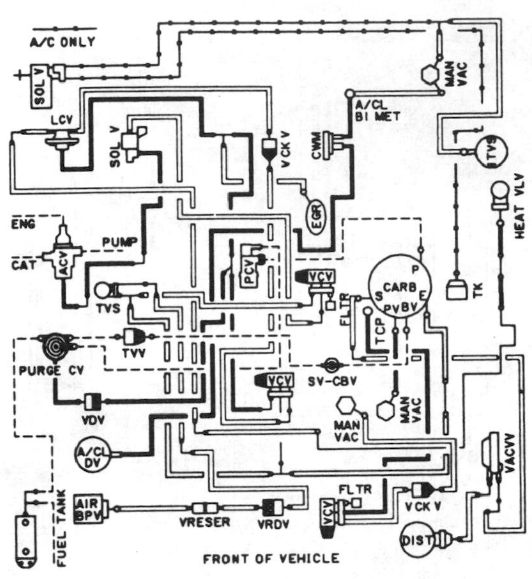

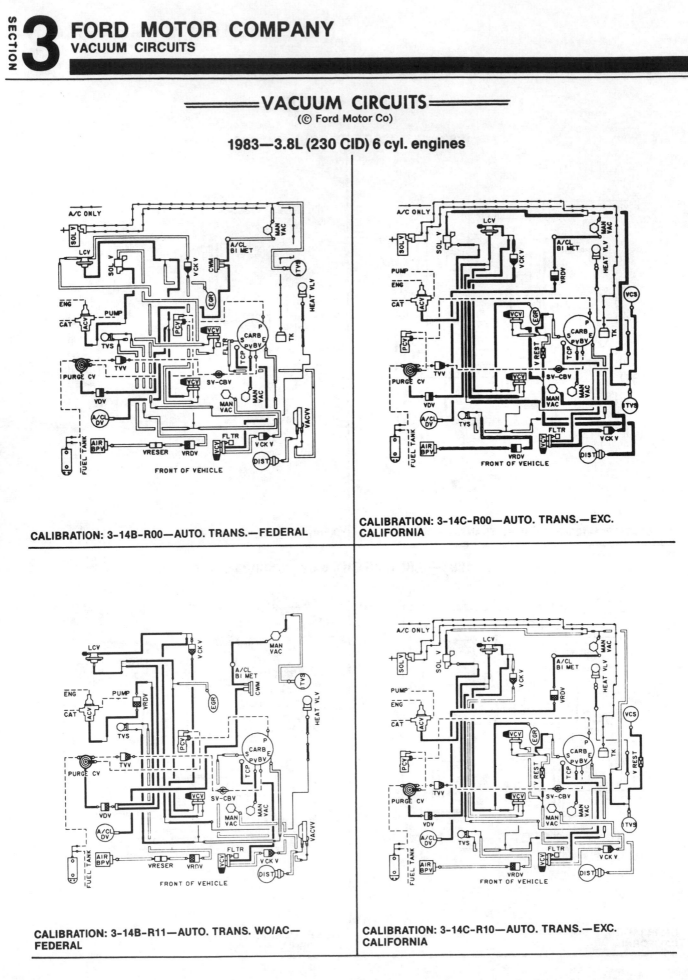

CALIBRATION: 3-14B-R00—AUTO. TRANS.—FEDERAL

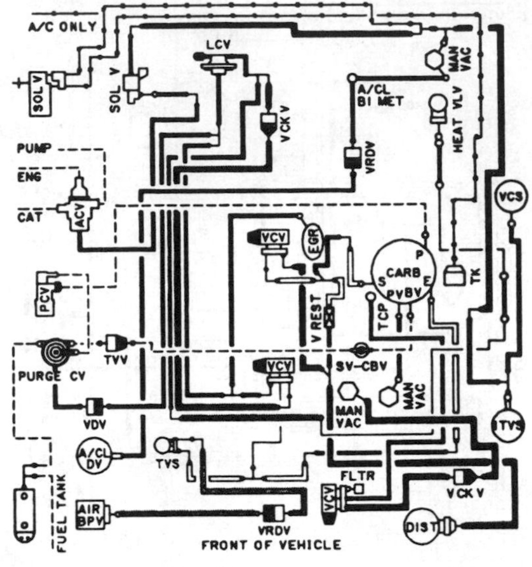

CALIBRATION: 3-14C-R00—AUTO. TRANS.—EXC. CALIFORNIA

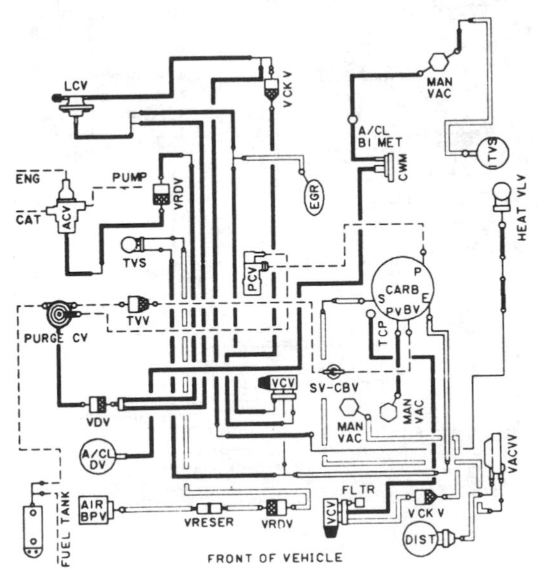

CALIBRATION: 3-14B-R11—AUTO. TRANS. WO/AC—FEDERAL

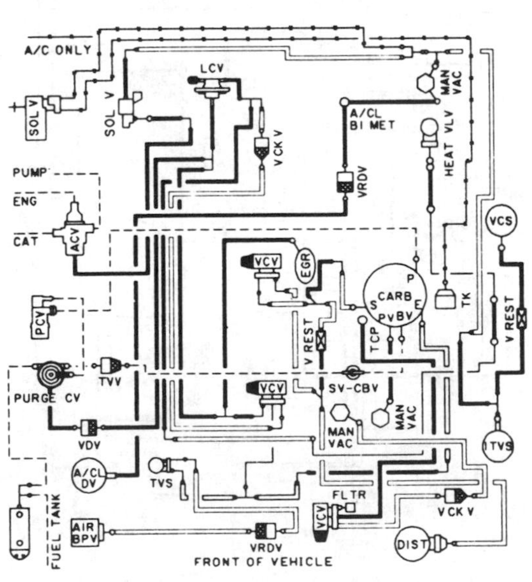

CALIBRATION: 3-14C-R10—AUTO. TRANS.—EXC. CALIFORNIA

VACUUM CIRCUITS

(© Ford Motor Co.)

1983—3.3L (200 CID) 6 cyl. engines

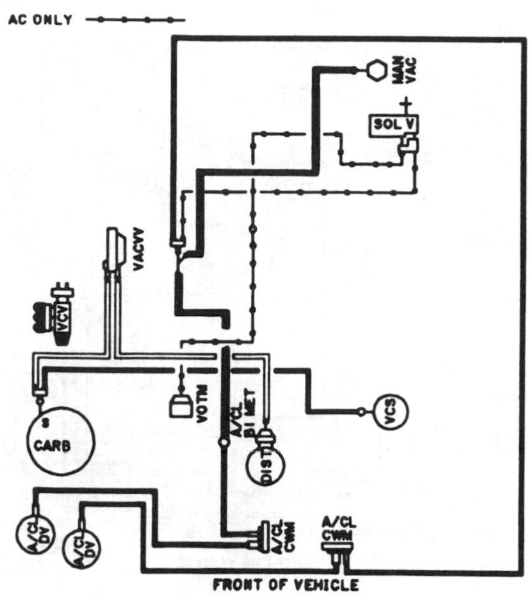

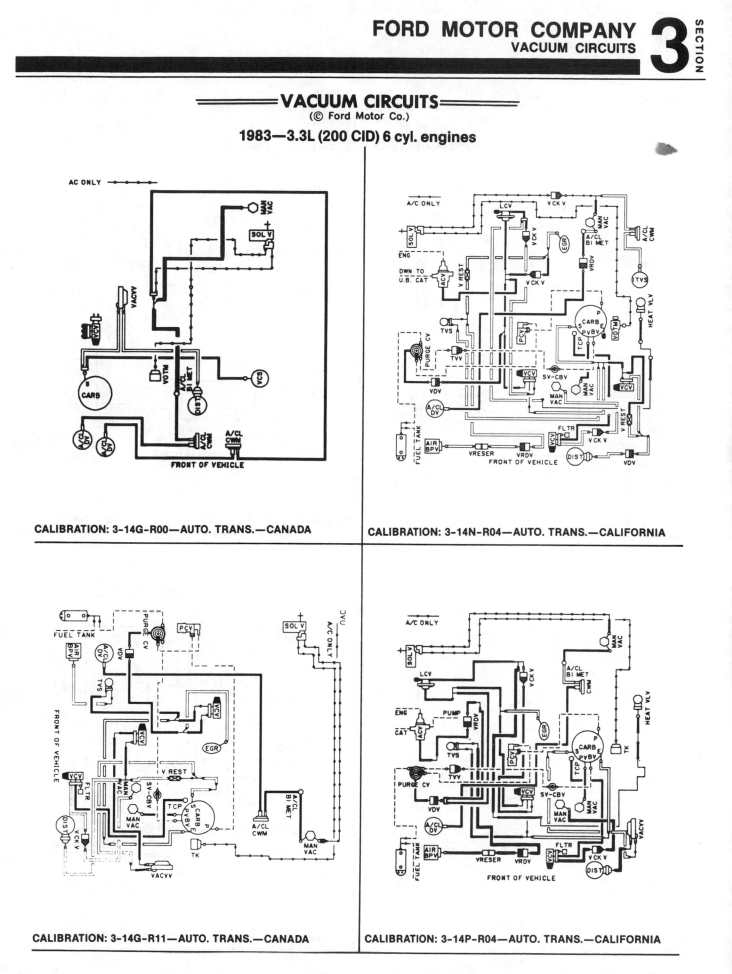

CALIBRATION: 3-14G-R00—AUTO. TRANS.—CANADA

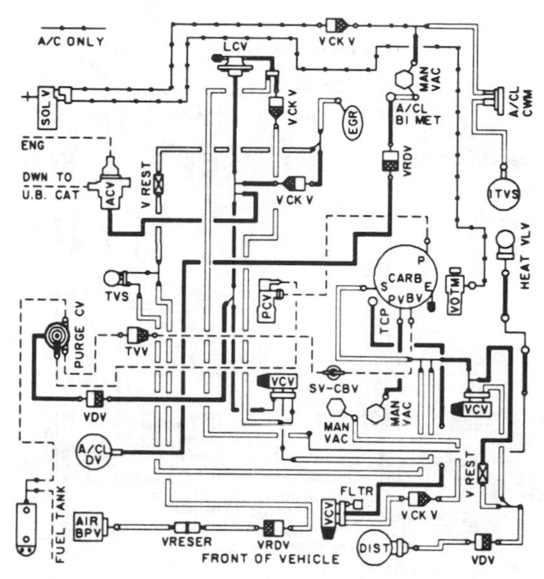

CALIBRATION: 3-14N-R04—AUTO. TRANS.—CALIFORNIA

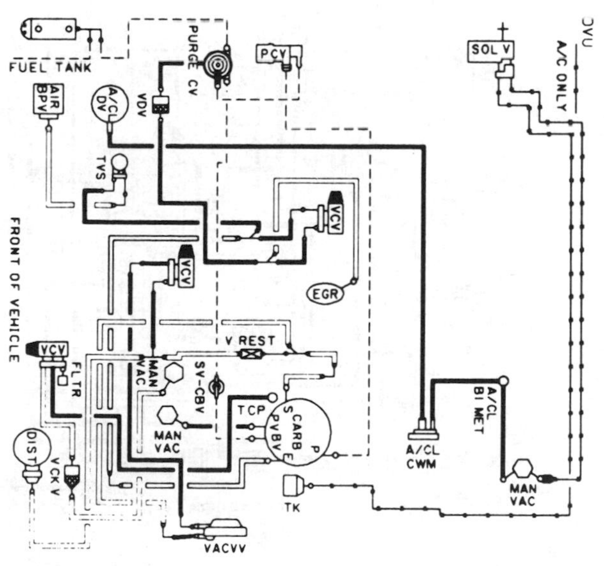

CALIBRATION: 3-14G-R11—AUTO. TRANS.—CANADA

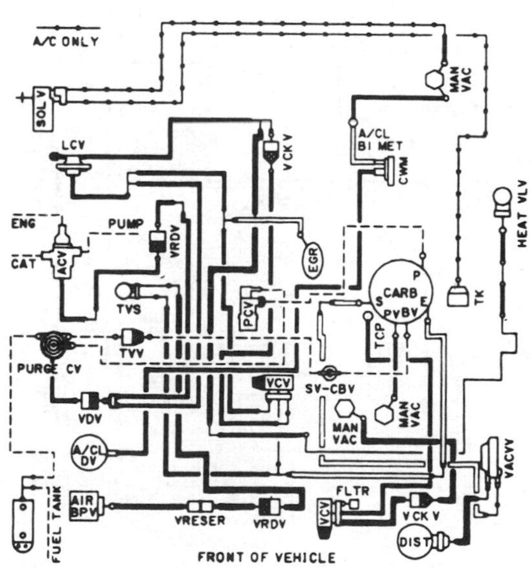

CALIBRATION: 3-14P-R04—AUTO. TRANS.—CALIFORNIA

VACUUM CIRCUITS

(© Ford Motor Co.)

1983—3.8L (230 CID) 6 cyl. engines

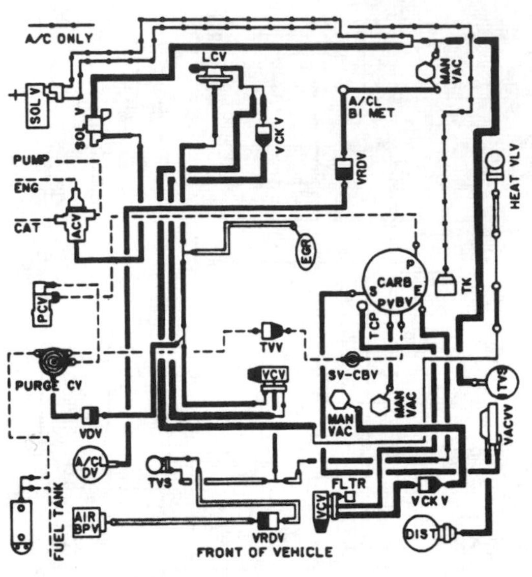

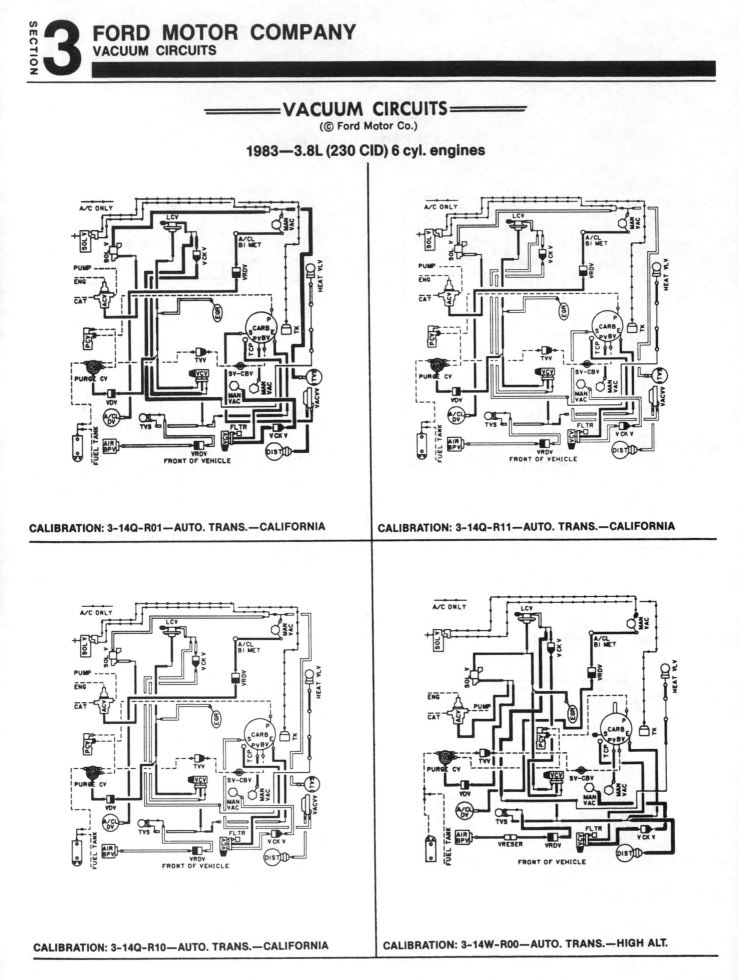

CALIBRATION: 3-14Q-R01—AUTO. TRANS.—CALIFORNIA

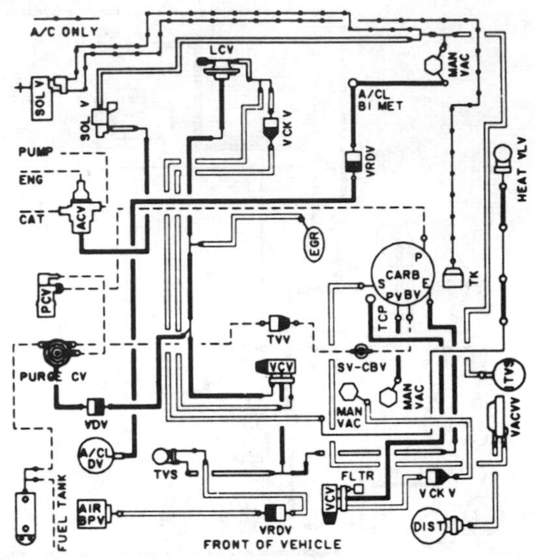

CALIBRATION: 3-14Q-R11—AUTO. TRANS.—CALIFORNIA

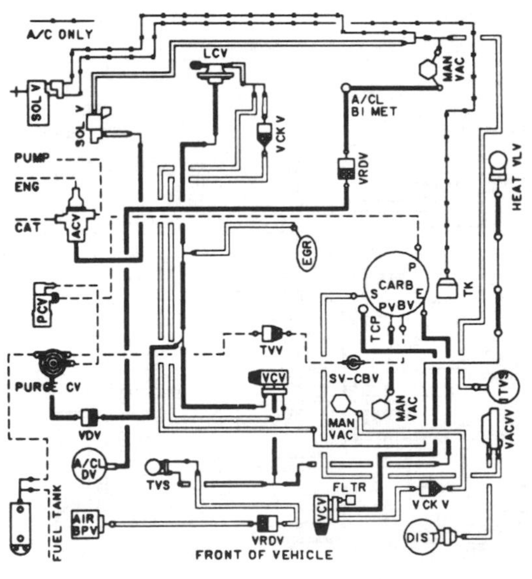

CALIBRATION: 3-14Q-R10—AUTO. TRANS.—CALIFORNIA

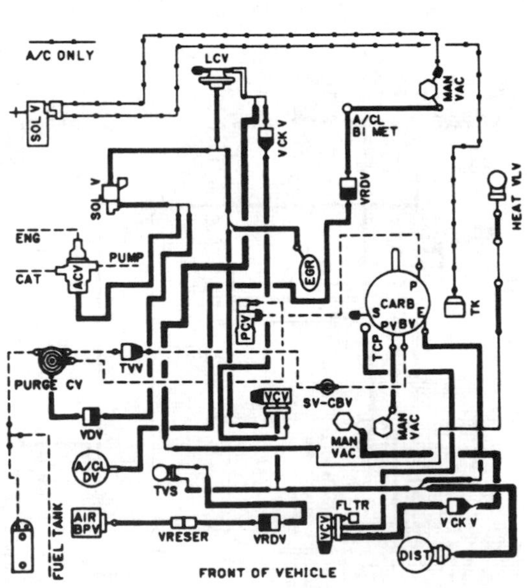

CALIBRATION: 3-14W-R00—AUTO. TRANS.—HIGH ALT.

VACUUM CIRCUITS

(© Ford Motor Co.)

1983—3.8L (230 CID) 6 cyl. engines

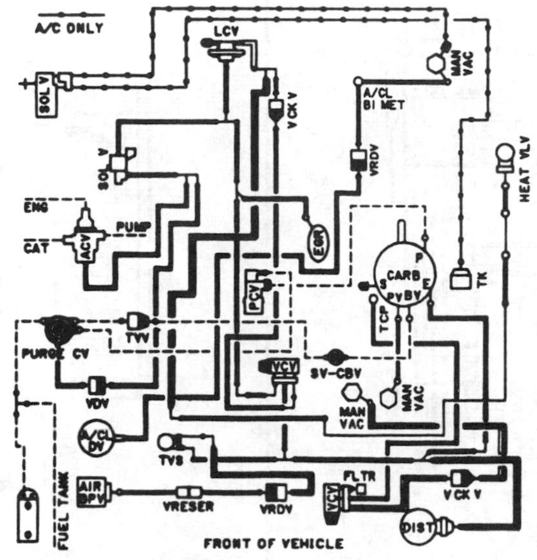

CALIBRATION: 3-14X-R00—AUTO. TRANS.—HIGH ALT.

1983—5.0L (302 CID) V8 Engines

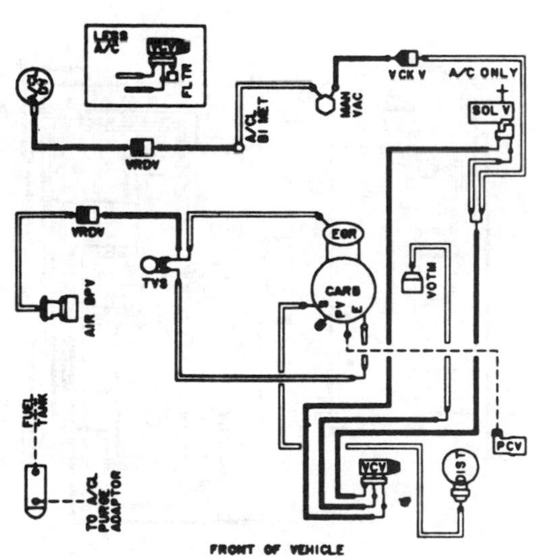

CALIBRATION: 2-20H-R00—AUTO. TRANS.—CANADA

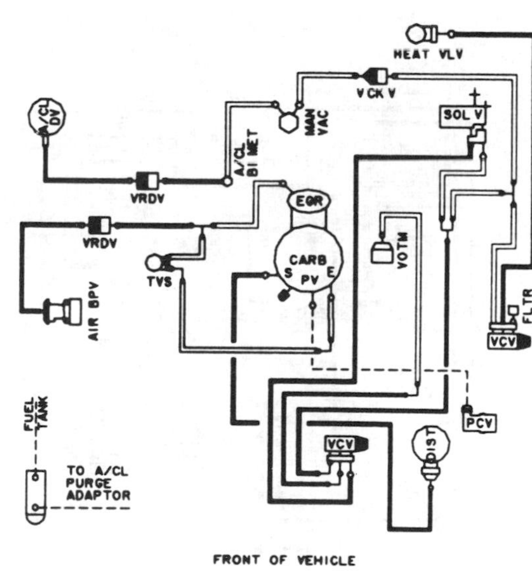

CALIBRATION: 2-20J-R11—AUTO. TRANS.—CANADA

VACUUM CIRCUITS

(© Ford Motor Co.)

1983—5.0L (302 CID) V8 Engines

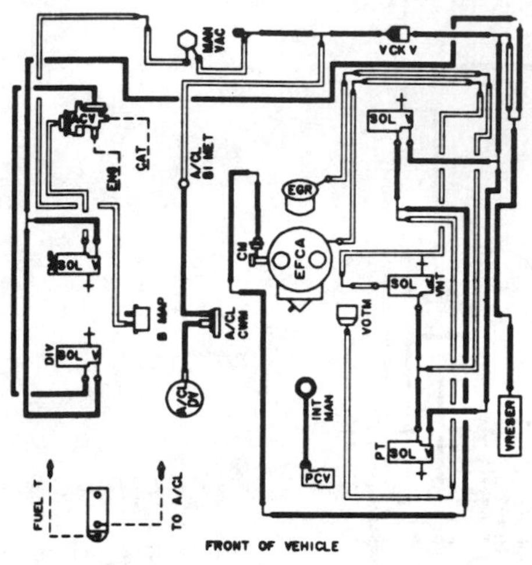

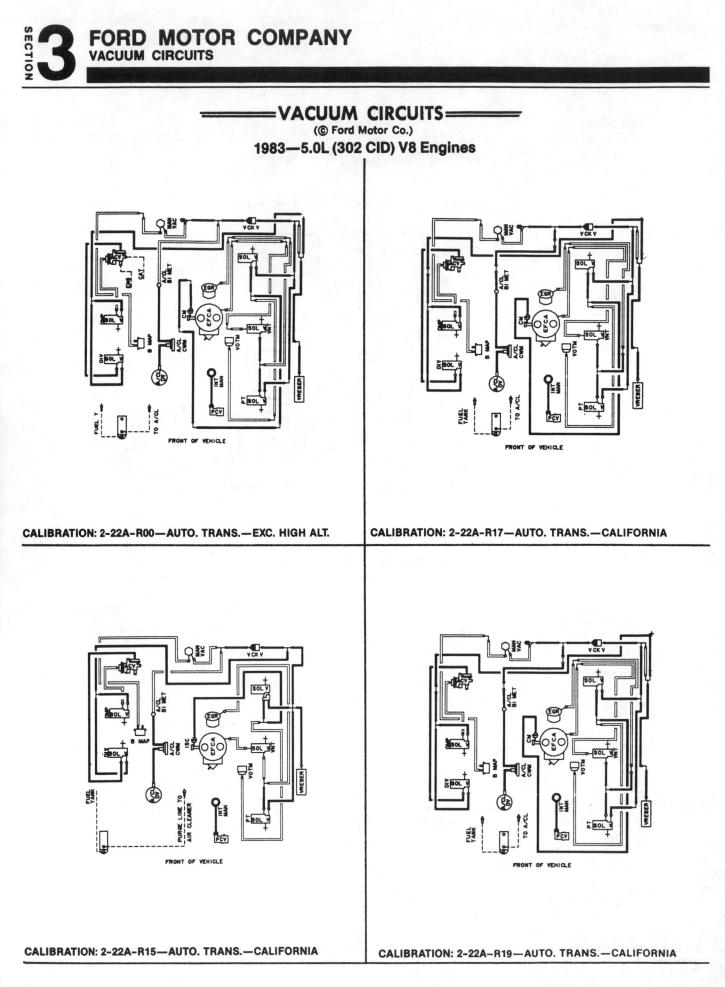

CALIBRATION: 2-22A-R00—AUTO. TRANS.—EXC. HIGH ALT.

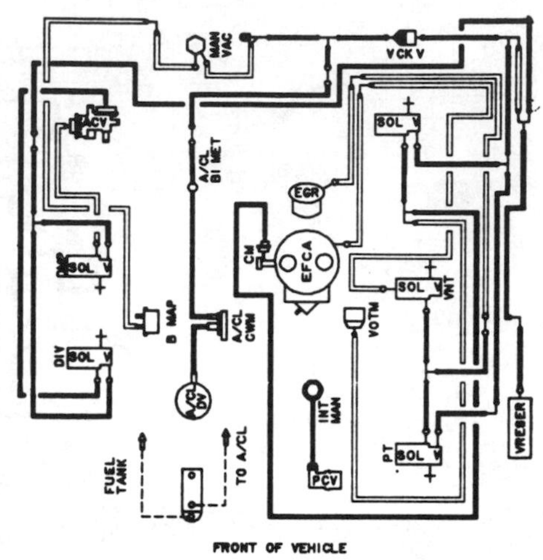

CALIBRATION: 2-22A-R17—AUTO. TRANS.—CALIFORNIA

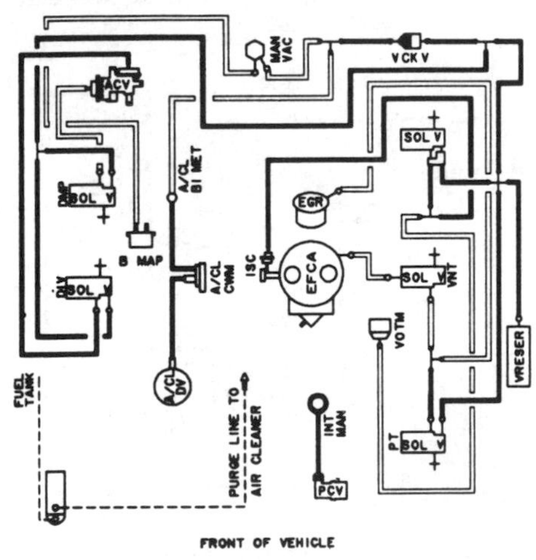

CALIBRATION: 2-22A-R15—AUTO. TRANS.—CALIFORNIA

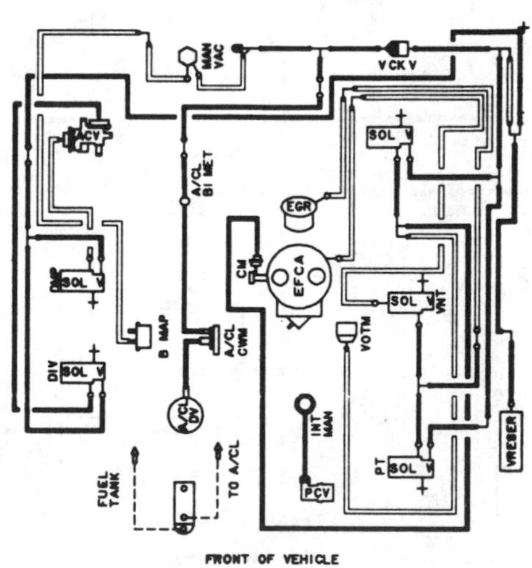

CALIBRATION: 2-22A-R19—AUTO. TRANS.—CALIFORNIA

VACUUM CIRCUITS

(© Ford Motor Co.)

1983—5.0L (302 CID) V8 Engines

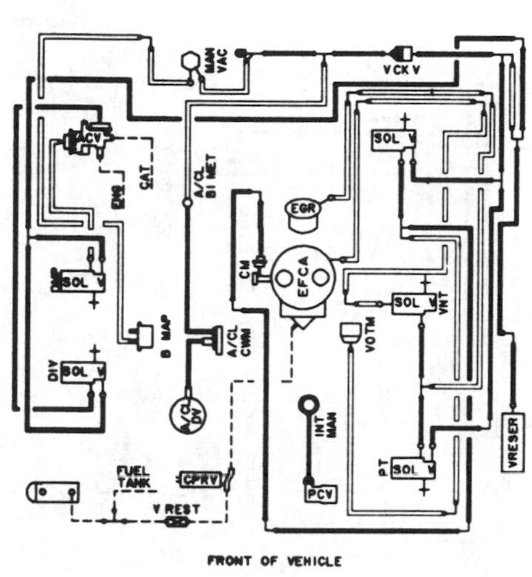

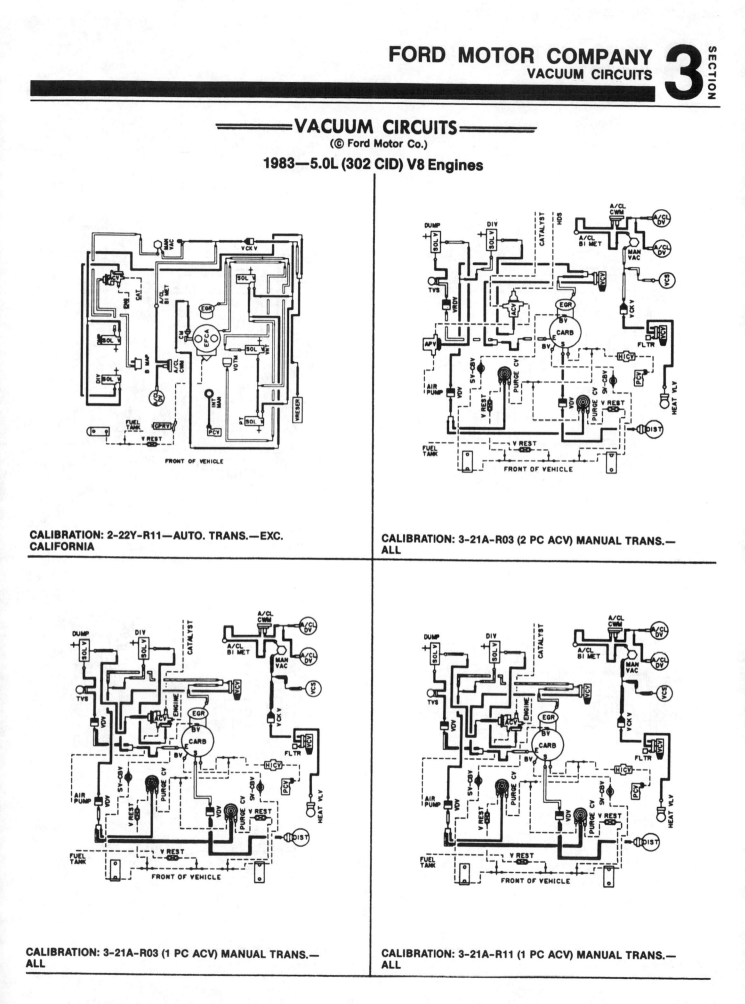

CALIBRATION: 2-22Y-R11—AUTO. TRANS.—EXC. CALIFORNIA

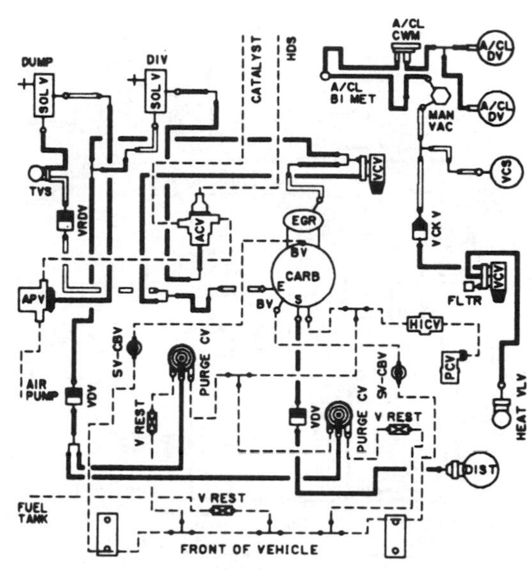

CALIBRATION: 3-21A-R03 (2 PC ACV) MANUAL TRANS.—ALL

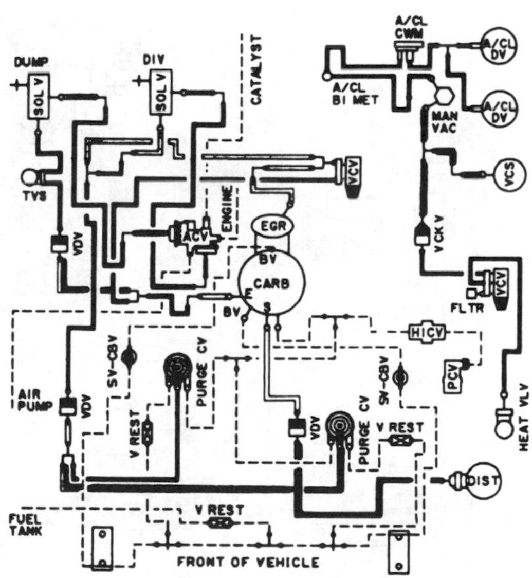

CALIBRATION: 3-21A-R03 (1 PC ACV) MANUAL TRANS.—ALL

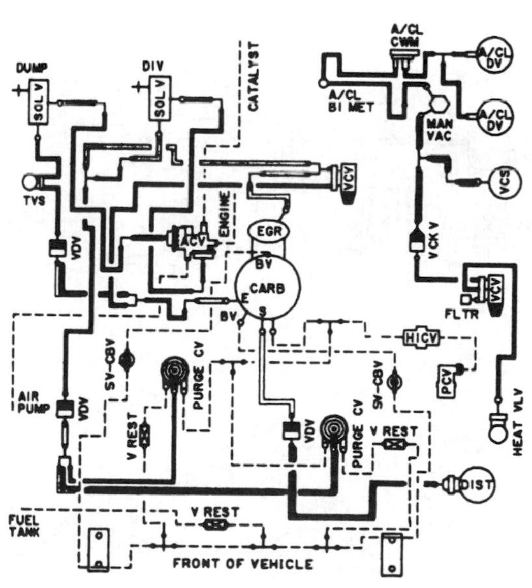

CALIBRATION: 3-21A-R11 (1 PC ACV) MANUAL TRANS.—ALL

VACUUM CIRCUITS

(© Ford Motor Co.)

1983—5.0L (302 CID) V8 Engines

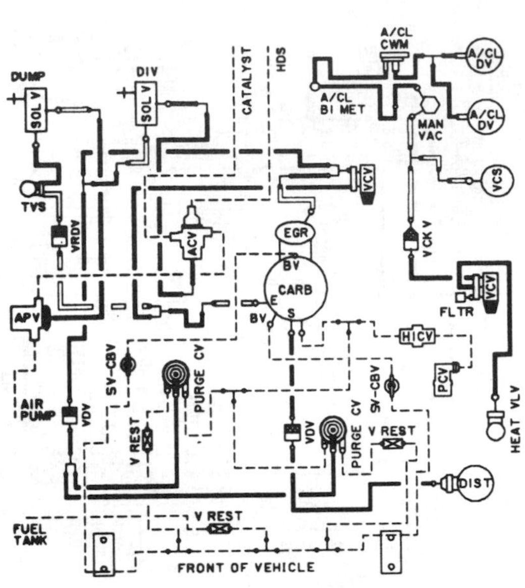

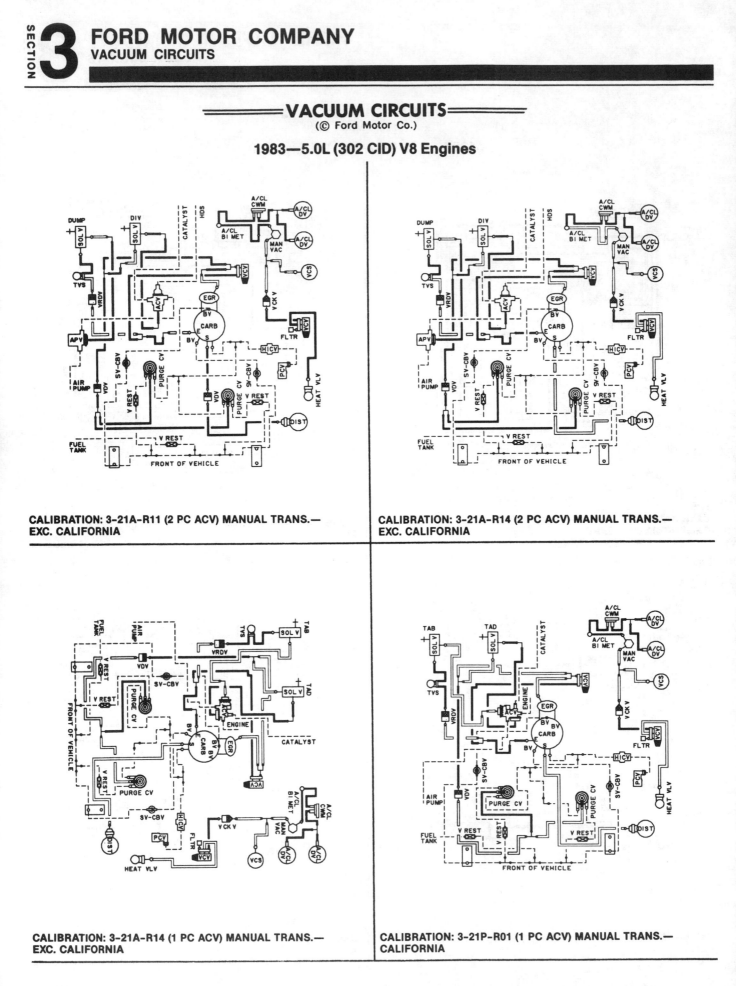

CALIBRATION: 3-21A-R11 (2 PC ACV) MANUAL TRANS.— EXC. CALIFORNIA

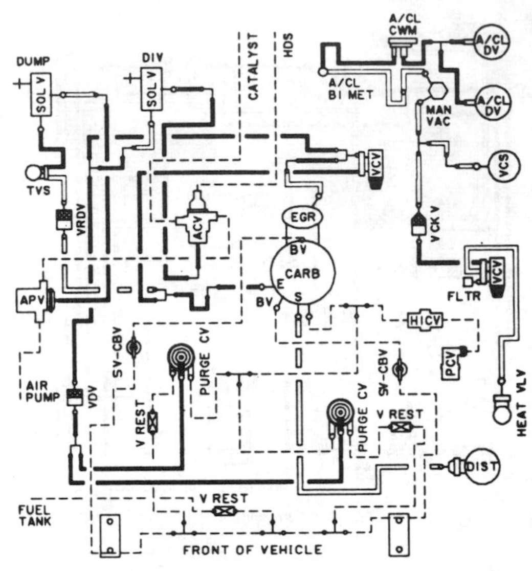

CALIBRATION: 3-21A-R14 (2 PC ACV) MANUAL TRANS.— EXC. CALIFORNIA

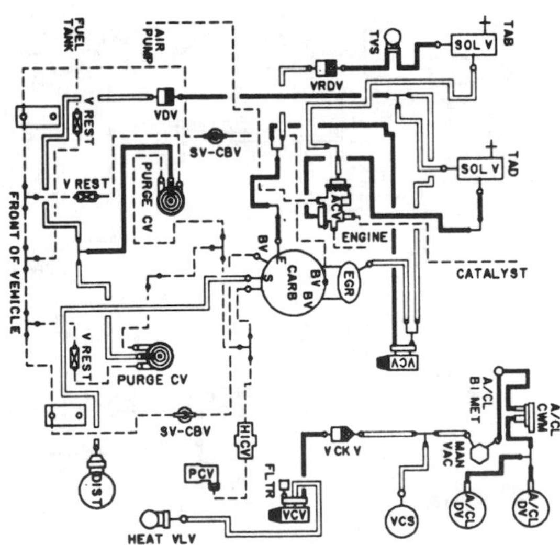

CALIBRATION: 3-21A-R14 (1 PC ACV) MANUAL TRANS.— EXC. CALIFORNIA

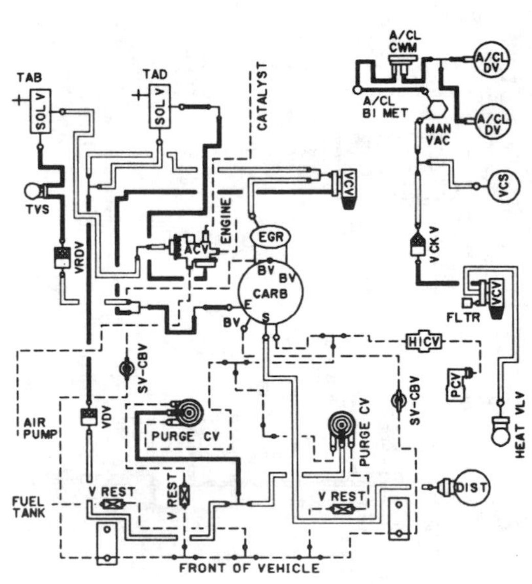

CALIBRATION: 3-21P-R01 (1 PC ACV) MANUAL TRANS.— CALIFORNIA

VACUUM CIRCUITS

(© Ford Motor Co)

1983—5.0L (302 CID) V8 Engines

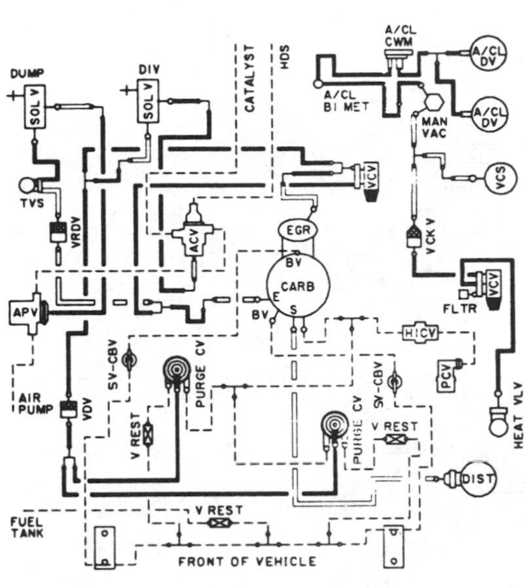

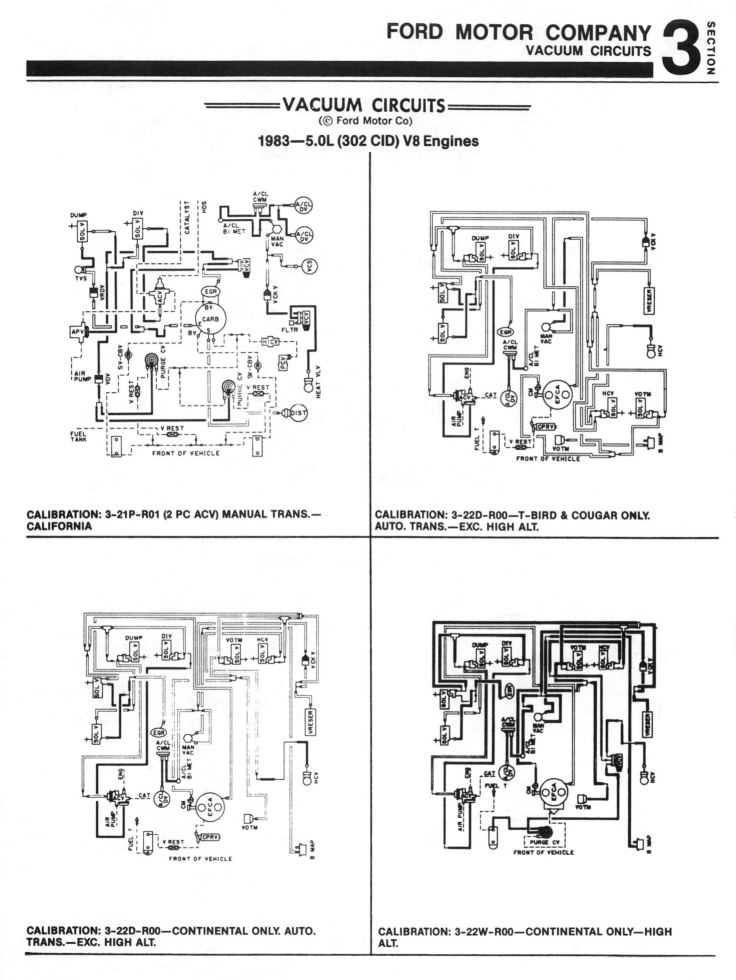

CALIBRATION: 3-21P-R01 (2 PC ACV) MANUAL TRANS.—CALIFORNIA

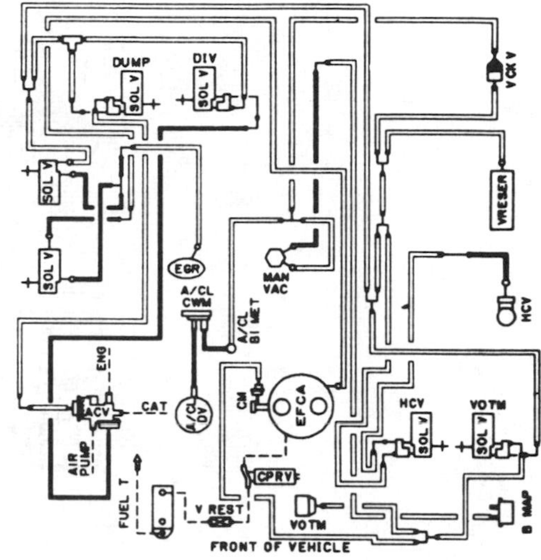

CALIBRATION: 3-22D-R00—T-BIRD & COUGAR ONLY. AUTO. TRANS.—EXC. HIGH ALT.

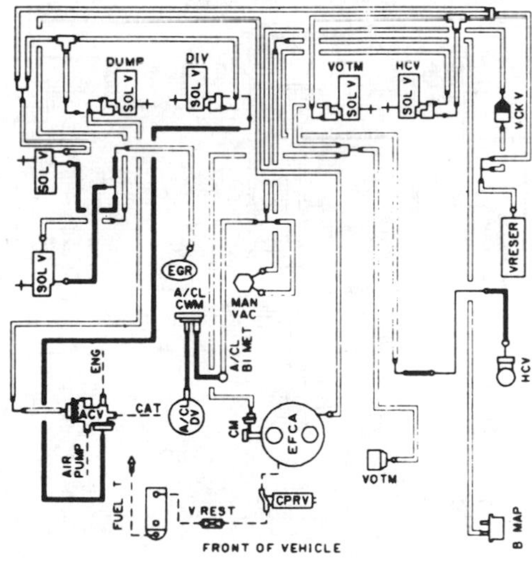

CALIBRATION: 3-22D-R00—CONTINENTAL ONLY. AUTO. TRANS.—EXC. HIGH ALT.

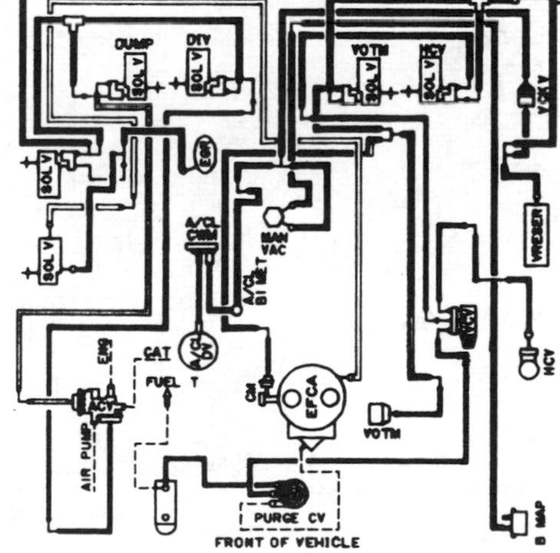

CALIBRATION: 3-22W-R00—CONTINENTAL ONLY—HIGH ALT.

VACUUM CIRCUITS

(© Ford Motor Co)

1983—5.0L (302 CID) V8 Engines

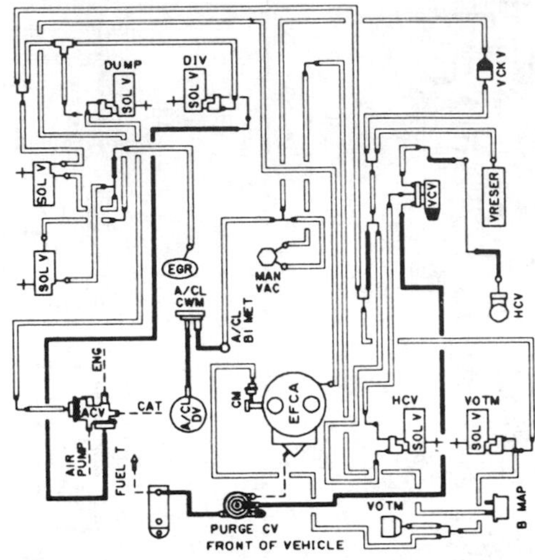

CALIBRATION: 3-22W-R00—T-BIRD & COUGAR ONLY.
AUTO. TRANS.—HIGH ALT.

1983—5.8L (351 CID) V8 Engines

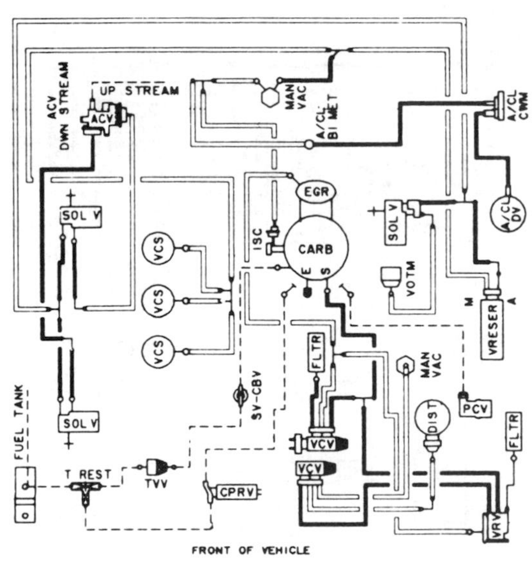

CALIBRATION: 2-24P-R10—AUTO. TRANS.—ALL

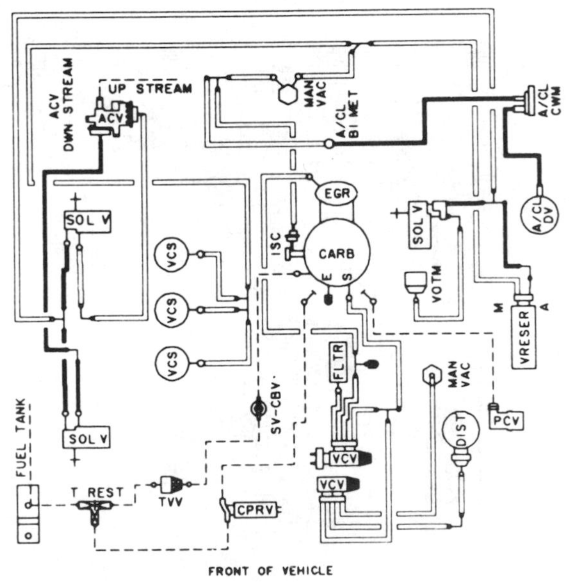

CALIBRATION: 2-24P-R11—AUTO. TRANS.—CANADA

Ford
Vacuum Circuits

INDEX

ENGINE AND CALIBRATION

VACUUM CIRCUITS

HOW TO DETERMINE VEHICLE CALIBRATION

Attached to all 1984 production engines is an Engine Code Information Label containing among other pertinent data—the engine calibration number. The size and shape of the decal is dependent upon the engine displacement. To determine the engine calibration, look at Step 1A. You do not have to refer to the Vehicle Emission decal for calibration numbers as in prior years.

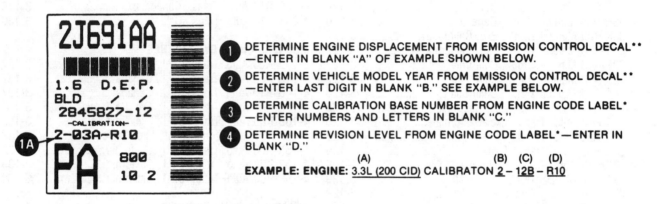

1. DETERMINE ENGINE DISPLACEMENT FROM EMISSION CONTROL DECAL** —ENTER IN BLANK "A" OF EXAMPLE SHOWN BELOW.
2. DETERMINE VEHICLE MODEL YEAR FROM EMISSION CONTROL DECAL** —ENTER LAST DIGIT IN BLANK "B." SEE EXAMPLE BELOW.
3. DETERMINE CALIBRATION BASE NUMBER FROM ENGINE CODE LABEL* —ENTER NUMBERS AND LETTERS IN BLANK "C."
4. DETERMINE REVISION LEVEL FROM ENGINE CODE LABEL*—ENTER IN BLANK "D."

EXAMPLE: ENGINE: (A) 3.3L (200 CID) CALIBRATON (B) 2 – (C) 12B – (D) R10

* Located on engine valve cover (Yellow or White).

** Located on engine valve cover, radiator support bracket, underside of hood or fresh air inlet tube (Gold or Silver), or windshield washer bottle.

VACUUM CIRCUITS

(© Ford Motor Co)

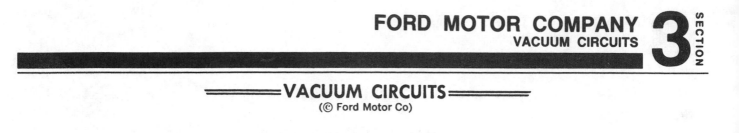

HOW TO READ VEHICLE EMISSION CONTROL INFORMATION (VECI) DECALS

The following callouts identify the various information areas of the VECI decals.

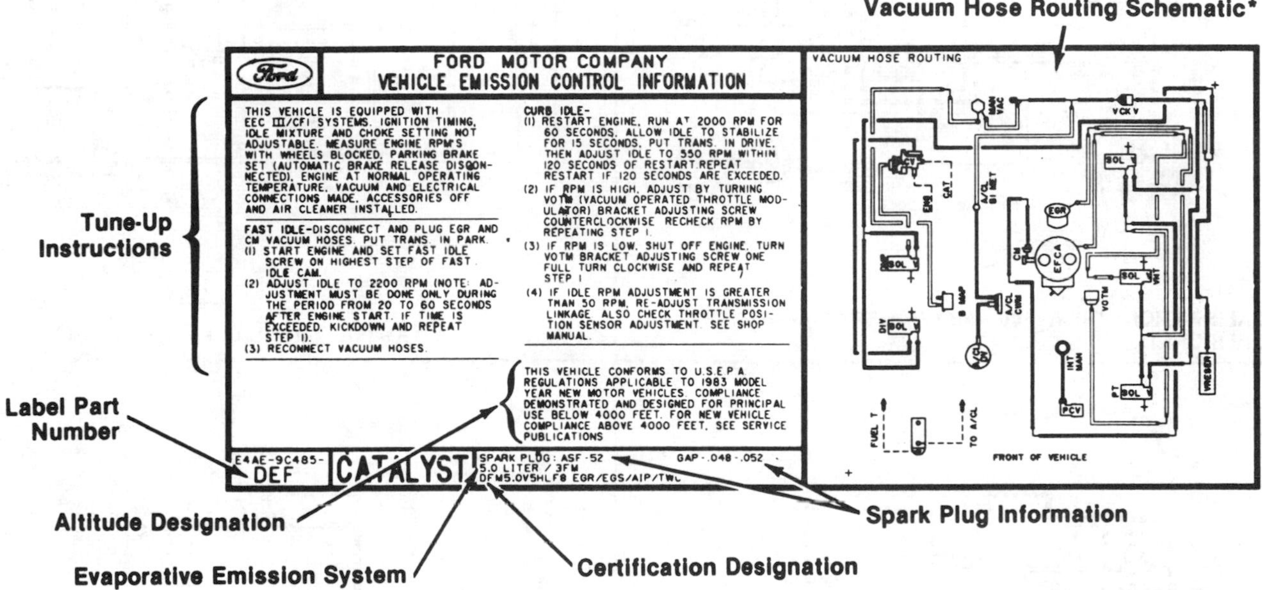

* The schematic refers only to emission systems (except exhaust). For vehicle vacuum systems, refer to Electrical and Vacuum Troubleshooting Manuals (EVTM).

VACUUM CIRCUITS

(© Ford Motor Co.)

1984—1.6L (98 CID) 4 cyl. engines

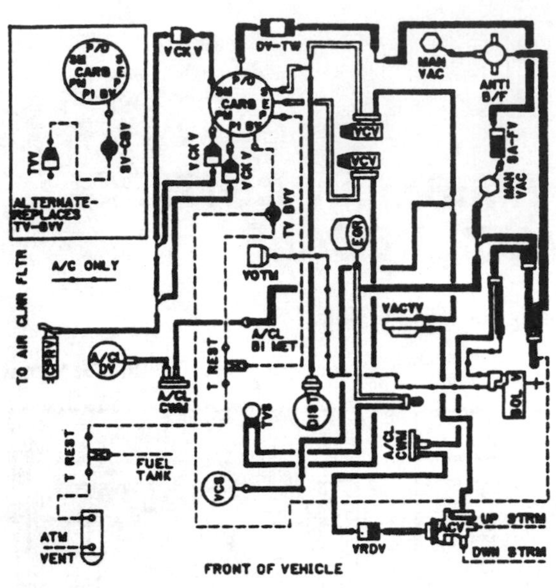

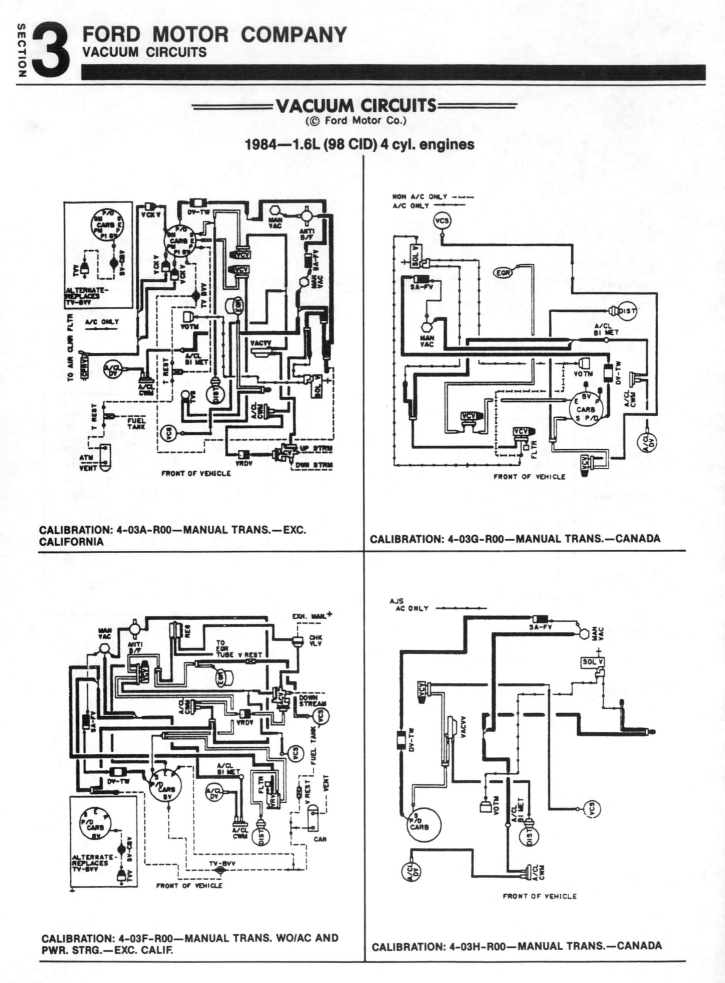

CALIBRATION: 4-03A-R00—MANUAL TRANS.—EXC. CALIFORNIA

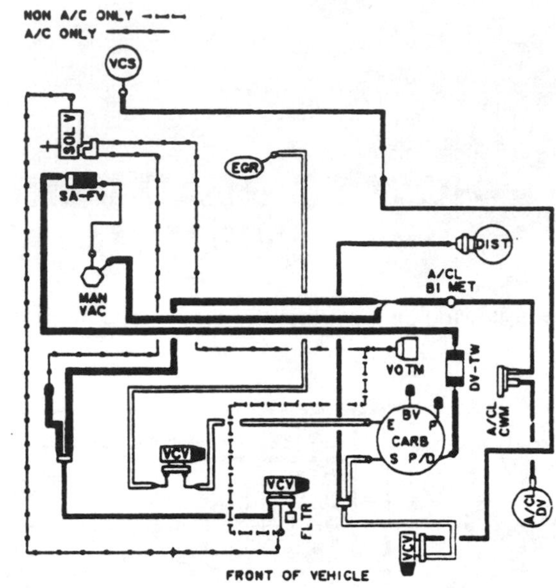

CALIBRATION: 4-03G-R00—MANUAL TRANS.—CANADA

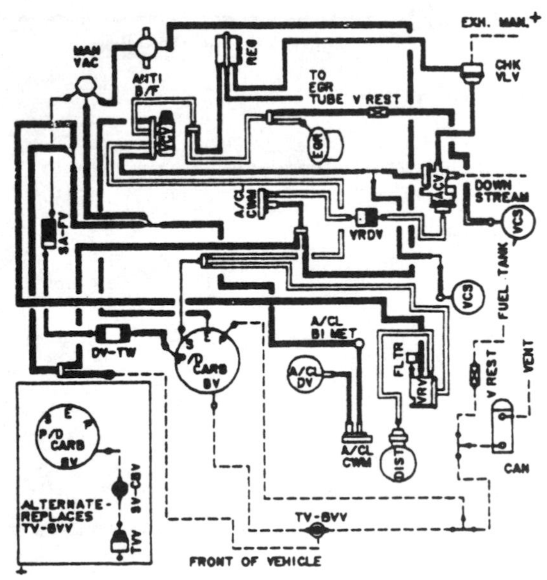

CALIBRATION: 4-03F-R00—MANUAL TRANS. WO/AC AND PWR. STRG.—EXC. CALIF.

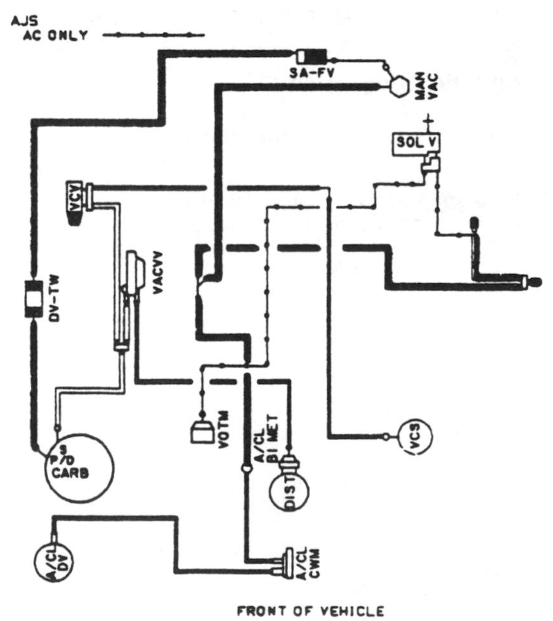

CALIBRATION: 4-03H-R00—MANUAL TRANS.—CANADA

VACUUM CIRCUITS

(© Ford Motor Co.)

1984—1.6L (98 CID) 4 cyl. engines

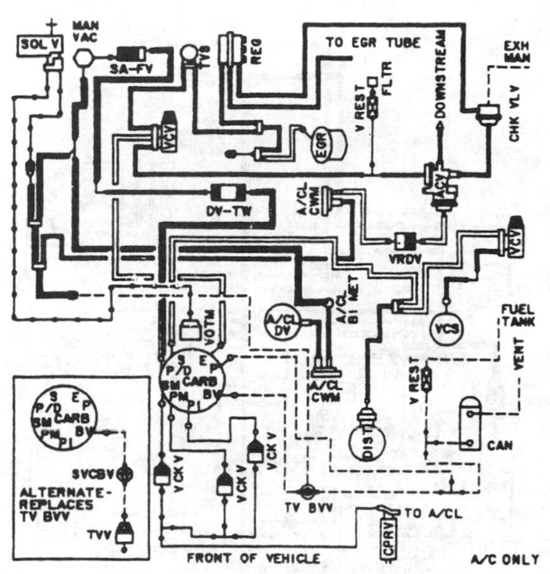

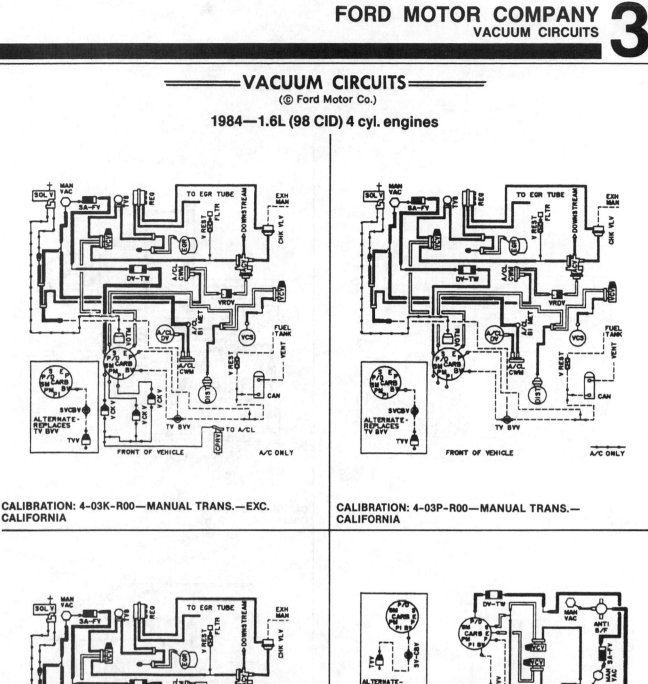

CALIBRATION: 4-03K-R00—MANUAL TRANS.—EXC. CALIFORNIA

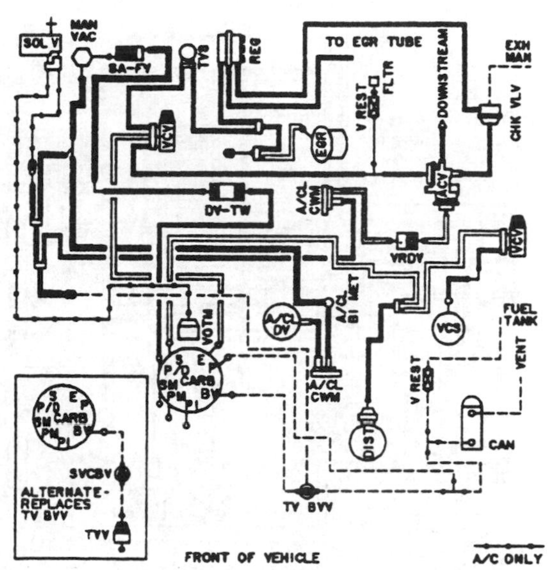

CALIBRATION: 4-03P-R00—MANUAL TRANS.—CALIFORNIA

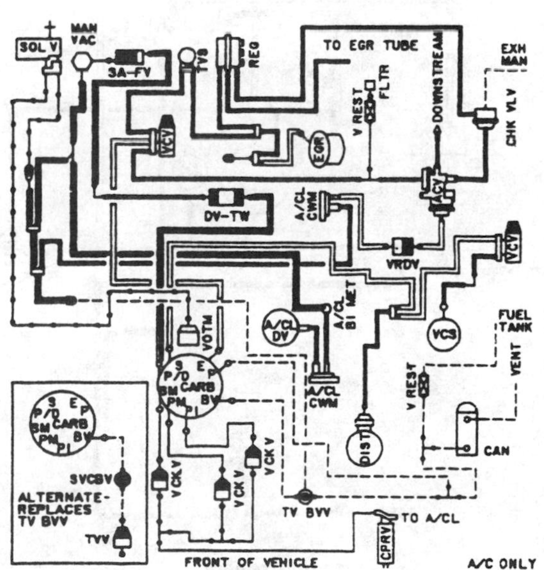

CALIBRATION: 4-03K-R10—MANUAL TRANS.—EXC. CALIFORNIA

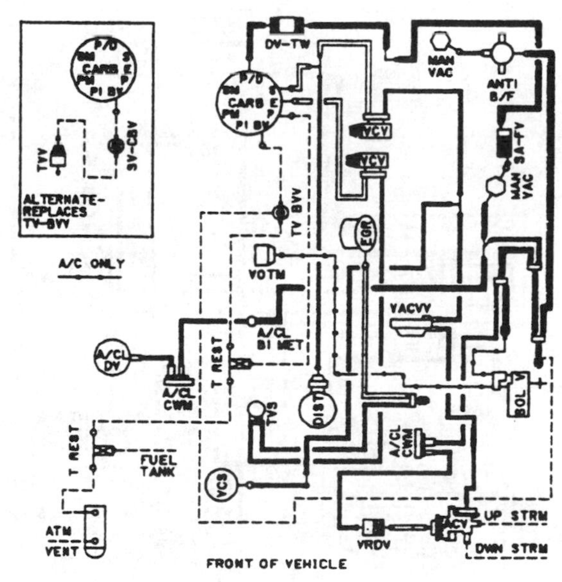

CALIBRATION: 4-03S-R00—MANUAL TRANS.—CALIFORNIA

VACUUM CIRCUITS

1984—1.6L (98 CID) 4 cyl. engines

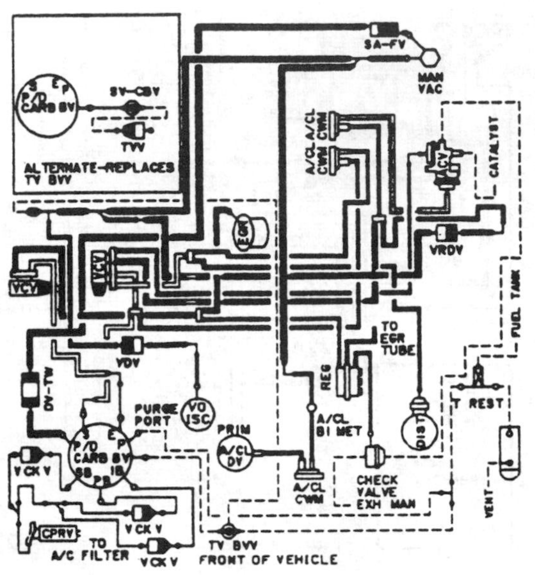

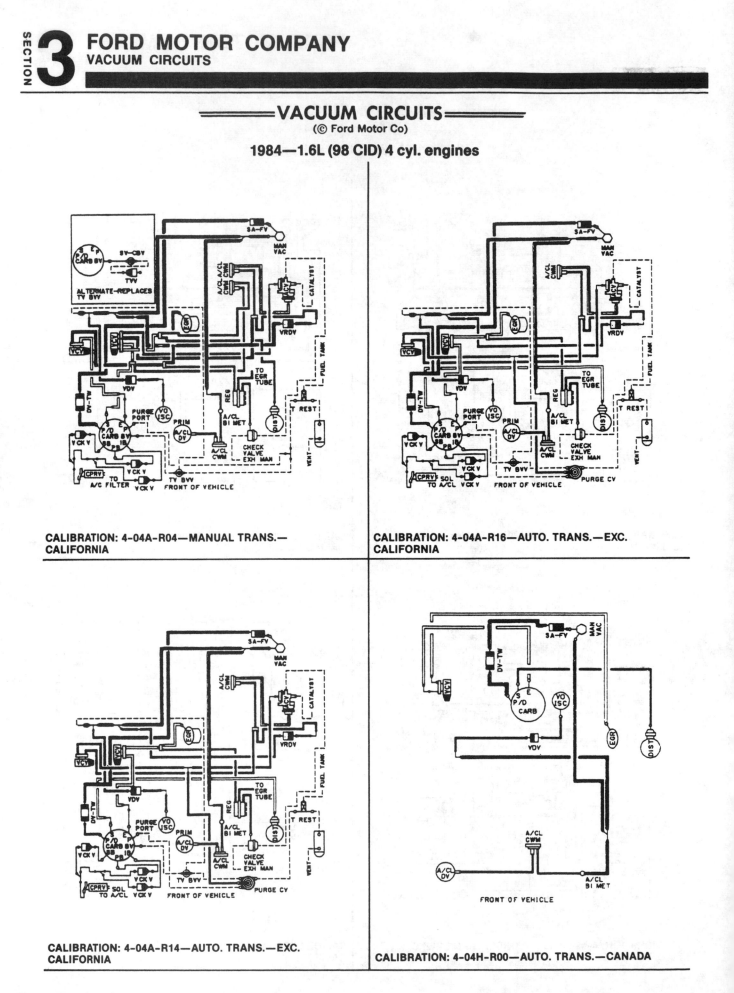

CALIBRATION: 4-04A-R04—MANUAL TRANS.—CALIFORNIA

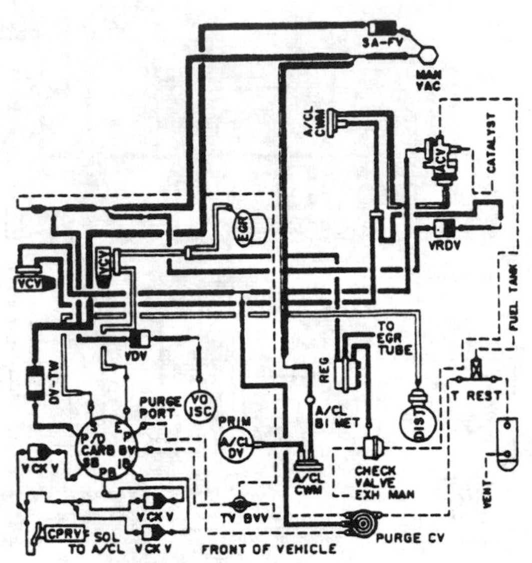

CALIBRATION: 4-04A-R16—AUTO. TRANS.—EXC. CALIFORNIA

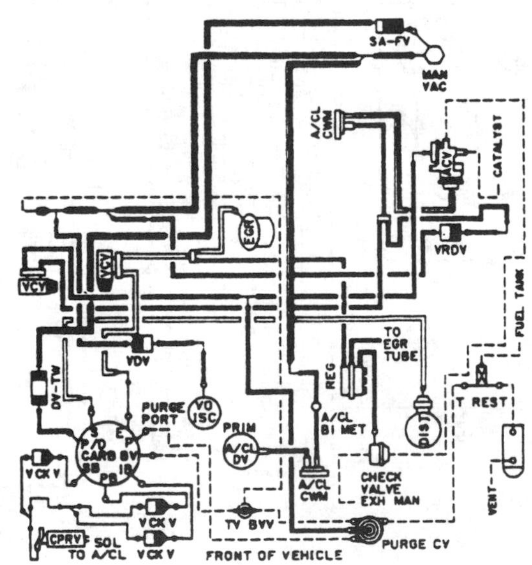

CALIBRATION: 4-04A-R14—AUTO. TRANS.—EXC. CALIFORNIA

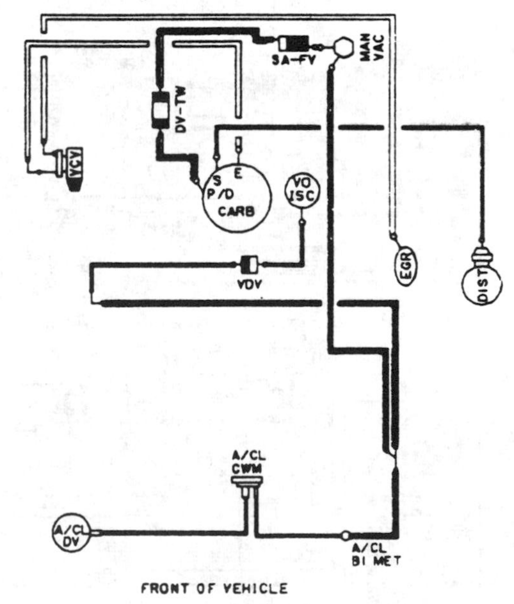

CALIBRATION: 4-04H-R00—AUTO. TRANS.—CANADA

VACUUM CIRCUITS

(© Ford Motor Co.)

1984—1.6L (98 CID) 4 cyl. engines

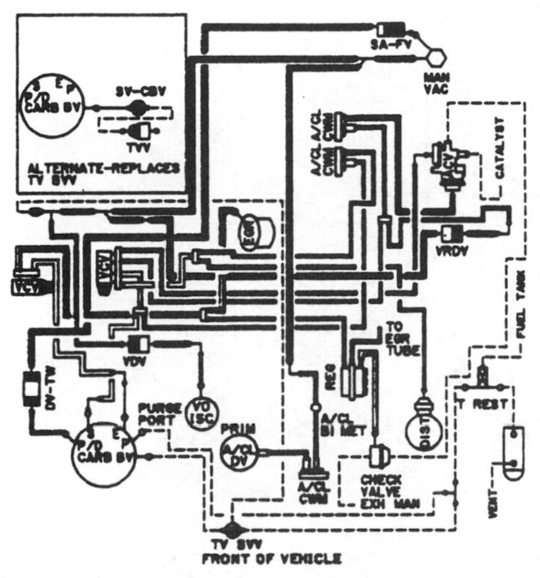

CALIBRATION: 4-04S-R06—AUTO. TRANS.—CANADA

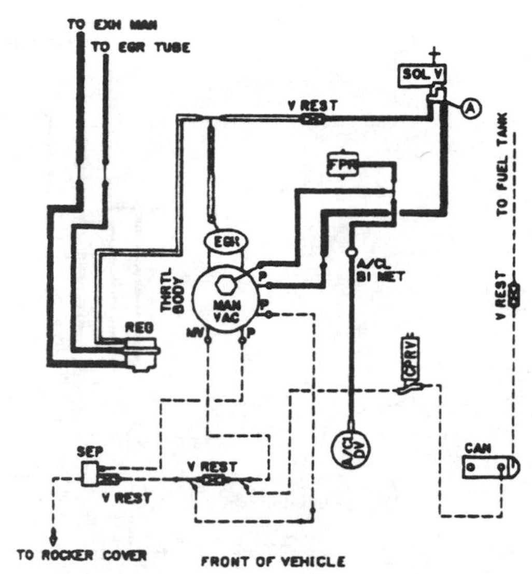

CALIBRATION: 4-27A-R00—MANUAL TRANS. E.F.I.—ALL

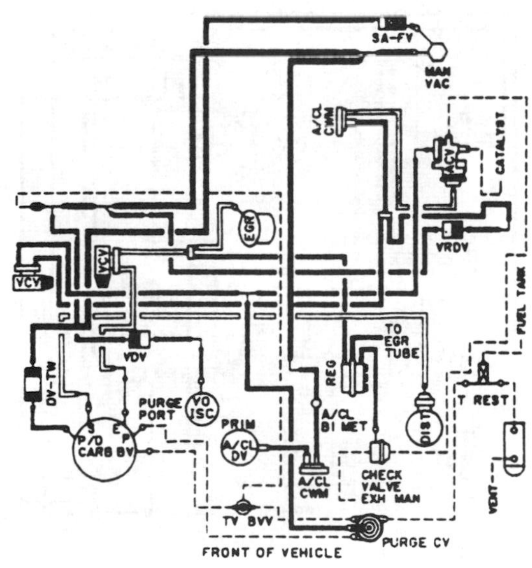

CALIBRATION: 4-04S-R12—AUTO. TRANS.—CALIFORNIA

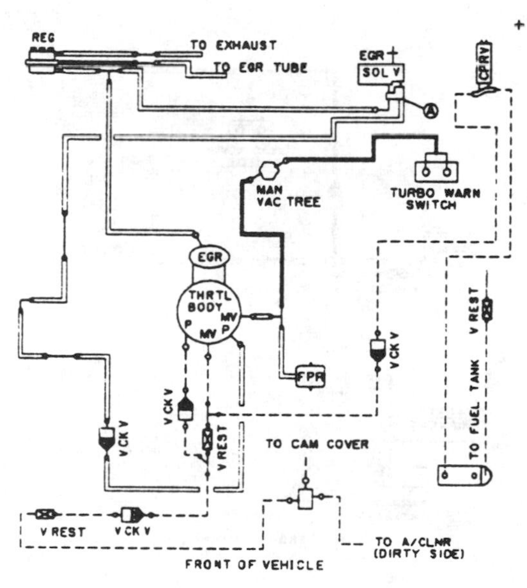

CALIBRATION: 4-27T-R00—MANUAL TRANS. E.F.I.—ALL

VACUUM CIRCUITS

(© Ford Motor Co)

1984—1.6L (98 CID) 4 cyl. engines

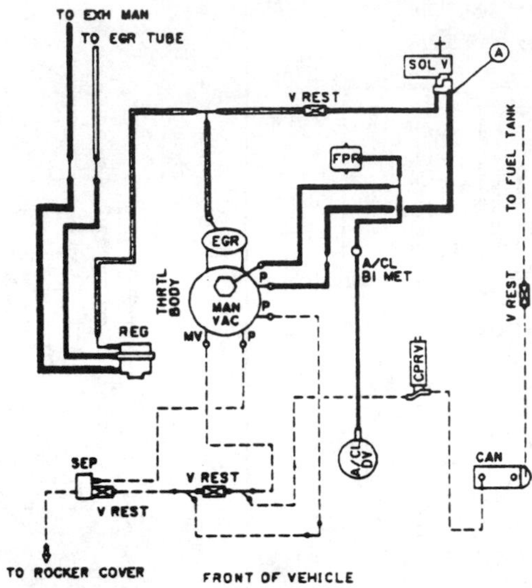

CALIBRATION: 4-28A-R00—AUTO. TRANS. E.F.I.—ALL

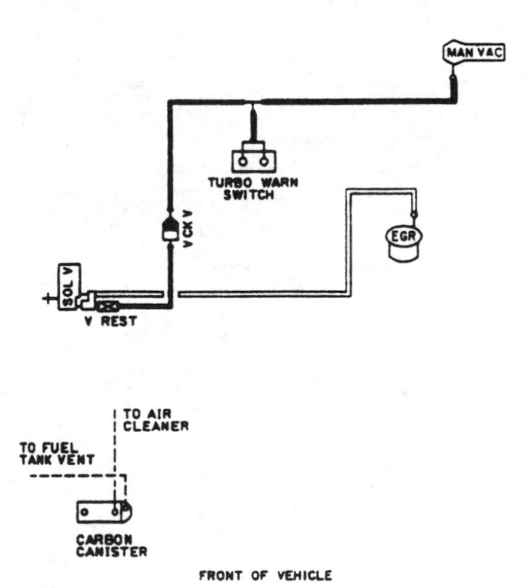

CALIBRATION: 3-05E-R12—MANUAL TRANS. E.F.I.—ALL

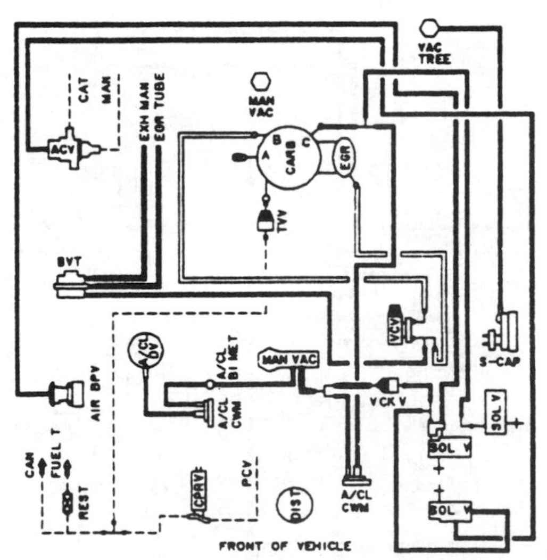

CALIBRATION: 4-05A-R00—MANUAL TRANS. EXC. E.F.I.—EXC. CALIFORNIA

VACUUM CIRCUITS

(© Ford Motor Co)

1984—2.3L (140 CID) 6 cyl. engines

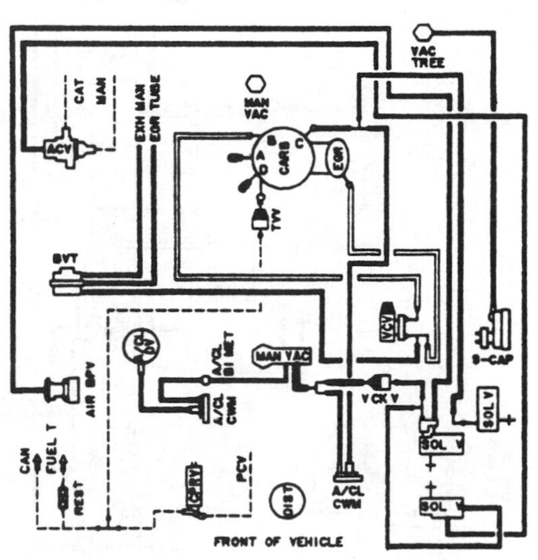

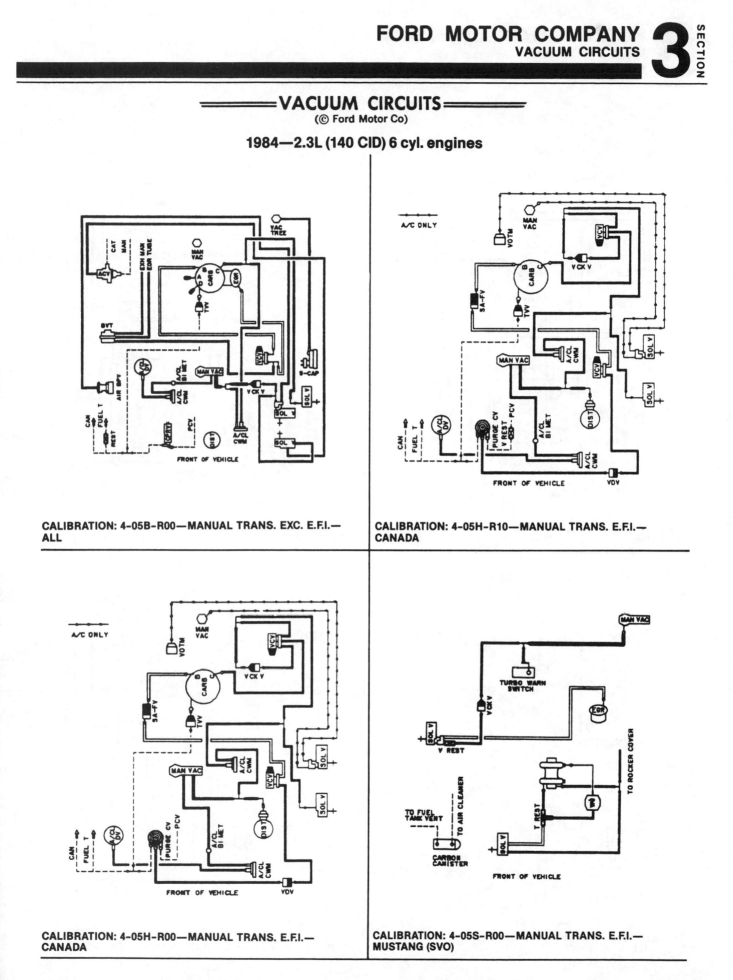

CALIBRATION: 4-05B-R00—MANUAL TRANS. EXC. E.F.I.—ALL

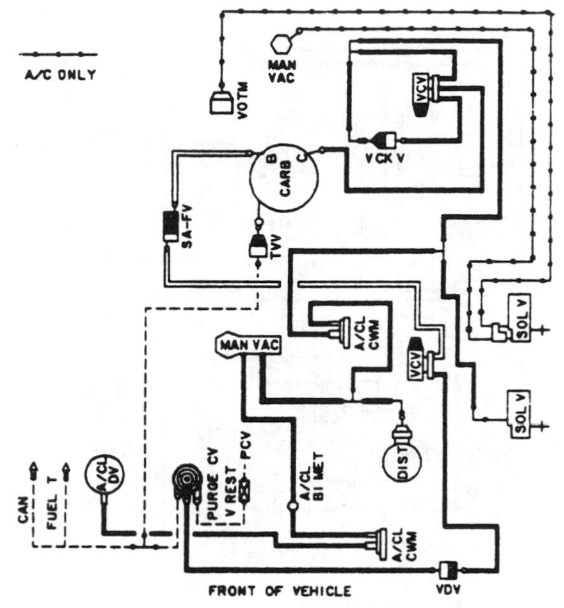

CALIBRATION: 4-05H-R10—MANUAL TRANS. E.F.I.—CANADA

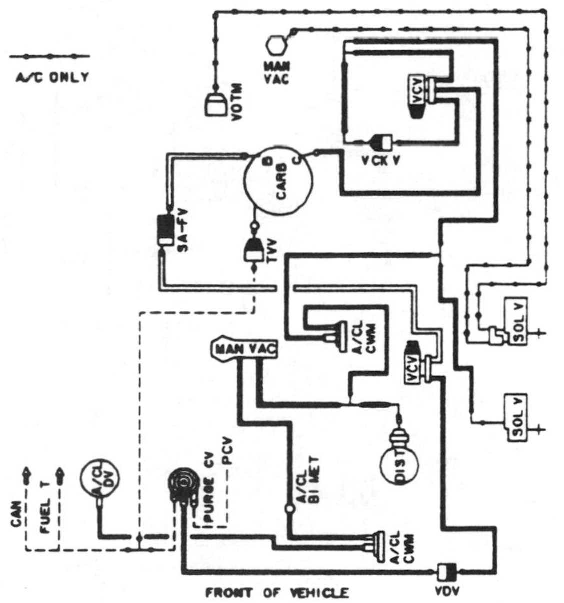

CALIBRATION: 4-05H-R00—MANUAL TRANS. E.F.I.—CANADA

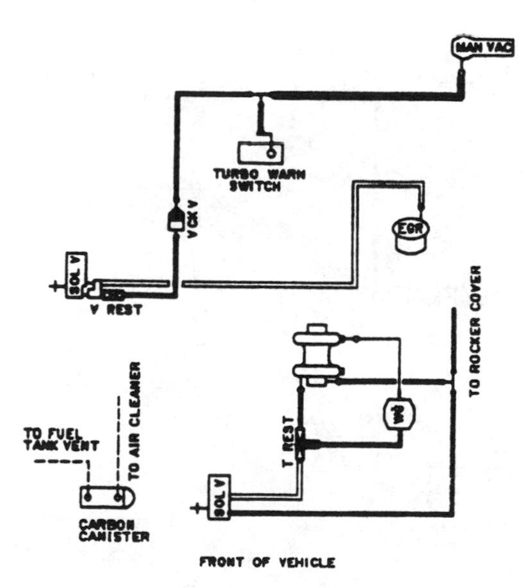

CALIBRATION: 4-05S-R00—MANUAL TRANS. E.F.I.—MUSTANG (SVO)

VACUUM CIRCUITS

(© Ford Motor Co.)

1984—2.3L (140 CID) 6 cyl. engines

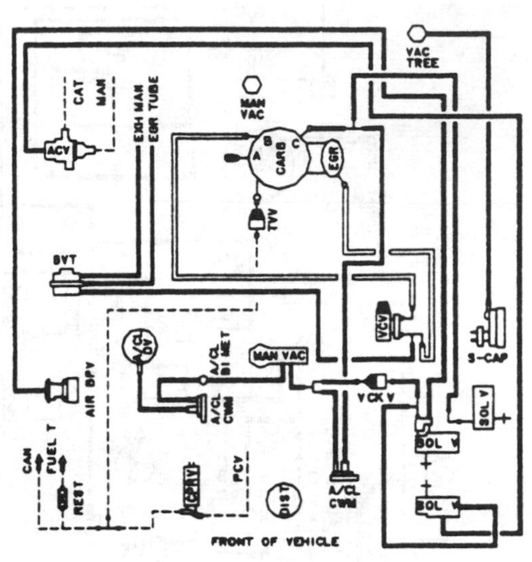

CALIBRATION: 4-06A-R10—AUTO. TRANS.—EXC. CALIFORNIA

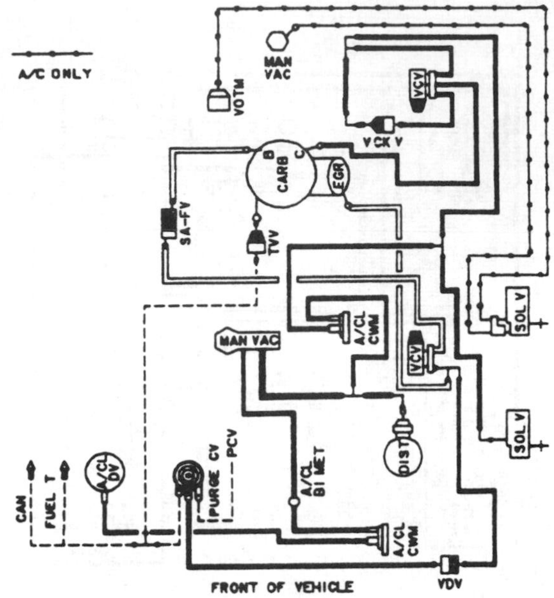

CALIBRATION: 4-06H-R00—AUTO. TRANS.—CANADA

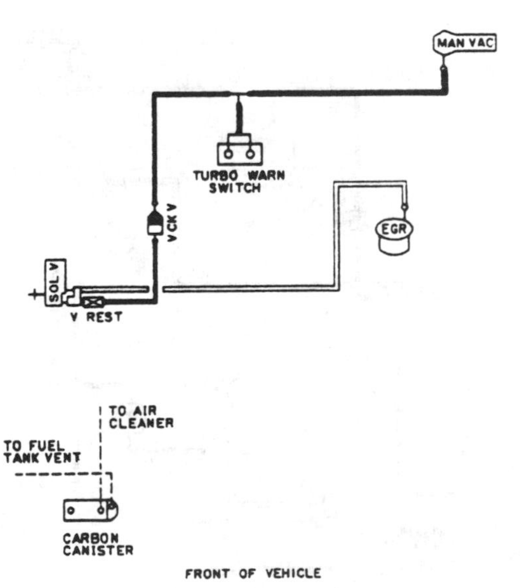

CALIBRATION: 4-06E-R00—MANUAL & AUTO. TRANS. E.F.I.—ALL

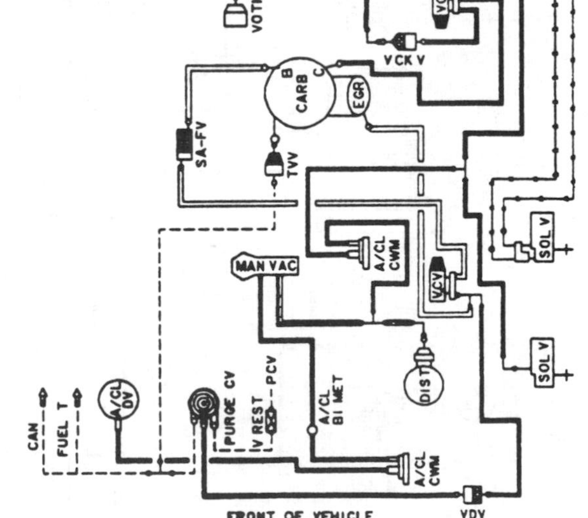

CALIBRATION: 4-06H-R10—AUTO. TRANS.—CANADA

VACUUM CIRCUITS

(© Ford Motor Co)

1984—2.3L (140 CID) 6 cyl. engines

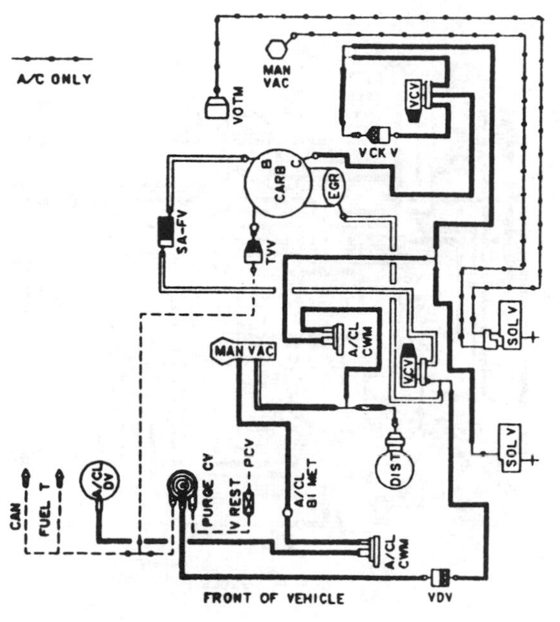

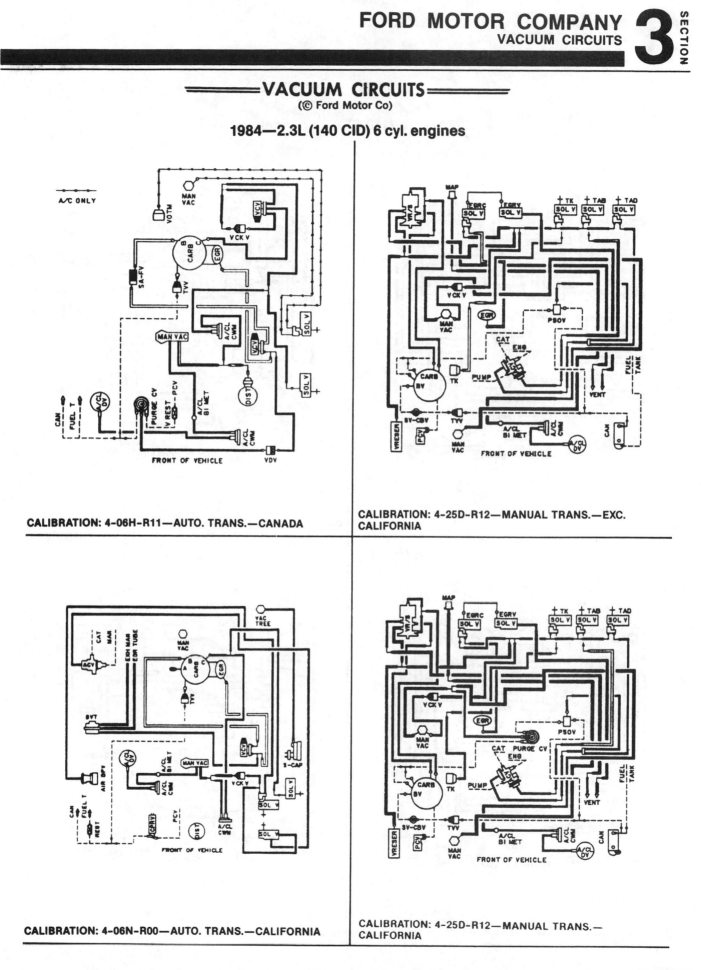

CALIBRATION: 4-06H-R11—AUTO. TRANS.—CANADA

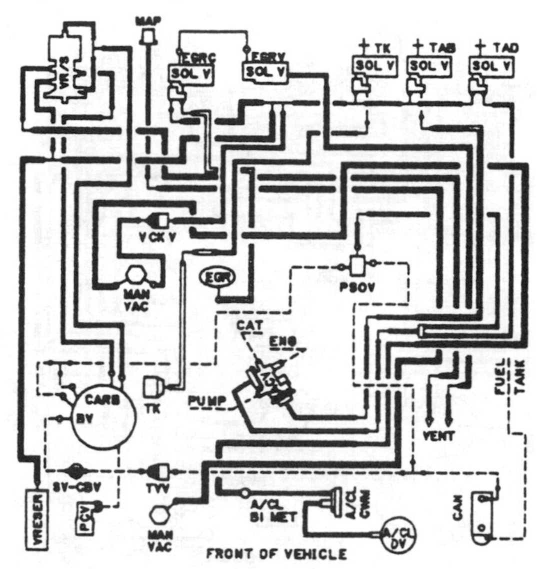

CALIBRATION: 4-25D-R12—MANUAL TRANS.—EXC. CALIFORNIA

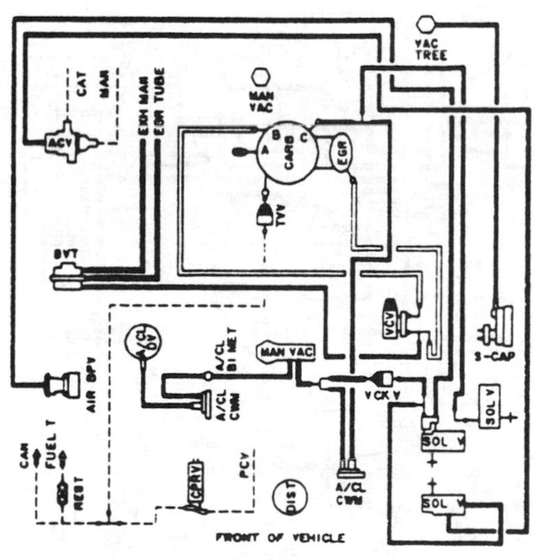

CALIBRATION: 4-06N-R00—AUTO. TRANS.—CALIFORNIA

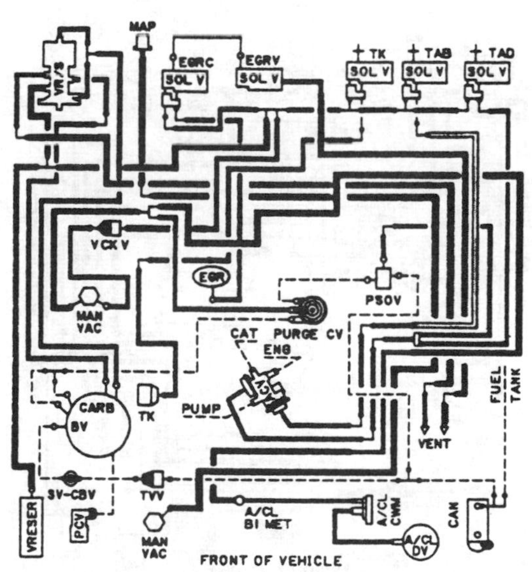

CALIBRATION: 4-25D-R12—MANUAL TRANS.—CALIFORNIA

VACUUM CIRCUITS

(© Ford Motor Co.)

1984—2.3L (140 CID) 6 cyl. engines

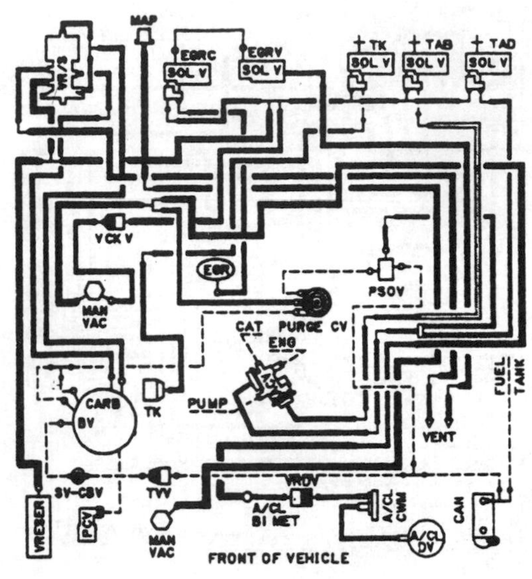

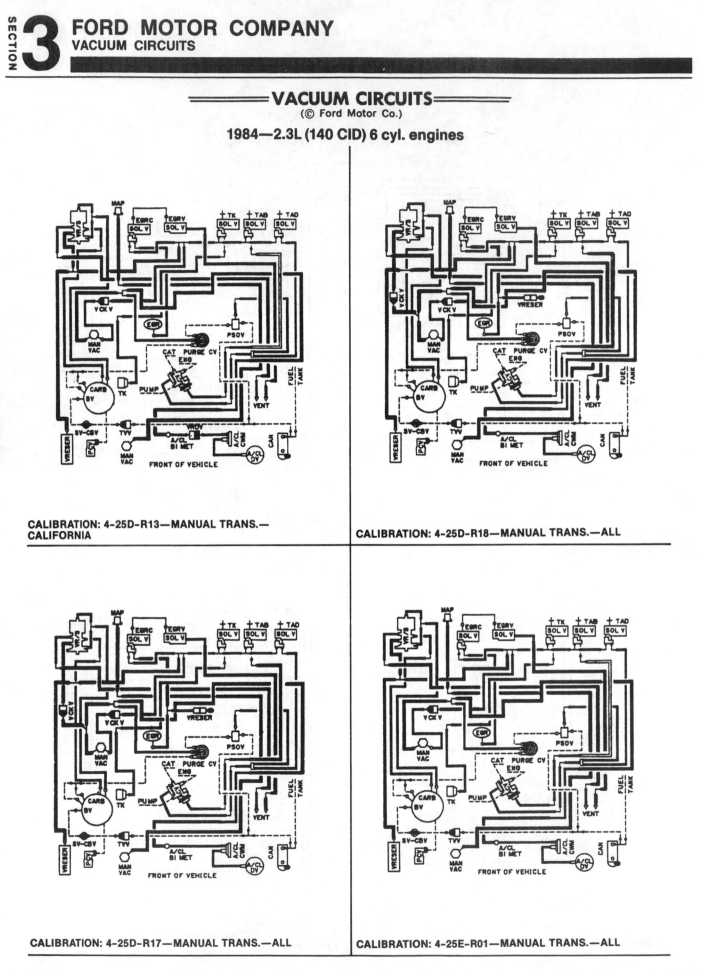

CALIBRATION: 4-25D-R13—MANUAL TRANS.—CALIFORNIA

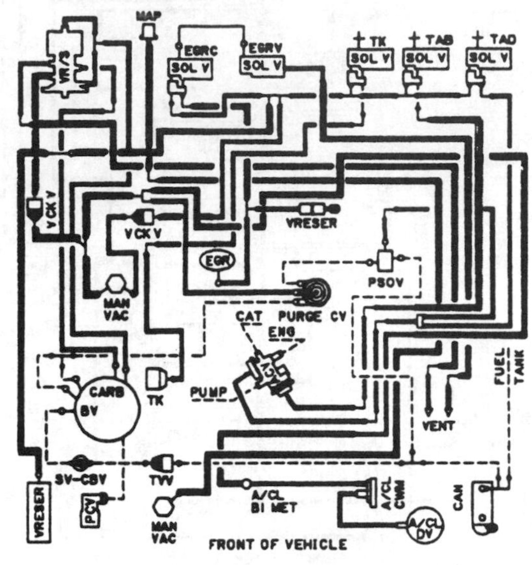

CALIBRATION: 4-25D-R18—MANUAL TRANS.—ALL

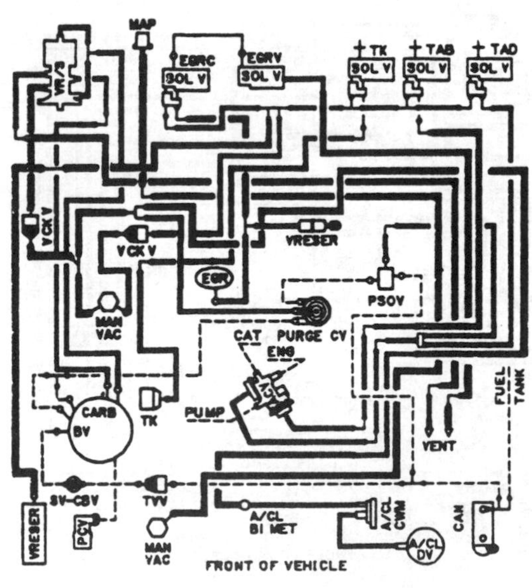

CALIBRATION: 4-25D-R17—MANUAL TRANS.—ALL

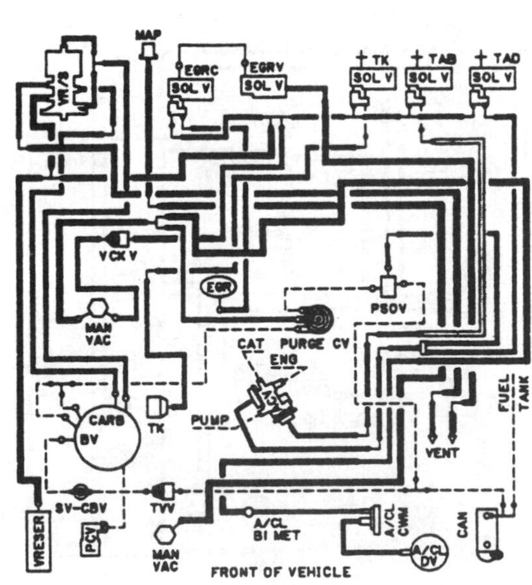

CALIBRATION: 4-25E-R01—MANUAL TRANS.—ALL

VACUUM CIRCUITS

(© Ford Motor Co)

1984—2.3L (140 CID) 6 cyl. engines

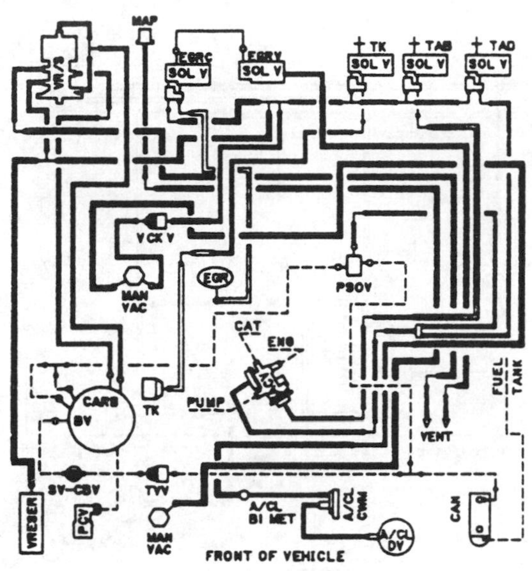

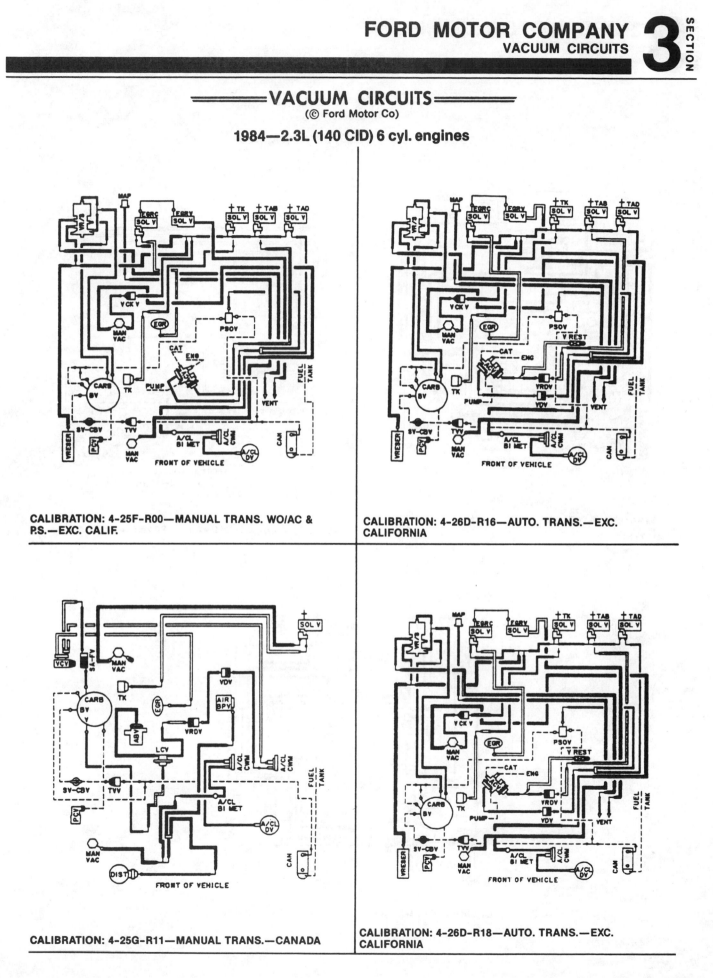

CALIBRATION: 4-25F-R00—MANUAL TRANS. WO/AC & P.S.—EXC. CALIF.

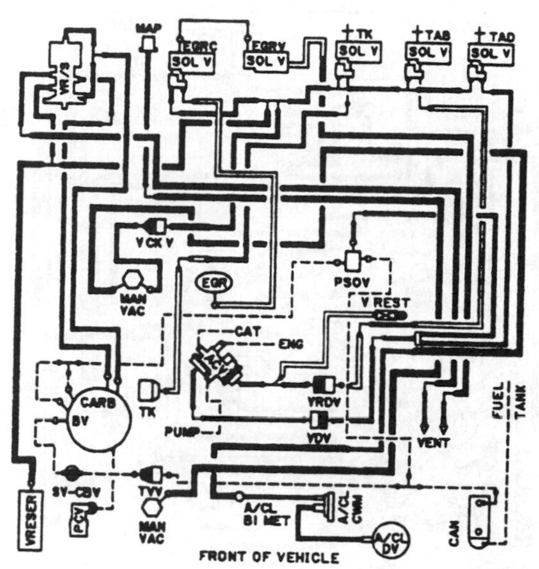

CALIBRATION: 4-26D-R16—AUTO. TRANS.—EXC. CALIFORNIA

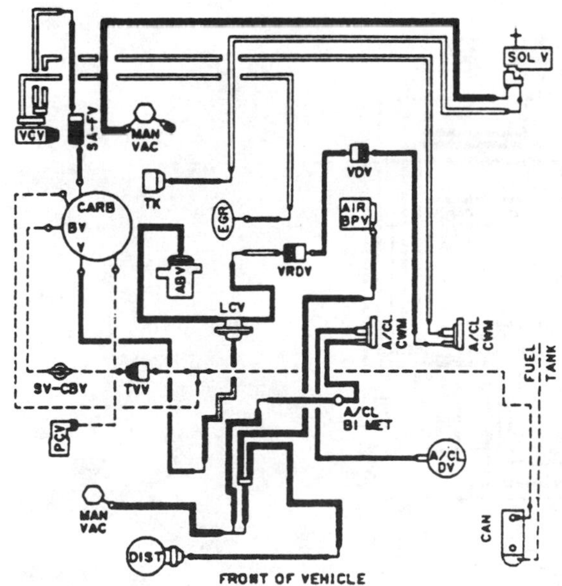

CALIBRATION: 4-25G-R11—MANUAL TRANS.—CANADA

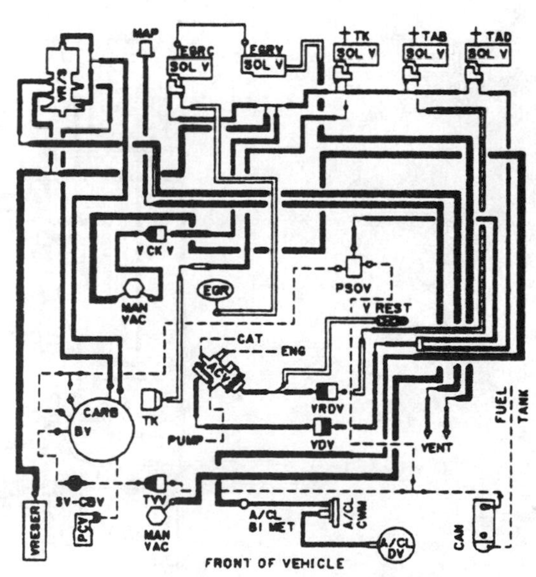

CALIBRATION: 4-26D-R18—AUTO. TRANS.—EXC. CALIFORNIA

VACUUM CIRCUITS

(© Ford Motor Co.)

1984—2.3L (140 CID) 6 cyl. engines

MAP
EGRC SOL V
EGRV SOL V
TK SOL V
TAB SOL V
TAD SOL V
VCK V
MAN VAC
EGR
PSOV
V REST
CAT
ENG
CARB
BV
TK
VRDV
PUMP
VDV
VENT
FUEL TANK
SV-CBV
TVV
VRESER
PCV
MAN VAC
A/CL BI MET
A/CL CWM
A/CL DV
CAN
FRONT OF VEHICLE

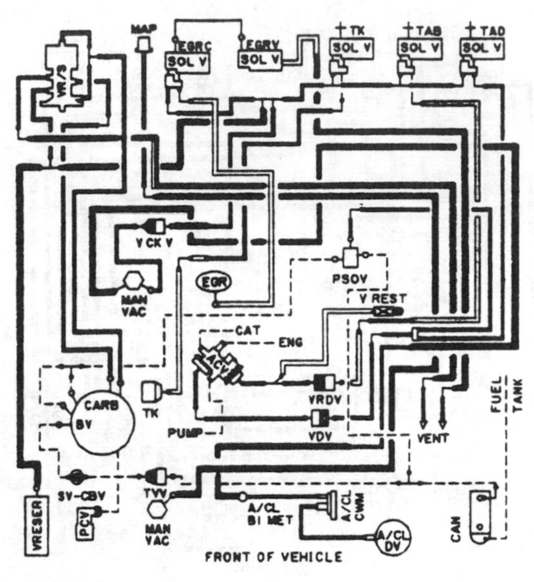

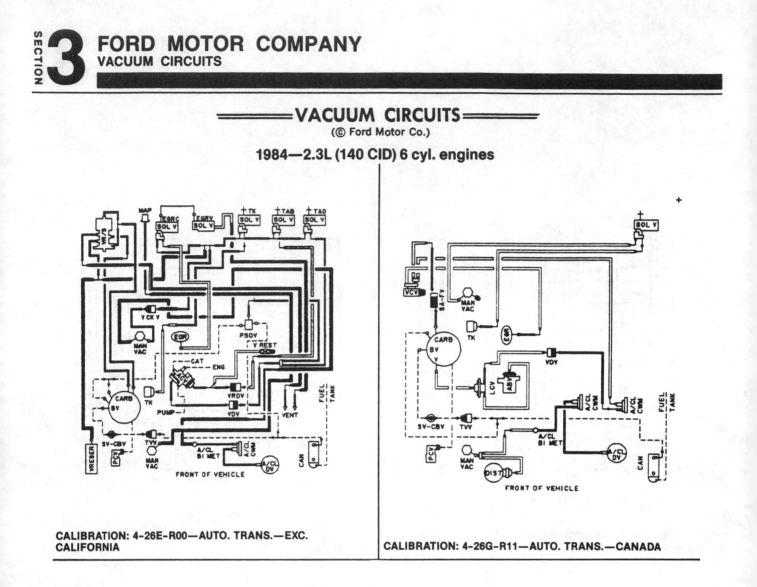

CALIBRATION: 4-26E-R00—AUTO. TRANS.—EXC. CALIFORNIA

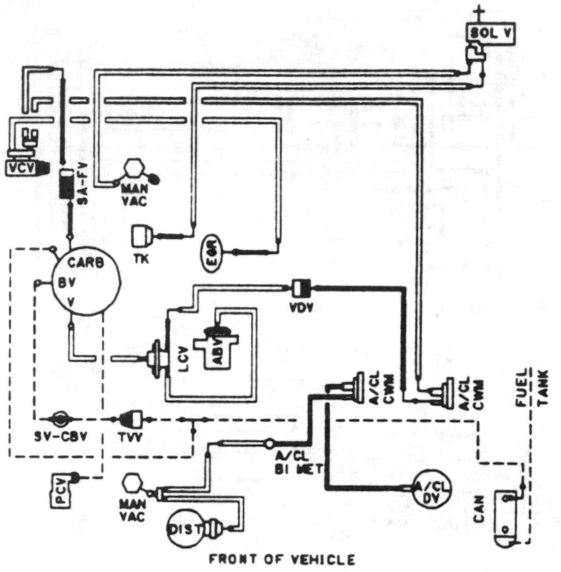

SOL V
VCV
SA-FV
MAN VAC
TK
EGR
CARB
BV
V
VDV
LCV
ABV
A/CL CWM
A/CL CWM
FUEL TANK
SV-CBV
TVV
A/CL BI MET
PCV
MAN VAC
A/CL DV
CAN
DIST
FRONT OF VEHICLE

CALIBRATION: 4-26G-R11—AUTO. TRANS.—CANADA

MAP
EGRC SOL V
EGRV SOL V
TK SOL V
TAB SOL V
TAD SOL V
VCK V
MAN VAC
EGR
PSOV
V REST
CAT
ENG
CARB
BV
TK
VRDV
PUMP
VDV
VENT
FUEL TANK
SV-CBV
TVV
VRESER
PCV
MAN VAC
A/CL BI MET
A/CL CWM
A/CL DV
CAN
FRONT OF VEHICLE

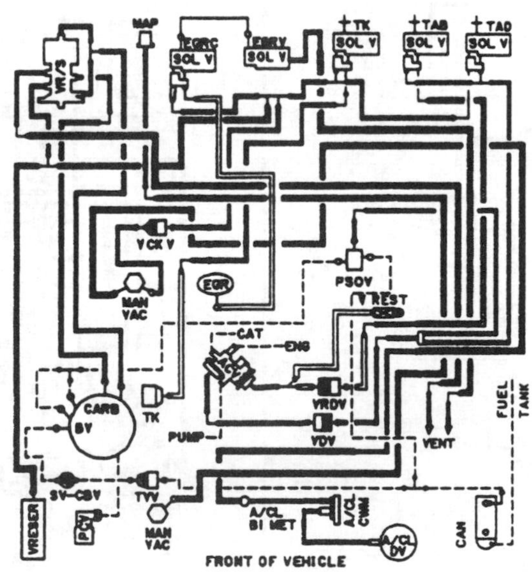

CALIBRATION: 4-26S-R13—AUTO. TRANS.—CALIFORNIA

VACUUM CIRCUITS

(© Ford Motor Co.)

1984—3.8L (230 CID) 6 cyl. engines

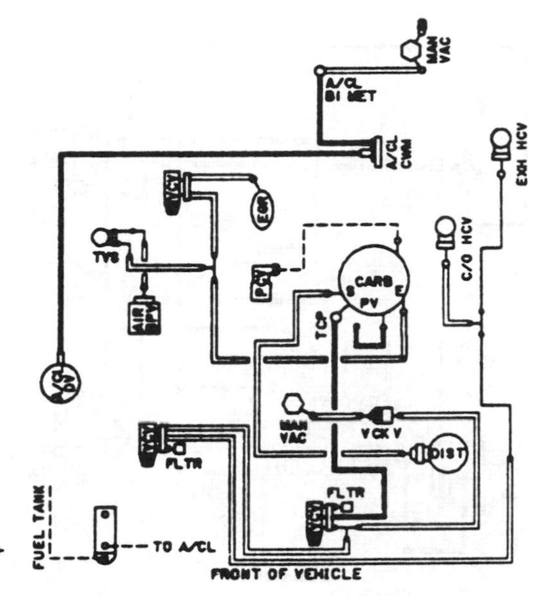

CALIBRATION: 4-14G-R00—AUTO. TRANS. C.F.I.—CANADA

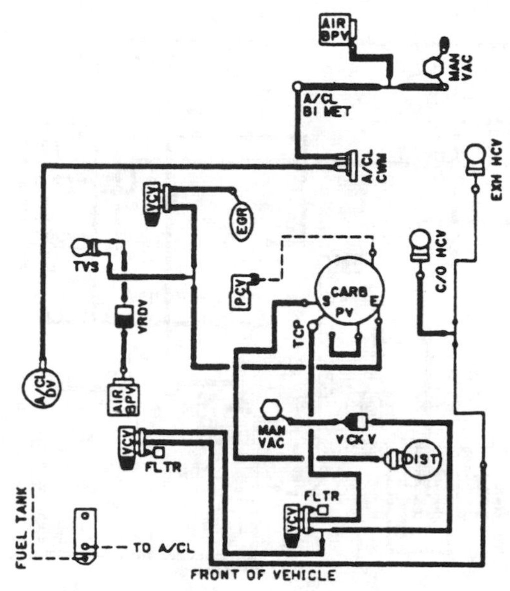

CALIBRATION: 4-14G-R10—AUTO. TRANS. C.F.I.—CANADA

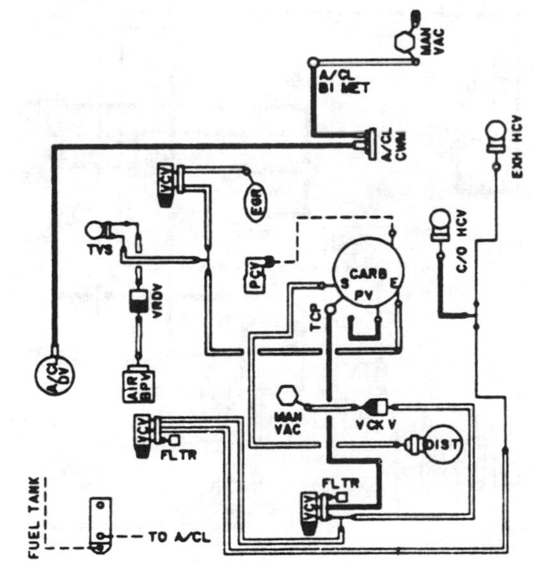

CALIBRATION: 4-14G-R03—AUTO. TRANS. C.F.I.—CANADA

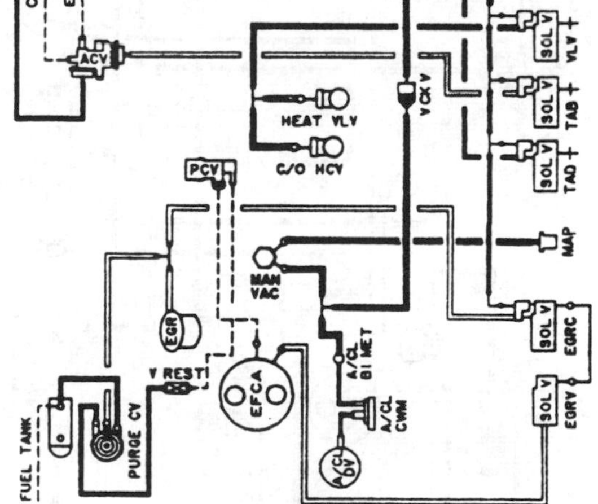

CALIBRATION: 4-16A-R00—AUTO. TRANS. C.F.I.—EXC. CALIFORNIA

VACUUM CIRCUITS

(© Ford Motor Co)

1984—3.8L (230 CID) 6 cyl. engines

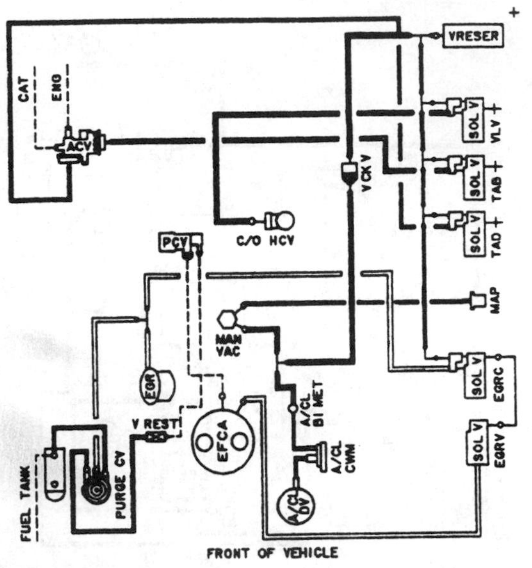

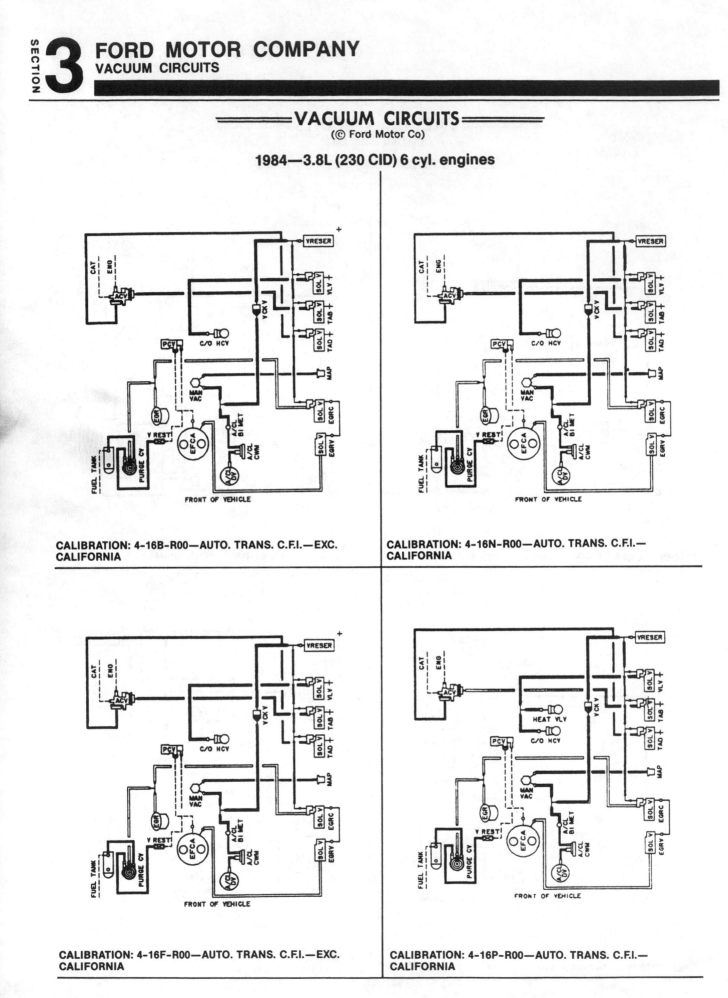

CALIBRATION: 4-16B-R00—AUTO. TRANS. C.F.I.—EXC. CALIFORNIA

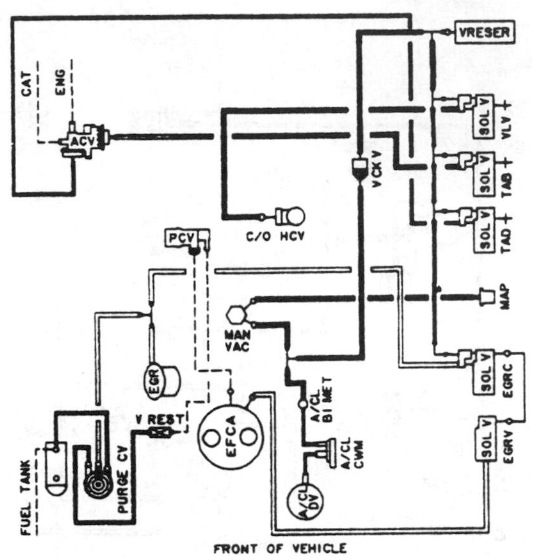

CALIBRATION: 4-16N-R00—AUTO. TRANS. C.F.I.—CALIFORNIA

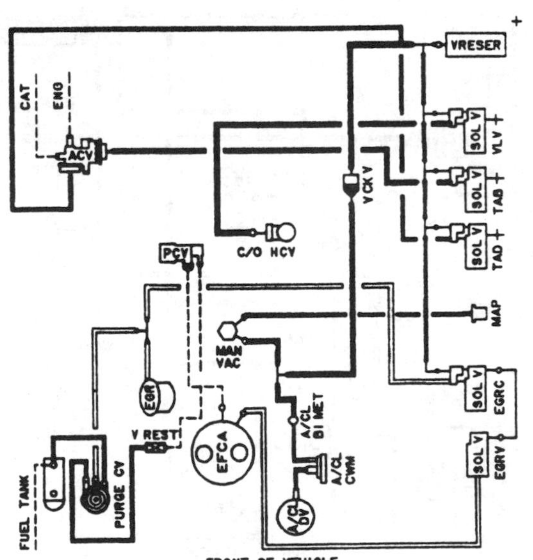

CALIBRATION: 4-16F-R00—AUTO. TRANS. C.F.I.—EXC. CALIFORNIA

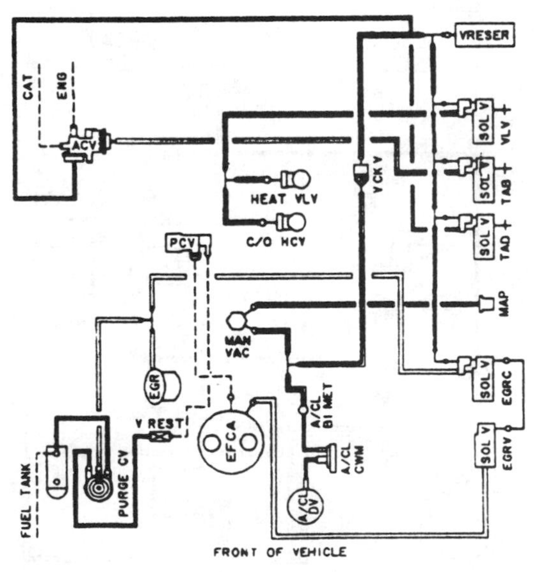

CALIBRATION: 4-16P-R00—AUTO. TRANS. C.F.I.—CALIFORNIA

VACUUM CIRCUITS

(© Ford Motor Co.)

1984—3.8L (230 CID) 6 cyl. engines

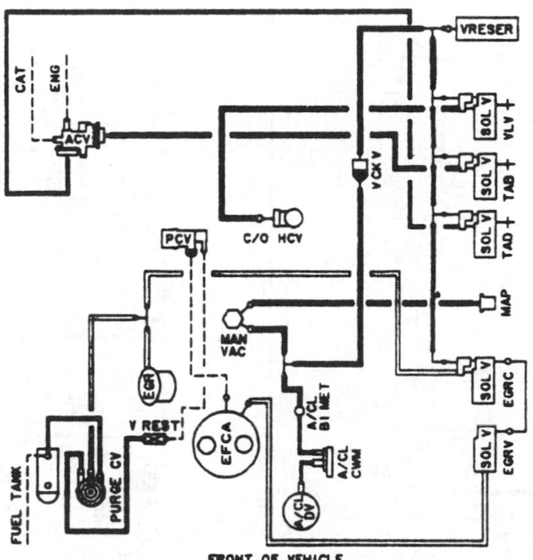

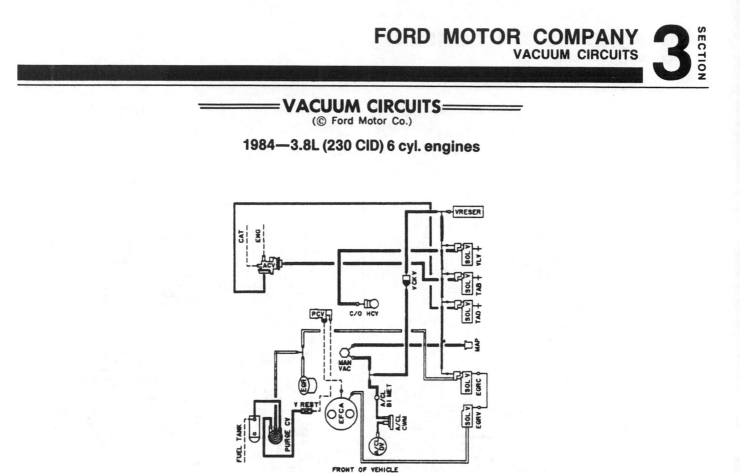

CALIBRATION: 4-16S-R00—AUTO. TRANS. C.F.I.—CALIFORNIA

1984—5.0L (302 CID) V8 engines

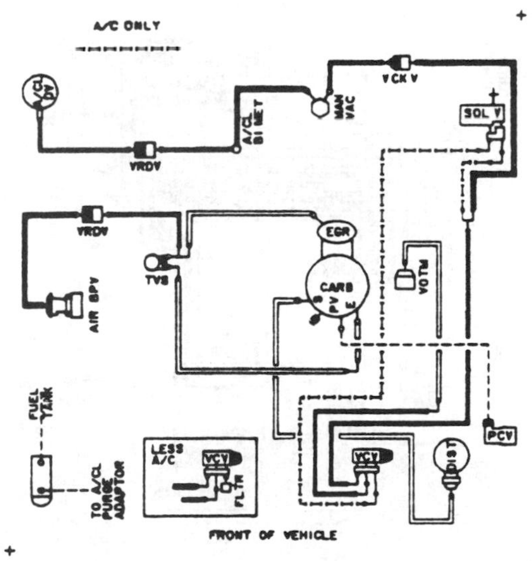

CALIBRATION: 2-20H-R00—AUTO. TRANS.—CANADA

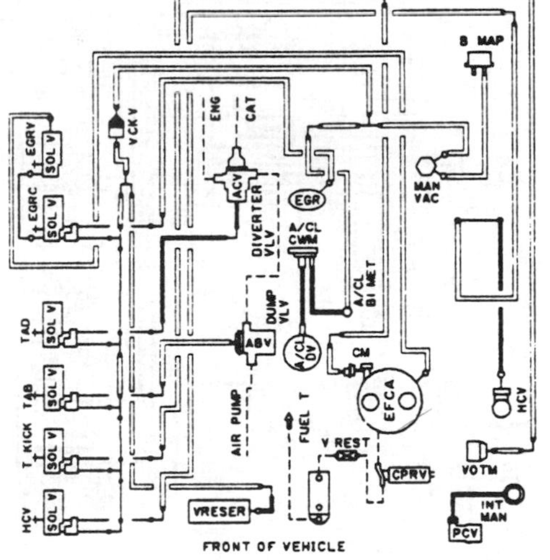

CALIBRATION: 3-22D-R00—CONTINENTAL/MARK VII, AUTO. TRANS.—CALIFORNIA

VACUUM CIRCUITS

(© Ford Motor Co)

1984—5.0L (302 CID) V8 engines

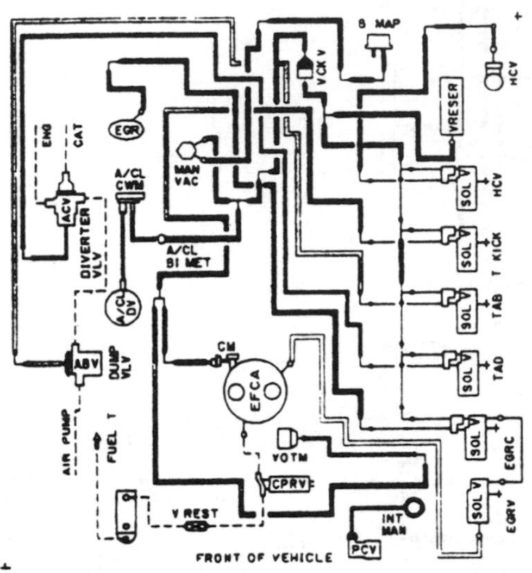

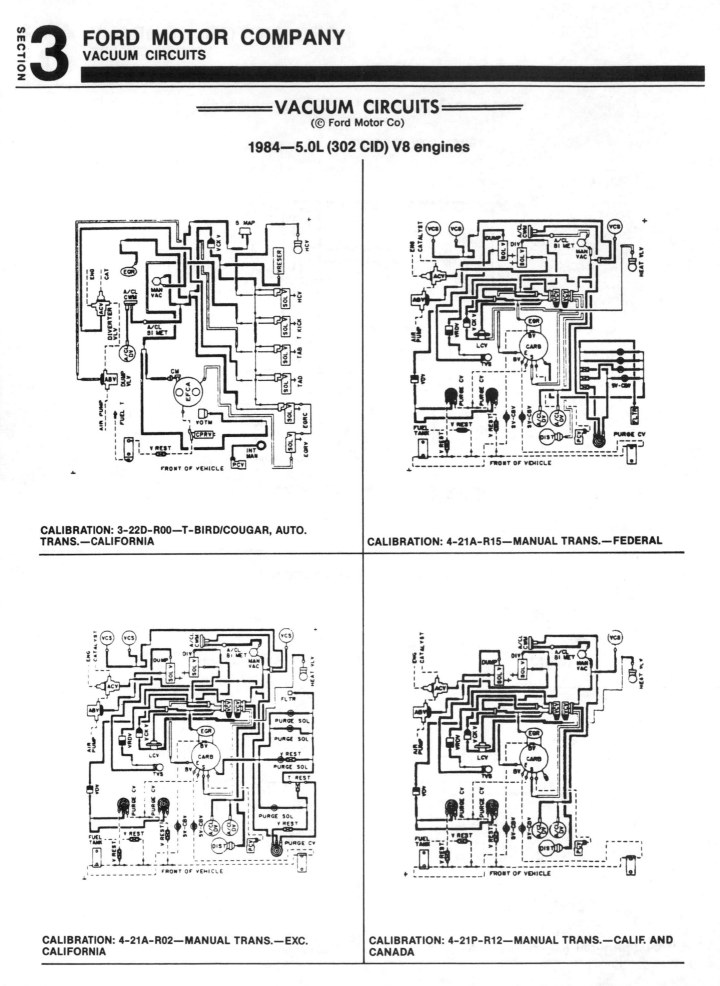

CALIBRATION: 3-22D-R00—T-BIRD/COUGAR, AUTO. TRANS.—CALIFORNIA

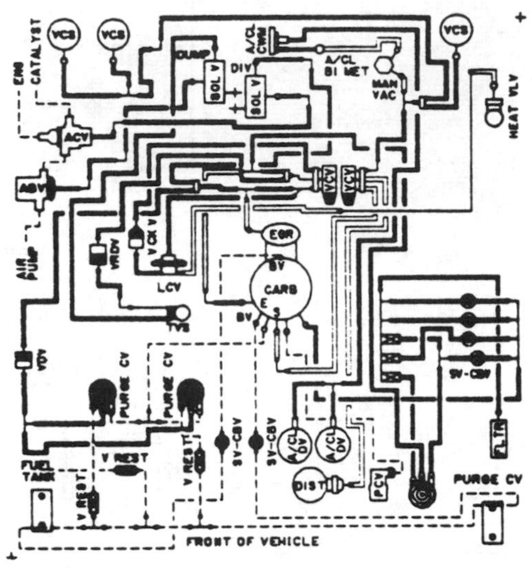

CALIBRATION: 4-21A-R15—MANUAL TRANS.—FEDERAL

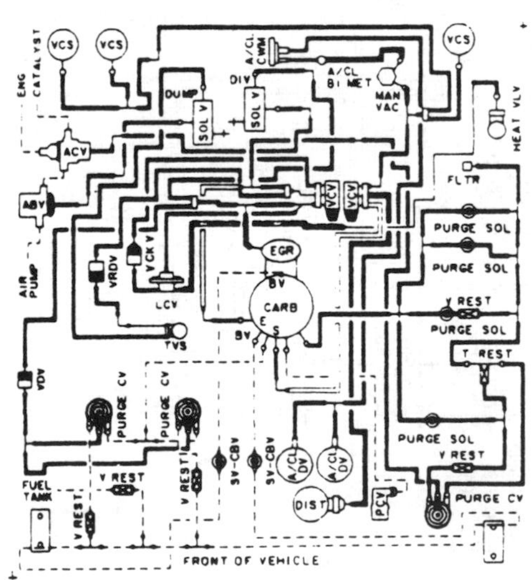

CALIBRATION: 4-21A-R02—MANUAL TRANS.—EXC. CALIFORNIA

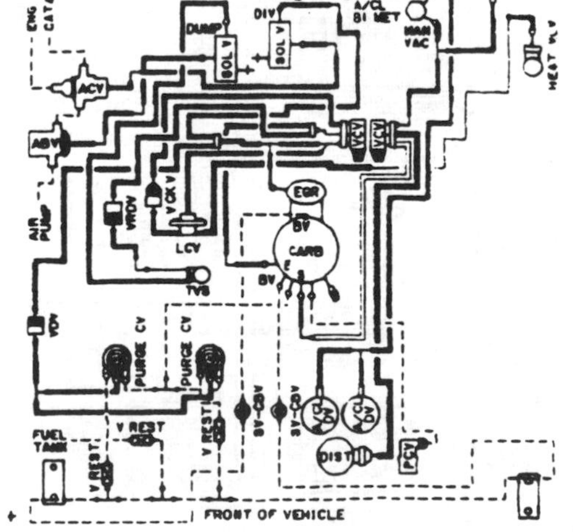

CALIBRATION: 4-21P-R12—MANUAL TRANS.—CALIF. AND CANADA

VACUUM CIRCUITS

(© Ford Motor Co)

1984—5.0L (302 CID) V8 engines

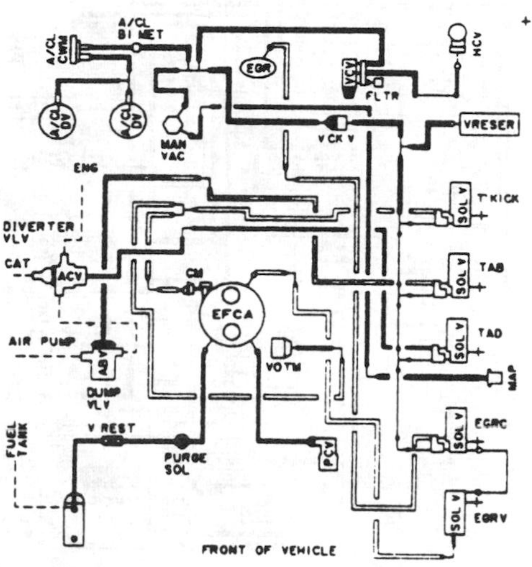

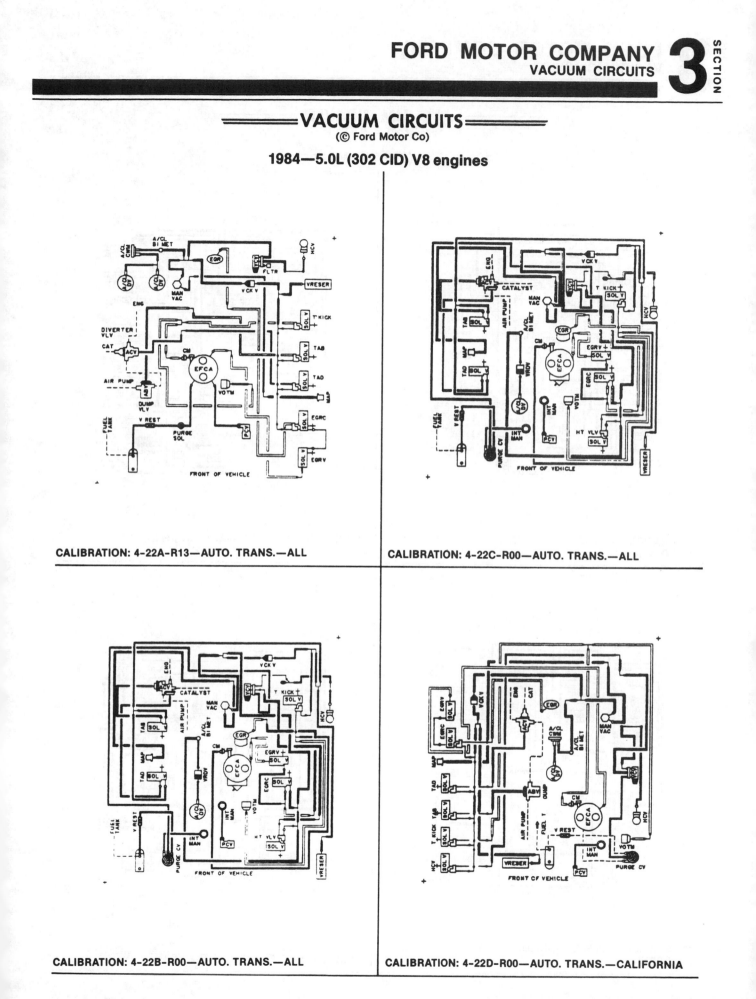

CALIBRATION: 4-22A-R13—AUTO. TRANS.—ALL

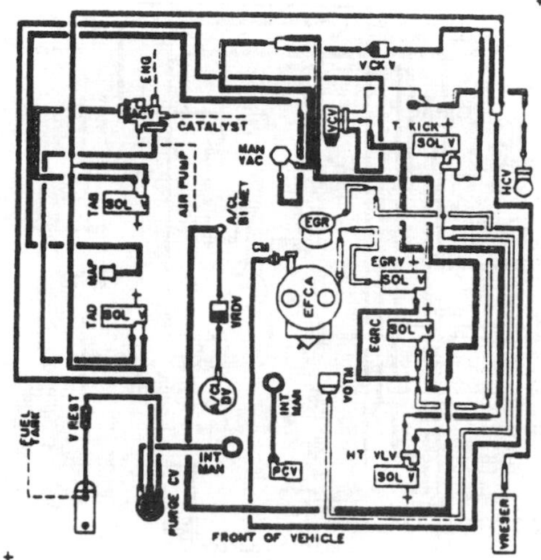

CALIBRATION: 4-22C-R00—AUTO. TRANS.—ALL

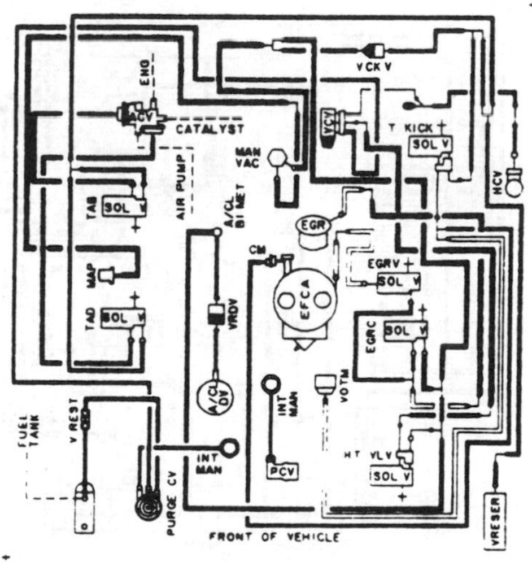

CALIBRATION: 4-22B-R00—AUTO. TRANS.—ALL

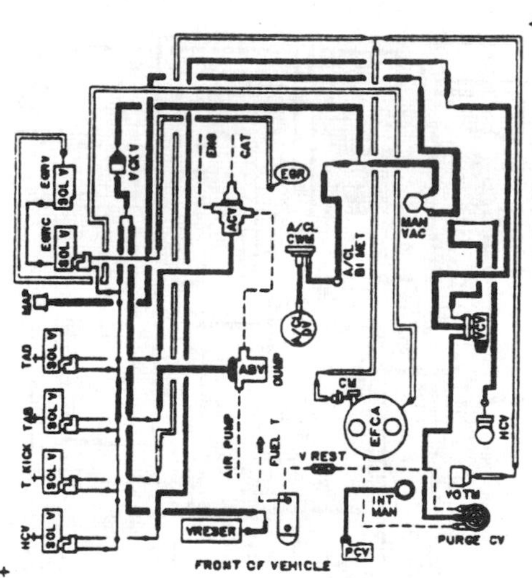

CALIBRATION: 4-22D-R00—AUTO. TRANS.—CALIFORNIA

VACUUM CIRCUITS

(© Ford Motor Co.)

1984—5.0L (302 CID) V8 engines

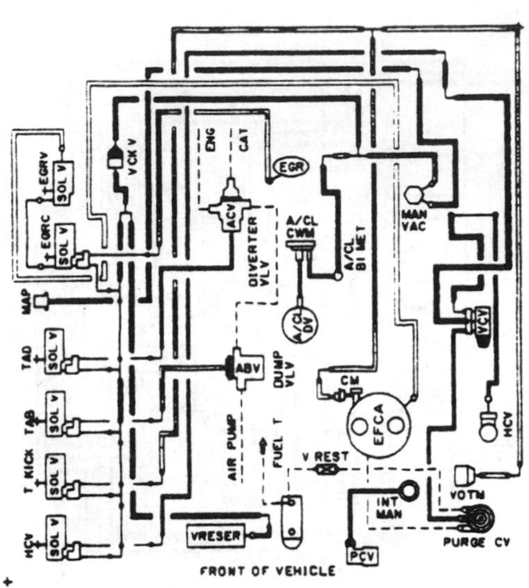

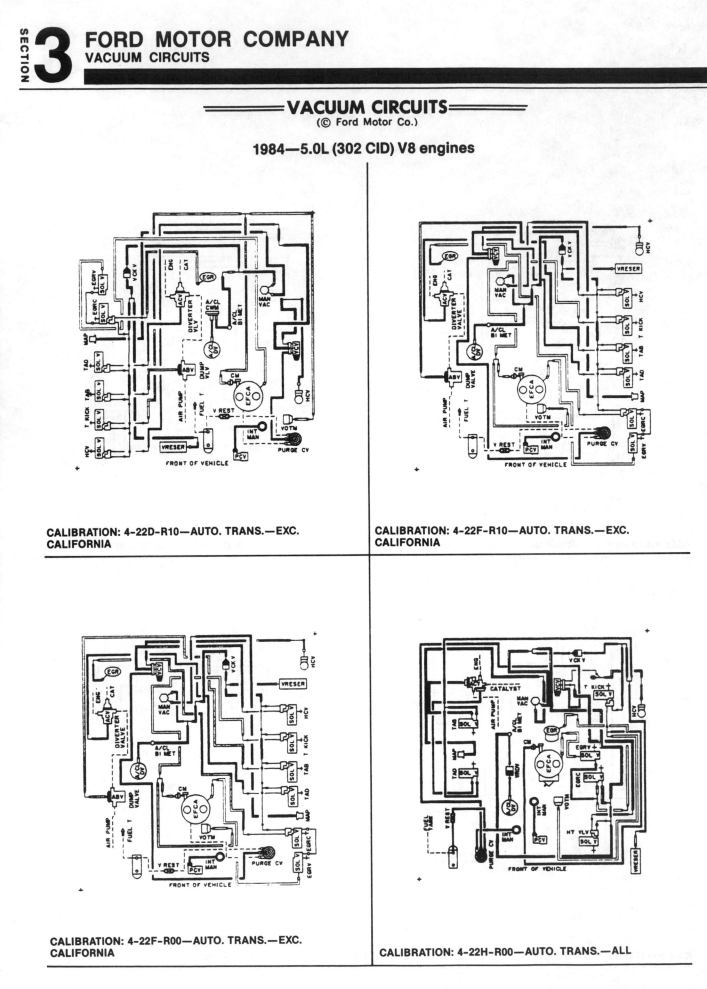

CALIBRATION: 4-22D-R10—AUTO. TRANS.—EXC. CALIFORNIA

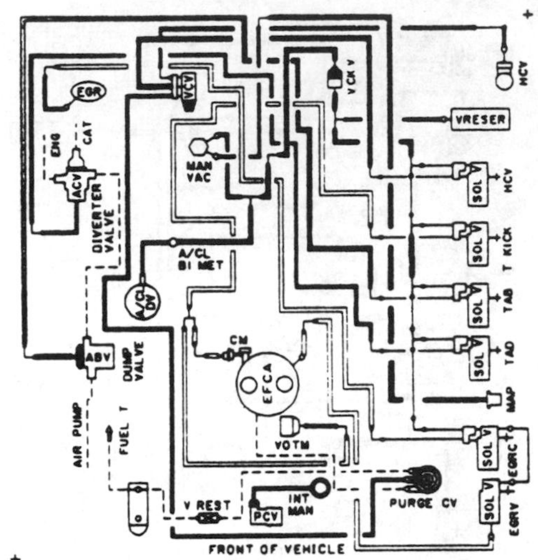

CALIBRATION: 4-22F-R10—AUTO. TRANS.—EXC. CALIFORNIA

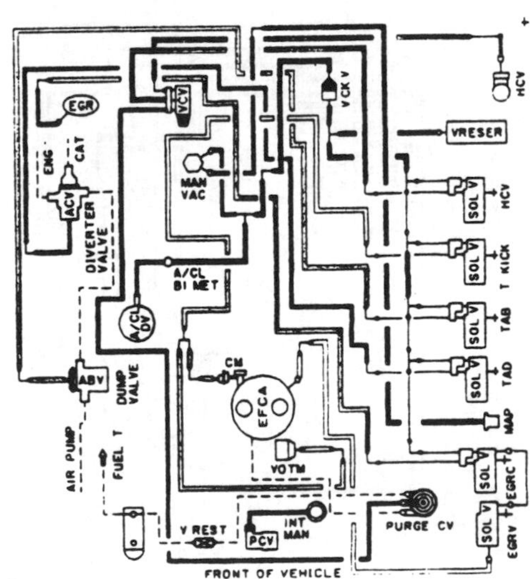

CALIBRATION: 4-22F-R00—AUTO. TRANS.—EXC. CALIFORNIA

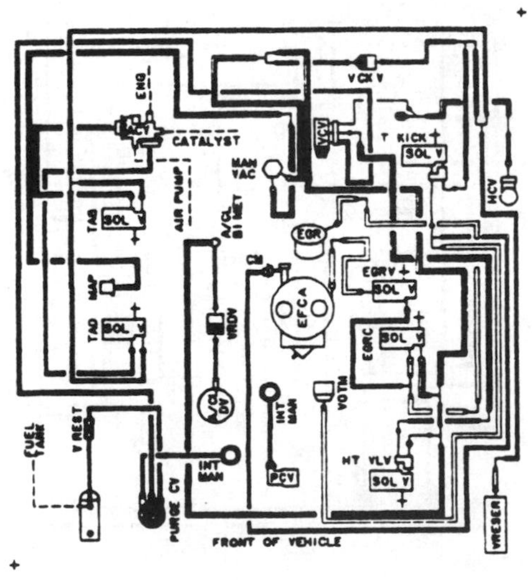

CALIBRATION: 4-22H-R00—AUTO. TRANS.—ALL

VACUUM CIRCUITS

1984—5.0L (302 CID) V8 engines

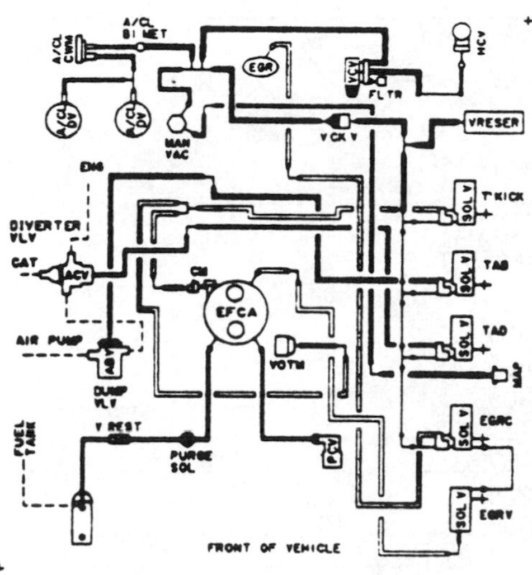

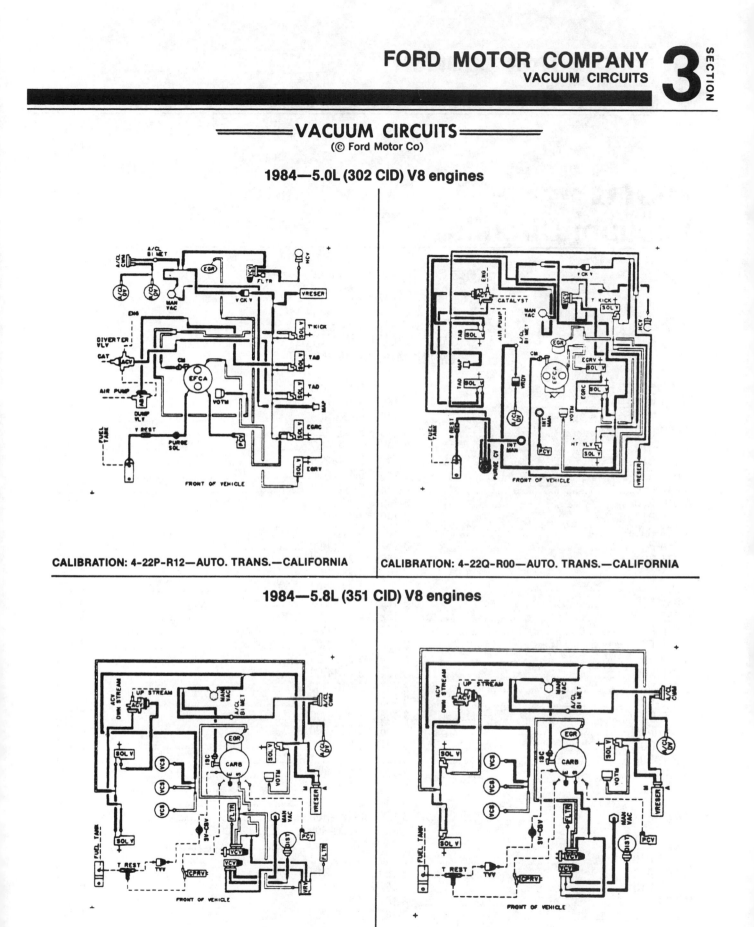

CALIBRATION: 4-22P-R12—AUTO. TRANS.—CALIFORNIA

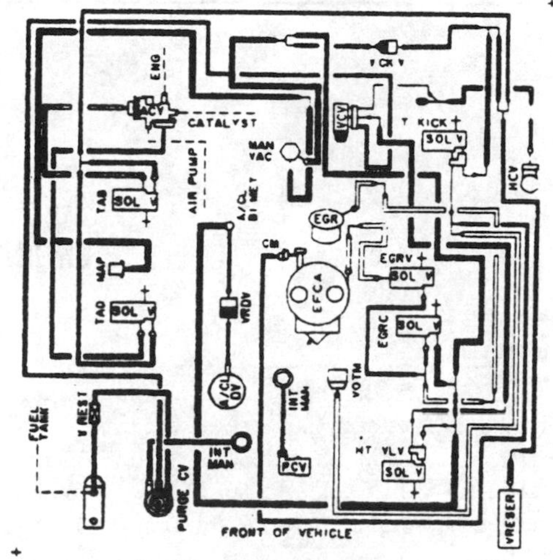

CALIBRATION: 4-22Q-R00—AUTO. TRANS.—CALIFORNIA

1984—5.8L (351 CID) V8 engines

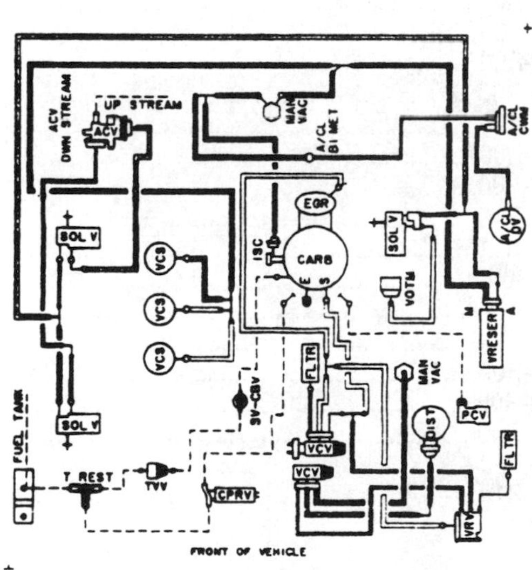

CALIBRATION: 4-24P-R00—AUTO. TRANS.—ALL

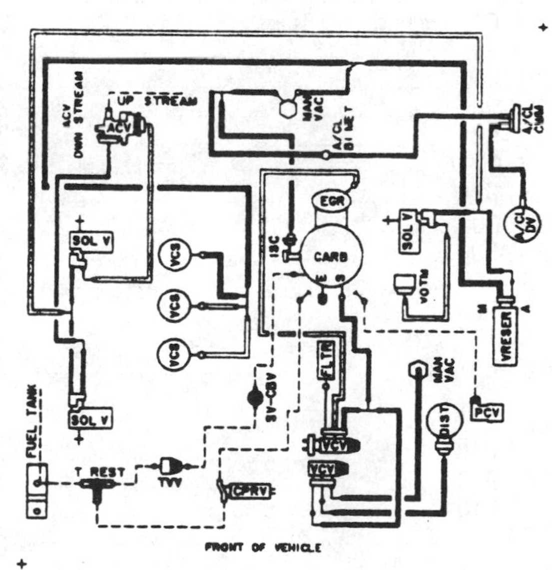

CALIBRATION: 4-24P-R11—AUTO. TRANS.—CANADA

Ford
Vacuum Circuits

INDEX

1985 VACUUM CIRCUITS
ENGINE AND CALIBRATION

VACUUM CIRCUITS

MODEL YEAR: 1985 **ENGINE: 1.6L**

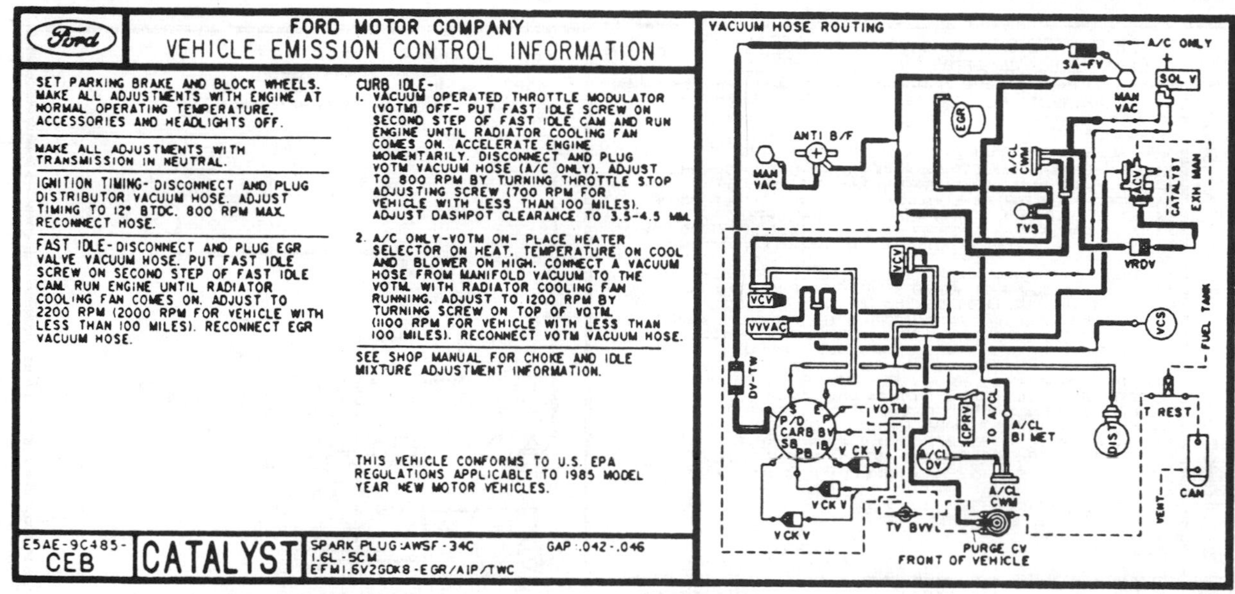

FORD MOTOR COMPANY
VEHICLE EMISSION CONTROL INFORMATION

SET PARKING BRAKE AND BLOCK WHEELS. MAKE ALL ADJUSTMENTS WITH ENGINE AT NORMAL OPERATING TEMPERATURE. ACCESSORIES AND HEADLIGHTS OFF.

MAKE ALL ADJUSTMENTS WITH TRANSMISSION IN NEUTRAL.

IGNITION TIMING- DISCONNECT AND PLUG DISTRIBUTOR VACUUM HOSE. ADJUST TIMING TO 12° BTDC. 800 RPM MAX. RECONNECT HOSE.

FAST IDLE- DISCONNECT AND PLUG EGR VALVE VACUUM HOSE. PUT FAST IDLE SCREW ON SECOND STEP OF FAST IDLE CAM. RUN ENGINE UNTIL RADIATOR COOLING FAN COMES ON. ADJUST TO 2200 RPM (2000 RPM FOR VEHICLE WITH LESS THAN 100 MILES). RECONNECT EGR VACUUM HOSE.

CURB IDLE-
1. VACUUM OPERATED THROTTLE MODULATOR (VOTM) OFF- PUT FAST IDLE SCREW ON SECOND STEP OF FAST IDLE CAM AND RUN ENGINE UNTIL RADIATOR COOLING FAN COMES ON. ACCELERATE ENGINE MOMENTARILY. DISCONNECT AND PLUG VOTM VACUUM HOSE (A/C ONLY). ADJUST TO 800 RPM BY TURNING THROTTLE STOP ADJUSTING SCREW (700 RPM FOR VEHICLE WITH LESS THAN 100 MILES). ADJUST DASHPOT CLEARANCE TO 3.5-4.5 MM.
2. A/C ONLY-VOTM ON- PLACE HEATER SELECTOR ON HEAT, TEMPERATURE ON COOL AND BLOWER ON HIGH. CONNECT A VACUUM HOSE FROM MANIFOLD VACUUM TO THE VOTM WITH RADIATOR COOLING FAN RUNNING. ADJUST TO 1200 RPM BY TURNING SCREW ON TOP OF VOTM. (1100 RPM FOR VEHICLE WITH LESS THAN 100 MILES). RECONNECT VOTM VACUUM HOSE.

SEE SHOP MANUAL FOR CHOKE AND IDLE MIXTURE ADJUSTMENT INFORMATION.

THIS VEHICLE CONFORMS TO U.S. EPA REGULATIONS APPLICABLE TO 1985 MODEL YEAR NEW MOTOR VEHICLES.

E5AE-9C485-CEB | CATALYST | SPARK PLUG: AWSF-34C GAP .042-.046 1.6L-5CM EFM1.6V2GDK8-EGR/AIP/TWC

CALIBRATION: 4—03A—R10

MODEL YEAR: 1985 **ENGINE: 1.6L**

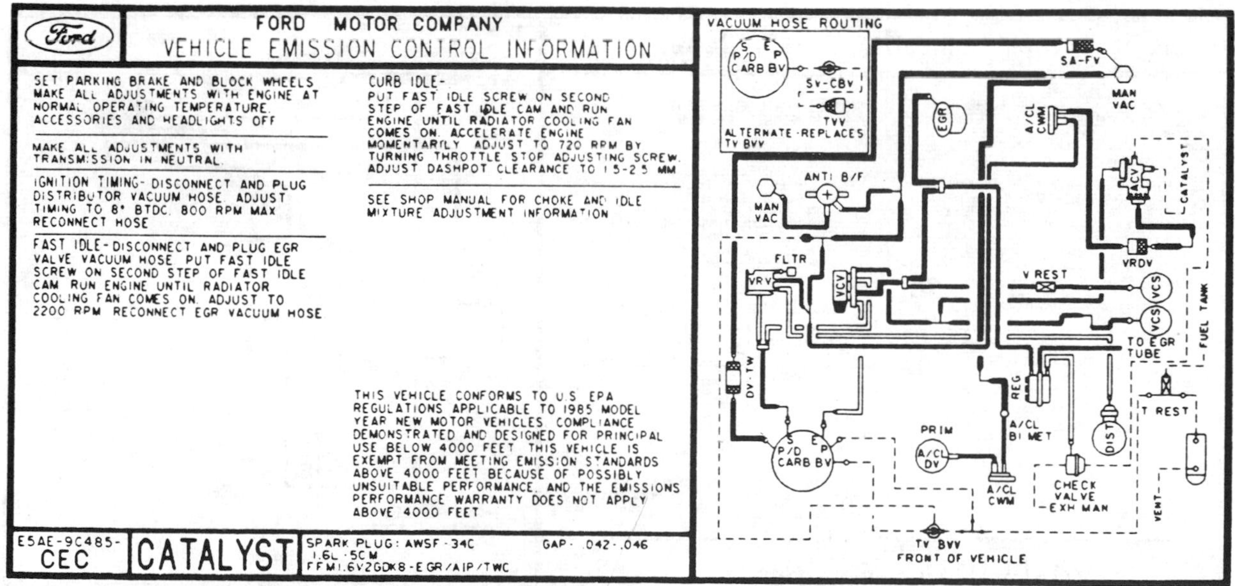

FORD MOTOR COMPANY
VEHICLE EMISSION CONTROL INFORMATION

SET PARKING BRAKE AND BLOCK WHEELS. MAKE ALL ADJUSTMENTS WITH ENGINE AT NORMAL OPERATING TEMPERATURE. ACCESSORIES AND HEADLIGHTS OFF

MAKE ALL ADJUSTMENTS WITH TRANSMISSION IN NEUTRAL.

IGNITION TIMING- DISCONNECT AND PLUG DISTRIBUTOR VACUUM HOSE. ADJUST TIMING TO 8° BTDC. 800 RPM MAX RECONNECT HOSE

FAST IDLE- DISCONNECT AND PLUG EGR VALVE VACUUM HOSE. PUT FAST IDLE SCREW ON SECOND STEP OF FAST IDLE CAM. RUN ENGINE UNTIL RADIATOR COOLING FAN COMES ON. ADJUST TO 2200 RPM. RECONNECT EGR VACUUM HOSE

CURB IDLE-
PUT FAST IDLE SCREW ON SECOND STEP OF FAST IDLE CAM AND RUN ENGINE UNTIL RADIATOR COOLING FAN COMES ON. ACCELERATE ENGINE MOMENTARILY. ADJUST TO 720 RPM BY TURNING THROTTLE STOP ADJUSTING SCREW. ADJUST DASHPOT CLEARANCE TO 1.5-2.5 MM

SEE SHOP MANUAL FOR CHOKE AND IDLE MIXTURE ADJUSTMENT INFORMATION

THIS VEHICLE CONFORMS TO U.S. EPA REGULATIONS APPLICABLE TO 1985 MODEL YEAR NEW MOTOR VEHICLES. COMPLIANCE DEMONSTRATED AND DESIGNED FOR PRINCIPAL USE BELOW 4000 FEET. THIS VEHICLE IS EXEMPT FROM MEETING EMISSION STANDARDS ABOVE 4000 FEET BECAUSE OF POSSIBLY UNSUITABLE PERFORMANCE, AND THE EMISSIONS PERFORMANCE WARRANTY DOES NOT APPLY ABOVE 4000 FEET

E5AE-9C485-CEC | CATALYST | SPARK PLUG: AWSF-34C GAP .042-.046 1.6L-5CM FFM1.6V2GDK8-EGR/AIP/TWC

CALIBRATION: 4—03F—R00

VACUUM CIRCUITS

(© Ford Motor Co.)

MODEL YEAR: 1985 **ENGINE: 1.6L**

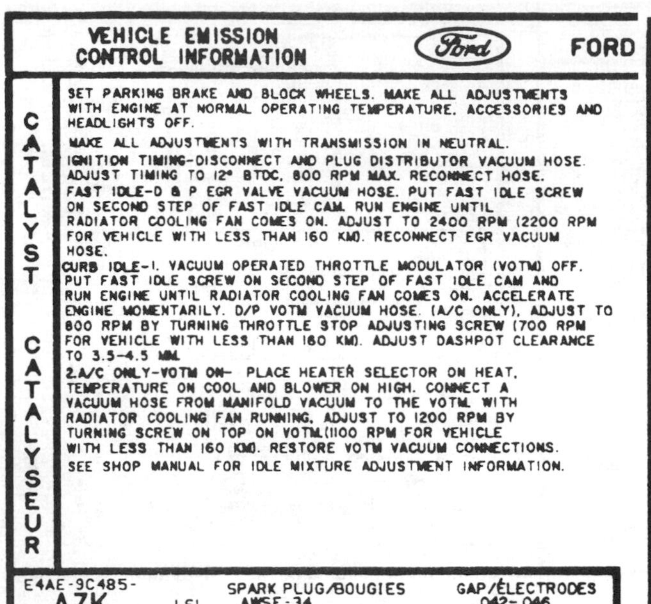

VEHICLE EMISSION CONTROL INFORMATION — FORD

CATALYST CATALYSEUR

SET PARKING BRAKE AND BLOCK WHEELS. MAKE ALL ADJUSTMENTS WITH ENGINE AT NORMAL OPERATING TEMPERATURE, ACCESSORIES AND HEADLIGHTS OFF.
MAKE ALL ADJUSTMENTS WITH TRANSMISSION IN NEUTRAL.
IGNITION TIMING-DISCONNECT AND PLUG DISTRIBUTOR VACUUM HOSE. ADJUST TIMING TO 12° BTDC, 800 RPM MAX. RECONNECT HOSE.
FAST IDLE-D & P EGR VALVE VACUUM HOSE. PUT FAST IDLE SCREW ON SECOND STEP OF FAST IDLE CAM. RUN ENGINE UNTIL RADIATOR COOLING FAN COMES ON. ADJUST TO 2400 RPM (2200 RPM FOR VEHICLE WITH LESS THAN 160 KM). RECONNECT EGR VACUUM HOSE.
CURB IDLE-1. VACUUM OPERATED THROTTLE MODULATOR (VOTM) OFF. PUT FAST IDLE SCREW ON SECOND STEP OF FAST IDLE CAM AND RUN ENGINE UNTIL RADIATOR COOLING FAN COMES ON. ACCELERATE ENGINE MOMENTARILY. D/P VOTM VACUUM HOSE. (A/C ONLY). ADJUST TO 800 RPM BY TURNING THROTTLE STOP ADJUSTING SCREW (700 RPM FOR VEHICLE WITH LESS THAN 160 KM). ADJUST DASHPOT CLEARANCE TO 3.5-4.5 MM.
2.A/C ONLY-VOTM ON- PLACE HEATER SELECTOR ON HEAT, TEMPERATURE ON COOL AND BLOWER ON HIGH. CONNECT A VACUUM HOSE FROM MANIFOLD VACUUM TO THE VOTM. WITH RADIATOR COOLING FAN RUNNING, ADJUST TO 1200 RPM BY TURNING SCREW ON TOP ON VOTM (1100 RPM FOR VEHICLE WITH LESS THAN 160 KM). RESTORE VOTM VACUUM CONNECTIONS.
SEE SHOP MANUAL FOR IDLE MIXTURE ADJUSTMENT INFORMATION.

E4AE-9C485-AZK 1.6L SPARK PLUG/BOUGIES AWSF-34 GAP/ÉLECTRODES .042-.046

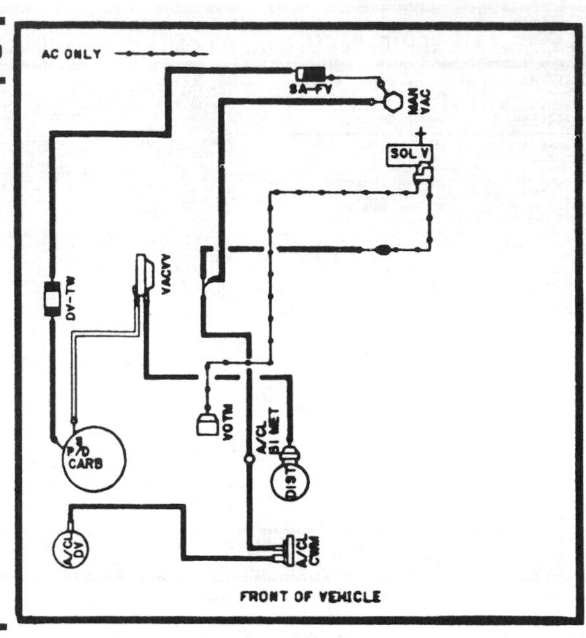

PROPANE SPECIFICATIONS
(Set in Neutral) Gain RPM: 40-110
Reset RPM: 70

CALIBRATION: 4—03H—R00

MODEL YEAR: 1985 **ENGINE: 1.6L**

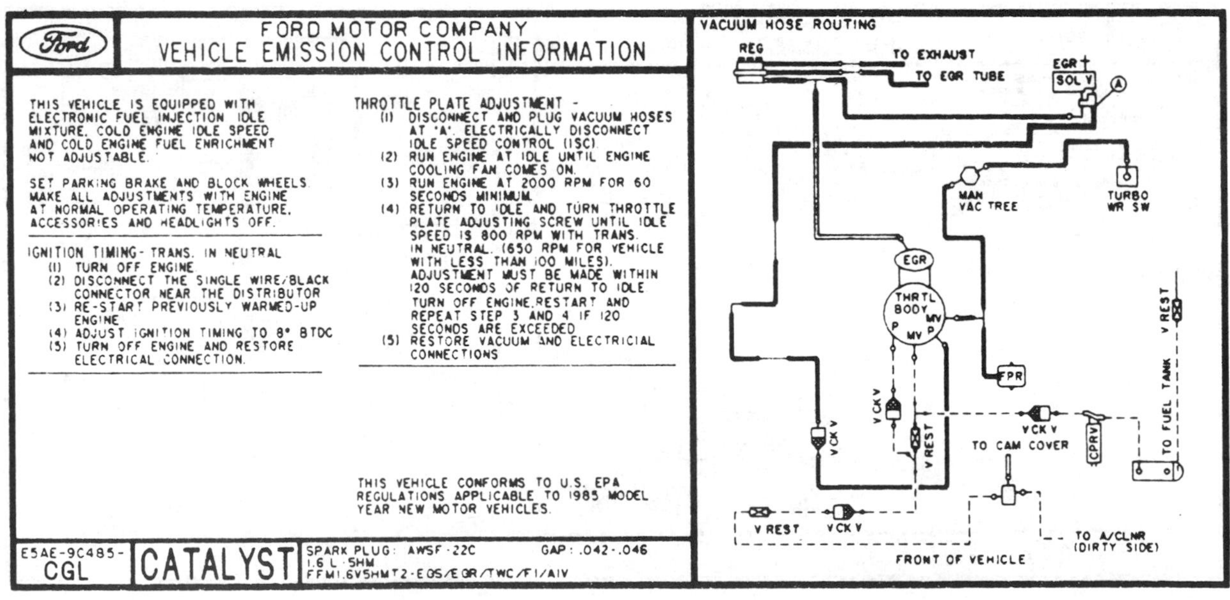

FORD MOTOR COMPANY
VEHICLE EMISSION CONTROL INFORMATION

THIS VEHICLE IS EQUIPPED WITH ELECTRONIC FUEL INJECTION. IDLE MIXTURE, COLD ENGINE IDLE SPEED AND COLD ENGINE FUEL ENRICHMENT NOT ADJUSTABLE.

SET PARKING BRAKE AND BLOCK WHEELS. MAKE ALL ADJUSTMENTS WITH ENGINE AT NORMAL OPERATING TEMPERATURE, ACCESSORIES AND HEADLIGHTS OFF.

IGNITION TIMING- TRANS. IN NEUTRAL
(1) TURN OFF ENGINE
(2) DISCONNECT THE SINGLE WIRE/BLACK CONNECTOR NEAR THE DISTRIBUTOR
(3) RE-START PREVIOUSLY WARMED-UP ENGINE
(4) ADJUST IGNITION TIMING TO 8° BTDC
(5) TURN OFF ENGINE AND RESTORE ELECTRICAL CONNECTION.

THROTTLE PLATE ADJUSTMENT -
(1) DISCONNECT AND PLUG VACUUM HOSES AT "A". ELECTRICALLY DISCONNECT IDLE SPEED CONTROL (ISC).
(2) RUN ENGINE AT IDLE UNTIL ENGINE COOLING FAN COMES ON.
(3) RUN ENGINE AT 2000 RPM FOR 60 SECONDS MINIMUM.
(4) RETURN TO IDLE AND TURN THROTTLE PLATE ADJUSTING SCREW UNTIL IDLE SPEED IS 800 RPM WITH TRANS. IN NEUTRAL. (650 RPM FOR VEHICLE WITH LESS THAN 100 MILES). ADJUSTMENT MUST BE MADE WITHIN 120 SECONDS OF RETURN TO IDLE. TURN OFF ENGINE, RESTART AND REPEAT STEP 3 AND 4 IF 120 SECONDS ARE EXCEEDED.
(5) RESTORE VACUUM AND ELECTRICIAL CONNECTIONS.

THIS VEHICLE CONFORMS TO U.S. EPA REGULATIONS APPLICABLE TO 1985 MODEL YEAR NEW MOTOR VEHICLES.

E5AE-9C485-CGL CATALYST SPARK PLUG: AWSF-22C GAP: .042-.046 1.6 L-5HM FFM1.6V5HMT2-EOS/EGR/TWC/FI/AIV

CALIBRATION: 5—27T—R00

VACUUM CIRCUITS

(© Ford Motor Co)

MODEL YEAR: 1985 **ENGINE: 1.6L**

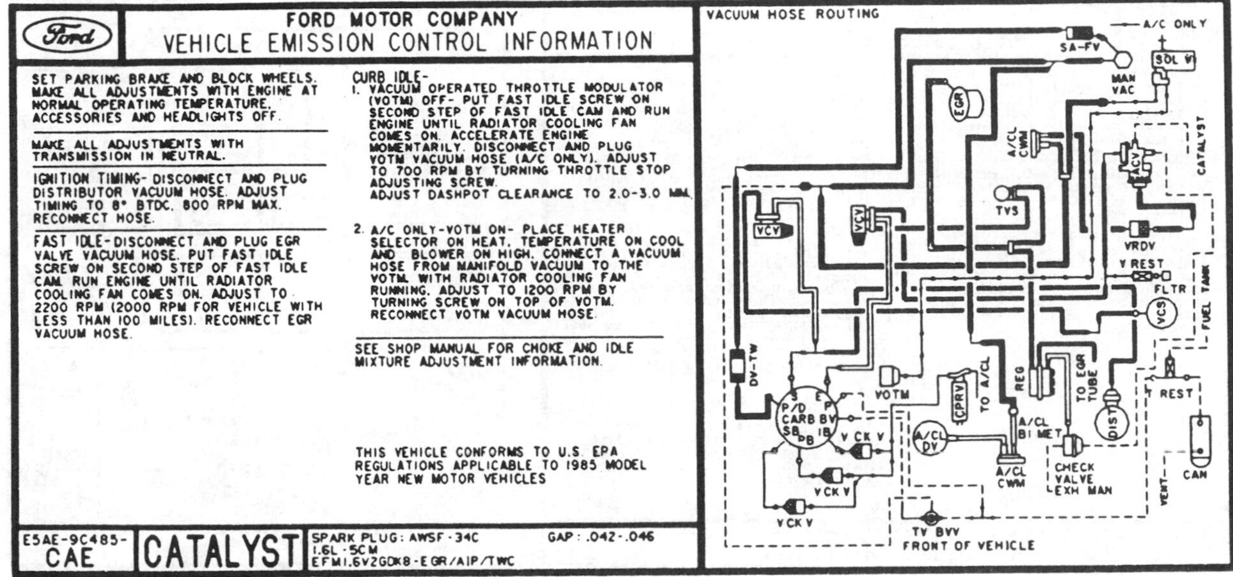

FORD MOTOR COMPANY
VEHICLE EMISSION CONTROL INFORMATION

SET PARKING BRAKE AND BLOCK WHEELS. MAKE ALL ADJUSTMENTS WITH ENGINE AT NORMAL OPERATING TEMPERATURE, ACCESSORIES AND HEADLIGHTS OFF.

MAKE ALL ADJUSTMENTS WITH TRANSMISSION IN NEUTRAL.

IGNITION TIMING- DISCONNECT AND PLUG DISTRIBUTOR VACUUM HOSE. ADJUST TIMING TO 8° BTDC. 800 RPM MAX. RECONNECT HOSE.

FAST IDLE- DISCONNECT AND PLUG EGR VALVE VACUUM HOSE. PUT FAST IDLE SCREW ON SECOND STEP OF FAST IDLE CAM. RUN ENGINE UNTIL RADIATOR COOLING FAN COMES ON. ADJUST TO 2200 RPM (2000 RPM FOR VEHICLE WITH LESS THAN 100 MILES). RECONNECT EGR VACUUM HOSE.

CURB IDLE-
1. VACUUM OPERATED THROTTLE MODULATOR (VOTM) OFF- PUT FAST IDLE SCREW ON SECOND STEP OF FAST IDLE CAM AND RUN ENGINE UNTIL RADIATOR COOLING FAN COMES ON. ACCELERATE ENGINE MOMENTARILY. DISCONNECT AND PLUG VOTM VACUUM HOSE (A/C ONLY). ADJUST TO 700 RPM BY TURNING THROTTLE STOP ADJUSTING SCREW.
ADJUST DASHPOT CLEARANCE TO 2.0-3.0 MM.
2. A/C ONLY-VOTM ON- PLACE HEATER SELECTOR ON HEAT, TEMPERATURE ON COOL AND BLOWER ON HIGH. CONNECT A VACUUM HOSE FROM MANIFOLD VACUUM TO THE VOTM WITH RADIATOR COOLING FAN RUNNING. ADJUST TO 1200 RPM BY TURNING SCREW ON TOP OF VOTM. RECONNECT VOTM VACUUM HOSE.

SEE SHOP MANUAL FOR CHOKE AND IDLE MIXTURE ADJUSTMENT INFORMATION.

THIS VEHICLE CONFORMS TO U.S. EPA REGULATIONS APPLICABLE TO 1985 MODEL YEAR NEW MOTOR VEHICLES

E5AE-9C485-CAE CATALYST SPARK PLUG: AWSF-34C GAP: .042-.046 1.6L-5CM EFM1.6V2GDK8-EGR/AIP/TWC

CALIBRATION: 4—03K—R10

MODEL YEAR: 1985 **ENGINE: 1.6L**

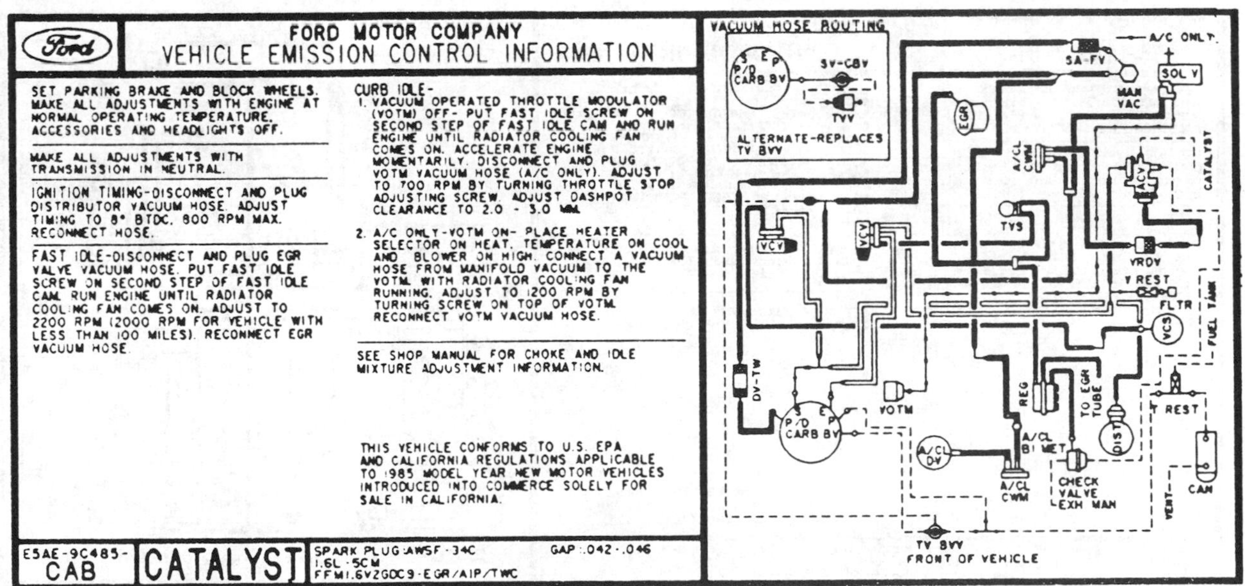

FORD MOTOR COMPANY
VEHICLE EMISSION CONTROL INFORMATION

SET PARKING BRAKE AND BLOCK WHEELS. MAKE ALL ADJUSTMENTS WITH ENGINE AT NORMAL OPERATING TEMPERATURE, ACCESSORIES AND HEADLIGHTS OFF.

MAKE ALL ADJUSTMENTS WITH TRANSMISSION IN NEUTRAL.

IGNITION TIMING-DISCONNECT AND PLUG DISTRIBUTOR VACUUM HOSE. ADJUST TIMING TO 8° BTDC. 800 RPM MAX. RECONNECT HOSE.

FAST IDLE-DISCONNECT AND PLUG EGR VALVE VACUUM HOSE. PUT FAST IDLE SCREW ON SECOND STEP OF FAST IDLE CAM. RUN ENGINE UNTIL RADIATOR COOLING FAN COMES ON. ADJUST TO 2200 RPM (2000 RPM FOR VEHICLE WITH LESS THAN 100 MILES). RECONNECT EGR VACUUM HOSE.

CURB IDLE-
1. VACUUM OPERATED THROTTLE MODULATOR (VOTM) OFF- PUT FAST IDLE SCREW ON SECOND STEP OF FAST IDLE CAM AND RUN ENGINE UNTIL RADIATOR COOLING FAN COMES ON. ACCELERATE ENGINE MOMENTARILY. DISCONNECT AND PLUG VOTM VACUUM HOSE (A/C ONLY). ADJUST TO 700 RPM BY TURNING THROTTLE STOP ADJUSTING SCREW. ADJUST DASHPOT CLEARANCE TO 2.0 - 3.0 MM.
2. A/C ONLY-VOTM ON- PLACE HEATER SELECTOR ON HEAT, TEMPERATURE ON COOL AND BLOWER ON HIGH. CONNECT A VACUUM HOSE FROM MANIFOLD VACUUM TO THE VOTM WITH RADIATOR COOLING FAN RUNNING. ADJUST TO 1200 RPM BY TURNING SCREW ON TOP OF VOTM. RECONNECT VOTM VACUUM HOSE.

SEE SHOP MANUAL FOR CHOKE AND IDLE MIXTURE ADJUSTMENT INFORMATION.

THIS VEHICLE CONFORMS TO U.S. EPA AND CALIFORNIA REGULATIONS APPLICABLE TO 1985 MODEL YEAR NEW MOTOR VEHICLES INTRODUCED INTO COMMERCE SOLELY FOR SALE IN CALIFORNIA.

E5AE-9C485-CAB CATALYST SPARK PLUG: AWSF-34C GAP: .042-.046 1.6L-5CM FFM1.6V2GDC9-EGR/AIP/TWC

CALIBRATION: 4—03P—R10

VACUUM CIRCUITS

(© Ford Motor Co.)

MODEL YEAR: 1985 — ENGINE: 1.6L

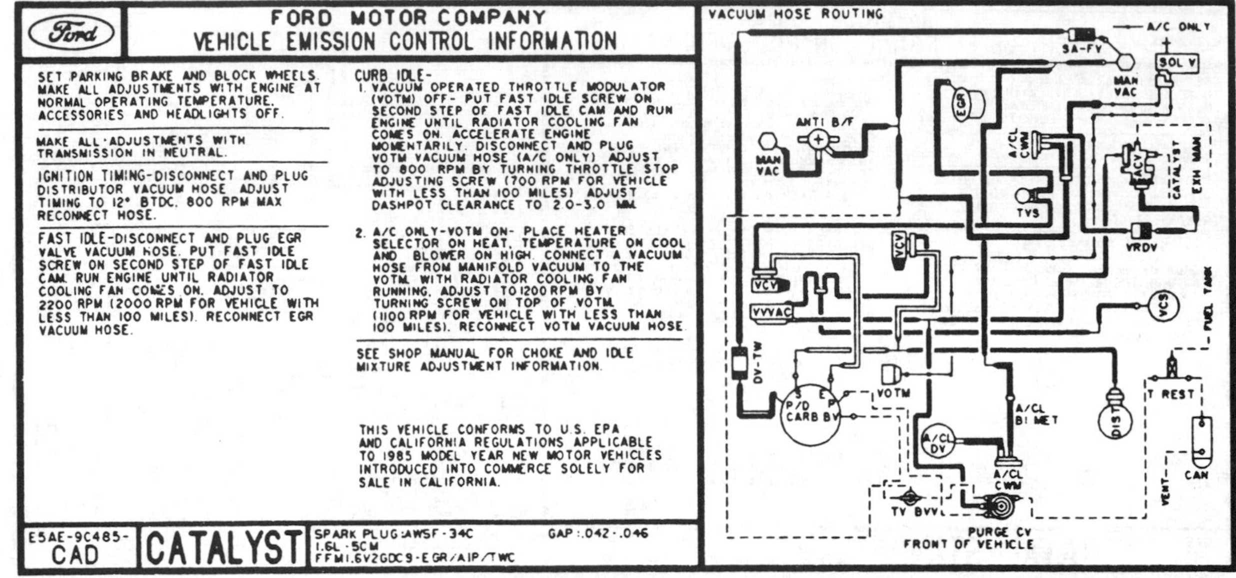

FORD MOTOR COMPANY
VEHICLE EMISSION CONTROL INFORMATION

SET PARKING BRAKE AND BLOCK WHEELS. MAKE ALL ADJUSTMENTS WITH ENGINE AT NORMAL OPERATING TEMPERATURE. ACCESSORIES AND HEADLIGHTS OFF.

MAKE ALL ADJUSTMENTS WITH TRANSMISSION IN NEUTRAL.

IGNITION TIMING-DISCONNECT AND PLUG DISTRIBUTOR VACUUM HOSE. ADJUST TIMING TO 12° BTDC. 800 RPM MAX RECONNECT HOSE.

FAST IDLE-DISCONNECT AND PLUG EGR VALVE VACUUM HOSE. PUT FAST IDLE SCREW ON SECOND STEP OF FAST IDLE CAM. RUN ENGINE UNTIL RADIATOR COOLING FAN COMES ON. ADJUST TO 2200 RPM (2000 RPM FOR VEHICLE WITH LESS THAN 100 MILES). RECONNECT EGR VACUUM HOSE.

CURB IDLE-

1. VACUUM OPERATED THROTTLE MODULATOR (VOTM) OFF- PUT FAST IDLE SCREW ON SECOND STEP OF FAST IDLE CAM AND RUN ENGINE UNTIL RADIATOR COOLING FAN COMES ON. ACCELERATE ENGINE MOMENTARILY. DISCONNECT AND PLUG VOTM VACUUM HOSE (A/C ONLY) ADJUST TO 800 RPM BY TURNING THROTTLE STOP ADJUSTING SCREW (700 RPM FOR VEHICLE WITH LESS THAN 100 MILES) ADJUST DASHPOT CLEARANCE TO 2.0-3.0 MM.
2. A/C ONLY-VOTM ON- PLACE HEATER SELECTOR ON HEAT, TEMPERATURE ON COOL AND BLOWER ON HIGH. CONNECT A VACUUM HOSE FROM MANIFOLD VACUUM TO THE VOTM WITH RADIATOR COOLING FAN RUNNING. ADJUST TO 1200 RPM BY TURNING SCREW ON TOP OF VOTM (1100 RPM FOR VEHICLE WITH LESS THAN 100 MILES). RECONNECT VOTM VACUUM HOSE.

SEE SHOP MANUAL FOR CHOKE AND IDLE MIXTURE ADJUSTMENT INFORMATION.

THIS VEHICLE CONFORMS TO U.S. EPA AND CALIFORNIA REGULATIONS APPLICABLE TO 1985 MODEL YEAR NEW MOTOR VEHICLES INTRODUCED INTO COMMERCE SOLELY FOR SALE IN CALIFORNIA.

E5AE-9C485-CAD | CATALYST | SPARK PLUG: AWSF-34C GAP: .042-.046 1.6L-5CM FFM1.6V2GDC9-EGR/AIP/TWC

CALIBRATION: 4—03S—R10

MODEL YEAR: 1985 — ENGINE: 1.6L

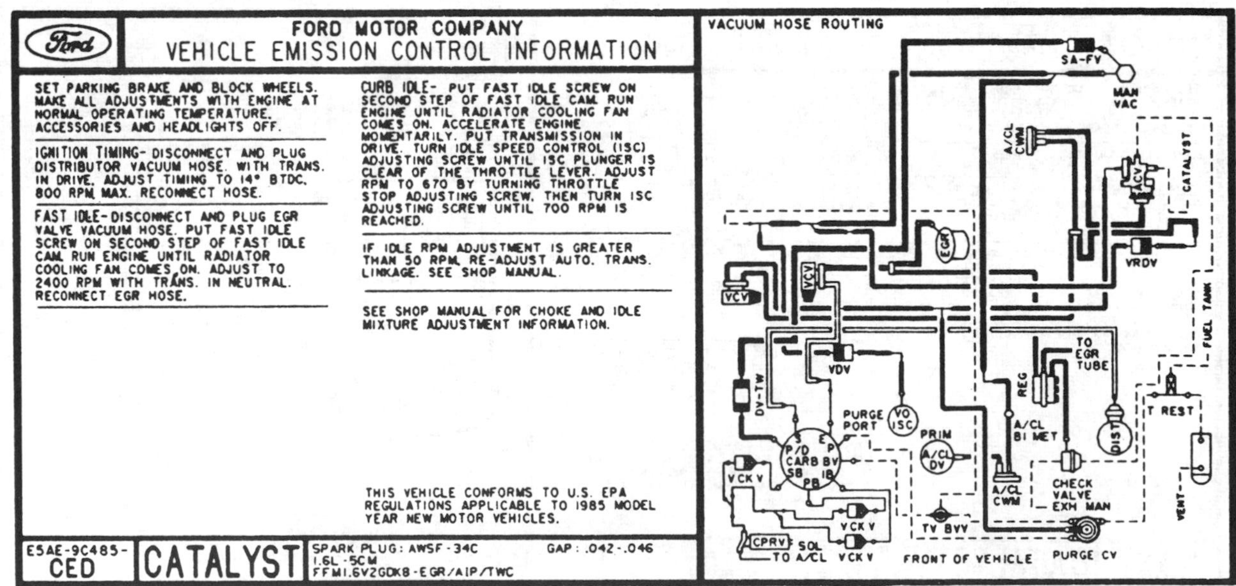

FORD MOTOR COMPANY
VEHICLE EMISSION CONTROL INFORMATION

SET PARKING BRAKE AND BLOCK WHEELS. MAKE ALL ADJUSTMENTS WITH ENGINE AT NORMAL OPERATING TEMPERATURE. ACCESSORIES AND HEADLIGHTS OFF.

IGNITION TIMING- DISCONNECT AND PLUG DISTRIBUTOR VACUUM HOSE. WITH TRANS. IN DRIVE, ADJUST TIMING TO 14° BTDC, 800 RPM MAX. RECONNECT HOSE.

FAST IDLE-DISCONNECT AND PLUG EGR VALVE VACUUM HOSE. PUT FAST IDLE SCREW ON SECOND STEP OF FAST IDLE CAM. RUN ENGINE UNTIL RADIATOR COOLING FAN COMES ON. ADJUST TO 2400 RPM WITH TRANS. IN NEUTRAL. RECONNECT EGR HOSE.

CURB IDLE- PUT FAST IDLE SCREW ON SECOND STEP OF FAST IDLE CAM. RUN ENGINE UNTIL RADIATOR COOLING FAN COMES ON. ACCELERATE ENGINE MOMENTARILY. PUT TRANSMISSION IN DRIVE. TURN IDLE SPEED CONTROL (ISC) ADJUSTING SCREW UNTIL ISC PLUNGER IS CLEAR OF THE THROTTLE LEVER. ADJUST RPM TO 670 BY TURNING THROTTLE STOP ADJUSTING SCREW, THEN TURN ISC ADJUSTING SCREW UNTIL 700 RPM IS REACHED.

IF IDLE RPM ADJUSTMENT IS GREATER THAN 50 RPM, RE-ADJUST AUTO. TRANS. LINKAGE. SEE SHOP MANUAL.

SEE SHOP MANUAL FOR CHOKE AND IDLE MIXTURE ADJUSTMENT INFORMATION.

THIS VEHICLE CONFORMS TO U.S. EPA REGULATIONS APPLICABLE TO 1985 MODEL YEAR NEW MOTOR VEHICLES.

E5AE-9C485-CED | CATALYST | SPARK PLUG: AWSF-34C GAP: .042-.046 1.6L-5CM FFM1.6V2GDK8-EGR/AIP/TWC

CALIBRATION: 4—04A—R16

VACUUM CIRCUITS

(© Ford Motor Co)

MODEL YEAR: 1985 **ENGINE: 1.6L**

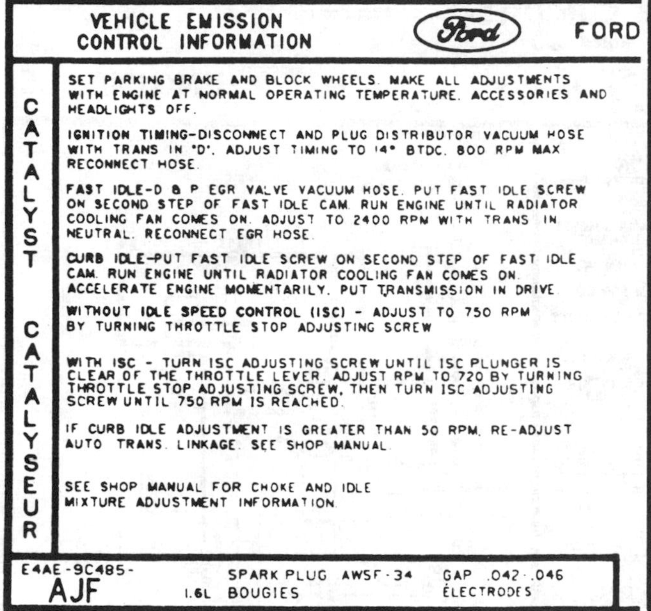

VEHICLE EMISSION CONTROL INFORMATION — FORD

CATALYST CATALYSEUR

SET PARKING BRAKE AND BLOCK WHEELS. MAKE ALL ADJUSTMENTS WITH ENGINE AT NORMAL OPERATING TEMPERATURE. ACCESSORIES AND HEADLIGHTS OFF.

IGNITION TIMING–DISCONNECT AND PLUG DISTRIBUTOR VACUUM HOSE WITH TRANS IN "D". ADJUST TIMING TO 14° BTDC, 800 RPM MAX. RECONNECT HOSE.

FAST IDLE–D & P EGR VALVE VACUUM HOSE. PUT FAST IDLE SCREW ON SECOND STEP OF FAST IDLE CAM. RUN ENGINE UNTIL RADIATOR COOLING FAN COMES ON. ADJUST TO 2400 RPM WITH TRANS IN NEUTRAL. RECONNECT EGR HOSE.

CURB IDLE–PUT FAST IDLE SCREW ON SECOND STEP OF FAST IDLE CAM. RUN ENGINE UNTIL RADIATOR COOLING FAN COMES ON. ACCELERATE ENGINE MOMENTARILY. PUT TRANSMISSION IN DRIVE

WITHOUT IDLE SPEED CONTROL (ISC) – ADJUST TO 750 RPM BY TURNING THROTTLE STOP ADJUSTING SCREW

WITH ISC – TURN ISC ADJUSTING SCREW UNTIL ISC PLUNGER IS CLEAR OF THE THROTTLE LEVER. ADJUST RPM TO 720 BY TURNING THROTTLE STOP ADJUSTING SCREW, THEN TURN ISC ADJUSTING SCREW UNTIL 750 RPM IS REACHED.

IF CURB IDLE ADJUSTMENT IS GREATER THAN 50 RPM, RE-ADJUST AUTO TRANS. LINKAGE. SEE SHOP MANUAL.

SEE SHOP MANUAL FOR CHOKE AND IDLE MIXTURE ADJUSTMENT INFORMATION.

E4AE-9C485-AJF — SPARK PLUG AWSF-34 — GAP .042-.046 — 1.6L BOUGIES — ÉLECTRODES

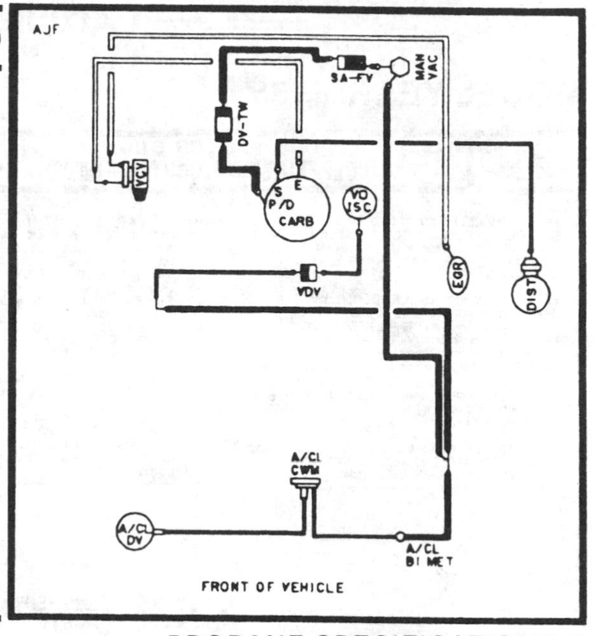

CALIBRATION: 4—04H—R00

PROPANE SPECIFICATIONS
(Set in Neutral) Gain RPM: 10-100
Reset RPM: 30

MODEL YEAR: 1985 **ENGINE: 1.6L**

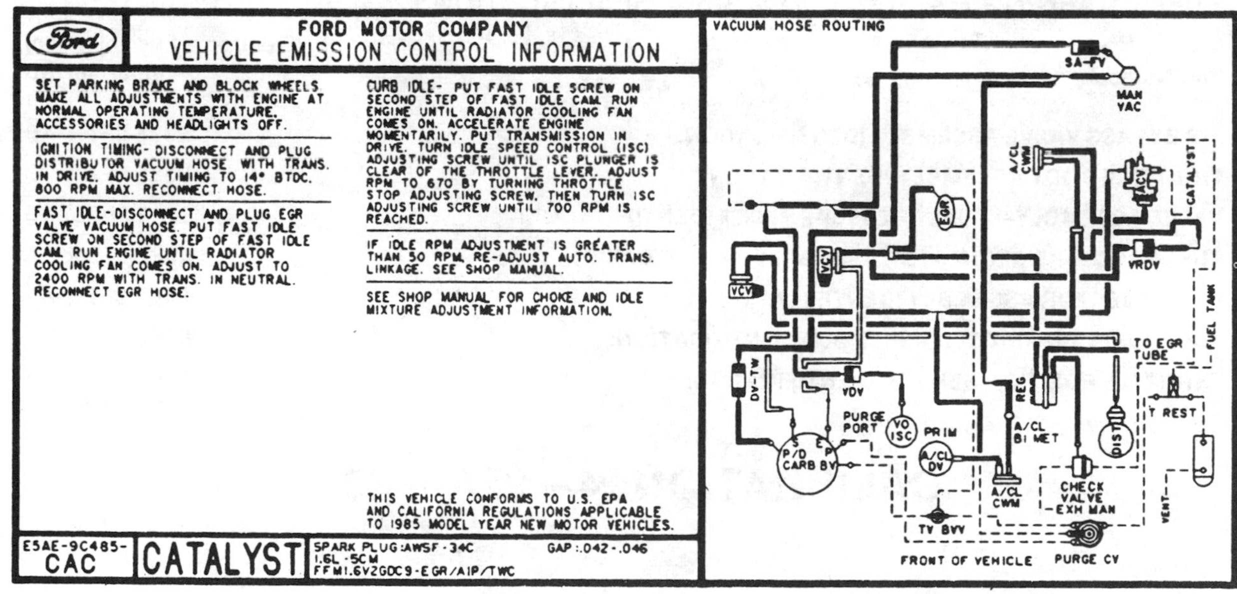

FORD MOTOR COMPANY
VEHICLE EMISSION CONTROL INFORMATION

SET PARKING BRAKE AND BLOCK WHEELS. MAKE ALL ADJUSTMENTS WITH ENGINE AT NORMAL OPERATING TEMPERATURE. ACCESSORIES AND HEADLIGHTS OFF.

IGNITION TIMING- DISCONNECT AND PLUG DISTRIBUTOR VACUUM HOSE WITH TRANS. IN DRIVE. ADJUST TIMING TO 14° BTDC. 800 RPM MAX. RECONNECT HOSE.

FAST IDLE- DISCONNECT AND PLUG EGR VALVE VACUUM HOSE. PUT FAST IDLE SCREW ON SECOND STEP OF FAST IDLE CAM. RUN ENGINE UNTIL RADIATOR COOLING FAN COMES ON. ADJUST TO 2400 RPM WITH TRANS. IN NEUTRAL. RECONNECT EGR HOSE.

CURB IDLE- PUT FAST IDLE SCREW ON SECOND STEP OF FAST IDLE CAM. RUN ENGINE UNTIL RADIATOR COOLING FAN COMES ON. ACCELERATE ENGINE MOMENTARILY. PUT TRANSMISSION IN DRIVE. TURN IDLE SPEED CONTROL (ISC) ADJUSTING SCREW UNTIL ISC PLUNGER IS CLEAR OF THE THROTTLE LEVER. ADJUST RPM TO 670 BY TURNING THROTTLE STOP ADJUSTING SCREW. THEN TURN ISC ADJUSTING SCREW UNTIL 700 RPM IS REACHED.

IF IDLE RPM ADJUSTMENT IS GREATER THAN 50 RPM, RE-ADJUST AUTO. TRANS. LINKAGE. SEE SHOP MANUAL.

SEE SHOP MANUAL FOR CHOKE AND IDLE MIXTURE ADJUSTMENT INFORMATION.

THIS VEHICLE CONFORMS TO U.S. EPA AND CALIFORNIA REGULATIONS APPLICABLE TO 1985 MODEL YEAR NEW MOTOR VEHICLES.

E5AE-9C485-CAC — CATALYST — SPARK PLUG:AWSF-34C — GAP:.042-.046 — 1.6L-5CM — FFM1.6V2GDC9-EGR/AIP/TWC

CALIBRATION: 4—04S—R12

VACUUM CIRCUITS

MODEL YEAR: 1985 **ENGINE: 1.6L**

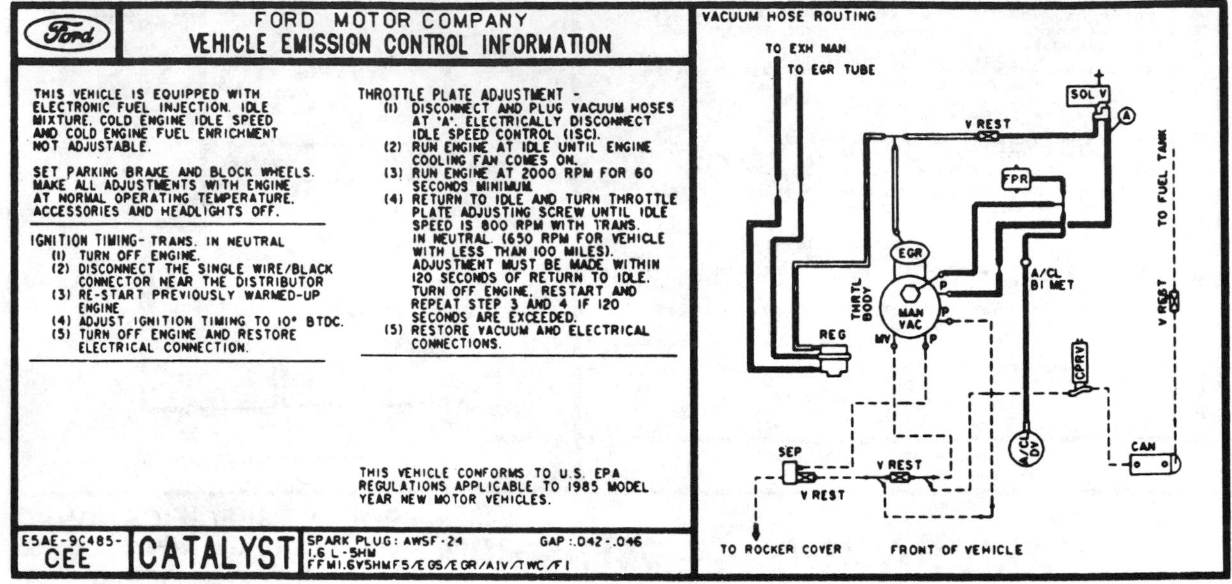

FORD MOTOR COMPANY
VEHICLE EMISSION CONTROL INFORMATION

THIS VEHICLE IS EQUIPPED WITH ELECTRONIC FUEL INJECTION. IDLE MIXTURE, COLD ENGINE IDLE SPEED AND COLD ENGINE FUEL ENRICHMENT NOT ADJUSTABLE.

SET PARKING BRAKE AND BLOCK WHEELS. MAKE ALL ADJUSTMENTS WITH ENGINE AT NORMAL OPERATING TEMPERATURE. ACCESSORIES AND HEADLIGHTS OFF.

IGNITION TIMING- TRANS. IN NEUTRAL
(1) TURN OFF ENGINE.
(2) DISCONNECT THE SINGLE WIRE/BLACK CONNECTOR NEAR THE DISTRIBUTOR
(3) RE-START PREVIOUSLY WARMED-UP ENGINE
(4) ADJUST IGNITION TIMING TO 10° BTDC.
(5) TURN OFF ENGINE AND RESTORE ELECTRICAL CONNECTION.

THROTTLE PLATE ADJUSTMENT -
(1) DISCONNECT AND PLUG VACUUM HOSES AT 'A'. ELECTRICALLY DISCONNECT IDLE SPEED CONTROL (ISC).
(2) RUN ENGINE AT IDLE UNTIL ENGINE COOLING FAN COMES ON.
(3) RUN ENGINE AT 2000 RPM FOR 60 SECONDS MINIMUM.
(4) RETURN TO IDLE AND TURN THROTTLE PLATE ADJUSTING SCREW UNTIL IDLE SPEED IS 800 RPM WITH TRANS. IN NEUTRAL. (650 RPM FOR VEHICLE WITH LESS THAN 100 MILES). ADJUSTMENT MUST BE MADE WITHIN 120 SECONDS OF RETURN TO IDLE. TURN OFF ENGINE, RESTART AND REPEAT STEP 3 AND 4 IF 120 SECONDS ARE EXCEEDED.
(5) RESTORE VACUUM AND ELECTRICAL CONNECTIONS.

THIS VEHICLE CONFORMS TO U.S. EPA REGULATIONS APPLICABLE TO 1985 MODEL YEAR NEW MOTOR VEHICLES.

E5AE-9C485-CEE | CATALYST | SPARK PLUG: AWSF-24 GAP: .042-.046
1.6 L-5HM
FFM1.6V5HMF5/EGS/EGR/AIV/TWC/FI

FUEL CHARGING MANIFOLD ASSEMBLY SPECIFICATIONS

Model:	Ford EFI (Electronic Fuel Injection)
Part Number:	E4EE-9H487-BB
AIR BYPASS VALVE ASSEMBLY IDENTIFICATION:	E3EE-9F715-AB
THROTTLE BODY ASSEMBLY IDENTIFICATION:	E3EE-9E926-DB
VALVE ASSEMBLY—FUEL PRESSURE RELIEF IDENTIFICATION:	E0AE-9H321-AB
FUEL PRESSURE REGULATOR ASSEMBLY IDENTIFICATION:	E3EE-9C968-AA
FUEL INJECTOR ASSEMBLY IDENTIFICATION (4):	E4EE-9F593-AA
FUEL SUPPLY MANIFOLD ASSEMBLY IDENTIFICATION:	E4EE-9D280-DA
THROTTLE POSITION SENSOR IDENTIFICATION:	E3EF-9B989-AA E5EF-9B989-CA (Alt.)

CALIBRATION: 4—27A—R02

VACUUM CIRCUITS

(© Ford Motor Co.)

MODEL YEAR: 1985 **ENGINE: 1.6L**

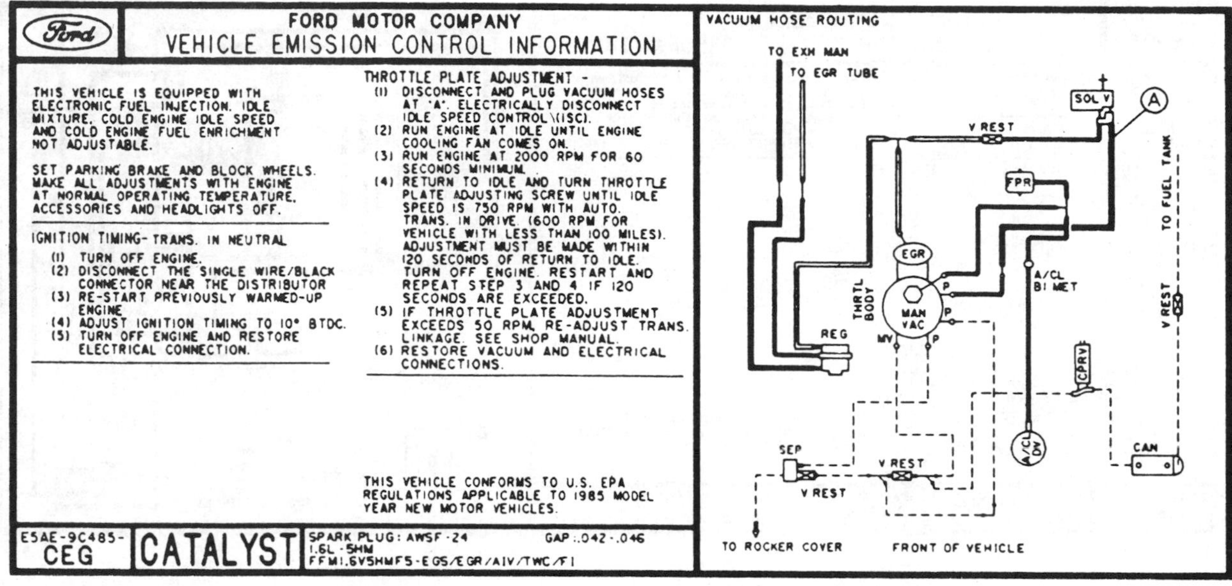

FORD MOTOR COMPANY
VEHICLE EMISSION CONTROL INFORMATION

THIS VEHICLE IS EQUIPPED WITH ELECTRONIC FUEL INJECTION. IDLE MIXTURE, COLD ENGINE IDLE SPEED AND COLD ENGINE FUEL ENRICHMENT NOT ADJUSTABLE.

SET PARKING BRAKE AND BLOCK WHEELS. MAKE ALL ADJUSTMENTS WITH ENGINE AT NORMAL OPERATING TEMPERATURE, ACCESSORIES AND HEADLIGHTS OFF.

IGNITION TIMING- TRANS. IN NEUTRAL
(1) TURN OFF ENGINE.
(2) DISCONNECT THE SINGLE WIRE/BLACK CONNECTOR NEAR THE DISTRIBUTOR
(3) RE-START PREVIOUSLY WARMED-UP ENGINE
(4) ADJUST IGNITION TIMING TO 10° BTDC.
(5) TURN OFF ENGINE AND RESTORE ELECTRICAL CONNECTION.

THROTTLE PLATE ADJUSTMENT -
(1) DISCONNECT AND PLUG VACUUM HOSES AT 'A'. ELECTRICALLY DISCONNECT IDLE SPEED CONTROL (ISC).
(2) RUN ENGINE AT IDLE UNTIL ENGINE COOLING FAN COMES ON.
(3) RUN ENGINE AT 2000 RPM FOR 60 SECONDS MINIMUM.
(4) RETURN TO IDLE AND TURN THROTTLE PLATE ADJUSTING SCREW UNTIL IDLE SPEED IS 750 RPM WITH AUTO. TRANS. IN DRIVE. (600 RPM FOR VEHICLE WITH LESS THAN 100 MILES). ADJUSTMENT MUST BE MADE WITHIN 120 SECONDS OF RETURN TO IDLE. TURN OFF ENGINE. RESTART AND REPEAT STEP 3 AND 4 IF 120 SECONDS ARE EXCEEDED.
(5) IF THROTTLE PLATE ADJUSTMENT EXCEEDS 50 RPM, RE-ADJUST TRANS. LINKAGE. SEE SHOP MANUAL.
(6) RESTORE VACUUM AND ELECTRICAL CONNECTIONS.

THIS VEHICLE CONFORMS TO U.S. EPA REGULATIONS APPLICABLE TO 1985 MODEL YEAR NEW MOTOR VEHICLES.

E5AE-9C485-CEG | CATALYST | SPARK PLUG: AWSF-24 GAP: .042-.046
1.6L-5HM
FFM1.6V5HMF5-EGS/EGR/AIV/TWC/FI

CALIBRATION: 4—28A—R00

MODEL YEAR: 1985 **ENGINE: 1.6L**

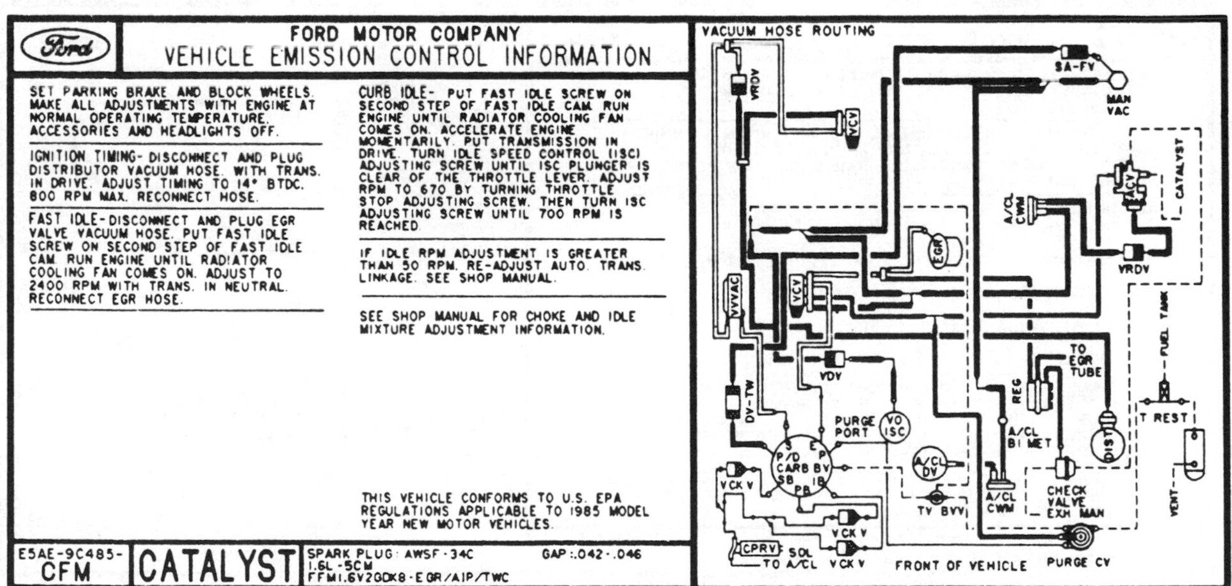

FORD MOTOR COMPANY
VEHICLE EMISSION CONTROL INFORMATION

SET PARKING BRAKE AND BLOCK WHEELS. MAKE ALL ADJUSTMENTS WITH ENGINE AT NORMAL OPERATING TEMPERATURE. ACCESSORIES AND HEADLIGHTS OFF.

IGNITION TIMING- DISCONNECT AND PLUG DISTRIBUTOR VACUUM HOSE. WITH TRANS. IN DRIVE. ADJUST TIMING TO 14° BTDC. 800 RPM MAX. RECONNECT HOSE.

FAST IDLE- DISCONNECT AND PLUG EGR VALVE VACUUM HOSE. PUT FAST IDLE SCREW ON SECOND STEP OF FAST IDLE CAM. RUN ENGINE UNTIL RADIATOR COOLING FAN COMES ON. ADJUST TO 2400 RPM WITH TRANS. IN NEUTRAL. RECONNECT EGR HOSE.

CURB IDLE- PUT FAST IDLE SCREW ON SECOND STEP OF FAST IDLE CAM. RUN ENGINE UNTIL RADIATOR COOLING FAN COMES ON. ACCELERATE ENGINE MOMENTARILY. PUT TRANSMISSION IN DRIVE. TURN IDLE SPEED CONTROL (ISC) ADJUSTING SCREW UNTIL ISC PLUNGER IS CLEAR OF THE THROTTLE LEVER. ADJUST RPM TO 670 BY TURNING THROTTLE STOP ADJUSTING SCREW. THEN TURN ISC ADJUSTING SCREW UNTIL 700 RPM IS REACHED.

IF IDLE RPM ADJUSTMENT IS GREATER THAN 50 RPM. RE-ADJUST AUTO. TRANS. LINKAGE. SEE SHOP MANUAL.

SEE SHOP MANUAL FOR CHOKE AND IDLE MIXTURE ADJUSTMENT INFORMATION.

THIS VEHICLE CONFORMS TO U.S. EPA REGULATIONS APPLICABLE TO 1985 MODEL YEAR NEW MOTOR VEHICLES.

E5AE-9C485-CFM | CATALYST | SPARK PLUG: AWSF-34C GAP: .042-.046
1.6L-5CM
FFM1.6V2GDK8-EGR/AIP/TWC

CALIBRATION: 5—04A—R12

VACUUM CIRCUITS

(© Ford Motor Co)

MODEL YEAR: 1985 ENGINE: 2.3L

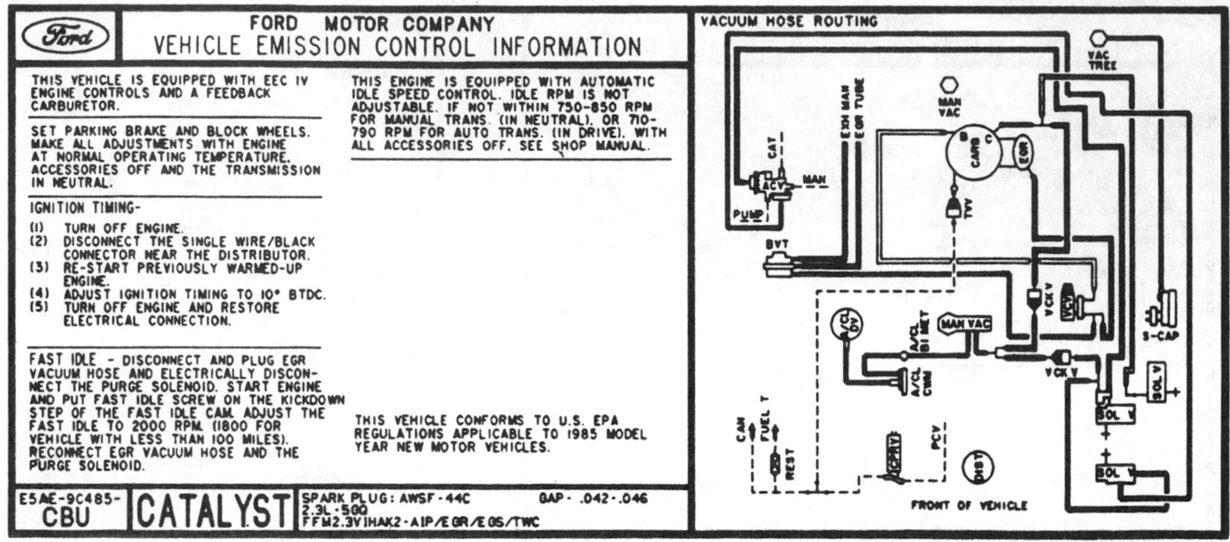

FORD MOTOR COMPANY
VEHICLE EMISSION CONTROL INFORMATION

THIS VEHICLE IS EQUIPPED WITH EEC IV ENGINE CONTROLS AND A FEEDBACK CARBURETOR.

SET PARKING BRAKE AND BLOCK WHEELS. MAKE ALL ADJUSTMENTS WITH ENGINE AT NORMAL OPERATING TEMPERATURE, ACCESSORIES OFF AND THE TRANSMISSION IN NEUTRAL.

IGNITION TIMING-

(1) TURN OFF ENGINE.
(2) DISCONNECT THE SINGLE WIRE/BLACK CONNECTOR NEAR THE DISTRIBUTOR.
(3) RE-START PREVIOUSLY WARMED-UP ENGINE.
(4) ADJUST IGNITION TIMING TO 10° BTDC.
(5) TURN OFF ENGINE AND RESTORE ELECTRICAL CONNECTION.

FAST IDLE - DISCONNECT AND PLUG EGR VACUUM HOSE AND ELECTRICALLY DISCONNECT THE PURGE SOLENOID. START ENGINE AND PUT FAST IDLE SCREW ON THE KICKDOWN STEP OF THE FAST IDLE CAM. ADJUST THE FAST IDLE TO 2000 RPM. (1800 FOR VEHICLE WITH LESS THAN 100 MILES). RECONNECT EGR VACUUM HOSE AND THE PURGE SOLENOID.

THIS ENGINE IS EQUIPPED WITH AUTOMATIC IDLE SPEED CONTROL. IDLE RPM IS NOT ADJUSTABLE. IF NOT WITHIN 750-850 RPM FOR MANUAL TRANS. (IN NEUTRAL), OR 710-790 RPM FOR AUTO TRANS. (IN DRIVE), WITH ALL ACCESSORIES OFF, SEE SHOP MANUAL.

THIS VEHICLE CONFORMS TO U.S. EPA REGULATIONS APPLICABLE TO 1985 MODEL YEAR NEW MOTOR VEHICLES.

E5AE-9C485-CBU CATALYST SPARK PLUG: AWSF-44C GAP-.042-.046
2.3L-5GQ
FFM2.3V1HAK2-AIP/EGR/EGS/TWC

CALIBRATION: 5—05A—R00

MODEL YEAR: 1985 ENGINE: 2.3L

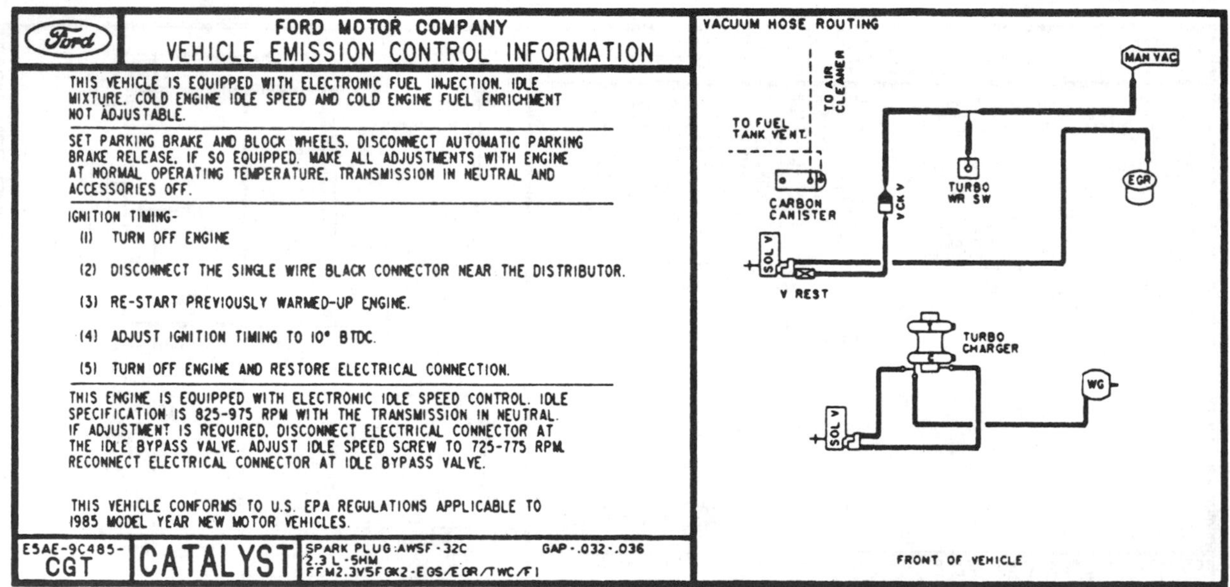

FORD MOTOR COMPANY
VEHICLE EMISSION CONTROL INFORMATION

THIS VEHICLE IS EQUIPPED WITH ELECTRONIC FUEL INJECTION. IDLE MIXTURE, COLD ENGINE IDLE SPEED AND COLD ENGINE FUEL ENRICHMENT NOT ADJUSTABLE.

SET PARKING BRAKE AND BLOCK WHEELS. DISCONNECT AUTOMATIC PARKING BRAKE RELEASE, IF SO EQUIPPED. MAKE ALL ADJUSTMENTS WITH ENGINE AT NORMAL OPERATING TEMPERATURE, TRANSMISSION IN NEUTRAL AND ACCESSORIES OFF.

IGNITION TIMING-

(1) TURN OFF ENGINE
(2) DISCONNECT THE SINGLE WIRE BLACK CONNECTOR NEAR THE DISTRIBUTOR.
(3) RE-START PREVIOUSLY WARMED-UP ENGINE.
(4) ADJUST IGNITION TIMING TO 10° BTDC.
(5) TURN OFF ENGINE AND RESTORE ELECTRICAL CONNECTION.

THIS ENGINE IS EQUIPPED WITH ELECTRONIC IDLE SPEED CONTROL. IDLE SPECIFICATION IS 825-975 RPM WITH THE TRANSMISSION IN NEUTRAL. IF ADJUSTMENT IS REQUIRED, DISCONNECT ELECTRICAL CONNECTOR AT THE IDLE BYPASS VALVE. ADJUST IDLE SPEED SCREW TO 725-775 RPM. RECONNECT ELECTRICAL CONNECTOR AT IDLE BYPASS VALVE.

THIS VEHICLE CONFORMS TO U.S. EPA REGULATIONS APPLICABLE TO 1985 MODEL YEAR NEW MOTOR VEHICLES.

E5AE-9C485-CGT CATALYST SPARK PLUG: AWSF-32C GAP-.032-.036
2.3L-5HM
FFM2.3V5FGK2-EGS/EGR/TWC/FI

CALIBRATION: 5—05D—R00

VACUUM CIRCUITS

(© Ford Motor Co)

MODEL YEAR: 1985 **ENGINE: 2.3L**

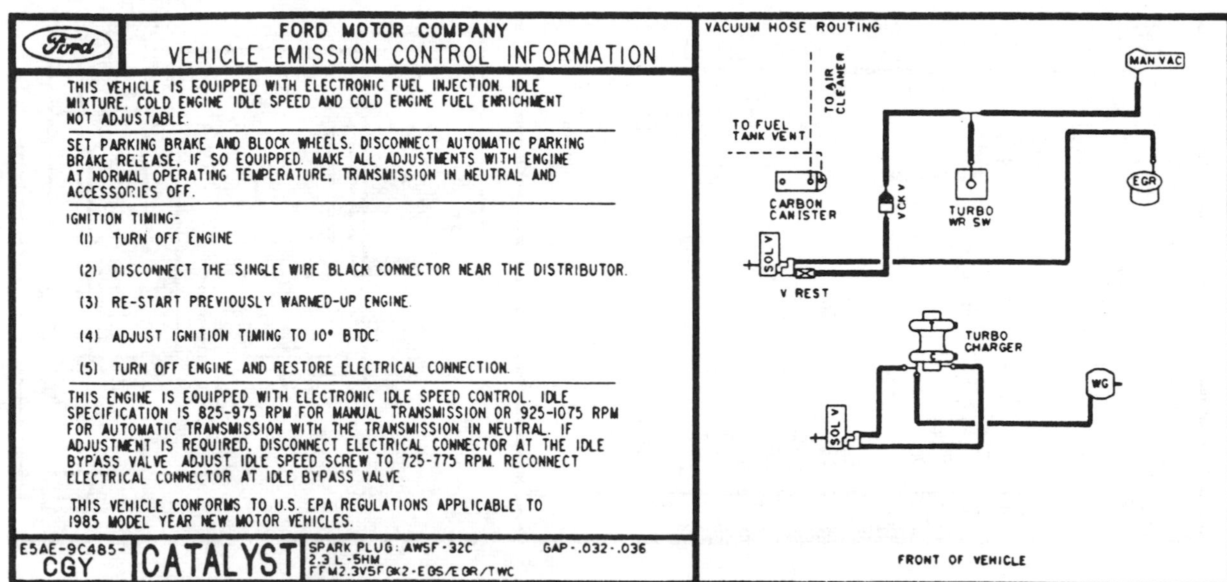

FORD MOTOR COMPANY
VEHICLE EMISSION CONTROL INFORMATION

THIS VEHICLE IS EQUIPPED WITH ELECTRONIC FUEL INJECTION. IDLE MIXTURE, COLD ENGINE IDLE SPEED AND COLD ENGINE FUEL ENRICHMENT NOT ADJUSTABLE.

SET PARKING BRAKE AND BLOCK WHEELS. DISCONNECT AUTOMATIC PARKING BRAKE RELEASE, IF SO EQUIPPED. MAKE ALL ADJUSTMENTS WITH ENGINE AT NORMAL OPERATING TEMPERATURE, TRANSMISSION IN NEUTRAL AND ACCESSORIES OFF.

IGNITION TIMING-
(1) TURN OFF ENGINE
(2) DISCONNECT THE SINGLE WIRE BLACK CONNECTOR NEAR THE DISTRIBUTOR.
(3) RE-START PREVIOUSLY WARMED-UP ENGINE.
(4) ADJUST IGNITION TIMING TO 10° BTDC.
(5) TURN OFF ENGINE AND RESTORE ELECTRICAL CONNECTION.

THIS ENGINE IS EQUIPPED WITH ELECTRONIC IDLE SPEED CONTROL. IDLE SPECIFICATION IS 825-975 RPM FOR MANUAL TRANSMISSION OR 925-1075 RPM FOR AUTOMATIC TRANSMISSION WITH THE TRANSMISSION IN NEUTRAL. IF ADJUSTMENT IS REQUIRED, DISCONNECT ELECTRICAL CONNECTOR AT THE IDLE BYPASS VALVE. ADJUST IDLE SPEED SCREW TO 725-775 RPM. RECONNECT ELECTRICAL CONNECTOR AT IDLE BYPASS VALVE.

THIS VEHICLE CONFORMS TO U.S. EPA REGULATIONS APPLICABLE TO 1985 MODEL YEAR NEW MOTOR VEHICLES.

E5AE-9C485-CGY CATALYST SPARK PLUG: AWSF-32C GAP-.032-.036
2.3 L-5HM
FFM2.3V5FGK2-EGS/EGR/TWC

CALIBRATION: 5—05E—R00

MODEL YEAR: 1985 **(Manual Transmission)** **ENGINE: 2.3L**

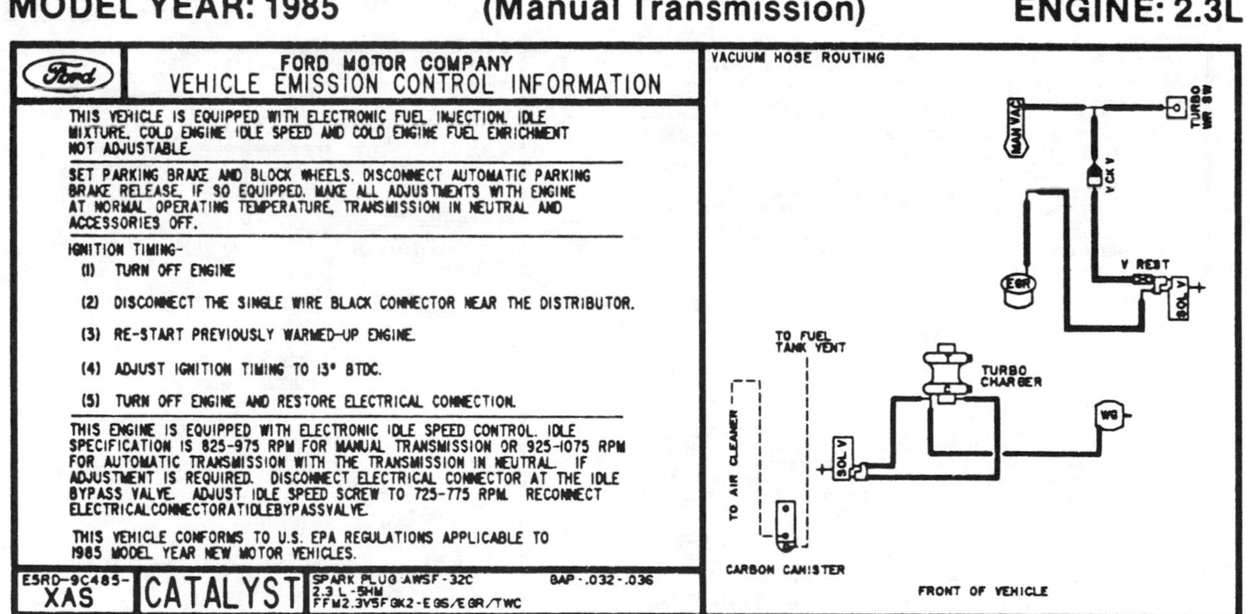

FORD MOTOR COMPANY
VEHICLE EMISSION CONTROL INFORMATION

THIS VEHICLE IS EQUIPPED WITH ELECTRONIC FUEL INJECTION. IDLE MIXTURE, COLD ENGINE IDLE SPEED AND COLD ENGINE FUEL ENRICHMENT NOT ADJUSTABLE.

SET PARKING BRAKE AND BLOCK WHEELS. DISCONNECT AUTOMATIC PARKING BRAKE RELEASE, IF SO EQUIPPED. MAKE ALL ADJUSTMENTS WITH ENGINE AT NORMAL OPERATING TEMPERATURE, TRANSMISSION IN NEUTRAL AND ACCESSORIES OFF.

IGNITION TIMING-
(1) TURN OFF ENGINE
(2) DISCONNECT THE SINGLE WIRE BLACK CONNECTOR NEAR THE DISTRIBUTOR.
(3) RE-START PREVIOUSLY WARMED-UP ENGINE.
(4) ADJUST IGNITION TIMING TO 13° BTDC.
(5) TURN OFF ENGINE AND RESTORE ELECTRICAL CONNECTION.

THIS ENGINE IS EQUIPPED WITH ELECTRONIC IDLE SPEED CONTROL. IDLE SPECIFICATION IS 825-975 RPM FOR MANUAL TRANSMISSION OR 925-1075 RPM FOR AUTOMATIC TRANSMISSION WITH THE TRANSMISSION IN NEUTRAL. IF ADJUSTMENT IS REQUIRED. DISCONNECT ELECTRICAL CONNECTOR AT THE IDLE BYPASS VALVE. ADJUST IDLE SPEED SCREW TO 725-775 RPM. RECONNECT ELECTRICALCONNECTORATIDLEBYPASSVALVE.

THIS VEHICLE CONFORMS TO U.S. EPA REGULATIONS APPLICABLE TO 1985 MODEL YEAR NEW MOTOR VEHICLES.

E5RD-9C485-XAS CATALYST SPARK PLUG AWSF-32C GAP-.032-.036
2.3 L-5HM
FFM2.3V5FGK2-EGS/EGR/TWC

CALIBRATION: 5—05R—R10

VACUUM CIRCUITS

(© Ford Motor Co)

MODEL YEAR: 1985 **ENGINE: 2.3L**

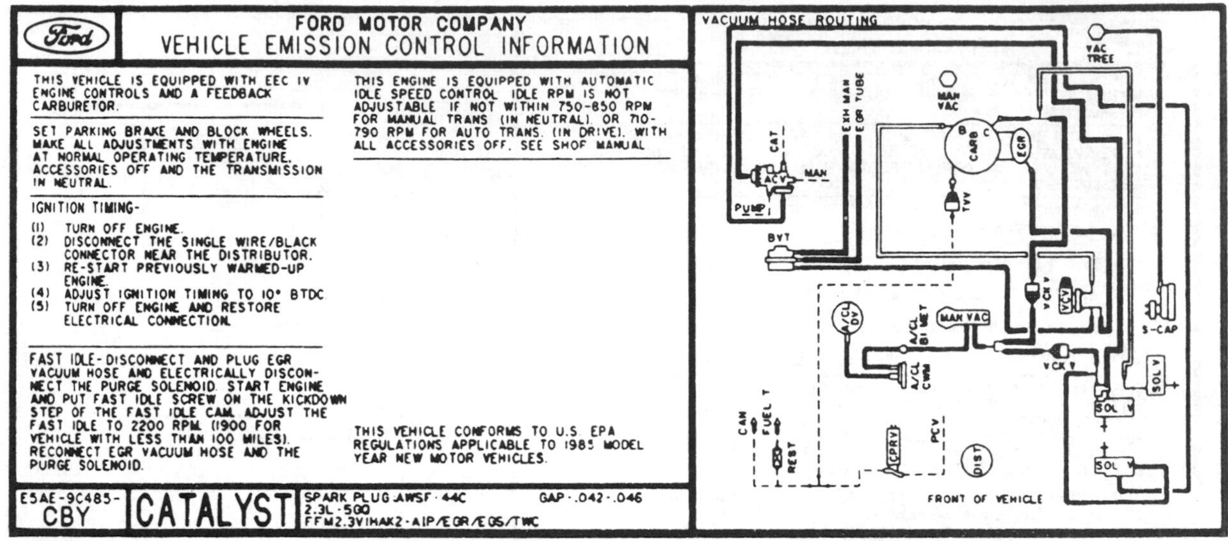

FORD MOTOR COMPANY
VEHICLE EMISSION CONTROL INFORMATION

THIS VEHICLE IS EQUIPPED WITH EEC IV ENGINE CONTROLS AND A FEEDBACK CARBURETOR.

SET PARKING BRAKE AND BLOCK WHEELS. MAKE ALL ADJUSTMENTS WITH ENGINE AT NORMAL OPERATING TEMPERATURE, ACCESSORIES OFF AND THE TRANSMISSION IN NEUTRAL.

IGNITION TIMING-

(1) TURN OFF ENGINE.
(2) DISCONNECT THE SINGLE WIRE/BLACK CONNECTOR NEAR THE DISTRIBUTOR.
(3) RE-START PREVIOUSLY WARMED-UP ENGINE.
(4) ADJUST IGNITION TIMING TO 10° BTDC
(5) TURN OFF ENGINE AND RESTORE ELECTRICAL CONNECTION.

FAST IDLE-DISCONNECT AND PLUG EGR VACUUM HOSE AND ELECTRICALLY DISCONNECT THE PURGE SOLENOID. START ENGINE AND PUT FAST IDLE SCREW ON THE KICKDOWN STEP OF THE FAST IDLE CAM. ADJUST THE FAST IDLE TO 2200 RPM. (1900 FOR VEHICLE WITH LESS THAN 100 MILES). RECONNECT EGR VACUUM HOSE AND THE PURGE SOLENOID.

THIS ENGINE IS EQUIPPED WITH AUTOMATIC IDLE SPEED CONTROL. IDLE RPM IS NOT ADJUSTABLE. IF NOT WITHIN 750-850 RPM FOR MANUAL TRANS (IN NEUTRAL), OR 710-790 RPM FOR AUTO TRANS. (IN DRIVE), WITH ALL ACCESSORIES OFF, SEE SHOP MANUAL.

THIS VEHICLE CONFORMS TO U.S. EPA REGULATIONS APPLICABLE TO 1985 MODEL YEAR NEW MOTOR VEHICLES.

E5AE-9C485-CBY | CATALYST | SPARK PLUG AWSF-44C GAP-.042-.046 2.3L-5G0 FFM2.3V1HAK2-AIP/EGR/EGS/TWC

CALIBRATION: 5—06A—R00

MODEL YEAR: 1985 **ENGINE: 2.3L**

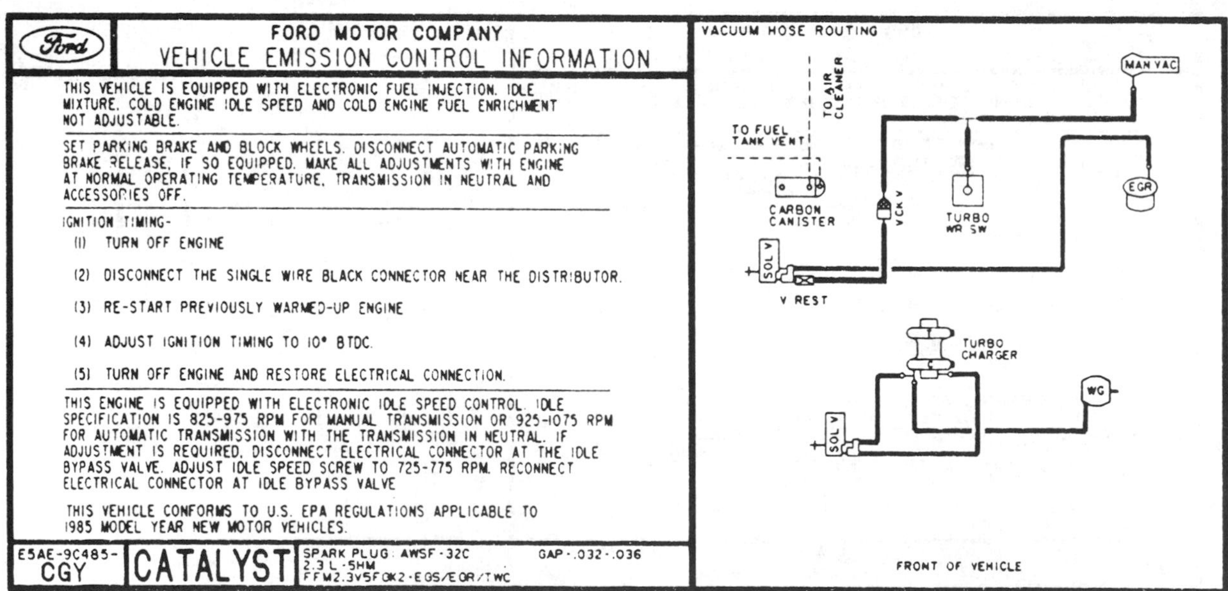

FORD MOTOR COMPANY
VEHICLE EMISSION CONTROL INFORMATION

THIS VEHICLE IS EQUIPPED WITH ELECTRONIC FUEL INJECTION. IDLE MIXTURE, COLD ENGINE IDLE SPEED AND COLD ENGINE FUEL ENRICHMENT NOT ADJUSTABLE.

SET PARKING BRAKE AND BLOCK WHEELS. DISCONNECT AUTOMATIC PARKING BRAKE RELEASE, IF SO EQUIPPED. MAKE ALL ADJUSTMENTS WITH ENGINE AT NORMAL OPERATING TEMPERATURE, TRANSMISSION IN NEUTRAL AND ACCESSORIES OFF.

IGNITION TIMING-

(1) TURN OFF ENGINE
(2) DISCONNECT THE SINGLE WIRE BLACK CONNECTOR NEAR THE DISTRIBUTOR.
(3) RE-START PREVIOUSLY WARMED-UP ENGINE
(4) ADJUST IGNITION TIMING TO 10° BTDC.
(5) TURN OFF ENGINE AND RESTORE ELECTRICAL CONNECTION.

THIS ENGINE IS EQUIPPED WITH ELECTRONIC IDLE SPEED CONTROL. IDLE SPECIFICATION IS 825-975 RPM FOR MANUAL TRANSMISSION OR 925-1075 RPM FOR AUTOMATIC TRANSMISSION WITH THE TRANSMISSION IN NEUTRAL. IF ADJUSTMENT IS REQUIRED, DISCONNECT ELECTRICAL CONNECTOR AT THE IDLE BYPASS VALVE. ADJUST IDLE SPEED SCREW TO 725-775 RPM. RECONNECT ELECTRICAL CONNECTOR AT IDLE BYPASS VALVE

THIS VEHICLE CONFORMS TO U.S. EPA REGULATIONS APPLICABLE TO 1985 MODEL YEAR NEW MOTOR VEHICLES.

E5AE-9C485-CGY | CATALYST | SPARK PLUG AWSF-32C GAP-.032-.036 2.3L-5HM FFM2.3V5FOK2-EGS/EGR/TWC

CALIBRATION: 5—06E—R00

VACUUM CIRCUITS

(© Ford Motor Co.)

MODEL YEAR: 1985 ENGINE: 2.3L

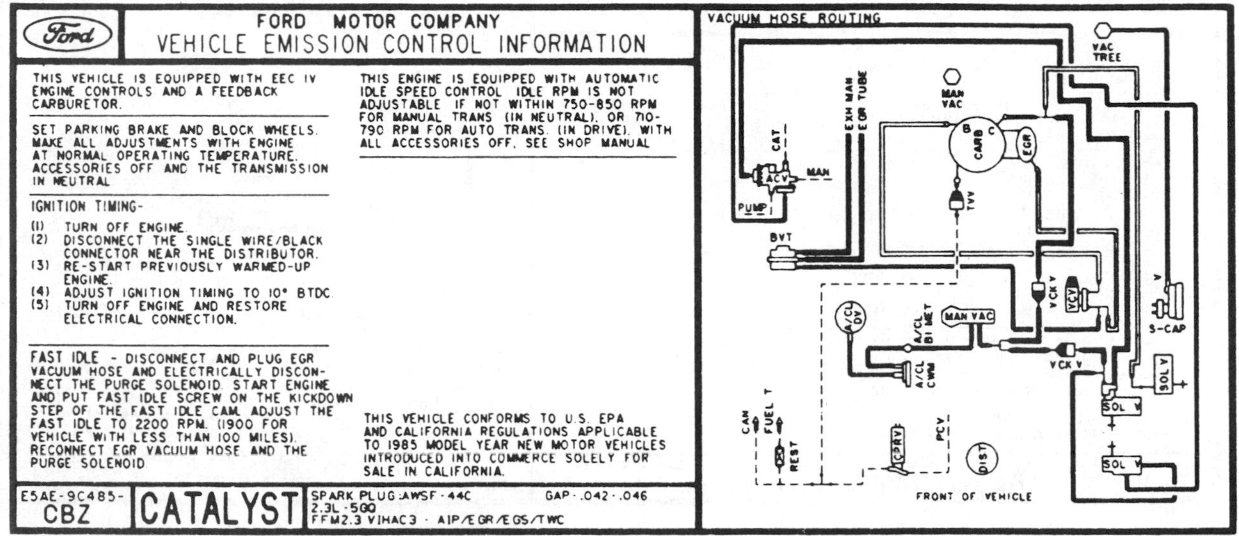

FORD MOTOR COMPANY
VEHICLE EMISSION CONTROL INFORMATION

THIS VEHICLE IS EQUIPPED WITH EEC IV ENGINE CONTROLS AND A FEEDBACK CARBURETOR.

SET PARKING BRAKE AND BLOCK WHEELS. MAKE ALL ADJUSTMENTS WITH ENGINE AT NORMAL OPERATING TEMPERATURE. ACCESSORIES OFF AND THE TRANSMISSION IN NEUTRAL

IGNITION TIMING-
(1) TURN OFF ENGINE.
(2) DISCONNECT THE SINGLE WIRE/BLACK CONNECTOR NEAR THE DISTRIBUTOR.
(3) RE-START PREVIOUSLY WARMED-UP ENGINE.
(4) ADJUST IGNITION TIMING TO 10° BTDC.
(5) TURN OFF ENGINE AND RESTORE ELECTRICAL CONNECTION.

FAST IDLE - DISCONNECT AND PLUG EGR VACUUM HOSE AND ELECTRICALLY DISCONNECT THE PURGE SOLENOID. START ENGINE AND PUT FAST IDLE SCREW ON THE KICKDOWN STEP OF THE FAST IDLE CAM. ADJUST THE FAST IDLE TO 2200 RPM. (1900 FOR VEHICLE WITH LESS THAN 100 MILES). RECONNECT EGR VACUUM HOSE AND THE PURGE SOLENOID.

THIS ENGINE IS EQUIPPED WITH AUTOMATIC IDLE SPEED CONTROL. IDLE RPM IS NOT ADJUSTABLE. IF NOT WITHIN 750-850 RPM FOR MANUAL TRANS (IN NEUTRAL), OR 710-790 RPM FOR AUTO TRANS. (IN DRIVE), WITH ALL ACCESSORIES OFF, SEE SHOP MANUAL

THIS VEHICLE CONFORMS TO U.S. EPA AND CALIFORNIA REGULATIONS APPLICABLE TO 1985 MODEL YEAR NEW MOTOR VEHICLES INTRODUCED INTO COMMERCE SOLELY FOR SALE IN CALIFORNIA.

E5AE-9C485-CBZ CATALYST SPARK PLUG:AWSF-44C GAP-.042-.046
2.3L-5GO
FFM2.3 V1HAC3 - AIP/EGR/EGS/TWC

CALIBRATION: 5—06N—R00

MODEL YEAR: 1985 ENGINE: 2.3L

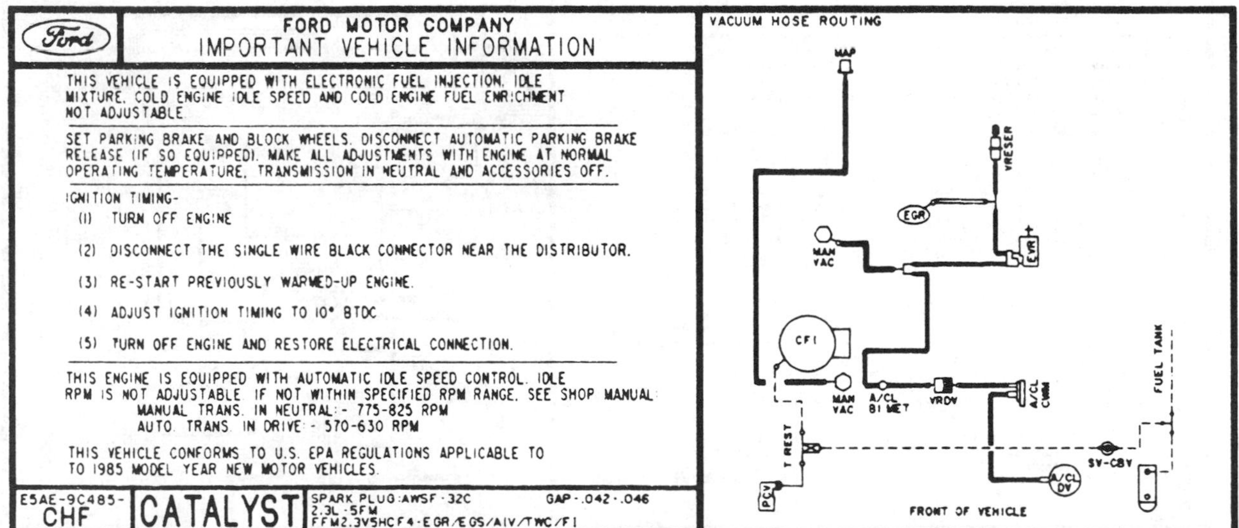

FORD MOTOR COMPANY
IMPORTANT VEHICLE INFORMATION

THIS VEHICLE IS EQUIPPED WITH ELECTRONIC FUEL INJECTION. IDLE MIXTURE, COLD ENGINE IDLE SPEED AND COLD ENGINE FUEL ENRICHMENT NOT ADJUSTABLE.

SET PARKING BRAKE AND BLOCK WHEELS. DISCONNECT AUTOMATIC PARKING BRAKE RELEASE (IF SO EQUIPPED). MAKE ALL ADJUSTMENTS WITH ENGINE AT NORMAL OPERATING TEMPERATURE, TRANSMISSION IN NEUTRAL AND ACCESSORIES OFF.

IGNITION TIMING-
(1) TURN OFF ENGINE
(2) DISCONNECT THE SINGLE WIRE BLACK CONNECTOR NEAR THE DISTRIBUTOR.
(3) RE-START PREVIOUSLY WARMED-UP ENGINE.
(4) ADJUST IGNITION TIMING TO 10° BTDC
(5) TURN OFF ENGINE AND RESTORE ELECTRICAL CONNECTION.

THIS ENGINE IS EQUIPPED WITH AUTOMATIC IDLE SPEED CONTROL. IDLE RPM IS NOT ADJUSTABLE. IF NOT WITHIN SPECIFIED RPM RANGE, SEE SHOP MANUAL.
MANUAL TRANS. IN NEUTRAL: - 775-825 RPM
AUTO. TRANS. IN DRIVE: - 570-630 RPM

THIS VEHICLE CONFORMS TO U.S. EPA REGULATIONS APPLICABLE TO TO 1985 MODEL YEAR NEW MOTOR VEHICLES.

E5AE-9C485-CHF CATALYST SPARK PLUG:AWSF-32C GAP-.042-.046
2.3L-5FM
FFM2.3V5HCF4-EGR/EGS/AIV/TWC/FI

CALIBRATION: 5—25C—R01

VACUUM CIRCUITS

(© Ford Motor Co.)

MODEL YEAR: 1985 ENGINE: 2.3L

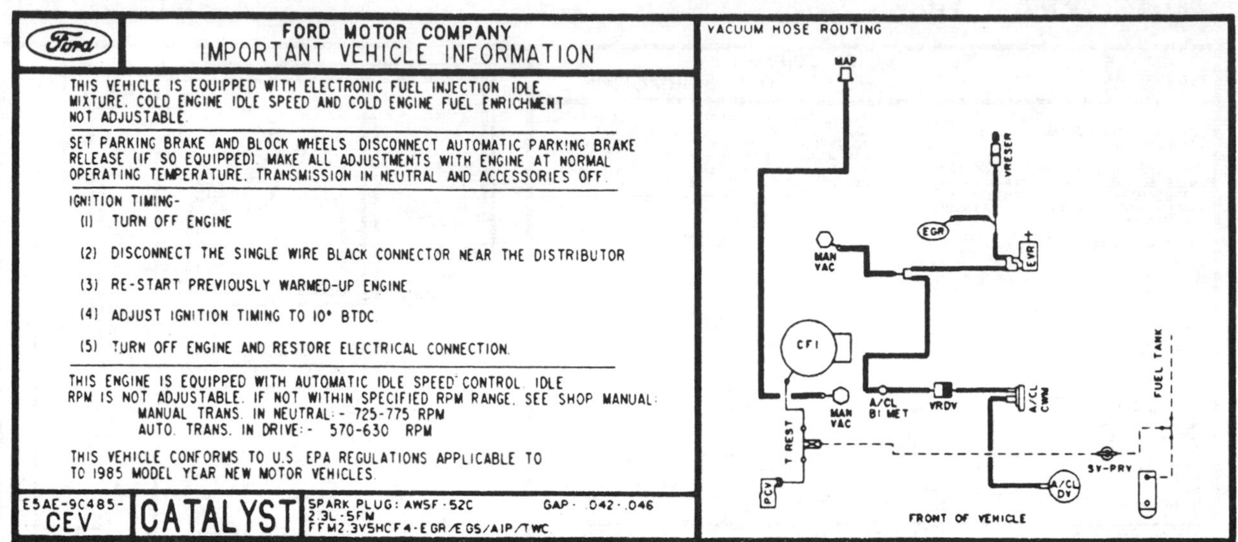

FORD MOTOR COMPANY
IMPORTANT VEHICLE INFORMATION

THIS VEHICLE IS EQUIPPED WITH ELECTRONIC FUEL INJECTION. IDLE MIXTURE, COLD ENGINE IDLE SPEED AND COLD ENGINE FUEL ENRICHMENT NOT ADJUSTABLE.

SET PARKING BRAKE AND BLOCK WHEELS. DISCONNECT AUTOMATIC PARKING BRAKE RELEASE (IF SO EQUIPPED). MAKE ALL ADJUSTMENTS WITH ENGINE AT NORMAL OPERATING TEMPERATURE, TRANSMISSION IN NEUTRAL AND ACCESSORIES OFF.

IGNITION TIMING-
(1) TURN OFF ENGINE
(2) DISCONNECT THE SINGLE WIRE BLACK CONNECTOR NEAR THE DISTRIBUTOR
(3) RE-START PREVIOUSLY WARMED-UP ENGINE
(4) ADJUST IGNITION TIMING TO 10° BTDC
(5) TURN OFF ENGINE AND RESTORE ELECTRICAL CONNECTION.

THIS ENGINE IS EQUIPPED WITH AUTOMATIC IDLE SPEED CONTROL. IDLE RPM IS NOT ADJUSTABLE. IF NOT WITHIN SPECIFIED RPM RANGE, SEE SHOP MANUAL:
MANUAL TRANS. IN NEUTRAL: - 725-775 RPM
AUTO. TRANS. IN DRIVE: - 570-630 RPM

THIS VEHICLE CONFORMS TO U.S. EPA REGULATIONS APPLICABLE TO 1985 MODEL YEAR NEW MOTOR VEHICLES.

E5AE-9C485- CEV | CATALYST | SPARK PLUG: AWSF-52C GAP: .042-.046 2.3L-5FM FFM2.3V5HCF4-EGR/EGS/AIP/TWC

CALIBRATION: 5—25F—R00

MODEL YEAR: 1985 ENGINE: 2.3L

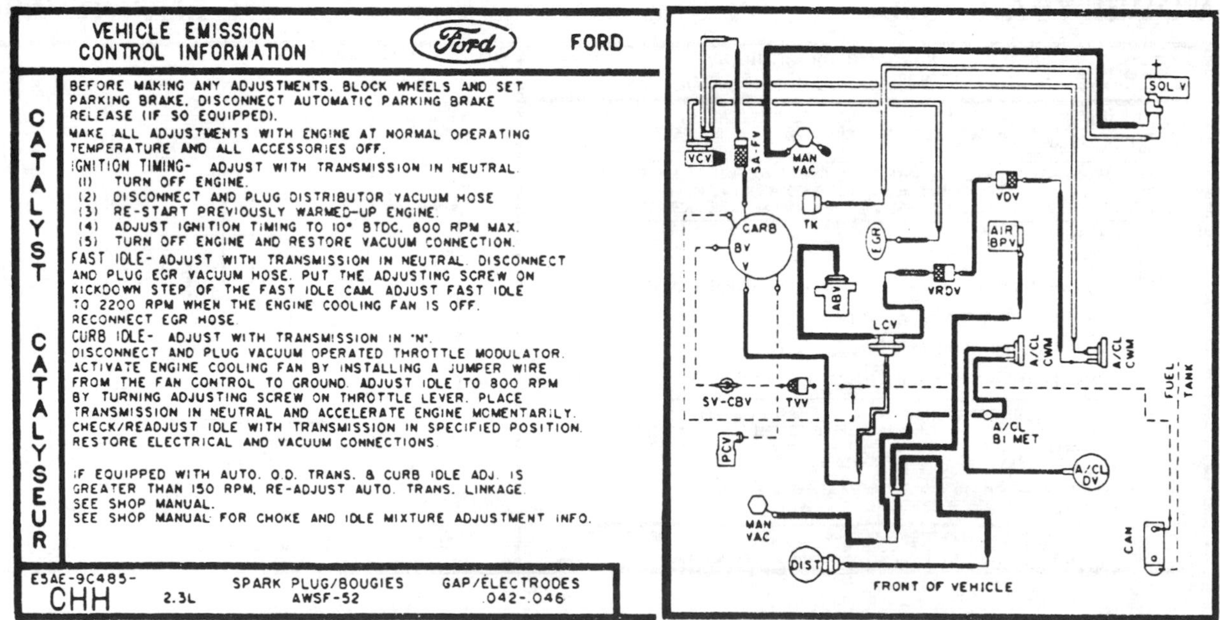

VEHICLE EMISSION CONTROL INFORMATION FORD

CATALYST CATALYSEUR

BEFORE MAKING ANY ADJUSTMENTS, BLOCK WHEELS AND SET PARKING BRAKE. DISCONNECT AUTOMATIC PARKING BRAKE RELEASE (IF SO EQUIPPED).
MAKE ALL ADJUSTMENTS WITH ENGINE AT NORMAL OPERATING TEMPERATURE AND ALL ACCESSORIES OFF.
IGNITION TIMING- ADJUST WITH TRANSMISSION IN NEUTRAL.
(1) TURN OFF ENGINE.
(2) DISCONNECT AND PLUG DISTRIBUTOR VACUUM HOSE.
(3) RE-START PREVIOUSLY WARMED-UP ENGINE.
(4) ADJUST IGNITION TIMING TO 10° BTDC. 800 RPM MAX.
(5) TURN OFF ENGINE AND RESTORE VACUUM CONNECTION.
FAST IDLE- ADJUST WITH TRANSMISSION IN NEUTRAL. DISCONNECT AND PLUG EGR VACUUM HOSE. PUT THE ADJUSTING SCREW ON KICKDOWN STEP OF THE FAST IDLE CAM. ADJUST FAST IDLE TO 2200 RPM WHEN THE ENGINE COOLING FAN IS OFF. RECONNECT EGR HOSE.
CURB IDLE- ADJUST WITH TRANSMISSION IN "N". DISCONNECT AND PLUG VACUUM OPERATED THROTTLE MODULATOR. ACTIVATE ENGINE COOLING FAN BY INSTALLING A JUMPER WIRE FROM THE FAN CONTROL TO GROUND. ADJUST IDLE TO 800 RPM BY TURNING ADJUSTING SCREW ON THROTTLE LEVER. PLACE TRANSMISSION IN NEUTRAL AND ACCELERATE ENGINE MOMENTARILY. CHECK/READJUST IDLE WITH TRANSMISSION IN SPECIFIED POSITION. RESTORE ELECTRICAL AND VACUUM CONNECTIONS.

IF EQUIPPED WITH AUTO. O.D. TRANS. & CURB IDLE ADJ. IS GREATER THAN 150 RPM, RE-ADJUST AUTO. TRANS. LINKAGE. SEE SHOP MANUAL.
SEE SHOP MANUAL FOR CHOKE AND IDLE MIXTURE ADJUSTMENT INFO.

E5AE-9C485- CHH | 2.3L | SPARK PLUG/BOUGIES AWSF-52 | GAP/ÉLECTRODES .042-.046

CALIBRATION: 5—25G—R00

VACUUM CIRCUITS

(© Ford Motor Co)

MODEL YEAR: 1985

ENGINE: 2.3L

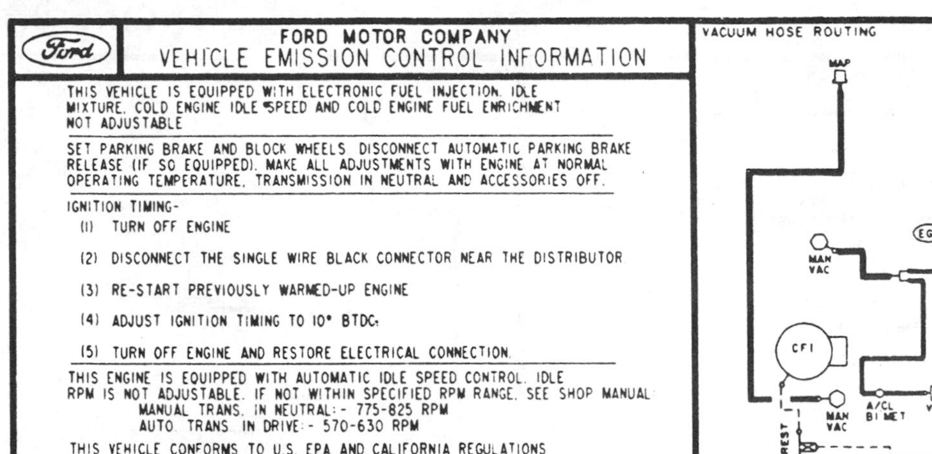

FORD MOTOR COMPANY
VEHICLE EMISSION CONTROL INFORMATION

THIS VEHICLE IS EQUIPPED WITH ELECTRONIC FUEL INJECTION. IDLE MIXTURE, COLD ENGINE IDLE SPEED AND COLD ENGINE FUEL ENRICHMENT NOT ADJUSTABLE

SET PARKING BRAKE AND BLOCK WHEELS. DISCONNECT AUTOMATIC PARKING BRAKE RELEASE (IF SO EQUIPPED). MAKE ALL ADJUSTMENTS WITH ENGINE AT NORMAL OPERATING TEMPERATURE, TRANSMISSION IN NEUTRAL AND ACCESSORIES OFF.

IGNITION TIMING-

(1) TURN OFF ENGINE

(2) DISCONNECT THE SINGLE WIRE BLACK CONNECTOR NEAR THE DISTRIBUTOR

(3) RE-START PREVIOUSLY WARMED-UP ENGINE

(4) ADJUST IGNITION TIMING TO 10° BTDC.

(5) TURN OFF ENGINE AND RESTORE ELECTRICAL CONNECTION.

THIS ENGINE IS EQUIPPED WITH AUTOMATIC IDLE SPEED CONTROL. IDLE RPM IS NOT ADJUSTABLE. IF NOT WITHIN SPECIFIED RPM RANGE, SEE SHOP MANUAL:
MANUAL TRANS. IN NEUTRAL:- 775-825 RPM
AUTO. TRANS. IN DRIVE:- 570-630 RPM

THIS VEHICLE CONFORMS TO U.S. EPA AND CALIFORNIA REGULATIONS APPLICABLE TO 1985 MODEL YEAR NEW MOTOR VEHICLES INTRODUCED INTO COMMERCE SOLELY FOR SALE IN CALIFORNIA.

E5AE-9C485-CHE | CATALYST | SPARK PLUG:AWSF-32C GAP-.042-.046 2.3L-5FM FFM2.3V5HCH6-EGR/EGS/AIV/TWC/FI

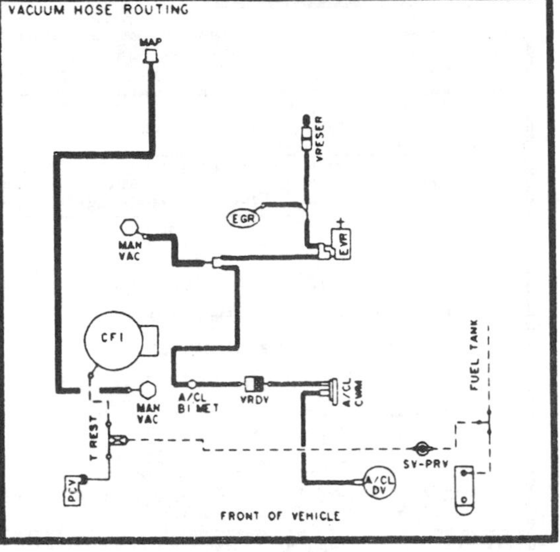

CALIBRATION: 5—25P—R00

MODEL YEAR: 1985

ENGINE: 2.3L

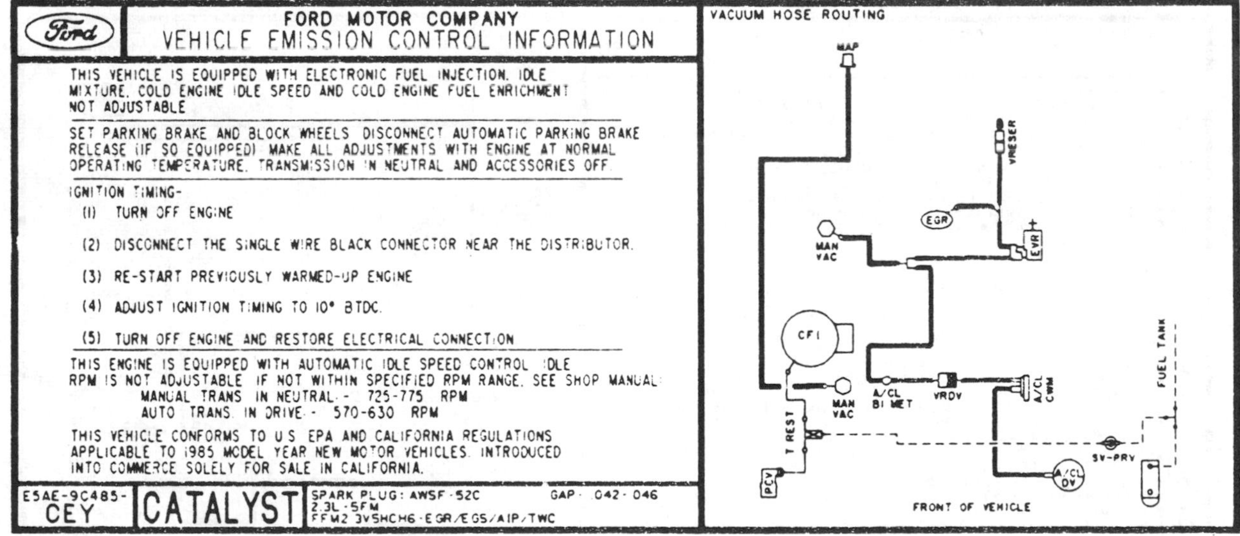

FORD MOTOR COMPANY
VEHICLE EMISSION CONTROL INFORMATION

THIS VEHICLE IS EQUIPPED WITH ELECTRONIC FUEL INJECTION. IDLE MIXTURE, COLD ENGINE IDLE SPEED AND COLD ENGINE FUEL ENRICHMENT NOT ADJUSTABLE

SET PARKING BRAKE AND BLOCK WHEELS. DISCONNECT AUTOMATIC PARKING BRAKE RELEASE (IF SO EQUIPPED). MAKE ALL ADJUSTMENTS WITH ENGINE AT NORMAL OPERATING TEMPERATURE, TRANSMISSION IN NEUTRAL AND ACCESSORIES OFF.

IGNITION TIMING-

(1) TURN OFF ENGINE

(2) DISCONNECT THE SINGLE WIRE BLACK CONNECTOR NEAR THE DISTRIBUTOR.

(3) RE-START PREVIOUSLY WARMED-UP ENGINE

(4) ADJUST IGNITION TIMING TO 10° BTDC.

(5) TURN OFF ENGINE AND RESTORE ELECTRICAL CONNECTION

THIS ENGINE IS EQUIPPED WITH AUTOMATIC IDLE SPEED CONTROL. IDLE RPM IS NOT ADJUSTABLE. IF NOT WITHIN SPECIFIED RPM RANGE, SEE SHOP MANUAL:
MANUAL TRANS. IN NEUTRAL:- 725-775 RPM
AUTO. TRANS. IN DRIVE:- 570-630 RPM

THIS VEHICLE CONFORMS TO U.S. EPA AND CALIFORNIA REGULATIONS APPLICABLE TO 1985 MODEL YEAR NEW MOTOR VEHICLES INTRODUCED INTO COMMERCE SOLELY FOR SALE IN CALIFORNIA.

E5AE-9C485-CEY | CATALYST | SPARK PLUG: AWSF-52C GAP-.042-.046 2.3L-5FM FFM2.3V5HCH6-EGR/EGS/AIP/TWC

CALIBRATION: 5—25Q—R00

VACUUM CIRCUITS

MODEL YEAR: 1985 **ENGINE: 2.3L**

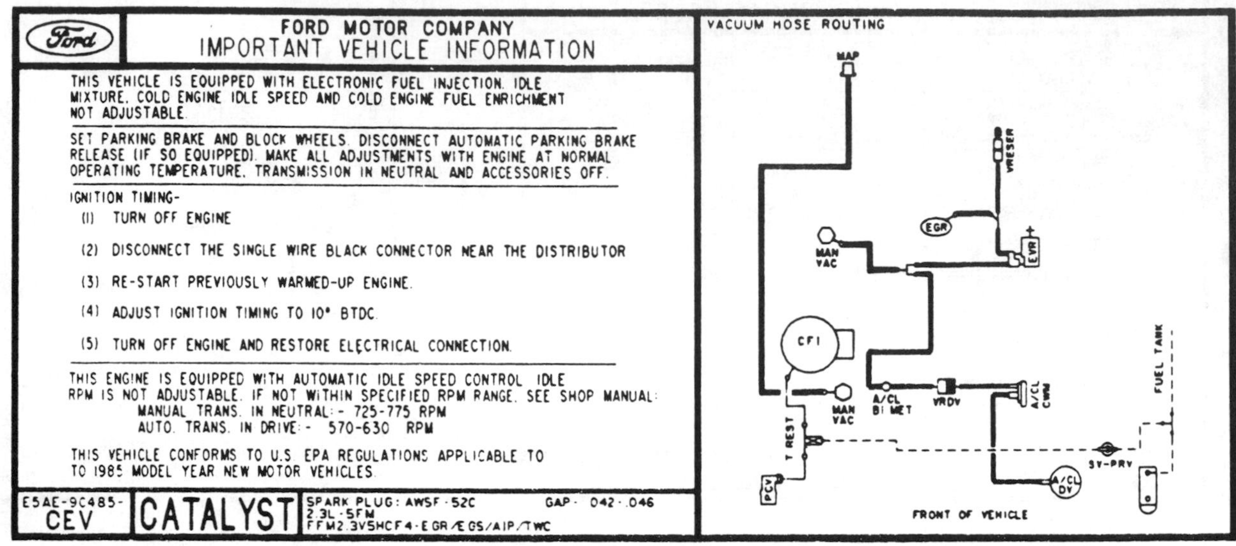

Ford

FORD MOTOR COMPANY
IMPORTANT VEHICLE INFORMATION

THIS VEHICLE IS EQUIPPED WITH ELECTRONIC FUEL INJECTION. IDLE MIXTURE, COLD ENGINE IDLE SPEED AND COLD ENGINE FUEL ENRICHMENT NOT ADJUSTABLE.

SET PARKING BRAKE AND BLOCK WHEELS. DISCONNECT AUTOMATIC PARKING BRAKE RELEASE (IF SO EQUIPPED). MAKE ALL ADJUSTMENTS WITH ENGINE AT NORMAL OPERATING TEMPERATURE, TRANSMISSION IN NEUTRAL AND ACCESSORIES OFF.

IGNITION TIMING-

(1) TURN OFF ENGINE

(2) DISCONNECT THE SINGLE WIRE BLACK CONNECTOR NEAR THE DISTRIBUTOR

(3) RE-START PREVIOUSLY WARMED-UP ENGINE.

(4) ADJUST IGNITION TIMING TO 10° BTDC.

(5) TURN OFF ENGINE AND RESTORE ELECTRICAL CONNECTION.

THIS ENGINE IS EQUIPPED WITH AUTOMATIC IDLE SPEED CONTROL. IDLE RPM IS NOT ADJUSTABLE. IF NOT WITHIN SPECIFIED RPM RANGE, SEE SHOP MANUAL:
MANUAL TRANS. IN NEUTRAL: - 725-775 RPM
AUTO. TRANS. IN DRIVE: - 570-630 RPM

THIS VEHICLE CONFORMS TO U.S. EPA REGULATIONS APPLICABLE TO 1985 MODEL YEAR NEW MOTOR VEHICLES.

E5AE-9C485-
CEV

CATALYST

SPARK PLUG: AWSF-52C GAP- .042-.046
2.3L-5FM
FFM2.3V5HCF4-EGR/EGS/AIP/TWC

CALIBRATION: 5—26E—R00

MODEL YEAR: 1985 **ENGINE: 2.3L**

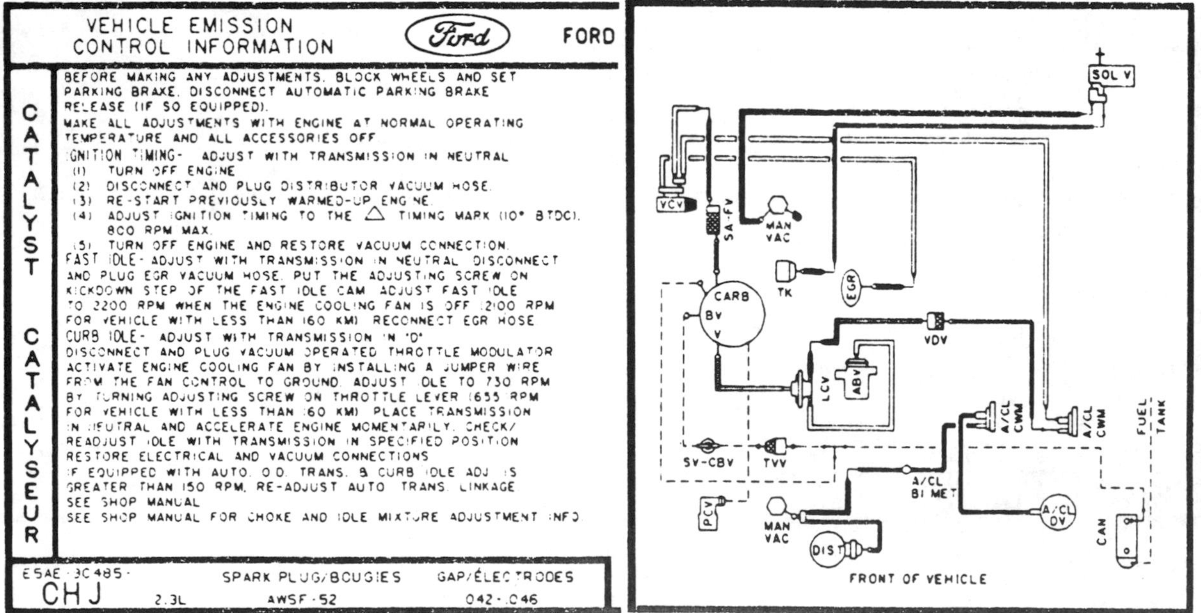

VEHICLE EMISSION
CONTROL INFORMATION
Ford
FORD

CATALYST CATALYSEUR

BEFORE MAKING ANY ADJUSTMENTS, BLOCK WHEELS AND SET PARKING BRAKE. DISCONNECT AUTOMATIC PARKING BRAKE RELEASE (IF SO EQUIPPED).
MAKE ALL ADJUSTMENTS WITH ENGINE AT NORMAL OPERATING TEMPERATURE AND ALL ACCESSORIES OFF.
IGNITION TIMING- ADJUST WITH TRANSMISSION IN NEUTRAL
(1) TURN OFF ENGINE
(2) DISCONNECT AND PLUG DISTRIBUTOR VACUUM HOSE.
(3) RE-START PREVIOUSLY WARMED-UP ENGINE.
(4) ADJUST IGNITION TIMING TO THE △ TIMING MARK (10° BTDC). 800 RPM MAX.
(5) TURN OFF ENGINE AND RESTORE VACUUM CONNECTION.
FAST IDLE- ADJUST WITH TRANSMISSION IN NEUTRAL. DISCONNECT AND PLUG EGR VACUUM HOSE. PUT THE ADJUSTING SCREW ON KICKDOWN STEP OF THE FAST IDLE CAM. ADJUST FAST IDLE TO 2200 RPM WHEN THE ENGINE COOLING FAN IS OFF (2100 RPM FOR VEHICLE WITH LESS THAN 160 KM). RECONNECT EGR HOSE.
CURB IDLE- ADJUST WITH TRANSMISSION IN "D"
DISCONNECT AND PLUG VACUUM OPERATED THROTTLE MODULATOR. ACTIVATE ENGINE COOLING FAN BY INSTALLING A JUMPER WIRE FROM THE FAN CONTROL TO GROUND. ADJUST IDLE TO 730 RPM BY TURNING ADJUSTING SCREW ON THROTTLE LEVER (655 RPM FOR VEHICLE WITH LESS THAN 160 KM). PLACE TRANSMISSION IN NEUTRAL AND ACCELERATE ENGINE MOMENTARILY. CHECK/READJUST IDLE WITH TRANSMISSION IN SPECIFIED POSITION. RESTORE ELECTRICAL AND VACUUM CONNECTIONS.
IF EQUIPPED WITH AUTO. O.D. TRANS. & CURB IDLE ADJ. IS GREATER THAN 150 RPM, RE-ADJUST AUTO. TRANS. LINKAGE. SEE SHOP MANUAL.
SEE SHOP MANUAL FOR CHOKE AND IDLE MIXTURE ADJUSTMENT INFO.

E5AE-9C485-
CHJ 2.3L

SPARK PLUG/BOUGIES AWSF-52
GAP/ÉLECTRODES .042-.046

CALIBRATION: 5—26G—R00

VACUUM CIRCUITS

(© Ford Motor Co)

MODEL YEAR: 1985 — ENGINE: 2.3L

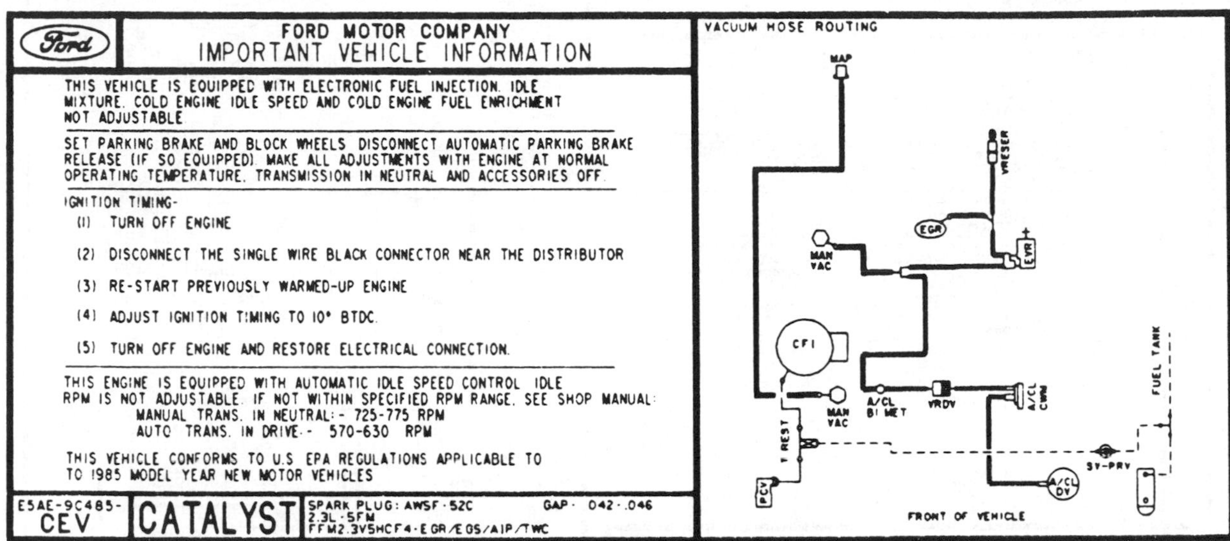

FORD MOTOR COMPANY
IMPORTANT VEHICLE INFORMATION

THIS VEHICLE IS EQUIPPED WITH ELECTRONIC FUEL INJECTION. IDLE MIXTURE. COLD ENGINE IDLE SPEED AND COLD ENGINE FUEL ENRICHMENT NOT ADJUSTABLE

SET PARKING BRAKE AND BLOCK WHEELS DISCONNECT AUTOMATIC PARKING BRAKE RELEASE (IF SO EQUIPPED) MAKE ALL ADJUSTMENTS WITH ENGINE AT NORMAL OPERATING TEMPERATURE. TRANSMISSION IN NEUTRAL AND ACCESSORIES OFF.

IGNITION TIMING-

(1) TURN OFF ENGINE

(2) DISCONNECT THE SINGLE WIRE BLACK CONNECTOR NEAR THE DISTRIBUTOR

(3) RE-START PREVIOUSLY WARMED-UP ENGINE

(4) ADJUST IGNITION TIMING TO 10° BTDC.

(5) TURN OFF ENGINE AND RESTORE ELECTRICAL CONNECTION.

THIS ENGINE IS EQUIPPED WITH AUTOMATIC IDLE SPEED CONTROL IDLE RPM IS NOT ADJUSTABLE. IF NOT WITHIN SPECIFIED RPM RANGE. SEE SHOP MANUAL:
MANUAL TRANS. IN NEUTRAL: - 725-775 RPM
AUTO TRANS. IN DRIVE - 570-630 RPM

THIS VEHICLE CONFORMS TO U.S EPA REGULATIONS APPLICABLE TO TO 1985 MODEL YEAR NEW MOTOR VEHICLES

E5AE-9C485-CEV | CATALYST | SPARK PLUG: AWSF-52C GAP-.042-.046 2.3L-5FM FFM2.3V5HCF4-EGR/EGS/AIP/TWC

CALIBRATION: 5—26J—R01

MODEL YEAR: 1985 — ENGINE: 2.3L

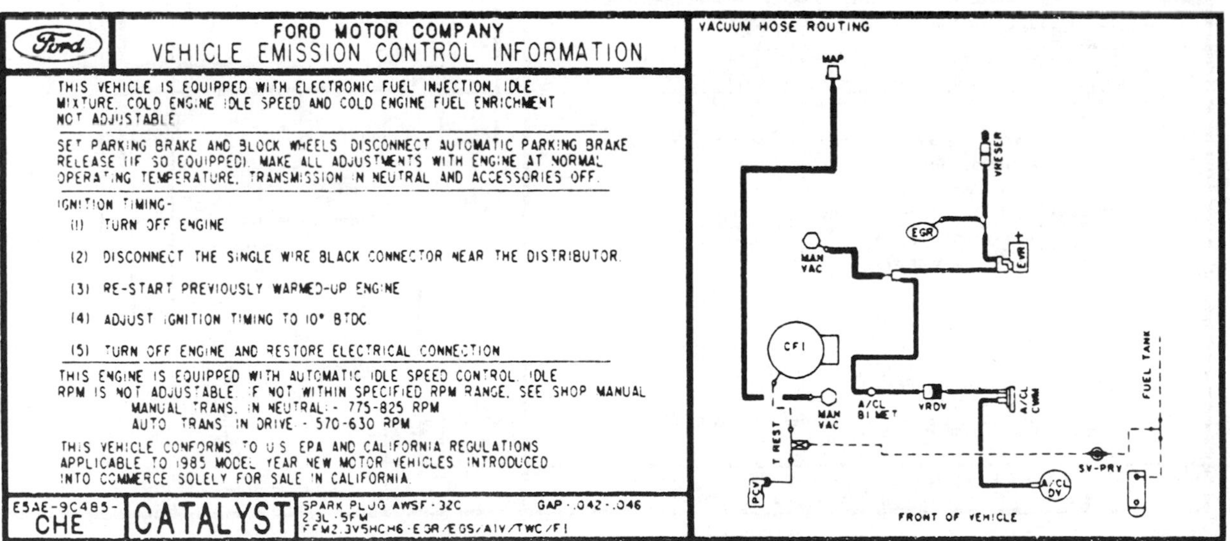

FORD MOTOR COMPANY
VEHICLE EMISSION CONTROL INFORMATION

THIS VEHICLE IS EQUIPPED WITH ELECTRONIC FUEL INJECTION. IDLE MIXTURE. COLD ENGINE IDLE SPEED AND COLD ENGINE FUEL ENRICHMENT NOT ADJUSTABLE

SET PARKING BRAKE AND BLOCK WHEELS DISCONNECT AUTOMATIC PARKING BRAKE RELEASE (IF SO EQUIPPED). MAKE ALL ADJUSTMENTS WITH ENGINE AT NORMAL OPERATING TEMPERATURE. TRANSMISSION IN NEUTRAL AND ACCESSORIES OFF.

IGNITION TIMING-

(1) TURN OFF ENGINE

(2) DISCONNECT THE SINGLE WIRE BLACK CONNECTOR NEAR THE DISTRIBUTOR.

(3) RE-START PREVIOUSLY WARMED-UP ENGINE

(4) ADJUST IGNITION TIMING TO 10° BTDC

(5) TURN OFF ENGINE AND RESTORE ELECTRICAL CONNECTION

THIS ENGINE IS EQUIPPED WITH AUTOMATIC IDLE SPEED CONTROL. IDLE RPM IS NOT ADJUSTABLE. IF NOT WITHIN SPECIFIED RPM RANGE. SEE SHOP MANUAL
MANUAL TRANS. IN NEUTRAL: - 775-825 RPM
AUTO TRANS IN DRIVE - 570-630 RPM

THIS VEHICLE CONFORMS TO U.S. EPA AND CALIFORNIA REGULATIONS APPLICABLE TO 1985 MODEL YEAR NEW MOTOR VEHICLES INTRODUCED INTO COMMERCE SOLELY FOR SALE IN CALIFORNIA.

E5AE-9C485-CHE | CATALYST | SPARK PLUG AWSF-32C GAP-.042-.046 2.3L-5FM FFM2.3V5HCH6-EGR/EGS/AIV/TWC/FI

CALIBRATION: 5—26R—R01

VACUUM CIRCUITS

(© Ford Motor Co)

MODEL YEAR: 1985 **ENGINE: 3.8L**

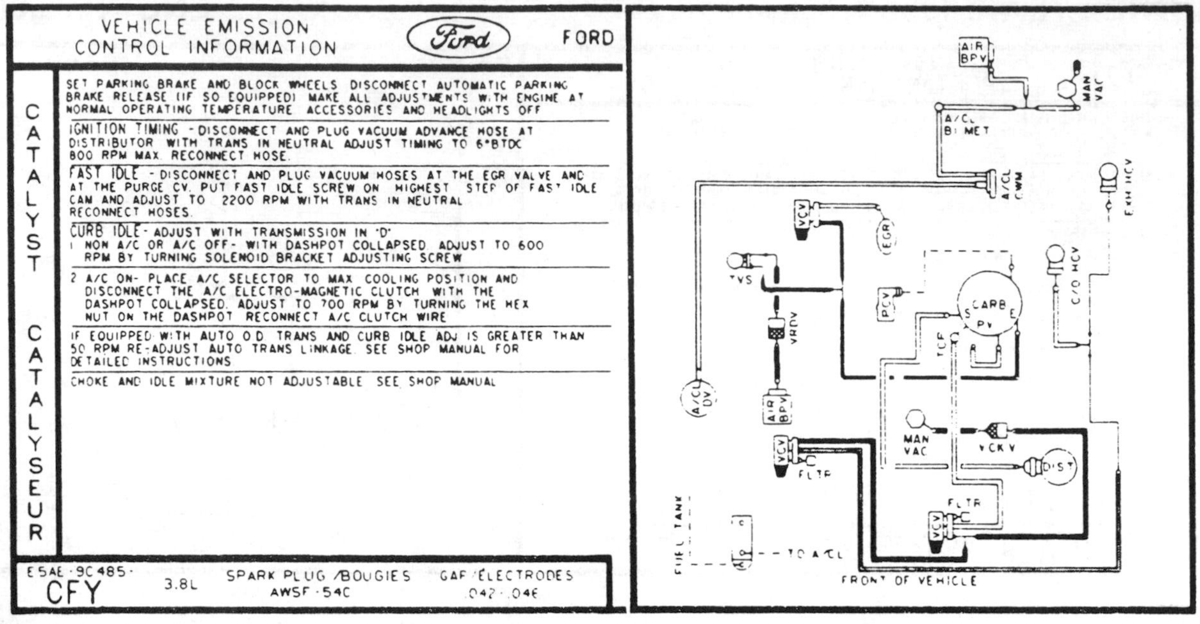

VEHICLE EMISSION CONTROL INFORMATION — Ford — FORD

CATALYST / CATALYSEUR

SET PARKING BRAKE AND BLOCK WHEELS DISCONNECT AUTOMATIC PARKING BRAKE RELEASE (IF SO EQUIPPED) MAKE ALL ADJUSTMENTS WITH ENGINE AT NORMAL OPERATING TEMPERATURE ACCESSORIES AND HEADLIGHTS OFF

IGNITION TIMING - DISCONNECT AND PLUG VACUUM ADVANCE HOSE AT DISTRIBUTOR WITH TRANS IN NEUTRAL ADJUST TIMING TO 6°BTDC 800 RPM MAX. RECONNECT HOSE.

FAST IDLE - DISCONNECT AND PLUG VACUUM HOSES AT THE EGR VALVE AND AT THE PURGE CV. PUT FAST IDLE SCREW ON HIGHEST STEP OF FAST IDLE CAM AND ADJUST TO 2200 RPM WITH TRANS IN NEUTRAL RECONNECT HOSES.

CURB IDLE - ADJUST WITH TRANSMISSION IN 'D'

1 NON A/C OR A/C OFF - WITH DASHPOT COLLAPSED. ADJUST TO 600 RPM BY TURNING SOLENOID BRACKET ADJUSTING SCREW

2 A/C ON - PLACE A/C SELECTOR TO MAX COOLING POSITION AND DISCONNECT THE A/C ELECTRO-MAGNETIC CLUTCH WITH THE DASHPOT COLLAPSED. ADJUST TO 700 RPM BY TURNING THE HEX NUT ON THE DASHPOT RECONNECT A/C CLUTCH WIRE

IF EQUIPPED WITH AUTO O.D. TRANS AND CURB IDLE ADJ IS GREATER THAN 50 RPM RE-ADJUST AUTO TRANS LINKAGE. SEE SHOP MANUAL FOR DETAILED INSTRUCTIONS

CHOKE AND IDLE MIXTURE NOT ADJUSTABLE SEE SHOP MANUAL

E5AE-9C485-CFY	3.8L	SPARK PLUG /BOUGIES AWSF-54C	GAP /ELECTRODES .042-.046

CALIBRATION: 4—14G—R10

MODEL YEAR: 1985 **ENGINE: 3.8L**

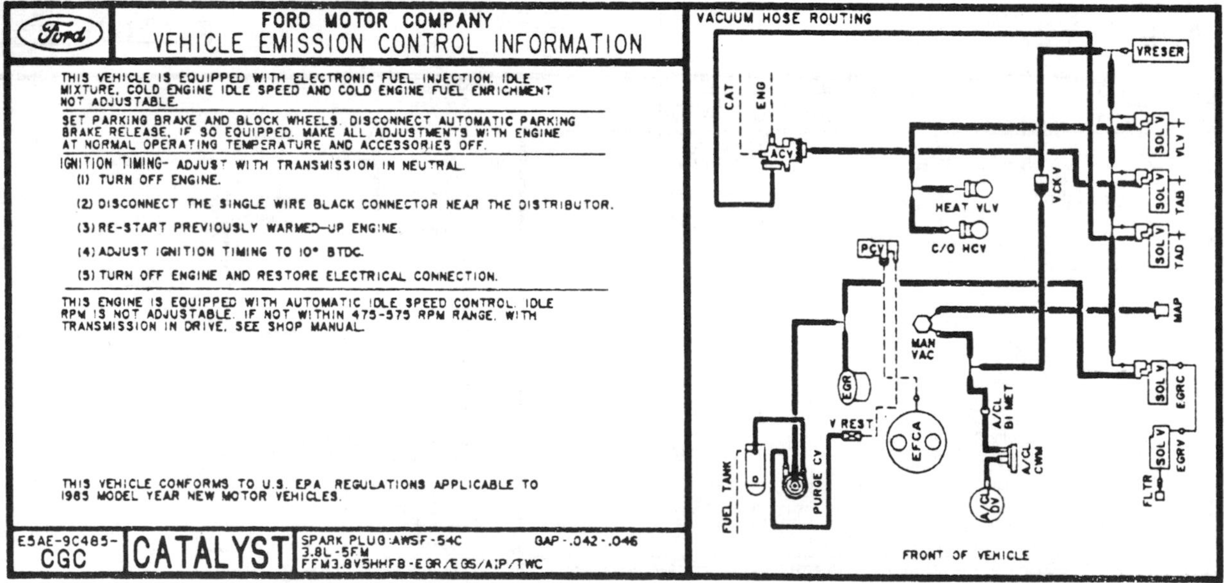

Ford — FORD MOTOR COMPANY
VEHICLE EMISSION CONTROL INFORMATION

THIS VEHICLE IS EQUIPPED WITH ELECTRONIC FUEL INJECTION. IDLE MIXTURE, COLD ENGINE IDLE SPEED AND COLD ENGINE FUEL ENRICHMENT NOT ADJUSTABLE.

SET PARKING BRAKE AND BLOCK WHEELS. DISCONNECT AUTOMATIC PARKING BRAKE RELEASE, IF SO EQUIPPED. MAKE ALL ADJUSTMENTS WITH ENGINE AT NORMAL OPERATING TEMPERATURE AND ACCESSORIES OFF.

IGNITION TIMING- ADJUST WITH TRANSMISSION IN NEUTRAL.

(1) TURN OFF ENGINE.

(2) DISCONNECT THE SINGLE WIRE BLACK CONNECTOR NEAR THE DISTRIBUTOR.

(3) RE-START PREVIOUSLY WARMED-UP ENGINE.

(4) ADJUST IGNITION TIMING TO 10° BTDC.

(5) TURN OFF ENGINE AND RESTORE ELECTRICAL CONNECTION.

THIS ENGINE IS EQUIPPED WITH AUTOMATIC IDLE SPEED CONTROL. IDLE RPM IS NOT ADJUSTABLE. IF NOT WITHIN 475-575 RPM RANGE, WITH TRANSMISSION IN DRIVE, SEE SHOP MANUAL.

THIS VEHICLE CONFORMS TO U.S. EPA REGULATIONS APPLICABLE TO 1985 MODEL YEAR NEW MOTOR VEHICLES.

E5AE-9C485-CGC	CATALYST	SPARK PLUG:AWSF-54C GAP-.042-.046 3.8L-5FM FFM3.8V5HHF8-EGR/EGS/AIP/TWC

CALIBRATION: 4—16A—R10

VACUUM CIRCUITS

(© Ford Motor Co)

MODEL YEAR: 1985

ENGINE: 3.8L

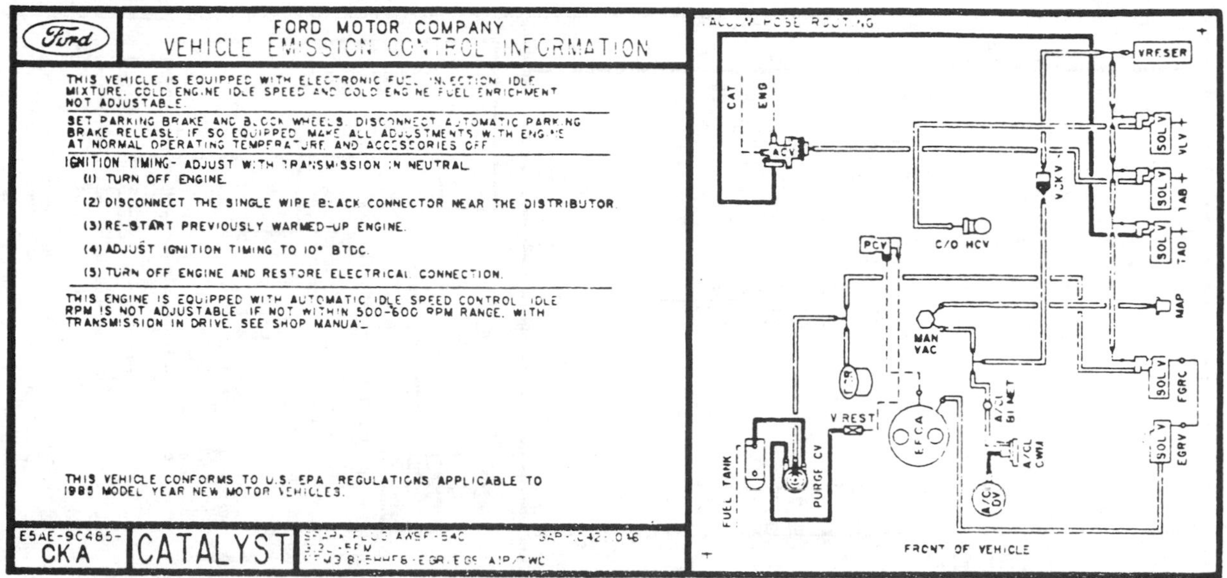

FORD MOTOR COMPANY
VEHICLE EMISSION CONTROL INFORMATION

THIS VEHICLE IS EQUIPPED WITH ELECTRONIC FUEL INJECTION. IDLE MIXTURE, COLD ENGINE IDLE SPEED AND COLD ENGINE FUEL ENRICHMENT NOT ADJUSTABLE.

SET PARKING BRAKE AND BLOCK WHEELS. DISCONNECT AUTOMATIC PARKING BRAKE RELEASE, IF SO EQUIPPED. MAKE ALL ADJUSTMENTS WITH ENGINE AT NORMAL OPERATING TEMPERATURE AND ACCESSORIES OFF.

IGNITION TIMING- ADJUST WITH TRANSMISSION IN NEUTRAL.

(1) TURN OFF ENGINE.

(2) DISCONNECT THE SINGLE WIRE BLACK CONNECTOR NEAR THE DISTRIBUTOR.

(3) RE-START PREVIOUSLY WARMED-UP ENGINE.

(4) ADJUST IGNITION TIMING TO 10° BTDC.

(5) TURN OFF ENGINE AND RESTORE ELECTRICAL CONNECTION.

THIS ENGINE IS EQUIPPED WITH AUTOMATIC IDLE SPEED CONTROL. IDLE RPM IS NOT ADJUSTABLE. IF NOT WITHIN 500-600 RPM RANGE, WITH TRANSMISSION IN DRIVE, SEE SHOP MANUAL.

THIS VEHICLE CONFORMS TO U.S. EPA REGULATIONS APPLICABLE TO 1985 MODEL YEAR NEW MOTOR VEHICLES.

E5AE-9C485-CKA | CATALYST | SPARK PLUG AWSF-54C GAP-.042-.046 3.8L-5FM FFM3.8 V5HHF6-EGR/EGS AIP/TWC

CALIBRATION: 4—16B—R10

MODEL YEAR: 1985

ENGINE: 3.8L

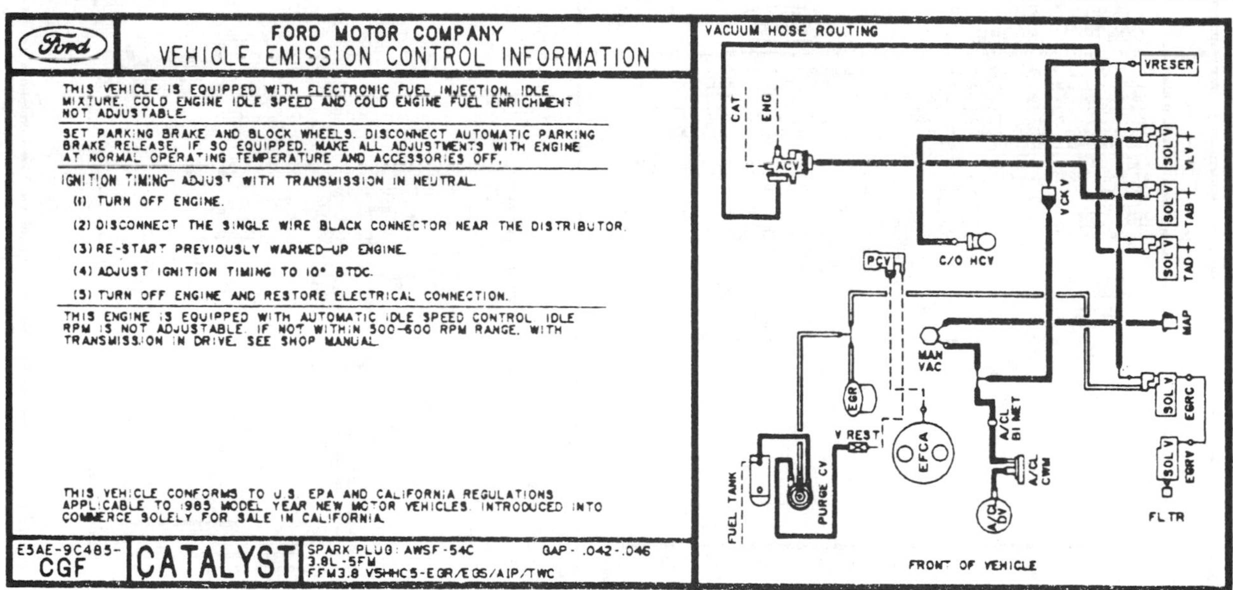

FORD MOTOR COMPANY
VEHICLE EMISSION CONTROL INFORMATION

THIS VEHICLE IS EQUIPPED WITH ELECTRONIC FUEL INJECTION. IDLE MIXTURE, COLD ENGINE IDLE SPEED AND COLD ENGINE FUEL ENRICHMENT NOT ADJUSTABLE.

SET PARKING BRAKE AND BLOCK WHEELS. DISCONNECT AUTOMATIC PARKING BRAKE RELEASE, IF SO EQUIPPED. MAKE ALL ADJUSTMENTS WITH ENGINE AT NORMAL OPERATING TEMPERATURE AND ACCESSORIES OFF.

IGNITION TIMING- ADJUST WITH TRANSMISSION IN NEUTRAL.

(1) TURN OFF ENGINE.

(2) DISCONNECT THE SINGLE WIRE BLACK CONNECTOR NEAR THE DISTRIBUTOR.

(3) RE-START PREVIOUSLY WARMED-UP ENGINE.

(4) ADJUST IGNITION TIMING TO 10° BTDC.

(5) TURN OFF ENGINE AND RESTORE ELECTRICAL CONNECTION.

THIS ENGINE IS EQUIPPED WITH AUTOMATIC IDLE SPEED CONTROL. IDLE RPM IS NOT ADJUSTABLE. IF NOT WITHIN 500-600 RPM RANGE, WITH TRANSMISSION IN DRIVE, SEE SHOP MANUAL.

THIS VEHICLE CONFORMS TO U.S. EPA AND CALIFORNIA REGULATIONS APPLICABLE TO 1985 MODEL YEAR NEW MOTOR VEHICLES. INTRODUCED INTO COMMERCE SOLELY FOR SALE IN CALIFORNIA.

E5AE-9C485-CGF | CATALYST | SPARK PLUG: AWSF-54C GAP-.042-.046 3.8L-5FM FFM3.8 V5HHC5-EGR/EGS/AIP/TWC

CALIBRATION: 4—16N—R10

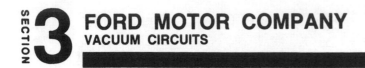

VACUUM CIRCUITS

MODEL YEAR: 1985 **ENGINE: 3.8L**

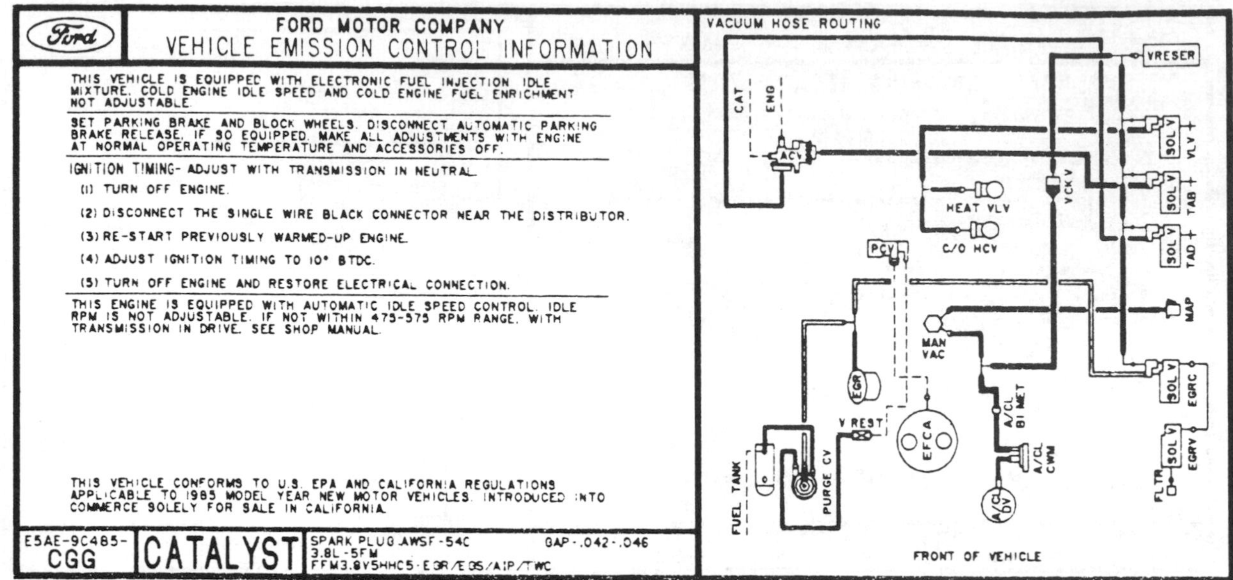

FORD MOTOR COMPANY
VEHICLE EMISSION CONTROL INFORMATION

THIS VEHICLE IS EQUIPPED WITH ELECTRONIC FUEL INJECTION IDLE MIXTURE. COLD ENGINE IDLE SPEED AND COLD ENGINE FUEL ENRICHMENT NOT ADJUSTABLE.

SET PARKING BRAKE AND BLOCK WHEELS. DISCONNECT AUTOMATIC PARKING BRAKE RELEASE, IF SO EQUIPPED. MAKE ALL ADJUSTMENTS WITH ENGINE AT NORMAL OPERATING TEMPERATURE AND ACCESSORIES OFF.

IGNITION TIMING- ADJUST WITH TRANSMISSION IN NEUTRAL.
(1) TURN OFF ENGINE.
(2) DISCONNECT THE SINGLE WIRE BLACK CONNECTOR NEAR THE DISTRIBUTOR.
(3) RE-START PREVIOUSLY WARMED-UP ENGINE.
(4) ADJUST IGNITION TIMING TO 10° BTDC.
(5) TURN OFF ENGINE AND RESTORE ELECTRICAL CONNECTION.

THIS ENGINE IS EQUIPPED WITH AUTOMATIC IDLE SPEED CONTROL. IDLE RPM IS NOT ADJUSTABLE. IF NOT WITHIN 475-575 RPM RANGE, WITH TRANSMISSION IN DRIVE. SEE SHOP MANUAL.

THIS VEHICLE CONFORMS TO U.S. EPA AND CALIFORNIA REGULATIONS APPLICABLE TO 1985 MODEL YEAR NEW MOTOR VEHICLES INTRODUCED INTO COMMERCE SOLELY FOR SALE IN CALIFORNIA.

E5AE-9C485-CGG | CATALYST | SPARK PLUG AWSF-54C GAP-.042-.046
3.8L-5FM
FFM3.8V5HHC5-EGR/EGS/AIP/TWC

CALIBRATION: 4—16P—R10

MODEL YEAR: 1985 **ENGINE: 3.8L**

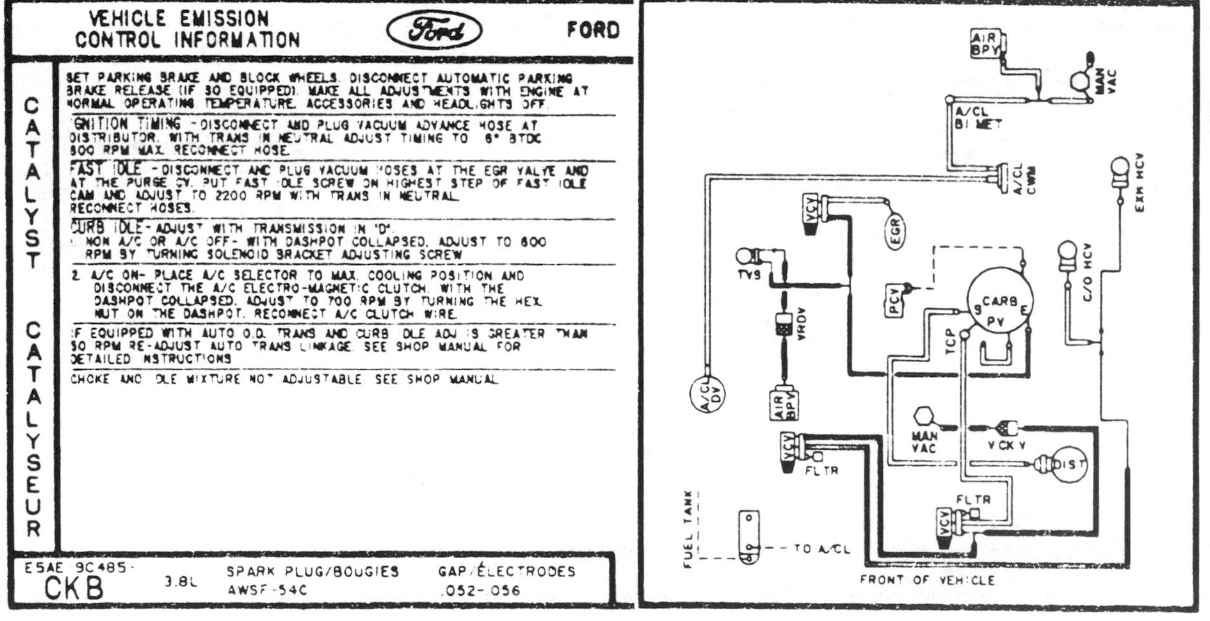

VEHICLE EMISSION CONTROL INFORMATION FORD

CATALYST CATALYSEUR

SET PARKING BRAKE AND BLOCK WHEELS. DISCONNECT AUTOMATIC PARKING BRAKE RELEASE (IF SO EQUIPPED). MAKE ALL ADJUSTMENTS WITH ENGINE AT NORMAL OPERATING TEMPERATURE, ACCESSORIES AND HEADLIGHTS OFF.

IGNITION TIMING - DISCONNECT AND PLUG VACUUM ADVANCE HOSE AT DISTRIBUTOR. WITH TRANS IN NEUTRAL ADJUST TIMING TO 6° BTDC 800 RPM MAX. RECONNECT HOSE.

FAST IDLE - DISCONNECT AND PLUG VACUUM HOSES AT THE EGR VALVE AND AT THE PURGE CV. PUT FAST IDLE SCREW ON HIGHEST STEP OF FAST IDLE CAM AND ADJUST TO 2200 RPM WITH TRANS IN NEUTRAL. RECONNECT HOSES.

CURB IDLE- ADJUST WITH TRANSMISSION IN 'D'.
1. NON A/C OR A/C OFF- WITH DASHPOT COLLAPSED, ADJUST TO 600 RPM BY TURNING SOLENOID BRACKET ADJUSTING SCREW.
2. A/C ON- PLACE A/C SELECTOR TO MAX. COOLING POSITION AND DISCONNECT THE A/C ELECTRO-MAGNETIC CLUTCH. WITH THE DASHPOT COLLAPSED, ADJUST TO 700 RPM BY TURNING THE HEX NUT ON THE DASHPOT. RECONNECT A/C CLUTCH WIRE.

IF EQUIPPED WITH AUTO O.D. TRANS AND CURB IDLE ADJ IS GREATER THAN 50 RPM RE-ADJUST AUTO TRANS LINKAGE. SEE SHOP MANUAL FOR DETAILED INSTRUCTIONS.

CHOKE AND IDLE MIXTURE NOT ADJUSTABLE. SEE SHOP MANUAL.

E5AE-9C485-CKB | 3.8L | SPARK PLUG/BOUGIES AWSF-54C | GAP/ÉLECTRODES .052-.056

CALIBRATION: 5—14G—R05

VACUUM CIRCUITS

(© Ford Motor Co.)

MODEL YEAR: 1985 **ENGINE: 3.8L**

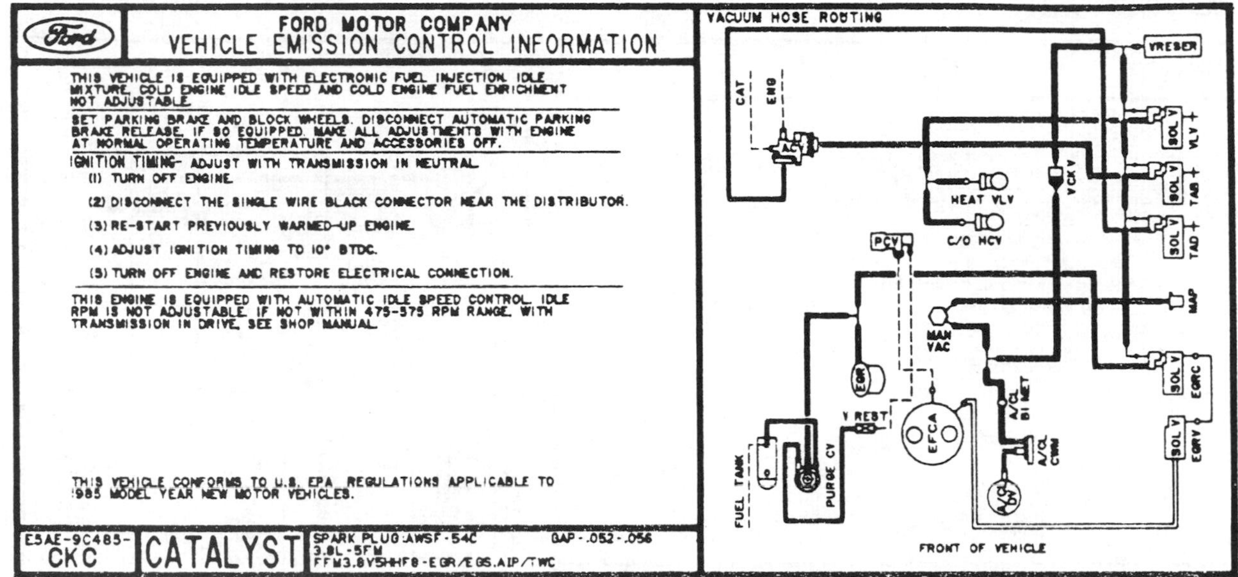

FORD MOTOR COMPANY
VEHICLE EMISSION CONTROL INFORMATION

THIS VEHICLE IS EQUIPPED WITH ELECTRONIC FUEL INJECTION. IDLE MIXTURE, COLD ENGINE IDLE SPEED AND COLD ENGINE FUEL ENRICHMENT NOT ADJUSTABLE.

SET PARKING BRAKE AND BLOCK WHEELS. DISCONNECT AUTOMATIC PARKING BRAKE RELEASE, IF SO EQUIPPED. MAKE ALL ADJUSTMENTS WITH ENGINE AT NORMAL OPERATING TEMPERATURE AND ACCESSORIES OFF.

IGNITION TIMING- ADJUST WITH TRANSMISSION IN NEUTRAL.

(1) TURN OFF ENGINE.

(2) DISCONNECT THE SINGLE WIRE BLACK CONNECTOR NEAR THE DISTRIBUTOR.

(3) RE-START PREVIOUSLY WARMED-UP ENGINE.

(4) ADJUST IGNITION TIMING TO 10° BTDC.

(5) TURN OFF ENGINE AND RESTORE ELECTRICAL CONNECTION.

THIS ENGINE IS EQUIPPED WITH AUTOMATIC IDLE SPEED CONTROL. IDLE RPM IS NOT ADJUSTABLE. IF NOT WITHIN 475-575 RPM RANGE, WITH TRANSMISSION IN DRIVE, SEE SHOP MANUAL.

THIS VEHICLE CONFORMS TO U.S. EPA REGULATIONS APPLICABLE TO 1985 MODEL YEAR NEW MOTOR VEHICLES.

E5AE-9C485-CKC CATALYST SPARK PLUG:AWSF-54C GAP-.052-.056
3.8L-5FM
FFM3.8V5HHF8-EGR/EGS.AIP/TWC

CALIBRATION: 5—16A—R05

MODEL YEAR: 1985 **ENGINE: 3.8L**

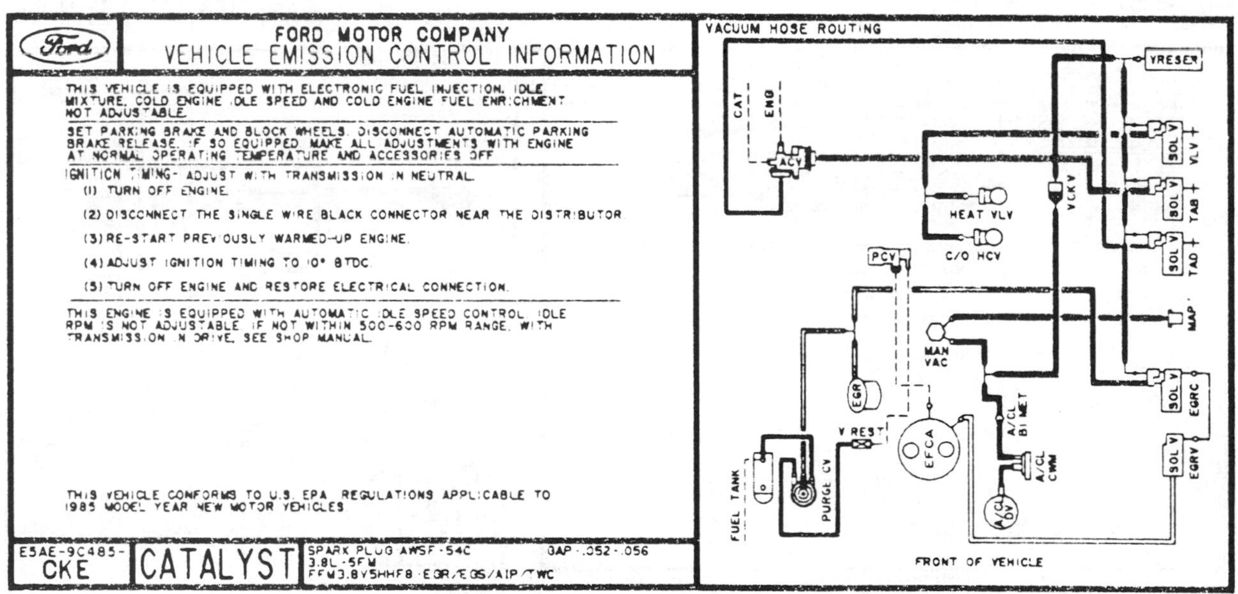

FORD MOTOR COMPANY
VEHICLE EMISSION CONTROL INFORMATION

THIS VEHICLE IS EQUIPPED WITH ELECTRONIC FUEL INJECTION. IDLE MIXTURE, COLD ENGINE IDLE SPEED AND COLD ENGINE FUEL ENRICHMENT NOT ADJUSTABLE.

SET PARKING BRAKE AND BLOCK WHEELS. DISCONNECT AUTOMATIC PARKING BRAKE RELEASE, IF SO EQUIPPED. MAKE ALL ADJUSTMENTS WITH ENGINE AT NORMAL OPERATING TEMPERATURE AND ACCESSORIES OFF.

IGNITION TIMING- ADJUST WITH TRANSMISSION IN NEUTRAL.

(1) TURN OFF ENGINE.

(2) DISCONNECT THE SINGLE WIRE BLACK CONNECTOR NEAR THE DISTRIBUTOR.

(3) RE-START PREVIOUSLY WARMED-UP ENGINE.

(4) ADJUST IGNITION TIMING TO 10° BTDC.

(5) TURN OFF ENGINE AND RESTORE ELECTRICAL CONNECTION.

THIS ENGINE IS EQUIPPED WITH AUTOMATIC IDLE SPEED CONTROL. IDLE RPM IS NOT ADJUSTABLE. IF NOT WITHIN 500-600 RPM RANGE, WITH TRANSMISSION IN DRIVE, SEE SHOP MANUAL.

THIS VEHICLE CONFORMS TO U.S. EPA REGULATIONS APPLICABLE TO 1985 MODEL YEAR NEW MOTOR VEHICLES.

E5AE-9C485-CKE CATALYST SPARK PLUG AWSF-54C GAP-.052-.056
3.8L-5FM
FFM3.8V5HHF8-EGR/EGS/AIP/TWC

CALIBRATION: 5—16B—R05

VACUUM CIRCUITS

(© Ford Motor Co.)

MODEL YEAR: 1985 **ENGINE: 3.8L**

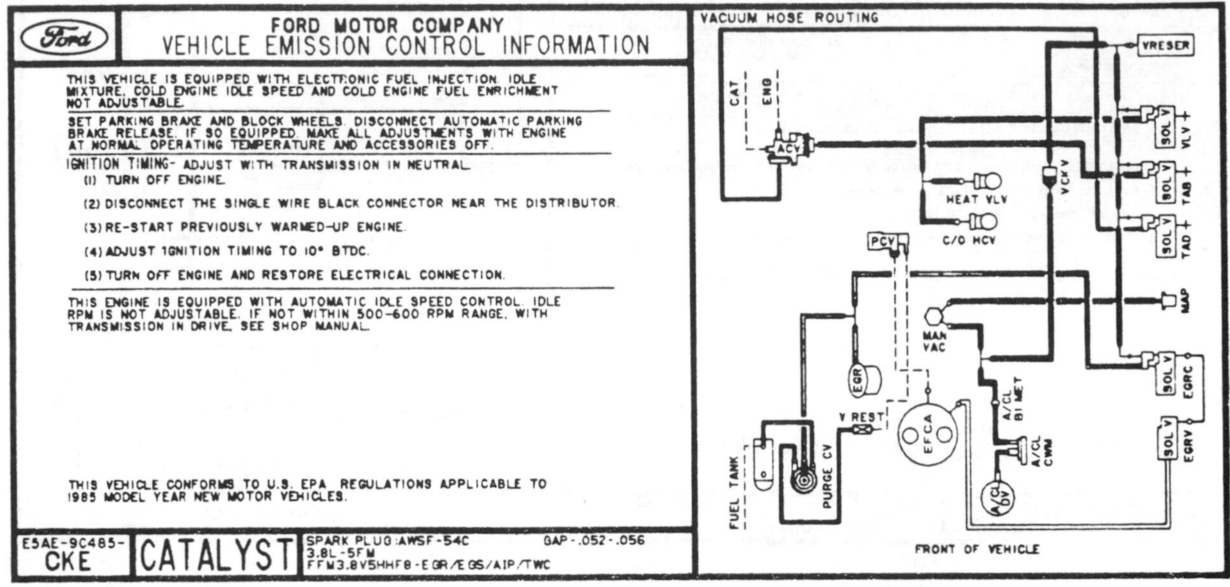

Ford

FORD MOTOR COMPANY
VEHICLE EMISSION CONTROL INFORMATION

THIS VEHICLE IS EQUIPPED WITH ELECTRONIC FUEL INJECTION. IDLE MIXTURE, COLD ENGINE IDLE SPEED AND COLD ENGINE FUEL ENRICHMENT NOT ADJUSTABLE.

SET PARKING BRAKE AND BLOCK WHEELS. DISCONNECT AUTOMATIC PARKING BRAKE RELEASE, IF SO EQUIPPED. MAKE ALL ADJUSTMENTS WITH ENGINE AT NORMAL OPERATING TEMPERATURE AND ACCESSORIES OFF.

IGNITION TIMING- ADJUST WITH TRANSMISSION IN NEUTRAL.
(1) TURN OFF ENGINE.
(2) DISCONNECT THE SINGLE WIRE BLACK CONNECTOR NEAR THE DISTRIBUTOR.
(3) RE-START PREVIOUSLY WARMED-UP ENGINE.
(4) ADJUST IGNITION TIMING TO 10° BTDC.
(5) TURN OFF ENGINE AND RESTORE ELECTRICAL CONNECTION.

THIS ENGINE IS EQUIPPED WITH AUTOMATIC IDLE SPEED CONTROL. IDLE RPM IS NOT ADJUSTABLE. IF NOT WITHIN 500-600 RPM RANGE, WITH TRANSMISSION IN DRIVE, SEE SHOP MANUAL.

THIS VEHICLE CONFORMS TO U.S. EPA REGULATIONS APPLICABLE TO 1985 MODEL YEAR NEW MOTOR VEHICLES.

E5AE-9C485-CKE | CATALYST | SPARK PLUG:AWSF-54C GAP-.052-.056 3.8L-5FM FFM3.8V5HHF8-EGR/EGS/AIP/TWC

CALIBRATION: 5—16F—R05

MODEL YEAR: 1985 **ENGINE: 3.8L**

Ford

FORD MOTOR COMPANY
VEHICLE EMISSION CONTROL INFORMATION

THIS VEHICLE IS EQUIPPED WITH ELECTRONIC FUEL INJECTION. IDLE MIXTURE, COLD ENGINE IDLE SPEED AND COLD ENGINE FUEL ENRICHMENT NOT ADJUSTABLE.

SET PARKING BRAKE AND BLOCK WHEELS. DISCONNECT AUTOMATIC PARKING BRAKE RELEASE, IF SO EQUIPPED. MAKE ALL ADJUSTMENTS WITH ENGINE AT NORMAL OPERATING TEMPERATURE AND ACCESSORIES OFF.

IGNITION TIMING- ADJUST WITH TRANSMISSION IN NEUTRAL.
(1) TURN OFF ENGINE.
(2) DISCONNECT THE SINGLE WIRE BLACK CONNECTOR NEAR THE DISTRIBUTOR.
(3) RE-START PREVIOUSLY WARMED-UP ENGINE.
(4) ADJUST IGNITION TIMING TO 10° BTDC.
(5) TURN OFF ENGINE AND RESTORE ELECTRICAL CONNECTION.

THIS ENGINE IS EQUIPPED WITH AUTOMATIC IDLE SPEED CONTROL. IDLE RPM IS NOT ADJUSTABLE. IF NOT WITHIN 500-600 RPM RANGE, WITH TRANSMISSION IN DRIVE, SEE SHOP MANUAL.

THIS VEHICLE CONFORMS TO U.S. EPA REGULATIONS APPLICABLE TO 1985 MODEL YEAR NEW MOTOR VEHICLES.

E5AE-9C485-CFZ | CATALYST | SPARK PLUG AWSF-54C GAP-.042-.046 3.8L-5FM FFM3.8V5HHF8-EGR/EGS/AIP/TWC

VACUUM HOSE ROUTING
VRESER
CAT
ENG
ACV
SOL V
VLV
SOL V
TAB
SOL V
TAD
VCK V
C/O HCV
PCV
MAP
MAN VAC
EGR
SOL V
EGRC
V REST
EFCA
A/CL BI MET
A/CL CWM
SOL V
EGRV
FLTR
FUEL TANK
PURGE CV
A/CL DV
FRONT OF VEHICLE

CALIBRATION: 5—16F—R00

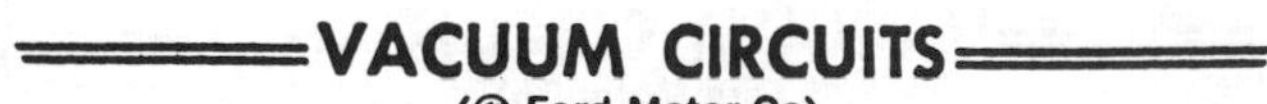

VACUUM CIRCUITS

(© Ford Motor Co)

MODEL YEAR: 1985 **ENGINE: 3.8L**

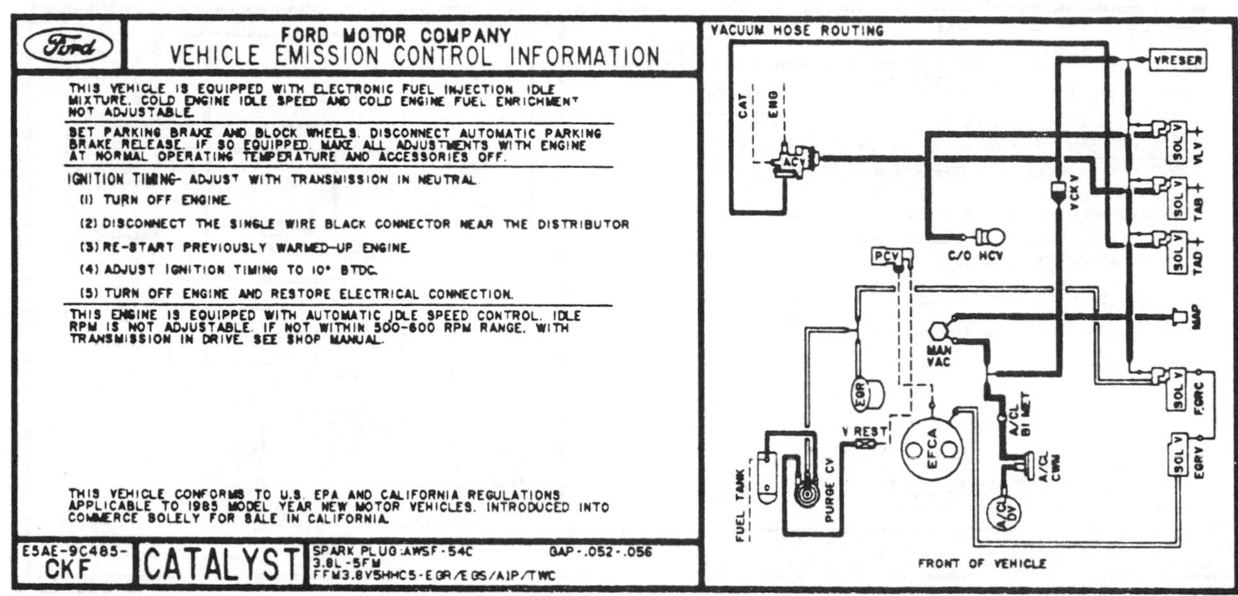

FORD MOTOR COMPANY
VEHICLE EMISSION CONTROL INFORMATION

THIS VEHICLE IS EQUIPPED WITH ELECTRONIC FUEL INJECTION. IDLE MIXTURE, COLD ENGINE IDLE SPEED AND COLD ENGINE FUEL ENRICHMENT NOT ADJUSTABLE.

SET PARKING BRAKE AND BLOCK WHEELS. DISCONNECT AUTOMATIC PARKING BRAKE RELEASE, IF SO EQUIPPED. MAKE ALL ADJUSTMENTS WITH ENGINE AT NORMAL OPERATING TEMPERATURE AND ACCESSORIES OFF.

IGNITION TIMING- ADJUST WITH TRANSMISSION IN NEUTRAL
(1) TURN OFF ENGINE.
(2) DISCONNECT THE SINGLE WIRE BLACK CONNECTOR NEAR THE DISTRIBUTOR
(3) RE-START PREVIOUSLY WARMED-UP ENGINE.
(4) ADJUST IGNITION TIMING TO 10° BTDC.
(5) TURN OFF ENGINE AND RESTORE ELECTRICAL CONNECTION.

THIS ENGINE IS EQUIPPED WITH AUTOMATIC IDLE SPEED CONTROL. IDLE RPM IS NOT ADJUSTABLE. IF NOT WITHIN 500-600 RPM RANGE, WITH TRANSMISSION IN DRIVE. SEE SHOP MANUAL.

THIS VEHICLE CONFORMS TO U.S. EPA AND CALIFORNIA REGULATIONS APPLICABLE TO 1985 MODEL YEAR NEW MOTOR VEHICLES, INTRODUCED INTO COMMERCE SOLELY FOR SALE IN CALIFORNIA.

E5AE-9C485-CKF | CATALYST | SPARK PLUG:AWSF-54C GAP-.052-.056 3.8L-5FM FFM3.8V5HHC5-EGR/EGS/AIP/TWC

CALIBRATION: 5—16N—R05

MODEL YEAR: 1985 **ENGINE: 3.8L**

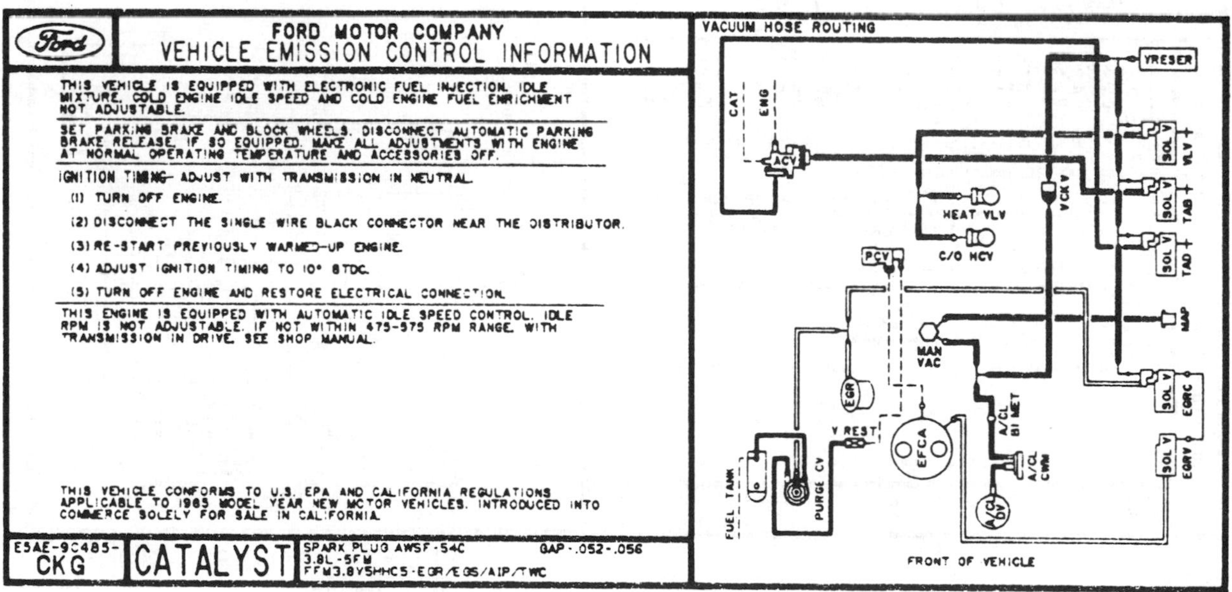

FORD MOTOR COMPANY
VEHICLE EMISSION CONTROL INFORMATION

THIS VEHICLE IS EQUIPPED WITH ELECTRONIC FUEL INJECTION. IDLE MIXTURE, COLD ENGINE IDLE SPEED AND COLD ENGINE FUEL ENRICHMENT NOT ADJUSTABLE.

SET PARKING BRAKE AND BLOCK WHEELS. DISCONNECT AUTOMATIC PARKING BRAKE RELEASE, IF SO EQUIPPED. MAKE ALL ADJUSTMENTS WITH ENGINE AT NORMAL OPERATING TEMPERATURE AND ACCESSORIES OFF.

IGNITION TIMING- ADJUST WITH TRANSMISSION IN NEUTRAL.
(1) TURN OFF ENGINE.
(2) DISCONNECT THE SINGLE WIRE BLACK CONNECTOR NEAR THE DISTRIBUTOR.
(3) RE-START PREVIOUSLY WARMED-UP ENGINE.
(4) ADJUST IGNITION TIMING TO 10° BTDC.
(5) TURN OFF ENGINE AND RESTORE ELECTRICAL CONNECTION.

THIS ENGINE IS EQUIPPED WITH AUTOMATIC IDLE SPEED CONTROL. IDLE RPM IS NOT ADJUSTABLE. IF NOT WITHIN 475-575 RPM RANGE, WITH TRANSMISSION IN DRIVE. SEE SHOP MANUAL.

THIS VEHICLE CONFORMS TO U.S. EPA AND CALIFORNIA REGULATIONS APPLICABLE TO 1985 MODEL YEAR NEW MOTOR VEHICLES, INTRODUCED INTO COMMERCE SOLELY FOR SALE IN CALIFORNIA.

E5AE-9C485-CKG | CATALYST | SPARK PLUG AWSF-54C GAP-.052-.056 3.8L-5FM FFM3.8V5HHC5-EGR/EGS/AIP/TWC

CALIBRATION: 5—16P—R05

VACUUM CIRCUITS

(© Ford Motor Co)

MODEL YEAR: 1985 **ENGINE: 3.8L**

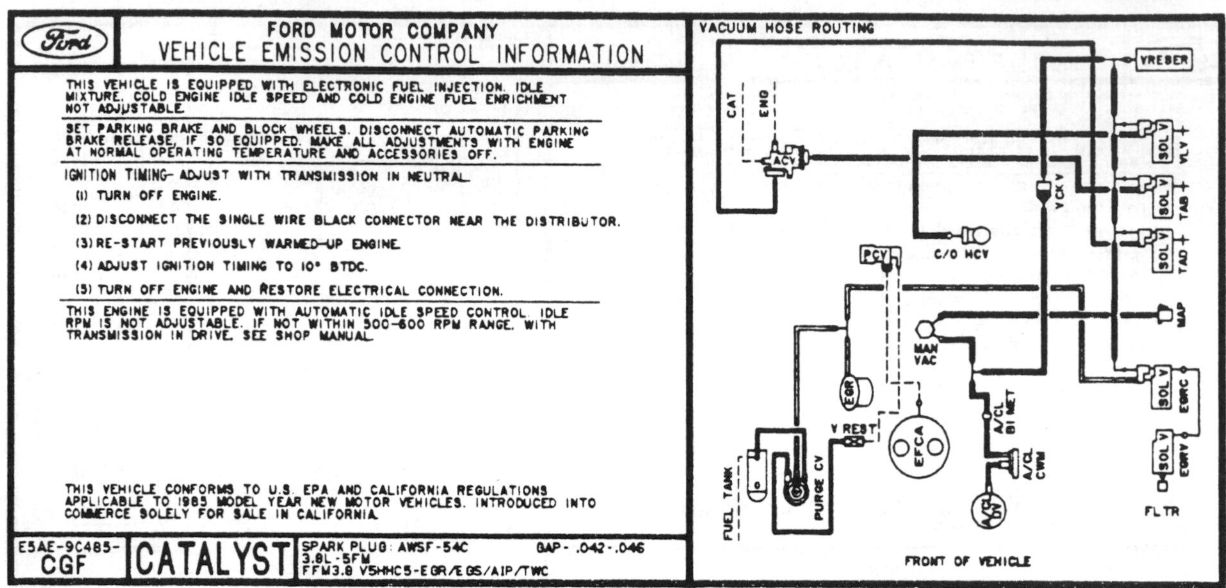

FORD MOTOR COMPANY
VEHICLE EMISSION CONTROL INFORMATION

THIS VEHICLE IS EQUIPPED WITH ELECTRONIC FUEL INJECTION. IDLE MIXTURE, COLD ENGINE IDLE SPEED AND COLD ENGINE FUEL ENRICHMENT NOT ADJUSTABLE.

SET PARKING BRAKE AND BLOCK WHEELS. DISCONNECT AUTOMATIC PARKING BRAKE RELEASE, IF SO EQUIPPED. MAKE ALL ADJUSTMENTS WITH ENGINE AT NORMAL OPERATING TEMPERATURE AND ACCESSORIES OFF.

IGNITION TIMING- ADJUST WITH TRANSMISSION IN NEUTRAL.

(1) TURN OFF ENGINE.

(2) DISCONNECT THE SINGLE WIRE BLACK CONNECTOR NEAR THE DISTRIBUTOR.

(3) RE-START PREVIOUSLY WARMED-UP ENGINE.

(4) ADJUST IGNITION TIMING TO 10° BTDC.

(5) TURN OFF ENGINE AND RESTORE ELECTRICAL CONNECTION.

THIS ENGINE IS EQUIPPED WITH AUTOMATIC IDLE SPEED CONTROL. IDLE RPM IS NOT ADJUSTABLE. IF NOT WITHIN 500-600 RPM RANGE, WITH TRANSMISSION IN DRIVE, SEE SHOP MANUAL.

THIS VEHICLE CONFORMS TO U.S. EPA AND CALIFORNIA REGULATIONS APPLICABLE TO 1985 MODEL YEAR NEW MOTOR VEHICLES. INTRODUCED INTO COMMERCE SOLELY FOR SALE IN CALIFORNIA.

E5AE-9C485-CGF CATALYST SPARK PLUG: AWSF-54C GAP-.042-.046
3.8L-5FM
FFM3.8 V5HHC5-EGR/EGS/AIP/TWC

CALIBRATION: 5—16S—R00

MODEL YEAR: 1985 **ENGINE: 5.0L**

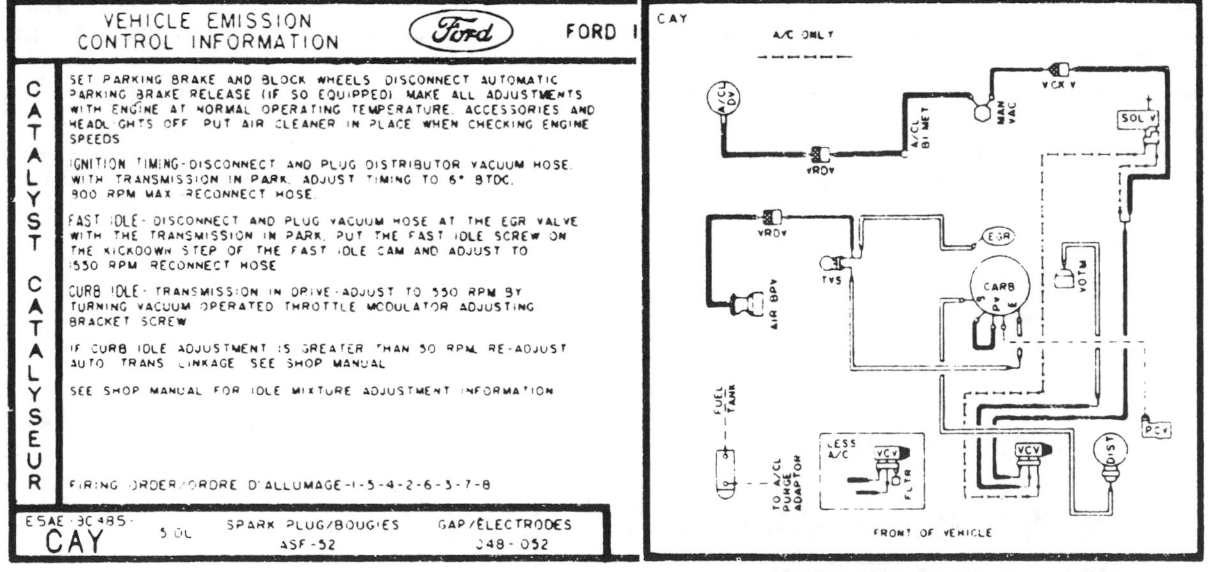

VEHICLE EMISSION CONTROL INFORMATION FORD

CATALYST CATALYSEUR

SET PARKING BRAKE AND BLOCK WHEELS DISCONNECT AUTOMATIC PARKING BRAKE RELEASE (IF SO EQUIPPED) MAKE ALL ADJUSTMENTS WITH ENGINE AT NORMAL OPERATING TEMPERATURE. ACCESSORIES AND HEADLIGHTS OFF PUT AIR CLEANER IN PLACE WHEN CHECKING ENGINE SPEEDS

IGNITION TIMING-DISCONNECT AND PLUG DISTRIBUTOR VACUUM HOSE. WITH TRANSMISSION IN PARK, ADJUST TIMING TO 6° BTDC. 900 RPM MAX RECONNECT HOSE.

FAST IDLE- DISCONNECT AND PLUG VACUUM HOSE AT THE EGR VALVE WITH THE TRANSMISSION IN PARK, PUT THE FAST IDLE SCREW ON THE KICKDOWN STEP OF THE FAST IDLE CAM AND ADJUST TO 1550 RPM RECONNECT HOSE

CURB IDLE- TRANSMISSION IN DRIVE-ADJUST TO 550 RPM BY TURNING VACUUM OPERATED THROTTLE MODULATOR ADJUSTING BRACKET SCREW

IF CURB IDLE ADJUSTMENT IS GREATER THAN 50 RPM RE-ADJUST AUTO TRANS LINKAGE SEE SHOP MANUAL

SEE SHOP MANUAL FOR IDLE MIXTURE ADJUSTMENT INFORMATION

FIRING ORDER/ORDRE D'ALLUMAGE-1-5-4-2-6-3-7-8

E5AE-9C485-CAY 5.0L SPARK PLUG/BOUGIES ASF-52 GAP/ÉLECTRODES .048-.052

CALIBRATION: 2—20H—R00

PROPANE SPECIFICATIONS
(Set in Neutral)

Gain RPM:	10-70
Reset RPM:	20

VACUUM CIRCUITS

MODEL YEAR: 1985 **ENGINE: 5.0L**

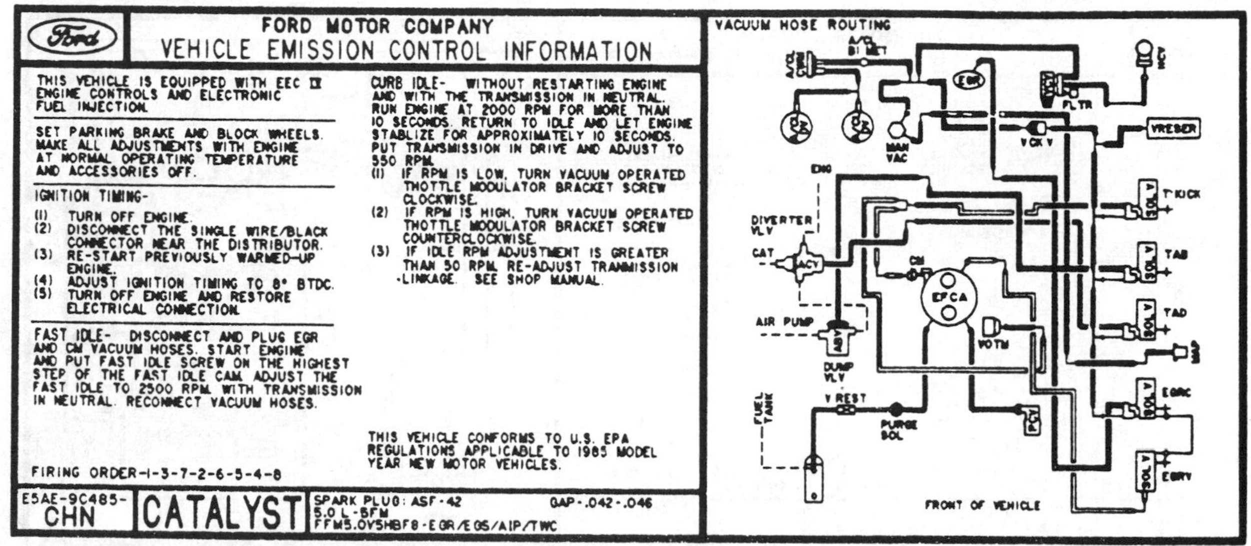

FORD MOTOR COMPANY
VEHICLE EMISSION CONTROL INFORMATION

THIS VEHICLE IS EQUIPPED WITH EEC IV ENGINE CONTROLS AND ELECTRONIC FUEL INJECTION.

SET PARKING BRAKE AND BLOCK WHEELS. MAKE ALL ADJUSTMENTS WITH ENGINE AT NORMAL OPERATING TEMPERATURE AND ACCESSORIES OFF.

IGNITION TIMING-
(1) TURN OFF ENGINE.
(2) DISCONNECT THE SINGLE WIRE/BLACK CONNECTOR NEAR THE DISTRIBUTOR.
(3) RE-START PREVIOUSLY WARMED-UP ENGINE.
(4) ADJUST IGNITION TIMING TO 8° BTDC.
(5) TURN OFF ENGINE AND RESTORE ELECTRICAL CONNECTION.

FAST IDLE- DISCONNECT AND PLUG EGR AND CM VACUUM HOSES. START ENGINE AND PUT FAST IDLE SCREW ON THE HIGHEST STEP OF THE FAST IDLE CAM. ADJUST THE FAST IDLE TO 2500 RPM. WITH TRANSMISSION IN NEUTRAL. RECONNECT VACUUM HOSES.

FIRING ORDER-1-3-7-2-6-5-4-8

CURB IDLE- WITHOUT RESTARTING ENGINE AND WITH THE TRANSMISSION IN NEUTRAL. RUN ENGINE AT 2000 RPM FOR MORE THAN 10 SECONDS. RETURN TO IDLE AND LET ENGINE STABLIZE FOR APPROXIMATELY 10 SECONDS. PUT TRANSMISSION IN DRIVE AND ADJUST TO 550 RPM.
(1) IF RPM IS LOW, TURN VACUUM OPERATED THOTTLE MODULATOR BRACKET SCREW CLOCKWISE.
(2) IF RPM IS HIGH, TURN VACUUM OPERATED THOTTLE MODULATOR BRACKET SCREW COUNTERCLOCKWISE.
(3) IF IDLE RPM ADJUSTMENT IS GREATER THAN 50 RPM. RE-ADJUST TRANMISSION LINKAGE. SEE SHOP MANUAL.

THIS VEHICLE CONFORMS TO U.S. EPA REGULATIONS APPLICABLE TO 1985 MODEL YEAR NEW MOTOR VEHICLES.

E5AE-9C485-CHN | CATALYST | SPARK PLUG: ASF-42 GAP-.042-.046 5.0 L-5FM FFM5.0V5HBF8-EGR/EGS/AIP/TWC

CALIBRATION: 4—22A—R13

MODEL YEAR: 1985 **ENGINE: 5.0L**

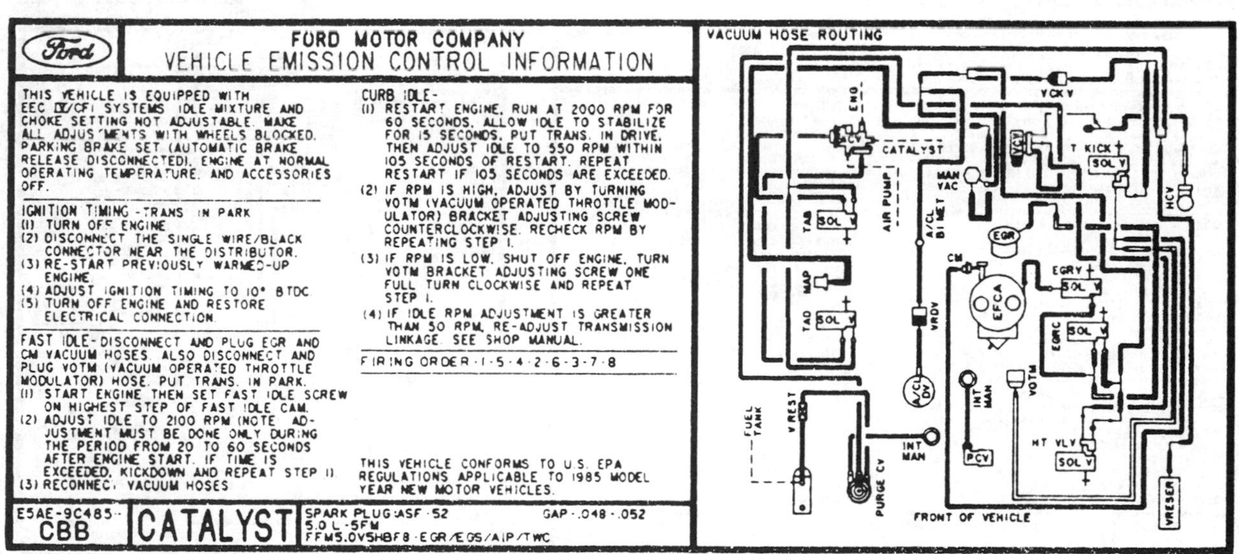

FORD MOTOR COMPANY
VEHICLE EMISSION CONTROL INFORMATION

THIS VEHICLE IS EQUIPPED WITH EEC IV/CFI SYSTEMS IDLE MIXTURE AND CHOKE SETTING NOT ADJUSTABLE. MAKE ALL ADJUSTMENTS WITH WHEELS BLOCKED, PARKING BRAKE SET (AUTOMATIC BRAKE RELEASE DISCONNECTED), ENGINE AT NORMAL OPERATING TEMPERATURE, AND ACCESSORIES OFF.

IGNITION TIMING - TRANS IN PARK
(1) TURN OFF ENGINE
(2) DISCONNECT THE SINGLE WIRE/BLACK CONNECTOR NEAR THE DISTRIBUTOR.
(3) RE-START PREVIOUSLY WARMED-UP ENGINE.
(4) ADJUST IGNITION TIMING TO 10° BTDC.
(5) TURN OFF ENGINE AND RESTORE ELECTRICAL CONNECTION.

FAST IDLE- DISCONNECT AND PLUG EGR AND CM VACUUM HOSES. ALSO DISCONNECT AND PLUG VOTM (VACUUM OPERATED THROTTLE MODULATOR) HOSE. PUT TRANS. IN PARK.
(1) START ENGINE THEN SET FAST IDLE SCREW ON HIGHEST STEP OF FAST IDLE CAM.
(2) ADJUST IDLE TO 2100 RPM (NOTE: ADJUSTMENT MUST BE DONE ONLY DURING THE PERIOD FROM 20 TO 60 SECONDS AFTER ENGINE START. IF TIME IS EXCEEDED, KICKDOWN AND REPEAT STEP 1).
(3) RECONNECT VACUUM HOSES

CURB IDLE-
(1) RESTART ENGINE, RUN AT 2000 RPM FOR 60 SECONDS, ALLOW IDLE TO STABILIZE FOR 15 SECONDS, PUT TRANS. IN DRIVE. THEN ADJUST IDLE TO 550 RPM WITHIN 105 SECONDS OF RESTART. REPEAT RESTART IF 105 SECONDS ARE EXCEEDED.
(2) IF RPM IS HIGH, ADJUST BY TURNING VOTM (VACUUM OPERATED THROTTLE MODULATOR) BRACKET ADJUSTING SCREW COUNTERCLOCKWISE. RECHECK RPM BY REPEATING STEP 1.
(3) IF RPM IS LOW, SHUT OFF ENGINE, TURN VOTM BRACKET ADJUSTING SCREW ONE FULL TURN CLOCKWISE AND REPEAT STEP 1.
(4) IF IDLE RPM ADJUSTMENT IS GREATER THAN 50 RPM, RE-ADJUST TRANSMISSION LINKAGE. SEE SHOP MANUAL.

FIRING ORDER-1-5-4-2-6-3-7-8

THIS VEHICLE CONFORMS TO U.S. EPA REGULATIONS APPLICABLE TO 1985 MODEL YEAR NEW MOTOR VEHICLES.

E5AE-9C485-CBB | CATALYST | SPARK PLUG: ASF-52 GAP-.048-.052 5.0 L-5FM FFM5.0V5HBF8-EGR/EGS/AIP/TWC

CALIBRATION: 4—22B—R00

VACUUM CIRCUITS

(© Ford Motor Co.)

MODEL YEAR: 1985 **ENGINE: 5.0L**

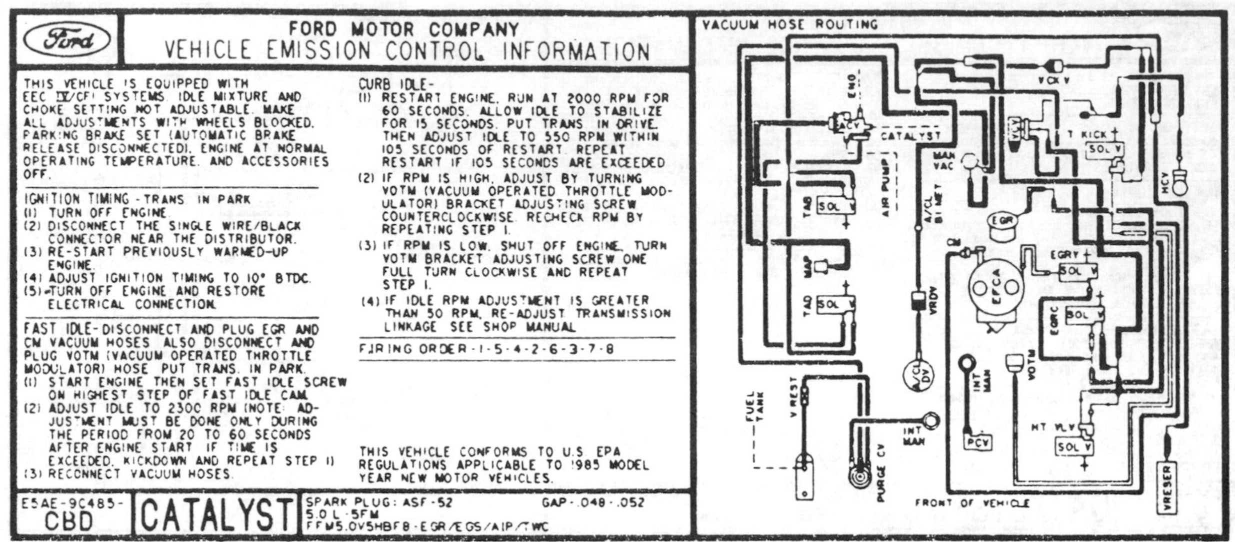

FORD MOTOR COMPANY
VEHICLE EMISSION CONTROL INFORMATION

THIS VEHICLE IS EQUIPPED WITH EEC IV/CFI SYSTEMS. IDLE MIXTURE AND CHOKE SETTING NOT ADJUSTABLE. MAKE ALL ADJUSTMENTS WITH WHEELS BLOCKED, PARKING BRAKE SET (AUTOMATIC BRAKE RELEASE DISCONNECTED), ENGINE AT NORMAL OPERATING TEMPERATURE, AND ACCESSORIES OFF.

IGNITION TIMING - TRANS. IN PARK
(1) TURN OFF ENGINE.
(2) DISCONNECT THE SINGLE WIRE/BLACK CONNECTOR NEAR THE DISTRIBUTOR.
(3) RE-START PREVIOUSLY WARMED-UP ENGINE.
(4) ADJUST IGNITION TIMING TO 10° BTDC.
(5) TURN OFF ENGINE AND RESTORE ELECTRICAL CONNECTION.

FAST IDLE - DISCONNECT AND PLUG EGR AND CM VACUUM HOSES. ALSO DISCONNECT AND PLUG VOTM (VACUUM OPERATED THROTTLE MODULATOR) HOSE. PUT TRANS. IN PARK.
(1) START ENGINE THEN SET FAST IDLE SCREW ON HIGHEST STEP OF FAST IDLE CAM.
(2) ADJUST IDLE TO 2300 RPM (NOTE: ADJUSTMENT MUST BE DONE ONLY DURING THE PERIOD FROM 20 TO 60 SECONDS AFTER ENGINE START. IF TIME IS EXCEEDED, KICKDOWN AND REPEAT STEP 1)
(3) RECONNECT VACUUM HOSES.

CURB IDLE -
(1) RESTART ENGINE. RUN AT 2000 RPM FOR 60 SECONDS. ALLOW IDLE TO STABILIZE FOR 15 SECONDS. PUT TRANS. IN DRIVE. THEN ADJUST IDLE TO 550 RPM WITHIN 105 SECONDS OF RESTART. REPEAT RESTART IF 105 SECONDS ARE EXCEEDED.
(2) IF RPM IS HIGH, ADJUST BY TURNING VOTM (VACUUM OPERATED THROTTLE MODULATOR) BRACKET ADJUSTING SCREW COUNTERCLOCKWISE. RECHECK RPM BY REPEATING STEP 1.
(3) IF RPM IS LOW, SHUT OFF ENGINE. TURN VOTM BRACKET ADJUSTING SCREW ONE FULL TURN CLOCKWISE AND REPEAT STEP 1.
(4) IF IDLE RPM ADJUSTMENT IS GREATER THAN 50 RPM, RE-ADJUST TRANSMISSION LINKAGE. SEE SHOP MANUAL.

FIRING ORDER - 1-5-4-2-6-3-7-8

THIS VEHICLE CONFORMS TO U.S. EPA REGULATIONS APPLICABLE TO 1985 MODEL YEAR NEW MOTOR VEHICLES.

E5AE-9C485-CBD | CATALYST | SPARK PLUG: ASF-52 GAP-.048-.052
5.0L-5FM
FFM5.0V5HBF8-EGR/EGS/AIP/TWC

CALIBRATION: 4—22C—R00

MODEL YEAR: 1985 **ENGINE: 5.0L**

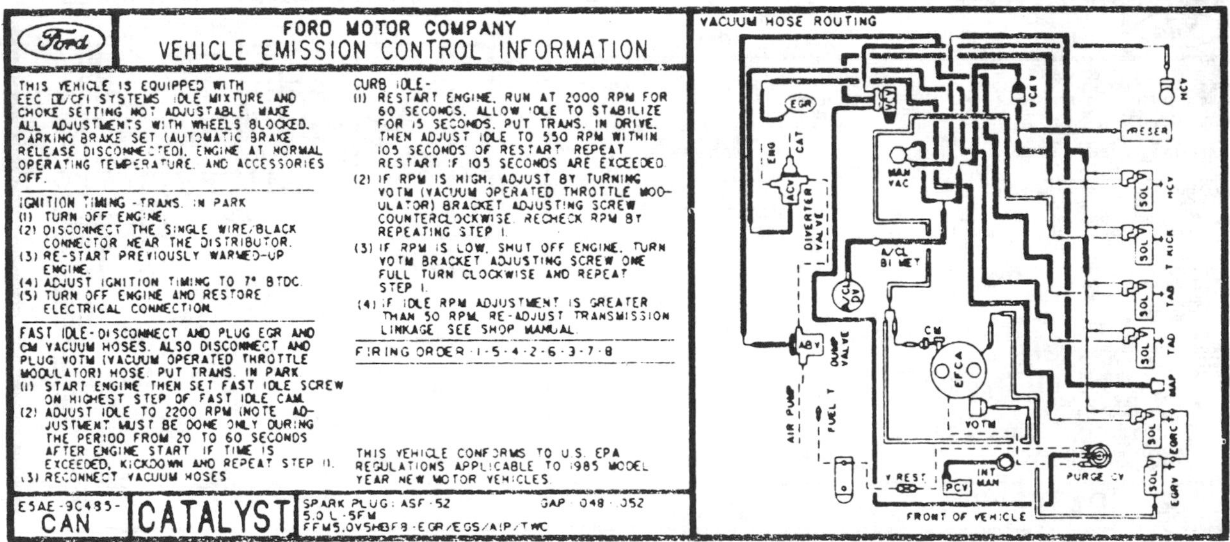

FORD MOTOR COMPANY
VEHICLE EMISSION CONTROL INFORMATION

THIS VEHICLE IS EQUIPPED WITH EEC IV/CFI SYSTEMS. IDLE MIXTURE AND CHOKE SETTING NOT ADJUSTABLE. MAKE ALL ADJUSTMENTS WITH WHEELS BLOCKED, PARKING BRAKE SET (AUTOMATIC BRAKE RELEASE DISCONNECTED), ENGINE AT NORMAL OPERATING TEMPERATURE, AND ACCESSORIES OFF.

IGNITION TIMING - TRANS. IN PARK
(1) TURN OFF ENGINE.
(2) DISCONNECT THE SINGLE WIRE/BLACK CONNECTOR NEAR THE DISTRIBUTOR.
(3) RE-START PREVIOUSLY WARMED-UP ENGINE.
(4) ADJUST IGNITION TIMING TO 7° BTDC.
(5) TURN OFF ENGINE AND RESTORE ELECTRICAL CONNECTION.

FAST IDLE - DISCONNECT AND PLUG EGR AND CM VACUUM HOSES. ALSO DISCONNECT AND PLUG VOTM (VACUUM OPERATED THROTTLE MODULATOR) HOSE. PUT TRANS. IN PARK.
(1) START ENGINE THEN SET FAST IDLE SCREW ON HIGHEST STEP OF FAST IDLE CAM.
(2) ADJUST IDLE TO 2200 RPM (NOTE: ADJUSTMENT MUST BE DONE ONLY DURING THE PERIOD FROM 20 TO 60 SECONDS AFTER ENGINE START. IF TIME IS EXCEEDED, KICKDOWN AND REPEAT STEP 1)
(3) RECONNECT VACUUM HOSES.

CURB IDLE -
(1) RESTART ENGINE. RUN AT 2000 RPM FOR 60 SECONDS. ALLOW IDLE TO STABILIZE FOR 15 SECONDS. PUT TRANS. IN DRIVE. THEN ADJUST IDLE TO 550 RPM WITHIN 105 SECONDS OF RESTART. REPEAT RESTART IF 105 SECONDS ARE EXCEEDED.
(2) IF RPM IS HIGH, ADJUST BY TURNING VOTM (VACUUM OPERATED THROTTLE MODULATOR) BRACKET ADJUSTING SCREW COUNTERCLOCKWISE. RECHECK RPM BY REPEATING STEP 1.
(3) IF RPM IS LOW, SHUT OFF ENGINE. TURN VOTM BRACKET ADJUSTING SCREW ONE FULL TURN CLOCKWISE AND REPEAT STEP 1.
(4) IF IDLE RPM ADJUSTMENT IS GREATER THAN 50 RPM, RE-ADJUST TRANSMISSION LINKAGE. SEE SHOP MANUAL.

FIRING ORDER - 1-5-4-2-6-3-7-8

THIS VEHICLE CONFORMS TO U.S. EPA REGULATIONS APPLICABLE TO 1985 MODEL YEAR NEW MOTOR VEHICLES.

E5AE-9C485-CAN | CATALYST | SPARK PLUG: ASF-52 GAP .048-.052
5.0L-5FM
FFM5.0V5HBF8-EGR/EGS/AIP/TWC

CALIBRATION: 4—22F—R10

VACUUM CIRCUITS

(© Ford Motor Co.)

MODEL YEAR: 1985

ENGINE: 5.0L

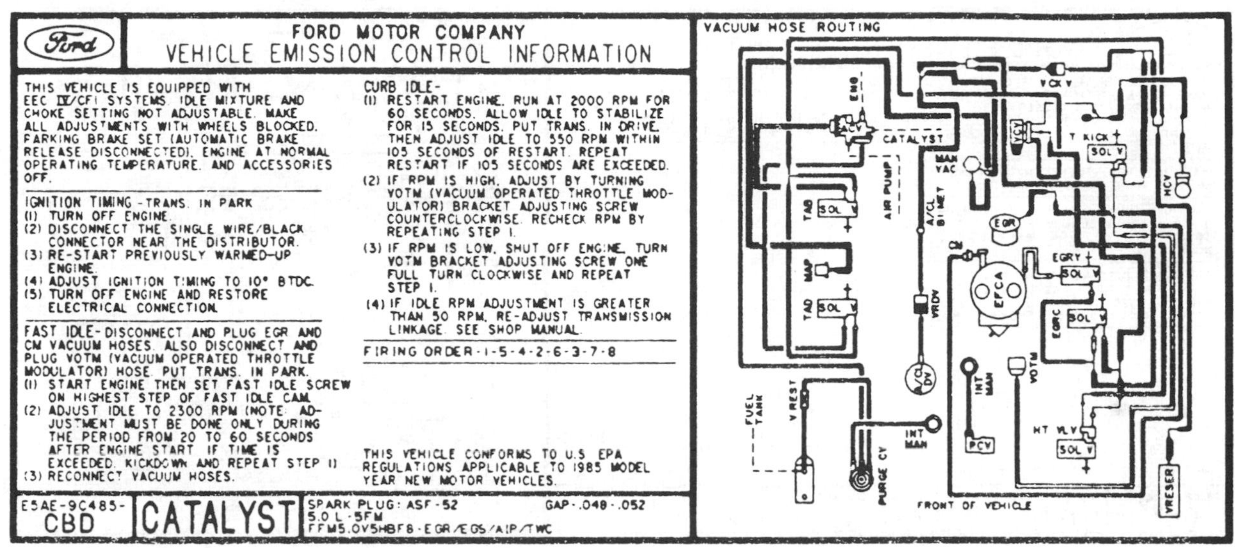

FORD MOTOR COMPANY
VEHICLE EMISSION CONTROL INFORMATION

THIS VEHICLE IS EQUIPPED WITH EEC IV/CFI SYSTEMS. IDLE MIXTURE AND CHOKE SETTING NOT ADJUSTABLE. MAKE ALL ADJUSTMENTS WITH WHEELS BLOCKED. PARKING BRAKE SET (AUTOMATIC BRAKE RELEASE DISCONNECTED), ENGINE AT NORMAL OPERATING TEMPERATURE, AND ACCESSORIES OFF.

IGNITION TIMING - TRANS. IN PARK
(1) TURN OFF ENGINE.
(2) DISCONNECT THE SINGLE WIRE/BLACK CONNECTOR NEAR THE DISTRIBUTOR.
(3) RE-START PREVIOUSLY WARMED-UP ENGINE.
(4) ADJUST IGNITION TIMING TO 10° BTDC.
(5) TURN OFF ENGINE AND RESTORE ELECTRICAL CONNECTION.

FAST IDLE - DISCONNECT AND PLUG EGR AND CM VACUUM HOSES. ALSO DISCONNECT AND PLUG VOTM (VACUUM OPERATED THROTTLE MODULATOR) HOSE. PUT TRANS. IN PARK.
(1) START ENGINE THEN SET FAST IDLE SCREW ON HIGHEST STEP OF FAST IDLE CAM.
(2) ADJUST IDLE TO 2300 RPM (NOTE: ADJUSTMENT MUST BE DONE ONLY DURING THE PERIOD FROM 20 TO 60 SECONDS AFTER ENGINE START. IF TIME IS EXCEEDED, KICKDOWN AND REPEAT STEP 1)
(3) RECONNECT VACUUM HOSES.

CURB IDLE -
(1) RESTART ENGINE. RUN AT 2000 RPM FOR 60 SECONDS. ALLOW IDLE TO STABILIZE FOR 15 SECONDS. PUT TRANS. IN DRIVE. THEN ADJUST IDLE TO 550 RPM WITHIN 105 SECONDS OF RESTART. REPEAT RESTART IF 105 SECONDS ARE EXCEEDED.
(2) IF RPM IS HIGH, ADJUST BY TURNING VOTM (VACUUM OPERATED THROTTLE MODULATOR) BRACKET ADJUSTING SCREW COUNTERCLOCKWISE. RECHECK RPM BY REPEATING STEP 1.
(3) IF RPM IS LOW, SHUT OFF ENGINE. TURN VOTM BRACKET ADJUSTING SCREW ONE FULL TURN CLOCKWISE AND REPEAT STEP 1.
(4) IF IDLE RPM ADJUSTMENT IS GREATER THAN 50 RPM, RE-ADJUST TRANSMISSION LINKAGE. SEE SHOP MANUAL.

FIRING ORDER - 1-5-4-2-6-3-7-8

THIS VEHICLE CONFORMS TO U.S. EPA REGULATIONS APPLICABLE TO 1985 MODEL YEAR NEW MOTOR VEHICLES.

E5AE-9C485-CBD | CATALYST | SPARK PLUG: ASF-52 GAP-.048-.052
5.0 L-5FM
FFM5.0V5HBF8-EGR/EGS/AIP/TWC

CALIBRATION: 4—22H—R00

MODEL YEAR: 1985

ENGINE: 5.0L

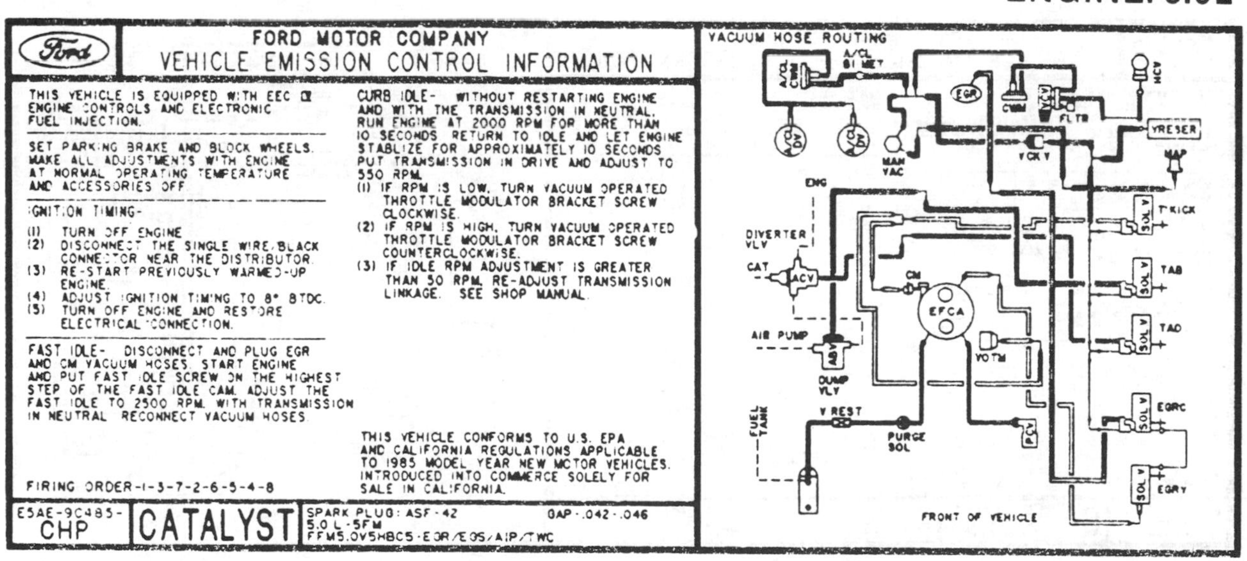

FORD MOTOR COMPANY
VEHICLE EMISSION CONTROL INFORMATION

THIS VEHICLE IS EQUIPPED WITH EEC IV ENGINE CONTROLS AND ELECTRONIC FUEL INJECTION.

SET PARKING BRAKE AND BLOCK WHEELS. MAKE ALL ADJUSTMENTS WITH ENGINE AT NORMAL OPERATING TEMPERATURE AND ACCESSORIES OFF.

IGNITION TIMING -
(1) TURN OFF ENGINE.
(2) DISCONNECT THE SINGLE WIRE/BLACK CONNECTOR NEAR THE DISTRIBUTOR.
(3) RE-START PREVIOUSLY WARMED-UP ENGINE.
(4) ADJUST IGNITION TIMING TO 8° BTDC.
(5) TURN OFF ENGINE AND RESTORE ELECTRICAL CONNECTION.

FAST IDLE - DISCONNECT AND PLUG EGR AND CM VACUUM HOSES. START ENGINE AND PUT FAST IDLE SCREW ON THE HIGHEST STEP OF THE FAST IDLE CAM. ADJUST THE FAST IDLE TO 2500 RPM, WITH TRANSMISSION IN NEUTRAL. RECONNECT VACUUM HOSES.

FIRING ORDER - 1-3-7-2-6-5-4-8

CURB IDLE - WITHOUT RESTARTING ENGINE AND WITH THE TRANSMISSION IN NEUTRAL, RUN ENGINE AT 2000 RPM FOR MORE THAN 10 SECONDS. RETURN TO IDLE AND LET ENGINE STABLIZE FOR APPROXIMATELY 10 SECONDS. PUT TRANSMISSION IN DRIVE AND ADJUST TO 550 RPM.
(1) IF RPM IS LOW, TURN VACUUM OPERATED THROTTLE MODULATOR BRACKET SCREW CLOCKWISE.
(2) IF RPM IS HIGH, TURN VACUUM OPERATED THROTTLE MODULATOR BRACKET SCREW COUNTERCLOCKWISE.
(3) IF IDLE RPM ADJUSTMENT IS GREATER THAN 50 RPM, RE-ADJUST TRANSMISSION LINKAGE. SEE SHOP MANUAL.

THIS VEHICLE CONFORMS TO U.S. EPA AND CALIFORNIA REGULATIONS APPLICABLE TO 1985 MODEL YEAR NEW MOTOR VEHICLES. INTRODUCED INTO COMMERCE SOLELY FOR SALE IN CALIFORNIA.

E5AE-9C485-CHP | CATALYST | SPARK PLUG: ASF-42 GAP-.042-.046
5.0 L-5FM
FFM5.0V5HBC5-EGR/EGS/AIP/TWC

CALIBRATION: 4—22P—R12

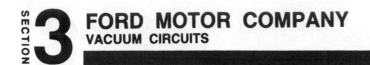

VACUUM CIRCUITS
(© Ford Motor Co.)

MODEL YEAR: 1985 **ENGINE: 5.0L**

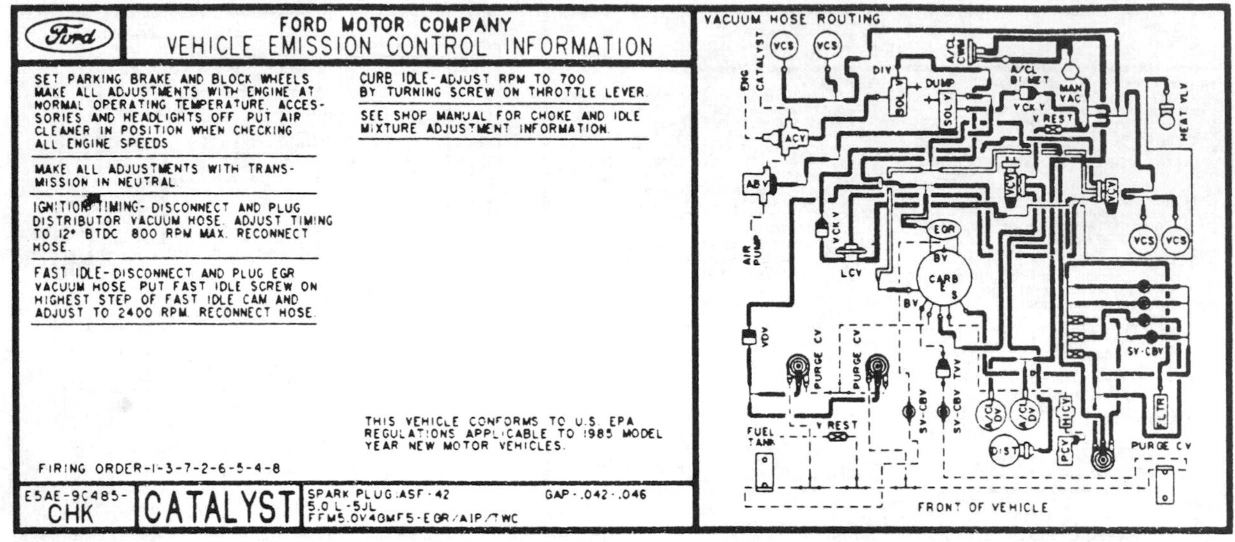
FORD MOTOR COMPANY
VEHICLE EMISSION CONTROL INFORMATION

SET PARKING BRAKE AND BLOCK WHEELS. MAKE ALL ADJUSTMENTS WITH ENGINE AT NORMAL OPERATING TEMPERATURE, ACCESSORIES AND HEADLIGHTS OFF. PUT AIR CLEANER IN POSITION WHEN CHECKING ALL ENGINE SPEEDS

MAKE ALL ADJUSTMENTS WITH TRANSMISSION IN NEUTRAL

IGNITION TIMING- DISCONNECT AND PLUG DISTRIBUTOR VACUUM HOSE. ADJUST TIMING TO 12° BTDC. 800 RPM MAX. RECONNECT HOSE.

FAST IDLE- DISCONNECT AND PLUG EGR VACUUM HOSE. PUT FAST IDLE SCREW ON HIGHEST STEP OF FAST IDLE CAM AND ADJUST TO 2400 RPM. RECONNECT HOSE.

CURB IDLE- ADJUST RPM TO 700 BY TURNING SCREW ON THROTTLE LEVER.

SEE SHOP MANUAL FOR CHOKE AND IDLE MIXTURE ADJUSTMENT INFORMATION.

THIS VEHICLE CONFORMS TO U.S. EPA REGULATIONS APPLICABLE TO 1985 MODEL YEAR NEW MOTOR VEHICLES.

FIRING ORDER-1-3-7-2-6-5-4-8

E5AE-9C485-CHK CATALYST SPARK PLUG: ASF-42 GAP-.042-.046
5.0 L-5JL
FFM5.0V4GMF5-EGR/AIP/TWC

CALIBRATION: 5—21A—R01

MODEL YEAR: 1985 **ENGINE: 5.0L**

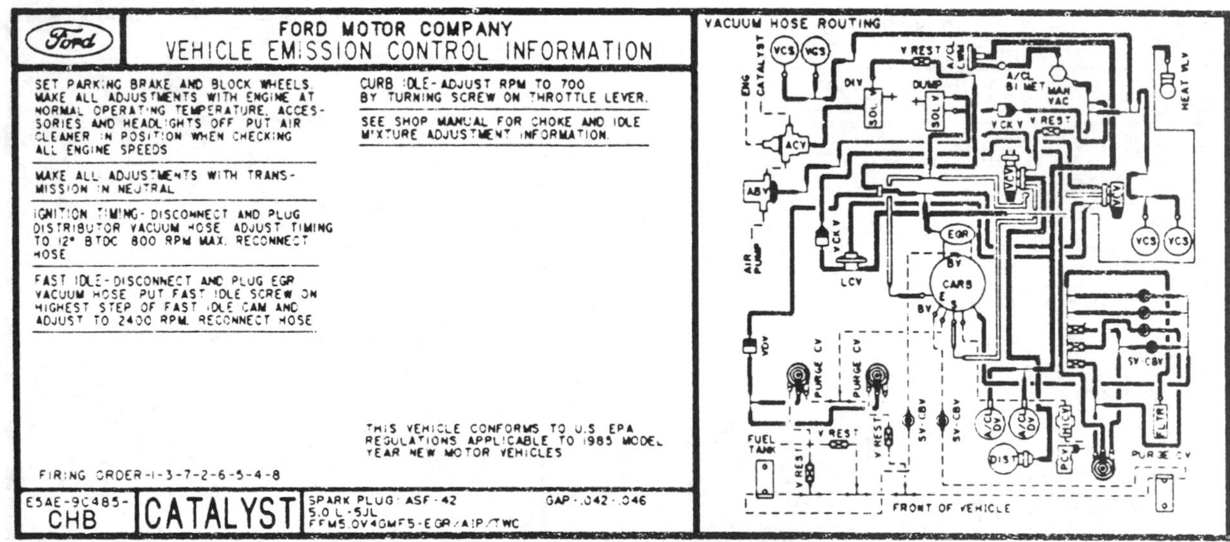
FORD MOTOR COMPANY
VEHICLE EMISSION CONTROL INFORMATION

SET PARKING BRAKE AND BLOCK WHEELS. MAKE ALL ADJUSTMENTS WITH ENGINE AT NORMAL OPERATING TEMPERATURE, ACCESSORIES AND HEADLIGHTS OFF. PUT AIR CLEANER IN POSITION WHEN CHECKING ALL ENGINE SPEEDS

MAKE ALL ADJUSTMENTS WITH TRANSMISSION IN NEUTRAL

IGNITION TIMING- DISCONNECT AND PLUG DISTRIBUTOR VACUUM HOSE. ADJUST TIMING TO 12° BTDC. 800 RPM MAX. RECONNECT HOSE

FAST IDLE- DISCONNECT AND PLUG EGR VACUUM HOSE. PUT FAST IDLE SCREW ON HIGHEST STEP OF FAST IDLE CAM AND ADJUST TO 2400 RPM. RECONNECT HOSE

CURB IDLE- ADJUST RPM TO 700 BY TURNING SCREW ON THROTTLE LEVER.

SEE SHOP MANUAL FOR CHOKE AND IDLE MIXTURE ADJUSTMENT INFORMATION.

THIS VEHICLE CONFORMS TO U.S. EPA REGULATIONS APPLICABLE TO 1985 MODEL YEAR NEW MOTOR VEHICLES

FIRING ORDER-1-3-7-2-6-5-4-8

E5AE-9C485-CHB CATALYST SPARK PLUG: ASF-42 GAP-.042-.046
5.0 L-5JL
FFM5.0V4GMF5-EGR/AIP/TWC

CALIBRATION: 5—21B—R00

VACUUM CIRCUITS

(© Ford Motor Co)

MODEL YEAR: 1985 **ENGINE: 5.0L**

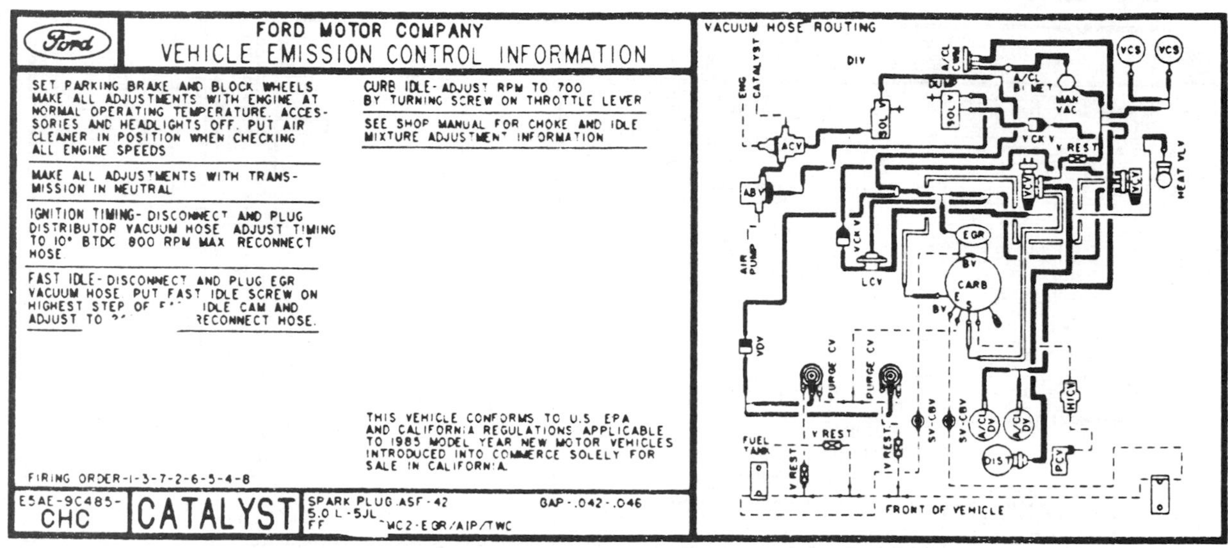

FORD MOTOR COMPANY
VEHICLE EMISSION CONTROL INFORMATION

SET PARKING BRAKE AND BLOCK WHEELS. MAKE ALL ADJUSTMENTS WITH ENGINE AT NORMAL OPERATING TEMPERATURE, ACCESSORIES AND HEADLIGHTS OFF, PUT AIR CLEANER IN POSITION WHEN CHECKING ALL ENGINE SPEEDS.

MAKE ALL ADJUSTMENTS WITH TRANSMISSION IN NEUTRAL.

IGNITION TIMING- DISCONNECT AND PLUG DISTRIBUTOR VACUUM HOSE. ADJUST TIMING TO 10° BTDC 800 RPM MAX. RECONNECT HOSE.

FAST IDLE- DISCONNECT AND PLUG EGR VACUUM HOSE. PUT FAST IDLE SCREW ON HIGHEST STEP OF F... IDLE CAM AND ADJUST TO ... RECONNECT HOSE.

CURB IDLE- ADJUST RPM TO 700 BY TURNING SCREW ON THROTTLE LEVER.

SEE SHOP MANUAL FOR CHOKE AND IDLE MIXTURE ADJUSTMENT INFORMATION.

THIS VEHICLE CONFORMS TO U.S. EPA AND CALIFORNIA REGULATIONS APPLICABLE TO 1985 MODEL YEAR NEW MOTOR VEHICLES INTRODUCED INTO COMMERCE SOLELY FOR SALE IN CALIFORNIA.

FIRING ORDER-1-3-7-2-6-5-4-8

E5AE-9C485-CHC | CATALYST | SPARK PLUG ASF-42 GAP-.042-.046 5.0L-5JL FF... MC2-EGR/AIP/TWC

CALIBRATION: 5—21P—R00

MODEL YEAR: 1985 **ENGINE: 5.0L**

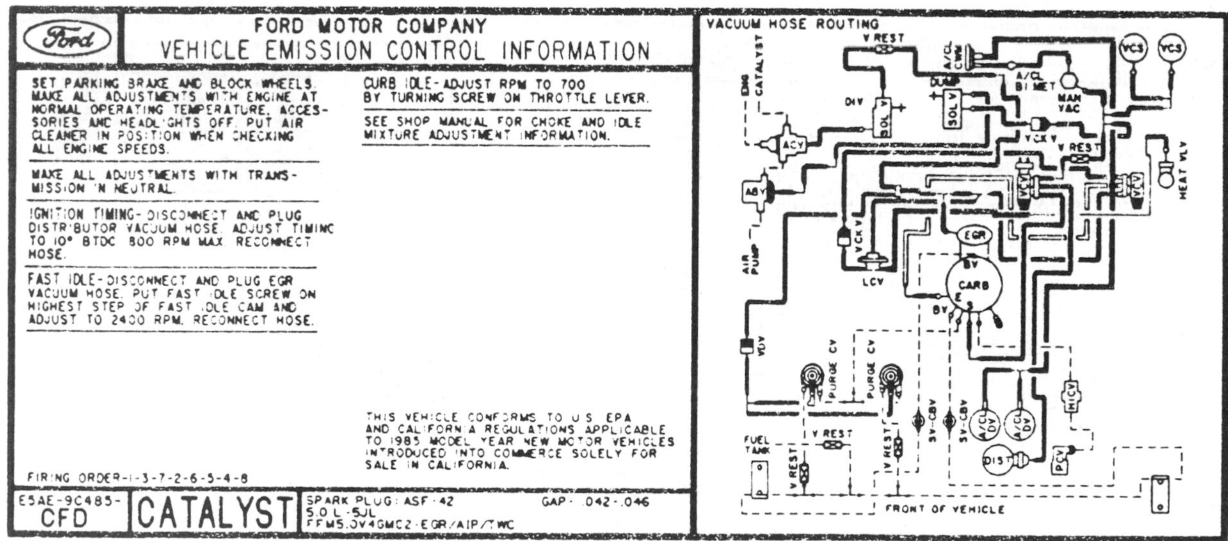

FORD MOTOR COMPANY
VEHICLE EMISSION CONTROL INFORMATION

SET PARKING BRAKE AND BLOCK WHEELS. MAKE ALL ADJUSTMENTS WITH ENGINE AT NORMAL OPERATING TEMPERATURE, ACCESSORIES AND HEADLIGHTS OFF, PUT AIR CLEANER IN POSITION WHEN CHECKING ALL ENGINE SPEEDS.

MAKE ALL ADJUSTMENTS WITH TRANSMISSION IN NEUTRAL.

IGNITION TIMING- DISCONNECT AND PLUG DISTRIBUTOR VACUUM HOSE. ADJUST TIMING TO 10° BTDC 800 RPM MAX. RECONNECT HOSE.

FAST IDLE- DISCONNECT AND PLUG EGR VACUUM HOSE. PUT FAST IDLE SCREW ON HIGHEST STEP OF FAST IDLE CAM AND ADJUST TO 2400 RPM. RECONNECT HOSE.

CURB IDLE- ADJUST RPM TO 700 BY TURNING SCREW ON THROTTLE LEVER.

SEE SHOP MANUAL FOR CHOKE AND IDLE MIXTURE ADJUSTMENT INFORMATION.

THIS VEHICLE CONFORMS TO U.S. EPA AND CALIFORNIA REGULATIONS APPLICABLE TO 1985 MODEL YEAR NEW MOTOR VEHICLES INTRODUCED INTO COMMERCE SOLELY FOR SALE IN CALIFORNIA.

FIRING ORDER-1-3-7-2-6-5-4-8

E5AE-9C485-CFD | CATALYST | SPARK PLUG ASF-42 GAP-.042-.046 5.0L-5JL FFM5.0V4GMC2-EGR/AIP/TWC

CALIBRATION: 5—21Q—R00

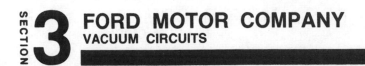

VACUUM CIRCUITS

(© Ford Motor Co.)

MODEL YEAR: 1985 — ENGINE: 5.0L

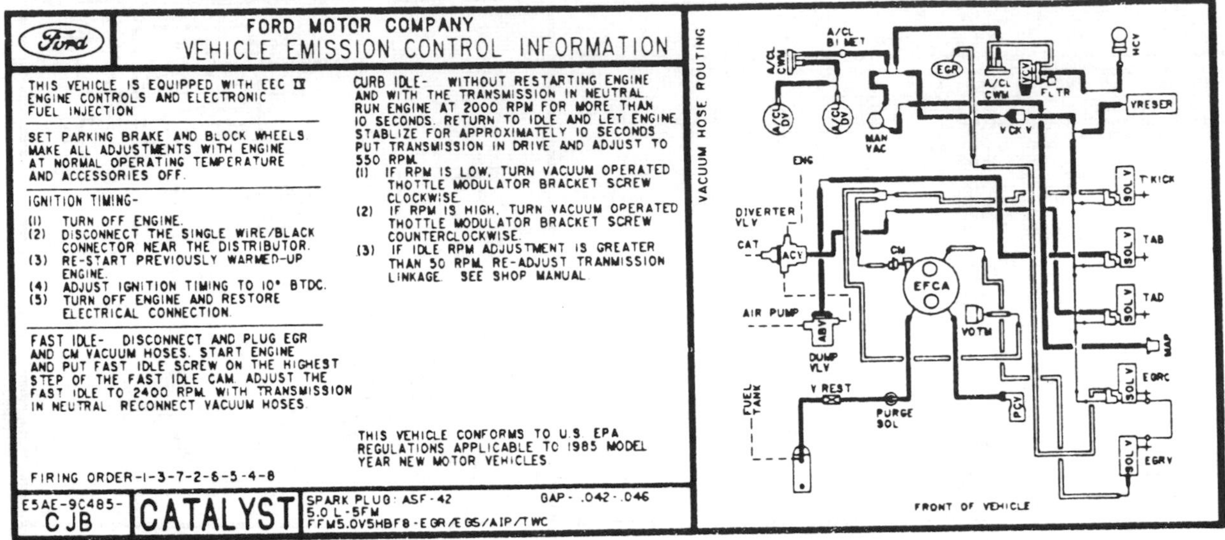

FORD MOTOR COMPANY
VEHICLE EMISSION CONTROL INFORMATION

THIS VEHICLE IS EQUIPPED WITH EEC IV ENGINE CONTROLS AND ELECTRONIC FUEL INJECTION

SET PARKING BRAKE AND BLOCK WHEELS MAKE ALL ADJUSTMENTS WITH ENGINE AT NORMAL OPERATING TEMPERATURE AND ACCESSORIES OFF.

IGNITION TIMING-
(1) TURN OFF ENGINE.
(2) DISCONNECT THE SINGLE WIRE/BLACK CONNECTOR NEAR THE DISTRIBUTOR.
(3) RE-START PREVIOUSLY WARMED-UP ENGINE.
(4) ADJUST IGNITION TIMING TO 10° BTDC.
(5) TURN OFF ENGINE AND RESTORE ELECTRICAL CONNECTION.

FAST IDLE- DISCONNECT AND PLUG EGR AND CM VACUUM HOSES. START ENGINE AND PUT FAST IDLE SCREW ON THE HIGHEST STEP OF THE FAST IDLE CAM. ADJUST THE FAST IDLE TO 2400 RPM WITH TRANSMISSION IN NEUTRAL. RECONNECT VACUUM HOSES.

CURB IDLE- WITHOUT RESTARTING ENGINE AND WITH THE TRANSMISSION IN NEUTRAL RUN ENGINE AT 2000 RPM FOR MORE THAN 10 SECONDS. RETURN TO IDLE AND LET ENGINE STABLIZE FOR APPROXIMATELY 10 SECONDS PUT TRANSMISSION IN DRIVE AND ADJUST TO 550 RPM.
(1) IF RPM IS LOW, TURN VACUUM OPERATED THOTTLE MODULATOR BRACKET SCREW CLOCKWISE.
(2) IF RPM IS HIGH, TURN VACUUM OPERATED THOTTLE MODULATOR BRACKET SCREW COUNTERCLOCKWISE.
(3) IF IDLE RPM ADJUSTMENT IS GREATER THAN 50 RPM, RE-ADJUST TRANMISSION LINKAGE. SEE SHOP MANUAL.

THIS VEHICLE CONFORMS TO U.S. EPA REGULATIONS APPLICABLE TO 1985 MODEL YEAR NEW MOTOR VEHICLES.

FIRING ORDER-1-3-7-2-6-5-4-8

E5AE-9C485-CJB | CATALYST | SPARK PLUG: ASF-42 GAP-.042-.046 5.0L-5FM FFM5.0V5HBF8-EGR/EGS/AIP/TWC

CALIBRATION: 5—22A—R00

MODEL YEAR: 1985 — ENGINE: 5.0L

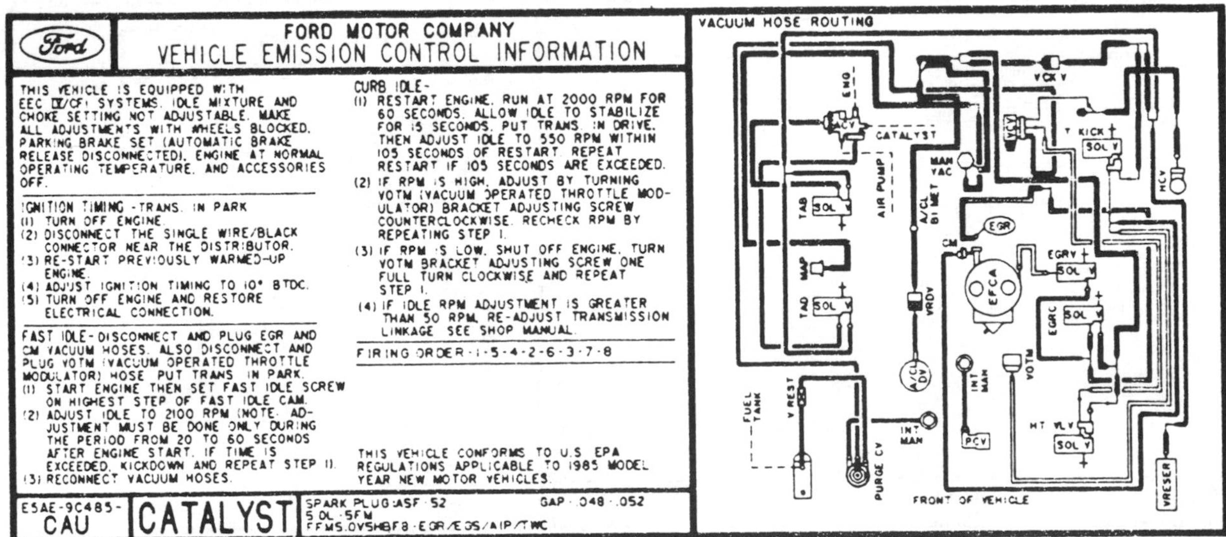

FORD MOTOR COMPANY
VEHICLE EMISSION CONTROL INFORMATION

THIS VEHICLE IS EQUIPPED WITH EEC IV/CFI SYSTEMS. IDLE MIXTURE AND CHOKE SETTING NOT ADJUSTABLE. MAKE ALL ADJUSTMENTS WITH WHEELS BLOCKED, PARKING BRAKE SET (AUTOMATIC BRAKE RELEASE DISCONNECTED), ENGINE AT NORMAL OPERATING TEMPERATURE, AND ACCESSORIES OFF.

IGNITION TIMING -TRANS. IN PARK
(1) TURN OFF ENGINE.
(2) DISCONNECT THE SINGLE WIRE/BLACK CONNECTOR NEAR THE DISTRIBUTOR.
(3) RE-START PREVIOUSLY WARMED-UP ENGINE.
(4) ADJUST IGNITION TIMING TO 10° BTDC.
(5) TURN OFF ENGINE AND RESTORE ELECTRICAL CONNECTION.

FAST IDLE-DISCONNECT AND PLUG EGR AND CM VACUUM HOSES. ALSO DISCONNECT AND PLUG VOTM (VACUUM OPERATED THROTTLE MODULATOR) HOSE. PUT TRANS. IN PARK.
(1) START ENGINE THEN SET FAST IDLE SCREW ON HIGHEST STEP OF FAST IDLE CAM.
(2) ADJUST IDLE TO 2100 RPM (NOTE: ADJUSTMENT MUST BE DONE ONLY DURING THE PERIOD FROM 20 TO 60 SECONDS AFTER ENGINE START. IF TIME IS EXCEEDED, KICKDOWN AND REPEAT STEP 1).
(3) RECONNECT VACUUM HOSES.

CURB IDLE-
(1) RESTART ENGINE, RUN AT 2000 RPM FOR 60 SECONDS. ALLOW IDLE TO STABILIZE FOR 15 SECONDS. PUT TRANS. IN DRIVE. THEN ADJUST IDLE TO 550 RPM WITHIN 105 SECONDS OF RESTART. REPEAT RESTART IF 105 SECONDS ARE EXCEEDED.
(2) IF RPM IS HIGH, ADJUST BY TURNING VOTM (VACUUM OPERATED THROTTLE MODULATOR) BRACKET ADJUSTING SCREW COUNTERCLOCKWISE. RECHECK RPM BY REPEATING STEP 1.
(3) IF RPM IS LOW, SHUT OFF ENGINE. TURN VOTM BRACKET ADJUSTING SCREW ONE FULL TURN CLOCKWISE AND REPEAT STEP 1.
(4) IF IDLE RPM ADJUSTMENT IS GREATER THAN 50 RPM, RE-ADJUST TRANSMISSION LINKAGE. SEE SHOP MANUAL.

FIRING ORDER-1-5-4-2-6-3-7-8

THIS VEHICLE CONFORMS TO U.S. EPA REGULATIONS APPLICABLE TO 1985 MODEL YEAR NEW MOTOR VEHICLES.

E5AE-9C485-CAU | CATALYST | SPARK PLUG: ASF-52 GAP-.048-.052 5.0L-5FM FFM5.0V5HBF8-EGR/EGS/AIP/TWC

CALIBRATION: 5—22B—R00

VACUUM CIRCUITS

(© Ford Motor Co.)

MODEL YEAR: 1985 **ENGINE: 5.0L**

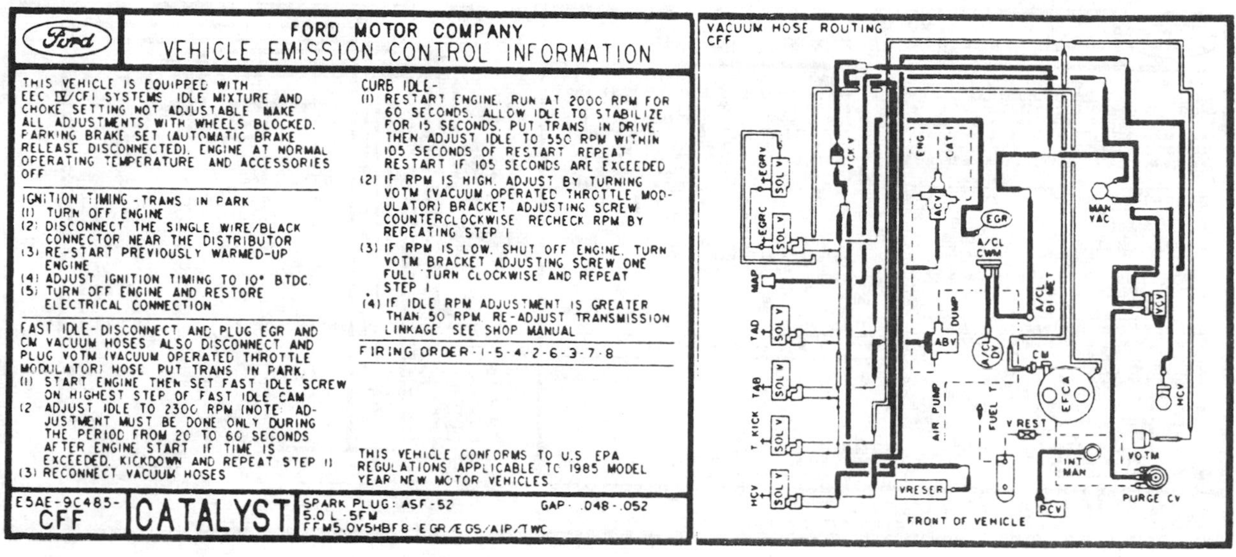

FORD MOTOR COMPANY
VEHICLE EMISSION CONTROL INFORMATION

THIS VEHICLE IS EQUIPPED WITH EEC IV/CFI SYSTEMS. IDLE MIXTURE AND CHOKE SETTING NOT ADJUSTABLE. MAKE ALL ADJUSTMENTS WITH WHEELS BLOCKED, PARKING BRAKE SET (AUTOMATIC BRAKE RELEASE DISCONNECTED), ENGINE AT NORMAL OPERATING TEMPERATURE AND ACCESSORIES OFF.

IGNITION TIMING - TRANS. IN PARK
(1) TURN OFF ENGINE
(2) DISCONNECT THE SINGLE WIRE/BLACK CONNECTOR NEAR THE DISTRIBUTOR
(3) RE-START PREVIOUSLY WARMED-UP ENGINE
(4) ADJUST IGNITION TIMING TO 10° BTDC
(5) TURN OFF ENGINE AND RESTORE ELECTRICAL CONNECTION

FAST IDLE - DISCONNECT AND PLUG EGR AND CM VACUUM HOSES. ALSO DISCONNECT AND PLUG VOTM (VACUUM OPERATED THROTTLE MODULATOR) HOSE. PUT TRANS. IN PARK.
(1) START ENGINE THEN SET FAST IDLE SCREW ON HIGHEST STEP OF FAST IDLE CAM.
(2) ADJUST IDLE TO 2300 RPM (NOTE: ADJUSTMENT MUST BE DONE ONLY DURING THE PERIOD FROM 20 TO 60 SECONDS AFTER ENGINE START. IF TIME IS EXCEEDED, KICKDOWN AND REPEAT STEP 1)
(3) RECONNECT VACUUM HOSES

CURB IDLE -
(1) RESTART ENGINE. RUN AT 2000 RPM FOR 60 SECONDS. ALLOW IDLE TO STABILIZE FOR 15 SECONDS. PUT TRANS. IN DRIVE. THEN ADJUST IDLE TO 550 RPM WITHIN 105 SECONDS OF RESTART. REPEAT RESTART IF 105 SECONDS ARE EXCEEDED.
(2) IF RPM IS HIGH, ADJUST BY TURNING VOTM (VACUUM OPERATED THROTTLE MODULATOR) BRACKET ADJUSTING SCREW COUNTERCLOCKWISE. RECHECK RPM BY REPEATING STEP 1.
(3) IF RPM IS LOW, SHUT OFF ENGINE. TURN VOTM BRACKET ADJUSTING SCREW ONE FULL TURN CLOCKWISE AND REPEAT STEP 1.
(4) IF IDLE RPM ADJUSTMENT IS GREATER THAN 50 RPM, RE-ADJUST TRANSMISSION LINKAGE. SEE SHOP MANUAL.

FIRING ORDER - 1-5-4-2-6-3-7-8

THIS VEHICLE CONFORMS TO U.S. EPA REGULATIONS APPLICABLE TO 1985 MODEL YEAR NEW MOTOR VEHICLES.

E5AE-9C485-CFF | CATALYST | SPARK PLUG: ASF-52 GAP: .048-.052
5.0L-5FM
FFM5.0V5HBF8-EGR/EGS/AIP/TWC

CALIBRATION: 5—22D—R00

MODEL YEAR: 1985 **ENGINE: 5.0L**

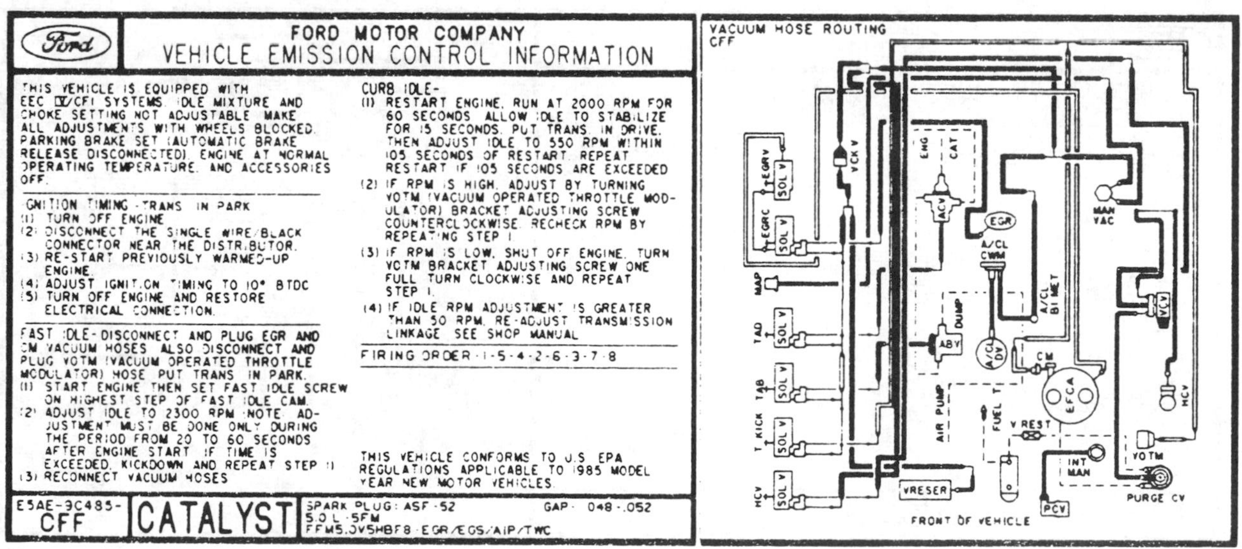

FORD MOTOR COMPANY
VEHICLE EMISSION CONTROL INFORMATION

THIS VEHICLE IS EQUIPPED WITH EEC IV/CFI SYSTEMS. IDLE MIXTURE AND CHOKE SETTING NOT ADJUSTABLE. MAKE ALL ADJUSTMENTS WITH WHEELS BLOCKED, PARKING BRAKE SET (AUTOMATIC BRAKE RELEASE DISCONNECTED), ENGINE AT NORMAL OPERATING TEMPERATURE AND ACCESSORIES OFF.

IGNITION TIMING - TRANS. IN PARK
(1) TURN OFF ENGINE
(2) DISCONNECT THE SINGLE WIRE/BLACK CONNECTOR NEAR THE DISTRIBUTOR
(3) RE-START PREVIOUSLY WARMED-UP ENGINE
(4) ADJUST IGNITION TIMING TO 10° BTDC
(5) TURN OFF ENGINE AND RESTORE ELECTRICAL CONNECTION

FAST IDLE - DISCONNECT AND PLUG EGR AND CM VACUUM HOSES. ALSO DISCONNECT AND PLUG VOTM (VACUUM OPERATED THROTTLE MODULATOR) HOSE. PUT TRANS. IN PARK.
(1) START ENGINE THEN SET FAST IDLE SCREW ON HIGHEST STEP OF FAST IDLE CAM.
(2) ADJUST IDLE TO 2300 RPM (NOTE: ADJUSTMENT MUST BE DONE ONLY DURING THE PERIOD FROM 20 TO 60 SECONDS AFTER ENGINE START. IF TIME IS EXCEEDED, KICKDOWN AND REPEAT STEP 1)
(3) RECONNECT VACUUM HOSES

CURB IDLE -
(1) RESTART ENGINE. RUN AT 2000 RPM FOR 60 SECONDS. ALLOW IDLE TO STABILIZE FOR 15 SECONDS. PUT TRANS. IN DRIVE. THEN ADJUST IDLE TO 550 RPM WITHIN 105 SECONDS OF RESTART. REPEAT RESTART IF 105 SECONDS ARE EXCEEDED.
(2) IF RPM IS HIGH, ADJUST BY TURNING VOTM (VACUUM OPERATED THROTTLE MODULATOR) BRACKET ADJUSTING SCREW COUNTERCLOCKWISE. RECHECK RPM BY REPEATING STEP 1.
(3) IF RPM IS LOW, SHUT OFF ENGINE. TURN VOTM BRACKET ADJUSTING SCREW ONE FULL TURN CLOCKWISE AND REPEAT STEP 1.
(4) IF IDLE RPM ADJUSTMENT IS GREATER THAN 50 RPM, RE-ADJUST TRANSMISSION LINKAGE. SEE SHOP MANUAL.

FIRING ORDER - 1-5-4-2-6-3-7-8

THIS VEHICLE CONFORMS TO U.S. EPA REGULATIONS APPLICABLE TO 1985 MODEL YEAR NEW MOTOR VEHICLES.

E5AE-9C485-CFF | CATALYST | SPARK PLUG: ASF-52 GAP: .048-.052
5.0L-5FM
FFM5.0V5HBF8-EGR/EGS/AIP/TWC

CALIBRATION: 5—22D—R10

VACUUM CIRCUITS

(© Ford Motor Co)

MODEL YEAR: 1985 **ENGINE: 5.0L**

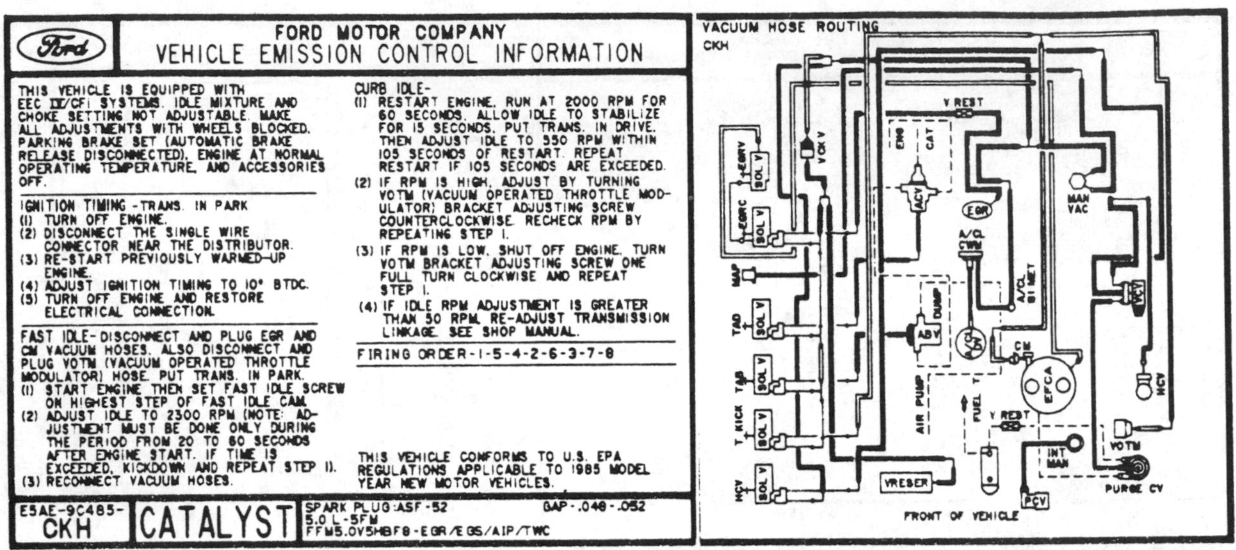

FORD MOTOR COMPANY
VEHICLE EMISSION CONTROL INFORMATION

THIS VEHICLE IS EQUIPPED WITH EEC IV/CFI SYSTEMS. IDLE MIXTURE AND CHOKE SETTING NOT ADJUSTABLE. MAKE ALL ADJUSTMENTS WITH WHEELS BLOCKED, PARKING BRAKE SET (AUTOMATIC BRAKE RELEASE DISCONNECTED), ENGINE AT NORMAL OPERATING TEMPERATURE, AND ACCESSORIES OFF.

IGNITION TIMING - TRANS. IN PARK
(1) TURN OFF ENGINE.
(2) DISCONNECT THE SINGLE WIRE CONNECTOR NEAR THE DISTRIBUTOR.
(3) RE-START PREVIOUSLY WARMED-UP ENGINE.
(4) ADJUST IGNITION TIMING TO 10° BTDC.
(5) TURN OFF ENGINE AND RESTORE ELECTRICAL CONNECTION.

FAST IDLE - DISCONNECT AND PLUG EGR AND CM VACUUM HOSES. ALSO DISCONNECT AND PLUG VOTM (VACUUM OPERATED THROTTLE MODULATOR) HOSE. PUT TRANS. IN PARK.
(1) START ENGINE THEN SET FAST IDLE SCREW ON HIGHEST STEP OF FAST IDLE CAM.
(2) ADJUST IDLE TO 2300 RPM (NOTE: ADJUSTMENT MUST BE DONE ONLY DURING THE PERIOD FROM 20 TO 60 SECONDS AFTER ENGINE START. IF TIME IS EXCEEDED, KICKDOWN AND REPEAT STEP 1).
(3) RECONNECT VACUUM HOSES.

CURB IDLE -
(1) RESTART ENGINE. RUN AT 2000 RPM FOR 60 SECONDS. ALLOW IDLE TO STABILIZE FOR 15 SECONDS. PUT TRANS. IN DRIVE. THEN ADJUST IDLE TO 550 RPM WITHIN 105 SECONDS OF RESTART. REPEAT RESTART IF 105 SECONDS ARE EXCEEDED.
(2) IF RPM IS HIGH, ADJUST BY TURNING VOTM (VACUUM OPERATED THROTTLE MODULATOR) BRACKET ADJUSTING SCREW COUNTERCLOCKWISE. RECHECK RPM BY REPEATING STEP 1.
(3) IF RPM IS LOW, SHUT OFF ENGINE. TURN VOTM BRACKET ADJUSTING SCREW ONE FULL TURN CLOCKWISE AND REPEAT STEP 1.
(4) IF IDLE RPM ADJUSTMENT IS GREATER THAN 50 RPM, RE-ADJUST TRANSMISSION LINKAGE. SEE SHOP MANUAL.

FIRING ORDER - 1-5-4-2-6-3-7-8

THIS VEHICLE CONFORMS TO U.S. EPA REGULATIONS APPLICABLE TO 1985 MODEL YEAR NEW MOTOR VEHICLES.

E5AE-9C485-CKH | CATALYST | SPARK PLUG: ASF-52 GAP-.048-.052
5.0 L-5FM
FFM5.0V5HBF8-EGR/EGS/AIP/TWC

CALIBRATION: 5—22D—R11

MODEL YEAR: 1985 **ENGINE: 5.0L**

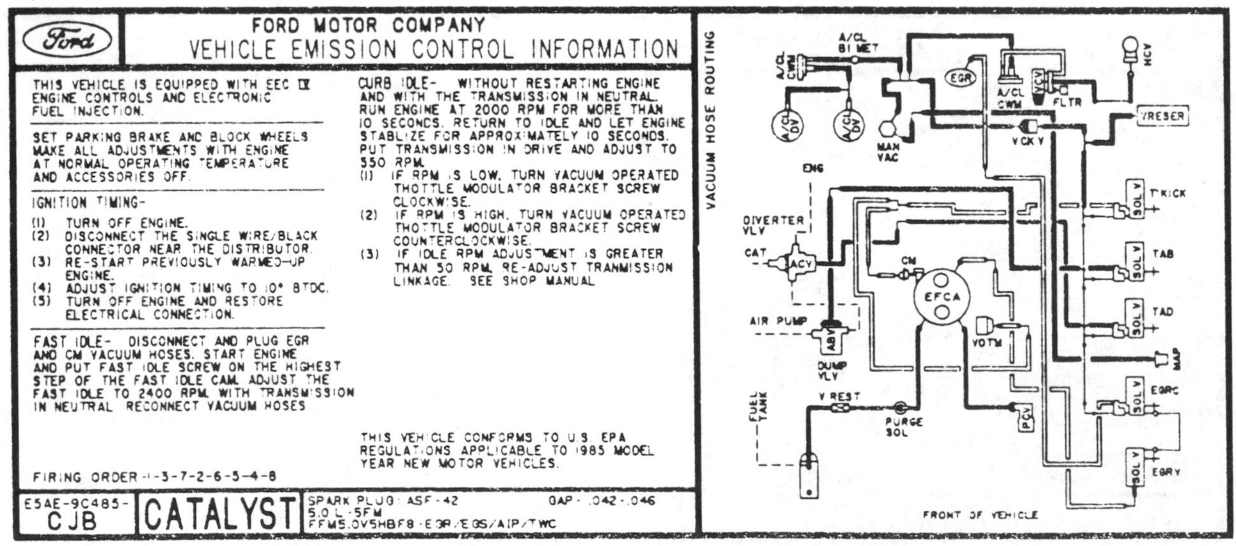

FORD MOTOR COMPANY
VEHICLE EMISSION CONTROL INFORMATION

THIS VEHICLE IS EQUIPPED WITH EEC IV ENGINE CONTROLS AND ELECTRONIC FUEL INJECTION.

SET PARKING BRAKE AND BLOCK WHEELS. MAKE ALL ADJUSTMENTS WITH ENGINE AT NORMAL OPERATING TEMPERATURE AND ACCESSORIES OFF.

IGNITION TIMING -
(1) TURN OFF ENGINE.
(2) DISCONNECT THE SINGLE WIRE/BLACK CONNECTOR NEAR THE DISTRIBUTOR.
(3) RE-START PREVIOUSLY WARMED-UP ENGINE.
(4) ADJUST IGNITION TIMING TO 10° BTDC.
(5) TURN OFF ENGINE AND RESTORE ELECTRICAL CONNECTION.

FAST IDLE - DISCONNECT AND PLUG EGR AND CM VACUUM HOSES. START ENGINE AND PUT FAST IDLE SCREW ON THE HIGHEST STEP OF THE FAST IDLE CAM. ADJUST THE FAST IDLE TO 2400 RPM WITH TRANSMISSION IN NEUTRAL. RECONNECT VACUUM HOSES

FIRING ORDER - 1-3-7-2-6-5-4-8

CURB IDLE - WITHOUT RESTARTING ENGINE AND WITH THE TRANSMISSION IN NEUTRAL. RUN ENGINE AT 2000 RPM FOR MORE THAN 10 SECONDS. RETURN TO IDLE AND LET ENGINE STABLIZE FOR APPROXIMATELY 10 SECONDS. PUT TRANSMISSION IN DRIVE AND ADJUST TO 550 RPM.
(1) IF RPM IS LOW, TURN VACUUM OPERATED THOTTLE MODULATOR BRACKET SCREW CLOCKWISE.
(2) IF RPM IS HIGH, TURN VACUUM OPERATED THOTTLE MODULATOR BRACKET SCREW COUNTERCLOCKWISE.
(3) IF IDLE RPM ADJUSTMENT IS GREATER THAN 50 RPM, RE-ADJUST TRANMISSION LINKAGE. SEE SHOP MANUAL

THIS VEHICLE CONFORMS TO U.S. EPA REGULATIONS APPLICABLE TO 1985 MODEL YEAR NEW MOTOR VEHICLES.

E5AE-9C485-CJB | CATALYST | SPARK PLUG: ASF-42 GAP-.042-.046
5.0 L-5FM
FFM5.0V5HBF8-EGR/EGS/AIP/TWC

CALIBRATION: 5—22I—R00

VACUUM CIRCUITS
(© Ford Motor Co.)

MODEL YEAR: 1985 **ENGINE: 5.0L**

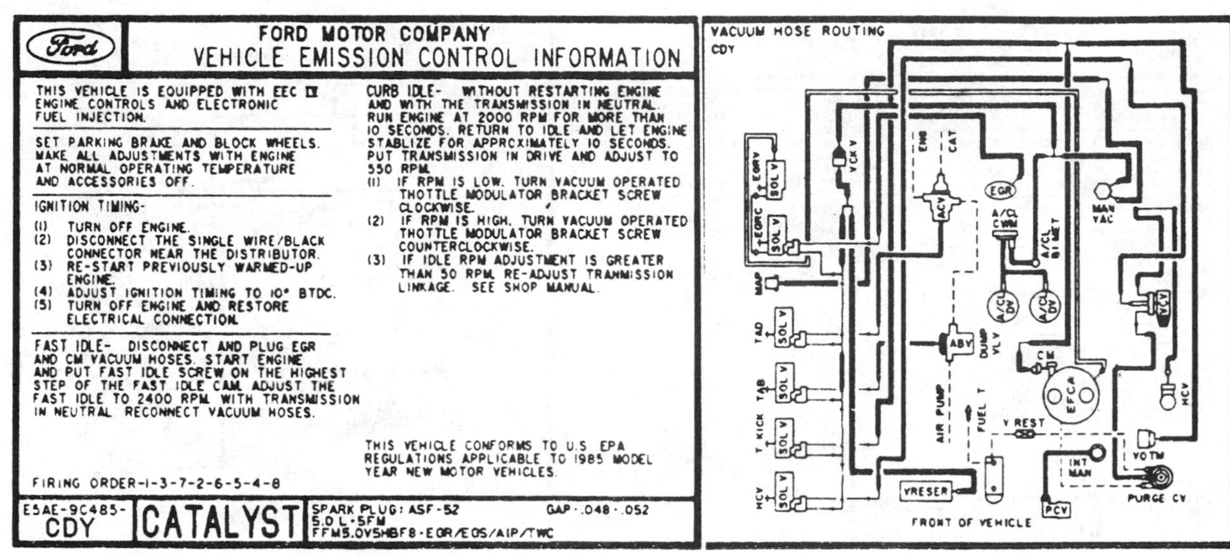

FORD MOTOR COMPANY
VEHICLE EMISSION CONTROL INFORMATION

THIS VEHICLE IS EQUIPPED WITH EEC IV ENGINE CONTROLS AND ELECTRONIC FUEL INJECTION.

SET PARKING BRAKE AND BLOCK WHEELS. MAKE ALL ADJUSTMENTS WITH ENGINE AT NORMAL OPERATING TEMPERATURE AND ACCESSORIES OFF.

IGNITION TIMING-
(1) TURN OFF ENGINE.
(2) DISCONNECT THE SINGLE WIRE/BLACK CONNECTOR NEAR THE DISTRIBUTOR.
(3) RE-START PREVIOUSLY WARMED-UP ENGINE.
(4) ADJUST IGNITION TIMING TO 10° BTDC.
(5) TURN OFF ENGINE AND RESTORE ELECTRICAL CONNECTION.

FAST IDLE- DISCONNECT AND PLUG EGR AND CM VACUUM HOSES. START ENGINE AND PUT FAST IDLE SCREW ON THE HIGHEST STEP OF THE FAST IDLE CAM. ADJUST THE FAST IDLE TO 2400 RPM WITH TRANSMISSION IN NEUTRAL. RECONNECT VACUUM HOSES.

CURB IDLE- WITHOUT RESTARTING ENGINE AND WITH THE TRANSMISSION IN NEUTRAL, RUN ENGINE AT 2000 RPM FOR MORE THAN 10 SECONDS. RETURN TO IDLE AND LET ENGINE STABLIZE FOR APPROXIMATELY 10 SECONDS. PUT TRANSMISSION IN DRIVE AND ADJUST TO 550 RPM.
(1) IF RPM IS LOW, TURN VACUUM OPERATED THOTTLE MODULATOR BRACKET SCREW CLOCKWISE.
(2) IF RPM IS HIGH, TURN VACUUM OPERATED THOTTLE MODULATOR BRACKET SCREW COUNTERCLOCKWISE.
(3) IF IDLE RPM ADJUSTMENT IS GREATER THAN 50 RPM, RE-ADJUST TRANMISSION LINKAGE. SEE SHOP MANUAL.

THIS VEHICLE CONFORMS TO U.S. EPA REGULATIONS APPLICABLE TO 1985 MODEL YEAR NEW MOTOR VEHICLES.

FIRING ORDER-1-3-7-2-6-5-4-8

E5AE-9C485- CDY | CATALYST | SPARK PLUG: ASF-52 GAP-.048-.052 5.0L-5FM FFM5.0V5HBF8-EGR/EOS/AIP/TWC

CALIBRATION: 5—22L—R00

MODEL YEAR: 1985 **ENGINE: 5.0L**

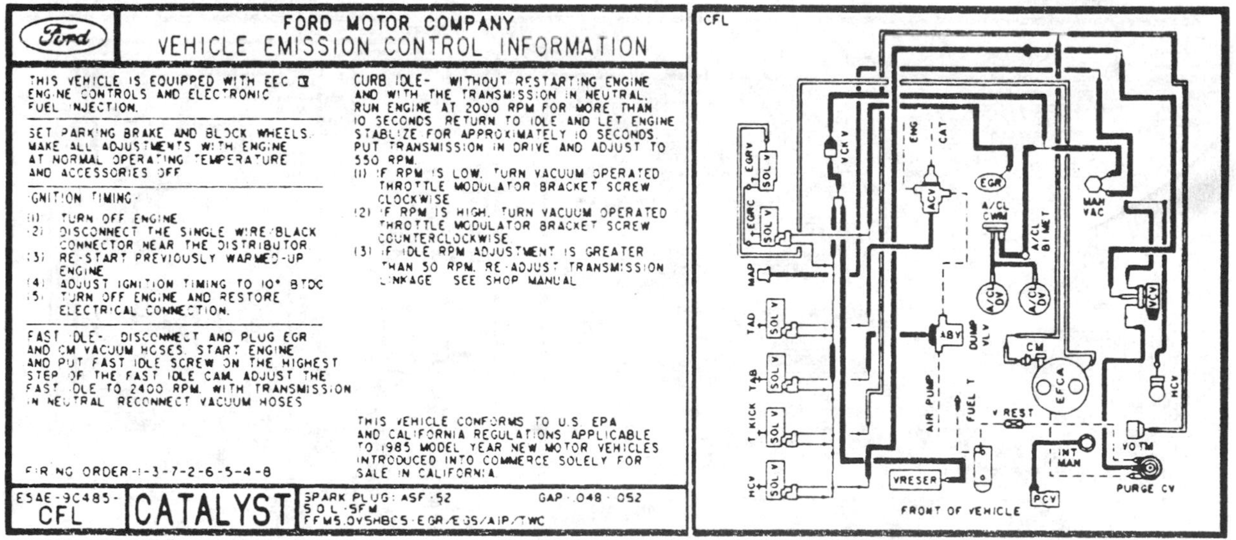

FORD MOTOR COMPANY
VEHICLE EMISSION CONTROL INFORMATION

THIS VEHICLE IS EQUIPPED WITH EEC IV ENGINE CONTROLS AND ELECTRONIC FUEL INJECTION.

SET PARKING BRAKE AND BLOCK WHEELS. MAKE ALL ADJUSTMENTS WITH ENGINE AT NORMAL OPERATING TEMPERATURE AND ACCESSORIES OFF.

IGNITION TIMING-
(1) TURN OFF ENGINE.
(2) DISCONNECT THE SINGLE WIRE/BLACK CONNECTOR NEAR THE DISTRIBUTOR.
(3) RE-START PREVIOUSLY WARMED-UP ENGINE.
(4) ADJUST IGNITION TIMING TO 10° BTDC.
(5) TURN OFF ENGINE AND RESTORE ELECTRICAL CONNECTION.

FAST IDLE- DISCONNECT AND PLUG EGR AND CM VACUUM HOSES. START ENGINE AND PUT FAST IDLE SCREW ON THE HIGHEST STEP OF THE FAST IDLE CAM. ADJUST THE FAST IDLE TO 2400 RPM WITH TRANSMISSION IN NEUTRAL. RECONNECT VACUUM HOSES.

CURB IDLE- WITHOUT RESTARTING ENGINE AND WITH THE TRANSMISSION IN NEUTRAL, RUN ENGINE AT 2000 RPM FOR MORE THAN 10 SECONDS. RETURN TO IDLE AND LET ENGINE STABLIZE FOR APPROXIMATELY 10 SECONDS. PUT TRANSMISSION IN DRIVE AND ADJUST TO 550 RPM.
(1) IF RPM IS LOW, TURN VACUUM OPERATED THROTTLE MODULATOR BRACKET SCREW CLOCKWISE.
(2) IF RPM IS HIGH, TURN VACUUM OPERATED THROTTLE MODULATOR BRACKET SCREW COUNTERCLOCKWISE.
(3) IF IDLE RPM ADJUSTMENT IS GREATER THAN 50 RPM, RE-ADJUST TRANSMISSION LINKAGE. SEE SHOP MANUAL.

THIS VEHICLE CONFORMS TO U.S. EPA AND CALIFORNIA REGULATIONS APPLICABLE TO 1985 MODEL YEAR NEW MOTOR VEHICLES INTRODUCED INTO COMMERCE SOLELY FOR SALE IN CALIFORNIA.

FIRING ORDER-1-3-7-2-6-5-4-8

E5AE-9C485- CFL | CATALYST | SPARK PLUG: ASF-52 GAP-.048-.052 5.0L-5FM FFM5.0V5HBC5-EGR/EOS/AIP/TWC

CALIBRATION: 5—22M—R01

VACUUM CIRCUITS

(© Ford Motor Co.)

MODEL YEAR: 1985 **ENGINE: 5.0L**

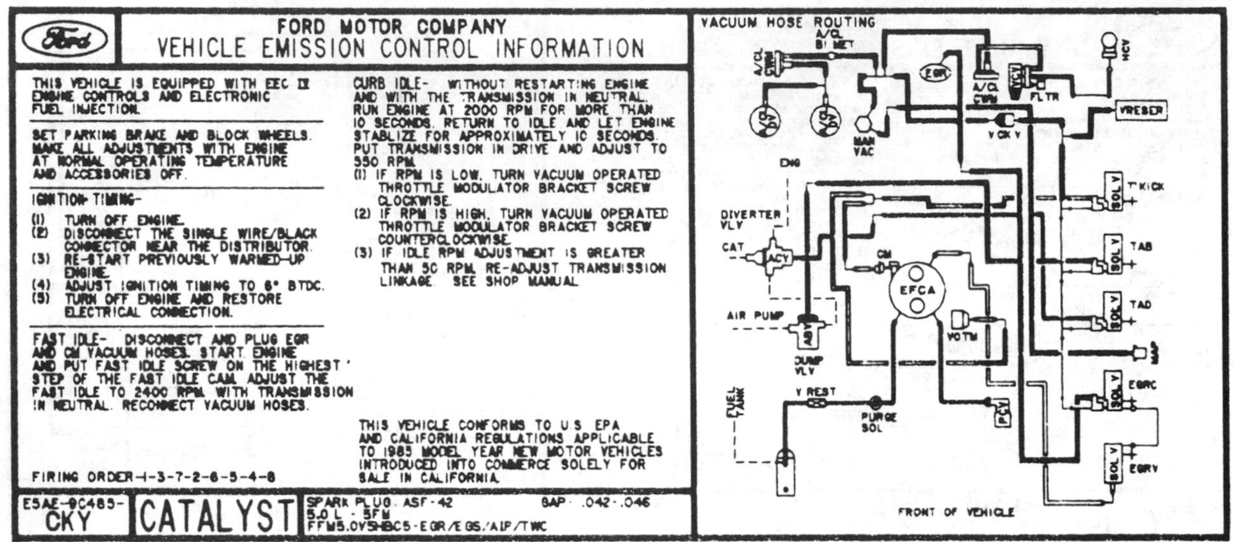

FORD MOTOR COMPANY
VEHICLE EMISSION CONTROL INFORMATION

THIS VEHICLE IS EQUIPPED WITH EEC IV ENGINE CONTROLS AND ELECTRONIC FUEL INJECTION.

SET PARKING BRAKE AND BLOCK WHEELS. MAKE ALL ADJUSTMENTS WITH ENGINE AT NORMAL OPERATING TEMPERATURE AND ACCESSORIES OFF.

IGNITION TIMING-
(1) TURN OFF ENGINE.
(2) DISCONNECT THE SINGLE WIRE/BLACK CONNECTOR NEAR THE DISTRIBUTOR.
(3) RE-START PREVIOUSLY WARMED-UP ENGINE.
(4) ADJUST IGNITION TIMING TO 6° BTDC.
(5) TURN OFF ENGINE AND RESTORE ELECTRICAL CONNECTION.

FAST IDLE- DISCONNECT AND PLUG EGR AND CM VACUUM HOSES. START ENGINE AND PUT FAST IDLE SCREW ON THE HIGHEST STEP OF THE FAST IDLE CAM. ADJUST THE FAST IDLE TO 2400 RPM WITH TRANSMISSION IN NEUTRAL. RECONNECT VACUUM HOSES.

FIRING ORDER-1-3-7-2-6-5-4-8

CURB IDLE- WITHOUT RESTARTING ENGINE AND WITH THE TRANSMISSION IN NEUTRAL. RUN ENGINE AT 2000 RPM FOR MORE THAN 10 SECONDS. RETURN TO IDLE AND LET ENGINE STABLIZE FOR APPROXIMATELY 10 SECONDS. PUT TRANSMISSION IN DRIVE AND ADJUST TO 550 RPM.
(1) IF RPM IS LOW, TURN VACUUM OPERATED THROTTLE MODULATOR BRACKET SCREW CLOCKWISE.
(2) IF RPM IS HIGH, TURN VACUUM OPERATED THROTTLE MODULATOR BRACKET SCREW COUNTERCLOCKWISE.
(3) IF IDLE RPM ADJUSTMENT IS GREATER THAN 50 RPM, RE-ADJUST TRANSMISSION LINKAGE. SEE SHOP MANUAL.

THIS VEHICLE CONFORMS TO U.S. EPA AND CALIFORNIA REGULATIONS APPLICABLE TO 1985 MODEL YEAR NEW MOTOR VEHICLES INTRODUCED INTO COMMERCE SOLELY FOR SALE IN CALIFORNIA.

E5AE-9C485-CKY | CATALYST | SPARK PLUG. ASF-42 GAP .042-.046
5.0 L - 5FM
FFM5.0V5HBC5-EGR/EGS/AIP/TWC

CALIBRATION: 5—22P—R00

MODEL YEAR: 1985 **ENGINE: 5.0L**

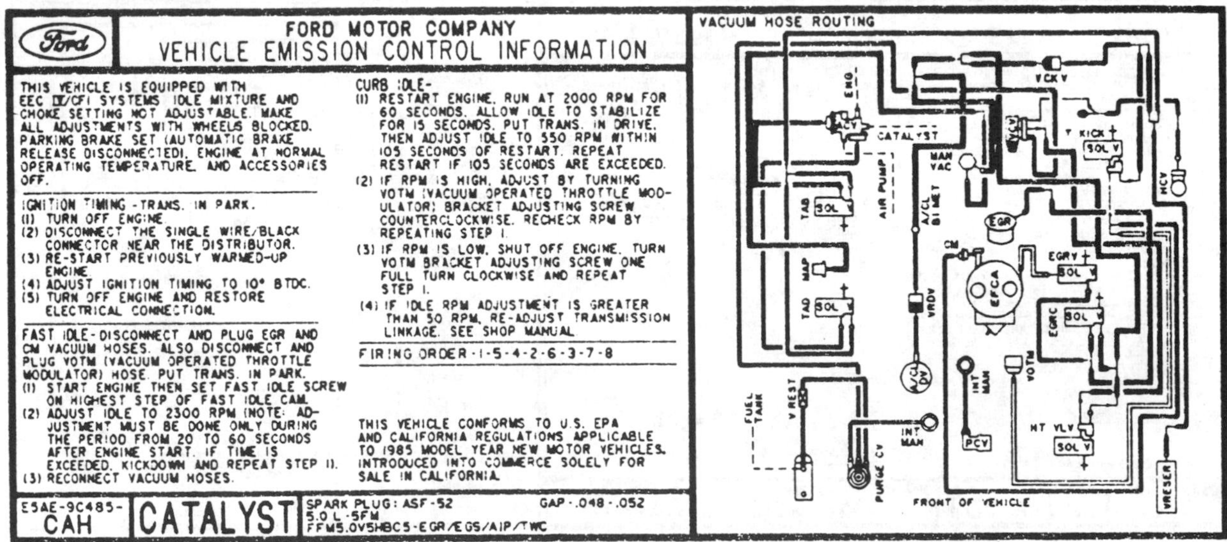

FORD MOTOR COMPANY
VEHICLE EMISSION CONTROL INFORMATION

THIS VEHICLE IS EQUIPPED WITH EEC IV/CFI SYSTEMS IDLE MIXTURE AND CHOKE SETTING NOT ADJUSTABLE. MAKE ALL ADJUSTMENTS WITH WHEELS BLOCKED. PARKING BRAKE SET (AUTOMATIC BRAKE RELEASE DISCONNECTED), ENGINE AT NORMAL OPERATING TEMPERATURE. AND ACCESSORIES OFF.

IGNITION TIMING -TRANS. IN PARK.
(1) TURN OFF ENGINE.
(2) DISCONNECT THE SINGLE WIRE/BLACK CONNECTOR NEAR THE DISTRIBUTOR.
(3) RE-START PREVIOUSLY WARMED-UP ENGINE.
(4) ADJUST IGNITION TIMING TO 10° BTDC.
(5) TURN OFF ENGINE AND RESTORE ELECTRICAL CONNECTION.

FAST IDLE-DISCONNECT AND PLUG EGR AND CM VACUUM HOSES. ALSO DISCONNECT AND PLUG VOTM (VACUUM OPERATED THROTTLE MODULATOR) HOSE. PUT TRANS. IN PARK.
(1) START ENGINE THEN SET FAST IDLE SCREW ON HIGHEST STEP OF FAST IDLE CAM.
(2) ADJUST IDLE TO 2300 RPM (NOTE: ADJUSTMENT MUST BE DONE ONLY DURING THE PERIOD FROM 20 TO 60 SECONDS AFTER ENGINE START. IF TIME IS EXCEEDED, KICKDOWN AND REPEAT STEP 1).
(3) RECONNECT VACUUM HOSES.

CURB IDLE-
(1) RESTART ENGINE, RUN AT 2000 RPM FOR 60 SECONDS. ALLOW IDLE TO STABILIZE FOR 15 SECONDS. PUT TRANS. IN DRIVE, THEN ADJUST IDLE TO 550 RPM WITHIN 105 SECONDS OF RESTART. REPEAT RESTART IF 105 SECONDS ARE EXCEEDED.
(2) IF RPM IS HIGH, ADJUST BY TURNING VOTM (VACUUM OPERATED THROTTLE MODULATOR) BRACKET ADJUSTING SCREW COUNTERCLOCKWISE. RECHECK RPM BY REPEATING STEP 1.
(3) IF RPM IS LOW, SHUT OFF ENGINE. TURN VOTM BRACKET ADJUSTING SCREW ONE FULL TURN CLOCKWISE AND REPEAT STEP 1.
(4) IF IDLE RPM ADJUSTMENT IS GREATER THAN 50 RPM, RE-ADJUST TRANSMISSION LINKAGE. SEE SHOP MANUAL.

FIRING ORDER-1-5-4-2-6-3-7-8

THIS VEHICLE CONFORMS TO U.S. EPA AND CALIFORNIA REGULATIONS APPLICABLE TO 1985 MODEL YEAR NEW MOTOR VEHICLES. INTRODUCED INTO COMMERCE SOLELY FOR SALE IN CALIFORNIA.

E5AE-9C485-CAH | CATALYST | SPARK PLUG: ASF-52 GAP-.048-.052
5.0 L-5FM
FFM5.0V5HBC5-EGR/EGS/AIP/TWC

CALIBRATION: 5—22Q—R00

VACUUM CIRCUITS

(© Ford Motor Co)

MODEL YEAR: 1985 — ENGINE: 5.0L

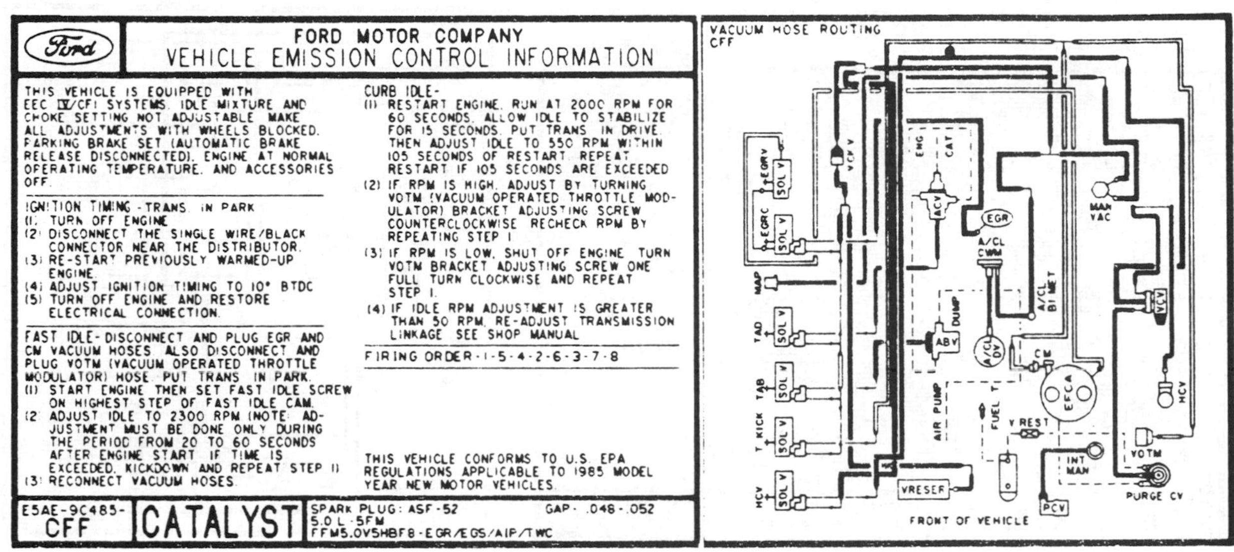

FORD MOTOR COMPANY
VEHICLE EMISSION CONTROL INFORMATION

THIS VEHICLE IS EQUIPPED WITH EEC IV/CFI SYSTEMS. IDLE MIXTURE AND CHOKE SETTING NOT ADJUSTABLE. MAKE ALL ADJUSTMENTS WITH WHEELS BLOCKED. PARKING BRAKE SET (AUTOMATIC BRAKE RELEASE DISCONNECTED). ENGINE AT NORMAL OPERATING TEMPERATURE. AND ACCESSORIES OFF.

IGNITION TIMING - TRANS. IN PARK
(1) TURN OFF ENGINE
(2) DISCONNECT THE SINGLE WIRE/BLACK CONNECTOR NEAR THE DISTRIBUTOR.
(3) RE-START PREVIOUSLY WARMED-UP ENGINE.
(4) ADJUST IGNITION TIMING TO 10° BTDC
(5) TURN OFF ENGINE AND RESTORE ELECTRICAL CONNECTION.

FAST IDLE - DISCONNECT AND PLUG EGR AND CM VACUUM HOSES. ALSO DISCONNECT AND PLUG VOTM (VACUUM OPERATED THROTTLE MODULATOR) HOSE. PUT TRANS. IN PARK.
(1) START ENGINE THEN SET FAST IDLE SCREW ON HIGHEST STEP OF FAST IDLE CAM.
(2) ADJUST IDLE TO 2300 RPM (NOTE: ADJUSTMENT MUST BE DONE ONLY DURING THE PERIOD FROM 20 TO 60 SECONDS AFTER ENGINE START. IF TIME IS EXCEEDED, KICKDOWN AND REPEAT STEP 1)
(3) RECONNECT VACUUM HOSES.

CURB IDLE-
(1) RESTART ENGINE. RUN AT 2000 RPM FOR 60 SECONDS. ALLOW IDLE TO STABILIZE FOR 15 SECONDS. PUT TRANS. IN DRIVE. THEN ADJUST IDLE TO 550 RPM WITHIN 105 SECONDS OF RESTART. REPEAT RESTART IF 105 SECONDS ARE EXCEEDED
(2) IF RPM IS HIGH, ADJUST BY TURNING VOTM (VACUUM OPERATED THROTTLE MODULATOR) BRACKET ADJUSTING SCREW COUNTERCLOCKWISE. RECHECK RPM BY REPEATING STEP 1
(3) IF RPM IS LOW, SHUT OFF ENGINE. TURN VOTM BRACKET ADJUSTING SCREW ONE FULL TURN CLOCKWISE AND REPEAT STEP 1.
(4) IF IDLE RPM ADJUSTMENT IS GREATER THAN 50 RPM, RE-ADJUST TRANSMISSION LINKAGE. SEE SHOP MANUAL

FIRING ORDER - 1 - 5 - 4 - 2 - 6 - 3 - 7 - 8

THIS VEHICLE CONFORMS TO U.S. EPA REGULATIONS APPLICABLE TO 1985 MODEL YEAR NEW MOTOR VEHICLES.

E5AE-9C485-CFF | CATALYST | SPARK PLUG: ASF-52 GAP - .048 - .052
5.0L-5FM
FFM5.0V5HBF8-EGR/EGS/AIP/TWC

CALIBRATION: 5—22R—R00

MODEL YEAR: 1985 — ENGINE: 5.0L

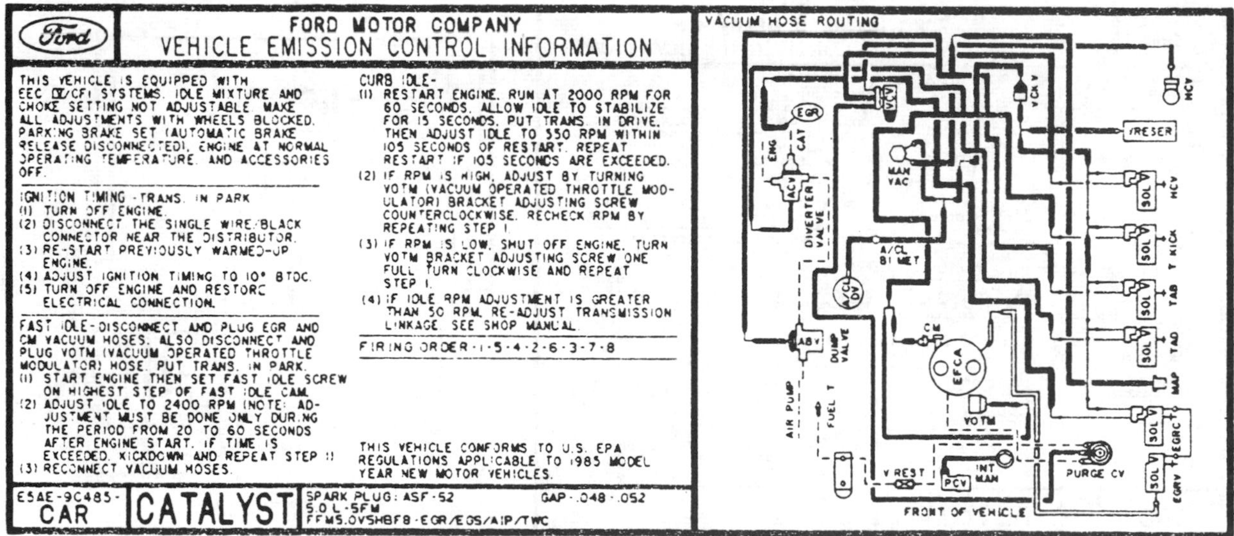

FORD MOTOR COMPANY
VEHICLE EMISSION CONTROL INFORMATION

THIS VEHICLE IS EQUIPPED WITH EEC IV/CFI SYSTEMS. IDLE MIXTURE AND CHOKE SETTING NOT ADJUSTABLE. MAKE ALL ADJUSTMENTS WITH WHEELS BLOCKED. PARKING BRAKE SET (AUTOMATIC BRAKE RELEASE DISCONNECTED). ENGINE AT NORMAL OPERATING TEMPERATURE. AND ACCESSORIES OFF.

IGNITION TIMING - TRANS. IN PARK
(1) TURN OFF ENGINE.
(2) DISCONNECT THE SINGLE WIRE/BLACK CONNECTOR NEAR THE DISTRIBUTOR.
(3) RE-START PREVIOUSLY WARMED-UP ENGINE.
(4) ADJUST IGNITION TIMING TO 10° BTDC.
(5) TURN OFF ENGINE AND RESTORE ELECTRICAL CONNECTION.

FAST IDLE - DISCONNECT AND PLUG EGR AND CM VACUUM HOSES. ALSO DISCONNECT AND PLUG VOTM (VACUUM OPERATED THROTTLE MODULATOR) HOSE. PUT TRANS. IN PARK.
(1) START ENGINE THEN SET FAST IDLE SCREW ON HIGHEST STEP OF FAST IDLE CAM.
(2) ADJUST IDLE TO 2400 RPM (NOTE: ADJUSTMENT MUST BE DONE ONLY DURING THE PERIOD FROM 20 TO 60 SECONDS AFTER ENGINE START. IF TIME IS EXCEEDED, KICKDOWN AND REPEAT STEP 1)
(3) RECONNECT VACUUM HOSES.

CURB IDLE-
(1) RESTART ENGINE. RUN AT 2000 RPM FOR 60 SECONDS. ALLOW IDLE TO STABILIZE FOR 15 SECONDS. PUT TRANS. IN DRIVE. THEN ADJUST IDLE TO 550 RPM WITHIN 105 SECONDS OF RESTART. REPEAT RESTART IF 105 SECONDS ARE EXCEEDED.
(2) IF RPM IS HIGH, ADJUST BY TURNING VOTM (VACUUM OPERATED THROTTLE MODULATOR) BRACKET ADJUSTING SCREW COUNTERCLOCKWISE. RECHECK RPM BY REPEATING STEP 1
(3) IF RPM IS LOW, SHUT OFF ENGINE. TURN VOTM BRACKET ADJUSTING SCREW ONE FULL TURN CLOCKWISE AND REPEAT STEP 1.
(4) IF IDLE RPM ADJUSTMENT IS GREATER THAN 50 RPM, RE-ADJUST TRANSMISSION LINKAGE. SEE SHOP MANUAL.

FIRING ORDER - 1 - 5 - 4 - 2 - 6 - 3 - 7 - 8

THIS VEHICLE CONFORMS TO U.S. EPA REGULATIONS APPLICABLE TO 1985 MODEL YEAR NEW MOTOR VEHICLES.

E5AE-9C485-CAR | CATALYST | SPARK PLUG: ASF-52 GAP - .048 - .052
5.0L-5FM
FFM5.0V5HBF8-EGR/EGS/AIP/TWC

CALIBRATION: 5—22T—R00

VACUUM CIRCUITS

MODEL YEAR: 1985 — ENGINE: 5.0L

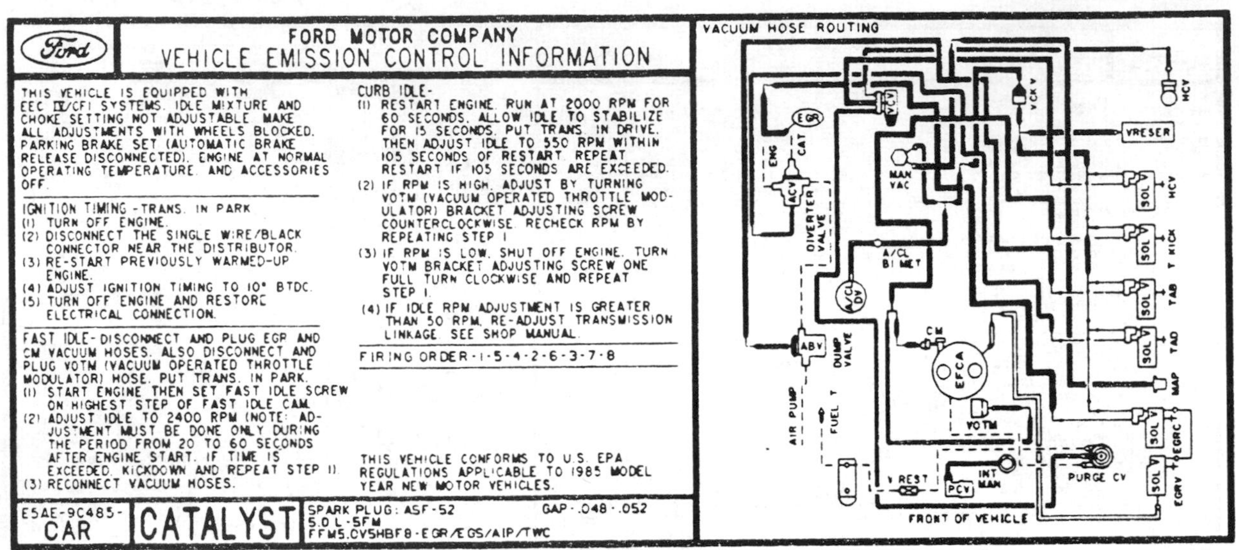

FORD MOTOR COMPANY
VEHICLE EMISSION CONTROL INFORMATION

THIS VEHICLE IS EQUIPPED WITH EEC IV/CFI SYSTEMS. IDLE MIXTURE AND CHOKE SETTING NOT ADJUSTABLE. MAKE ALL ADJUSTMENTS WITH WHEELS BLOCKED. PARKING BRAKE SET (AUTOMATIC BRAKE RELEASE DISCONNECTED). ENGINE AT NORMAL OPERATING TEMPERATURE. AND ACCESSORIES OFF.

IGNITION TIMING - TRANS. IN PARK
(1) TURN OFF ENGINE.
(2) DISCONNECT THE SINGLE WIRE/BLACK CONNECTOR NEAR THE DISTRIBUTOR.
(3) RE-START PREVIOUSLY WARMED-UP ENGINE.
(4) ADJUST IGNITION TIMING TO 10° BTDC.
(5) TURN OFF ENGINE AND RESTORE ELECTRICAL CONNECTION.

FAST IDLE - DISCONNECT AND PLUG EGR AND CM VACUUM HOSES. ALSO DISCONNECT AND PLUG VOTM (VACUUM OPERATED THROTTLE MODULATOR) HOSE. PUT TRANS. IN PARK.
(1) START ENGINE THEN SET FAST IDLE SCREW ON HIGHEST STEP OF FAST IDLE CAM.
(2) ADJUST IDLE TO 2400 RPM (NOTE: ADJUSTMENT MUST BE DONE ONLY DURING THE PERIOD FROM 20 TO 60 SECONDS AFTER ENGINE START. IF TIME IS EXCEEDED. KICKDOWN AND REPEAT STEP 1).
(3) RECONNECT VACUUM HOSES.

CURB IDLE -
(1) RESTART ENGINE. RUN AT 2000 RPM FOR 60 SECONDS. ALLOW IDLE TO STABILIZE FOR 15 SECONDS. PUT TRANS. IN DRIVE. THEN ADJUST IDLE TO 550 RPM WITHIN 105 SECONDS OF RESTART. REPEAT RESTART IF 105 SECONDS ARE EXCEEDED.
(2) IF RPM IS HIGH. ADJUST BY TURNING VOTM (VACUUM OPERATED THROTTLE MODULATOR) BRACKET ADJUSTING SCREW COUNTERCLOCKWISE. RECHECK RPM BY REPEATING STEP 1.
(3) IF RPM IS LOW. SHUT OFF ENGINE. TURN VOTM BRACKET ADJUSTING SCREW ONE FULL TURN CLOCKWISE AND REPEAT STEP 1.
(4) IF IDLE RPM ADJUSTMENT IS GREATER THAN 50 RPM. RE-ADJUST TRANSMISSION LINKAGE. SEE SHOP MANUAL.

FIRING ORDER - 1 - 5 - 4 - 2 - 6 - 3 - 7 - 8

THIS VEHICLE CONFORMS TO U.S. EPA REGULATIONS APPLICABLE TO 1985 MODEL YEAR NEW MOTOR VEHICLES.

E5AE-9C485-CAR | CATALYST | SPARK PLUG: ASF-52 GAP - .048 - .052
5.0L-5FM
FFM5.0V5HBF8-EGR/EGS/AIP/TWC

CALIBRATION: 5—22T—R10

MODEL YEAR: 1985 — ENGINE: 5.8L

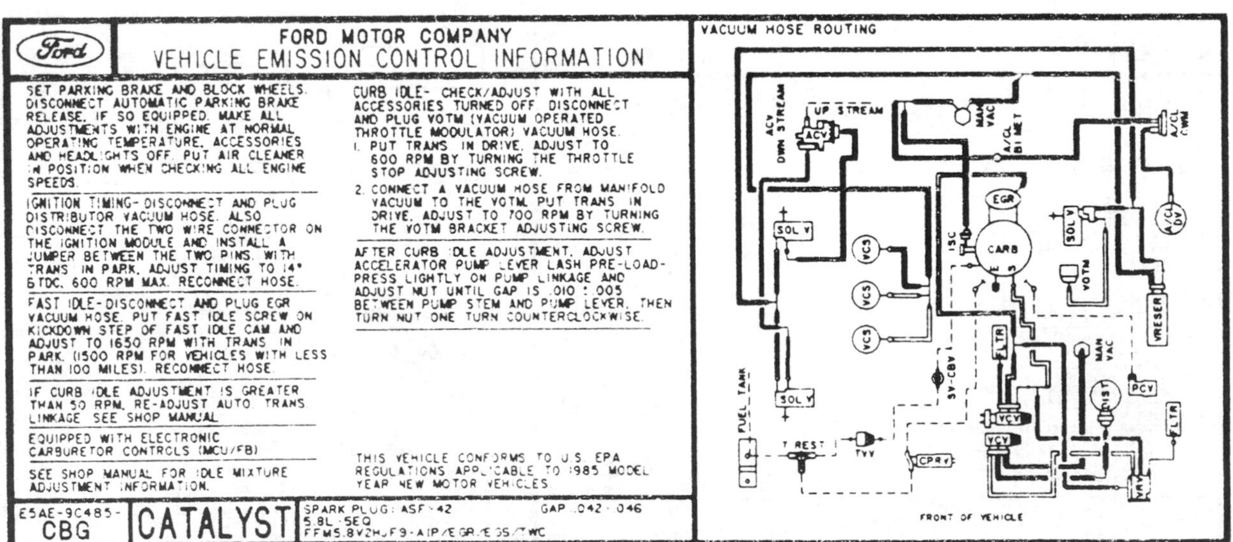

FORD MOTOR COMPANY
VEHICLE EMISSION CONTROL INFORMATION

SET PARKING BRAKE AND BLOCK WHEELS. DISCONNECT AUTOMATIC PARKING BRAKE RELEASE. IF SO EQUIPPED. MAKE ALL ADJUSTMENTS WITH ENGINE AT NORMAL OPERATING TEMPERATURE. ACCESSORIES AND HEADLIGHTS OFF. PUT AIR CLEANER IN POSITION WHEN CHECKING ALL ENGINE SPEEDS.

IGNITION TIMING - DISCONNECT AND PLUG DISTRIBUTOR VACUUM HOSE. ALSO DISCONNECT THE TWO WIRE CONNECTOR ON THE IGNITION MODULE AND INSTALL A JUMPER BETWEEN THE TWO PINS. WITH TRANS. IN PARK. ADJUST TIMING TO 14° BTDC. 600 RPM MAX. RECONNECT HOSE.

FAST IDLE - DISCONNECT AND PLUG EGR VACUUM HOSE. PUT FAST IDLE SCREW ON KICKDOWN STEP OF FAST IDLE CAM AND ADJUST TO 1650 RPM WITH TRANS. IN PARK. (1500 RPM FOR VEHICLES WITH LESS THAN 100 MILES). RECONNECT HOSE.

IF CURB IDLE ADJUSTMENT IS GREATER THAN 50 RPM. RE-ADJUST AUTO. TRANS. LINKAGE. SEE SHOP MANUAL.

EQUIPPED WITH ELECTRONIC CARBURETOR CONTROLS (MCU/FB)

SEE SHOP MANUAL FOR IDLE MIXTURE ADJUSTMENT INFORMATION.

CURB IDLE - CHECK/ADJUST WITH ALL ACCESSORIES TURNED OFF. DISCONNECT AND PLUG VOTM (VACUUM OPERATED THROTTLE MODULATOR) VACUUM HOSE.
1. PUT TRANS. IN DRIVE. ADJUST TO 600 RPM BY TURNING THE THROTTLE STOP ADJUSTING SCREW.
2. CONNECT A VACUUM HOSE FROM MANIFOLD VACUUM TO THE VOTM. PUT TRANS. IN DRIVE. ADJUST TO 700 RPM BY TURNING THE VOTM BRACKET ADJUSTING SCREW.

AFTER CURB IDLE ADJUSTMENT. ADJUST ACCELERATOR PUMP LEVER LASH PRE-LOAD-PRESS LIGHTLY ON PUMP LINKAGE AND ADJUST NUT UNTIL GAP IS .010 ± .005 BETWEEN PUMP STEM AND PUMP LEVER. THEN TURN NUT ONE TURN COUNTERCLOCKWISE.

THIS VEHICLE CONFORMS TO U.S. EPA REGULATIONS APPLICABLE TO 1985 MODEL YEAR NEW MOTOR VEHICLES.

E5AE-9C485-CBG | CATALYST | SPARK PLUG: ASF-42 GAP .042 - .046
5.8L-5EQ
FFM5.8V2HJF9-AIP/EGR/EGS/TWC

CALIBRATION: 4—24P—R00

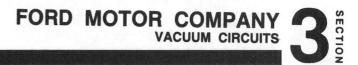

Ford
Vacuum Circuits

INDEX

1986 VACUUM CIRCUITS
ENGINE AND CALIBRATION

VACUUM CIRCUITS

CALIBRATION: 6—07A—R12

MODEL YEAR: 1986 **ENGINE: 1.9L**

FORD MOTOR COMPANY
VEHICLE EMISSION CONTROL INFORMATION

SET PARKING BRAKE AND BLOCK WHEELS. MAKE ALL ADJUSTMENTS WITH ENGINE AT NORMAL OPERATING TEMPERATURE, ACCESSORIES AND HEADLIGHTS OFF.

IGNITION TIMING-DISCONNECT AND PLUG DISTRIBUTOR VACUUM HOSE. WITH TRANS. IN NEUTRAL , ADJUST TIMING TO 10 ° BTDC, 750 RPM MAX. RECONNECT HOSE.

FAST [illegible]E-DISCONNECT AND PLUG EGR [illegible]AL. [illegible] HOSE. PUT FAST IDLE [illegible]CREW ON SECO[illegible] STEP OF FAST IDLE CAM. RUN ENGINE UNTIL RADIATOR COOLING FAN CO[illegible] ON. ADJUST TO 2200 RPM WITH TRANS. IN NEUTRAL. RECONNE[illegible]T EGR HOSE.

CURB IDLE-DISCONNECT AND PLUG THERMACTOR BY-PASS VALVE VACUUM HOSE. PUT FAST IDLE SCREW ON SECOND HIGHEST STEP OF FAST IDLE CAM. RUN ENGINE UNTIL RADIATOR FAN COMES ON. FAN MUST BE OPERATING DURING ALL RPM CHECKS AND RPM ADJUSTMENTS.
1. ACCELERATE ENGINE MOMENTARILY. DISCONNECT AND PLUG ISC (IDLE SPEED CONTROL) VACUUM HOSE. PUT TRANS. IN NEUTRAL. ADJUST TO 1350 RPM BY TURNING FULL STROKE ADJUSTING SCREW ON THROTTLE LEVER.
2. CONNECT A VACUUM PUMP TO ISC AND APPLY ENOUGH VACUUM TO RETRACT THE PLUNGER FROM THE FULL STROKE ADJUSTING SCREW. PUT TRANS IN NEUTRAL. ADJUST TO 720 RPM BY TURNING THROTTLE STOP ADJUSTING SCREW IN THROTTLE BODY. ALSO ADJUST DASHPOT CLEARANCE TO 1.0 - 2.0 MM. RESTORE ISC VACUUM CONNECTION.
3. PUT TRANS. IN NEUTRAL. ADJUST TO 750 RPM BY TURNING SCREW ON BACK OF ISC. RECONNECT THERMACTOR BY-PASS VALVE VACUUM HOSE.

IF IDLE RPM ADJUSTMENT IS GREATER THAN 50 RPM, RE-ADJUST AUTO. TRANS. LINKAGE. SEE SHOP MANUAL.

SEE SHOP MANUAL FOR CHOKE AND IDLE MIXTURE ADJUSTMENT INFORMATION.

USE SAE 5W-30 OIL - API CATEGORY SF, SF/CC OR SF/CD.

THIS VEHICLE CONFORMS TO U.S. EPA REGULATIONS APPLICABLE TO 1986 MODEL YEAR NEW MOTOR VEHICLES.

E6AE-9C485-AGL | CATALYST | SPARK PLUG: AWSF-34C GAP: /.042-.046
1.9L-6CM
GFM1.9V2GDF9-EGR/AIP/TWC

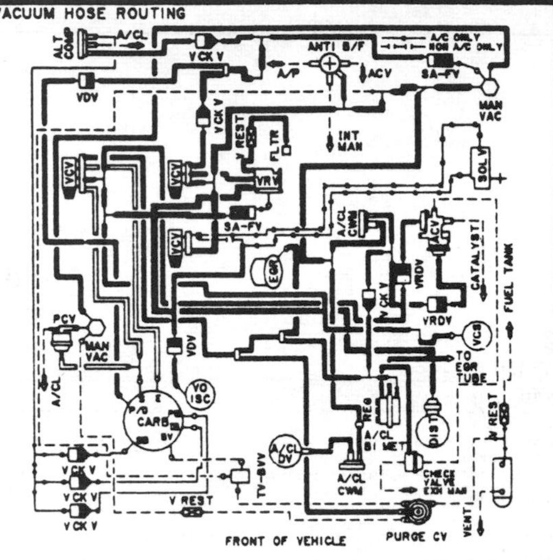

CALIBRATION: 6—07E—R00

MODEL YEAR: 1986 **ENGINE: 1.9L**

FORD MOTOR COMPANY
VEHICLE EMISSION CONTROL INFORMATION

THIS VEHICLE IS EQUIPPED WITH ELECTRONIC FUEL INJECTION. IDLE MIXTURE, COLD ENGINE IDLE SPEED AND COLD ENGINE FUEL ENRICHMENT NOT ADJUSTABLE.

SET PARKING BRAKE AND BLOCK WHEELS. MAKE ALL ADJUSTMENTS WITH ENGINE AT NORMAL OPERATING TEMPERATURE, ACCESSORIES AND HEADLIGHTS OFF.

IGNITION TIMING- TRANS. IN NEUTRAL
(1) TURN OFF ENGINE.
(2) DISCONNECT THE IN-LINE SPOUT CONNECTOR (-□□- OR =□d).
(3) RE-START PREVIOUSLY WARMED-UP ENGINE
(4) ADJUST IGNITION TIMING TO 10° BTDC.
(5) TURN OFF ENGINE AND RESTORE ELECTRICAL CONNECTION.

USE SAE 5W-30 OIL - API CATEGORY SF, SF/CC OR SF/CD.

THROTTLE PLATE ADJUSTMENT -
(1) ELECTRICALLY DISCONNECT IDLE SPEED CONTROL (ISC).
(2) RUN ENGINE AT IDLE UNTIL ENGINE COOLING FAN COMES ON.
(3) RUN ENGINE AT 2000 RPM FOR 60 SECONDS MINIMUM.
(4) RETURN TO IDLE AND TURN THROTTLE PLATE ADJUSTING SCREW UNTIL IDLE SPEED IS 900 RPM WITH TRANS. IN NEUTRAL AND FAN ON. (800 RPM FOR VEHICLE WITH LESS THAN 100 MILES). ADJUSTMENT MUST BE MADE WITHIN 120 SECONDS OF RETURN TO IDLE. TURN OFF ENGINE, RESTART AND REPEAT STEP 3 AND 4 IF 120 SECONDS ARE EXCEEDED.
(5) RESTORE VACUUM AND ELECTRICAL CONNECTIONS.

THIS VEHICLE CONFORMS TO U.S. EPA REGULATIONS APPLICABLE TO 1986 MODEL YEAR NEW MOTOR VEHICLES.

E6AE-9C485-AKM | CATALYST | SPARK PLUG: AWSF-24C GAP: .042-.046
1.9 L-6HM
GFM1.9V5HMK9-EGS/EGR/AIV/TWC/FI

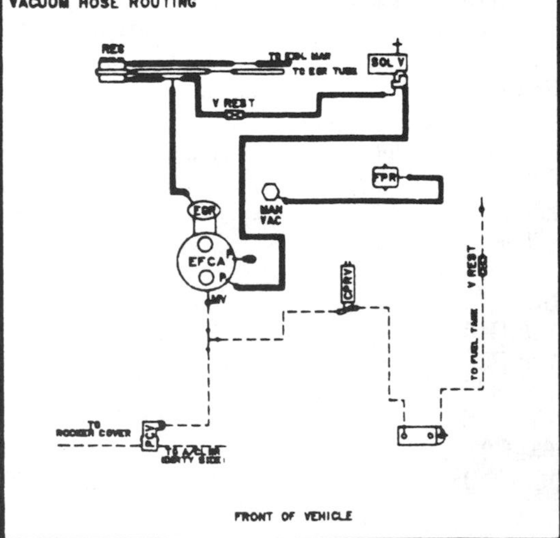

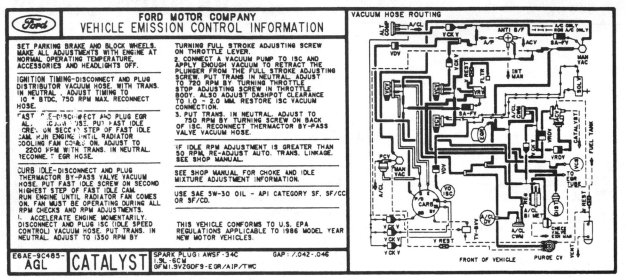

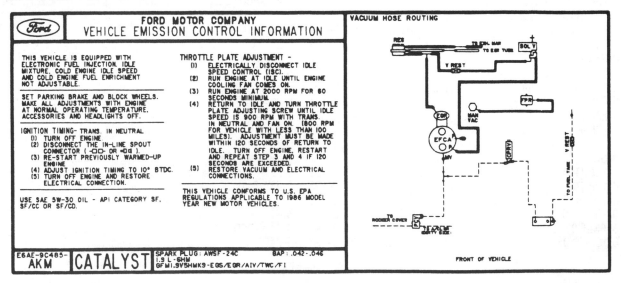

VACUUM CIRCUITS

(© Ford Motor Co)

CALIBRATION: 6—07F—R13

MODEL YEAR: 1986 **ENGINE: 1.9L**

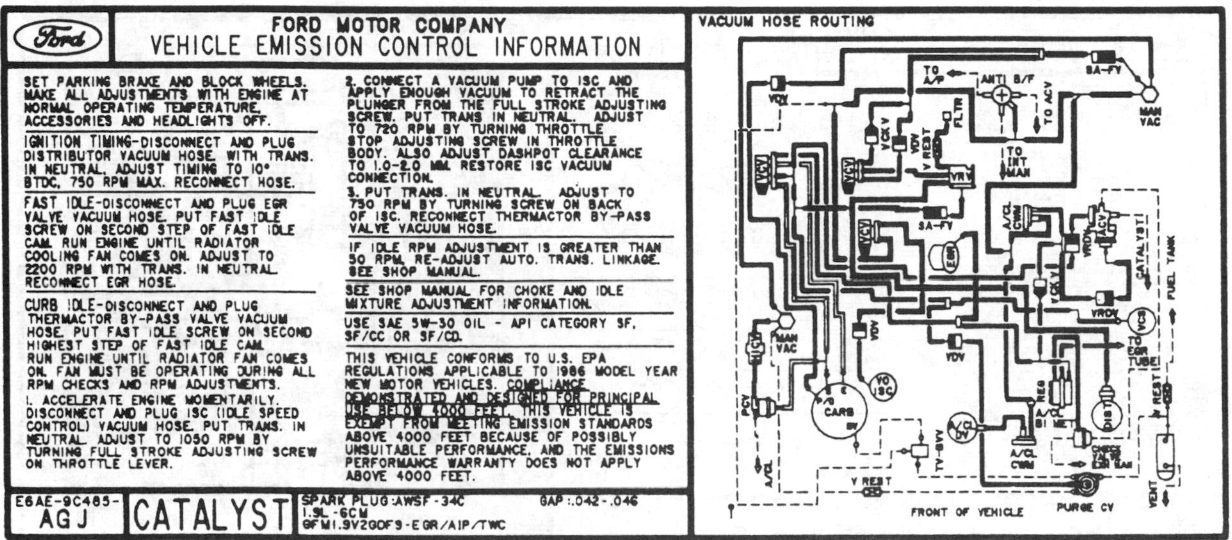

FORD MOTOR COMPANY
VEHICLE EMISSION CONTROL INFORMATION

SET PARKING BRAKE AND BLOCK WHEELS. MAKE ALL ADJUSTMENTS WITH ENGINE AT NORMAL OPERATING TEMPERATURE, ACCESSORIES AND HEADLIGHTS OFF.

IGNITION TIMING-DISCONNECT AND PLUG DISTRIBUTOR VACUUM HOSE. WITH TRANS. IN NEUTRAL, ADJUST TIMING TO 10° BTDC, 750 RPM MAX. RECONNECT HOSE.

FAST IDLE-DISCONNECT AND PLUG EGR VALVE VACUUM HOSE. PUT FAST IDLE SCREW ON SECOND STEP OF FAST IDLE CAM. RUN ENGINE UNTIL RADIATOR COOLING FAN COMES ON. ADJUST TO 2200 RPM WITH TRANS. IN NEUTRAL. RECONNECT EGR HOSE.

CURB IDLE-DISCONNECT AND PLUG THERMACTOR BY-PASS VALVE VACUUM HOSE. PUT FAST IDLE SCREW ON SECOND HIGHEST STEP OF FAST IDLE CAM. RUN ENGINE UNTIL RADIATOR FAN COMES ON. FAN MUST BE OPERATING DURING ALL RPM CHECKS AND RPM ADJUSTMENTS.
1. ACCELERATE ENGINE MOMENTARILY. DISCONNECT AND PLUG ISC (IDLE SPEED CONTROL) VACUUM HOSE. PUT TRANS. IN NEUTRAL. ADJUST TO 1050 RPM BY TURNING FULL STROKE ADJUSTING SCREW ON THROTTLE LEVER.
2. CONNECT A VACUUM PUMP TO ISC AND APPLY ENOUGH VACUUM TO RETRACT THE PLUNGER FROM THE FULL STROKE ADJUSTING SCREW. PUT TRANS IN NEUTRAL. ADJUST TO 720 RPM BY TURNING THROTTLE STOP ADJUSTING SCREW IN THROTTLE BODY. ALSO ADJUST DASHPOT CLEARANCE TO 1.0-2.0 MM. RESTORE ISC VACUUM CONNECTION.
3. PUT TRANS. IN NEUTRAL. ADJUST TO 750 RPM BY TURNING SCREW ON BACK OF ISC. RECONNECT THERMACTOR BY-PASS VALVE VACUUM HOSE.

IF IDLE RPM ADJUSTMENT IS GREATER THAN 50 RPM, RE-ADJUST AUTO. TRANS. LINKAGE. SEE SHOP MANUAL.

SEE SHOP MANUAL FOR CHOKE AND IDLE MIXTURE ADJUSTMENT INFORMATION.

USE SAE 5W-30 OIL - API CATEGORY SF, SF/CC OR SF/CD.

THIS VEHICLE CONFORMS TO U.S. EPA REGULATIONS APPLICABLE TO 1986 MODEL YEAR NEW MOTOR VEHICLES. COMPLIANCE DEMONSTRATED AND DESIGNED FOR PRINCIPAL USE BELOW 4000 FEET. THIS VEHICLE IS EXEMPT FROM MEETING EMISSION STANDARDS ABOVE 4000 FEET BECAUSE OF POSSIBLY UNSUITABLE PERFORMANCE, AND THE EMISSIONS PERFORMANCE WARRANTY DOES NOT APPLY ABOVE 4000 FEET.

E6AE-9C485-AGJ | CATALYST | SPARK PLUG:AWSF-34C GAP:.042-.046
1.9L-6CM
GFM1.9V2GDF9-EGR/AIP/TWC

CALIBRATION: 6—07S—R12

MODEL YEAR: 1986 **ENGINE: 1.9L**

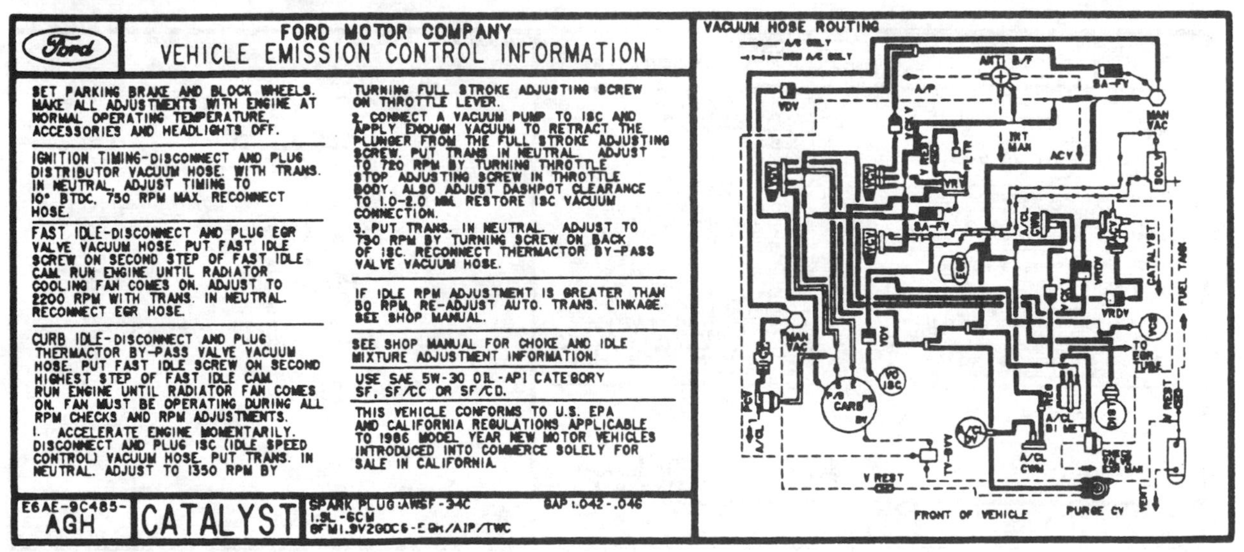

FORD MOTOR COMPANY
VEHICLE EMISSION CONTROL INFORMATION

SET PARKING BRAKE AND BLOCK WHEELS. MAKE ALL ADJUSTMENTS WITH ENGINE AT NORMAL OPERATING TEMPERATURE, ACCESSORIES AND HEADLIGHTS OFF.

IGNITION TIMING-DISCONNECT AND PLUG DISTRIBUTOR VACUUM HOSE. WITH TRANS. IN NEUTRAL, ADJUST TIMING TO 10° BTDC, 750 RPM MAX. RECONNECT HOSE.

FAST IDLE-DISCONNECT AND PLUG EGR VALVE VACUUM HOSE. PUT FAST IDLE SCREW ON SECOND STEP OF FAST IDLE CAM. RUN ENGINE UNTIL RADIATOR COOLING FAN COMES ON. ADJUST TO 2200 RPM WITH TRANS. IN NEUTRAL. RECONNECT EGR HOSE.

CURB IDLE-DISCONNECT AND PLUG THERMACTOR BY-PASS VALVE VACUUM HOSE. PUT FAST IDLE SCREW ON SECOND HIGHEST STEP OF FAST IDLE CAM. RUN ENGINE UNTIL RADIATOR FAN COMES ON. FAN MUST BE OPERATING DURING ALL RPM CHECKS AND RPM ADJUSTMENTS.
1. ACCELERATE ENGINE MOMENTARILY. DISCONNECT AND PLUG ISC (IDLE SPEED CONTROL) VACUUM HOSE. PUT TRANS. IN NEUTRAL. ADJUST TO 1350 RPM BY TURNING FULL STROKE ADJUSTING SCREW ON THROTTLE LEVER.
2. CONNECT A VACUUM PUMP TO ISC AND APPLY ENOUGH VACUUM TO RETRACT THE PLUNGER FROM THE FULL STROKE ADJUSTING SCREW. PUT TRANS IN NEUTRAL. ADJUST TO 720 RPM BY TURNING THROTTLE STOP ADJUSTING SCREW IN THROTTLE BODY. ALSO ADJUST DASHPOT CLEARANCE TO 1.0-2.0 MM. RESTORE ISC VACUUM CONNECTION.
3. PUT TRANS. IN NEUTRAL. ADJUST TO 750 RPM BY TURNING SCREW ON BACK OF ISC. RECONNECT THERMACTOR BY-PASS VALVE VACUUM HOSE.

IF IDLE RPM ADJUSTMENT IS GREATER THAN 50 RPM, RE-ADJUST AUTO. TRANS. LINKAGE. SEE SHOP MANUAL.

SEE SHOP MANUAL FOR CHOKE AND IDLE MIXTURE ADJUSTMENT INFORMATION.

USE SAE 5W-30 OIL-API CATEGORY SF, SF/CC OR SF/CD.

THIS VEHICLE CONFORMS TO U.S. EPA AND CALIFORNIA REGULATIONS APPLICABLE TO 1986 MODEL YEAR NEW MOTOR VEHICLES INTRODUCED INTO COMMERCE SOLELY FOR SALE IN CALIFORNIA.

E6AE-9C485-AGH | CATALYST | SPARK PLUG:AWSF-34C GAP:.042-.046
1.9L-6CM
GFM1.9V2GDC6-EGR/AIP/TWC

VACUUM CIRCUITS

(© Ford Motor Co)

CALIBRATION: 6—08A—R17

MODEL YEAR: 1986 **ENGINE: 1.9L**

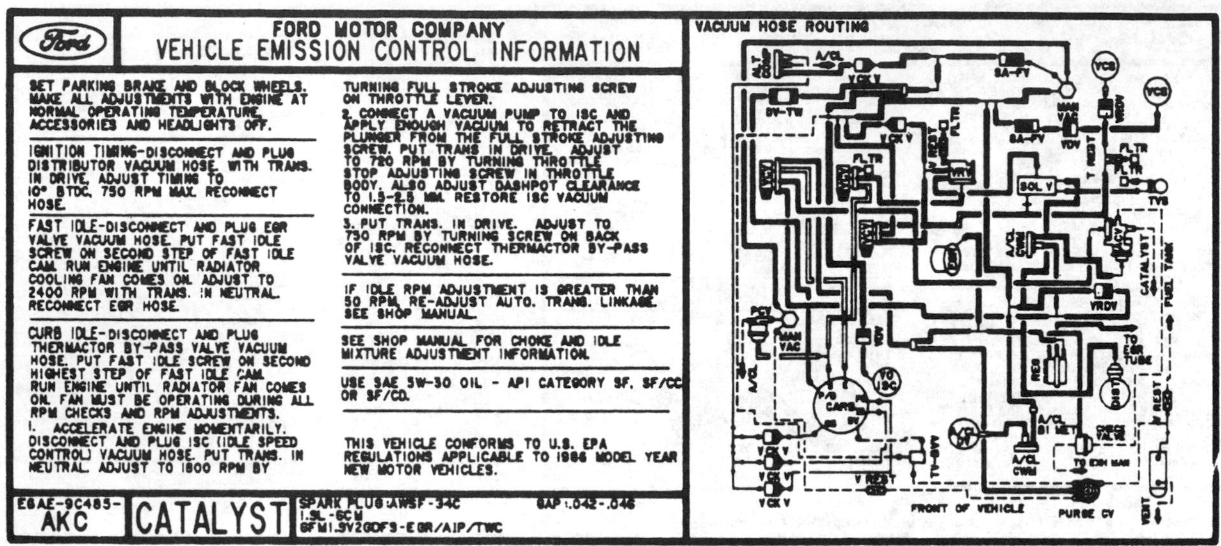

FORD MOTOR COMPANY
VEHICLE EMISSION CONTROL INFORMATION

SET PARKING BRAKE AND BLOCK WHEELS. MAKE ALL ADJUSTMENTS WITH ENGINE AT NORMAL OPERATING TEMPERATURE. ACCESSORIES AND HEADLIGHTS OFF.

IGNITION TIMING-DISCONNECT AND PLUG DISTRIBUTOR VACUUM HOSE. WITH TRANS. IN DRIVE, ADJUST TIMING TO 10° BTDC, 750 RPM MAX. RECONNECT HOSE.

FAST IDLE-DISCONNECT AND PLUG EGR VALVE VACUUM HOSE. PUT FAST IDLE SCREW ON SECOND STEP OF FAST IDLE CAM. RUN ENGINE UNTIL RADIATOR COOLING FAN COMES ON. ADJUST TO 2400 RPM WITH TRANS. IN NEUTRAL. RECONNECT EGR HOSE.

CURB IDLE-DISCONNECT AND PLUG THERMACTOR BY-PASS VALVE VACUUM HOSE. PUT FAST IDLE SCREW ON SECOND HIGHEST STEP OF FAST IDLE CAM. RUN ENGINE UNTIL RADIATOR FAN COMES ON. FAN MUST BE OPERATING DURING ALL RPM CHECKS AND RPM ADJUSTMENTS.
1. ACCELERATE ENGINE MOMENTARILY. DISCONNECT AND PLUG ISC (IDLE SPEED CONTROL) VACUUM HOSE. PUT TRANS. IN NEUTRAL. ADJUST TO 1800 RPM BY TURNING FULL STROKE ADJUSTING SCREW ON THROTTLE LEVER.
2. CONNECT A VACUUM PUMP TO ISC AND APPLY ENOUGH VACUUM TO RETRACT THE PLUNGER FROM THE FULL STROKE ADJUSTING SCREW. PUT TRANS IN DRIVE. ADJUST TO 720 RPM BY TURNING THROTTLE STOP ADJUSTING SCREW IN THROTTLE BODY. ALSO ADJUST DASHPOT CLEARANCE TO 1.5-2.5 MM. RESTORE ISC VACUUM CONNECTION.
3. PUT TRANS. IN DRIVE. ADJUST TO 730 RPM BY TURNING SCREW ON BACK OF ISC. RECONNECT THERMACTOR BY-PASS VALVE VACUUM HOSE.

IF IDLE RPM ADJUSTMENT IS GREATER THAN 50 RPM, RE-ADJUST AUTO. TRANS. LINKAGE. SEE SHOP MANUAL.

SEE SHOP MANUAL FOR CHOKE AND IDLE MIXTURE ADJUSTMENT INFORMATION.

USE SAE 5W-30 OIL - API CATEGORY SF, SF/CC OR SF/CD.

THIS VEHICLE CONFORMS TO U.S. EPA REGULATIONS APPLICABLE TO 1986 MODEL YEAR NEW MOTOR VEHICLES.

E6AE-9C485-AKC | CATALYST | SPARK PLUG:AWSF-34C GAP:.042-.046 1.9L-6CM 6FM1.9V2GDF9-EGR/AIP/TWC

CALIBRATION: 6—08E—R00

MODEL YEAR: 1986 **ENGINE: 1.9L**

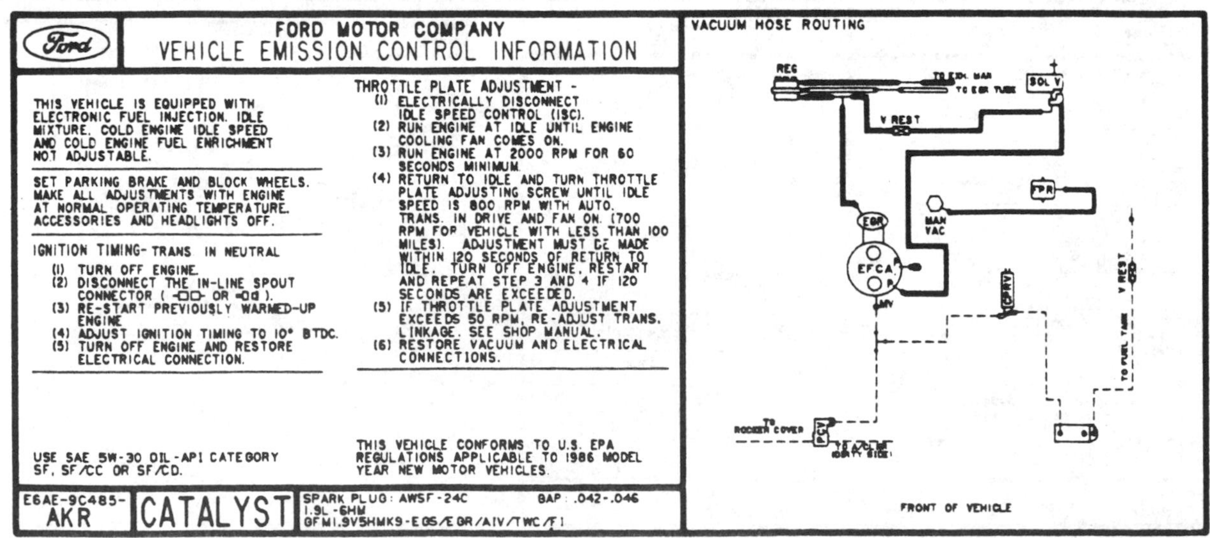

FORD MOTOR COMPANY
VEHICLE EMISSION CONTROL INFORMATION

THIS VEHICLE IS EQUIPPED WITH ELECTRONIC FUEL INJECTION. IDLE MIXTURE, COLD ENGINE IDLE SPEED AND COLD ENGINE FUEL ENRICHMENT NOT ADJUSTABLE.

SET PARKING BRAKE AND BLOCK WHEELS. MAKE ALL ADJUSTMENTS WITH ENGINE AT NORMAL OPERATING TEMPERATURE. ACCESSORIES AND HEADLIGHTS OFF.

IGNITION TIMING- TRANS IN NEUTRAL
(1) TURN OFF ENGINE.
(2) DISCONNECT THE IN-LINE SPOUT CONNECTOR (-□□- OR -□d).
(3) RE-START PREVIOUSLY WARMED-UP ENGINE
(4) ADJUST IGNITION TIMING TO 10° BTDC.
(5) TURN OFF ENGINE AND RESTORE ELECTRICAL CONNECTION.

THROTTLE PLATE ADJUSTMENT -
(1) ELECTRICALLY DISCONNECT IDLE SPEED CONTROL (ISC).
(2) RUN ENGINE AT IDLE UNTIL ENGINE COOLING FAN COMES ON.
(3) RUN ENGINE AT 2000 RPM FOR 60 SECONDS MINIMUM.
(4) RETURN TO IDLE AND TURN THROTTLE PLATE ADJUSTING SCREW UNTIL IDLE SPEED IS 800 RPM WITH AUTO. TRANS. IN DRIVE AND FAN ON. (700 RPM FOR VEHICLE WITH LESS THAN 100 MILES). ADJUSTMENT MUST BE MADE WITHIN 120 SECONDS OF RETURN TO IDLE. TURN OFF ENGINE, RESTART AND REPEAT STEP 3 AND 4 IF 120 SECONDS ARE EXCEEDED.
(5) IF THROTTLE PLATE ADJUSTMENT EXCEEDS 50 RPM, RE-ADJUST TRANS. LINKAGE. SEE SHOP MANUAL.
(6) RESTORE VACUUM AND ELECTRICAL CONNECTIONS.

USE SAE 5W-30 OIL-API CATEGORY SF, SF/CC OR SF/CD.

THIS VEHICLE CONFORMS TO U.S. EPA REGULATIONS APPLICABLE TO 1986 MODEL YEAR NEW MOTOR VEHICLES.

E6AE-9C485-AKR | CATALYST | SPARK PLUG: AWSF-24C GAP: .042-.046 1.9L-6HM GFM1.9V5HMK9-EGS/EGR/AIV/TWC/FI

VACUUM CIRCUITS

(© Ford Motor Co)

CALIBRATION: 6—08S—R21

MODEL YEAR: 1986 **ENGINE: 1.9L**

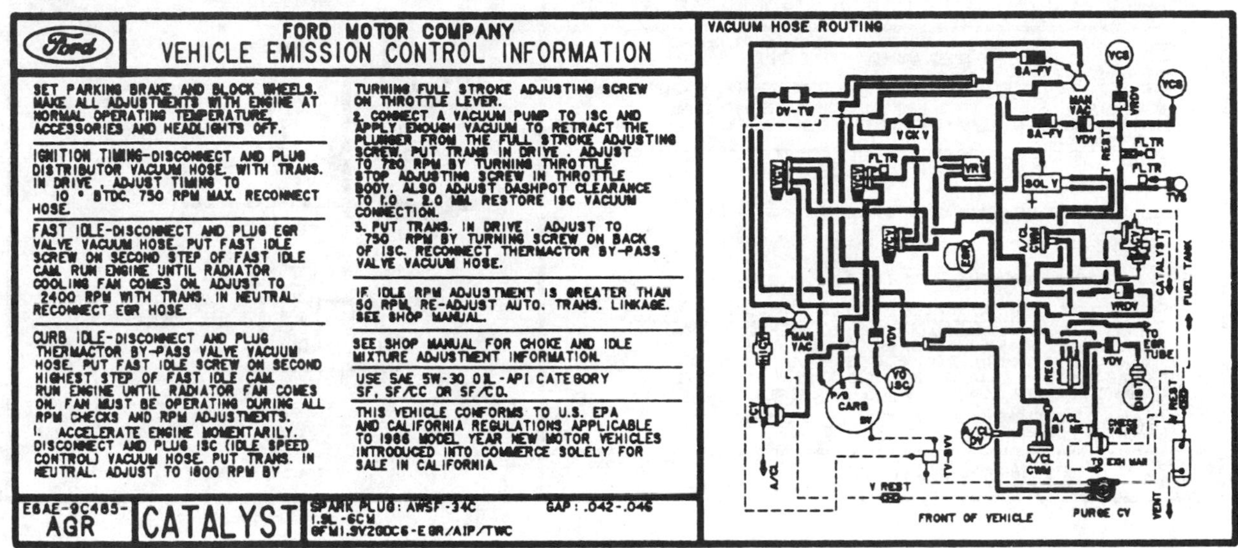

FORD MOTOR COMPANY
VEHICLE EMISSION CONTROL INFORMATION

SET PARKING BRAKE AND BLOCK WHEELS. MAKE ALL ADJUSTMENTS WITH ENGINE AT NORMAL OPERATING TEMPERATURE, ACCESSORIES AND HEADLIGHTS OFF.

IGNITION TIMING-DISCONNECT AND PLUG DISTRIBUTOR VACUUM HOSE. WITH TRANS. IN DRIVE , ADJUST TIMING TO 10 ° BTDC, 750 RPM MAX. RECONNECT HOSE.

FAST IDLE-DISCONNECT AND PLUG EGR VALVE VACUUM HOSE. PUT FAST IDLE SCREW ON SECOND STEP OF FAST IDLE CAM. RUN ENGINE UNTIL RADIATOR COOLING FAN COMES ON. ADJUST TO 2400 RPM WITH TRANS. IN NEUTRAL. RECONNECT EGR HOSE.

CURB IDLE-DISCONNECT AND PLUG THERMACTOR BY-PASS VALVE VACUUM HOSE. PUT FAST IDLE SCREW ON SECOND HIGHEST STEP OF FAST IDLE CAM. RUN ENGINE UNTIL RADIATOR FAN COMES ON. FAN MUST BE OPERATING DURING ALL RPM CHECKS AND RPM ADJUSTMENTS.
1. ACCELERATE ENGINE MOMENTARILY. DISCONNECT AND PLUG ISC (IDLE SPEED CONTROL) VACUUM HOSE. PUT TRANS. IN NEUTRAL. ADJUST TO 1800 RPM BY TURNING FULL STROKE ADJUSTING SCREW ON THROTTLE LEVER.
2. CONNECT A VACUUM PUMP TO ISC AND APPLY ENOUGH VACUUM TO RETRACT THE PLUNGER FROM THE FULL STROKE ADJUSTING SCREW. PUT TRANS IN DRIVE . ADJUST TO 720 RPM BY TURNING THROTTLE STOP ADJUSTING SCREW IN THROTTLE BODY. ALSO ADJUST DASHPOT CLEARANCE TO 1.0 - 2.0 MM. RESTORE ISC VACUUM CONNECTION.
3. PUT TRANS. IN DRIVE . ADJUST TO 750 RPM BY TURNING SCREW ON BACK OF ISC. RECONNECT THERMACTOR BY-PASS VALVE VACUUM HOSE.

IF IDLE RPM ADJUSTMENT IS GREATER THAN 50 RPM, RE-ADJUST AUTO. TRANS. LINKAGE. SEE SHOP MANUAL.

SEE SHOP MANUAL FOR CHOKE AND IDLE MIXTURE ADJUSTMENT INFORMATION.

USE SAE 5W-30 OIL-API CATEGORY SF, SF/CC OR SF/CD.

THIS VEHICLE CONFORMS TO U.S. EPA AND CALIFORNIA REGULATIONS APPLICABLE TO 1986 MODEL YEAR NEW MOTOR VEHICLES INTRODUCED INTO COMMERCE SOLELY FOR SALE IN CALIFORNIA.

E6AE-9C485-AGR | CATALYST | SPARK PLUG: AWSF-34C GAP: .042-.046
1.9L-6CM
GFM1.9V2GDC6-EGR/AIP/TWC

CALIBRATION: 5—05A—R00

MODEL YEAR: 1986 **ENGINE: 2.3L**

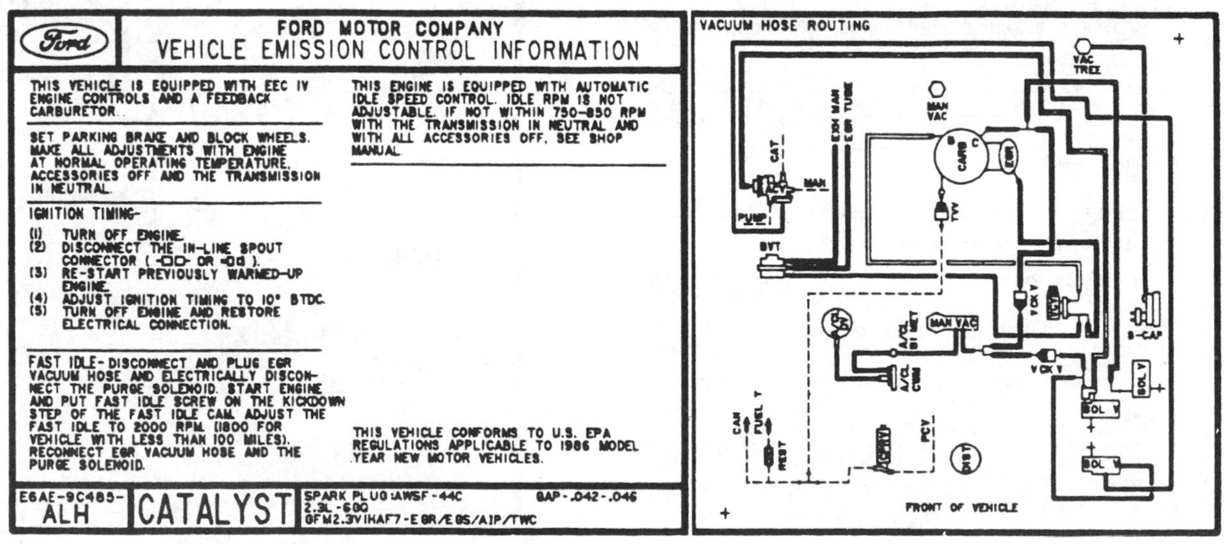

FORD MOTOR COMPANY
VEHICLE EMISSION CONTROL INFORMATION

THIS VEHICLE IS EQUIPPED WITH EEC IV ENGINE CONTROLS AND A FEEDBACK CARBURETOR.

SET PARKING BRAKE AND BLOCK WHEELS. MAKE ALL ADJUSTMENTS WITH ENGINE AT NORMAL OPERATING TEMPERATURE, ACCESSORIES OFF AND THE TRANSMISSION IN NEUTRAL.

IGNITION TIMING-
(1) TURN OFF ENGINE.
(2) DISCONNECT THE IN-LINE SPOUT CONNECTOR (-□□- OR =□□).
(3) RE-START PREVIOUSLY WARMED-UP ENGINE.
(4) ADJUST IGNITION TIMING TO 10° BTDC.
(5) TURN OFF ENGINE AND RESTORE ELECTRICAL CONNECTION.

FAST IDLE- DISCONNECT AND PLUG EGR VACUUM HOSE AND ELECTRICALLY DISCONNECT THE PURGE SOLENOID. START ENGINE AND PUT FAST IDLE SCREW ON THE KICKDOWN STEP OF THE FAST IDLE CAM. ADJUST THE FAST IDLE TO 2000 RPM. (1800 FOR VEHICLE WITH LESS THAN 100 MILES). RECONNECT EGR VACUUM HOSE AND THE PURGE SOLENOID.

THIS ENGINE IS EQUIPPED WITH AUTOMATIC IDLE SPEED CONTROL. IDLE RPM IS NOT ADJUSTABLE. IF NOT WITHIN 750-850 RPM WITH THE TRANSMISSION IN NEUTRAL AND WITH ALL ACCESSORIES OFF, SEE SHOP MANUAL.

THIS VEHICLE CONFORMS TO U.S. EPA REGULATIONS APPLICABLE TO 1986 MODEL YEAR NEW MOTOR VEHICLES.

E6AE-9C485-ALH | CATALYST | SPARK PLUG:AWSF-44C GAP-.042-.046
2.3L-6GQ
GFM2.3V1HAF7-EGR/EGS/AIP/TWC

VACUUM CIRCUITS

(© Ford Motor Co)

CALIBRATION: 5—05E—R00

MODEL YEAR: 1986 **ENGINE: 2.3L T/C**

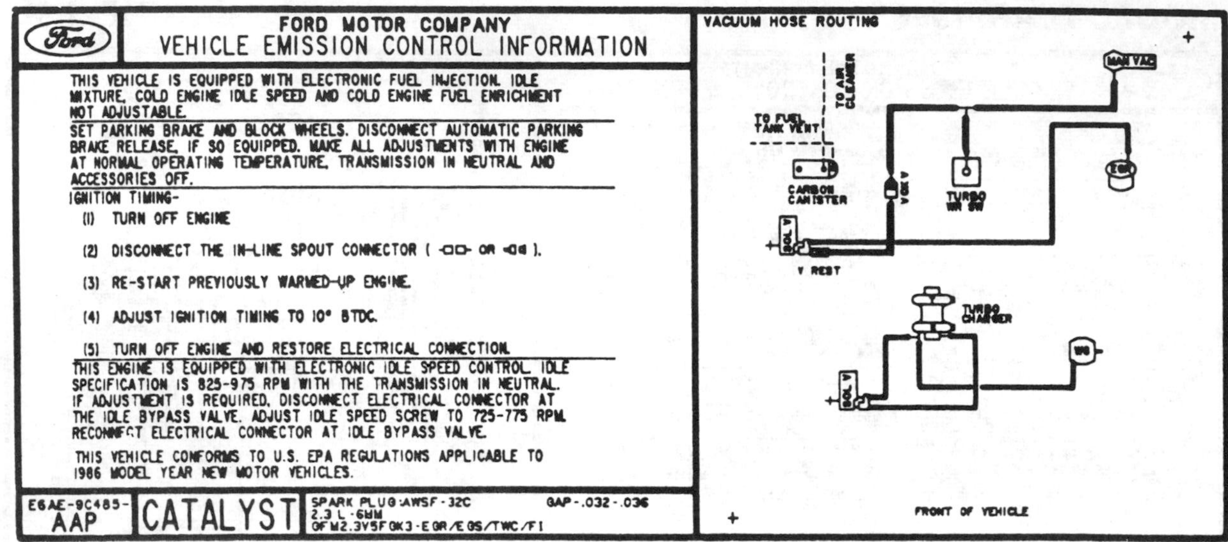

FORD MOTOR COMPANY
VEHICLE EMISSION CONTROL INFORMATION

THIS VEHICLE IS EQUIPPED WITH ELECTRONIC FUEL INJECTION. IDLE MIXTURE, COLD ENGINE IDLE SPEED AND COLD ENGINE FUEL ENRICHMENT NOT ADJUSTABLE.

SET PARKING BRAKE AND BLOCK WHEELS. DISCONNECT AUTOMATIC PARKING BRAKE RELEASE, IF SO EQUIPPED. MAKE ALL ADJUSTMENTS WITH ENGINE AT NORMAL OPERATING TEMPERATURE, TRANSMISSION IN NEUTRAL AND ACCESSORIES OFF.

IGNITION TIMING-

(1) TURN OFF ENGINE

(2) DISCONNECT THE IN-LINE SPOUT CONNECTOR (-▭▭- OR -◁◁).

(3) RE-START PREVIOUSLY WARMED-UP ENGINE.

(4) ADJUST IGNITION TIMING TO 10° BTDC.

(5) TURN OFF ENGINE AND RESTORE ELECTRICAL CONNECTION.

THIS ENGINE IS EQUIPPED WITH ELECTRONIC IDLE SPEED CONTROL. IDLE SPECIFICATION IS 825-975 RPM WITH THE TRANSMISSION IN NEUTRAL. IF ADJUSTMENT IS REQUIRED, DISCONNECT ELECTRICAL CONNECTOR AT THE IDLE BYPASS VALVE. ADJUST IDLE SPEED SCREW TO 725-775 RPM. RECONNECT ELECTRICAL CONNECTOR AT IDLE BYPASS VALVE.

THIS VEHICLE CONFORMS TO U.S. EPA REGULATIONS APPLICABLE TO 1986 MODEL YEAR NEW MOTOR VEHICLES.

E6AE-9C485-AAP CATALYST SPARK PLUG:AWSF-32C GAP-.032-.036 2.3L-6MM GFM2.3V5FGK3-EGR/EGS/TWC/FI

CALIBRATION: 5—05R—R10

MODEL YEAR: 1986 **ENGINE: 2.3L T/C**

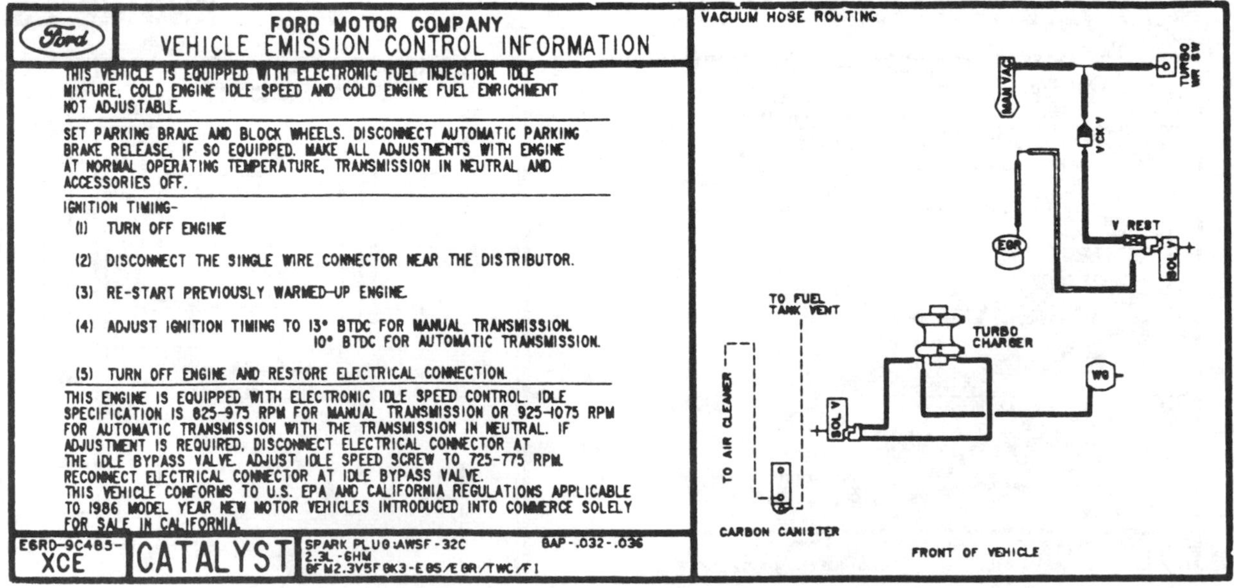

FORD MOTOR COMPANY
VEHICLE EMISSION CONTROL INFORMATION

THIS VEHICLE IS EQUIPPED WITH ELECTRONIC FUEL INJECTION. IDLE MIXTURE, COLD ENGINE IDLE SPEED AND COLD ENGINE FUEL ENRICHMENT NOT ADJUSTABLE.

SET PARKING BRAKE AND BLOCK WHEELS. DISCONNECT AUTOMATIC PARKING BRAKE RELEASE, IF SO EQUIPPED. MAKE ALL ADJUSTMENTS WITH ENGINE AT NORMAL OPERATING TEMPERATURE, TRANSMISSION IN NEUTRAL AND ACCESSORIES OFF.

IGNITION TIMING-

(1) TURN OFF ENGINE

(2) DISCONNECT THE SINGLE WIRE CONNECTOR NEAR THE DISTRIBUTOR.

(3) RE-START PREVIOUSLY WARMED-UP ENGINE.

(4) ADJUST IGNITION TIMING TO 13° BTDC FOR MANUAL TRANSMISSION.
10° BTDC FOR AUTOMATIC TRANSMISSION.

(5) TURN OFF ENGINE AND RESTORE ELECTRICAL CONNECTION.

THIS ENGINE IS EQUIPPED WITH ELECTRONIC IDLE SPEED CONTROL. IDLE SPECIFICATION IS 825-975 RPM FOR MANUAL TRANSMISSION OR 925-1075 RPM FOR AUTOMATIC TRANSMISSION WITH THE TRANSMISSION IN NEUTRAL. IF ADJUSTMENT IS REQUIRED, DISCONNECT ELECTRICAL CONNECTOR AT THE IDLE BYPASS VALVE. ADJUST IDLE SPEED SCREW TO 725-775 RPM. RECONNECT ELECTRICAL CONNECTOR AT IDLE BYPASS VALVE.
THIS VEHICLE CONFORMS TO U.S. EPA AND CALIFORNIA REGULATIONS APPLICABLE TO 1986 MODEL YEAR NEW MOTOR VEHICLES INTRODUCED INTO COMMERCE SOLELY FOR SALE IN CALIFORNIA.

E6RD-9C485-XCE CATALYST SPARK PLUG:AWSF-32C GAP-.032-.036 2.3L-6MM GFM2.3V5FGK3-EGS/EGR/TWC/FI

VACUUM CIRCUITS

(© Ford Motor Co)

CALIBRATION: 5—05S—R01

MODEL YEAR: 1986 **ENGINE: 2.3L T/C**

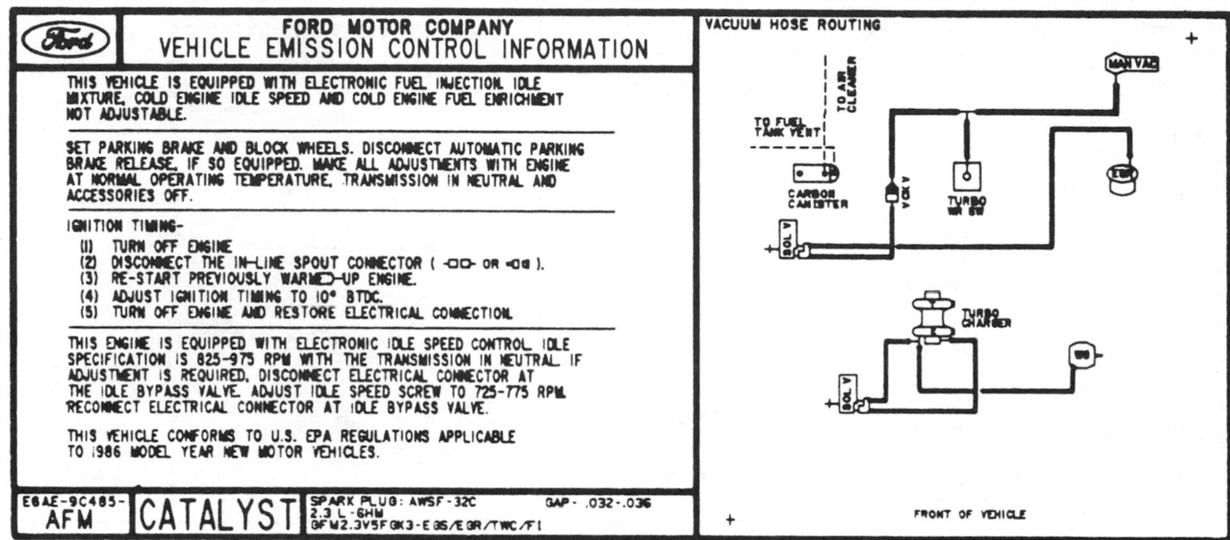

FORD MOTOR COMPANY
VEHICLE EMISSION CONTROL INFORMATION

THIS VEHICLE IS EQUIPPED WITH ELECTRONIC FUEL INJECTION. IDLE MIXTURE, COLD ENGINE IDLE SPEED AND COLD ENGINE FUEL ENRICHMENT NOT ADJUSTABLE.

SET PARKING BRAKE AND BLOCK WHEELS. DISCONNECT AUTOMATIC PARKING BRAKE RELEASE, IF SO EQUIPPED. MAKE ALL ADJUSTMENTS WITH ENGINE AT NORMAL OPERATING TEMPERATURE, TRANSMISSION IN NEUTRAL AND ACCESSORIES OFF.

IGNITION TIMING-

(1) TURN OFF ENGINE
(2) DISCONNECT THE IN-LINE SPOUT CONNECTOR (-□□- OR -□◘).
(3) RE-START PREVIOUSLY WARMED-UP ENGINE.
(4) ADJUST IGNITION TIMING TO 10° BTDC.
(5) TURN OFF ENGINE AND RESTORE ELECTRICAL CONNECTION.

THIS ENGINE IS EQUIPPED WITH ELECTRONIC IDLE SPEED CONTROL. IDLE SPECIFICATION IS 825-975 RPM WITH THE TRANSMISSION IN NEUTRAL. IF ADJUSTMENT IS REQUIRED, DISCONNECT ELECTRICAL CONNECTOR AT THE IDLE BYPASS VALVE. ADJUST IDLE SPEED SCREW TO 725-775 RPM. RECONNECT ELECTRICAL CONNECTOR AT IDLE BYPASS VALVE.

THIS VEHICLE CONFORMS TO U.S. EPA REGULATIONS APPLICABLE TO 1986 MODEL YEAR NEW MOTOR VEHICLES.

E6AE-9C485-AFM | CATALYST | SPARK PLUG: AWSF-32C GAP-.032-.036 2.3 L-6HM GFM2.3V5FGK3-EGS/EGR/TWC/FI

CALIBRATION: 5—06A—R00

MODEL YEAR: 1986 **ENGINE: 2.3L**

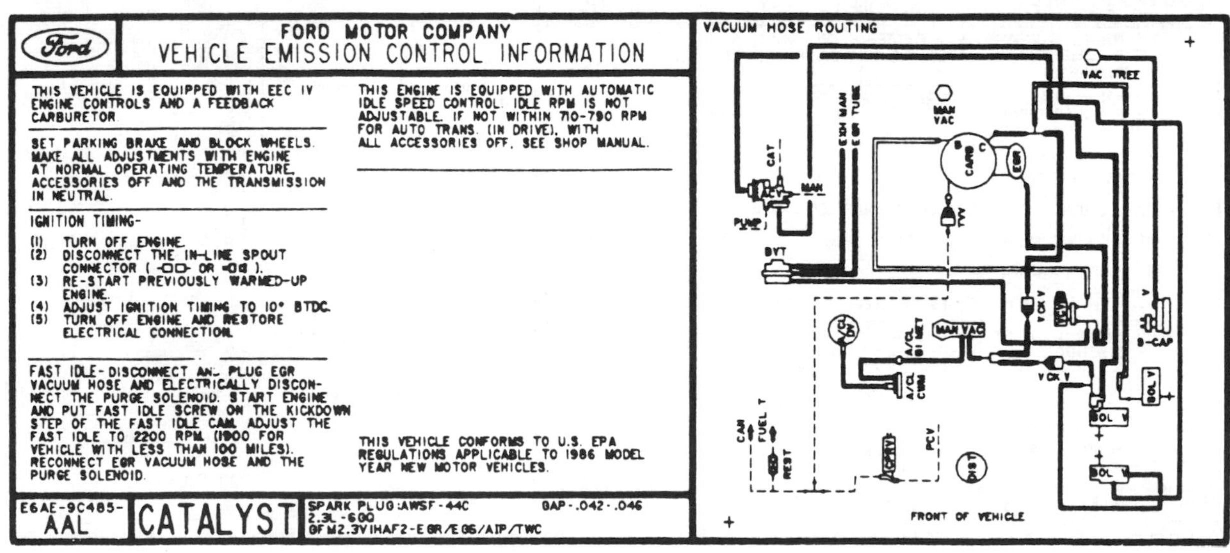

FORD MOTOR COMPANY
VEHICLE EMISSION CONTROL INFORMATION

THIS VEHICLE IS EQUIPPED WITH EEC IV ENGINE CONTROLS AND A FEEDBACK CARBURETOR.

SET PARKING BRAKE AND BLOCK WHEELS. MAKE ALL ADJUSTMENTS WITH ENGINE AT NORMAL OPERATING TEMPERATURE, ACCESSORIES OFF AND THE TRANSMISSION IN NEUTRAL.

IGNITION TIMING-

(1) TURN OFF ENGINE.
(2) DISCONNECT THE IN-LINE SPOUT CONNECTOR (-□□- OR -□◘).
(3) RE-START PREVIOUSLY WARMED-UP ENGINE.
(4) ADJUST IGNITION TIMING TO 10° BTDC.
(5) TURN OFF ENGINE AND RESTORE ELECTRICAL CONNECTION.

FAST IDLE-DISCONNECT AND PLUG EGR VACUUM HOSE AND ELECTRICALLY DISCONNECT THE PURGE SOLENOID. START ENGINE AND PUT FAST IDLE SCREW ON THE KICKDOWN STEP OF THE FAST IDLE CAM. ADJUST THE FAST IDLE TO 2200 RPM. (1900 FOR VEHICLE WITH LESS THAN 100 MILES). RECONNECT EGR VACUUM HOSE AND THE PURGE SOLENOID.

THIS ENGINE IS EQUIPPED WITH AUTOMATIC IDLE SPEED CONTROL. IDLE RPM IS NOT ADJUSTABLE. IF NOT WITHIN 710-790 RPM FOR AUTO TRANS. (IN DRIVE), WITH ALL ACCESSORIES OFF, SEE SHOP MANUAL.

THIS VEHICLE CONFORMS TO U.S. EPA REGULATIONS APPLICABLE TO 1986 MODEL YEAR NEW MOTOR VEHICLES.

E6AE-9C485-AAL | CATALYST | SPARK PLUG: AWSF-44C GAP-.042-.046 2.3L-6GQ GFM2.3V1HAF2-EGR/EGS/AIP/TWC

VACUUM CIRCUITS

(© Ford Motor Co)

CALIBRATION: 5—06N—R00

MODEL YEAR: 1986 **ENGINE: 2.3L**

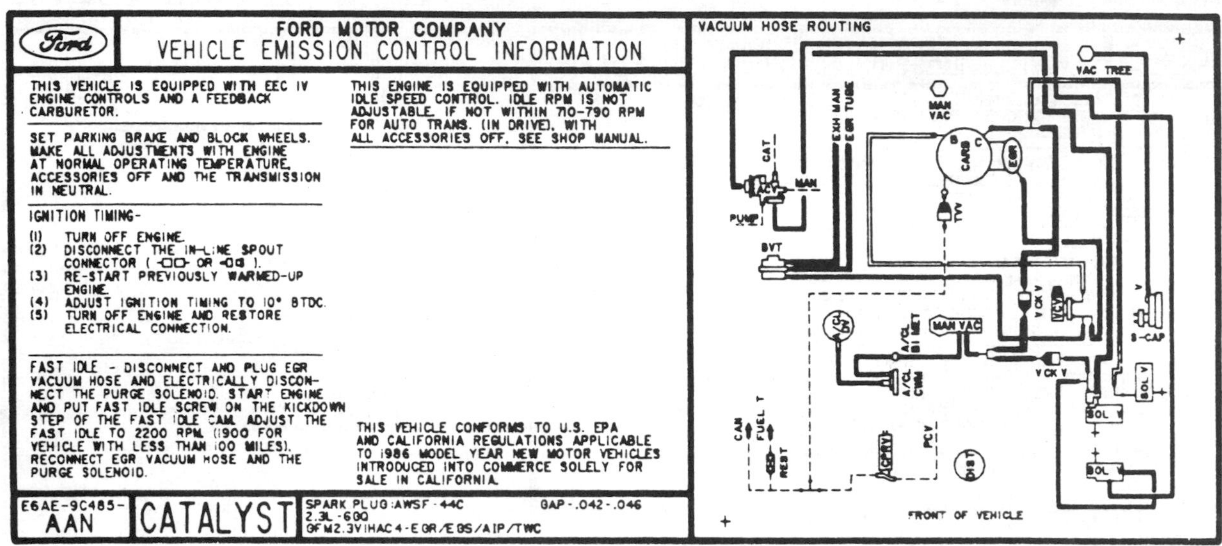

Ford

FORD MOTOR COMPANY
VEHICLE EMISSION CONTROL INFORMATION

THIS VEHICLE IS EQUIPPED WITH EEC IV ENGINE CONTROLS AND A FEEDBACK CARBURETOR.

SET PARKING BRAKE AND BLOCK WHEELS. MAKE ALL ADJUSTMENTS WITH ENGINE AT NORMAL OPERATING TEMPERATURE, ACCESSORIES OFF AND THE TRANSMISSION IN NEUTRAL.

IGNITION TIMING-

(1) TURN OFF ENGINE.
(2) DISCONNECT THE IN-LINE SPOUT CONNECTOR (-□□- OR -□□).
(3) RE-START PREVIOUSLY WARMED-UP ENGINE.
(4) ADJUST IGNITION TIMING TO 10° BTDC.
(5) TURN OFF ENGINE AND RESTORE ELECTRICAL CONNECTION.

FAST IDLE - DISCONNECT AND PLUG EGR VACUUM HOSE AND ELECTRICALLY DISCONNECT THE PURGE SOLENOID. START ENGINE AND PUT FAST IDLE SCREW ON THE KICKDOWN STEP OF THE FAST IDLE CAM. ADJUST THE FAST IDLE TO 2200 RPM. (1900 FOR VEHICLE WITH LESS THAN 100 MILES). RECONNECT EGR VACUUM HOSE AND THE PURGE SOLENOID.

THIS ENGINE IS EQUIPPED WITH AUTOMATIC IDLE SPEED CONTROL. IDLE RPM IS NOT ADJUSTABLE. IF NOT WITHIN 710-790 RPM FOR AUTO TRANS. (IN DRIVE), WITH ALL ACCESSORIES OFF. SEE SHOP MANUAL.

THIS VEHICLE CONFORMS TO U.S. EPA AND CALIFORNIA REGULATIONS APPLICABLE TO 1986 MODEL YEAR NEW MOTOR VEHICLES INTRODUCED INTO COMMERCE SOLELY FOR SALE IN CALIFORNIA.

E6AE-9C485-AAN | CATALYST | SPARK PLUG:AWSF-44C GAP-.042-.046
2.3L-60Q
GFM2.3V1HAC4-EGR/EGS/AIP/TWC

CALIBRATION: 5—25C—R01

MODEL YEAR: 1986 **ENGINE: 2.3L**

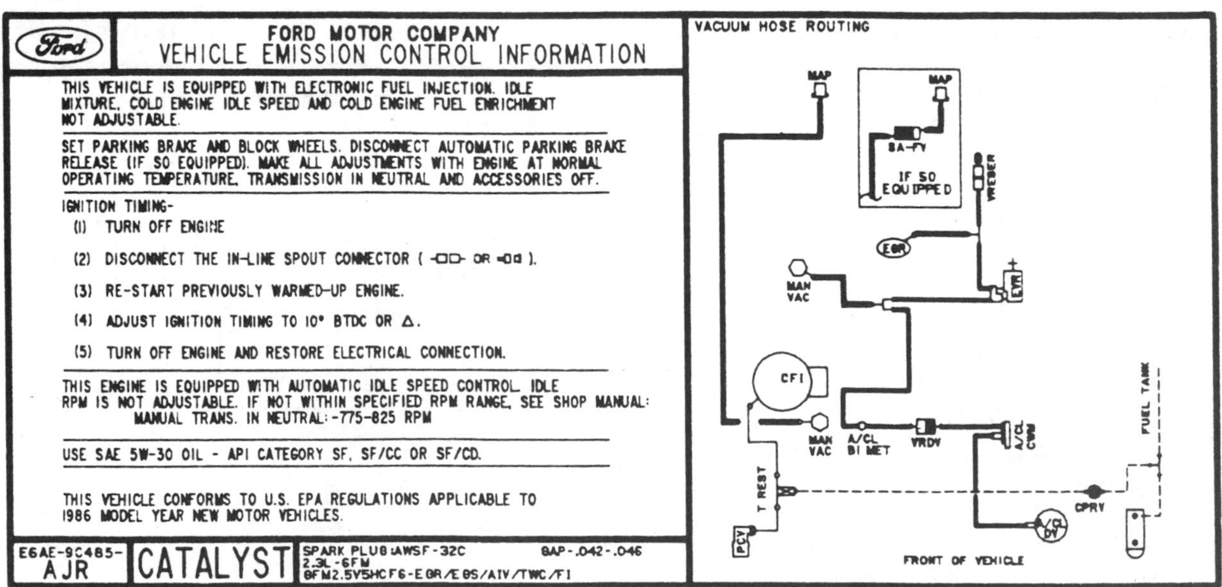

Ford

FORD MOTOR COMPANY
VEHICLE EMISSION CONTROL INFORMATION

THIS VEHICLE IS EQUIPPED WITH ELECTRONIC FUEL INJECTION. IDLE MIXTURE, COLD ENGINE IDLE SPEED AND COLD ENGINE FUEL ENRICHMENT NOT ADJUSTABLE.

SET PARKING BRAKE AND BLOCK WHEELS. DISCONNECT AUTOMATIC PARKING BRAKE RELEASE (IF SO EQUIPPED). MAKE ALL ADJUSTMENTS WITH ENGINE AT NORMAL OPERATING TEMPERATURE, TRANSMISSION IN NEUTRAL AND ACCESSORIES OFF.

IGNITION TIMING-

(1) TURN OFF ENGINE
(2) DISCONNECT THE IN-LINE SPOUT CONNECTOR (-□□- OR -□□).
(3) RE-START PREVIOUSLY WARMED-UP ENGINE.
(4) ADJUST IGNITION TIMING TO 10° BTDC OR △.
(5) TURN OFF ENGINE AND RESTORE ELECTRICAL CONNECTION.

THIS ENGINE IS EQUIPPED WITH AUTOMATIC IDLE SPEED CONTROL. IDLE RPM IS NOT ADJUSTABLE. IF NOT WITHIN SPECIFIED RPM RANGE, SEE SHOP MANUAL: MANUAL TRANS. IN NEUTRAL:-775-825 RPM

USE SAE 5W-30 OIL - API CATEGORY SF, SF/CC OR SF/CD.

THIS VEHICLE CONFORMS TO U.S. EPA REGULATIONS APPLICABLE TO 1986 MODEL YEAR NEW MOTOR VEHICLES.

E6AE-9C485-AJR | CATALYST | SPARK PLUG:AWSF-32C GAP-.042-.046
2.3L-6FM
GFM2.5Y5HCF6-EGR/EGS/AIV/TWC/FI

VACUUM CIRCUITS

(© Ford Motor Co)

CALIBRATION: 5—25F—R10

MODEL YEAR: 1986 **ENGINE: 2.3L**

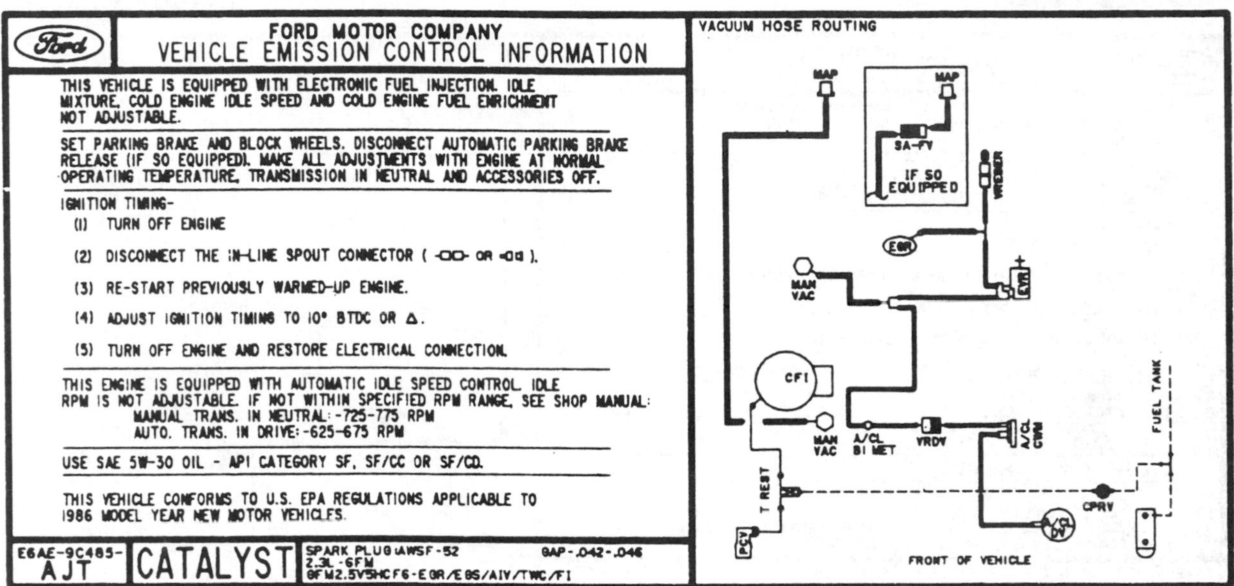

FORD MOTOR COMPANY
VEHICLE EMISSION CONTROL INFORMATION

THIS VEHICLE IS EQUIPPED WITH ELECTRONIC FUEL INJECTION. IDLE MIXTURE, COLD ENGINE IDLE SPEED AND COLD ENGINE FUEL ENRICHMENT NOT ADJUSTABLE.

SET PARKING BRAKE AND BLOCK WHEELS. DISCONNECT AUTOMATIC PARKING BRAKE RELEASE (IF SO EQUIPPED). MAKE ALL ADJUSTMENTS WITH ENGINE AT NORMAL OPERATING TEMPERATURE, TRANSMISSION IN NEUTRAL AND ACCESSORIES OFF.

IGNITION TIMING-
(1) TURN OFF ENGINE
(2) DISCONNECT THE IN-LINE SPOUT CONNECTOR (-○○- OR -○d).
(3) RE-START PREVIOUSLY WARMED-UP ENGINE.
(4) ADJUST IGNITION TIMING TO 10° BTDC OR Δ.
(5) TURN OFF ENGINE AND RESTORE ELECTRICAL CONNECTION.

THIS ENGINE IS EQUIPPED WITH AUTOMATIC IDLE SPEED CONTROL. IDLE RPM IS NOT ADJUSTABLE. IF NOT WITHIN SPECIFIED RPM RANGE, SEE SHOP MANUAL:
MANUAL TRANS. IN NEUTRAL:-725-775 RPM
AUTO. TRANS. IN DRIVE:-625-675 RPM

USE SAE 5W-30 OIL - API CATEGORY SF, SF/CC OR SF/CD.

THIS VEHICLE CONFORMS TO U.S. EPA REGULATIONS APPLICABLE TO 1986 MODEL YEAR NEW MOTOR VEHICLES.

E6AE-9C485-AJT CATALYST SPARK PLUG:AWSF-52 GAP-.042-.046
2.3L-6FM
GFM2.5V5HCF6-EGR/EGS/AIV/TWC/FI

CALIBRATION: 5—25P—R00

MODEL YEAR: 1986 **ENGINE: 2.3L**

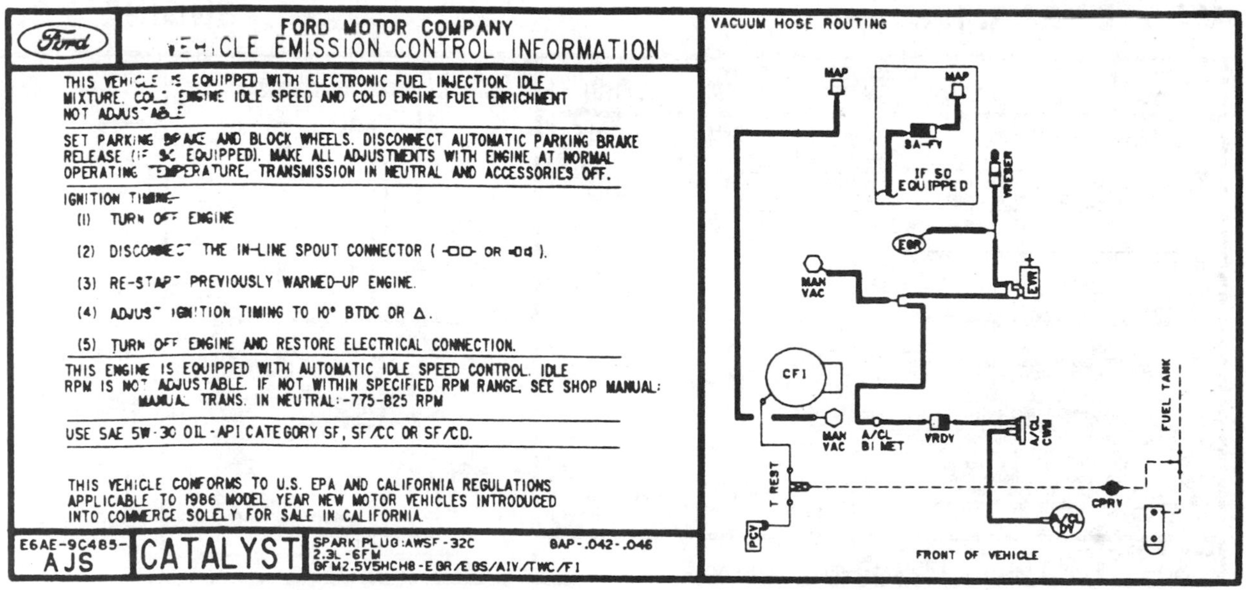

FORD MOTOR COMPANY
VEHICLE EMISSION CONTROL INFORMATION

THIS VEHICLE IS EQUIPPED WITH ELECTRONIC FUEL INJECTION. IDLE MIXTURE, COLD ENGINE IDLE SPEED AND COLD ENGINE FUEL ENRICHMENT NOT ADJUSTABLE.

SET PARKING BRAKE AND BLOCK WHEELS. DISCONNECT AUTOMATIC PARKING BRAKE RELEASE (IF SO EQUIPPED). MAKE ALL ADJUSTMENTS WITH ENGINE AT NORMAL OPERATING TEMPERATURE, TRANSMISSION IN NEUTRAL AND ACCESSORIES OFF.

IGNITION TIMING-
(1) TURN OFF ENGINE
(2) DISCONNECT THE IN-LINE SPOUT CONNECTOR (-○○- OR -○d).
(3) RE-START PREVIOUSLY WARMED-UP ENGINE.
(4) ADJUST IGNITION TIMING TO 10° BTDC OR Δ.
(5) TURN OFF ENGINE AND RESTORE ELECTRICAL CONNECTION.

THIS ENGINE IS EQUIPPED WITH AUTOMATIC IDLE SPEED CONTROL. IDLE RPM IS NOT ADJUSTABLE. IF NOT WITHIN SPECIFIED RPM RANGE, SEE SHOP MANUAL:
MANUAL TRANS. IN NEUTRAL:-775-825 RPM

USE SAE 5W-30 OIL-API CATEGORY SF, SF/CC OR SF/CD.

THIS VEHICLE CONFORMS TO U.S. EPA AND CALIFORNIA REGULATIONS APPLICABLE TO 1986 MODEL YEAR NEW MOTOR VEHICLES INTRODUCED INTO COMMERCE SOLELY FOR SALE IN CALIFORNIA.

E6AE-9C485-AJS CATALYST SPARK PLUG:AWSF-32C GAP-.042-.046
2.3L-6FM
GFM2.5V5HCH8-EGR/EGS/AIV/TWC/FI

VACUUM CIRCUITS

(© Ford Motor Co)

CALIBRATION: 5—25Q—R10

MODEL YEAR: 1986 **ENGINE: 2.3L**

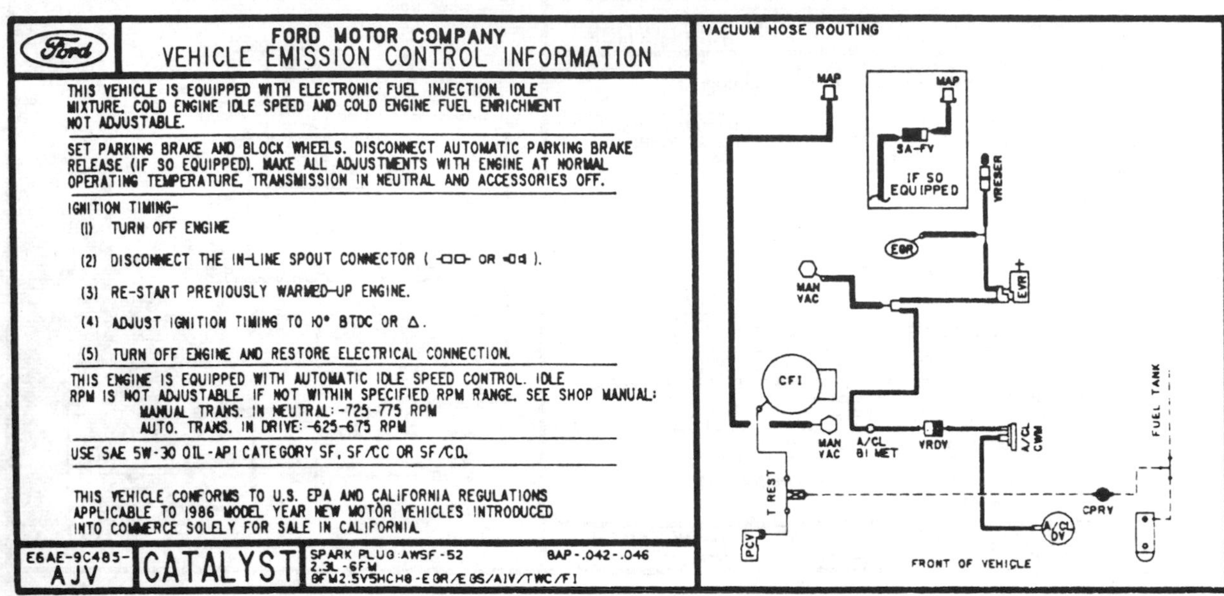

FORD MOTOR COMPANY
VEHICLE EMISSION CONTROL INFORMATION

THIS VEHICLE IS EQUIPPED WITH ELECTRONIC FUEL INJECTION. IDLE MIXTURE, COLD ENGINE IDLE SPEED AND COLD ENGINE FUEL ENRICHMENT NOT ADJUSTABLE.

SET PARKING BRAKE AND BLOCK WHEELS. DISCONNECT AUTOMATIC PARKING BRAKE RELEASE (IF SO EQUIPPED). MAKE ALL ADJUSTMENTS WITH ENGINE AT NORMAL OPERATING TEMPERATURE, TRANSMISSION IN NEUTRAL AND ACCESSORIES OFF.

IGNITION TIMING-
(1) TURN OFF ENGINE
(2) DISCONNECT THE IN-LINE SPOUT CONNECTOR (-□□- OR ⊲□⊲).
(3) RE-START PREVIOUSLY WARMED-UP ENGINE.
(4) ADJUST IGNITION TIMING TO 10° BTDC OR Δ.
(5) TURN OFF ENGINE AND RESTORE ELECTRICAL CONNECTION.

THIS ENGINE IS EQUIPPED WITH AUTOMATIC IDLE SPEED CONTROL. IDLE RPM IS NOT ADJUSTABLE. IF NOT WITHIN SPECIFIED RPM RANGE, SEE SHOP MANUAL:
MANUAL TRANS. IN NEUTRAL: -725-775 RPM
AUTO. TRANS. IN DRIVE: -625-675 RPM

USE SAE 5W-30 OIL-API CATEGORY SF, SF/CC OR SF/CD.

THIS VEHICLE CONFORMS TO U.S. EPA AND CALIFORNIA REGULATIONS APPLICABLE TO 1986 MODEL YEAR NEW MOTOR VEHICLES INTRODUCED INTO COMMERCE SOLELY FOR SALE IN CALIFORNIA.

E6AE-9C485- AJV | CATALYST | SPARK PLUG:AWSF-52 GAP-.042-.046
2.3L-6FM
6FM2.5V5HCH8-EGR/EGS/AIV/TWC/FI

CALIBRATION: 6—05A—R11

MODEL YEAR: 1986 **ENGINE: 2.3L**

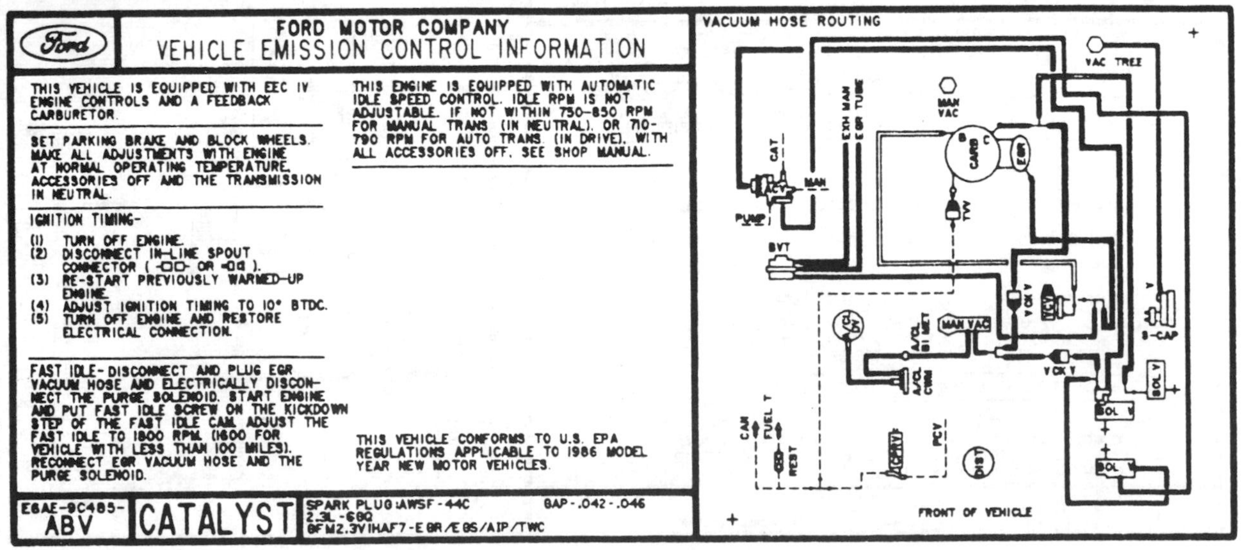

FORD MOTOR COMPANY
VEHICLE EMISSION CONTROL INFORMATION

THIS VEHICLE IS EQUIPPED WITH EEC IV ENGINE CONTROLS AND A FEEDBACK CARBURETOR.

SET PARKING BRAKE AND BLOCK WHEELS. MAKE ALL ADJUSTMENTS WITH ENGINE AT NORMAL OPERATING TEMPERATURE, ACCESSORIES OFF AND THE TRANSMISSION IN NEUTRAL.

IGNITION TIMING-
(1) TURN OFF ENGINE.
(2) DISCONNECT IN-LINE SPOUT CONNECTOR (-□□- OR ⊲□⊲).
(3) RE-START PREVIOUSLY WARMED-UP ENGINE.
(4) ADJUST IGNITION TIMING TO 10° BTDC.
(5) TURN OFF ENGINE AND RESTORE ELECTRICAL CONNECTION.

FAST IDLE-DISCONNECT AND PLUG EGR VACUUM HOSE AND ELECTRICALLY DISCONNECT THE PURGE SOLENOID. START ENGINE AND PUT FAST IDLE SCREW ON THE KICKDOWN STEP OF THE FAST IDLE CAM. ADJUST THE FAST IDLE TO 1800 RPM. (1600 FOR VEHICLE WITH LESS THAN 100 MILES). RECONNECT EGR VACUUM HOSE AND THE PURGE SOLENOID.

THIS ENGINE IS EQUIPPED WITH AUTOMATIC IDLE SPEED CONTROL. IDLE RPM IS NOT ADJUSTABLE. IF NOT WITHIN 750-850 RPM FOR MANUAL TRANS (IN NEUTRAL), OR 710-790 RPM FOR AUTO TRANS. (IN DRIVE), WITH ALL ACCESSORIES OFF. SEE SHOP MANUAL.

THIS VEHICLE CONFORMS TO U.S. EPA REGULATIONS APPLICABLE TO 1986 MODEL YEAR NEW MOTOR VEHICLES.

E6AE-9C485- ABV | CATALYST | SPARK PLUG:AWSF-44C GAP-.042-.046
2.3L-68Q
6FM2.3V1HAF7-EGR/EGS/AIP/TWC

VACUUM CIRCUITS

(© Ford Motor Co)

CALIBRATION: 6—05R—R00

MODEL YEAR: 1986 **ENGINE: 2.3L T/C**

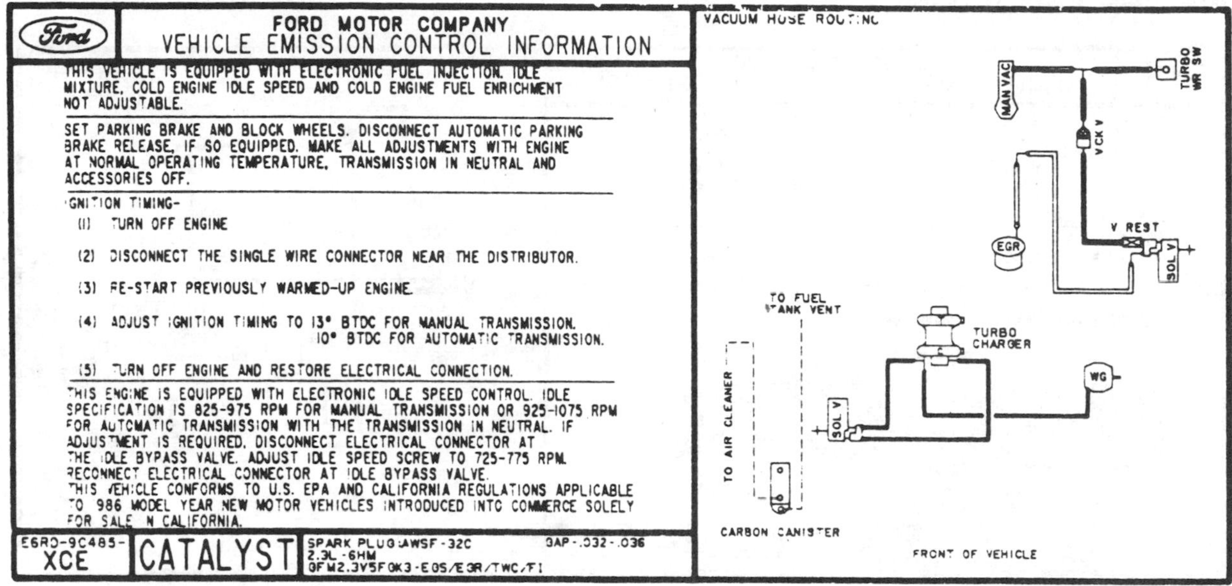

FORD MOTOR COMPANY
VEHICLE EMISSION CONTROL INFORMATION

THIS VEHICLE IS EQUIPPED WITH ELECTRONIC FUEL INJECTION. IDLE MIXTURE, COLD ENGINE IDLE SPEED AND COLD ENGINE FUEL ENRICHMENT NOT ADJUSTABLE.

SET PARKING BRAKE AND BLOCK WHEELS. DISCONNECT AUTOMATIC PARKING BRAKE RELEASE, IF SO EQUIPPED. MAKE ALL ADJUSTMENTS WITH ENGINE AT NORMAL OPERATING TEMPERATURE, TRANSMISSION IN NEUTRAL AND ACCESSORIES OFF.

IGNITION TIMING-

(1) TURN OFF ENGINE

(2) DISCONNECT THE SINGLE WIRE CONNECTOR NEAR THE DISTRIBUTOR.

(3) RE-START PREVIOUSLY WARMED-UP ENGINE.

(4) ADJUST IGNITION TIMING TO 13° BTDC FOR MANUAL TRANSMISSION.
10° BTDC FOR AUTOMATIC TRANSMISSION.

(5) TURN OFF ENGINE AND RESTORE ELECTRICAL CONNECTION.

THIS ENGINE IS EQUIPPED WITH ELECTRONIC IDLE SPEED CONTROL. IDLE SPECIFICATION IS 825-975 RPM FOR MANUAL TRANSMISSION OR 925-1075 RPM FOR AUTOMATIC TRANSMISSION WITH THE TRANSMISSION IN NEUTRAL. IF ADJUSTMENT IS REQUIRED, DISCONNECT ELECTRICAL CONNECTOR AT THE IDLE BYPASS VALVE. ADJUST IDLE SPEED SCREW TO 725-775 RPM. RECONNECT ELECTRICAL CONNECTOR AT IDLE BYPASS VALVE.
THIS VEHICLE CONFORMS TO U.S. EPA AND CALIFORNIA REGULATIONS APPLICABLE TO 1986 MODEL YEAR NEW MOTOR VEHICLES INTRODUCED INTO COMMERCE SOLELY FOR SALE IN CALIFORNIA.

E6RD-9C485-XCE | CATALYST | SPARK PLUG:AWSF-32C GAP-.032-.036
2.3L-6HM
GFM2.3V5FGK3-EGS/EGR/TWC/FI

CALIBRATION: 6—06A—R10

MODEL YEAR: 1986 **ENGINE: 2.3L**

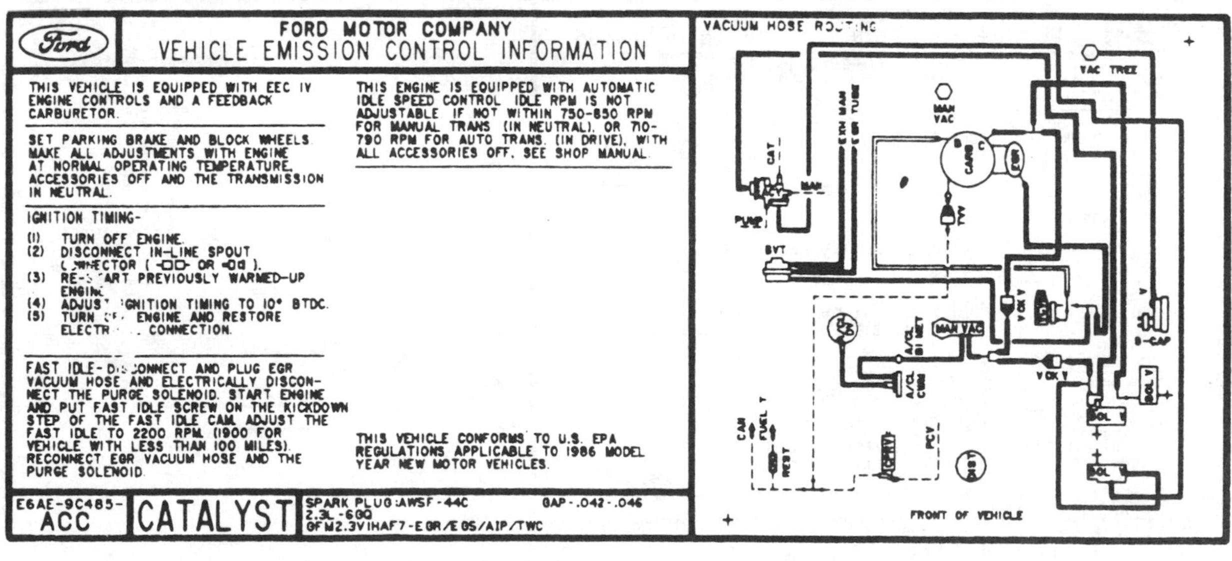

FORD MOTOR COMPANY
VEHICLE EMISSION CONTROL INFORMATION

THIS VEHICLE IS EQUIPPED WITH EEC IV ENGINE CONTROLS AND A FEEDBACK CARBURETOR.

SET PARKING BRAKE AND BLOCK WHEELS. MAKE ALL ADJUSTMENTS WITH ENGINE AT NORMAL OPERATING TEMPERATURE, ACCESSORIES OFF AND THE TRANSMISSION IN NEUTRAL.

IGNITION TIMING-

(1) TURN OFF ENGINE.
(2) DISCONNECT IN-LINE SPOUT CONNECTOR (-□□- OR -□α).
(3) RE-START PREVIOUSLY WARMED-UP ENGINE.
(4) ADJUST IGNITION TIMING TO 10° BTDC.
(5) TURN OFF ENGINE AND RESTORE ELECTRICAL CONNECTION.

FAST IDLE- DISCONNECT AND PLUG EGR VACUUM HOSE AND ELECTRICALLY DISCONNECT THE PURGE SOLENOID. START ENGINE AND PUT FAST IDLE SCREW ON THE KICKDOWN STEP OF THE FAST IDLE CAM. ADJUST THE FAST IDLE TO 2200 RPM. (1900 FOR VEHICLE WITH LESS THAN 100 MILES). RECONNECT EGR VACUUM HOSE AND THE PURGE SOLENOID.

THIS ENGINE IS EQUIPPED WITH AUTOMATIC IDLE SPEED CONTROL. IDLE RPM IS NOT ADJUSTABLE. IF NOT WITHIN 750-850 RPM FOR MANUAL TRANS (IN NEUTRAL), OR 710-790 RPM FOR AUTO TRANS. (IN DRIVE), WITH ALL ACCESSORIES OFF. SEE SHOP MANUAL.

THIS VEHICLE CONFORMS TO U.S. EPA REGULATIONS APPLICABLE TO 1986 MODEL YEAR NEW MOTOR VEHICLES.

E6AE-9C485-ACC | CATALYST | SPARK PLUG:AWSF-44C GAP-.042-.046
2.3L-6GQ
GFM2.3V1HAF7-EGR/EGS/AIP/TWC

VACUUM CIRCUITS

CALIBRATION: 6—06E—R01

(T-Bird/Cougar)

MODEL YEAR: 1986 **ENGINE: 2.3L T/C**

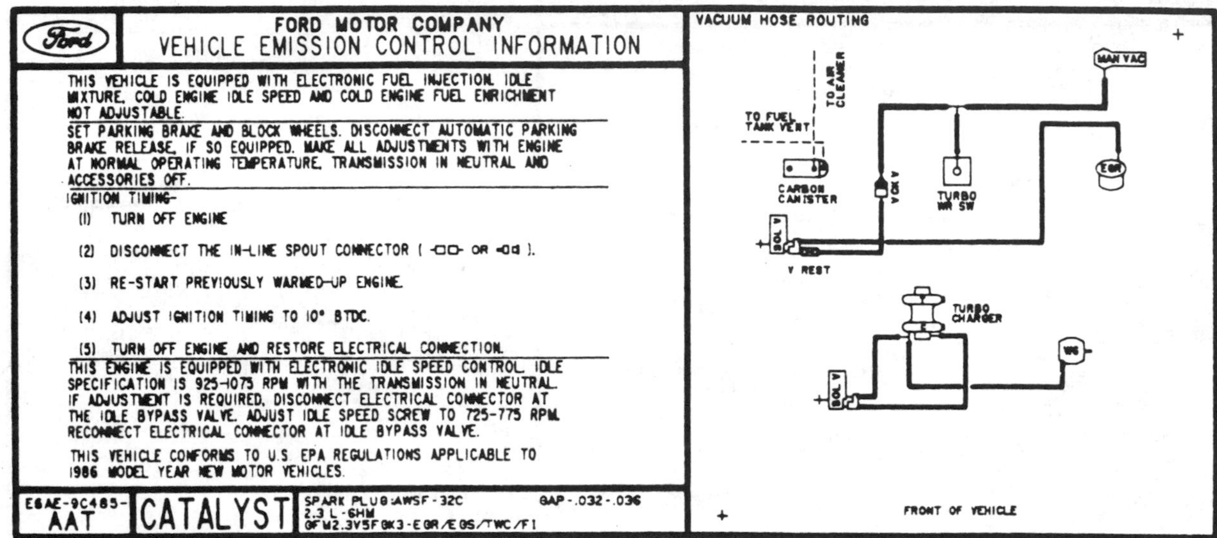

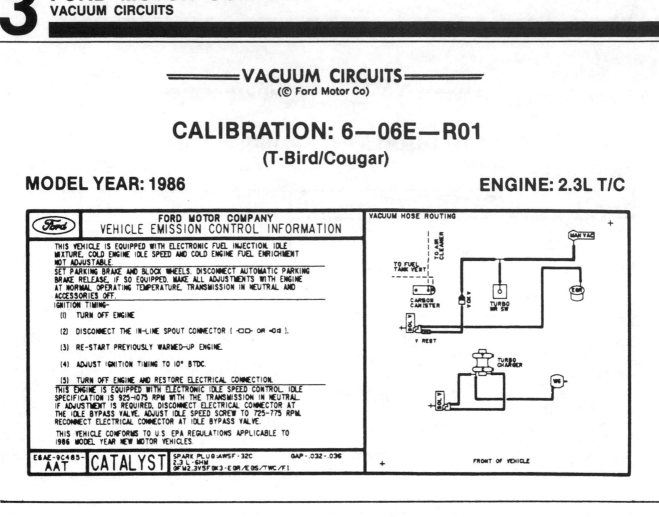

FORD MOTOR COMPANY
VEHICLE EMISSION CONTROL INFORMATION

THIS VEHICLE IS EQUIPPED WITH ELECTRONIC FUEL INJECTION. IDLE MIXTURE, COLD ENGINE IDLE SPEED AND COLD ENGINE FUEL ENRICHMENT NOT ADJUSTABLE.

SET PARKING BRAKE AND BLOCK WHEELS. DISCONNECT AUTOMATIC PARKING BRAKE RELEASE, IF SO EQUIPPED. MAKE ALL ADJUSTMENTS WITH ENGINE AT NORMAL OPERATING TEMPERATURE, TRANSMISSION IN NEUTRAL AND ACCESSORIES OFF.

IGNITION TIMING-

(1) TURN OFF ENGINE

(2) DISCONNECT THE IN-LINE SPOUT CONNECTOR (-□□- OR -□◁).

(3) RE-START PREVIOUSLY WARMED-UP ENGINE.

(4) ADJUST IGNITION TIMING TO 10° BTDC.

(5) TURN OFF ENGINE AND RESTORE ELECTRICAL CONNECTION.

THIS ENGINE IS EQUIPPED WITH ELECTRONIC IDLE SPEED CONTROL. IDLE SPECIFICATION IS 925-1075 RPM WITH THE TRANSMISSION IN NEUTRAL. IF ADJUSTMENT IS REQUIRED, DISCONNECT ELECTRICAL CONNECTOR AT THE IDLE BYPASS VALVE. ADJUST IDLE SPEED SCREW TO 725-775 RPM. RECONNECT ELECTRICAL CONNECTOR AT IDLE BYPASS VALVE.

THIS VEHICLE CONFORMS TO U.S. EPA REGULATIONS APPLICABLE TO 1986 MODEL YEAR NEW MOTOR VEHICLES.

E6AE-9C485-AAT | CATALYST | SPARK PLUG:AWSF-32C GAP-.032-.036 2.3L-6HM GFM2.3V5FGK3-EGR/EGS/TWC/FI

CALIBRATION: 6—06E—R01

(Merkur XR4Ti)

MODEL YEAR: 1986 **ENGINE: 2.3L T/C**

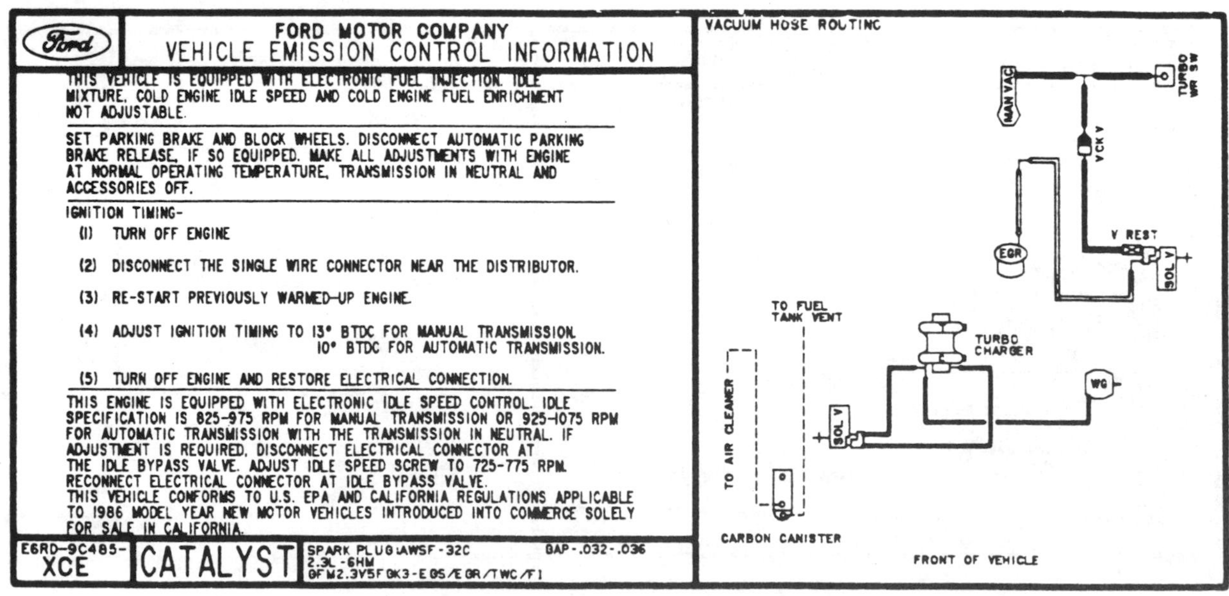

FORD MOTOR COMPANY
VEHICLE EMISSION CONTROL INFORMATION

THIS VEHICLE IS EQUIPPED WITH ELECTRONIC FUEL INJECTION. IDLE MIXTURE, COLD ENGINE IDLE SPEED AND COLD ENGINE FUEL ENRICHMENT NOT ADJUSTABLE.

SET PARKING BRAKE AND BLOCK WHEELS. DISCONNECT AUTOMATIC PARKING BRAKE RELEASE, IF SO EQUIPPED. MAKE ALL ADJUSTMENTS WITH ENGINE AT NORMAL OPERATING TEMPERATURE, TRANSMISSION IN NEUTRAL AND ACCESSORIES OFF.

IGNITION TIMING-

(1) TURN OFF ENGINE

(2) DISCONNECT THE SINGLE WIRE CONNECTOR NEAR THE DISTRIBUTOR.

(3) RE-START PREVIOUSLY WARMED-UP ENGINE.

(4) ADJUST IGNITION TIMING TO 13° BTDC FOR MANUAL TRANSMISSION.
10° BTDC FOR AUTOMATIC TRANSMISSION.

(5) TURN OFF ENGINE AND RESTORE ELECTRICAL CONNECTION.

THIS ENGINE IS EQUIPPED WITH ELECTRONIC IDLE SPEED CONTROL. IDLE SPECIFICATION IS 825-975 RPM FOR MANUAL TRANSMISSION OR 925-1075 RPM FOR AUTOMATIC TRANSMISSION WITH THE TRANSMISSION IN NEUTRAL. IF ADJUSTMENT IS REQUIRED, DISCONNECT ELECTRICAL CONNECTOR AT THE IDLE BYPASS VALVE. ADJUST IDLE SPEED SCREW TO 725-775 RPM. RECONNECT ELECTRICAL CONNECTOR AT IDLE BYPASS VALVE.
THIS VEHICLE CONFORMS TO U.S. EPA AND CALIFORNIA REGULATIONS APPLICABLE TO 1986 MODEL YEAR NEW MOTOR VEHICLES INTRODUCED INTO COMMERCE SOLELY FOR SALE IN CALIFORNIA.

E6RD-9C485-XCE | CATALYST | SPARK PLUG:AWSF-32C GAP-.032-.036 2.3L-6HM GFM2.3V5FGK3-EGS/EGR/TWC/FI

VACUUM CIRCUITS

CALIBRATION: 6—06N—R10

MODEL YEAR: 1986 **ENGINE: 2.3L**

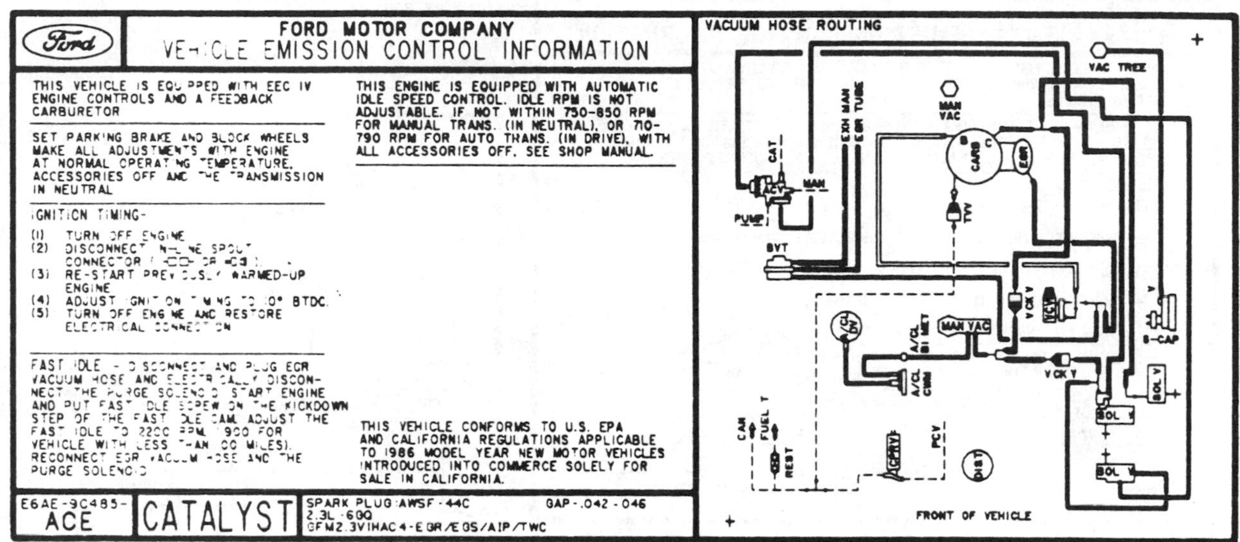

FORD MOTOR COMPANY
VEHICLE EMISSION CONTROL INFORMATION

THIS VEHICLE IS EQUIPPED WITH EEC IV ENGINE CONTROLS AND A FEEDBACK CARBURETOR

SET PARKING BRAKE AND BLOCK WHEELS MAKE ALL ADJUSTMENTS WITH ENGINE AT NORMAL OPERATING TEMPERATURE, ACCESSORIES OFF AND THE TRANSMISSION IN NEUTRAL

IGNITION TIMING-

(1) TURN OFF ENGINE
(2) DISCONNECT IN-LINE SPOUT CONNECTOR (-▭▭- OR -▭▭).
(3) RE-START PREVIOUSLY WARMED-UP ENGINE
(4) ADJUST IGNITION TIMING TO 10° BTDC.
(5) TURN OFF ENGINE AND RESTORE ELECTRICAL CONNECTION

FAST IDLE - DISCONNECT AND PLUG EGR VACUUM HOSE AND ELECTRICALLY DISCONNECT THE PURGE SOLENOID. START ENGINE AND PUT FAST IDLE SCREW ON THE KICKDOWN STEP OF THE FAST IDLE CAM. ADJUST THE FAST IDLE TO 2200 RPM. (1900 FOR VEHICLE WITH LESS THAN 100 MILES). RECONNECT EGR VACUUM HOSE AND THE PURGE SOLENOID

THIS ENGINE IS EQUIPPED WITH AUTOMATIC IDLE SPEED CONTROL. IDLE RPM IS NOT ADJUSTABLE. IF NOT WITHIN 750-850 RPM FOR MANUAL TRANS. (IN NEUTRAL), OR 710-790 RPM FOR AUTO TRANS. (IN DRIVE), WITH ALL ACCESSORIES OFF. SEE SHOP MANUAL.

THIS VEHICLE CONFORMS TO U.S. EPA AND CALIFORNIA REGULATIONS APPLICABLE TO 1986 MODEL YEAR NEW MOTOR VEHICLES INTRODUCED INTO COMMERCE SOLELY FOR SALE IN CALIFORNIA.

E6AE-9C485-ACE | CATALYST | SPARK PLUG:AWSF-44C GAP-.042-.046
2.3L-6GQ
GFM2.3V1HAC4-EGR/EGS/AIP/TWC

CALIBRATION: 6—26E—R00

MODEL YEAR: 1986 **ENGINE: 2.3L**

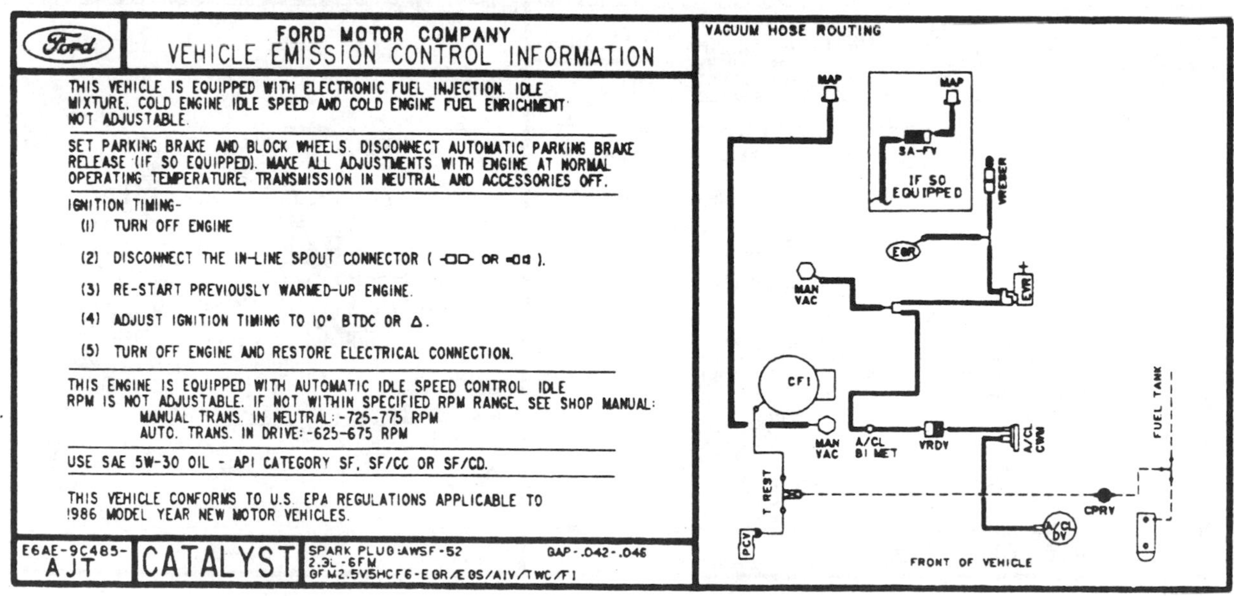

FORD MOTOR COMPANY
VEHICLE EMISSION CONTROL INFORMATION

THIS VEHICLE IS EQUIPPED WITH ELECTRONIC FUEL INJECTION. IDLE MIXTURE, COLD ENGINE IDLE SPEED AND COLD ENGINE FUEL ENRICHMENT NOT ADJUSTABLE.

SET PARKING BRAKE AND BLOCK WHEELS. DISCONNECT AUTOMATIC PARKING BRAKE RELEASE (IF SO EQUIPPED). MAKE ALL ADJUSTMENTS WITH ENGINE AT NORMAL OPERATING TEMPERATURE, TRANSMISSION IN NEUTRAL AND ACCESSORIES OFF.

IGNITION TIMING-

(1) TURN OFF ENGINE
(2) DISCONNECT THE IN-LINE SPOUT CONNECTOR (-▭▭- OR -▭▭).
(3) RE-START PREVIOUSLY WARMED-UP ENGINE.
(4) ADJUST IGNITION TIMING TO 10° BTDC OR Δ.
(5) TURN OFF ENGINE AND RESTORE ELECTRICAL CONNECTION.

THIS ENGINE IS EQUIPPED WITH AUTOMATIC IDLE SPEED CONTROL. IDLE RPM IS NOT ADJUSTABLE. IF NOT WITHIN SPECIFIED RPM RANGE, SEE SHOP MANUAL:
MANUAL TRANS. IN NEUTRAL:-725-775 RPM
AUTO. TRANS. IN DRIVE:-625-675 RPM

USE SAE 5W-30 OIL - API CATEGORY SF, SF/CC OR SF/CD.

THIS VEHICLE CONFORMS TO U.S. EPA REGULATIONS APPLICABLE TO 1986 MODEL YEAR NEW MOTOR VEHICLES.

E6AE-9C485-AJT | CATALYST | SPARK PLUG:AWSF-52 GAP-.042-.046
2.3L-6FM
GFM2.5V5HCF6-EGR/EGS/AIV/TWC/FI

VACUUM CIRCUITS

(© Ford Motor Co)

CALIBRATION: 6—26R—R10

MODEL YEAR: 1986 **ENGINE: 2.3L**

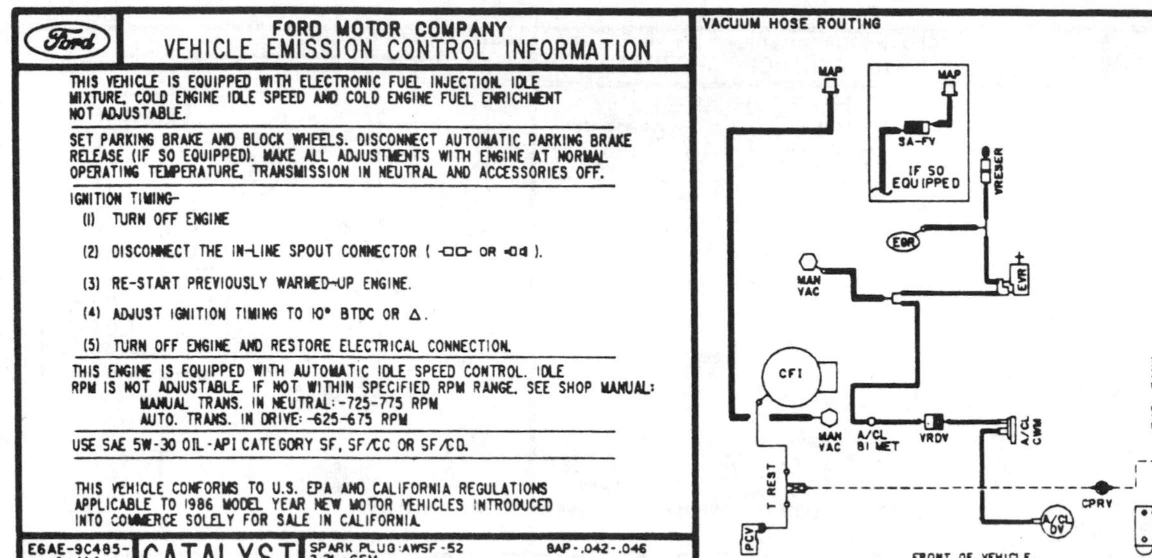

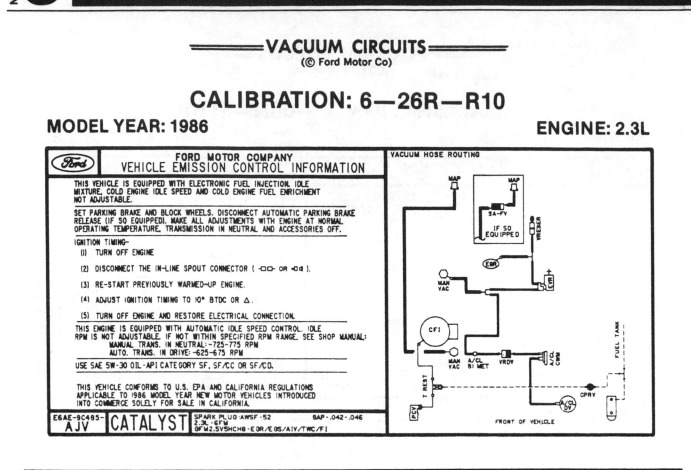

FORD MOTOR COMPANY
VEHICLE EMISSION CONTROL INFORMATION

THIS VEHICLE IS EQUIPPED WITH ELECTRONIC FUEL INJECTION. IDLE MIXTURE, COLD ENGINE IDLE SPEED AND COLD ENGINE FUEL ENRICHMENT NOT ADJUSTABLE.

SET PARKING BRAKE AND BLOCK WHEELS. DISCONNECT AUTOMATIC PARKING BRAKE RELEASE (IF SO EQUIPPED). MAKE ALL ADJUSTMENTS WITH ENGINE AT NORMAL OPERATING TEMPERATURE, TRANSMISSION IN NEUTRAL AND ACCESSORIES OFF.

IGNITION TIMING—
(1) TURN OFF ENGINE
(2) DISCONNECT THE IN-LINE SPOUT CONNECTOR (-□□- OR ≡□◁).
(3) RE-START PREVIOUSLY WARMED-UP ENGINE.
(4) ADJUST IGNITION TIMING TO 10° BTDC OR △.
(5) TURN OFF ENGINE AND RESTORE ELECTRICAL CONNECTION.

THIS ENGINE IS EQUIPPED WITH AUTOMATIC IDLE SPEED CONTROL. IDLE RPM IS NOT ADJUSTABLE. IF NOT WITHIN SPECIFIED RPM RANGE, SEE SHOP MANUAL:
MANUAL TRANS. IN NEUTRAL:-725-775 RPM
AUTO. TRANS. IN DRIVE:-625-675 RPM

USE SAE 5W-30 OIL-API CATEGORY SF, SF/CC OR SF/CD.

THIS VEHICLE CONFORMS TO U.S. EPA AND CALIFORNIA REGULATIONS APPLICABLE TO 1986 MODEL YEAR NEW MOTOR VEHICLES INTRODUCED INTO COMMERCE SOLELY FOR SALE IN CALIFORNIA.

E6AE-9C485-AJV CATALYST SPARK PLUG:AWSF-52 GAP-.042-.046
2.3L-6FM
GFM2.5V5HCH8-EGR/EGS/AIV/TWC/FI

CALIBRATION: 6—10A—R10

MODEL YEAR: 1986 **ENGINE: 3.0L**

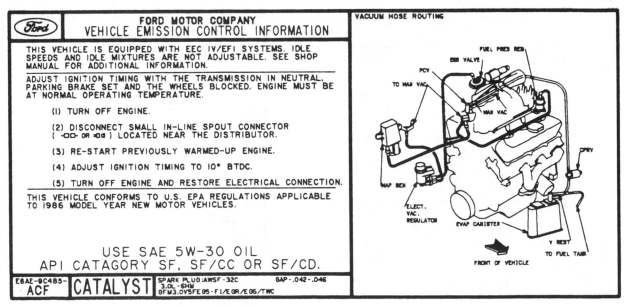

FORD MOTOR COMPANY
VEHICLE EMISSION CONTROL INFORMATION

THIS VEHICLE IS EQUIPPED WITH EEC IV/EFI SYSTEMS. IDLE SPEEDS AND IDLE MIXTURES ARE NOT ADJUSTABLE. SEE SHOP MANUAL FOR ADDITIONAL INFORMATION.

ADJUST IGNITION TIMING WITH THE TRANSMISSION IN NEUTRAL, PARKING BRAKE SET AND THE WHEELS BLOCKED. ENGINE MUST BE AT NORMAL OPERATING TEMPERATURE.

(1) TURN OFF ENGINE.
(2) DISCONNECT SMALL IN-LINE SPOUT CONNECTOR (-□□- OR ≡□◁) LOCATED NEAR THE DISTRIBUTOR.
(3) RE-START PREVIOUSLY WARMED-UP ENGINE.
(4) ADJUST IGNITION TIMING TO 10° BTDC.
(5) TURN OFF ENGINE AND RESTORE ELECTRICAL CONNECTION.

THIS VEHICLE CONFORMS TO U.S. EPA REGULATIONS APPLICABLE TO 1986 MODEL YEAR NEW MOTOR VEHICLES.

USE SAE 5W-30 OIL
API CATAGORY SF, SF/CC OR SF/CD.

E6AE-9C485-ACF CATALYST SPARK PLUG:AWSF-32C GAP-.042-.046
3.0L-6HM
GFM3.0V5FE05-FI/EGR/EGS/TWC

VACUUM HOSE ROUTING

FUEL PRES REG
EGR VALVE
PCV
TO MAN VAC
MAN VAC
CPRV
MAP SEN
ELECT. VAC. REGULATOR
EVAP CANISTER
V REST
TO FUEL TANK
FRONT OF VEHICLE

VACUUM CIRCUITS

(© Ford Motor Co)

CALIBRATION: 6—10B—R05

MODEL YEAR: 1986 **ENGINE: 3.0L**

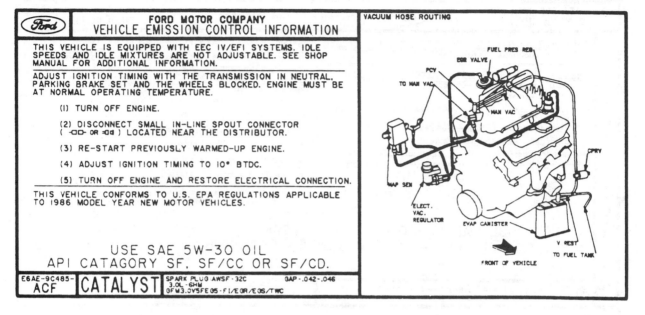

FORD MOTOR COMPANY
VEHICLE EMISSION CONTROL INFORMATION

THIS VEHICLE IS EQUIPPED WITH EEC IV/EFI SYSTEMS. IDLE SPEEDS AND IDLE MIXTURES ARE NOT ADJUSTABLE. SEE SHOP MANUAL FOR ADDITIONAL INFORMATION.

ADJUST IGNITION TIMING WITH THE TRANSMISSION IN NEUTRAL, PARKING BRAKE SET AND THE WHEELS BLOCKED. ENGINE MUST BE AT NORMAL OPERATING TEMPERATURE.

(1) TURN OFF ENGINE.

(2) DISCONNECT SMALL IN-LINE SPOUT CONNECTOR (-□□- OR =□d) LOCATED NEAR THE DISTRIBUTOR.

(3) RE-START PREVIOUSLY WARMED-UP ENGINE.

(4) ADJUST IGNITION TIMING TO 10° BTDC.

(5) TURN OFF ENGINE AND RESTORE ELECTRICAL CONNECTION.

THIS VEHICLE CONFORMS TO U.S. EPA REGULATIONS APPLICABLE TO 1986 MODEL YEAR NEW MOTOR VEHICLES.

USE SAE 5W-30 OIL
API CATAGORY SF, SF/CC OR SF/CD.

E6AE-9C485-ACF | CATALYST | SPARK PLUG:AWSF-32C 3.0L-6HM GAP-.042-.046 GFM3.0V5FE05-FI/EGR/EGS/TWC

VACUUM HOSE ROUTING

FUEL PRES REG
EGR VALVE
PCV
TO MAN VAC
MAN VAC
CPRV
MAP SEN
ELECT. VAC. REGULATOR
EVAP CANISTER
V REST
TO FUEL TANK
FRONT OF VEHICLE

CALIBRATION: 6—10C—R05

MODEL YEAR: 1986 **ENGINE: 3.0L**

FORD MOTOR COMPANY
VEHICLE EMISSION CONTROL INFORMATION

THIS VEHICLE IS EQUIPPED WITH EEC IV/EFI SYSTEMS. IDLE SPEEDS AND IDLE MIXTURES ARE NOT ADJUSTABLE. SEE SHOP MANUAL FOR ADDITIONAL INFORMATION.

ADJUST IGNITION TIMING WITH THE TRANSMISSION IN NEUTRAL, PARKING BRAKE SET AND THE WHEELS BLOCKED. ENGINE MUST BE AT NORMAL OPERATING TEMPERATURE.

(1) TURN OFF ENGINE.

(2) DISCONNECT SMALL IN-LINE SPOUT CONNECTOR (-□□- OR =□d) LOCATED NEAR THE DISTRIBUTOR.

(3) RE-START PREVIOUSLY WARMED-UP ENGINE.

(4) ADJUST IGNITION TIMING TO 10° BTDC.

(5) TURN OFF ENGINE AND RESTORE ELECTRICAL CONNECTION.

THIS VEHICLE CONFORMS TO U.S. EPA REGULATIONS APPLICABLE TO 1986 MODEL YEAR NEW MOTOR VEHICLES.

USE SAE 5W-30 OIL
API CATAGORY SF, SF/CC OR SF/CD.

E6AE-9C485-ACF | CATALYST | SPARK PLUG:AWSF-32C 3.0L-6HM GAP-.042-.046 GFM3.0V5FE05-FI/EGR/EGS/TWC

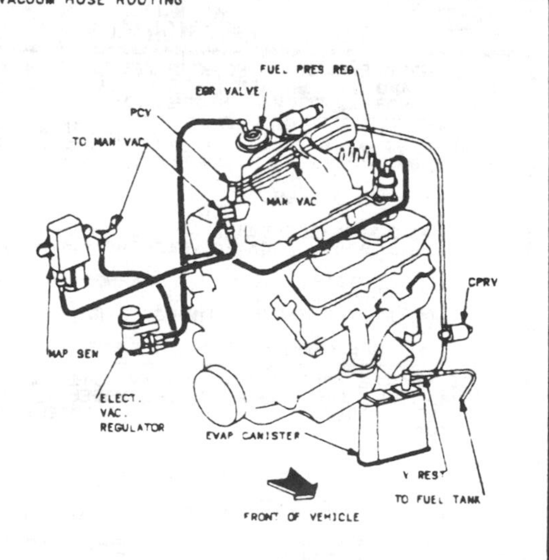

VACUUM CIRCUITS

(© Ford Motor Co)

CALIBRATION: 6—10H—R05

MODEL YEAR: 1986 **ENGINE: 3.0L**

VEHICLE EMISSION CONTROL INFORMATION  FORD MOTOR COMPANY CONTRÔLE DES ÉMISSIONS DU VÉHICULE

CATALYST CATALYSEUR

THIS VEHICLE IS EQUIPPED WITH EEC IV/EFI SYSTEMS. IDLE SPEED AND IDLE MIXTURES ARE NOT ADJUSTABLE. SEE SHOP MANUAL FOR ADDITIONAL INFORMATION.

ADJUST IGNITION TIMING WITH THE TRANSMISSION IN NEUTRAL, PARKING BRAKE SET AND THE WHEELS BLOCKED. ENGINE MUST BE AT NORMAL OPERATING TEMPERATURE.

(1) TURN OFF ENGINE.

(2) DISCONNECT SMALL IN-LINE SPOUT CONNECTOR (-□□- OR =□◁) LOCATED NEAR THE DISTRIBUTOR.

(3) RE-START PREVIOUSLY WARMED-UP ENGINE.

(4) ADJUST IGNITION TIMING TO 10° BTDC.

(5) TURN OFF ENGINE AND RESTORE ELECTRICAL CONNECTION.

CE VÉHICULE EST MUNI DES SYSTÈMES EEC IV ET EFI.• LE RÉGIME DE RALENTI ET LE MÉLANGE DE RALENTI NE SONT PAS RÉGLABLES. POUR DE PLUS AMPLES DÉTAILS, CONSULTEZ LE MANUEL DE RÉPARATION.

POUR LE CALAGE DE L'ALLUMAGE, PLACEZ LE LEVIER DE VITESSE EN POSITION "N". SERREZ LE FREIN DE STATIONNEMENT ET BLOQUEZ LES ROUES. LE MOTEUR DOIT ÊTRE NORMALEMENT CHAUD.

(1) ARRÊTEZ LE MOTEUR.

(2) DÉBRANCHEZ LE PETIT CONNECTEUR (-□□- OR =□◁) (DU CIRCUIT DE DÉCLENCHEMENT DE L'ETINCELLE) MONTÉ À MÊME LE FIL PRÈS DE L'ALLUMEUR.

(3) REDÉMARREZ LE MOTEUR PRÉALABLEMENT RÉCHAUFFÉ.

(4) CALEZ L'ALLUMAGE À 10° AVPMH.

(5) ARRÊTEZ LE MOTEUR ET REBRANCHEZ LE CONNECTEUR.

• EEC IV = COMMANDE ÉLECTRONIQUE DU MOTEUR, VERSION IV
EFI = INJECTION ÉLECTRONIQUE

USE SAE 5W-30 OIL-API CATEGORY SF, SF/CC OR SF/CD.
EMPLOYER L'HUILE SAE 5W-30, CLASSIFICATION API: SF, SF/CC OU SF/CD.

E6AE-9C485- ACH	3.0 L	SPARK PLUG/BOUGIES AWSF-32C	GAP/ÉLECTRODES .042-.046

CALIBRATION: 6—10M—R05

MODEL YEAR: 1986 **ENGINE: 3.0L**

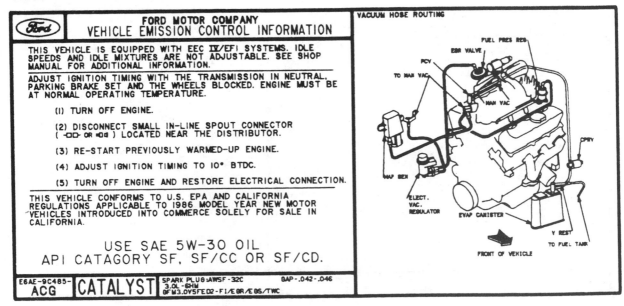

 FORD MOTOR COMPANY
VEHICLE EMISSION CONTROL INFORMATION

THIS VEHICLE IS EQUIPPED WITH EEC IV/EFI SYSTEMS. IDLE SPEEDS AND IDLE MIXTURES ARE NOT ADJUSTABLE. SEE SHOP MANUAL FOR ADDITIONAL INFORMATION.

ADJUST IGNITION TIMING WITH THE TRANSMISSION IN NEUTRAL, PARKING BRAKE SET AND THE WHEELS BLOCKED. ENGINE MUST BE AT NORMAL OPERATING TEMPERATURE.

(1) TURN OFF ENGINE.

(2) DISCONNECT SMALL IN-LINE SPOUT CONNECTOR (-□□- OR =□◁) LOCATED NEAR THE DISTRIBUTOR.

(3) RE-START PREVIOUSLY WARMED-UP ENGINE.

(4) ADJUST IGNITION TIMING TO 10° BTDC.

(5) TURN OFF ENGINE AND RESTORE ELECTRICAL CONNECTION.

THIS VEHICLE CONFORMS TO U.S. EPA AND CALIFORNIA REGULATIONS APPLICABLE TO 1986 MODEL YEAR NEW MOTOR VEHICLES INTRODUCED INTO COMMERCE SOLELY FOR SALE IN CALIFORNIA.

USE SAE 5W-30 OIL
API CATAGORY SF, SF/CC OR SF/CD.

E6AE-9C485- ACG	CATALYST	SPARK PLUG:AWSF-32C 3.0L-6HM GFM3.0V5FED2-FI/EGR/EGS/TWC	GAP-.042-.046

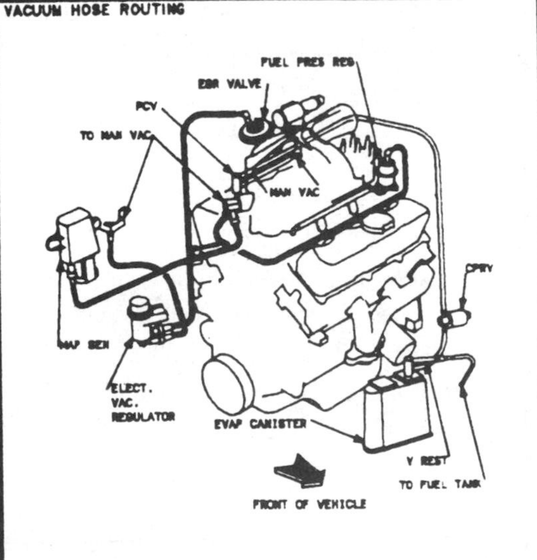

VACUUM CIRCUITS

(© Ford Motor Co)

CALIBRATION: 6—10S—R10

MODEL YEAR: 1986 **ENGINE: 3.0L**

FORD MOTOR COMPANY
VEHICLE EMISSION CONTROL INFORMATION

THIS VEHICLE IS EQUIPPED WITH EEC IV/EFI SYSTEMS. IDLE SPEEDS AND IDLE MIXTURES ARE NOT ADJUSTABLE. SEE SHOP MANUAL FOR ADDITIONAL INFORMATION.

ADJUST IGNITION TIMING WITH THE TRANSMISSION IN NEUTRAL, PARKING BRAKE SET AND THE WHEELS BLOCKED. ENGINE MUST BE AT NORMAL OPERATING TEMPERATURE.

(1) TURN OFF ENGINE.

(2) DISCONNECT SMALL IN-LINE SPOUT CONNECTOR (-□□- OR =□□) LOCATED NEAR THE DISTRIBUTOR.

(3) RE-START PREVIOUSLY WARMED-UP ENGINE.

(4) ADJUST IGNITION TIMING TO 10° BTDC.

(5) TURN OFF ENGINE AND RESTORE ELECTRICAL CONNECTION.

THIS VEHICLE CONFORMS TO U.S. EPA AND CALIFORNIA REGULATIONS APPLICABLE TO 1986 MODEL YEAR NEW MOTOR VEHICLES INTRODUCED INTO COMMERCE SOLELY FOR SALE IN CALIFORNIA.

USE SAE 5W-30 OIL
API CATAGORY SF, SF/CC OR SF/CD.

E6AE-9C485-ACG | CATALYST | SPARK PLUG:AWSF-32C GAP-.042-.046 3.0L-6HM GFM3.0V5FED2-FI/EGR/EGS/TWC

VACUUM HOSE ROUTING

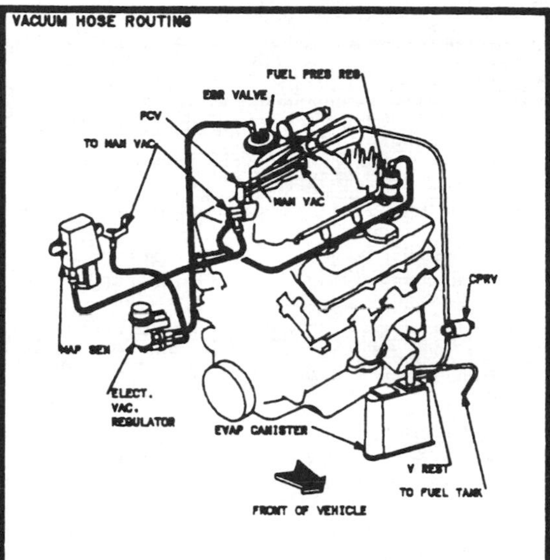

CALIBRATION: 6—16B—R00

MODEL YEAR: 1986 **ENGINE: 3.8L**

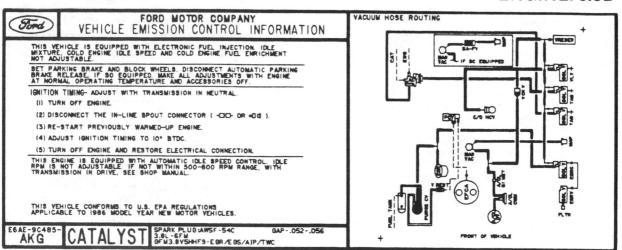

FORD MOTOR COMPANY
VEHICLE EMISSION CONTROL INFORMATION

THIS VEHICLE IS EQUIPPED WITH ELECTRONIC FUEL INJECTION. IDLE MIXTURE, COLD ENGINE IDLE SPEED AND COLD ENGINE FUEL ENRICHMENT NOT ADJUSTABLE.

SET PARKING BRAKE AND BLOCK WHEELS. DISCONNECT AUTOMATIC PARKING BRAKE RELEASE, IF SO EQUIPPED. MAKE ALL ADJUSTMENTS WITH ENGINE AT NORMAL OPERATING TEMPERATURE AND ACCESSORIES OFF.

IGNITION TIMING- ADJUST WITH TRANSMISSION IN NEUTRAL.

(1) TURN OFF ENGINE.

(2) DISCONNECT THE IN-LINE SPOUT CONNECTOR (-□□- OR =□□).

(3) RE-START PREVIOUSLY WARMED-UP ENGINE.

(4) ADJUST IGNITION TIMING TO 10° BTDC.

(5) TURN OFF ENGINE AND RESTORE ELECTRICAL CONNECTION.

THIS ENGINE IS EQUIPPED WITH AUTOMATIC IDLE SPEED CONTROL. IDLE RPM IS NOT ADJUSTABLE. IF NOT WITHIN 500-600 RPM RANGE, WITH TRANSMISSION IN DRIVE, SEE SHOP MANUAL.

THIS VEHICLE CONFORMS TO U.S. EPA REGULATIONS APPLICABLE TO 1986 MODEL YEAR NEW MOTOR VEHICLES.

E6AE-9C485-AKG | CATALYST | SPARK PLUG:AWSF-54C GAP-.052-.056 3.8L-6FM GFM3.8V5HHF9-EGR/EGS/AIP/TWC

VACUUM HOSE ROUTING

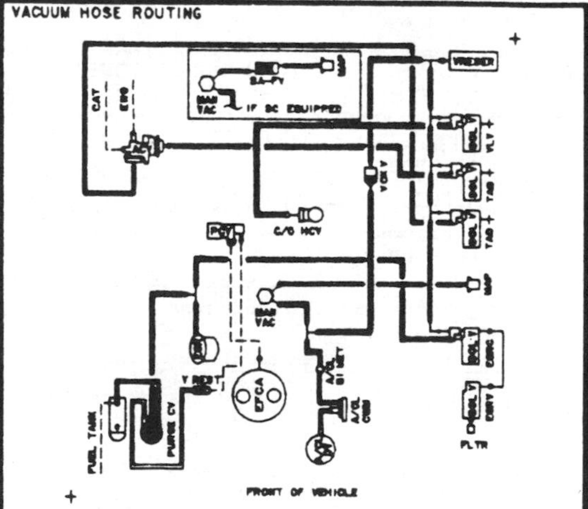

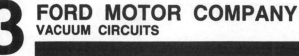

VACUUM CIRCUITS

(© Ford Motor Co)

CALIBRATION: 6—16B—R02

MODEL YEAR: 1986 **ENGINE: 3.8L**

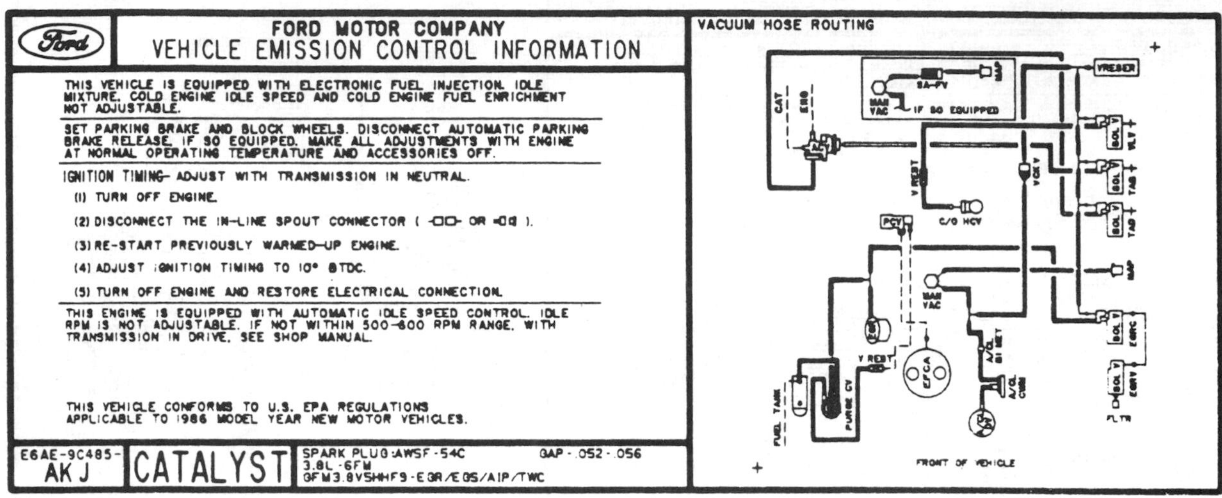

Ford

FORD MOTOR COMPANY
VEHICLE EMISSION CONTROL INFORMATION

THIS VEHICLE IS EQUIPPED WITH ELECTRONIC FUEL INJECTION. IDLE MIXTURE, COLD ENGINE IDLE SPEED AND COLD ENGINE FUEL ENRICHMENT NOT ADJUSTABLE.

SET PARKING BRAKE AND BLOCK WHEELS. DISCONNECT AUTOMATIC PARKING BRAKE RELEASE, IF SO EQUIPPED. MAKE ALL ADJUSTMENTS WITH ENGINE AT NORMAL OPERATING TEMPERATURE AND ACCESSORIES OFF.

IGNITION TIMING- ADJUST WITH TRANSMISSION IN NEUTRAL.

(1) TURN OFF ENGINE.

(2) DISCONNECT THE IN-LINE SPOUT CONNECTOR (-□□- OR =□□).

(3) RE-START PREVIOUSLY WARMED-UP ENGINE.

(4) ADJUST IGNITION TIMING TO 10° BTDC.

(5) TURN OFF ENGINE AND RESTORE ELECTRICAL CONNECTION.

THIS ENGINE IS EQUIPPED WITH AUTOMATIC IDLE SPEED CONTROL. IDLE RPM IS NOT ADJUSTABLE. IF NOT WITHIN 500-600 RPM RANGE, WITH TRANSMISSION IN DRIVE, SEE SHOP MANUAL.

THIS VEHICLE CONFORMS TO U.S. EPA REGULATIONS APPLICABLE TO 1986 MODEL YEAR NEW MOTOR VEHICLES.

E6AE-9C485-AKJ | CATALYST | SPARK PLUG:AWSF-54C GAP-.052-.056
3.8L-6FM
GFM3.8V5HHF9-EGR/EOS/AIP/TWC

CALIBRATION: 6—16N—R02

MODEL YEAR: 1986 **ENGINE: 3.8L**

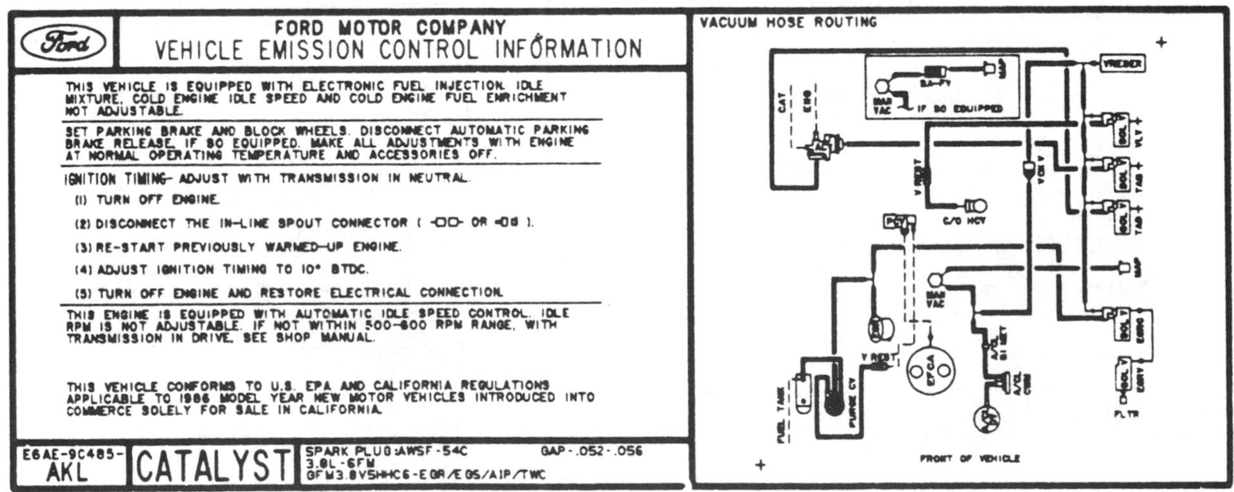

Ford

FORD MOTOR COMPANY
VEHICLE EMISSION CONTROL INFORMATION

THIS VEHICLE IS EQUIPPED WITH ELECTRONIC FUEL INJECTION. IDLE MIXTURE, COLD ENGINE IDLE SPEED AND COLD ENGINE FUEL ENRICHMENT NOT ADJUSTABLE.

SET PARKING BRAKE AND BLOCK WHEELS. DISCONNECT AUTOMATIC PARKING BRAKE RELEASE, IF SO EQUIPPED. MAKE ALL ADJUSTMENTS WITH ENGINE AT NORMAL OPERATING TEMPERATURE AND ACCESSORIES OFF.

IGNITION TIMING- ADJUST WITH TRANSMISSION IN NEUTRAL.

(1) TURN OFF ENGINE.

(2) DISCONNECT THE IN-LINE SPOUT CONNECTOR (-□□- OR =□□).

(3) RE-START PREVIOUSLY WARMED-UP ENGINE.

(4) ADJUST IGNITION TIMING TO 10° BTDC.

(5) TURN OFF ENGINE AND RESTORE ELECTRICAL CONNECTION.

THIS ENGINE IS EQUIPPED WITH AUTOMATIC IDLE SPEED CONTROL. IDLE RPM IS NOT ADJUSTABLE. IF NOT WITHIN 500-600 RPM RANGE, WITH TRANSMISSION IN DRIVE, SEE SHOP MANUAL.

THIS VEHICLE CONFORMS TO U.S. EPA AND CALIFORNIA REGULATIONS APPLICABLE TO 1986 MODEL YEAR NEW MOTOR VEHICLES INTRODUCED INTO COMMERCE SOLELY FOR SALE IN CALIFORNIA.

E6AE-9C485-AKL | CATALYST | SPARK PLUG:AWSF-54C GAP-.052-.056
3.8L-6FM
GFM3.8V5HHC6-EGR/EOS/AIP/TWC

VACUUM CIRCUITS
(© Ford Motor Co)

CALIBRATION: 6—21A—R00

MODEL YEAR: 1986 **ENGINE: 5.0L**

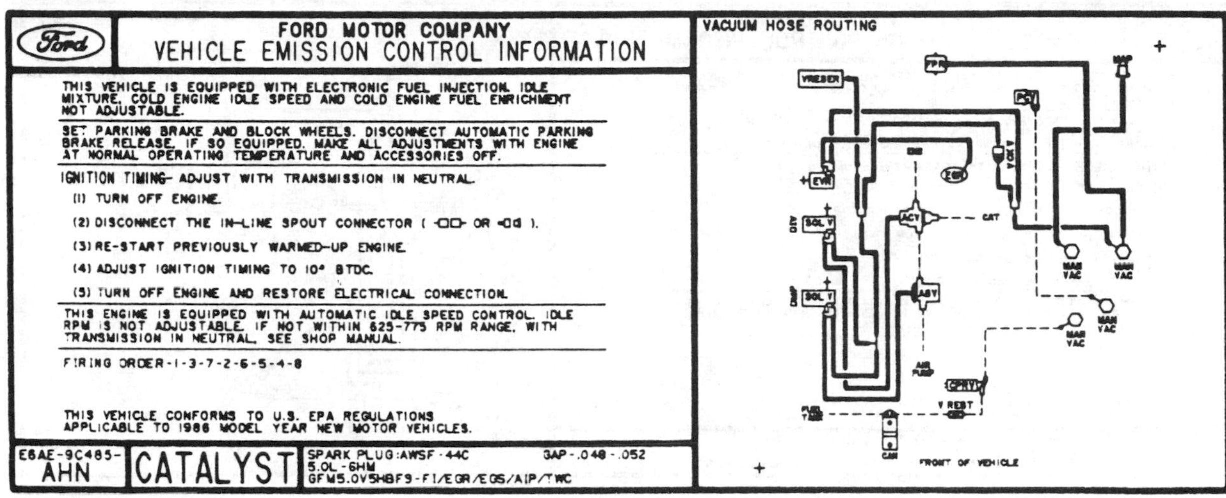

FORD MOTOR COMPANY
VEHICLE EMISSION CONTROL INFORMATION

THIS VEHICLE IS EQUIPPED WITH ELECTRONIC FUEL INJECTION. IDLE MIXTURE, COLD ENGINE IDLE SPEED AND COLD ENGINE FUEL ENRICHMENT NOT ADJUSTABLE.

SET PARKING BRAKE AND BLOCK WHEELS. DISCONNECT AUTOMATIC PARKING BRAKE RELEASE, IF SO EQUIPPED. MAKE ALL ADJUSTMENTS WITH ENGINE AT NORMAL OPERATING TEMPERATURE AND ACCESSORIES OFF.

IGNITION TIMING- ADJUST WITH TRANSMISSION IN NEUTRAL.
(1) TURN OFF ENGINE.
(2) DISCONNECT THE IN-LINE SPOUT CONNECTOR (-□□- OR =□□).
(3) RE-START PREVIOUSLY WARMED-UP ENGINE.
(4) ADJUST IGNITION TIMING TO 10° BTDC.
(5) TURN OFF ENGINE AND RESTORE ELECTRICAL CONNECTION.

THIS ENGINE IS EQUIPPED WITH AUTOMATIC IDLE SPEED CONTROL. IDLE RPM IS NOT ADJUSTABLE. IF NOT WITHIN 625-775 RPM RANGE, WITH TRANSMISSION IN NEUTRAL, SEE SHOP MANUAL.

FIRING ORDER-1-3-7-2-6-5-4-8

THIS VEHICLE CONFORMS TO U.S. EPA REGULATIONS APPLICABLE TO 1986 MODEL YEAR NEW MOTOR VEHICLES.

E6AE-9C485-AHN | CATALYST | SPARK PLUG:AWSF-44C GAP-.048-.052 5.0L-6HM GFM5.0V5HBF9-FI/EGR/EGS/AIP/TWC

CALIBRATION: 6—21B—R00

MODEL YEAR: 1986 **ENGINE: 5.0L**

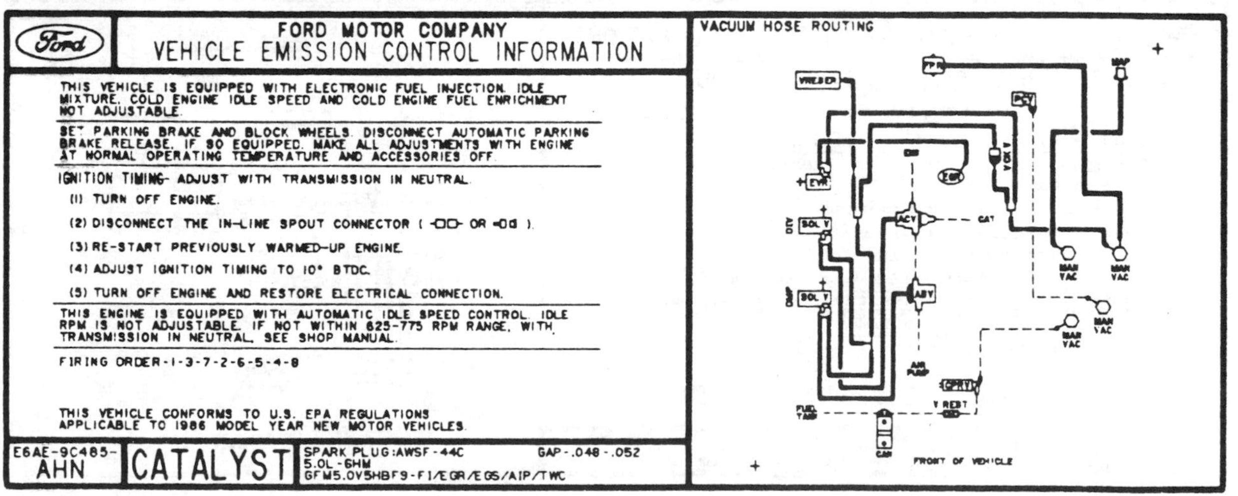

FORD MOTOR COMPANY
VEHICLE EMISSION CONTROL INFORMATION

THIS VEHICLE IS EQUIPPED WITH ELECTRONIC FUEL INJECTION. IDLE MIXTURE, COLD ENGINE IDLE SPEED AND COLD ENGINE FUEL ENRICHMENT NOT ADJUSTABLE.

SET PARKING BRAKE AND BLOCK WHEELS. DISCONNECT AUTOMATIC PARKING BRAKE RELEASE, IF SO EQUIPPED. MAKE ALL ADJUSTMENTS WITH ENGINE AT NORMAL OPERATING TEMPERATURE AND ACCESSORIES OFF.

IGNITION TIMING- ADJUST WITH TRANSMISSION IN NEUTRAL.
(1) TURN OFF ENGINE.
(2) DISCONNECT THE IN-LINE SPOUT CONNECTOR (-□□- OR =□□).
(3) RE-START PREVIOUSLY WARMED-UP ENGINE.
(4) ADJUST IGNITION TIMING TO 10° BTDC.
(5) TURN OFF ENGINE AND RESTORE ELECTRICAL CONNECTION.

THIS ENGINE IS EQUIPPED WITH AUTOMATIC IDLE SPEED CONTROL. IDLE RPM IS NOT ADJUSTABLE. IF NOT WITHIN 625-775 RPM RANGE, WITH TRANSMISSION IN NEUTRAL, SEE SHOP MANUAL.

FIRING ORDER-1-3-7-2-6-5-4-8

THIS VEHICLE CONFORMS TO U.S. EPA REGULATIONS APPLICABLE TO 1986 MODEL YEAR NEW MOTOR VEHICLES.

E6AE-9C485-AHN | CATALYST | SPARK PLUG:AWSF-44C GAP-.048-.052 5.0L-6HM GFM5.0V5HBF9-FI/EGR/EGS/AIP/TWC

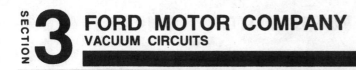

VACUUM CIRCUITS

(© Ford Motor Co)

CALIBRATION: 6—21C—R00

MODEL YEAR: 1986 **ENGINE: 5.0L**

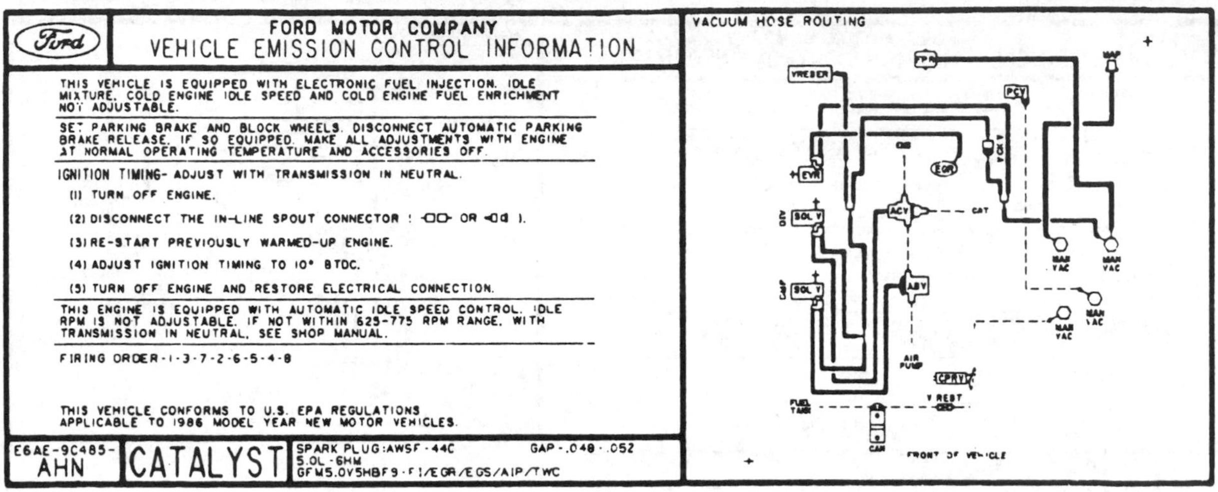

FORD MOTOR COMPANY
VEHICLE EMISSION CONTROL INFORMATION

THIS VEHICLE IS EQUIPPED WITH ELECTRONIC FUEL INJECTION. IDLE MIXTURE, COLD ENGINE IDLE SPEED AND COLD ENGINE FUEL ENRICHMENT NOT ADJUSTABLE.

SET PARKING BRAKE AND BLOCK WHEELS. DISCONNECT AUTOMATIC PARKING BRAKE RELEASE, IF SO EQUIPPED. MAKE ALL ADJUSTMENTS WITH ENGINE AT NORMAL OPERATING TEMPERATURE AND ACCESSORIES OFF.

IGNITION TIMING- ADJUST WITH TRANSMISSION IN NEUTRAL.

(1) TURN OFF ENGINE.
(2) DISCONNECT THE IN-LINE SPOUT CONNECTOR (-□□- OR -□◘).
(3) RE-START PREVIOUSLY WARMED-UP ENGINE.
(4) ADJUST IGNITION TIMING TO 10° BTDC.
(5) TURN OFF ENGINE AND RESTORE ELECTRICAL CONNECTION.

THIS ENGINE IS EQUIPPED WITH AUTOMATIC IDLE SPEED CONTROL. IDLE RPM IS NOT ADJUSTABLE. IF NOT WITHIN 625-775 RPM RANGE, WITH TRANSMISSION IN NEUTRAL, SEE SHOP MANUAL.

FIRING ORDER-1-3-7-2-6-5-4-8

THIS VEHICLE CONFORMS TO U.S. EPA REGULATIONS APPLICABLE TO 1986 MODEL YEAR NEW MOTOR VEHICLES.

E6AE-9C485-AHN CATALYST SPARK PLUG:AWSF-44C GAP-.048-.052
5.0L-GHM
GFM5.0V5HBF9-FI/EGR/EGS/AIP/TWC

CALIBRATION: 6—21P—R00

MODEL YEAR: 1986 **ENGINE: 5.0L**

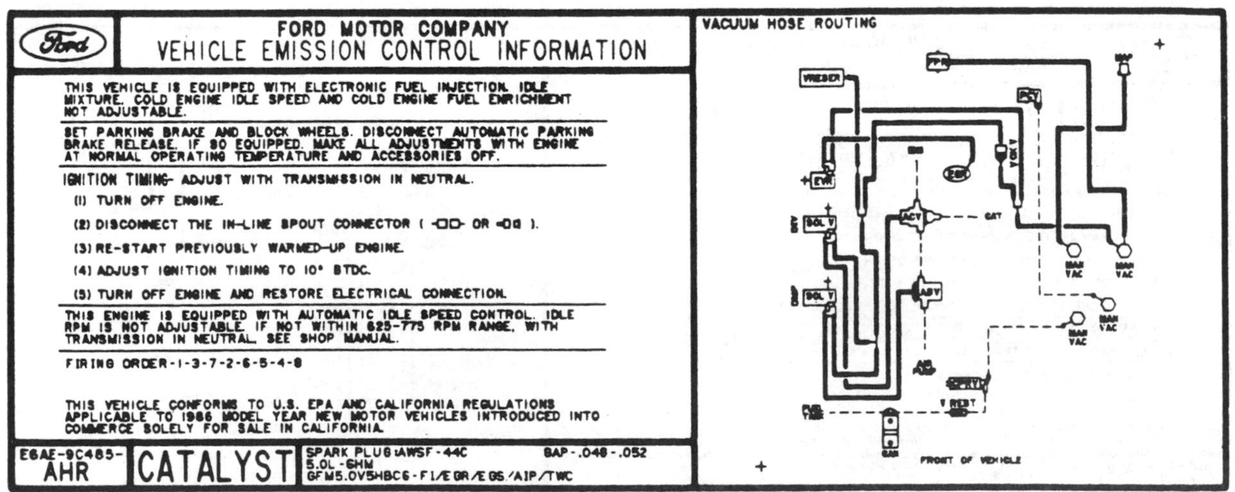

FORD MOTOR COMPANY
VEHICLE EMISSION CONTROL INFORMATION

THIS VEHICLE IS EQUIPPED WITH ELECTRONIC FUEL INJECTION. IDLE MIXTURE, COLD ENGINE IDLE SPEED AND COLD ENGINE FUEL ENRICHMENT NOT ADJUSTABLE.

SET PARKING BRAKE AND BLOCK WHEELS. DISCONNECT AUTOMATIC PARKING BRAKE RELEASE, IF SO EQUIPPED. MAKE ALL ADJUSTMENTS WITH ENGINE AT NORMAL OPERATING TEMPERATURE AND ACCESSORIES OFF.

IGNITION TIMING- ADJUST WITH TRANSMISSION IN NEUTRAL.

(1) TURN OFF ENGINE.
(2) DISCONNECT THE IN-LINE SPOUT CONNECTOR (-□□- OR -□◘).
(3) RE-START PREVIOUSLY WARMED-UP ENGINE.
(4) ADJUST IGNITION TIMING TO 10° BTDC.
(5) TURN OFF ENGINE AND RESTORE ELECTRICAL CONNECTION.

THIS ENGINE IS EQUIPPED WITH AUTOMATIC IDLE SPEED CONTROL. IDLE RPM IS NOT ADJUSTABLE. IF NOT WITHIN 625-775 RPM RANGE, WITH TRANSMISSION IN NEUTRAL, SEE SHOP MANUAL.

FIRING ORDER-1-3-7-2-6-5-4-8

THIS VEHICLE CONFORMS TO U.S. EPA AND CALIFORNIA REGULATIONS APPLICABLE TO 1986 MODEL YEAR NEW MOTOR VEHICLES INTRODUCED INTO COMMERCE SOLELY FOR SALE IN CALIFORNIA.

E6AE-9C485-AHR CATALYST SPARK PLUG:AWSF-44C GAP-.048-.052
5.0L-GHM
GFM5.0V5HBC6-FI/EGR/EGS/AIP/TWC

VACUUM CIRCUITS
(© Ford Motor Co)

CALIBRATION: 6—22A—R00

MODEL YEAR: 1986 **ENGINE: 5.0L**

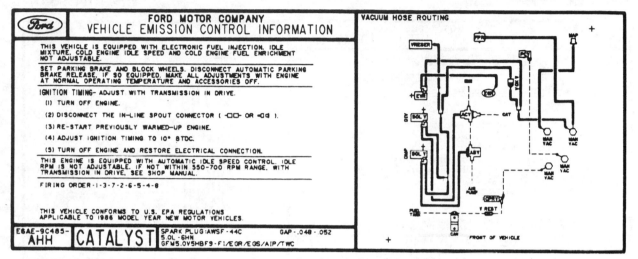

FORD MOTOR COMPANY
VEHICLE EMISSION CONTROL INFORMATION

THIS VEHICLE IS EQUIPPED WITH ELECTRONIC FUEL INJECTION. IDLE MIXTURE, COLD ENGINE IDLE SPEED AND COLD ENGINE FUEL ENRICHMENT NOT ADJUSTABLE.

SET PARKING BRAKE AND BLOCK WHEELS. DISCONNECT AUTOMATIC PARKING BRAKE RELEASE, IF SO EQUIPPED. MAKE ALL ADJUSTMENTS WITH ENGINE AT NORMAL OPERATING TEMPERATURE AND ACCESSORIES OFF.

IGNITION TIMING- ADJUST WITH TRANSMISSION IN DRIVE.
(1) TURN OFF ENGINE.
(2) DISCONNECT THE IN-LINE SPOUT CONNECTOR (-□□- OR =□⊐).
(3) RE-START PREVIOUSLY WARMED-UP ENGINE.
(4) ADJUST IGNITION TIMING TO 10° BTDC.
(5) TURN OFF ENGINE AND RESTORE ELECTRICAL CONNECTION.

THIS ENGINE IS EQUIPPED WITH AUTOMATIC IDLE SPEED CONTROL. IDLE RPM IS NOT ADJUSTABLE. IF NOT WITHIN 550-700 RPM RANGE, WITH TRANSMISSION IN DRIVE, SEE SHOP MANUAL.

FIRING ORDER-1-3-7-2-6-5-4-8

THIS VEHICLE CONFORMS TO U.S. EPA REGULATIONS APPLICABLE TO 1986 MODEL YEAR NEW MOTOR VEHICLES.

E6AE-9C485-AHH CATALYST SPARK PLUG:AWSF-44C GAP-.048-.052
5.0L-6HN
GFM5.0V5HBF9-FI/EGR/EGS/AIP/TWC

VACUUM HOSE ROUTING

FPR MAP VREGEN PCV VCKV EVR EGR ACV CAT SOL V DIV SOL V DMP ABV MAN VAC AIR PUMP CPRV FUEL TANK V REST CAN FRONT OF VEHICLE

CALIBRATION: 6—22B—R00

MODEL YEAR: 1986 **ENGINE: 5.0L**

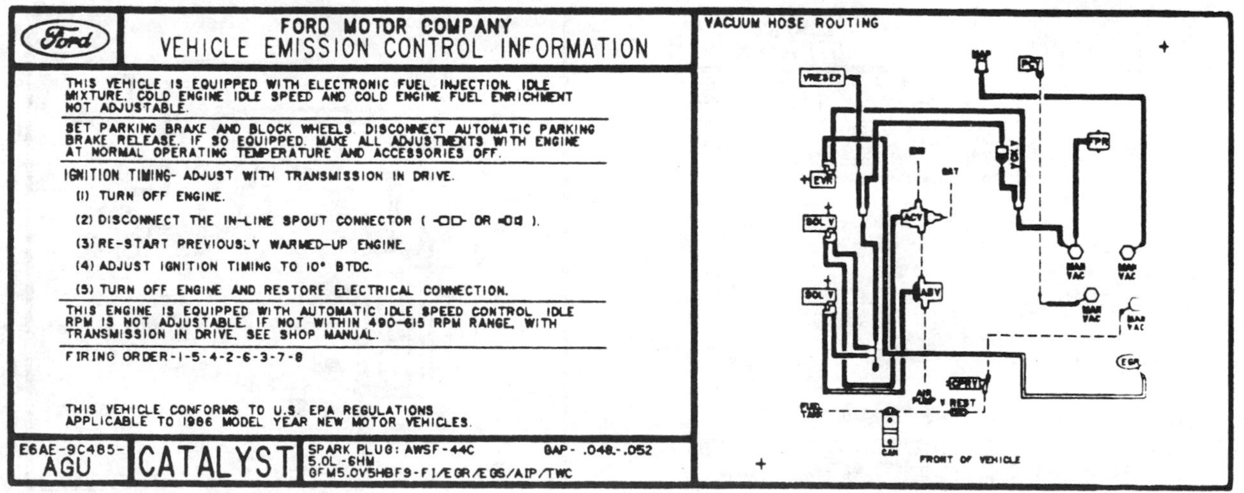

FORD MOTOR COMPANY
VEHICLE EMISSION CONTROL INFORMATION

THIS VEHICLE IS EQUIPPED WITH ELECTRONIC FUEL INJECTION. IDLE MIXTURE, COLD ENGINE IDLE SPEED AND COLD ENGINE FUEL ENRICHMENT NOT ADJUSTABLE.

SET PARKING BRAKE AND BLOCK WHEELS. DISCONNECT AUTOMATIC PARKING BRAKE RELEASE, IF SO EQUIPPED. MAKE ALL ADJUSTMENTS WITH ENGINE AT NORMAL OPERATING TEMPERATURE AND ACCESSORIES OFF.

IGNITION TIMING- ADJUST WITH TRANSMISSION IN DRIVE.
(1) TURN OFF ENGINE.
(2) DISCONNECT THE IN-LINE SPOUT CONNECTOR (-□□- OR =□⊐).
(3) RE-START PREVIOUSLY WARMED-UP ENGINE.
(4) ADJUST IGNITION TIMING TO 10° BTDC.
(5) TURN OFF ENGINE AND RESTORE ELECTRICAL CONNECTION.

THIS ENGINE IS EQUIPPED WITH AUTOMATIC IDLE SPEED CONTROL. IDLE RPM IS NOT ADJUSTABLE. IF NOT WITHIN 490-615 RPM RANGE, WITH TRANSMISSION IN DRIVE, SEE SHOP MANUAL.

FIRING ORDER-1-5-4-2-6-3-7-8

THIS VEHICLE CONFORMS TO U.S. EPA REGULATIONS APPLICABLE TO 1986 MODEL YEAR NEW MOTOR VEHICLES.

E6AE-9C485-AGU CATALYST SPARK PLUG: AWSF-44C GAP-.048-.052
5.0L-6HM
GFM5.0V5HBF9-FI/EGR/EGS/AIP/TWC

VACUUM CIRCUITS
(© Ford Motor Co)

CALIBRATION: 6—22C—R00

MODEL YEAR: 1986 **ENGINE: 5.0L**

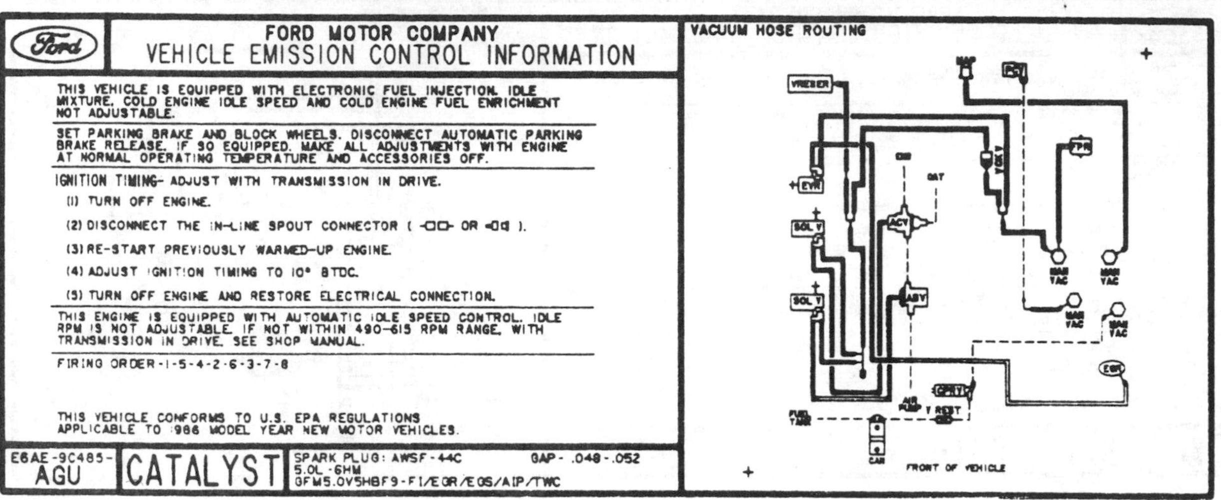

FORD MOTOR COMPANY
VEHICLE EMISSION CONTROL INFORMATION

THIS VEHICLE IS EQUIPPED WITH ELECTRONIC FUEL INJECTION. IDLE MIXTURE, COLD ENGINE IDLE SPEED AND COLD ENGINE FUEL ENRICHMENT NOT ADJUSTABLE.

SET PARKING BRAKE AND BLOCK WHEELS. DISCONNECT AUTOMATIC PARKING BRAKE RELEASE, IF SO EQUIPPED. MAKE ALL ADJUSTMENTS WITH ENGINE AT NORMAL OPERATING TEMPERATURE AND ACCESSORIES OFF.

IGNITION TIMING- ADJUST WITH TRANSMISSION IN DRIVE.
(1) TURN OFF ENGINE.
(2) DISCONNECT THE IN-LINE SPOUT CONNECTOR (-□□- OR =□□).
(3) RE-START PREVIOUSLY WARMED-UP ENGINE.
(4) ADJUST IGNITION TIMING TO 10° BTDC.
(5) TURN OFF ENGINE AND RESTORE ELECTRICAL CONNECTION.

THIS ENGINE IS EQUIPPED WITH AUTOMATIC IDLE SPEED CONTROL. IDLE RPM IS NOT ADJUSTABLE. IF NOT WITHIN 490-615 RPM RANGE, WITH TRANSMISSION IN DRIVE, SEE SHOP MANUAL.

FIRING ORDER-1-5-4-2-6-3-7-8

THIS VEHICLE CONFORMS TO U.S. EPA REGULATIONS APPLICABLE TO 1986 MODEL YEAR NEW MOTOR VEHICLES.

E6AE-9C485-AGU | CATALYST | SPARK PLUG: AWSF-44C GAP- .048-.052 5.0L-6HM GFM5.0V5HBF9-FI/EGR/EGS/AIP/TWC

CALIBRATION: 6—22D—R22
(T-Bird/Cougar)

MODEL YEAR: 1986 **ENGINE: 5.0L**

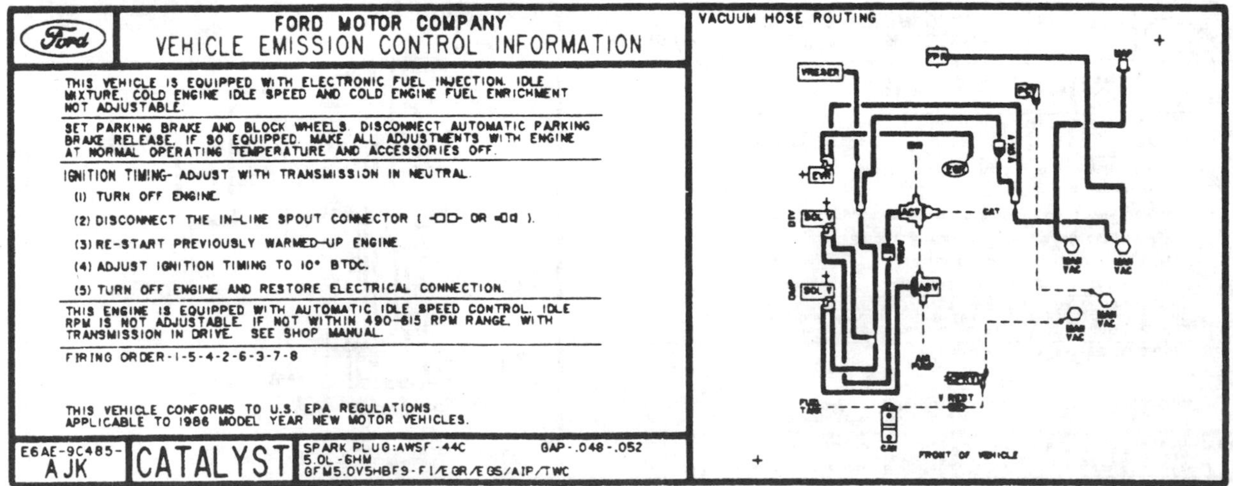

FORD MOTOR COMPANY
VEHICLE EMISSION CONTROL INFORMATION

THIS VEHICLE IS EQUIPPED WITH ELECTRONIC FUEL INJECTION. IDLE MIXTURE, COLD ENGINE IDLE SPEED AND COLD ENGINE FUEL ENRICHMENT NOT ADJUSTABLE.

SET PARKING BRAKE AND BLOCK WHEELS. DISCONNECT AUTOMATIC PARKING BRAKE RELEASE, IF SO EQUIPPED. MAKE ALL ADJUSTMENTS WITH ENGINE AT NORMAL OPERATING TEMPERATURE AND ACCESSORIES OFF.

IGNITION TIMING- ADJUST WITH TRANSMISSION IN NEUTRAL.
(1) TURN OFF ENGINE.
(2) DISCONNECT THE IN-LINE SPOUT CONNECTOR (-□□- OR =□□).
(3) RE-START PREVIOUSLY WARMED-UP ENGINE.
(4) ADJUST IGNITION TIMING TO 10° BTDC.
(5) TURN OFF ENGINE AND RESTORE ELECTRICAL CONNECTION.

THIS ENGINE IS EQUIPPED WITH AUTOMATIC IDLE SPEED CONTROL. IDLE RPM IS NOT ADJUSTABLE. IF NOT WITHIN 490-615 RPM RANGE, WITH TRANSMISSION IN DRIVE. SEE SHOP MANUAL.

FIRING ORDER-1-5-4-2-6-3-7-8

THIS VEHICLE CONFORMS TO U.S. EPA REGULATIONS APPLICABLE TO 1986 MODEL YEAR NEW MOTOR VEHICLES.

E6AE-9C485-AJK | CATALYST | SPARK PLUG: AWSF-44C GAP-.048-.052 5.0L-6HM GFM5.0V5HBF9-FI/EGR/EGS/AIP/TWC

VACUUM CIRCUITS

(© Ford Motor Co)

CALIBRATION: 6—22D—R22

(Cont./Mark—2.73 Axle)

MODEL YEAR: 1986 **ENGINE: 5.0L**

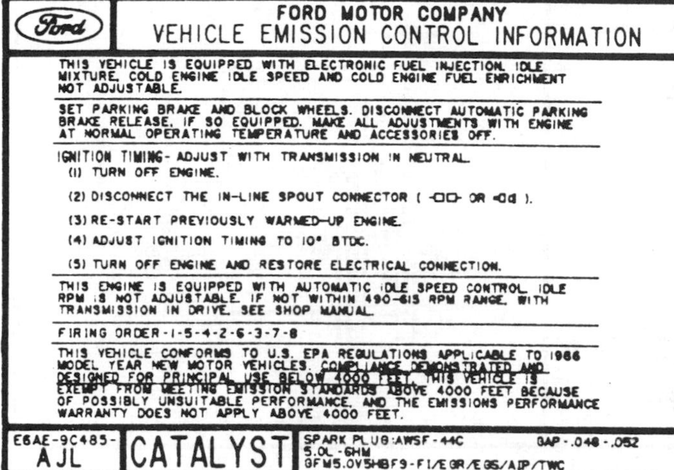

Ford

FORD MOTOR COMPANY
VEHICLE EMISSION CONTROL INFORMATION

THIS VEHICLE IS EQUIPPED WITH ELECTRONIC FUEL INJECTION. IDLE MIXTURE, COLD ENGINE IDLE SPEED AND COLD ENGINE FUEL ENRICHMENT NOT ADJUSTABLE.

SET PARKING BRAKE AND BLOCK WHEELS. DISCONNECT AUTOMATIC PARKING BRAKE RELEASE, IF SO EQUIPPED. MAKE ALL ADJUSTMENTS WITH ENGINE AT NORMAL OPERATING TEMPERATURE AND ACCESSORIES OFF.

IGNITION TIMING- ADJUST WITH TRANSMISSION IN NEUTRAL.
(1) TURN OFF ENGINE.
(2) DISCONNECT THE IN-LINE SPOUT CONNECTOR (-OD- OR -Od).
(3) RE-START PREVIOUSLY WARMED-UP ENGINE.
(4) ADJUST IGNITION TIMING TO 10° BTDC.
(5) TURN OFF ENGINE AND RESTORE ELECTRICAL CONNECTION.

THIS ENGINE IS EQUIPPED WITH AUTOMATIC IDLE SPEED CONTROL. IDLE RPM IS NOT ADJUSTABLE. IF NOT WITHIN 490-615 RPM RANGE, WITH TRANSMISSION IN DRIVE, SEE SHOP MANUAL.

FIRING ORDER-1-5-4-2-6-3-7-8

THIS VEHICLE CONFORMS TO U.S. EPA REGULATIONS APPLICABLE TO 1986 MODEL YEAR NEW MOTOR VEHICLES. COMPLIANCE DEMONSTRATED AND DESIGNED FOR PRINCIPAL USE BELOW 4000 FEET. THIS VEHICLE IS EXEMPT FROM MEETING EMISSION STANDARDS ABOVE 4000 FEET BECAUSE OF POSSIBLY UNSUITABLE PERFORMANCE, AND THE EMISSIONS PERFORMANCE WARRANTY DOES NOT APPLY ABOVE 4000 FEET.

E6AE-9C485-AJL | CATALYST | SPARK PLUG:AWSF-44C GAP-.048-.052 5.0L-6HM GFM5.0V5HBF9-FI/EGR/EGS/AIP/TWC

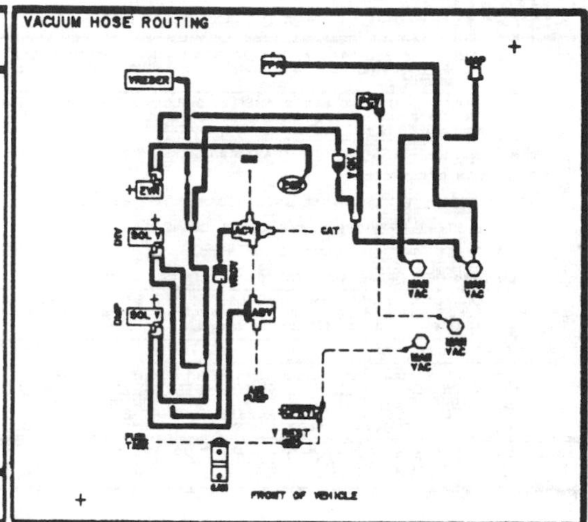

CALIBRATION: 6—22D—R22

(Cont./Mark—3.08 Axle)

MODEL YEAR: 1986 **ENGINE: 5.0L**

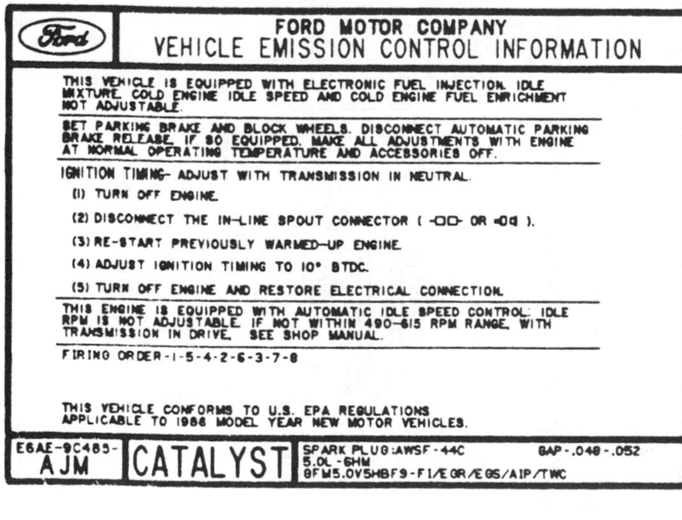

Ford

FORD MOTOR COMPANY
VEHICLE EMISSION CONTROL INFORMATION

THIS VEHICLE IS EQUIPPED WITH ELECTRONIC FUEL INJECTION. IDLE MIXTURE, COLD ENGINE IDLE SPEED AND COLD ENGINE FUEL ENRICHMENT NOT ADJUSTABLE.

SET PARKING BRAKE AND BLOCK WHEELS. DISCONNECT AUTOMATIC PARKING BRAKE RELEASE, IF SO EQUIPPED. MAKE ALL ADJUSTMENTS WITH ENGINE AT NORMAL OPERATING TEMPERATURE AND ACCESSORIES OFF.

IGNITION TIMING- ADJUST WITH TRANSMISSION IN NEUTRAL.
(1) TURN OFF ENGINE.
(2) DISCONNECT THE IN-LINE SPOUT CONNECTOR (-OD- OR -Od).
(3) RE-START PREVIOUSLY WARMED-UP ENGINE.
(4) ADJUST IGNITION TIMING TO 10° BTDC.
(5) TURN OFF ENGINE AND RESTORE ELECTRICAL CONNECTION.

THIS ENGINE IS EQUIPPED WITH AUTOMATIC IDLE SPEED CONTROL. IDLE RPM IS NOT ADJUSTABLE. IF NOT WITHIN 490-615 RPM RANGE, WITH TRANSMISSION IN DRIVE, SEE SHOP MANUAL.

FIRING ORDER-1-5-4-2-6-3-7-8

THIS VEHICLE CONFORMS TO U.S. EPA REGULATIONS APPLICABLE TO 1986 MODEL YEAR NEW MOTOR VEHICLES.

E6AE-9C485-AJM | CATALYST | SPARK PLUG:AWSF-44C GAP-.048-.052 5.0L-6HM GFM5.0V5HBF9-FI/EGR/EGS/AIP/TWC

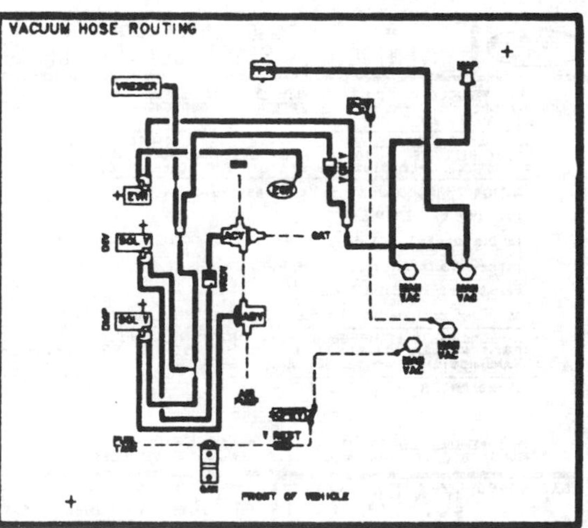

VACUUM CIRCUITS

(© Ford Motor Co)

CALIBRATION: 6—22E—R00

MODEL YEAR: 1986 **ENGINE: 5.0L**

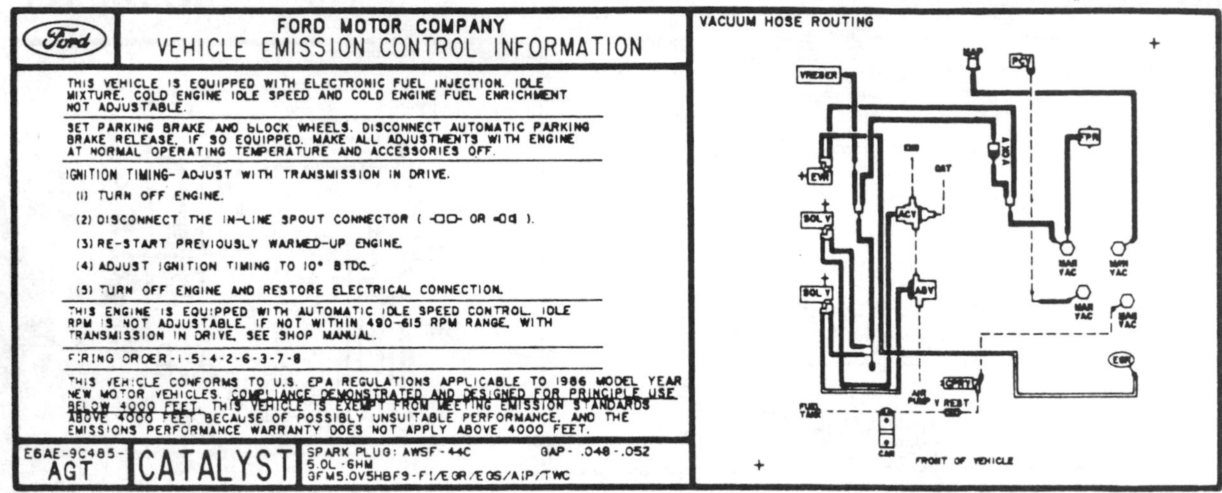

FORD MOTOR COMPANY
VEHICLE EMISSION CONTROL INFORMATION

THIS VEHICLE IS EQUIPPED WITH ELECTRONIC FUEL INJECTION. IDLE MIXTURE, COLD ENGINE IDLE SPEED AND COLD ENGINE FUEL ENRICHMENT NOT ADJUSTABLE.

SET PARKING BRAKE AND BLOCK WHEELS. DISCONNECT AUTOMATIC PARKING BRAKE RELEASE, IF SO EQUIPPED. MAKE ALL ADJUSTMENTS WITH ENGINE AT NORMAL OPERATING TEMPERATURE AND ACCESSORIES OFF.

IGNITION TIMING- ADJUST WITH TRANSMISSION IN DRIVE.

(1) TURN OFF ENGINE.

(2) DISCONNECT THE IN-LINE SPOUT CONNECTOR (-▭▭- OR =▭◘).

(3) RE-START PREVIOUSLY WARMED-UP ENGINE.

(4) ADJUST IGNITION TIMING TO 10° BTDC.

(5) TURN OFF ENGINE AND RESTORE ELECTRICAL CONNECTION.

THIS ENGINE IS EQUIPPED WITH AUTOMATIC IDLE SPEED CONTROL. IDLE RPM IS NOT ADJUSTABLE. IF NOT WITHIN 490-615 RPM RANGE, WITH TRANSMISSION IN DRIVE, SEE SHOP MANUAL.

FIRING ORDER-1-5-4-2-6-3-7-8

THIS VEHICLE CONFORMS TO U.S. EPA REGULATIONS APPLICABLE TO 1986 MODEL YEAR NEW MOTOR VEHICLES. COMPLIANCE DEMONSTRATED AND DESIGNED FOR PRINCIPLE USE BELOW 4000 FEET. THIS VEHICLE IS EXEMPT FROM MEETING EMISSION STANDARDS ABOVE 4000 FEET BECAUSE OF POSSIBLY UNSUITABLE PERFORMANCE, AND THE EMISSIONS PERFORMANCE WARRANTY DOES NOT APPLY ABOVE 4000 FEET.

E6AE-9C485-AGT | CATALYST | SPARK PLUG: AWSF-44C GAP- .048-.052
5.0L-6HM
GFM5.0V5HBF9-FI/EGR/EGS/AIP/TWC

CALIBRATION: 6—22F—R00

MODEL YEAR: 1986 **ENGINE: 5.0L**

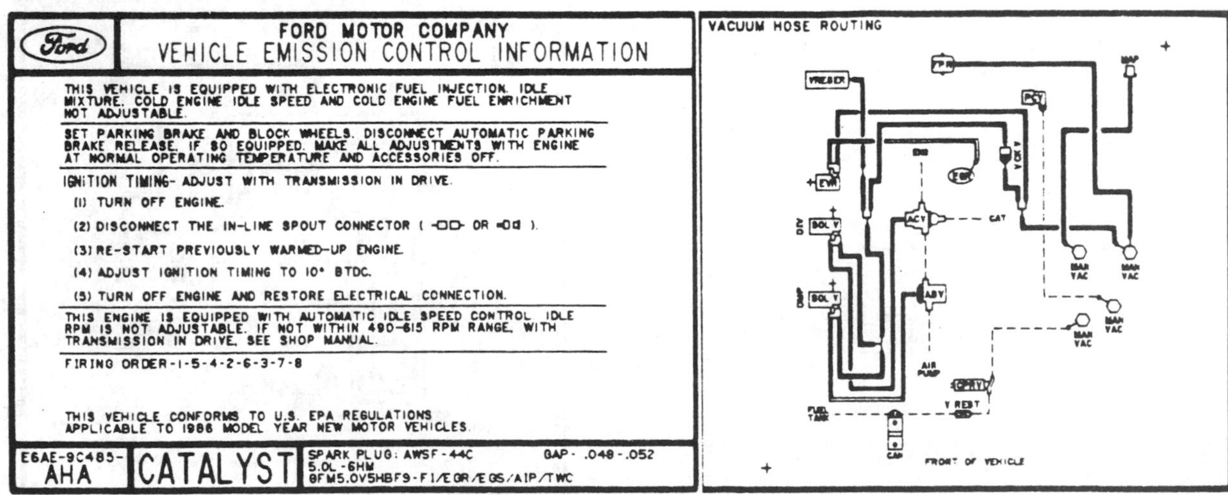

FORD MOTOR COMPANY
VEHICLE EMISSION CONTROL INFORMATION

THIS VEHICLE IS EQUIPPED WITH ELECTRONIC FUEL INJECTION. IDLE MIXTURE, COLD ENGINE IDLE SPEED AND COLD ENGINE FUEL ENRICHMENT NOT ADJUSTABLE.

SET PARKING BRAKE AND BLOCK WHEELS. DISCONNECT AUTOMATIC PARKING BRAKE RELEASE, IF SO EQUIPPED. MAKE ALL ADJUSTMENTS WITH ENGINE AT NORMAL OPERATING TEMPERATURE AND ACCESSORIES OFF.

IGNITION TIMING- ADJUST WITH TRANSMISSION IN DRIVE.

(1) TURN OFF ENGINE.

(2) DISCONNECT THE IN-LINE SPOUT CONNECTOR (-▭▭- OR =▭◘).

(3) RE-START PREVIOUSLY WARMED-UP ENGINE.

(4) ADJUST IGNITION TIMING TO 10° BTDC.

(5) TURN OFF ENGINE AND RESTORE ELECTRICAL CONNECTION.

THIS ENGINE IS EQUIPPED WITH AUTOMATIC IDLE SPEED CONTROL. IDLE RPM IS NOT ADJUSTABLE. IF NOT WITHIN 490-615 RPM RANGE, WITH TRANSMISSION IN DRIVE, SEE SHOP MANUAL.

FIRING ORDER-1-5-4-2-6-3-7-8

THIS VEHICLE CONFORMS TO U.S. EPA REGULATIONS APPLICABLE TO 1986 MODEL YEAR NEW MOTOR VEHICLES.

E6AE-9C485-AHA | CATALYST | SPARK PLUG: AWSF-44C GAP- .048-.052
5.0L-6HM
GFM5.0V5HBF9-FI/EGR/EGS/AIP/TWC

VACUUM CIRCUITS

(© Ford Motor Co)

CALIBRATION: 6—22G—R00

MODEL YEAR: 1986 **ENGINE: 5.0L**

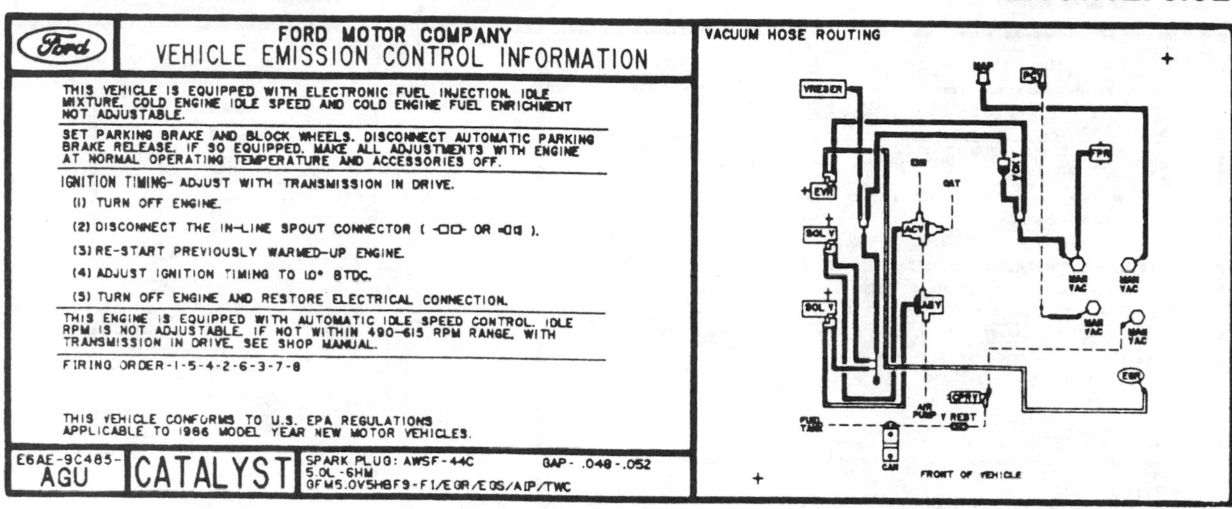

FORD MOTOR COMPANY
VEHICLE EMISSION CONTROL INFORMATION

THIS VEHICLE IS EQUIPPED WITH ELECTRONIC FUEL INJECTION. IDLE MIXTURE, COLD ENGINE IDLE SPEED AND COLD ENGINE FUEL ENRICHMENT NOT ADJUSTABLE.

SET PARKING BRAKE AND BLOCK WHEELS. DISCONNECT AUTOMATIC PARKING BRAKE RELEASE, IF SO EQUIPPED. MAKE ALL ADJUSTMENTS WITH ENGINE AT NORMAL OPERATING TEMPERATURE AND ACCESSORIES OFF.

IGNITION TIMING- ADJUST WITH TRANSMISSION IN DRIVE.

(1) TURN OFF ENGINE.
(2) DISCONNECT THE IN-LINE SPOUT CONNECTOR (-□□- OR =□□).
(3) RE-START PREVIOUSLY WARMED-UP ENGINE.
(4) ADJUST IGNITION TIMING TO 10° BTDC.
(5) TURN OFF ENGINE AND RESTORE ELECTRICAL CONNECTION.

THIS ENGINE IS EQUIPPED WITH AUTOMATIC IDLE SPEED CONTROL. IDLE RPM IS NOT ADJUSTABLE. IF NOT WITHIN 490-615 RPM RANGE, WITH TRANSMISSION IN DRIVE, SEE SHOP MANUAL.

FIRING ORDER-1-5-4-2-6-3-7-8

THIS VEHICLE CONFORMS TO U.S. EPA REGULATIONS APPLICABLE TO 1986 MODEL YEAR NEW MOTOR VEHICLES.

E6AE-9C485-AGU | CATALYST | SPARK PLUG: AWSF-44C GAP- .048-.052 5.0L-6HM GFM5.0V5HBF9-F1/EGR/EGS/AIP/TWC

CALIBRATION: 6—22H—R00

MODEL YEAR: 1986 **ENGINE: 5.0L**

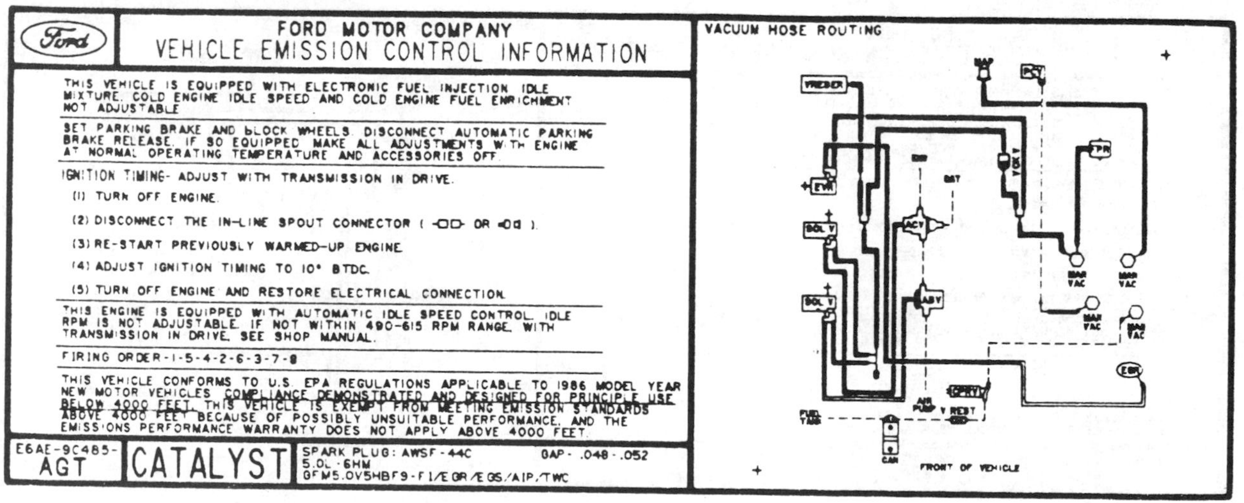

FORD MOTOR COMPANY
VEHICLE EMISSION CONTROL INFORMATION

THIS VEHICLE IS EQUIPPED WITH ELECTRONIC FUEL INJECTION. IDLE MIXTURE, COLD ENGINE IDLE SPEED AND COLD ENGINE FUEL ENRICHMENT NOT ADJUSTABLE.

SET PARKING BRAKE AND BLOCK WHEELS. DISCONNECT AUTOMATIC PARKING BRAKE RELEASE, IF SO EQUIPPED. MAKE ALL ADJUSTMENTS WITH ENGINE AT NORMAL OPERATING TEMPERATURE AND ACCESSORIES OFF.

IGNITION TIMING- ADJUST WITH TRANSMISSION IN DRIVE.

(1) TURN OFF ENGINE.
(2) DISCONNECT THE IN-LINE SPOUT CONNECTOR (-□□- OR =□□).
(3) RE-START PREVIOUSLY WARMED-UP ENGINE.
(4) ADJUST IGNITION TIMING TO 10° BTDC.
(5) TURN OFF ENGINE AND RESTORE ELECTRICAL CONNECTION.

THIS ENGINE IS EQUIPPED WITH AUTOMATIC IDLE SPEED CONTROL. IDLE RPM IS NOT ADJUSTABLE. IF NOT WITHIN 490-615 RPM RANGE, WITH TRANSMISSION IN DRIVE, SEE SHOP MANUAL.

FIRING ORDER-1-5-4-2-6-3-7-8

THIS VEHICLE CONFORMS TO U.S. EPA REGULATIONS APPLICABLE TO 1986 MODEL YEAR NEW MOTOR VEHICLES. COMPLIANCE DEMONSTRATED AND DESIGNED FOR PRINCIPLE USE BELOW 4000 FEET. THIS VEHICLE IS EXEMPT FROM MEETING EMISSION STANDARDS ABOVE 4000 FEET BECAUSE OF POSSIBLY UNSUITABLE PERFORMANCE, AND THE EMISSIONS PERFORMANCE WARRANTY DOES NOT APPLY ABOVE 4000 FEET.

E6AE-9C485-AGT | CATALYST | SPARK PLUG: AWSF-44C GAP- .048-.052 5.0L-6HM GFM5.0V5HBF9-F1/EGR/EGS/AIP/TWC

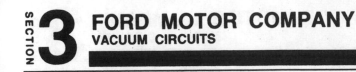

VACUUM CIRCUITS

(© Ford Motor Co)

CALIBRATION: 6—22I—R00

MODEL YEAR: 1986 **ENGINE: 5.0L**

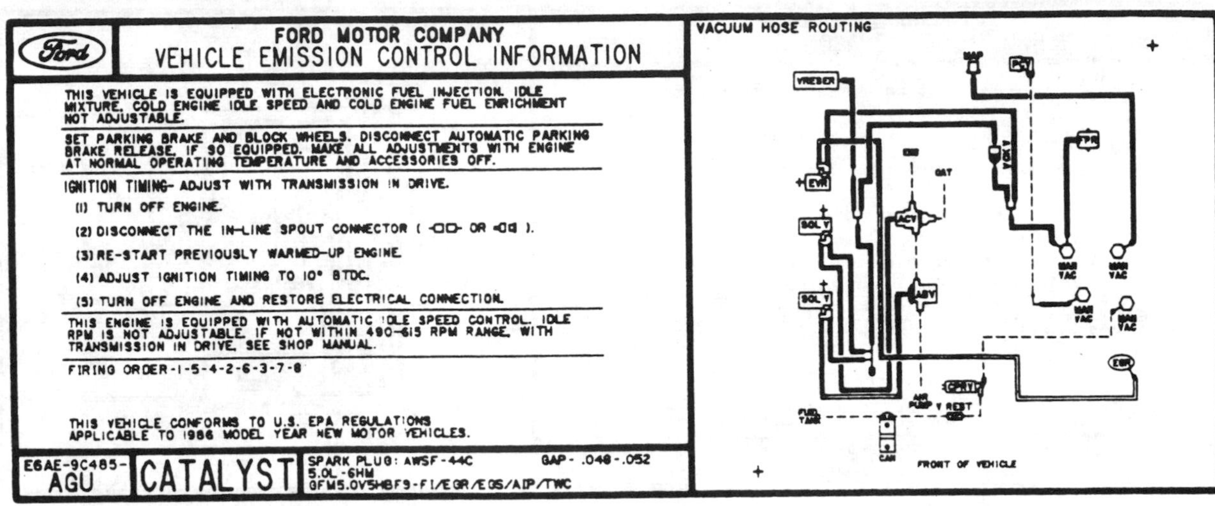

FORD MOTOR COMPANY
VEHICLE EMISSION CONTROL INFORMATION

THIS VEHICLE IS EQUIPPED WITH ELECTRONIC FUEL INJECTION. IDLE MIXTURE, COLD ENGINE IDLE SPEED AND COLD ENGINE FUEL ENRICHMENT NOT ADJUSTABLE.

SET PARKING BRAKE AND BLOCK WHEELS. DISCONNECT AUTOMATIC PARKING BRAKE RELEASE, IF SO EQUIPPED. MAKE ALL ADJUSTMENTS WITH ENGINE AT NORMAL OPERATING TEMPERATURE AND ACCESSORIES OFF.

IGNITION TIMING- ADJUST WITH TRANSMISSION IN DRIVE.

(1) TURN OFF ENGINE.
(2) DISCONNECT THE IN-LINE SPOUT CONNECTOR (-□□- OR =□□).
(3) RE-START PREVIOUSLY WARMED-UP ENGINE.
(4) ADJUST IGNITION TIMING TO 10° BTDC.
(5) TURN OFF ENGINE AND RESTORE ELECTRICAL CONNECTION.

THIS ENGINE IS EQUIPPED WITH AUTOMATIC IDLE SPEED CONTROL. IDLE RPM IS NOT ADJUSTABLE. IF NOT WITHIN 490-615 RPM RANGE, WITH TRANSMISSION IN DRIVE, SEE SHOP MANUAL.

FIRING ORDER-1-5-4-2-6-3-7-8

THIS VEHICLE CONFORMS TO U.S. EPA REGULATIONS APPLICABLE TO 1986 MODEL YEAR NEW MOTOR VEHICLES.

E6AE-9C485-AGU CATALYST SPARK PLUG: AWSF-44C GAP- .048-.052
5.0L-6HM
GFM5.0V5HBF9-FI/EGR/EGS/AIP/TWC

CALIBRATION: 6—22J—R00

MODEL YEAR: 1986 **ENGINE: 5.0L**

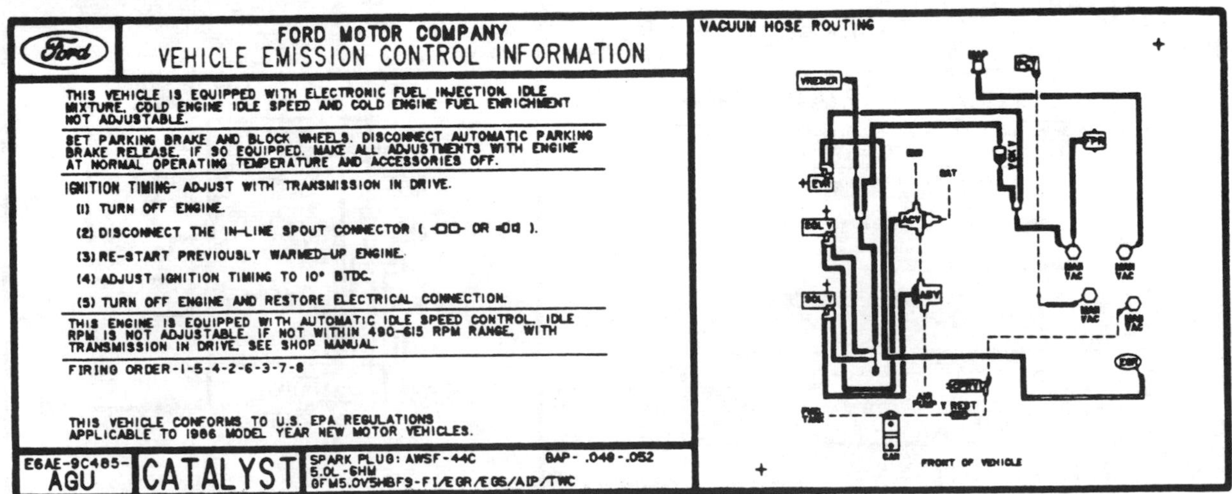

FORD MOTOR COMPANY
VEHICLE EMISSION CONTROL INFORMATION

THIS VEHICLE IS EQUIPPED WITH ELECTRONIC FUEL INJECTION. IDLE MIXTURE, COLD ENGINE IDLE SPEED AND COLD ENGINE FUEL ENRICHMENT NOT ADJUSTABLE.

SET PARKING BRAKE AND BLOCK WHEELS. DISCONNECT AUTOMATIC PARKING BRAKE RELEASE, IF SO EQUIPPED. MAKE ALL ADJUSTMENTS WITH ENGINE AT NORMAL OPERATING TEMPERATURE AND ACCESSORIES OFF.

IGNITION TIMING- ADJUST WITH TRANSMISSION IN DRIVE.

(1) TURN OFF ENGINE.
(2) DISCONNECT THE IN-LINE SPOUT CONNECTOR (-□□- OR =□□).
(3) RE-START PREVIOUSLY WARMED-UP ENGINE.
(4) ADJUST IGNITION TIMING TO 10° BTDC.
(5) TURN OFF ENGINE AND RESTORE ELECTRICAL CONNECTION.

THIS ENGINE IS EQUIPPED WITH AUTOMATIC IDLE SPEED CONTROL. IDLE RPM IS NOT ADJUSTABLE. IF NOT WITHIN 490-615 RPM RANGE, WITH TRANSMISSION IN DRIVE, SEE SHOP MANUAL.

FIRING ORDER-1-5-4-2-6-3-7-8

THIS VEHICLE CONFORMS TO U.S. EPA REGULATIONS APPLICABLE TO 1986 MODEL YEAR NEW MOTOR VEHICLES.

E6AE-9C485-AGU CATALYST SPARK PLUG: AWSF-44C GAP- .048-.052
5.0L-6HM
GFM5.0V5HBF9-FI/EGR/EGS/AIP/TWC

VACUUM CIRCUITS

(© Ford Motor Co)

CALIBRATION: 6—22K—R00

MODEL YEAR: 1986 **ENGINE: 5.0L**

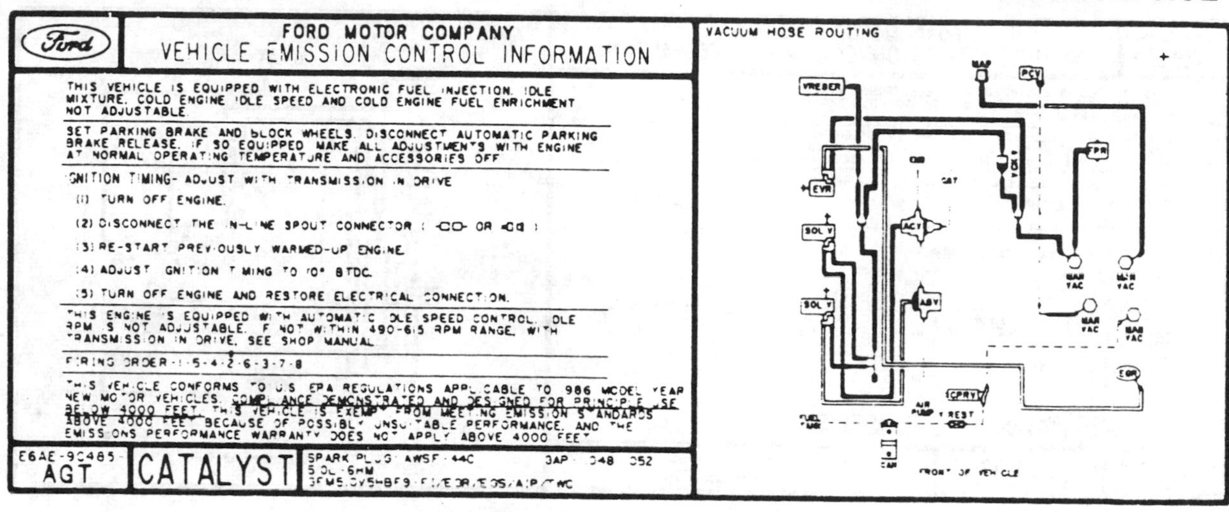

FORD MOTOR COMPANY
VEHICLE EMISSION CONTROL INFORMATION

THIS VEHICLE IS EQUIPPED WITH ELECTRONIC FUEL INJECTION. IDLE MIXTURE, COLD ENGINE IDLE SPEED AND COLD ENGINE FUEL ENRICHMENT NOT ADJUSTABLE.

SET PARKING BRAKE AND BLOCK WHEELS. DISCONNECT AUTOMATIC PARKING BRAKE RELEASE, IF SO EQUIPPED. MAKE ALL ADJUSTMENTS WITH ENGINE AT NORMAL OPERATING TEMPERATURE AND ACCESSORIES OFF.

IGNITION TIMING- ADJUST WITH TRANSMISSION IN DRIVE

(1) TURN OFF ENGINE.

(2) DISCONNECT THE IN-LINE SPOUT CONNECTOR (-□□- OR =□□).

(3) RE-START PREVIOUSLY WARMED-UP ENGINE.

(4) ADJUST IGNITION TIMING TO 10° BTDC.

(5) TURN OFF ENGINE AND RESTORE ELECTRICAL CONNECTION.

THIS ENGINE IS EQUIPPED WITH AUTOMATIC IDLE SPEED CONTROL. IDLE RPM IS NOT ADJUSTABLE. IF NOT WITHIN 490-615 RPM RANGE, WITH TRANSMISSION IN DRIVE, SEE SHOP MANUAL.

FIRING ORDER-1-5-4-2-6-3-7-8

THIS VEHICLE CONFORMS TO U.S. EPA REGULATIONS APPLICABLE TO 1986 MODEL YEAR NEW MOTOR VEHICLES. COMPLIANCE DEMONSTRATED AND DESIGNED FOR PRINCIPLE USE BELOW 4000 FEET. THIS VEHICLE IS EXEMPT FROM MEETING EMISSION STANDARDS ABOVE 4000 FEET BECAUSE OF POSSIBLY UNSUITABLE PERFORMANCE, AND THE EMISSIONS PERFORMANCE WARRANTY DOES NOT APPLY ABOVE 4000 FEET.

E6AE-9C485-AGT | CATALYST | SPARK PLUG: AWSF-44C GAP-.048-.052
5.0L-6HM
GFM5.0V5HBF9-FI/EGR/EGS/AIP/TWC

CALIBRATION: 6—22L—R00

MODEL YEAR: 1986 **ENGINE: 5.0L**

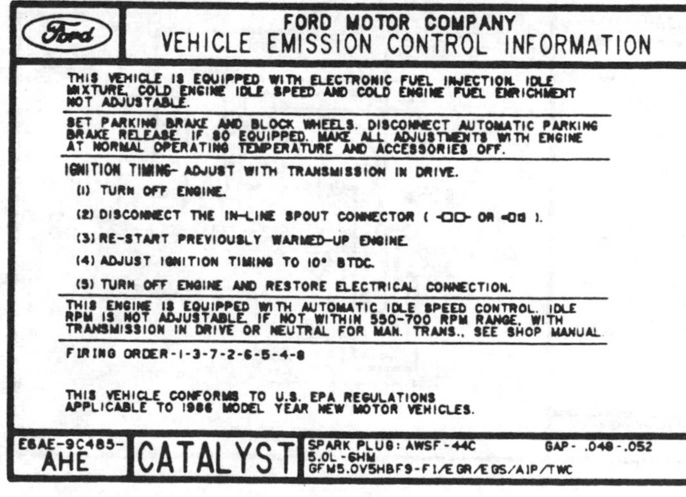

FORD MOTOR COMPANY
VEHICLE EMISSION CONTROL INFORMATION

THIS VEHICLE IS EQUIPPED WITH ELECTRONIC FUEL INJECTION. IDLE MIXTURE, COLD ENGINE IDLE SPEED AND COLD ENGINE FUEL ENRICHMENT NOT ADJUSTABLE.

SET PARKING BRAKE AND BLOCK WHEELS. DISCONNECT AUTOMATIC PARKING BRAKE RELEASE, IF SO EQUIPPED. MAKE ALL ADJUSTMENTS WITH ENGINE AT NORMAL OPERATING TEMPERATURE AND ACCESSORIES OFF.

IGNITION TIMING- ADJUST WITH TRANSMISSION IN DRIVE.

(1) TURN OFF ENGINE.

(2) DISCONNECT THE IN-LINE SPOUT CONNECTOR (-□□- OR =□□).

(3) RE-START PREVIOUSLY WARMED-UP ENGINE.

(4) ADJUST IGNITION TIMING TO 10° BTDC.

(5) TURN OFF ENGINE AND RESTORE ELECTRICAL CONNECTION.

THIS ENGINE IS EQUIPPED WITH AUTOMATIC IDLE SPEED CONTROL. IDLE RPM IS NOT ADJUSTABLE. IF NOT WITHIN 550-700 RPM RANGE, WITH TRANSMISSION IN DRIVE OR NEUTRAL FOR MAN. TRANS., SEE SHOP MANUAL.

FIRING ORDER-1-3-7-2-6-5-4-8

THIS VEHICLE CONFORMS TO U.S. EPA REGULATIONS APPLICABLE TO 1986 MODEL YEAR NEW MOTOR VEHICLES.

E6AE-9C485-AHE | CATALYST | SPARK PLUG: AWSF-44C GAP-.048-.052
5.0L-6HM
GFM5.0V5HBF9-FI/EGR/EGS/AIP/TWC

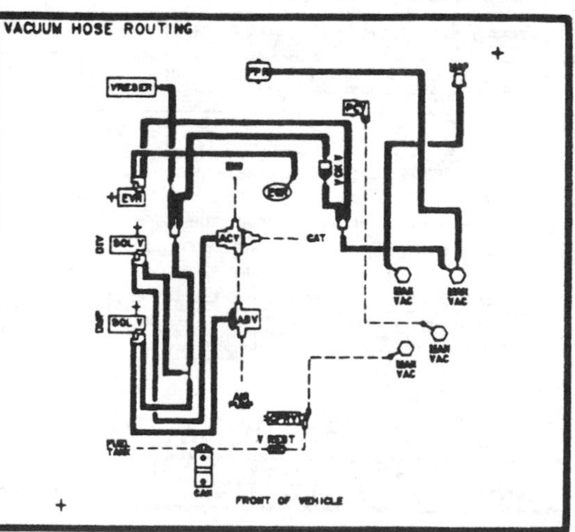

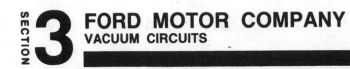

VACUUM CIRCUITS

CALIBRATION: 6—22M—R00

MODEL YEAR: 1986 **ENGINE: 5.0L**

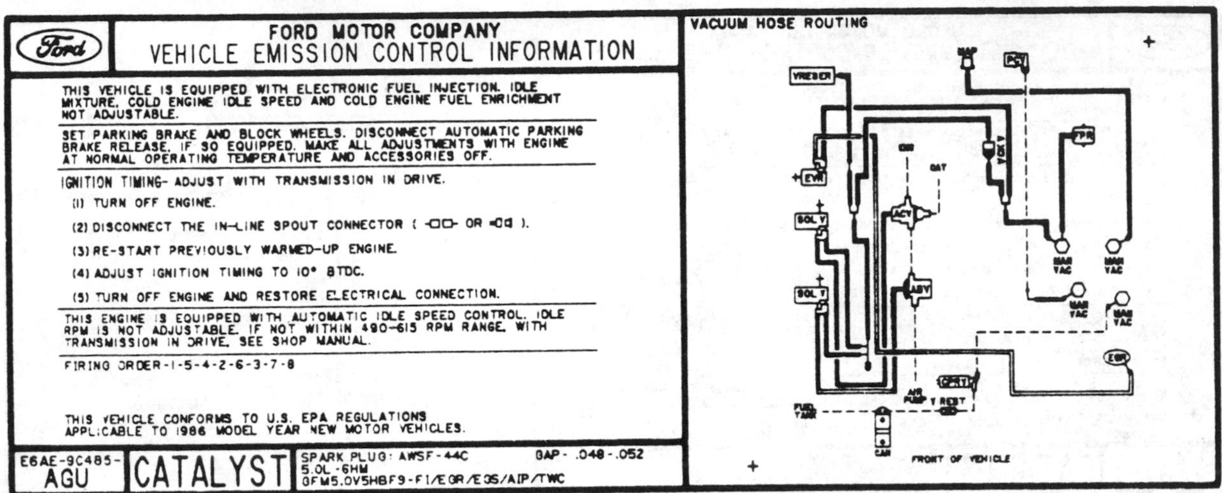

FORD MOTOR COMPANY
VEHICLE EMISSION CONTROL INFORMATION

THIS VEHICLE IS EQUIPPED WITH ELECTRONIC FUEL INJECTION. IDLE MIXTURE, COLD ENGINE IDLE SPEED AND COLD ENGINE FUEL ENRICHMENT NOT ADJUSTABLE.

SET PARKING BRAKE AND BLOCK WHEELS. DISCONNECT AUTOMATIC PARKING BRAKE RELEASE, IF SO EQUIPPED. MAKE ALL ADJUSTMENTS WITH ENGINE AT NORMAL OPERATING TEMPERATURE AND ACCESSORIES OFF.

IGNITION TIMING- ADJUST WITH TRANSMISSION IN DRIVE.
(1) TURN OFF ENGINE.
(2) DISCONNECT THE IN-LINE SPOUT CONNECTOR (-□□- OR =□□).
(3) RE-START PREVIOUSLY WARMED-UP ENGINE.
(4) ADJUST IGNITION TIMING TO 10° BTDC.
(5) TURN OFF ENGINE AND RESTORE ELECTRICAL CONNECTION.

THIS ENGINE IS EQUIPPED WITH AUTOMATIC IDLE SPEED CONTROL. IDLE RPM IS NOT ADJUSTABLE. IF NOT WITHIN 490-615 RPM RANGE, WITH TRANSMISSION IN DRIVE, SEE SHOP MANUAL.

FIRING ORDER-1-5-4-2-6-3-7-8

THIS VEHICLE CONFORMS TO U.S. EPA REGULATIONS APPLICABLE TO 1986 MODEL YEAR NEW MOTOR VEHICLES.

E6AE-9C485-AGU | CATALYST | SPARK PLUG: AWSF-44C GAP-.048-.052 5.0L-6HM GFM5.0V5HBF9-FI/EGR/EGS/AIP/TWC

CALIBRATION: 6—22P—R00

MODEL YEAR: 1986 **ENGINE: 5.0L**

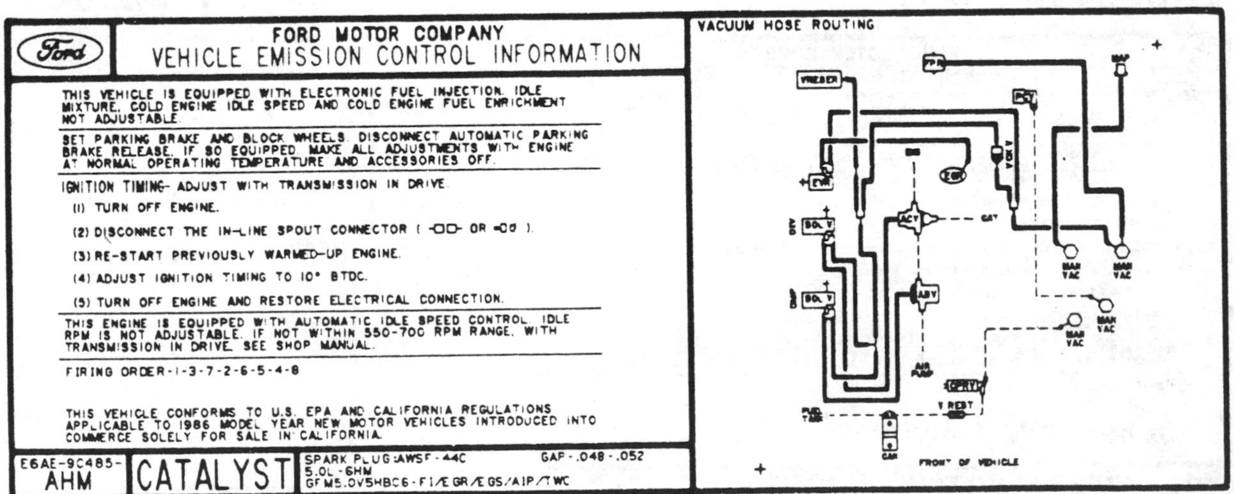

FORD MOTOR COMPANY
VEHICLE EMISSION CONTROL INFORMATION

THIS VEHICLE IS EQUIPPED WITH ELECTRONIC FUEL INJECTION. IDLE MIXTURE, COLD ENGINE IDLE SPEED AND COLD ENGINE FUEL ENRICHMENT NOT ADJUSTABLE.

SET PARKING BRAKE AND BLOCK WHEELS. DISCONNECT AUTOMATIC PARKING BRAKE RELEASE, IF SO EQUIPPED. MAKE ALL ADJUSTMENTS WITH ENGINE AT NORMAL OPERATING TEMPERATURE AND ACCESSORIES OFF.

IGNITION TIMING- ADJUST WITH TRANSMISSION IN DRIVE.
(1) TURN OFF ENGINE.
(2) DISCONNECT THE IN-LINE SPOUT CONNECTOR (-□□- OR =□□).
(3) RE-START PREVIOUSLY WARMED-UP ENGINE.
(4) ADJUST IGNITION TIMING TO 10° BTDC.
(5) TURN OFF ENGINE AND RESTORE ELECTRICAL CONNECTION.

THIS ENGINE IS EQUIPPED WITH AUTOMATIC IDLE SPEED CONTROL. IDLE RPM IS NOT ADJUSTABLE. IF NOT WITHIN 550-700 RPM RANGE, WITH TRANSMISSION IN DRIVE, SEE SHOP MANUAL.

FIRING ORDER-1-3-7-2-6-5-4-8

THIS VEHICLE CONFORMS TO U.S. EPA AND CALIFORNIA REGULATIONS APPLICABLE TO 1986 MODEL YEAR NEW MOTOR VEHICLES INTRODUCED INTO COMMERCE SOLELY FOR SALE IN CALIFORNIA.

E6AE-9C485-AHM | CATALYST | SPARK PLUG: AWSF-44C GAP-.048-.052 5.0L-6HM GFM5.0V5HBC6-FI/EGR/EGS/AIP/TWC

VACUUM CIRCUITS

(© Ford Motor Co)

CALIBRATION: 6—22Q—R00

MODEL YEAR: 1986 **ENGINE: 5.0L**

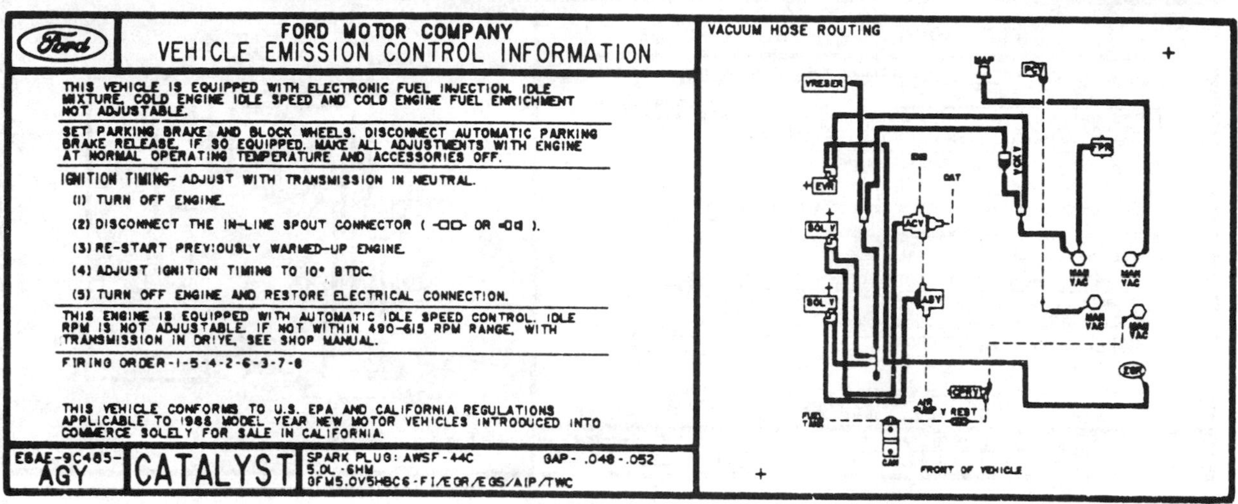
FORD MOTOR COMPANY
VEHICLE EMISSION CONTROL INFORMATION

THIS VEHICLE IS EQUIPPED WITH ELECTRONIC FUEL INJECTION. IDLE MIXTURE, COLD ENGINE IDLE SPEED AND COLD ENGINE FUEL ENRICHMENT NOT ADJUSTABLE.

SET PARKING BRAKE AND BLOCK WHEELS. DISCONNECT AUTOMATIC PARKING BRAKE RELEASE, IF SO EQUIPPED. MAKE ALL ADJUSTMENTS WITH ENGINE AT NORMAL OPERATING TEMPERATURE AND ACCESSORIES OFF.

IGNITION TIMING- ADJUST WITH TRANSMISSION IN NEUTRAL.

(1) TURN OFF ENGINE.

(2) DISCONNECT THE IN-LINE SPOUT CONNECTOR (-□□- OR =□□).

(3) RE-START PREVIOUSLY WARMED-UP ENGINE.

(4) ADJUST IGNITION TIMING TO 10° BTDC.

(5) TURN OFF ENGINE AND RESTORE ELECTRICAL CONNECTION.

THIS ENGINE IS EQUIPPED WITH AUTOMATIC IDLE SPEED CONTROL. IDLE RPM IS NOT ADJUSTABLE. IF NOT WITHIN 490-615 RPM RANGE, WITH TRANSMISSION IN DRIVE, SEE SHOP MANUAL.

FIRING ORDER-1-5-4-2-6-3-7-8

THIS VEHICLE CONFORMS TO U.S. EPA AND CALIFORNIA REGULATIONS APPLICABLE TO 1986 MODEL YEAR NEW MOTOR VEHICLES INTRODUCED INTO COMMERCE SOLELY FOR SALE IN CALIFORNIA.

E6AE-9C485-AGY | CATALYST | SPARK PLUG: AWSF-44C GAP-.048-.052 5.0L-6HM GFM5.0V5HBC6-FI/EGR/EGS/AIP/TWC

CALIBRATION: 6—22R—R00

MODEL YEAR: 1986 **ENGINE: 5.0L**

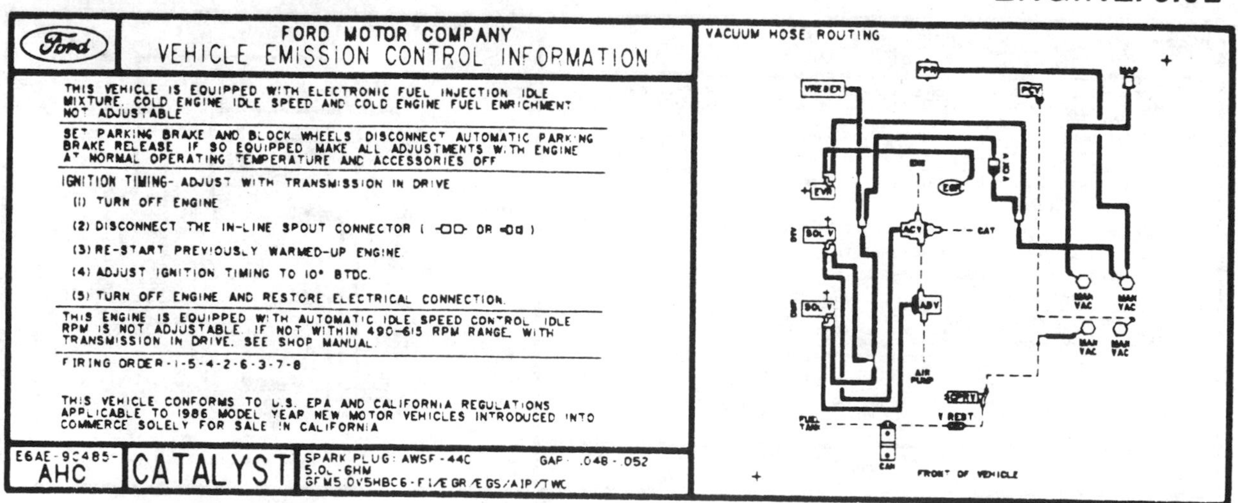
FORD MOTOR COMPANY
VEHICLE EMISSION CONTROL INFORMATION

THIS VEHICLE IS EQUIPPED WITH ELECTRONIC FUEL INJECTION. IDLE MIXTURE, COLD ENGINE IDLE SPEED AND COLD ENGINE FUEL ENRICHMENT NOT ADJUSTABLE

SET PARKING BRAKE AND BLOCK WHEELS. DISCONNECT AUTOMATIC PARKING BRAKE RELEASE IF SO EQUIPPED MAKE ALL ADJUSTMENTS WITH ENGINE AT NORMAL OPERATING TEMPERATURE AND ACCESSORIES OFF

IGNITION TIMING- ADJUST WITH TRANSMISSION IN DRIVE

(1) TURN OFF ENGINE

(2) DISCONNECT THE IN-LINE SPOUT CONNECTOR (-□□- OR =□□)

(3) RE-START PREVIOUSLY WARMED-UP ENGINE

(4) ADJUST IGNITION TIMING TO 10° BTDC.

(5) TURN OFF ENGINE AND RESTORE ELECTRICAL CONNECTION.

THIS ENGINE IS EQUIPPED WITH AUTOMATIC IDLE SPEED CONTROL. IDLE RPM IS NOT ADJUSTABLE. IF NOT WITHIN 490-615 RPM RANGE, WITH TRANSMISSION IN DRIVE. SEE SHOP MANUAL.

FIRING ORDER-1-5-4-2-6-3-7-8

THIS VEHICLE CONFORMS TO U.S. EPA AND CALIFORNIA REGULATIONS APPLICABLE TO 1986 MODEL YEAR NEW MOTOR VEHICLES INTRODUCED INTO COMMERCE SOLELY FOR SALE IN CALIFORNIA

E6AE-9C485-AHC | CATALYST | SPARK PLUG: AWSF-44C GAP .048-.052 5.0L-6HM GFM5.0V5HBC6-FI/EGR/EGS/AIP/TWC

VACUUM CIRCUITS

(© Ford Motor Co)

CALIBRATION: 6—22S—R00

MODEL YEAR: 1986 **ENGINE: 5.0L**

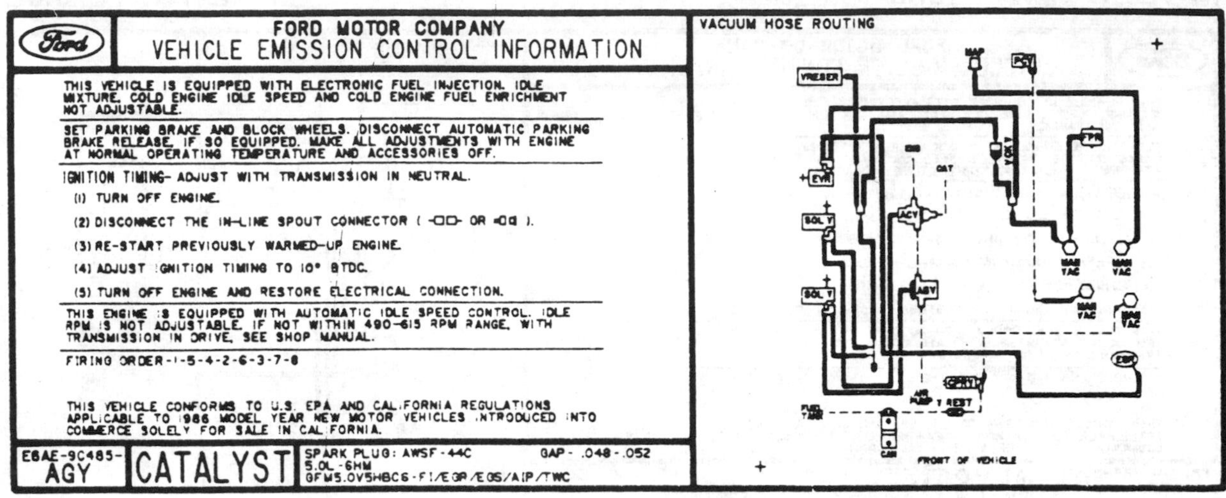
FORD MOTOR COMPANY
VEHICLE EMISSION CONTROL INFORMATION

THIS VEHICLE IS EQUIPPED WITH ELECTRONIC FUEL INJECTION. IDLE MIXTURE, COLD ENGINE IDLE SPEED AND COLD ENGINE FUEL ENRICHMENT NOT ADJUSTABLE.

SET PARKING BRAKE AND BLOCK WHEELS. DISCONNECT AUTOMATIC PARKING BRAKE RELEASE, IF SO EQUIPPED. MAKE ALL ADJUSTMENTS WITH ENGINE AT NORMAL OPERATING TEMPERATURE AND ACCESSORIES OFF.

IGNITION TIMING- ADJUST WITH TRANSMISSION IN NEUTRAL.

(1) TURN OFF ENGINE.

(2) DISCONNECT THE IN-LINE SPOUT CONNECTOR (-▭▭- OR =▭▭).

(3) RE-START PREVIOUSLY WARMED-UP ENGINE.

(4) ADJUST IGNITION TIMING TO 10° BTDC.

(5) TURN OFF ENGINE AND RESTORE ELECTRICAL CONNECTION.

THIS ENGINE IS EQUIPPED WITH AUTOMATIC IDLE SPEED CONTROL. IDLE RPM IS NOT ADJUSTABLE. IF NOT WITHIN 490-615 RPM RANGE, WITH TRANSMISSION IN DRIVE, SEE SHOP MANUAL.

FIRING ORDER-1-5-4-2-6-3-7-8

THIS VEHICLE CONFORMS TO U.S. EPA AND CALIFORNIA REGULATIONS APPLICABLE TO 1986 MODEL YEAR NEW MOTOR VEHICLES INTRODUCED INTO COMMERCE SOLELY FOR SALE IN CALIFORNIA.

E6AE-9C485- AGY | CATALYST | SPARK PLUG: AWSF-44C GAP- .048-.052
5.0L-6HM
GFM5.0V5HBC6-FI/EGR/EGS/AIP/TWC

CALIBRATION: 6—22T—R00

MODEL YEAR: 1986 **ENGINE: 5.0L**

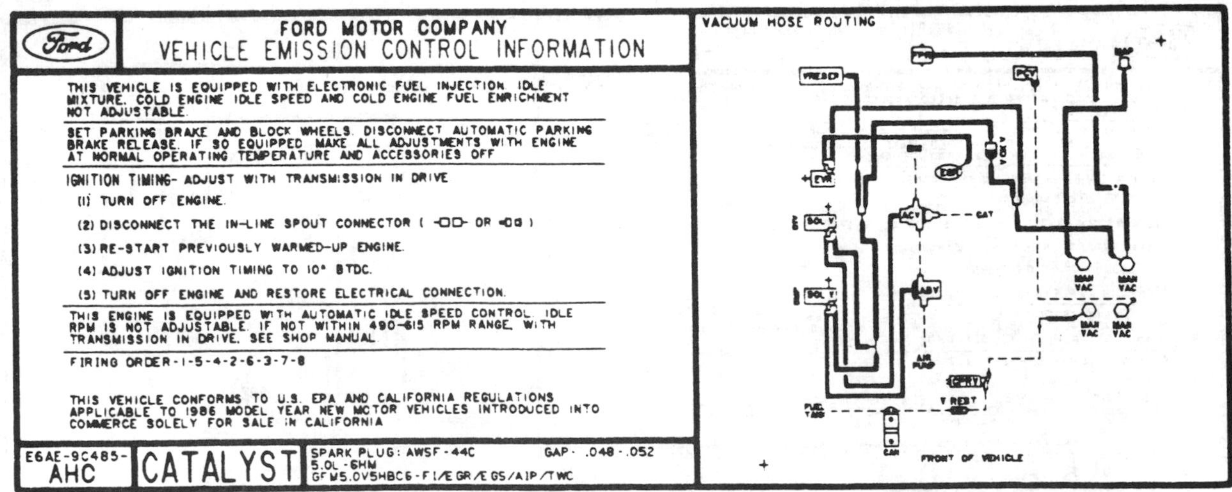
FORD MOTOR COMPANY
VEHICLE EMISSION CONTROL INFORMATION

THIS VEHICLE IS EQUIPPED WITH ELECTRONIC FUEL INJECTION. IDLE MIXTURE, COLD ENGINE IDLE SPEED AND COLD ENGINE FUEL ENRICHMENT NOT ADJUSTABLE.

SET PARKING BRAKE AND BLOCK WHEELS. DISCONNECT AUTOMATIC PARKING BRAKE RELEASE. IF SO EQUIPPED. MAKE ALL ADJUSTMENTS WITH ENGINE AT NORMAL OPERATING TEMPERATURE AND ACCESSORIES OFF

IGNITION TIMING- ADJUST WITH TRANSMISSION IN DRIVE

(1) TURN OFF ENGINE.

(2) DISCONNECT THE IN-LINE SPOUT CONNECTOR (-▭▭- OR =▭▭)

(3) RE-START PREVIOUSLY WARMED-UP ENGINE.

(4) ADJUST IGNITION TIMING TO 10° BTDC.

(5) TURN OFF ENGINE AND RESTORE ELECTRICAL CONNECTION.

THIS ENGINE IS EQUIPPED WITH AUTOMATIC IDLE SPEED CONTROL. IDLE RPM IS NOT ADJUSTABLE. IF NOT WITHIN 490-615 RPM RANGE, WITH TRANSMISSION IN DRIVE. SEE SHOP MANUAL

FIRING ORDER-1-5-4-2-6-3-7-8

THIS VEHICLE CONFORMS TO U.S. EPA AND CALIFORNIA REGULATIONS APPLICABLE TO 1986 MODEL YEAR NEW MOTOR VEHICLES INTRODUCED INTO COMMERCE SOLELY FOR SALE IN CALIFORNIA

E6AE-9C485- AHC | CATALYST | SPARK PLUG: AWSF-44C GAP- .048-.052
5.0L-6HM
GFM5.0V5HBC6-FI/EGR/EGS/AIP/TWC

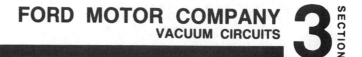

VACUUM CIRCUITS

(© Ford Motor Co)

CALIBRATION: 6—23A—R00

MODEL YEAR: 1986

ENGINE: 5.0L

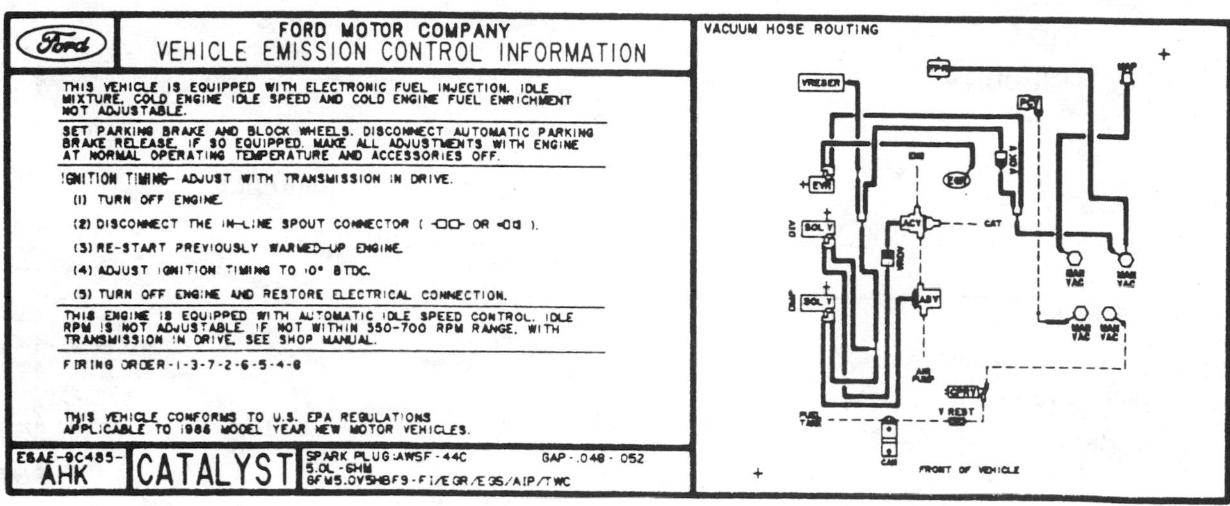

FORD MOTOR COMPANY
VEHICLE EMISSION CONTROL INFORMATION

THIS VEHICLE IS EQUIPPED WITH ELECTRONIC FUEL INJECTION. IDLE MIXTURE, COLD ENGINE IDLE SPEED AND COLD ENGINE FUEL ENRICHMENT NOT ADJUSTABLE.

SET PARKING BRAKE AND BLOCK WHEELS. DISCONNECT AUTOMATIC PARKING BRAKE RELEASE, IF SO EQUIPPED. MAKE ALL ADJUSTMENTS WITH ENGINE AT NORMAL OPERATING TEMPERATURE AND ACCESSORIES OFF.

IGNITION TIMING- ADJUST WITH TRANSMISSION IN DRIVE.

(1) TURN OFF ENGINE.

(2) DISCONNECT THE IN-LINE SPOUT CONNECTOR (-□□- OR -□□).

(3) RE-START PREVIOUSLY WARMED-UP ENGINE.

(4) ADJUST IGNITION TIMING TO 10° BTDC.

(5) TURN OFF ENGINE AND RESTORE ELECTRICAL CONNECTION.

THIS ENGINE IS EQUIPPED WITH AUTOMATIC IDLE SPEED CONTROL. IDLE RPM IS NOT ADJUSTABLE. IF NOT WITHIN 550-700 RPM RANGE, WITH TRANSMISSION IN DRIVE, SEE SHOP MANUAL.

FIRING ORDER-1-3-7-2-6-5-4-8

THIS VEHICLE CONFORMS TO U.S. EPA REGULATIONS APPLICABLE TO 1986 MODEL YEAR NEW MOTOR VEHICLES.

E6AE-9C485-AHK | CATALYST | SPARK PLUG:AWSF-44C GAP-.048-052 5.0L-6HM 6FM5.0V5HBF9-FI/EGR/EGS/AIP/TWC

ENGINE: 5.8L

CALIBRATION: 4—24P—R00

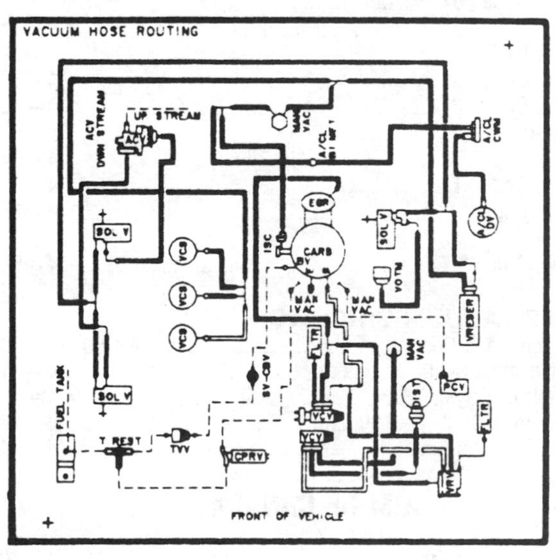

CALIBRATION: 4—24P—R11

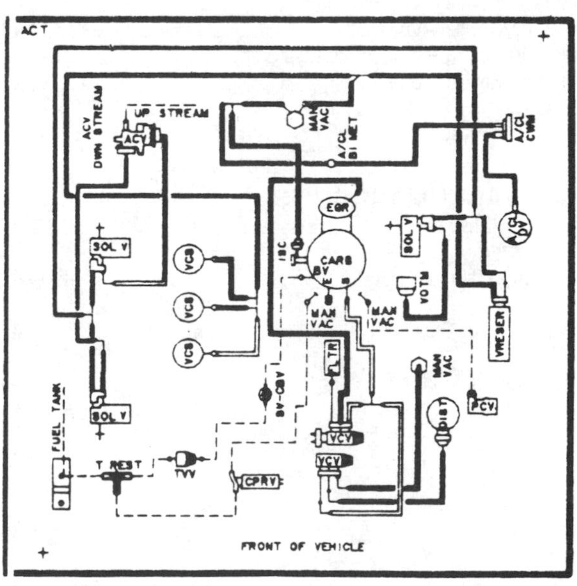

General Motors 1980-82
Vacuum Circuits

INDEX

VACUUM CIRCUITS

(© Buick Div., G.M. Corp.)

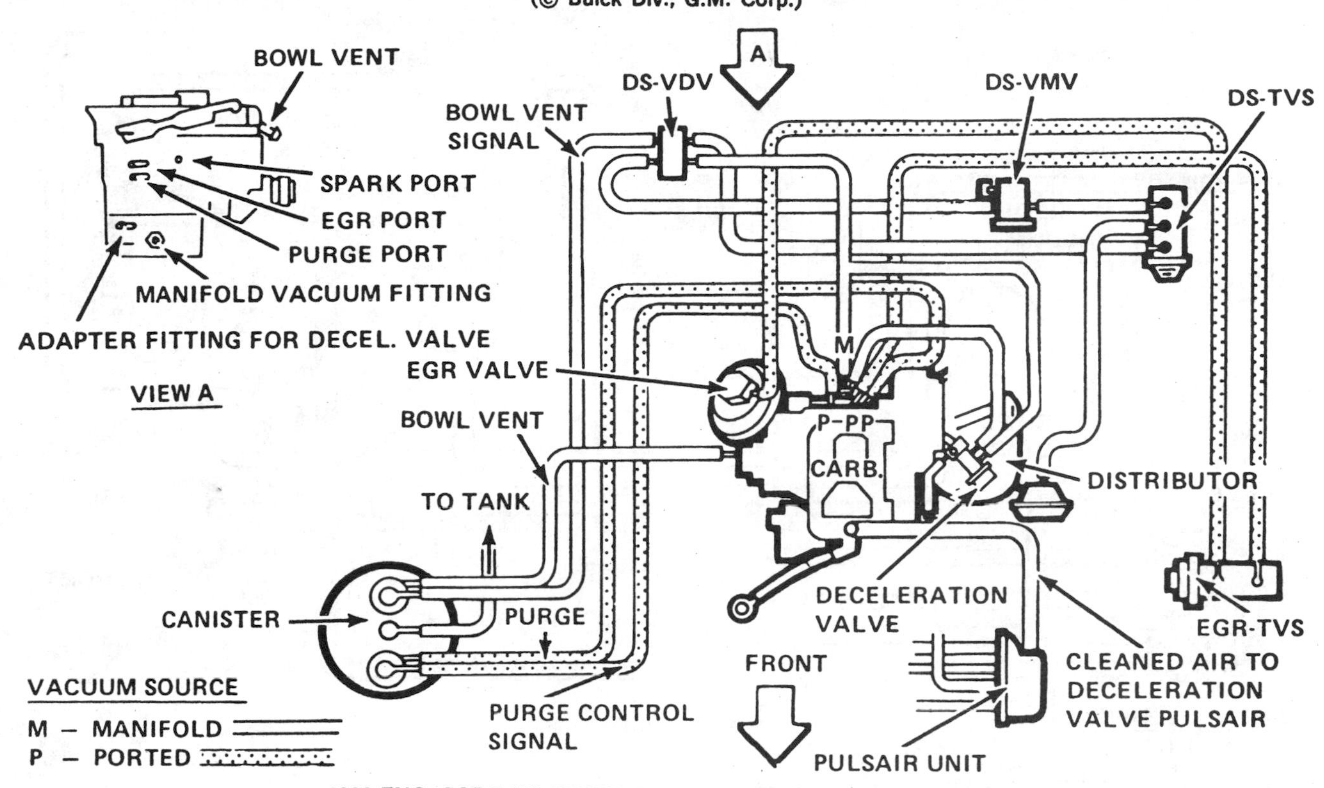

1980 ENG. CODE "5"-W/MAN. TRANS.-SKYLARK-LOW ALTITUDE

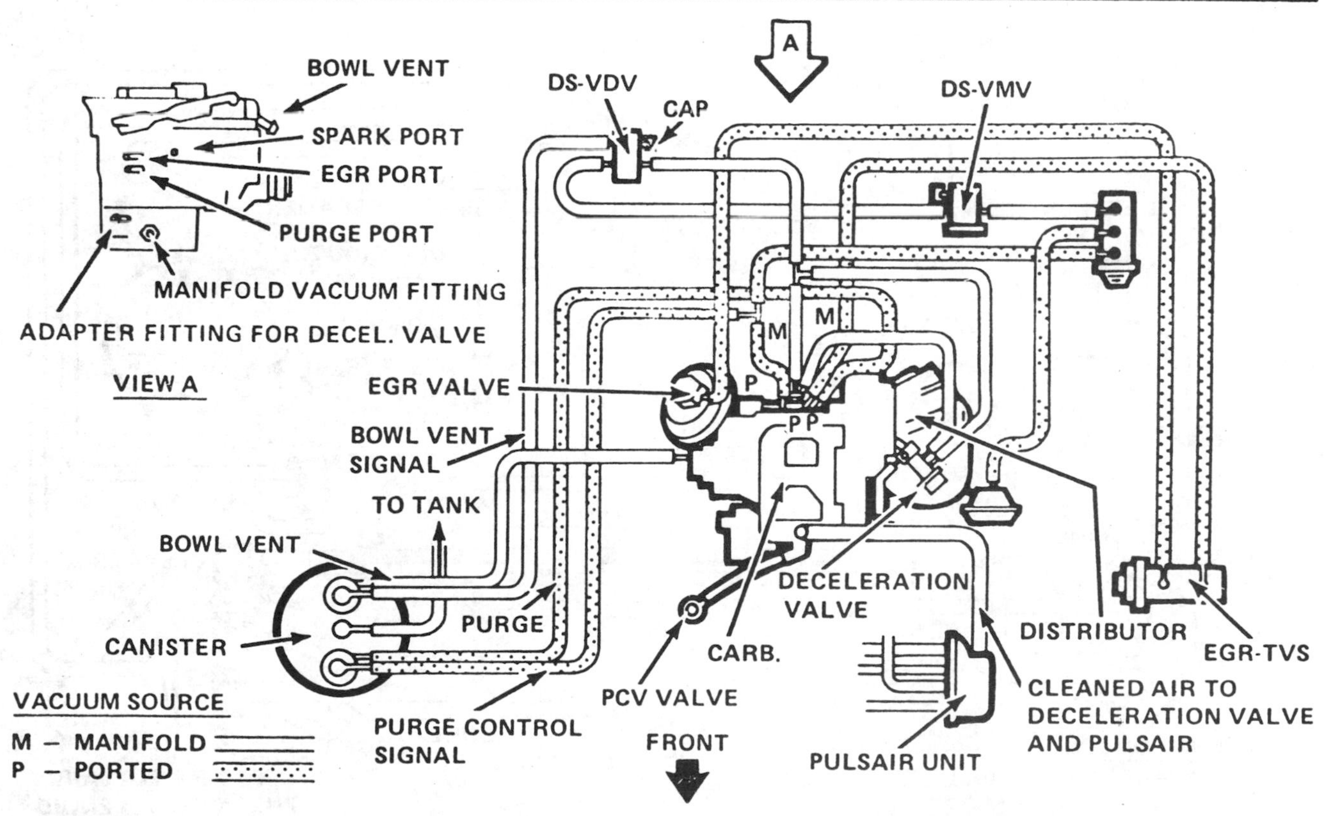

1980 ENG. CODE "5"-W/AUTO. TRANS.-SKYLARK-LOW ALTITUDE

VACUUM CIRCUITS

(© Buick Div., G.M. Corp.)

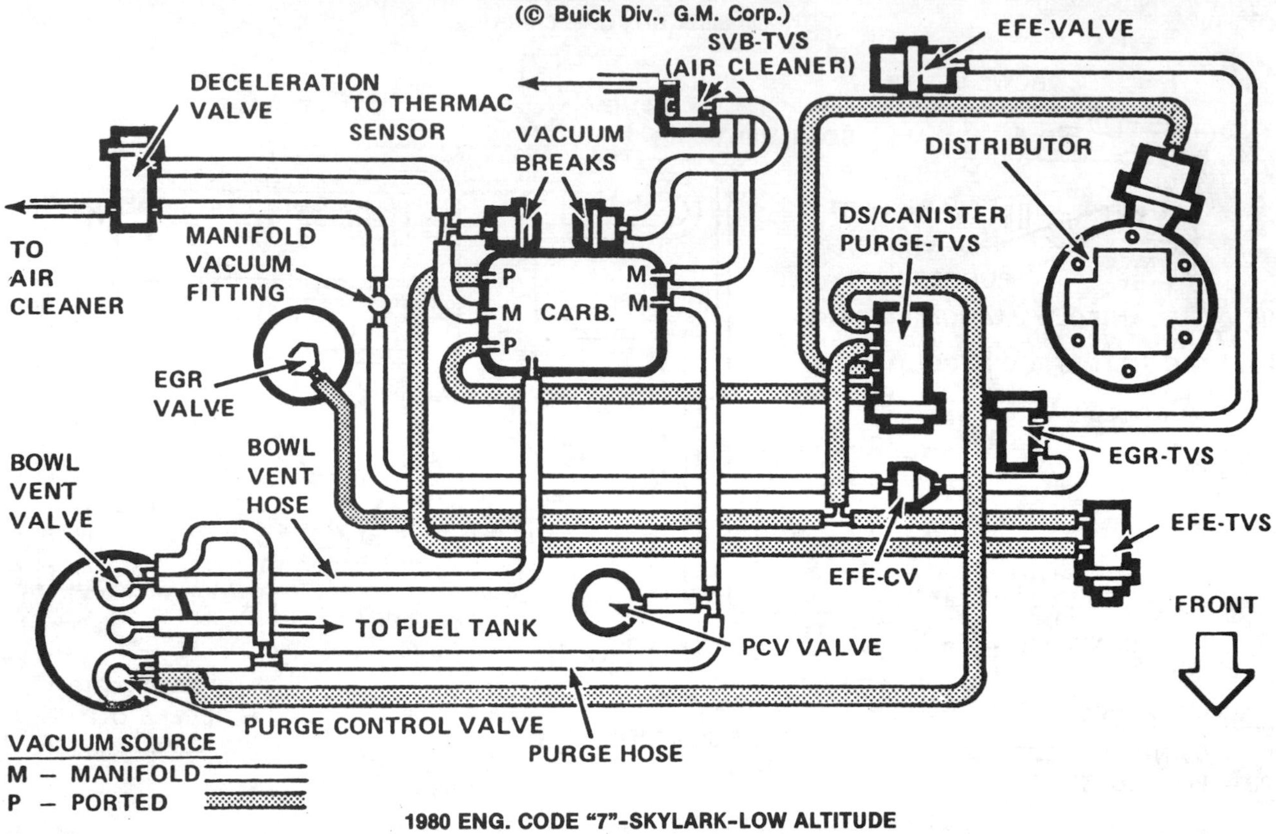

SVB-TVS
(AIR CLEANER)
EFE-VALVE
DECELERATION
VALVE
TO THERMAC
SENSOR
VACUUM
BREAKS
DISTRIBUTOR
TO
AIR
CLEANER
MANIFOLD
VACUUM
FITTING
DS/CANISTER
PURGE-TVS
P
M
P
M
M
CARB.
EGR
VALVE
BOWL
VENT
VALVE
BOWL
VENT
HOSE
EGR-TVS
EFE-TVS
EFE-CV
TO FUEL TANK
PCV VALVE
FRONT
PURGE CONTROL VALVE
PURGE HOSE
VACUUM SOURCE
M – MANIFOLD
P – PORTED

1980 ENG. CODE "7"-SKYLARK-LOW ALTITUDE

RIGHT PULSAIR
EFE VALVE
LEAN
AUTHORITY
SWITCH
TO THERMAL
SENSOR
SVB-TVS
(AIR CLEANER)
DECELERATION
VALVE
VACUUM
BREAK
DISTRIBUTOR
VACUUM
DS-VRDV
MANIFOLD
VACUUM
FITTING
P
M
P
M
M
CARB.
TO AIR
CLEANER
EGR
VALVE
DISTRIBUTOR
EFE-CV
EFE-TVS
BOWL VENT
LINE
BOWL VENT
VALVE
TO AIR
CLEANER
PCV-VALVE
EGR-TVS
TO FUEL TANK
PURGE HOSE
TANK PRESSURE
CONTROL VALVE
PURGE CONTROL
VALVE
FRONT
LEFT
PULSAIR
PULSAIR
SOLENOID
VACUUM SOURCE
M – MANIFOLD VACUUM
P – PORTED VACUUM

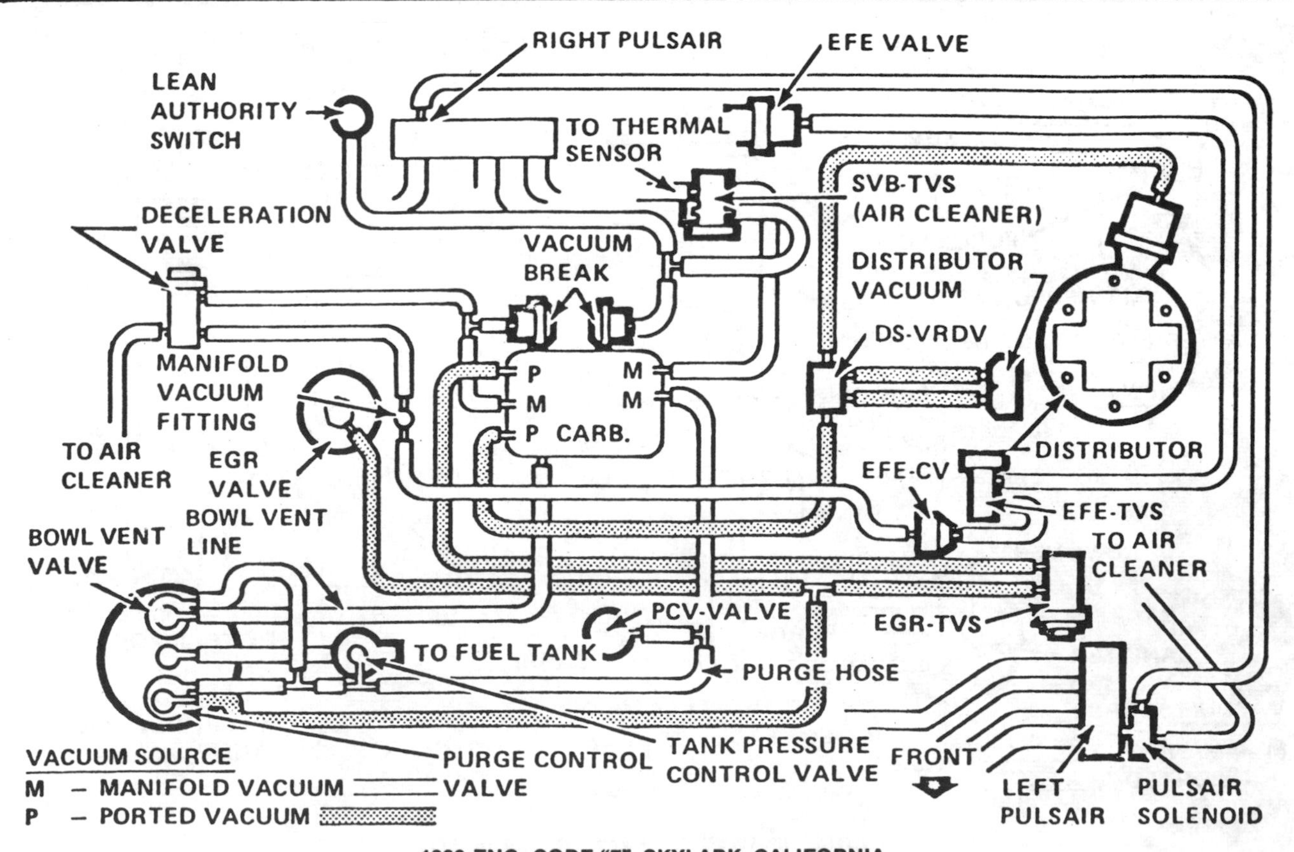

1980 ENG. CODE "7"-SKYLARK-CALIFORNIA

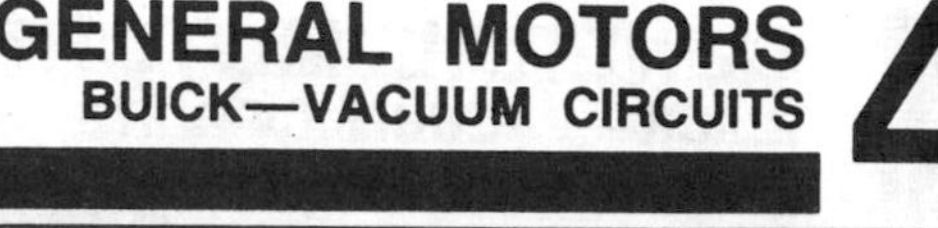

VACUUM CIRCUITS

(© Buick Div., G.M. Corp.)

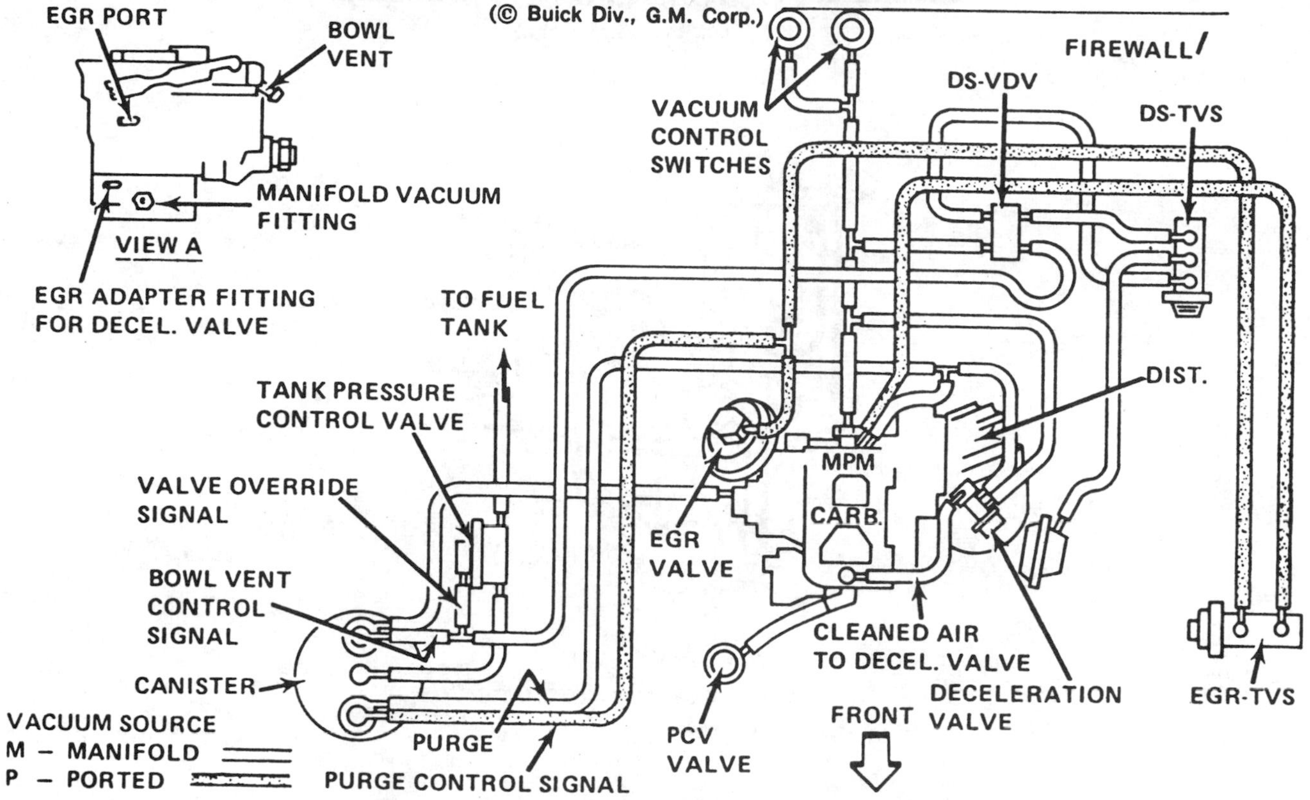

EGR PORT
BOWL VENT
MANIFOLD VACUUM FITTING
VIEW A
EGR ADAPTER FITTING FOR DECEL. VALVE
VACUUM CONTROL SWITCHES
FIREWALL
DS-VDV
DS-TVS
TO FUEL TANK
TANK PRESSURE CONTROL VALVE
VALVE OVERRIDE SIGNAL
BOWL VENT CONTROL SIGNAL
CANISTER
VACUUM SOURCE
M – MANIFOLD
P – PORTED
PURGE
PURGE CONTROL SIGNAL
MPM
CARB.
EGR VALVE
DIST.
CLEANED AIR TO DECEL. VALVE
DECELERATION VALVE
PCV VALVE
FRONT
EGR-TVS

1980 ENG. CODE "5"-W/MAN. TRANS.-SKYLARK-CALIFORNIA

SPARK PORT
BOWL VENT
FIREWALL
VACUUM CONTROL SWITCHES
A
DS-VDV
DS-TVS
EGR PORT
MANIFOLD VACUUM FITTING
VIEW A
EGR ADAPTER FITTING FOR DECEL. VALVE
TO FUEL TANK
TANK PRESSURE CONTROL VALVE
VENT OVERRIDE SIGNAL
CARB. BOWL VENT
BOWL VENT CONTROL SIGNAL
CANISTER
VACUUM SOURCE
M – MANIFOLD
P – PORTED
PURGE
PURGE CONTROL SIGNAL
P M P M
CARB.
EGR VALVE
DIST.
PCV VALVE
CLEANED AIR TO DECEL. VALVE
FRONT
DECELERATION VALVE
EGR/CAN. PURGE-TVS

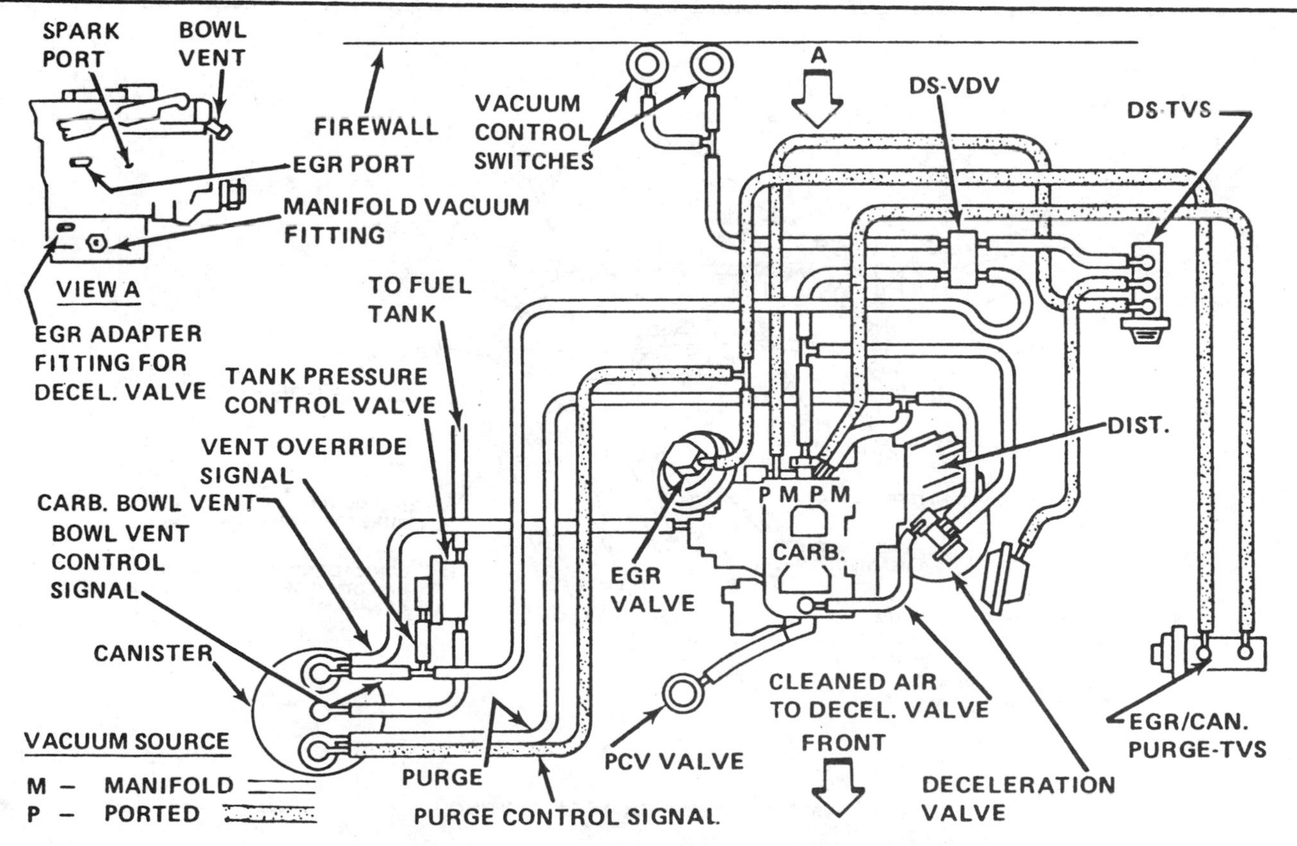

1980 ENG. CODE "5"-W/AUTO. TRANS.-SKYLARK-CALIFORNIA

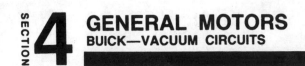

VACUUM CIRCUITS

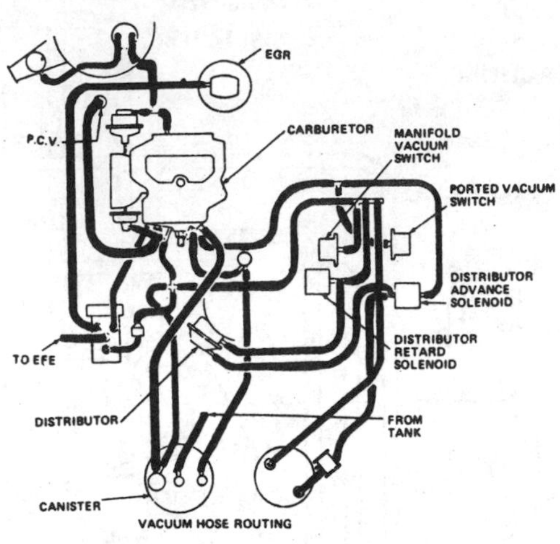

1979-80 C-4 SYSTEM-SCHEMATIC

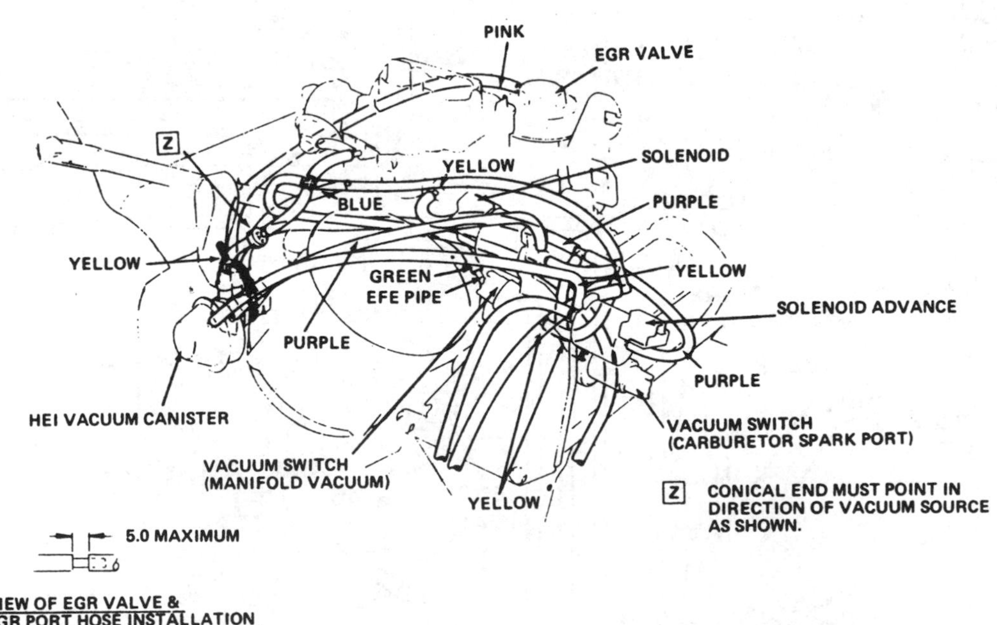

1979-80 C-4 SYSTEM-HOSE ROUTING

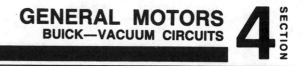

COMPUTER COMMAND CONTROL SYSTEM

The Computer Command Control System (the "System"), is an electronically controlled emission system that can monitor up to fifteen (15) various engine/vehicle operating conditions and using this information can control as many as nine (9) engine related systems (Fig. 6E-2 and 6E-3). The "System" is thereby making constant adjustments to maintain good vehicle performance under all normal driving conditions while allowing the catalytic converter to effectively control Oxides of Nitrogen (NOX), Hydrocarbons (DC)

The "System" has a built-in diagnostic system that recognizes and identifies possible operational problems and alerts the driver through a "CHECK ENGINE" light in the instrument panel. The light will remain "ON" until the problem is corrected. This same light is used, with the built-in diagnostic system, to aid the technician in locating and repairing the problem by flashing a stored code or codes which identifies the possible problem areas.

The "System" also has built-in back-up systems that in most cases of an operation problem will allow continued operation of the vehicle in a near normal manner until repairs can be made.

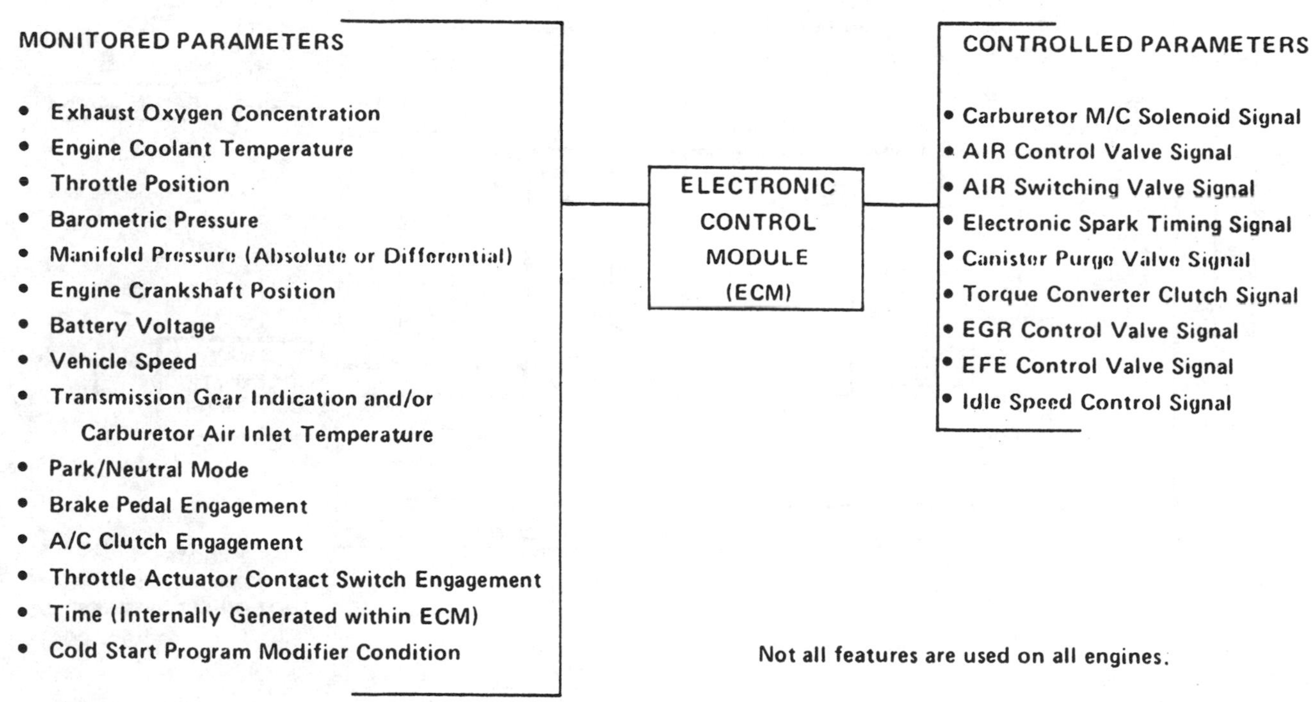

System Parameters

1981-82 COMPUTER COMMAND CONTROL SYSTEM

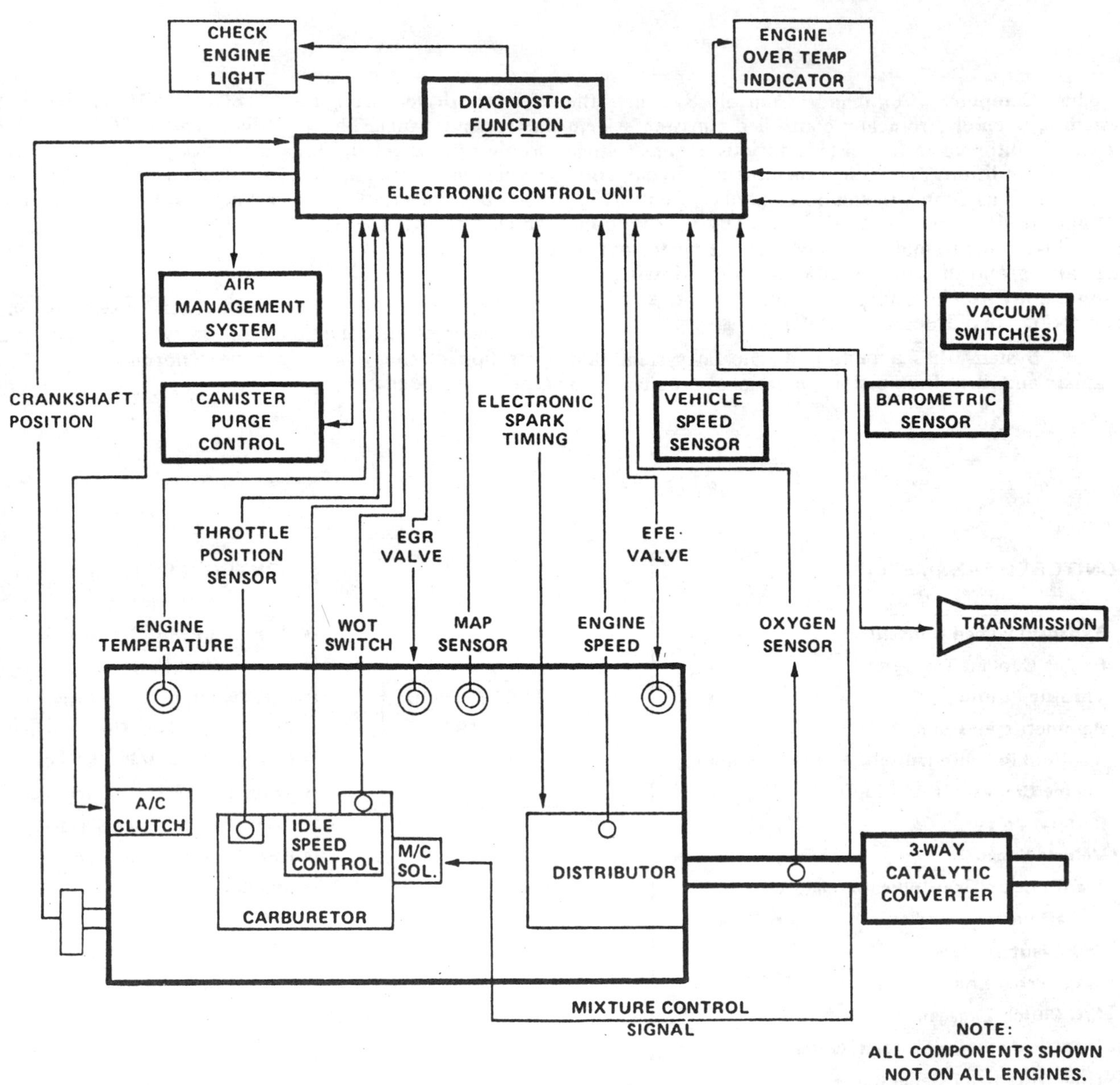
CHECK ENGINE LIGHT
ENGINE OVER TEMP INDICATOR
DIAGNOSTIC FUNCTION
ELECTRONIC CONTROL UNIT
AIR MANAGEMENT SYSTEM
VACUUM SWITCH(ES)
CRANKSHAFT POSITION
CANISTER PURGE CONTROL
ELECTRONIC SPARK TIMING
VEHICLE SPEED SENSOR
BAROMETRIC SENSOR
THROTTLE POSITION SENSOR
EGR VALVE
EFE VALVE
ENGINE TEMPERATURE
WOT SWITCH
MAP SENSOR
ENGINE SPEED
OXYGEN SENSOR
TRANSMISSION
A/C CLUTCH
IDLE SPEED CONTROL
M/C SOL.
CARBURETOR
DISTRIBUTOR
3-WAY CATALYTIC CONVERTER
MIXTURE CONTROL SIGNAL
NOTE: ALL COMPONENTS SHOWN NOT ON ALL ENGINES.

1981-82 COMPUTER COMMAND CONTROL SYSTEM

VACUUM CIRCUITS

(© Buick Div., G.M. Corp.)

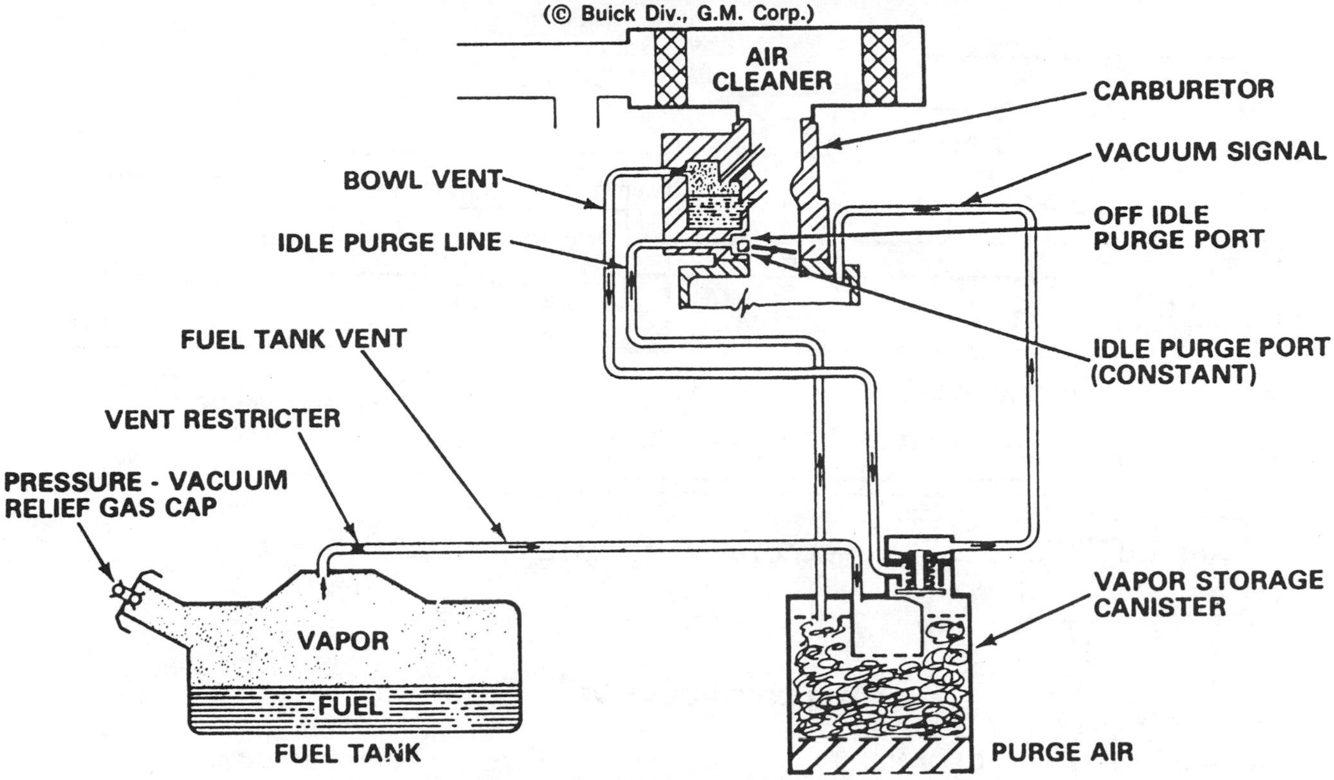

1981-82 EVAPORATIVE EMISSION CONTROL SYSTEM (EECS)-OPEN CANISTER

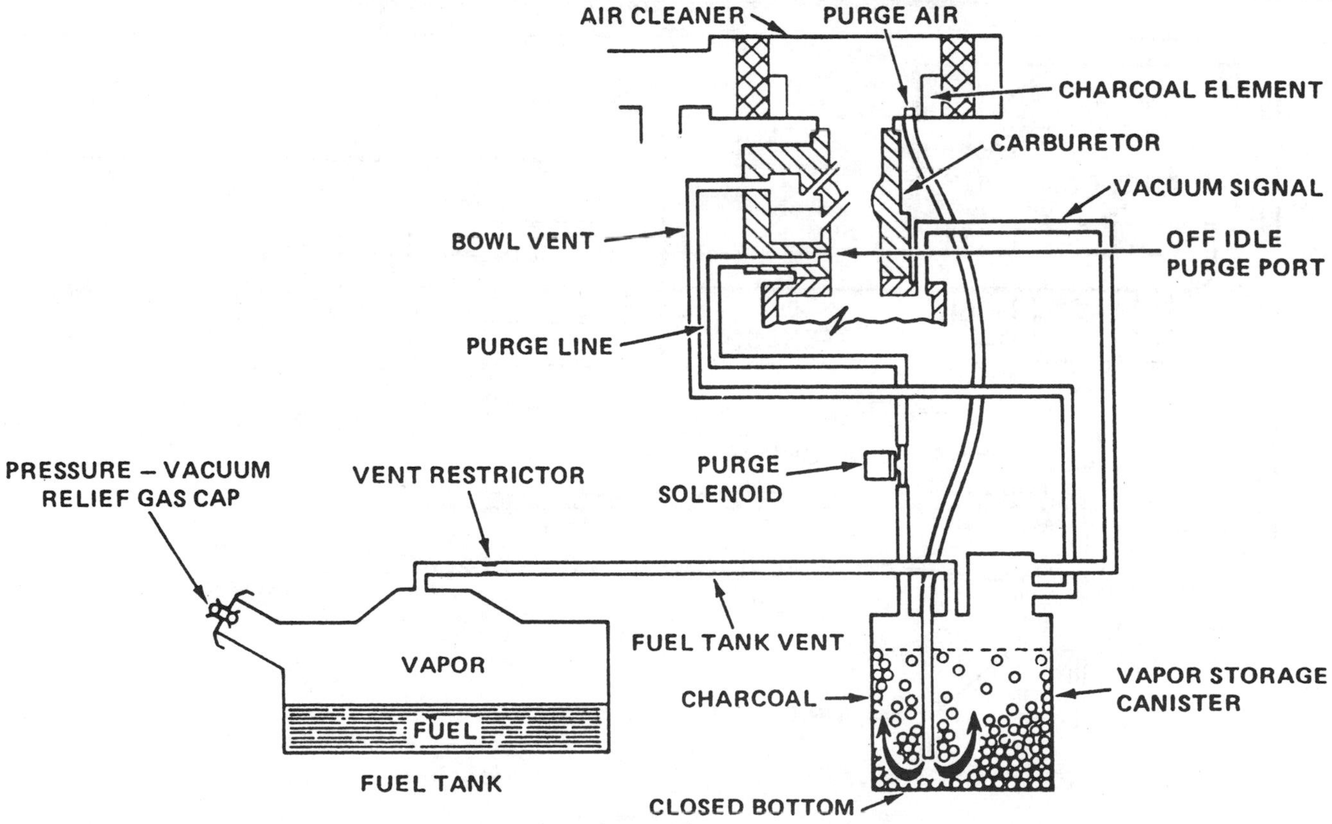

1981-82 EVAPORATIVE EMISSION CONTROL SYSTEM-(EECS)-CLOSED CANISTER

VACUUM CIRCUITS

(© Buick Div., G.M. Corp.)

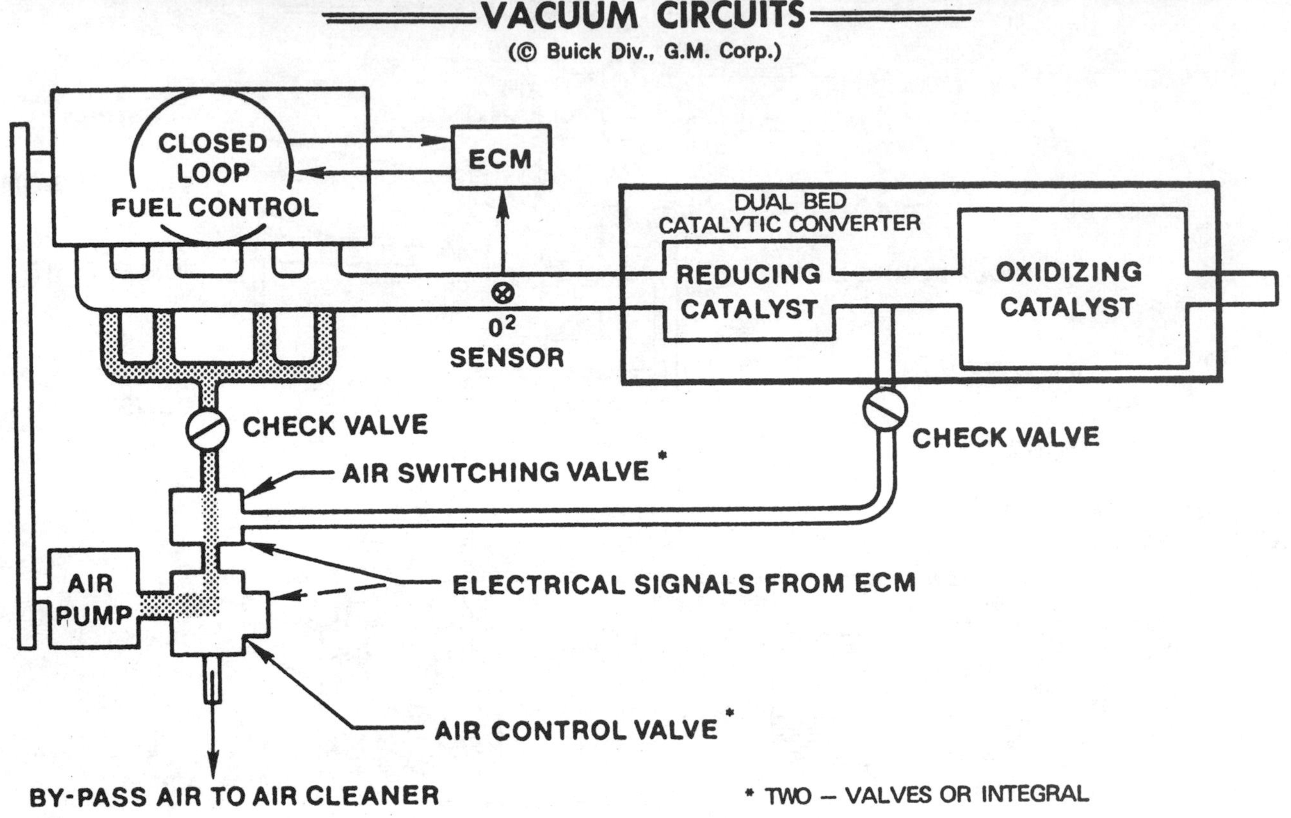

1981-82 AIR MANAGEMENT SYSTEM-COLD ENGINE MODE

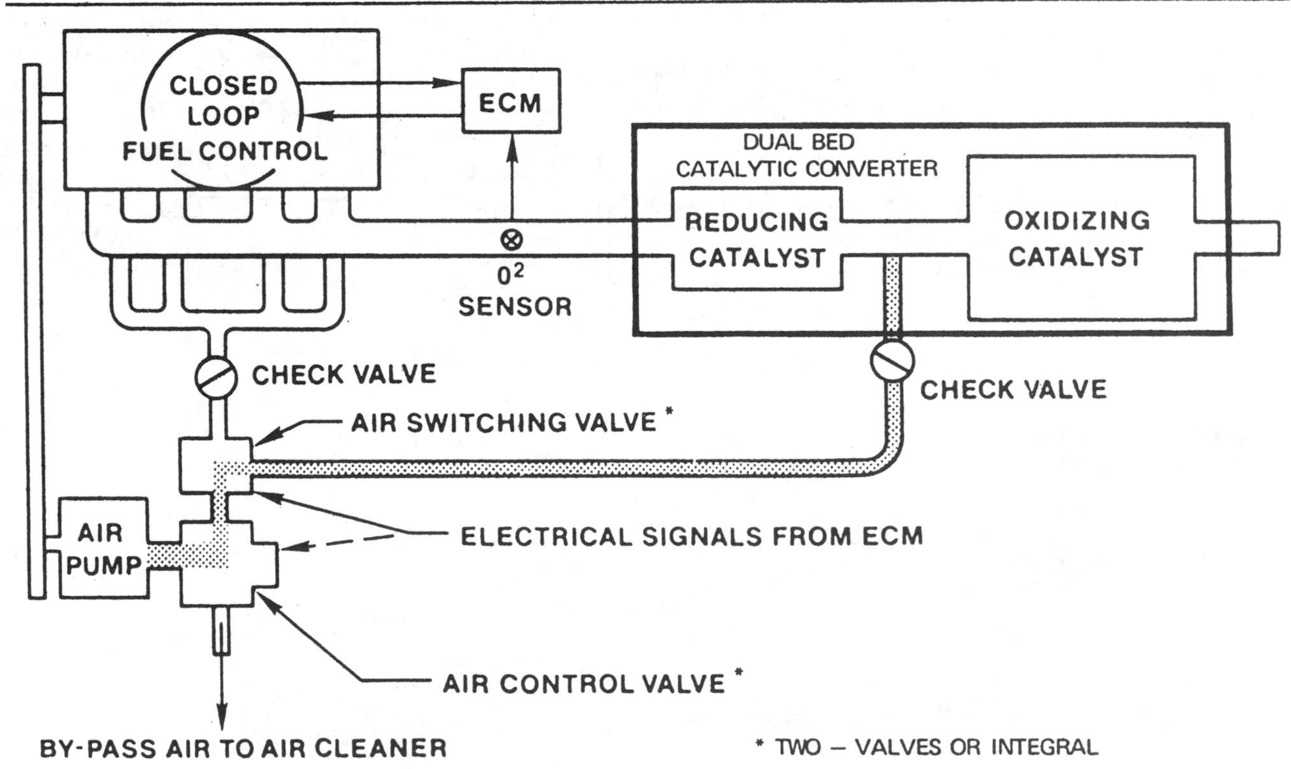

1981-82 AIR MANAGEMENT SYSTEM-WARM ENGINE MODE

VACUUM CIRCUITS

(© Cadillac Div., G.M. Corp.)

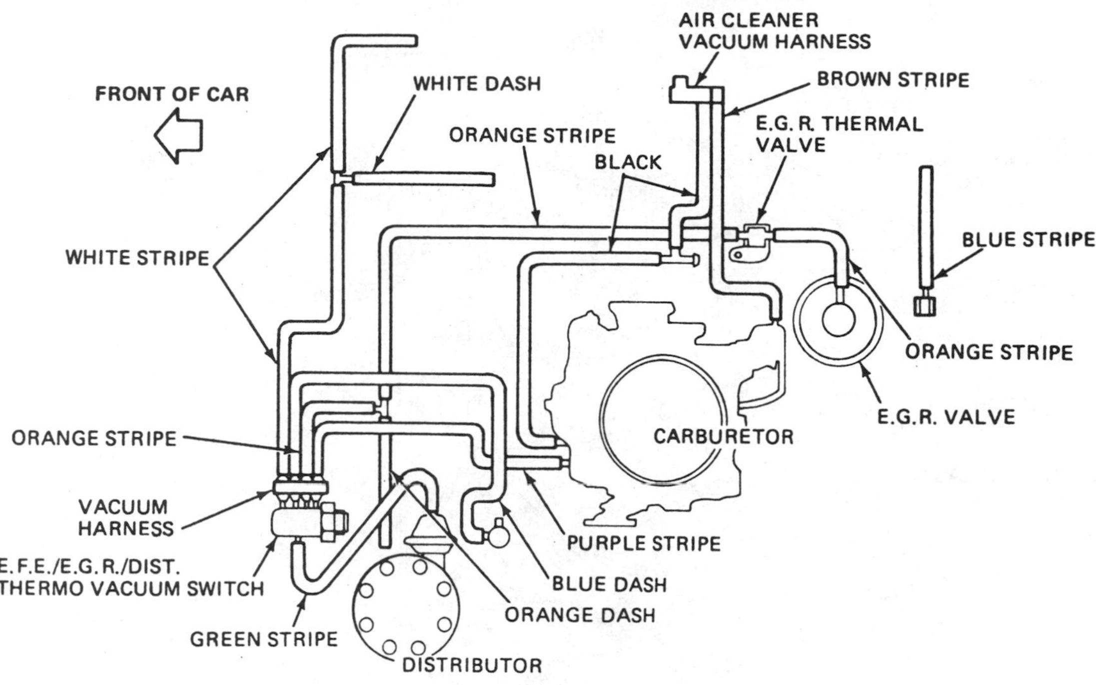

1980 FLEETWOOD & DeVILLE-CALIFORNIA

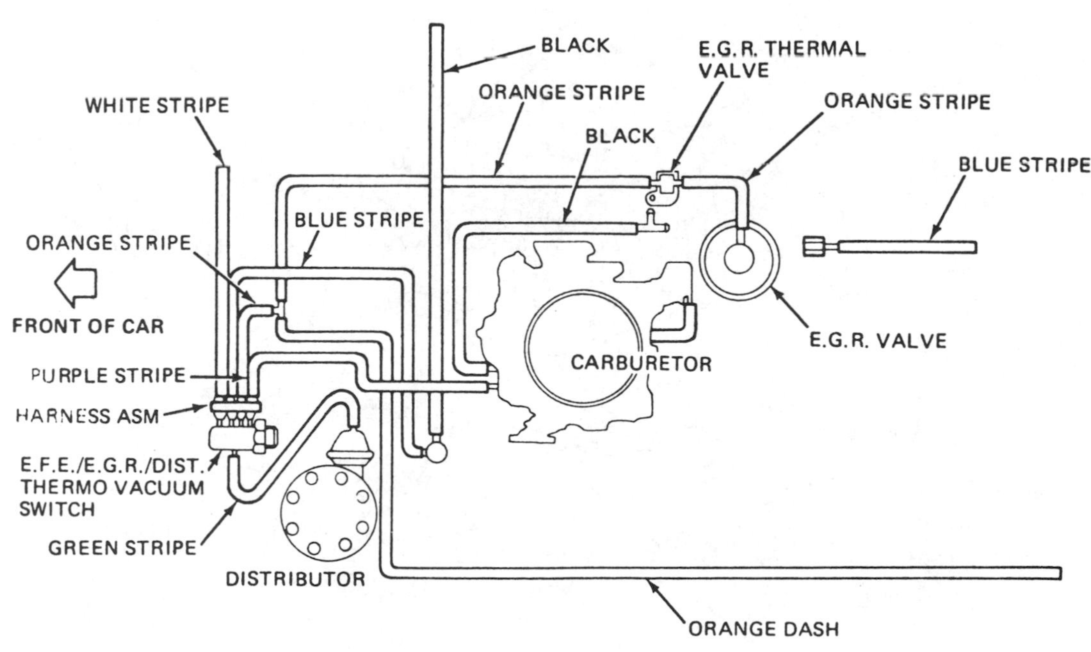

1980 FLEETWOOD & DeVILLE-EXC. CALIFORNIA

VACUUM CIRCUITS

(© Cadillac Div., G.M. Corp.)

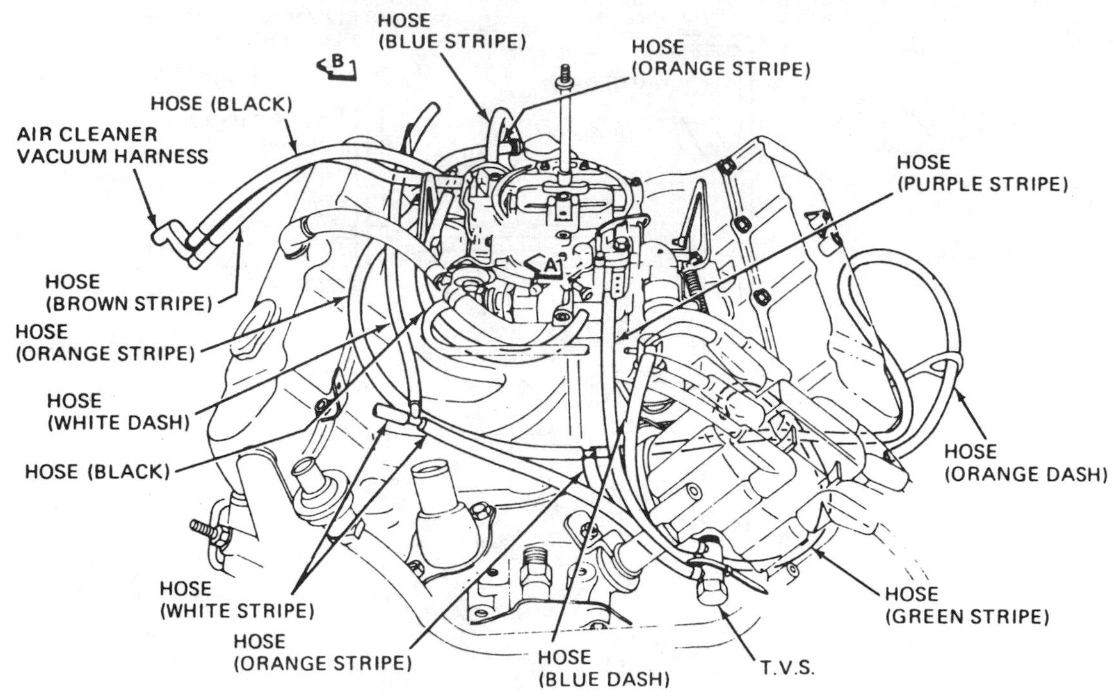

1980 FLEETWOOD & DeVILLE-CALIFORNIA

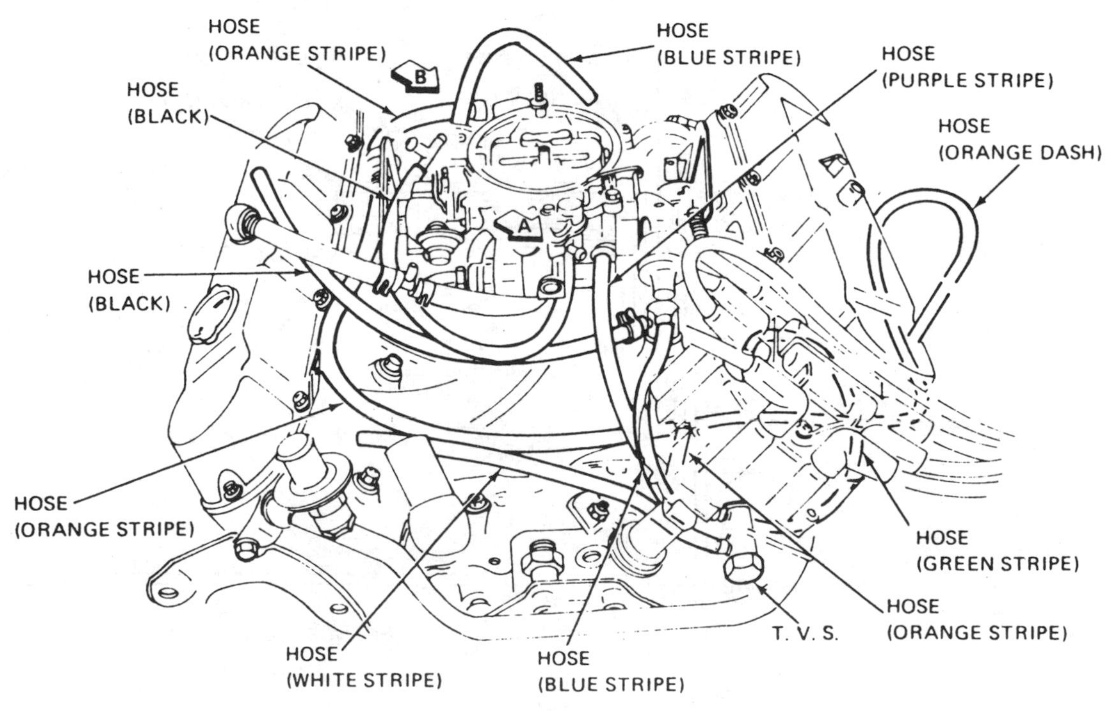

1980 FLEETWOOD & DeVILLE-EXC. CALIFORNIA

VACUUM CIRCUITS

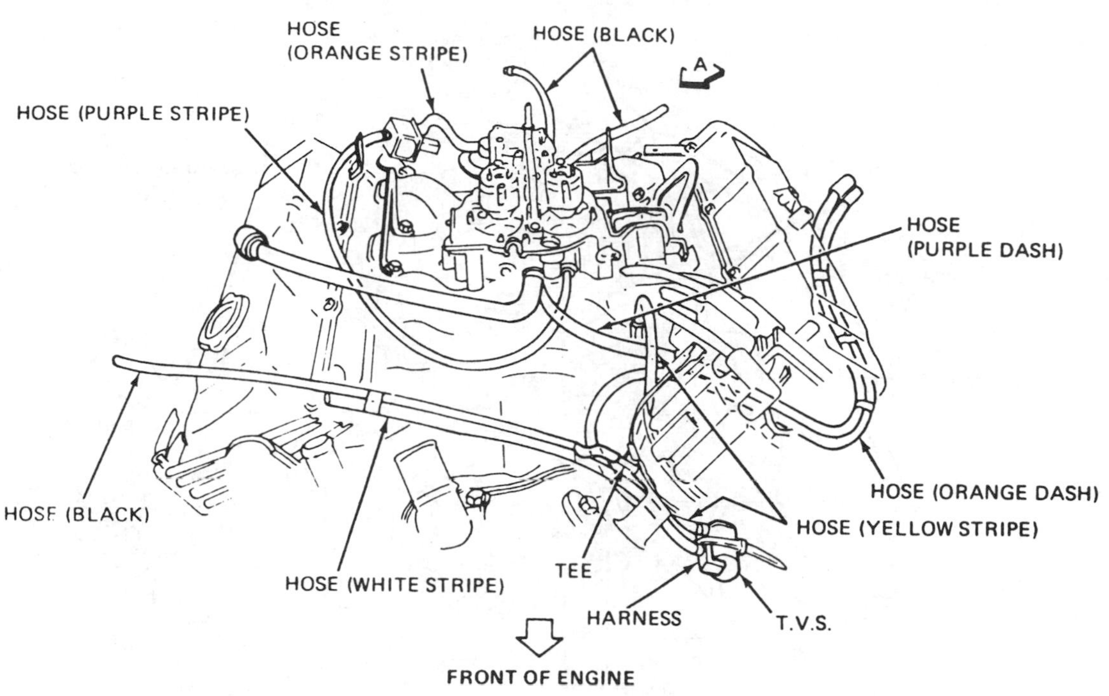

1980 ELDORADO & SEVILLE (DEFI)

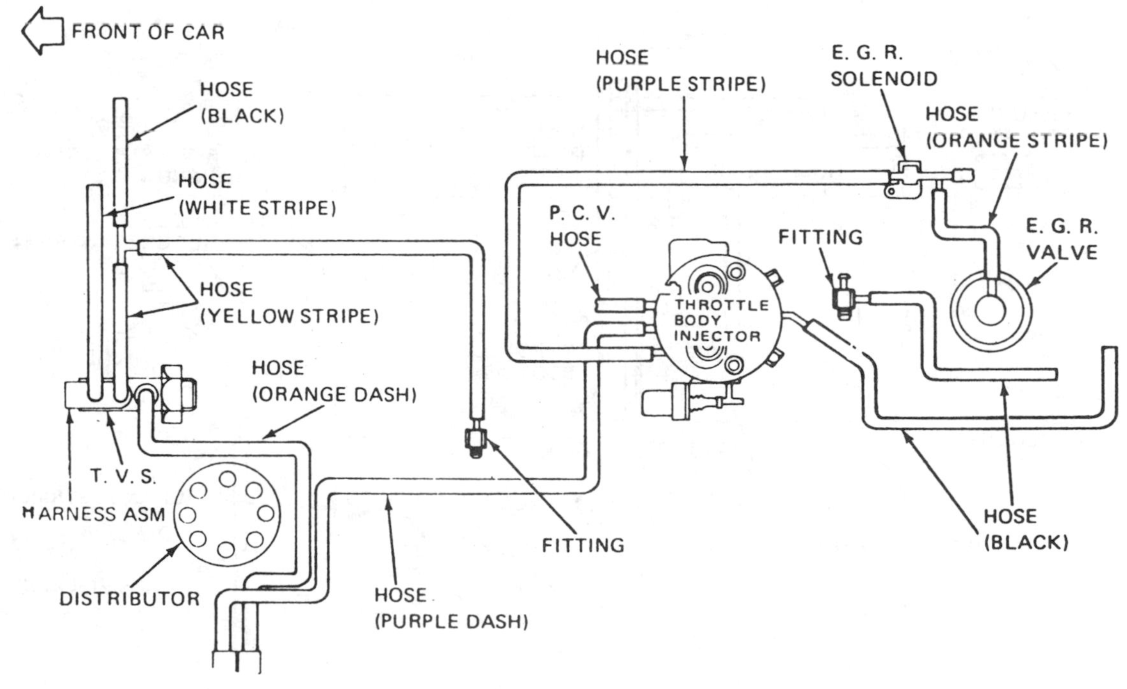

1980 ELDORADO & SEVILLE (DEFI)

VACUUM CIRCUITS

(© Cadillac Div., G.M. Corp.)

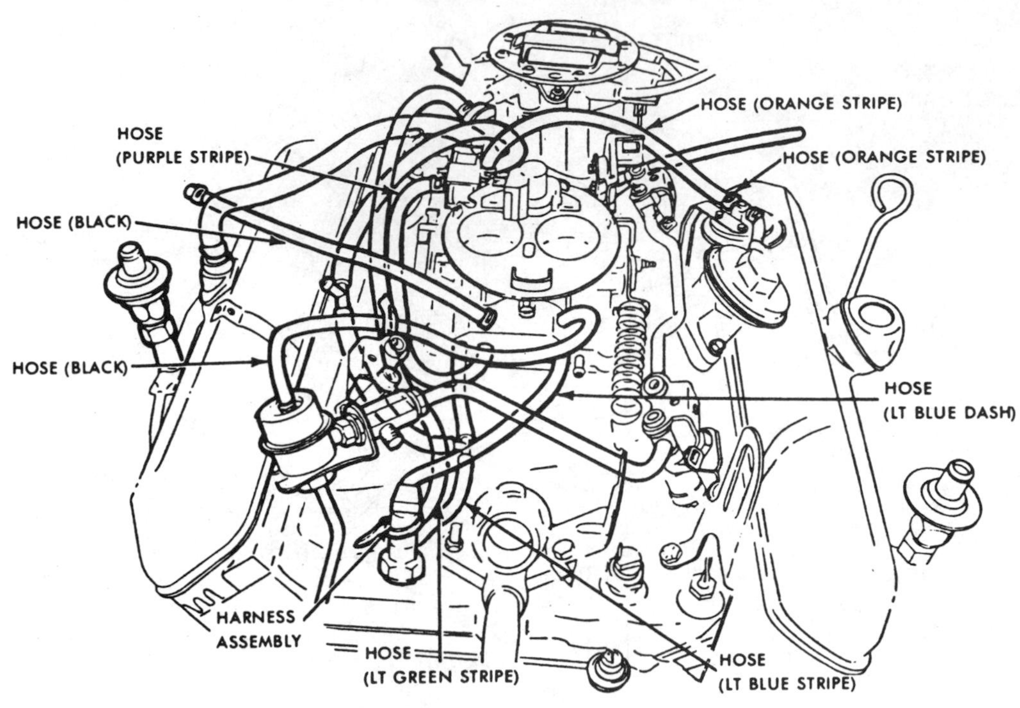

1980 ELDORADO & SEVILLE (EFI)

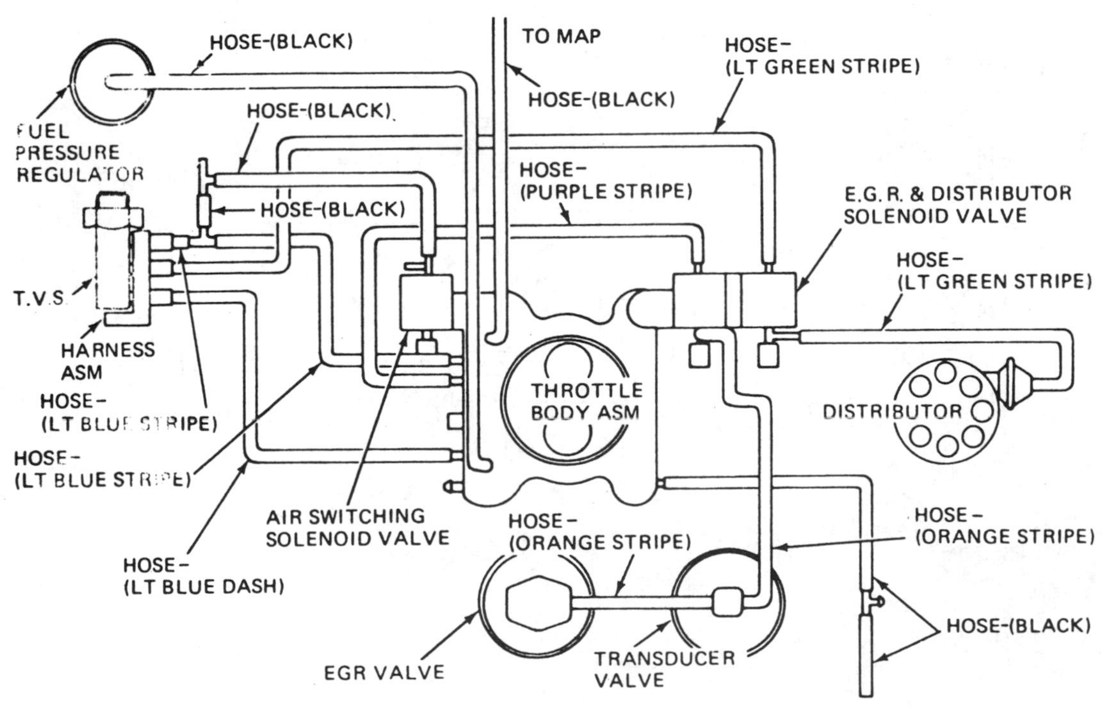

1980 ELDORADO & SEVILLE (EFI)

VACUUM CIRCUITS

(© Cadillac Div., G.M. Corp.)

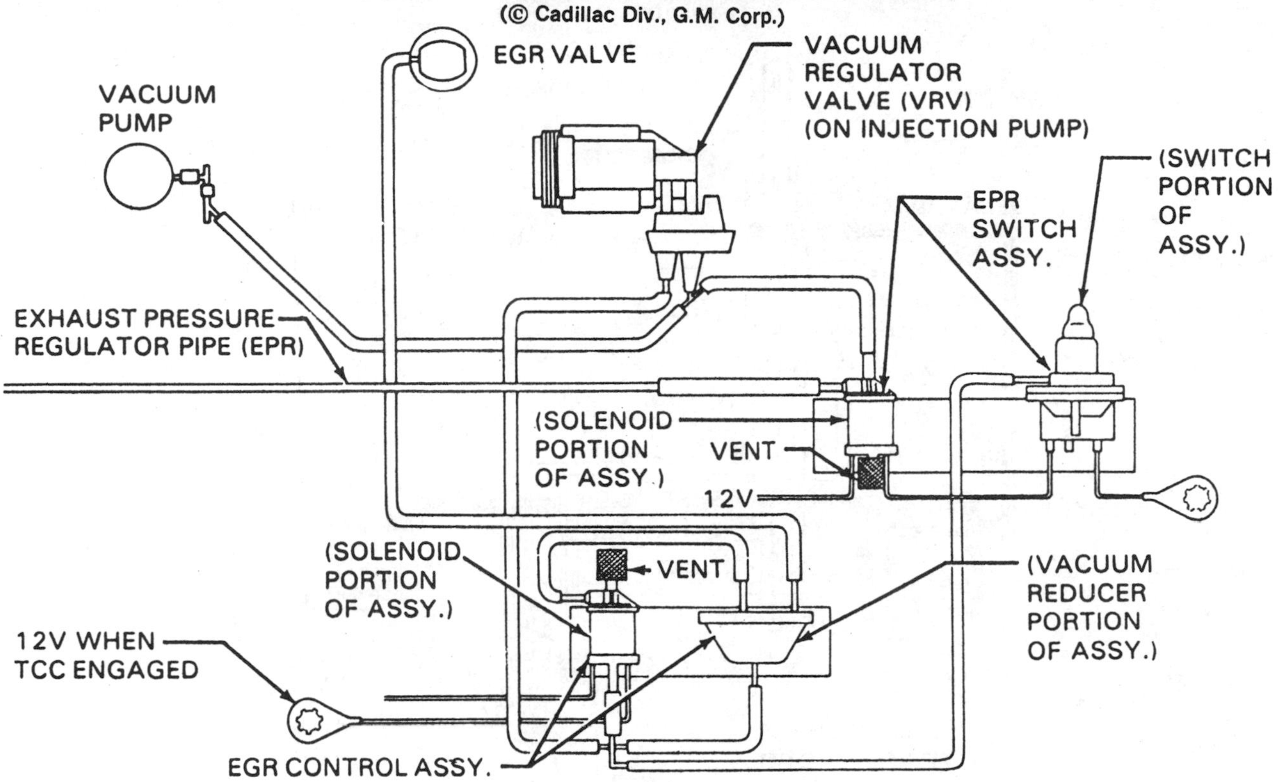

EGR VALVE
VACUUM REGULATOR VALVE (VRV) (ON INJECTION PUMP)
VACUUM PUMP
EPR SWITCH ASSY.
(SWITCH PORTION OF ASSY.)
EXHAUST PRESSURE REGULATOR PIPE (EPR)
(SOLENOID PORTION OF ASSY.)
VENT
12V
(SOLENOID PORTION OF ASSY.)
VENT
(VACUUM REDUCER PORTION OF ASSY.)
12V WHEN TCC ENGAGED
EGR CONTROL ASSY.

1981-82 FLEETWOOD & DeVILLE-DIESEL-EGR/EPR-EXC. CALIFORNIA

EGR VALVE
VACUUM PUMP
VACUUM REGULATOR VALVE (VRV) (ON INJECTION PUMP)
EPR SWITCH ASSY.
(SWITCH PORTION OF ASSY.)
EXHAUST PRESSURE REGULATOR PIPE (EPR)
(SOLENOID PORTION OF ASSY.)
12V
VENT

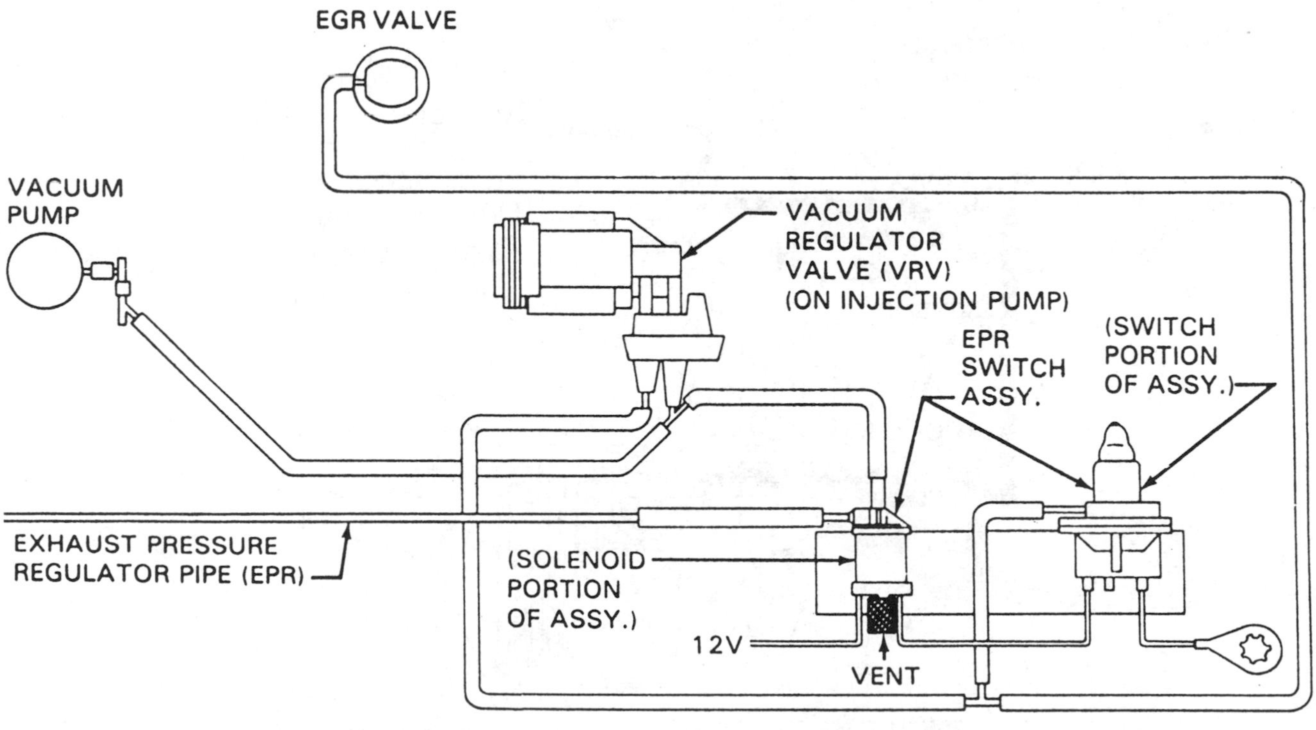

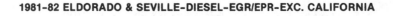

1981-82 ELDORADO & SEVILLE-DIESEL-EGR/EPR-EXC. CALIFORNIA

VACUUM CIRCUITS

(© Cadillac Div., G.M. Corp.)

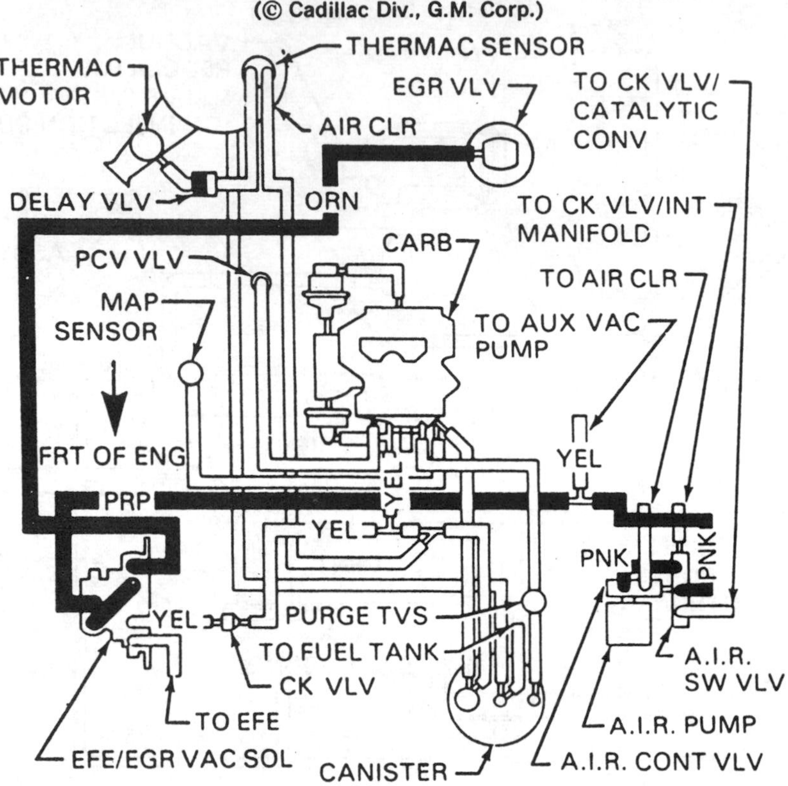

1981-82 FLEETWOOD & DeVILLE V6 ENG.-EXC. CALIFORNIA

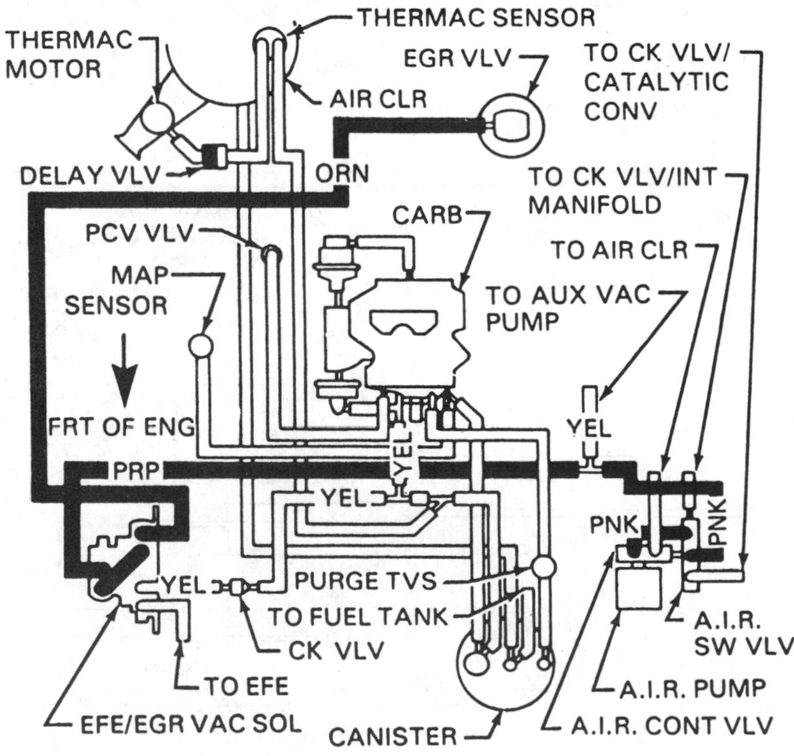

1981-82 FLEETWOOD & DeVILLE-V6 ENG.-CALIFORNIA

VACUUM CIRCUITS

(© Cadillac Div., G.M. Corp.)

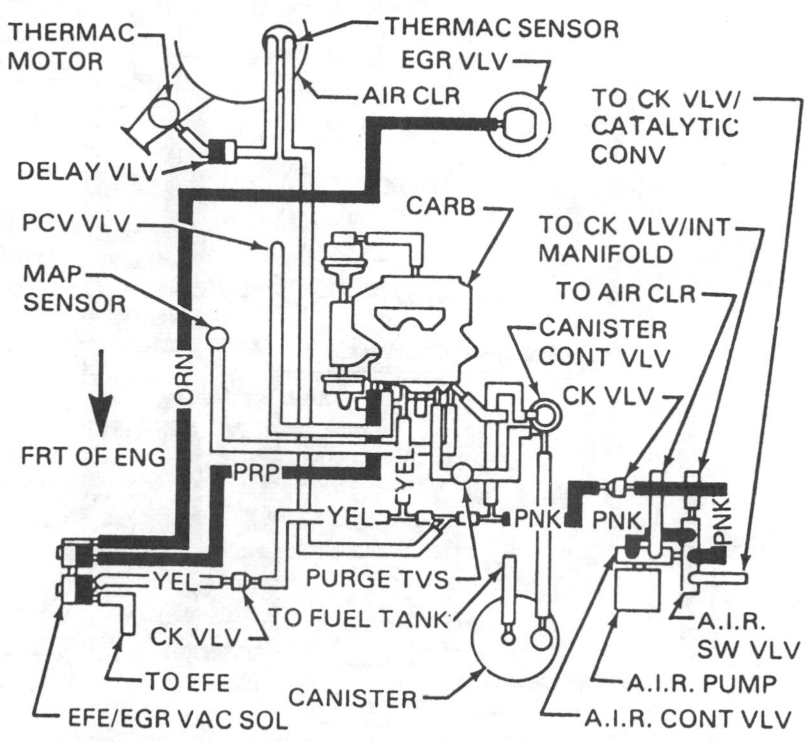

1981-82 ELDORADO & SEVILLE-V6 ENG.-EXC. CALIFORNIA

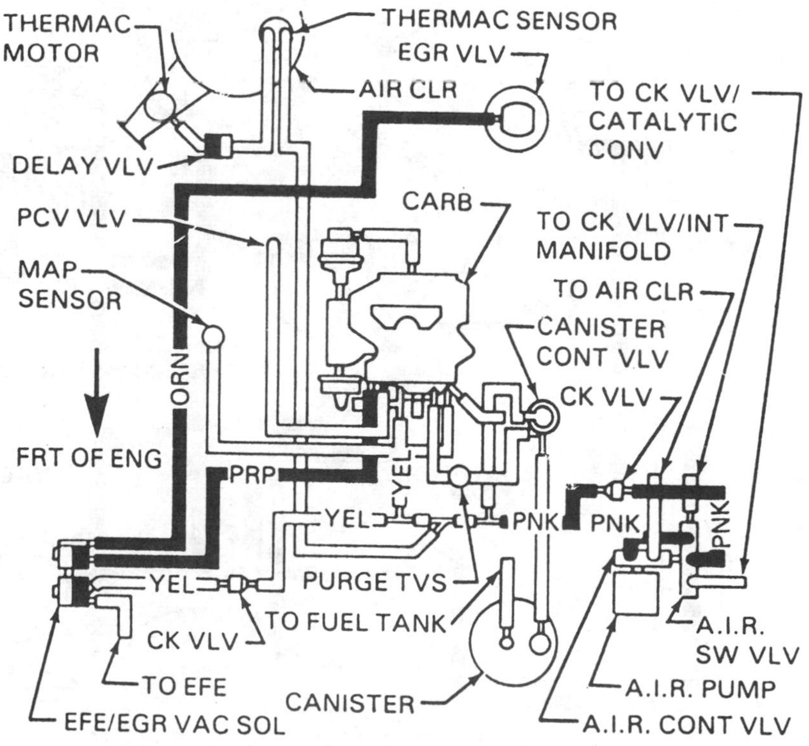

1981-82 ELDORADO & SEVILLE-V6 ENG.-CALIFORNIA

VACUUM CIRCUITS

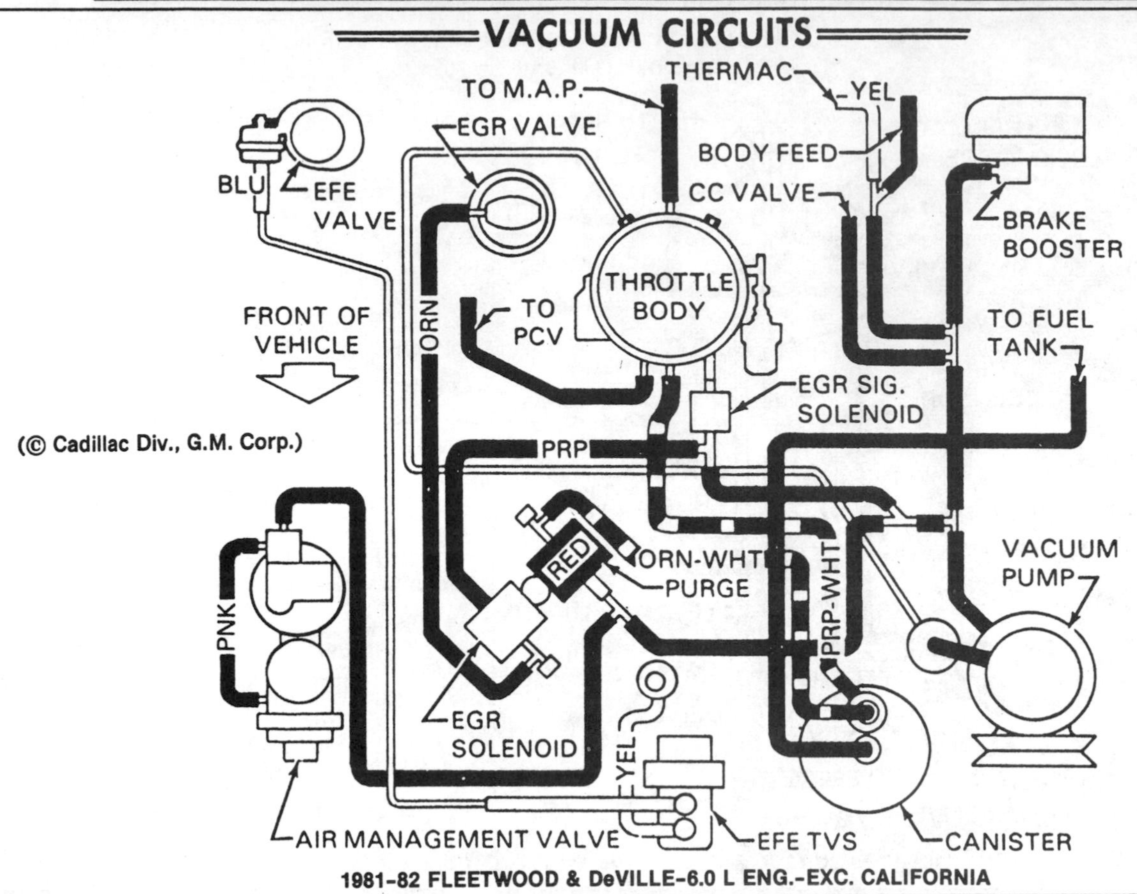

(© Cadillac Div., G.M. Corp.)

1981-82 FLEETWOOD & DeVILLE-6.0 L ENG.-EXC. CALIFORNIA

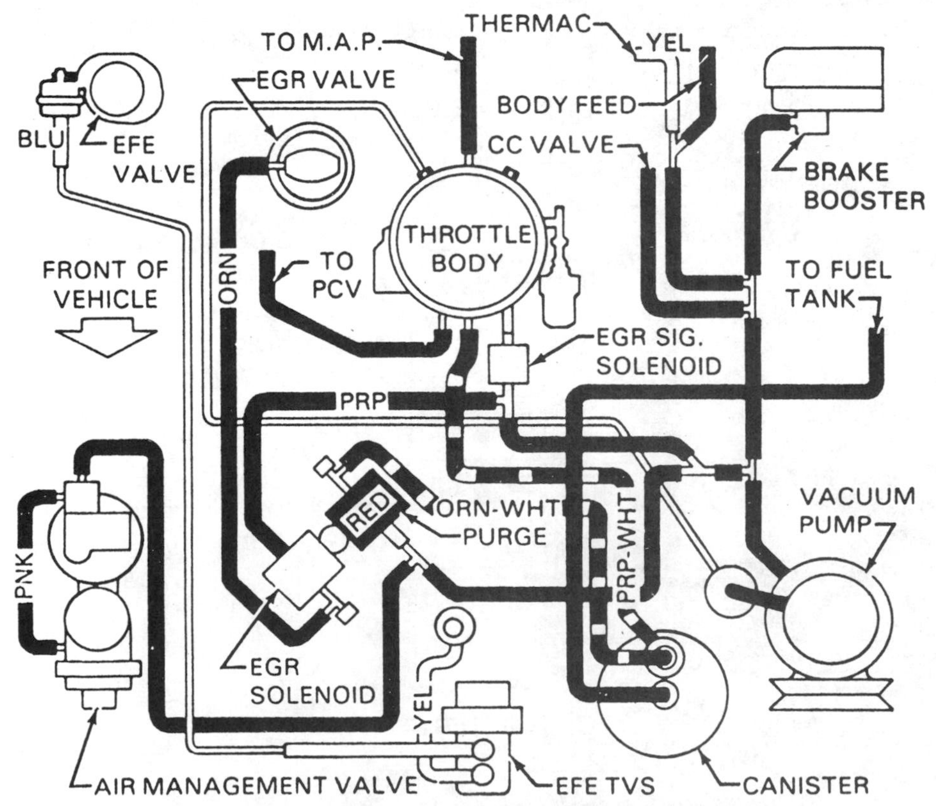

1981-82 FLEETWOOD & DeVILLE-6.0 L ENG.-CALIFORNIA

VACUUM CIRCUITS

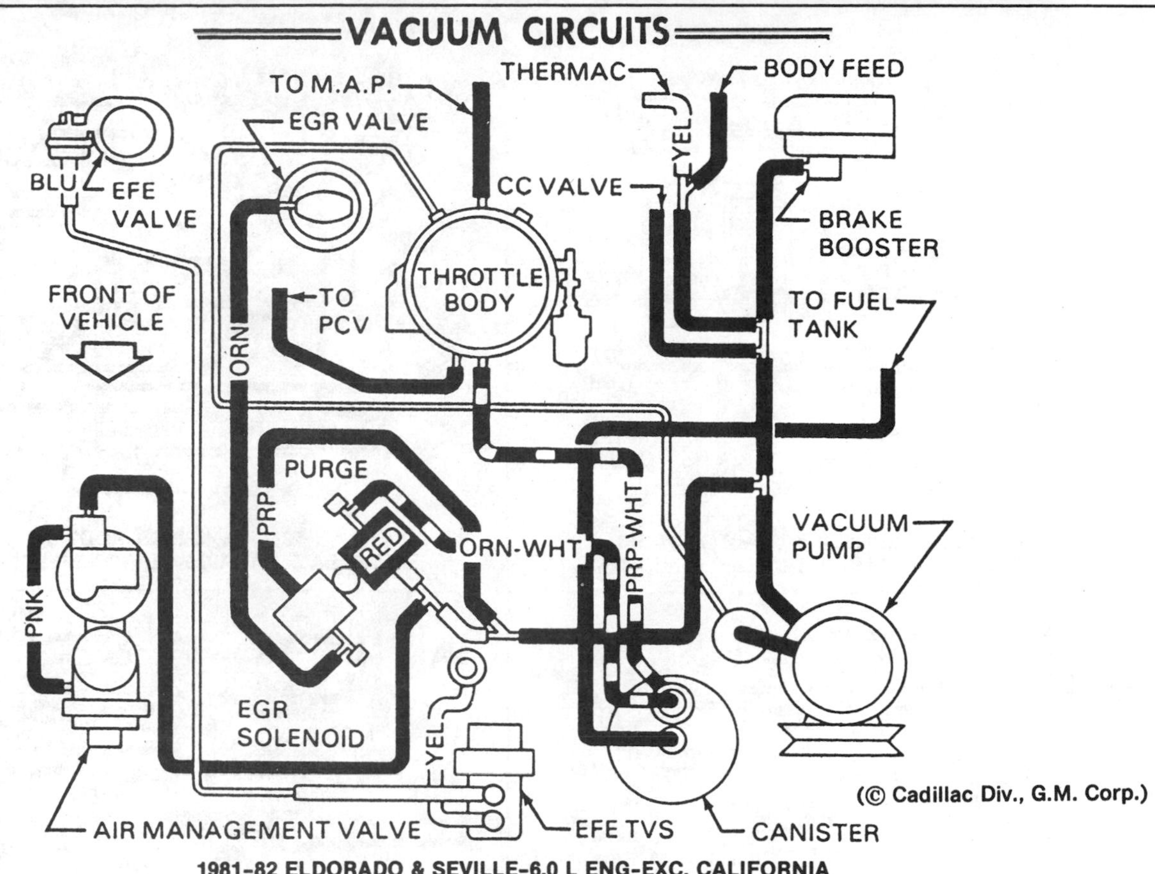

(© Cadillac Div., G.M. Corp.)

1981-82 ELDORADO & SEVILLE-6.0 L ENG-EXC. CALIFORNIA

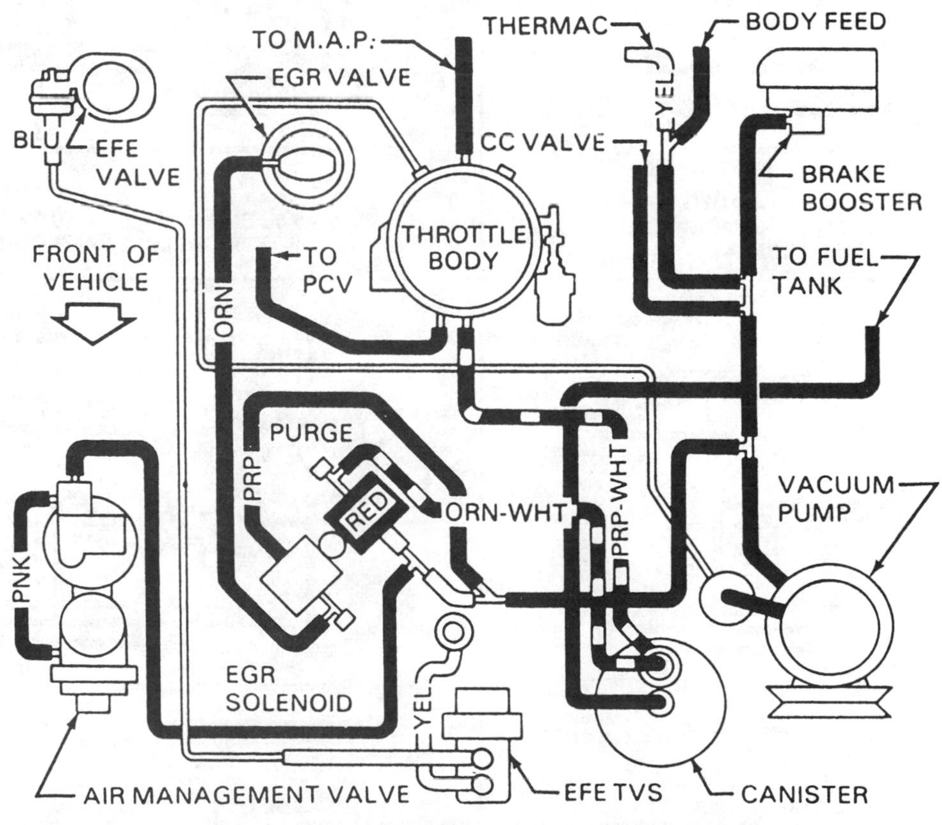

1981-82 ELDORADO & SEVILLE-6.0 L ENG-CALIFORNIA

VACUUM CIRCUITS

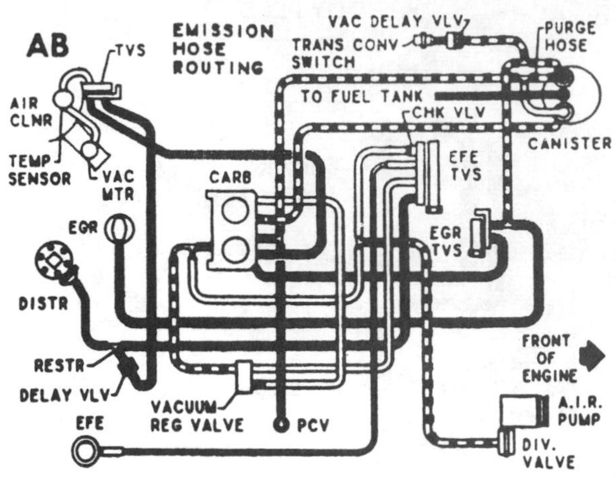

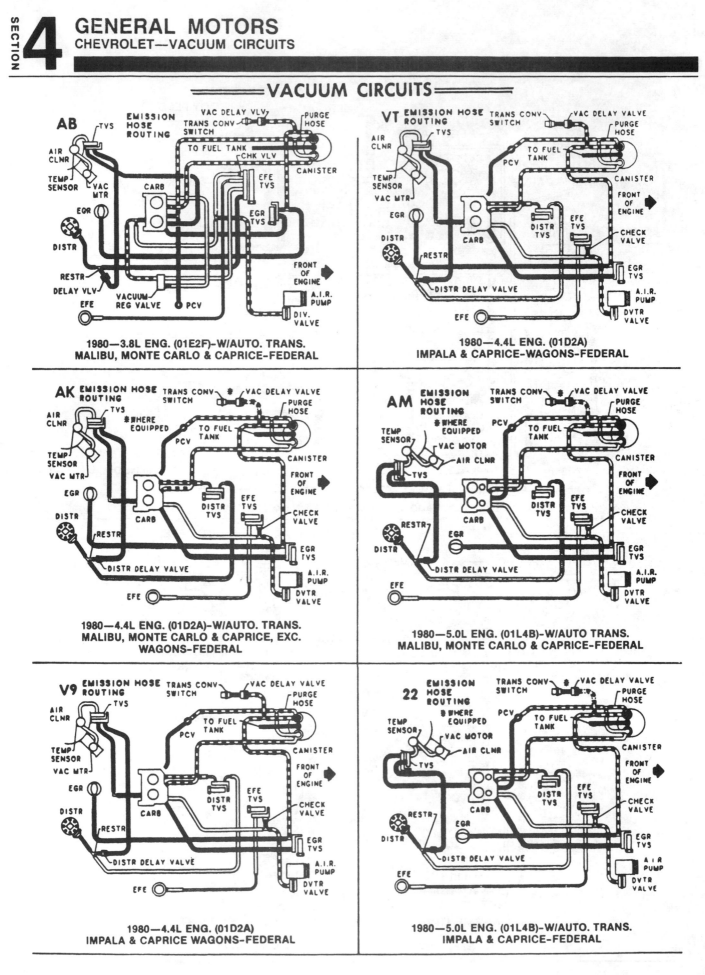

1980—3.8L ENG. (01E2F)-W/AUTO. TRANS.
MALIBU, MONTE CARLO & CAPRICE-FEDERAL

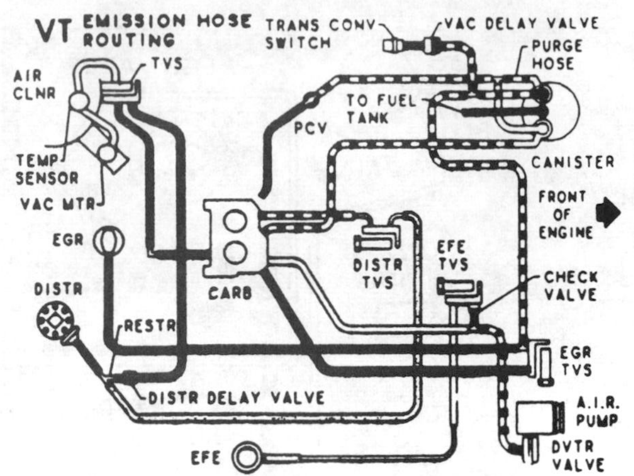

1980—4.4L ENG. (01D2A)
IMPALA & CAPRICE-WAGONS-FEDERAL

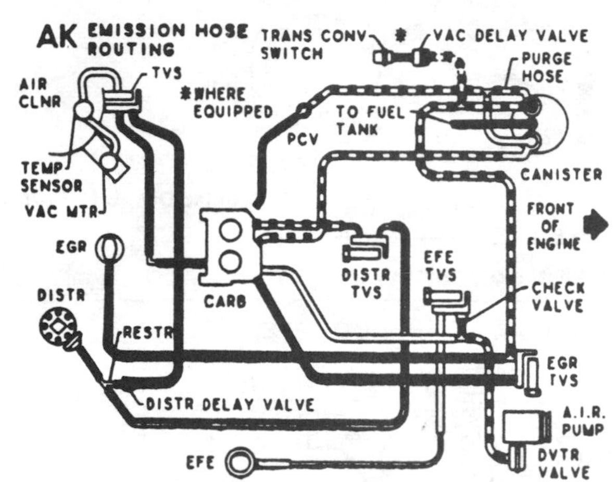

1980—4.4L ENG. (01D2A)-W/AUTO. TRANS.
MALIBU, MONTE CARLO & CAPRICE, EXC.
WAGONS-FEDERAL

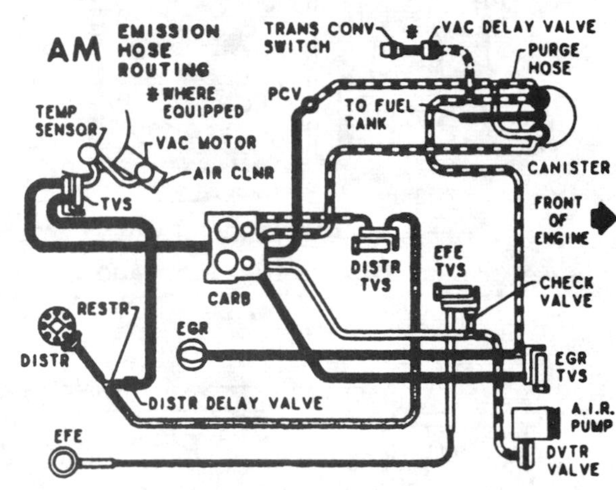

1980—5.0L ENG. (01L4B)-W/AUTO TRANS.
MALIBU, MONTE CARLO & CAPRICE-FEDERAL

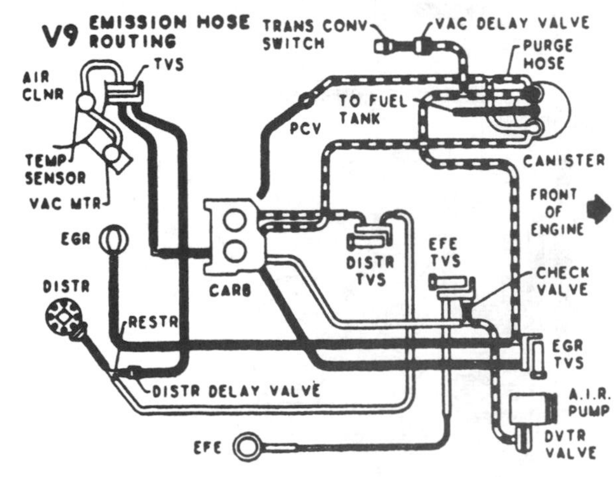

1980—4.4L ENG. (01D2A)
IMPALA & CAPRICE WAGONS-FEDERAL

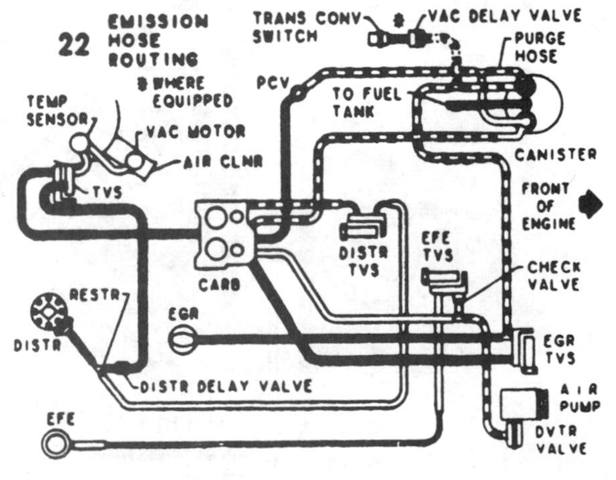

1980—5.0L ENG. (01L4B)-W/AUTO. TRANS.
IMPALA & CAPRICE-FEDERAL

VACUUM CIRCUITS

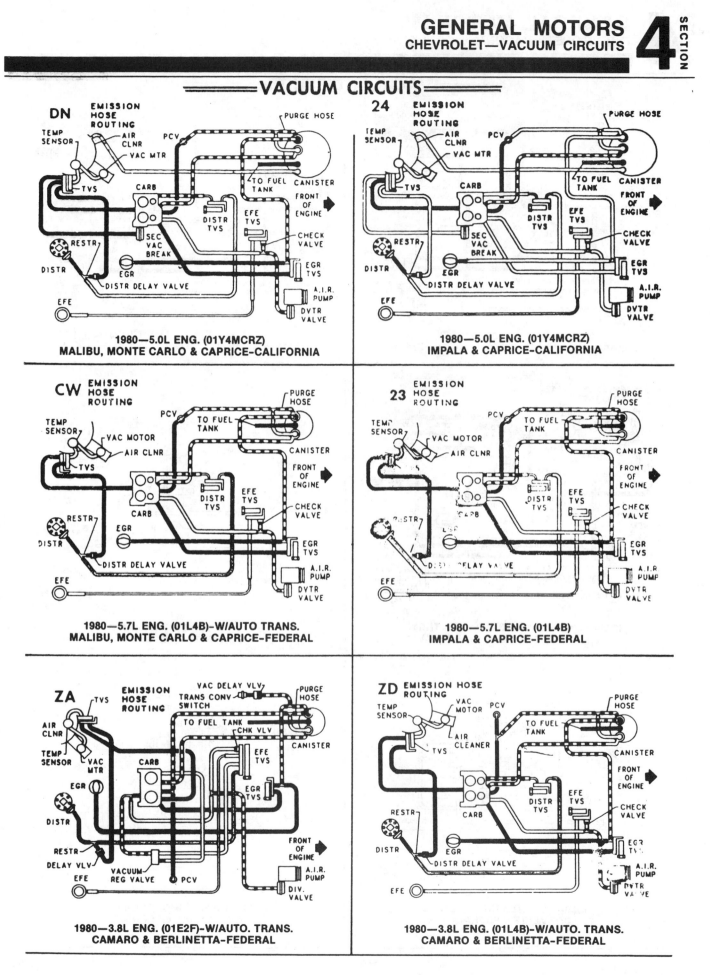

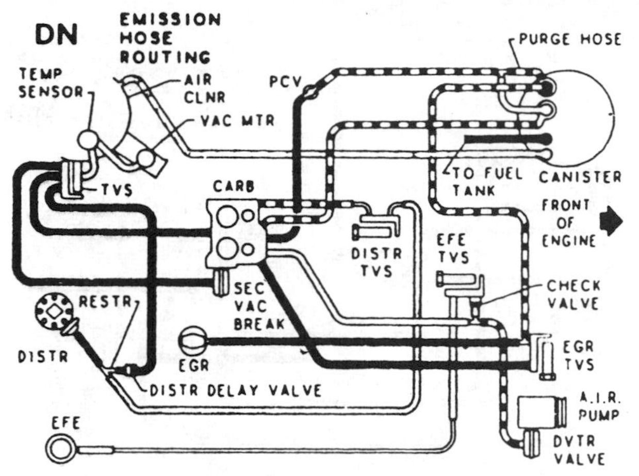

1980—5.0L ENG. (01Y4MCRZ)
MALIBU, MONTE CARLO & CAPRICE-CALIFORNIA

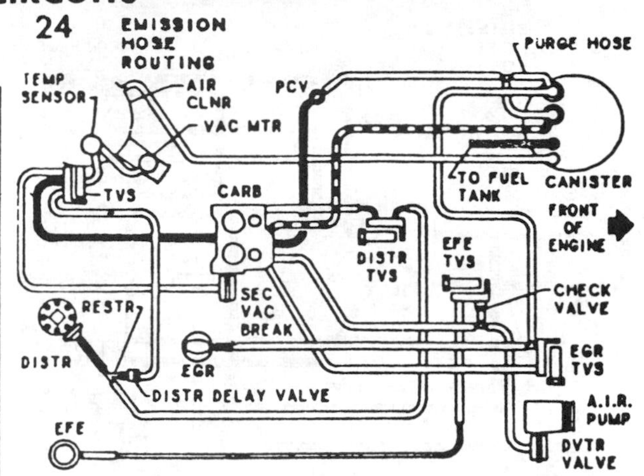

1980—5.0L ENG. (01Y4MCRZ)
IMPALA & CAPRICE-CALIFORNIA

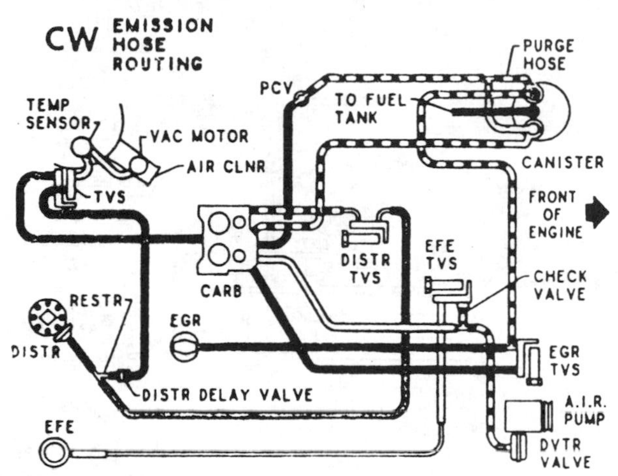

1980—5.7L ENG. (01L4B)-W/AUTO TRANS.
MALIBU, MONTE CARLO & CAPRICE-FEDERAL

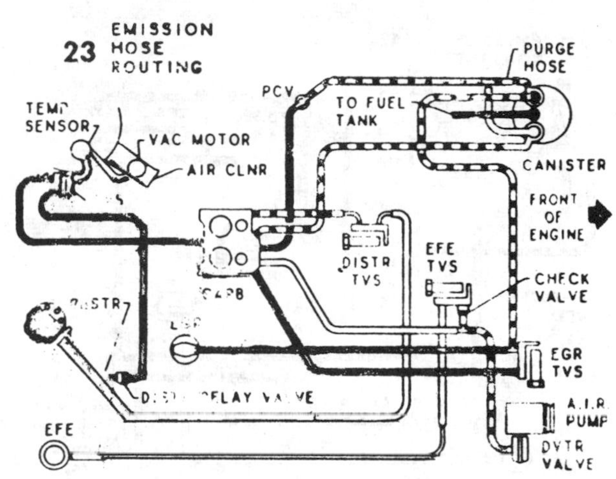

1980—5.7L ENG. (01L4B)
IMPALA & CAPRICE-FEDERAL

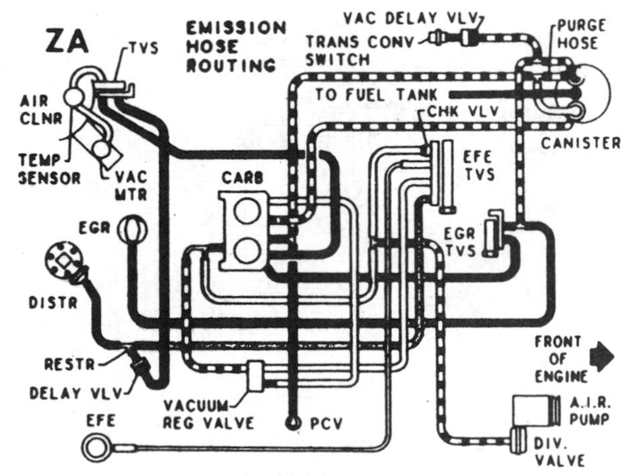

1980—3.8L ENG. (01E2F)-W/AUTO. TRANS.
CAMARO & BERLINETTA-FEDERAL

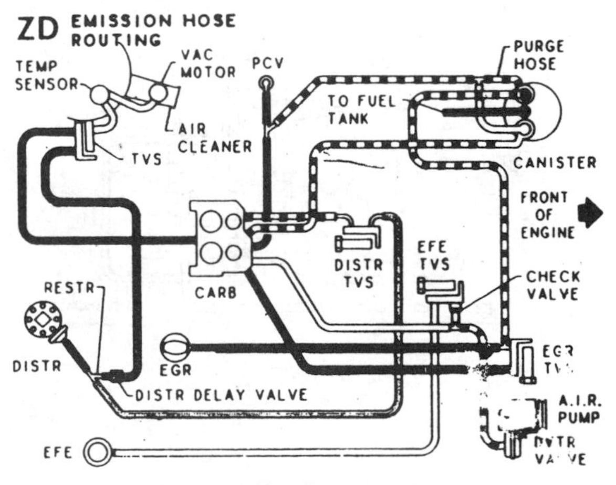

1980—3.8L ENG. (01L4B)-W/AUTO. TRANS.
CAMARO & BERLINETTA-FEDERAL

VACUUM CIRCUITS

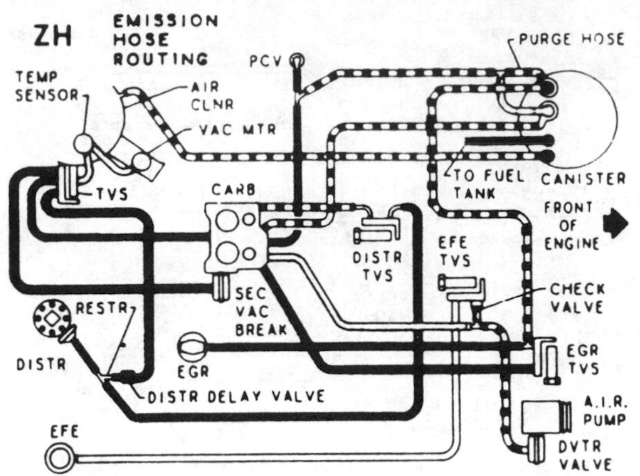

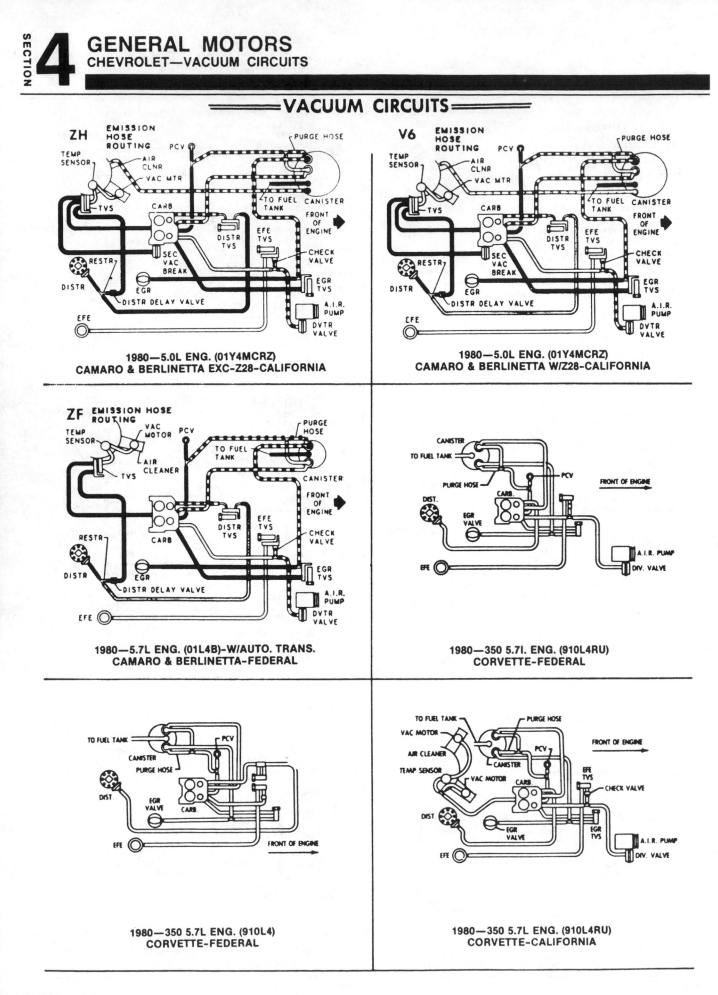

1980—5.0L ENG. (01Y4MCRZ)
CAMARO & BERLINETTA EXC-Z28-CALIFORNIA

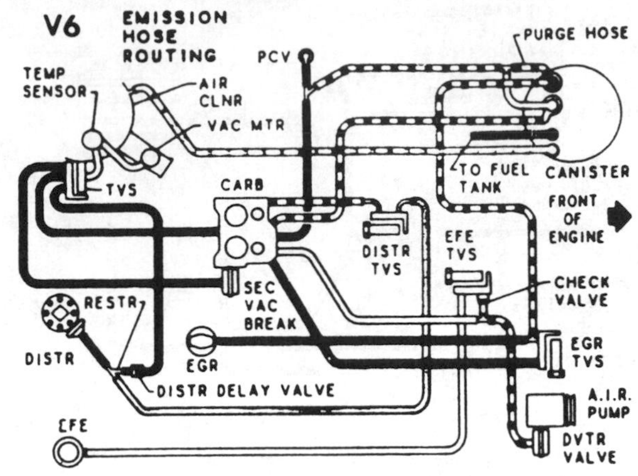

1980—5.0L ENG. (01Y4MCRZ)
CAMARO & BERLINETTA W/Z28-CALIFORNIA

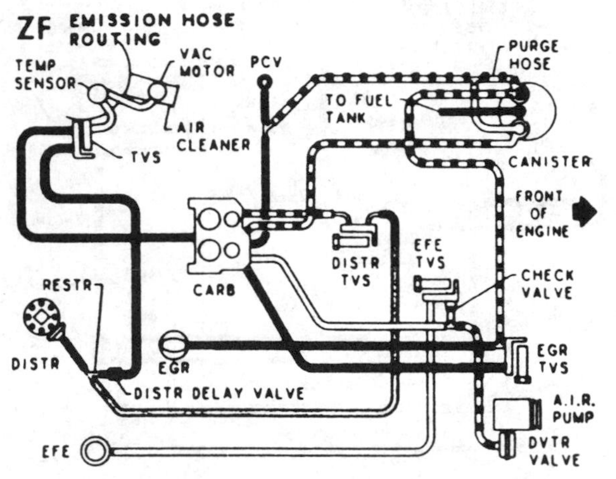

1980—5.7L ENG. (01L4B)-W/AUTO. TRANS.
CAMARO & BERLINETTA-FEDERAL

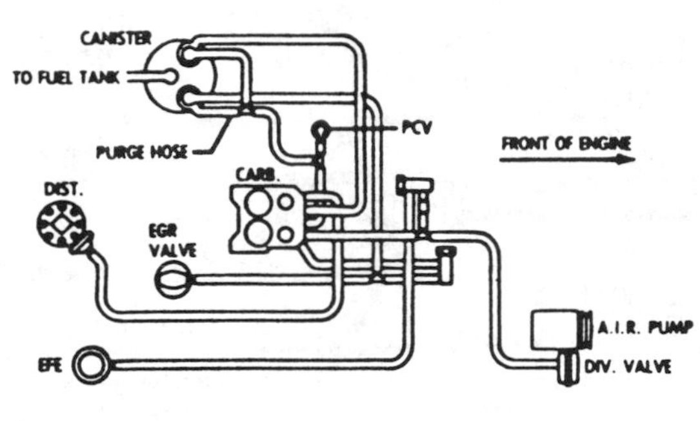

1980—350 5.7l. ENG. (910L4RU)
CORVETTE-FEDERAL

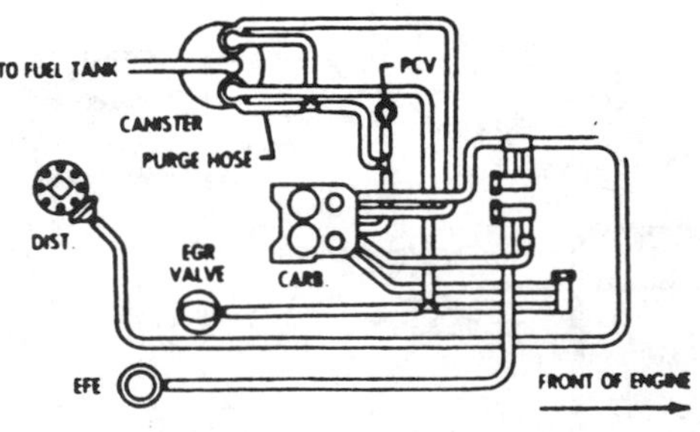

1980—350 5.7L ENG. (910L4)
CORVETTE-FEDERAL

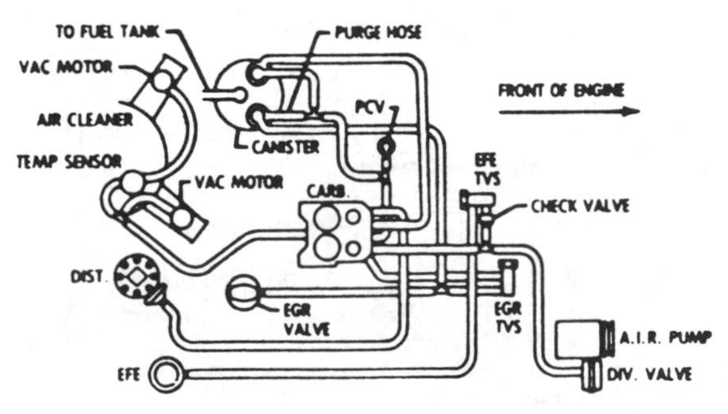

1980—350 5.7L ENG. (910L4RU)
CORVETTE-CALIFORNIA

VACUUM CIRCUITS

(© Chevrolet Div., G.M. Corp.)

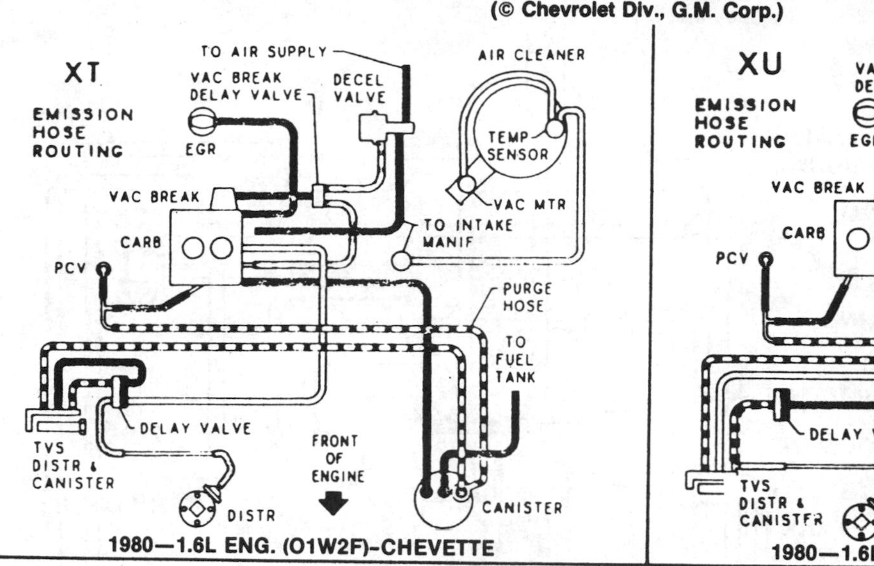

1980—1.6L ENG. (O1W2F)-CHEVETTE

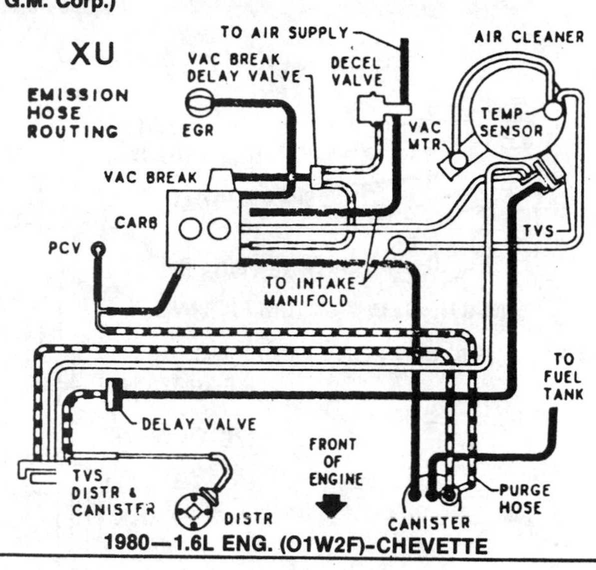

1980—1.6L ENG. (O1W2F)-CHEVETTE

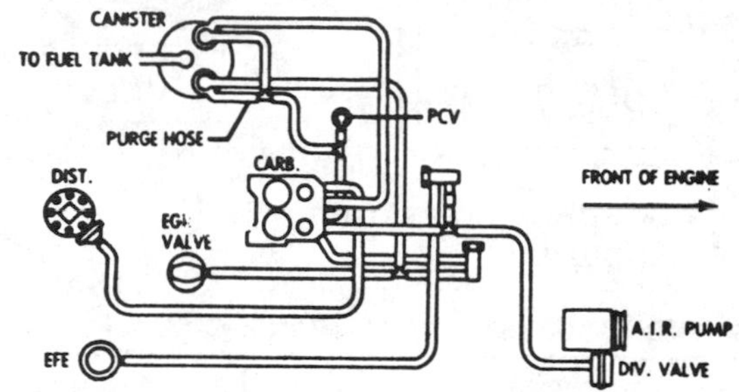

1980—350 5.7L ENG. (910L4RU)-CORVETTE-HIGH ALTITUDE

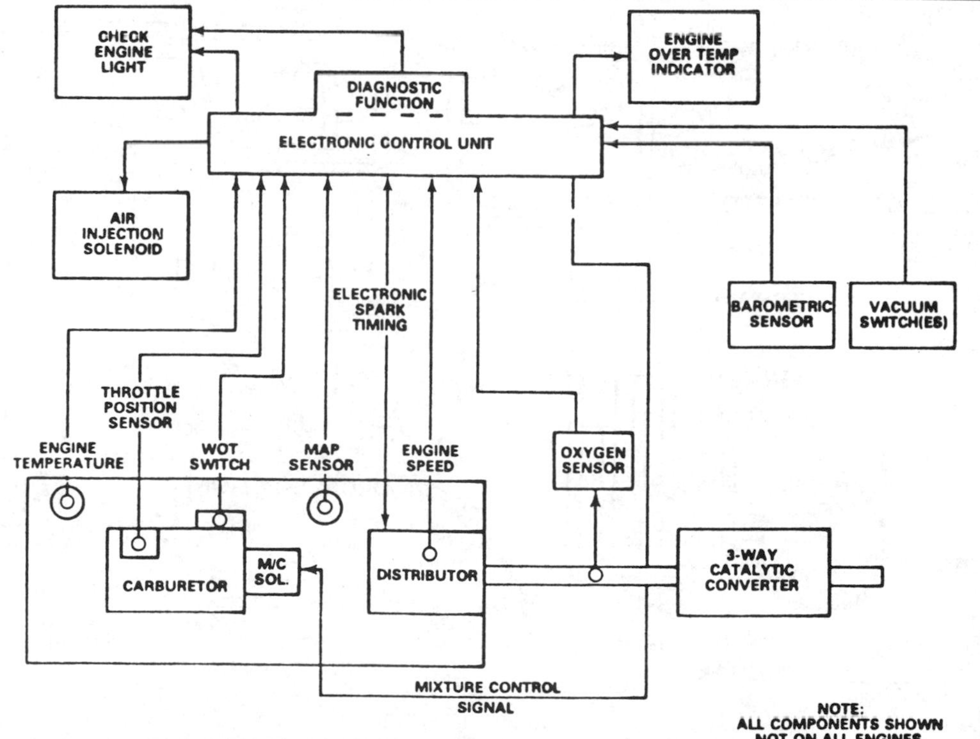

1980 MONZA-COMPUTER CONTROLLED CATALYTIC CONVERTER SYSTEM

VACUUM CIRCUITS

(© Chevrolet Div., G.M. Corp.)

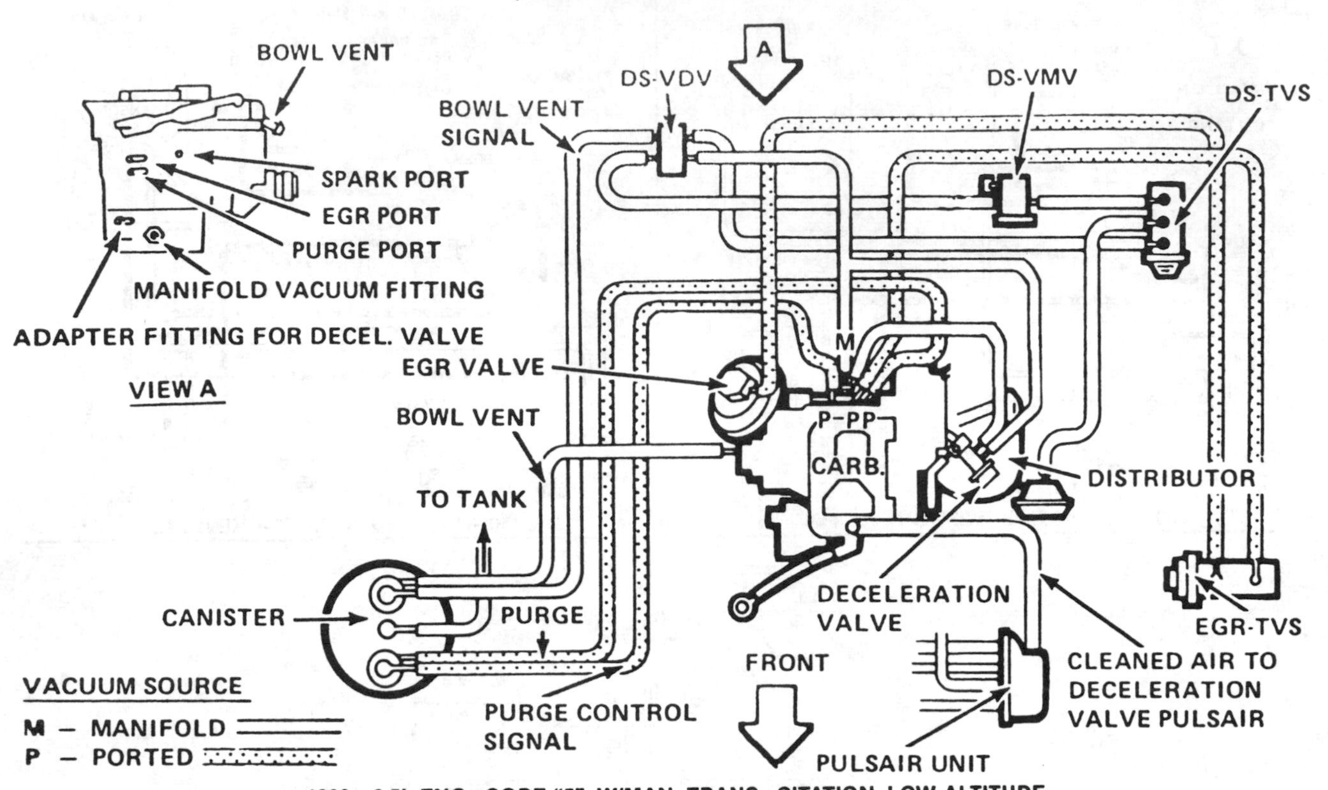

1980—2.5L ENG.-CODE "5"-W/MAN. TRANS.-CITATION-LOW ALTITUDE

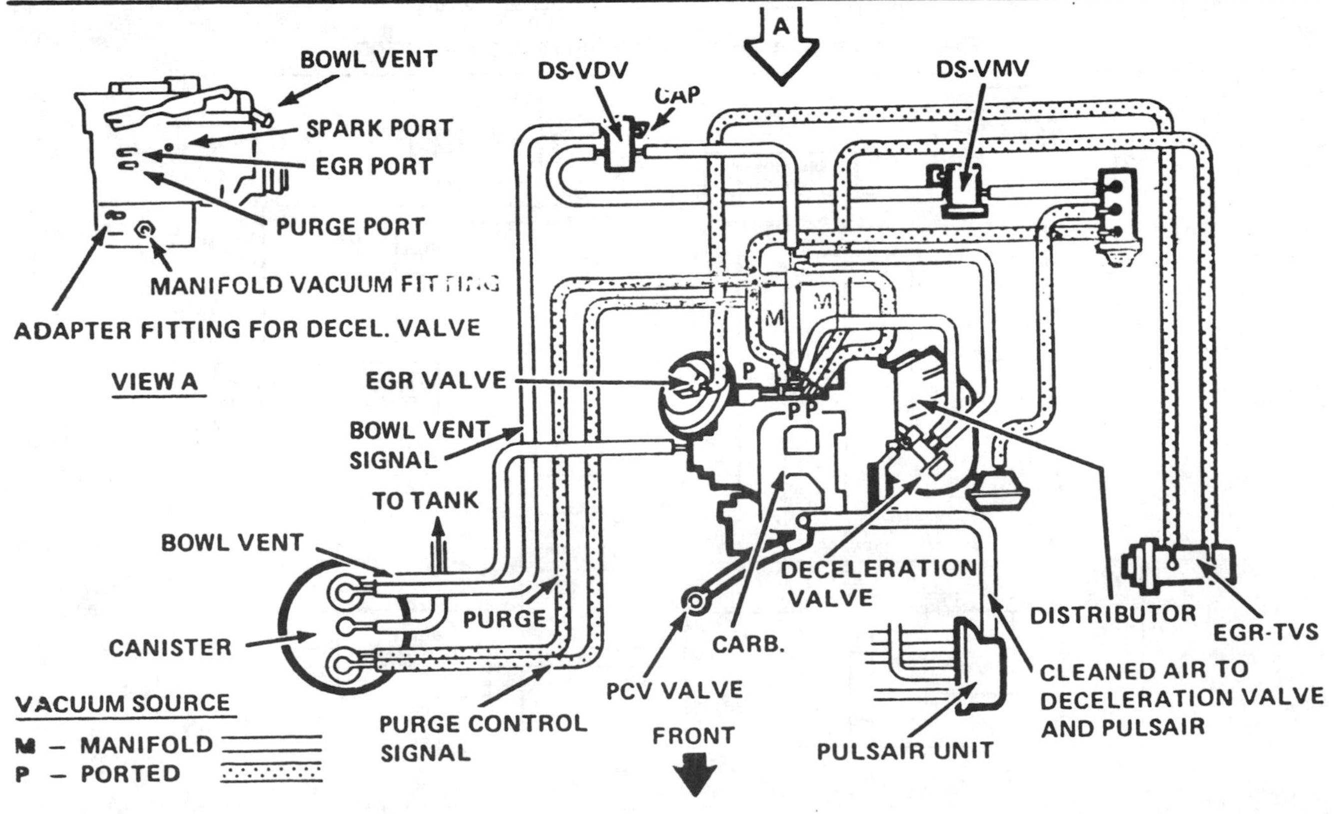

1980—2.5L ENG.-CODE "5"-W/AUTO. TRANS.-CITATION-LOW ALTITUDE

VACUUM CIRCUITS

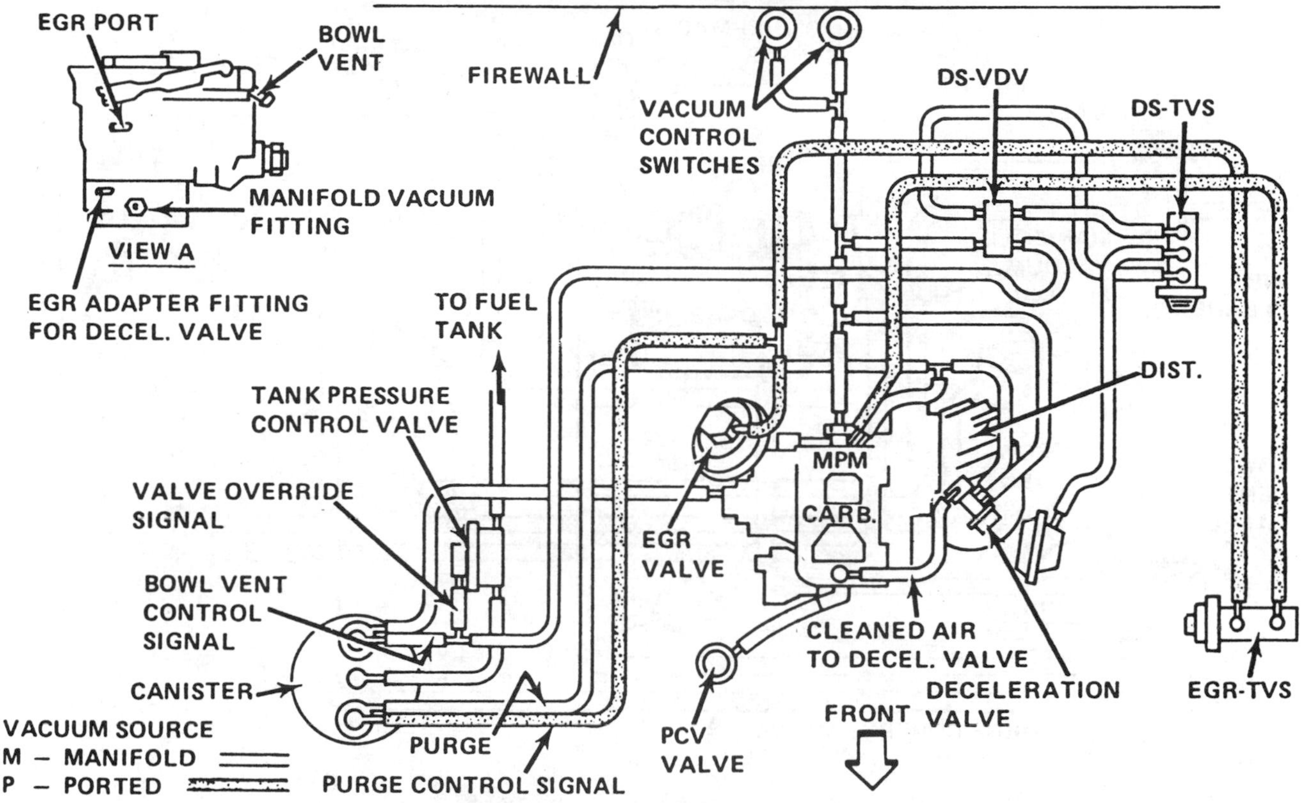

1980—2.5L ENG.-CODE "5"-W/MAN. TRANS.-CITATION-CALIFORNIA

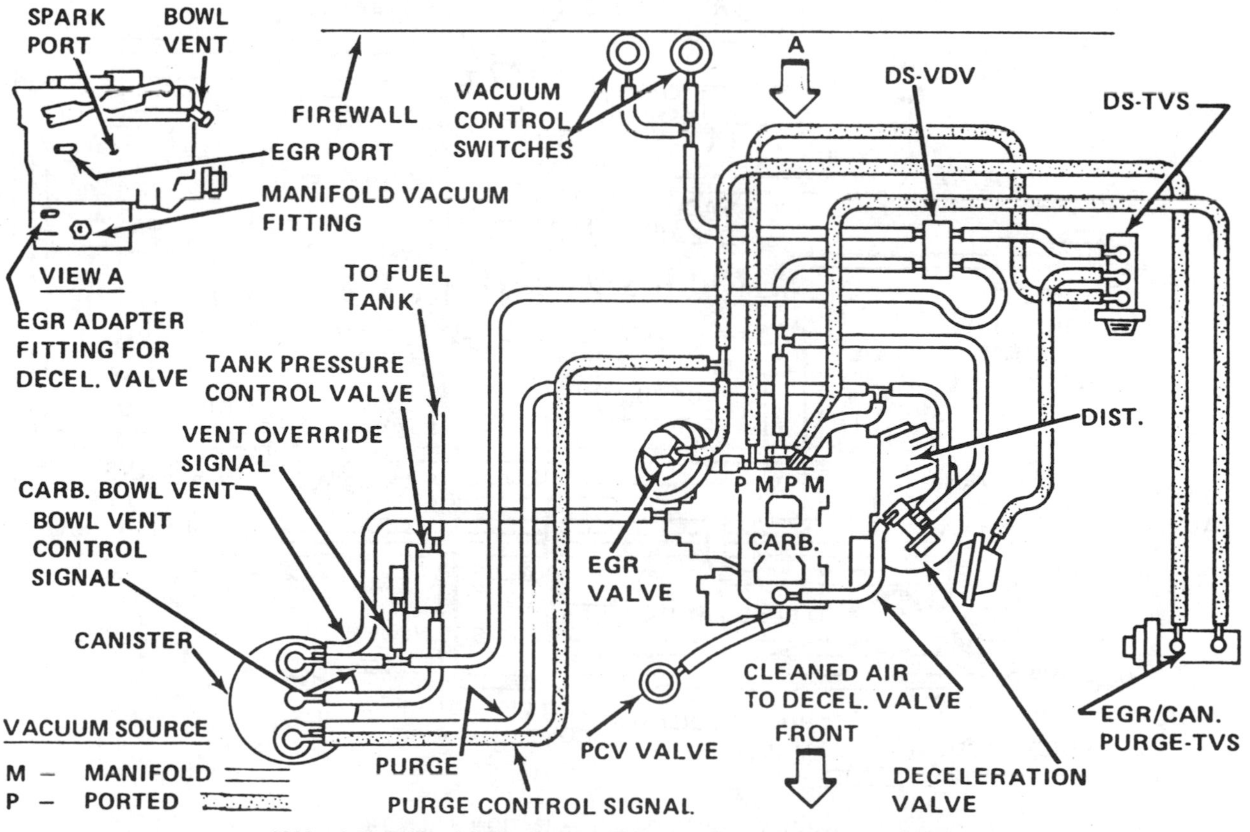

1980—2.5L ENG.-CODE "5"-W/AUTO. TRANS.-CITATION-CALIFORNIA

VACUUM CIRCUITS

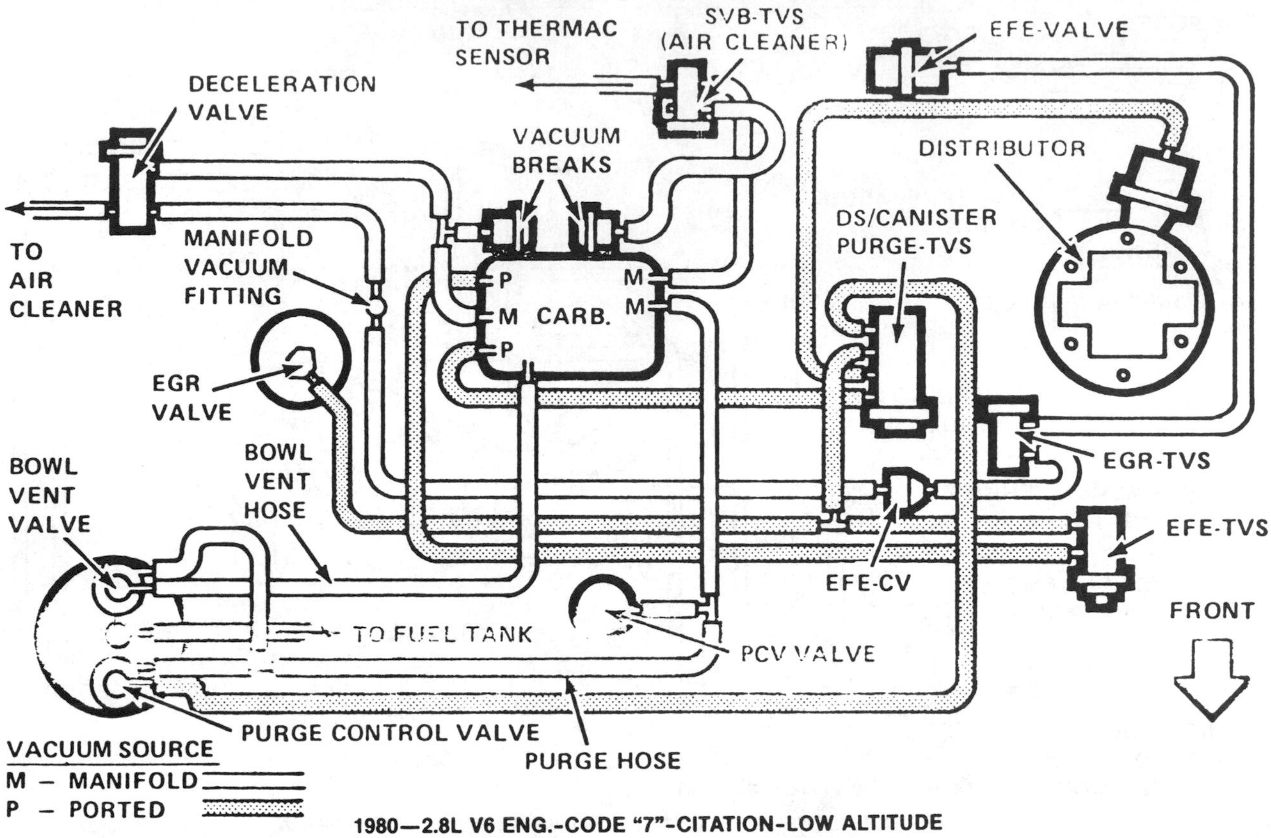

TO THERMAC SENSOR
SVB-TVS (AIR CLEANER)
EFE-VALVE
DECELERATION VALVE
VACUUM BREAKS
DISTRIBUTOR
TO AIR CLEANER
MANIFOLD VACUUM FITTING
DS/CANISTER PURGE-TVS
P
M CARB.
P
M
M
EGR VALVE
BOWL VENT VALVE
BOWL VENT HOSE
EGR-TVS
EFE-TVS
EFE-CV
TO FUEL TANK
PCV VALVE
FRONT
PURGE CONTROL VALVE
PURGE HOSE

VACUUM SOURCE
M – MANIFOLD
P – PORTED

1980—2.8L V6 ENG.-CODE "7"-CITATION-LOW ALTITUDE

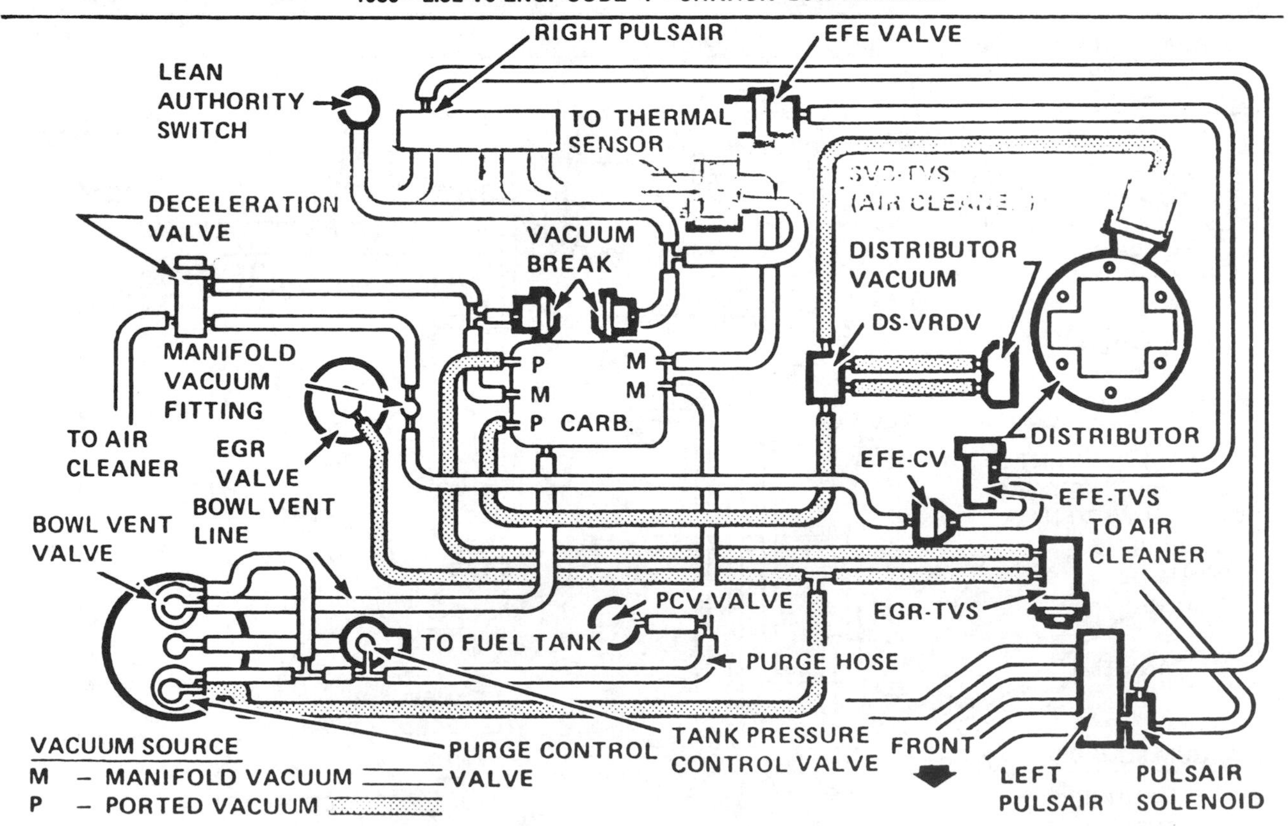

1980—2.8L V6 ENG.-CODE "7"-CITATION-CALIFORNIA

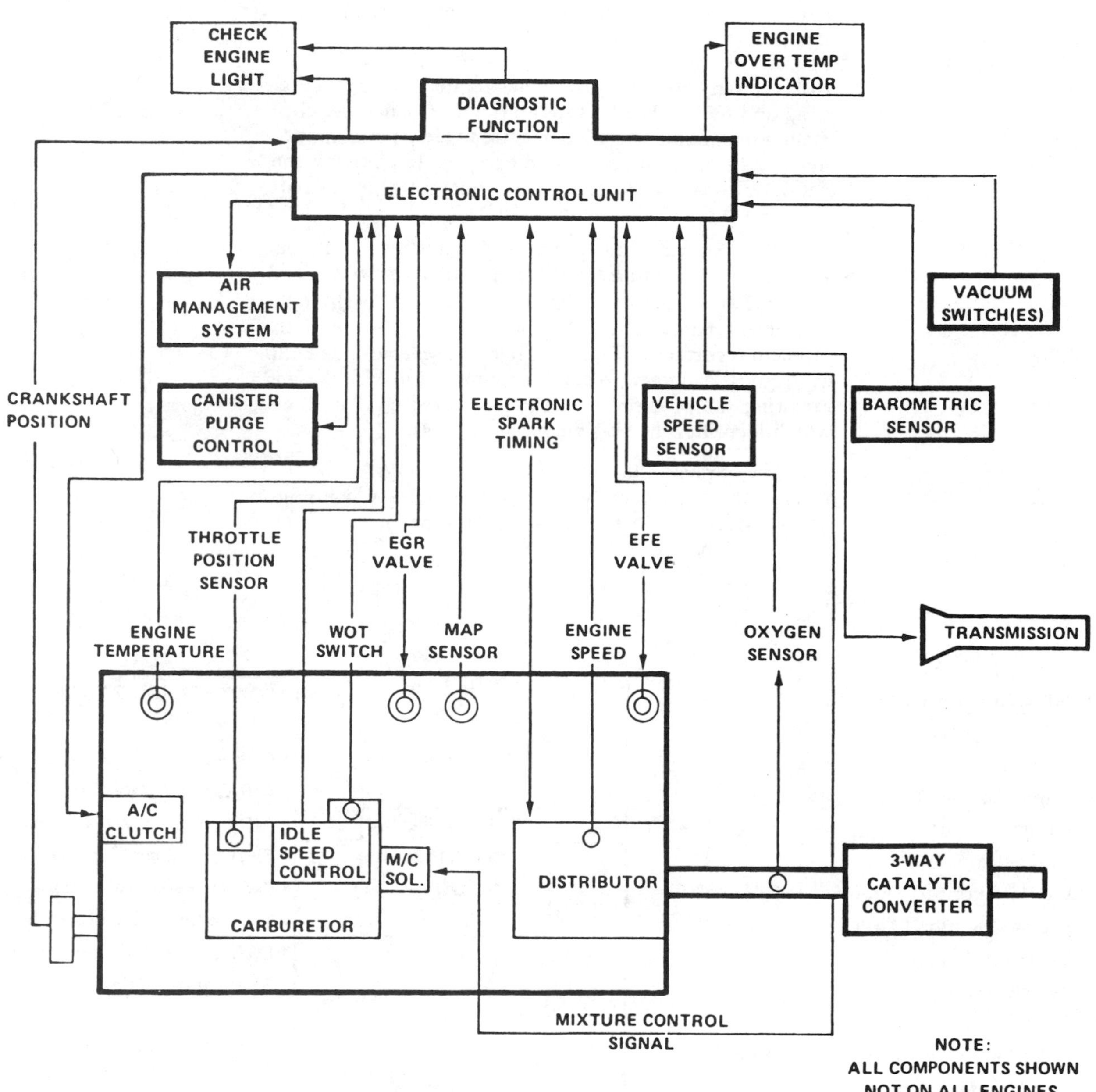

1981-82—COMPUTER COMMAND CONTROL SYSTEM

COMPUTER COMMAND CONTROL

The Computer Command Control System (the "System"), is an electronically controlled emission system that can monitor up to fifteen (15) various engine/vehicle operating conditions and using this information can control as many as nine (9) engine related systems (Fig. 6E-2 and 6E-3). The "System" is thereby making constant adjustments to maintain good vehicle performance under all normal driving conditions while allowing the catalytic converter to effectively control Oxides of Nitrogen (NOX), Hydrocarbons (DC)

The "System" has a built-in diagnostic system that recognizes and identifies possible operational problems and alerts the driver through a "CHECK ENGINE" light in the instrument panel. The light will remain "ON" until the problem is corrected. This same light is used, with the built-in diagnostic system, to aid the technician in locating and repairing the problem by flashing a stored code or codes which identifies the possible problem areas.

The "System" also has built-in back-up systems that in most cases of an operation problem will allow continued operation of the vehicle in a near normal manner until repairs can be made.

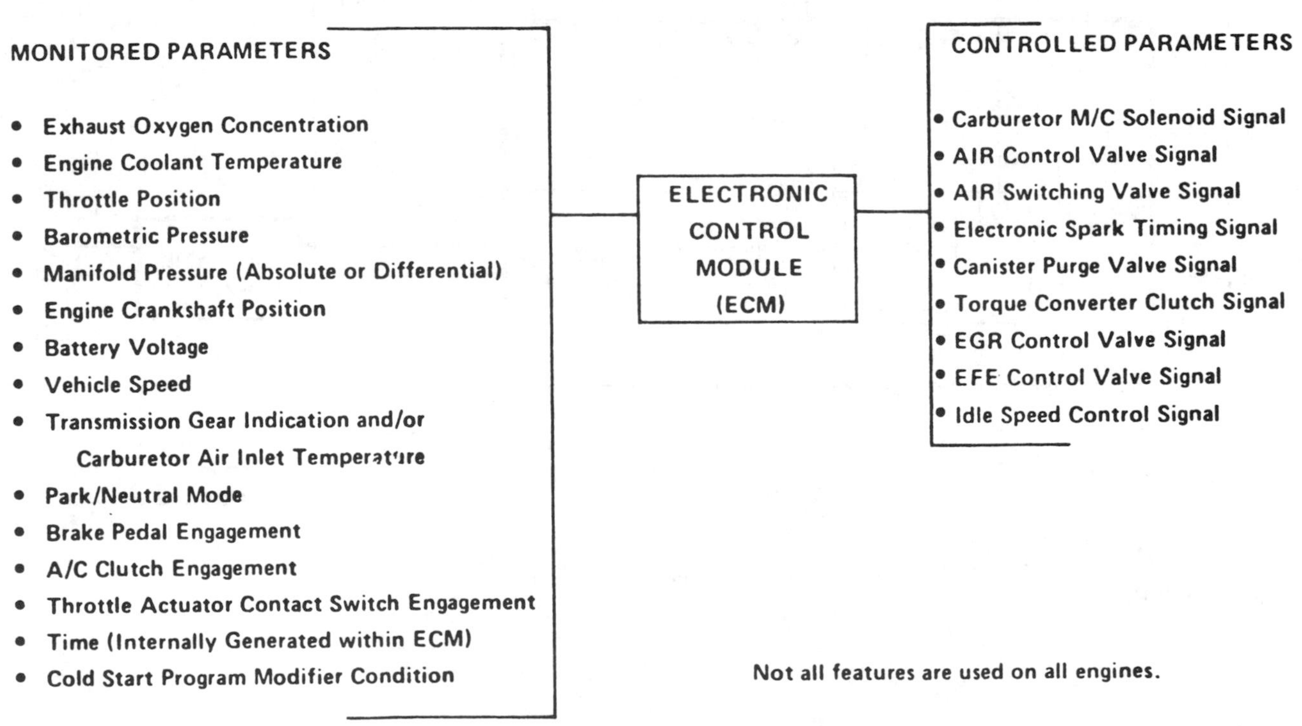

System Parameters

1981-82—COMPUTER COMMAND CONTROL SYSTEM

VEHICLE EMISSION CONTROL INFORMATION LABEL

The Vehicle Emission Control Information Label is located in the engine compartment. The label contains important emission specifications and setting procedures, as well as a vacuum hose schematic with emission components identified.

When servicing the engine or emission systems, the Vehicle Emission Control Information Label should be checked for up-to-date information.

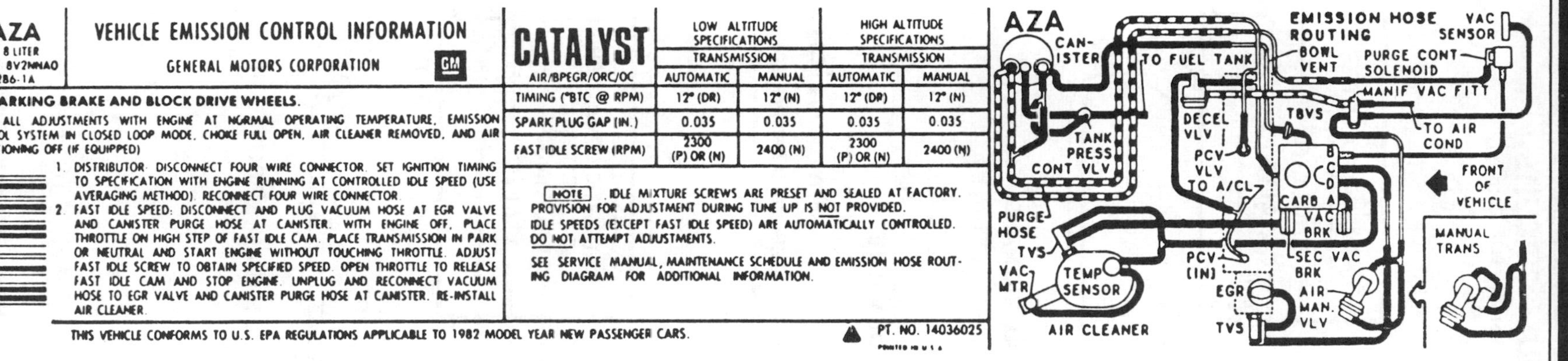

AZA 1.8 LITER ★C1G1 8V2NNAO 2B6-1A

VEHICLE EMISSION CONTROL INFORMATION
GENERAL MOTORS CORPORATION GM

SET PARKING BRAKE AND BLOCK DRIVE WHEELS.

MAKE ALL ADJUSTMENTS WITH ENGINE AT NORMAL OPERATING TEMPERATURE, EMISSION CONTROL SYSTEM IN CLOSED LOOP MODE, CHOKE FULL OPEN, AIR CLEANER REMOVED, AND AIR CONDITIONING OFF (IF EQUIPPED)

1. DISTRIBUTOR: DISCONNECT FOUR WIRE CONNECTOR. SET IGNITION TIMING TO SPECIFICATION WITH ENGINE RUNNING AT CONTROLLED IDLE SPEED (USE AVERAGING METHOD). RECONNECT FOUR WIRE CONNECTOR.
2. FAST IDLE SPEED: DISCONNECT AND PLUG VACUUM HOSE AT EGR VALVE AND CANISTER PURGE HOSE AT CANISTER. WITH ENGINE OFF, PLACE THROTTLE ON HIGH STEP OF FAST IDLE CAM. PLACE TRANSMISSION IN PARK OR NEUTRAL AND START ENGINE WITHOUT TOUCHING THROTTLE. ADJUST FAST IDLE SCREW TO OBTAIN SPECIFIED SPEED. OPEN THROTTLE TO RELEASE FAST IDLE CAM AND STOP ENGINE. UNPLUG AND RECONNECT VACUUM HOSE TO EGR VALVE AND CANISTER PURGE HOSE AT CANISTER. RE-INSTALL AIR CLEANER.

CATALYST AIR/BPEGR/ORC/OC

	LOW ALTITUDE SPECIFICATIONS TRANSMISSION		HIGH ALTITUDE SPECIFICATIONS TRANSMISSION	
	AUTOMATIC	MANUAL	AUTOMATIC	MANUAL
TIMING (°BTC @ RPM)	12° (DR)	12° (N)	12° (DP)	12° (N)
SPARK PLUG GAP (IN.)	0.035	0.035	0.035	0.035
FAST IDLE SCREW (RPM)	2300 (P) OR (N)	2400 (N)	2300 (P) OR (N)	2400 (N)

NOTE: IDLE MIXTURE SCREWS ARE PRESET AND SEALED AT FACTORY. PROVISION FOR ADJUSTMENT DURING TUNE UP IS NOT PROVIDED. IDLE SPEEDS (EXCEPT FAST IDLE SPEED) ARE AUTOMATICALLY CONTROLLED. DO NOT ATTEMPT ADJUSTMENTS.

SEE SERVICE MANUAL, MAINTENANCE SCHEDULE AND EMISSION HOSE ROUTING DIAGRAM FOR ADDITIONAL INFORMATION.

THIS VEHICLE CONFORMS TO U.S. EPA REGULATIONS APPLICABLE TO 1982 MODEL YEAR NEW PASSENGER CARS.

PT. NO. 14036025 PRINTED IN U.S.A.

AZB 1.8 LITER ★C1G1 8V2NNAO 2B6-1A

VEHICLE EMISSION CONTROL INFORMATION
GENERAL MOTORS CORPORATION GM

SET PARKING BRAKE AND BLOCK DRIVE WHEELS.

MAKE ALL ADJUSTMENTS WITH ENGINE AT NORMAL OPERATING TEMPERATURE, EMISSION CONTROL SYSTEM IN CLOSED LOOP MODE, CHOKE FULL OPEN, AIR CLEANER REMOVED, AND AIR CONDITIONING OFF (IF EQUIPPED)

1. DISTRIBUTOR: DISCONNECT FOUR WIRE CONNECTOR. SET IGNITION TIMING TO SPECIFICATION WITH ENGINE RUNNING AT CONTROLLED IDLE SPEED (USE AVERAGING METHOD). RECONNECT FOUR WIRE CONNECTOR.
2. FAST IDLE SPEED: DISCONNECT AND PLUG VACUUM HOSE AT EGR VALVE AND CANISTER PURGE HOSE AT CANISTER. WITH ENGINE OFF, PLACE THROTTLE ON HIGH STEP OF FAST IDLE CAM. PLACE TRANSMISSION IN PARK OR NEUTRAL AND START ENGINE WITHOUT TOUCHING THROTTLE. ADJUST FAST IDLE SCREW TO OBTAIN SPECIFIED SPEED. OPEN THROTTLE TO RELEASE FAST IDLE CAM AND STOP ENGINE. UNPLUG AND RECONNECT VACUUM HOSE TO EGR VALVE AND CANISTER PURGE HOSE AT CANISTER. RE-INSTALL AIR CLEANER.

CATALYST AIR/BPEGR/ORC/OC

	LOW ALTITUDE SPECIFICATIONS TRANSMISSION		HIGH ALTITUDE SPECIFICATIONS TRANSMISSION	
	AUTOMATIC	MANUAL	AUTOMATIC	MANUAL
TIMING (°BTC @ RPM)	12° (DR)	12° (N)	12° (DR)	12° (N)
SPARK PLUG GAP (IN.)	0.035	0.035	0.035	0.035
FAST IDLE SCREW (RPM)	2300 (P) OR (N)	2400 (N)	2300 (P) OR (N)	2400 (N)

NOTE: IDLE MIXTURE SCREWS ARE PRESET AND SEALED AT FACTORY. PROVISION FOR ADJUSTMENT DURING TUNE UP IS NOT PROVIDED. IDLE SPEEDS (EXCEPT FAST IDLE SPEED) ARE AUTOMATICALLY CONTROLLED. DO NOT ATTEMPT ADJUSTMENTS.

SEE SERVICE MANUAL, MAINTENANCE SCHEDULE AND EMISSION HOSE ROUTING DIAGRAM FOR ADDITIONAL INFORMATION.

THIS VEHICLE CONFORMS TO U.S. EPA AND CALIFORNIA REGULATIONS APPLICABLE TO 1982 MODEL YEAR NEW PASSENGER CARS.

PT NO 1403€026 PRINTED IN U.S.A.

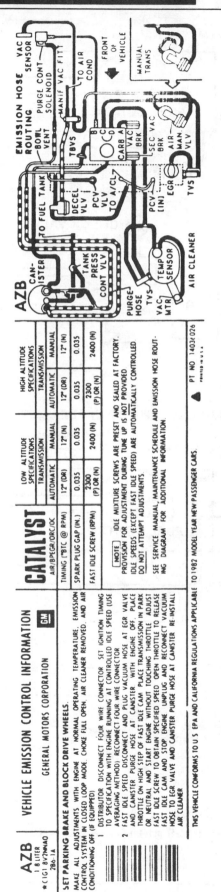

Vehicle Emission Control Information Label

VACUUM CIRCUITS

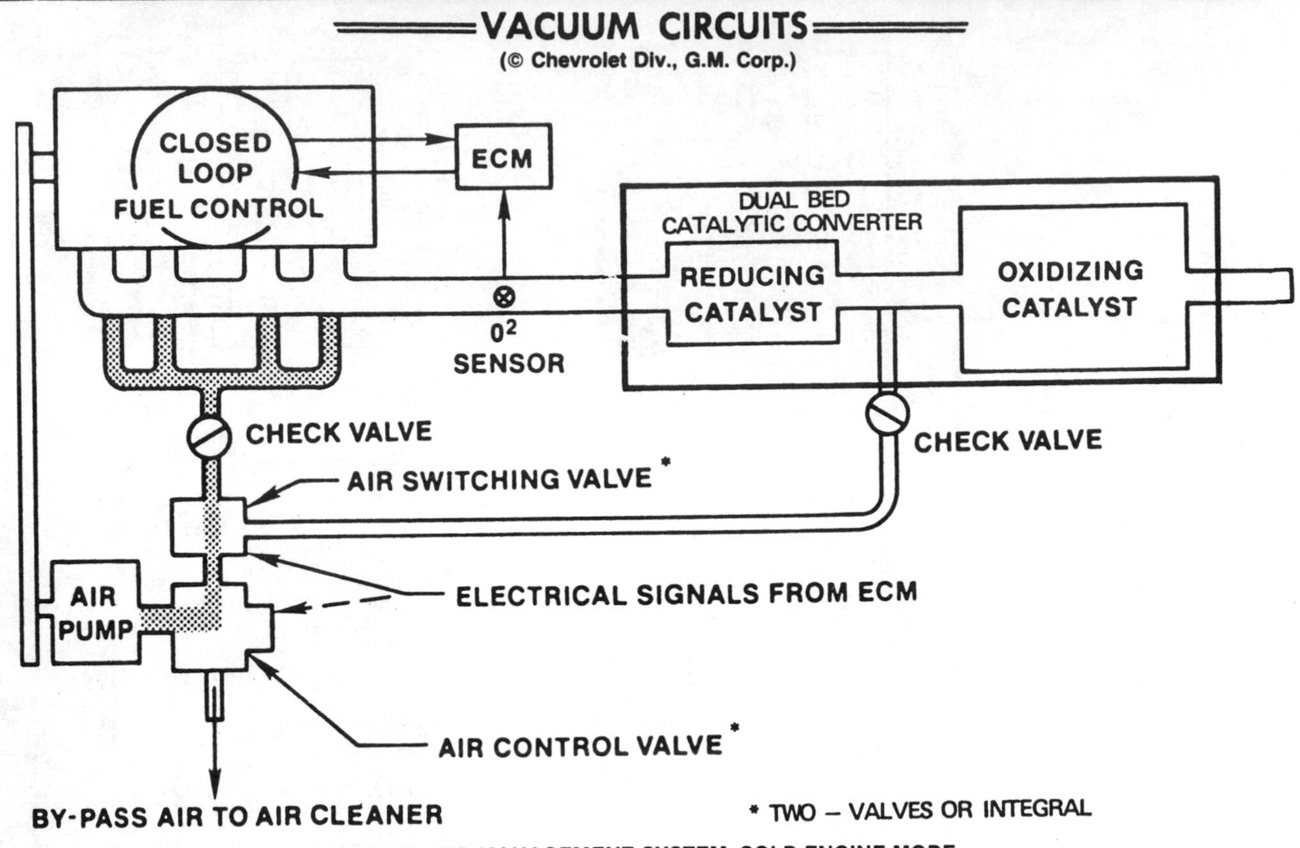

1981-82—AIR MANAGEMENT SYSTEM-COLD ENGINE MODE

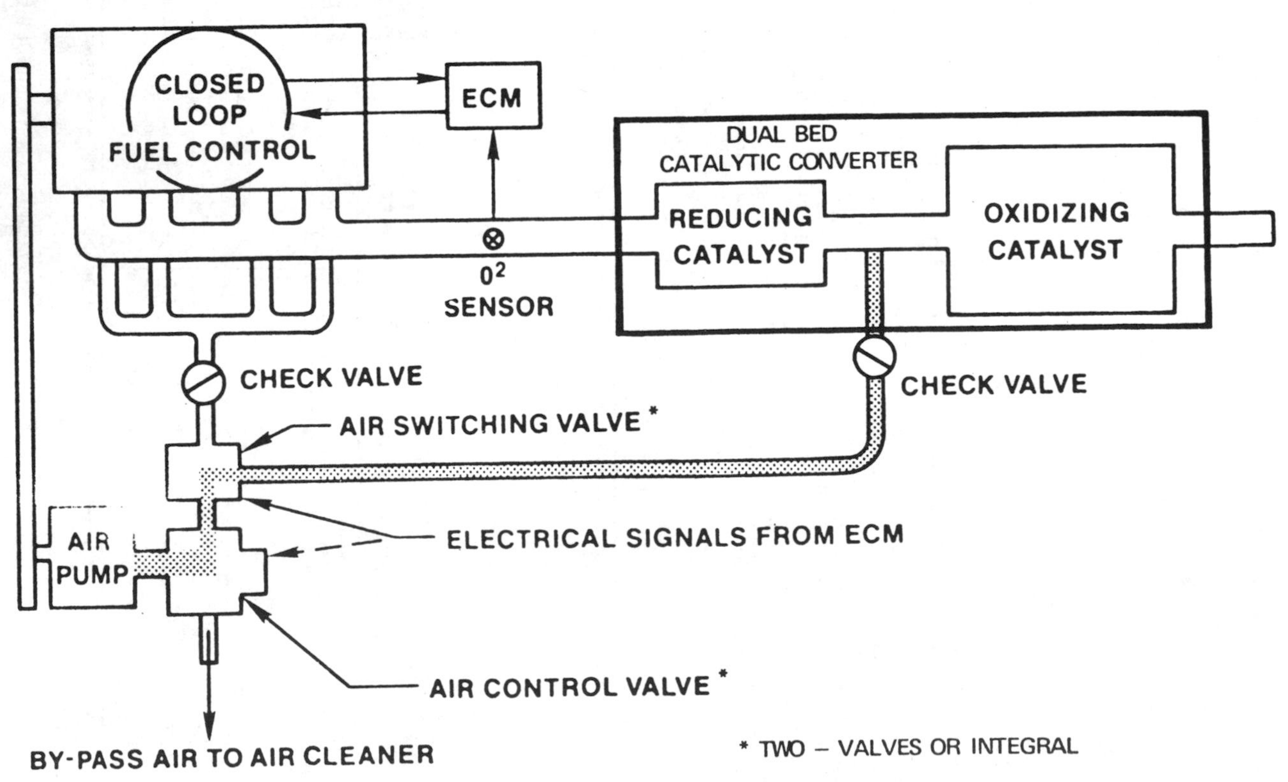

1981-82—AIR MANAGEMENT SYSTEM-WARM ENGINE MODE

VACUUM CIRCUITS

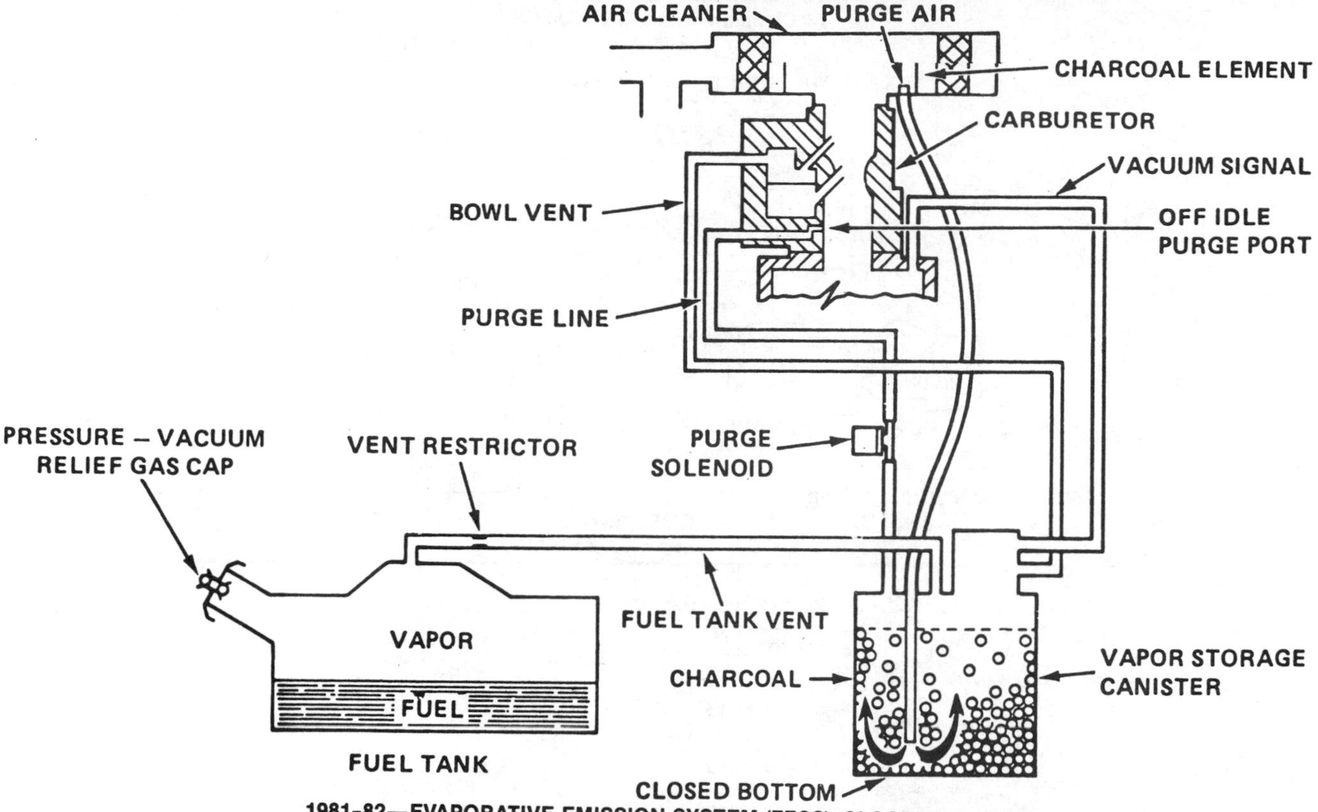

AIR CLEANER
PURGE AIR
CHARCOAL ELEMENT
CARBURETOR
VACUUM SIGNAL
BOWL VENT
OFF IDLE PURGE PORT
PURGE LINE
PRESSURE – VACUUM RELIEF GAS CAP
VENT RESTRICTOR
PURGE SOLENOID
FUEL TANK VENT
VAPOR
FUEL
CHARCOAL
VAPOR STORAGE CANISTER
FUEL TANK
CLOSED BOTTOM

1981-82—EVAPORATIVE EMISSION SYSTEM (EECS)-CLOSED CANISTER

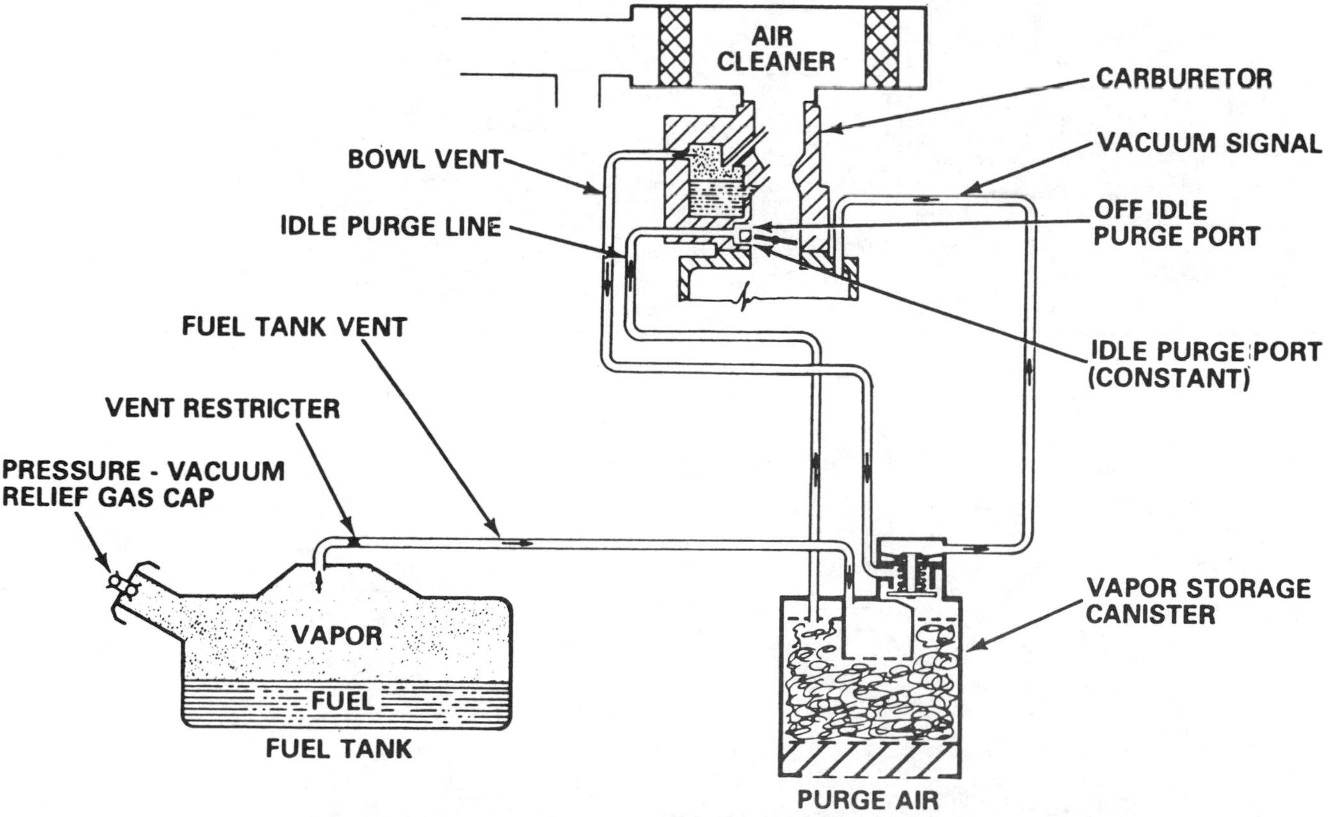

AIR CLEANER
CARBURETOR
BOWL VENT
VACUUM SIGNAL
IDLE PURGE LINE
OFF IDLE PURGE PORT
FUEL TANK VENT
IDLE PURGE PORT (CONSTANT)
VENT RESTRICTER
PRESSURE - VACUUM RELIEF GAS CAP
VAPOR STORAGE CANISTER
VAPOR
FUEL
FUEL TANK
PURGE AIR

1981-82—EVAPORATIVE EMISSION SYSTEM (EECS)-OPEN CANISTER

VACUUM CIRCUITS

(© Oldsmobile Div., G.M. Corp.)

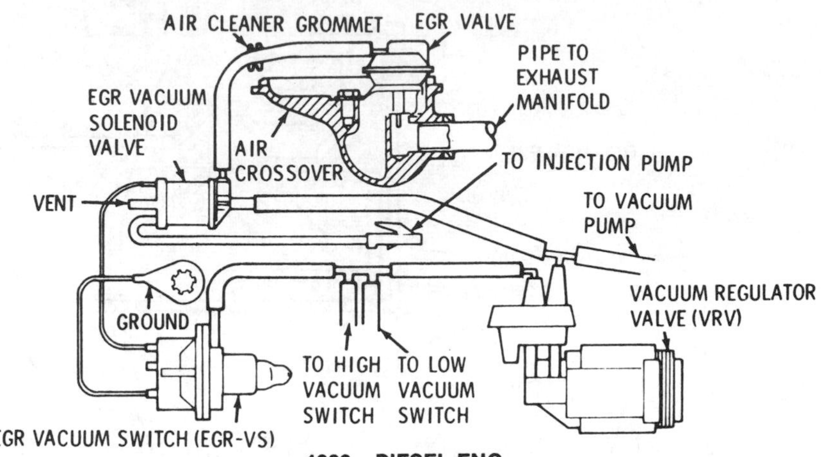

**1980—DIESEL ENG.
CUTLASS, 88 & 98 EXC. 88 WAG.**

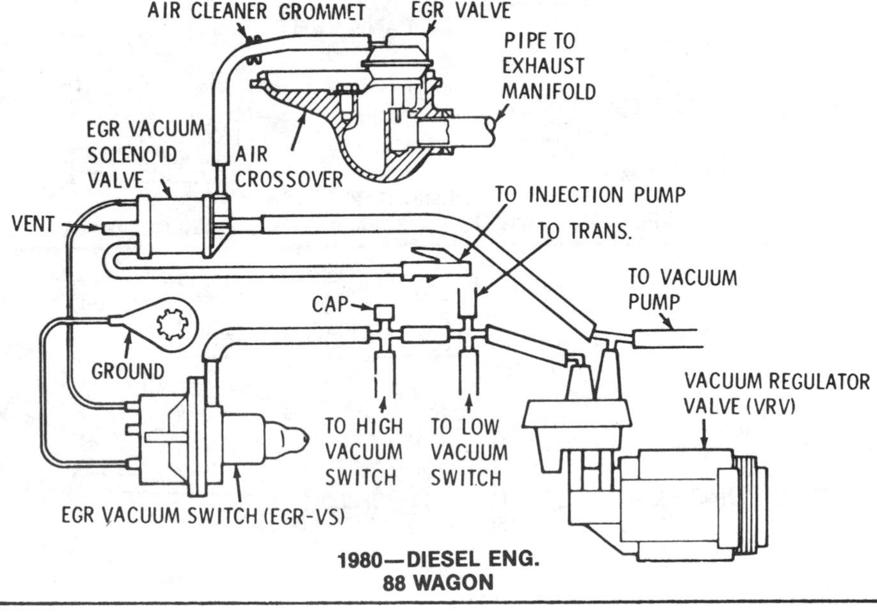

**1980—DIESEL ENG.
88 WAGON**

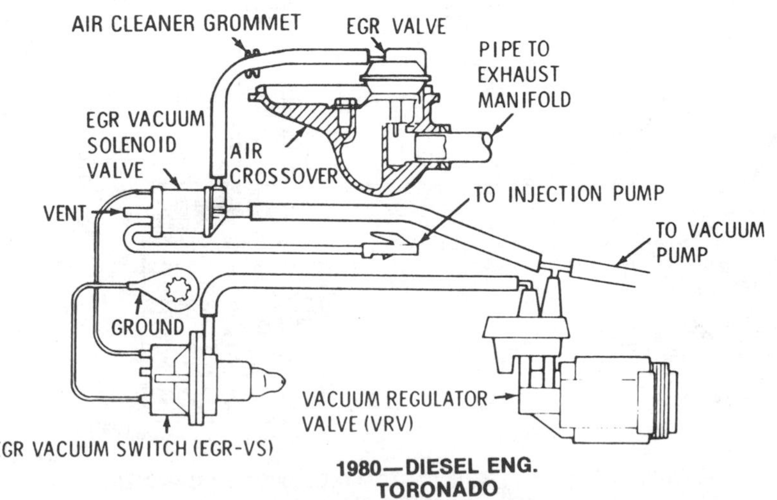

**1980—DIESEL ENG.
TORONADO**

VEHICLE EMISSION CONTROL LABELS AND VACUUM HOSE ROUTINGS

1980

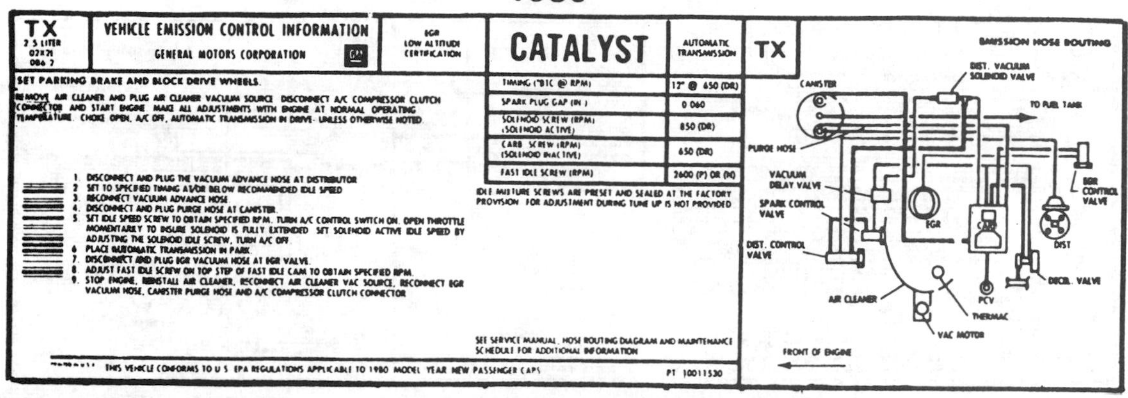

TX
2.5 LITER
02X21
086 2

VEHICLE EMISSION CONTROL INFORMATION
GENERAL MOTORS CORPORATION

EGR LOW ALTITUDE CERTIFICATION

CATALYST

AUTOMATIC TRANSMISSION

TX

SET PARKING BRAKE AND BLOCK DRIVE WHEELS.

REMOVE AIR CLEANER AND PLUG AIR CLEANER VACUUM SOURCE. DISCONNECT A/C COMPRESSOR CLUTCH CONNECTOR AND START ENGINE. MAKE ALL ADJUSTMENTS WITH ENGINE AT NORMAL OPERATING TEMPERATURE. CHOKE OPEN, A/C OFF, AUTOMATIC TRANSMISSION IN DRIVE- UNLESS OTHERWISE NOTED.

1. DISCONNECT AND PLUG THE VACUUM ADVANCE HOSE AT DISTRIBUTOR
2. SET TO SPECIFIED TIMING AT/OR BELOW RECOMMENDED IDLE SPEED
3. RECONNECT VACUUM ADVANCE HOSE.
4. DISCONNECT AND PLUG PURGE HOSE AT CANISTER.
5. SET IDLE SPEED SCREW TO OBTAIN SPECIFIED RPM. TURN A/C CONTROL SWITCH ON. OPEN THROTTLE MOMENTARILY TO INSURE SOLENOID IS FULLY EXTENDED. SET SOLENOID ACTIVE IDLE SPEED BY ADJUSTING THE SOLENOID IDLE SCREW. TURN A/C OFF.
6. PLACE AUTOMATIC TRANSMISSION IN PARK.
7. DISCONNECT AND PLUG EGR VACUUM HOSE AT EGR VALVE.
8. ADJUST FAST IDLE SCREW ON TOP STEP OF FAST IDLE CAM TO OBTAIN SPECIFIED RPM.
9. STOP ENGINE, REINSTALL AIR CLEANER, RECONNECT AIR CLEANER VAC SOURCE, RECONNECT EGR VACUUM HOSE, CANISTER PURGE HOSE AND A/C COMPRESSOR CLUTCH CONNECTOR

TIMING (°BTC @ RPM)	12° @ 650 (DR)
SPARK PLUG GAP (IN.)	0.060
SOLENOID SCREW (RPM) (SOLENOID ACTIVE)	850 (DR)
CARB SCREW (RPM) (SOLENOID INACTIVE)	650 (DR)
FAST IDLE SCREW (RPM)	2600 (P) OR (N)

IDLE MIXTURE SCREWS ARE PRESET AND SEALED AT THE FACTORY. PROVISION FOR ADJUSTMENT DURING TUNE UP IS NOT PROVIDED.

SEE SERVICE MANUAL, HOSE ROUTING DIAGRAM AND MAINTENANCE SCHEDULE FOR ADDITIONAL INFORMATION

THIS VEHICLE CONFORMS TO U.S. EPA REGULATIONS APPLICABLE TO 1980 MODEL YEAR NEW PASSENGER CARS

PT 10011530

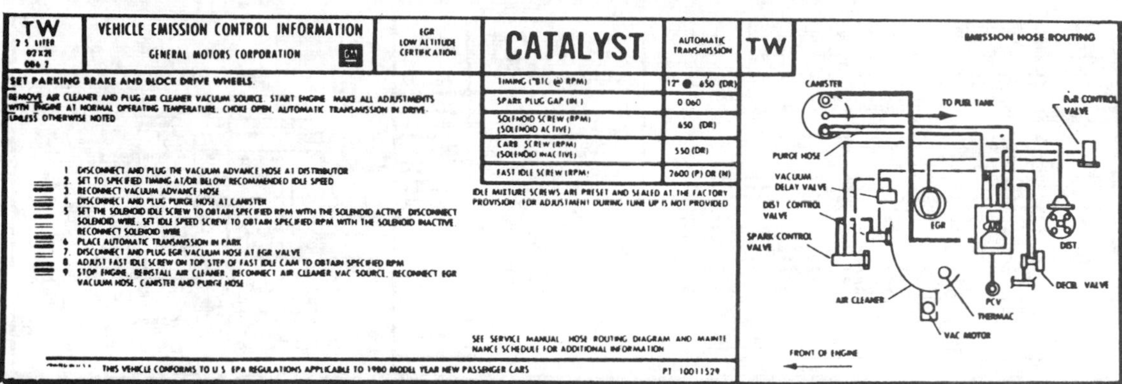

TW
2.5 LITER
02X21
086 2

VEHICLE EMISSION CONTROL INFORMATION
GENERAL MOTORS CORPORATION

EGR LOW ALTITUDE CERTIFICATION

CATALYST

AUTOMATIC TRANSMISSION

TW

SET PARKING BRAKE AND BLOCK DRIVE WHEELS.

REMOVE AIR CLEANER AND PLUG AIR CLEANER VACUUM SOURCE. START ENGINE. MAKE ALL ADJUSTMENTS WITH ENGINE AT NORMAL OPERATING TEMPERATURE, CHOKE OPEN, AUTOMATIC TRANSMISSION IN DRIVE- UNLESS OTHERWISE NOTED

1. DISCONNECT AND PLUG THE VACUUM ADVANCE HOSE AT DISTRIBUTOR
2. SET TO SPECIFIED TIMING AT/OR BELOW RECOMMENDED IDLE SPEED
3. RECONNECT VACUUM ADVANCE HOSE
4. DISCONNECT AND PLUG PURGE HOSE AT CANISTER
5. SET THE SOLENOID IDLE SCREW TO OBTAIN SPECIFIED RPM WITH THE SOLENOID ACTIVE. DISCONNECT SOLENOID WIRE. SET IDLE SPEED SCREW TO OBTAIN SPECIFIED RPM WITH THE SOLENOID INACTIVE. RECONNECT SOLENOID WIRE
6. PLACE AUTOMATIC TRANSMISSION IN PARK
7. DISCONNECT AND PLUG EGR VACUUM HOSE AT EGR VALVE
8. ADJUST FAST IDLE SCREW ON TOP STEP OF FAST IDLE CAM TO OBTAIN SPECIFIED RPM
9. STOP ENGINE, REINSTALL AIR CLEANER, RECONNECT AIR CLEANER VAC SOURCE, RECONNECT EGR VACUUM HOSE, CANISTER AND PURGE HOSE

TIMING (°BTC @ RPM)	12° @ 650 (DR)
SPARK PLUG GAP (IN.)	0.060
SOLENOID SCREW (RPM) (SOLENOID ACTIVE)	650 (DR)
CARB SCREW (RPM) (SOLENOID INACTIVE)	550 (DR)
FAST IDLE SCREW (RPM)	2600 (P) OR (N)

IDLE MIXTURE SCREWS ARE PRESET AND SEALED AT THE FACTORY. PROVISION FOR ADJUSTMENT DURING TUNE UP IS NOT PROVIDED

SEE SERVICE MANUAL, HOSE ROUTING DIAGRAM AND MAINTENANCE SCHEDULE FOR ADDITIONAL INFORMATION

THIS VEHICLE CONFORMS TO U.S. EPA REGULATIONS APPLICABLE TO 1980 MODEL YEAR NEW PASSENGER CARS

PT 10011529

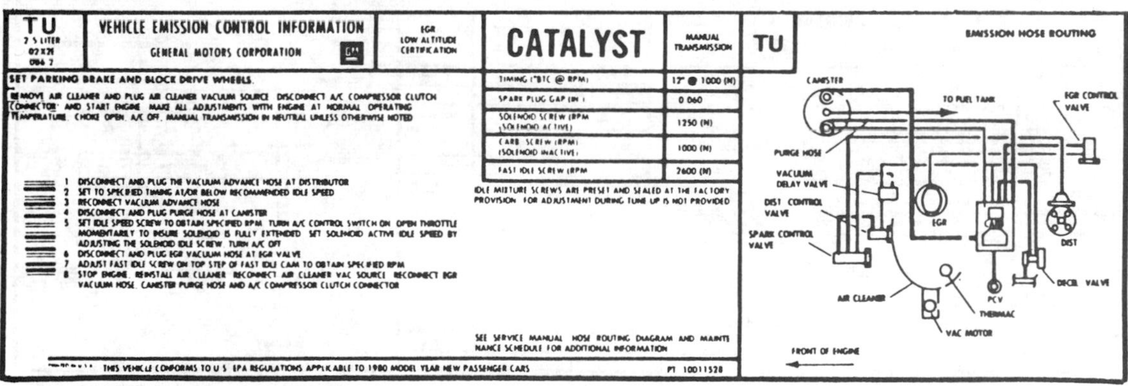

TU
2.5 LITER
02X21
086 2

VEHICLE EMISSION CONTROL INFORMATION
GENERAL MOTORS CORPORATION

EGR LOW ALTITUDE CERTIFICATION

CATALYST

MANUAL TRANSMISSION

TU

SET PARKING BRAKE AND BLOCK DRIVE WHEELS.

REMOVE AIR CLEANER AND PLUG AIR CLEANER VACUUM SOURCE. DISCONNECT A/C COMPRESSOR CLUTCH CONNECTOR AND START ENGINE. MAKE ALL ADJUSTMENTS WITH ENGINE AT NORMAL OPERATING TEMPERATURE, CHOKE OPEN, A/C OFF, MANUAL TRANSMISSION IN NEUTRAL UNLESS OTHERWISE NOTED

1. DISCONNECT AND PLUG THE VACUUM ADVANCE HOSE AT DISTRIBUTOR
2. SET TO SPECIFIED TIMING AT/OR BELOW RECOMMENDED IDLE SPEED
3. RECONNECT VACUUM ADVANCE HOSE
4. DISCONNECT AND PLUG PURGE HOSE AT CANISTER
5. SET IDLE SPEED SCREW TO OBTAIN SPECIFIED RPM. TURN A/C CONTROL SWITCH ON. OPEN THROTTLE MOMENTARILY TO INSURE SOLENOID IS FULLY EXTENDED. SET SOLENOID ACTIVE IDLE SPEED BY ADJUSTING THE SOLENOID IDLE SCREW. TURN A/C OFF
6. DISCONNECT AND PLUG EGR VACUUM HOSE AT EGR VALVE
7. ADJUST FAST IDLE SCREW ON TOP STEP OF FAST IDLE CAM TO OBTAIN SPECIFIED RPM
8. STOP ENGINE, REINSTALL AIR CLEANER, RECONNECT AIR CLEANER VAC SOURCE, RECONNECT EGR VACUUM HOSE, CANISTER PURGE HOSE AND A/C COMPRESSOR CLUTCH CONNECTOR

TIMING (°BTC @ RPM)	12° @ 1000 (N)
SPARK PLUG GAP (IN.)	0.060
SOLENOID SCREW (RPM (SOLENOID ACTIVE)	1250 (N)
CARB SCREW (RPM) (SOLENOID INACTIVE)	1000 (N)
FAST IDLE SCREW (RPM	2600 (N)

IDLE MIXTURE SCREWS ARE PRESET AND SEALED AT THE FACTORY. PROVISION FOR ADJUSTMENT DURING TUNE UP IS NOT PROVIDED

SEE SERVICE MANUAL, HOSE ROUTING DIAGRAM AND MAINTENANCE SCHEDULE FOR ADDITIONAL INFORMATION

THIS VEHICLE CONFORMS TO U.S. EPA REGULATIONS APPLICABLE TO 1980 MODEL YEAR NEW PASSENGER CARS

PT 10011528

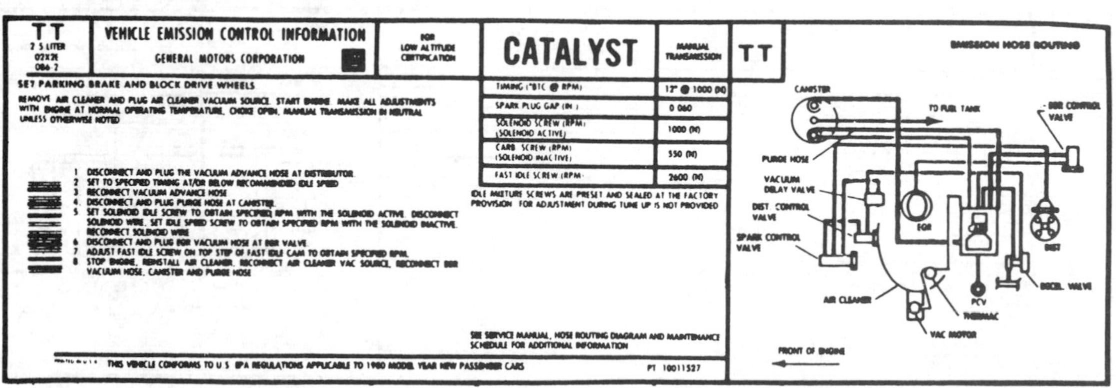

TT
2.5 LITER
02X21
086 2

VEHICLE EMISSION CONTROL INFORMATION
GENERAL MOTORS CORPORATION

EGR LOW ALTITUDE CERTIFICATION

CATALYST

MANUAL TRANSMISSION

TT

SET PARKING BRAKE AND BLOCK DRIVE WHEELS

REMOVE AIR CLEANER AND PLUG AIR CLEANER VACUUM SOURCE. START ENGINE. MAKE ALL ADJUSTMENTS WITH ENGINE AT NORMAL OPERATING TEMPERATURE, CHOKE OPEN, MANUAL TRANSMISSION IN NEUTRAL UNLESS OTHERWISE NOTED

1. DISCONNECT AND PLUG THE VACUUM ADVANCE HOSE AT DISTRIBUTOR.
2. SET TO SPECIFIED TIMING AT/OR BELOW RECOMMENDED IDLE SPEED
3. RECONNECT VACUUM ADVANCE HOSE
4. DISCONNECT AND PLUG PURGE HOSE AT CANISTER.
5. SET SOLENOID IDLE SCREW TO OBTAIN SPECIFIED RPM WITH THE SOLENOID ACTIVE. DISCONNECT SOLENOID WIRE. SET IDLE SPEED SCREW TO OBTAIN SPECIFIED RPM WITH THE SOLENOID INACTIVE. RECONNECT SOLENOID WIRE
6. DISCONNECT AND PLUG EGR VACUUM HOSE AT EGR VALVE.
7. ADJUST FAST IDLE SCREW ON TOP STEP OF FAST IDLE CAM TO OBTAIN SPECIFIED RPM.
8. STOP ENGINE, REINSTALL AIR CLEANER, RECONNECT AIR CLEANER VAC SOURCE, RECONNECT EGR VACUUM HOSE, CANISTER AND PURGE HOSE

TIMING (°BTC @ RPM)	12° @ 1000 (N)
SPARK PLUG GAP (IN.)	0.060
SOLENOID SCREW (RPM) (SOLENOID ACTIVE)	1000 (N)
CARB SCREW (RPM) (SOLENOID INACTIVE)	550 (N)
FAST IDLE SCREW (RPM	2600 (N)

IDLE MIXTURE SCREWS ARE PRESET AND SEALED AT THE FACTORY. PROVISION FOR ADJUSTMENT DURING TUNE UP IS NOT PROVIDED

SEE SERVICE MANUAL, HOSE ROUTING DIAGRAM AND MAINTENANCE SCHEDULE FOR ADDITIONAL INFORMATION

THIS VEHICLE CONFORMS TO U.S. EPA REGULATIONS APPLICABLE TO 1980 MODEL YEAR NEW PASSENGER CARS

PT 10011527

VEHICLE EMISSION CONTROL LABELS AND VACUUM HOSE ROUTINGS

1980

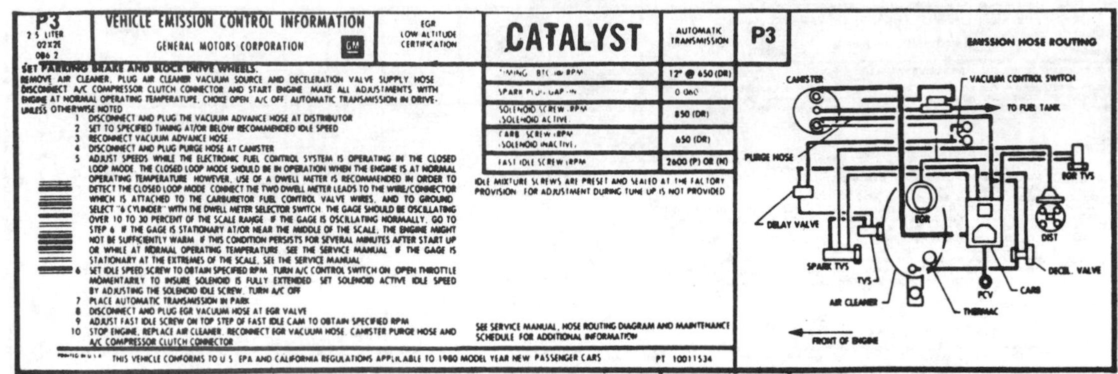

P3 — 2 5 LITER 02X2E 0B6 2

VEHICLE EMISSION CONTROL INFORMATION — GENERAL MOTORS CORPORATION — EGR LOW ALTITUDE CERTIFICATION — CATALYST — AUTOMATIC TRANSMISSION — P3

SET PARKING BRAKE AND BLOCK DRIVE WHEELS.
REMOVE AIR CLEANER. PLUG AIR CLEANER VACUUM SOURCE AND DECELERATION VALVE SUPPLY HOSE DISCONNECT A/C COMPRESSOR CLUTCH CONNECTOR AND START ENGINE MAKE ALL ADJUSTMENTS WITH ENGINE AT NORMAL OPERATING TEMPERATURE. CHOKE OPEN A/C OFF AUTOMATIC TRANSMISSION IN DRIVE-UNLESS OTHERWISE NOTED

1 DISCONNECT AND PLUG THE VACUUM ADVANCE HOSE AT DISTRIBUTOR
2 SET TO SPECIFIED TIMING AT/OR BELOW RECOMMENDED IDLE SPEED
3 RECONNECT VACUUM ADVANCE HOSE
4 DISCONNECT AND PLUG PURGE HOSE AT CANISTER
5 ADJUST SPEEDS WHILE THE ELECTRONIC FUEL CONTROL SYSTEM IS OPERATING IN THE CLOSED LOOP MODE THE CLOSED LOOP MODE SHOULD BE IN OPERATION WHEN THE ENGINE IS AT NORMAL OPERATING TEMPERATURE HOWEVER, USE OF A DWELL METER IS RECOMMENDED IN ORDER TO DETECT THE CLOSED LOOP MODE CONNECT THE TWO DWELL METER LEADS TO THE WIRE/CONNECTOR WHICH IS ATTACHED TO THE CARBURETOR FUEL CONTROL VALVE WIRES, AND TO GROUND SELECT "6 CYLINDER" WITH THE DWELL METER SELECTOR SWITCH THE GAGE SHOULD BE OSCILLATING OVER 10 TO 30 PERCENT OF THE SCALE RANGE IF THE GAGE IS OSCILLATING NORMALLY, GO TO STEP 6 IF THE GAGE IS STATIONARY AT/OR NEAR THE MIDDLE OF THE SCALE, THE ENGINE MIGHT NOT BE SUFFICIENTLY WARM IF THIS CONDITION PERSISTS FOR SEVERAL MINUTES AFTER START UP OR WHILE AT NORMAL OPERATING TEMPERATURE SEE THE SERVICE MANUAL IF THE GAGE IS STATIONARY AT THE EXTREMES OF THE SCALE, SEE THE SERVICE MANUAL
6 SET IDLE SPEED SCREW TO OBTAIN SPECIFIED RPM TURN A/C CONTROL SWITCH ON OPEN THROTTLE MOMENTARILY TO INSURE SOLENOID IS FULLY EXTENDED SET SOLENOID ACTIVE IDLE SPEED BY ADJUSTING THE SOLENOID IDLE SCREW. TURN A/C OFF
7 PLACE AUTOMATIC TRANSMISSION IN PARK
8 DISCONNECT AND PLUG EGR VACUUM HOSE AT EGR VALVE
9 ADJUST FAST IDLE SCREW ON TOP STEP OF FAST IDLE CAM TO OBTAIN SPECIFIED RPM
10 STOP ENGINE, REPLACE AIR CLEANER. RECONNECT EGR VACUUM HOSE. CANISTER PURGE HOSE AND A/C COMPRESSOR CLUTCH CONNECTOR

	AUTOMATIC TRANSMISSION
TIMING (°BTC @ RPM)	12° @ 650 (DR)
SPARK PLUG GAP (IN)	0 060
SOLENOID SCREW (RPM) (SOLENOID ACTIVE)	850 (DR)
CARB SCREW (RPM) (SOLENOID INACTIVE)	650 (DR)
FAST IDLE SCREW (RPM)	2600 (P) OR (N)

IDLE MIXTURE SCREWS ARE PRESET AND SEALED AT THE FACTORY PROVISION FOR ADJUSTMENT DURING TUNE UP IS NOT PROVIDED

SEE SERVICE MANUAL, HOSE ROUTING DIAGRAM AND MAINTENANCE SCHEDULE FOR ADDITIONAL INFORMATION

PRINTED IN U S A THIS VEHICLE CONFORMS TO U S EPA AND CALIFORNIA REGULATIONS APPLICABLE TO 1980 MODEL YEAR NEW PASSENGER CARS PT 10011534

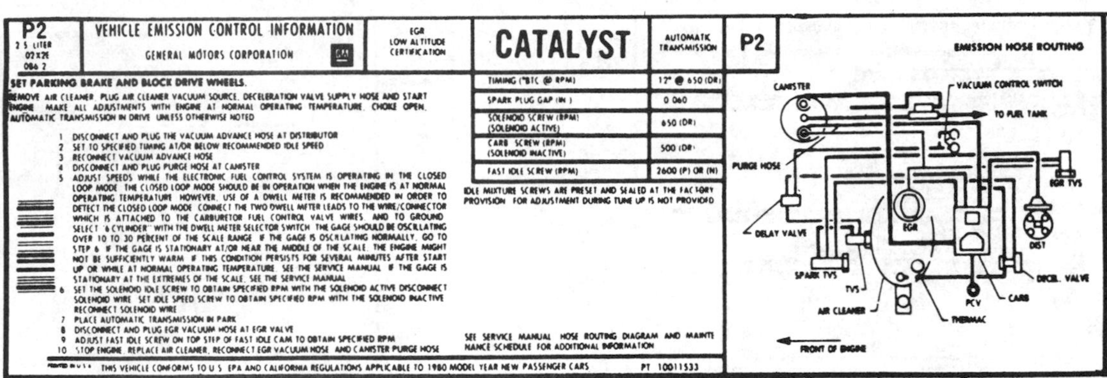

P2 — 2 5 LITER 02X2E 0B6 2

VEHICLE EMISSION CONTROL INFORMATION — GENERAL MOTORS CORPORATION — EGR LOW ALTITUDE CERTIFICATION — CATALYST — AUTOMATIC TRANSMISSION — P2

SET PARKING BRAKE AND BLOCK DRIVE WHEELS.
REMOVE AIR CLEANER. PLUG AIR CLEANER VACUUM SOURCE. DECELERATION VALVE SUPPLY HOSE AND START ENGINE MAKE ALL ADJUSTMENTS WITH ENGINE AT NORMAL OPERATING TEMPERATURE. CHOKE OPEN. AUTOMATIC TRANSMISSION IN DRIVE UNLESS OTHERWISE NOTED

1 DISCONNECT AND PLUG THE VACUUM ADVANCE HOSE AT DISTRIBUTOR
2 SET TO SPECIFIED TIMING AT/OR BELOW RECOMMENDED IDLE SPEED
3 RECONNECT VACUUM ADVANCE HOSE
4 DISCONNECT AND PLUG PURGE HOSE AT CANISTER
5 ADJUST SPEEDS WHILE THE ELECTRONIC FUEL CONTROL SYSTEM IS OPERATING IN THE CLOSED LOOP MODE THE CLOSED LOOP MODE SHOULD BE IN OPERATION WHEN THE ENGINE IS AT NORMAL OPERATING TEMPERATURE HOWEVER, USE OF A DWELL METER IS RECOMMENDED IN ORDER TO DETECT THE CLOSED LOOP MODE CONNECT THE TWO DWELL METER LEADS TO THE WIRE/CONNECTOR WHICH IS ATTACHED TO THE CARBURETOR FUEL CONTROL VALVE WIRES. AND TO GROUND SELECT "6 CYLINDER" WITH THE DWELL METER SELECTOR SWITCH THE GAGE SHOULD BE OSCILLATING OVER 10 TO 30 PERCENT OF THE SCALE RANGE IF THE GAGE IS OSCILLATING NORMALLY, GO TO STEP 6 IF THE GAGE IS STATIONARY AT/OR NEAR THE MIDDLE OF THE SCALE. THE ENGINE MIGHT NOT BE SUFFICIENTLY WARM IF THIS CONDITION PERSISTS FOR SEVERAL MINUTES AFTER START UP OR WHILE AT NORMAL OPERATING TEMPERATURE. SEE THE SERVICE MANUAL IF THE GAGE IS STATIONARY AT THE EXTREMES OF THE SCALE. SEE THE SERVICE MANUAL
6 SET THE SOLENOID IDLE SCREW TO OBTAIN SPECIFIED RPM WITH THE SOLENOID ACTIVE DISCONNECT SOLENOID WIRE SET IDLE SPEED SCREW TO OBTAIN SPECIFIED RPM WITH THE SOLENOID INACTIVE RECONNECT SOLENOID WIRE
7 PLACE AUTOMATIC TRANSMISSION IN PARK
8 DISCONNECT AND PLUG EGR VACUUM HOSE AT EGR VALVE
9 ADJUST FAST IDLE SCREW ON TOP STEP OF FAST IDLE CAM TO OBTAIN SPECIFIED RPM
10 STOP ENGINE. REPLACE AIR CLEANER. RECONNECT EGR VACUUM HOSE AND CANISTER PURGE HOSE

	AUTOMATIC TRANSMISSION
TIMING (°BTC @ RPM)	12° @ 650 (DR)
SPARK PLUG GAP (IN)	0 060
SOLENOID SCREW (RPM) (SOLENOID ACTIVE)	650 (DR)
CARB SCREW (RPM) (SOLENOID INACTIVE)	500 (DR)
FAST IDLE SCREW (RPM)	2600 (P) OR (N)

IDLE MIXTURE SCREWS ARE PRESET AND SEALED AT THE FACTORY PROVISION FOR ADJUSTMENT DURING TUNE UP IS NOT PROVIDED

SEE SERVICE MANUAL HOSE ROUTING DIAGRAM AND MAINTENANCE SCHEDULE FOR ADDITIONAL INFORMATION

PRINTED IN U S A THIS VEHICLE CONFORMS TO U S EPA AND CALIFORNIA REGULATIONS APPLICABLE TO 1980 MODEL YEAR NEW PASSENGER CARS PT 10011533

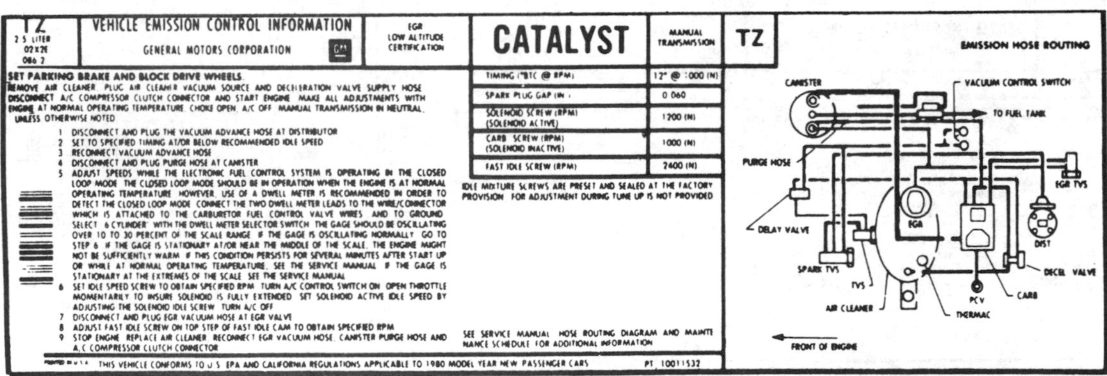

TZ — 2 5 LITER 02X2E 0B6 2

VEHICLE EMISSION CONTROL INFORMATION — GENERAL MOTORS CORPORATION — EGR LOW ALTITUDE CERTIFICATION — CATALYST — MANUAL TRANSMISSION — TZ

SET PARKING BRAKE AND BLOCK DRIVE WHEELS.
REMOVE AIR CLEANER. PLUG AIR CLEANER VACUUM SOURCE AND DECELERATION VALVE SUPPLY HOSE DISCONNECT A/C COMPRESSOR CLUTCH CONNECTOR AND START ENGINE MAKE ALL ADJUSTMENTS WITH ENGINE AT NORMAL OPERATING TEMPERATURE. CHOKE OPEN. A/C OFF MANUAL TRANSMISSION IN NEUTRAL. UNLESS OTHERWISE NOTED

1 DISCONNECT AND PLUG THE VACUUM ADVANCE HOSE AT DISTRIBUTOR
2 SET TO SPECIFIED TIMING AT/OR BELOW RECOMMENDED IDLE SPEED
3 RECONNECT VACUUM ADVANCE HOSE
4 DISCONNECT AND PLUG PURGE HOSE AT CANISTER
5 ADJUST SPEEDS WHILE THE ELECTRONIC FUEL CONTROL SYSTEM IS OPERATING IN THE CLOSED LOOP MODE THE CLOSED LOOP MODE SHOULD BE IN OPERATION WHEN THE ENGINE IS AT NORMAL OPERATING TEMPERATURE HOWEVER, USE OF A DWELL METER IS RECOMMENDED IN ORDER TO DETECT THE CLOSED LOOP MODE CONNECT THE TWO DWELL METER LEADS TO THE WIRE/CONNECTOR WHICH IS ATTACHED TO THE CARBURETOR FUEL CONTROL VALVE WIRES AND TO GROUND SELECT 6 CYLINDER WITH THE DWELL METER SELECTOR SWITCH THE GAGE SHOULD BE OSCILLATING OVER 10 TO 30 PERCENT OF THE SCALE RANGE IF THE GAGE IS OSCILLATING NORMALLY GO TO STEP 6 IF THE GAGE IS STATIONARY AT/OR NEAR THE MIDDLE OF THE SCALE. THE ENGINE MIGHT NOT BE SUFFICIENTLY WARM IF THIS CONDITION PERSISTS FOR SEVERAL MINUTES AFTER START UP OR WHILE AT NORMAL OPERATING TEMPERATURE. SEE THE SERVICE MANUAL IF THE GAGE IS STATIONARY AT THE EXTREMES OF THE SCALE SEE THE SERVICE MANUAL
6 SET IDLE SPEED SCREW TO OBTAIN SPECIFIED RPM TURN A/C CONTROL SWITCH ON OPEN THROTTLE MOMENTARILY TO INSURE SOLENOID IS FULLY EXTENDED SET SOLENOID ACTIVE IDLE SPEED BY ADJUSTING THE SOLENOID IDLE SCREW TURN A/C OFF
7 DISCONNECT AND PLUG EGR VACUUM HOSE AT EGR VALVE
8 ADJUST FAST IDLE SCREW ON TOP STEP OF FAST IDLE CAM TO OBTAIN SPECIFIED RPM
9 STOP ENGINE REPLACE AIR CLEANER RECONNECT EGR VACUUM HOSE. CANISTER PURGE HOSE AND A.C COMPRESSOR CLUTCH CONNECTOR

	MANUAL TRANSMISSION
TIMING (°BTC @ RPM)	12° @ 1000 (N)
SPARK PLUG GAP (IN)	0 060
SOLENOID SCREW (RPM) (SOLENOID ACTIVE)	1200 (N)
CARB SCREW (RPM) (SOLENOID INACTIVE)	1000 (N)
FAST IDLE SCREW (RPM)	2400 (N)

IDLE MIXTURE SCREWS ARE PRESET AND SEALED AT THE FACTORY PROVISION FOR ADJUSTMENT DURING TUNE UP IS NOT PROVIDED

SEE SERVICE MANUAL HOSE ROUTING DIAGRAM AND MAINTENANCE SCHEDULE FOR ADDITIONAL INFORMATION

PRINTED IN U S A THIS VEHICLE CONFORMS TO U S EPA AND CALIFORNIA REGULATIONS APPLICABLE TO 1980 MODEL YEAR NEW PASSENGER CARS PT 10011532

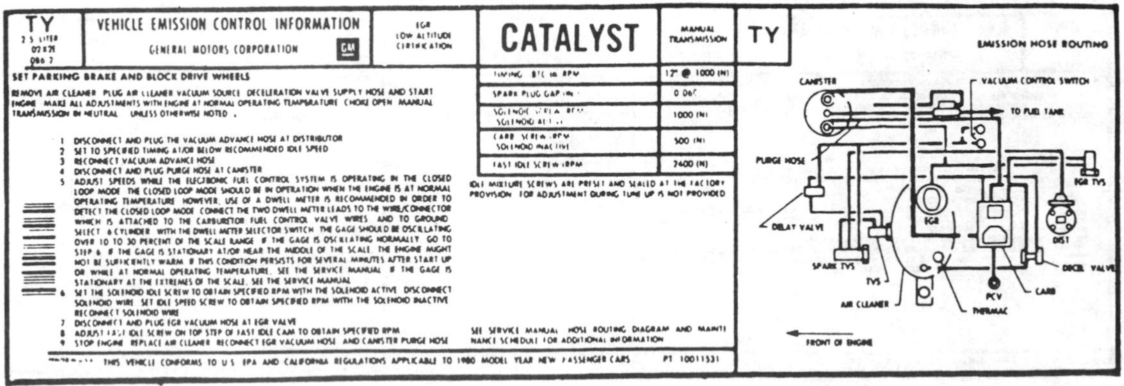

TY — 2 5 LITER 02X2E 0B6 2

VEHICLE EMISSION CONTROL INFORMATION — GENERAL MOTORS CORPORATION — EGR LOW ALTITUDE CERTIFICATION — CATALYST — MANUAL TRANSMISSION — TY

SET PARKING BRAKE AND BLOCK DRIVE WHEELS
REMOVE AIR CLEANER PLUG AIR CLEANER VACUUM SOURCE DECELERATION VALVE SUPPLY HOSE AND START ENGINE MAKE ALL ADJUSTMENTS WITH ENGINE AT NORMAL OPERATING TEMPERATURE CHOKE OPEN MANUAL TRANSMISSION IN NEUTRAL UNLESS OTHERWISE NOTED .

1 DISCONNECT AND PLUG THE VACUUM ADVANCE HOSE AT DISTRIBUTOR
2 SET TO SPECIFIED TIMING AT/OR BELOW RECOMMENDED IDLE SPEED
3 RECONNECT VACUUM ADVANCE HOSE
4 DISCONNECT AND PLUG PURGE HOSE AT CANISTER
5 ADJUST SPEEDS WHILE THE ELECTRONIC FUEL CONTROL SYSTEM IS OPERATING IN THE CLOSED LOOP MODE THE CLOSED LOOP MODE SHOULD BE IN OPERATION WHEN THE ENGINE IS AT NORMAL OPERATING TEMPERATURE HOWEVER, USE OF A DWELL METER IS RECOMMENDED IN ORDER TO DETECT THE CLOSED LOOP MODE CONNECT THE TWO DWELL METER LEADS TO THE WIRE/CONNECTOR WHICH IS ATTACHED TO THE CARBURETOR FUEL CONTROL VALVE WIRES AND TO GROUND SELECT 6 CYLINDER WITH THE DWELL METER SELECTOR SWITCH THE GAGE SHOULD BE OSCILLATING OVER 10 TO 30 PERCENT OF THE SCALE RANGE IF THE GAGE IS OSCILLATING NORMALLY GO TO STEP 6 IF THE GAGE IS STATIONARY AT/OR NEAR THE MIDDLE OF THE SCALE THE ENGINE MIGHT NOT BE SUFFICIENTLY WARM IF THIS CONDITION PERSISTS FOR SEVERAL MINUTES AFTER START UP OR WHILE AT NORMAL OPERATING TEMPERATURE. SEE THE SERVICE MANUAL IF THE GAGE IS STATIONARY AT THE EXTREMES OF THE SCALE. SEE THE SERVICE MANUAL
6 SET THE SOLENOID IDLE SCREW TO OBTAIN SPECIFIED RPM WITH THE SOLENOID ACTIVE DISCONNECT SOLENOID WIRE SET IDLE SPEED SCREW TO OBTAIN SPECIFIED RPM WITH THE SOLENOID INACTIVE RECONNECT SOLENOID WIRE
7 DISCONNECT AND PLUG EGR VACUUM HOSE AT EGR VALVE
8 ADJUST FAST IDLE SCREW ON TOP STEP OF FAST IDLE CAM TO OBTAIN SPECIFIED RPM
9 STOP ENGINE REPLACE AIR CLEANER RECONNECT EGR VACUUM HOSE AND CANISTER PURGE HOSE

	MANUAL TRANSMISSION
TIMING (°BTC @ RPM)	12° @ 1000 (N)
SPARK PLUG GAP (IN)	0 060
SOLENOID SCREW (RPM) (SOLENOID ACTIVE)	1000 (N)
CARB SCREW (RPM) (SOLENOID INACTIVE)	500 (N)
FAST IDLE SCREW (RPM)	2400 (N)

IDLE MIXTURE SCREWS ARE PRESET AND SEALED AT THE FACTORY PROVISION FOR ADJUSTMENT DURING TUNE UP IS NOT PROVIDED

SEE SERVICE MANUAL HOSE ROUTING DIAGRAM AND MAINTENANCE SCHEDULE FOR ADDITIONAL INFORMATION

PRINTED IN U S A THIS VEHICLE CONFORMS TO U S EPA AND CALIFORNIA REGULATIONS APPLICABLE TO 1980 MODEL YEAR NEW PASSENGER CARS PT 10011531

VEHICLE EMISSION CONTROL LABELS AND VACUUM HOSE ROUTINGS

1980

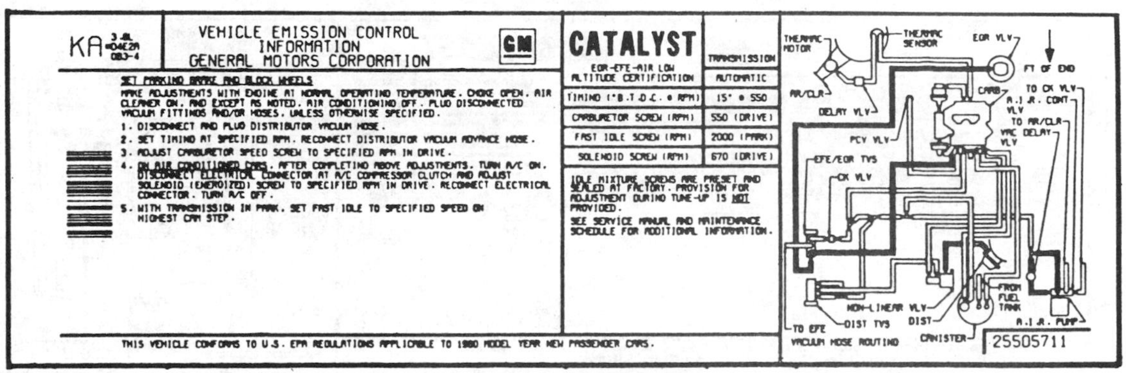

KA 3.8L =04E2A 083-4

VEHICLE EMISSION CONTROL INFORMATION
GENERAL MOTORS CORPORATION
GM

SET PARKING BRAKE AND BLOCK WHEELS
MAKE ADJUSTMENTS WITH ENGINE AT NORMAL OPERATING TEMPERATURE. CHOKE OPEN. AIR CLEANER ON. AND EXCEPT AS NOTED. AIR CONDITIONING OFF. PLUG DISCONNECTED VACUUM FITTINGS AND/OR HOSES. UNLESS OTHERWISE SPECIFIED.

1. DISCONNECT AND PLUG DISTRIBUTOR VACUUM HOSE.
2. SET TIMING AT SPECIFIED RPM. RECONNECT DISTRIBUTOR VACUUM ADVANCE HOSE.
3. ADJUST CARBURETOR SPEED SCREW TO SPECIFIED RPM IN DRIVE.
4. ON AIR CONDITIONED CARS. AFTER COMPLETING ABOVE ADJUSTMENTS. TURN A/C ON. DISCONNECT ELECTRICAL CONNECTOR AT A/C COMPRESSOR CLUTCH AND ADJUST SOLENOID (ENERGIZED) SCREW TO SPECIFIED RPM IN DRIVE. RECONNECT ELECTRICAL CONNECTOR. TURN A/C OFF.
5. WITH TRANSMISSION IN PARK. SET FAST IDLE TO SPECIFIED SPEED ON HIGHEST CAM STEP.

CATALYST EGR-EFE-AIR LOW ALTITUDE CERTIFICATION	TRANSMISSION AUTOMATIC
TIMING (°B.T.D.C. @ RPM)	15° @ 550
CARBURETOR SCREW (RPM)	550 (DRIVE)
FAST IDLE SCREW (RPM)	2000 (PARK)
SOLENOID SCREW (RPM)	670 (DRIVE)

IDLE MIXTURE SCREWS ARE PRESET AND SEALED AT FACTORY. PROVISION FOR ADJUSTMENT DURING TUNE-UP IS NOT PROVIDED.

SEE SERVICE MANUAL AND MAINTENANCE SCHEDULE FOR ADDITIONAL INFORMATION.

THIS VEHICLE CONFORMS TO U.S. EPA REGULATIONS APPLICABLE TO 1980 MODEL YEAR NEW PASSENGER CARS.

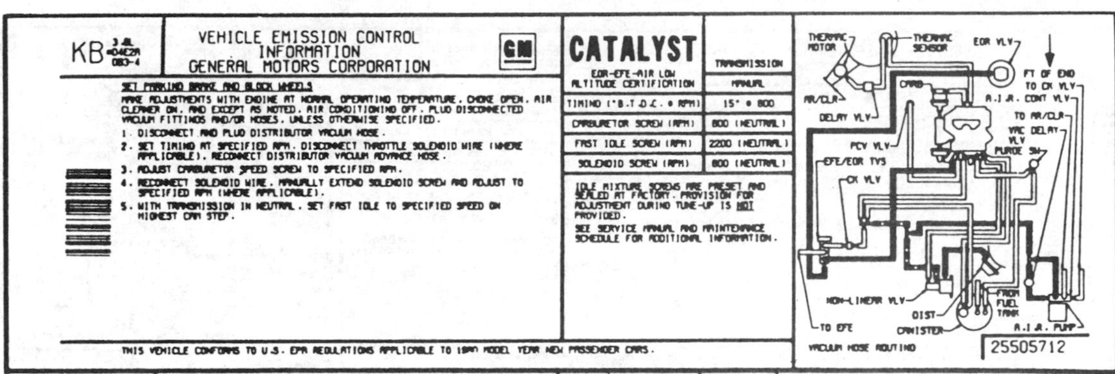

KB 3.8L =04E2A 083-4

VEHICLE EMISSION CONTROL INFORMATION
GENERAL MOTORS CORPORATION
GM

SET PARKING BRAKE AND BLOCK WHEELS
MAKE ADJUSTMENTS WITH ENGINE AT NORMAL OPERATING TEMPERATURE. CHOKE OPEN. AIR CLEANER ON. AND EXCEPT AS NOTED. AIR CONDITIONING OFF. PLUG DISCONNECTED VACUUM FITTINGS AND/OR HOSES. UNLESS OTHERWISE SPECIFIED.

1. DISCONNECT AND PLUG DISTRIBUTOR VACUUM HOSE.
2. SET TIMING AT SPECIFIED RPM. DISCONNECT THROTTLE SOLENOID WIRE (WHERE APPLICABLE). RECONNECT DISTRIBUTOR VACUUM ADVANCE HOSE.
3. ADJUST CARBURETOR SPEED SCREW TO SPECIFIED RPM.
4. RECONNECT SOLENOID WIRE. MANUALLY EXTEND SOLENOID SCREW AND ADJUST TO SPECIFIED RPM (WHERE APPLICABLE).
5. WITH TRANSMISSION IN NEUTRAL. SET FAST IDLE TO SPECIFIED SPEED ON HIGHEST CAM STEP.

CATALYST EGR-EFE-AIR LOW ALTITUDE CERTIFICATION	TRANSMISSION MANUAL
TIMING (°B.T.D.C. @ RPM)	15° @ 800
CARBURETOR SCREW (RPM)	800 (NEUTRAL)
FAST IDLE SCREW (RPM)	2200 (NEUTRAL)
SOLENOID SCREW (RPM)	800 (NEUTRAL)

IDLE MIXTURE SCREWS ARE PRESET AND SEALED AT FACTORY. PROVISION FOR ADJUSTMENT DURING TUNE-UP IS NOT PROVIDED.

SEE SERVICE MANUAL AND MAINTENANCE SCHEDULE FOR ADDITIONAL INFORMATION.

THIS VEHICLE CONFORMS TO U.S. EPA REGULATIONS APPLICABLE TO 1980 MODEL YEAR NEW PASSENGER CARS.

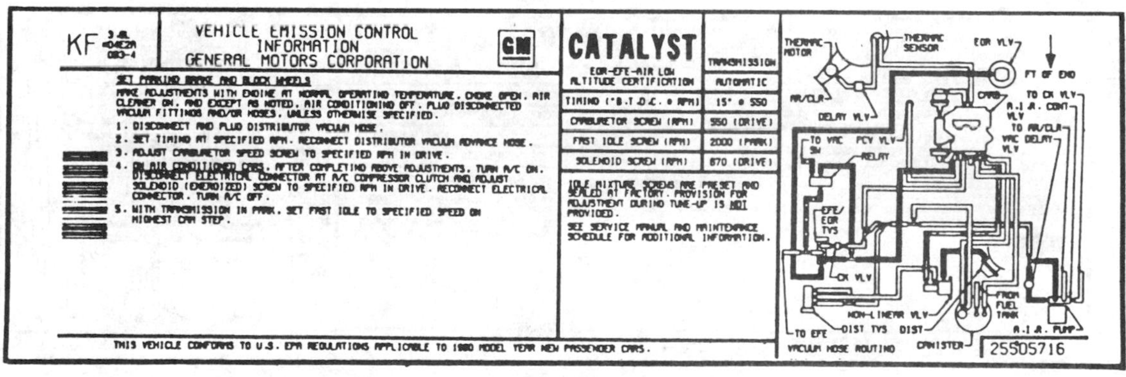

KF 3.8L =04E2A 083-4

VEHICLE EMISSION CONTROL INFORMATION
GENERAL MOTORS CORPORATION
GM

SET PARKING BRAKE AND BLOCK WHEELS
MAKE ADJUSTMENTS WITH ENGINE AT NORMAL OPERATING TEMPERATURE. CHOKE OPEN. AIR CLEANER ON. AND EXCEPT AS NOTED. AIR CONDITIONING OFF. PLUG DISCONNECTED VACUUM FITTINGS AND/OR HOSES. UNLESS OTHERWISE SPECIFIED.

1. DISCONNECT AND PLUG DISTRIBUTOR VACUUM HOSE.
2. SET TIMING AT SPECIFIED RPM. RECONNECT DISTRIBUTOR VACUUM ADVANCE HOSE.
3. ADJUST CARBURETOR SPEED SCREW TO SPECIFIED RPM IN DRIVE.
4. ON AIR CONDITIONED CARS. AFTER COMPLETING ABOVE ADJUSTMENTS. TURN A/C ON. DISCONNECT ELECTRICAL CONNECTOR AT A/C COMPRESSOR CLUTCH AND ADJUST SOLENOID (ENERGIZED) SCREW TO SPECIFIED RPM IN DRIVE. RECONNECT ELECTRICAL CONNECTOR. TURN A/C OFF.
5. WITH TRANSMISSION IN PARK. SET FAST IDLE TO SPECIFIED SPEED ON HIGHEST CAM STEP.

CATALYST EGR-EFE-AIR LOW ALTITUDE CERTIFICATION	TRANSMISSION AUTOMATIC
TIMING (°B.T.D.C. @ RPM)	15° @ 550
CARBURETOR SCREW (RPM)	550 (DRIVE)
FAST IDLE SCREW (RPM)	2000 (PARK)
SOLENOID SCREW (RPM)	670 (DRIVE)

IDLE MIXTURE SCREWS ARE PRESET AND SEALED AT FACTORY. PROVISION FOR ADJUSTMENT DURING TUNE-UP IS NOT PROVIDED.

SEE SERVICE MANUAL AND MAINTENANCE SCHEDULE FOR ADDITIONAL INFORMATION.

THIS VEHICLE CONFORMS TO U.S. EPA REGULATIONS APPLICABLE TO 1980 MODEL YEAR NEW PASSENGER CARS.

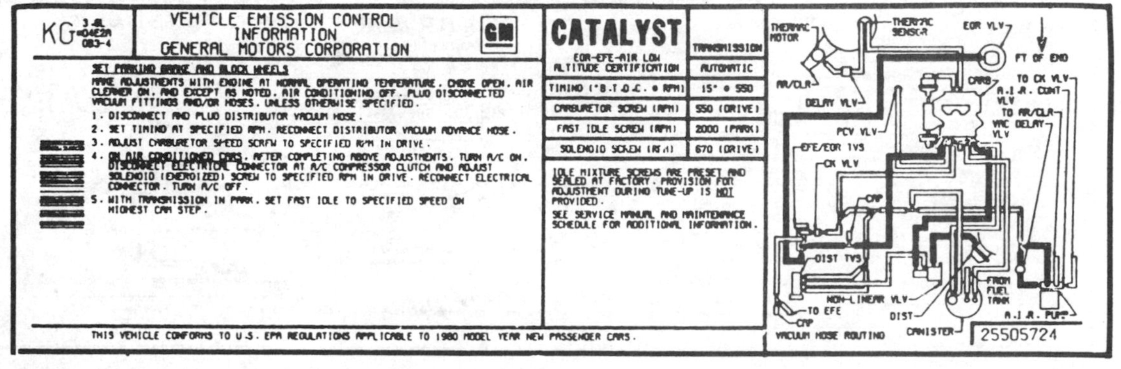

KG 3.8L =04E2A 083-4

VEHICLE EMISSION CONTROL INFORMATION
GENERAL MOTORS CORPORATION
GM

SET PARKING BRAKE AND BLOCK WHEELS
MAKE ADJUSTMENTS WITH ENGINE AT NORMAL OPERATING TEMPERATURE. CHOKE OPEN. AIR CLEANER ON. AND EXCEPT AS NOTED. AIR CONDITIONING OFF. PLUG DISCONNECTED VACUUM FITTINGS AND/OR HOSES. UNLESS OTHERWISE SPECIFIED.

1. DISCONNECT AND PLUG DISTRIBUTOR VACUUM HOSE.
2. SET TIMING AT SPECIFIED RPM. RECONNECT DISTRIBUTOR VACUUM ADVANCE HOSE.
3. ADJUST CARBURETOR SPEED SCREW TO SPECIFIED RPM IN DRIVE.
4. ON AIR CONDITIONED CARS. AFTER COMPLETING ABOVE ADJUSTMENTS. TURN A/C ON. DISCONNECT ELECTRICAL CONNECTOR AT A/C COMPRESSOR CLUTCH AND ADJUST SOLENOID (ENERGIZED) SCREW TO SPECIFIED RPM IN DRIVE. RECONNECT ELECTRICAL CONNECTOR. TURN A/C OFF.
5. WITH TRANSMISSION IN PARK. SET FAST IDLE TO SPECIFIED SPEED ON HIGHEST CAM STEP.

CATALYST EGR-EFE-AIR LOW ALTITUDE CERTIFICATION	TRANSMISSION AUTOMATIC
TIMING (°B.T.D.C. @ RPM)	15° @ 550
CARBURETOR SCREW (RPM)	550 (DRIVE)
FAST IDLE SCREW (RPM)	2000 (PARK)
SOLENOID SCREW (RPM)	670 (DRIVE)

IDLE MIXTURE SCREWS ARE PRESET AND SEALED AT FACTORY. PROVISION FOR ADJUSTMENT DURING TUNE-UP IS NOT PROVIDED.

SEE SERVICE MANUAL AND MAINTENANCE SCHEDULE FOR ADDITIONAL INFORMATION.

THIS VEHICLE CONFORMS TO U.S. EPA REGULATIONS APPLICABLE TO 1980 MODEL YEAR NEW PASSENGER CARS.

VEHICLE EMISSION CONTROL LABELS AND VACUUM HOSE ROUTINGS

1980

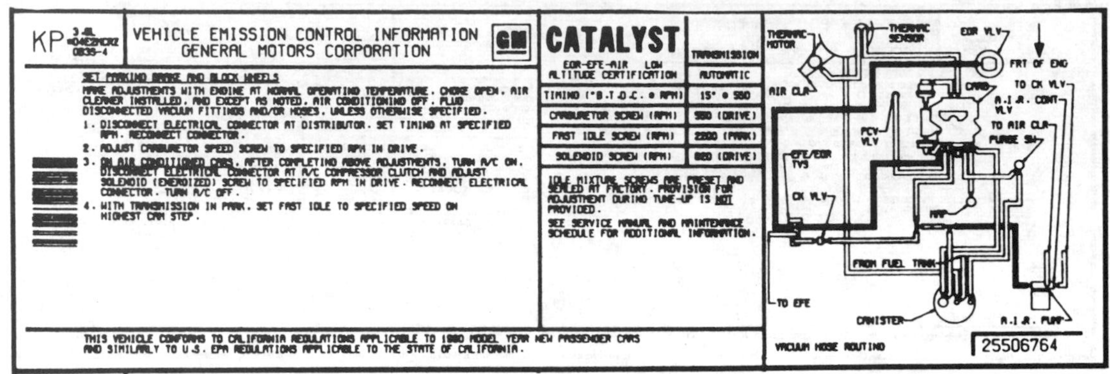

KP 3.8L 0835-4 VEHICLE EMISSION CONTROL INFORMATION GENERAL MOTORS CORPORATION

SET PARKING BRAKE AND BLOCK WHEELS

MAKE ADJUSTMENTS WITH ENGINE AT NORMAL OPERATING TEMPERATURE. CHOKE OPEN. AIR CLEANER INSTALLED. AND EXCEPT AS NOTED. AIR CONDITIONING OFF. PLUG DISCONNECTED VACUUM FITTINGS AND/OR HOSES. UNLESS OTHERWISE SPECIFIED.

1. DISCONNECT ELECTRICAL CONNECTOR AT DISTRIBUTOR. SET TIMING AT SPECIFIED RPM. RECONNECT CONNECTOR.
2. ADJUST CARBURETOR SPEED SCREW TO SPECIFIED RPM IN DRIVE.
3. ON AIR CONDITIONED CARS. AFTER COMPLETING ABOVE ADJUSTMENTS. TURN A/C ON. DISCONNECT ELECTRICAL CONNECTOR AT A/C COMPRESSOR CLUTCH AND ADJUST SOLENOID (ENERGIZED) SCREW TO SPECIFIED RPM IN DRIVE. RECONNECT ELECTRICAL CONNECTOR. TURN A/C OFF.
4. WITH TRANSMISSION IN PARK. SET FAST IDLE TO SPECIFIED SPEED ON HIGHEST CAM STEP.

CATALYST

EGR-EFE-AIR LOW ALTITUDE CERTIFICATION	TRANSMISSION AUTOMATIC
TIMING (°B.T.D.C. @ RPM)	15° @ 550
CARBURETOR SCREW (RPM)	550 (DRIVE)
FAST IDLE SCREW (RPM)	2200 (PARK)
SOLENOID SCREW (RPM)	620 (DRIVE)

IDLE MIXTURE SCREWS ARE PRESET AND SEALED AT FACTORY. PROVISION FOR ADJUSTMENT DURING TUNE-UP IS NOT PROVIDED.

SEE SERVICE MANUAL AND MAINTENANCE SCHEDULE FOR ADDITIONAL INFORMATION.

THIS VEHICLE CONFORMS TO CALIFORNIA REGULATIONS APPLICABLE TO 1980 MODEL YEAR NEW PASSENGER CARS AND SIMILARLY TO U.S. EPA REGULATIONS APPLICABLE TO THE STATE OF CALIFORNIA.

VACUUM HOSE ROUTING 25506764

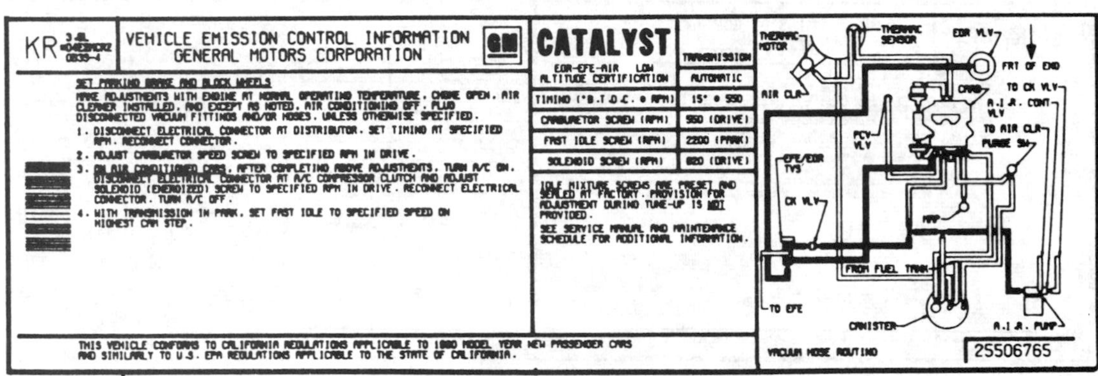

KR 3.8L 0835-4 VEHICLE EMISSION CONTROL INFORMATION GENERAL MOTORS CORPORATION

SET PARKING BRAKE AND BLOCK WHEELS

MAKE ADJUSTMENTS WITH ENGINE AT NORMAL OPERATING TEMPERATURE. CHOKE OPEN. AIR CLEANER INSTALLED. AND EXCEPT AS NOTED. AIR CONDITIONING OFF. PLUG DISCONNECTED VACUUM FITTINGS AND/OR HOSES. UNLESS OTHERWISE SPECIFIED.

1. DISCONNECT ELECTRICAL CONNECTOR AT DISTRIBUTOR. SET TIMING AT SPECIFIED RPM. RECONNECT CONNECTOR.
2. ADJUST CARBURETOR SPEED SCREW TO SPECIFIED RPM IN DRIVE.
3. ON AIR CONDITIONED CARS. AFTER COMPLETING ABOVE ADJUSTMENTS. TURN A/C ON. DISCONNECT ELECTRICAL CONNECTOR AT A/C COMPRESSOR CLUTCH AND ADJUST SOLENOID (ENERGIZED) SCREW TO SPECIFIED RPM IN DRIVE. RECONNECT ELECTRICAL CONNECTOR. TURN A/C OFF.
4. WITH TRANSMISSION IN PARK. SET FAST IDLE TO SPECIFIED SPEED ON HIGHEST CAM STEP.

CATALYST

EGR-EFE-AIR LOW ALTITUDE CERTIFICATION	TRANSMISSION AUTOMATIC
TIMING (°B.T.D.C. @ RPM)	15° @ 550
CARBURETOR SCREW (RPM)	550 (DRIVE)
FAST IDLE SCREW (RPM)	2200 (PARK)
SOLENOID SCREW (RPM)	620 (DRIVE)

IDLE MIXTURE SCREWS ARE PRESET AND SEALED AT FACTORY. PROVISION FOR ADJUSTMENT DURING TUNE-UP IS NOT PROVIDED.

SEE SERVICE MANUAL AND MAINTENANCE SCHEDULE FOR ADDITIONAL INFORMATION.

THIS VEHICLE CONFORMS TO CALIFORNIA REGULATIONS APPLICABLE TO 1980 MODEL YEAR NEW PASSENGER CARS AND SIMILARLY TO U.S. EPA REGULATIONS APPLICABLE TO THE STATE OF CALIFORNIA.

VACUUM HOSE ROUTING 25506765

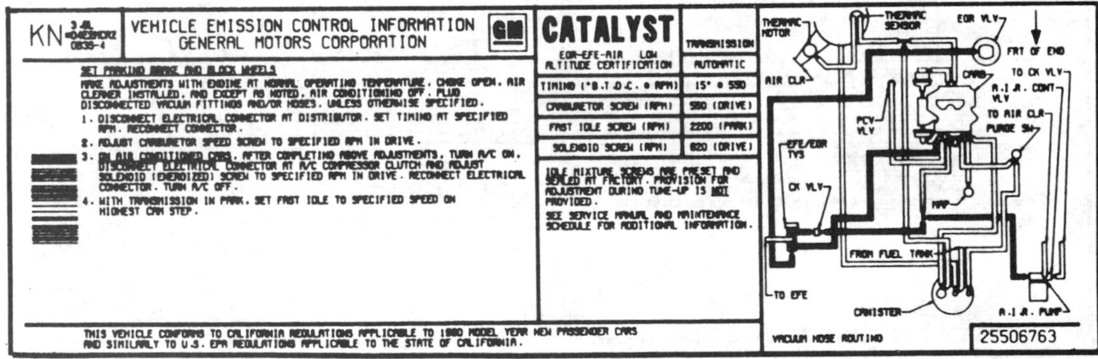

KN 3.8L 0835-4 VEHICLE EMISSION CONTROL INFORMATION GENERAL MOTORS CORPORATION

SET PARKING BRAKE AND BLOCK WHEELS

MAKE ADJUSTMENTS WITH ENGINE AT NORMAL OPERATING TEMPERATURE. CHOKE OPEN. AIR CLEANER INSTALLED. AND EXCEPT AS NOTED. AIR CONDITIONING OFF. PLUG DISCONNECTED VACUUM FITTINGS AND/OR HOSES. UNLESS OTHERWISE SPECIFIED.

1. DISCONNECT ELECTRICAL CONNECTOR AT DISTRIBUTOR. SET TIMING AT SPECIFIED RPM. RECONNECT CONNECTOR.
2. ADJUST CARBURETOR SPEED SCREW TO SPECIFIED RPM IN DRIVE.
3. ON AIR CONDITIONED CARS. AFTER COMPLETING ABOVE ADJUSTMENTS. TURN A/C ON. DISCONNECT ELECTRICAL CONNECTOR AT A/C COMPRESSOR CLUTCH AND ADJUST SOLENOID (ENERGIZED) SCREW TO SPECIFIED RPM IN DRIVE. RECONNECT ELECTRICAL CONNECTOR. TURN A/C OFF.
4. WITH TRANSMISSION IN PARK. SET FAST IDLE TO SPECIFIED SPEED ON HIGHEST CAM STEP.

CATALYST

EGR-EFE-AIR LOW ALTITUDE CERTIFICATION	TRANSMISSION AUTOMATIC
TIMING (°B.T.D.C. @ RPM)	15° @ 550
CARBURETOR SCREW (RPM)	550 (DRIVE)
FAST IDLE SCREW (RPM)	2200 (PARK)
SOLENOID SCREW (RPM)	620 (DRIVE)

IDLE MIXTURE SCREWS ARE PRESET AND SEALED AT FACTORY. PROVISION FOR ADJUSTMENT DURING TUNE-UP IS NOT PROVIDED.

SEE SERVICE MANUAL AND MAINTENANCE SCHEDULE FOR ADDITIONAL INFORMATION.

THIS VEHICLE CONFORMS TO CALIFORNIA REGULATIONS APPLICABLE TO 1980 MODEL YEAR NEW PASSENGER CARS AND SIMILARLY TO U.S. EPA REGULATIONS APPLICABLE TO THE STATE OF CALIFORNIA.

VACUUM HOSE ROUTING 25506763

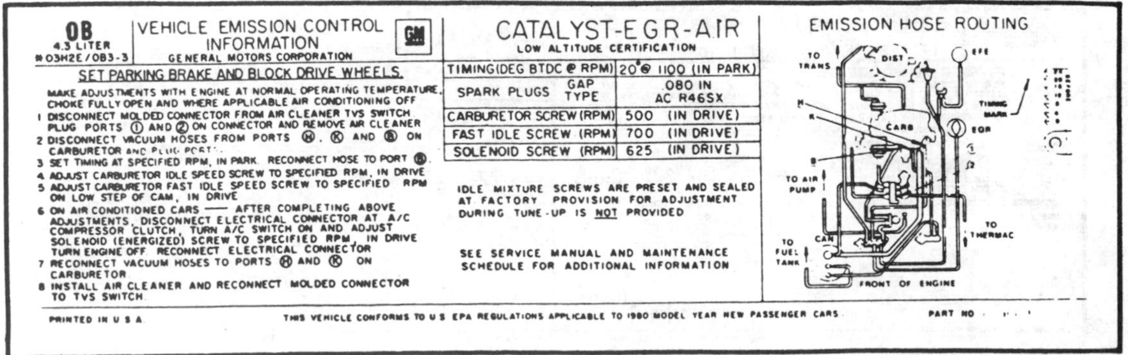

OB 4.3 LITER #03H2E/OB3-3 VEHICLE EMISSION CONTROL INFORMATION GENERAL MOTORS CORPORATION

SET PARKING BRAKE AND BLOCK DRIVE WHEELS.

MAKE ADJUSTMENTS WITH ENGINE AT NORMAL OPERATING TEMPERATURE, CHOKE FULLY OPEN AND WHERE APPLICABLE AIR CONDITIONING OFF

1. DISCONNECT MOLDED CONNECTOR FROM AIR CLEANER TVS SWITCH PLUG PORTS ① AND ② ON CONNECTOR AND REMOVE AIR CLEANER
2. DISCONNECT VACUUM HOSES FROM PORTS Ⓗ, Ⓚ AND Ⓑ ON CARBURETOR AND PLUG PORTS
3. SET TIMING AT SPECIFIED RPM, IN PARK. RECONNECT HOSE TO PORT Ⓑ.
4. ADJUST CARBURETOR IDLE SPEED SCREW TO SPECIFIED RPM, IN DRIVE
5. ADJUST CARBURETOR FAST IDLE SPEED SCREW TO SPECIFIED RPM ON LOW STEP OF CAM, IN DRIVE
6. ON AIR CONDITIONED CARS —— AFTER COMPLETING ABOVE ADJUSTMENTS, DISCONNECT ELECTRICAL CONNECTOR AT A/C COMPRESSOR CLUTCH, TURN A/C SWITCH ON AND ADJUST SOLENOID (ENERGIZED) SCREW TO SPECIFIED RPM, IN DRIVE TURN ENGINE OFF RECONNECT ELECTRICAL CONNECTOR
7. RECONNECT VACUUM HOSES TO PORTS Ⓗ AND Ⓚ ON CARBURETOR
8. INSTALL AIR CLEANER AND RECONNECT MOLDED CONNECTOR TO TVS SWITCH

CATALYST-EGR-AIR
LOW ALTITUDE CERTIFICATION

TIMING (DEG BTDC @ RPM)	20° @ 1100 (IN PARK)
SPARK PLUGS GAP / TYPE	.080 IN / AC R46SX
CARBURETOR SCREW (RPM)	500 (IN DRIVE)
FAST IDLE SCREW (RPM)	700 (IN DRIVE)
SOLENOID SCREW (RPM)	625 (IN DRIVE)

IDLE MIXTURE SCREWS ARE PRESET AND SEALED AT FACTORY PROVISION FOR ADJUSTMENT DURING TUNE-UP IS NOT PROVIDED

SEE SERVICE MANUAL AND MAINTENANCE SCHEDULE FOR ADDITIONAL INFORMATION

EMISSION HOSE ROUTING

PRINTED IN U S A

THIS VEHICLE CONFORMS TO U S EPA REGULATIONS APPLICABLE TO 1980 MODEL YEAR NEW PASSENGER CARS

PART NO

VEHICLE EMISSION CONTROL LABELS AND VACUUM HOSE ROUTINGS

1980

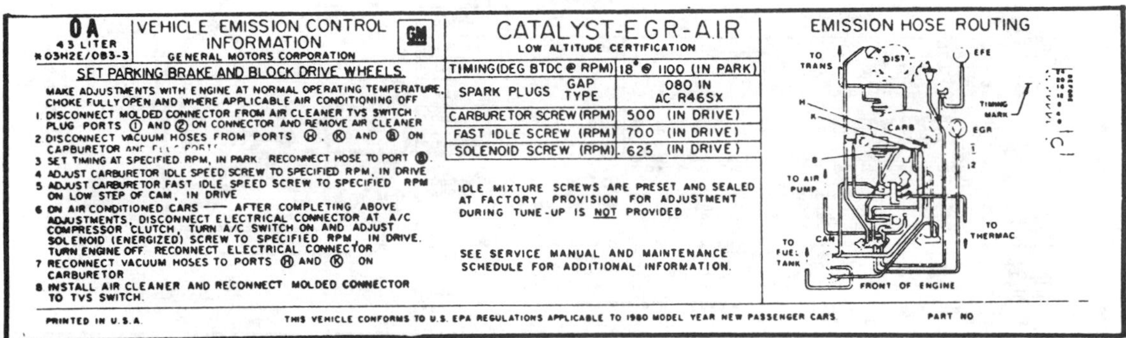

0A 4.3 LITER #03H2E/0B3-3

VEHICLE EMISSION CONTROL INFORMATION
GENERAL MOTORS CORPORATION

SET PARKING BRAKE AND BLOCK DRIVE WHEELS.

MAKE ADJUSTMENTS WITH ENGINE AT NORMAL OPERATING TEMPERATURE, CHOKE FULLY OPEN AND WHERE APPLICABLE AIR CONDITIONING OFF

1 DISCONNECT MOLDED CONNECTOR FROM AIR CLEANER TVS SWITCH. PLUG PORTS ① AND ② ON CONNECTOR AND REMOVE AIR CLEANER
2 DISCONNECT VACUUM HOSES FROM PORTS Ⓗ, Ⓚ AND Ⓢ ON CARBURETOR AND PLUG PORTS.
3 SET TIMING AT SPECIFIED RPM, IN PARK. RECONNECT HOSE TO PORT Ⓢ.
4 ADJUST CARBURETOR IDLE SPEED SCREW TO SPECIFIED RPM, IN DRIVE
5 ADJUST CARBURETOR FAST IDLE SPEED SCREW TO SPECIFIED RPM ON LOW STEP OF CAM, IN DRIVE
6 ON AIR CONDITIONED CARS — AFTER COMPLETING ABOVE ADJUSTMENTS, DISCONNECT ELECTRICAL CONNECTOR AT A/C COMPRESSOR CLUTCH, TURN A/C SWITCH ON AND ADJUST SOLENOID (ENERGIZED) SCREW TO SPECIFIED RPM, IN DRIVE. TURN ENGINE OFF. RECONNECT ELECTRICAL CONNECTOR
7 RECONNECT VACUUM HOSES TO PORTS Ⓗ AND Ⓚ ON CARBURETOR
8 INSTALL AIR CLEANER AND RECONNECT MOLDED CONNECTOR TO TVS SWITCH.

CATALYST-EGR-AIR
LOW ALTITUDE CERTIFICATION

TIMING(DEG BTDC @ RPM)	18° @ 1100 (IN PARK)
SPARK PLUGS GAP / TYPE	.080 IN / AC R46SX
CARBURETOR SCREW (RPM)	500 (IN DRIVE)
FAST IDLE SCREW (RPM)	700 (IN DRIVE)
SOLENOID SCREW (RPM)	625 (IN DRIVE)

IDLE MIXTURE SCREWS ARE PRESET AND SEALED AT FACTORY. PROVISION FOR ADJUSTMENT DURING TUNE-UP IS NOT PROVIDED

SEE SERVICE MANUAL AND MAINTENANCE SCHEDULE FOR ADDITIONAL INFORMATION.

EMISSION HOSE ROUTING

PRINTED IN U.S.A. THIS VEHICLE CONFORMS TO U.S. EPA REGULATIONS APPLICABLE TO 1980 MODEL YEAR NEW PASSENGER CARS PART NO

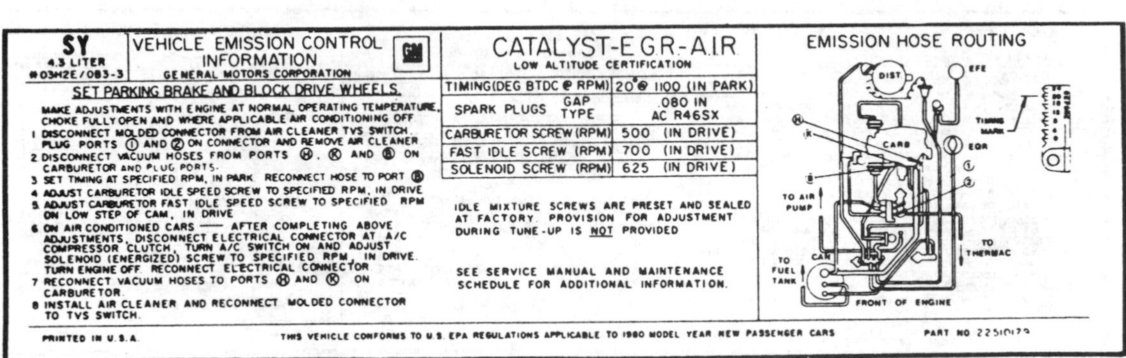

SY 4.3 LITER #03H2E/0B3-3

VEHICLE EMISSION CONTROL INFORMATION
GENERAL MOTORS CORPORATION

SET PARKING BRAKE AND BLOCK DRIVE WHEELS.

MAKE ADJUSTMENTS WITH ENGINE AT NORMAL OPERATING TEMPERATURE, CHOKE FULLY OPEN AND WHERE APPLICABLE AIR CONDITIONING OFF

1 DISCONNECT MOLDED CONNECTOR FROM AIR CLEANER TVS SWITCH. PLUG PORTS ① AND ② ON CONNECTOR AND REMOVE AIR CLEANER
2 DISCONNECT VACUUM HOSES FROM PORTS Ⓗ, Ⓚ AND Ⓢ ON CARBURETOR AND PLUG PORTS.
3 SET TIMING AT SPECIFIED RPM, IN PARK. RECONNECT HOSE TO PORT Ⓢ.
4 ADJUST CARBURETOR IDLE SPEED SCREW TO SPECIFIED RPM, IN DRIVE
5 ADJUST CARBURETOR FAST IDLE SPEED SCREW TO SPECIFIED RPM ON LOW STEP OF CAM, IN DRIVE
6 ON AIR CONDITIONED CARS — AFTER COMPLETING ABOVE ADJUSTMENTS, DISCONNECT ELECTRICAL CONNECTOR AT A/C COMPRESSOR CLUTCH, TURN A/C SWITCH ON AND ADJUST SOLENOID (ENERGIZED) SCREW TO SPECIFIED RPM, IN DRIVE. TURN ENGINE OFF. RECONNECT ELECTRICAL CONNECTOR
7 RECONNECT VACUUM HOSES TO PORTS Ⓗ AND Ⓚ ON CARBURETOR
8 INSTALL AIR CLEANER AND RECONNECT MOLDED CONNECTOR TO TVS SWITCH.

CATALYST-EGR-AIR
LOW ALTITUDE CERTIFICATION

TIMING(DEG BTDC @ RPM)	20° @ 1100 (IN PARK)
SPARK PLUGS GAP / TYPE	.080 IN / AC R46SX
CARBURETOR SCREW (RPM)	500 (IN DRIVE)
FAST IDLE SCREW (RPM)	700 (IN DRIVE)
SOLENOID SCREW (RPM)	625 (IN DRIVE)

IDLE MIXTURE SCREWS ARE PRESET AND SEALED AT FACTORY. PROVISION FOR ADJUSTMENT DURING TUNE-UP IS NOT PROVIDED

SEE SERVICE MANUAL AND MAINTENANCE SCHEDULE FOR ADDITIONAL INFORMATION.

EMISSION HOSE ROUTING

PRINTED IN U.S.A. THIS VEHICLE CONFORMS TO U.S. EPA REGULATIONS APPLICABLE TO 1980 MODEL YEAR NEW PASSENGER CARS PART NO 22510173

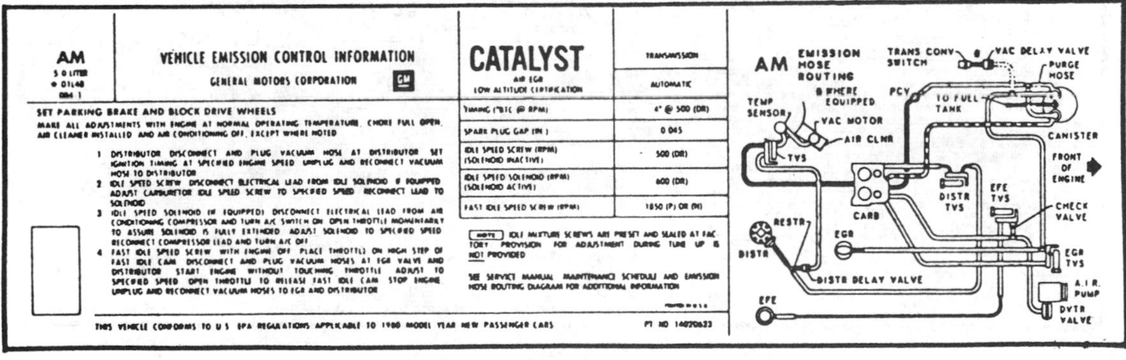

AM 5.0 LITER ● 01L48 0M-1

VEHICLE EMISSION CONTROL INFORMATION
GENERAL MOTORS CORPORATION

SET PARKING BRAKE AND BLOCK DRIVE WHEELS

MAKE ALL ADJUSTMENTS WITH ENGINE AT NORMAL OPERATING TEMPERATURE, CHOKE FULL OPEN, AIR CLEANER INSTALLED AND AIR CONDITIONING OFF, EXCEPT WHERE NOTED

1 DISTRIBUTOR: DISCONNECT AND PLUG VACUUM HOSE AT DISTRIBUTOR. SET IGNITION TIMING AT SPECIFIED ENGINE SPEED. UNPLUG AND RECONNECT VACUUM HOSE TO DISTRIBUTOR
2 IDLE SPEED SCREW: DISCONNECT ELECTRICAL LEAD FROM IDLE SOLENOID IF EQUIPPED. ADJUST CARBURETOR IDLE SPEED SCREW TO SPECIFIED SPEED. RECONNECT LEAD TO SOLENOID
3 IDLE SPEED SOLENOID (IF EQUIPPED): DISCONNECT ELECTRICAL LEAD FROM AIR CONDITIONING COMPRESSOR AND TURN A/C SWITCH ON. OPEN THROTTLE MOMENTARILY TO ASSURE SOLENOID IS FULLY EXTENDED. ADJUST SOLENOID TO SPECIFIED SPEED. RECONNECT COMPRESSOR LEAD AND TURN A/C OFF
4 FAST IDLE SPEED SCREW: WITH ENGINE OFF, PLACE THROTTLE ON HIGH STEP OF FAST IDLE CAM. DISCONNECT AND PLUG VACUUM HOSES AT EGR VALVE AND DISTRIBUTOR. START ENGINE WITHOUT TOUCHING THROTTLE. ADJUST TO SPECIFIED SPEED. OPEN THROTTLE TO RELEASE FAST IDLE CAM. STOP ENGINE. UNPLUG AND RECONNECT VACUUM HOSES TO EGR AND DISTRIBUTOR

CATALYST
AIR EGR
LOW ALTITUDE CERTIFICATION

	TRANSMISSION
	AUTOMATIC
TIMING (°BTC @ RPM)	4° @ 500 (DR)
SPARK PLUG GAP (IN.)	0.045
IDLE SPEED SCREW (RPM) (SOLENOID INACTIVE)	500 (DR)
IDLE SPEED SOLENOID (RPM) (SOLENOID ACTIVE)	600 (DR)
FAST IDLE SPEED SCREW (RPM)	1850 (P) OR (N)

NOTE: IDLE MIXTURE SCREWS ARE PRESET AND SEALED AT FACTORY. PROVISION FOR ADJUSTMENT DURING TUNE UP IS NOT PROVIDED

SEE SERVICE MANUAL, MAINTENANCE SCHEDULE AND EMISSION HOSE ROUTING DIAGRAM FOR ADDITIONAL INFORMATION

PRINTED IN U.S.A.

THIS VEHICLE CONFORMS TO U.S. EPA REGULATIONS APPLICABLE TO 1980 MODEL YEAR NEW PASSENGER CARS PT NO 14070623

AM EMISSION HOSE ROUTING

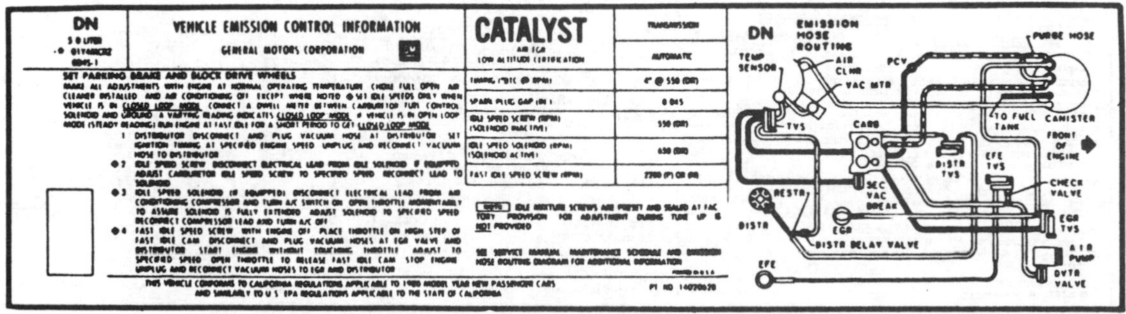

DN 5.0 LITER ● 01Y4MCR2 0D45-1

VEHICLE EMISSION CONTROL INFORMATION
GENERAL MOTORS CORPORATION

SET PARKING BRAKE AND BLOCK DRIVE WHEELS

MAKE ALL ADJUSTMENTS WITH ENGINE AT NORMAL OPERATING TEMPERATURE, CHOKE FULL OPEN, AIR CLEANER INSTALLED AND AIR CONDITIONING OFF, EXCEPT WHERE NOTED. ● SET IDLE SPEEDS ONLY WHEN VEHICLE IS IN CLOSED LOOP MODE. CONNECT A DWELL METER BETWEEN CARBURETOR FUEL CONTROL SOLENOID AND GROUND. A VARYING READING INDICATES CLOSED LOOP MODE. IF VEHICLE IS IN OPEN LOOP MODE (STEADY READING) RUN ENGINE AT FAST IDLE FOR A SHORT PERIOD TO GET CLOSED LOOP MODE.

1 DISTRIBUTOR: DISCONNECT AND PLUG VACUUM HOSE AT DISTRIBUTOR. SET IGNITION TIMING AT SPECIFIED ENGINE SPEED. UNPLUG AND RECONNECT VACUUM HOSE TO DISTRIBUTOR
●2 IDLE SPEED SCREW: DISCONNECT ELECTRICAL LEAD FROM IDLE SOLENOID IF EQUIPPED. ADJUST CARBURETOR IDLE SPEED SCREW TO SPECIFIED SPEED. RECONNECT LEAD TO SOLENOID
●3 IDLE SPEED SOLENOID (IF EQUIPPED): DISCONNECT ELECTRICAL LEAD FROM AIR CONDITIONING COMPRESSOR AND TURN A/C SWITCH ON. OPEN THROTTLE MOMENTARILY TO ASSURE SOLENOID IS FULLY EXTENDED. ADJUST SOLENOID TO SPECIFIED SPEED. RECONNECT COMPRESSOR LEAD AND TURN A/C OFF
●4 FAST IDLE SPEED SCREW: WITH ENGINE OFF, PLACE THROTTLE ON HIGH STEP OF FAST IDLE CAM. DISCONNECT AND PLUG VACUUM HOSES AT EGR VALVE AND DISTRIBUTOR. START ENGINE WITHOUT TOUCHING THROTTLE. ADJUST TO SPECIFIED SPEED. OPEN THROTTLE TO RELEASE FAST IDLE CAM. STOP ENGINE. UNPLUG AND RECONNECT VACUUM HOSES TO EGR AND DISTRIBUTOR

CATALYST
AIR EGR
LOW ALTITUDE CERTIFICATION

	TRANSMISSION
	AUTOMATIC
TIMING (°BTC @ RPM)	4° @ 550 (DR)
SPARK PLUG GAP (IN.)	0.045
IDLE SPEED SCREW (RPM) (SOLENOID INACTIVE)	550 (DR)
IDLE SPEED SOLENOID (RPM) (SOLENOID ACTIVE)	650 (DR)
FAST IDLE SPEED SCREW (RPM)	2200 (P) OR (N)

NOTE: IDLE MIXTURE SCREWS ARE PRESET AND SEALED AT FACTORY. PROVISION FOR ADJUSTMENT DURING TUNE UP IS NOT PROVIDED

SEE SERVICE MANUAL, MAINTENANCE SCHEDULE AND EMISSION HOSE ROUTING DIAGRAM FOR ADDITIONAL INFORMATION

PRINTED IN U.S.A.

THIS VEHICLE CONFORMS TO CALIFORNIA REGULATIONS APPLICABLE TO 1980 MODEL YEAR NEW PASSENGER CARS AND SIMILARLY TO U.S. EPA REGULATIONS APPLICABLE TO THE STATE OF CALIFORNIA PT NO 14070620

DN EMISSION HOSE ROUTING

VEHICLE EMISSION CONTROL LABELS AND VACUUM HOSE ROUTINGS 1980

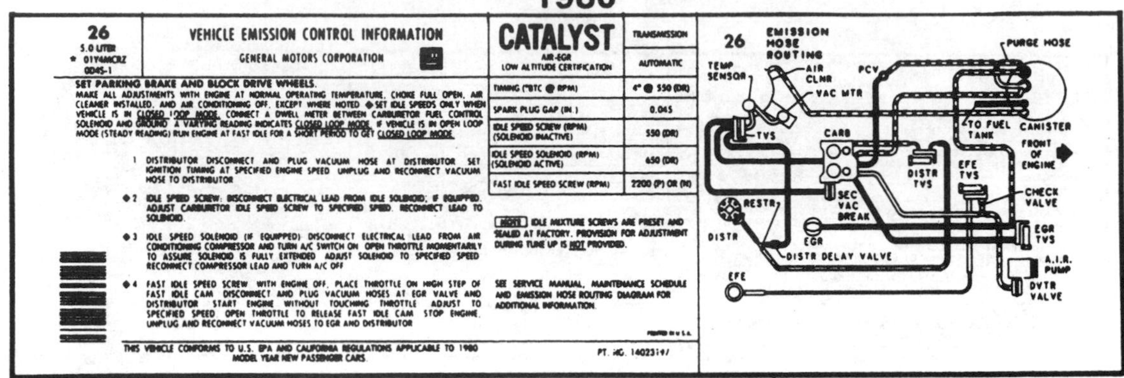

26
5.0 LITER
* 01Y4MCRZ
0D45-1

VEHICLE EMISSION CONTROL INFORMATION
GENERAL MOTORS CORPORATION

SET PARKING BRAKE AND BLOCK DRIVE WHEELS.
MAKE ALL ADJUSTMENTS WITH ENGINE AT NORMAL OPERATING TEMPERATURE, CHOKE FULL OPEN, AIR CLEANER INSTALLED, AND AIR CONDITIONING OFF, EXCEPT WHERE NOTED ◆ SET IDLE SPEEDS ONLY WHEN VEHICLE IS IN CLOSED LOOP MODE. CONNECT A DWELL METER BETWEEN CARBURETOR FUEL CONTROL SOLENOID AND GROUND. A VARYING READING INDICATES CLOSED LOOP MODE. IF VEHICLE IS IN OPEN LOOP MODE (STEADY READING) RUN ENGINE AT FAST IDLE FOR A SHORT PERIOD TO GET CLOSED LOOP MODE.

1 DISTRIBUTOR: DISCONNECT AND PLUG VACUUM HOSE AT DISTRIBUTOR. SET IGNITION TIMING AT SPECIFIED ENGINE SPEED. UNPLUG AND RECONNECT VACUUM HOSE TO DISTRIBUTOR.

◆2 IDLE SPEED SCREW: DISCONNECT ELECTRICAL LEAD FROM IDLE SOLENOID, IF EQUIPPED. ADJUST CARBURETOR IDLE SPEED SCREW TO SPECIFIED SPEED. RECONNECT LEAD TO SOLENOID.

◆3 IDLE SPEED SOLENOID (IF EQUIPPED): DISCONNECT ELECTRICAL LEAD FROM AIR CONDITIONING COMPRESSOR AND TURN A/C SWITCH ON. OPEN THROTTLE MOMENTARILY TO ASSURE SOLENOID IS FULLY EXTENDED. ADJUST SOLENOID TO SPECIFIED SPEED. RECONNECT COMPRESSOR LEAD AND TURN A/C OFF.

◆4 FAST IDLE SPEED SCREW: WITH ENGINE OFF, PLACE THROTTLE ON HIGH STEP OF FAST IDLE CAM. DISCONNECT AND PLUG VACUUM HOSES AT EGR VALVE AND DISTRIBUTOR. START ENGINE WITHOUT TOUCHING THROTTLE. ADJUST TO SPECIFIED SPEED. OPEN THROTTLE TO RELEASE FAST IDLE CAM. STOP ENGINE. UNPLUG AND RECONNECT VACUUM HOSES TO EGR AND DISTRIBUTOR.

THIS VEHICLE CONFORMS TO U.S. EPA AND CALIFORNIA REGULATIONS APPLICABLE TO 1980 MODEL YEAR NEW PASSENGER CARS.

CATALYST
AIR-EGR
LOW ALTITUDE CERTIFICATION

	TRANSMISSION
	AUTOMATIC
TIMING (°BTC @ RPM)	4° @ 550 (DR)
SPARK PLUG GAP (IN.)	0.045
IDLE SPEED SCREW (RPM) (SOLENOID INACTIVE)	550 (DR)
IDLE SPEED SOLENOID (RPM) (SOLENOID ACTIVE)	650 (DR)
FAST IDLE SPEED SCREW (RPM)	2200 (P) OR (N)

NOTE IDLE MIXTURE SCREWS ARE PRESET AND SEALED AT FACTORY. PROVISION FOR ADJUSTMENT DURING TUNE UP IS NOT PROVIDED.

SEE SERVICE MANUAL, MAINTENANCE SCHEDULE AND EMISSION HOSE ROUTING DIAGRAM FOR ADDITIONAL INFORMATION.

PRINTED IN U.S.A.
PT. NO. 14023197

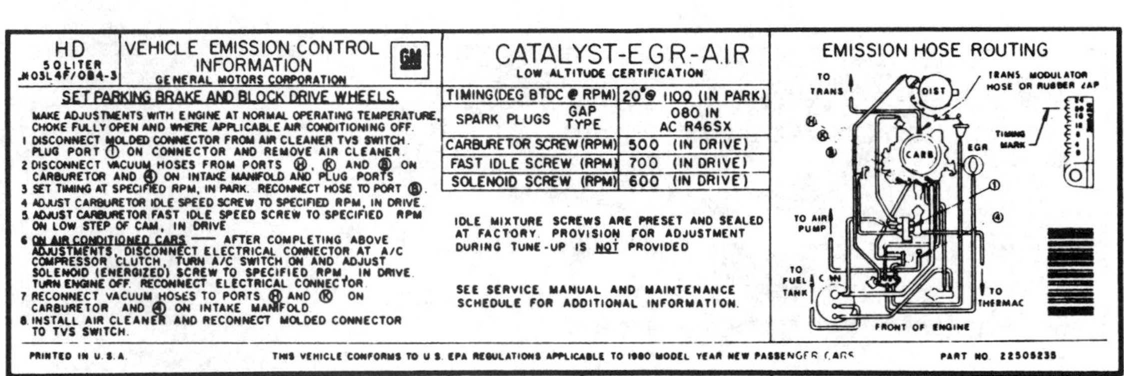

HD
5.0 LITER
N03L4F/0B4-3

VEHICLE EMISSION CONTROL INFORMATION
GENERAL MOTORS CORPORATION
GM

SET PARKING BRAKE AND BLOCK DRIVE WHEELS.
MAKE ADJUSTMENTS WITH ENGINE AT NORMAL OPERATING TEMPERATURE, CHOKE FULLY OPEN AND WHERE APPLICABLE AIR CONDITIONING OFF.
1 DISCONNECT MOLDED CONNECTOR FROM AIR CLEANER TVS SWITCH. PLUG PORT Ⓘ ON CONNECTOR AND REMOVE AIR CLEANER.
2 DISCONNECT VACUUM HOSES FROM PORTS Ⓗ, Ⓚ AND Ⓑ ON CARBURETOR AND Ⓓ ON INTAKE MANIFOLD AND PLUG PORTS.
3 SET TIMING AT SPECIFIED RPM, IN PARK. RECONNECT HOSE TO PORT Ⓑ.
4 ADJUST CARBURETOR IDLE SPEED SCREW TO SPECIFIED RPM, IN DRIVE.
5 ADJUST CARBURETOR FAST IDLE SPEED SCREW TO SPECIFIED RPM ON LOW STEP OF CAM, IN DRIVE.
6 ON AIR CONDITIONED CARS —— AFTER COMPLETING ABOVE ADJUSTMENTS, DISCONNECT ELECTRICAL CONNECTOR AT A/C COMPRESSOR CLUTCH, TURN A/C SWITCH ON AND ADJUST SOLENOID (ENERGIZED) SCREW TO SPECIFIED RPM, IN DRIVE. TURN ENGINE OFF. RECONNECT ELECTRICAL CONNECTOR.
7 RECONNECT VACUUM HOSES TO PORTS Ⓗ AND Ⓚ ON CARBURETOR AND Ⓓ ON INTAKE MANIFOLD.
8 INSTALL AIR CLEANER AND RECONNECT MOLDED CONNECTOR TO TVS SWITCH.

CATALYST-EGR-AIR
LOW ALTITUDE CERTIFICATION

TIMING (DEG BTDC @ RPM)	20° @ 1100 (IN PARK)	
SPARK PLUGS	GAP	.080 IN
	TYPE	AC R46SX
CARBURETOR SCREW (RPM)	500 (IN DRIVE)	
FAST IDLE SCREW (RPM)	700 (IN DRIVE)	
SOLENOID SCREW (RPM)	600 (IN DRIVE)	

IDLE MIXTURE SCREWS ARE PRESET AND SEALED AT FACTORY. PROVISION FOR ADJUSTMENT DURING TUNE-UP IS NOT PROVIDED.

SEE SERVICE MANUAL AND MAINTENANCE SCHEDULE FOR ADDITIONAL INFORMATION.

PRINTED IN U.S.A. THIS VEHICLE CONFORMS TO U.S. EPA REGULATIONS APPLICABLE TO 1980 MODEL YEAR NEW PASSENGER CARS PART NO. 22505235

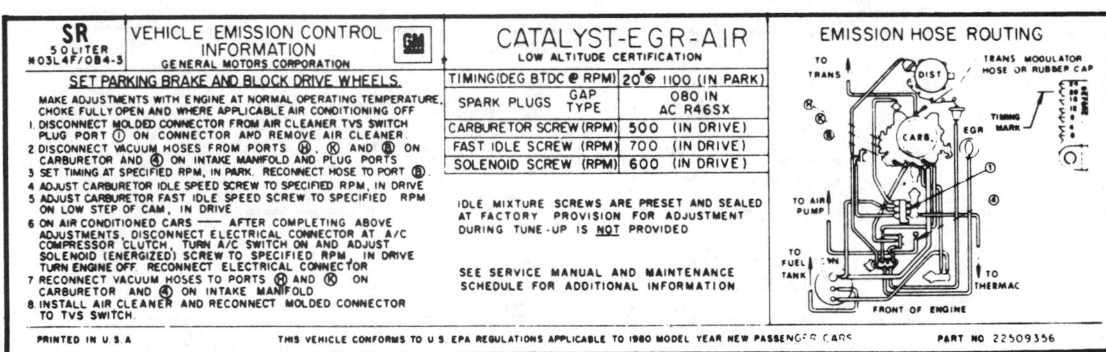

SR
5.0 LITER
N03L4F/0B4-3

VEHICLE EMISSION CONTROL INFORMATION
GENERAL MOTORS CORPORATION
GM

SET PARKING BRAKE AND BLOCK DRIVE WHEELS.
MAKE ADJUSTMENTS WITH ENGINE AT NORMAL OPERATING TEMPERATURE, CHOKE FULLY OPEN AND WHERE APPLICABLE AIR CONDITIONING OFF.
1. DISCONNECT MOLDED CONNECTOR FROM AIR CLEANER TVS SWITCH. PLUG PORT Ⓘ ON CONNECTOR AND REMOVE AIR CLEANER.
2 DISCONNECT VACUUM HOSES FROM PORTS Ⓗ, Ⓚ AND Ⓑ ON CARBURETOR AND Ⓓ ON INTAKE MANIFOLD AND PLUG PORTS.
3 SET TIMING AT SPECIFIED RPM, IN PARK. RECONNECT HOSE TO PORT Ⓑ.
4 ADJUST CARBURETOR IDLE SPEED SCREW TO SPECIFIED RPM, IN DRIVE.
5 ADJUST CARBURETOR FAST IDLE SPEED SCREW TO SPECIFIED RPM ON LOW STEP OF CAM, IN DRIVE.
6 ON AIR CONDITIONED CARS —— AFTER COMPLETING ABOVE ADJUSTMENTS, DISCONNECT ELECTRICAL CONNECTOR AT A/C COMPRESSOR CLUTCH, TURN A/C SWITCH ON AND ADJUST SOLENOID (ENERGIZED) SCREW TO SPECIFIED RPM, IN DRIVE. TURN ENGINE OFF. RECONNECT ELECTRICAL CONNECTOR.
7 RECONNECT VACUUM HOSES TO PORTS Ⓗ AND Ⓚ ON CARBURETOR AND Ⓓ ON INTAKE MANIFOLD.
8. INSTALL AIR CLEANER AND RECONNECT MOLDED CONNECTOR TO TVS SWITCH.

CATALYST-EGR-AIR
LOW ALTITUDE CERTIFICATION

TIMING (DEG BTDC @ RPM)	20° @ 1100 (IN PARK)	
SPARK PLUGS	GAP	.080 IN
	TYPE	AC R46SX
CARBURETOR SCREW (RPM)	500 (IN DRIVE)	
FAST IDLE SCREW (RPM)	700 (IN DRIVE)	
SOLENOID SCREW (RPM)	600 (IN DRIVE)	

IDLE MIXTURE SCREWS ARE PRESET AND SEALED AT FACTORY. PROVISION FOR ADJUSTMENT DURING TUNE-UP IS NOT PROVIDED.

SEE SERVICE MANUAL AND MAINTENANCE SCHEDULE FOR ADDITIONAL INFORMATION.

PRINTED IN U.S.A. THIS VEHICLE CONFORMS TO U.S. EPA REGULATIONS APPLICABLE TO 1980 MODEL YEAR NEW PASSENGER CARS PART NO. 22509356

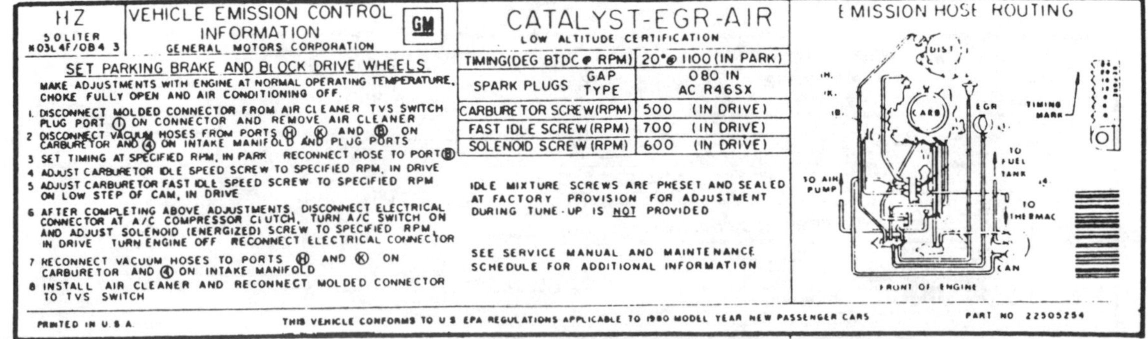

HZ
5.0 LITER
N03L4F/0B4 3

VEHICLE EMISSION CONTROL INFORMATION
GENERAL MOTORS CORPORATION
GM

SET PARKING BRAKE AND BLOCK DRIVE WHEELS.
MAKE ADJUSTMENTS WITH ENGINE AT NORMAL OPERATING TEMPERATURE, CHOKE FULLY OPEN AND AIR CONDITIONING OFF.
1. DISCONNECT MOLDED CONNECTOR FROM AIR CLEANER TVS SWITCH. PLUG PORT Ⓘ ON CONNECTOR AND REMOVE AIR CLEANER.
2 DISCONNECT VACUUM HOSES FROM PORTS Ⓗ, Ⓚ AND Ⓑ ON CARBURETOR AND Ⓓ ON INTAKE MANIFOLD AND PLUG PORTS.
3 SET TIMING AT SPECIFIED RPM, IN PARK. RECONNECT HOSE TO PORT Ⓑ.
4 ADJUST CARBURETOR IDLE SPEED SCREW TO SPECIFIED RPM, IN DRIVE.
5 ADJUST CARBURETOR FAST IDLE SPEED SCREW TO SPECIFIED RPM ON LOW STEP OF CAM, IN DRIVE.
6 AFTER COMPLETING ABOVE ADJUSTMENTS, DISCONNECT ELECTRICAL CONNECTOR AT A/C COMPRESSOR CLUTCH, TURN A/C SWITCH ON AND ADJUST SOLENOID (ENERGIZED) SCREW TO SPECIFIED RPM, IN DRIVE. TURN ENGINE OFF. RECONNECT ELECTRICAL CONNECTOR.
7 RECONNECT VACUUM HOSES TO PORTS Ⓗ AND Ⓚ ON CARBURETOR AND Ⓓ ON INTAKE MANIFOLD.
8 INSTALL AIR CLEANER AND RECONNECT MOLDED CONNECTOR TO TVS SWITCH.

CATALYST-EGR-AIR
LOW ALTITUDE CERTIFICATION

TIMING (DEG BTDC @ RPM)	20° @ 1100 (IN PARK)	
SPARK PLUGS	GAP	.080 IN
	TYPE	AC R46SX
CARBURETOR SCREW (RPM)	500 (IN DRIVE)	
FAST IDLE SCREW (RPM)	700 (IN DRIVE)	
SOLENOID SCREW (RPM)	600 (IN DRIVE)	

IDLE MIXTURE SCREWS ARE PRESET AND SEALED AT FACTORY. PROVISION FOR ADJUSTMENT DURING TUNE-UP IS NOT PROVIDED.

SEE SERVICE MANUAL AND MAINTENANCE SCHEDULE FOR ADDITIONAL INFORMATION.

PRINTED IN U.S.A. THIS VEHICLE CONFORMS TO U.S. EPA REGULATIONS APPLICABLE TO 1980 MODEL YEAR NEW PASSENGER CARS PART NO. 22505254

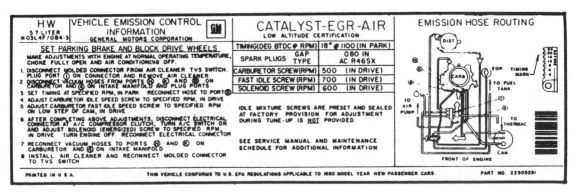

VEHICLE EMISSION CONTROL LABELS AND VACUUM HOSE ROUTINGS

1980

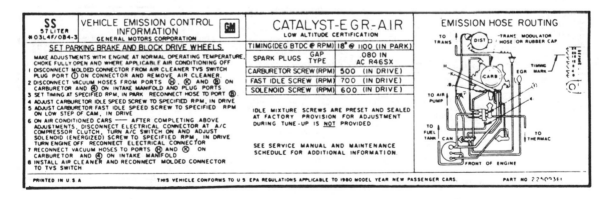

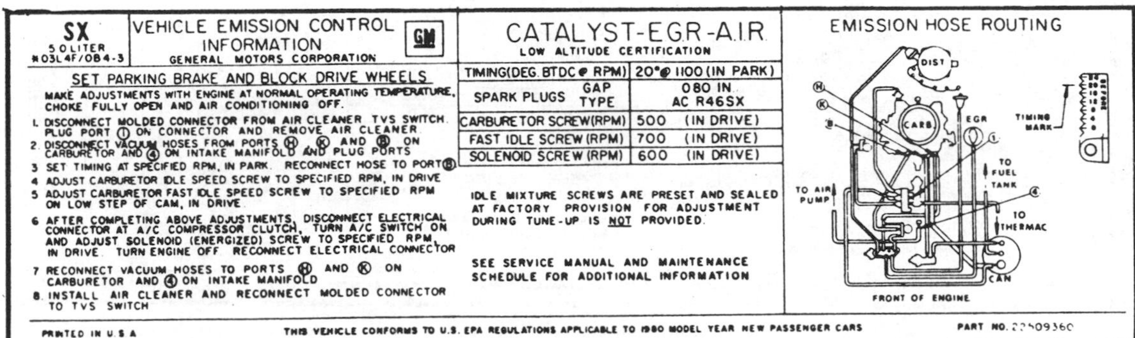

SX
5.0 LITER
#03L4F/0B4-3

VEHICLE EMISSION CONTROL INFORMATION
GENERAL MOTORS CORPORATION

GM

SET PARKING BRAKE AND BLOCK DRIVE WHEELS

MAKE ADJUSTMENTS WITH ENGINE AT NORMAL OPERATING TEMPERATURE, CHOKE FULLY OPEN AND AIR CONDITIONING OFF.

1. DISCONNECT MOLDED CONNECTOR FROM AIR CLEANER TVS SWITCH. PLUG PORT (1) ON CONNECTOR AND REMOVE AIR CLEANER
2. DISCONNECT VACUUM HOSES FROM PORTS (H), (K) AND (B) ON CARBURETOR AND (4) ON INTAKE MANIFOLD AND PLUG PORTS
3. SET TIMING AT SPECIFIED RPM, IN PARK. RECONNECT HOSE TO PORT (B)
4. ADJUST CARBURETOR IDLE SPEED SCREW TO SPECIFIED RPM, IN DRIVE
5. ADJUST CARBURETOR FAST IDLE SPEED SCREW TO SPECIFIED RPM ON LOW STEP OF CAM, IN DRIVE
6. AFTER COMPLETING ABOVE ADJUSTMENTS, DISCONNECT ELECTRICAL CONNECTOR AT A/C COMPRESSOR CLUTCH, TURN A/C SWITCH ON AND ADJUST SOLENOID (ENERGIZED) SCREW TO SPECIFIED RPM, IN DRIVE. TURN ENGINE OFF. RECONNECT ELECTRICAL CONNECTOR
7. RECONNECT VACUUM HOSES TO PORTS (H) AND (K) ON CARBURETOR AND (4) ON INTAKE MANIFOLD
8. INSTALL AIR CLEANER AND RECONNECT MOLDED CONNECTOR TO TVS SWITCH

CATALYST-EGR-AIR
LOW ALTITUDE CERTIFICATION

TIMING (DEG. BTDC @ RPM)		20° @ 1100 (IN PARK)
SPARK PLUGS	GAP	.080 IN
	TYPE	AC R46SX
CARBURETOR SCREW (RPM)		500 (IN DRIVE)
FAST IDLE SCREW (RPM)		700 (IN DRIVE)
SOLENOID SCREW (RPM)		600 (IN DRIVE)

IDLE MIXTURE SCREWS ARE PRESET AND SEALED AT FACTORY. PROVISION FOR ADJUSTMENT DURING TUNE-UP IS NOT PROVIDED.

SEE SERVICE MANUAL AND MAINTENANCE SCHEDULE FOR ADDITIONAL INFORMATION

EMISSION HOSE ROUTING

PRINTED IN U.S.A.

THIS VEHICLE CONFORMS TO U.S. EPA REGULATIONS APPLICABLE TO 1980 MODEL YEAR NEW PASSENGER CARS

PART NO. 22509360

HM
5.7 LITER
#03L4F/0B4-3

VEHICLE EMISSION CONTROL INFORMATION
GENERAL MOTORS CORPORATION

GM

SET PARKING BRAKE AND BLOCK DRIVE WHEELS

MAKE ADJUSTMENTS WITH ENGINE AT NORMAL OPERATING TEMPERATURE, CHOKE FULLY OPEN AND WHERE APPLICABLE AIR CONDITIONING OFF

1 DISCONNECT MOLDED CONNECTOR FROM AIR CLEANER TVS SWITCH. PLUG PORT (1) ON CONNECTOR AND REMOVE AIR CLEANER.
2 DISCONNECT VACUUM HOSES FROM PORTS (H), (K) AND (B) ON CARBURETOR AND (4) ON INTAKE MANIFOLD AND PLUG PORTS
3 SET TIMING AT SPECIFIED RPM, IN PARK. RECONNECT HOSE TO PORT (B).
4 ADJUST CARBURETOR IDLE SPEED SCREW TO SPECIFIED RPM, IN DRIVE
5 ADJUST CARBURETOR FAST IDLE SPEED SCREW TO SPECIFIED RPM ON LOW STEP OF CAM, IN DRIVE
6 ON AIR CONDITIONED CARS — AFTER COMPLETING ABOVE ADJUSTMENTS, DISCONNECT ELECTRICAL CONNECTOR AT A/C COMPRESSOR CLUTCH, TURN A/C SWITCH ON AND ADJUST SOLENOID (ENERGIZED) SCREW TO SPECIFIED RPM, IN DRIVE. TURN ENGINE OFF. RECONNECT ELECTRICAL CONNECTOR
7 RECONNECT VACUUM HOSES TO PORTS (H) AND (K) ON CARBURETOR AND (4) ON INTAKE MANIFOLD
8 INSTALL AIR CLEANER AND RECONNECT MOLDED CONNECTOR TO TVS SWITCH.

CATALYST-EGR-AIR
LOW ALTITUDE CERTIFICATION

TIMING (DEG BTDC @ RPM)		18° @ 1100 (IN PARK)
SPARK PLUGS	GAP	.080 IN
	TYPE	AC R46SX
CARBURETOR SCREW (RPM)		500 (IN DRIVE)
FAST IDLE SCREW (RPM)		700 (IN DRIVE)
SOLENOID SCREW (RPM)		600 (IN DRIVE)

IDLE MIXTURE SCREWS ARE PRESET AND SEALED AT FACTORY. PROVISION FOR ADJUSTMENT DURING TUNE-UP IS NOT PROVIDED

SEE SERVICE MANUAL AND MAINTENANCE SCHEDULE FOR ADDITIONAL INFORMATION.

EMISSION HOSE ROUTING

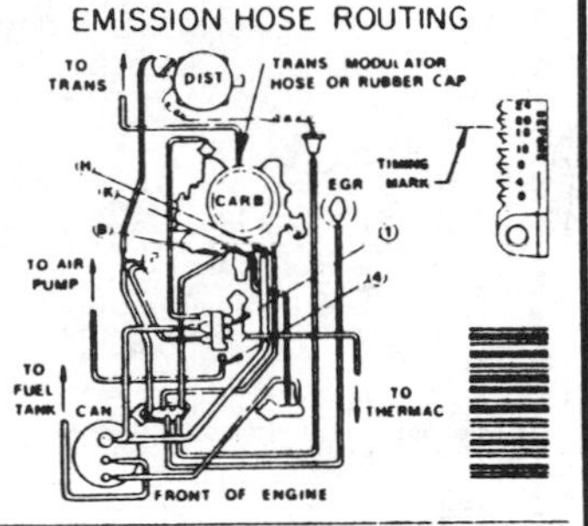

PRINTED IN U.S.A.

THIS VEHICLE CONFORMS TO U.S. EPA REGULATIONS APPLICABLE TO 1980 MODEL YEAR NEW PASSENGER CARS.

PART NO 22505241

SS
5.7 LITER
#03L4F/0B4-3

VEHICLE EMISSION CONTROL INFORMATION
GENERAL MOTORS CORPORATION

GM

SET PARKING BRAKE AND BLOCK DRIVE WHEELS.

MAKE ADJUSTMENTS WITH ENGINE AT NORMAL OPERATING TEMPERATURE, CHOKE FULLY OPEN AND WHERE APPLICABLE AIR CONDITIONING OFF

1 DISCONNECT MOLDED CONNECTOR FROM AIR CLEANER TVS SWITCH PLUG PORT (1) ON CONNECTOR AND REMOVE AIR CLEANER.
2 DISCONNECT VACUUM HOSES FROM PORTS (H), (K) AND (B) ON CARBURETOR AND (4) ON INTAKE MANIFOLD AND PLUG PORTS
3 SET TIMING AT SPECIFIED RPM, IN PARK. RECONNECT HOSE TO PORT (B).
4 ADJUST CARBURETOR IDLE SPEED SCREW TO SPECIFIED RPM, IN DRIVE
5 ADJUST CARBURETOR FAST IDLE SPEED SCREW TO SPECIFIED RPM ON LOW STEP OF CAM, IN DRIVE
6 ON AIR CONDITIONED CARS — AFTER COMPLETING ABOVE ADJUSTMENTS, DISCONNECT ELECTRICAL CONNECTOR AT A/C COMPRESSOR CLUTCH, TURN A/C SWITCH ON AND ADJUST SOLENOID (ENERGIZED) SCREW TO SPECIFIED RPM, IN DRIVE TURN ENGINE OFF RECONNECT ELECTRICAL CONNECTOR
7 RECONNECT VACUUM HOSES TO PORTS (H) AND (K) ON CARBURETOR AND (4) ON INTAKE MANIFOLD
8 INSTALL AIR CLEANER AND RECONNECT MOLDED CONNECTOR TO TVS SWITCH

CATALYST-EGR-AIR
LOW ALTITUDE CERTIFICATION

TIMING (DEG BTDC @ RPM)		18° @ 1100 (IN PARK)
SPARK PLUGS	GAP	.080 IN
	TYPE	AC R46SX
CARBURETOR SCREW (RPM)		500 (IN DRIVE)
FAST IDLE SCREW (RPM)		700 (IN DRIVE)
SOLENOID SCREW (RPM)		600 (IN DRIVE)

IDLE MIXTURE SCREWS ARE PRESET AND SEALED AT FACTORY. PROVISION FOR ADJUSTMENT DURING TUNE-UP IS NOT PROVIDED

SEE SERVICE MANUAL AND MAINTENANCE SCHEDULE FOR ADDITIONAL INFORMATION.

EMISSION HOSE ROUTING

TO TRANS
DIST
TRANS MODULATOR HOSE OR RUBBER CAP
H
K
B
CARB
EGR
TIMING MARK
(1)
TO AIR PUMP
4
TO FUEL TANK
CAN
TO THERMAC
FRONT OF ENGINE

PRINTED IN U.S.A.

THIS VEHICLE CONFORMS TO U.S. EPA REGULATIONS APPLICABLE TO 1980 MODEL YEAR NEW PASSENGER CARS.

PART NO 22509361

HW
5.7 LITER
#03L4F/0B4-3

VEHICLE EMISSION CONTROL INFORMATION
GENERAL MOTORS CORPORATION

GM

SET PARKING BRAKE AND BLOCK DRIVE WHEELS

MAKE ADJUSTMENTS WITH ENGINE AT NORMAL OPERATING TEMPERATURE, CHOKE FULLY OPEN AND AIR CONDITIONING OFF.

1. DISCONNECT MOLDED CONNECTOR FROM AIR CLEANER TVS SWITCH. PLUG PORT (1) ON CONNECTOR AND REMOVE AIR CLEANER.
2. DISCONNECT VACUUM HOSES FROM PORTS (H), (K) AND (B) ON CARBURETOR AND (4) ON INTAKE MANIFOLD AND PLUG PORTS
3. SET TIMING AT SPECIFIED RPM, IN PARK. RECONNECT HOSE TO PORT (B)
4 ADJUST CARBURETOR IDLE SPEED SCREW TO SPECIFIED RPM, IN DRIVE
5. ADJUST CARBURETOR FAST IDLE SPEED SCREW TO SPECIFIED RPM ON LOW STEP OF CAM, IN DRIVE.
6 AFTER COMPLETING ABOVE ADJUSTMENTS, DISCONNECT ELECTRICAL CONNECTOR AT A/C COMPRESSOR CLUTCH, TURN A/C SWITCH ON AND ADJUST SOLENOID (ENERGIZED) SCREW TO SPECIFIED RPM, IN DRIVE. TURN ENGINE OFF. RECONNECT ELECTRICAL CONNECTOR
7 RECONNECT VACUUM HOSES TO PORTS (H) AND (K) ON CARBURETOR AND (4) ON INTAKE MANIFOLD
8. INSTALL AIR CLEANER AND RECONNECT MOLDED CONNECTOR TO TVS SWITCH

CATALYST-EGR-AIR
LOW ALTITUDE CERTIFICATION

TIMING (DEG. BTDC @ RPM)		18° @ 1100 (IN PARK)
SPARK PLUGS	GAP	.080 IN.
	TYPE	AC R46SX
CARBURETOR SCREW (RPM)		500 (IN DRIVE)
FAST IDLE SCREW (RPM)		700 (IN DRIVE)
SOLENOID SCREW (RPM)		600 (IN DRIVE)

IDLE MIXTURE SCREWS ARE PRESET AND SEALED AT FACTORY. PROVISION FOR ADJUSTMENT DURING TUNE-UP IS NOT PROVIDED

SEE SERVICE MANUAL AND MAINTENANCE SCHEDULE FOR ADDITIONAL INFORMATION

EMISSION HOSE ROUTING

DIST
(H)
(K)
(B)
CARB
EGR
TIMING MARK
TO FUEL TANK
(1)
TO AIR PUMP
(4)
TO THERMAC
CAN
FRONT OF ENGINE

PRINTED IN U.S.A.

THIS VEHICLE CONFORMS TO U.S. EPA REGULATIONS APPLICABLE TO 1980 MODEL YEAR NEW PASSENGER CARS.

PART NO. 22505251

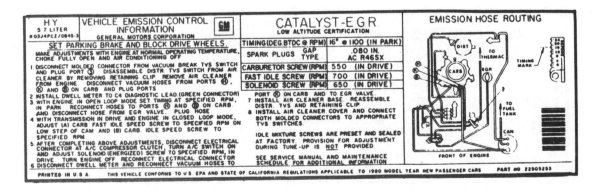

VEHICLE EMISSION CONTROL LABELS AND VACUUM HOSE ROUTINGS

1980

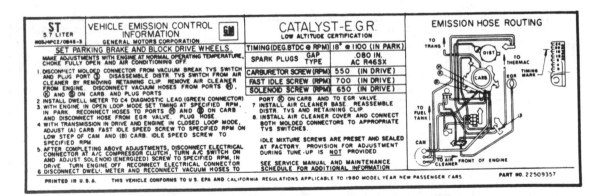

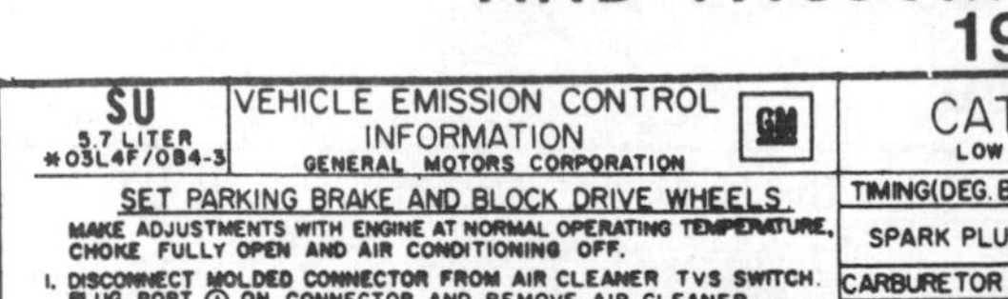

SU 5.7 LITER #03L4F/0B4-3 — VEHICLE EMISSION CONTROL INFORMATION — GENERAL MOTORS CORPORATION — GM

SET PARKING BRAKE AND BLOCK DRIVE WHEELS.

MAKE ADJUSTMENTS WITH ENGINE AT NORMAL OPERATING TEMPERATURE, CHOKE FULLY OPEN AND AIR CONDITIONING OFF.

1. DISCONNECT MOLDED CONNECTOR FROM AIR CLEANER TVS SWITCH. PLUG PORT (1) ON CONNECTOR AND REMOVE AIR CLEANER.
2. DISCONNECT VACUUM HOSES FROM PORTS (H), (K) AND (B) ON CARBURETOR AND (4) ON INTAKE MANIFOLD AND PLUG PORTS.
3. SET TIMING AT SPECIFIED RPM, IN PARK. RECONNECT HOSE TO PORT (B)
4. ADJUST CARBURETOR IDLE SPEED SCREW TO SPECIFIED RPM, IN DRIVE.
5. ADJUST CARBURETOR FAST IDLE SPEED SCREW TO SPECIFIED RPM ON LOW STEP OF CAM, IN DRIVE.
6. AFTER COMPLETING ABOVE ADJUSTMENTS, DISCONNECT ELECTRICAL CONNECTOR AT A/C COMPRESSOR CLUTCH, TURN A/C SWITCH ON AND ADJUST SOLENOID (ENERGIZED) SCREW TO SPECIFIED RPM, IN DRIVE. TURN ENGINE OFF. RECONNECT ELECTRICAL CONNECTOR
7. RECONNECT VACUUM HOSES TO PORTS (H) AND (K) ON CARBURETOR AND (4) ON INTAKE MANIFOLD.
8. INSTALL AIR CLEANER AND RECONNECT MOLDED CONNECTOR TO TVS SWITCH

CATALYST-E.G.R.-A.I.R.
LOW ALTITUDE CERTIFICATION

TIMING (DEG. BTDC @ RPM)	18° @ 1100 (IN PARK)
SPARK PLUGS GAP / TYPE	.080 IN. / AC R46SX
CARBURETOR SCREW (RPM)	500 (IN DRIVE)
FAST IDLE SCREW (RPM)	700 (IN DRIVE)
SOLENOID SCREW (RPM)	600 (IN DRIVE)

IDLE MIXTURE SCREWS ARE PRESET AND SEALED AT FACTORY. PROVISION FOR ADJUSTMENT DURING TUNE-UP IS NOT PROVIDED.

SEE SERVICE MANUAL AND MAINTENANCE SCHEDULE FOR ADDITIONAL INFORMATION.

EMISSION HOSE ROUTING

DIST
CARB.
EGR
TIMING MARK
TO FUEL TANK
TO AIR PUMP
TO THERMAC
CAN
FRONT OF ENGINE

PRINTED IN U.S.A. THIS VEHICLE CONFORMS TO U.S. EPA REGULATIONS APPLICABLE TO 1980 MODEL YEAR NEW PASSENGER CARS. PART NO. 22509358

HN 5.7 LITER #03J4PCZ/0848-3 — VEHICLE EMISSION CONTROL INFORMATION — GENERAL MOTORS CORPORATION — GM

SET PARKING BRAKE AND BLOCK DRIVE WHEELS.

MAKE ADJUSTMENTS WITH ENGINE AT NORMAL OPERATING TEMPERATURE, CHOKE FULLY OPEN AND AIR CONDITIONING OFF.

1. DISCONNECT MOLDED CONNECTOR FROM VACUUM BREAK TVS SWITCH AND PLUG PORT (3) DISASSEMBLE DISTR. TVS SWITCH FROM AIR CLEANER BY REMOVING RETAINING CLIP. REMOVE AIR CLEANER FROM ENGINE. DISCONNECT VACUUM HOSES FROM PORTS (H), (K) AND (B) ON CARB. AND PLUG PORTS
2. INSTALL DWELL METER TO C4 DIAGNOSTIC LEAD (GREEN CONNECTOR)
3. WITH ENGINE IN OPEN LOOP MODE SET TIMING AT SPECIFIED RPM, IN PARK. RECONNECT HOSES TO PORTS (H) AND (B) ON CARB. AND DISCONNECT HOSE FROM EGR VALVE. PLUG HOSE.
4. WITH TRANSMISSION IN DRIVE AND ENGINE IN CLOSED LOOP MODE, ADJUST (A) CARB FAST IDLE SPEED SCREW TO SPECIFIED RPM ON LOW STEP OF CAM AND (B) CARB. IDLE SPEED SCREW TO SPECIFIED RPM.
5. AFTER COMPLETING ABOVE ADJUSTMENTS, DISCONNECT ELECTRICAL CONNECTOR AT A/C COMPRESSOR CLUTCH, TURN A/C SWITCH ON AND ADJUST SOLENOID (ENERGIZED) SCREW TO SPECIFIED RPM, IN DRIVE. TURN ENGINE OFF RECONNECT ELECTRICAL CONNECTOR
6. DISCONNECT DWELL METER AND RECONNECT VACUUM HOSES TO PORT (K) ON CARB. AND TO EGR VALVE
7. INSTALL AIR CLEANER BASE. REASSEMBLE DISTR. TVS AND RETAINING CLIP.
8. INSTALL AIR CLEANER COVER AND CONNECT BOTH MOLDED CONNECTORS TO APPROPRIATE TVS SWITCHES.

CATALYST-E.G.R.
LOW ALTITUDE CERTIFICATION

TIMING (DEG. BTDC @ RPM)	18° @ 1100 (IN PARK)
SPARK PLUGS GAP / TYPE	.080 IN. / AC R46SX
CARBURETOR SCREW (RPM)	550 (IN DRIVE)
FAST IDLE SCREW (RPM)	700 (IN DRIVE)
SOLENOID SCREW (RPM)	650 (IN DRIVE)

IDLE MIXTURE SCREWS ARE PRESET AND SEALED AT FACTORY. PROVISION FOR ADJUSTMENT DURING TUNE-UP IS NOT PROVIDED.

SEE SERVICE MANUAL AND MAINTENANCE SCHEDULE FOR ADDITIONAL INFORMATION

EMISSION HOSE ROUTING

TO TRANS
DIST.
TO THERMAC
TIMING MARK
CARB.
EGR
TO FUEL TANK
CAN
FRONT OF ENGINE

PRINTED IN U.S.A. THIS VEHICLE CONFORMS TO U.S. EPA AND STATE OF CALIFORNIA REGULATIONS APPLICABLE TO 1980 MODEL YEAR NEW PASSENGER CARS PART NO. 22505242

ST 5.7 LITER #03J4PCZ/0848-3 — VEHICLE EMISSION CONTROL INFORMATION — GENERAL MOTORS CORPORATION — GM

SET PARKING BRAKE AND BLOCK DRIVE WHEELS.

MAKE ADJUSTMENTS WITH ENGINE AT NORMAL OPERATING TEMPERATURE, CHOKE FULLY OPEN AND AIR CONDITIONING OFF.

1. DISCONNECT MOLDED CONNECTOR FROM VACUUM BREAK TVS SWITCH AND PLUG PORT (3) DISASSEMBLE DISTR. TVS SWITCH FROM AIR CLEANER BY REMOVING RETAINING CLIP. REMOVE AIR CLEANER FROM ENGINE. DISCONNECT VACUUM HOSES FROM PORTS (H), (K) AND (B) ON CARB. AND PLUG PORTS
2. INSTALL DWELL METER TO C4 DIAGNOSTIC LEAD (GREEN CONNECTOR)
3. WITH ENGINE IN OPEN LOOP MODE SET TIMING AT SPECIFIED RPM, IN PARK. RECONNECT HOSES TO PORTS (H) AND (B) ON CARB AND DISCONNECT HOSE FROM EGR VALVE. PLUG HOSE.
4. WITH TRANSMISSION IN DRIVE AND ENGINE IN CLOSED LOOP MODE, ADJUST (A) CARB. FAST IDLE SPEED SCREW TO SPECIFIED RPM ON LOW STEP OF CAM AND (B) CARB. IDLE SPEED SCREW TO SPECIFIED RPM
5. AFTER COMPLETING ABOVE ADJUSTMENTS, DISCONNECT ELECTRICAL CONNECTOR AT A/C COMPRESSOR CLUTCH, TURN A/C SWITCH ON AND ADJUST SOLENOID (ENERGIZED) SCREW TO SPECIFIED RPM, IN DRIVE. TURN ENGINE OFF RECONNECT ELECTRICAL CONNECTOR
6. DISCONNECT DWELL METER AND RECONNECT VACUUM HOSES TO PORT (K) ON CARB AND TO EGR VALVE
7. INSTALL AIR CLEANER BASE. REASSEMBLE DISTR. TVS AND RETAINING CLIP.
8. INSTALL AIR CLEANER COVER AND CONNECT BOTH MOLDED CONNECTORS TO APPROPRIATE TVS SWITCHES.

CATALYST-E.G.R.
LOW ALTITUDE CERTIFICATION

TIMING (DEG. BTDC @ RPM)	18° @ 1100 (IN PARK)
SPARK PLUGS GAP / TYPE	.080 IN. / AC R46SX
CARBURETOR SCREW (RPM)	550 (IN DRIVE)
FAST IDLE SCREW (RPM)	700 (IN DRIVE)
SOLENOID SCREW (RPM)	650 (IN DRIVE)

IDLE MIXTURE SCREWS ARE PRESET AND SEALED AT FACTORY. PROVISION FOR ADJUSTMENT DURING TUNE-UP IS NOT PROVIDED

SEE SERVICE MANUAL AND MAINTENANCE SCHEDULE FOR ADDITIONAL INFORMATION

EMISSION HOSE ROUTING

TO TRANS
DIST
TO THERMAC
TIMING MARK
CARB
EGR
TO FUEL TANK
CAN
TO AIR CLEANER
FRONT OF ENGINE

PRINTED IN U.S.A. THIS VEHICLE CONFORMS TO U.S. EPA AND CALIFORNIA REGULATIONS APPLICABLE TO 1980 MODEL YEAR NEW PASSENGER CARS. PART NO. 22509357

HY 5.7 LITER #03J4PCZ/0845-3 — VEHICLE EMISSION CONTROL INFORMATION — GENERAL MOTORS CORPORATION — GM

SET PARKING BRAKE AND BLOCK DRIVE WHEELS.

MAKE ADJUSTMENTS WITH ENGINE AT NORMAL OPERATING TEMPERATURE, CHOKE FULLY OPEN AND AIR CONDITIONING OFF

1. DISCONNECT MOLDED CONNECTOR FROM VACUUM BREAK TVS SWITCH AND PLUG PORT (3) DISASSEMBLE DISTR TVS SWITCH FROM AIR CLEANER BY REMOVING RETAINING CLIP REMOVE AIR CLEANER FROM ENGINE. DISCONNECT VACUUM HOSES FROM PORTS (H), (K) AND (B) ON CARB AND PLUG PORTS
2. INSTALL DWELL METER TO C4 DIAGNOSTIC LEAD (GREEN CONNECTOR)
3. WITH ENGINE IN OPEN LOOP MODE SET TIMING AT SPECIFIED RPM, IN PARK RECONNECT HOSES TO PORTS (H) AND (B) ON CARB. AND DISCONNECT HOSE FROM EGR VALVE. PLUG HOSE
4. WITH TRANSMISSION IN DRIVE AND ENGINE IN CLOSED LOOP MODE, ADJUST (A) CARB FAST IDLE SPEED SCREW TO SPECIFIED RPM ON LOW STEP OF CAM AND (B) CARB. IDLE SPEED SCREW TO SPECIFIED RPM
5. AFTER COMPLETING ABOVE ADJUSTMENTS, DISCONNECT ELECTRICAL CONNECTOR AT A/C COMPRESSOR CLUTCH, TURN A/C SWITCH ON AND ADJUST SOLENOID (ENERGIZED) SCREW TO SPECIFIED RPM, IN DRIVE TURN ENGINE OFF RECONNECT ELECTRICAL CONNECTOR
6. DISCONNECT DWELL METER AND RECONNECT VACUUM HOSES TO PORT (K) ON CARB AND TO EGR VALVE.
7. INSTALL AIR CLEANER BASE. REASSEMBLE DISTR. TVS AND RETAINING CLIP
8. INSTALL AIR CLEANER COVER AND CONNECT BOTH MOLDED CONNECTORS TO APPROPRIATE TVS SWITCHES.

CATALYST-E.G.R.
LOW ALTITUDE CERTIFICATION

TIMING (DEG. BTDC @ RPM)	16° @ 1100 (IN PARK)
SPARK PLUGS GAP / TYPE	.080 IN. / AC R46SX
CARBURETOR SCREW (RPM)	550 (IN DRIVE)
FAST IDLE SCREW (RPM)	700 (IN DRIVE)
SOLENOID SCREW (RPM)	650 (IN DRIVE)

IDLE MIXTURE SCREWS ARE PRESET AND SEALED AT FACTORY. PROVISION FOR ADJUSTMENT DURING TUNE-UP IS NOT PROVIDED.

SEE SERVICE MANUAL AND MAINTENANCE SCHEDULE FOR ADDITIONAL INFORMATION

EMISSION HOSE ROUTING

DIST
TO THERMAC
TIMING MARK
CARB
EGR
TO FUEL TANK
CAN
FRONT OF ENGINE

PRINTED IN U.S.A. THIS VEHICLE CONFORMS TO U.S. EPA AND STATE OF CALIFORNIA REGULATIONS APPLICABLE TO 1980 MODEL YEAR NEW PASSENGER CARS PART NO. 22505253

VEHICLE EMISSION CONTROL LABELS AND VACUUM HOSE ROUTINGS

1980

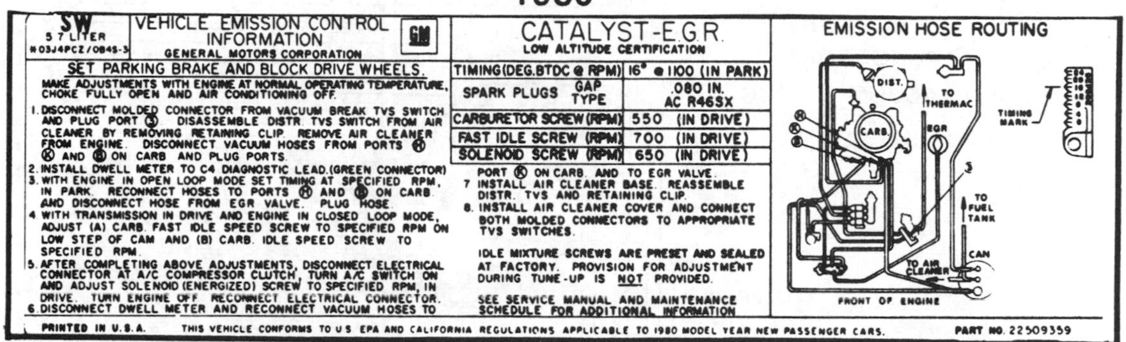

SW
5.7 LITER
#03J4PCZ/0845-3

VEHICLE EMISSION CONTROL INFORMATION
GENERAL MOTORS CORPORATION
GM

SET PARKING BRAKE AND BLOCK DRIVE WHEELS.

MAKE ADJUSTMENTS WITH ENGINE AT NORMAL OPERATING TEMPERATURE, CHOKE FULLY OPEN AND AIR CONDITIONING OFF.

1. DISCONNECT MOLDED CONNECTOR FROM VACUUM BREAK TVS SWITCH AND PLUG PORT (S). DISASSEMBLE DISTR. TVS SWITCH FROM AIR CLEANER BY REMOVING RETAINING CLIP. REMOVE AIR CLEANER FROM ENGINE. DISCONNECT VACUUM HOSES FROM PORTS (H) (K) AND (B) ON CARB. AND PLUG PORTS.
2. INSTALL DWELL METER TO C4 DIAGNOSTIC LEAD. (GREEN CONNECTOR)
3. WITH ENGINE IN OPEN LOOP MODE SET TIMING AT SPECIFIED RPM, IN PARK. RECONNECT HOSES TO PORTS (H) AND (B) ON CARB. AND DISCONNECT HOSE FROM EGR VALVE. PLUG HOSE.
4. WITH TRANSMISSION IN DRIVE AND ENGINE IN CLOSED LOOP MODE, ADJUST (A) CARB. FAST IDLE SPEED SCREW TO SPECIFIED RPM ON LOW STEP OF CAM AND (B) CARB. IDLE SPEED SCREW TO SPECIFIED RPM.
5. AFTER COMPLETING ABOVE ADJUSTMENTS, DISCONNECT ELECTRICAL CONNECTOR AT A/C COMPRESSOR CLUTCH, TURN A/C SWITCH ON AND ADJUST SOLENOID (ENERGIZED) SCREW TO SPECIFIED RPM, IN DRIVE. TURN ENGINE OFF. RECONNECT ELECTRICAL CONNECTOR.
6. DISCONNECT DWELL METER AND RECONNECT VACUUM HOSES TO PORT (K) ON CARB. AND TO EGR VALVE.
7. INSTALL AIR CLEANER BASE. REASSEMBLE DISTR. TVS AND RETAINING CLIP.
8. INSTALL AIR CLEANER COVER AND CONNECT BOTH MOLDED CONNECTORS TO APPROPRIATE TVS SWITCHES.

CATALYST-E.G.R.
LOW ALTITUDE CERTIFICATION

TIMING (DEG. BTDC @ RPM)	16° @ 1100 (IN PARK)
SPARK PLUGS GAP / TYPE	.080 IN. / AC R46SX
CARBURETOR SCREW (RPM)	550 (IN DRIVE)
FAST IDLE SCREW (RPM)	700 (IN DRIVE)
SOLENOID SCREW (RPM)	650 (IN DRIVE)

IDLE MIXTURE SCREWS ARE PRESET AND SEALED AT FACTORY. PROVISION FOR ADJUSTMENT DURING TUNE-UP IS NOT PROVIDED.

SEE SERVICE MANUAL AND MAINTENANCE SCHEDULE FOR ADDITIONAL INFORMATION

EMISSION HOSE ROUTING

PRINTED IN U.S.A. THIS VEHICLE CONFORMS TO U.S. EPA AND CALIFORNIA REGULATIONS APPLICABLE TO 1980 MODEL YEAR NEW PASSENGER CARS. PART NO. 22509359

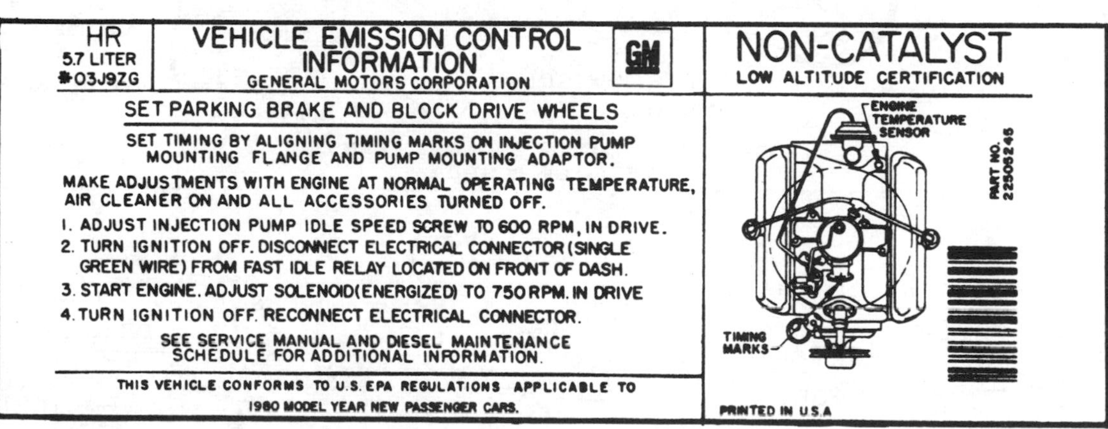

HR
5.7 LITER
#03J9ZG

VEHICLE EMISSION CONTROL INFORMATION
GENERAL MOTORS CORPORATION
GM

SET PARKING BRAKE AND BLOCK DRIVE WHEELS

SET TIMING BY ALIGNING TIMING MARKS ON INJECTION PUMP MOUNTING FLANGE AND PUMP MOUNTING ADAPTOR.

MAKE ADJUSTMENTS WITH ENGINE AT NORMAL OPERATING TEMPERATURE, AIR CLEANER ON AND ALL ACCESSORIES TURNED OFF.

1. ADJUST INJECTION PUMP IDLE SPEED SCREW TO 600 RPM, IN DRIVE.
2. TURN IGNITION OFF. DISCONNECT ELECTRICAL CONNECTOR (SINGLE GREEN WIRE) FROM FAST IDLE RELAY LOCATED ON FRONT OF DASH.
3. START ENGINE. ADJUST SOLENOID (ENERGIZED) TO 750 RPM. IN DRIVE
4. TURN IGNITION OFF. RECONNECT ELECTRICAL CONNECTOR.

SEE SERVICE MANUAL AND DIESEL MAINTENANCE SCHEDULE FOR ADDITIONAL INFORMATION.

THIS VEHICLE CONFORMS TO U.S. EPA REGULATIONS APPLICABLE TO 1980 MODEL YEAR NEW PASSENGER CARS.

NON-CATALYST
LOW ALTITUDE CERTIFICATION

PRINTED IN U.S.A.

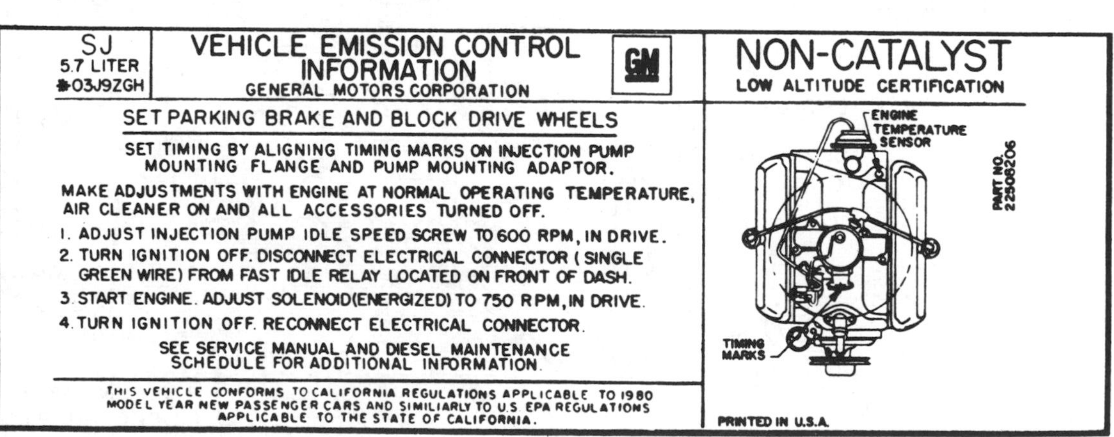

SJ
5.7 LITER
#03J9ZGH

VEHICLE EMISSION CONTROL INFORMATION
GENERAL MOTORS CORPORATION
GM

SET PARKING BRAKE AND BLOCK DRIVE WHEELS

SET TIMING BY ALIGNING TIMING MARKS ON INJECTION PUMP MOUNTING FLANGE AND PUMP MOUNTING ADAPTOR.

MAKE ADJUSTMENTS WITH ENGINE AT NORMAL OPERATING TEMPERATURE, AIR CLEANER ON AND ALL ACCESSORIES TURNED OFF.

1. ADJUST INJECTION PUMP IDLE SPEED SCREW TO 600 RPM, IN DRIVE.
2. TURN IGNITION OFF. DISCONNECT ELECTRICAL CONNECTOR (SINGLE GREEN WIRE) FROM FAST IDLE RELAY LOCATED ON FRONT OF DASH.
3. START ENGINE. ADJUST SOLENOID (ENERGIZED) TO 750 RPM, IN DRIVE.
4. TURN IGNITION OFF. RECONNECT ELECTRICAL CONNECTOR.

SEE SERVICE MANUAL AND DIESEL MAINTENANCE SCHEDULE FOR ADDITIONAL INFORMATION.

THIS VEHICLE CONFORMS TO CALIFORNIA REGULATIONS APPLICABLE TO 1980 MODEL YEAR NEW PASSENGER CARS AND SIMILIARLY TO U.S. EPA REGULATIONS APPLICABLE TO THE STATE OF CALIFORNIA.

NON-CATALYST
LOW ALTITUDE CERTIFICATION

PRINTED IN U.S.A.

VACUUM CIRCUITS

(© Oldsmobile Div., G.M. Corp.)

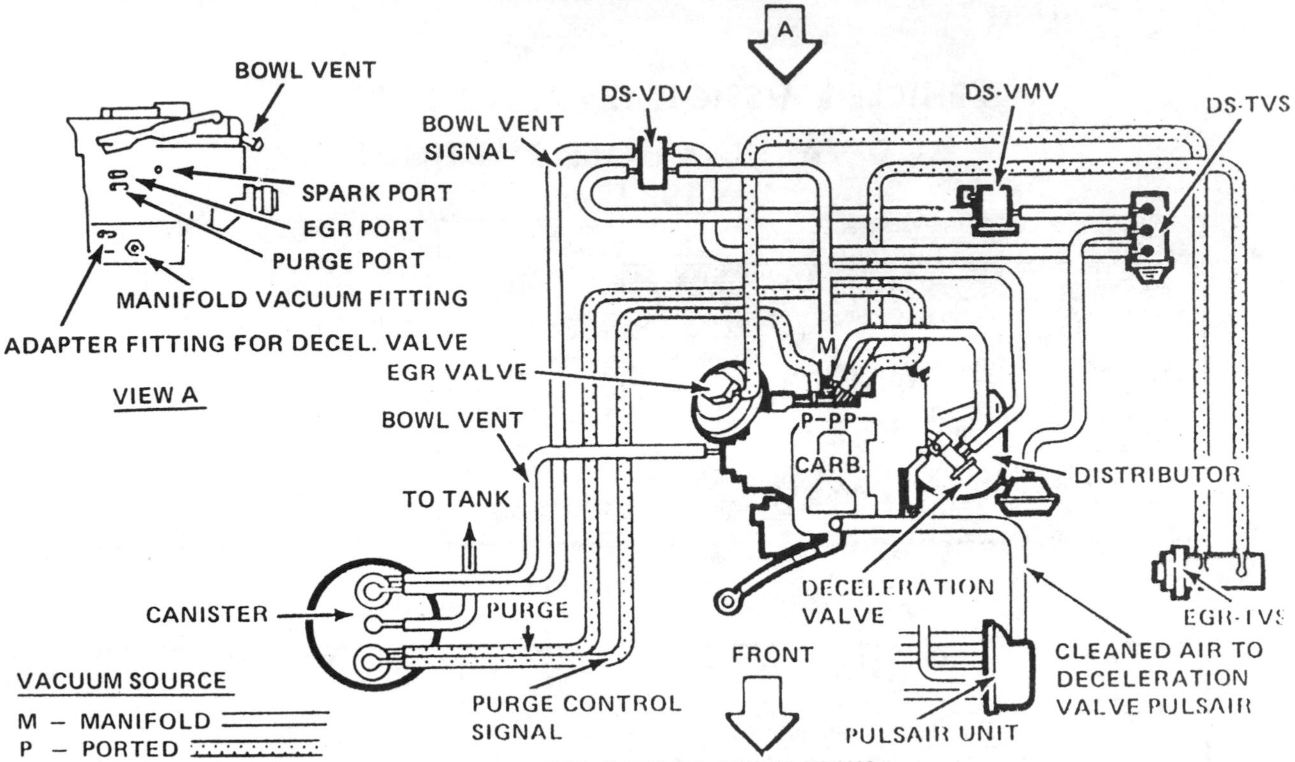

1980—2.5L ENG.-CODE "5"-W/MAN. TRANS.
OMEGA-LOW ALTITUDE

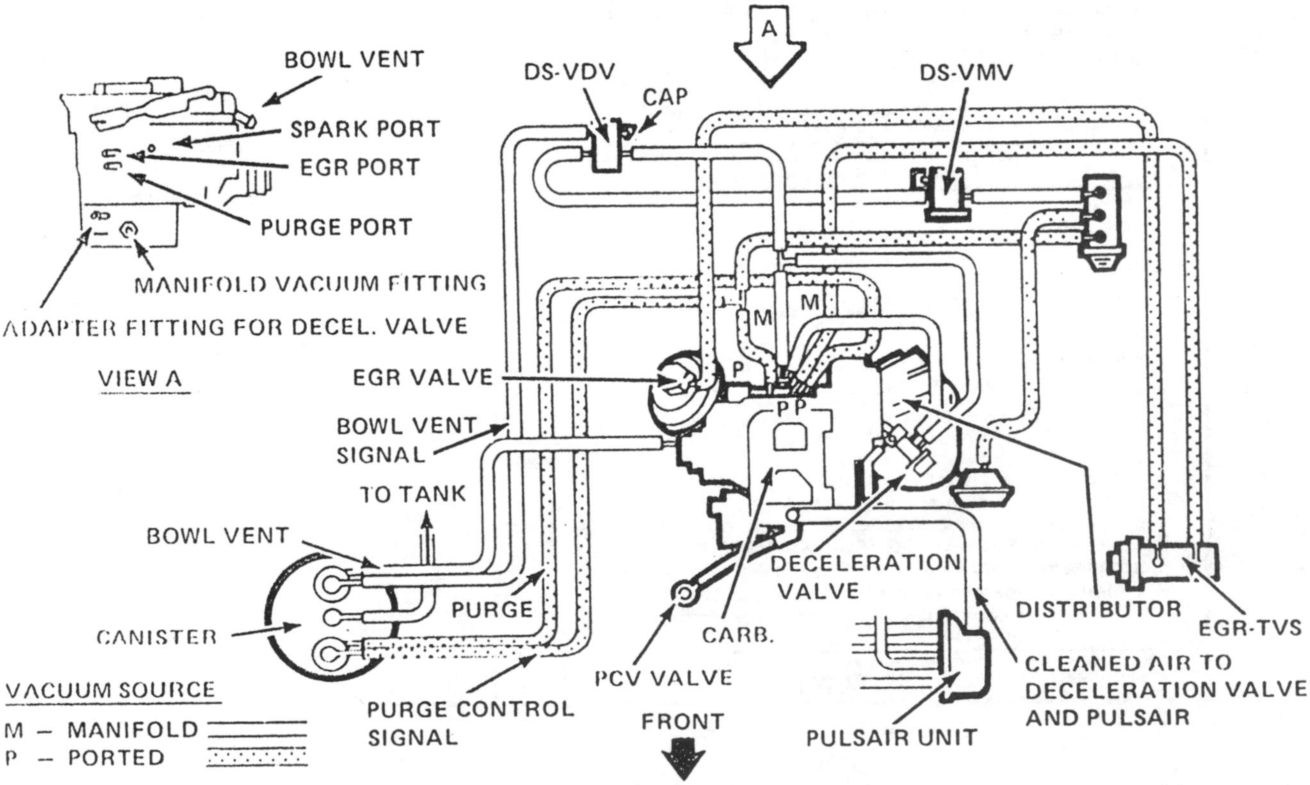

1980—2.5L ENG.-CODE "5"-W/AUTO. TRANS.
OMEGA-LOW ALTITUDE

VACUUM CIRCUITS

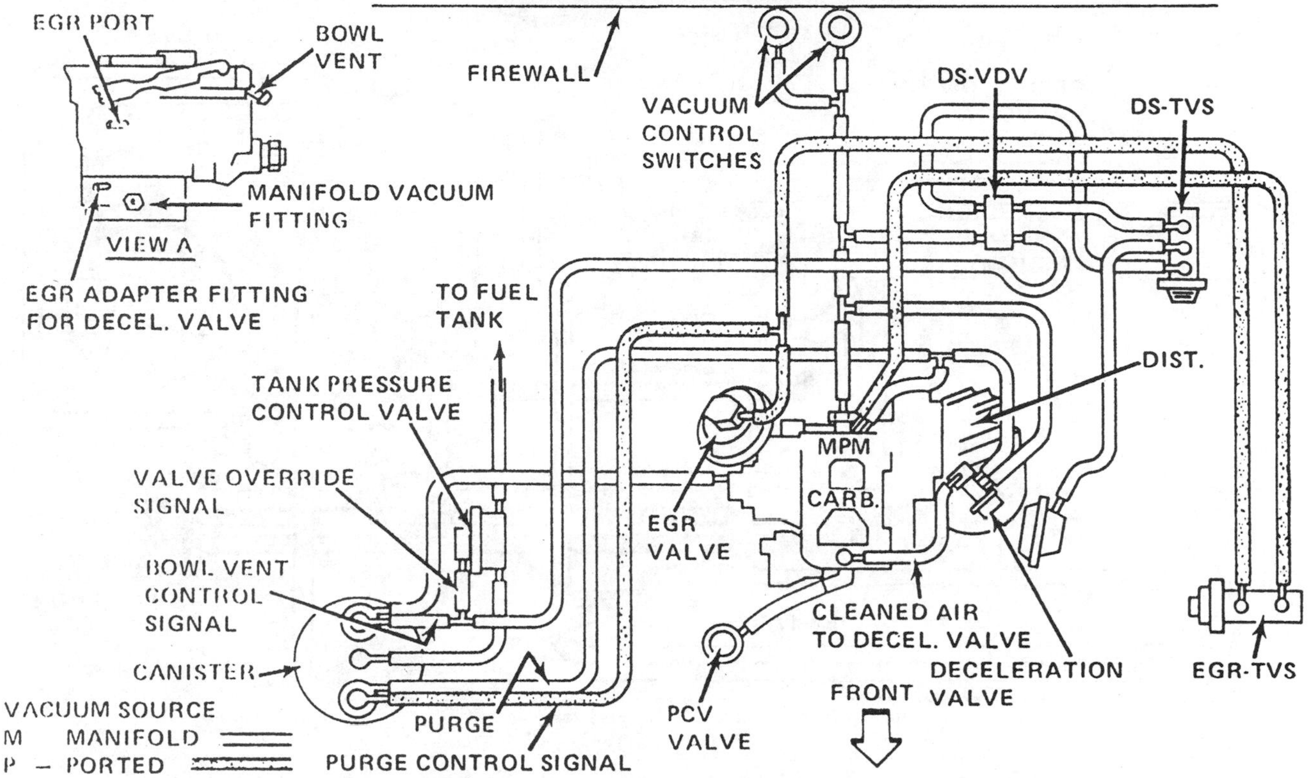

EGR PORT
BOWL VENT
FIREWALL
VACUUM CONTROL SWITCHES
DS-VDV
DS-TVS
MANIFOLD VACUUM FITTING
VIEW A
EGR ADAPTER FITTING FOR DECEL. VALVE
TO FUEL TANK
DIST.
TANK PRESSURE CONTROL VALVE
MPM
CARB.
VALVE OVERRIDE SIGNAL
EGR VALVE
BOWL VENT CONTROL SIGNAL
CLEANED AIR TO DECEL. VALVE
CANISTER
DECELERATION VALVE
EGR-TVS
FRONT
VACUUM SOURCE
M MANIFOLD
P – PORTED
PURGE
PCV VALVE
PURGE CONTROL SIGNAL

**1980—2.5L ENG.-CODE "5"-W/MAN. TRANS.
OMEGA-CALIF.**

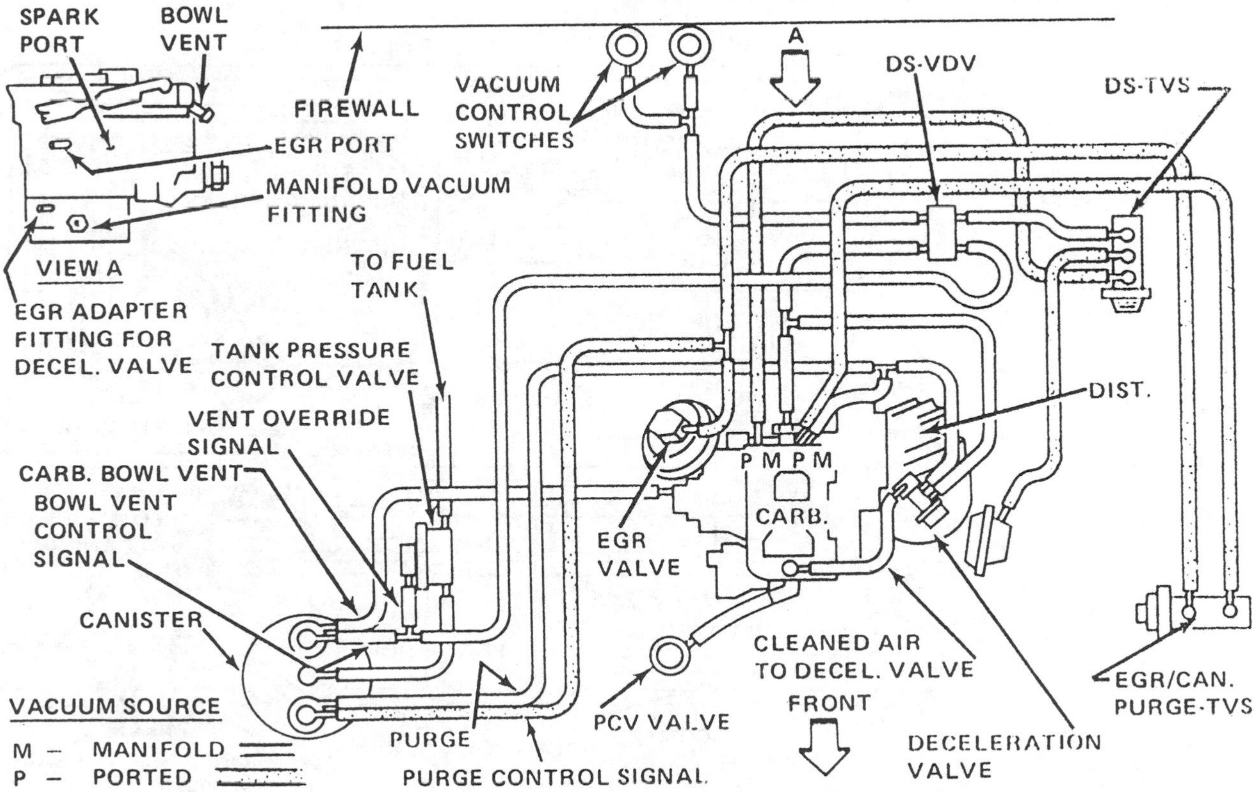

SPARK PORT
BOWL VENT
FIREWALL
VACUUM CONTROL SWITCHES
A
DS-VDV
DS-TVS
EGR PORT
MANIFOLD VACUUM FITTING
VIEW A
TO FUEL TANK
EGR ADAPTER FITTING FOR DECEL. VALVE
TANK PRESSURE CONTROL VALVE
DIST.
VENT OVERRIDE SIGNAL
P M P M
CARB. BOWL VENT
BOWL VENT CONTROL SIGNAL
CARB.
EGR VALVE
CANISTER
CLEANED AIR TO DECEL. VALVE
EGR/CAN. PURGE-TVS
VACUUM SOURCE
FRONT
PCV VALVE
M – MANIFOLD
P – PORTED
PURGE
DECELERATION VALVE
PURGE CONTROL SIGNAL

**1980—2.5L ENG.-CODE"5"-W/AUTO. TRANS.
OMEGA-CALIF.**

VACUUM CIRCUITS

(© Oldsmobile Div., G.M. Corp.)

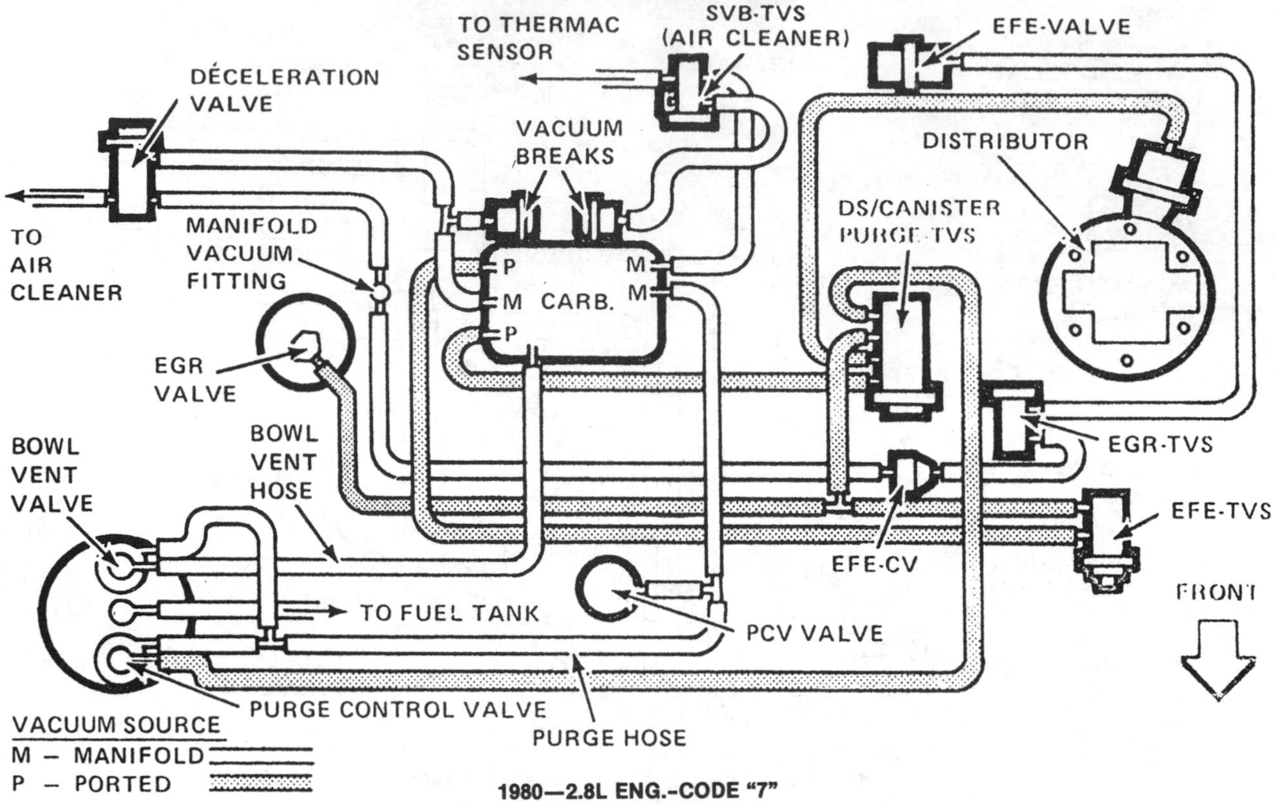

TO THERMAC SENSOR
SVB-TVS (AIR CLEANER)
EFE-VALVE
DECELERATION VALVE
VACUUM BREAKS
DISTRIBUTOR
DS/CANISTER PURGE-TVS
TO AIR CLEANER
MANIFOLD VACUUM FITTING
P
M
P
CARB.
M
M
EGR VALVE
EGR-TVS
BOWL VENT VALVE
BOWL VENT HOSE
EFE-TVS
EFE-CV
FRONT
TO FUEL TANK
PCV VALVE
PURGE CONTROL VALVE
PURGE HOSE
VACUUM SOURCE
M – MANIFOLD
P – PORTED

1980—2.8L ENG.-CODE "7"
OMEGA-LOW ALT.

RIGHT PULSAIR
EFE VALVE
LEAN AUTHORITY SWITCH
TO THERMAL SENSOR
SVB-TVS (AIR CLEANER)
DECELERATION VALVE
VACUUM BREAK
DISTRIBUTOR VACUUM
DS-VRDV
MANIFOLD VACUUM FITTING
P
M
P
CARB.
M
M
TO AIR CLEANER
EGR VALVE
DISTRIBUTOR
EFE-CV
EFE-TVS
BOWL VENT LINE
BOWL VENT VALVE
TO AIR CLEANER
PCV-VALVE
EGR-TVS
TO FUEL TANK
PURGE HOSE
VACUUM SOURCE
M – MANIFOLD VACUUM
P PORTED VACUUM
PURGE CONTROL VALVE
TANK PRESSURE CONTROL VALVE
FRONT
LEFT PULSAIR
PULSAIR SOLENOID

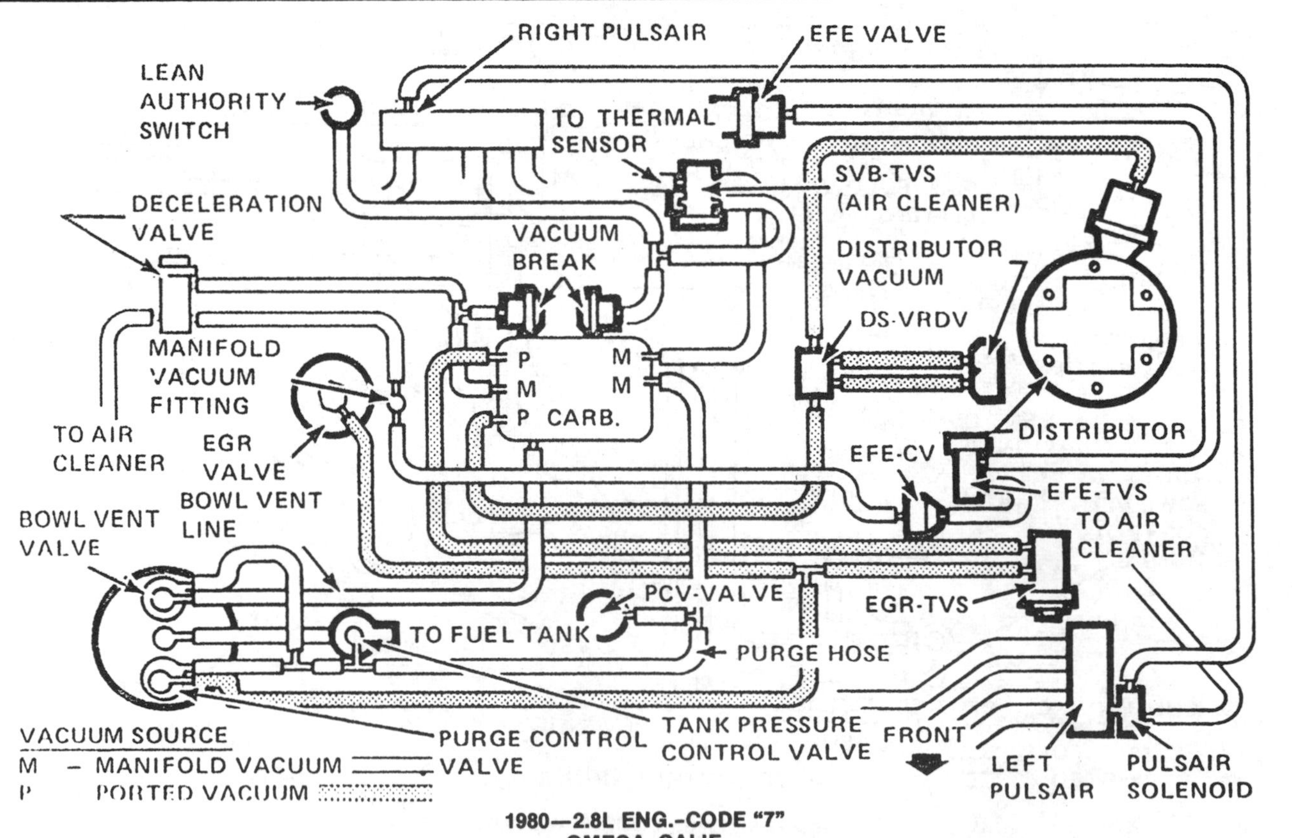

1980—2.8L ENG.-CODE "7"
OMEGA-CALIF.

VACUUM CIRCUITS

(© Oldsmobile Div., G.M. Corp.)

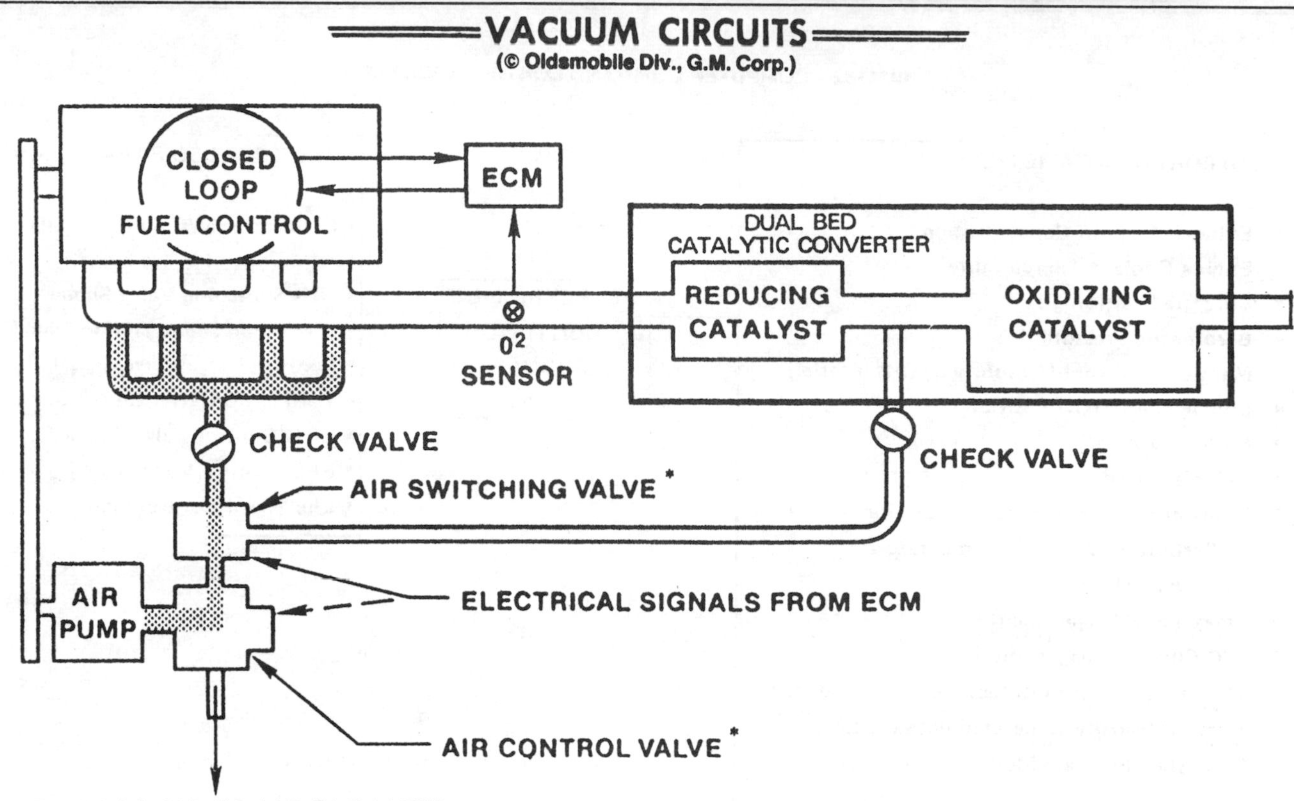

1981-82 AIR MANAGEMENT SYSTEM-COLD ENGINE MODE

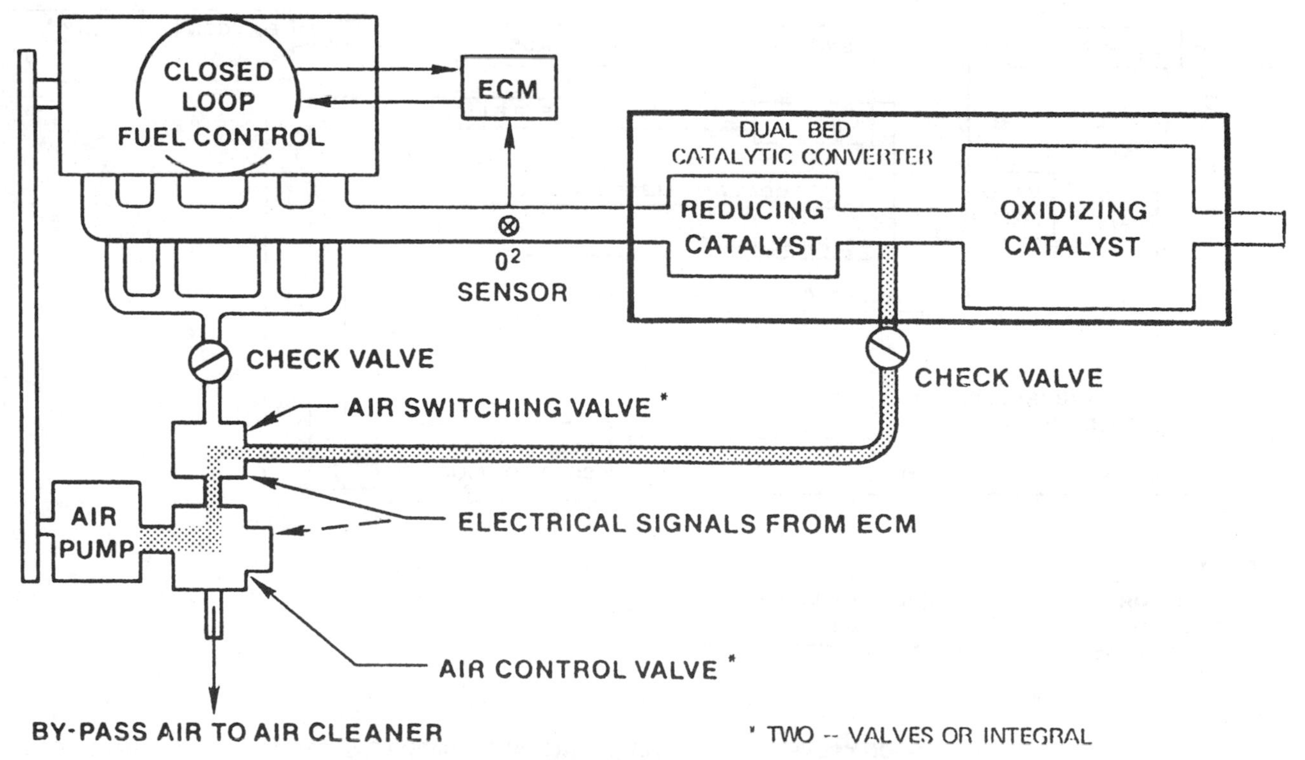

1981-82 AIR MANAGEMENT SYSTEM-WARM ENGINE MODE

1981-82—COMPUTER COMMAND CONTROL SYSTEM

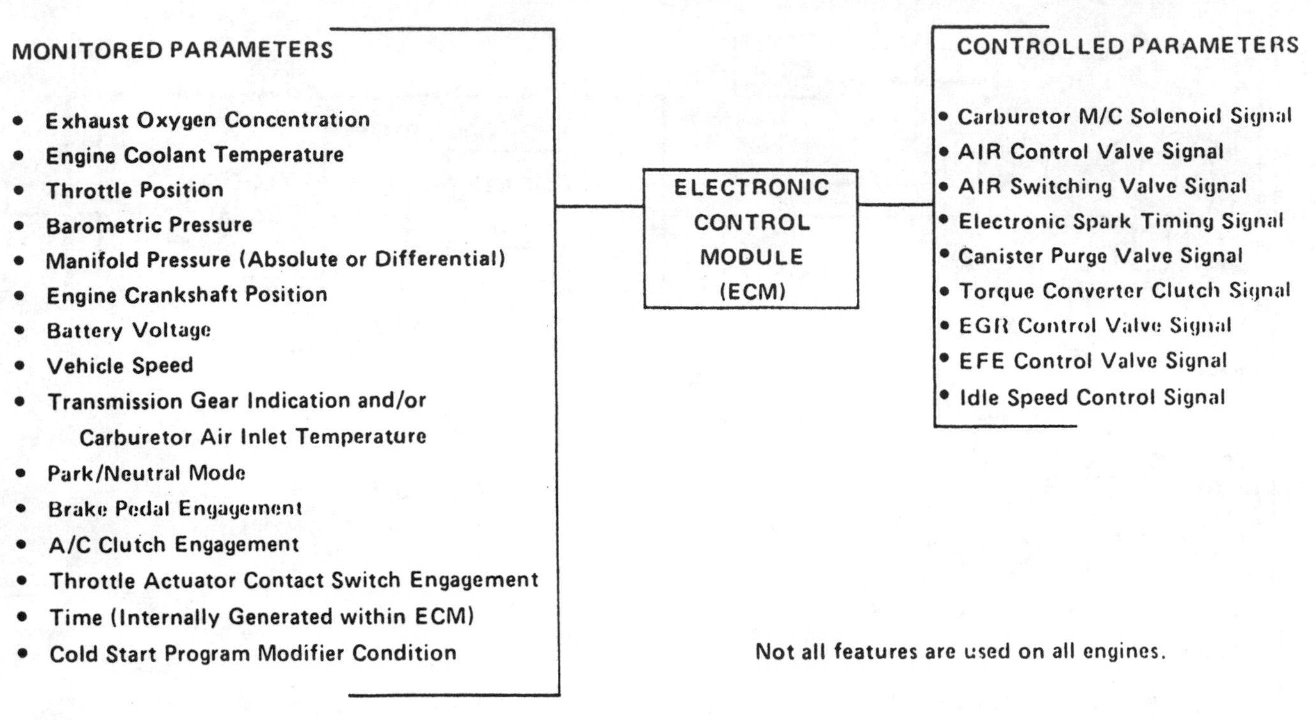
MONITORED PARAMETERS
• Exhaust Oxygen Concentration
• Engine Coolant Temperature
• Throttle Position
• Barometric Pressure
• Manifold Pressure (Absolute or Differential)
• Engine Crankshaft Position
• Battery Voltage
• Vehicle Speed
• Transmission Gear Indication and/or Carburetor Air Inlet Temperature
• Park/Neutral Mode
• Brake Pedal Engagement
• A/C Clutch Engagement
• Throttle Actuator Contact Switch Engagement
• Time (Internally Generated within ECM)
• Cold Start Program Modifier Condition
ELECTRONIC CONTROL MODULE (ECM)
CONTROLLED PARAMETERS
• Carburetor M/C Solenoid Signal
• AIR Control Valve Signal
• AIR Switching Valve Signal
• Electronic Spark Timing Signal
• Canister Purge Valve Signal
• Torque Converter Clutch Signal
• EGR Control Valve Signal
• EFE Control Valve Signal
• Idle Speed Control Signal
Not all features are used on all engines.

System Parameters

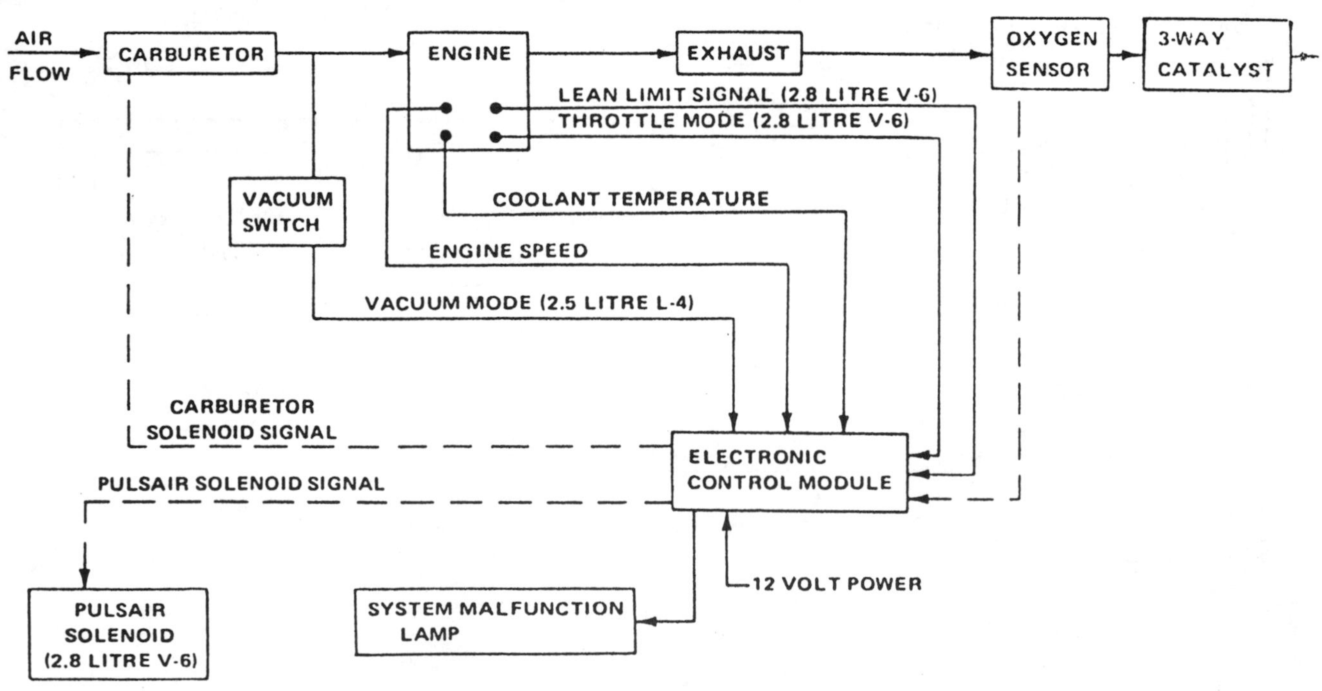
AIR FLOW
CARBURETOR
ENGINE
EXHAUST
OXYGEN SENSOR
3-WAY CATALYST
LEAN LIMIT SIGNAL (2.8 LITRE V-6)
THROTTLE MODE (2.8 LITRE V-6)
VACUUM SWITCH
COOLANT TEMPERATURE
ENGINE SPEED
VACUUM MODE (2.5 LITRE L-4)
CARBURETOR SOLENOID SIGNAL
ELECTRONIC CONTROL MODULE
PULSAIR SOLENOID SIGNAL
12 VOLT POWER
PULSAIR SOLENOID (2.8 LITRE V-6)
SYSTEM MALFUNCTION LAMP

1981-82—COMPUTER CONTROLLED CATALYTIC CONVERTER SYSTEM

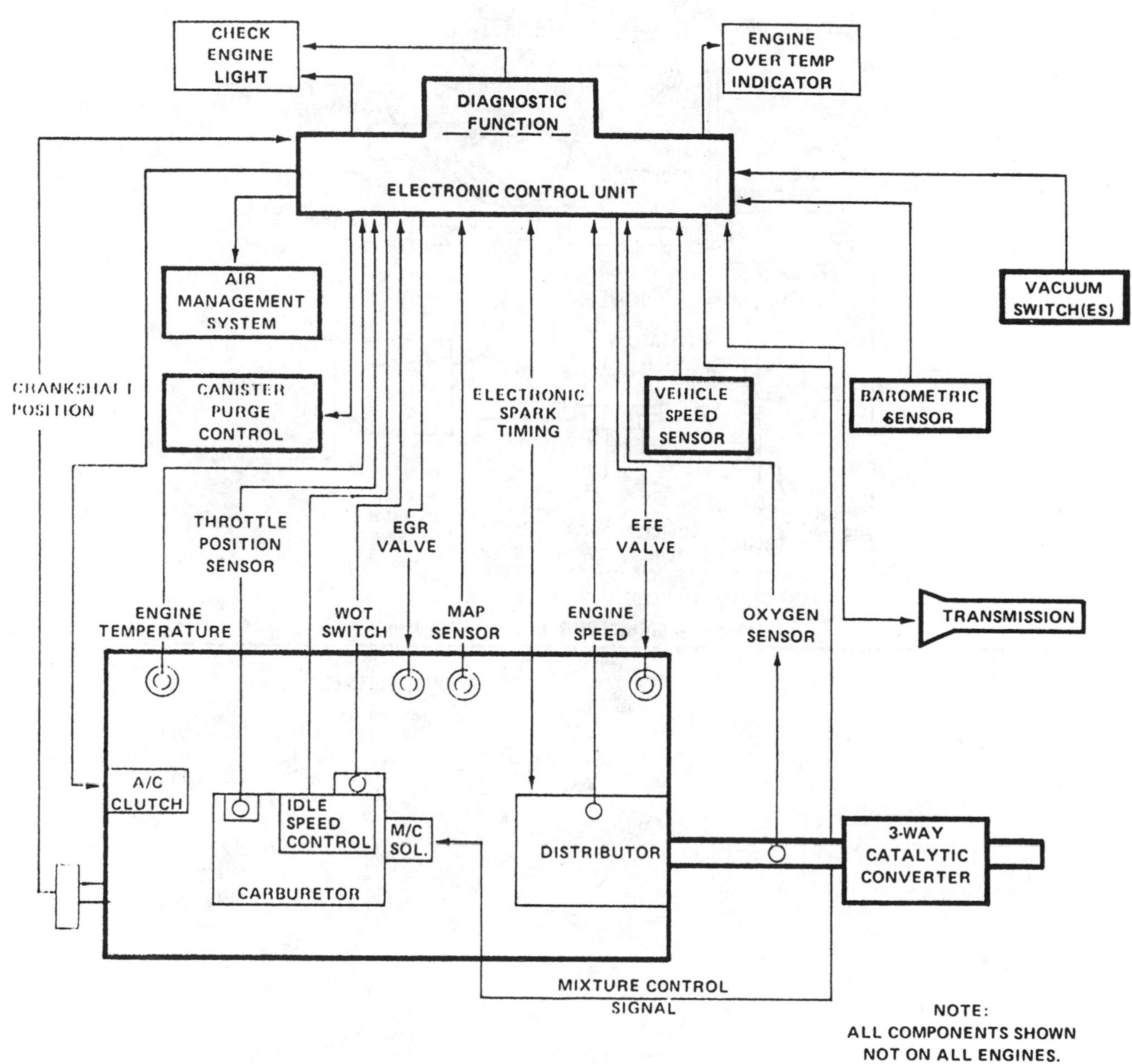

1981-82—COMPUTER COMMAND CONTROL SYSTEM

The Vehicle Emission Control Information Label is located in the engine compartment (fan shroud, radiator support, hood underside, etc.) of every vehicle produced by General Motors Corporation. The label contains important emission specifications and setting procedures, as well as a vacuum hose schematic with emission components identified.

When servicing the engine or emission systems, the Vehicle Emission Control Information Label should be checked for up-to-date information.

VACUUM CIRCUITS

(© Oldsmobile Div., G.M. Corp.)

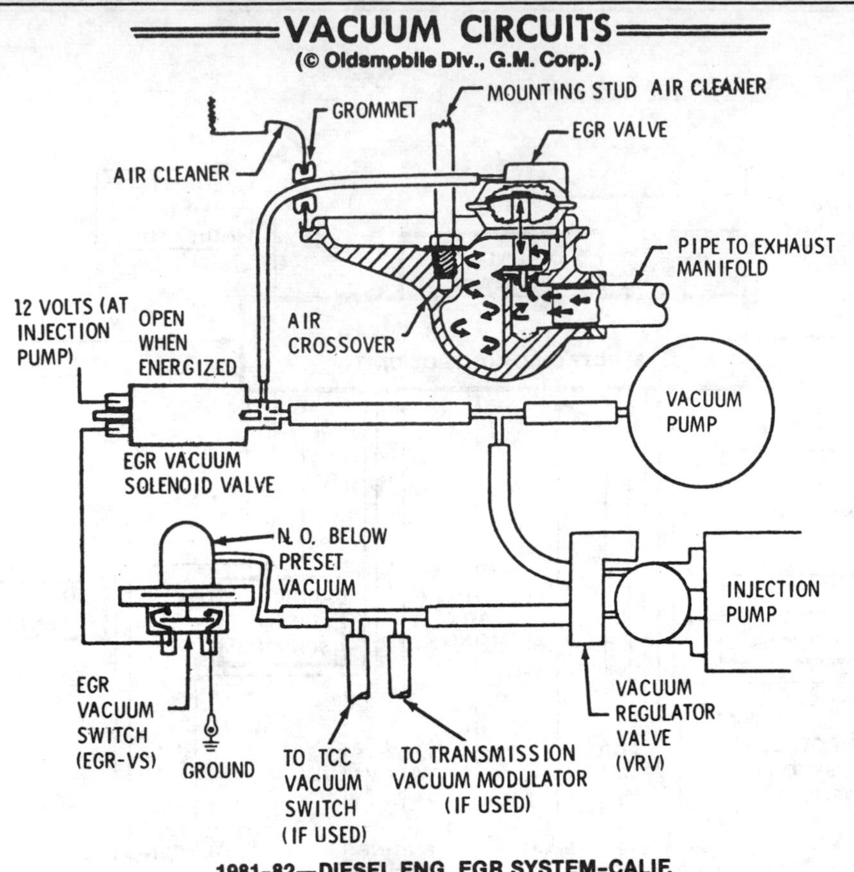

1981-82—DIESEL ENG. EGR SYSTEM-CALIF.

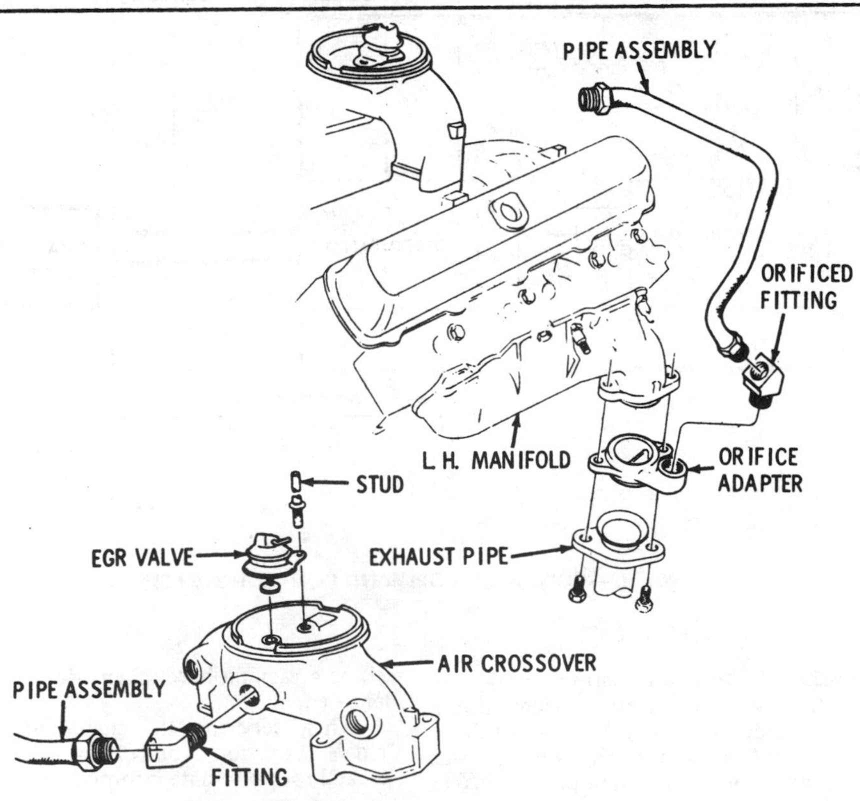

1981-82—DIESEL ENG. EGR ORIFICE SYSTEM-CALIF.

VACUUM CIRCUITS

(© Oldsmobile Div., G.M. Corp.)

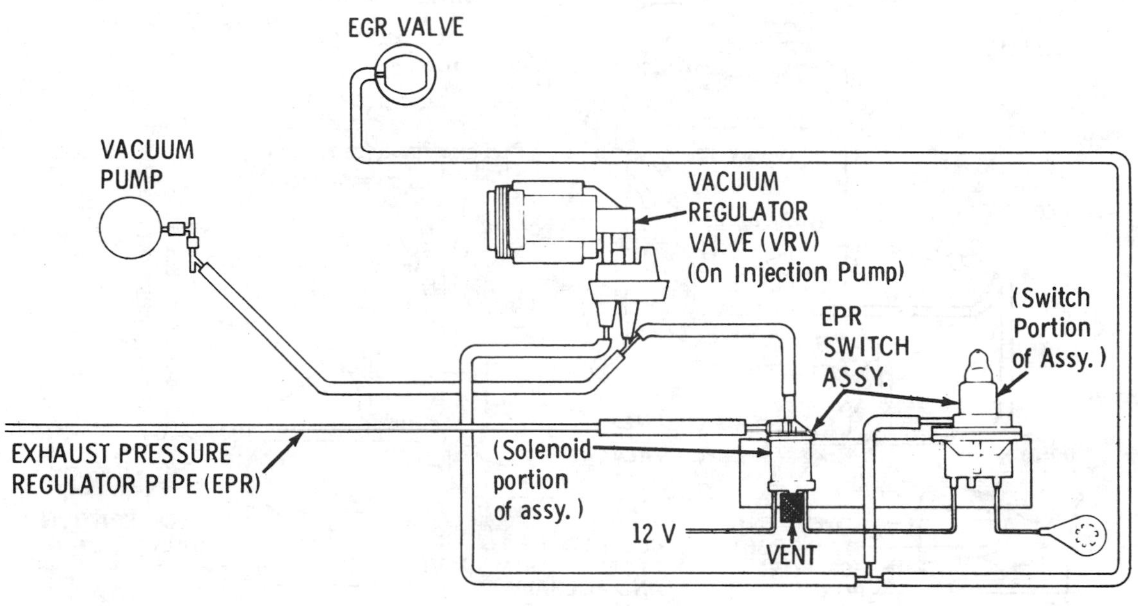

1981-82—DIESEL ENG. EGR SYSTEM-TORONADO-EXC. CALIF.

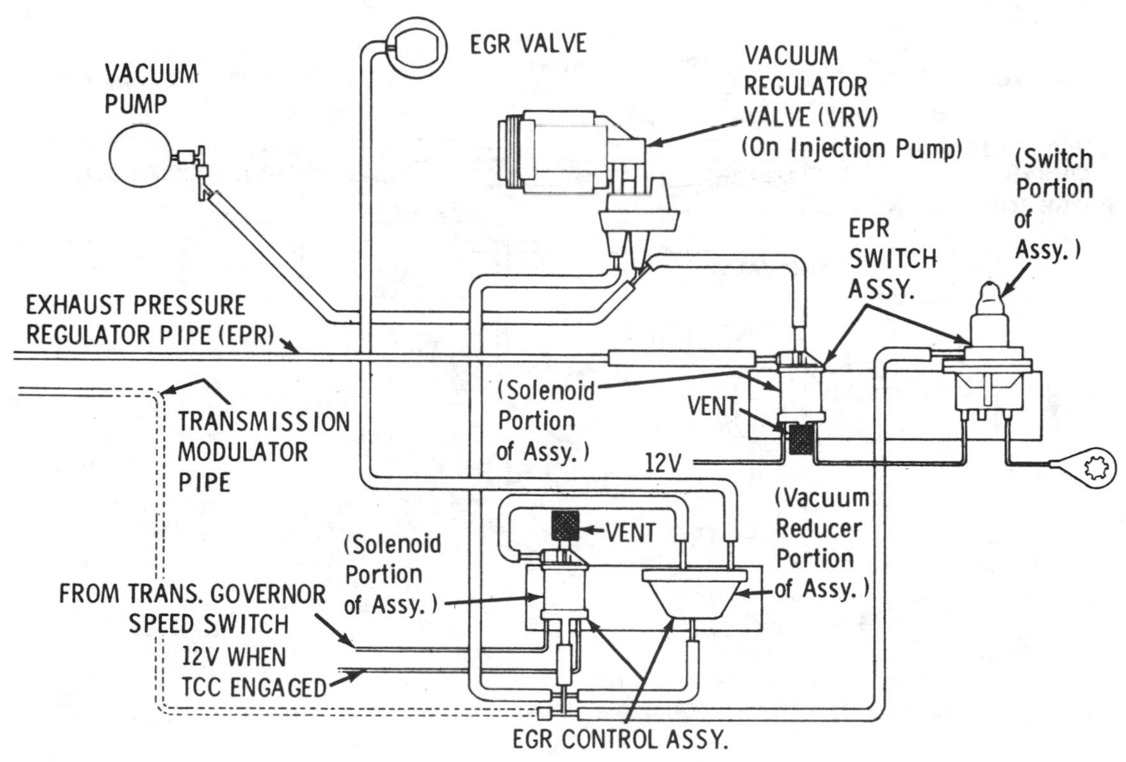

1981-82—DIESEL ENG. EGR SYSTEM-EXC. TORONADO & CALIF.

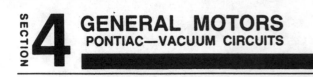

VACUUM CIRCUITS

(© Pontiac Div., G.M. Corp.)

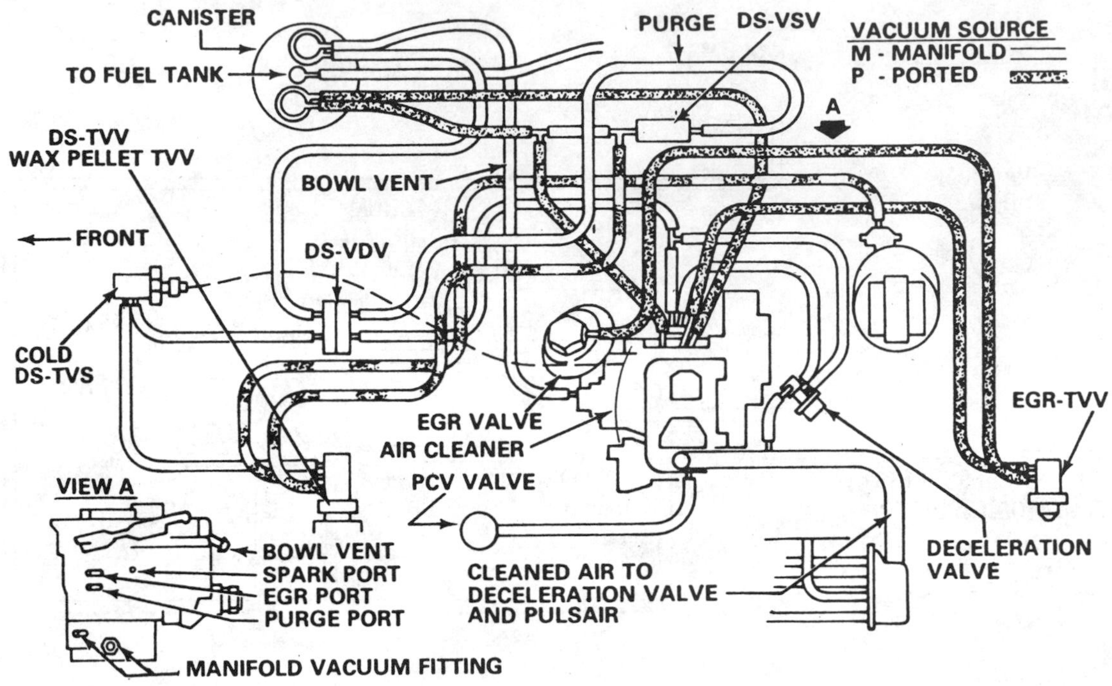

1980—2.5L 4 CYL. ENG.-CODE "V"-WO/C-4 AUTO. TRANS.-W/AIR COND.

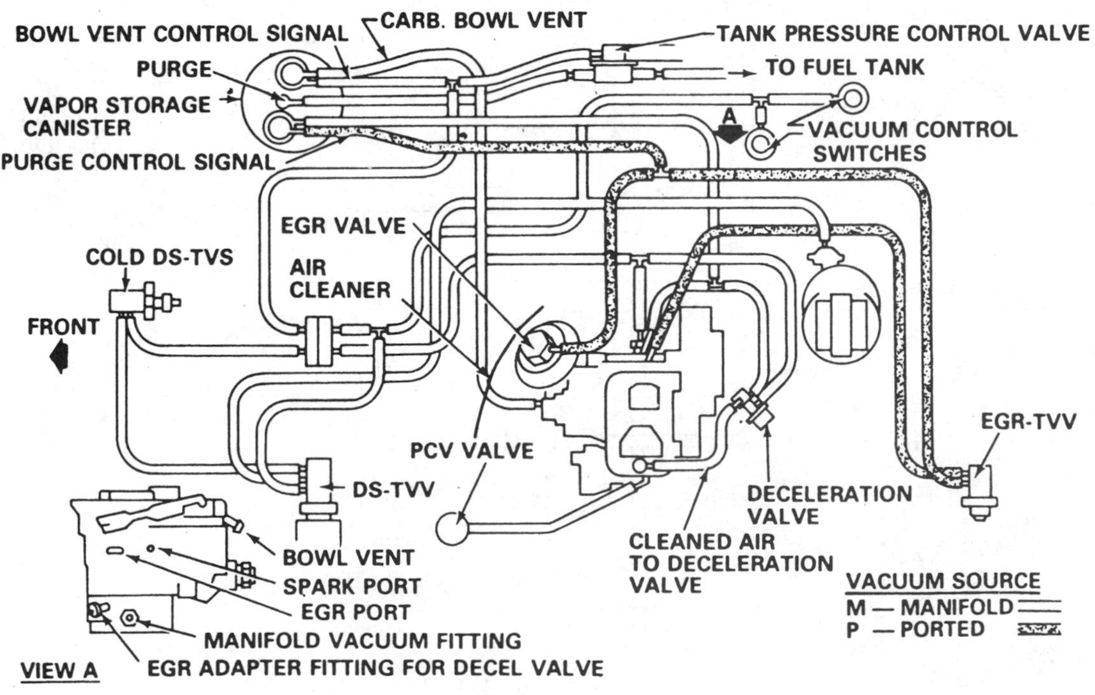

1980—2.5L 4 CYL. ENG.-CODE "V"-W/C-4 AUTO. TRANS. & MAN. TRANS.

VACUUM CIRCUITS

(© Pontiac Div., G.M. Corp.)

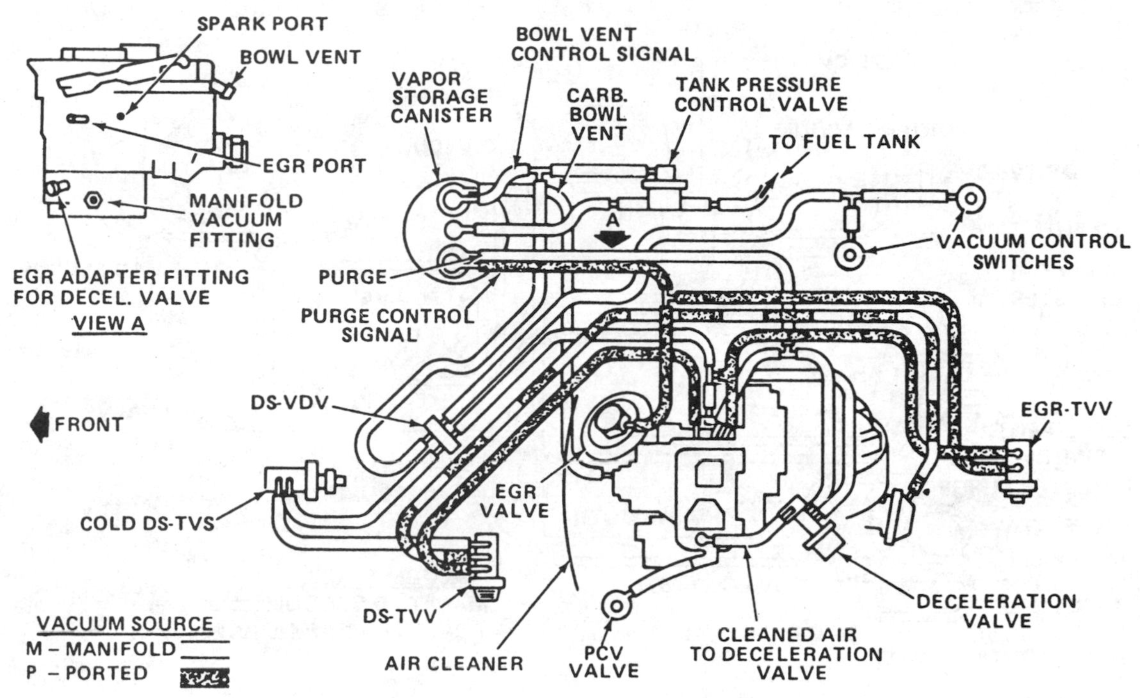

1980—2.5L 4 CYL. ENG.-CODE "V"-W/C-4 AUTO. TRANS.

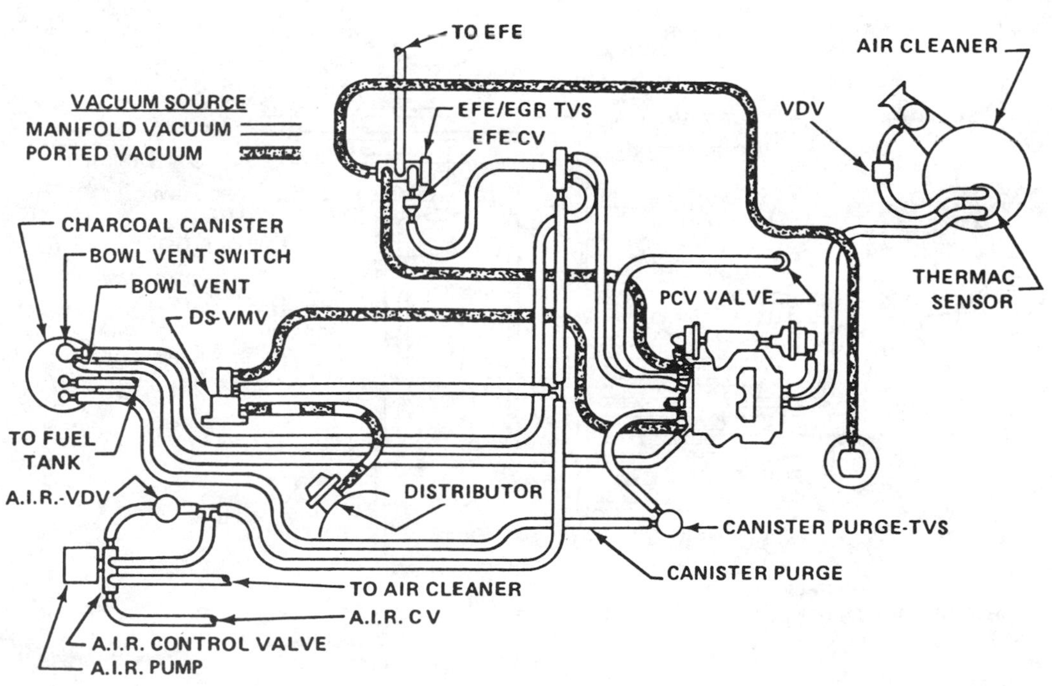

1980—3.8L V6 ENG.-CODE "A"-WO/C-4 AUTO. TRANS.-W/MAN. TRANS.

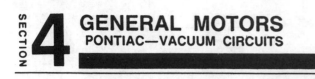

VACUUM CIRCUITS

(© Pontiac Div., G.M. Corp.)

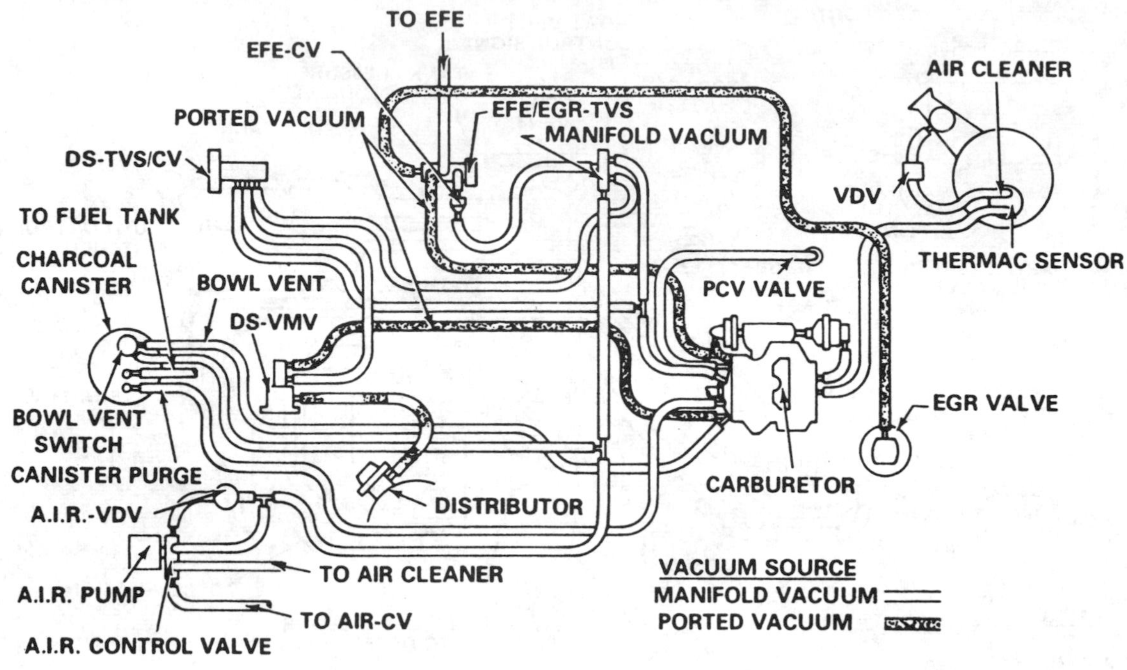

1980—3.8L V6 ENG.-CODE "A"-WO/C-4 AUTO. TRANS.

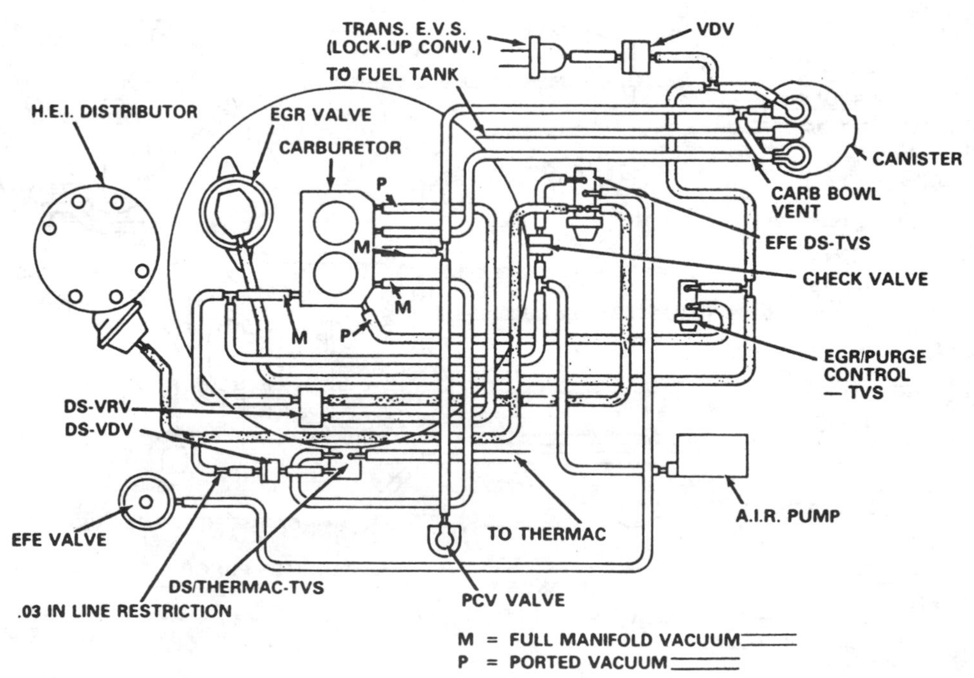

1980—3.8L V6 ENG.-CODE "K"

VACUUM CIRCUITS

(© Pontiac Div., G.M. Corp.)

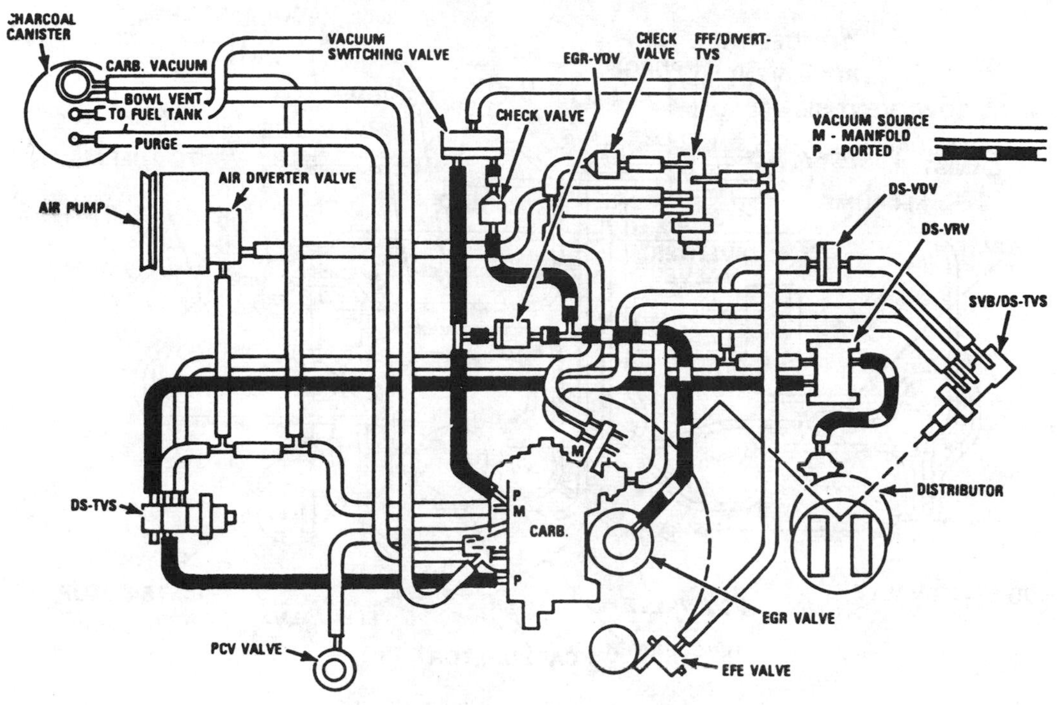

1980—3.8 L V6 ENG.-CODE "S"

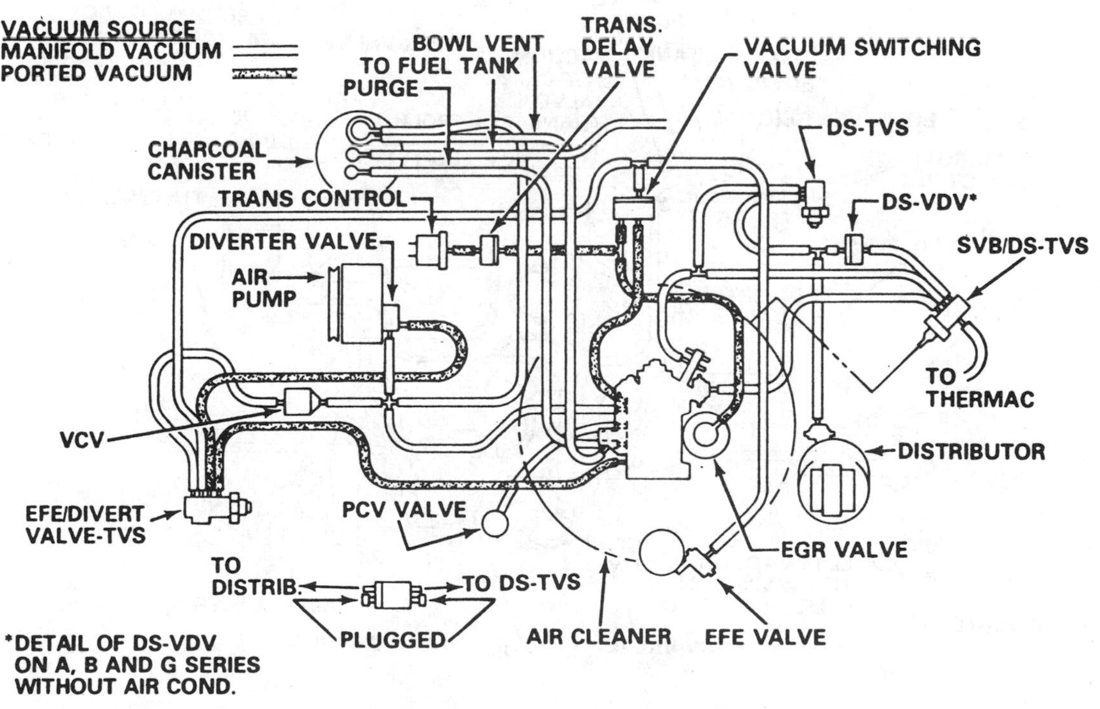

1980—4.3L V8 ENG.-CODE "S"-W/LOCK-UP TORQUE CONVERTER

VACUUM CIRCUITS

(© Pontiac Div., G.M. Corp.)

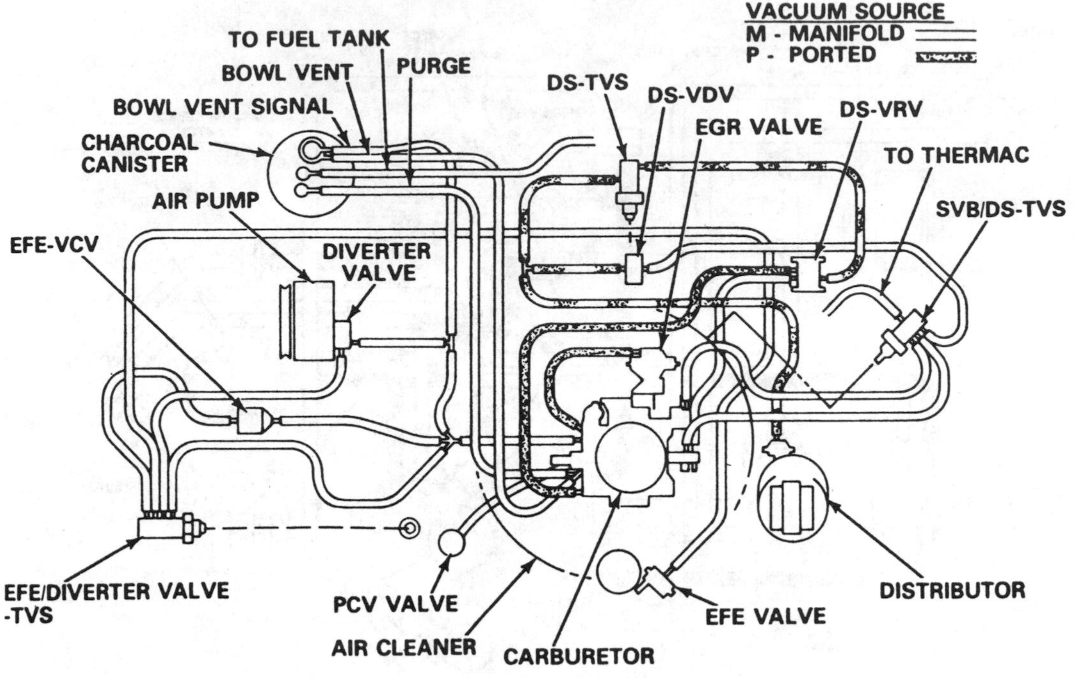

1980—4.9L V8 ENG.-CODE "W"

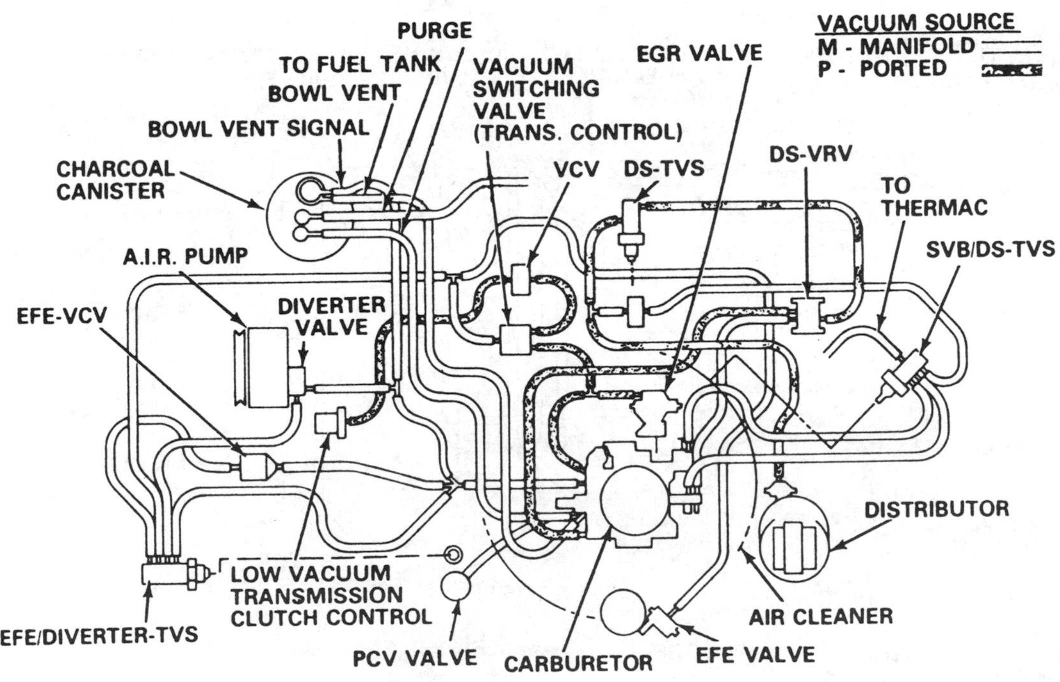

1980—4.9L V8 ENG.-CODE "W"-W/LOCK-UP TORQUE CONVERTER

VACUUM CIRCUITS

(© Pontiac Div., G.M. Corp.)

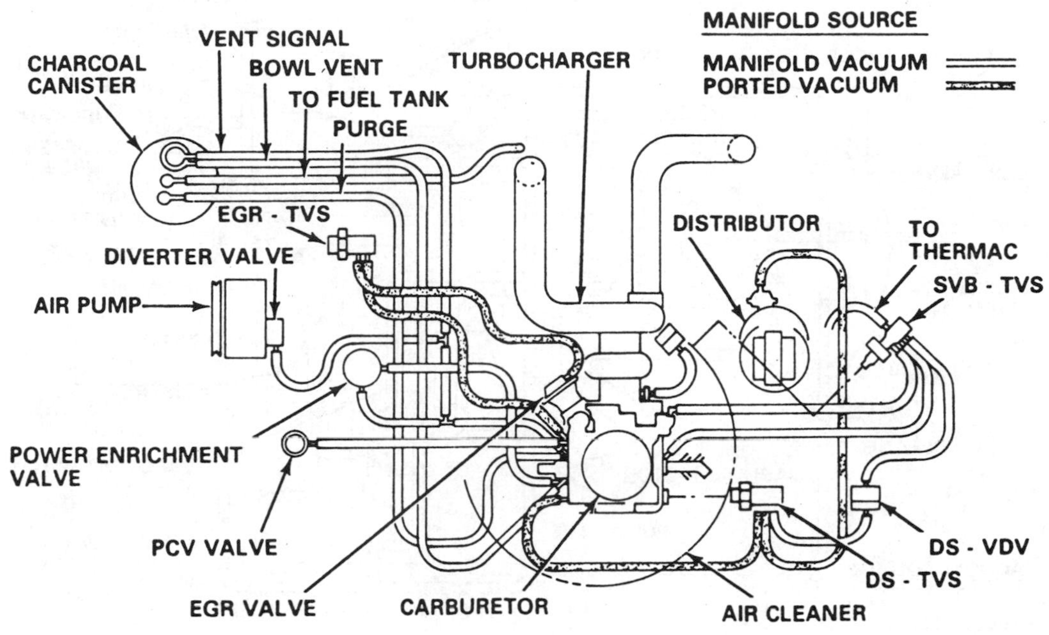

1980—4.9L V8 ENG.-CODE "T"

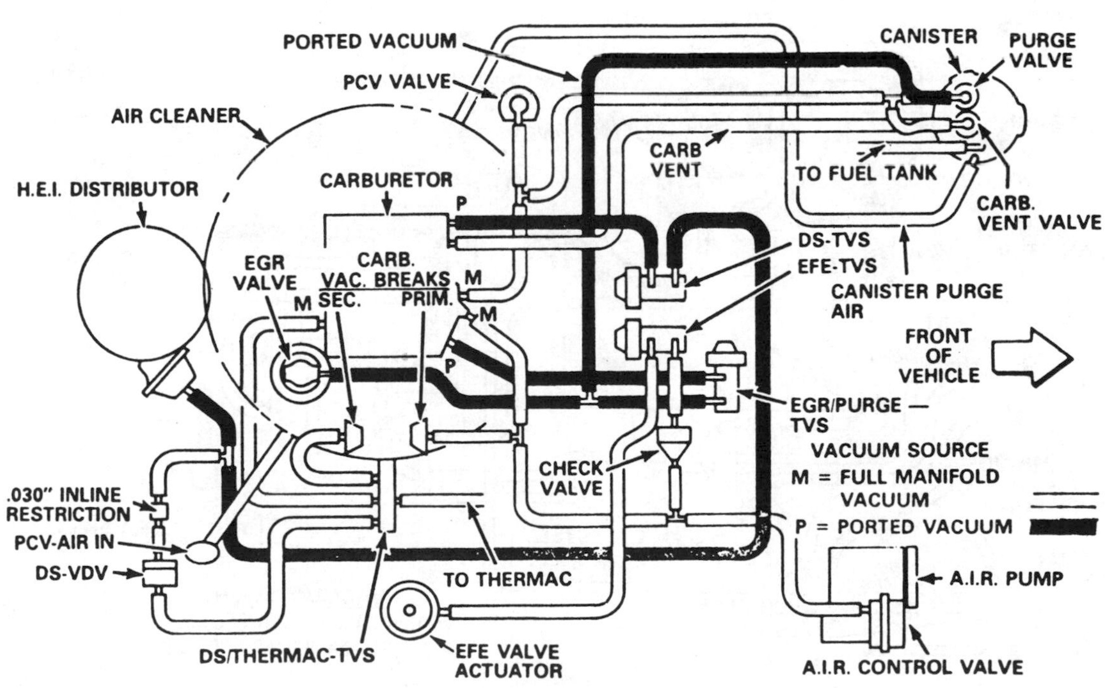

1980—4.9L V8 ENG.-CODE "H"

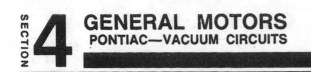

VACUUM CIRCUITS

(© Pontiac Div., G.M. Corp.)

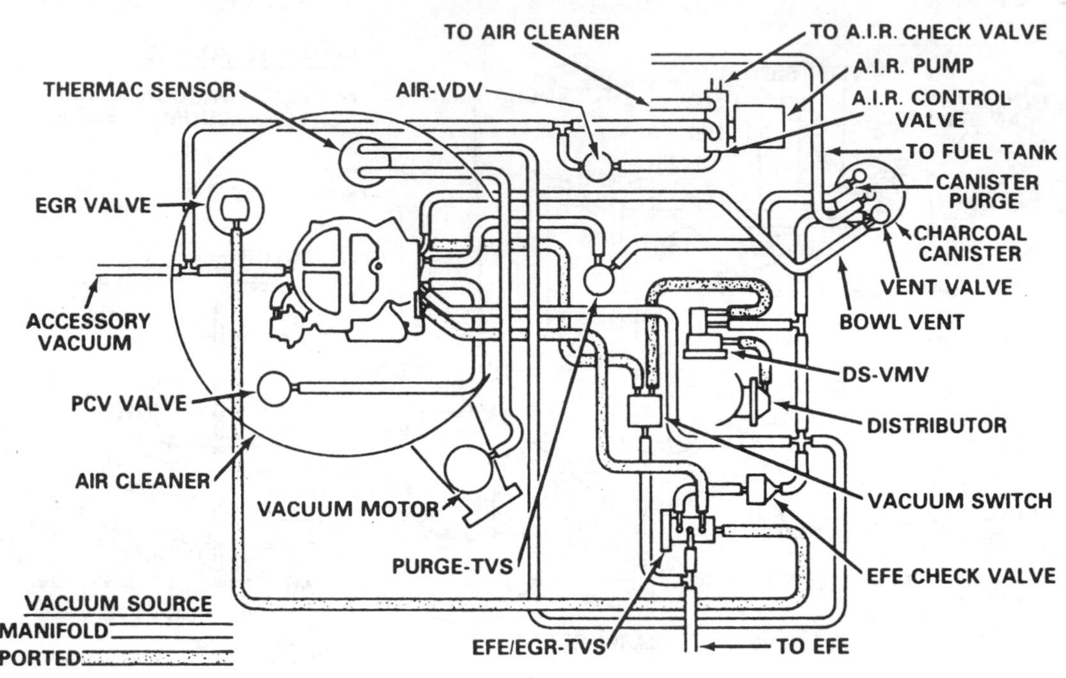

1980—5.7L V8 ENG.-CODE "X"

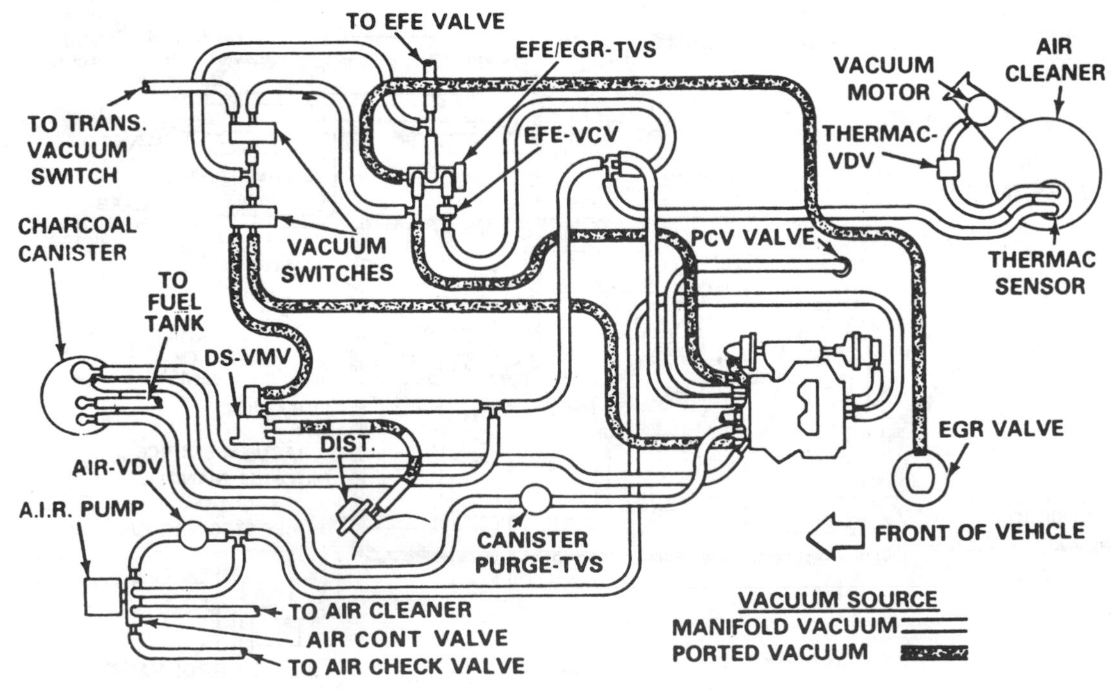

1980—5.7L V8 ENG.-CODE "X"-W/LOCK-UP TORQUE CONVERTER

VACUUM CIRCUITS

(© Pontiac Div., G.M. Corp.)

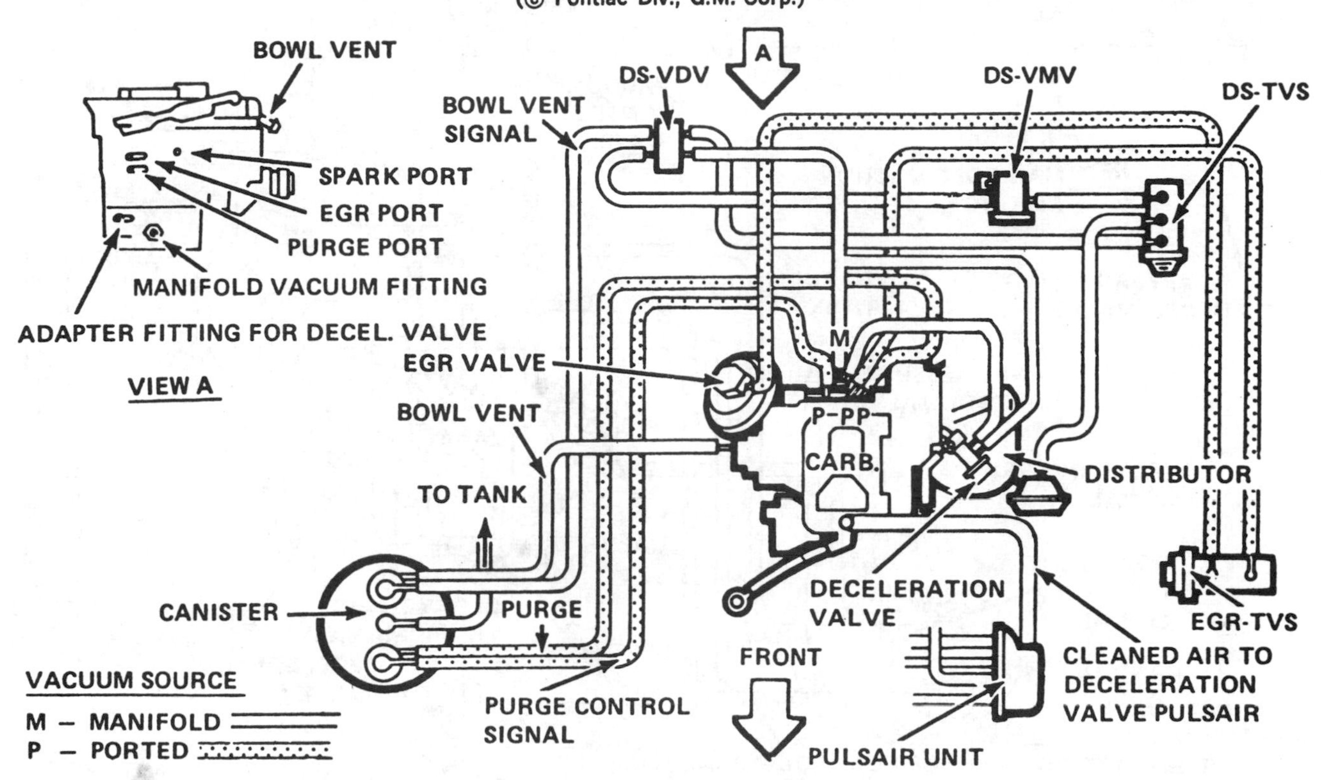

1980—2.5L 4 CYL. ENG.-CODE "5"-PHOENIX-W/MAN. TRANS.-LOW ALT.

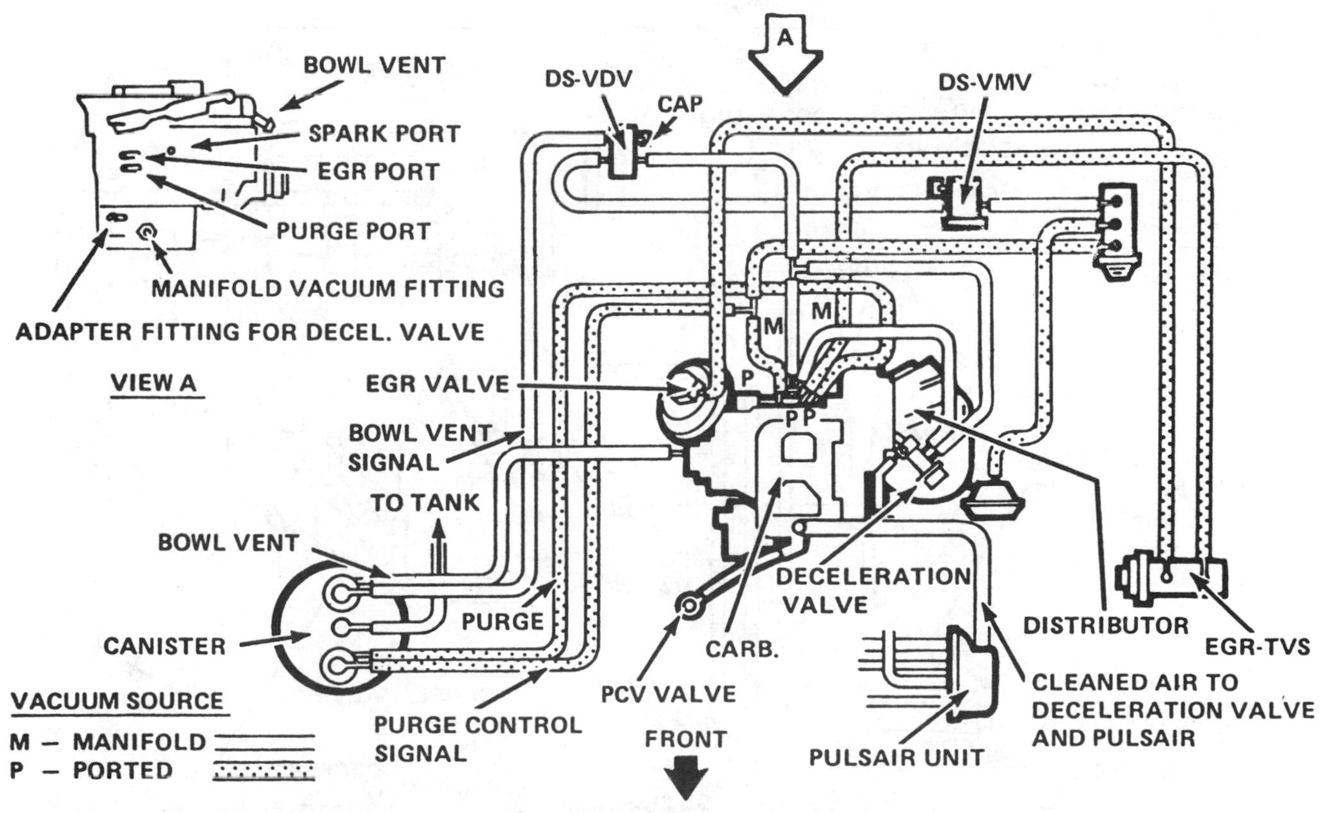

1980—2.5L 4 CYL. ENG.-CODE "5"-PHOENIX-W/AUTO. TRANS.-LOW ALT.

VACUUM CIRCUITS

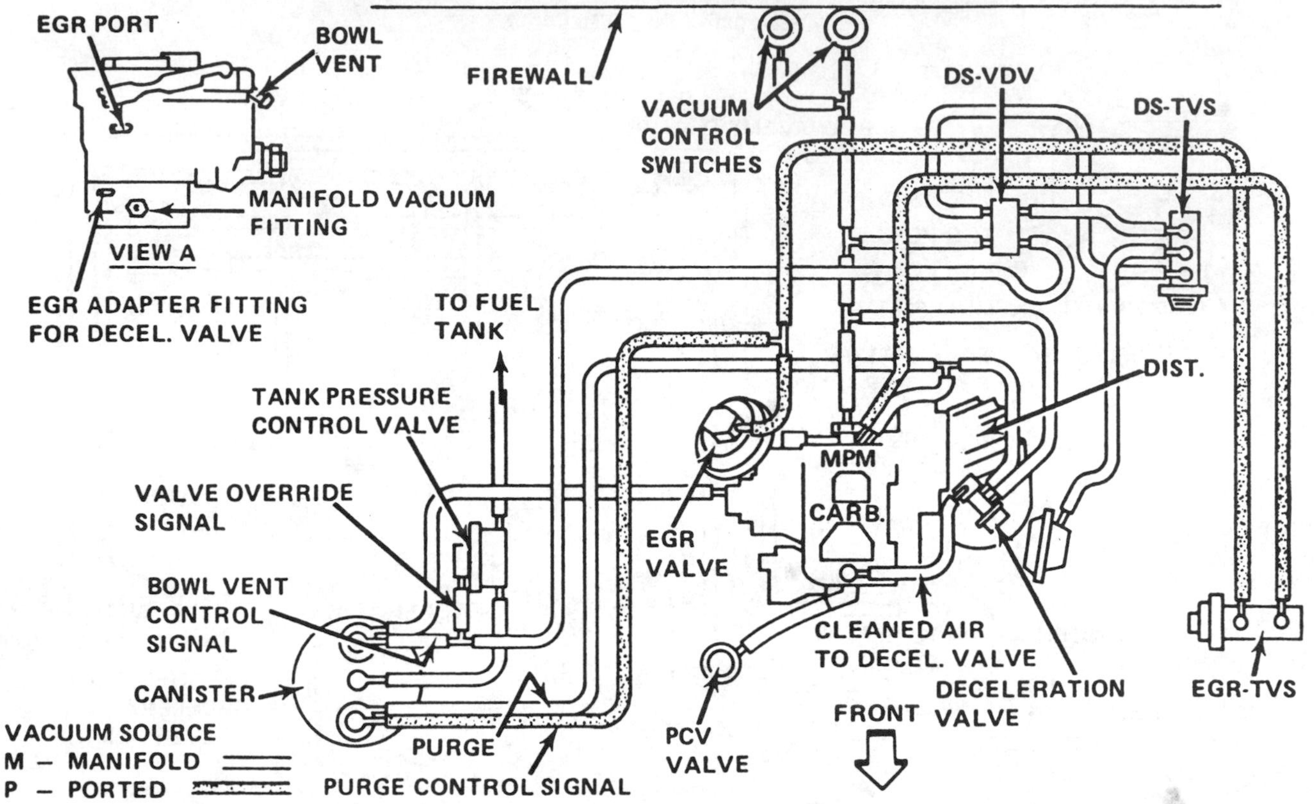

EGR PORT
BOWL VENT
FIREWALL
VACUUM CONTROL SWITCHES
DS-VDV
DS-TVS
MANIFOLD VACUUM FITTING
VIEW A
EGR ADAPTER FITTING FOR DECEL. VALVE
TO FUEL TANK
TANK PRESSURE CONTROL VALVE
DIST.
MPM
CARB.
VALVE OVERRIDE SIGNAL
EGR VALVE
BOWL VENT CONTROL SIGNAL
CLEANED AIR TO DECEL. VALVE
CANISTER
DECELERATION VALVE
EGR-TVS
FRONT
VACUUM SOURCE
M – MANIFOLD
P – PORTED
PURGE
PCV VALVE
PURGE CONTROL SIGNAL

1980—2.5L 4 CYL. ENG.-CODE "5"-PHOENIX-W/MAN. TRANS.-CALIF.

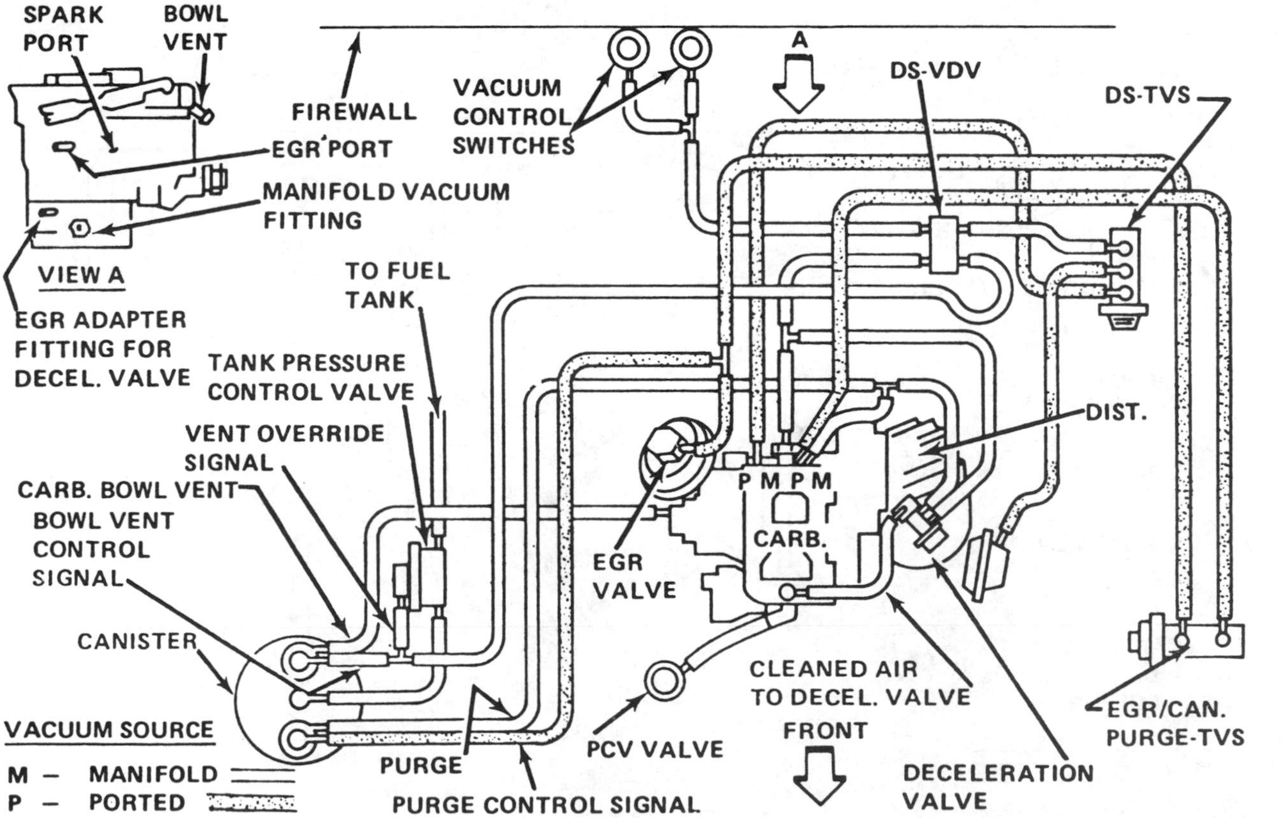

SPARK PORT
BOWL VENT
FIREWALL
VACUUM CONTROL SWITCHES
A
DS-VDV
DS-TVS
EGR PORT
MANIFOLD VACUUM FITTING
VIEW A
TO FUEL TANK
EGR ADAPTER FITTING FOR DECEL. VALVE
TANK PRESSURE CONTROL VALVE
DIST.
VENT OVERRIDE SIGNAL
CARB. BOWL VENT
P M P M
BOWL VENT CONTROL SIGNAL
CARB.
EGR VALVE
CANISTER
CLEANED AIR TO DECEL. VALVE
EGR/CAN. PURGE-TVS
VACUUM SOURCE
FRONT
PCV VALVE
M – MANIFOLD
P – PORTED
PURGE
DECELERATION VALVE
PURGE CONTROL SIGNAL

1980—2.5L 4 CYL. ENG.-CODE "5"-PHOENIX-W/AUTO. TRANS.-CALIF.

VEHICLE EMISSION CONTROL INFORMATION LABEL

The Vehicle Emission Control Information Label is located in the engine compartment (fan shroud, radiator support, hood underside, etc.) of every vehicle produced by General Motors Corporation. The label contains important emission specifications and setting procedures, as well as a vacuum hose schematic with emission components identified.

When servicing the engine or emission systems, the Vehicle Emission Control Information Label should be checked for up-to-date information.

COMPUTER COMMAND CONTROL SYSTEM

The Computer Command Control System (the "System"), is an electronically controlled emission system that can monitor up to fifteen (15) various engine/vehicle operating conditions and using this information can control as many as nine (9) engine related systems and The "System" is thereby making constant adjustments to maintain good vehicle performance under all normal driving conditions while allowing the catalytic converter to effectively control Oxides of Nitrogen (NOX), Hydrocarbons (HC) and Carbon Monoxide (CO).

The "System" has a built-in diagnostic system that recognizes and identifies possible operational problems and alerts the driver through a "CHECK ENGINE" light in the instrument panel. The light will remain "ON" until the problem is corrected. This same light is used, with the built-in diagnostic system, to aid the technician in locating and repairing the problem by flashing a stored code or codes which identifies the possible problem areas.

The "System" also has built-in back-up systems that in most cases of an operation problem will allow continued operation of the vehicle in a near normal manner until repairs can be made.

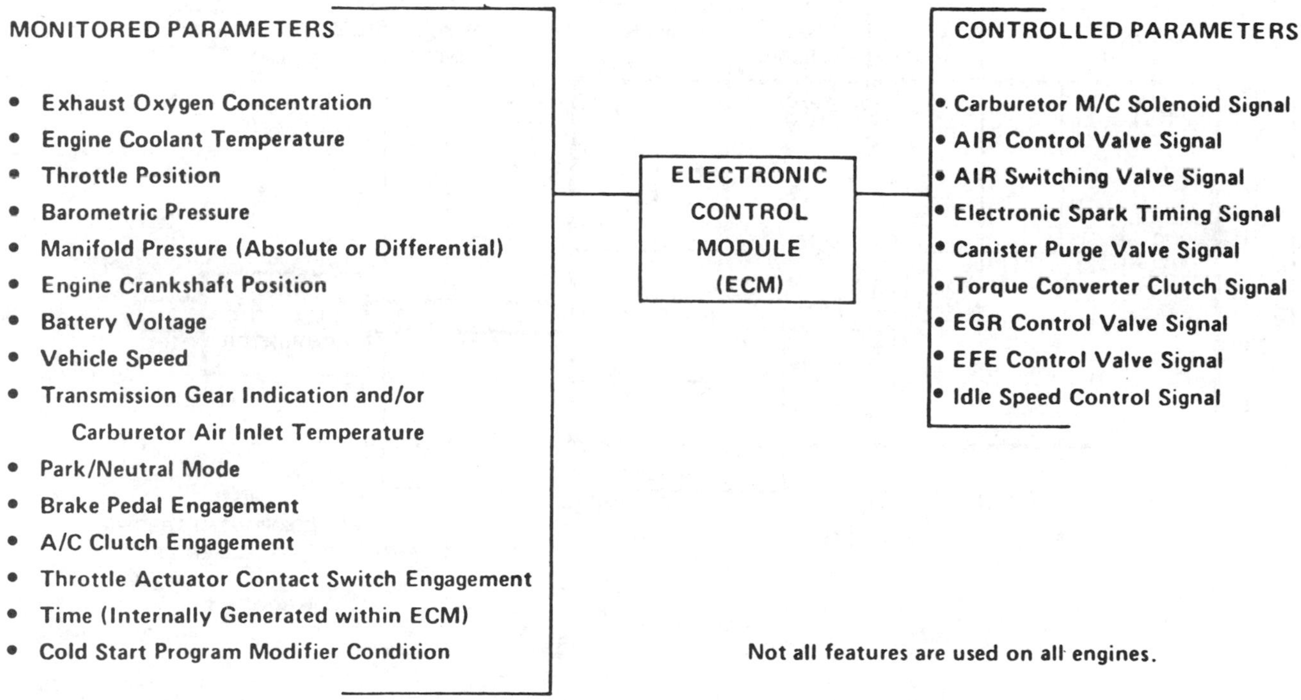

MONITORED PARAMETERS

- Exhaust Oxygen Concentration
- Engine Coolant Temperature
- Throttle Position
- Barometric Pressure
- Manifold Pressure (Absolute or Differential)
- Engine Crankshaft Position
- Battery Voltage
- Vehicle Speed
- Transmission Gear Indication and/or Carburetor Air Inlet Temperature
- Park/Neutral Mode
- Brake Pedal Engagement
- A/C Clutch Engagement
- Throttle Actuator Contact Switch Engagement
- Time (Internally Generated within ECM)
- Cold Start Program Modifier Condition

ELECTRONIC CONTROL MODULE (ECM)

CONTROLLED PARAMETERS

- Carburetor M/C Solenoid Signal
- AIR Control Valve Signal
- AIR Switching Valve Signal
- Electronic Spark Timing Signal
- Canister Purge Valve Signal
- Torque Converter Clutch Signal
- EGR Control Valve Signal
- EFE Control Valve Signal
- Idle Speed Control Signal

Not all features are used on all engines.

System Parameters

1981-82 COMPUTER COMMAND CONTROL SYSTEM

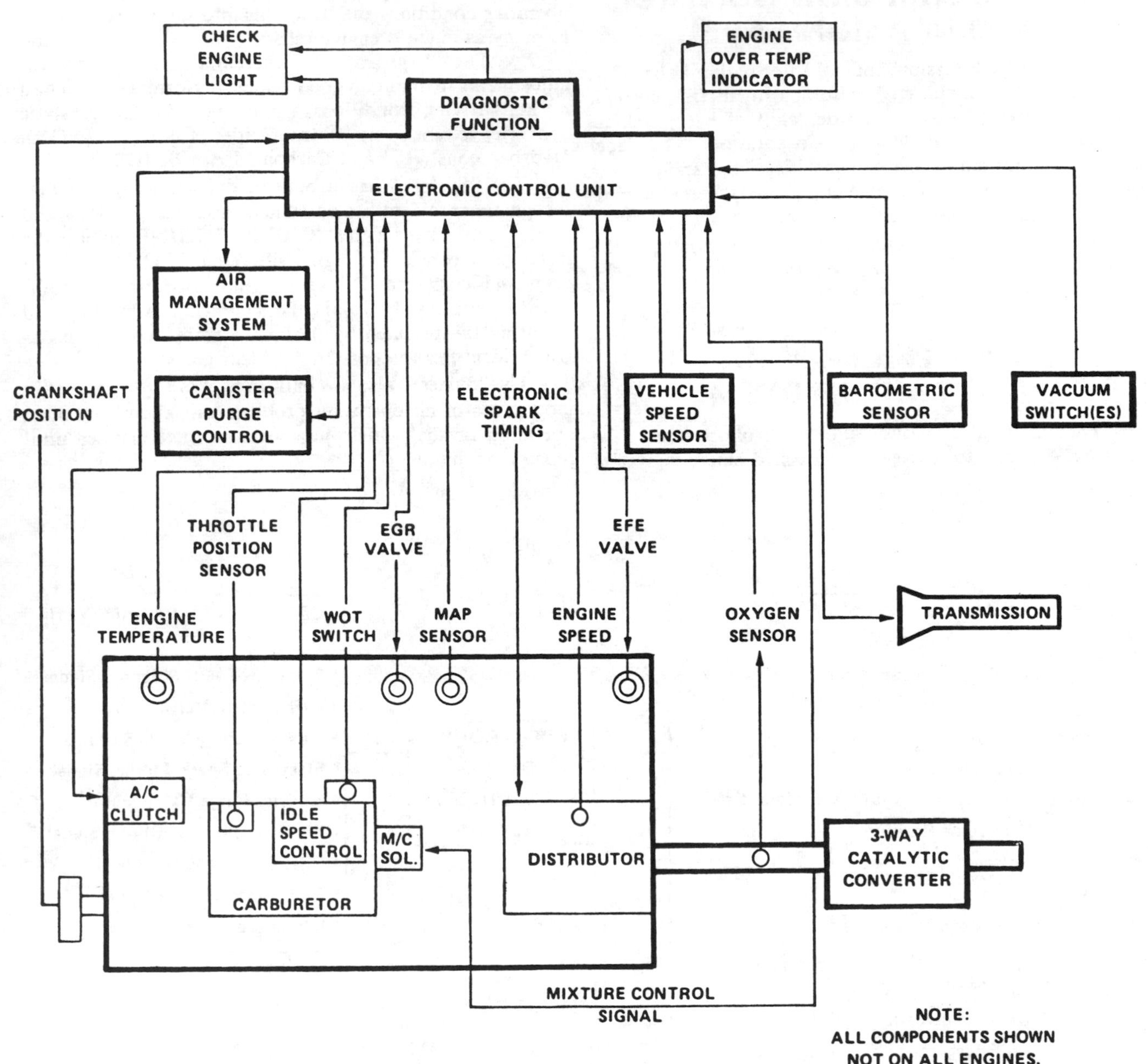

1981–82 COMPUTER COMMAND CONTROL SYSTEM

VACUUM CIRCUITS

(© Pontiac Div., G.M. Corp.)

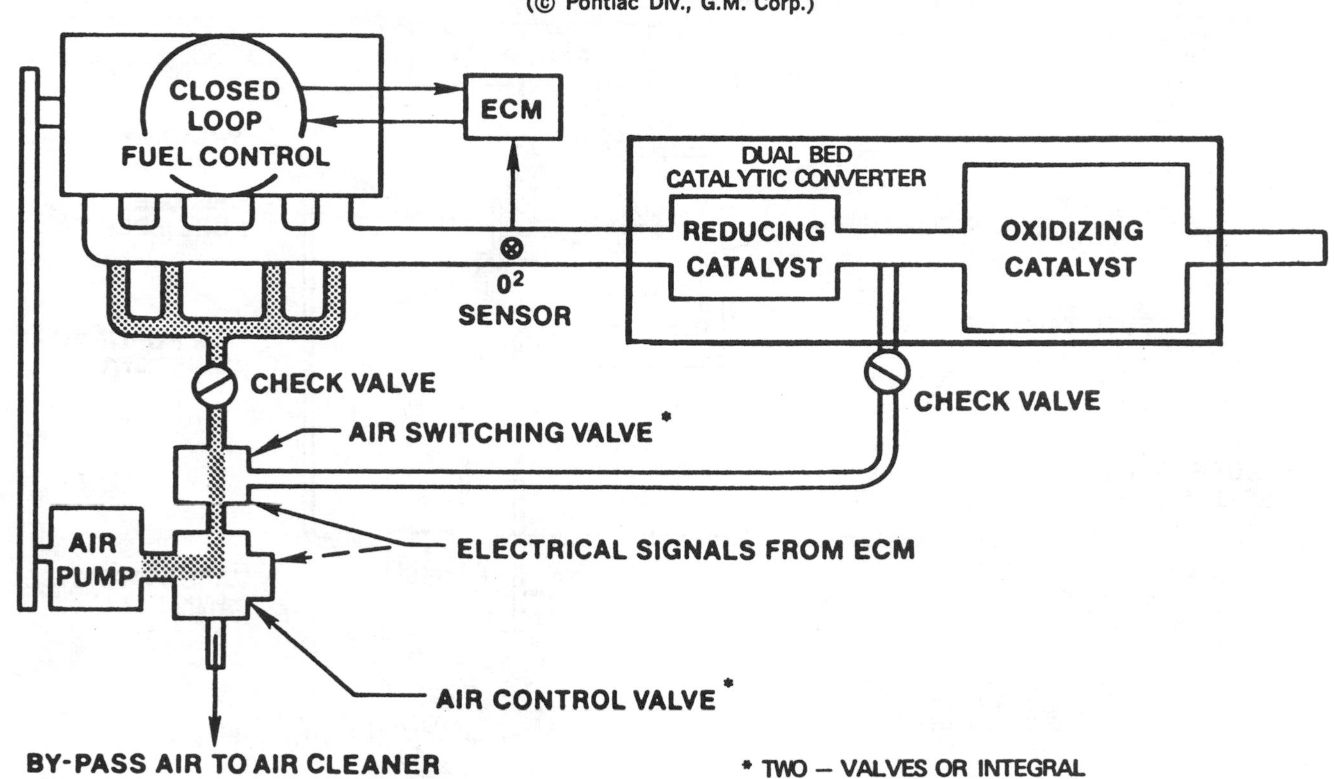

1981-82 AIR MANAGEMENT SYSTEM-COLD ENGINE MODE

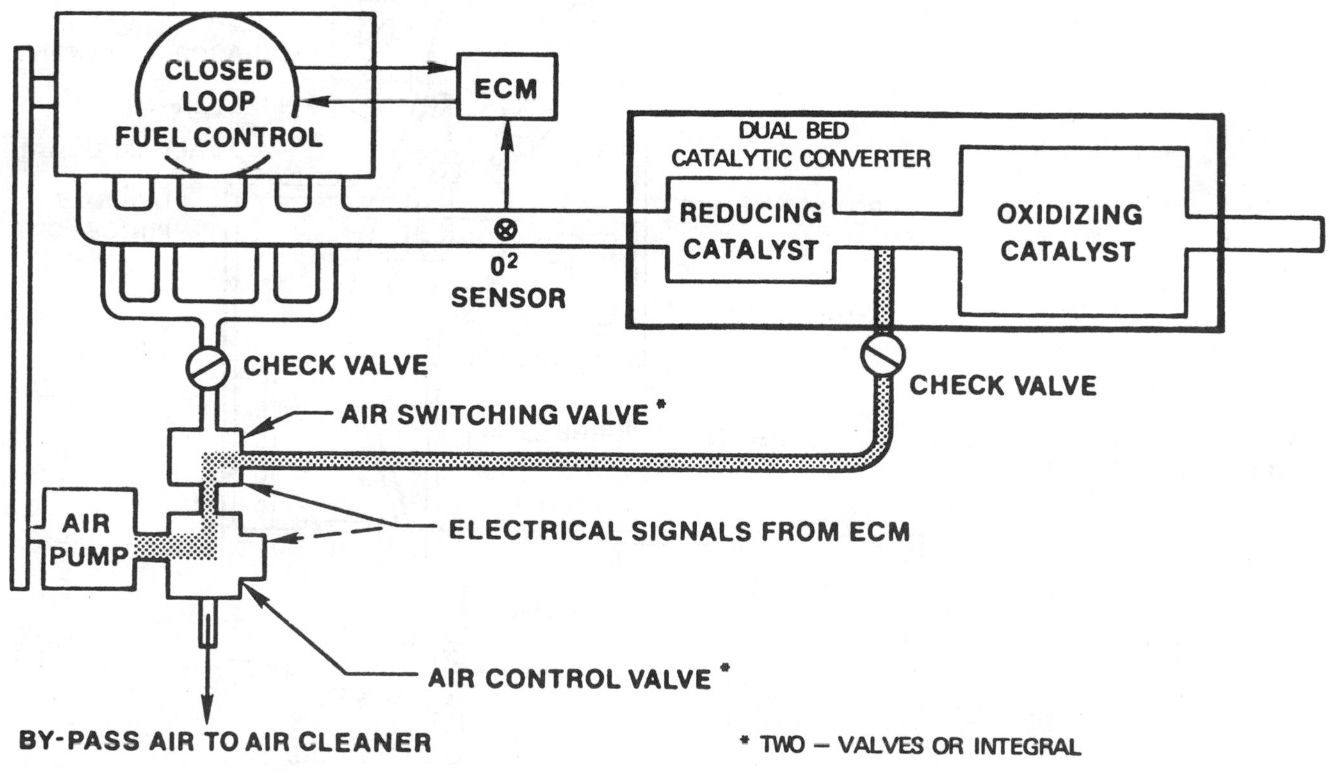

1981-82 AIR MANAGEMENT SYSTEM-WARM ENGINE MODE

VACUUM CIRCUITS

(© Pontiac Div., G.M. Corp.)

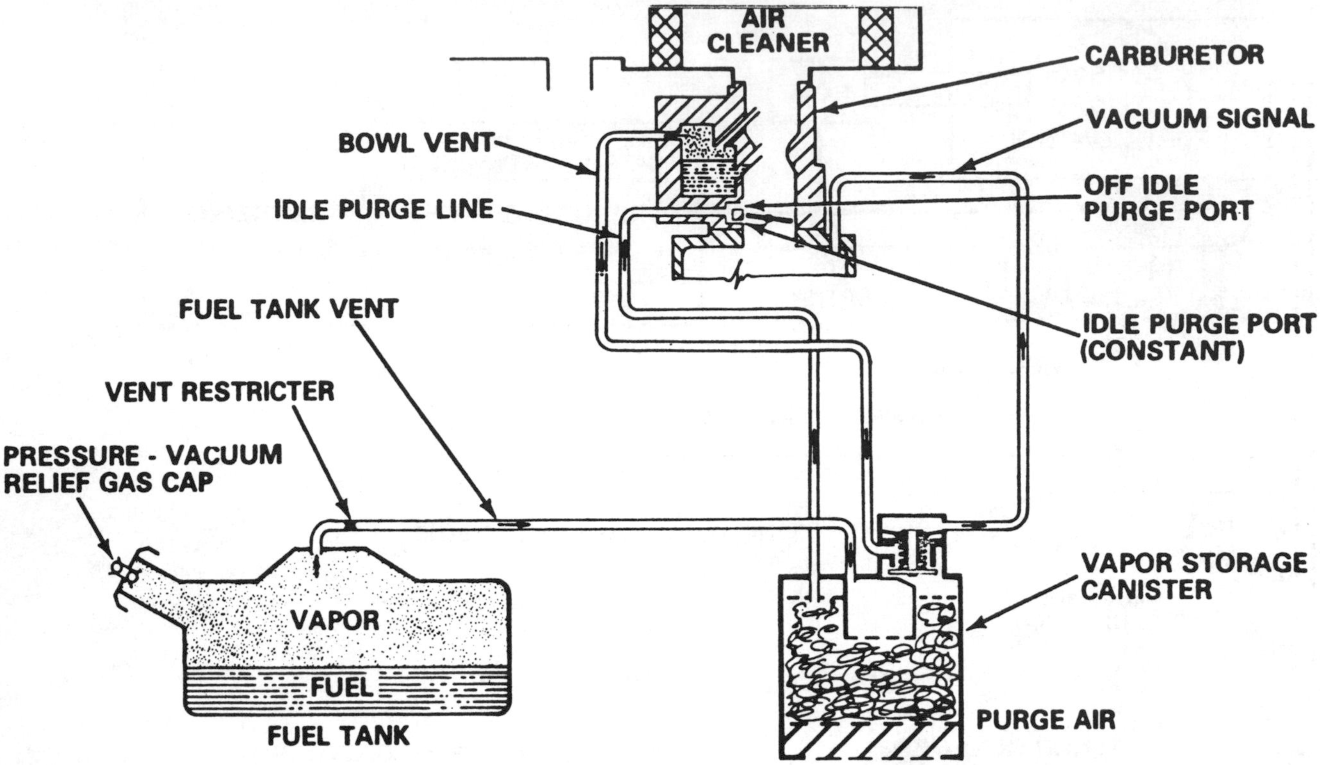

1981-82 EVAPORATIVE EMISSION CONTROL SYSTEM (EECS)-OPEN CANISTER

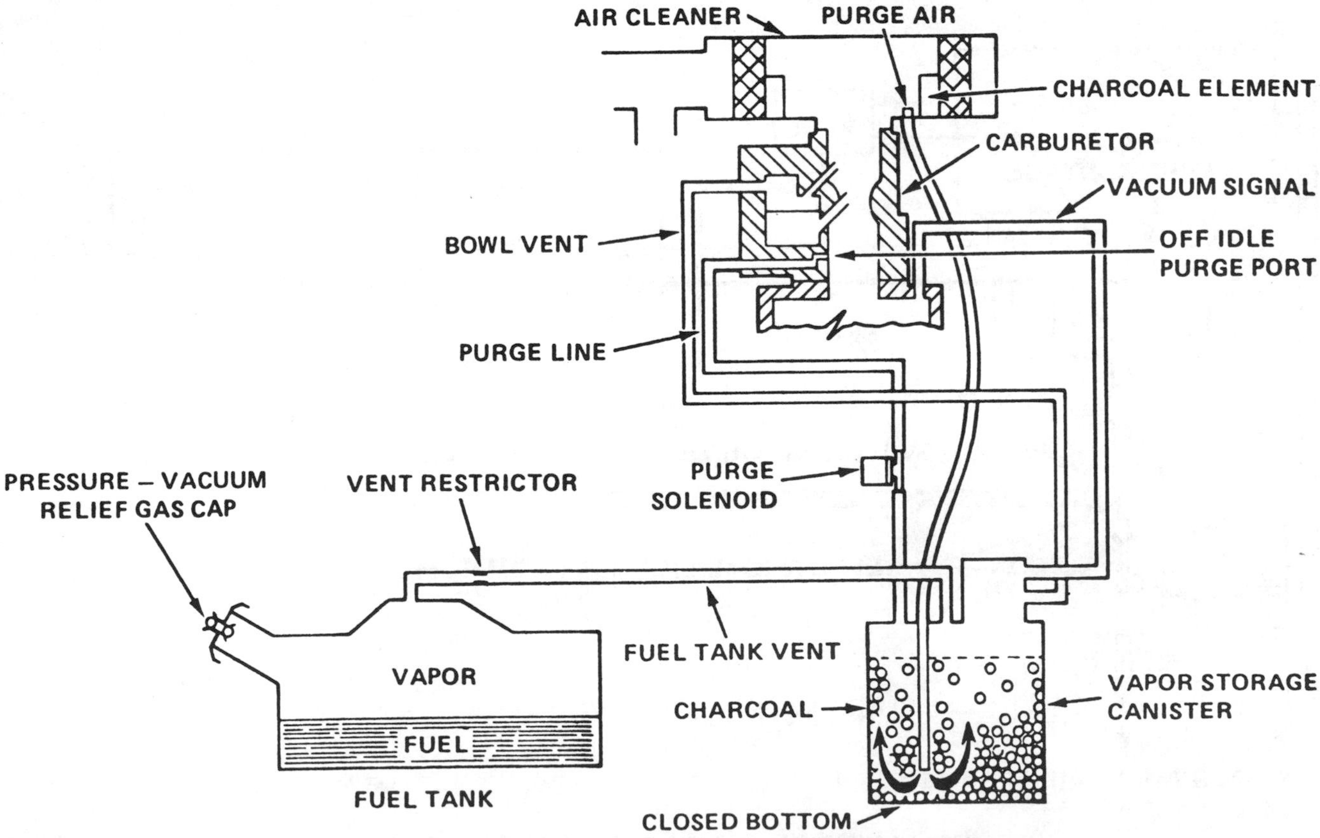

1981-82 EVAPORATIVE EMISSION CONTROL SYSTEM (EECS)-CLOSED CANISTER

COMPUTER COMMAND CONTROL

SYSTEM DESCRIPTION

The Computer Command Control system is an electronically controlled exhaust emission system that monitors up to fifteen (15) different engine/vehicle functions and can control as many as nine (9) different operations including the transmission converter clutch.

The system has back-up programs in the event of a failure to alert or instruct the operator through a "CHECK ENGINE" lamp on the instrument panel. This lamp will light indicating a fault in the system and will remain "on" until the problem is corrected. This same lamp through an integral diagnostic system, will aid the technician in locating the cause of the problem.

The system helps to lower exhaust emissions while maintaining good fuel economy and driveability. The system controls the following operations:

- Fuel Control System
- Electronic Spark Timing (EST)
- Air Management
- Evaporative Emission Control System
- Early Fuel Evaporation

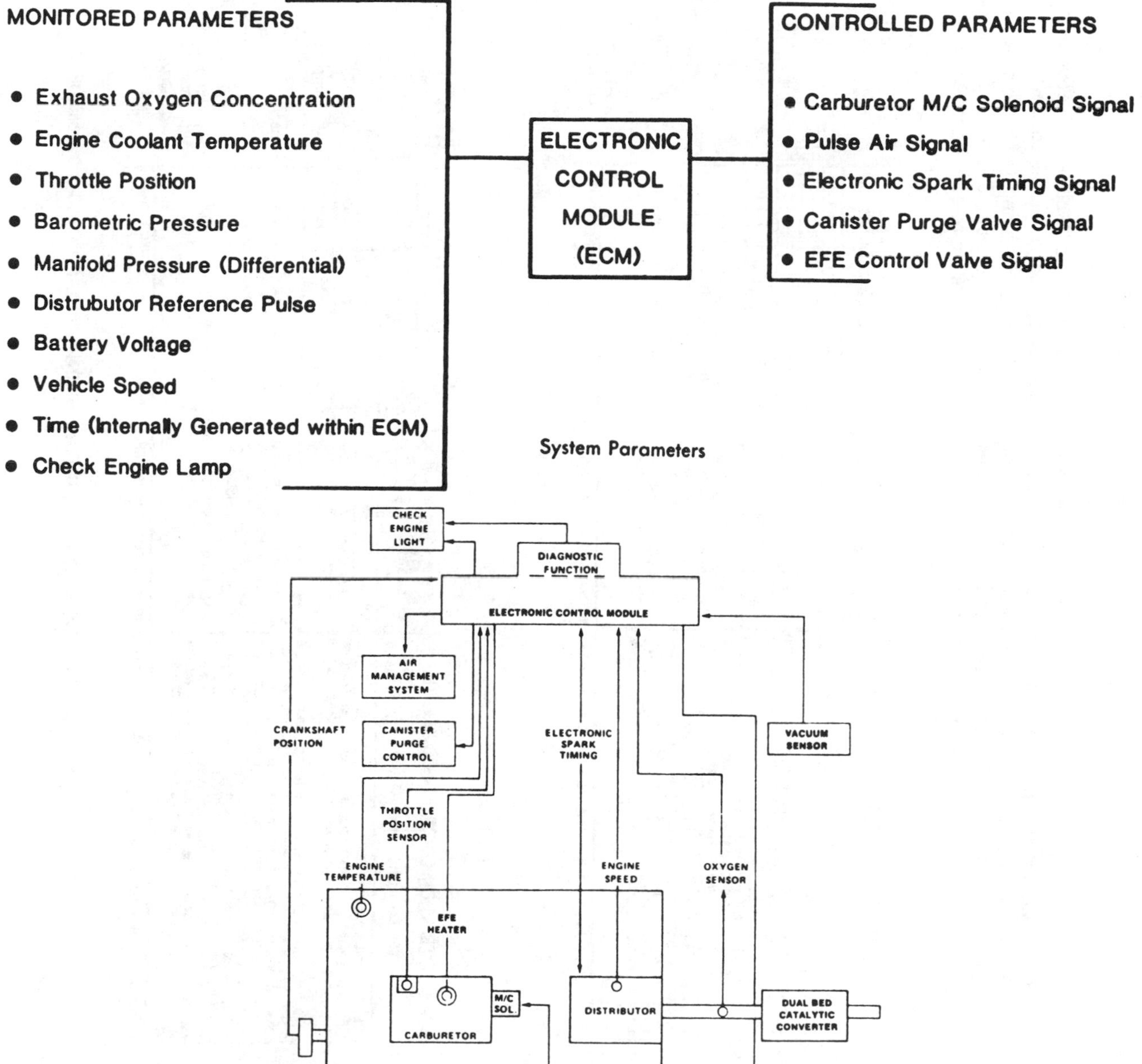

System Parameters

COMPUTER COMMAND CONTROL SYSTEM-PONTIAC T1000 MODELS

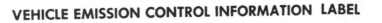

VEHICLE EMISSION CONTROL INFORMATION LABEL

The Vehicle Emission Control Information Label (Fig. 6E-1) is located in the engine compartment (fan shroud, radiator support, hood underside, etc.) of every vehicle produced by General Motors Corporation. The label contains important emission specifications and setting procedures, as well as a vacuum hose schematic with emission components identified.

When servicing the engine or emission systems, the Vehicle Emission Control Information Label should be checked for up-to-date information.

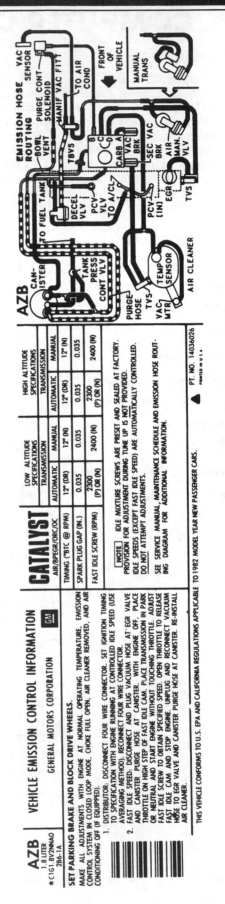

AZA
1.8 LITER
★C1G1.8V2NNAO
2B6-1A

VEHICLE EMISSION CONTROL INFORMATION
GENERAL MOTORS CORPORATION GM

SET PARKING BRAKE AND BLOCK DRIVE WHEELS.

MAKE ALL ADJUSTMENTS WITH ENGINE AT NORMAL OPERATING TEMPERATURE, EMISSION CONTROL SYSTEM IN CLOSED LOOP MODE, CHOKE FULL OPEN, AIR CLEANER REMOVED, AND AIR CONDITIONING OFF (IF EQUIPPED).

1. DISTRIBUTOR: DISCONNECT FOUR WIRE CONNECTOR. SET IGNITION TIMING TO SPECIFICATION WITH ENGINE RUNNING AT CONTROLLED IDLE SPEED (USE AVERAGING METHOD). RECONNECT FOUR WIRE CONNECTOR.
2. FAST IDLE SPEED: DISCONNECT AND PLUG VACUUM HOSE AT EGR VALVE AND CANISTER PURGE HOSE AT CANISTER. WITH ENGINE OFF, PLACE THROTTLE ON HIGH STEP OF FAST IDLE CAM. PLACE TRANSMISSION IN PARK OR NEUTRAL AND START ENGINE WITHOUT TOUCHING THROTTLE. ADJUST FAST IDLE SCREW TO OBTAIN SPECIFIED SPEED. OPEN THROTTLE TO RELEASE FAST IDLE CAM AND STOP ENGINE. UNPLUG AND RECONNECT VACUUM HOSE TO EGR VALVE AND CANISTER PURGE HOSE AT CANISTER. RE-INSTALL AIR CLEANER.

CATALYST	LOW ALTITUDE SPECIFICATIONS TRANSMISSION		HIGH ALTITUDE SPECIFICATIONS TRANSMISSION	
AIR/BPEGR/ORC/OC	AUTOMATIC	MANUAL	AUTOMATIC	MANUAL
TIMING (°BTC @ RPM)	12° (DR)	12° (N)	12° (DR)	12° (N)
SPARK PLUG GAP (IN.)	0.035	0.035	0.035	0.035
FAST IDLE SCREW (RPM)	2300 (P) OR (N)	2400 (N)	2300 (P) OR (N)	2400 (N)

NOTE IDLE MIXTURE SCREWS ARE PRESET AND SEALED AT FACTORY. PROVISION FOR ADJUSTMENT DURING TUNE UP IS NOT PROVIDED.
IDLE SPEEDS (EXCEPT FAST IDLE SPEED) ARE AUTOMATICALLY CONTROLLED.
DO NOT ATTEMPT ADJUSTMENTS.

SEE SERVICE MANUAL, MAINTENANCE SCHEDULE AND EMISSION HOSE ROUTING DIAGRAM FOR ADDITIONAL INFORMATION.

PT. NO. 14036025
PRINTED IN U.S.A.

THIS VEHICLE CONFORMS TO U.S. EPA REGULATIONS APPLICABLE TO 1982 MODEL YEAR NEW PASSENGER CARS.

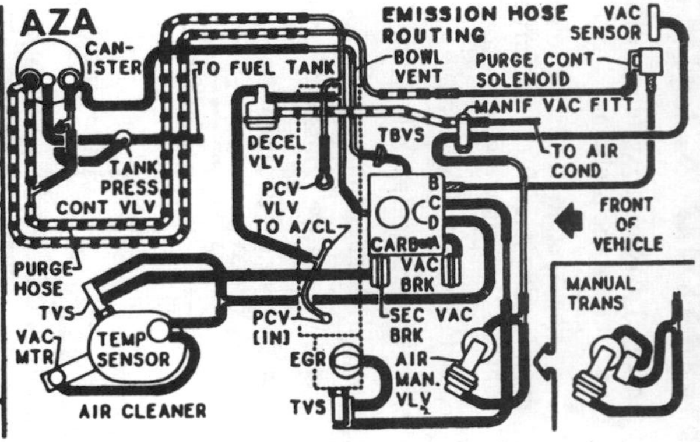

AZA
EMISSION HOSE ROUTING
CAN-ISTER
TO FUEL TANK
BOWL VENT
PURGE CONT SOLENOID
VAC SENSOR
MANIF VAC FITT
TO AIR COND
DECEL VLV
TBVS
TANK PRESS CONT VLV
PCV VLV
TO A/CL
B C D
CARB
VAC BRK
FRONT OF VEHICLE
PURGE HOSE
TVS
VAC MTR
TEMP SENSOR
PCV (IN)
SEC VAC BRK
MANUAL TRANS
EGR
AIR MAN. VLV
AIR CLEANER
TVS

AZB
1.8 LITER
★C1G1.8V2NNAO
2B6-1A

VEHICLE EMISSION CONTROL INFORMATION
GENERAL MOTORS CORPORATION GM

SET PARKING BRAKE AND BLOCK DRIVE WHEELS.

MAKE ALL ADJUSTMENTS WITH ENGINE AT NORMAL OPERATING TEMPERATURE, EMISSION CONTROL SYSTEM IN CLOSED LOOP MODE, CHOKE FULL OPEN, AIR CLEANER REMOVED, AND AIR CONDITIONING OFF (IF EQUIPPED).

1. DISTRIBUTOR: DISCONNECT FOUR WIRE CONNECTOR. SET IGNITION TIMING TO SPECIFICATION WITH ENGINE RUNNING AT CONTROLLED IDLE SPEED (USE AVERAGING METHOD). RECONNECT FOUR WIRE CONNECTOR.
2. FAST IDLE SPEED: DISCONNECT AND PLUG VACUUM HOSE AT EGR VALVE AND CANISTER PURGE HOSE AT CANISTER. WITH ENGINE OFF, PLACE THROTTLE ON HIGH STEP OF FAST IDLE CAM. PLACE TRANSMISSION IN PARK OR NEUTRAL AND START ENGINE WITHOUT TOUCHING THROTTLE. ADJUST FAST IDLE SCREW TO OBTAIN SPECIFIED SPEED. OPEN THROTTLE TO RELEASE FAST IDLE CAM AND STOP ENGINE. UNPLUG AND RECONNECT VACUUM HOSE TO EGR VALVE AND CANISTER PURGE HOSE AT CANISTER. RE-INSTALL AIR CLEANER.

CATALYST	LOW ALTITUDE SPECIFICATIONS TRANSMISSION		HIGH ALTITUDE SPECIFICATIONS TRANSMISSION	
AIR/BPEGR/ORC/OC	AUTOMATIC	MANUAL	AUTOMATIC	MANUAL
TIMING (°BTC @ RPM)	12° (DR)	12° (N)	12° (DR)	12° (N)
SPARK PLUG GAP (IN.)	0.035	0.035	0.035	0.035
FAST IDLE SCREW (RPM)	2300 (P) OR (N)	2400 (N)	2300 (P) OR (N)	2400 (N)

NOTE IDLE MIXTURE SCREWS ARE PRESET AND SEALED AT FACTORY. PROVISION FOR ADJUSTMENT DURING TUNE UP IS NOT PROVIDED.
IDLE SPEEDS (EXCEPT FAST IDLE SPEED) ARE AUTOMATICALLY CONTROLLED.
DO NOT ATTEMPT ADJUSTMENTS.

SEE SERVICE MANUAL, MAINTENANCE SCHEDULE AND EMISSION HOSE ROUTING DIAGRAM FOR ADDITIONAL INFORMATION.

PT. NO. 14036026
PRINTED IN U.S.A.

THIS VEHICLE CONFORMS TO U.S. EPA AND CALIFORNIA REGULATIONS APPLICABLE TO 1982 MODEL YEAR NEW PASSENGER CARS.

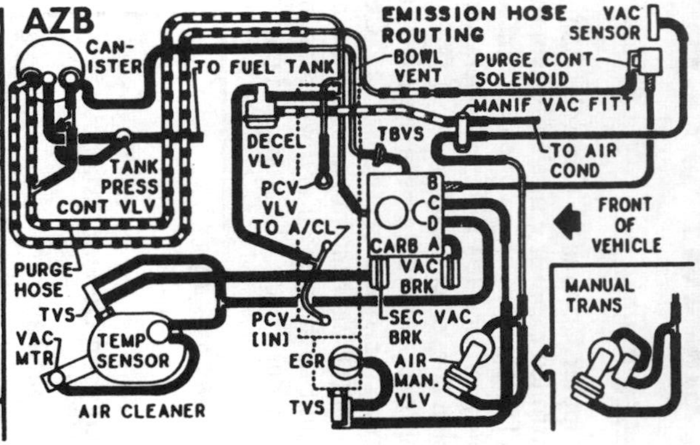

AZB
EMISSION HOSE ROUTING
CAN-ISTER
TO FUEL TANK
BOWL VENT
PURGE CONT SOLENOID
VAC SENSOR
MANIF VAC FITT
TO AIR COND
DECEL VLV
TBVS
TANK PRESS CONT VLV
PCV VLV
TO A/CL
B C D
CARB A
VAC BRK
FRONT OF VEHICLE
PURGE HOSE
TVS
VAC MTR
TEMP SENSOR
PCV (IN)
SEC VAC BRK
MANUAL TRANS
EGR
AIR MAN. VLV
AIR CLEANER
TVS

VEHICLE EMISSION CONTROL INFORMATION LABEL-PONTIAC J2000 MODELS

VACUUM CIRCUITS

1981

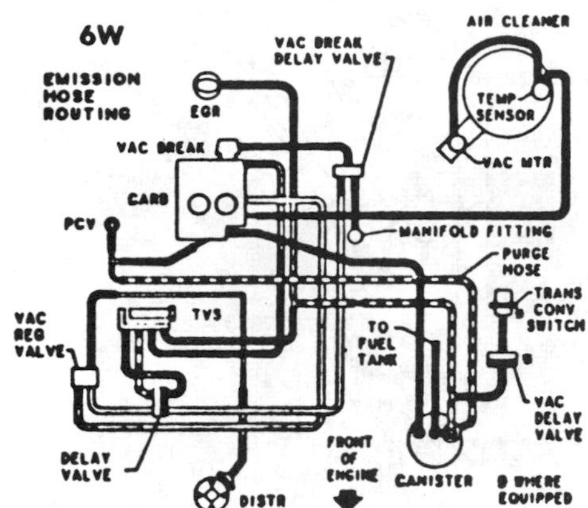

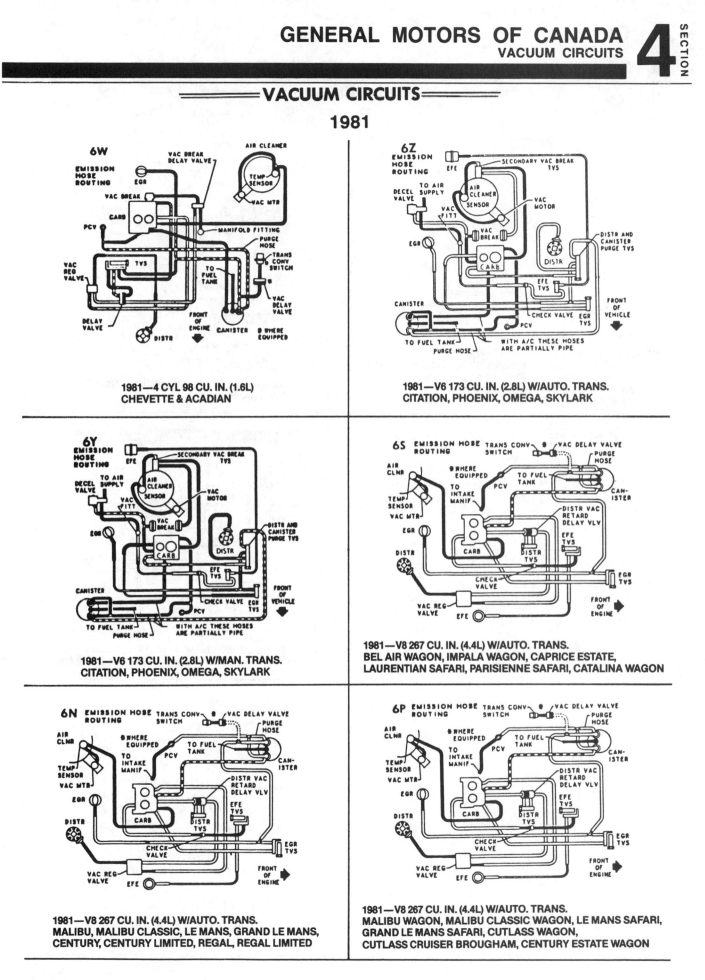

1981—4 CYL 98 CU. IN. (1.6L)
CHEVETTE & ACADIAN

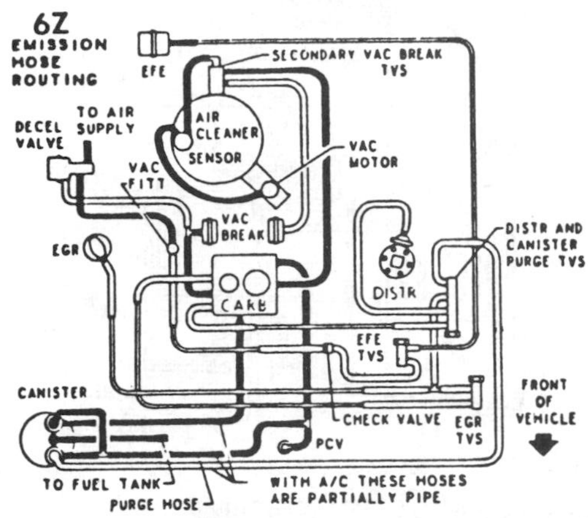

1981—V6 173 CU. IN. (2.8L) W/AUTO. TRANS.
CITATION, PHOENIX, OMEGA, SKYLARK

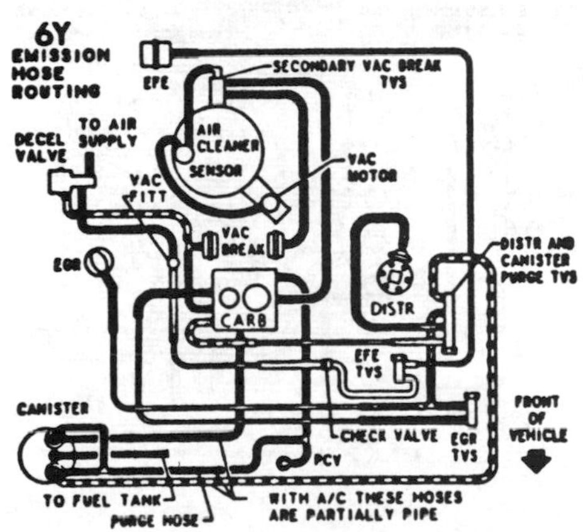

1981—V6 173 CU. IN. (2.8L) W/MAN. TRANS.
CITATION, PHOENIX, OMEGA, SKYLARK

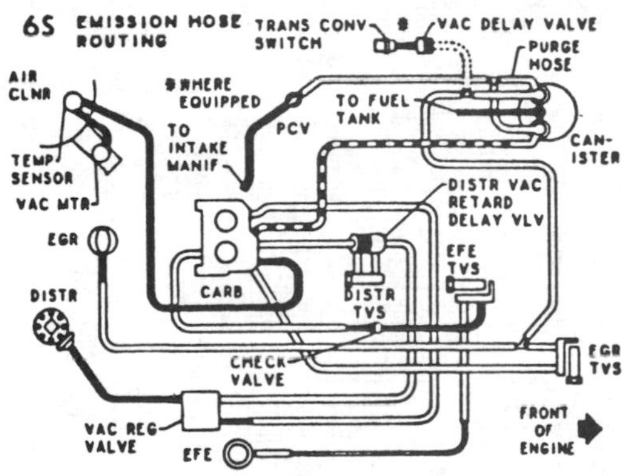

1981—V8 267 CU. IN. (4.4L) W/AUTO. TRANS.
BEL AIR WAGON, IMPALA WAGON, CAPRICE ESTATE,
LAURENTIAN SAFARI, PARISIENNE SAFARI, CATALINA WAGON

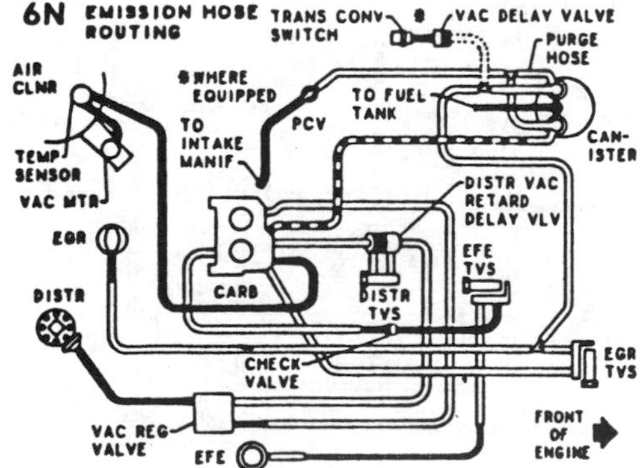

1981—V8 267 CU. IN. (4.4L) W/AUTO. TRANS.
MALIBU, MALIBU CLASSIC, LE MANS, GRAND LE MANS,
CENTURY, CENTURY LIMITED, REGAL, REGAL LIMITED

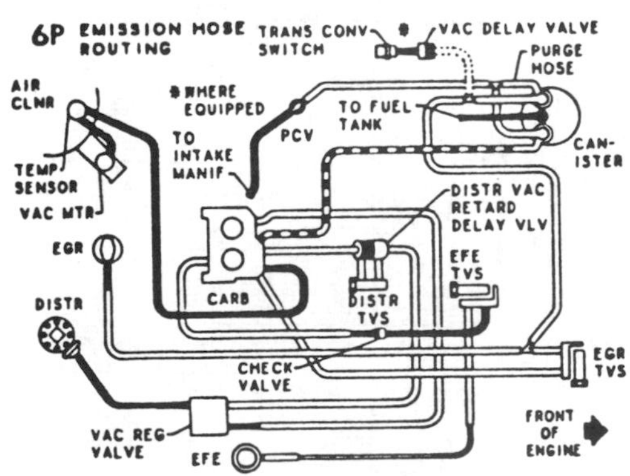

1981—V8 267 CU. IN. (4.4L) W/AUTO. TRANS.
MALIBU WAGON, MALIBU CLASSIC WAGON, LE MANS SAFARI,
GRAND LE MANS SAFARI, CUTLASS WAGON,
CUTLASS CRUISER BROUGHAM, CENTURY ESTATE WAGON

VACUUM CIRCUITS

1981

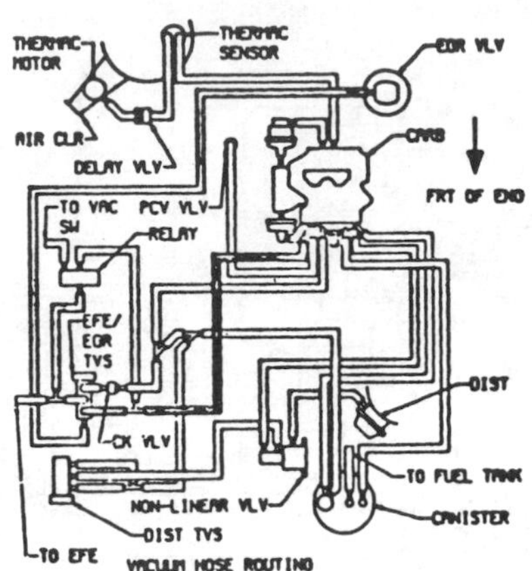

1981—V6 231 CU. IN. (3.8L) W/AUTO. TRANS.
LE MANS SAFARI, GRAND LE MANS SAFARI, CENTURY LIMITED, CENTURY, REGAL, REGAL LIMITED, CENTURY WAGON, CENTURY ESTATE, CUTLASS CRUISER BROUGHAM, CUTLASS WAGON

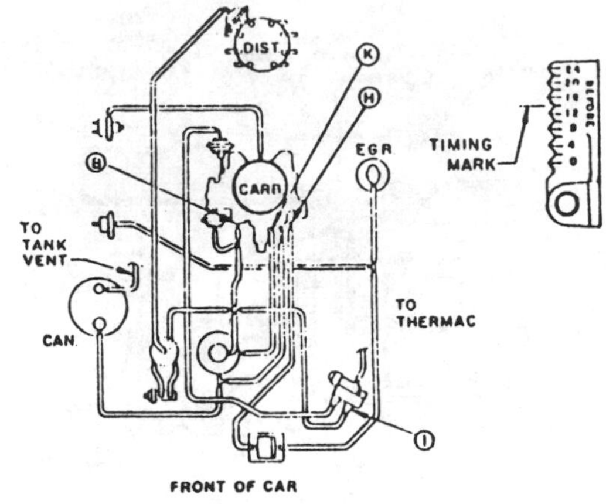

1981—V8 307 CU. IN. (5.0L) W/AUTO. TRANS.
ELECTRA, 98

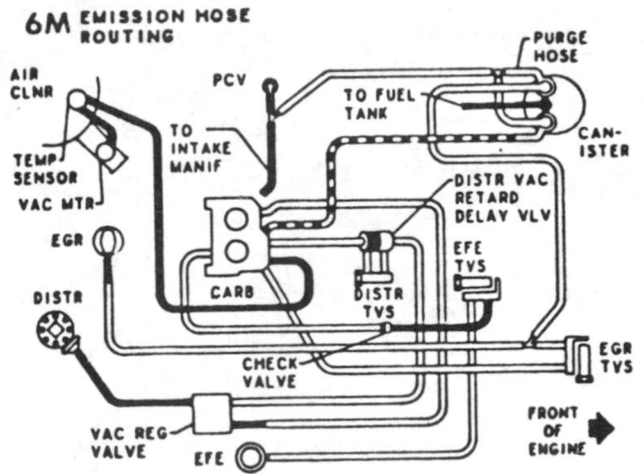

1981—V8 267 CU. IN. (4.4L) W/AUTO. TRANS.
CAMARO, FIREBIRD, FORMULA

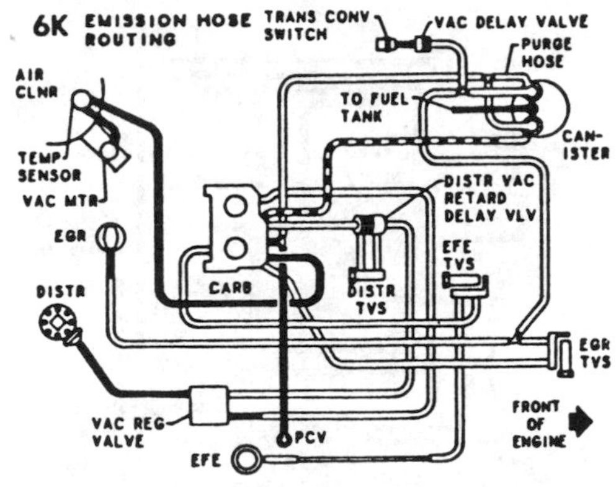

1981—V6 229 CU. IN. (3.8L) W/AUTO. TRANS.
MALIBU, MALIBU CLASSIC, LE MANS, GRAND LE MANS

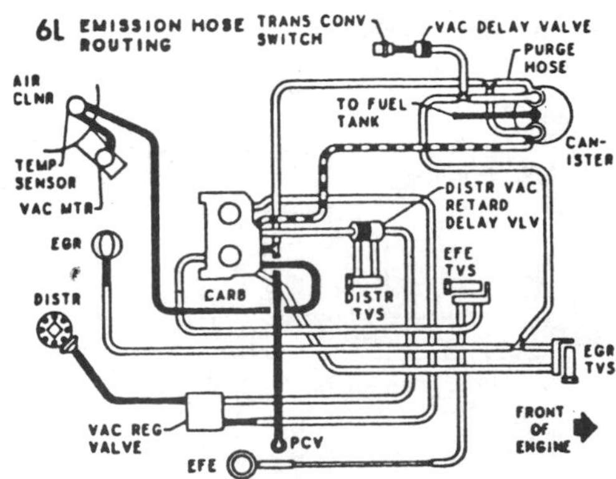

1981—V6 229 CU. IN. (3.8L) W/AUTO. TRANS.
MALIBU WAGON, MALIBU CLASSIC WAGON, EL CAMINO, CABALLERO

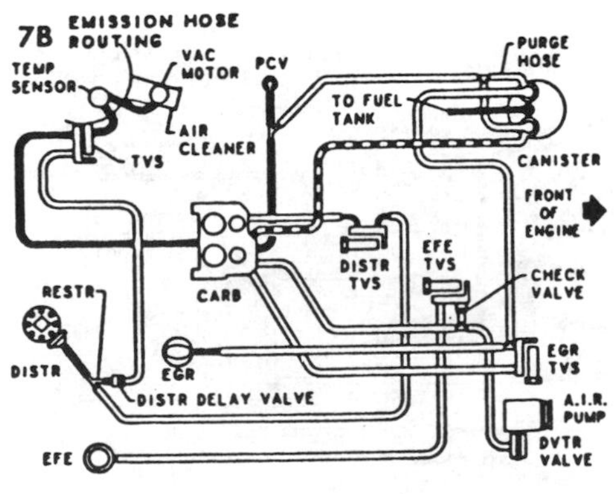

1981—V8 350 CU. IN. (5.7L) W/AUTO. TRANS.
Z28

VACUUM CIRCUITS

1981

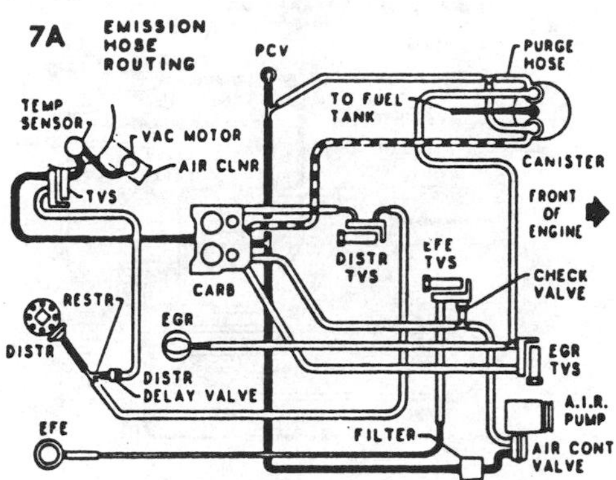

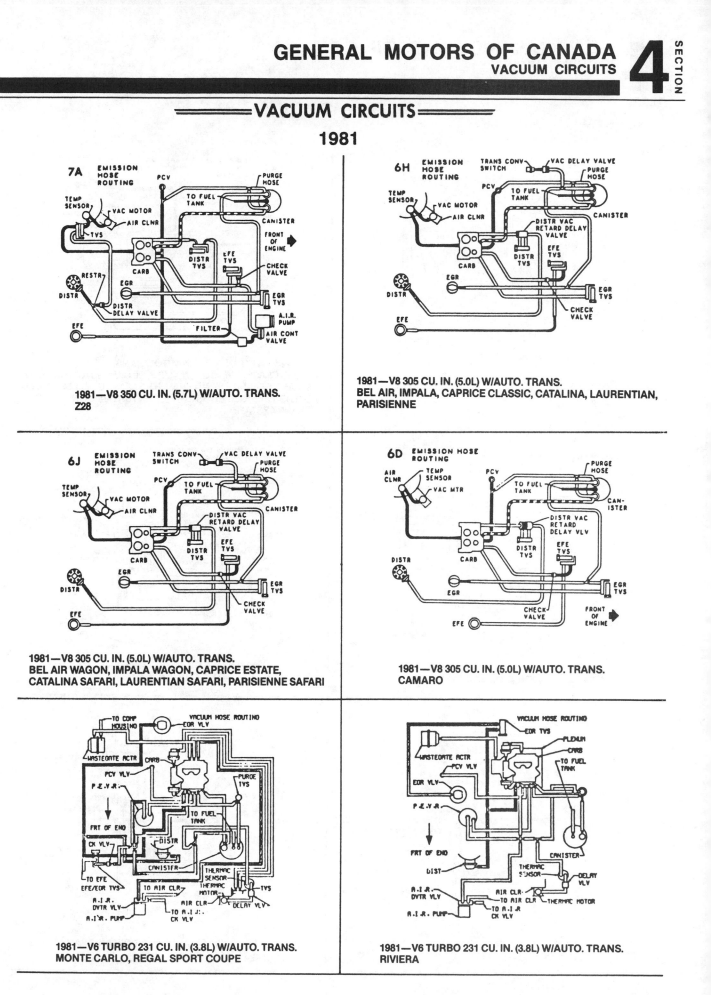

1981—V8 350 CU. IN. (5.7L) W/AUTO. TRANS.
Z28

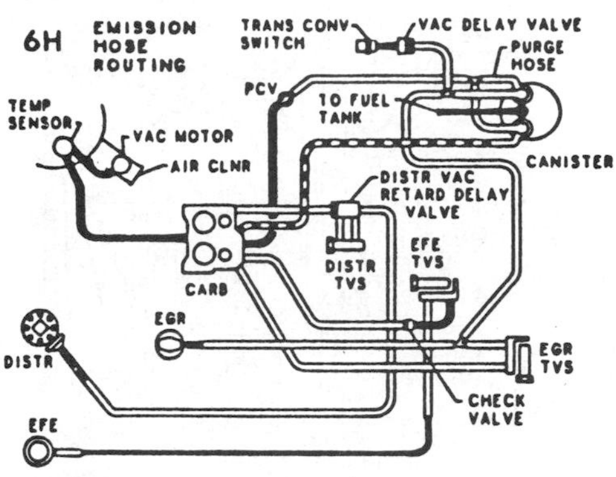

1981—V8 305 CU. IN. (5.0L) W/AUTO. TRANS.
BEL AIR, IMPALA, CAPRICE CLASSIC, CATALINA, LAURENTIAN, PARISIENNE

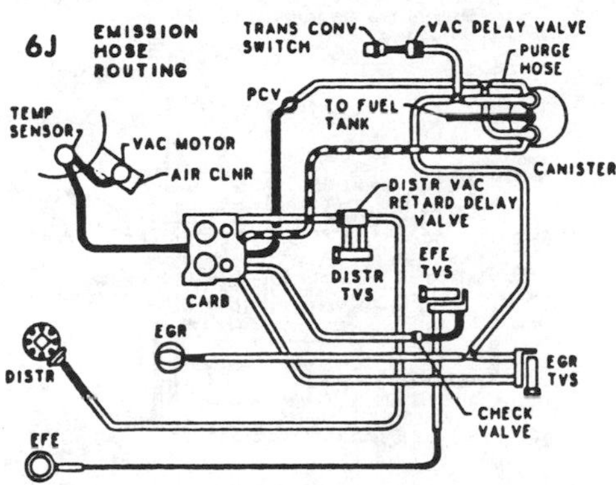

1981—V8 305 CU. IN. (5.0L) W/AUTO. TRANS.
BEL AIR WAGON, IMPALA WAGON, CAPRICE ESTATE, CATALINA SAFARI, LAURENTIAN SAFARI, PARISIENNE SAFARI

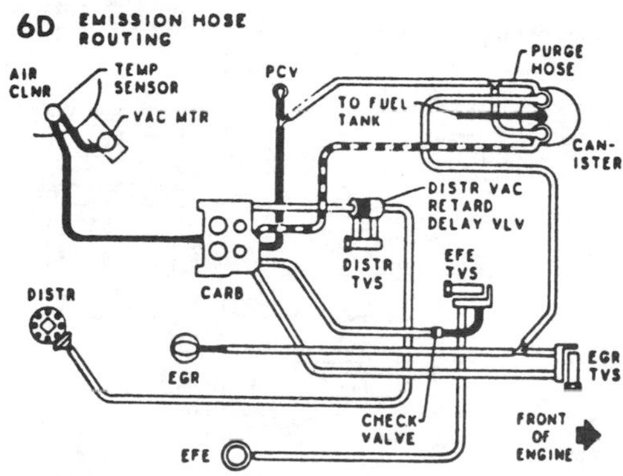

1981—V8 305 CU. IN. (5.0L) W/AUTO. TRANS.
CAMARO

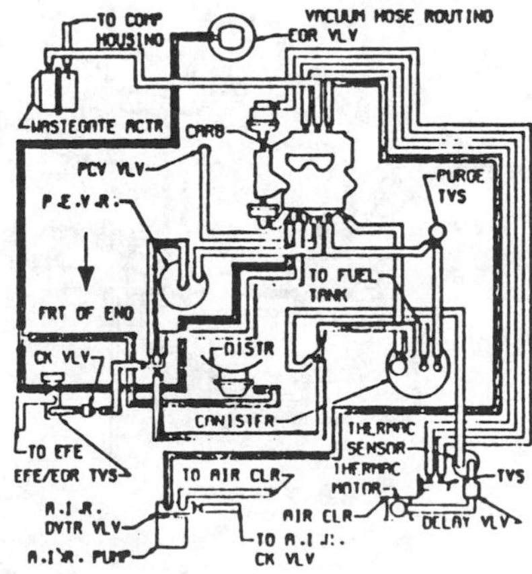

1981—V6 TURBO 231 CU. IN. (3.8L) W/AUTO. TRANS.
MONTE CARLO, REGAL SPORT COUPE

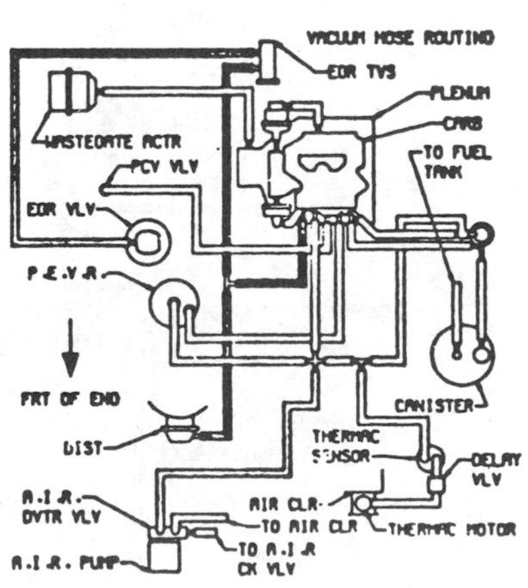

1981—V6 TURBO 231 CU. IN. (3.8L) W/AUTO. TRANS.
RIVIERA

VACUUM CIRCUITS

1981

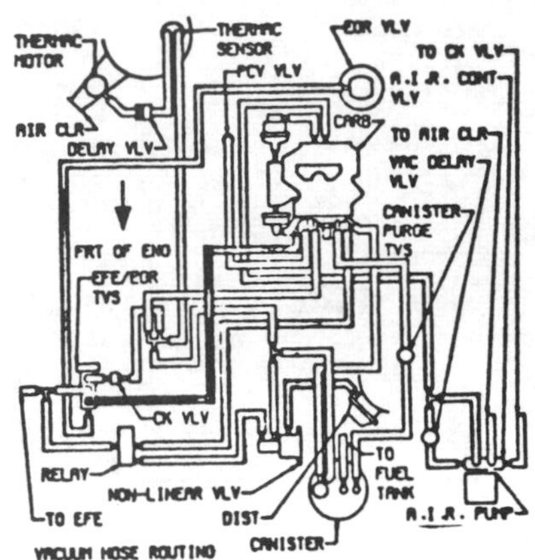

1981—V8 350 CU. IN. (5.7L) W/AUTO. TRANS.
ELECTRA

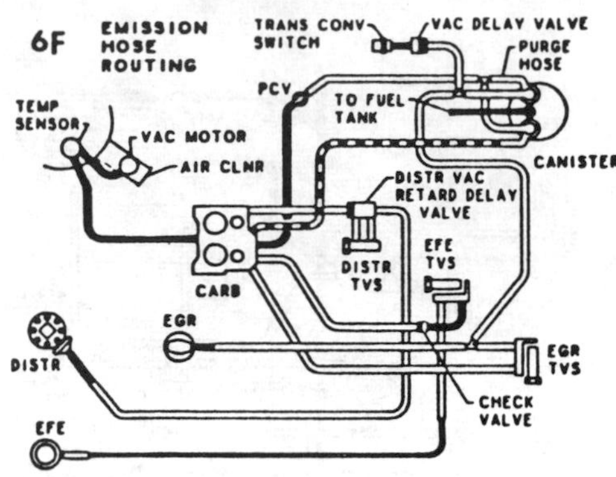

1981—V8 305 CU. IN. (5.0L) W/AUTO. TRANS.
MALIBU CLASSIC WAGON, MALIBU, EL CAMINO, CABALLERO, LE MANS SAFARI, GRAND LE MANS SAFARI, CUTLASS WAGON, CUTLASS CRUISER BROUGHAM, CENTURY WAGON, CENTURY ESTATE

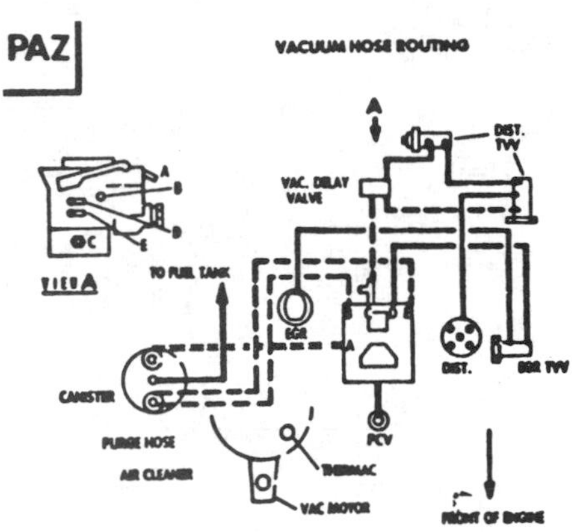

1981—4 CYL. 151 CU. IN. (2.5L) W/MAN. TRANS.
CITATION, PHOENIX, OMEGA, SKYLARK

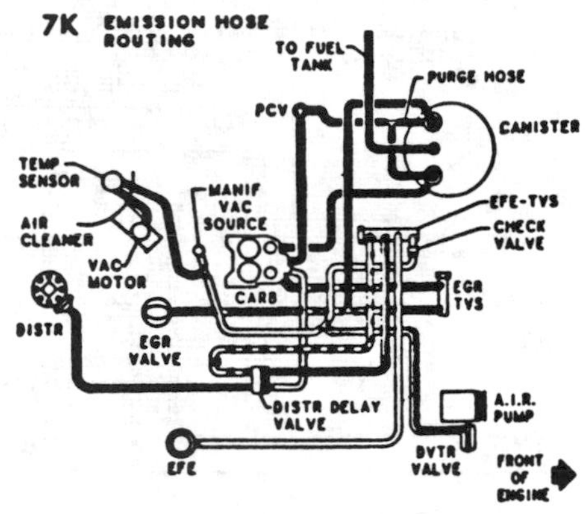

1981—V8 350 CU. IN. (5.7L) W/MAN. TRANS.
½ TON CONV. CAB (4WD)

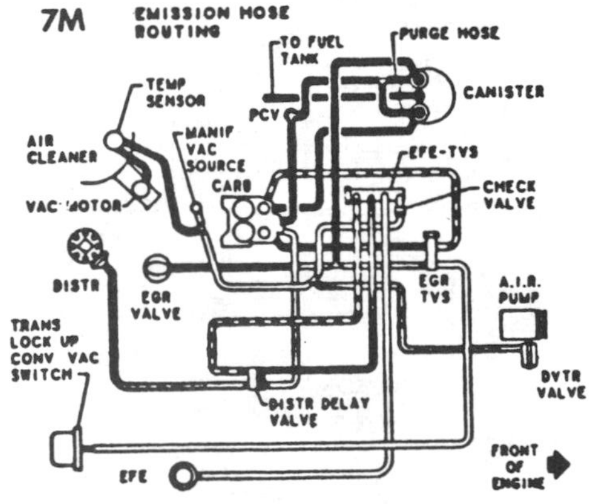

1981—V8 350 CU. IN. (5.7L) W/AUTO. TRANS.
½ TON CONV. CAB (4WD)

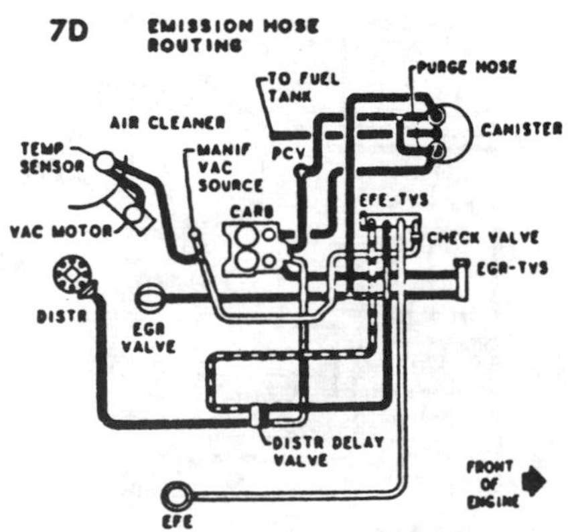

1981—V8 350 CU. IN. (5.7L) W/MAN. TRANS.
½ TON CONV. CAB (2WD)

VACUUM CIRCUITS

1981

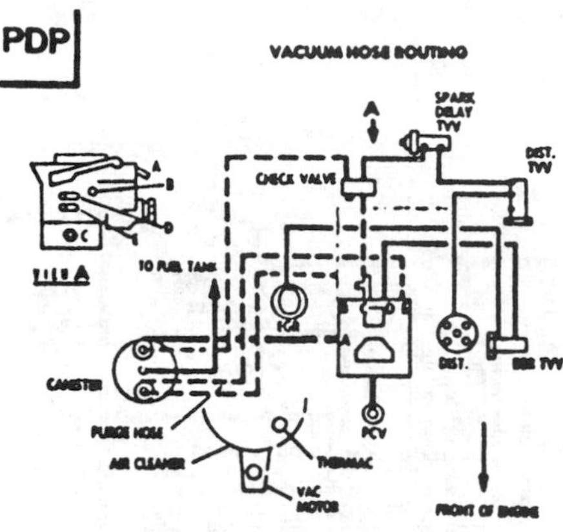

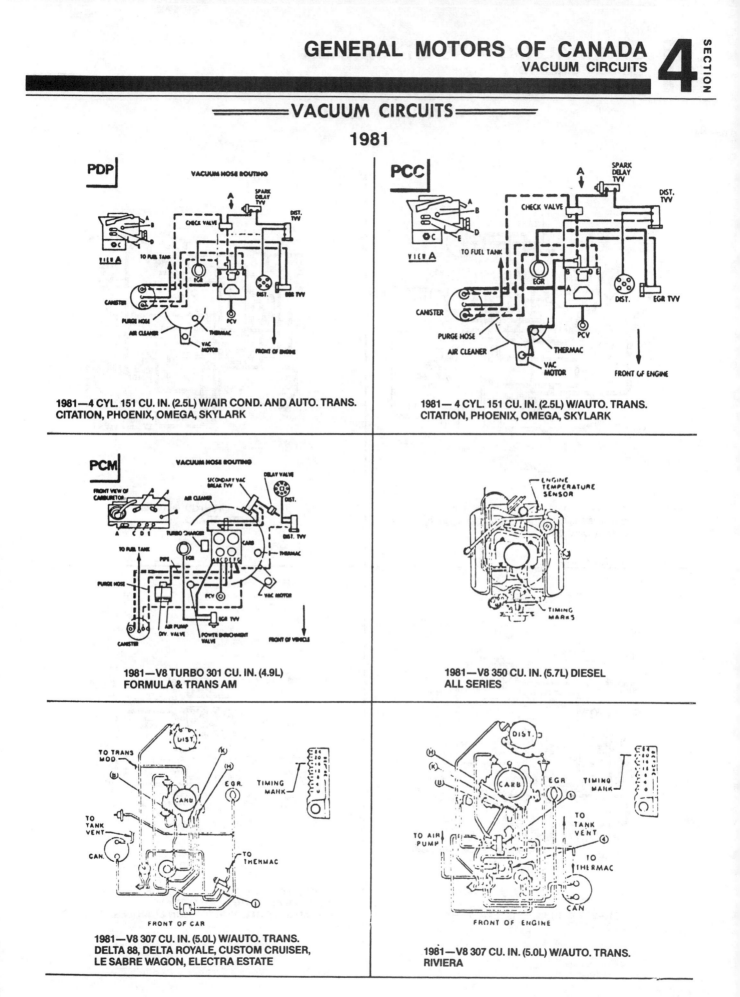

1981—4 CYL. 151 CU. IN. (2.5L) W/AIR COND. AND AUTO. TRANS.
CITATION, PHOENIX, OMEGA, SKYLARK

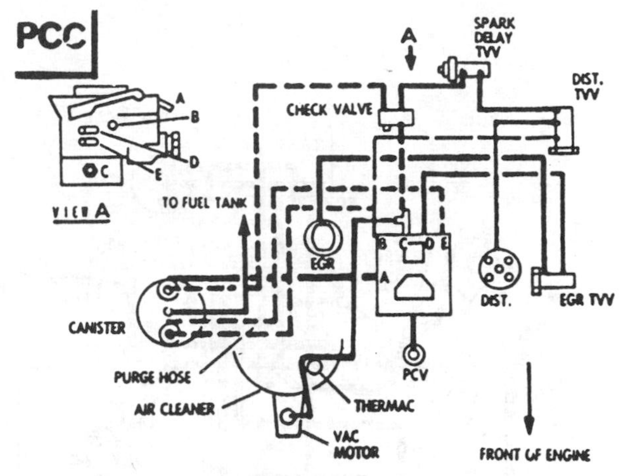

1981— 4 CYL. 151 CU. IN. (2.5L) W/AUTO. TRANS.
CITATION, PHOENIX, OMEGA, SKYLARK

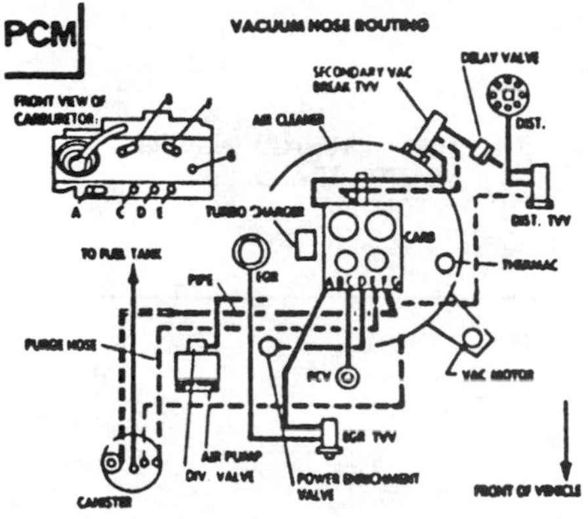

1981—V8 TURBO 301 CU. IN. (4.9L)
FORMULA & TRANS AM

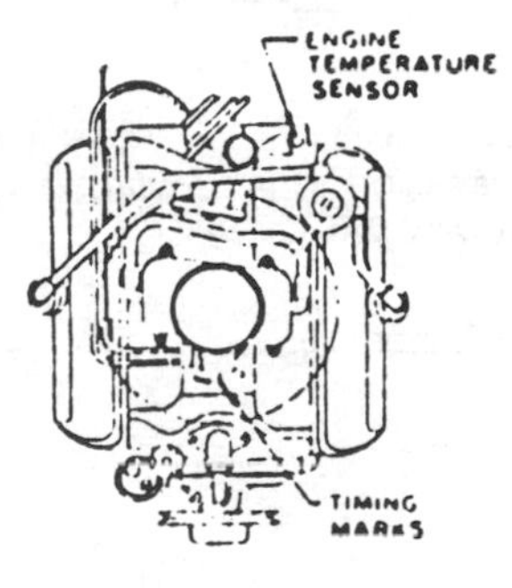

1981—V8 350 CU. IN. (5.7L) DIESEL
ALL SERIES

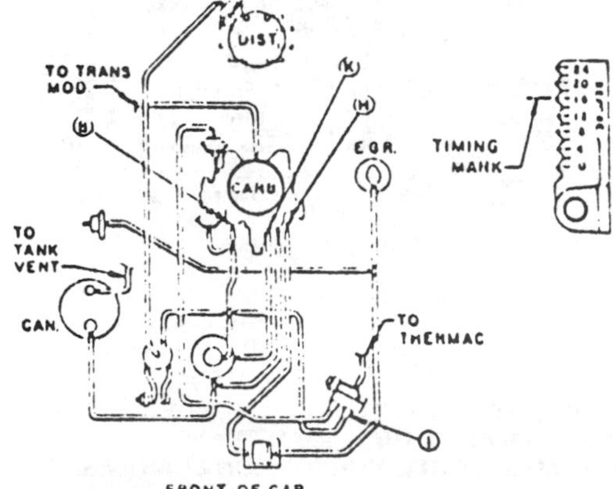

1981—V8 307 CU. IN. (5.0L) W/AUTO. TRANS.
DELTA 88, DELTA ROYALE, CUSTOM CRUISER,
LE SABRE WAGON, ELECTRA ESTATE

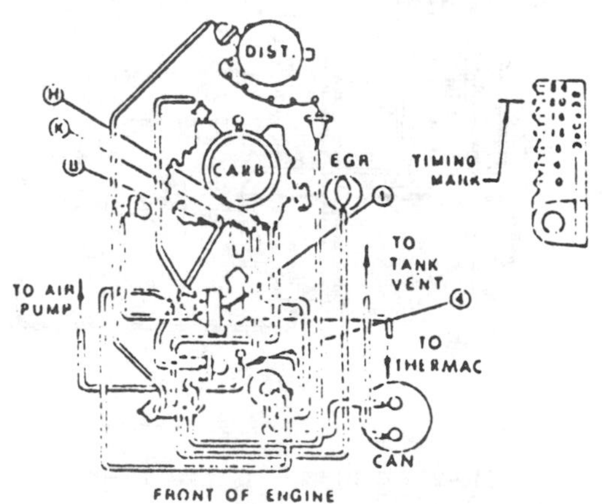

1981—V8 307 CU. IN. (5.0L) W/AUTO. TRANS.
RIVIERA

VACUUM CIRCUITS

1981

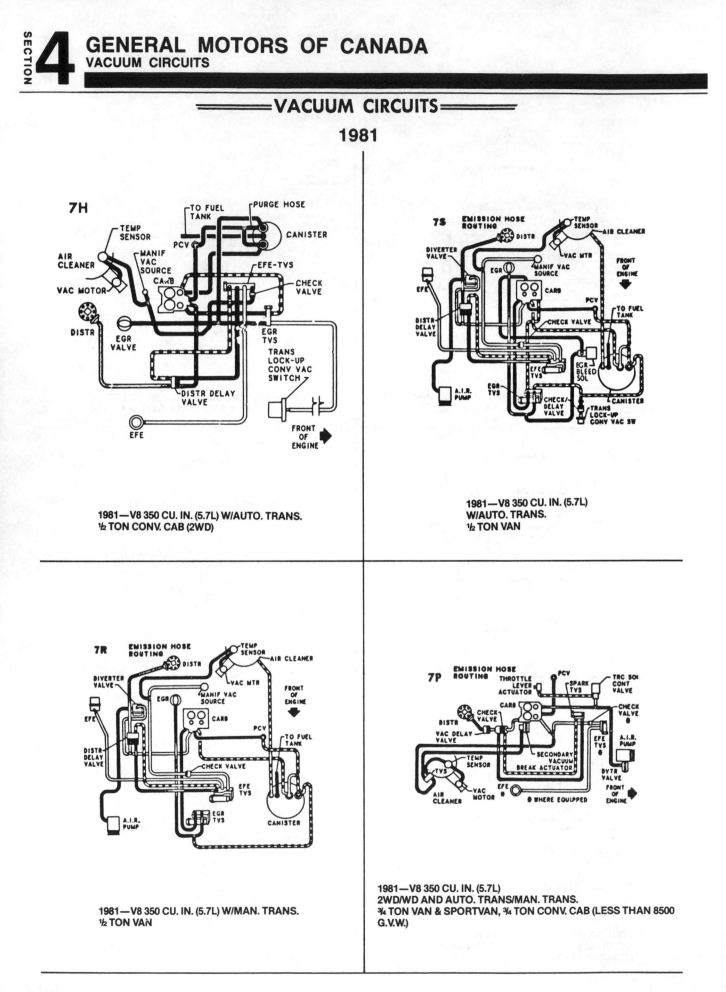

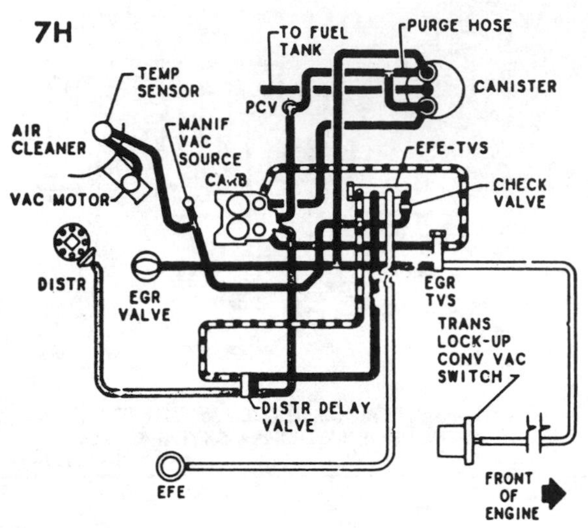

1981—V8 350 CU. IN. (5.7L) W/AUTO. TRANS.
½ TON CONV. CAB (2WD)

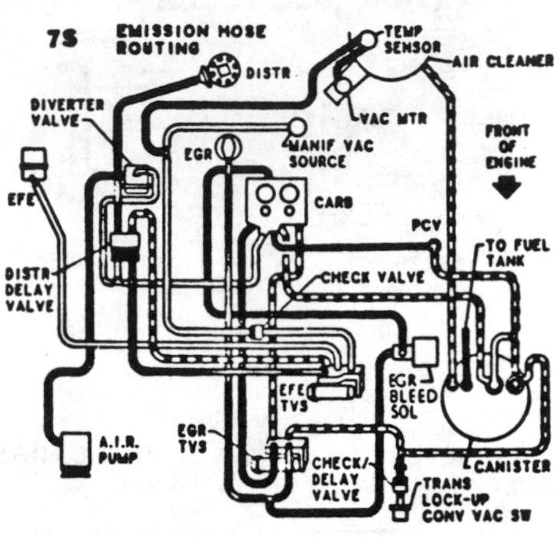

1981—V8 350 CU. IN. (5.7L)
W/AUTO. TRANS.
½ TON VAN

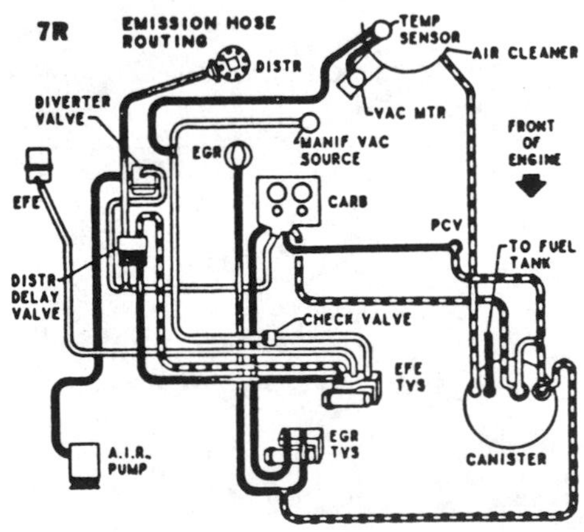

1981—V8 350 CU. IN. (5.7L) W/MAN. TRANS.
½ TON VAN

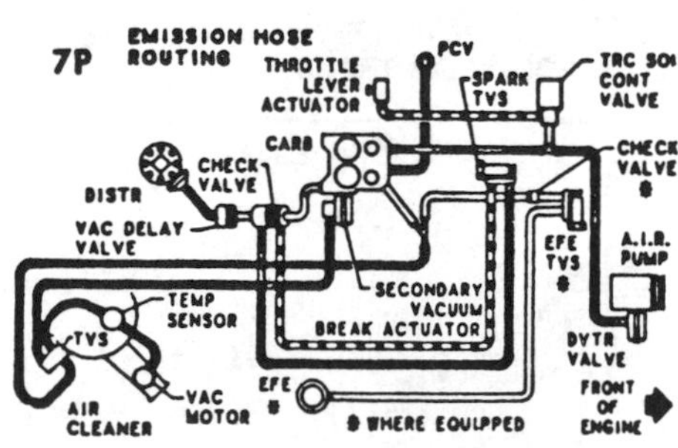

1981—V8 350 CU. IN. (5.7L)
2WD/WD AND AUTO. TRANS/MAN. TRANS.
¾ TON VAN & SPORTVAN, ¾ TON CONV. CAB (LESS THAN 8500 G.V.W.)

VACUUM CIRCUITS

1982

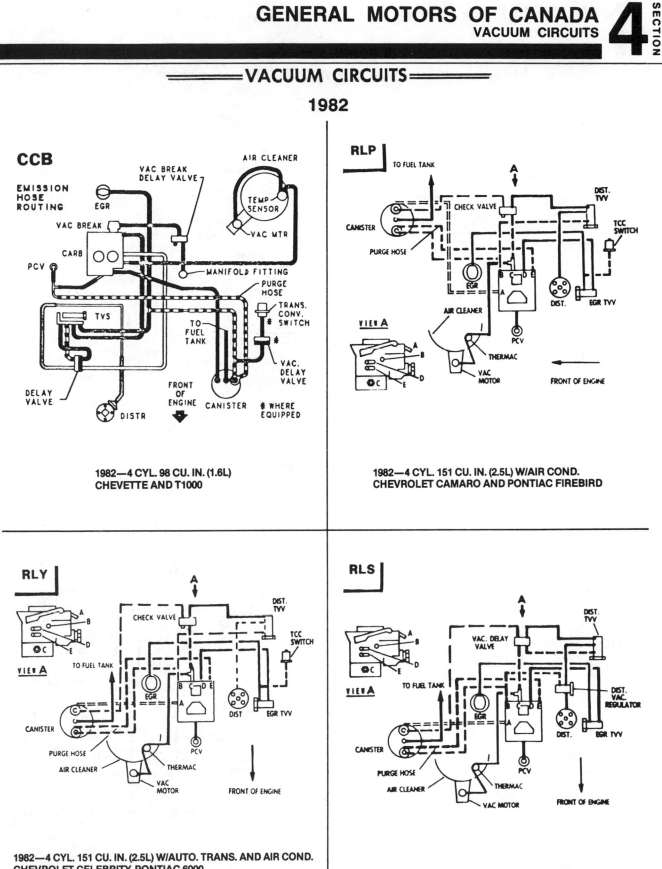

CCB

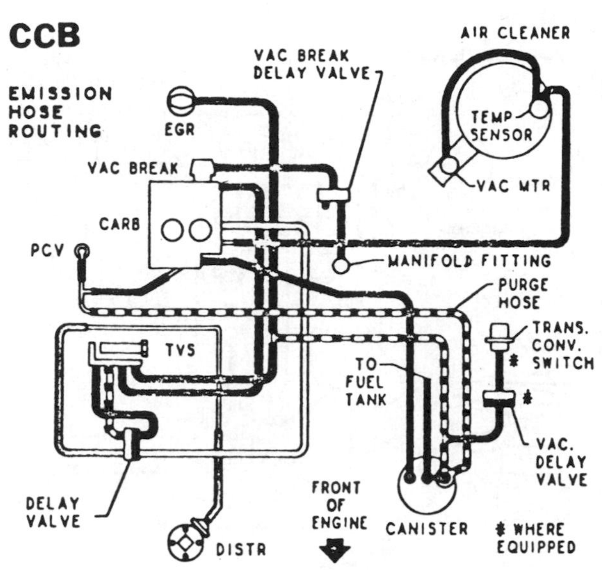

1982—4 CYL. 98 CU. IN. (1.6L)
CHEVETTE AND T1000

RLP

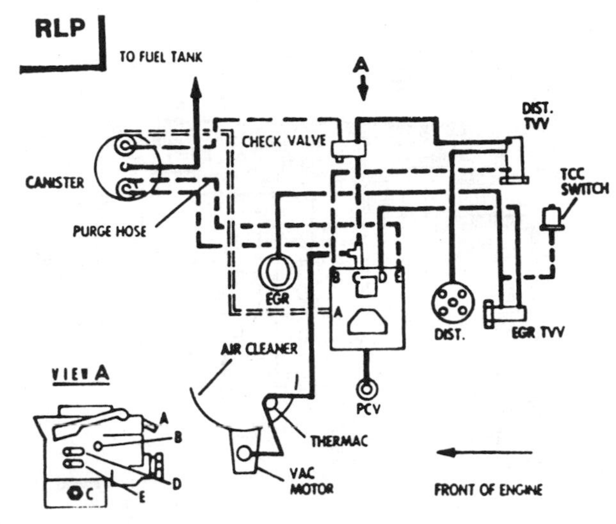

1982—4 CYL. 151 CU. IN. (2.5L) W/AIR COND.
CHEVROLET CAMARO AND PONTIAC FIREBIRD

RLY

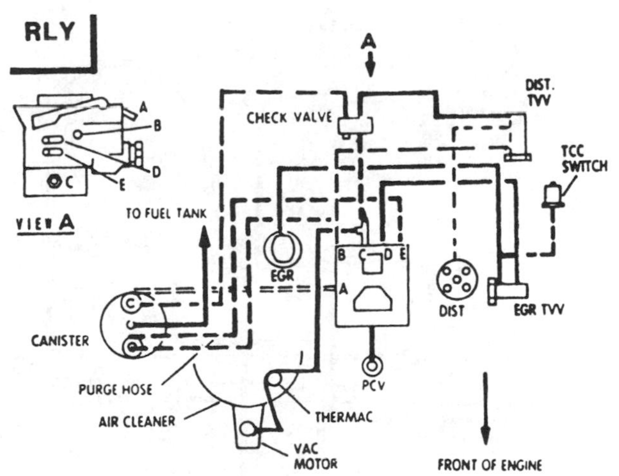

1982—4 CYL. 151 CU. IN. (2.5L) W/AUTO. TRANS. AND AIR COND.
CHEVROLET CELEBRITY, PONTIAC 6000,
OLDSMOBILE CUTLASS CIERA, BUICK CENTURY,
CHEVROLET CITATION, PONTIAC PHOENIX, OLDSMOBILE
OMEGA, BUICK SKYLARK

RLS

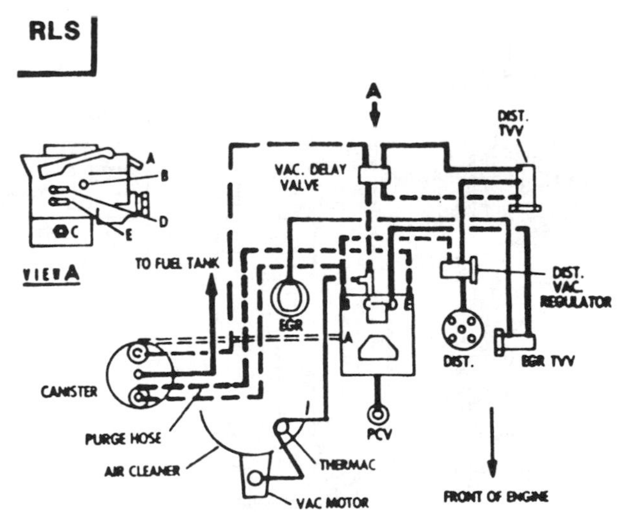

1982—4 CYL. 151 CU. IN. (2.5L) W/MAN. TRANS.
CHEVROLET CITATION, PONTIAC PHOENIX,
OLDSMOBILE OMEGA, BUICK SKYLARK

VACUUM CIRCUITS

1982

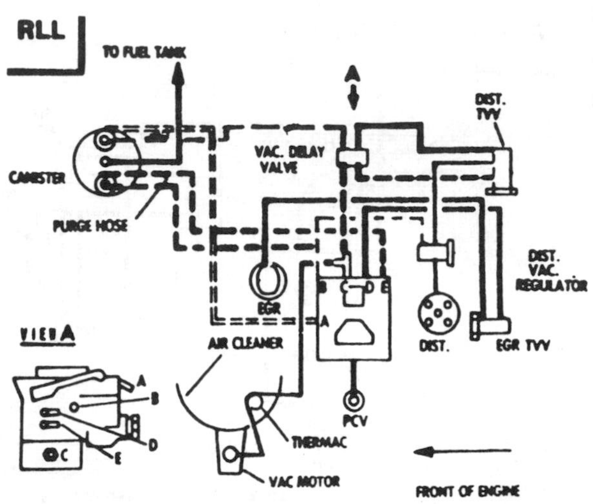

1982—4 CYL. 151 CU. IN. (2.5L) W/MAN. TRANS. AND NO AIR
CHEVROLET CAMARO AND PONTIAC FIREBIRD

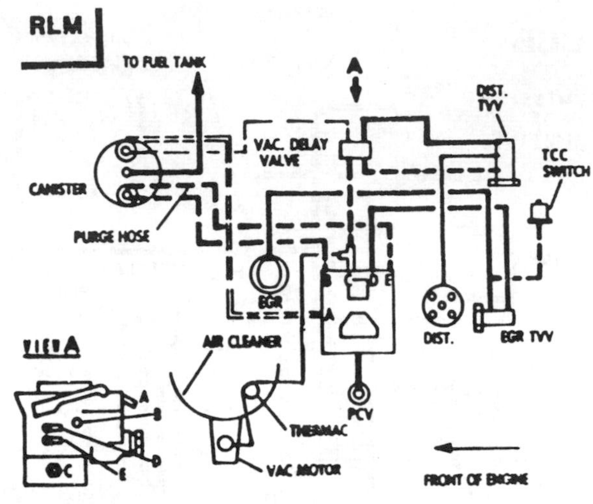

1982—4 CYL. 151 CU. IN. (2.5L) W/AUTO. TRANS. AND NO AIR
CHEVROLET CAMARO AND PONTIAC FIREBIRD

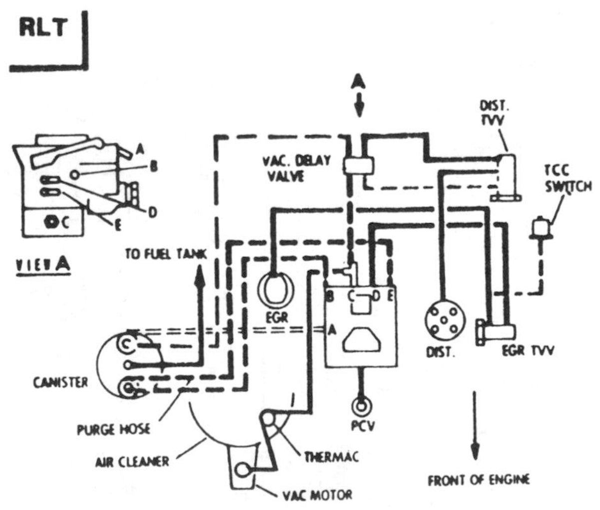

1982—4 CYL. 151 CU. IN. (2.5L) W/AUTO. TRANS. AND NO AIR
CHEVROLET CELEBRITY, PONTIAC 6000,
OLDSMOBILE CUTLASS CIERA, BUICK CENTURY,
CHEVROLET CITATION, PONTIAC PHOENIX, OLDSMOBILE
OMEGA, BUICK SKYLARK

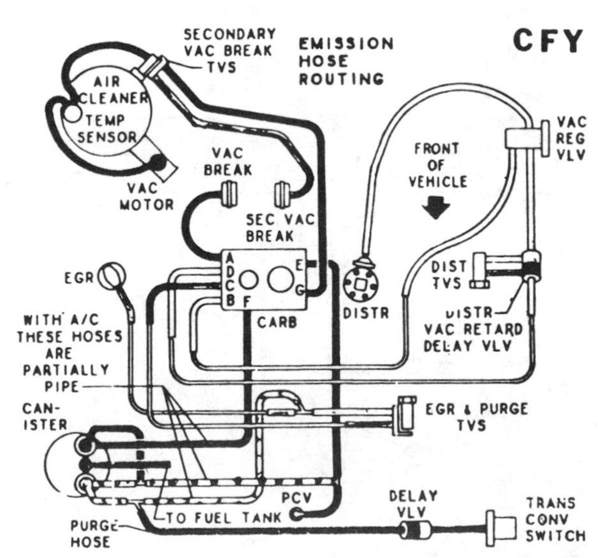

1982—V6 173 CU. IN. (2.8L)
CHEVROLET CELEBRITY, PONTIAC 6000, OLDSMOBILE OMEGA,
BUICK CENTURY

VACUUM CIRCUITS

1982

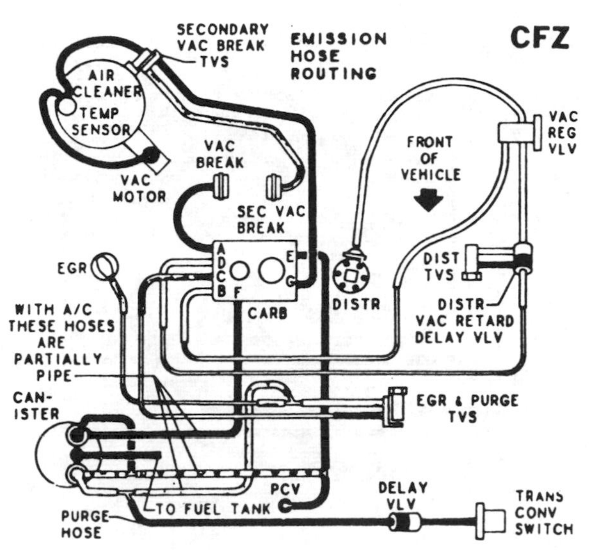

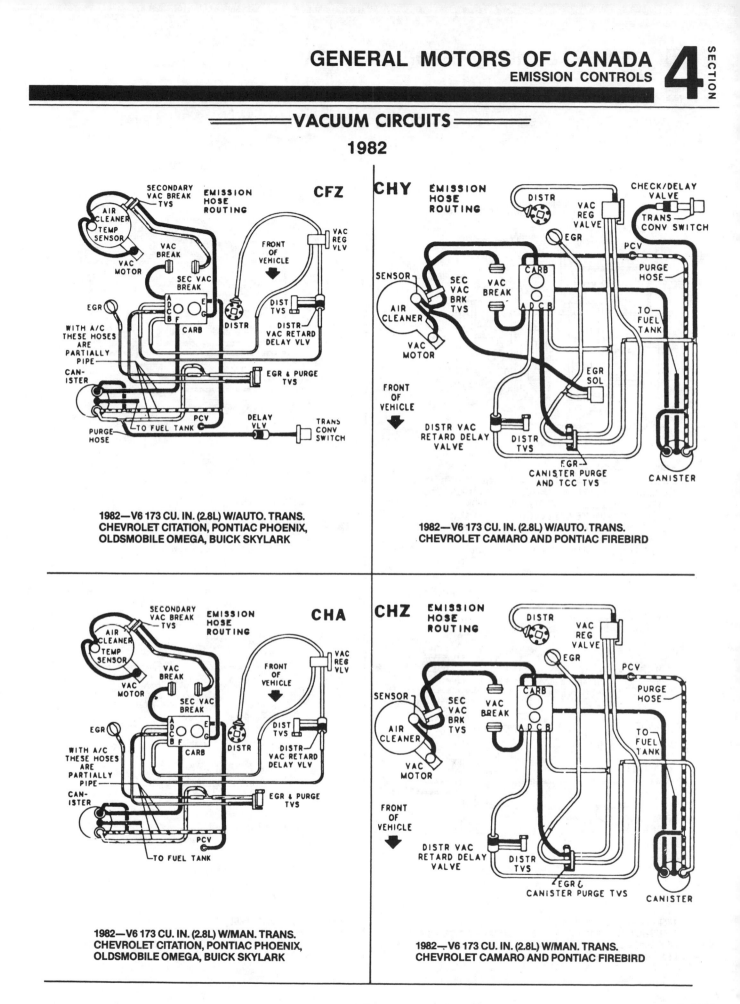

1982—V6 173 CU. IN. (2.8L) W/AUTO. TRANS.
CHEVROLET CITATION, PONTIAC PHOENIX,
OLDSMOBILE OMEGA, BUICK SKYLARK

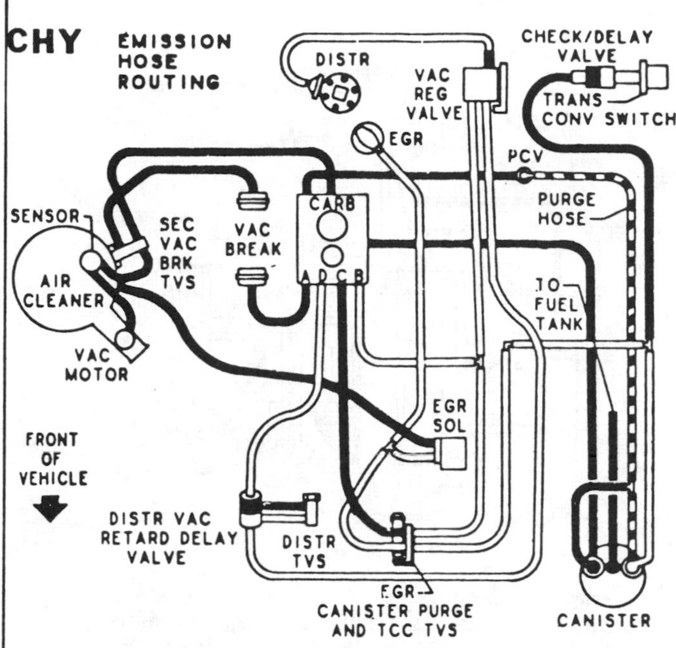

1982—V6 173 CU. IN. (2.8L) W/AUTO. TRANS.
CHEVROLET CAMARO AND PONTIAC FIREBIRD

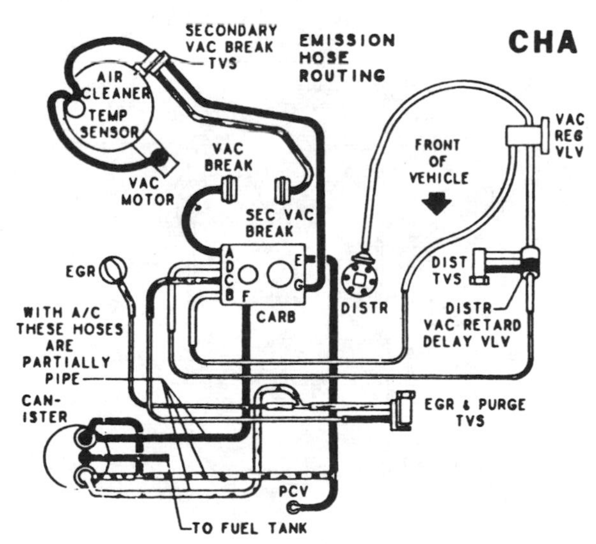

1982—V6 173 CU. IN. (2.8L) W/MAN. TRANS.
CHEVROLET CITATION, PONTIAC PHOENIX,
OLDSMOBILE OMEGA, BUICK SKYLARK

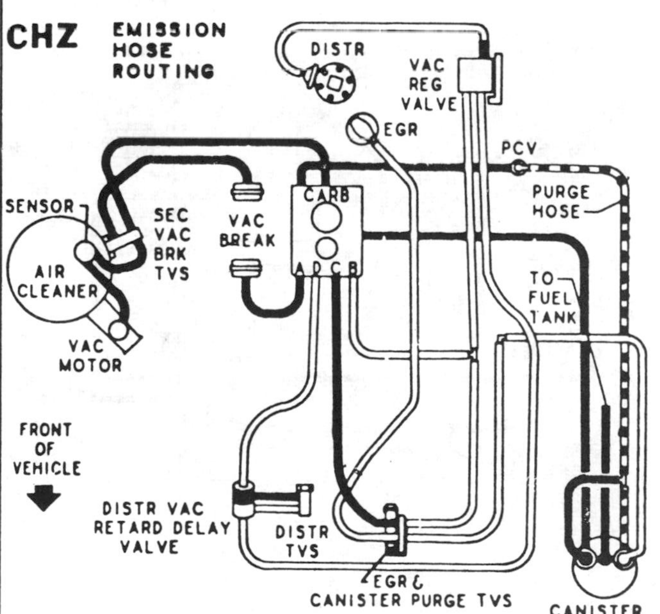

1982—V6 173 CU. IN. (2.8L) W/MAN. TRANS.
CHEVROLET CAMARO AND PONTIAC FIREBIRD

VACUUM CIRCUITS

1982

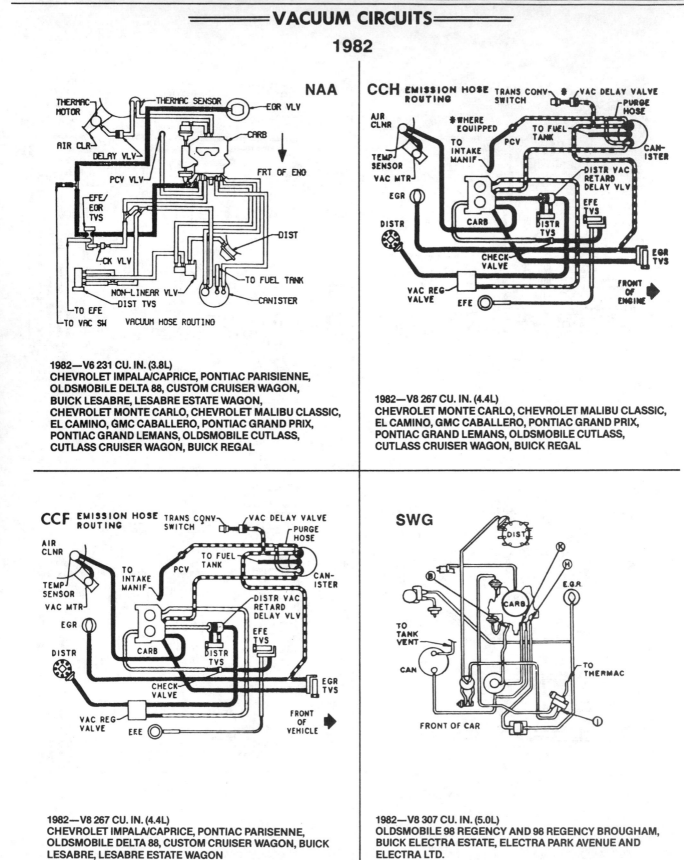

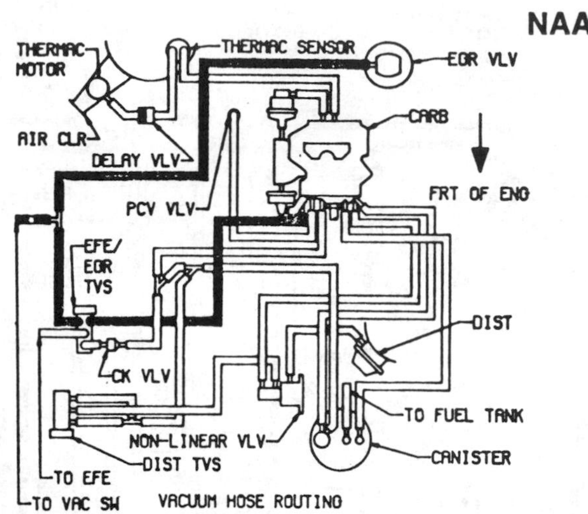

1982—V6 231 CU. IN. (3.8L)
CHEVROLET IMPALA/CAPRICE, PONTIAC PARISIENNE, OLDSMOBILE DELTA 88, CUSTOM CRUISER WAGON, BUICK LESABRE, LESABRE ESTATE WAGON, CHEVROLET MONTE CARLO, CHEVROLET MALIBU CLASSIC, EL CAMINO, GMC CABALLERO, PONTIAC GRAND PRIX, PONTIAC GRAND LEMANS, OLDSMOBILE CUTLASS, CUTLASS CRUISER WAGON, BUICK REGAL

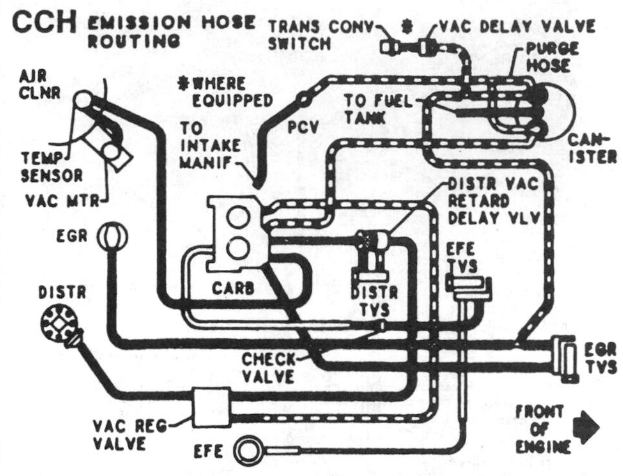

1982—V8 267 CU. IN. (4.4L)
CHEVROLET MONTE CARLO, CHEVROLET MALIBU CLASSIC, EL CAMINO, GMC CABALLERO, PONTIAC GRAND PRIX, PONTIAC GRAND LEMANS, OLDSMOBILE CUTLASS, CUTLASS CRUISER WAGON, BUICK REGAL

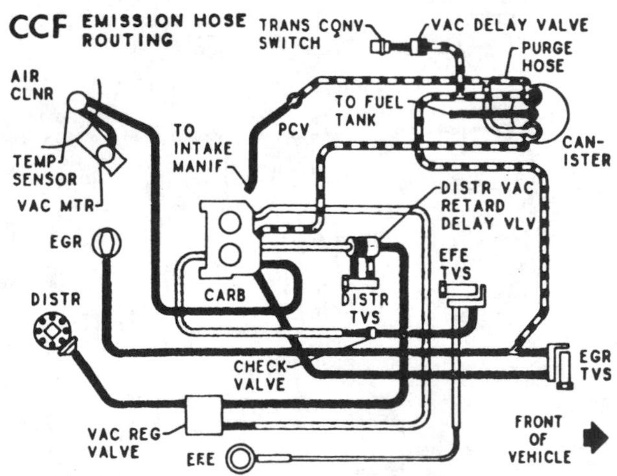

1982—V8 267 CU. IN. (4.4L)
CHEVROLET IMPALA/CAPRICE, PONTIAC PARISENNE, OLDSMOBILE DELTA 88, CUSTOM CRUISER WAGON, BUICK LESABRE, LESABRE ESTATE WAGON

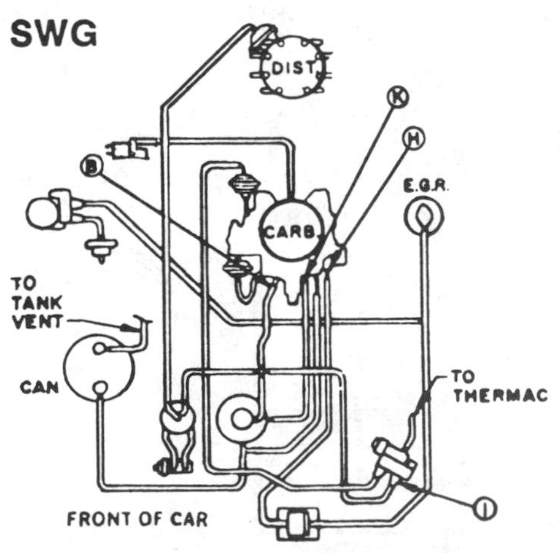

1982—V8 307 CU. IN. (5.0L)
OLDSMOBILE 98 REGENCY AND 98 REGENCY BROUGHAM, BUICK ELECTRA ESTATE, ELECTRA PARK AVENUE AND ELECTRA LTD.

VACUUM CIRCUITS

1982

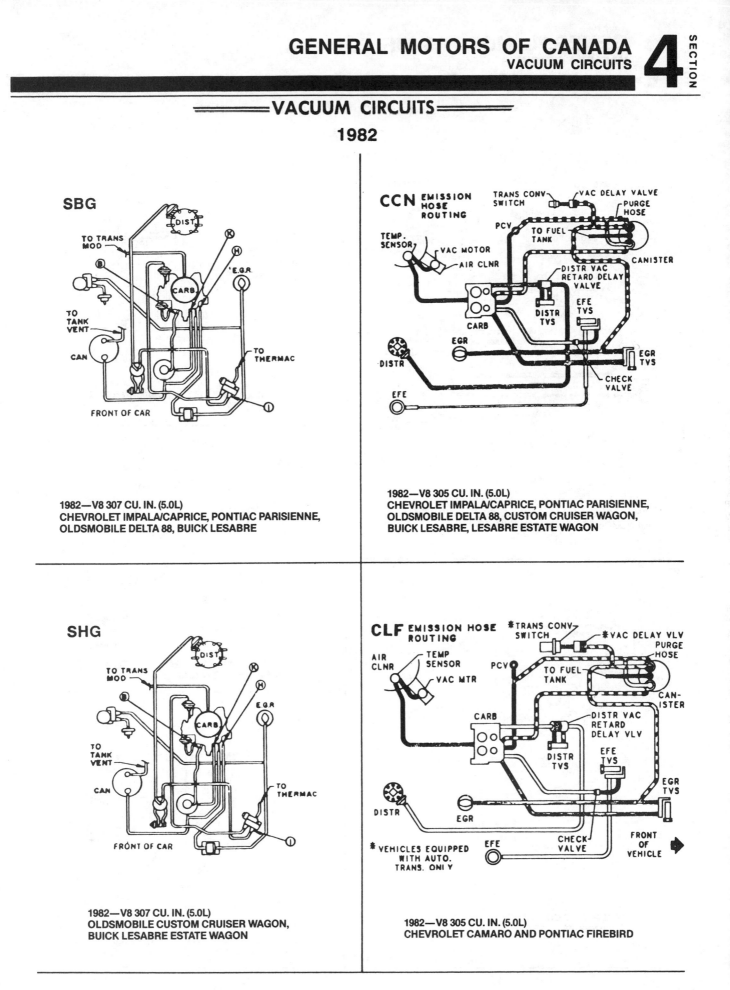

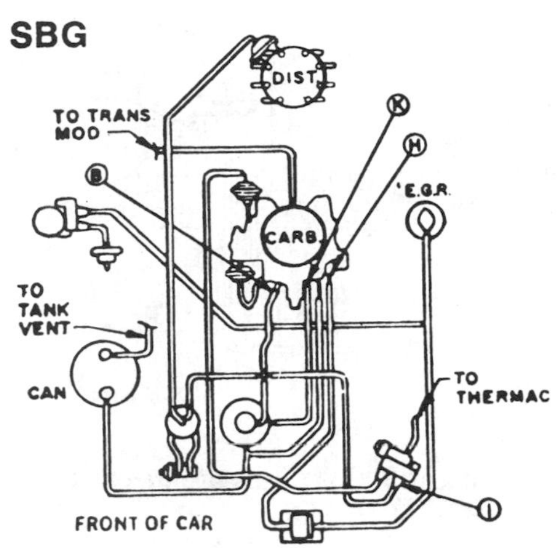

1982—V8 307 CU. IN. (5.0L)
CHEVROLET IMPALA/CAPRICE, PONTIAC PARISIENNE, OLDSMOBILE DELTA 88, BUICK LESABRE

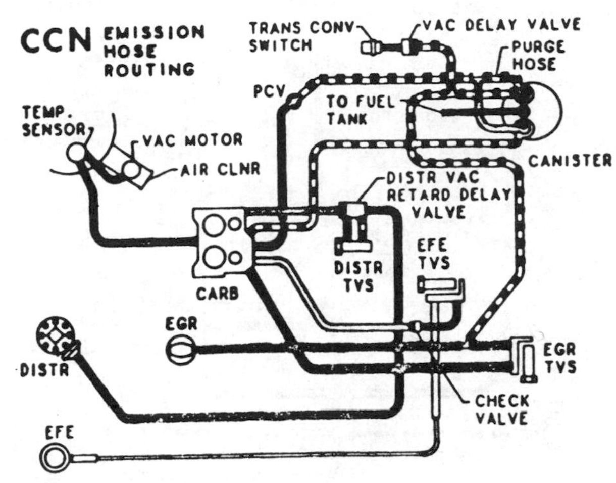

1982—V8 305 CU. IN. (5.0L)
CHEVROLET IMPALA/CAPRICE, PONTIAC PARISIENNE, OLDSMOBILE DELTA 88, CUSTOM CRUISER WAGON, BUICK LESABRE, LESABRE ESTATE WAGON

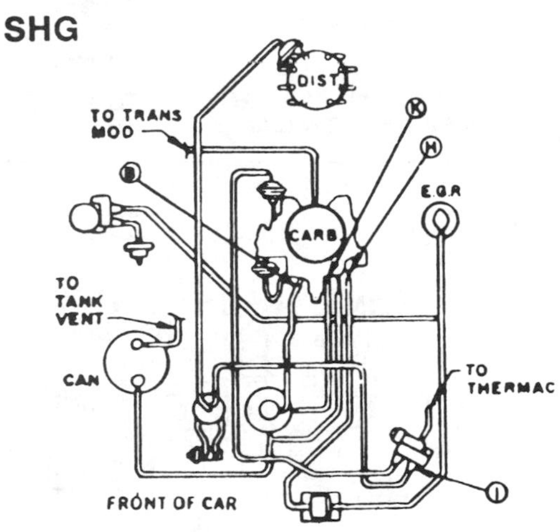

1982—V8 307 CU. IN. (5.0L)
OLDSMOBILE CUSTOM CRUISER WAGON, BUICK LESABRE ESTATE WAGON

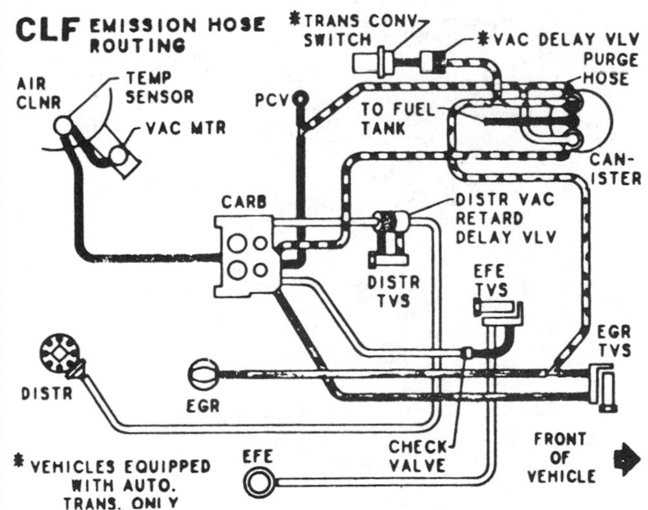

1982—V8 305 CU. IN. (5.0L)
CHEVROLET CAMARO AND PONTIAC FIREBIRD

VACUUM CIRCUITS

1982

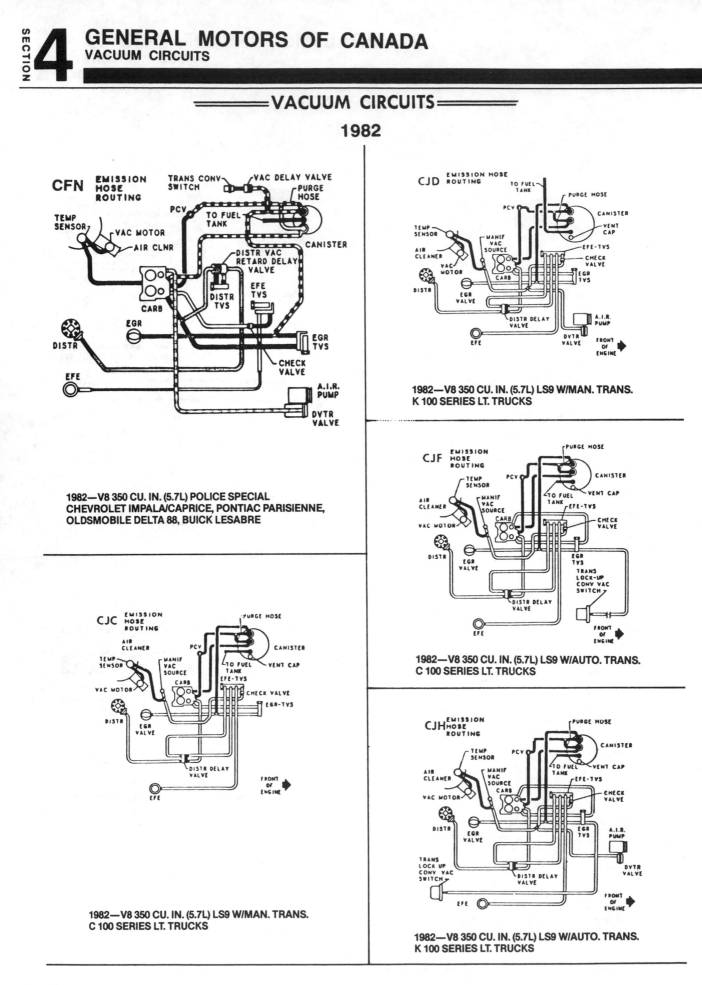

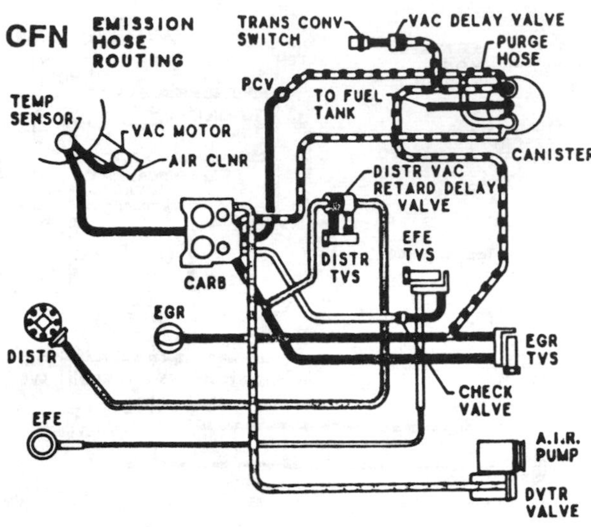

1982—V8 350 CU. IN. (5.7L) POLICE SPECIAL
CHEVROLET IMPALA/CAPRICE, PONTIAC PARISIENNE,
OLDSMOBILE DELTA 88, BUICK LESABRE

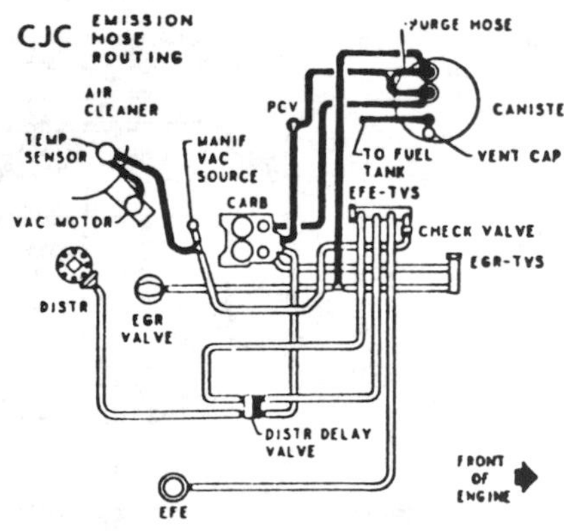

1982—V8 350 CU. IN. (5.7L) LS9 W/MAN. TRANS.
C 100 SERIES LT. TRUCKS

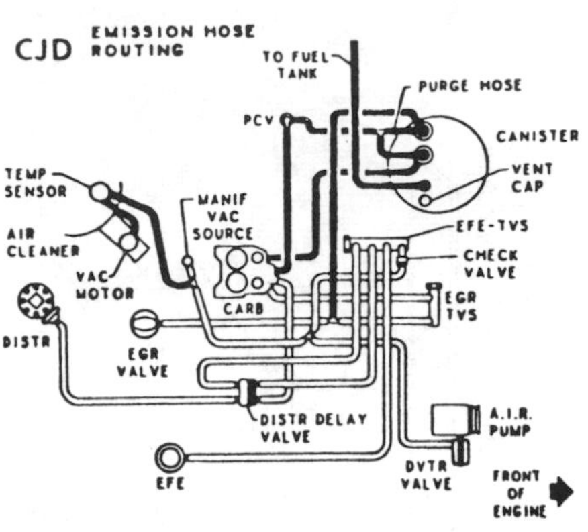

1982—V8 350 CU. IN. (5.7L) LS9 W/MAN. TRANS.
K 100 SERIES LT. TRUCKS

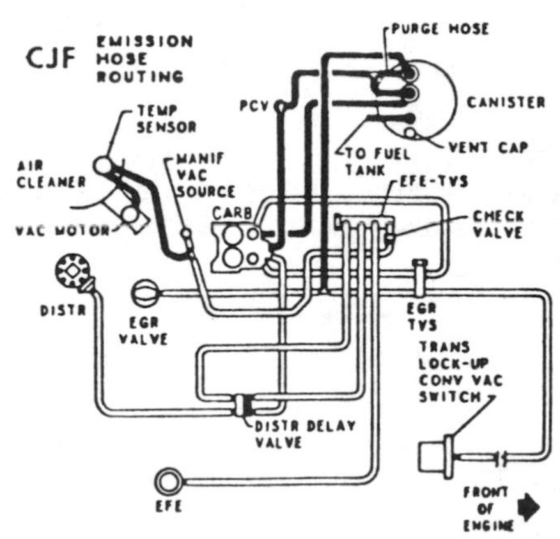

1982—V8 350 CU. IN. (5.7L) LS9 W/AUTO. TRANS.
C 100 SERIES LT. TRUCKS

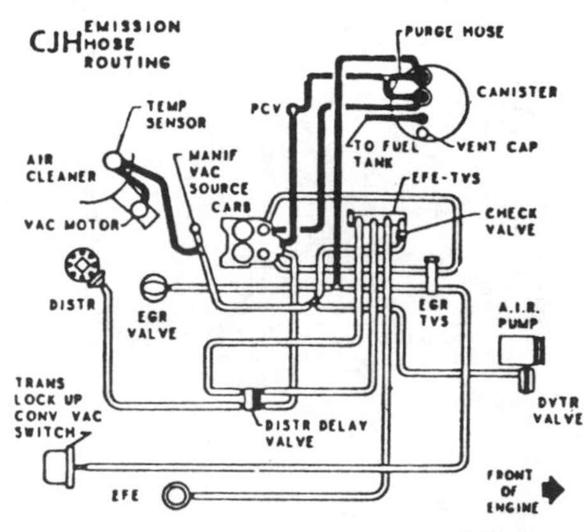

1982—V8 350 CU. IN. (5.7L) LS9 W/AUTO. TRANS.
K 100 SERIES LT. TRUCKS

VACUUM CIRCUITS

1982

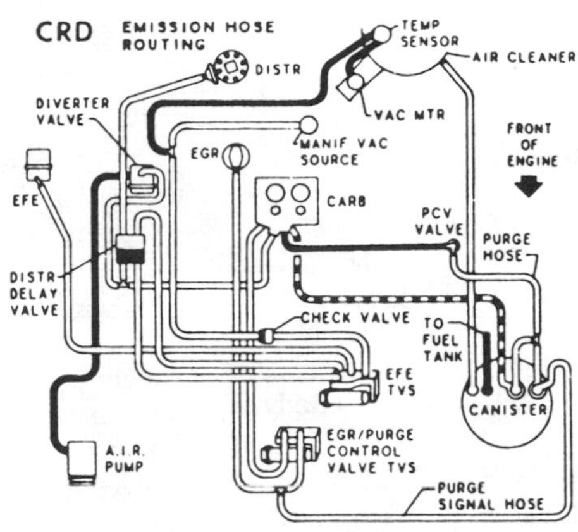

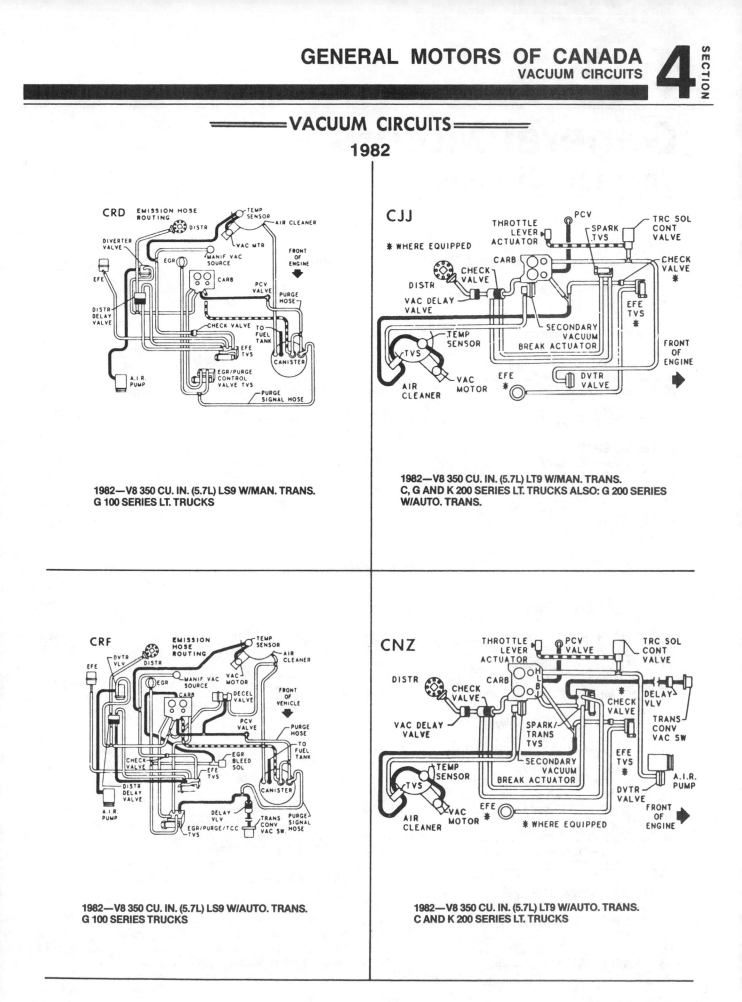

1982—V8 350 CU. IN. (5.7L) LS9 W/MAN. TRANS. G 100 SERIES LT. TRUCKS

CJJ
THROTTLE LEVER ACTUATOR
PCV
SPARK TVS
TRC SOL CONT VALVE
✱ WHERE EQUIPPED
CARB
CHECK VALVE
CHECK VALVE ✱
DISTR
VAC DELAY VALVE
EFE TVS ✱
SECONDARY VACUUM BREAK ACTUATOR
TEMP SENSOR
TVS
FRONT OF ENGINE
EFE ✱
DVTR VALVE
AIR CLEANER
VAC MOTOR

1982—V8 350 CU. IN. (5.7L) LT9 W/MAN. TRANS. C, G AND K 200 SERIES LT. TRUCKS ALSO: G 200 SERIES W/AUTO. TRANS.

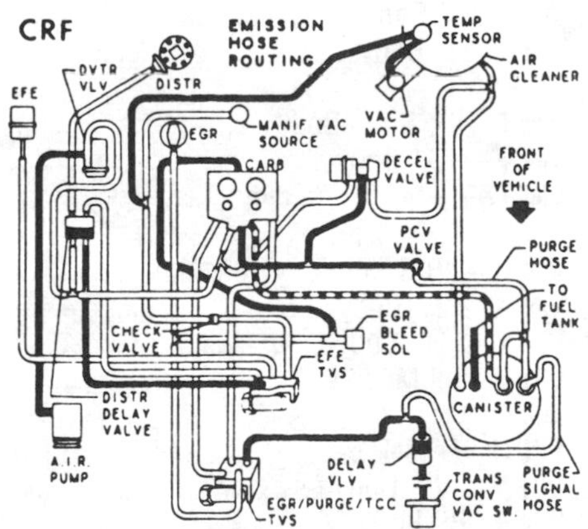

1982—V8 350 CU. IN. (5.7L) LS9 W/AUTO. TRANS. G 100 SERIES TRUCKS

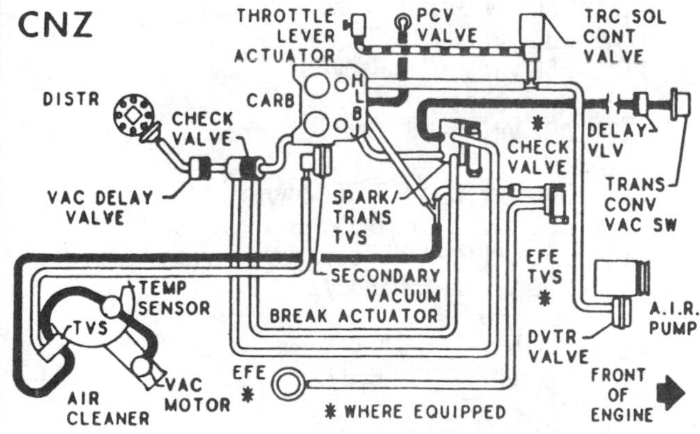

1982—V8 350 CU. IN. (5.7L) LT9 W/AUTO. TRANS. C AND K 200 SERIES LT. TRUCKS

General Motors
Vacuum Circuits

INDEX

1983 VACUUM CIRCUITS

1984 VACUUM CIRCUITS

GENERAL MOTORS BODY CODE INDENTIFICATION

Reference will be made to body style codes for GM cars in the vacuum diagrams index. Refer to the table below for identification of these body style codes.

Body Code	GM Division	Model Name
A	Buick	Century
A	Chevrolet	Celebrity
A	Oldsmobile	Ciera
A	Pontiac	6000
B	Buick	Electra, LeSabre
B	Chevrolet	Caprice Classic, Impala
B	Oldsmobile	Delta 88
B	Pontiac	Parisienne
C	Buick	Electra
C	Cadillac	DeVille, Fleetwood
C	Oldsmobile	(84) Ninety-Eight
D	Cadillac	Fleetwood Limousine DeVille, Olds 98 (83)
E	Buick	Riviera
E	Cadillac	Eldorado
E	Oldsmobile	Toronado
F	Chevrolet	Camaro
F	Pontiac	Firebird
G	Buick	Regal
G	Chevrolet	Malibu, Monte Carlo El Camino
G	Oldsmobile	Cutlass
G	Pontiac	Bonneville, Grand Prix
H	Buick	Skyhawk
H	Chevrolet	Monza
H	Oldsmobile	Starfire
H	Pontiac	Sunbird
J	Buick	Skyhawk
J	Cadillac	Cimarron
J	Chevrolet	Cavalier
J	Oldsmobile	Firenza
J	Pontiac	Sunbird, 2000
K	Cadillac	Seville 83–84 Eldorado
P	Pontiac	Fiero
T	Chevrolet	Chevette
T	Pontiac	1000
X	Buick	Skylark
X	Chevrolet	Citation
X	Oldsmobile	Omega
X	Pontiac	Phoenix
Y	Chevrolet	Corvette

VACUUM CIRCUITS

1983

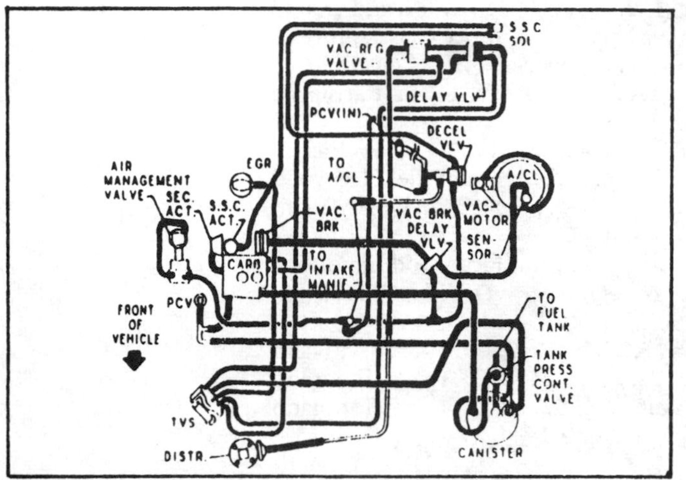

1983—4 CYL.-1.6L VIN C 2 BBL., MANUAL TRANS., "T" BODY—FED. & HIGH ALT.

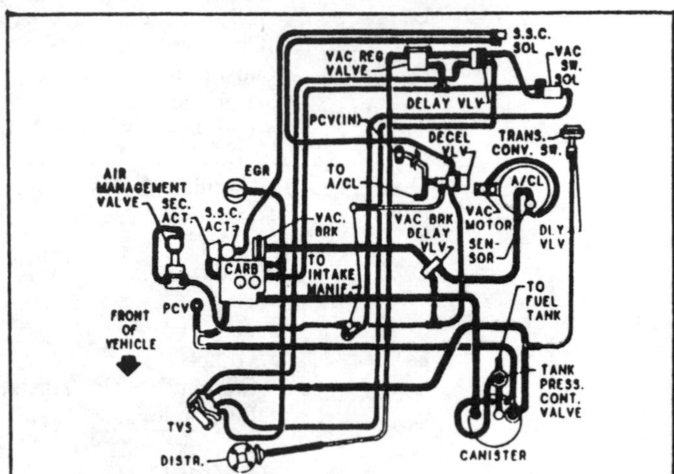

1983—4 CYL.-1.6L VIN C 2 BBL., AUTO. TRANS. & PWR. STRG.—"T" BODY—ALL

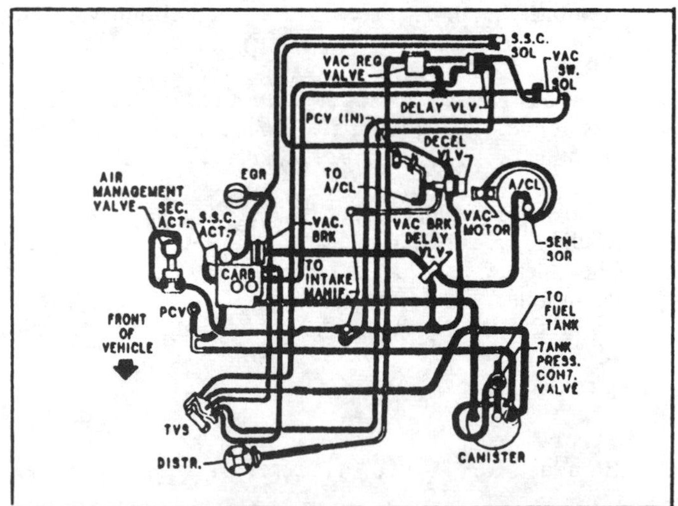

1983—4 CYL.-1.6L VIN C 2 BBL., MANUAL TRANS. & A/C—"T" BODY—FED. & HIGH ALT.

VACUUM CIRCUITS

1983

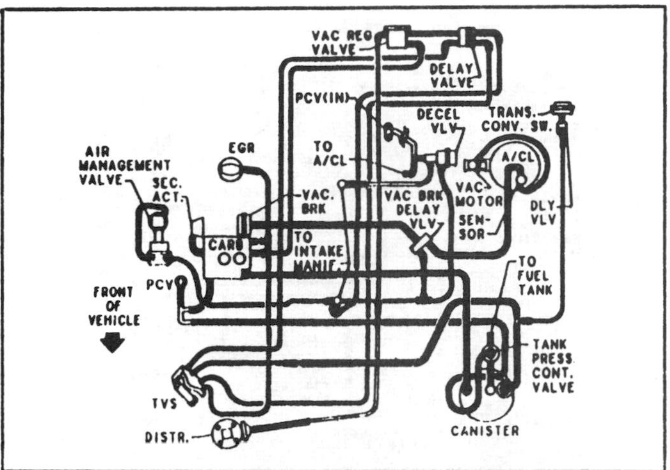

1983—4 CYL.-1.6L VIN C 2 BBL., AUTO. TRANS., WO/PWR. STRG.—"T" BODY—ALL

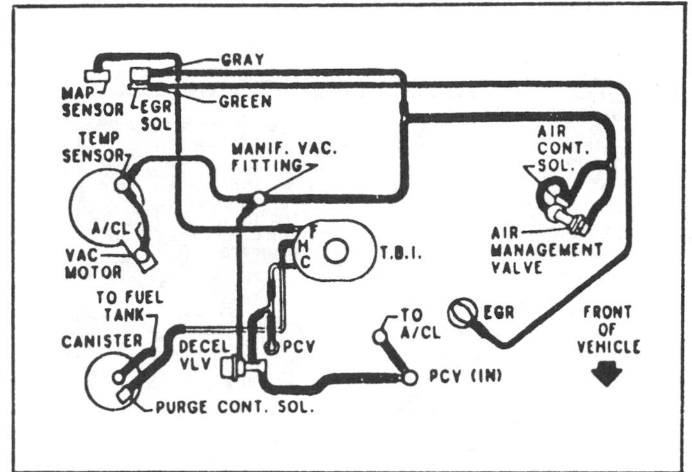

1983—4 CYL.-2.0L VIN P T.B.I., MANUAL TRANS., W/AMV—"J" BODY—FEDERAL

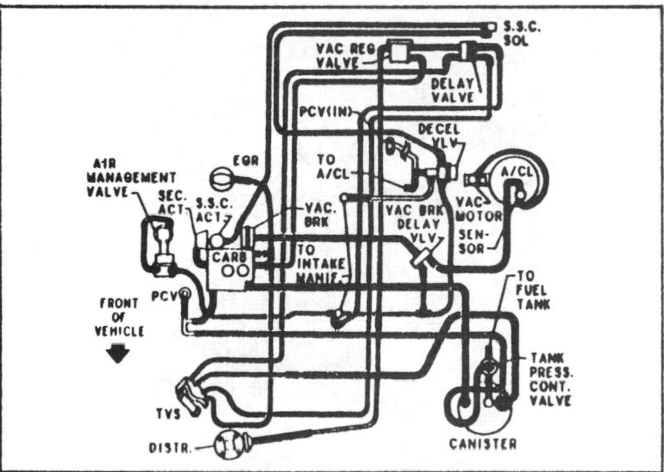

1983—4 CYL.-1.6L VIN C 2 BBL., MANUAL TRANS.—"T" BODY—CALIFORNIA

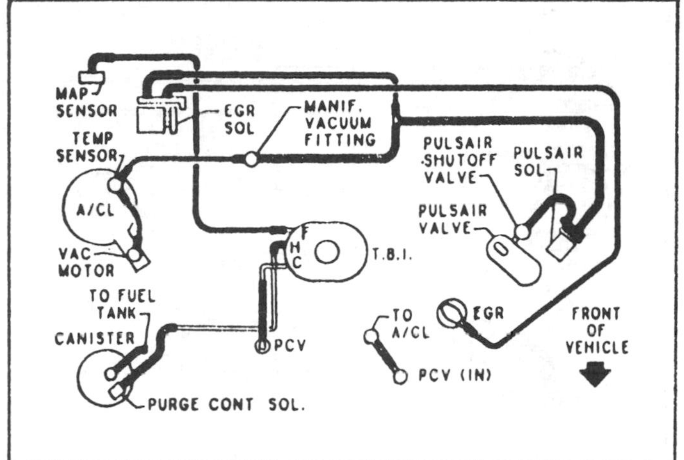

1983—4 CYL.-2.0L VIN P T.B.I., AUTO. TRANS., WO/CRUISE CONT.—"J" BODY—ALL

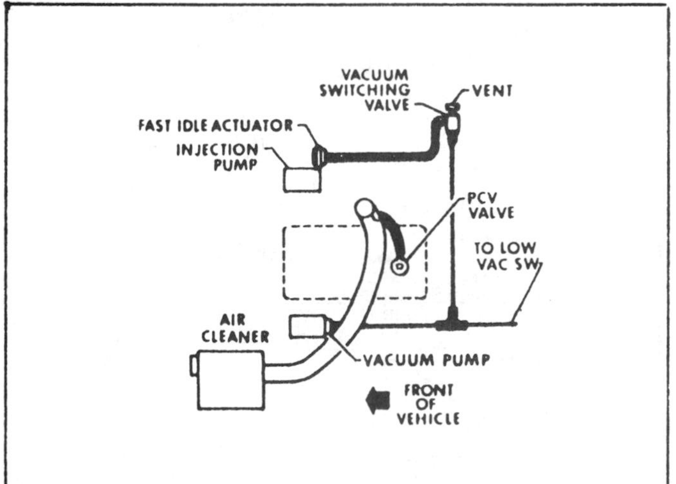

1983—4 CYL.-1.8L VIN D Diesel—"T" BODY—ALL

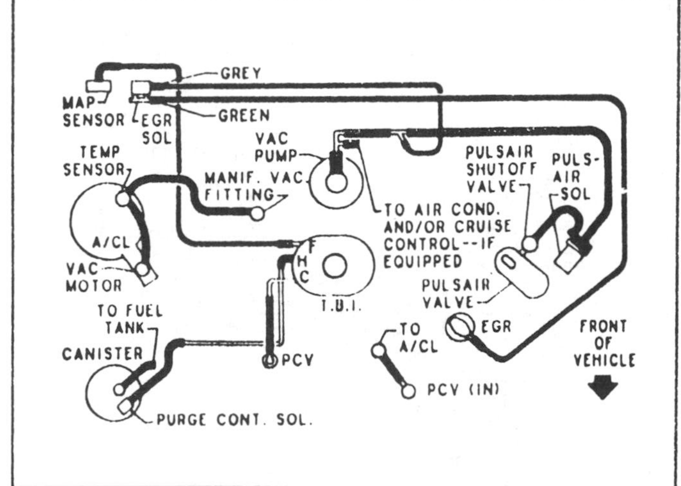

1983—4 CYL.-2.0L VIN P T.B.I., AUTO. TRANS., W/CRUISE CONT.—"J" BODY—ALL

VACUUM CIRCUITS

1983

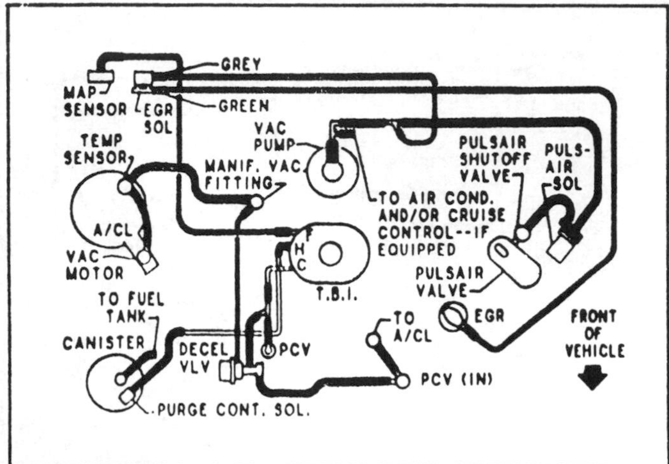

1983—4 CYL.-2.0L VIN P T.B.I., MANUAL TRANS., W/CRUISE CONT.—"J" BODY—ALL

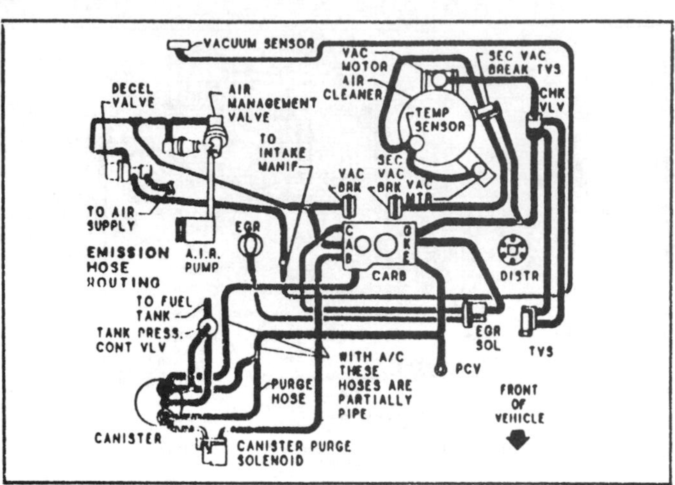

1983—V6-2.8L VIN Z 2 BBL., W/FRESH AIR OPTION—"X" BODY—ALL

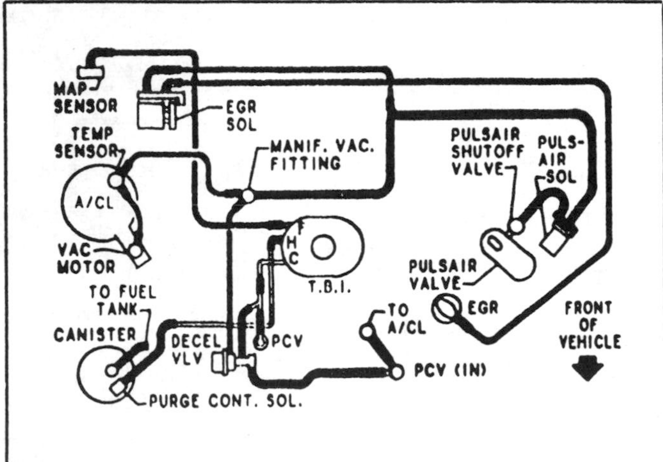

1983—4 CYL.-2.0L VIN P T.B.I., MANUAL TRANS., WO/ CRUISE CONT.—"J" BODY—ALL

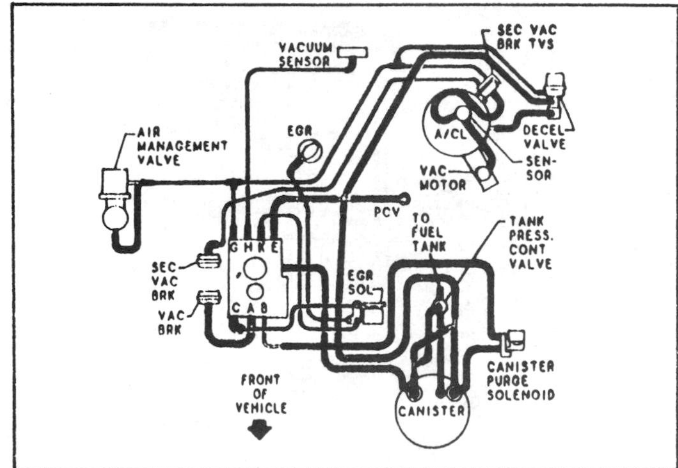

1983—V6-2.8L VIN 1 2 BBL.—"F" BODY—ALL

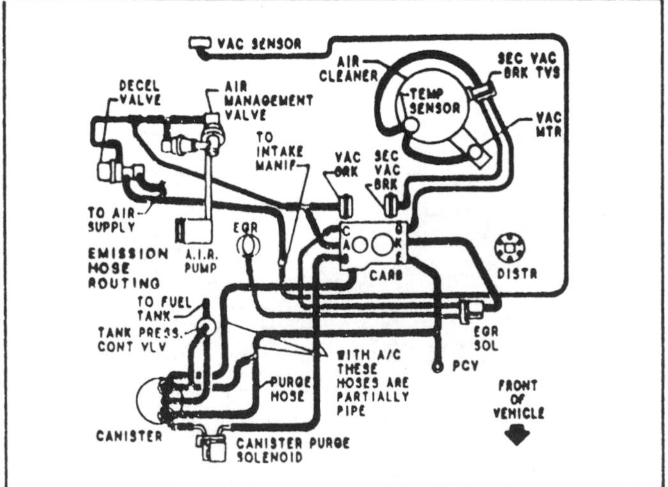

1983—V6-2.8L VIN X & Z 2 BBL., WO/FRESH AIR OPTION—"A" & "X" BODY—ALL

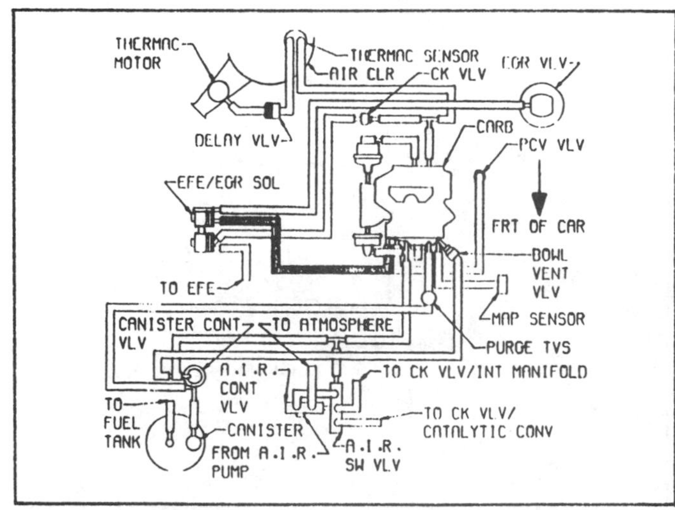

1983—V6-3.0L VIN E 2 BBL., MANUAL TRANS.—"A" BODY—ALL

VACUUM CIRCUITS

1983

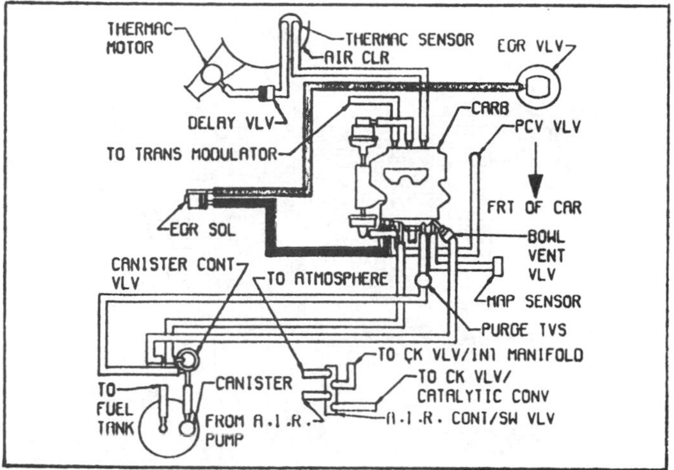

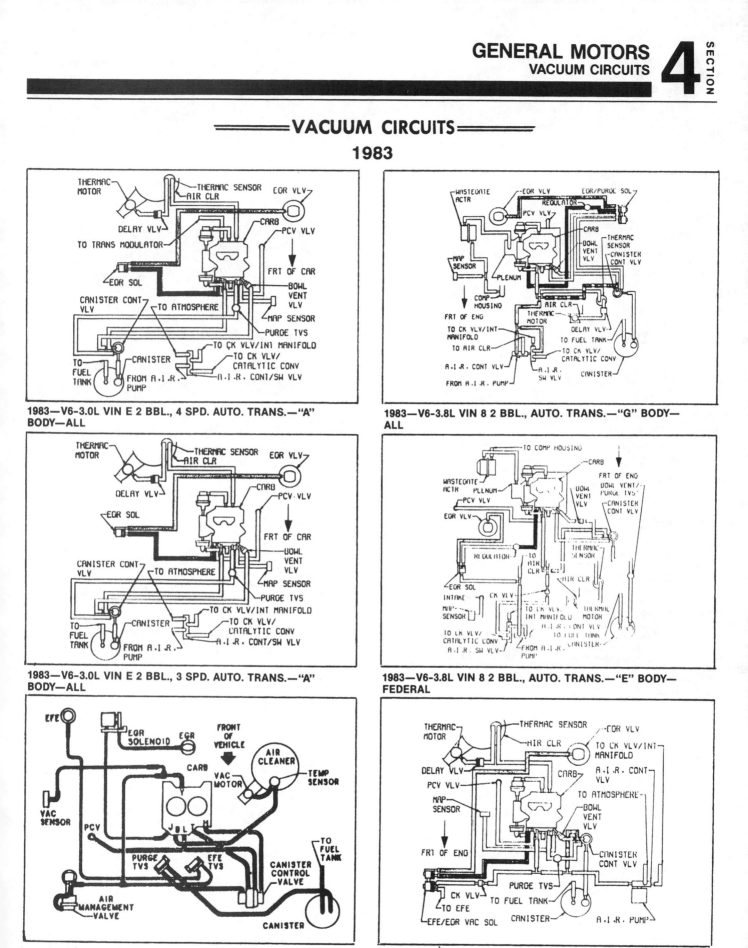

1983—V6-3.0L VIN E 2 BBL., 4 SPD. AUTO. TRANS.—"A" BODY—ALL

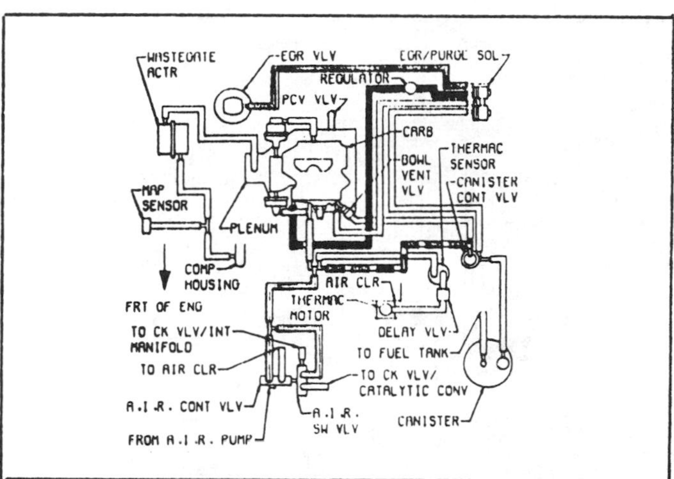

1983—V6-3.8L VIN 8 2 BBL., AUTO. TRANS.—"G" BODY—ALL

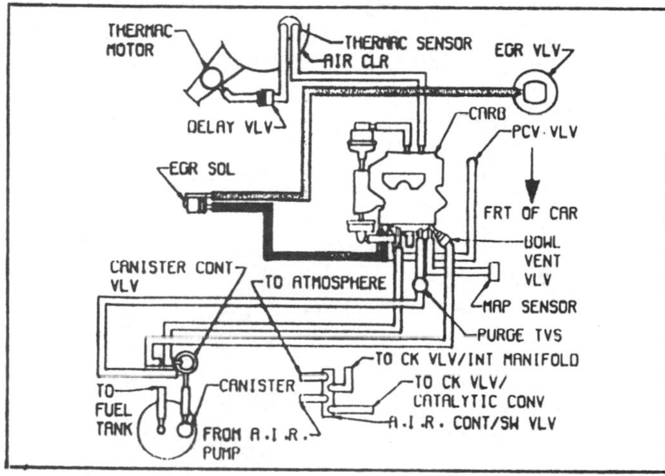

1983—V6-3.0L VIN E 2 BBL., 3 SPD. AUTO. TRANS.—"A" BODY—ALL

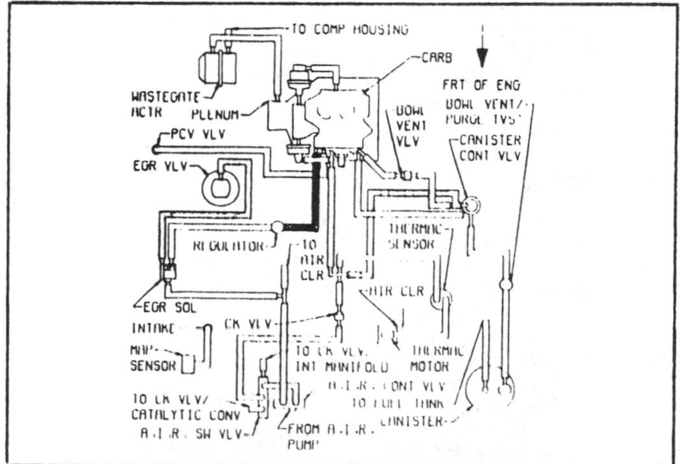

1983—V6-3.8L VIN 8 2 BBL., AUTO. TRANS.—"E" BODY—FEDERAL

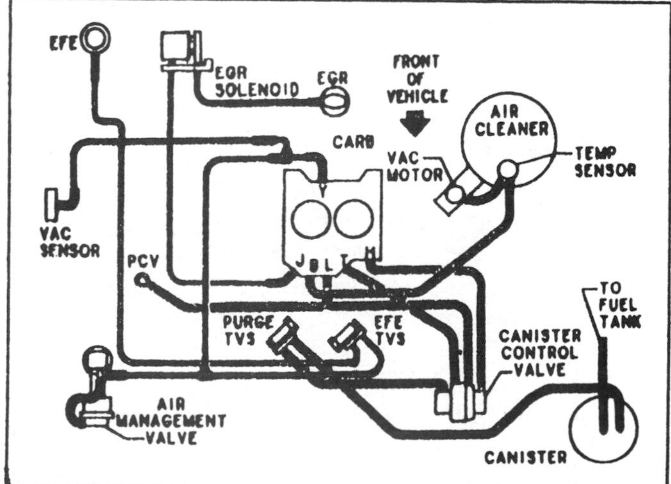

1983—V6-3.8L VIN 9 2 BBL., AUTO. TRANS.—"B" & "G" BODY—FEDERAL & HIGH ALT.

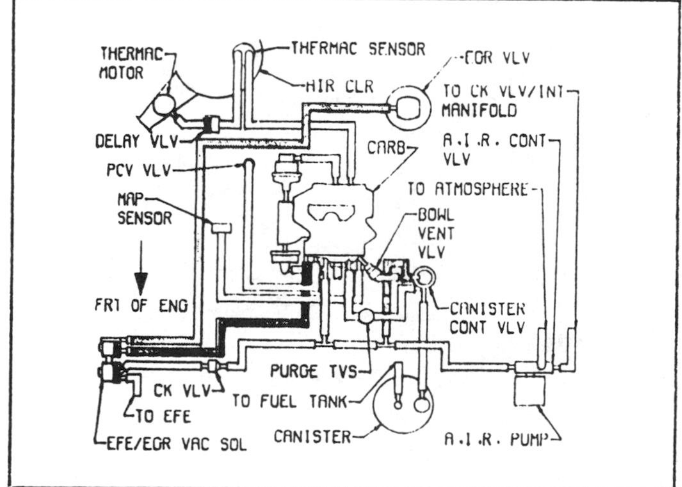

1983—V6-3.8L VIN A 2 BBL., AUTO. TRANS.—"B" & "G" BODY—CALIF.

VACUUM CIRCUITS

1983

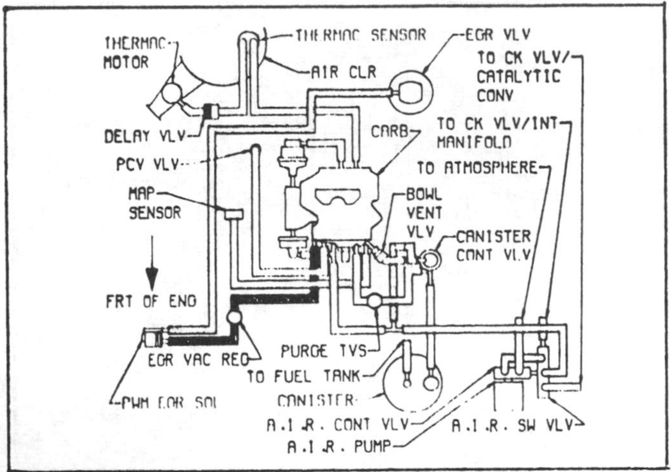

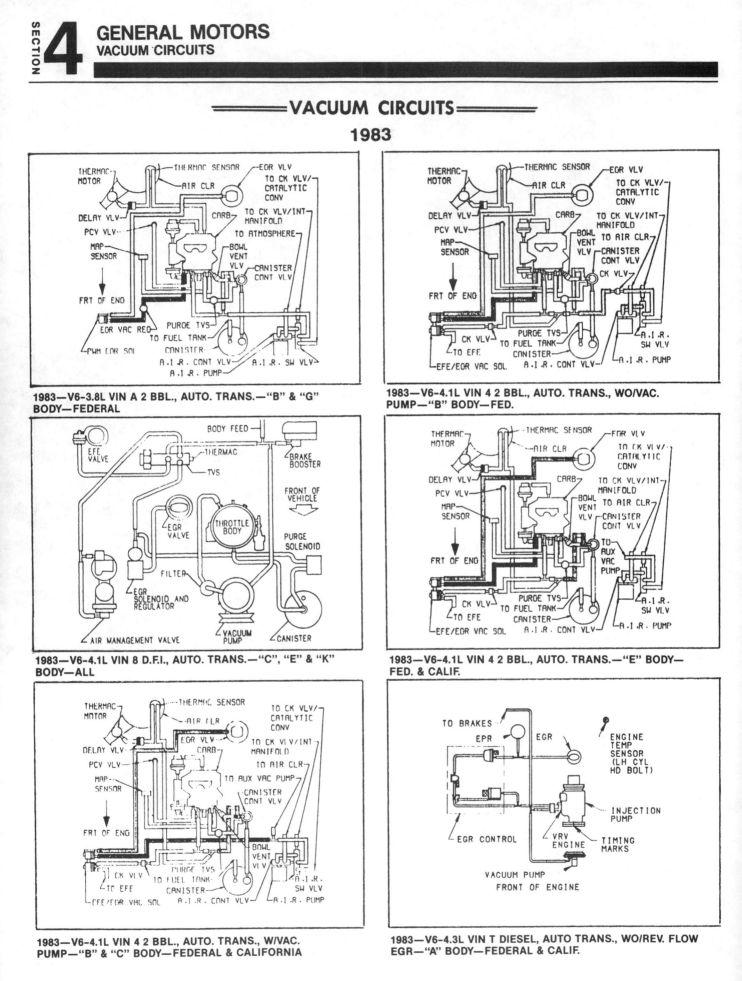

1983—V6-3.8L VIN A 2 BBL., AUTO. TRANS.—"B" & "G" BODY—FEDERAL

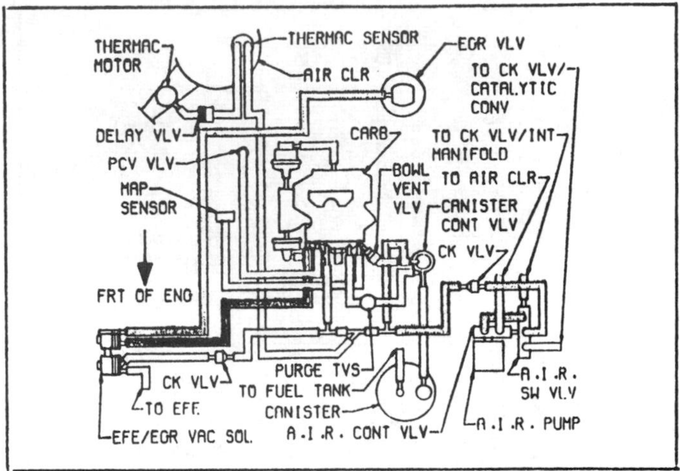

1983—V6-4.1L VIN 4 2 BBL., AUTO. TRANS., WO/VAC. PUMP—"B" BODY—FED.

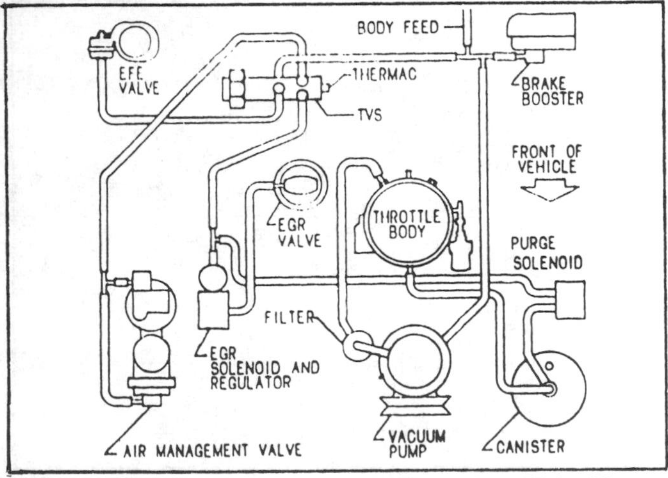

1983—V6-4.1L VIN 8 D.F.I., AUTO. TRANS.—"C", "E" & "K" BODY—ALL

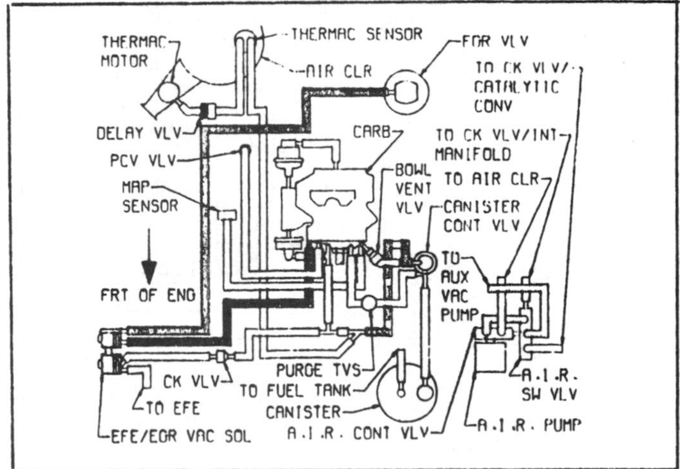

1983—V6-4.1L VIN 4 2 BBL., AUTO. TRANS.—"E" BODY—FED. & CALIF.

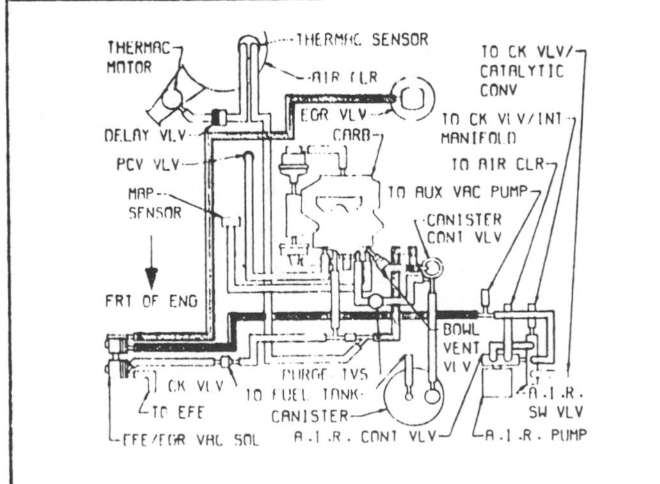

1983—V6-4.1L VIN 4 2 BBL., AUTO. TRANS., W/VAC. PUMP—"B" & "C" BODY—FEDERAL & CALIFORNIA

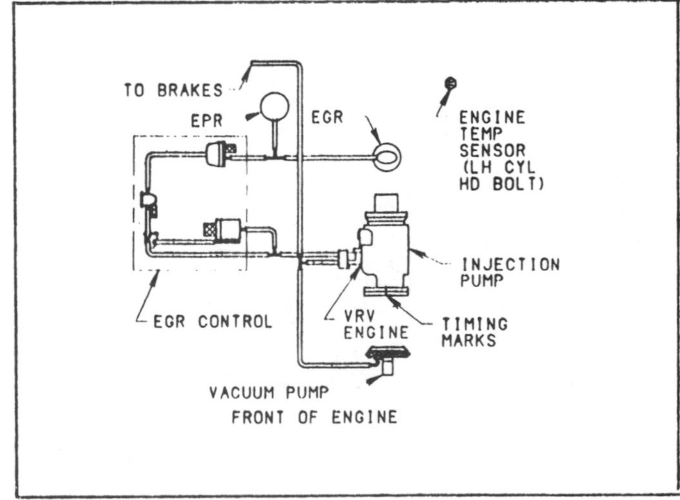

1983—V6-4.3L VIN T DIESEL, AUTO TRANS., WO/REV. FLOW EGR—"A" BODY—FEDERAL & CALIF.

VACUUM CIRCUITS

1983

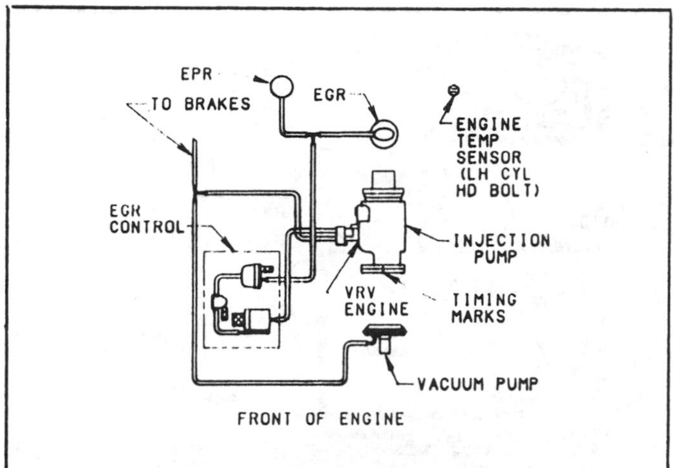

1983—V6-4.3L VIN T & V DIESEL, AUTO TRANS., W/REV. FLOW EGR—"A" BODY—FEDERAL & CALIF.

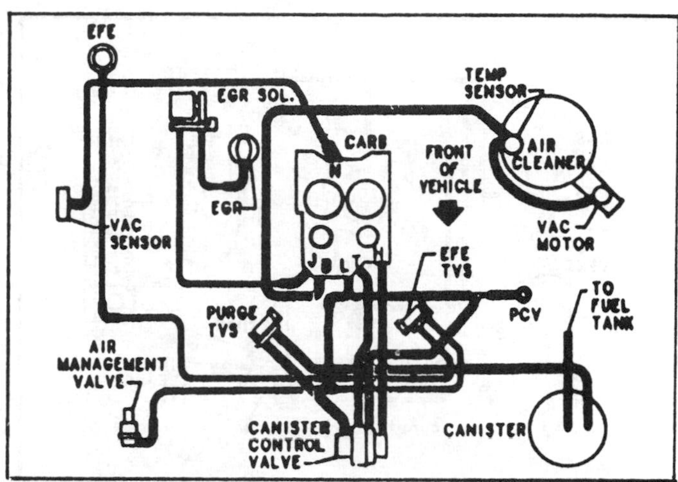

1983—V8-5.0L VIN H 4 BBL., AUTO. TRANS.—"B" & "G" BODY—CALIF.

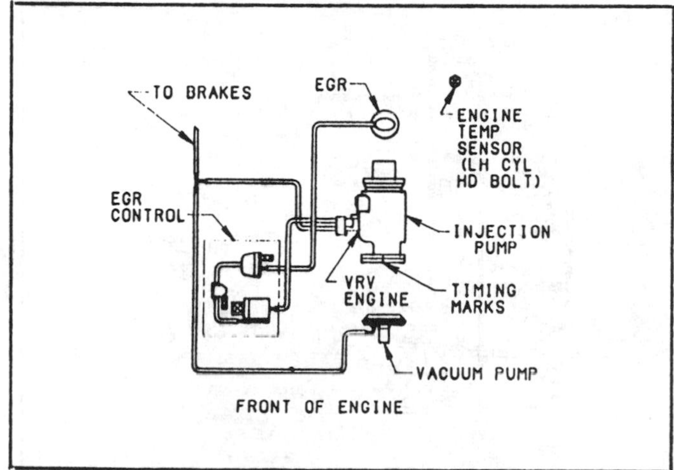

1983—V6-4.3L VIN V DIESEL, AUTO TRANS.—"A" & "G" BODY—FED. & HIGH ALT

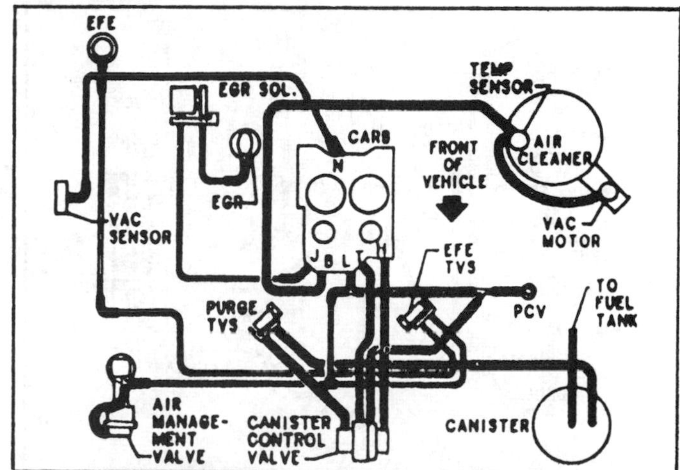

1983—V8-5.0L VIN H & 7 4 BBL., AUTO. TRANS.—"B" & "G" BODY—EXC. CALIF.

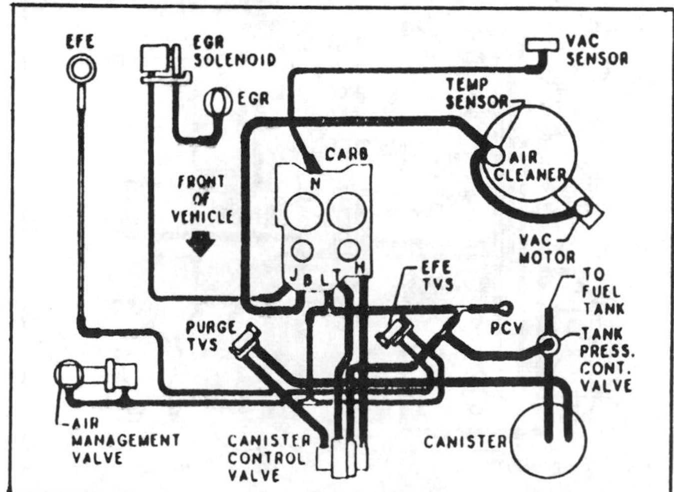

1983—V8-5.0L VIN H 2 BBL., MAN. & AUTO. TRANS.—"F" BODY—ALL

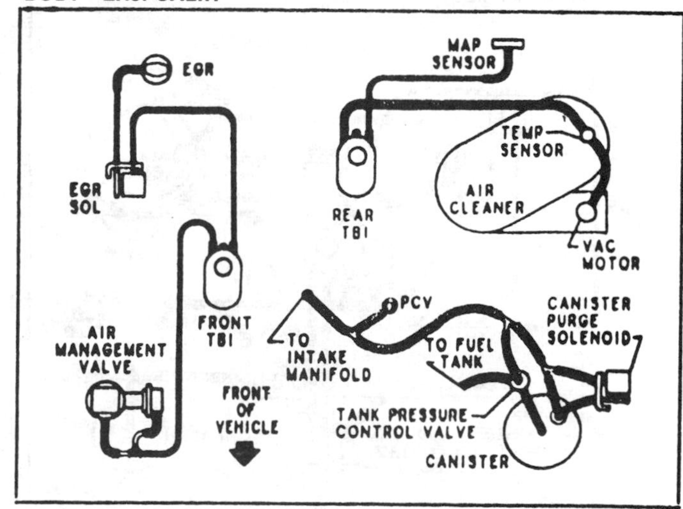

1983—V8-5.0L VIN S T.B.I., AUTO. TRANS.—"F" BODY—ALL

VACUUM CIRCUITS

1983

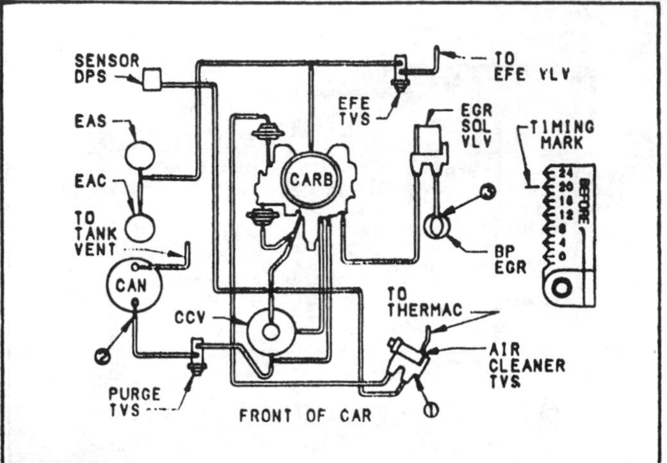

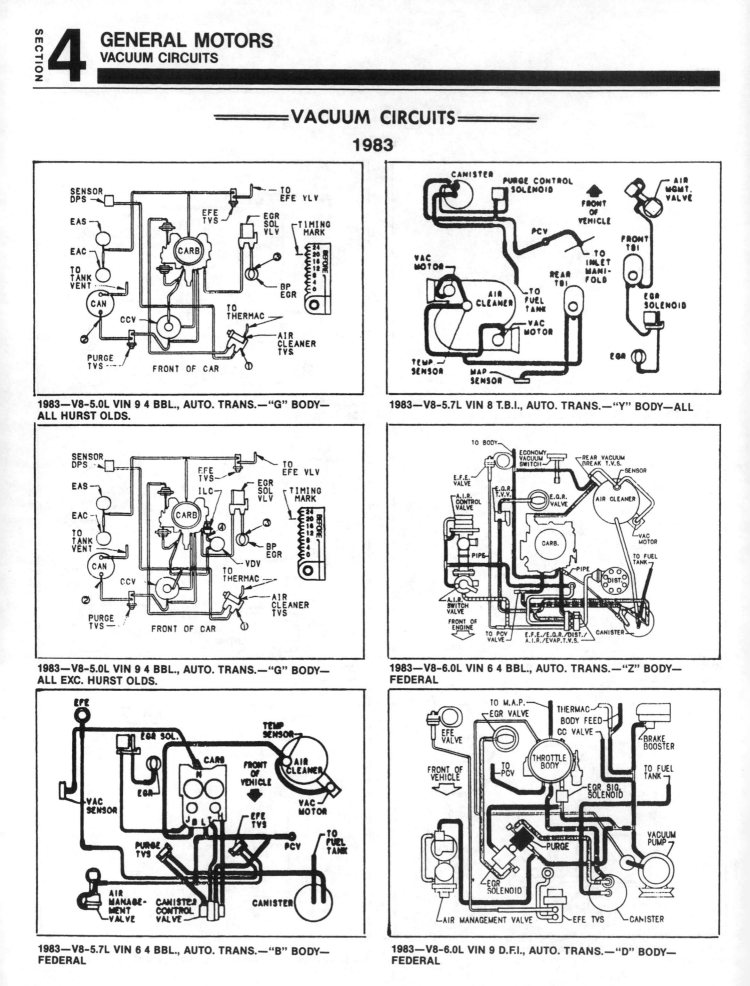

1983—V8-5.0L VIN 9 4 BBL., AUTO. TRANS.—"G" BODY—ALL HURST OLDS.

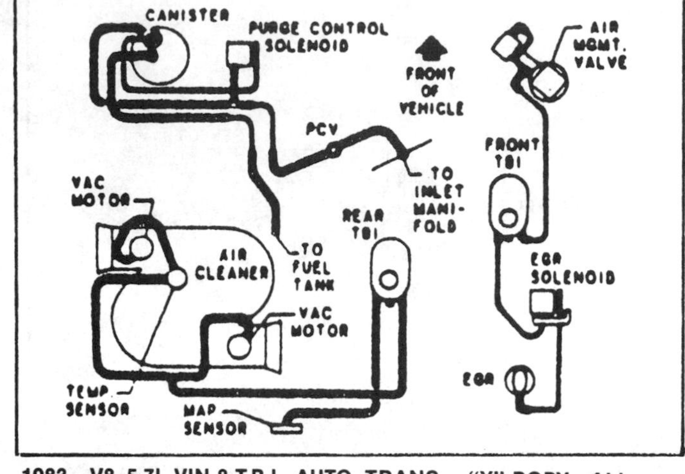

1983—V8-5.7L VIN 8 T.B.I., AUTO. TRANS.—"Y" BODY—ALL

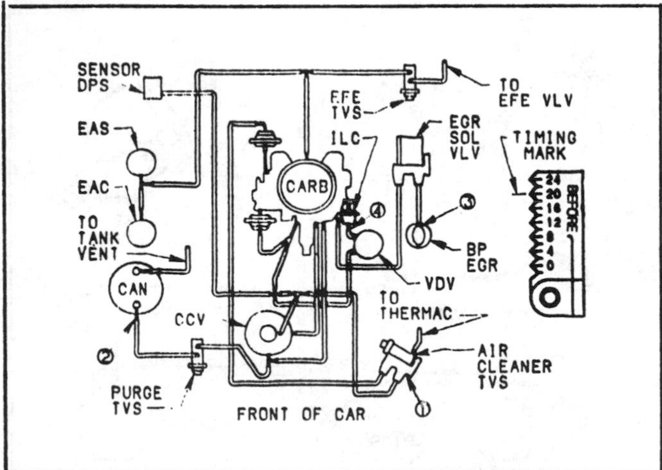

1983—V8-5.0L VIN 9 4 BBL., AUTO. TRANS.—"G" BODY—ALL EXC. HURST OLDS.

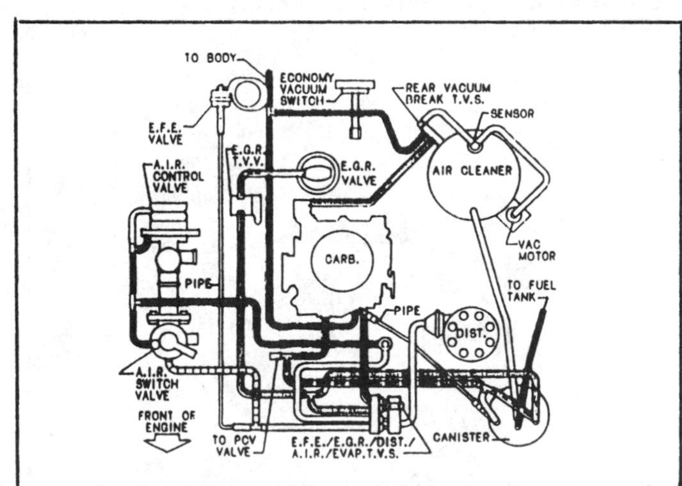

1983—V8-6.0L VIN 6 4 BBL., AUTO. TRANS.—"Z" BODY—FEDERAL

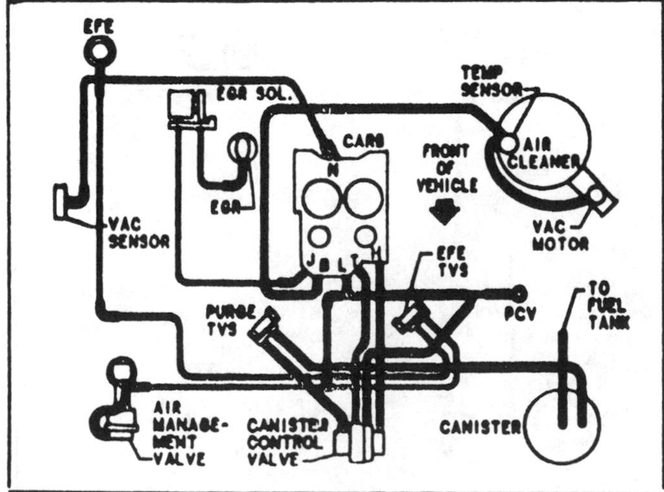

1983—V8-5.7L VIN 6 4 BBL., AUTO. TRANS.—"B" BODY—FEDERAL

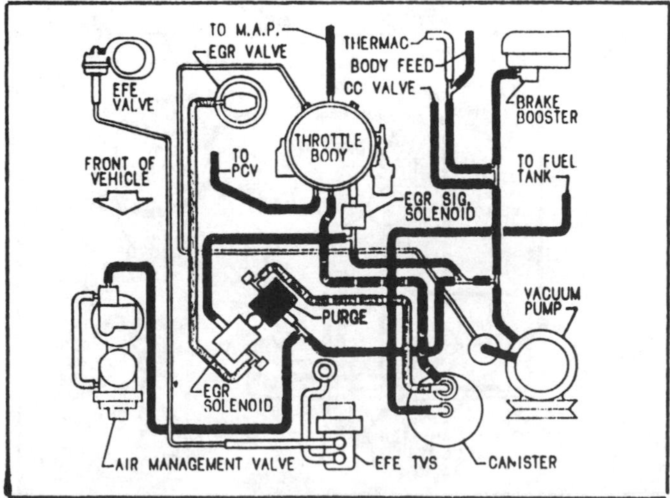

1983—V8-6.0L VIN 9 D.F.I., AUTO. TRANS.—"D" BODY—FEDERAL

VACUUM CIRCUITS

1984

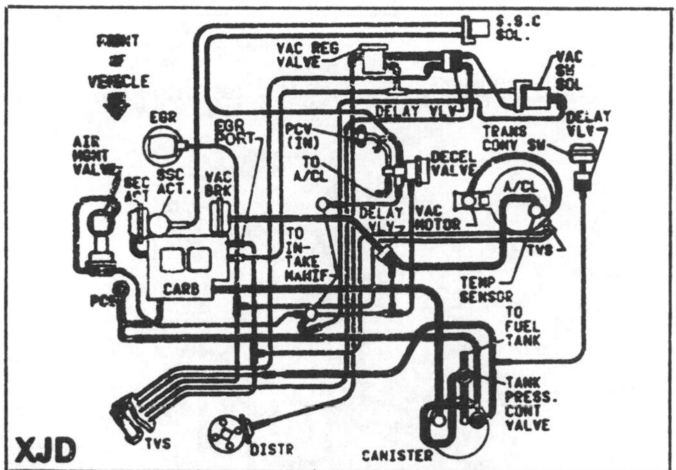

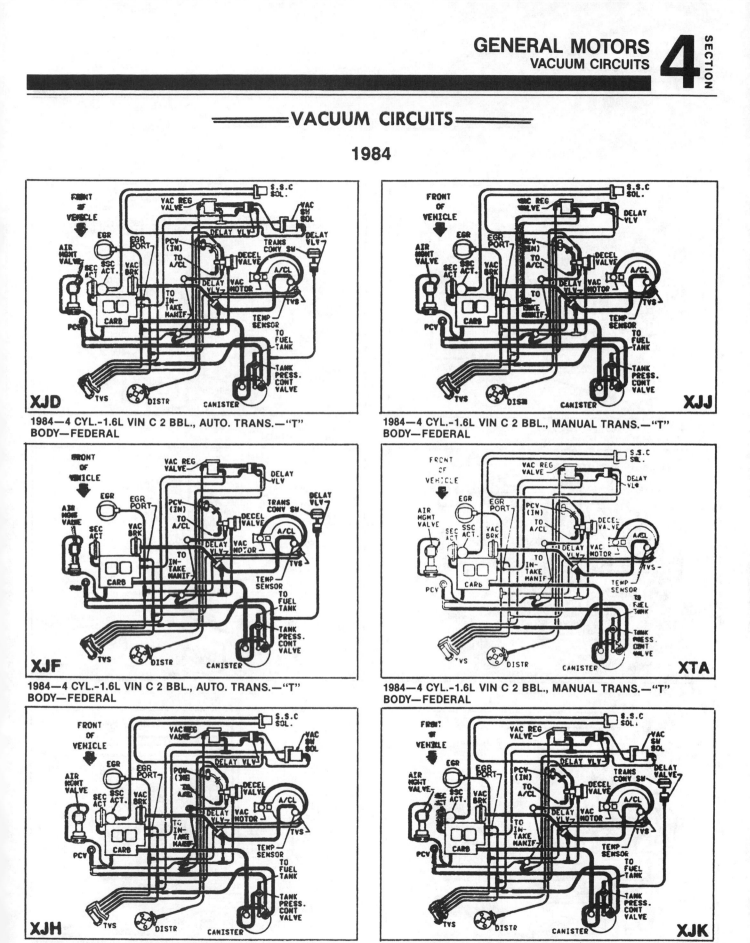

1984—4 CYL.-1.6L VIN C 2 BBL., AUTO. TRANS.—"T" BODY—FEDERAL

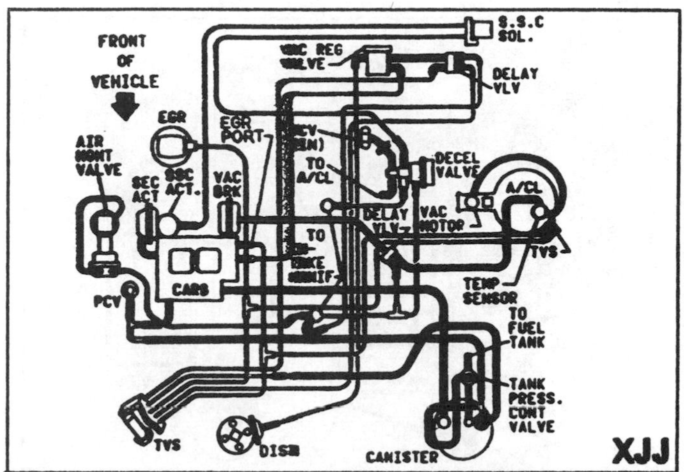

1984—4 CYL.-1.6L VIN C 2 BBL., MANUAL TRANS.—"T" BODY—FEDERAL

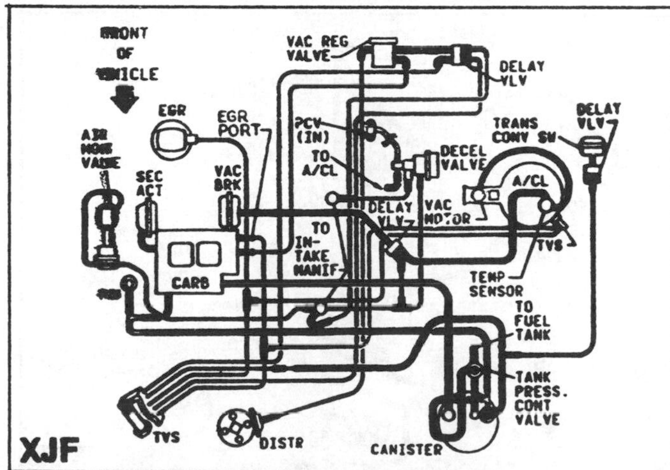

1984—4 CYL.-1.6L VIN C 2 BBL., AUTO. TRANS.—"T" BODY—FEDERAL

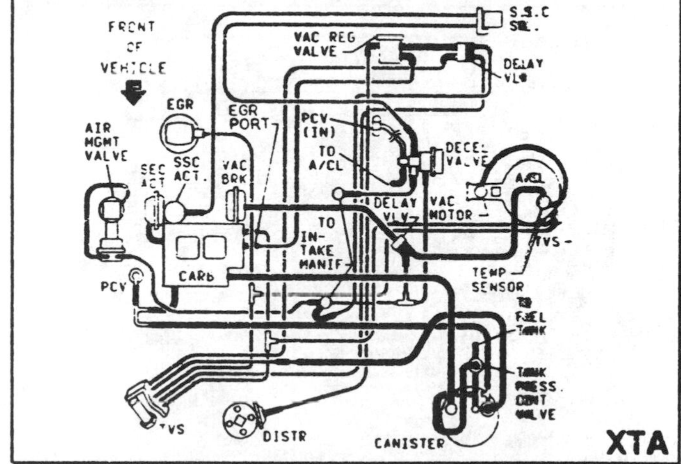

1984—4 CYL.-1.6L VIN C 2 BBL., MANUAL TRANS.—"T" BODY—FEDERAL

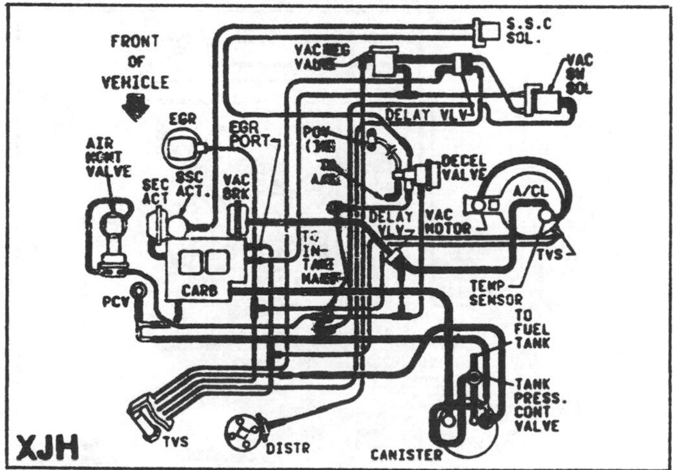

1984—4 CYL.-1.6L VIN C 2 BBL., MANUAL TRANS.—"T" BODY—FEDERAL

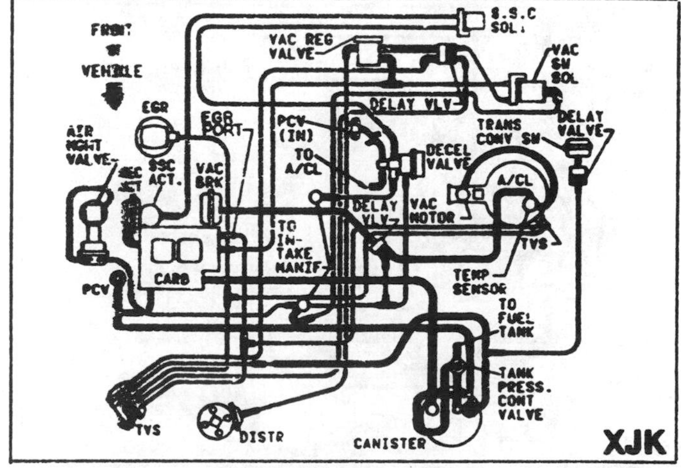

1984—4 CYL.-1.6L VIN C 2 BBL., AUTO. TRANS.—"T" BODY—LOW ALT. & CALIF.

VACUUM CIRCUITS

1984

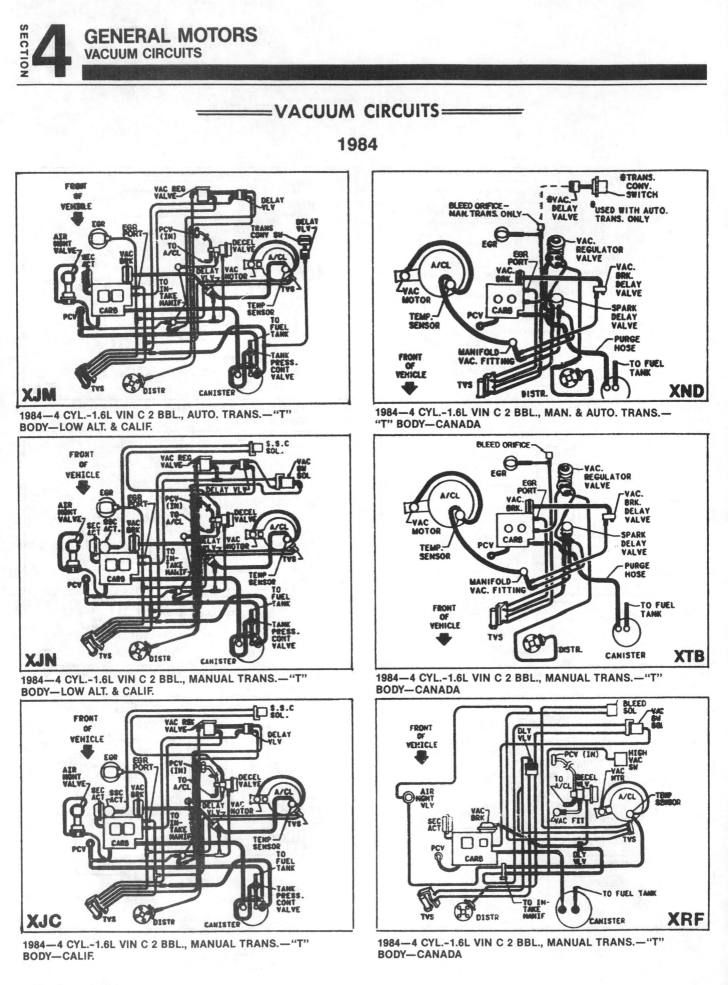

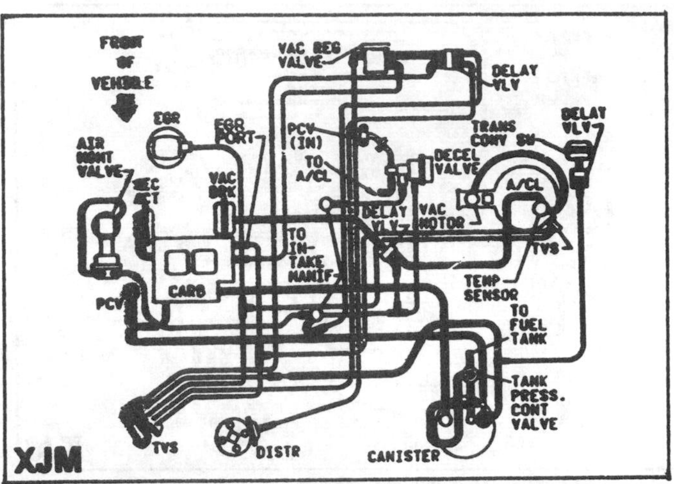

1984—4 CYL.-1.6L VIN C 2 BBL., AUTO. TRANS.—"T" BODY—LOW ALT. & CALIF.

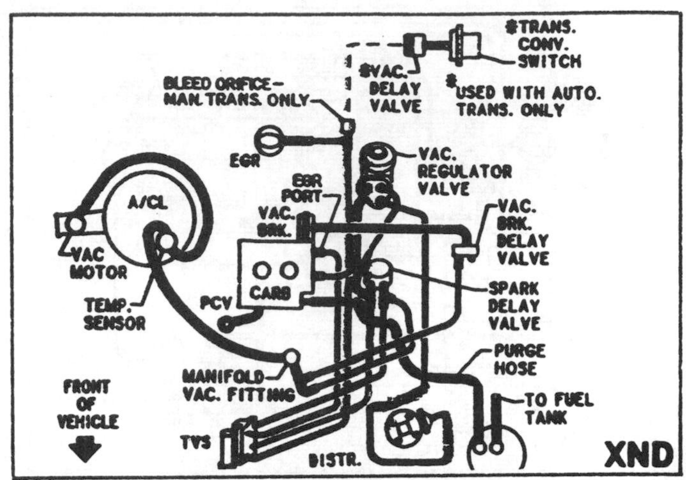

1984—4 CYL.-1.6L VIN C 2 BBL., MAN. & AUTO. TRANS.—"T" BODY—CANADA

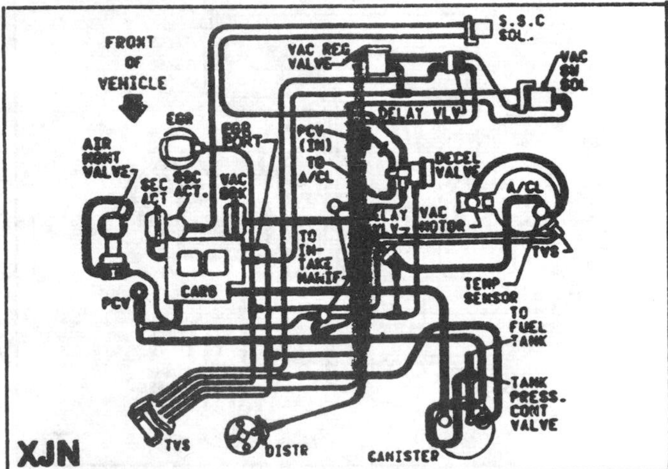

1984—4 CYL.-1.6L VIN C 2 BBL., MANUAL TRANS.—"T" BODY—LOW ALT. & CALIF.

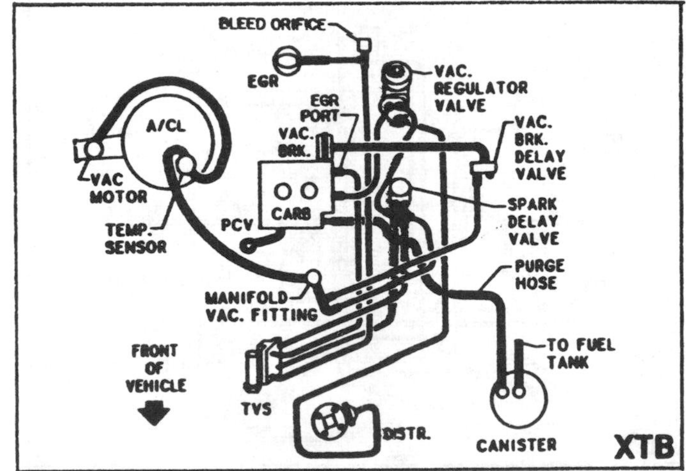

1984—4 CYL.-1.6L VIN C 2 BBL., MANUAL TRANS.—"T" BODY—CANADA

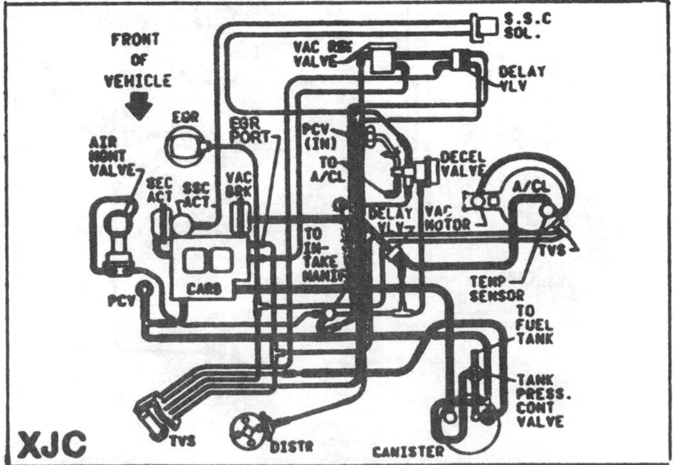

1984—4 CYL.-1.6L VIN C 2 BBL., MANUAL TRANS.—"T" BODY—CALIF.

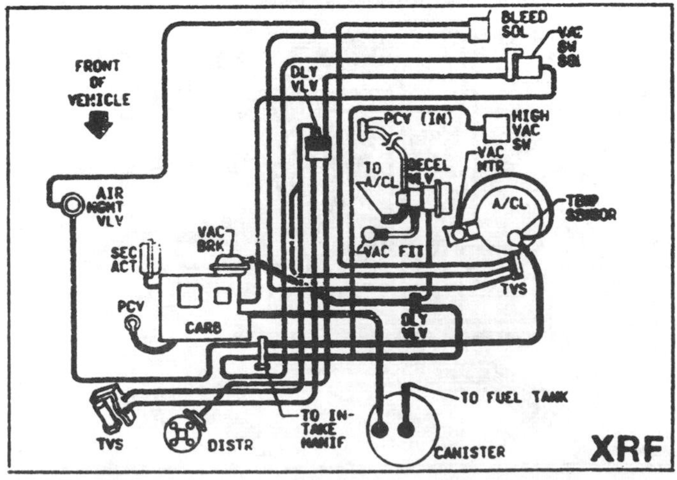

1984—4 CYL.-1.6L VIN C 2 BBL., MANUAL TRANS.—"T" BODY—CANADA

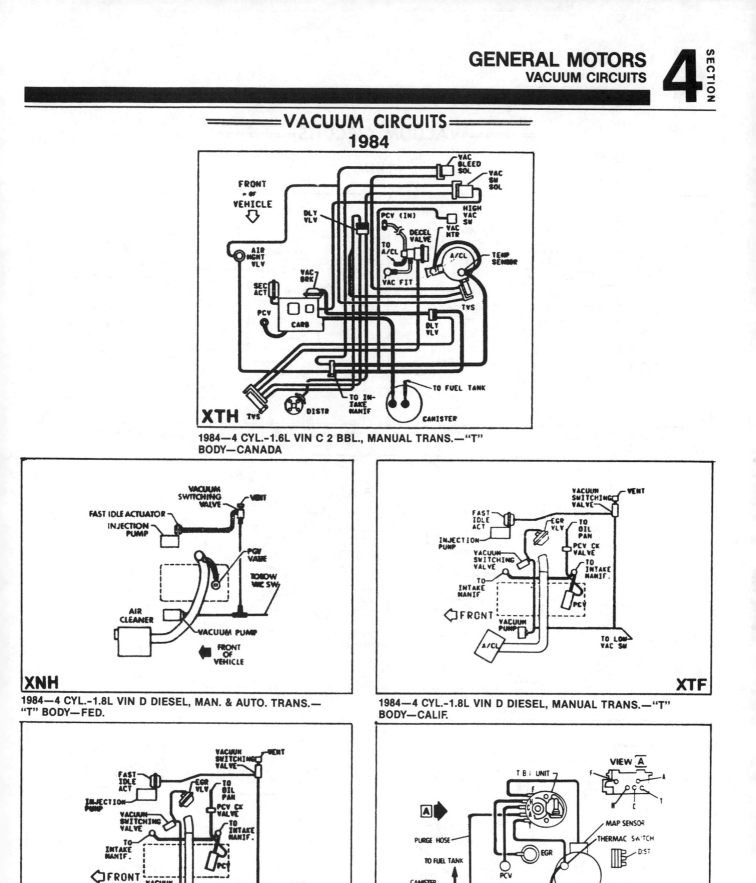

VACUUM CIRCUITS
1984

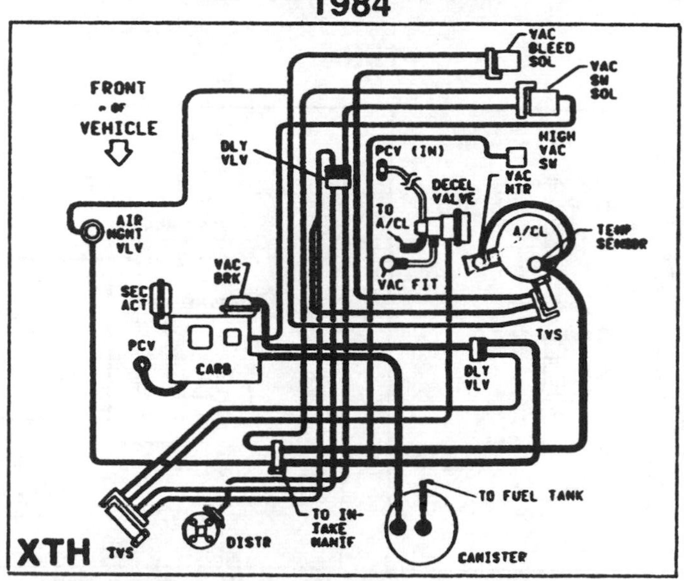

1984—4 CYL.-1.6L VIN C 2 BBL., MANUAL TRANS.—"T" BODY—CANADA

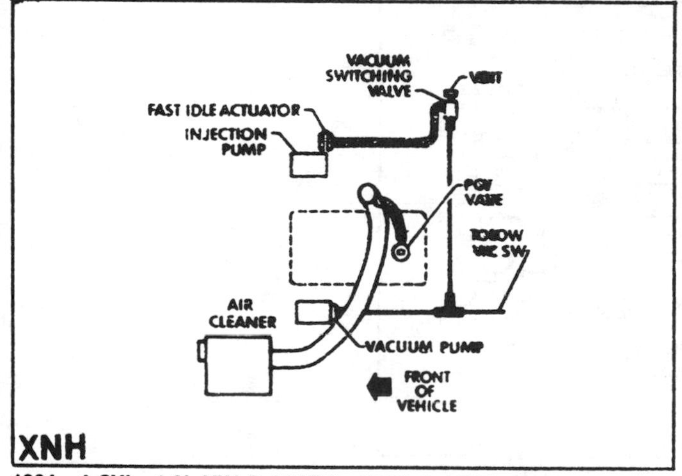

1984—4 CYL.-1.8L VIN D DIESEL, MAN. & AUTO. TRANS.—"T" BODY—FED.

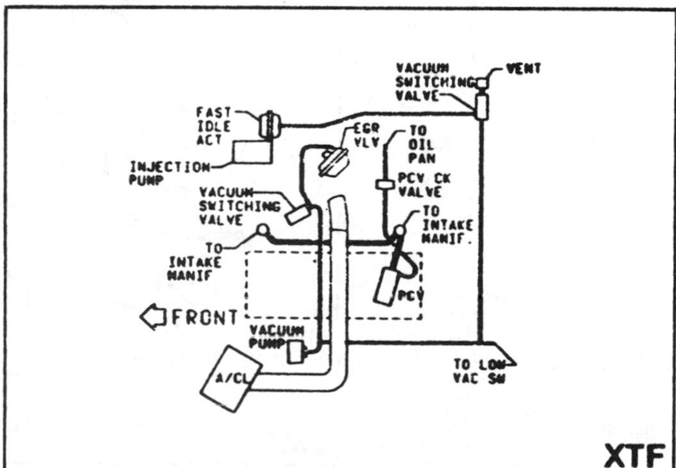

1984—4 CYL.-1.8L VIN D DIESEL, MANUAL TRANS.—"T" BODY—CALIF.

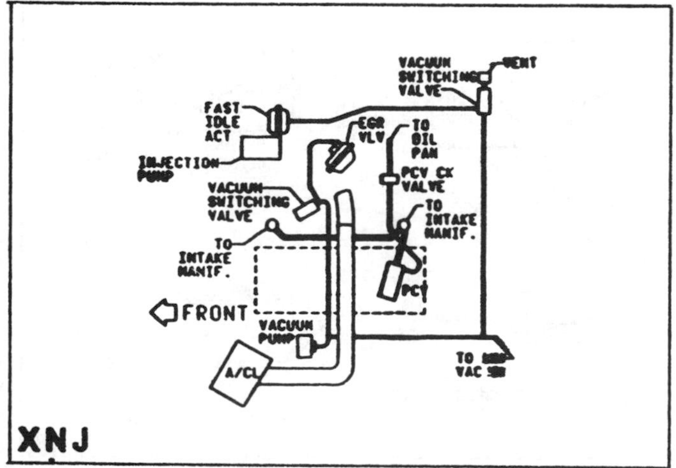

1984—4 CYL.-1.8L VIN D DIESEL, MANUAL TRANS.—"T" BODY—CALIF.

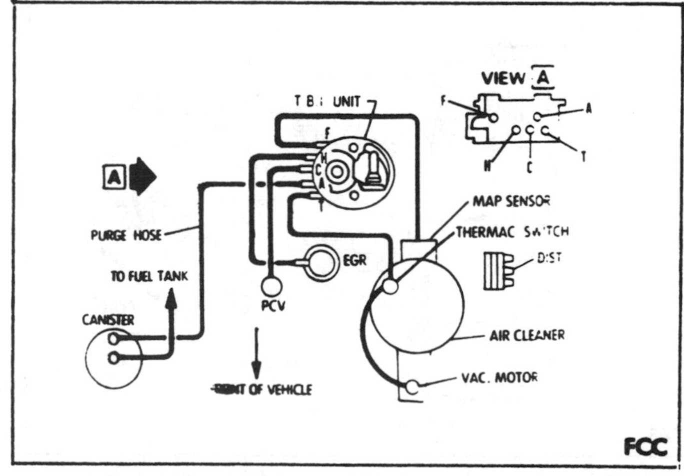

1984—4 CYL.-1.8L VIN O T.B.I., MAN. & AUTO. TRANS.—"J" BODY—FEDERAL

VACUUM CIRCUITS

1984

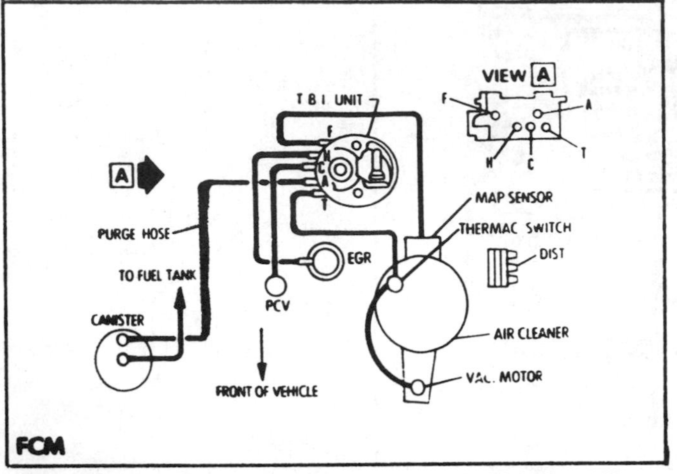

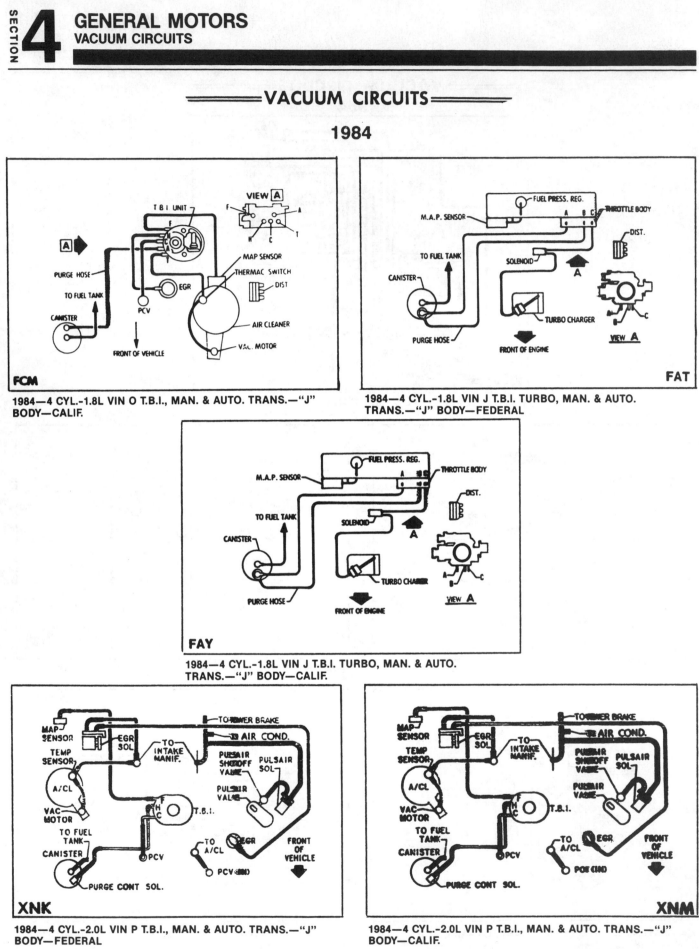

1984—4 CYL.-1.8L VIN O T.B.I., MAN. & AUTO. TRANS.—"J" BODY—CALIF.

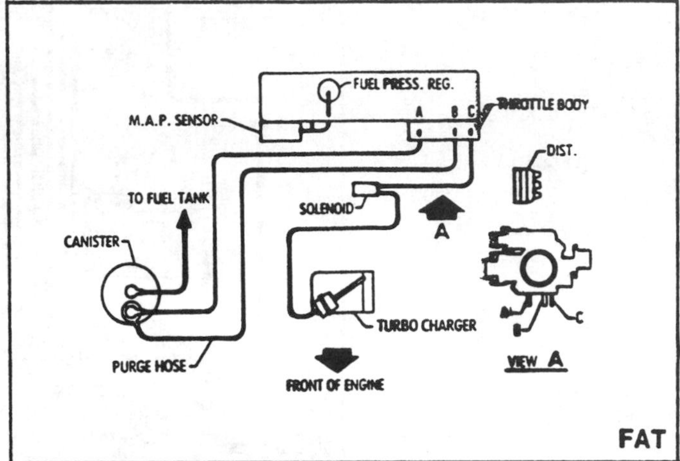

1984—4 CYL.-1.8L VIN J T.B.I. TURBO, MAN. & AUTO. TRANS.—"J" BODY—FEDERAL

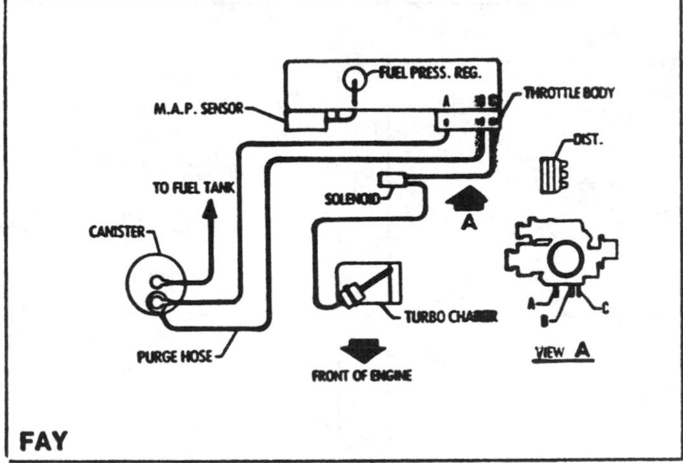

1984—4 CYL.-1.8L VIN J T.B.I. TURBO, MAN. & AUTO. TRANS.—"J" BODY—CALIF.

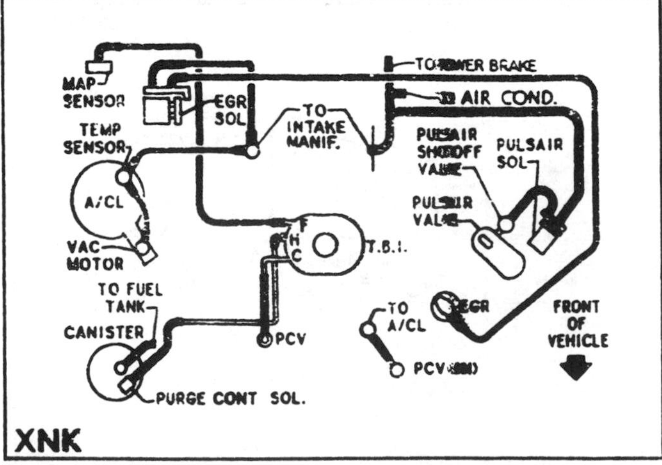

1984—4 CYL.-2.0L VIN P T.B.I., MAN. & AUTO. TRANS.—"J" BODY—FEDERAL

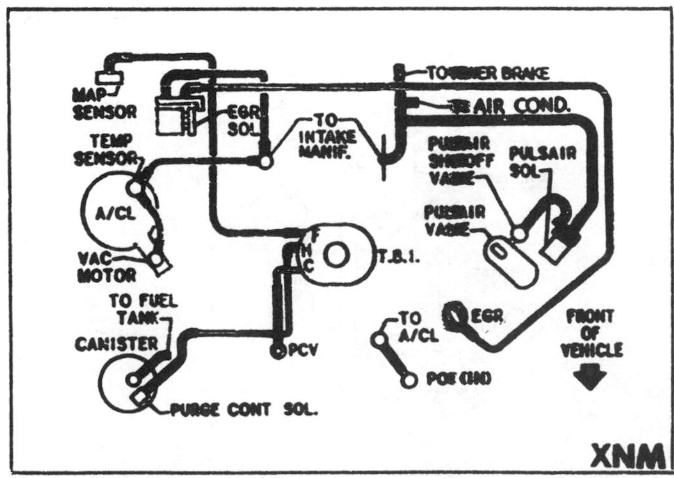

1984—4 CYL.-2.0L VIN P T.B.I., MAN. & AUTO. TRANS.—"J" BODY—CALIF.

VACUUM CIRCUITS

1984

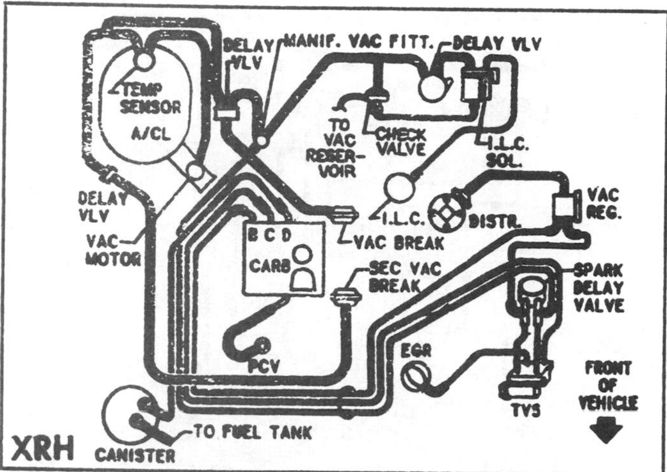

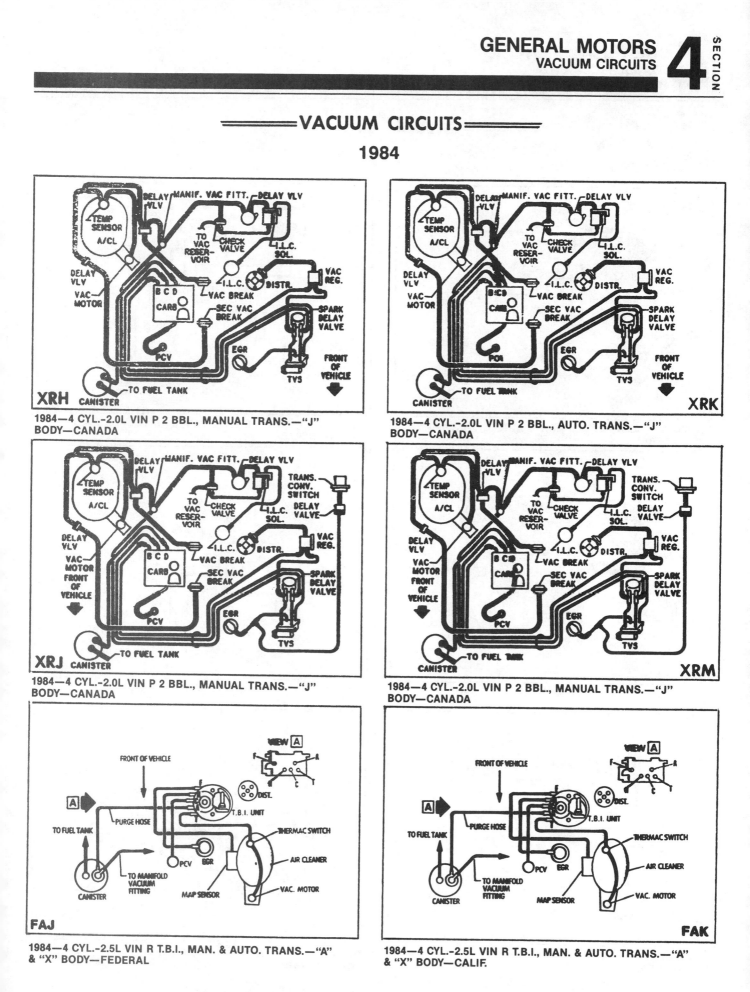

1984—4 CYL.-2.0L VIN P 2 BBL., MANUAL TRANS.—"J" BODY—CANADA

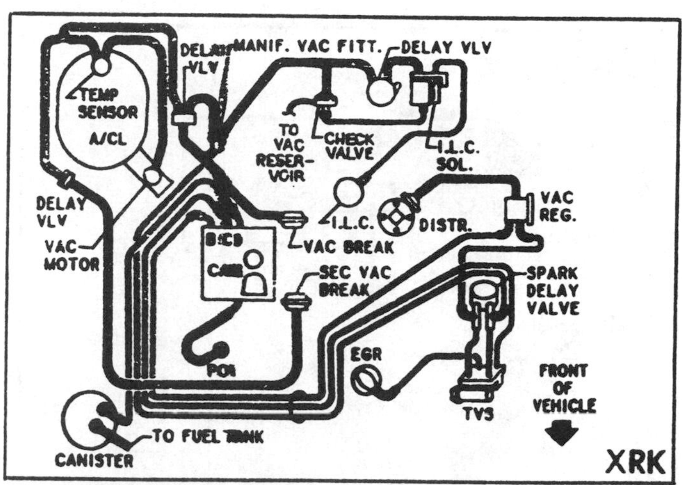

1984—4 CYL.-2.0L VIN P 2 BBL., AUTO. TRANS.—"J" BODY—CANADA

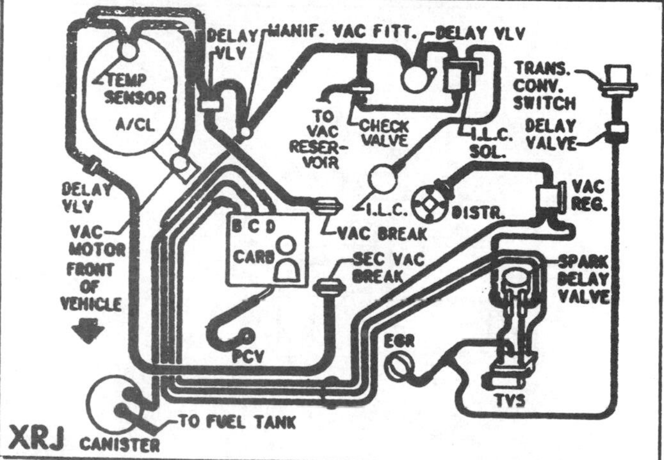

1984—4 CYL.-2.0L VIN P 2 BBL., MANUAL TRANS.—"J" BODY—CANADA

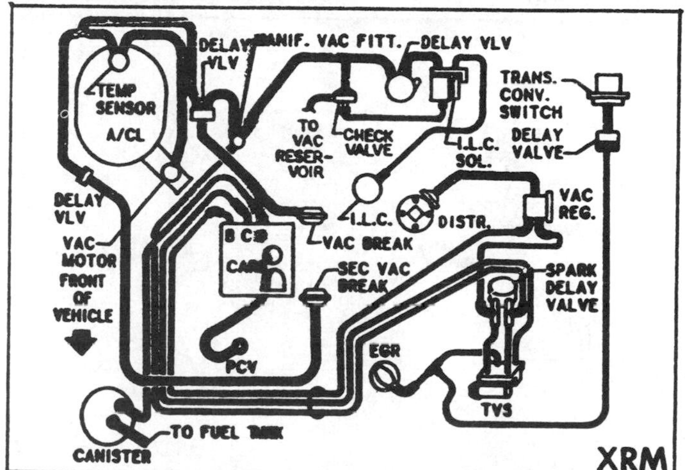

1984—4 CYL.-2.0L VIN P 2 BBL., MANUAL TRANS.—"J" BODY—CANADA

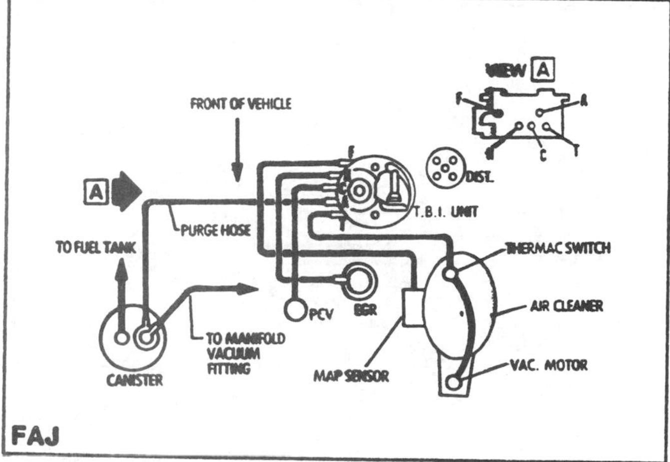

1984—4 CYL.-2.5L VIN R T.B.I., MAN. & AUTO. TRANS.—"A" & "X" BODY—FEDERAL

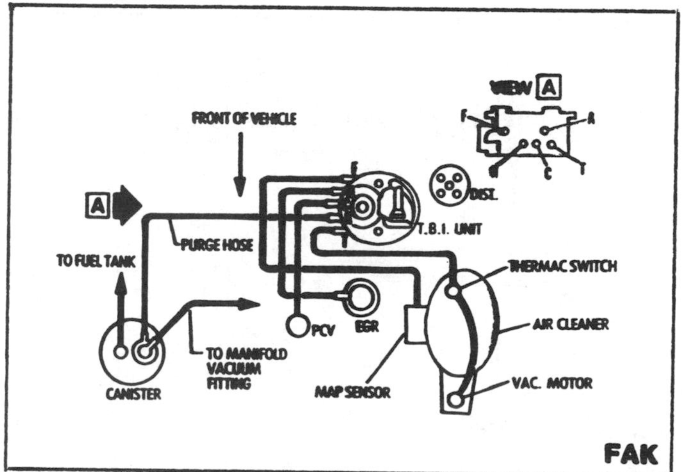

1984—4 CYL.-2.5L VIN R T.B.I., MAN. & AUTO. TRANS.—"A" & "X" BODY—CALIF.

VACUUM CIRCUITS

1984

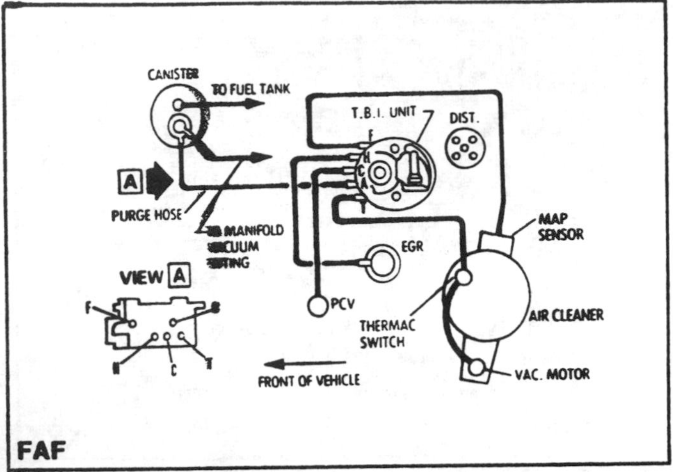

1984—4 CYL.-2.5L VIN 2 T.B.I., MAN. & AUTO. TRANS.—"F" BODY—FEDERAL

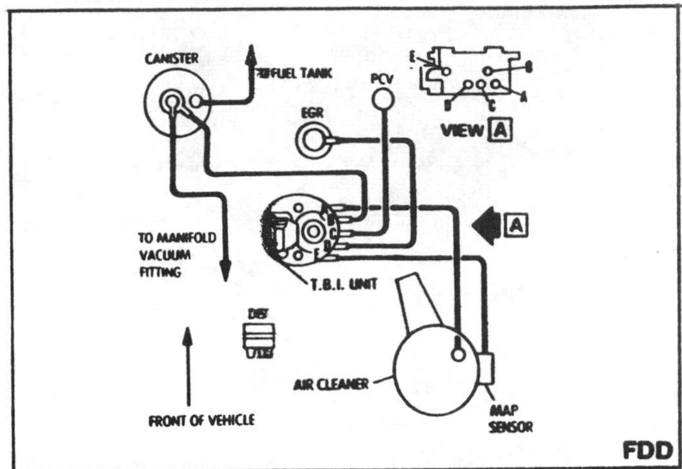

1984—4 CYL.-2.5L VIN R T.B.I., MAN. & AUTO. TRANS.—"P" BODY—FEDERAL

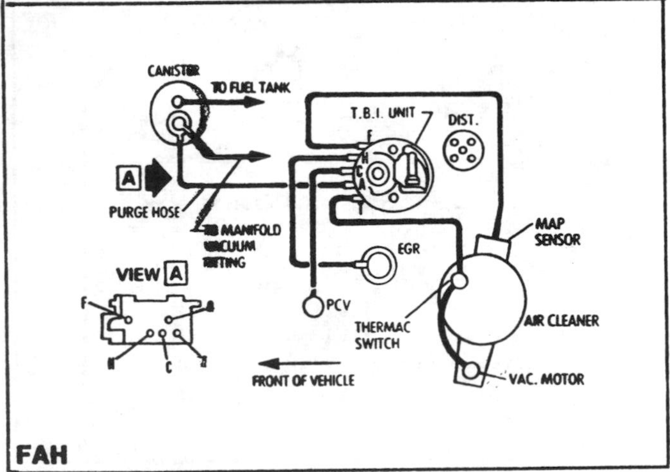

1984—4 CYL.-2.5L VIN 2 T.B.I., MAN. & AUTO. TRANS.—"F" BODY—CALIF.

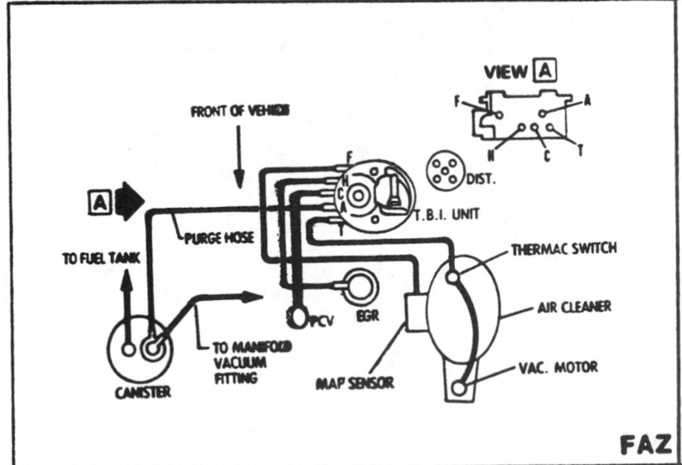

1984—4 CYL.-2.5L VIN R T.B.I., AUTO. TRANS.—"X" BODY—CANADA

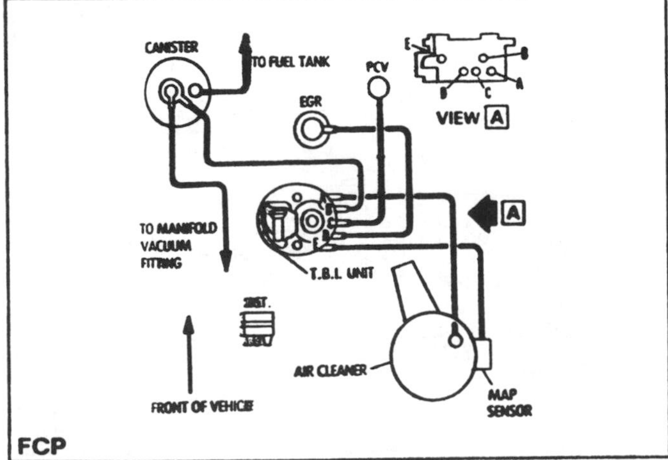

1984—4 CYL.-2.5L VIN R T.B.I., MAN. & AUTO. TRANS.—"P" BODY—FEDERAL

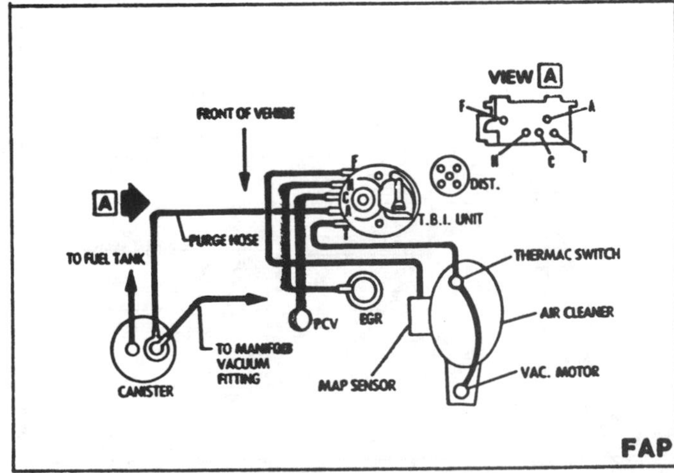

1984—4 CYL.-2.5L VIN R T.B.I., AUTO. TRANS.—"A" & "X" BODY—CANADA

VACUUM CIRCUITS

1984

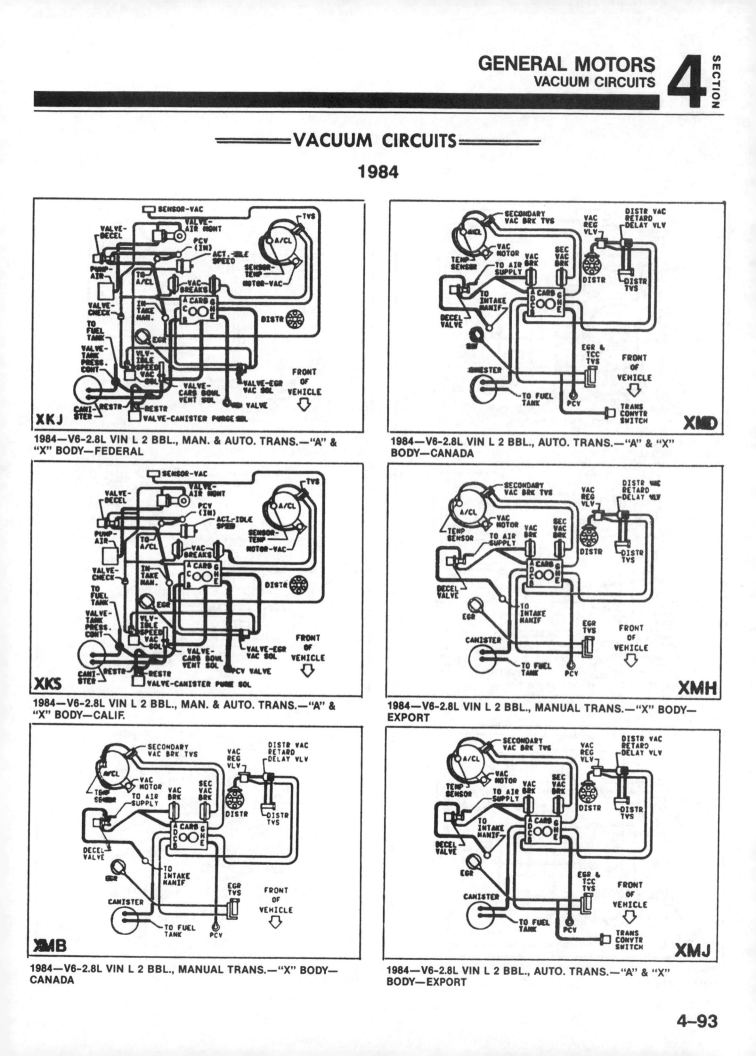

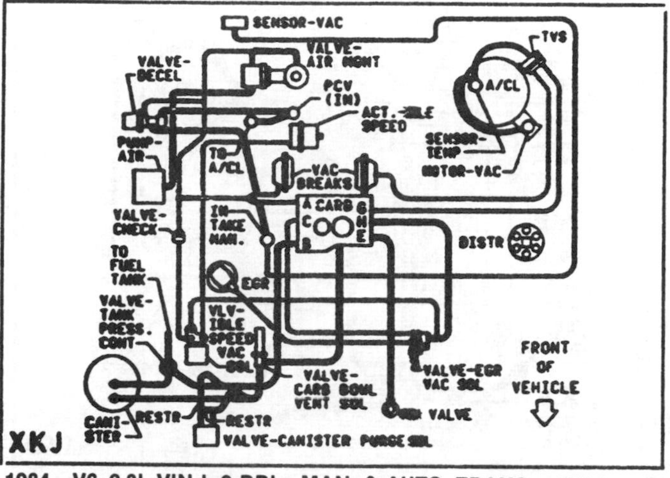

1984—V6-2.8L VIN L 2 BBL., MAN. & AUTO. TRANS.—"A" & "X" BODY—FEDERAL

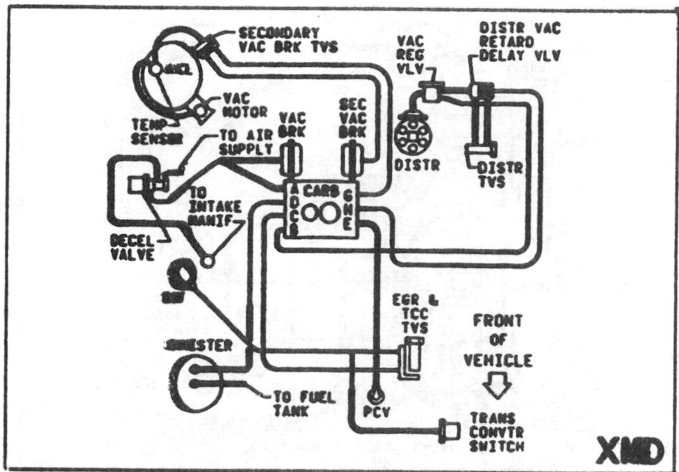

1984—V6-2.8L VIN L 2 BBL., AUTO. TRANS.—"A" & "X" BODY—CANADA

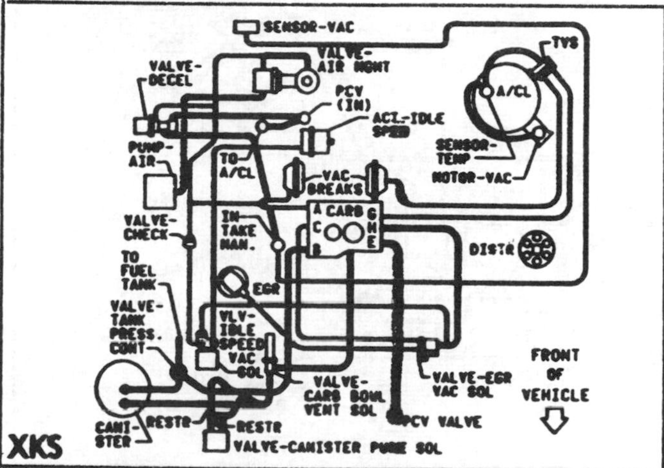

1984—V6-2.8L VIN L 2 BBL., MAN. & AUTO. TRANS.—"A" & "X" BODY—CALIF.

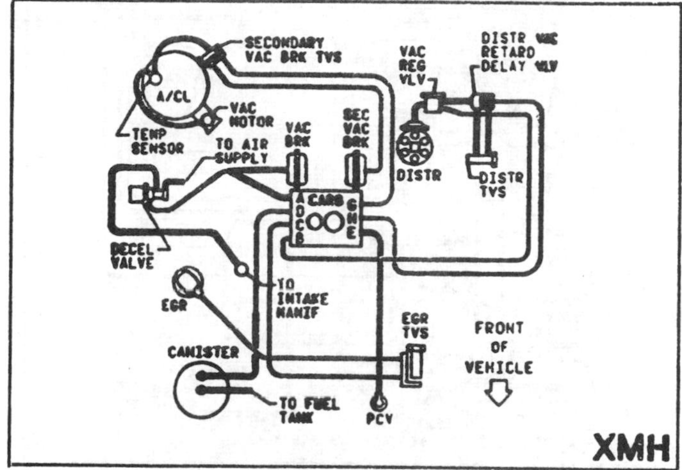

1984—V6-2.8L VIN L 2 BBL., MANUAL TRANS.—"X" BODY—EXPORT

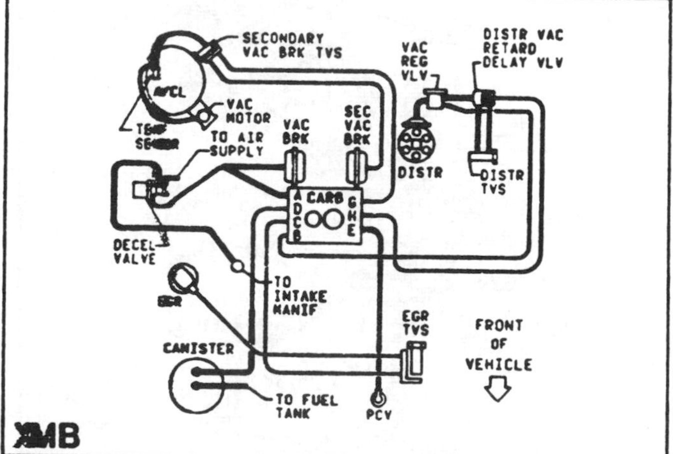

1984—V6-2.8L VIN L 2 BBL., MANUAL TRANS.—"X" BODY—CANADA

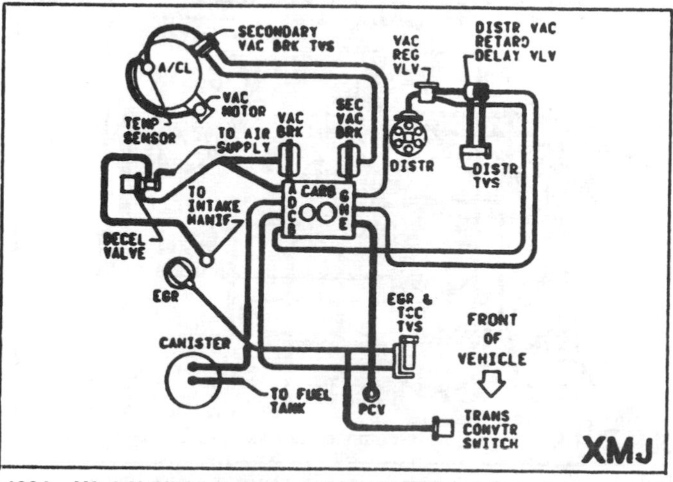

1984—V6-2.8L VIN L 2 BBL., AUTO. TRANS.—"A" & "X" BODY—EXPORT

VACUUM CIRCUITS

1984

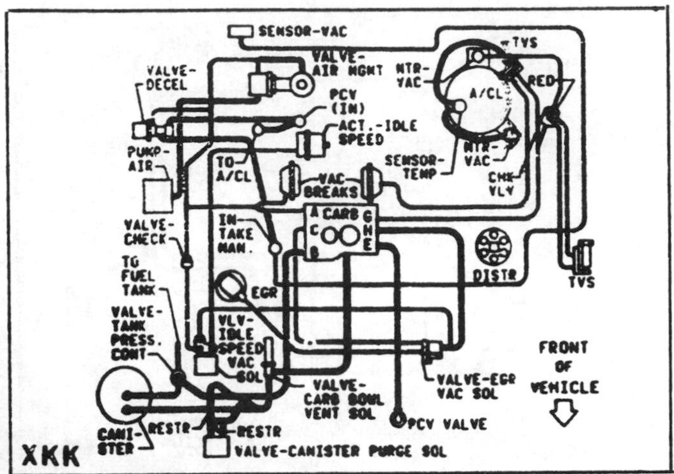

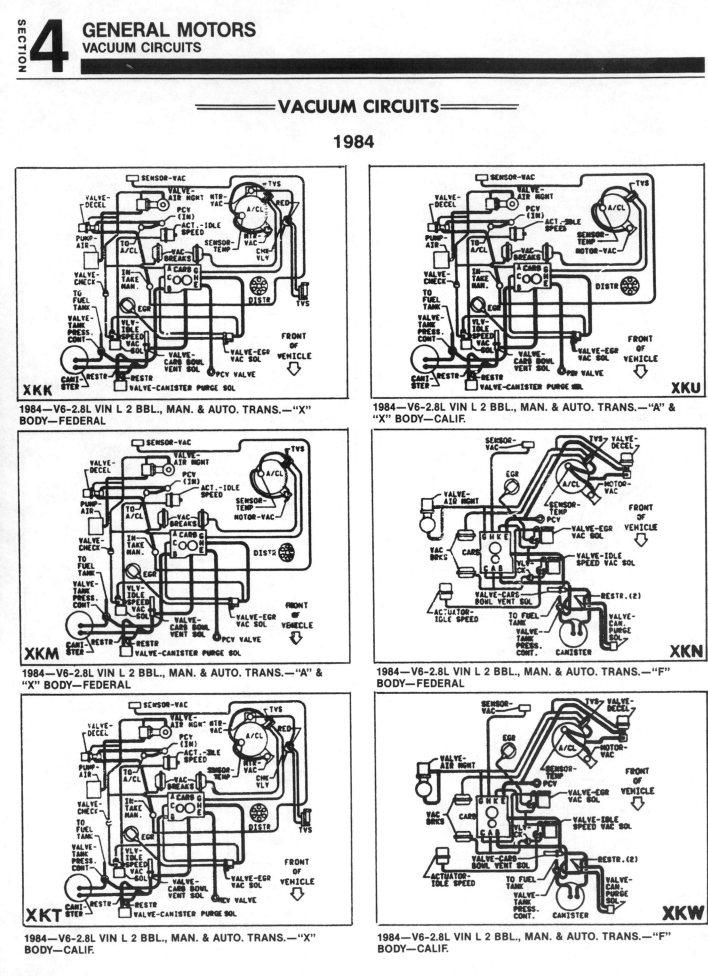

1984—V6-2.8L VIN L 2 BBL., MAN. & AUTO. TRANS.—"X" BODY—FEDERAL

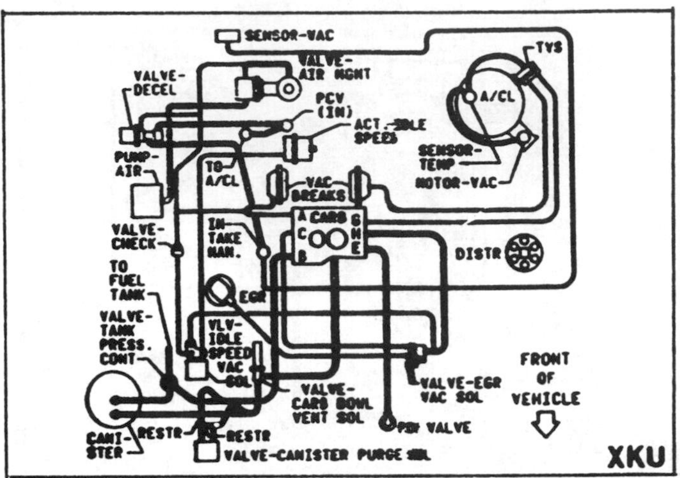

1984—V6-2.8L VIN L 2 BBL., MAN. & AUTO. TRANS.—"A" & "X" BODY—CALIF.

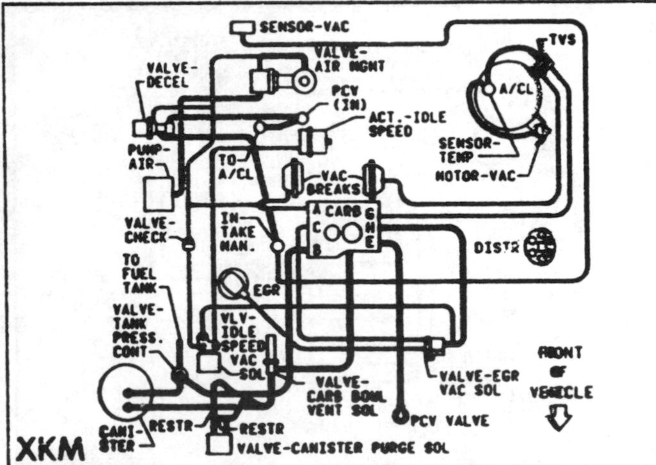

1984—V6-2.8L VIN L 2 BBL., MAN. & AUTO. TRANS.—"A" & "X" BODY—FEDERAL

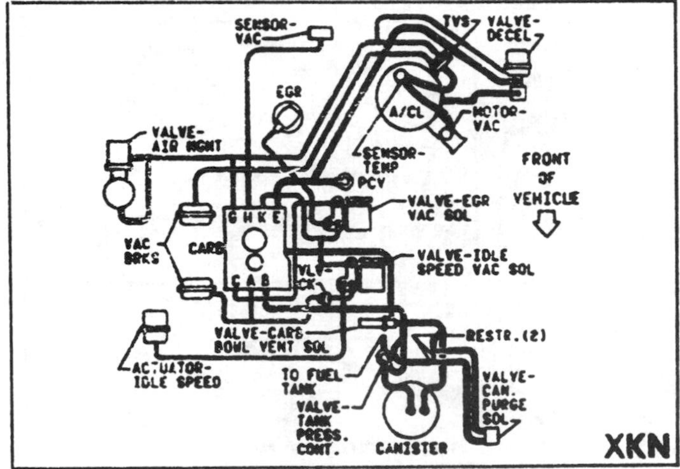

1984—V6-2.8L VIN L 2 BBL., MAN. & AUTO. TRANS.—"F" BODY—FEDERAL

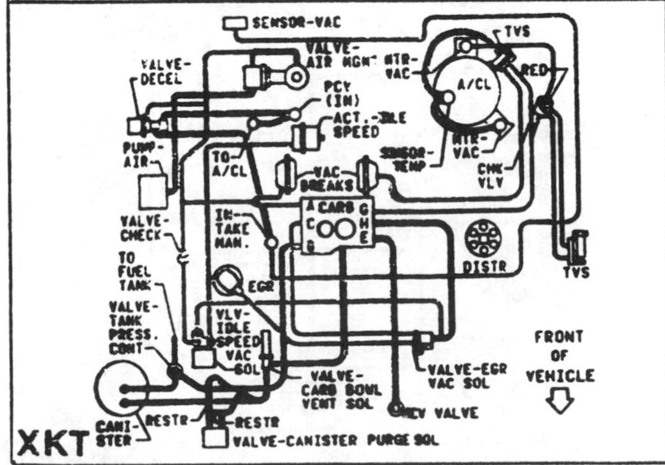

1984—V6-2.8L VIN L 2 BBL., MAN. & AUTO. TRANS.—"X" BODY—CALIF.

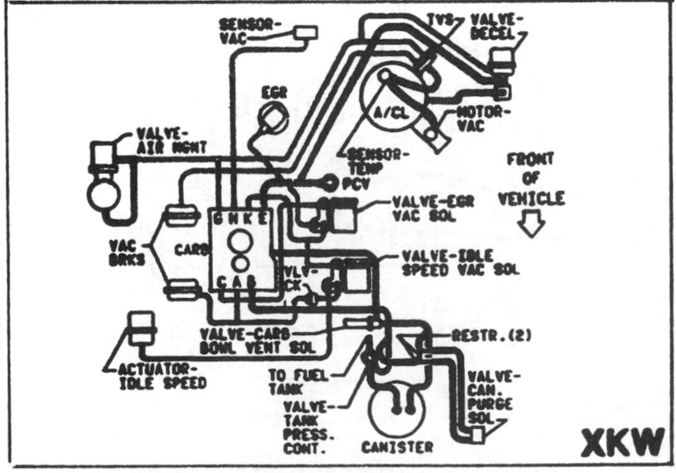

1984—V6-2.8L VIN L 2 BBL., MAN. & AUTO. TRANS.—"F" BODY—CALIF.

VACUUM CIRCUITS

1984

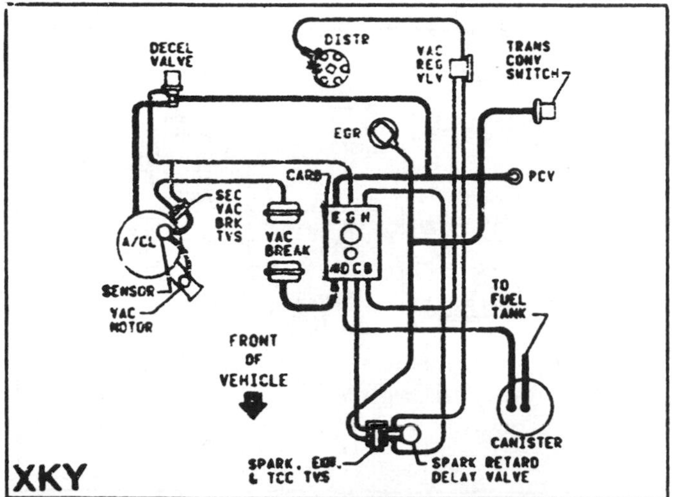

1984—V6-2.8L VIN L 2 BBL., AUTO. TRANS.—"F" BODY—CANADA

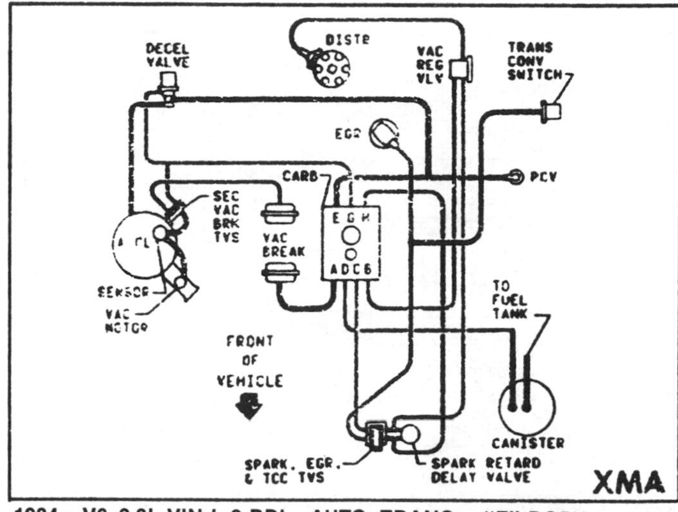

1984—V6-2.8L VIN L 2 BBL., AUTO. TRANS.—"F" BODY—EXPORT

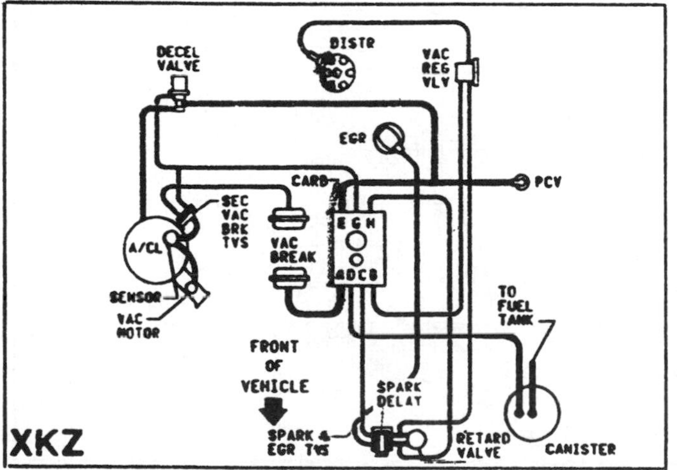

1984—V6-2.8L VIN L 2 BBL., MANUAL TRANS.—"F" BODY—CANADA

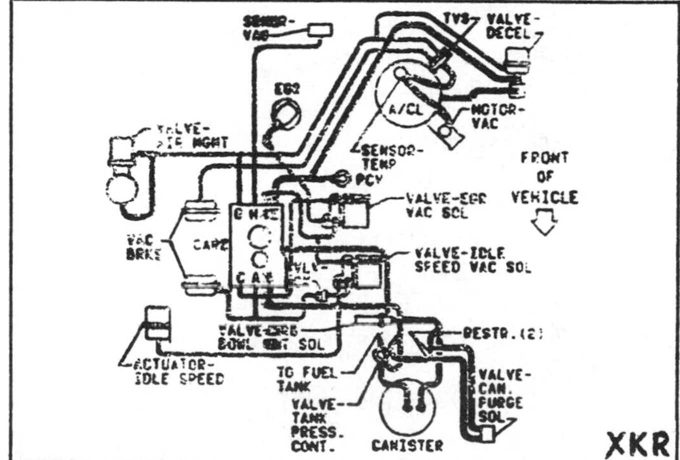

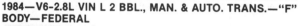

1984—V6-2.8L VIN L 2 BBL., MAN. & AUTO. TRANS.—"F" BODY—FEDERAL

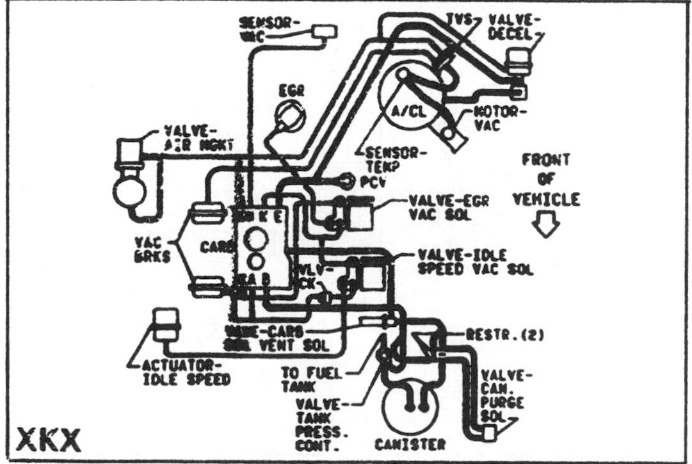

1984—V6-2.8L VIN L 2 BBL., AUTO. TRANS.—"F" BODY—CALIF.

VACUUM CIRCUITS

1984

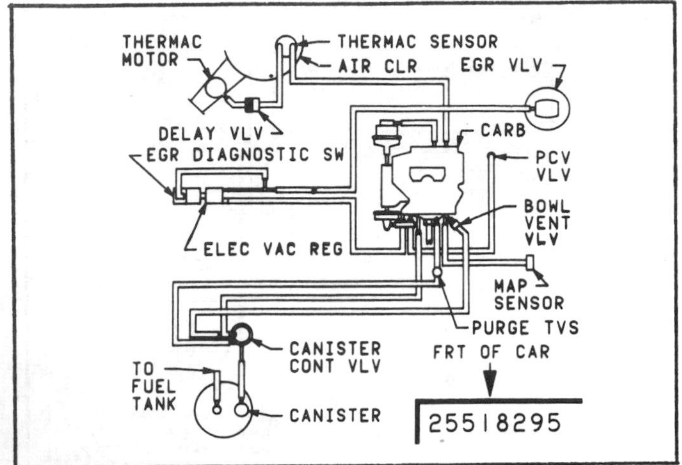

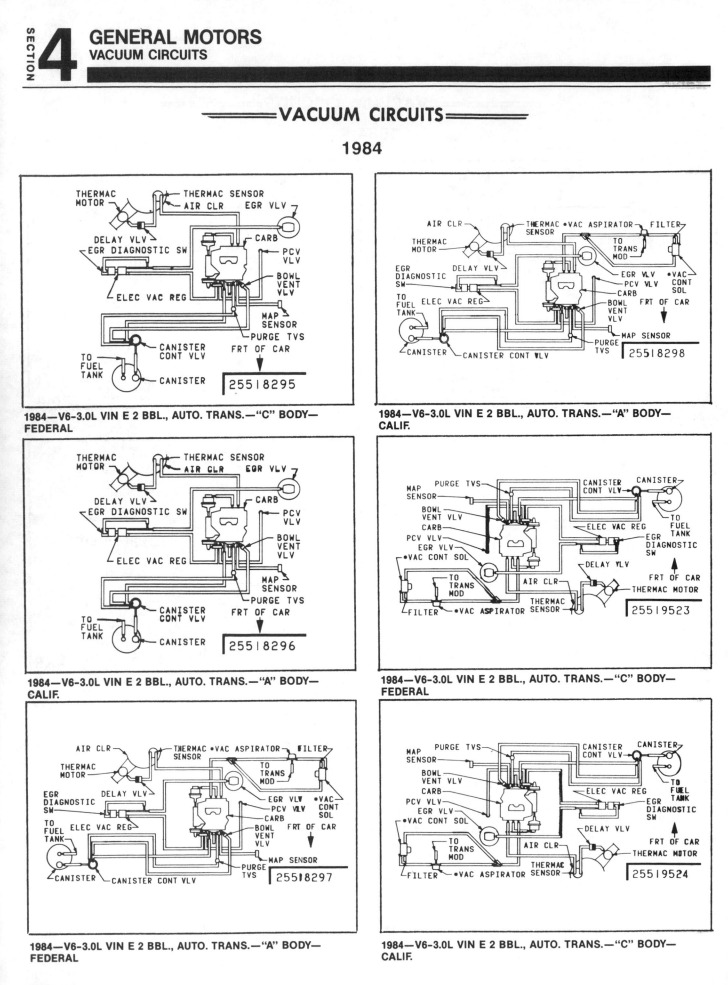

1984—V6-3.0L VIN E 2 BBL., AUTO. TRANS.—"C" BODY—FEDERAL

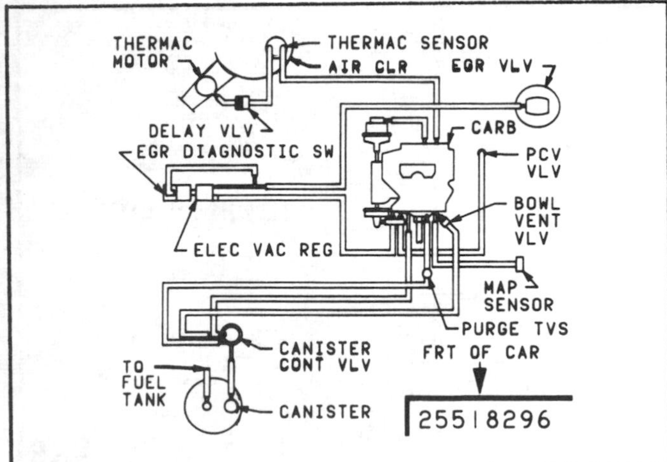

1984—V6-3.0L VIN E 2 BBL., AUTO. TRANS.—"A" BODY—CALIF.

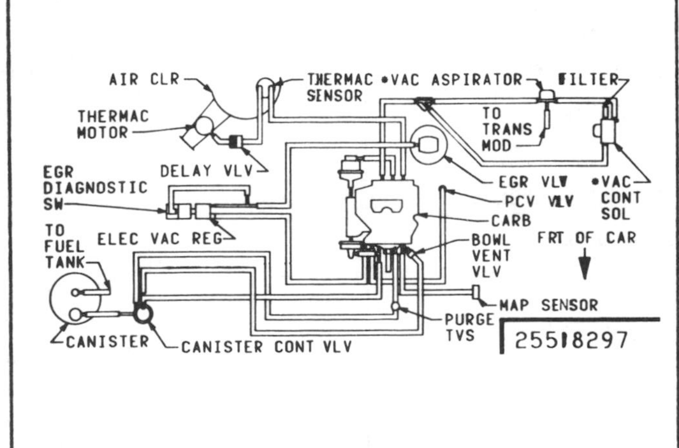

1984—V6-3.0L VIN E 2 BBL., AUTO. TRANS.—"A" BODY—FEDERAL

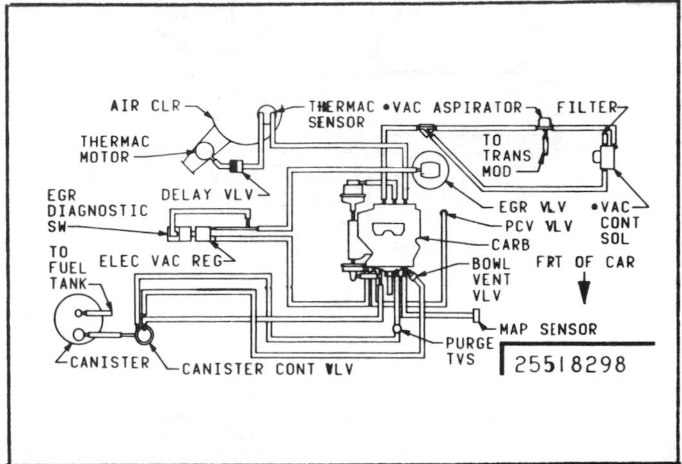

1984—V6-3.0L VIN E 2 BBL., AUTO. TRANS.—"A" BODY—CALIF.

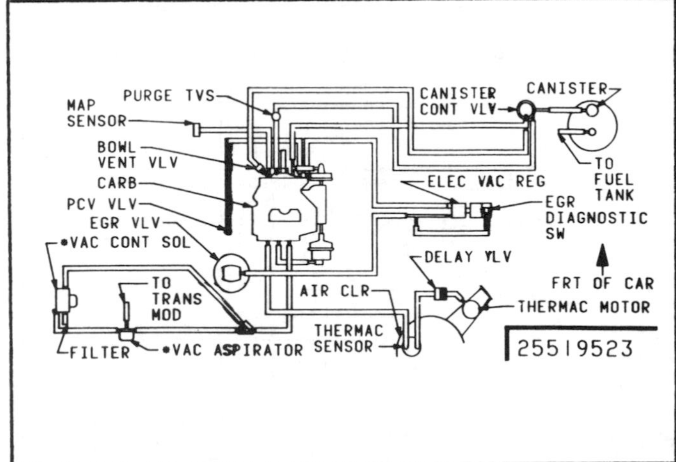

1984—V6-3.0L VIN E 2 BBL., AUTO. TRANS.—"C" BODY—FEDERAL

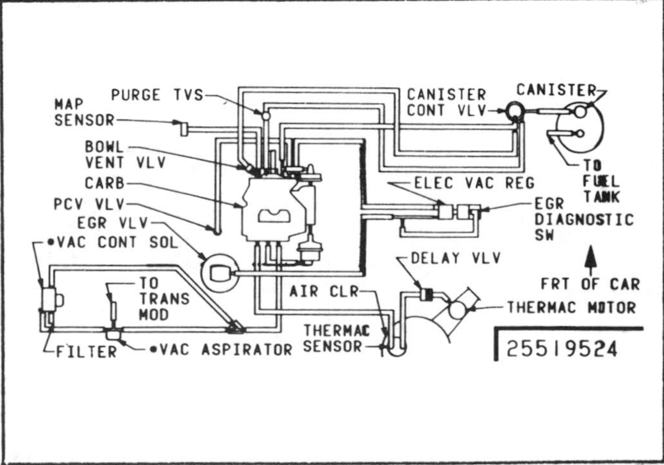

1984—V6-3.0L VIN E 2 BBL., AUTO. TRANS.—"C" BODY—CALIF.

VACUUM CIRCUITS

1984

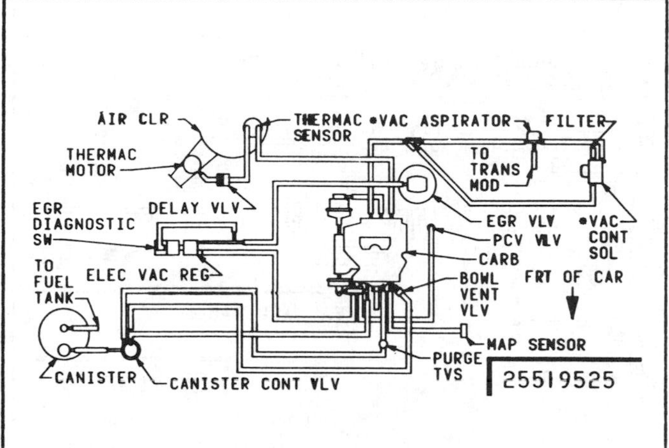

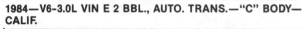

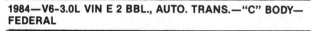

1984—V6-3.0L VIN E 2 BBL., AUTO. TRANS.—"C" BODY—FEDERAL

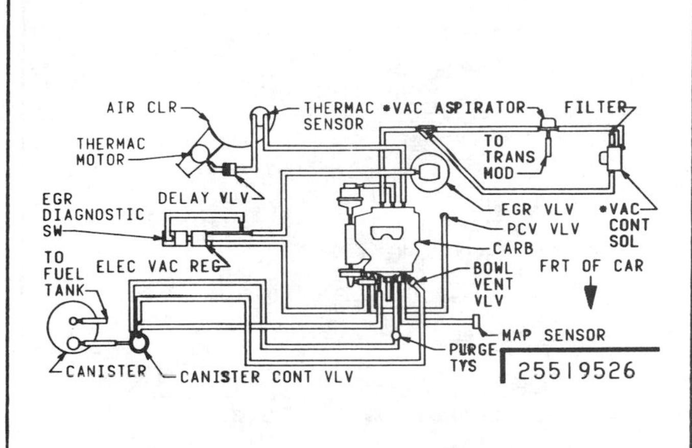

1984—V6-3.0L VIN E 2 BBL., AUTO. TRANS.—"C" BODY—CALIF.

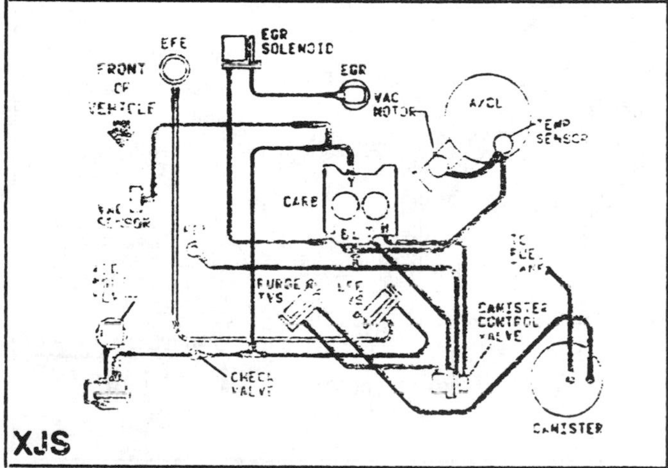

1984—V6-3.8L VIN A 2 BBL., AUTO. TRANS.—"B" & "G" BODY—FEDERAL

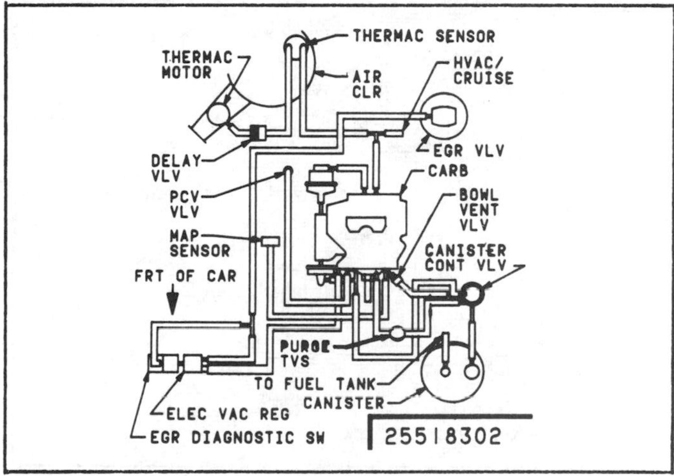

1984—V6-3.8L VIN A 2 BBL., AUTO. TRANS.—"B" BODY—CALIF.

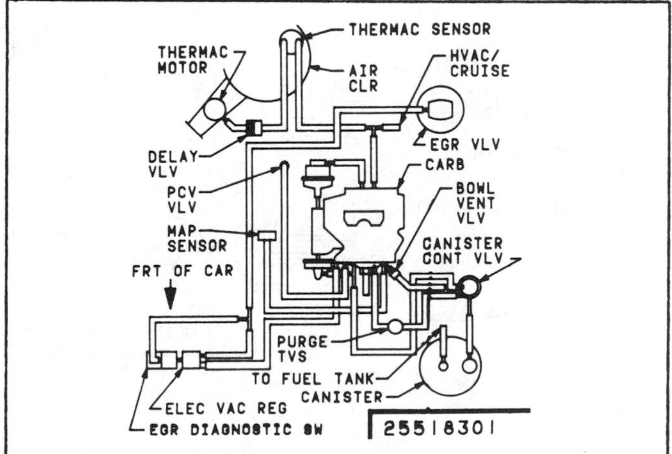

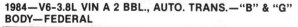

1984—V6-3.8L VIN A 2 BBL., AUTO. TRANS.—"B" BODY—FEDERAL

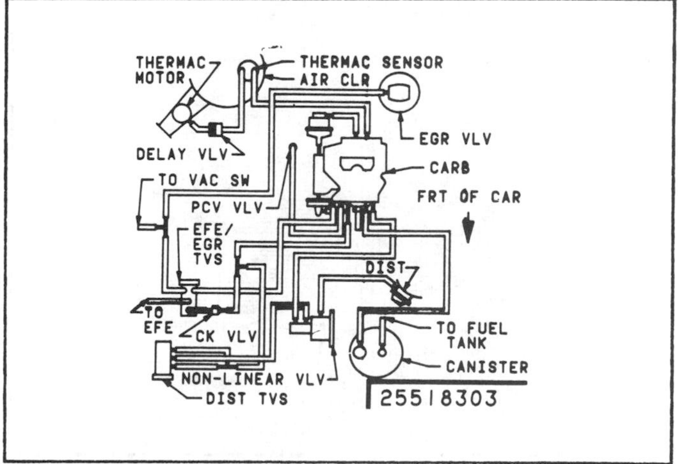

1984—V6-3.8L VIN A 2 BBL., AUTO. TRANS.—"G" BODY—FEDERAL

VACUUM CIRCUITS

1984

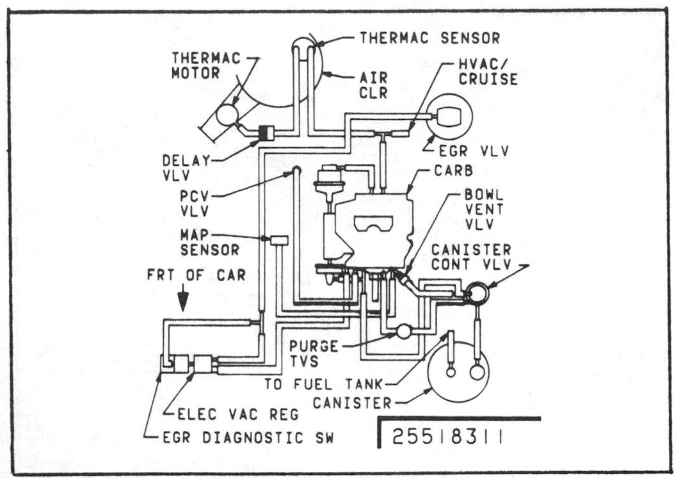

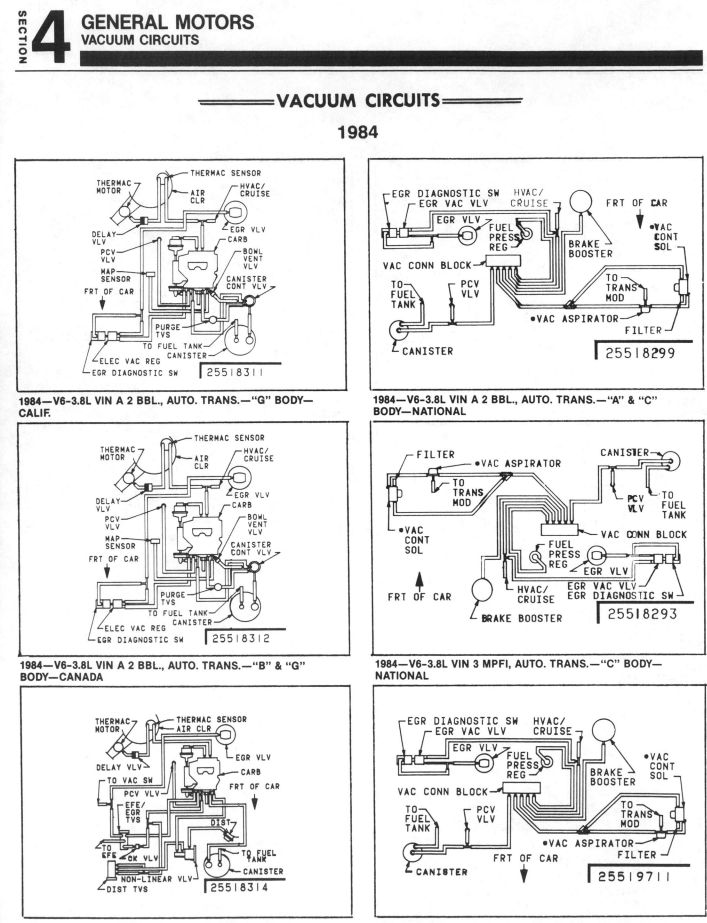

1984—V6-3.8L VIN A 2 BBL., AUTO. TRANS.—"G" BODY—CALIF.

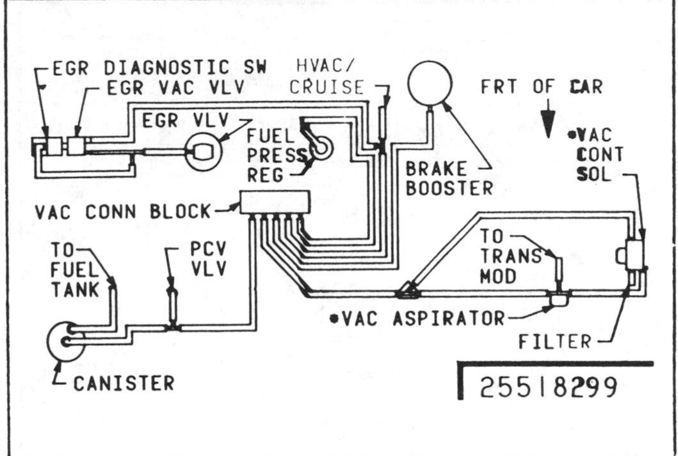

1984—V6-3.8L VIN A 2 BBL., AUTO. TRANS.—"A" & "C" BODY—NATIONAL

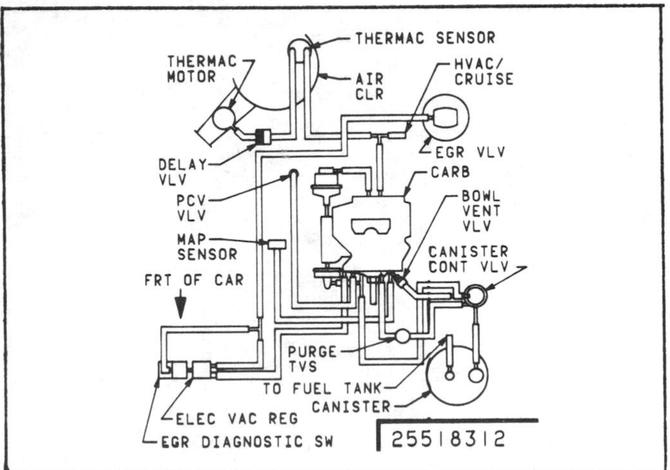

1984—V6-3.8L VIN A 2 BBL., AUTO. TRANS.—"B" & "G" BODY—CANADA

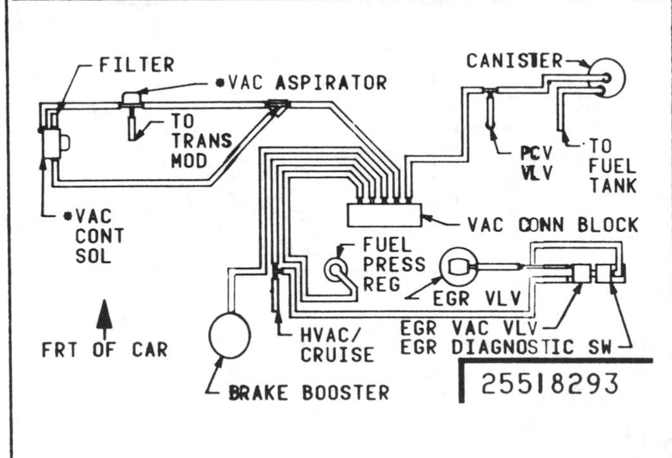

1984—V6-3.8L VIN 3 MPFI, AUTO. TRANS.—"C" BODY—NATIONAL

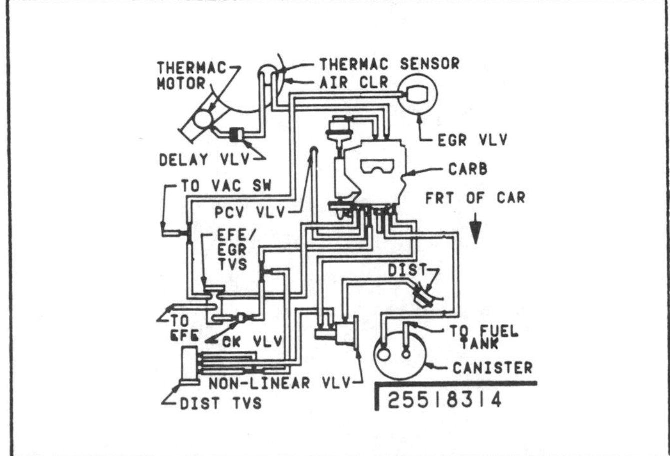

1984—V6-3.8L VIN A 2 BBL., AUTO. TRANS.—"B" & "G" BODY—EXPORT

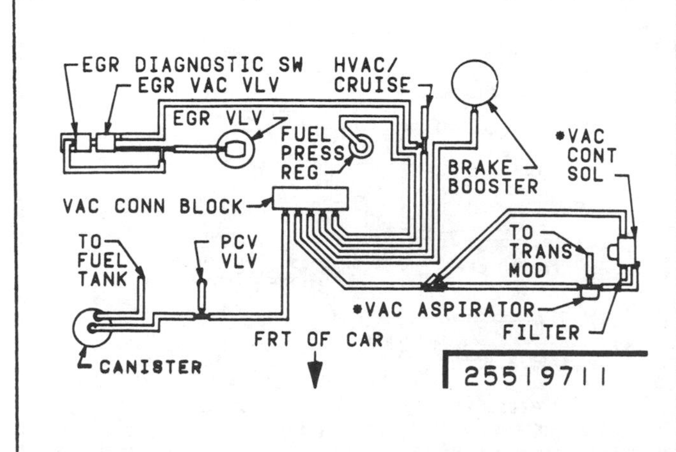

1984—V6-3.8L VIN 3 MPFI, AUTO. TRANS.—"A" & "C" BODY—EXPORT

VACUUM CIRCUITS

1984

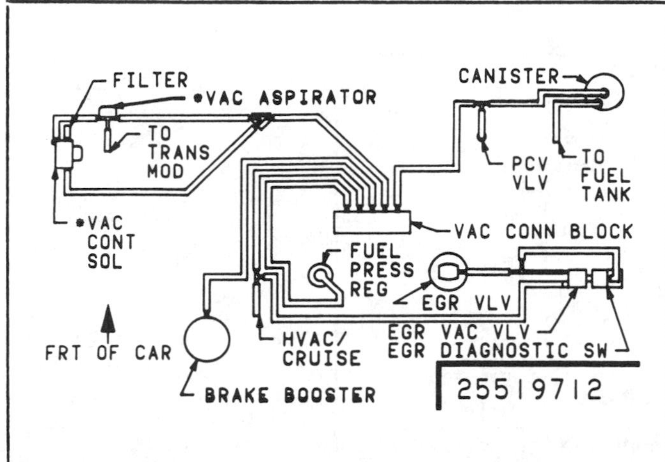

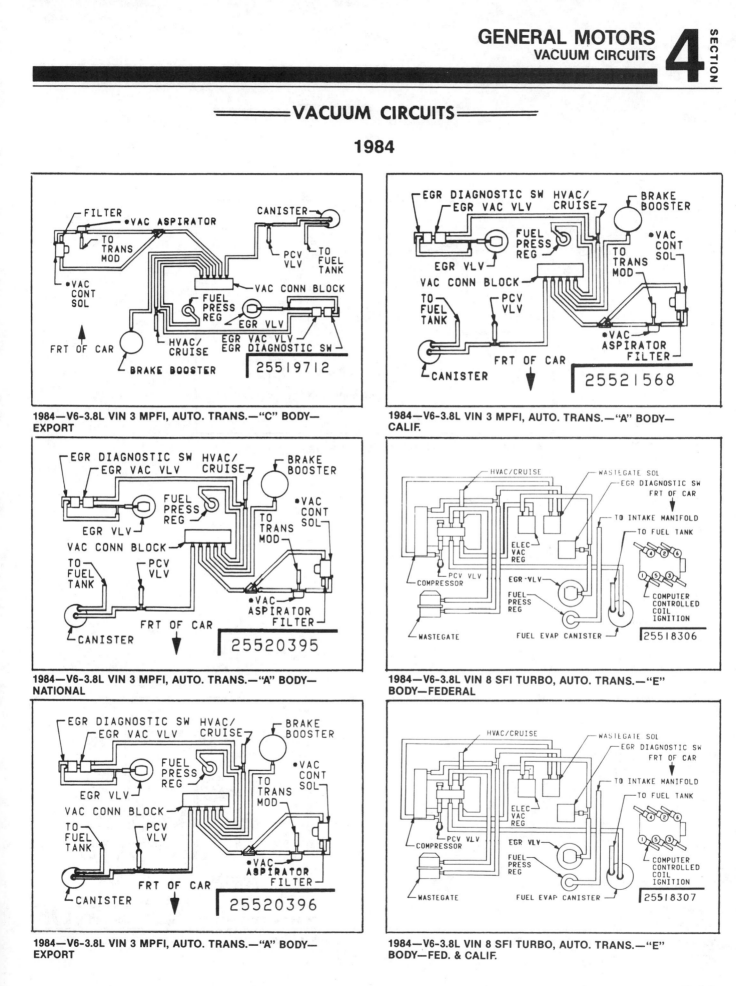

1984—V6-3.8L VIN 3 MPFI, AUTO. TRANS.—"C" BODY—EXPORT

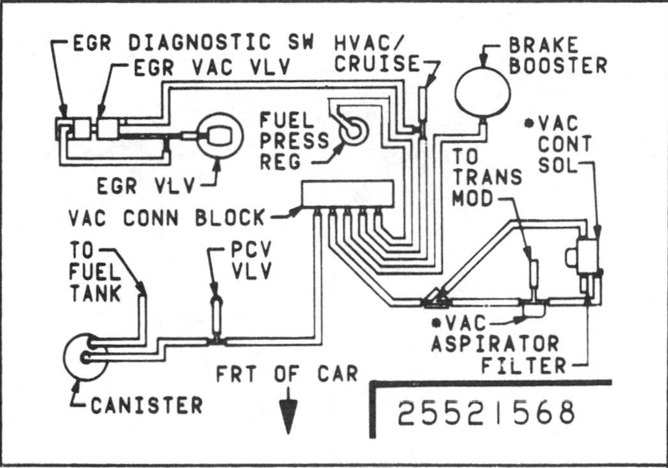

1984—V6-3.8L VIN 3 MPFI, AUTO. TRANS.—"A" BODY—CALIF.

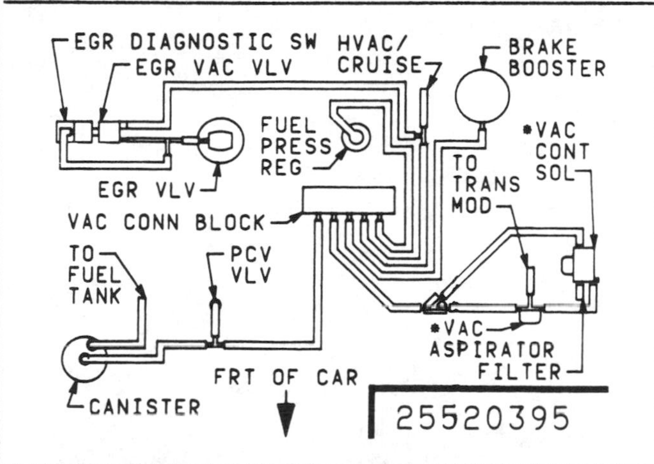

1984—V6-3.8L VIN 3 MPFI, AUTO. TRANS.—"A" BODY—NATIONAL

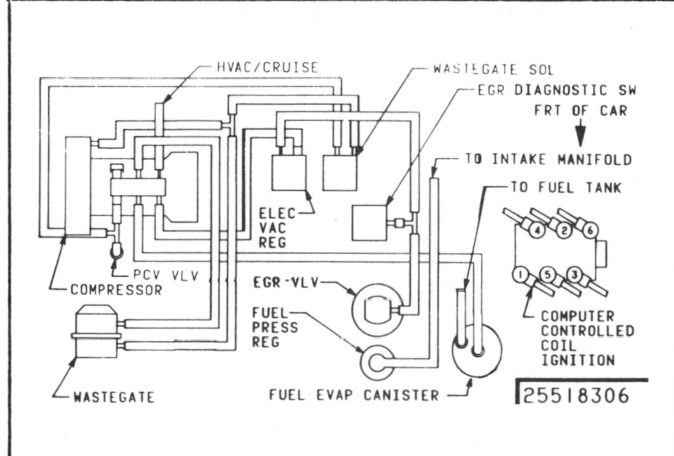

1984—V6-3.8L VIN 8 SFI TURBO, AUTO. TRANS.—"E" BODY—FEDERAL

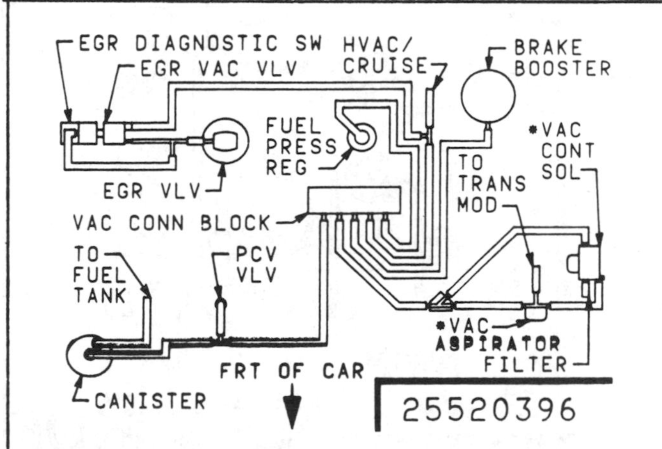

1984—V6-3.8L VIN 3 MPFI, AUTO. TRANS.—"A" BODY—EXPORT

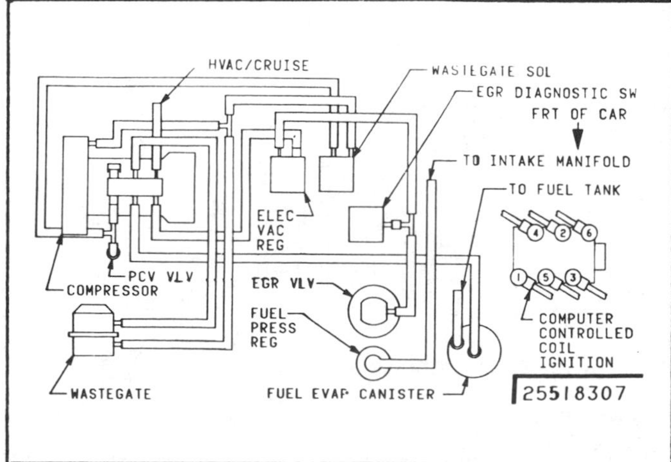

1984—V6-3.8L VIN 8 SFI TURBO, AUTO. TRANS.—"E" BODY—FED. & CALIF.

VACUUM CIRCUITS

1984

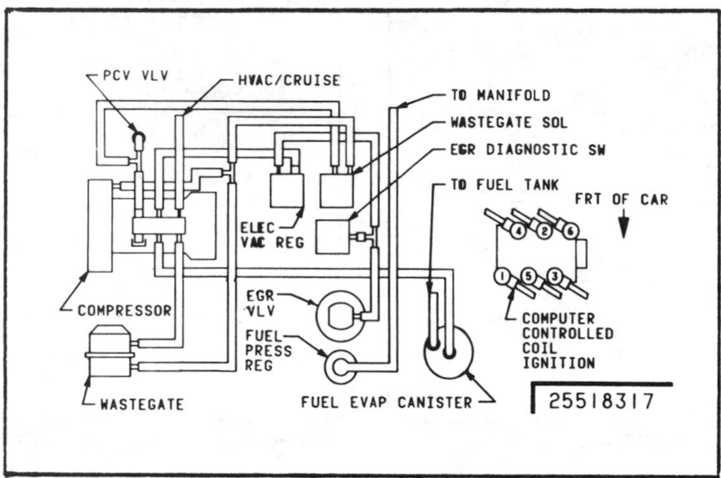

1984—V6-3.8L VIN 8 SFI TURBO, AUTO. TRANS.—"G" BODY—NATIONAL

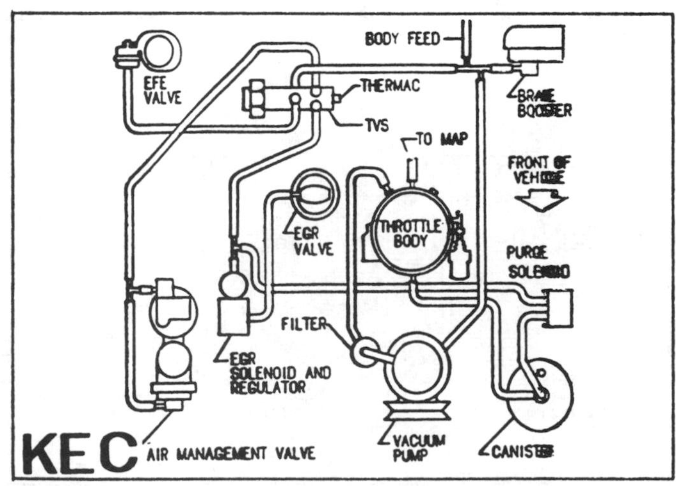

1984—V6-4.1L VIN 8 DFI, AUTO. TRANS.—CADILLAC—CALIF.

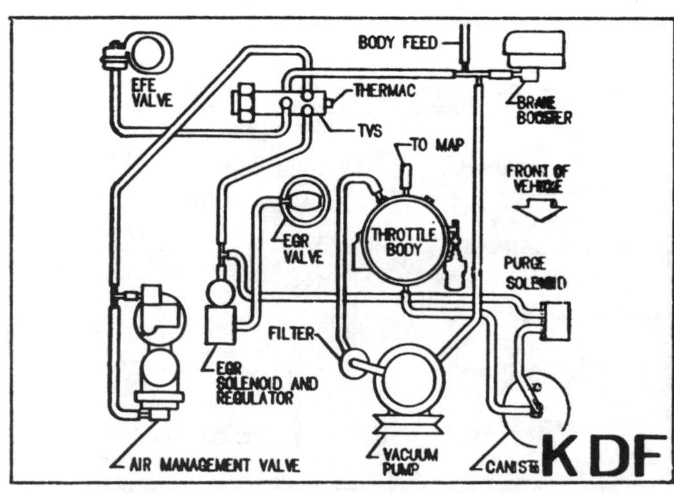

1984—V6-4.1L VIN 8 DFI, AUTO. TRANS.—CADILLAC—FED. & HIGH ALT.

VACUUM CIRCUITS

1984

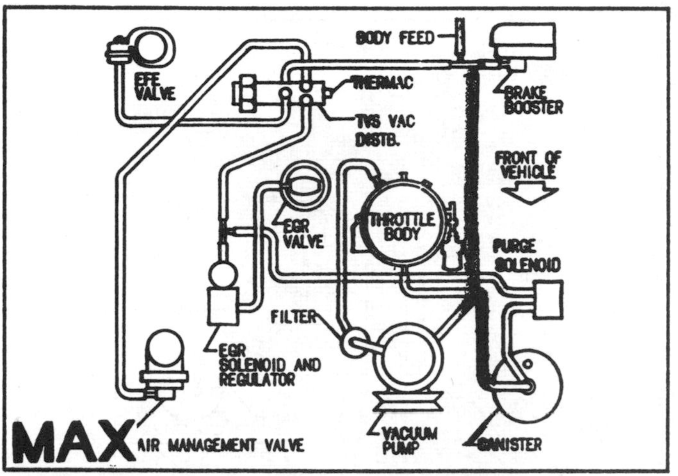

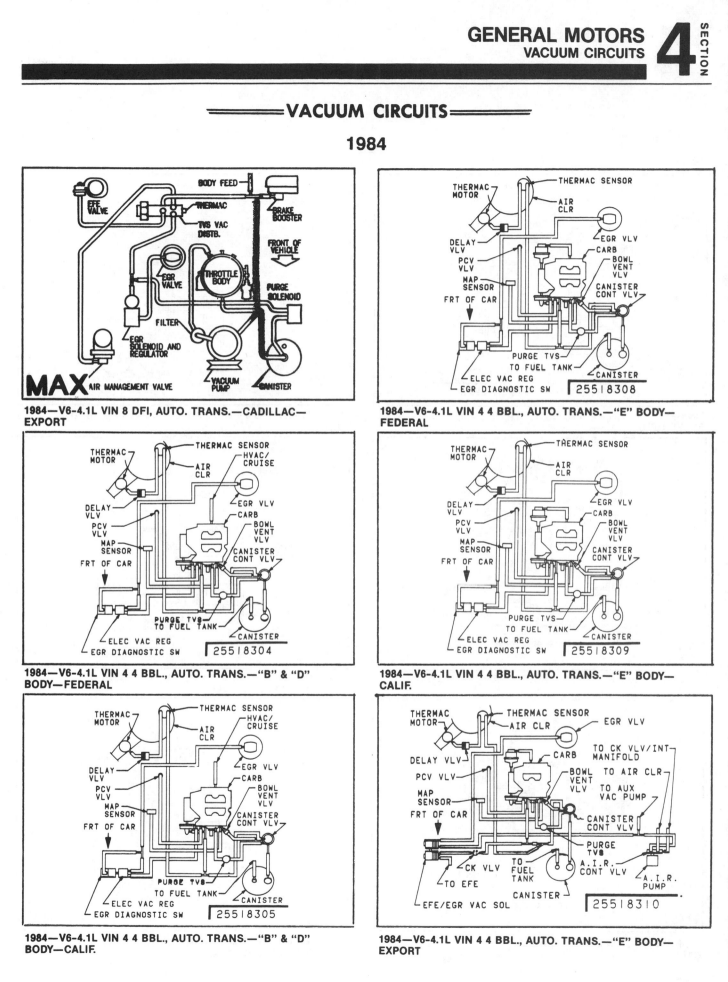

1984—V6-4.1L VIN 8 DFI, AUTO. TRANS.—CADILLAC—EXPORT

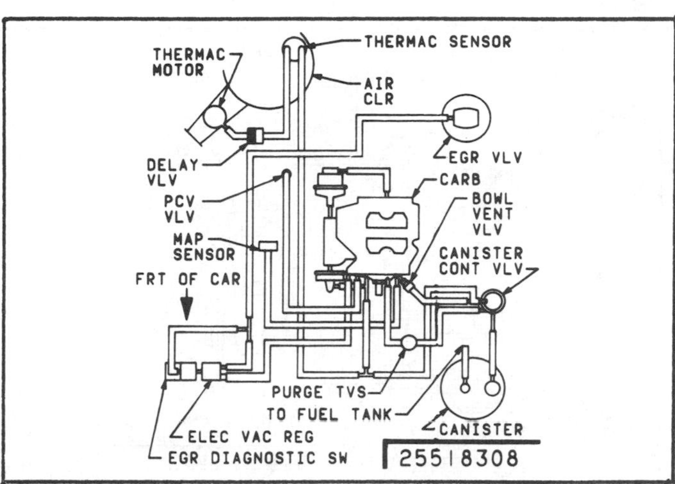

1984—V6-4.1L VIN 4 4 BBL., AUTO. TRANS.—"E" BODY—FEDERAL

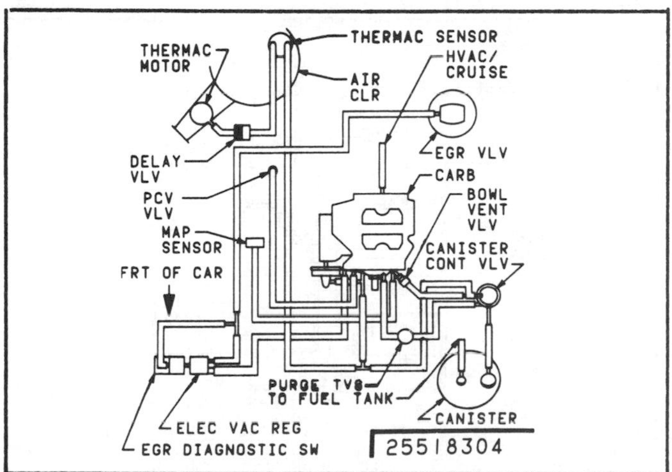

1984—V6-4.1L VIN 4 4 BBL., AUTO. TRANS.—"B" & "D" BODY—FEDERAL

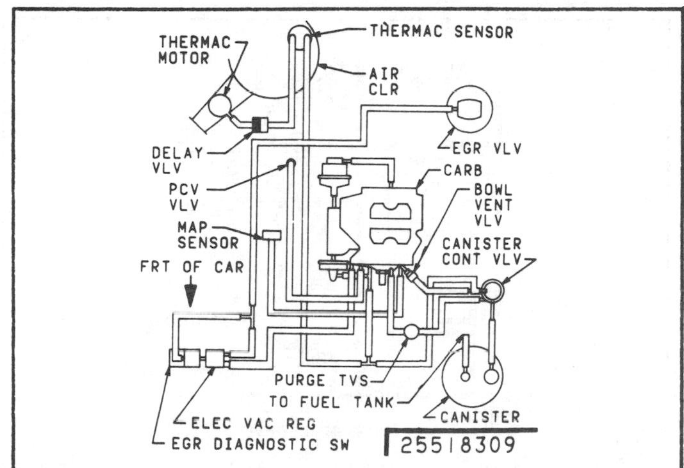

1984—V6-4.1L VIN 4 4 BBL., AUTO. TRANS.—"E" BODY—CALIF.

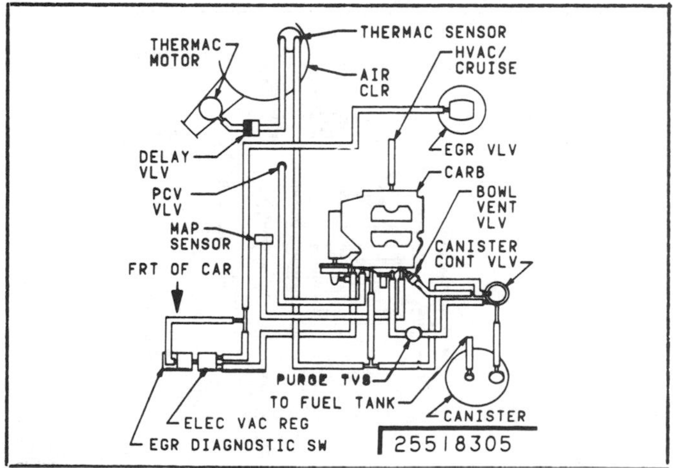

1984—V6-4.1L VIN 4 4 BBL., AUTO. TRANS.—"B" & "D" BODY—CALIF.

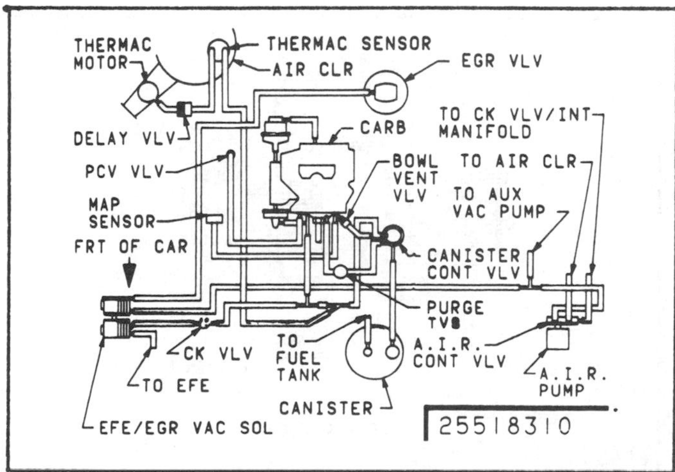

1984—V6-4.1L VIN 4 4 BBL., AUTO. TRANS.—"E" BODY—EXPORT

VACUUM CIRCUITS

1984

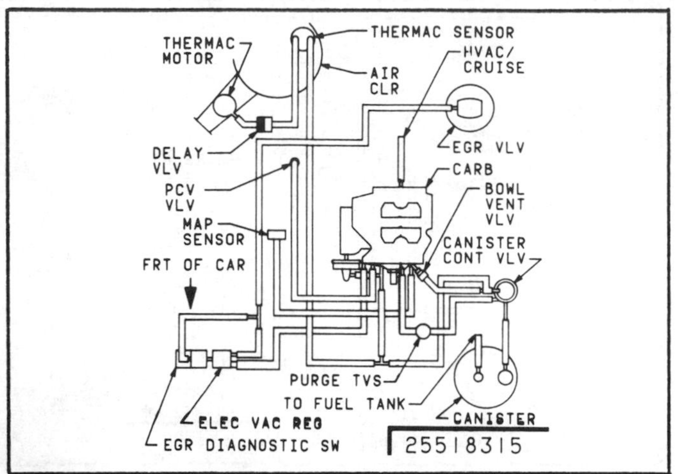

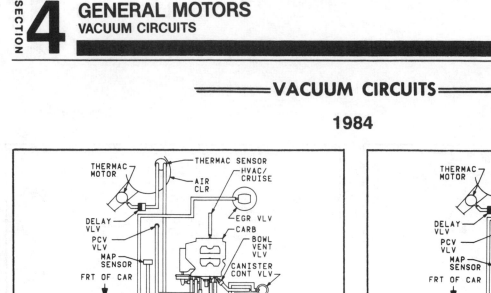

1984—V6-4.1L VIN 4 4 BBL., AUTO. TRANS.—"G" BODY—FEDERAL

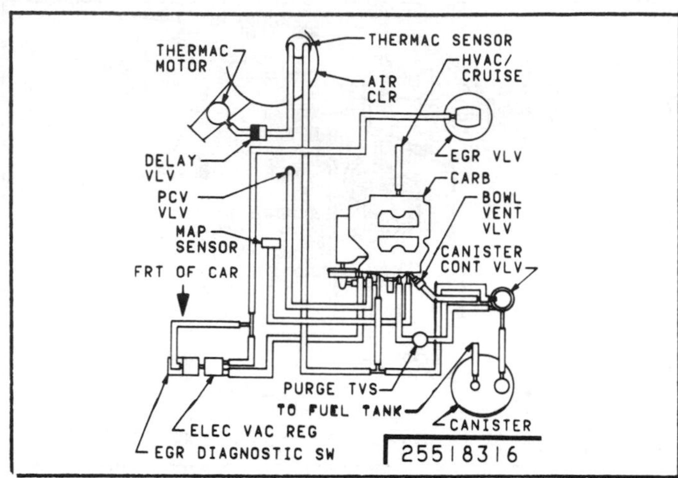

1984—V6-4.1L VIN 4 4 BBL., AUTO. TRANS.—"G" BODY—CALIF.

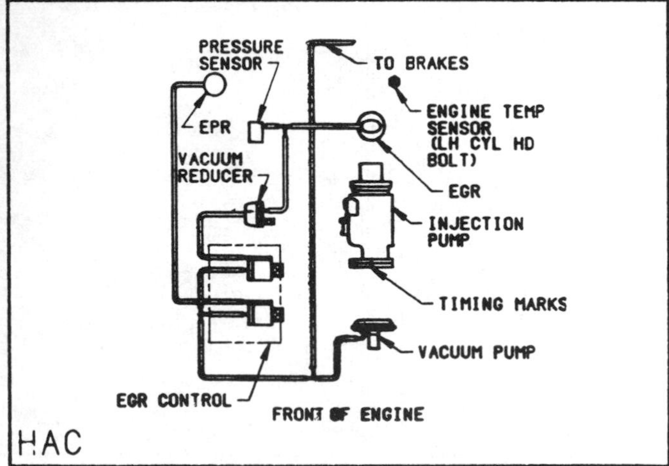

1984—V6-4.3L VIN T DIESEL, AUTO. TRANS.—"A" BODY—CALIF.

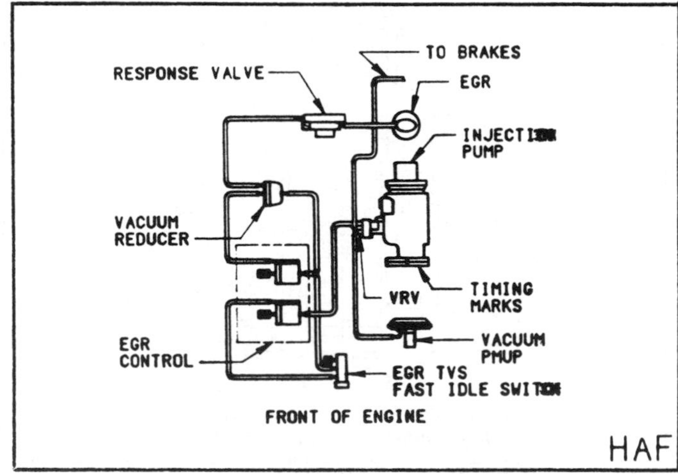

1984—V6-4.3L VIN T & V DIESEL, AUTO. TRANS.—"A" & "G" BODY—FED. LOW ALT.

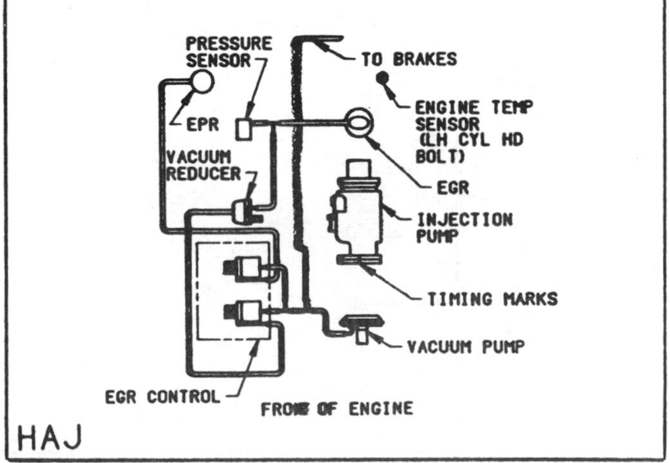

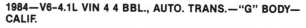

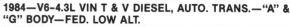

1984—V6-4.3L VIN V DIESEL, AUTO. TRANS.—"G" BODY—CALIF.

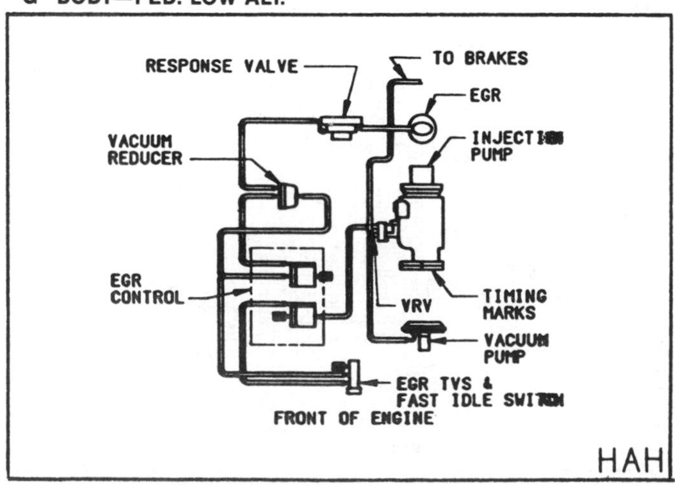

1984—V6-4.3L VIN T & V DIESEL, AUTO. TRANS.—"A" & "G" BODY—FED. HIGH ALT.

VACUUM CIRCUITS

1984

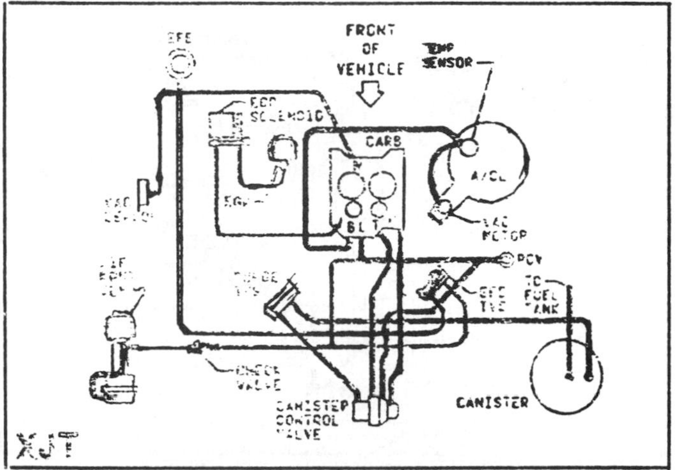

1984—V8-5.0L VIN H 4 BBL., AUTO. TRANS.—"B" & "G" BODY—FEDERAL

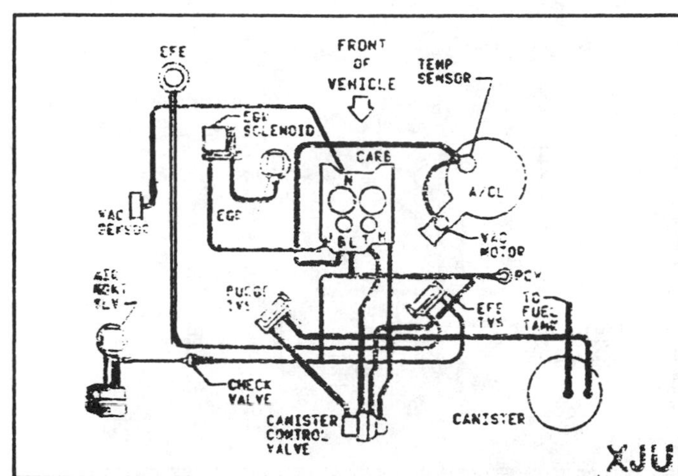

1984—V8-5.0L VIN H 4 BBL., AUTO. TRANS.—"B" & "G" BODY—CALIF.

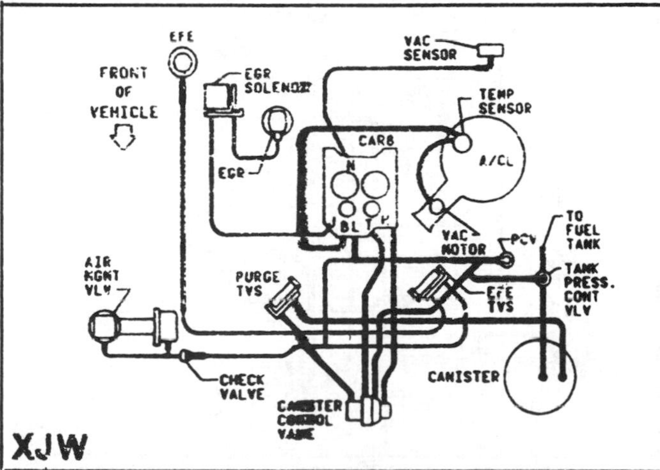

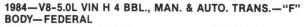

1984—V8-5.0L VIN H 4 BBL., MAN. & AUTO. TRANS.—"F" BODY—FEDERAL

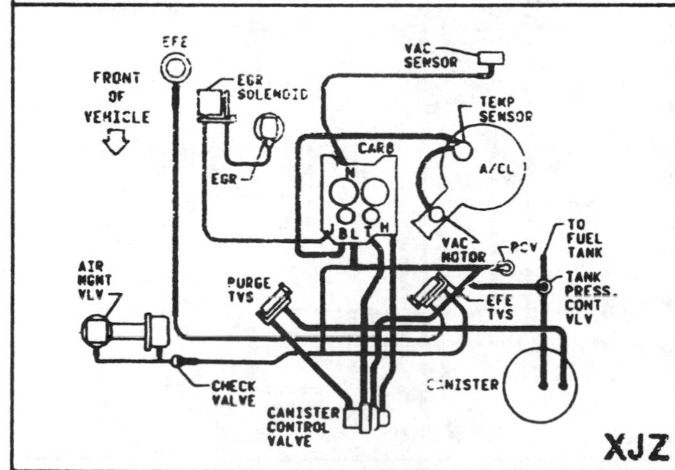

1984—V8-5.0L VIN H 4 BBL., MAN. & AUTO. TRANS.—"F" BODY—CALIF.

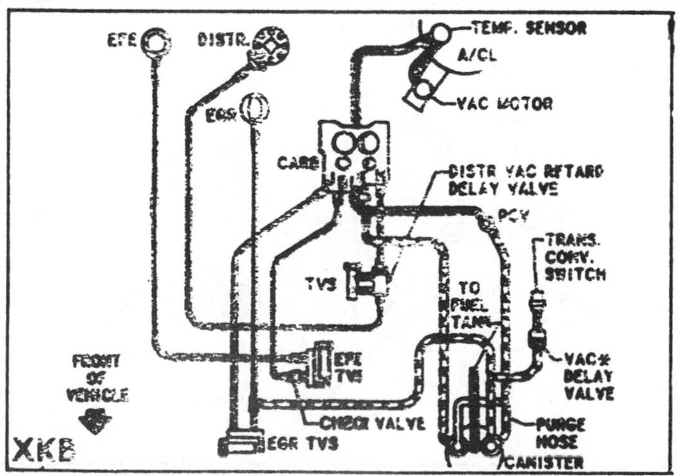

1984—V8-5.0L VIN H 4 BBL., AUTO. TRANS.—"B" & "G" BODY—CANADA

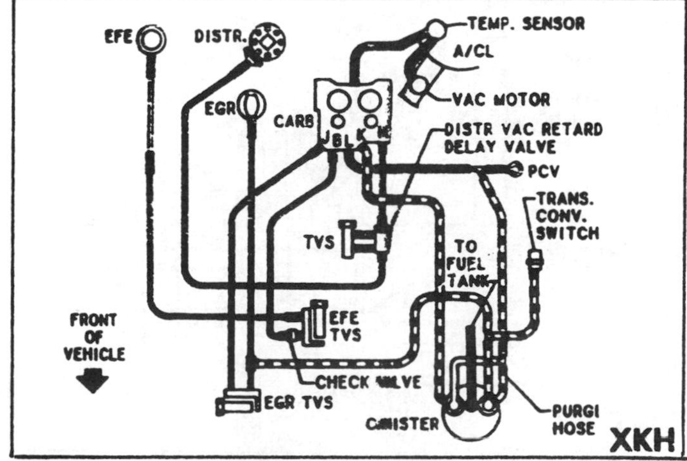

1984—V8-5.0L VIN H 4 BBL., AUTO. TRANS.—"F" BODY—EXPORT

VACUUM CIRCUITS

1984

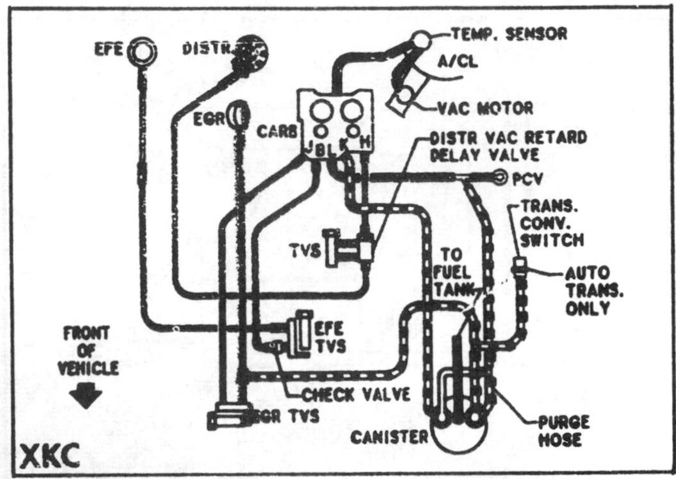

1984—V8-5.0L VIN H 4 BBL., MAN. & AUTO. TRANS.—"F" BODY—CANADA

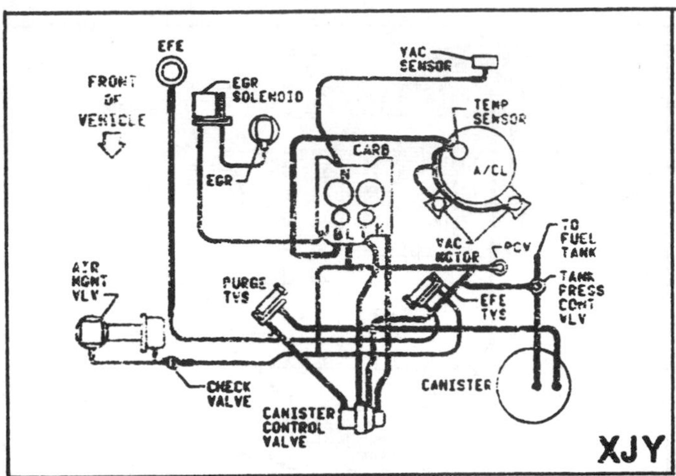

1984—V8-5.0L VIN H 4 BBL., MAN. & AUTO. TRANS.—"F" BODY—FEDERAL

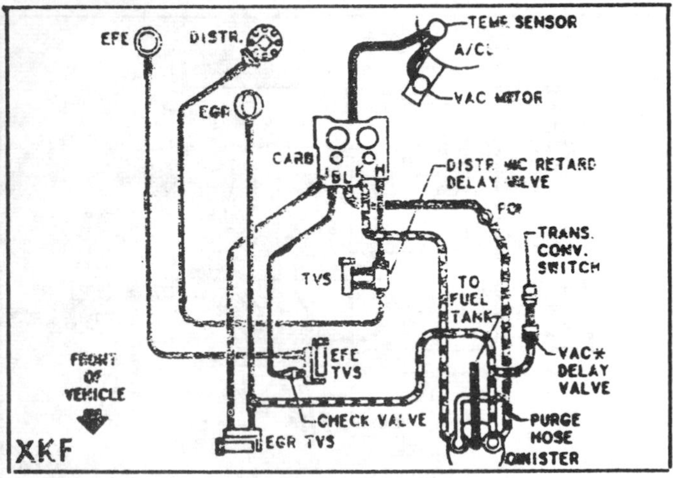

1984—V8-5.0L VIN H 4 BBL., AUTO. TRANS.—"B" & "G" BODY—EXPORT

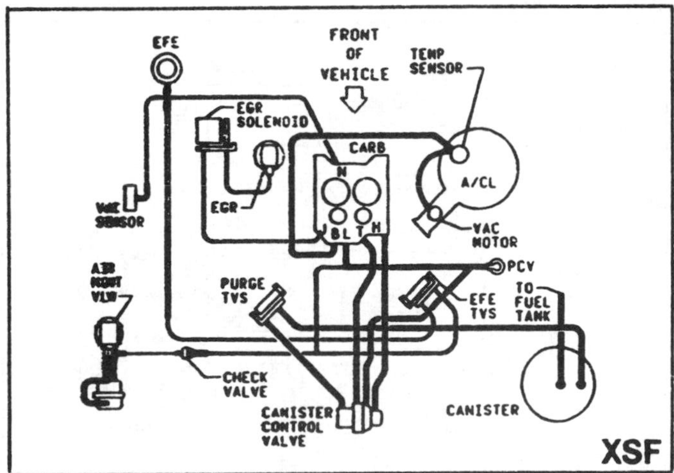

1984—V8-5.0L VIN H 4 BBL., AUTO. TRANS.—"G" BODY—FEDERAL

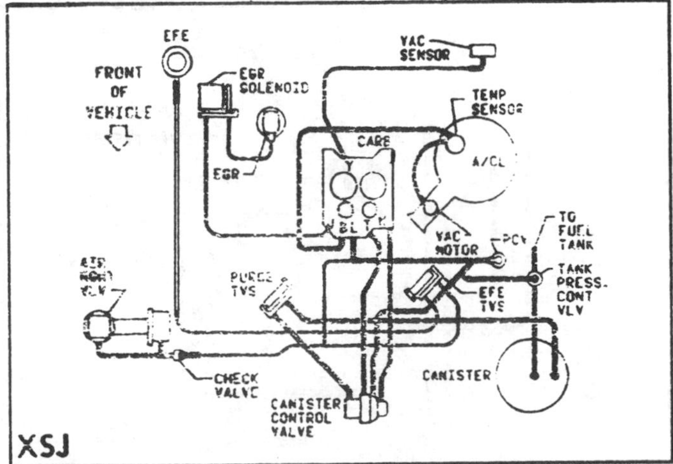

1984—V8-5.0L VIN H 4 BBL., MAN. & AUTO. TRANS.—"F" BODY—FEDERAL

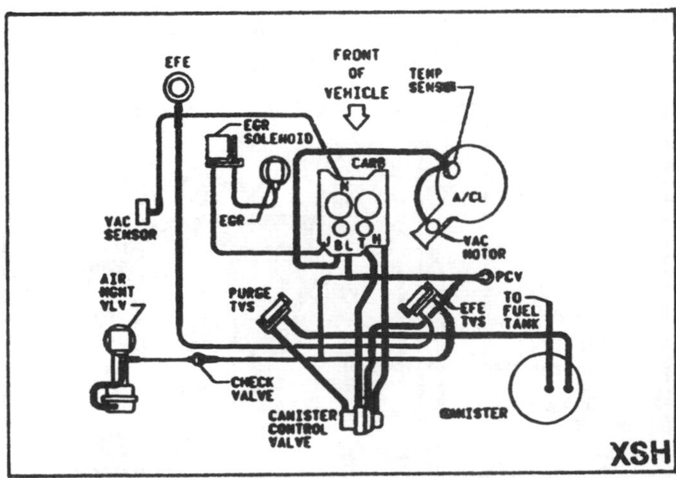

1984—V8-5.0L VIN H 4 BBL., AUTO. TRANS.—"G" BODY—FEDERAL

VACUUM CIRCUITS

1984

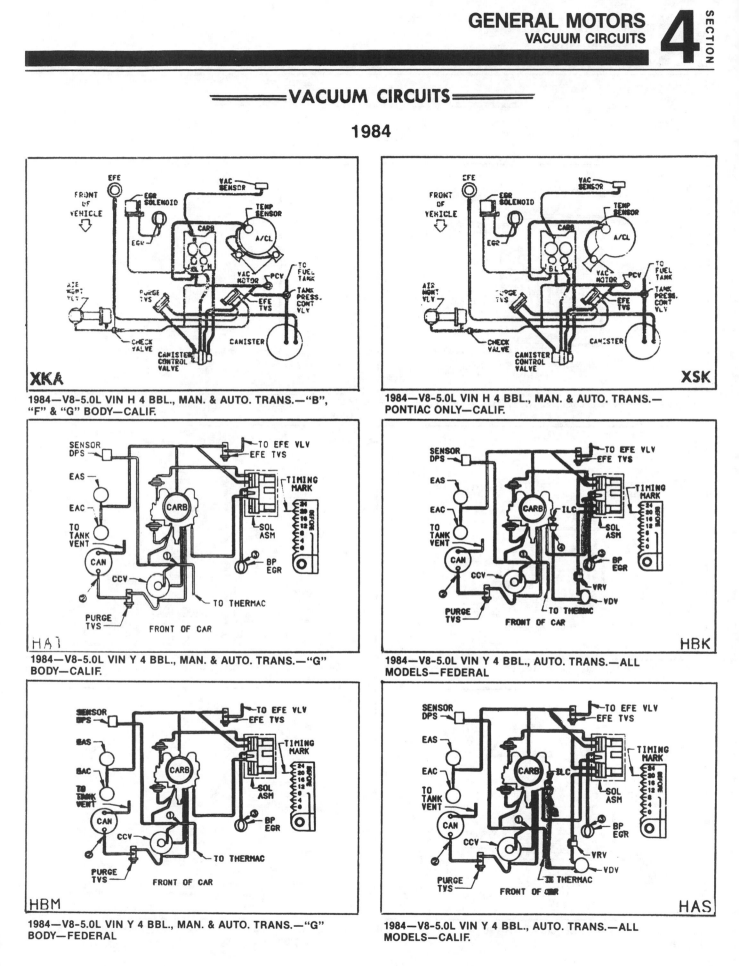

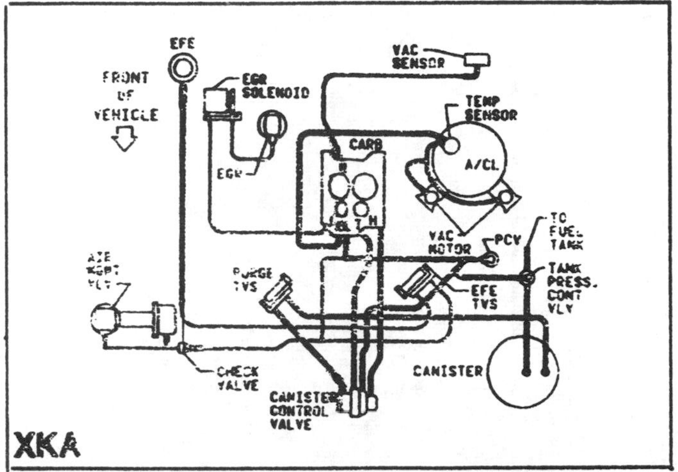

1984—V8-5.0L VIN H 4 BBL., MAN. & AUTO. TRANS.—"B", "F" & "G" BODY—CALIF.

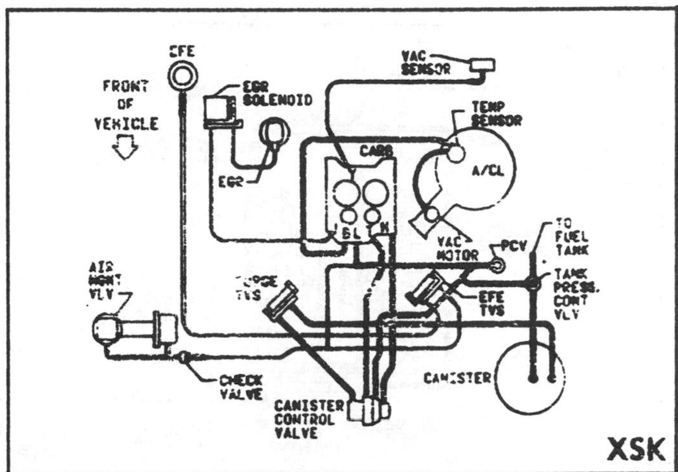

1984—V8-5.0L VIN H 4 BBL., MAN. & AUTO. TRANS.—PONTIAC ONLY—CALIF.

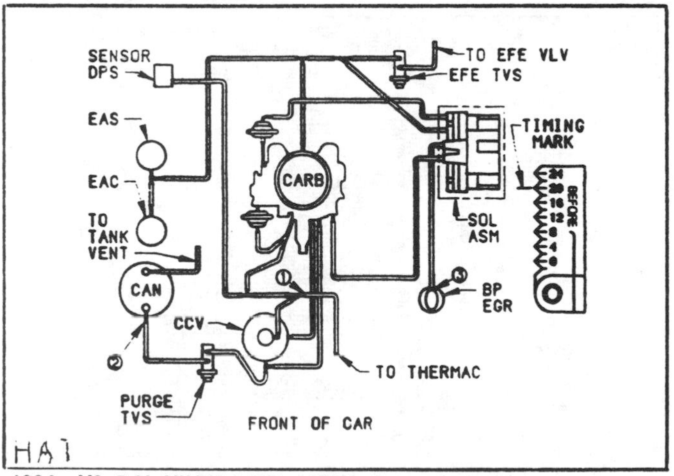

1984—V8-5.0L VIN Y 4 BBL., MAN. & AUTO. TRANS.—"G" BODY—CALIF.

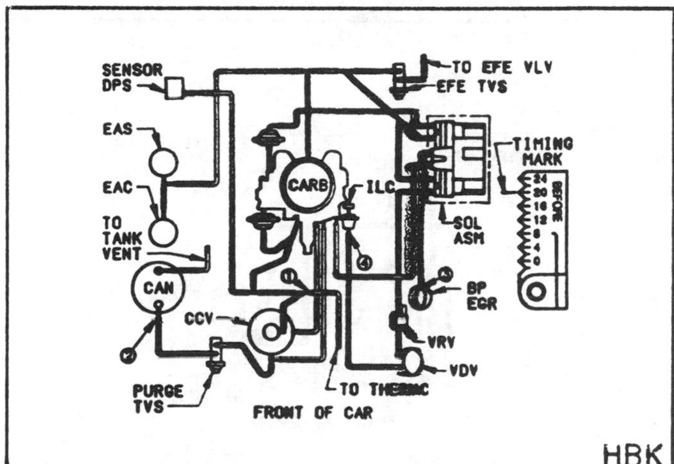

1984—V8-5.0L VIN Y 4 BBL., AUTO. TRANS.—ALL MODELS—FEDERAL

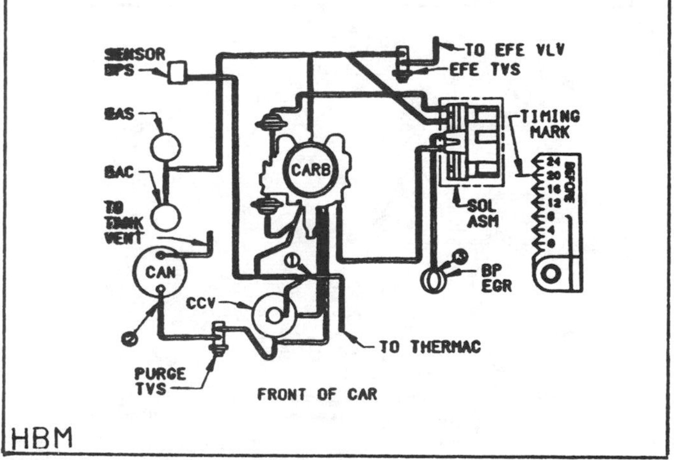

1984—V8-5.0L VIN Y 4 BBL., MAN. & AUTO. TRANS.—"G" BODY—FEDERAL

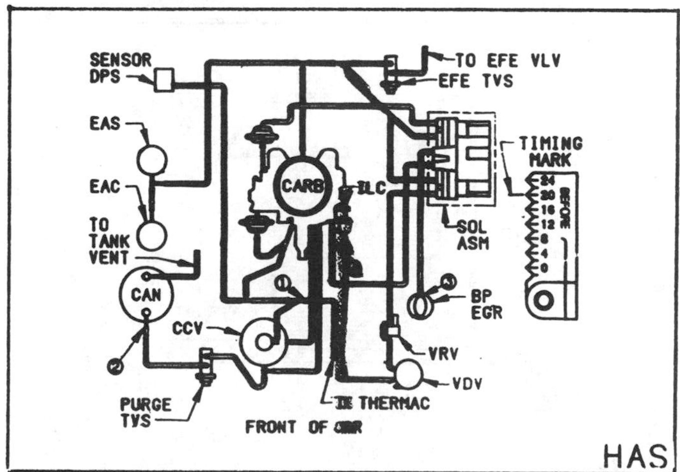

1984—V8-5.0L VIN Y 4 BBL., AUTO. TRANS.—ALL MODELS—CALIF.

VACUUM CIRCUITS

1984

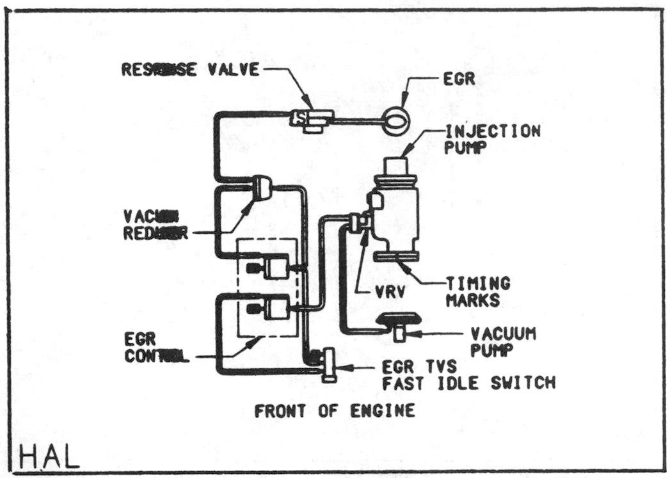

1984—V8-5.7L VIN N DIESEL, AUTO. TRANS.—FED. LOW ALT.

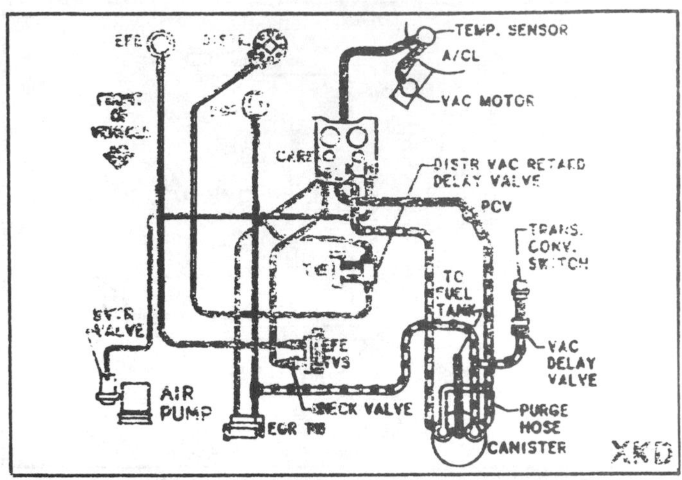

1984—V8-5.7L VIN N 4 BBL., AUTO. TRANS.—"B" BODY—CANADA

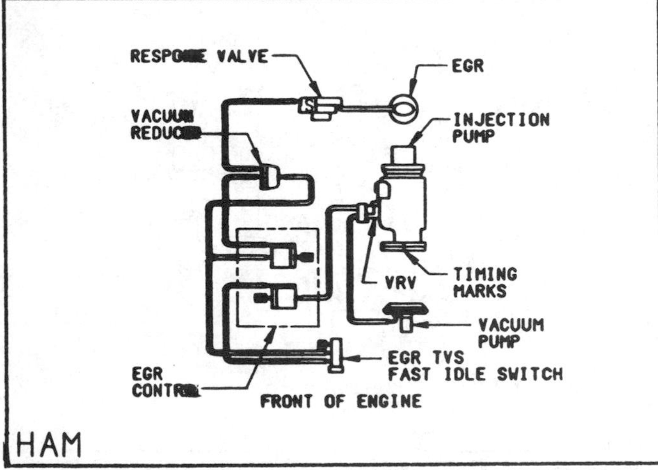

1984—V8-5.7L VIN N DIESEL, AUTO. TRANS.—FED. HIGH ALT.

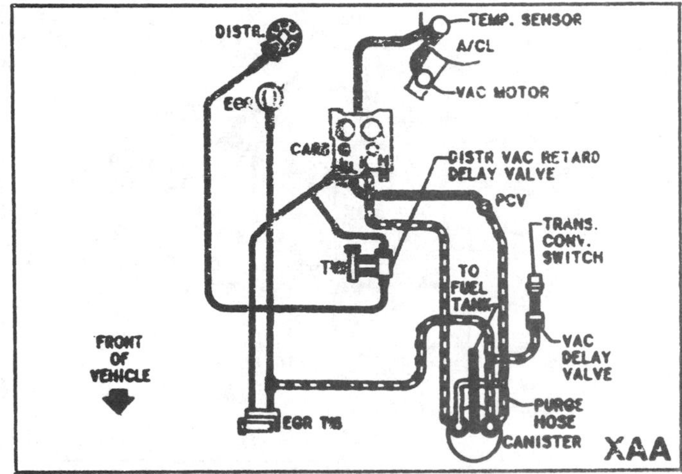

1984—V8-5.7L VIN 4 BBL., AUTO. TRANS.—"B" BODY—CANADA

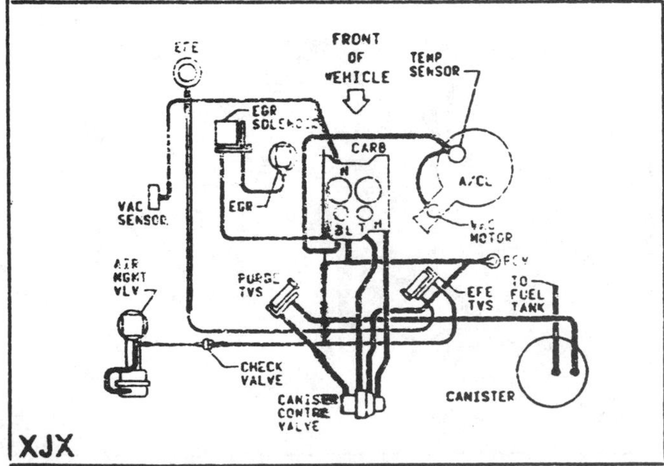

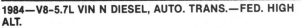

1984—V8-5.7L VIN N 4 BBL., AUTO. TRANS.—"B" BODY—FEDERAL

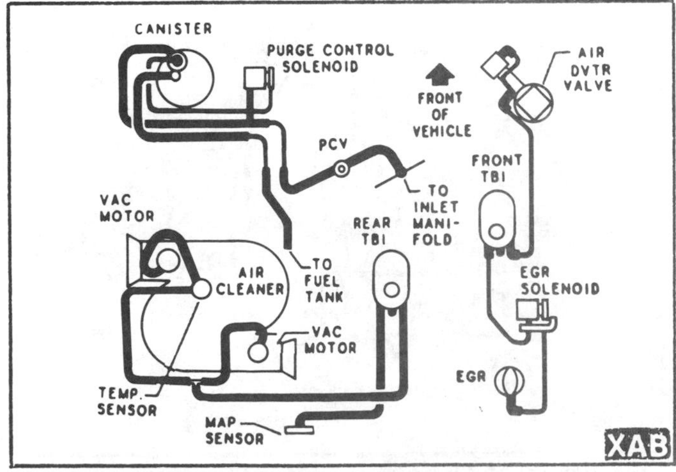

1984—V8-5.7L VIN 8 T.B.I., MAN. TRANS. W/OD—"Y" BODY—FEDERAL

VACUUM CIRCUITS

1984

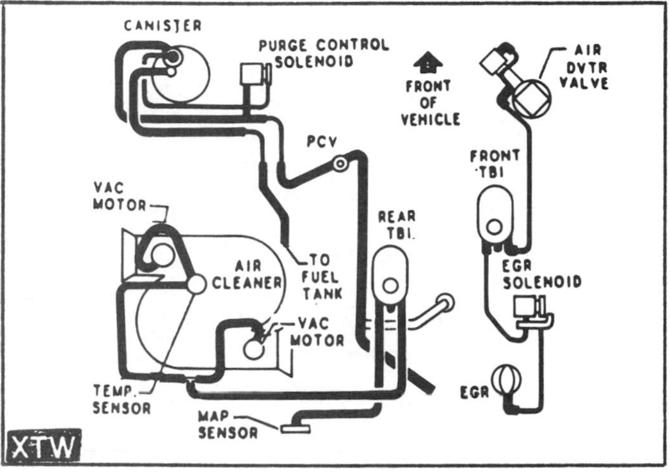

1984—V8-5.7L VIN 8 T.B.I., MAN. TRANS. W/OD—"Y" BODY—FEDERAL

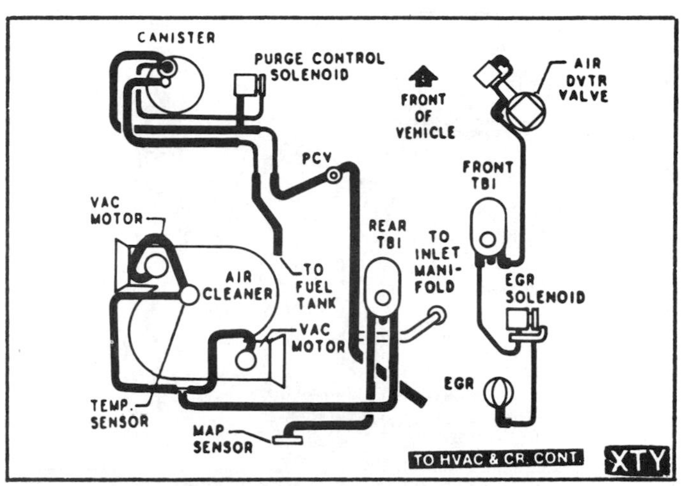

1984—V8-5.7L VIN 8 T.B.I., MAN. TRANS. W/OD—"Y" BODY—CALIF.

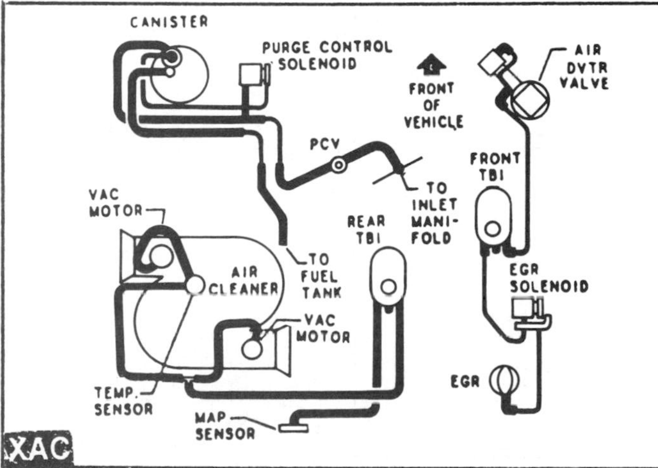

1984—V8-5.7L VIN 8 T.B.I., AUTO. TRANS.—"Y" BODY—FEDERAL

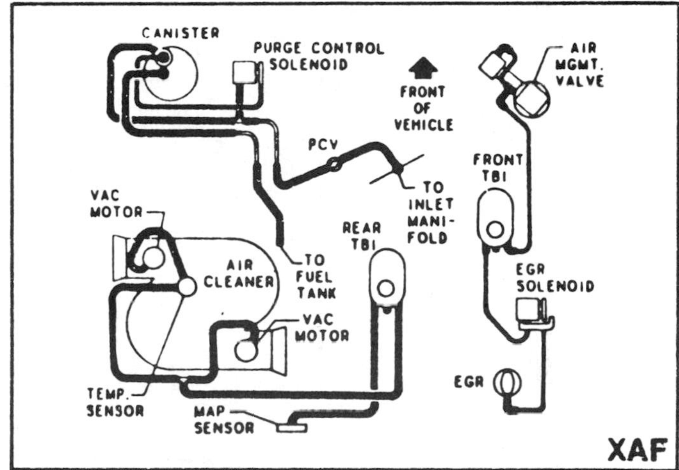

1984—V8-5.7L VIN 8 T.B.I., AUTO. TRANS.—"Y" BODY—CALIF.

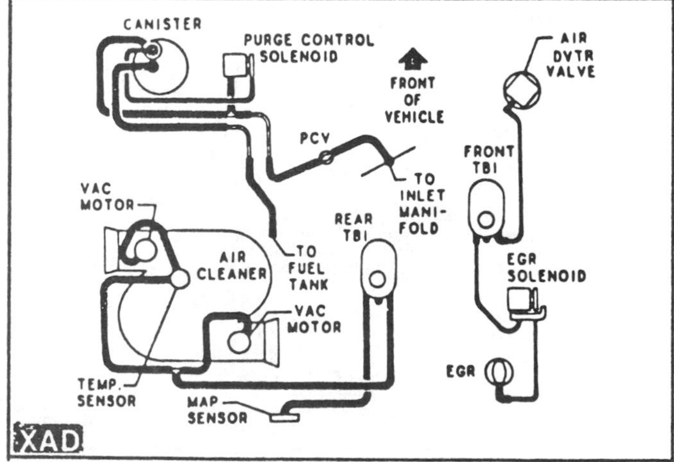

1984—V8-5.7L VIN 8 T.B.I., MAN. TRANS. W/OD—"Y" BODY—CALIF.

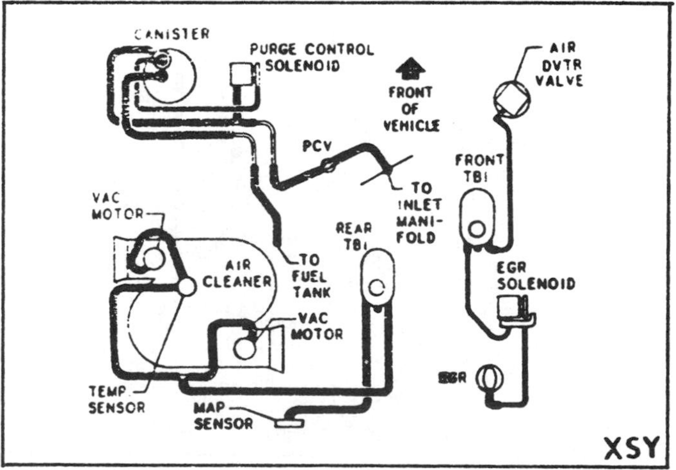

1984—V8-5.7L VIN 8 T.B.I., MAN. TRANS. W/OD—"Y" BODY—EXPORT

VACUUM CIRCUITS

1984

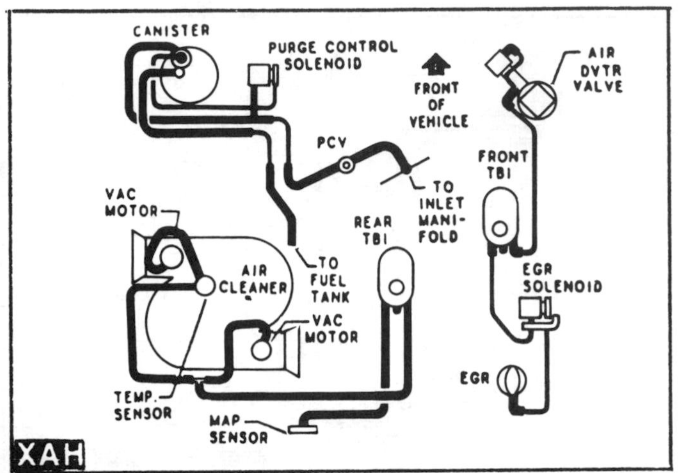

1984—V8-5.7L VIN 8 T.B.I., AUTO. TRANS.—"Y" BODY—EXPORT

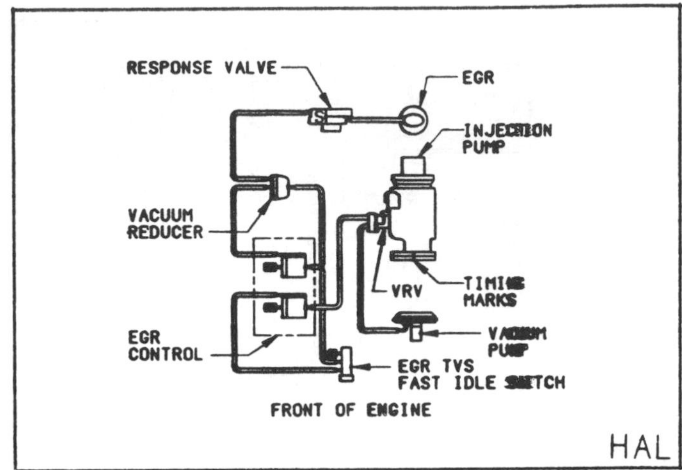

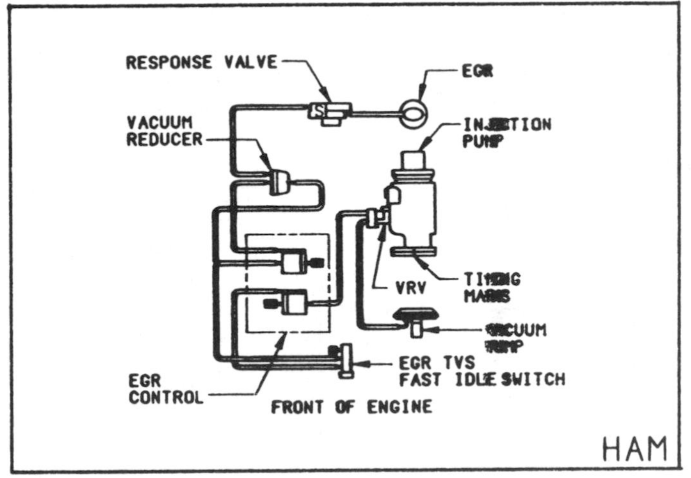

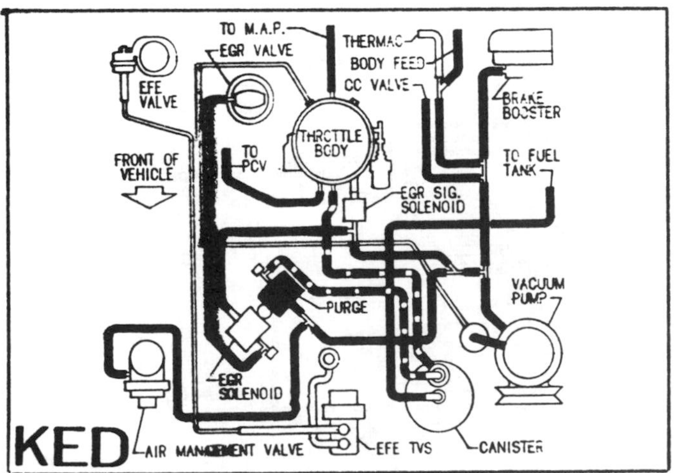

1984—V8-6.0L VIN 9 DFI, AUTO TRANS.—CAD. LIMO.—EXPORT

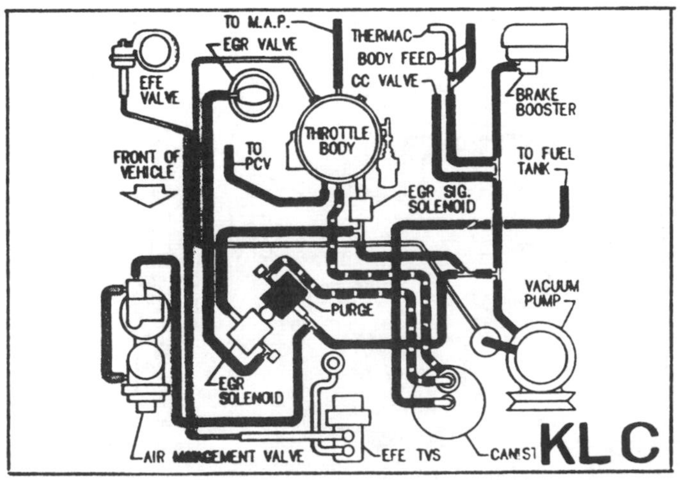

1984—V8-6.0L VIN 9 DFI, AUTO TRANS.—CAD. LIMO.—CALIF.

VACUUM CIRCUITS

1984

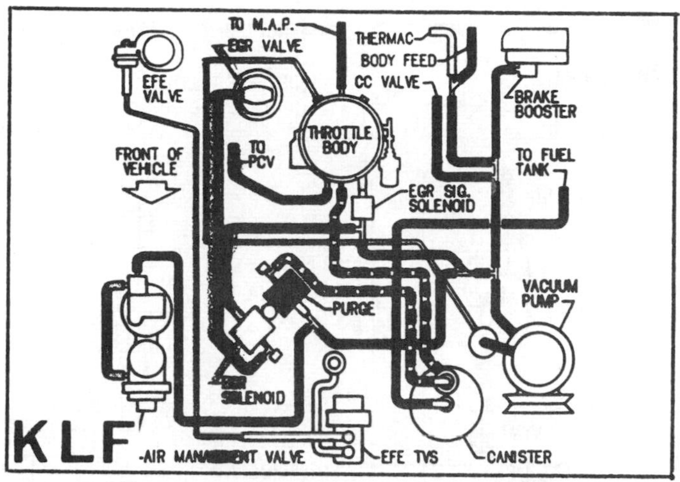

1984—V8-6.0L VIN 9 DFI, AUTO TRANS.—CAD. LIMO.—FEDERAL

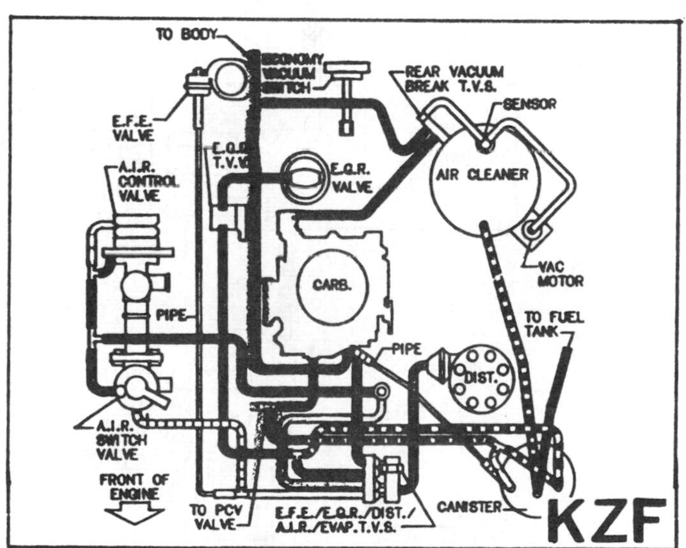

1984—V8-6.0L VIN 9 4 BBL., AUTO TRANS.—CAD. COM. CHASSIS—FEDERAL

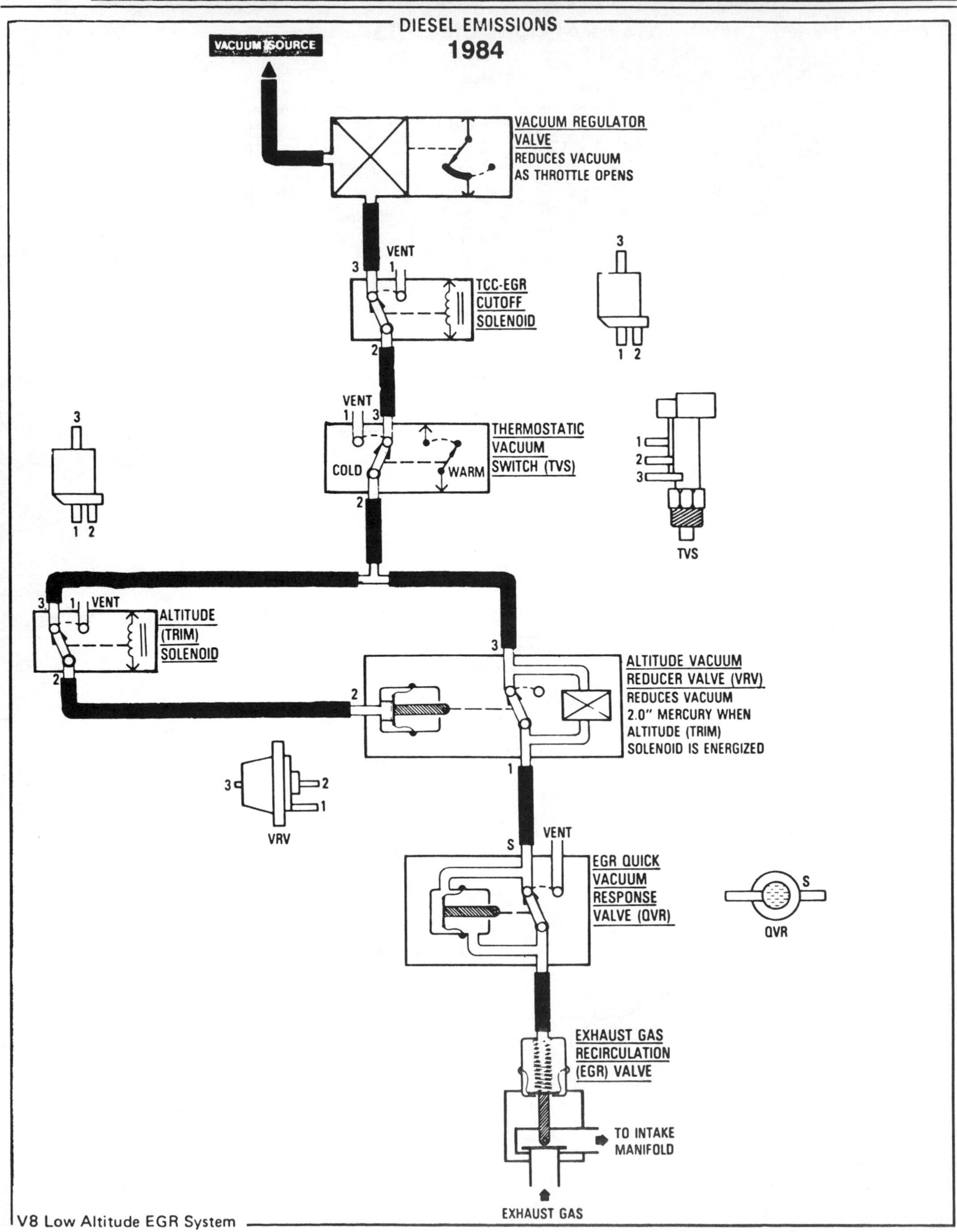
DIESEL EMISSIONS
1984
VACUUM SOURCE
VACUUM REGULATOR VALVE
REDUCES VACUUM AS THROTTLE OPENS
VENT
3
1
2
TCC-EGR CUTOFF SOLENOID
THERMOSTATIC VACUUM SWITCH (TVS)
COLD
WARM
TVS
ALTITUDE (TRIM) SOLENOID
ALTITUDE VACUUM REDUCER VALVE (VRV)
REDUCES VACUUM 2.0" MERCURY WHEN ALTITUDE (TRIM) SOLENOID IS ENERGIZED
VRV
S
EGR QUICK VACUUM RESPONSE VALVE (QVR)
QVR
EXHAUST GAS RECIRCULATION (EGR) VALVE
TO INTAKE MANIFOLD
EXHAUST GAS
V8 Low Altitude EGR System

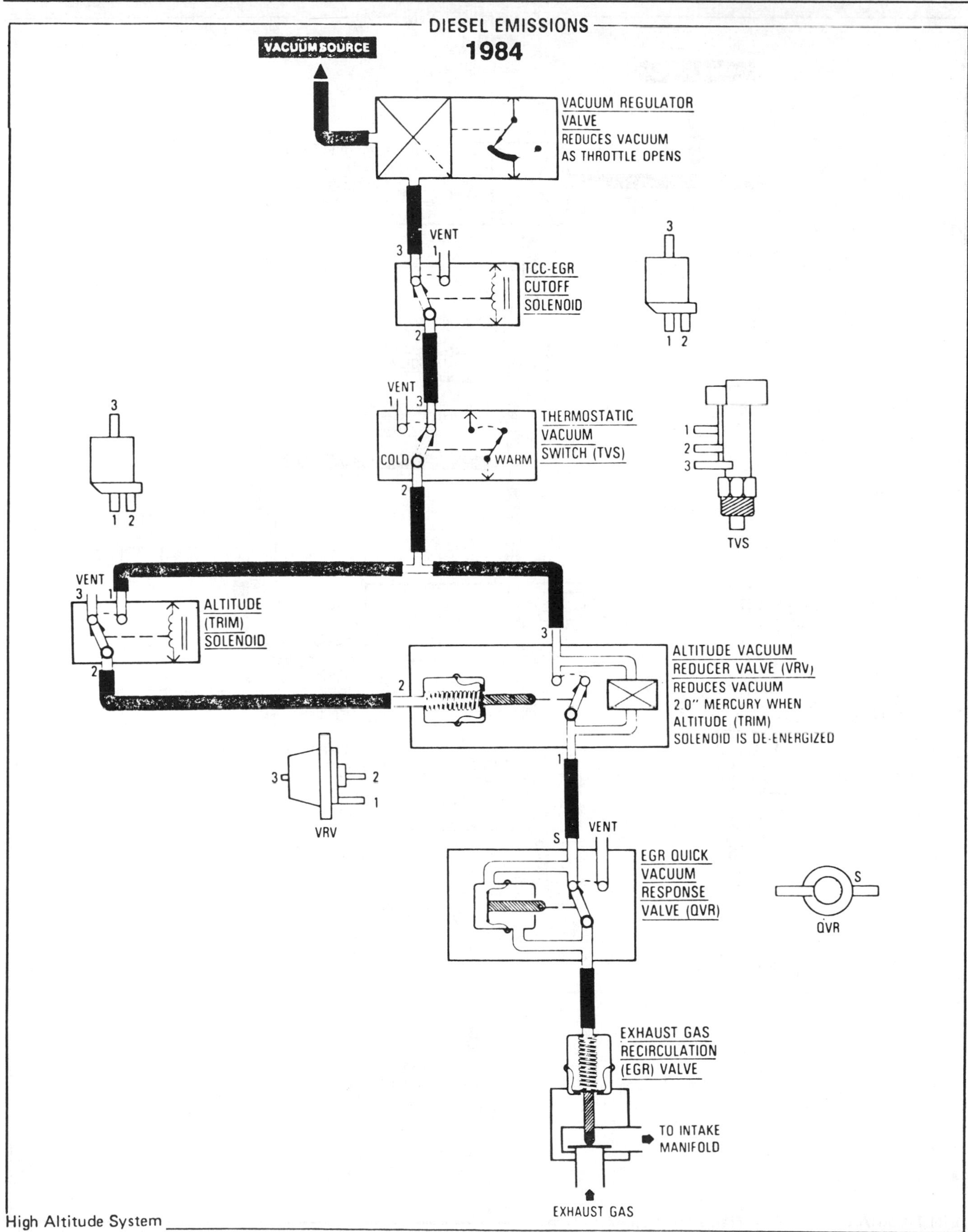

High Altitude System

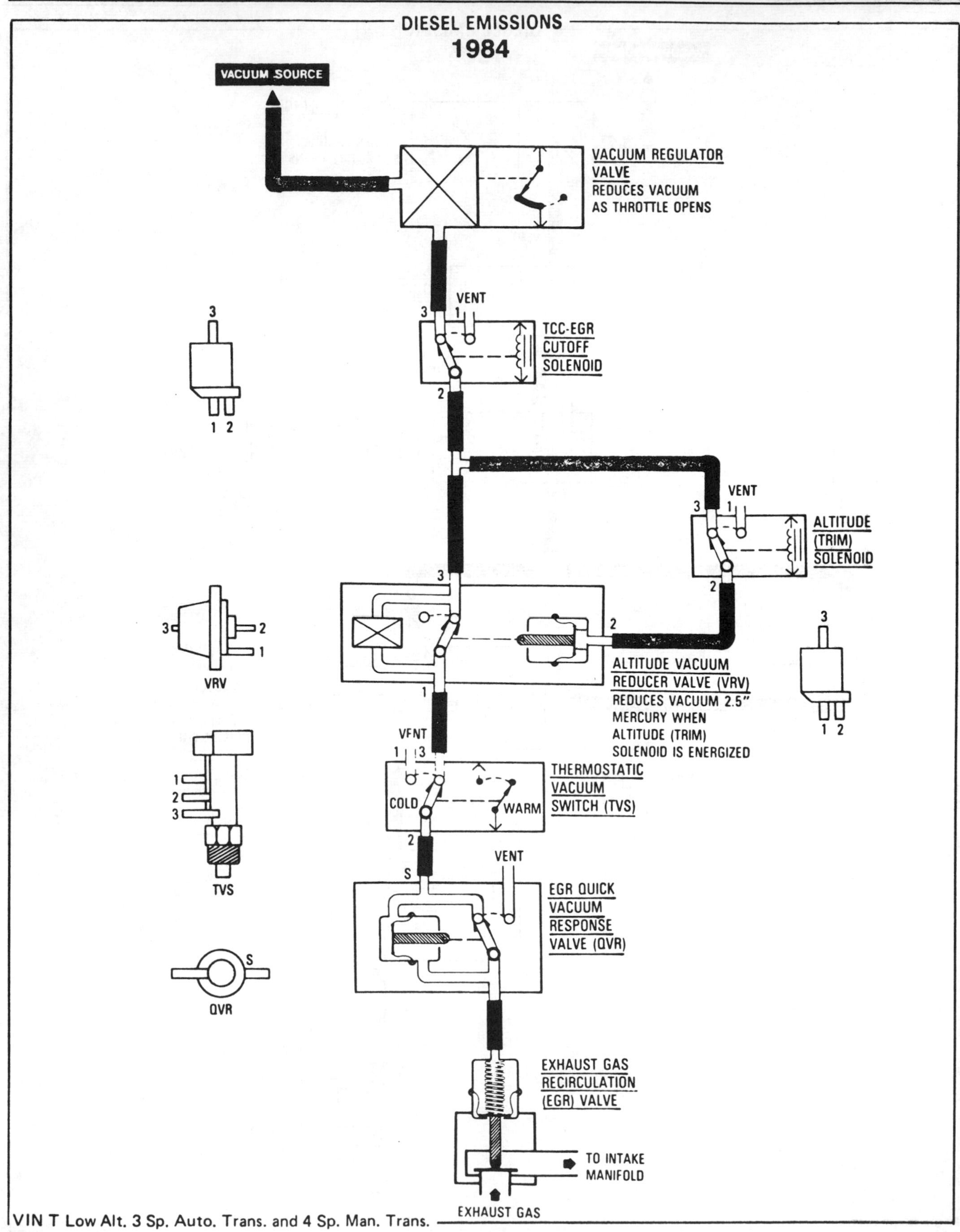

VIN T Low Alt. 3 Sp. Auto. Trans. and 4 Sp. Man. Trans.

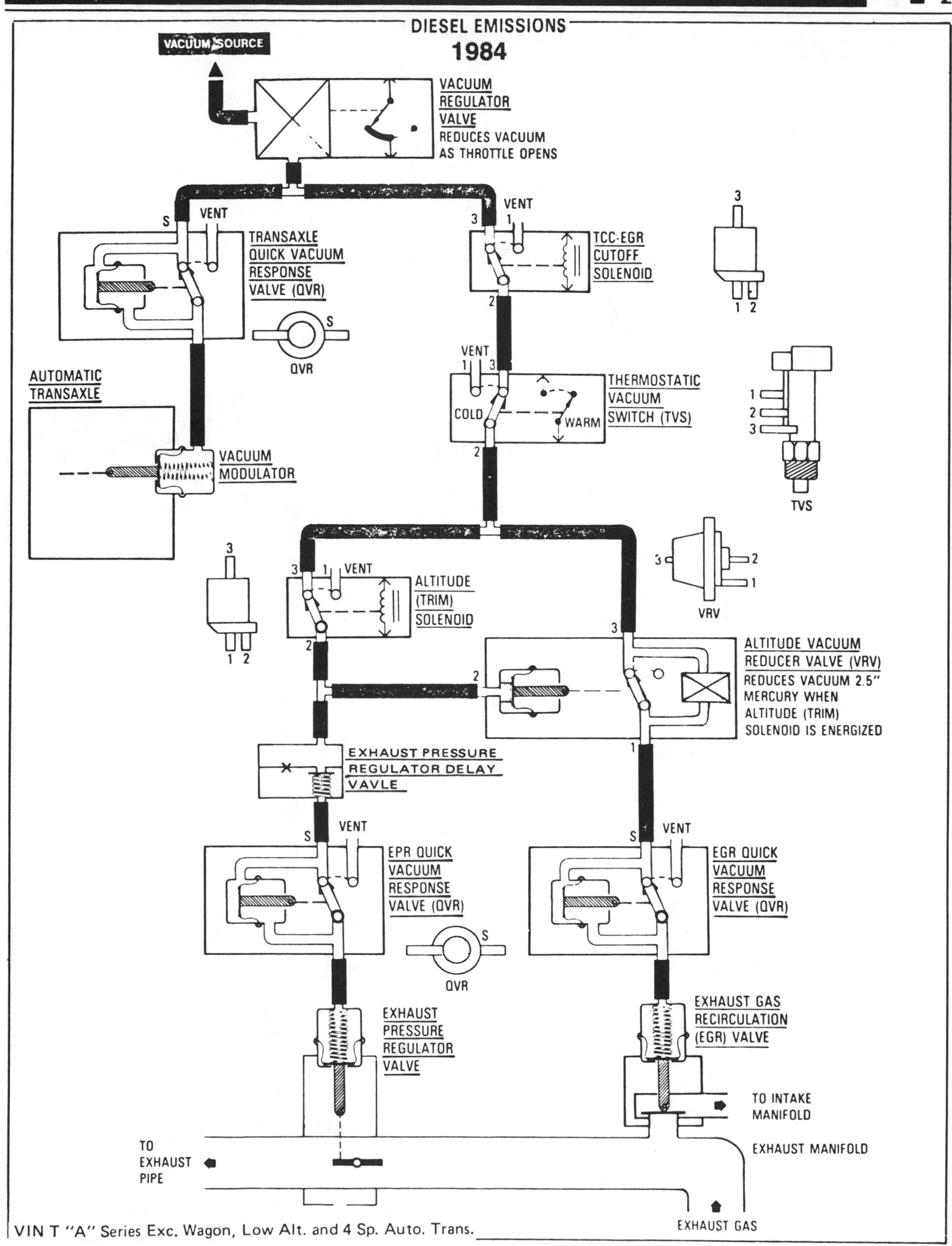
DIESEL EMISSIONS
1984
VACUUM SOURCE
VACUUM REGULATOR VALVE
REDUCES VACUUM AS THROTTLE OPENS
S
VENT
TRANSAXLE QUICK VACUUM RESPONSE VALVE (QVR)
3
VENT
1
TCC-EGR CUTOFF SOLENOID
2
3
1 2
S
QVR
AUTOMATIC TRANSAXLE
VENT
1 3
THERMOSTATIC VACUUM SWITCH (TVS)
COLD
WARM
2
1
2
3
TVS
VACUUM MODULATOR
3
VRV
3
1 VENT
ALTITUDE (TRIM) SOLENOID
2
1 2
3
ALTITUDE VACUUM REDUCER VALVE (VRV)
REDUCES VACUUM 2.5" MERCURY WHEN ALTITUDE (TRIM) SOLENOID IS ENERGIZED
2
1
EXHAUST PRESSURE REGULATOR DELAY VAVLE
S
VENT
EPR QUICK VACUUM RESPONSE VALVE (QVR)
S
VENT
EGR QUICK VACUUM RESPONSE VALVE (QVR)
S
QVR
EXHAUST PRESSURE REGULATOR VALVE
EXHAUST GAS RECIRCULATION (EGR) VALVE
TO INTAKE MANIFOLD
EXHAUST MANIFOLD
TO EXHAUST PIPE
EXHAUST GAS

VIN T "A" Series Exc. Wagon, Low Alt. and 4 Sp. Auto. Trans.

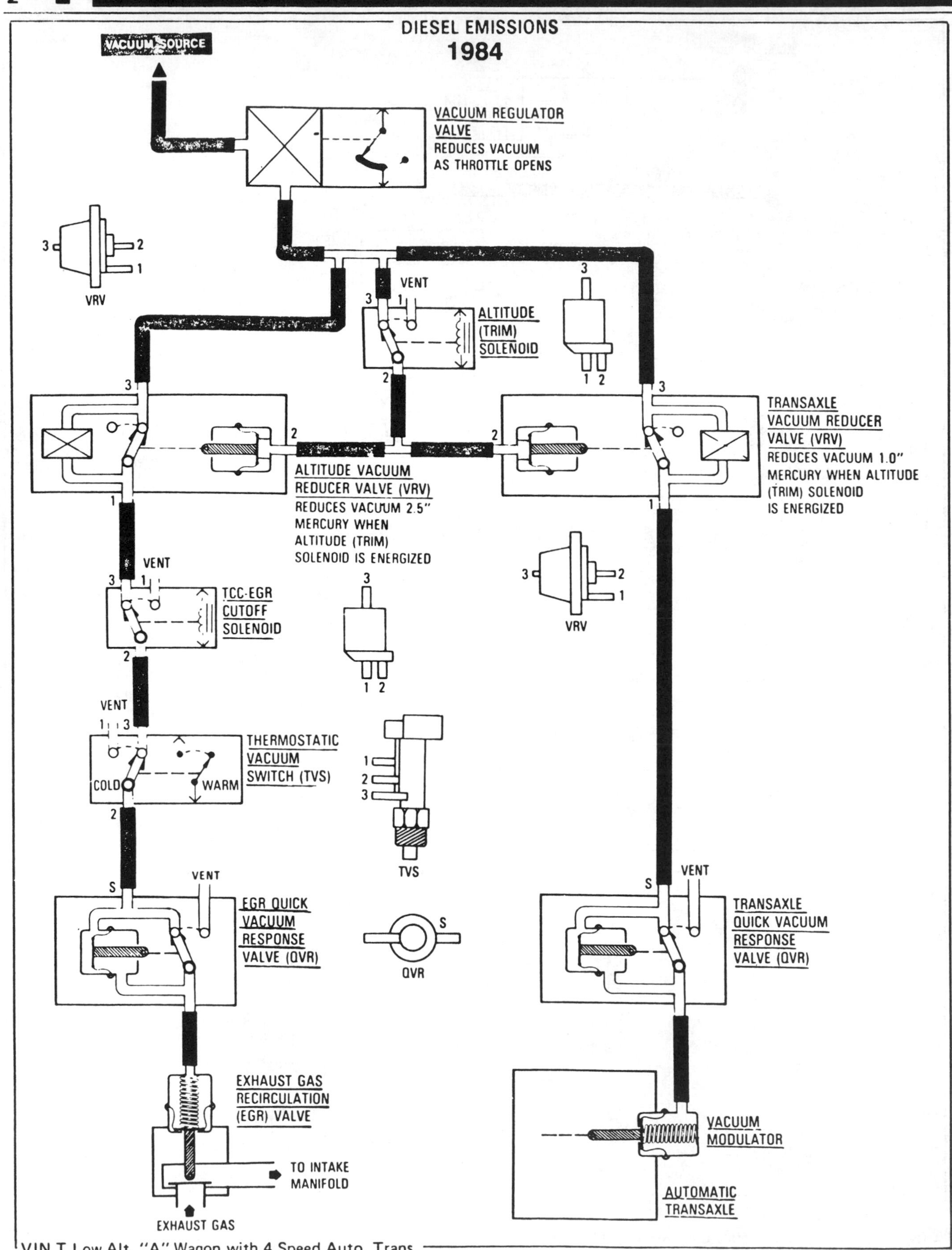

VIN T Low Alt. "A" Wagon with 4 Speed Auto. Trans.

DIESEL EMISSIONS
1984

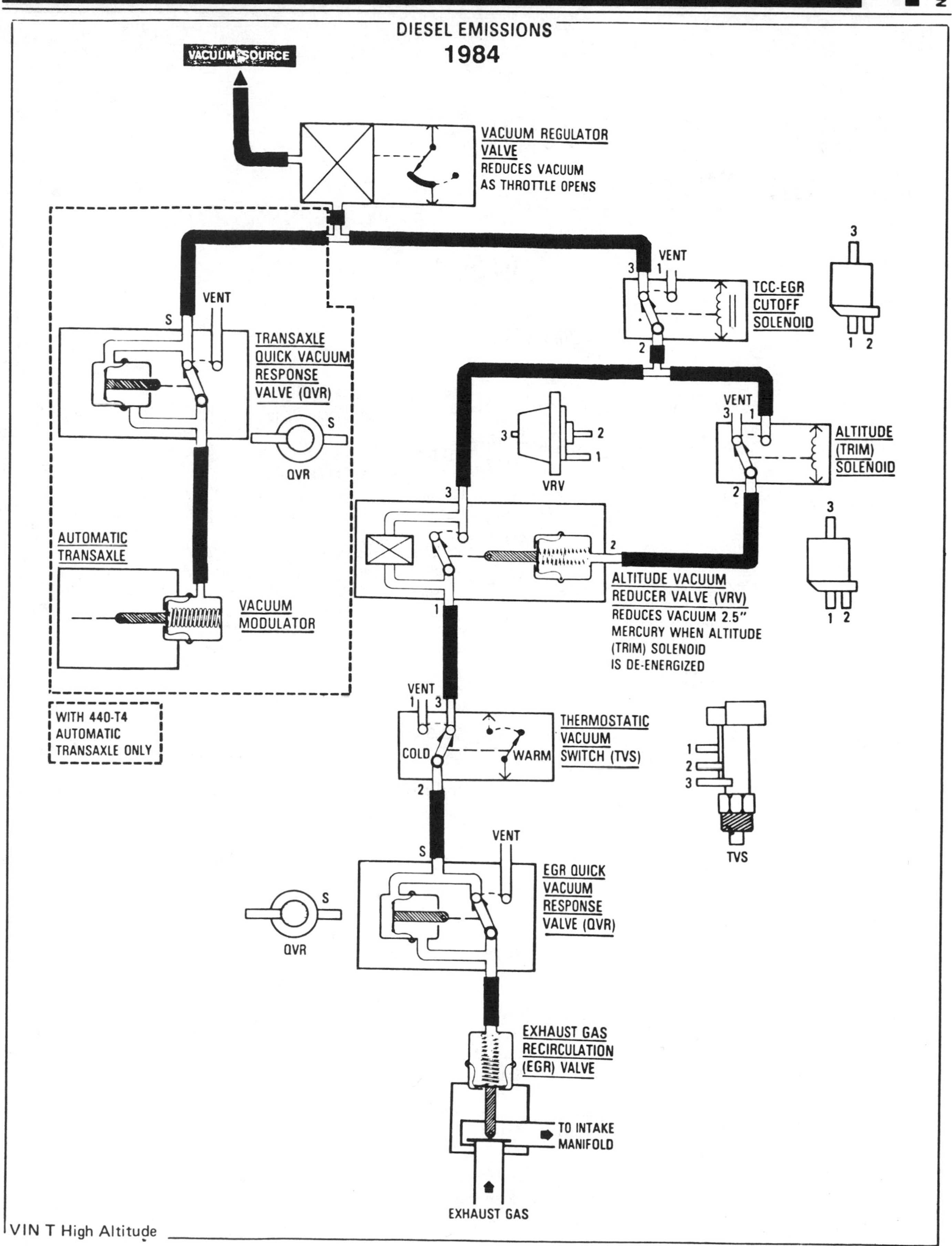

VIN T High Altitude

General Motors
Vacuum Circuits

INDEX

1985 VACUUM CIRCUITS

BUICK

CADILLAC

CHEVROLET

OLDSMOBILE

PONTIAC

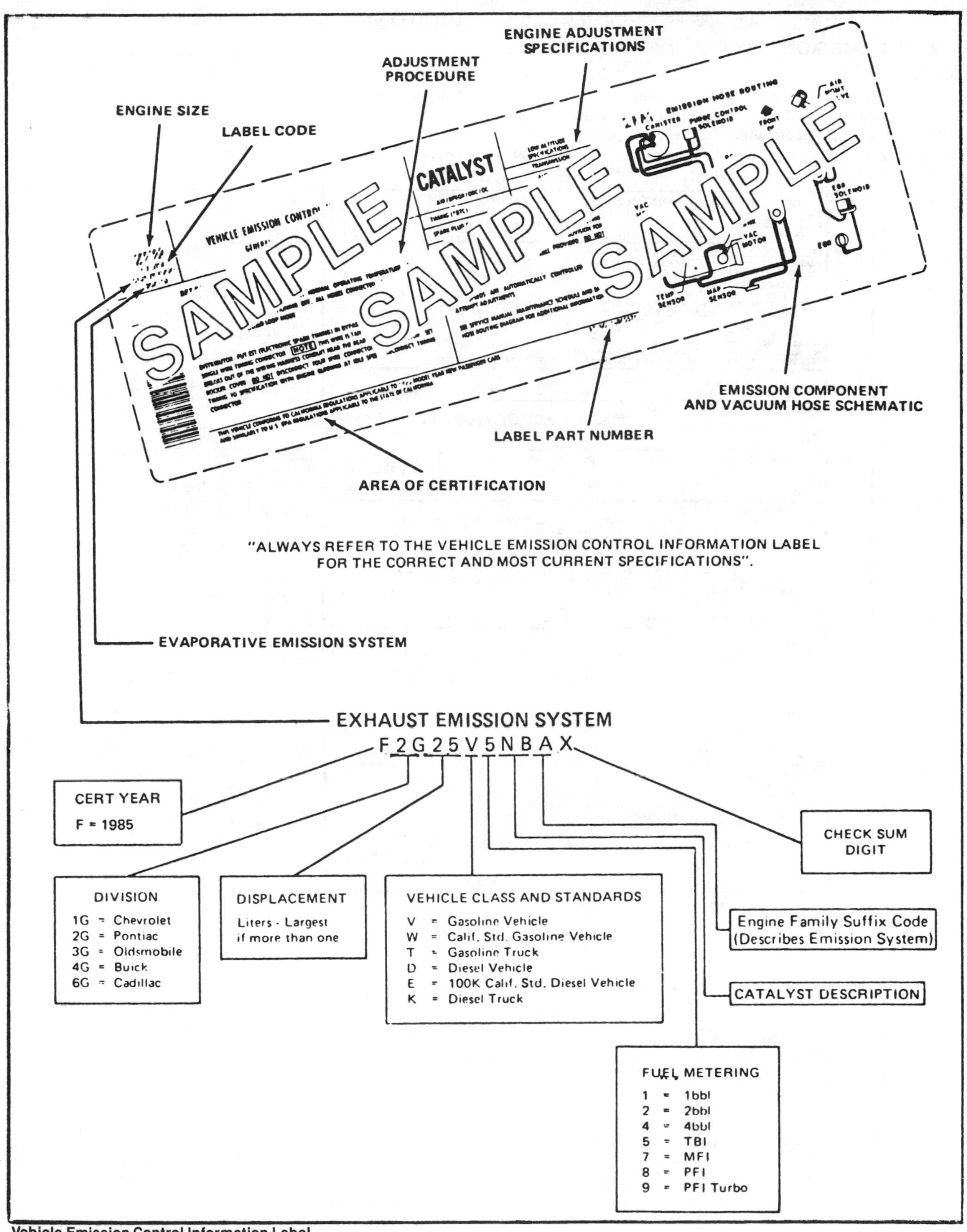
ENGINE ADJUSTMENT SPECIFICATIONS
ADJUSTMENT PROCEDURE
ENGINE SIZE
LABEL CODE
CATALYST
SAMPLE
SAMPLE
SAMPLE
EMISSION COMPONENT AND VACUUM HOSE SCHEMATIC
LABEL PART NUMBER
AREA OF CERTIFICATION
"ALWAYS REFER TO THE VEHICLE EMISSION CONTROL INFORMATION LABEL FOR THE CORRECT AND MOST CURRENT SPECIFICATIONS".
EVAPORATIVE EMISSION SYSTEM
EXHAUST EMISSION SYSTEM
F2G25V5NBAX
CERT YEAR
F = 1985
CHECK SUM DIGIT
DIVISION
1G = Chevrolet
2G = Pontiac
3G = Oldsmobile
4G = Buick
6G = Cadillac
DISPLACEMENT
Liters - Largest if more than one
VEHICLE CLASS AND STANDARDS
V = Gasoline Vehicle
W = Calif. Std. Gasoline Vehicle
T = Gasoline Truck
D = Diesel Vehicle
E = 100K Calif. Std. Diesel Vehicle
K = Diesel Truck
Engine Family Suffix Code (Describes Emission System)
CATALYST DESCRIPTION
FUEL METERING
1 = 1bbl
2 = 2bbl
4 = 4bbl
5 = TBI
7 = MFI
8 = PFI
9 = PFI Turbo
Vehicle Emission Control Information Label

VACUUM CIRCUITS

VEHICLE EMISSION CONTROL INFORMATION LABEL

The Vehicle Emission Control Information label is located in the engine compartment (fan shroud, radiator support, hood underside, etc.) of every vehicle produced by General Motors Corporation. The label contains important emission specifications and setting procedures, as well as a vacuum hose schematic with emission components identified.

When servicing the engine or emission systems, the Vehicle Emission Control Information label should be checked for up-to date information.

1985 BUICK 3.0L

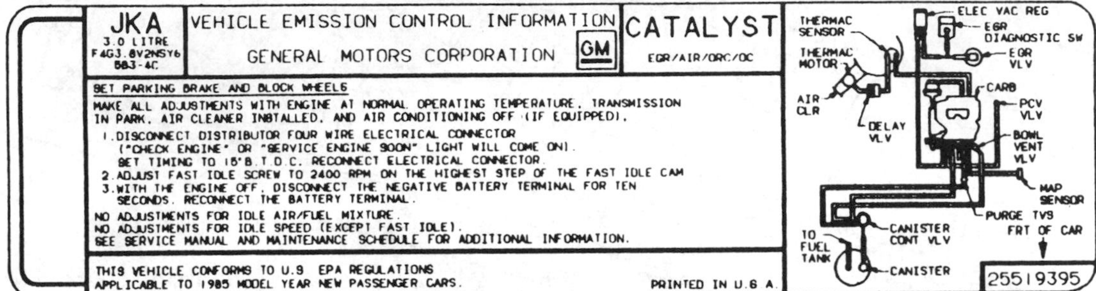

JKA
3.0 LITRE
F4G3.8V2NSY6
5B3-4C

VEHICLE EMISSION CONTROL INFORMATION
GENERAL MOTORS CORPORATION GM

CATALYST
EGR/AIR/ORC/OC

SET PARKING BRAKE AND BLOCK WHEELS
MAKE ALL ADJUSTMENTS WITH ENGINE AT NORMAL OPERATING TEMPERATURE, TRANSMISSION IN PARK, AIR CLEANER INSTALLED, AND AIR CONDITIONING OFF (IF EQUIPPED).
1. DISCONNECT DISTRIBUTOR FOUR WIRE ELECTRICAL CONNECTOR ("CHECK ENGINE" OR "SERVICE ENGINE SOON" LIGHT WILL COME ON). SET TIMING TO 15°B.T.D.C. RECONNECT ELECTRICAL CONNECTOR.
2. ADJUST FAST IDLE SCREW TO 2400 RPM ON THE HIGHEST STEP OF THE FAST IDLE CAM
3. WITH THE ENGINE OFF, DISCONNECT THE NEGATIVE BATTERY TERMINAL FOR TEN SECONDS. RECONNECT THE BATTERY TERMINAL.
NO ADJUSTMENTS FOR IDLE AIR/FUEL MIXTURE.
NO ADJUSTMENTS FOR IDLE SPEED (EXCEPT FAST IDLE).
SEE SERVICE MANUAL AND MAINTENANCE SCHEDULE FOR ADDITIONAL INFORMATION.

THIS VEHICLE CONFORMS TO U.S. EPA REGULATIONS APPLICABLE TO 1985 MODEL YEAR NEW PASSENGER CARS.
PRINTED IN U.S.A.

1985-V6-3.0L CENTURY—FEDERAL

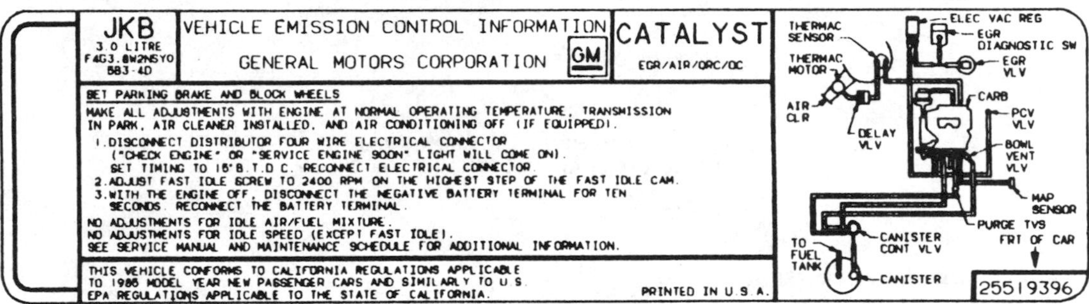

JKB
3.0 LITRE
F4G3.8W2NSYO
5B3-4D

VEHICLE EMISSION CONTROL INFORMATION
GENERAL MOTORS CORPORATION GM

CATALYST
EGR/AIR/ORC/OC

SET PARKING BRAKE AND BLOCK WHEELS
MAKE ALL ADJUSTMENTS WITH ENGINE AT NORMAL OPERATING TEMPERATURE, TRANSMISSION IN PARK, AIR CLEANER INSTALLED, AND AIR CONDITIONING OFF (IF EQUIPPED).
1. DISCONNECT DISTRIBUTOR FOUR WIRE ELECTRICAL CONNECTOR ("CHECK ENGINE" OR "SERVICE ENGINE SOON" LIGHT WILL COME ON). SET TIMING TO 15°B.T.D.C. RECONNECT ELECTRICAL CONNECTOR.
2. ADJUST FAST IDLE SCREW TO 2400 RPM ON THE HIGHEST STEP OF THE FAST IDLE CAM.
3. WITH THE ENGINE OFF, DISCONNECT THE NEGATIVE BATTERY TERMINAL FOR TEN SECONDS. RECONNECT THE BATTERY TERMINAL.
NO ADJUSTMENTS FOR IDLE AIR/FUEL MIXTURE.
NO ADJUSTMENTS FOR IDLE SPEED (EXCEPT FAST IDLE).
SEE SERVICE MANUAL AND MAINTENANCE SCHEDULE FOR ADDITIONAL INFORMATION.

THIS VEHICLE CONFORMS TO CALIFORNIA REGULATIONS APPLICABLE TO 1985 MODEL YEAR NEW PASSENGER CARS AND SIMILARLY TO U.S. EPA REGULATIONS APPLICABLE TO THE STATE OF CALIFORNIA.
PRINTED IN U.S.A.

1985-V6-3.0L CENTURY—CALIFORNIA

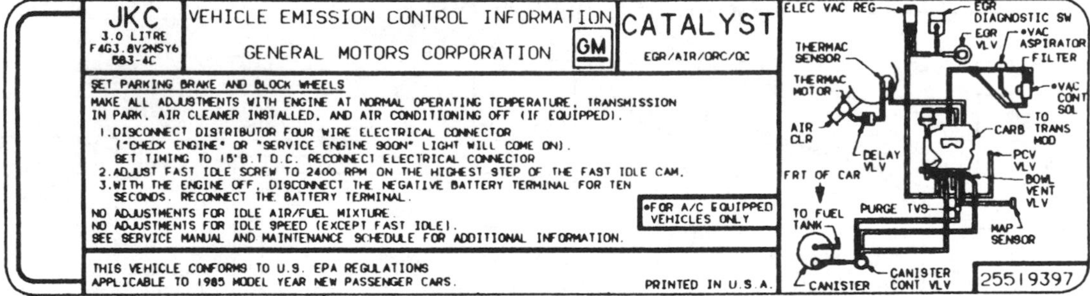

JKC
3.0 LITRE
F4G3.8V2NSY6
5B3-4C

VEHICLE EMISSION CONTROL INFORMATION
GENERAL MOTORS CORPORATION GM

CATALYST
EGR/AIR/ORC/OC

SET PARKING BRAKE AND BLOCK WHEELS
MAKE ALL ADJUSTMENTS WITH ENGINE AT NORMAL OPERATING TEMPERATURE, TRANSMISSION IN PARK, AIR CLEANER INSTALLED, AND AIR CONDITIONING OFF (IF EQUIPPED).
1. DISCONNECT DISTRIBUTOR FOUR WIRE ELECTRICAL CONNECTOR ("CHECK ENGINE" OR "SERVICE ENGINE SOON" LIGHT WILL COME ON). SET TIMING TO 15°B.T.D.C. RECONNECT ELECTRICAL CONNECTOR
2. ADJUST FAST IDLE SCREW TO 2400 RPM ON THE HIGHEST STEP OF THE FAST IDLE CAM.
3. WITH THE ENGINE OFF, DISCONNECT THE NEGATIVE BATTERY TERMINAL FOR TEN SECONDS. RECONNECT THE BATTERY TERMINAL.
NO ADJUSTMENTS FOR IDLE AIR/FUEL MIXTURE.
NO ADJUSTMENTS FOR IDLE SPEED (EXCEPT FAST IDLE).
SEE SERVICE MANUAL AND MAINTENANCE SCHEDULE FOR ADDITIONAL INFORMATION.

•FOR A/C EQUIPPED VEHICLES ONLY

THIS VEHICLE CONFORMS TO U.S. EPA REGULATIONS APPLICABLE TO 1985 MODEL YEAR NEW PASSENGER CARS.
PRINTED IN U.S.A.

1985-V6-3.0L CENTURY—FEDERAL

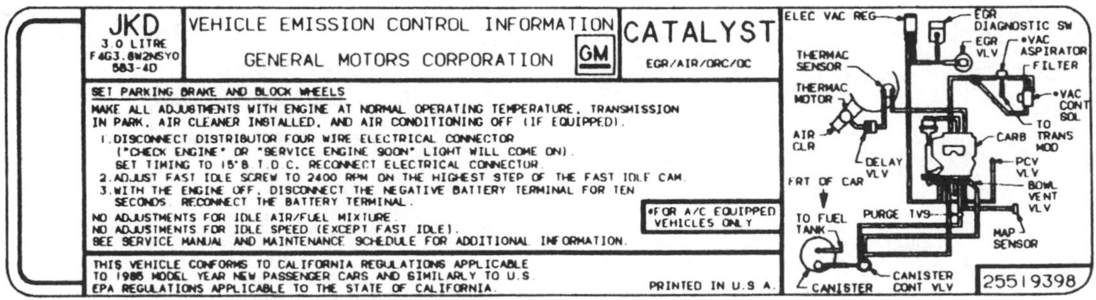

JKD
3.0 LITRE
F4G3.8W2NSYO
5B3-4D

VEHICLE EMISSION CONTROL INFORMATION
GENERAL MOTORS CORPORATION GM

CATALYST
EGR/AIR/ORC/OC

SET PARKING BRAKE AND BLOCK WHEELS
MAKE ALL ADJUSTMENTS WITH ENGINE AT NORMAL OPERATING TEMPERATURE, TRANSMISSION IN PARK, AIR CLEANER INSTALLED, AND AIR CONDITIONING OFF (IF EQUIPPED).
1. DISCONNECT DISTRIBUTOR FOUR WIRE ELECTRICAL CONNECTOR ("CHECK ENGINE" OR "SERVICE ENGINE SOON" LIGHT WILL COME ON). SET TIMING TO 15°B.T.D.C. RECONNECT ELECTRICAL CONNECTOR.
2. ADJUST FAST IDLE SCREW TO 2400 RPM ON THE HIGHEST STEP OF THE FAST IDLE CAM.
3. WITH THE ENGINE OFF, DISCONNECT THE NEGATIVE BATTERY TERMINAL FOR TEN SECONDS. RECONNECT THE BATTERY TERMINAL.
NO ADJUSTMENTS FOR IDLE AIR/FUEL MIXTURE.
NO ADJUSTMENTS FOR IDLE SPEED (EXCEPT FAST IDLE).
SEE SERVICE MANUAL AND MAINTENANCE SCHEDULE FOR ADDITIONAL INFORMATION.

•FOR A/C EQUIPPED VEHICLES ONLY

THIS VEHICLE CONFORMS TO CALIFORNIA REGULATIONS APPLICABLE TO 1985 MODEL YEAR NEW PASSENGER CARS AND SIMILARLY TO U.S. EPA REGULATIONS APPLICABLE TO THE STATE OF CALIFORNIA.
PRINTED IN U.S.A.

1985-V6-3.0L CENTURY—CALIFORNIA

VACUUM CIRCUITS

1985 BUICK 3.0L

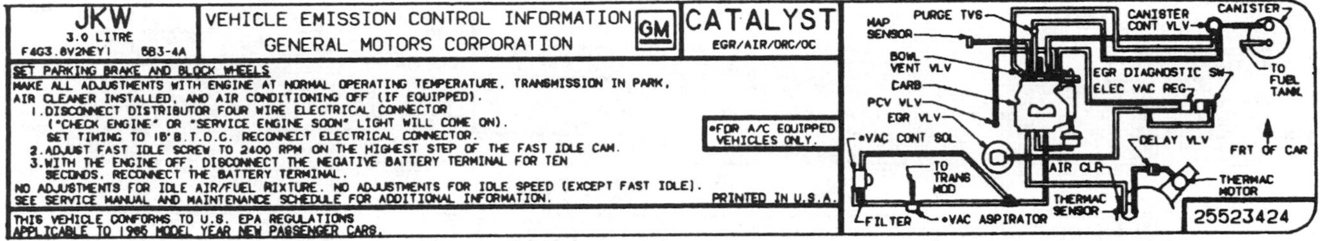
JKW
3.0 LITRE
F4G3.8V2NEY1 5B3-4A

VEHICLE EMISSION CONTROL INFORMATION
GENERAL MOTORS CORPORATION
GM

CATALYST
EGR/AIR/ORC/OC

SET PARKING BRAKE AND BLOCK WHEELS
MAKE ALL ADJUSTMENTS WITH ENGINE AT NORMAL OPERATING TEMPERATURE, TRANSMISSION IN PARK, AIR CLEANER INSTALLED, AND AIR CONDITIONING OFF (IF EQUIPPED).
1. DISCONNECT DISTRIBUTOR FOUR WIRE ELECTRICAL CONNECTOR ("CHECK ENGINE" OR "SERVICE ENGINE SOON" LIGHT WILL COME ON). SET TIMING TO 18°B.T.D.C. RECONNECT ELECTRICAL CONNECTOR.
2. ADJUST FAST IDLE SCREW TO 2400 RPM ON THE HIGHEST STEP OF THE FAST IDLE CAM.
3. WITH THE ENGINE OFF, DISCONNECT THE NEGATIVE BATTERY TERMINAL FOR TEN SECONDS. RECONNECT THE BATTERY TERMINAL.
NO ADJUSTMENTS FOR IDLE AIR/FUEL MIXTURE. NO ADJUSTMENTS FOR IDLE SPEED (EXCEPT FAST IDLE).
SEE SERVICE MANUAL AND MAINTENANCE SCHEDULE FOR ADDITIONAL INFORMATION.

•FOR A/C EQUIPPED VEHICLES ONLY.

PRINTED IN U.S.A.

THIS VEHICLE CONFORMS TO U.S. EPA REGULATIONS APPLICABLE TO 1985 MODEL YEAR NEW PASSENGER CARS.

1985-V6-3.0L ELECTRA—FEDERAL

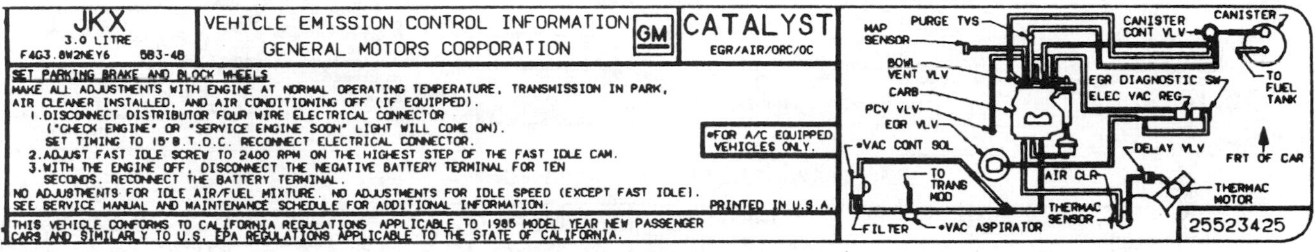
JKX
3.0 LITRE
F4G3.8W2NEY6 5B3-4B

VEHICLE EMISSION CONTROL INFORMATION
GENERAL MOTORS CORPORATION
GM

CATALYST
EGR/AIR/ORC/OC

SET PARKING BRAKE AND BLOCK WHEELS
MAKE ALL ADJUSTMENTS WITH ENGINE AT NORMAL OPERATING TEMPERATURE, TRANSMISSION IN PARK, AIR CLEANER INSTALLED, AND AIR CONDITIONING OFF (IF EQUIPPED).
1. DISCONNECT DISTRIBUTOR FOUR WIRE ELECTRICAL CONNECTOR ("CHECK ENGINE" OR "SERVICE ENGINE SOON" LIGHT WILL COME ON). SET TIMING TO 18°B.T.D.C. RECONNECT ELECTRICAL CONNECTOR.
2. ADJUST FAST IDLE SCREW TO 2400 RPM ON THE HIGHEST STEP OF THE FAST IDLE CAM.
3. WITH THE ENGINE OFF, DISCONNECT THE NEGATIVE BATTERY TERMINAL FOR TEN SECONDS. RECONNECT THE BATTERY TERMINAL.
NO ADJUSTMENTS FOR IDLE AIR/FUEL MIXTURE. NO ADJUSTMENTS FOR IDLE SPEED (EXCEPT FAST IDLE).
SEE SERVICE MANUAL AND MAINTENANCE SCHEDULE FOR ADDITIONAL INFORMATION.

•FOR A/C EQUIPPED VEHICLES ONLY.

PRINTED IN U.S.A.

THIS VEHICLE CONFORMS TO CALIFORNIA REGULATIONS APPLICABLE TO 1985 MODEL YEAR NEW PASSENGER CARS AND SIMILARLY TO U.S. EPA REGULATIONS APPLICABLE TO THE STATE OF CALIFORNIA.

1985-V6-3.0L ELECTRA—CALIFORNIA

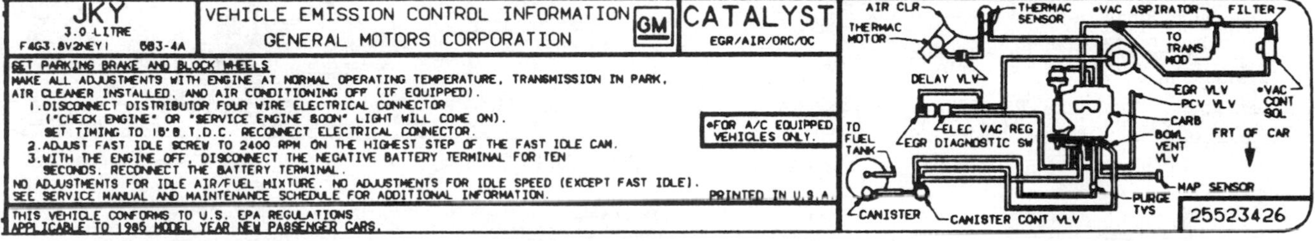
JKY
3.0 LITRE
F4G3.8V2NEY1 5B3-4A

VEHICLE EMISSION CONTROL INFORMATION
GENERAL MOTORS CORPORATION
GM

CATALYST
EGR/AIR/ORC/OC

SET PARKING BRAKE AND BLOCK WHEELS
MAKE ALL ADJUSTMENTS WITH ENGINE AT NORMAL OPERATING TEMPERATURE, TRANSMISSION IN PARK, AIR CLEANER INSTALLED, AND AIR CONDITIONING OFF (IF EQUIPPED).
1. DISCONNECT DISTRIBUTOR FOUR WIRE ELECTRICAL CONNECTOR ("CHECK ENGINE" OR "SERVICE ENGINE SOON" LIGHT WILL COME ON). SET TIMING TO 18°B.T.D.C. RECONNECT ELECTRICAL CONNECTOR.
2. ADJUST FAST IDLE SCREW TO 2400 RPM ON THE HIGHEST STEP OF THE FAST IDLE CAM.
3. WITH THE ENGINE OFF, DISCONNECT THE NEGATIVE BATTERY TERMINAL FOR TEN SECONDS. RECONNECT THE BATTERY TERMINAL.
NO ADJUSTMENTS FOR IDLE AIR/FUEL MIXTURE. NO ADJUSTMENTS FOR IDLE SPEED (EXCEPT FAST IDLE).
SEE SERVICE MANUAL AND MAINTENANCE SCHEDULE FOR ADDITIONAL INFORMATION.

•FOR A/C EQUIPPED VEHICLES ONLY.

PRINTED IN U.S.A.

THIS VEHICLE CONFORMS TO U.S. EPA REGULATIONS APPLICABLE TO 1985 MODEL YEAR NEW PASSENGER CARS.

1985-V6-3.0L ELECTRA—FEDERAL

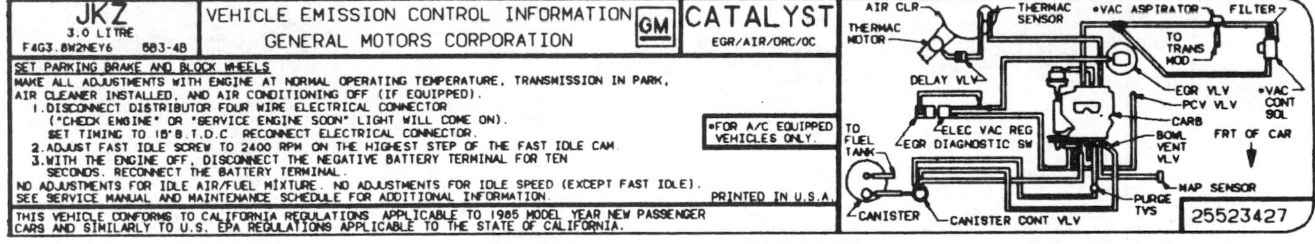
JKZ
3.0 LITRE
F4G3.8W2NEY6 5B3-4B

VEHICLE EMISSION CONTROL INFORMATION
GENERAL MOTORS CORPORATION
GM

CATALYST
EGR/AIR/ORC/OC

SET PARKING BRAKE AND BLOCK WHEELS
MAKE ALL ADJUSTMENTS WITH ENGINE AT NORMAL OPERATING TEMPERATURE, TRANSMISSION IN PARK, AIR CLEANER INSTALLED, AND AIR CONDITIONING OFF (IF EQUIPPED).
1. DISCONNECT DISTRIBUTOR FOUR WIRE ELECTRICAL CONNECTOR ("CHECK ENGINE" OR "SERVICE ENGINE SOON" LIGHT WILL COME ON). SET TIMING TO 18°B.T.D.C. RECONNECT ELECTRICAL CONNECTOR.
2. ADJUST FAST IDLE SCREW TO 2400 RPM ON THE HIGHEST STEP OF THE FAST IDLE CAM.
3. WITH THE ENGINE OFF, DISCONNECT THE NEGATIVE BATTERY TERMINAL FOR TEN SECONDS. RECONNECT THE BATTERY TERMINAL.
NO ADJUSTMENTS FOR IDLE AIR/FUEL MIXTURE. NO ADJUSTMENTS FOR IDLE SPEED (EXCEPT FAST IDLE).
SEE SERVICE MANUAL AND MAINTENANCE SCHEDULE FOR ADDITIONAL INFORMATION.

•FOR A/C EQUIPPED VEHICLES ONLY.

PRINTED IN U.S.A.

THIS VEHICLE CONFORMS TO CALIFORNIA REGULATIONS APPLICABLE TO 1985 MODEL YEAR NEW PASSENGER CARS AND SIMILARLY TO U.S. EPA REGULATIONS APPLICABLE TO THE STATE OF CALIFORNIA.

1985-V6-3.0L ELECTRA—CALIFORNIA

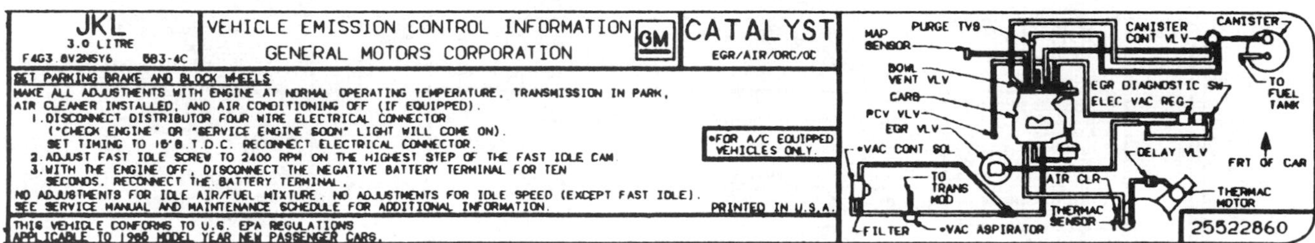
JKL
3.0 LITRE
F4G3 8V2NSY6 5B3-4C

VEHICLE EMISSION CONTROL INFORMATION
GENERAL MOTORS CORPORATION
GM

CATALYST
EGR/AIR/ORC/OC

SET PARKING BRAKE AND BLOCK WHEELS
MAKE ALL ADJUSTMENTS WITH ENGINE AT NORMAL OPERATING TEMPERATURE, TRANSMISSION IN PARK, AIR CLEANER INSTALLED, AND AIR CONDITIONING OFF (IF EQUIPPED).
1. DISCONNECT DISTRIBUTOR FOUR WIRE ELECTRICAL CONNECTOR ("CHECK ENGINE" OR "SERVICE ENGINE SOON" LIGHT WILL COME ON). SET TIMING TO 18°B.T.D.C. RECONNECT ELECTRICAL CONNECTOR.
2. ADJUST FAST IDLE SCREW TO 2400 RPM ON THE HIGHEST STEP OF THE FAST IDLE CAM
3. WITH THE ENGINE OFF, DISCONNECT THE NEGATIVE BATTERY TERMINAL FOR TEN SECONDS. RECONNECT THE BATTERY TERMINAL.
NO ADJUSTMENTS FOR IDLE AIR/FUEL MIXTURE. NO ADJUSTMENTS FOR IDLE SPEED (EXCEPT FAST IDLE).
SEE SERVICE MANUAL AND MAINTENANCE SCHEDULE FOR ADDITIONAL INFORMATION.

•FOR A/C EQUIPPED VEHICLES ONLY.

PRINTED IN U.S.A.

THIS VEHICLE CONFORMS TO U.S. EPA REGULATIONS APPLICABLE TO 1985 MODEL YEAR NEW PASSENGER CARS.

1985-V6-3.0L ELECTRA—FEDERAL

VACUUM CIRCUITS

1985 BUICK 3.0L

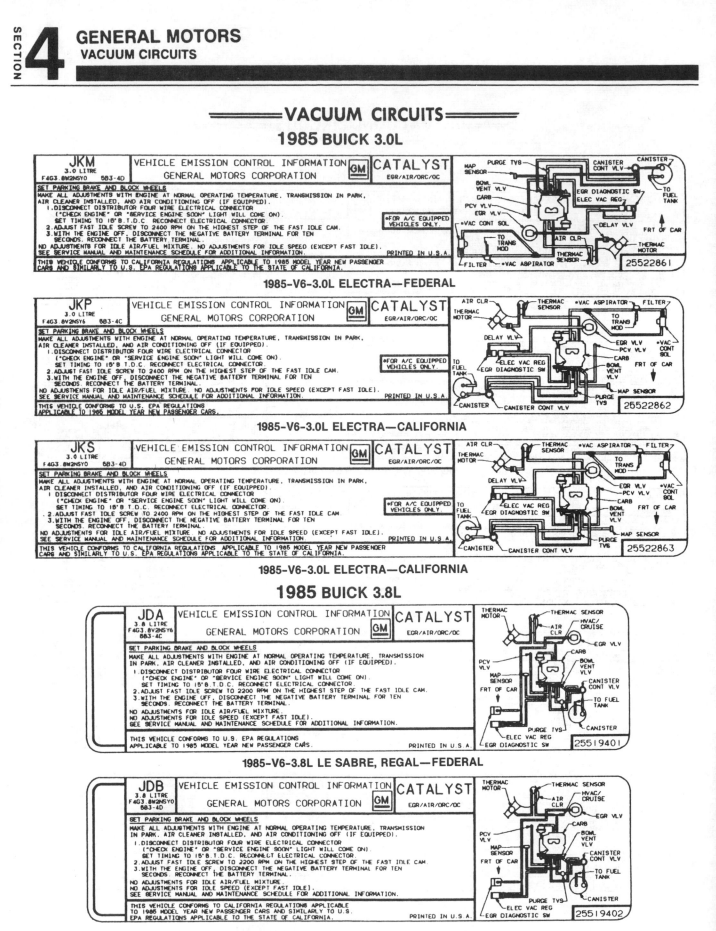

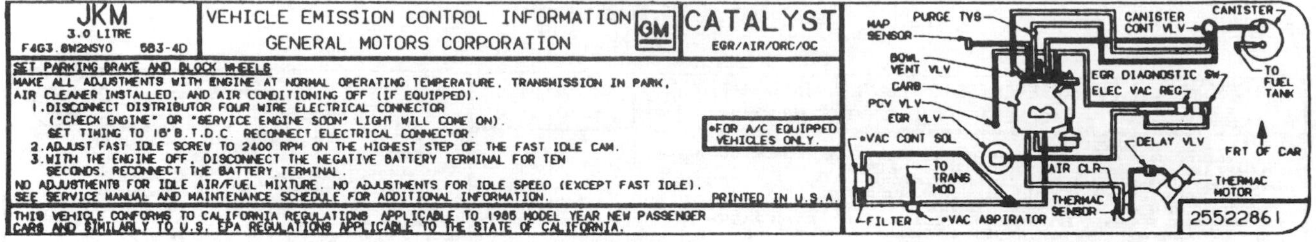
JKM
3.0 LITRE
F4G3.8W2NSY0 5B3-4D

VEHICLE EMISSION CONTROL INFORMATION
GENERAL MOTORS CORPORATION
GM
CATALYST
EGR/AIR/ORC/OC

SET PARKING BRAKE AND BLOCK WHEELS
MAKE ALL ADJUSTMENTS WITH ENGINE AT NORMAL OPERATING TEMPERATURE, TRANSMISSION IN PARK, AIR CLEANER INSTALLED, AND AIR CONDITIONING OFF (IF EQUIPPED).
1. DISCONNECT DISTRIBUTOR FOUR WIRE ELECTRICAL CONNECTOR ("CHECK ENGINE" OR "SERVICE ENGINE SOON" LIGHT WILL COME ON). SET TIMING TO 15°B.T.D.C. RECONNECT ELECTRICAL CONNECTOR.
2. ADJUST FAST IDLE SCREW TO 2400 RPM ON THE HIGHEST STEP OF THE FAST IDLE CAM.
3. WITH THE ENGINE OFF, DISCONNECT THE NEGATIVE BATTERY TERMINAL FOR TEN SECONDS. RECONNECT THE BATTERY TERMINAL.
NO ADJUSTMENTS FOR IDLE AIR/FUEL MIXTURE. NO ADJUSTMENTS FOR IDLE SPEED (EXCEPT FAST IDLE).
SEE SERVICE MANUAL AND MAINTENANCE SCHEDULE FOR ADDITIONAL INFORMATION.

•FOR A/C EQUIPPED VEHICLES ONLY.

PRINTED IN U.S.A.

THIS VEHICLE CONFORMS TO CALIFORNIA REGULATIONS APPLICABLE TO 1985 MODEL YEAR NEW PASSENGER CARS AND SIMILARLY TO U.S. EPA REGULATIONS APPLICABLE TO THE STATE OF CALIFORNIA.

1985-V6-3.0L ELECTRA—FEDERAL

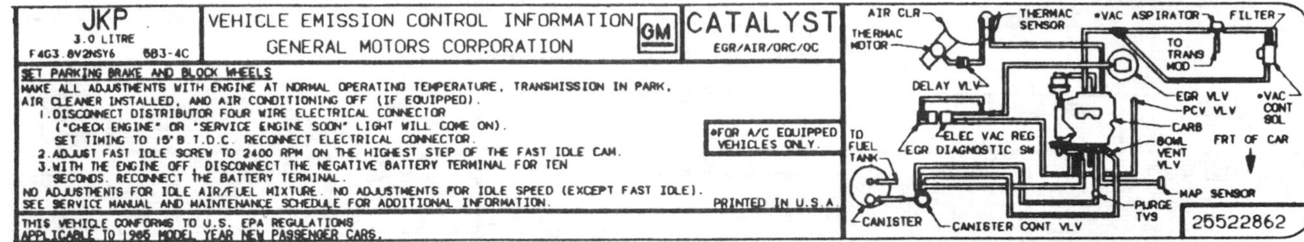
JKP
3.0 LITRE
F4G3.8V2NSY6 5B3-4C

VEHICLE EMISSION CONTROL INFORMATION
GENERAL MOTORS CORPORATION
GM
CATALYST
EGR/AIR/ORC/OC

SET PARKING BRAKE AND BLOCK WHEELS
MAKE ALL ADJUSTMENTS WITH ENGINE AT NORMAL OPERATING TEMPERATURE, TRANSMISSION IN PARK, AIR CLEANER INSTALLED, AND AIR CONDITIONING OFF (IF EQUIPPED).
1. DISCONNECT DISTRIBUTOR FOUR WIRE ELECTRICAL CONNECTOR ("CHECK ENGINE" OR "SERVICE ENGINE SOON" LIGHT WILL COME ON). SET TIMING TO 15°B.T.D.C. RECONNECT ELECTRICAL CONNECTOR.
2. ADJUST FAST IDLE SCREW TO 2400 RPM ON THE HIGHEST STEP OF THE FAST IDLE CAM.
3. WITH THE ENGINE OFF, DISCONNECT THE NEGATIVE BATTERY TERMINAL FOR TEN SECONDS. RECONNECT THE BATTERY TERMINAL.
NO ADJUSTMENTS FOR IDLE AIR/FUEL MIXTURE. NO ADJUSTMENTS FOR IDLE SPEED (EXCEPT FAST IDLE).
SEE SERVICE MANUAL AND MAINTENANCE SCHEDULE FOR ADDITIONAL INFORMATION.

•FOR A/C EQUIPPED VEHICLES ONLY.

PRINTED IN U.S.A.

THIS VEHICLE CONFORMS TO U.S. EPA REGULATIONS APPLICABLE TO 1985 MODEL YEAR NEW PASSENGER CARS.

1985-V6-3.0L ELECTRA—CALIFORNIA

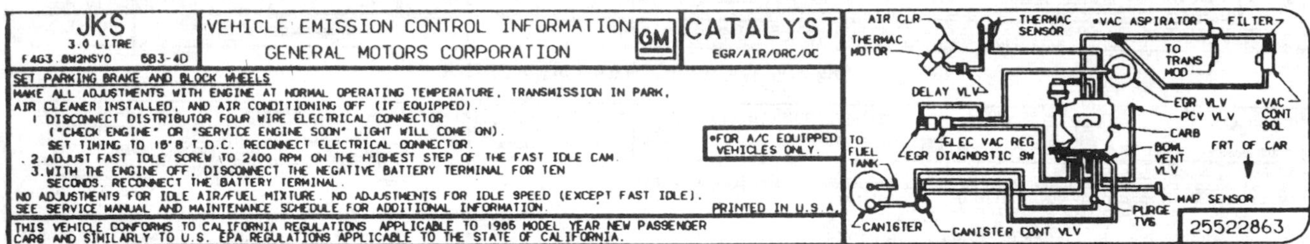
JKS
3.0 LITRE
F4G3.8W2NSY0 5B3-4D

VEHICLE EMISSION CONTROL INFORMATION
GENERAL MOTORS CORPORATION
GM
CATALYST
EGR/AIR/ORC/OC

SET PARKING BRAKE AND BLOCK WHEELS
MAKE ALL ADJUSTMENTS WITH ENGINE AT NORMAL OPERATING TEMPERATURE, TRANSMISSION IN PARK, AIR CLEANER INSTALLED, AND AIR CONDITIONING OFF (IF EQUIPPED).
1. DISCONNECT DISTRIBUTOR FOUR WIRE ELECTRICAL CONNECTOR ("CHECK ENGINE" OR "SERVICE ENGINE SOON" LIGHT WILL COME ON). SET TIMING TO 15°B.T.D.C. RECONNECT ELECTRICAL CONNECTOR.
2. ADJUST FAST IDLE SCREW TO 2400 RPM ON THE HIGHEST STEP OF THE FAST IDLE CAM.
3. WITH THE ENGINE OFF, DISCONNECT THE NEGATIVE BATTERY TERMINAL FOR TEN SECONDS. RECONNECT THE BATTERY TERMINAL.
NO ADJUSTMENTS FOR IDLE AIR/FUEL MIXTURE. NO ADJUSTMENTS FOR IDLE SPEED (EXCEPT FAST IDLE).
SEE SERVICE MANUAL AND MAINTENANCE SCHEDULE FOR ADDITIONAL INFORMATION.

•FOR A/C EQUIPPED VEHICLES ONLY.

PRINTED IN U.S.A.

THIS VEHICLE CONFORMS TO CALIFORNIA REGULATIONS APPLICABLE TO 1985 MODEL YEAR NEW PASSENGER CARS AND SIMILARLY TO U.S. EPA REGULATIONS APPLICABLE TO THE STATE OF CALIFORNIA.

1985-V6-3.0L ELECTRA—CALIFORNIA

1985 BUICK 3.8L

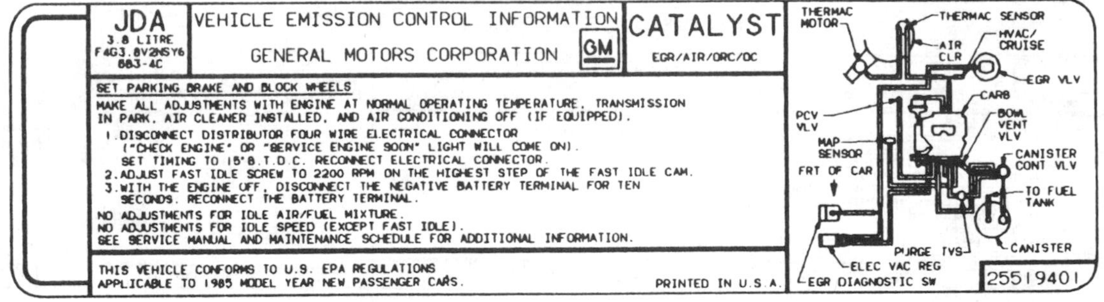
JDA
3.8 LITRE
F4G3.8V2NSY6
5B3-4C

VEHICLE EMISSION CONTROL INFORMATION
GENERAL MOTORS CORPORATION
GM
CATALYST
EGR/AIR/ORC/OC

SET PARKING BRAKE AND BLOCK WHEELS
MAKE ALL ADJUSTMENTS WITH ENGINE AT NORMAL OPERATING TEMPERATURE, TRANSMISSION IN PARK, AIR CLEANER INSTALLED, AND AIR CONDITIONING OFF (IF EQUIPPED).
1. DISCONNECT DISTRIBUTOR FOUR WIRE ELECTRICAL CONNECTOR ("CHECK ENGINE" OR "SERVICE ENGINE SOON" LIGHT WILL COME ON). SET TIMING TO 15°B.T.D.C. RECONNECT ELECTRICAL CONNECTOR.
2. ADJUST FAST IDLE SCREW TO 2200 RPM ON THE HIGHEST STEP OF THE FAST IDLE CAM.
3. WITH THE ENGINE OFF, DISCONNECT THE NEGATIVE BATTERY TERMINAL FOR TEN SECONDS. RECONNECT THE BATTERY TERMINAL.
NO ADJUSTMENTS FOR IDLE AIR/FUEL MIXTURE.
NO ADJUSTMENTS FOR IDLE SPEED (EXCEPT FAST IDLE).
SEE SERVICE MANUAL AND MAINTENANCE SCHEDULE FOR ADDITIONAL INFORMATION.

THIS VEHICLE CONFORMS TO U.S. EPA REGULATIONS APPLICABLE TO 1985 MODEL YEAR NEW PASSENGER CARS.

PRINTED IN U.S.A.

1985-V6-3.8L LE SABRE, REGAL—FEDERAL

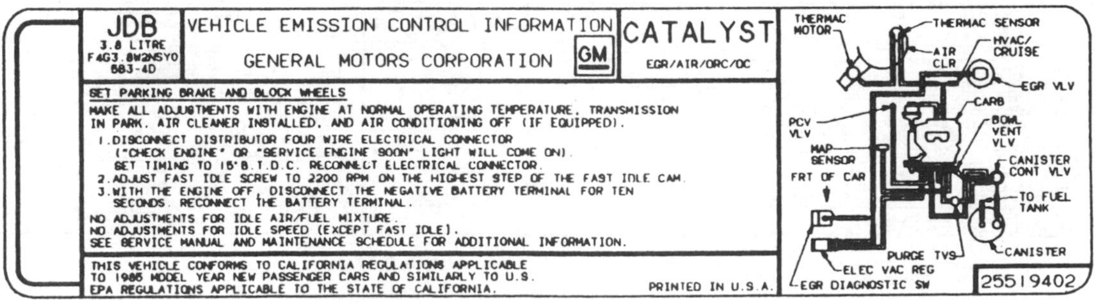
JDB
3.8 LITRE
F4G3.8W2NSY0
5B3-4D

VEHICLE EMISSION CONTROL INFORMATION
GENERAL MOTORS CORPORATION
GM
CATALYST
EGR/AIR/ORC/OC

SET PARKING BRAKE AND BLOCK WHEELS
MAKE ALL ADJUSTMENTS WITH ENGINE AT NORMAL OPERATING TEMPERATURE, TRANSMISSION IN PARK, AIR CLEANER INSTALLED, AND AIR CONDITIONING OFF (IF EQUIPPED).
1. DISCONNECT DISTRIBUTOR FOUR WIRE ELECTRICAL CONNECTOR ("CHECK ENGINE" OR "SERVICE ENGINE SOON" LIGHT WILL COME ON). SET TIMING TO 15°B.T.D.C. RECONNECT ELECTRICAL CONNECTOR.
2. ADJUST FAST IDLE SCREW TO 2200 RPM ON THE HIGHEST STEP OF THE FAST IDLE CAM.
3. WITH THE ENGINE OFF, DISCONNECT THE NEGATIVE BATTERY TERMINAL FOR TEN SECONDS. RECONNECT THE BATTERY TERMINAL.
NO ADJUSTMENTS FOR IDLE AIR/FUEL MIXTURE.
NO ADJUSTMENTS FOR IDLE SPEED (EXCEPT FAST IDLE).
SEE SERVICE MANUAL AND MAINTENANCE SCHEDULE FOR ADDITIONAL INFORMATION.

THIS VEHICLE CONFORMS TO CALIFORNIA REGULATIONS APPLICABLE TO 1985 MODEL YEAR NEW PASSENGER CARS AND SIMILARLY TO U.S. EPA REGULATIONS APPLICABLE TO THE STATE OF CALIFORNIA.

PRINTED IN U.S.A.

1985-V6-3.8L LE SABRE, REGAL—CALIFORNIA

VACUUM CIRCUITS

1985 BUICK 3.8L

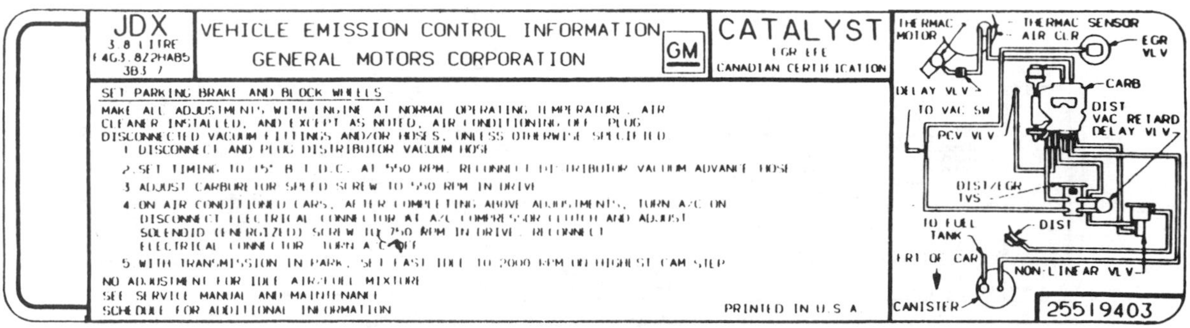

JDX
3.8 LITRE
F4G3.8Z2HAB5
3B3 7

VEHICLE EMISSION CONTROL INFORMATION
GENERAL MOTORS CORPORATION
GM

CATALYST
EGR EFE
CANADIAN CERTIFICATION

SET PARKING BRAKE AND BLOCK WHEELS
MAKE ALL ADJUSTMENTS WITH ENGINE AT NORMAL OPERATING TEMPERATURE, AIR CLEANER INSTALLED, AND EXCEPT AS NOTED, AIR CONDITIONING OFF. PLUG DISCONNECTED VACUUM FITTINGS AND/OR HOSES, UNLESS OTHERWISE SPECIFIED.
1. DISCONNECT AND PLUG DISTRIBUTOR VACUUM HOSE
2. SET TIMING TO 15° B.T.D.C. AT 550 RPM. RECONNECT DISTRIBUTOR VACUUM ADVANCE HOSE
3. ADJUST CARBURETOR SPEED SCREW TO 550 RPM IN DRIVE
4. ON AIR CONDITIONED CARS, AFTER COMPLETING ABOVE ADJUSTMENTS, TURN A/C ON. DISCONNECT ELECTRICAL CONNECTOR AT A/C COMPRESSOR CLUTCH AND ADJUST SOLENOID (ENERGIZED) SCREW TO 750 RPM IN DRIVE. RECONNECT ELECTRICAL CONNECTOR. TURN A/C OFF
5. WITH TRANSMISSION IN PARK, SET FAST IDLE TO 2000 RPM ON HIGHEST CAM STEP
NO ADJUSTMENT FOR IDLE AIR/FUEL MIXTURE
SEE SERVICE MANUAL AND MAINTENANCE SCHEDULE FOR ADDITIONAL INFORMATION
PRINTED IN U.S.A.

1985-V6-3.8L LE SABRE, REGAL—CANADA

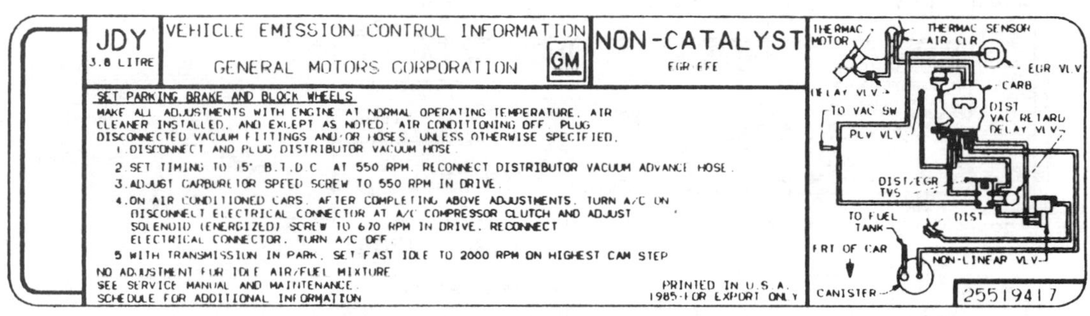

JDY
3.8 LITRE

VEHICLE EMISSION CONTROL INFORMATION
GENERAL MOTORS CORPORATION
GM

NON-CATALYST
EGR EFE

SET PARKING BRAKE AND BLOCK WHEELS
MAKE ALL ADJUSTMENTS WITH ENGINE AT NORMAL OPERATING TEMPERATURE, AIR CLEANER INSTALLED, AND EXCEPT AS NOTED, AIR CONDITIONING OFF. PLUG DISCONNECTED VACUUM FITTINGS AND/OR HOSES, UNLESS OTHERWISE SPECIFIED.
1. DISCONNECT AND PLUG DISTRIBUTOR VACUUM HOSE.
2. SET TIMING TO 15° B.T.D.C. AT 550 RPM. RECONNECT DISTRIBUTOR VACUUM ADVANCE HOSE.
3. ADJUST CARBURETOR SPEED SCREW TO 550 RPM IN DRIVE.
4. ON AIR CONDITIONED CARS, AFTER COMPLETING ABOVE ADJUSTMENTS, TURN A/C ON. DISCONNECT ELECTRICAL CONNECTOR AT A/C COMPRESSOR CLUTCH AND ADJUST SOLENOID (ENERGIZED) SCREW TO 670 RPM IN DRIVE. RECONNECT ELECTRICAL CONNECTOR. TURN A/C OFF.
5. WITH TRANSMISSION IN PARK, SET FAST IDLE TO 2000 RPM ON HIGHEST CAM STEP
NO ADJUSTMENT FOR IDLE AIR/FUEL MIXTURE
SEE SERVICE MANUAL AND MAINTENANCE SCHEDULE FOR ADDITIONAL INFORMATION
PRINTED IN U.S.A.
1985-FOR EXPORT ONLY

1985-V6-3.8L LE SABRE, REGAL—EXPORT

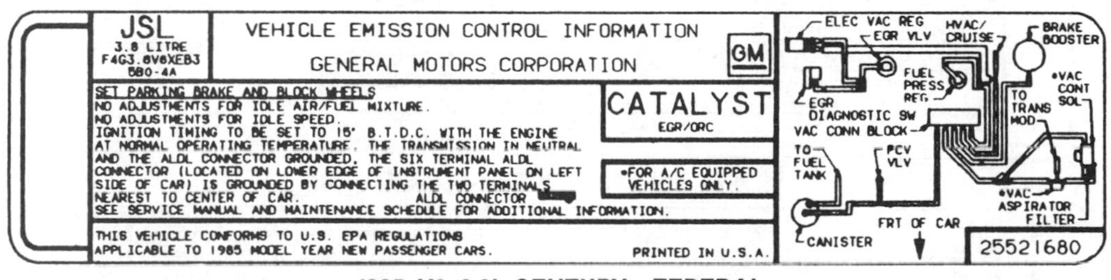

JSL
3.8 LITRE
F4G3.8V6XEB3
5B0-4A

VEHICLE EMISSION CONTROL INFORMATION
GENERAL MOTORS CORPORATION
GM

SET PARKING BRAKE AND BLOCK WHEELS
NO ADJUSTMENTS FOR IDLE AIR/FUEL MIXTURE.
NO ADJUSTMENTS FOR IDLE SPEED.
IGNITION TIMING TO BE SET TO 15° B.T.D.C. WITH THE ENGINE AT NORMAL OPERATING TEMPERATURE, THE TRANSMISSION IN NEUTRAL AND THE ALDL CONNECTOR GROUNDED. THE SIX TERMINAL ALDL CONNECTOR (LOCATED ON LOWER EDGE OF INSTRUMENT PANEL ON LEFT SIDE OF CAR) IS GROUNDED BY CONNECTING THE TWO TERMINALS NEAREST TO CENTER OF CAR. ALDL CONNECTOR
SEE SERVICE MANUAL AND MAINTENANCE SCHEDULE FOR ADDITIONAL INFORMATION.

CATALYST
EGR/ORC

•FOR A/C EQUIPPED VEHICLES ONLY.

THIS VEHICLE CONFORMS TO U.S. EPA REGULATIONS APPLICABLE TO 1985 MODEL YEAR NEW PASSENGER CARS.
PRINTED IN U.S.A.

1985-V6-3.8L CENTURY—FEDERAL

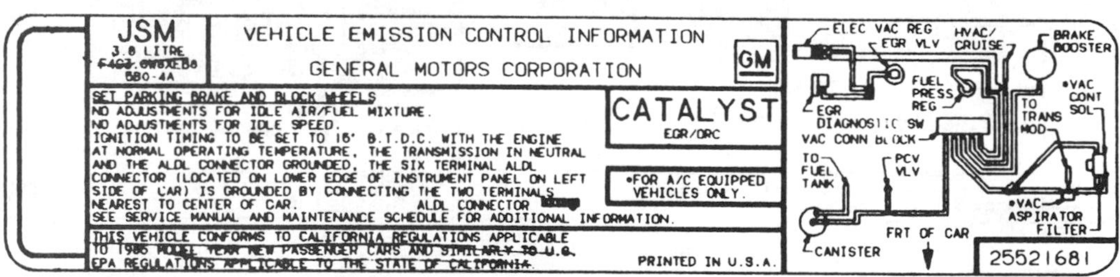

JSM
3.8 LITRE
F4G3.8V8XEB8
5B0-4A

VEHICLE EMISSION CONTROL INFORMATION
GENERAL MOTORS CORPORATION
GM

SET PARKING BRAKE AND BLOCK WHEELS
NO ADJUSTMENTS FOR IDLE AIR/FUEL MIXTURE.
NO ADJUSTMENTS FOR IDLE SPEED.
IGNITION TIMING TO BE SET TO 15° B.T.D.C. WITH THE ENGINE AT NORMAL OPERATING TEMPERATURE, THE TRANSMISSION IN NEUTRAL AND THE ALDL CONNECTOR GROUNDED. THE SIX TERMINAL ALDL CONNECTOR (LOCATED ON LOWER EDGE OF INSTRUMENT PANEL ON LEFT SIDE OF CAR) IS GROUNDED BY CONNECTING THE TWO TERMINALS NEAREST TO CENTER OF CAR. ALDL CONNECTOR
SEE SERVICE MANUAL AND MAINTENANCE SCHEDULE FOR ADDITIONAL INFORMATION.

CATALYST
EGR/ORC

•FOR A/C EQUIPPED VEHICLES ONLY.

THIS VEHICLE CONFORMS TO CALIFORNIA REGULATIONS APPLICABLE TO 1985 MODEL YEAR NEW PASSENGER CARS AND SIMILARLY TO U.S. EPA REGULATIONS APPLICABLE TO THE STATE OF CALIFORNIA.
PRINTED IN U.S.A.

1985-V6-3.8L CENTURY—CALIFORNIA

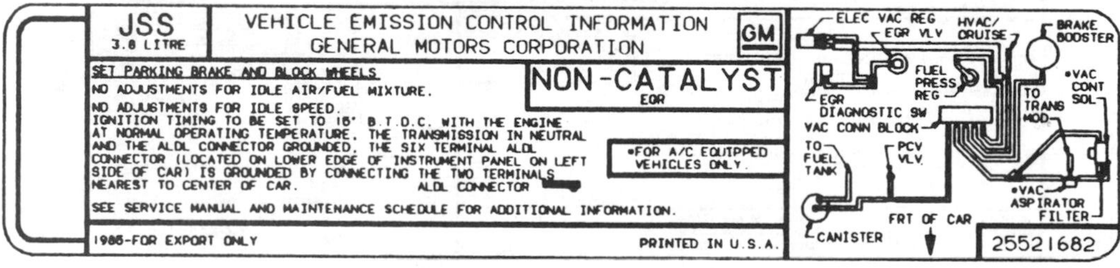

JSS
3.8 LITRE

VEHICLE EMISSION CONTROL INFORMATION
GENERAL MOTORS CORPORATION
GM

SET PARKING BRAKE AND BLOCK WHEELS
NO ADJUSTMENTS FOR IDLE AIR/FUEL MIXTURE.
NO ADJUSTMENTS FOR IDLE SPEED.
IGNITION TIMING TO BE SET TO 15° B.T.D.C. WITH THE ENGINE AT NORMAL OPERATING TEMPERATURE, THE TRANSMISSION IN NEUTRAL AND THE ALDL CONNECTOR GROUNDED. THE SIX TERMINAL ALDL CONNECTOR (LOCATED ON LOWER EDGE OF INSTRUMENT PANEL ON LEFT SIDE OF CAR) IS GROUNDED BY CONNECTING THE TWO TERMINALS NEAREST TO CENTER OF CAR. ALDL CONNECTOR
SEE SERVICE MANUAL AND MAINTENANCE SCHEDULE FOR ADDITIONAL INFORMATION.

NON-CATALYST
EGR

•FOR A/C EQUIPPED VEHICLES ONLY.

1985-FOR EXPORT ONLY
PRINTED IN U.S.A.

1985-V6-3.8L CENTURY—EXPORT

VACUUM CIRCUITS

1985 BUICK 3.8L

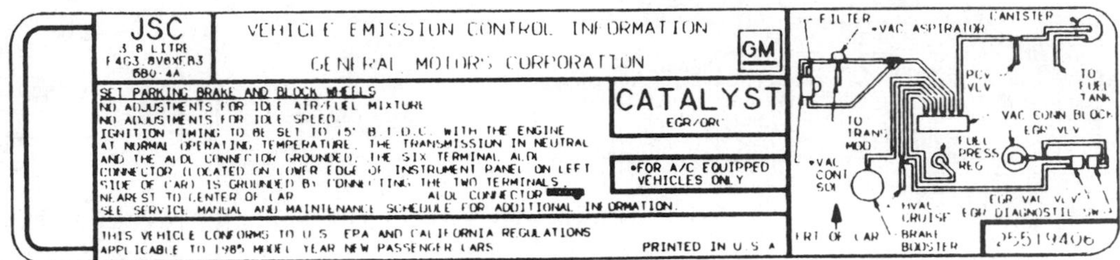

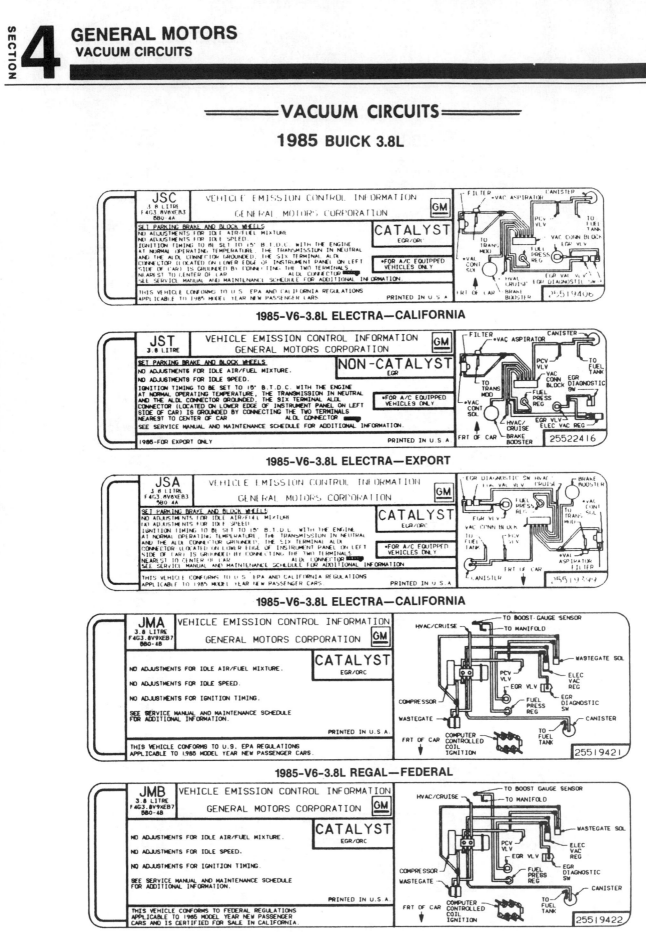

1985–V6–3.8L ELECTRA—CALIFORNIA

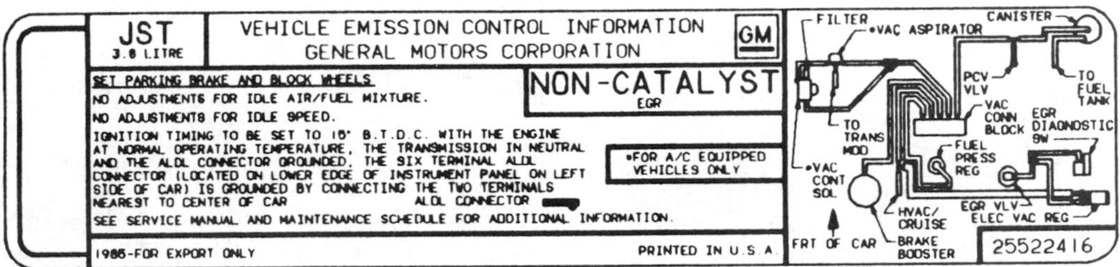

1985–V6–3.8L ELECTRA—EXPORT

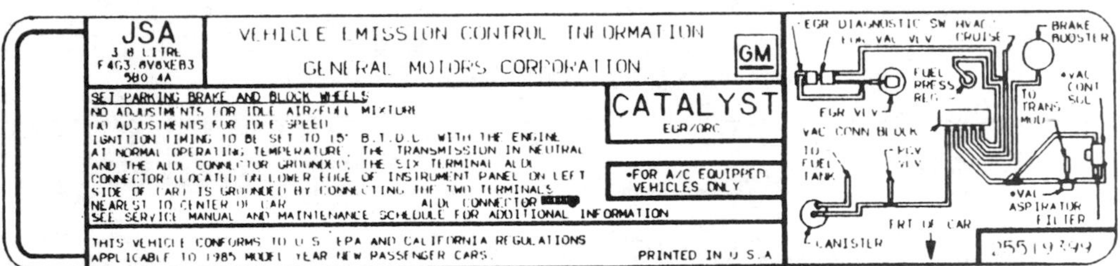

1985–V6–3.8L ELECTRA—CALIFORNIA

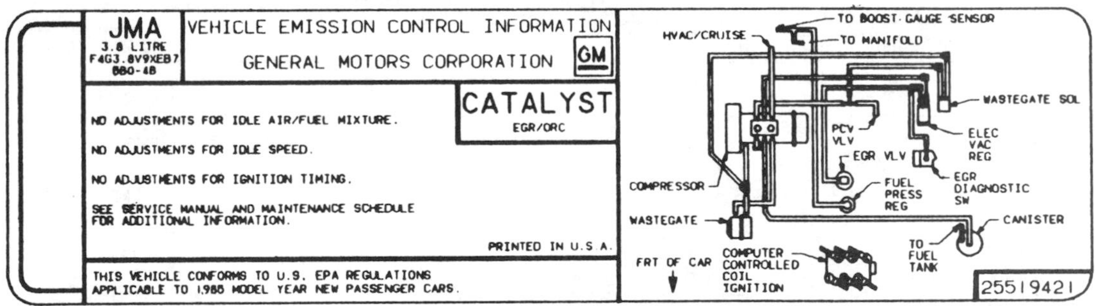

1985–V6–3.8L REGAL—FEDERAL

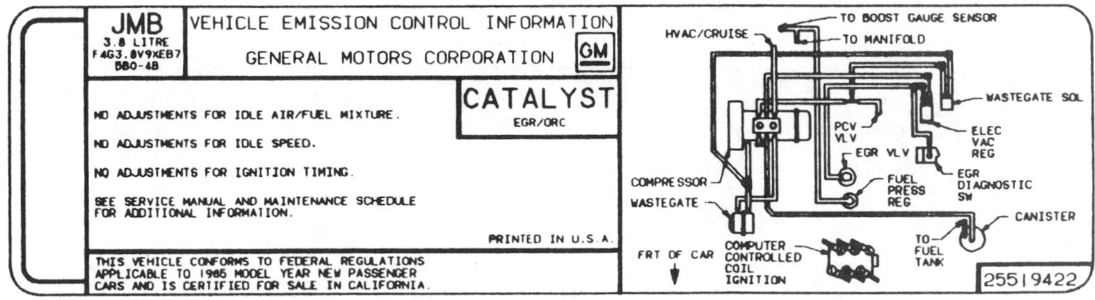

JMB
3.8 LITRE
F4G3.8V9XEB7
VEHICLE EMISSION CONTROL INFORMATION
GENERAL MOTORS CORPORATION
GM
CATALYST
EGR/ORC
NO ADJUSTMENTS FOR IDLE AIR/FUEL MIXTURE.
NO ADJUSTMENTS FOR IDLE SPEED.
NO ADJUSTMENTS FOR IGNITION TIMING.
SEE SERVICE MANUAL AND MAINTENANCE SCHEDULE FOR ADDITIONAL INFORMATION.
PRINTED IN U.S.A.
THIS VEHICLE CONFORMS TO FEDERAL REGULATIONS APPLICABLE TO 1985 MODEL YEAR NEW PASSENGER CARS AND IS CERTIFIED FOR SALE IN CALIFORNIA.
TO BOOST GAUGE SENSOR
HVAC/CRUISE
TO MANIFOLD
WASTEGATE SOL
PCV VLV
ELEC VAC REG
EGR VLV
EGR DIAGNOSTIC SW
COMPRESSOR
FUEL PRESS REG
WASTEGATE
CANISTER
FRT OF CAR
COMPUTER CONTROLLED COIL IGNITION
TO FUEL TANK
25519422

1985–V6–3.8L REGAL—FEDERAL

VACUUM CIRCUITS

1985 BUICK 3.8L

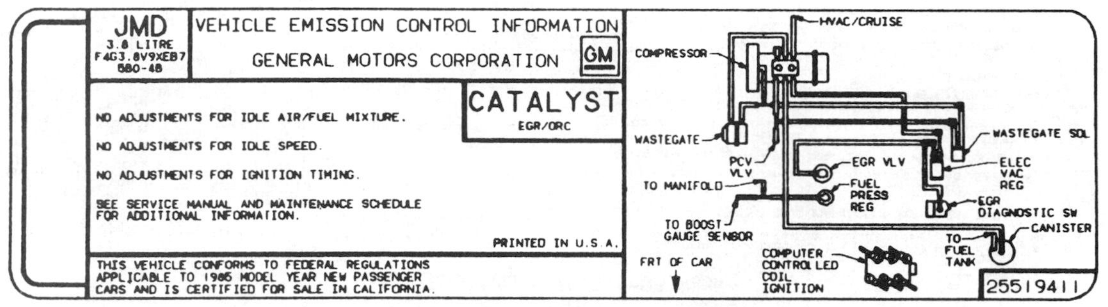

1985-V6-3.8L RIVIERA—FEDERAL

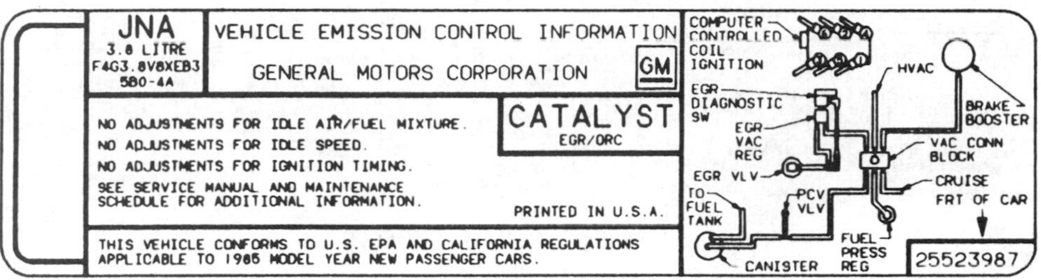

1985-V6-3.8L SOMERSET REGAL—NATIONAL

1985 CADILLAC 4.1L

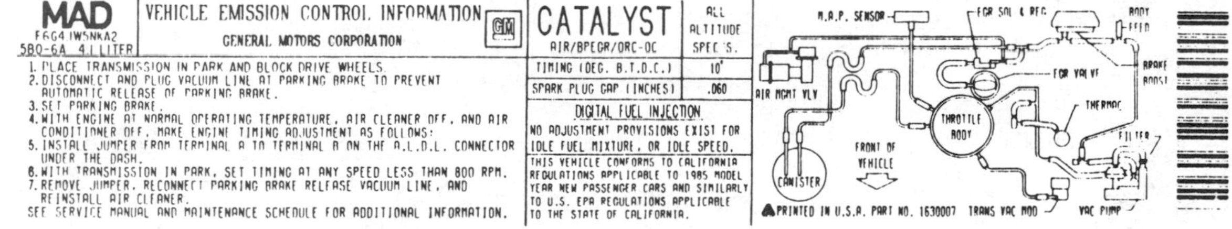

1985-V8-4.1L DFI—CALIFORNIA

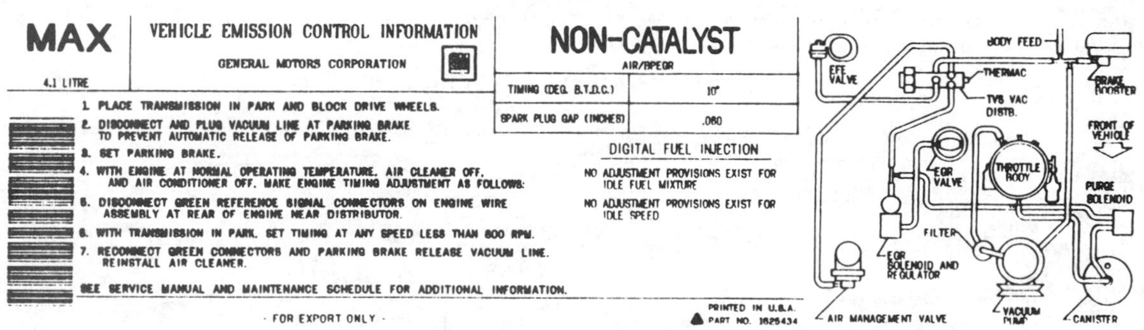

1985-V8-4.1L DFI—EXPORT

VACUUM CIRCUITS

1985 CADILLAC 4.1L

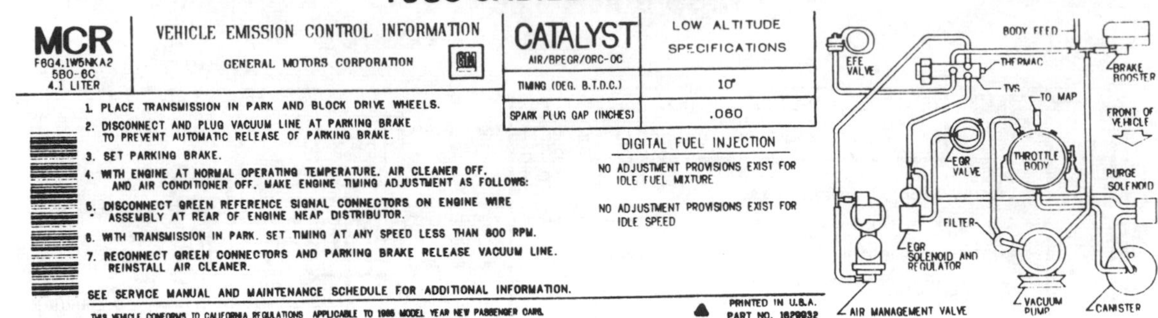

1985-V8-4.1L DFI—CALIFORNIA

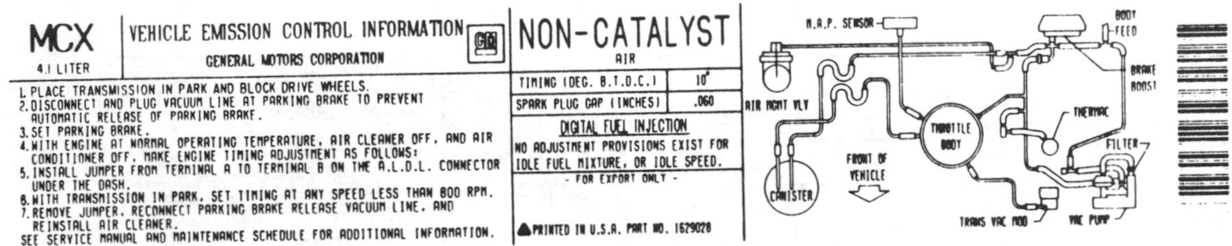

1985-V8-4.1L DFI—EXPORT

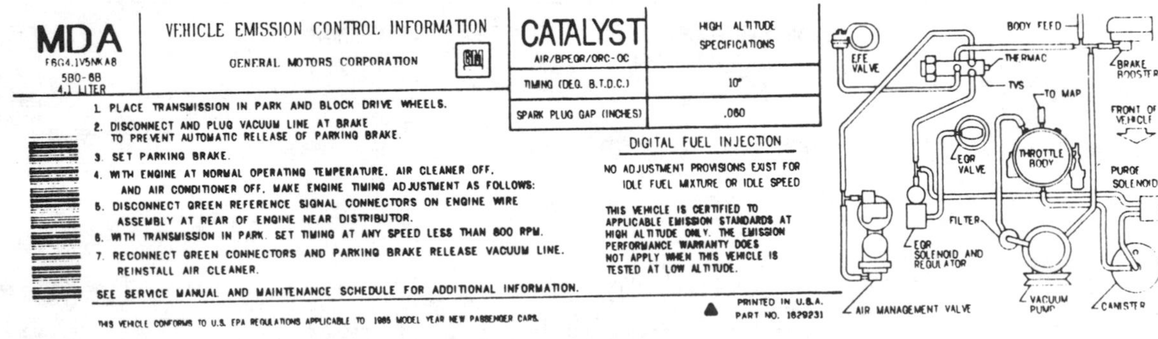

1985-V8-4.1L DFI—HIGH ALT.

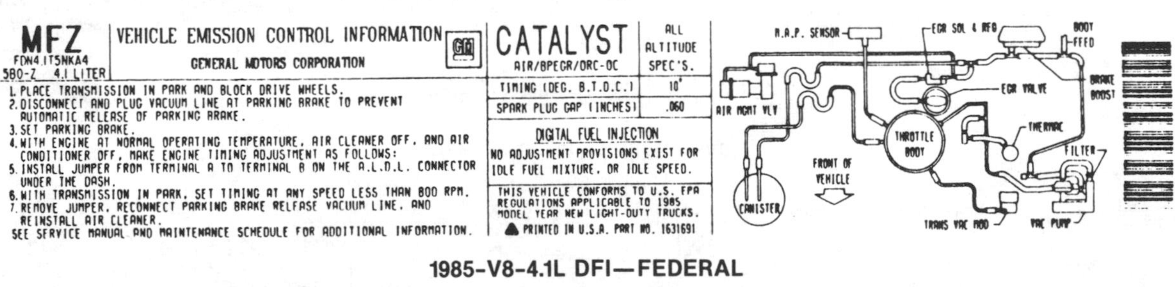

1985-V8-4.1L DFI—FEDERAL

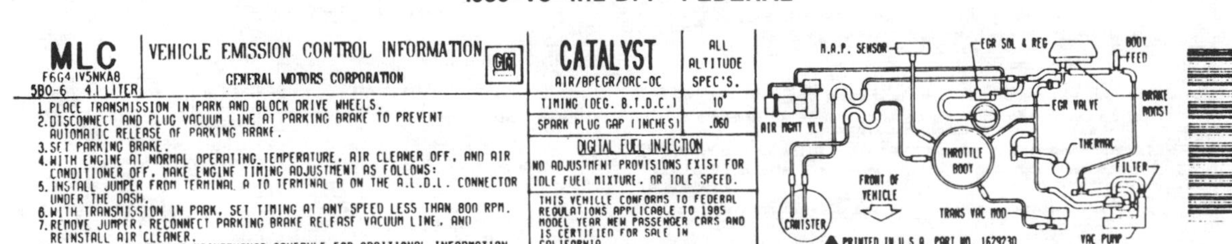

1985-V8-4.1L DFI—CALIFORNIA

VACUUM CIRCUITS

1985 CADILLAC 4.1L

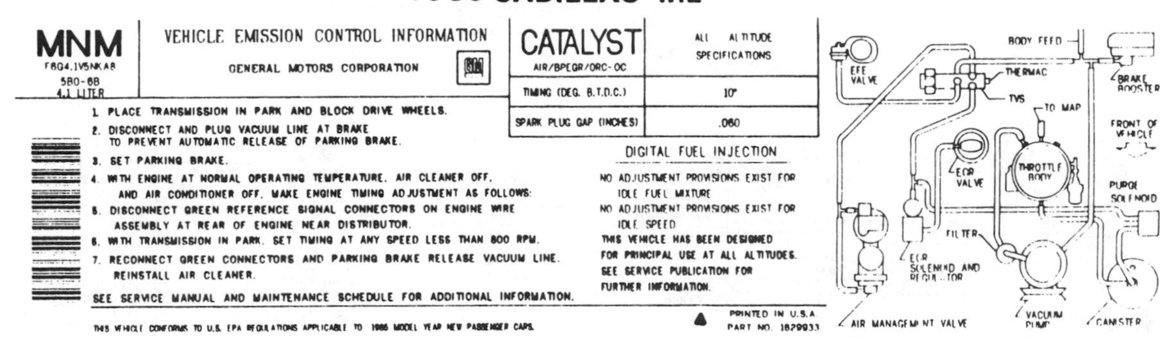

MNM
F8G4.1V5NKA8
5B0-6B
4.1 LITER

VEHICLE EMISSION CONTROL INFORMATION
GENERAL MOTORS CORPORATION

1. PLACE TRANSMISSION IN PARK AND BLOCK DRIVE WHEELS.
2. DISCONNECT AND PLUG VACUUM LINE AT BRAKE TO PREVENT AUTOMATIC RELEASE OF PARKING BRAKE.
3. SET PARKING BRAKE.
4. WITH ENGINE AT NORMAL OPERATING TEMPERATURE, AIR CLEANER OFF, AND AIR CONDITIONER OFF, MAKE ENGINE TIMING ADJUSTMENT AS FOLLOWS:
5. DISCONNECT GREEN REFERENCE SIGNAL CONNECTORS ON ENGINE WIRE ASSEMBLY AT REAR OF ENGINE NEAR DISTRIBUTOR.
6. WITH TRANSMISSION IN PARK, SET TIMING AT ANY SPEED LESS THAN 800 RPM.
7. RECONNECT GREEN CONNECTORS AND PARKING BRAKE RELEASE VACUUM LINE. REINSTALL AIR CLEANER.

SEE SERVICE MANUAL AND MAINTENANCE SCHEDULE FOR ADDITIONAL INFORMATION.

THIS VEHICLE CONFORMS TO U.S. EPA REGULATIONS APPLICABLE TO 1985 MODEL YEAR NEW PASSENGER CARS.

CATALYST AIR/BPEGR/ORC-OC	ALL ALTITUDE SPECIFICATIONS
TIMING (DEG. B.T.D.C.)	10°
SPARK PLUG GAP (INCHES)	.060

DIGITAL FUEL INJECTION

NO ADJUSTMENT PROVISIONS EXIST FOR IDLE FUEL MIXTURE
NO ADJUSTMENT PROVISIONS EXIST FOR IDLE SPEED
THIS VEHICLE HAS BEEN DESIGNED FOR PRINCIPAL USE AT ALL ALTITUDES.
SEE SERVICE PUBLICATION FOR FURTHER INFORMATION.

PRINTED IN U.S.A. PART NO. 1829933

1985-V8-4.1L DFI—FEDERAL

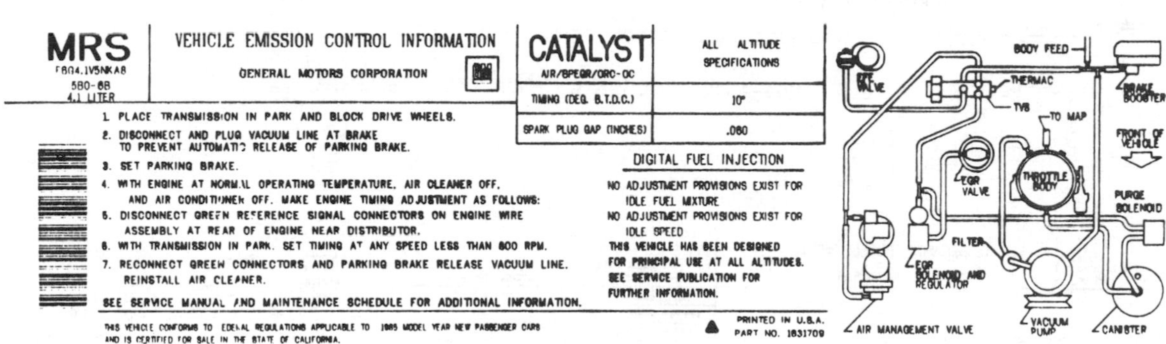

MRS
F8G4.1V5NKA8
5B0-6B
4.1 LITER

VEHICLE EMISSION CONTROL INFORMATION
GENERAL MOTORS CORPORATION

1. PLACE TRANSMISSION IN PARK AND BLOCK DRIVE WHEELS.
2. DISCONNECT AND PLUG VACUUM LINE AT BRAKE TO PREVENT AUTOMATIC RELEASE OF PARKING BRAKE.
3. SET PARKING BRAKE.
4. WITH ENGINE AT NORMAL OPERATING TEMPERATURE, AIR CLEANER OFF, AND AIR CONDITIONER OFF, MAKE ENGINE TIMING ADJUSTMENT AS FOLLOWS:
5. DISCONNECT GREEN REFERENCE SIGNAL CONNECTORS ON ENGINE WIRE ASSEMBLY AT REAR OF ENGINE NEAR DISTRIBUTOR.
6. WITH TRANSMISSION IN PARK, SET TIMING AT ANY SPEED LESS THAN 800 RPM.
7. RECONNECT GREEN CONNECTORS AND PARKING BRAKE RELEASE VACUUM LINE. REINSTALL AIR CLEANER.

SEE SERVICE MANUAL AND MAINTENANCE SCHEDULE FOR ADDITIONAL INFORMATION.

THIS VEHICLE CONFORMS TO FEDERAL REGULATIONS APPLICABLE TO 1985 MODEL YEAR NEW PASSENGER CARS AND IS CERTIFIED FOR SALE IN THE STATE OF CALIFORNIA.

CATALYST AIR/BPEGR/ORC-OC	ALL ALTITUDE SPECIFICATIONS
TIMING (DEG. B.T.D.C.)	10°
SPARK PLUG GAP (INCHES)	.060

DIGITAL FUEL INJECTION

NO ADJUSTMENT PROVISIONS EXIST FOR IDLE FUEL MIXTURE
NO ADJUSTMENT PROVISIONS EXIST FOR IDLE SPEED
THIS VEHICLE HAS BEEN DESIGNED FOR PRINCIPAL USE AT ALL ALTITUDES.
SEE SERVICE PUBLICATION FOR FURTHER INFORMATION.

PRINTED IN U.S.A. PART NO. 1831709

1985-V8-4.1L DFI—FEDERAL

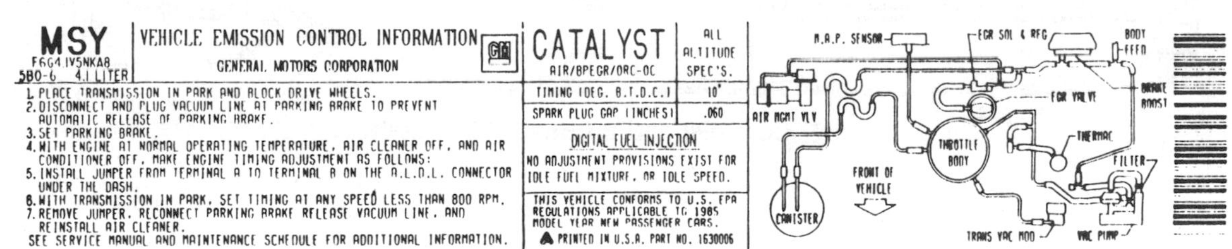

MSY
F6G4.1V5NKA8
5B0-6 4.1 LITER

VEHICLE EMISSION CONTROL INFORMATION
GENERAL MOTORS CORPORATION

1. PLACE TRANSMISSION IN PARK AND BLOCK DRIVE WHEELS.
2. DISCONNECT AND PLUG VACUUM LINE AT PARKING BRAKE TO PREVENT AUTOMATIC RELEASE OF PARKING BRAKE.
3. SET PARKING BRAKE.
4. WITH ENGINE AT NORMAL OPERATING TEMPERATURE, AIR CLEANER OFF, AND AIR CONDITIONER OFF, MAKE ENGINE TIMING ADJUSTMENT AS FOLLOWS:
5. INSTALL JUMPER FROM TERMINAL A TO TERMINAL B ON THE A.L.D.L. CONNECTOR UNDER THE DASH.
6. WITH TRANSMISSION IN PARK, SET TIMING AT ANY SPEED LESS THAN 800 RPM.
7. REMOVE JUMPER. RECONNECT PARKING BRAKE RELEASE VACUUM LINE, AND REINSTALL AIR CLEANER.

SEE SERVICE MANUAL AND MAINTENANCE SCHEDULE FOR ADDITIONAL INFORMATION.

CATALYST AIR/BPEGR/ORC-OC	ALL ALTITUDE SPEC'S.
TIMING (DEG. B.T.D.C.)	10°
SPARK PLUG GAP (INCHES)	.060

DIGITAL FUEL INJECTION

NO ADJUSTMENT PROVISIONS EXIST FOR IDLE FUEL MIXTURE, OR IDLE SPEED.

THIS VEHICLE CONFORMS TO U.S. EPA REGULATIONS APPLICABLE TO 1985 MODEL YEAR NEW PASSENGER CARS.

PRINTED IN U.S.A. PART NO. 1630006

1985-V8-4.1L DFI—FEDERAL

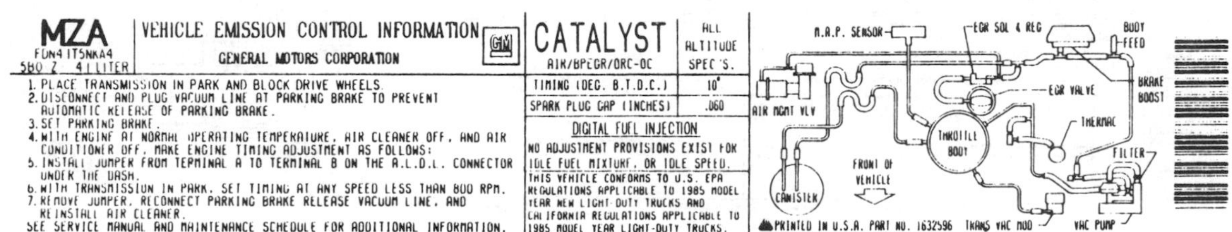

MZA
FDN4.1T5NKA4
5B0-2 4.1 LITER

VEHICLE EMISSION CONTROL INFORMATION
GENERAL MOTORS CORPORATION

1. PLACE TRANSMISSION IN PARK AND BLOCK DRIVE WHEELS.
2. DISCONNECT AND PLUG VACUUM LINE AT PARKING BRAKE TO PREVENT AUTOMATIC RELEASE OF PARKING BRAKE.
3. SET PARKING BRAKE.
4. WITH ENGINE AT NORMAL OPERATING TEMPERATURE, AIR CLEANER OFF, AND AIR CONDITIONER OFF, MAKE ENGINE TIMING ADJUSTMENT AS FOLLOWS:
5. INSTALL JUMPER FROM TERMINAL A TO TERMINAL B ON THE A.L.D.L. CONNECTOR UNDER THE DASH.
6. WITH TRANSMISSION IN PARK, SET TIMING AT ANY SPEED LESS THAN 800 RPM.
7. REMOVE JUMPER. RECONNECT PARKING BRAKE RELEASE VACUUM LINE, AND REINSTALL AIR CLEANER.

SEE SERVICE MANUAL AND MAINTENANCE SCHEDULE FOR ADDITIONAL INFORMATION.

CATALYST AIR/BPEGR/ORC-OC	ALL ALTITUDE SPEC'S.
TIMING (DEG. B.T.D.C.)	10°
SPARK PLUG GAP (INCHES)	.060

DIGITAL FUEL INJECTION

NO ADJUSTMENT PROVISIONS EXIST FOR IDLE FUEL MIXTURE, OR IDLE SPEED.

THIS VEHICLE CONFORMS TO U.S. EPA REGULATIONS APPLICABLE TO 1985 MODEL YEAR NEW LIGHT-DUTY TRUCKS AND CALIFORNIA REGULATIONS APPLICABLE TO 1985 MODEL YEAR LIGHT-DUTY TRUCKS.

PRINTED IN U.S.A. PART NO. 1632596

1985-V8-4.1L DFI—CALIFORNIA

VACUUM CIRCUITS

1985 CHEVROLET 1.6L

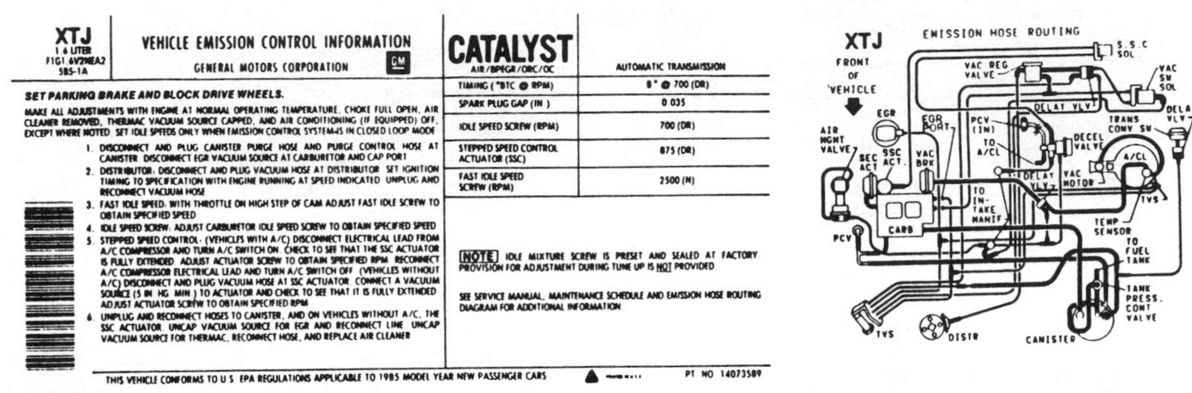

XTJ
1.6 LITER
F1G1.6V2NEA2
5B5-1A

VEHICLE EMISSION CONTROL INFORMATION
GENERAL MOTORS CORPORATION

CATALYST
AIR/BPEGR/ORC/OC

	AUTOMATIC TRANSMISSION
TIMING (°BTC @ RPM)	8° @ 700 (DR)
SPARK PLUG GAP (IN.)	0.035
IDLE SPEED SCREW (RPM)	700 (DR)
STEPPED SPEED CONTROL ACTUATOR (SSC)	875 (DR)
FAST IDLE SPEED SCREW (RPM)	2500 (N)

SET PARKING BRAKE AND BLOCK DRIVE WHEELS.

MAKE ALL ADJUSTMENTS WITH ENGINE AT NORMAL OPERATING TEMPERATURE, CHOKE FULL OPEN, AIR CLEANER REMOVED, THERMAC VACUUM SOURCE CAPPED, AND AIR CONDITIONING (IF EQUIPPED) OFF, EXCEPT WHERE NOTED. SET IDLE SPEEDS ONLY WHEN EMISSION CONTROL SYSTEM IS IN CLOSED LOOP MODE

1. DISCONNECT AND PLUG CANISTER PURGE HOSE AND PURGE CONTROL HOSE AT CANISTER. DISCONNECT EGR VACUUM SOURCE AT CARBURETOR AND CAP PORT
2. DISTRIBUTOR: DISCONNECT AND PLUG VACUUM HOSE AT DISTRIBUTOR. SET IGNITION TIMING TO SPECIFICATION WITH ENGINE RUNNING AT SPEED INDICATED. UNPLUG AND RECONNECT VACUUM HOSE
3. FAST IDLE SPEED: WITH THROTTLE ON HIGH STEP OF CAM ADJUST FAST IDLE SCREW TO OBTAIN SPECIFIED SPEED
4. IDLE SPEED SCREW: ADJUST CARBURETOR IDLE SPEED SCREW TO OBTAIN SPECIFIED SPEED
5. STEPPED SPEED CONTROL: (VEHICLES WITH A/C) DISCONNECT ELECTRICAL LEAD FROM A/C COMPRESSOR AND TURN A/C SWITCH ON. CHECK TO SEE THAT THE SSC ACTUATOR IS FULLY EXTENDED. ADJUST ACTUATOR SCREW TO OBTAIN SPECIFIED RPM. RECONNECT A/C COMPRESSOR ELECTRICAL LEAD AND TURN A/C SWITCH OFF. (VEHICLES WITHOUT A/C) DISCONNECT AND PLUG VACUUM HOSE AT SSC ACTUATOR. CONNECT A VACUUM SOURCE (5 IN. HG. MIN.) TO ACTUATOR AND CHECK TO SEE THAT IT IS FULLY EXTENDED. ADJUST ACTUATOR SCREW TO OBTAIN SPECIFIED RPM
6. UNPLUG AND RECONNECT HOSES TO CANISTER, AND ON VEHICLES WITHOUT A/C, THE SSC ACTUATOR. UNCAP VACUUM SOURCE FOR EGR AND RECONNECT LINE. UNCAP VACUUM SOURCE FOR THERMAC, RECONNECT HOSE, AND REPLACE AIR CLEANER

NOTE IDLE MIXTURE SCREW IS PRESET AND SEALED AT FACTORY. PROVISION FOR ADJUSTMENT DURING TUNE UP IS NOT PROVIDED.

SEE SERVICE MANUAL, MAINTENANCE SCHEDULE AND EMISSION HOSE ROUTING DIAGRAM FOR ADDITIONAL INFORMATION.

THIS VEHICLE CONFORMS TO U.S. EPA REGULATIONS APPLICABLE TO 1985 MODEL YEAR NEW PASSENGER CARS

PT. NO. 14073589

1985–4 CYL–1.6L VIN C CHEVETTE—FEDERAL

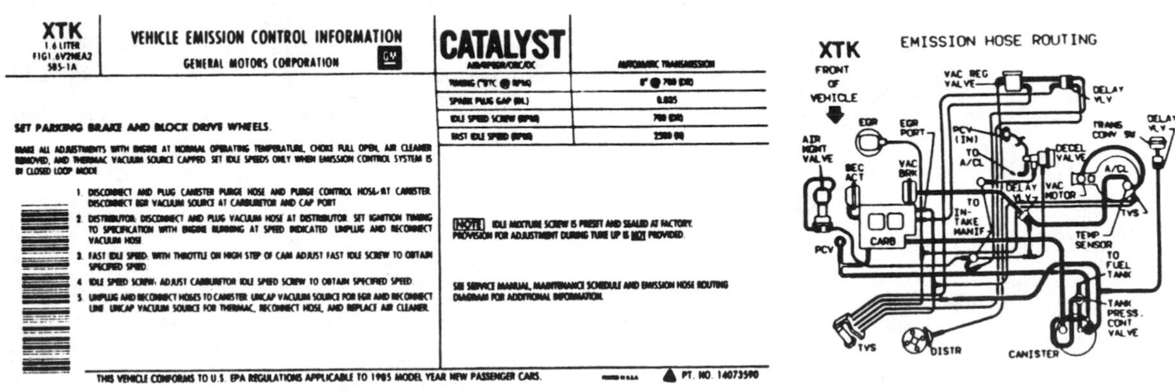

XTK
1.6 LITER
F1G1.6V2NEA2
5B5-1A

VEHICLE EMISSION CONTROL INFORMATION
GENERAL MOTORS CORPORATION

CATALYST
AIR/BPEGR/ORC/OC

	AUTOMATIC TRANSMISSION
TIMING (°BTC @ RPM)	8° @ 700 (DR)
SPARK PLUG GAP (IN.)	0.035
IDLE SPEED SCREW (RPM)	700 (DR)
FAST IDLE SPEED (RPM)	2500 (N)

SET PARKING BRAKE AND BLOCK DRIVE WHEELS.

MAKE ALL ADJUSTMENTS WITH ENGINE AT NORMAL OPERATING TEMPERATURE, CHOKE FULL OPEN, AIR CLEANER REMOVED, AND THERMAC VACUUM SOURCE CAPPED. SET IDLE SPEEDS ONLY WHEN EMISSION CONTROL SYSTEM IS IN CLOSED LOOP MODE

1. DISCONNECT AND PLUG CANISTER PURGE HOSE AND PURGE CONTROL HOSE AT CANISTER. DISCONNECT EGR VACUUM SOURCE AT CARBURETOR AND CAP PORT
2. DISTRIBUTOR: DISCONNECT AND PLUG VACUUM HOSE AT DISTRIBUTOR. SET IGNITION TIMING TO SPECIFICATION WITH ENGINE RUNNING AT SPEED INDICATED. UNPLUG AND RECONNECT VACUUM HOSE
3. FAST IDLE SPEED: WITH THROTTLE ON HIGH STEP OF CAM ADJUST FAST IDLE SCREW TO OBTAIN SPECIFIED SPEED
4. IDLE SPEED SCREW: ADJUST CARBURETOR IDLE SPEED SCREW TO OBTAIN SPECIFIED SPEED.
5. UNPLUG AND RECONNECT HOSES TO CANISTER. UNCAP VACUUM SOURCE FOR EGR AND RECONNECT LINE. UNCAP VACUUM SOURCE FOR THERMAC, RECONNECT HOSE, AND REPLACE AIR CLEANER.

NOTE IDLE MIXTURE SCREW IS PRESET AND SEALED AT FACTORY. PROVISION FOR ADJUSTMENT DURING TUNE UP IS NOT PROVIDED.

SEE SERVICE MANUAL, MAINTENANCE SCHEDULE AND EMISSION HOSE ROUTING DIAGRAM FOR ADDITIONAL INFORMATION.

THIS VEHICLE CONFORMS TO U.S. EPA REGULATIONS APPLICABLE TO 1985 MODEL YEAR NEW PASSENGER CARS.

PT. NO. 14073590

1985–4 CYL–1.6L VIN C CHEVETTE—FEDERAL

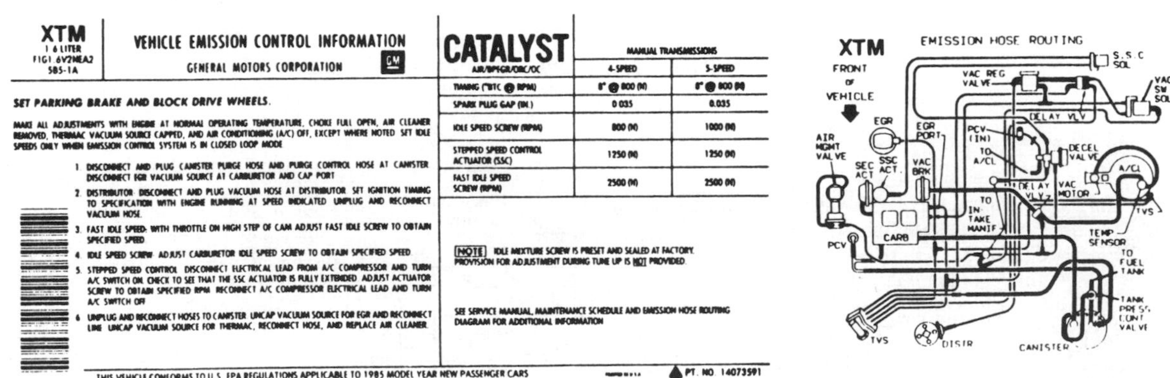

XTM
1.6 LITER
F1G1.6V2NEA2
5B5-1A

VEHICLE EMISSION CONTROL INFORMATION
GENERAL MOTORS CORPORATION

CATALYST
AIR/BPEGR/ORC/OC

	MANUAL TRANSMISSIONS	
	4-SPEED	5-SPEED
TIMING (°BTC @ RPM)	8° @ 800 (N)	8° @ 800 (N)
SPARK PLUG GAP (IN.)	0.035	0.035
IDLE SPEED SCREW (RPM)	800 (N)	1000 (N)
STEPPED SPEED CONTROL ACTUATOR (SSC)	1250 (N)	1250 (N)
FAST IDLE SPEED SCREW (RPM)	2500 (N)	2500 (N)

SET PARKING BRAKE AND BLOCK DRIVE WHEELS.

MAKE ALL ADJUSTMENTS WITH ENGINE AT NORMAL OPERATING TEMPERATURE, CHOKE FULL OPEN, AIR CLEANER REMOVED, THERMAC VACUUM SOURCE CAPPED, AND AIR CONDITIONING (A/C) OFF, EXCEPT WHERE NOTED. SET IDLE SPEEDS ONLY WHEN EMISSION CONTROL SYSTEM IS IN CLOSED LOOP MODE

1. DISCONNECT AND PLUG CANISTER PURGE HOSE AND PURGE CONTROL HOSE AT CANISTER. DISCONNECT EGR VACUUM SOURCE AT CARBURETOR AND CAP PORT
2. DISTRIBUTOR: DISCONNECT AND PLUG VACUUM HOSE AT DISTRIBUTOR. SET IGNITION TIMING TO SPECIFICATION WITH ENGINE RUNNING AT SPEED INDICATED. UNPLUG AND RECONNECT VACUUM HOSE.
3. FAST IDLE SPEED: WITH THROTTLE ON HIGH STEP OF CAM ADJUST FAST IDLE SCREW TO OBTAIN SPECIFIED SPEED
4. IDLE SPEED SCREW: ADJUST CARBURETOR IDLE SPEED SCREW TO OBTAIN SPECIFIED SPEED
5. STEPPED SPEED CONTROL: DISCONNECT ELECTRICAL LEAD FROM A/C COMPRESSOR AND TURN A/C SWITCH ON. CHECK TO SEE THAT THE SSC ACTUATOR IS FULLY EXTENDED. ADJUST ACTUATOR SCREW TO OBTAIN SPECIFIED RPM. RECONNECT A/C COMPRESSOR ELECTRICAL LEAD AND TURN A/C SWITCH OFF
6. UNPLUG AND RECONNECT HOSES TO CANISTER. UNCAP VACUUM SOURCE FOR EGR AND RECONNECT LINE. UNCAP VACUUM SOURCE FOR THERMAC, RECONNECT HOSE, AND REPLACE AIR CLEANER.

NOTE IDLE MIXTURE SCREW IS PRESET AND SEALED AT FACTORY. PROVISION FOR ADJUSTMENT DURING TUNE UP IS NOT PROVIDED.

SEE SERVICE MANUAL, MAINTENANCE SCHEDULE AND EMISSION HOSE ROUTING DIAGRAM FOR ADDITIONAL INFORMATION

THIS VEHICLE CONFORMS TO U.S. EPA REGULATIONS APPLICABLE TO 1985 MODEL YEAR NEW PASSENGER CARS

PT. NO. 14073591

1985–4 CYL–1.6L VIN C CHEVETTE—FEDERAL

VACUUM CIRCUITS

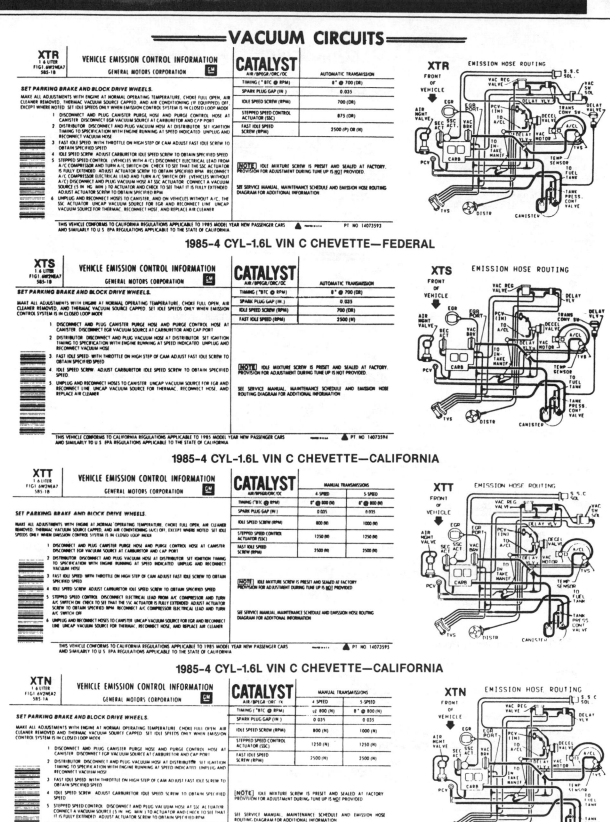

XTR
1 6 LITER
FIG1.6W2NEA7
5B5-1B

VEHICLE EMISSION CONTROL INFORMATION
GENERAL MOTORS CORPORATION GM

CATALYST AIR/BPEGR/ORC/OC	AUTOMATIC TRANSMISSION
TIMING (° BTC @ RPM)	8° @ 700 (DR)
SPARK PLUG GAP (IN.)	0.035
IDLE SPEED SCREW (RPM)	700 (DR)
STEPPED SPEED CONTROL ACTUATOR (SSC)	875 (DR)
FAST IDLE SPEED SCREW (RPM)	2500 (P) OR (N)

SET PARKING BRAKE AND BLOCK DRIVE WHEELS.

MAKE ALL ADJUSTMENTS WITH ENGINE AT NORMAL OPERATING TEMPERATURE, CHOKE FULL OPEN, AIR CLEANER REMOVED, THERMAC VACUUM SOURCE CAPPED, AND AIR CONDITIONING (IF EQUIPPED) OFF, EXCEPT WHERE NOTED SET IDLE SPEEDS ONLY WHEN EMISSION CONTROL SYSTEM IS IN CLOSED LOOP MODE

1 DISCONNECT AND PLUG CANISTER PURGE HOSE AND PURGE CONTROL HOSE AT CANISTER DISCONNECT EGR VACUUM SOURCE AT CARBURETOR AND C/P PORT
2 DISTRIBUTOR DISCONNECT AND PLUG VACUUM HOSE AT DISTRIBUTOR SET IGNITION TIMING TO SPECIFICATION WITH ENGINE RUNNING AT SPEED INDICATED UNPLUG AND RECONNECT VACUUM HOSE
3 FAST IDLE SPEED WITH THROTTLE ON HIGH STEP OF CAM ADJUST FAST IDLE SCREW TO OBTAIN SPECIFIED SPEED
4 IDLE SPEED SCREW ADJUST CARBURETOR IDLE SPEED SCREW TO OBTAIN SPECIFIED SPEED
5 STEPPED SPEED CONTROL (VEHICLES WITH A/C) DISCONNECT ELECTRICAL LEAD FROM A/C COMPRESSOR AND TURN A/C SWITCH ON CHECK TO SEE THAT THE SSC ACTUATOR IS FULLY EXTENDED ADJUST ACTUATOR SCREW TO OBTAIN SPECIFIED RPM RECONNECT A/C COMPRESSOR ELECTRICAL LEAD AND TURN A/C SWITCH OFF (VEHICLES WITHOUT A/C) DISCONNECT AND PLUG VACUUM HOSE AT SSC ACTUATOR CONNECT A VACUUM SOURCE (5 IN HG MIN) TO ACTUATOR AND CHECK TO SEE THAT IT IS FULLY EXTENDED ADJUST ACTUATOR SCREW TO OBTAIN SPECIFIED RPM
6 UNPLUG AND RECONNECT HOSES TO CANISTER, AND ON VEHICLES WITHOUT A/C, THE SSC ACTUATOR UNCAP VACUUM SOURCE FOR EGR AND RECONNECT LINE UNCAP VACUUM SOURCE FOR THERMAC, RECONNECT HOSE, AND REPLACE AIR CLEANER

NOTE IDLE MIXTURE SCREW IS PRESET AND SEALED AT FACTORY. PROVISION FOR ADJUSTMENT DURING TUNE UP IS NOT PROVIDED.

SEE SERVICE MANUAL, MAINTENANCE SCHEDULE AND EMISSION HOSE ROUTING DIAGRAM FOR ADDITIONAL INFORMATION

THIS VEHICLE CONFORMS TO CALIFORNIA REGULATIONS APPLICABLE TO 1985 MODEL YEAR NEW PASSENGER CARS AND SIMILARLY TO U S EPA REGULATIONS APPLICABLE TO THE STATE OF CALIFORNIA PT NO 14073593

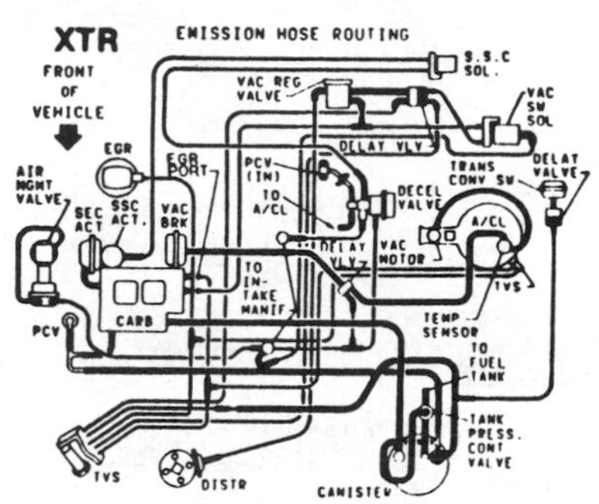

1985–4 CYL–1.6L VIN C CHEVETTE—FEDERAL

XTS
1.6 LITER
FIG1.6W2NEA7
5B5-1B

VEHICLE EMISSION CONTROL INFORMATION
GENERAL MOTORS CORPORATION GM

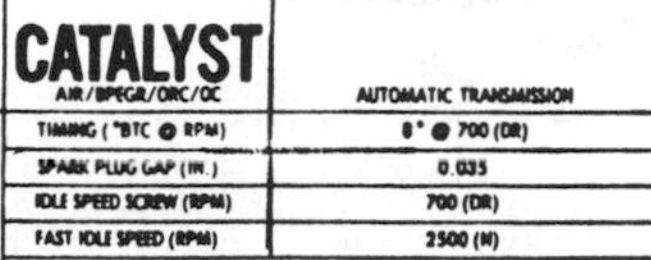

CATALYST AIR/BPEGR/ORC/OC	AUTOMATIC TRANSMISSION
TIMING (° BTC @ RPM)	8° @ 700 (DR)
SPARK PLUG GAP (IN.)	0.035
IDLE SPEED SCREW (RPM)	700 (DR)
FAST IDLE SPEED (RPM)	2500 (N)

SET PARKING BRAKE AND BLOCK DRIVE WHEELS.

MAKE ALL ADJUSTMENTS WITH ENGINE AT NORMAL OPERATING TEMPERATURE, CHOKE FULL OPEN, AIR CLEANER REMOVED, AND THERMAC VACUUM SOURCE CAPPED. SET IDLE SPEEDS ONLY WHEN EMISSION CONTROL SYSTEM IS IN CLOSED LOOP MODE

1. DISCONNECT AND PLUG CANISTER PURGE HOSE AND PURGE CONTROL HOSE AT CANISTER. DISCONNECT EGR VACUUM SOURCE AT CARBURETOR AND CAP PORT
2. DISTRIBUTOR DISCONNECT AND PLUG VACUUM HOSE AT DISTRIBUTOR SET IGNITION TIMING TO SPECIFICATION WITH ENGINE RUNNING AT SPEED INDICATED UNPLUG AND RECONNECT VACUUM HOSE
3. FAST IDLE SPEED: WITH THROTTLE ON HIGH STEP OF CAM ADJUST FAST IDLE SCREW TO OBTAIN SPECIFIED SPEED
4. IDLE SPEED SCREW: ADJUST CARBURETOR IDLE SPEED SCREW TO OBTAIN SPECIFIED SPEED.
5. UNPLUG AND RECONNECT HOSES TO CANISTER UNCAP VACUUM SOURCE FOR EGR AND RECONNECT LINE. UNCAP VACUUM SOURCE FOR THERMAC, RECONNECT HOSE, AND REPLACE AIR CLEANER

NOTE IDLE MIXTURE SCREW IS PRESET AND SEALED AT FACTORY. PROVISION FOR ADJUSTMENT DURING TUNE UP IS NOT PROVIDED.

SEE SERVICE MANUAL, MAINTENANCE SCHEDULE AND EMISSION HOSE ROUTING DIAGRAM FOR ADDITIONAL INFORMATION

THIS VEHICLE CONFORMS TO CALIFORNIA REGULATIONS APPLICABLE TO 1985 MODEL YEAR NEW PASSENGER CARS AND SIMILARLY TO U S EPA REGULATIONS APPLICABLE TO THE STATE OF CALIFORNIA PT. NO 14073594

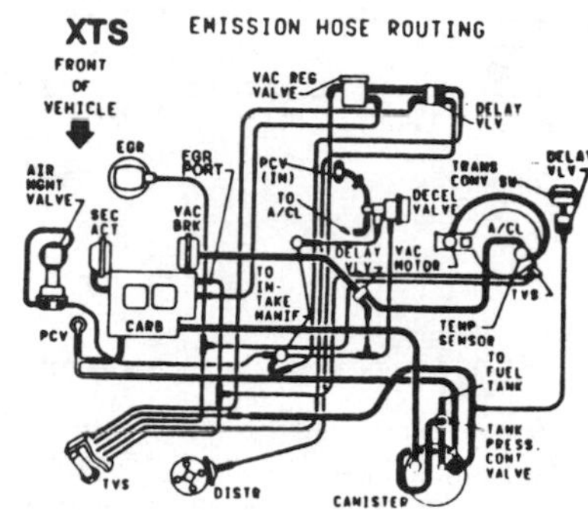

1985–4 CYL–1.6L VIN C CHEVETTE—CALIFORNIA

XTT
1 6 LITER
FIG1 6W2NEA7
5B5-1B

VEHICLE EMISSION CONTROL INFORMATION
GENERAL MOTORS CORPORATION GM

CATALYST AIR/BPEGR/ORC/OC	MANUAL TRANSMISSIONS 4-SPEED	5-SPEED
TIMING (°BTC @ RPM)	8° @ 800 (N)	8° @ 800 (N)
SPARK PLUG GAP (IN)	0.035	0.035
IDLE SPEED SCREW (RPM)	800 (N)	1000 (N)
STEPPED SPEED CONTROL ACTUATOR (SSC)	1250 (N)	1250 (N)
FAST IDLE SPEED SCREW (RPM)	2500 (N)	2500 (N)

SET PARKING BRAKE AND BLOCK DRIVE WHEELS.

MAKE ALL ADJUSTMENTS WITH ENGINE AT NORMAL OPERATING TEMPERATURE, CHOKE FULL OPEN, AIR CLEANER REMOVED, THERMAC VACUUM SOURCE CAPPED, AND AIR CONDITIONING (A/C) OFF, EXCEPT WHERE NOTED SET IDLE SPEEDS ONLY WHEN EMISSION CONTROL SYSTEM IS IN CLOSED LOOP MODE

1 DISCONNECT AND PLUG CANISTER PURGE HOSE AND PURGE CONTROL HOSE AT CANISTER DISCONNECT EGR VACUUM SOURCE AT CARBURETOR AND CAP PORT
2 DISTRIBUTOR DISCONNECT AND PLUG VACUUM HOSE AT DISTRIBUTOR SET IGNITION TIMING TO SPECIFICATION WITH ENGINE RUNNING AT SPEED INDICATED UNPLUG AND RECONNECT VACUUM HOSE
3 FAST IDLE SPEED WITH THROTTLE ON HIGH STEP OF CAM ADJUST FAST IDLE SCREW TO OBTAIN SPECIFIED SPEED
4 IDLE SPEED SCREW ADJUST CARBURETOR IDLE SPEED SCREW TO OBTAIN SPECIFIED SPEED
5 STEPPED SPEED CONTROL DISCONNECT ELECTRICAL LEAD FROM A/C COMPRESSOR AND TURN A/C SWITCH ON CHECK TO SEE THAT THE SSC ACTUATOR IS FULLY EXTENDED ADJUST ACTUATOR SCREW TO OBTAIN SPECIFIED RPM RECONNECT A/C COMPRESSOR ELECTRICAL LEAD AND TURN A/C SWITCH OFF
6 UNPLUG AND RECONNECT HOSES TO CANISTER UNCAP VACUUM SOURCE FOR EGR AND RECONNECT LINE UNCAP VACUUM SOURCE FOR THERMAC, RECONNECT HOSE, AND REPLACE AIR CLEANER

NOTE IDLE MIXTURE SCREW IS PRESET AND SEALED AT FACTORY PROVISION FOR ADJUSTMENT DURING TUNE UP IS NOT PROVIDED

SEE SERVICE MANUAL, MAINTENANCE SCHEDULE AND EMISSION HOSE ROUTING DIAGRAM FOR ADDITIONAL INFORMATION

THIS VEHICLE CONFORMS TO CALIFORNIA REGULATIONS APPLICABLE TO 1985 MODEL YEAR NEW PASSENGER CARS AND SIMILARLY TO U S EPA REGULATIONS APPLICABLE TO THE STATE OF CALIFORNIA PT NO 14073595

XTT
EMISSION HOSE ROUTING
FRONT OF VEHICLE
S.S.C SOL
VAC REG VALVE
VAC SW SOL
DELAY VLV
EGR
EGR PORT
AIR MGMT VALVE
PCV (IN)
TO A/CL
DECEL VALVE
A/CL
SEC ACT
SSC ACT
VAC BRK
DELAY VLV
VAC MOTOR
TVS
CARB
TO IN TAKE MANIF
PCV
TEMP SENSOR
TO FUEL TANK
TANK PRESS CONT VALVE
TVS
DISTR
CANISTER

1985–4 CYL–1.6L VIN C CHEVETTE—CALIFORNIA

XTN
1 6 LITER
FIG1 6V2NEA2
5B5-1A

VEHICLE EMISSION CONTROL INFORMATION
GENERAL MOTORS CORPORATION GM

CATALYST AIR/BPEGR/ORC/OC	MANUAL TRANSMISSIONS 4 SPEED	5-SPEED
TIMING (°BTC @ RPM)	8° @ 800 (N)	8° @ 800 (N)
SPARK PLUG GAP (IN)	0 035	0 035
IDLE SPEED SCREW (RPM)	800 (N)	1000 (N)
STEPPED SPEED CONTROL ACTUATOR (SSC)	1250 (N)	1250 (N)
FAST IDLE SPEED SCREW (RPM)	2500 (N)	2500 (N)

SET PARKING BRAKE AND BLOCK DRIVE WHEELS.

MAKE ALL ADJUSTMENTS WITH ENGINE AT NORMAL OPERATING TEMPERATURE CHOKE FULL OPEN AIR CLEANER REMOVED AND THERMAC VACUUM SOURCE CAPPED SET IDLE SPEEDS ONLY WHEN EMISSION CONTROL SYSTEM IS IN CLOSED LOOP MODE

1 DISCONNECT AND PLUG CANISTER PURGE HOSE AND PURGE CONTROL HOSE AT CANISTER DISCONNECT EGR VACUUM SOURCE AT CARBURETOR AND CAP PORT
2 DISTRIBUTOR DISCONNECT AND PLUG VACUUM HOSE AT DISTRIBUTOR SET IGNITION TIMING TO SPECIFICATION WITH ENGINE RUNNING AT SPEED INDICATED UNPLUG AND RECONNECT VACUUM HOSE
3 FAST IDLE SPEED WITH THROTTLE ON HIGH STEP OF CAM ADJUST FAST IDLE SCREW TO OBTAIN SPECIFIED SPEED
4 IDLE SPEED SCREW ADJUST CARBURETOR IDLE SPEED SCREW TO OBTAIN SPECIFIED SPEED
5 STEPPED SPEED CONTROL DISCONNECT AND PLUG VACUUM HOSE AT SSC ACTUATOR CONNECT A VACUUM SOURCE (5 IN HG MIN) TO ACTUATOR AND CHECK TO SEE THAT IT IS FULLY EXTENDED ADJUST ACTUATOR SCREW TO OBTAIN SPECIFIED RPM
6 UNPLUG AND RECONNECT HOSES TO CANISTER AND SSC ACTUATOR UNCAP VACUUM SOURCE FOR EGR AND RECONNECT LINE UNCAP VACUUM SOURCE FOR THERMAC RECONNECT HOSE AND REPLACE AIR CLEANER

NOTE IDLE MIXTURE SCREW IS PRESET AND SEALED AT FACTORY PROVISION FOR ADJUSTMENT DURING TUNE UP IS NOT PROVIDED

SEE SERVICE MANUAL, MAINTENANCE SCHEDULE AND EMISSION HOSE ROUTING DIAGRAM FOR ADDITIONAL INFORMATION

THIS VEHICLE CONFORMS TO U S EPA REGULATIONS APPLICABLE TO 1985 MODEL YEAR NEW PASSENGER CARS PT NO 14073592

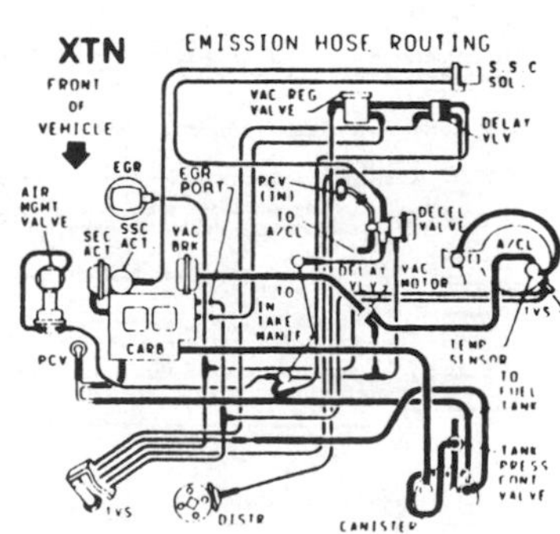

1985–4 CYL–1.6L VIN C CHEVETTE—CALIFORNIA

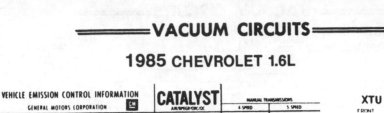

VACUUM CIRCUITS

1985 CHEVROLET 1.6L

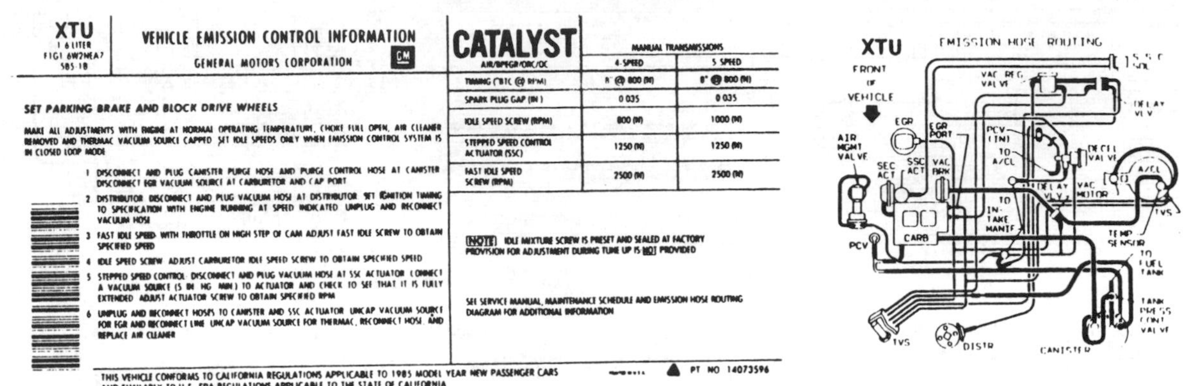

XTU
1 6 LITER
FIGI 6W2NEA7
5B5-1B

VEHICLE EMISSION CONTROL INFORMATION
GENERAL MOTORS CORPORATION

SET PARKING BRAKE AND BLOCK DRIVE WHEELS

MAKE ALL ADJUSTMENTS WITH ENGINE AT NORMAL OPERATING TEMPERATURE, CHOKE FULL OPEN, AIR CLEANER REMOVED AND THERMAC VACUUM SOURCE CAPPED SET IDLE SPEEDS ONLY WHEN EMISSION CONTROL SYSTEM IS IN CLOSED LOOP MODE

1 DISCONNECT AND PLUG CANISTER PURGE HOSE AND PURGE CONTROL HOSE AT CANISTER DISCONNECT EGR VACUUM SOURCE AT CARBURETOR AND CAP PORT
2 DISTRIBUTOR DISCONNECT AND PLUG VACUUM HOSE AT DISTRIBUTOR SET IGNITION TIMING TO SPECIFICATION WITH ENGINE RUNNING AT SPEED INDICATED UNPLUG AND RECONNECT VACUUM HOSE
3 FAST IDLE SPEED WITH THROTTLE ON HIGH STEP OF CAM ADJUST FAST IDLE SCREW TO OBTAIN SPECIFIED SPEED
4 IDLE SPEED SCREW ADJUST CARBURETOR IDLE SPEED SCREW TO OBTAIN SPECIFIED SPEED
5 STEPPED SPEED CONTROL DISCONNECT AND PLUG VACUUM HOSE AT SSC ACTUATOR CONNECT A VACUUM SOURCE (5 IN HG MIN) TO ACTUATOR AND CHECK TO SEE THAT IT IS FULLY EXTENDED ADJUST ACTUATOR SCREW TO OBTAIN SPECIFIED RPM
6 UNPLUG AND RECONNECT HOSES TO CANISTER AND SSC ACTUATOR UNCAP VACUUM SOURCE FOR EGR AND RECONNECT LINE UNCAP VACUUM SOURCE FOR THERMAC, RECONNECT HOSE AND REPLACE AIR CLEANER

CATALYST
AIR/BPEGR/ORC/OC

	MANUAL TRANSMISSIONS	
	4-SPEED	5 SPEED
TIMING (°BTC @ RPM)	8° @ 800 (N)	8° @ 800 (N)
SPARK PLUG GAP (IN)	0 035	0 035
IDLE SPEED SCREW (RPM)	800 (N)	1000 (N)
STEPPED SPEED CONTROL ACTUATOR (SSC)	1250 (N)	1250 (N)
FAST IDLE SPEED SCREW (RPM)	2500 (N)	2500 (N)

NOTE IDLE MIXTURE SCREW IS PRESET AND SEALED AT FACTORY PROVISION FOR ADJUSTMENT DURING TUNE UP IS NOT PROVIDED

SEE SERVICE MANUAL, MAINTENANCE SCHEDULE AND EMISSION HOSE ROUTING DIAGRAM FOR ADDITIONAL INFORMATION

THIS VEHICLE CONFORMS TO CALIFORNIA REGULATIONS APPLICABLE TO 1985 MODEL YEAR NEW PASSENGER CARS AND SIMILARLY TO U S EPA REGULATIONS APPLICABLE TO THE STATE OF CALIFORNIA

PT NO 14073596

1985-4 CYL-1.6L VIN C CHEVETTE—CALIFORNIA

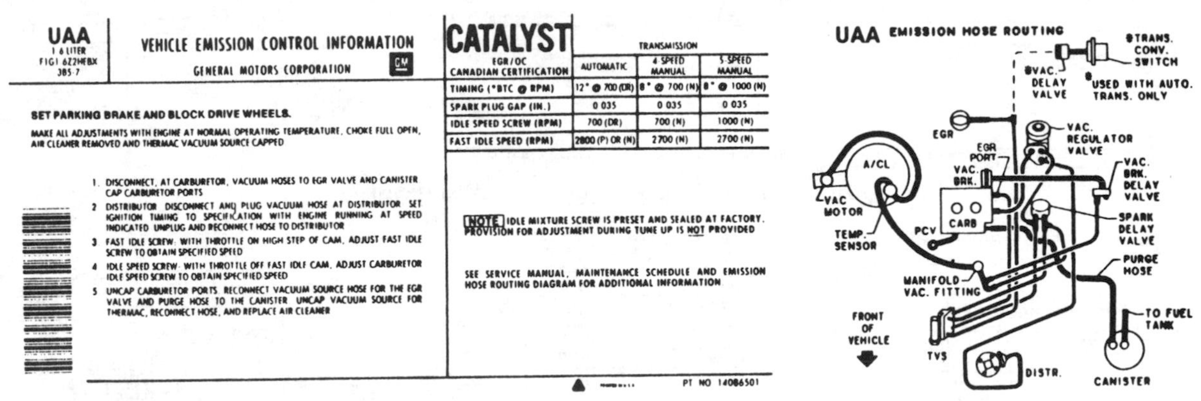

UAA
1 6 LITER
FIGI 6Z2HEBX
3B5-7

VEHICLE EMISSION CONTROL INFORMATION
GENERAL MOTORS CORPORATION

SET PARKING BRAKE AND BLOCK DRIVE WHEELS.

MAKE ALL ADJUSTMENTS WITH ENGINE AT NORMAL OPERATING TEMPERATURE, CHOKE FULL OPEN, AIR CLEANER REMOVED AND THERMAC VACUUM SOURCE CAPPED

1. DISCONNECT, AT CARBURETOR, VACUUM HOSES TO EGR VALVE AND CANISTER CAP CARBURETOR PORTS
2 DISTRIBUTOR DISCONNECT AND PLUG VACUUM HOSE AT DISTRIBUTOR SET IGNITION TIMING TO SPECIFICATION WITH ENGINE RUNNING AT SPEED INDICATED UNPLUG AND RECONNECT HOSE TO DISTRIBUTOR
3 FAST IDLE SCREW: WITH THROTTLE ON HIGH STEP OF CAM, ADJUST FAST IDLE SCREW TO OBTAIN SPECIFIED SPEED
4 IDLE SPEED SCREW WITH THROTTLE OFF FAST IDLE CAM, ADJUST CARBURETOR IDLE SPEED SCREW TO OBTAIN SPECIFIED SPEED
5 UNCAP CARBURETOR PORTS RECONNECT VACUUM SOURCE HOSE FOR THE EGR VALVE AND PURGE HOSE TO THE CANISTER UNCAP VACUUM SOURCE FOR THERMAC, RECONNECT HOSE, AND REPLACE AIR CLEANER

CATALYST
EGR/OC
CANADIAN CERTIFICATION

	TRANSMISSION		
	AUTOMATIC	4-SPEED MANUAL	5-SPEED MANUAL
TIMING (°BTC @ RPM)	12° @ 700 (DR)	8° @ 700 (N)	8° @ 1000 (N)
SPARK PLUG GAP (IN.)	0 035	0 035	0 035
IDLE SPEED SCREW (RPM)	700 (DR)	700 (N)	1000 (N)
FAST IDLE SPEED (RPM)	2800 (P) OR (N)	2700 (N)	2700 (N)

NOTE IDLE MIXTURE SCREW IS PRESET AND SEALED AT FACTORY. PROVISION FOR ADJUSTMENT DURING TUNE UP IS NOT PROVIDED

SEE SERVICE MANUAL, MAINTENANCE SCHEDULE AND EMISSION HOSE ROUTING DIAGRAM FOR ADDITIONAL INFORMATION

PT NO 14086501

1985-4 CYL-1.6L VIN C CHEVETTE—CANADA

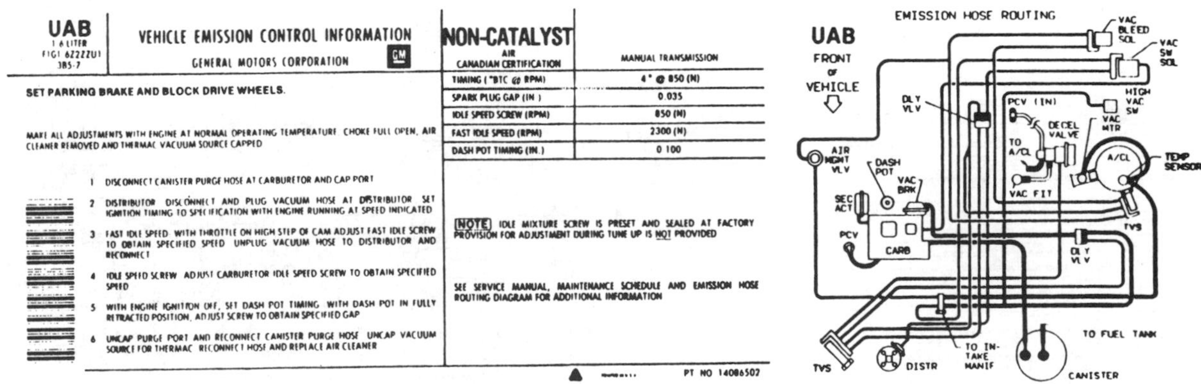

UAB
1 6 LITER
FIGI 6ZZZZU1
3B5-7

VEHICLE EMISSION CONTROL INFORMATION
GENERAL MOTORS CORPORATION

SET PARKING BRAKE AND BLOCK DRIVE WHEELS.

MAKE ALL ADJUSTMENTS WITH ENGINE AT NORMAL OPERATING TEMPERATURE CHOKE FULL OPEN, AIR CLEANER REMOVED AND THERMAC VACUUM SOURCE CAPPED

1 DISCONNECT CANISTER PURGE HOSE AT CARBURETOR AND CAP PORT
2 DISTRIBUTOR DISCONNECT AND PLUG VACUUM HOSE AT DISTRIBUTOR SET IGNITION TIMING TO SPECIFICATION WITH ENGINE RUNNING AT SPEED INDICATED
3 FAST IDLE SPEED WITH THROTTLE ON HIGH STEP OF CAM ADJUST FAST IDLE SCREW TO OBTAIN SPECIFIED SPEED UNPLUG VACUUM HOSE TO DISTRIBUTOR AND RECONNECT
4 IDLE SPEED SCREW ADJUST CARBURETOR IDLE SPEED SCREW TO OBTAIN SPECIFIED SPEED
5 WITH ENGINE IGNITION OFF, SET DASH POT TIMING WITH DASH POT IN FULLY RETRACTED POSITION, ADJUST SCREW TO OBTAIN SPECIFIED GAP
6 UNCAP PURGE PORT AND RECONNECT CANISTER PURGE HOSE UNCAP VACUUM SOURCE FOR THERMAC RECONNECT HOSE AND REPLACE AIR CLEANER

NON-CATALYST
AIR
CANADIAN CERTIFICATION

	MANUAL TRANSMISSION
TIMING (°BTC @ RPM)	4° @ 850 (N)
SPARK PLUG GAP (IN)	0.035
IDLE SPEED SCREW (RPM)	850 (N)
FAST IDLE SPEED (RPM)	2300 (N)
DASH POT TIMING (IN.)	0 100

NOTE IDLE MIXTURE SCREW IS PRESET AND SEALED AT FACTORY PROVISION FOR ADJUSTMENT DURING TUNE UP IS NOT PROVIDED

SEE SERVICE MANUAL, MAINTENANCE SCHEDULE AND EMISSION HOSE ROUTING DIAGRAM FOR ADDITIONAL INFORMATION

PT NO 14086502

1985-4 CYL-1.6L VIN C CHEVETTE—CANADA

VACUUM CIRCUITS

1985 CHEVROLET 1.8L

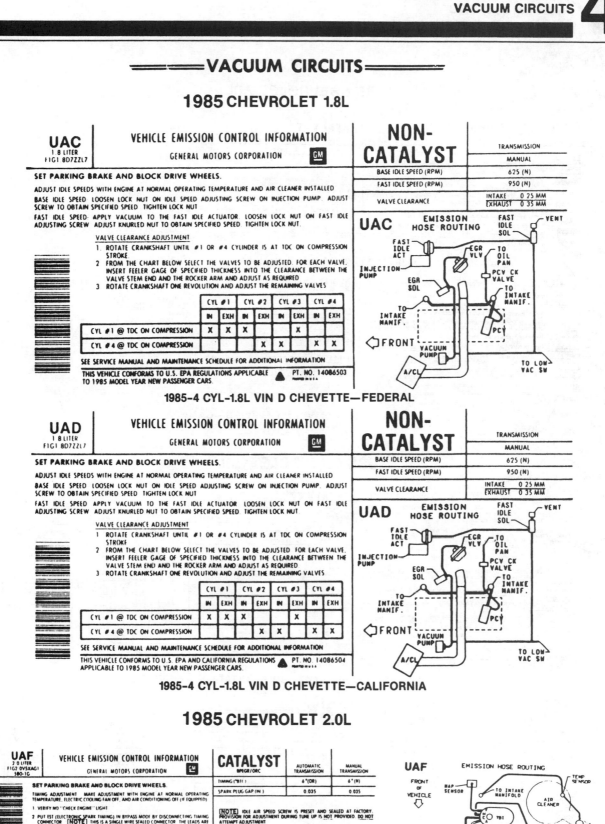

UAC
1.8 LITER
FIG1 8D7ZZL7

VEHICLE EMISSION CONTROL INFORMATION
GENERAL MOTORS CORPORATION GM

NON-CATALYST

	TRANSMISSION
	MANUAL
BASE IDLE SPEED (RPM)	625 (N)
FAST IDLE SPEED (RPM)	950 (N)
VALVE CLEARANCE	INTAKE 0.25 MM EXHAUST 0.35 MM

SET PARKING BRAKE AND BLOCK DRIVE WHEELS.

ADJUST IDLE SPEEDS WITH ENGINE AT NORMAL OPERATING TEMPERATURE AND AIR CLEANER INSTALLED

BASE IDLE SPEED: LOOSEN LOCK NUT ON IDLE SPEED ADJUSTING SCREW ON INJECTION PUMP. ADJUST SCREW TO OBTAIN SPECIFIED SPEED. TIGHTEN LOCK NUT.

FAST IDLE SPEED: APPLY VACUUM TO THE FAST IDLE ACTUATOR. LOOSEN LOCK NUT ON FAST IDLE ADJUSTING SCREW. ADJUST KNURLED NUT TO OBTAIN SPECIFIED SPEED. TIGHTEN LOCK NUT.

VALVE CLEARANCE ADJUSTMENT

1. ROTATE CRANKSHAFT UNTIL #1 OR #4 CYLINDER IS AT TDC ON COMPRESSION STROKE.
2. FROM THE CHART BELOW SELECT THE VALVES TO BE ADJUSTED. FOR EACH VALVE, INSERT FEELER GAGE OF SPECIFIED THICKNESS INTO THE CLEARANCE BETWEEN THE VALVE STEM END AND THE ROCKER ARM AND ADJUST AS REQUIRED.
3. ROTATE CRANKSHAFT ONE REVOLUTION AND ADJUST THE REMAINING VALVES.

	CYL #1		CYL #2		CYL #3		CYL #4	
	IN	EXH	IN	EXH	IN	EXH	IN	EXH
CYL #1 @ TDC ON COMPRESSION	X	X	X			X		
CYL #4 @ TDC ON COMPRESSION				X	X		X	X

SEE SERVICE MANUAL AND MAINTENANCE SCHEDULE FOR ADDITIONAL INFORMATION

THIS VEHICLE CONFORMS TO U.S. EPA REGULATIONS APPLICABLE TO 1985 MODEL YEAR NEW PASSENGER CARS. PT. NO. 14086503

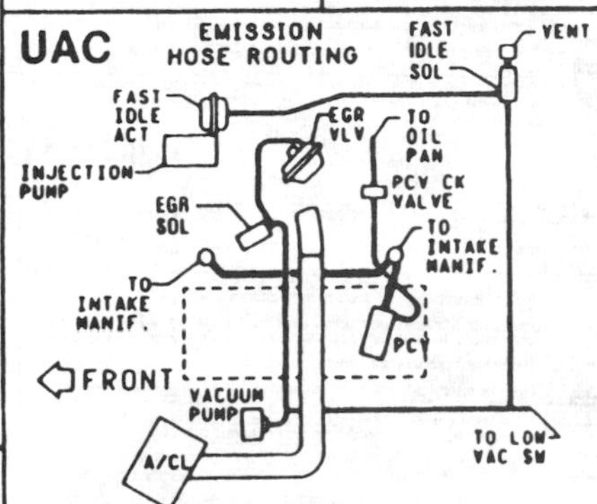

1985-4 CYL-1.8L VIN D CHEVETTE—FEDERAL

UAD
1.8 LITER
FIG1 8D7ZZL7

VEHICLE EMISSION CONTROL INFORMATION
GENERAL MOTORS CORPORATION GM

NON-CATALYST

	TRANSMISSION
	MANUAL
BASE IDLE SPEED (RPM)	625 (N)
FAST IDLE SPEED (RPM)	950 (N)
VALVE CLEARANCE	INTAKE 0.25 MM EXHAUST 0.35 MM

SET PARKING BRAKE AND BLOCK DRIVE WHEELS.

ADJUST IDLE SPEEDS WITH ENGINE AT NORMAL OPERATING TEMPERATURE AND AIR CLEANER INSTALLED

BASE IDLE SPEED: LOOSEN LOCK NUT ON IDLE SPEED ADJUSTING SCREW ON INJECTION PUMP. ADJUST SCREW TO OBTAIN SPECIFIED SPEED. TIGHTEN LOCK NUT.

FAST IDLE SPEED: APPLY VACUUM TO THE FAST IDLE ACTUATOR. LOOSEN LOCK NUT ON FAST IDLE ADJUSTING SCREW. ADJUST KNURLED NUT TO OBTAIN SPECIFIED SPEED. TIGHTEN LOCK NUT.

VALVE CLEARANCE ADJUSTMENT

1. ROTATE CRANKSHAFT UNTIL #1 OR #4 CYLINDER IS AT TDC ON COMPRESSION STROKE.
2. FROM THE CHART BELOW SELECT THE VALVES TO BE ADJUSTED. FOR EACH VALVE, INSERT FEELER GAGE OF SPECIFIED THICKNESS INTO THE CLEARANCE BETWEEN THE VALVE STEM END AND THE ROCKER ARM AND ADJUST AS REQUIRED.
3. ROTATE CRANKSHAFT ONE REVOLUTION AND ADJUST THE REMAINING VALVES.

	CYL #1		CYL #2		CYL #3		CYL #4	
	IN	EXH	IN	EXH	IN	EXH	IN	EXH
CYL #1 @ TDC ON COMPRESSION	X	X	X			X		
CYL #4 @ TDC ON COMPRESSION				X	X		X	X

SEE SERVICE MANUAL AND MAINTENANCE SCHEDULE FOR ADDITIONAL INFORMATION

THIS VEHICLE CONFORMS TO U.S. EPA AND CALIFORNIA REGULATIONS APPLICABLE TO 1985 MODEL YEAR NEW PASSENGER CARS. PT. NO. 14086504

UAD
EMISSION HOSE ROUTING
FAST IDLE SOL
VENT
FAST IDLE ACT
EGR VLV
TO OIL PAN
INJECTION PUMP
PCV CK VALVE
EGR SOL
TO INTAKE MANIF.
TO INTAKE MANIF.
PCV
FRONT
VACUUM PUMP
A/C
TO LOW VAC SW

1985-4 CYL-1.8L VIN D CHEVETTE—CALIFORNIA

1985 CHEVROLET 2.0L

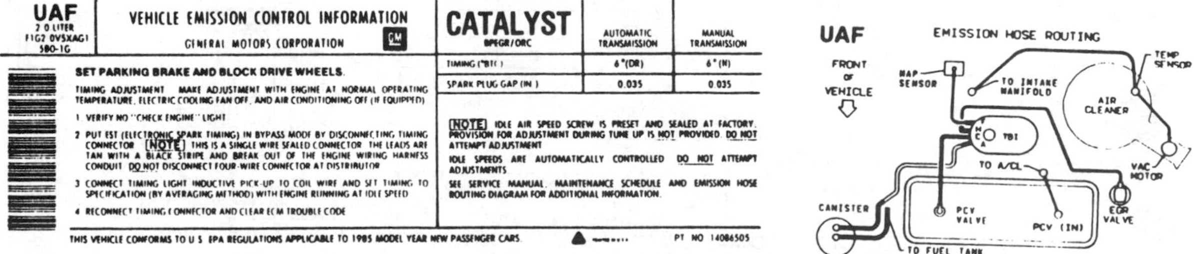

UAF
2.0 LITER
FIG2 0VSXAG1
580-1G

VEHICLE EMISSION CONTROL INFORMATION
GENERAL MOTORS CORPORATION GM

CATALYST
BPEGR/ORC

	AUTOMATIC TRANSMISSION	MANUAL TRANSMISSION
TIMING (°BTC)	6°(DR)	6°(N)
SPARK PLUG GAP (IN.)	0.035	0.035

SET PARKING BRAKE AND BLOCK DRIVE WHEELS.

TIMING ADJUSTMENT: MAKE ADJUSTMENT WITH ENGINE AT NORMAL OPERATING TEMPERATURE, ELECTRIC COOLING FAN OFF, AND AIR CONDITIONING OFF (IF EQUIPPED)

1. VERIFY NO "CHECK ENGINE" LIGHT
2. PUT EST (ELECTRONIC SPARK TIMING) IN BYPASS MODE BY DISCONNECTING TIMING CONNECTOR. NOTE: THIS IS A SINGLE WIRE SEALED CONNECTOR. THE LEADS ARE TAN WITH A BLACK STRIPE AND BREAK OUT OF THE ENGINE WIRING HARNESS CONDUIT. DO NOT DISCONNECT FOUR-WIRE CONNECTOR AT DISTRIBUTOR
3. CONNECT TIMING LIGHT INDUCTIVE PICK-UP TO COIL WIRE AND SET TIMING TO SPECIFICATION (BY AVERAGING METHOD) WITH ENGINE RUNNING AT IDLE SPEED
4. RECONNECT TIMING CONNECTOR AND CLEAR ECM TROUBLE CODE

NOTE: IDLE AIR SPEED SCREW IS PRESET AND SEALED AT FACTORY. PROVISION FOR ADJUSTMENT DURING TUNE UP IS NOT PROVIDED. DO NOT ATTEMPT ADJUSTMENT

IDLE SPEEDS ARE AUTOMATICALLY CONTROLLED. DO NOT ATTEMPT ADJUSTMENTS

SEE SERVICE MANUAL, MAINTENANCE SCHEDULE AND EMISSION HOSE ROUTING DIAGRAM FOR ADDITIONAL INFORMATION.

THIS VEHICLE CONFORMS TO U.S. EPA REGULATIONS APPLICABLE TO 1985 MODEL YEAR NEW PASSENGER CARS. PT. NO. 14086505

1985-4 CYL-2.0L VIN P CAVALIER—FEDERAL

VACUUM CIRCUITS

1985 CHEVROLET 2.0L

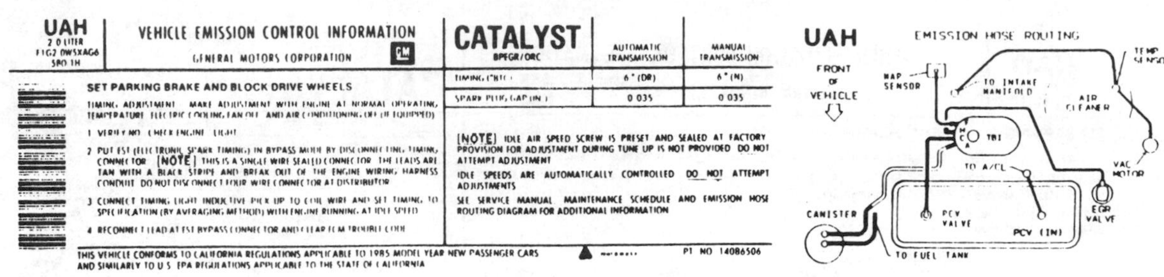

UAH
2.0 LITER
F1G2 0WSXAG6
5B0 1H

VEHICLE EMISSION CONTROL INFORMATION
GENERAL MOTORS CORPORATION

SET PARKING BRAKE AND BLOCK DRIVE WHEELS

TIMING ADJUSTMENT: MAKE ADJUSTMENT WITH ENGINE AT NORMAL OPERATING TEMPERATURE, ELECTRIC COOLING FAN OFF, AND AIR CONDITIONING OFF (IF EQUIPPED)

1. VERIFY NO "CHECK ENGINE" LIGHT
2. PUT EST (ELECTRONIC SPARK TIMING) IN BYPASS MODE BY DISCONNECTING TIMING CONNECTOR. [NOTE] THIS IS A SINGLE WIRE SEALED CONNECTOR. THE LEADS ARE TAN WITH A BLACK STRIPE AND BREAK OUT OF THE ENGINE WIRING HARNESS CONDUIT. DO NOT DISCONNECT FOUR WIRE CONNECTOR AT DISTRIBUTOR
3. CONNECT TIMING LIGHT INDUCTIVE PICK UP TO COIL WIRE AND SET TIMING TO SPECIFICATION (BY AVERAGING METHOD) WITH ENGINE RUNNING AT IDLE SPEED
4. RECONNECT LEAD AT EST BYPASS CONNECTOR AND CLEAR ECM TROUBLE CODE

CATALYST
BPEGR/ORC

	AUTOMATIC TRANSMISSION	MANUAL TRANSMISSION
TIMING (°BTC)	6° (DR)	6° (N)
SPARK PLUG GAP (IN.)	0.035	0.035

[NOTE] IDLE AIR SPEED SCREW IS PRESET AND SEALED AT FACTORY. PROVISION FOR ADJUSTMENT DURING TUNE UP IS NOT PROVIDED. DO NOT ATTEMPT ADJUSTMENT

IDLE SPEEDS ARE AUTOMATICALLY CONTROLLED. DO NOT ATTEMPT ADJUSTMENTS

SEE SERVICE MANUAL, MAINTENANCE SCHEDULE AND EMISSION HOSE ROUTING DIAGRAM FOR ADDITIONAL INFORMATION

THIS VEHICLE CONFORMS TO CALIFORNIA REGULATIONS APPLICABLE TO 1985 MODEL YEAR NEW PASSENGER CARS AND SIMILARLY TO U.S. EPA REGULATIONS APPLICABLE TO THE STATE OF CALIFORNIA

PT. NO. 14086506

1985-4 CYL-2.0L VIN P CAVALIER—CALIFORNIA

ULY
2.0 LITER
F1G2 0Z5ZZGX
5B0-1G

VEHICLE EMISSION CONTROL INFORMATION
GENERAL MOTORS CORPORATION

SET PARKING BRAKE AND BLOCK DRIVE WHEELS.

TIMING ADJUSTMENT: MAKE ADJUSTMENT WITH ENGINE AT NORMAL OPERATING TEMPERATURE, ELECTRIC COOLING FAN OFF, AND AIR CONDITIONING OFF (IF EQUIPPED)

1. VERIFY NO "CHECK ENGINE" LIGHT
2. PUT EST (ELECTRONIC SPARK TIMING) IN BYPASS MODE BY DISCONNECTING TIMING CONNECTOR. [NOTE] THIS IS A SINGLE WIRE SEALED CONNECTOR. THE LEADS ARE TAN WITH A BLACK STRIPE AND BREAK OUT OF THE ENGINE WIRING HARNESS CONDUIT. DO NOT DISCONNECT FOUR WIRE CONNECTOR AT DISTRIBUTOR
3. CONNECT TIMING LIGHT INDUCTIVE PICK-UP TO COIL WIRE AND SET TIMING TO SPECIFICATION (BY AVERAGING METHOD) WITH ENGINE RUNNING AT IDLE SPEED
4. RECONNECT LEAD AT EST BYPASS CONNECTOR AND CLEAR ECM TROUBLE CODE

NON-CATALYST
BPEGR/OI
CANADIAN CERTIFICATION

	AUTOMATIC TRANSMISSION
TIMING (°BTC)	6° (DR)
SPARK PLUG GAP (IN.)	0.035

[NOTE] IDLE AIR SPEED SCREW IS PRESET AND SEALED AT FACTORY. PROVISION FOR ADJUSTMENT DURING TUNE UP IS NOT PROVIDED. DO NOT ATTEMPT ADJUSTMENT

IDLE SPEEDS ARE AUTOMATICALLY CONTROLLED. DO NOT ATTEMPT ADJUSTMENTS

SEE SERVICE MANUAL, MAINTENANCE SCHEDULE AND EMISSION HOSE ROUTING DIAGRAM FOR ADDITIONAL INFORMATION

PT. NO. 14086571

1985-4 CYL-2.0L VIN P CAVALIER—CANADA

1985 CHEVROLET 2.8L

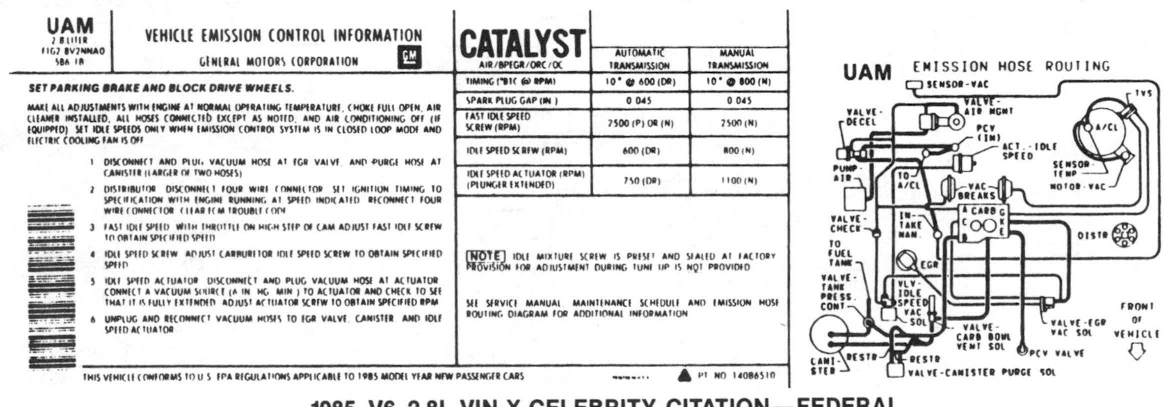

UAM
2.8 LITER
F1G2 8V2NNA0
5B6 1B

VEHICLE EMISSION CONTROL INFORMATION
GENERAL MOTORS CORPORATION

SET PARKING BRAKE AND BLOCK DRIVE WHEELS.

MAKE ALL ADJUSTMENTS WITH ENGINE AT NORMAL OPERATING TEMPERATURE, CHOKE FULL OPEN, AIR CLEANER INSTALLED, ALL HOSES CONNECTED EXCEPT AS NOTED, AND AIR CONDITIONING OFF (IF EQUIPPED). SET IDLE SPEEDS ONLY WHEN EMISSION CONTROL SYSTEM IS IN CLOSED LOOP MODE AND ELECTRIC COOLING FAN IS OFF

1. DISCONNECT AND PLUG VACUUM HOSE AT EGR VALVE, AND PURGE HOSE AT CANISTER (LARGER OF TWO HOSES)
2. DISTRIBUTOR: DISCONNECT FOUR WIRE CONNECTOR. SET IGNITION TIMING TO SPECIFICATION WITH ENGINE RUNNING AT SPEED INDICATED. RECONNECT FOUR WIRE CONNECTOR. CLEAR ECM TROUBLE CODE
3. FAST IDLE SPEED: WITH THROTTLE ON HIGH STEP OF CAM ADJUST FAST IDLE SCREW TO OBTAIN SPECIFIED SPEED
4. IDLE SPEED SCREW: ADJUST CARBURETOR IDLE SPEED SCREW TO OBTAIN SPECIFIED SPEED
5. IDLE SPEED ACTUATOR: DISCONNECT AND PLUG VACUUM HOSE AT ACTUATOR. CONNECT A VACUUM SOURCE (6 IN. HG. MIN.) TO ACTUATOR AND CHECK TO SEE THAT IT IS FULLY EXTENDED. ADJUST ACTUATOR SCREW TO OBTAIN SPECIFIED RPM
6. UNPLUG AND RECONNECT VACUUM HOSES TO EGR VALVE, CANISTER, AND IDLE SPEED ACTUATOR

CATALYST
AIR/BPEGR/ORC/OC

	AUTOMATIC TRANSMISSION	MANUAL TRANSMISSION
TIMING (°BTC @ RPM)	10° @ 600 (DR)	10° @ 800 (N)
SPARK PLUG GAP (IN.)	0.045	0.045
FAST IDLE SPEED SCREW (RPM)	2500 (P) OR (N)	2500 (N)
IDLE SPEED SCREW (RPM)	600 (DR)	800 (N)
IDLE SPEED ACTUATOR (RPM) (PLUNGER EXTENDED)	750 (DR)	1100 (N)

[NOTE] IDLE MIXTURE SCREW IS PRESET AND SEALED AT FACTORY. PROVISION FOR ADJUSTMENT DURING TUNE UP IS NOT PROVIDED

SEE SERVICE MANUAL, MAINTENANCE SCHEDULE AND EMISSION HOSE ROUTING DIAGRAM FOR ADDITIONAL INFORMATION

THIS VEHICLE CONFORMS TO U.S. EPA REGULATIONS APPLICABLE TO 1985 MODEL YEAR NEW PASSENGER CARS

PT. NO. 14086510

1985-V6-2.8L VIN X CELEBRITY, CITATION—FEDERAL

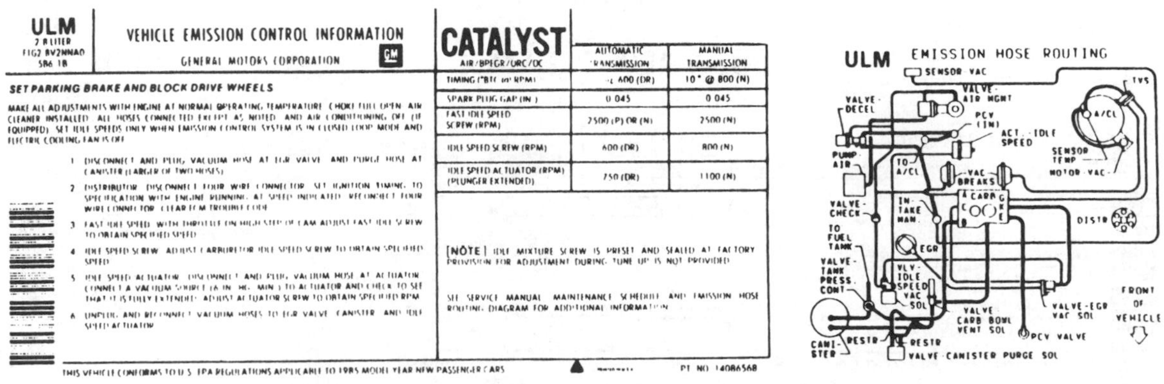

ULM
2.8 LITER
F1G2 8V2NNA0
5B6 1B

VEHICLE EMISSION CONTROL INFORMATION
GENERAL MOTORS CORPORATION

SET PARKING BRAKE AND BLOCK DRIVE WHEELS

MAKE ALL ADJUSTMENTS WITH ENGINE AT NORMAL OPERATING TEMPERATURE, CHOKE FULL OPEN, AIR CLEANER INSTALLED, ALL HOSES CONNECTED EXCEPT AS NOTED, AND AIR CONDITIONING OFF (IF EQUIPPED). SET IDLE SPEEDS ONLY WHEN EMISSION CONTROL SYSTEM IS IN CLOSED LOOP MODE AND ELECTRIC COOLING FAN IS OFF

1. DISCONNECT AND PLUG VACUUM HOSE AT EGR VALVE, AND PURGE HOSE AT CANISTER (LARGER OF TWO HOSES)
2. DISTRIBUTOR: DISCONNECT FOUR WIRE CONNECTOR. SET IGNITION TIMING TO SPECIFICATION WITH ENGINE RUNNING AT SPEED INDICATED. RECONNECT FOUR WIRE CONNECTOR. CLEAR ECM TROUBLE CODE
3. FAST IDLE SPEED: WITH THROTTLE ON HIGH STEP OF CAM ADJUST FAST IDLE SCREW TO OBTAIN SPECIFIED SPEED
4. IDLE SPEED SCREW: ADJUST CARBURETOR IDLE SPEED SCREW TO OBTAIN SPECIFIED SPEED
5. IDLE SPEED ACTUATOR: DISCONNECT AND PLUG VACUUM HOSE AT ACTUATOR. CONNECT A VACUUM SOURCE (6 IN. HG. MIN.) TO ACTUATOR AND CHECK TO SEE THAT IT IS FULLY EXTENDED. ADJUST ACTUATOR SCREW TO OBTAIN SPECIFIED RPM
6. UNPLUG AND RECONNECT VACUUM HOSES TO EGR VALVE, CANISTER, AND IDLE SPEED ACTUATOR

CATALYST
AIR/BPEGR/ORC/OC

	AUTOMATIC TRANSMISSION	MANUAL TRANSMISSION
TIMING (°BTC @ RPM)	[illegible] 600 (DR)	10° @ 800 (N)
SPARK PLUG GAP (IN.)	0.045	0.045
FAST IDLE SPEED SCREW (RPM)	2500 (P) OR (N)	2500 (N)
IDLE SPEED SCREW (RPM)	600 (DR)	800 (N)
IDLE SPEED ACTUATOR (RPM) (PLUNGER EXTENDED)	750 (DR)	1100 (N)

[NOTE] IDLE MIXTURE SCREW IS PRESET AND SEALED AT FACTORY. PROVISION FOR ADJUSTMENT DURING TUNE UP IS NOT PROVIDED

SEE SERVICE MANUAL, MAINTENANCE SCHEDULE AND EMISSION HOSE ROUTING DIAGRAM FOR ADDITIONAL INFORMATION

THIS VEHICLE CONFORMS TO U.S. EPA REGULATIONS APPLICABLE TO 1985 MODEL YEAR NEW PASSENGER CARS

PT. NO. 14086568

1985-V6-2.8L VIN X CELEBRITY—FEDERAL (POLICE, TAXI)

VACUUM CIRCUITS

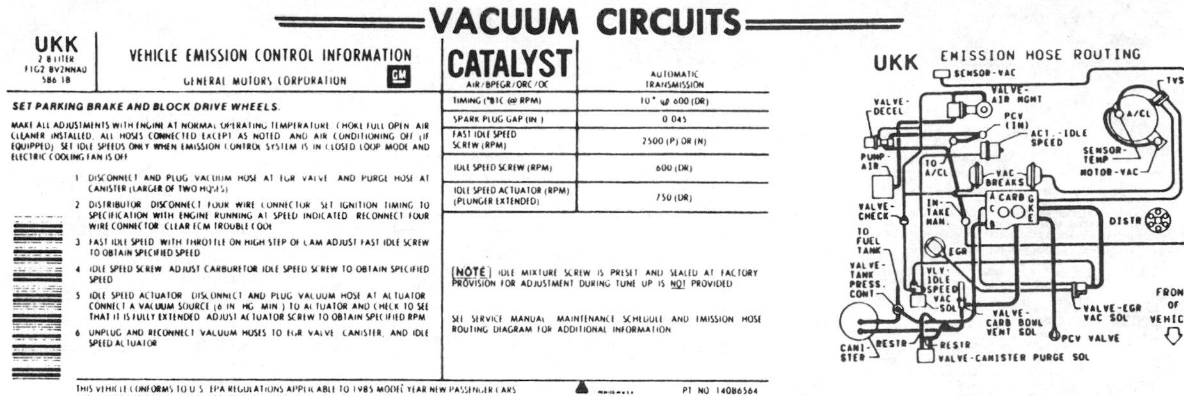

UKK
2 8 LITER
F1G2 8V2NNA0
5B6 1B

VEHICLE EMISSION CONTROL INFORMATION
GENERAL MOTORS CORPORATION GM

SET PARKING BRAKE AND BLOCK DRIVE WHEELS.

MAKE ALL ADJUSTMENTS WITH ENGINE AT NORMAL OPERATING TEMPERATURE CHOKE FULL OPEN AIR CLEANER INSTALLED, ALL HOSES CONNECTED EXCEPT AS NOTED AND AIR CONDITIONING OFF (IF EQUIPPED) SET IDLE SPEEDS ONLY WHEN EMISSION CONTROL SYSTEM IS IN CLOSED LOOP MODE AND ELECTRIC COOLING FAN IS OFF

1 DISCONNECT AND PLUG VACUUM HOSE AT EGR VALVE AND PURGE HOSE AT CANISTER (LARGER OF TWO HOSES)
2 DISTRIBUTOR DISCONNECT FOUR WIRE CONNECTOR SET IGNITION TIMING TO SPECIFICATION WITH ENGINE RUNNING AT SPEED INDICATED RECONNECT FOUR WIRE CONNECTOR CLEAR ECM TROUBLE CODE
3 FAST IDLE SPEED WITH THROTTLE ON HIGH STEP OF CAM ADJUST FAST IDLE SCREW TO OBTAIN SPECIFIED SPEED
4 IDLE SPEED SCREW ADJUST CARBURETOR IDLE SPEED SCREW TO OBTAIN SPECIFIED SPEED
5 IDLE SPEED ACTUATOR DISCONNECT AND PLUG VACUUM HOSE AT ACTUATOR CONNECT A VACUUM SOURCE (6 IN HG MIN) TO ACTUATOR AND CHECK TO SEE THAT IT IS FULLY EXTENDED ADJUST ACTUATOR SCREW TO OBTAIN SPECIFIED RPM
6 UNPLUG AND RECONNECT VACUUM HOSES TO EGR VALVE CANISTER, AND IDLE SPEED ACTUATOR

CATALYST AIR/BPEGR/ORC/OC	AUTOMATIC TRANSMISSION
TIMING (°BTC @ RPM)	10° @ 600 (DR)
SPARK PLUG GAP (IN)	0 045
FAST IDLE SPEED SCREW (RPM)	2500 (P) OR (N)
IDLE SPEED SCREW (RPM)	600 (DR)
IDLE SPEED ACTUATOR (RPM) (PLUNGER EXTENDED)	750 (DR)

[NOTE] IDLE MIXTURE SCREW IS PRESET AND SEALED AT FACTORY PROVISION FOR ADJUSTMENT DURING TUNE UP IS NOT PROVIDED

SEE SERVICE MANUAL MAINTENANCE SCHEDULE AND EMISSION HOSE ROUTING DIAGRAM FOR ADDITIONAL INFORMATION

THIS VEHICLE CONFORMS TO U S EPA REGULATIONS APPLICABLE TO 1985 MODEL YEAR NEW PASSENGER CARS AND IS CERTIFIED FOR SALE IN THE STATE OF CALIFORNIA

PT NO 14086564

1985-V6-2.8L VIN X CELEBRITY—CALIFORNIA (TAXI)

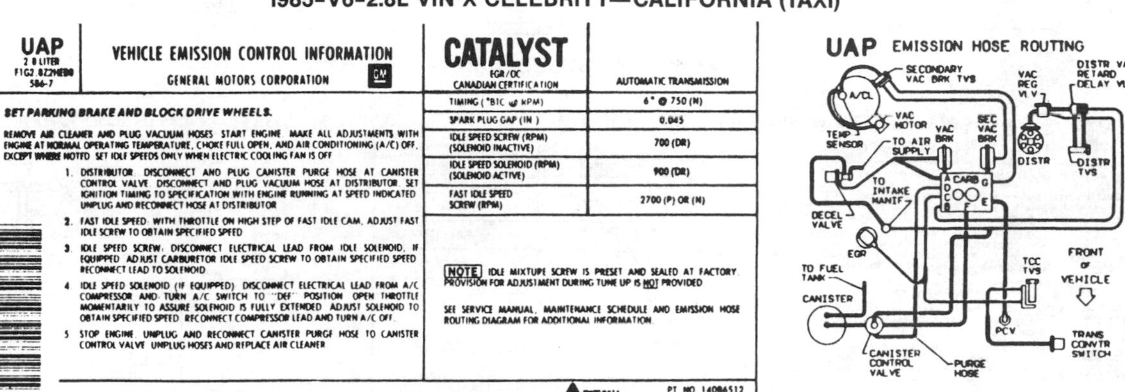

UAP
2 8 LITER
F1G2 8Z2HEB8
5B6-7

VEHICLE EMISSION CONTROL INFORMATION
GENERAL MOTORS CORPORATION GM

SET PARKING BRAKE AND BLOCK DRIVE WHEELS.

REMOVE AIR CLEANER AND PLUG VACUUM HOSES START ENGINE MAKE ALL ADJUSTMENTS WITH ENGINE AT NORMAL OPERATING TEMPERATURE, CHOKE FULL OPEN, AND AIR CONDITIONING (A/C) OFF, EXCEPT WHERE NOTED SET IDLE SPEEDS ONLY WHEN ELECTRIC COOLING FAN IS OFF

1. DISTRIBUTOR DISCONNECT AND PLUG CANISTER PURGE HOSE AT CANISTER CONTROL VALVE DISCONNECT AND PLUG VACUUM HOSE AT DISTRIBUTOR. SET IGNITION TIMING TO SPECIFICATION WITH ENGINE RUNNING AT SPEED INDICATED UNPLUG AND RECONNECT HOSE AT DISTRIBUTOR
2. FAST IDLE SPEED: WITH THROTTLE ON HIGH STEP OF FAST IDLE CAM, ADJUST FAST IDLE SCREW TO OBTAIN SPECIFIED SPEED
3. IDLE SPEED SCREW: DISCONNECT ELECTRICAL LEAD FROM IDLE SOLENOID, IF EQUIPPED ADJUST CARBURETOR IDLE SPEED SCREW TO OBTAIN SPECIFIED SPEED RECONNECT LEAD TO SOLENOID
4 IDLE SPEED SOLENOID (IF EQUIPPED) DISCONNECT ELECTRICAL LEAD FROM A/C COMPRESSOR AND TURN A/C SWITCH TO "DEF" POSITION OPEN THROTTLE MOMENTARILY TO ASSURE SOLENOID IS FULLY EXTENDED ADJUST SOLENOID TO OBTAIN SPECIFIED SPEED RECONNECT COMPRESSOR LEAD AND TURN A/C OFF.
5 STOP ENGINE UNPLUG AND RECONNECT CANISTER PURGE HOSE TO CANISTER CONTROL VALVE UNPLUG HOSES AND REPLACE AIR CLEANER

CATALYST EGR/OC CANADIAN CERTIFICATION	AUTOMATIC TRANSMISSION
TIMING (°BTC @ RPM)	6° @ 750 (N)
SPARK PLUG GAP (IN)	0.045
IDLE SPEED SCREW (RPM) (SOLENOID INACTIVE)	700 (DR)
IDLE SPEED SOLENOID (RPM) (SOLENOID ACTIVE)	900 (DR)
FAST IDLE SPEED SCREW (RPM)	2700 (P) OR (N)

[NOTE] IDLE MIXTURE SCREW IS PRESET AND SEALED AT FACTORY. PROVISION FOR ADJUSTMENT DURING TUNE UP IS NOT PROVIDED

SEE SERVICE MANUAL, MAINTENANCE SCHEDULE AND EMISSION HOSE ROUTING DIAGRAM FOR ADDITIONAL INFORMATION.

PT NO 14086512

1985-V6-2.8L VIN X CELEBRITY, CITATION—CANADA

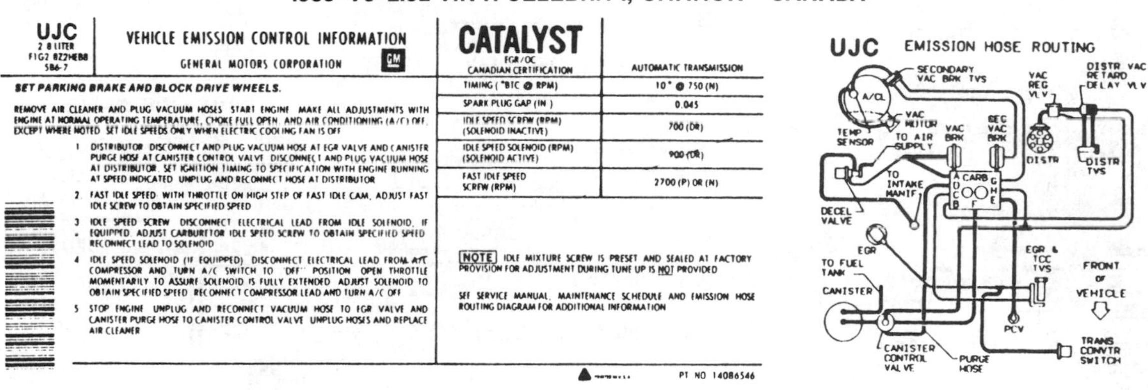

UJC
2 8 LITER
F1G2 8Z2HEB8
5B6-7

VEHICLE EMISSION CONTROL INFORMATION
GENERAL MOTORS CORPORATION GM

SET PARKING BRAKE AND BLOCK DRIVE WHEELS.

REMOVE AIR CLEANER AND PLUG VACUUM HOSES START ENGINE MAKE ALL ADJUSTMENTS WITH ENGINE AT NORMAL OPERATING TEMPERATURE, CHOKE FULL OPEN AND AIR CONDITIONING (A/C) OFF, EXCEPT WHERE NOTED SET IDLE SPEEDS ONLY WHEN ELECTRIC COOLING FAN IS OFF

1 DISTRIBUTOR DISCONNECT AND PLUG VACUUM HOSE AT EGR VALVE AND CANISTER PURGE HOSE AT CANISTER CONTROL VALVE DISCONNECT AND PLUG VACUUM HOSE AT DISTRIBUTOR. SET IGNITION TIMING TO SPECIFICATION WITH ENGINE RUNNING AT SPEED INDICATED UNPLUG AND RECONNECT HOSE AT DISTRIBUTOR
2. FAST IDLE SPEED WITH THROTTLE ON HIGH STEP OF FAST IDLE CAM, ADJUST FAST IDLE SCREW TO OBTAIN SPECIFIED SPEED
3 IDLE SPEED SCREW DISCONNECT ELECTRICAL LEAD FROM IDLE SOLENOID, IF EQUIPPED ADJUST CARBURETOR IDLE SPEED SCREW TO OBTAIN SPECIFIED SPEED RECONNECT LEAD TO SOLENOID
4 IDLE SPEED SOLENOID (IF EQUIPPED) DISCONNECT ELECTRICAL LEAD FROM A/C COMPRESSOR AND TURN A/C SWITCH TO "DEF" POSITION OPEN THROTTLE MOMENTARILY TO ASSURE SOLENOID IS FULLY EXTENDED ADJUST SOLENOID TO OBTAIN SPECIFIED SPEED RECONNECT COMPRESSOR LEAD AND TURN A/C OFF
5 STOP ENGINE UNPLUG AND RECONNECT VACUUM HOSE TO EGR VALVE AND CANISTER PURGE HOSE TO CANISTER CONTROL VALVE UNPLUG HOSES AND REPLACE AIR CLEANER

CATALYST EGR/OC CANADIAN CERTIFICATION	AUTOMATIC TRANSMISSION
TIMING (°BTC @ RPM)	10° @ 750 (N)
SPARK PLUG GAP (IN)	0.045
IDLE SPEED SCREW (RPM) (SOLENOID INACTIVE)	700 (DR)
IDLE SPEED SOLENOID (RPM) (SOLENOID ACTIVE)	900 (DR)
FAST IDLE SPEED SCREW (RPM)	2700 (P) OR (N)

[NOTE] IDLE MIXTURE SCREW IS PRESET AND SEALED AT FACTORY PROVISION FOR ADJUSTMENT DURING TUNE UP IS NOT PROVIDED

SEE SERVICE MANUAL, MAINTENANCE SCHEDULE AND EMISSION HOSE ROUTING DIAGRAM FOR ADDITIONAL INFORMATION

PT NO 14086546

1985-V6-2.8L VIN X CELEBRITY—CANADA

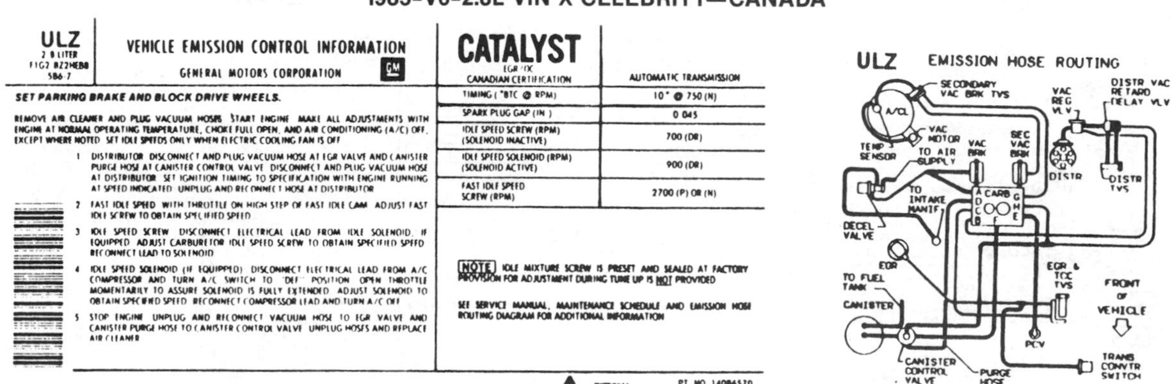

ULZ
2 8 LITER
F1G2 8Z2HEB8
5B6-7

VEHICLE EMISSION CONTROL INFORMATION
GENERAL MOTORS CORPORATION GM

SET PARKING BRAKE AND BLOCK DRIVE WHEELS.

REMOVE AIR CLEANER AND PLUG VACUUM HOSES START ENGINE MAKE ALL ADJUSTMENTS WITH ENGINE AT NORMAL OPERATING TEMPERATURE, CHOKE FULL OPEN, AND AIR CONDITIONING (A/C) OFF, EXCEPT WHERE NOTED SET IDLE SPEEDS ONLY WHEN ELECTRIC COOLING FAN IS OFF

1 DISTRIBUTOR DISCONNECT AND PLUG VACUUM HOSE AT EGR VALVE AND CANISTER PURGE HOSE AT CANISTER CONTROL VALVE DISCONNECT AND PLUG VACUUM HOSE AT DISTRIBUTOR SET IGNITION TIMING TO SPECIFICATION WITH ENGINE RUNNING AT SPEED INDICATED UNPLUG AND RECONNECT HOSE AT DISTRIBUTOR
2 FAST IDLE SPEED WITH THROTTLE ON HIGH STEP OF FAST IDLE CAM ADJUST FAST IDLE SCREW TO OBTAIN SPECIFIED SPEED
3 IDLE SPEED SCREW DISCONNECT ELECTRICAL LEAD FROM IDLE SOLENOID, IF EQUIPPED ADJUST CARBURETOR IDLE SPEED SCREW TO OBTAIN SPECIFIED SPEED RECONNECT LEAD TO SOLENOID
4 IDLE SPEED SOLENOID (IF EQUIPPED) DISCONNECT ELECTRICAL LEAD FROM A/C COMPRESSOR AND TURN A/C SWITCH TO "DEF" POSITION OPEN THROTTLE MOMENTARILY TO ASSURE SOLENOID IS FULLY EXTENDED ADJUST SOLENOID TO OBTAIN SPECIFIED SPEED RECONNECT COMPRESSOR LEAD AND TURN A/C OFF
5 STOP ENGINE UNPLUG AND RECONNECT VACUUM HOSE TO EGR VALVE AND CANISTER PURGE HOSE TO CANISTER CONTROL VALVE UNPLUG HOSES AND REPLACE AIR CLEANER

CATALYST EGR/OC CANADIAN CERTIFICATION	AUTOMATIC TRANSMISSION
TIMING (°BTC @ RPM)	10° @ 750 (N)
SPARK PLUG GAP (IN)	0 045
IDLE SPEED SCREW (RPM) (SOLENOID INACTIVE)	700 (DR)
IDLE SPEED SOLENOID (RPM) (SOLENOID ACTIVE)	900 (DR)
FAST IDLE SPEED SCREW (RPM)	2700 (P) OR (N)

[NOTE] IDLE MIXTURE SCREW IS PRESET AND SEALED AT FACTORY PROVISION FOR ADJUSTMENT DURING TUNE UP IS NOT PROVIDED

SEE SERVICE MANUAL, MAINTENANCE SCHEDULE AND EMISSION HOSE ROUTING DIAGRAM FOR ADDITIONAL INFORMATION

PT NO 14086570

1985-V6-2.8L VIN X CELEBRITY—CANADA

VACUUM CIRCUITS

1985 CHEVROLET 2.8L

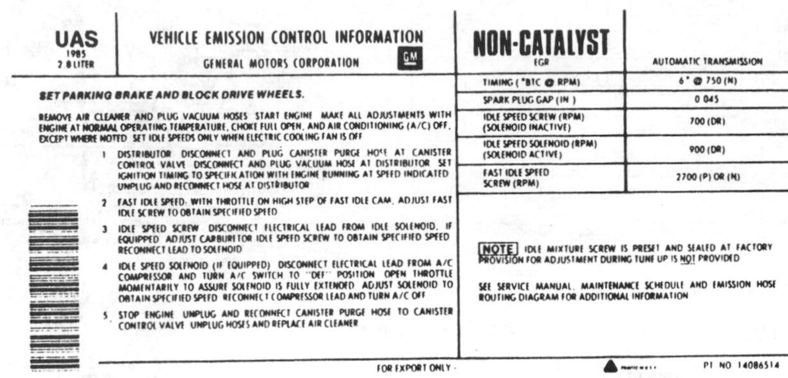

UAS 1985 2.8 LITER — VEHICLE EMISSION CONTROL INFORMATION — GENERAL MOTORS CORPORATION

NON-CATALYST EGR

	AUTOMATIC TRANSMISSION
TIMING (°BTC @ RPM)	6° @ 750 (N)
SPARK PLUG GAP (IN.)	0.045
IDLE SPEED SCREW (RPM) (SOLENOID INACTIVE)	700 (DR)
IDLE SPEED SOLENOID (RPM) (SOLENOID ACTIVE)	900 (DR)
FAST IDLE SPEED SCREW (RPM)	2700 (P) OR (N)

SET PARKING BRAKE AND BLOCK DRIVE WHEELS.

REMOVE AIR CLEANER AND PLUG VACUUM HOSES. START ENGINE. MAKE ALL ADJUSTMENTS WITH ENGINE AT NORMAL OPERATING TEMPERATURE, CHOKE FULL OPEN, AND AIR CONDITIONING (A/C) OFF, EXCEPT WHERE NOTED. SET IDLE SPEEDS ONLY WHEN ELECTRIC COOLING FAN IS OFF.

1. DISTRIBUTOR: DISCONNECT AND PLUG CANISTER PURGE HOSE AT CANISTER CONTROL VALVE. DISCONNECT AND PLUG VACUUM HOSE AT DISTRIBUTOR. SET IGNITION TIMING TO SPECIFICATION WITH ENGINE RUNNING AT SPEED INDICATED. UNPLUG AND RECONNECT HOSE AT DISTRIBUTOR.
2. FAST IDLE SPEED: WITH THROTTLE ON HIGH STEP OF FAST IDLE CAM, ADJUST FAST IDLE SCREW TO OBTAIN SPECIFIED SPEED.
3. IDLE SPEED SCREW: DISCONNECT ELECTRICAL LEAD FROM IDLE SOLENOID, IF EQUIPPED. ADJUST CARBURETOR IDLE SPEED SCREW TO OBTAIN SPECIFIED SPEED. RECONNECT LEAD TO SOLENOID.
4. IDLE SPEED SOLENOID (IF EQUIPPED): DISCONNECT ELECTRICAL LEAD FROM A/C COMPRESSOR AND TURN A/C SWITCH TO "DEF" POSITION. OPEN THROTTLE MOMENTARILY TO ASSURE SOLENOID IS FULLY EXTENDED. ADJUST SOLENOID TO OBTAIN SPECIFIED SPEED. RECONNECT COMPRESSOR LEAD AND TURN A/C OFF.
5. STOP ENGINE. UNPLUG AND RECONNECT CANISTER PURGE HOSE TO CANISTER CONTROL VALVE. UNPLUG HOSES AND REPLACE AIR CLEANER.

NOTE: IDLE MIXTURE SCREW IS PRESET AND SEALED AT FACTORY. PROVISION FOR ADJUSTMENT DURING TUNE UP IS NOT PROVIDED.

SEE SERVICE MANUAL, MAINTENANCE SCHEDULE AND EMISSION HOSE ROUTING DIAGRAM FOR ADDITIONAL INFORMATION.

FOR EXPORT ONLY — PT. NO. 14086514

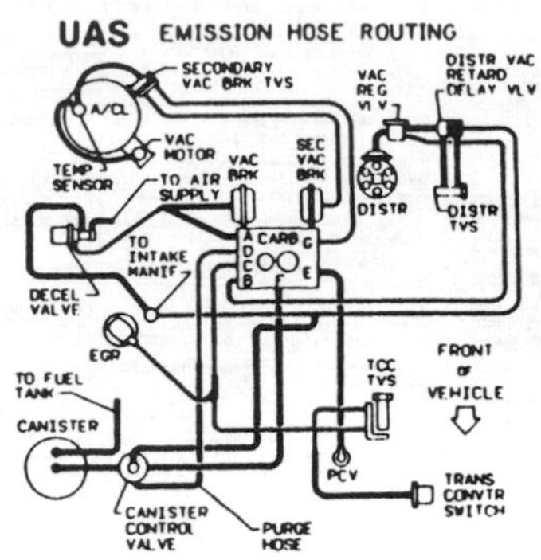

1985-V6-2.8L VIN X CELEBRITY—EXPORT

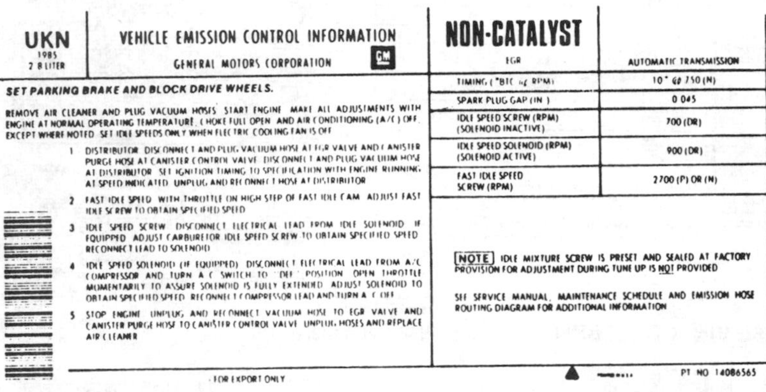

UKN 1985 2.8 LITER — VEHICLE EMISSION CONTROL INFORMATION — GENERAL MOTORS CORPORATION

NON-CATALYST EGR

	AUTOMATIC TRANSMISSION
TIMING (°BTC @ RPM)	10° @ 750 (N)
SPARK PLUG GAP (IN.)	0.045
IDLE SPEED SCREW (RPM) (SOLENOID INACTIVE)	700 (DR)
IDLE SPEED SOLENOID (RPM) (SOLENOID ACTIVE)	900 (DR)
FAST IDLE SPEED SCREW (RPM)	2700 (P) OR (N)

SET PARKING BRAKE AND BLOCK DRIVE WHEELS.

REMOVE AIR CLEANER AND PLUG VACUUM HOSES. START ENGINE. MAKE ALL ADJUSTMENTS WITH ENGINE AT NORMAL OPERATING TEMPERATURE, CHOKE FULL OPEN, AND AIR CONDITIONING (A/C) OFF, EXCEPT WHERE NOTED. SET IDLE SPEEDS ONLY WHEN ELECTRIC COOLING FAN IS OFF.

1. DISTRIBUTOR: DISCONNECT AND PLUG VACUUM HOSE AT EGR VALVE AND CANISTER PURGE HOSE AT CANISTER CONTROL VALVE. DISCONNECT AND PLUG VACUUM HOSE AT DISTRIBUTOR. SET IGNITION TIMING TO SPECIFICATION WITH ENGINE RUNNING AT SPEED INDICATED. UNPLUG AND RECONNECT HOSE AT DISTRIBUTOR.
2. FAST IDLE SPEED: WITH THROTTLE ON HIGH STEP OF FAST IDLE CAM, ADJUST FAST IDLE SCREW TO OBTAIN SPECIFIED SPEED.
3. IDLE SPEED SCREW: DISCONNECT ELECTRICAL LEAD FROM IDLE SOLENOID, IF EQUIPPED. ADJUST CARBURETOR IDLE SPEED SCREW TO OBTAIN SPECIFIED SPEED. RECONNECT LEAD TO SOLENOID.
4. IDLE SPEED SOLENOID (IF EQUIPPED): DISCONNECT ELECTRICAL LEAD FROM A/C COMPRESSOR AND TURN A/C SWITCH TO "DEF" POSITION. OPEN THROTTLE MOMENTARILY TO ASSURE SOLENOID IS FULLY EXTENDED. ADJUST SOLENOID TO OBTAIN SPECIFIED SPEED. RECONNECT COMPRESSOR LEAD AND TURN A/C OFF.
5. STOP ENGINE. UNPLUG AND RECONNECT VACUUM HOSE TO EGR VALVE AND CANISTER PURGE HOSE TO CANISTER CONTROL VALVE. UNPLUG HOSES AND REPLACE AIR CLEANER.

NOTE: IDLE MIXTURE SCREW IS PRESET AND SEALED AT FACTORY. PROVISION FOR ADJUSTMENT DURING TUNE UP IS NOT PROVIDED.

SEE SERVICE MANUAL, MAINTENANCE SCHEDULE AND EMISSION HOSE ROUTING DIAGRAM FOR ADDITIONAL INFORMATION.

FOR EXPORT ONLY — PT. NO. 14086565

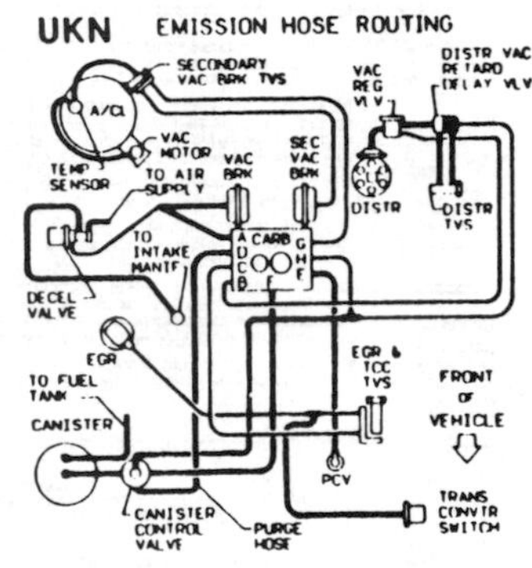

1985-V6-2.8L VIN X CELEBRITY—EXPORT

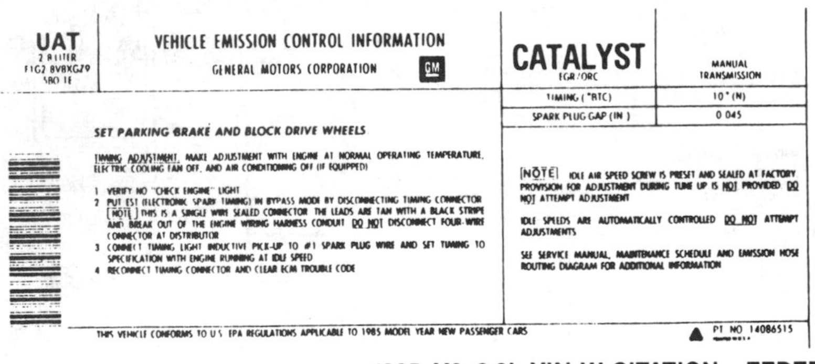

UAT 2.8 LITER F1G2 8V8XGZ9 5BO 1E — VEHICLE EMISSION CONTROL INFORMATION — GENERAL MOTORS CORPORATION

CATALYST EGR/ORC

	MANUAL TRANSMISSION
TIMING (°BTC)	10° (N)
SPARK PLUG GAP (IN.)	0.045

SET PARKING BRAKE AND BLOCK DRIVE WHEELS.

TIMING ADJUSTMENT: MAKE ADJUSTMENT WITH ENGINE AT NORMAL OPERATING TEMPERATURE, ELECTRIC COOLING FAN OFF, AND AIR CONDITIONING OFF (IF EQUIPPED).

1. VERIFY NO "CHECK ENGINE" LIGHT.
2. PUT EST (ELECTRONIC SPARK TIMING) IN BYPASS MODE BY DISCONNECTING TIMING CONNECTOR. NOTE: THIS IS A SINGLE WIRE SEALED CONNECTOR. THE LEADS ARE TAN WITH A BLACK STRIPE AND BREAK OUT OF THE ENGINE WIRING HARNESS CONDUIT. DO NOT DISCONNECT FOUR-WIRE CONNECTOR AT DISTRIBUTOR.
3. CONNECT TIMING LIGHT INDUCTIVE PICK-UP TO #1 SPARK PLUG WIRE AND SET TIMING TO SPECIFICATION WITH ENGINE RUNNING AT IDLE SPEED.
4. RECONNECT TIMING CONNECTOR AND CLEAR ECM TROUBLE CODE.

NOTE: IDLE AIR SPEED SCREW IS PRESET AND SEALED AT FACTORY. PROVISION FOR ADJUSTMENT DURING TUNE UP IS NOT PROVIDED. DO NOT ATTEMPT ADJUSTMENT.

IDLE SPEEDS ARE AUTOMATICALLY CONTROLLED. DO NOT ATTEMPT ADJUSTMENTS.

SEE SERVICE MANUAL, MAINTENANCE SCHEDULE AND EMISSION HOSE ROUTING DIAGRAM FOR ADDITIONAL INFORMATION.

THIS VEHICLE CONFORMS TO U.S. EPA REGULATIONS APPLICABLE TO 1985 MODEL YEAR NEW PASSENGER CARS. — PT. NO. 14086515

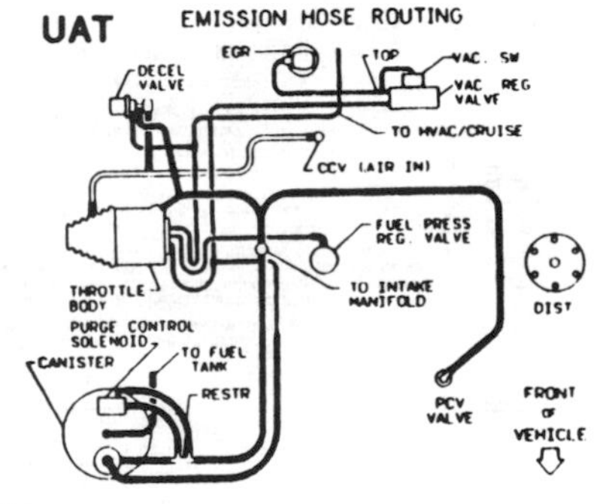

1985-V6-2.8L VIN W CITATION—FEDERAL

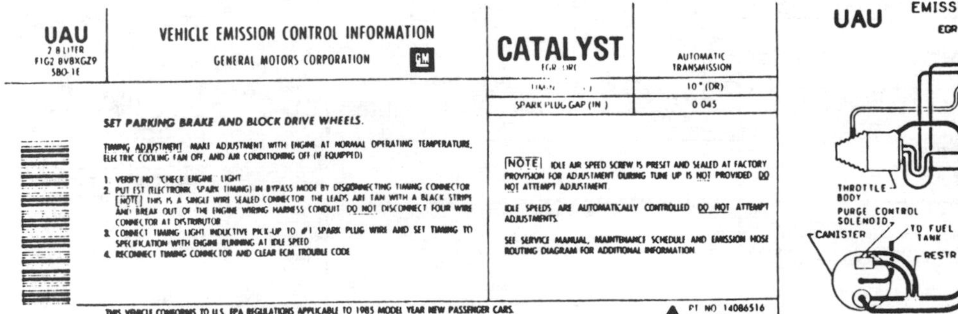

UAU 2.8 LITER F1G2 8V8XGZ9 5BO 1E — VEHICLE EMISSION CONTROL INFORMATION — GENERAL MOTORS CORPORATION

CATALYST EGR/ORC

	AUTOMATIC TRANSMISSION
TIMING (°BTC)	10° (DR)
SPARK PLUG GAP (IN.)	0.045

SET PARKING BRAKE AND BLOCK DRIVE WHEELS.

TIMING ADJUSTMENT: MAKE ADJUSTMENT WITH ENGINE AT NORMAL OPERATING TEMPERATURE, ELECTRIC COOLING FAN OFF, AND AIR CONDITIONING OFF (IF EQUIPPED).

1. VERIFY NO "CHECK ENGINE" LIGHT.
2. PUT EST (ELECTRONIC SPARK TIMING) IN BYPASS MODE BY DISCONNECTING TIMING CONNECTOR. NOTE: THIS IS A SINGLE WIRE SEALED CONNECTOR. THE LEADS ARE TAN WITH A BLACK STRIPE AND BREAK OUT OF THE ENGINE WIRING HARNESS CONDUIT. DO NOT DISCONNECT FOUR-WIRE CONNECTOR AT DISTRIBUTOR.
3. CONNECT TIMING LIGHT INDUCTIVE PICK-UP TO #1 SPARK PLUG WIRE AND SET TIMING TO SPECIFICATION WITH ENGINE RUNNING AT IDLE SPEED.
4. RECONNECT TIMING CONNECTOR AND CLEAR ECM TROUBLE CODE.

NOTE: IDLE AIR SPEED SCREW IS PRESET AND SEALED AT FACTORY. PROVISION FOR ADJUSTMENT DURING TUNE UP IS NOT PROVIDED. DO NOT ATTEMPT ADJUSTMENT.

IDLE SPEEDS ARE AUTOMATICALLY CONTROLLED. DO NOT ATTEMPT ADJUSTMENTS.

SEE SERVICE MANUAL, MAINTENANCE SCHEDULE AND EMISSION HOSE ROUTING DIAGRAM FOR ADDITIONAL INFORMATION.

THIS VEHICLE CONFORMS TO U.S. EPA REGULATIONS APPLICABLE TO 1985 MODEL YEAR NEW PASSENGER CARS. — PT. NO. 14086516

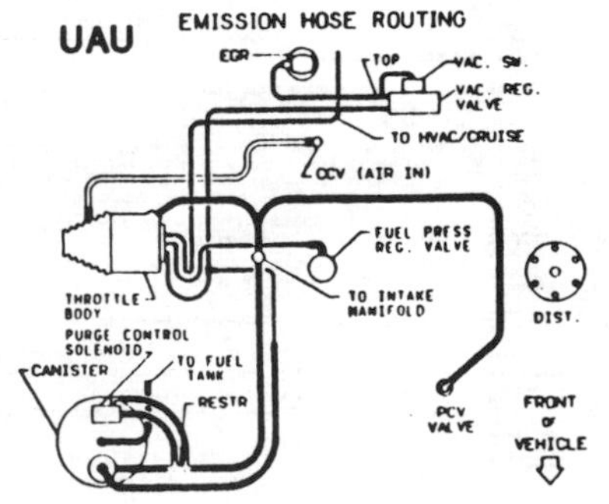

1985-V6-2.8L VIN W CELEBRITY, CITATION—FEDERAL

VACUUM CIRCUITS

1985 CHEVROLET 2.8L

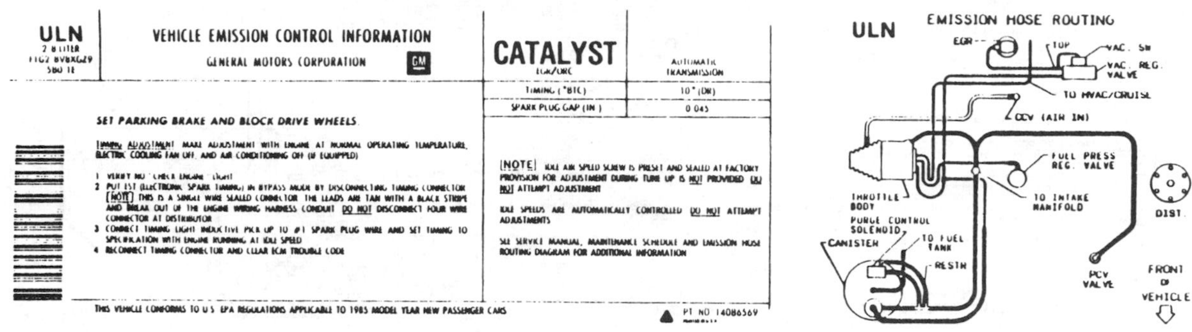

ULN
2.8 LITER
FIG2 8V8XGZ9
5B0-1E

VEHICLE EMISSION CONTROL INFORMATION
GENERAL MOTORS CORPORATION GM

CATALYST EGR/ORC	AUTOMATIC TRANSMISSION
TIMING (°BTC)	10° (DR)
SPARK PLUG GAP (IN.)	0.045

SET PARKING BRAKE AND BLOCK DRIVE WHEELS.

TIMING ADJUSTMENT: MAKE ADJUSTMENT WITH ENGINE AT NORMAL OPERATING TEMPERATURE, ELECTRIC COOLING FAN OFF, AND AIR CONDITIONING OFF (IF EQUIPPED)

1. VERIFY NO "CHECK ENGINE" LIGHT
2. PUT EST (ELECTRONIC SPARK TIMING) IN BYPASS MODE BY DISCONNECTING TIMING CONNECTOR. [NOTE] THIS IS A SINGLE WIRE SEALED CONNECTOR. THE LEADS ARE TAN WITH A BLACK STRIPE AND BREAK OUT OF THE ENGINE WIRING HARNESS CONDUIT. DO NOT DISCONNECT FOUR WIRE CONNECTOR AT DISTRIBUTOR
3. CONNECT TIMING LIGHT INDUCTIVE PICK-UP TO #1 SPARK PLUG WIRE AND SET TIMING TO SPECIFICATION WITH ENGINE RUNNING AT IDLE SPEED
4. RECONNECT TIMING CONNECTOR AND CLEAR ECM TROUBLE CODE

[NOTE] IDLE AIR SPEED SCREW IS PRESET AND SEALED AT FACTORY. PROVISION FOR ADJUSTMENT DURING TUNE UP IS NOT PROVIDED. DO NOT ATTEMPT ADJUSTMENT

IDLE SPEEDS ARE AUTOMATICALLY CONTROLLED. DO NOT ATTEMPT ADJUSTMENTS

SEE SERVICE MANUAL, MAINTENANCE SCHEDULE AND EMISSION HOSE ROUTING DIAGRAM FOR ADDITIONAL INFORMATION

THIS VEHICLE CONFORMS TO U.S. EPA REGULATIONS APPLICABLE TO 1985 MODEL YEAR NEW PASSENGER CARS

PT NO 14086569

1985-V6-2.8L VIN W CELEBRITY—FEDERAL

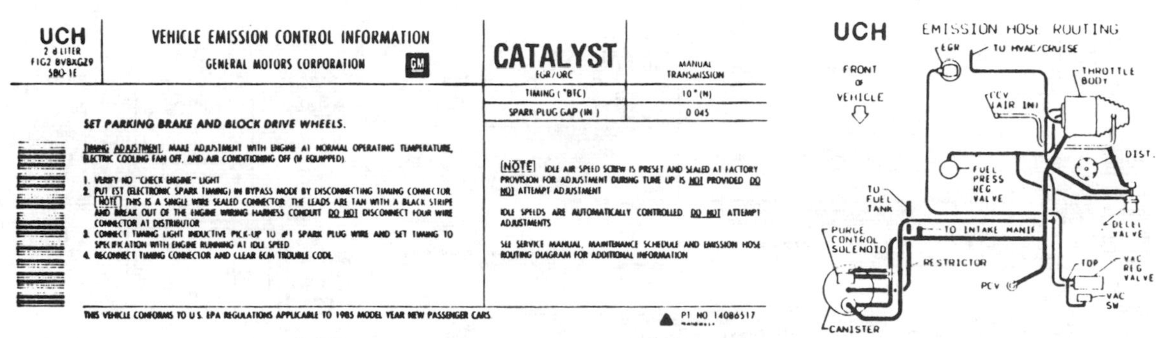

UCH
2.8 LITER
FIG2 8V8XGZ9
5B0-1E

VEHICLE EMISSION CONTROL INFORMATION
GENERAL MOTORS CORPORATION GM

CATALYST EGR/ORC	MANUAL TRANSMISSION
TIMING (°BTC)	10° (N)
SPARK PLUG GAP (IN.)	0.045

SET PARKING BRAKE AND BLOCK DRIVE WHEELS.

TIMING ADJUSTMENT: MAKE ADJUSTMENT WITH ENGINE AT NORMAL OPERATING TEMPERATURE, ELECTRIC COOLING FAN OFF, AND AIR CONDITIONING OFF (IF EQUIPPED)

1. VERIFY NO "CHECK ENGINE" LIGHT
2. PUT EST (ELECTRONIC SPARK TIMING) IN BYPASS MODE BY DISCONNECTING TIMING CONNECTOR. [NOTE] THIS IS A SINGLE WIRE SEALED CONNECTOR. THE LEADS ARE TAN WITH A BLACK STRIPE AND BREAK OUT OF THE ENGINE WIRING HARNESS CONDUIT. DO NOT DISCONNECT FOUR WIRE CONNECTOR AT DISTRIBUTOR
3. CONNECT TIMING LIGHT INDUCTIVE PICK-UP TO #1 SPARK PLUG WIRE AND SET TIMING TO SPECIFICATION WITH ENGINE RUNNING AT IDLE SPEED
4. RECONNECT TIMING CONNECTOR AND CLEAR ECM TROUBLE CODE

[NOTE] IDLE AIR SPEED SCREW IS PRESET AND SEALED AT FACTORY. PROVISION FOR ADJUSTMENT DURING TUNE UP IS NOT PROVIDED. DO NOT ATTEMPT ADJUSTMENT

IDLE SPEEDS ARE AUTOMATICALLY CONTROLLED. DO NOT ATTEMPT ADJUSTMENTS

SEE SERVICE MANUAL, MAINTENANCE SCHEDULE AND EMISSION HOSE ROUTING DIAGRAM FOR ADDITIONAL INFORMATION

THIS VEHICLE CONFORMS TO U.S. EPA REGULATIONS APPLICABLE TO 1985 MODEL YEAR NEW PASSENGER CARS

PT NO 14086517

1985-V6-2.8L VIN W CAVALIER—FEDERAL

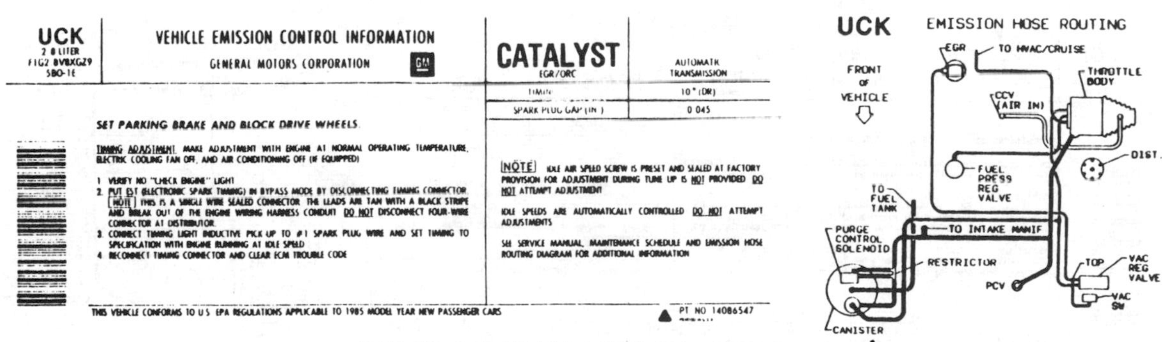

UCK
2.8 LITER
FIG2 8V8XGZ9
5B0-1E

VEHICLE EMISSION CONTROL INFORMATION
GENERAL MOTORS CORPORATION GM

CATALYST EGR/ORC	AUTOMATIC TRANSMISSION
TIMING	10° (DR)
SPARK PLUG GAP (IN.)	0.045

SET PARKING BRAKE AND BLOCK DRIVE WHEELS.

TIMING ADJUSTMENT: MAKE ADJUSTMENT WITH ENGINE AT NORMAL OPERATING TEMPERATURE, ELECTRIC COOLING FAN OFF, AND AIR CONDITIONING OFF (IF EQUIPPED)

1. VERIFY NO "CHECK ENGINE" LIGHT
2. PUT EST (ELECTRONIC SPARK TIMING) IN BYPASS MODE BY DISCONNECTING TIMING CONNECTOR. [NOTE] THIS IS A SINGLE WIRE SEALED CONNECTOR. THE LEADS ARE TAN WITH A BLACK STRIPE AND BREAK OUT OF THE ENGINE WIRING HARNESS CONDUIT. DO NOT DISCONNECT FOUR WIRE CONNECTOR AT DISTRIBUTOR
3. CONNECT TIMING LIGHT INDUCTIVE PICK-UP TO #1 SPARK PLUG WIRE AND SET TIMING TO SPECIFICATION WITH ENGINE RUNNING AT IDLE SPEED
4. RECONNECT TIMING CONNECTOR AND CLEAR ECM TROUBLE CODE

[NOTE] IDLE AIR SPEED SCREW IS PRESET AND SEALED AT FACTORY. PROVISION FOR ADJUSTMENT DURING TUNE UP IS NOT PROVIDED. DO NOT ATTEMPT ADJUSTMENT

IDLE SPEEDS ARE AUTOMATICALLY CONTROLLED. DO NOT ATTEMPT ADJUSTMENTS

SEE SERVICE MANUAL, MAINTENANCE SCHEDULE AND EMISSION HOSE ROUTING DIAGRAM FOR ADDITIONAL INFORMATION

THIS VEHICLE CONFORMS TO U.S. EPA REGULATIONS APPLICABLE TO 1985 MODEL YEAR NEW PASSENGER CARS

PT NO 14086547

1985-V6-2.8L VIN W CAVALIER—FEDERAL

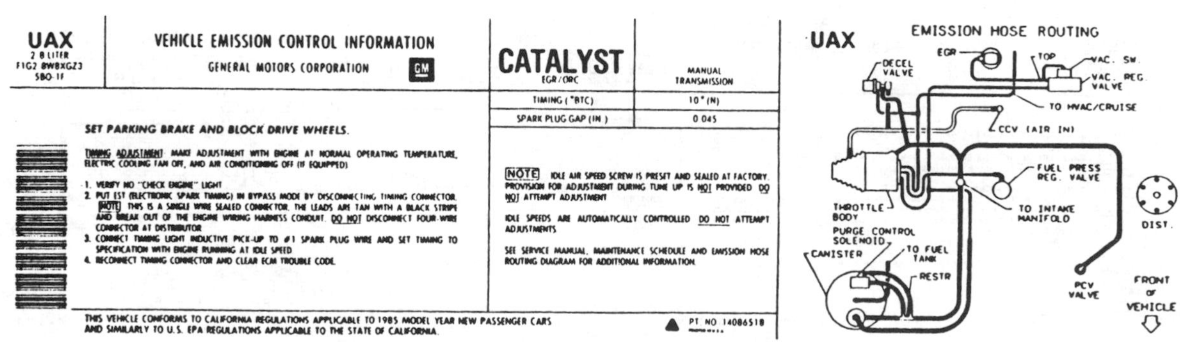

UAX
2.8 LITER
FIG2 8W8XGZ3
5B0-1F

VEHICLE EMISSION CONTROL INFORMATION
GENERAL MOTORS CORPORATION GM

CATALYST EGR/ORC	MANUAL TRANSMISSION
TIMING (°BTC)	10° (N)
SPARK PLUG GAP (IN.)	0.045

SET PARKING BRAKE AND BLOCK DRIVE WHEELS.

TIMING ADJUSTMENT: MAKE ADJUSTMENT WITH ENGINE AT NORMAL OPERATING TEMPERATURE, ELECTRIC COOLING FAN OFF, AND AIR CONDITIONING OFF (IF EQUIPPED)

1. VERIFY NO "CHECK ENGINE" LIGHT
2. PUT EST (ELECTRONIC SPARK TIMING) IN BYPASS MODE BY DISCONNECTING TIMING CONNECTOR. [NOTE] THIS IS A SINGLE WIRE SEALED CONNECTOR. THE LEADS ARE TAN WITH A BLACK STRIPE AND BREAK OUT OF THE ENGINE WIRING HARNESS CONDUIT. DO NOT DISCONNECT FOUR WIRE CONNECTOR AT DISTRIBUTOR
3. CONNECT TIMING LIGHT INDUCTIVE PICK-UP TO #1 SPARK PLUG WIRE AND SET TIMING TO SPECIFICATION WITH ENGINE RUNNING AT IDLE SPEED
4. RECONNECT TIMING CONNECTOR AND CLEAR ECM TROUBLE CODE

[NOTE] IDLE AIR SPEED SCREW IS PRESET AND SEALED AT FACTORY. PROVISION FOR ADJUSTMENT DURING TUNE UP IS NOT PROVIDED. DO NOT ATTEMPT ADJUSTMENT

IDLE SPEEDS ARE AUTOMATICALLY CONTROLLED. DO NOT ATTEMPT ADJUSTMENTS

SEE SERVICE MANUAL, MAINTENANCE SCHEDULE AND EMISSION HOSE ROUTING DIAGRAM FOR ADDITIONAL INFORMATION

THIS VEHICLE CONFORMS TO CALIFORNIA REGULATIONS APPLICABLE TO 1985 MODEL YEAR NEW PASSENGER CARS AND SIMILARLY TO U.S. EPA REGULATIONS APPLICABLE TO THE STATE OF CALIFORNIA

PT NO 14086518

1985-V6-2.8L VIN W CITATION—CALIFORNIA

VACUUM CIRCUITS

1985 CHEVROLET 2.8L

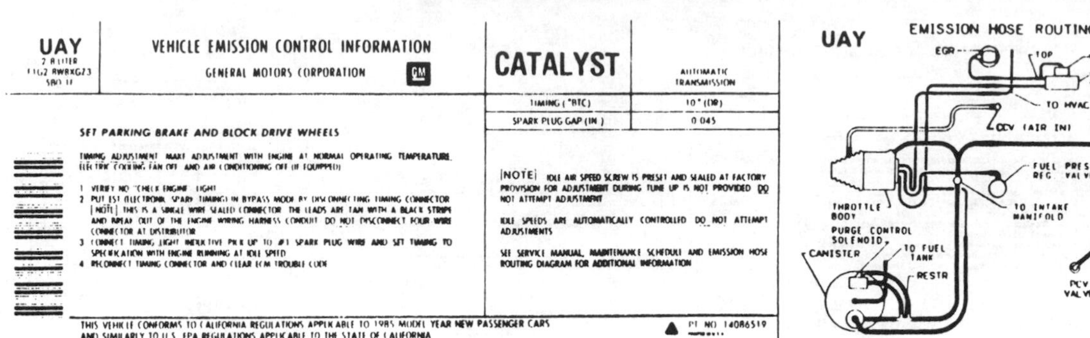

UAY
2.8 LITER
FIG2 8W8XGZ3
5B0-1F

VEHICLE EMISSION CONTROL INFORMATION
GENERAL MOTORS CORPORATION GM

CATALYST

AUTOMATIC TRANSMISSION

TIMING (°BTC)	10° (DR)
SPARK PLUG GAP (IN.)	0.045

SET PARKING BRAKE AND BLOCK DRIVE WHEELS.

TIMING ADJUSTMENT: MAKE ADJUSTMENT WITH ENGINE AT NORMAL OPERATING TEMPERATURE, ELECTRIC COOLING FAN OFF, AND AIR CONDITIONING OFF (IF EQUIPPED).

1. VERIFY NO "CHECK ENGINE" LIGHT.
2. PUT EST (ELECTRONIC SPARK TIMING) IN BYPASS MODE BY DISCONNECTING TIMING CONNECTOR. [NOTE] THIS IS A SINGLE WIRE SEALED CONNECTOR. THE LEADS ARE TAN WITH A BLACK STRIPE AND BREAK OUT OF THE ENGINE WIRING HARNESS CONDUIT. DO NOT DISCONNECT FOUR WIRE CONNECTOR AT DISTRIBUTOR.
3. CONNECT TIMING LIGHT INDUCTIVE PICK-UP TO #1 SPARK PLUG WIRE AND SET TIMING TO SPECIFICATION WITH ENGINE RUNNING AT IDLE SPEED.
4. RECONNECT TIMING CONNECTOR AND CLEAR ECM TROUBLE CODE.

[NOTE] IDLE AIR SPEED SCREW IS PRESET AND SEALED AT FACTORY. PROVISION FOR ADJUSTMENT DURING TUNE UP IS NOT PROVIDED. DO NOT ATTEMPT ADJUSTMENT.

IDLE SPEEDS ARE AUTOMATICALLY CONTROLLED. DO NOT ATTEMPT ADJUSTMENTS.

SEE SERVICE MANUAL, MAINTENANCE SCHEDULE AND EMISSION HOSE ROUTING DIAGRAM FOR ADDITIONAL INFORMATION.

THIS VEHICLE CONFORMS TO CALIFORNIA REGULATIONS APPLICABLE TO 1985 MODEL YEAR NEW PASSENGER CARS AND SIMILARLY TO U.S. EPA REGULATIONS APPLICABLE TO THE STATE OF CALIFORNIA.

PT NO 14086519

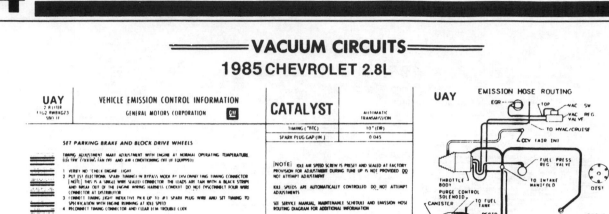

1985-V6-2.8L VIN W CELEBRITY, CITATION—CALIFORNIA

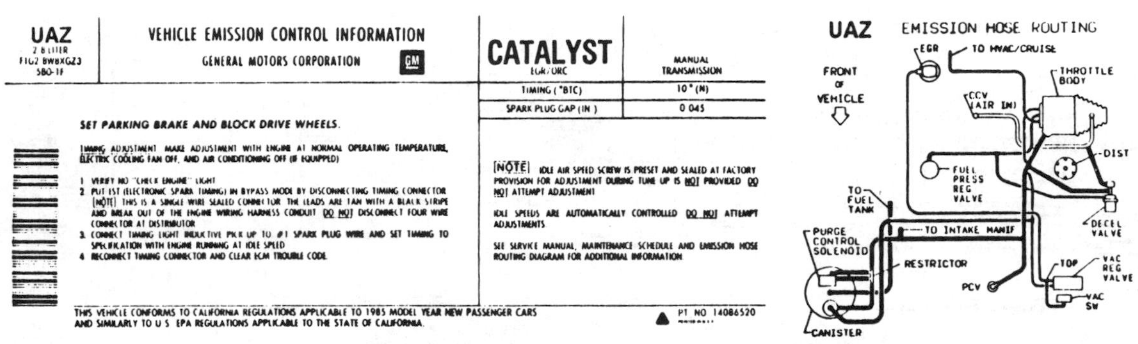

UAZ
2.8 LITER
FIG2 8W8XGZ3
5B0-1F

VEHICLE EMISSION CONTROL INFORMATION
GENERAL MOTORS CORPORATION GM

CATALYST
EGR/ORC

MANUAL TRANSMISSION

TIMING (°BTC)	10° (N)
SPARK PLUG GAP (IN.)	0.045

SET PARKING BRAKE AND BLOCK DRIVE WHEELS.

TIMING ADJUSTMENT: MAKE ADJUSTMENT WITH ENGINE AT NORMAL OPERATING TEMPERATURE, ELECTRIC COOLING FAN OFF, AND AIR CONDITIONING OFF (IF EQUIPPED).

1. VERIFY NO "CHECK ENGINE" LIGHT.
2. PUT EST (ELECTRONIC SPARK TIMING) IN BYPASS MODE BY DISCONNECTING TIMING CONNECTOR. [NOTE] THIS IS A SINGLE WIRE SEALED CONNECTOR. THE LEADS ARE TAN WITH A BLACK STRIPE AND BREAK OUT OF THE ENGINE WIRING HARNESS CONDUIT. DO NOT DISCONNECT FOUR WIRE CONNECTOR AT DISTRIBUTOR.
3. CONNECT TIMING LIGHT INDUCTIVE PICK-UP TO #1 SPARK PLUG WIRE AND SET TIMING TO SPECIFICATION WITH ENGINE RUNNING AT IDLE SPEED.
4. RECONNECT TIMING CONNECTOR AND CLEAR ECM TROUBLE CODE.

[NOTE] IDLE AIR SPEED SCREW IS PRESET AND SEALED AT FACTORY. PROVISION FOR ADJUSTMENT DURING TUNE UP IS NOT PROVIDED. DO NOT ATTEMPT ADJUSTMENT.

IDLE SPEEDS ARE AUTOMATICALLY CONTROLLED. DO NOT ATTEMPT ADJUSTMENTS.

SEE SERVICE MANUAL, MAINTENANCE SCHEDULE AND EMISSION HOSE ROUTING DIAGRAM FOR ADDITIONAL INFORMATION.

THIS VEHICLE CONFORMS TO CALIFORNIA REGULATIONS APPLICABLE TO 1985 MODEL YEAR NEW PASSENGER CARS AND SIMILARLY TO U.S. EPA REGULATIONS APPLICABLE TO THE STATE OF CALIFORNIA.

PT NO 14086520

1985-V6-2.8L VIN W CAVALIER—CALIFORNIA

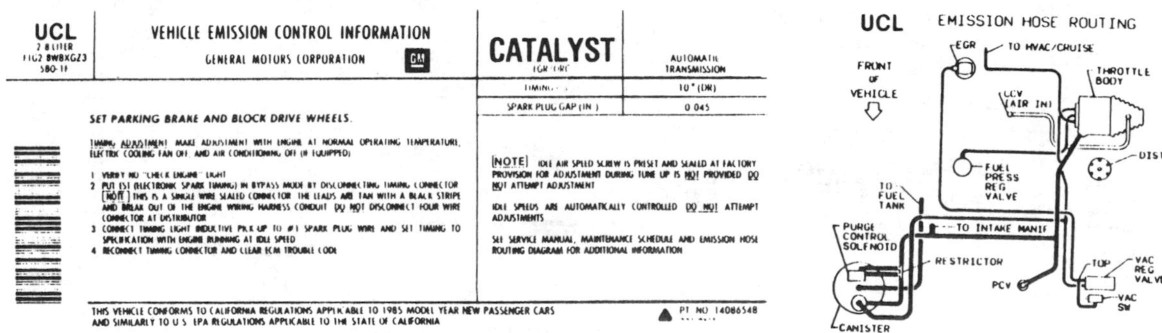

UCL
2.8 LITER
FIG2 8W8XGZ3
5B0-1F

VEHICLE EMISSION CONTROL INFORMATION
GENERAL MOTORS CORPORATION GM

CATALYST
EGR/ORC

AUTOMATIC TRANSMISSION

TIMING (°BTC)	10° (DR)
SPARK PLUG GAP (IN.)	0.045

SET PARKING BRAKE AND BLOCK DRIVE WHEELS.

TIMING ADJUSTMENT: MAKE ADJUSTMENT WITH ENGINE AT NORMAL OPERATING TEMPERATURE, ELECTRIC COOLING FAN OFF, AND AIR CONDITIONING OFF (IF EQUIPPED).

1. VERIFY NO "CHECK ENGINE" LIGHT.
2. PUT EST (ELECTRONIC SPARK TIMING) IN BYPASS MODE BY DISCONNECTING TIMING CONNECTOR. [NOTE] THIS IS A SINGLE WIRE SEALED CONNECTOR. THE LEADS ARE TAN WITH A BLACK STRIPE AND BREAK OUT OF THE ENGINE WIRING HARNESS CONDUIT. DO NOT DISCONNECT FOUR WIRE CONNECTOR AT DISTRIBUTOR.
3. CONNECT TIMING LIGHT INDUCTIVE PICK-UP TO #1 SPARK PLUG WIRE AND SET TIMING TO SPECIFICATION WITH ENGINE RUNNING AT IDLE SPEED.
4. RECONNECT TIMING CONNECTOR AND CLEAR ECM TROUBLE CODE.

[NOTE] IDLE AIR SPEED SCREW IS PRESET AND SEALED AT FACTORY. PROVISION FOR ADJUSTMENT DURING TUNE UP IS NOT PROVIDED. DO NOT ATTEMPT ADJUSTMENT.

IDLE SPEEDS ARE AUTOMATICALLY CONTROLLED. DO NOT ATTEMPT ADJUSTMENTS.

SEE SERVICE MANUAL, MAINTENANCE SCHEDULE AND EMISSION HOSE ROUTING DIAGRAM FOR ADDITIONAL INFORMATION.

THIS VEHICLE CONFORMS TO CALIFORNIA REGULATIONS APPLICABLE TO 1985 MODEL YEAR NEW PASSENGER CARS AND SIMILARLY TO U.S. EPA REGULATIONS APPLICABLE TO THE STATE OF CALIFORNIA.

PT NO 14086548

1985-V6-2.8L VIN W CAVALIER—CALIFORNIA

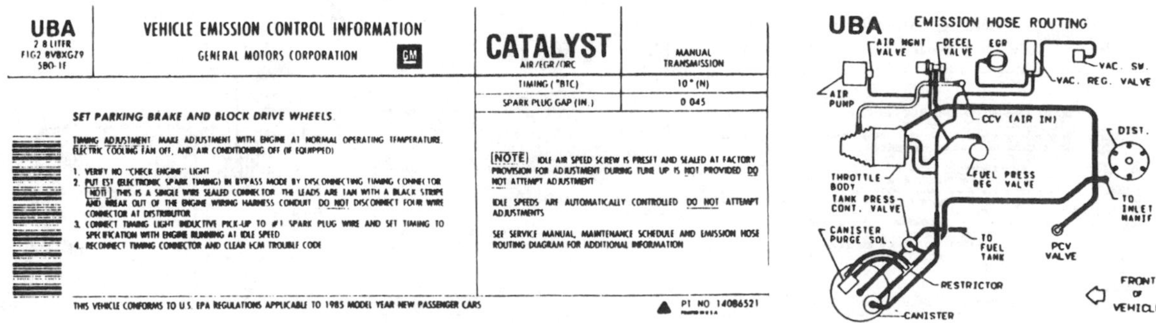

UBA
2.8 LITER
FIG2 8V8XGZ9
5B0-1F

VEHICLE EMISSION CONTROL INFORMATION
GENERAL MOTORS CORPORATION GM

CATALYST
AIR/EGR/ORC

MANUAL TRANSMISSION

TIMING (°BTC)	10° (N)
SPARK PLUG GAP (IN.)	0.045

SET PARKING BRAKE AND BLOCK DRIVE WHEELS.

TIMING ADJUSTMENT: MAKE ADJUSTMENT WITH ENGINE AT NORMAL OPERATING TEMPERATURE, ELECTRIC COOLING FAN OFF, AND AIR CONDITIONING OFF (IF EQUIPPED).

1. VERIFY NO "CHECK ENGINE" LIGHT.
2. PUT EST (ELECTRONIC SPARK TIMING) IN BYPASS MODE BY DISCONNECTING TIMING CONNECTOR. [NOTE] THIS IS A SINGLE WIRE SEALED CONNECTOR. THE LEADS ARE TAN WITH A BLACK STRIPE AND BREAK OUT OF THE ENGINE WIRING HARNESS CONDUIT. DO NOT DISCONNECT FOUR WIRE CONNECTOR AT DISTRIBUTOR.
3. CONNECT TIMING LIGHT INDUCTIVE PICK-UP TO #1 SPARK PLUG WIRE AND SET TIMING TO SPECIFICATION WITH ENGINE RUNNING AT IDLE SPEED.
4. RECONNECT TIMING CONNECTOR AND CLEAR ECM TROUBLE CODE.

[NOTE] IDLE AIR SPEED SCREW IS PRESET AND SEALED AT FACTORY. PROVISION FOR ADJUSTMENT DURING TUNE UP IS NOT PROVIDED. DO NOT ATTEMPT ADJUSTMENT.

IDLE SPEEDS ARE AUTOMATICALLY CONTROLLED. DO NOT ATTEMPT ADJUSTMENTS.

SEE SERVICE MANUAL, MAINTENANCE SCHEDULE AND EMISSION HOSE ROUTING DIAGRAM FOR ADDITIONAL INFORMATION.

THIS VEHICLE CONFORMS TO U.S. EPA REGULATIONS APPLICABLE TO 1985 MODEL YEAR NEW PASSENGER CARS.

PT NO 14086521

1985-V6-2.8L VIN S CHEV.—FEDERAL

VACUUM CIRCUITS

1985 CHEVROLET 2.8L

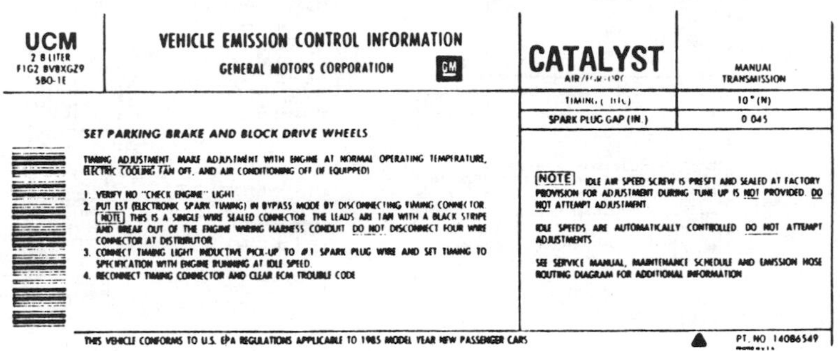
UCM
2 8 LITER
F1G2 8V8XGZ9
5BO-1E

VEHICLE EMISSION CONTROL INFORMATION
GENERAL MOTORS CORPORATION GM

CATALYST
AIR/EGR/ORC

	MANUAL TRANSMISSION
TIMING (°BTC)	10° (N)
SPARK PLUG GAP (IN.)	0.045

SET PARKING BRAKE AND BLOCK DRIVE WHEELS

TIMING ADJUSTMENT: MAKE ADJUSTMENT WITH ENGINE AT NORMAL OPERATING TEMPERATURE, ELECTRIC COOLING FAN OFF, AND AIR CONDITIONING OFF (IF EQUIPPED)

1. VERIFY NO "CHECK ENGINE" LIGHT
2. PUT EST (ELECTRONIC SPARK TIMING) IN BYPASS MODE BY DISCONNECTING TIMING CONNECTOR. [NOTE] THIS IS A SINGLE WIRE SEALED CONNECTOR. THE LEADS ARE TAN WITH A BLACK STRIPE AND BREAK OUT OF THE ENGINE WIRING HARNESS CONDUIT. DO NOT DISCONNECT FOUR WIRE CONNECTOR AT DISTRIBUTOR
3. CONNECT TIMING LIGHT INDUCTIVE PICK-UP TO #1 SPARK PLUG WIRE AND SET TIMING TO SPECIFICATION WITH ENGINE RUNNING AT IDLE SPEED.
4. RECONNECT TIMING CONNECTOR AND CLEAR ECM TROUBLE CODE

[NOTE] IDLE AIR SPEED SCREW IS PRESET AND SEALED AT FACTORY. PROVISION FOR ADJUSTMENT DURING TUNE UP IS NOT PROVIDED. DO NOT ATTEMPT ADJUSTMENT

IDLE SPEEDS ARE AUTOMATICALLY CONTROLLED. DO NOT ATTEMPT ADJUSTMENTS

SEE SERVICE MANUAL, MAINTENANCE SCHEDULE AND EMISSION HOSE ROUTING DIAGRAM FOR ADDITIONAL INFORMATION

THIS VEHICLE CONFORMS TO U.S. EPA REGULATIONS APPLICABLE TO 1985 MODEL YEAR NEW PASSENGER CARS

PT. NO 14086549

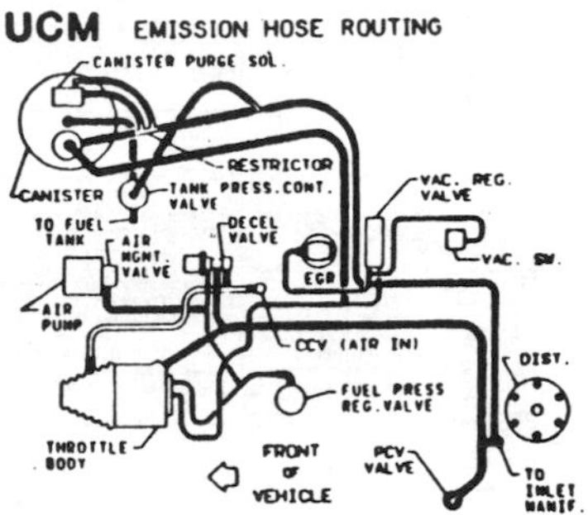

● 1985-V6-2.8L VIN S PONT.—FEDERAL

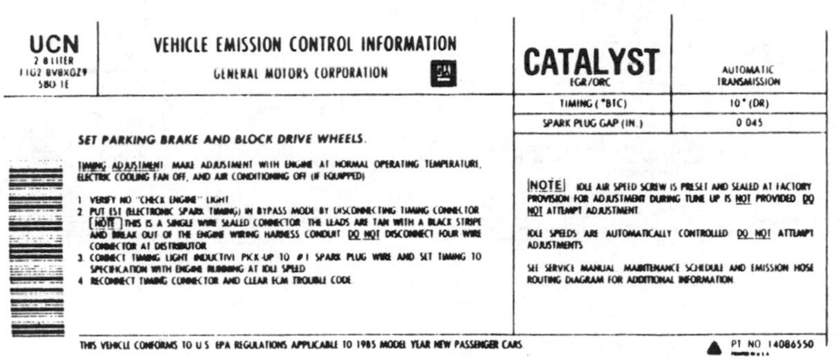
UCN
2 8 LITER
F1G2 8V8XGZ9
5BO-1E

VEHICLE EMISSION CONTROL INFORMATION
GENERAL MOTORS CORPORATION GM

CATALYST
EGR/ORC

	AUTOMATIC TRANSMISSION
TIMING (°BTC)	10° (DR)
SPARK PLUG GAP (IN.)	0.045

SET PARKING BRAKE AND BLOCK DRIVE WHEELS.

TIMING ADJUSTMENT: MAKE ADJUSTMENT WITH ENGINE AT NORMAL OPERATING TEMPERATURE, ELECTRIC COOLING FAN OFF, AND AIR CONDITIONING OFF (IF EQUIPPED)

1. VERIFY NO "CHECK ENGINE" LIGHT
2. PUT EST (ELECTRONIC SPARK TIMING) IN BYPASS MODE BY DISCONNECTING TIMING CONNECTOR. [NOTE] THIS IS A SINGLE WIRE SEALED CONNECTOR. THE LEADS ARE TAN WITH A BLACK STRIPE AND BREAK OUT OF THE ENGINE WIRING HARNESS CONDUIT. DO NOT DISCONNECT FOUR WIRE CONNECTOR AT DISTRIBUTOR
3. CONNECT TIMING LIGHT INDUCTIVE PICK-UP TO #1 SPARK PLUG WIRE AND SET TIMING TO SPECIFICATION WITH ENGINE RUNNING AT IDLE SPEED
4. RECONNECT TIMING CONNECTOR AND CLEAR ECM TROUBLE CODE

[NOTE] IDLE AIR SPEED SCREW IS PRESET AND SEALED AT FACTORY. PROVISION FOR ADJUSTMENT DURING TUNE UP IS NOT PROVIDED. DO NOT ATTEMPT ADJUSTMENT

IDLE SPEEDS ARE AUTOMATICALLY CONTROLLED. DO NOT ATTEMPT ADJUSTMENTS

SEE SERVICE MANUAL, MAINTENANCE SCHEDULE AND EMISSION HOSE ROUTING DIAGRAM FOR ADDITIONAL INFORMATION

THIS VEHICLE CONFORMS TO U.S. EPA REGULATIONS APPLICABLE TO 1985 MODEL YEAR NEW PASSENGER CARS

PT. NO 14086550

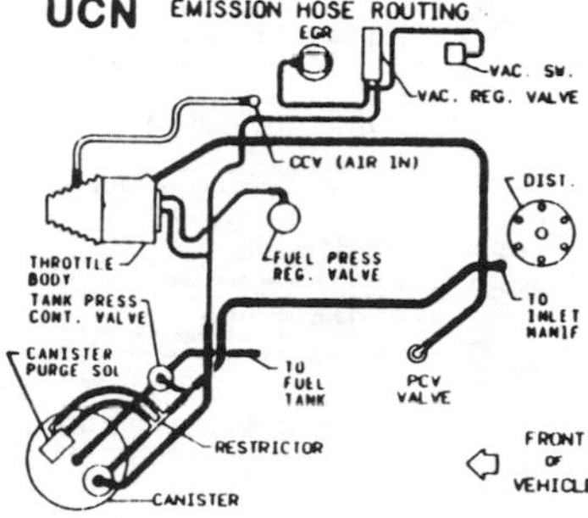

1985-V6-2.8L VIN S CHEV.—FEDERAL

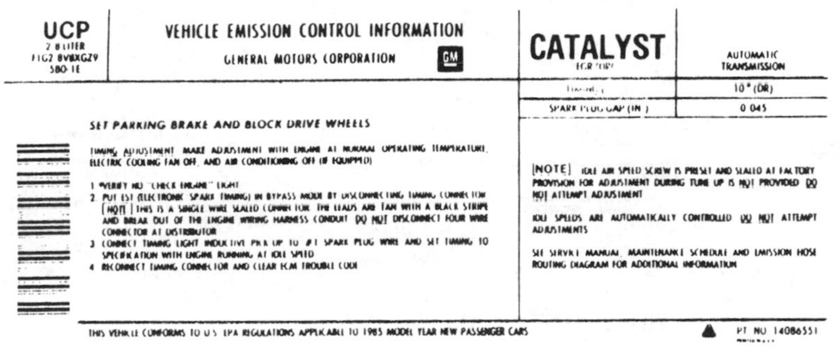
UCP
2 8 LITER
F1G2 8V8XGZ9
5BO-1E

VEHICLE EMISSION CONTROL INFORMATION
GENERAL MOTORS CORPORATION GM

CATALYST
EGR/ORC

	AUTOMATIC TRANSMISSION
TIMING (°BTC)	10° (DR)
SPARK PLUG GAP (IN.)	0.045

SET PARKING BRAKE AND BLOCK DRIVE WHEELS

TIMING ADJUSTMENT: MAKE ADJUSTMENT WITH ENGINE AT NORMAL OPERATING TEMPERATURE, ELECTRIC COOLING FAN OFF, AND AIR CONDITIONING OFF (IF EQUIPPED)

1. VERIFY NO "CHECK ENGINE" LIGHT
2. PUT EST (ELECTRONIC SPARK TIMING) IN BYPASS MODE BY DISCONNECTING TIMING CONNECTOR. [NOTE] THIS IS A SINGLE WIRE SEALED CONNECTOR. THE LEADS ARE TAN WITH A BLACK STRIPE AND BREAK OUT OF THE ENGINE WIRING HARNESS CONDUIT. DO NOT DISCONNECT FOUR WIRE CONNECTOR AT DISTRIBUTOR
3. CONNECT TIMING LIGHT INDUCTIVE PICK-UP TO #1 SPARK PLUG WIRE AND SET TIMING TO SPECIFICATION WITH ENGINE RUNNING AT IDLE SPEED
4. RECONNECT TIMING CONNECTOR AND CLEAR ECM TROUBLE CODE

[NOTE] IDLE AIR SPEED SCREW IS PRESET AND SEALED AT FACTORY. PROVISION FOR ADJUSTMENT DURING TUNE UP IS NOT PROVIDED. DO NOT ATTEMPT ADJUSTMENT

IDLE SPEEDS ARE AUTOMATICALLY CONTROLLED. DO NOT ATTEMPT ADJUSTMENTS

SEE SERVICE MANUAL, MAINTENANCE SCHEDULE AND EMISSION HOSE ROUTING DIAGRAM FOR ADDITIONAL INFORMATION

THIS VEHICLE CONFORMS TO U.S. EPA REGULATIONS APPLICABLE TO 1985 MODEL YEAR NEW PASSENGER CARS

PT. NO 14086551

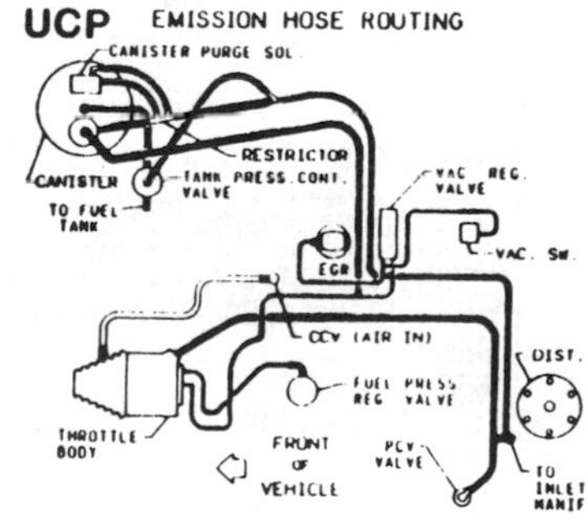

1985-V6-2.8L VIN S PONT.—FEDERAL

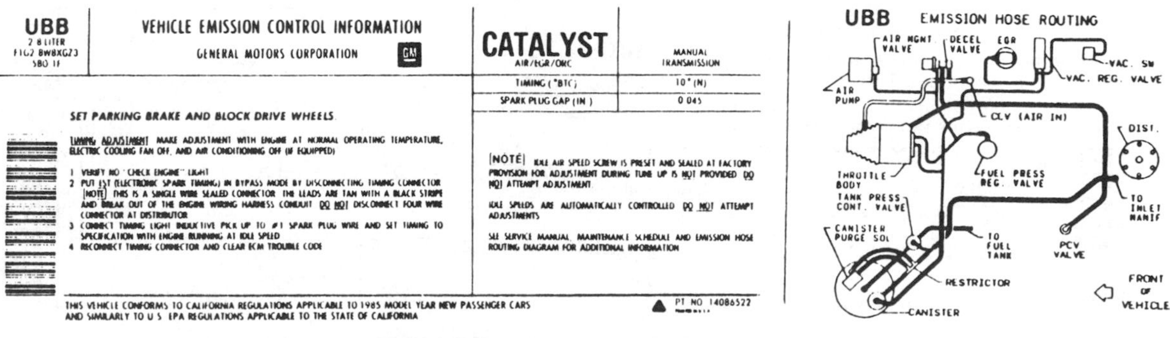
UBB
2 8 LITER
F1G2 8W8XGZ3
5BO-1F

VEHICLE EMISSION CONTROL INFORMATION
GENERAL MOTORS CORPORATION GM

CATALYST
AIR/EGR/ORC

	MANUAL TRANSMISSION
TIMING (°BTC)	10° (N)
SPARK PLUG GAP (IN.)	0.045

SET PARKING BRAKE AND BLOCK DRIVE WHEELS

TIMING ADJUSTMENT: MAKE ADJUSTMENT WITH ENGINE AT NORMAL OPERATING TEMPERATURE, ELECTRIC COOLING FAN OFF, AND AIR CONDITIONING OFF (IF EQUIPPED)

1. VERIFY NO "CHECK ENGINE" LIGHT
2. PUT EST (ELECTRONIC SPARK TIMING) IN BYPASS MODE BY DISCONNECTING TIMING CONNECTOR. [NOTE] THIS IS A SINGLE WIRE SEALED CONNECTOR. THE LEADS ARE TAN WITH A BLACK STRIPE AND BREAK OUT OF THE ENGINE WIRING HARNESS CONDUIT. DO NOT DISCONNECT FOUR WIRE CONNECTOR AT DISTRIBUTOR
3. CONNECT TIMING LIGHT INDUCTIVE PICK-UP TO #1 SPARK PLUG WIRE AND SET TIMING TO SPECIFICATION WITH ENGINE RUNNING AT IDLE SPEED
4. RECONNECT TIMING CONNECTOR AND CLEAR ECM TROUBLE CODE

[NOTE] IDLE AIR SPEED SCREW IS PRESET AND SEALED AT FACTORY. PROVISION FOR ADJUSTMENT DURING TUNE UP IS NOT PROVIDED. DO NOT ATTEMPT ADJUSTMENT

IDLE SPEEDS ARE AUTOMATICALLY CONTROLLED. DO NOT ATTEMPT ADJUSTMENTS

SEE SERVICE MANUAL, MAINTENANCE SCHEDULE AND EMISSION HOSE ROUTING DIAGRAM FOR ADDITIONAL INFORMATION

THIS VEHICLE CONFORMS TO CALIFORNIA REGULATIONS APPLICABLE TO 1985 MODEL YEAR NEW PASSENGER CARS AND SIMILARLY TO U.S. EPA REGULATIONS APPLICABLE TO THE STATE OF CALIFORNIA

PT. NO 14086522

1985-V6-2.8L VIN S CHEV.—CALIFORNIA

VACUUM CIRCUITS

1985 CHEVROLET 2.8L

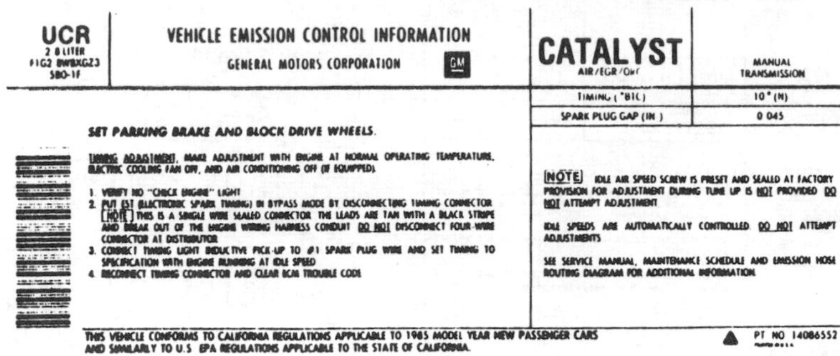

UCR
2.8 LITER
FIG2 8WBXGZ3
5B0-1F

VEHICLE EMISSION CONTROL INFORMATION
GENERAL MOTORS CORPORATION GM

CATALYST
AIR/EGR/ORC

	MANUAL TRANSMISSION
TIMING (°BTC)	10° (N)
SPARK PLUG GAP (IN.)	0.045

SET PARKING BRAKE AND BLOCK DRIVE WHEELS.

TIMING ADJUSTMENT. MAKE ADJUSTMENT WITH ENGINE AT NORMAL OPERATING TEMPERATURE, ELECTRIC COOLING FAN OFF, AND AIR CONDITIONING OFF (IF EQUIPPED).

1. VERIFY NO "CHECK ENGINE" LIGHT
2. PUT EST (ELECTRONIC SPARK TIMING) IN BYPASS MODE BY DISCONNECTING TIMING CONNECTOR. [NOTE] THIS IS A SINGLE WIRE SEALED CONNECTOR. THE LEADS ARE TAN WITH A BLACK STRIPE AND BREAK OUT OF THE ENGINE WIRING HARNESS CONDUIT. DO NOT DISCONNECT FOUR-WIRE CONNECTOR AT DISTRIBUTOR.
3. CONNECT TIMING LIGHT INDUCTIVE PICK-UP TO #1 SPARK PLUG WIRE AND SET TIMING TO SPECIFICATION WITH ENGINE RUNNING AT IDLE SPEED.
4. RECONNECT TIMING CONNECTOR AND CLEAR ECM TROUBLE CODE.

[NOTE] IDLE AIR SPEED SCREW IS PRESET AND SEALED AT FACTORY. PROVISION FOR ADJUSTMENT DURING TUNE UP IS NOT PROVIDED. DO NOT ATTEMPT ADJUSTMENT.

IDLE SPEEDS ARE AUTOMATICALLY CONTROLLED. DO NOT ATTEMPT ADJUSTMENTS.

SEE SERVICE MANUAL, MAINTENANCE SCHEDULE AND EMISSION HOSE ROUTING DIAGRAM FOR ADDITIONAL INFORMATION.

THIS VEHICLE CONFORMS TO CALIFORNIA REGULATIONS APPLICABLE TO 1985 MODEL YEAR NEW PASSENGER CARS AND SIMILARLY TO U.S. EPA REGULATIONS APPLICABLE TO THE STATE OF CALIFORNIA.

PT NO 14086552

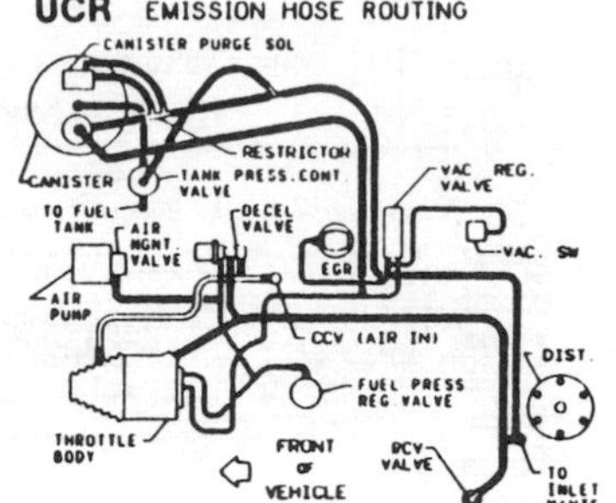

1985-V6-2.8L VIN S PONT.—CALIFORNIA

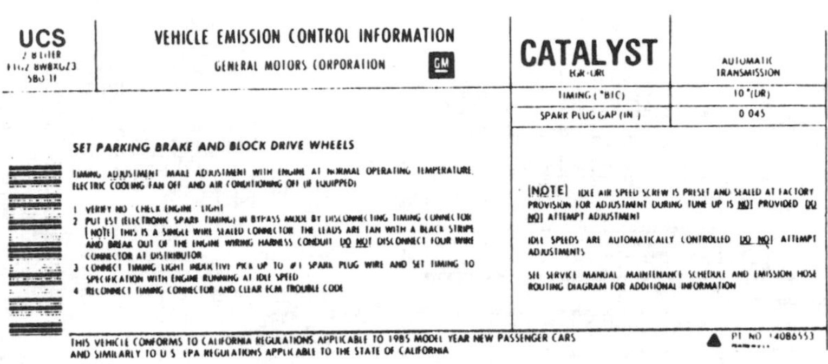

UCS
2.8 LITER
FIG2 8WBXGZ3
5B0-1F

VEHICLE EMISSION CONTROL INFORMATION
GENERAL MOTORS CORPORATION GM

CATALYST
EGR/ORC

	AUTOMATIC TRANSMISSION
TIMING (°BTC)	10° (DR)
SPARK PLUG GAP (IN.)	0.045

SET PARKING BRAKE AND BLOCK DRIVE WHEELS.

TIMING ADJUSTMENT. MAKE ADJUSTMENT WITH ENGINE AT NORMAL OPERATING TEMPERATURE, ELECTRIC COOLING FAN OFF, AND AIR CONDITIONING OFF (IF EQUIPPED).

1. VERIFY NO "CHECK ENGINE" LIGHT
2. PUT EST (ELECTRONIC SPARK TIMING) IN BYPASS MODE BY DISCONNECTING TIMING CONNECTOR. [NOTE] THIS IS A SINGLE WIRE SEALED CONNECTOR. THE LEADS ARE TAN WITH A BLACK STRIPE AND BREAK OUT OF THE ENGINE WIRING HARNESS CONDUIT. DO NOT DISCONNECT FOUR-WIRE CONNECTOR AT DISTRIBUTOR.
3. CONNECT TIMING LIGHT INDUCTIVE PICK-UP TO #1 SPARK PLUG WIRE AND SET TIMING TO SPECIFICATION WITH ENGINE RUNNING AT IDLE SPEED.
4. RECONNECT TIMING CONNECTOR AND CLEAR ECM TROUBLE CODE.

[NOTE] IDLE AIR SPEED SCREW IS PRESET AND SEALED AT FACTORY. PROVISION FOR ADJUSTMENT DURING TUNE UP IS NOT PROVIDED. DO NOT ATTEMPT ADJUSTMENT.

IDLE SPEEDS ARE AUTOMATICALLY CONTROLLED. DO NOT ATTEMPT ADJUSTMENTS.

SEE SERVICE MANUAL, MAINTENANCE SCHEDULE AND EMISSION HOSE ROUTING DIAGRAM FOR ADDITIONAL INFORMATION.

THIS VEHICLE CONFORMS TO CALIFORNIA REGULATIONS APPLICABLE TO 1985 MODEL YEAR NEW PASSENGER CARS AND SIMILARLY TO U.S. EPA REGULATIONS APPLICABLE TO THE STATE OF CALIFORNIA.

PT NO 14086553

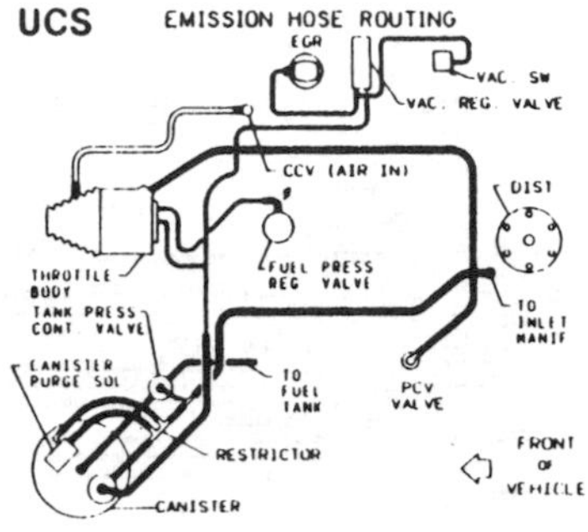

1985-V6-2.8L VIN S CHEV.—CALIFORNIA

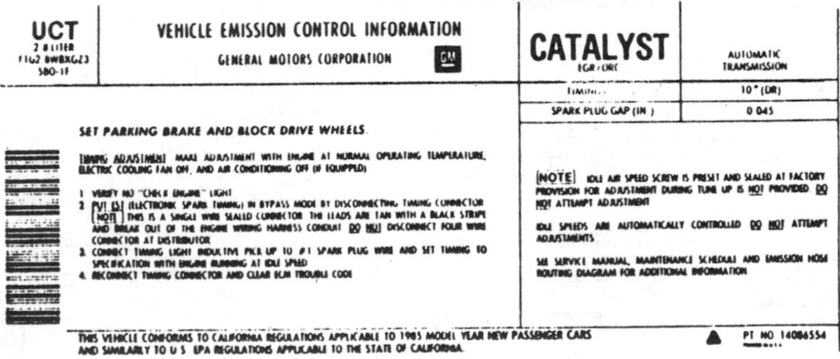

UCT
2.8 LITER
FIG2 8WBXGZ3
5B0-1F

VEHICLE EMISSION CONTROL INFORMATION
GENERAL MOTORS CORPORATION GM

CATALYST
EGR/ORC

	AUTOMATIC TRANSMISSION
TIMING (°BTC)	10° (DR)
SPARK PLUG GAP (IN.)	0.045

SET PARKING BRAKE AND BLOCK DRIVE WHEELS.

TIMING ADJUSTMENT. MAKE ADJUSTMENT WITH ENGINE AT NORMAL OPERATING TEMPERATURE, ELECTRIC COOLING FAN OFF, AND AIR CONDITIONING OFF (IF EQUIPPED).

1. VERIFY NO "CHECK ENGINE" LIGHT
2. PUT EST (ELECTRONIC SPARK TIMING) IN BYPASS MODE BY DISCONNECTING TIMING CONNECTOR. [NOTE] THIS IS A SINGLE WIRE SEALED CONNECTOR. THE LEADS ARE TAN WITH A BLACK STRIPE AND BREAK OUT OF THE ENGINE WIRING HARNESS CONDUIT. DO NOT DISCONNECT FOUR-WIRE CONNECTOR AT DISTRIBUTOR.
3. CONNECT TIMING LIGHT INDUCTIVE PICK-UP TO #1 SPARK PLUG WIRE AND SET TIMING TO SPECIFICATION WITH ENGINE RUNNING AT IDLE SPEED.
4. RECONNECT TIMING CONNECTOR AND CLEAR ECM TROUBLE CODE.

[NOTE] IDLE AIR SPEED SCREW IS PRESET AND SEALED AT FACTORY. PROVISION FOR ADJUSTMENT DURING TUNE UP IS NOT PROVIDED. DO NOT ATTEMPT ADJUSTMENT.

IDLE SPEEDS ARE AUTOMATICALLY CONTROLLED. DO NOT ATTEMPT ADJUSTMENTS.

SEE SERVICE MANUAL, MAINTENANCE SCHEDULE AND EMISSION HOSE ROUTING DIAGRAM FOR ADDITIONAL INFORMATION.

THIS VEHICLE CONFORMS TO CALIFORNIA REGULATIONS APPLICABLE TO 1985 MODEL YEAR NEW PASSENGER CARS AND SIMILARLY TO U.S. EPA REGULATIONS APPLICABLE TO THE STATE OF CALIFORNIA.

PT NO 14086554

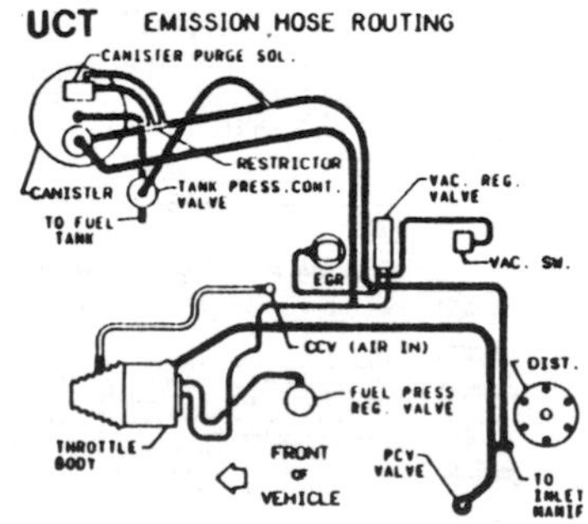

1985-V6-2.8L VIN S PONT.—CALIFORNIA

1985 CHEVROLET 4.3L

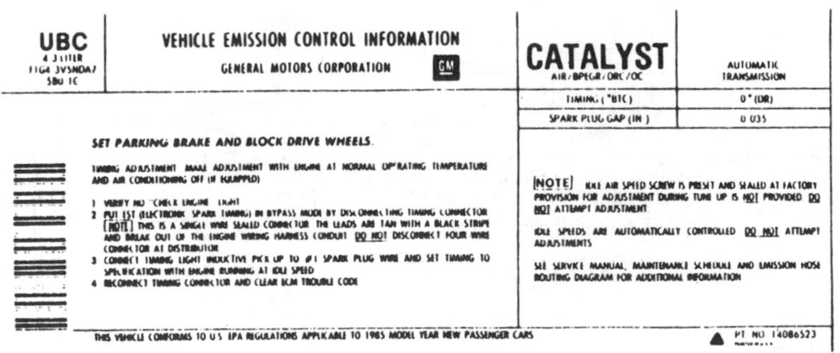

UBC
4.3 LITER
FIG4 3VSNDA7
5B0-1C

VEHICLE EMISSION CONTROL INFORMATION
GENERAL MOTORS CORPORATION GM

CATALYST
AIR/BPEGR/ORC/OC

	AUTOMATIC TRANSMISSION
TIMING (°BTC)	0° (DR)
SPARK PLUG GAP (IN.)	0.035

SET PARKING BRAKE AND BLOCK DRIVE WHEELS.

TIMING ADJUSTMENT. MAKE ADJUSTMENT WITH ENGINE AT NORMAL OPERATING TEMPERATURE AND AIR CONDITIONING OFF (IF EQUIPPED).

1. VERIFY NO "CHECK ENGINE" LIGHT
2. PUT EST (ELECTRONIC SPARK TIMING) IN BYPASS MODE BY DISCONNECTING TIMING CONNECTOR. [NOTE] THIS IS A SINGLE WIRE SEALED CONNECTOR. THE LEADS ARE TAN WITH A BLACK STRIPE AND BREAK OUT OF THE ENGINE WIRING HARNESS CONDUIT. DO NOT DISCONNECT FOUR WIRE CONNECTOR AT DISTRIBUTOR.
3. CONNECT TIMING LIGHT INDUCTIVE PICK UP TO #1 SPARK PLUG WIRE AND SET TIMING TO SPECIFICATION WITH ENGINE RUNNING AT IDLE SPEED.
4. RECONNECT TIMING CONNECTOR AND CLEAR ECM TROUBLE CODE.

[NOTE] IDLE AIR SPEED SCREW IS PRESET AND SEALED AT FACTORY. PROVISION FOR ADJUSTMENT DURING TUNE UP IS NOT PROVIDED. DO NOT ATTEMPT ADJUSTMENT.

IDLE SPEEDS ARE AUTOMATICALLY CONTROLLED. DO NOT ATTEMPT ADJUSTMENTS.

SEE SERVICE MANUAL, MAINTENANCE SCHEDULE AND EMISSION HOSE ROUTING DIAGRAM FOR ADDITIONAL INFORMATION.

THIS VEHICLE CONFORMS TO U.S. EPA REGULATIONS APPLICABLE TO 1985 MODEL YEAR NEW PASSENGER CARS.

PT NO 14086523

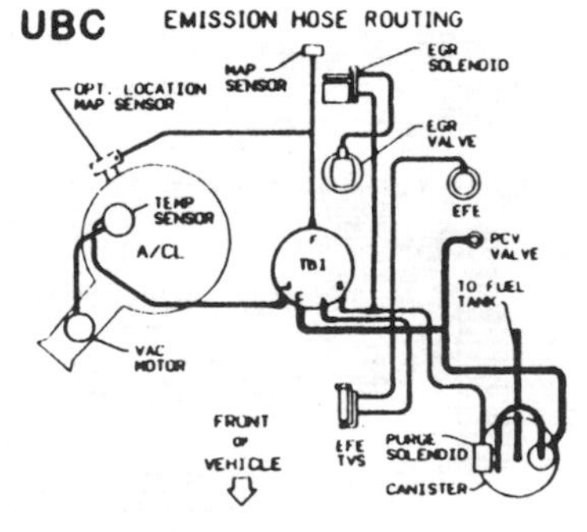

1985-V6-4.3L VIN V IMPALA, CAPRICE, MONTE CARLO—FEDERAL

VACUUM CIRCUITS

1985 CHEVROLET 4.3L

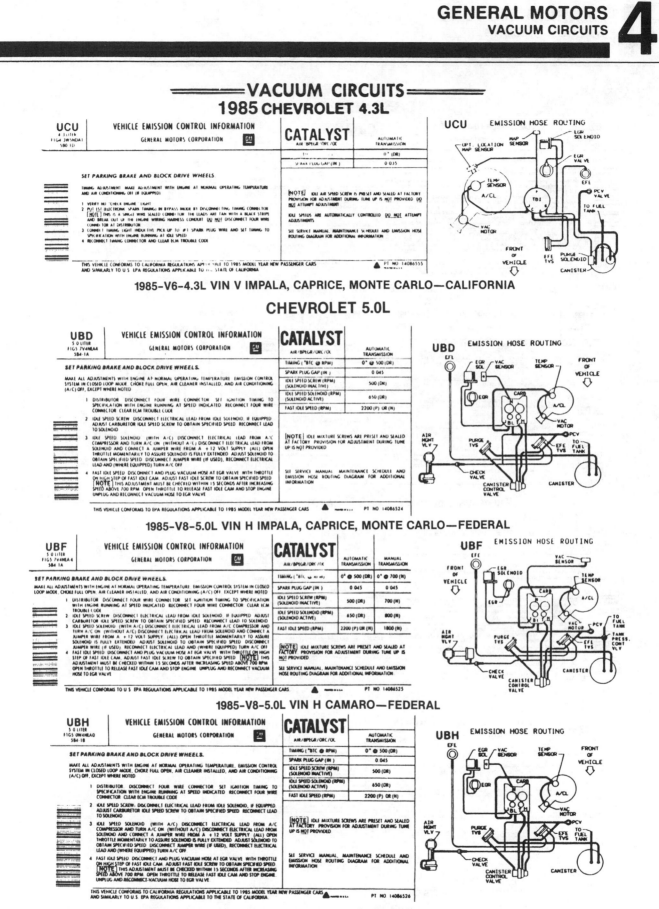

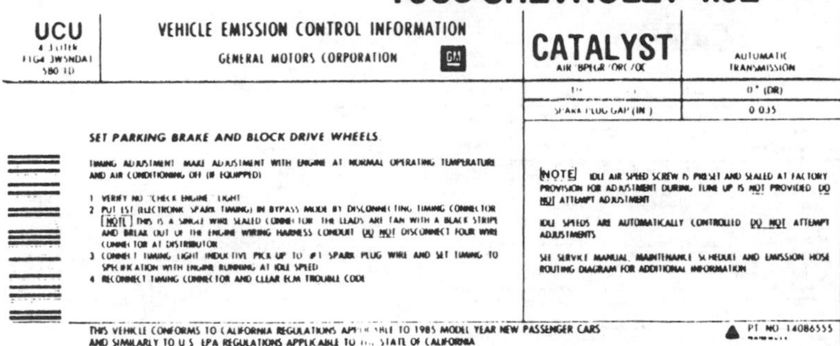

UCU
4.3 LITER
FIG4 3W5NDA1
5B0-1D

VEHICLE EMISSION CONTROL INFORMATION
GENERAL MOTORS CORPORATION

CATALYST
AIR/BPEGR/ORC/OC

	AUTOMATIC TRANSMISSION
TI[illegible]	0° (DR)
SPARK PLUG GAP (IN.)	0.035

SET PARKING BRAKE AND BLOCK DRIVE WHEELS.

TIMING ADJUSTMENT: MAKE ADJUSTMENT WITH ENGINE AT NORMAL OPERATING TEMPERATURE AND AIR CONDITIONING OFF (IF EQUIPPED)

1. VERIFY NO "CHECK ENGINE" LIGHT
2. PUT EST (ELECTRONIC SPARK TIMING) IN BYPASS MODE BY DISCONNECTING TIMING CONNECTOR [NOTE] THIS IS A SINGLE WIRE SEALED CONNECTOR. THE LEADS ARE TAN WITH A BLACK STRIPE AND BREAK OUT OF THE ENGINE WIRING HARNESS CONDUIT. DO NOT DISCONNECT FOUR WIRE CONNECTOR AT DISTRIBUTOR
3. CONNECT TIMING LIGHT INDUCTIVE PICK UP TO #1 SPARK PLUG WIRE AND SET TIMING TO SPECIFICATION WITH ENGINE RUNNING AT IDLE SPEED
4. RECONNECT TIMING CONNECTOR AND CLEAR ECM TROUBLE CODE

[NOTE] IDLE AIR SPEED SCREW IS PRESET AND SEALED AT FACTORY. PROVISION FOR ADJUSTMENT DURING TUNE UP IS NOT PROVIDED. DO NOT ATTEMPT ADJUSTMENT

IDLE SPEEDS ARE AUTOMATICALLY CONTROLLED. DO NOT ATTEMPT ADJUSTMENTS

SEE SERVICE MANUAL, MAINTENANCE SCHEDULE AND EMISSION HOSE ROUTING DIAGRAM FOR ADDITIONAL INFORMATION

THIS VEHICLE CONFORMS TO CALIFORNIA REGULATIONS APPLICABLE TO 1985 MODEL YEAR NEW PASSENGER CARS AND SIMILARLY TO U.S. EPA REGULATIONS APPLICABLE TO THE STATE OF CALIFORNIA

PT NO 14086555

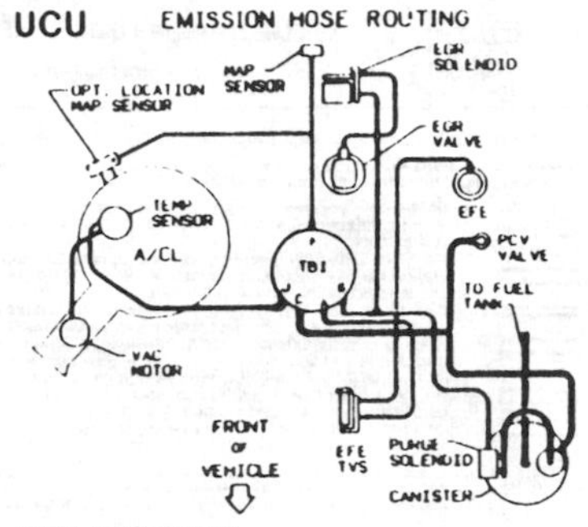

1985-V6-4.3L VIN V IMPALA, CAPRICE, MONTE CARLO—CALIFORNIA

CHEVROLET 5.0L

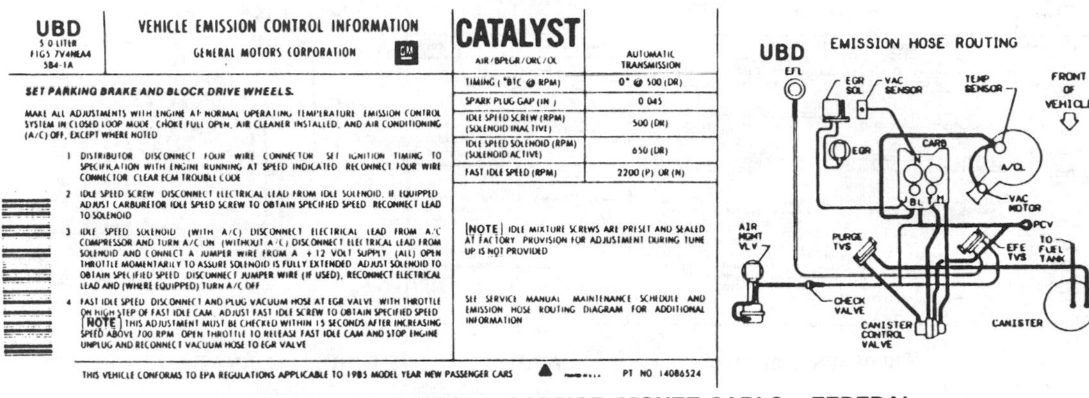

UBD
5.0 LITER
FIGS 7V4NEA4
5B4-1A

VEHICLE EMISSION CONTROL INFORMATION
GENERAL MOTORS CORPORATION

CATALYST
AIR/BPEGR/ORC/OC

	AUTOMATIC TRANSMISSION
TIMING (°BTC @ RPM)	0° @ 500 (DR)
SPARK PLUG GAP (IN.)	0.045
IDLE SPEED SCREW (RPM) (SOLENOID INACTIVE)	500 (DR)
IDLE SPEED SOLENOID (RPM) (SOLENOID ACTIVE)	650 (DR)
FAST IDLE SPEED (RPM)	2200 (P) OR (N)

SET PARKING BRAKE AND BLOCK DRIVE WHEELS.

MAKE ALL ADJUSTMENTS WITH ENGINE AT NORMAL OPERATING TEMPERATURE. EMISSION CONTROL SYSTEM IN CLOSED LOOP MODE, CHOKE FULL OPEN, AIR CLEANER INSTALLED, AND AIR CONDITIONING (A/C) OFF, EXCEPT WHERE NOTED

1. DISTRIBUTOR: DISCONNECT FOUR WIRE CONNECTOR. SET IGNITION TIMING TO SPECIFICATION WITH ENGINE RUNNING AT SPEED INDICATED. RECONNECT FOUR WIRE CONNECTOR. CLEAR ECM TROUBLE CODE
2. IDLE SPEED SCREW: DISCONNECT ELECTRICAL LEAD FROM IDLE SOLENOID, IF EQUIPPED. ADJUST CARBURETOR IDLE SPEED SCREW TO OBTAIN SPECIFIED SPEED. RECONNECT LEAD TO SOLENOID
3. IDLE SPEED SOLENOID: (WITH A/C) DISCONNECT ELECTRICAL LEAD FROM A/C COMPRESSOR AND TURN A/C ON. (WITHOUT A/C) DISCONNECT ELECTRICAL LEAD FROM SOLENOID AND CONNECT A JUMPER WIRE FROM A +12 VOLT SUPPLY. (ALL) OPEN THROTTLE MOMENTARILY TO ASSURE SOLENOID IS FULLY EXTENDED. ADJUST SOLENOID TO OBTAIN SPECIFIED SPEED. DISCONNECT JUMPER WIRE (IF USED), RECONNECT ELECTRICAL LEAD AND (WHERE EQUIPPED) TURN A/C OFF
4. FAST IDLE SPEED: DISCONNECT AND PLUG VACUUM HOSE AT EGR VALVE. WITH THROTTLE ON HIGH STEP OF FAST IDLE CAM, ADJUST FAST IDLE SCREW TO OBTAIN SPECIFIED SPEED. [NOTE] THIS ADJUSTMENT MUST BE CHECKED WITHIN 15 SECONDS AFTER INCREASING SPEED ABOVE 700 RPM. OPEN THROTTLE TO RELEASE FAST IDLE CAM AND STOP ENGINE. UNPLUG AND RECONNECT VACUUM HOSE TO EGR VALVE

[NOTE] IDLE MIXTURE SCREWS ARE PRESET AND SEALED AT FACTORY. PROVISION FOR ADJUSTMENT DURING TUNE UP IS NOT PROVIDED

SEE SERVICE MANUAL, MAINTENANCE SCHEDULE AND EMISSION HOSE ROUTING DIAGRAM FOR ADDITIONAL INFORMATION

THIS VEHICLE CONFORMS TO EPA REGULATIONS APPLICABLE TO 1985 MODEL YEAR NEW PASSENGER CARS

PT NO 14086524

1985-V8-5.0L VIN H IMPALA, CAPRICE, MONTE CARLO—FEDERAL

UBF
5.0 LITER
FIGS 7V4NEA4
5B4-1A

VEHICLE EMISSION CONTROL INFORMATION
GENERAL MOTORS CORPORATION

CATALYST
AIR/BPEGR/ORC/OC

	AUTOMATIC TRANSMISSION	MANUAL TRANSMISSION
TIMING (°BTC @ RPM)	0° @ 500 (DR)	0° @ 700 (N)
SPARK PLUG GAP (IN.)	0.045	0.045
IDLE SPEED SCREW (RPM) (SOLENOID INACTIVE)	500 (DR)	700 (N)
IDLE SPEED SOLENOID (RPM) (SOLENOID ACTIVE)	650 (DR)	800 (N)
FAST IDLE SPEED (RPM)	2200 (P) OR (N)	1800 (N)

SET PARKING BRAKE AND BLOCK DRIVE WHEELS.

MAKE ALL ADJUSTMENTS WITH ENGINE AT NORMAL OPERATING TEMPERATURE. EMISSION CONTROL SYSTEM IN CLOSED LOOP MODE, CHOKE FULL OPEN, AIR CLEANER INSTALLED, AND AIR CONDITIONING (A/C) OFF. EXCEPT WHERE NOTED

1. DISTRIBUTOR: DISCONNECT FOUR WIRE CONNECTOR. SET IGNITION TIMING TO SPECIFICATION WITH ENGINE RUNNING AT SPEED INDICATED. RECONNECT FOUR WIRE CONNECTOR. CLEAR ECM TROUBLE CODE
2. IDLE SPEED SCREW: DISCONNECT ELECTRICAL LEAD FROM IDLE SOLENOID, IF EQUIPPED. ADJUST CARBURETOR IDLE SPEED SCREW TO OBTAIN SPECIFIED SPEED. RECONNECT LEAD TO SOLENOID
3. IDLE SPEED SOLENOID: (WITH A/C) DISCONNECT ELECTRICAL LEAD FROM A/C COMPRESSOR AND TURN A/C ON. (WITHOUT A/C) DISCONNECT ELECTRICAL LEAD FROM SOLENOID AND CONNECT A JUMPER WIRE FROM A +12 VOLT SUPPLY. (ALL) OPEN THROTTLE MOMENTARILY TO ASSURE SOLENOID IS FULLY EXTENDED. ADJUST SOLENOID TO OBTAIN SPECIFIED SPEED. DISCONNECT JUMPER WIRE (IF USED). RECONNECT ELECTRICAL LEAD AND (WHERE EQUIPPED) TURN A/C OFF
4. FAST IDLE SPEED: DISCONNECT AND PLUG VACUUM HOSE AT EGR VALVE. WITH THROTTLE ON HIGH STEP OF FAST IDLE CAM, ADJUST FAST IDLE SCREW TO OBTAIN SPECIFIED SPEED. [NOTE] THIS ADJUSTMENT MUST BE CHECKED WITHIN 15 SECONDS AFTER INCREASING SPEED ABOVE 700 RPM. OPEN THROTTLE TO RELEASE FAST IDLE CAM AND STOP ENGINE. UNPLUG AND RECONNECT VACUUM HOSE TO EGR VALVE

[NOTE] IDLE MIXTURE SCREWS ARE PRESET AND SEALED AT FACTORY. PROVISION FOR ADJUSTMENT DURING TUNE UP IS NOT PROVIDED.

SEE SERVICE MANUAL, MAINTENANCE SCHEDULE AND EMISSION HOSE ROUTING DIAGRAM FOR ADDITIONAL INFORMATION.

THIS VEHICLE CONFORMS TO U.S. EPA REGULATIONS APPLICABLE TO 1985 MODEL YEAR NEW PASSENGER CARS.

PT NO 14086525

UBF
EMISSION HOSE ROUTING
EFE
FRONT OF VEHICLE
EGR SOLENOID
VAC SENSOR
TEMP SENSOR
EGR
CARB
A/CL
VAC MOTOR
PCV
TO FUEL TANK
AIR MGMT VLV
PURGE TVS
EFE TVS
TANK PRESS. CONT VLV
CHECK VALVE
CANISTER CONTROL VALVE
CANISTER

1985-V8-5.0L VIN H CAMARO—FEDERAL

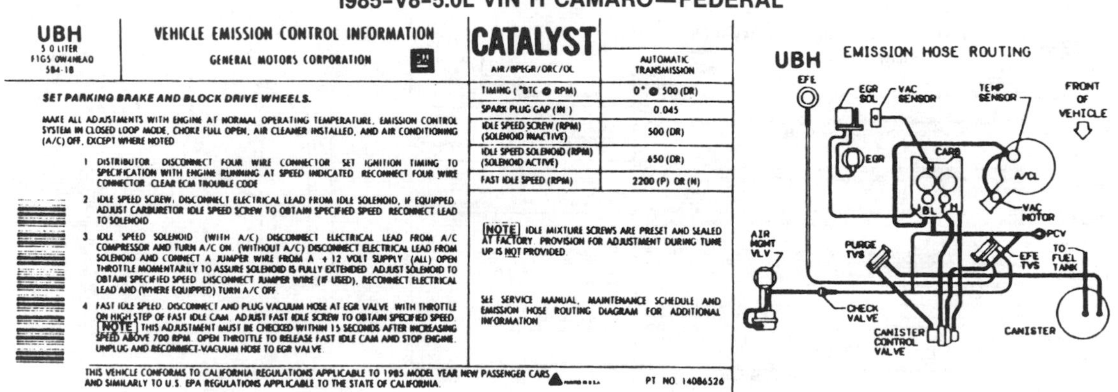

UBH
5.0 LITER
FIGS 0W4NEA0
5B4-1B

VEHICLE EMISSION CONTROL INFORMATION
GENERAL MOTORS CORPORATION

CATALYST
AIR/BPEGR/ORC/OC

	AUTOMATIC TRANSMISSION
TIMING (°BTC @ RPM)	0° @ 500 (DR)
SPARK PLUG GAP (IN.)	0.045
IDLE SPEED SCREW (RPM) (SOLENOID INACTIVE)	500 (DR)
IDLE SPEED SOLENOID (RPM) (SOLENOID ACTIVE)	650 (DR)
FAST IDLE SPEED (RPM)	2200 (P) OR (N)

SET PARKING BRAKE AND BLOCK DRIVE WHEELS.

MAKE ALL ADJUSTMENTS WITH ENGINE AT NORMAL OPERATING TEMPERATURE, EMISSION CONTROL SYSTEM IN CLOSED LOOP MODE, CHOKE FULL OPEN, AIR CLEANER INSTALLED, AND AIR CONDITIONING (A/C) OFF, EXCEPT WHERE NOTED

1. DISTRIBUTOR: DISCONNECT FOUR WIRE CONNECTOR. SET IGNITION TIMING TO SPECIFICATION WITH ENGINE RUNNING AT SPEED INDICATED. RECONNECT FOUR WIRE CONNECTOR. CLEAR ECM TROUBLE CODE
2. IDLE SPEED SCREW: DISCONNECT ELECTRICAL LEAD FROM IDLE SOLENOID, IF EQUIPPED. ADJUST CARBURETOR IDLE SPEED SCREW TO OBTAIN SPECIFIED SPEED. RECONNECT LEAD TO SOLENOID
3. IDLE SPEED SOLENOID: (WITH A/C) DISCONNECT ELECTRICAL LEAD FROM A/C COMPRESSOR AND TURN A/C ON. (WITHOUT A/C) DISCONNECT ELECTRICAL LEAD FROM SOLENOID AND CONNECT A JUMPER WIRE FROM A +12 VOLT SUPPLY. (ALL) OPEN THROTTLE MOMENTARILY TO ASSURE SOLENOID IS FULLY EXTENDED. ADJUST SOLENOID TO OBTAIN SPECIFIED SPEED. DISCONNECT JUMPER WIRE (IF USED), RECONNECT ELECTRICAL LEAD AND (WHERE EQUIPPED) TURN A/C OFF
4. FAST IDLE SPEED: DISCONNECT AND PLUG VACUUM HOSE AT EGR VALVE. WITH THROTTLE ON HIGH STEP OF FAST IDLE CAM, ADJUST FAST IDLE SCREW TO OBTAIN SPECIFIED SPEED. [NOTE] THIS ADJUSTMENT MUST BE CHECKED WITHIN 15 SECONDS AFTER INCREASING SPEED ABOVE 700 RPM. OPEN THROTTLE TO RELEASE FAST IDLE CAM AND STOP ENGINE. UNPLUG AND RECONNECT VACUUM HOSE TO EGR VALVE.

[NOTE] IDLE MIXTURE SCREWS ARE PRESET AND SEALED AT FACTORY. PROVISION FOR ADJUSTMENT DURING TUNE UP IS NOT PROVIDED.

SEE SERVICE MANUAL, MAINTENANCE SCHEDULE AND EMISSION HOSE ROUTING DIAGRAM FOR ADDITIONAL INFORMATION

THIS VEHICLE CONFORMS TO CALIFORNIA REGULATIONS APPLICABLE TO 1985 MODEL YEAR NEW PASSENGER CARS AND SIMILARLY TO U.S. EPA REGULATIONS APPLICABLE TO THE STATE OF CALIFORNIA.

PT NO 14086526

1985-V8-5.0L VIN H IMPALA, CAPRICE, MONTE CARLO—CALIFORNIA

VACUUM CIRCUITS
1985 CHEVROLET 5.0L

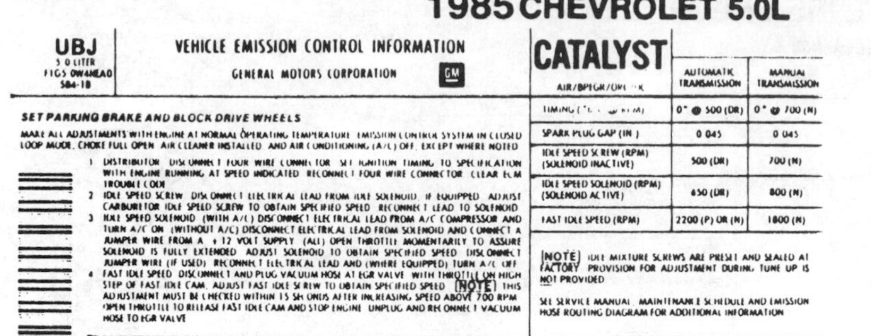

UBJ 5.0 LITER FIGS 0W4NEA0 584-1B

VEHICLE EMISSION CONTROL INFORMATION
GENERAL MOTORS CORPORATION GM

SET PARKING BRAKE AND BLOCK DRIVE WHEELS

MAKE ALL ADJUSTMENTS WITH ENGINE AT NORMAL OPERATING TEMPERATURE, EMISSION CONTROL SYSTEM IN CLOSED LOOP MODE, CHOKE FULL OPEN, AIR CLEANER INSTALLED, AND AIR CONDITIONING (A/C) OFF, EXCEPT WHERE NOTED

1. DISTRIBUTOR: DISCONNECT FOUR WIRE CONNECTOR. SET IGNITION TIMING TO SPECIFICATION WITH ENGINE RUNNING AT SPEED INDICATED. RECONNECT FOUR WIRE CONNECTOR. CLEAR ECM TROUBLE CODE
2. IDLE SPEED SCREW: DISCONNECT ELECTRICAL LEAD FROM IDLE SOLENOID, IF EQUIPPED. ADJUST CARBURETOR IDLE SPEED SCREW TO OBTAIN SPECIFIED SPEED. RECONNECT LEAD TO SOLENOID
3. IDLE SPEED SOLENOID: (WITH A/C) DISCONNECT ELECTRICAL LEAD FROM A/C COMPRESSOR AND TURN A/C ON. (WITHOUT A/C) DISCONNECT ELECTRICAL LEAD FROM SOLENOID AND CONNECT A JUMPER WIRE FROM A +12 VOLT SUPPLY. (ALL) OPEN THROTTLE MOMENTARILY TO ASSURE SOLENOID IS FULLY EXTENDED. ADJUST SOLENOID TO OBTAIN SPECIFIED SPEED. DISCONNECT JUMPER WIRE (IF USED). RECONNECT ELECTRICAL LEAD AND (WHERE EQUIPPED) TURN A/C OFF
4. FAST IDLE SPEED: DISCONNECT AND PLUG VACUUM HOSE AT EGR VALVE. WITH THROTTLE ON HIGH STEP OF FAST IDLE CAM, ADJUST FAST IDLE SCREW TO OBTAIN SPECIFIED SPEED. [NOTE] THIS ADJUSTMENT MUST BE CHECKED WITHIN 15 SECONDS AFTER INCREASING SPEED ABOVE 700 RPM. OPEN THROTTLE TO RELEASE FAST IDLE CAM AND STOP ENGINE. UNPLUG AND RECONNECT VACUUM HOSE TO EGR VALVE

CATALYST AIR/BPEGR/OPEN LOOP	AUTOMATIC TRANSMISSION	MANUAL TRANSMISSION
TIMING (°BTC @ RPM)	0° @ 500 (DR)	0° @ 700 (N)
SPARK PLUG GAP (IN.)	0.045	0.045
IDLE SPEED SCREW (RPM) (SOLENOID INACTIVE)	500 (DR)	700 (N)
IDLE SPEED SOLENOID (RPM) (SOLENOID ACTIVE)	650 (DR)	800 (N)
FAST IDLE SPEED (RPM)	2200 (P) OR (N)	1800 (N)

[NOTE] IDLE MIXTURE SCREWS ARE PRESET AND SEALED AT FACTORY. PROVISION FOR ADJUSTMENT DURING TUNE UP IS NOT PROVIDED

SEE SERVICE MANUAL, MAINTENANCE SCHEDULE AND EMISSION HOSE ROUTING DIAGRAM FOR ADDITIONAL INFORMATION

THIS VEHICLE CONFORMS TO CALIFORNIA REGULATIONS APPLICABLE TO 1985 MODEL YEAR NEW PASSENGER CARS AND SIMILARLY TO U.S. EPA REGULATIONS APPLICABLE TO THE STATE OF CALIFORNIA

PT NO 14086527

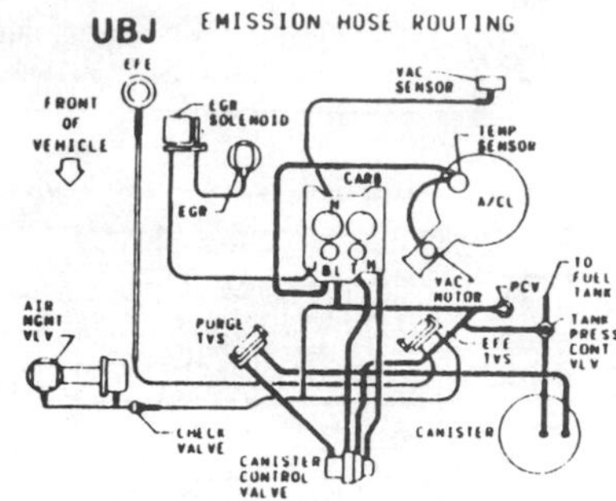

1985-V8-5.0L VIN H CAMARO—CALIFORNIA

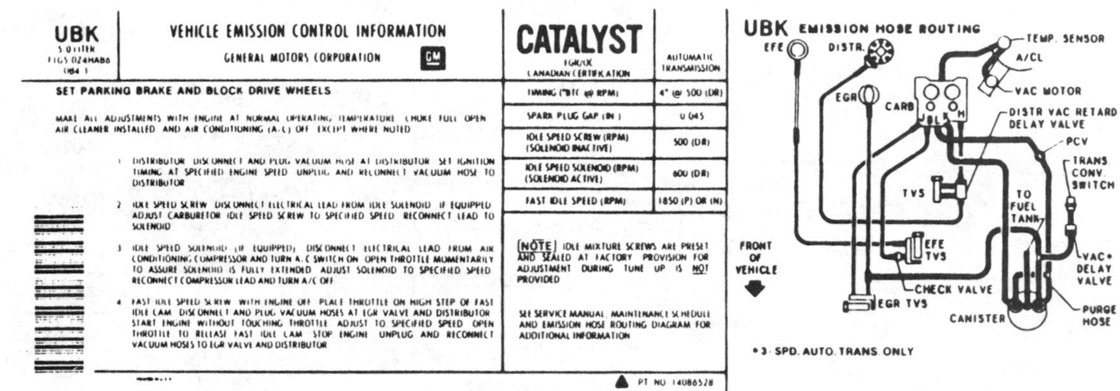

UBK 5.0 LITER FIGS 0Z4HAB6 084 1

VEHICLE EMISSION CONTROL INFORMATION
GENERAL MOTORS CORPORATION GM

SET PARKING BRAKE AND BLOCK DRIVE WHEELS

MAKE ALL ADJUSTMENTS WITH ENGINE AT NORMAL OPERATING TEMPERATURE, CHOKE FULL OPEN, AIR CLEANER INSTALLED, AND AIR CONDITIONING (A/C) OFF, EXCEPT WHERE NOTED

1. DISTRIBUTOR: DISCONNECT AND PLUG VACUUM HOSE AT DISTRIBUTOR. SET IGNITION TIMING AT SPECIFIED ENGINE SPEED. UNPLUG AND RECONNECT VACUUM HOSE TO DISTRIBUTOR
2. IDLE SPEED SCREW: DISCONNECT ELECTRICAL LEAD FROM IDLE SOLENOID, IF EQUIPPED. ADJUST CARBURETOR IDLE SPEED SCREW TO SPECIFIED SPEED. RECONNECT LEAD TO SOLENOID
3. IDLE SPEED SOLENOID (IF EQUIPPED): DISCONNECT ELECTRICAL LEAD FROM AIR CONDITIONING COMPRESSOR AND TURN A/C SWITCH ON. OPEN THROTTLE MOMENTARILY TO ASSURE SOLENOID IS FULLY EXTENDED. ADJUST SOLENOID TO SPECIFIED SPEED. RECONNECT COMPRESSOR LEAD AND TURN A/C OFF
4. FAST IDLE SPEED SCREW: WITH ENGINE OFF, PLACE THROTTLE ON HIGH STEP OF FAST IDLE CAM. DISCONNECT AND PLUG VACUUM HOSES AT EGR VALVE AND DISTRIBUTOR. START ENGINE WITHOUT TOUCHING THROTTLE. ADJUST TO SPECIFIED SPEED. OPEN THROTTLE TO RELEASE FAST IDLE CAM. STOP ENGINE. UNPLUG AND RECONNECT VACUUM HOSES TO EGR VALVE AND DISTRIBUTOR

CATALYST EGR/OC CANADIAN CERTIFICATION	AUTOMATIC TRANSMISSION
TIMING (°BTC @ RPM)	4° @ 500 (DR)
SPARK PLUG GAP (IN.)	0.045
IDLE SPEED SCREW (RPM) (SOLENOID INACTIVE)	500 (DR)
IDLE SPEED SOLENOID (RPM) (SOLENOID ACTIVE)	600 (DR)
FAST IDLE SPEED (RPM)	1850 (P) OR (N)

[NOTE] IDLE MIXTURE SCREWS ARE PRESET AND SEALED AT FACTORY. PROVISION FOR ADJUSTMENT DURING TUNE UP IS NOT PROVIDED

SEE SERVICE MANUAL, MAINTENANCE SCHEDULE AND EMISSION HOSE ROUTING DIAGRAM FOR ADDITIONAL INFORMATION

PT NO 14086528

1985-V8-5.0L VIN H IMPALA, CAPRICE, MONTE CARLO—CANADA

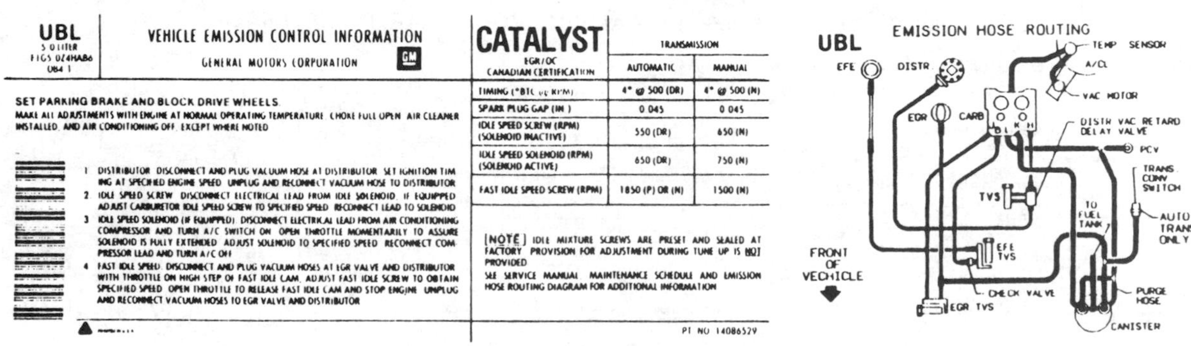

UBL 5.0 LITER FIGS 0Z4HAB6 084 1

VEHICLE EMISSION CONTROL INFORMATION
GENERAL MOTORS CORPORATION GM

SET PARKING BRAKE AND BLOCK DRIVE WHEELS.

MAKE ALL ADJUSTMENTS WITH ENGINE AT NORMAL OPERATING TEMPERATURE, CHOKE FULL OPEN, AIR CLEANER INSTALLED, AND AIR CONDITIONING OFF, EXCEPT WHERE NOTED

1. DISTRIBUTOR: DISCONNECT AND PLUG VACUUM HOSE AT DISTRIBUTOR. SET IGNITION TIMING AT SPECIFIED ENGINE SPEED. UNPLUG AND RECONNECT VACUUM HOSE TO DISTRIBUTOR
2. IDLE SPEED SCREW: DISCONNECT ELECTRICAL LEAD FROM IDLE SOLENOID, IF EQUIPPED. ADJUST CARBURETOR IDLE SPEED SCREW TO SPECIFIED SPEED. RECONNECT LEAD TO SOLENOID
3. IDLE SPEED SOLENOID (IF EQUIPPED): DISCONNECT ELECTRICAL LEAD FROM AIR CONDITIONING COMPRESSOR AND TURN A/C SWITCH ON. OPEN THROTTLE MOMENTARILY TO ASSURE SOLENOID IS FULLY EXTENDED. ADJUST SOLENOID TO SPECIFIED SPEED. RECONNECT COMPRESSOR LEAD AND TURN A/C OFF
4. FAST IDLE SPEED: DISCONNECT AND PLUG VACUUM HOSES AT EGR VALVE AND DISTRIBUTOR. WITH THROTTLE ON HIGH STEP OF FAST IDLE CAM, ADJUST FAST IDLE SCREW TO OBTAIN SPECIFIED SPEED. OPEN THROTTLE TO RELEASE FAST IDLE CAM AND STOP ENGINE. UNPLUG AND RECONNECT VACUUM HOSES TO EGR VALVE AND DISTRIBUTOR

CATALYST EGR/OC CANADIAN CERTIFICATION	TRANSMISSION AUTOMATIC	TRANSMISSION MANUAL
TIMING (°BTC @ RPM)	4° @ 500 (DR)	4° @ 500 (N)
SPARK PLUG GAP (IN.)	0.045	0.045
IDLE SPEED SCREW (RPM) (SOLENOID INACTIVE)	550 (DR)	650 (N)
IDLE SPEED SOLENOID (RPM) (SOLENOID ACTIVE)	650 (DR)	750 (N)
FAST IDLE SPEED SCREW (RPM)	1850 (P) OR (N)	1500 (N)

[NOTE] IDLE MIXTURE SCREWS ARE PRESET AND SEALED AT FACTORY. PROVISION FOR ADJUSTMENT DURING TUNE UP IS NOT PROVIDED

SEE SERVICE MANUAL, MAINTENANCE SCHEDULE AND EMISSION HOSE ROUTING DIAGRAM FOR ADDITIONAL INFORMATION

PT NO 14086529

1985-V8-5.0L VIN H CAMARO—CANADA

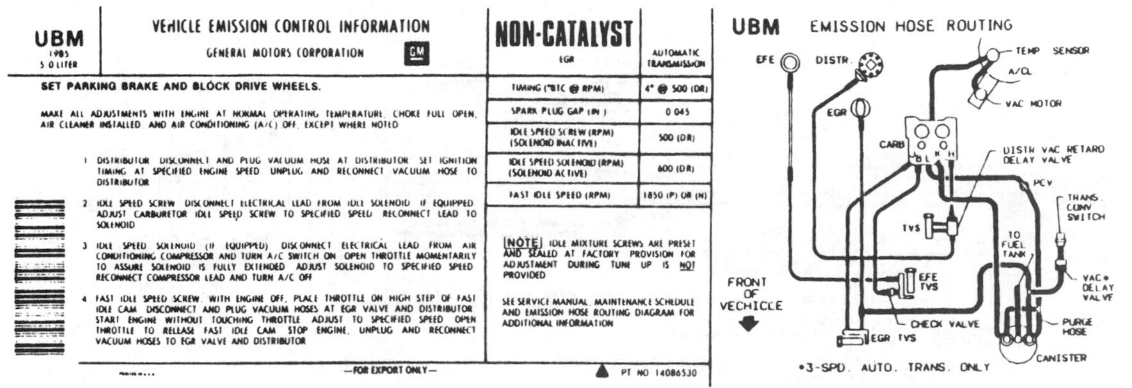

UBM 1985 5.0 LITER

VEHICLE EMISSION CONTROL INFORMATION
GENERAL MOTORS CORPORATION GM

SET PARKING BRAKE AND BLOCK DRIVE WHEELS.

MAKE ALL ADJUSTMENTS WITH ENGINE AT NORMAL OPERATING TEMPERATURE, CHOKE FULL OPEN, AIR CLEANER INSTALLED, AND AIR CONDITIONING (A/C) OFF, EXCEPT WHERE NOTED

1. DISTRIBUTOR: DISCONNECT AND PLUG VACUUM HOSE AT DISTRIBUTOR. SET IGNITION TIMING AT SPECIFIED ENGINE SPEED. UNPLUG AND RECONNECT VACUUM HOSE TO DISTRIBUTOR
2. IDLE SPEED SCREW: DISCONNECT ELECTRICAL LEAD FROM IDLE SOLENOID, IF EQUIPPED. ADJUST CARBURETOR IDLE SPEED SCREW TO SPECIFIED SPEED. RECONNECT LEAD TO SOLENOID
3. IDLE SPEED SOLENOID (IF EQUIPPED): DISCONNECT ELECTRICAL LEAD FROM AIR CONDITIONING COMPRESSOR AND TURN A/C SWITCH ON. OPEN THROTTLE MOMENTARILY TO ASSURE SOLENOID IS FULLY EXTENDED. ADJUST SOLENOID TO SPECIFIED SPEED. RECONNECT COMPRESSOR LEAD AND TURN A/C OFF
4. FAST IDLE SPEED SCREW: WITH ENGINE OFF, PLACE THROTTLE ON HIGH STEP OF FAST IDLE CAM. DISCONNECT AND PLUG VACUUM HOSES AT EGR VALVE AND DISTRIBUTOR. START ENGINE WITHOUT TOUCHING THROTTLE. ADJUST TO SPECIFIED SPEED. OPEN THROTTLE TO RELEASE FAST IDLE CAM. STOP ENGINE, UNPLUG AND RECONNECT VACUUM HOSES TO EGR VALVE AND DISTRIBUTOR

NON-CATALYST EGR	AUTOMATIC TRANSMISSION
TIMING (°BTC @ RPM)	4° @ 500 (DR)
SPARK PLUG GAP (IN.)	0.045
IDLE SPEED SCREW (RPM) (SOLENOID INACTIVE)	500 (DR)
IDLE SPEED SOLENOID (RPM) (SOLENOID ACTIVE)	600 (DR)
FAST IDLE SPEED (RPM)	1850 (P) OR (N)

[NOTE] IDLE MIXTURE SCREWS ARE PRESET AND SEALED AT FACTORY. PROVISION FOR ADJUSTMENT DURING TUNE UP IS NOT PROVIDED

SEE SERVICE MANUAL, MAINTENANCE SCHEDULE AND EMISSION HOSE ROUTING DIAGRAM FOR ADDITIONAL INFORMATION

—FOR EXPORT ONLY—

PT NO 14086530

1985-V8-5.0L VIN H IMPALA, CAPRICE, MONTE CARLO—EXPORT

VACUUM CIRCUITS
1985 CHEVROLET 5.0L

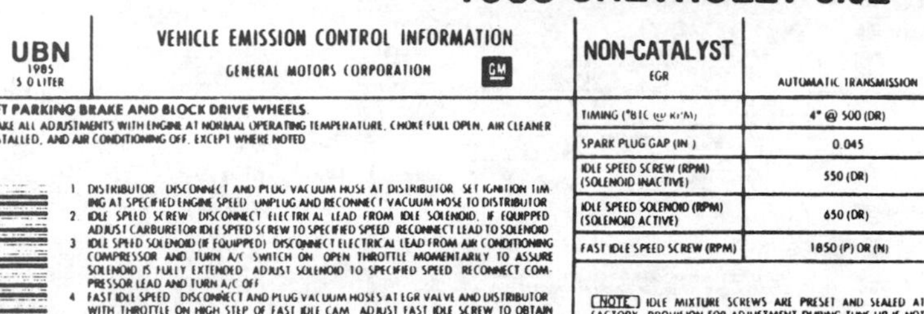

UBN 1985 5.0 LITER — VEHICLE EMISSION CONTROL INFORMATION — GENERAL MOTORS CORPORATION — GM

NON-CATALYST — EGR

SET PARKING BRAKE AND BLOCK DRIVE WHEELS.
MAKE ALL ADJUSTMENTS WITH ENGINE AT NORMAL OPERATING TEMPERATURE, CHOKE FULL OPEN, AIR CLEANER INSTALLED, AND AIR CONDITIONING OFF, EXCEPT WHERE NOTED

1. DISTRIBUTOR: DISCONNECT AND PLUG VACUUM HOSE AT DISTRIBUTOR. SET IGNITION TIMING AT SPECIFIED ENGINE SPEED. UNPLUG AND RECONNECT VACUUM HOSE TO DISTRIBUTOR
2. IDLE SPEED SCREW: DISCONNECT ELECTRICAL LEAD FROM IDLE SOLENOID, IF EQUIPPED. ADJUST CARBURETOR IDLE SPEED SCREW TO SPECIFIED SPEED. RECONNECT LEAD TO SOLENOID
3. IDLE SPEED SOLENOID (IF EQUIPPED): DISCONNECT ELECTRICAL LEAD FROM AIR CONDITIONING COMPRESSOR AND TURN A/C SWITCH ON. OPEN THROTTLE MOMENTARILY TO ASSURE SOLENOID IS FULLY EXTENDED. ADJUST SOLENOID TO SPECIFIED SPEED. RECONNECT COMPRESSOR LEAD AND TURN A/C OFF
4. FAST IDLE SPEED: DISCONNECT AND PLUG VACUUM HOSES AT EGR VALVE AND DISTRIBUTOR WITH THROTTLE ON HIGH STEP OF FAST IDLE CAM. ADJUST FAST IDLE SCREW TO OBTAIN SPECIFIED SPEED. OPEN THROTTLE TO RELEASE FAST IDLE CAM AND STOP ENGINE. UNPLUG AND RECONNECT VACUUM HOSES TO EGR VALVE AND DISTRIBUTOR

	AUTOMATIC TRANSMISSION
TIMING (°BTC @ RPM)	4° @ 500 (DR)
SPARK PLUG GAP (IN.)	0.045
IDLE SPEED SCREW (RPM) (SOLENOID INACTIVE)	550 (DR)
IDLE SPEED SOLENOID (RPM) (SOLENOID ACTIVE)	650 (DR)
FAST IDLE SPEED SCREW (RPM)	1850 (P) OR (N)

NOTE: IDLE MIXTURE SCREWS ARE PRESET AND SEALED AT FACTORY. PROVISION FOR ADJUSTMENT DURING TUNE UP IS NOT PROVIDED
SEE SERVICE MANUAL, MAINTENANCE SCHEDULE AND EMISSION HOSE ROUTING DIAGRAM FOR ADDITIONAL INFORMATION

— FOR EXPORT ONLY —
PT NO 14086531

1985-V8-5.0L VIN H CAMARO—EXPORT

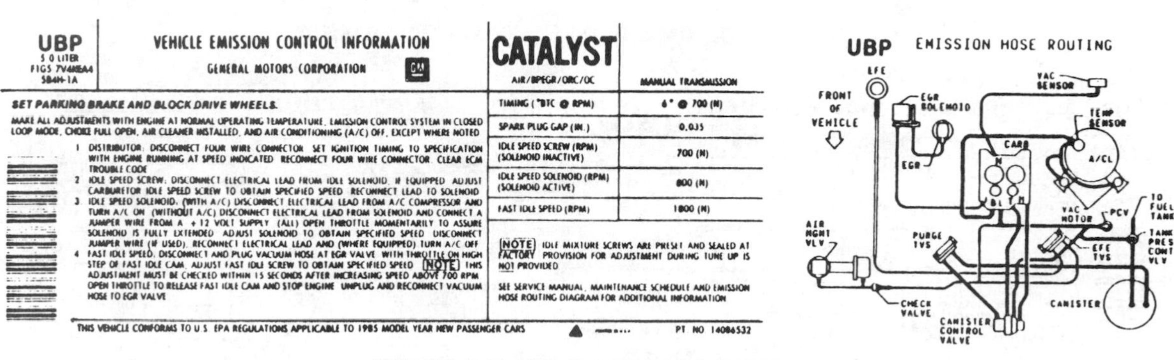

UBP 5.0 LITER FIGS 7V4NEA4 5B4H-1A — VEHICLE EMISSION CONTROL INFORMATION — GENERAL MOTORS CORPORATION — GM

CATALYST — AIR/BPEGR/ORC/OC

SET PARKING BRAKE AND BLOCK DRIVE WHEELS.
MAKE ALL ADJUSTMENTS WITH ENGINE AT NORMAL OPERATING TEMPERATURE, EMISSION CONTROL SYSTEM IN CLOSED LOOP MODE, CHOKE FULL OPEN, AIR CLEANER INSTALLED, AND AIR CONDITIONING (A/C) OFF, EXCEPT WHERE NOTED

1. DISTRIBUTOR: DISCONNECT FOUR WIRE CONNECTOR. SET IGNITION TIMING TO SPECIFICATION WITH ENGINE RUNNING AT SPEED INDICATED. RECONNECT FOUR WIRE CONNECTOR. CLEAR ECM TROUBLE CODE
2. IDLE SPEED SCREW: DISCONNECT ELECTRICAL LEAD FROM IDLE SOLENOID, IF EQUIPPED. ADJUST CARBURETOR IDLE SPEED SCREW TO OBTAIN SPECIFIED SPEED. RECONNECT LEAD TO SOLENOID
3. IDLE SPEED SOLENOID: (WITH A/C) DISCONNECT ELECTRICAL LEAD FROM A/C COMPRESSOR AND TURN A/C ON. (WITHOUT A/C) DISCONNECT ELECTRICAL LEAD FROM SOLENOID AND CONNECT A JUMPER WIRE FROM A +12 VOLT SUPPLY. (ALL) OPEN THROTTLE MOMENTARILY TO ASSURE SOLENOID IS FULLY EXTENDED. ADJUST SOLENOID TO OBTAIN SPECIFIED SPEED. DISCONNECT JUMPER WIRE (IF USED). RECONNECT ELECTRICAL LEAD AND (WHERE EQUIPPED) TURN A/C OFF
4. FAST IDLE SPEED: DISCONNECT AND PLUG VACUUM HOSE AT EGR VALVE. WITH THROTTLE ON HIGH STEP OF FAST IDLE CAM, ADJUST FAST IDLE SCREW TO OBTAIN SPECIFIED SPEED. NOTE: THIS ADJUSTMENT MUST BE CHECKED WITHIN 15 SECONDS AFTER INCREASING SPEED ABOVE 700 RPM. OPEN THROTTLE TO RELEASE FAST IDLE CAM AND STOP ENGINE. UNPLUG AND RECONNECT VACUUM HOSE TO EGR VALVE

	MANUAL TRANSMISSION
TIMING (°BTC @ RPM)	6° @ 700 (N)
SPARK PLUG GAP (IN.)	0.035
IDLE SPEED SCREW (RPM) (SOLENOID INACTIVE)	700 (N)
IDLE SPEED SOLENOID (RPM) (SOLENOID ACTIVE)	800 (N)
FAST IDLE SPEED (RPM)	1800 (N)

NOTE: IDLE MIXTURE SCREWS ARE PRESET AND SEALED AT FACTORY. PROVISION FOR ADJUSTMENT DURING TUNE UP IS NOT PROVIDED
SEE SERVICE MANUAL, MAINTENANCE SCHEDULE AND EMISSION HOSE ROUTING DIAGRAM FOR ADDITIONAL INFORMATION

THIS VEHICLE CONFORMS TO U.S. EPA REGULATIONS APPLICABLE TO 1985 MODEL YEAR NEW PASSENGER CARS
PT. NO 14086532

1985-V8-5.0L VIN G CAMARO—FEDERAL

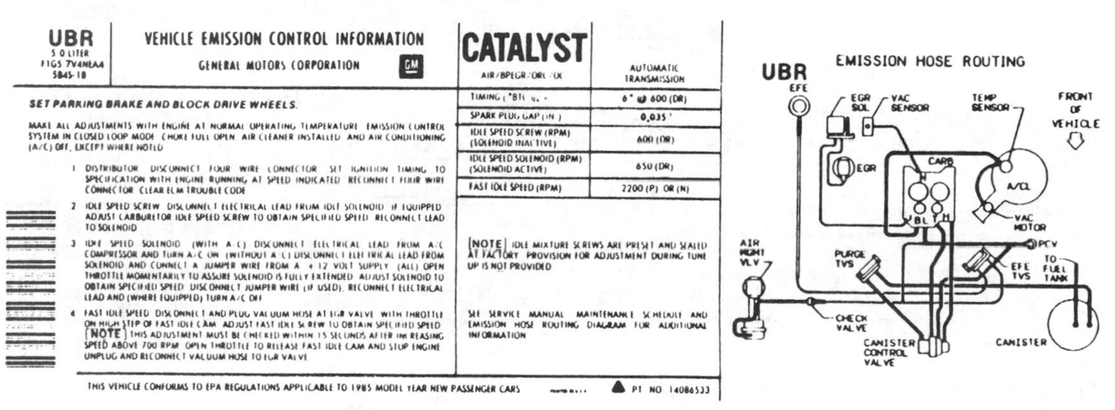

UBR 5.0 LITER FIGS 7V4NEA4 5B4S-1B — VEHICLE EMISSION CONTROL INFORMATION — GENERAL MOTORS CORPORATION — GM

CATALYST — AIR/BPEGR/ORC/OC

SET PARKING BRAKE AND BLOCK DRIVE WHEELS.
MAKE ALL ADJUSTMENTS WITH ENGINE AT NORMAL OPERATING TEMPERATURE, EMISSION CONTROL SYSTEM IN CLOSED LOOP MODE, CHOKE FULL OPEN, AIR CLEANER INSTALLED, AND AIR CONDITIONING (A/C) OFF, EXCEPT WHERE NOTED

1. DISTRIBUTOR: DISCONNECT FOUR WIRE CONNECTOR. SET IGNITION TIMING TO SPECIFICATION WITH ENGINE RUNNING AT SPEED INDICATED. RECONNECT FOUR WIRE CONNECTOR. CLEAR ECM TROUBLE CODE
2. IDLE SPEED SCREW: DISCONNECT ELECTRICAL LEAD FROM IDLE SOLENOID, IF EQUIPPED. ADJUST CARBURETOR IDLE SPEED SCREW TO OBTAIN SPECIFIED SPEED. RECONNECT LEAD TO SOLENOID
3. IDLE SPEED SOLENOID: (WITH A/C) DISCONNECT ELECTRICAL LEAD FROM A/C COMPRESSOR AND TURN A/C ON. (WITHOUT A/C) DISCONNECT ELECTRICAL LEAD FROM SOLENOID AND CONNECT A JUMPER WIRE FROM A +12 VOLT SUPPLY. (ALL) OPEN THROTTLE MOMENTARILY TO ASSURE SOLENOID IS FULLY EXTENDED. ADJUST SOLENOID TO OBTAIN SPECIFIED SPEED. DISCONNECT JUMPER WIRE (IF USED). RECONNECT ELECTRICAL LEAD AND (WHERE EQUIPPED) TURN A/C OFF
4. FAST IDLE SPEED: DISCONNECT AND PLUG VACUUM HOSE AT EGR VALVE. WITH THROTTLE ON HIGH STEP OF FAST IDLE CAM, ADJUST FAST IDLE SCREW TO OBTAIN SPECIFIED SPEED. NOTE: THIS ADJUSTMENT MUST BE CHECKED WITHIN 15 SECONDS AFTER INCREASING SPEED ABOVE 700 RPM. OPEN THROTTLE TO RELEASE FAST IDLE CAM AND STOP ENGINE. UNPLUG AND RECONNECT VACUUM HOSE TO EGR VALVE

	AUTOMATIC TRANSMISSION
TIMING (°BTC @ RPM)	6° @ 600 (DR)
SPARK PLUG GAP (IN.)	0.035
IDLE SPEED SCREW (RPM) (SOLENOID INACTIVE)	600 (DR)
IDLE SPEED SOLENOID (RPM) (SOLENOID ACTIVE)	650 (DR)
FAST IDLE SPEED (RPM)	2200 (P) OR (N)

NOTE: IDLE MIXTURE SCREWS ARE PRESET AND SEALED AT FACTORY. PROVISION FOR ADJUSTMENT DURING TUNE UP IS NOT PROVIDED
SEE SERVICE MANUAL, MAINTENANCE SCHEDULE AND EMISSION HOSE ROUTING DIAGRAM FOR ADDITIONAL INFORMATION

THIS VEHICLE CONFORMS TO EPA REGULATIONS APPLICABLE TO 1985 MODEL YEAR NEW PASSENGER CARS
PT NO 14086533

1985-V8-5.0L VIN G MONTE CARLO—FEDERAL

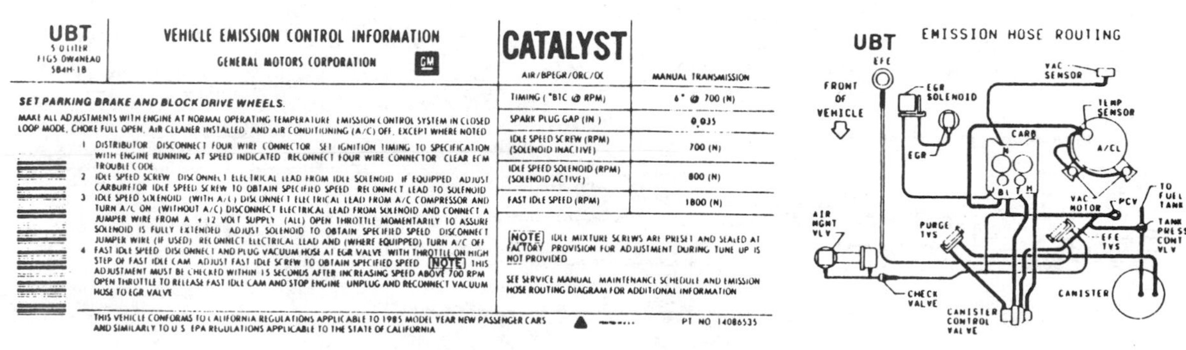

UBT 5.0 LITER FIGS 0W4NEA0 5B4H-1B — VEHICLE EMISSION CONTROL INFORMATION — GENERAL MOTORS CORPORATION — GM

CATALYST — AIR/BPEGR/ORC/OC

SET PARKING BRAKE AND BLOCK DRIVE WHEELS.
MAKE ALL ADJUSTMENTS WITH ENGINE AT NORMAL OPERATING TEMPERATURE, EMISSION CONTROL SYSTEM IN CLOSED LOOP MODE, CHOKE FULL OPEN, AIR CLEANER INSTALLED, AND AIR CONDITIONING (A/C) OFF, EXCEPT WHERE NOTED

1. DISTRIBUTOR: DISCONNECT FOUR WIRE CONNECTOR. SET IGNITION TIMING TO SPECIFICATION WITH ENGINE RUNNING AT SPEED INDICATED. RECONNECT FOUR WIRE CONNECTOR. CLEAR ECM TROUBLE CODE
2. IDLE SPEED SCREW: DISCONNECT ELECTRICAL LEAD FROM IDLE SOLENOID, IF EQUIPPED. ADJUST CARBURETOR IDLE SPEED SCREW TO OBTAIN SPECIFIED SPEED. RECONNECT LEAD TO SOLENOID
3. IDLE SPEED SOLENOID: (WITH A/C) DISCONNECT ELECTRICAL LEAD FROM A/C COMPRESSOR AND TURN A/C ON. (WITHOUT A/C) DISCONNECT ELECTRICAL LEAD FROM SOLENOID AND CONNECT A JUMPER WIRE FROM A +12 VOLT SUPPLY. (ALL) OPEN THROTTLE MOMENTARILY TO ASSURE SOLENOID IS FULLY EXTENDED. ADJUST SOLENOID TO OBTAIN SPECIFIED SPEED. DISCONNECT JUMPER WIRE (IF USED). RECONNECT ELECTRICAL LEAD AND (WHERE EQUIPPED) TURN A/C OFF
4. FAST IDLE SPEED: DISCONNECT AND PLUG VACUUM HOSE AT EGR VALVE. WITH THROTTLE ON HIGH STEP OF FAST IDLE CAM, ADJUST FAST IDLE SCREW TO OBTAIN SPECIFIED SPEED. NOTE: THIS ADJUSTMENT MUST BE CHECKED WITHIN 15 SECONDS AFTER INCREASING SPEED ABOVE 700 RPM. OPEN THROTTLE TO RELEASE FAST IDLE CAM AND STOP ENGINE. UNPLUG AND RECONNECT VACUUM HOSE TO EGR VALVE

	MANUAL TRANSMISSION
TIMING (°BTC @ RPM)	6° @ 700 (N)
SPARK PLUG GAP (IN.)	0.035
IDLE SPEED SCREW (RPM) (SOLENOID INACTIVE)	700 (N)
IDLE SPEED SOLENOID (RPM) (SOLENOID ACTIVE)	800 (N)
FAST IDLE SPEED (RPM)	1800 (N)

NOTE: IDLE MIXTURE SCREWS ARE PRESET AND SEALED AT FACTORY. PROVISION FOR ADJUSTMENT DURING TUNE UP IS NOT PROVIDED
SEE SERVICE MANUAL, MAINTENANCE SCHEDULE AND EMISSION HOSE ROUTING DIAGRAM FOR ADDITIONAL INFORMATION

THIS VEHICLE CONFORMS TO CALIFORNIA REGULATIONS APPLICABLE TO 1985 MODEL YEAR NEW PASSENGER CARS AND SIMILARLY TO U.S. EPA REGULATIONS APPLICABLE TO THE STATE OF CALIFORNIA
PT NO 14086535

1985-V8-5.0L VIN G CAMARO—CALIFORNIA

VACUUM CIRCUITS

1985 CHEVROLET 5.0L

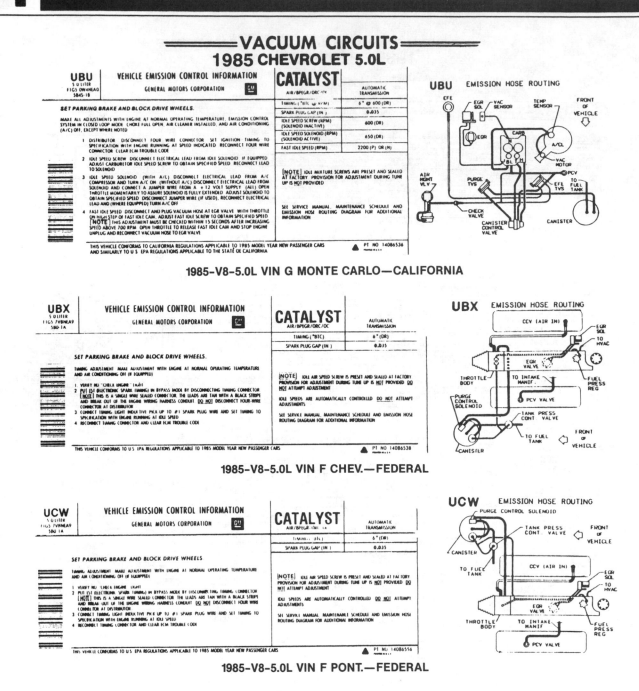

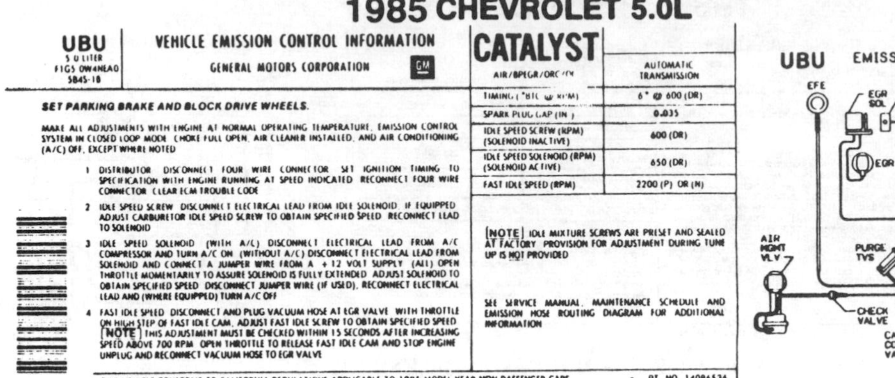

UBU
5.0 LITER
FIGS 0W4NEA0
5B45-1B

VEHICLE EMISSION CONTROL INFORMATION
GENERAL MOTORS CORPORATION

CATALYST
AIR/BPEGR/ORC/OC

	AUTOMATIC TRANSMISSION
TIMING (°BTC @ RPM)	6° @ 600 (DR)
SPARK PLUG GAP (IN.)	0.035
IDLE SPEED SCREW (RPM) (SOLENOID INACTIVE)	600 (DR)
IDLE SPEED SOLENOID (RPM) (SOLENOID ACTIVE)	650 (DR)
FAST IDLE SPEED (RPM)	2200 (P) OR (N)

SET PARKING BRAKE AND BLOCK DRIVE WHEELS.

MAKE ALL ADJUSTMENTS WITH ENGINE AT NORMAL OPERATING TEMPERATURE, EMISSION CONTROL SYSTEM IN CLOSED LOOP MODE, CHOKE FULL OPEN, AIR CLEANER INSTALLED, AND AIR CONDITIONING (A/C) OFF, EXCEPT WHERE NOTED.

1. DISTRIBUTOR: DISCONNECT FOUR WIRE CONNECTOR. SET IGNITION TIMING TO SPECIFICATION WITH ENGINE RUNNING AT SPEED INDICATED. RECONNECT FOUR WIRE CONNECTOR. CLEAR ECM TROUBLE CODE.
2. IDLE SPEED SCREW: DISCONNECT ELECTRICAL LEAD FROM IDLE SOLENOID, IF EQUIPPED. ADJUST CARBURETOR IDLE SPEED SCREW TO OBTAIN SPECIFIED SPEED. RECONNECT LEAD TO SOLENOID.
3. IDLE SPEED SOLENOID: (WITH A/C) DISCONNECT ELECTRICAL LEAD FROM A/C COMPRESSOR AND TURN A/C ON. (WITHOUT A/C) DISCONNECT ELECTRICAL LEAD FROM SOLENOID AND CONNECT A JUMPER WIRE FROM A +12 VOLT SUPPLY. (ALL) OPEN THROTTLE MOMENTARILY TO ASSURE SOLENOID IS FULLY EXTENDED. ADJUST SOLENOID TO OBTAIN SPECIFIED SPEED. DISCONNECT JUMPER WIRE (IF USED), RECONNECT ELECTRICAL LEAD AND (WHERE EQUIPPED) TURN A/C OFF.
4. FAST IDLE SPEED: DISCONNECT AND PLUG VACUUM HOSE AT EGR VALVE. WITH THROTTLE ON HIGH STEP OF FAST IDLE CAM, ADJUST FAST IDLE SCREW TO OBTAIN SPECIFIED SPEED. [NOTE] THIS ADJUSTMENT MUST BE CHECKED WITHIN 15 SECONDS AFTER INCREASING SPEED ABOVE 700 RPM. OPEN THROTTLE TO RELEASE FAST IDLE CAM AND STOP ENGINE. UNPLUG AND RECONNECT VACUUM HOSE TO EGR VALVE.

[NOTE] IDLE MIXTURE SCREWS ARE PRESET AND SEALED AT FACTORY. PROVISION FOR ADJUSTMENT DURING TUNE UP IS NOT PROVIDED.

SEE SERVICE MANUAL, MAINTENANCE SCHEDULE AND EMISSION HOSE ROUTING DIAGRAM FOR ADDITIONAL INFORMATION.

THIS VEHICLE CONFORMS TO CALIFORNIA REGULATIONS APPLICABLE TO 1985 MODEL YEAR NEW PASSENGER CARS AND SIMILARLY TO U.S. EPA REGULATIONS APPLICABLE TO THE STATE OF CALIFORNIA.

PT. NO. 14086536

1985–V8–5.0L VIN G MONTE CARLO—CALIFORNIA

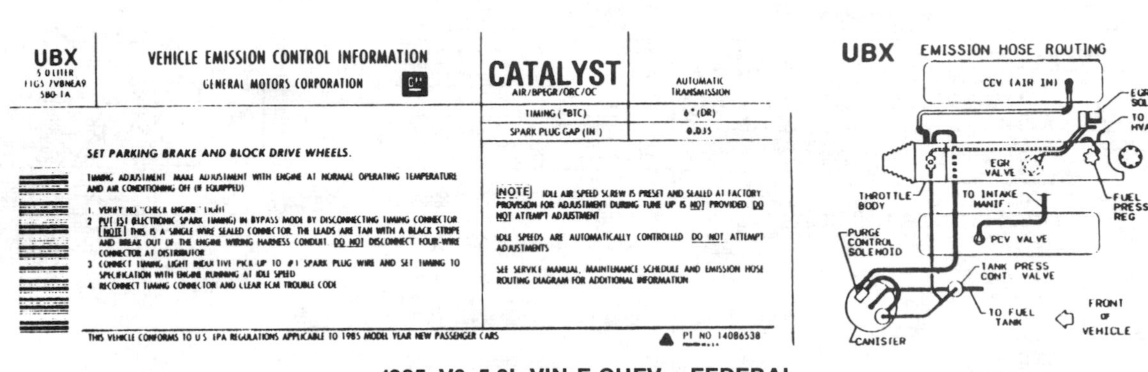

UBX
5.0 LITER
FIGS 7V8NEA9
5B0-1A

VEHICLE EMISSION CONTROL INFORMATION
GENERAL MOTORS CORPORATION

CATALYST
AIR/BPEGR/ORC/OC

	AUTOMATIC TRANSMISSION
TIMING (°BTC)	6° (DR)
SPARK PLUG GAP (IN.)	0.035

SET PARKING BRAKE AND BLOCK DRIVE WHEELS.

TIMING ADJUSTMENT: MAKE ADJUSTMENT WITH ENGINE AT NORMAL OPERATING TEMPERATURE AND AIR CONDITIONING OFF (IF EQUIPPED).

1. VERIFY NO "CHECK ENGINE" LIGHT.
2. PUT EST (ELECTRONIC SPARK TIMING) IN BYPASS MODE BY DISCONNECTING TIMING CONNECTOR. [NOTE] THIS IS A SINGLE WIRE SEALED CONNECTOR. THE LEADS ARE TAN WITH A BLACK STRIPE AND BREAK OUT OF THE ENGINE WIRING HARNESS CONDUIT. DO NOT DISCONNECT FOUR-WIRE CONNECTOR AT DISTRIBUTOR.
3. CONNECT TIMING LIGHT INDUCTIVE PICK UP TO #1 SPARK PLUG WIRE AND SET TIMING TO SPECIFICATION WITH ENGINE RUNNING AT IDLE SPEED.
4. RECONNECT TIMING CONNECTOR AND CLEAR ECM TROUBLE CODE.

[NOTE] IDLE AIR SPEED SCREW IS PRESET AND SEALED AT FACTORY. PROVISION FOR ADJUSTMENT DURING TUNE UP IS NOT PROVIDED. DO NOT ATTEMPT ADJUSTMENT.

IDLE SPEEDS ARE AUTOMATICALLY CONTROLLED. DO NOT ATTEMPT ADJUSTMENTS.

SEE SERVICE MANUAL, MAINTENANCE SCHEDULE AND EMISSION HOSE ROUTING DIAGRAM FOR ADDITIONAL INFORMATION.

THIS VEHICLE CONFORMS TO U.S. EPA REGULATIONS APPLICABLE TO 1985 MODEL YEAR NEW PASSENGER CARS.

PT. NO. 14086538

1985–V8–5.0L VIN F CHEV.—FEDERAL

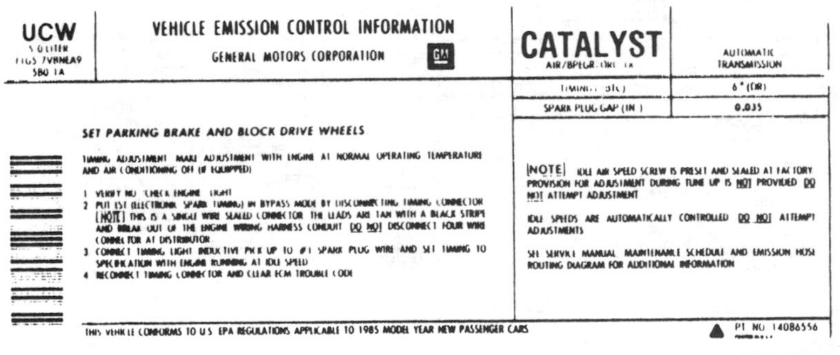

UCW
5.0 LITER
FIGS 7V8NEA9
5B0-1A

VEHICLE EMISSION CONTROL INFORMATION
GENERAL MOTORS CORPORATION

CATALYST
AIR/BPEGR/ORC/OC

	AUTOMATIC TRANSMISSION
TIMING (°BTC)	6° (DR)
SPARK PLUG GAP (IN.)	0.035

SET PARKING BRAKE AND BLOCK DRIVE WHEELS.

TIMING ADJUSTMENT: MAKE ADJUSTMENT WITH ENGINE AT NORMAL OPERATING TEMPERATURE AND AIR CONDITIONING OFF (IF EQUIPPED).

1. VERIFY NO "CHECK ENGINE" LIGHT.
2. PUT EST (ELECTRONIC SPARK TIMING) IN BYPASS MODE BY DISCONNECTING TIMING CONNECTOR. [NOTE] THIS IS A SINGLE WIRE SEALED CONNECTOR. THE LEADS ARE TAN WITH A BLACK STRIPE AND BREAK OUT OF THE ENGINE WIRING HARNESS CONDUIT. DO NOT DISCONNECT FOUR-WIRE CONNECTOR AT DISTRIBUTOR.
3. CONNECT TIMING LIGHT INDUCTIVE PICK UP TO #1 SPARK PLUG WIRE AND SET TIMING TO SPECIFICATION WITH ENGINE RUNNING AT IDLE SPEED.
4. RECONNECT TIMING CONNECTOR AND CLEAR ECM TROUBLE CODE.

[NOTE] IDLE AIR SPEED SCREW IS PRESET AND SEALED AT FACTORY. PROVISION FOR ADJUSTMENT DURING TUNE UP IS NOT PROVIDED. DO NOT ATTEMPT ADJUSTMENT.

IDLE SPEEDS ARE AUTOMATICALLY CONTROLLED. DO NOT ATTEMPT ADJUSTMENTS.

SEE SERVICE MANUAL, MAINTENANCE SCHEDULE AND EMISSION HOSE ROUTING DIAGRAM FOR ADDITIONAL INFORMATION.

THIS VEHICLE CONFORMS TO U.S. EPA REGULATIONS APPLICABLE TO 1985 MODEL YEAR NEW PASSENGER CARS.

PT. NO. 14086556

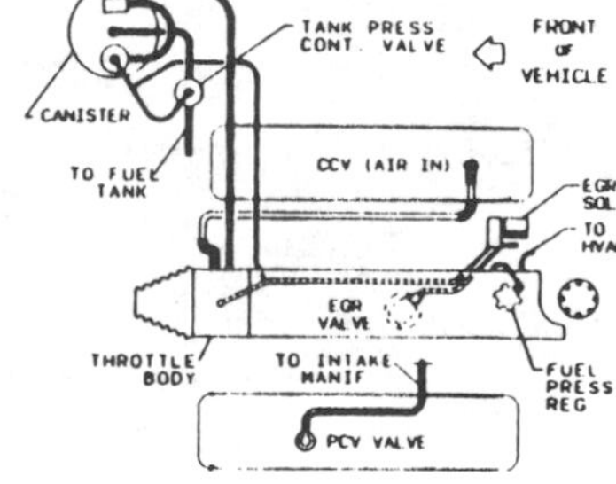

1985–V8–5.0L VIN F PONT.—FEDERAL

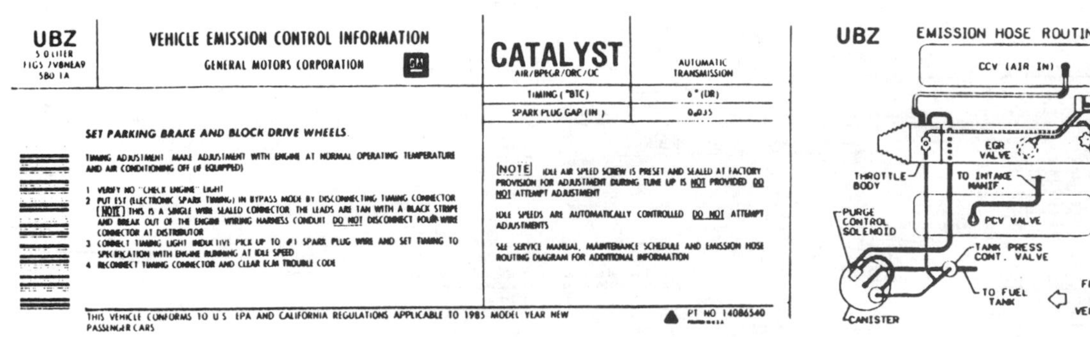

UBZ
5.0 LITER
FIGS 7V8NEA9
5B0-1A

VEHICLE EMISSION CONTROL INFORMATION
GENERAL MOTORS CORPORATION

CATALYST
AIR/BPEGR/ORC/OC

	AUTOMATIC TRANSMISSION
TIMING (°BTC)	6° (DR)
SPARK PLUG GAP (IN.)	0.035

SET PARKING BRAKE AND BLOCK DRIVE WHEELS.

TIMING ADJUSTMENT: MAKE ADJUSTMENT WITH ENGINE AT NORMAL OPERATING TEMPERATURE AND AIR CONDITIONING OFF (IF EQUIPPED).

1. VERIFY NO "CHECK ENGINE" LIGHT.
2. PUT EST (ELECTRONIC SPARK TIMING) IN BYPASS MODE BY DISCONNECTING TIMING CONNECTOR. [NOTE] THIS IS A SINGLE WIRE SEALED CONNECTOR. THE LEADS ARE TAN WITH A BLACK STRIPE AND BREAK OUT OF THE ENGINE WIRING HARNESS CONDUIT. DO NOT DISCONNECT FOUR-WIRE CONNECTOR AT DISTRIBUTOR.
3. CONNECT TIMING LIGHT INDUCTIVE PICK UP TO #1 SPARK PLUG WIRE AND SET TIMING TO SPECIFICATION WITH ENGINE RUNNING AT IDLE SPEED.
4. RECONNECT TIMING CONNECTOR AND CLEAR ECM TROUBLE CODE.

[NOTE] IDLE AIR SPEED SCREW IS PRESET AND SEALED AT FACTORY. PROVISION FOR ADJUSTMENT DURING TUNE UP IS NOT PROVIDED. DO NOT ATTEMPT ADJUSTMENT.

IDLE SPEEDS ARE AUTOMATICALLY CONTROLLED. DO NOT ATTEMPT ADJUSTMENTS.

SEE SERVICE MANUAL, MAINTENANCE SCHEDULE AND EMISSION HOSE ROUTING DIAGRAM FOR ADDITIONAL INFORMATION.

THIS VEHICLE CONFORMS TO U.S. EPA AND CALIFORNIA REGULATIONS APPLICABLE TO 1985 MODEL YEAR NEW PASSENGER CARS.

PT. NO. 14086540

1985–V8–5.0L VIN F CHEV.—CALIFORNIA

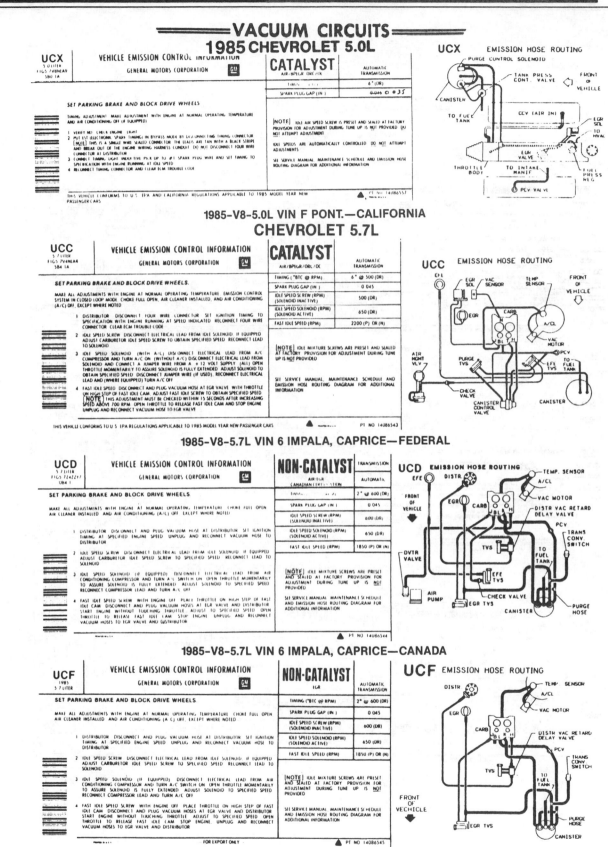

VACUUM CIRCUITS

1985 CHEVROLET 5.0L

UCX 5.0 LITER FIGS 7V8NEA9 5B0 1A

VEHICLE EMISSION CONTROL INFORMATION
GENERAL MOTORS CORPORATION GM

CATALYST AIR/BPEGR ORC/OX	AUTOMATIC TRANSMISSION
TIMING [illegible]	6° (DR)
SPARK PLUG GAP (IN.)	~~0.045~~ 0.035

SET PARKING BRAKE AND BLOCK DRIVE WHEELS

TIMING ADJUSTMENT. MAKE ADJUSTMENT WITH ENGINE AT NORMAL OPERATING TEMPERATURE AND AIR CONDITIONING OFF (IF EQUIPPED)

1. VERIFY NO "CHECK ENGINE" LIGHT
2. PUT EST (ELECTRONIC SPARK TIMING) IN BYPASS MODE BY DISCONNECTING TIMING CONNECTOR [NOTE] THIS IS A SINGLE WIRE SEALED CONNECTOR. THE LEADS ARE TAN WITH A BLACK STRIPE AND BREAK OUT OF THE ENGINE WIRING HARNESS CONDUIT. DO NOT DISCONNECT FOUR WIRE CONNECTOR AT DISTRIBUTOR
3. CONNECT TIMING LIGHT INDUCTIVE PICK UP TO #1 SPARK PLUG WIRE AND SET TIMING TO SPECIFICATION WITH ENGINE RUNNING AT IDLE SPEED
4. RECONNECT TIMING CONNECTOR AND CLEAR ECM TROUBLE CODE

[NOTE] IDLE AIR SPEED SCREW IS PRESET AND SEALED AT FACTORY. PROVISION FOR ADJUSTMENT DURING TUNE UP IS NOT PROVIDED. DO NOT ATTEMPT ADJUSTMENT

IDLE SPEEDS ARE AUTOMATICALLY CONTROLLED. DO NOT ATTEMPT ADJUSTMENTS

SEE SERVICE MANUAL, MAINTENANCE SCHEDULE AND EMISSION HOSE ROUTING DIAGRAM FOR ADDITIONAL INFORMATION

THIS VEHICLE CONFORMS TO U.S. EPA AND CALIFORNIA REGULATIONS APPLICABLE TO 1985 MODEL YEAR NEW PASSENGER CARS

PT NO 14086557

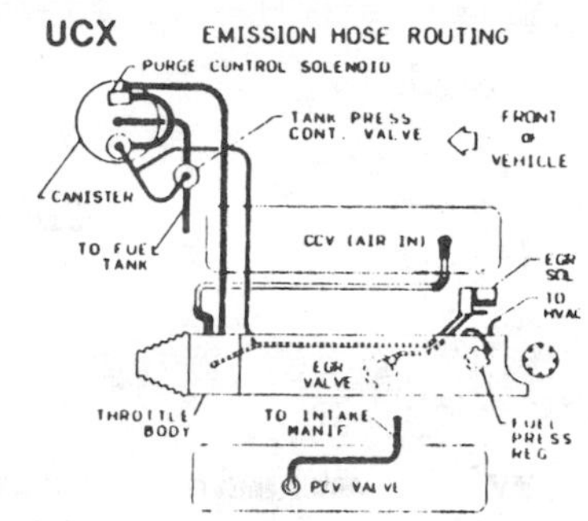

1985–V8–5.0L VIN F PONT.—CALIFORNIA

CHEVROLET 5.7L

UCC 5.7 LITER FIGS 7V4NEA4 5B4 1A

VEHICLE EMISSION CONTROL INFORMATION
GENERAL MOTORS CORPORATION GM

CATALYST AIR/BPEGR/ORC/OC	AUTOMATIC TRANSMISSION
TIMING (°BTC @ RPM)	6° @ 500 (DR)
SPARK PLUG GAP (IN.)	0.045
IDLE SPEED SCREW (RPM) (SOLENOID INACTIVE)	500 (DR)
IDLE SPEED SOLENOID (RPM) (SOLENOID ACTIVE)	650 (DR)
FAST IDLE SPEED (RPM)	2200 (P) OR (N)

SET PARKING BRAKE AND BLOCK DRIVE WHEELS.

MAKE ALL ADJUSTMENTS WITH ENGINE AT NORMAL OPERATING TEMPERATURE. EMISSION CONTROL SYSTEM IN CLOSED LOOP MODE, CHOKE FULL OPEN, AIR CLEANER INSTALLED, AND AIR CONDITIONING (A/C) OFF, EXCEPT WHERE NOTED

1. DISTRIBUTOR. DISCONNECT FOUR WIRE CONNECTOR. SET IGNITION TIMING TO SPECIFICATION WITH ENGINE RUNNING AT SPEED INDICATED. RECONNECT FOUR WIRE CONNECTOR. CLEAR ECM TROUBLE CODE
2. IDLE SPEED SCREW. DISCONNECT ELECTRICAL LEAD FROM IDLE SOLENOID, IF EQUIPPED. ADJUST CARBURETOR IDLE SPEED SCREW TO OBTAIN SPECIFIED SPEED. RECONNECT LEAD TO SOLENOID
3. IDLE SPEED SOLENOID. (WITH A/C) DISCONNECT ELECTRICAL LEAD FROM A/C COMPRESSOR AND TURN A/C ON. (WITHOUT A/C) DISCONNECT ELECTRICAL LEAD FROM SOLENOID AND CONNECT A JUMPER WIRE FROM A +12 VOLT SUPPLY. (ALL) OPEN THROTTLE MOMENTARILY TO ASSURE SOLENOID IS FULLY EXTENDED. ADJUST SOLENOID TO OBTAIN SPECIFIED SPEED. DISCONNECT JUMPER WIRE (IF USED), RECONNECT ELECTRICAL LEAD AND (WHERE EQUIPPED) TURN A/C OFF
4. FAST IDLE SPEED. DISCONNECT AND PLUG VACUUM HOSE AT EGR VALVE. WITH THROTTLE ON HIGH STEP OF FAST IDLE CAM, ADJUST FAST IDLE SCREW TO OBTAIN SPECIFIED SPEED. [NOTE] THIS ADJUSTMENT MUST BE CHECKED WITHIN 15 SECONDS AFTER INCREASING SPEED ABOVE 700 RPM. OPEN THROTTLE TO RELEASE FAST IDLE CAM AND STOP ENGINE. UNPLUG AND RECONNECT VACUUM HOSE TO EGR VALVE

[NOTE] IDLE MIXTURE SCREWS ARE PRESET AND SEALED AT FACTORY. PROVISION FOR ADJUSTMENT DURING TUNE UP IS NOT PROVIDED

SEE SERVICE MANUAL, MAINTENANCE SCHEDULE AND EMISSION HOSE ROUTING DIAGRAM FOR ADDITIONAL INFORMATION

THIS VEHICLE CONFORMS TO U.S. EPA REGULATIONS APPLICABLE TO 1985 MODEL YEAR NEW PASSENGER CARS

PT NO 14086543

UCC EMISSION HOSE ROUTING
EFE
EGR SOL
VAC SENSOR
TEMP SENSOR
FRONT OF VEHICLE
EGR
CARB
A/CL
VAC MOTOR
PCV
AIR MGMT VLV
PURGE TVS
EFE TVS
TO FUEL TANK
CHECK VALVE
CANISTER CONTROL VALVE
CANISTER

1985–V8–5.7L VIN 6 IMPALA, CAPRICE—FEDERAL

UCD 5.7 LITER FIGS 7242ZY7 0B4 1

VEHICLE EMISSION CONTROL INFORMATION
GENERAL MOTORS CORPORATION GM

NON-CATALYST AIR-EGR CANADIAN CERTIFICATION	TRANSMISSION AUTOMATIC
TIMING [illegible]	2° @ 600 (DR)
SPARK PLUG GAP (IN.)	0.045
IDLE SPEED SCREW (RPM) (SOLENOID INACTIVE)	600 (DR)
IDLE SPEED SOLENOID (RPM) (SOLENOID ACTIVE)	650 (DR)
FAST IDLE SPEED (RPM)	1850 (P) OR (N)

SET PARKING BRAKE AND BLOCK DRIVE WHEELS.

MAKE ALL ADJUSTMENTS WITH ENGINE AT NORMAL OPERATING TEMPERATURE, CHOKE FULL OPEN, AIR CLEANER INSTALLED AND AIR CONDITIONING (A/C) OFF, EXCEPT WHERE NOTED

1. DISTRIBUTOR. DISCONNECT AND PLUG VACUUM HOSE AT DISTRIBUTOR. SET IGNITION TIMING AT SPECIFIED ENGINE SPEED. UNPLUG AND RECONNECT VACUUM HOSE TO DISTRIBUTOR
2. IDLE SPEED SCREW. DISCONNECT ELECTRICAL LEAD FROM IDLE SOLENOID, IF EQUIPPED. ADJUST CARBURETOR IDLE SPEED SCREW TO SPECIFIED SPEED. RECONNECT LEAD TO SOLENOID
3. IDLE SPEED SOLENOID (IF EQUIPPED). DISCONNECT ELECTRICAL LEAD FROM AIR CONDITIONING COMPRESSOR AND TURN A/C SWITCH ON. OPEN THROTTLE MOMENTARILY TO ASSURE SOLENOID IS FULLY EXTENDED. ADJUST SOLENOID TO SPECIFIED SPEED. RECONNECT COMPRESSOR LEAD AND TURN A/C OFF
4. FAST IDLE SPEED SCREW. WITH ENGINE OFF, PLACE THROTTLE ON HIGH STEP OF FAST IDLE CAM. DISCONNECT AND PLUG VACUUM HOSES AT EGR VALVE AND DISTRIBUTOR. START ENGINE WITHOUT TOUCHING THROTTLE. ADJUST TO SPECIFIED SPEED. OPEN THROTTLE TO RELEASE FAST IDLE CAM. STOP ENGINE. UNPLUG AND RECONNECT VACUUM HOSES TO EGR VALVE AND DISTRIBUTOR

[NOTE] IDLE MIXTURE SCREWS ARE PRESET AND SEALED AT FACTORY. PROVISION FOR ADJUSTMENT DURING TUNE UP IS NOT PROVIDED

SEE SERVICE MANUAL, MAINTENANCE SCHEDULE AND EMISSION HOSE ROUTING DIAGRAM FOR ADDITIONAL INFORMATION

PT NO 14086544

UCD EMISSION HOSE ROUTING
EFE
DISTR.
TEMP. SENSOR
A/CL
VAC MOTOR
FRONT OF VEHICLE
EGR
CARB
DISTR VAC RETARD DELAY VALVE
PCV
TRANS CONV. SWITCH
TVS
TO FUEL TANK
DVTR VALVE
EFE TVS
AIR PUMP
CHECK VALVE
EGR TVS
CANISTER
PURGE HOSE

1985–V8–5.7L VIN 6 IMPALA, CAPRICE—CANADA

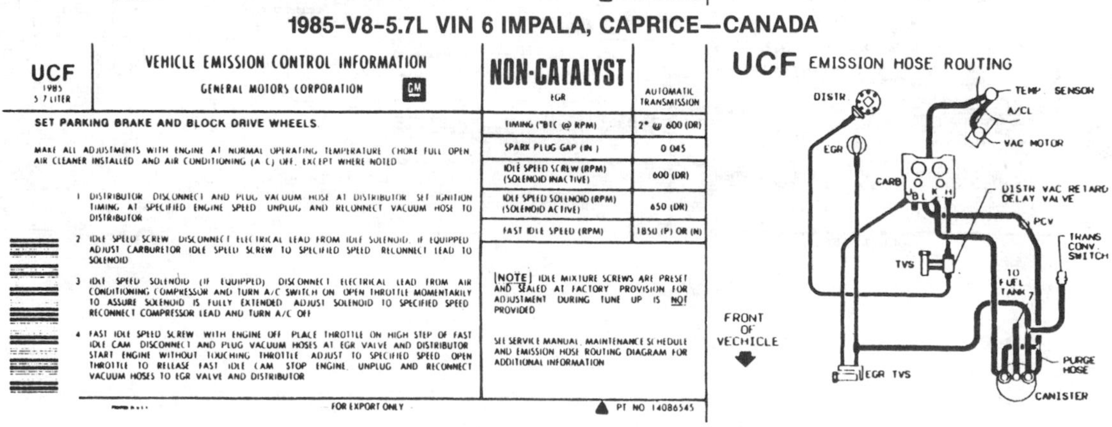

UCF 1985 5.7 LITER

VEHICLE EMISSION CONTROL INFORMATION
GENERAL MOTORS CORPORATION GM

NON-CATALYST EGR	AUTOMATIC TRANSMISSION
TIMING (°BTC @ RPM)	2° @ 600 (DR)
SPARK PLUG GAP (IN.)	0.045
IDLE SPEED SCREW (RPM) (SOLENOID INACTIVE)	600 (DR)
IDLE SPEED SOLENOID (RPM) (SOLENOID ACTIVE)	650 (DR)
FAST IDLE SPEED (RPM)	1850 (P) OR (N)

SET PARKING BRAKE AND BLOCK DRIVE WHEELS.

MAKE ALL ADJUSTMENTS WITH ENGINE AT NORMAL OPERATING TEMPERATURE, CHOKE FULL OPEN, AIR CLEANER INSTALLED AND AIR CONDITIONING (A/C) OFF, EXCEPT WHERE NOTED

1. DISTRIBUTOR. DISCONNECT AND PLUG VACUUM HOSE AT DISTRIBUTOR. SET IGNITION TIMING AT SPECIFIED ENGINE SPEED. UNPLUG AND RECONNECT VACUUM HOSE TO DISTRIBUTOR
2. IDLE SPEED SCREW. DISCONNECT ELECTRICAL LEAD FROM IDLE SOLENOID, IF EQUIPPED. ADJUST CARBURETOR IDLE SPEED SCREW TO SPECIFIED SPEED. RECONNECT LEAD TO SOLENOID
3. IDLE SPEED SOLENOID (IF EQUIPPED). DISCONNECT ELECTRICAL LEAD FROM AIR CONDITIONING COMPRESSOR AND TURN A/C SWITCH ON. OPEN THROTTLE MOMENTARILY TO ASSURE SOLENOID IS FULLY EXTENDED. ADJUST SOLENOID TO SPECIFIED SPEED. RECONNECT COMPRESSOR LEAD AND TURN A/C OFF
4. FAST IDLE SPEED SCREW. WITH ENGINE OFF, PLACE THROTTLE ON HIGH STEP OF FAST IDLE CAM. DISCONNECT AND PLUG VACUUM HOSES AT EGR VALVE AND DISTRIBUTOR. START ENGINE WITHOUT TOUCHING THROTTLE. ADJUST TO SPECIFIED SPEED. OPEN THROTTLE TO RELEASE FAST IDLE CAM. STOP ENGINE. UNPLUG AND RECONNECT VACUUM HOSES TO EGR VALVE AND DISTRIBUTOR

[NOTE] IDLE MIXTURE SCREWS ARE PRESET AND SEALED AT FACTORY. PROVISION FOR ADJUSTMENT DURING TUNE UP IS NOT PROVIDED

SEE SERVICE MANUAL, MAINTENANCE SCHEDULE AND EMISSION HOSE ROUTING DIAGRAM FOR ADDITIONAL INFORMATION

FOR EXPORT ONLY

PT NO 14086545

1985–V8–5.7L VIN 6 IMPALA, CAPRICE—EXPORT

VACUUM CIRCUITS

1985 CHEVROLET 5.7L

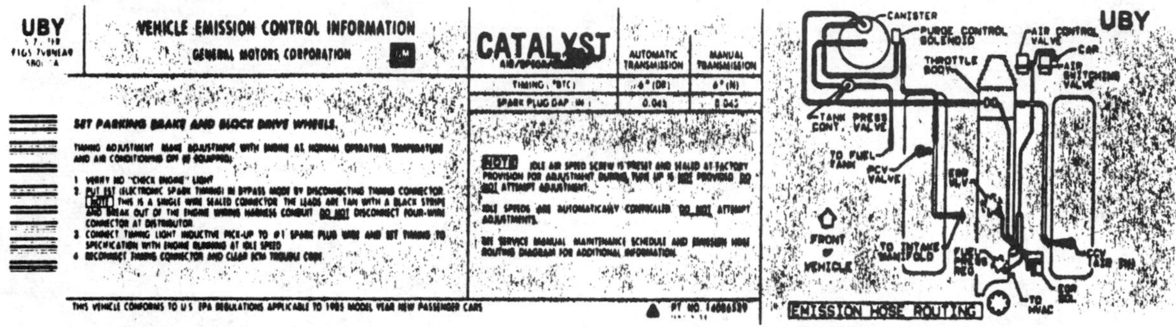

UBY

VEHICLE EMISSION CONTROL INFORMATION

GENERAL MOTORS CORPORATION

CATALYST

	AUTOMATIC TRANSMISSION	MANUAL TRANSMISSION
TIMING (°BTC)	6° (DR)	6° (N)
SPARK PLUG GAP (IN.)	0.045	0.045

SET PARKING BRAKE AND BLOCK DRIVE WHEELS.

TIMING ADJUSTMENT: MAKE ADJUSTMENT WITH ENGINE AT NORMAL OPERATING TEMPERATURE AND AIR CONDITIONING OFF (IF EQUIPPED).

1. VERIFY NO "CHECK ENGINE" LIGHT
2. PUT EST (ELECTRONIC SPARK TIMING) IN BYPASS MODE BY DISCONNECTING TIMING CONNECTOR. NOTE: THIS IS A SINGLE WIRE SEALED CONNECTOR. THE LEADS ARE TAN WITH A BLACK STRIPE AND BREAK OUT OF THE ENGINE WIRING HARNESS CONDUIT. DO NOT DISCONNECT FOUR-WIRE CONNECTOR AT DISTRIBUTOR.
3. CONNECT TIMING LIGHT INDUCTIVE PICK-UP TO #1 SPARK PLUG WIRE AND SET TIMING TO SPECIFICATION WITH ENGINE RUNNING AT IDLE SPEED.
4. RECONNECT TIMING CONNECTOR AND CLEAR ECM TROUBLE CODE.

NOTE: IDLE AIR SPEED SCREW IS PRESET AND SEALED AT FACTORY. PROVISION FOR ADJUSTMENT DURING TUNE UP IS NOT PROVIDED. DO NOT ATTEMPT ADJUSTMENT.

IDLE SPEEDS ARE AUTOMATICALLY CONTROLLED. DO NOT ATTEMPT ADJUSTMENTS.

SEE SERVICE MANUAL, MAINTENANCE SCHEDULE AND EMISSION HOSE ROUTING DIAGRAM FOR ADDITIONAL INFORMATION.

THIS VEHICLE CONFORMS TO U.S. EPA REGULATIONS APPLICABLE TO 1985 MODEL YEAR NEW PASSENGER CARS

PT. NO. 14086529

1985–V8–5.7L VIN 8 CORVETTE—FEDERAL

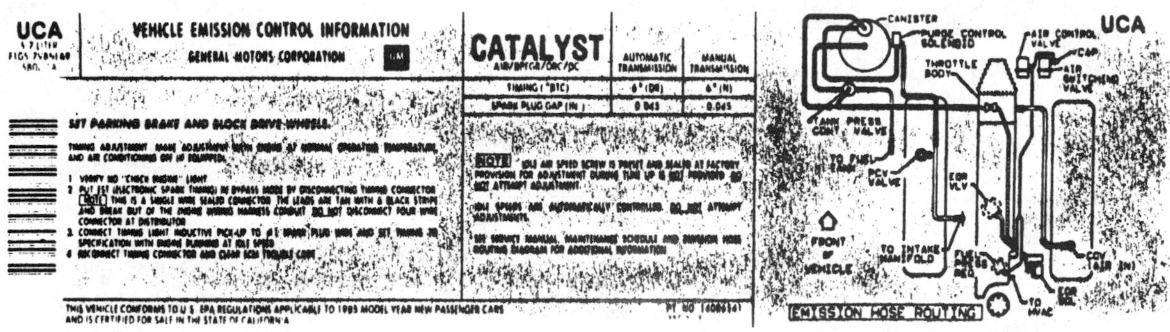

UCA

VEHICLE EMISSION CONTROL INFORMATION

GENERAL MOTORS CORPORATION

CATALYST

	AUTOMATIC TRANSMISSION	MANUAL TRANSMISSION
TIMING (°BTC)	6° (DR)	6° (N)
SPARK PLUG GAP (IN.)	0.045	0.045

SET PARKING BRAKE AND BLOCK DRIVE WHEELS.

TIMING ADJUSTMENT: MAKE ADJUSTMENT WITH ENGINE AT NORMAL OPERATING TEMPERATURE AND AIR CONDITIONING OFF (IF EQUIPPED).

1. VERIFY NO "CHECK ENGINE" LIGHT
2. PUT EST (ELECTRONIC SPARK TIMING) IN BYPASS MODE BY DISCONNECTING TIMING CONNECTOR. NOTE: THIS IS A SINGLE WIRE SEALED CONNECTOR. THE LEADS ARE TAN WITH A BLACK STRIPE AND BREAK OUT OF THE ENGINE WIRING HARNESS CONDUIT. DO NOT DISCONNECT FOUR-WIRE CONNECTOR AT DISTRIBUTOR.
3. CONNECT TIMING LIGHT INDUCTIVE PICK-UP TO #1 SPARK PLUG WIRE AND SET TIMING TO SPECIFICATION WITH ENGINE RUNNING AT IDLE SPEED.
4. RECONNECT TIMING CONNECTOR AND CLEAR ECM TROUBLE CODE.

NOTE: IDLE AIR SPEED SCREW IS PRESET AND SEALED AT FACTORY. PROVISION FOR ADJUSTMENT DURING TUNE UP IS NOT PROVIDED. DO NOT ATTEMPT ADJUSTMENT.

IDLE SPEEDS ARE AUTOMATICALLY CONTROLLED. DO NOT ATTEMPT ADJUSTMENTS.

SEE SERVICE MANUAL, MAINTENANCE SCHEDULE AND EMISSION HOSE ROUTING DIAGRAM FOR ADDITIONAL INFORMATION.

THIS VEHICLE CONFORMS TO U.S. EPA REGULATIONS APPLICABLE TO 1985 MODEL YEAR NEW PASSENGER CARS AND IS CERTIFIED FOR SALE IN THE STATE OF CALIFORNIA.

1985–V8–5.7L VIN 8 CORVETTE—FEDERAL

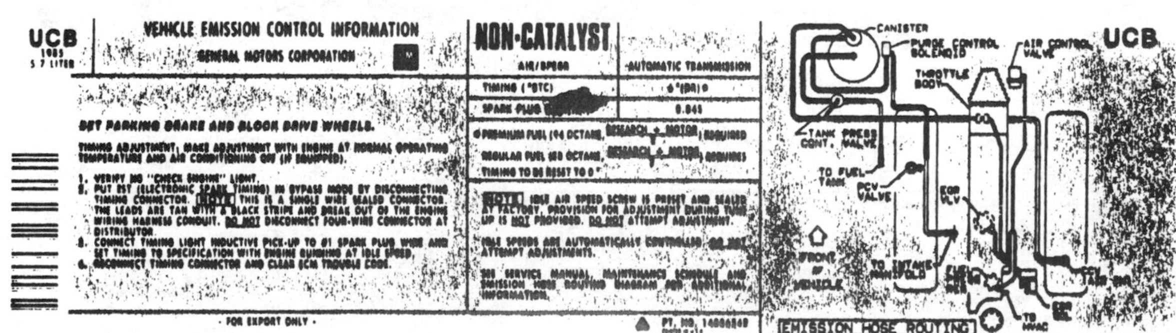

UCB
1985
5.7 LITER

VEHICLE EMISSION CONTROL INFORMATION

GENERAL MOTORS CORPORATION

NON-CATALYST

AIR/SPEGR	AUTOMATIC TRANSMISSION
TIMING (°BTC)	6° (DR) ⊕
SPARK PLUG [illegible]	0.045

⊕ PREMIUM FUEL (94 OCTANE, RESEARCH + MOTOR / 2) REQUIRED. REGULAR FUEL (88 OCTANE, RESEARCH + MOTOR / 2) REQUIRES TIMING TO BE RESET TO 0°.

SET PARKING BRAKE AND BLOCK DRIVE WHEELS.

TIMING ADJUSTMENT: MAKE ADJUSTMENT WITH ENGINE AT NORMAL OPERATING TEMPERATURE AND AIR CONDITIONING OFF (IF EQUIPPED).

1. VERIFY NO "CHECK ENGINE" LIGHT.
2. PUT EST (ELECTRONIC SPARK TIMING) IN BYPASS MODE BY DISCONNECTING TIMING CONNECTOR. NOTE: THIS IS A SINGLE WIRE SEALED CONNECTOR. THE LEADS ARE TAN WITH A BLACK STRIPE AND BREAK OUT OF THE ENGINE WIRING HARNESS CONDUIT. DO NOT DISCONNECT FOUR-WIRE CONNECTOR AT DISTRIBUTOR.
3. CONNECT TIMING LIGHT INDUCTIVE PICK-UP TO #1 SPARK PLUG WIRE AND SET TIMING TO SPECIFICATION WITH ENGINE RUNNING AT IDLE SPEED.
4. RECONNECT TIMING CONNECTOR AND CLEAR ECM TROUBLE CODE.

NOTE: IDLE AIR SPEED SCREW IS PRESET AND SEALED AT FACTORY. PROVISION FOR ADJUSTMENT DURING TUNE UP IS NOT PROVIDED. DO NOT ATTEMPT ADJUSTMENT.

IDLE SPEEDS ARE AUTOMATICALLY CONTROLLED. DO NOT ATTEMPT ADJUSTMENTS.

SEE SERVICE MANUAL, MAINTENANCE SCHEDULE AND EMISSION HOSE ROUTING DIAGRAM FOR ADDITIONAL INFORMATION.

· FOR EXPORT ONLY ·

1985–V8–5.7L VIN 8 CORVETTE—EXPORT

VACUUM CIRCUITS

1985 OLDSMOBILE 4.3L

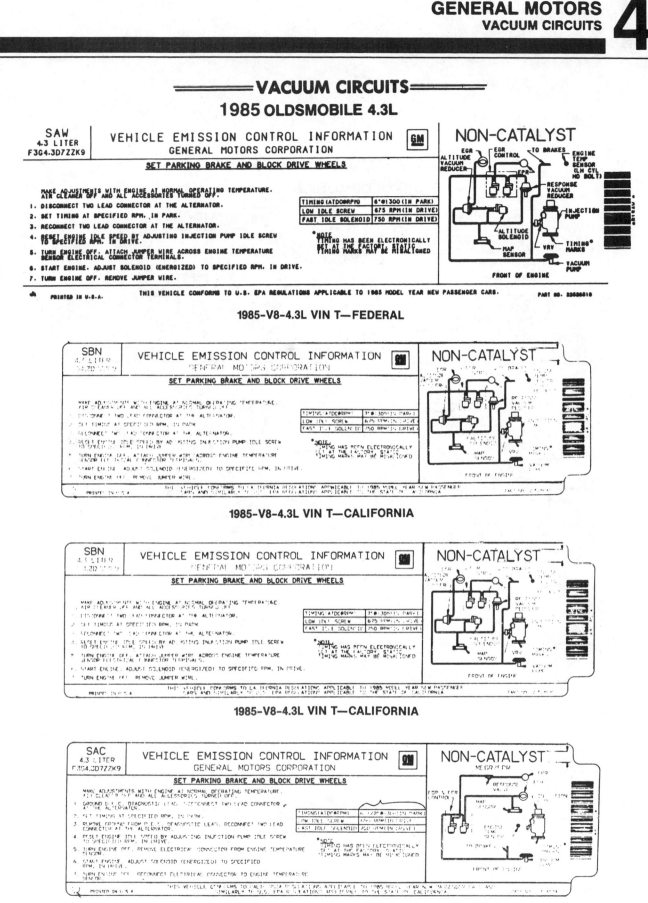

SAW
4.3 LITER
F3G4.3D7ZZK9

VEHICLE EMISSION CONTROL INFORMATION
GENERAL MOTORS CORPORATION
GM

NON-CATALYST

SET PARKING BRAKE AND BLOCK DRIVE WHEELS

MAKE ADJUSTMENTS WITH ENGINE AT NORMAL OPERATING TEMPERATURE. AIR CLEANER OFF AND ALL ACCESSORIES TURNED OFF.

1. DISCONNECT TWO LEAD CONNECTOR AT THE ALTERNATOR.
2. SET TIMING AT SPECIFIED RPM. IN PARK.
3. RECONNECT TWO LEAD CONNECTOR AT THE ALTERNATOR.
4. RESET ENGINE IDLE SPEED BY ADJUSTING INJECTION PUMP IDLE SCREW TO SPECIFIED RPM. IN DRIVE.
5. TURN ENGINE OFF. ATTACH JUMPER WIRE ACROSS ENGINE TEMPERATURE SENSOR ELECTRICAL CONNECTOR TERMINALS.
6. START ENGINE. ADJUST SOLENOID (ENERGIZED) TO SPECIFIED RPM. IN DRIVE.
7. TURN ENGINE OFF. REMOVE JUMPER WIRE.

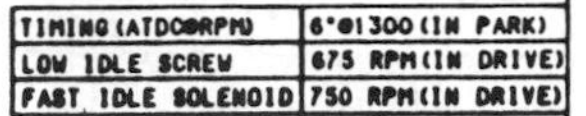

TIMING (ATDC@RPM)	6°@1300 (IN PARK)
LOW IDLE SCREW	675 RPM (IN DRIVE)
FAST IDLE SOLENOID	750 RPM (IN DRIVE)

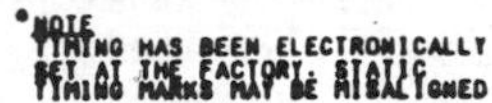

*NOTE
TIMING HAS BEEN ELECTRONICALLY SET AT THE FACTORY. STATIC TIMING MARKS MAY BE MISALIGNED

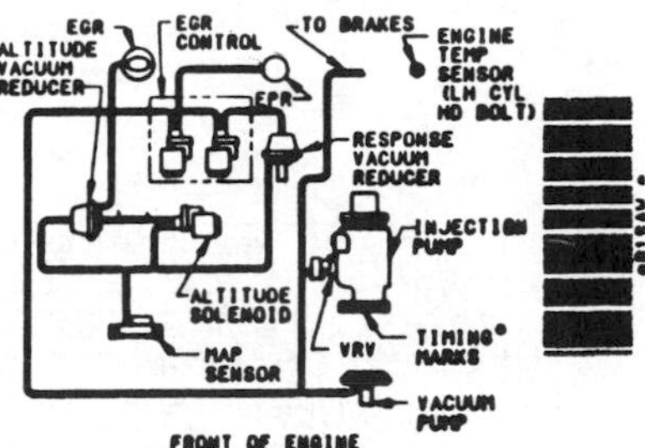

PRINTED IN U.S.A.
THIS VEHICLE CONFORMS TO U.S. EPA REGULATIONS APPLICABLE TO 1985 MODEL YEAR NEW PASSENGER CARS.
PART NO. 22526610

1985-V8-4.3L VIN T—FEDERAL

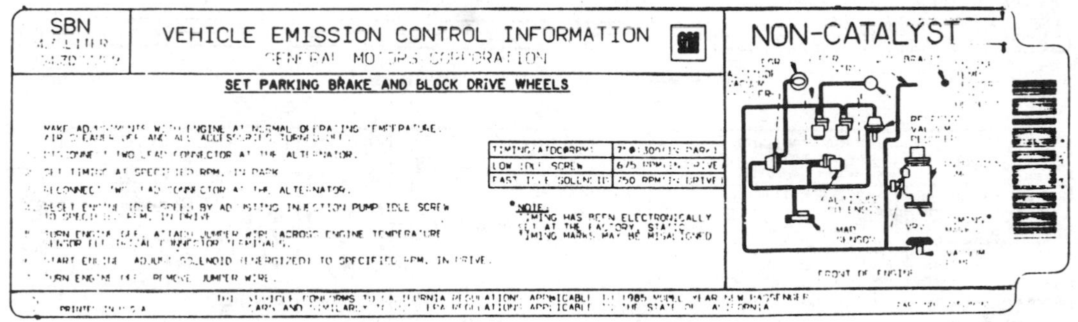

SBN

VEHICLE EMISSION CONTROL INFORMATION

NON-CATALYST

SET PARKING BRAKE AND BLOCK DRIVE WHEELS

1985-V8-4.3L VIN T—CALIFORNIA

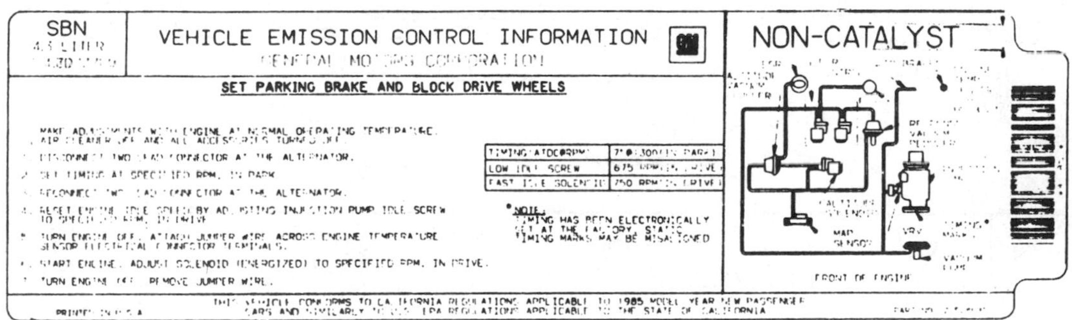

SBN

VEHICLE EMISSION CONTROL INFORMATION

NON-CATALYST

SET PARKING BRAKE AND BLOCK DRIVE WHEELS

1985-V8-4.3L VIN T—CALIFORNIA

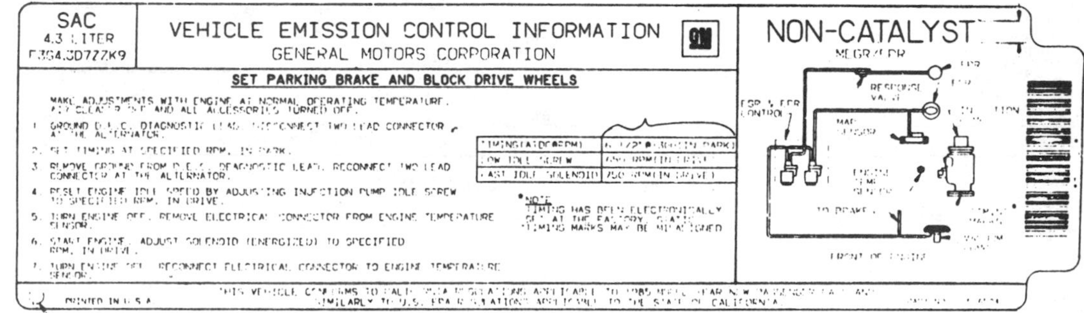

SAC
4.3 LITER
F3G4.3D7ZZK9

VEHICLE EMISSION CONTROL INFORMATION
GENERAL MOTORS CORPORATION

NON-CATALYST

SET PARKING BRAKE AND BLOCK DRIVE WHEELS

1985-V8-4.3L DIESEL—CALIFORNIA

VACUUM CIRCUITS

1985 OLDSMOBILE 5.0L

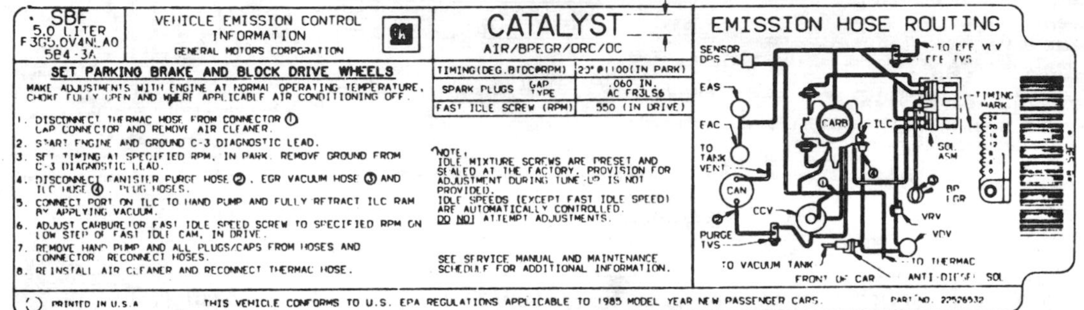

SBF
5.0 LITER
F3G5.0V4NLA0
5B4-3A

VEHICLE EMISSION CONTROL INFORMATION
GENERAL MOTORS CORPORATION

CATALYST
AIR/BPEGR/ORC/OC

TIMING (DEG. BTDC@RPM)	20°@1100 (IN PARK)
SPARK PLUGS GAP / TYPE	.060 IN. / AC FR3LS6
FAST IDLE SCREW (RPM)	550 (IN DRIVE)

SET PARKING BRAKE AND BLOCK DRIVE WHEELS

MAKE ADJUSTMENTS WITH ENGINE AT NORMAL OPERATING TEMPERATURE, CHOKE FULLY OPEN AND WHERE APPLICABLE AIR CONDITIONING OFF.

1. DISCONNECT THERMAC HOSE FROM CONNECTOR ①. CAP CONNECTOR AND REMOVE AIR CLEANER.
2. START ENGINE AND GROUND C-3 DIAGNOSTIC LEAD.
3. SET TIMING AT SPECIFIED RPM, IN PARK. REMOVE GROUND FROM C-3 DIAGNOSTIC LEAD.
4. DISCONNECT CANISTER PURGE HOSE ②, EGR VACUUM HOSE ③ AND ILC HOSE ④. PLUG HOSES.
5. CONNECT PORT ON ILC TO HAND PUMP AND FULLY RETRACT ILC RAM BY APPLYING VACUUM.
6. ADJUST CARBURETOR FAST IDLE SPEED SCREW TO SPECIFIED RPM ON LOW STEP OF FAST IDLE CAM, IN DRIVE.
7. REMOVE HAND PUMP AND ALL PLUGS/CAPS FROM HOSES AND CONNECTOR. RECONNECT HOSES.
8. REINSTALL AIR CLEANER AND RECONNECT THERMAC HOSE.

NOTE: IDLE MIXTURE SCREWS ARE PRESET AND SEALED AT THE FACTORY. PROVISION FOR ADJUSTMENT DURING TUNE-UP IS NOT PROVIDED. IDLE SPEEDS (EXCEPT FAST IDLE SPEED) ARE AUTOMATICALLY CONTROLLED. DO NOT ATTEMPT ADJUSTMENTS.

SEE SERVICE MANUAL AND MAINTENANCE SCHEDULE FOR ADDITIONAL INFORMATION.

EMISSION HOSE ROUTING

PRINTED IN U.S.A. THIS VEHICLE CONFORMS TO U.S. EPA REGULATIONS APPLICABLE TO 1985 MODEL YEAR NEW PASSENGER CARS. PART NO. 22526532

1985-V8-5.0L EXC LG8—FEDERAL

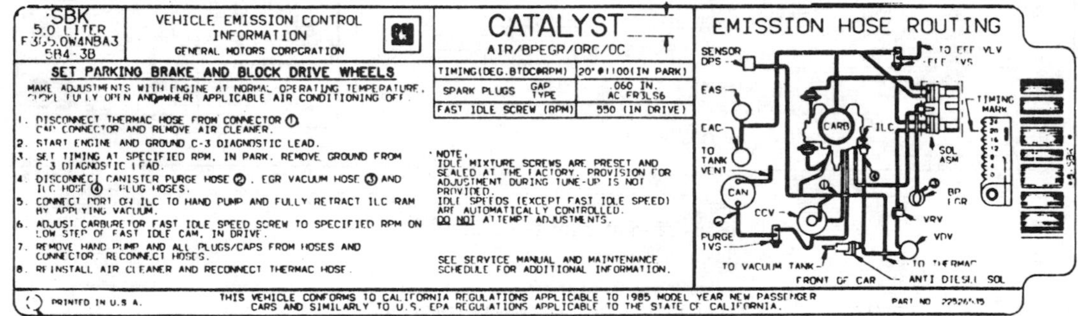

SBK
5.0 LITER
F3G5.0W4NBA3
5B4-3B

VEHICLE EMISSION CONTROL INFORMATION
GENERAL MOTORS CORPORATION

CATALYST
AIR/BPEGR/ORC/OC

TIMING (DEG. BTDC@RPM)	20°@1100 (IN PARK)
SPARK PLUGS GAP / TYPE	.060 IN. / AC FR3LS6
FAST IDLE SCREW (RPM)	550 (IN DRIVE)

SET PARKING BRAKE AND BLOCK DRIVE WHEELS

MAKE ADJUSTMENTS WITH ENGINE AT NORMAL OPERATING TEMPERATURE, CHOKE FULLY OPEN AND WHERE APPLICABLE AIR CONDITIONING OFF.

1. DISCONNECT THERMAC HOSE FROM CONNECTOR ①. CAP CONNECTOR AND REMOVE AIR CLEANER.
2. START ENGINE AND GROUND C-3 DIAGNOSTIC LEAD.
3. SET TIMING AT SPECIFIED RPM, IN PARK. REMOVE GROUND FROM C-3 DIAGNOSTIC LEAD.
4. DISCONNECT CANISTER PURGE HOSE ②, EGR VACUUM HOSE ③ AND ILC HOSE ④. PLUG HOSES.
5. CONNECT PORT ON ILC TO HAND PUMP AND FULLY RETRACT ILC RAM BY APPLYING VACUUM.
6. ADJUST CARBURETOR FAST IDLE SPEED SCREW TO SPECIFIED RPM ON LOW STEP OF FAST IDLE CAM, IN DRIVE.
7. REMOVE HAND PUMP AND ALL PLUGS/CAPS FROM HOSES AND CONNECTOR. RECONNECT HOSES.
8. REINSTALL AIR CLEANER AND RECONNECT THERMAC HOSE.

NOTE: IDLE MIXTURE SCREWS ARE PRESET AND SEALED AT THE FACTORY. PROVISION FOR ADJUSTMENT DURING TUNE-UP IS NOT PROVIDED. IDLE SPEEDS (EXCEPT FAST IDLE SPEED) ARE AUTOMATICALLY CONTROLLED. DO NOT ATTEMPT ADJUSTMENTS.

SEE SERVICE MANUAL AND MAINTENANCE SCHEDULE FOR ADDITIONAL INFORMATION.

EMISSION HOSE ROUTING

PRINTED IN U.S.A. THIS VEHICLE CONFORMS TO CALIFORNIA REGULATIONS APPLICABLE TO 1985 MODEL YEAR NEW PASSENGER CARS AND SIMILARLY TO U.S. EPA REGULATIONS APPLICABLE TO THE STATE OF CALIFORNIA. PART NO. 22526535

1985-V8-5.0L EXC LG8—CALIFORNIA

OLDSMOBILE 5.7L

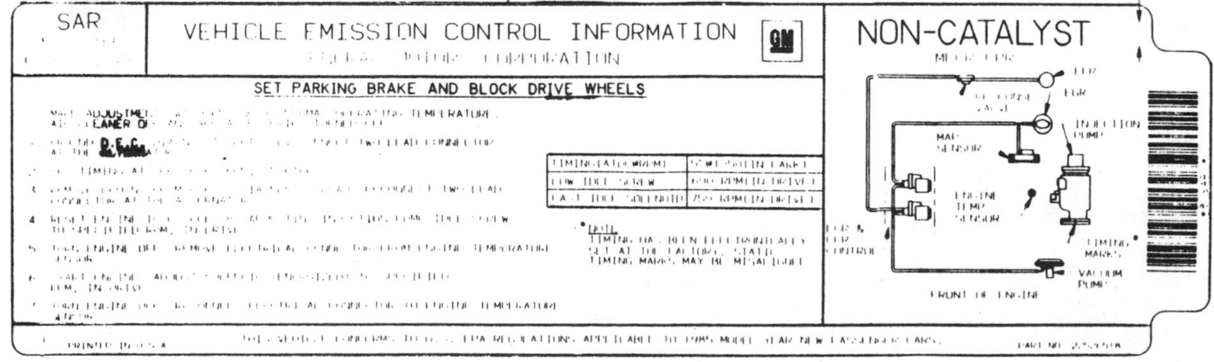

SAR

VEHICLE EMISSION CONTROL INFORMATION

NON-CATALYST

SET PARKING BRAKE AND BLOCK DRIVE WHEELS

1985-V8-5.7L DIESEL—FEDERAL

VACUUM CIRCUITS

1985 PONTIAC 1.8L

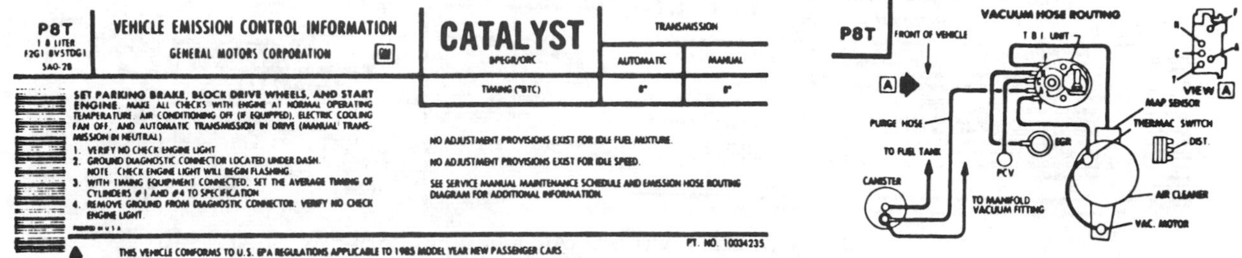

P8T
1.8 LITER
F2G1.8V5TDG1
5A0-2B

VEHICLE EMISSION CONTROL INFORMATION
GENERAL MOTORS CORPORATION

CATALYST BPEGR/ORC	TRANSMISSION AUTOMATIC	TRANSMISSION MANUAL
TIMING (°BTC)	8°	8°

SET PARKING BRAKE, BLOCK DRIVE WHEELS, AND START ENGINE. MAKE ALL CHECKS WITH ENGINE AT NORMAL OPERATING TEMPERATURE, AIR CONDITIONING OFF (IF EQUIPPED), ELECTRIC COOLING FAN OFF, AND AUTOMATIC TRANSMISSION IN DRIVE (MANUAL TRANSMISSION IN NEUTRAL).

1. VERIFY NO CHECK ENGINE LIGHT.
2. GROUND DIAGNOSTIC CONNECTOR LOCATED UNDER DASH. NOTE: CHECK ENGINE LIGHT WILL BEGIN FLASHING.
3. WITH TIMING EQUIPMENT CONNECTED, SET THE AVERAGE TIMING OF CYLINDERS #1 AND #4 TO SPECIFICATION.
4. REMOVE GROUND FROM DIAGNOSTIC CONNECTOR. VERIFY NO CHECK ENGINE LIGHT.

PRINTED IN U.S.A.

NO ADJUSTMENT PROVISIONS EXIST FOR IDLE FUEL MIXTURE.

NO ADJUSTMENT PROVISIONS EXIST FOR IDLE SPEED.

SEE SERVICE MANUAL MAINTENANCE SCHEDULE AND EMISSION HOSE ROUTING DIAGRAM FOR ADDITIONAL INFORMATION.

THIS VEHICLE CONFORMS TO U.S. EPA REGULATIONS APPLICABLE TO 1985 MODEL YEAR NEW PASSENGER CARS.

PT. NO. 10034235

1985-4 CYL-1.8L VIN O PONT 2000—FEDERAL

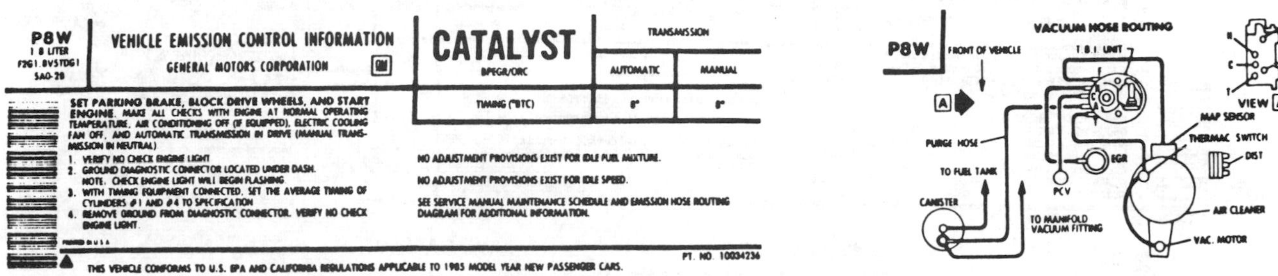

P8W
1.8 LITER
F2G1.8V5TDG1
5A0-2B

VEHICLE EMISSION CONTROL INFORMATION
GENERAL MOTORS CORPORATION

CATALYST BPEGR/ORC	TRANSMISSION AUTOMATIC	TRANSMISSION MANUAL
TIMING (°BTC)	8°	8°

SET PARKING BRAKE, BLOCK DRIVE WHEELS, AND START ENGINE. MAKE ALL CHECKS WITH ENGINE AT NORMAL OPERATING TEMPERATURE, AIR CONDITIONING OFF (IF EQUIPPED), ELECTRIC COOLING FAN OFF, AND AUTOMATIC TRANSMISSION IN DRIVE (MANUAL TRANSMISSION IN NEUTRAL).

1. VERIFY NO CHECK ENGINE LIGHT.
2. GROUND DIAGNOSTIC CONNECTOR LOCATED UNDER DASH. NOTE: CHECK ENGINE LIGHT WILL BEGIN FLASHING.
3. WITH TIMING EQUIPMENT CONNECTED, SET THE AVERAGE TIMING OF CYLINDERS #1 AND #4 TO SPECIFICATION.
4. REMOVE GROUND FROM DIAGNOSTIC CONNECTOR. VERIFY NO CHECK ENGINE LIGHT.

PRINTED IN U.S.A.

NO ADJUSTMENT PROVISIONS EXIST FOR IDLE FUEL MIXTURE.

NO ADJUSTMENT PROVISIONS EXIST FOR IDLE SPEED.

SEE SERVICE MANUAL MAINTENANCE SCHEDULE AND EMISSION HOSE ROUTING DIAGRAM FOR ADDITIONAL INFORMATION.

THIS VEHICLE CONFORMS TO U.S. EPA AND CALIFORNIA REGULATIONS APPLICABLE TO 1985 MODEL YEAR NEW PASSENGER CARS.

PT. NO. 10034236

1985-4 CYL-1.8L VIN O PONT 2000—CALIFORNIA

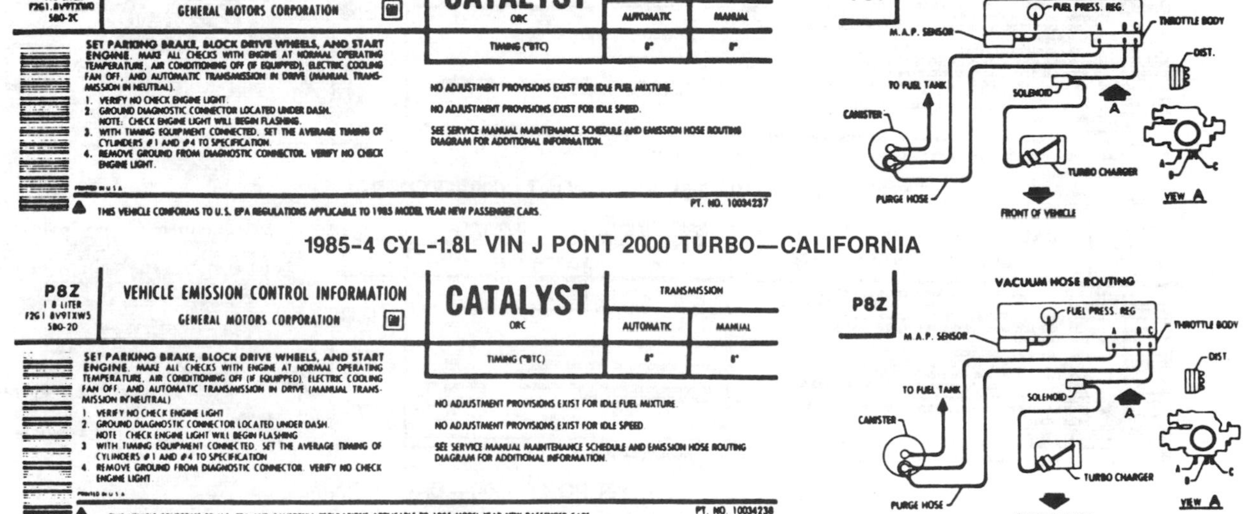

P8Y
1.8 LITER
F2G1.8V9TXWD
5B0-2C

VEHICLE EMISSION CONTROL INFORMATION
GENERAL MOTORS CORPORATION

CATALYST ORC	TRANSMISSION AUTOMATIC	TRANSMISSION MANUAL
TIMING (°BTC)	8°	8°

SET PARKING BRAKE, BLOCK DRIVE WHEELS, AND START ENGINE. MAKE ALL CHECKS WITH ENGINE AT NORMAL OPERATING TEMPERATURE, AIR CONDITIONING OFF (IF EQUIPPED), ELECTRIC COOLING FAN OFF, AND AUTOMATIC TRANSMISSION IN DRIVE (MANUAL TRANSMISSION IN NEUTRAL).

1. VERIFY NO CHECK ENGINE LIGHT.
2. GROUND DIAGNOSTIC CONNECTOR LOCATED UNDER DASH. NOTE: CHECK ENGINE LIGHT WILL BEGIN FLASHING.
3. WITH TIMING EQUIPMENT CONNECTED, SET THE AVERAGE TIMING OF CYLINDERS #1 AND #4 TO SPECIFICATION.
4. REMOVE GROUND FROM DIAGNOSTIC CONNECTOR. VERIFY NO CHECK ENGINE LIGHT.

PRINTED IN U.S.A.

NO ADJUSTMENT PROVISIONS EXIST FOR IDLE FUEL MIXTURE.

NO ADJUSTMENT PROVISIONS EXIST FOR IDLE SPEED.

SEE SERVICE MANUAL MAINTENANCE SCHEDULE AND EMISSION HOSE ROUTING DIAGRAM FOR ADDITIONAL INFORMATION.

THIS VEHICLE CONFORMS TO U.S. EPA REGULATIONS APPLICABLE TO 1985 MODEL YEAR NEW PASSENGER CARS.

PT. NO. 10034237

1985-4 CYL-1.8L VIN J PONT 2000 TURBO—CALIFORNIA

P8Z
1.8 LITER
F2G1.8V9TXWS
5B0-2D

VEHICLE EMISSION CONTROL INFORMATION
GENERAL MOTORS CORPORATION

CATALYST ORC	TRANSMISSION AUTOMATIC	TRANSMISSION MANUAL
TIMING (°BTC)	8°	8°

SET PARKING BRAKE, BLOCK DRIVE WHEELS, AND START ENGINE. MAKE ALL CHECKS WITH ENGINE AT NORMAL OPERATING TEMPERATURE, AIR CONDITIONING OFF (IF EQUIPPED), ELECTRIC COOLING FAN OFF, AND AUTOMATIC TRANSMISSION IN DRIVE (MANUAL TRANSMISSION IN NEUTRAL).

1. VERIFY NO CHECK ENGINE LIGHT.
2. GROUND DIAGNOSTIC CONNECTOR LOCATED UNDER DASH. NOTE: CHECK ENGINE LIGHT WILL BEGIN FLASHING.
3. WITH TIMING EQUIPMENT CONNECTED, SET THE AVERAGE TIMING OF CYLINDERS #1 AND #4 TO SPECIFICATION.
4. REMOVE GROUND FROM DIAGNOSTIC CONNECTOR. VERIFY NO CHECK ENGINE LIGHT.

PRINTED IN U.S.A.

NO ADJUSTMENT PROVISIONS EXIST FOR IDLE FUEL MIXTURE.

NO ADJUSTMENT PROVISIONS EXIST FOR IDLE SPEED.

SEE SERVICE MANUAL MAINTENANCE SCHEDULE AND EMISSION HOSE ROUTING DIAGRAM FOR ADDITIONAL INFORMATION.

THIS VEHICLE CONFORMS TO U.S. EPA AND CALIFORNIA REGULATIONS APPLICABLE TO 1985 MODEL YEAR NEW PASSENGER CARS.

PT. NO. 10034238

1985-4 CYL-1.8L VIN J PONT 2000 TURBO—CALIFORNIA

1985 PONTIAC 2.5L

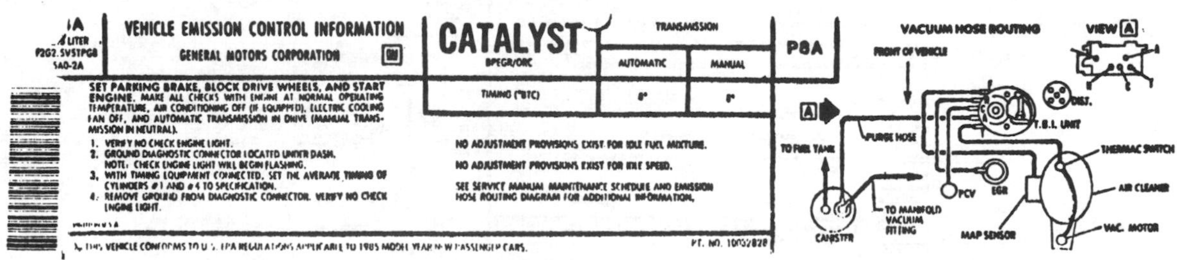

P8A
2.5 LITER
F2G2.5V5TPG8
5A0-2A

VEHICLE EMISSION CONTROL INFORMATION
GENERAL MOTORS CORPORATION

CATALYST BPEGR/ORC	TRANSMISSION AUTOMATIC	TRANSMISSION MANUAL
TIMING (°BTC)	8°	8°

SET PARKING BRAKE, BLOCK DRIVE WHEELS, AND START ENGINE. MAKE ALL CHECKS WITH ENGINE AT NORMAL OPERATING TEMPERATURE, AIR CONDITIONING OFF (IF EQUIPPED), ELECTRIC COOLING FAN OFF, AND AUTOMATIC TRANSMISSION IN DRIVE (MANUAL TRANSMISSION IN NEUTRAL).

1. VERIFY NO CHECK ENGINE LIGHT.
2. GROUND DIAGNOSTIC CONNECTOR LOCATED UNDER DASH. NOTE: CHECK ENGINE LIGHT WILL BEGIN FLASHING.
3. WITH TIMING EQUIPMENT CONNECTED, SET THE AVERAGE TIMING OF CYLINDERS #1 AND #4 TO SPECIFICATION.
4. REMOVE GROUND FROM DIAGNOSTIC CONNECTOR. VERIFY NO CHECK ENGINE LIGHT.

PRINTED IN U.S.A.

NO ADJUSTMENT PROVISIONS EXIST FOR IDLE FUEL MIXTURE.

NO ADJUSTMENT PROVISIONS EXIST FOR IDLE SPEED.

SEE SERVICE MANUAL MAINTENANCE SCHEDULE AND EMISSION HOSE ROUTING DIAGRAM FOR ADDITIONAL INFORMATION.

THIS VEHICLE CONFORMS TO U.S. EPA REGULATIONS APPLICABLE TO 1985 MODEL YEAR NEW PASSENGER CARS.

PT. NO. 10032828

1985-4 CYL-2.5L VIN U GRAND AM—FEDERAL

VACUUM CIRCUITS
1985 PONTIAC 2.5L

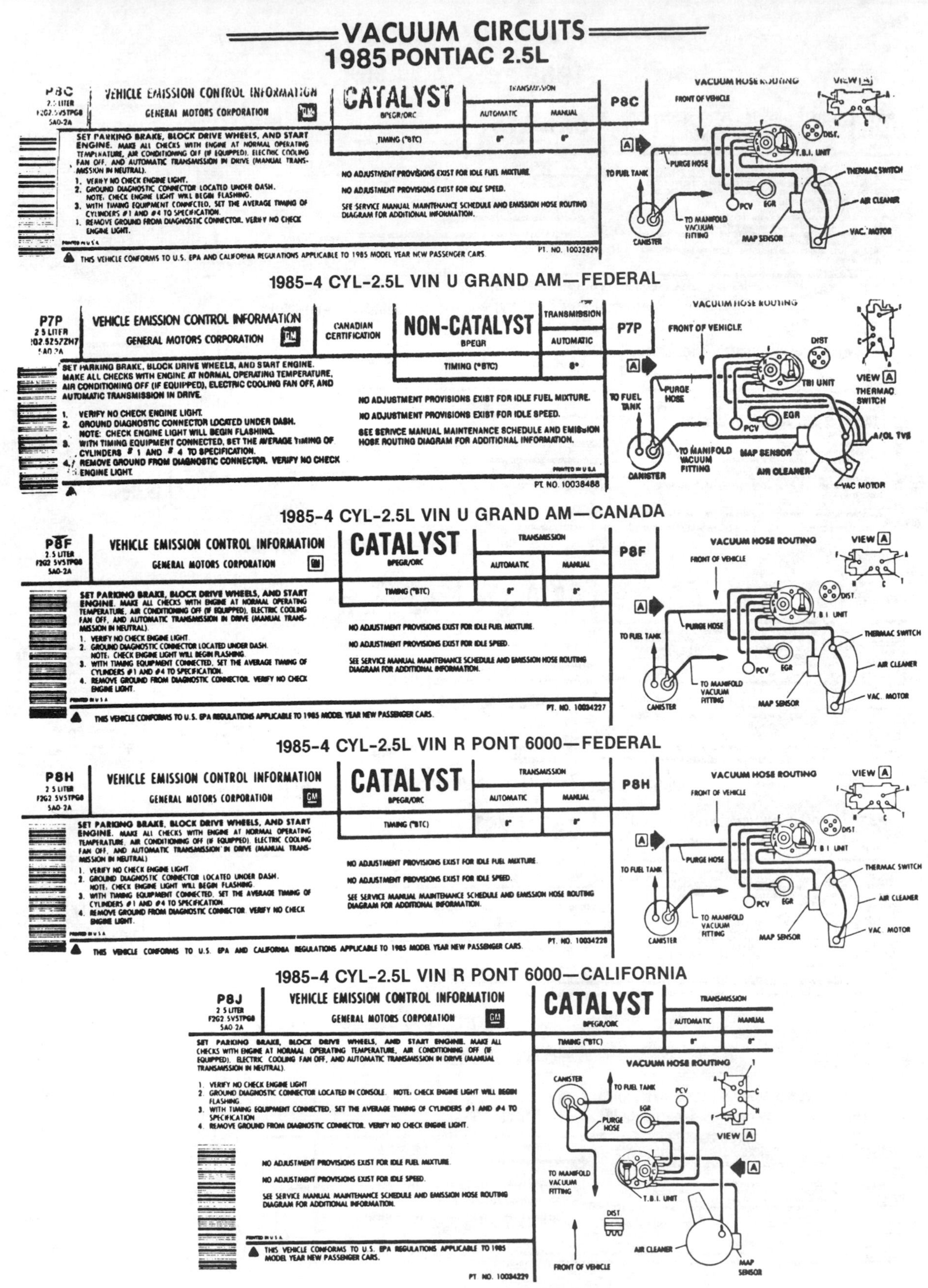

P8C 2.5 LITER F2G2.5V5TPG8 5A0-2A

VEHICLE EMISSION CONTROL INFORMATION
GENERAL MOTORS CORPORATION

CATALYST
BPEGR/ORC

	TRANSMISSION	
	AUTOMATIC	MANUAL
TIMING (°BTC)	8°	8°

SET PARKING BRAKE, BLOCK DRIVE WHEELS, AND START ENGINE. MAKE ALL CHECKS WITH ENGINE AT NORMAL OPERATING TEMPERATURE, AIR CONDITIONING OFF (IF EQUIPPED), ELECTRIC COOLING FAN OFF, AND AUTOMATIC TRANSMISSION IN DRIVE (MANUAL TRANSMISSION IN NEUTRAL).

1. VERIFY NO CHECK ENGINE LIGHT.
2. GROUND DIAGNOSTIC CONNECTOR LOCATED UNDER DASH. NOTE: CHECK ENGINE LIGHT WILL BEGIN FLASHING.
3. WITH TIMING EQUIPMENT CONNECTED, SET THE AVERAGE TIMING OF CYLINDERS #1 AND #4 TO SPECIFICATION.
4. REMOVE GROUND FROM DIAGNOSTIC CONNECTOR. VERIFY NO CHECK ENGINE LIGHT.

NO ADJUSTMENT PROVISIONS EXIST FOR IDLE FUEL MIXTURE.
NO ADJUSTMENT PROVISIONS EXIST FOR IDLE SPEED.
SEE SERVICE MANUAL MAINTENANCE SCHEDULE AND EMISSION HOSE ROUTING DIAGRAM FOR ADDITIONAL INFORMATION.

PRINTED IN U.S.A.

THIS VEHICLE CONFORMS TO U.S. EPA AND CALIFORNIA REGULATIONS APPLICABLE TO 1985 MODEL YEAR NEW PASSENGER CARS.

PT. NO. 10032829

1985–4 CYL–2.5L VIN U GRAND AM—FEDERAL

P7P 2.5 LITER F02.5Z5Z7H7 5A0-2A

VEHICLE EMISSION CONTROL INFORMATION
GENERAL MOTORS CORPORATION

CANADIAN CERTIFICATION

NON-CATALYST
BPEGR

	TRANSMISSION
	AUTOMATIC
TIMING (°BTC)	8°

SET PARKING BRAKE, BLOCK DRIVE WHEELS, AND START ENGINE. MAKE ALL CHECKS WITH ENGINE AT NORMAL OPERATING TEMPERATURE, AIR CONDITIONING OFF (IF EQUIPPED), ELECTRIC COOLING FAN OFF, AND AUTOMATIC TRANSMISSION IN DRIVE.

1. VERIFY NO CHECK ENGINE LIGHT.
2. GROUND DIAGNOSTIC CONNECTOR LOCATED UNDER DASH. NOTE: CHECK ENGINE LIGHT WILL BEGIN FLASHING.
3. WITH TIMING EQUIPMENT CONNECTED, SET THE AVERAGE TIMING OF CYLINDERS # 1 AND # 4 TO SPECIFICATION.
4. REMOVE GROUND FROM DIAGNOSTIC CONNECTOR. VERIFY NO CHECK ENGINE LIGHT.

NO ADJUSTMENT PROVISIONS EXIST FOR IDLE FUEL MIXTURE.
NO ADJUSTMENT PROVISIONS EXIST FOR IDLE SPEED.
SEE SERVICE MANUAL MAINTENANCE SCHEDULE AND EMISSION HOSE ROUTING DIAGRAM FOR ADDITIONAL INFORMATION.

PRINTED IN U.S.A.

PT. NO. 10038488

1985–4 CYL–2.5L VIN U GRAND AM—CANADA

P8F 2.5 LITER F2G2.5V5TPG8 5A0-2A

VEHICLE EMISSION CONTROL INFORMATION
GENERAL MOTORS CORPORATION

CATALYST
BPEGR/ORC

	TRANSMISSION	
	AUTOMATIC	MANUAL
TIMING (°BTC)	8°	8°

SET PARKING BRAKE, BLOCK DRIVE WHEELS, AND START ENGINE. MAKE ALL CHECKS WITH ENGINE AT NORMAL OPERATING TEMPERATURE, AIR CONDITIONING OFF (IF EQUIPPED), ELECTRIC COOLING FAN OFF, AND AUTOMATIC TRANSMISSION IN DRIVE (MANUAL TRANSMISSION IN NEUTRAL).

1. VERIFY NO CHECK ENGINE LIGHT.
2. GROUND DIAGNOSTIC CONNECTOR LOCATED UNDER DASH. NOTE: CHECK ENGINE LIGHT WILL BEGIN FLASHING.
3. WITH TIMING EQUIPMENT CONNECTED, SET THE AVERAGE TIMING OF CYLINDERS #1 AND #4 TO SPECIFICATION.
4. REMOVE GROUND FROM DIAGNOSTIC CONNECTOR. VERIFY NO CHECK ENGINE LIGHT.

NO ADJUSTMENT PROVISIONS EXIST FOR IDLE FUEL MIXTURE.
NO ADJUSTMENT PROVISIONS EXIST FOR IDLE SPEED.
SEE SERVICE MANUAL MAINTENANCE SCHEDULE AND EMISSION HOSE ROUTING DIAGRAM FOR ADDITIONAL INFORMATION.

PRINTED IN U.S.A.

THIS VEHICLE CONFORMS TO U.S. EPA REGULATIONS APPLICABLE TO 1985 MODEL YEAR NEW PASSENGER CARS.

PT. NO. 10034227

1985–4 CYL–2.5L VIN R PONT 6000—FEDERAL

P8H 2.5 LITER F2G2.5V5TPG8 5A0-2A

VEHICLE EMISSION CONTROL INFORMATION
GENERAL MOTORS CORPORATION

CATALYST
BPEGR/ORC

	TRANSMISSION	
	AUTOMATIC	MANUAL
TIMING (°BTC)	8°	8°

SET PARKING BRAKE, BLOCK DRIVE WHEELS, AND START ENGINE. MAKE ALL CHECKS WITH ENGINE AT NORMAL OPERATING TEMPERATURE, AIR CONDITIONING OFF (IF EQUIPPED), ELECTRIC COOLING FAN OFF, AND AUTOMATIC TRANSMISSION IN DRIVE (MANUAL TRANSMISSION IN NEUTRAL).

1. VERIFY NO CHECK ENGINE LIGHT.
2. GROUND DIAGNOSTIC CONNECTOR LOCATED UNDER DASH. NOTE: CHECK ENGINE LIGHT WILL BEGIN FLASHING.
3. WITH TIMING EQUIPMENT CONNECTED, SET THE AVERAGE TIMING OF CYLINDERS #1 AND #4 TO SPECIFICATION.
4. REMOVE GROUND FROM DIAGNOSTIC CONNECTOR. VERIFY NO CHECK ENGINE LIGHT.

NO ADJUSTMENT PROVISIONS EXIST FOR IDLE FUEL MIXTURE.
NO ADJUSTMENT PROVISIONS EXIST FOR IDLE SPEED.
SEE SERVICE MANUAL MAINTENANCE SCHEDULE AND EMISSION HOSE ROUTING DIAGRAM FOR ADDITIONAL INFORMATION.

PRINTED IN U.S.A.

THIS VEHICLE CONFORMS TO U.S. EPA AND CALIFORNIA REGULATIONS APPLICABLE TO 1985 MODEL YEAR NEW PASSENGER CARS.

PT. NO. 10034228

1985–4 CYL–2.5L VIN R PONT 6000—CALIFORNIA

P8J 2.5 LITER F2G2.5V5TPG8 5A0-2A

VEHICLE EMISSION CONTROL INFORMATION
GENERAL MOTORS CORPORATION

CATALYST
BPEGR/ORC

	TRANSMISSION	
	AUTOMATIC	MANUAL
TIMING (°BTC)	8°	8°

SET PARKING BRAKE, BLOCK DRIVE WHEELS, AND START ENGINE. MAKE ALL CHECKS WITH ENGINE AT NORMAL OPERATING TEMPERATURE, AIR CONDITIONING OFF (IF EQUIPPED), ELECTRIC COOLING FAN OFF, AND AUTOMATIC TRANSMISSION IN DRIVE (MANUAL TRANSMISSION IN NEUTRAL).

1. VERIFY NO CHECK ENGINE LIGHT.
2. GROUND DIAGNOSTIC CONNECTOR LOCATED IN CONSOLE. NOTE: CHECK ENGINE LIGHT WILL BEGIN FLASHING.
3. WITH TIMING EQUIPMENT CONNECTED, SET THE AVERAGE TIMING OF CYLINDERS #1 AND #4 TO SPECIFICATION.
4. REMOVE GROUND FROM DIAGNOSTIC CONNECTOR. VERIFY NO CHECK ENGINE LIGHT.

NO ADJUSTMENT PROVISIONS EXIST FOR IDLE FUEL MIXTURE.
NO ADJUSTMENT PROVISIONS EXIST FOR IDLE SPEED.
SEE SERVICE MANUAL MAINTENANCE SCHEDULE AND EMISSION HOSE ROUTING DIAGRAM FOR ADDITIONAL INFORMATION.

PRINTED IN U.S.A.

THIS VEHICLE CONFORMS TO U.S. EPA REGULATIONS APPLICABLE TO 1985 MODEL YEAR NEW PASSENGER CARS.

PT. NO. 10034229

1985–4 CYL–2.5L VIN R FIERO—FEDERAL

VACUUM CIRCUITS

1985 PONTIAC 2.5L

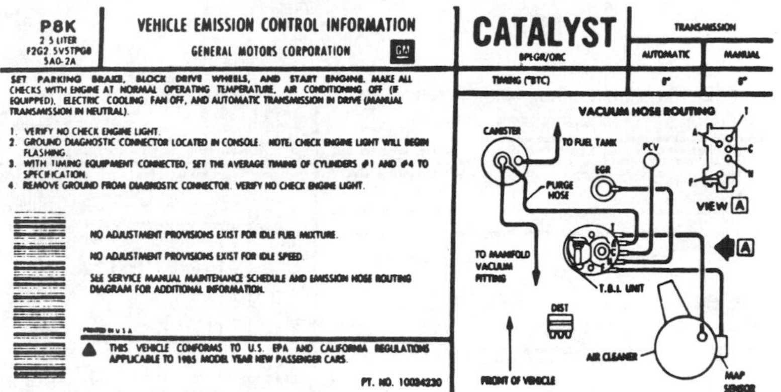

1985-4 CYL-2.5L VIN R FIERO—CALIFORNIA

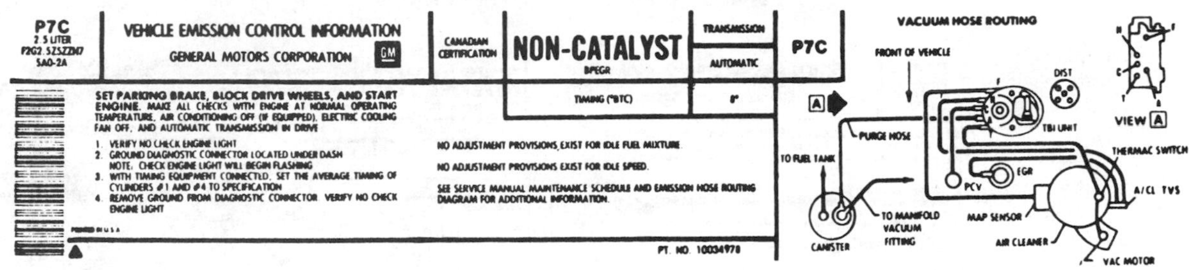

1985-4 CYL-2.5L VIN R "X" BODY-TBI—CANADA

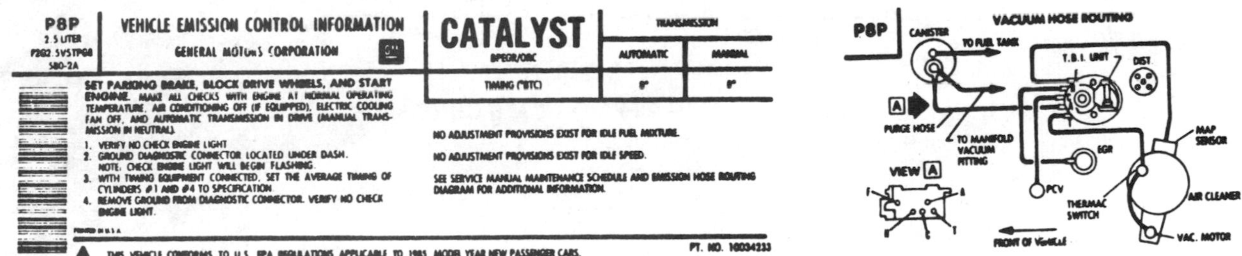

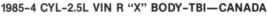

1985-4 CYL-2.5L VIN 2 FIREBIRD—FEDERAL

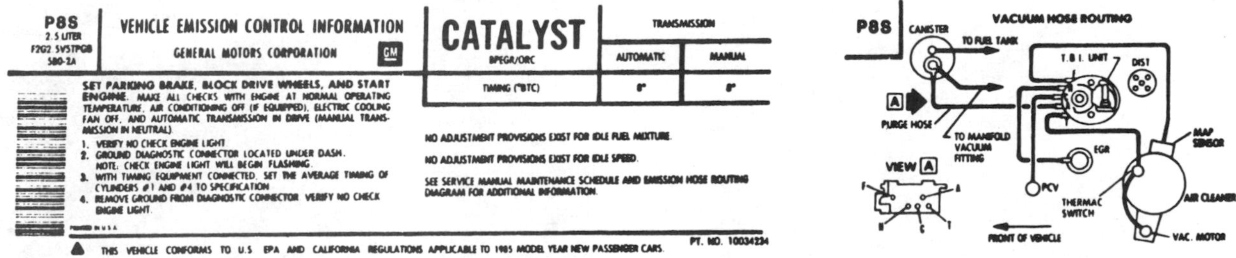

1985-4 CYL-2.5L VIN 2 FIREBIRD—CALIFORNIA

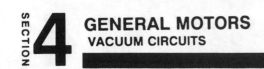

VACUUM CIRCUITS

1985 PONTIAC 2.8L

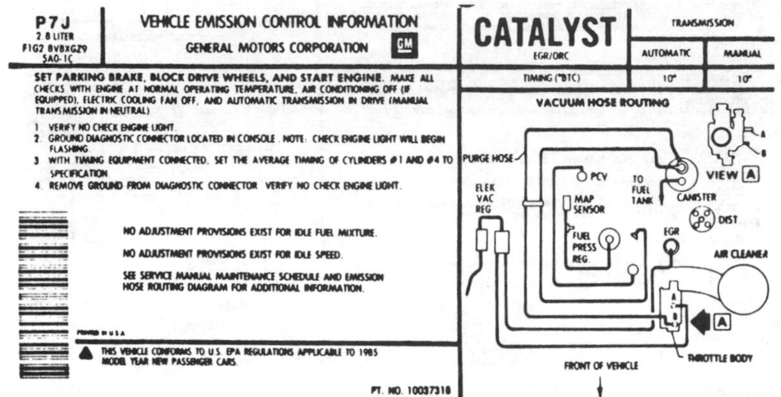

1985-V6-2.8L VIN 9 FIERO—FEDERAL

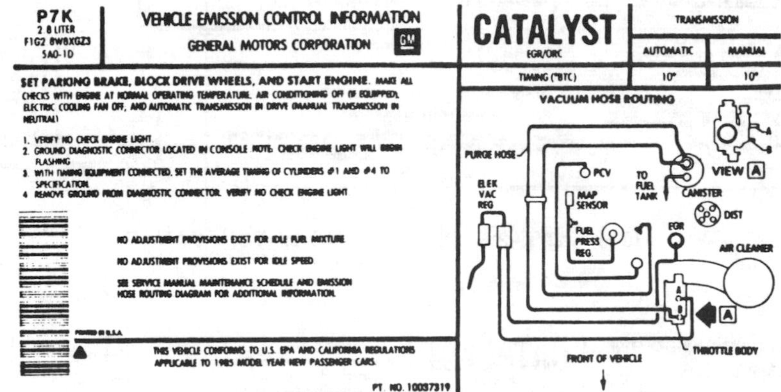

1985-V6-2.8L VIN 9 FIERO—CALIFORNIA

General Motors
Vacuum Circuits

INDEX

1986 VACUUM CIRCUITS

VACUUM CIRCUITS

(© G.M. Corp.)

1986 BUICK 3.0L

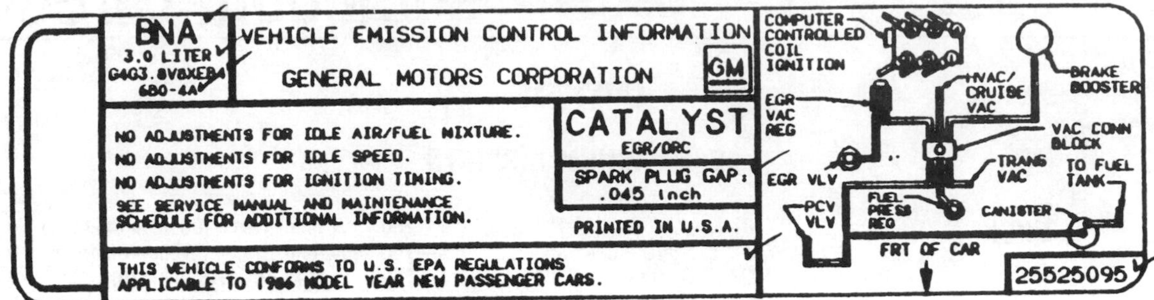

3.0L W/AUTO. TRANS – FEDERAL

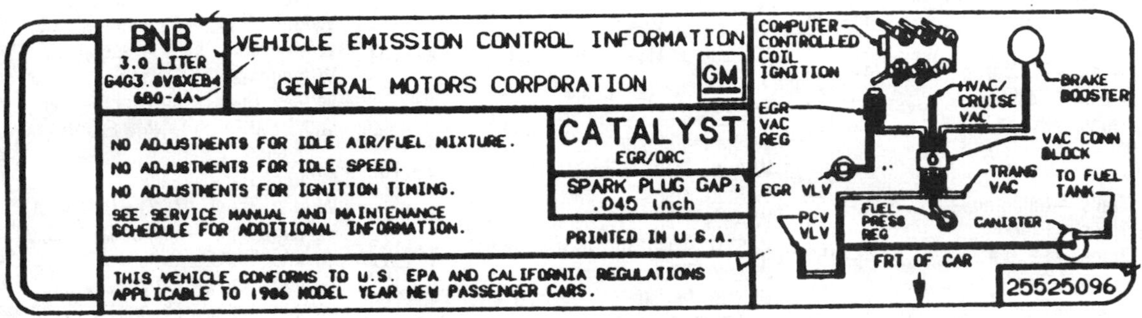

3.0L W/AUTO. TRANS. – CALIFORNIA

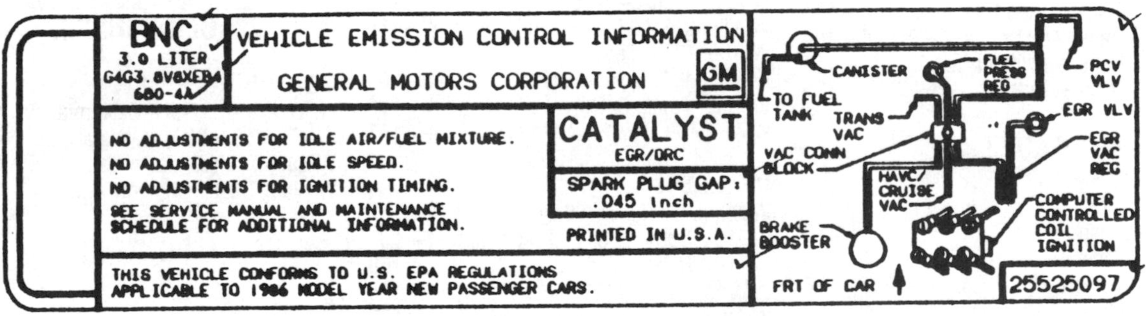

3.0L W/AUTO. TRANS. – FEDERAL

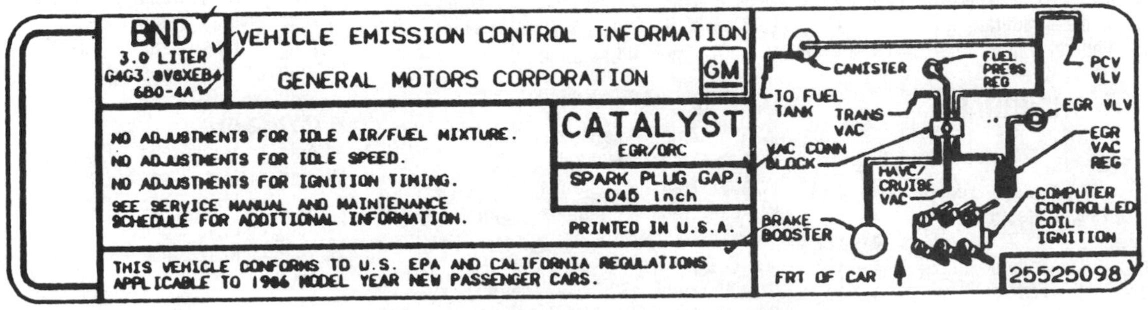

3.0L W/AUTO. TRANS. – CALIFORNIA

VACUUM CIRCUITS

(© G.M. Corp.)

1986 BUICK 3.0L

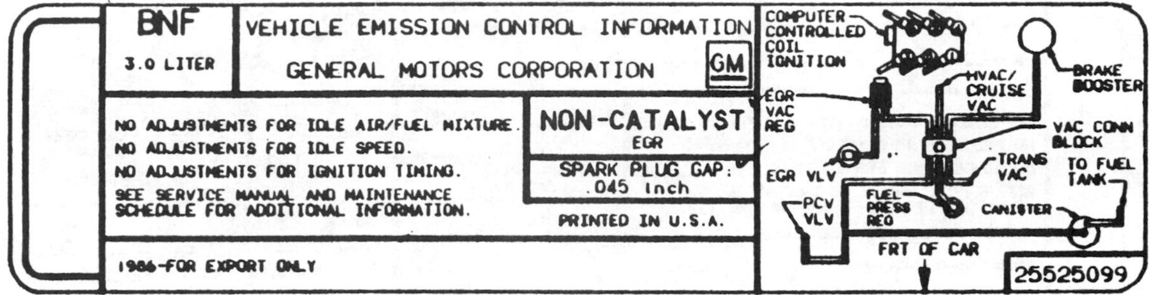

3.0L W/AUTO. TRANS.—EXPORT

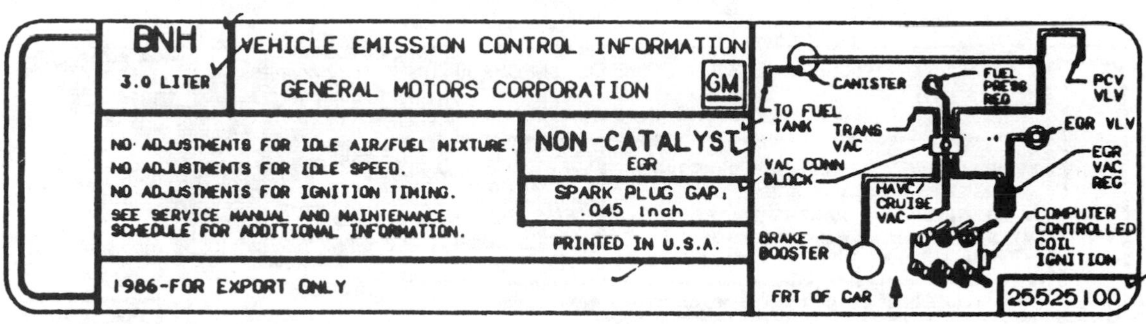

3.0L W/AUTO. TRANS.—EXPORT

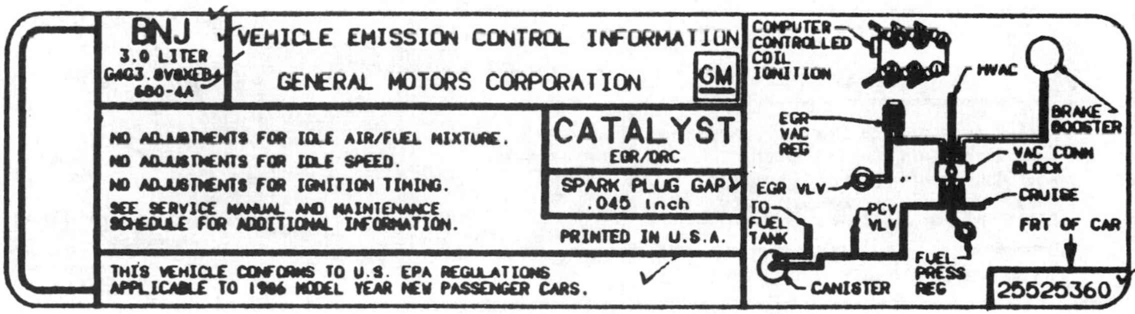

3.0L W/AUTO. TRANS.—FEDERAL

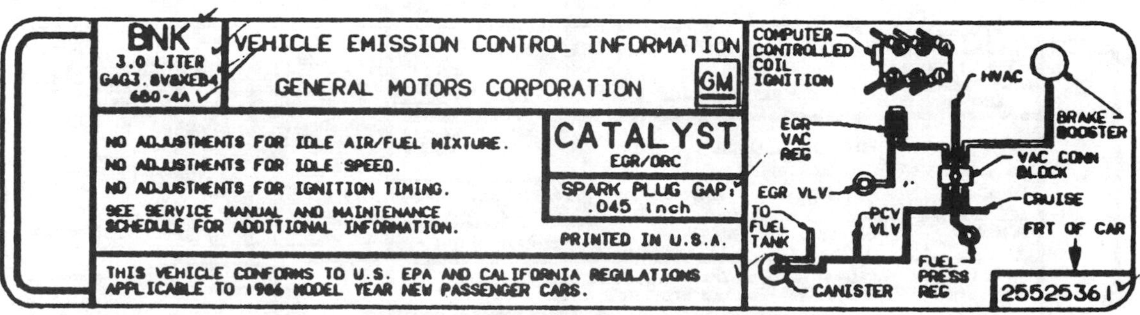

3.0L W/AUTO. TRANS.—CALIFORNIA

VACUUM CIRCUITS

(© G.M. Corp.)

1986 BUICK 3.0L

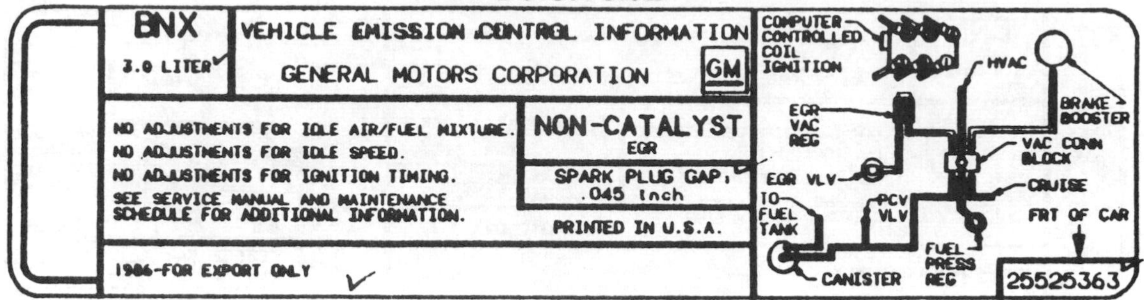

3.0L W/AUTO. TRANS.—EXPORT

1986 BUICK 3.8L

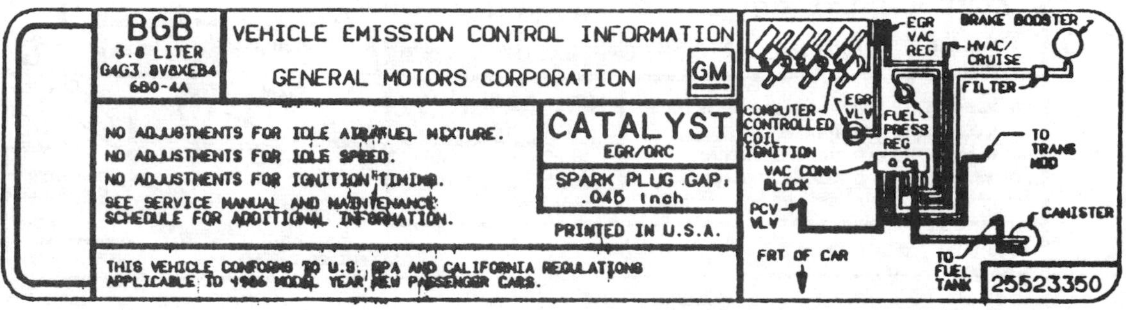

3.8L W/AUTO. TRANS.—FEDERAL

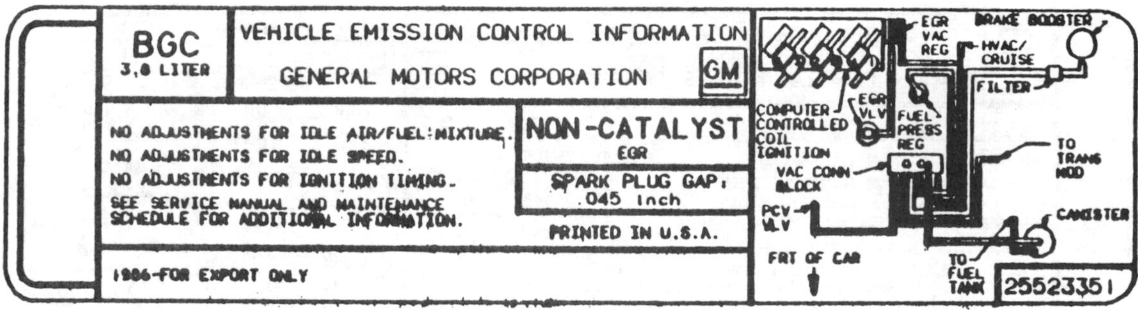

3.8L W/AUTO. TRANS.—EXPORT

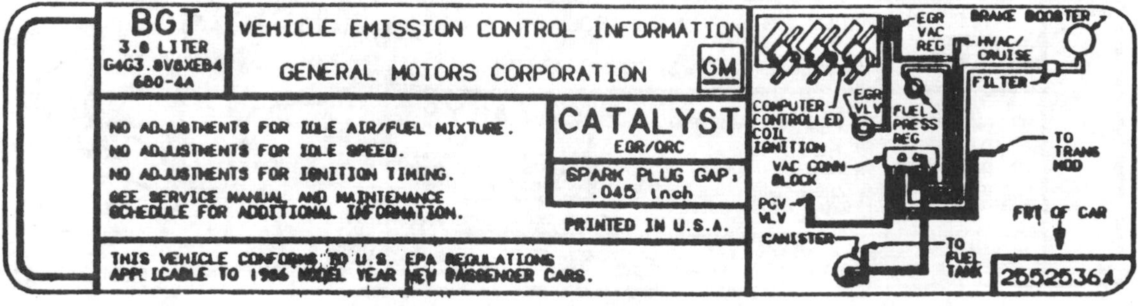

3.8L W/AUTO. TRANS.—FEDERAL

VACUUM CIRCUITS

(© G.M. Corp.)

1986 BUICK 3.8L

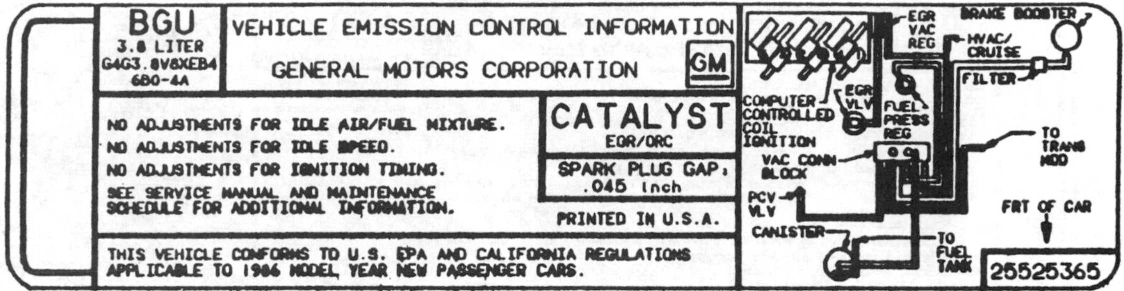
BGU
3.8 LITER
G4G3.8V8XEB4
6B0-4A

VEHICLE EMISSION CONTROL INFORMATION
GENERAL MOTORS CORPORATION GM

NO ADJUSTMENTS FOR IDLE AIR/FUEL MIXTURE.
NO ADJUSTMENTS FOR IDLE SPEED.
NO ADJUSTMENTS FOR IGNITION TIMING.
SEE SERVICE MANUAL AND MAINTENANCE SCHEDULE FOR ADDITIONAL INFORMATION.

CATALYST
EGR/ORC

SPARK PLUG GAP: .045 Inch

PRINTED IN U.S.A.

THIS VEHICLE CONFORMS TO U.S. EPA AND CALIFORNIA REGULATIONS APPLICABLE TO 1986 MODEL YEAR NEW PASSENGER CARS.

3.8L W/AUTO. TRANS.—CALIFORNIA

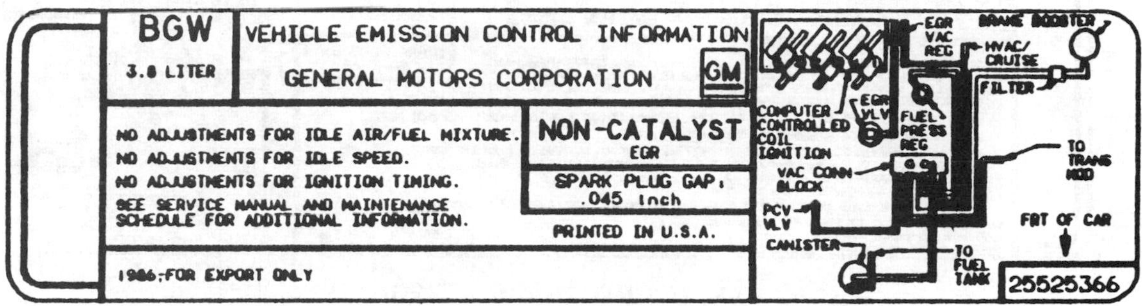
BGW
3.8 LITER

VEHICLE EMISSION CONTROL INFORMATION
GENERAL MOTORS CORPORATION GM

NO ADJUSTMENTS FOR IDLE AIR/FUEL MIXTURE.
NO ADJUSTMENTS FOR IDLE SPEED.
NO ADJUSTMENTS FOR IGNITION TIMING.
SEE SERVICE MANUAL AND MAINTENANCE SCHEDULE FOR ADDITIONAL INFORMATION.

NON-CATALYST
EGR

SPARK PLUG GAP: .045 Inch

PRINTED IN U.S.A.

1986-FOR EXPORT ONLY

3.8L W/AUTO. TRANS.—EXPORT

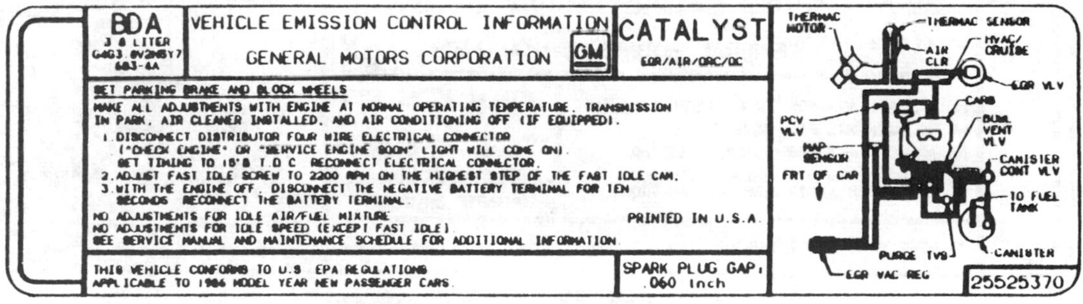
BDA
3.8 LITER
G4G3.8V2NSY7
6B3-4A

VEHICLE EMISSION CONTROL INFORMATION
GENERAL MOTORS CORPORATION GM

CATALYST
EGR/AIR/ORC/OC

SET PARKING BRAKE AND BLOCK WHEELS

MAKE ALL ADJUSTMENTS WITH ENGINE AT NORMAL OPERATING TEMPERATURE. TRANSMISSION IN PARK, AIR CLEANER INSTALLED, AND AIR CONDITIONING OFF (IF EQUIPPED).

1. DISCONNECT DISTRIBUTOR FOUR WIRE ELECTRICAL CONNECTOR ("CHECK ENGINE" OR "SERVICE ENGINE SOON" LIGHT WILL COME ON). SET TIMING TO 15°B T.D.C. RECONNECT ELECTRICAL CONNECTOR.
2. ADJUST FAST IDLE SCREW TO 2200 RPM ON THE HIGHEST STEP OF THE FAST IDLE CAM.
3. WITH THE ENGINE OFF, DISCONNECT THE NEGATIVE BATTERY TERMINAL FOR TEN SECONDS. RECONNECT THE BATTERY TERMINAL.

NO ADJUSTMENTS FOR IDLE AIR/FUEL MIXTURE
NO ADJUSTMENTS FOR IDLE SPEED (EXCEPT FAST IDLE)
SEE SERVICE MANUAL AND MAINTENANCE SCHEDULE FOR ADDITIONAL INFORMATION.

PRINTED IN U.S.A.

THIS VEHICLE CONFORMS TO U.S. EPA REGULATIONS APPLICABLE TO 1986 MODEL YEAR NEW PASSENGER CARS.

SPARK PLUG GAP: .060 Inch

3.8L W/AUTO. TRANS.—FEDERAL

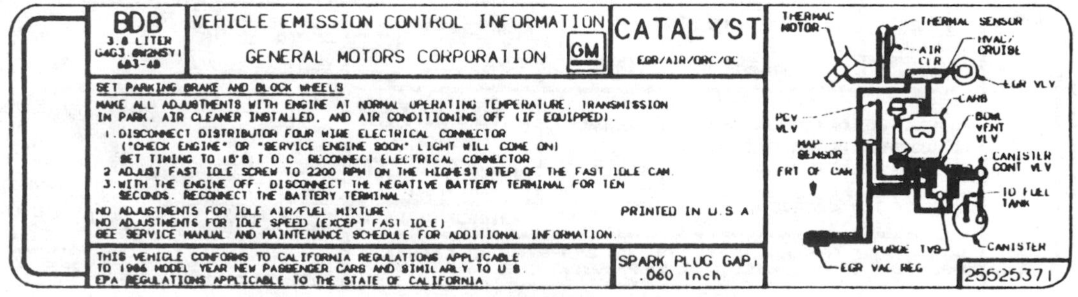
BDB
3.8 LITER
G4G3.8W2NSY1
6B3-4B

VEHICLE EMISSION CONTROL INFORMATION
GENERAL MOTORS CORPORATION GM

CATALYST
EGR/AIR/ORC/OC

SET PARKING BRAKE AND BLOCK WHEELS

MAKE ALL ADJUSTMENTS WITH ENGINE AT NORMAL OPERATING TEMPERATURE. TRANSMISSION IN PARK, AIR CLEANER INSTALLED, AND AIR CONDITIONING OFF (IF EQUIPPED).

1. DISCONNECT DISTRIBUTOR FOUR WIRE ELECTRICAL CONNECTOR ("CHECK ENGINE" OR "SERVICE ENGINE SOON" LIGHT WILL COME ON) SET TIMING TO 15°B T.D.C. RECONNECT ELECTRICAL CONNECTOR
2. ADJUST FAST IDLE SCREW TO 2200 RPM ON THE HIGHEST STEP OF THE FAST IDLE CAM.
3. WITH THE ENGINE OFF, DISCONNECT THE NEGATIVE BATTERY TERMINAL FOR TEN SECONDS. RECONNECT THE BATTERY TERMINAL

NO ADJUSTMENTS FOR IDLE AIR/FUEL MIXTURE
NO ADJUSTMENTS FOR IDLE SPEED (EXCEPT FAST IDLE)
SEE SERVICE MANUAL AND MAINTENANCE SCHEDULE FOR ADDITIONAL INFORMATION.

PRINTED IN U.S.A.

THIS VEHICLE CONFORMS TO CALIFORNIA REGULATIONS APPLICABLE TO 1986 MODEL YEAR NEW PASSENGER CARS AND SIMILARLY TO U.S. EPA REGULATIONS APPLICABLE TO THE STATE OF CALIFORNIA

SPARK PLUG GAP: .060 Inch

3.8L W/AUTO. TRANS.—CALIFORNIA

VACUUM CIRCUITS

(© G.M. Corp.)

1986 BUICK 3.8L

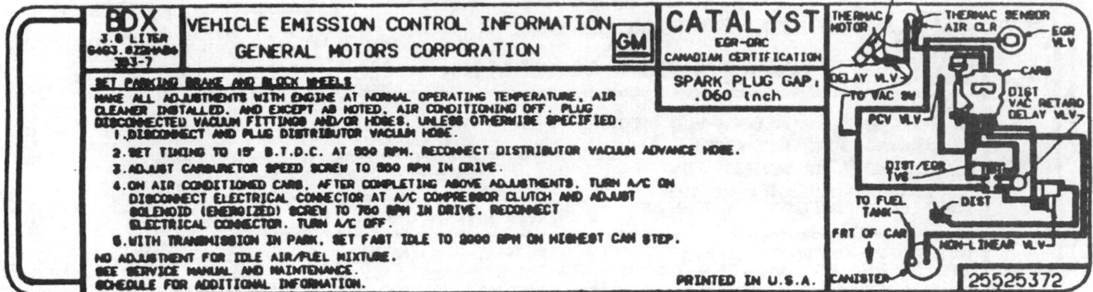

3.8L W/AUTO. TRANS.—CANADA

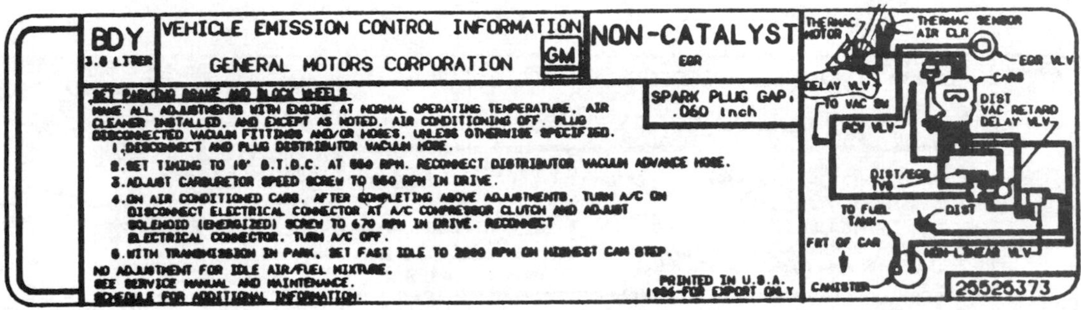

3.8L W/AUTO. TRANS.—EXPORT

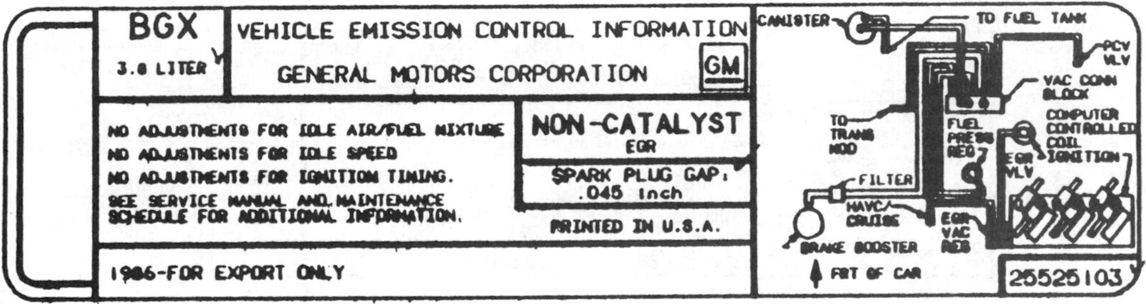

3.8L W/AUTO. TRANS.—EXPORT

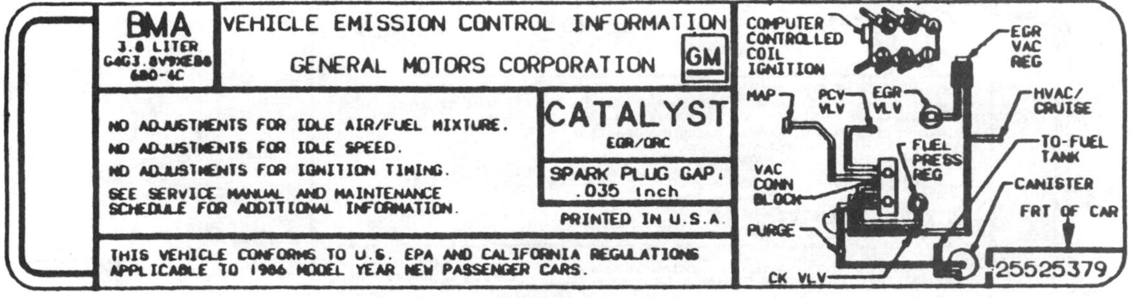

3.8L W/AUTO. TRANS.—FEDERAL

VACUUM CIRCUITS

(© G.M. Corp.)

1986 BUICK 3.8L

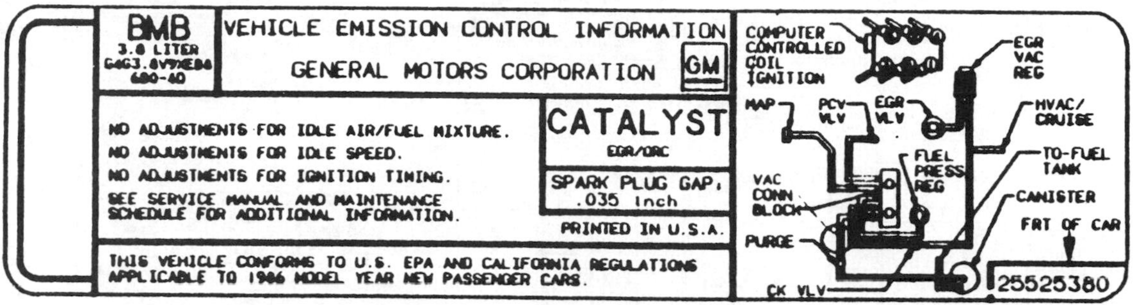

3.8L W/AUTO. TRANS.—CALIFORNIA

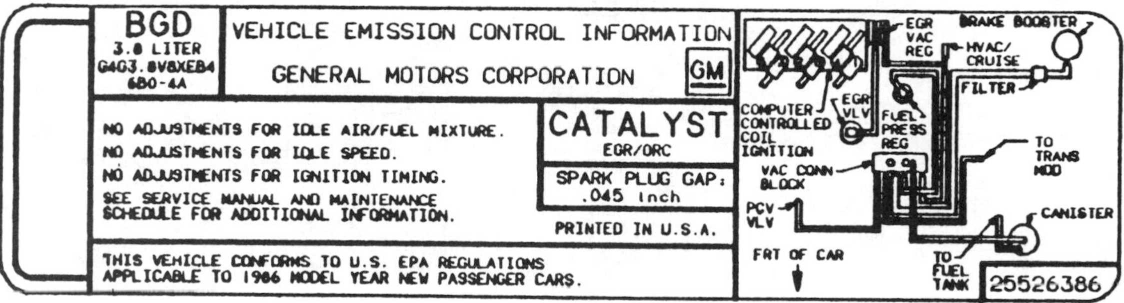

3.8L W/AUTO. TRANS.—FEDERAL

VACUUM CIRCUITS

(© G.M. Corp.)

1986 CADILLAC 4.1L

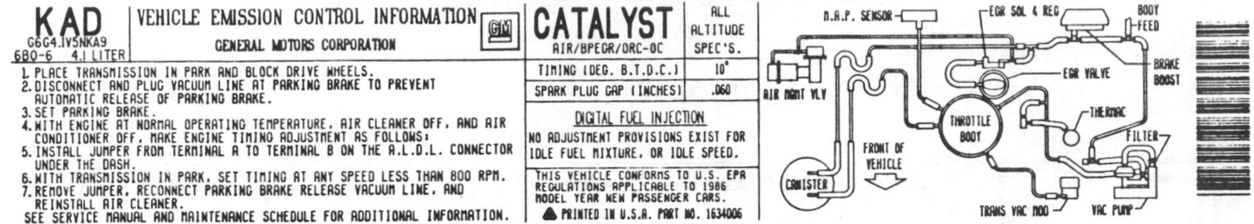
KAD
G6G4.1V5NKA9
6B0-6 4.1 LITER

VEHICLE EMISSION CONTROL INFORMATION
GENERAL MOTORS CORPORATION

1. PLACE TRANSMISSION IN PARK AND BLOCK DRIVE WHEELS.
2. DISCONNECT AND PLUG VACUUM LINE AT PARKING BRAKE TO PREVENT AUTOMATIC RELEASE OF PARKING BRAKE.
3. SET PARKING BRAKE.
4. WITH ENGINE AT NORMAL OPERATING TEMPERATURE, AIR CLEANER OFF, AND AIR CONDITIONER OFF, MAKE ENGINE TIMING ADJUSTMENT AS FOLLOWS:
5. INSTALL JUMPER FROM TERMINAL A TO TERMINAL B ON THE A.L.D.L. CONNECTOR UNDER THE DASH.
6. WITH TRANSMISSION IN PARK, SET TIMING AT ANY SPEED LESS THAN 800 RPM.
7. REMOVE JUMPER, RECONNECT PARKING BRAKE RELEASE VACUUM LINE, AND REINSTALL AIR CLEANER.

SEE SERVICE MANUAL AND MAINTENANCE SCHEDULE FOR ADDITIONAL INFORMATION.

CATALYST AIR/BPEGR/ORC-OC	ALL ALTITUDE SPEC'S.
TIMING (DEG. B.T.D.C.)	10°
SPARK PLUG GAP (INCHES)	.060

DIGITAL FUEL INJECTION
NO ADJUSTMENT PROVISIONS EXIST FOR IDLE FUEL MIXTURE, OR IDLE SPEED.

THIS VEHICLE CONFORMS TO U.S. EPA REGULATIONS APPLICABLE TO 1986 MODEL YEAR NEW PASSENGER CARS.

▲ PRINTED IN U.S.A. PART NO. 1634006

4.1L—DeVILLE, LIMOS—FEDERAL

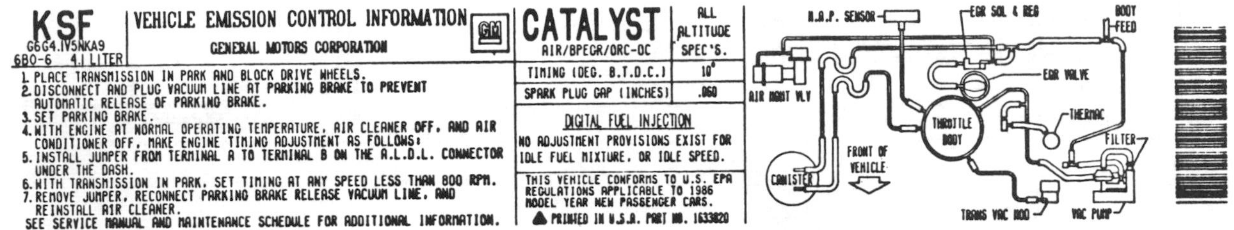
KSF
G6G4.1V5NKA9
6B0-6 4.1 LITER

VEHICLE EMISSION CONTROL INFORMATION
GENERAL MOTORS CORPORATION

1. PLACE TRANSMISSION IN PARK AND BLOCK DRIVE WHEELS.
2. DISCONNECT AND PLUG VACUUM LINE AT PARKING BRAKE TO PREVENT AUTOMATIC RELEASE OF PARKING BRAKE.
3. SET PARKING BRAKE.
4. WITH ENGINE AT NORMAL OPERATING TEMPERATURE, AIR CLEANER OFF, AND AIR CONDITIONER OFF, MAKE ENGINE TIMING ADJUSTMENT AS FOLLOWS:
5. INSTALL JUMPER FROM TERMINAL A TO TERMINAL B ON THE A.L.D.L. CONNECTOR UNDER THE DASH.
6. WITH TRANSMISSION IN PARK, SET TIMING AT ANY SPEED LESS THAN 800 RPM.
7. REMOVE JUMPER, RECONNECT PARKING BRAKE RELEASE VACUUM LINE, AND REINSTALL AIR CLEANER.

SEE SERVICE MANUAL AND MAINTENANCE SCHEDULE FOR ADDITIONAL INFORMATION.

CATALYST AIR/BPEGR/ORC-OC	ALL ALTITUDE SPEC'S.
TIMING (DEG. B.T.D.C.)	10°
SPARK PLUG GAP (INCHES)	.060

DIGITAL FUEL INJECTION
NO ADJUSTMENT PROVISIONS EXIST FOR IDLE FUEL MIXTURE, OR IDLE SPEED.

THIS VEHICLE CONFORMS TO U.S. EPA REGULATIONS APPLICABLE TO 1986 MODEL YEAR NEW PASSENGER CARS.

▲ PRINTED IN U.S.A. PART NO. 1633820

4.1L—DeVILLE—FEDERAL

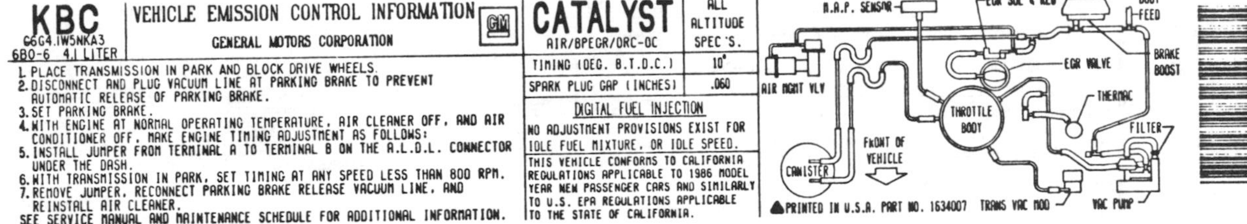
KBC
G6G4.1W5NKA3
6B0-6 4.1 LITER

VEHICLE EMISSION CONTROL INFORMATION
GENERAL MOTORS CORPORATION

1. PLACE TRANSMISSION IN PARK AND BLOCK DRIVE WHEELS.
2. DISCONNECT AND PLUG VACUUM LINE AT PARKING BRAKE TO PREVENT AUTOMATIC RELEASE OF PARKING BRAKE.
3. SET PARKING BRAKE.
4. WITH ENGINE AT NORMAL OPERATING TEMPERATURE, AIR CLEANER OFF, AND AIR CONDITIONER OFF, MAKE ENGINE TIMING ADJUSTMENT AS FOLLOWS:
5. INSTALL JUMPER FROM TERMINAL A TO TERMINAL B ON THE A.L.D.L. CONNECTOR UNDER THE DASH.
6. WITH TRANSMISSION IN PARK, SET TIMING AT ANY SPEED LESS THAN 800 RPM.
7. REMOVE JUMPER, RECONNECT PARKING BRAKE RELEASE VACUUM LINE, AND REINSTALL AIR CLEANER.

SEE SERVICE MANUAL AND MAINTENANCE SCHEDULE FOR ADDITIONAL INFORMATION.

CATALYST AIR/BPEGR/ORC-OC	ALL ALTITUDE SPEC'S.
TIMING (DEG. B.T.D.C.)	10°
SPARK PLUG GAP (INCHES)	.060

DIGITAL FUEL INJECTION
NO ADJUSTMENT PROVISIONS EXIST FOR IDLE FUEL MIXTURE, OR IDLE SPEED.

THIS VEHICLE CONFORMS TO CALIFORNIA REGULATIONS APPLICABLE TO 1986 MODEL YEAR NEW PASSENGER CARS AND SIMILARLY TO U.S. EPA REGULATIONS APPLICABLE TO THE STATE OF CALIFORNIA.

▲ PRINTED IN U.S.A. PART NO. 1634007

4.1L—DeVILLE—CALIFORNIA

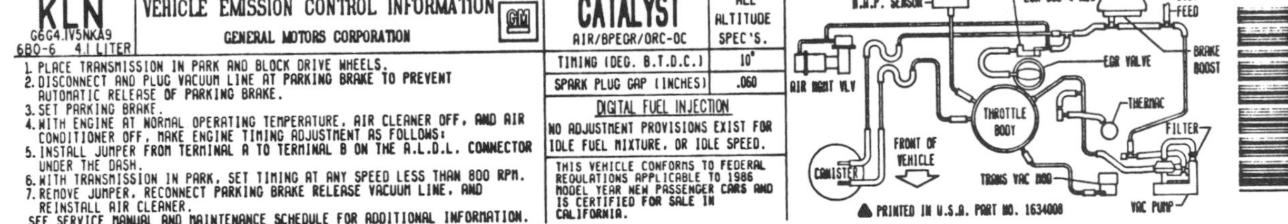
KLN
G6G4.1V5NKA9
6B0-6 4.1 LITER

VEHICLE EMISSION CONTROL INFORMATION
GENERAL MOTORS CORPORATION

1. PLACE TRANSMISSION IN PARK AND BLOCK DRIVE WHEELS.
2. DISCONNECT AND PLUG VACUUM LINE AT PARKING BRAKE TO PREVENT AUTOMATIC RELEASE OF PARKING BRAKE.
3. SET PARKING BRAKE.
4. WITH ENGINE AT NORMAL OPERATING TEMPERATURE, AIR CLEANER OFF, AND AIR CONDITIONER OFF, MAKE ENGINE TIMING ADJUSTMENT AS FOLLOWS:
5. INSTALL JUMPER FROM TERMINAL A TO TERMINAL B ON THE A.L.D.L. CONNECTOR UNDER THE DASH.
6. WITH TRANSMISSION IN PARK, SET TIMING AT ANY SPEED LESS THAN 800 RPM.
7. REMOVE JUMPER, RECONNECT PARKING BRAKE RELEASE VACUUM LINE, AND REINSTALL AIR CLEANER.

SEE SERVICE MANUAL AND MAINTENANCE SCHEDULE FOR ADDITIONAL INFORMATION.

CATALYST AIR/BPEGR/ORC-OC	ALL ALTITUDE SPEC'S.
TIMING (DEG. B.T.D.C.)	10°
SPARK PLUG GAP (INCHES)	.060

DIGITAL FUEL INJECTION
NO ADJUSTMENT PROVISIONS EXIST FOR IDLE FUEL MIXTURE, OR IDLE SPEED.

THIS VEHICLE CONFORMS TO FEDERAL REGULATIONS APPLICABLE TO 1986 MODEL YEAR NEW PASSENGER CARS AND IS CERTIFIED FOR SALE IN CALIFORNIA.

▲ PRINTED IN U.S.A. PART NO. 1634008

4.1L—LIMO—FEDERAL

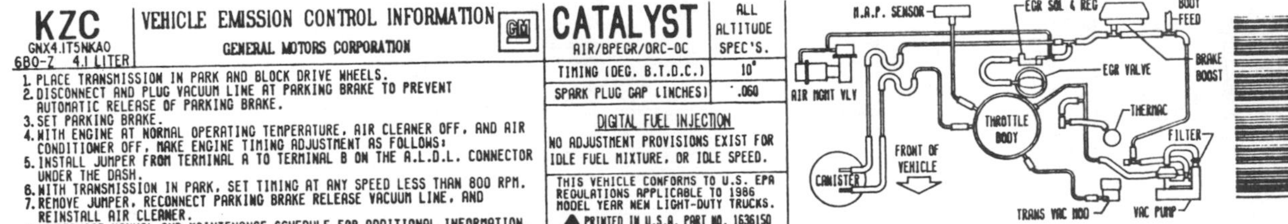
KZC
GNX4.1T5NKA0
6B0-Z 4.1 LITER

VEHICLE EMISSION CONTROL INFORMATION
GENERAL MOTORS CORPORATION

1. PLACE TRANSMISSION IN PARK AND BLOCK DRIVE WHEELS.
2. DISCONNECT AND PLUG VACUUM LINE AT PARKING BRAKE TO PREVENT AUTOMATIC RELEASE OF PARKING BRAKE.
3. SET PARKING BRAKE.
4. WITH ENGINE AT NORMAL OPERATING TEMPERATURE, AIR CLEANER OFF, AND AIR CONDITIONER OFF, MAKE ENGINE TIMING ADJUSTMENT AS FOLLOWS:
5. INSTALL JUMPER FROM TERMINAL A TO TERMINAL B ON THE A.L.D.L. CONNECTOR UNDER THE DASH.
6. WITH TRANSMISSION IN PARK, SET TIMING AT ANY SPEED LESS THAN 800 RPM.
7. REMOVE JUMPER, RECONNECT PARKING BRAKE RELEASE VACUUM LINE, AND REINSTALL AIR CLEANER.

SEE SERVICE MANUAL AND MAINTENANCE SCHEDULE FOR ADDITIONAL INFORMATION.

CATALYST AIR/BPEGR/ORC-OC	ALL ALTITUDE SPEC'S.
TIMING (DEG. B.T.D.C.)	10°
SPARK PLUG GAP (INCHES)	.060

DIGITAL FUEL INJECTION
NO ADJUSTMENT PROVISIONS EXIST FOR IDLE FUEL MIXTURE, OR IDLE SPEED.

THIS VEHICLE CONFORMS TO U.S. EPA REGULATIONS APPLICABLE TO 1986 MODEL YEAR NEW LIGHT-DUTY TRUCKS.

▲ PRINTED IN U.S.A. PART NO. 1636150

4.1L—COMMERCIAL CHASSIS—FEDERAL

VACUUM CIRCUITS

(© G.M. Corp.)

1986 CADILLAC 4.1L

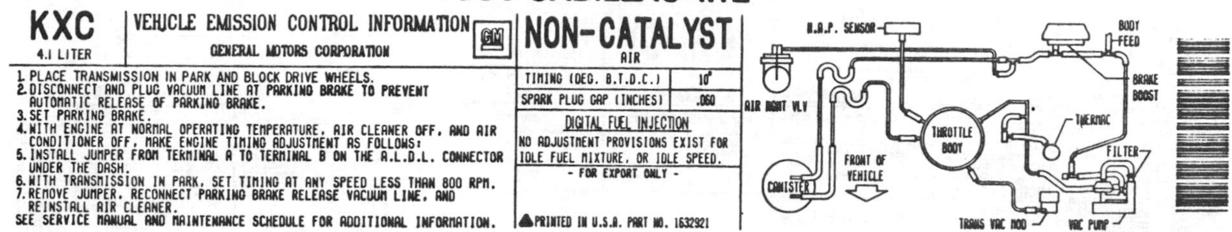

KXC
4.1 LITER
VEHICLE EMISSION CONTROL INFORMATION
GENERAL MOTORS CORPORATION

1. PLACE TRANSMISSION IN PARK AND BLOCK DRIVE WHEELS.
2. DISCONNECT AND PLUG VACUUM LINE AT PARKING BRAKE TO PREVENT AUTOMATIC RELEASE OF PARKING BRAKE.
3. SET PARKING BRAKE.
4. WITH ENGINE AT NORMAL OPERATING TEMPERATURE, AIR CLEANER OFF, AND AIR CONDITIONER OFF, MAKE ENGINE TIMING ADJUSTMENT AS FOLLOWS:
5. INSTALL JUMPER FROM TERMINAL A TO TERMINAL B ON THE A.L.D.L. CONNECTOR UNDER THE DASH.
6. WITH TRANSMISSION IN PARK, SET TIMING AT ANY SPEED LESS THAN 800 RPM.
7. REMOVE JUMPER, RECONNECT PARKING BRAKE RELEASE VACUUM LINE, AND REINSTALL AIR CLEANER.

SEE SERVICE MANUAL AND MAINTENANCE SCHEDULE FOR ADDITIONAL INFORMATION.

NON-CATALYST
AIR

TIMING (DEG. B.T.D.C.)	10°
SPARK PLUG GAP (INCHES)	.060

DIGITAL FUEL INJECTION
NO ADJUSTMENT PROVISIONS EXIST FOR IDLE FUEL MIXTURE, OR IDLE SPEED.
- FOR EXPORT ONLY -
PRINTED IN U.S.A. PART NO. 1632921

4.1L—DeVILLE, LIMO—EXPORT

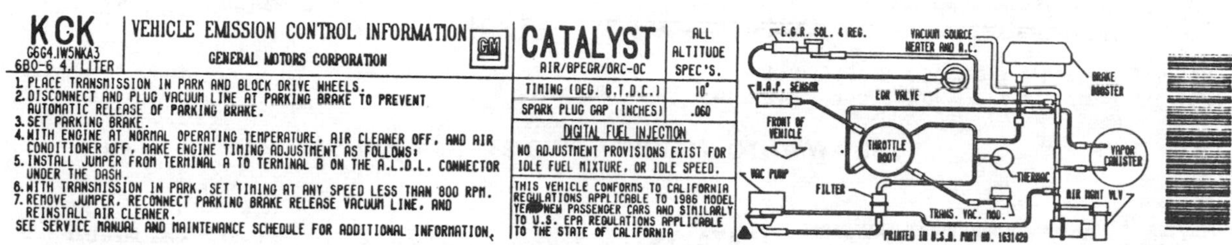

KCK
G6G4.1W5NKA3
6B0-6 4.1 LITER
VEHICLE EMISSION CONTROL INFORMATION
GENERAL MOTORS CORPORATION

1. PLACE TRANSMISSION IN PARK AND BLOCK DRIVE WHEELS.
2. DISCONNECT AND PLUG VACUUM LINE AT PARKING BRAKE TO PREVENT AUTOMATIC RELEASE OF PARKING BRAKE.
3. SET PARKING BRAKE.
4. WITH ENGINE AT NORMAL OPERATING TEMPERATURE, AIR CLEANER OFF, AND AIR CONDITIONER OFF, MAKE ENGINE TIMING ADJUSTMENT AS FOLLOWS:
5. INSTALL JUMPER FROM TERMINAL A TO TERMINAL B ON THE A.L.D.L. CONNECTOR UNDER THE DASH.
6. WITH TRANSMISSION IN PARK, SET TIMING AT ANY SPEED LESS THAN 800 RPM.
7. REMOVE JUMPER, RECONNECT PARKING BRAKE RELEASE VACUUM LINE, AND REINSTALL AIR CLEANER.

SEE SERVICE MANUAL AND MAINTENANCE SCHEDULE FOR ADDITIONAL INFORMATION.

CATALYST
AIR/BPEGR/ORC-OC
ALL ALTITUDE SPEC'S.

TIMING (DEG. B.T.D.C.)	10°
SPARK PLUG GAP (INCHES)	.060

DIGITAL FUEL INJECTION
NO ADJUSTMENT PROVISIONS EXIST FOR IDLE FUEL MIXTURE, OR IDLE SPEED.
THIS VEHICLE CONFORMS TO CALIFORNIA REGULATIONS APPLICABLE TO 1986 MODEL YEAR NEW PASSENGER CARS AND SIMILARLY TO U.S. EPA REGULATIONS APPLICABLE TO THE STATE OF CALIFORNIA
PRINTED IN U.S.A. PART NO. 1631420

4.1L—ELDORADO, SEVILLE—CALIFORNIA

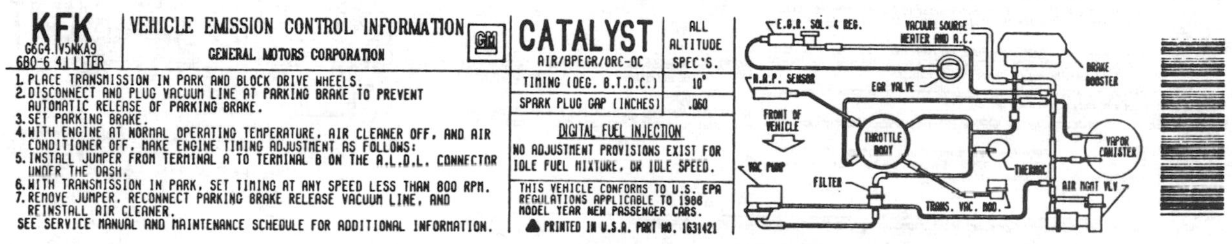

KFK
G6G4.1V5NKA9
6B0-6 4.1 LITER
VEHICLE EMISSION CONTROL INFORMATION
GENERAL MOTORS CORPORATION

1. PLACE TRANSMISSION IN PARK AND BLOCK DRIVE WHEELS.
2. DISCONNECT AND PLUG VACUUM LINE AT PARKING BRAKE TO PREVENT AUTOMATIC RELEASE OF PARKING BRAKE.
3. SET PARKING BRAKE.
4. WITH ENGINE AT NORMAL OPERATING TEMPERATURE, AIR CLEANER OFF, AND AIR CONDITIONER OFF, MAKE ENGINE TIMING ADJUSTMENT AS FOLLOWS:
5. INSTALL JUMPER FROM TERMINAL A TO TERMINAL B ON THE A.L.D.L. CONNECTOR UNDER THE DASH.
6. WITH TRANSMISSION IN PARK, SET TIMING AT ANY SPEED LESS THAN 800 RPM.
7. REMOVE JUMPER, RECONNECT PARKING BRAKE RELEASE VACUUM LINE, AND REINSTALL AIR CLEANER.

SEE SERVICE MANUAL AND MAINTENANCE SCHEDULE FOR ADDITIONAL INFORMATION.

CATALYST
AIR/BPEGR/ORC-OC
ALL ALTITUDE SPEC'S.

TIMING (DEG. B.T.D.C.)	10°
SPARK PLUG GAP (INCHES)	.060

DIGITAL FUEL INJECTION
NO ADJUSTMENT PROVISIONS EXIST FOR IDLE FUEL MIXTURE, OR IDLE SPEED.
THIS VEHICLE CONFORMS TO U.S. EPA REGULATIONS APPLICABLE TO 1986 MODEL YEAR NEW PASSENGER CARS.
PRINTED IN U.S.A. PART NO. 1631421

4.1L—ELDORADO, SEVILLE—FEDERAL

1986 ELDORADO 4.1L

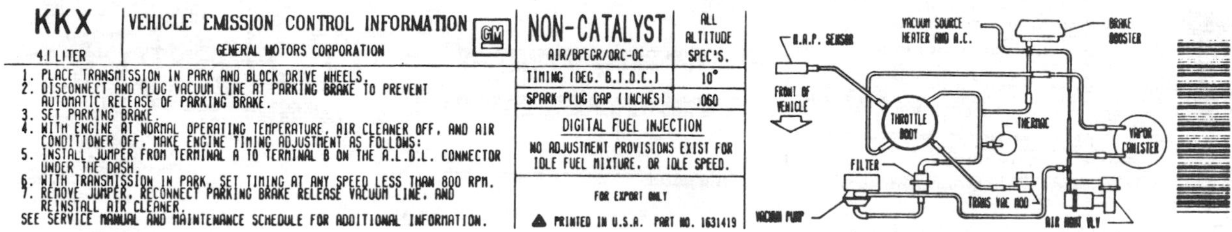

KKX
4.1 LITER
VEHICLE EMISSION CONTROL INFORMATION
GENERAL MOTORS CORPORATION

1. PLACE TRANSMISSION IN PARK AND BLOCK DRIVE WHEELS.
2. DISCONNECT AND PLUG VACUUM LINE AT PARKING BRAKE TO PREVENT AUTOMATIC RELEASE OF PARKING BRAKE.
3. SET PARKING BRAKE.
4. WITH ENGINE AT NORMAL OPERATING TEMPERATURE, AIR CLEANER OFF, AND AIR CONDITIONER OFF, MAKE ENGINE TIMING ADJUSTMENT AS FOLLOWS:
5. INSTALL JUMPER FROM TERMINAL A TO TERMINAL B ON THE A.L.D.L. CONNECTOR UNDER THE DASH.
6. WITH TRANSMISSION IN PARK, SET TIMING AT ANY SPEED LESS THAN 800 RPM.
7. REMOVE JUMPER, RECONNECT PARKING BRAKE RELEASE VACUUM LINE, AND REINSTALL AIR CLEANER.

SEE SERVICE MANUAL AND MAINTENANCE SCHEDULE FOR ADDITIONAL INFORMATION.

NON-CATALYST
AIR/BPEGR/ORC-OC
ALL ALTITUDE SPEC'S.

TIMING (DEG. B.T.D.C.)	10°
SPARK PLUG GAP (INCHES)	.060

DIGITAL FUEL INJECTION
NO ADJUSTMENT PROVISIONS EXIST FOR IDLE FUEL MIXTURE, OR IDLE SPEED.
FOR EXPORT ONLY
PRINTED IN U.S.A. PART NO. 1631419

4.1L—ELDORADO—FEDERAL

VACUUM CIRCUITS

(© G.M. Corp.)

1986 CHEVROLET 1.6L

YAR
1.6 LITER
G1G1.6V2HEA3
685-1A

VEHICLE EMISSION CONTROL INFORMATION
GENERAL MOTORS CORPORATION

SET PARKING BRAKE AND BLOCK DRIVE WHEELS.

MAKE ALL ADJUSTMENTS WITH ENGINE AT NORMAL OPERATING TEMPERATURE, CHOKE FULL OPEN, AIR CLEANER REMOVED, THERMAC VACUUM SOURCE CAPPED, AND AIR CONDITIONING (IF EQUIPPED) OFF, EXCEPT WHERE NOTED. SET IDLE SPEEDS ONLY WHEN EMISSION CONTROL SYSTEM IS IN CLOSED LOOP MODE.

1. DISCONNECT AND PLUG CANISTER PURGE HOSE AND PURGE CONTROL HOSE AT CANISTER. DISCONNECT EGR VACUUM SOURCE AT CARBURETOR AND CAP PORT.
2. DISTRIBUTOR: DISCONNECT AND PLUG VACUUM HOSE AT DISTRIBUTOR. SET IGNITION TIMING TO SPECIFICATION WITH ENGINE RUNNING AT SPEED INDICATED. UNPLUG AND RECONNECT VACUUM HOSE.
3. FAST IDLE SPEED: WITH THROTTLE ON HIGH STEP OF CAM ADJUST FAST IDLE SCREW TO OBTAIN SPECIFIED SPEED.
4. IDLE SPEED SCREW: ADJUST CARBURETOR IDLE SPEED SCREW TO OBTAIN SPECIFIED SPEED.
5. STEPPED SPEED CONTROL: (VEHICLES WITH A/C) DISCONNECT ELECTRICAL LEAD FROM A/C COMPRESSOR AND TURN A/C SWITCH ON. CHECK TO SEE THAT THE SSC ACTUATOR IS FULLY EXTENDED. ADJUST ACTUATOR SCREW TO OBTAIN SPECIFIED RPM. RECONNECT A/C COMPRESSOR ELECTRICAL LEAD AND TURN A/C SWITCH OFF. (VEHICLES WITHOUT A/C) DISCONNECT AND PLUG VACUUM HOSE AT SSC ACTUATOR. CONNECT A VACUUM SOURCE (5 IN. HG. MIN.) TO ACTUATOR AND CHECK TO SEE THAT IT IS FULLY EXTENDED. ADJUST ACTUATOR SCREW TO OBTAIN SPECIFIED RPM.
6. UNPLUG AND RECONNECT HOSES TO CANISTER, AND ON VEHICLES WITHOUT A/C, THE SSC ACTUATOR. UNCAP VACUUM SOURCE FOR EGR AND RECONNECT LINE. UNCAP VACUUM SOURCE FOR THERMAC, RECONNECT HOSE, AND REPLACE AIR CLEANER.

CATALYST
AIR/DPEGR/OBC/OC

	AUTOMATIC TRANSMISSION
TIMING (°BTC @ RPM)	8° @ 700 (DR)
SPARK PLUG GAP (IN.)	0.035
IDLE SPEED SCREW (RPM)	700 (DR)
STEPPED SPEED CONTROL ACTUATOR (SSC)	875 (DR)
FAST IDLE SPEED SCREW (RPM)	2500 (N)

NOTE: IDLE MIXTURE SCREW IS PRESET AND SEALED AT FACTORY. PROVISION FOR ADJUSTMENT DURING TUNE UP IS NOT PROVIDED.

SEE SERVICE MANUAL, MAINTENANCE SCHEDULE AND EMISSION HOSE ROUTING DIAGRAM FOR ADDITIONAL INFORMATION.

THIS VEHICLE CONFORMS TO U.S. EPA REGULATIONS APPLICABLE TO 1986 MODEL YEAR NEW PASSENGER CARS.

PT. NO. 14086596
PRINTED IN U.S.A.

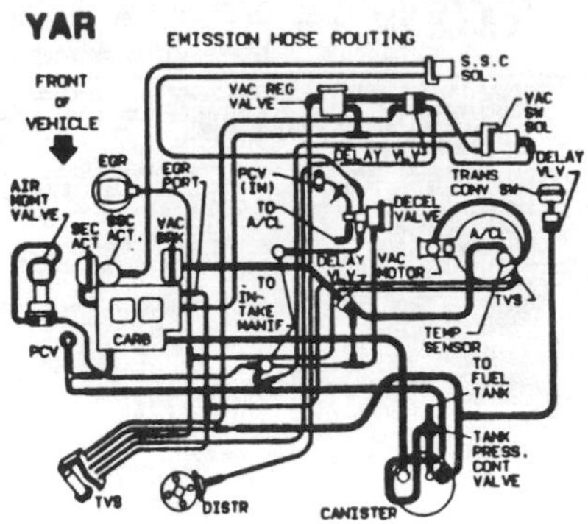

USAGE: 1-2TA00 & L17 & MD2 & C60 & NA5
1-2TA00 & L17 & MD2 & C41 & N41 & NA5

1.6L—W/AUTO. TRANS. & PWR STRG.—FEDERAL

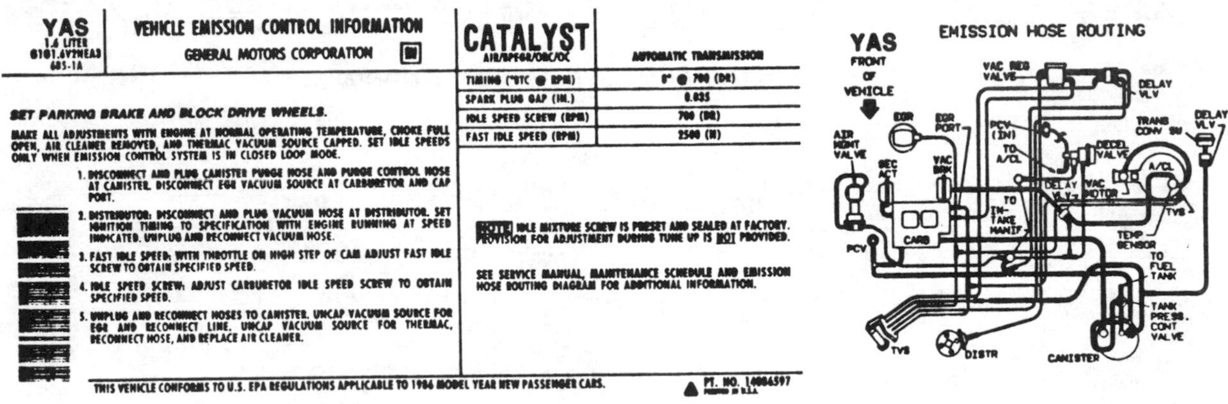

YAS
1.6 LITER
G1G1.6V2HEA3
685-1A

VEHICLE EMISSION CONTROL INFORMATION
GENERAL MOTORS CORPORATION

SET PARKING BRAKE AND BLOCK DRIVE WHEELS.

MAKE ALL ADJUSTMENTS WITH ENGINE AT NORMAL OPERATING TEMPERATURE, CHOKE FULL OPEN, AIR CLEANER REMOVED, AND THERMAC VACUUM SOURCE CAPPED. SET IDLE SPEEDS ONLY WHEN EMISSION CONTROL SYSTEM IS IN CLOSED LOOP MODE.

1. DISCONNECT AND PLUG CANISTER PURGE HOSE AND PURGE CONTROL HOSE AT CANISTER. DISCONNECT EGR VACUUM SOURCE AT CARBURETOR AND CAP PORT.
2. DISTRIBUTOR: DISCONNECT AND PLUG VACUUM HOSE AT DISTRIBUTOR. SET IGNITION TIMING TO SPECIFICATION WITH ENGINE RUNNING AT SPEED INDICATED. UNPLUG AND RECONNECT VACUUM HOSE.
3. FAST IDLE SPEED: WITH THROTTLE ON HIGH STEP OF CAM ADJUST FAST IDLE SCREW TO OBTAIN SPECIFIED SPEED.
4. IDLE SPEED SCREW: ADJUST CARBURETOR IDLE SPEED SCREW TO OBTAIN SPECIFIED SPEED.
5. UNPLUG AND RECONNECT HOSES TO CANISTER. UNCAP VACUUM SOURCE FOR EGR AND RECONNECT LINE. UNCAP VACUUM SOURCE FOR THERMAC, RECONNECT HOSE, AND REPLACE AIR CLEANER.

CATALYST
AIR/DPEGR/OBC/OC

	AUTOMATIC TRANSMISSION
TIMING (°BTC @ RPM)	8° @ 700 (DR)
SPARK PLUG GAP (IN.)	0.035
IDLE SPEED SCREW (RPM)	700 (DR)
FAST IDLE SPEED (RPM)	2500 (N)

NOTE: IDLE MIXTURE SCREW IS PRESET AND SEALED AT FACTORY. PROVISION FOR ADJUSTMENT DURING TUNE UP IS NOT PROVIDED.

SEE SERVICE MANUAL, MAINTENANCE SCHEDULE AND EMISSION HOSE ROUTING DIAGRAM FOR ADDITIONAL INFORMATION.

THIS VEHICLE CONFORMS TO U.S. EPA REGULATIONS APPLICABLE TO 1986 MODEL YEAR NEW PASSENGER CARS.

PT. NO. 14086597
PRINTED IN U.S.A.

USAGE: 1-2TA00 & L17 & MD2 & C41 & N51 & NA5

1.6L—W/AUTO. TRANS. , EXC PWR. STRG.—FEDERAL

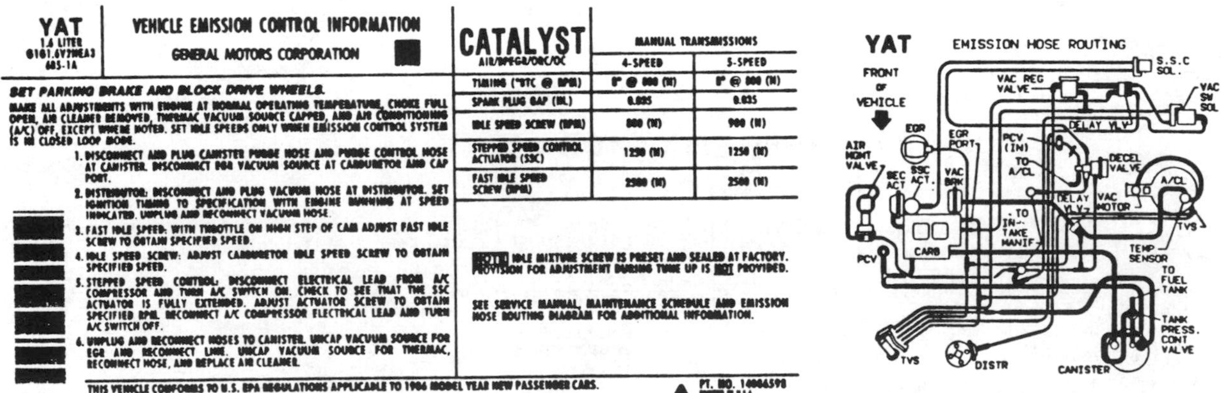

YAT
1.6 LITER
G1G1.6V2HEA3
685-1A

VEHICLE EMISSION CONTROL INFORMATION
GENERAL MOTORS CORPORATION

SET PARKING BRAKE AND BLOCK DRIVE WHEELS.

MAKE ALL ADJUSTMENTS WITH ENGINE AT NORMAL OPERATING TEMPERATURE, CHOKE FULL OPEN, AIR CLEANER REMOVED, THERMAC VACUUM SOURCE CAPPED, AND AIR CONDITIONING (A/C) OFF, EXCEPT WHERE NOTED. SET IDLE SPEEDS ONLY WHEN EMISSION CONTROL SYSTEM IS IN CLOSED LOOP MODE.

1. DISCONNECT AND PLUG CANISTER PURGE HOSE AND PURGE CONTROL HOSE AT CANISTER. DISCONNECT EGR VACUUM SOURCE AT CARBURETOR AND CAP PORT.
2. DISTRIBUTOR: DISCONNECT AND PLUG VACUUM HOSE AT DISTRIBUTOR. SET IGNITION TIMING TO SPECIFICATION WITH ENGINE RUNNING AT SPEED INDICATED. UNPLUG AND RECONNECT VACUUM HOSE.
3. FAST IDLE SPEED: WITH THROTTLE ON HIGH STEP OF CAM ADJUST FAST IDLE SCREW TO OBTAIN SPECIFIED SPEED.
4. IDLE SPEED SCREW: ADJUST CARBURETOR IDLE SPEED SCREW TO OBTAIN SPECIFIED SPEED.
5. STEPPED SPEED CONTROL: DISCONNECT ELECTRICAL LEAD FROM A/C COMPRESSOR AND TURN A/C SWITCH ON. CHECK TO SEE THAT THE SSC ACTUATOR IS FULLY EXTENDED. ADJUST ACTUATOR SCREW TO OBTAIN SPECIFIED RPM. RECONNECT A/C COMPRESSOR ELECTRICAL LEAD AND TURN A/C SWITCH OFF.
6. UNPLUG AND RECONNECT HOSES TO CANISTER. UNCAP VACUUM SOURCE FOR EGR AND RECONNECT LINE. UNCAP VACUUM SOURCE FOR THERMAC, RECONNECT HOSE, AND REPLACE AIR CLEANER.

CATALYST
AIR/DPEGR/OBC/OC

	MANUAL TRANSMISSIONS	
	4-SPEED	5-SPEED
TIMING (°BTC @ RPM)	8° @ 800 (N)	8° @ 800 (N)
SPARK PLUG GAP (IN.)	0.035	0.035
IDLE SPEED SCREW (RPM)	800 (N)	900 (N)
STEPPED SPEED CONTROL ACTUATOR (SSC)	1250 (N)	1250 (N)
FAST IDLE SPEED SCREW (RPM)	2500 (N)	2500 (N)

NOTE: IDLE MIXTURE SCREW IS PRESET AND SEALED AT FACTORY. PROVISION FOR ADJUSTMENT DURING TUNE UP IS NOT PROVIDED.

SEE SERVICE MANUAL, MAINTENANCE SCHEDULE AND EMISSION HOSE ROUTING DIAGRAM FOR ADDITIONAL INFORMATION.

THIS VEHICLE CONFORMS TO U.S. EPA REGULATIONS APPLICABLE TO 1986 MODEL YEAR NEW PASSENGER CARS.

PT. NO. 14086598
PRINTED IN U.S.A.

USAGE: 1-2TA00 & L17 & MY1/MB4 & C60 & NA5

1.6L—W/AUTO. TRANS. , EXC. AIR. COND.—FEDERAL

VACUUM CIRCUITS

(© G.M. Corp.)

1986 CHEVROLET 1.6L

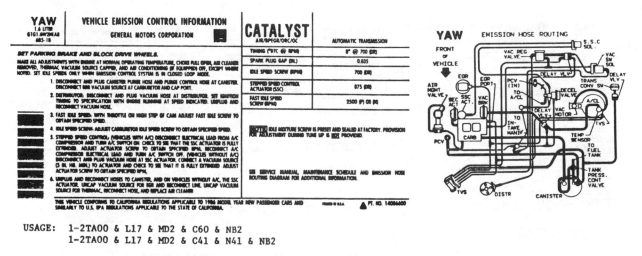

YAU
1.6 LITER
G1G1.6V2NEA3
6B5-1A

VEHICLE EMISSION CONTROL INFORMATION
GENERAL MOTORS CORPORATION

CATALYST
AIR/DPEGR/ORC/OC

SET PARKING BRAKE AND BLOCK DRIVE WHEELS.

MAKE ALL ADJUSTMENTS WITH ENGINE AT NORMAL OPERATING TEMPERATURE, CHOKE FULL OPEN, AIR CLEANER REMOVED AND THERMAC VACUUM SOURCE CAPPED. SET IDLE SPEEDS ONLY WHEN EMISSION CONTROL SYSTEM IS IN CLOSED LOOP MODE.

1. DISCONNECT AND PLUG CANISTER PURGE HOSE AND PURGE CONTROL HOSE AT CANISTER. DISCONNECT EGR VACUUM SOURCE AT CARBURETOR AND CAP PORT.
2. DISTRIBUTOR: DISCONNECT AND PLUG VACUUM HOSE AT DISTRIBUTOR. SET IGNITION TIMING TO SPECIFICATION WITH ENGINE RUNNING AT SPEED INDICATED. UNPLUG AND RECONNECT VACUUM HOSE.
3. FAST IDLE SPEED: WITH THROTTLE ON HIGH STEP OF CAM ADJUST FAST IDLE SCREW TO OBTAIN SPECIFIED SPEED.
4. IDLE SPEED SCREW: ADJUST CARBURETOR IDLE SPEED SCREW TO OBTAIN SPECIFIED SPEED.
5. STEPPED SPEED CONTROL: DISCONNECT AND PLUG VACUUM HOSE AT SSC ACTUATOR. CONNECT A VACUUM SOURCE (5 IN. HG. MIN.) TO ACTUATOR AND CHECK TO SEE THAT IT IS FULLY EXTENDED. ADJUST ACTUATOR SCREW TO OBTAIN SPECIFIED RPM.
6. UNPLUG AND RECONNECT HOSES TO CANISTER AND SSC ACTUATOR. UNCAP VACUUM SOURCE FOR EGR AND RECONNECT LINE. UNCAP VACUUM SOURCE FOR THERMAC, RECONNECT HOSE, AND REPLACE AIR CLEANER.

	MANUAL TRANSMISSIONS	
	4-SPEED	5-SPEED
TIMING (°BTC @ RPM)	8° @ 800 (N)	8° @ 800 (N)
SPARK PLUG GAP (IN.)	0.035	0.035
IDLE SPEED SCREW (RPM)	800 (N)	900 (N)
STEPPED SPEED CONTROL ACTUATOR (SSC)	1250 (N)	1250 (N)
FAST IDLE SPEED SCREW (RPM)	2500 (N)	2500 (N)

NOTE: IDLE MIXTURE SCREW IS PRESET AND SEALED AT FACTORY. PROVISION FOR ADJUSTMENT DURING TUNE UP IS NOT PROVIDED.

SEE SERVICE MANUAL, MAINTENANCE SCHEDULE AND EMISSION HOSE ROUTING DIAGRAM FOR ADDITIONAL INFORMATION.

THIS VEHICLE CONFORMS TO U.S. EPA REGULATIONS APPLICABLE TO 1986 MODEL YEAR NEW PASSENGER CARS.

PT. NO. 14086599
PRINTED IN U.S.A.

YAU EMISSION HOSE ROUTING
FRONT OF VEHICLE
S.S.C SOL.
VAC REG VALVE
DELAY VLV
EGR
EGR PORT
PCV (IN)
TO A/CL
DECEL VALVE
AIR MGMT VALVE
SEC ACT
SSC ACT
VAC BRK
A/CL
DELAY VLV
VAC MOTOR
TVS
TO IN-TAKE MANIF
CARB
PCV
TEMP SENSOR
TO FUEL TANK
TANK PRESS. CONT VALVE
TVS
DISTR
CANISTER

USAGE: 1-2TA00 & L17 & MY1/MB4 & C41 & NA5

1.6l—W/AUTO. TRANS. , EXC. AIR COND.—FEDERAL

YAW
1.6 LITER
G1G1.6W2NEA8
6B5-1B

VEHICLE EMISSION CONTROL INFORMATION
GENERAL MOTORS CORPORATION

CATALYST
AIR/BPEGR/ORC/OC

SET PARKING BRAKE AND BLOCK DRIVE WHEELS.

MAKE ALL ADJUSTMENTS WITH ENGINE AT NORMAL OPERATING TEMPERATURE, CHOKE FULL OPEN, AIR CLEANER REMOVED, THERMAC VACUUM SOURCE CAPPED, AND AIR CONDITIONING (IF EQUIPPED) OFF, EXCEPT WHERE NOTED. SET IDLE SPEEDS ONLY WHEN EMISSION CONTROL SYSTEM IS IN CLOSED LOOP MODE.

1. DISCONNECT AND PLUG CANISTER PURGE HOSE AND PURGE CONTROL HOSE AT CANISTER. DISCONNECT EGR VACUUM SOURCE AT CARBURETOR AND CAP PORT.
2. DISTRIBUTOR: DISCONNECT AND PLUG VACUUM HOSE AT DISTRIBUTOR. SET IGNITION TIMING TO SPECIFICATION WITH ENGINE RUNNING AT SPEED INDICATED. UNPLUG AND RECONNECT VACUUM HOSE.
3. FAST IDLE SPEED: WITH THROTTLE ON HIGH STEP OF CAM ADJUST FAST IDLE SCREW TO OBTAIN SPECIFIED SPEED.
4. IDLE SPEED SCREW: ADJUST CARBURETOR IDLE SPEED SCREW TO OBTAIN SPECIFIED SPEED.
5. STEPPED SPEED CONTROL: (VEHICLES WITH A/C) DISCONNECT ELECTRICAL LEAD FROM A/C COMPRESSOR AND TURN A/C SWITCH ON. CHECK TO SEE THAT THE SSC ACTUATOR IS FULLY EXTENDED. ADJUST ACTUATOR SCREW TO OBTAIN SPECIFIED RPM. RECONNECT A/C COMPRESSOR ELECTRICAL LEAD AND TURN A/C SWITCH OFF. (VEHICLES WITHOUT A/C) DISCONNECT AND PLUG VACUUM HOSE AT SSC ACTUATOR. CONNECT A VACUUM SOURCE (5 IN. HG. MIN.) TO ACTUATOR AND CHECK TO SEE THAT IT IS FULLY EXTENDED. ADJUST ACTUATOR SCREW TO OBTAIN SPECIFIED RPM.
6. UNPLUG AND RECONNECT HOSES TO CANISTER, AND ON VEHICLES WITHOUT A/C, THE SSC ACTUATOR. UNCAP VACUUM SOURCE FOR EGR AND RECONNECT LINE. UNCAP VACUUM SOURCE FOR THERMAC, RECONNECT HOSE, AND REPLACE AIR CLEANER.

	AUTOMATIC TRANSMISSION
TIMING (°BTC @ RPM)	8° @ 700 (DR)
SPARK PLUG GAP (IN.)	0.035
IDLE SPEED SCREW (RPM)	700 (DR)
STEPPED SPEED CONTROL ACTUATOR (SSC)	875 (DR)
FAST IDLE SPEED SCREW (RPM)	2500 (P) OR (N)

NOTE: IDLE MIXTURE SCREW IS PRESET AND SEALED AT FACTORY. PROVISION FOR ADJUSTMENT DURING TUNE UP IS NOT PROVIDED.

SEE SERVICE MANUAL, MAINTENANCE SCHEDULE AND EMISSION HOSE ROUTING DIAGRAM FOR ADDITIONAL INFORMATION.

THIS VEHICLE CONFORMS TO CALIFORNIA REGULATIONS APPLICABLE TO 1986 MODEL YEAR NEW PASSENGER CARS AND SIMILARLY TO U.S. EPA REGULATIONS APPLICABLE TO THE STATE OF CALIFORNIA.

PRINTED IN U.S.A.
PT. NO. 14086600

YAW EMISSION HOSE ROUTING
FRONT OF VEHICLE
S.S.C SOL.
VAC REG VALVE
VAC SW SOL
DELAY VLV
DELAY VLV
EGR
EGR PORT
PCV (IN)
TO A/CL
TRANS CONV SW
DECEL VALVE
AIR MGMT VALVE
SEC ACT
SSC ACT.
VAC BRK
DELAY VLV
VAC MOTOR
TVS
TO IN-TAKE MANIF
CARB
PCV
TEMP SENSOR
TO FUEL TANK
TANK PRESS. CONT VALVE
TVS
DISTR
CANISTER

USAGE: 1-2TA00 & L17 & MD2 & C60 & NB2
1-2TA00 & L17 & MD2 & C41 & N41 & NB2

1.6L—W/AUTO. TRANS. & PWR. STRG. EXC. AIR. COND.—CALIFORNIA

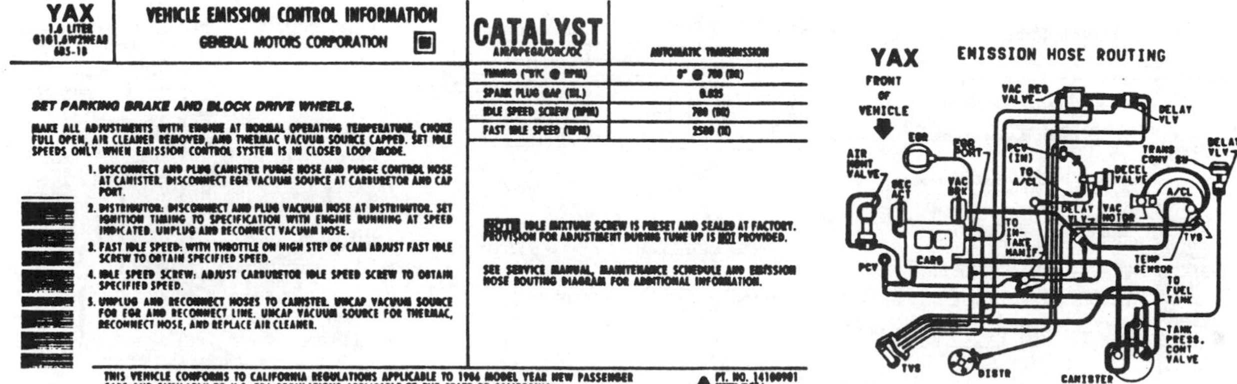

YAX
1.6 LITER
G1G1.6W2NEA8
6B5-1B

VEHICLE EMISSION CONTROL INFORMATION
GENERAL MOTORS CORPORATION

CATALYST
AIR/BPEGR/ORC/OC

SET PARKING BRAKE AND BLOCK DRIVE WHEELS.

MAKE ALL ADJUSTMENTS WITH ENGINE AT NORMAL OPERATING TEMPERATURE, CHOKE FULL OPEN, AIR CLEANER REMOVED, AND THERMAC VACUUM SOURCE CAPPED. SET IDLE SPEEDS ONLY WHEN EMISSION CONTROL SYSTEM IS IN CLOSED LOOP MODE.

1. DISCONNECT AND PLUG CANISTER PURGE HOSE AND PURGE CONTROL HOSE AT CANISTER. DISCONNECT EGR VACUUM SOURCE AT CARBURETOR AND CAP PORT.
2. DISTRIBUTOR: DISCONNECT AND PLUG VACUUM HOSE AT DISTRIBUTOR. SET IGNITION TIMING TO SPECIFICATION WITH ENGINE RUNNING AT SPEED INDICATED. UNPLUG AND RECONNECT VACUUM HOSE.
3. FAST IDLE SPEED: WITH THROTTLE ON HIGH STEP OF CAM ADJUST FAST IDLE SCREW TO OBTAIN SPECIFIED SPEED.
4. IDLE SPEED SCREW: ADJUST CARBURETOR IDLE SPEED SCREW TO OBTAIN SPECIFIED SPEED.
5. UNPLUG AND RECONNECT HOSES TO CANISTER. UNCAP VACUUM SOURCE FOR EGR AND RECONNECT LINE. UNCAP VACUUM SOURCE FOR THERMAC, RECONNECT HOSE, AND REPLACE AIR CLEANER.

	AUTOMATIC TRANSMISSION
TIMING (°BTC @ RPM)	8° @ 700 (DR)
SPARK PLUG GAP (IN.)	0.035
IDLE SPEED SCREW (RPM)	700 (DR)
FAST IDLE SPEED (RPM)	2500 (N)

NOTE: IDLE MIXTURE SCREW IS PRESET AND SEALED AT FACTORY. PROVISION FOR ADJUSTMENT DURING TUNE UP IS NOT PROVIDED.

SEE SERVICE MANUAL, MAINTENANCE SCHEDULE AND EMISSION HOSE ROUTING DIAGRAM FOR ADDITIONAL INFORMATION.

THIS VEHICLE CONFORMS TO CALIFORNIA REGULATIONS APPLICABLE TO 1986 MODEL YEAR NEW PASSENGER CARS AND SIMILARLY TO U.S. EPA REGULATIONS APPLICABLE TO THE STATE OF CALIFORNIA.

PT. NO. 14100901
PRINTED IN U.S.A.

USAGE: 1-2TA00 & L17 & MD2 & C41 & N51 & NB2

1.6L—W/AUTO. TRANS. , EXC. AIR COND. & PWR. STRG.—CALIFORNIA

VACUUM CIRCUITS

(© G.M. Corp.)

1986 CHEVROLET 1.6L

YAY 1.6 LITER 6161.6W2HEA8 685-18

VEHICLE EMISSION CONTROL INFORMATION
GENERAL MOTORS CORPORATION

SET PARKING BRAKE AND BLOCK DRIVE WHEELS.

MAKE ALL ADJUSTMENTS WITH ENGINE AT NORMAL OPERATING TEMPERATURE, CHOKE FULL OPEN, AIR CLEANER REMOVED, THERMAC VACUUM SOURCE CAPPED AND AIR CONDITIONING (A/C) OFF, EXCEPT WHERE NOTED. SET IDLE SPEEDS ONLY WHEN EMISSION CONTROL SYSTEM IS IN CLOSED LOOP MODE.

1. DISCONNECT AND PLUG CANISTER PURGE HOSE AND PURGE CONTROL HOSE AT CANISTER. DISCONNECT EGR VACUUM SOURCE AT CARBURETOR AND CAP PORT.
2. DISTRIBUTOR: DISCONNECT AND PLUG VACUUM HOSE AT DISTRIBUTOR. SET IGNITION TIMING TO SPECIFICATION WITH ENGINE RUNNING AT SPEED INDICATED. UNPLUG AND RECONNECT VACUUM HOSE.
3. FAST IDLE SPEED: WITH THROTTLE ON HIGH STEP OF CAM ADJUST FAST IDLE SCREW TO OBTAIN SPECIFIED SPEED.
4. IDLE SPEED SCREW: ADJUST CARBURETOR IDLE SPEED SCREW TO OBTAIN SPECIFIED SPEED.
5. STEPPED SPEED CONTROL: DISCONNECT ELECTRICAL LEAD FROM A/C COMPRESSOR AND TURN A/C SWITCH ON. CHECK TO SEE THAT THE SSC ACTUATOR IS FULLY EXTENDED. ADJUST ACTUATOR SCREW TO OBTAIN SPECIFIED RPM. RECONNECT A/C COMPRESSOR ELECTRICAL LEAD AND TURN A/C SWITCH OFF.
6. UNPLUG AND RECONNECT HOSES TO CANISTER. UNCAP VACUUM SOURCE FOR EGR AND RECONNECT LINE. UNCAP VACUUM SOURCE FOR THERMAC, RECONNECT HOSE, AND REPLACE AIR CLEANER.

CATALYST AIR/DPEGR/ORC/OC

	MANUAL TRANSMISSIONS	
	4-SPEED	5-SPEED
TIMING (°BTC @ RPM)	8° @ 800 (N)	8° @ 800 (N)
SPARK PLUG GAP (IN.)	0.035	0.035
IDLE SPEED SCREW (RPM)	800 (N)	900 (N)
STEPPED SPEED CONTROL ACTUATOR (SSC)	1250 (N)	1250 (N)
FAST IDLE SPEED SCREW (RPM)	2500 (N)	2500 (N)

NOTE: IDLE MIXTURE SCREW IS PRESET AND SEALED AT FACTORY. PROVISION FOR ADJUSTMENT DURING TUNE UP IS NOT PROVIDED.

SEE SERVICE MANUAL, MAINTENANCE SCHEDULE AND EMISSION HOSE ROUTING DIAGRAM FOR ADDITIONAL INFORMATION.

THIS VEHICLE CONFORMS TO CALIFORNIA REGULATIONS APPLICABLE TO 1986 MODEL YEAR NEW PASSENGER CARS AND SIMILARLY TO U.S. EPA REGULATIONS APPLICABLE TO THE STATE OF CALIFORNIA. PRINTED IN U.S.A. PT. NO. 14100902

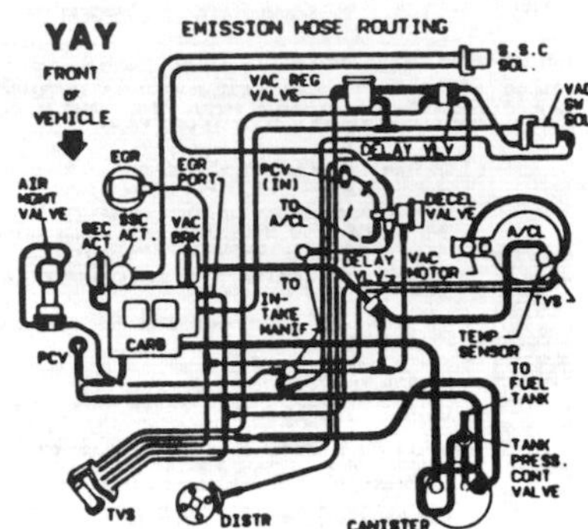

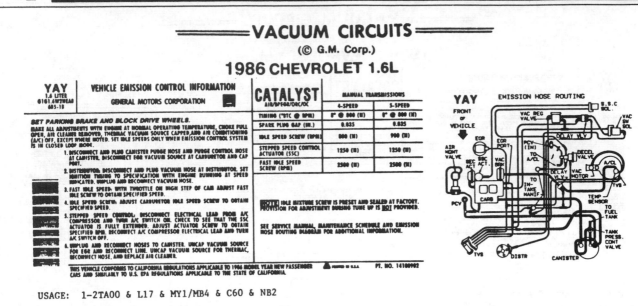

USAGE: 1-2TA00 & L17 & MY1/MB4 & C60 & NB2

1.6L—W/AUTO. TRANS. & AIR COND. (C60)—CALIFORNIA

YAZ 1.6 LITER 6161.6W2HEA8 685-18

VEHICLE EMISSION CONTROL INFORMATION
GENERAL MOTORS CORPORATION

SET PARKING BRAKE AND BLOCK DRIVE WHEELS.

MAKE ALL ADJUSTMENTS WITH ENGINE AT NORMAL OPERATING TEMPERATURE, CHOKE FULL OPEN, AIR CLEANER REMOVED AND THERMAC VACUUM SOURCE CAPPED. SET IDLE SPEEDS ONLY WHEN EMISSION CONTROL SYSTEM IS IN CLOSED LOOP MODE.

1. DISCONNECT AND PLUG CANISTER PURGE HOSE AND PURGE CONTROL HOSE AT CANISTER. DISCONNECT EGR VACUUM SOURCE AT CARBURETOR AND CAP PORT.
2. DISTRIBUTOR: DISCONNECT AND PLUG VACUUM HOSE AT DISTRIBUTOR. SET IGNITION TIMING TO SPECIFICATION WITH ENGINE RUNNING AT SPEED INDICATED. UNPLUG AND RECONNECT VACUUM HOSE.
3. FAST IDLE SPEED: WITH THROTTLE ON HIGH STEP OF CAM ADJUST FAST IDLE SCREW TO OBTAIN SPECIFIED SPEED.
4. IDLE SPEED SCREW: ADJUST CARBURETOR IDLE SPEED SCREW TO OBTAIN SPECIFIED SPEED.
5. STEPPED SPEED CONTROL: DISCONNECT AND PLUG VACUUM HOSE AT SSC ACTUATOR. CONNECT A VACUUM SOURCE (5 IN. HG. MIN.) TO ACTUATOR AND CHECK TO SEE THAT IT IS FULLY EXTENDED. ADJUST ACTUATOR SCREW TO OBTAIN SPECIFIED RPM.
6. UNPLUG AND RECONNECT HOSES TO CANISTER AND SSC ACTUATOR. UNCAP VACUUM SOURCE FOR EGR AND RECONNECT LINE. UNCAP VACUUM SOURCE FOR THERMAC, RECONNECT HOSE, AND REPLACE AIR CLEANER.

CATALYST AIR/DPEGR/ORC/OC

	MANUAL TRANSMISSIONS	
	4-SPEED	5-SPEED
TIMING (°BTC @ RPM).	8° @ 800 (N)	8° @ 800 (N)
SPARK PLUG GAP (IN.)	0.035	0.035
IDLE SPEED SCREW (RPM)	800 (N)	900 (N)
STEPPED SPEED CONTROL ACTUATOR (SSC)	1250 (N)	1250 (N)
FAST IDLE SPEED SCREW (RPM)	2500 (N)	2500 (N)

NOTE: IDLE MIXTURE SCREW IS PRESET AND SEALED AT FACTORY. PROVISION FOR ADJUSTMENT DURING TUNE UP IS NOT PROVIDED.

SEE SERVICE MANUAL, MAINTENANCE SCHEDULE AND EMISSION HOSE ROUTING DIAGRAM FOR ADDITIONAL INFORMATION.

THIS VEHICLE CONFORMS TO CALIFORNIA REGULATIONS APPLICABLE TO 1986 MODEL YEAR NEW PASSENGER CARS AND SIMILARLY TO U.S. EPA REGULATIONS APPLICABLE TO THE STATE OF CALIFORNIA. PT. NO. 14100903 PRINTED IN U.S.A.

YAZ EMISSION HOSE ROUTING

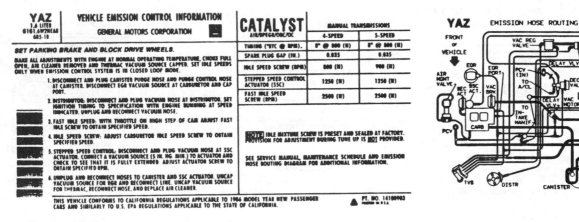

USAGE: 1-2TA00 & MY1/MB4 & C41 & NB2

1.6L—W/AUTO. TRANS., EXC. AIR COND.—CALIFORNIA

YBA 1.6 LITER G1G1.6Z2HED0 385-7

VEHICLE EMISSION CONTROL INFORMATION
GENERAL MOTORS CORPORATION

SET PARKING BRAKE AND BLOCK DRIVE WHEELS.

MAKE ALL ADJUSTMENTS WITH ENGINE AT NORMAL OPERATING TEMPERATURE, CHOKE FULL OPEN, AIR CLEANER REMOVED AND THERMAC VACUUM SOURCE CAPPED.

1. DISCONNECT, AT CARBURETOR, VACUUM HOSES TO EGR VALVE AND CANISTER. CAP CARBURETOR PORTS.
2. DISTRIBUTOR: DISCONNECT AND PLUG VACUUM HOSE AT DISTRIBUTOR. SET IGNITION TIMING TO SPECIFICATION WITH ENGINE RUNNING AT SPEED INDICATED. UNPLUG AND RECONNECT HOSE TO DISTRIBUTOR.
3. FAST IDLE SCREW: WITH THROTTLE ON HIGH STEP OF CAM, ADJUST FAST IDLE SCREW TO OBTAIN SPECIFIED SPEED.
4. IDLE SPEED SCREW: WITH THROTTLE OFF FAST IDLE CAM, ADJUST CARBURETOR IDLE SPEED SCREW TO OBTAIN SPECIFIED SPEED.
5. UNCAP CARBURETOR PORTS. RECONNECT VACUUM SOURCE HOSE FOR THE EGR VALVE AND PURGE HOSE TO THE CANISTER. UNCAP VACUUM SOURCE FOR THERMAC, RECONNECT HOSE, AND REPLACE AIR CLEANER.

CATALYST EGR/OC CANADIAN CERTIFICATION

	TRANSMISSION		
	AUTOMATIC	4-SPEED MANUAL	5-SPEED MANUAL
TIMING (°BTC @ RPM)	12° @ 700 (DR)	8° @ 700 (N)	8° @ 900 (N)
SPARK PLUG GAP (IN.)	0.035	0.035	0.035
IDLE SPEED SCREW (RPM)	700 (DR)	700 (N)	900 (N)
FAST IDLE SPEED (RPM)	2000 (P) OR (N)	2700 (N)	2700 (N)

NOTE: IDLE MIXTURE SCREW IS PRESET AND SEALED AT FACTORY. PROVISION FOR ADJUSTMENT DURING TUNE UP IS NOT PROVIDED.

SEE SERVICE MANUAL, MAINTENANCE SCHEDULE AND EMISSION HOSE ROUTING DIAGRAM FOR ADDITIONAL INFORMATION.

PT. NO. 14100904 PRINTED IN U.S.A.

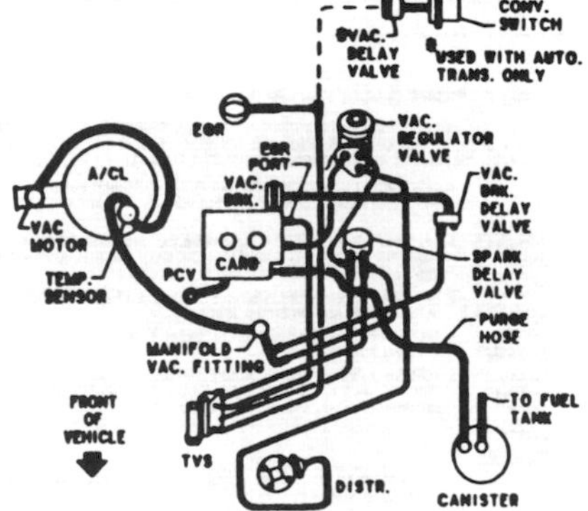

USAGE: 1-2-7TA00 & L17 & NM5

1.6L—W/NM5—CANADA

VACUUM CIRCUITS

(© G.M. Corp.)

1986 CHEVROLET 1.8L

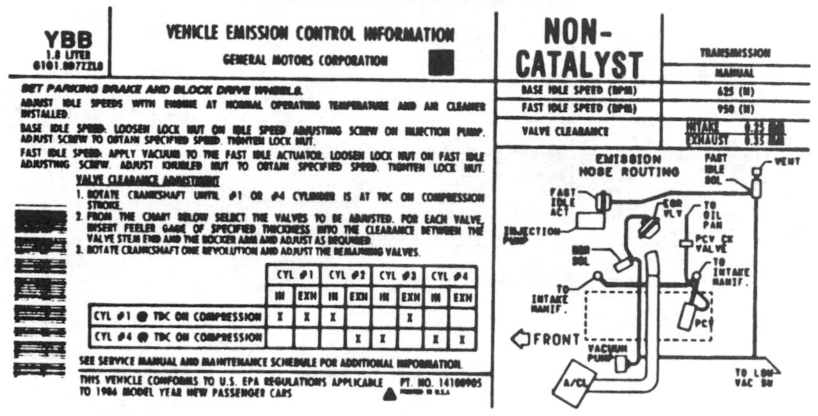

YBB
VEHICLE EMISSION CONTROL INFORMATION
GENERAL MOTORS CORPORATION
NON-CATALYST
TRANSMISSION
MANUAL
BASE IDLE SPEED (RPM)
FAST IDLE SPEED (RPM)
VALVE CLEARANCE
EMISSION HOSE ROUTING
FRONT
SEE SERVICE MANUAL AND MAINTENANCE SCHEDULE FOR ADDITIONAL INFORMATION.

USAGE: 1-7TA & LJ5 & NA5

1.8L – W/MAN. TRANS. – FEDERAL

1986 CHEVROLET 2.0L

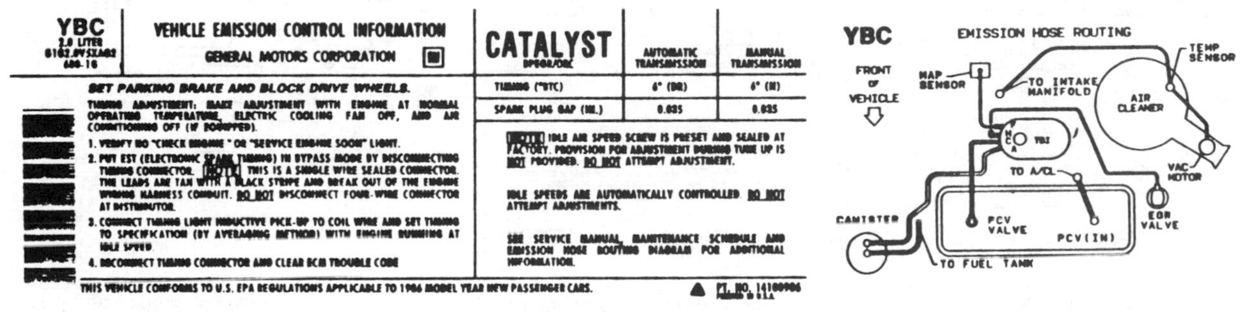

USAGE: 1-2JA00 & LQ5 & NA5

2.0L – W/AUTO. TRANS. OR MAN. TRANS. – FEDERAL

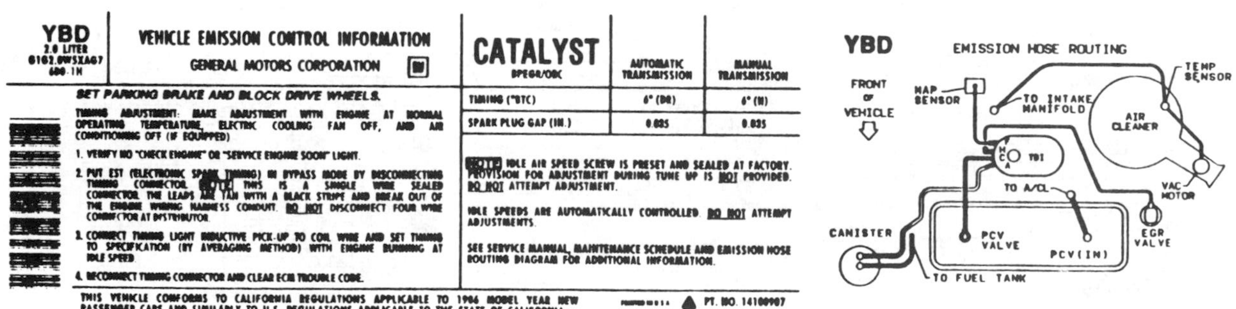

USAGE: 1-2JA00 & LQ5 & NB2

2.0L – W/AUTO. TRANS. OR MAN. TRANS. – CALIFORNIA

VACUUM CIRCUITS

(© G.M. Corp.)

1986 CHEVROLET 2.8L

YAC
2.8 LITER
G1G2.8Z5ZZG0
6D0-1G

VEHICLE EMISSION CONTROL INFORMATION
GENERAL MOTORS CORPORATION

NON-CATALYST
BPEGR/OL
CANADIAN CERTIFICATION

SET PARKING BRAKE AND BLOCK DRIVE WHEELS.

TIMING ADJUSTMENT: MAKE ADJUSTMENT WITH ENGINE AT NORMAL OPERATING TEMPERATURE, ELECTRIC COOLING FAN OFF, AND AIR CONDITIONING OFF (IF EQUIPPED).

1. VERIFY NO "CHECK ENGINE" OR "SERVICE ENGINE SOON" LIGHT.
2. PUT EST (ELECTRONIC SPARK TIMING) IN BYPASS MODE BY DISCONNECTING TIMING CONNECTOR. NOTE: THIS IS A SINGLE WIRE SEALED CONNECTOR. THE LEADS ARE TAN WITH A BLACK STRIPE AND BREAK OUT OF THE ENGINE WIRING HARNESS CONDUIT. DO NOT DISCONNECT FOUR-WIRE CONNECTOR AT DISTRIBUTOR.
3. CONNECT TIMING LIGHT INDUCTIVE PICK-UP TO COIL WIRE AND SET TIMING TO SPECIFICATION (BY AVERAGING METHOD) WITH ENGINE RUNNING AT IDLE SPEED.
4. RECONNECT LEAD AT EST BYPASS CONNECTOR AND CLEAR ECM TROUBLE CODE.

	AUTOMATIC TRANSMISSION
TIMING (°BTC)	6° (DR)
SPARK PLUG GAP (IN.)	0.035

NOTE: IDLE AIR SPEED SCREW IS PRESET AND SEALED AT FACTORY. PROVISION FOR ADJUSTMENT DURING TUNE UP IS NOT PROVIDED. DO NOT ATTEMPT ADJUSTMENT.

IDLE SPEEDS ARE AUTOMATICALLY CONTROLLED. DO NOT ATTEMPT ADJUSTMENTS.

SEE SERVICE MANUAL, MAINTENANCE SCHEDULE AND EMISSION HOSE ROUTING DIAGRAM FOR ADDITIONAL INFORMATION.

PRINTED IN U.S.A. PT. NO. 14086572

YAC EMISSION HOSE ROUTING

FRONT OF VEHICLE
MAP SENSOR
TO INTAKE MANIFOLD
TEMP SENSOR
AIR CLEANER
TBI
TO A/CL
VAC MOTOR
CANISTER
PCV VALVE
PCV(IN)
EGR VALVE
TO FUEL TANK

USAGE: 1-2JA00 & LQ5 & NM5

2.8L – W/NM5 – CANADA

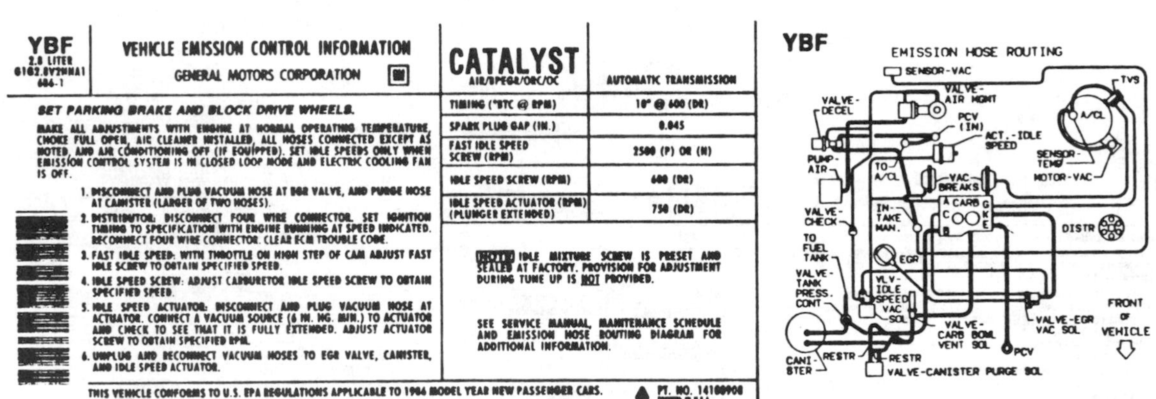

YBF
2.8 LITER
G1G2.8V2NNA1
6B6-1

VEHICLE EMISSION CONTROL INFORMATION
GENERAL MOTORS CORPORATION

CATALYST
AIR/BPEGR/ORC/OC

SET PARKING BRAKE AND BLOCK DRIVE WHEELS.

MAKE ALL ADJUSTMENTS WITH ENGINE AT NORMAL OPERATING TEMPERATURE, CHOKE FULL OPEN, AIR CLEANER INSTALLED, ALL HOSES CONNECTED EXCEPT AS NOTED, AND AIR CONDITIONING OFF (IF EQUIPPED). SET IDLE SPEEDS ONLY WHEN EMISSION CONTROL SYSTEM IS IN CLOSED LOOP MODE AND ELECTRIC COOLING FAN IS OFF.

1. DISCONNECT AND PLUG VACUUM HOSE AT EGR VALVE, AND PURGE HOSE AT CANISTER (LARGER OF TWO HOSES).
2. DISTRIBUTOR: DISCONNECT FOUR WIRE CONNECTOR. SET IGNITION TIMING TO SPECIFICATION WITH ENGINE RUNNING AT SPEED INDICATED. RECONNECT FOUR WIRE CONNECTOR. CLEAR ECM TROUBLE CODE.
3. FAST IDLE SPEED: WITH THROTTLE ON HIGH STEP OF CAM ADJUST FAST IDLE SCREW TO OBTAIN SPECIFIED SPEED.
4. IDLE SPEED SCREW: ADJUST CARBURETOR IDLE SPEED SCREW TO OBTAIN SPECIFIED SPEED.
5. IDLE SPEED ACTUATOR: DISCONNECT AND PLUG VACUUM HOSE AT ACTUATOR. CONNECT A VACUUM SOURCE (6 IN. HG. MIN.) TO ACTUATOR AND CHECK TO SEE THAT IT IS FULLY EXTENDED. ADJUST ACTUATOR SCREW TO OBTAIN SPECIFIED RPM.
6. UNPLUG AND RECONNECT VACUUM HOSES TO EGR VALVE, CANISTER, AND IDLE SPEED ACTUATOR.

	AUTOMATIC TRANSMISSION
TIMING (°BTC @ RPM)	10° @ 600 (DR)
SPARK PLUG GAP (IN.)	0.045
FAST IDLE SPEED SCREW (RPM)	2500 (P) OR (N)
IDLE SPEED SCREW (RPM)	600 (DR)
IDLE SPEED ACTUATOR (RPM) (PLUNGER EXTENDED)	750 (DR)

NOTE: IDLE MIXTURE SCREW IS PRESET AND SEALED AT FACTORY. PROVISION FOR ADJUSTMENT DURING TUNE UP IS NOT PROVIDED.

SEE SERVICE MANUAL, MAINTENANCE SCHEDULE AND EMISSION HOSE ROUTING DIAGRAM FOR ADDITIONAL INFORMATION.

THIS VEHICLE CONFORMS TO U.S. EPA REGULATIONS APPLICABLE TO 1986 MODEL YEAR NEW PASSENGER CARS.

PT. NO. 14100900
PRINTED IN U.S.A.

USAGE: 1-2AA00 & LE2 & NA5
<u>NOTE</u>: This label contains an error in the hose routing diagram. It has been cancelled and replaced by Label "YMD" below (ECA 46578). All production and service stock should be disposed of. This label was used for early production without diagram (PAA 16-A-5903)

2.8L – W/AUTO. TRANS. – FEDERAL

YMD
2.8 LITER
G1G2.8V2NNA1
6B6-1

VEHICLE EMISSION CONTROL INFORMATION
GENERAL MOTORS CORPORATION

CATALYST
AIR/BPEGR/ORC/OC

SET PARKING BRAKE AND BLOCK DRIVE WHEELS.

MAKE ALL ADJUSTMENTS WITH ENGINE AT NORMAL OPERATING TEMPERATURE, CHOKE FULL OPEN, AIR CLEANER INSTALLED, ALL HOSES CONNECTED EXCEPT AS NOTED, AND AIR CONDITIONING OFF (IF EQUIPPED). SET IDLE SPEEDS ONLY WHEN EMISSION CONTROL SYSTEM IS IN CLOSED LOOP MODE AND ELECTRIC COOLING FAN IS OFF.

1. DISCONNECT AND PLUG VACUUM HOSE AT EGR VALVE, AND PURGE HOSE AT CANISTER (LARGER OF TWO HOSES).
2. DISTRIBUTOR: DISCONNECT FOUR WIRE CONNECTOR. SET IGNITION TIMING TO SPECIFICATION WITH ENGINE RUNNING AT SPEED INDICATED. RECONNECT FOUR WIRE CONNECTOR. CLEAR ECM TROUBLE CODE.
3. FAST IDLE SPEED: WITH THROTTLE ON HIGH STEP OF CAM ADJUST FAST IDLE SCREW TO OBTAIN SPECIFIED SPEED.
4. IDLE SPEED SCREW: ADJUST CARBURETOR IDLE SPEED SCREW TO OBTAIN SPECIFIED SPEED.
5. IDLE SPEED ACTUATOR: DISCONNECT AND PLUG VACUUM HOSE AT ACTUATOR. CONNECT A VACUUM SOURCE (6 IN. HG. MIN.) TO ACTUATOR AND CHECK TO SEE THAT IT IS FULLY EXTENDED. ADJUST ACTUATOR SCREW TO OBTAIN SPECIFIED RPM.
6. UNPLUG AND RECONNECT VACUUM HOSES TO EGR VALVE, CANISTER, AND IDLE SPEED ACTUATOR.

	AUTOMATIC TRANSMISSION
TIMING (°BTC @ RPM)	10° @ 600 (DR)
SPARK PLUG GAP (IN.)	0.045
FAST IDLE SPEED SCREW (RPM)	2500 (P) OR (N)
IDLE SPEED SCREW (RPM)	600 (DR)
IDLE SPEED ACTUATOR (RPM) (PLUNGER EXTENDED)	750 (DR)

NOTE: IDLE MIXTURE SCREW IS PRESET AND SEALED AT FACTORY. PROVISION FOR ADJUSTMENT DURING TUNE UP IS NOT PROVIDED.

SEE SERVICE MANUAL, MAINTENANCE SCHEDULE AND EMISSION HOSE ROUTING DIAGRAM FOR ADDITIONAL INFORMATION.

THIS VEHICLE CONFORMS TO U.S. EPA REGULATIONS APPLICABLE TO 1986 MODEL YEAR NEW PASSENGER CARS.

PT. NO. 14100960
PRINTED IN U.S.A.

YMD EMISSION HOSE ROUTING

SENSOR-VAC
VALVE-AIR MGMT
TVS
VALVE-DECEL
PCV (IN)
ACT.-IDLE SPEED
A/CL
PUMP-AIR
SENSOR-TEMP
MOTOR-VAC
TO A/CL
VAC BREAKS
CARB
VALVE-CHECK
IN-TAKE MAN.
DISTR
TO FUEL TANK
EGR
VALVE-TANK PRESS. CONT
VLV-IDLE SPEED VAC SOL
VALVE-CARB BOWL VENT SOL
VALVE-EGR VAC SOL
FRONT OF VEHICLE
CANI-STER
RESTR
RESTR
PCV
VALVE-CANISTER PURGE SOL

USAGE: 1-2AA00 & LE2 & NA5
<u>NOTE</u>: Replaces "YBF" for all uses (ECA 46578)

2.8L – W/AUTO. TRANS. – FEDERAL

VACUUM CIRCUITS

(© G.M. Corp.)

1986 CHEVROLET 2.8L

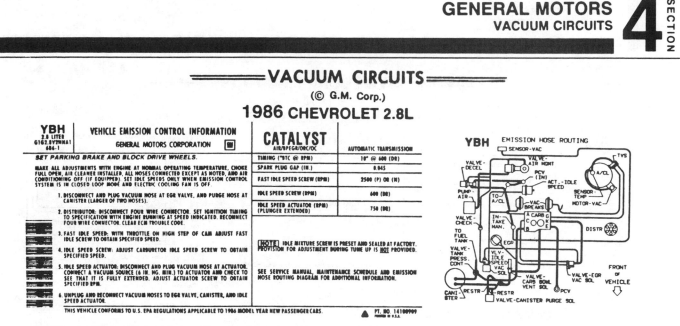

YBH
2.8 LITER
G1G2.8V2NNA1
686-1

VEHICLE EMISSION CONTROL INFORMATION
GENERAL MOTORS CORPORATION

SET PARKING BRAKE AND BLOCK DRIVE WHEELS.

MAKE ALL ADJUSTMENTS WITH ENGINE AT NORMAL OPERATING TEMPERATURE, CHOKE FULL OPEN, AIR CLEANER INSTALLED, ALL HOSES CONNECTED EXCEPT AS NOTED, AND AIR CONDITIONING OFF (IF EQUIPPED). SET IDLE SPEEDS ONLY WHEN EMISSION CONTROL SYSTEM IS IN CLOSED LOOP MODE AND ELECTRIC COOLING FAN IS OFF.

1. DISCONNECT AND PLUG VACUUM HOSE AT EGR VALVE, AND PURGE HOSE AT CANISTER (LARGER OF TWO HOSES).
2. DISTRIBUTOR: DISCONNECT FOUR WIRE CONNECTOR. SET IGNITION TIMING TO SPECIFICATION WITH ENGINE RUNNING AT SPEED INDICATED. RECONNECT FOUR WIRE CONNECTOR. CLEAR ECM TROUBLE CODE.
3. FAST IDLE SPEED: WITH THROTTLE ON HIGH STEP OF CAM ADJUST FAST IDLE SCREW TO OBTAIN SPECIFIED SPEED.
4. IDLE SPEED SCREW: ADJUST CARBURETOR IDLE SPEED SCREW TO OBTAIN SPECIFIED SPEED.
5. IDLE SPEED ACTUATOR: DISCONNECT AND PLUG VACUUM HOSE AT ACTUATOR. CONNECT A VACUUM SOURCE (6 IN. HG. MIN.) TO ACTUATOR AND CHECK TO SEE THAT IT IS FULLY EXTENDED. ADJUST ACTUATOR SCREW TO OBTAIN SPECIFIED RPM.
6. UNPLUG AND RECONNECT VACUUM HOSES TO EGR VALVE, CANISTER, AND IDLE SPEED ACTUATOR.

CATALYST AIR/BPEGR/ORC/OC	AUTOMATIC TRANSMISSION
TIMING (°BTC @ RPM)	10° @ 600 (DR)
SPARK PLUG GAP (IN.)	0.045
FAST IDLE SPEED SCREW (RPM)	2500 (P) OR (N)
IDLE SPEED SCREW (RPM)	600 (DR)
IDLE SPEED ACTUATOR (RPM) (PLUNGER EXTENDED)	750 (DR)

NOTE IDLE MIXTURE SCREW IS PRESET AND SEALED AT FACTORY. PROVISION FOR ADJUSTMENT DURING TUNE UP IS NOT PROVIDED.

SEE SERVICE MANUAL, MAINTENANCE SCHEDULE AND EMISSION HOSE ROUTING DIAGRAM FOR ADDITIONAL INFORMATION.

THIS VEHICLE CONFORMS TO U.S. EPA REGULATIONS APPLICABLE TO 1986 MODEL YEAR NEW PASSENGER CARS.

PT. NO. 14100909
PRINTED IN U.S.A.

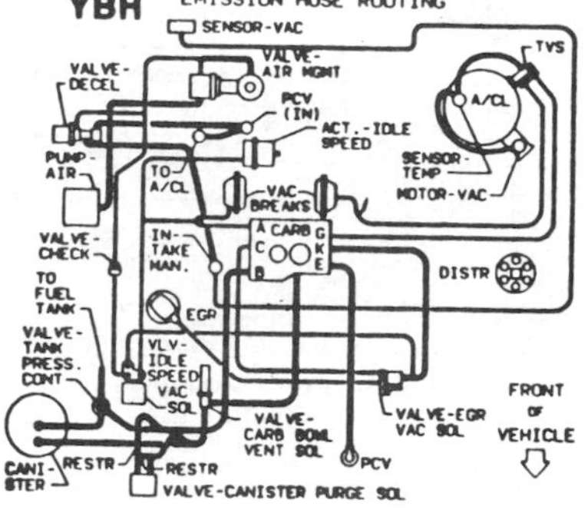

USAGE: 1AA00 & LE2 & NA5 (-NN5) & 9C1/9C6 (police-taxi only)
NOTE: This label contains an error in the hose routing diagram. It has been cancelled and replaced by Label "YMF" below (ECA 46578). All production and service stock should be disposed of.

2.8L – W/AUTO. TRANS. – FEDERAL

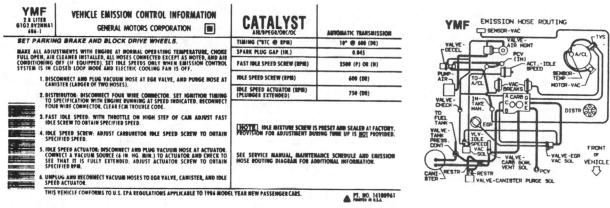

YMF
2.8 LITER
G1G2.8V2NNA1
686-1

VEHICLE EMISSION CONTROL INFORMATION
GENERAL MOTORS CORPORATION

SET PARKING BRAKE AND BLOCK DRIVE WHEELS.

MAKE ALL ADJUSTMENTS WITH ENGINE AT NORMAL OPERATING TEMPERATURE, CHOKE FULL OPEN, AIR CLEANER INSTALLED, ALL HOSES CONNECTED EXCEPT AS NOTED, AND AIR CONDITIONING OFF (IF EQUIPPED). SET IDLE SPEEDS ONLY WHEN EMISSION CONTROL SYSTEM IS IN CLOSED LOOP MODE AND ELECTRIC COOLING FAN IS OFF.

1. DISCONNECT AND PLUG VACUUM HOSE AT EGR VALVE, AND PURGE HOSE AT CANISTER (LARGER OF TWO HOSES).
2. DISTRIBUTOR: DISCONNECT FOUR WIRE CONNECTOR. SET IGNITION TIMING TO SPECIFICATION WITH ENGINE RUNNING AT SPEED INDICATED. RECONNECT FOUR WIRE CONNECTOR. CLEAR ECM TROUBLE CODE.
3. FAST IDLE SPEED: WITH THROTTLE ON HIGH STEP OF CAM ADJUST FAST IDLE SCREW TO OBTAIN SPECIFIED SPEED.
4. IDLE SPEED SCREW: ADJUST CARBURETOR IDLE SPEED SCREW TO OBTAIN SPECIFIED SPEED.
5. IDLE SPEED ACTUATOR: DISCONNECT AND PLUG VACUUM HOSE AT ACTUATOR. CONNECT A VACUUM SOURCE (6 IN. HG. MIN.) TO ACTUATOR AND CHECK TO SEE THAT IT IS FULLY EXTENDED. ADJUST ACTUATOR SCREW TO OBTAIN SPECIFIED RPM.
6. UNPLUG AND RECONNECT VACUUM HOSES TO EGR VALVE, CANISTER, AND IDLE SPEED ACTUATOR.

CATALYST AIR/BPEGR/ORC/OC	AUTOMATIC TRANSMISSION
TIMING (°BTC @ RPM)	10° @ 600 (DR)
SPARK PLUG GAP (IN.)	0.045
FAST IDLE SPEED SCREW (RPM)	2500 (P) OR (N)
IDLE SPEED SCREW (RPM)	600 (DR)
IDLE SPEED ACTUATOR (RPM) (PLUNGER EXTENDED)	750 (DR)

NOTE IDLE MIXTURE SCREW IS PRESET AND SEALED AT FACTORY. PROVISION FOR ADJUSTMENT DURING TUNE UP IS NOT PROVIDED.

SEE SERVICE MANUAL, MAINTENANCE SCHEDULE AND EMISSION HOSE ROUTING DIAGRAM FOR ADDITIONAL INFORMATION.

THIS VEHICLE CONFORMS TO U.S. EPA REGULATIONS APPLICABLE TO 1986 MODEL YEAR NEW PASSENGER CARS.

PT. NO. 14100961
PRINTED IN U.S.A.

YMF EMISSION HOSE ROUTING

USAGE: 1AA00 & LE2 & NA5 (-NN5) & 9C1/9C6 (police-taxi only)
NOTE: Replaces "YBH" for all uses (ECA 46578)

2.8L – W/SEO (9CI) (EXC. NN5) – FEDERAL

YBJ
2.8 LITER
G1G2.8V2NNA1
686-1

VEHICLE EMISSION CONTROL INFORMATION
GENERAL MOTORS CORPORATION

SET PARKING BRAKE AND BLOCK DRIVE WHEELS.

MAKE ALL ADJUSTMENTS WITH ENGINE AT NORMAL OPERATING TEMPERATURE, CHOKE FULL OPEN, AIR CLEANER INSTALLED, ALL HOSES CONNECTED EXCEPT AS NOTED, AND AIR CONDITIONING OFF (IF EQUIPPED). SET IDLE SPEEDS ONLY WHEN EMISSION CONTROL SYSTEM IS IN CLOSED LOOP MODE AND ELECTRIC COOLING FAN IS OFF.

1. DISCONNECT AND PLUG VACUUM HOSE AT EGR VALVE, AND PURGE HOSE AT CANISTER (LARGER OF TWO HOSES).
2. DISTRIBUTOR: DISCONNECT FOUR WIRE CONNECTOR. SET IGNITION TIMING TO SPECIFICATION WITH ENGINE RUNNING AT SPEED INDICATED. RECONNECT FOUR WIRE CONNECTOR. CLEAR ECM TROUBLE CODE.
3. FAST IDLE SPEED: WITH THROTTLE ON HIGH STEP OF CAM ADJUST FAST IDLE SCREW TO OBTAIN SPECIFIED SPEED.
4. IDLE SPEED SCREW: ADJUST CARBURETOR IDLE SPEED SCREW TO OBTAIN SPECIFIED SPEED.
5. IDLE SPEED ACTUATOR: DISCONNECT AND PLUG VACUUM HOSE AT ACTUATOR. CONNECT A VACUUM SOURCE (6 IN. HG. MIN.) TO ACTUATOR AND CHECK TO SEE THAT IT IS FULLY EXTENDED. ADJUST ACTUATOR SCREW TO OBTAIN SPECIFIED RPM.
6. UNPLUG AND RECONNECT VACUUM HOSES TO EGR VALVE, CANISTER, AND IDLE SPEED ACTUATOR.

CATALYST AIR/BPEGR/ORC/OC	AUTOMATIC TRANSMISSION
TIMING (°BTC @ RPM)	10° @ 600 (DR)
SPARK PLUG GAP (IN.)	0.045
FAST IDLE SPEED SCREW (RPM)	2500 (P) OR (N)
IDLE SPEED SCREW (RPM)	600 (DR)
IDLE SPEED ACTUATOR (RPM) (PLUNGER EXTENDED)	750 (DR)

NOTE IDLE MIXTURE SCREW IS PRESET AND SEALED AT FACTORY. PROVISION FOR ADJUSTMENT DURING TUNE UP IS NOT PROVIDED.

SEE SERVICE MANUAL, MAINTENANCE SCHEDULE AND EMISSION HOSE ROUTING DIAGRAM FOR ADDITIONAL INFORMATION.

THIS VEHICLE CONFORMS TO U.S. EPA REGULATIONS APPLICABLE TO 1986 MODEL YEAR NEW PASSENGER CARS AND IS CERTIFIED FOR SALE IN THE STATE OF CALIFORNIA.

PRINTED IN U.S.A. PT. NO. 14100910

YBJ EMISSION HOSE ROUTING

USAGE: 1AA00 & LE2 & NA5 & NN5 (Calif. taxi only)
NOTE: This label contains an error in the hose routing diagram. It has been cancelled and replaced by Label "YMH" below (ECA 46578). All production and service stock should be disposed of.

2.8L – W/AUTO. TRANS. – FEDERAL

VACUUM CIRCUITS

(© G.M. Corp.)

1986 CHEVROLET 2.8L

YMH 2.8 LITER 6G12.8V2NNA1 606-1

VEHICLE EMISSION CONTROL INFORMATION
GENERAL MOTORS CORPORATION

SET PARKING BRAKE AND BLOCK DRIVE WHEELS.

MAKE ALL ADJUSTMENTS WITH ENGINE AT NORMAL OPERATING TEMPERATURE, CHOKE FULL OPEN, AIR CLEANER INSTALLED, ALL HOSES CONNECTED EXCEPT AS NOTED, AND AIR CONDITIONING OFF (IF EQUIPPED). SET IDLE SPEEDS ONLY WHEN EMISSION CONTROL SYSTEM IS IN CLOSED LOOP MODE AND ELECTRIC COOLING FAN IS OFF.

1. DISCONNECT AND PLUG VACUUM HOSE AT EGR VALVE, AND PURGE HOSE AT CANISTER (LARGER OF TWO HOSES).
2. DISTRIBUTOR: DISCONNECT FOUR WIRE CONNECTOR. SET IGNITION TIMING TO SPECIFICATION WITH ENGINE RUNNING AT SPEED INDICATED. RECONNECT FOUR WIRE CONNECTOR. CLEAR ECM TROUBLE CODE.
3. FAST IDLE SPEED: WITH THROTTLE ON HIGH STEP OF CAM ADJUST FAST IDLE SCREW TO OBTAIN SPECIFIED SPEED
4. IDLE SPEED SCREW: ADJUST CARBURETOR IDLE SPEED SCREW TO OBTAIN SPECIFIED SPEED
5. IDLE SPEED ACTUATOR: DISCONNECT AND PLUG VACUUM HOSE AT ACTUATOR. CONNECT A VACUUM SOURCE (6 IN. HG. MIN.) TO ACTUATOR AND CHECK TO SEE THAT IT IS FULLY EXTENDED. ADJUST ACTUATOR SCREW TO OBTAIN SPECIFIED RPM
6. UNPLUG AND RECONNECT VACUUM HOSES TO EGR VALVE, CANISTER, AND IDLE SPEED ACTUATOR.

CATALYST AIR/BPEGR/ORC/OC	AUTOMATIC TRANSMISSION
TIMING (°BTC @ RPM)	10° @ 600 (DR)
SPARK PLUG GAP (IN.)	0.045
FAST IDLE SPEED SCREW (RPM)	2500 (P) OR (N)
IDLE SPEED SCREW (RPM)	600 (DR)
IDLE SPEED ACTUATOR (RPM) (PLUNGER EXTENDED)	750 (DR)

NOTE IDLE MIXTURE SCREW IS PRESET AND SEALED AT FACTORY. PROVISION FOR ADJUSTMENT DURING TUNE UP IS NOT PROVIDED

SEE SERVICE MANUAL, MAINTENANCE SCHEDULE AND EMISSION HOSE ROUTING DIAGRAM FOR ADDITIONAL INFORMATION.

THIS VEHICLE CONFORMS TO U.S. EPA REGULATIONS APPLICABLE TO 1986 MODEL YEAR NEW PASSENGER CARS AND IS CERTIFIED FOR SALE IN THE STATE OF CALIFORNIA.

PT. NO. 14100962 PRINTED IN U.S.A.

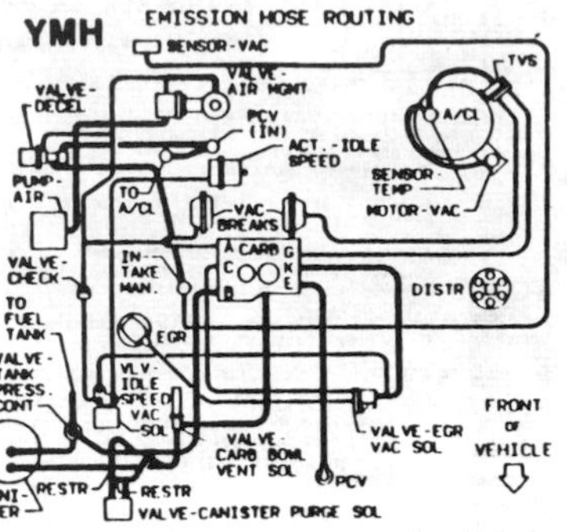

YMH EMISSION HOSE ROUTING

USAGE: 1AA00 & LE2 & NA5 & NN5 (Calif. taxi only)
NOTE: Replaces "YBJ" for all uses (ECA 46578)

2.8L—W/NN5, SEO (9C6)—FEDERAL

YAH 2.8 LITER G1G2.8Z2HED9 586-7

VEHICLE EMISSION CONTROL INFORMATION
GENERAL MOTORS CORPORATION

SET PARKING BRAKE AND BLOCK DRIVE WHEELS.

REMOVE AIR CLEANER AND PLUG VACUUM HOSES. START ENGINE. MAKE ALL ADJUSTMENTS WITH ENGINE AT NORMAL OPERATING TEMPERATURE, CHOKE FULL OPEN, AND AIR CONDITIONING (A/C) OFF, EXCEPT WHERE NOTED. SET IDLE SPEEDS ONLY WHEN ELECTRIC COOLING FAN IS OFF.

1. DISTRIBUTOR: DISCONNECT AND PLUG CANISTER PURGE HOSE AT CANISTER CONTROL VALVE. DISCONNECT AND PLUG VACUUM HOSE AT DISTRIBUTOR. SET IGNITION TIMING TO SPECIFICATION WITH ENGINE RUNNING AT SPEED INDICATED. UNPLUG AND RECONNECT HOSE AT DISTRIBUTOR.
2. FAST IDLE SPEED: WITH THROTTLE ON HIGH STEP OF FAST IDLE CAM, ADJUST FAST IDLE SCREW TO OBTAIN SPECIFIED SPEED.
3. IDLE SPEED SCREW: DISCONNECT ELECTRICAL LEAD FROM IDLE SOLENOID, IF EQUIPPED. ADJUST CARBURETOR IDLE SPEED SCREW TO OBTAIN SPECIFIED SPEED. RECONNECT LEAD TO SOLENOID.
4. IDLE SPEED SOLENOID (IF EQUIPPED): DISCONNECT ELECTRICAL LEAD FROM A/C COMPRESSOR AND TURN A/C SWITCH TO "DEF" POSITION. OPEN THROTTLE MOMENTARILY TO ASSURE SOLENOID IS FULLY EXTENDED. ADJUST SOLENOID TO OBTAIN SPECIFIED SPEED. RECONNECT COMPRESSOR LEAD AND TURN A/C OFF.
5. STOP ENGINE. UNPLUG AND RECONNECT CANISTER PURGE HOSE TO CANISTER CONTROL VALVE. UNPLUG HOSES AND REPLACE AIR CLEANER.

CATALYST EGR/OC CANADIAN CERTIFICATION	AUTOMATIC TRANSMISSION
TIMING (°BTC @ RPM)	6° @ 750 (N)
SPARK PLUG GAP (IN.)	0.045
IDLE SPEED SCREW (RPM) (SOLENOID INACTIVE)	700 (DR)
IDLE SPEED SOLENOID (RPM) (SOLENOID ACTIVE)	900 (DR)
FAST IDLE SPEED SCREW (RPM)	2700 (P) OR (N)

NOTE IDLE MIXTURE SCREW IS PRESET AND SEALED AT FACTORY. PROVISION FOR ADJUSTMENT DURING TUNE UP IS NOT PROVIDED.

SEE SERVICE MANUAL, MAINTENANCE SCHEDULE AND EMISSION HOSE ROUTING DIAGRAM FOR ADDITIONAL INFORMATION.

PRINTED IN U.S.A. PT. NO. 14004575

YAH EMISSION HOSE ROUTING

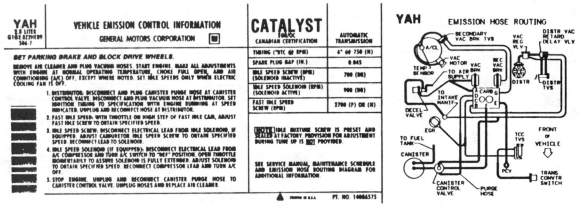

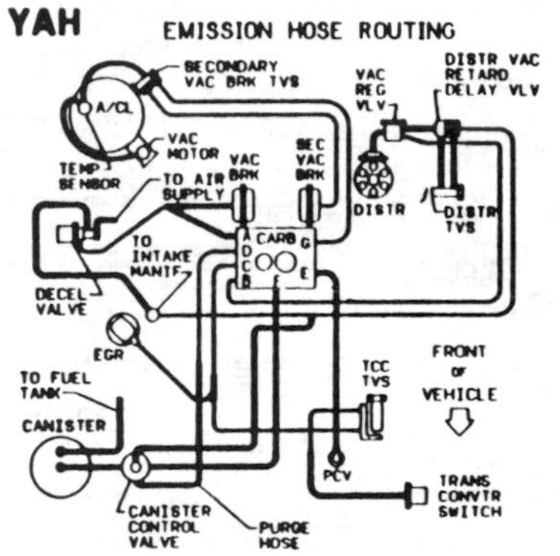

USAGE: 1-2AA00 & LE2 & NM5 (-9C1) (Except taxi)

2.8L—W/AUTO. TRANS. (MD9), NM5 (CANADA), SEO (9CI), EXC. CANADA

YAJ 2.8 LITER G1G2.8Z2HED9 586-7

VEHICLE EMISSION CONTROL INFORMATION
GENERAL MOTORS CORPORATION

SET PARKING BRAKE AND BLOCK DRIVE WHEELS.

REMOVE AIR CLEANER AND PLUG VACUUM HOSES. START ENGINE. MAKE ALL ADJUSTMENTS WITH ENGINE AT NORMAL OPERATING TEMPERATURE, CHOKE FULL OPEN, AND AIR CONDITIONING (A/C) OFF, EXCEPT WHERE NOTED. SET IDLE SPEEDS ONLY WHEN ELECTRIC COOLING FAN IS OFF.

1. DISTRIBUTOR: DISCONNECT AND PLUG VACUUM HOSE AT EGR VALVE AND CANISTER PURGE HOSE AT CANISTER CONTROL VALVE. DISCONNECT AND PLUG VACUUM HOSE AT DISTRIBUTOR. SET IGNITION TIMING TO SPECIFICATION WITH ENGINE RUNNING AT SPEED INDICATED. UNPLUG AND RECONNECT HOSE AT DISTRIBUTOR.
2. FAST IDLE SPEED: WITH THROTTLE ON HIGH STEP OF FAST IDLE CAM, ADJUST FAST IDLE SCREW TO OBTAIN SPECIFIED SPEED.
3. IDLE SPEED SCREW: DISCONNECT ELECTRICAL LEAD FROM IDLE SOLENOID, IF EQUIPPED. ADJUST CARBURETOR IDLE SPEED SCREW TO OBTAIN SPECIFIED SPEED. RECONNECT LEAD TO SOLENOID.
4. IDLE SPEED SOLENOID (IF EQUIPPED): DISCONNECT ELECTRICAL LEAD FROM A/C COMPRESSOR AND TURN A/C SWITCH TO "DEF" POSITION. OPEN THROTTLE MOMENTARILY TO ASSURE SOLENOID IS FULLY EXTENDED. ADJUST SOLENOID TO OBTAIN SPECIFIED SPEED. RECONNECT COMPRESSOR LEAD AND TURN A/C OFF.
5. STOP ENGINE. UNPLUG AND RECONNECT VACUUM HOSE TO EGR VALVE AND CANISTER PURGE HOSE TO CANISTER CONTROL VALVE. UNPLUG HOSES AND REPLACE AIR CLEANER.

CATALYST EGR/OC CANADIAN CERTIFICATION	AUTOMATIC TRANSMISSION
TIMING (°BTC @ RPM)	10° @ 750 (N)
SPARK PLUG GAP (IN.)	0.045
IDLE SPEED SCREW (RPM) (SOLENOID INACTIVE)	700 (DR)
IDLE SPEED SOLENOID (RPM) (SOLENOID ACTIVE)	900 (DR)
FAST IDLE SPEED SCREW (RPM)	2700 (P) OR (N)

NOTE IDLE MIXTURE SCREW IS PRESET AND SEALED AT FACTORY. PROVISION FOR ADJUSTMENT DURING TUNE UP IS NOT PROVIDED.

SEE SERVICE MANUAL, MAINTENANCE SCHEDULE AND EMISSION HOSE ROUTING DIAGRAM FOR ADDITIONAL INFORMATION.

PRINTED IN U.S.A. PT. NO. 14004576

YAJ EMISSION HOSE ROUTING

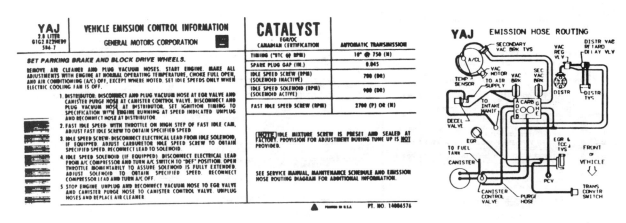

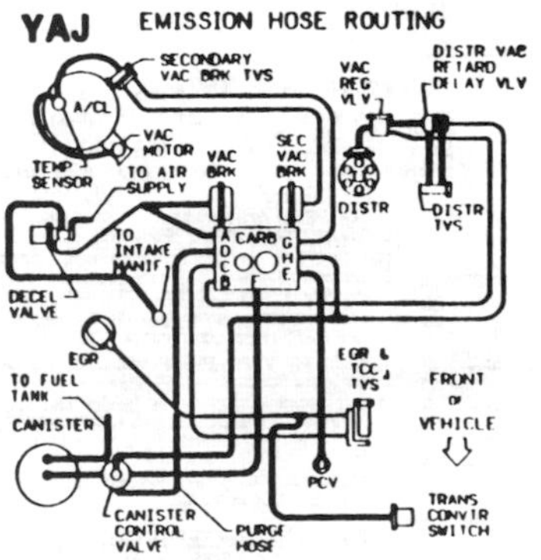

USAGE: 1AA00 & LE2 & NM5 & 9C1 (taxi only)

2.8L—W/AUTO. TRANS.—CANADA

VACUUM CIRCUITS

(© G.M. Corp.)

1986 CHEVROLET 2.8L

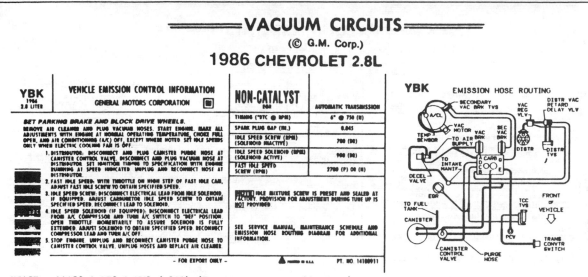

YBK 1986 2.8 LITER

VEHICLE EMISSION CONTROL INFORMATION
GENERAL MOTORS CORPORATION

SET PARKING BRAKE AND BLOCK DRIVE WHEELS.

REMOVE AIR CLEANER AND PLUG VACUUM HOSES. START ENGINE. MAKE ALL ADJUSTMENTS WITH ENGINE AT NORMAL OPERATING TEMPERATURE, CHOKE FULL OPEN, AND AIR CONDITIONING (A/C) OFF, EXCEPT WHERE NOTED. SET IDLE SPEEDS ONLY WHEN ELECTRIC COOLING FAN IS OFF.

1. DISTRIBUTOR: DISCONNECT AND PLUG CANISTER PURGE HOSE AT CANISTER CONTROL VALVE. DISCONNECT AND PLUG VACUUM HOSE AT DISTRIBUTOR. SET IGNITION TIMING TO SPECIFICATION WITH ENGINE RUNNING AT SPEED INDICATED. UNPLUG AND RECONNECT HOSE AT DISTRIBUTOR.
2. FAST IDLE SPEED: WITH THROTTLE ON HIGH STEP OF FAST IDLE CAM, ADJUST FAST IDLE SCREW TO OBTAIN SPECIFIED SPEED.
3. IDLE SPEED SCREW: DISCONNECT ELECTRICAL LEAD FROM IDLE SOLENOID, IF EQUIPPED. ADJUST CARBURETOR IDLE SPEED SCREW TO OBTAIN SPECIFIED SPEED. RECONNECT LEAD TO SOLENOID.
4. IDLE SPEED SOLENOID (IF EQUIPPED): DISCONNECT ELECTRICAL LEAD FROM A/C COMPRESSOR AND TURN A/C SWITCH TO "DEF" POSITION. OPEN THROTTLE MOMENTARILY TO ASSURE SOLENOID IS FULLY EXTENDED. ADJUST SOLENOID TO OBTAIN SPECIFIED SPEED. RECONNECT COMPRESSOR LEAD AND TURN A/C OFF.
5. STOP ENGINE. UNPLUG AND RECONNECT CANISTER PURGE HOSE TO CANISTER CONTROL VALVE. UNPLUG HOSES AND REPLACE AIR CLEANER.

NON-CATALYST EGR	AUTOMATIC TRANSMISSION
TIMING (°BTC @ RPM)	6° @ 750 (N)
SPARK PLUG GAP (IN.)	0.045
IDLE SPEED SCREW (RPM) (SOLENOID INACTIVE)	700 (DR)
IDLE SPEED SOLENOID (RPM) (SOLENOID ACTIVE)	900 (DR)
FAST IDLE SPEED SCREW (RPM)	2700 (P) OR (N)

NOTE: IDLE MIXTURE SCREW IS PRESET AND SEALED AT FACTORY. PROVISION FOR ADJUSTMENT DURING TUNE UP IS NOT PROVIDED.

SEE SERVICE MANUAL, MAINTENANCE SCHEDULE AND EMISSION HOSE ROUTING DIAGRAM FOR ADDITIONAL INFORMATION.

- FOR EXPORT ONLY -   PRINTED IN U.S.A.   PT. NO. 14100911

YBK EMISSION HOSE ROUTING

SECONDARY VAC BRK TVS; VAC REG VLV; DISTR VAC RETARD DELAY VLV; A/CL; VAC MOTOR; TEMP SENSOR; TO AIR SUPPLY; VAC BRK; SEC VAC BRK; DISTR; DISTR TVS; TO INTAKE MANIF.; CARB; DECEL VALVE; EGR; TO FUEL TANK; TCC TVS; FRONT OF VEHICLE; CANISTER; PCV; TRANS CONVTR SWITCH; CANISTER CONTROL VALVE; PURGE HOSE

USAGE: 1AA00 & LE2 & NM8 (-9C1) (Export - except police car)

2.8L—W/AUTO. TRANS. (MD9, ME9), NM8—EXPORT

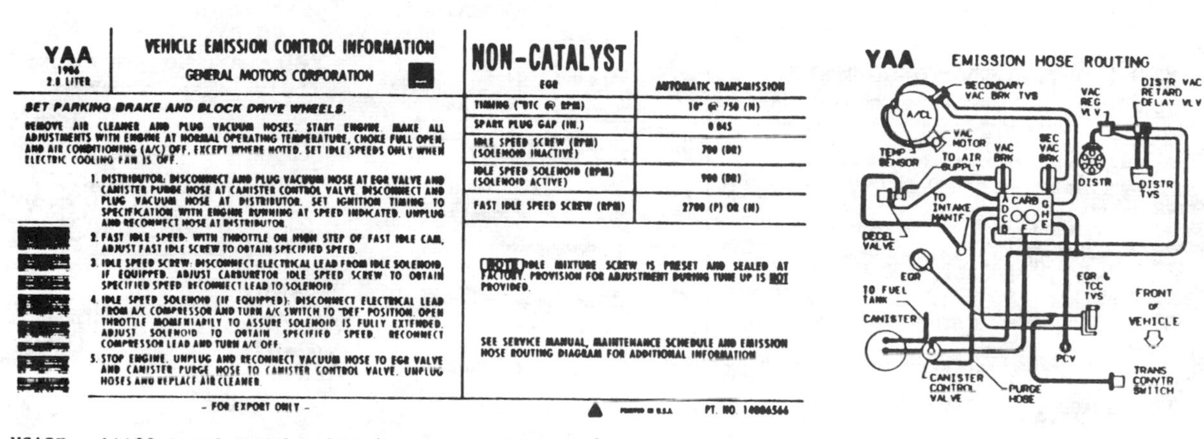
YAA 1986 2.8 LITER

VEHICLE EMISSION CONTROL INFORMATION
GENERAL MOTORS CORPORATION

SET PARKING BRAKE AND BLOCK DRIVE WHEELS.

REMOVE AIR CLEANER AND PLUG VACUUM HOSES. START ENGINE. MAKE ALL ADJUSTMENTS WITH ENGINE AT NORMAL OPERATING TEMPERATURE, CHOKE FULL OPEN, AND AIR CONDITIONING (A/C) OFF, EXCEPT WHERE NOTED. SET IDLE SPEEDS ONLY WHEN ELECTRIC COOLING FAN IS OFF.

1. DISTRIBUTOR: DISCONNECT AND PLUG VACUUM HOSE AT EGR VALVE AND CANISTER PURGE HOSE AT CANISTER CONTROL VALVE. DISCONNECT AND PLUG VACUUM HOSE AT DISTRIBUTOR. SET IGNITION TIMING TO SPECIFICATION WITH ENGINE RUNNING AT SPEED INDICATED. UNPLUG AND RECONNECT HOSE AT DISTRIBUTOR.
2. FAST IDLE SPEED: WITH THROTTLE ON HIGH STEP OF FAST IDLE CAM, ADJUST FAST IDLE SCREW TO OBTAIN SPECIFIED SPEED.
3. IDLE SPEED SCREW: DISCONNECT ELECTRICAL LEAD FROM IDLE SOLENOID, IF EQUIPPED. ADJUST CARBURETOR IDLE SPEED SCREW TO OBTAIN SPECIFIED SPEED. RECONNECT LEAD TO SOLENOID.
4. IDLE SPEED SOLENOID (IF EQUIPPED): DISCONNECT ELECTRICAL LEAD FROM A/C COMPRESSOR AND TURN A/C SWITCH TO "DEF" POSITION. OPEN THROTTLE MOMENTARILY TO ASSURE SOLENOID IS FULLY EXTENDED. ADJUST SOLENOID TO OBTAIN SPECIFIED SPEED. RECONNECT COMPRESSOR LEAD AND TURN A/C OFF.
5. STOP ENGINE. UNPLUG AND RECONNECT VACUUM HOSE TO EGR VALVE AND CANISTER PURGE HOSE TO CANISTER CONTROL VALVE. UNPLUG HOSES AND REPLACE AIR CLEANER.

NON-CATALYST EGR	AUTOMATIC TRANSMISSION
TIMING (°BTC @ RPM)	10° @ 750 (N)
SPARK PLUG GAP (IN.)	0.045
IDLE SPEED SCREW (RPM) (SOLENOID INACTIVE)	700 (DR)
IDLE SPEED SOLENOID (RPM) (SOLENOID ACTIVE)	900 (DR)
FAST IDLE SPEED SCREW (RPM)	2700 (P) OR (N)

NOTE: IDLE MIXTURE SCREW IS PRESET AND SEALED AT FACTORY. PROVISION FOR ADJUSTMENT DURING TUNE UP IS NOT PROVIDED.

SEE SERVICE MANUAL, MAINTENANCE SCHEDULE AND EMISSION HOSE ROUTING DIAGRAM FOR ADDITIONAL INFORMATION

- FOR EXPORT ONLY -   PRINTED IN U.S.A.   PT. NO. 14086566

USAGE: 1AA00 & LE2 & NM8 & 9C1 (Export - police car)

2.8L—W/AUTO. TRANS.—EXPORT

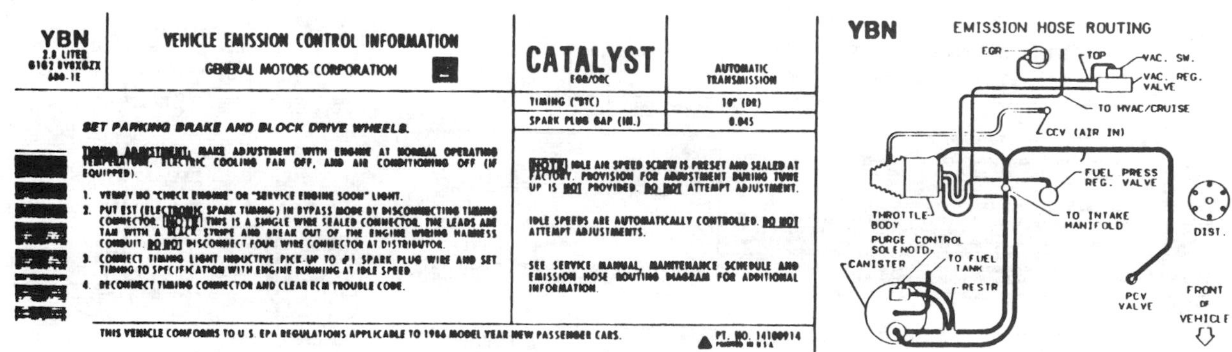
YBN 2.8 LITER 6162 8V8XG2X 600-1E

VEHICLE EMISSION CONTROL INFORMATION
GENERAL MOTORS CORPORATION

SET PARKING BRAKE AND BLOCK DRIVE WHEELS.

TIMING ADJUSTMENT: MAKE ADJUSTMENT WITH ENGINE AT NORMAL OPERATING TEMPERATURE, ELECTRIC COOLING FAN OFF, AND AIR CONDITIONING OFF (IF EQUIPPED).

1. VERIFY NO "CHECK ENGINE" OR "SERVICE ENGINE SOON" LIGHT.
2. PUT EST (ELECTRONIC SPARK TIMING) IN BYPASS MODE BY DISCONNECTING TIMING CONNECTOR. NOTE: THIS IS A SINGLE WIRE SEALED CONNECTOR. THE LEADS ARE TAN WITH A BLACK STRIPE AND BREAK OUT OF THE ENGINE WIRING HARNESS CONDUIT. DO NOT DISCONNECT FOUR WIRE CONNECTOR AT DISTRIBUTOR.
3. CONNECT TIMING LIGHT INDUCTIVE PICK-UP TO #1 SPARK PLUG WIRE AND SET TIMING TO SPECIFICATION WITH ENGINE RUNNING AT IDLE SPEED.
4. RECONNECT TIMING CONNECTOR AND CLEAR ECM TROUBLE CODE.

CATALYST EGR/ORC	AUTOMATIC TRANSMISSION
TIMING (°BTC)	10° (DR)
SPARK PLUG GAP (IN.)	0.045

NOTE: IDLE AIR SPEED SCREW IS PRESET AND SEALED AT FACTORY. PROVISION FOR ADJUSTMENT DURING TUNE UP IS NOT PROVIDED. DO NOT ATTEMPT ADJUSTMENT.

IDLE SPEEDS ARE AUTOMATICALLY CONTROLLED. DO NOT ATTEMPT ADJUSTMENTS.

SEE SERVICE MANUAL, MAINTENANCE SCHEDULE AND EMISSION HOSE ROUTING DIAGRAM FOR ADDITIONAL INFORMATION.

THIS VEHICLE CONFORMS TO U.S. EPA REGULATIONS APPLICABLE TO 1986 MODEL YEAR NEW PASSENGER CARS.   PT. NO. 14100914 PRINTED IN U.S.A.

USAGE: 1-2AA00 & LB6 & MD9/ME9 & NA5 (-9C1) (Except police car)

2.8L—W/AUTO. TRANS. (MD9, ME9)—FEDERAL

VACUUM CIRCUITS

(© G.M. Corp.)

1986 CHEVROLET 2.8L

YBP
2.8 LITER
G102.8V8XGZX
600-1E

VEHICLE EMISSION CONTROL INFORMATION
GENERAL MOTORS CORPORATION

CATALYST
EGR/ORC

	AUTOMATIC TRANSMISSION
TIMING (°BTC)	10° (DR)
SPARK PLUG GAP (IN.)	0.045

SET PARKING BRAKE AND BLOCK DRIVE WHEELS.

TIMING ADJUSTMENT, MAKE ADJUSTMENT WITH ENGINE AT NORMAL OPERATING TEMPERATURE, ELECTRIC COOLING FAN OFF, AND AIR CONDITIONING OFF (IF EQUIPPED).

1. VERIFY NO "CHECK ENGINE" OR "SERVICE ENGINE SOON" LIGHT.
2. PUT EST (ELECTRONIC SPARK TIMING) IN BYPASS MODE BY DISCONNECTING TIMING CONNECTOR. NOTE: THIS IS A SINGLE WIRE SEALED CONNECTOR. THE LEADS ARE TAN WITH A BLACK STRIPE AND BREAK OUT OF THE ENGINE WIRING HARNESS CONDUIT. DO NOT DISCONNECT FOUR-WIRE CONNECTOR AT DISTRIBUTOR.
3. CONNECT TIMING LIGHT INDUCTIVE PICK-UP TO #1 SPARK PLUG WIRE AND SET TIMING TO SPECIFICATION WITH ENGINE RUNNING AT IDLE SPEED.
4. RECONNECT TIMING CONNECTOR AND CLEAR ECM TROUBLE CODE.

NOTE: IDLE AIR SPEED SCREW IS PRESET AND SEALED AT FACTORY. PROVISION FOR ADJUSTMENT DURING TUNE UP IS NOT PROVIDED. DO NOT ATTEMPT ADJUSTMENT.

IDLE SPEEDS ARE AUTOMATICALLY CONTROLLED. DO NOT ATTEMPT ADJUSTMENTS.

SEE SERVICE MANUAL, MAINTENANCE SCHEDULE AND EMISSION HOSE ROUTING DIAGRAM FOR ADDITIONAL INFORMATION.

THIS VEHICLE CONFORMS TO U.S. EPA REGULATIONS APPLICABLE TO 1986 MODEL YEAR NEW PASSENGER CARS.

PT. NO. 14100915
PRINTED IN U.S.A.

YBP EMISSION HOSE ROUTING
EGR
TOP
VAC. SW.
VAC. REG. VALVE
TO HVAC/CRUISE
CCV (AIR IN)
FUEL PRESS REG. VALVE
THROTTLE BODY
TO INTAKE MANIFOLD
DIST.
PURGE CONTROL SOLENOID
CANISTER
TO FUEL TANK
RESTR
PCV VALVE
FRONT OF VEHICLE

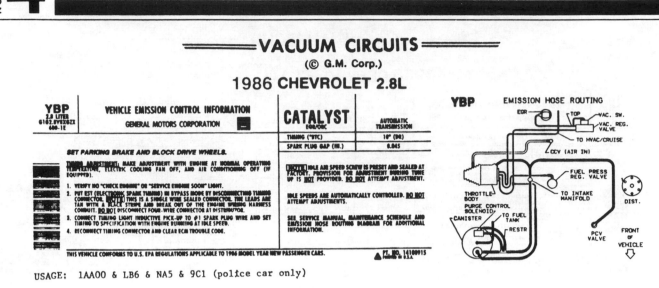

USAGE: 1AA00 & LB6 & NA5 & 9C1 (police car only)

2.8L—W/SEO (9CI)—FEDERAL

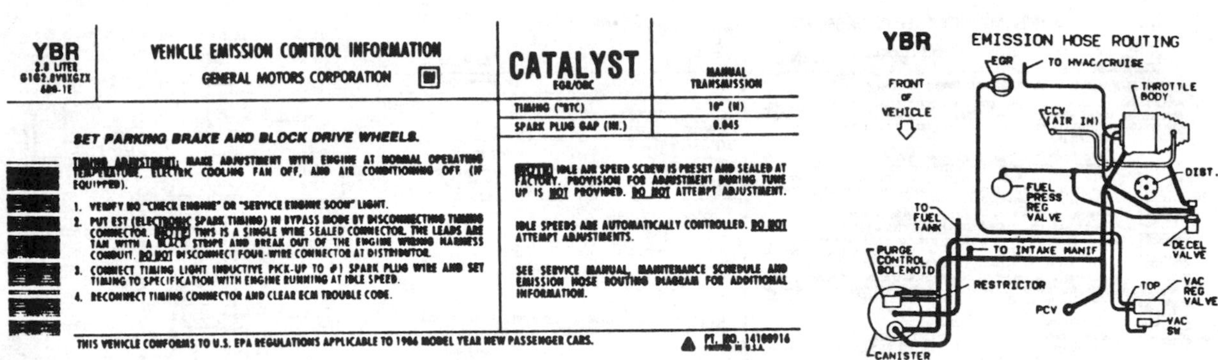

YBR
2.8 LITER
G102.8V8XGZX
600-1E

VEHICLE EMISSION CONTROL INFORMATION
GENERAL MOTORS CORPORATION

CATALYST
EGR/ORC

	MANUAL TRANSMISSION
TIMING (°BTC)	10° (N)
SPARK PLUG GAP (IN.)	0.045

SET PARKING BRAKE AND BLOCK DRIVE WHEELS.

TIMING ADJUSTMENT, MAKE ADJUSTMENT WITH ENGINE AT NORMAL OPERATING TEMPERATURE, ELECTRIC COOLING FAN OFF, AND AIR CONDITIONING OFF (IF EQUIPPED).

1. VERIFY NO "CHECK ENGINE" OR "SERVICE ENGINE SOON" LIGHT.
2. PUT EST (ELECTRONIC SPARK TIMING) IN BYPASS MODE BY DISCONNECTING TIMING CONNECTOR. NOTE: THIS IS A SINGLE WIRE SEALED CONNECTOR. THE LEADS ARE TAN WITH A BLACK STRIPE AND BREAK OUT OF THE ENGINE WIRING HARNESS CONDUIT. DO NOT DISCONNECT FOUR-WIRE CONNECTOR AT DISTRIBUTOR.
3. CONNECT TIMING LIGHT INDUCTIVE PICK-UP TO #1 SPARK PLUG WIRE AND SET TIMING TO SPECIFICATION WITH ENGINE RUNNING AT IDLE SPEED.
4. RECONNECT TIMING CONNECTOR AND CLEAR ECM TROUBLE CODE.

NOTE: IDLE AIR SPEED SCREW IS PRESET AND SEALED AT FACTORY. PROVISION FOR ADJUSTMENT DURING TUNE UP IS NOT PROVIDED. DO NOT ATTEMPT ADJUSTMENT.

IDLE SPEEDS ARE AUTOMATICALLY CONTROLLED. DO NOT ATTEMPT ADJUSTMENTS.

SEE SERVICE MANUAL, MAINTENANCE SCHEDULE AND EMISSION HOSE ROUTING DIAGRAM FOR ADDITIONAL INFORMATION.

THIS VEHICLE CONFORMS TO U.S. EPA REGULATIONS APPLICABLE TO 1986 MODEL YEAR NEW PASSENGER CARS.

PT. NO. 14100916
PRINTED IN U.S.A.

USAGE: 1JA00 & LB6 & MX6/MG2 & NA5

2.8L—W/MAN. TRANS. (MG2, MG6)—FEDERAL

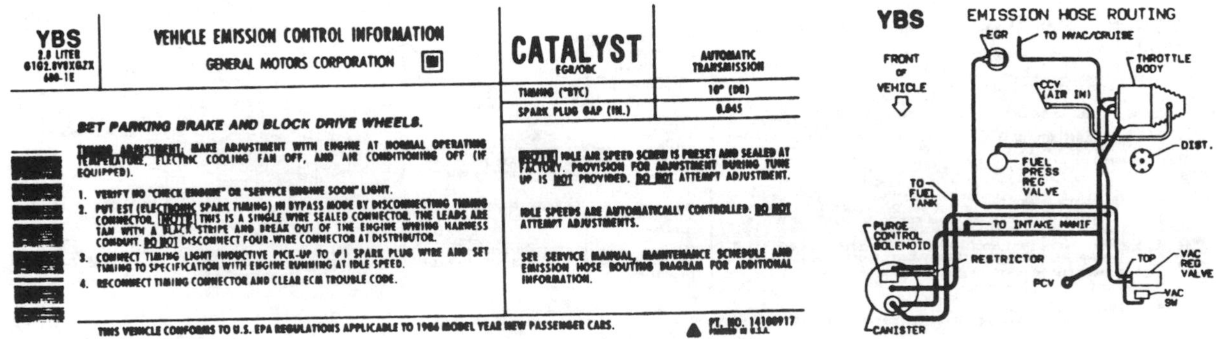

YBS
2.8 LITER
G102.8V8XGZX
600-1E

VEHICLE EMISSION CONTROL INFORMATION
GENERAL MOTORS CORPORATION

CATALYST
EGR/ORC

	AUTOMATIC TRANSMISSION
TIMING (°BTC)	10° (DR)
SPARK PLUG GAP (IN.)	0.045

SET PARKING BRAKE AND BLOCK DRIVE WHEELS.

TIMING ADJUSTMENT, MAKE ADJUSTMENT WITH ENGINE AT NORMAL OPERATING TEMPERATURE, ELECTRIC COOLING FAN OFF, AND AIR CONDITIONING OFF (IF EQUIPPED).

1. VERIFY NO "CHECK ENGINE" OR "SERVICE ENGINE SOON" LIGHT.
2. PUT EST (ELECTRONIC SPARK TIMING) IN BYPASS MODE BY DISCONNECTING TIMING CONNECTOR. NOTE: THIS IS A SINGLE WIRE SEALED CONNECTOR. THE LEADS ARE TAN WITH A BLACK STRIPE AND BREAK OUT OF THE ENGINE WIRING HARNESS CONDUIT. DO NOT DISCONNECT FOUR-WIRE CONNECTOR AT DISTRIBUTOR.
3. CONNECT TIMING LIGHT INDUCTIVE PICK-UP TO #1 SPARK PLUG WIRE AND SET TIMING TO SPECIFICATION WITH ENGINE RUNNING AT IDLE SPEED.
4. RECONNECT TIMING CONNECTOR AND CLEAR ECM TROUBLE CODE.

NOTE: IDLE AIR SPEED SCREW IS PRESET AND SEALED AT FACTORY. PROVISION FOR ADJUSTMENT DURING TUNE UP IS NOT PROVIDED. DO NOT ATTEMPT ADJUSTMENT.

IDLE SPEEDS ARE AUTOMATICALLY CONTROLLED. DO NOT ATTEMPT ADJUSTMENTS.

SEE SERVICE MANUAL, MAINTENANCE SCHEDULE AND EMISSION HOSE ROUTING DIAGRAM FOR ADDITIONAL INFORMATION.

THIS VEHICLE CONFORMS TO U.S. EPA REGULATIONS APPLICABLE TO 1986 MODEL YEAR NEW PASSENGER CARS.

PT. NO. 14100917
PRINTED IN U.S.A.

USAGE: 1JA00 & LB6 & MD9 & NA5

2.8L— W/AUTO. TRANS. (MD9)—FEDERAL

VACUUM CIRCUITS

(© G.M. Corp.)

1986 CHEVROLET 2.8L

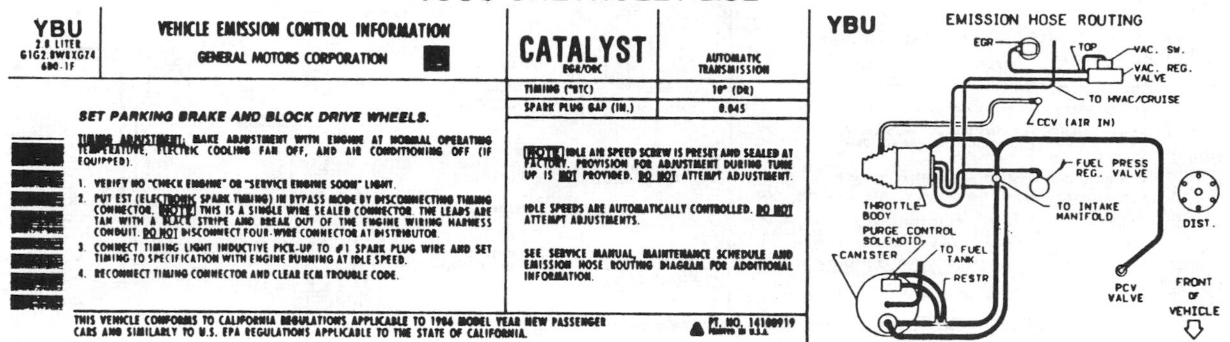

YBU
2.8 LITER
G1G2.8W8XGZ4
6D0-1F

VEHICLE EMISSION CONTROL INFORMATION
GENERAL MOTORS CORPORATION

CATALYST
EGR/ORC

AUTOMATIC TRANSMISSION

TIMING (°BTC)	10° (DR)
SPARK PLUG GAP (IN.)	0.045

SET PARKING BRAKE AND BLOCK DRIVE WHEELS.

TIMING ADJUSTMENT: MAKE ADJUSTMENT WITH ENGINE AT NORMAL OPERATING TEMPERATURE, ELECTRIC COOLING FAN OFF, AND AIR CONDITIONING OFF (IF EQUIPPED).

1. VERIFY NO "CHECK ENGINE" OR "SERVICE ENGINE SOON" LIGHT.
2. PUT EST (ELECTRONIC SPARK TIMING) IN BYPASS MODE BY DISCONNECTING TIMING CONNECTOR. NOTE: THIS IS A SINGLE WIRE SEALED CONNECTOR. THE LEADS ARE TAN WITH A BLACK STRIPE AND BREAK OUT OF THE ENGINE WIRING HARNESS CONDUIT. DO NOT DISCONNECT FOUR-WIRE CONNECTOR AT DISTRIBUTOR.
3. CONNECT TIMING LIGHT INDUCTIVE PICK-UP TO #1 SPARK PLUG WIRE AND SET TIMING TO SPECIFICATION WITH ENGINE RUNNING AT IDLE SPEED.
4. RECONNECT TIMING CONNECTOR AND CLEAR ECM TROUBLE CODE.

NOTE: IDLE AIR SPEED SCREW IS PRESET AND SEALED AT FACTORY. PROVISION FOR ADJUSTMENT DURING TUNE UP IS NOT PROVIDED. DO NOT ATTEMPT ADJUSTMENT.

IDLE SPEEDS ARE AUTOMATICALLY CONTROLLED. DO NOT ATTEMPT ADJUSTMENTS.

SEE SERVICE MANUAL, MAINTENANCE SCHEDULE AND EMISSION HOSE ROUTING DIAGRAM FOR ADDITIONAL INFORMATION.

THIS VEHICLE CONFORMS TO CALIFORNIA REGULATIONS APPLICABLE TO 1986 MODEL YEAR NEW PASSENGER CARS AND SIMILARLY TO U.S. EPA REGULATIONS APPLICABLE TO THE STATE OF CALIFORNIA.

PT. NO. 14100919
PRINTED IN U.S.A.

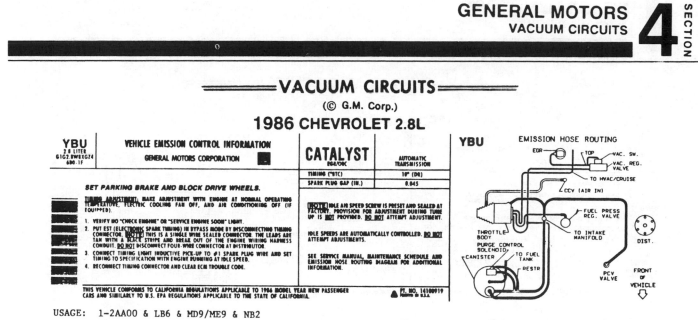

USAGE: 1-2AA00 & LB6 & MD9/ME9 & NB2

2.8L – W/AUTO. TRANS. (MD9, ME9) – CALIFORNIA

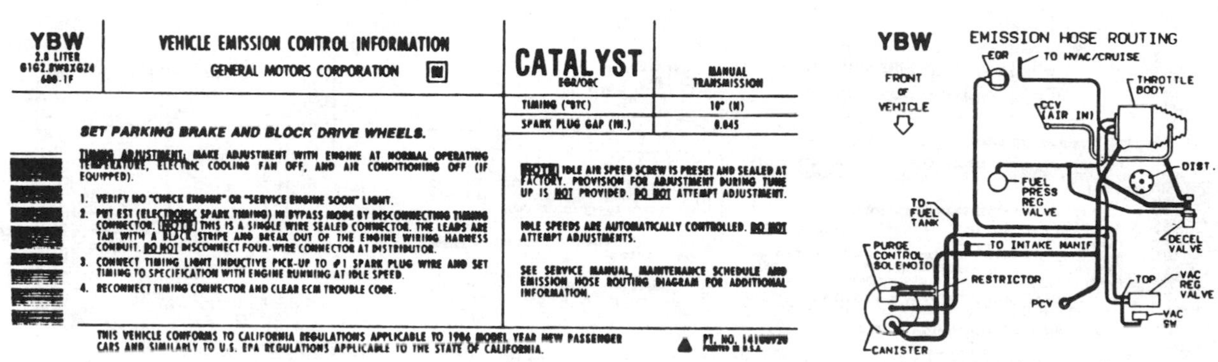

YBW
2.8 LITER
G1G2.8W8XGZ4
6D0-1F

VEHICLE EMISSION CONTROL INFORMATION
GENERAL MOTORS CORPORATION

CATALYST
EGR/ORC

MANUAL TRANSMISSION

TIMING (°BTC)	10° (N)
SPARK PLUG GAP (IN.)	0.045

SET PARKING BRAKE AND BLOCK DRIVE WHEELS.

TIMING ADJUSTMENT: MAKE ADJUSTMENT WITH ENGINE AT NORMAL OPERATING TEMPERATURE, ELECTRIC COOLING FAN OFF, AND AIR CONDITIONING OFF (IF EQUIPPED).

1. VERIFY NO "CHECK ENGINE" OR "SERVICE ENGINE SOON" LIGHT.
2. PUT EST (ELECTRONIC SPARK TIMING) IN BYPASS MODE BY DISCONNECTING TIMING CONNECTOR. NOTE: THIS IS A SINGLE WIRE SEALED CONNECTOR. THE LEADS ARE TAN WITH A BLACK STRIPE AND BREAK OUT OF THE ENGINE WIRING HARNESS CONDUIT. DO NOT DISCONNECT FOUR-WIRE CONNECTOR AT DISTRIBUTOR.
3. CONNECT TIMING LIGHT INDUCTIVE PICK-UP TO #1 SPARK PLUG WIRE AND SET TIMING TO SPECIFICATION WITH ENGINE RUNNING AT IDLE SPEED.
4. RECONNECT TIMING CONNECTOR AND CLEAR ECM TROUBLE CODE.

NOTE: IDLE AIR SPEED SCREW IS PRESET AND SEALED AT FACTORY. PROVISION FOR ADJUSTMENT DURING TUNE UP IS NOT PROVIDED. DO NOT ATTEMPT ADJUSTMENT.

IDLE SPEEDS ARE AUTOMATICALLY CONTROLLED. DO NOT ATTEMPT ADJUSTMENTS.

SEE SERVICE MANUAL, MAINTENANCE SCHEDULE AND EMISSION HOSE ROUTING DIAGRAM FOR ADDITIONAL INFORMATION.

THIS VEHICLE CONFORMS TO CALIFORNIA REGULATIONS APPLICABLE TO 1986 MODEL YEAR NEW PASSENGER CARS AND SIMILARLY TO U.S. EPA REGULATIONS APPLICABLE TO THE STATE OF CALIFORNIA.

PT. NO. 14100920
PRINTED IN U.S.A.

USAGE: 1JA00 & LB6 MX6/MG2 & NB2

2.8L – W/MAN. TRANS. (MG2, MX6) – CALIFORNIA

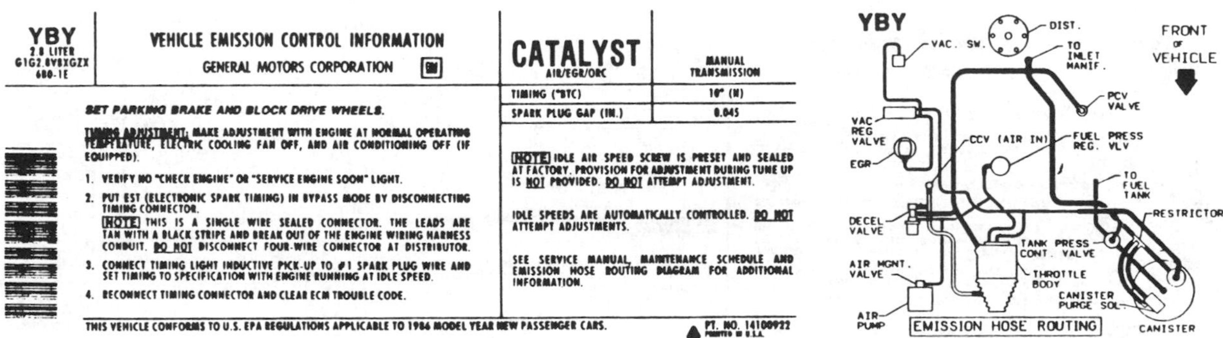

YBY
2.8 LITER
G1G2.8V8XGZX
6D0-1E

VEHICLE EMISSION CONTROL INFORMATION
GENERAL MOTORS CORPORATION

CATALYST
AIR/EGR/ORC

MANUAL TRANSMISSION

TIMING (°BTC)	10° (N)
SPARK PLUG GAP (IN.)	0.045

SET PARKING BRAKE AND BLOCK DRIVE WHEELS.

TIMING ADJUSTMENT: MAKE ADJUSTMENT WITH ENGINE AT NORMAL OPERATING TEMPERATURE, ELECTRIC COOLING FAN OFF, AND AIR CONDITIONING OFF (IF EQUIPPED).

1. VERIFY NO "CHECK ENGINE" OR "SERVICE ENGINE SOON" LIGHT.
2. PUT EST (ELECTRONIC SPARK TIMING) IN BYPASS MODE BY DISCONNECTING TIMING CONNECTOR.
NOTE THIS IS A SINGLE WIRE SEALED CONNECTOR. THE LEADS ARE TAN WITH A BLACK STRIPE AND BREAK OUT OF THE ENGINE WIRING HARNESS CONDUIT. DO NOT DISCONNECT FOUR-WIRE CONNECTOR AT DISTRIBUTOR.
3. CONNECT TIMING LIGHT INDUCTIVE PICK-UP TO #1 SPARK PLUG WIRE AND SET TIMING TO SPECIFICATION WITH ENGINE RUNNING AT IDLE SPEED.
4. RECONNECT TIMING CONNECTOR AND CLEAR ECM TROUBLE CODE.

NOTE IDLE AIR SPEED SCREW IS PRESET AND SEALED AT FACTORY. PROVISION FOR ADJUSTMENT DURING TUNE UP IS NOT PROVIDED. DO NOT ATTEMPT ADJUSTMENT.

IDLE SPEEDS ARE AUTOMATICALLY CONTROLLED. DO NOT ATTEMPT ADJUSTMENTS.

SEE SERVICE MANUAL, MAINTENANCE SCHEDULE AND EMISSION HOSE ROUTING DIAGRAM FOR ADDITIONAL INFORMATION.

THIS VEHICLE CONFORMS TO U.S. EPA REGULATIONS APPLICABLE TO 1986 MODEL YEAR NEW PASSENGER CARS.

PT. NO. 14100922
PRINTED IN U.S.A.

USAGE: 1FA00 & LB8 & MB1 & NA5 (Chev. only)

2.8L – W/MAN. TRANS. (MB1) – FEDERAL

VACUUM CIRCUITS

(© G.M. Corp.)

1986 CHEVROLET 2.8L

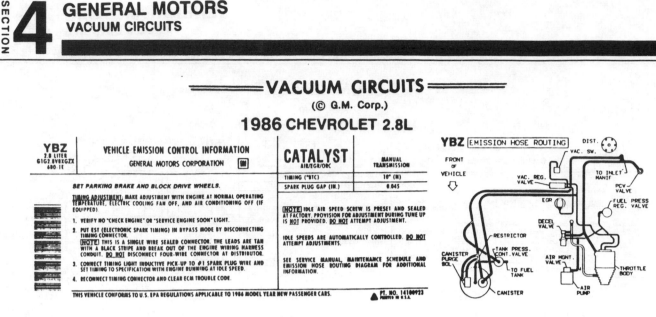

YBZ
2.8 LITER
G1G2.8V8XGZX
6B0-1E

VEHICLE EMISSION CONTROL INFORMATION
GENERAL MOTORS CORPORATION

CATALYST
AIR/EGR/ORC

	MANUAL TRANSMISSION
TIMING (°BTC)	10° (N)
SPARK PLUG GAP (IN.)	0.045

SET PARKING BRAKE AND BLOCK DRIVE WHEELS.

TIMING ADJUSTMENT: MAKE ADJUSTMENT WITH ENGINE AT NORMAL OPERATING TEMPERATURE, ELECTRIC COOLING FAN OFF, AND AIR CONDITIONING OFF (IF EQUIPPED).

1. VERIFY NO "CHECK ENGINE" OR "SERVICE ENGINE SOON" LIGHT.
2. PUT EST (ELECTRONIC SPARK TIMING) IN BYPASS MODE BY DISCONNECTING TIMING CONNECTOR.
   NOTE THIS IS A SINGLE WIRE SEALED CONNECTOR. THE LEADS ARE TAN WITH A BLACK STRIPE AND BREAK OUT OF THE ENGINE WIRING HARNESS CONDUIT. DO NOT DISCONNECT FOUR-WIRE CONNECTOR AT DISTRIBUTOR.
3. CONNECT TIMING LIGHT INDUCTIVE PICK-UP TO #1 SPARK PLUG WIRE AND SET TIMING TO SPECIFICATION WITH ENGINE RUNNING AT IDLE SPEED.
4. RECONNECT TIMING CONNECTOR AND CLEAR ECM TROUBLE CODE.

NOTE IDLE AIR SPEED SCREW IS PRESET AND SEALED AT FACTORY. PROVISION FOR ADJUSTMENT DURING TUNE UP IS NOT PROVIDED. DO NOT ATTEMPT ADJUSTMENT.

IDLE SPEEDS ARE AUTOMATICALLY CONTROLLED. DO NOT ATTEMPT ADJUSTMENTS.

SEE SERVICE MANUAL, MAINTENANCE SCHEDULE AND EMISSION HOSE ROUTING DIAGRAM FOR ADDITIONAL INFORMATION.

THIS VEHICLE CONFORMS TO U.S. EPA REGULATIONS APPLICABLE TO 1986 MODEL YEAR NEW PASSENGER CARS.

PT. NO. 14100923
PRINTED IN U.S.A.

YBZ EMISSION HOSE ROUTING
DIST.
VAC. SW.
FRONT OF VEHICLE
VAC. REG. VALVE
TO INLET MANIF
PCV VALVE
EGR
FUEL PRESS REG. VALVE
DECEL VALVE
RESTRICTOR
TANK PRESS. CONT. VALVE
CANISTER PURGE SOL
TO FUEL TANK
AIR MGNT. VALVE
THROTTLE BODY
CANISTER
AIR PUMP

USAGE: 2FA00 & LB8 & MB1 & NA5 (Pontiac only)

2.8L – W/MAN. TRANS. – FEDERAL

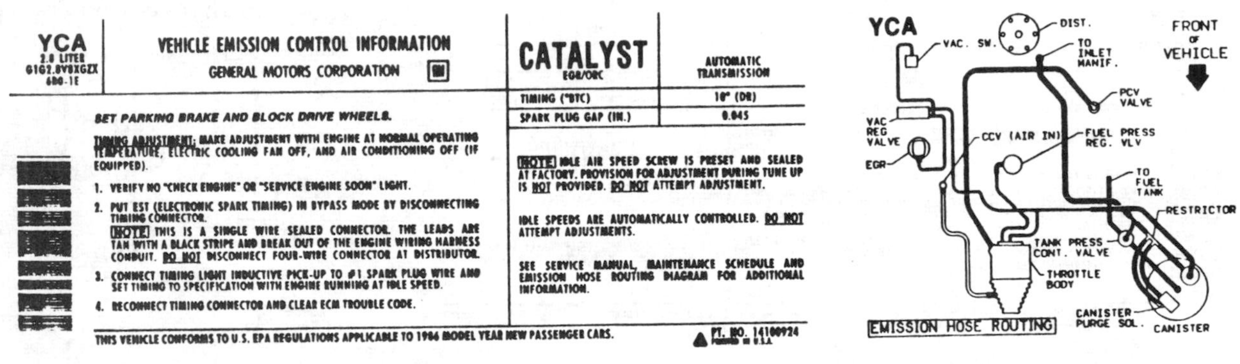

YCA
2.8 LITER
G1G2.8V8XGZX
6B0-1E

VEHICLE EMISSION CONTROL INFORMATION
GENERAL MOTORS CORPORATION

CATALYST
EGR/ORC

	AUTOMATIC TRANSMISSION
TIMING (°BTC)	10° (DR)
SPARK PLUG GAP (IN.)	0.045

SET PARKING BRAKE AND BLOCK DRIVE WHEELS.

TIMING ADJUSTMENT: MAKE ADJUSTMENT WITH ENGINE AT NORMAL OPERATING TEMPERATURE, ELECTRIC COOLING FAN OFF, AND AIR CONDITIONING OFF (IF EQUIPPED).

1. VERIFY NO "CHECK ENGINE" OR "SERVICE ENGINE SOON" LIGHT.
2. PUT EST (ELECTRONIC SPARK TIMING) IN BYPASS MODE BY DISCONNECTING TIMING CONNECTOR.
   NOTE THIS IS A SINGLE WIRE SEALED CONNECTOR. THE LEADS ARE TAN WITH A BLACK STRIPE AND BREAK OUT OF THE ENGINE WIRING HARNESS CONDUIT. DO NOT DISCONNECT FOUR-WIRE CONNECTOR AT DISTRIBUTOR.
3. CONNECT TIMING LIGHT INDUCTIVE PICK-UP TO #1 SPARK PLUG WIRE AND SET TIMING TO SPECIFICATION WITH ENGINE RUNNING AT IDLE SPEED.
4. RECONNECT TIMING CONNECTOR AND CLEAR ECM TROUBLE CODE.

NOTE IDLE AIR SPEED SCREW IS PRESET AND SEALED AT FACTORY. PROVISION FOR ADJUSTMENT DURING TUNE UP IS NOT PROVIDED. DO NOT ATTEMPT ADJUSTMENT.

IDLE SPEEDS ARE AUTOMATICALLY CONTROLLED. DO NOT ATTEMPT ADJUSTMENTS.

SEE SERVICE MANUAL, MAINTENANCE SCHEDULE AND EMISSION HOSE ROUTING DIAGRAM FOR ADDITIONAL INFORMATION.

THIS VEHICLE CONFORMS TO U.S. EPA REGULATIONS APPLICABLE TO 1986 MODEL YEAR NEW PASSENGER CARS.

PT. NO. 14100924
PRINTED IN U.S.A.

USAGE: 1FA00 & LB8 & MD8 & NA5 (Chev. only)

2.8L – W/AUTO. TRANS. (MD8) – FEDERAL

YCB
2.8 LITER
G1G2.8V8XGZX
6B0-1E

VEHICLE EMISSION CONTROL INFORMATION
GENERAL MOTORS CORPORATION

CATALYST
EGR/ORC

	AUTOMATIC TRANSMISSION
TIMING (°BTC)	10° (DR)
SPARK PLUG GAP (IN.)	0.045

SET PARKING BRAKE AND BLOCK DRIVE WHEELS.

TIMING ADJUSTMENT: MAKE ADJUSTMENT WITH ENGINE AT NORMAL OPERATING TEMPERATURE, ELECTRIC COOLING FAN OFF, AND AIR CONDITIONING OFF (IF EQUIPPED).

1. VERIFY NO "CHECK ENGINE" OR "SERVICE ENGINE SOON" LIGHT.
2. PUT EST (ELECTRONIC SPARK TIMING) IN BYPASS MODE BY DISCONNECTING TIMING CONNECTOR.
   NOTE THIS IS A SINGLE WIRE SEALED CONNECTOR. THE LEADS ARE TAN WITH A BLACK STRIPE AND BREAK OUT OF THE ENGINE WIRING HARNESS CONDUIT. DO NOT DISCONNECT FOUR-WIRE CONNECTOR AT DISTRIBUTOR.
3. CONNECT TIMING LIGHT INDUCTIVE PICK-UP TO #1 SPARK PLUG WIRE AND SET TIMING TO SPECIFICATION WITH ENGINE RUNNING AT IDLE SPEED.
4. RECONNECT TIMING CONNECTOR AND CLEAR ECM TROUBLE CODE.

NOTE IDLE AIR SPEED SCREW IS PRESET AND SEALED AT FACTORY. PROVISION FOR ADJUSTMENT DURING TUNE UP IS NOT PROVIDED. DO NOT ATTEMPT ADJUSTMENT.

IDLE SPEEDS ARE AUTOMATICALLY CONTROLLED. DO NOT ATTEMPT ADJUSTMENTS.

SEE SERVICE MANUAL, MAINTENANCE SCHEDULE AND EMISSION HOSE ROUTING DIAGRAM FOR ADDITIONAL INFORMATION.

THIS VEHICLE CONFORMS TO U S. EPA REGULATIONS APPLICABLE TO 1986 MODEL YEAR NEW PASSENGER CARS.

PT. NO. 14100925
PRINTED IN U S A

USAGE: 2FA00 & LB8 & MD8 & NA5 (Pontiac only)

2.8L – W/AUTO. TRANS. – FEDERAL

VACUUM CIRCUITS

(© G.M. Corp.)

1986 CHEVROLET 2.8L

YCC
2.8 LITER
G1G2.8W8XGZ4
680-1F

VEHICLE EMISSION CONTROL INFORMATION
GENERAL MOTORS CORPORATION

CATALYST
AIR/EGR/ORC

MANUAL TRANSMISSION

TIMING (°BTC)	10° (N)
SPARK PLUG GAP (IN.)	0.045

SET PARKING BRAKE AND BLOCK DRIVE WHEELS.

TIMING ADJUSTMENT: MAKE ADJUSTMENT WITH ENGINE AT NORMAL OPERATING TEMPERATURE, ELECTRIC COOLING FAN OFF, AND AIR CONDITIONING OFF (IF EQUIPPED).

1. VERIFY NO "CHECK ENGINE" OR "SERVICE ENGINE SOON" LIGHT.
2. PUT EST (ELECTRONIC SPARK TIMING) IN BYPASS MODE BY DISCONNECTING TIMING CONNECTOR.
   NOTE THIS IS A SINGLE WIRE SEALED CONNECTOR. THE LEADS ARE TAN WITH A BLACK STRIPE AND BREAK OUT OF THE ENGINE WIRING HARNESS CONDUIT. DO NOT DISCONNECT FOUR-WIRE CONNECTOR AT DISTRIBUTOR.
3. CONNECT TIMING LIGHT INDUCTIVE PICK-UP TO #1 SPARK PLUG WIRE AND SET TIMING TO SPECIFICATION WITH ENGINE RUNNING AT IDLE SPEED.
4. RECONNECT TIMING CONNECTOR AND CLEAR ECM TROUBLE CODE.

NOTE IDLE AIR SPEED SCREW IS PRESET AND SEALED AT FACTORY. PROVISION FOR ADJUSTMENT DURING TUNE UP IS NOT PROVIDED. DO NOT ATTEMPT ADJUSTMENT.

IDLE SPEEDS ARE AUTOMATICALLY CONTROLLED. DO NOT ATTEMPT ADJUSTMENTS.

SEE SERVICE MANUAL, MAINTENANCE SCHEDULE AND EMISSION HOSE ROUTING DIAGRAM FOR ADDITIONAL INFORMATION.

THIS VEHICLE CONFORMS TO CALIFORNIA REGULATIONS APPLICABLE TO 1986 MODEL YEAR NEW PASSENGER CARS AND SIMILARLY TO U.S. EPA REGULATIONS APPLICABLE TO THE STATE OF CALIFORNIA.

PT. NO. 14100926
PRINTED IN U.S.A.

YCC
DIST.
VAC. SW.
TO INLET MANIF
FRONT OF VEHICLE
PCV VALVE
VAC REG VALVE
CCV (AIR IN)
FUEL PRESS REG. VLV
EGR
TO FUEL TANK
RESTRICTOR
DECEL VALVE
TANK PRESS CONT. VALVE
AIR MGNT. VALVE
THROTTLE BODY
CANISTER PURGE SOL.
AIR PUMP
EMISSION HOSE ROUTING
CANISTER

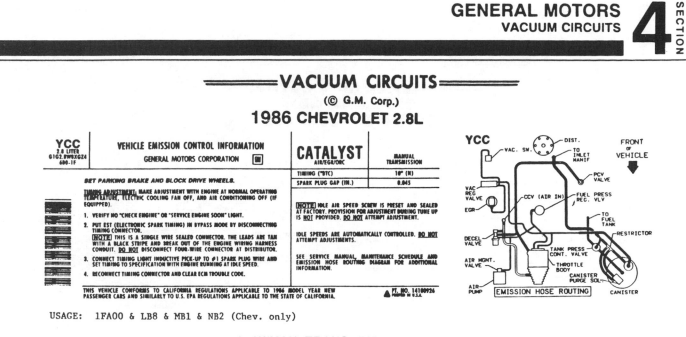

USAGE: 1FA00 & LB8 & MB1 & NB2 (Chev. only)

2.8l – W/MAN. TRANS. (MB1) – CALIFORNIA

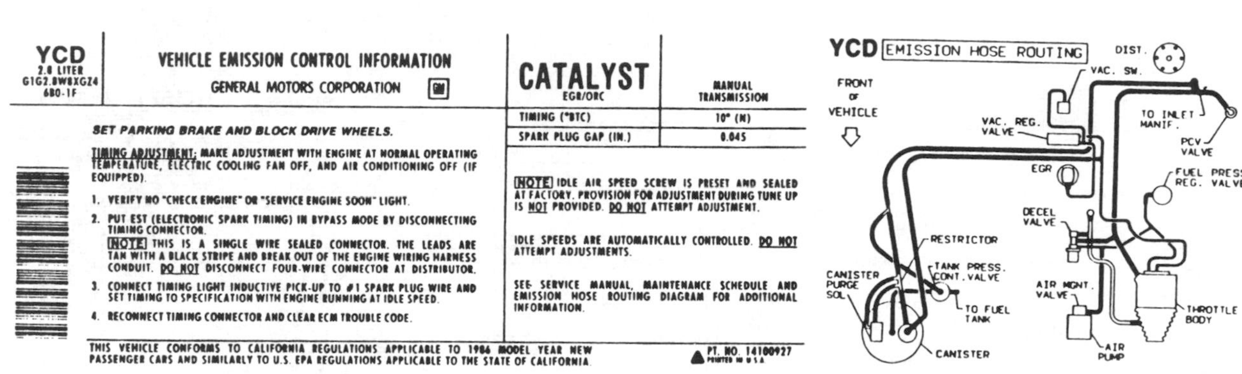

YCD
2.8 LITER
G1G2.8W8XGZ4
680-1F

VEHICLE EMISSION CONTROL INFORMATION
GENERAL MOTORS CORPORATION

CATALYST
EGR/ORC

MANUAL TRANSMISSION

TIMING (°BTC)	10° (N)
SPARK PLUG GAP (IN.)	0.045

SET PARKING BRAKE AND BLOCK DRIVE WHEELS.

TIMING ADJUSTMENT: MAKE ADJUSTMENT WITH ENGINE AT NORMAL OPERATING TEMPERATURE, ELECTRIC COOLING FAN OFF, AND AIR CONDITIONING OFF (IF EQUIPPED).

1. VERIFY NO "CHECK ENGINE" OR "SERVICE ENGINE SOON" LIGHT.
2. PUT EST (ELECTRONIC SPARK TIMING) IN BYPASS MODE BY DISCONNECTING TIMING CONNECTOR.
   NOTE THIS IS A SINGLE WIRE SEALED CONNECTOR. THE LEADS ARE TAN WITH A BLACK STRIPE AND BREAK OUT OF THE ENGINE WIRING HARNESS CONDUIT. DO NOT DISCONNECT FOUR-WIRE CONNECTOR AT DISTRIBUTOR.
3. CONNECT TIMING LIGHT INDUCTIVE PICK-UP TO #1 SPARK PLUG WIRE AND SET TIMING TO SPECIFICATION WITH ENGINE RUNNING AT IDLE SPEED.
4. RECONNECT TIMING CONNECTOR AND CLEAR ECM TROUBLE CODE.

NOTE IDLE AIR SPEED SCREW IS PRESET AND SEALED AT FACTORY. PROVISION FOR ADJUSTMENT DURING TUNE UP IS NOT PROVIDED. DO NOT ATTEMPT ADJUSTMENT.

IDLE SPEEDS ARE AUTOMATICALLY CONTROLLED. DO NOT ATTEMPT ADJUSTMENTS.

SEE SERVICE MANUAL, MAINTENANCE SCHEDULE AND EMISSION HOSE ROUTING DIAGRAM FOR ADDITIONAL INFORMATION.

THIS VEHICLE CONFORMS TO CALIFORNIA REGULATIONS APPLICABLE TO 1986 MODEL YEAR NEW PASSENGER CARS AND SIMILARLY TO U.S. EPA REGULATIONS APPLICABLE TO THE STATE OF CALIFORNIA.

PT. NO. 14100927
PRINTED IN U.S.A.

USAGE: 2FA00 & LB8 & MB1 &NB2 (Pontiac only)

NOTE: This label was cancelled and replaced by label "YMH" to correct emission system type from "EGR/ORC" to "AIR/EGR/ORC" (ECA 46589). All production and service stock should be disposed of.

2.8L – W/MAN. TRANS. – CALIFORNIA

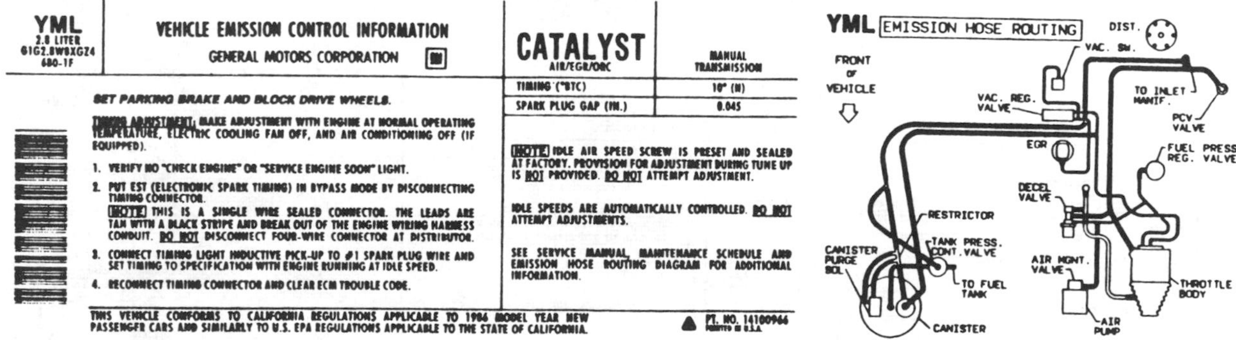

YML
2.8 LITER
G1G2.8W8XGZ4
680-1F

VEHICLE EMISSION CONTROL INFORMATION
GENERAL MOTORS CORPORATION

CATALYST
AIR/EGR/ORC

MANUAL TRANSMISSION

TIMING (°BTC)	10° (N)
SPARK PLUG GAP (IN.)	0.045

SET PARKING BRAKE AND BLOCK DRIVE WHEELS.

TIMING ADJUSTMENT: MAKE ADJUSTMENT WITH ENGINE AT NORMAL OPERATING TEMPERATURE, ELECTRIC COOLING FAN OFF, AND AIR CONDITIONING OFF (IF EQUIPPED).

1. VERIFY NO "CHECK ENGINE" OR "SERVICE ENGINE SOON" LIGHT.
2. PUT EST (ELECTRONIC SPARK TIMING) IN BYPASS MODE BY DISCONNECTING TIMING CONNECTOR.
   NOTE THIS IS A SINGLE WIRE SEALED CONNECTOR. THE LEADS ARE TAN WITH A BLACK STRIPE AND BREAK OUT OF THE ENGINE WIRING HARNESS CONDUIT. DO NOT DISCONNECT FOUR-WIRE CONNECTOR AT DISTRIBUTOR.
3. CONNECT TIMING LIGHT INDUCTIVE PICK-UP TO #1 SPARK PLUG WIRE AND SET TIMING TO SPECIFICATION WITH ENGINE RUNNING AT IDLE SPEED.
4. RECONNECT TIMING CONNECTOR AND CLEAR ECM TROUBLE CODE.

NOTE IDLE AIR SPEED SCREW IS PRESET AND SEALED AT FACTORY. PROVISION FOR ADJUSTMENT DURING TUNE UP IS NOT PROVIDED. DO NOT ATTEMPT ADJUSTMENT.

IDLE SPEEDS ARE AUTOMATICALLY CONTROLLED. DO NOT ATTEMPT ADJUSTMENTS.

SEE SERVICE MANUAL, MAINTENANCE SCHEDULE AND EMISSION HOSE ROUTING DIAGRAM FOR ADDITIONAL INFORMATION.

THIS VEHICLE CONFORMS TO CALIFORNIA REGULATIONS APPLICABLE TO 1986 MODEL YEAR NEW PASSENGER CARS AND SIMILARLY TO U.S. EPA REGULATIONS APPLICABLE TO THE STATE OF CALIFORNIA.

PT. NO. 14100966
PRINTED IN U.S.A.

USAGE: 2FA00 & LB8 & MB1 & NB2 (Pontiac only)

NOTE: This label replaces label "YCD" for all production and service uses to provide a label with correct emission system type (ECA 46589).

2.8L – W/MAN. TRANS. – CALIFORNIA

VACUUM CIRCUITS

(© G.M. Corp.)

1986 CHEVROLET 2.8L

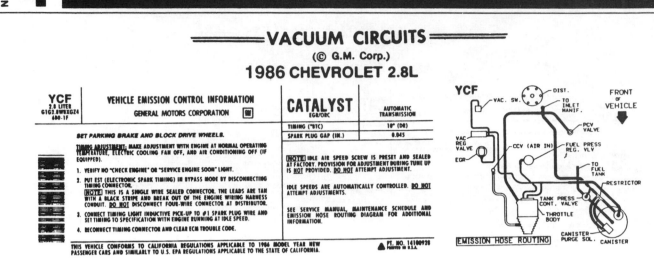

YCF
2.8 LITER
G1G2.8W8XGZ4
680-1F

VEHICLE EMISSION CONTROL INFORMATION
GENERAL MOTORS CORPORATION

CATALYST
EGR/ORC

	AUTOMATIC TRANSMISSION
TIMING (°BTC)	10° (DR)
SPARK PLUG GAP (IN.)	0.045

SET PARKING BRAKE AND BLOCK DRIVE WHEELS.

TIMING ADJUSTMENT: MAKE ADJUSTMENT WITH ENGINE AT NORMAL OPERATING TEMPERATURE, ELECTRIC COOLING FAN OFF, AND AIR CONDITIONING OFF (IF EQUIPPED).

1. VERIFY NO "CHECK ENGINE" OR "SERVICE ENGINE SOON" LIGHT.
2. PUT EST (ELECTRONIC SPARK TIMING) IN BYPASS MODE BY DISCONNECTING TIMING CONNECTOR.
   NOTE THIS IS A SINGLE WIRE SEALED CONNECTOR. THE LEADS ARE TAN WITH A BLACK STRIPE AND BREAK OUT OF THE ENGINE WIRING HARNESS CONDUIT. DO NOT DISCONNECT FOUR-WIRE CONNECTOR AT DISTRIBUTOR.
3. CONNECT TIMING LIGHT INDUCTIVE PICK-UP TO #1 SPARK PLUG WIRE AND SET TIMING TO SPECIFICATION WITH ENGINE RUNNING AT IDLE SPEED.
4. RECONNECT TIMING CONNECTOR AND CLEAR ECM TROUBLE CODE.

NOTE IDLE AIR SPEED SCREW IS PRESET AND SEALED AT FACTORY. PROVISION FOR ADJUSTMENT DURING TUNE UP IS NOT PROVIDED. DO NOT ATTEMPT ADJUSTMENT.

IDLE SPEEDS ARE AUTOMATICALLY CONTROLLED. DO NOT ATTEMPT ADJUSTMENTS.

SEE SERVICE MANUAL, MAINTENANCE SCHEDULE AND EMISSION HOSE ROUTING DIAGRAM FOR ADDITIONAL INFORMATION.

THIS VEHICLE CONFORMS TO CALIFORNIA REGULATIONS APPLICABLE TO 1986 MODEL YEAR NEW PASSENGER CARS AND SIMILARLY TO U.S. EPA REGULATIONS APPLICABLE TO THE STATE OF CALIFORNIA.

PT. NO. 14100928
PRINTED IN U.S.A.

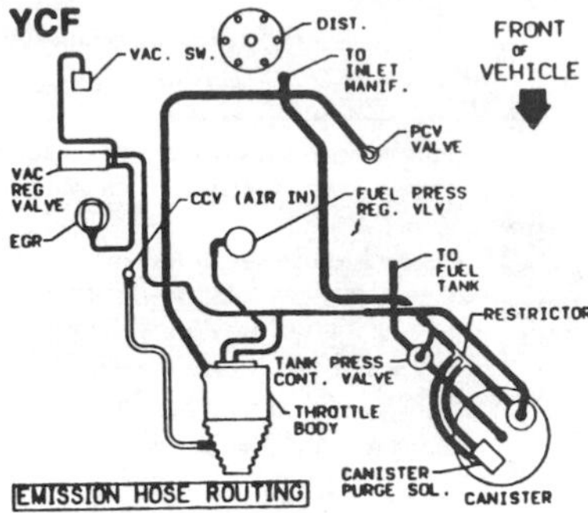

USAGE: 1FA00 & LB8 & MD8 & NB2 (Chev. only)

2.8L—W/AUTO. TRANS. (MD8)—CALIFORNIA

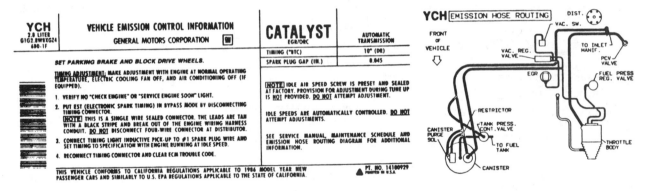

YCH
2.8 LITER
G1G2.8W8XGZ4
680-1F

VEHICLE EMISSION CONTROL INFORMATION
GENERAL MOTORS CORPORATION

CATALYST
EGR/ORC

	AUTOMATIC TRANSMISSION
TIMING (°BTC)	10° (DR)
SPARK PLUG GAP (IN.)	0.045

SET PARKING BRAKE AND BLOCK DRIVE WHEELS.

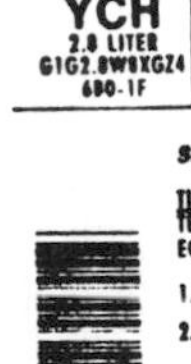

TIMING ADJUSTMENT: MAKE ADJUSTMENT WITH ENGINE AT NORMAL OPERATING TEMPERATURE, ELECTRIC COOLING FAN OFF, AND AIR CONDITIONING OFF (IF EQUIPPED).

1. VERIFY NO "CHECK ENGINE" OR "SERVICE ENGINE SOON" LIGHT.
2. PUT EST (ELECTRONIC SPARK TIMING) IN BYPASS MODE BY DISCONNECTING TIMING CONNECTOR.
   NOTE THIS IS A SINGLE WIRE SEALED CONNECTOR. THE LEADS ARE TAN WITH A BLACK STRIPE AND BREAK OUT OF THE ENGINE WIRING HARNESS CONDUIT. DO NOT DISCONNECT FOUR-WIRE CONNECTOR AT DISTRIBUTOR.
3. CONNECT TIMING LIGHT INDUCTIVE PICK-UP TO #1 SPARK PLUG WIRE AND SET TIMING TO SPECIFICATION WITH ENGINE RUNNING AT IDLE SPEED.
4. RECONNECT TIMING CONNECTOR AND CLEAR ECM TROUBLE CODE.

NOTE IDLE AIR SPEED SCREW IS PRESET AND SEALED AT FACTORY. PROVISION FOR ADJUSTMENT DURING TUNE UP IS NOT PROVIDED. DO NOT ATTEMPT ADJUSTMENT.

IDLE SPEEDS ARE AUTOMATICALLY CONTROLLED. DO NOT ATTEMPT ADJUSTMENTS.

SEE SERVICE MANUAL, MAINTENANCE SCHEDULE AND EMISSION HOSE ROUTING DIAGRAM FOR ADDITIONAL INFORMATION.

THIS VEHICLE CONFORMS TO CALIFORNIA REGULATIONS APPLICABLE TO 1986 MODEL YEAR NEW PASSENGER CARS AND SIMILARLY TO U.S. EPA REGULATIONS APPLICABLE TO THE STATE OF CALIFORNIA.

PT. NO. 14100929
PRINTED IN U.S.A.

YCH EMISSION HOSE ROUTING
DIST.
VAC. SW.
FRONT OF VEHICLE
VAC. REG. VALVE
TO INLET MANIF.
PCV VALVE
EGR
FUEL PRESS REG. VALVE
RESTRICTOR
TANK PRESS. CONT. VALVE
CANISTER PURGE SOL.
TO FUEL TANK
THROTTLE BODY
CANISTER

USAGE: 2FA00 & LB8 & MD8 & NB2 (Pontiac only)

2.8L—W/AUTO. TRANS.—CALIFORNIA

1986 CHEVROLET 4.3L

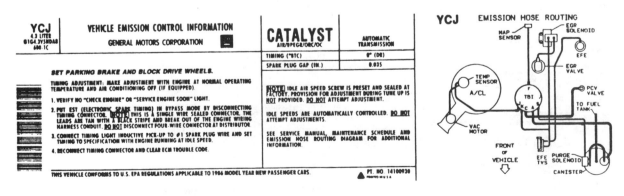

YCJ
4.3 LITER
G1G4.3V5NDA8
680-1C

VEHICLE EMISSION CONTROL INFORMATION
GENERAL MOTORS CORPORATION

CATALYST
AIR/BPEGR/ORC/OC

	AUTOMATIC TRANSMISSION
TIMING (°BTC)	0° (DR)
SPARK PLUG GAP (IN.)	0.035

SET PARKING BRAKE AND BLOCK DRIVE WHEELS.

TIMING ADJUSTMENT: MAKE ADJUSTMENT WITH ENGINE AT NORMAL OPERATING TEMPERATURE AND AIR CONDITIONING OFF (IF EQUIPPED).

1. VERIFY NO "CHECK ENGINE" OR "SERVICE ENGINE SOON" LIGHT.
2. PUT EST (ELECTRONIC SPARK TIMING) IN BYPASS MODE BY DISCONNECTING TIMING CONNECTOR. NOTE THIS IS A SINGLE WIRE SEALED CONNECTOR. THE LEADS ARE TAN WITH A BLACK STRIPE AND BREAK OUT OF THE ENGINE WIRING HARNESS CONDUIT. DO NOT DISCONNECT FOUR-WIRE CONNECTOR AT DISTRIBUTOR.
3. CONNECT TIMING LIGHT INDUCTIVE PICK-UP TO #1 SPARK PLUG WIRE AND SET TIMING TO SPECIFICATION WITH ENGINE RUNNING AT IDLE SPEED.
4. RECONNECT TIMING CONNECTOR AND CLEAR ECM TROUBLE CODE.

NOTE IDLE AIR SPEED SCREW IS PRESET AND SEALED AT FACTORY. PROVISION FOR ADJUSTMENT DURING TUNE UP IS NOT PROVIDED. DO NOT ATTEMPT ADJUSTMENT.

IDLE SPEEDS ARE AUTOMATICALLY CONTROLLED. DO NOT ATTEMPT ADJUSTMENTS.

SEE SERVICE MANUAL, MAINTENANCE SCHEDULE AND EMISSION HOSE ROUTING DIAGRAM FOR ADDITIONAL INFORMATION.

THIS VEHICLE CONFORMS TO U.S. EPA REGULATIONS APPLICABLE TO 1986 MODEL YEAR NEW PASSENGER CARS.

PT. NO. 14100930
PRINTED IN U.S.A.

YCJ EMISSION HOSE ROUTING
MAP SENSOR
EGR SOLENOID
EFE
EGR VALVE
TEMP SENSOR
A/CL
TBI
PCV VALVE
TO FUEL TANK
VAC MOTOR
FRONT OF VEHICLE
EFE TVS
PURGE SOLENOID
CANISTER

USAGE: 1-2BA00 & LB4 & NA5

4.3L—W/AUTO. TRANS.—FEDERAL

VACUUM CIRCUITS

(© G.M. Corp.)

1986 CHEVROLET 4.3L

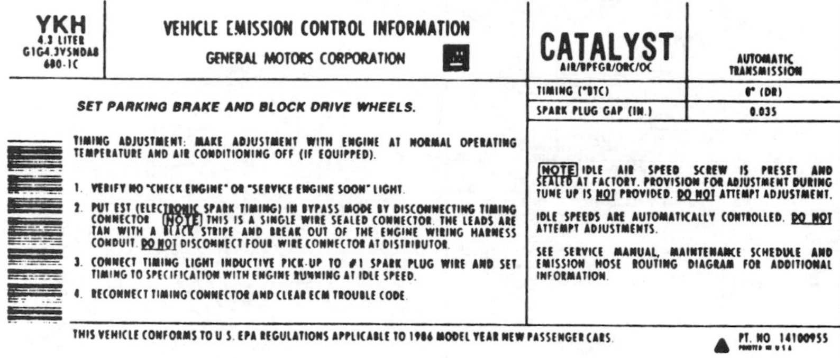

YKH
4.3 LITER
G1G4.3VSNDA8
680-1C

VEHICLE EMISSION CONTROL INFORMATION
GENERAL MOTORS CORPORATION

CATALYST AIR/BPEGR/ORC/OC	AUTOMATIC TRANSMISSION
TIMING (°BTC)	0° (DR)
SPARK PLUG GAP (IN.)	0.035

SET PARKING BRAKE AND BLOCK DRIVE WHEELS.

TIMING ADJUSTMENT: MAKE ADJUSTMENT WITH ENGINE AT NORMAL OPERATING TEMPERATURE AND AIR CONDITIONING OFF (IF EQUIPPED).

1. VERIFY NO "CHECK ENGINE" OR "SERVICE ENGINE SOON" LIGHT.
2. PUT EST (ELECTRONIC SPARK TIMING) IN BYPASS MODE BY DISCONNECTING TIMING CONNECTOR. NOTE THIS IS A SINGLE WIRE SEALED CONNECTOR. THE LEADS ARE TAN WITH A BLACK STRIPE AND BREAK OUT OF THE ENGINE WIRING HARNESS CONDUIT. DO NOT DISCONNECT FOUR WIRE CONNECTOR AT DISTRIBUTOR.
3. CONNECT TIMING LIGHT INDUCTIVE PICK-UP TO #1 SPARK PLUG WIRE AND SET TIMING TO SPECIFICATION WITH ENGINE RUNNING AT IDLE SPEED.
4. RECONNECT TIMING CONNECTOR AND CLEAR ECM TROUBLE CODE.

NOTE IDLE AIR SPEED SCREW IS PRESET AND SEALED AT FACTORY. PROVISION FOR ADJUSTMENT DURING TUNE UP IS NOT PROVIDED. DO NOT ATTEMPT ADJUSTMENT.

IDLE SPEEDS ARE AUTOMATICALLY CONTROLLED. DO NOT ATTEMPT ADJUSTMENTS.

SEE SERVICE MANUAL, MAINTENANCE SCHEDULE AND EMISSION HOSE ROUTING DIAGRAM FOR ADDITIONAL INFORMATION.

THIS VEHICLE CONFORMS TO U.S. EPA REGULATIONS APPLICABLE TO 1986 MODEL YEAR NEW PASSENGER CARS.

PT. NO. 14100955
PRINTED IN U.S.A.

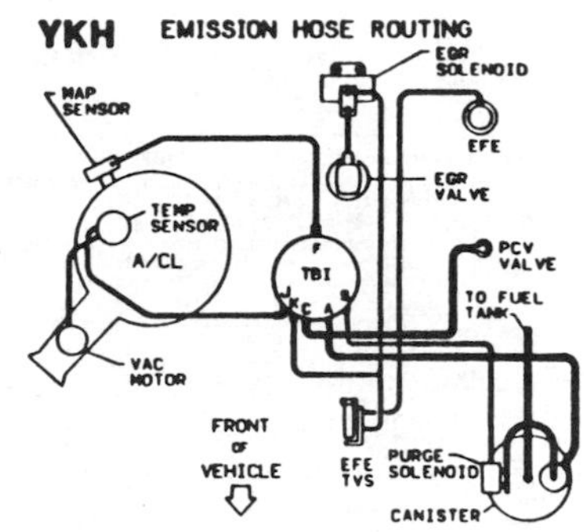

USAGE: 1-2GA00 & LB4 & NA5

4.3L—W/AUTO. TRANS.—FEDERAL

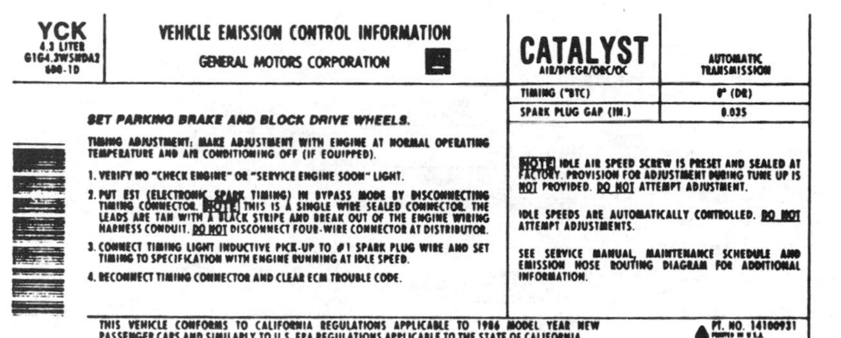

YCK
4.3 LITER
G1G4.3WSNDA2
680-1D

VEHICLE EMISSION CONTROL INFORMATION
GENERAL MOTORS CORPORATION

CATALYST AIR/BPEGR/ORC/OC	AUTOMATIC TRANSMISSION
TIMING (°BTC)	0° (DR)
SPARK PLUG GAP (IN.)	0.035

SET PARKING BRAKE AND BLOCK DRIVE WHEELS.

TIMING ADJUSTMENT: MAKE ADJUSTMENT WITH ENGINE AT NORMAL OPERATING TEMPERATURE AND AIR CONDITIONING OFF (IF EQUIPPED).

1. VERIFY NO "CHECK ENGINE" OR "SERVICE ENGINE SOON" LIGHT.
2. PUT EST (ELECTRONIC SPARK TIMING) IN BYPASS MODE BY DISCONNECTING TIMING CONNECTOR. NOTE THIS IS A SINGLE WIRE SEALED CONNECTOR. THE LEADS ARE TAN WITH A BLACK STRIPE AND BREAK OUT OF THE ENGINE WIRING HARNESS CONDUIT. DO NOT DISCONNECT FOUR-WIRE CONNECTOR AT DISTRIBUTOR.
3. CONNECT TIMING LIGHT INDUCTIVE PICK-UP TO #1 SPARK PLUG WIRE AND SET TIMING TO SPECIFICATION WITH ENGINE RUNNING AT IDLE SPEED.
4. RECONNECT TIMING CONNECTOR AND CLEAR ECM TROUBLE CODE.

NOTE IDLE AIR SPEED SCREW IS PRESET AND SEALED AT FACTORY. PROVISION FOR ADJUSTMENT DURING TUNE UP IS NOT PROVIDED. DO NOT ATTEMPT ADJUSTMENT.

IDLE SPEEDS ARE AUTOMATICALLY CONTROLLED. DO NOT ATTEMPT ADJUSTMENTS.

SEE SERVICE MANUAL, MAINTENANCE SCHEDULE AND EMISSION HOSE ROUTING DIAGRAM FOR ADDITIONAL INFORMATION.

THIS VEHICLE CONFORMS TO CALIFORNIA REGULATIONS APPLICABLE TO 1986 MODEL YEAR NEW PASSENGER CARS AND SIMILARLY TO U.S. EPA REGULATIONS APPLICABLE TO THE STATE OF CALIFORNIA.

PT. NO. 14100931
PRINTED IN U.S.A.

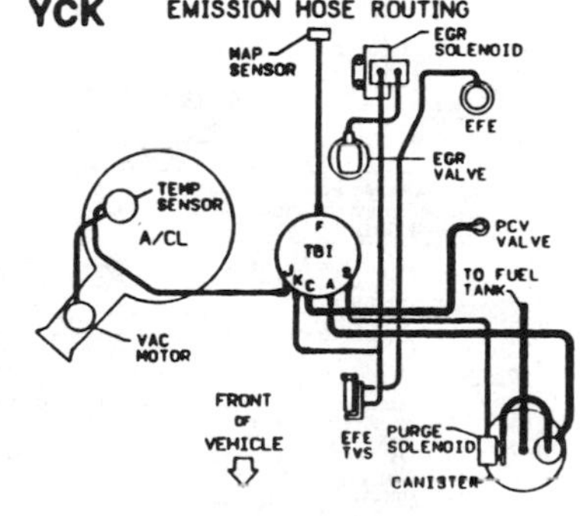

USAGE: 1-2BA00 & LB4 & NB2

4.3L—W/AUTO. TRANS.—CALIFORNIA

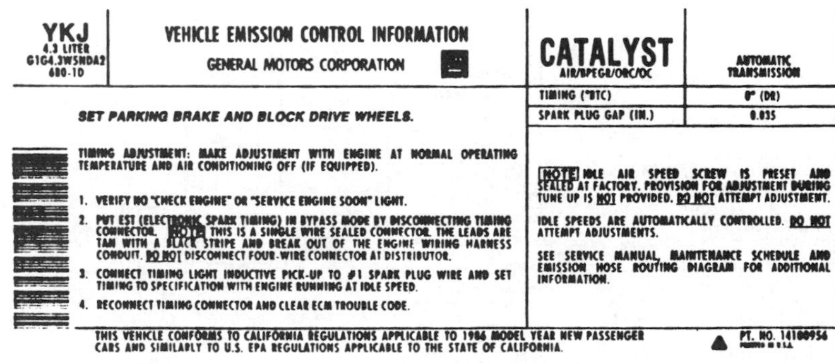

YKJ
4.3 LITER
G1G4.3WSNDA2
680-1D

VEHICLE EMISSION CONTROL INFORMATION
GENERAL MOTORS CORPORATION

CATALYST AIR/BPEGR/ORC/OC	AUTOMATIC TRANSMISSION
TIMING (°BTC)	0° (DR)
SPARK PLUG GAP (IN.)	0.035

SET PARKING BRAKE AND BLOCK DRIVE WHEELS.

TIMING ADJUSTMENT: MAKE ADJUSTMENT WITH ENGINE AT NORMAL OPERATING TEMPERATURE AND AIR CONDITIONING OFF (IF EQUIPPED).

1. VERIFY NO "CHECK ENGINE" OR "SERVICE ENGINE SOON" LIGHT.
2. PUT EST (ELECTRONIC SPARK TIMING) IN BYPASS MODE BY DISCONNECTING TIMING CONNECTOR. NOTE THIS IS A SINGLE WIRE SEALED CONNECTOR. THE LEADS ARE TAN WITH A BLACK STRIPE AND BREAK OUT OF THE ENGINE WIRING HARNESS CONDUIT. DO NOT DISCONNECT FOUR-WIRE CONNECTOR AT DISTRIBUTOR.
3. CONNECT TIMING LIGHT INDUCTIVE PICK-UP TO #1 SPARK PLUG WIRE AND SET TIMING TO SPECIFICATION WITH ENGINE RUNNING AT IDLE SPEED.
4. RECONNECT TIMING CONNECTOR AND CLEAR ECM TROUBLE CODE.

NOTE IDLE AIR SPEED SCREW IS PRESET AND SEALED AT FACTORY. PROVISION FOR ADJUSTMENT DURING TUNE UP IS NOT PROVIDED. DO NOT ATTEMPT ADJUSTMENT.

IDLE SPEEDS ARE AUTOMATICALLY CONTROLLED. DO NOT ATTEMPT ADJUSTMENTS.

SEE SERVICE MANUAL, MAINTENANCE SCHEDULE AND EMISSION HOSE ROUTING DIAGRAM FOR ADDITIONAL INFORMATION.

THIS VEHICLE CONFORMS TO CALIFORNIA REGULATIONS APPLICABLE TO 1986 MODEL YEAR NEW PASSENGER CARS AND SIMILARLY TO U.S. EPA REGULATIONS APPLICABLE TO THE STATE OF CALIFORNIA.

PT. NO. 14100956
PRINTED IN U.S.A.

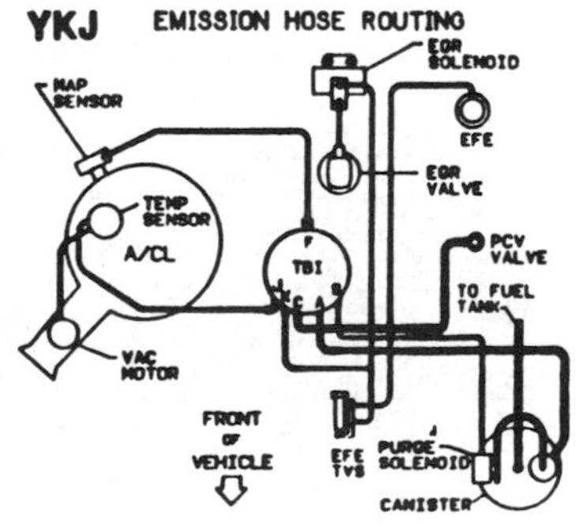

USAGE: 1-2GA00 & LB4 & NB2

4.3—W/AUTO. TRANS.—CALIFORNIA

VACUUM CIRCUITS

(© G.M. Corp.)

1986 CHEVROLET 5.0L

YCL
5.0 LITER
G1G5 7V4NEA5
6B4H 1A

VEHICLE EMISSION CONTROL INFORMATION
GENERAL MOTORS CORPORATION

SET PARKING BRAKE AND BLOCK DRIVE WHEELS.

MAKE ALL ADJUSTMENTS WITH ENGINE AT NORMAL OPERATING TEMPERATURE, EMISSION CONTROL SYSTEM IN CLOSED LOOP MODE, CHOKE FULL OPEN, AIR CLEANER INSTALLED, AND AIR CONDITIONING (A/C) OFF, EXCEPT WHERE NOTED

1. DISTRIBUTOR DISCONNECT FOUR WIRE CONNECTOR SET IGNITION TIMING TO SPECIFICATION WITH ENGINE RUNNING AT SPEED INDICATED RECONNECT FOUR WIRE CONNECTOR CLEAR ECM TROUBLE CODE
2. IDLE SPEED SCREW DISCONNECT ELECTRICAL LEAD FROM IDLE SOLENOID, IF EQUIPPED ADJUST CARBURETOR IDLE SPEED SCREW TO OBTAIN SPECIFIED SPEED RECONNECT LEAD TO SOLENOID
3. IDLE SPEED SOLENOID (WITH A/C) DISCONNECT ELECTRICAL LEAD FROM A/C COMPRESSOR AND TURN A/C ON (WITHOUT A/C) DISCONNECT ELECTRICAL LEAD FROM SOLENOID AND CONNECT A JUMPER WIRE FROM A + 12 VOLT SUPPLY (ALL) OPEN THROTTLE MOMENTARILY TO ASSURE SOLENOID IS FULLY EXTENDED ADJUST SOLENOID TO OBTAIN SPECIFIED SPEED DISCONNECT JUMPER WIRE (IF USED), RECONNECT ELECTRICAL LEAD AND (WHERE EQUIPPED) TURN A/C OFF
4. FAST IDLE SPEED DISCONNECT AND PLUG VACUUM HOSE AT EGR VALVE WITH THROTTLE ON HIGH STEP OF FAST IDLE CAM, ADJUST FAST IDLE SCREW TO OBTAIN SPECIFIED SPEED NOTE THIS ADJUSTMENT MUST BE CHECKED WITHIN 15 SECONDS AFTER INCREASING SPEED ABOVE 700 RPM OPEN THROTTLE TO RELEASE FAST IDLE CAM AND STOP ENGINE UNPLUG AND RECONNECT VACUUM HOSE TO EGR VALVE

CATALYST AIR/BPEGR/ORC/OC	AUTOMATIC TRANSMISSION
TIMING (°BTC @ RPM)	0° @ 500 (DR)
SPARK PLUG GAP (IN.)	0.035
IDLE SPEED SCREW (RPM) (SOLENOID INACTIVE)	500 (DR)
IDLE SPEED SOLENOID (RPM) (SOLENOID ACTIVE)	650 (DR)
FAST IDLE SPEED (RPM)	2200 (P) OR (N)

NOTE IDLE MIXTURE SCREWS ARE PRESET AND SEALED AT FACTORY PROVISION FOR ADJUSTMENT DURING TUNE UP IS NOT PROVIDED

SEE SERVICE MANUAL, MAINTENANCE SCHEDULE AND EMISSION HOSE ROUTING DIAGRAM FOR ADDITIONAL INFORMATION

THIS VEHICLE CONFORMS TO U S EPA REGULATIONS APPLICABLE TO 1986 MODEL YEAR NEW PASSENGER CARS. PT. NO 14100932

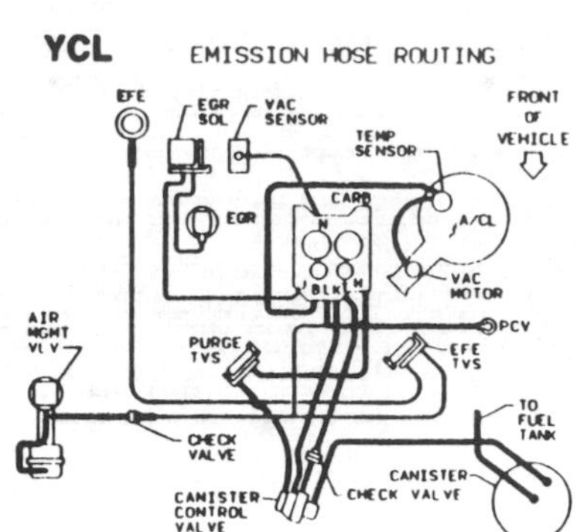

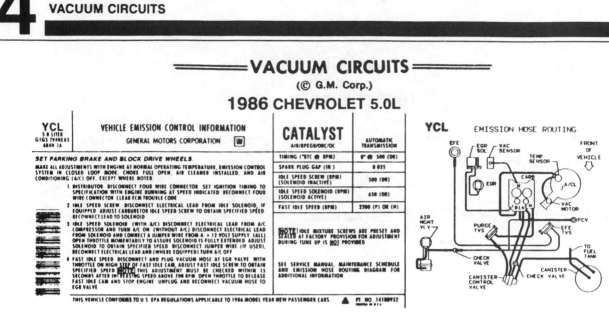

USAGE: 1-2GA00/1-2BA00 & LG4 & NA5

5.0L—W/AUTO. TRANS. (305H)—FEDERAL

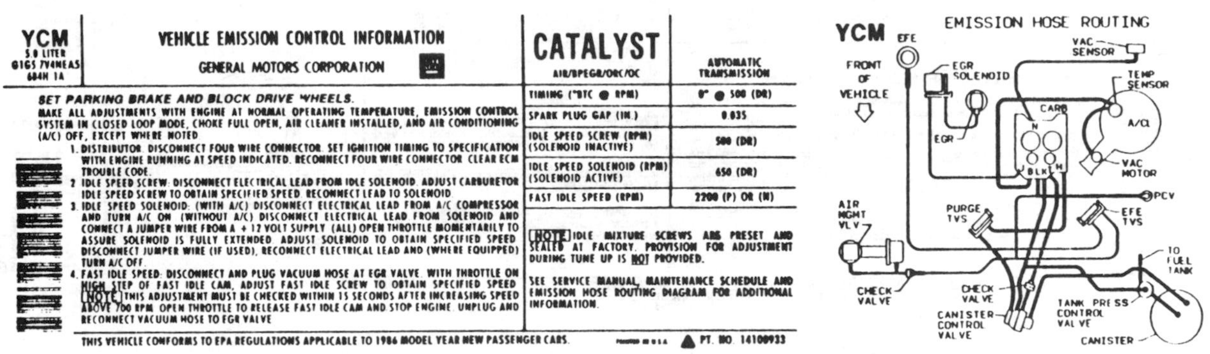

YCM
5.0 LITER
G1G5 7V4NEA5
6B4H 1A

VEHICLE EMISSION CONTROL INFORMATION
GENERAL MOTORS CORPORATION

SET PARKING BRAKE AND BLOCK DRIVE WHEELS.

MAKE ALL ADJUSTMENTS WITH ENGINE AT NORMAL OPERATING TEMPERATURE, EMISSION CONTROL SYSTEM IN CLOSED LOOP MODE, CHOKE FULL OPEN, AIR CLEANER INSTALLED, AND AIR CONDITIONING (A/C) OFF, EXCEPT WHERE NOTED

1. DISTRIBUTOR. DISCONNECT FOUR WIRE CONNECTOR. SET IGNITION TIMING TO SPECIFICATION WITH ENGINE RUNNING AT SPEED INDICATED. RECONNECT FOUR WIRE CONNECTOR CLEAR ECM TROUBLE CODE.
2. IDLE SPEED SCREW: DISCONNECT ELECTRICAL LEAD FROM IDLE SOLENOID. ADJUST CARBURETOR IDLE SPEED SCREW TO OBTAIN SPECIFIED SPEED. RECONNECT LEAD TO SOLENOID.
3. IDLE SPEED SOLENOID: (WITH A/C) DISCONNECT ELECTRICAL LEAD FROM A/C COMPRESSOR AND TURN A/C ON (WITHOUT A/C) DISCONNECT ELECTRICAL LEAD FROM SOLENOID AND CONNECT A JUMPER WIRE FROM A + 12 VOLT SUPPLY (ALL) OPEN THROTTLE MOMENTARILY TO ASSURE SOLENOID IS FULLY EXTENDED. ADJUST SOLENOID TO OBTAIN SPECIFIED SPEED. DISCONNECT JUMPER WIRE (IF USED), RECONNECT ELECTRICAL LEAD AND (WHERE EQUIPPED) TURN A/C OFF.
4. FAST IDLE SPEED: DISCONNECT AND PLUG VACUUM HOSE AT EGR VALVE. WITH THROTTLE ON HIGH STEP OF FAST IDLE CAM, ADJUST FAST IDLE SCREW TO OBTAIN SPECIFIED SPEED. NOTE THIS ADJUSTMENT MUST BE CHECKED WITHIN 15 SECONDS AFTER INCREASING SPEED ABOVE 700 RPM. OPEN THROTTLE TO RELEASE FAST IDLE CAM AND STOP ENGINE. UNPLUG AND RECONNECT VACUUM HOSE TO EGR VALVE

CATALYST AIR/BPEGR/ORC/OC	AUTOMATIC TRANSMISSION
TIMING (°BTC @ RPM)	0° @ 500 (DR)
SPARK PLUG GAP (IN.)	0.035
IDLE SPEED SCREW (RPM) (SOLENOID INACTIVE)	500 (DR)
IDLE SPEED SOLENOID (RPM) (SOLENOID ACTIVE)	650 (DR)
FAST IDLE SPEED (RPM)	2200 (P) OR (N)

NOTE IDLE MIXTURE SCREWS ARE PRESET AND SEALED AT FACTORY. PROVISION FOR ADJUSTMENT DURING TUNE UP IS NOT PROVIDED.

SEE SERVICE MANUAL, MAINTENANCE SCHEDULE AND EMISSION HOSE ROUTING DIAGRAM FOR ADDITIONAL INFORMATION.

THIS VEHICLE CONFORMS TO EPA REGULATIONS APPLICABLE TO 1986 MODEL YEAR NEW PASSENGER CARS. PT. NO. 14100933

USAGE: 1-2FA00 & LG4 & MD8 & NA5

5.0L—W/AUTO. TRANS. (MD8)—FEDERAL

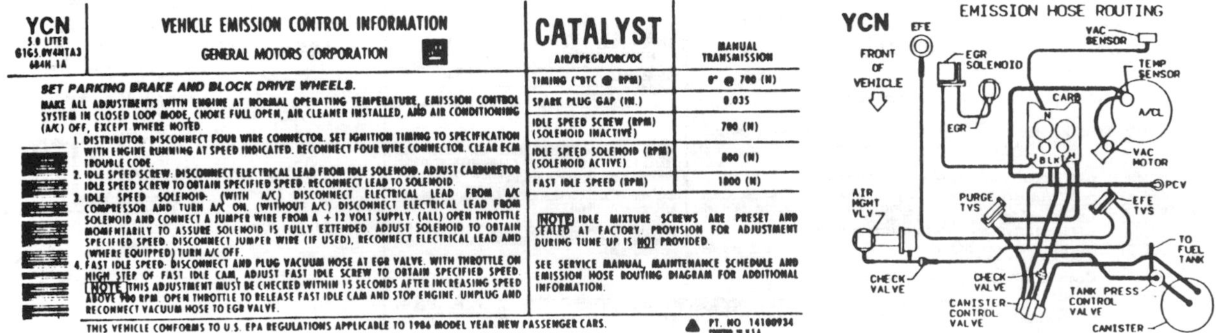

YCN
5.0 LITER
G1G5.8V4NTA3
6B4H 1A

VEHICLE EMISSION CONTROL INFORMATION
GENERAL MOTORS CORPORATION

SET PARKING BRAKE AND BLOCK DRIVE WHEELS.

MAKE ALL ADJUSTMENTS WITH ENGINE AT NORMAL OPERATING TEMPERATURE, EMISSION CONTROL SYSTEM IN CLOSED LOOP MODE, CHOKE FULL OPEN, AIR CLEANER INSTALLED, AND AIR CONDITIONING (A/C) OFF, EXCEPT WHERE NOTED.

1. DISTRIBUTOR. DISCONNECT FOUR WIRE CONNECTOR. SET IGNITION TIMING TO SPECIFICATION WITH ENGINE RUNNING AT SPEED INDICATED. RECONNECT FOUR WIRE CONNECTOR. CLEAR ECM TROUBLE CODE.
2. IDLE SPEED SCREW: DISCONNECT ELECTRICAL LEAD FROM IDLE SOLENOID. ADJUST CARBURETOR IDLE SPEED SCREW TO OBTAIN SPECIFIED SPEED. RECONNECT LEAD TO SOLENOID.
3. IDLE SPEED SOLENOID: (WITH A/C) DISCONNECT ELECTRICAL LEAD FROM A/C COMPRESSOR AND TURN A/C ON. (WITHOUT A/C) DISCONNECT ELECTRICAL LEAD FROM SOLENOID AND CONNECT A JUMPER WIRE FROM A + 12 VOLT SUPPLY. (ALL) OPEN THROTTLE MOMENTARILY TO ASSURE SOLENOID IS FULLY EXTENDED. ADJUST SOLENOID TO OBTAIN SPECIFIED SPEED. DISCONNECT JUMPER WIRE (IF USED), RECONNECT ELECTRICAL LEAD AND (WHERE EQUIPPED) TURN A/C OFF.
4. FAST IDLE SPEED: DISCONNECT AND PLUG VACUUM HOSE AT EGR VALVE. WITH THROTTLE ON HIGH STEP OF FAST IDLE CAM, ADJUST FAST IDLE SCREW TO OBTAIN SPECIFIED SPEED. NOTE THIS ADJUSTMENT MUST BE CHECKED WITHIN 15 SECONDS AFTER INCREASING SPEED ABOVE 900 RPM. OPEN THROTTLE TO RELEASE FAST IDLE CAM AND STOP ENGINE. UNPLUG AND RECONNECT VACUUM HOSE TO EGR VALVE.

CATALYST AIR/BPEGR/ORC/OC	MANUAL TRANSMISSION
TIMING (°BTC @ RPM)	0° @ 700 (N)
SPARK PLUG GAP (IN.)	0.035
IDLE SPEED SCREW (RPM) (SOLENOID INACTIVE)	700 (N)
IDLE SPEED SOLENOID (RPM) (SOLENOID ACTIVE)	800 (N)
FAST IDLE SPEED (RPM)	1800 (N)

NOTE IDLE MIXTURE SCREWS ARE PRESET AND SEALED AT FACTORY. PROVISION FOR ADJUSTMENT DURING TUNE UP IS NOT PROVIDED.

SEE SERVICE MANUAL, MAINTENANCE SCHEDULE AND EMISSION HOSE ROUTING DIAGRAM FOR ADDITIONAL INFORMATION.

THIS VEHICLE CONFORMS TO U.S. EPA REGULATIONS APPLICABLE TO 1986 MODEL YEAR NEW PASSENGER CARS. PT. NO 14100934

USAGE: 1-2FA00 & LG4 & MC4/M39 & NA5

5.0L—W/AUTO. TRANS. (MC4, M39)—FEDERAL

VACUUM CIRCUITS

1986 CHEVROLET 5.0L

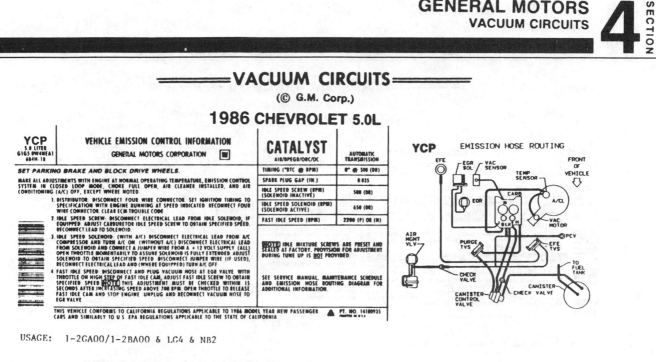

YCP
5.0 LITER
G1G5.0W4NEA1
6B4H-1B

VEHICLE EMISSION CONTROL INFORMATION
GENERAL MOTORS CORPORATION

SET PARKING BRAKE AND BLOCK DRIVE WHEELS.

MAKE ALL ADJUSTMENTS WITH ENGINE AT NORMAL OPERATING TEMPERATURE, EMISSION CONTROL SYSTEM IN CLOSED LOOP MODE, CHOKE FULL OPEN, AIR CLEANER INSTALLED, AND AIR CONDITIONING (A/C) OFF, EXCEPT WHERE NOTED

1. DISTRIBUTOR: DISCONNECT FOUR WIRE CONNECTOR. SET IGNITION TIMING TO SPECIFICATION WITH ENGINE RUNNING AT SPEED INDICATED. RECONNECT FOUR WIRE CONNECTOR. CLEAR ECM TROUBLE CODE
2. IDLE SPEED SCREW: DISCONNECT ELECTRICAL LEAD FROM IDLE SOLENOID, IF EQUIPPED. ADJUST CARBURETOR IDLE SPEED SCREW TO OBTAIN SPECIFIED SPEED. RECONNECT LEAD TO SOLENOID.
3. IDLE SPEED SOLENOID: (WITH A/C) DISCONNECT ELECTRICAL LEAD FROM A/C COMPRESSOR AND TURN A/C ON. (WITHOUT A/C) DISCONNECT ELECTRICAL LEAD FROM SOLENOID AND CONNECT A JUMPER WIRE FROM A + 12 VOLT SUPPLY. (ALL) OPEN THROTTLE MOMENTARILY TO ASSURE SOLENOID IS FULLY EXTENDED. ADJUST SOLENOID TO OBTAIN SPECIFIED SPEED. DISCONNECT JUMPER WIRE (IF USED), RECONNECT ELECTRICAL LEAD AND (WHERE EQUIPPED) TURN A/C OFF
4. FAST IDLE SPEED: DISCONNECT AND PLUG VACUUM HOSE AT EGR VALVE. WITH THROTTLE ON HIGH STEP OF FAST IDLE CAM, ADJUST FAST IDLE SCREW TO OBTAIN SPECIFIED SPEED. NOTE THIS ADJUSTMENT MUST BE CHECKED WITHIN 15 SECONDS AFTER INCREASING SPEED ABOVE 700 RPM. OPEN THROTTLE TO RELEASE FAST IDLE CAM AND STOP ENGINE. UNPLUG AND RECONNECT VACUUM HOSE TO EGR VALVE.

CATALYST AIR/BPEGR/ORC/OC	AUTOMATIC TRANSMISSION
TIMING (°BTC @ RPM)	0° @ 500 (DR)
SPARK PLUG GAP (IN.)	0.035
IDLE SPEED SCREW (RPM) (SOLENOID INACTIVE)	500 (DR)
IDLE SPEED SOLENOID (RPM) (SOLENOID ACTIVE)	650 (DR)
FAST IDLE SPEED (RPM)	2200 (P) OR (N)

NOTE IDLE MIXTURE SCREWS ARE PRESET AND SEALED AT FACTORY. PROVISION FOR ADJUSTMENT DURING TUNE UP IS NOT PROVIDED.

SEE SERVICE MANUAL, MAINTENANCE SCHEDULE AND EMISSION HOSE ROUTING DIAGRAM FOR ADDITIONAL INFORMATION.

THIS VEHICLE CONFORMS TO CALIFORNIA REGULATIONS APPLICABLE TO 1986 MODEL YEAR NEW PASSENGER CARS AND SIMILARLY TO U.S. EPA REGULATIONS APPLICABLE TO THE STATE OF CALIFORNIA

PT. NO. 14100935
PRINTED IN U.S.A.

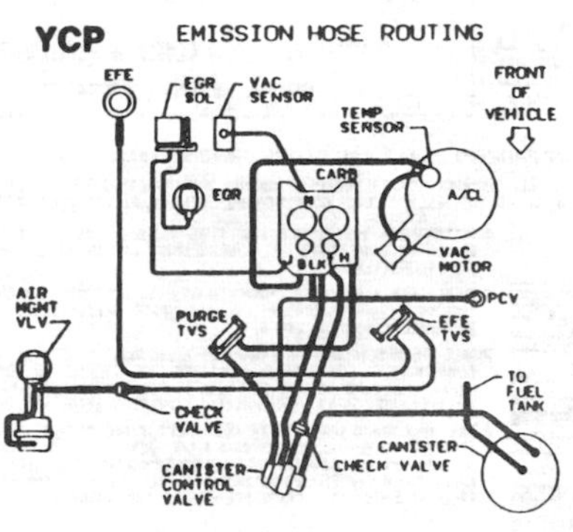

USAGE: 1-2GA00/1-2BA00 & LG4 & NB2

5.0L—W/AUTO. TRANS. (305H)—CALIFORNIA

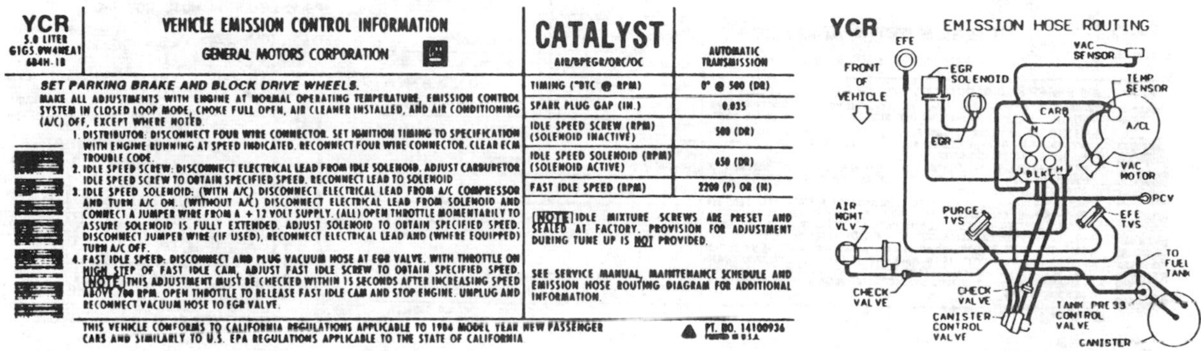

YCR
5.0 LITER
G1G5.0W4NEA1
6B4H-1B

VEHICLE EMISSION CONTROL INFORMATION
GENERAL MOTORS CORPORATION

SET PARKING BRAKE AND BLOCK DRIVE WHEELS.

MAKE ALL ADJUSTMENTS WITH ENGINE AT NORMAL OPERATING TEMPERATURE, EMISSION CONTROL SYSTEM IN CLOSED LOOP MODE, CHOKE FULL OPEN, AIR CLEANER INSTALLED, AND AIR CONDITIONING (A/C) OFF, EXCEPT WHERE NOTED.

1. DISTRIBUTOR: DISCONNECT FOUR WIRE CONNECTOR. SET IGNITION TIMING TO SPECIFICATION WITH ENGINE RUNNING AT SPEED INDICATED. RECONNECT FOUR WIRE CONNECTOR. CLEAR ECM TROUBLE CODE.
2. IDLE SPEED SCREW: DISCONNECT ELECTRICAL LEAD FROM IDLE SOLENOID. ADJUST CARBURETOR IDLE SPEED SCREW TO OBTAIN SPECIFIED SPEED. RECONNECT LEAD TO SOLENOID
3. IDLE SPEED SOLENOID: (WITH A/C) DISCONNECT ELECTRICAL LEAD FROM A/C COMPRESSOR AND TURN A/C ON. (WITHOUT A/C) DISCONNECT ELECTRICAL LEAD FROM SOLENOID AND CONNECT A JUMPER WIRE FROM A + 12 VOLT SUPPLY. (ALL) OPEN THROTTLE MOMENTARILY TO ASSURE SOLENOID IS FULLY EXTENDED. ADJUST SOLENOID TO OBTAIN SPECIFIED SPEED. DISCONNECT JUMPER WIRE (IF USED), RECONNECT ELECTRICAL LEAD AND (WHERE EQUIPPED) TURN A/C OFF.
4. FAST IDLE SPEED: DISCONNECT AND PLUG VACUUM HOSE AT EGR VALVE. WITH THROTTLE ON HIGH STEP OF FAST IDLE CAM, ADJUST FAST IDLE SCREW TO OBTAIN SPECIFIED SPEED. NOTE THIS ADJUSTMENT MUST BE CHECKED WITHIN 15 SECONDS AFTER INCREASING SPEED ABOVE 700 RPM. OPEN THROTTLE TO RELEASE FAST IDLE CAM AND STOP ENGINE. UNPLUG AND RECONNECT VACUUM HOSE TO EGR VALVE.

CATALYST AIR/BPEGR/ORC/OC	AUTOMATIC TRANSMISSION
TIMING (°BTC @ RPM)	0° @ 500 (DR)
SPARK PLUG GAP (IN.)	0.035
IDLE SPEED SCREW (RPM) (SOLENOID INACTIVE)	500 (DR)
IDLE SPEED SOLENOID (RPM) (SOLENOID ACTIVE)	650 (DR)
FAST IDLE SPEED (RPM)	2200 (P) OR (N)

NOTE IDLE MIXTURE SCREWS ARE PRESET AND SEALED AT FACTORY. PROVISION FOR ADJUSTMENT DURING TUNE UP IS NOT PROVIDED.

SEE SERVICE MANUAL, MAINTENANCE SCHEDULE AND EMISSION HOSE ROUTING DIAGRAM FOR ADDITIONAL INFORMATION.

THIS VEHICLE CONFORMS TO CALIFORNIA REGULATIONS APPLICABLE TO 1986 MODEL YEAR NEW PASSENGER CARS AND SIMILARLY TO U.S. EPA REGULATIONS APPLICABLE TO THE STATE OF CALIFORNIA

PT. NO. 14100936
PRINTED IN U.S.A.

USAGE: 1-2FA00 & LG4 & MD8 & NB2

5.0L—W/AUTO. TRANS. (MD8)—CALIFORNIA

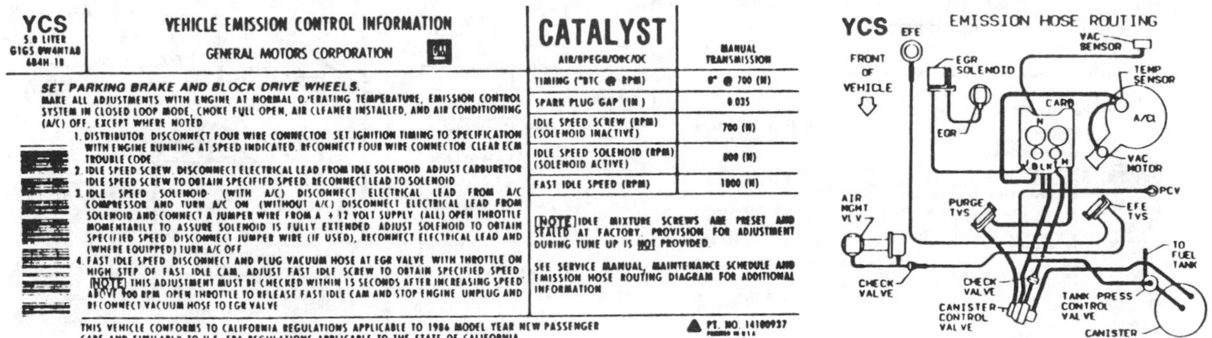

YCS
5.0 LITER
G1G5.0W4NTA8
6B4H 1B

VEHICLE EMISSION CONTROL INFORMATION
GENERAL MOTORS CORPORATION

SET PARKING BRAKE AND BLOCK DRIVE WHEELS.

MAKE ALL ADJUSTMENTS WITH ENGINE AT NORMAL OPERATING TEMPERATURE, EMISSION CONTROL SYSTEM IN CLOSED LOOP MODE, CHOKE FULL OPEN, AIR CLEANER INSTALLED, AND AIR CONDITIONING (A/C) OFF, EXCEPT WHERE NOTED

1. DISTRIBUTOR DISCONNECT FOUR WIRE CONNECTOR SET IGNITION TIMING TO SPECIFICATION WITH ENGINE RUNNING AT SPEED INDICATED. RECONNECT FOUR WIRE CONNECTOR CLEAR ECM TROUBLE CODE
2. IDLE SPEED SCREW. DISCONNECT ELECTRICAL LEAD FROM IDLE SOLENOID ADJUST CARBURETOR IDLE SPEED SCREW TO OBTAIN SPECIFIED SPEED. RECONNECT LEAD TO SOLENOID
3. IDLE SPEED SOLENOID (WITH A/C) DISCONNECT ELECTRICAL LEAD FROM A/C COMPRESSOR AND TURN A/C ON (WITHOUT A/C) DISCONNECT ELECTRICAL LEAD FROM SOLENOID AND CONNECT A JUMPER WIRE FROM A + 12 VOLT SUPPLY (ALL) OPEN THROTTLE MOMENTARILY TO ASSURE SOLENOID IS FULLY EXTENDED ADJUST SOLENOID TO OBTAIN SPECIFIED SPEED DISCONNECT JUMPER WIRE (IF USED), RECONNECT ELECTRICAL LEAD AND (WHERE EQUIPPED) TURN A/C OFF
4. FAST IDLE SPEED DISCONNECT AND PLUG VACUUM HOSE AT EGR VALVE WITH THROTTLE ON HIGH STEP OF FAST IDLE CAM, ADJUST FAST IDLE SCREW TO OBTAIN SPECIFIED SPEED NOTE THIS ADJUSTMENT MUST BE CHECKED WITHIN 15 SECONDS AFTER INCREASING SPEED ABOVE 900 RPM OPEN THROTTLE TO RELEASE FAST IDLE CAM AND STOP ENGINE UNPLUG AND RECONNECT VACUUM HOSE TO EGR VALVE

CATALYST AIR/BPEGR/ORC/OC	MANUAL TRANSMISSION
TIMING (°BTC @ RPM)	0° @ 700 (N)
SPARK PLUG GAP (IN.)	0.035
IDLE SPEED SCREW (RPM) (SOLENOID INACTIVE)	700 (N)
IDLE SPEED SOLENOID (RPM) (SOLENOID ACTIVE)	800 (N)
FAST IDLE SPEED (RPM)	1000 (N)

NOTE IDLE MIXTURE SCREWS ARE PRESET AND SEALED AT FACTORY. PROVISION FOR ADJUSTMENT DURING TUNE UP IS NOT PROVIDED.

SEE SERVICE MANUAL, MAINTENANCE SCHEDULE AND EMISSION HOSE ROUTING DIAGRAM FOR ADDITIONAL INFORMATION.

THIS VEHICLE CONFORMS TO CALIFORNIA REGULATIONS APPLICABLE TO 1986 MODEL YEAR NEW PASSENGER CARS AND SIMILARLY TO U.S. EPA REGULATIONS APPLICABLE TO THE STATE OF CALIFORNIA

PT. NO. 14100937
PRINTED IN U.S.A.

USAGE: 1-2FA00 & LG4 & MC4/M39 & NB2

5.0L—W/MAN. TRANS. (MC4, M39)—CALIFORNIA

VACUUM CIRCUITS

(© G.M. Corp.)

1986 CHEVROLET 5.0L

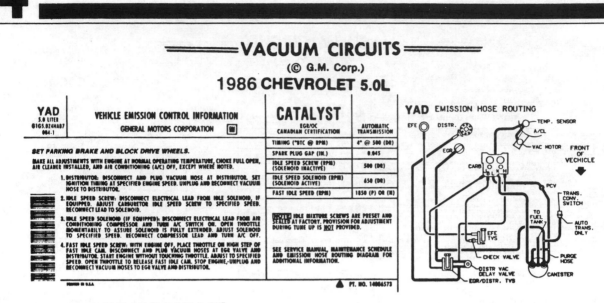

YAD
5.0 LITER
01G5.0Z4HAB7
004-1

VEHICLE EMISSION CONTROL INFORMATION
GENERAL MOTORS CORPORATION

SET PARKING BRAKE AND BLOCK DRIVE WHEELS.

MAKE ALL ADJUSTMENTS WITH ENGINE AT NORMAL OPERATING TEMPERATURE, CHOKE FULL OPEN, AIR CLEANER INSTALLED, AND AIR CONDITIONING (A/C) OFF, EXCEPT WHERE NOTED.

1. DISTRIBUTOR: DISCONNECT AND PLUG VACUUM HOSE AT DISTRIBUTOR. SET IGNITION TIMING AT SPECIFIED ENGINE SPEED. UNPLUG AND RECONNECT VACUUM HOSE TO DISTRIBUTOR.
2. IDLE SPEED SCREW: DISCONNECT ELECTRICAL LEAD FROM IDLE SOLENOID, IF EQUIPPED. ADJUST CARBURETOR IDLE SPEED SCREW TO SPECIFIED SPEED. RECONNECT LEAD TO SOLENOID.
3. IDLE SPEED SOLENOID (IF EQUIPPED): DISCONNECT ELECTRICAL LEAD FROM AIR CONDITIONING COMPRESSOR AND TURN A/C SWITCH ON. OPEN THROTTLE MOMENTARILY TO ASSURE SOLENOID IS FULLY EXTENDED. ADJUST SOLENOID TO SPECIFIED SPEED. RECONNECT COMPRESSOR LEAD AND TURN A/C OFF.
4. FAST IDLE SPEED SCREW: WITH ENGINE OFF, PLACE THROTTLE ON HIGH STEP OF FAST IDLE CAM. DISCONNECT AND PLUG VACUUM HOSES AT EGR VALVE AND DISTRIBUTOR. START ENGINE WITHOUT TOUCHING THROTTLE. ADJUST TO SPECIFIED SPEED. OPEN THROTTLE TO RELEASE FAST IDLE CAM. STOP ENGINE, UNPLUG AND RECONNECT VACUUM HOSES TO EGR VALVE AND DISTRIBUTOR.

PRINTED IN U.S.A.

CATALYST EGR/OC CANADIAN CERTIFICATION	AUTOMATIC TRANSMISSION
TIMING (°BTC @ RPM)	4° @ 500 (DR)
SPARK PLUG GAP (IN.)	0.045
IDLE SPEED SCREW (RPM) (SOLENOID INACTIVE)	500 (DR)
IDLE SPEED SOLENOID (RPM) (SOLENOID ACTIVE)	650 (DR)
FAST IDLE SPEED (RPM)	1850 (P) OR (N)

NOTE IDLE MIXTURE SCREWS ARE PRESET AND SEALED AT FACTORY. PROVISION FOR ADJUSTMENT DURING TUNE UP IS NOT PROVIDED.

SEE SERVICE MANUAL, MAINTENANCE SCHEDULE AND EMISSION HOSE ROUTING DIAGRAM FOR ADDITIONAL INFORMATION.

PT. NO. 14086573

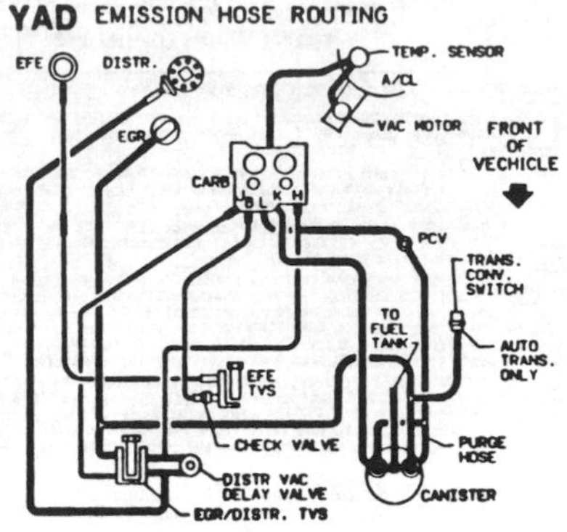

USAGE: 1-2BA00/1-2GA00 & LG4 & NM5

5.0L—W/AUTO. TRANS. (NM5 EXPORT, CANADA), EXC. NM8 EXPORT

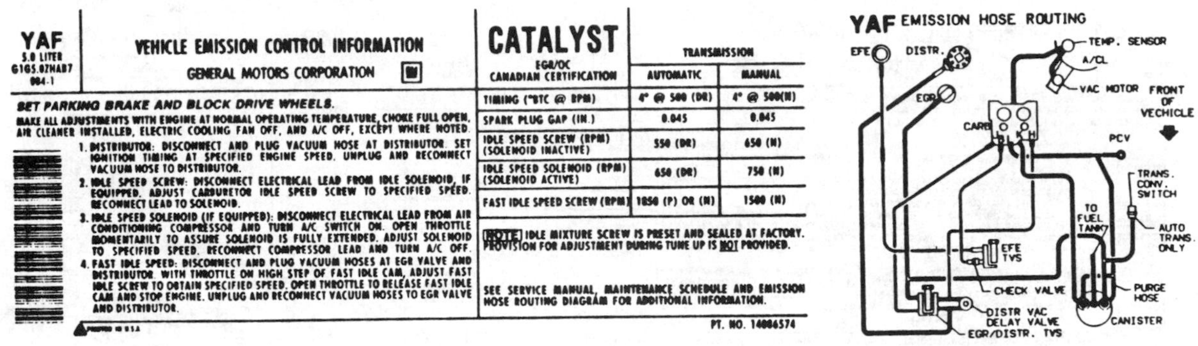

YAF
5.0 LITER
G1G5.0ZHAB7
004-1

VEHICLE EMISSION CONTROL INFORMATION
GENERAL MOTORS CORPORATION

SET PARKING BRAKE AND BLOCK DRIVE WHEELS.

MAKE ALL ADJUSTMENTS WITH ENGINE AT NORMAL OPERATING TEMPERATURE, CHOKE FULL OPEN, AIR CLEANER INSTALLED, ELECTRIC COOLING FAN OFF, AND A/C OFF, EXCEPT WHERE NOTED.

1. DISTRIBUTOR: DISCONNECT AND PLUG VACUUM HOSE AT DISTRIBUTOR. SET IGNITION TIMING AT SPECIFIED ENGINE SPEED. UNPLUG AND RECONNECT VACUUM HOSE TO DISTRIBUTOR.
2. IDLE SPEED SCREW: DISCONNECT ELECTRICAL LEAD FROM IDLE SOLENOID, IF EQUIPPED. ADJUST CARBURETOR IDLE SPEED SCREW TO SPECIFIED SPEED. RECONNECT LEAD TO SOLENOID.
3. IDLE SPEED SOLENOID (IF EQUIPPED): DISCONNECT ELECTRICAL LEAD FROM AIR CONDITIONING COMPRESSOR AND TURN A/C SWITCH ON. OPEN THROTTLE MOMENTARILY TO ASSURE SOLENOID IS FULLY EXTENDED. ADJUST SOLENOID TO SPECIFIED SPEED. RECONNECT COMPRESSOR LEAD AND TURN A/C OFF.
4. FAST IDLE SPEED: DISCONNECT AND PLUG VACUUM HOSES AT EGR VALVE AND DISTRIBUTOR. WITH THROTTLE ON HIGH STEP OF FAST IDLE CAM, ADJUST FAST IDLE SCREW TO OBTAIN SPECIFIED SPEED. OPEN THROTTLE TO RELEASE FAST IDLE CAM AND STOP ENGINE. UNPLUG AND RECONNECT VACUUM HOSES TO EGR VALVE AND DISTRIBUTOR.

PRINTED IN U.S.A.

CATALYST EGR/OC CANADIAN CERTIFICATION	TRANSMISSION AUTOMATIC	 MANUAL
TIMING (°BTC @ RPM)	4° @ 500 (DR)	4° @ 500(N)
SPARK PLUG GAP (IN.)	0.045	0.045
IDLE SPEED SCREW (RPM) (SOLENOID INACTIVE)	550 (DR)	650 (N)
IDLE SPEED SOLENOID (RPM) (SOLENOID ACTIVE)	650 (DR)	750 (N)
FAST IDLE SPEED SCREW (RPM)	1850 (P) OR (N)	1500 (N)

NOTE IDLE MIXTURE SCREW IS PRESET AND SEALED AT FACTORY. PROVISION FOR ADJUSTMENT DURING TUNE UP IS NOT PROVIDED.

SEE SERVICE MANUAL, MAINTENANCE SCHEDULE AND EMISSION HOSE ROUTING DIAGRAM FOR ADDITIONAL INFORMATION.

PT. NO. 14086574

USAGE: 1-2FA00 & LG4 & NM5

5.0L—W/AUTO. TRANS. & MAN. TRANS.

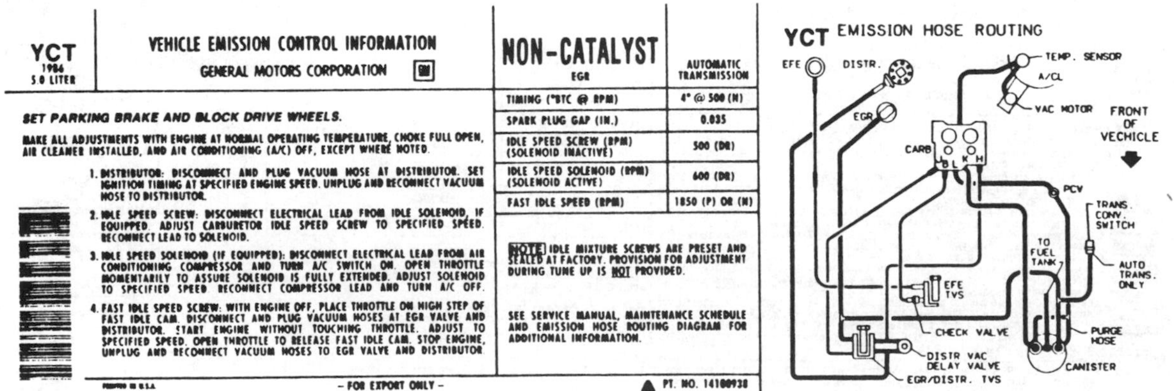

YCT
1986
5.0 LITER

VEHICLE EMISSION CONTROL INFORMATION
GENERAL MOTORS CORPORATION

SET PARKING BRAKE AND BLOCK DRIVE WHEELS.

MAKE ALL ADJUSTMENTS WITH ENGINE AT NORMAL OPERATING TEMPERATURE, CHOKE FULL OPEN, AIR CLEANER INSTALLED, AND AIR CONDITIONING (A/C) OFF, EXCEPT WHERE NOTED.

1. DISTRIBUTOR: DISCONNECT AND PLUG VACUUM HOSE AT DISTRIBUTOR. SET IGNITION TIMING AT SPECIFIED ENGINE SPEED. UNPLUG AND RECONNECT VACUUM HOSE TO DISTRIBUTOR.
2. IDLE SPEED SCREW: DISCONNECT ELECTRICAL LEAD FROM IDLE SOLENOID, IF EQUIPPED. ADJUST CARBURETOR IDLE SPEED SCREW TO SPECIFIED SPEED. RECONNECT LEAD TO SOLENOID.
3. IDLE SPEED SOLENOID (IF EQUIPPED): DISCONNECT ELECTRICAL LEAD FROM AIR CONDITIONING COMPRESSOR AND TURN A/C SWITCH ON. OPEN THROTTLE MOMENTARILY TO ASSURE SOLENOID IS FULLY EXTENDED. ADJUST SOLENOID TO SPECIFIED SPEED. RECONNECT COMPRESSOR LEAD AND TURN A/C OFF.
4. FAST IDLE SPEED SCREW: WITH ENGINE OFF, PLACE THROTTLE ON HIGH STEP OF FAST IDLE CAM. DISCONNECT AND PLUG VACUUM HOSES AT EGR VALVE AND DISTRIBUTOR. START ENGINE WITHOUT TOUCHING THROTTLE. ADJUST TO SPECIFIED SPEED. OPEN THROTTLE TO RELEASE FAST IDLE CAM. STOP ENGINE, UNPLUG AND RECONNECT VACUUM HOSES TO EGR VALVE AND DISTRIBUTOR.

PRINTED IN U.S.A. — FOR EXPORT ONLY —

NON-CATALYST EGR	AUTOMATIC TRANSMISSION
TIMING (°BTC @ RPM)	4° @ 500 (N)
SPARK PLUG GAP (IN.)	0.035
IDLE SPEED SCREW (RPM) (SOLENOID INACTIVE)	500 (DR)
IDLE SPEED SOLENOID (RPM) (SOLENOID ACTIVE)	600 (DR)
FAST IDLE SPEED (RPM)	1850 (P) OR (N)

NOTE IDLE MIXTURE SCREWS ARE PRESET AND SEALED AT FACTORY. PROVISION FOR ADJUSTMENT DURING TUNE UP IS NOT PROVIDED.

SEE SERVICE MANUAL, MAINTENANCE SCHEDULE AND EMISSION HOSE ROUTING DIAGRAM FOR ADDITIONAL INFORMATION.

PT. NO. 14100938

USAGE: 1-2BA00/1-2GA00 & LG4 & NM8

5.0L—W/AUTO. TRANS. (NM5 EXPORT, CANADA) EXC. NM8 EXPORT

VACUUM CIRCUITS

(© G.M. Corp.)

1986 CHEVROLET 5.0L

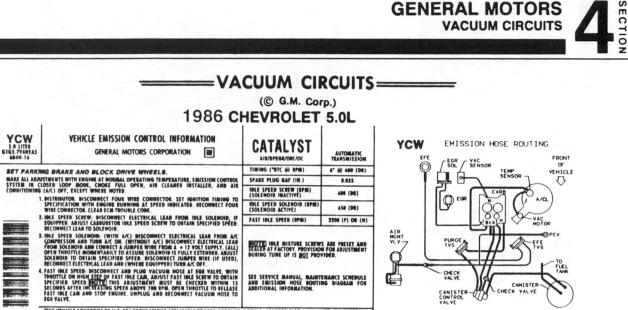

YCW
5.0 LITER
G1G5.7V4NEA5
6B4H-1A

VEHICLE EMISSION CONTROL INFORMATION
GENERAL MOTORS CORPORATION

CATALYST
AIR/DPEGR/ORC/OC

SET PARKING BRAKE AND BLOCK DRIVE WHEELS.

MAKE ALL ADJUSTMENTS WITH ENGINE AT NORMAL OPERATING TEMPERATURE, EMISSION CONTROL SYSTEM IN CLOSED LOOP MODE, CHOKE FULL OPEN, AIR CLEANER INSTALLED, AND AIR CONDITIONING (A/C) OFF, EXCEPT WHERE NOTED.

1. DISTRIBUTOR: DISCONNECT FOUR WIRE CONNECTOR. SET IGNITION TIMING TO SPECIFICATION WITH ENGINE RUNNING AT SPEED INDICATED. RECONNECT FOUR WIRE CONNECTOR. CLEAR ECM TROUBLE CODE.
2. IDLE SPEED SCREW: DISCONNECT ELECTRICAL LEAD FROM IDLE SOLENOID, IF EQUIPPED. ADJUST CARBURETOR IDLE SPEED SCREW TO OBTAIN SPECIFIED SPEED. RECONNECT LEAD TO SOLENOID.
3. IDLE SPEED SOLENOID: (WITH A/C) DISCONNECT ELECTRICAL LEAD FROM A/C COMPRESSOR AND TURN A/C ON. (WITHOUT A/C) DISCONNECT ELECTRICAL LEAD FROM SOLENOID AND CONNECT A JUMPER WIRE FROM A + 12 VOLT SUPPLY. (ALL) OPEN THROTTLE MOMENTARILY TO ASSURE SOLENOID IS FULLY EXTENDED. ADJUST SOLENOID TO OBTAIN SPECIFIED SPEED. DISCONNECT JUMPER WIRE (IF USED), RECONNECT ELECTRICAL LEAD AND (WHERE EQUIPPED) TURN A/C OFF.
4. FAST IDLE SPEED: DISCONNECT AND PLUG VACUUM HOSE AT EGR VALVE. WITH THROTTLE ON HIGH STEP OF FAST IDLE CAM, ADJUST FAST IDLE SCREW TO OBTAIN SPECIFIED SPEED. NOTE THIS ADJUSTMENT MUST BE CHECKED WITHIN 15 SECONDS AFTER INCREASING SPEED ABOVE 700 RPM. OPEN THROTTLE TO RELEASE FAST IDLE CAM AND STOP ENGINE. UNPLUG AND RECONNECT VACUUM HOSE TO EGR VALVE.

	AUTOMATIC TRANSMISSION
TIMING (°BTC @ RPM)	6° @ 600 (DR)
SPARK PLUG GAP (IN.)	0.035
IDLE SPEED SCREW (RPM) (SOLENOID INACTIVE)	600 (DR)
IDLE SPEED SOLENOID (RPM) (SOLENOID ACTIVE)	650 (DR)
FAST IDLE SPEED (RPM)	2200 (P) OR (N)

NOTE IDLE MIXTURE SCREWS ARE PRESET AND SEALED AT FACTORY. PROVISION FOR ADJUSTMENT DURING TUNE UP IS NOT PROVIDED.

SEE SERVICE MANUAL, MAINTENANCE SCHEDULE AND EMISSION HOSE ROUTING DIAGRAM FOR ADDITIONAL INFORMATION.

THIS VEHICLE CONFORMS TO U.S. EPA REGULATIONS APPLICABLE TO 1986 MODEL YEAR NEW PASSENGER CARS. PT. NO. 14100940 PRINTED IN U.S.A.

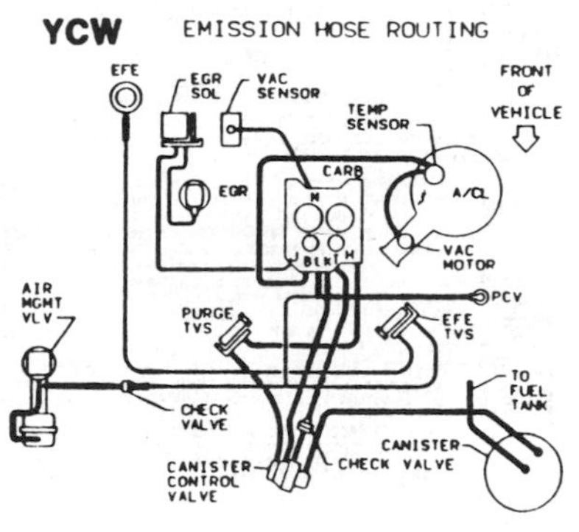

USAGE: 1GA00 & L69 & MW9 & NA5

5.0L—W/AUTO. TRANS.—FEDERAL

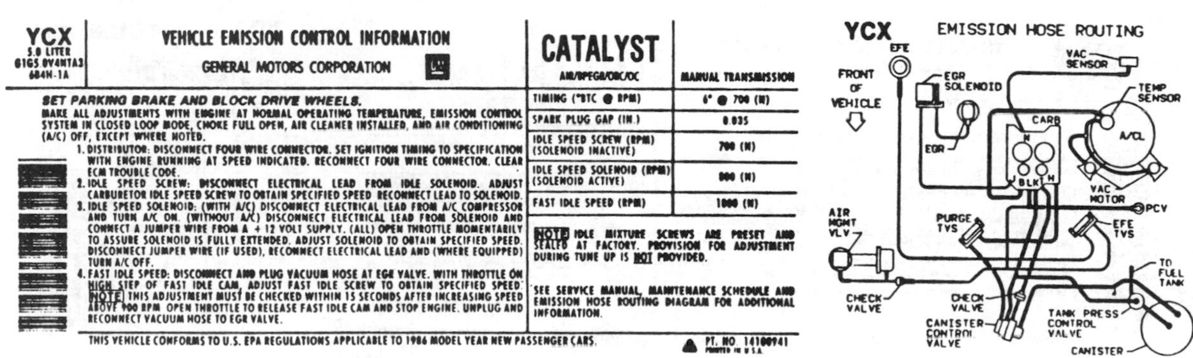

YCX
5.0 LITER
G1G5.0V4NTA3
6B4H-1A

VEHICLE EMISSION CONTROL INFORMATION
GENERAL MOTORS CORPORATION

CATALYST
AIR/DPEGR/ORC/OC

SET PARKING BRAKE AND BLOCK DRIVE WHEELS.

MAKE ALL ADJUSTMENTS WITH ENGINE AT NORMAL OPERATING TEMPERATURE, EMISSION CONTROL SYSTEM IN CLOSED LOOP MODE, CHOKE FULL OPEN, AIR CLEANER INSTALLED, AND AIR CONDITIONING (A/C) OFF, EXCEPT WHERE NOTED.

1. DISTRIBUTOR: DISCONNECT FOUR WIRE CONNECTOR. SET IGNITION TIMING TO SPECIFICATION WITH ENGINE RUNNING AT SPEED INDICATED. RECONNECT FOUR WIRE CONNECTOR. CLEAR ECM TROUBLE CODE.
2. IDLE SPEED SCREW: DISCONNECT ELECTRICAL LEAD FROM IDLE SOLENOID. ADJUST CARBURETOR IDLE SPEED SCREW TO OBTAIN SPECIFIED SPEED RECONNECT LEAD TO SOLENOID.
3. IDLE SPEED SOLENOID: (WITH A/C) DISCONNECT ELECTRICAL LEAD FROM A/C COMPRESSOR AND TURN A/C ON. (WITHOUT A/C) DISCONNECT ELECTRICAL LEAD FROM SOLENOID AND CONNECT A JUMPER WIRE FROM A + 12 VOLT SUPPLY. (ALL) OPEN THROTTLE MOMENTARILY TO ASSURE SOLENOID IS FULLY EXTENDED. ADJUST SOLENOID TO OBTAIN SPECIFIED SPEED. DISCONNECT JUMPER WIRE (IF USED), RECONNECT ELECTRICAL LEAD AND (WHERE EQUIPPED) TURN A/C OFF.
4. FAST IDLE SPEED: DISCONNECT AND PLUG VACUUM HOSE AT EGR VALVE. WITH THROTTLE ON HIGH STEP OF FAST IDLE CAM, ADJUST FAST IDLE SCREW TO OBTAIN SPECIFIED SPEED. NOTE THIS ADJUSTMENT MUST BE CHECKED WITHIN 15 SECONDS AFTER INCREASING SPEED ABOVE 900 RPM OPEN THROTTLE TO RELEASE FAST IDLE CAM AND STOP ENGINE. UNPLUG AND RECONNECT VACUUM HOSE TO EGR VALVE.

	MANUAL TRANSMISSION
TIMING (°BTC @ RPM)	6° @ 700 (N)
SPARK PLUG GAP (IN.)	0.035
IDLE SPEED SCREW (RPM) (SOLENOID INACTIVE)	700 (N)
IDLE SPEED SOLENOID (RPM) (SOLENOID ACTIVE)	800 (N)
FAST IDLE SPEED (RPM)	1000 (N)

NOTE IDLE MIXTURE SCREWS ARE PRESET AND SEALED AT FACTORY. PROVISION FOR ADJUSTMENT DURING TUNE UP IS NOT PROVIDED.

SEE SERVICE MANUAL, MAINTENANCE SCHEDULE AND EMISSION HOSE ROUTING DIAGRAM FOR ADDITIONAL INFORMATION.

THIS VEHICLE CONFORMS TO U.S. EPA REGULATIONS APPLICABLE TO 1986 MODEL YEAR NEW PASSENGER CARS. PT. NO. 14100941 PRINTED IN U.S.A.

USAGE: 1-2FA00 & L69 & M39/MC4 & NA5

5.0L—W/MAN. TRANS. (MC4, M39)—FEDERAL

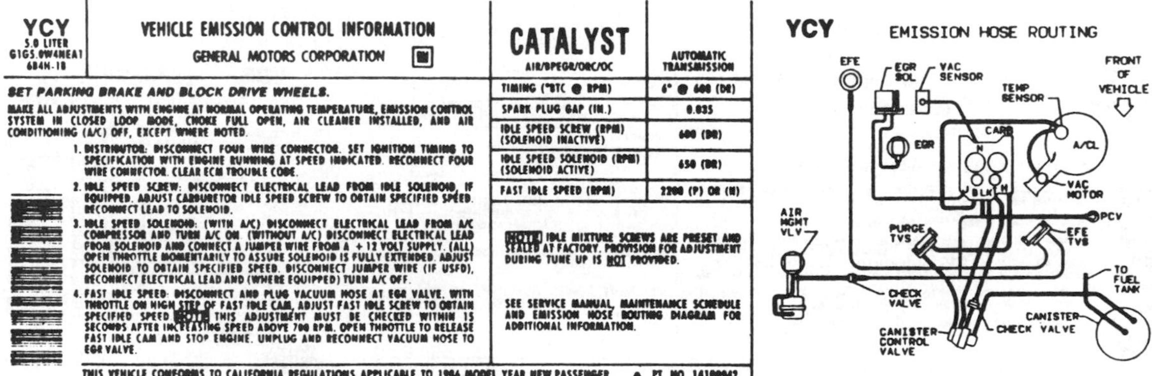

YCY
5.0 LITER
G1G5.0W4NEA1
6B4H-1B

VEHICLE EMISSION CONTROL INFORMATION
GENERAL MOTORS CORPORATION

CATALYST
AIR/DPEGR/ORC/OC

SET PARKING BRAKE AND BLOCK DRIVE WHEELS.

MAKE ALL ADJUSTMENTS WITH ENGINE AT NORMAL OPERATING TEMPERATURE, EMISSION CONTROL SYSTEM IN CLOSED LOOP MODE, CHOKE FULL OPEN, AIR CLEANER INSTALLED, AND AIR CONDITIONING (A/C) OFF, EXCEPT WHERE NOTED.

1. DISTRIBUTOR: DISCONNECT FOUR WIRE CONNECTOR. SET IGNITION TIMING TO SPECIFICATION WITH ENGINE RUNNING AT SPEED INDICATED. RECONNECT FOUR WIRE CONNECTOR. CLEAR ECM TROUBLE CODE.
2. IDLE SPEED SCREW: DISCONNECT ELECTRICAL LEAD FROM IDLE SOLENOID, IF EQUIPPED. ADJUST CARBURETOR IDLE SPEED SCREW TO OBTAIN SPECIFIED SPEED. RECONNECT LEAD TO SOLENOID.
3. IDLE SPEED SOLENOID: (WITH A/C) DISCONNECT ELECTRICAL LEAD FROM A/C COMPRESSOR AND TURN A/C ON. (WITHOUT A/C) DISCONNECT ELECTRICAL LEAD FROM SOLENOID AND CONNECT A JUMPER WIRE FROM A + 12 VOLT SUPPLY. (ALL) OPEN THROTTLE MOMENTARILY TO ASSURE SOLENOID IS FULLY EXTENDED. ADJUST SOLENOID TO OBTAIN SPECIFIED SPEED. DISCONNECT JUMPER WIRE (IF USED), RECONNECT ELECTRICAL LEAD AND (WHERE EQUIPPED) TURN A/C OFF.
4. FAST IDLE SPEED: DISCONNECT AND PLUG VACUUM HOSE AT EGR VALVE. WITH THROTTLE ON HIGH STEP OF FAST IDLE CAM, ADJUST FAST IDLE SCREW TO OBTAIN SPECIFIED SPEED NOTE THIS ADJUSTMENT MUST BE CHECKED WITHIN 15 SECONDS AFTER INCREASING SPEED ABOVE 700 RPM. OPEN THROTTLE TO RELEASE FAST IDLE CAM AND STOP ENGINE. UNPLUG AND RECONNECT VACUUM HOSE TO EGR VALVE.

	AUTOMATIC TRANSMISSION
TIMING (°BTC @ RPM)	6° @ 600 (DR)
SPARK PLUG GAP (IN.)	0.035
IDLE SPEED SCREW (RPM) (SOLENOID INACTIVE)	600 (DR)
IDLE SPEED SOLENOID (RPM) (SOLENOID ACTIVE)	650 (DR)
FAST IDLE SPEED (RPM)	2200 (P) OR (N)

NOTE IDLE MIXTURE SCREWS ARE PRESET AND SEALED AT FACTORY. PROVISION FOR ADJUSTMENT DURING TUNE UP IS NOT PROVIDED.

SEE SERVICE MANUAL, MAINTENANCE SCHEDULE AND EMISSION HOSE ROUTING DIAGRAM FOR ADDITIONAL INFORMATION.

THIS VEHICLE CONFORMS TO CALIFORNIA REGULATIONS APPLICABLE TO 1986 MODEL YEAR NEW PASSENGER CARS AND SIMILARLY TO U.S. EPA REGULATIONS APPLICABLE TO THE STATE OF CALIFORNIA. PT. NO. 14100942 PRINTED IN U.S.A.

USAGE: 1GA00 & L69 & MW9 & NB2

5.0L—W/AUTO. TRANS.—CALIFORNIA

VACUUM CIRCUITS

(© G.M. Corp.)

1986 CHEVROLET 5.0L

YCZ
5.0 LITER
G1GS.0W4NTA8
6BAM 1B

VEHICLE EMISSION CONTROL INFORMATION
GENERAL MOTORS CORPORATION

SET PARKING BRAKE AND BLOCK DRIVE WHEELS.

MAKE ALL ADJUSTMENTS WITH ENGINE AT NORMAL OPERATING TEMPERATURE, EMISSION CONTROL SYSTEM IN CLOSED LOOP MODE, CHOKE FULL OPEN, AIR CLEANER INSTALLED, AND AIR CONDITIONING (A/C) OFF, EXCEPT WHERE NOTED.

1. DISTRIBUTOR: DISCONNECT FOUR WIRE CONNECTOR. SET IGNITION TIMING TO SPECIFICATION WITH ENGINE RUNNING AT SPEED INDICATED. RECONNECT FOUR WIRE CONNECTOR. CLEAR ECM TROUBLE CODE.
2. IDLE SPEED SCREW: DISCONNECT ELECTRICAL LEAD FROM IDLE SOLENOID ADJUST CARBURETOR IDLE SPEED SCREW TO OBTAIN SPECIFIED SPEED. RECONNECT LEAD TO SOLENOID
3. IDLE SPEED SOLENOID: (WITH A/C) DISCONNECT ELECTRICAL LEAD FROM A/C COMPRESSOR AND TURN A/C ON. (WITHOUT A/C) DISCONNECT ELECTRICAL LEAD FROM SOLENOID AND CONNECT A JUMPER WIRE FROM A +12 VOLT SUPPLY. (ALL) OPEN THROTTLE MOMENTARILY TO ASSURE SOLENOID IS FULLY EXTENDED. ADJUST SOLENOID TO OBTAIN SPECIFIED SPEED. DISCONNECT JUMPER WIRE (IF USED), RECONNECT ELECTRICAL LEAD AND (WHERE EQUIPPED) TURN A/C OFF.
4. FAST IDLE SPEED: DISCONNECT AND PLUG VACUUM HOSE AT EGR VALVE. WITH THROTTLE ON HIGH STEP OF FAST IDLE CAM, ADJUST FAST IDLE SCREW TO OBTAIN SPECIFIED SPEED. NOTE: THIS ADJUSTMENT MUST BE CHECKED WITHIN 15 SECONDS AFTER INCREASING SPEED ABOVE 900 RPM. OPEN THROTTLE TO RELEASE FAST IDLE CAM AND STOP ENGINE. UNPLUG AND RECONNECT VACUUM HOSE TO EGR VALVE.

CATALYST
AIR/DPEGR/ORC/OC

	MANUAL TRANSMISSION
TIMING (°BTC @ RPM)	6° @ 700 (N)
SPARK PLUG GAP (IN.)	0.035
IDLE SPEED SCREW (RPM) (SOLENOID INACTIVE)	700 (N)
IDLE SPEED SOLENOID (RPM) (SOLENOID ACTIVE)	800 (N)
FAST IDLE SPEED (RPM)	1000 (N)

NOTE: IDLE MIXTURE SCREWS ARE PRESET AND SEALED AT FACTORY. PROVISION FOR ADJUSTMENT DURING TUNE UP IS NOT PROVIDED.

SEE SERVICE MANUAL, MAINTENANCE SCHEDULE AND EMISSION HOSE ROUTING DIAGRAM FOR ADDITIONAL INFORMATION.

THIS VEHICLE CONFORMS TO CALIFORNIA REGULATIONS APPLICABLE TO 1986 MODEL YEAR NEW PASSENGER CARS AND SIMILARLY TO U.S. EPA REGULATIONS APPLICABLE TO THE STATE OF CALIFORNIA. PT. NO. 14100943

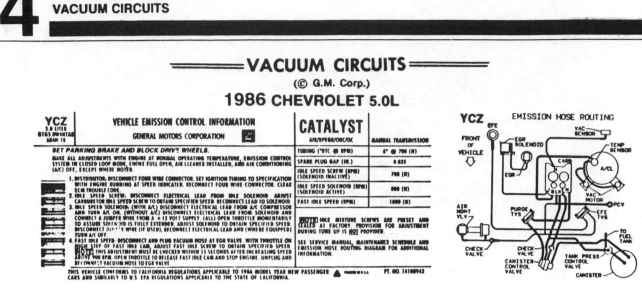

USAGE: 1-2FA00 & L69 & M39/MC4 & NB2

5.0L – W/MAN. TRANS. (MC4, M39) – CALIFORNIA

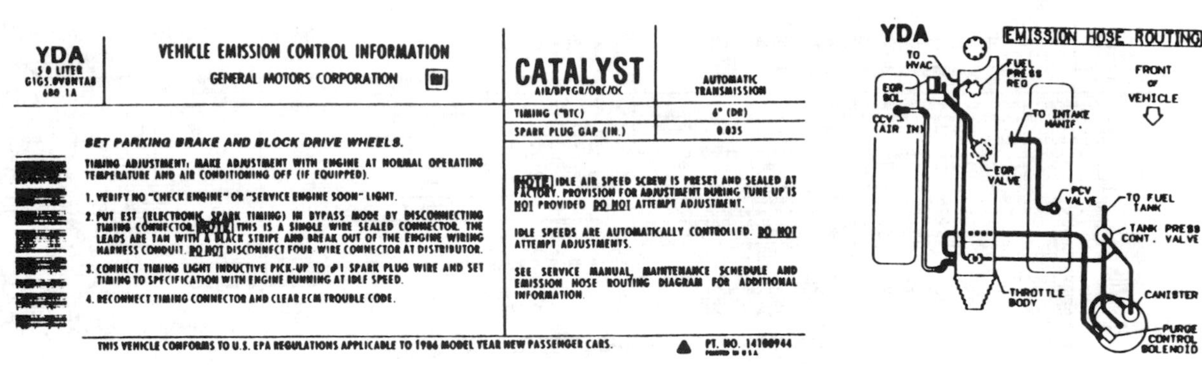

YDA
5.0 LITER
G1GS.0V8NTA8
6B0 1A

VEHICLE EMISSION CONTROL INFORMATION
GENERAL MOTORS CORPORATION

SET PARKING BRAKE AND BLOCK DRIVE WHEELS.

TIMING ADJUSTMENT: MAKE ADJUSTMENT WITH ENGINE AT NORMAL OPERATING TEMPERATURE AND AIR CONDITIONING OFF (IF EQUIPPED).

1. VERIFY NO "CHECK ENGINE" OR "SERVICE ENGINE SOON" LIGHT.
2. PUT EST (ELECTRONIC SPARK TIMING) IN BYPASS MODE BY DISCONNECTING TIMING CONNECTOR. NOTE: THIS IS A SINGLE WIRE SEALED CONNECTOR. THE LEADS ARE TAN WITH A BLACK STRIPE AND BREAK OUT OF THE ENGINE WIRING HARNESS CONDUIT. DO NOT DISCONNECT FOUR WIRE CONNECTOR AT DISTRIBUTOR.
3. CONNECT TIMING LIGHT INDUCTIVE PICK-UP TO #1 SPARK PLUG WIRE AND SET TIMING TO SPECIFICATION WITH ENGINE RUNNING AT IDLE SPEED.
4. RECONNECT TIMING CONNECTOR AND CLEAR ECM TROUBLE CODE.

CATALYST
AIR/DPEGR/ORC/OC

	AUTOMATIC TRANSMISSION
TIMING (°BTC)	6° (DR)
SPARK PLUG GAP (IN.)	0.035

NOTE: IDLE AIR SPEED SCREW IS PRESET AND SEALED AT FACTORY. PROVISION FOR ADJUSTMENT DURING TUNE UP IS NOT PROVIDED. DO NOT ATTEMPT ADJUSTMENT.

IDLE SPEEDS ARE AUTOMATICALLY CONTROLLED. DO NOT ATTEMPT ADJUSTMENTS.

SEE SERVICE MANUAL, MAINTENANCE SCHEDULE AND EMISSION HOSE ROUTING DIAGRAM FOR ADDITIONAL INFORMATION.

THIS VEHICLE CONFORMS TO U.S. EPA REGULATIONS APPLICABLE TO 1986 MODEL YEAR NEW PASSENGER CARS. PT. NO. 14100944

USAGE: 1FA00 & LB9 & NA5 (Chev. only)

5.0L – W/AUTO. TRANS. (MD8), EXC NM8 EXPORT – FEDERAL

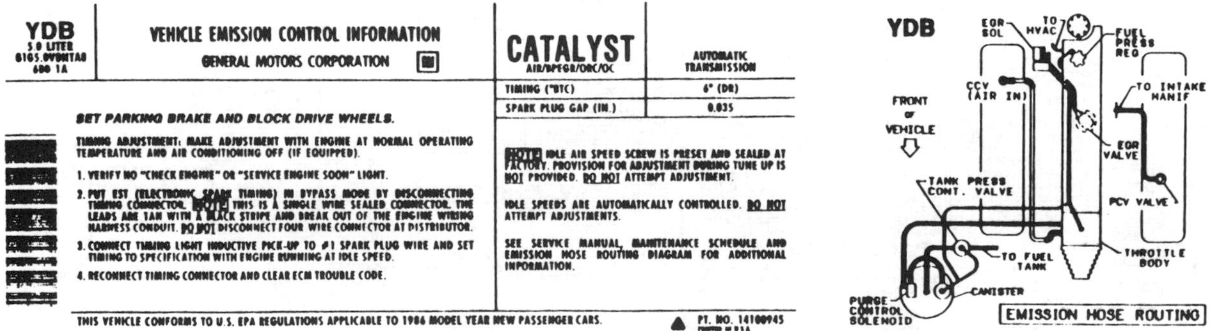

YDB
5.0 LITER
G1GS.0V8NTA8
6B0 1A

VEHICLE EMISSION CONTROL INFORMATION
GENERAL MOTORS CORPORATION

SET PARKING BRAKE AND BLOCK DRIVE WHEELS.

TIMING ADJUSTMENT: MAKE ADJUSTMENT WITH ENGINE AT NORMAL OPERATING TEMPERATURE AND AIR CONDITIONING OFF (IF EQUIPPED).

1. VERIFY NO "CHECK ENGINE" OR "SERVICE ENGINE SOON" LIGHT.
2. PUT EST (ELECTRONIC SPARK TIMING) IN BYPASS MODE BY DISCONNECTING TIMING CONNECTOR. NOTE: THIS IS A SINGLE WIRE SEALED CONNECTOR. THE LEADS ARE TAN WITH A BLACK STRIPE AND BREAK OUT OF THE ENGINE WIRING HARNESS CONDUIT. DO NOT DISCONNECT FOUR WIRE CONNECTOR AT DISTRIBUTOR.
3. CONNECT TIMING LIGHT INDUCTIVE PICK-UP TO #1 SPARK PLUG WIRE AND SET TIMING TO SPECIFICATION WITH ENGINE RUNNING AT IDLE SPEED.
4. RECONNECT TIMING CONNECTOR AND CLEAR ECM TROUBLE CODE.

CATALYST
AIR/DPEGR/ORC/OC

	AUTOMATIC TRANSMISSION
TIMING (°BTC)	6° (DR)
SPARK PLUG GAP (IN.)	0.035

NOTE: IDLE AIR SPEED SCREW IS PRESET AND SEALED AT FACTORY. PROVISION FOR ADJUSTMENT DURING TUNE UP IS NOT PROVIDED. DO NOT ATTEMPT ADJUSTMENT.

IDLE SPEEDS ARE AUTOMATICALLY CONTROLLED. DO NOT ATTEMPT ADJUSTMENTS.

SEE SERVICE MANUAL, MAINTENANCE SCHEDULE AND EMISSION HOSE ROUTING DIAGRAM FOR ADDITIONAL INFORMATION.

THIS VEHICLE CONFORMS TO U.S. EPA REGULATIONS APPLICABLE TO 1986 MODEL YEAR NEW PASSENGER CARS. PT. NO. 14100945

USAGE: 2FA00 & LB9 & NA5 (Pontiac only)

5.0L – W/AUTO. TRANS. – FEDERAL

VACUUM CIRCUITS

(© G.M. Corp.)

1986 CHEVROLET 5.0L

YDC
5.0 LITER
6165.0W8NTA2
600-1B

VEHICLE EMISSION CONTROL INFORMATION
GENERAL MOTORS CORPORATION

CATALYST
AIR/DPEGR/ORC/OC

AUTOMATIC TRANSMISSION

	AUTOMATIC TRANSMISSION
TIMING (°BTC)	6° (DR)
SPARK PLUG GAP (IN.)	0.035

SET PARKING BRAKE AND BLOCK DRIVE WHEELS.

TIMING ADJUSTMENT: MAKE ADJUSTMENT WITH ENGINE AT NORMAL OPERATING TEMPERATURE AND AIR CONDITIONING OFF (IF EQUIPPED).

1. VERIFY NO "CHECK ENGINE" OR "SERVICE ENGINE SOON" LIGHT.
2. PUT EST (ELECTRONIC SPARK TIMING) IN BYPASS MODE BY DISCONNECTING TIMING CONNECTOR. NOTE: THIS IS A SINGLE WIRE SEALED CONNECTOR. THE LEADS ARE TAN WITH A BLACK STRIPE AND BREAK OUT OF THE ENGINE WIRING HARNESS CONDUIT. DO NOT DISCONNECT FOUR-WIRE CONNECTOR AT DISTRIBUTOR.
3. CONNECT TIMING LIGHT INDUCTIVE PICK-UP TO #1 SPARK PLUG WIRE AND SET TIMING TO SPECIFICATION WITH ENGINE RUNNING AT IDLE SPEED.
4. RECONNECT TIMING CONNECTOR AND CLEAR ECM TROUBLE CODE.

NOTE: IDLE AIR SPEED SCREW IS PRESET AND SEALED AT FACTORY. PROVISION FOR ADJUSTMENT DURING TUNE UP IS NOT PROVIDED. DO NOT ATTEMPT ADJUSTMENT.

IDLE SPEEDS ARE AUTOMATICALLY CONTROLLED. DO NOT ATTEMPT ADJUSTMENTS.

SEE SERVICE MANUAL, MAINTENANCE SCHEDULE AND EMISSION HOSE ROUTING DIAGRAM FOR ADDITIONAL INFORMATION.

THIS VEHICLE CONFORMS TO CALIFORNIA REGULATIONS APPLICABLE TO 1986 MODEL YEAR NEW PASSENGER CARS AND SIMILARLY TO U.S. EPA REGULATIONS APPLICABLE TO THE STATE OF CALIFORNIA.

PT. NO. 14100946
PRINTED IN U.S.A.

YDC
EMISSION HOSE ROUTING
TO HVAC
FUEL PRESS REG
EGR SOL
CCV (AIR IN)
TO INTAKE MANIF.
FRONT OF VEHICLE
EGR VALVE
PCV VALVE
TO FUEL TANK
TANK PRESS CONT. VALVE
CANISTER
THROTTLE BODY
PURGE CONTROL SOLENOID

USAGE: 1FA00 & LB9 & NB2 (Chev. only)

5.0L—W/AUTO. TRANS. (MD8)—CALIFORNIA

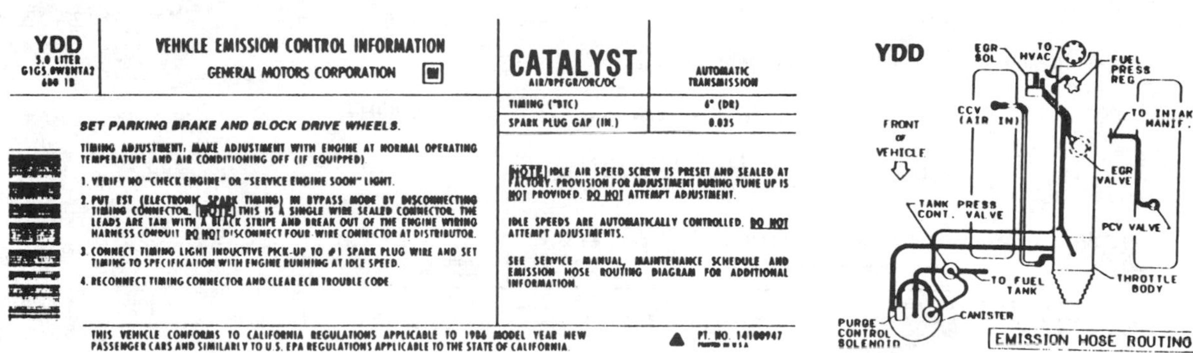

YDD
5.0 LITER
G1G5.0W8NTA2
600 1B

VEHICLE EMISSION CONTROL INFORMATION
GENERAL MOTORS CORPORATION

CATALYST
AIR/DPEGR/ORC/OC

	AUTOMATIC TRANSMISSION
TIMING (°BTC)	6° (DR)
SPARK PLUG GAP (IN.)	0.035

SET PARKING BRAKE AND BLOCK DRIVE WHEELS.

TIMING ADJUSTMENT: MAKE ADJUSTMENT WITH ENGINE AT NORMAL OPERATING TEMPERATURE AND AIR CONDITIONING OFF (IF EQUIPPED).

1. VERIFY NO "CHECK ENGINE" OR "SERVICE ENGINE SOON" LIGHT.
2. PUT EST (ELECTRONIC SPARK TIMING) IN BYPASS MODE BY DISCONNECTING TIMING CONNECTOR. NOTE: THIS IS A SINGLE WIRE SEALED CONNECTOR. THE LEADS ARE TAN WITH A BLACK STRIPE AND BREAK OUT OF THE ENGINE WIRING HARNESS CONDUIT. DO NOT DISCONNECT FOUR-WIRE CONNECTOR AT DISTRIBUTOR.
3. CONNECT TIMING LIGHT INDUCTIVE PICK-UP TO #1 SPARK PLUG WIRE AND SET TIMING TO SPECIFICATION WITH ENGINE RUNNING AT IDLE SPEED.
4. RECONNECT TIMING CONNECTOR AND CLEAR ECM TROUBLE CODE.

NOTE: IDLE AIR SPEED SCREW IS PRESET AND SEALED AT FACTORY. PROVISION FOR ADJUSTMENT DURING TUNE UP IS NOT PROVIDED. DO NOT ATTEMPT ADJUSTMENT.

IDLE SPEEDS ARE AUTOMATICALLY CONTROLLED. DO NOT ATTEMPT ADJUSTMENTS.

SEE SERVICE MANUAL, MAINTENANCE SCHEDULE AND EMISSION HOSE ROUTING DIAGRAM FOR ADDITIONAL INFORMATION.

THIS VEHICLE CONFORMS TO CALIFORNIA REGULATIONS APPLICABLE TO 1986 MODEL YEAR NEW PASSENGER CARS AND SIMILARLY TO U.S. EPA REGULATIONS APPLICABLE TO THE STATE OF CALIFORNIA.

PT. NO. 14100947
PRINTED IN U.S.A.

USAGE: 2FA00 & LB9 & NB2 (Pontiac only)

5.0L—W/AUTO. TRANS.—CALIFORNIA

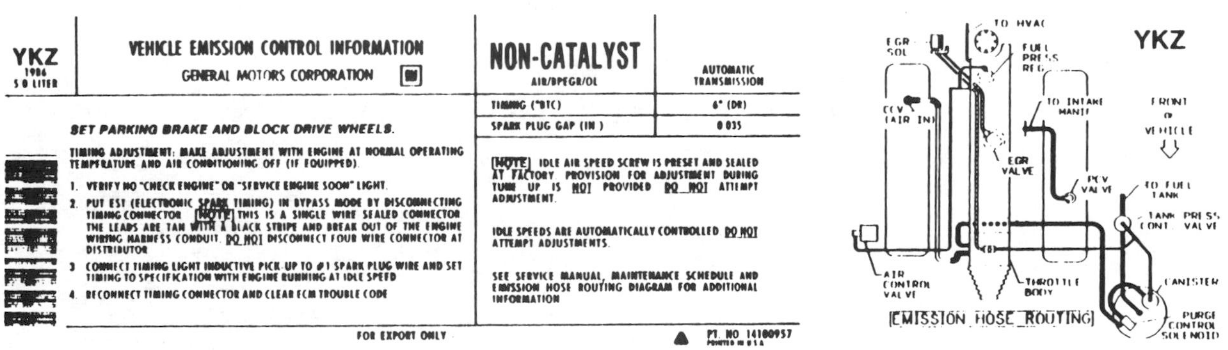

YKZ
1986
5.0 LITER

VEHICLE EMISSION CONTROL INFORMATION
GENERAL MOTORS CORPORATION

NON-CATALYST
AIR/DPEGR/OL

	AUTOMATIC TRANSMISSION
TIMING (°BTC)	6° (DR)
SPARK PLUG GAP (IN.)	0.035

SET PARKING BRAKE AND BLOCK DRIVE WHEELS.

TIMING ADJUSTMENT: MAKE ADJUSTMENT WITH ENGINE AT NORMAL OPERATING TEMPERATURE AND AIR CONDITIONING OFF (IF EQUIPPED).

1. VERIFY NO "CHECK ENGINE" OR "SERVICE ENGINE SOON" LIGHT.
2. PUT EST (ELECTRONIC SPARK TIMING) IN BYPASS MODE BY DISCONNECTING TIMING CONNECTOR. NOTE: THIS IS A SINGLE WIRE SEALED CONNECTOR. THE LEADS ARE TAN WITH A BLACK STRIPE AND BREAK OUT OF THE ENGINE WIRING HARNESS CONDUIT. DO NOT DISCONNECT FOUR WIRE CONNECTOR AT DISTRIBUTOR.
3. CONNECT TIMING LIGHT INDUCTIVE PICK-UP TO #1 SPARK PLUG WIRE AND SET TIMING TO SPECIFICATION WITH ENGINE RUNNING AT IDLE SPEED.
4. RECONNECT TIMING CONNECTOR AND CLEAR ECM TROUBLE CODE.

NOTE: IDLE AIR SPEED SCREW IS PRESET AND SEALED AT FACTORY. PROVISION FOR ADJUSTMENT DURING TUNE UP IS NOT PROVIDED. DO NOT ATTEMPT ADJUSTMENT.

IDLE SPEEDS ARE AUTOMATICALLY CONTROLLED. DO NOT ATTEMPT ADJUSTMENTS.

SEE SERVICE MANUAL, MAINTENANCE SCHEDULE AND EMISSION HOSE ROUTING DIAGRAM FOR ADDITIONAL INFORMATION.

FOR EXPORT ONLY

PT. NO. 14100957
PRINTED IN U.S.A.

USAGE: 1FA00 & LB9 & NM8 (Chev. only)

5.0L—W/AUTO. TRANS.—EXPORT

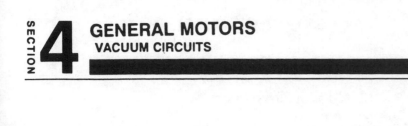

VACUUM CIRCUITS

(© G.M. Corp.)

1986 CHEVROLET 5.0L

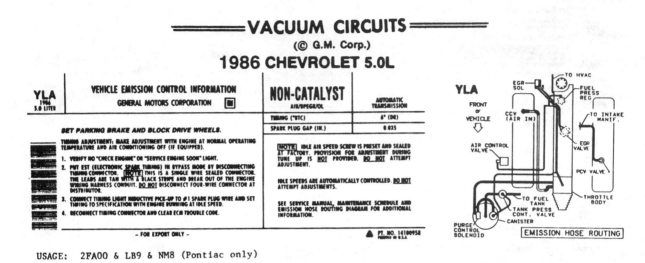

YLA
1986
5.0 LITER

VEHICLE EMISSION CONTROL INFORMATION
GENERAL MOTORS CORPORATION

NON-CATALYST AIR/BPEGR/OL	AUTOMATIC TRANSMISSION
TIMING (°BTC)	6° (DR)
SPARK PLUG GAP (IN.)	0.035

SET PARKING BRAKE AND BLOCK DRIVE WHEELS.

TIMING ADJUSTMENT: MAKE ADJUSTMENT WITH ENGINE AT NORMAL OPERATING TEMPERATURE AND AIR CONDITIONING OFF (IF EQUIPPED).

1. VERIFY NO "CHECK ENGINE" OR "SERVICE ENGINE SOON" LIGHT.
2. PUT EST (ELECTRONIC SPARK TIMING) IN BYPASS MODE BY DISCONNECTING TIMING CONNECTOR. NOTE THIS IS A SINGLE WIRE SEALED CONNECTOR. THE LEADS ARE TAN WITH A BLACK STRIPE AND BREAK OUT OF THE ENGINE WIRING HARNESS CONDUIT. DO NOT DISCONNECT FOUR-WIRE CONNECTOR AT DISTRIBUTOR.
3. CONNECT TIMING LIGHT INDUCTIVE PICK-UP TO #1 SPARK PLUG WIRE AND SET TIMING TO SPECIFICATION WITH ENGINE RUNNING AT IDLE SPEED.
4. RECONNECT TIMING CONNECTOR AND CLEAR ECM TROUBLE CODE.

NOTE IDLE AIR SPEED SCREW IS PRESET AND SEALED AT FACTORY. PROVISION FOR ADJUSTMENT DURING TUNE UP IS NOT PROVIDED. DO NOT ATTEMPT ADJUSTMENT.

IDLE SPEEDS ARE AUTOMATICALLY CONTROLLED. DO NOT ATTEMPT ADJUSTMENTS.

SEE SERVICE MANUAL, MAINTENANCE SCHEDULE AND EMISSION HOSE ROUTING DIAGRAM FOR ADDITIONAL INFORMATION.

- FOR EXPORT ONLY -

PT. NO. 14100958
PRINTED IN U.S.A.

YLA

FRONT OF VEHICLE
EGR SOL
TO HVAC
FUEL PRESS REG
CCV (AIR IN)
TO INTAKE MANIF.
AIR CONTROL VALVE
EGR VALVE
PCV VALVE
THROTTLE BODY
TO FUEL TANK
TANK PRESS CONT. VALVE
CANISTER
PURGE CONTROL SOLENOID

EMISSION HOSE ROUTING

USAGE: 2FA00 & LB9 & NM8 (Pontiac only)

5.0L—W/AUTO. TRANS.—EXPORT

1986 CHEVROLET 5.7L

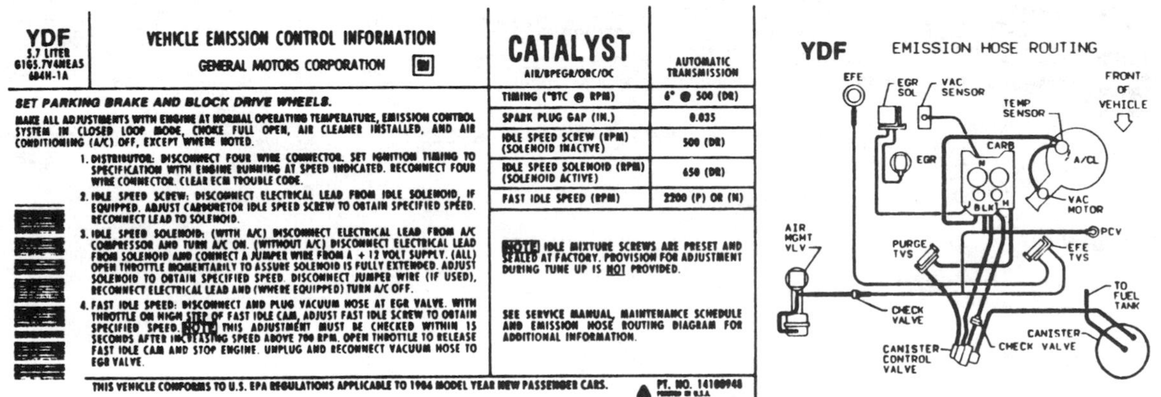
YDF
5.7 LITER
6G5.7V4NEA5
684N-1A

VEHICLE EMISSION CONTROL INFORMATION
GENERAL MOTORS CORPORATION

CATALYST AIR/BPEGR/ORC/OC	AUTOMATIC TRANSMISSION
TIMING (°BTC @ RPM)	6° @ 500 (DR)
SPARK PLUG GAP (IN.)	0.035
IDLE SPEED SCREW (RPM) (SOLENOID INACTVE)	500 (DR)
IDLE SPEED SOLENOID (RPM) (SOLENOID ACTIVE)	650 (DR)
FAST IDLE SPEED (RPM)	2200 (P) OR (N)

SET PARKING BRAKE AND BLOCK DRIVE WHEELS.

MAKE ALL ADJUSTMENTS WITH ENGINE AT NORMAL OPERATING TEMPERATURE, EMISSION CONTROL SYSTEM IN CLOSED LOOP MODE, CHOKE FULL OPEN, AIR CLEANER INSTALLED, AND AIR CONDITIONING (A/C) OFF, EXCEPT WHERE NOTED.

1. DISTRIBUTOR: DISCONNECT FOUR WIRE CONNECTOR. SET IGNITION TIMING TO SPECIFICATION WITH ENGINE RUNNING AT SPEED INDICATED. RECONNECT FOUR WIRE CONNECTOR. CLEAR ECM TROUBLE CODE.
2. IDLE SPEED SCREW: DISCONNECT ELECTRICAL LEAD FROM IDLE SOLENOID, IF EQUIPPED. ADJUST CARBURETOR IDLE SPEED SCREW TO OBTAIN SPECIFIED SPEED. RECONNECT LEAD TO SOLENOID.
3. IDLE SPEED SOLENOID: (WITH A/C) DISCONNECT ELECTRICAL LEAD FROM A/C COMPRESSOR AND TURN A/C ON. (WITHOUT A/C) DISCONNECT ELECTRICAL LEAD FROM SOLENOID AND CONNECT A JUMPER WIRE FROM A + 12 VOLT SUPPLY. (ALL) OPEN THROTTLE MOMENTARILY TO ASSURE SOLENOID IS FULLY EXTENDED. ADJUST SOLENOID TO OBTAIN SPECIFIED SPEED. DISCONNECT JUMPER WIRE (IF USED), RECONNECT ELECTRICAL LEAD AND (WHERE EQUIPPED) TURN A/C OFF.
4. FAST IDLE SPEED: DISCONNECT AND PLUG VACUUM HOSE AT EGR VALVE. WITH THROTTLE ON HIGH STEP OF FAST IDLE CAM, ADJUST FAST IDLE SCREW TO OBTAIN SPECIFIED SPEED. NOTE THIS ADJUSTMENT MUST BE CHECKED WITHIN 15 SECONDS AFTER INCREASING SPEED ABOVE 700 RPM. OPEN THROTTLE TO RELEASE FAST IDLE CAM AND STOP ENGINE. UNPLUG AND RECONNECT VACUUM HOSE TO EGR VALVE.

NOTE IDLE MIXTURE SCREWS ARE PRESET AND SEALED AT FACTORY. PROVISION FOR ADJUSTMENT DURING TUNE UP IS NOT PROVIDED.

SEE SERVICE MANUAL, MAINTENANCE SCHEDULE AND EMISSION HOSE ROUTING DIAGRAM FOR ADDITIONAL INFORMATION.

THIS VEHICLE CONFORMS TO U.S. EPA REGULATIONS APPLICABLE TO 1986 MODEL YEAR NEW PASSENGER CARS.

PT. NO. 14100948
PRINTED IN U.S.A.

USAGE: 1BA00 & LM1 & NA5 & 9C1 (police car only)

5.7L—W/AUTO. TRANS.—FEDERAL

VACUUM CIRCUITS

(© G.M. Corp.)

1986 CHEVROLET 5.7L

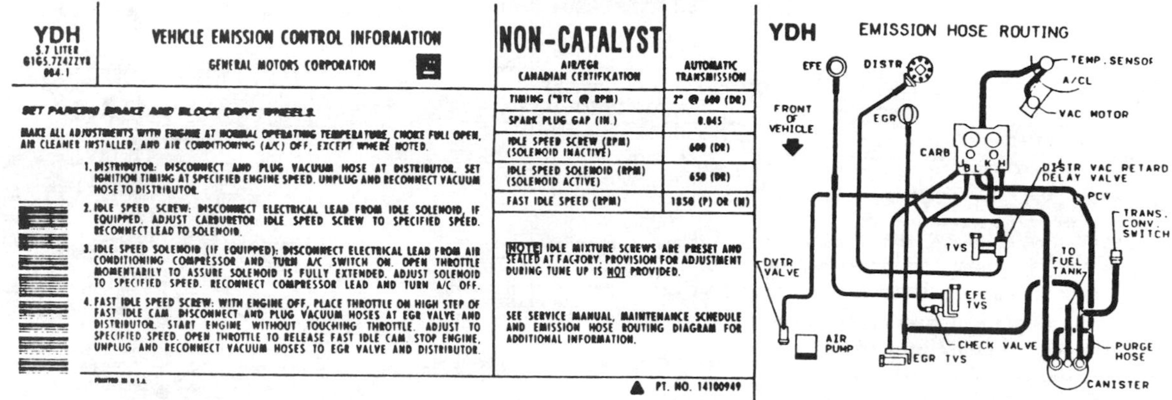

YDH
5.7 LITER
G1G5.7Z4ZZY8
004.1

VEHICLE EMISSION CONTROL INFORMATION
GENERAL MOTORS CORPORATION

SET PARKING BRAKE AND BLOCK DRIVE WHEELS.

MAKE ALL ADJUSTMENTS WITH ENGINE AT NORMAL OPERATING TEMPERATURE, CHOKE FULL OPEN, AIR CLEANER INSTALLED, AND AIR CONDITIONING (A/C) OFF, EXCEPT WHERE NOTED.

1. DISTRIBUTOR: DISCONNECT AND PLUG VACUUM HOSE AT DISTRIBUTOR. SET IGNITION TIMING AT SPECIFIED ENGINE SPEED. UNPLUG AND RECONNECT VACUUM HOSE TO DISTRIBUTOR.
2. IDLE SPEED SCREW: DISCONNECT ELECTRICAL LEAD FROM IDLE SOLENOID, IF EQUIPPED. ADJUST CARBURETOR IDLE SPEED SCREW TO SPECIFIED SPEED. RECONNECT LEAD TO SOLENOID.
3. IDLE SPEED SOLENOID (IF EQUIPPED): DISCONNECT ELECTRICAL LEAD FROM AIR CONDITIONING COMPRESSOR AND TURN A/C SWITCH ON. OPEN THROTTLE MOMENTARILY TO ASSURE SOLENOID IS FULLY EXTENDED. ADJUST SOLENOID TO SPECIFIED SPEED. RECONNECT COMPRESSOR LEAD AND TURN A/C OFF.
4. FAST IDLE SPEED SCREW: WITH ENGINE OFF, PLACE THROTTLE ON HIGH STEP OF FAST IDLE CAM. DISCONNECT AND PLUG VACUUM HOSES AT EGR VALVE AND DISTRIBUTOR. START ENGINE WITHOUT TOUCHING THROTTLE. ADJUST TO SPECIFIED SPEED. OPEN THROTTLE TO RELEASE FAST IDLE CAM. STOP ENGINE, UNPLUG AND RECONNECT VACUUM HOSES TO EGR VALVE AND DISTRIBUTOR.

PRINTED IN U.S.A.

NON-CATALYST AIR/EGR CANADIAN CERTIFICATION	AUTOMATIC TRANSMISSION
TIMING (°BTC @ RPM)	2° @ 600 (DR)
SPARK PLUG GAP (IN.)	0.045
IDLE SPEED SCREW (RPM) (SOLENOID INACTIVE)	600 (DR)
IDLE SPEED SOLENOID (RPM) (SOLENOID ACTIVE)	650 (DR)
FAST IDLE SPEED (RPM)	1850 (P) OR (N)

NOTE IDLE MIXTURE SCREWS ARE PRESET AND SEALED AT FACTORY. PROVISION FOR ADJUSTMENT DURING TUNE UP IS NOT PROVIDED.

SEE SERVICE MANUAL, MAINTENANCE SCHEDULE AND EMISSION HOSE ROUTING DIAGRAM FOR ADDITIONAL INFORMATION.

PT. NO. 14100949

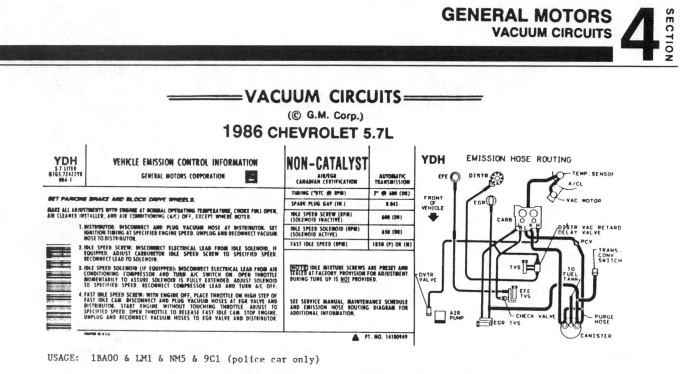

USAGE: 1BA00 & LM1 & NM5 & 9C1 (police car only)

5.7L—W/AUTO. TRANS. (NM5 EXPORT, CANADA) EXC NM8 EXPORT

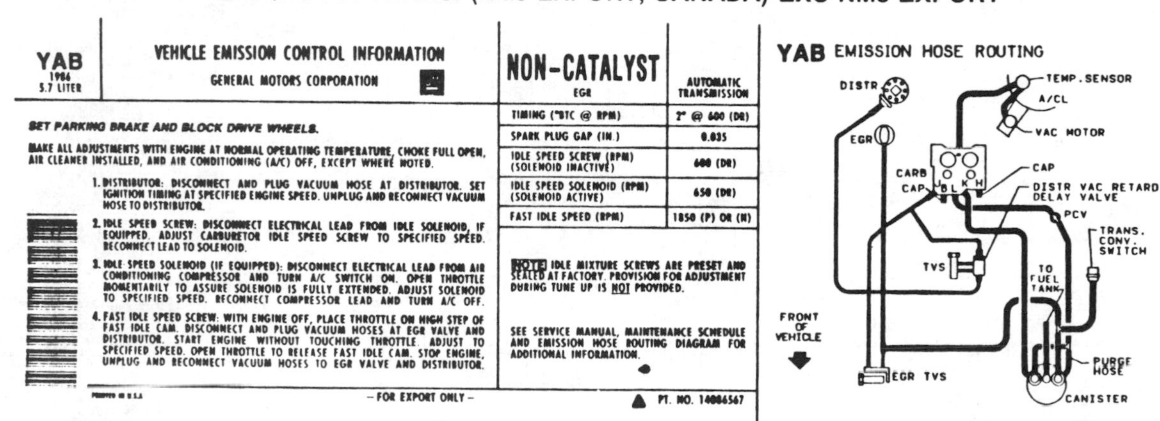

YAB
1986
5.7 LITER

VEHICLE EMISSION CONTROL INFORMATION
GENERAL MOTORS CORPORATION

SET PARKING BRAKE AND BLOCK DRIVE WHEELS.

MAKE ALL ADJUSTMENTS WITH ENGINE AT NORMAL OPERATING TEMPERATURE, CHOKE FULL OPEN, AIR CLEANER INSTALLED, AND AIR CONDITIONING (A/C) OFF, EXCEPT WHERE NOTED.

1. DISTRIBUTOR: DISCONNECT AND PLUG VACUUM HOSE AT DISTRIBUTOR. SET IGNITION TIMING AT SPECIFIED ENGINE SPEED. UNPLUG AND RECONNECT VACUUM HOSE TO DISTRIBUTOR.
2. IDLE SPEED SCREW: DISCONNECT ELECTRICAL LEAD FROM IDLE SOLENOID, IF EQUIPPED. ADJUST CARBURETOR IDLE SPEED SCREW TO SPECIFIED SPEED. RECONNECT LEAD TO SOLENOID.
3. IDLE SPEED SOLENOID (IF EQUIPPED): DISCONNECT ELECTRICAL LEAD FROM AIR CONDITIONING COMPRESSOR AND TURN A/C SWITCH ON. OPEN THROTTLE MOMENTARILY TO ASSURE SOLENOID IS FULLY EXTENDED. ADJUST SOLENOID TO SPECIFIED SPEED. RECONNECT COMPRESSOR LEAD AND TURN A/C OFF.
4. FAST IDLE SPEED SCREW: WITH ENGINE OFF, PLACE THROTTLE ON HIGH STEP OF FAST IDLE CAM. DISCONNECT AND PLUG VACUUM HOSES AT EGR VALVE AND DISTRIBUTOR. START ENGINE WITHOUT TOUCHING THROTTLE. ADJUST TO SPECIFIED SPEED. OPEN THROTTLE TO RELEASE FAST IDLE CAM. STOP ENGINE, UNPLUG AND RECONNECT VACUUM HOSES TO EGR VALVE AND DISTRIBUTOR.

PRINTED IN U.S.A. — FOR EXPORT ONLY —

NON-CATALYST EGR	AUTOMATIC TRANSMISSION
TIMING (°BTC @ RPM)	2° @ 600 (DR)
SPARK PLUG GAP (IN.)	0.035
IDLE SPEED SCREW (RPM) (SOLENOID INACTIVE)	600 (DR)
IDLE SPEED SOLENOID (RPM) (SOLENOID ACTIVE)	650 (DR)
FAST IDLE SPEED (RPM)	1850 (P) OR (N)

NOTE IDLE MIXTURE SCREWS ARE PRESET AND SEALED AT FACTORY. PROVISION FOR ADJUSTMENT DURING TUNE UP IS NOT PROVIDED.

SEE SERVICE MANUAL, MAINTENANCE SCHEDULE AND EMISSION HOSE ROUTING DIAGRAM FOR ADDITIONAL INFORMATION.

PT. NO. 14086567

USAGE: 1BA00 & LM1 & NM8 (export)

5.7L—W/AUTO. TRANS. (NM8)—EXPORT

1986 OLDSMOBILE 5.0L

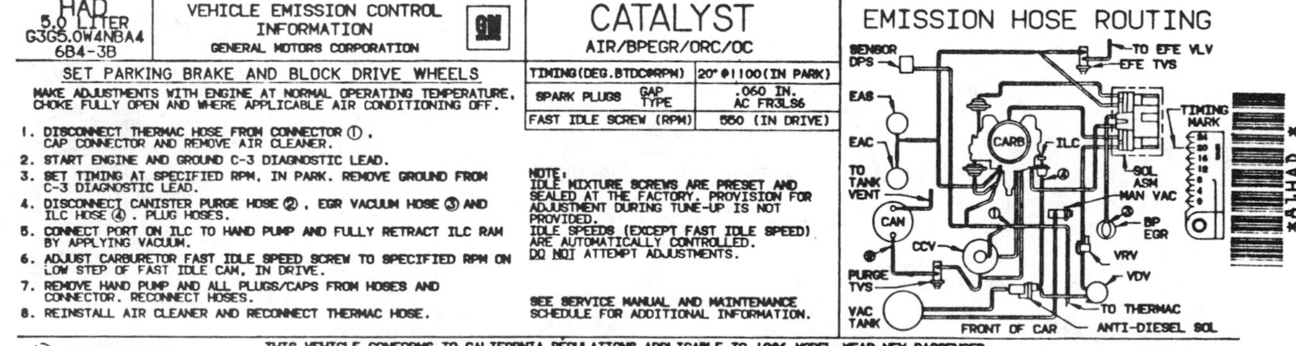

HAD
5.0 LITER
G3G5.0W4NBA4
6B4-3B

VEHICLE EMISSION CONTROL INFORMATION
GENERAL MOTORS CORPORATION

SET PARKING BRAKE AND BLOCK DRIVE WHEELS

MAKE ADJUSTMENTS WITH ENGINE AT NORMAL OPERATING TEMPERATURE, CHOKE FULLY OPEN AND WHERE APPLICABLE AIR CONDITIONING OFF.

1. DISCONNECT THERMAC HOSE FROM CONNECTOR ①. CAP CONNECTOR AND REMOVE AIR CLEANER.
2. START ENGINE AND GROUND C-3 DIAGNOSTIC LEAD.
3. SET TIMING AT SPECIFIED RPM, IN PARK. REMOVE GROUND FROM C-3 DIAGNOSTIC LEAD.
4. DISCONNECT CANISTER PURGE HOSE ②, EGR VACUUM HOSE ③ AND ILC HOSE ④. PLUG HOSES.
5. CONNECT PORT ON ILC TO HAND PUMP AND FULLY RETRACT ILC RAM BY APPLYING VACUUM.
6. ADJUST CARBURETOR FAST IDLE SPEED SCREW TO SPECIFIED RPM ON LOW STEP OF FAST IDLE CAM, IN DRIVE.
7. REMOVE HAND PUMP AND ALL PLUGS/CAPS FROM HOSES AND CONNECTOR. RECONNECT HOSES.
8. REINSTALL AIR CLEANER AND RECONNECT THERMAC HOSE.

CATALYST
AIR/BPEGR/ORC/OC

TIMING(DEG.BTDC@RPM)	20° @1100(IN PARK)
SPARK PLUGS GAP / TYPE	.060 IN. / AC FR3LS6
FAST IDLE SCREW (RPM)	550 (IN DRIVE)

NOTE:
IDLE MIXTURE SCREWS ARE PRESET AND SEALED AT THE FACTORY. PROVISION FOR ADJUSTMENT DURING TUNE-UP IS NOT PROVIDED.
IDLE SPEEDS (EXCEPT FAST IDLE SPEED) ARE AUTOMATICALLY CONTROLLED.
DO NOT ATTEMPT ADJUSTMENTS.

SEE SERVICE MANUAL AND MAINTENANCE SCHEDULE FOR ADDITIONAL INFORMATION.

(K) PRINTED IN U.S.A.

THIS VEHICLE CONFORMS TO CALIFORNIA REGULATIONS APPLICABLE TO 1986 MODEL YEAR NEW PASSENGER CARS AND SIMILARLY TO U.S. EPA REGULATIONS APPLICABLE TO THE STATE OF CALIFORNIA.

PART NO. 22530564

5.0L—W/AUTO. TRANS. (MV9, MW9)—CALIFORNIA

VACUUM CIRCUITS

(© G.M. Corp.)

1986 OLDSMOBILE 5.0L

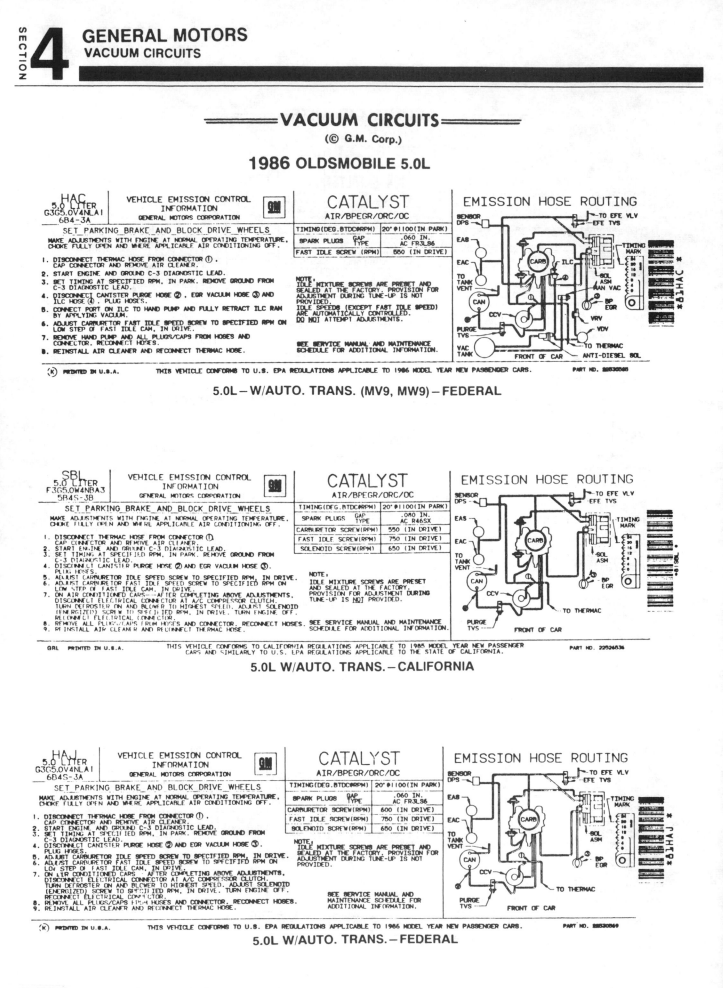

HAC
5.0 LITER
G3G5.0V4NLA1
6B4-3A

VEHICLE EMISSION CONTROL INFORMATION
GENERAL MOTORS CORPORATION
GM

SET PARKING BRAKE AND BLOCK DRIVE WHEELS

MAKE ADJUSTMENTS WITH ENGINE AT NORMAL OPERATING TEMPERATURE. CHOKE FULLY OPEN AND WHERE APPLICABLE AIR CONDITIONING OFF.

1. DISCONNECT THERMAC HOSE FROM CONNECTOR ①. CAP CONNECTOR AND REMOVE AIR CLEANER.
2. START ENGINE AND GROUND C-3 DIAGNOSTIC LEAD.
3. SET TIMING AT SPECIFIED RPM, IN PARK. REMOVE GROUND FROM C-3 DIAGNOSTIC LEAD.
4. DISCONNECT CANISTER PURGE HOSE ②, EGR VACUUM HOSE ③ AND ILC HOSE ④. PLUG HOSES.
5. CONNECT PORT ON ILC TO HAND PUMP AND FULLY RETRACT ILC RAM BY APPLYING VACUUM.
6. ADJUST CARBURETOR FAST IDLE SPEED SCREW TO SPECIFIED RPM ON LOW STEP OF FAST IDLE CAM, IN DRIVE.
7. REMOVE HAND PUMP AND ALL PLUGS/CAPS FROM HOSES AND CONNECTOR. RECONNECT HOSES.
8. REINSTALL AIR CLEANER AND RECONNECT THERMAC HOSE.

CATALYST
AIR/BPEGR/ORC/OC

TIMING(DEG.BTDC@RPM)	20° @1100(IN PARK)
SPARK PLUGS GAP TYPE	.060 IN. AC FR3LS6
FAST IDLE SCREW (RPM)	550 (IN DRIVE)

NOTE:
IDLE MIXTURE SCREWS ARE PRESET AND SEALED AT THE FACTORY. PROVISION FOR ADJUSTMENT DURING TUNE-UP IS NOT PROVIDED.
IDLE SPEEDS (EXCEPT FAST IDLE SPEED) ARE AUTOMATICALLY CONTROLLED. DO NOT ATTEMPT ADJUSTMENTS.

SEE SERVICE MANUAL AND MAINTENANCE SCHEDULE FOR ADDITIONAL INFORMATION.

EMISSION HOSE ROUTING

SENSOR DPS
TO EFE VLV
EFE TVS
EAS
TIMING MARK
EAC
CARB
ILC
TO TANK VENT
SOL ASM
MAN VAC
CAN
BP EGR
CCV
VRV
PURGE TVS
VDV
VAC TANK
TO THERMAC
FRONT OF CAR
ANTI-DIESEL SOL
81HAC

(K) PRINTED IN U.S.A. THIS VEHICLE CONFORMS TO U.S. EPA REGULATIONS APPLICABLE TO 1986 MODEL YEAR NEW PASSENGER CARS. PART NO. [illegible]

5.0L—W/AUTO. TRANS. (MV9, MW9)—FEDERAL

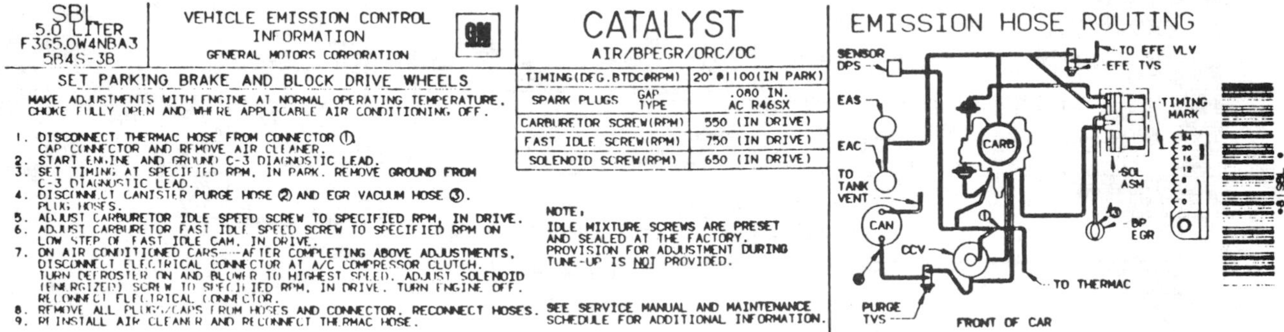
SBL
5.0 LITER
F3G5.0W4NBA3
5B4S-3B

VEHICLE EMISSION CONTROL INFORMATION
GENERAL MOTORS CORPORATION
GM

SET PARKING BRAKE AND BLOCK DRIVE WHEELS

MAKE ADJUSTMENTS WITH ENGINE AT NORMAL OPERATING TEMPERATURE. CHOKE FULLY OPEN AND WHERE APPLICABLE AIR CONDITIONING OFF.

1. DISCONNECT THERMAC HOSE FROM CONNECTOR ①. CAP CONNECTOR AND REMOVE AIR CLEANER.
2. START ENGINE AND GROUND C-3 DIAGNOSTIC LEAD.
3. SET TIMING AT SPECIFIED RPM, IN PARK. REMOVE GROUND FROM C-3 DIAGNOSTIC LEAD.
4. DISCONNECT CANISTER PURGE HOSE ② AND EGR VACUUM HOSE ③. PLUG HOSES.
5. ADJUST CARBURETOR IDLE SPEED SCREW TO SPECIFIED RPM, IN DRIVE.
6. ADJUST CARBURETOR FAST IDLE SPEED SCREW TO SPECIFIED RPM ON LOW STEP OF FAST IDLE CAM, IN DRIVE.
7. ON AIR CONDITIONED CARS—AFTER COMPLETING ABOVE ADJUSTMENTS, DISCONNECT ELECTRICAL CONNECTOR AT A/C COMPRESSOR CLUTCH. TURN DEFROSTER ON AND BLOWER TO HIGHEST SPEED. ADJUST SOLENOID (ENERGIZED) SCREW TO SPECIFIED RPM, IN DRIVE. TURN ENGINE OFF. RECONNECT ELECTRICAL CONNECTOR.
8. REMOVE ALL PLUGS/CAPS FROM HOSES AND CONNECTOR. RECONNECT HOSES.
9. REINSTALL AIR CLEANER AND RECONNECT THERMAC HOSE.

CATALYST
AIR/BPEGR/ORC/OC

TIMING(DEG.BTDC@RPM)	20° @1100(IN PARK)
SPARK PLUGS GAP TYPE	.080 IN. AC R46SX
CARBURETOR SCREW(RPM)	550 (IN DRIVE)
FAST IDLE SCREW(RPM)	750 (IN DRIVE)
SOLENOID SCREW(RPM)	650 (IN DRIVE)

NOTE:
IDLE MIXTURE SCREWS ARE PRESET AND SEALED AT THE FACTORY. PROVISION FOR ADJUSTMENT DURING TUNE-UP IS NOT PROVIDED.

SEE SERVICE MANUAL AND MAINTENANCE SCHEDULE FOR ADDITIONAL INFORMATION.

GRL PRINTED IN U.S.A. THIS VEHICLE CONFORMS TO CALIFORNIA REGULATIONS APPLICABLE TO 1985 MODEL YEAR NEW PASSENGER CARS AND SIMILARLY TO U.S. EPA REGULATIONS APPLICABLE TO THE STATE OF CALIFORNIA. PART NO. 22526836

5.0L W/AUTO. TRANS.—CALIFORNIA

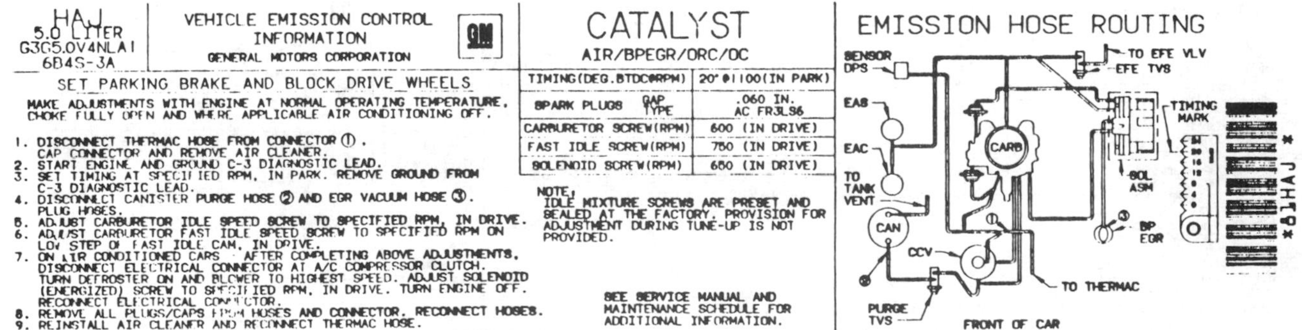
HAJ
5.0 LITER
G3G5.0V4NLA1
6B4S-3A

VEHICLE EMISSION CONTROL INFORMATION
GENERAL MOTORS CORPORATION
GM

SET PARKING BRAKE AND BLOCK DRIVE WHEELS

MAKE ADJUSTMENTS WITH ENGINE AT NORMAL OPERATING TEMPERATURE. CHOKE FULLY OPEN AND WHERE APPLICABLE AIR CONDITIONING OFF.

1. DISCONNECT THERMAC HOSE FROM CONNECTOR ①. CAP CONNECTOR AND REMOVE AIR CLEANER.
2. START ENGINE AND GROUND C-3 DIAGNOSTIC LEAD.
3. SET TIMING AT SPECIFIED RPM, IN PARK. REMOVE GROUND FROM C-3 DIAGNOSTIC LEAD.
4. DISCONNECT CANISTER PURGE HOSE ② AND EGR VACUUM HOSE ③. PLUG HOSES.
5. ADJUST CARBURETOR IDLE SPEED SCREW TO SPECIFIED RPM, IN DRIVE.
6. ADJUST CARBURETOR FAST IDLE SPEED SCREW TO SPECIFIED RPM ON LOW STEP OF FAST IDLE CAM, IN DRIVE.
7. ON AIR CONDITIONED CARS—AFTER COMPLETING ABOVE ADJUSTMENTS, DISCONNECT ELECTRICAL CONNECTOR AT A/C COMPRESSOR CLUTCH. TURN DEFROSTER ON AND BLOWER TO HIGHEST SPEED. ADJUST SOLENOID (ENERGIZED) SCREW TO SPECIFIED RPM, IN DRIVE. TURN ENGINE OFF. RECONNECT ELECTRICAL CONNECTOR.
8. REMOVE ALL PLUGS/CAPS FROM HOSES AND CONNECTOR. RECONNECT HOSES.
9. REINSTALL AIR CLEANER AND RECONNECT THERMAC HOSE.

CATALYST
AIR/BPEGR/ORC/OC

TIMING(DEG.BTDC@RPM)	20° @1100(IN PARK)
SPARK PLUGS GAP TYPE	.060 IN. AC FR3LS6
CARBURETOR SCREW(RPM)	600 (IN DRIVE)
FAST IDLE SCREW(RPM)	750 (IN DRIVE)
SOLENOID SCREW(RPM)	650 (IN DRIVE)

NOTE:
IDLE MIXTURE SCREWS ARE PRESET AND SEALED AT THE FACTORY. PROVISION FOR ADJUSTMENT DURING TUNE-UP IS NOT PROVIDED.

SEE SERVICE MANUAL AND MAINTENANCE SCHEDULE FOR ADDITIONAL INFORMATION.

(K) PRINTED IN U.S.A. THIS VEHICLE CONFORMS TO U.S. EPA REGULATIONS APPLICABLE TO 1986 MODEL YEAR NEW PASSENGER CARS. PART NO. [illegible]

5.0L W/AUTO. TRANS.—FEDERAL

VACUUM CIRCUITS

(© G.M. Corp.)

1986 PONTIAC 1.8L

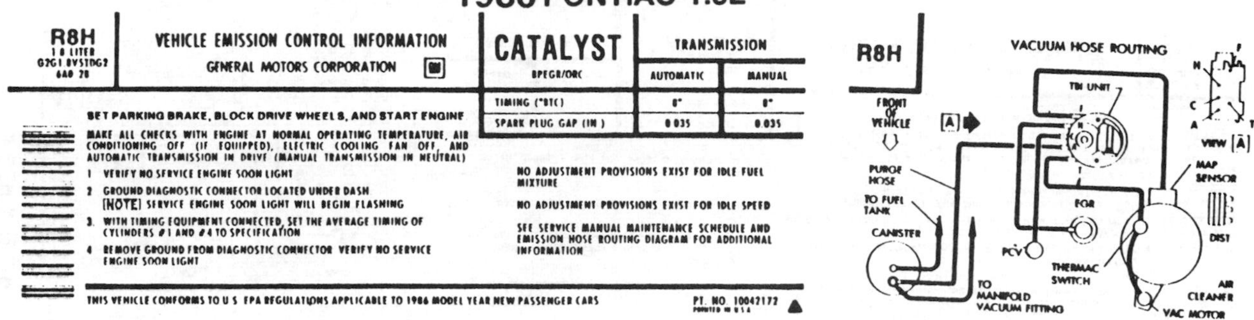

USAGE: J CAR 1.8L LH8 & NA5

1.8L – W/AUTO. TRANS. OR MAN. TRANS. – FEDERAL

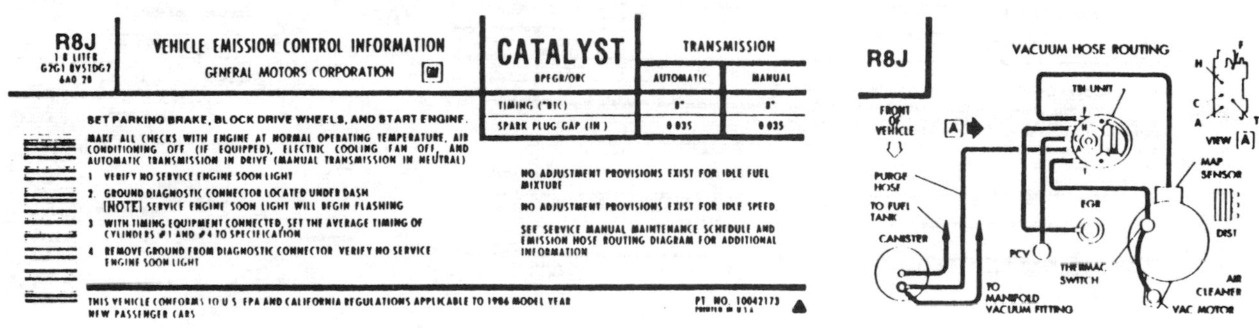

USAGE: J CAR 1.8L LH8 & NB2

1.8L – W/AUTO. TRANS. OR MAN. TRANS. – CALIFORNIA

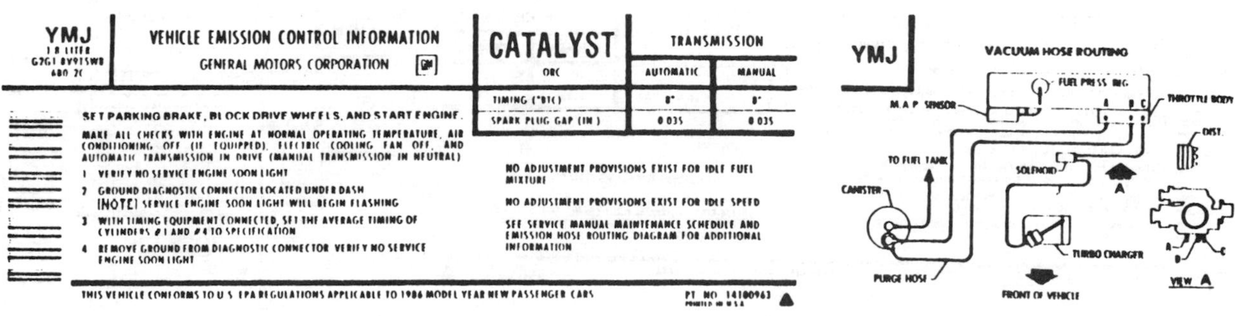

USAGE: J CAR 1.8L LA5 & NA5

1.8L – W/AUTO. TRANS. OR MAN. TRANS. – FEDERAL

VACUUM CIRCUITS

(© G.M. Corp.)

1986 PONTIAC 1.8L

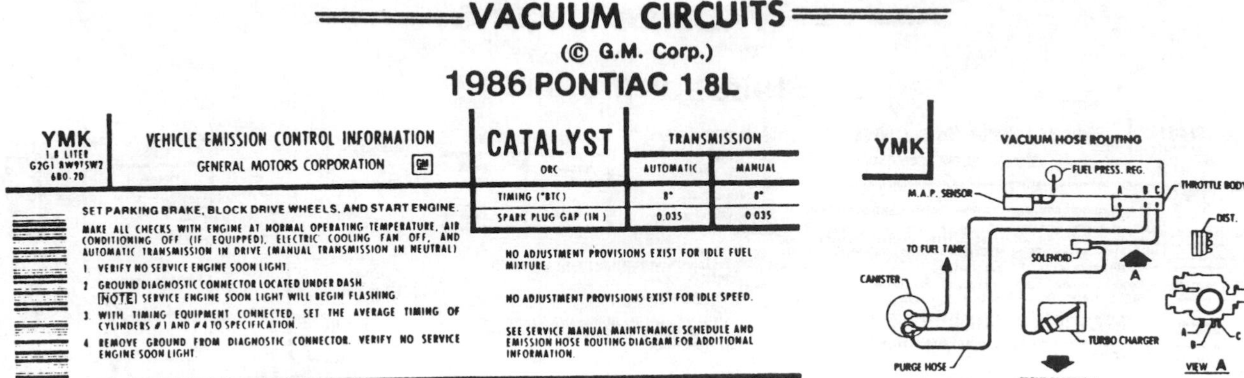

YMK
1.8 LITER
G2G1.8W9TSW2
6B0-7D

VEHICLE EMISSION CONTROL INFORMATION
GENERAL MOTORS CORPORATION

CATALYST
ORC

TRANSMISSION	AUTOMATIC	MANUAL
TIMING (°BTC)	8°	8°
SPARK PLUG GAP (IN.)	0.035	0.035

SET PARKING BRAKE, BLOCK DRIVE WHEELS, AND START ENGINE.

MAKE ALL CHECKS WITH ENGINE AT NORMAL OPERATING TEMPERATURE, AIR CONDITIONING OFF (IF EQUIPPED), ELECTRIC COOLING FAN OFF, AND AUTOMATIC TRANSMISSION IN DRIVE (MANUAL TRANSMISSION IN NEUTRAL)

1. VERIFY NO SERVICE ENGINE SOON LIGHT.
2. GROUND DIAGNOSTIC CONNECTOR LOCATED UNDER DASH. [NOTE] SERVICE ENGINE SOON LIGHT WILL BEGIN FLASHING.
3. WITH TIMING EQUIPMENT CONNECTED, SET THE AVERAGE TIMING OF CYLINDERS #1 AND #4 TO SPECIFICATION.
4. REMOVE GROUND FROM DIAGNOSTIC CONNECTOR. VERIFY NO SERVICE ENGINE SOON LIGHT.

NO ADJUSTMENT PROVISIONS EXIST FOR IDLE FUEL MIXTURE.

NO ADJUSTMENT PROVISIONS EXIST FOR IDLE SPEED.

SEE SERVICE MANUAL MAINTENANCE SCHEDULE AND EMISSION HOSE ROUTING DIAGRAM FOR ADDITIONAL INFORMATION.

THIS VEHICLE CONFORMS TO CALIFORNIA REGULATIONS APPLICABLE TO 1986 MODEL YEAR NEW PASSENGER CARS AND SIMILARLY TO U.S. EPA REGULATIONS APPLICABLE TO THE STATE OF CALIFORNIA.

PT. NO. 14100964
PRINTED IN U.S.A.

USAGE: J CAR 1.8L LA5 & NB2

1.8L—W/AUTO. TRANS. OR MAN. TRANS.—CALIFORNIA

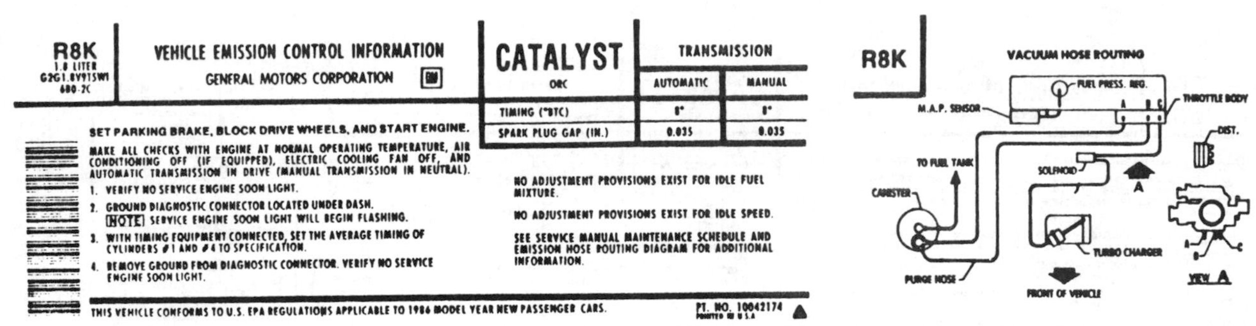

R8K
1.8 LITER
G2G1.8V9TSW1
6B0-2C

VEHICLE EMISSION CONTROL INFORMATION
GENERAL MOTORS CORPORATION

CATALYST
ORC

TRANSMISSION	AUTOMATIC	MANUAL
TIMING (°BTC)	8°	8°
SPARK PLUG GAP (IN.)	0.035	0.035

SET PARKING BRAKE, BLOCK DRIVE WHEELS, AND START ENGINE.

MAKE ALL CHECKS WITH ENGINE AT NORMAL OPERATING TEMPERATURE, AIR CONDITIONING OFF (IF EQUIPPED), ELECTRIC COOLING FAN OFF, AND AUTOMATIC TRANSMISSION IN DRIVE (MANUAL TRANSMISSION IN NEUTRAL).

1. VERIFY NO SERVICE ENGINE SOON LIGHT.
2. GROUND DIAGNOSTIC CONNECTOR LOCATED UNDER DASH. [NOTE] SERVICE ENGINE SOON LIGHT WILL BEGIN FLASHING.
3. WITH TIMING EQUIPMENT CONNECTED, SET THE AVERAGE TIMING OF CYLINDERS #1 AND #4 TO SPECIFICATION.
4. REMOVE GROUND FROM DIAGNOSTIC CONNECTOR. VERIFY NO SERVICE ENGINE SOON LIGHT.

NO ADJUSTMENT PROVISIONS EXIST FOR IDLE FUEL MIXTURE.

NO ADJUSTMENT PROVISIONS EXIST FOR IDLE SPEED.

SEE SERVICE MANUAL MAINTENANCE SCHEDULE AND EMISSION HOSE ROUTING DIAGRAM FOR ADDITIONAL INFORMATION.

THIS VEHICLE CONFORMS TO U.S. EPA REGULATIONS APPLICABLE TO 1986 MODEL YEAR NEW PASSENGER CARS.

PT. NO. 10042174
PRINTED IN U.S.A.

USAGE: J CAR 1.8L LA5 & NA5
CANCELLED - REPLACED WITH 14100963

1.8L—W/AUTO. TRANS. OR MAN. TRANS.—FEDERAL

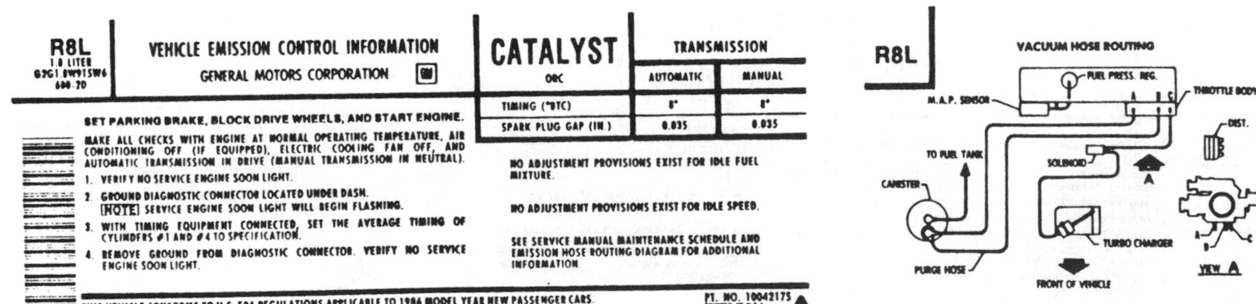

R8L
1.8 LITER
G2G1.8W9TSW6
6B0-7D

VEHICLE EMISSION CONTROL INFORMATION
GENERAL MOTORS CORPORATION

CATALYST
ORC

TRANSMISSION	AUTOMATIC	MANUAL
TIMING (°BTC)	8°	8°
SPARK PLUG GAP (IN.)	0.035	0.035

SET PARKING BRAKE, BLOCK DRIVE WHEELS, AND START ENGINE.

MAKE ALL CHECKS WITH ENGINE AT NORMAL OPERATING TEMPERATURE, AIR CONDITIONING OFF (IF EQUIPPED), ELECTRIC COOLING FAN OFF, AND AUTOMATIC TRANSMISSION IN DRIVE (MANUAL TRANSMISSION IN NEUTRAL).

1. VERIFY NO SERVICE ENGINE SOON LIGHT.
2. GROUND DIAGNOSTIC CONNECTOR LOCATED UNDER DASH. [NOTE] SERVICE ENGINE SOON LIGHT WILL BEGIN FLASHING.
3. WITH TIMING EQUIPMENT CONNECTED, SET THE AVERAGE TIMING OF CYLINDERS #1 AND #4 TO SPECIFICATION.
4. REMOVE GROUND FROM DIAGNOSTIC CONNECTOR. VERIFY NO SERVICE ENGINE SOON LIGHT.

NO ADJUSTMENT PROVISIONS EXIST FOR IDLE FUEL MIXTURE.

NO ADJUSTMENT PROVISIONS EXIST FOR IDLE SPEED.

SEE SERVICE MANUAL MAINTENANCE SCHEDULE AND EMISSION HOSE ROUTING DIAGRAM FOR ADDITIONAL INFORMATION.

THIS VEHICLE CONFORMS TO U.S. EPA REGULATIONS APPLICABLE TO 1986 MODEL YEAR NEW PASSENGER CARS.

PT. NO. 10042175
PRINTED IN U.S.A.

USAGE: J CAR 1.8L LA5 & NB2
CANCELLED - REPLACED WITH 14100964

1.8L—W/AUTO. TRANS. OR MAN. TRANS.—CALIFORNIA

VACUUM CIRCUITS

(© G.M. Corp.)

1986 PONTIAC 2.5L

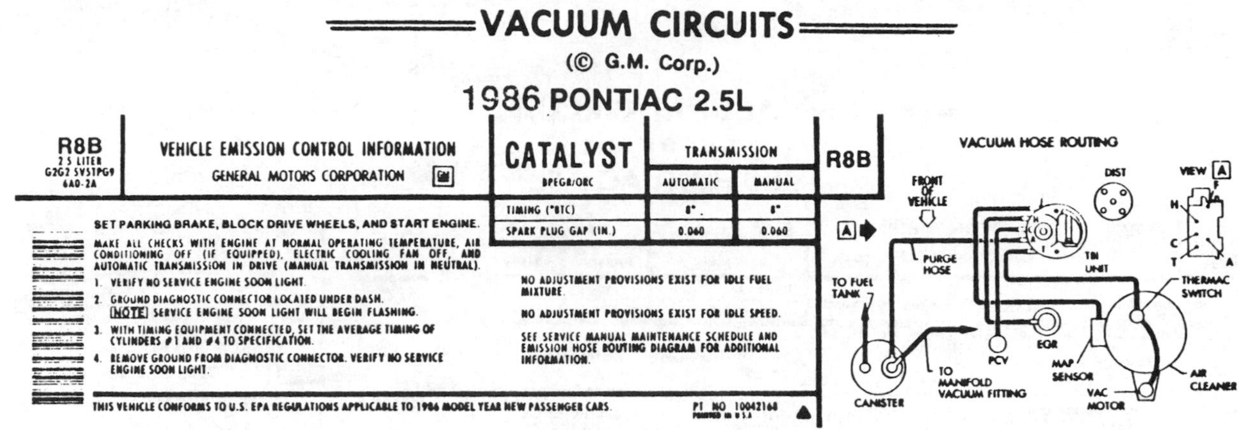

USAGE: A CAR 2.5L LR8 & NA5

2.5L—W/AUTO. TRANS. OR MAN. TRANS.—FEDERAL

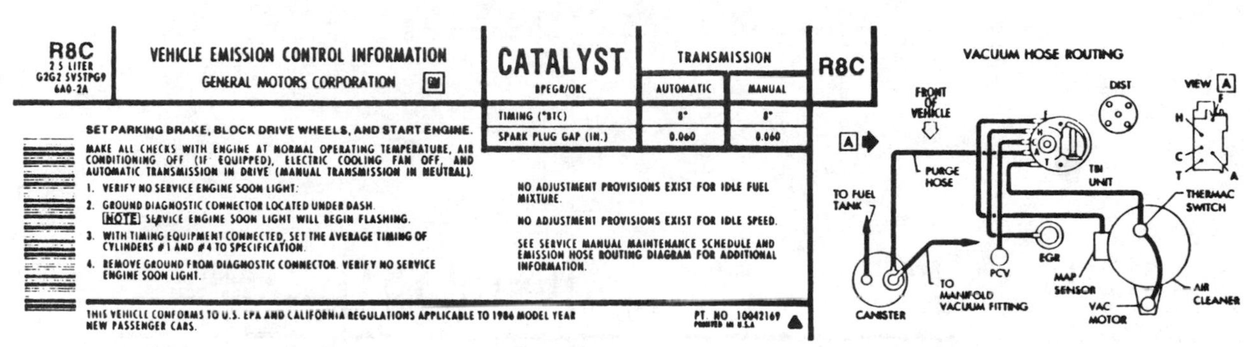

USAGE: A CAR 2.5L LR8 & NB2

2.5L—W/AUTO. TRANS. OR MAN. TRANS.—CALIFORNIA

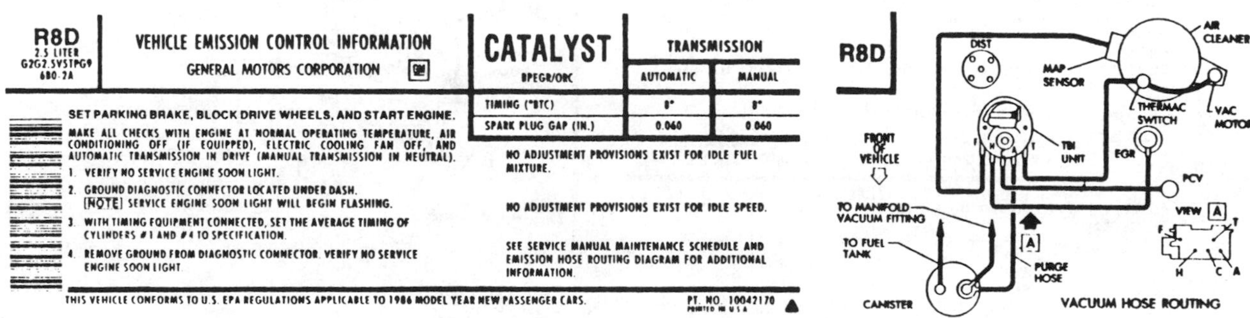

USAGE: F CAR 2.5L LQ9 & NA5

2.5L—W/AUTO. TRANS. OR MAN. TRANS.—FEDERAL

VACUUM CIRCUITS

(© G.M. Corp.)

1986 PONTIAC 2.5L

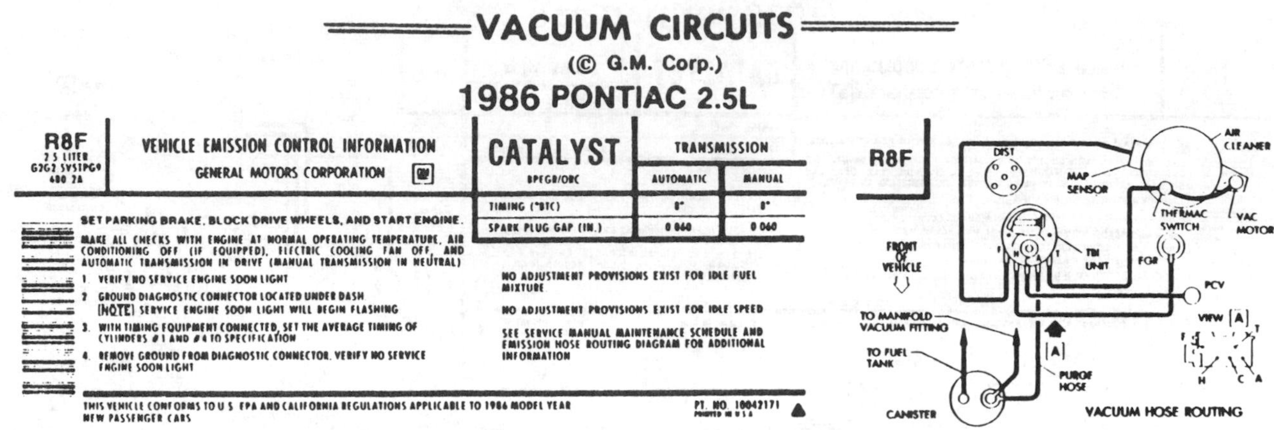

R8F
2 5 LITER
G2G2 5V5TPG9
680 2A

VEHICLE EMISSION CONTROL INFORMATION
GENERAL MOTORS CORPORATION

CATALYST BPEGR/ORC	TRANSMISSION AUTOMATIC	MANUAL
TIMING (°BTC)	8°	8°
SPARK PLUG GAP (IN.)	0 060	0 060

SET PARKING BRAKE, BLOCK DRIVE WHEELS, AND START ENGINE.

MAKE ALL CHECKS WITH ENGINE AT NORMAL OPERATING TEMPERATURE, AIR CONDITIONING OFF (IF EQUIPPED), ELECTRIC COOLING FAN OFF, AND AUTOMATIC TRANSMISSION IN DRIVE (MANUAL TRANSMISSION IN NEUTRAL)

1. VERIFY NO SERVICE ENGINE SOON LIGHT
2. GROUND DIAGNOSTIC CONNECTOR LOCATED UNDER DASH
   NOTE SERVICE ENGINE SOON LIGHT WILL BEGIN FLASHING
3. WITH TIMING EQUIPMENT CONNECTED, SET THE AVERAGE TIMING OF CYLINDERS #1 AND #4 TO SPECIFICATION
4. REMOVE GROUND FROM DIAGNOSTIC CONNECTOR. VERIFY NO SERVICE ENGINE SOON LIGHT

NO ADJUSTMENT PROVISIONS EXIST FOR IDLE FUEL MIXTURE

NO ADJUSTMENT PROVISIONS EXIST FOR IDLE SPEED

SEE SERVICE MANUAL MAINTENANCE SCHEDULE AND EMISSION HOSE ROUTING DIAGRAM FOR ADDITIONAL INFORMATION

THIS VEHICLE CONFORMS TO U S EPA AND CALIFORNIA REGULATIONS APPLICABLE TO 1986 MODEL YEAR NEW PASSENGER CARS

PT. NO. 10042171
PRINTED IN U S A

USAGE: F CAR 2.5L LQ9 & NB2

2.5L–W/AUTO. TRANS. OR MAN. TRANS.–CALIFORNIA

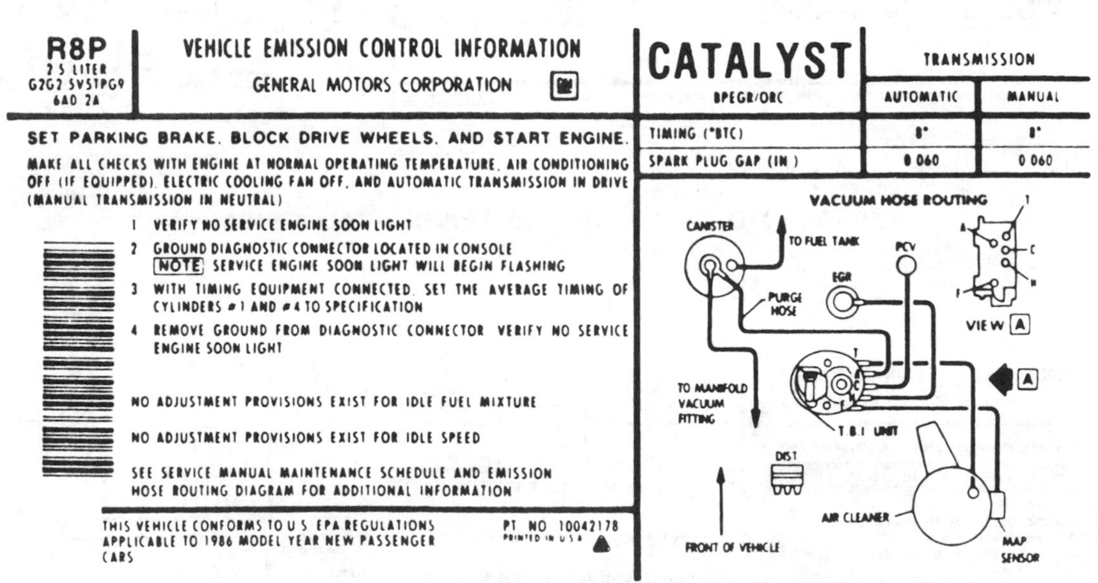

R8P
2 5 LITER
G2G2 5V5TPG9
6A0 2A

VEHICLE EMISSION CONTROL INFORMATION
GENERAL MOTORS CORPORATION

CATALYST BPEGR/ORC	TRANSMISSION AUTOMATIC	MANUAL
TIMING (°BTC)	8°	8°
SPARK PLUG GAP (IN.)	0 060	0 060

SET PARKING BRAKE, BLOCK DRIVE WHEELS, AND START ENGINE.

MAKE ALL CHECKS WITH ENGINE AT NORMAL OPERATING TEMPERATURE, AIR CONDITIONING OFF (IF EQUIPPED), ELECTRIC COOLING FAN OFF, AND AUTOMATIC TRANSMISSION IN DRIVE (MANUAL TRANSMISSION IN NEUTRAL)

1. VERIFY NO SERVICE ENGINE SOON LIGHT
2. GROUND DIAGNOSTIC CONNECTOR LOCATED IN CONSOLE
   NOTE SERVICE ENGINE SOON LIGHT WILL BEGIN FLASHING
3. WITH TIMING EQUIPMENT CONNECTED, SET THE AVERAGE TIMING OF CYLINDERS #1 AND #4 TO SPECIFICATION
4. REMOVE GROUND FROM DIAGNOSTIC CONNECTOR VERIFY NO SERVICE ENGINE SOON LIGHT

NO ADJUSTMENT PROVISIONS EXIST FOR IDLE FUEL MIXTURE

NO ADJUSTMENT PROVISIONS EXIST FOR IDLE SPEED

SEE SERVICE MANUAL MAINTENANCE SCHEDULE AND EMISSION HOSE ROUTING DIAGRAM FOR ADDITIONAL INFORMATION

THIS VEHICLE CONFORMS TO U S EPA REGULATIONS APPLICABLE TO 1986 MODEL YEAR NEW PASSENGER CARS

PT NO 10042178
PRINTED IN U S A

USAGE: P CAR 2.5L LR8 & NA5

2.5L–W/AUTO. TRANS. OR MAN. TRANS.–FEDERAL

VACUUM CIRCUITS

(© G.M. Corp.)

1986 PONTIAC 2.5L

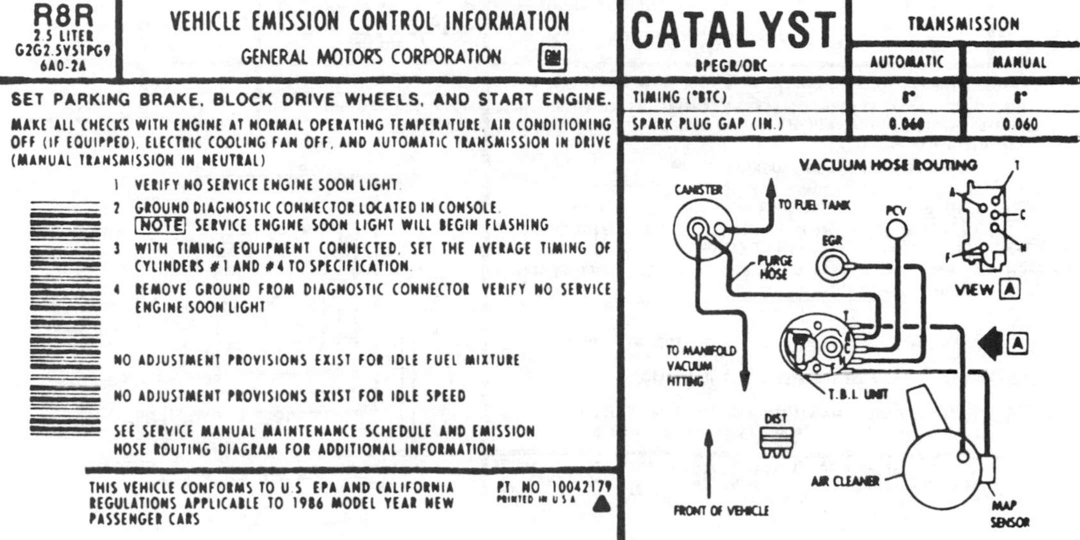

R8R
2.5 LITER
G2G2.5V5IPG9
6A0-2A

VEHICLE EMISSION CONTROL INFORMATION
GENERAL MOTORS CORPORATION

CATALYST
BPEGR/ORC

	TRANSMISSION	
	AUTOMATIC	MANUAL
TIMING (°BTC)	8°	8°
SPARK PLUG GAP (IN.)	0.060	0.060

SET PARKING BRAKE, BLOCK DRIVE WHEELS, AND START ENGINE.

MAKE ALL CHECKS WITH ENGINE AT NORMAL OPERATING TEMPERATURE, AIR CONDITIONING OFF (IF EQUIPPED), ELECTRIC COOLING FAN OFF, AND AUTOMATIC TRANSMISSION IN DRIVE (MANUAL TRANSMISSION IN NEUTRAL)

1. VERIFY NO SERVICE ENGINE SOON LIGHT.
2. GROUND DIAGNOSTIC CONNECTOR LOCATED IN CONSOLE. NOTE SERVICE ENGINE SOON LIGHT WILL BEGIN FLASHING
3. WITH TIMING EQUIPMENT CONNECTED, SET THE AVERAGE TIMING OF CYLINDERS #1 AND #4 TO SPECIFICATION.
4. REMOVE GROUND FROM DIAGNOSTIC CONNECTOR. VERIFY NO SERVICE ENGINE SOON LIGHT

NO ADJUSTMENT PROVISIONS EXIST FOR IDLE FUEL MIXTURE

NO ADJUSTMENT PROVISIONS EXIST FOR IDLE SPEED

SEE SERVICE MANUAL MAINTENANCE SCHEDULE AND EMISSION HOSE ROUTING DIAGRAM FOR ADDITIONAL INFORMATION

THIS VEHICLE CONFORMS TO U.S. EPA AND CALIFORNIA REGULATIONS APPLICABLE TO 1986 MODEL YEAR NEW PASSENGER CARS

PT. NO. 10042179
PRINTED IN U.S.A.

USAGE: P CAR 2.5L LR8 & NB2

2.5L—W/AUTO. TRANS. OR MAN. TRANS.—CALIFORNIA

1986 PONTIAC 2.8L

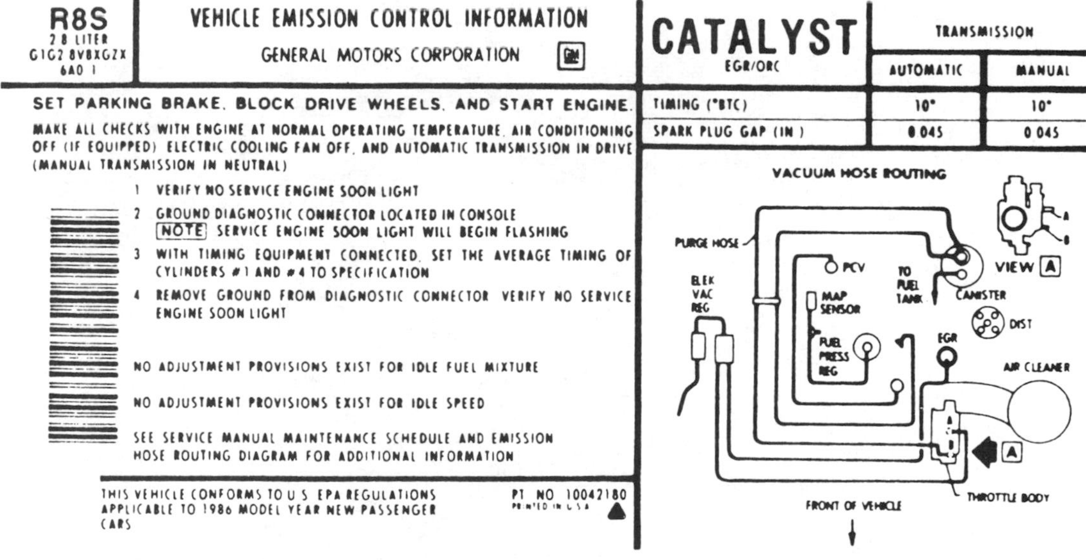

R8S
2.8 LITER
G1G2.8V8XG2X
6A0-1

VEHICLE EMISSION CONTROL INFORMATION
GENERAL MOTORS CORPORATION

CATALYST
EGR/ORC

	TRANSMISSION	
	AUTOMATIC	MANUAL
TIMING (°BTC)	10°	10°
SPARK PLUG GAP (IN.)	0.045	0.045

SET PARKING BRAKE, BLOCK DRIVE WHEELS, AND START ENGINE.

MAKE ALL CHECKS WITH ENGINE AT NORMAL OPERATING TEMPERATURE, AIR CONDITIONING OFF (IF EQUIPPED), ELECTRIC COOLING FAN OFF, AND AUTOMATIC TRANSMISSION IN DRIVE (MANUAL TRANSMISSION IN NEUTRAL)

1. VERIFY NO SERVICE ENGINE SOON LIGHT
2. GROUND DIAGNOSTIC CONNECTOR LOCATED IN CONSOLE. NOTE SERVICE ENGINE SOON LIGHT WILL BEGIN FLASHING
3. WITH TIMING EQUIPMENT CONNECTED, SET THE AVERAGE TIMING OF CYLINDERS #1 AND #4 TO SPECIFICATION
4. REMOVE GROUND FROM DIAGNOSTIC CONNECTOR. VERIFY NO SERVICE ENGINE SOON LIGHT

NO ADJUSTMENT PROVISIONS EXIST FOR IDLE FUEL MIXTURE

NO ADJUSTMENT PROVISIONS EXIST FOR IDLE SPEED

SEE SERVICE MANUAL MAINTENANCE SCHEDULE AND EMISSION HOSE ROUTING DIAGRAM FOR ADDITIONAL INFORMATION

THIS VEHICLE CONFORMS TO U.S. EPA REGULATIONS APPLICABLE TO 1986 MODEL YEAR NEW PASSENGER CARS

PT. NO. 10042180
PRINTED IN U.S.A.

USAGE: P CAR 2.8L L44 & NA5

2.8L—W/AUTO. TRANS. OR MAN. TRANS.—FEDERAL

VACUUM CIRCUITS

(© G.M. Corp.)

1986 PONTIAC 2.8L

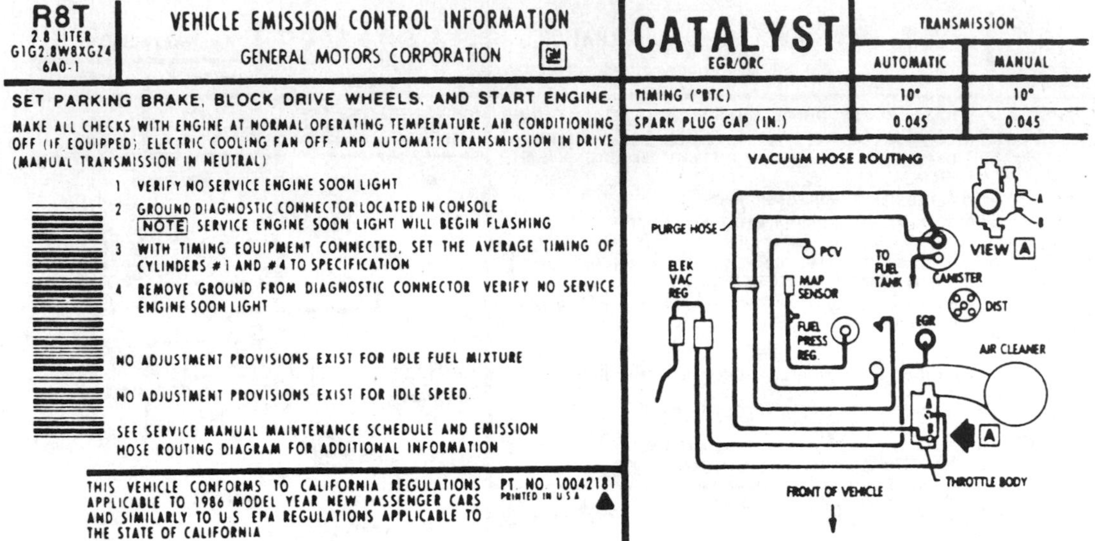

R8T
2.8 LITER
G1G2.8W8XGZ4
6A0-1

VEHICLE EMISSION CONTROL INFORMATION
GENERAL MOTORS CORPORATION

CATALYST
EGR/ORC

	TRANSMISSION	
	AUTOMATIC	MANUAL
TIMING (°BTC)	10°	10°
SPARK PLUG GAP (IN.)	0.045	0.045

SET PARKING BRAKE, BLOCK DRIVE WHEELS, AND START ENGINE.

MAKE ALL CHECKS WITH ENGINE AT NORMAL OPERATING TEMPERATURE, AIR CONDITIONING OFF (IF EQUIPPED), ELECTRIC COOLING FAN OFF, AND AUTOMATIC TRANSMISSION IN DRIVE (MANUAL TRANSMISSION IN NEUTRAL)

1. VERIFY NO SERVICE ENGINE SOON LIGHT
2. GROUND DIAGNOSTIC CONNECTOR LOCATED IN CONSOLE
   NOTE SERVICE ENGINE SOON LIGHT WILL BEGIN FLASHING
3. WITH TIMING EQUIPMENT CONNECTED, SET THE AVERAGE TIMING OF CYLINDERS #1 AND #4 TO SPECIFICATION
4. REMOVE GROUND FROM DIAGNOSTIC CONNECTOR VERIFY NO SERVICE ENGINE SOON LIGHT

NO ADJUSTMENT PROVISIONS EXIST FOR IDLE FUEL MIXTURE

NO ADJUSTMENT PROVISIONS EXIST FOR IDLE SPEED

SEE SERVICE MANUAL MAINTENANCE SCHEDULE AND EMISSION HOSE ROUTING DIAGRAM FOR ADDITIONAL INFORMATION

THIS VEHICLE CONFORMS TO CALIFORNIA REGULATIONS APPLICABLE TO 1986 MODEL YEAR NEW PASSENGER CARS AND SIMILARLY TO U.S. EPA REGULATIONS APPLICABLE TO THE STATE OF CALIFORNIA

PT. NO. 10042181
PRINTED IN U.S.A.

VACUUM HOSE ROUTING

USAGE: P CAR 2.8: L44 & NB2

2.8L—W/AUTO. TRANS. OR MAN. TRANS.—CALIFORNIA